Metals Handbook Ninth Edition

Volume 6 Welding, Brazing, and Soldering

Prepared under the direction of the ASM Handbook Committee

Planned, prepared, organized, and reviewed by the ASM Joining Division Council

Coordinator

Ernest F. Nippes

William H. Cubberly, Director of Publications
Robert L. Stedfeld, Assistant Director of Reference Publications
Kathleen Mills, Managing Editor
Joseph R. Davis, Technical Editor
Bonnie R. Sanders, Production Editor

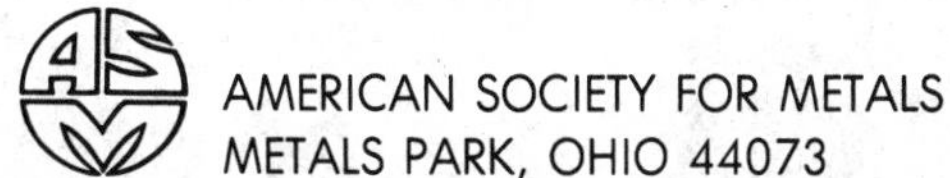

AMERICAN SOCIETY FOR METALS
METALS PARK, OHIO 44073

First printing, August 1983

Metals Handbook is a collective effort involving thousands of technical specialists. It brings together in one book a wealth of information from worldwide sources to help scientists, engineers, and technicians solve current and long-range problems.

Great care is taken in the compilation and production of this volume, but it should be made clear that no warranties, express or implied, are given in connection with the accuracy or completeness of this publication, and no responsibility can be taken for any claims that may arise.

Nothing contained in the Metals Handbook shall be construed as a grant of any right of manufacture, sale, use, or reproduction, in connection with any method, process, apparatus, product, composition, or system, whether or not covered by letters patent, copyright, or trademark, and nothing contained in the Metals Handbook shall be construed as a defense against any alleged infringement of letters patent, copyright, or trademark, or as a defense against any liability for such infringement.

Comments, criticisms, and suggestions are invited, and should be forwarded to the American Society for Metals.

Library of Congress Cataloging in Publication Data

American Society for Metals

Metals handbook.

Includes bibliographical references and indexes.
Contents: v. 1. Properties and selection—v. 2.
Properties and selection—nonferrous alloys and pure
metals—[etc.]—v. 6. Welding, brazing, and soldering.

1. Metals—Handbooks, manuals, etc. I. American
Society for Metals. Handbook Committee.
TA459.M43 1978 669 78-14934
ISBN 0-87170-007-7 (v. 1)

SAN 204-7586

Printed in the United States of America

Foreword

Exactly 50 years ago, the first article on welding was published in *Metals Handbook*. The short piece—equal in size to about five of today's *Metals Handbook* pages—was a small chapter in the 1933 Edition of the *National Metals Handbook* published by the American Society for Steel Treating, ASM's predecessor. Brazing and soldering were disposed of in one brief sentence.

By the time Volume 6 of the 8th Edition was published in 1971, the small chapter had grown into a completely separate volume of 734 pages on welding and brazing.

This pattern of growth, both in size and scope, has continued into the 9th Edition. In the present volume, the number of pages needed to span the areas of welding, brazing, and soldering has increased by 25%. The welding section has grown substantially, the amount of information about brazing has also increased, and an entirely new section on soldering has been added.

The task of researching, writing, reviewing, editing, and producing a volume of this size is a huge one. The book is larger than any *Metals Handbook* volume published during the past 21 years (Volume 1 of the 8th Edition, published in 1961, still stands as the all-time record-holder).

To achieve this growth, over 200 contributors and reviewers gave freely of their time, effort, and expertise. We particularly appreciate the enthusiastic leadership of Peter Patriarca, Chairman of ASM's Joining Division, and the capable direction of Dr. Ernest F. Nippes, who steered this volume to its successful conclusion. Our profound thanks also goes to the many members of the Joining Division who participated, to the ASM Handbook Committee, and to the editorial staff of *Metals Handbook*.

The thought and effort that have gone into this volume show clearly. We are proud to present this practical and authoritative reference work to the technical community.

George H. Bodeen
President

Allan Ray Putnam
Managing Director

The Ninth Edition of Metals Handbook
is dedicated to the memory of
TAYLOR LYMAN, A.B. (Eng.), S.M., Ph.D.
(1917-1973)
Editor, Metals Handbook, 1945-1973

Preface

Many developments have occurred in the joining industry since the publication of "Welding and Brazing" as Volume 6 of the 8th Edition of *Metals Handbook* in 1971.

Volume 6 of the 9th Edition, entitled "Welding, Brazing, and Soldering," contains substantially more information than the volume it replaces. In addition to incorporating a complete revision of the material contained in the prior volume, as well as an entirely new section on soldering, this volume places expanded emphasis on the metallurgical aspects of joining. New articles have been included which cover the principles of joining metallurgy; joining processes and their selection; joint design and preparation; codes, specifications, and inspection; weld discontinuities; and residual stresses and distortion.

In view of the advancement of joining technology, articles have been added to provide information on additional joining-related processes, such as thermal cutting, laser beam welding, solid-state welding, thermit welding, explosion welding, ultrasonic welding, high frequency welding, underwater welding and cutting, surfacing, and laser brazing.

Reflecting the development of joining processes for new materials, appropriate articles have been added on arc welding of reactive metals and refractory metals, brazing of reactive metals and refractory metals, and brazing of carbon and graphite.

Thus, this volume constitutes a substantial contribution to the welding, brazing, and soldering communities. The book provides up-to-date, reliable information and reference material for the producer and user of welded products.

The planning and organization of this volume were accomplished through the efforts of the membership of the ASM Joining Division Council, under the leadership of Peter Patriarca. I would like to thank them for their tremendous help, and I would also like to thank the American Welding Society for its cooperation and generosity.

The many contributors to this volume include representatives of all fields of welding and joining: industry, private and governmental laboratories, universities, and consulting firms. ASM expresses appreciation to them. Their knowledge, expertise, efforts, perseverance, and perspicacity have produced this prestigious volume.

Dr. Ernest F. Nippes, FASM
Coordinator
Metals Handbook Activity
Joining Division

Handbook Committee, Officers and Trustees

Members of the ASM Handbook Committee (1982-1983)

Officers and Trustees of the American Society for Metals

Previous Chairmen of the ASM Handbook Committee

R.S. Archer
(1940-1942) (Member, 1937-1942)

L.B. Case
(1931-1933) (Member, 1927-1933)

E.O. Dixon
(1952-1954) (Member, 1947-1955)

R.L. Dowdell
(1938-1939) (Member, 1935-1939)

J.P. Gill
(1937) (Member, 1934-1937)

J.D. Graham
(1965-1968) (Member, 1961-1970)

J.F. Harper
(1923-1926) (Member, 1923-1926)

C.H. Herty, Jr.
(1934-1936) (Member, 1930-1936)

J.B. Johnson
(1948-1951) (Member, 1944-1951)

R.W.E. Leiter
(1962-1963) (Member, 1955-1958, 1960-1964)

G.V. Luerssen
(1943-1947) (Member, 1942-1947)

Gunvant N. Maniar
(1978-1980) (Member, 1974-1980)

James L. McCall
(1982-1983) (Member, 1977-1983)

W.J. Merten
(1927-1930) (Member, 1923-1933)

N.E. Promisel
(1955-1961) (Member, 1954-1963)

G.J. Shubat
(1972-1975) (Member, 1966-1975)

W.A. Stadtler
(1968-1972) (Member, 1962-1972)

Raymond Ward
(1976-1978) (Member, 1972-1978)

Martin G.H. Wells
(1980-1982) (Member, 1976-1982)

D.J. Wright
(1964-1965) (Member, 1959-1967)

ASM Joining Division Council

Peter Patriarca
Chairman
Manager, Breeder Reactor Materials Program
Oak Ridge National Laboratory

Rosalie Brosilow
Editor
Welding Design & Fabrication
Penton/IPC Publications

Stan A. David
Group Leader, Welding and Brazing
Metals and Ceramics Division
Oak Ridge National Laboratory

Gene M. Goodwin
Metallurgist
Metals and Ceramics Division
Oak Ridge National Laboratory

Daniel Hauser
Senior Research Metallurgist
Battelle Columbus Laboratories

Tom Hikido
Vice President
Pyromet Industries

Wartan A. Jemian
Professor and Chairman
Department of Materials Engineering
Auburn University

J. W. Morris, Jr.
Professor
Department of Materials Science and Mineral Engineering
University of California at Berkeley

Ernest F. Nippes
Professor
Department of Materials Engineering
Rensselaer Polytechnic Institute

David LeRoy Olson
Professor
Department of Metallurgical Engineering
Colorado School of Mines

Glenn W. Oyler
Executive Director
Welding Research Council

Paul W. Ramsey
Executive Director
American Welding Society

R. David Thomas, Jr.
President
R.D. Thomas & Co., Inc.

James C. Williams
Professor
Department of Metallurgy and Materials Science
Carnegie-Mellon University

William G.G. Winship
Director
Welding Institute of Canada

John H. Zirnhelt
General Manager
Canadian Welding Bureau

Author/Reviewer Committees

Introduction to Joining

John Bartley
Vice President—Marketing
Nu-Weld/MCI
Divisions of Dimetrics, Inc.

Omer W. Blodgett
Design Consultant
The Lincoln Electric Co.

Jack H. Devletian
Associate Professor
Oregon Graduate Center

Issac S. Goodman
Consultant

Edwin J. Hemzy
Construction Manager
Commonwealth Edison Co.

L.G. Kvidahl
Chief Welding Engineer
Ingalls Shipbuilding Co.

Kenneth E. Richter
President
Nu-Weld/MCI
Divisions of Dimetrics, Inc.

Edwin Shifrin
Welding Engineer
Detroit Edison Co.

William E. Wood
Professor and Chairman
Oregon Graduate Center

Resistance Welding

John M. Gerken
Chairman
Staff Engineer
TRW Inc.

Robert S. Brown
Senior Metallurgist
Stainless Alloy Metallurgy
Carpenter Technology Corp.

Daniel A. DeAntonio
Associate Metallurgist
Corrosion and Welding Research
Carpenter Technology Corp.

David W. Dickinson
Supervisor, Welding and Flat Rolled Research & Development
Republic Steel Corp.

Robert H. Foxall
Assistant Sales Manager
Wean United, Inc.

I.A. Oehler
Retired

L. McDonald Schetky
Technical Director—Metallurgy
International Copper Research Association, Inc.

Michael J. Shields
Associate Metallurgist
Corrosion and Welding Research
Carpenter Technology Corp.

Thomas A. Siewert
Manager, Development
Alloy Rods

Lawrence W. Weller, Jr.
Engineering Supervisor of Welded Rings
The American Welding & Manufacturing Co.

Oxyfuel Gas Welding

Gene Meyer
Chairman
Manager, Product Distribution and Warranty Service
Victor Equipment

Joseph E. McQuillen
Product Manager
Air Products and Chemicals Inc.

Clarence N. Vaughn
Corporate Burning Manager
Chicago Bridge & Iron Co.

Arc Welding of Carbon Steels

Daniel Hauser
Chairman
Senior Research Metallurgist
Battelle Columbus Laboratories

Dennis R. Amos
Welding Engineer
Westinghouse Turbine Plant

Arno R. Bebernitz
Development Engineer
Caterpillar Tractor Co.

Charles W. Bosworth
Senior Specialist Engineer
Boeing Engineering & Construction Co.

Elwood Chaney
Staff Engineer
Caterpillar Tractor Co.

George E. Cook
Manager, Research & Development
CRC Welding Systems, Inc.

Craig B. Dallam
Research Assistant
Colorado School of Mines

Emet C. Dunn, Jr.
Senior Principal Engineer
Rockwell International

Glen R. Edwards
Professor of Metallurgical Engineering
Colorado School of Mines

Stanley E. Ferree
Research Engineer
Alloy Rods Division
Allegheny Ludlum Industries

G.A. Gix
Chief Welding Engineer
American Bridge Division
U.S. Steel Corp.

James R. Hannahs
President
Midwest Testing Laboratories, Inc.

R.W. Heid
Supervisor, Welding Engineering
Newport News Shipbuilding and Drydock Co.

Richard L. Hellner
Welding Engineer
Public Service Company of Colorado

Raymond T. Hemzacek
Manager, Welding Technical Services
International Harvester Co.

Franklin W. Henderson
Manager—Engineering
Richlin Equipment, Inc.

Jerald E. Jones
Assistant Professor
Department of Metallurgical Engineering
Colorado School of Mines

L.G. Kvidahl
Chief Welding Engineer
Ingalls Shipbuilding Co.

Guy A. Leclair
Manager—Welding Development Laboratory
Foster Wheeler Energy Corp.

A. Lesnewich
Director, Research & Development
Filler Metals Division
Airco Welding Products

Peter C. Levick
Manager, Process & Testing
CRC Welding Sytems, Inc.

Stephen Liu
Research Assistant
Colorado School of Mines

Kevin A. Lyttle
Laboratory Division Head
Linde Division
Union Carbide Corp.

Thomas A. Nevitt, II
Division Welding Engineer
McDermott International
Fabricators Division

Thomas H. North
Senior Research Associate
Stelco Inc.

Daniel O'Hara
Vice President—Engineering
Thermal Dynamics Corp.

David LeRoy Olson
Professor of Metallurgical Engineering
Colorado School of Mines

Abe Pollack
Head, Joining Branch
David Taylor Naval Ship Research & Development Center

Richard S. Sabo
Manager, Publicity and Educational Services
The Lincoln Electric Co.

Gregory L. Serangeli
Manager, Industrial Technology
Caterpillar Tractor Co.

Robert C. Shutt
Vice President
The Lincoln Electric Co.

James E. Sims
Manager, Welding Research & Development
Chicago Bridge & Iron Co.

John Springer
Principal Engineer
Boeing Commercial Airplane Co.

R. David Thomas, Jr.
President
R.D. Thomas & Co., Inc.

Albert E. Wiehe
Senior Research Scientist
Hobart Brothers Co.

Arc Welding of Low-Alloy Steels and Other Ferrous Metals

David LeRoy Olson
Chairman
Professor of Metallurgical Engineering
Colorado School of Mines

Robert S. Brown
Senior Metallurgist
Stainless Alloy Metallurgy
Carpenter Technology Corp.

Jon R. Bryant
Manager, Metallurgy & Welding
Otis Engineering Corp.

Charles H. Cadden
Metallurgist
Cabot Corp.

Harold Conaway
Vice President
Major Tool & Machine, Inc.

Daniel A. DeAntonio
Associate Metallurgist
Corrosion and Welding Research
Carpenter Technology Corp.

Anthony Di Nardo
Vice President—Marketing
Materials Development Corp.

Robert J. Dybas
Manager, Welding & Flame Spray
General Electric Co.

Harry W. Ebert
Engineering Associate
Exxon Research & Engineering Co.

Glen Edwards
Professor of Metallurgical Engineering
Colorado School of Mines

Richard L. Hellner
Welding Engineer
Public Service Company of Colorado

James F. Key
Group Leader
Materials Joining Section
EG&G Idaho, Inc.

S.D. Kiser
Product Manager
Welding Products Co.
Huntington Alloys, Inc.

Damian J. Kotecki
Associate Director of Research
Welding Products Division
Teledyne McKay

L.G. Kvidahl
Chief Welding Engineer
Ingalls Shipbuilding Co.

Robert B. Lazor
Group Leader, Materials Technology
Welding Institute of Canada

Leroy W. Myers, Jr.
Welding Engineer
Dresser Clark Division
Dresser Industries

William S. Ricci
Welding Engineer/Metallurgist
U.S. Army Materials and Mechanics Research Center

James M. Sawhill, Jr.
Engineer, Product Metallurgy
Homer Research Laboratories
Bethlehem Steel Corp.

Michael J. Shields
Associate Metallurgist
Corrosion and Welding Research
Carpenter Technology Corp.

Lewis E. Shoemaker
Metallurgist—Advanced
Huntington Alloys, Inc.

Thomas A. Siewert
Manager, Development
Alloy Rods

C.E. Spaeder, Jr.
Associate Research Consultant
Materials Technology Division
United States Steel Corp.

Charles J. Sponaugle
Fabrication Manager
Technical and Product Services
Cabot Corp.

Robert A. Swift
Supervisor of Metallurgical Development
Lukens Steel Co.

R. Vasudevan
Project Engineer
Linde Division
Union Carbide Corp.

Donald E. Wenschhof, Jr.
Corporate Marketing Manager
High Technology Products
A.M. Castle Co.

James T. Worman
Development Engineer
Linde Division
Union Carbide Corp.

Arc Welding of Nonferrous Metals and Alloys

William A. Baeslack
Assistant Professor
Department of Welding Engineering
Ohio State University

Donald W. Becker
Manager, C & M Laboratory
Kohler Co.

Alan H. Braun
Director of Engineering
Wellman Dynamics Corp.

Anthony J. Bryhan
Senior Research Associate
Climax Molybdenum Company of Michigan

Paul M. Chewey
Design Specialist
Lockheed Missiles & Space Co.

C.C. Clark
Manager of Market Development
Tungsten Division
AMAX of Michigan, Inc.

Paul B. Dickerson
Technical Specialist
Product Engineering Division
Alcoa Laboratories
Aluminum Company of America

John M. Gerken
Staff Engineer
TRW Inc.

William C. Hagel
Manager, High Temperature Materials Development
Climax Molybdenum Company of Michigan

Dennis F. Hasson
Associate Professor
Department of Mechanical Engineering
U.S. Naval Academy

V.L. Hill
Project Leader
Technical Service and Development
Dow Chemical U.S.A.

James L. Jellison
Supervisor, Process Metallurgy
Sandia National Laboratories

S.D. Kiser
Product Manager
Welding Products Co.
Huntington Alloys, Inc.

Arthur H. Lentz
Welding Engineer
Reynolds Metals Co.

G.K. Mathew
Engineer
Bettis Atomic Power Laboratory

Herbert Nagler
Research Engineer
Rohr Industries, Inc.

I.B. Robinson
Head, Joining Section
Kaiser Aluminum & Chemical Corp.

H.L. Saunders
Welding Engineer
Alcan International Ltd.

L. McDonald Schetky
Technical Director—Metallurgy
International Copper Research Association, Inc.

Frank Sheara
Executive Vice President
Magnesium Elektron, Inc.

Joseph H. Waibel
Group Leader
Dow Chemical U.S.A.

R. Terrence Webster
Principal Metallurgical Engineer
Teledyne Wah Chang Albany

Haskell Weiss
Welding Engineer
Lawrence Livermore National Laboratory

Charles S. Young
Manager, Metallurgical Development
Titanium Industries

H.G. Ziegenfuss
Manager, Welding and Cutting Applications
Bettis Atomic Power Laboratory

Special Welding Processes

Charles G. Albright
Assistant Professor
Department of Welding Engineering
Ohio State University

Robert Bakish
President
Bakish Materials Corp.

Conrad M. Banas
Chief, Industrial Laser Processing
United Technologies Research Center

John T. Berry
Professor
School of Mechanical Engineering
Georgia Institute of Technology

Carl B. Biro
Senior Project Engineer
Bulova Systems & Instruments Corp.

Hugh Casey
Associate Group Leader
Materials Technology
Los Alamos National Laboratory

Janet Devine
Vice President and Technical Director
Sonobond Corp.

Jeff Flynn
Materials Project Engineer
Pratt & Whitney Aircraft

Hans D. Fricke
President
U.S. Thermit Inc.

Daniel S. Gnanamuthu
Member of Technical Staff
Rockwell International

Gene M. Goodwin
Metallurgist
Metals and Ceramics Division
Oak Ridge National Laboratory

Hans Guntermann
President
Elektro-Thermit GmbH

Richard H. Jackson
Program Development Manager
Advanced Manufacturing Technology
Mechanical Technology, Inc.

James L. Jellison
Supervisor, Process Metallurgy
Sandia National Laboratories

Stanley E. Knaus
Senior Research Engineer
Rockwell International

Matthias J. Kotowski
Safety Engineer
Lawrence Livermore Laboratory

Akira Kubota
Manager, Technical Department
Explosives Division
Asahi Chemical Industry Co., Ltd.

Vonne D. Linse
Associate Section Manager
Metals Joining & Nondestructive Testing
Battelle Columbus Laboratories

Kim W. Mahin
Welding Metallurgist
Lawrence Livermore Laboratory

Edward A. Metzbower
Supervisory Metallurgist
Naval Research Laboratory

Marc A. Meyers
Associate Professor
Department of Metallurgical Engineering
New Mexico Institute of Mining and Technology

John Nurminen
Manager, Metals Joining Research
Westinghouse Research & Development Center

Edgar D. Oppenheimer
Consulting Engineer

Daniel F. Paulonis
Materials Project Engineer
Pratt & Whitney Aircraft

Charles C. Pease
Chief Metallurgist
KSM Division
Omark Industries

Andrew Pocalyko
Technical Superintendent
Detaclad Operations
E.I. Du Pont de Nemours & Co., Inc.

Donald E. Powers
Manager—Advanced Product Development
Leybold-Heraeus Vacuum Systems Inc.

Stanley L. Ream
Consultant
Laser Manufacturing Technologies Inc.

T. Renshaw
Senior Engineer
Manufacturing Technology
Fairchild Republic Co.

Wallace C. Rudd
Consulting Engineer
Thermatool Corp.

Joseph Wayne Schroeder
Senior Engineer
Materials Technology
Foster Wheeler Development Corp.

Larry C. Schroeder
Supervisor, Physical Test Laboratory
The Atchinson, Topeka, and Santa Fe Railway Co.

Thomas E. Shoup
Manager, Product Development
Nelson Division
TRW Inc.

Lyle B. Spiegel
Manager, Applications Development
Avco Everett Metalworking Lasers

Dietmar E. Spindler
General Manager
Manufacturing Technology, Inc.

Mikio Tamiyasu
Director of Engineering
Erico-Jones Co., Inc.

J. Paul Thorne
Vice President—Research & Development
Bay City Division
Newcor, Inc.

C.A. Tudbury
Consultant
Thermatool Corp.

Humfrey N. Udall
Director of Research & Development
Thermatool Corp.

Robert G. Vollmer
Chief, Production Technology Branch
Army Aviation Research & Development Command

A.S. Wadleigh
President
Interface Welding

Frank J. Wallace
Chief, Welding Development Programs
Pratt & Whitney Aircraft

Frank J. Zanner
Member of the Technical Staff
Sandia National Laboratories

Surfacing

Kenneth C. Antony
Manager, Technical Services
Wear Technology Division
Cabot Corp.

J.D. Ayers
Metallurgist
Naval Research Laboratory

J.J. Barger
Supervisor, Prototype Equipment
Combustion Engineering, Inc.

Kirit J. Bhansali
Group Leader, Chemical Metallurgy
National Bureau of Standards

Howard N. Farmer
Director, Metallurgical Services
Stoody Co.

Daniel S. Gnanamuthu
Member of Technical Staff
Rockwell International

Jerald E. Jones
Assistant Professor
Department of Metallurgical Engineering
Colorado School of Mines

William E. Layo
Manager of Technical Services
Sandvik, Inc.

Steven J. Matthews
Group Leader—Welding
Wear Technology Division
Cabot Corp.

Robert W. Messler, Jr.
Director of Welding Research & Development
Eutectic Corp.

John J. Meyer
Chief Welding Engineer
Nooter Corp.

Albert E. Miller
Professor of Metallurgical Engineering and Materials Science
University of Notre Dame

Marianne O. Price
Senior Project Engineer
Engineering Products Division
Union Carbide Corp.

Louis E. Stark
Welding Engineer
Babcock & Wilcox

R. David Thomas, Jr.
President
R.D. Thomas & Co., Inc.

R.C. Tucker, Jr.
Associate Director, Materials Development
Coatings Service Department
Union Carbide Corp.

Other Welding Topics

Michael F. Ahern
Engineer
Northeast Utilities

Donald A. Bolstad
Section Chief, Materials Engineering
Martin Marietta Aerospace

Rosalie Brosilow
Editor
Welding Design & Fabrication
Penton/IPC Publications

Alfred E. Burnell
Development Engineer
Harris Calorific

Spencer H. Bush
Senior Staff Consultant
Battelle Pacific Northwest Laboratory

John C. Duke, Jr.
Associate Professor
Engineering Science and Mechanics
Virginia Polytechnic Institute and State University

Richard W. Jacobus
Manager of Marketing Communications
Hypertherm Inc.

Ben Jezek
Product Planning Manager
Victor Equipment

Maurice B. Kasen
Metallurgist
Fracture and Deformation Division
National Bureau of Standards

John Koziol
Manager, Systems Materials
Nuclear Power Systems Division
Combustion Engineering, Inc.

L.G. Kvidahl
Chief Welding Engineer
Ingalls Shipbuilding Co.

Koichi Masubuchi
Professor of Ocean Engineering and Materials Science
Massachusetts Institute of Technology

William H. Munse
Professor Emeritus of Civil Engineering
University of Illinois

Alan W. Pense
Professor of Metallurgy and Materials Engineering
Lehigh University

Clayton O. Ruud
Senior Research Associate
Pennsylvania State University

L.W. Sandor
Manager of Materials Technology
Franklin Research Center
Professor of Materials Engineering
Widener University

E.A. Silva
Division Leader
Technology & Support Division
Office of Naval Research

Brazing

Robert L. Peaslee
Chairman
Vice President
Wall Colmonoy Corp.

G.A. Andreano
Manufacturing Technology Engineer
Garrett Turbine Engine Co.

Pat Capolongo
Manager, Technical Sales
Lepel Corp.

Edward J. Cove
Senior Joining & Coatings Engineer
General Electric Co.

Austin Dixon
Consultant, Process Design
Metals Joining Technology

Jeff Flynn
Materials Project Engineer
Pratt & Whitney Aircraft

Gary W. Gaines
Project Engineer
Linde Division
Union Carbide Corp.

John M. Gerken
Staff Engineer
TRW Inc.

Tom Hikido
Vice President
Pyromet Industries

Thomas J. Kelly
Senior Metallurgist
International Nickel Co.

C.R. Kennedy
Research Staff Member
Oak Ridge National Laboratory

Arthur H. Lentz
Welding Engineer
Reynolds Metals Co.

Joseph A. Lincoln
Vice President—Technology
Atmosphere Furnace Co.

Quentin D. Mehrkam
Senior Vice President
Ajax Electric Co.

A.J. Moorhead
Group Leader
Oak Ridge National Laboratories

C. Philp
Product Engineer—Brazing
Handy & Harman

Martin Prager
Consultant

Edward J. Ryan
Senior Metallurgical Engineer
Pratt & Whitney Aircraft

Mel Schwartz
Manager, Manufacturing Technology
Sikorsky Aircraft

G.M. Slaughter
Manager, Engineering Materials Section
Oak Ridge National Laboratory

Charles J. Sponaugle
Fabrication Manager
Technical and Product Services
Cabot Corp.

C. Van Dyke
Brazing Consultant
Lucas-Milhaupt, Inc.

S.J. Whalen
President
Aerobraze Corp.

G.F. Willhauck
Vice President—Marketing
Atmosphere Furnace Co.

Charles E. Witherell
Leader, Metal Fabrication
Consulting Engineer
Lawrence Livermore National Laboratory

William G. Wood
Vice President—Technology/Research & Development
Kolene Corp.

Soldering

Roy E. Beal
Chairman
President
Amalgamated Technologies, Inc.

Paul J. Bud
President
Electrovert Consulting Services

William B. Hampshire
Assistant Manager
Tin Research Institute, Inc.

Jerome F. Smith
Vice President
Lead Industries Association, Inc.

C.E.T. White
Executive Vice President
The Indium Corporation of America

Contents

Glossary

A

abrasion soldering. A soldering process variation in which the faying surface of the base metal is mechanically abraded during soldering.

abrasive. Material such as sand, crushed chilled cast iron, crushed steel grit, aluminum oxide, silicon carbide, flint, garnet, or crushed slag used for cleaning or surface roughening.

abrasive blasting. See preferred term *blasting*.

absorptive lens (eye protection). A filter lens whose physical properties are designed to attenuate the effects of glare and reflected and stray light. See also *filter plate*.

accelerating potential. The potential in electron beam welding that imparts the velocity to the electrons, thus giving them energy.

acceptable weld. A weld that meets all the requirements and the acceptance criteria prescribed by the welding specifications.

acid core solder. See *cored solder*.

activated rosin flux. A rosin or resin base flux containing an additive which increases wetting by the solder.

actual throat. See *throat of a fillet weld*.

adhesive bond. Attractive forces, generally physical in character, between an adhesive and the base materials. Two principal interactions that contribute to the adhesion are van der Waals bonds and dipole bonds. See *van der Waals bond* and *permanent dipole bond*.

adhesive bonding (ABD). A materials joining process in which an adhesive, placed between the faying surfaces, solidifies to produce an adhesive bond. See *adhesive bond*.

air acetylene welding (AAW). A fuel gas welding process in which coalescence is produced by heating with a gas flame or flames obtained from the combustion of acetylene with air, without the application of pressure, and with or without the use of filler metal.

air cap (thermal spraying). A device for forming, shaping, and directing an air pattern for the atomization of wire or ceramic rod.

air carbon arc cutting (AAC). An arc cutting process in which metals to be cut are melted by the heat of a carbon arc and the molten metal is removed by a blast of air.

air feed. A thermal spraying process variation in which an air stream carries the powdered material to be sprayed through the gun and into the heat source.

all-weld-metal test specimen. A test specimen with the reduced section composed wholly of weld metal.

alternate polarity operation. A resistance welding process variation in which succeeding welds are made with pulses of alternating polarity.

angle of bevel. See preferred term *bevel angle*.

arc blow. The deflection of an electric arc from its normal path because of magnetic forces.

arc brazing (AB). A brazing process in which the heat required is obtained from an electric arc. See *twin carbon arc brazing*.

arc cutting (AC). A group of cutting processes which melts the metals to be cut with the heat of an arc between an electrode and the base metal. See *carbon arc cutting, metal arc cutting, gas metal arc cutting, gas tungsten arc cutting, plasma arc cutting,* and *air carbon arc cutting*. Compare with *oxygen arc cutting*.

arc force. The axial force developed by a plasma.

arc gouging. An arc cutting process variation used to form a bevel or groove.

arc oxygen cutting. See preferred term *oxygen arc cutting*.

arc plasma. See *plasma*.

arc seam weld. A seam weld made by an arc welding process.

arc spot weld. A spot weld made by an arc welding process.

arc strike. A discontinuity consisting of any localized remelted metal, heat-affected metal, or change in the surface profile of any part of a weld or base metal resulting from an arc.

arc time. The time during which an arc is maintained in making an arc weld.

arc voltage. The voltage across the welding arc.

arc welding (AW). A group of welding processes which produces coalescence of metals by heating them with an arc, with or without the application of pressure, and with or without the use of filler metal.

arc welding electrode. A component of the welding circuit through which current is conducted between the electrode holder and the arc. See *arc welding*.

arc welding gun. A device used in semiautomatic, machine, and automatic arc welding to transfer current, guide the consumable electrode, and direct shielding gas when used.

arm (resistance welding). A projecting beam extending from the frame of a resistance welding machine which transmits the electrode force and may conduct the welding current.

as-brazed. The condition of brazements after brazing, prior to any subsequent thermal, mechanical, or chemical treatments.

as-welded. The condition of weld metal, welded joints, and weldments after welding but prior to any subsequent thermal, mechanical, or chemical treatments.

atomic hydrogen welding (AHW). An arc welding process which produces coalescence of metals by heating them with an electric arc maintained between two metal electrodes in an atmosphere of hydrogen. Shielding is obtained from the hydrogen. Pressure may or may not be used and filler metal may or may not be used.

atomization (thermal spraying). The division of molten material at the end of the wire or rod into fine particles.

autogenous weld. A fusion weld made without the addition of filler metal.

automatic brazing. Brazing with equipment which performs the brazing operation without constant observation and adjustment by a brazing operator. The equipment may or may not perform the loading and unloading of the work. See *machine brazing*.

automatic gas cutting. See preferred term *automatic oxygen cutting*.

automatic oxygen cutting. Oxygen cutting with equipment which performs the cutting operation without constant observation and adjustment of the controls by an operator. The equipment may or may not perform loading and unloading of the work. See *machine oxygen cutting*.

automatic welding. Welding with equipment which performs the welding operation without adjustment of the controls by a welding operator. The equipment may or may not perform the loading and unloading of the work. See *machine welding*.

auxiliary magnifier or enlarger (eye protection). An additional lens or plate, associated with eye protection equipment, used to magnify or enlarge the field of vision.

axis of a weld. A line through the length of a weld, perpendicular to and at the geometric center of its cross section.

B

back bead. See preferred term *back weld*.

backfire. The momentary recession of the flame into the welding tip or cutting tip followed by immediate reappearance or complete extinction of the flame.

back gouging. The removal of weld metal and base metal from the other side of a partially welded joint to ensure complete penetration upon subsequent welding from that side.

backhand welding. A welding technique in which the welding torch or gun is directed opposite to the progress of welding. Sometimes referred to as the "pull gun technique" in GMAW and FCAW. See *travel angle, work angle*, and *drag angle*.

backing. A material (base metal, weld metal, carbon, or granular material) placed at the root of a weld joint for the purpose of supporting molten weld metal.

backing bead. See preferred term *backing weld*.

backing filler metal. See *consumable insert*.

backing pass. A pass made to deposit a backing weld.

backing ring. Backing in the form of a ring, generally used in the welding of piping.

backing-split pipe. See *split pipe backing*.

backing strap. See preferred term *backing strip*.

backing strip. Backing in the form of a strip.

backing weld. Backing in the form of a weld.

backstep sequence. A longitudinal sequence in which the weld bead increments are deposited in the direction opposite to the progress of welding the joint. See *block sequence, cascade sequence, continuous sequence, joint building sequence*, and *longitudinal sequence*.

backup (flash and upset welding). A locator used to transmit all or a portion of the upsetting force to the workpieces or to aid in preventing the workpieces from slipping during upsetting.

back weld. A weld deposited at the back of a single groove weld.

balling up. The formation of globules of molten brazing filler metal or flux due to lack of wetting of the base metal.

bare electrode. A filler metal electrode consisting of a single metal or alloy that has been produced into a wire, strip, or bar form and that has had no coating or covering applied to it other than that which was incidental to its manufacture or preservation.

bare metal arc welding (BMAW). An arc welding process which produces coalescence of metals by heating them with an electric arc between a bare or lightly coated metal electrode and the work. Neither shielding nor pressure is used and filler metal is obtained from the electrode.

base material. The material to be welded, brazed, soldered, or cut. See also *base metal* and *substrate*.

base metal. The metal to be welded, brazed, soldered, or cut. The use of this term implies that materials other than metals are also referred to, where this is appropriate. See also *base material* and *substrate*.

base metal test specimen. A test specimen composed wholly of base metal.

bead. See preferred term *weld bead*.

bead weld. See preferred term *surfacing weld*.

bevel. An angular type of edge preparation.

bevel angle. The angle formed between the prepared edge of a member and a plane perpendicular to the surface of the member.

bevel groove. See *groove weld*.

bit (soldering). That part of the soldering iron, usually made of copper, which actually transfers heat (and sometimes solder) to the joint.

bit soldering. See preferred term *iron soldering*.

blacksmith welding. See preferred term *forge welding*.

blasting. A method of cleaning or surface roughening by a forcibly projected stream of sharp angular abrasive.

blind joint. A joint, no portion of which is visible.

block brazing (BB). A brazing process in which the heat required is obtained from heated blocks applied to the parts to be joined.

block sequence. A combined longitudinal and buildup sequence for a continuous multiple-pass weld in which separated lengths are completely or partially built up in cross section before intervening lengths are deposited. See also *backstep sequence, longitudinal sequence*, etc.

blowhole. See preferred term *porosity*.

blowpipe.

welding and cutting blowpipe. See preferred term *welding torch* or *cutting torch*.

brazing and soldering blowpipe. A device used to obtain a small, accurately directed flame for fine work, such as in the dental and jewelry trades. Any flame may be used, a portion of it being blown to the desired location for the required time by the blowpipe which is usually mouth operated.

bond. See *adhesive bond, mechanical bond*, and *metallic bond*.

bond coat (thermal spraying). A preliminary (or prime) coat of material which improves adherence of the subsequent thermal spray deposit.

bonding force. The force that holds two atoms together; it results from a decrease in energy as two atoms are brought closer to one another.

bond line. The cross section of the interface between thermal spray deposits and substrate, or the interface between adhesive and adherend in an adhesive bonded joint.

bottle. See preferred term *cylinder*.

boxing. The continuation of a fillet weld

around a corner of a member as an extension of the principal weld.

braze. A weld produced by heating an assembly to suitable temperatures and by using a filler metal having a liquidus above 840 °F and below the solidus of the base metal. The filler metal is distributed between the closely fitted faying surfaces of the joint by capillary action.

brazeability. The capacity of a metal to be brazed under the fabrication conditions imposed into a specific suitably designed structure and to perform satisfactorily in the intended service.

braze interface. See *weld interface*.

brazement. An assembly whose component parts are joined by brazing.

brazer. One who performs a manual or semi-automatic brazing operation.

braze welding. A welding process variation in which a filler metal, having a liquidus above 840 °F and below the solidus of the base metal, is used. Unlike brazing, in braze welding the filler metal is not distributed in the joint by capillary action.

brazing (B). A group of welding processes which produces coalescence of materials by heating them to a suitable temperature and by using a filler metal having a liquidus above 840 °F and below the solidus of the base metal. The filler metal is distributed between the closely fitted faying surfaces of the joint by capillary action.

brazing alloy. See preferred term *brazing filler metal*.

brazing filler metal. The metal which fills the capillary gap and has a liquidus above 840 °F but below the solidus of the base materials.

brazing operator. One who operates machine or automatic brazing equipment.

brazing procedure. The detailed methods and practices including all joint brazing procedures involved in the production of a brazement. See *joint brazing procedure*.

brazing sheet. Brazing filler metal in sheet form.

brazing technique. The details of a brazing operation which, within the limitations of the prescribed brazing procedure, are controlled by the brazer or the brazing operator.

brazing temperature. The temperature to which the base metal is heated to enable the filler metal to wet the base metal and form a brazed joint.

brazing temperature range. The temperature range within which brazing can be conducted.

bridge size (eye protection). The distance between lenses on the nose side of each eye, expressed in millimeters.

bronze welding. A term erroneously used to denote braze welding. See *braze welding*.

buildup sequence. See *joint buildup sequence*.

burnback time. See preferred term *meltback time*.

burner. See preferred term *oxygen cutter*.

burning. See preferred term *oxygen cutting*.

burning in. See preferred term *flow welding*.

burnoff rate. See preferred term *melting rate*.

burn-thru. A term erroneously used to denote excessive melt-thru or a hole. See *melt-thru*.

burn-thru weld. A term erroneously used to denote a seam weld or spot weld.

buttering. A surfacing variation in which one or more layers of weld metal are deposited on the groove face of one member (for example, a high alloy weld deposit on steel base metal which is to be welded to a dissimilar base metal). The buttering provides a suitable transition weld deposit for subsequent completion of the butt joint.

butt joint. A joint between two members aligned approximately in the same plane.

button. That part of a weld, including all or part of the nugget, which tears out in the destructive testing of spot, seam, or projection welded specimens.

butt weld. An erroneous term for a weld in a butt joint. See *butt joint*.

C

capillary action. The force by which liquid, in contact with a solid, is distributed between closely fitted faying surfaces of the joint to be brazed or soldered.

carbon arc cutting (CAC). An arc cutting process in which metals are severed by melting them with the heat of an arc between a carbon electrode and the base metal.

carbon arc welding (CAW). An arc welding process which produces coalescence of metals by heating them with an arc between a carbon electrode and the work. No shielding is used. Pressure and filler metal may or may not be used.

carbon electrode. A non-filler material electrode used in arc welding or cutting, consisting of a carbon or graphite rod, which may be coated with copper or other coatings.

carbonizing flame. See preferred term *reducing flame*.

carburizing flame. See preferred term *reducing flame*.

carrier gas (thermal spraying). The gas used to carry powdered materials from the powder feeder or hopper to the gun.

cascade sequence. A combined longitudinal and buildup sequence during which weld beads are deposited in overlapping layers. See also *backstep sequence, block sequence, buildup sequence,* and *longitudinal sequence*.

caulk weld. See preferred term *seal weld*.

ceramic rod flame spraying. A thermal spraying process variation in which the material to be sprayed is in ceramic rod form. See *flame spraying (FLSP)*.

chain intermittent welds. Intermittent welds on both sides of a joint in which the weld increments on one side are approximately opposite those on the other side.

chamfer. See preferred term *bevel*.

chemical flux cutting (FOC). An oxygen cutting process in which metals are severed using a chemical flux to facilitate cutting.

chill ring. See preferred term *backing ring*.

chill time. See preferred term *quench time*.

circular electrode. See *electrode—resistance welding electrode*.

circular resistance seam welding. See preferred term *transverse resistance seam welding*.

circumferential resistance seam welding. See preferred term *transverse resistance seam welding*.

clad brazing sheet. A metal sheet on which one or both sides are clad with brazing filler metal.

clad metal. A composite metal containing two or three layers that have been welded together. The welding may have been accomplished by roll welding, arc welding, casting, heavy chemical deposition, or heavy electroplating. See *cladding* and *surfacing*.

cladding. A relatively thick layer (>0.04 in.) of material applied by surfacing for the purpose of improved corrosion resistance or other properties (see *coating, surfacing,* and *hardfacing*).

coalescence. The growing together or growth into one body of the materials being welded.

coated electrode. See preferred terms *covered electrode* and *lightly coated electrode*. See *electrode*.

coating. A relatively thin layer (<0.04 in.) of material applied by surfacing for the purpose of corrosion prevention, resistance to high temperature scaling, wear resistance, lubrication, or other purposes. See *cladding, surfacing,* and *hardfacing*.

coating density (thermal spraying). The ratio of the determined density of a thermal sprayed coating to the theoretical density of the material used in the coating process. Usually expressed as percent of theoretical density.

coextrusion welding (CEW). A solid-state welding process which produces coalescence of the faying surfaces by heating and forcing materials through an extrusion die.

coil without support. A filler metal package type consisting of a continuous length of electrode in coil form without an internal support. It is appropriately bound to maintain its shape.

coil with support. A filler metal package type consisting of a continuous length of electrode in coil form wound on an internal support which is a simple cylindrical section without flanges.

cold soldered joint. A joint with incomplete coalescence caused by insufficient application of heat to the base metal during soldering.

cold welding (CW). A solid-state welding process in which pressure is used at room temperature to produce coalescence of metals with substantial deformation at the weld. Compare *hot pressure welding, diffusion welding,* and *forge welding*.

collar. The reinforcing metal of a nonpressure thermit weld.

collaring (thermal spraying). Adding a shoulder to a shaft or similar component as a protective confining wall for the thermal spray deposit.

commutator-controlled welding (resistance welding). The making of a number of spot or projection welds. Several electrodes, in simultaneous contact with the work, function progressively under the control of an electrical commutating device.

complete fusion. Fusion which has occurred over the entire base material surfaces intended for welding and between all layers and weld beads.

complete joint penetration. Joint penetration in which the weld metal completely fills the groove and is fused to the base metal throughout its total thickness.

complete penetration. See preferred term *complete joint penetration*.

composite electrode. Any of a number of multicomponent filler metal electrodes in various physical forms such as stranded wires, tubes, and covered wire. See *covered electrode, flux cored electrode, metal cored electrode,* and *stranded electrode*.

composite joint. A joint produced by welding used in conjunction with a nonwelding process. See *weldbonding*.

concave fillet weld. A fillet weld having a concave face.

concave root surface. A root surface which is concave.

concavity. The maximum distance from the face of a concave fillet weld perpendicular to a line joining the toes.

concurrent heating. The application of supplemental heat to a structure during a welding or cutting operation.

cone. The conical part of an oxyfuel gas flame next to the orifice of the tip.

constricted arc (plasma arc welding and cutting). A plasma arc column that is shaped by a constricting nozzle orifice.

constricting nozzle (plasma arc welding and cutting). A water-cooled copper nozzle surrounding the electrode and containing the constricting orifice.

constricting orifice (plasma arc welding and cutting). The hole in the constricting nozzle through which the arc passes.

consumable guide electroslag welding. An electroslag welding process variation in which filler metal is supplied by an electrode and its guiding member. See *electroslag welding (ESW)*.

consumable insert. Preplaced filler metal which is completely fused into the root of the joint and becomes part of the weld.

contact tube. A device which transfers current to a continuous electrode.

continuous sequence. A longitudinal sequence in which each pass is made continuously from one end of the joint to the other. See *backstep sequence, longitudinal sequence,* etc.

continuous weld. A weld which extends continuously from one end of a joint to the other. Where the joint is essentially circular, it extends completely around the joint.

convex fillet weld. A fillet weld having a convex face.

convex root surface. A root surface which is convex.

convexity. The maximum distance from the face of a convex fillet weld perpendicular to a line joining the toes.

cool time (resistance welding). The time interval between successive heat times in multiple-impulse welding or in the making of seam welds.

copper brazing. A term improperly used to denote brazing with a copper filler metal. See preferred terms *furnace brazing* and *braze welding*.

cored solder. A solder wire or bar containing flux as a core.

corner-flange weld. A flange weld with only one member flanged at the location of welding.

corner joint. A joint between two members located approximately at right angles to each other.

corona (resistance welding). The area sometimes surrounding the nugget of a spot weld at the faying surfaces which provides a degree of solid-state welding.

corrective lens (eye protection). A lens ground to the wearer's individual corrective prescription.

corrosive flux. A flux with a residue that chemically attacks the base metal. It may be composed of inorganic salts and acids, organic salts and acids, or activated rosins or resins.

cosmetic pass. A weld pass made primarily for the purpose of enhancing appearance.

CO_2 welding. See preferred term *gas metal arc welding*.

covalent bond. A primary bond arising from the reduction in energy associated with overlapping half-filled orbitals of two atoms.

covered electrode. A composite filler metal electrode consisting of a core of a bare electrode or metal cored electrode to which a covering sufficient to provide a slag layer on the weld metal has been applied. The covering may contain materials providing such functions as shielding from the atmosphere, deoxidation, and arc stabilization and can serve as a source of metallic additions to the weld.

cover lens (eye protection). A round cover plate.

cover plate (eye protection). A removable pane of colorless glass, plastic-coated glass, or plastic that covers the filter plate and protects it from weld spatter, pitting, or scratching when used in a helmet, hood, or goggles.

crack. A fracture type discontinuity characterized by a sharp tip and high ratio of length and width to opening displacement.

crater. In arc welding, a depression at the termination of a weld bead or in the molten weld pool.

crater crack. A crack in the crater of a weld bead.

crater fill current. The current value during crater fill time.

crater fill time. The time interval following weld time but prior to meltback time during which arc voltage or current reaches a preset value greater or less than welding values. Weld travel may or may not stop at this point.

crater fill voltage. The arc voltage value during crater fill time.

cross wire weld. A weld made between crossed wires or bars.

cup. See preferred term *nozzle*.

cutting attachment. A device for converting an oxyfuel gas welding torch into an oxygen cutting torch.

cutting head. The part of a cutting machine or automatic cutting equipment in which a cutting torch or tip is incorporated.

cutting nozzle. See preferred term *cutting tip*.

cutting process. A process which brings about the severing or removal of metals. See *arc cutting* and *oxygen cutting*.

cutting tip. That part of an oxygen cutting torch from which the gases issue.

cutting torch (oxyfuel gas). A device used for directing the preheating flame produced by the controlled combustion of fuel gases and to direct and control the cutting oxygen.

cutting torch (plasma arc). A device used for plasma arc cutting to control the position of the electrode, to transfer current to the arc, and to direct the flow of plasma and shielding gas.

cylinder. A portable container used for transportation and storage of a compressed gas.

cylinder manifold. See preferred term *manifold*.

D

defect. A discontinuity or discontinuities which by nature or accumulated effect (for example, total crack length) render a part or product unable to meet minimum applicable acceptance standards or specifications. This term designates rejectability. See *discontinuity* and *flaw*.

defective weld. A weld containing one or more defects.

deposit. (thermal spraying). See preferred term *spray deposit*.

deposited metal. Filler metal that has been added during a welding operation.

deposition efficiency (arc welding). The ratio of the weight of deposited metal to the net weight of filler metal consumed, exclusive of stubs.

deposition efficiency (thermal spraying). The ratio, usually expressed in percent, of the weight of spray deposit to the weight of the material sprayed.

deposition rate. The weight of material deposited in a unit of time. It is usually expressed as pounds per hour (lb/h).

deposition sequence. The order in which the increments of weld metal are deposited. See *longitudinal sequence*, *buildup sequence*, and *pass sequence*.

deposit sequence. See preferred term *deposition sequence*.

depth of fusion. The distance that fusion extends into the base metal or previous pass from the surface melted during welding.

detonation flame spraying. A thermal spraying process variation in which the controlled explosion of a mixture of fuel gas, oxygen, and powdered coating material is utilized to melt and propel the material to the workpiece.

die.

resistance welding die. A member usually shaped to the work contour to clamp the parts being welded and to conduct the welding current.

forge welding die. A device used in forge welding primarily to form the work while hot and apply the necessary pressure.

die welding. See preferred terms *forge welding* and *cold welding*.

diffusion aid. A solid filler metal applied to the faying surfaces to assist in diffusion welding.

diffusion bonding. See preferred terms *diffusion brazing* and *diffusion welding*.

diffusion brazing (DFB). A brazing process which produces coalescence of metals by heating them to suitable temperatures and by using a filler metal or an *in situ* liquid phase. The filler metal may be distributed by capillary action or may be placed or formed at the faying surfaces. The filler metal is diffused with the base metal to the extent that the joint properties have been changed to approach those of the base metal. Pressure may or may not be applied.

diffusion welding (DFW). A solid-state welding process which produces coalescence of the faying surfaces by the application of pressure at elevated temperature. The process does not involve macroscopic deformation, melting, or relative motion of parts. A solid filler metal (diffusion aid) may or may not be inserted between the faying surfaces. See also *forge welding, hot pressure welding*, and *cold welding*.

dilution. The change in chemical composition of a welding filler metal caused by the admixture of the base metal or previously deposited weld metal in the deposited weld bead. It is normally measured by the percentage of base metal or previously deposited weld metal in the weld bead.

dip brazing (DB). A brazing process in which the heat required is furnished by a molten chemical or metal bath. When a molten chemical bath is used, the bath may act as a flux. When a molten metal bath is used, the bath provides the filler metal.

dip soldering (DS). A soldering process in which the heat required is furnished by a molten metal bath which provides the solder filler metal.

direct current electrode negative (DCEN). The arrangement of direct current arc welding leads in which the work is the positive pole and the electrode is the negative pole of the welding arc. See also *straight polarity*.

direct current electrode positive (DCEP). The arrangement of direct current arc welding leads in which the work is the negative pole and the electrode is the positive pole of the welding arc. See also *reverse polarity*.

direct current reverse polarity (DCRP). See *reverse polarity* and *direct current electrode positive*.

direct current straight polarity (DCSP). See *straight polarity* and *direct current electrode negative*.

discontinuity. An interruption of the typical structure of a weldment, such as a lack of homogeneity in the mechanical, metallurgical, or physical characteristics of the material or weldment. A discontinuity is not necessarily a defect. See *defect* and *flaw*.

doped solder. A solder containing a small amount of an element intentionally added to ensure retention of one or more characteristics of the materials on which it is used.

double arcing (plasma arc welding and cutting). A condition in which the main arc does not pass through the constricting orifice but transfers to the inside surface of the nozzle. A secondary arc is simultaneously established between the outside surface of the nozzle and the workpiece. Double arcing usually damages the nozzle.

double-welded joint. In arc and oxyfuel gas welding, any joint welded from both sides.

dovetailing (thermal spraying). A method of surface roughening involving angular undercutting to interlock the spray deposit.

downhand. See preferred term *flat position*.

downslope time (resistance welding). The time during which the welding current is continuously decreased.

downslope time (automatic arc welding). The time during which the current is changed continuously from final taper current or welding current to final current.

drag (thermal cutting). The offset distance between the actual and the theoretical exit points of the cutting oxygen stream measured on the exit surface of the material.

drag angle. The travel angle when the electrode is pointing backward. See also *backhand welding*. Note: This angle can be used to define the position of welding guns, welding torches, high energy beams, welding rods, and thermal cutting and thermal spraying guns.

drop-thru. An undesirable sagging or surface irregularity, usually encountered when brazing or welding near the solidus of the base metal, caused by overheating with rapid diffusion or alloying between the filler metal and the base metal.

drum. A filler metal package type consisting of a continuous length of electrode wound or coiled within an enclosed cylindrical container.

duty cycle. The percentage of time during an arbitrary test period, usually 10 min, during which a power supply can be operated at its rated output without overloading.

dynamic electrode force. See *electrode force*.

E

edge-flange weld. A flange weld with two members flanged at the location of welding.

edge joint. A joint between the edges of two or more parallel or nearly parallel members.

edge preparation. The surface prepared on the edge of a member for welding.

edge weld. A weld in an edge joint.

effective length of weld. The length of weld throughout which the correctly proportioned cross section exists. In a curved weld, it shall be measured along the axis of the weld.

effective throat. The minimum distance from the root of a weld to its face less any reinforcement. See also *joint penetration*.

electric arc spraying (EASP) (thermal spraying). A process using as a heat source an electric arc between two consumable electrodes of a coating material and a compressed gas which is used to atomize and propel the material to the substrate.

electric bonding (thermal spraying). See preferred term *surfacing*.

electric brazing. See preferred terms *resistance brazing* and *arc brazing*.

electrode. A component of the welding circuit through which current is conducted to the arc, molten slag, or base metal. See *arc welding electrode, bare electrode, carbon electrode, composite electrode, covered electrode, electroslag welding electrode, emissive electrode, flux cored electrode, lightly coated electrode, metal cored electrode, metal electrode, resistance welding electrode, stranded electrode,* and *tungsten electrode*.

electrode extension (GMAW, FCAW, SAW). The length of unmelted electrode extending beyond the end of the contact tube during welding.

electrode force.

dynamic electrode force (resistance welding). The force in newtons (pounds force) between the electrodes during the actual welding cycle in making spot, seam, or projection welds by resistance welding.

static electrode force (resistance welding). The force between the electrodes in making spot, seam, or projection welds by resistance welding under welding conditions, but with no current flowing and no movement in the welding machine.

theoretical electrode force (resistance welding). The force, neglecting friction and inertia, in making spot, seam, or projection welds by resistance welding, available at the electrodes of a resistance welding machine by virtue of the initial force application and the theoretical mechanical advantage of the system.

electrode holder. A device used for mechanically holding the electrode while conducting current to it.

electrode lead. The electrical conductor between the source of arc welding current and the electrode holder.

electrode setback (plasma arc welding and cutting). The distance the electrode is recessed behind the constricting orifice measured from the outer face of the nozzle.

electrode skid (resistance welding). The sliding of an electrode along the surface of the work during the making of spot, seam, or projection welds by resistance welding.

electrogas welding (EGW). An arc welding process which produces coalescence of metals by heating them with an arc between a continuous filler metal (consumable) electrode and the work. Molding shoe(s) are used to confine the molten weld metal for vertical position welding. The electrodes may be either flux cored or solid. Shielding may or may not be obtained from an externally supplied gas or mixture.

electron beam cutting (EBC). A cutting process which uses the heat obtained from a concentrated beam composed primarily of high velocity electrons which impinge upon the workpieces to be cut; it may or may not use an externally supplied gas.

electron beam gun. A device for producing and accelerating electrons. Typical components include the emitter (also called the filament or cathode) which is heated to produce electrons via thermionic emission, a cup (also called the grid or grid cup), and the anode.

electron beam gun column. The electron beam gun plus auxiliary mechanical and electrical components which may include beam alignment, focus, and deflection coils.

electron beam welding (EBW). A welding process which produces coalescence of metals with the heat obtained from a concentrated beam composed primarily of high velocity electrons impinging upon the joint. See *electron beam welding—high vacuum, electron beam welding—medium vacuum,* and *electron beam welding—nonvacuum*.

electron beam welding—high vacuum (EBW-HV). An electron beam welding process variation in which welding is accomplished at a pressure of approximately 10^{-6} to 10^{-3} torr.

electron beam welding—medium vacuum (EBW-MV). An electron beam welding process variation in which welding is accomplished at a pressure of approximately 10^{-3} to 25 torr. The term "medium vacuum" encompasses the range of pressures often referred to as "soft," and "partial vacuum."

electron beam welding—nonvacuum (EBW-NV). An electron beam welding process variation in which welding is accomplished at atmospheric pressure.

electronic heat control (resistance welding). A device for adjusting the heating value (rms value) of the current in making a resistance weld by controlling the ignition or firing of the electronic devices in an electronic contactor. The current is initiated each half-cycle at an adjustable time with respect to the zero point on the voltage wave.

electroslag welding (ESW). A welding process producing coalescence of metals with molten slag which melts the filler metal and the surfaces of the work to be welded. The molten weld pool is shielded by this slag which moves along the full cross section of the joint as welding progresses. The process is initiated by an arc which heats the slag. The arc is then extinguished and the conductive slag is maintained in a molten condition by its resistance to electric current passing between the electrode and the work. See *electroslag welding electrode* and *consumable guide electroslag welding*.

electroslag welding electrode. A filler metal component of the welding circuit through which current is conducted between the electrode guiding member and the molten slag.

emissive electrode. A filler metal electrode consisting of a core of a bare electrode or a composite electrode to which a very light coating has been applied to produce a stable arc.

end return.. See preferred term *boxing*.

erosion (brazing). A condition caused by dissolution of the base metal by molten filler metal resulting in a postbraze reduction in the thickness of the base metal.

exhaust booth. A mechanically ventilated, semi-enclosed area in which an air flow across the work area is used to remove fumes, gases, and material particles.

explosion welding (EXW). A solid-state welding process in which coalescence is effected by high-velocity movement together of parts to be joined produced by a controlled detonation.

eye size (eye protection). The nominal size of the lens-holding section of an eye frame expressed in millimeters.

F

face feed. The application of filler metal to the joint, usually by hand, during brazing and soldering.

face of weld. The exposed surface of a weld on the side from which welding was done.

face reinforcement. Reinforcement of weld at the side of the joint from which welding was done. See also *root reinforcement*.

face shield (eye protection). A device positioned in front of the eyes and a portion of, or all of, the face, whose predominant function is protection of the eyes and face. See also *hand shield* and *helmet*.

faying surface. That mating surface of a member which is in contact or in close proximity to another member to which it is to be joined.

feed rate (thermal spraying). The rate at which material passes through the gun in a unit of time. A synonym for *spray rate*.

ferrite number. An arbitrary, standardized value designating the ferrite content of an austenitic stainless steel weld metal. It should be used in place of percent ferrite or volume percent ferrite on a direct one-to-one replacement basis. See the latest edition of AWS A4.2, Standard Procedures for Calibrating Magnetic Instruments to Measure the Delta Ferrite Content of Austenitic Stainless Steel Weld Metal.

filler metal. The metal to be added in making a welded, brazed, or soldered joint. See *electrode, welding rod, backing filler metal, brazing filler metal, diffusion aid, solder*, and *spray deposit*.

filler metal start delay time. The time interval from arc initiation to the start of filler metal feeding.

filler metal stop delay time. The time delay interval from beginning of downslope time to the stop of filler metal.

fillet weld. A weld of approximately triangular cross section joining two surfaces approximately at right angles to each other in a lap joint, T-joint, or corner joint.

fillet weld size. See preferred term *size of weld*.

filter glass. See preferred term *filter plate*.

filter lens (eye protection). A round filter plate.

filter plate (eye protection). An optical material which protects the eyes against excessive ultraviolet, infrared, and visible radiation.

final current. The current after downslope but prior to current shut-off.

final taper current. The current at the end of the taper interval prior to downslope. A changing current is sometimes used to compensate for heat buildup or for change in section thickness.

fines. Any or all material finer than a particular mesh under consideration.

firecracker welding. A variation of the shielded metal arc welding process in which a length of covered electrode is placed along the joint in contact with the parts to be welded; during the welding operation, the stationary electrode is consumed as the arc travels the length of the electrode.

fisheye. A discontinuity found on the fracture surface of a weld in steel that consists of a small pore or inclusion surrounded by an approximately round, bright area.

fissure. A small crack-like discontinuity with only slight separation (opening displacement) of the fracture surfaces. The prefixes macro or micro indicate relative size.

fixture. A device designed to hold parts to be joined in proper relation to each other.

flame cutting. See preferred term *oxygen cutting*.

flame spraying (FLSP). A thermal spraying process in which an oxyfuel gas flame is the source of heat for melting the coating material. Compressed gas may or may not be used for atomizing and propelling the material to the substrate.

flange weld.* A weld made on the edges of two or more members to be joined, usually light gage metal, at least one of the members being flanged.

flare-bevel groove weld.* A weld in a groove formed by a member with a curved surface in contact with a planar member.

flare-V-groove weld.* A weld in a groove formed by two members with curved surfaces.

flash. The material which is expelled or squeezed out of a weld joint and which forms around the weld.

flashback. A recession of the flame into or back of the mixing chamber of the torch.

flashback arrester. A device to limit damage from a flashback by preventing propagation of the flame front beyond the point at which the arrester is installed.

flash coat. A thin coating usually less than 0.002 in. in thickness.

flash butt welding. See preferred term *flash welding*.

flashing time. The time during which the flashing action is taking place in flash welding.

flash-off time. See preferred term *flashing time*.

flash weld. A weld made by flash welding.

flash welding (FW). A resistance welding process which produces coalescence simultaneously over the entire area of abutting surfaces, by the heat obtained from resistance to electric current between the two surfaces, and by the application of pressure after heating is substantially completed. Flashing and upsetting are accompanied by expulsion of metal from the joint.

flat position. The welding position used to weld from the upper side of the joint; the face of the weld is approximately horizontal.

flaw. A near synonym for discontinuity but with an undesirable connotation. See *defect* and *discontinuity*.

*Flange weld, flare-bevel, and flare-V-groove welds may be confused because they have similar geometry before welding. A flange is welded on the edge and a flare is welded in the groove.

flowability. The ability of molten filler metal to flow or spread over a metal surface.

flow brazing (FLB). A brazing process which produces coalescence of metals by heating them with molten nonferrous filler metal poured over the joint until brazing temperature is attained. The filler metal is distributed in the joint by capillary action.

flow brightening (soldering). Fusion (melting) of a chemically or mechanically deposited metallic coating on a substrate.

flow welding (FLOW). A welding process which produces coalescence of metals by heating them with molten filler metal poured over the surfaces to be welded until the welding temperature is attained and until the required filler metal has been added. The filler metal is not distributed in the joint by capillary action.

flux. Material used to prevent, dissolve, or facilitate removal of oxides and other undesirable surface substances.

flux cored arc welding (FCAW). An arc welding process which produces coalescence of metals by heating them with an arc between a continuous filler metal (consumable) electrode and the work. Shielding is provided by a flux contained within the tubular electrode. Additional shielding may or may not be obtained from an externally supplied gas or gas mixture. See *flux cored electrode*.

flux cored electrode. A composite filler metal electrode consisting of a metal tube or other hollow configuration containing ingredients to provide such functions as shielding atmosphere, deoxidation, arc stabilization, and slag formation. Alloying materials may be included in the core. External shielding may or may not be used.

flux cover. In metal bath dip brazing and dip soldering, a cover of flux over the molten filler metal bath.

flux oxygen cutting. See preferred term *chemical flux cutting*.

focal point. See preferred term *focal spot*.

focal spot (EBW and LBW). A spot at which an energy beam has the most concentrated energy level and the smallest cross-sectional area.

forehand welding. A welding technique in which the welding torch or gun is directed toward the progress of welding. See also *travel angle, work angle,* and *push angle*.

forge-delay time (resistance welding). The time elapsing between the beginning of weld time or weld interval and the instant of application of forging force to the electrodes.

forge welding (FOW). A solid-state welding process which produces coalescence of metals by heating them in air in a forge and by applying pressure or blows sufficient to cause permanent deformation at the interface. Compare *cold welding, roll welding, diffusion welding,* and *hot pressure welding*.

freezing point. See preferred terms *liquidus* and *solidus*.

friction soldering. See preferred term *abrasion soldering*.

friction welding (FRW). A solid-state welding process which produces coalescence of metals by the heat obtained from a mechanically induced sliding motion between rubbing surfaces. The work parts are held together under pressure.

fuel gases. Gases usually used with oxygen for heating such as acetylene, natural gas, hydrogen, propane, methylacetylene propadiene stablized, and other synthetic fuels and hydrocarbons.

full fillet weld. A fillet weld whose size is equal to the thickness of the thinner member joined.

furnace brazing (FB). A brazing process in which the parts to be joined are placed in a furnace heated to a suitable temperature.

furnace soldering (FS). A soldering process in which the parts to be joined are placed in a furnace heated to a suitable temperature.

fused spray deposit (thermal spraying). A self-fluxing spray deposit which is subsequently heated to coalescence within itself and with the substrate.

fused zone. See preferred terms *fusion zone, nugget,* and *weld interface*.

fusing (thermal spraying). See preferred term *fusion*.

fusion. The melting together of filler metal and base metal (substrate), or of base metal only, which results in coalescence. See *depth of fusion*.

fusion face. A surface of the base metal which will be melted during welding.

fusion welding. Any welding process or method which uses fusion to complete the weld.

fusion zone. The area of base metal melted as determined on the cross section of a weld.

G

gas brazing. See preferred term *torch brazing*.

gas carbon arc welding (CAW-G). A carbon arc welding process variation which produces coalescence of metals by heating them with an electric arc between a single carbon electrode and the work. Shielding is obtained from a gas or gas mixture.

gas cutter. See preferred term *oxygen cutter*.

gas cutting. See preferred term *oxygen cutting*.

gas gouging. See preferred term *oxygen gouging*.

gas metal arc cutting (GMAC). An arc cutting process used to sever metals by melting them with the heat of an arc between a continuous metal (consumable) electrode and the work. Shielding is obtained entirely from an externally supplied gas or gas mixture.

gas metal arc welding (GMAW). An arc welding process which produces coalescence of metals by heating them with an arc between a continuous filler metal (consumable) electrode and the work. Shielding is obtained entirely from an externally supplied gas or gas mixture. Some variations of this process are called MIG or CO_2 welding (nonpreferred terms).

gas metal arc welding—pulsed arc (GMAW-P). A gas metal arc welding process variation in which the current is pulsed. See also *pulsed power welding*.

gas metal arc welding—short circuit arc (GMAW-S). A gas metal arc welding process variation in which the consumable electrode is deposited during repeated short circuits. Sometimes this process is referred to as MIG or CO_2 welding (nonpreferred terms).

gas pocket. See preferred term *porosity*.

gas regulator. See preferred term *regulator*.

gas shielded arc welding. A general term used to describe gas metal arc welding, gas tungsten arc welding, and flux cored arc welding when gas shielding is employed.

gas shielded stud welding. See *stud arc welding*.

gas torch. See preferred terms *welding torch* and *cutting torch*.

gas tungsten arc cutting (GTAC). An arc cutting process in which metals are severed by melting them with an arc between a single tungsten (nonconsumable) electrode and the work. Shielding is obtained from a gas or gas mixture.

gas tungsten arc welding (GTAW). An arc welding process which produces coalescence of metals by heating them with an arc between a tungsten (nonconsumable) electrode and the work. Shielding is obtained from a gas or gas mixture.

Pressure may or may not be used and filler metal may or may not be used. (This process has sometimes been called TIG welding, a nonpreferred term.)

gas tungsten arc welding—pulsed arc (GTAW-P). A gas tungsten arc welding process variation in which the current is pulsed. See also *pulsed power welding*.

gas welding. See preferred term *oxyfuel gas welding (OFW)*.

globular transfer (arc welding). The transfer of molten metal from a consumable electrode across the arc in large droplets.

gouging. The forming of a bevel or groove by material removal. See also *back gouging*, *arc gouging*, and *oxygen gouging*.

gradated coating. A thermal sprayed deposit composed of mixed materials in successive layers which progressively change in composition from the constituent material of the substrate to the surface of the sprayed deposit.

grit. See preferred term *abrasive*.

groove. An opening or channel in the surface of a part or between two components which provides space to contain a weld.

groove and rotary roughening (thermal spraying). A method of surface roughening in which grooves are made and the original surface roughened and spread.

groove angle. The total included angle of the groove between parts to be joined by a groove weld.

groove face. That surface of a member included in the groove.

groove radius. The radius used to form the shape of a J- or U-groove weld joint.

groove type. The geometric configuration of a groove.

groove weld. A weld made in the groove between two members to be joined. The standard types of groove welds are as follows:

double-bevel-groove weld
double-flare-bevel-groove weld
double-flare-V-groove weld
double-J-groove weld
double-U-groove weld
double-V-groove weld
single-bevel-groove weld
single-flare-bevel-groove weld
single-flare-V-groove weld
single-J-groove weld
single-U-groove weld
single-V-groove weld
square-groove weld

ground connection. An electrical connection of the welding machine frame to the earth for safety. See also *work connection* and *work lead*.

ground lead. See preferred term *work lead*.

gun. See preferred terms *arc welding gun*, *electron beam gun*, *resistance welding gun*, *soldering gun*, and *thermal spraying gun*.

gun extension (thermal spraying). The extension tube attached in front of the thermal spraying device to permit spraying within confined areas or deep recesses.

H

hammer welding. See preferred terms *forge welding* and *cold welding*.

hand shield. A protective device, used in arc welding, for shielding the eyes, face, and neck. A hand shield is equipped with a suitable filter plate and is designed to be held by hand.

hardfacing. A particular form of surfacing in which a coating or cladding is applied to a substrate for the main purpose of reducing wear or loss of material by abrasion, impact, erosion, galling, and cavitation. See *coating*, *cladding*, and *surfacing*.

hard solder. A term erroneously used to denote silver-base brazing filler metals.

hard surfacing. See preferred terms *surfacing* or *hardfacing*.

head. See *welding head* and *cutting head*.

heat-affected zone. That portion of the base metal which has not been melted, but whose mechanical properties or microstructure have been altered by the heat of welding, brazing, soldering, or cutting.

heating gate. The opening in a thermit mold through which the parts to be welded are preheated.

heat time (resistance welding). The duration of current flow during any one impulse in multiple-impulse welding, or the duration of current flow when making a simple welding cycle.

helmet (eye protection). A protection device used in arc welding for shielding the eyes, face, and neck. A helmet is equipped with a suitable filter plate and is designed to be worn on the head.

high frequency resistance welding (HFRW). A resistance welding process which produces coalescence of metals with the heat generated from the resistance of the workpieces to a high frequency alternating current in the 10 to 500 kHz range and the rapid application of an upsetting force after heating is substantially completed. The path of the current in the workpiece is controlled by the use of the proximity effect (the feed current follows closely the return current conductor).

high pulse current. The current levels during the high pulse time which produces the high heat level.

high pulse time. The duration of the high current pulse time.

hold time (resistance welding). The duration of force application at the point of welding after the last impulse of current ceases.

holding time. In brazing and soldering, the amount of time a joint is held within a specified temperature range.

horizontal fixed position (pipe welding). The position of pipe joint in which the axis of the pipe is approximately horizontal and the pipe is not rotated during welding.

horizontal position (fillet weld). The position in which welding is performed on the upper side of an approximately horizontal surface and against an approximately vertical surface.

horizontal position (groove weld). The position of welding in which the axis of the weld lies in an approximately horizontal plane and the face of the weld lies in an approximately vertical plane.

horizontal rolled position (pipe welding). The position of a pipe joint in which the axis of the pipe is approximately horizontal, and welding is performed in the flat position by rotating the pipe.

horn (resistance welding). An essentially cylindrical arm or extension of an arm of a resistance welding machine which transmits the electrode force and usually conducts the welding current. See *arm*.

horn spacing (resistance welding). The distance between adjacent surfaces of the horns of a resistance welding machine.

hot isostatic pressure welding. A diffusion welding process variation which produces coalescence of materials by heating and applying hot inert gas under pressure.

hot pressure welding (HPW). A solid-state welding process which produces coalescence of metals with heat and application of pressure sufficient to produce macrodeformation of the base metal. Vacuum or other shielding media may be used. See also *forge welding* and *diffusion welding*.

hot start current. A very brief current pulse at arc initiation to stabilize the arc quickly.

hot wire welding. A variation of arc welding processes (GTAW, PAW, SAW) in which a filler metal wire is resistance heated as it is fed into the molten weld pool.

hydrogen brazing. A term erroneously used to denote any brazing process which takes place in a hydrogen or hydrogen-containing atmosphere.

hydromatic welding. See preferred term *pressure-controlled welding*.

I

impulse (resistance welding). An impulse of welding current consisting of a single pulse or a series of pulses, separated only by an interpulse time.

inadequate joint penetration. Joint penetration which is less than that specified.

inclined position. The position of a pipe joint in which the axis of the pipe is approximately at an angle of 45° to the horizontal and the pipe is not rotated during welding.

inclined position (with restriction ring). The position of a pipe joint in which the axis of the pipe is approximately at an angle of 45° to the horizontal and a restriction ring is located near the joint. The pipe is not rotated during welding.

included angle. See preferred term *groove angle*.

incomplete fusion. Fusion which is less than complete.

indentation. In a spot, seam, or projection weld, the depression on the exterior surface or surfaces of the base metal.

induction brazing (IB). A brazing process in which the heat required is obtained from the resistance of the work to induced electric current.

induction soldering (IS). A soldering process in which the heat required is obtained from the resistance of the work to induced electric current.

induction welding (IW). A welding process which produces coalescence of metals through the heat obtained from resistance of the work to induced electric current, with or without the application of pressure.

induction work coil. See preferred term *work coil*.

inert gas. A gas which does not normally combine chemically with the base metal or filler metal. See also *protective atmosphere*.

inert gas metal arc welding. See preferred term *gas metal arc welding*.

inert gas tungsten arc welding. See preferred term *gas tungsten arc welding*.

infrared brazing (IRB). A brazing process in which the heat required is furnished by infrared radiation.

infrared radiation. Electromagnetic energy with wavelengths from 770 to 12 000 nanometers.

infrared soldering (IRS). A soldering process in which the heat required is furnished by infrared radiation.

initial current. The current after starting, but before establishment of welding current.

intergranular penetration. The penetration of a filler metal along the grain boundaries of a base metal.

intermediate flux. A soldering flux with a residue that generally does not attack the base metal. The original composition may be corrosive.

intermittent weld. A weld in which the continuity is broken by recurring unwelded spaces.

interpass temperature. In a multiple-pass weld, the temperature (minimum or maximum as specified) of the deposited weld metal before the next pass is started.

interpulse time (resistance welding). The time between successive pulses of current within the same impulse.

interrupted spot welding. See preferred term *multiple-impulse welding*.

ionic bond. A primary bond arising from the electrostatic attraction between two oppositely charged ions.

iron soldering (INS). A soldering process in which the heat required is obtained from a soldering iron.

J

joint. The junction of members or the edges of members which are to be joined or have been joined.

joint brazing procedure. The materials, detailed methods, and practices employed in the brazing of a particular joint.

joint buildup sequence. The order in which the weld beads of a multiple-pass weld are deposited with respect to the cross section of the joint. See also *block sequence* and *longitudinal sequence*.

joint clearance. The distance between the faying surfaces of a joint. In brazing, this distance is referred to as that which is present before brazing, at the brazing temperature, or after brazing is completed.

joint design. The joint geometry together with the required dimensions of the welded joint.

joint efficiency. The ratio of the strength of a joint to the strength of the base metal (expressed in percent).

joint geometry. The shape and dimensions of a joint in cross section prior to welding.

joint penetration. The minimum depth a groove or flange weld extends from its face into a joint, exclusive of reinforcement. Joint penetration may include root penetration. See also *complete joint penetration, root penetration*, and *effective throat*.

joint welding sequence. See preferred term *joint buildup sequence*.

K

kerf. The width of the cut produced during a cutting process.

keyhole. A technique of welding in which a concentrated heat source penetrates completely through a workpiece forming a hole at the leading edge of the molten weld metal. As the heat source progresses, the molten metal fills in behind the hole to form the weld bead.

keying (thermal spraying). See preferred term *mechanical bond*.

knee (resistance welding). The lower arm supporting structure in a resistance welding machine.

knurling (thermal spraying). See preferred terms *rotary roughening, groove and rotary roughening*, and *threading and knurling*.

L

lack of fusion. See preferred term *incomplete fusion*.

lack of joint penetration. See preferred term *inadequate joint penetration*.

land. See preferred term *root face*.

lap joint. A joint between two overlapping members.

laser beam cutting (LBC). A cutting process which severs materials with the heat obtained from the application of a concentrated coherent light beam impinging upon the workpiece to be cut. The process can be used with or without an externally supplied gas.

laser beam welding (LBW). A welding process which produces coalescence of materials with the heat obtained from the application of a concentrated coherent light beam impinging upon the members to be joined.

layer. A stratum of weld metal or surfacing material. The layer may consist of one or more weld beads laid side by side.

layer level wound. See preferred term *level wound*.

layer wound. See preferred term *level wound*.

lead angle. See preferred term *travel angle*.

lead burning. An erroneous term used to denote the welding of lead.

leg of a fillet weld. The distance from the root of the joint to the toe of the fillet weld.

lens. See preferred term *filter lens*.

level wound. Spooled or coiled filler metal that has been wound in distinct layers such that adjacent turns touch.

lightly coated electrode. A filler metal electrode consisting of a metal wire with a light coating applied subsequent to the drawing operation, primarily for stabilizing the arc.

liquation. The separation of a low melting constituent of an alloy from the remaining constituents, usually apparent in alloys having a wide melting range.

liquidus. The lowest temperature at which a metal or an alloy is completely liquid.

local preheating. Preheating a specific portion of a structure.

local stress relief heat treatment. Stress relief heat treatment of a specific portion of a structure.

locked-up stress. See preferred term *residual stress*.

longitudinal resistance seam welding. The making of a resistance seam weld in a direction essentially parallel to the throat depth of a resistance welding machine.

longitudinal sequence. The order in which the increments of a continuous weld are deposited with respect to its length. See *backstep sequence, block sequence,* etc.

low frequency cycle (resistance welding). One positive and one negative pulse of current within the same weld or heat time at a frequency lower than the power supply frequency from which it is obtained.

low pulse current. The current levels during the low pulse time which produce the low heat levels.

low pulse time. The duration of the low current pulse time.

M

machine brazing. Brazing with equipment which performs the brazing operation under the constant observation and control of a brazing operator. The equipment may or may not perform the loading and unloading of the work. See *automatic brazing*.

machine oxygen cutting. Oxygen cutting with equipment which performs the cutting operation under the constant observation and control of an oxygen cutting operator. The equipment may or may not perform the loading and unloading of the work. See *automatic oxygen cutting*.

machine welding. Welding with equipment which performs the welding operation under the constant observation and control of a welding operator. The equipment may or may not perform the loading and unloading of the work. See *automatic welding*.

manifold. A multiple header for interconnection of gas or fluid sources with distribution points.

manual brazing. A brazing operation performed and controlled completely by hand. See *automatic brazing* and *machine brazing*.

manual oxygen cutting. A cutting operation performed and controlled completely by hand. See *automatic oxygen cutting* and *machine oxygen cutting*.

manual welding. A welding operation performed and controlled completely by hand. See *automatic welding, machine welding*, and *semiautomatic arc welding*.

mash resistance seam welding. A resistance seam welding process variation in which a seam weld is made in a lap joint primarily by high temperature plastic working and diffusion as opposed to melting and solidification. The joint thickness after welding is less than the original assembled thickness.

mask (thermal spraying). A device for protecting a surface from the effects of blasting or coating adherence or coalescence with the substrate. Masks are generally of two types: reusable or disposable.

matrix (thermal spraying). The major continuous substance of a thermal sprayed coating as opposed to inclusions or particles of materials having dissimilar characteristics.

mechanical bond (thermal spraying). The adherence of a thermal sprayed deposit to a roughened surface by the mechanism of particle interlocking.

meltback time. The time interval at the end of crater fill time to arc outage during which electrode feed is stopped. Arc voltage and arc length increase and current decreases to zero to prevent the electrode from freezing in the weld deposit.

melting range. The temperature range between solidus and liquidus.

melting rate. The weight or length of electrode melted in a unit of time.

melt-thru. Complete joint penetration for a joint welded from one side. Visible root reinforcement is produced.

metal arc cutting (MAC). Any of a group of arc cutting processes which severs metals by melting them with the heat of an arc between a metal electrode and the base metal. See *shielded metal arc cutting* and *gas metal arc cutting*.

metal arc welding. See *shielded metal arc welding, flux cored arc welding, gas metal arc welding, gas tungsten arc welding, submerged arc welding, plasma arc welding,* and *stud arc welding*.

metal cored electrode. A composite filler metal electrode consisting of a metal tube or other hollow configuration containing alloying ingredients. Minor amounts of ingredients providing such functions as arc stabilization and fluxing of oxides may be included. External shielding gas may or may not be used.

metal electrode. A filler or non-filler metal electrode used in arc welding or cutting which consists of a metal wire or rod that has been manufactured by any method and that is either bare or covered with a suitable covering or coating.

metallic bond. The principal bond which holds metals together and which is formed between base metals and filler metals in all welding processes. This is a primary bond arising from the increased spatial extension of the valence electron wave functions when an aggregate of metal atoms is brought close together. See *bonding force, covalent bond*, and *ionic bond*.

metallizing. See preferred term *thermal spraying*.

metallurgical bond. See preferred term *metallic bond*.

metal powder cutting (POC). An oxygen cutting process which severs metals through the use of powder, such as iron, to facilitate cutting.

MIG welding. See preferred terms *gas metal arc welding* and *flux cored arc welding*.

mixing chamber. That part of a welding or cutting torch in which a fuel gas and oxygen are mixed.

molten chemical-bath dip brazing. A dip brazing process variation. See *dip brazing (DB)*.

molten metal-bath dip brazing. A dip brazing process variation. See *dip brazing (DB)*.

molten metal flame spraying. A thermal spraying process variation in which the metallic material to be sprayed is in the molten condition. See *flame spraying (FLSP)*.

molten weld pool. The liquid state of a weld prior to solidification as weld metal.

multiple-impulse welding. A resistance welding process variation in which welds are made by more than one impulse of current.

multiple-impulse weld timer (resistance welding). A device for multiple-impulse welding which controls only the heat time, the cool time, and the weld interval or the number of heat times.

multiport nozzle (plasma arc welding and cutting). A constricting nozzle containing two or more orifices located in a configuration to achieve a degree of control over the arc shape.

N

neutral flame. An oxyfuel gas flame in which the portion used is neither oxidizing nor reducing.

noncorrosive flux. A soldering flux which in neither its original nor residual form chemically attacks the base metal. It usually is composed of rosin or resin-base materials.

nonsynchronous initiation (resistance welding). The initiation or termination of the welding transformer primary current at any random time with respect to the voltage wave.

nonsynchronous timing. See preferred term *nonsynchronous initiation*.

nontransferred arc (plasma arc welding and cutting, and thermal spraying). An arc established between the electrode and the constricting nozzle. The workpiece is not in the electrical circuit. See *transferred arc*.

nozzle. A device which directs shielding media.

nugget (resistance welding). The weld metal joining the parts in spot, roll spot, seam, or projection welds.

nugget size. The diameter or width of the nugget measured in the plane of the interface between the pieces joined.

O

off time (resistance welding). The time during which the electrodes are off the work. This term is generally used when the welding cycle is repetitive.

open-circuit voltage. The voltage between the output terminals of the welding machine when no current is flowing in the welding circuit.

orifice gas (plasma arc welding and cutting). The gas that is directed into the torch to surround the electrode. It becomes ionized in the arc to form the plasma, and issues from the orifice in the torch nozzle as the plasma jet.

orifice throat length (plasma arc welding and cutting). The length of the constricting orifice.

oven soldering. See preferred term *furnace soldering*.

overhead position. The position in which welding is performed from the underside of the joint.

overlap. The protrusion of weld metal beyond the toe, face, or root of the weld; in resistance seam welding, the area in the preceding weld remelted by the succeeding weld.

overlaying. See preferred term *surfacing*.

oxidizing flame. An oxyfuel gas flame having an oxidizing effect (excess oxygen).

oxyacetylene cutting (OFC-A). An oxyfuel gas cutting process used to sever metals by means of the chemical reaction of oxygen with the base metal at elevated temperatures. The necessary temperature is maintained by gas flames resulting from the combustion of acetylene with oxygen.

oxyacetylene welding (OAW). An oxyfuel gas welding process which produces coalescence of metals by heating them with a gas flame or flames obtained from the combustion of acetylene with oxygen. The process may be used with or without the application of pressure and with or without the use of filler metal.

oxyfuel gas cutting (OFC). A group of cutting processes used to sever metals by means of the chemical reaction of oxygen with the base metal at elevated temperatures. The necessary temperature is maintained by means of gas flames obtained from the combustion of a specified fuel gas and oxygen. See *oxygen cutting, oxyacetylene cutting, oxyhydrogen cutting, oxynatural gas cutting*, and *oxypropane cutting*.

oxyfuel gas welding (OFW). A group of welding processes which produces coalescence by heating materials with an oxyfuel gas flame or flames, with or without the application of pressure and with or without the use of filler metal.

oxyfuel gas spraying. See preferred term *flame spraying*.

oxygas cutting. See preferred term *oxygen cutting*.

oxygen arc cutting (AOC). An oxygen cutting process used to sever metals by means of the chemical reaction of oxygen with the base metal at elevated temperatures. The necessary temperature is maintained by an arc between a consumable tubular electrode and the base metal.

oxygen cutter. One who performs a manual oxygen cutting operation.

oxygen cutting (OC). A group of cutting processes used to sever or remove metals by means of the chemical reaction of oxygen with the base metal at elevated temperatures. In the case of oxidation-resistant metals, the reaction is facilitated by the use of a chemical flux or metal powder. See *oxygen arc cutting, oxyfuel gas cutting, oxygen lance cutting, chemical flux cutting*, and *metal powder cutting*.

oxygen cutting operator. One who operates machine or automatic oxygen cutting equipment.

oxygen gouging. An application of oxygen cutting in which a bevel or groove is formed.

oxygen grooving. See preferred term *oxygen gouging*.

oxygen lance. A length of pipe used to convey oxygen to the point of cutting in oxygen lance cutting.

oxygen lance cutting (LOC). An oxygen cutting process used to sever metals with oxygen supplied through a consumable lance; the preheat to start the cutting is obtained by other means.

oxygen lancing. See preferred term *oxygen lance cutting*.

oxyhydrogen cutting (OFC-H). An oxyfuel gas cutting process used to sever metals by means of the chemical reaction of oxygen with the base metal at elevated temperatures. The necessary temperature is maintained by gas flames resulting from the combustion of hydrogen with oxygen.

oxyhydrogen welding (OHW). An oxyfuel gas welding process which produces coalescence of materials by heating them with a gas flame or flames obtained from the combustion of hydrogen with oxygen, without the application of pressure and with or without the use of filler metal.

oxynatural gas cutting (OFC-N). An oxyfuel gas cutting process used to sever metals by means of the chemical reaction of oxygen with the base metal at elevated temperatures. The necessary temperature is maintained by gas flames resulting from the combustion of natural gas with oxygen.

oxypropane cutting (OFC-P). An oxyfuel gas cutting process used to sever metals by means of the chemical reaction of oxygen with the base metal at elevated temperatures. The necessary temperature is maintained by gas flames resulting from the combustion of propane with oxygen.

P

parent metal. See preferred term *base metal*.

partial joint penetration. Joint penetration which is less than complete. See also *complete joint penetration*.

pass. See preferred term *weld pass*.

pass sequence. See *deposition sequence*.

paste brazing filler metal. A mixture of finely divided brazing filler metal with an organic or inorganic flux or neutral vehicle or carrier.

paste soldering filler metal. A mixture of finely divided metallic solder with an organic or inorganic flux or neutral vehicle or carrier.

peel test. A destructive method of inspection which mechanically separates a lap joint by peeling.

peening. The mechanical working of metals using impact blows.

penetration. See preferred terms *joint penetration* and *root penetration*.

percent ferrite. See preferred term *ferrite number*.

percussion weld. A weld made by percussion welding.

percussion welding (PEW). A resistance welding process which produces coalescence of the abutting members using heat from an arc produced by a rapid discharge of electrical energy. Pressure is applied percussively during or immediately following the electrical discharge.

permanent dipole bond. A secondary bond arising from the attraction between dipoles, the oppositely charged ends of which are electronegative and electropositive atoms.

pilot arc (plasma arc welding). A low current continuous arc between the electrode and the constricting nozzle to ionize the gas and facilitate the start of the main welding arc.

plano lens (eye protection). A lens which does not incorporate correction.

plasma. A gas that has been heated to an at least partially ionized condition, enabling it to conduct an electric current.

plasma arc cutting (PAC). An arc cutting process which severs metal by melting a localized area with a constricted arc and removing the molten material with a high velocity jet of hot, ionized gas issuing from the orifice.

plasma arc welding (PAW). An arc welding process which produces coalescence of metals by heating them with a constricted arc between an electrode and the workpiece (transferred arc) or the electrode and the constricting nozzle (nontransferred arc). Shielding is obtained from the hot, ionized gas issuing from the orifice which may be supplemented by an auxiliary source of shielding gas. Shielding gas may be an inert gas or a mixture of gases. Pressure may or may not be used, and filler metal may or may not be supplied.

plasma metallizing. See preferred term *plasma spraying*.

plasma spraying (PSP). A thermal spraying process in which a nontransferred arc is utilized as the source of heat for melting and propelling the coating material to the workpiece.

platen (resistance welding). A member with a substantially flat surface to which dies, fixtures, backups, or electrode holders are attached, and which transmits the electrode force or upsetting force. One platen usually is fixed and the other moveable.

platen force (resistance welding). The force available at the moveable platen to cause upsetting in flash or upset welding. This force may be dynamic, theoretical, or static.

platen spacing. The distance between adjacent surfaces of the platens in a resistance welding machine.

plenum chamber. See *plenum*.

plenum (plasma arc welding and cutting, and thermal spraying). The space between the inside wall of the constricting nozzle and the electrode.

plug weld. A circular weld made through a hole in one member of a lap or T-joint fusing that member to the other. The walls of the hole may or may not be parallel and the hole may be partially or completely filled with weld metal. (A fillet welded hole or a spot weld should not be construed as conforming to this definition.)

poke weld. See preferred term *push weld*.

poke welding. See preferred term *push welding*.

polarity. See *direct current electrode negative, direct current electrode positive, straight polarity,* and *reverse polarity*.

porosity. Cavity type discontinuities formed by gas entrapment during solidification.

positioned weld. A weld made in a joint which has been so placed as to facilitate making the weld.

position of welding. See *flat position, horizontal position, horizontal fixed position, horizontal rolled position, inclined position, overhead position,* and *vertical position*.

postflow time. The time interval from current shut-off to shielding gas and/or cooling water shut-off.

postheat current (resistance welding). The current through the welding circuit during postheat time in resistance welding.

postheating. The application of heat to an assembly after a welding, brazing, soldering, thermal spraying, or cutting operation. See *postweld heat treatment*.

postheat time (resistance welding). The time from the end of weld heat time to the end of weld time.

postweld heat treatment. Any heat treatment subsequent to welding.

postweld interval (resistance welding). The total time elapsing between the end of welding current and the start of hold time.

powder cutting. See preferred terms *chemical flux cutting* and *metal powder cutting*.

powder feeder (thermal spraying). A device for supplying powdered materials to thermal spraying equipment.

powder flame spraying. A thermal spraying process variation in which the material to be sprayed is in powder form. See *flame spraying (FLSP)*.

powder metallizing. See preferred term *powder flame spraying*.

precoating. Coating the base metal in the joint by dipping, electroplating, or other applicable means prior to soldering or brazing.

preflow time. The time interval between start of shielding gas flow and arc starting.

preform. Brazing or soldering filler metal fabricated in a shape or form for a specific application.

preheat. See preferred term *preheat temperature*.

preheat current (resistance welding). An impulse or series of current impulses which occurs prior to and is separated from welding current.

preheating. The application of heat to the base metal immediately before welding, brazing, soldering, thermal spraying, or cutting.

preheat temperature. A specified temperature that the base metal must attain in the welding, brazing, soldering, thermal spraying, or cutting area immediately before these operations are performed.

preheat time (resistance welding). A portion of the preweld interval during which preheat current occurs.

pressure-controlled welding. A resistance welding process variation in which a number of spot or projection welds are made with several electrodes functioning progressively under the control of a pressure-sequencing device.

pressure gas welding (PGW). An oxyfuel gas welding process which produces coalescence simultaneously over the entire area of abutting surfaces by heating them with gas flames obtained from the combustion of a fuel gas with oxygen and by the application of pressure, without the use of filler metal.

pressure welding. See preferred terms *solid-state welding, hot pressure welding, forge welding, diffusion welding, pressure gas welding,* and *cold welding.*

pretinning. See preferred term *precoating.*

preweld interval (resistance welding). The time elapsing between the end of squeeze time and the beginning of welding current in making spot welds and in projection or upset welding. In flash welding, it is the time during which the material is preheated.

procedure. The detailed elements (with prescribed values or ranges of values) of a process or method used to produce a specific result.

procedure qualification. The demonstration that welds made by a specific procedure can meet prescribed standards.

procedure qualification record (PQR). A document providing the actual welding variables used to produce an acceptable test weld and the results of tests conducted on the weld for the purpose of qualifying a welding procedure specification.

progressive block sequence. A block sequence during which successive blocks are completed progressively along the joint, either from one end to the other or from the center of the joint toward either end.

projection weld. A weld made by projection welding.

projection welding (RPW). A resistance welding process which produces coalescence of metals with the heat obtained from resistance to electric current through the work parts held together under pressure by electrodes. The resulting welds are localized at predetermined points by projections, embossments, or intersections.

protective atmosphere. A gas envelope surrounding the part to be brazed, welded, or thermal sprayed, with the gas composition controlled with respect to chemical composition, dew point, pressure, flow rate, etc. Examples are inert gases, combusted fuel gases, hydrogen, and vacuum.

puddle. See preferred term *molten weld pool.*

pull gun technique. See *backhand welding.*

pulsation welding. See preferred term *multiple-impulse welding.*

pulsation weld timer. See preferred term *multiple-impulse weld timer.*

pulse (resistance welding). A current of controlled duration through a welding circuit.

pulsed power welding. An arc welding process variation in which the power is cyclically programmed to pulse so that effective but short duration values of a parameter can be utilized. Such short duration values are significantly different from the average value of the parameter. Equivalent terms are pulsed voltage or pulsed current welding; see also *pulsed spray welding.*

pulsed spray welding. An arc welding process variation in which the current is pulsed to utilize the advantages of the spray mode of metal transfer at average currents equal to or less than the globular to spray transition current.

pulse start delay time. The time interval from current initiation to the beginning of current pulsation, if pulsation is used.

pulse time (resistance welding). The duration of a pulse.

push angle. The travel angle when the electrode is pointing forward. See also *forehand welding.* Note: This angle can be used to define the position of welding guns, welding torches, high energy beams, welding rods, thermal cutting and thermal spraying torches, and thermal spraying guns.

push weld (resistance welding). A spot or projection weld made by push welding.

push welding. A resistance welding process variation in which spot or projection welds are made by manually applying force to one electrode and using the work or a backing as the other electrode.

Q

qualification. See preferred terms *welder performance qualification* and *procedure qualification.*

quench time (resistance welding). The time from the end of weld time to the beginning of temper time.

R

random intermittent welds. Intermittent welds on one or both sides of a joint in which the weld increments are deposited without regard to spacing.

random sequence. A longitudinal sequence in which the weld bead increments are deposited at random.

random wound. Spooled or coiled filler metal that has not been wound in distinct layers.

rate of deposition. See *deposition rate.*

rate of flame propagation. The speed at which a flame travels through a mixture of gases.

reaction flux (soldering). A flux composition in which one or more of the ingredients reacts with a base metal upon heating to deposit one or more metals.

reaction soldering. A soldering process variation in which a reaction flux is used.

reaction stress. The residual stress which could not otherwise exist if the members or parts being welded were isolated as free bodies without connection to other parts of the structure.

reactor (arc welding). A device used in arc welding circuits for the purpose of minimizing irregularities in the flow of welding current.

reducing atmosphere. A chemically active protective atmosphere which at elevated temperature will reduce metal oxides to their metallic state. (Reducing atmosphere is a relative term and such an atmosphere may be reducing to one oxide but not to another oxide.)

reducing flame. A gas flame having a reducing effect. (Excess fuel gas.)

reflow soldering. A nonpreferred term used to describe a soldering process variation in which preplaced solder is melted to produce a soldered joint or coated surface.

reflowing (soldering). See preferred term *flow brightening.*

regulator. A device for controlling the delivery of gas at some substantially constant pressure.

reinforcement of weld. Weld metal in excess of the quantity required to fill a joint. See *face reinforcement* and *root reinforcement.*

residual stress. Stress remaining in a structure or member as a result of thermal or mechanical treatment or both. Stress arises in fusion welding primarily because the weld metal contracts on cooling from the solidus to room temperature.

resistance brazing (RB). A brazing process in which the heat required is obtained from the resistance to electric current in a circuit of which the work is a part.

resistance butt weld. See preferred terms *upset weld* and *flash weld.*

resistance butt welding. See preferred terms *upset welding* and *flash welding.*

resistance seam weld timer. In resistance seam welding, a device which controls the heat times and the cool times.

resistance seam welding (RSEW). A resistance welding process which produces coalescence at the faying surfaces by the heat obtained from resistance to electric current through the work parts held together under pressure by electrodes. The resulting weld is a series of overlapping resistance spot welds made progressively along a joint by rotating the electrodes.

resistance soldering (RS). A soldering process in which the heat required is obtained from the resistance to electric current in a circuit of which the work is a part.

resistance spot welding (RSW). A resistance welding process which produces coalescence at the faying surfaces in one spot by the heat obtained from the resistance to electric current through the work parts held together under pressure by electrodes. The size and shape of the individually formed welds are limited primarily by the size and contour of the electrodes.

resistance welding (RW). A group of welding processes which produces coalescence of metals with the heat obtained from resistance of the work to electric current in a circuit of which the work is a part, and by the application of pressure. See *resistance welding electrode*.

resistance welding electrode. The part or parts of a resistance welding machine through which the welding current and, in most cases, pressure are applied directly to the work. The electrode may be in the form of a rotating wheel, rotating roll, bar, cylinder, plate, clamp, chuck, or modification thereof. See *resistance welding*.

resistance welding gun. A manipulating device to transfer current and provide electrode force to the weld area (usually in reference to a portable gun).

reverse polarity. The arrangement of direct current arc welding leads with the work as the negative pole and the electrode as the positive pole of the welding arc. A synonym for direct current electrode positive.

roll resistance spot welding. A resistance welding process variation in which separated spot welds are made with one or more rotating circular electrodes. The rotation of the electrodes may or may not be stopped during the making of a weld.

roll welding (ROW). A solid-state welding process which produces coalescence of metals by heating and by applying pressure with rolls sufficient to cause deformation at the faying surfaces. See also *forge welding*.

root. See preferred terms *root of joint* and *root of weld*.

root head. A weld deposit that extends into or includes part or all of the root of the joint.

root crack. A crack in the weld or heat-affected zone occurring at the root of a weld.

root edge. A root face of zero width. See *root face*.

root face. That portion of the groove face adjacent to the root of the joint.

root gap. See preferred term *root opening*.

root of joint. That portion of a joint to be welded where the members approach closest to each other. In cross section, the root of the joint may be either a point, a line, or an area.

root of weld. The points, as shown in cross section, at which the back of the weld intersects the base metal surfaces.

root opening. The separation between the members to be joined at the root of the joint.

root penetration. The depth that a weld extends into the root of a joint measured on the centerline of the root cross section.

root radius. See preferred term *groove radius*.

root reinforcement. Reinforcement of weld at the side other than that from which welding was done.

root surface. The exposed surface of a weld on the side other than that from which welding was done.

rotary roughening (thermal spraying). A method of surface roughening in which a revolving roughening tool is pressed against the surface being prepared, while either the work or the tool, or both, move.

rough threading (thermal spraying). A method of surface roughening which consists of cutting threads with the sides and tops of the threads jagged and torn.

S

salt-bath dip brazing. See preferred term *molten chemical-bath dip brazing*.

sandwich braze. A brazed assembly of dissimilar materials using a preplaced shim, other than the filler metals, as a transition layer to minimize thermal stresses.

scarf. See preferred term *edge preparation*.

scarf joint. A form of butt joint.

seal coat (thermal spraying). Material applied to infiltrate the pores of a thermal spray deposit.

seal weld. Any weld designed primarily to provide a specific degree of tightness against leakage.

seam weld. A continuous weld made between or upon overlapping members, in which coalescence may start and occur on the faying surfaces, or may have proceeded from the surface of one member. The continuous weld may consist of a single weld bead or a series of overlapping spot welds.

seam welding. The making of seam welds.

secondary circuit. That portion of a welding machine which conducts the secondary current between the secondary terminals of the welding transformer and the electrodes, or electrode and work.

selective block sequence. A block sequence in which successive blocks are completed in a certain order selected to create a predetermined stress pattern.

self-fluxing alloys (thermal spraying). Certain materials that "wet" the substrate and coalesce when heated to their melting point, without the addition of a fluxing agent.

semiautomatic arc welding. Arc welding with equipment which controls only the filler metal feed. The advance of the welding is manually controlled.

semiautomatic brazing. Brazing with equipment which controls only the brazing filler metal feed. The advance of the brazing is manually controlled.

semiblind joint. A joint in which one extremity of the joint is not visible.

sequence time (automatic arc welding). See preferred term *weld cycle*.

sequence timer (resistance welding). A device for controlling the sequence and duration of any or all of the elements of a complete welding cycle except weld time or heat time.

sequence weld timer (resistance welding). A device which performs the functions of a sequence timer and the functions of either a weld timer or a multiple-impulse weld timer.

series submerged arc welding (SAW-S). A submerged arc welding process variation in which electric current is established between two (consumable) electrodes which meet just above the surface of the work. The work is not in the electrical circuit.

series welding (resistance welding). A resistance welding process variation in which two spot or seam welds or two or more projection welds are made simultaneously with three electrodes, forming a series circuit.

shadow mask (thermal spraying). A thermal spraying process variation in which an area is partially shielded during the thermal spraying operation, thus permitting some overspray to produce a feathering at the coating edge.

sheet separation (resistance welding). The gap surrounding the weld, between faying surfaces, after the joint has been welded in spot, seam, or projection welding.

shielded carbon arc welding (CAW-S). A carbon arc welding process variation which produces coalescence of metals by heating them with an electric arc between a carbon electrode and the work. Shielding is obtained from the combustion of a solid material fed into the arc or from a blanket of flux on the work, or both. Pressure may or may not be used and filler metal may or may not be used.

shielded metal arc cutting (SMAC). A metal arc cutting process in which metals are severed by melting them with the heat of an arc between a covered metal electrode and the base metal.

shielded metal arc welding (SMAW). An arc welding process which produces coalescence of metals by heating them with an arc between a covered metal electrode and the work. Shielding is obtained from decomposition of the electrode covering. Pressure is not used and filler metal is obtained from the electrode.

shielding gas. Protective gas used to prevent atmospheric contamination.

short circuiting arc welding. See *gas metal arc welding—short circuit arc (GMAW-S)*.

short circuiting transfer (arc welding). Metal transfer in which molten metal from a consumable electrode is deposited during repeated short circuits.

shoulder. See preferred term *root face*.

shrinkage stress. See preferred term *residual stress*.

shrinkage void. A cavity type discontinuity normally formed by shrinkage during solidification.

sieve analysis. A method of determining particle size distribution, usually expressed as the weight percentage retained upon each of a series of standard screens of decreasing mesh size.

silver soldering, silver alloy brazing. Nonpreferred terms used to denote brazing with a silver-base filler metal. See preferred terms *furnace brazing, induction brazing,* and *torch brazing*.

single-impulse welding (resistance welding). A resistance welding process variation in which spot, projection, or upset welds are made by a single impulse of current. When alternating current is used, an impulse may consist of a fraction of a cycle or a number of cycles.

single-port nozzle. A constricting nozzle containing one orifice, located below and concentric with the electrode.

single-welded joint. In arc and gas welding, any joint welded from one side only.

size of weld.

groove weld. The joint penetration (depth of bevel plus the root penetration when specified). The size of a groove weld and its effective throat are one and the same.

fillet weld. For equal leg fillet welds, the leg lengths of the largest isosceles right triangle which can be inscribed within the fillet weld cross section. For unequal leg fillet welds, the leg lengths of the largest right triangle which can be inscribed within the fillet weld cross section. Note: When one member makes an angle with the other member greater than 105 degrees, the leg length (size) is of less significance than the effective throat which is the controlling factor for the strength of a weld.

flange weld. The weld metal thickness measured at the root of the weld.

skip weld. See preferred term *intermittent weld*.

skull. The unmelted residue from a liquated filler metal.

slag inclusion. Nonmetallic solid material entrapped in weld metal or between weld metal and base metal.

slot weld. A weld made in an elongated hole in one member of a lap or T-joint joining that member to that portion of the surface of the other member which is exposed through the hole. The hole may be open at one end and may be partially or completely filled with weld metal. (A fillet welded slot should not be construed as conforming to this definition.)

slugging. The act of adding a separate piece or pieces of material in a joint before or during welding that results in a welded joint not complying with design, drawing, or specification requirements.

soft solder. See preferred term *solder*.

solder. A filler metal used in soldering which has a liquidus not exceeding 850 °F.

solderability. The capacity of a material to be soldered under the fabrication conditions imposed upon a specific, suitably designed structure.

soldering (S). A group of welding processes which produces coalescence of materials by heating them to a suitable temperature and by using a filler metal having a liquidus not exceeding 840 °F and below the solidus of the base metals. The filler metal is distributed between the closely fitted faying surfaces of the joint by capillary action.

soldering gun. An electrical soldering iron with a pistol grip and a quick heating, relatively small bit.

solder interface. See *weld interface*.

soldering iron. A soldering tool having an internally or externally heated metal bit usually made of copper.

solid-state welding (SSW). A group of welding processes which produces coalescence at temperatures essentially below the melting point of the base metal being joined, without the addition of a brazing filler metal. Pressure may or may not be used.

solidus. The highest temperature at which a metal or alloy is completely solid.

spacer strip. A metal strip or bar prepared for a groove weld and inserted in the root of a joint to serve as a backing and to maintain root opening during welding. It can also bridge an exceptionally wide gap due to poor fitup.

spatter. The metal particles expelled during welding and which do not form a part of the weld.

spatter loss. Metal lost due to spatter.

spiking (electron beam welding and laser welding). A condition where the depth of penetration is nonuniform and changes abruptly over the length of the weld.

spit. See preferred term *flash*.

split pipe backing. Backing in the form of a pipe segment used for welding round bars.

spool. A type of filler metal package consisting of a continuous length of electrode wound on a cylinder (called the barrel) which is flanged at both ends. The flange extends below the inside diameter of the barrel and contains a spindle hole.

spot weld. A weld made between or upon overlapping members in which coalescence may start and occur on the faying surfaces or may proceed from the surface of one member. The weld cross section (plan view) is approximately circular. See also *arc spot weld* and *resistance spot welding*.

spot welding. The making of spot welds.

spray deposit (thermal spraying). The coating applied by a thermal spraying process.

spraying sequence (thermal spraying). The order in which different layers of similar or different materials are applied in a planned relationship, such as overlapped, superimposed, or at certain angles.

spray rate. A synonym for feed rate. See *feed rate*.

spray transfer (arc welding). Metal transfer in which molten metal from a consumable electrode is propelled axially across the arc in small droplets.

squeeze time (resistance welding). The time interval between the initial application of the electrode force on the work and the first application of current in making spot and seam welds and in projection or upset welding.

stack cutting. Thermal cutting of stacked metal plates arranged so that all the plates are severed by a single cut.

staggered intermittent welds. Intermittent welds on both sides of a joint in which the weld increments on one side are alternated with respect to those on the other side.

standoff distance. The distance between a nozzle and the base metal.

start current. The current value during start time interval.

start time. The time interval prior to weld time during which arc voltage and current reach a preset value greater or less than welding values.

start voltage. The arc voltage during the start time.

static electrode force. See *electrode force*.

stepback sequence. See preferred term *backstep sequence*.

step brazing. The brazing of successive joints on a given part with filler metals of successively lower brazing temperatures so as to accomplish the joining without disturbing the joints previously brazed. A similar result can be achieved at a single brazing temperature if the remelt temperature of prior joints is increased by metallurgical interaction.

step soldering. The soldering of successive joints on a given part with solders of successively lower soldering temperatures so as to accomplish the joining without disturbing the joints previously soldered.

stick electrode. See *covered electrode*.

stick electrode welding. See preferred term *shielded metal arc welding*.

stickout. See preferred term *electrode extension*.

stitch weld. See preferred term *intermittent weld*.

stopoff. A material used on the surfaces adjacent to the joint to limit the spread of soldering or brazing filler metal.

stored energy welding (resistance welding). A resistance welding process variation in which welds are made with electrical energy accumulated electrostatically, electromagnetically, or electrochemically at a relatively low rate and made available at the required welding rate.

straight polarity. The arrangement of direct current arc welding leads in which the work is the positive pole and the electrode is the negative pole of the welding arc. A synonym for *direct current electrode negative*.

stranded electrode. A composite filler metal electrode consisting of stranded wires which may mechanically enclose materials to improve properties, stabilize the arc, or provide shielding.

stress corrosion cracking. Failure of metals by cracking under combined action of corrosion and stress, residual or applied. In brazing, the term applies to the cracking of stressed base metal due to the presence of a liquid filler metal.

stress relief cracking. Intergranular cracking in the heat-affected zone or weld metal that occurs during the exposure of weldments to elevated temperatures during postweld heat treatment or high temperature service.

stress relief heat treatment. Uniform heating of a structure or a portion thereof to a sufficient temperature to relieve the major portion of the residual stresses, followed by uniform cooling.

stringer bead. A type of weld bead made without appreciable weaving motion. See also *weave bead*.

stud arc welding (SW). An arc welding process which produces coalescence of metals by heating them with an arc between a metal stud, or similar part, and the other work part. When the surfaces to be joined are properly heated, they are brought together under pressure. Partial shielding may be obtained by the use of a ceramic ferrule surrounding the stud. Shielding gas or flux may or may not be used.

stud welding. A general term for the joining of a metal stud or similar part to a workpiece. Welding may be accomplished by arc, resistance, friction, or other suitable process with or without external gas shielding.

submerged arc welding (SAW). An arc welding process which produces coalescence of metals by heating them with an arc or arcs between a bare metal electrode or electrodes and the work. The arc and molten metal are shielded by a blanket of granular, fusible material on the work. Pressure is not used, and filler metal is obtained from the electrode and sometimes from a supplemental source(s) (welding rod, flux, or metal granules).

substrate. Any base material to which a thermal sprayed coating or surfacing weld is applied.

suck-back. See preferred term *concave root surface*.

surface preparation. The operations necessary to produce a desired or specified surface condition.

surface roughening. A group of procedures for producing irregularities on a surface. See *blasting, dovetailing, groove and rotary roughening, rotary roughening, rough threading*, and *threading and knurling*.

surfacing. The deposition of filler metal (material) on a base metal (substrate) to obtain desired properties or dimensions. See also *buttering, cladding, coating*, and *hardfacing*.

surfacing weld. A type of weld composed of one or more stringer or weave beads deposited on an unbroken surface to obtain desired properties or dimensions.

sweat soldering. A soldering process variation in which two or more parts which have been precoated with solder are reheated and assembled into a joint without the use of additional solder.

synchronous initiation (resistance welding). The initiation and termination of each half-cycle of welding transformer primary current so that all half-cycles of such current are identical in making spot and seam welds or in making projection welds.

synchronous timing. See preferred term *synchronous initiation*.

T

tack weld. A weld made to hold parts of a weldment in proper alignment until the final welds are made.

taper delay time. The time interval after upslope during which the current is constant.

taper time. The time interval when current increases or decreases continuously from the welding current to final taper current.

taps. Connections to a transformer winding which are used to vary the transformer turns ratio, thereby controlling welding voltage and current.

temper time (resistance welding). That part of the postweld interval following quench time to the beginning of hold time in resistance welding.

temporary weld. A weld made to attach a piece or pieces to a weldment for temporary use in handling, shipping, or working on the weldment.

theoretical electrode force. See *electrode force*.

theoretical throat. See *throat of a fillet weld*.

thermal cutting. A group of cutting processes which melts the metal (material) to be cut. See *arc cutting, oxygen cutting, electron beam cutting*, and *laser beam cutting*.

thermal spraying (THSP). A group of welding or allied processes in which finely divided metallic or nonmetallic materials are deposited in a molten or semi-molten condition to form a coating. The coating material may be in the form of powder, ceramic rod, wire, or molten materials. See *flame spraying, plasma spraying,* and *electric arc spraying*.

thermal spraying gun. A device for heating, feeding, and directing the flow of a thermal spraying material.

thermal stresses. Stresses in metal resulting from nonuniform temperature distributions.

thermit crucible. The vessel in which the thermit reaction takes place.

thermit mixture. A mixture of metal oxide and finely divided aluminum with the addition of alloying metals as required.

thermit mold. A mold formed around the parts to be welded to receive the molten metal.

thermit reaction. The chemical reaction between metal oxide and aluminum which produces superheated molten metal and aluminum oxide slag.

thermit welding (TW). A welding process which produces coalescence of metals by heating them with superheated liquid metal from a chemical reaction between a metal oxide and aluminum, with or without the application of pressure. Filler metal, when used, is obtained from the liquid metal.

thermocompression bonding. See preferred term *hot pressure welding*.

threading and knurling (thermal spraying). A method of surface roughening in which spiral threads are prepared, followed by upsetting with a knurling tool.

throat depth (resistance welding). The distance from the center line of the electrodes or platens to the nearest point of interference for flat sheets in a resistance welding machine. In the case of a resistance seam welding machine with a universal head, the throat depth is measured with the machine arranged for transverse welding.

throat height (resistance welding). The unobstructed dimension between arms throughout the throat depth in a resistance welding machine.

throat of a fillet weld.

theoretical throat. The distance from the beginning of the root of the joint perpendicular to the hypotenuse of the largest right triangle that can be inscribed within the fillet weld cross section. This dimension is based on the assumption that the root opening is equal to zero.

actual throat. The shortest distance from the root of weld to its face.

effective throat. The minimum distance minus any reinforcement from the root of weld to its face.

throat of a groove weld. See preferred term *size of weld*.

throat opening. See preferred term *horn spacing*.

TIG welding. See preferred term *gas tungsten arc welding*.

tinning. See preferred term *precoating*.

tip skid. See preferred term *electrode skid*.

T-joint. A joint between two members located approximately at right angles to each other in the form of a T.

toe crack. A crack in the base metal occurring at the toe of a weld.

toe of weld. The junction between the face of a weld and the base metal.

torch. See preferred terms *cutting torch* and *welding torch*.

torch brazing (TB). A brazing process in which the heat required is furnished by a fuel gas flame.

torch soldering (TS). A soldering process in which the heat required is furnished by a fuel gas flame.

torch tip. See preferred terms *welding tip* or *cutting tip*.

transferred arc (plasma arc welding). A plasma arc established between the electrode and the workpiece.

transverse resistance seam welding. The making of a resistance seam weld in a direction essentially at right angles to the throat depth of a resistance seam welding machine.

travel angle. The angle that the electrode makes with a reference line perpendicular to the axis of the weld in the plane of the weld axis. See also *drag angle* and *push angle*. Note: This angle can be used to define the position of welding guns, welding torches, high energy beams, welding rods, thermal cutting and thermal spraying torches, and thermal spraying guns.

travel angle (pipe). The angle that the electrode makes with a reference line extending from the center of the pipe through the molten weld pool in the plane of the weld axis. Note: This angle can be used to define the position of welding guns, welding torches, high energy beams, welding rods, thermal cutting and thermal spraying torches, and thermal spraying guns.

travel start delay time. The time interval from arc initiation to the start of work or torch travel.

travel stop delay time. The time interval from beginning of downslope time or crater fill time to shut-off of torch or work travel.

tungsten electrode. A non-filler metal electrode used in arc welding or cutting, made principally of tungsten.

twin carbon arc brazing (TCAB). A brazing process which produces coalescence of metals by heating them with an electric arc between two carbon electrodes. The filler metal is distributed in the joint by capillary attraction.

twin carbon arc welding (CAW-T). A carbon arc welding process variation which produces coalescence of metals by heating them with an electric arc between two carbon electrodes. No shielding is used. Pressure and filler metal may or may not be used.

U

ultrasonic coupler (ultrasonic soldering and ultrasonic welding). Elements through which ultrasonic vibration is transmitted from the transducer to the tip.

ultrasonic soldering. A soldering process variation in which high frequency vibratory energy is transmitted through molten solder to remove undesirable surface films and thereby promote wetting of the base metal. This operation is usually accomplished without a flux.

ultrasonic welding (USW). A solid-state welding process which produces coalescence of materials by the local application of high frequency vibratory energy as the work parts are held together under pressure.

ultra-speed welding. See preferred term *commutator-controlled welding*.

underbead crack. A crack in the heat-affected zone generally not extending to the surface of the base metal.

undercut. A groove melted into the base metal adjacent to the toe or root of a weld and left unfilled by weld metal.

underfill. A depression on the face of the weld or root surface extending below the surface of the adjacent base metal.

unipolarity operation. A resistance welding process variation in which succeeding welds are made with pulses of the same polarity.

upset. Bulk deformation resulting from the application of pressure in welding. The upset may be measured as a percent increase in interfacial area, a reduction in length, or a percent reduction in thickness (for lap joints).

upset butt welding. See preferred term *upset welding*.

upsetting force. The force exerted at the faying surfaces during upsetting.

upsetting time. The time during upsetting.

upset weld. A weld made by upset welding.

upset welding (UW). A resistance welding process which produces coalescence simultaneously over the entire area of abutting surfaces or progressively along a joint by the heat obtained from resistance to electric current through the area where those surfaces are in contact. Pressure is applied before heating is started and is maintained throughout the heating period.

upslope time (resistance welding). The time during which the welding current continuously increases from the beginning of welding current.

upslope time (automatic arc welding). The time during which the current changes continuously from initial current value to the welding value.

V

van der Waals bond. A secondary bond arising from the fluctuating dipole nature of an atom with all occupied electron shells filled.

vacuum brazing. A nonpreferred term used to denote various brazing processes which take place in a chamber or retort below atmospheric pressure.

vertical position. The position of welding in which the axis of the weld is approximately vertical.

vertical position (pipe welding). The position of a pipe joint in which welding is performed in the horizontal position and the pipe may or may not be rotated.

voltage regulator. An automatic electrical control device for maintaining a constant voltage supply to the primary of a welding transformer.

W

water wash (thermal spraying). The forcing of exhaust air and fumes from a spray booth through water so that the vented air is free of thermal sprayed particles or fumes.

wave soldering (WS). An automatic soldering process where work parts are automatically passed through a wave of molten solder. See also *dip soldering*.

wax pattern (thermit welding). Wax molded around the parts to be welded to the form desired for the completed weld.

weave bead. A type of weld bead made with transverse oscillation.

weld. A localized coalescence of metals or nonmetals produced either by heating the materials to suitable temperatures, with or without the application of pressure, or by the application of pressure alone and with or without the use of filler material.

weldability. The capacity of a material to be welded under the fabrication conditions imposed into a specific, suitably designed structure and to perform satisfactorily in the intended service.

weld bead. A weld deposit resulting from a pass. See *stringer bead* and *weave bead*.

weldbonding. A joining method which combines resistance spot welding or resistance seam welding with adhesive bonding. The adhesive may be applied to a faying surface before welding or may be applied to the areas of sheet separation after welding.

weld brazing. A joining method which combines resistance welding with brazing.

weld crack. A crack in weld metal.

weld-delay time (resistance welding). The amount of time the beginning of welding current is delayed with respect to the initiation of the forge-delay timer in order to synchronize the forging force with welding current flow.

welder. One who performs a manual or semiautomatic welding operation. (Sometimes erroneously used to denote a welding machine.)

welder certification. Certification in writing that a welder has produced welds meeting prescribed standards.

welder performance qualification. The demonstration of a welder's ability to produce welds meeting prescribed standards.

welder registration. The act of registering a welder certification or a photostatic copy thereof.

weld gage. A device designed for checking the shape and size of welds.

weld-heat time (resistance welding). The time from the beginning of welding current to the beginning of post-heat time.

welding. A materials joining process used in making welds.

welding current. The current in the welding circuit during the making of a weld.

welding current (automatic arc welding). The current in the welding circuit during the making of a weld, but excluding upslope, downslope, start, and crater fill current.

welding current (resistance welding). The current in the welding circuit during the making of a weld, but excluding preweld or postweld current.

welding cycle. The complete series of events involved in the making of a weld.

welding electrode. See preferred term *electrode*.

welding force. See preferred terms *electrode force* and *platen force*.

welding generator. A generator used for supplying current for welding.

welding ground. See preferred term *work connection*.

welding head. The part of a welding machine or automatic welding equipment in which a welding gun or torch is incorporated.

welding leads. The work lead and electrode lead of an arc welding circuit.

welding machine. Equipment used to perform the welding operation. For example, spot welding machine, arc welding machine, seam welding machine, etc.

welding operator. One who operates machine or automatic welding equipment.

welding position. See *flat position, horizontal position, horizontal fixed position, horizontal rolled position, inclined position, overhead position,* and *vertical position*.

welding pressure. The pressure exerted during the welding operation on the parts being welded. See also *electrode force* and *platen force*.

welding procedure. The detailed methods and practices including all welding procedure specifications involved in the production of a weldment. See *welding procedure specification*.

welding procedure specification (WPS). A document providing in detail the required variables for a specific application to assure repeatability by properly trained welders and welding operators.

welding process. A materials joining process which produces coalescence of materials by heating them to suitable temperatures, with or without the application of pressure or by the application of pressure alone, and with or without the use of filler metal.

welding rectifier. A device in a welding machine for converting alternating current to direct current.

welding rod. A form of filler metal used for welding or brazing which does not conduct the electrical current.

welding sequence. The order of making the welds in a weldment.

welding technique. The details of a welding procedure which are controlled by the welder or welding operator.

welding tip. A welding torch tip designed for welding.

welding torch (arc). A device used in the gas tungsten and plasma arc welding processes to control the position of the electrode, to transfer current to the arc, and to direct the flow of shielding and plasma gas.

welding torch (oxyfuel gas). A device used in oxyfuel gas welding, torch brazing, and torch soldering for directing the heating flame produced by the controlled combustion of fuel gases.

welding transformer. A transformer used for supplying current for welding. See also *reactor* (arc welding).

welding voltage. See *arc voltage*.

welding wheel. See preferred term *electrode*.

welding wire. See preferred terms *electrode* and *welding rod*.

weld interface. The interface between weld metal and base metal in a fusion weld, between base metals in a solid-state weld without filler metal, or between filler metal and base metal in a solid-state weld with filler metal and in a braze.

weld interval (resistance welding). The total of all heat and cool time when making one multiple-impulse weld.

weld interval timer (resistance welding). A device which controls heat and cool times and weld interval when making multiple-impulse welds singly or simultaneously.

weld length. See preferred term *effective length of weld*.

weld line. See preferred term *weld interface*.

weldment. An assembly whose component parts are joined by welding.

weld metal. That portion of a weld which has been melted during welding.

weld metal area. The area of the weld metal as measured on the cross section of a weld.

weldor. See preferred term *welder*.

weld pass. A single progression of a welding or surfacing operation along a joint, weld deposit, or substrate. The result of a pass is a weld bead, layer, or spray deposit.

weld penetration. See preferred terms *joint penetration* and *root penetration*.

weld size. See preferred term *size of weld*.

weld tab. Additional material on which the weld may be initiated or terminated.

weld time (automatic arc welding). The time interval from the end of start time or end of upslope to beginning of crater fill time or beginning of downslope.

weld time (resistance welding). The time that welding current is applied to the work in making a weld by single-impulse welding or flash welding.

weld timer. A device which controls only the weld time in resistance welding.

weld voltage. See *arc voltage*.

wetting. The phenomenon whereby a liquid filler metal or flux spreads and adheres in a thin continuous layer on a solid base metal.

wiped joint. A joint made with solder having a wide melting range and with the heat supplied by the molten solder poured onto the joint. The solder is manipulated with a hand-held cloth or paddle so as to obtain the required size and contour.

wire flame spraying. A thermal spraying process variation in which the material to be sprayed is in wire or rod form. See *flame spraying (FLSP)*.

wire feed speed. The rate of speed in mm/s or in./min at which a filler metal is consumed in arc welding or thermal spraying.

wire straightener. A device used for controlling the cast of coiled wire to enable it to be easily fed into the gun.

work angle. The angle that the electrode makes with the referenced plane or surface of the base metal in a plane perpendicular to the axis of the weld. See also *drag angle* and *push angle*. Note: This angle can be used to define the position of welding guns, welding torches, high energy beams, welding rods, thermal cutting and thermal spraying torches, and thermal spraying guns.

work angle (pipe). The angle that the electrode makes with the referenced plane extending from the center of the pipe through the molten weld pool. Note: This angle can be used to define the position of welding guns, welding torches, high energy beams, welding rods, thermal cutting and thermal spraying torches, and thermal spraying guns.

work coil. The inductor used when welding, brazing, or soldering with induction heating equipment.

work connection. The connection of the work lead to the work.

work lead. The electric conductor between the source of arc welding current and the work.

Principles of Joining Metallurgy

By Jack H. Devletian
Associate Professor
Oregon Graduate Center
and
William E. Wood
Professor and Chairman
Oregon Graduate Center

JOINING METALLURGY has assumed an even greater role in the fabrication of metals within the last two decades, largely because of the development of new alloys with tremendously increased strength and toughness. Therefore, a working knowledge of metallurgy is essential to understanding current engineering structures and, in particular, the mechanisms that control weldment performance. Fundamental to joining metallurgy are the microstructures of a weld joint, which determine the mechanical properties, and welding variables such as weld thermal cycle, chemical reactions in the molten pool, alloying, flux composition, and contaminants, which significantly affect the weld and heat-affected zone (HAZ) microstructures.

In this article, a general introduction to metallurgy provides a background of terminology and fundamentals. Because of their importance in welding modern alloys, several metallurgical phenomena are discussed, including cold working, recrystallization, grain growth, aging, and tempering, as well as the utilization of equilibrium phase diagrams and nonequilibrium time-temperature transformation (TTT) and continuous cooling transformation (CCT) diagrams. More detailed explanations are given for the following welding subjects: solidification, heat flow, microstructure, mechanical properties of welds, preheating and postweld heat treatment, and causes and remedies of common welding flaws. Although efforts have been made to use English units of measure throughout this article, some metric notations have been used in heat flow calculations.

General Metallurgy

Phases in Metals. Single-phase metals, which are very prevalent in structural metallurgical systems, retain the same crystal structure at all temperatures up to the melting temperature. Most pure metals are single phase. They include copper, nickel, aluminum, lead, platinum, gold, and silver, which are face-centered cubic (fcc); chromium, niobium, molybdenum, tungsten, and vanadium, which are body-centered cubic (bcc); and beryllium, cadmium, magnesium, zinc, and rhenium, which are hexagonal close-packed (hcp). Metals exhibiting more than one crystal structure are allotropic or polymorphic in behavior. Titanium, for example, is hcp below 1621 °F and bcc above 1621 °F. Iron is bcc below 1674 °F and above 2541 °F, but fcc between that temperature range. Other metals exhibiting allotropy are cobalt, zirconium, tin, and uranium.

Combinations of different single-phase metals can be melted together also to form a single-phase alloy system, such as in the copper-nickel system shown in Fig. 1. This isomorphous phase diagram defines the regions of temperature and composition in which the liquid phase and the α solid-solution phase occur in an alloy system under conditions of constant atmospheric pressure. Figure 1 also demonstrates several interesting phenomena. Although the pure elements have discrete melting points, such as nickel at 2647 °F and copper at 1981 °F, any alloy of nickel and copper melts and, conversely, solidifies over a range of temperatures. For example, when nickel and copper mixtures are melted together, a single liquid phase (L) will exist at temperatures above the line indicating liquidus, while a single solid fcc phase (α) will exist for all temperatures below the line indicating solidus. Unlike the pure elements, a two-phase region of liquid and solid exists in the temperature and compositional range between the liquidus and solidus for a single-phase alloy system.

To exhibit unlimited solid-solution formation, an alloy system must satisfy the rules of Hume-Rothery, which require that the two elements have:

- Atomic radii that do not differ by more than 15%
- The same crystal structure
- Similar electronegativity values (elements should be close to one another in the periodic table)
- The same valency

There are many commercially available single-phase nickel-copper alloys ranging from the nickel-rich Monels to the high-conductivity copper-rich alloys. Additionally, large numbers of isomorphous alloy systems are based on iron-chromium, iron-vanadium, tungsten-molybdenum, chromium-molybdenum, and others. Although isomorphous alloys are not limited to two-component systems, they are much more common.

Eutectic binary alloy systems have components that do not conform with these Hume-Rothery rules. Therefore, these alloys do not exhibit unlimited solid solubility at room temperature, existing instead as a mixture of two different solid phases. Some examples of such alloy systems include aluminum-copper, aluminum-silicon, aluminum-magnesium, lead-tin, and nickel-titanium.

Fig. 1 Isomorphous phase diagram of the copper-nickel alloy system. Source: Ref 3

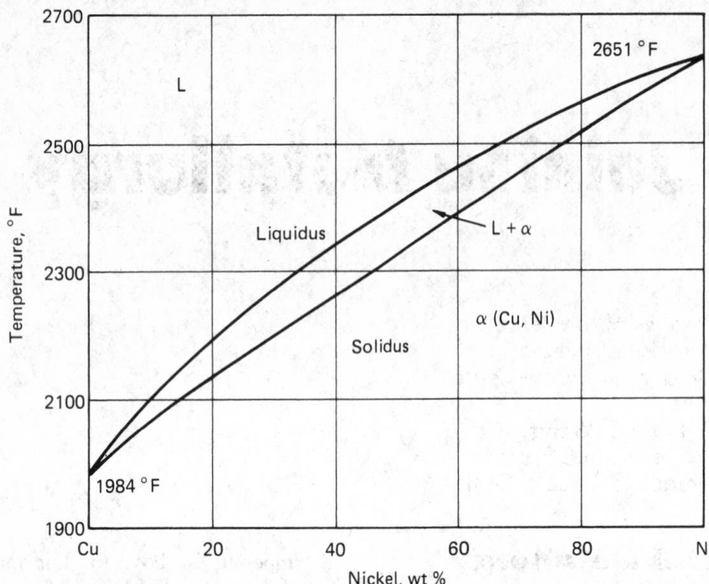

Fig. 2 Eutectic phase diagram of the aluminum-copper alloy system. Source: Ref 3

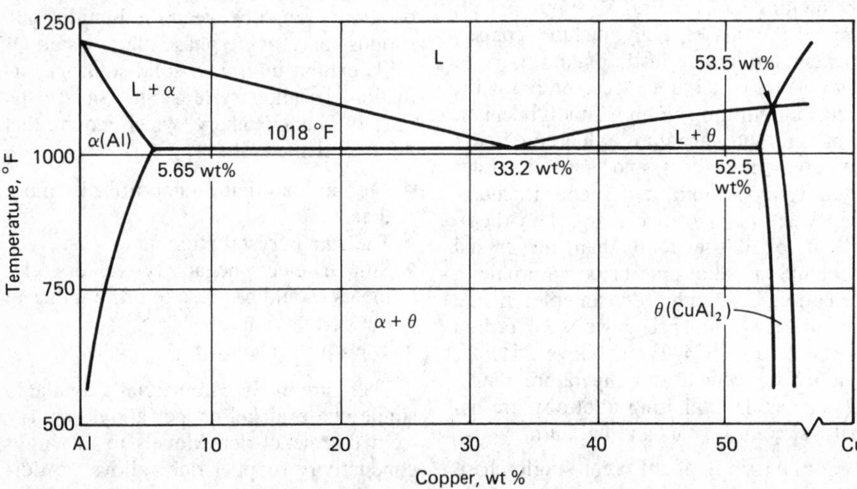

In the aluminum-copper phase diagram (Fig. 2), the liquidus temperature decreases to a minimum at 33.2% Cu where solidification occurs at a single temperature (1018 °F) as though the alloy were a pure metal. The eutectic reaction that occurs on heating or cooling across the eutectic isotherm (horizontal line in Fig. 2) at 1018 °F between the compositional limits of 5.65 and 52.5% Cu is:

$$\text{L}(33.2\%\ \text{Cu}) \underset{\text{heating}}{\overset{\text{cooling}}{\rightleftarrows}} \alpha(5.65\%\ \text{Cu}) + \Theta(52.5\%\ \text{Cu})$$

The resulting solid alloy consists of a mixture of two distinct phases (α Θ), where α is fcc and Θ is body-centered tetragonal (bct). At all compositions in the aluminum-copper system except pure aluminum, Θ at 53.5 and 33.2% Cu, the liquid solidifies over a range of temperatures. The low-melting alloy at 33.2% Cu is called the eutectic composition.

Another example of a eutectic system can be seen in the iron-carbon diagram (Fig. 3) at 2098 °F between the compositional limits of 2.11 and 6.69% C where:

$$\text{L}(4.3\%\ \text{C}) \underset{\text{heating}}{\overset{\text{cooling}}{\rightleftarrows}} \gamma(2.11\%\ \text{C}) + \text{Fe}_3\text{C}(6.69\%\ \text{C})$$

The resulting eutectic mixture at room temperature is called ledeburite. Eutectic alloys generally have broad applications in casting of complex shapes.

The peritectic reaction occurs in many engineering alloy systems and is characterized by the simultaneous solidification of a liquid and solid to form a new solid. The reaction occurring at 2723 °F between 0.09 to 0.53% C in the iron-carbon system (Fig. 4) is the peritectic type from which austenite (γ) in steel is formed from liquid and δ ferrite on cooling or:

$$\delta(0.09\%\ \text{C}) + \text{L}(0.53\%\ \text{C}) \underset{\text{heating}}{\overset{\text{cooling}}{\rightleftarrows}} \gamma(0.17\%\ \text{C})$$

Many binary alloy systems, such as iron-carbon, iron-nickel, iron-manganese, copper-zinc, and silver-platinum, contain peritectic reactions.

In several systems, the solid-state transformation reaction occurs where a single-phase solid of a particular composition transforms to a mixture of two other phases on cooling. This phase transformation is called the eutectoid reaction and is involved in all compositions of steel. Significantly, the iron-carbon alloy system exhibits all three types of phase transformation—peritectic, eutectic, and eutectoid reactions. This system has the most widely studied eutectoid reaction, which occurs at 1340 °F from 0.0218 to 6.69% C where upon slow cooling:

$$\text{austenite}\ (0.77\%\ \text{C}) \rightarrow \text{ferrite}\ (0.0218\%\ \text{C}) + \text{Fe}_3\text{C}(6.69\%\ \text{C})$$

The lamellar eutectoid (ferrite and Fe_3C) occurring at 0.77% C is called pearlite.

Figure 5 illustrates the microstructures obtained when 0, 0.2, and 0.77% C steels are slowly cooled from the austenite range. The iron (0% C), containing absolutely no carbon, is pure bcc ferrite. The 0.2% C steel contains approximately 25% pearlite (dark) and 75% ferrite (light). The 0.77% C steel contains 100% pearlite.

Cold Working. Fabrication processes that reduce the cross section of a metal, strengthen a metal by a phenomenon called work hardening (also known as strain hardening), and typically reduce the elongation of a metal are considered cold working processes. Processes such as ex-

Fig. 3 Peritectic, eutectic, and eutectoid reaction isotherms for the iron-carbon system. Source: Ref 3

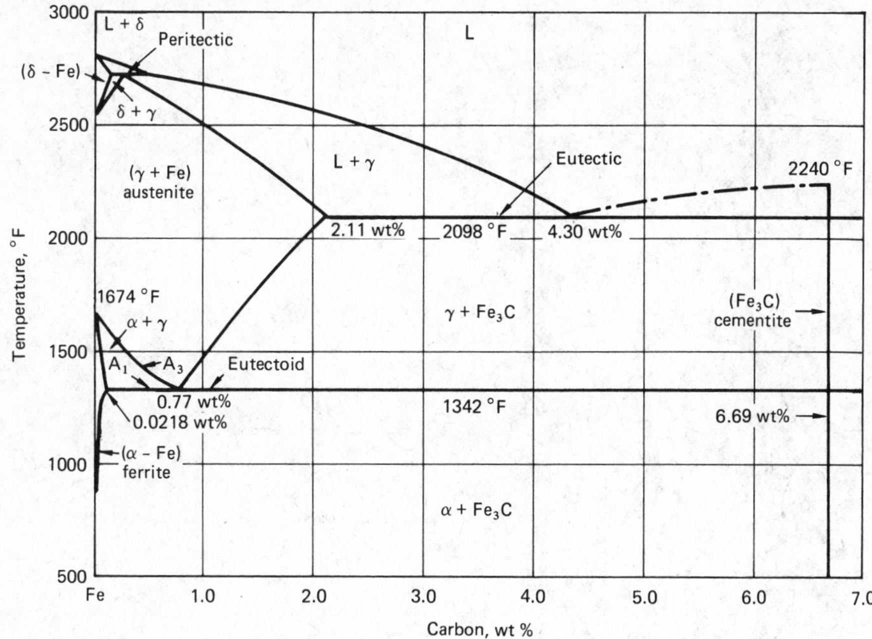

Fig. 4 Peritectic portion of the iron-carbon system

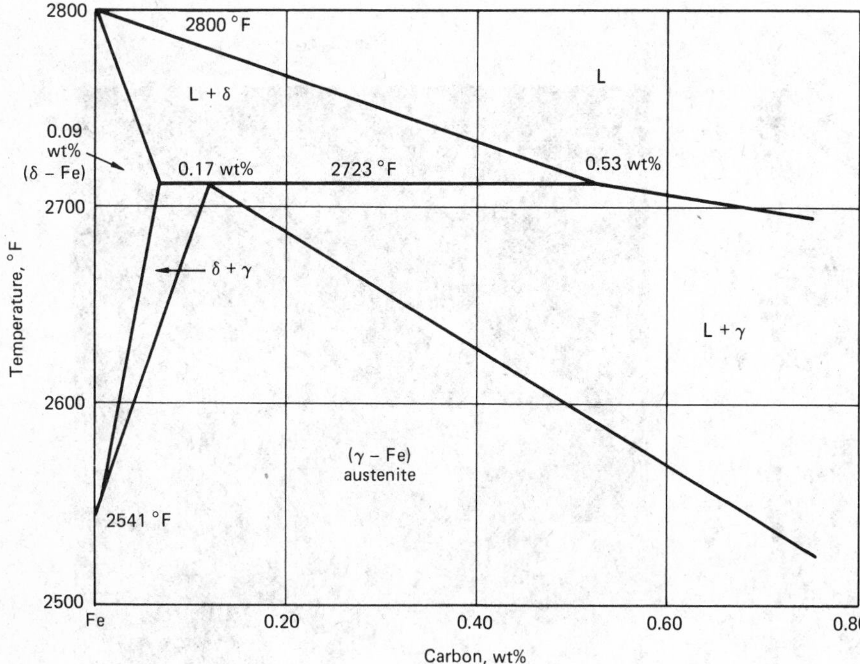

trusion, forging, rolling, and swaging strengthen as well as fabricate the material. Some of these processes employ elevated temperatures, while others are conducted at ambient temperatures. The term "cold working" is used generically to indicate the mechanical (fabrication) processes, provided that the temperature of working does not alter the structural changes produced by the process.

The purpose of cold working is to strengthen metals that are inherently very weak and ductile, such as pure aluminum and copper. The process of cold working produces material with grains that are distorted in comparison to grains in the annealed condition. The severity of distortion is directly proportional to the percent reduction of cross-sectional area of the workpiece. Severe cold working, however, often leads to a reduction in ductility. For restored ductility, the cold worked material is heated to an elevated temperature to permit microstructural changes. The effects of temperature on cold worked (material) structure can be divided into two stages: recrystallization and grain growth.

Recrystallization. The process of recrystallization begins when a cold worked material is heated to a suitable temperature—the recrystallization temperature. In recrystallization, the distorted grains are replaced by new, small, stress-free, equiaxed grains having greater ductility. As the degree of force exerted in the initial cold work increases, the recrystallization temperature becomes lower and the resulting grain size becomes smaller. As the recrystallization temperature decreases, the recrystallization time increases until a limiting temperature is reached below which recrystallization will not occur.

Grain Growth. Holding the material at a temperature substantially above the minimum recrystallization temperature results in grain growth, which occurs because the grains have a tendency to decrease their surface energy. Large grains have smaller grain-boundary area per unit volume and, hence, smaller surface energy. Because the larger grains grow at the expense of smaller grains, energy is conserved. Significant grain growth occurs at temperatures above $T_m/2$, where T_m is the melting temperature of the metal in degrees kelvin; for example, $T_m/2$ for aluminum is 872 °F. Annealing of metals is performed above $T_m/2$ to obtain excellent ductility.

Age Hardening and Tempering Precipitation. Although the procedures for both are essentially the same and involve heating a supersaturated solid-solution alloy to a carefully selected temperature until the desired precipitation reaction occurs, the objectives of age hardening (precipitation hardening) and tempering precipitation are totally opposite. The best age-hardening effects occur for alloy compositions on the phase diagram where maximum solid solubility occurs. For ex-

Fig. 5 Typical microstructures of fully annealed iron and steel. (a) Pure iron. (b) 0.2% C steel. (c) 0.77% C steel (pearlitic structure). Magnification: (a) ×180; (b) ×90; (c) ×900. Etchant: 2% nital

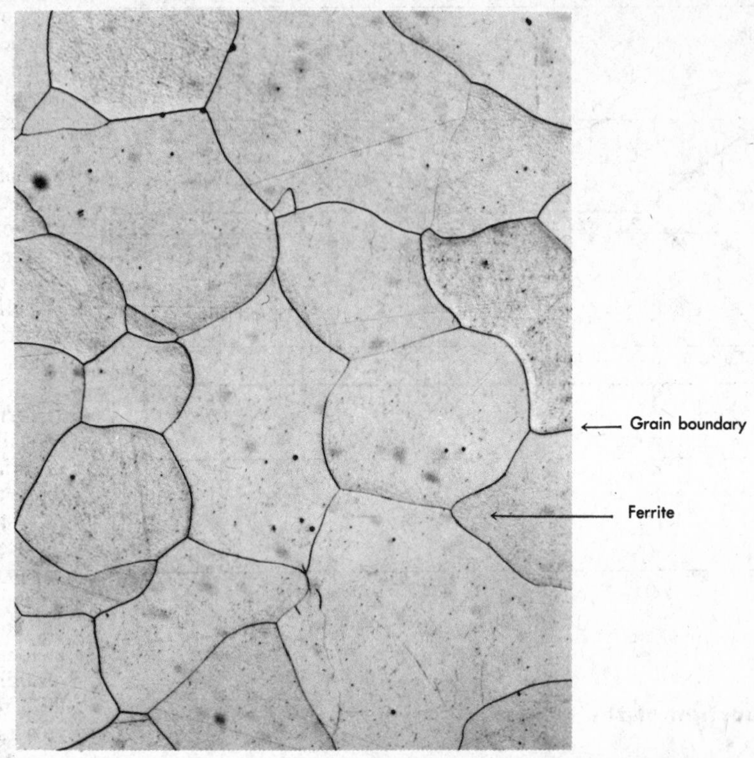

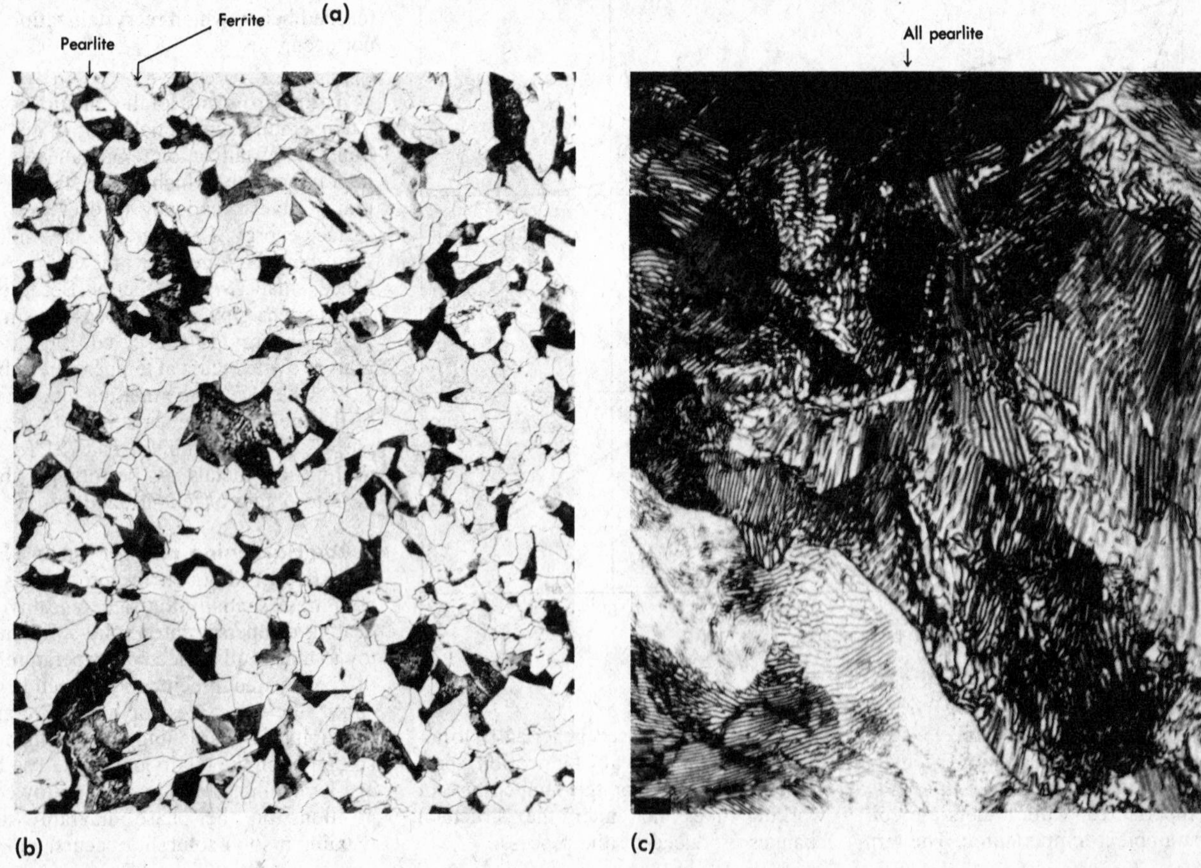

ample, in Fig. 2, the composition for maximum precipitation hardening occurs at approximately 4 to 6% Cu. Such an alloy is normally heated to a temperature just below the eutectic reaction temperature to about 950 °F to dissolve all (or nearly all) of the copper in solid solution. This alloy is then rapidly quenched in water to obtain a supersaturated solid solution at room temperature. In the as-quenched condition, the alloy is soft and ductile. When the alloy is then aged at an intermediate temperature, such as 375 °F, strength increases rapidly with aging time because of the precipitation of microscopic particles approaching the Θ composition. After a maximum strength level is achieved, further aging results in a reduced strength level, called overaging, because of the excess precipitation of large-sized Θ phase.

Tempering, in contrast, only applies to metals that transform upon quenching to martensitic structures. Although the martensitic reaction occurs in a number of alloy systems, only in steel is the greatest commercial benefit derived. Martensite does not appear on the iron-carbon diagram because it is a nonequilibrium phase that forms only upon rapid cooling from the austenite range. Because as-quenched martensite is usually too brittle for engineering structures, tempering at a temperature between 390 and 1200 °F provides an ideal means to obtain the desired strength, ductility, and toughness for a given application. As the tempering temperature increases, the strength decreases and generally the ductility and toughness improve, except for possible tempered martensite embrittlement and temper embrittlement, which can range from about 570 to 750 °F and 750 to 1000 °F, respectively. The heat treatment of steel has become so essential to all types of structural uses that additional diagrams, such as TTT and CCT diagrams, have been developed to more fully utilize the properties of steels through nonequilibrium phase transformations.

Time-Temperature Transformation Diagrams. The iron-carbon phase diagram (Fig. 3) is valid only under equilibrium conditions, when heating or cooling is done at a very slow rate and usually held for a long time at the desired temperature range. The diagram does not take into account the kinetics of transformation reactions, such as the austenite decomposition. The formation of nonequilibrium (phases) microstructures, such as bainite and martensite, is the basis for nearly all alloy systems. Isothermal transformation or TTT graphs are used to predict the decomposition kinetics of austenite (γ) transformation to pearlite or bainite, as well as to proeutectoid ferrite or cementite.

Figure 6 shows one curve for a 0.77% C eutectoid steel. The beginning and the end of austenite decomposition as a function of temperature are shown by the curves labeled start and finish. These curves are experimentally determined by rapidly quenching austenitized samples of a given steel to temperatures below the eutectoid temperature, holding for various times, and then quenching to room temperature or below. Metallographic examination of these samples reveals the extent to which austenite has transformed just before the final quench. The type of microstructure obtained after γ decomposition depends on the intermediate holding temperature. Just below the eutectoid temperature, coarse pearlite with wide lamellar spacing is usually developed. As the isothermal transformation temperature is decreased, the pearlite becomes finer. At still lower temperatures (around 930 °F), an entirely new microstructure called bainite results. Bainite is a mixture of extremely fine carbides dispersed in a ferrite matrix. At temperatures below M_s (martensite start temperature), austenite directly transforms to martensite, normally by shear transformation.

Continuous Cooling Transformation Diagrams. Although a necessary aid to complement the equilibrium phase diagram, the TTT diagram does not represent most heat treatment conditions under which continuous cooling occurs and more complex microstructures are developed. Thus, CCT diagrams were developed for various steel compositions to characterize the microstructures present as a function of cooling time or cooling rate from the austenitizing temperature. Figure 7 shows CCT curves for a eutectoid steel.

Figure 7 also includes qualitative representations of water, oil, and air quenches with different cooling rates that result in varying microstructures and hardness. For example, consider a 1/2-in.-diam eutectoid (0.77% C) steel bar that is heated and held at a suitable austenitizing temperature of about 1600 °F, followed by water quenching. According to the CCT diagram in Fig. 7, this steel has the martensitic microstructure shown in Fig. 8(a) and a hardness of 840 HV. If the same bar is air cooled, the microstructure may be entirely pearlitic, as shown in Fig. 5(c), with a hardness of only 270 HV. Finally, consider a 0.060-in.-diam wire that is austenitized and air cooled to room temperature. This microstructure (Fig. 8b) contains both bainite (dark) and martensite (light) and has a hardness of 560 HV. Thus, a steel of a given composition can be heat treated

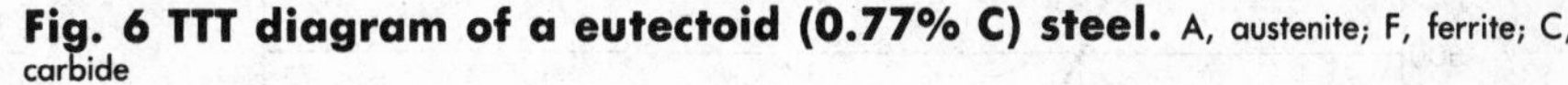

Fig. 6 TTT diagram of a eutectoid (0.77% C) steel. A, austenite; F, ferrite; C, carbide

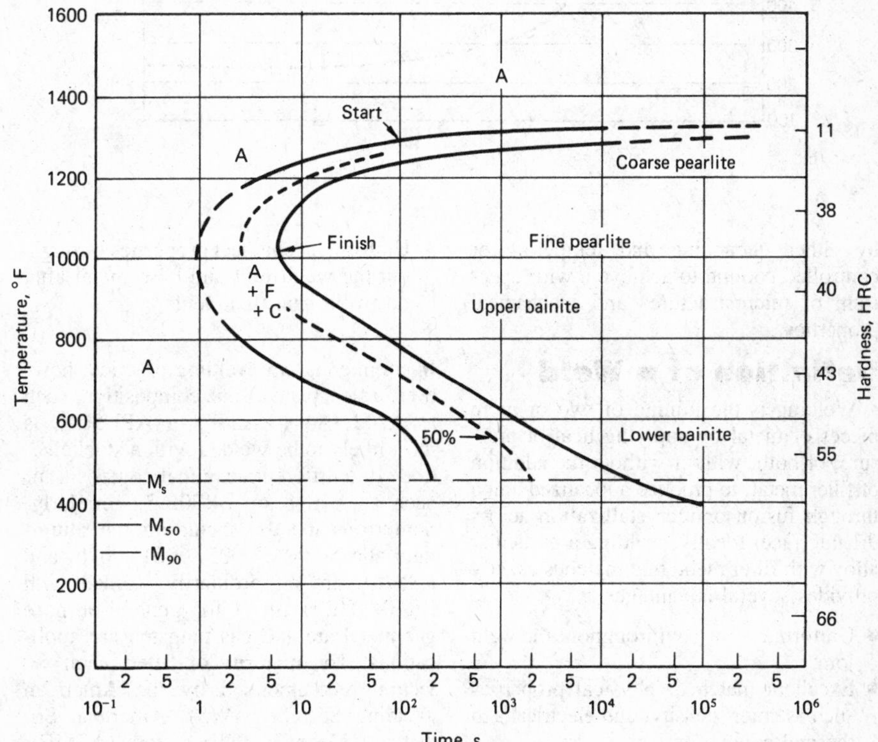

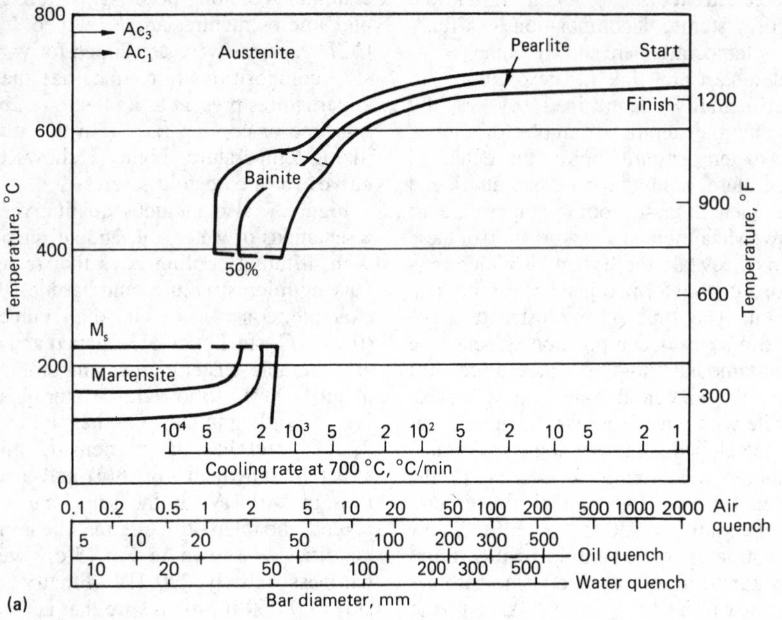

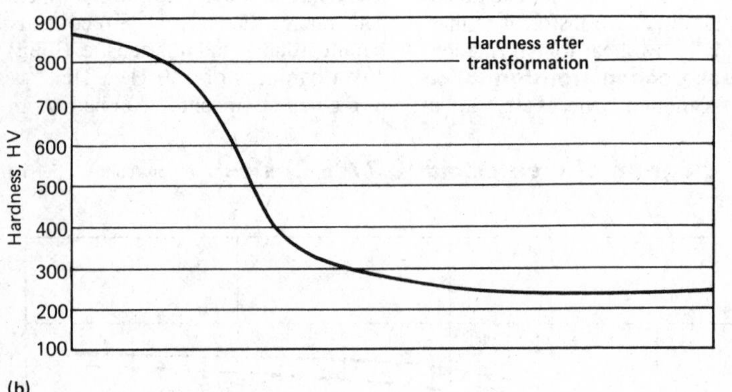

Fig. 7 CCT diagram of a eutectoid steel and accompanying hardness values. (a) CCT diagram. (b) Effect of the bar diameter on hardness after transformation (Ref 1). For conversion data, see the "Appendix on Metric and Conversion Data for Welding" in this Volume.

by either quenching and tempering or controlled cooling to achieve a wide spectrum of microstructures and mechanical properties.

Definition of a Weld

Welding is the joining of two or more pieces of metal by applying heat or pressure, or both, with or without the addition of filler metal, to produce a localized union through fusion or recrystallization across the interface. Ideally, welding a particular alloy with filler metal that matches exactly provides several advantages:

- Uniform chemistry throughout the weld joint
- Excellent match of physical properties such as color, density, and electrical and thermal conductivities
- Uniform mechanical properties throughout the weld joint and base metal after postweld heat treatment

In commercial arc welding practice, however, a steel plate of one composition, such as A242, A441, A588, or API-5LX, is most likely to be welded with a steel electrode of a different chemical composition, such as E7018 or ER70S-3. Similarly, nonferrous metals, including the aluminum alloys 3004, 5005, 6061, 6070, and A357.0, are all ordinarily welded with ER4043 filler metal for general-purpose gas metal arc and gas tungsten arc applications. The majority of filler metal selection recommended by the American Welding Society (AWS), American Society of Mechanical Engineers (ASME), American Petroleum Institute (API), and military welding codes is based on providing crack-free welds and closely matching the tensile properties of the as-deposited filler metal with those of the base metal. The composition match, although important, is the secondary consideration.

As a result of the nonmatching filler metal and heat distribution characteristics, the weld joint is usually a chemically heterogeneous composite consisting of as many as six metallurgically distinct regions. Based on the work by Savage, Nippes, and Szekeres (Ref 2), a typical weld consists of (1) the composite zone, (2) the unmixed zone, (3) the weld interface, (4) the partially melted zone, (5) the heat-affected zone (HAZ), and (6) the unaffected base metal. These zones are illustrated in Fig. 9.

Composite Zone. The admixture of filler metal and melted base metal comprises a completely melted and homogeneous weld fusion zone in the composite zone or region. For example, when gray cast iron is welded with a nickel electrode, the composite region contains a homogeneous molten pool of nickel filler metal diluted with melted gray iron-based metal. Similarly, when E10018 electrodes are used to weld HY-80 steel, the chemical composition of the composite zone is the weighted average of the elements (i.e., carbon, nickel, or manganese) from both the filler metal and the melted base metal. Even totally dissimilar metals such as copper and nickel can be welded autogenously (no filler metal) to each other by gas tungsten arc welding (GTAW), and the bulk composition throughout the composite zone will be surprisingly uniform. Thorough mixing is promoted by forced convection in the molten pool combined with the substantial reduction of free energy contributed by the great increase in mixing entropy.

Unmixed Zone. The narrow region surrounding the bulk composite zone is the unmixed zone (Fig. 9), which consists of a boundary layer of melted base metal that froze before experiencing any mixing in the molten composite zone. This layer at the extremities of the weld pool is characterized by a composition essentially identical to the base metal with a typical thickness of about 0.05 to 0.10 in., depending on welding process and weld cooling rate. Although the unmixed zone is present in all fusion welds, it is readily visible only in those welds utilizing a filler-metal alloy of substantially different composition than the base metal.

Fig. 8 Microstructure of eutectoid steel. (a) Martensite. (b) Martensite (light) and bainite (dark) formed under continuous cooling condition. Magnification: ×700. Hardness: (a) 840 HV, (b) 600 HV. Etchant: 2% nital

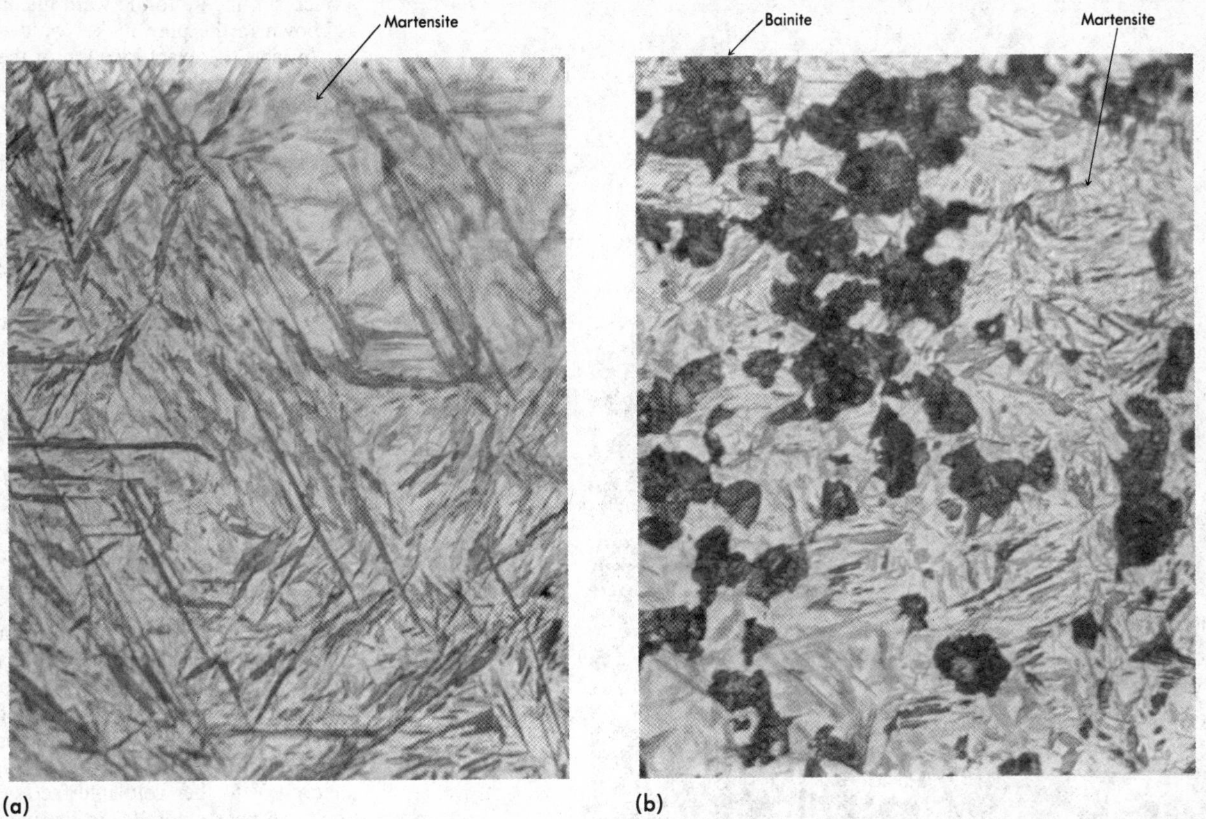

Fig. 9 Metallurgical zones developed in a typical weld. Source: Ref 2

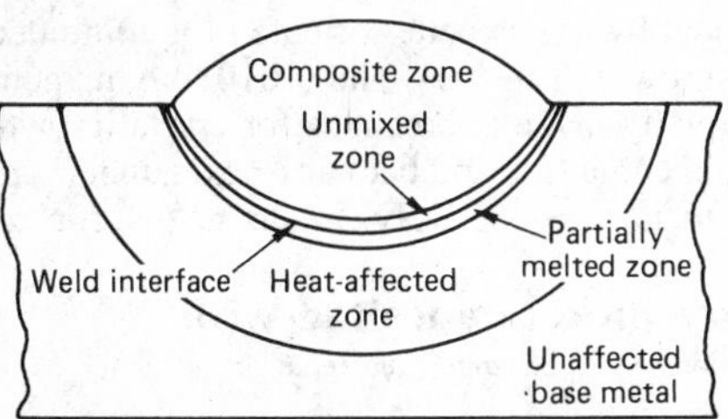

Consider, for example, the shielded metal arc welding (SMAW) of class 30 gray cast iron with nickel-rich filler metal. The unmixed zone is clearly visible (Fig. 10) because the molten gray iron solidifies as a white iron (with eutectic of Fe_3C plus γ) structure, while the composite zone, containing a majority of nickel filler metal, solidifies as austenite. Conversely, consider the gas tungsten arc welding of pure nickel deposited with a nickel filler in Fig. 11, in which no unmixed zone is visible, because the composite zone liquid composition and cooling conditions are identical to the liquid in the unmixed zone.

Weld Interface. The third region defined in a weldment is the weld interface. This surface clearly delineates the boundary between the unmelted base metal on one side and the solidified weld metal on the other side. Often in pure metals or very dilute alloys using matching filler metal, the transition from base metal to weld metal is difficult to observe metallographically, but can be revealed through alloy-sensitive etching of the solidification substructure. Generally, as the alloy content and the solidification range between liquidus and solidus of a given weld increase, the solidification structure is more easily revealed by etching.

Partially Melted Zone. In the base metal immediately adjacent to the weld interface where some localized melting may occur, the partially melted zone can be observed. In many alloys that contain low-melting inclusions and impurity or alloy segregation at grain boundaries, liquation of these low-melting microscopic regions may occur and extend from the weld interface into the partially melted zone. The depth to which a liquated region penetrates into the base metal depends on the solidus temperature of the liquated matter. In steels, the classic example of constitutional liquation in the partially melted zone occurs in HY-80 steel weldments. Typically, the liquation of manganese sulfide inclusions results in a hot crack or microfissure which extends from the unmixed zone, across the weld interface, and into the partially melted zone.

Heat-Affected Zone. The true HAZ (Fig. 9) is the portion of the weld joint which has experienced peak temperatures high enough to produce solid-state microstructural changes but too low to cause any melting. For example, this zone in single-phase wrought alloy is characterized by a steadily increasing grain size from the outer extremity of the HAZ to a maximum grain size at the weld interface, as shown in Fig. 11.

Unaffected Base Metal. Finally, that part of the workpiece that has not undergone any metallurgical change is the unaffected base metal. Although metallurgi-

Fig. 10 Weld deposited on gray iron with nickel filler metal. Composite zone is austenite. Unmixed zone is white iron. HAZ is martensite and undissolved graphite. Magnification: ×160. Etchant: 2% nital

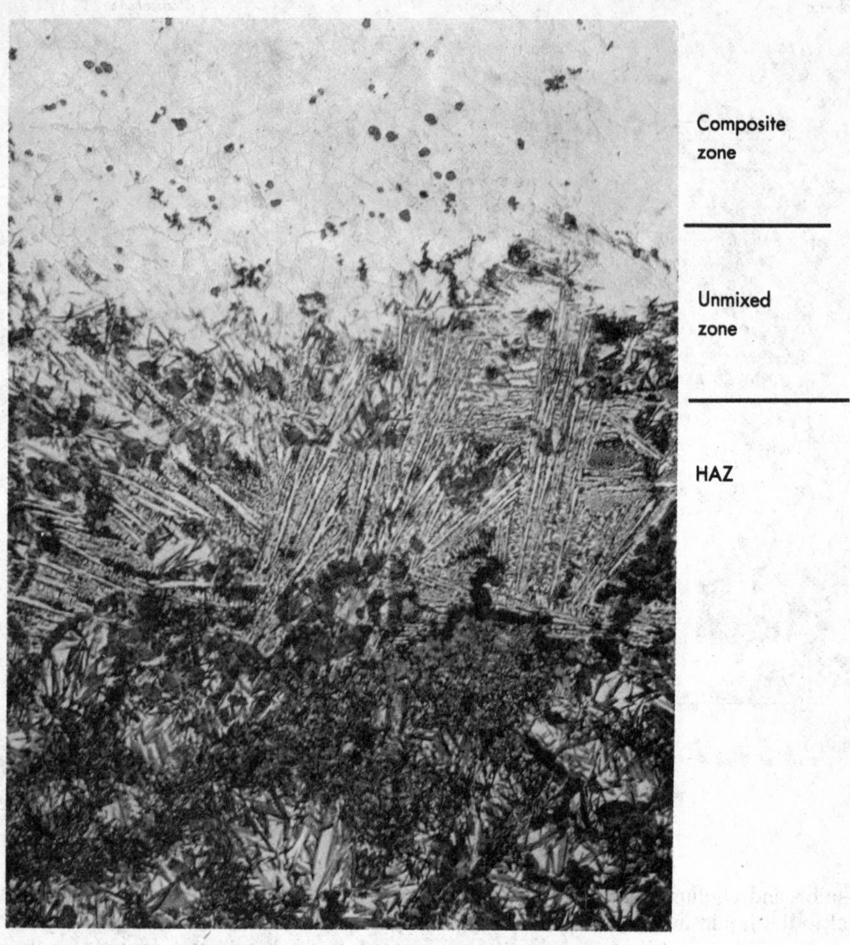

cally unchanged, the unaffected base metal, as well as the entire weld joint, is likely to be in a state of high residual transverse and longitudinal shrinkage stress, depending on the degree of restraint imposed on the weld.

Solidification of Welds

Epitaxial Growth. Fundamental solidification mechanics developed primarily for cast metals have been successfully applied to the solidification of welds. The outstanding difference between the solidification of a casting and that of a weld (aside from the relative size and cooling rates) is the phenomenon of epitaxial growth in welds. In castings, formation of solid crystals from the melt requires heterogeneous nucleation of solid particles, principally on the mold walls, followed by grain growth. In contrast, the nucleation event in welds is eliminated during the initial stages of solidification because of the mechanism of epitaxial growth wherein atoms from the molten weld pool are rapidly deposited on pre-existing lattice sites in the adjacent solid base metal. As a result, the structure and crystallographic orientation of the HAZ grains at the weld interface continue into the weld fusion zone as shown in the pure nickel weld in Fig. 11. In fact, the exact location of the weld interface is very difficult to determine in any weld deposited on pure metals using matching filler metal. Even microstructural features, such as annealing twins located in the HAZ weld joints, will continue to grow epitaxially into the weld during solidification. Similarly, nonmatching filler metals will also solidify epitaxially, particularly if the filler metal and base metal have the same crystal structure upon solidification, e.g., welding Monel (fcc) with nickel (fcc) filler metal.

Weld Pool Shape. Because it controls the grain structure of the weld, weld pool shape is an important factor in welding. For example, if a single-phase metal is gas tungsten arc welded at a low velocity, the weld pool is elliptical (nearly circular), as shown in Fig. 12(a). The columnar grains grow in the direction of the thermal gradient produced by the moving heat source (arc). The grains grow epitaxially from the base metal toward the arc. Because the direction of maximum temperature gradient is constantly changing from approximately 90° to the weld interface at position A to nearly parallel to the weld axis at position B, the grains must grow from position A and continuously turn toward the position of the moving arc. The process of "competitive growth" provides a means whereby grains less favorably oriented for growth are pinched off or crowded out by grains better oriented for continued growth. The $\langle 001 \rangle$ and $\langle 10\bar{1}0 \rangle$ are the generally favored directions for crystal growth in cubic (fcc and bcc) and hexagonal (hcp) metals, respectively. In fcc metals, for ex-

Fig. 11 Gas tungsten arc weld of pure nickel deposited with matching nickel filler metal. Magnification: ×20. Etchant: aqua regia

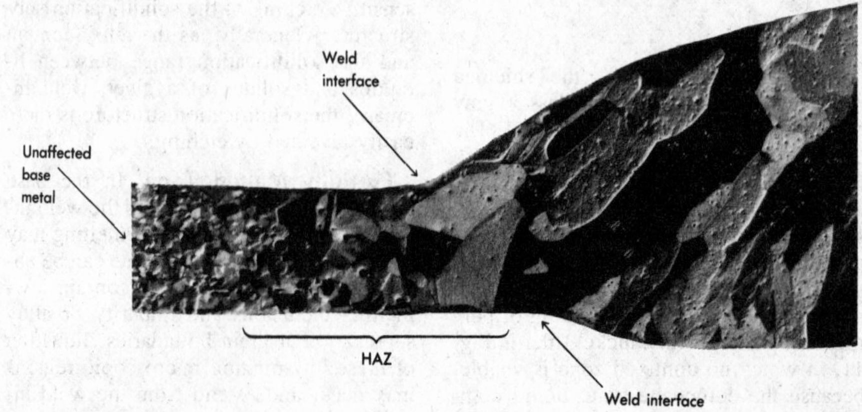

Fig. 12 Comparison of weld pool shapes. Travel speeds: (a) slow, (b) intermediate, (c) fast

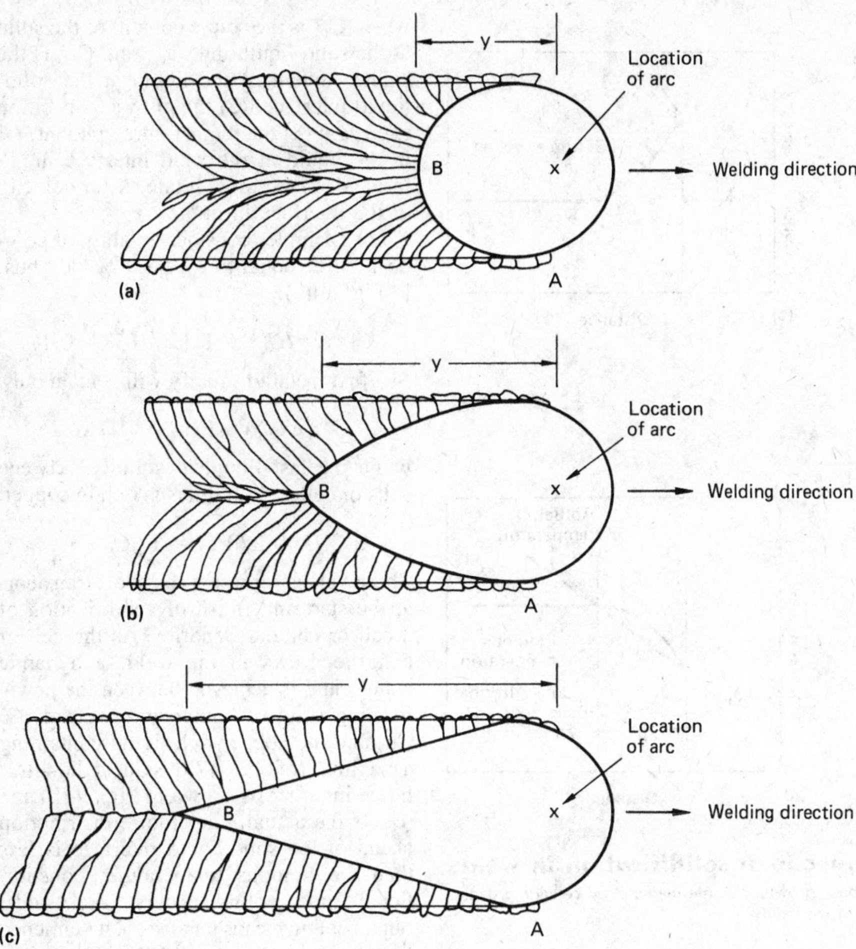

ample, the ⟨001⟩ most favored direction leads each solidifying grain because the four close-packed {111} planes symmetrically located around the ⟨001⟩ axes require the greatest time to solidify and, therefore, serve both to drag and guide the growth of solidifying grains.

The shape of the weld pool tends to become more elongated with increasing welding speed. In Fig. 12(b), the direction of maximum temperature gradient is perpendicular to the weld interface at positions A and B, but because the weld pool is trailing a greater distance behind the arc, the temperature gradient at position B is no longer strongly directed toward the electrode. Therefore, the columnar grains do not turn as much as in the case of a nearly circular weld pool.

Finally, the weld takes on a teardrop shape at the fast welding speeds that are usually encountered in commercial welding practice. The weld pool is elongated so far behind the welding arc that the directions of the maximum temperature gradient at position A and B in Fig. 12(c) have changed only slightly. As a result, the grains grow from the base metal and converge abruptly at the centerline of the weld with little change in direction. Welds that solidify in a teardrop shape have the poorest resistance to centerline hot cracking because low-melting impurities and other low-melting constituents tend to segregate at the centerline. Unfortunately, this solidification geometry occurs most frequently in commercial welding applications, because high heat input and fast travel speeds produce the most cost-effective method of welding.

Cells, Dendrites, and Microsegregation. Each columnar weld-metal grain may contain a solidification substructure. No substructures are visible metallographically in the pure nickel weld shown in Fig. 11, but the weld in Fig. 13 shows a definite substructure within each grain. Although the bulk weld-metal composition is homogeneous, cells or cellular dendrites represent a commonly observed pattern of microsegregation that is developed during nonequilibrium solidification of a weld or casting. Microsegregation is characterized by a compositional difference between the cores and peripheries of individual cells and cellular dendrites. Cells are microscopic pencil-shaped protrusions of solid metal that freeze ahead of the solid-liquid interface in the weld. Cellular dendrites are more developed than cells and appear to have a "tree-like" shape; the main stalk is called the "primary dendrite arm," and the orthogonal branches are called the "secondary dendrite arms." The cores of the cells and dendrite arms have a higher solidus temperature and contain less solute than the intercellular and interdendritic regions. In actual welding practice, cellular or dendritic microsegregation is virtually impossible to avoid unless the metal being welded is a pure element.

Generally, the important parameters controlling the cellular or cellular dendritic substructures in welds are (1) the equilibrium partition ratio, K, which is an in-

Fig. 13 Cellular dendritic substructure in Hadfield steel weld metal deposited with the GTAW process. Matching filler metal. Arrows indicate grain boundaries and dark etching regions are cores of cellular dendrites. Magnification: ×65. Etchant: Stead's reagent

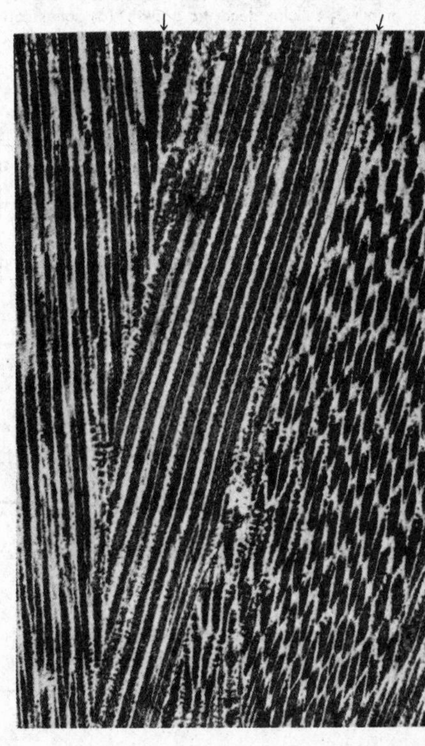

Fig. 14 Solidification of dendrites in a weld. (a) Solidification of 3% Cu-Al alloy by the growth of dendrites, (b). (c) Solute redistribution occurring ahead of the solid/liquid interface. (d) Constitutional supercooling develops when the actual temperature of liquid in the copper-rich zone is greater than the liquidus temperature.

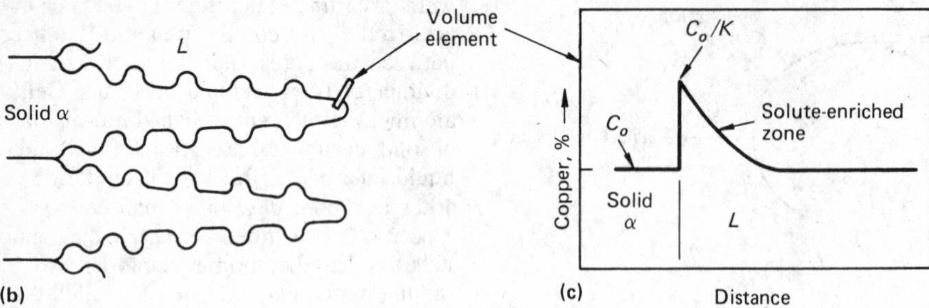

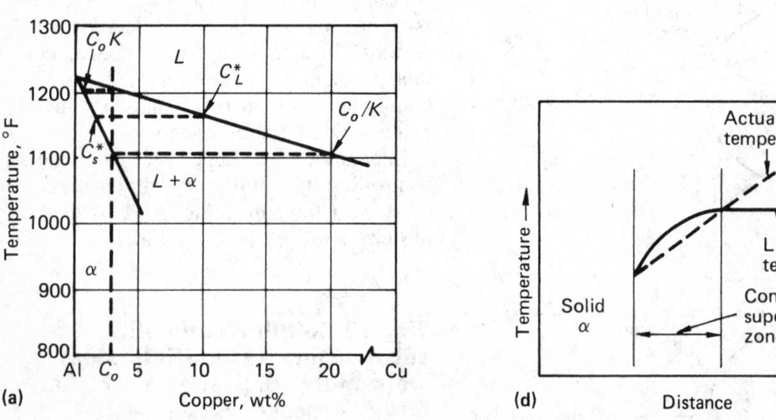

Fig. 15 Effect of thermal gradient on mode of solidification in welds for constant growth rate. (a) Steep G_1 planar growth. (b) Intermediate G_2 cellular growth. (c) Small G_3 cellular dendritic growth. (d) Solidification of the weld

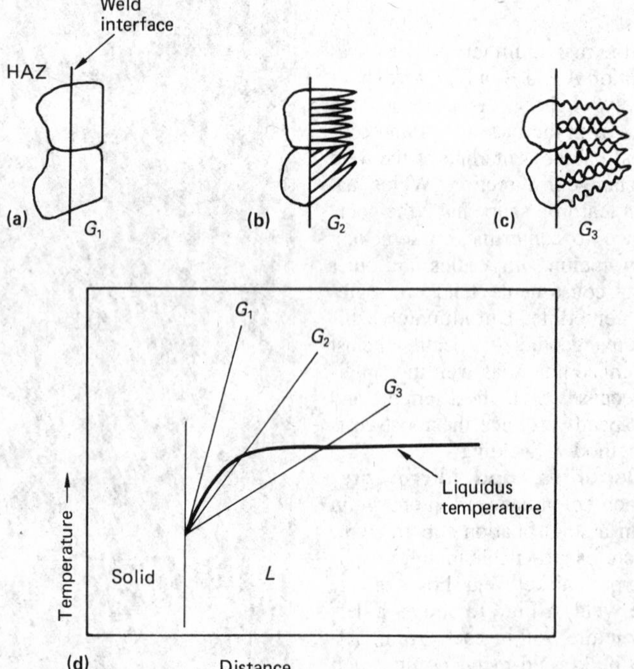

dex of the segregation potential of an alloy:

$$K = C_s^*/C_L^* \quad \text{(Eq 1)}$$

where C_s^* is the solute content of the solid at the solid-liquid interface and C_L^* is the solute content of the liquid at the solid-liquid interface; (2) the alloy composition itself, C_o; (3) the temperature gradient, G, in the liquid at the weld interface in °F/in.; and (4) the growth rate, R, or velocity of the interface in in./s.

For example, consider an alloy of composition C_o equal to 3%Cu-97%Al. Thus, in Fig. 14(a):

$$K = C_s^*/C_L^* = 1.7\%/10\% = 0.17$$

The first metal to solidify will contain only:

$$C_oK = (3)\,0.17 = 0.51\%\ \text{Cu}$$

while the last liquid to solidify between cells or cellular dendrites is rich in copper:

$$C_o/K = 3/0.17 = 17.6\%\ \text{Cu}$$

These values represent the short transients at the start and finish of solidification of a cell or cellular dendrite. As the cell or dendrite grows in the weld, a dynamic equilibrium is achieved between the newly forming solid of composition, C_o = 3% Cu, and the copper-rich liquid containing a maximum of C_o/K 17.6% Cu, at the solid/liquid interface as shown in Fig. 14(b) and (c). If the actual temperature distribution ahead of the solid/liquid interface is less than the liquidus temperature, constitutional supercooling occurs (Fig. 14d). Supercooling means that the solute-enriched liquid ahead of the solid-liquid interface has been cooled below its equilibrium freezing temperature, and constitutional indicates that the supercooling originated from an enrichment in composition rather than temperature.

Microsegregation results when the copper-rich liquid at the solid/liquid interface solidifies between the cellular dendrites. The interdendritic regions are so segregated with copper (solute) that a small amount of eutectic (α Θ) is frequently observed. Eutectic structures can only occur when the composition of solidifying metal exceeds the maximum solid solubility of 5.65% Cu in α (Fig. 2).

Whether or not a planar, cellular, or dendritic substructure occurs upon solidification is largely determined by G and R, which control the amount of constitutional supercooling. If a weld is deposited at a constant travel speed, R becomes fixed. By inducing an extremely steep temperature gradient G_1 (Fig. 15a), no constitutional supercooling occurs and the solidified weld-metal grain structure is planar.

For example, the epitaxially grown columnar grains in the pure nickel weld (Fig. 11) contain no solidification substructure. Therefore, the location of the weld/base metal interface is extremely difficult to distinguish.

When the gradient is decreased slightly to G_2 (Fig. 15b), any protuberance of solid metal on the interface will grow faster than the remaining flat interface because the solid is growing into supercooled liquid; that is, the solid protuberance exists at a temperature below that of the liquidus for that alloy. As a result, a cellular substructure develops in each epitaxially grown grain. The liquid ahead and alongside each cell contains greater solute content than the cell core.

If the value of the temperature gradient is decreased further to G_3 (Fig. 15c), constitutional supercooling becomes so extensive that secondary arms form and cellular dendritic growth is observed. The greatest degree of microsegregation occurs during columnar dendritic solidification, while no measurable segregation is encountered in planar growth. Whether planar, cellular, or cellular dendritic, growth is always anisotropic.

Investigators have found that these solidification substructures can be characterized by the combined parameter G/R. Figure 16 shows that a large value of G/R combined with a very dilute alloy will result in a planar solidification structure, while a low G/R and high solute concentration will produce a heavily segregated columnar dendritic structure. Both columnar dendritic and equiaxed dendritic structures, although common in large castings, are not frequently encountered in welds. In practice, cellular and cellular dendritic substructures are most frequently observed in welds. The difference between the cellular dendritic and columnar dendritic structures is related to the length of the constitutionally supercooled zone ahead of the solid-liquid interface. This zone is typically much smaller for cellular dendritic than for columnar dendritic solidification. Therefore, each grain will contain many cellular dendrites (Fig. 13), whereas only one columnar dendrite occupies one grain. Unfortunately, it is very difficult to control G and R independently in welding practice. As a general rule, a fast welding speed (R) will produce a steep G. The relative values of G and R, however, determine the solidification morphology for a given alloy of fixed C_o and K.

Solidification Rate. While G/R controls the mode of solidification, the weld cooling rate, in terms of the parameter GR

Fig. 16 Dependence of mode of solidification on G/R parameter for different solute concentrations (C_o)

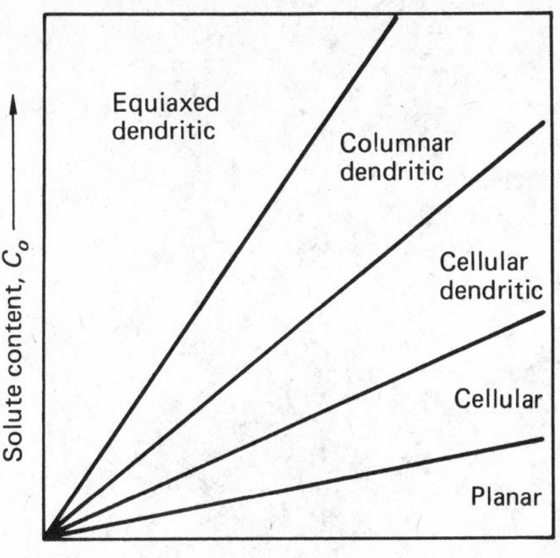

(solidification rate in units of °F/s), determines both the size and spacing of cells and dendrites. Flemings and others have demonstrated that the effect of solidification rate on the dendrite arm spacing (d) is:

$$d = a(GR)^{-n} \quad \text{(Eq 2)}$$

where a is a constant and n is approximately $^1/_2$ for primary arms and between $^1/_3$ and $^1/_2$ for secondary dendrite arms. The dendrite arm spacing of stainless steel in an electroslag weld is often several hundred times greater than that found in a rapidly cooled laser weld.

Solute Banding. The phenomenon of solute banding occurs to some degree in all alloy welds. The formation of ripples on the weld surface and solute banding within the weld are both caused by the discontinuous nature of weld-metal solidification and occurs in manual as well as in automatic welds where the travel speed is mechanically constant. During weld-metal solidification, however, R fluctuates cyclically above and below a mean value of growth rate that is determined by the weld travel speed. Fluctuations in R result in not only ripple formation, but also solute banding. Because an abrupt increase in R causes a reduction in the amount of solute that can be held in the solute-enriched liquid (Fig. 14c), excess solute is dumped and appears as a solute-rich band. Similarly, a sudden decrease in R produces a solute-poor band. Solute banding lines (Fig. 17) are very helpful in welding research because they always outline the weld-pool shape at a given instant during solidification. For example, the form factor (ratio of width to depth of weld pool), which is so important in electroslag welding (ESW), can be easily measured metallographically using solute band lines.

Basic Heat Flow Calculations

Heat Input. In arc welding, the total heat energy, H, generated by the power source is given by:

$$H = \frac{EI}{v} \quad \text{(Eq 3)}$$

where E is the voltage, I is the current, and v is the travel speed in mm/s (in./min). Due to small electrical losses in the arc, the total heat does not reach the workpiece. The heat actually transferred to the workpiece in units of J/mm (J/in.) is defined as the net heat input, H_n:

$$H_n = \frac{f_1 EI}{v} \quad \text{(Eq 4)}$$

where f_1 is the heat transfer efficiency. The value of f_1 for most arc welding processes is generally between 0.7 and 1.0. The submerged arc welding (SAW) process is the most efficient, with f_1 approaching 1.0, and the GTAW process is the least efficient, with f_1 equaling approximately 0.7. Other arc welding processes, such as gas metal arc welding (GMAW), flux cored arc welding (FCAW), and SMAW, have g_1 values approximately equal to 0.9.

Of the heat energy (H_n) transferred to the workpiece, only a portion is actually used to melt metal. For example, at low arc energy inputs, the gas tungsten arc can be maintained on a massive block of copper for several hours without melting any of the copper workpiece. Therefore, the melting efficiency, f_2, would be zero, where f_2 is defined as the ratio of the minimum heat input necessary to cause melting for a given H_n delivered to the workpiece, or:

$$f_2 = \frac{[H_f + C(T_m - T_o)]W}{(H_n)(L)} \quad \text{(Eq 5)}$$

where H_f is the heat of fusion; L is the length of weld metal deposited; T_m is the melting or liquidus temperature of the base metal in °C; T_o is the temperature of the base metal prior to welding in °C; C is the specific heat; and W is the weight of the deposited weld metal.

For example, calculate the melting efficiency of this weld. A gas metal arc weld is deposited on steel plate using current of 300 A, voltage of 29 V, and travel speed of 7.6 mm/s (18 in./min), with f_1 equal to 0.9, and the weld deposit measuring 250 mm (10 in.) in length. The melting point (T_m) of the steel is 1510 °C (2750 °F), and

Fig. 17 Solute banding lines in electroslag weld metal of 6% Ni steel. (a) Etched in nital. (b) Etched in solute-sensitive Stead's reagent; solute banding lines are indicated by arrows. Magnification: ×25

(a)

(b)

the temperature of the base plate prior to welding (T_o) is 25 °C (77 °F). From *Metals Handbook* (Ref 4), the heat of fusion (H_f) is 274 000 J/kg (125 000 J/lb). From Ref 5, an approximate value of the specific heat (C) is 670 J/kg · °C (169 J/lb · °F).

The weight of deposited weld metal can either be weighed directly or calculated by:

$$W = AL\rho$$

where W is the weight in kilograms of weld metal; A is the cross-sectional area of the weld bead determined by metallographic sectioning, given in this example to be 34 mm² (0.053 in.²); and ρ is density, which from Ref 3 is 7.8 × 10^{-6} kg/mm³ (0.283 lb/in.³). Therefore:

$$W = (34)(250)(7.8 \times 10^{-6})$$

$$W = 0.066 \text{ kg}$$

$$H_n = \frac{f_1 EI}{v} = \frac{(0.9)(29)(300)}{7.6}$$

$$H_n = 1.03 \text{ kJ/mm} = 1030 \text{ J/mm}$$

or in English units:

$$W = (0.053)(10)(0.283)$$

$$W = 0.15 \text{ lb}$$

$$H_n = \frac{f_1 EI}{v} = \frac{(0.9)(29)(300)}{0.3}$$

$$H_n = 26\ 100 \text{ J/in.}$$

Substituting into Eq 5:

$$f_2 = \frac{[274\ 000 + 670\ (1510 - 25)]\ 0.066}{(1030)\ 250}$$

$$f_2 = 0.33$$

or in English units:

$$f_2 = \left[\frac{125\ 000 + 169\ (2750 - 77)}{26\ 100\ (10)}\right] 0.15$$

$$f_2 = 0.33$$

In most arc welding processes, the amount of heat energy actually causing melting is significantly less than the net heat input.

As a general rule, the combination of a high-intensity heat source for welding and a low-conductivity workpiece results in nearly 100% melting efficiency. Finely focused laser or electron beams are examples of high-power density sources that can deliver up to 10^{12} W/m² (6.5 × 10^8 W/in.²). The melting efficiencies of laser and electron beam sources of heat approach f_2 = 1. When the intensity is greater than approximately 10^9 W/m² (6.5 × 10^5 W/in.²), selective vaporization of low-melting elements in the molten pool can occur. Typical arc welding techniques, such as SAW and GMAW, can only produce up to approximately 5 × 10^8 W/m² (3.2 × 10^5 W/in.²). Therefore, selective evaporation is not a serious problem, although it occurs. The oxyacetylene process has a melting efficiency so low that it is not economically feasible to weld large structural components in production.

Cooling Rates in Fusion Zone. Determination of the cooling rate of the weld fusion zone is important to (1) avoid a martensitic transformation or possible cold cracking; (2) establish a relationship to cooling rates for wrought steels, which are recorded in the Jominy hardenability tests or CCT diagrams; (3) identify any rate-sensitive metallurgical reaction occurring in the weld and HAZ; and (4) prevent excessive dendrite coarsening in weld metal.

Typically, weld-metal cooling rates decrease rapidly as welds cool to room temperature. Therefore, weld cooling rates are usually measured experimentally as the slope of the cooling curve (temperature versus time) at a particular temperature of interest. Temperatures of interest that are used frequently to calculate or measure the cooling rates of steels are 550 °C (1020 °F) and 700 °C (1300 °F). Once established, cooling rate data from sample to sample for a temperature of interest can be effectively evaluated or compared.

Adams (Ref 6, 7) has shown that the cooling rate, S, at the weld centerline of a thin plate workpiece, which requires less than four passes for full penetration, is:

$$S = 2\pi K\rho C\left(\frac{t}{H_n}\right)^2 (T_i - T_o)^3 \quad \text{(Eq 6)}$$

where T_i is the temperature of interest and t is the plate thickness. This equation applies to two-dimensional heat flow situations where the actual thickness of thin plate being welded is relative. For example, 25-mm (1-in.) thick steel plate is easily welded to full penetration with the SAW process. Because fewer than approximately four passes were used to complete this weld, it fulfills the thin plate conditions. Similarly, a 250-mm (10-in.) thick electroslag weld is also considered as meeting thin plate conditions because this full-penetration weld is deposited in a single pass.

If, however, a 25-mm (1-in.) thick plate is welded in 25 passes by GMAW, the thin plate criteria cannot be used; heat flow is now three-dimensional. Adams derived a thick plate cooling rate equation for such cases:

$$S = \frac{2\pi K(T_i - T_o)^2}{H_n} \quad \text{(Eq 7)}$$

Here, the cooling rate is maximum under normal welding conditions. Values for plate thickness and specific heat are not included in this thick plate equation. Whenever the size of the weld bead of one pass is small compared to the thickness of plate, the thick plate equation is used.

In some cases, it is unclear whether thin or thick plate criteria should be used. Consider welding 3-, 6-, 12-, and 25-mm (1/8-, 1/4-, 1/2-, and 1-in.) thick steel by GMAW. Calculate the cooling rates at 550 °C (1020 °F) using both thick and thin plate equations and compare the results.

For the 3-mm (0.118-in.) thick plate, the thin plate calculation is:

$$H_n = 800 \text{ J/mm}$$

$$K = 0.028 \text{ J/mm} \cdot \text{s} \cdot {}^\circ\text{C}$$

$$\rho C = 0.0044 \text{ J/mm}^3 \cdot {}^\circ\text{C}$$

$$S = 2\pi(0.028)(0.0044)(3/800)^2 (550 - 25)^3$$

$$S = 1.6\ {}^\circ\text{C/s at } 550\ {}^\circ\text{C}$$

or in English units:

$$H_n = 20\,320 \text{ J/in.}$$

$$K = 0.395 \text{ J/in.} \cdot \text{s} \cdot {}^\circ\text{F}$$

$$\rho C = 40 \text{ J/in.}^3 \cdot {}^\circ\text{F}$$

$$S = 2\pi(0.395)(40)(0.118/20\,320)^2 (1020 - 77)^3$$

$$S = 2.9\ {}^\circ\text{F/s at } 1020\ {}^\circ\text{F}$$

Similarly, S for 6-, 12-, and 25-mm (1/4-, 1/2-, and 1-in.) thicknesses is 6.3, 25, and 109 °C/s (11, 45, and 196 °F/s), respectively. The thin plate equation, however, is open ended; an unrealistically high cooling rate value results with thicknesses increasing beyond the limits of the thin plate equation. Therefore, the thick plate equation is used to calculate the maximum cooling rate as a function of H_n and T_i. Using the thick plate equation, the 3-, 6-, 12-, and 25-mm (1/8-, 1/4-, 1/2-, and 1-in.) thick welds result in the same cooling rate:

$$S = \frac{2\pi(0.028)(550 - 25)^2}{800}$$

$$S = 61\ {}^\circ\text{C/s at } 550\ {}^\circ\text{C}$$

or in English units:

$$S = \frac{2\pi(0.395)(1020 - 77)^2}{20\,320}$$

$$S = 108.6\ {}^\circ\text{F/s at } 1020\ {}^\circ\text{F}$$

Comparison of the results obtained with both thin and thick equations shows that the thin plate equation applies to 3-, 6-, and 12-mm (1/8-, 1/4-, and 1/2-in.) plate thicknesses because the cooling rates determined by the thin plate equation cannot exceed that value determined by the thick plate equation, which consequently applies only to the 25-mm (1-in.) thick weld.

To simplify the choice between the thin and thick plate equations, Adams developed the relative plate thickness equation, τ:

$$\tau = t\left(\frac{\rho C(T_i - T_o)}{H_n}\right)^{1/2} \quad \text{(Eq 8)}$$

According to Adams, the thin plate equation should be applied when $\tau < 0.75$ and the thick plate equation for $\tau > 0.75$. For example, using the 3-mm (1/8-in.) thickness:

$$\tau = 3\left(\frac{0.0044\,(550 - 25)}{800}\right)^{1/2}$$

$$\tau = 0.16$$

and for:

$$t = 6 \text{ mm}, \tau = 0.32$$

$$t = 12 \text{ mm}, \tau = 0.64$$

$$t = 25 \text{ mm}, \tau = 1.3$$

or in English units:

$$\tau = 0.125\left(\frac{40\,(1020 - 77)^{1/2}}{20\,320}\right)^{1/2}$$

$$\tau = 0.17$$

and for:

$$t = 1/4 \text{ in.}, \tau = 0.34$$

$$t = 1/2 \text{ in.}, \tau = 0.68$$

$$t = 1 \text{ in.}, \tau = 1.36$$

Thus, the thin plate equation applies to the 3-, 6-, and 12-mm (1/8-, 1/4-, and 1/2-in.) thick welded plates, while the thick plate equation applies to the 25-mm (1-in.) thick welded plate. Note that the values of τ depend on welding conditions incorporated into H_n.

Frequently, in solidification work, the most reliable temperature of interest used to calculate weld cooling rates is the liquidus temperature of the molten admixture. During solidification, if the thermal gradient ahead of the solid-liquid interface (on the liquid side) and growth rate of the interface (R) can be determined, the cooling rate during solidification can also be calculated. Many methods have been used to calculate the value of G (temperature gradient) for an arc weld. Using the method of Lancaster, the gradient in arc welding was derived as:

$$G = T_m/y \quad \text{(Eq 9)}$$

where y is the distance between the heat source (Fig. 12) and the rear of the molten weld pool. The value of y is a function of H_n and will increase accordingly. The average growth rate (R) has been well established as the velocity of the solidification front at the rear of the weld pool, or:

$$R = v \sin \phi \quad \text{(Eq 10)}$$

where ϕ is the angle between the welding

direction and the tangent of the weld pool boundary. At the rear of the weld pool, $R = v$ and:

$$GR = \frac{T_m R}{y} \quad \text{(Eq 11)}$$

where the product GR has units of cooling rate in °C/s (°F/s).

For example, calculate the solidification rate for a gas metal arc weld deposited on steel with a value of y equal to 11 mm (0.43 in.) at a welding speed of 4 mm/s (0.15666 in./s or 9.4 in./min):

$$GR = \frac{(1510)(4)}{11}$$

$$GR = 549\ °\text{C/s at } 1510\ °\text{C}$$

or in English units:

$$GR = \frac{(2750)(0.15666)}{(0.43)}$$

$$GR = 1000\ °\text{F/s at } 2750\ °\text{F}$$

This weld has a cooling rate measured at 550 °C (1020 °F) of approximately 69 °C/s (125 °F/s) using thick plate criteria or a value less than 69 °C/s (125 °F/s) using the thin plate equation, depending on workpiece thickness.

Heat-Affected Zone Thermal Cycle. The unfused base metal adjacent to the weld bead experiences localized peak temperatures up to the melting point or solidus temperature of the base metal and forms the HAZ of the weld joint. The width of this zone depends upon the definition of the HAZ. The extremity of the HAZ is generally associated with a minimum peak temperature that causes an observable microstructural change or disruption in the metallurgical structure of the base metal. For example, in normalized steels, the extremity of the HAZ may be the temperature where pearlite is slightly disturbed by a brief excursion into the austenite range. In single-phase alloys that have been cold worked, the minimum HAZ peak temperature may be that required for the first visible indication of recrystallization of the cold worked structure.

In quantitative terms, the peak temperature distribution in the HAZ of a single-pass weld is a function of several variables, including net heat input and plate thickness.

According to Adams (Ref 6, 7), the peak temperature, T_p, experienced in the HAZ at a distance y from the fusion line is:

$$\frac{1}{T_p - T_o} = \frac{4.13\ \rho C t y}{H_n} + \frac{1}{T_m - T_o} \quad \text{(Eq 12)}$$

Typical HAZ thermal cycles for different values of y are shown in Fig. 18. Consider, for example, a shielded metal arc weld deposited on ASTM A36 steel plate where:

$$t = 12\ \text{mm}$$

$$T_m = 1510\ °\text{C}$$

$$T_o = 25\ °\text{C}$$

$$\rho C = 0.0044\ \text{J/mm}^3 \cdot °\text{C}$$

$$H_n = 2\ \text{kJ/mm}$$

or in English units:

$$t = 0.5\ \text{in.}$$

$$T_m = 2750\ °\text{F}$$

$$T_o = 77\ °\text{F}$$

$$\rho C = 40\ \text{J/in.}^3 \cdot °\text{F}$$

$$H_n = 50.8\ \text{kJ/in. or } 50\ 800\ \text{J/in.}$$

Calculate the peak HAZ temperature occurring 1, 5, and 10 mm (0.04, 0.20, and 0.40 in.) from the weld fusion line. The solution for y equaling 1 mm (0.04 in.) is:

$$\frac{1}{T_p - 25} = \frac{4.13\ (0.0044)(12)(1)}{2000} + \frac{1}{1510 - 25}$$

$$T_p = 1300\ °\text{C}$$

Similarly, for:

$$y = 5\ \text{mm},\ T_p = 846\ °\text{C}$$

$$y = 10\ \text{mm},\ T_p = 592\ °\text{C}$$

or in English units:

$$\frac{1}{T_p - 77} = \frac{4.13\ (40)(0.5)(0.04)}{5} + \frac{1}{2750 - 77}$$

$$T_p = 2272\ °\text{F}$$

$$y = 0.20\ \text{in.},\ T_p = 1506\ °\text{F}$$

$$y = 0.40\ \text{in.},\ T_p = 1052\ °\text{F}$$

As these calculations show, the peak temperature experienced in the HAZ of any weld decreases with increased distance from the weld interface.

Fig. 18 Typical thermal cycles experienced at various locations in HAZ of electroslag welds as a function of time from the fusion line

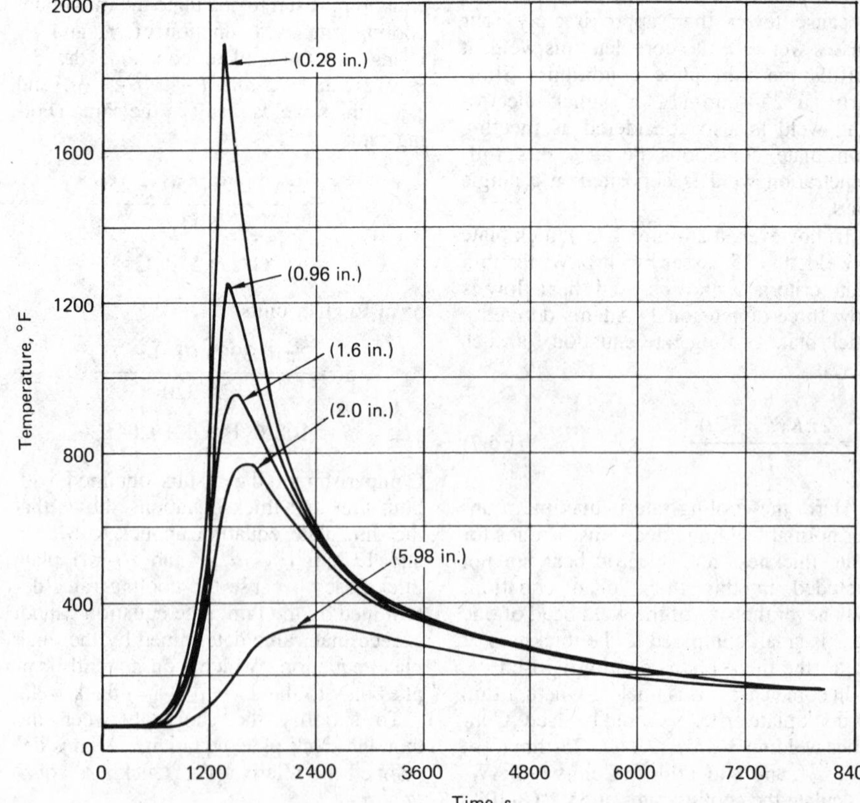

Microstructure of the Weld and Heat-Affected Zone

Through the process of epitaxial growth, the initial columnar grain width of the fusion zone is determined by the size of the base-metal grains adjacent to the weld interface. Because the peak HAZ temperature increases with decreasing distance from the weld interface (Fig. 18) and grain growth is a function of temperature, the maximum grain size in the HAZ always occurs along the weld interface. It is this maximum grain size that is transmitted into the weld fusion zone.

Grain Size. The relationship used to calculate the grain size in the HAZ is:

$$D - D_o = be^{-Q/2RT_P}(t')^n \quad \text{(Eq 13)}$$

where D is final grain diameter; D_o is the original grain diameter; e, which equals 2.718, is the natural base for logarithms; T_p is the peak temperature, which would be the solidus temperature at the fusion line; t' is the time at temperature; Q is the activation energy for grain growth of the alloy; R is the universal gas constant; and b and n are constants determined by the materials. Both temperature and time at temperature produce grain growth in the HAZ, and as stated previously, the maximum grain size always occurs immediately adjacent to the weld interface. Because all welds experience the same spectrum of peak temperature from T_o to the solidus temperature, the only significant variable in Eq 13 is the residence time, t'. As the cooling rate decreases, residence time increases, substantially coarsening the maximum HAZ grain size.

The process of competitive grain growth may cause further lateral growth of the weld-metal grain size. In pure nickel welds (Fig. 11), columnar grains emanating from the HAZ continue to widen as they grow into the weld fusion zone. Similarly, in steels, the columnar grain width in the weld will be several times greater than the maximum grain diameter in the HAZ, as shown in Fig. 19.

The maximum columnar grain width in the weld metal is limited only by the physical size of the weld bead and the arc energy input. For example, it is virtually impossible for a gas tungsten arc weld deposited on a coarse-grained copper casting to exhibit any distinguishable HAZ (Fig. 20) because the residence time is insufficient to cause noticeable grain growth; for example, D in Eq 13 is insignificantly greater than D_o. Furthermore, because the large base-metal grains that grow epitaxially into the weld must squeeze into a bead of limited volume, lateral growth of the columnar grains in the weld is not possible.

Many metals that cannot be strengthened by heat treating are often strength-

Fig. 19 Weld-metal, HAZ, and unaffected base-metal microstructures of A588 steel welded by ESW. Microstructures: (a) weld joint; (b) centerline of weld; (c) columnar portion of weld; (d) coarse-grained HAZ; (e) fine-grained HAZ; (f) disturbed pearlite; and (g) unaffected base metal. Magnification: (a) ×0.4; (b), (c), (d), (e), (f), and (g) ×50. Etchant: 2% nital

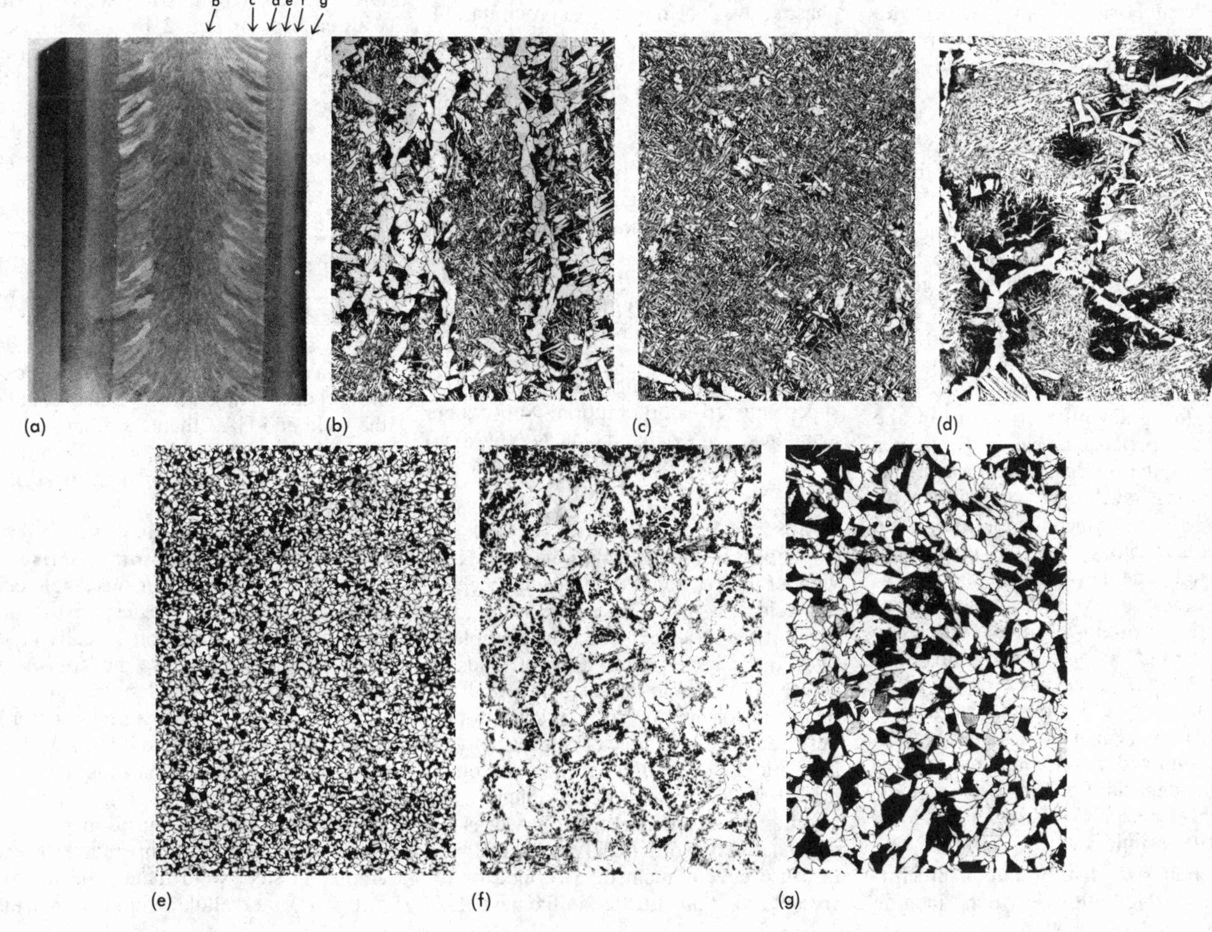

Fig. 20 Gas tungsten arc weld deposited on coarse-grained copper casting using matching filler metal. Dotted line indicates weld interface. Magnification: ×4. Etchant: hydrogen peroxide

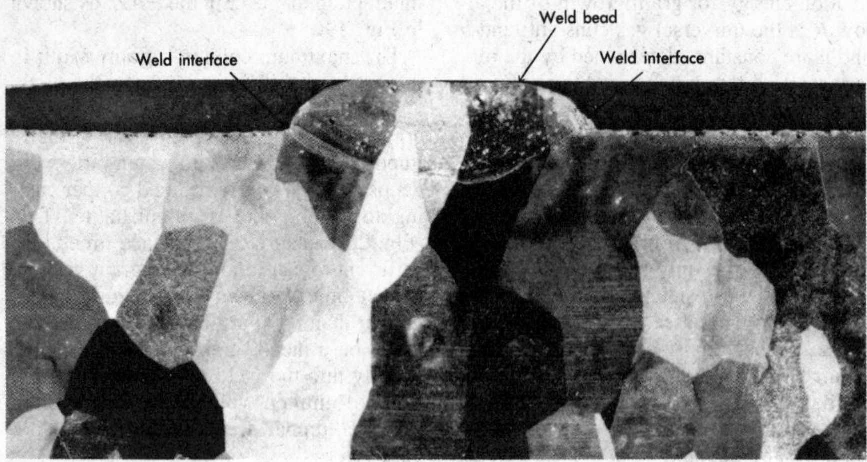

ened by cold rolling, including 1*xxx*, 3*xxx*, and 5*xxx* series aluminum alloys, copper alloys, and others. When these alloys are welded, the HAZ experiences both recrystallization and grain growth reactions. The hardness and strength properties of the recrystallized HAZ lose the benefits derived by cold working, and joint strength approaches that of an annealed alloy. Although weld-metal properties can always be controlled by judicious alloying, HAZ properties can only be controlled by regulating heat input or by changing the base-metal composition.

In steels and other metals that undergo allotropic phase transformation, the HAZ is conveniently divided into two regions: the grain growth region, which lies adjacent to the weld interface (Fig. 19d), and the grain-refined region, which is farther away from the weld interface (Fig. 19e). Because the grain growth region of the HAZ has experienced peak temperatures approaching the solidus of the base metal, coarse grains develop in accordance with Eq 13. The grain-refined region of the HAZ has been thermally cycled only briefly into the low-temperature portion of the austenite region, resulting in significant grain refinement. This grain-refining reaction occurs by the nucleation of new grains each time the A_1 and A_3 lines are crossed, either upon heating or weld cooling. The general structure of a steel weld will always appear fine-grained when compared to similar welds deposited on single-phase metals, such as pure nickel, copper, α brass, and ferritic stainless steel.

The grain size distribution in precipitation-hardening alloys—which include maraging steels; precipitation-hardening stainless steels; 2*xxx*, 6*xxx*, and 7*xxx* series aluminum alloys; cobalt- or nickel-based superalloys; copper, titanium, and magnesium alloys; and many others—is generally similar to that of the single-phase alloys. The majority of precipitation-hardening alloys develop coarse grain structures in both the weld and HAZ, and the small amount of second-phase transformation is insufficient to produce any grain refinement. For example, welding and slow cooling a typical nickel-based superalloy containing small additions of titanium and aluminum result in a coarse-grained weld and HAZ structure with small amounts of $Ni_3(Al,Ti)$ phase along the γ (nickel solid solution) grain boundaries. If the weld cooling rate is fast, as in an electron beam weld, the $Ni_3(Al,Ti)$ does not form at all on cooling, but remains in a supersaturated solid solution. Subsequent aging treatments only precipitate $Ni_3(Al,Ti)$ as microscopic particles throughout the weld and the HAZ.

Multiple-Pass Welds. Grains of single-phase metals continue to grow without obstruction through each succeeding weld pass of multiple-pass welds, until all of the required passes are complete. Such interpass epitaxial growth leads to coarse columnar grain structures and extreme anisotropy of mechanical properties. Peening or cold working each weld pass prior to deposition of the subsequent pass helps mitigate the problem. The peening action sufficiently cold works the columnar grains of a freshly deposited pass to cause development of a refined or recrystallized grain structure in the new HAZ of this weld pass. Through epitaxial growth, these refined grains grow into the weld. By interpass peening, columnar grains are restricted to growth only within each weld pass, thus greatly reducing the overall grain size and anisotropy of multiple-pass welds of single-phase alloys. Peening is not recommended by most welding codes for the first and last (surface) passes because of the likelihood of fracturing the first pass and heavily distorting the surface of the last pass.

In multiple-pass welds deposited on allotropic metals, such as steel, substantial interpass grain refinement occurs in the weld fusion zone. Figure 21(a) shows how a portion of each weld pass becomes the HAZ of the subsequent pass. Each interpass HAZ recrystallizes into a fine-grained structure that effectively prohibits the uninterrupted growth of large columnar grains from one weld pass to the next, as occurs in multiple-pass welding of single-phase alloys. Grain refinement is achieved each time a portion of the steel in the weld transforms to austenite upon heating followed by a transformation back to ferrite plus bainite or pearlite upon cooling. A thick band of fine, equiaxed grains separates the coarse columnar grains deposited with each pass (Fig. 21b).

Each transformation from one phase to another is accompanied by the nucleation and growth of very fine grains which completely replace the coarse columnar grain structure from the previous weld pass. When the HAZ is heated into the austenite region, growth of the newly formed grains may take place rapidly if the residence time at the peak temperature is long, or the weld cooling rate is slow. Fast cooling of a weld produces fine grain structures. For example, a weld deposited on 2-in.-thick steel plate in 50 passes with the SMAW process will produce a much finer grain size in both the weld and HAZ than a similar weld deposited in 1 pass with the ESW process, which is an extremely high heat input welding process.

Influence of Solidification Structure on Solid-State Transformations. The microstructure always remains single phase in single-phase alloys, despite the presence of microsegregation in cells or cellular dendrites, because no transformations occur in the solid state. For alloys that experience solid-state transformations, such as the eutectoid transformation in steels, the solidification structure of cells or dendrites affects to some extent any subsequent solid-state transformation as the weld cools to room temperature. For example, in steel weld metal, the interdendritic or intercellular regions are more

Fig. 21 Grain refinement of shielded metal arc, multiple-pass weld of low-carbon steel. (a) Two weld passes and grain-refined HAZ between passes. (b) Equiaxed grain structure in the interpass HAZ. Magnification: (a) ×15, (b) ×125. Etchant: 2% nital

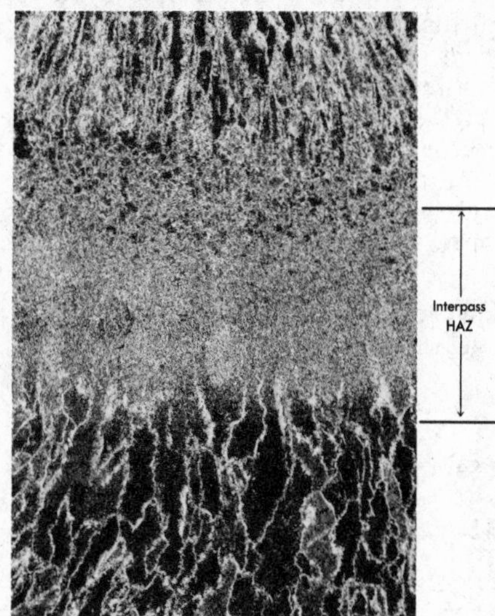

(a)

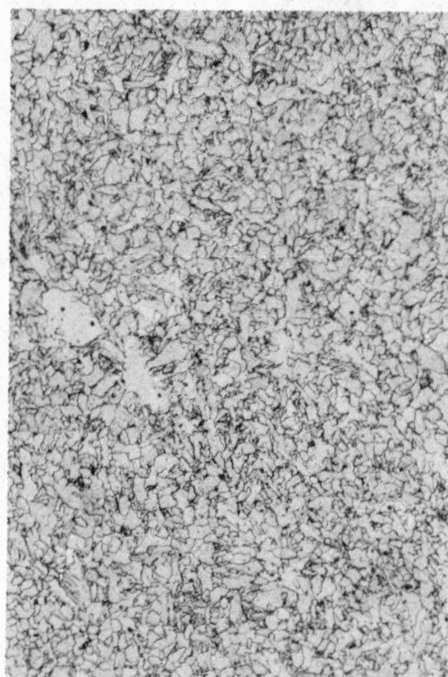

(b)

hardenable than the cores because these regions contain substantially higher percentages of alloying elements such as nickel, manganese, and molybdenum. Generally, as the alloy content increases, the compositional differences that may be developed between the cores and interdendritic regions are more significant.

In plain-carbon or low-alloy steel weldments, the cellular dendritic solidification that normally occurs under most welding conditions cannot be seen metallographically unless specially prepared solute-sensitive etchants are used, because the strong etching characteristics of proeutectoid ferrite and the eutectoid transformation products completely mask any metallographic effects caused by the solidification structure. No evidence of any solidification structure is apparent unless the weld is etched with a solute-sensitive reagent such as Stead's reagent. Stead's selectively deposits copper on regions of high-phosphorus segregation. Because phosphorus segregates intercellularly or interdendritically in steels, the solidification structure of the weld metal can be revealed (Fig. 17). In low-carbon steels, the degree of alloy segregation, although appreciable, is not sufficient to significantly change any solid-state transformation.

In high-alloy steels, the solidification of 304 stainless steel is an excellent example of the influence of the solidification structure on the resulting solid-state transformation. When wrought 304 stainless plate is examined metallographically and magnetically (with the ferrite gage), the microstructure is fully austenitic. When this alloy is autogenously gas tungsten arc welded (without filler metal), however, the weld metal is no longer fully austenitic, but now contains about 5% ferrite. During solidification of the weld, the first solid to form is δ ferrite dendrites containing an enriched chromium content at the core of the dendrite arms. Upon further cooling, the δ transforms almost entirely to austenite, except in the dendrite core where the concentration of chromium is sufficiently high to stabilize approximately 5% of the δ ferrite (Fig. 22). If this weld metal is chemically homogenized by hot working and annealing, the δ ferrite disappears, and the weld microstructure becomes fully austenitic again.

Fig. 22 Weld microstructure of 304 stainless steel containing austenite with approximately 5% δ ferrite. Magnification: ×100. Etchant: nitric and hydrochloric acid in glycerol

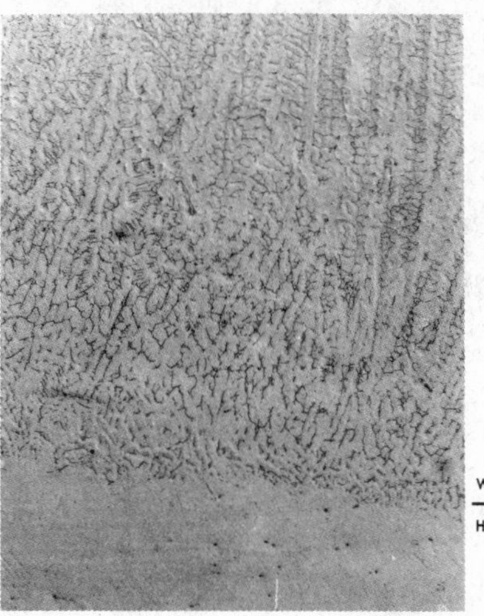

In weld metal of high-alloy steels, such as 18% Ni maraging steel, the segregated solidification structure is directly responsible for formation of small pools of unwanted retained austenite. When 18% Ni maraging steel is gas metal arc welded with nearly matching filler metal, the resulting dendritic structure is so segregated that the nickel-enriched interdendritic regions do not transform fully to martensite upon cooling. The resulting weld-metal microstructure is not 100% martensite as intended, but is a duplex structure of martensite with about 8% interdendritic retained austenite that is detrimental to mechanical properties.

In summary, the solidification structure generally has little or no effect on subsequent solid-state transformations when the degree of segregation is small, as in dilute alloys which have undergone little constitutional supercooling, because the parameter G/R was high. Heavily alloyed metals, however, tend to solidify with a segregated dendritic structure where the dendrite core is rich in the high melting point element(s) and the interdendritic areas are rich in the low melting point element(s). Because the dendrite core and interdendritic regions are virtually two entirely different alloys, substantial deviations from the normal solid-state transformations can be expected.

Effects of Welding on Microstructure

In welding more complex metals that undergo solid-state phase transformations, alloy composition is a major factor in determining the final microstructure and, hence, mechanical properties of a weld joint. For example, when welding plain-carbon and low-alloy steels, a weld metal carbon content below 0.1%, or as low as possible, should be maintained to achieve excellent weld toughness and maximum resistance to hot cracking from liquid sulfide films (discussed in the hot cracking section in this article). To counteract this hot cracking tendency, manganese is usually added to steel; as the carbon content of the weld metal increases, the manganese-to-sulfur ratio becomes larger, as shown in Fig. 23. Fortunately, most commercial filler metals contain minimal carbon and sulfur levels, while alloy content is raised to achieve specified strength requirements. Most structural steels, such as ASTM A36 and A588 or AISI-SAE

Fig. 23 Relationship of manganese-to-sulfur ratio, carbon content, and hot cracking susceptibility in welds

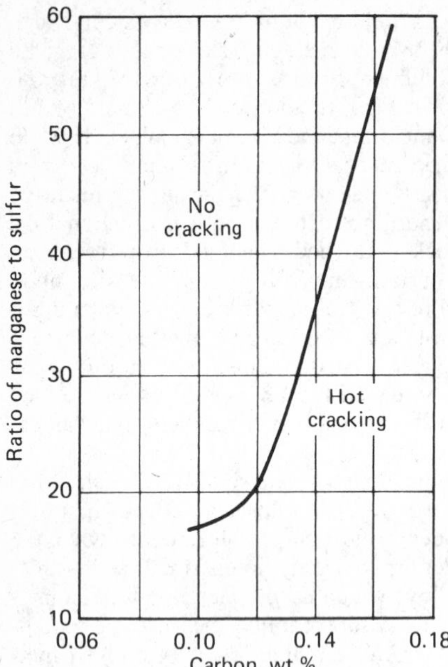

1018, 1020, and 1045, however, contain well over 0.10% C primarily because carbon is a far less expensive strengthener than alloying additions such as manganese, chromium, molybdenum, and nickel.

Effects of Cooling Rates. Consider SMAW of 0.18% C, low-alloy steel (ASTM A588) with an E7018 electrode. Temporarily neglecting the small amounts of alloying elements in the weld, development of the microstructure of this weld can be followed using the iron-carbon diagram or a CCT diagram if the cooling rate is known. Assuming the weld metal admixture contains 0.12% C, the iron-carbon diagram (Fig. 3) shows that the first solid to freeze is δ ferrite cells or dendrites, depending on G/R and C_o. At 2725 °F, the δ ferrite begins to transform by a peritectic reaction to austenite through epitaxial growth from the HAZ following the solidification front. Although some nucleation of austenite occurs at δ ferrite columnar grain boundaries, the majority of austenitic growth occurs epitaxially in a direction parallel to the maximum temperature gradient. The growing austenite grains replace the primary δ ferritic structure virtually grain-upon-grain with little change in the columnar nature of the weld structure. Despite the transformation of δ to austenite, austenite grain boundaries can occupy the same positions that the original δ ferrite grains occupied during solidification with little deviation. Therefore, the original δ ferrite solidification structure, although replaced by austenite, remains essentially intact.

The austenitic weld metal may transform to a variety of proeutectoid and eutectoid products, depending upon weld cooling rate and alloy composition. At a normal or intermediate cooling rate, as in a typical shielded metal arc weld, proeutectoid ferrite nucleates at the austenite grain boundaries as the weld cools below the A_3 temperature (Fig. 3). This continuous network or veining of ferrite grains outlines the prior austenite grain boundaries and is often called "grain-boundary ferrite" (Fig. 24). Within each prior austenite grain, the structure is normally "acicular ferrite," which is characterized by fine elongated platelets arranged in an interlocking path. At high magnification, the acicular ferrite grains, typically 40 to 80 μin. wide, are separated by local regions of carbon-enriched austenite which transform to carbide in the form of bainite. This microstructure is characterized by excellent strength and fracture toughness.

At slow weld cooling rates associated with high preheating temperatures or heat input, the amount of acicular ferrite

Fig. 24 Microstructures of A588 steel weld metal deposited by SMAW. Cooling rates: (a) normal, (b) slow, (c) fast. Magnification: (a) ×320, (b) ×125, (c) ×320. Etchant: 2% nital

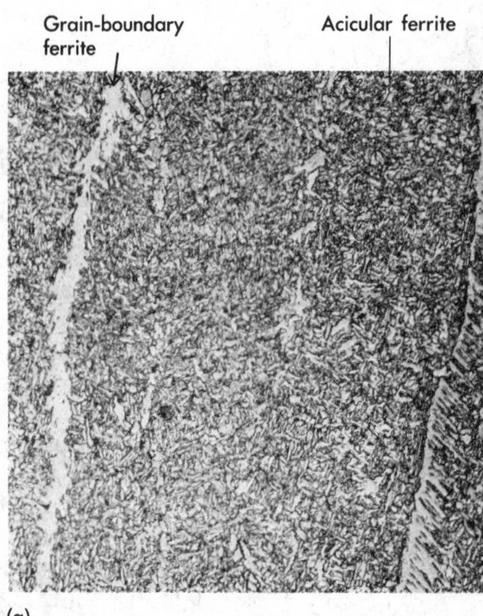

(a)

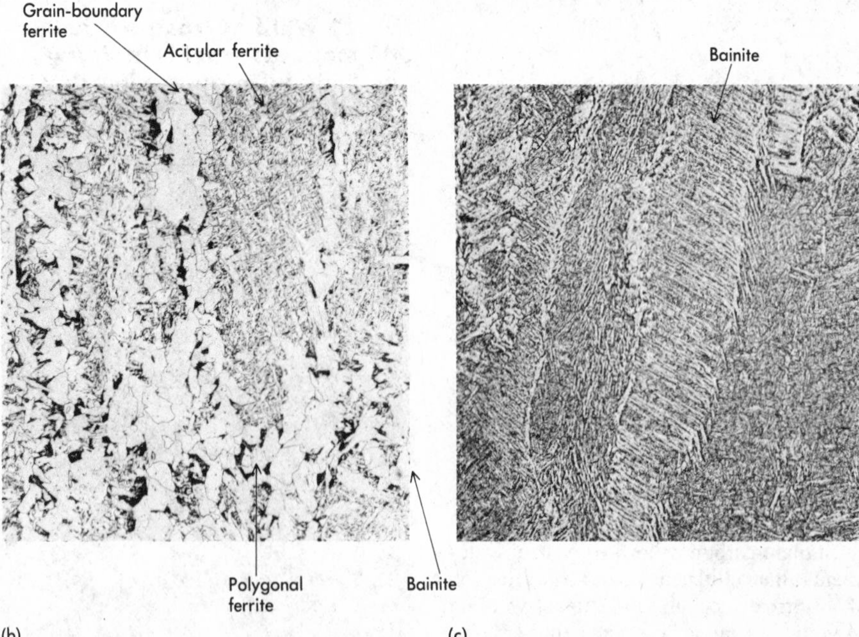

(b) (c)

Fig. 25a CCT diagram for a low-alloy steel. Source: Ref 8

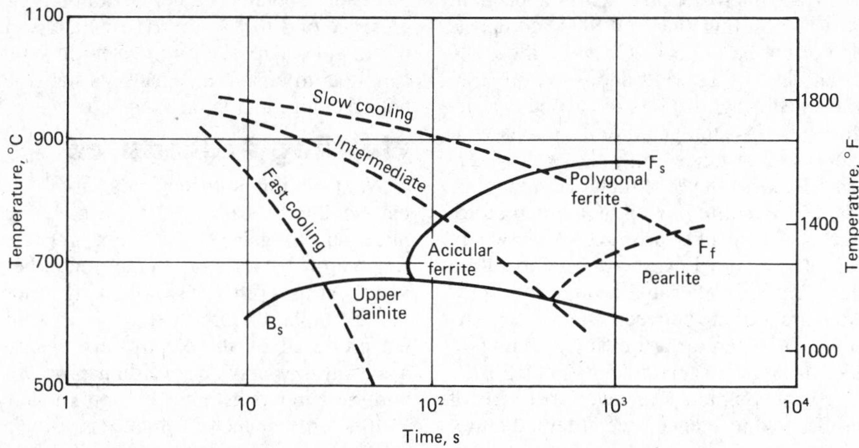

Fig. 25b CCT diagram for 1035 steel. Source: Ref 1

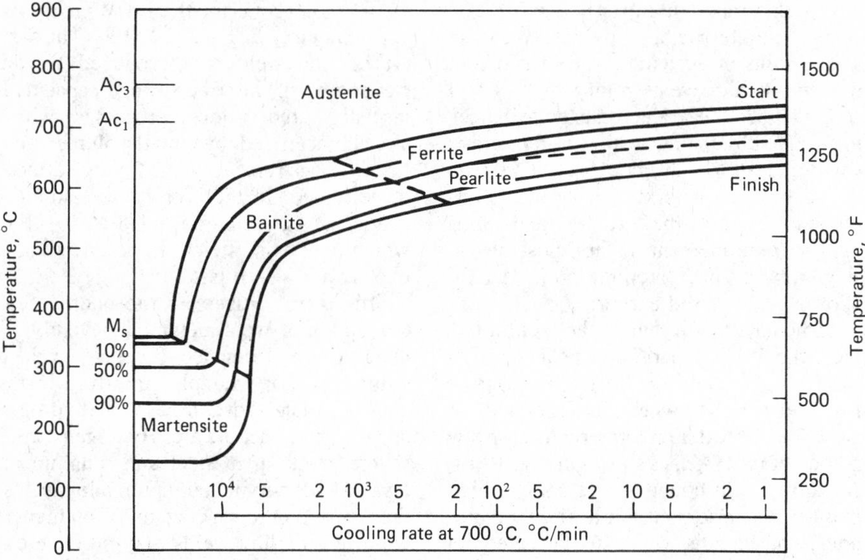

decreases substantially and is replaced by a coarse structure of additional grain-boundary ferrite and a second form of proeutectoid ferrite called "polygonal" or blocky ferrite shown in Fig. 24(b). The CCT schematic in Fig. 25(a) shows that the cooling path for a slowly cooled weld passes through the polygonal ferrite and pearlite fields. Pearlite appears as both large colonies and small islands between the acicular ferrite platelets in this microstructure, which is characterized by generally low toughness because of the extremely coarse and continuous ferrite grain structure.

At an optimum cooling rate, a maximum amount of acicular ferrite (as much as 90% of the weld) is obtained and any further increase in cooling rate tends to decrease its presence. For example, at a fast weld cooling rate, acicular ferrite is replaced by various forms of bainitic structures (Fig. 24c), which usually exhibit Widmanstatten or side-plate ferrite. Ferrite side plates grow from the prior austenite grain boundary into the grain interior. At these fast cooling rates, the bainitic structure may also be accompanied by regions of martensite and some retained austenite. The side-plate ferrite, bainitic, and martensitic structures are generally undesirable because of their low ductility and poor fracture toughness.

Weld-Metal Composition. The alloy composition of the weld metal also influences the resulting microstructure. Increasing the hardenability of the weld metal by the simple addition of alloying elements such as chromium, manganese, molybdenum, or nickel to the filler metal substantially decreases the amount of grain-boundary or polygonal ferrite in slowly cooled welds to the point of extinction. Because it offers a continuous crack-propagation path through a weld, grain-boundary ferrite is generally undesirable.

Unlike grain-boundary ferrite, acicular ferrite is believed to form by shear and diffusion modes in a temperature range on the CCT diagram (Fig. 25) above the bainite reaction and at a cooling rate faster than required for polygonal or grain-boundary ferrite. Certain alloying elements, such as manganese, molybdenum, nickel, silicon, chromium, boron, niobium, vanadium, and titanium, are known not only to increase the amount of acicular ferrite but also to strengthen it. Manganese, molybdenum, nickel, silicon, and chromium promote acicular ferrite at the expense of grain-boundary and polygonal ferrite and strengthen the ferrite through solid-solution strengthening. Boron, niobium, vanadium, and titanium increase acicular ferrite and increase strength through precipitation hardening; for example, carbonitrides of niobium are precipitated in the acicular ferrite matrix.

There is an optimum amount of one or more of these alloy additions that will achieve maximum toughness through the formation of acicular ferrite. Excess alloying lowers toughness due to overstrengthening the weld metal either by hardening of the acicular ferrite or the replacement of acicular ferrite with bainitic and martensitic structures. For example, increasing the manganese content in low-carbon steel welds promotes acicular ferrite formation, which increases toughness. Maximum toughness is achieved at an optimum concentration of 1.4% Mn, but increasing manganese content further serves only to continue strengthening the acicular ferrite while toughness deteriorates. Under ideal conditions of favorable alloying and cooling rate, the maximum amount of acicular ferrite attainable in a weld is about 90%; the remaining phases are grain-boundary ferrite and bainitic carbides. For example, the coarse columnar zone of the A588 weld in Fig. 19(c) clearly exhibits approximately 90% acicular ferrite.

The weld-metal and HAZ microstructures can be predicted for numerous steel compositions if appropriate CCT diagrams are available. This method of predicting microstructures can be applied effectively to any fusion welding process and follows a three-step process depending on (1) calculation of the maximum weld cooling rate, (2) procurement or availability of the appropriate CCT diagrams for the weld admixture and base plate compositions, and (3) comparison of the calculated cooling rate with that given in the CCT diagrams to obtain the microstructural phases present.

For example, the weld-metal and HAZ microstructures may be predicted for a shielded metal arc weld that is deposited on 12-mm (0.5-in.) thick 1080 steel using E7018 electrodes and a net arc energy input of 800 J/mm (20 320 J/in.) without preheating, and with a preheating temperature of 325 °C (615 °F). Knowing the groove geometry and chemical compositions of both electrode and base metal, an approximate composition of the composite zone admixture can be calculated. In this example, assume the weld-metal composition was found to be essentially equivalent to a 1035 steel. Weld cooling rate from Eq 6 for thin plate without preheating is:

$$S = 2\pi K\rho C(t/H_n)^2(T_i - T_o)^3$$

$$S = 2\pi\,(0.028)(0.0044)(13/800)^2 (700 - 25)^3$$

$$S = 63\ ^\circ\text{C/s} = 3800\ ^\circ\text{C/min}$$

or in English units:

$$S = 2\pi(0.395)(40)(0.5/20\,320)^2 (1292 - 77)^3$$

$$S = 108\ ^\circ\text{F/s} = 6473\ ^\circ\text{F/min}$$

From the CCT diagram in Fig. 7, the HAZ microstructure is essentially martensitic, and from the CCT diagram in Fig. 25(b), the weld microstructure is entirely bainitic.

Next, taking the effect of 325 °C (615 °F) preheating into account and recalculating using S for T_o equal to 325 °C (615 °F):

$$S = 10.8\ ^\circ\text{C/s} = 647\ ^\circ\text{C/min}$$

$$S = 19.4\ ^\circ\text{F/s} = 1165\ ^\circ\text{F/min}$$

With preheating, the HAZ microstructure contains pearlite plus bainite, while the microstructure of the weld metal contains grain-boundary ferrite plus bainite. Unfortunately, CCT diagrams do not presently differentiate among the various forms of ferrite (i.e., acicular or polygonal). In addition to eliminating martensite from the weld joint, the slower cooling rates attained with preheating provide additional benefits, such as greatly decreased susceptibility to hydrogen-assisted cold cracking and reduced residual stresses and distortion.

Distance From the Weld Interface. The HAZ microstructure also is dependent on weld cooling rate and alloy composition of the base metal. Consider the HAZ of an electroslag weld deposited on low-alloy steel (Fig. 19). As discussed earlier, the peak HAZ temperature decreases with increasing distance from the weld interface. Because the base metal is in the normalized condition, having a microstructure of ferrite plus pearlite, as shown in Fig. 19(g), the lowest peak temperature that causes a noticeable change in microstructure occurs between the A_1 and A_3 (Fig. 3), where pearlite is beginning to transform to austenite. This transformation is incomplete, however, because of insufficient residence time at temperature. The resulting microstructure, shown in Fig. 19(f), exhibits disturbed pearlite colonies, while the ferrite is virtually unaltered.

Closer to the weld interface, the localized peak temperature briefly exceeds A_3, where both ferrite and pearlite nucleate small grains of austenite, and insufficient residence time prevents grain growth. Upon cooling, the fine-grained austenite is further refined by the nucleation of ferrite and bainite, as shown in Fig. 19(e). This area of the weld joint is known as the grain-refined portion of the HAZ of steel. Similar grain refinement is not possible in single-phase or precipitation-hardening alloys, as discussed earlier.

Immediately adjacent to the weld interface, the HAZ experiences peak temperatures approaching the peritectic reaction isotherm (Fig. 4) where maximum grain growth occurs. The microstructure of this portion of the HAZ, as shown in Fig. 19(d), contains grain-boundary ferrite, which outlines the prior austenite grain boundaries, and bainite, which is the eutectoid transformation product.

Cooling rate and base plate alloy content determine the structure of the HAZ. Increasing either the weld cooling rate or the hardenability (by alloy additions) of the base plate decreases the amount of proeutectoid ferrite and increases the tendency of the HAZ to form lower transformation products such as martensite.

Stainless Steel Welds. The microstructure of as-deposited stainless steel welds can also be predicted using the Schaeffler diagram and, more recently, the DeLong diagram. With these diagrams, the weighted average compositions of the stainless steel base metal and filler metal are grouped in terms of equivalent chromium and nickel percentages. By plotting the chromium and nickel equivalents for a given weld composition on a DeLong diagram, the percentage of ferrite and austenite or the presence of martensite is immediately obtained. For example, the presence of 3 to 8% δ ferrite, which is vital in welds deposited on austenitic stainless steel to ensure adequate resistance to hot cracking, can be ascertained.

Welding Procedures

Weld cooling rate depends largely on the welding process used for a given application. In general, welding processes capable of delivering extremely high heat inputs per pass produce slow cooling welds. For example, the slowest cooling welds are produced by the electroslag, electrogas, and oxyacetylene welding processes with heat inputs as high as approximately 1140 kJ/in. per inch of thickness. Single-electrode or multiple-electrode submerged arc welds cool faster than electroslag welds but slower than welds deposited by GTAW, GMAW, SMAW, and FCAW. The typical weld cooling rate for the ESW process is approximately 5 °F/s at 1290 °F. The fastest cooling welds are produced by the electron beam, laser beam, and capacitive-discharge stud welding processes, which are characterized by small-volume weld fusion zones relative to the faying surfaces to be joined. In these processes, solidification rates can approach 1.8×10^6 °F/s, which may be sufficient to produce amorphous weld deposits.

Shielding Gases. Atmospheric contamination of weld metal can severely reduce the performance level of welded structures. For example, moisture in the air immediately decomposes to hydrogen and oxygen under the arc. Hydrogen causes severe cracking in steel and titanium alloys and porosity in aluminum alloy welds. Oxygen and nitrogen contamination in most engineering alloy welds may produce oxide and nitride inclusions and porosity. In reactive metals, such as titanium, a small concentration of oxygen (as low as 200 ppm) can so impair ductility of titanium weld metal that cracking in heavy sections is virtually inevitable. In austenitic stainless steel welds, nitrogen not only forms detrimental inclusions, but also acts as a potent austenite former, which may reduce the δ ferrite content needed in the weld to prevent hot cracking.

The welding process itself has a substantial influence on the ultimate purity of the weld deposit. The most contamination-free welds are deposited by the electron beam welding process, primarily because welding is conventionally performed in a vacuum of up to 10^{-6} torr. However, the great majority of fusion welding processes require either a protective shielding gas or flux to prevent oxygen, nitrogen, and water vapor in the air from contaminating the weld deposit. The most common shielding gases used in welding are argon, helium, and carbon dioxide. Argon and helium, which are totally inert, are used primarily for the GMAW, GTAW, and plasma arc welding (PAW) processes. When adequately dispensed, shielding gas completely excludes air from the molten weld pool. Because helium has a higher ionization potential than argon, it is a hotter gas—suitable for welding thick metals and highly conductive metals, such as copper and aluminum alloys. Argon is more dense and cost-effective than helium for welding applications. Deep-penetrating spray transfer is more easily achieved at lower current levels with argon.

In GMAW of ferrous metals, small quantities of reactive gases, such as 1 to 5% oxygen and up to 25% carbon dioxide, can be added to argon to improve metal transfer, arc stability, and overall economy. Filler metals used with gas shielding containing oxygen or carbon dioxide must be alloyed with deoxidizers, such as aluminum, titanium, or zirconium, to prevent porosity and the oxidation of alloying elements in the weld admixture. Mixtures of argon and approximately 25% carbon dioxide work well for the short circuiting type of transfer when welding steels. Because of high susceptibility to porosity, aluminum alloy welds can only be shielded with argon and/or helium.

In GTAW, argon provides the best arc stability and cleaning action for alternating current (ac) welding of aluminum and magnesium. Direct current (dc) is used to weld all other metals because the alternating current cleaning action works effectively only for aluminum and magnesium. Reactive gases cannot be used with either GTAW or PAW, because the tungsten electrodes oxidize rapidly and become useless for welding. The only exception is the use of argon-hydrogen mixtures to weld austenitic stainless steels and nickel-based alloys. Small additions of hydrogen to helium or argon greatly increase arc voltage and welding heat. Hydrogen is a reducing agent that does not attack the tungsten electrode. Except for stainless steel and nickel-based alloys, hydrogen additions to the shielding gas result in welding defects of varying severity, depending on the metal welded.

Fluxes. The use of fluxes can provide a protective slag covering over the molten weld pool to effectively exclude oxygen and nitrogen contamination in processes such as ESW, SAW, FCAW, and SMAW.

Fig. 26 Effect of basicity index of SAW fluxes on resulting weld-metal oxygen content. ○ Tuliani *et al.*; △ Garland *et al.*; □ commercial data (Ref 9)

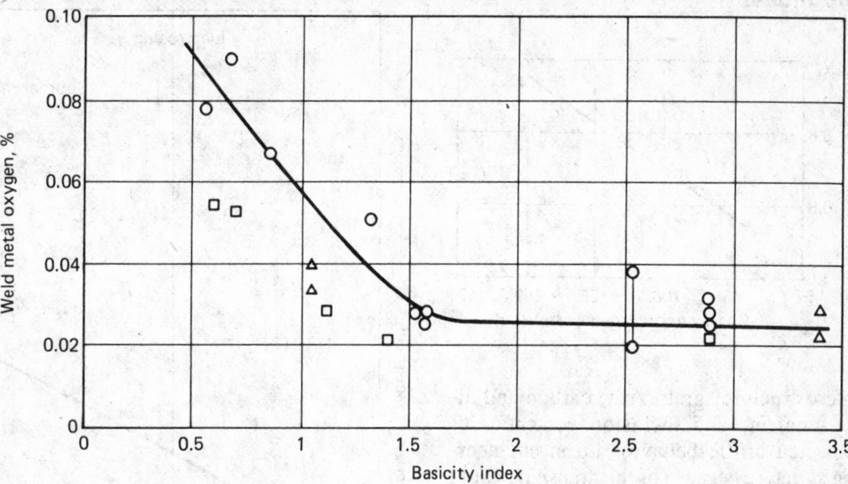

The density and melting point of fluxes are always substantially lower than those of the molten weld metal to allow liquid slag to float and solidify over the weld metal for protection against atmospheric contamination.

The flux influences both the chemical composition and microstructure of the as-deposited weld metal. In welding with fluxes, weld-metal oxygen, sulfur, manganese, and silicon contents may be affected by the basicity of the flux. The Tuliani equation (see below), which uses weight percentage units, is often used to measure the basicity index, *BI*. The ranges of *BI* are acid, <1; neutral, 1 to 1.5; semibasic, 1.5 to 2.5; and basic, >2.5. Increasing flux basicity reduces weld-metal oxygen content to a minimum for *BI* values greater than 1.5 (Fig. 26). As oxygen levels decrease, the number of inclusions in the weld tends to decrease. The nucleation of ferrite during weld cooling often occurs at inclusion sites as well as austenite grain boundaries. Weld-metal grain size is also influenced by the inclusion content of the weld, because inclusions often act as barriers to grain coarsening by pinning the movement of grain boundaries.

Sulfur content in the weld pool is also reduced as *BI* is increased; phosphorus, however, is generally unaffected. In addition, the degree of basicity influences partitioning of manganese and silicon in the weld. As *BI* increases, silicon levels in the weld decrease while manganese increases. Decreased silicon and increased manganese in the weld yield a larger *BI* or basic flux; conversely, increased silicon and decreased manganese in the weld yield a smaller *BI* or acid flux. Thus, a neutral flux generally does not alter the chemical content of the weld metal significantly, whereas basic and acidic fluxes do.

An active flux contains metallic deoxidants such as silicon, aluminum, titanium, or zirconium, which alter not only the weld-metal composition, but also the degree of change affected by welding variables such as voltage and heat input.

In summary, acidic fluxes have excellent welding characteristics, arc stability, and good detachability, but also produce a high inclusion concentration in the weld and poor impact toughness. Basic fluxes produce welds with excellent impact toughness but poor welding characteristics. Because they provide good welding characteristics and high impact toughness, semibasic fluxes are gaining popularity as working compromises which combine the advantages of the acid and basic fluxes.

Controlling Toughness in the Heat-Affected Zone

Unlike fcc metals which are ductile at all temperatures, bcc metals such as steel undergo a ductile-to-brittle transition at a temperature that is substantially influenced by metallurgical factors, including

$$BI = \frac{CaO + MgO + BaO + SrO + Na_2O + K_2O + Li_2O + CaF_2 + {}^{1}/_{2}(MnO + FeO)}{SiO_2 + {}^{1}/_{2}(Al_2O_3 + TiO_2 + ZrO_2)}$$

Fig. 27 Damaging effect of inclusion content on weld-metal toughness

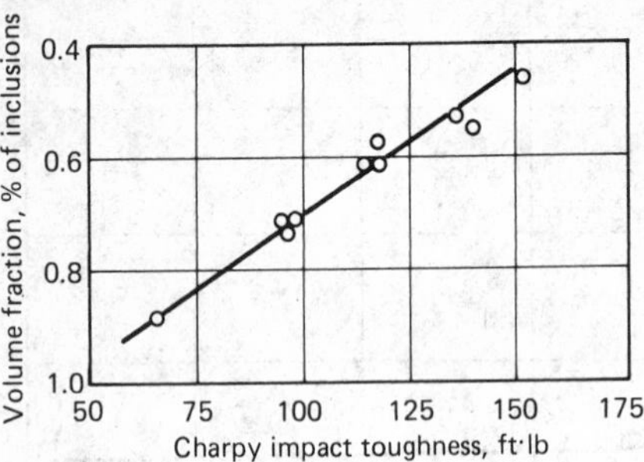

Fig. 28 Effect of welding process on weld-metal toughness for low-alloy steels. Source: Ref 11

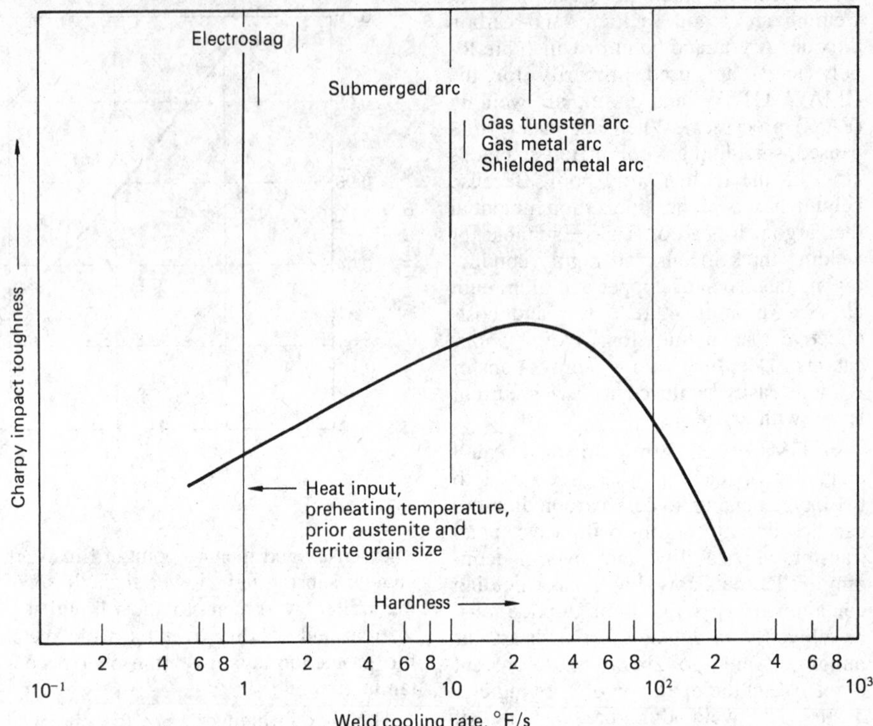

microstructure, grain size, carbon and alloy content, and inclusion content. Unexpected brittle behavior in an engineering structure because of this transition could cause a catastrophic failure.

Microstructure. Excellent toughness can be achieved in steel weld metal if the microstructure is essentially acicular ferrite with only a trace of grain-boundary ferrite, minimal bainite, and no martensite. Unless the carbon content is extremely low, fully bainitic and/or martensitic structures must be avoided. The grain size and oxide inclusion content should be as minimal as welding conditions permit. Figure 27 shows that reduced inclusion content of a weld increases the weld-metal toughness.

Welding Process. Primarily because the prior austenite and acicular ferrite grain sizes are small and the inclusion content is negligible in welds deposited by the GTAW, SMAW, and GMAW processes, these methods provide the best means of obtaining welds with excellent toughness in the as-welded condition (Fig. 28). The HAZ toughness values of the resulting welds are generally good because of the fine HAZ grain size characteristic of low heat input welding methods.

Filler Metal. Alloying elements in the filler metal, such as manganese, nickel, molybdenum, chromium, and vanadium, are very effective in promoting the formation of acicular ferrite over a greater range of weld cooling rates. An optimum alloy addition, however, is usually required to provide maximum weld-metal toughness (Fig. 29); ductility is adversely affected by quantities of the elements larger than optimum composition.

Means of Improving Toughness. The most serious problem confronting the use of cost-effective, high heat input welding processes, such as electroslag and submerged arc, is the severe loss of toughness (Fig. 28) in the weld centerline and the coarse-grained HAZ at the weld interface. Several means of increasing weld and HAZ toughness levels have been identified.

Weld-metal toughness levels can be improved by (1) judicious alloying and low carbon content, (2) special fluxes, (3) use of a high convection producing device, (4) vibration, (5) narrow-gap techniques, and (6) supplemental powdered metal additions. The most practical methods of improving weld-metal toughness employ combinations of low-carbon, alloyed filler metal to promote acicular ferrite while retarding polygonal and grain-boundary ferrite and the low inclusion content provided by semibasic or basic fluxes.

The HAZ toughness is much more difficult to control because it is part of the base metal. Small improvements in HAZ toughness have been realized by utilizing low heat input welding techniques, including the use of narrow gaps, high welding speeds, and powder metal additions. The greatest potential to increase HAZ toughness lies in the use of the new high heat input steels (Ref 10). These special alloy steels, which contain titanium and nitrogen additions, are designed to resist grain coarsening in the HAZ.

Postweld normalizing of submerged arc and electroslag welds offers the greatest improvement in toughness in both the weld and HAZ. Normalizing produces a uniform, fine-grained structure throughout the weld joint.

Preheating

Preheating of joints to be welded is an extremely effective method ordinarily used to reduce (1) the cooling rates of the weld and HAZ, (2) the magnitude of distortion and residual shrinkage stress, and (3) the arc energy input required to deposit a given weld. The first two factors are essential to

Fig. 29 Relationship between toughness and manganese content in low-alloy steel welds. Source: Ref 12

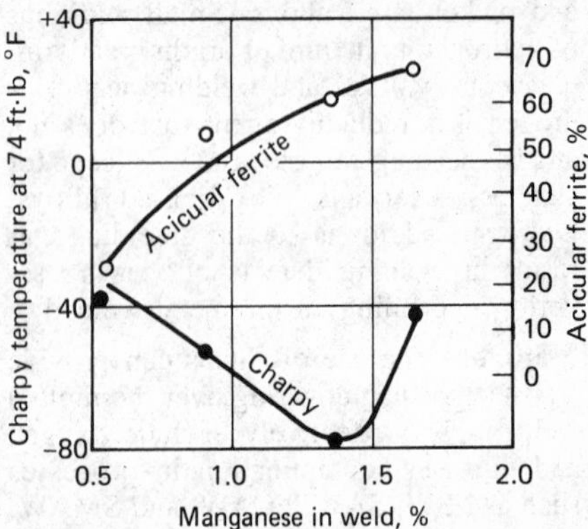

prevent cracking in hardenable steels. The third is often necessary to weld thick sections of high-conductive metals, such as copper or aluminum.

From the Adams cooling rate and peak HAZ temperature equations (Ref 6), it is clear that preheating the workpiece can significantly reduce weld cooling rate and increase the width of the HAZ. The accompanying changes in microstructure and hardness of the weld joint can be substantial. Previously in this article, adequate preheating of a high-carbon 1080 steel was shown to produce a crack-free pearlitic microstructure in the HAZ, while a similar weld without preheating resulted in a brittle martensitic structure in the HAZ.

Calculation of Preheat Temperatures. Methods used to determine the proper preheating temperature for welding a given plain or low-alloy steel are (1) consultation of the ASME Boiler Code (Section IX), AWS D1.1 Structural Welding Code, API or American Association of State Highway and Transportation Officials (AASHTO) codes, or other recognized welding codes; (2) carbon equivalent calculations; (3) reference to published literature; and (4) use of published CCT curves. When welding to a code, preheating temperatures are mandated for a particular grade of steel and thickness to be welded. For example, the AWS D1.1 code specifies that 2-in.-thick A588 steel must be submerged arc welded with a minimum preheat of 150 °F.

As the carbon content, alloy contents, and plate thickness increase, the need for preheating becomes essential to prevent cracking. For steels of a given composition, preheating temperature can be calculated on the basis of carbon equivalent, *CE*, by the following empirical equation:

$$CE = \%C + \frac{\%Mn}{6} + \frac{\%Ni}{15} + \frac{\%Mo}{4} + \frac{\%Cr}{4} + \frac{\%Cu}{13}$$

Using values derived from the equation, the requirements for preheating temperature ranges necessary to prevent cracking are $CE < 0.45\%$, optional preheating; $CE > 0.45\%$ or $< 0.60\%$, 200 to 400 °F; and $CE > 0.60\%$, 400 to 700 °F.

Published literature can also be consulted to determine preheating temperatures. *Weldability of Steels,* published by the Welding Research Council, is an excellent source for recommended temperatures.

Provided that CCT diagrams are available for both the steel plate to be welded and the anticipated admixture compositions, this method of calculating preheating temperatures is very useful and can be applied to virtually any alloy steel composition. In general, structural steels should be welded with sufficient preheating to prevent the formation of embrittling martensite. By studying the CCT diagrams for the plate and weld-metal compositions, the maximum permissible cooling rate can be determined. This cooling rate is then substituted into Eq 6 or 7. The solution of the equation yields the preheating temperature, T_o. If the solution provides a value of T_o less than ambient temperature, no preheating is required.

Reduction of Distortion and Residual Stress. The second purpose of preheating both ferrous and nonferrous metals is to reduce distortion and residual stress. As a weld cools through the austenite range in steels and through an elevated temperature range in nonferrous alloys, the metal has little strength and good plasticity. Therefore, the weld metal and HAZ deform plastically to accommodate the change in dimensions imposed by the shrinking weld. Upon cooling to room temperature, residual stresses build up because of continued shrinkage, but not to the extent experienced in a similar weld without preheating. The degree of reduction of distortion and residual stress is difficult to predict in practical welding applications, because it depends on many variables, including magnitude of restraint, preheating temperature, groove preparations, and heat input.

Postweld Heat Treatment

A wide variety of metallurgical objectives may be accomplished through postweld heat treatments, including stress relief, dimensional stability, resistance to stress corrosion, and occasionally, improved toughness and mechanical properties. The most common postweld heat treatments for steels are subcritical stress relieving, normalizing, and quench and tempering. Typical treatments for nonferrous metals, such as aluminum alloys, include postweld stress relieving, full solution heat treatment and aging, aging only, and annealing.

Stress relieving is probably the most frequently used treatment to reduce the residual welding stresses in welds that are heavily restrained or susceptible to cracking. The dominant mechanisms of stress relief are relaxation of stress and tempering of martensite or overaging of precipitation-hardening alloys. In steels, the stress-relief temperature ranges from 895 to 1240 °F, which is below the eutectoid transformation, for a minimum of 1 h per inch of thickness. Most often, a weld deposited on a high hardenability steel, such as 4130, is put into the stress-relieving furnace before having a chance to cool below the preheating or interpass temperature. Thus, the microstructure does not contain any martensite, because any austenite remaining after welding is transformed to bainite during stress relieving in accordance with the TTT diagram for 4130 steel. If the steel weld does form martensite after welding because of insufficient preheating, the stress-relieving operation tempers the martensite to a lower value of hardness, but improves toughness and ductility.

Postweld stress relieving can virtually eliminate caustic stress corrosion cracking occurring in the HAZ of ASTM A516, grade 70 steel, which is ordinarily used in the pulp and paper industry. The combined reduction of residual stress and galvanic differences among the weld metal, the HAZ, and the base metal undoubtedly contributes significantly to the improved stress corrosion resistance of these weldments.

Postweld normalizing applies primarily to steels. This treatment has generally the same function in welding as in the casting industry, promoting toughness and eliminating the coarse grain structure. Therefore, the ESW process especially benefits from normalizing. Because of the high heat input, 2286 kJ/in. for a 2-in.-thick weld, the large HAZ and weld-metal grain sizes result in a severe loss of Charpy V-notch toughness. Several welding codes permit electroslag welds in primary tension members only if the welds are normalized. Unfortunately, normalizing is a far more expensive operation than stress relieving. Although stress relieving can be conducted in the field, normalizing requires such high temperatures, 1600 to 1705 °F for 1 h per inch of thickness, that the entire welded assembly must be transported to a large furnace.

Benefits of postweld normalizing include the elimination of both the coarse columnar grain structure in the weld and the large grain size in the HAZ and substantial improvement in toughness at the weld centerline and HAZ. The microstructure of typical normalized weld joints contains a fine-grained mixture of pearlite and polygonal ferrite. For example, the microstructural changes induced by normalizing electroslag welds deposited on A588 steel will raise the CVN toughness of the weld centerline from approximately 7 ft · lb at 0 °F to more than 50 ft · lb.

Normalizing is generally not as beneficial for gas tungsten arc, shielded metal arc, or gas metal arc welds as for electroslag or submerged arc welds, because grain structures in the weld and HAZ are already refined by virtue of their comparatively low heat input and small bead size. Nonetheless, normalizing can be applied to any weld and will replace the weld and HAZ structures with a uniform structure of polygonal ferrite plus pearlite.

Quenching and tempering of welded joints is an expensive operation and is reserved only for welds deposited on heat treatable steels such as 4130, 4140, 4340, H-11, and other steels used for high-strength, high-hardness applications. Unlike the quenched and tempered steels, which include A514, A517 (T-1), and A508 (HY-80), heat treatable steel welds contain high carbon levels and alloying elements to match the hardness and hardenability of the base metal after postweld quenching and tempering.

Solution treating and aging are postweld heat treatments used with precipitation-hardening alloys for uniform strength of the weld, HAZ, and base metal. If the weld cooling is rapid, as in electron beam welding, aging alone can yield significant strengthening. In the latter case, the weld and the large-grained portion of the HAZ are considered after welding to be essentially solution treated.

Welding Flaws

Porosity, or fine holes or pores within the weld metal, can occur by absorption of evolved gases and chemical reaction. Metals susceptible to porosity are those which can dissolve large quantities of gas contaminants (hydrogen, oxygen, nitrogen, etc.) in the molten weld pool and subsequently reject most of the gas during solidification. Aluminum alloys are more susceptible to porosity than any other structural material.

The mechanism by which porosity forms in welds is inextricably related to the solidification kinetics and morphologies of each alloy. The most common types of porosity are either interdendritic or spherical. Variations in solidification mode (planar, cellular, or cellular dendritic) will affect not only the resulting size, shape, and distribution of hydrogen pores, but also the ability of pores to detach from the interdendritic regions to be free floating. The process by which gas bubbles grow in the weld pool is coalescence of smaller bubbles in accordance with Stokes law, and a strong convective fluid flow.

Weld cooling rate substantially affects the volume of porosity retained in a gas-contaminated weld. At fast cooling rates, the level of porosity is low, as the nucleation and growth of bubbles in the liquid are severely suppressed. Similarly, at very slow cooling rates, porosity is minimal because bubbles have ample time to coalesce, float, and escape from the weld pool. At intermediate cooling rates, the greatest volume of porosity in a weld is observed, as conditions are optimum for both formation and entrapment of virtually all of the evolved gases in the weld.

The sources of porosity are contaminants, including moisture, oils, paints, rust, mill scale, and oxygen and nitrogen in the air. The heat from the welding arc decomposes moisture, oils, and paints into hydrogen and other gaseous products. Hydrogen may produce severe porosity in aluminum alloy welds, as well as welds deposited on copper, magnesium, niobium, and titanium alloys. Because hydrogen is not easily nucleated in steels during solidification, the majority of hydrogen contamination remains in supersaturated solid solution after welding is complete.

The base metal itself may contain gas in appreciable amounts, as in rimmed steels or volatile coatings, such as zinc (from galvanizing) and cadmium. Volatile ingredients, including sulfur, lead, and selenium, are required for free-machining applications. However, during welding, these additions vaporize and produce excessive porosity. In addition, the evolution of gases in a weld may cause wormholes and oxide inclusions, as well as porosity.

Porosity is also caused to a lesser extent by chemical reactions between absorbed active gases and easily reducible ingredients in the weld pool. For example, in steel, absorption of oxygen gas causes decarburization:

$$C + O \rightarrow CO \text{ gas}$$

In copper alloys that almost always contain some oxygen impurity, absorbed hydrogen reacts as follows:

$$2H + O \rightarrow H_2O \text{ gas}$$

Hydrogen accidentally introduced into a welding atmosphere tends to reduce most metal oxides:

$$4H + 2Cu_2O \rightarrow 4Cu + 2H_2O \text{ gas}$$

$$2H + FeO \rightarrow Fe + H_2O \text{ gas}$$

Reduction reactions occurring in the weld pool can be approximated by using simple thermodynamic calculations or functional charts, such as the Ellingham diagram.

Because the production of porous-free welds is not cost-effective in commercial welding applications, limited amounts and sizes of porosity are permitted by all welding codes. The best prevention for porosity in welds is utilization of acceptable standards of workmanship, as specified in various structural welding codes. If parts to be welded and consumables are cleaned and dried, a great percentage of gaseous porosity is eliminated. If the rust and light mill scale on steel plates cannot be removed before welding, filler metal or fluxes containing strong deoxidizers, such as aluminum, titanium, and zirconium, should be used. These deoxidizers are effective in reducing porosity to acceptable levels. Only hydrogen contamination cannot be removed by deoxidizers.

Hydrogen-Induced Cold Cracking. A major source of underbead cracking in welds deposited on low-alloy and other hardenable steels is cold cracking (also called delayed cracking). Cold cracks may form within minutes, hours, or days after welding and can result in catastrophic failures of welded structures. Factors required for cold cracking to occur are (1) a crack-sensitive microstructure, usually martensitic; (2) sufficient hydrogen concentration in the weld; (3) rigid tensile restraint; and (4) a temperature between approximately 300 to −150 °F. Elimination of one or more of these factors greatly reduces crack susceptibility.

Hydrogen, when entering a steel weld, diffuses and segregates at pores, discontinuities, inclusions, and other microscopic flaws. These flaws are effective traps and can severely reduce the diffusion coefficient of hydrogen (Fig. 30). General belief, however, is that the remaining diffusible hydrogen is responsible for the cold cracking problems in welds. The mechanism of hydrogen-induced cold cracking has been widely studied.

This phenomenon is known to be diffusion-controlled, time-dependent, and either transgranular or intergranular. Several theories explain why cold cracking is time-dependent. Generally, a pre-existing microcrack or discontinuity acts as a stress-concentration site. When a tensile stress is applied, hydrogen diffuses at room temperature to the regions of greatest tensile strain. After the concentration of hydrogen at or near the tip of the discontinuity has accumulated to a critical value, which depends on the magnitude of externally applied tensile stresses or residual stresses, the hydrogen is believed to cause severe reduction in the cohesive bonding energy between iron atoms ahead of the discontinuity, and cracking initiates. Propagation of the crack occurs in discrete bursts or steps, which are repeated as fresh hydrogen diffuses ahead of the crack tip. At low

Fig. 30 Diffusion coefficient of hydrogen in iron alloys. Source: Ref 13

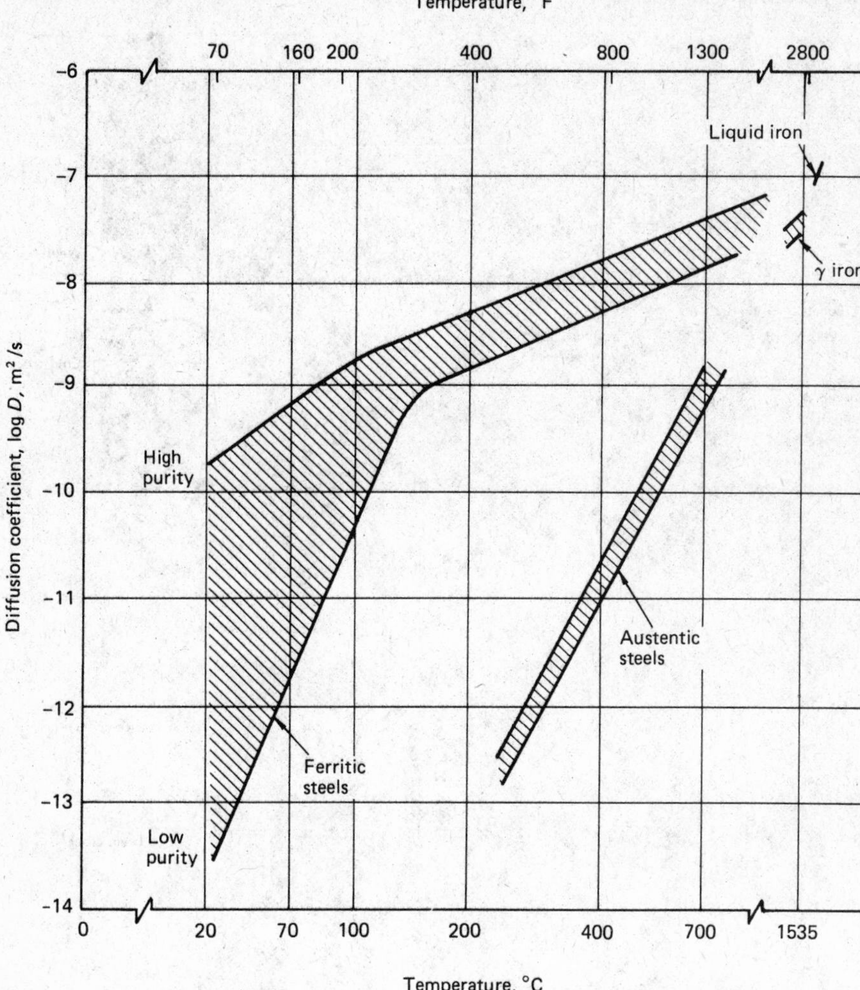

stress intensity values, cracking is likely to follow an intergranular path between prior austenite grains which have transformed to martensite, while at high stress intensities, the fracture could be transgranular.

In welding, the combination of tensile shrinkage stresses and hydrogen contamination may cause microcracks to occur in both the weld fusion zone and HAZ. In fact, cold cracking occurs more commonly in the HAZ because the hydrogen contamination entering the molten weld pool diffuses rapidly into the HAZ and most steel filler metals have less carbon than the base metal for good weldability, making the HAZ microstructure more susceptible. Typical cold cracks are shown in Fig. 31(a) and (b).

The cold cracking susceptibility of a given composition of steel is related to the Dearden and O'Neill equation for *CE*:

$$CE = \%C + \frac{\%Mn}{6} + \frac{\%Cu + \%Ni}{15} + \frac{\%Cr + \%V + \%Mo}{5}$$

This formula was derived for plain-carbon and low-alloy steels containing 0.12% C or more. Weight percentages are used in the calculation. For low-carbon steels in the range from 0.07 to 0.22% C, the Ito and Bessyo equation can be used:

$$CE = C + \frac{Si}{30} + \frac{Mn + Cu + Cr}{20} + \frac{Ni}{60} + \frac{Mo}{15} + \frac{V}{10} + 5B$$

Generally, most users of the *CE* equation agree that a value of *CE* >0.35 to 0.40 (depending on plate thickness and the degree of restraint) indicates that a given steel composition will be susceptible to cold cracking in the HAZ unless steps are taken to reduce the amount of hydrogen contamination entering the molten weld pool.

For example, consider the SMAW of ASTM A572 grade 65 with a *CE* of 0.44 to 0.48%. Although susceptible to cold cracking, this grade of steel is welded routinely without cracking if proper welding procedure is maintained. Structural welding codes, such as AWS D1.1, require that A572 grade 65 be welded with low-hydrogen electrodes E8015, 8016, or 8018; adequate preheating temperature, depending on thickness; and a suitable welding procedure to ensure low-hydrogen welding conditions.

Because hydrogen enters the weld pool by the dissociation of moisture and hydrocarbons in the arc, consumables that are inherently high in moisture should not be used to weld steels having greater than either a 300 HB hardness value or 0.40% *CE*. In SMAW, cellulosic and rutile electrodes, such as E6010, E6011, E6012, E6013, E7014, and E7024, contain organic substances which evolve abundant amounts of hydrogen during welding. Similarly, in FCAW, an E70T-1 filler metal contains a rutile base flux and may evolve substantial hydrogen, particularly if not dried by baking prior to welding. In SAW, fused fluxes contain no water, but agglomerated fluxes may have substantial amounts of residual moisture because these fluxes are hygroscopic and should be baked at a prescribed temperature immediately prior to welding.

To ensure crack-free welding of hardenable steels, low-hydrogen covered electrodes, such as E7015, E7016, and E7018 for SMAW, must be baked at 450 to 500 °F for at least 2 h prior to welding. Higher strength electrodes, such as E8018, E9018, and E10018, must be dried at 700 to 800 °F for crack prevention. Fused fluxes for SAW may be dried at 250 °F to eliminate moisture.

Preheating is essential to prevent the occurrence of cold cracking. The diffusible hydrogen content of the weld is substantially reduced by preheating and maintaining an interpass temperature such as 480 °F. At approximately 480 °F, no hydrogen-induced cold cracking is possible because hydrogen diffuses so rapidly that it will not segregate at the tips of discontinuities or stress concentrations. Eventually, the rapidly diffusing hydrogen atoms escape from the weld surface.

The level of diffusible hydrogen in a given weld can be measured by standard techniques that are specified in the military specification MIL-E-24403A(SH),

International Standard ISO-3690, and several other specifications. Once the hydrogen content is determined, an upper limit on the permitted hydrogen level in the weld joint can be set, based on actual experience. The International Institute of Welding's general terminology for diffusible hydrogen concentrations in welds is:

Term	Hydrogen concentration, mL/100 g(a)
Very low	0-5
Low	5-10
Medium	10-15
High	>15

(a) Millilitres of hydrogen evolved per 100 g of weld metal deposited

Ferrite Vein Cracking. Another form of hydrogen-induced cracking has been observed in electroslag weld metal. Unlike cold cracking of high-strength steels (discussed in the previous section), ferrite vein cracking occurs along the grain-boundary ferrite in low-carbon steels that were believed "immune" to hydrogen-induced cracking such as A36 and A588. A typical vein crack, shown in Fig. 31(b), is usually less than 1/4 in. long and always occurs along the grain-boundary ferrite. Although the mechanism of cracking is unknown, the cause has been definitely attributed to the presence of hydrogen-bearing contaminants in the weld pool. The cure is simply to ensure that all fluxes and consumables are sufficiently free of moisture, grease, oil, and excessive oxide scale.

Lamellar Tearing. This type of cracking, which occurs in the base metal or HAZ of restrained weld joints, results from inadequate ductility in the through-thickness direction of steel plate. Susceptibility to lamellar tearing depends on the joint geometry, oxide and sulfide inclusion content, and the extent to which these inclusions are elongated or flattened in the rolling direction to form parallel planes of weakness. The reduction of area values from tensile tests on steels taken in the through-thickness direction indicates susceptibility to lamellar tearing. Steel plates exhibiting reduction of area values less than 10% will be sensitive to cracking. Hydrogen is known to be preferentially trapped at inclusions and tends to accelerate the occurrence of lamellar tearing. Therefore, welds on steels deposited by low-hydrogen processes are more resistant to lamellar tearing than similar welds deposited with methods other than low-hydrogen welding.

Preheating and buttering can be effective in reducing susceptibility of steel to lamellar tearing. Preheating decreases both the magnitude of residual tensile stresses acting in the through-thickness direction and the severity of embrittling effects caused by hydrogen. Because lamellar tearing usually occurs within 0.1 to 0.2 in. from the weld interface in the HAZ, susceptible plates can be buttered with crack-resistant weld metal. When joining two buttered plates, the major tensile stresses act on the relatively immune buttered regions of the joint. The most economical preventative measure, when possible, is to assemble the plates to be welded with the rolling direction of the plate perpendicular to the weld axis.

Fig. 31 Hydrogen-induced cold cracking. (a) Underbead cold cracking in the HAZ of a shielded metal arc welded 1080 steel. (b) Ferrite vein cold cracking in an electroslag welded A588 steel. Magnification: (a) ×65, (b) ×65

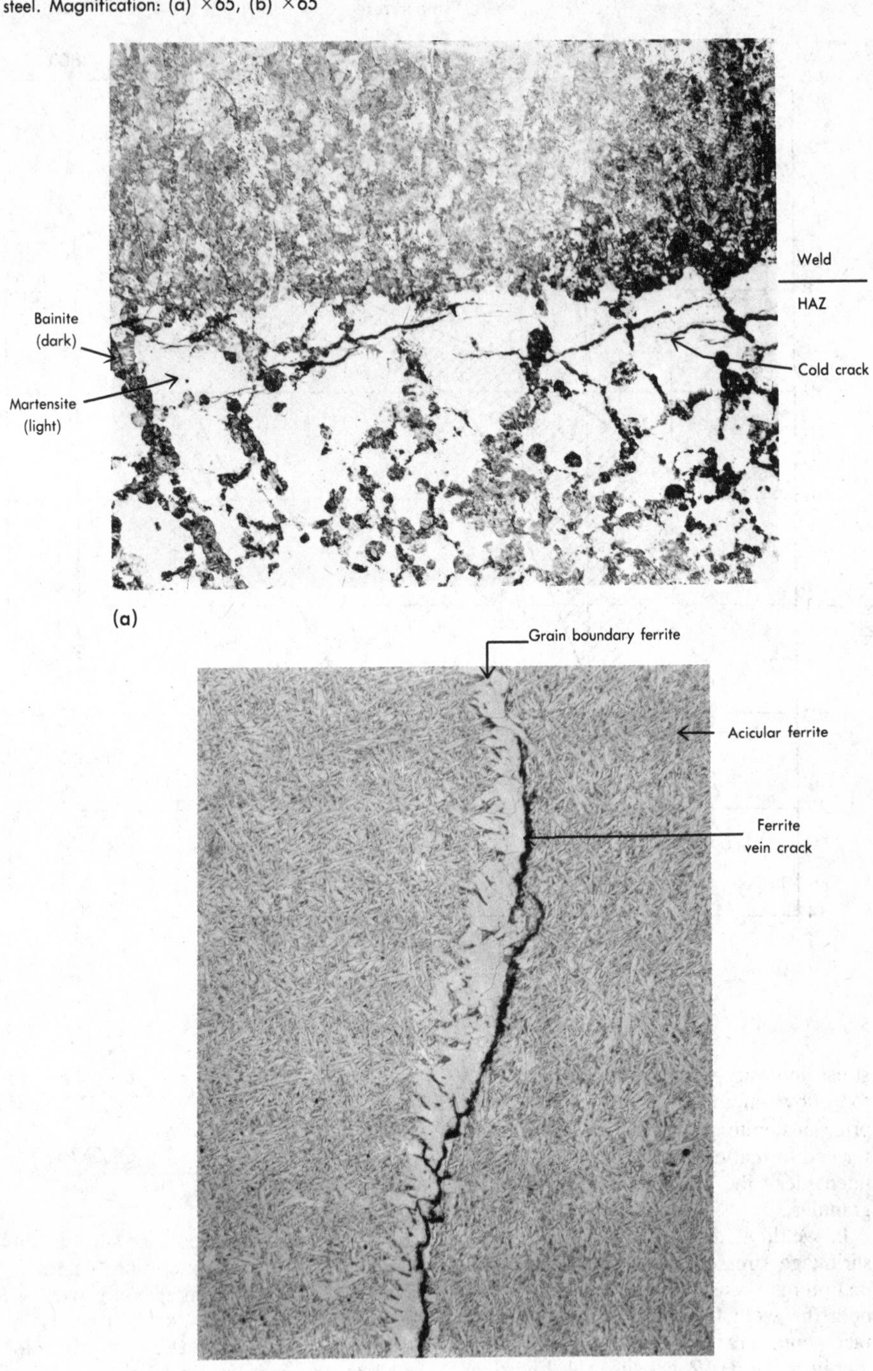

Reheat Cracking. In welding high-strength low-alloy steels that have been alloyed with vanadium, boron, niobium, molybdenum, or titanium to promote the formation of acicular ferrite for maximum toughness, postweld stress relief is often prohibited by reheat cracking in the HAZ. Reheat cracking is entirely intergranular, and although not definitely established, the mechanism is believed to be a result of precipitation hardening within the HAZ grains while the grain boundaries are left in a weakened condition. If the residual stress and restraint are great, brittle intergranular cracking occurs. The relative propensity for a steel to be subject to reheat cracking can be qualitatively measured by the Ito equation, which is calculated using weight percentages:

$$C = 10V + 7Nb + 5Ti + Cr + Cu - 2$$

If carbon (C) is equal to or greater than zero, the steel may be susceptible to reheat cracking. Although boron is not considered in the Ito equation, steels containing boron, particularly in combination with 1/2% Mo, have been suspected of contributing indirectly to reheat cracking. In addition, the presence of nitrogen, sulfur, phosphorus, arsenic, antimony, and aluminum is suspected of contributing indirectly to the reheat cracking problem. The best prevention for reheat cracking is to use a nonsusceptible alloy. If this is not possible, the stress relief anneal must not be performed, at least not within the temperature range of maximum susceptibility, which is usually between 1020 to 1200 °F. Finally, reducing external restraint and minimizing stress concentrations through proper design and good welding practice are helpful in moderating susceptibility to reheat cracking.

Hot Cracking in the Weld Metal and Heat-Affected Zone. Another type of cracking, which results from internal stress developed on cooling following solidification, is hot cracking. This defect occurs at a temperature above the solidus of an alloy. As discussed earlier, dendritically solidified weld metal is characterized by compositional differences between the dendrite core and the solute-rich, lower melting interdendritic regions. During the last stages of solidification, a semirigid network of mechanically interlocked dendrites forms with a small amount of low-melting liquid interdispersed. As the weld shrinks during solidification, tensile stresses tend to pull the loosely bonded dendrites apart. This separation is always interdendritic or intercellular and results in hot cracked weld metal (Fig. 32).

Fig. 32 Hot cracking in a gas tungsten arc weld of Fe-12%Mn alloy. Magnification: ×200. Etchant: three acids

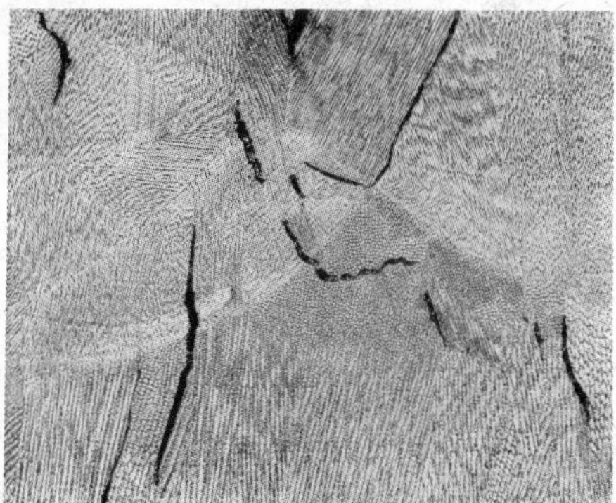

The amount of interdispersed liquid is a critical factor in determining the susceptibility of an alloy to hot cracking. In the eutectic system shown in Fig. 33, an alloy weld of composition C_1 solidifies in the portion of the phase diagram that is most susceptible to hot cracking, because the greatest solid and liquid temperature ranges occur at C_1, and because the amount of liquid remaining after an interdendritic separation or crack has taken place is not sufficient to fill or heal the crack as a riser would in a casting. If an alloy of composition C_2 is welded, the solidification temperature range decreases and, more importantly, a plentiful amount of liquid of eutectic composition is available to heal any interdendritic crack that may form. Finally, the alloy of composition C_3 is virtually immune to hot cracking under normal welding practice because each dendrite is surrounded by eutectic liquid that solidifies at one temperature in a planar mode and provides ample eutectic healing of any dendritic separation.

Very often in complex alloys, hot cracking is caused by impurity liquid films such as the low-melting sulfide film in steels, stainless steels, and nickel alloys. In welding low-alloy steels, for example, weld-metal carbon content should be kept below 0.1% so the weld solidifies as δ ferrite and avoids the peritectic reaction shown in Fig. 4. Sulfur dissolves readily in δ ferrite but is relatively insoluble in austenite. As a result, austenite rejects sulfur to cell or dendrite boundaries and forms an intergranular low-melting film, which is partially responsible for hot cracking during solidification of many iron- and nickel-based alloys. Fortunately, increasing the manganese content, as shown in Fig. 23, helps reduce hot cracking sensitivity by raising the melting temperature of the sulfides.

Other elements, such as phosphorus and niobium, can be identified as promoters of

Fig. 33 Effect of composition on hot cracking susceptibility of welds in a eutectic system. Regions of hot crack susceptibility: A, no cracking; B, liquid healing is possible; C, hot crack sensitive. (Source: Ref 14)

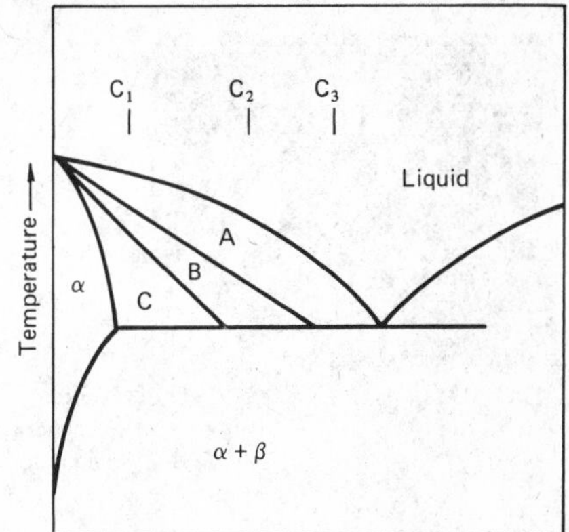

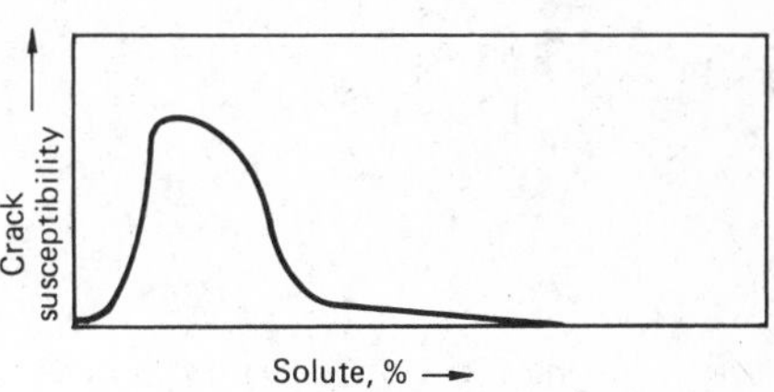

hot cracking in steels using this formula developed by Bailey and Jones:

$$UCS = 230\%\ C^* + 190\%\ S + 75\%\ P + 45\%\ Nb - 12.3\%\ Si - 5.4\%\ Mn - 1$$

If $C < 0.08\%$, C^* is taken to be 0.08% C. If units of crack susceptibility, *UCS*, exceed 19 for T-fillet welds and 25 for butt welds, the chances for hot cracking become an important consideration in material, design, and production choices. Methods to minimize hot cracking in weld metal include: (1) maintenance of adequate manganese-to-sulfur ratio; (2) reduction of sulfur, phosphorus, carbon, and niobium to minimal amounts; (3) production of weld bead convexes, such as slightly overfilling the weld groove or fillet; and (4) reduction of the tensile restraint exerted on the weld.

Hot cracking may also occur in the partially melted zone by the liquation of low-melting inclusions or second-phase segregates and grain-boundary melting. The width of the partially melted zone depends on the nature of the segregates and the thermal gradient perpendicular to the weld interface. The liquation of large manganese sulfide inclusions was described in the section on definition of a weld in this

Fig. 34 Hot cracking in a stainless steel weld joint. Copper-contaminated weld deposited by GTAW on 304 stainless steel. Severe hot cracking in weld metal with grain-boundary liquation in the partially melted zone. Magnification: ×100. Etchant: nitric and hydrochloric acid in glycerol

article. If the grain boundary in the HAZ is segregated with a low melting phase, cracking can be extensive. For example, welding of 304 stainless steel with 308 filler metal is occasionally contaminated with copper from various types of tooling or other sources. The accidental addition of copper to the weld pool causes not only severe hot cracking of the weld metal, but also extensive grain-boundary melting or hot cracking in the partially melted zone by a liquid-metal embrittlement mechanism. Figure 34 illustrates the severity of hot cracking that is possible in stainless steel weld joints.

Chevron cracking, also referred to as staircase cracking, can occur in the fusion zones of both submerged arc welds using agglomerated fluxes and shielded metal arc welds deposited on medium-strength steels. This type of cracking is characterized by many parallel cracks angled approximately 45° from the plane of the plates in a butt joint. The multiple cracks are best viewed by cutting a longitudinal section of a weld joint. A chevron pattern is produced by the 90° intersections of these cracks, which are partially intercolumnar and partially transcolumnar within the weld fusion zone. Caused by the presence of hydrogen, chevron cracking is apparently a form of hydrogen-induced cold cracking. The exact mechanism of cracking is unexplained, although low-hydrogen welding practice, particularly in SAW and SMAW, is known to eliminate chevron cracking.

Ductility-Dip Cracking. Many alloys, including cupronickels, austenitic chromium-nickel steels, some nickel-based alloys, and aluminum bronzes, exhibit a temperature range over which ductility and tensile strength drop sharply. These alloys are susceptible to HAZ and weld-metal cracking when welded under excessive restraint. Above and below the ductility-dip embrittlement range, fracture occurs by microvoid coalescence resulting in high values of strength and ductility. Within the embrittlement range, fracture is intergranular and brittle.

The actual embrittling mechanism is not well understood, but is believed in most instances to be caused by the grain-boundary segregation of phosphorus and sulfur in austenitic chromium-nickel steels, nickel in cupronickels, and aluminum and nickel in aluminum bronzes. Crack susceptibility is extremely composition-dependent. For example, increasing nickel content in cupronickel alloys to above approximately 18% tends to decrease ductility at 1290 °F. Similarly, aluminum bronzes containing less than approximately 11% Al and an increased amount of nickel tend to exhibit reduced ductility in the temperature range of approximately 930 to 1110 °F. In some cases, cracking may be initiated by hot cracking followed by ductility-dip cracking when the segregating constituents not only embrittle grain-boundary regions in the ductility-dip temperature range, but also substantially depress the melting temperature at the grain boundaries.

Fortunately, ductility-dip embrittlement is not a widespread problem and occurs in only a few structural alloys. This form of cracking is best prevented by avoiding alloys known to exhibit ductility-dip problems. If such an alloy must be welded, cracking will be difficult to avoid. Therefore, welding of susceptible alloys should be limited to thin sections, and the welding conditions should minimize shrinkage stresses during cooling.

Intergranular Corrosion of Austenitic Stainless Steels. A recurring problem in the welding of austenitic stainless steels is the accelerated corrosion that takes place in the HAZ. When stainless steels such as 304 and 308 are heated into or very slowly cooled through the sensitizing temperature range of 930 to 1470 °F, chromium carbide precipitation occurs along the grain boundaries. The narrow region adjacent to the carbide-decorated boundaries is depleted of chromium, making this region susceptible to corrosion by numerous media. Corrosion is always intergranular and can be so severe that grains may actually dislodge from the metal. In welding stainless steels, sensitization is particularly severe in portions of the HAZ that have experienced peak temperatures in the sensitizing range for a time sufficient to precipitate intergranular chromium carbides.

Intergranular corrosion can be avoided by utilizing any of three practical solutions. The first and most common solution is to only weld plate that contains extra-low carbon, as designated by the suffix L, such as types 304L and 308L. Filler metals should also contain low carbon, because without adequate carbon, the precipitation of damaging intergranular carbides is essentially eliminated.

Intergranular corrosion can also be avoided by using stainless steel base metal that has been stabilized with small additions of titanium or niobium. These elements form tenacious carbides that are nearly insoluble at elevated temperatures. As a result, carbon is tied up as stable compounds (titanium or niobium carbides) and is no longer free to precipitate chromium carbides when exposed to the sensitizing temperature range. Typical stabilized stainless steels include types 321, 347, and 348.

In addition, a postweld solution heat treatment can be applied to a sensitized weld joint to prevent intergranular corrosion. By heating the entire welded assembly to approximately 1650 to 2100 °F for a time sufficient to dissolve all chromium carbides and quenching to room temperature, corrosion resistance can be fully restored.

For more information on welding flaws, see the article "Weld Discontinuities" in this Volume.

REFERENCES

1. Atkins, M., *Atlas of Continuous Cooling Transformation Diagrams for Engineering Steels,* American Society for Metals and British Steel Corporation, 1980
2. Savage, W.F., Nippes, E.F., and Szekeres, E.S., A Study of Weld Interface Phenomena in a Low Alloy Steel, *Welding Journal,* Sept 1976, p 260s-268s
3. Lyman, T., Ed., *Metals Handbook,* 8th ed., Vol 8, Metallography, Structures and Phase Diagrams, American Society for Metals, 1973
4. Lyman, T., Ed., *Metals Handbook,* 8th ed., Vol 1, Properties and Selection, American Society for Metals, 1961, p 46, 52

5. Flemings, M.C., *Solidification Processing*, McGraw-Hill, New York, 1974
6. Adams, C.M., Jr., Cooling Rates and Peak Temperatures in Fusion Welding, *Welding Journal*, Vol 37 (No. 5), 1958, p 210s-215s
7. Weisman, C., Fundamentals of Welding, *Welding Handbook*, 7th ed., Vol 1, American Welding Society, Miami, 1976
8. Glover, A.G., *et al.*, The Influence of Cooling Rate and Composition on Weld Metal Microstructure in a C/Mn and HSLA Steel, *Welding Journal*, Vol 56 (No. 9), 1977, p 267s-273s
9. Eager, T.W., Sources of Weld Metal Oxygen Contamination During Submerged Arc Welding, *Welding Journal*, Vol 57 (No. 3), 1978, p 76s-80s
10. Kanazawa, S., *et al.*, Improvement of Weld Fusion Zone Toughness by Fine TiN, *Transactions of Iron and Steel Institute of Japan*, Vol 16, 1976, p 487-495
11. Dolby, R.E., Factors Controlling HAZ and Weld Metal Toughness in C-Mn Steels, *Proceedings of First National Conference on Fracture*, Johannesburg, South Africa, 1979, Contained in Garrett, G.G. and Marriott, D.L., *Engineering Applications of Fracture Analysis*, Pergamon Press, p 117-134
12. Evans, G.M., The Effect of Heat Input on the Microstructure and Properties of C-Mn All-Weld-Metal Deposits, *Welding Journal*, Vol 61 (No. 4), 1982, p 125s-132s
13. Coe, F.R., *Welding in the World*, Vol 14 (1-2), 1976, p 1-7
14. Barland, J.C., *British Welding Journal*, Vol 7, 1960, p 508

SELECTED REFERENCES

- Gray, T.G.F. and Spence, J., *Rational Welding Design*, 2nd ed., Butterworths, London, 1982
- Lancaster, J.F., *Metallurgy of Welding*, 3rd ed., George Allen & Unwin, Boston, 1980
- Linnert, G.E., *Welding Metallurgy*, Vol 2, American Welding Society, New York, 1967
- Stout, R.D. and Doty, W.D., *Weldability of Steels*, 3rd ed., Welding Research Council, New York, 1978

Introduction to Joining Processes

By the ASM Committee on Joining Processes*

SELECTION OF A JOINING PROCESS requires a basic knowledge of the various processes and of their relationships to such variables as joint design, base-metal properties, equipment cost, and welder skill. Discussion of these variables in this article will help the reader to attain the level of understanding necessary to adequately select the correct process for a specific application. For more in-depth discussion, see the articles on individual processes in this Volume.

Process Selection

For increased economy, automatic welding modes should be considered in process selection. Most basic welding processes generally can be automated through the use of robotics, mechanical indexing, and positioning systems. The success of any automated mode depends on the adaptability of the item to be welded and the time devoted to the proper design of equipment.

Welding processes that use an electric arc are the most widely used in industry. The arc may be established between an electrode and the base material, as in shielded metal arc welding (SMAW) and gas tungsten arc welding (GTAW), or the arc may occur within the welding heat source, as in plasma arc welding (PAW). Furthermore, the arc and molten metal may be protected by an inert gas, granulated flux, or gaseous slag products of a consumable electrode.

Other welding processes include (1) oxyfuel gas welding (OFW), in which a combustible gas is burned with additions of oxygen to produce a high-temperature flame; (2) resistance welding, in which high current density is introduced to create a high metal temperature and pressure is applied to produce a weld; (3) flash welding, in which an arc is created and followed by instantaneous force to bring the parts being welded together; (4) diffusion welding (DFW), in which clean metallic parts are brought together with high force to create bonding through diffusion; (5) friction welding (FRW), in which two parts to be welded are brought together with force and movement at high speed to create high temperature and bonding; (6) electron beam welding (EBW), in which a focused stream of electrons produces melting and joining; (7) laser beam welding (LBW), in which a coherent light beam is focused on the workpiece to create melting for welding or cutting; (8) ultrasonic welding (USW), in which a concentrated beam of sound waves is used; and (9) explosion welding (EXW), in which a high-energy explosive is used to create very high forces between two workpieces, thus bonding them.

The OFW process is the oldest in use, but it is usually limited to manual techniques for braze welding or as a heat source for silver brazing and soldering. Fusion welding is possible with OFW, but the high heat input may create metallurgical problems.

The SMAW process (stick) is widely used because of its versatility, portability, and low cost, which make it useful for field fabrication and installation. Although individual weld joint costs may be high compared to automated processes, initial equipment cost and portability often are deciding factors in selection.

The shielded GTAW process (heliarc) is a strong competitor with SMAW because of its adaptability to certain materials, such as titanium, zirconium, stainless steel, and aluminum, as well as its capability to produce high-quality welds. Shielded GTAW is well suited to automated welding modes.

Submerged arc welding (SAW) is a high-production process that can be used for shop, field, and semiautomated applications. Submerged arc welding has certain limitations for weld position requirements.

Plasma arc welding is a high-energy heat source application that is particularly adaptable to automated welding techniques. It has been used advantageously for hardfacing with special metal alloys for wear and abrasion applications.

The LBW, EBW, DFW, EXW, FRW, USW, and flash welding processes are somewhat specialized and limited in application. Resistance welding is a low-cost, high-production process for use in industrial applications. It is an excellent substitute for riveted construction of thin metal members.

Fusion and Nonfusion Processes

In selecting a welding process and determining the welding procedure, consideration must be given not only to the nature of the base metal, but also to the effects of the joining process on the properties of the weldment. These effects are greater if the base metal is partially melted than if no fusion occurs. Usually, changes in base-metal properties occur in the areas adjacent to the weld because of the heat gen-

*Kenneth E. Richter, *Chairman,* President, Nu-Weld/MCI, Divisions of Dimetrics, Inc.; John Bartley, Vice President—Marketing, Nu-Weld/MCI, Divisions of Dimetrics, Inc.; Isaac S. Goodman, Consultant; Edwin J. Hemzy, Construction Manager, Commonwealth Edison Co.; Edwin Shifrin, Welding Engineer, Detroit Edison Co.

erated during the joining process. The width of these areas is called the heat-affected zone (HAZ).

Most welding processes involve partial melting of the base metal. In some cases, the molten product is largely excluded from the completed joint, as in friction and flash resistance welding. Nonfusion processes involve little or no melting of the base metal. These include soldering, brazing, USW, EXW, and DFW.

For joining to occur, there must be sufficient compatibility of the metals involved to create strong atom-to-atom forces once the metals are brought into intimate contact. This requirement determines the nature of the welding processes and filler metals (if any) that may be used. It also significantly affects some details of the welding procedure; for example, use of a nonfusion process may be required. When metals do not join to each other, they sometimes are said not to "wet." Such nonfusion may be the result of the composition of the metals or of the presence of foreign substances which prevent intimate contact between the workpieces.

Sometimes it is possible to overcome this type of incompatibility by using an intermediate material capable of joining both of the incompatible metals. This can be accomplished by weld cladding, by sandwiching the intermediate material between the base metals, or by using a filler metal whose composition is different from either of the base metals.

When fusion occurs, some of the base metal mixes with the weld metal. The properties of the resultant mixture must be considered when selecting the process and filler metal to be used. The amount of weld-metal dilution by the base metal depends on the welding process and welding variables, such as current density, travel speed, welding technique, and rate of cooling by the base metal. In thicker materials where weld grooves are used, the amount of dilution by the base metal decreases as the weld proceeds from root to crown.

If there is to be fusion, elevated temperatures must be used; the method of raising the temperature also may introduce complications. When the process involves production of an arc, all constituents of the electrode may not be transferred uniformly across the arc to the base metal. Some elements may be lost through vaporization or chemical combination, which produces a product that may be lost in the slag. It may or may not be possible to compensate for such losses. If the welding process includes both a heat source and a separate filler metal (such as GTAW or PAW), such problems can be minimized or eliminated.

When molten filler metals are present, their fluidity must be considered in the welding procedure. With low fluidity, the metal must be directed to the required location, as it may not flow into narrow openings or fuse to the side of the weld groove. Furthermore, with low fluidity, there is an increased likelihood of incomplete penetration and slag entrapment. Conversely, high fluidity may limit the choice of welding position; very fluid weld metal may not stay in place in the weld groove in the vertical, overhead, and horizontal positions.

Welds produced by a fusion-type process may or may not require the addition of filler metal, depending on the material being welded. For example, filler metal is required in the case of some of the austenitic stainless steels and high-copper alloys, because welds made without it are crack sensitive. When filler metal is required, it may not always be possible to purchase the exact composition desired due to insufficient demand or the relatively new introduction of the material into the marketplace. In such cases, a study of alternative filler metals that are metallurgically compatible with the base metal must be made.

If manufacturers of filler metals are not able to purchase wire of a desired composition, they add alloys to wire of lower composition in order to achieve the required composition. This is done in flux-coated electrodes for SMAW by adding alloy to the flux coating. In filler wire for flux cored arc welding (FCAW), alloy can be added to the flux core of the wire. The addition of alloying agents to the core of a coiled wire also may be required to produce a ductile material if the desired composition is too "stiff" or brittle to permit coiling.

In fusion processes, the metal that does melt solidifies in a relatively short period and generally can be compared to a casting made with a metal mold. The type of cast structure in a weld deposit can affect the tendency to crack. For example, a dendritic structure is more apt to crack than a cellular-type structure. The nature of the weld-deposit structure can be affected by the composition of the weld metal and the rate of cooling.

Joint Design and Preparation

Joint design and preparation and selection of welding process are interdependent.

Workpiece Thickness. The joining of thin materials precludes the use of processes with high heat input and deep penetration, such as SAW, electroslag welding (ESW), and thermit welding. Conversely, the joining of thick materials precludes the use of processes with low heat input. Rapid solidification of molten metal increases the likelihood of incomplete penetration, incomplete fusion, and the entrapment of slag and/or gases. For a detailed list of joining processes and recommended joining thicknesses for several alloys, see Table 1.

Welding Position. The position of the weldment significantly affects the selection of welding process. For example, welding in the overhead position cannot be accomplished with processes that produce large volumes of molten metal, such as SAW, ESW, SMAW, or thermit welding, with electrodes designed for use in the flat position. In FRW applications, at least one of the parts to be joined must be rotatable. With SAW, if the part is rotated, the molten metal must solidify before the unfused flux falls off and eliminates its protective layer.

Groove Size. There must be a correlation between the size of the groove opening and angle and the welding process if full penetration is to be achieved. Relatively narrow openings and angles can be used with processes such as EBW and LBW. Submerged arc welding does not require as much of an angle as does SMAW or oxyacetylene welding (OAW). With ESW, the opening is usually quite wide, and beveling is not required.

Solidification of molten metal begins at the coolest surfaces (the base metals) and proceeds toward the center of the weld. Impurities and lower melting constituents tend to move toward the center of the weld-metal solidification. Thus, the center of a weld not only contains the most crack-sensitive parts, but also is subject to greater stress as the adjacent, already solidified metal begins to cool and contract. If the volume of the as-deposited molten metal is sufficiently large, cracking may result if the ratio of depth to width of the deposit is too great. However, with processes that produce relatively small volumes of molten metal, such as EBW and LBW, there is no such limitation, and the weld can be very long and narrow without cracking.

Cleanliness is an important factor in achieving high weld quality. If the welding process does not include a flux that removes impurities, any foreign constituent or other impurity present on the surface can potentially enter the weld and affect its quality. Even if a flux is used, there is always the possibility that it will not be able to remove all of the impurities. Therefore, successful welding procedures require thorough cleaning of all surfaces involved in the joining process, including the filler metals. When the welding process does not include fluxing action and relies on cover gases for protection from the atmosphere, such as GTAW and gas metal arc welding (GMAW), care must be taken that the protective gases are not disturbed by air movement in the vicinity of the weld.

Discontinuities may be present in the base metal because of the fabrication process, and care should be taken to compensate for this condition. Lamellar tearing in steels can be avoided by eliminating residual welding stresses applied perpendicular to laminations. Castings are more porous than rolled or forged parts, and special cleaning or nondestructive examination may be necessary for preparing these surfaces for welding.

Heat Factors. The amount of preheat required varies with the metals involved and the heat input from the welding process. Preheating reduces residual stresses, drives off moisture, reduces quench hardening, and gives gases and slag (if any) more time to reach the surface of the weld before solidification occurs. In some instances, however, preheating may be detrimental. If quenched and tempered steels are heated above the tempering temperature, they become softened and may affect the required mechnical properties. Therefore, in welding quenched and tempered materials, care must be taken to restrict the heat input from welding.

A temperature limitation during welding may also be required if harmful precipitations can occur or if existing precipitates are caused to increase in size to an undesirable degree (overaging). In such cases, the temperature reached during welding must be kept low by variations in the welding procedure and/or by restricting the interpass temperature. Interpass temperature is the temperature the base metal reaches between passes of a multiple-pass weld. To limit this temperature, the base metal must be cooled before welding resumes.

Base-Metal Properties

The condition and form of the materials to be joined may affect the choice of welding process. Likewise, the welding process can have various effects on the base metals. For example, high heat input may affect the mechanical properties of the base metal adversely. The following factors should be considered in the selection of a welding process when the base metals are known.

Physical Properties. Weldability generally is inversely proportional to both electrical and thermal conductivity. If electrical conductivity is high, methods of joining that use a heat source that depends on resistance to an electric current become less useful. For example, good contact resistance is essential for resistance spot welding (RSW) or other resistance heating methods to be effective, and induction welding requires resistance to the electric current induced in the base metals. If thermal conductivity is high, more preheat and/or greater heat input must be used. The coefficient of thermal expansion has a significant effect on the fit-up, the required allowance for distortion and shrinkage, and the amount of residual stress produced.

The metal liquidus and solidus temperatures, as well as the metal condition between these temperatures (generally referred to as the "mushy range"), are significant. If the mushy range is large, cracking is more likely to occur, and the procedure used must either eliminate melting or make solidification occur quickly. The relative temperature of melting helps in the selection of a welding process.

Most base metals used in welding are alloys, and the various constituents in them have different melting and freezing points. Cracking may occur when films of molten metal exist around grains of solidified weld or base metal. Stresses are set up when solidified metal begins to cool, but the metal is restrained from contraction by the surrounding cold metal. Should such stresses be applied to partially solidified material, hot tearing is likely to occur. Lower melting constituents that produce films around grains may be present in the metals to be joined or may have been introduced through the use of cutting fluids, paint, crayon markings, or other foreign material that remains on the surfaces to be welded because of improper cleaning.

In some alloys, such as brasses, some constituents may have very low melting points, which causes them to vaporize when the temperature is raised sufficiently to melt the remainder of the base metal. For example, zinc vaporization may produce porosity in the weld joint and, furthermore, is a health hazard to the welder.

Mechanical Properties. Cracking occurs when a material is unable to resist the stresses that are applied to it. The level of applied stress varies with the welding process. In turn, the ductility of the base metal determines its ability to resist stresses that are introduced by the joining process, without cracking. Base metals that have high hardness values or low yield strengths are more difficult to join. This type of base metal may require heat treatment prior to welding or the use of specific welding processes.

The joining process may change the mechanical properties of the base metal; consequently, this factor must be considered in conjunction with usefulness after joining. The weld or HAZ may be different from the base metal in terms of hardness, strength, impact resistance, creep strength, and wear resistance.

Chemical Composition. For joining to occur, the atoms on the surface of the workpieces must be in intimate contact. Foreign matter, such as oxides or dirt, on the surface make fusion impossible or, at best, difficult. Also, highly reactive metals such as titanium and zirconium are more difficult to join. Surface compounds, such as oxides, must be removed prior to joining. Changes in the chemical composition of the base metal at the weld joint may be produced during the joining process; also, weld metals often have chemical compositions different from the base metals.

Effect of Fabrication. Many problems in joining are the result of materials or conditions that preexist in the base metal. Castings may contain one or more of the following: sand inclusions, hot tears, gas holes, shrinkage cavities, inclusions, precipitates, and films. Wrought and forged products may contain one or more of the following: laminations, laps, scabs, seams, cracks, and surface contaminants. Any of the above can affect the resultant weldment and must be considered in the determination of the welding process.

All of the joining processes discussed in this Volume produce or require heating of the base metal to some degree. Heating is greater in fusion than in nonfusion processes. It may have one or more of the following effects:

- Residual stresses may be reduced by base-metal or weld-metal yielding, or increased because of contraction stresses as the molten filler metal solidifies and cools. Either of these conditions may affect the geometry of the final weldment or produce hardening, softening, or cracking of the base metal.
- Grain growth and/or recrystallization into finer grains may occur.
- Changes may take place in the metallurgical phases present in the HAZ.
- Precipitation of particles or an increase in the size of particles already precipitated in the base-metal HAZ may occur.

Table 1 Recommended joining processes for various metal groups

Joining process(a)	Recommended thickness (in.) for: Carbon steels	Low-alloy steels	Stainless steels	Cast irons
Arc welding				
AHW	Up to 1/4	Up to 1/4	Up to 1/8	1/8 and up
BMAW	Up to 1/4	...	...	...
CAW	Up to 1/4	Up to 1/4	...	1/8 to 3/4
CAW-G	Up to 1/4	Up to 1/4	...	1/8 to 3/4
CAW-S	Up to 1/4	Up to 1/4	...	1/8 to 3/4
CAW-T	Up to 1/4	Up to 1/4	...	1/8 to 3/4
EGW	1/4 and up	1/4 and up	1/4 and up(b)	...
FCAW	1/8 and up	1/8 and up	1/8 and up	1/8 to 3/4
GMAW	1/8 and up	1/8 and up	1/8 and up	1/8 to 3/4
GMAW-P	All thicknesses	All thicknesses	All thicknesses	1/8 and up
GMAW-S	Up to 1/4	Up to 1/4	Up to 1/4	...
GTAW	Up to 1/4	Up to 1/4	Up to 1/4	...
GTAW-P	Up to 1/4	Up to 1/4	Up to 1/4	...
PAW	...	...	Up to 3/4	...
SAW	All thicknesses	All thicknesses	All thicknesses	1/4 and up
SAW-S	1/4 and up	1/4 and up	1/4 and up	...
SMAW	All thicknesses	All thicknesses	All thicknesses	All thicknesses
SW	All thicknesses	All thicknesses	All thicknesses	...
Resistance welding				
FW	All thicknesses	All thicknesses	All thicknesses	...
HFRW	Up to 1/4	Up to 1/4	Up to 1/4	...
PEW	Up to 1/4	Up to 1/4	Up to 1/4	...
RPW	Up to 1/4	Up to 1/4	Up to 1/4	...
RSEW	Up to 1/4	Up to 1/4	Up to 1/4	...
RSW	Up to 1/4	Up to 1/4	Up to 1/4	...
UW	Up to 1/4	Up to 1/4	Up to 1/4	...
Solid-state welding				
CW	1/4 and up	...	...	...
DFW	...	All thicknesses	All thicknesses	...
EXW	Up to 3/4	Up to 3/4	Up to 3/4	...
FOW	All thicknesses	...	...	...
FRW	1/8 and up	1/8 and up	1/8 and up	...
HPW	1/8 and up	1/8 and up	1/8 and up	...
USW	Up to 1/8	Up to 1/8	Up to 1/8	...
Oxyfuel gas welding				
AAW	Up to 1/8	Up to 1/8	Up to 1/8	...
OAW	Up to 3/4	Up to 1/8	Up to 1/8	All thicknesses
OHW	Up to 1/4	Up to 1/8	Up to 1/8	Up to 1/4
Other welding processes				
EBW	All thicknesses	All thicknesses	All thicknesses	...
ESW	3/4 and up	3/4 and up	3/4 and up	...
IW	Up to 1/8	...	...	...
LBW	Up to 3/4	Up to 3/4	Up to 3/4	...
Brazing				
AB	Up to 1/4	Up to 1/4	Up to 1/4	Up to 3/4
DFB	All thicknesses	All thicknesses	All thicknesses	All thicknesses
DB	Up to 1/4	Up to 1/8	Up to 1/8	...
FB	All thicknesses	All thicknesses	All thicknesses	All thicknesses
IB	Up to 3/4	Up to 3/4	Up to 3/4	Up to 1/4
IRB	Up to 1/8	Up to 1/8	Up to 1/8	...
LB	Up to 1/8	Up to 1/8	Up to 1/8	...
RB	Up to 1/4	Up to 1/8	Up to 1/8	...
TB	Up to 3/4	Up to 3/4	Up to 3/4	Up to 1/4
TCAB	Up to 1/8	Up to 1/8	Up to 1/8	Up to 1/8
Soldering				
DS	Up to 1/8	Up to 1/8	Up to 1/8	...
FS	Up to 1/8	Up to 1/8	Up to 1/8	...
IS	Up to 1/8	Up to 1/8	Up to 1/8	...
IRS	Up to 1/8	Up to 1/8	Up to 1/8	...
INS	Up to 1/8	Up to 1/8	Up to 1/8	...
RS	Up to 1/8	Up to 1/8	Up to 1/8	...
TS	Up to 1/8	Up to 1/8	Up to 1/8	...
WS	Up to 1/8	Up to 1/8	Up to 1/8	...

(continued)

(a) All abbreviations in this table are listed in the section on Abbreviations and Symbols at the end of this Volume. (b) Applicable to EGW using solid electrode wire

Table 1 (continued)

Joining process(a)	Recommended thickness (in.) for: Nickel and nickel alloys	Aluminum and aluminum alloys	Titanium and titanium alloys	Copper and copper alloys	Magnesium and magnesium alloys	Refractory metals and alloys
Arc welding						
AHW	Up to 1/8	Up to 1/8	...	Up to 1/8	Up to 1/8	...
BMAW	...	...	...	...	...	...
CAW	...	...	...	...	...	...
CAW-G	...	...	...	...	...	...
CAW-S	...	...	...	...	...	...
CAW-T	...	...	...	...	...	...
EGW	1/4 and up (b)	...	...	...	...	...
FCAW	Up to 3/4	...	...	...	...	...
GMAW	All thicknesses	Up to 3/4	Up to 3/4	Up to 3/4	Up to 3/4	1/8 to 1/4
GMAW-P	All thicknesses	Up to 1/4	All thicknesses	Up to 1/4	All thicknesses	1/8 to 3/4
GMAW-S	Up to 1/4	...	...	...	...	...
GTAW	Up to 1/4	Up to 3/4	Up to 3/4	Up to 1/8	Up to 1/4	Up to 1/8
GTAW-P	Up to 1/4	Up to 3/4	Up to 3/4	Up to 1/8	Up to 1/4	Up to 1/8
PAW	Up to 3/4	Up to 1/8	Up to 3/4	Up to 1/4	...	Up to 1/4
SAW	1/4 and up	...	...	...	...	...
SAW-S	3/4 and up	...	...	...	...	...
SMAW	All thicknesses	...	...	...	...	...
SW	...	All thicknesses	...	...	All thicknesses	...
Resistance welding						
FW	All thicknesses	All thicknesses	All thicknesses	All thicknesses	1/8 and up	1/8 and up
HFRW	Up to 1/8	Up to 1/4	Up to 1/4	Up to 1/4	Up to 1/4	...
PEW	Up to 1/4	Up to 1/4	Up to 1/4	Up to 1/4	Up to 1/4	...
RPW	Up to 1/4	Up to 1/4	Up to 1/4	Up to 1/4	Up to 1/4	...
RSEW	Up to 1/4	Up to 1/4	Up to 1/4	Up to 1/4	Up to 1/4	...
RSW	Up to 1/4	Up to 1/4	Up to 1/4	Up to 1/4	Up to 1/4	...
UW	Up to 1/4	Up to 1/4	Up to 1/4	Up to 1/4	Up to 1/4	...
Solid-state welding						
CW	...	Up to 3/4	...	Up to 1/4	Up to 1/4	...
DFW	...	Up to 1/4	All thicknesses	...	...	...
EXW	Up to 3/4	All thicknesses	All thicknesses	All thicknesses	All thicknesses	Up to 3/4
FOW	...	...	...	...	...	1/8 and up
FRW	1/8 and up	1/8 and up	1/8 and up	1/8 and up	1/8 and up	...
HPW	1/8 and up	...	...	1/8 and up	...	...
USW	Up to 1/8	Up to 1/4	Up to 1/8	Up to 1/8	Up to 1/8	Up to 1/8
Oxyfuel gas welding						
AAW	...	...	...	...	...	...
OAW	Up to 1/8	Up to 1/8	...	...	...	...
OHW	Up to 1/8	Up to 1/8	...	...	...	...
Other welding processes						
EBW	All thicknesses	All thicknesses	All thicknesses	All thicknesses	All thicknesses	Up to 1/4
ESW	3/4 and up	...	...	...	...	...
IW	...	...	...	...	...	...
LBW	Up to 3/4	Up to 1/4	Up to 3/4	...	Up to 3/4	...
Brazing						
AB	Up to 1/4	Up to 3/4	Up to 1/8	Up to 3/4	Up to 3/4	Up to 1/8
DFB	All thicknesses	All thicknesses	All thicknesses	All thicknesses	Up to 3/4	Up to 1/4
DB	Up to 1/8	Up to 3/4	...	...	Up to 1/4	...
FB	All thicknesses	Up to 3/4	All thicknesses	All thicknesses	Up to 3/4	Up to 1/4
IB	Up to 1/4	Up to 1/8	Up to 1/8	Up to 1/8	...	Up to 1/8
IRB	Up to 1/8	Up to 1/8	Up to 1/8	...	...	Up to 1/8
LB	Up to 1/8	Up to 1/8	Up to 1/8	Up to 1/8	...	Up to 1/8
RB	Up to 1/8	...	...	Up to 1/4	...	...
TB	Up to 3/4	Up to 3/4	...	Up to 3/4	Up to 1/4	Up to 1/4
TCAB	Up to 1/8	...	...	Up to 1/8	...	...
Soldering						
DS	Up to 1/8	Up to 1/8	...	Up to 1/8	...	...
FS	Up to 1/8	Up to 1/8	...	Up to 1/8	...	...
IS	Up to 1/8	Up to 1/8	...	Up to 1/8	...	...
IRS	Up to 1/8	Up to 1/8	...	Up to 1/8	...	...
INS	Up to 1/8	Up to 1/8	...	Up to 1/8	...	...
RS	Up to 1/8	Up to 1/8	...	Up to 1/8	...	...
TS	Up to 1/8	Up to 1/8	...	Up to 1/8	...	...
WS	Up to 1/8	Up to 1/8	...	Up to 1/8	...	...

(a) All abbreviations in this table are listed in the section on Abbreviations and Symbols at the end of this Volume. (b) Applicable to EGW using solid electrode wire

- Dissolved gases may increase in mobility, which may be beneficial or detrimental.
- Local melting and/or partial liquation around solid grain boundaries may take place. This is the major reason for hot shortness, or metal cracking at elevated temperature.

The processes involving the least amount of heat input to the base metal are: resistance welding, DFW, and USW. Soldering and brazing usually occur at lower temperatures than the fusion processes. When the minimum width of HAZ is required, processes that produce rapid solidification are desirable. These include GMAW, LBW, EBW, induction welding, and flash welding.

In considering the effect of the welding process on the base metal, it must be remembered that there are several distinct zones in a weldment. In fusion processes, the weld metal contains some melted base metal, with the highest concentration adjacent to the base metal that has been melted. The remainder of the HAZ contains metal that has gone through a range of temperatures, from the melting point of the weld to the lowest temperature of the base metal during the welding process. Accordingly, the nonuniformity of the effects of the process on the base metal must be taken into consideration.

The effect of the welding process on the base metal may include factors beyond those required for mechanical usefulness. There may be surface or visual requirements that must be met, such as color match between weld and base metal, and surface appearance. The latter requirement can eliminate the use of processes which result in projections, indentations, or flash. Also to be considered are the visual effects of the mechanical means which must be used to achieve an acceptable surface appearance.

Work Place Location

Welding is a versatile technology for industry use. For this reason, a great variety of product applications exist that use all available processes and in a wide variety of locations. However, every process cannot be used for every welding application, as all have some limitations. Certain welding processes are very versatile and adaptable and can be used in almost any location and environment. There are also welding processes whose use is limited because of complicated installation requirements or equipment size. For these reasons, it is necessary to weigh the desire or need to use a specific welding process against an understanding of the limitations or impractical aspects of that process. Environment, portability, access to necessary electrical power, and availability of auxiliary supplies such as water, air, and other support services directly affect process selection for a given location.

Shielded metal arc welding continues to be the simplest and most versatile arc welding process. Because of its simplicity and portability, SMAW has many advantages. An electrical power source (either alternating current or direct current), welding cables, and an electrode holder are generally the only equipment needed for manual welding. The power source can be either a primary line or fuel-powered generator. Use of a fuel-powered generator may be an important factor where portability or a remote location is a consideration. The SMAW process is used extensively for maintenance and field construction work.

Submerged Arc Welding. For many years, SAW was a complementary welding process to SMAW for shop and field applications where mechanized and semiautomatic welding techniques could be applied for cost advantages. In recent years, GMAW and FCAW have replaced SAW for some of these applications. Properly trained, an operator with minimal manual ability can produce consistent high-quality welded joints using the submerged arc process.

Gas Tungsten Arc Welding. In recent years, GTAW has evolved as a complementary process to SMAW. Although best suited for applications involving aluminum, magnesium, and titanium or other refractory metals, GTAW also has found favor where high-quality weld joints are required. The GTAW process requires auxiliary supplies, including inert gas cylinders with pressure regulators and flowmeters, and for some applications a water supply.

Gas metal arc welding and flux cored arc welding use a continuous-wire filler metal. The required equipment for these processes is relatively simple and is considered portable even though it is somewhat more cumbersome compared to SMAW equipment. The GMAW and FCAW processes can be used to join virtually the same metals as SMAW and SAW in any position. When shielding gases are used with these processes, the effect from drafts must be considered. In most cases, protective shields alleviate this problem.

Plasma arc welding has some advantages, because of greater energy concentration, arc stabiliy, and higher attainable weld travel speeds. Plasma arc welding equipment includes a control/power source, weld torch, welding cables, inert gas cylinders with pressure regulator and flowmeter, shielding gas hose, and orifice gas hose. The equipment is considered semiportable.

Resistance welding is limited to fixed locations, primarily because of the high electrical power input required. Also, the mass of the equipment and controls is better suited to a shop application. Equipment setup by qualified personnel is required, but welding operators need only minimal skills.

Electrogas welding (EGW) and electroslag welding normally are limited to joining plate materials edge to edge. Equipment and installation generally are limited to fixed locations, such as fabricating shops and production plants. However, these processes have been applied to field erection (storage tanks); application is somewhat limited by plate thickness because of equipment size and controls.

Brazing and soldering lend themselves to a wide range of applications—from low-volume manual applications using oxyfuel gas equipment to sophisticated furnace installations with conveyor belts for high-volume production. Manual methods involve equipment that is portable. Furnace and other mechanized installations are fixed in a shop environment.

Electron beam welding and laser beam welding involve fairly sophisticated equipment installation and are fixed in a shop because of the equipment mass and precise tooling required. These processes are relatively new, but are finding a market where precise tolerance control of the weldment is required along with a high degree of quality. Both of these processes can be performed by unskilled operators, but require qualified technicians and engineers for setup and maintenance.

Flash, friction, and diffusion welding are best suited to special applications, depending on the design and type of parts to be joined. Equipment for these processes is normally nonportable. Flash and friction welding are used for moderate to high-volume production, whereas DFW generally is applied to low-volume production.

Welder Skill

A major factor to be considered in selecting a welding process is the level of welder skill required to operate the equipment. Many welding applications are required to conform to specific standards or codes. In this case, both the welder and welding process must be qualified.

Over 100 different occupational titles are listed under "Welding" in the Dictionary of Occupational Titles published by the U.S. Department of Labor. The types of skills include welder helper, welding machine operator, welder, welder technician, welding inspector, welding supervisor, and welding engineer. The American Welding Society (AWS) has numerous publications referencing welding processes and labor skill requirements. Another domestic source on the availability of skilled welder labor is state employment agencies. Local colleges, trade and vocational schools, and union training centers are excellent sources for information on welder availability.

Equipment Costs

Welding equipment and systems vary in cost from a few hundred dollars for a simple manual SMAW or OFW setup to well over a million dollars for a sophisticated fully automatic LBW installation. When selecting a process, all aspects of the job need to be analyzed, and equipment costs play an important part in the evaluation. Aside from the cost of the basic welding equipment, there are auxiliary equipment costs such as special tooling and fixtures that may be required. Equipment consumables, such as nozzles, cups, and wire guides, also need to be accounted for during the selection of a process.

Brazing processes using oxyacetylene may cost as little as a hundred dollars for a manual setup. Furnace brazing setups, however, may cost over a million dollars for an installation that requires a very large furnace and an inert gas purge system.

Industrial Usage

The industrial usage of a welding process depends to a great extent on the following considerations:

- Material and its weldability
- Production requirements
- Design specifications and intended service
- Size and complexity of weldment
- Fabrication site—shop or field
- Cost of welding equipment
- Welder skill and training required

Welding processes enjoying the greatest industrial usage are the manual welding processes—SMAW, OFW, GTAW, GMAW, and FCAW. The mechanized processes used most frequently in industry are GTAW, GMAW, FCAW, and SAW.

Industrial usage of a welding process depends, for the most part, on the material to be welded. Carbon steel, the most widely used material, can be welded with most manual or automatic welding processes. Aluminum, on the other hand, frequently is welded with an inert gas process, such as GTAW or GMAW.

At the other end of the spectrum are the more sophisticated processes such as EBW and LBW. These processes generally are used for more specialized applications.

Applications

The choice of welding process used on any construction assignment depends, to a great extent, on the type of job involved. Applicable codes and standards and the location where the welding is to be carried out have a direct bearing on the choice of process. Although welding can be applied to a broad spectrum of manufacturing and construction activities, there are certain applications that encompass the major volume of welding in industry. These are discussed below, with specific attention to the welding processes involved. For additional information on applicable codes and standards for the welding industry, see the article "Codes, Standards, and Inspection" in this Volume.

For structural welding, there are two categories of welding activity—buildings and bridges—that normally are recognized. Although there are different design considerations and requirements for these two categories of structures, they both require the use of AWS D1.1, "Structural Welding Code." This code covers qualification of procedures and welders for building and structural bridge fabrication and erection. One of the options suggested in AWS D1.1 is the use of prequalified procedures involving four processes (SMAW, SAW, GMAW, and FCAW). Three other processes (ESW, EGW, and stud welding) are permitted but must be qualified for the specific application. It is the responsibility of the designer to specify the type of joint required; the fabricator then must select the welding process to meet joint-design requirements. Because the selection of base metals is limited to those permitted in AWS D1.1, certain variables are decided upon before a fabricator is contacted.

Piping, pressure vessel, boiler, and storage tank construction involves a large portion of current welding applications that encompasses the petroleum, petrochemical, chemical, power generation (utility), and transmission pipeline industries. Codes and standards pertaining to the application include the following:

- ASME, "Boiler and Pressure Vessel Code"
- ANSI B31.1, "Pressure Piping Code"
- API standard 1104
- API standard 620
- API standard 650
- American Water Works Association (AWWA) standard D100 (AWS D5.2)

Power piping, pressure vessel, and boiler welding construction normally is governed by Section IX of the ASME "Boiler and Pressure Vessel Code" for procedure and operator qualification. The manufacturer or fabricator is responsible for qualifying the process used with the procedures to be followed in construction. The only constraints are processes that may not be permitted by governing codes or standards of job specifications. Processes that are impractical for use at the location of the construction are also eliminated. The most widely used processes in field erection/construction are SMAW and GTAW, with limited use of SAW, GMAW, and FCAW. For shop fabrication, the only limiting factors to process use are design, equipment cost, or governing codes and standards.

Transmission piping construction normally is controlled through API standard 1104. Welding processes generally are limited by field conditions. Although SMAW has dominated transmission piping applications for many years, automation has led to the widespread use of the GMAW and FCAW processes.

Storage tank construction generally is governed by API standard 620, API standard 650, or AWWA standard D100 (AWS D5.2), depending on the type of storage tank. A considerable amount of storage tank construction is performed with the SMAW process, but automated SAW also is used to a great extent. For storage tanks constructed with aluminum-based metal, GTAW and GMAW are used extensively.

Shipbuilding construction is governed by American Bureau of Shipping (ABS) requirements. The major welding process utilized in shipbuilding is SMAW, but there is increased development in the use of SAW, GMAW, and FCAW. Some applications are well suited to the use of ESW for joining thick plate.

Aircraft and aerospace welded construction is governed by military specifications, and welding processes are governed by material, production, and quality considerations. Gas tungsten arc welding continues to be the dominant process used, but SMAW, GMAW, EBW, PAW, resistance welding, stud welding, and brazing also are used for aircraft or aerospace construction because of the flexibility of these processes.

Automotive and railroad industries use almost every welding process available because of the many types of materials and applications encountered. Resistance welding is the dominant process employed on automotive assembly lines. Other processes used in the automotive industry include FRW, GMAW, EBW, brazing, soldering, and flash welding. The railroad industry utilizes flash welding, SMAW, GMAW, FCAW, SAW, and thermit welding. Although there are no established codes or standards that specifically cover welding procedures and operator qualifications for these industries, guidelines set forth in ASME Section IX and AWS D1.1 should be followed extensively.

Weld Joint Properties

Evaluation of weld joint properties by the designer and fabricator is an important consideration in process selection. The environment in which a weldment must perform may be the deciding factor as to whether the weld joint will fail or provide reliable service. Factors such as temperature, vibration, earthquakes, and corrosion must be evaluated to determine if the welding procedure, including process, filler metal, and heat treatment selection, is correct.

Temperature. When a weldment is subjected to service temperatures lower than the nil-ductility-transition temperature (NDTT) of the base-metal HAZ or weld metal, brittle fracture can occur. The NDTT point is the temperature below which the behavior of the metal becomes brittle rather than ductile. When embrittlement is a possibility, procedure qualification should include tests such as the Charpy V-notch test to determine the NDTT of the weld joint. Brittle failure can occur during hydrostatic testing of pressure piping and vessels when proper control of the test fluid is not maintained.

Vibration. When service environment involves vibration, the weldment must be evaluated for points of stress concentration, such as thickness transition, to ensure that fatigue failure will not occur. The study should include a check of weldment hardness to ensure that potential metallurgical notches do not exist, which could lead to failure. In such cases, repair, redesign, or postweld heat treatment may be necessary.

Earthquake. When the possibility of an earthquake exists, the potential stresses on the weldment must be analyzed to ensure that design factors are adequate to prevent failure.

Corrosion. The possible effect of a corrosive environment must be evaluated to determine if the resultant weldment may be susceptible to failure in service. Intergranular stress corrosion cracking of stainless steel weld joints in nuclear power plants are an example of such corrosive environments that create the need for added precautions. High oxygen levels in the water have caused cracking to occur in the weld-sensitized base metal for this type of application.

In addition to the above, any possible effect of welding on base-metal properties should be studied, including decreased mechanical properties, increased hardness, or weld-metal dilution.

Quality Requirements

Assurance that imposed quality requirements will be attained and maintained can be obtained by a controlled program or system. Such a system should impose the following minimum controls:

- Full testing of base metals and filler metals for chemical, mechanical, and physical properties, including traceability through construction
- Qualification of procedures and welders
- In-process controls to ensure that specified filler metal is used
- Control of filler metal and flux storage to maintain their quality
- Fit-up and cleanliness inspection
- Nondestructive examination by qualified personnel
- A record for each weld joint, recording all pertinent data, tests, and examinations

The sophistication of such a system depends on the desired and required quality. For example, the extent of documentation, testing, and nondestructive examination for nuclear power plants is greater than that required for transmission pipelines.

Safety

The utilization of welding in manufacturing, construction, and maintenance activities involves many considerations. Of greatest importance, however, is the need to be attentive to safety. Factors involved in personnel safety include protection against eye and ear damage, burns, radiation, respiratory damage, and crushed or broken limbs. Loss and damage to equipment and building are generally the result of fire or explosion.

The following publications and documents are helpful to gain an understanding of the precautions that should be taken to meet minimum levels of safety:

- American National Standards Institute (ANSI) Standard Z49.1, "Safety in Welding and Cutting"
- ANSI/NFPA No. 70, "National Electric Code"
- Occupational Safety and Health Administration (OSHA), various publications and documents
- AWS 6.3, "Recommended Safe Practices for Plasma Arc Cutting"
- AWS F2.1, "Recommended Safe Practices for Electron Beam Welding and Cutting"
- ANSI Z136.1, "Safe Use of Lasers"
- Compressed Gas Association (CGA) P-1, "Safe Handling of Compressed Gases"
- ANSI Z87.1, "Practice for Occupational and Educational Eye and Face Protection"
- ANSI Z88.2, "Practices for Respiratory Protection"
- ANSI Z89.1, "Safety Requirements for Industrial Head Protection"

Each welding process and its application determines the requirements for protection of personnel. General requirements for protection of personnel involved in welding are covered by ANSI Z49.1.

Eyes and Face. The eyes, face, and neck should be protected at all times through use of helmets, face shields, goggles, or hand-held shields. The shade of the glass lens used in helmets, goggles, or shields depends on the intensity of the arc. Recommended lens shades for welding current ranges to be used with arc welding processes are:

Shade No.	Welding current, A
6	Up to 30
8	30 to 75
10	75 to 200

For LBW and cutting, special safety glasses are available to filter out the infrared wavelengths that cause retinal damage. The lenses in goggles used for oxyfuel processes are lighter shades than those used for arc welding.

Hearing. For some welding processes, such as EXW and metal spraying, precautionary measures are needed to protect the hearing of personnel. In such cases, the use of ear protection or properly fitted ear plugs should be worn to protect personnel from the high-intensity sound. The type of protective device chosen should be capable of reducing the sound level below 80 decibels. Governing codes should be reviewed for sound-level protection requirements.

Clothing. To properly protect the arms, body, and legs, special protective clothing is available. In many cases, clothes made of high-grade denim provide sufficient protection against ultraviolet or infrared radiation and occasional spatter or sparks. However, where the work is performed in a very confined space or the application involves high temperatures, heavy spatter, and/or large molten weld pools, the use of flame- and temperature-resistant clothing material, such as leather, should be considered. In addition to shirt and pants, aprons, gauntlets, and leggings may be needed. High-top shoes should be worn with cuffless pants that cover shoetops. Loose pocket flaps and open shirt collars present a potential hazard.

Respiratory. To prevent the inhalation of fumes, gases, and particulate matter, a positive ventilation system must be installed. Local codes and OSHA rules generally govern the requirements. The nature, type, and magnitude of fume and gas exposure will, for the most part, determine the type, location, and volume of air removal needed.

For certain welding processes, such as metal spraying, additional consideration must be given to the use of respiratory protection devices. Selection of these devices should be in accordance with ANSI Z88.2, which contains descriptions, limitations, operational procedures, and maintenance requirements for standard respiratory protective devices. All such devices should be of a type approved by the U.S. Bureau of Mines, OSHA, state and local codes, or other regulating agencies.

Toxic Materials. The vapors of some cleaning solvents break down in the presence of a welding arc or burning flame and can cause dizziness, nausea, or danger to life. Consequently, precautions should be taken to ensure the use and storage of these materials in a safe manner. Manufacturers of such materials should be consulted for needed precaution in their use. Protective clothing, such as rubber gloves, should be worn while using any materials, substances, or liquids that could cause irritation or allergic reaction.

Training. Personnel should be trained in the need and use of safe practices that are necessary for their protection. Safe practices should include:

- Anchoring of compressed gas cylinders
- Insulation of electrical circuits
- Monitoring of molten metal (sparks or discharge) to prevent fire
- Protection from arc exposure to eyes and skin
- Elimination of fumes to prevent their inhalation
- Use of protective clothing to prevent skin burns

Fire Protection. All flammable materials, such as solvents and cleaning agents, should be stored in tightly sealed drums and issued in suitably labeled safety containers to prevent fires during storage and use. Solvents and flammable liquids should not be used in poorly ventilated or confined areas. When solvents are used in trays, safety lids should be provided. Flames, sparking, or spark-producing equipment must not be permitted in the area where flammable materials are being used.

When welding or cutting operations produce heavy spatter or sparking, precautions must be taken to protect or remove combustible materials from the immediate vicinity to eliminate potential fire hazards or danger to personnel. Personnel should be trained properly in fire protection action, and firefighting devices, such as fire extinguishers, should be readily available.

Gas cylinders are pressurized and as such must be handled with care. They should be kept upright at all times and securely anchored by chain, rope, or cable to a wall or rack. The cylinder can be a hazard to personnel, equipment, or building if tipped or dropped. If a cylinder valve is broken while the cylinder is fully pressurized, it could become a missile, causing serious injury, loss of life, or damage to equipment and buildings.

Oxygen should never be used to operate pneumatic tools, to dust clothing, to blow away debris, or to pressurize tanks or piping systems. It can react violently with oil or grease. Therefore, all fittings, gages, hoses, and valves should be checked thoroughly for cleanliness prior to installation and use.

Liquefied gas cylinders should be of double-walled construction, with a vacuum between the inner and outer shell. These cylinders should be handled with extreme care to prevent damage to the internal piping and loss of vacuum.

Acetylene in contact with copper, mercury, or silver may form acetylides, especially if impurities are present. These compounds are violently explosive and can be detonated by the slightest shock or the application of heat. Alloys containing more than 67% Cu should not be used in acetylene systems unless such alloys have proven safe in a specific application by experimental testing.

Electrical installations for equipment and controls should be manufactured and installed in accordance with the appropriate codes and standards. High-voltage parts must be insulated suitably and enclosed with access doors and panels that are locked to prevent access by unauthorized persons. All electrical equipment must be grounded suitably or provided with equivalent protection. External weld-initiating control circuits should operate at low voltage for portable equipment.

Precautions for Specific Joining Processes

Listed below are important safety precautions that should be followed during welding and related processes. No attempt has been made to list all required safety measures.

Shielded Metal Arc Welding. The welder should be properly protected from the arc. This requires suitable spatter- and spark-resistant clothing, a welding helmet, and gloves.

Gas Tungsten Arc Welding. The same precautions apply to GTAW as for SMAW. A darker shade of welding lens may be necessary, because the gas tungsten arc is more intensive.

Submerged Arc Welding. No protective shield or helmet is necessary in SAW, but safety glasses and gloves should be used for routine protection. Safety glasses should be tinted for protection against flash when the arc is inadvertently exposed.

Oxyfuel Gas Welding. Protection is required in OFW from the glare of the flame and molten metal. Goggles or eyeshields and suitable gloves must be worn.

Plasma Arc Cutting. Operators and persons in the vicinity of plasma arc cutting (PAC) equipment must be fully protected from strong arc glare, spatter, fumes, and noise when the unit is in operation.

Resistance Welding. Operators should wear eye and face protection against the ejection of molten metal and sparks during welding. Equipment should be designed to prevent crushing of hands or other parts of the body.

Thermit Welding. Personnel should wear protective clothing to shield against hot metal particles or sparks. Full face shields and safety boots also should be worn.

Brazing. Personnel and property should be adequately protected from hot materials, gases, and fumes. Adequate ventilation is needed to prevent the inhalation of gases and fumes, as some metals and fluxes contain toxic materials. In furnace brazing, furnaces or retorts must be purged if the brazing atmosphere is flammable. Also, before personnel enter brazing furnaces or chambers for maintenance or cleaning, a complete air purge is needed.

Flash, Upset, and Percussion Welding. Some type of suitable fire-resistant shielding should be provided to protect the operator from molten metal particles. Eye protection with suitable shaded lenses also should be worn.

High Frequency Welding. No special protective clothing is needed in high frequency welding (HFW) applications. However, proper instruction and training of personnel is needed to ensure an understanding of safety precautions to prevent injuries from the high frequency power source. Electrical aspects of equipment installation are extremely important.

Electron Beam Welding. Operator protection from x-ray radiation, visible radiation, and electric shock is necessary in EBW applications. Although protection generally is designed into the equipment, a complete x-ray radiation survey should be made at installation and at regular intervals thereafter to ensure continued protection.

Laser Beam Welding and Cutting. The primary hazards in LBW are eye damage, skin burns, respiratory system damage, and electric shock. Proper eye shielding from beam exposure and fume elimination systems are also important factors to be considered.

Friction Welding. Welding personnel should wear appropriate eye protection and clothing commonly used with machine tool operations. Machines should be equipped with appropriate mechanical guards and shields and should be designed to prevent operation when the work area, rotating drive, or force system is accessible to the operator.

Explosion Welding. Ear protection is required in EXW application. Only highly trained and competent personnel should operate EXW equipment in order to prevent serious injury or damage to property.

Ultrasonic welding presents no unusual hazards to operating personnel. However, because of the high voltages involved, proper installation must be made to protect the operator. Proper controls must be incorporated into the installation to prevent operator injury from clamping fixtures.

Joint Design and Preparation

By O.W. Blodgett
Design Consultant
The Lincoln Electric Company

A WELD JOINT serves to transfer the stresses between the joined members and throughout the welded assembly. Forces and loads, which are introduced at different points, are transmitted to different areas throughout the weldment. The type of loading and service of the weldment influences the selection of a joint design. The names of joint types describe how the members meet. They are butt, T, corner, lap, and edge.

The weld selected may or may not require preparation in the form of a groove to permit proper access to the root of the joint. Several methods—machining, chipping, shearing, grinding, gas cutting, gas gouging, or air carbon arc gouging—may be used to create single- or double-bevel, V-, J-, or U-grooves.

No preparation is needed for a square groove weld. The members can be abutted in the same plane. If there is zero root opening, the weld arc essentially fuses the members together. Usually, very little or no weld metal bridges the weld members, although some weld metal or reinforcement may be left on the top and bottom surfaces. Similarly, fillet welds, which join pieces perpendicular to each other, can be used on T-joints without preparation. The welds and the fusion into both members join the two pieces.

Groove preparations are often necessary, however, for successful welding of various corner, T-, and butt joint applications. Frequently required with larger metal thicknesses, these weld preparations ensure that the welding heat and weld metal reach and fuse the root of the joint.

Nomenclature

Terminology describing various joints and welds is often confused. Types of welds and joints recognized by the American Welding Society (AWS) are illustrated in Fig. 1. Note that joints are the junctions where the members join and welds are the mechanisms (groove or preparation) which complete joints. Figure 2 shows applications of some single-bevel and single-V welds on T- and corner joints.

Although precise nomenclature to differentiate between welds and joints is desirable, engineers will probably continue to use the terms interchangeably. For example, although Fig. 19 is technically a joint made with two partial-penetration double-bevel groove welds, the joint is referred to as a partial-penetration double-bevel groove joint. Thus, in this article, as well as other practical applications, the terms joint and weld are used interchangeably to describe how a joint is constructed.

Fig. 1 Types of joints and welds

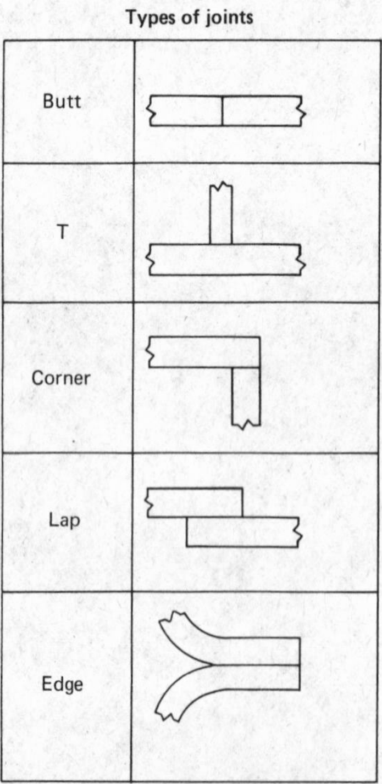

Types of Joints

Butt joints are joints between two abutting members lying approximately in the same plane. Although they may be welded with no preparation, butt joints are often grooved, especially with thicker and heavier plate.

Butt joints are preferable in uses where continuity of section is desired, such as

Fig. 2 Welds in T- and corner joints. (a) and (b) single-bevel welds. (c) Single-V-groove weld

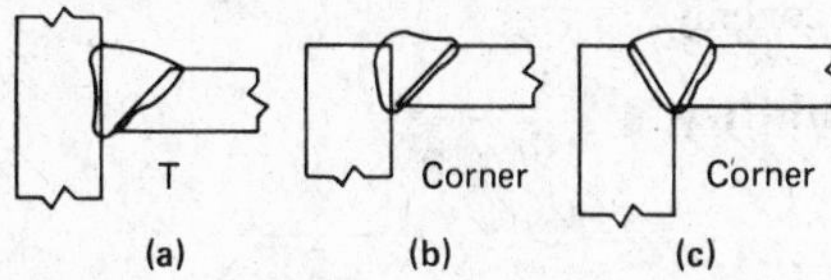

fusion-welded joints in sheet or plate for pressure vessels, pipelines, decking, tanks, and similar construction applications. Generally, when the metal thickness exceeds approximately $^3/_{16}$ in., the butt joint requires grooving to prevent a lack of fusion from weakening the weld.

T-joints are joints in which the members are positioned in the form of a T. They are well suited for no-preparation fillet welds, but in thick material the preparation welds, such as the single-bevel in Fig. 2, may be necessary.

T-joints are frequently used in production of fabricated weldments, welded machinery components, attachment of flanges and plate stiffeners to webs of girders, and assembly of rolled section and tubular building structures. Fillet-welded T-joints should not be used for fusion welding with ordinary gas or arc welding of containers because corrosive solids may be trapped.

Corner joints are joints between members located at approximately right angles to each other that meet in the form of an L. Typical applications are the same as those for T-joints.

Figure 3 shows various corner joints with fillet and preparation welds. The corner-to-corner joint (Fig. 3a) is difficult to assemble because neither plate can be supported by the other. A small electrode with low welding current must be used to prevent the first welding pass from melting through. This joint requires a large amount of weld metal. In contrast, the corner joint in Fig. 3(b) is simple to assemble, does not melt through easily, and requires half the amount of weld metal. Half the weld size is also used when welding the identical joint with two welds, one outside and the other inside (Fig. 3c), a design that has the same total throat as Fig. 3(a).

Fig. 3 Corner joints. (a), (b), (c), and (f) fillet welds. (d) V-groove weld. (e) J-groove weld

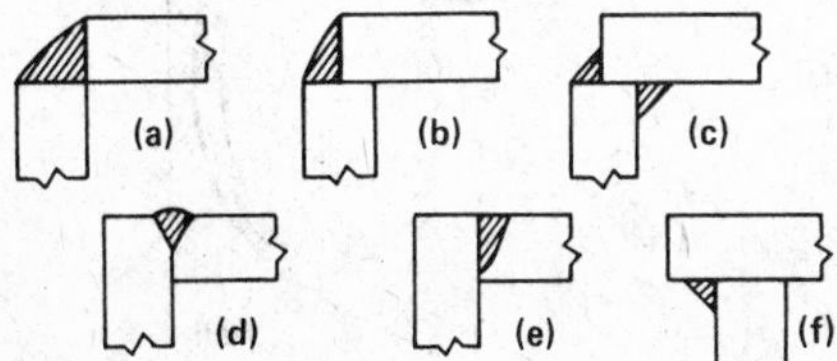

With thick plates, a partial-penetration groove weld (Fig. 3d) is often used in a corner joint. This weld requires beveling. For a deeper weld, a J-groove preparation may be chosen instead of a bevel (Fig. 3e). Another possibility, the fillet weld in Fig. 3(f), is inconspicuous and makes a neat and economical corner.

Lap joints are joints made with two overlapping members. They do not require edge preparation. Lap joints are necessary for use in seam and spot welding of sheet metals.

Edge joints are joints between the edges of two or more parallel or nearly parallel members. They may not require the addition of filler metal for welding and are frequently used to join thin turned-out edges, such as those formed where the sides of a container meet.

Types of Welds

Fillet welds are welds approximately triangular in cross section, joining two surfaces essentially at right angles to each other in a lap, T-, or corner joint. Basically, they fill in a corner. Fillet welds require no preparation and are the most common type of weld used in structural work.

Square groove welds are welds in which the abutting surfaces are square (zero degree included angle). Because welding from one side makes penetration difficult, the double-square weld is used frequently to ensure the strength of the weld. Sometimes the root of the weld is opened and a backing bar is used. The purpose of a backing material, which is placed at the root of the weld joint, is to support the molten weld metal.

Bevel groove welds are welds in which only one member is beveled. The groove between the two members is measured between the prepared edge of a member and the surface of the nonprepared member. The bevel angle (material removed by cutting) is measured between the prepared edge of a member and a plane perpendicular to the surface of the other member. Single-bevel groove preparations are widely used because they are easily prepared by oxyfuel gas cutting. They are well suited for corner and T-joints (Fig. 2), as well as butt joints $^1/_4$ in. in thickness and greater. Double-bevel grooves are recommended, if welding from both sides is possible, when metal is $^3/_4$ in. or thicker, because these grooves produce less distortion of the welded parts and reduce weld-metal requirements by about half.

V-groove welds are welds in which the total included angle of the groove between the members to be joined is the sum of the two bevel angles. Similar to single-bevel groove welds, single-V-groove welds are widely used because they are easily prepared by oxyfuel gas cutting. Figure 2 shows a single-V weld in a corner joint. These groove preparations are also used extensively in edge preparations for butt joints $^1/_4$ in. in thickness and greater. The advantages of double-V bevel grooves are the same as those for double-bevel grooves.

J-groove welds are welds made in J-shaped grooves between the two members to be welded. Single-J-grooves are well suited for butt corner (Fig. 3e) and T-joints. The advantages of double-J-groove preparations are the same as for double-bevel grooves.

U-groove welds are welds made in U-shaped grooves between two members to be welded. Because of the rounded base, larger electrodes can be used with narrower groove angles than for V-groove welds. The U-groove preparations can be prepared by air carbon arc gouging or by machining, which yields more uniform grooves. Because J-grooves and U-grooves require at least $^1/_8$-in. root face and $^1/_4$-in. radius at the root of the joint, the material must be more than $^3/_8$ in. thick for this joint to be used. The advantages of double-U-groove welds are the same as those for double-bevel welds.

Weld Joint Design

The first consideration in the design of a weld joint is its ability to transfer load; the second is cost. The ideal weld joint is one that can handle the loads imposed, usually with a substantial safety margin, and still be produced at minimal cost.

Therefore, once the type of joint has been selected primarily on the basis of load requirements, the choice of weld to complete the joint should be determined by the effects of the structural design and layout on weld metal, accessibility, and preparation requirements—variables that directly influence the cost of the weld joint.

Weld Metal. In general, joint and weld types specified should require the least amount of filler metal to avoid unnecessary expense. The size of the weld should always be designed with reference to the size of the thinner member. The joint cannot be made any stronger by matching the weld size to the thicker member, which

Fig. 4 Determination of weld size

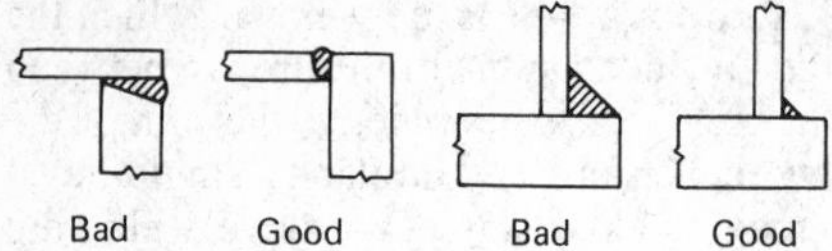

would require a greater amount of weld metal (Fig. 4).

In the design stage, joints that create extremely deep grooves should be avoided, because these require more filler metal. Deep-penetration welding processes, such as submerged arc welding, especially automatic welding, reduce the volume of weld metal needed. Because greater thickness of plate can be welded, use of these methods may make the additional cost of groove preparations unnecessary.

If groove welds are required, minimum root openings and included angle should be used to reduce the filler metal required. The use of double- instead of single-groove welds on thick plate where welding from both sides is feasible decreases the total filler metal.

To prevent waste of weld metal, welding specifications should avoid the unnecessary use of the all-around welding symbol (Fig. 5). Welding everything that touches has no engineering advantages, yet increases the amount of weld metal used.

When joining high-strength materials, cost can be minimized by specifying the use of high-strength weld metal only where required for primary load-carrying welds.

Accessibility. An important factor in the design of a weld joint is the accessibility of the members to be welded. The welder needs space to manipulate the electrode when making the weld. Frequently, what appears straightforward on the drawing board may be impractical in the shop or very costly to produce. Figure 5 illustrates examples of joint placement that are often difficult to weld.

Fillet Welds

A fillet weld is measured by the leg size (ω) of the largest right triangle that may be inscribed within the cross-sectional area (Fig. 6). Although the actual leg of the fillet weld is defined by the distance from the root of the joint (the point where the members are closest before welding) to the toe, which is the junction of the weld face and base metal, the leg size of the weld may be shorter than the leg in a concave fillet, as shown in Fig. 6. The leg size (ω) is equal to the side of a right triangle, but does not extend from root to toe of the weld. Unequal leg size should be avoided, as the longer leg adds nothing to strength but does add to cost.

For efficient weld filler-metal use, the convexity of the fillet weld should be minimized. The 45° flat fillet, very slightly convex, achieves maximum strength and economy.

Fig. 5 Effect of design on joint accessibility

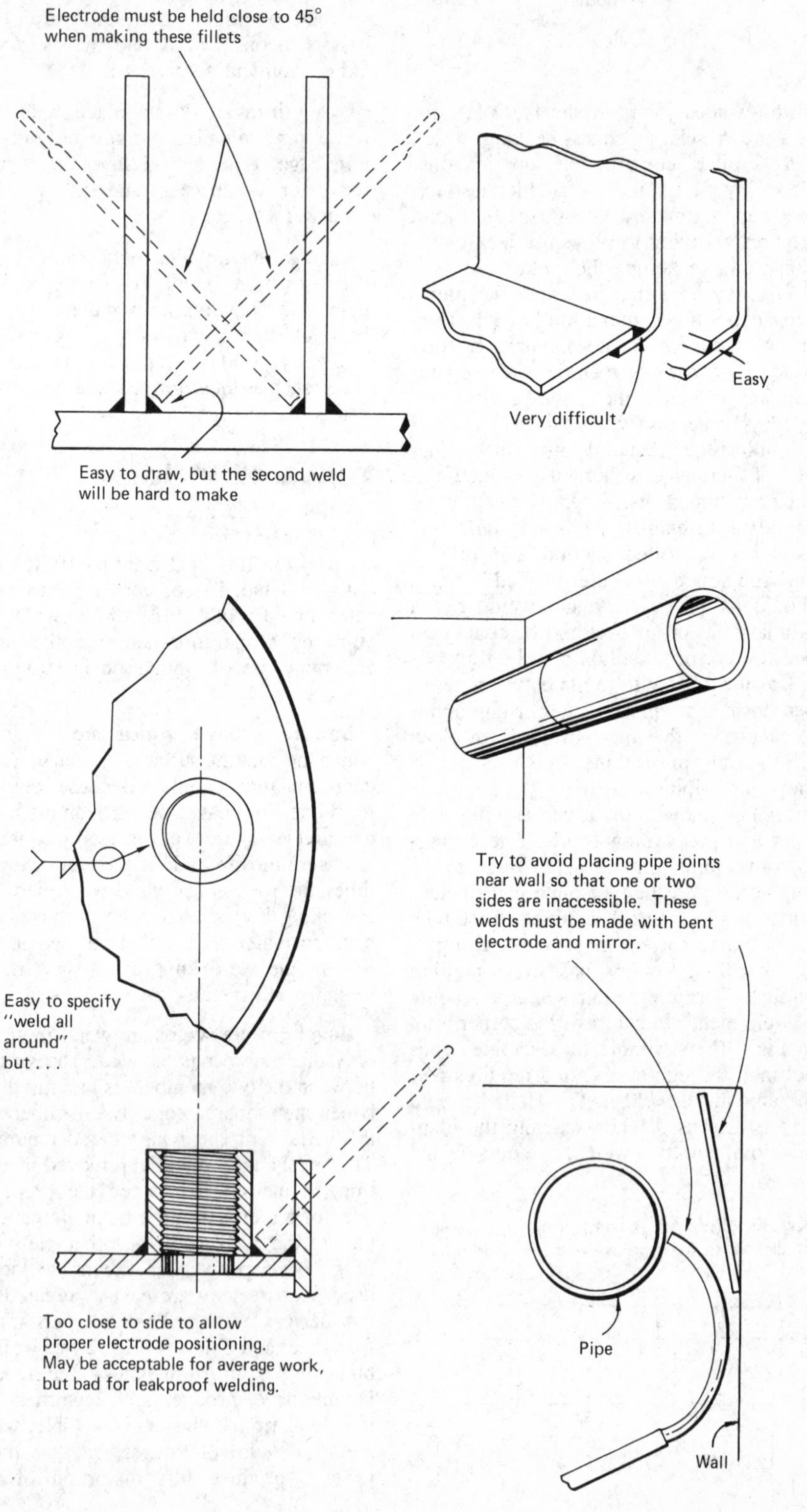

Electrode must be held close to 45° when making these fillets

Easy to draw, but the second weld will be hard to make

Very difficult

Easy

Easy to specify "weld all around" but . . .

Too close to side to allow proper electrode positioning. May be acceptable for average work, but bad for leakproof welding.

Try to avoid placing pipe joints near wall so that one or two sides are inaccessible. These welds must be made with bent electrode and mirror.

Pipe

Wall

Fig. 6 Leg size (ω) of a fillet weld. *t* is throat.

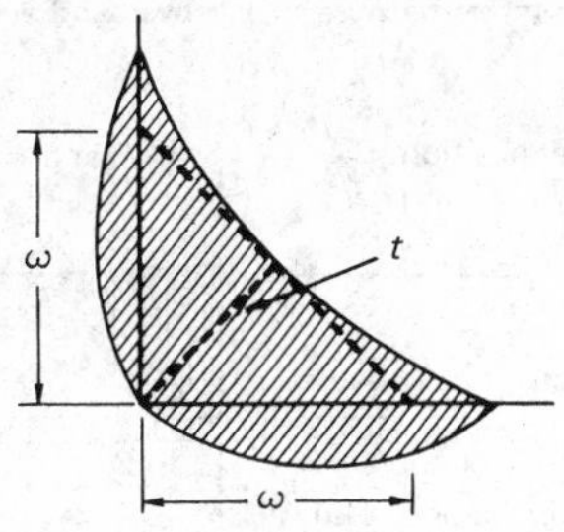

Throat Size. The shortest distance between the root of the joint and the face of the diagrammatical weld, or the throat (Fig. 6), is a better index of strength of a weld than leg size. It is along this throat (t) that the allowable stress is applied. The leg size (ω) of a fillet weld simply provides this throat.

Weld Size. When sizing fillet welds, the direction of loading should be considered. Extensive test data have shown a one third increase in weld strength under transverse versus parallel loading. Therefore, weld size may be reduced under transverse loading conditions. Table 1 gives the allowable unit forces on fillets of varying sizes under both loading conditions.

At one time, two 45° fillet welds that had legs equal to three fourths of the plate thickness were considered to develop the full strength of the plate for either transverse or parallel loading, assuming that the weld metal was equivalent to the base metal and average penetration was obtained. Originally, calculations for full-strength welds were based on thinner plate welded on both sides, with the welds extending full length. Although this ratio of 0.75 is true for the old values of E60 weld metal and A7 low-carbon steel prior to 1969, these values are not necessarily applicable when working with the wide range of steels and weld-metal strength levels available today. Table 2 shows how the full-strength factor varies with weld-metal and plate strength levels for a number of steel grades. Suggested factors for the weld size in relation to plate thickness in rigidity or non-full-strength designs are also given.

Weld Placement. In addition to sizing, the placement of welds in the most effective load-carrying positions can conserve weld metal. Although the welds in Fig. 7 are both simple fillets requiring the same amount of weld metal, when section

Table 2 Weld size factors for full-strength and rigidity fillet welds

ASTM No.	Yield strength, ksi	Weld-metal strength level and factor: E60	E70	E80	E90	E100
Full-strength weld size factor(a)						
A36	36	0.65	0.56	...	...	...
A441	50	...	0.76	...	...	...
A572-65	65	...	1.00	0.86	0.77	...
A514	100	...	1.52	1.33	1.18	1.06
Rigidity design (non-full-strength) weld size factor(a)(b)						
A36	36	0.22-0.32	0.19-0.28	...	...	...
A441	50	...	0.25-0.38	...	...	...
A572-65	65	...	0.33-0.49	0.29-0.43	0.26-0.38	...
A514	100	...	0.51-0.76	0.44-0.66	0.39-0.59	0.35-0.53

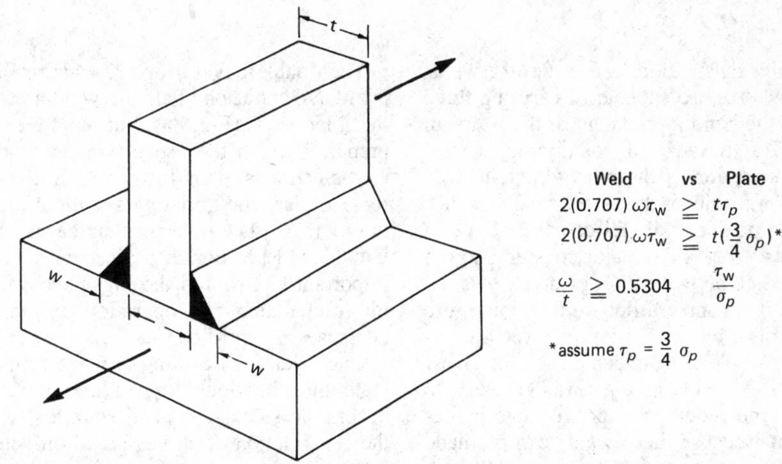

Note: τ_w is allowable shear stress on throat of weld; τ_p is allowable shear stress on plate; σ_p is allowable tensile stress on plate. (a) ω/t; where ω is leg size and t is thickness of thinner plate. (b) First ω/t value will develop one third of plate allowable; second value will develop one half of plate allowable.

Table 1 Allowable unit forces on fillet welds

Leg size, in.	E60 or F-60 weld metal: Parallel load, 10^3 lbf/in.	E60 or F-60 weld metal: Transverse load(a), 10^3 lbf/in.	E70 or F-70 weld metal: Parallel load, 10^3 lbf/in.	E70 or F-70 weld metal: Transverse load(a), 10^3 lbf/in.
1/16	0.80	0.93	0.93	1.09
1/8	1.59	1.86	1.86	2.17
3/16	2.39	2.80	2.78	3.26
1/4	3.18	3.73	3.71	4.35
5/16	3.98	4.66	4.64	5.44
3/8	4.77	5.59	5.57	6.52
7/16	5.57	6.52	6.50	7.61
1/2	6.36	7.46	7.42	8.70
5/8	7.95	9.32	9.28	10.87
3/4	9.54	11.16	11.14	13.05
7/8	11.14	13.05	13.00	15.22
1	12.73	14.91	14.85	17.40

(a) Non-code work

Fig. 7 Comparison of the effect of weld position on bending moment using weld section moduli. M is bending moment (in. · lb); b is width of welded connection (in.); d is depth of welded connection (in.); S_w is section modulus of welded connection (in.^{2}). If b and d each equal 4 in., then S_w equals $5^1/_3$ in.^{2} for (a) and 16 in.^{2} for (b), showing that (b) can resist three times the moment of (a).

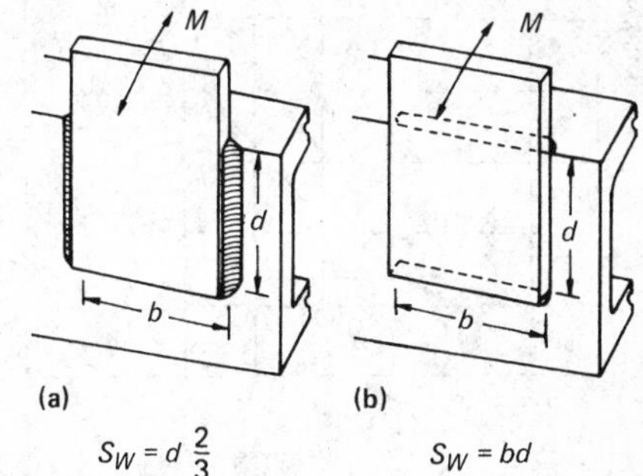

Fig. 8 Comparison of fillet and bevel-groove welds. ω is the leg size of the fillet weld in inches = ¾ in.; A is the cross-sectional area of the weld in square inches = ½ ω^2; t is the plate thickness in inches = 1.0 in. (a) Fillet welds. (b) Double-bevel groove weld. (c) Single-bevel groove weld

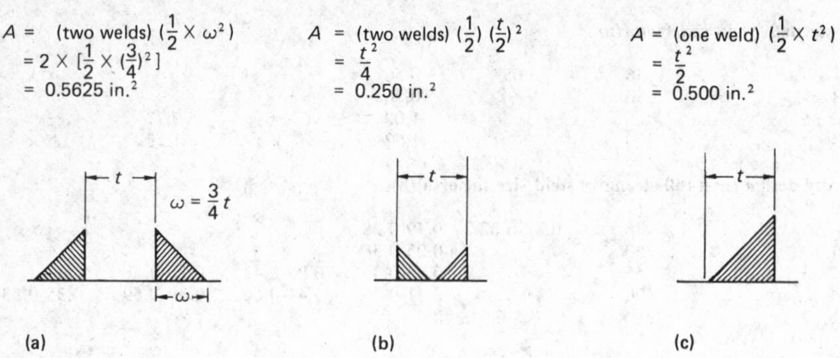

Fig. 10 Comparison of welds in the flat position in a T-joint. Fillet welds (a) are more expensive than a single-bevel groove joint (b) because of the overhead weld required.

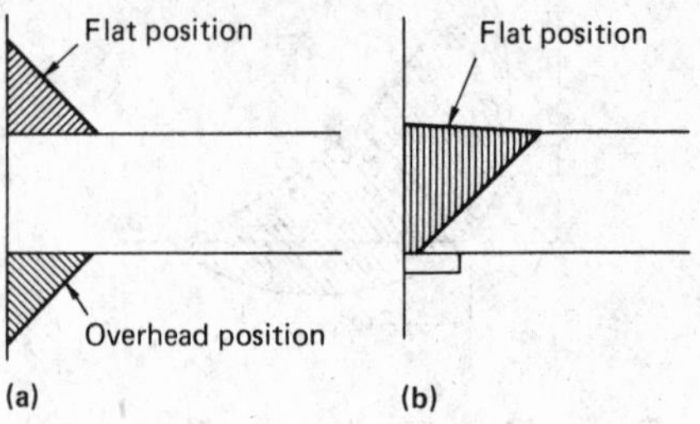

modulus calculations are made, the welds in Fig. 7(b) are capable of carrying three times the bending moment of the welds in Fig. 7(a) because of positioning. Therefore, selection of the most effective weld position facilitates the use of smaller welds.

Comparison of Fillet and Groove Welds. Cost is the major consideration in the choice between fillet or groove welds. Although simple fillet-welded joints are the easiest to make, they may require excessive weld filler metal for larger sizes. The fillet welds in Fig. 8(a) are easy to apply and require no special plate preparation. Because these welds can be made with large-diameter electrodes having high welding currents, the deposition rate is high. However, the amount of weld metal increases in relation to the square of the leg size.

Fig. 9 Relative cost of welds having the full strength of the plate. (a) Fillet welds. (b) 45° double-bevel groove welds. (c) 60° double-bevel groove welds

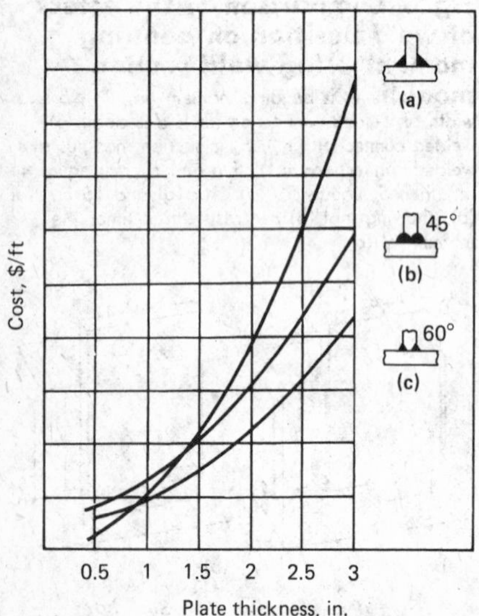

The double-bevel groove welds in Fig. 8(b) have about one half the weld area of the fillet welds (Fig. 8a), but require extra preparation and the use of smaller diameter electrodes with lower welding currents to place the initial pass without melt-through. As plate thickness increases, this initial low-deposition region becomes a less important factor, and the higher cost factor (preparation and operation expenses) decreases in significance.

One means of deciding at what point in plate thickness double-bevel groove welds become less costly than fillet welds is by the construction of curves based on determination of the cost of welding, cutting, and assembling. By reading upward from the plate thickness in Fig. 9, the relative costs of fillet welds or double-bevel welds of 45 or 60° can be determined. Thus, for smaller plate sizes the 45° double-bevel groove weld is more expensive, but as plate thickness increases, the cost of fillet welds is higher. The accuracy of this device depends on the accuracy of the cost data used in constructing the curves.

The single-bevel groove weld in Fig. 8(c) requires about the same amount of weld metal as the fillet welds in Fig. 8(a). Thus, there is no apparent economic advantage. There are some disadvantages, however. The single-bevel joint requires bevel preparation and, initially, a lower deposition rate at the root of the joint. From a design standpoint, the single-bevel groove weld offers a more direct transfer of force through the joint, which provides better service under fatigue loading.

Although the full-strength fillet welds (Fig. 8a) would be sufficient, some codes have lower allowable limits for fillet welds and may require a leg size equal to the plate thickness. In those cases, the cost of the fillet-welded joint may exceed the cost of the single-bevel groove in thicker plates. Also, if the joint is positioned so that the weld can be made in the flat position, a single-bevel groove weld would be less expensive than fillet welds, because one of the fillets would have to be made in the overhead position, a costly operation (Fig. 10).

Groove Preparation Welds

The important design considerations for groove-weld selection are the included angle, root opening, root face, and radius at root. The included angle, which is the angle of the groove weld, and the root opening, which is the portion of the joint before welding where the weld members are closest, are directly related. The root opening should be increased as the included angle decreases (Fig. 11) to allow for electrode access. The root face of a groove preparation, which helps to prevent melt-through at the root opening, is the portion of a weld groove face that is adjacent to the root of the joint. The radius at root (groove radius) is the radius

Fig. 11 Relationship of included angle size to the root opening

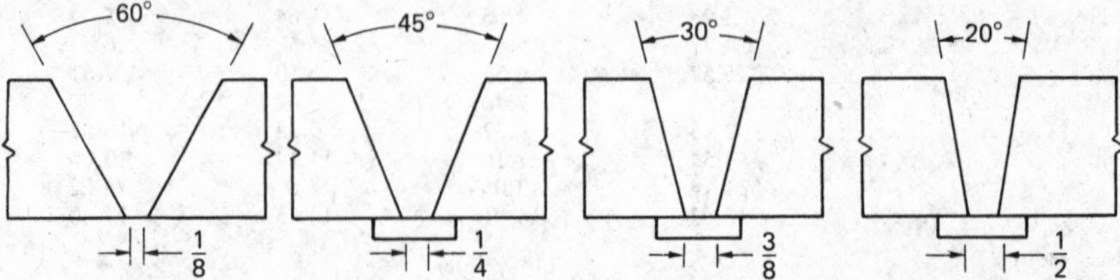

Table 3 Effect of included angle size on weld requirements

Plate thickness, in.	Weld metal needed for a root opening (r) of: $^{1}/_{8}$ in. lb/ft	$^{1}/_{4}$ in. lb/ft	$^{3}/_{8}$ in. lb/ft	$^{1}/_{2}$ in. lb/ft
$^{1}/_{2}$	0.84	0.90	0.99	1.13
$^{5}/_{8}$	1.21	1.23	1.29	1.44
$^{3}/_{4}$	1.64	1.61	1.62	1.76
1	2.69	2.50	2.38	2.47
$1^{1}/_{2}$	5.55	4.81	4.23	4.12
2	9.43	7.85	6.54	6.07
3	20.21	16.08	12.57	10.88
4	35.03	27.19	20.44	16.91
6	76.71	58.01	41.72	32.58
8	134.43	100.29	70.37	53.10
10	208.14	154.01	106.40	78.45

60° 45° 30° 20°
r r r r

used to form the shape of a J- or U-groove joint.

Included angles vary from 20 to 60° (Fig. 11). Generally, the smaller the included angle, the less weld metal required. The included angle and root opening must be sufficient to permit the electrode access to the root of the joint and to ensure good fusion to side walls with multiple passes. All four preparations in Fig. 11 are acceptable; all are conducive to good welding procedure and good weld quality. Therefore, selection usually depends on root opening and joint preparation, which directly affect weld cost (pounds of metal required).

Although the smaller included angle apparently requires less weld metal, it also requires a wider root opening. The wider root opening may defeat any savings achieved by the need for less metal in a narrower included angle, particularly in thinner plates. Thus, with $^{5}/_{8}$-in. plate, a 60° included angle in a double-bevel joint, with its $^{1}/_{8}$-in. root opening, is more economical than a 20° included angle with a $^{1}/_{2}$-in. root opening.

Fig. 12 Typical root openings in groove joints. *R* is the root opening.

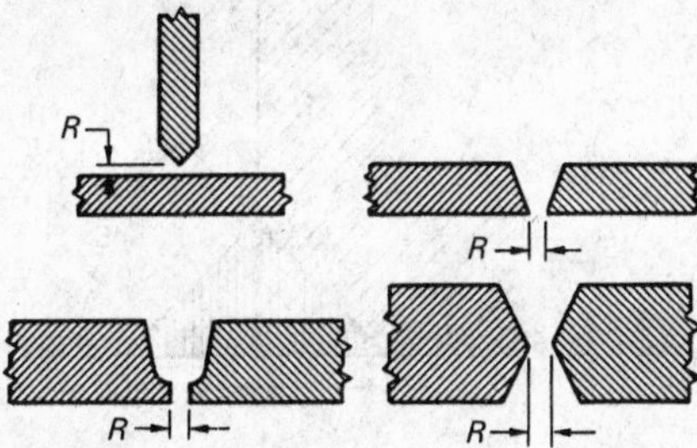

Table 3 shows that a 60° included angle requires 1.21 lb of weld metal per foot, while the 20° angle requires 1.44 lb/ft. With 2-in. plate, however, the 20° included angle is the most economical choice, requiring 6.07 lb/ft of weld metal compared to 9.43 lb/ft for the 60° angle.

Root openings (Fig. 12) are the separations between the members to be joined and provide electrode access to the root of the joint. The smaller the angle of bevel, the larger the root opening must be to obtain good fusion at the root.

If the root opening is too small, root fusion is more difficult to obtain, and smaller electrodes must be used, thus slowing the welding process. If the root opening is too large, weld quality does not suffer with use

Fig. 13 Effect of bevel angle and root opening on weld joint accessibility. θ is the included angle.

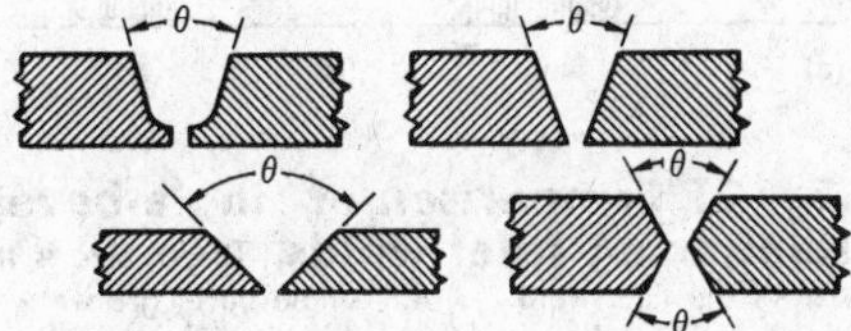

Fig. 14 Relationship of electrode angle on bevel angle

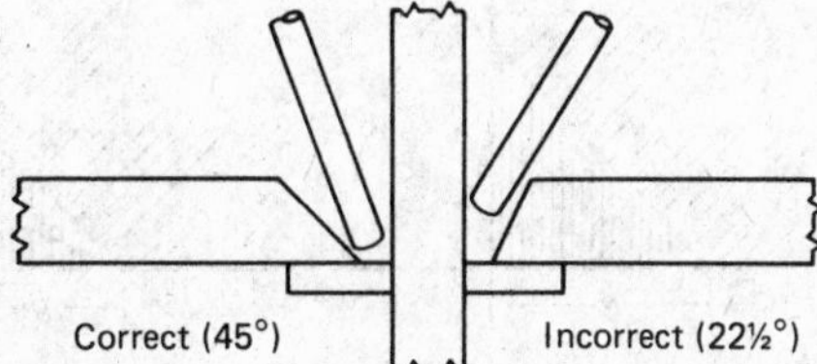

Fig. 15 Effect of included angle and root opening on correct welding. (a) Improper; gap is too small. (b) Proper joint preparation. (c) Improper; melt-through results from too large root opening.

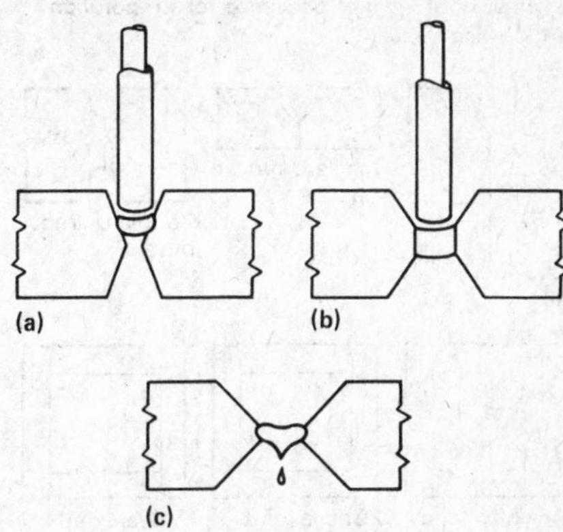

of a backing bar, but more weld metal is required, thus increasing welding cost and tending to increase distortion.

Bevel angles, because they affect the size of the root opening, also affect accessibility to all parts of the joint and the quality of fusion throughout the entire weld cross section. Accessibility can be improved by compromising between maximum bevel and minimum root opening (Fig. 13). As in Fig. 14, the importance of maintaining proper electrode angle in confined quarters may affect the angle of bevel. The minimum recommended bevel for the conditions in Fig. 14 is 45°.

Figure 15 illustrates an improperly sized included angle and root opening. In Fig. 15(a), the bevel and root opening are too small; the first weld pass bridges the gap, but may deposit slag at the root, which would require excessive back gouging. In Fig. 15(c), the angle and root opening are too large; melt-through may occur, resulting in weld-metal drop-through. In Fig. 15(b), a proper balance that permits good fusion of the joint has been obtained between included angle and root opening.

Double-groove weld joints reduce the amount of weld filler metal required for single-groove preparations by about half (Fig. 16). The decreased welding reduces distortion and facilitates alternating weld

Fig. 16 Comparison of weld metal required for single- and double-V-groove weld joints

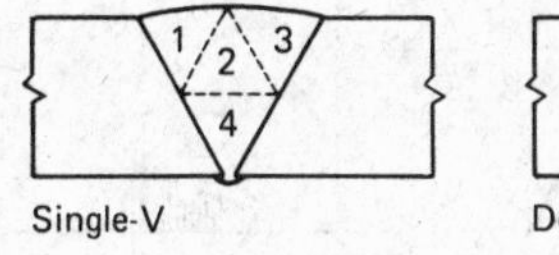

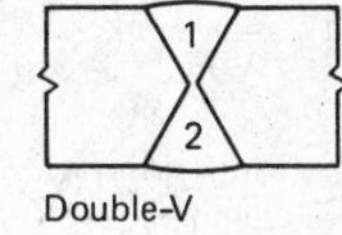

Fig. 17 Partial-penetration U-groove welds suitable for joint preparation after assembly and recommended procedure. (a) U-groove preparations suitable for completion prior to or after fitting. (b) Sequence for preparation after fitting

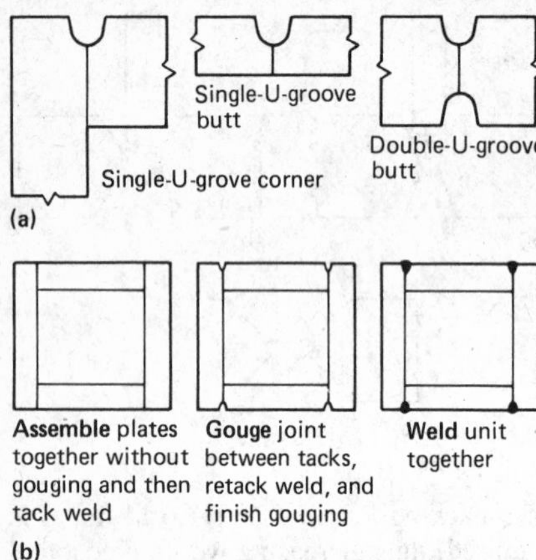

passes on each side of the joint, which further reduces distortion.

Joint preparation after assembly has not always been considered acceptable. This practice—assembly and tackup before gouge preparation of U-grooves for welding—is practical and less costly in certain applications. Now a sanctioned alternative, the AWS Structural Welding Codes address the procedures of joint preparation after assembly. U-groove joints for complete- and partial-penetration welds may, according to AWS, be made prior to or after fitting. Figure 17 shows joints that are adaptable to gouge preparations of U-grooves after fit-up assembly, as well as the sequence for fabricating preparation welds that are made after assembly.

Groove- and Fillet-Weld Combinations

Combinations of partial-penetration groove welds and fillet welds, such as those shown in Fig. 18, are used for many joints. The AWS prequalified, single-bevel groove T-joint, for example, is shown reinforced with a fillet weld.

Combination Double-Bevel Groove and Fillet Welds. The combination weld joint in Fig. 19 is welded using a partial-penetration double-bevel groove weld and two fillet welds. The plates are beveled to 60° on both sides to give a bevel whose depth is at least 29% of the thickness of the plate (0.29*t*). After the grooves are filled, they are reinforced with fillet welds of equal cross-sectional area and shape. This joint has a strength equal to that of the plate. These partial-penetration double-bevel groove joints have 57.8% of the weld metal of full-strength fillet welds. Although they require joint preparation, the 60° angles allow the use of large electrodes and high welding current.

Fig. 18 Combination groove- and fillet-welded joints

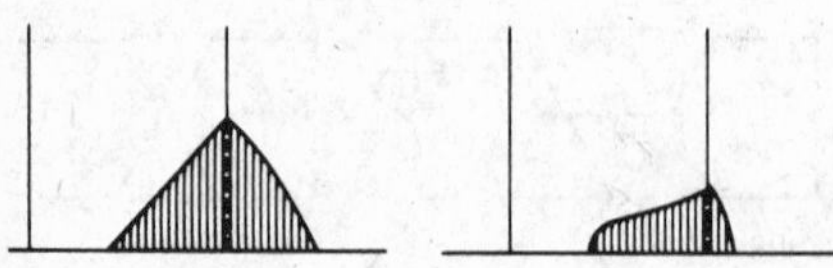

Fig. 19 Partial-penetration double-bevel groove joint. *t* is plate thickness. Depth of bevel is 29% of plate thickness.

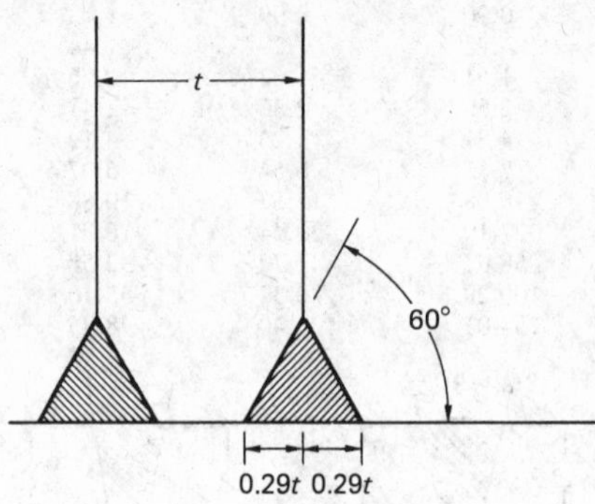

Comparison of Single-Bevel Groove, Fillet, and Combination Welds. When full-strength fillet welds are not required in the design, savings can often be achieved by using partial-penetration groove welds. As shown in Fig. 20, a 45° partial-penetration single-bevel groove weld with a 1-in. throat (Fig. 20b) requires only one half the weld area needed for a fillet weld (Fig. 20a). For smaller welds, however, this weld may not be as economical as the same strength fillet weld, because of the cost of edge preparation and the need to

Fig. 20 Comparison of weld joints wth equal throat size. *A* is the cross-sectional area of the weld; *t* is throat size; ω is the leg size of the fillet weld. (a) Fillet weld. (b) Single-bevel groove weld. (c) 45° single-bevel groove weld reinforced with an equal-leg fillet weld. (d) 60° single-bevel groove weld reinforced with an equal fillet weld

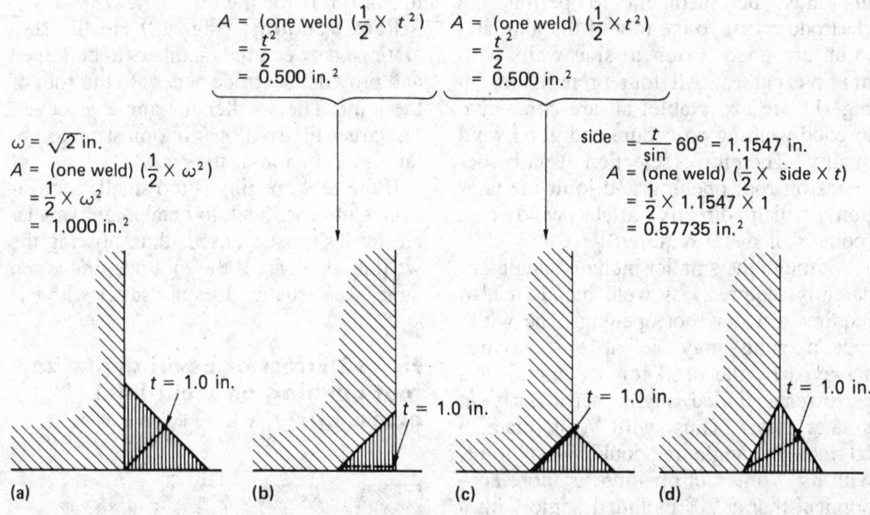

Fig. 21 Comparison of single-bevel joints with and without reinforcing fillet welds. Throat size is the same in each. *p* is depth of bevel; ω is the leg size of the fillet weld. (a) 45° single-bevel groove weld. (b) 45° single-bevel groove weld with less than equal sized reinforcing fillet weld. (c) 45° single-bevel groove weld with equal, full 45° fillet weld

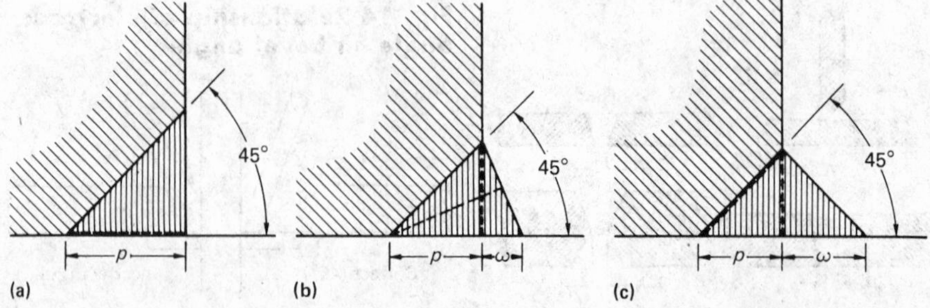

Fig. 22 Determination of minimum throat

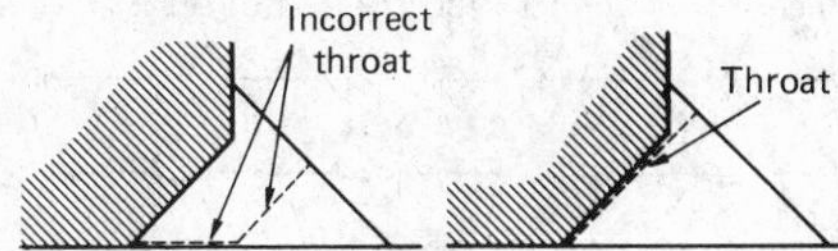

use a smaller electrode and lower current on the initial pass.

The welds in Fig. 20 all have equal throat sizes. As defined earlier in this article, the throat is the shortest distance between the root of the joint and the face of the diagrammatical weld. Thus, Fig. 20(c) shows that when a single-bevel groove joint is reinforced with an equal-leg fillet weld, the cross-sectional area for the same throat size is still one half the area of the fillet and requires less beveling. The single-bevel 60° groove joint with an equal fillet weld reinforcement for the same throat size (Fig. 20d) has an area 57.8% of the simple fillet weld. This joint has the benefit of smaller cross-sectional area, while the 60° included angle allows the use of higher welding current and larger electrodes. The only disadvantage is extra preparation cost.

Throat Size. The three parts of Fig. 21 illustrate a single-bevel 45° joint used alone and in combination with reinforcing fillet welds. In each weld, the throat size is equal.

Similar to the use of the minimum throat for fillet or partial-penetration groove welds, the minimum throat is used for designing a partial-penetration combination groove weld. As Fig. 22 shows, the allowable unit force for a combination weld is not the sum of the allowable unit forces for each portion of the combination weld, which would result in a total throat larger than the actual. Figure 23 illustrates the faulty calculations which result when the incorrect throat is used to determine the allowable unit force on a combination weld.

Edge Preparation

When considering whether to construct the weld joint using any of the groove welds—bevel, V-, J-, or U-groove—the cost of preparing the plate and weldment handling must be weighed against the savings in weld metal or other advantages of the groove-preparation welds.

Methods of cutting the groove-weld preparations include machining, chipping,

Fig. 23 Determination of allowable load on a combination weld. (a) Weld allowable load incorrectly figured adding each weld separately. (b) Weld allowable load correctly figured using minimum throat

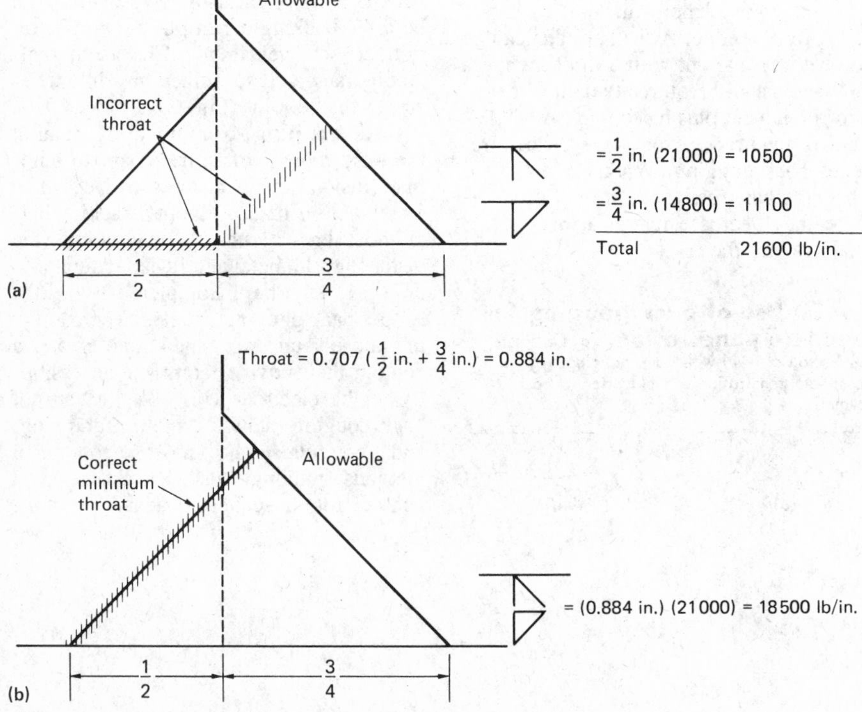

Fig. 24 Comparison of welding preparations

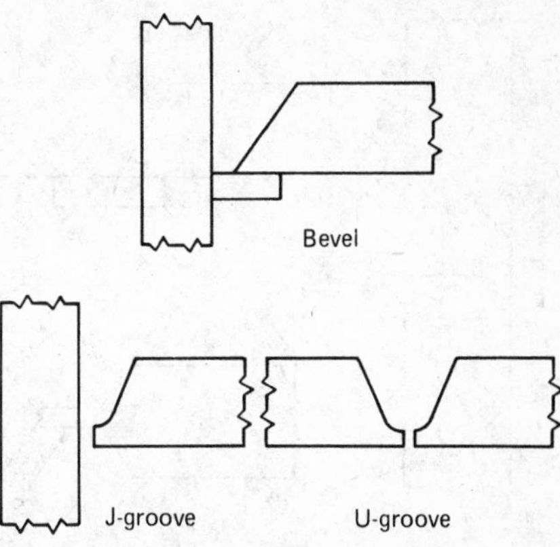

Fig. 25 Typical root faces on edge preparations

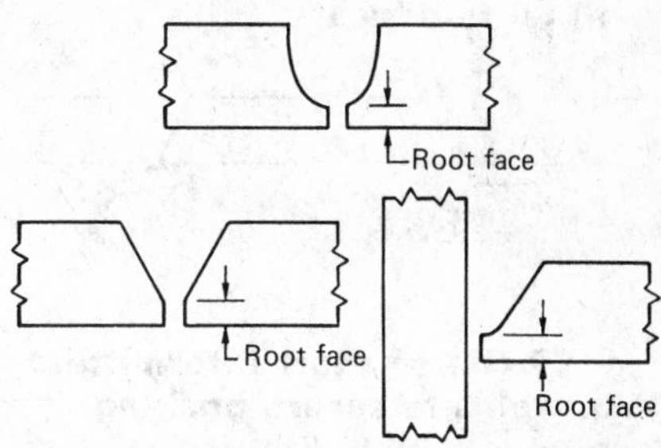

Fig. 26 Comparison of feather-edge and root-face preparations. Feather edge (a) is more prone to melt-through than root face (b).

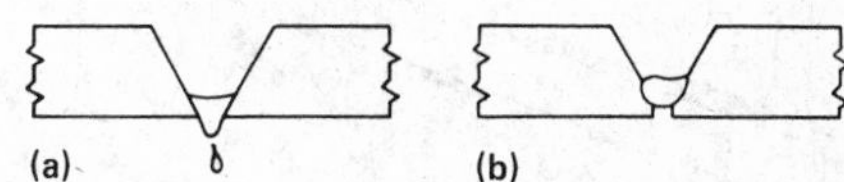

shearing, grinding, gas cutting, gas gouging, and air carbon arc gouging. The choice of the most economical means depends on factors such as the type of material, section characteristics, quality required, and available equipment.

Edge preparations are cut into the weld member using flame cutting, shearing, sawing, punch-press blanking, nibbling, or lathe cutoff (for bar and tube stock). In addition to cost, selection of the appropriate cutting operation depends on the quality of the edge for fit-up and whether a bevel is needed.

Because they are easily prepared by gas cutting, bevel and V-groove welds are more widely used. Although they offer the advantages of less weld filler metal and a more accessible work area, J- and U-groove preparations are not as applicable because they usually require machining or air car-

Fig. 27 Welding edge preparations for wide root openings. (a), (b), and (c) backing bars. (d) Spacer bar

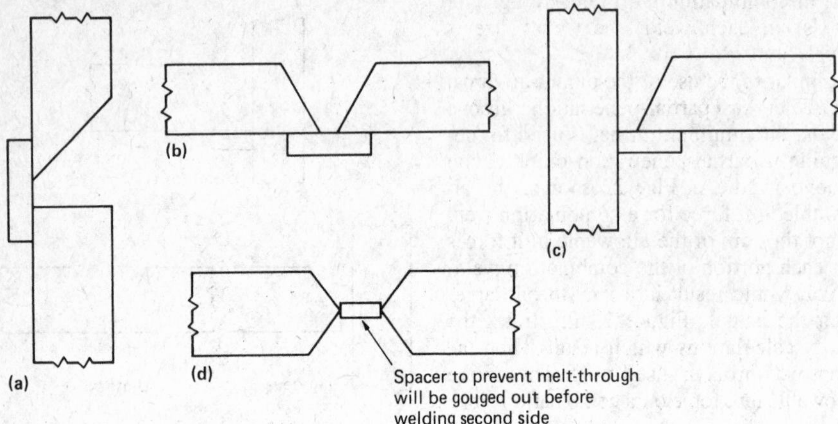

Fig. 28 Contact of backing bar with plate edges

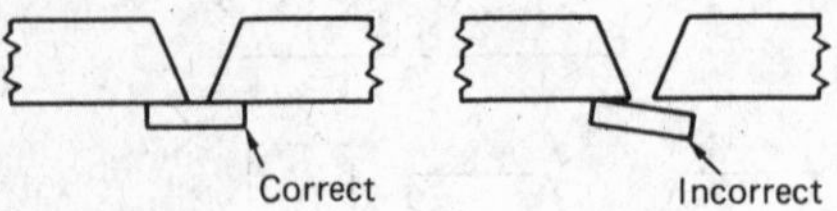

Fig. 29 Use of short intermittent tack welds to secure backing bars

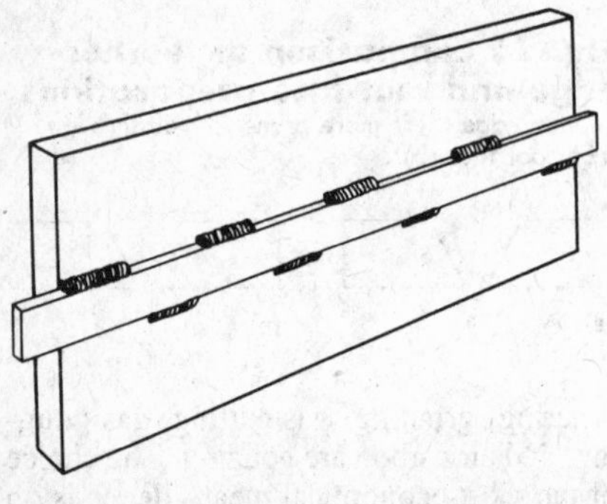

bon arc gouging, more costly methods of edge preparation. Thus, the bevel-groove preparation with a backing bar, shown in Fig. 24, may be more economical than a J- or U-groove. If a plate planer is available, however, J- or U-groove preparations are often specified because of reduced weld-metal requirements.

For single-bevel and single-V-groove preparations, single-tip flame-cutting torches are used because only one cut is necessary. Double-bevel or double-V-groove preparations are usually made with multiple-tip flame-cutting torches, which facilitate completion of the cut in one pass of the cutting machine.

Root faces on edge preparations (Fig. 25) provide additional thicknesses of metal to minimize melt-through, the tendency of the weld metal to pass through the root opening during welding. Feather-edge preparations (Fig. 26a), which provide only minimal thickness at the root opening, are more prone to melt-through than weld joints with root faces (Fig. 26b), especially if the gap becomes too large, and they require more weld filler metal to fill the joint.

A root face is more difficult to obtain than a feather edge. A feather edge can be achieved by one cut with a torch, while a root face usually requires two cuts or possibly a torch cut plus machining. If a 100% weld is required, a root face usually requires back gouging. When welding into a backing bar, root faces are not used because they decrease fusion at the root of the joint into the bar.

Fig. 30 Use of back gouging for complete penetration. (a) Complete penetration of weld with back gouging. (b) Incomplete penetration of weld prior to back gouging

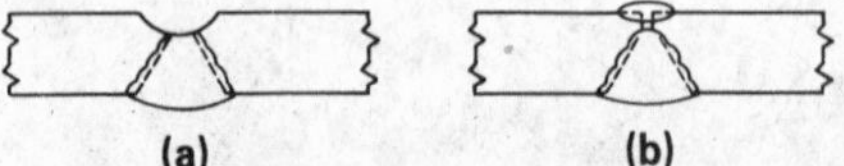

Fig. 31 Depth and contour of back gouging

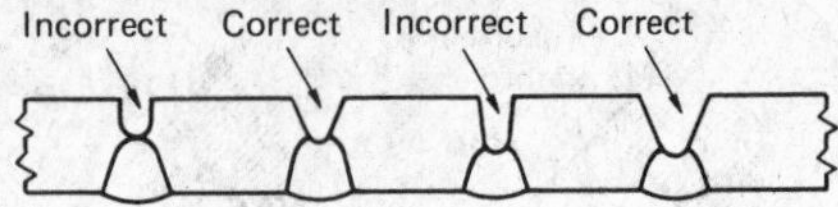

Spacer bars may be required to prevent melt-through when very large root openings are used on double grooved joints (Fig. 27d). A wide gap may be unavoidable for practical reasons, such as difficulty in pulling the parts together for desired fit-up.

The metal spacer bar serves as a backing and maintains root opening throughout the course of the welding operation. When using a spacer bar, the joint must be back gouged to sound metal before the second side is welded.

Backing bars, another edge preparation frequently required when the root opening is excessively large, are also used when welding must be done from one side (Fig. 27). Composed of metal, weld metal, or nonmetal, a backing bar is placed under or behind a joint to enhance the quality of the weld at the root. The bar usually remains in place after welding, becoming an integral part of the joint.

Steel backing bar material should conform to the base metal, maintaining a close contact with both edges to avoid trapped slag at the root of the weld joint (Fig. 28). Feather edges are recommended on the weld members for best performance of backing bars in ensuring weld quality. To hold the backing bar in place, short intermittent tack welds should be used, preferably staggered to reduce any initial restraint of the joint (Fig. 29).

Back gouging, or removal of metal at the weld root, is often necessary to eliminate fusion defects at the root face when a butt weld is made without a backing bar. Without back gouging, penetration (the distance weld metal and fusion extend into a joint) may be incomplete (Fig. 30). Proper back gouging extends deep enough to expose sound weld metal and creates a contour that provides complete accessibility to the electrode (Fig. 31). Means of back gouging include grinding, chipping, and gouging. The most economical method, gouging, also leaves an ideal contour for subsequent beads.

Appendix
Recommended Proportions of Grooves for Arc Welding

Fig. 1 Recommended proportions of grooves for butt joints. Made by shielded metal arc welding, gas metal arc welding, gas tungsten arc welding, flux cored arc welding, and oxyfuel gas welding (except pressure gas welding). Dimensions that apply to gas metal arc welding only are noted.

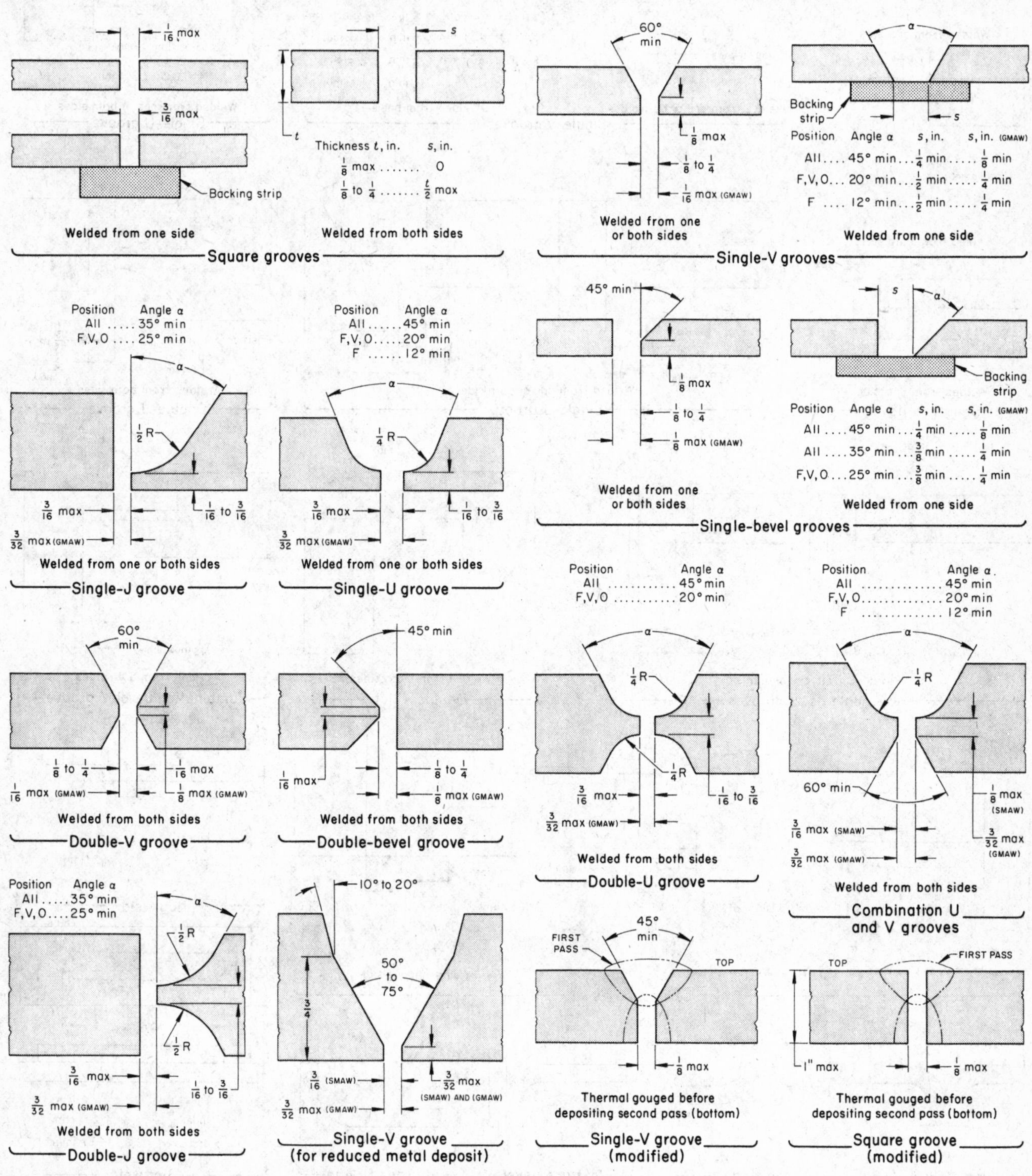

Fig. 2 Recommended proportions of grooves for corner and flange joints and plug welds. Made by shielded metal arc welding, gas metal arc welding, gas tungsten arc welding, flux cored arc welding, and oxyfuel gas welding (except pressure gas welding). Dimensions that apply to gas metal arc welding only are noted.

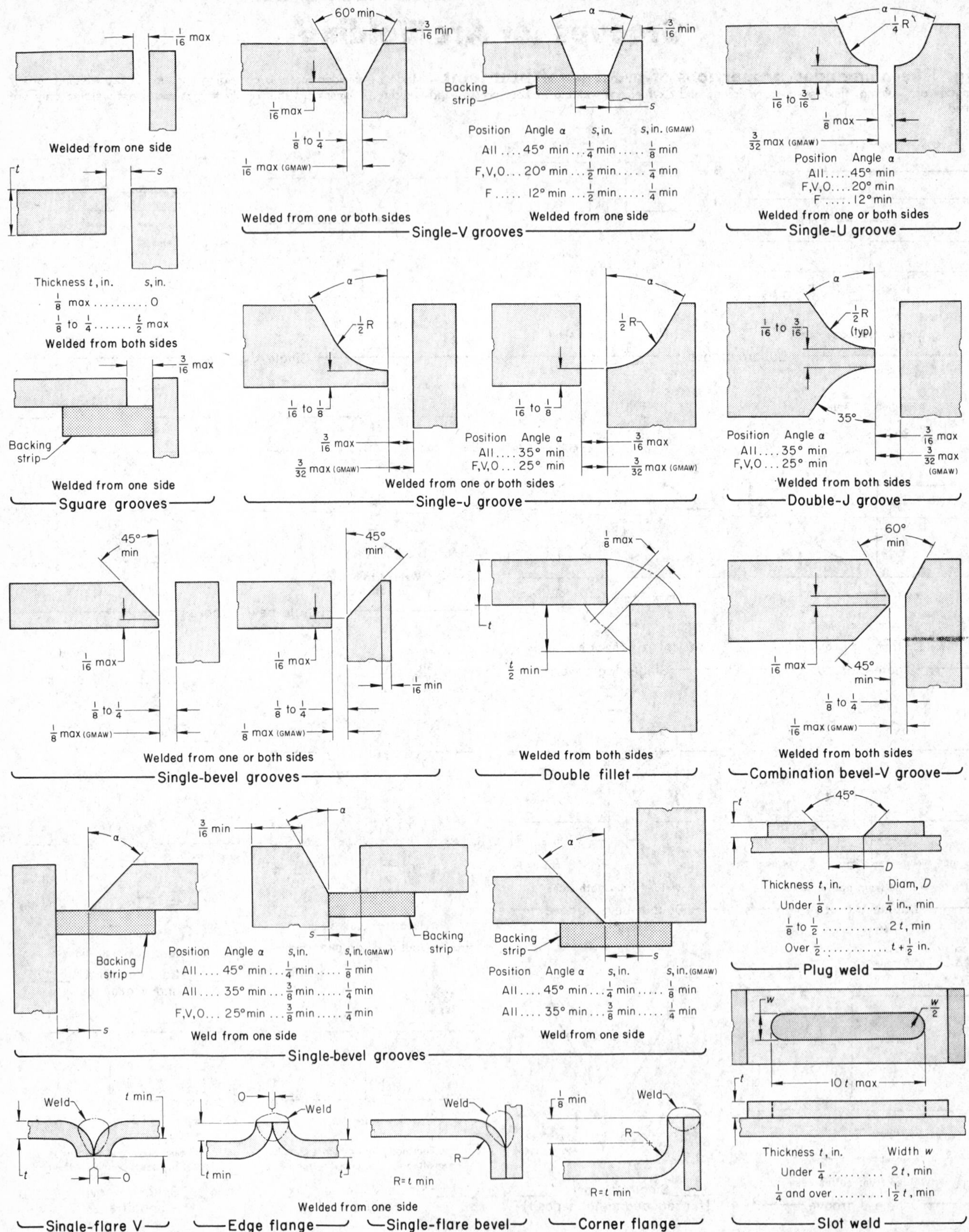

Fig. 3 Recommended proportions of grooves for T-joints and joints for specific applications. Made by shielded metal arc welding, gas metal arc welding, gas tungsten arc welding, flux cored arc welding, and oxyfuel gas welding (except pressure gas welding). Dimensions for specific welding processes are noted.

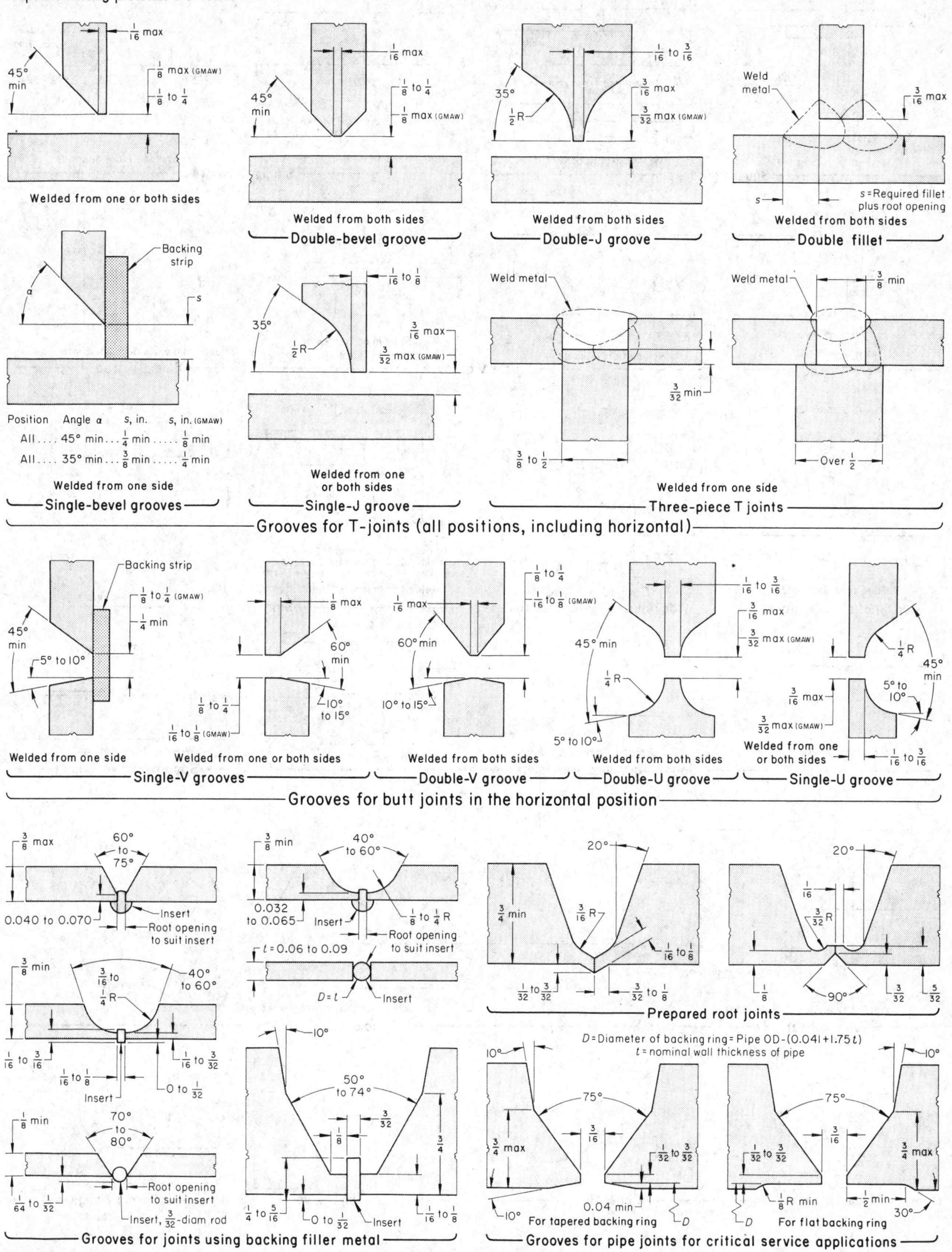

Fig. 4 Recommended proportions of grooves for joints made by submerged arc welding

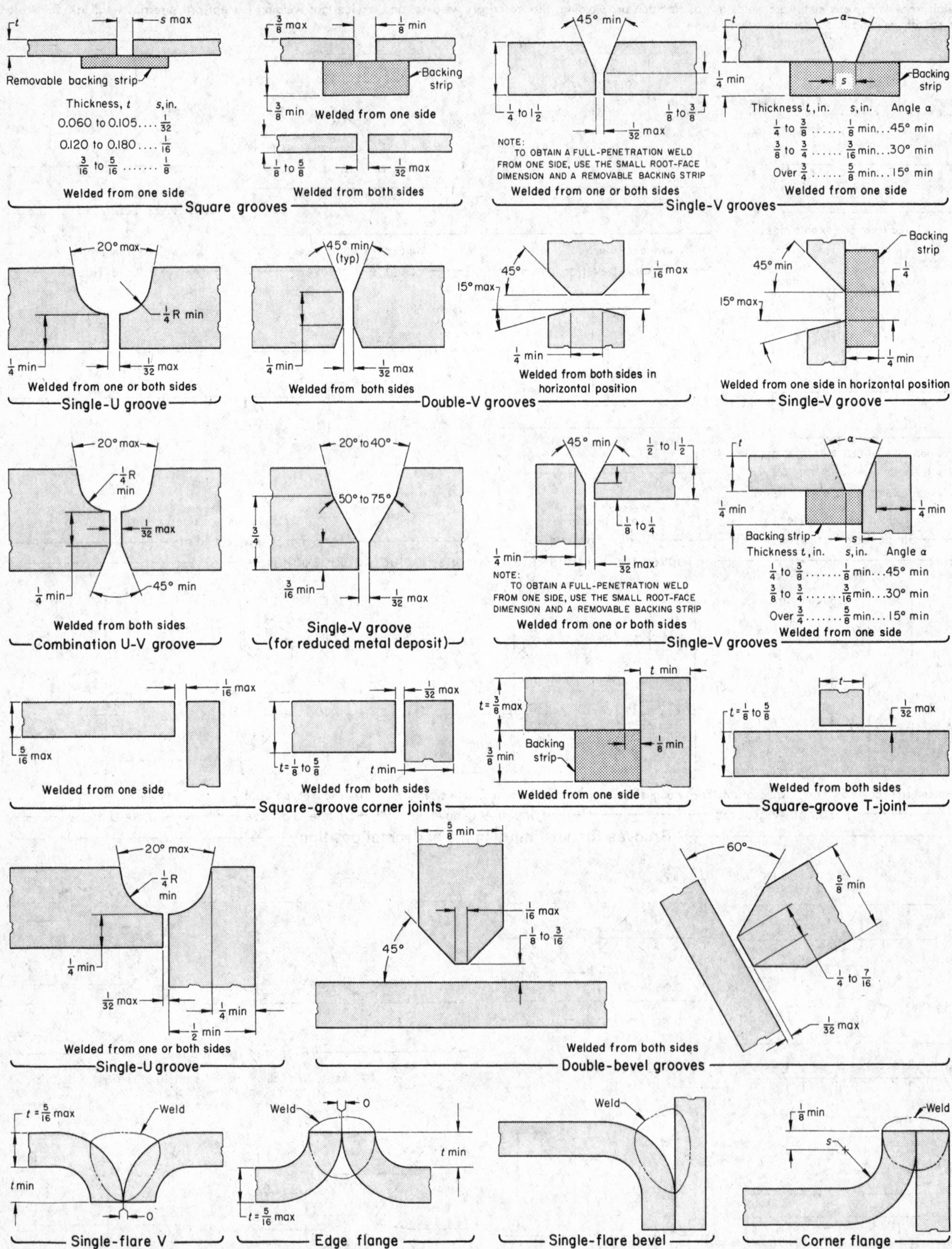

Arc Welding of Carbon Steels

Shielded Metal Arc Welding

By the ASM Committee on Shielded Metal Arc Welding*

SHIELDED METAL ARC WELDING (SMAW) is a manual arc welding process in which the heat for welding is generated by an arc established between a flux-covered consumable electrode and a workpiece. The electrode tip, molten weld pool, arc, and adjacent areas of the workpiece are protected from atmospheric contamination by a gaseous shield obtained from the combustion and decomposition of the electrode covering. Additional shielding is provided for the molten metal in the weld pool by a covering of molten flux or slag. Filler metal is supplied by the core of the consumable electrode and from metal powder mixed with the electrode covering of certain electrodes. Shielded metal arc welding is often referred to as arc welding with stick electrodes, manual metal arc welding, and stick welding.

Process Capabilities

Shielded metal arc welding is the most widely used welding process for joining metal parts because of its versatility and its less complex, more portable, and less costly equipment.

Versatility. Shielded metal arc welding can be done indoors or outdoors. Joints in any position that can be reached with an electrode (i.e., overhead joints and vertical joints) can be welded. By the use of bent electrodes, joints in blind areas can be welded, including the back sides of pipes in restricted areas—inaccessible for most welding processes.

The power supply leads can be extended for relatively long distances, and no hoses are required for shielding gas or water cooling. Shielded metal arc welding has been used for storage tanks, ship structures, and bridges. Machinery in manufacturing plants and in remote areas can be repaired using SMAW, because the welding equipment is light and portable.

Shielded metal arc welding is useful for joining the components of complex structural assemblies because it is adapted to multiposition welding in difficult locations. In assembling pipe and coupling connections, joints prepared for automatic or semiautomatic equipment may not fit. When this occurs, the assembly may be cut apart, adjusted in place to fit, tack welded, and then the assembly may be welded by SMAW. Welded joints can be cut apart on location and rebuilt with added structural members, as may be necessary for well drilling, quarrying, and mining.

Joint Quality and Strength. The quality and strength of shielded metal arc welded joints can be controlled as easily as the quality and strength of joints welded by other manual methods that use consumable electrodes. Shielded metal arc welding electrode materials are available to match the properties of most ferrous base metals, allowing the properties of a joint to match those of the alloys joined.

Metals commonly welded by SMAW are carbon and low-alloy steels, stainless steels, and heat-resistant alloys. Cast irons and the high-strength and hardenable steels can also be shielded metal arc welded, but procedures that include preheating, postheating, or both, may be needed. Electrode selection and care in following the welding schedule are more critical for welding hardenable steels. Nickel alloys are often welded by SMAW, although gas metal arc welding (GMAW) and gas tungsten arc welding (GTAW) are usually preferred and are more widely used for joining nickel alloys. A few aluminum and copper alloys are shielded metal arc welded. The softer metals such as zinc, lead, and tin, which have low melting and boiling temperatures, and the refractory and reactive metals, are not amenable to SMAW.

Limitations of SMAW when compared with other arc welding processes, such as GMAW and submerged arc welding (SAW), are related to metal deposition rate and deposition efficiency. Electrodes used in SMAW have fixed lengths (usually 18 in. or less), and welding must be stopped after each electrode is consumed. With arc welding processes that use continuously fed electrode wires or no filler metal, welding can be done with fewer interruptions. Deslagging is required after each pass to remove the slag covering that forms on the weld, in contrast to GMAW, where multiple passes are made without stopping for slag removal because no flux is used. Operator skill is a limiting factor in SMAW. Less training is needed by both GMAW and cored wire welders.

Principles of Operation

An adequate power supply is required for SMAW. Suitable cables are used to attach one terminal of the power supply to the electrode holder and the other terminal to a ground clamp (Fig. 1).

To start welding, an arc is struck by briefly touching the workpiece with the tip of the electrode. The welder guides the electrode by hand in welding a joint, and controls its position, direction, travel speed, and arc length (the distance between the end of the electrode and the work surface). In many applications in which electrodes with heavy coverings are used, the welder actually drags the electrode in the joint or on the work and uses the electrode

*Gregory L. Serangeli, *Chairman,* Manager, Industrial Technology, Caterpillar Tractor Co.; Arno R. Bebernitz, Development Engineer, Caterpillar Tractor Co.; Raymond T. Hemzacek, Manager, Welding Technical Services, International Harvester Co.; A. Lesnewich, Director, Research & Development, Filler Metals Division, Airco Welding Products

Fig. 1 Shielded metal arc welding. Setup and fundamentals of operation

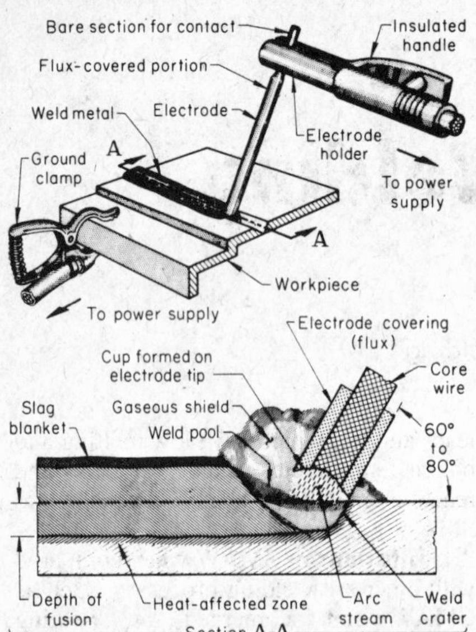

angle to control arc length. Electrodes must be discarded at a length of about 2 in.

A power supply with a drooping volt-ampere characteristic is required. With it, the current decreases as the arc becomes longer and increases as the arc becomes shorter. However, the basically simple concept is complicated by the effect of metal transfer across the arc, which can short circuit the power supply. As a result, the dynamic characteristics of the power supply are all-important; the reactance of the circuit that controls the speed of current response to short circuits affects the drop size and stability of the arc.

Electrode Coverings. The chemical and metallurgical properties of a weld depend on the electrode to be used and its covering. Atmospheric oxygen and nitrogen cause excessive porosity and poor ductility in the welded joint unless excluded from the molten weld pool. In SMAW, combustion and decomposition of the electrode covering from the heat of the welding arc produce a gaseous shield that excludes the atmosphere from the weld area. The molten metal is further protected by materials in the electrode covering that form a molten slag, which acts like a protective blanket until the metal has solidified (Fig. 1). Most electrode coverings also contain deoxidizers.

The electrode covering includes materials such as sodium and potassium that are readily ionized when heated by the arc. These help keep the gap between the end of the electrode and the workpiece conductive and stabilize the arc.

The electrode covering may also be used to introduce alloying additions into the weld. Most electrodes used for welding low-carbon steel have either a rimmed or a semikilled steel core. Alloying elements are added from the covering. Most coverings also contain iron powder, which increases metal deposition rates.

The heavier electrode coverings provide a cuplike shape at the end of the electrode (Fig. 1). This cup permits electrode drag and acts like a nozzle, increasing thermal efficiency and providing arc stream direction, which helps the welder direct the transfer of metal from the electrode into the molten weld pool.

The usability characteristics of an electrode, including speed of deposition, variety of possible welding positions, shape of weld bead, ease of slag removal, and weld properties, are controlled by the chemical composition of the electrode covering.

Welding Positions. The usual welding positions are flat, horizontal, vertical, and overhead, as shown for groove and fillet welds in Fig. 2. In some welding, the positions of the hours on a clock are used as reference locations. The flat position is the easiest for welding. In this position, weld quality benefits from the force of gravity, and maximum deposition rates are obtained. Next in ease of welding is the horizontal fillet position, in which the force of gravity helps to some extent. For welding in these positions, the joint should be level, or nearly so, when possible. Welding in positions other than flat (referred to as out-of-position welding) requires the use of manipulative techniques and electrodes designed for faster freezing of the molten metal and slag to counteract the effect of gravity.

Power Supplies

Many types and sizes of power supply are used for SMAW. When alternating current (ac) is used, high-voltage input power is transformed or stepped down through a transformer to a voltage low enough for safe use. When direct current (dc) is used, a motor-generator or a transformer-rectifier unit is used. A transformer-rectifier consists of a step-down voltage transformer with means to rectify alternating current to direct current.

Both alternating current and direct current produce acceptable results in welding low-carbon steel. Although each has distinct advantages, the choice usually depends on cost, availability of equipment, and the electrode used.

Combination ac/dc power supplies are widely used and are versatile for general-purpose applications. They consist of a transformer and a rectifier in combination and are capable of supplying either alternating current or direct current. When direct current is used, either straight or reverse polarity is available. Output ratings

Fig. 2 Welding positions

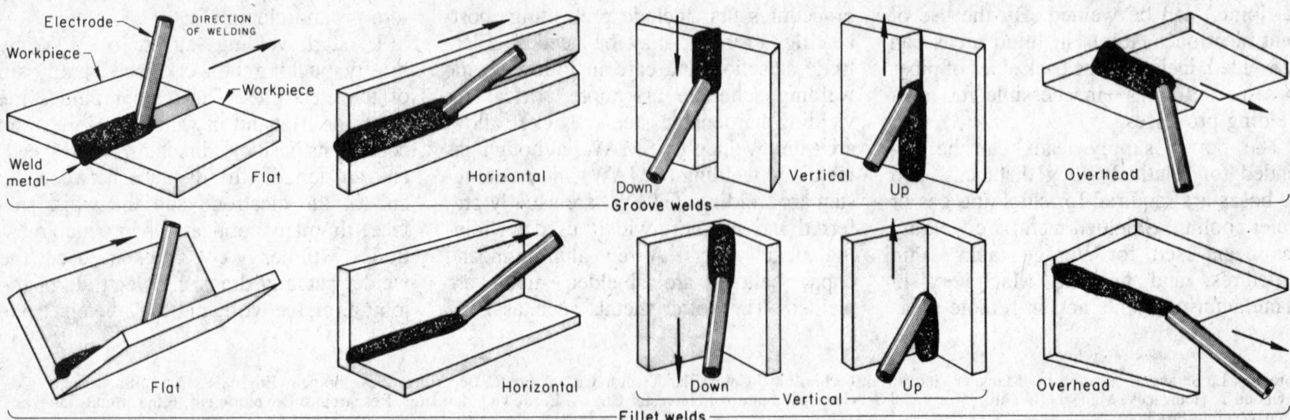

Table 1 Output ratings of power supplies used in SMAW

Output current, A, at load, V — Rated (60% duty cycle)(a)	Maximum (35% duty cycle)(b)
Constant-current motor-generators (dc)	
300 at 32	375 at 35
400 at 36	500 at 40
500 at 40	625 at 44
600 at 44	750 at 44
Transformer-rectifiers (dc)	
400 at 36	500 at 40
500 at 40	625 at 44
600 at 44	750 at 44
800 at 44	1000 at 44
Transformers (ac)	
400 at 36	500 at 40
500 at 40	625 at 44
600 at 44	750 at 44
(c)	(b)
750 at 44	925 at 44
1000 at 44	1250 at 44
1500 at 44	1875 at 44

(a) Rated current can be delivered continuously for 6 min out of every 10 min. (b) Maximum current can be delivered continuously for 3½ min out of every 10 min. (c) One-hour duty rating. Rated current can be delivered continuously for 1 h, then for 45 min of every hour for the next 3 h.

for various power supplies are given in Table 1.

Constant-Current Output. Power supplies with constant-current output and a means of controlling current are required for SMAW. Constant-current output is obtained with a drooping volt-ampere characteristic; voltage within a specified range is reduced as current increases. Constant-current output is preferred because small variations in arc lengths causing small variations in voltage do not significantly affect current output and deposition rate. A 3-V drop in voltage results in only about a 10-A increase in current output compared to hundreds of amperes with a constant voltage machine.

Selection Factors. Factors that influence the selection of power supply include: (1) available input power, (2) available floor space, (3) initial costs, (4) location of the operation (in a plant or in the field), (5) personnel available for maintenance, (6) versatility, (7) required output, (8) duty cycle, (9) efficiency, (10) need to minimize arc blow, and (11) safety. Table 2 lists space requirements and outputs of power supplies. The deposition rate desired determines the output required from a power supply to a great extent. The deposition rate obtained, however, is related to the duty cycle, output, and efficiency of a power supply.

The duty cycle of a power supply is the percentage of time, calculated on several successive 10-min periods, during which it can operate at rated capacity without exceeding recommended maximum operating temperature. If a power supply is operated at its rated capacity but beyond its rated duty cycle, overheating occurs, resulting either in automatic shutoff or, if the power supply is an unprotected rectifier, in burning out of the diode plates. However, short periods of overloading are not usually harmful to constant-current power supplies; i.e., a duty cycle of 76% can be tolerated for short periods on a power supply rated at 60%.

Transformers for SMAW with alternating current generally have constant-current (drooping) volt-ampere characteristics. Although heavy-duty transformers with outputs up to 1500 A are available, transformers with ratings of 200 to 500 A are used in most industrial welding applications. Transformers with ratings of less than 200 A are used extensively in farm and small job shop applications, because they can be operated on 220-V power and

Table 2 Space requirements and output ranges of ac and dc power supplies for SMAW

Floor space required, in.	Output range, A
Light-duty transformers (ac)	
14 × 14	30-180
	35-295
Heavy-duty transformers (ac)	
22 × 29	50-375
	50-625
28 × 45	100-1300
Light-duty motor-generators (dc)	
18 × 26	40-260
Heavy-duty motor-generators (dc)	
19 × 36	30-450
19 × 39	40-600
19 × 42	50-800
Transformer-rectifiers (dc)	
21 × 39	25-425
24 × 44	25-525
Heavy-duty transformer-rectifiers (dc)	
22 × 39	40-375
	50-625
Gasoline-engine generators (dc)	
31 × 75 (52 hp)	30-450
(69 hp)	30-600
(85 hp)	40-800

usually have sufficient output. Transformers have lower initial costs than direct current power supplies. They are also more economical in terms of cost of input power.

Poor power factor is inherent in the use of constant-current transformers. Welding characteristics are not affected by a poor power factor, but power cost is increased. For this reason, correcting capacitors can be used to reduce operating costs.

Some plants have large synchronous electric motors that operate air compressors and other equipment. These motors have a favorable influence on power factor and help to keep the plant power factor within acceptable limits. Under these conditions, welding transformers without power factor correction are satisfactory. However, in plants where welding transformers consume the major portion of the electricity, or where the reserve capacity is inadequate, transformers with power factor correction should be chosen.

Motor-generators and transformer-rectifiers are power supplies for welding with direct current. Most of these direct current machines have constant-current (drooping) volt-ampere characteristics. A constant-voltage power supply, one with nearly constant voltage output with increasing current output, is not well suited to SMAW, because the volt-ampere curve does not have sufficient droop to compensate for variations in arc voltage.

Motor-generators used in industry for SMAW have outputs of 200 to 600 A. However, units having much larger capacities are available. Engine-driven generators can be used in the shop or field where no line power supply is available.

Maintenance costs for motor-generators are higher than for transformer-rectifiers. Motor-generators have some moving parts that require periodic repair or replacement, while engine-driven generators have more moving parts and high initial cost.

Most motor-generators and transformer-rectifiers are available with a switch that permits current to be supplied with either direct current electrode negative (DCEN) or direct current electrode positive (DCEP), without disconnecting and reconnecting cables. This facility is useful in applications that involve a wide range of work metals and electrode types.

Efficiency of Power Supplies

Efficiencies of power supplies can be compared in terms of cost of input power for current output, melting rate, and deposition rate.

Cost of electric power is significant in welding, but usually is a relatively small fraction of the total direct cost. A motor-generator is the least efficient power supply, because a large mass is continually rotated, even when welding power is not being used. Transformer-rectifiers and transformers operate at higher efficiencies, because they have no moving parts except a cooling fan, and their no-load power loss is negligible. Their higher efficiency is especially significant when duty cycles are low. The efficiency of transformer-rectifiers or transformers at high duty cycles remains greater than that of motor-generators.

Efficiencies are always measured at rated load with arc or resistance loads and represent actual operating conditions. Figure 3 compares efficiencies of a transformer, a transformer-rectifier, and a motor-generator, operating at 20 to 120% of rated load.

Although direct current power supplies are less efficient than alternating current supplies, direct current is usually the better choice when power is needed for many different base metals and electrodes.

Melting rate, or burnoff rate, is the rate at which an electrode of a specific type and size is melted by a specific welding current. It is usually expressed in inches per minute. The melting rate of an electrode increases as current is increased. Figure 4 shows the effect of current supplied by a transformer and by a transformer-rectifier on the melting rate of E6012 electrodes. The melting rate increases rapidly as the current is increased, especially for smaller diameter electrodes. When welding current is too high, however, electrode deposition efficiency decreases rapidly because of arc blow, weld spatter, and excessive heating of the electrode.

Deposition rate, usually expressed in pounds per hour or minute, is a direct measure of the amount of weld metal produced under a given set of conditions.

Fig. 3 Effect of operating load on efficiency. Operating load shown as a percentage of rated load

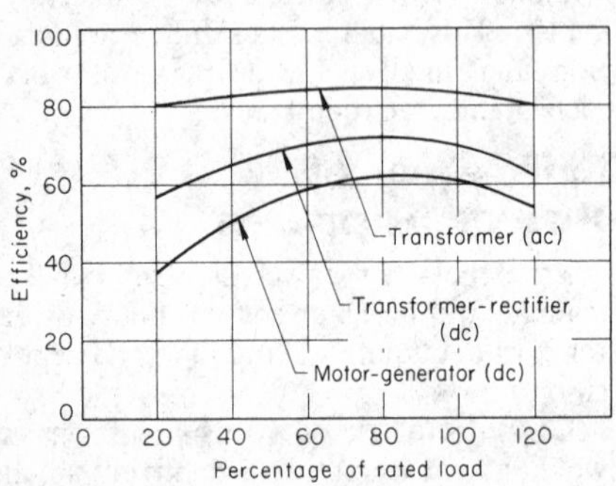

Fig. 4 Level of welding current and type of power supply versus melting rate. E6012 electrodes

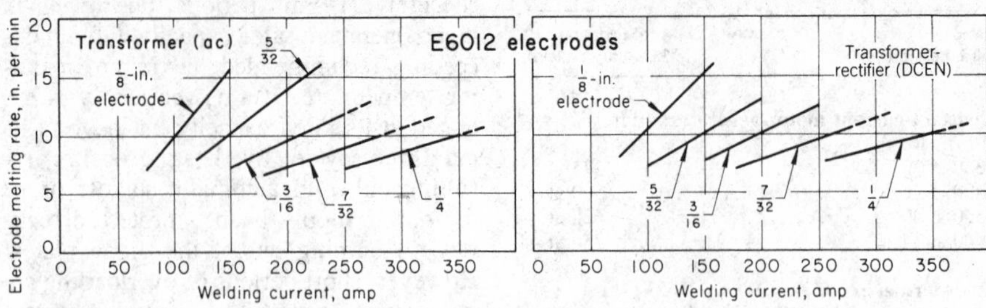

Deposition rate of an electrode is always less than melting rate because of losses by spatter, slag, and fume. The ratio of deposited weight to melted weight, times 100, is the electrode efficiency. Arc blow sometimes offsets the efficient operation of motor-generators at high welding currents. To overcome arc blow, the current must sometimes be decreased.

Variations in performance among different power supplies of the same general type must be considered when efficiencies are evaluated. For example, performance tests have shown that deposition rates at given currents for transformer-rectifiers from five different manufacturers varied 11% and that melting rates for the same equipment varied 9%.

Direct Current

Direct current flows continuously in one direction through the welding circuit. Whether the current is uniform or fluctuating does not affect its direction for any given welding setup. Because the current flow is continuous, the welding arc is relatively steady and smooth.

Voltage Drop in Cables. Welding cables should be as short as possible. Cable length is more critical for direct current than for alternating current. The voltage drop in long cables, added to that occurring at the arc, can either overload the power supply or reduce the available voltage needed for a proper welding arc.

Low Currents. Direct current surpasses alternating current for use at low amperages with small-diameter electrodes.

Electrodes. All classes of covered electrodes can be used with direct current.

Arc starting is generally easier with direct current than with alternating current, particularly with small-diameter electrodes.

Maintaining a short arc when the electrode must be close to the molten pool is easier with direct current than with alternating current.

Arc Blow. Direct current is highly susceptible to arc blow, particularly when welding is being done near the ends of joints, in corners, or on small, complex structures composed of a number of pieces. Welding on massive structures with high currents, or where fit-up is poor, also encourages arc blow. Arc blow causes excessive weld spatter and weld defects.

Welding Positions. Direct current is somewhat easier to use for out-of-position welding on thicker sections than alternating current, because lower currents can be used. Experienced welders usually can produce equal results with both types of current.

Welding of Sheet Metal. Because of the steady, easily started arc, direct current is preferable to alternating current for welding of sheet metal.

Polarity

Polarity, the direction of current flow, is important when direct current is used for welding. Connections for DCEN (straight polarity) and DCEP (reverse polarity) are illustrated in Fig. 5. Some direct current power supplies have switch-

Fig. 5 Connections for DCEN (straight polarity) and DCEP (reverse polarity)

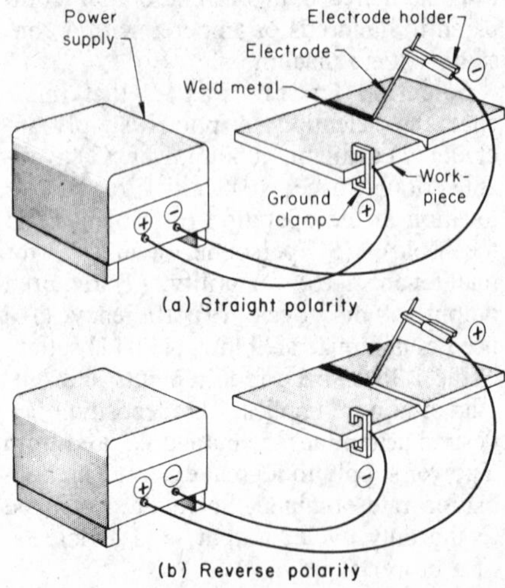

Fig. 6 Relative depths of penetration for different current characteristics

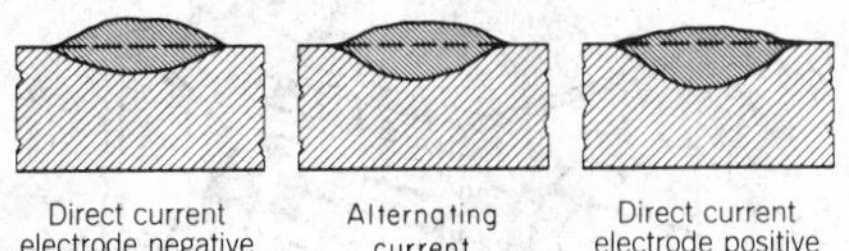

ing arrangements for reversing the polarity of the welding current whenever required.

Electrode type, metal being welded, and required penetration are important factors in the choice of polarity. Some electrodes are specially made for deep penetration and work better with reverse polarity. Other electrodes are designed for shallower penetration and work better with straight polarity.

Direct current electrode negative can be used for SMAW of all steels. Melting and deposition rates are higher than with DCEP, while penetration is shallower and narrower (Fig. 6). Contractional stresses are less severe, and restraint cracking is less likely. Also, because more of the heat is concentrated on the electrode, welding is more rapid with DCEN, and the workpiece is less susceptible to distortion. In addition, higher welding speed is usually possible with DCEN.

Direct current electrode negative is preferred for welding sheet metal, because the shallow penetration minimizes melt-through in thin sections. It is also preferred for welding joints with excessively wide gaps or root openings, as well as for buildup.

Direct current electrode positive produces maximum penetration for a given set of welding conditions. Although welding current is the main determinant of the extent of penetration, electrodes operating on DCEP provide deeper penetration than those operating on either DCEN or alternating current (Fig. 6). This makes DCEP the better choice for both root passes in groove welds made with the use of backing bars or strips and out-of-position welds.

Alternating Current

Alternating current combines both reverse and straight polarity alternately in regular cycles. In each cycle, the current starts at zero, builds up to its maximum value in one direction, decays to zero, builds up to its maximum value in the opposite direction, and decays to zero again. Cycles are repeated continuously while the welding arc is maintained. For 60-cycle alternating current, the polarity and therefore the direction of the current change 120 times each second. These changes produce a rapidly pulsating arc that is somewhat harsh and less stable when compared with a direct current arc.

Deposition rates and depth of penetration obtained with alternating current fall between those obtained with DCEP and DCEN, when welding at the same current rating. All electrodes that operate well on alternating current operate on direct current with either reverse or straight polarity, although one polarity is usually preferred.

Voltage Drop in Cables. Alternating current is preferred to direct current for welding that must be done at considerable distances from the power supply, because the voltage drop is less than for direct current in long cables. Long cables should not, however, be coiled either on the floor or over hooks, because the inductance set up by the coils reduces the output of the power supply and may overload the transformer. Cables should never be longer than required for the job.

Low Currents. Alternating current is less suited than direct current for use at low amperages with small-diameter electrodes.

Electrodes. Only ac/dc electrodes with coverings specifically designed for use with alternating current should be used. Because the current reverses, coverings must contain arc stabilizers to re-establish the arc immediately after the current decays to zero during each cycle.

Arc starting with small-diameter electrodes is more difficult with alternating current than with direct current. When the arc is struck using low current, electrodes may stick or freeze unless designed specifically for alternating current welding with low open-circuit voltages.

Maintaining a short arc (arc crowding) is more difficult with alternating current than with direct current except when iron-powder electrodes are used.

Arc blow is rarely a problem with alternating current. Reversing current causes the magnetic field in the work to build up and decay alternately in opposite directions and neutralizes the magnetic field.

Weld Spatter. Somewhat more weld spatter is produced with alternating current than with direct current.

Welding Positions. With the use of suitable electrodes, satisfactory welds can be made in all positions.

Welding of Sheet Metal. Alternating current, because of the difficulty in arc starting, is generally less desirable than direct current for the welding of sheet metal.

Welding of Thick Sections. Alternating current can be used to weld thick sections successfully by using large-diameter electrodes and maximum currents, because arc blow is seldom serious.

Electrode Holder

The electrode holder is a simple clamping device that holds the electrode and allows the welder to provide control during welding. The welding current is conducted from the welding cable through the jaws of the holder to the electrode. An insulated handle on the holder separates the hand of the welder from the welding circuit. The holder should be designed to provide good electrical contact and hold the electrode securely in position while allowing electrodes to be changed quickly and easily.

Electrode holders vary from basic to relatively large, heavily insulated holders used for long duty cycles and high welding current. Electrode holders are designed to accommodate a range of electrode diameters and to carry welding current required for the largest diameter electrode possible with the particular holder. The lightest and simplest electrode holder that can do the job without overheating should be used.

The jaws of an electrode holder should be kept in good condition to ensure good electrical contact and to minimize heating of the electrode holder. Poor contact and overheating can result in poor performance and low-quality welds. Electrode holders should never be cooled by immersion in water.

Ground Clamps

A ground clamp connects the ground cable to the workpiece. It should furnish a substantial mechanical and electrical connection and should be able to be readily attached to and removed from the work. For light duty, a spring-loaded clamp may be suitable, whereas for high currents, a screw clamp may be needed to provide a good connection and prevent overheating.

Jigs, Fixtures, and Positioners

Jigs, fixtures, and positioners are used to (1) minimize distortion caused by the heat of welding, (2) permit welding in a more convenient position, (3) increase welding efficiency, and (4) minimize fit-up problems. With a welding jig or fixture, components of a weldment can be assembled and held securely in proper relationship and with correct fit-up for positioning and welding. The time required to assemble parts to be welded (setup and fit-up time) is often a large percentage of total production time. If assembly follows a predetermined sequence of controlled fit-up and alignment conditions, welding efficiency is increased.

For making a single weldment or a few weldments of the same size and shape, temporary tooling can be used. For quantity production, designing and constructing accurate, durable jigs and fixtures is economical. The increased production that is possible through use of jigs, fixtures, and positioners depends on the proportion of arc time to handling and setup time and varies widely for different weldments.

Desirable features for fixture design include:

- The fixture should be designed so that all joints in an assembly are convenient and accessible for welding without removing the assembly from the fixture. Slots, or other means of access to joints on the reverse side of the weldment, may be provided.
- The fixture should be strong and light, but rigid enough to ensure accurate alignment.
- The fixture should permit quick and easy manipulation to position (by one hand, if possible). Balancing of the fixture may be necessary.
- The use of light alloys in the design of moving parts reduces weight. Air or electric motors should be used for revolving and air or hydraulic rams or racks for tilting the fixtured assembly.
- The accuracy and complexity of a fixture should be no greater than required.
- The fixture should permit clamping and location of components in position so that assembly, tacking, and welding can be done in one fixture.
- The fixture should permit freedom of movement in one plane to minimize residual stress in the completed weldment. Design should permit heat distortion to release, rather than bind, the assembly being welded.
- Floating rather than rigid anchoring of parts is recommended.
- Fixtures should be kept cool enough to handle; air, water, fins, or insulated handles can be used.
- Jigs or fixtures can be mounted on wheels or used in conjunction with floor-mounted or overhead conveyors. They are also located in a production line to facilitate shop flow.
- Indexing arrangements are helpful in providing accurate and quick positioning of work.
- Clamps must operate quickly. Screws and moving parts should be protected against weld spatter. Fusion to a fixture or to clamps can be avoided by the use of slots or copper backing. If possible, clamps and locating devices should be integral with, and hinged to, the fixture.
- Use of nuts and bolts, wrenches, C-clamps, wedges and hammers, and hand screws should be minimal. Eccentrics and cams, cranks, pinions and racks, air or hydraulic rams, solenoids, or magnetic clamps are preferred.
- The fixture must be designed to permit quick and easy removal of the weldment.

Fig. 7 Reversing pass direction minimizes distortion without clamping

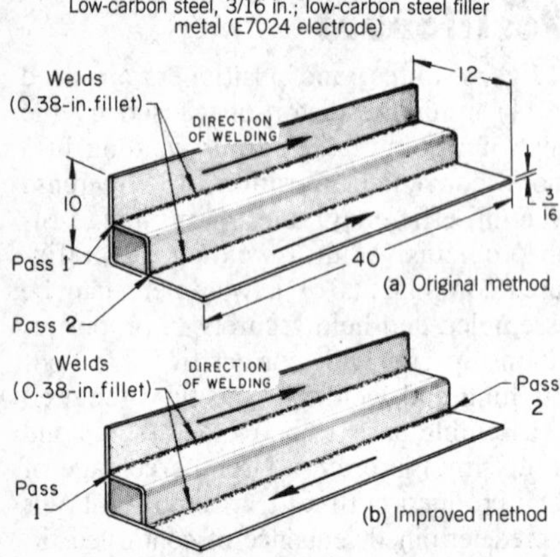

Minimizing Fixturing. Sometimes the need for fixturing or clamping to prevent excessive distortion can be minimized by a change in welding technique or weld sequencing. The change in welding technique need not be complicated and may consist merely of a change in welding direction.

Example 1. Changing Pass Direction to Minimize Distortion. The $^3/_{16}$-in.-thick low-carbon steel weldment shown in Fig. 7 was initially produced by welding both joints in the same direction (Fig. 7a). With this technique, the assembly had to be clamped securely to a table during welding and cooling to keep distortion within acceptable limits. When the two joints were welded in opposite directions (Fig. 7b), distortion was within acceptable limits without clamping. The weldment was made by first tack welding, using a $^5/_{32}$-in.-diam E7014 electrode and 200-A current, and then depositing the 0.38-in. fillet welds with $^1/_4$-in.-diam E7024 electrodes at 405 A.

Positioning is sometimes needed to make joints accessible, because assemblies usually require more than one weld. The main object of positioning, however, is to increase welding speed by putting the joint in a flat (or more favorable) position. Increases in speed are often considerable. Changing from a vertical or overhead position to a flat or horizontal position can increase welding speed up to 400%.

Fig. 8 Headstock-and-tailstock positioner

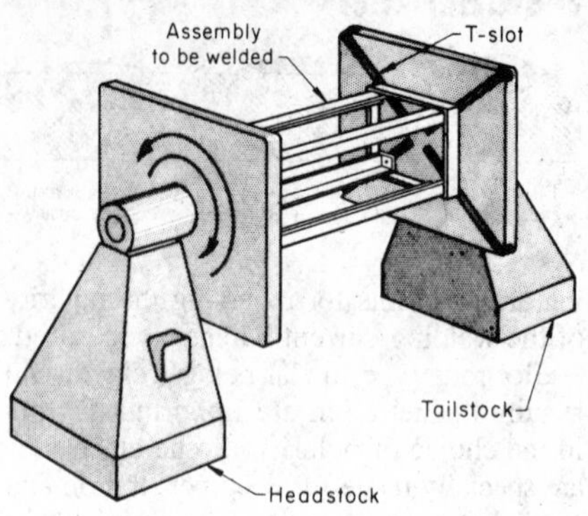

In flat position welding, deposition is assisted by gravity. Electrodes with coverings containing iron powder, which have a deposition rate up to 50% greater than that of other electrodes having the same core wire size, can be used at maximum efficiency when welding in the flat position.

Positioners. Commercial positioners of many types and sizes are readily available. The headstock-and-tailstock positioner (Fig. 8) is used for positioning large, bulky assemblies. This positioner is extremely flexible, because the headstock and the tailstock are separate units. Figure 9 shows a positioner that allows the workpiece to be rotated and tilted by using manual or power drives. This positioner is available in sizes ranging from small bench models to large floor models with capacities exceeding 100 tons. It can be modified as needed. For welding cylindrical workpieces such as tanks, a roll-type positioner (Fig. 10) is used. The speed of the

Fig. 9 Positioner that rotates and tilts workpieces

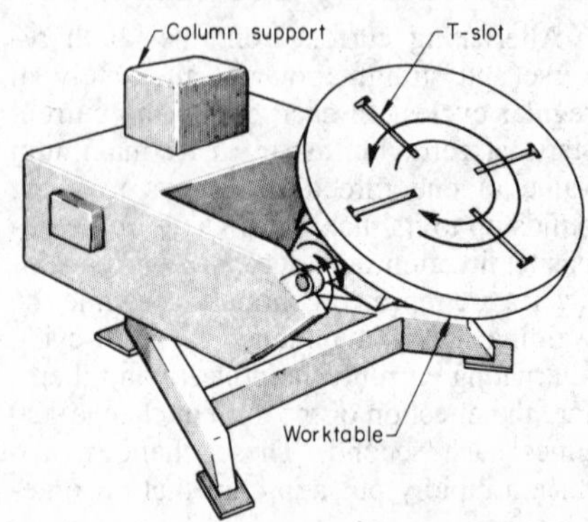

Fig. 10 Roll-type positioner

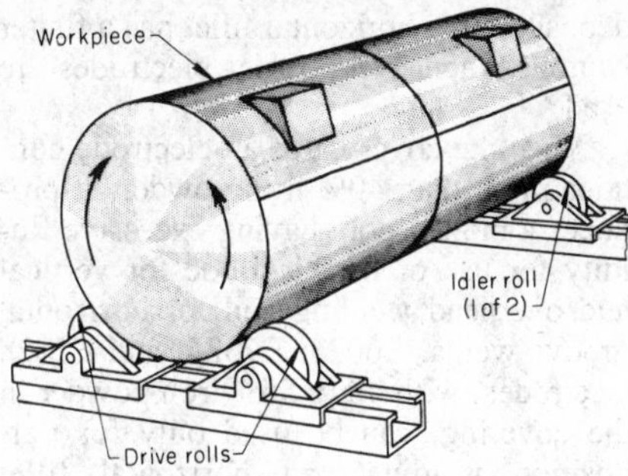

drive rolls can be regulated as needed, and rotation can be constant or intermittent in either direction. Positioning devices discussed in the articles "Gas Metal Arc Welding" and "Gas Tungsten Arc Welding" in this Volume are also applicable to SMAW.

Accessory Equipment. A steel wire brush, hammer, chisel, and chipping hammer are basic accessory tools used in conjunction with SMAW. These tools are used to remove dirt, rust, slag, and other foreign material from the weld area to provide a relatively clean surface for the deposition of weld metal, to cut tack welds, and to chip slag from the previous weld bead.

Electrodes

Electrodes used in SMAW have many different compositions of core wire and a wide variety of types and weights of flux covering. Standard electrode diameters (diameters of the core wire) range from $^1/_{16}$ to $^5/_{16}$ in. Length is usually 9 to 18 in., although electrodes up to 36 in. long have been made for special applications. A bare (uncoated) end of the electrode, standardized at a length of $^3/_4$ to $1^1/_2$ in., is provided for making electrical contact with the electrode holder.

Table 3 Coverings and currents indicated by fourth digit in AWS classifications of low-carbon steel covered arc welding electrodes

Fourth digit	Covering	Current(a)
0	High cellulose, sodium(b)	DCEP(b)
	High iron oxide(c)	ac or dc(c)(d)
1	High cellulose, potassium	ac or DCEP
2	High titania, sodium(e)	ac or dc(f)
3	High titania, potassium	ac or dc(f)
4	Iron powder, titania	ac or dc(f)
5	Low hydrogen, sodium	DCEP
6	Low hydrogen, potassium	ac or DCEP
7	Iron powder, iron oxide	ac or dc(d)
8	Iron powder, low hydrogen	ac or DCEP

(a) ac, alternating current; dc, direct current; DCEP, direct current electrode positive. (b) When third digit is 1. (c) When third digit is 2. (d) Either polarity for flat welds; DCEN for horizontal fillet welds. (e) Can also imply high iron oxide, ac or dc, either polarity. (f) Either polarity

Classification of low-carbon steel covered electrodes according to the system devised by the American Welding Society (AWS) is generally used throughout industry. In this system, designations consist of the letter E (for electrode) and four digits (five digits for weld-metal strength of 100 ksi or more). For electrodes producing under 100 ksi tensile strength, the first two digits indicate minimum tensile strength, in ksi, of deposited weld metal in the as-welded condition. The third digit indicates welding positions for which the electrode can be used successfully. For example, E*xx*1*x* indicates all positions; E*xx*2*x*, flat welds and horizontal fillet welds only; and E*xx*4*x*, specific design for vertical-down welds. The fourth digit indicates the type of covering and suitable current characteristics (Table 3). For example, E6011 is an electrode that deposits weld metal with a minimum tensile strength of 60 ksi (first two digits); can be used for welds in all positions (third digit); has a high-cellulose, potassium covering and can be used either with alternating current or with DCEP (fourth digit). A suffix added to the classification number indicates chemical composition of the deposited metal as specified in AWS specification A5.5.

Electrode Coverings. The composition of the electrode covering largely determines the performance of an electrode and the soundness of the weld. Table 4 lists materials used in making electrode coverings. More than 18 can be used; 12 or more may be included in a specific covering. Each material is often used for more than one purpose. Table 4 gives compositions of electrode coverings for classes of electrodes used in welding low-carbon steel and lists the primary and secondary functions of the constituents of the coverings. The thickness of the covering varies from 10 to 55% of the total diameter of a covered electrode, depending on the type of covering. Coverings are usually

Table 4 Functions and composition ranges of constituents of coverings on low-carbon steel arc welding electrodes

Constituent of covering	Function of constituent: Primary	Function of constituent: Secondary	E6010, E6011	E6012, E6013	E6020, E6022	E6027	E7014	E7016	E7018, E7048	E7024	E7028
			Composition range in covering on electrode of class, %								
Cellulose	Shielding gas	...	25-40	2-12	1-5	0-5	2-6	...	...	1-5	...
Calcium carbonate	Shielding gas	Fluxing agent	...	0-5	0-5	0-5	0-5	15-30	15-30	0-5	0-5
Fluorspar	Slag former	Fluxing agent	...	...	...	...	...	15-30	15-30	...	5-10
Dolomite	Shielding gas	Fluxing agent	...	...	...	...	...	...	...	...	5-10
Titanium dioxide (rutile)	Slag former	Arc stabilizer	10-20	30-55	0-5	0-5	20-35	15-30	0-5	20-35	10-20
Potassium titanate	Arc stabilizer	Slag former	(a)	(a)	...	...	...	...	0-5	...	0-5
Feldspar	Slag former	Stabilizer	...	0-20	5-20	0-5	0-5	0-5	0-5	...	0-5
Mica	Extrusion	Stabilizer	...	0-15	0-10	...	0-5	...	...	0-5	...
Clay	Extrusion	Slag former	...	0-10	0-5	0-5	0-5	...	...	...	...
Silica	Slag former	...	...	...	5-20	...	...	...	...	...	...
Manganese oxide	Slag former	Alloying	...	...	0-20	0-15	...	...	...	...	...
Iron oxide	Slag former	...	...	...	15-45	5-20	...	...	...	...	...
Iron powder	Deposition rate	Contact welding	...	...	...	40-55	25-40	...	25-40	40-55	40-55
Ferrosilicon	Deoxidizer	...	...	...	0-5	0-10	0-5	5-10	5-10	0-5	2-6
Ferromanganese	Alloying	Deoxidizer	5-10	5-10	5-20	5-15	5-10	2-6	2-6	5-10	2-6
Sodium silicate	Binder	Fluxing agent	20-30	5-10	5-15	5-10	0-10	0-5	0-5	0-10	0-5
Potassium silicate	Arc stabilizer	Binder	(a)	5-15(a)	0-5	0-5	5-10	5-10	5-10	0-10	0-5

(a) Used (in place of constituent on line above) in E6011 and E6013 electrodes to permit welding with alternating current.

applied by extrusion of the flux onto the core wire.

Characteristics of Electrode Classes

E6010 and E6011 electrodes give a deep-penetrating, forceful spray arc and can be used in all welding positions. These electrodes develop a low volume of slag that is easily removed. The deposits usually have good mechanical properties and are radiographically acceptable. The main constituent of the covering is cellulose, which decomposes during welding to provide a gas shield. The gases formed from the decomposition of cellulose and the high moisture content (up to 5%) of the electrodes produce the arc characteristics. The covering of E6011 electrodes contains potassium to assist in maintaining the arc when alternating current is used.

E6012 and E6013 electrodes provide a medium-penetrating arc. They yield a semiglobular to globular viscous slag that permits welding of joints with poor fit-up. The contour of horizontal fillet welds obtained using these electrodes varies from convex with E6012 electrodes to nearly flat with E6013 electrodes. Both can be used for out-of-position welding, and most E6013 electrodes operate successfully in the vertical-down position. E6012 electrodes can be used at relatively high welding current, because the coverings contain only a small proportion of cellulose and a large proportion of refractory material. Coverings of E6013 electrodes contain more potassium than those of E6012 electrodes and give a quieter arc with less penetration. The high potassium content of the coverings of some E6013 electrodes permits the use of low open-circuit voltage. In small sizes, E6013 electrodes are used for welding of sheet metal.

E6020 electrodes provide a spray arc with medium-to-deep penetration. These electrodes produce a heavy honeycombed slag that is easily removed. Coverings for these electrodes consist mainly of oxides of iron and manganese. Protection is provided by the volume of slag. The molten weld metal from these electrodes is too fluid for welding other than horizontal fillets or joints in the flat position. These electrodes have higher deposition rates than other conventional electrodes of equivalent sizes and produce welds of equivalent strength and soundness.

E6022 electrodes are designed for single-pass, high-speed, high-current, flat-groove and horizontal-lap and fillet welds in sheet metal. Because of high welding speeds, the profiles of the weld beads tend to be convex.

E7016 electrodes have low-hydrogen coverings containing little or no hydrogen-bearing materials, such as cellulose and clays. These electrodes are baked at a relatively high temperature (500 to 600 °F) to minimize the retention of water from the silicate binder. Because of the low hydrogen content of the covering, the weld metal deposited is also low in hydrogen and free of porosity. The carbon dioxide from the calcium carbonate and the silicon fluoride (SiF_4) (formed from the fluorspar reaction with silicon dioxide) provide the shielding gas. The high proportions of titanium dioxide and potassium silicate in the covering permit the use of E7016 electrodes with alternating current as well as DCEP.

E7015 electrodes are low-hydrogen electrodes used with DCEP. Although included in AWS A5.1, E7015 electrodes are no longer generally manufactured.

E7048 electrodes are low-hydrogen electrodes that have similar usability, composition, and design characteristics of E7018 electrodes, but are designed specifically for vertical-down welds.

Iron-Powder Electrodes. As shown in Table 4, the coverings of E6027, E7014, E7018, E7024, E7028, and E7048 electrodes contain iron powder in addition to the constituents present in coverings on several classes of conventional electrodes. Except for the iron powder, the constituents of E7014 and E7024 coverings resemble those of E6012 and E6013, and the constituents of E7018, E7028, and E7048 resemble those of E7016. In general, covering thickness increases as the content of iron powder increases. The iron powder and additional covering thickness permit higher welding currents and higher deposition rates than are possible with electrodes having coverings containing similar constituents without iron powder. Thicker coverings also provide a deeper shield and permit the use of the drag technique in the flat position. In addition, the weld bead deposited in a horizontal fillet has a flatter contour when iron-powder electrodes are used.

When the covering of an electrode contains more than 40% iron powder, it produces a molten pool having excessive fluidity for use of the electrode for vertical and overhead welding and for horizontal groove welds. E6027, E7024, and E7028 electrodes, with 40 to 55% iron powder in the covering, can be used only for flat-position welding and horizontal fillet welds. Welding positions for E6027 electrodes are also restricted by the fluidity of the covering.

Effect of Moisture in Electrode Coverings

Moisture is not as harmful in the coverings of some low-carbon steel electrodes as has been assumed. Precautions should be taken to store electrodes in dry places. However, redrying, which is often done after prolonged storage, can impair both quality and operation of electrodes with cellulose coverings, especially E6010 and E6011 electrodes.

Table 5 shows recommended moisture contents of coverings and storage and redrying conditions for different classes of electrodes. Some brands of E6010 and E6011 electrodes operate satisfactorily and produce satisfactory weld-metal deposits when the moisture content of the covering is above the recommended range. All other electrodes usually operate best when the moisture content is lowest.

The redrying temperature depends on the composition and thickness of the covering. Coverings containing organic material are usually redried at temperatures below the charring point (250 °F), whereas inorganic coverings, such as the low-hydrogen types, are redried at temperatures up to 800 °F. The drying procedure prescribed by the manufacturer must be followed precisely for specific electrodes,

Table 5 Recommended moisture content of coverings and storage and redrying conditions, for low-carbon steel covered arc welding electrodes

Electrode class	Recommended moisture content of covering, %	Relative humidity, %(a)	Temperature of holding oven, °F	Redrying temperature, °F(b)
E6010	3.0-5.0	20-60	(c)	(c)
E6011	2.0-4.0	20-60	(c)	(c)
E6012, E6013, E6020, E6022	Less than 1	60 max	100-120	275 ± 25
E6027, E7014, E7024	Less than 0.5	60 max	100-120	275 ± 25
E7015, E7016	Less than 0.4	50 max	130-330	650 ± 50
E7018, E7028, E7048	Less than 0.4	50 max	130-330	750 ± 50

(a) For storage at normal temperature of 80 °F ± 20 °F. (b) 1 h at temperature. (c) Follow manufacturer's recommendation.

or the electrodes may become unusable. Holding oven temperatures shown in Table 5 should be maintained after redrying. Permissible electrode exposure to the workplace atmosphere varies with the coating. Manufacturers' recommendations regarding exposure and redrying should be followed.

Determination of Moisture Content. The moisture content of coverings on all except low-hydrogen electrodes can be determined by weighing an approximately 1-g sample of the covering before and after redrying for 1 h in an electric oven at 220 °F. The weight loss calculated as percentage of predrying weight is the moisture content.

Moisture content for low-hydrogen electrodes is usually determined at temperatures between 1650 and 1800 °F, using a special apparatus and technique as described in AWS A5.1 and A5.5 specifications.

E6010 and E6011 Electrodes. Manufacturers of E6010 and E6011 electrodes control moisture content so that it is between 3 and 5% for E6010 and between 2 and 4% for E6011. E6010 and E6011 electrodes may perform satisfactorily when the moisture content in the covering exceeds optimum values, but only if the covering does not blister during welding and does not interfere with operation of the electrode.

Ordinarily, the soundness of welds made with E6010 or E6011 electrodes is not impaired by excess moisture if the operation of the electrode remains satisfactory. When most brands of E6010 and E6011 electrodes have a moisture content much below 2.0%, more weld spatter and increased probability of porosity in the weld metal are encountered. Arc control may also be impaired. The moisture content of E6010 electrodes can accidentally be decreased to values lower than the recommended level when a container is left open and electrodes are exposed to a hot, dry atmosphere.

Low-Hydrogen Coverings. The moisture content of low-hydrogen coverings (E7015, E7016, E7018, and E7028 electrodes) should be kept below the values in Table 5 (or preferably below 0.3%). If moisture content is much above these values, underbead cracking is likely. Low-hydrogen electrodes should not be allowed to remain in open boxes or bins for 2 to 8 h, depending on the humidity.

After a working-shift exposure, all unused low-hydrogen electrodes should be returned to a redrying oven maintained at 250 to 350 °F for at least 8 h before reissuing them.

Fig. 11 Effects of humidity and redrying temperature on moisture in electrode coverings

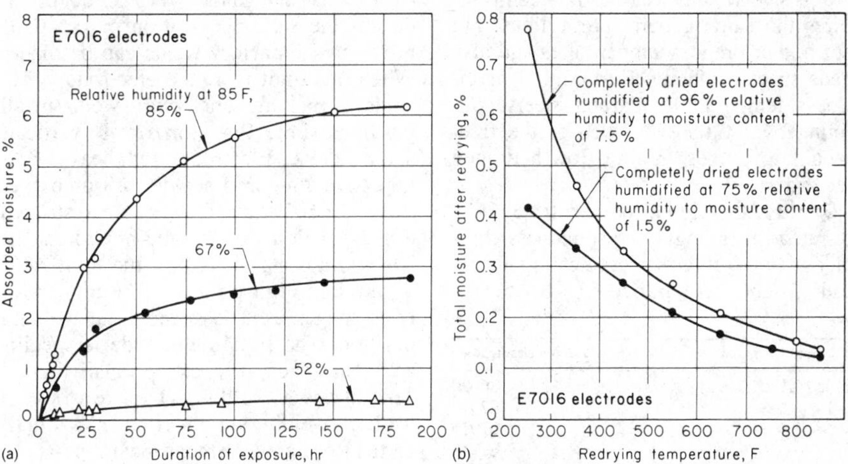

Figure 11 shows amounts of moisture absorbed by the covering on E7016 electrodes exposed for various periods of time to a humid condition. Figure 11(b) shows the total percentages of moisture contained in coverings on E7016 electrodes after redrying the electrodes at about 250 to 850 °F in a well-ventilated oven. Values shown in Fig. 11 are typical; they vary considerably among different brands.

Times and temperatures required for redrying E7016 electrodes to their original usability differ for electrodes from different manufacturers. A redrying procedure for one commercial brand of E7016 electrodes entails redrying for approximately 1 h in a well-ventilated oven by slowly increasing the temperature to 550 °F ± 50 °F. Prolonged holding periods (weeks or months) at either lower or higher temperatures can cause the coverings of some brands of E7016 electrodes to become brittle and crack.

Selection of Electrode Class

The selection of welding electrodes is based on compatibility between base metals to be joined and the service requirements of the weldment. Carbon steels can be welded with essentially any classification of covered carbon steel electrodes (AWS A5.1, "Specification for Covered Carbon Steel Electrodes") within certain limitations and service requirements. These limitations are the carbon and strength levels of the base metal, the restraint conditions, and the ambient temperature. The essential factors in the selection of electrodes for the carbon steels are: (1) mechanical properties, (2) materials composition, (3) welding current, (4) quality, (5) position of welding, and (6) cost.

Mechanical Properties. Carbon steel electrodes classified in the type E60*xx* series may be used for welding lower carbon content steels. Where higher strengths are required, electrodes possessing higher weld-metal deposit strength, i.e., type E70*xx* series, are used. For full-penetration welds, an electrode must be selected to ensure compatibility with the minimum strength level of the base metal. In addition, where notch toughness is a factor, an electrode possessing the proper level of impact criteria should be selected. For fillet welds and partial-penetration groove welds, the matching of minimum strength properties may be compromised to the extent of modifying the weld size, if needed. Steels having a yield strength above 60 ksi become increasingly difficult to weld, particularly as the conditions of restraint (based on mass or design) or low ambient temperature are encountered. Under these conditions, certain precautions must be observed, such as the use of electrodes in a low-hydrogen classification and perhaps the use of preheating and possibly postweld heat treatments.

Material Composition. Carbon steels which possess less than 0.30% C (low-carbon steel) are readily weldable with any class of low-carbon steel electrodes. Steels which possess a carbon content in the range of 0.30 to 0.60% C (medium-carbon steel) have restrictions in the applied welding procedure and electrode selection. In this range, low-hydrogen electrodes are necessary and, depending on carbon content, condition of restraint, and ambient temperature, may require the welding procedure to include preheat, and possibly postheat, depending on service requirements. With carbon content above 0.60% C (high-carbon steels), the welding procedure requires the same general precautions, but uses higher preheat temperatures and also needs stricter adherence to control of interpass temperature. Carbon steels containing over 0.05% S, such as the free-machining grades, require low-hydrogen electrodes.

Quality. Covered electrodes for SMAW are rated for radiographic soundness characteristics in three categories: grade 1, grade 2, and not required (AWS A5.1).

Radiographic standard	AWS classification
Grade 1(a)	E6020
	E7015
	E7016
	E7018
	E7048
Grade 2(a)	E6010
	E6011
	E6013
	E7014
	E7024
	E6027
	E7027
	E7028
Not required	E6012
	E6022

(a) Per porosity and inclusion standards of AWS A5.1

In almost all classifications, the electrodes specified in AWS A5.1 are capable of meeting most radiographic soundness requirements. Correct application is an important factor. Certain welding conditions, such as long, continuous welds, open root conditions, and thickness of base metal, provide optimum soundness characteristics for particular classifications. Selection for soundness should be based on production conditions and weldment design. Based on arc characteristics, these electrode classifications also produce a varying degree of penetration into the root of the weld. The following table compares the depths of penetration normally obtained with various classes of low-carbon steel covered electrodes. Penetration is rated as deep, medium, or shallow:

Electrode class	Penetration
E6010	Deep
E6011	Deep
E6012	Medium
E6013	Less than E6012
E6020	Medium(a)
E6022	Medium
E6027	Medium
E7014	Medium
E7016	Medium
E7018	Shallow
E7024	Shallow
E7028	Shallow
E7048	Shallow

(a) Deep with high current

Position of Welding. In the covered carbon steel classification system, the third digit of the designation (e.g., E6010) indicates the welding position or positions in which satisfactory welds can be made. When this digit is 1, the electrode is capable of making satisfactory welds in all positions, i.e., flat, horizontal, vertical, and overhead. Electrodes with heavy coatings containing iron powder or iron oxide, such as E6020 and E7028, are restricted to flat-position welding and horizontal fillets and are designated by the numeral 2 in the third digit position. The numeral 3 is no longer used to specify flat welding position. The E7048 electrode is specifically designed for exceptionally good vertical-down welding characteristics as well as welding in the flat, horizontal, vertical-up, and overhead positions.

Cost. The principal factor when considering welding cost is deposition rate. This influences overall welding cost in two areas: cost of labor and productivity. The higher the deposition rate, the greater the number of inches of weld that can be produced per unit of time and the more weldments that can be completed per unit of time. Deposition rate, however, with its effect upon welding travel speed, should not exceed the maximum travel speed capability required to maintain quality welds.

The deposition rate for a given electrode diameter is highest for those covered electrodes that contain iron powder or iron oxide in their coatings (Table 6). Electrodes containing the greatest amount of these additives in their coatings and offering the best potential for highest deposition rate are designed for flat-position and horizontal-fillet welding. To maximize productivity with electrode selection, the weldment should be positioned by using a positioner. The use of a positioner enables processing for maximum productivity with any electrode classification by minimizing the effect of gravity on the molten weld pool and allowing maximum deposition rates.

Selection of Electrode Size

Selection of electrode size directly influences both weld quality and welding costs. The electrode size (diameter of the core wire) must be compatible with joint design and the position of welding to ensure that the fusion characteristics, weld-bead profile, and degree of internal discontinuities are optimized. In addition, the electrode size should be maximized for any given application to obtain the highest deposition rate possible for the most efficient productivity. The principal considerations in selecting electrode size are joint design, welding position, thickness of weld layer, and permissible heat input. Table 7 shows ranges of welding current and welding travel speed for several sizes and classifications of carbon steel covered electrodes for fillet and groove welds of various sizes and positions.

Dimensions for weld joint design are established by specification or through the development of a welding procedure that has been qualified through testing. The number of weld passes required depends on weld joint design, electrode size, base-metal thickness, and welding position.

All classifications of carbon steel covered electrodes, except type E6022, are designed for multiple-pass welding. Type

Table 6 Electrode performance and mechanical properties of weld metal

Electrode class	Electrode size, in.	Current(a), A	Deposition rate(b), oz/h	Deposition efficiency, %	Properties of weld metal: Tensile strength, ksi	Yield strength, ksi	Elongation in 2 in., %
E6011	3/16	200 max	68	76	70	60	25
	1/4	300 max	112	72	...	...	...
E6012	3/16	225	66	68	79	66	20
	1/4	380	120	78	71	58	15
	5/16	475	178	78	...	...	...
E6013	3/16	225	67	67	76	67	22
E6020	3/16	225	92	68	69	59	31
	1/4	380	177	69	65	52	29
	5/16	450	215	69	63	50	28
E7014	3/16	260	87	68	74	68	23
	1/4	340	119	69	73	67	25
E7016	3/16	225	63	70	87	75	29
E7018	3/16	240	83	69	80	67	29
E7024	3/16	270	134	68	90	80	22
	1/4	360	186	69	89	77	22
	5/16	475	250	73	83	74	22
E7028	3/16	300	125	68	90	81	25
	1/4	390	192	70	88	80	25

(a) Alternating current. (b) 100% arc time

Table 7 Operating conditions and speeds for electrodes in shielded metal arc fillet and butt welding

Electrode size, in.	Current, A	Weld size, in.	Speed for welding positions, in./min: Flat	Horizontal	Downhill, 30°	Vertical	Overhead
E6011 electrodes							
1/8	100-120	1/8	8-10	8-10	10-12	...	...
	110-120	1/8	...	...	...	7-8	7-8
5/32	120-140	5/32	8-10	8-10	10-12	...	...
	130-140	5/32	...	...	...	7-9	7-9
3/16	150-165	3/16	...	...	...	6-7	6-7
	160-175	3/16	9-11	9-11	11-13	...	...
E6012 electrodes							
1/8	120-140	1/8	13-15	12-14	14-18	4-5	4-5
5/32	150-170	1/8	15-16	14-16	13-17	...	...
	150-170	5/32	...	...	...	3-4	3-4
3/16	190-210	5/32	13-15	12-14	12-15	...	...
	190-210	3/16	...	...	...	3-4	3-4
	200-220	3/16	10-12	9-11	11-14	...	...
E7014 electrodes							
5/32	180-200	5/32	10-11	10-11	12-13	...	...
3/16	230-250	3/16	11-12	11-12	12-13	...	...
7/32	280-310	1/4	10-11	10-11	11-12	...	...
1/4	340-370	5/16	8-9	8-9	9-10	...	...
E7018 electrodes							
1/8	120-140	1/8	8-10	8-10	...	4-6	7-9
3/16	200-225	3/16	...	...	...	4-6	7-9
	220-240	5/32	13-14	12-13	...	...	...
	220-240	3/16	10-13	8-12	...	...	...
7/32	250-275	3/16	12-13	10-11	...	...	...
1/4	320-350	1/4	8-9	8-9	...	...	...
	320-350	5/16	6-7	6-7	...	...	...
E7024 and E6027 electrodes							
1/8	160-170	5/32	15-16	14-15	...	...	...
5/32	215-225	3/16	15-16	14-15	...	...	...
3/16	265-275	1/4	12-13	11-12	...	...	...
7/32	330-360	1/4	14-15	13-14	...	...	...
1/4	370-400	5/16	11-12	10-11	...	...	...

E6022 electrodes are designed for single-pass welding of sheet metal. Electrode sizes suited for specific applications are:

- *For the first pass in an open root joint design in pipe or plate*: 1/8- or 5/32-in.-diam electrode; for the remainder of the weld, 5/32- or 3/16-in.-diam electrode can be used in all welding positions, or when welding in the flat position, an electrode 3/16 in. in diameter or larger.
- *For flat-position welding of double-groove weld joints (with back gouging), or single-groove weld joints with backing strip*: 3/16-in.-diam electrode for the root pass, with 3/16-in.-diam or larger electrode for the remainder.
- *For fillet welds in the flat position*: Electrode size should not exceed the size of the fillet weld. Electrodes 3/16, 7/32, or 1/4 in. in diameter are satisfactory. In heavier weldments, larger electrode sizes may be used where the mass can tolerate the additional heat.
- *For out-of-position welding*: 3/16-in.-diam electrode and under, except for classifications E7014, E7015, E7016, and E7018, which should not have an electrode exceeding 5/32 in. in diameter.

Electrode Deposition Rates and Properties of Welds

Electrode deposition rate is the weight of weld metal deposited in a unit of arc time and is generally expressed as pounds per hour. Table 6 shows deposition rates obtained with covered carbon steel electrodes of various sizes and classifications. Table 6 also shows deposition efficiencies and mechanical properties of weld metal deposited under these conditions. Deposition efficiency is the ratio, expressed as a percentage of deposited metal to electrode net weight (excluding stubs), used in an operation. These data, under normal conditions and within the filler-metal specification tolerances allowed (AWS A5.1), can be expected to vary by as much as ±7%. For any specific class of carbon steel covered electrode, deposition rate is primarily a function of electrode size and welding current (melting rate). Manufacturers of welding electrodes provide recommended current ranges for each welding position in which their product may be used (Table 7). Figure 12 depicts the influence of welding current on deposition rate and deposition efficiency for a 1/4-in.-diam type E6011 electrode.

Welding Speed

Welding speed is the rate of electrode travel in relation to the workpiece. The optimum travel speed is the highest speed

Fig. 12 Effect of current on deposition rate, deposition efficiency, spatter loss, and mechanical properties of weld metal

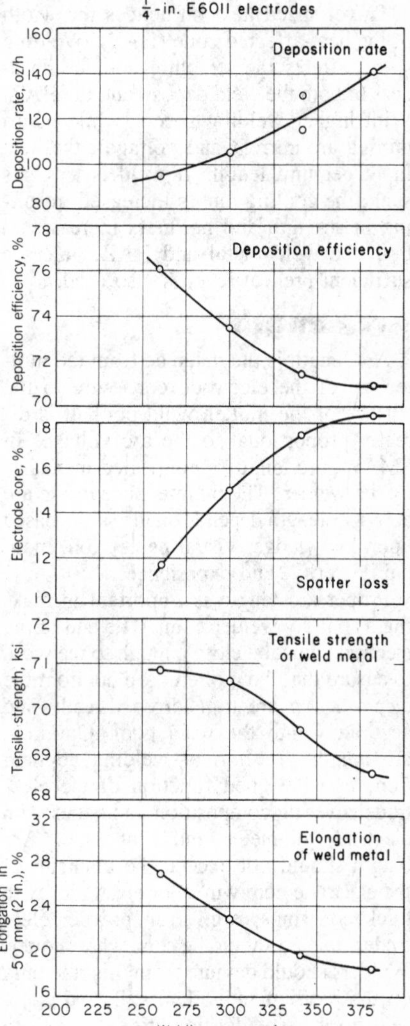

at which a weld bead of proper contour, fusion characteristics, and appearance is produced. Welding travel speed depends on two primary factors: electrode melting rate (magnitude of welding current) and the weld-bead size. Several additional factors influence these primary factors directly: (1) type of welding current and polarity, (2) position of welding, (3) weld joint design, (4) weld joint fit-up, (5) thickness of material, (6) surface condition of the base metal, and (7) electrode manipulation.

Welding travel speed must be high enough to permit the arc to lead the molten weld pool slightly. Up to a point, increasing welding speed (while maintaining arc current and voltage) reduces the width of the weld bead and increases penetration. Beyond this, increasing the travel speed results in decreased penetration, poor appearance, undercutting, poor slag removal, and entrapped gases, which create weld-metal porosity. With low travel speed, the weld bead tends to be wide and convex in profile with shallow penetration.

Travel speed also influences the width of the heat-affected zone (HAZ). Welding speed affects the structure and the hardness of both the weld deposit and the HAZ. With higher welding speed (while maintaining arc current and voltage), the heat input per unit length of weld is less. As such, the cooling rate is increased, resulting in strength and hardness increases in both the weld metal and HAZ, unless a sufficient preheat level is also used.

Arc Length

Arc length is the distance from the molten tip of the electrode core wire to the surface of the molten weld pool. It is directly proportional to the arc voltage. In SMAW, arc length is controlled manually by the welder. The maintenance of the arc at proper length depends on his skill—based upon knowledge, visual perception, manual dexterity, and experience.

Proper arc length is important in making a soundly welded joint. The end of the electrode must be close enough to the work to ensure that the molten droplets from the electrode tip are transferred directly and accurately into the weld pool. Optimum arc length, for normal welding application, is a designed function of the electrode covering composition and varies with electrode diameter and amperage. Arc length should not exceed the diameter of the electrode core wire. For electrodes with thick coverings, e.g., iron powder electrodes, the arc should be somewhat shorter. Welders should deviate from this rule only on the basis of individual skill and experience for particular conditions.

Arc length largely controls arc voltage and directly affects welding speed and efficiency. Shorter arcs cause an increase in current, which increases the rate of deposition and welding speed. When an arc is too long, heat is dissipated into the air; the stream of molten metal from the electrode to the work is scattered in the form of weld spatter, and deposition rate is reduced. In addition, susceptibility to arc blow and porosity from loss of shielding increase as length of arc increases. In welding with direct current, the shortest possible arc is used to minimize arc blow and contamination by the air. Control of arc length in vertical and overhead welding demands greater attention from the welder and more skill than in flat-position welding. In overhead welding, only certain types of electrodes can be used, and the welder must adjust the arc length during deposition to retain control of the weld pool. For fillet welds and for root passes in properly prepared joints, the arc can easily be crowded into the joint for maximum speed and penetration.

Example 2. Control of Arc Length to Reduce Rejection for Porosity. Porosity in welded pipe joints was the cause of rejections that averaged 14.8% over a 19-week period with a high of 40% (Fig. 13a). Investigation showed that the porosity resulted from failure to maintain a short arc length in welding. When arc length was brought under control, average rejection rate over an 18-week period dropped to 3.2%, with no rejections in 12 of the 18 weeks (Fig. 13b).

Striking, Maintaining, and Breaking the Welding Arc

The welding arc in SMAW is initiated by tapping or scratching the work with the tip of the electrode and then quickly withdrawing it a short distance to produce an arc of the proper length. When the electrode tip touches the work, the two tend to freeze or stick together. The technique of striking the arc by tapping or scratching requires some development of skill.

The manner of restriking the welding arc (i.e., after once initiating and subsequently breaking the welding arc) varies with electrode type. Generally, the projecting covering on the electrode tip remains conductive while still hot and assists in restriking the arc. This is particularly true for electrodes containing iron powder in their coverings, which may also be conductive when cold. However, when using heavily coated electrodes, such as type E6020, or low-hydrogen electrodes, the projecting cup of the electrode covering may have to be broken off to expose the core wire to effect restriking.

The arc length is maintained by providing uniform movement of the electrode toward the weld pool to compensate for the portion of the electrode that has been melted and deposited in the weld pool. The welder must coordinate this movement with progressive advancement in the direction of welding.

Several techniques may be used to break the arc at the completion of a weld. In one technique, the electrode may be drawn to the end of the joint, quickening the travel speed to make a somewhat lighter deposit, and then reversing the travel direction to complete the weld size over this portion before withdrawing the electrode. In another technique, the electrode is held essentially stationary at the end of the weld long enough to fill the weld crater before withdrawing the electrode.

To break the welding arc at some point in the weld length (i.e., allow for electrode change) where welding is to be continued from the crater, the arc is shortened and then broken by quickly moving the

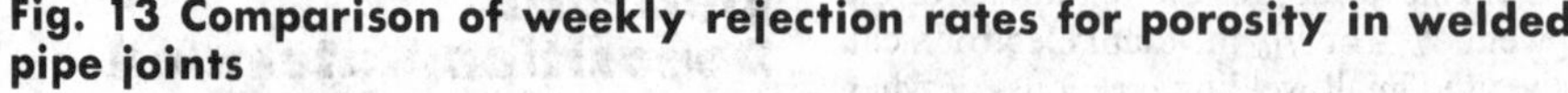

Fig. 13 Comparison of weekly rejection rates for porosity in welded pipe joints

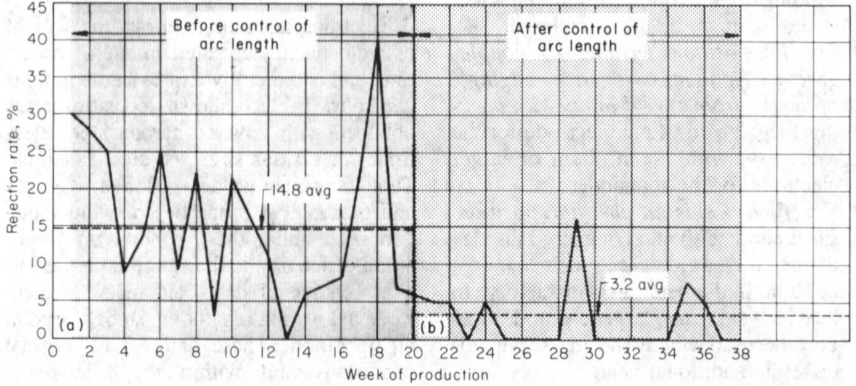

electrode sideways out of the crater. When the arc is re-established in the crater, it should be struck at the forward or cold end of the crater, moved backward over the crater until the weld is reached, and then forward to continue the weld. If the slag has cooled below a red heat, it should be removed before restriking the arc. With this procedure, the crater is filled, and porosity and trapping of slag are avoided. This procedure should always be used in welding with low-hydrogen electrodes.

Electrode Orientation and Manipulation

The angular position of the electrode in relation to the weld joint and the work (electrode orientation) is important to the quality of a weld. Correct positioning of the electrode influences fusion characteristics, weld-bead contour, and weld quality. Slag entrapment, porosity, and undercutting can result from improper positioning. Electrode orientation is specified by the travel angle and the work angle. Travel angle is the angle that the electrode makes with a reference line perpendicular to the axis of the weld in the plane of the weld axis (Fig. 14). When the electrode is pointing forward in the direction of travel, the travel angle is referred to as the push angle, and the welding technique is referred to as forehand welding. When the electrode is pointing backward, the travel angle is referred to as the drag angle, and the welding technique is referred to as backhand welding.

Electrode manipulation techniques for specific welding positions and conditions (e.g., poor fit-up) vary considerably among welders. The mastering of these techniques requires the development of considerable skill. Regardless of the techniques used, the objective is a quality weld. Two fundamental rules that apply to all techniques are: radical motions of the electrode should not be used, and the arc must not be broken if the electrode is manipulated from one side of the weld pool to the other. Manipulation should be such that the slag does not run ahead of the weld pool and become entrapped in the weld, nor should it be allowed to run off the weld metal during out-of-position welding. The manipulation technique may need modification when the electrode type is changed.

Fig. 14 Electrode orientation. Source: AWS standard A3.0

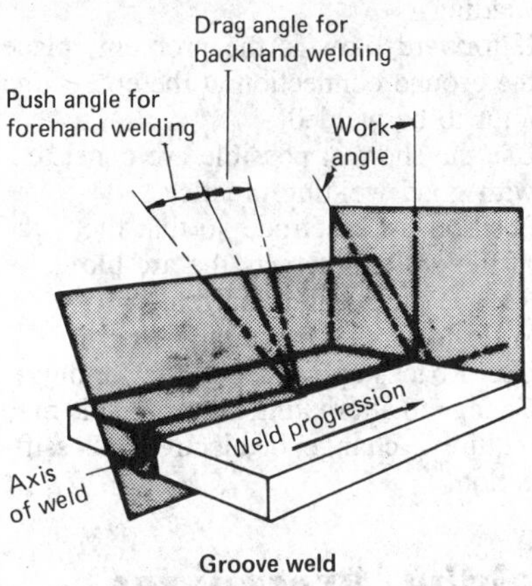

Groove weld

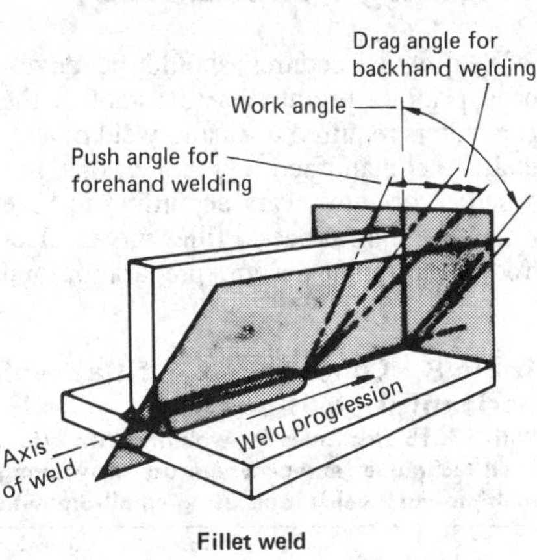

Fillet weld

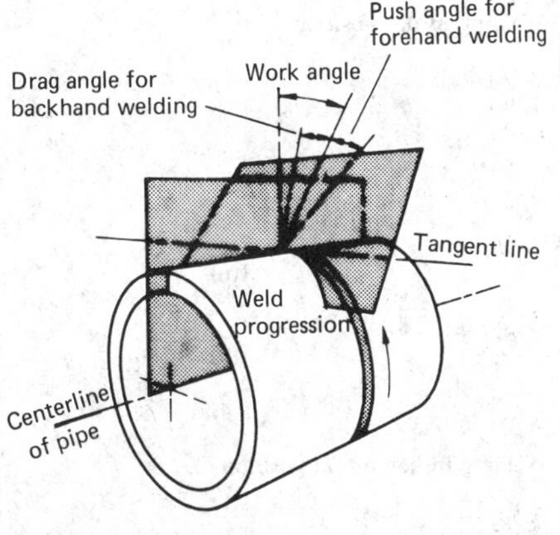

Pipe weld

Accessibility

The joint to be welded should have clearance between the parts to be joined and the electrode. Figure 15 shows minimum dimensions for clearance that avoid interference and provide visibility to the bottom or end of the joint. Where accessibility is limited, the electrode can be bent slightly to reach the weld joint. When this is done, the electrode covering must remain intact to provide protection for the arc and the weld pool. Not all electrodes bend and still retain their covering intact. Correct moisture content in the covering is extremely important for satisfactory bending; if a covering is too dry, it can crack and spall when bent. Heating the electrode by shorting it out may help in bending.

Arc Blow

The electric current flowing through the electrode, workpiece, and ground cable sets up a circular magnetic field around the current path. When the fields around the workpiece or around the electrode are unbalanced, the arc bends away from the greater concentration of the magnetic field. This arc deflection from the intended path is known as arc blow. Arc blow, when it occurs, is usually encountered when using direct current, because the induced magnetic fields are constant in direction. It occurs only to a minor degree with alternating current, because the induced magnetic field that builds up collapses as soon as the current reverses. Figure 16(a) shows magnetic fields around the electrode that result in forward blow (deflection in the direction of electrode travel) and backward blow (deflection opposite the direction of electrode travel). Deflection may occur to one side when the concentration of magnetic field is greater on one side of the arc than on the other, but usually deflection is in the direction of electrode travel or opposite to it.

Backward blow is encountered when welding toward the ground connection, near the end of the joint, or into a corner. Forward blow is encountered when welding away from the ground connection or at the start of a joint. The conditions may become so severe that a satisfactory weld cannot be made because of incomplete fusion and excessive weld spatter. This is especially true for electrode coverings containing large amounts of iron powder or other coverings which produce a large amount of slag. The difficulty arises be-

Fig. 15 Electrode clearances. Used to avoid interference and provide welder visibility

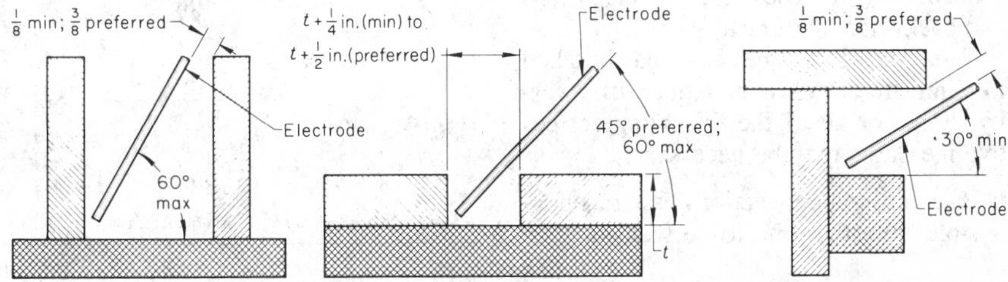

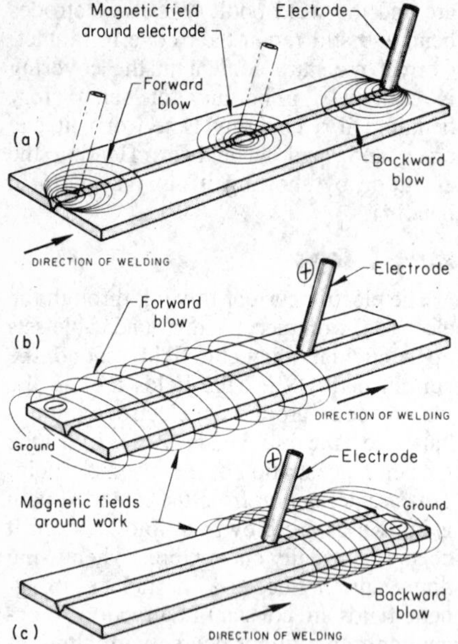

Fig. 16 Magnetic fields that result in forward or backward arc blow. (a) Around the electrode. (b) Around the work. (c) Around the work

cause the forward fanning-out of the arc permits the heavy slag deposit in the crater to run under and ahead of the arc.

The forward blow at the start of a weld deposit is only momentary, because the magnetic field soon finds an easy path through the weld metal being deposited behind the arc. Because the field that follows is in the workpiece and the weld, a slight backward blow is created for the remainder of the weld. At the end of a weld, the magnetic field ahead of the electrode becomes crowded and may cause a problem. As the end approaches, the backward blow increases and may become severe.

The welding current passing through the work also causes the work to act as a conductor surrounded by a magnetic field. The circles of the magnetic field around the current path are in planes perpendicular to the workpiece and are present from the arc to the point at which the workpiece is grounded. This condition is most likely to occur in narrow plates. In welding away from the ground connection, the magnetic field perpendicular to the workpiece is behind the electrode (Fig. 16b) and forward blow results. In welding toward the ground connection (Fig. 16c), the reverse is true and backward blow results.

Certain corrective measures may be taken to eliminate arc blow or reduce its severity. Some or all of the following corrective measures may be necessary:

- Place ground connections as far as possible from the joint to be welded.
- If backward blow is the problem, place the ground connection at the start of welding.
- If forward blow is the problem, place the ground connection at the end of the joint to be welded.
- Use the shortest possible arc consistent with good welding practice.
- Position the electrode so that the force of the arc counteracts the arc blow.
- Reduce the welding current.
- Weld toward a heavy tack weld.
- Use a backstep sequence for welding.
- Change to alternating current. This may require a change in electrode classification.

Welding Procedures

Welding procedures should be developed prior to production to establish the parameters required to ensure weld quality under the manufacturing conditions imposed. Procedures vary according to base metal, welding process, filler metal, electrode class, joint design, preparation and position, thickness of sections being welded, and variations of thickness among sections being welded. Additional information dealing with joint design, edge preparation, weld location, and fit-up can be found in the article "Joint Design and Preparation" in this Volume.

Fillet Welds

Fillet welds as large as 1 in. require six to eight passes and usually have poor appearance. Where strength greater than a 3/8-in. fillet weld can provide is required, edge preparation such as a 1/4-in. bevel in the joint should be considered before enlarging the fillet size. Welding conditions for various sizes of fillet welds are given in Table 8.

Single-Pass Fillet Welds. The maximum size of a sound fillet weld that can be made in a single pass depends on the position of welding, electrode size, electrode type, amperage, and work-metal thickness. Attempts to make fillet welds that are too large for one pass can result in undercutting, unequal legs, lack of penetration, and slag entrapment. The largest

Table 8 Conditions for fillet welding in flat and horizontal positions

With E7018 electrodes for welding with alternating current or DCEP, using the stringer-bead technique (except where use of weaving is footnoted to pass number), chipping for multiple-pass welds, and using small-diameter electrodes to tie in ends and corners of joints

Fillet size, in.	Pass No.	Electrode size, in.	Current, A: Range	Current, A: Optimum	Melting rate, in./min
Welding in flat position					
1/8-5/32	1	1/8	120-160	140	...
1/8-3/16	1	5/32	150-220	170	...
5/32-3/16	1	3/16	240-330	260	11.60
3/16-5/16	1	7/32	280-350	300	...
1/4-3/8	1	1/4	320-400	350	9.54
3/8	1(a)	5/16	420-540	460	...
7/16-5/8	1	7/32	280-350	300	...
	1(b)	1/4	320-400	350	9.54
	2(a)	5/16	420-540	460	...
11/16-3/4	1	7/32	280-350	300	...
	1(b)	1/4	320-400	350	9.54
	2	5/16	420-540	460	...
	3,4(a)	5/16	420-540	460	...
Welding in horizontal position					
1/8-5/32	1	1/8	120-160	140	...
1/8-3/16	1	5/32	150-220	170	...
5/32-1/4	1	3/16	240-300	260	11.60
3/16-5/16	1	7/32	280-350	300	...
1/4-5/16	1	1/4	320-400	350	9.54
3/8-7/16	1	7/32	280-350	300	...
	1(b)	1/4	320-400	350	9.54
	2	1/4	320-400	350	9.54
1/2-5/8	1	7/32	280-350	300	9.54
	1(b)	1/4	320-400	350	9.54
	2,3	1/4	320-400	350	9.54
11/16-3/4	1	7/32	280-350	300	...
	1(b)	1/4	320-400	350	9.54
	2-6	1/4	320-400	350	9.54

(a) Using weaving technique. (b) Alternative first pass

Table 9 Classes and sizes of electrodes for single-pass horizontal fillet welds

Electrode class	Electrode size for fillet size, in. 1/8	3/16	1/4	5/16	3/8(a)
E6012, E6013	5/32	3/16	1/4	1/4	1/4
E7014	1/8	5/32	7/32	1/4	1/4
E7018	1/8	5/32	3/16, 7/32	1/4	1/4
E7024	1/8	5/32	3/16, 7/32	7/32	...

(a) Usually made in two passes; great operator skill is required for making a 3/8-in. fillet weld in one pass.

size of a single-pass fillet weld is 3/8 in. in the flat position, 5/16 in. in the horizontal and overhead positions, and 1/2 in. in the vertical position. Sizes of various classes of electrodes for single-pass horizontal fillet welds are given in Table 9.

Multiple-Pass Fillet Welds. For larger fillets that require multiple-pass welding, horizontal fillets usually require more passes than flat fillets.

Effect of Work-Metal Thickness. For a given thickness of work metal, a fillet weld should be no larger than is required for adequate strength. Needlessly large welds add to cost by increasing deposition time and wasting metal. Minimum sizes of fillet welds for adequate strength, as well as for avoiding cracking caused by joint stresses during cooling, are:

Thickness of thicker member, in.	Minimum size of fillet weld, in.
Up to 1/4	1/8
1/4-1/2	3/16
1/2-3/4	1/4
3/4-1 1/2	5/16
1 1/2-2 1/4	3/8
2 1/4-6	1/2
Over 6	5/8

Intermittent Welds. An intermittent fillet weld, one of broken continuity, is often used when the strength of a continuous fillet weld of minimum size exceeds requirements (unless a continuous weld is required for appearance or because the welded joint must be leak resistant or weather resistant). For adequate strength, the minimum length of an intermittent weld should be at least four times the size of the fillet but not less than 1 1/2 in.

The maximum center-to-center spacing of intermittent welds should not exceed 16 times the thickness of the thinner member of joints that are to be loaded in compression, or 32 times the thickness of the thinner member of joints to be subjected to other types of loading. The length of the unwelded spaces between continuous sections of weld should not, however, exceed 12 in.

Groove Welds

For making a full-penetration square-groove weld, the electrode must be of a size that ensures fusion of the weld root and gives proper arc length. The amount of penetration obtained varies with the welding technique used. The welder should maintain a short arc and the correct electrode angle. This automatically forces the weld pool away from under the electrode tip, especially when travel speed is high. Even a very shallow weld pool under the arc forms an insulating layer and decreases the penetrating power of the arc. When the electrode is raised to keep the tip out of the molten metal, the resulting longer arc dissipates more of its heat into the air and, in flaring out, widens the weld bead. With heavily covered electrodes, the correct arc length is obtained when the electrode is dragged along the joint with the covering touching the workpieces. Smooth square-groove welds are readily produced on thicknesses of 0.060 to 3/16 in., but for thicknesses of 3/16 to 1/4 in. weld-bead surface roughness is often a problem.

The service requirements of a weldment often influence joint design and welding procedure for groove welds. For example, the weld type (joint design), electrode class and size, number of passes, and the direction of welding for vertical seams in storage tanks usually depend on whether the tanks will be subjected to extremely low temperatures or to normal temperatures in service.

Example 3. Use of Different Steels, Joint Grooves, and Welding Procedures. Vertical joints in tanks intended for storage of anhydrous ammonia at −28 °F were designed and welded differently from those in tanks designed for service at normal temperatures (ranging from outdoor ambient to that of hot asphalt). Figure 17 compares the two joint designs and shows the number and sequence of passes

Fig. 17 Groove designs and buildup sequences for vertical butt welds. Storage tanks for normal and low-temperature (−28 °F) service

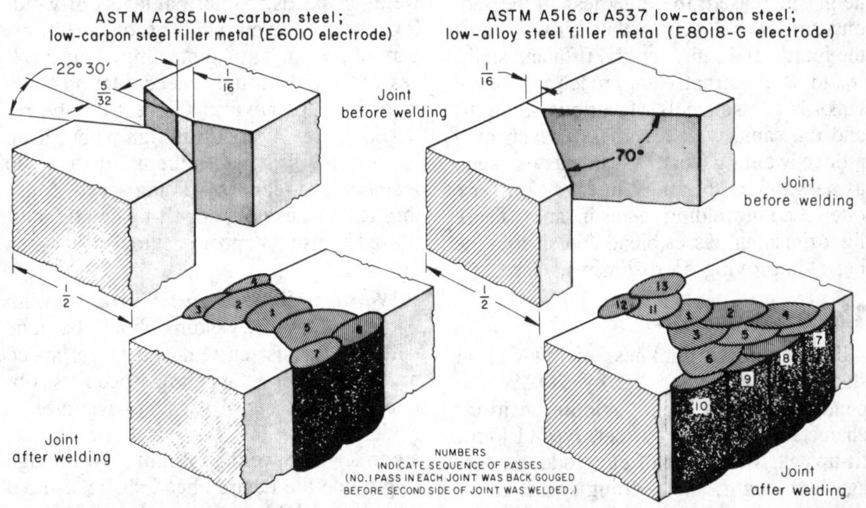

(a) Joint in tank for normal service (double-V-groove) (b) Joint in tank for low-temperature service (single-V-groove)

Edge preparation .... Gas cut and descale
Preheat .... None for workpieces above 32 °F(a)
Electrode class and size:
Double-V-groove weld .. E6010, 5/32 and 3/16 in.
Single-V-groove weld .... E8018-G, 1/8 in.
Welding positions:
5/32-in. E6010 .... Vertical-up
3/16-in. E6010 and 1/8-in. E8018-G .... Vertical-down
Voltage and current (DCEP)(b):
5/32-in. E6010 .... 26-29 V, 120-130 A
3/16-in. E6010 .... 26-29 V, 150-170 A
1/8-in. E8018-G .... 20-22 V, 125-135 A
Deposition rate, lb/h:
5/32-in. E6010 .... 2.54
3/16-in. E6010 .... 3.66
1/8-in. E8018-G 2.52

(a) Workpieces below 32 °F were heated with a gas burner until they were warm to the hand. (b) Supplied by a motor-generator or a transformer-rectifier

made in welding each. E6010 electrodes were used because of their low cost. As shown in Fig. 17(a), the joint was welded in seven passes. Beads in passes 1, 2, and 5 were made with $^{5}/_{32}$-in.-diam electrodes in the vertical-up position. Beads in passes 3, 4, 6, and 7 (wash passes) were made with $^{3}/_{16}$-in.-diam electrodes in the vertical-down position. Pass 1 was back gouged (air carbon arc) before welding the second side (pass 5). Back gouging was done with a $^{5}/_{16}$-in.-diam carbon electrode at 300 to 350 A.

For welding tanks for low-temperature service, a single-V-groove joint (Fig. 17b) was selected to provide good access and room for electrode manipulation for the $^{1}/_{2}$-in. plate thickness. The joint was welded in 13 passes (10 on one side and 3 on the other); pass 1 was back gouged before pass 11 was made. Beads in all passes were made with $^{1}/_{8}$-in.-diam E8018-G low-alloy steel electrodes (iron-powder, low-hydrogen covering) in the vertical-down position to minimize heat input and obtain the greatest tempering effect on previous beads.

For many large storage tanks, plates are not all of the same thickness. For instance, the lowest ring of a tank that was 120 ft in diameter and 48 ft high had a plate thickness of 0.83 in. The plate thickness of the other rings decreased as tank height increased; the thickness of the second ring was $^{11}/_{16}$ in.; the third, $^{9}/_{16}$ in.; the fourth, 0.41 in.; and the fifth and sixth, $^{5}/_{16}$ in. An alternative approach in storage tanks is to use steels of various strengths and the same wall thickness, which may reduce welding cost. Wash passes, such as passes 3, 4, 6, and 7 in Fig. 17(a), are often used in welding seams in tanks. These light finishing passes blend and smooth the bead, improving corrosion resistance.

Thin Sections

The minimum thickness of low-carbon steel that can be welded by SMAW depends on welder skill, welding position, characteristics of the current, type of joint, fit-up, class and size of electrode, amperage, arc length, and welding speed. Low-carbon steel sheet as thin as 0.036 in. has been welded successfully with flux-covered electrodes. Welding sheet this thin, however, requires exceptional skill and the use of electrodes smaller than $^{3}/_{32}$ in. in diameter. Fit-up and position must also be favorable. Generally, the minimum practical sheet thickness for an average welding unit is 0.060 in. A procedure for welding steel sheet 0.060 in. thick is:

- *Electrode*: Use $^{3}/_{32}$-in.-diam E6012 or E6013.
- *Current*: Direct current electrode negative, 50 to 60 A at start; vary amperage as required. E6013 electrodes require slightly higher current than E6012 electrodes.
- *Welding position*: Keep longitudinal axis of joint within 45 to 60° from horizontal. Increasing the slope of the joint decreases both the penetration and the size of the weld; decreasing the slope does the opposite.
- *Technique*: Weld from top to bottom, using a light drag technique with sufficient speed of travel to lead the weld pool and slag. Electrode lead angle should be 10 to 30°.

Electrodes should be $^{3}/_{32}$-in.- or $^{1}/_{16}$-in.-diam E6012 or E6013, preferably DCEN. These electrodes can be used in all positions. If welds are vertical, however, the weld metal is deposited from top to bottom (vertical-down). Of the two electrodes, E6012 produces a smaller bead with moderate penetration. E6013 electrodes, although depositing a larger bead, provide a very soft, low-penetrating arc and are preferred for welding thin sheet. When only alternating current is available, E6013 electrodes are preferred for easier starting and arc maintenance.

Fit-Up. Tight-fitting joints are essential for producing good welds at maximum welding speeds. At normal levels of welding current, even slight root openings, particularly in butt joints, may cause excessive melt-through, necessitating considerable repair work. Current can be reduced to prevent melt-through when fit-up is poor, but this makes the arc difficult to maintain and reduces welding speed. Melt-through because of poor fit-up can be minimized by using copper backing strips when practical.

Welding Speed and Current. With tight-fitting joints, welding should be done with the highest current and travel speed practical. Proper welding speed is obtained for any set of variables when electrode travel is fast enough to permit the arc to lead the weld pool, but slow enough to provide the desired bead shape without undercut. Welding speeds for 0.060-in.-thick steel sheet are so high when compared with speeds for thicker sections that welders inexperienced with thin sheet have difficulty keeping the arc on the joint. To concentrate the arc on the joint and minimize heat input, welding should be done with a slight drag technique. The tip of the flux covering should lightly touch the work as the electrode is advanced. Positioning the work with the axis of the joint to be welded 45 to 60° from horizontal enables maximum efficiency with good control over the weld deposit, permits the use of maximum welding current and travel speed, and minimizes melt-through.

Fig. 18 Sequence of buildups in welding thick sections with a double-U-groove. Uses root gouge between the first and second buildups

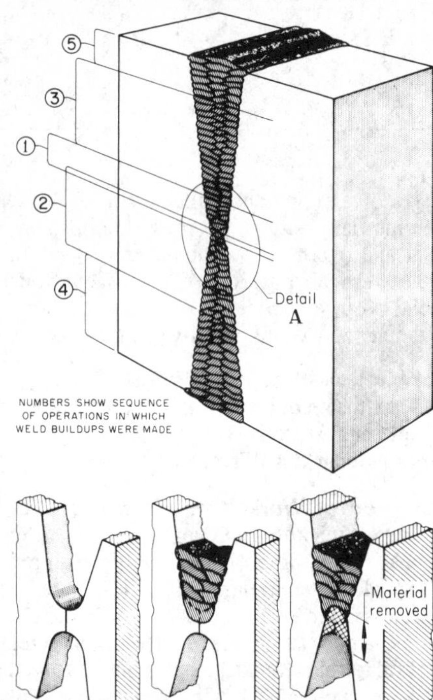

Tack welds should be as small as possible, because weld beads pile up when deposited over them. When uniform welds are required, tack welds should be ground before welding or deposited on the opposite side of the joint from which welding is to be done. Tack welds should be close enough to one another to prevent sheets from separating ahead of the arc. Separation of sheets is most prevalent in lap joints.

Thick Sections

All commercial thicknesses of low-carbon steel plate can be welded successfully by SMAW, provided that the root of the joint can be reached with the electrode. Beveling is used on thicknesses above $^{3}/_{16}$ or $^{1}/_{4}$ in. A single groove is used on thicknesses up to about $^{3}/_{4}$ in.; a double groove is needed on thicker sections.

Figure 18 depicts a sequence for welding thick sections with a double-U-groove using the root-gouging technique. In root gouging, the first buildup is made with a series of passes. The root of the weld is then gouged, usually by oxyfuel gas cutting or air carbon arc cutting. The major difficulties that can occur in welding of thick sections are slag entrapment, porosity, transverse shrinkage, angular distortion, and cracking.

Transverse shrinkage is affected by restraint, volume of deposited weld metal, joint design, and welding procedure. Transverse shrinkage, which can be $^1/_4$ in. or more in full-section welds in plate 6 in. and upward in thickness, cannot be prevented, but it can be controlled or minimized by using one or more of these design or welding practices:

- Design groove joints for minimum volume of weld metal.
- Weld temporary strongbacks across the joints before welding.
- Use a block sequence in welding.
- Make allowances in design to compensate for estimated shrinkage.
- Peen all but the root and cover passes, using a blunt-nose tool. Peening should flatten the surface of the weld bead, but not cut into it. To retain notch toughness, do not peen at 500 to 900 °F.

Fig. 19 Thick tube sheet with machined groove. Minimizes heat sink differential during welding of thin-walled heat-exchanger tube to the tube sheet

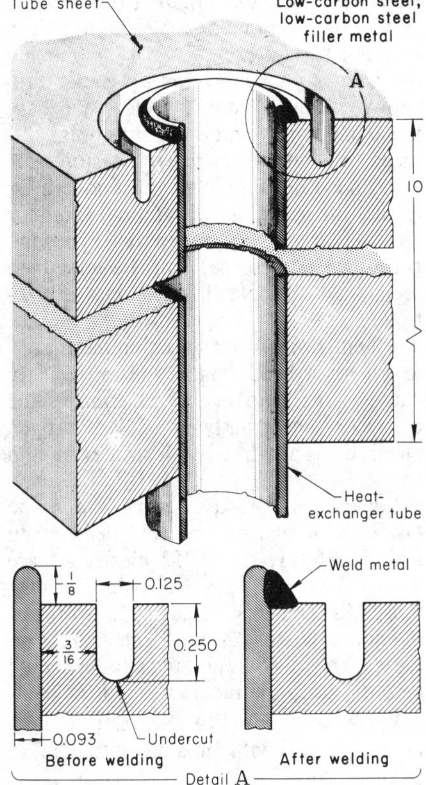

Fig. 20 Use of a copper backing block as a chill. Minimizes heat sink differential

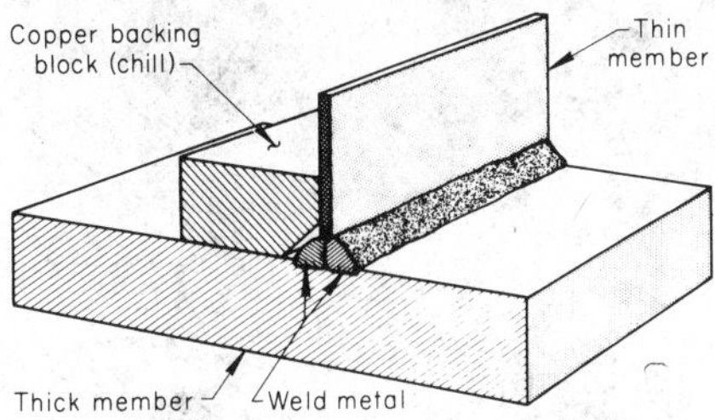

Sections of Unequal Thickness

In SMAW, as in other welding processes, special techniques are required when the components to be welded have a large heat sink differential (difference in heat-dissipating capacities). When a thick member is joined to a thin member, the welding current needed to obtain good penetration into the thick member is sometimes too much for the thin member and results in undercutting of the thin member and a poor weld. If the proper amount of current for the thin member is used, the heat is not sufficient to provide adequate fusion in the thick member, and again a poor weld results.

An application involving components of unequal section thicknesses is the welding of heat-exchanger tubes having 0.093-in. wall thickness to a tube sheet as thick as 10 in. The usual method of avoiding difficulty is to cut a $^1/_4$-in.-deep circular groove in the upper surface of the tube sheet around the opening for the tube (Fig. 19). By restricting heat transfer, this groove minimizes the heat sink differential between the thin tube wall and thick tube sheet.

A more widely applicable method of minimizing heat sink differential is to place a copper backing block against the thin member during welding (Fig. 20). The block serves as a chill, or heat sink, for the thin member. The block can be beveled along one edge so that it can be used when horizontal fillet welds are deposited on both sides of a thin member. Copper backing bars or strips are made in a variety of shapes and sizes to dissipate heat as needed. Often some experimentation is required to obtain the optimum backing location and design. Another way to obtain equalized heating and obtain smooth transfer of stress where unequal section thicknesses are being welded is to taper one or both members to obtain a common width or thickness at the joint. If thicknesses are not greatly different, directing the arc toward the thicker member may produce correct penetration.

Distortion

Welding of any structure or member results in nonuniform heating and cooling at the point of welding. Welding is always accompanied by some distortion, and some unbalanced residual stress remains. In practice, the amount of distortion from welding depends on several factors. In some weldments, distortion may be so slight that it may be disregarded, while in others it may be so great that special precautions must be taken before and during welding, or straightening must be done after welding. In many welded structures, residual stresses have little influence on the service life of the structure. On the other hand, in structures subjected to dynamic loading or those that must retain their shape during service, residual stresses are important.

Use of the following practices may assist in control of detrimental stresses and distortion which could impair the usability of a weldment:

- The forming and preparation of steel sheet and plate induce stresses within the material. These may be released by the heat of welding and cause substantial distortion. Stress relieving of the components before welding is sometimes required.
- On multiple-pass welds, angular distortion across the welded joint increases with the number of passes and joint volume.
- A more uniform distribution of heat in a longitudinal seam can be obtained by welding with a backstep sequence or a wandering (skip) sequence. This practice also gives greater rigidity to the seam and results in less distortion.
- Clamping does not entirely eliminate warping, but is more likely to be effective if the weld is permitted to remain in the clamps until cooled.
- Peening the weld metal while it is hot has been used to reduce stresses and minimize distortion or warping. Peening should be avoided in the temperature range of 500 to 900 °F, because peening in this range may result in loss of notch toughness.
- Stresses may be substantially reduced if welding direction is away from the point of greatest restraint.
- Changing the speed of welding may either increase or decrease distortion of the weldment. Experimentation may be needed to determine the optimum welding speed for specific joints where freedom from distortion is critical.

- Residual stresses may be minimized by postweld heat treatment of the completed weldment. Proper support of the weldment is required to prevent warpage and distortion from occurring during stress relief.
- Welding of parallel joints in opposite directions can sometimes minimize distortion.
- Stresses can be balanced by using a double groove or by alternating the welding from one side of the joint to the other.

Distortion in Thin Sections. Distortion or buckling is more often encountered in the welding of thin sheet. To minimize distortion, beads should be small and should be deposited as rapidly as possible. Intermittent welds, clamping fixtures, and copper backing bars minimize distortion.

Distortion in Thick Sections. Welding of thick sections is accompanied by angular distortion, which is caused by unequal shrinkage across the welded joint. Unequal shrinkage occurs because the welds are deposited in successive layers and because, in most joint designs, the width of the layers increases as the joint is filled.

Eliminating angular distortion in the welding of thick sections is virtually impossible when all welding is from one side. Temporary strongbacks, built-in ribbing, block sequence welding, preheating, and peening help reduce angular distortion.

When individual plates a few inches thick are to be butt welded, presetting the plates opposite to the anticipated direction of angular distortion can offset most of the distortion; during welding, the plates move into position as a result of weld shrinkage. However, presetting is seldom possible when the thick sections to be welded are integral parts of structures. Also, presetting is impractical when sections with thicknesses of 6 in. or more are welded from one side, because angular movement resulting from contraction of the weld metal tears the joint apart at the weld root.

Whenever possible, heavy sections should be welded from both sides. Joints should be designed so that both sides require the same volume of weld metal, and a balanced welding technique should be used. Balancing the welding on both sides of the joint results in good control over angular distortion, but when welding is done in the flat position, the workpiece must be turned over frequently, unless restraining devices are used. Joint design is important for control of distortion. Groove angles should be the minimum size that allows root penetration and accessibility for remaining passes.

Fig. 21 Weld defects

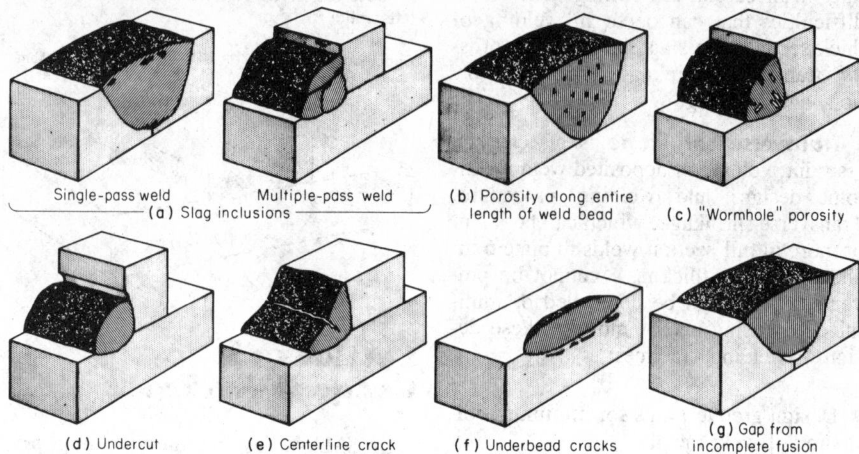

Sequencing. Establishing the order of deposit in which the several passes in a weldment should be made is called sequencing. It may be extremely important to the final usability of the weldment that the welds be deposited in a precise order (buildup sequence). For pilot models of critical weldments, the sequence may require change several times to obtain the optimum order of passes.

Causes and Prevention of Weld Defects

In SMAW of steel, slag inclusions and porosity are common weld defects. Other defects that often occur are undercuts, longitudinal (centerline) cracks, underbead cracks, and gaps resulting from incomplete fusion (Fig. 21).

Slag inclusions (Fig. 21a) may be the result of (1) incomplete deslagging of a previous pass; (2) wide weaving, which permits slag to solidify at the sides of the bead; (3) erratic progression of travel; (4) excessive amounts of slag ahead of the arc, particularly in deep grooves; and (5) use of electrodes that are too large. Preventive measures are: (1) deslag deposits thoroughly by chipping, wire brushing, or grinding before a subsequent weld bead is deposited; (2) restrict the width of weaving so that the entire width of the slag immediately behind the weld pool remains molten; (3) use a uniform travel speed; (4) keep slag behind the arc by shortening the arc, increasing the electrode angle, or increasing travel speed; and (5) use a smaller electrode.

Wagon Tracks. Linear slag inclusions along the axis of the weld are sometimes called wagon tracks. Wagon tracks ordinarily result from failing to remove slag remaining from previous passes, or allowing slag to run ahead of the weld pool. Wagon tracks above the root pass have the same effect on weld quality as other inclusions in the weld metal or base metal.

Porosity that is scattered along the entire length of a weld bead (Fig. 21b) can be caused by (1) impurities, such as sulfur or phosphorus, in the base metal; (2) contamination of the surface of the base metal by rust, grease, moisture, or dirt; (3) hydrogen from excessive moisture in electrode coverings; (4) nitrogen entrapment from improper arc length; (5) excessive current; (6) welding at a speed too high to permit gases to escape; and (7) freezing the weld pool before gases escape. Preventive measures are: (1) clean the base metal and remove moisture from joint surfaces; (2) redry electrodes to restore recommended moisture content of coverings; (3) use the proper length of arc; (4) reduce welding current; (5) reduce travel speed to permit gases to escape; (6) preheat the base metal; and (7) use a different class of electrode.

When porosity occurs within the first $1/4$ to $1/2$ in. of bead length, the most likely cause is moisture in the covering of the electrode, particularly if a low-hydrogen electrode is used. Dry electrodes should be used.

Sometimes porosity occurs within the last inch or two of the weld bead when E6010, E6011, or E6012 electrodes are used. The usual cause is excessive current, which results in electrode overheating and excessive drying of the flux covering. When this occurs, the current should be reduced to the recommended range.

Wormhole porosity (Fig. 21c) is usually associated with moisture entrapped in the joint. When visible at the surface, wormhole porosity is sometimes referred to as gas shoots. Sulfur in the base metal is also a contributing factor. One means of preventing wormhole porosity is to reduce travel speed to permit gases to escape before the metal freezes.

Undercuts (Fig. 21d) are usually caused by excessive welding current, arc length, and weaving speed. In horizontal or vertical welding, additional causes are excessive electrode size and incorrect electrode angles. Travel speed should allow the deposited weld metal to fill all melted-out portions of the base metal completely. When the weaving technique is used, the welder should pause briefly at each side of the weld. The arc should be as short as possible without shorting, and current should be appropriate for the electrode size and type and for the welding position.

Hot cracking occurs at elevated temperature, generally just after the weld starts to solidify. Hot cracks, which are for the most part intergranular, can be identified by a coating of oxide on their surfaces. Depending on the magnitude of strain, hot cracks vary from microfissures to readily visible cracks.

Hot cracks are most likely to occur in the root-pass weld bead, because of the small cross section of the bead in comparison to the mass of material being welded. Hot cracking often occurs in deep-penetrating welds and in welds in free-machining steels. If the initial crack is not repaired, it usually continues through successive layers as they are deposited. Hot cracking of the root bead may be minimized or prevented by preheating to modify strain, increasing the cross-sectional area of the root bead, using low-hydrogen electrodes, or changing the contour or the composition of the weld bead.

Cold cracking occurs in the HAZ after the weld cools to ambient temperature. Its occurrence may be delayed for as long as several days. Cold cracking is recognized as being caused by the presence of hydrogen in hardenable steels, excessive restraint of the joint, or by martensite formation resulting from rapid cooling. Dry low-hydrogen electrodes are used extensively to overcome cold cracking; preheating is also helpful. Although a stress-relief heat treatment cannot heal cracks that have already formed, it does reduce the residual stress sensitivity from welding and the susceptibility to fracture.

Centerline cracks (Fig. 21e) are cold cracks that often occur in single-pass concave fillet welds. The usual causes are an incorrect relationship between the size of the weld and the thickness of the base metal, poor fit-up, and overly rigid fit-up. The usual ways of preventing centerline cracks are: (1) positioning the joint slightly uphill to produce flat or slightly convex bead contours; (2) increasing bead size; (3) decreasing gap width or filling one side of the joint before welding the parts together; and (4) providing a small gap to allow movement during cooling.

Centerline cracks can occur as extensions of cracks in weld craters or root-bead cracks. The first extension of the crack may occur after the weld is completed and is cooling.

In one application of multiple-pass welding, centerline cracking occurred when an E6011 electrode was used for the root pass, an E6012 for the intermediate passes, and an E6020 for the final pass. The cracking was eliminated by reducing the number of passes with the E6012 electrode and increasing the depth of deposit with the E6020 electrode in the final pass.

Underbead cracks (Fig. 21f) are cold cracks that are most frequently encountered when welding a hardenable base metal. Excessive joint restraint and the presence of hydrogen are contributing causes. Underbead cracking can be minimized or prevented by using low-hydrogen electrodes or by preheating joints to the temperature range of 250 to 400 °F.

Base-metal cracking usually originates in the HAZ of the base metal as cold cracks. Cracks often extend into and through the weld metal. A reduction in welding stresses by using weld metal with high ductility, for instance, may alleviate the problem. The presence of a small slag inclusion or mass of porosity can increase the probability of cracking.

Cold cracks may originate at the toe of a weld and propagate randomly. Increasing restraint of the base metal, the thickness of the base metal, or the length of the weld increases the likelihood of cold cracking. The probability of cold cracking can be decreased by preheating the base metal, postweld heating, depositing the weld in an intermittent pattern, or welding with higher currents.

Microfissuring can be detected only by the use of a microscope. It is associated with either hot or cold cracking. Extremely small cracks are not detrimental for many applications, but they make the weldment unacceptable for applications involving fatigue.

Weld Craters. Weld craters are a depression of the weld surface caused by solidification of the molten weld pool after the arc has been extinguished. Weld crater cracks are often the origin of linear cracking. They are usually removed by chipping or grinding and filling the depression with a small deposit of filler metal. A backstep or reverse travel technique to fill each crater before breaking the arc helps prevent crater cracking.

Arc Strikes. Striking the arc on the base metal outside the weld joint can result in hardened spots. Failures can occur because of the notch effect of the arc strike or the crater formed. Welders should avoid indiscriminate marking of the base-metal surface with arc strikes. It is unnecessary and may be harmful. Many codes prohibit striking the arc on workpiece surfaces.

Gaps from incomplete fusion may occur between the weld metal and the base metal (Fig. 21g) or between the weld beads of a multiple-pass weld. Failure to obtain complete fusion can be caused by excessive travel speed, bridging, excessive electrode size, insufficient current, or poor joint preparation. Gaps usually can be prevented by reducing travel speed, improving joint preparation, or increasing the current.

Oxidation. Surface oxidation occurs when the base metal or weld metal has been inadequately protected from the atmosphere. Severe drafts, which disrupt the protection offered by the gas shield, must be avoided so that the weld metal and adjacent base metal are shielded until they have cooled enough to prevent harmful oxidation.

Sink or concavity is produced by surface tension on the surface of the weld pool, which pulls the molten metal up into the joint. If the condition is severe, root-bead cracking can result.

Weld Reinforcement. The reinforcement (crown) left on the weld can have a significant effect on the fatigue strength of the weldment at that point. On highly stressed weldments, the reinforcement is removed by grinding the weld until it is flush with the adjacent base metal. Use of appropriate welding techniques results in a weld reinforcement that is smoothly blended into the adjacent surface.

Overlapping, or protrusion of the weld metal beyond the toe or root of the weld, is most often caused by (1) insufficient travel speed, which permits the weld pool to get ahead of the electrode and cushion the arc; (2) incorrect electrode angle, which allows the force of the arc to push molten metal over unfused portions of the base metal; or (3) welding away from the ground with large electrodes having very fluid weld pools, such as E6020, E6027, E7024, and E7028.

To prevent overlapping, the travel speed should be such that the arc leads the weld pool and the electrode angle should be such that the force of the arc does not push the molten metal out of the pool and over the cold base metal. Overlapping is corrected by grinding off the excess weld metal.

Excessive weld spatter, if coarse, is often caused by an excessive arc length; if the spatter is fine, excessive current is a likely cause. Weld spatter can be minimized by keeping the arc length at the minimum that does not result in shorting out. A slight drag technique is recommended when using electrodes with iron-powder coverings. Current should be kept within the range recommended for the specific electrode. To improve the surface appearance of the weld, spatter can be removed by grinding.

Effect of Number of Passes. The use of multiple passes often prevents weld cracking. Multiple passes can be effective in fillet welding when cracking has resulted from single-pass welding, because the carbon equivalent of the steel was borderline or attempts to minimize distortion caused severe joint restraint in cooling.

The use of multiple passes (preferably a minimum of three) can sometimes eliminate the need for preheating or postheating with borderline conditions. In multiple-pass welding, the first pass provides some preheating for the second pass, and the third pass provides a certain amount of postheating for the previous two passes. Welding should be done quickly to minimize cooling between passes. Also, the first-pass weld bead must be deep-penetrating and of sufficient size to be strong enough to resist cracking. If the first-pass bead cracks, the crack usually propagates through subsequent beads.

Multiple passes are particularly adaptable for repair welding in the field and have proved useful in shop welding applications. However, this technique cannot always be considered a substitute for preheating and postheating. In shop practice, weld cracking often has two or more causes rather than a single cause.

Example 6. Changes in Welding Procedure to Prevent Cracking. Figure 22 shows a four-piece 840-lb truck axle weldment that required 16 fillet welds. In the original procedure, all of the $^3/_8$-in. fillet welds were deposited in two passes with E7024 electrodes. Although the high degree of restraint during cooling of a 2-in.-thick plate would normally indicate the need for preheating to at least 200 °F and possibly postheating for stress relief, welding trials made without heating appeared to be satisfactory; no cracking was detected. However, when the truck axles were put in service, cracks appeared in some of the welds. Because heating facilities were not available at the plant, welding procedures were changed in an effort to prevent cracking.

Fig. 22 Truck axle weldment joined with 16 welds. Cracking was prevented by changes in welding procedure.

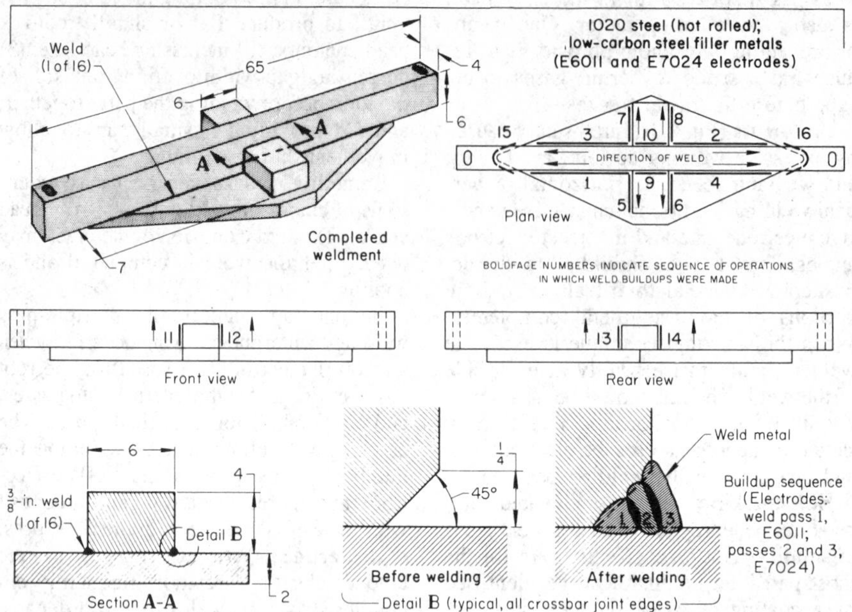

In the first revision, the size of the fillet welds was increased to $^1/_2$ in. This reduced the cracking, but did not eliminate it. In the second revision, fillet size was increased to $^3/_4$ in. This reduced cracking further, but magnetic-particle inspection showed occasional cracks in the corners and in the top fillet welds of the two crossbars. In addition, cost and appearance of the $^3/_4$-in. fillet welds were objectionable. The third revision involved welding in the directions of decreasing restraint. This also reduced cracking, but did not eliminate it. In a fourth revision, the class of electrodes was changed—first to low-hydrogen types with tensile strengths of 70 to 110 ksi and then, because of their greater ductility, to austenitic stainless steel electrodes—but neither change resulted in a satisfactory solution to the problem.

Finally, the welds were made in three passes rather than the two passes used in all previous procedures. A deep-penetrating root pass with a $^5/_{32}$-in.-diam E6011 electrode was followed immediately by two passes with $^3/_{16}$-in.-diam E7024 electrodes. This procedure greatly reduced the amount of cracking, largely because of the preheating effect of the first and second passes. A further improvement was obtained by cutting a $^1/_4$-in. bevel on the joint edges of the crossbars to increase penetration. Weld size was changed back to $^3/_8$ in. The three-pass welding and the edge beveling, combined with the use of the directional welding sequence, eliminated weld cracking.

Rating of Weldability

Some factors that determine whether a steel of a specific composition can be welded by conventional procedures include flexibility or rigidity of the structure, cooling conditions, and weld quality requirements. Welds good enough for many commercial applications can be made in low-carbon steel with an abnormally high content of inclusions, but these welds would not be adequate for severe service conditions. A steel containing a high carbon content or alloying elements may be unsuitable for welding into complex structures, but completely satisfactory for simple structures. Resulfurized steels are highly susceptible to underbead cracking, porosity, and hot cracking. In some applications, however, resulfurized steels can be welded successfully with low-hydrogen electrodes.

The weldability of a steel in terms of its susceptibility to cracking can be roughly estimated by using a carbon equivalent (CE), which can be calculated using one of several formulas:

$$CE = \%C + \frac{\%Mn}{6} + \frac{\%Cr}{5} + \frac{\%Mo}{4}$$

With this equation, if the carbon equivalent is not more than about 0.45%, the steel is considered weldable without any special precautions, such as preheating, postheating, or using low-hydrogen electrodes or additional passes. However, weldability depends on section thickness. Most high-strength low-alloy (HSLA) structural steels do not present serious problems in welding, even if their alloy content is high enough to raise their carbon equivalent above 0.45%.

Effect of Surface Condition on Weld Quality

Many satisfactory welds are made without special surface preparation, but quality is not usually high in such welds. The most common surface contaminants that adversely affect weld quality are carburization (resulting from improper gas or arc cutting), oxides and mill scale, shop dirt (oil and grease), paint, and moisture.

Surface carburization can cause cracking, because in effect, high-carbon steel is being welded. Carburized surfaces can be removed by mechanical means such as grinding. The other contaminants can result in weld porosity or microporosity, which usually cannot be detected without magnification. Oil, grease, and moisture can cause cold cracking by supplying hydrogen during welding.

Abrasive blasting is widely used for removing mill scale, rust, or other oxides in production welding. Grinding is used for odd jobs or for weldments that cannot be blasted conveniently. Oil and grease can be removed by wiping with a clean solvent and then washing and drying if needed. Paint must be removed from surfaces to be welded.

Moisture is frequently present on components that are to be welded outdoors or that are brought indoors just before welding. A humid atmosphere can cause a layer of moisture to form on workpiece surfaces. Surfaces should always be dry before welding. Moderate preheating with a gas torch may be used to remove surface moisture.

Cost

Many cost studies made to compare SMAW with SAW and GMAW have shown that when welding conditions (such as welding position, accessibility, and the amount of welding done without changing classes or sizes of electrodes) are favorable, SMAW is the most costly of the three processes. Cost studies have also shown that if SMAW is used in a production application and if electrodes of the same class and size are used for more than half the welding time, the use of a process that employs a continuously fed electrode should be considered. Conversely, in applications where welding must be done in a variety of positions, in areas difficult to reach, or when specifications require frequent changes in class and size of electrode, SMAW often is the lowest cost process and may well be the only practical one.

Shielded Metal Arc Welding Versus Alternative Welding Processes

When more than one welding process can produce acceptable results, the choice depends primarily on the equipment available, the skill of the welder, the number of similar weldments to be produced, and cost. Shielded metal arc welding is the most versatile arc welding process, but it is not always the lowest in cost or the fastest and does not produce the highest quality welds.

Although several welding processes may be suitable alternatives to SMAW in many applications, GMAW and flux cored arc welding (FCAW) compete closely for most applications. Both deposit metal faster than SMAW. Because both are semiautomatic processes with filler metal supplied at an established rate from a coil, arc time can be a higher percentage of total time than in SMAW. In addition, both welding processes can be fully automated. However, GMAW and FCAW are less versatile than SMAW.

Safety

To protect the welder from the direct rays of the welding arc and weld spatter, a helmet and protective clothing are used. To provide vision through the helmet, a filter plate is inserted into the helmet, which absorbs essentially all ultraviolet and infrared arc rays and most of the visible light spectrum emanating from the welding arc. To protect the filter plate surface from molten spatter, it is placed between expendable clear cover glass plates in the helmet. Recommended filter plate shades for SMAW are:

Electrode size	Filter plate
$^5/_{32}$ in. diam, or less	10
$^3/_{16}$-$^1/_4$ in. diam	12
Over $^1/_4$ in. diam	14

Protective clothing is used to protect otherwise exposed skin from burns caused by weld spatter and arc radiation (similar to sunburn). For this purpose, the welder wears gloves, an apron, jacket, and spats of flame-resistant material, as required. Cuffless pants and high-top work shoes or boots are recommended. Further information on safety may be found in ANSI Z49.1, "Safety in Welding and Cutting."

Flux Cored Arc Welding

By the ASM Committee on Flux Cored Arc Welding*

FLUX CORED ARC WELDING (FCAW) is a process in which the heat for welding is produced by an arc between a tubular consumable electrode wire and the work metal, with shielding provided by gas evolved during combustion and decomposition of a flux contained within the tubular electrode wire, or by the flux gas plus an auxiliary shielding gas. Thus, there are two major versions of the process: one that uses both an auxiliary shielding gas (usually carbon dioxide or a mixture of argon and carbon dioxide) and shielding obtained from the flux core of the electrode; and the self-shielding method, which depends on combustion and decomposition of flux-core compounds for shielding.

Both methods of FCAW are closely related to other arc welding processes. The method that uses an auxiliary gas shield is similar to gas metal arc welding (GMAW), which employs a solid consumable electrode and depends on an externally applied gas shield for protecting the arc and molten metal from contamination by the atmosphere. The self-shielding method is more closely related to shielded metal arc welding (SMAW), which also depends on the combustion and decomposition of a solid flux to provide the gaseous shield. In SMAW, the flux is on the outside of the electrode, which limits the form of the electrode to a straight length (a stick electrode), whereas in FCAW, the flux is inside a tubular electrode, which can be coiled and supplied to the arc as a continuous wire. For a size comparison, see the bottom row in Fig. 5.

Applicability

Applications of the two methods of FCAW overlap considerably. However, the specific characteristics of each process make it suitable for different operating conditions.

Flux cored arc welding with auxiliary gas shielding is used mainly for welding carbon, low-alloy steels, and some stainless steels. The method is applicable to semiautomatic work (manually manipulated electrode holder) and to the various machine and automatic welding procedures (in which the electrode holder is held mechanically). It is also adaptable to arc spot welding.

Originally, FCAW with auxiliary gas shielding was restricted to welding in the flat and horizontal positions, because tubular electrode wires were available only in relatively large diameters. With these large-diameter wires, the welder was unable to control the weld pool when welding in the vertical and overhead positions. This position restriction no longer holds, partly because small-diameter electrode wires (as small as 0.035 in. in diameter) have become available and partly because electrode-wire compositions that are better suited to out-of-position welding have been developed. Flux cored arc welding with auxiliary gas shielding is now an all-position process. It is applicable to a wide range of work-metal thicknesses, beginning with metal as thin as $^{1}/_{16}$ in.

Because of the protection offered by auxiliary gas shielding, a spray-type arc at high current density can be used, which results in maximum joint penetration. An arc producing globular transfer and a short circuiting arc can also be used. The latter types of arcs are produced at lower current density, result in shallower penetration, and are therefore better suited to the welding of thin sections.

Joints made by FCAW with auxiliary gas shielding meet the quality requirements of many codes; radiographs of welds show that sound deposits that meet exacting face and root bend tests can be produced. In many applications, FCAW with an auxiliary gas shield is competitive with GMAW and submerged arc welding (SAW) processes.

A solid layer of slag covers the weld deposit produced during FCAW, much the same as, although thinner than, the deposit made by SMAW. This slag is usually removed between welding passes. Consistency and adherence of the slag depend on the composition of the electrode wire. Some slag coverings crack during cooling and can be easily removed with a wire brush; others adhere firmly and must be broken up with a deslagging hammer to permit removal by wire brushing.

Flux Cored Arc Welding With Self-Shielding. This method of FCAW has become popular for many applications, mainly because of the simplicity of operation that results from the absence of the equipment necessary for gas shielding. In addition, the electrode holder is simpler than that required for use with auxiliary shielding gas. Because the self-shielding method does not penetrate as deeply as the auxiliary gas shielded method, it can be used under conditions of poor fit-up to better advantage. The self-shielding method is used mainly for welding carbon steels, but has been used successfully for welding some low-alloy steels, and several flux cored electrode wires have been developed for welding austenitic stainless steels.

The self-shielding method can be used for out-of-position welding by selection of a small-diameter electrode wire of suitable composition, just as for the auxiliary gas shielded method. The range of work-metal thickness is the same for both methods.

*Gregory L. Serangeli, *Chariman*, Manager, Industrial Technology, Caterpillar Tractor Co.; Stanley E. Ferree, Research Engineer, Alloy Rods Division, Allegheny Ludlum Industries; R.W. Heid, Supervisor, Welding Engineering, Newport News Shipbuilding and Drydock Co.; Albert E. Wiehe, Senior Research Scientist, Hobart Brothers Co.

Fig. 1 (a) Operating principles for FCAW with auxiliary gas shielding; (b) and (c) nozzles for auxiliary gas shielded and for self-shielding methods of FCAW. Note difference in lengths of electrode extension.

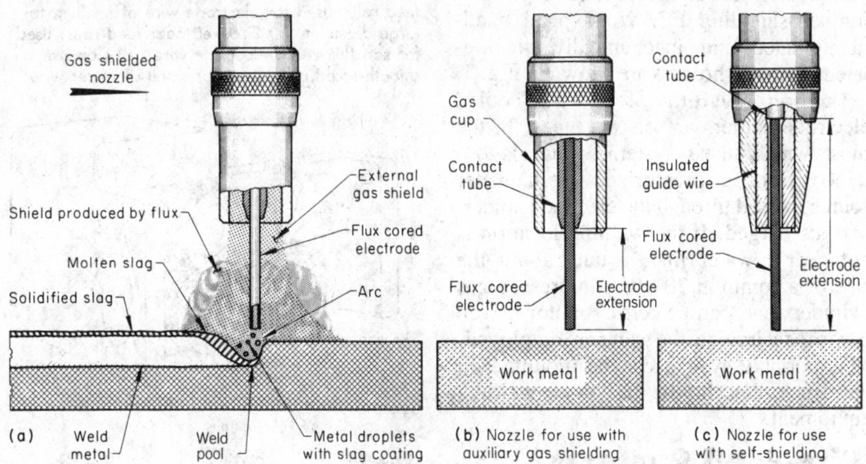

Flux cored arc welding without an auxiliary gas shield is not suited to a spray-type arc. Because metal is transferred from the outside of the electrode to the weld pool, protection from oxidation in this area is minimal without an auxiliary gas shield. When a fine spray is developed, the total surface area of the metal particles is large, which results in excessive oxidation. Therefore, the self-shielding method should be operated with globular or short circuiting metal transfer.

The quality of welds made by self-shielding FCAW is generally lower than that of welds made with auxiliary gas shielding. The main cause of lower weld quality is the greater contamination of the weld deposit by the atmosphere during welding; another cause is the presence of gas-forming and deoxidizing elements in the flux core of the electrode wire. The mechanical properties adversely affected are ductility and impact toughness, particularly at low temperature. Some codes do not permit the use of self-shielding FCAW for steels having a yield strength greater than 42 ksi.

An advantage of the self-shielding method is that it can be used in a draft, as can SMAW, although such practice is not recommended for either process.

Fundamentals of the Process

Basic equipment requirements for FCAW are essentially the same as for GMAW: a power supply, a wire feeder to advance the electrode wire as it melts to form the weld deposit, an electrode holder, and, when appropriate, a means of supplying auxiliary shielding gas. These components are shown in the article "Gas Metal Arc Welding" in this Volume.

Likewise, the arc characteristics that can be obtained in FCAW are essentially the same as those in GMAW. The spray arc, the globular arc, and the short circuiting arc also are shown in the article "Gas Metal Arc Welding" in this Volume.

Principles of Operation. The flux cored wire is the main difference between GMAW and FCAW with an auxiliary gas shield (Fig. 1a). The flux core provides a molten slag that covers the weld metal and a gas that assists in shielding of the arc. The necessary equipment, including the electrode holder, is essentially the same for both of these processes.

Aside from the use or nonuse of auxiliary shielding gas, the self-shielding and auxiliary gas shielded methods differ mainly in the type of electrode holder used and in the length of electrode extension. As illustrated in Fig. 1(b), with the type of electrode holder used with the auxiliary gas shielded process, the contact tube extends nearly to the end of the gas cup, so that the electrode extension (distance from the end of the contact tube to the weld pool) is nearly the same as the visible extension (distance from the end of the gas cup to the weld pool). Electrode extension is generally what is referred to when the word "stickout" is used without qualification. In many electrode holders, the distance from the end of the contact tube to the end of the gas cup is $^3/_8$ in. or more, about the same as for GMAW. For some applications, the extension may be as much as 1 in.

For the self-shielding method, a much greater electrode extension is used ($2^1/_2$ in. or more), as indicated in Fig. 1(c). Because no provision need be made for shielding gas, the space at the end of the electrode holder can be occupied by an insulated wire guide, which prevents the wire from touching the end or nozzle portion of the electrode holder. The use of a longer electrode extension allows a higher deposition rate, as the electrode wire is preheated more because of the greater distance along which the welding current flows through the electrode.

Power Supply

Alternating current is seldom used for welding with flux cored electrode wires. Direct current supplied by a rectifier, motor-generator, or engine-driven generator and operated with reverse polarity (direct current electrode positive, or DCEP) is generally used; straight polarity (direct current electrode negative, or DCEN) is also used, although only to a limited extent.

Either of two general types of power supply can be employed: the constant-current type (drooping voltage characteristic), such as is commonly used for SMAW, or the constant-voltage type, originally developed for use with GMAW. When a constant-current power supply is employed, it is usually matched to an arc-voltage-sensing electrode-wire feeder. A constant-voltage power supply, which is the type most often used for FCAW, should be employed only for continuous-feed electrode wire processes. When a constant-voltage power supply is used for FCAW, the electrode wire is fed into the arc at a specific rate, and it automatically draws the amount of current from the constant-voltage power supply required to maintain the preset arc voltage. If wire-feed rate is increased, current increases and deposition rate increases. This system is simple to control. Figure 2 shows the interrelation of wire-feed rate and

Fig. 2 Interrelation of wire-feed rate and welding current for four sizes of electrode wire, using carbon dioxide shielding gas

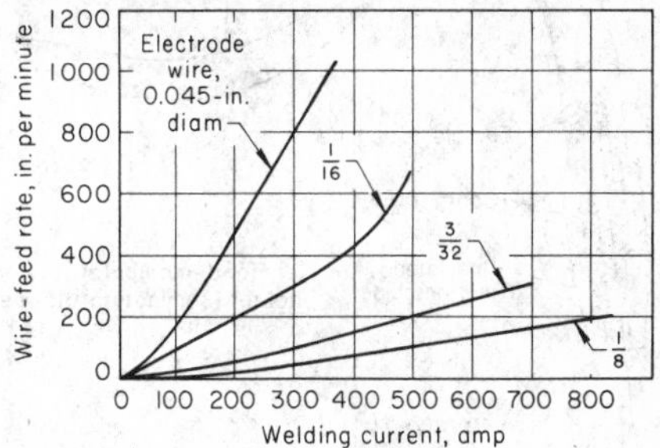

welding current for four common sizes of flux cored electrode wire.

Various models and sizes of constant-voltage power supplies are available, most of which incorporate slope and inductance control. Power supplies that can be varied in slope and inductance are more versatile and can be used for a wide variety of conditions. For further discussion of power supplies suitable for FCAW, see the article "Gas Metal Arc Welding" in this Volume.

Electrode Holders

Electrode holders for auxiliary gas shielded FCAW are similar to those used for GMAW. Various sizes, ratings, and styles, suitable for both semiautomatic and automatic welding, are available. Both air-cooled and water-cooled electrode holders are made. Air-cooled holders depend on radiation of heat to the surrounding air for cooling, although the shielding gas, which is quite cold as it passes through the holders, helps to cool them. Choice of holder depends largely on the welding current and shielding gas used. When current is 500 A or more, a water-cooled electrode holder is usually employed. Some welders prefer water-cooled holders when using welding currents of less than 500 A.

Three types of electrode holders are shown in the article "Gas Metal Arc Welding" in this Volume, together with additional discussion of electrode holders for use with a shielding gas.

Electrode Holders for the Self-Shielding Method. The same electrode holder can be used for self-shielding FCAW as is used for the auxiliary gas shielded method. Ordinarily, the electrode holder requires minor modification for use without auxiliary gas shielding—for example, a longer electrode extension is used (Fig. 1). Actually, it is uncommon to use the same electrode holder for self-shielding welding as for welding with an auxiliary gas shield. Electrode holders of various sizes and styles are made especially for self-shielding FCAW. Typical hand-manipulated and mechanically manipulated electrode holders are shown in Fig. 3.

Cooling systems for water-cooled electrode holders are of two types. In the most widely used system, water is obtained from a pressurized fresh water source, passed through the electrode holder, and discharged. If the welding location is not near a water line, a tank about the size of a common 200-ft^3 compressed-gas cylinder and an electric motor-driven pump to recirculate the water are employed. This portable system can be mounted on the truck that holds the other welding equipment.

Wire-Feed Systems

Constant-voltage power supplies are used most often for both methods of FCAW, which require a constant wire-feed system. A push-type wire-feed system is universally employed for both methods, for three reasons: (1) electrode-wire diameter is fairly large—the smallest size made is 0.035 in. in diameter and the smallest size commonly used is 0.045 in. in diameter; (2) all electrode wires are made of steel and are stiff, so feeding is greatly facilitated with a push-type feeder; and (3) the feed mechanism is not located in the electrode holder, so the holder weighs less and is easier to manipulate. The system used for feeding flux cored wire is similar to that used for feeding solid wire (see the section dealing with wire-feed systems in the article "Gas Metal Arc Welding" in this Volume.)

Push-Type Systems. In a push-type wire-feed system, the electrode wire is pulled from a reel by feed rolls and pushed through a flexible wire conduit to the electrode holder and then into the arc. A constant rate of feed is mandatory in a constant-voltage system, but the feed rate must be capable of adjustment to obtain the required welding current.

A four-roll feeder is shown in Fig. 4(a). In this feeder, all four rolls are driven. Push-type feeding systems vary considerably among manufacturers. Some have only two rolls—one driven roll and one pressure roll. The design must ensure that the roll speed and pressure can be varied and that the rolls can be changed quickly.

Push-type wire-feed systems are coupled electrically to the power supply. For convenience, wire feeders are often mounted on the power supply, although they have been separated from the power supply by distances up to 200 ft. They can be mounted on overhead jib cranes or booms, thus allowing the welder to cover a large area, and they can be located on the shop floor, on the ground, or on overhead scaffolding.

A given joint may require the use of electrodes of two different sizes. A typical application would call for a root pass to be made by GMAW, employing a very small-diameter solid wire and for the weld to be completed by FCAW, using a larger

Fig. 4 Electrode wire-feed system and three designs of feed rolls. (a) Four-roll push-type feeder, in which all rolls are driven. (b) V-groove knurled feed rolls, used for electrode wire of medium to large diameter. (c) Grooved-gear feed rolls, used for soft flux cored electrode wire. (d) Concave, smooth-faced rolls, used for small-diameter wire

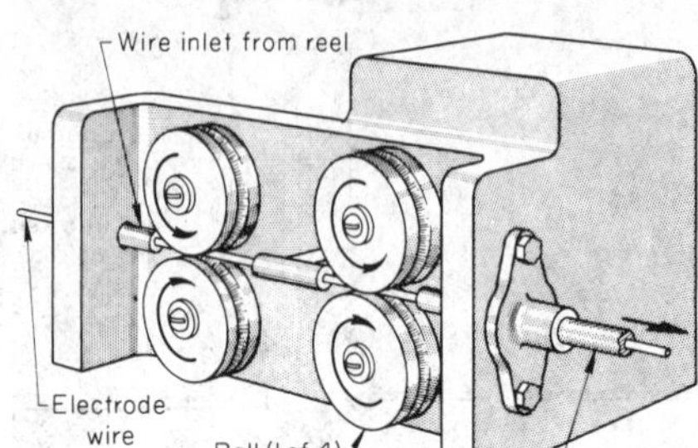

(a) Four-roll push-type feeder

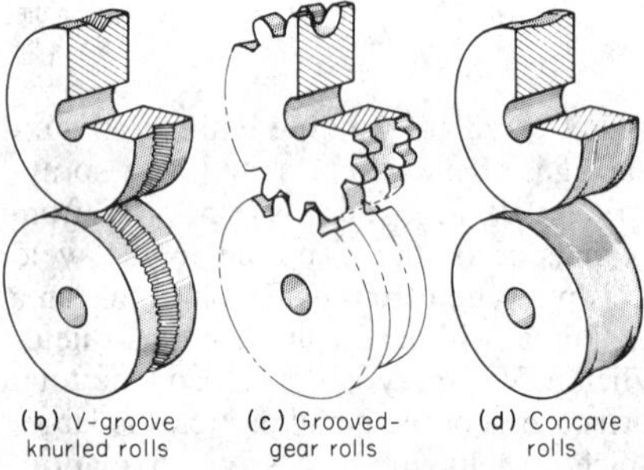

Fig. 3 Typical electrode holders for self-shielding FCAW

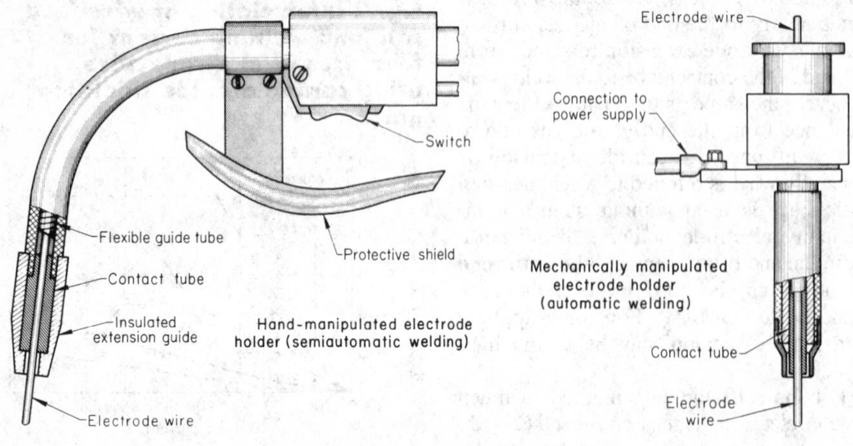

diameter wire. A wire-feed system capable of supplying electrode wire to either of two electrode holders is available for such applications. This type of wire feeder has two electrode-holder-and-cable assemblies and two sources of wire. By actuating a switch and picking up the appropriate electrode holder, the welder can select GMAW or FCAW. A single power supply is used. With some models, two motors are employed for driving the two wire feeders, and in other models, the two wire feeders are driven by a single 1/4-hp motor and a dual drive.

Feed Rolls. Because of their tubular construction, flux cored electrode wires are easily flattened. As a result, the design of the feed rolls is critical. Various types of grooved, grooved and knurled, and concave rolls have been used successfully (Fig. 4).

Rolls that are compatible with the wire size should be selected; rolls must be changed when there is an appreciable change in wire diameter. The V-groove knurled rolls shown in Fig. 4(b) are commonly used for feeding wires with a diameter greater than about 1/16 in. This type of roll is lightly knurled so that the wire can be fed without slipping and without the use of excessive pressure.

Another type of feed roll, shown in Fig. 4(c), is used in a two-roll wire feeder. Each roll is a gear having a round groove with the same radius as the wire to be fed. The groove is cut into both rolls but is cut only part way through the tooth. The grasp of the wire is by the relatively flat surface of the groove. The advantage in this type of roll is that it does not mar the surface of soft flux cored wires. Small-diameter wires (for example, 0.045 in.) are less easily flattened than are larger diameter wires, and simple concave rolls without knurls, as shown in Fig. 4(d), are usually satisfactory.

Maintenance of Feed Systems. Wire-feeding mechanisms require scheduled regular maintenance to ensure a smooth and constant delivery of wire to the arc. The guides, feed rolls, and reels must be properly aligned and adjusted, and the wire conduit must be cleaned at regular intervals, normally before a fresh reel of wire is put into service. This should be done by removing the wire and blowing clean air or shielding gas through the conduit.

Flux Cored Electrode Wires

Flux cored electrode wire consists of a low-carbon steel sheath surrounding a core of fluxing and alloying material.

Manufacture of flux cored electrode wire is a specialized and precise operation. Most flux cored electrode wire is made by passing low-carbon steel strip through contour-forming rolls that bend the strip into a U-shaped cross section. The U-shaped product is then filled with a measured amount of granular core material (flux) by passing it through a filling device. Next, the flux-filled U-shaped strip passes through closing rolls that form it into a tube and tightly compress the core materials. The tube is then pulled through drawing dies that reduce its diameter and further compress the core materials. The drawing operation secures the core materials inside the tube. The electrode may or may not be baked during or between drawing operations, depending on its type.

Additional drawing operations are performed to produce different sizes of electrode wire. The standard sizes are 0.045, 1/16, 5/64, 3/32, 7/64, 1/8, and 5/32 in. in diameter (0.035-in.-diam wire is available in limited quantities). The 3/32-in.- and 5/64-in.-diam sizes are most widely used, except for out-of-position work, where 0.052- and 0.062-in.-diam wire is usually used.

The finished electrode wire is wound into a continuous coil or onto spools, as required. Various standardized methods of packaging are in use; most electrode wire is wound into 25-lb and 50-lb spools and 60-lb coils, but other sizes and forms are used on occasion. The spools and coils are placed in moistureproof plastic bags and then in shipping boxes, to ensure that deterioration does not occur.

Functions of the compounds contained in the core are similar to those in the covering on the stick electrodes used for SMAW, which are to:

- Act as deoxidizers or scavengers to help purify the weld metal and produce a sound deposit
- Form slag to float on the molten weld metal and protect it from the atmosphere during solidification
- Act as arc stabilizers to produce a smooth welding arc and reduce weld spatter
- Add alloying elements to the weld metal to increase weld strength and to provide other required weld-metal properties
- Provide shielding gas

Construction. Several styles of flux cored electrodes are shown in cross section in Fig. 5. In the three styles shown in the top row of the illustration, the steel portion of the electrode comprises about 75 to 85% of the total weight and about 75% of the cross-sectional area of the electrode.

The amount of flux contained in the core of a flux cored electrode is less than is used on flux-covered SMAW electrodes of comparable size (see bottom row in Fig. 5), because the covering on SMAW electrodes must contain binders to keep the covering adherent and continuous and constituents that enable it to be extruded. Comparison of a typical flux-covered electrode with a typical flux cored electrode shows the percentage of steel in each to be as follows:

Steel	Flux-covered E7016, %	Flux cored E70T-1, %
By area	45	75
By weight	76	85

Metal transfer from consumable electrodes across an arc is by the spray, globular, or short circuiting mode.

On cored electrodes, the molten droplets form on the periphery (or sheath) of the electrode. A droplet forms and is transferred, and then another droplet forms at another location on the metal sheath and is transferred. The core material appears to transfer to the weld-pool surface independently.

At low current densities, the droplets are larger than at high current densities. For instance, the transfer from a 3/32-in.-diam flux cored electrode was observed at 350, 425, and 550 A, respectively. At 550 A, it appeared that some metal was being transferred by spray. Large droplets, which formed at lower current densities, caused splashing as they entered the weld pool, thus increasing the amount of visible weld spatter. This explains why there is less visible spatter, the arc appears smoother, and deposition efficiency is higher when current density is high.

Classification of Electrodes. The various types of flux cored electrodes are classified by the American Welding Society (AWS). The classification for flux cored electrodes follows the standard pattern used in other AWS specifications. The AWS specification A5.20-79 covers carbon steel flux cored electrodes for welding carbon and low-alloy steels. Figure 6 illustrates this classification system, which can be explained further by considering a typical designation such as E70T-1:

- The prefix "E" indicates an electrode.
- The first digit, 7, indicates a minimum tensile strength of 70 ksi.
- The second digit, 0, indicates that the electrode can be welded in the flat and horizontal positions.
- The "T" indicates a tubular or flux cored electrode.

Fig. 5 Several styles of flux cored electrodes, and size comparison of flux-covered, flux cored, and solid metal electrodes for use at the same nominal welding current

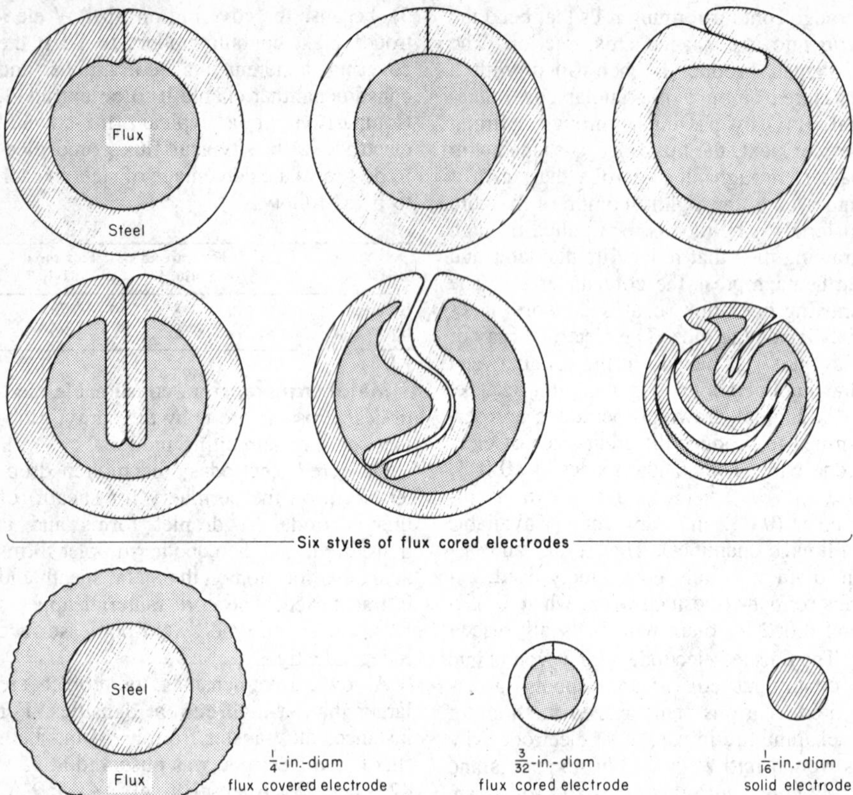

- The last digit, 1, indicates the chemistry of the deposited weld metal, method of shielding, usability, and performance characteristics of the electrode wire. More detailed information on each suffix, such as T-1 or T-2, can be found in the section on carbon steel flux cored electrodes in this article.

The AWS specification A5.29-80 covers low-alloy flux cored electrodes for welding carbon and low-alloy steels. Figure 7 illustrates this classification system, which can be further explained by considering a typical designation such as E80T5-Ni1:

- The prefix "E" indicates an electrode.
- The first digit, 8, indicates a minimum tensile strength of 80 ksi.
- The second digit, 0, indicates that the electrode can be welded in the flat and horizontal positions.
- The "T" indicates a tubular or flux cored electrode.
- The third digit, 5, indicates the chemistry of the deposited weld metal, method of shielding, usability, and performance characteristics of the electrode wire.
- The ending suffix, Ni1, designates this electrode as a 1% Ni alloy.

Flux cored electrodes for welding corrosion-resistant chromium and chromium-nickel steels are covered by AWS A5.22-80. Figure 8 illustrates this classification system, which can be further explained by considering a typical designation such as E308T-3:

- The prefix "E" indicates an electrode.
- The number 308 designates the American Iron and Steel Institute (AISI) alloy code number.
- The "T" indicates that this is a tubular or flux cored electrode.
- The final digit or suffix, 3, indicates that the electrode is self-shielding and that the current and polarity is DCEP (reverse polarity). More detailed information on shielding-gas requirements and current type for stainless steel flux cored electrodes can be found in the section on stainless steel flux cored electrodes in this article.

Carbon Steel Flux Cored Electrodes. Characteristics of specific types of carbon steel flux cored electrodes covered by the AWS A5.20-79 specification are summarized below. Each suffix (T-1, T-2, and so forth) indicates a general grouping of electrodes that produce similar usability and performance characteristics and contain similar flux ingredients:

- *T-1 electrodes* are used with auxiliary gas shielding and are designed for single- and multiple-pass welding. The larger diameters (5/64 in. and larger) are used for welding in the flat and horizontal positions. The smaller diameters (1/16 in. and smaller) are used in all positions. A quiet arc, high deposition rate, low spatter loss, flat to slightly convex bead, and easily controlled and removed slag are characteristics of these electrodes. Weld deposits have good impact properties. Electrodes in this group have a rutile base slag system.
- *T-2 electrodes* are essentially T-1 electrodes with higher levels of manganese or silicon, or both, in the core of the electrode. The higher levels of deoxidizers allow for welding material that

Fig. 6 Classification system for carbon steel flux cored electrodes, AWS A5.20-79. The letter X as used in this figure and in electrode classification designations in this specification substitutes for specific designations indicated by this figure.

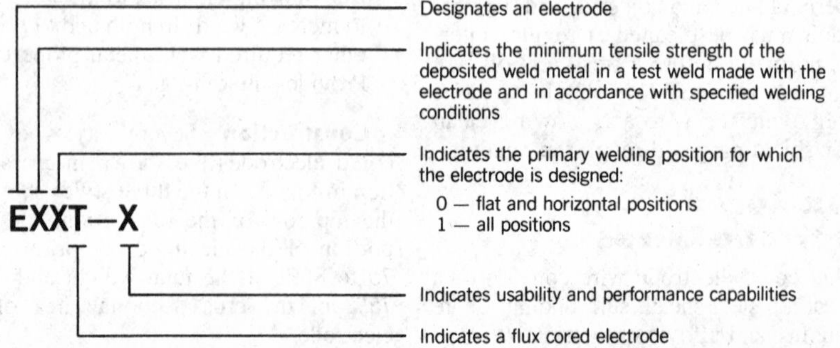

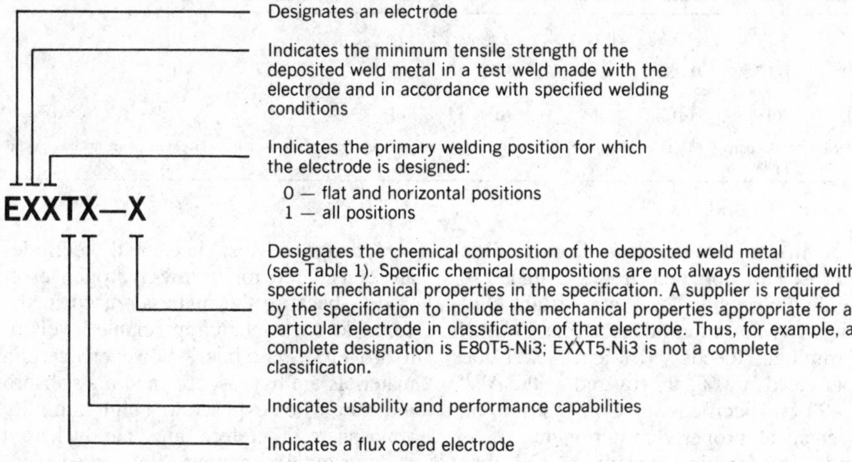

Fig. 7 Classification system for low-alloy flux cored electrodes, AWS A5.29-80. The letter X as used in this figure and in electrode classification designations in this specification substitutes for specific designations indicated by this figure.

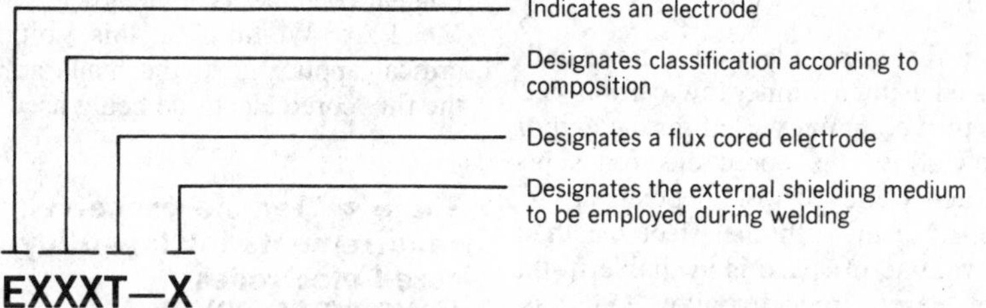

Fig. 8 Classification system for stainless steel flux cored electrodes, AWS A5.22-80

has heavier mill scale, rust, or other foreign materials on its surface than can be tolerated by electrodes of the T-1 classification and still produce welds of radiographic quality. Although the T-2 electrodes are primarily designed for single-pass welding in the flat position and horizontal fillet welds, multiple-pass welds can be made when the weld beads are heavy and appreciable admixture of base and filler metal occurs.

- *T-3 electrodes* are used without auxiliary gas shielding and have a spray-type transfer. They are primarily intended for depositing single-pass welds at high speeds in the flat, horizontal, and downhill (up to 20°) positions on relatively thin steels ($^3/_{16}$ in. thick).
- *T-4 electrodes* are used without auxiliary gas shielding for single-pass and multiple-pass welding in the flat and horizontal positions. The slag system is designed to desulfurize the weld metal and produce deposits that have low sensitivity to cracking. A globular transfer, low penetration, and high deposition rates are other characteristics of these types.
- *T-5 electrodes* are used with auxiliary gas shielding in the flat and horizontal positions for single- and multiple-pass applications. Typical characteristics are a globular transfer, convex bead, shallow penetration, and a thin slag that may not completely cover the weld bead. Excellent impact properties and resistance to cracking are characteristic of these electrodes because of their lime-fluoride basic slag system.
- *T-6 electrodes* are designed to be used without auxiliary gas shielding for single- and multiple-pass welding in the flat and horizontal positions. The slag system is designed to give good impact properties and excellent deep groove slag removal.
- *T-7 electrodes* are used without auxiliary gas shielding for single- and multiple-pass welding with DCEN. Weld deposits have low sensitivity to cracking and the smaller diameter electrodes can be used in all positions.
- *T-8 electrodes* are used without auxiliary gas shielding and operate on DCEN. Single- and multiple-pass welds with good impact properties can be produced in all positions.
- *T-10 electrodes* are used without auxiliary gas shielding and also operate on DCEN. High travel speeds can be used to make single-pass welds on material of $^3/_4$-in. thickness in the flat, horizontal, and downhill (up to 20°) positions.
- *T-11 electrodes* are used without auxiliary gas shielding, operate on DCEN, and produce a spray-type arc. They are used as general-purpose electrodes for single- and multiple-pass welding in all positions.
- *T-G electrodes* (EXXT-G) include new multiple-pass electrodes that are not covered by the presently defined classifications. The slag system, arc characteristics, weld appearance, and polarity are not defined.
- *T-GS electrodes* (EXXT-GS) are new single-pass electrodes that have not been previously defined.

The composition requirements for weld metal deposited from carbon steel flux cored electrodes of the AWS A5.20-79 specification are given in Table 1. Note that no carbon content is specified. The carbon content of an actual weld depends largely on the carbon in the base metal, somewhat on the carbon in the electrode, and to a slight extent on the shielding gas, if carbon dioxide is used. Additional information on shielding gas, current type, and mechanical properties is given in Tables 2 and 3.

Low-Alloy Flux Cored Electrodes. Various low-alloy flux cored electrodes are available for welding high-strength structural steels, high-strength alloy steels, and other types of low-alloy steels. The AWS A5.29-80 specification covers the four general classifications (T1, T4, T5, and T8) of low-alloy flux cored electrodes. The usability characteristics, shielding gas, and current requirements of these electrodes are the same as the carbon steel flux cored electrodes of the equivalent classification. The major difference is that the cores of the low-alloy flux cored electrodes carry alloying elements, in addition to the normal gas formers, fluxing elements, arc stabilizers, and deoxidizers. The alloying elements in the core melt in the arc and mix with the molten low-carbon steel outer sheath to produce the required composition of deposited weld metal.

One major advantage of flux cored electrodes over solid wire electrodes is that the cored wires can readily be made to match virtually any base-metal steel composition. There are 23 alloy grades covered by the AWS A5.29-80 specification,

Table 1 Composition requirements for undiluted weld metal from carbon steel flux cored electrodes (AWS A5.20-79)

AWS classification	Weld-metal composition, max %									
	Carbon	Phosphorus	Sulfur	Vanadium(a)	Silicon	Nickel(a)	Chromium(a)	Molybdenum(a)	Manganese	Aluminum(b)
EXXT-1, EXXT-4, EXXT-5, EXXT-6, EXXT-7, EXXT-8, EXXT-11, EXXT-G	(c)	0.04	0.03	0.08	0.90	0.50	0.20	0.30	1.75	1.8
EXXT-GS, EXXT-2, EXXT-3, EXXT-10	(d)	(d)	(d)	(d)	(d)	(d)	(d)	(d)	(d)	(d)

(a) These elements shall be reported if added intentionally. (b) For self-shielded electrodes only. (c) This element shall be determined. (d) No requirements. Analyses of undiluted weld metal are not meaningful, because these classifications are intended for single-pass welding with resultant high dilution.

Table 2 Shielding gas and current type for carbon steel flux cored electrodes (AWS A5.20-79)

AWS classification	Shielding gas	Current and polarity
EXXT-1	CO_2	DCEP
EXXT-2	CO_2	DCEP
EXXT-3	None	DCEP
EXXT-4	None	DCEP
EXXT-5	CO_2	DCEP
EXXT-6	None	DCEP
EXXT-7	None	DCEN
EXXT-8	None	DCEN
EXXT-10	None	DCEN
EXXT-11	None	DCEN
EXXT-G	Not defined	Not defined
EXXT-GS	Not defined	Not defined

which can be classified into five major groups: carbon-molybdenum, chromium-molybdenum, nickel, manganese-molybdenum, and other low-alloy grades. Depending on the alloy grade, impact requirements range from 20 ft · lb at 0 °F to 20 ft · lb at −100 °F. Information on the tensile property requirements is given in Table 4.

Stainless Steel Flux Cored Electrodes. Flux cored electrodes designed for welding corrosion- or heat-resistant chromium and chromium-nickel steels, in which chromium exceeds 4.0% and nickel does not exceed 50.0%, are covered by the AWS A5.22-80 specification. Composition and mechanical property requirements of 21 grades of ferritic, austenitic, and martensitic alloys are included in this specification. Information on shielding gas requirements and current type is given in Table 5.

These flux cored electrodes are usually fabricated using a thinner low-carbon steel sheath than commonly used for the carbon and low-alloy flux cored electrodes because higher alloyed weld deposits are required. By using a thinner steel sheath, a larger volume of space is available in the core of the electrode for alloy additions. Sometimes an austenitic or a ferritic grade stainless steel sheath may be used to fabricate special high-alloy grades or to improve drawing characteristics during manufacturing of small-diameter ($^1/_{16}$ in. and smaller) electrodes.

Hydrogen. Most flux cored electrodes are considered to be low-hydrogen electrodes, because the materials used in the cores do not contain appreciable levels of hydrogen. Sometimes, however, certain materials are hygroscopic and may absorb moisture when exposed to a high humidity atmosphere. Therefore, after removal from their original container, flux cored electrodes should be treated in the same manner as low-hydrogen flux-covered electrodes are treated after removal from their original container (see the article "Shielded Metal Arc Welding" in this Volume). In critical applications, the manufacturer of the flux cored electrode being used should

Table 3 Mechanical property requirements for carbon steel flux cored electrodes (AWS A5.20-79)(a)

AWS classification	Tensile strength, ksi	Yield strength, ksi	Elongation in 2 in., %	Charpy V-notch impact
E6XT-1	62	50	22	20 ft · lb at 0 °F
E6XT-4	62	50	22	Not required
E6XT-5	62	50	22	20 ft · lb at −20 °F
E6XT-6	62	50	22	20 ft · lb at −20 °F
E6XT-7	62	50	22	Not required
E6XT-8	62	50	22	20 ft · lb at −20 °F
E6XT-11	62	50	22	Not required
E6XT-G	62	50	22	Not required
E6XT-GS	62	Not required	Not required	Not required
E7XT-1	72	60	22	20 ft · lb at 0 °F
E7XT-4	72	60	22	Not required
E7XT-5	72	60	22	20 ft · lb at −20 °F
E7XT-6	72	60	22	20 ft · lb at −20 °F
E7XT-7	72	60	22	Not required
E7XT-8	72	60	22	20 ft · lb at −20 °F
E7XT-11	72	60	22	Not required
E7XT-G	72	60	22	Not required
E7XT-2, E7XT-3, E7XT-10, E7XT-GS	72	Not required	Not required	Not required

(a) All values are minimums in the as-welded condition.

Table 4 Tensile property requirements for low-alloy flux cored electrodes (AWS A5.29-80)(a)

AWS classification	Tensile strength, ksi	Yield strength (min), ksi	Elongation in 2 in., min %
E6XTX-X	60-80	50	22
E7XTX-X	70-90	58	20
E8XTX-X	80-100	68	19
E9XTX-X	90-110	78	17
E10XTX-X	100-120	88	16
E11XTX-X	110-130	98	15
E12XTX-X	120-140	108	14
EXXXTX-G	(b)	(b)	(b)

(a) Properties of electrodes that use external gas shielding (T1 and T5 types) may vary with gas mixtures. Therefore, electrodes should not be used with gases other than those recommended by the manufacturer of the electrode. (b) Properties as agreed upon between supplier and purchaser

Table 5 Shielding gas and current type for stainless steel flux cored electrodes (AWS A5.22-80)

AWS classification(a)	Shielding gas(b)	Current and polarity
EXXXT-1	CO_2	DCEP
EXXXT-2	Ar + 2% O_2	DCEP
EXXXT-3	None	DCEP
EXXXT-G	Not specified	Not specified

(a) The letters "XXX" stand for the AISI code number; such as 308 and 309. (b) The requirement for the use of specified shielding gases for classification purposes shall not be construed to restrict the use of other gases for industrial use as recommended by the manufacturer.

be consulted for recommended storage and rebaking procedures.

Fluxes and Slags. The American Welding Society has no classification system for the fluxes used in FCAW. As a result, electrodes may vary in characteristics even when made by a single manufacturer. Electrode users must rely on "trade name" electrodes for consistent electrode performance. Fluxes generally are considered to be proprietary products, each formulated to produce a specific result. The flux core consists primarily of gas formers, fluxing agents, arc stabilizers, deoxidizers, denitrifiers, and alloying elements to suit the specific application and metal alloy being welded. The elements present in a flux may exist in several usable forms, and therefore the raw materials used in fluxes that produce similar results may vary considerably.

The flux core of the wire serves several purposes. It may provide alloy material that improves the properties of the weld metal by becoming part of the weld, or it may provide "scavengers" that improve the weld by removing undesirable oxygen, nitrogen, or sulfur from the molten weld metal. The slag produced provides protection to the weld metal from the air while it is molten, solidifying, or solid. Molten metal retention and pool shape are influenced by the slag covering.

When alloying elements are used, they may be present in the flux core as ferroalloys (chromium, manganese, silicon, molybdenum, and titanium); as metal powders (nickel, zirconium, vanadium, manganese, chromium, and iron); or as oxides such as vanadium and titanium, which are present in rutile (TiO_2). Table 6 lists the elements found in flux cored electrodes, including their form and function.

Shielding Gases

Gases used for auxiliary gas shielded FCAW can be the same as those used for GMAW, which include carbon dioxide, 98% argon with 2% oxygen, and 75% argon with 25% carbon dioxide. The chemical reactions of carbon dioxide with the carbon steel base metal and with the cored electrodes make it desirable as a shielding gas. In addition, carbon dioxide is less costly than the mixtures containing argon, based on the cost of gas per pound of weld metal deposited.

Characteristics of Carbon Dioxide. At room temperature, dry carbon dioxide is an inactive gas that has no adverse effect on the metals with which it comes in contact; but at the high temperature of the welding arc, it dissociates, in the following reaction:

$$2CO_2 \leftrightarrows 2CO + O_2$$

This dissociation leaves considerable oxygen available in the arc to oxidize metallic elements. Molten iron, for example, reacts with carbon dioxide to produce iron oxide and carbon monoxide, as follows:

$$Fe + CO_2 \leftrightarrows FeO + CO$$

The oxidizing characteristics of carbon dioxide have been considered in the development of flux cored electrodes, and deoxidizing materials are added to the flux core to compensate for it. Some of the carbon monoxide produced during the reaction between iron and carbon dioxide may dissociate to carbon and oxygen. The carbon released by this reaction is available to dissolve in the weld metal, thereby increasing the carbon content.

Depending on the carbon content of the base metal and of the electrode, a carbon dioxide atmosphere may be either carburizing or decarburizing. Carbon content in the weld deposit usually ranges between 0.05 and 0.12%. If the carbon content of the electrode metal and of the base metal is lower than 0.05%, the weld metal will pick up carbon from the carbon dioxide shielding atmosphere. Conversely, if the carbon content of the electrode metal and of the base metal is greater than 0.12%, the weld metal may lose carbon. This loss of carbon can occur as follows:

$$FeO + C \leftrightarrows Fe + CO$$

When the above reaction occurs, carbon monoxide gas can be trapped in the weld deposit, where it may produce porosity. The reaction is avoided by keeping the level of the deoxidizing elements in the core of the electrode sufficiently high.

In general, joint penetration with FCAW with carbon dioxide shielding is much greater than with SMAW using low-hydrogen iron powder electrodes, but not as deep as with GMAW using solid electrode wire and carbon dioxide shielding, operating in the same range of welding current.

Containers for Carbon Dioxide. Carbon dioxide is available in cylinders or in bulk. The most widely used container is the Interstate Commerce Commission-approved high-pressure steel cylinder. Cylinders containing carbon dioxide are always labeled "CO_2," and for welding applications they should be labeled "Welding Grade."

A cylinder from which no gas has been drawn contains, at 70 °F, gaseous carbon dioxide and liquid carbon dioxide. The gas occupies about one third of the volume of the cylinder, and when gas is withdrawn, liquid vaporizes and maintains the gas pressure. When all the liquid carbon dioxide has been vaporized, the pressure starts to fall. Originally, the weight of liquid carbon dioxide is about 90% of the weight of the contents of a cylinder. In order to determine how much remains in a partly used cylinder, the cylinder should be weighed (pressure is not a reliable indication of cylinder content). The tare weight of the (empty) cylinder is usually stenciled

Table 6 Elements found in flux cored electrodes

Element	Form	Function
Aluminum	Powder	Deoxidizer; denitrifier
Calcium	Present in fluorspar (CaF_2) and calcium carbonate ($CaCO_3$)	Slag former; provides shielding
Chromium	Ferroalloy or powder	Alloying element to improve strength and corrosion resistance
Iron	Powder and present in ferroalloys	Alloy matrix in iron-based deposits(a)
Manganese	Ferroalloy or powder	Deoxidizer; slag former; increases strength and hardness(b)
Molybdenum	Ferroalloy	Alloying element to increase hardness and strength; increases corrosion resistance in austenitic stainless steels
Nickel	Powder	Alloying element to improve hardness, strength, toughness and corrosion resistance
Potassium	Present in feldspars and silicates	Arc stabilizer; slag former
Silicon	Ferroalloy or present in silicates and feldspars	Deoxidizer; slag former
Sodium	Present in feldspars and silicates	Arc stabilizer; slag former
Titanium	Ferroalloy and present in rutile (TiO_2)	Deoxidizer and denitrifier; slag former(c)
Zirconium	Powder or oxide	Deoxidizer and denitrifier
Vanadium	Powder or oxide	Alloying element to increase strength

Note: Carbon, which increases hardness and strength in the weld, is present in various ferroalloys, e.g., ferromanganese.
(a) Iron serves as an alloying element in nickel-based deposits. (b) Manganese combines with sulfur to form manganese sulfide (MnS), which helps prevent hot shortness (high-temperature brittleness). (c) Titanium also stabilizes carbon in some stainless steels.

on the cylinder neck. At 70 °F, 8.47 ft^3 of carbon dioxide weighs 1 lb at standard pressure.

As the pressure drops from cylinder pressure to discharge pressure in the regulator, carbon dioxide absorbs a considerable amount of heat. If the flow rate is too high, the heat absorption can cause freezing of the regulator and flowmeter, which interrupts gas flow. Heaters are available that prevent regulator freeze-up. Excessive rates of gas flow also can result in the withdrawal of liquid carbon dioxide from the cylinder.

A positive pressure should always be maintained in the cylinder (the valve should be closed before the cylinder is removed from service) to keep moisture or other contaminants from backing into the cylinder. Carbon dioxide cylinders should always be kept in an upright position and held firmly in place when in use at the welding station. If it is placed in a horizontal position, carbon dioxide will be drawn off as liquid instead of as gas.

Siphoned-tube cylinders require a heater to transform the liquid to gas. Nonsiphoned-tube cylinders operating at flow rates greater than 35 ft^3/h should be manifolded to reduce flow rate and prevent freezing.

Flow rate for carbon dioxide shielding gas should be at least 30 ft^3/h for most welding applications. When welding outdoors or at a drafty site, the operation should be protected by a windshield, and the flow rate should be increased to 50 ft^3/h.

Purity. Welding-grade carbon dioxide is required. The purity of the gas is based on the percentage of moisture present, which is indicated by a dew point temperature. At −40 °F dew point, the percentage of moisture by weight is 0.0066, or 66 ppm.

Equipment Installations

Support equipment required for FCAW is about the same as for GMAW. Less equipment may be needed if no auxiliary gas shielding is applied. It can range from completely portable equipment to large, highly mechanized, stationary setups. Support equipment is dealt with in the section on equipment installations in the article "Gas Metal Arc Welding" in this Volume.

Holding and Handling of Workpieces

Jigging and fixturing requirements for FCAW are the same as those for GMAW. Holding and handling devices are discussed and illustrated in the article "Gas Metal Arc Welding" in this Volume. Applicable techniques are discussed also in the section on accessory equipment in the article "Gas Tungsten Arc Welding," and in the section on jigs, fixtures, and positioners in the article "Shielded Metal Arc Welding," both in this Volume.

The importance of positioning workpieces so that they can be welded in as flat a position as possible is no less, and may be greater, than when welding by GMAW. In most arc welding methods, maximum efficiency is obtained when welding is done in the flat position. Although FCAW can be done in all positions, out-of-position welding increases production time, restricts electrode selection, and requires greater operator skill.

Deposition Rate

As is true for all arc welding processes, the deposition rate for FCAW depends on the welding current and the electrode diameter. Typical relations between current and deposition rate for four different electrode-wire diameters are given in Fig. 9, which indicates for each electrode-wire diameter the current range that is normally appropriate. Deposition efficiency in FCAW is generally high, 70 to 85%, but occasionally as high as 92%.

Fig. 9 Effect of welding current and electrode-wire diameter on deposition rate. The solid part of each curve indicates the normal current range for that size of electrode.

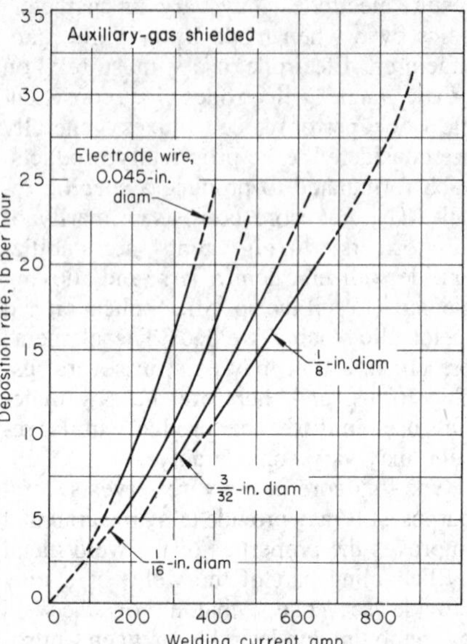

Effects of Operating Variables

The principal operating variables that must be controlled are arc voltage, current, travel speed, and electrode extension. The effects of changes in these variables are summarized in the paragraphs that follow.

Arc voltage variations have the following effects:

- Excessive arc voltage results in heavy spatter, porosity, and undercutting.
- Increasing the arc voltage flattens and widens the weld bead.
- Decreasing the voltage may cause a convex bead having a ropey appearance.
- Extremely low voltage causes the electrode to "stub" on the workpiece. The electrode dives through the molten weld pool and strikes the unmelted base metal at the bottom of the pool.
- With higher current, higher voltage can be used without causing porosity. Using the highest voltage possible (without causing porosity) will result in a weld-bead shape that is satisfactory for most applications.

Current variations have the following effects when arc voltage, travel speed, and electrode extension are all held constant:

- Excessive current produces convex weld beads, which result in waste of weld metal and poor appearance.
- Melting rate, deposition rate, and penetration are increased by increasing the current.
- Large-droplet transfer results from current that is too low, causing difficulty in maintaining a uniform weld bead.
- Increasing the current also increases the maximum voltage that can be used without causing porosity.

Travel speed variations have the following major effects when arc voltage, current, and electrode extension are held constant:

- Convexity of the weld bead, with uneven edges, and shallower penetration result from travel speed that is too high.
- Slag interference and slag inclusions, together with a rough, uneven weld bead, result from travel speed that is too slow.

To obtain weld beads of even contour, maintenance of uniform travel speed is essential. The welder should maintain a uniform distance between the end of the electrode and the molten slag behind the electrode.

As in all other welding processes in which the molten weld pool is protected

Table 7 Typical conditions for flat-position FCAW of square grooves and 60° single-V-grooves in low-carbon steel, using backing and auxiliary gas shielding (CO_2 at 35 ft^3/h)

Steel thickness (t), in.	No. of passes	Electrode diameter, in.	Current (DCEP), A	Voltage, V	Travel speed, in./min
Square-groove weld					
1/8(a)	1	3/32	325	24-26	56
3/16(b)	1	3/32	350	24-26	48
60° single-V-groove weld					
1/4	1	3/32	375	25-27	41
3/8	1	1/8	500	25-27	24
1/2	2	1/8	550	27-30	18
5/8	2	1/8	550	27-30	18
3/4	3	1/8	550	27-30	18
7/8	4	1/8	550	27-30	11
1	6	1/8	550	27-30	11

Square-groove weld

60° single-V-groove weld

(a) Root opening (w), 1/32 in. (b) Root opening (w), 1/16 in.

Table 8 Typical conditions for flat-position FCAW of 30 and 40° single-V-grooves in low-carbon steel, using backing and auxiliary gas shielding (CO_2 at 35 ft^3/h)

Steel thickness (t), in.	Pass No.	Electrode diameter, in.	Current (DCEP), A	Voltage, V	Travel speed, in./min
30° single-V-groove weld					
5/8	1	1/8	575	31	14
	2	1/8	600	32.5	16
3/4	1	1/8	575	32.5	19
	2	1/8	600	32.5	18
	3	1/8	600	32.5	15
1	1	1/8	575	31.5	21
	2	1/8	575	32	11
	3	1/8	575	32	15
	4	1/8	575	32	12
40° single-V-groove weld					
5/8	1	1/8	575	32	16
	2	1/8	600	32	13
3/4	1	1/8	575	32	23
	2	1/8	575	32	18
	3	1/8	600	32	15
1	1	1/8	575	31	15
	2	1/8	575	31	13
	3	1/8	575	31	15
	4	1/8	600	32	12

30° single-V-groove weld

40° single-V-groove weld

by slag or flux, a travel speed should be used that will produce the desired weld size and appearance by maintaining a proper relationship between the positions of the molten weld pool and the slag that protects it.

Electrode extension (stickout) is the distance between the electrode nozzle contact tip and the workpiece (Fig. 1). If voltage, current setting, and travel speed are held constant, variations in stickout have the following major effects:

- Increasing stickout decreases the welding current; decreasing stickout increases current.
- When stickout is increased, actual arc voltage is lowered. Lower arc voltage increases weld-bead convexity and reduces the likelihood of porosity.
- When stickout is excessive, spatter and irregular arc action will result.
- Short stickout gives greater penetration than long stickout.
- When stickout is too short, spatter will build up on the nozzle and contact tube.

Groove Welding

Welding position, work-metal thickness, shape of the groove, and use or nonuse of backing affect procedures and techniques employed in making groove welds. Typical welding conditions for several types of grooves welded in various positions are given in Tables 7 to 11. Specific applications may necessitate considerable deviation from the practices suggested in these tables; for instance, conditions given in the examples later in this article do not always

Table 9 Typical conditions for flat-position FCAW of a 45° single-bevel groove in 1-in.-thick low-carbon steel, using backing and auxiliary gas shielding (CO_2 at 35 ft^3/h)

Pass No.	Electrode diameter, in.	Current (DCEP), A	Voltage, V	Travel speed, in./min
1	1/8	600	32	17
2	1/8	600	32	24
3	1/8	600	32	18
4	1/8	600	32	15
5	1/8	600	32	16
6	1/8	600	32	21
7	1/8	600	32	21
8	1/8	600	32	18

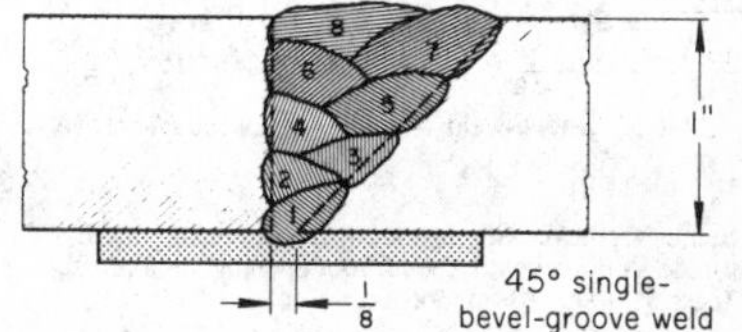

45° single-bevel-groove weld

Table 10 Typical conditions for horizontal-position FCAW of a single-bevel groove in 1-in.-thick low-carbon steel, using backing and auxiliary gas shielding (CO_2 at 35 ft³/h)

Pass No.	Electrode diameter, in.	Current (DCEP), A	Voltage, V	Travel speed, in./min
1	3/32	450	27	14
2	3/32	450	27	14
3-18	3/32	380	27	18

Using backing and auxiliary gas shielding (CO_2 at 35 ft³/h)

Single-bevel-groove weld, horizontal position

Table 11 Typical conditions for vertical-position FCAW of 60° V-grooves in low-carbon steel
Using auxiliary shielding with CO_2 at 35 ft³/h, DCEP, and 0.045-in.-diam electrodes

3/8-in. steel(a)

Current .... 180 A
Voltage .... 22 V

Travel speed, in./min:
- Pass 1 .... 13 down
- Pass 2 ..... 7.7 up
- Pass 3 .... 5 up

1-in. steel(b)

Current .... 180 A
Voltage .... 22 V
Travel speed, in./min:
- Pass 1 .... 13 down
- Pass 2 ..... 1.4 up
- Pass 3 ..... 2.3 up
- Pass 4 ..... 1.6 up
- Pass 5 .... 11 down

2-in. steel

(Double-V-groove; 0- to 1/16-in. root opening)
Current ... 190-200 A
Voltage .... 22 V
Travel speed, in./min:
- Pass 1 .... 11 down
- Pass 2 .... 3 up
- Pass 3 ..... 3.5 up
- Pass 4 ..... 2.1 up
- Pass 5 ..... 2.7 up
- Pass 6 .... 2 up
- Pass 7 ..... 1.8 up
- Pass 8 ..... 1.4 up
- Pass 9 ..... 1.3 up

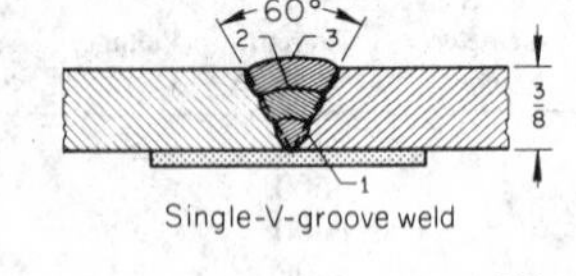

Single-V-groove weld

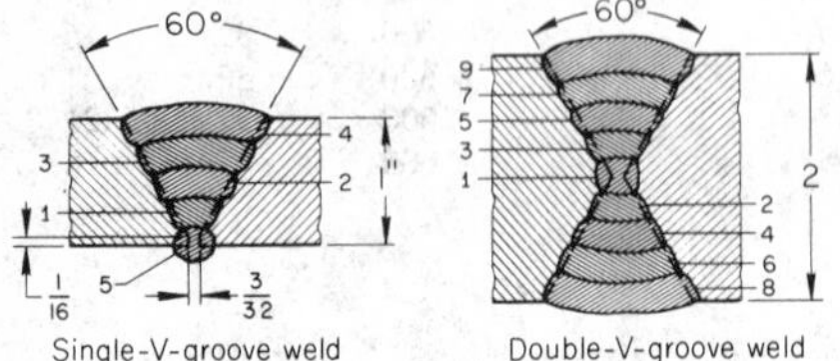

Single-V-groove weld   Double-V-groove weld

(a) Single-V-groove with no root opening; backing used.
(b) Single-V-groove with 3/32-in. root opening. Before back welding (pass 5), joint is back gouged.

agree with those given in Tables 7 to 11. However, the data shown in these tables are based on extensive development work and should provide useful guidelines for establishing welding procedures for specific applications.

Fillet Welding

As with groove welding, procedures for fillet welding vary widely. The welding conditions shown in Tables 12 and 13 provide guidelines for several sets of conditions. Welding position has a major influence on technique. For example, Table 12 shows that a 1/2-in. fillet weld can be made in two passes in the flat position, whereas Table 13 indicates that three passes are required for a 1/2-in. fillet weld in the horizontal position.

Angle of the electrode wire to the joint has a considerable effect on the appearance of the completed weld deposit. The electrode wire should point at the bottom plate close to the corner of the joint if best bead shape is desired on single-pass 5/16- and 3/8-in. horizontal fillet welds. The angle between the electrode wire and the bottom plate should be less than 45°, because with this position the molten metal is caused to wash up onto the vertical member. If root porosity occurs, it may be decreased by pointing the electrode directly into the joint and using an angle of 45 to 55°, but this may cause some weld spatter, as well as a convex weld bead. For 1/4-in. and smaller fillet welds, the electrode wire should be pointed directly into the joint and at an angle to the joint about 40° above horizontal.

Multiple-Position Welding

The development of small-diameter electrode wire (0.062 in. or less) has made it possible to employ FCAW in the vertical position, thereby extending the use of the process to building construction and tank fabrication. The desirability of using FCAW for structural members helped to establish conditions that meet code requirements. Typical electrode sizes and amperage and voltage ranges are given in Table 14. These conditions apply to both groove and fillet welding. With a choice of 1/16-in.- and 5/64-in.-diam electrode wires, and using the appropriate type of flux and suitable process adjustments, welding can be done in any position.

Downhill and vertical-down welding can be done in low-cost, single-pass welds. About a 60° downhill angle and 5/64- or 3/32-in. electrode wires result in maximum deposit speed, although the smaller 1/16-in. electrode can also be used. To make such deposits, stringer beads and currents in the middle to high part of the amperage range for the electrode diameter should be used. If the electrode holder is tipped in

Table 12 Typical conditions for flat-position flux cored arc fillet welding of low-carbon steel, using auxiliary gas shielding (CO_2 at 35 ft³/h)(a)

Steel thickness (*t*), in.	Electrode diameter, in.	No. of passes	Current (DCEP), A	Voltage, V	Travel speed, in./min
1/8	3/32	1	300	24-26	53
3/16	3/32	1	350	24-26	41
	1/8	1	450	24-26	40
1/4	3/32	1	400	24-26	24
	1/8	1	500	25-27	25
5/16	3/32	1	550	28-30	22
	1/8	1	460	26-28	20
3/8	3/32	1	575	29-31	20
	1/8	1	575	29-31	20
1/2	3/32	2	525	30-32	16
	1/8	2	525	30-32	16
5/8	3/32	3	475	29-31	12
	1/8	3	450	27-29	14
3/4	3/32	3	500	29-31	13
	1/8	3	500	28-30	12

Corner joint   T-joint   Lap joint

(a) Weld size (*s*) is usually the same as steel thickness (*t*) for the range of steel thicknesses given here.

Table 13 Typical conditions for flux cored arc fillet welding in the horizontal and vertical positions, using auxiliary gas shielding (CO_2 at 35 ft^3/h)(a)

Steel thickness (t), in.	Electrode diameter, in.	No. of passes	Current (DCEP), A	Voltage, V	Travel speed, in./min
Horizontal position					
1/8	3/32	1	350	24-26	60
3/16	3/32	1	400	24-26	41
	1/8	1	425	24-26	32
1/4	3/32	1	400	24-26	24
	1/8	1	450	25-27	25
5/16	3/32	1	440	25-27	20
	1/8	1	460	26-28	20
3/8	3/32	1	475	26-28	15
	1/8	1	500	28-80	14
Horizontal position (continued)					
1/2	3/32	3	400	24-26	20
	1/8	3	450	25-27	18
5/8	3/32	3	450	26-28	14
	1/8	3	450	27-29	14
3/4	3/32	6	470	28-30	20
	1/8	6	470	28-30	20
Vertical position					
3/8	0.045	1	180	21	3-4

s
t
Lap joints
T-joint
T-joint (multiple-pass weld)
Welding in horizontal position
Corner joint
T-joint
Lap joint
Welding in vertical position

(a) Weld size (s) usually equals steel thickness (t) for the range of thicknesses given here.

Table 14 Typical conditions for multiple-position FCAW of grooves and fillets, using various sizes of electrodes(a)

	Amperage (DCEP) and voltage(b)					
	Flat		Horizontal		Vertical	
Electrode diameter, in.	Amperes	Volts	Amperes	Volts	Amperes	Volts
0.045	150-225	22-27	150-225	22-26	125-220	22-25
1/16	175-300	24-29	175-275	25-28	150-200	24-27
5/64	200-400	25-30	200-375	26-30	175-225	25-29
3/32	300-500	25-32	300-450	25-30	...	...
7/64	400-525	26-33	...	...	...	...
1/8	450-650	28-34	...	...	...	...

(a) Flow rate of shielding gas is 30 to 45 ft^3/h, depending on electrode size. (b) Current ranges can be expanded. Higher amperage can be used with automatic travel. Voltage will increase when longer electrical extension (stickout) is used.

the direction of travel so that the arc force tends to hold the molten metal in the joint, the deposits will appear much more uniform. Welding should be done as fast as is possible for the desired size and shape of weld bead.

Out-of-position welding can be done if a few basic rules are followed: (1) do not whip or move the arc around rapidly, (2) do not break the arc too rapidly, (3) do not move out of the weld pool too fast, (4) do not move too fast in any direction, and (5) use current in the low part of the amperage range.

Thick Sections

The high deposition rates and deep joint penetration obtainable with auxiliary gas shielded FCAW have proved the process to be economical for welding sections more than 1/2 in. thick.

Sections 1/2 to 3/4 in. Thick (Comparison of Processes). The two examples that follow describe the practice used for, and indicate specific advantages of, auxiliary gas shielded FCAW as compared with SMAW for sections between 1/2 and 3/4 in. thick.

Advantages of FCAW Over SMAW

Example 1. Increased Deposition—Smaller Fillets. The weldment shown in Fig. 10, consisting of two flat plates 1/2 in. thick and one formed plate 3/4 in. thick, was originally joined by SMAW, using 3/8-in. fillet welds. The deposition rate was 6 to 10 lb of weld metal per hour. By changing to auxiliary gas shielded FCAW, the deposition rate was increased to 12 to 18 lb per hour. In addition, because of the deeper penetration obtained in auxiliary gas shielded FCAW, it was possible to use a 1/4-in. fillet weld. The strength of the joint was equal to that of the joint made with the 3/8-in. fillet weld deposited by SMAW.

The higher deposition rate and smaller fillet weld made FCAW more economical. Welding details for the flux cored arc process are given in the table accompanying Fig. 10.

Fig. 10 Weldment of thick plates, for which process was changed from SMAW to FCAW. When SMAW was used, 3/8-in. fillet welds were required and deposition rate was only 6 to 10 lb/h.

Low-carbon steel; low-carbon steel filler metal

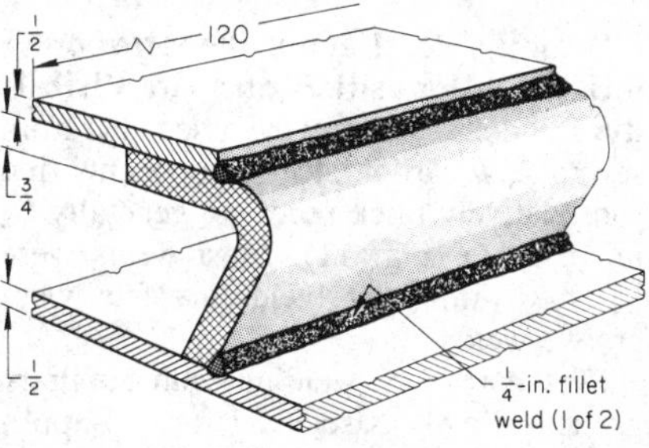

Auxiliary gas shielded FCAW

Joint type	Lap and modified T
Weld type	1/4-in. fillet
Welding position	Horizontal
Number of passes	1
Shielding gas	Carbon dioxide at 35 ft^3/h
Electrode	3/32-in.-diam flux cored wire
Electrode feed	160-175 in./min
Current	420-450 A
Voltage	29-31 V
Travel speed	14-15 in./min
Deposition rate	12-18 lb/h

Fig. 11 Bulldozer blade and comparison of joint penetration (and actual throat depth) of fillet welds made by SMAW and by auxiliary gas shielded FCAW

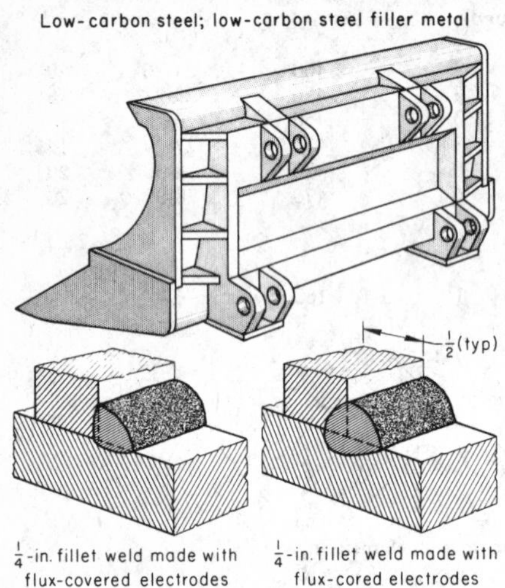

Auxiliary gas shielded FCAW

Joint type	Corner
Weld type	Fillet; some groove
Weld size	$^1/_4$ to $^1/_2$ in.
Welding position	Flat; horizontal
Number of passes for fillet welds:	
For $^1/_4$ and $^5/_{16}$ in., flat position	1
For $^3/_8$ in., horizontal position	2
For $^7/_{16}$ and $^1/_2$ in., flat position	1
For $^7/_{16}$ and $^1/_2$ in., horizontal position	3
Shielding gas	Carbon dioxide at 35 ft^3/h
Electrode	$^3/_{32}$-in.-diam flux cored wire
Current	450 A
Voltage	30-32 V
Electrode feed	206 in./min

Example 2. Deeper Penetration—Increased Deposition and Arc Visibility. Bulldozer blades were assembled from several low-carbon steel components that had relatively thick sections, generally $^1/_2$ in. or more (Fig. 11). Most welds were $^1/_4$- to $^1/_2$-in. fillet welds. A few were groove welds.

Flux cored arc welding with auxiliary gas shielding was selected for this application in preference to SMAW for three reasons: (1) deeper joint penetration, permitting the use of smaller fillets without decreasing the strength of the joint; (2) higher deposition rate; and (3) greater visibility of the arc to the welder, resulting in a better weld. The difference in joint penetration for the two processes is shown in Fig. 11. Details for FCAW are given in the table accompanying Fig. 11.

Sections 3 to 15 in. Thick. Thicknesses much greater than those discussed in Examples 1 and 2 are often welded by auxiliary gas shielded flux cored arc methods. The two examples that follow describe applications of this process for such sections.

Example 3. Use of Auxiliary Gas Shielded FCAW and a High-Strength Electrode Wire for Welds Subjected to Heavy Loads. The shank portion of the weldment shown in Fig. 12, a component of an earthmoving machine, was subjected to heavy loads in service. Auxiliary gas shielded FCAW with a high-strength electrode was selected to join the 3-in. section to the $3^1/_2$-in. section, because it provided the required strength and weld quality and was faster than SMAW. Welding details, including the sequence of passes, are given in the table accompanying Fig. 12.

Example 4. Welding 4-in.-Thick Plates for I-Beam Flange Sections. The plates shown in Fig. 13 were successfully butt welded by FCAW with auxiliary gas shielding, using a double-V-groove. Runoff tabs attached to both ends of the joint, after fit-up but prior to welding, provided a means for starting and stopping the weld. After welding, the runoff tabs were removed, and the edges of the plates were ground smooth.

Welding was started on the side with the 60° groove (side A in Fig. 13) and was continued until that groove was about one-quarter full. The plates were then turned over, and the groove area was back gouged until sound weld metal was reached. Welding was then continued on the side with the 80° groove until this groove was about half full. The plates were turned again, and welding was completed on the 60°-groove side. The plates were turned once more to the 80°-groove side, and

Fig. 12 Thick-section earthmover part that was flux cored arc welded

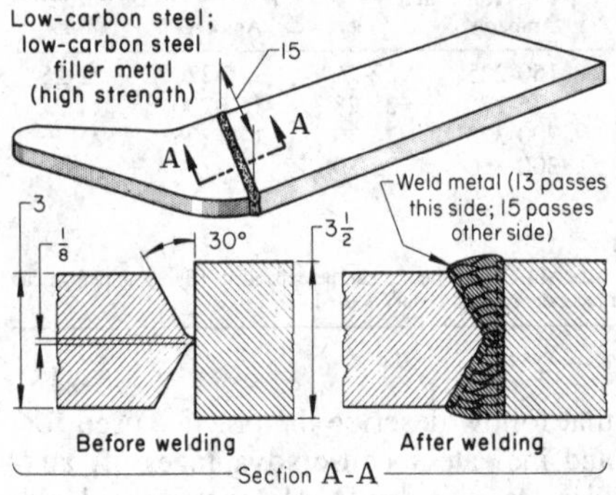

Auxiliary gas shielded FCAW

Joint type	Butt
Weld type	Double-J-groove
Welding position	Flat
Shielding gas	Carbon dioxide at 35 ft^3/h
Electrode	$^3/_{32}$-in.-diam high-strength flux cored wire
Current	450 A (DCEP)
Voltage	30-32 V
Electrode feed	206 in./min

Sequence of passes

First side. Make four passes without weaving. Turn to second side.
Second side. Back gouge to obtain complete joint penetration and weld four passes as on the first side.
Alternate sides. Make two passes per side until 13 passes have been made on the first side and 15 passes on the second side. Weld from center to end; deslag after each pass; use weaving technique. Use copper dams for the last three passes on each side.

Fig. 13 Joint and welding sequence used for joining plates 4 in. thick by 26 in. wide by FCAW with auxiliary gas shielding. Numbers shown on weld metal represent sequence of welding, not number of passes; many small passes were used.

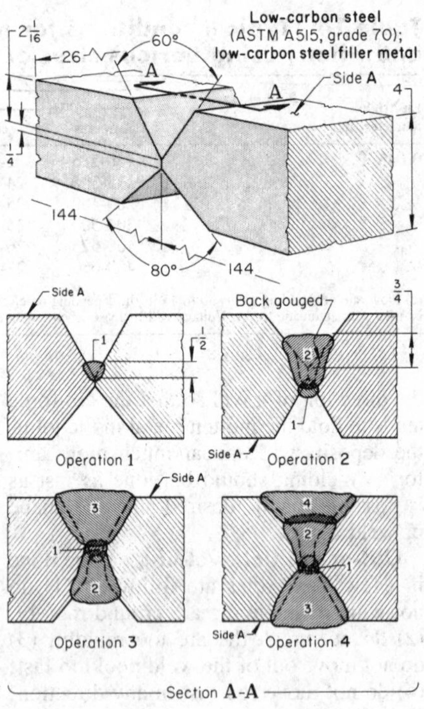

Auxiliary gas shielded FCAW

Joint type	Butt
Weld type	Double-V-groove
Welding position	Flat
Shielding gas	Carbon dioxide
Electrode	$^3/_{32}$-in.-diam flux cored wire
Power supply	600-A motor-generator, drooping output
Deposition rate	7.1 lb/h

Fig. 14 Joint design and pass sequence for fixed-position welding of 1/2-in.-wall pipe with a 0.045-in.-diam electrode wire

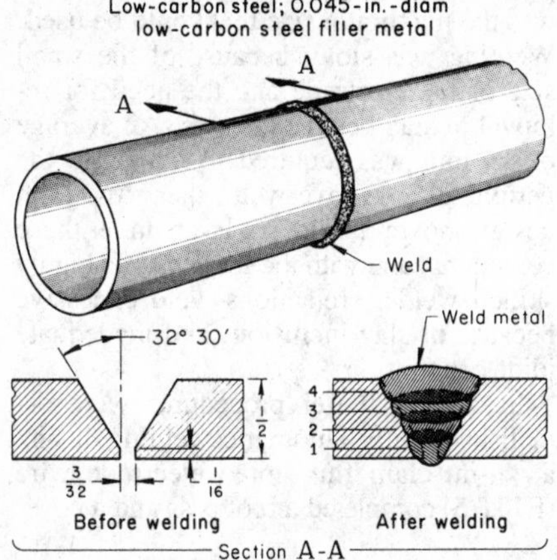

Pass	Current (DCEP), A	Voltage, V	Gas (CO_2) flow, ft^3/min	Travel direction
1	100	20	25	Down
2, 3, 4	120	20	25	Up

Note: Interpass temperature 300 °F; pipe axis horizontal

welding was completed. Turning the plates during welding minimized distortion.

Weldments were essentially free of defects, as determined by radiographic examination and testing of side-bend specimens. Tests of transverse tensile specimens showed that the tensile strength of the welds in the stress-relieved condition ranged from 75 to 80 ksi. Welding details are given in the table accompanying Fig. 13.

Pipe Welding

When pipe sections can be rotated during welding, the operation is greatly simplified, because all of the welding can be done in the most favorable position, the downhand position. However, when elbows or other irregular sections are involved, welding must be done with the workpieces in a fixed position unless special manipulating fixtures are provided.

Fixed-Position Welding. Although the larger sizes of flux cored electrode wires are not suitable for welding pipe in the fixed position, development of the smaller (0.045- to 1/16-in.-diam) flux cored electrode wire for welding with auxiliary gas shielding enables this to be done on low-carbon and low-alloy steel pipe sections to a quality standard that satisfies requirements of the American Society of Mechanical Engineers (ASME) Boiler and Pressure Vessel Code.

Joint designs for welding with small-diameter flux cored electrodes are the same as those used for SMAW. Joint design and pass sequence for fixed-position welding of pipe having a wall thickness of 1/2 in. are shown in Fig. 14. Typical welding conditions are given in the table that accompanies Fig. 14.

Rotation of the workpieces, which permits all welding to be performed in the downhand position, is greatly preferred for pipe welding and is done whenever possible. In the example that follows, thick-walled pipe and tube were rotated during welding.

Example 5. Butt Welding of Large-Diameter Thick-Walled Pipe and Tube. Steel pipe and tube, ranging in outside diameter from 8 to 36 in. and in wall thickness from 1/2 to 3 in., were butt welded in the horizontal-rolled position. The groove design used is shown in Fig. 15.

The root pass was made by GMAW, using a 0.035-in.-diam solid electrode and argon-carbon dioxide shielding. Pipe was preheated to 70 °F, 125 °F, or 200 °F, depending on wall thickness (see the table accompanying Fig. 15). After preheating, maximum root opening was 3/32 in. The root pass was made with a hand-held electrode holder positioned 45° above the horizontal centerline as the pipe rotated away from the welder.

For subsequent passes (3 to 17, depending on wall thickness), FCAW was used, with a 7/64-in.-diam flux cored electrode wire and carbon dioxide shielding. To control the weld pool during these passes (because a large-diameter electrode wire was employed), the electrode holder was held 1/2 to 3/4 in. below the vertical centerline, with the pipe rotating away from the welder. Maximum width of weave was 1 1/4 in. When the groove width exceeded 1 1/4 in., the split-weave technique was used. For the first few passes that followed the root pass, the voltage and amperage were near the low end of the "remaining passes" ranges in the table with Fig. 15. As the depth of weld buildup increased, lessening the possibility of melt-through, voltage and amperage were increased to the high end of these ranges. Travel speed varied with pipe diameter; it was controlled to ensure that slag did not run ahead of the arc. Slag was removed between passes. Deposition rate for FCAW ranged from 15 to 25 lb/h.

Typical mechanical properties of the weld metal were 76 ksi tensile strength, 67 ksi yield strength, 32% elongation, and 68% reduction of area. Charpy V-notch impact values of the weld metal at selected temperatures were as follows: 92 ft · lb at 75 °F, 75 ft · lb at 0 °F, 48 ft · lb at −40 °F, and 30 ft · lb at −75 °F. The

Fig. 15 Circumferential butt weld in pipe or tubing 8 to 36 in. in outside diameter and 1/2 to 3 in. in wall thickness, and joint design. Horizontal-rolled position was used in making the weld.

Joint type	Circumferential butt
Weld type	Modified U-groove
Process	GMAW and FCAW(a)
Welding position	Horizontal rolled
Fixture	Variable-speed turning rolls
Preheating:	
1/2- to 1 1/2-in. wall thickness	70 °F
1 1/2- to 2 1/4-in. wall thickness	125 °F
2 1/4- to 3-in. wall thickness	200 °F
Shielding gas:	
Root pass	75% Ar-25% CO_2, at 15-20 ft^3/h
Remaining passes	CO_2, at 40-45 ft^3/h
Electrode:	
Root pass	Solid wire(b)
Remaining passes	Flux cored wire(c)
Power supply	500 A constant-voltage three-phase rectifier, with slope control
Current and voltage:	
Root pass	90-110 A, 17-19 V
Remaining passes	300-475 A, 24-28 V
Wire feeder	Constant-feed, either two standard units or a special unit featuring dual-drive mechanism

(a) Root pass was made by GMAW and remaining passes by FCAW with auxiliary gas shielding. (b) Composition, 0.06% C max, 1.20% Mn, 0.50% Si, 0.05 to 0.15% Al, 0.02 to 0.12% Zr, 0.05 to 0.15% Ti. (c) Composition, 0.09% C, 1.00% Mn, 0.45% Si

joint and procedures were qualified by tests made according to the ASME Boiler and Pressure Vessel Code, American National Standard Code for Pressure Piping, and U.S. Navy specifications.

Arc Spot Welding

In arc spot welding, a weld is made in a lap joint through one piece of metal into the other piece. There is no travel of the electrode holder or workpiece. For thin sheets, the weld is a melt-through weld, made without preparing a hole in the top sheet.

Thickness and Position Limitations. Flux cored arc spot welding is used to make lap joints in low-carbon steel $^1/_{16}$ to $^1/_4$ in. thick. Sheet or plate of the same or of different thicknesses can be welded together. If stock of different thicknesses is being welded, the weld should be made from the side of the thinner member. Arc spot welding is best done in the flat position, although thin sheet can be welded in other positions.

Typical Procedures and Conditions. In flux cored arc spot welding, an air-cooled electrode holder equipped with a special nozzle is held against the top member of a lap joint. The arc, which is maintained by the continuously fed electrode, melts through the top sheet into the bottom sheet and fuses the two sheets together. The electrode continues to feed for a preset arc time and produces a slightly convex spot on the upper surface of the top sheet.

Most flux cored arc spot welding is done with carbon dioxide shielding, because this shielding provides deeper penetration. However, the self-shielding method can be used for arc spot welding of thin sheet (up to $^1/_8$ in. thick). Typical welding conditions for flux cored arc spot welding of steel from $^1/_{16}$ to $^1/_4$ in. thick are given in Table 15.

Weld Characteristics. The strength of arc spot welds produced by FCAW is about the same as that produced by resistance welding in steel of the same thickness. Typical values of shear strength of arc spot welds are given in Table 15.

Arc spot welding with a flux cored electrode offers several advantages over arc spot welding with a solid electrode. Generally, the weld nugget is larger at the interface when welding through metal of the same thickness, which results in higher strength per spot weld. Also, with a flux cored electrode, there is a slag cover on the surface of the weld, which results in a smoother surface and, for some applications, eliminates the necessity for a finishing operation.

A flux cored electrode has been found to be better than a solid electrode for spot welding galvanized steel; it provides smoother surfaces and deeper penetration. Also, weld spatter from flux cored electrodes does not adhere to galvanized surfaces. For spot welding of galvanized steel, the current is reduced slightly and the arc time is increased, compared with spot welding of uncoated sheet of the same thickness.

Table 15 Typical conditions for flux cored arc spot welding and shear strength per spot

Steel thickness, in.	Current (DCEP), A	Voltage, V	Arc time, s	Shear strength per spot, lb
With carbon dioxide shielding(a)				
$^1/_{16}$	400	30	0.6	2 550
$^1/_8$	500	34	0.8	3 400
$^3/_{16}$	650	38	1.6	7 050
$^1/_4$	750	40	2.2	10 300
Self-shielding(b)				
$^1/_{16}$	500	29	0.5	2 250
$^1/_8$	500	29	1.5	2 450

(a) $^3/_{32}$-in.-diam E70T-2 electrode; electrode extension, $^7/_8$ in.; flow rate of shielding gas, 35 ft^3/h. (b) $^3/_{32}$-in.-diam E70T-4 electrode; electrode extension, $1^1/_2$ in.

Automatic Welding

Flux cored arc welding can be done automatically, using essentially the same techniques as for GMAW. Electrode holders used for automatic GMAW can be used also for automatic FCAW (see the section on electrode holders in this article). Power supply, wire feeders, and other basic equipment for automatic welding are described in that section.

The quantity of similar weldments required determines whether or not automatic welding is feasible and the degree of automation that is appropriate. For some applications, quality requirements are so high that they can be met economically only by automatic welding. Automatic welding is most advantageously applied to long in-line welds such as long seam welds and girth welds on large pipes. Tooling cost for an application of this type is generally moderate, but increases as weld complexity increases.

In some applications, tooling can be simplified by moving the components to be welded while the electrode holder remains stationary. If the workpiece is large or of complex shape, moving it past the electrode holder may not be practical, and the alternative of moving the electrode holder will have to be adopted.

Circumferential welding is usually easy to automate, because round workpieces of all sizes can almost always be held and rotated without difficulty. In practice, the electrode holder remains stationary and makes the weld as the workpiece slowly rotates. The two examples that follow describe applications in which automatic welding was advantageous.

Example 6. Change from SMAW to Automatic FCAW of Heat Exchanger Components. A shell and a fitting for a heat exchanger were originally joined by SMAW. Because of the relatively small diameter of the workpiece (3-in. schedule 40 pipe) and the requirement for 100% joint penetration, a $^3/_{32}$-in.-diam electrode wire was the maximum size that could be used. Welding was slow, because of the small size of the electrode and the need for removal of slag between passes. An average of 13 min was required to complete the peripheral V-groove weld, using the joint design shown in Fig. 16. Even under these conditions, and with the services of a highly skilled welder, rejections were excessive because of slag inclusions and inadequate joint penetration.

In contrast, after procedures were established, fully automatic welding, using a $^7/_{64}$-in.-diam flux cored electrode wire (E70T-5) completed a root pass and a cover

Fig. 16 Heat exchanger shell and fitting, and joint designs for SMAW and automatic FCAW

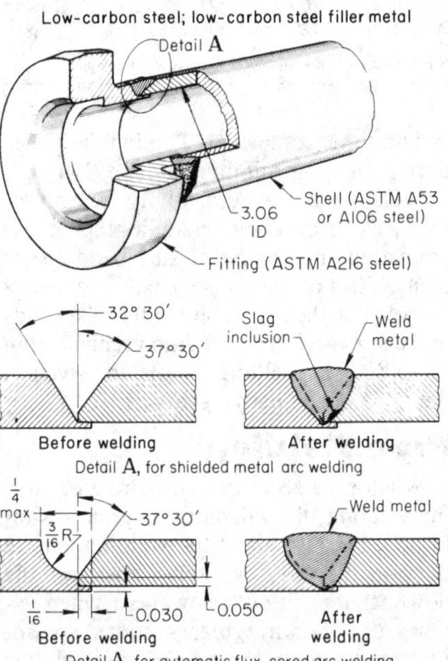

Automatic FCAW

Joint type	Modified lap and butt (see figure)
Weld type	Groove (see figure)
Welding position	Horizontal rolled
Preheat	None
Shielding gas	Carbon dioxide at 30 ft^3/h
Electrode	$^7/_{64}$-in.-diam E70T-5
Electrode holder	Air cooled, mechanically held
Power supply	500-A constant-voltage three-phase rectifier
Current	430 A (DCEP)
Voltage	28 V
Wire feeder	Constant feed
Welding time:	
Root pass	16 s
Cover pass	22 s

pass in 38 s. For automatic FCAW, the joint was redesigned as shown in Fig. 16, although the 37$^1/_2$° bevel angle was retained on the shell. This change in joint design prevented slag inclusions and lack of penetration. Because use of electrode wire E70T-5 results in a very thin friable slag layer, it was possible to make the two-pass weld without removing slag between passes. Even with a relatively unskilled operator, acceptable parts were produced more consistently.

Conditions for automatic FCAW are given in the table that accompanies Fig. 16. A mechanically held, retractable, air-cooled electrode holder was used. Other special tooling included an air-operated internal chucking device to pull the shell into intimate contact with the fitting, equipment for automatic loading and unloading of shells, and a mechanism for variable-speed rotation. It was required that the welding procedure be qualified by tests performed in accordance with the ASME Boiler and Pressure Vessel Code.

Example 7. Welding Long Channel Sections by Automatic FCAW. The fully automatic equipment for FCAW shown in Fig. 17 was developed to butt weld two structural channels into a box section for a hitch for a gang plow. Originally, the box sections had been produced by SMAW. Cost of welding was reduced about 40% by using the auxiliary gas shielded flux cored arc method. A principal factor in the decrease in cost was reduction of labor from two welders to one operator.

The channels were cut to length and all holes were punched in a progressive die to make a pair of channels, one a right-hand section and the other a left-hand section. Channel length was from 30 in. to 13$^1/_2$ ft for different designs.

The operator slid a pair of channels onto a roller conveyor, ahead of the entry end of the welding machine, and aligned the channels to form the box. The channels were then pushed into the welding machine on rollers, which became the bottom supporting surface for the workpiece during welding. A pin was inserted through holes in the frame of the machine and through matching holes in each end of the pair of channels, to hold hole alignment. Next, air cylinders were actuated by a foot switch to clamp the workpiece. The electrode holder was brought into line, the drive was locked, and the start button was pushed to begin welding on the first side. The locating pins were removed as the welding head progressed along the seam.

Fig. 17 Machine for automatic FCAW of long channels into box sections. Joint design and location for welds on the box section are shown at the left.

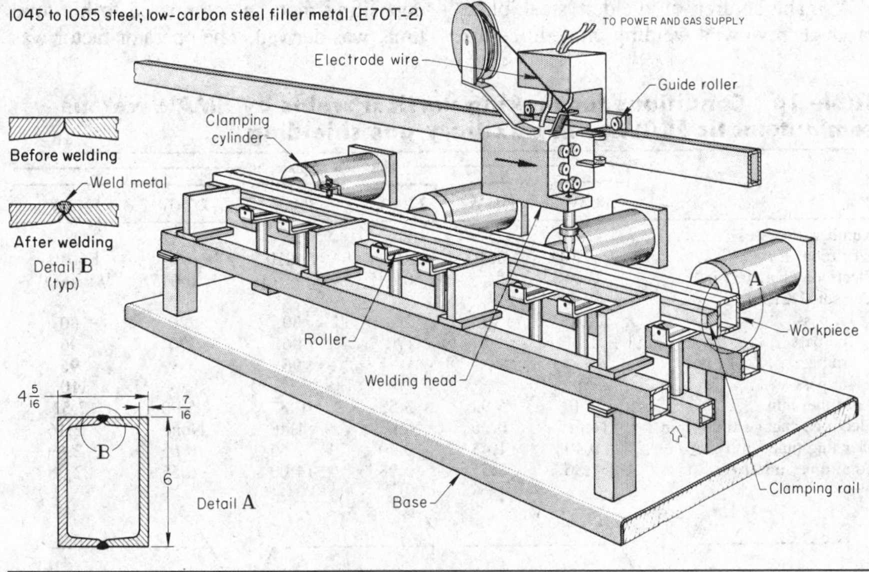

Number of passes	1	Electrode consumption	9 lb/h
Preheat and postheat	None	Power supply	600-A constant-voltage motor-generator
Shielding gas	Carbon dioxide at 45 ft^3/h	Current	290-300 A (DCEP)
Electrode	$^3/_{32}$-in.-diam E70T-2	Voltage	29-30 V
Electrode holder	Mechanically held, CO_2 cooled	Wire feeder	Constant feed
Electrode-feed rate	95 in./min	Travel speed	62-65 in./min
Electrode extension (stickout)	1$^1/_4$ in. max		

On completing the first seam, the welding head was returned to the starting position, the air clamps were released, and by means of an air-operated linkage, the rollers supporting the channels were raised above the machine frame so that the workpiece could be turned over to weld the second side. The rollers were then lowered, the workpiece was reclamped, and the second side was welded. On completion of the second weld, the air clamps were released, the weldment was struck six or more blows with a hammer to free flux deposits, and the workpiece was pushed out onto a conveyor table.

Because the equipment operated automatically, the operator was free to prepare the next pair of channels while the pair in the machine was being welded. While the second pair was being welded, the completed workpiece was removed from the exit conveyor table to a skid, by means of a bridge crane.

The welding machine was extremely versatile. Various sizes of channels were welded, ranging from 4 by 1$^3/_4$ in. to 6 by 3 in. and in length from 30 in. to 13$^1/_2$ ft. V-notches were added to the clamping jaws to make possible welding of straight-line corner joints joining pairs of angles into box sections, and the gas nozzle on the electrode holder was pointed 8 to 10° in the travel direction to expel dirt, rust, and oil from the joint area and to spread the weld deposit. Joint penetration was found to be 90 to 95%.

In the original method (SMAW), a $^5/_{32}$-in.-diam E7018 electrode was operated at 275 A DCEP, 30 to 31 V, at a travel speed of 10 to 12 in./min.

Submerged arc welding was considered as an alternative process for joining the structural channels, but the channels did not match up evenly enough for this type of welding. In addition, it was estimated that the removal of unfused flux would require too much time and that the consumption of flux would be excessive. Details for automatic FCAW of the channel sections are given in the table that accompanies Fig. 17.

Twin-Electrode (Twin-Wire) Welding. When automatic welding equipment is used, two electrode holders can be mounted on a single carriage for twin-electrode FCAW (Fig. 18). For example, the twin-electrode technique was used for depositing horizontal fillet welds $^1/_4$ to $^3/_8$ in. in size and flat fillet welds up to $^3/_4$ in. in size. Both of the electrode wires were $^1/_8$ in. in diameter, and they were in line along the weld joint, with one electrode trailing about 2 in. behind the other. Thus, there were two arcs and two separate weld

Fig. 18 Position of electrodes, and weld deposits made, for automatic twin-electrode welding in the flat position

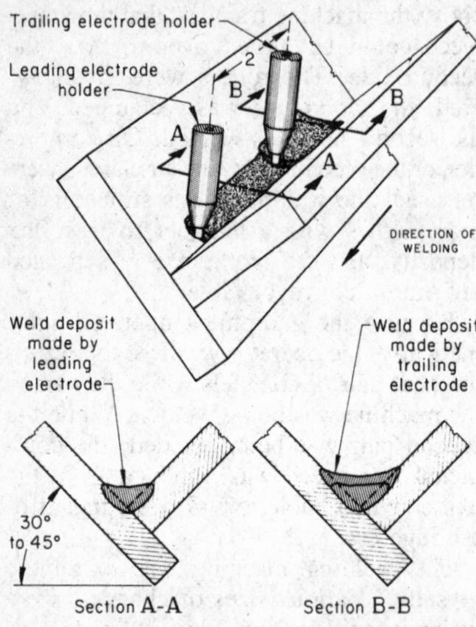

Fillet size, in.	Electrode diam, in.	Total current (DCEP), A	Voltage, V	Travel speed, in./min
1/2	1/8	1300	35-36	34
7/8	1/8	1300	35-36	13

pools. The two weld pools were smaller and were more easily controlled than the single, larger pool that would have resulted if a weld of the same size had been deposited from a single electrode in one pass. Electrode positions used for the twin-electrode technique are shown in Fig. 18, and typical welding conditions for depositing 1/2- and 7/8-in. fillet welds are given in the table that accompanies Fig. 18.

Flux Cored Arc Welding Versus Alternative Welding Processes

For applications in which two or more welding processes can produce acceptable results, the choice of process depends mainly on the equipment available, the skill of the welders, the number of similar weldments to be produced, and the cost. For some applications, SMAW, GMAW, and SAW compete with FCAW.

Flux Cored Arc Welding Versus Shielded Metal Arc Welding. Although FCAW is less versatile than SMAW, it is usually faster, because it uses a continuously fed electrode wire. Also, the flux cored method with auxiliary gas shielding results in deeper penetration, which often permits a smaller fillet to be used to achieve the required joint strength, thus requiring less filler metal.

Development of small-diameter flux cored electrode wires has increased the versatility of FCAW. Prior to this development, FCAW was generally restricted to welding in the flat and horizontal positions. The examples that follow compare FCAW with SMAW for making vertical groove welds and fillet welds.

Example 8. Auxiliary Gas Shielded FCAW Versus SMAW for Vertical Groove Welds. A study was conducted to compare FCAW with SMAW for making 1/4-in. vertical groove welds 12 in. long on 1/4-, 3/8-, and 1/2-in.-thick steel plates. The time required to make these welds is shown in Table 16. Flux cored arc welding proved to be three to four times faster than SMAW. Joint designs for the two processes are shown in the figure that accompanies Table 16. The 1/8-in. flux-covered electrode required a relatively wide root opening for the root pass, but a narrow root opening could be used for making the root pass with the 0.045-in. flux cored electrode. The lower total time for FCAW resulted from fewer passes (for the 3/8- and 1/2-in.-thick plates), less arc time, and less cleaning time.

Flux Cored Arc Welding Without Auxiliary Gas Shielding Versus SMAW for Fillet Welds

A manufacturer of welded steel plate products reviewed welding procedures for fillet welds. All fillet welds were being made by SMAW. Three types of semiautomatic welding were initially considered: gas metal arc, flux cored arc with carbon dioxide shielding, and flux cored arc without auxiliary gas shielding. After reviewing costs of equipment, shielding gas, welder training, and portability and operability in the field, it was decided to limit the comparative study to FCAW without auxiliary gas shielding and SMAW.

Two important objectives of the investigation were to determine the actual length of weld that could be deposited per hour and the amount of electrode consumed per foot of weld for the two processes. To do this, three types of joints commonly used in shop fabrication were selected. Joints of each type were welded by the two processes and data pertinent to the comparison were recorded. Shielded metal arc welding was done according to the procedures then in use in the shop. Flux cored arc welding was done using electrodes of the same size and type, and essentially the same current and electrode-wire feed rate, for all three joints.

Example 9. T-Joint. A T-joint in 1/4-in. plate was selected for this test. The requirement was to deposit a single 1/4-in. fillet weld in the flat position, as shown in the view for this example in Table 17. When arc-time studies were made, a special electric clock was hooked up with the welding circuit, and an "operator factor," which reflected the time a welder spent depositing metal as compared to his total time, was derived. The operator factor was

Table 16 Conditions for making vertical welds by SMAW versus semiautomatic FCAW with auxiliary gas shielding

Item	Thickness of plate, in.: 1/4 FCAW	1/4 SMAW	3/8 FCAW	3/8 SMAW	1/2 FCAW	1/2 SMAW
Number of passes	2	2	2	3	2	4
Electrode	E70T-1	E6010	E70T-1	E6010	E70T-1	E6010
Electrode diameter, in.	0.045	1/8	0.045	1/8; 5/32(a)	0.045	1/8; 5/32(a)
Current (DCEP), A:						
1st pass	220	80	220	80	220	80
2nd pass	220	80	170	96	170	96
3rd pass	...	...	...	96	...	96
4th pass	...	...	...	...	...	110
Arc time, min	1.46	5.92	3.58	10.87	5.90	17.53
Electrode-change time, min	None	0.96	None	1.48	None	2.16
Cleaning time, min	0.40	1.43	0.40	1.65	0.65	2.69
Total time, min(b)	1.86	8.31	3.98	14.00	6.55	22.38

60° — 1/8 — 1/8 — Shielded metal-arc welding

60° — 1/16 — 3/32 — Flux-cored arc welding

(a) 1/8 in. for first pass only. Remainder of passes, use 5/32-in. electrodes. (b) All times based on making a 12-in.-long weld in each thickness.

Table 17 Production rate, electrode consumption, and welding variables for fillet welds made by SMAW versus FCAW without auxiliary gas shielding(a)

Item	Example 9 — T-joint FCAW	Example 9 — T-joint SMAW	Example 10 — Corner joint FCAW	Example 10 — Corner joint SMAW	Example 11 — Corner joint FCAW	Example 11 — Corner joint SMAW
Welding position	Flat	Flat	Flat	Flat	Horizontal	Horizontal
Number of passes	1	1	1	1	1	3
Electrode type	E70T-4	E7014	E70T-4	E7024	E70T-4	E7018
Electrode diameter, in.	$^3/_{32}$	$^3/_{16}$	$^3/_{32}$	$^5/_{32}$	$^3/_{32}$	$^3/_{16}$
Electrode length, in.	Continuous	14	Continuous	24	Continuous	14
Current, A	350; dcep	215; ac	325; dcep	170; ac	350; dcep	225; dcep
Voltage, V	31	(b)	30	(b)	31	(b)
Electrode-feed rate, in./min	126	6.5	126	11	126	9.5
Length of weld per electrode, in.(c)	...	10.5	...	20.5	...	10.25
Operator factor(d), %	50	35	50	35	50	35
Welding speed, in./min	18	6.5	25.5	16.2	8.85	4(e)
Welding speed, ft^3/h (actual)	45	11.4	63.75	31.85	22.1	7(e)
Electrode consumption per foot of weld, lb	0.171	0.191	0.133	0.133	0.346	0.480

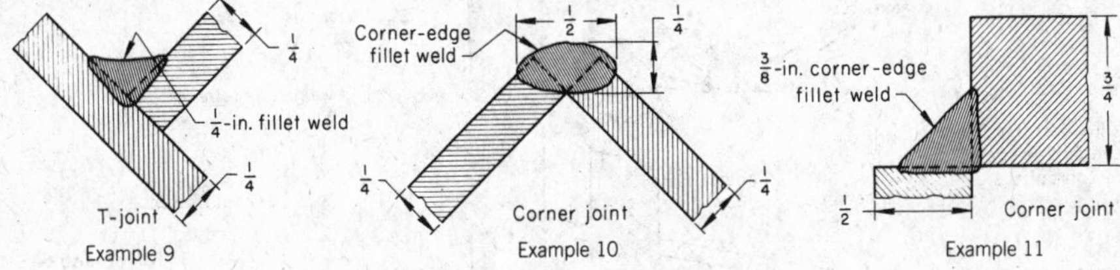

(a) Power consumption was not considered, because it was not significant. (b) Not metered. (c) For covered electrodes only. Takes into account average stub loss of 2 in. per electrode. (d) Average; arc time divided by man-hours multiplied by 100. (e) For three passes

important in this comparison, because when SMAW was used, considerable time was spent in changing covered electrodes and restriking the arc.

The actual number of feet welded per hour (on the average) was obtained by converting welding speed in inches per minute to feet per hour and multiplying by the operator factor. Flux cored arc welding produced more than three times as much weld per hour as SMAW.

For the flux-covered electrodes, electrode consumption in pounds per foot of weld was obtained by dividing the consumption in inches of weld per electrode by 12 to obtain feet of weld per electrode and dividing this figure into the weight of one electrode. For the continuous flux cored electrode wire, consumption in pounds of electrode per foot of weld was obtained by multiplying the number of pounds purchased by 85%, the electrode-deposition efficiency usually accepted for flux cored electrodes.

Example 10. Corner Joint. For this joint, strength was of minor importance. The joint, which is often used in box construction, consisted of two $^1/_4$-in. plates assembled at a right angle, with the inside-corner edges touching to form a 90° groove, which was filled by welding (Table 17).

Welding was done in the flat position. This type of weld is often used in low-stress applications. The iron-powder E7024 electrode selected for SMAW is intended for high-speed fillet welding. The data shown in Table 17 for this weld were obtained in the same manner as described in Example 9. The production rate for FCAW, expressed as actual feet welded per hour, was twice that obtained for SMAW.

Example 11. Corner Joint. The corner joint for this comparison called for a $^3/_8$-in. horizontal fillet weld as shown in the view for Example 11 in Table 17 and was made between plates $^1/_2$ and $^3/_4$ in. thick. For SMAW, the low-hydrogen iron-powder E7018 electrode was selected because of its resistance to underbead cracking in higher strength steels. Because of the thicker sections involved, cooling rate for the weld was higher than in the two previous examples. Three passes were required to fill the joint. The same type of electrode was used for each pass. The data shown for this weld were obtained in the manner described in Example 9. Production rate for FCAW was three times that obtained for SMAW. Because the types of joints described in Examples 10 and 11 are generally used in applications involving low stress, strength tests and quality inspection were not performed.

Flux Cored Arc Welding Versus Submerged Arc Welding. These welding processes are often competitive. Both are capable of high deposition rates and deep joint penetration, although FCAW with auxiliary gas shielding usually penetrates deeper. Weld position and length of weld often influence the choice of process.

Submerged Arc Welding

By the ASM Committee on Submerged Arc Welding*

SUBMERGED ARC WELDING (SAW) is an arc welding process in which the heat for welding is supplied by an arc (or arcs) developed between a bare metal (or flux cored) consumable electrode (or electrodes) and a workpiece. The arc is shielded by a layer of granular and fusible flux, which blankets the molten weld metal and the base metal near the joint and protects the molten weld metal from atmospheric contamination.

Principles of Operation

In the SAW process, electric current flows through the arc and the weld pool, which consists of molten flux and molten weld metal. The molten flux is usually highly conductive, although solid flux does not conduct electricity. In addition to acting as a protective shield, the flux cover may supply deoxidizers and scavengers that react chemically with the molten weld metal. Fluxes for SAW of alloy steel may also contain alloying ingredients that modify the composition of the molten weld metal.

Fig. 1 Cutaway view of a single-V-groove weld in a butt joint

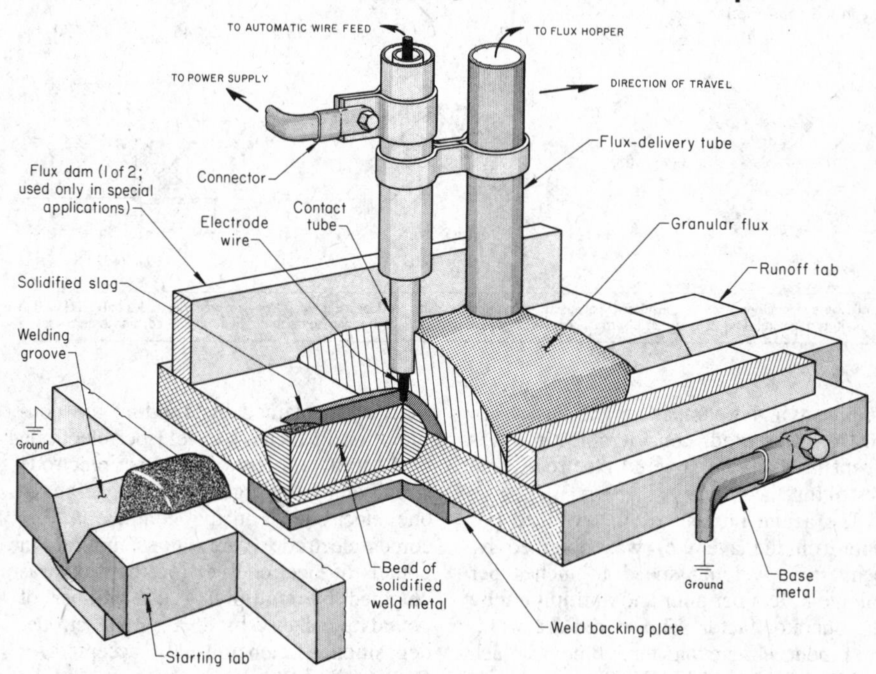

Figure 1 shows a SAW setup. Electric current from a power supply passes through the contact tube, and then through the electrode wire, to produce an arc between the electrode and the base metal. The heat of the arc melts the electrode, flux, and some base metal, forming a weld pool that fills the joint.

In all types of equipment, mechanically powered drive rolls continuously feed the consumable electrode wire through a contact tube (nozzle) and through the flux blanket to the joint being welded. The electrode wire is coiled on a reel or in a drum. The electrode wire melts off at the weld zone and is deposited along the joint. Granular flux is deposited ahead of the arc, and after the weld metal solidifies, unfused flux is removed by a vacuum pickup system to be screened and reused. In automatic welding, flux recovery may be an integral function of the equipment, as a flux-recovery tube follows directly behind the contact tube.

Submerged arc welding is adaptable to both semiautomatic and fully automatic operation, although the latter, because of inherent advantages, is more widely used. In semiautomatic welding, the welder manually controls the rate of travel by guiding a welding gun that feeds the flux and the electrode to the joint. In fully automatic welding, the equipment automatically feeds and guides the electrode and the flux along the joint and controls the rate of deposition. A typical machine for automatic SAW is shown in Fig. 2.

In some automatic SAW applications, two or more electrodes are fed simultaneously to the same joint. The electrodes can be side-by-side and fed into the same weld pool, or they can be spaced just far enough apart to permit two weld pools to

*David LeRoy Olson, *Chairman*, Professor of Metallurgical Engineering, Colorado School of Mines; Elwood Chaney, Staff Engineer, Caterpillar Tractor Co.; Craig B. Dallam, Research Assistant, Colorado School of Mines; Richard L. Hellner, Welding Engineer, Public Service Company of Colorado; Jerald E. Jones, Assistant Professor of Metallurgical Engineering, Colorado School of Mines; Stephen Liu, Research Assistant, Colorado School of Mines; Thomas H. North, Senior Research Associate, Stelco Inc.; Richard S. Sabo, Manager, Publicity and Educational Services, Lincoln Electric Co.; James E. Sims, Manager, Welding Research & Development, Chicago Bridge & Iron Co.

Fig. 2 Typical automatic SAW unit

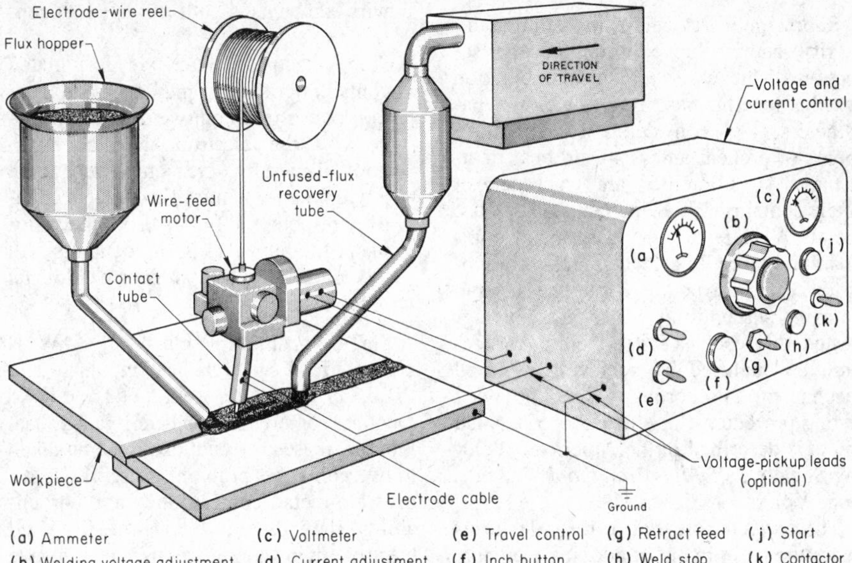

solidify independently. The latter technique, sometimes referred to as tandem arc welding, produces a multiple-pass weld in a single traverse of the joint, thereby increasing rate of deposition and welding speed. A variation of a dual-wire process, which uses two small-diameter electrodes fed from a common contact nozzle and supplied with current from one power source, is shown in Fig. 3. The two wires in close proximity result in an elongated pool that improves pool shape characteristics and enables fast travel speeds while maintaining the proper weld-bead shape.

Advantages and Limitations

Submerged arc welding, either semiautomatic or fully automatic, offers the following advantages over some other welding processes:

- Joints can be prepared with a shallow V-groove, resulting in less filler metal being used. In some applications, a groove is not required.
- The arc operates under the flux cover, thus eliminating weld spatter and arc flash.
- The process can be used at high welding speeds and deposition rates to weld flat plate or the surfaces of cylindrical shapes of virtually any size or thickness. It also can be used for hardfacing or weld overlay applications.
- The flux acts as a scavenger and deoxidizer to remove undesirable contaminants from the molten weld pool and to produce sound welds with good mechanical properties. The flux may, if required, supply alloying elements to the weld metal.
- For welding unalloyed low-carbon steels, inexpensive electrode wires can be used—usually carbon steel wire, either bare or flash plated with copper to improve electrical contact and protect against rusting. Flux cored electrodes also may be used, generally with neutral fluxes.
- The SAW process can be used for welding in exposed areas with relatively high winds; the granular flux shielding provides protection superior to that obtained from the electrode covering in shielded metal arc welding (SMAW) or the gas shielding in gas metal arc welding (GMAW).
- Low-hydrogen weld metal can be produced.

Fig. 3 Dual-wire SAW setup. Two small-diameter electrodes fed from a common nozzle are supplied with current from a single power source.

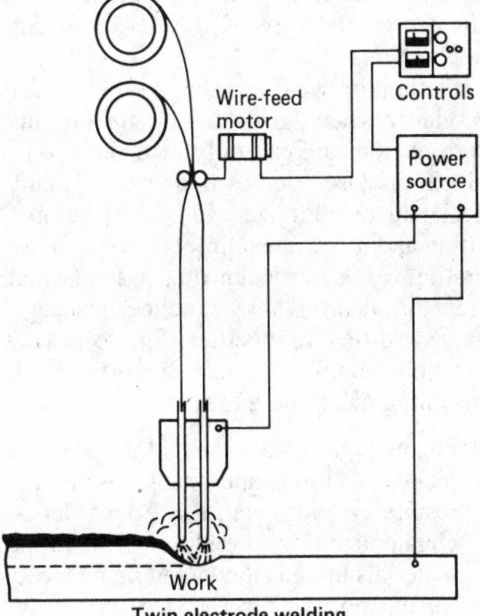

Twin electrode welding

Limitations of SAW include the following:

- Flux, flux-handling equipment, and workholding fixtures are required. Many joints also require the use of backing plates, strips, or rings.
- Flux is subject to contamination that may cause weld discontinuities.
- To obtain welds of good quality, the base metal must be homogeneous and essentially free of scale, rust, oil, and other contaminants.
- Slag (solidified residue from the fused flux) must be removed from the weld bead; this is sometimes difficult. In multiple-pass welding, slag must be removed after each pass to avoid entrapment in the weld metal.
- The process usually is not suitable for use on metal less than $^3/_{16}$ in. thick because melt-through is likely, unless backup methods are provided.
- Except for special applications, welding is largely restricted to the flat and horizontal positions in order to avoid runoff of flux.

Weldability of Steels

Steels that are suitable for welding by other major welding processes are equally well suited to welding by the SAW process. Submerged arc welding is used most widely in the production welding of unalloyed (plain) low-carbon steels containing less than 0.30% C and 0.05% S.

Manganese and silicon contents can be used up to 1.10 and 0.15%, respectively, without affecting the welding characteristics significantly. However, above 1.40% Mn, 0.35% C, or 0.30% Si, special welding procedures may be needed. For high sulfur content, manganese must be increased to compensate for the adverse effect of sulfur.

Submerged arc weldable steels, which range in tensile strength from about 45 to 90 ksi, usually are welded with a combination of flux and electrode wire included in American Welding Society (AWS) A5.17-80, "Specification for Carbon Steel Electrodes and Fluxes for Submerged Arc Welding."

Medium-carbon and low-alloy structural steels (high-strength low-alloy, or

HSLA, steels) are equally well suited to SAW. However, in such applications, preheating (and/or postweld heat treating) may be necessary. Where special notch toughness and/or strength are required, special electrode wires and fluxes are used. Submerged arc welding consumables are capable of producing high-strength weldments at high deposition rates.

Several structural steels, such as American Society for Testing and Materials (ASTM) grades A36 and A283, are usually welded without difficulty and without the use of special procedures. In practice, most steel produced to these specifications avoids the upper compositional limits, particularly those for sulfur and phosphorus. However, when these steels are purchased to ASTM specifications, there is no guarantee that composition will avoid the upper limits of those elements most likely to promote welding difficulties. Some ASTM specifications do not establish upper limits for certain elements. Thus, the weldability of these steels cannot be predicted on the basis of compliance with the material specification. Therefore, each heat of base material should be tested to verify weldability, or an acceptable chemistry range should be specified for each specific welding procedure. In general, the importance of the chemical composition of the steel, as it affects welding quality, increases directly with increases in joint thickness and joint restraint.

Stainless steel and hardenable carbon steel also are welded by SAW. Procedures for welding these steels are described in the articles "Arc Welding of Hardenable Carbon and Alloy Steels" and "Arc Welding of Stainless Steels" in this Volume. Submerged arc welding also is used to deposit buildup, corrosion-resistant, and abrasion-resistant coatings on steel surfaces that are subjected to wear.

Submerged arc welding is a high heat input and high deposition rate process. Some steels that can be submerged arc welded are more often welded by lower heat input welding processes, such as SMAW and GMAW, and multipass welding operations that give a narrower heat-affected zone (HAZ) and meet higher weld-metal toughness specifications. Problems may be encountered when welding special steels, e.g., high-carbon steels, maraging steels, and cast irons. In the case of high-carbon steels and in some cast irons, this would be due to HAZ and weld-metal cracking (hydrogen-induced cracking and/or solidification cracking). In the case of maraging steels, problems are due to reheat treatment of the maraging steel microstructure in the HAZ.

Metallurgical Considerations

Submerged arc weldments of high integrity can be achieved through proper selection of the wire and flux combination for the specific base metal, welding parameters, and consideration of requirements of preheat and postweld heat treatment. The submerged arc weld deposit often contains a high percentage of base metal. As a result, the base-metal composition markedly affects deposit composition, microstructure, toughness, and cracking susceptibility. It is strongly suggested that each specific welding procedure be qualified by achieving required mechanical properties. Recommended testing procedures for evaluation of welded joints is described in the American Welding Society's *Welding Handbook,* 7th edition, Volume 1, Chapter 5.

The oxygen potential of the flux influences the loss or gain of alloying elements during welding, the weld-deposit oxygen content, and the type, size, and distribution of oxide inclusions in the solidified weld metal. High-oxygen-potential fluxes (sometimes referred to as acidic) contain large amounts of SiO_2 and MnO and produce weld metals containing very high oxygen levels (approximately 600 to 1000 ppm of oxygen). Flux formulations high in SiO_2 have excellent operating characteristics, namely, good current/voltage stability and formation of desirable weld-bead profiles. Low-oxygen-potential fluxes (sometimes referred to as basic) contain increased contents of CaO, CaF_2, and Al_2O_3 and produce weld metals containing approximately 250 to 450 ppm of oxygen. Weld metals produced using low-oxygen-potential fluxes exhibit superior toughness properties. Generally, this requirement (in user specifications) has an overriding influence on the type of flux used for any application.

Optimum weld-metal properties are produced when the electrode wire and flux formulations are carefully matched. Some electrode-wire compositions are specially designed to optimize weld-metal composition and mechanical properties when operating with a particular flux. Use of other flux formulations may produce unacceptable weld-metal properties. The following examples emphasize the importance of matching electrode and flux materials:

- Fluxes high in SiO_2 and MnO promote pickup of silicon and manganese by the resulting weld metal; this effect can be counteracted by employing electrode materials low in silicon and manganese; for example, a 0.05% Si content wire can be used with a high SiO_2 content flux, and a 0.5% Mn content electrode wire can be used with a high MnO content flux.
- Some electrode wires are formulated containing titanium and boron additions to produce optimum weld-metal toughness. These electrode-wire compositions generally are employed with low-oxygen-potential fluxes. If a high-oxygen-potential flux is used, the weld-metal titanium and boron contents will be reduced, resulting in poor weld-metal toughness.

Single- or multiple-electrode SAW is characterized by high heat inputs (40 to 250 kJ/in.). Because weld-metal and HAZ toughness or strength can deteriorate as heat input increases, precautions must be taken to overcome this problem. Careful control of weld-metal composition and hardenability by judicious choice of flux and electrode-wire compositions can improve weld-metal toughness in high heat input welding. Tensile properties of the HAZ can be improved by selecting multipass welding procedures and by setting upper limits on the heat input used during welding. Where multipass welding is impractical (due to productivity limitations), HAZ toughness can be improved by modification of the base-metal composition. For example, in the manufacturing of linepipe, the two-pass seam welding operation precludes multipass or setting limiting heat input levels because inside and outside surface weld beads must overlap. In this application, small base-metal additions of titanium restrict austenite grain growth in the HAZ region immediately adjacent to the weld interface to improve toughness.

Weld-Metal Microstructures

Submerged arc weld-metal microstructures generally are composed of a complex mixture of microstructural constituents. Weld-deposit mechanical properties depend as much on the relative proportions and morphologies of different constituents that are formed as they do on the actual phases themselves. Submerged arc deposits in HSLA steels comprise the following microstructural constituents:

- Proeutectoid ferrite, either in a massive equiaxed form or as thin veins delineating prior austenite grain boundaries
- Sideplate Widmanstatten ferrite (parallel ferrite laths emanating from prior austenite grain boundaries)
- Acicular ferrite (a tough structure found within the body of prior austenite grains

Fig. 4 Weld-metal microstructure. (a) Proeutectoid ferrite. (b) Sideplate ferrite. (c) Acicular ferrite

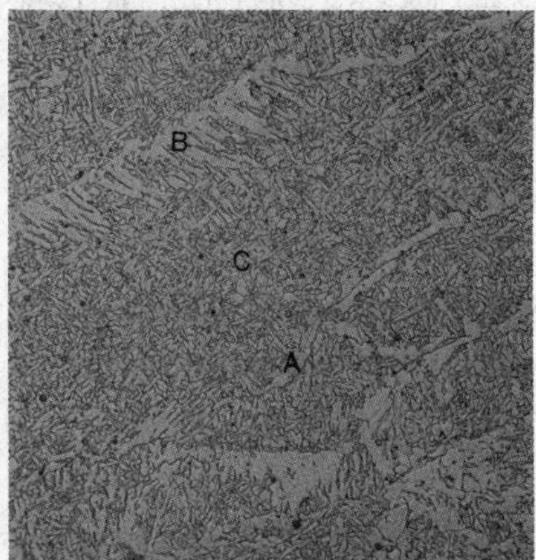

that is formed between 1065 to 1200 °F on cooling)

- Retained austenite and twinned or lath martensite (sometimes referred to as martensite-austenitic phases)
- Other products, including pearlite and bainite

Weld-metal ferrite microstructures are illustrated in Fig. 4. The microstructural products and their relative contents formed in the weld metal depend on a complex interplay of flux choice, electrode and base-metal composition, and on the heat input (cooling rate) occurring in welding operations.

Flux Formulation and Deposit Microstructure. Weld-metal oxygen content and the types, size, and distribution of oxide inclusions largely are determined by flux composition. Oxygen is almost completely insoluble in solidified weld metal, and the resulting inclusions range from approximately 0.1 to 5 μm in diameter. Phase transformations that occur as weld metal cools to room temperature are influenced by the presence of these oxide inclusions. In high-oxygen weld metal, austenite grain growth is limited through pinning of grain boundaries by preferentially segregated inclusions. Also, inclusions located at prior austenite grain boundaries provide energetically favorable sites for ferrite nucleation. Consequently, high-oxygen-content submerged arc weld deposits exhibit large amounts of proeutectoid and sideplate ferrite. In low-oxygen-content weld metal (250 to 450 ppm of oxygen), oxide inclusions nucleate acicular ferrite within austenite grains so that the weld-metal microstructure is comprised of acicular ferrite and possibly proeutectoid ferrite (at prior austenite grain boundaries).

Effect of Base-Metal and Electrode Compositions. Because dilution of base metal can be considerable (up to 60%) in SAW, base-metal composition plays an important part in determining weld-metal microstructure. Weld-metal microstructure can be related to weld-metal composition by using the following hardenability factor (HF) calculation:

HF =

$$1000\,\text{C} + \frac{\text{Mn}}{6} + \frac{\text{Cr} + \text{Mo}}{10} + \frac{\text{Ni}}{20} + \frac{\text{Cu}}{40}$$

The HF parameter predicts the complex interaction of several alloying elements as it relates to the formation of acicular ferrite in the weld deposit. Jominy hardenability predictions are quite different and relate to the formation of martensite in a quenched steel. Figure 5 shows the relation between the HF and toughness when welding with a single flux composition at a given heat input level. Increasing hardenability (due to increasing alloy element contributions by the base material and/or filler wire) can affect both weld-metal microstructure and toughness. The influence of base-material sulfur content on weld-metal toughness is apparent (increased toughness at low plate sulfur levels).

Effect of Weld-Metal Cooling Rate. The cooling rate after welding depends inversely on the heat input employed. Cooling rates are given in terms of the time for cooling between the the temperatures 1470 and 930 °F. Figure 6 shows a typical microalloyed HSLA steel continuous cooling transformation diagram for a weld-metal composition of 0.1% C, 1.37% Mn, 0.31% Si, 0.002% P, 0.020% S, 0.06% Cr, 0.0 to 2% Mo, 0.008% Ni, 0.03% Nb, 0.034% O, and 0.01% N. A change in the cooling time between 1470 and 930 °F (by means of heat input variation) has a marked influence on weld-metal microstructure. Notice the variations in type and amount of weld-metal microstructural constituents (Fig. 6) that are possible with changes in heat input.

Weld-Metal Microstructure and Toughness

Much research has been undertaken to establish a relationship between weld-metal microstructure and toughness. The highest resistance to cleavage fracture is obtained when there is a high proportion of fine-grained acicular ferrite. Alloying with elements such as manganese, molybdenum, and nickel promotes acicular ferrite formation and improves toughness, in spite of an increase in yield strength. However, the balance between increasing acicular ferrite content and increasing yield strength (due to alloy addition) becomes important

Fig. 5 Relationship between weld-deposit toughness and the hardenability factor (HF) using low-oxygen-potential flux

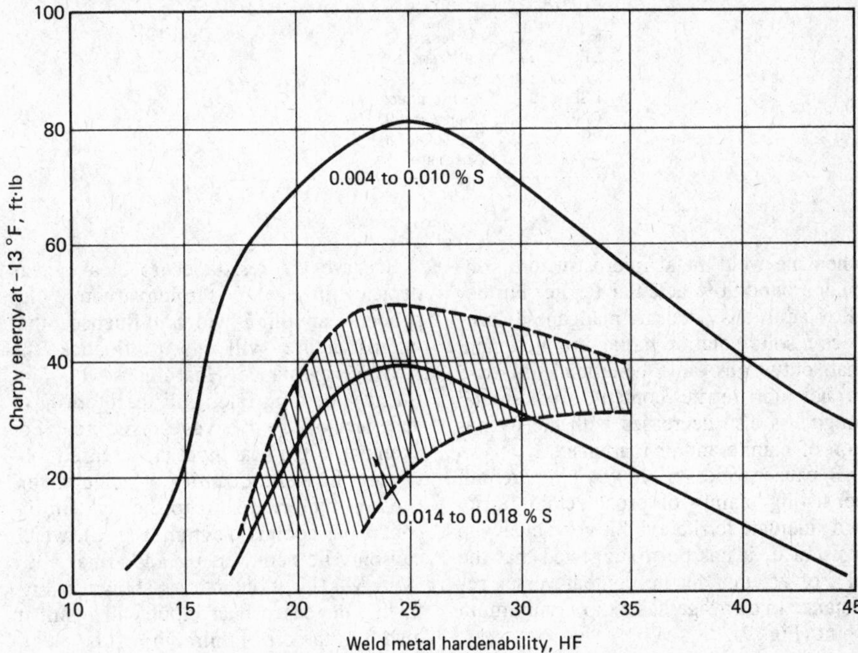

Fig. 6 Continuous cooling transformation diagram for a microalloyed HSLA steel. Submerged arc weld-metal composition: 0.1% C, 1.37% Mn, 0.31% Si, 0.002% P, 0.020% S, 0.06% Cr, 0.02% Mo, 0.08% Ni, 0.15% Cu, 0.03% Nb, 0.034% O, and 0.010% N

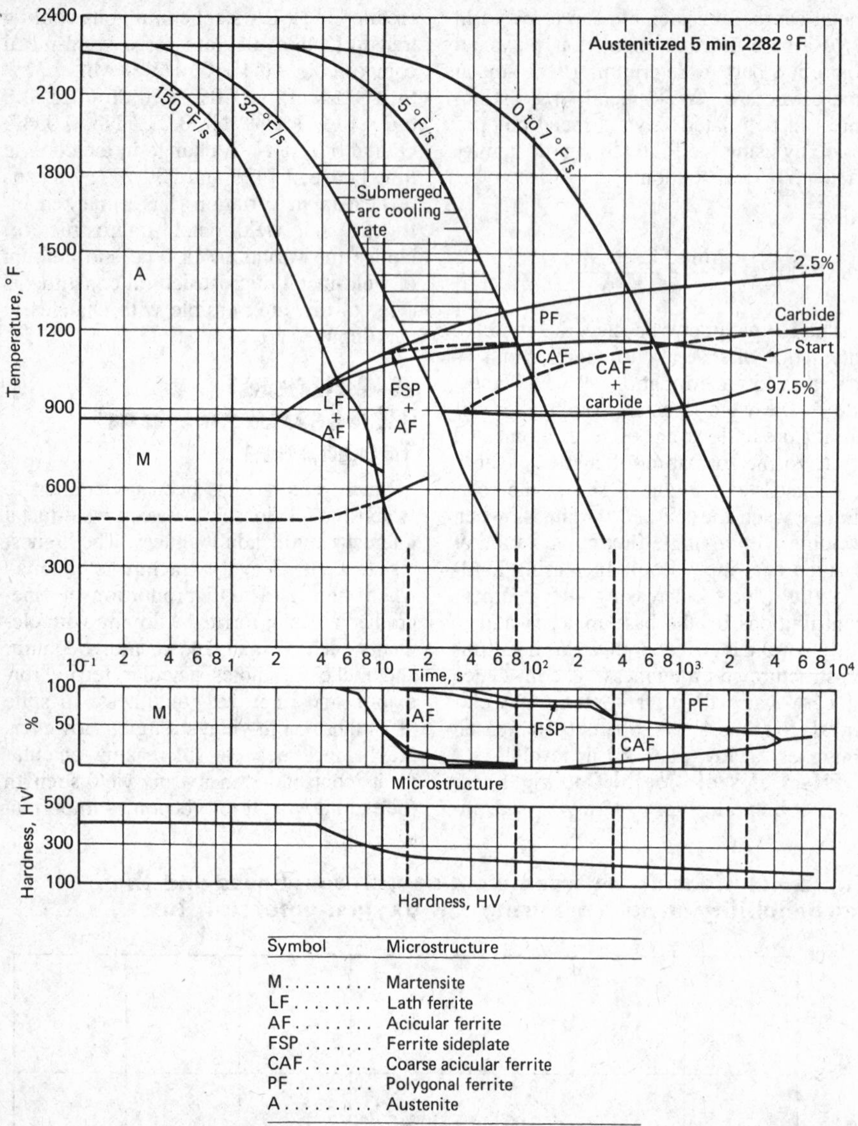

when the weld-metal microstructure contains around 90% acicular ferrite. Further alloy additions (such as manganese) produce a solid-solution hardening increment that outweighs any beneficial increase in acicular ferrite content. Weld-metal toughness also decreases with the formation of bainite and/or martensite.

Because welds of low yield strength consisting mainly of proeutectoid ferrite and sideplate ferrite exhibit good cleavage resistance, it has been suggested that the plot of acicular ferrite content versus resistance to cleavage has a dip or minimum point (Fig. 7).

Because the resistance to cleavage as depicted in Fig. 7 is microstructure dependent, anything which influences that microstructure will also influence the cleavage resistance. Heat input, which influences cooling rate, will therefore have an effect on the cleavage resistance. For example, if the heat input is increased, obtaining the same quantity of acicular ferrite will require a shift to the right in the continuous cooling diagram (Fig. 6), which can only be achieved by additional alloy (Mn, Mo, Ni) content. As shown in Fig. 8, this increased heat input will result in a shift in the dip (minimum) of the cleavage resistance curve due to higher alloy content.

In addition, as shown in Fig. 8, the resistance to cleavage of high heat input will be below that of a low heat input weld. This decrease in cleavage resistance is due to a general coarsening of the microstructure which results from the slower cooling rate. Thus, the grains which result from additional alloy (Mn, Mo, Ni) content do not completely offset the losses due to increased heat input.

Heat-Affected Zone Microstructure and Toughness

During welding, the region of the HAZ closest to the weld interface is heated above 2200 °F and grain coarsening occurs. Final grain size of this particular region depends on the heat input used during welding and can vary from around 100 μm at heat inputs of approximately 75 kJ/in. (typical of SAW) to over 300 μm for heat inputs around 500 kJ/in., which are typical of electroslag welding (ESW). The presence of AlN and Nb (C, N) particles in the base material (in aluminum-killed carbon-manganese-niobium HSLA steels, for example) does not alter grain size in the region close to the weld interface, but does narrow the width of the grain-

Fig. 7 Relationship among cleavage resistance, acicular ferrite content, and yield strength

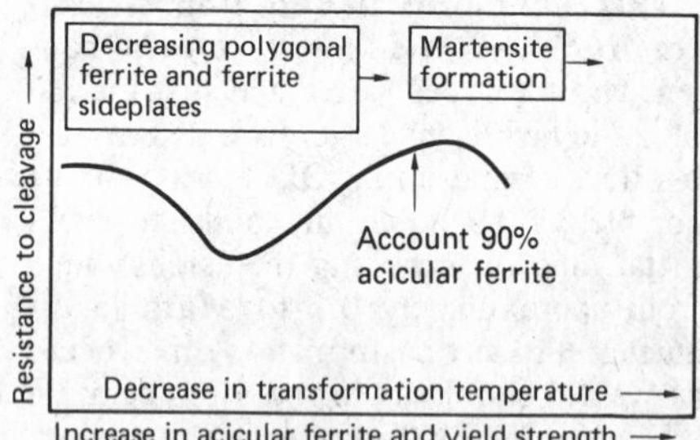

Fig. 8 Effect of heat input and alloy content on cleavage resistance

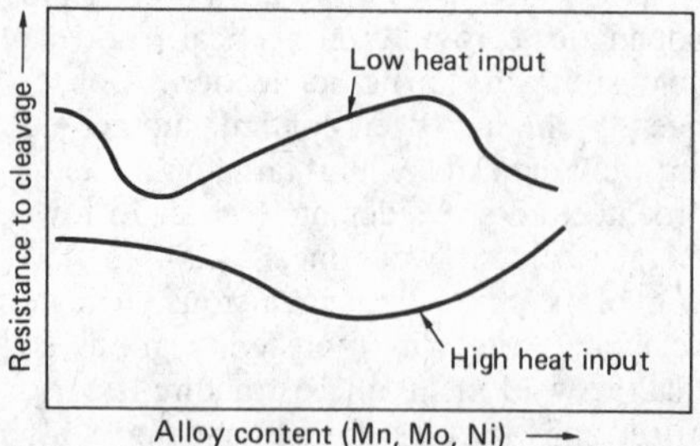

Fig. 9 Effect of arc energy on HAZ toughness for a carbon-manganese steel

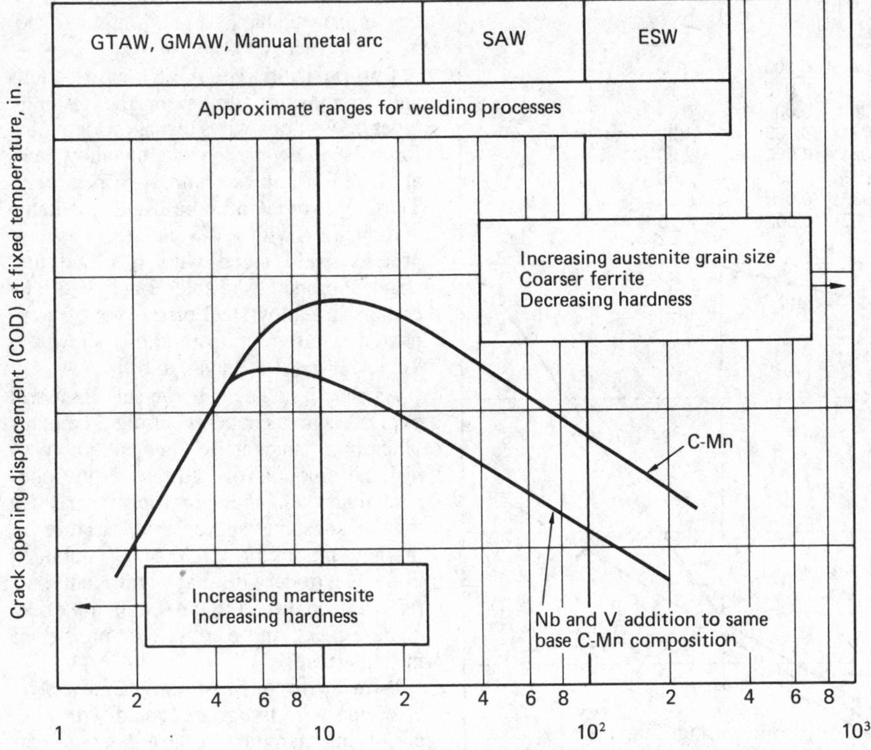

coarsened region by restricting grain growth at lower peak temperatures.

In carbon-manganese steels, the HAZ region closest to the weld interface transforms to a complex mixture of one or more of the following microstructures: (1) proeutectoid ferrite at prior austenite grain boundaries; (2) transgranular Widmanstatten ferrite; (3) high-carbide-content microstructures; such as pearlite; (4) upper bainite, and lower bainite; and (5) martensite.

Figure 9 shows the relationship between heat input (essentially cooling time between 1470 and 930 °F) and HAZ toughness for a 0.17%C-1.3%Mn base material. Maximum toughness occurs at a $t_{1470\text{-}930}$ of around 10 s (equivalent to a heat input of 63.5 kJ/in. when welding 1-in.-thick plate). Toughness decreases at higher cooling rates due to increased proportions of martensite in the HAZ microstructure. At very low cooling rates, toughness decreases due to increased contents of coarse ferrite structures. When niobium or vanadium is present, toughness decreases more rapidly than in carbon-manganese steels when welding at high heat input levels. This decrease in toughness is associated with niobium or vanadium depressing the austenite transformation temperature and promoting the formation of fine bainitic microstructures that have high hardness due to niobium precipitation clustering.

Improved HAZ microstructures and toughness when welding HSLA steels at high heat inputs can be obtained by titanium additions to the base-metal composition. Titanium nitride particles restrict austenite grain growth during welding (see Fig. 10). The beneficial influence of 0.005 to 0.011% Ti on the HAZ toughness of X-70 linepipe material is illustrated in Fig. 11. High titanium-to-nitrogen ratios exceeding 3 to 1 do not improve toughness, but actually cause deterioration of HAZ toughness.

Submerged Arc Welding Consumables

The filler metal and flux used during SAW are classified as consumables. The specific chemical and physical properties of the wire, base metal and flux, as well as process parameters, control weld-metal composition, microstructure, and properties. Because of the complex interactions of these factors, specific fluxes and wires must be combined to optimize weld-metal properties.

Electrode Wire

Solid electrode wire for SAW of steel is available commercially in sizes from 1/16 to 1/4 in. in diameter and 5/64 to 0.120 in. in tubular form. Electrode wire 3/8 in. in diameter is seldom used. Electrodes are produced in various ferrous alloy compositions, ranging from unalloyed low-carbon steel to high-alloy steel, and also in a variety of nonferrous compositions.

Packaging. Continuous unalloyed low-carbon steel electrode wire is either coiled or wound into drums. The weights of electrode coils vary slightly among manufacturers. Typically, electrode wire 1/8 in. in diameter or smaller is available in coils weighing 25 or 60 lb. Electrode wire 3/32 in. in diameter or larger is available in coils weighing 150 or 200 lb. The tolerance for all coils is ±10%. The smaller coils are used where weight reduction and compactness are essential, as in semiautomatic welding.

Drums of electrode wire are used primarily in high-production applications, where a large volume of wire is consumed

Fig. 10 Effect of titanium-to-nitrogen ratio on grain coarsening of as-cast steel. Partially acicular ferrite X-70 linepipe. Composition: 0.070% C, 1.95% Mn, 0.002% P, 0.004% S, 0.30% Si, 0.39% Mo, and 0.050% Nb

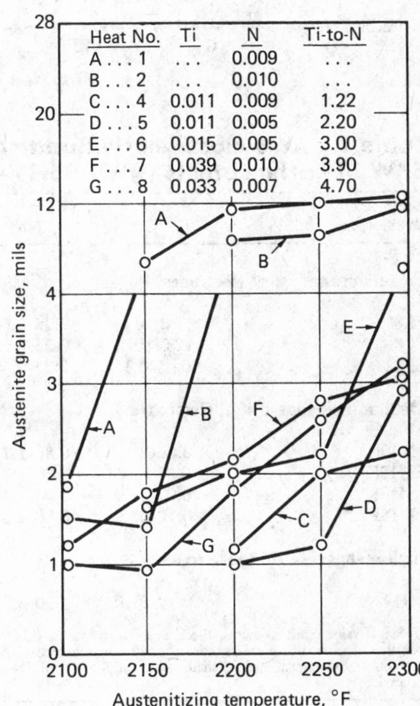

Fig. 11 Effect of titanium-to-nitrogen ratio on the COD toughness of coarse-grained HAZ material. Partially acicular ferrite X-70 linepipe. Composition: 0.07% C, 1.95% Mn, 0.002% P, 0.004% S, 0.30% Si, 0.30% Mo, and 0.05% Nb

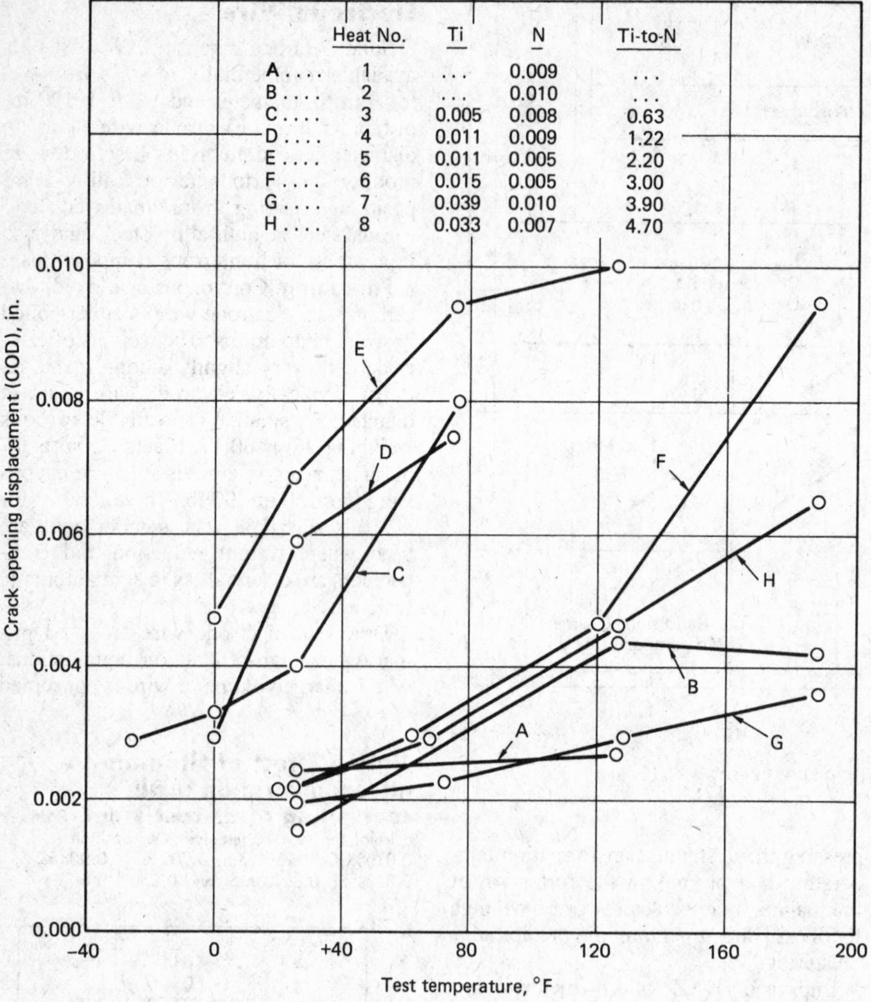

Table 1 AWS classification and composition limits for electrodes for SAW of carbon steels (AWS A5.17-80)

Electrode classification	Composition(a), wt% C	Mn	Si	S	P	Cu(b)
Low-manganese steel electrodes						
EL8	0.10	0.25-0.60	0.07	0.035	0.035	0.35
EL8K	0.10	0.25-0.60	0.10-0.25	0.035	0.035	0.35
EL12	0.05-0.15	0.25-0.60	0.07	0.035	0.035	0.35
Medium-manganese steel electrodes						
EM12	0.06-0.15	0.80-1.25	0.10	0.035	0.035	0.35
EM12K	0.05-0.15	0.80-1.25	0.10-0.35	0.035	0.035	0.35
EM13K	0.07-0.19	0.90-1.40	0.35-0.75	0.035	0.035	0.35
EM15K	0.10-0.20	0.80-1.25	0.10-0.35	0.035	0.035	0.35
High-manganese steel electrodes						
EH14	0.10-0.20	1.70-2.20	0.10	0.035	0.035	0.35

(a) Single values are maximums. Electrodes shall be analyzed for those elements for which specific values are shown. Elements other than those shown, which are intentionally added (except iron), shall also be reported. The total of these latter elements and all other elements not intentionally added shall not exceed 0.50%. (b) The copper limit includes any copper coating that may be applied to the electrode.

and where the welding equipment has been designed to support and channel the electrode wire from the drum into the welding head. Electrode wire $^1/_{16}$ to $^1/_4$ in. in diameter is available in drums weighing 250, 500, 750, and 1000 lb.

Composition. The AWS classifications and composition limits for the different types of electrode wires that are widely used for SAW of low-carbon and low-alloy steels are listed in Tables 1 and 2, respectively. These electrodes may be used in suitable combination with AWS classes of flux to produce weld metal with specified mechanical properties (Tables 3 and 4). Many carbon and alloy steel electrodes of compositions different from those shown in Tables 1 and 2 are available.

As shown in Table 1, the standard steel electrode wires contain controlled amounts of carbon, manganese, and silicon with residual amounts of sulfur and phosphorus. When all other significant variables remain constant, appreciable variation in the amounts of these elements contained in the electrode wire may significantly affect weld quality. See the section on effect of base-metal and electrode compositions in this article.

Wire Surface. Most non-stainless steel electrode wire is lightly coated with copper during manufacture. The copper coating provides some protection from rust and ensures good electrical contact between the electrode and the contact tube nozzle of the welding head. Good electrical contact is essential to maintain satisfactory arc characteristics.

Electrode wire should have a smooth, clean surface. Rust and other surface contaminants can produce weld porosity, cause excessive wear of contact tubes, and adversely affect arc characteristics. Some small-diameter electrode wire is coated with a small amount of special hydrogen-free lubricant. This lubricant helps to feed the wire through the welding cable conduit that carries the wire from the wire feeder to the welding gun contact tip.

Selection of wire size (diameter) depends on equipment capabilities and application. Small-diameter electrode wire ($^1/_{16}$ to $^3/_{32}$ in.) is used almost exclusively with semiautomatic welding equipment. The $^3/_{32}$-in.-diam electrode wire is used with either semiautomatic or fully automatic equipment. Larger sizes ($^1/_8$ in. diam and above) are used only with fully automatic equipment.

Small-diameter electrode wires ($^1/_{16}$ to $^5/_{64}$ in.) perform better at low currents than large-diameter wires of the same composition and type. Flexibility is essential to the use of semiautomatic SAW equip-

Table 2 Compositions of low-alloy steel electrodes for SAW composition requirements for solid electrodes (AWS 5.23-80)

Electrode classification	Composition(a), wt% C	Mn	Si	S	P	Cr	Ni	Mo	Cu(b)	V	Al	Ti	Zr
Carbon-molybdenum steels													
EA1	0.07-0.17	0.65-1.00	0.20	0.035	0.025	...	...	0.45-0.65	0.35	...	...	...	...
EA2	0.07-0.17	0.95-1.35	0.20	0.035	0.025	...	...	0.45-0.65	0.35	...	...	...	...
EA3	0.10-0.18	1.65-2.15	0.20	0.035	0.025	...	...	0.45-0.65	0.35	...	...	...	...
EA4	0.10-0.18	1.25-1.65	0.25	0.035	0.025	...	...	0.45-0.65	0.35	...	...	...	...
Chromium-molybdenum steels													
EB2	0.07-0.15	0.45-0.80	0.05-0.30	0.030	0.025	1.00-1.75	...	0.45-0.65	0.35	...	...	...	...
EB2H	0.28-0.33	0.45-0.65	0.55-0.75	0.015	0.015	1.00-1.50	...	0.40-0.65	0.30	0.20-0.30			
EB3	0.07-0.15	0.45-0.80	0.05-0.30	0.030	0.025	2.25-3.00	...	0.90-1.10	0.35	...	...	...	...
EB5	0.18-0.23	0.40-0.70	0.40-0.60	0.025	0.025	0.45-0.65	...	0.90-1.10	0.30	...	...	...	...
EB6(c)	0.10	0.40-0.65	0.20-0.50	0.025	0.025	4.50-6.00	...	0.45-0.65	0.35	...	...	...	...
EB6H	0.25-0.40	0.75-1.00	0.25-0.50	0.030	0.025	4.80-6.00	...	0.45-0.65	0.35	...	...	...	...
Nickel steel													
ENi1	0.10	0.75-1.25	0.05-0.25	0.010	0.010	0.15	0.80-1.20	0.30	0.35	...	...	...	...
ENi2	0.10	0.75-1.25	0.05-0.25	0.010	0.010	...	2.10-2.90	...	0.35	...	...	...	
ENi3	0.13	0.60-1.20	0.05-0.25	0.012	0.012	0.15	3.10-3.80	...	0.35	...	...	...	...
ENi4	0.12-0.19	0.60-1.00	0.10-0.30	0.020	0.015	...	1.60-2.10	0.10-0.30	0.35	...	...	...	...
Other low-alloy steels													
EF1	0.07-0.15	0.90-1.70	0.15-0.35	0.025	0.025	...	0.95-1.60	0.25-0.55	0.35	...	...	...	...
EF2	0.10-0.18	1.70-2.40	0.20	0.025	0.025	...	0.40-0.80	0.40-0.65	0.35	...	...	...	...
EF3	0.10-0.18	1.70-2.40	0.20	0.025	0.025	...	0.70-1.10	0.45-0.65	0.35	...	...	...	...
EF4	0.16-0.23	0.60-0.90	0.15-0.35	0.035	0.025	0.40-0.60	0.40-0.80	0.15-0.30	0.35	...	...	...	...
EF5	0.10-0.17	1.70-2.20	0.20	0.010	0.010	0.25-0.50	2.30-2.80	0.45-0.65	0.50	...	...	...	...
EF6	0.07-0.15	1.45-1.90	0.10-0.30	0.015	0.015	0.20-0.55	1.75-2.25	0.40-0.65	0.35	...	...	...	...
EM2(d)	0.10	1.25-1.80	0.20-0.60	0.010	0.010	0.30	1.40-2.10	0.25-0.55	0.25	0.05	0.10	0.10	0.10
EM3(d)	0.10	1.40-1.80	0.20-0.60	0.010	0.010	0.55	1.90-2.60	0.25-0.65	0.25	0.04	0.10	0.10	0.10
EM4(d)	0.10	1.40-1.80	0.20-0.60	0.010	0.010	0.60	2.00-2.80	0.30-0.65	0.25	0.03	0.10	0.10	0.10
EW	0.12	0.35-0.65	0.20-0.35	0.040	0.030	0.50-0.80	0.40-0.80	...	0.30-0.80	...	...	...	...
EG	No requirements specified												

(a) Single values are maximums. Electrodes shall be analyzed for those elements for which specific values are shown. Elements other than those shown, which are intentionally added (except iron), shall also be reported. The total of these latter elements and all other elements not intentionally added shall not exceed 0.50%. The letter "N" as a suffix to a classification indicates that the electrode is intended for welds in the core belt region of nuclear reactor vessels, as described in paragraph A2.2 of the Appendix to this specification. This suffix changes the limits on the phosphorus, vanadium, and copper, as follows: P, 0.012% max; V, 0.05% max; Cu, 0.08% max. "N" electrodes shall not be coated with copper or any material containing copper. The "EF5" and "EW" electrodes shall not be designated as "N" electrodes. (b) The copper limit includes any copper coating which may be applied to the electrode. (c) The EB6 classification is similar to, but not identical with, the ER502 classification in A5.9-77, "Specification for Corrosion-Resisting Chromium and Chromium-Nickel Steel Bare and Composite Metal Cored and Stranded Arc Welding Electrodes and Welding Rods." (d) The composition ranges of classifications with the "EM" prefix are intended to conform to the ranges for similar electrodes in the military specifications.

ment. If the electrode wire is too stiff, the welder has difficulty manipulating the welding gun and cable into the positions necessary to produce sound welds. A stiff welding cable also increases welder fatigue, thereby reducing arc time and efficiency.

Most semiautomatic SAW is done with electrode wire $^5/_{64}$ or $^3/_{32}$ in. in diameter. The use of wire less than $^5/_{64}$ in. in diameter is usually limited to applications requiring stable arc characteristics at low current—for example, for joining thin metal, for making circumferential welds on small-diameter assemblies, and for multiple-arc welding. In addition to their ability to develop stable arc characteristics, small-diameter wires provide better control of the molten weld pool and bead size and shape. They also provide a more uniform weld appearance.

Large-diameter electrode wires ($^5/_{32}$ to $^1/_4$ in.) are used to take advantage of higher current capacity for increased deposition rate. They occasionally are used to reduce depth of joint penetration and to increase bead width. Also, the large sizes make tightness of fit-up at the root of a joint less critical.

Current Range. In SAW, an electrode of specific diameter can operate within a wide current range, as shown in Table 5. The overlap of current ranges makes it possible to use any of several wire sizes at a particular welding-current setting. Changing to a smaller diameter electrode wire at a given current may serve to increase depth of fusion and reduce the width of the weld bead.

Table 3 Mechanical property requirements of flux types

Flux classification(a)	Tensile strength, ksi	Yield strength(b), ksi	Elongation(b), %
F6XX-EXXX	60-80	48	22
F7XX-EXXX	70-95	58	22

(a) The letter "X" stands for, respectively, the condition of heat treatment, toughness of the weld metal, and classification of the electrode. (b) Yield strength at 0.2% offset, and elongation in 2-in. gage length.

Table 4 Impact strength requirements of flux types

Digit	Test temperature, °F	Energy level, min, ft·lb
Z	...	No impact requirement
0	0	20
2	−20	20
4	−40	20
5	−50	20
6	−60	20
8	−80	20

Note: Based on the results of the impact tests of the weld metal, the manufacturer shall insert in the above flux classification the appropriate digit from the table above. Weld metal from a specific flux-electrode combination that meets impact requirements at a given temperature also meets the requirements at all higher temperatures in this table (i.e., weld metal meeting the requirements for digit 5 also meets the requirements for digits 4, 2, 0, and Z).

Table 5 Current ranges for electrode wires used in SAW

Wire diameter, in.	Current range(a), A	Wire diameter, in.	Current range(a), A
1/16	115-500	5/32	340-1100
5/64	125-600	3/16	400-1300
3/32	150-700	7/32	500-1400
1/8	220-1000	1/4	600-1600

(a) Upper and lower limits of ranges are extremes and are rarely used.

With small-diameter electrode wire, the arc is initiated more readily than with a wire of large diameter, and small-diameter wires are particularly well suited to hot starts and to high-frequency arc starting. When the welding current is on the low side of one of the current ranges given in Table 5, an electrode wire of the next smaller diameter (resulting in a higher current density) provides a more stable arc and a higher deposition rate.

In SAW, the current density in the electrode wire is several times greater than in SMAW, and the melting rate of the electrode wire is increased. Typical current densities for various electrode diameters are given in Fig. 12.

The melting rate of the electrode wire is sometimes expressed in inches or feet per minute, but a more useful unit is pounds per minute or pounds per hour. For comparing applications, the melting rate is sometimes expressed as pounds per minute per ampere, or as pounds per thousand amperes per minute. In SAW, the electrode melting rate increases as the current increases, as shown in Fig. 13.

The melting rate of a welding electrode is the sum of the rate due to arc melting and that due to resistance (I^2R) heating in the length of electrode that extends beyond the contact tube, or the electrode extension. The melting rate due to I^2R heating is dependent on current density and electrode extension (Fig. 14). Both the current density and the electrode extension depend on the electrode diameter. The melting rate of a practical electrode extension of eight times the electrode diameter is shown in Fig. 15.

Fig. 12 Typical current densities and current ranges for various diameters of electrode wire

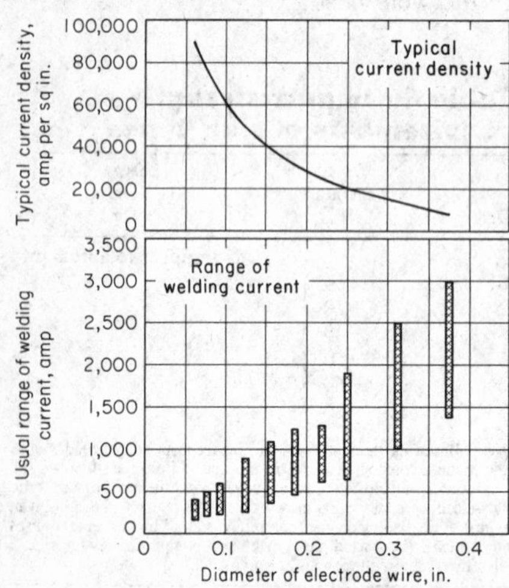

Fig. 13 Relationship of welding current to melting rate of electrodes for SAW

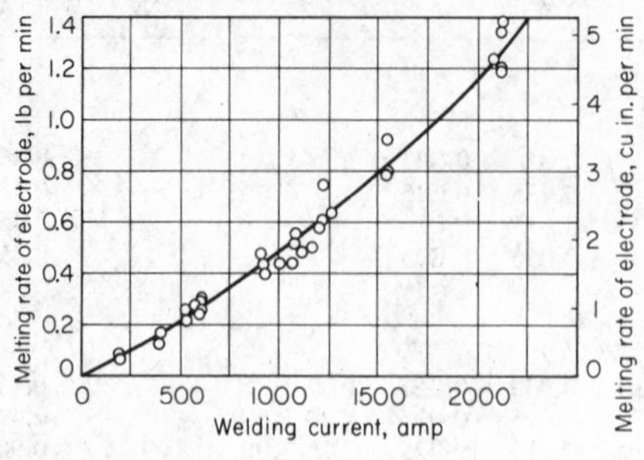

Fig. 14 Increase in electrode melting rate due to I^2R heating

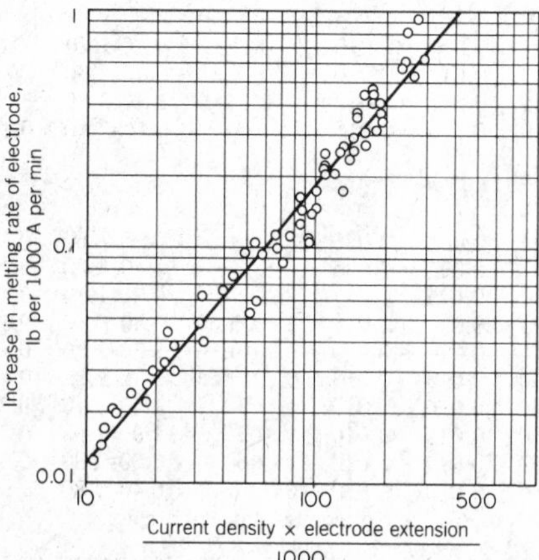

Fluxes

Fluxes used in SAW are granular, fusible mineral materials containing oxides of manganese, silicon, titanium, aluminum, calcium, zirconium, and magnesium and other compounds such as calcium fluoride. They are melted by the welding arc and, in the molten condition, blanket the weld metal and shield it from the atmosphere.

Fluxes are classified by AWS on the basis of the mechanical properties of a weld deposit. Figure 16 illustrates the classification system for fluxes used for SAW, which can be further explained by considering a typical designation such as F7A6-EM12K. Following the guidelines illustrated in Fig. 16, this designation refers to a flux that will produce weld metal which, in the as-welded condition, will have a tensile strength no lower than 70 ksi and Charpy V-notch impact strength of at least 20 ft · lb at −60 °F when deposited with an EM12K electrode. A flux used in combination with an electrode of any of the classes shown in Tables 1 or 2 must produce weld metal that conforms with the requirements in Tables 3 and 4. Submerged arc welding fluxes are produced in three forms: prefused, bonded, and agglomerated.

Prefused Fluxes. In the production of a prefused flux, the ingredients are dry mixed and then melted. Typical melting and pouring temperatures are between 2700 and 3100 °F. The molten flux may either be water shotted or poured on chill plates, then crushed and sized. With proper chilling, a glassy product is obtained. The product is passed over a series of screens that set upper and lower limits on the particle size—for example, through 12 mesh and on 200 mesh.

Advantages of prefused fluxes include:

- Chemical homogeneity is extremely good.
- Fines can be removed without changing the composition of the flux.
- The fluxes do not absorb moisture, thus simplifying storage problems.
- Nonconsumed or nonmelted flux can be recycled several times without significant change in average particle size or flux composition. Some codes, however, prohibit the recycling of flux.
- The fluxes are suitable for the highest travel speeds in welding operations.

The primary disadvantage of prefused fluxes is that deoxidizers and ferroalloys cannot be added without segregation or prohibitive losses during processing because of the high temperatures involved.

Bonded Fluxes. In the production of bonded flux, the raw materials are ground

Fig. 15 Effect of electrode diameter and electrode extension on melting rate

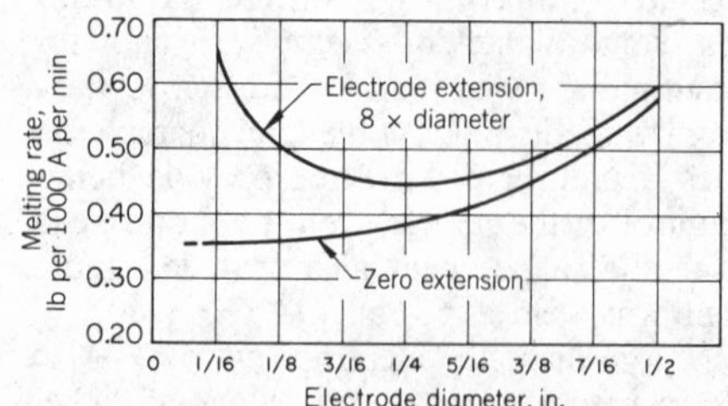

Fig. 16 AWS classification system for SAW fluxes

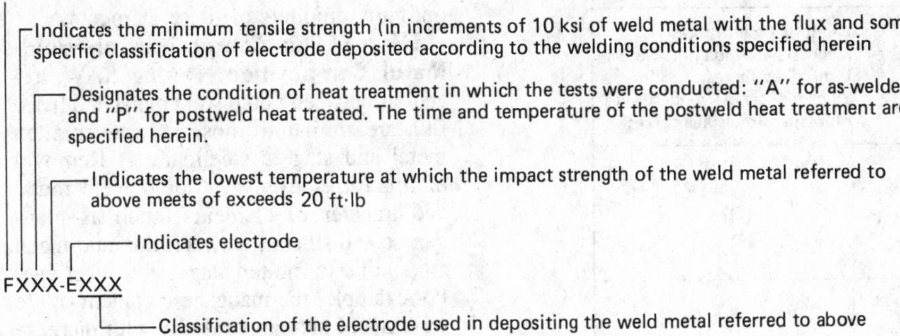

to approximately 100 mesh; they are dry mixed and then bonded with an addition of potassium silicate or sodium silicate. The resulting mixture is pelletized, dried at a relatively low temperature, broken up by mechanical means, and screened. Advantages of bonded fluxes are:

- Because of the low temperatures involved in the bonding process, metallic deoxidizers and ferroalloys can be included in the flux.
- The density of the flux is lower, which permits use of a thicker layer of flux in the weld zone.
- The solidified slag is readily detachable after welding.

One disadvantage of bonded fluxes is that fines cannot be removed without some alteration in the flux composition. Another disadvantage is that bonded fluxes are likely to absorb moisture, which can cause weld-metal porosity or hydrogen-induced cracking.

Agglomerated fluxes are similar to bonded fluxes except that a ceramic binder is used. The problem of moisture pickup by agglomerated or bonded fluxes can affect the operability of these particular formulations. For example, some manufacturers recommend use of controlled temperature/humidity storage environments. Also, careful flux-handling practices may be used, such as flux preheating prior to use and disposal of flux that has been exposed to high temperature/humidity environments for long periods. Redrying of large volumes of flux after exposure to moisture-absorbing conditions is difficult. In fact, recommended drying practices such as heating flux on trays for long periods at prescribed temperatures may be impractical when large volumes of flux are to be used.

Sizing of a SAW flux is important because the size of the particles and the distribution of particle sizes determine the level of welding current at which a flux performs best. For currents greater than 1500 A, the percentage of small particles must be increased. For bonded fluxes that are to be used at lower currents, sizing is less critical; these fluxes generally are available in only one sizing. The maximum welding current at which bonded fluxes can be used is generally 800 to 1000 A. Some prefused fluxes of the modified calcium silicate type can be used at more than 2000 A.

Handling and Storage. Commercial fluxes are thoroughly dry prior to shipping from the manufacturer and are packaged in moisture-resistant containers. Fluxes should be stored in dry areas at room temperature, 40 to 120 °F. Flux containers, after opening, should be kept in holding ovens at temperatures recommended by the manufacturer. If fluxes become damp or wet, they can be dried by rebaking at temperatures recommended by the manufacturer. Moist or damp flux causes porosity and cracking in weld metal. When flux is recycled, care should be taken to avoid contamination by rust, mill scale, and other foreign substances. Flux that has been contaminated by oil or other foreign material should be discarded.

Composition. In the development of the SAW process, prefused fluxes consisting of complex silicates were used. Formulations were chiefly alumina silicates of manganese, calcium, and magnesium. Manganese silicate compositions corresponding to the typical analyses shown in Table 6 have been extensively used. Flux formulations of this type generally have excellent operating characteristics, such as good current/voltage stability, and produce weld deposits with excellent bead shape.

The demand for submerged arc deposits having lower oxygen contents (in the range of 250 to 450 ppm of oxygen) and optimum toughness has been satisfied by the use of fluxes containing substantial contents of CaO, Al_2O_3, MgO, and CaF_2. Typical low-oxygen-potential flux compositions (sometimes referred to as basic) are presented in Table 7 (see the section of this article on the effect of flux composition on weld-metal composition for an explanation of the basicity index). Figure 17 shows the relationship between weld-metal toughness and weld-metal oxygen content for various types of flux. The various types of flux are classified as shown in Table 8.

Table 6 Typical compositions of manganese silicate fluxes for SAW

Substance	Prefused flux	Bonded flux
MnO	42.0%	36.5%
MnO_2	...	5.2(a)
SiO_2	45.0	38.0
CaF_2	6.9	3.9
CaO	1.2	0.8
MgO	0.3	2.7
BaO	0.1	0.3
Al_2O_3	2.0	1.1
FeO	1.5	...
Fe_2O_3	...	2.7
TiO_2	0.1	0.1
K_2O	0.4	...
Na_2O	0.4	1.5
PbO	0.1	0.1
FeSi (50%)	...	7.1(b)
Ratio MnO/SiO_2	$\frac{42.0}{45.0} = 0.93$	$\frac{40.7}{45.6} = 0.89$

(a) At welding temperatures, MnO_2 reacts with silicon in ferrosilicon to yield additional MnO equal to 4.2% of the total flux (0.815 × 5.2% = 4.2%). (b) In reaction with MnO_2 at welding temperatures, silicon in ferrosilicon forms additional SiO_2 equal to 7.6% of the total flux (2.14 × 7.1% ÷ 2 = 7.6%).

Table 7 Typical low-oxygen-potential fluxes

Flux	Composition, % Al_2O_3	SiO_2	TiO_2	MgO	CaF_2	CaO	MnO	Na_2O	K_2O	Basicity index (*BI*)
A	49.9	13.7	10.1	2.9	5.7	...	15.1	1.6	0.2	0.4
B	45.7	12.9	10.7	0.1	15.6	...	10.9	0.2	1.7	0.6
C	24.9	18.4	0.2	28.9	24.2	...	1.8	2.1	0.07	1.8
D	23.2	14.5	0.9	30.8	17.7	4.2	2.3	2.3	0.1	2.1
E	19.3	16.3	0.8	27.2	23.6	9.8	0.08	0.9	1.1	2.4
F	18.5	15.2	1.0	31.8	29.8	1.8	1.5	1.0	1.05	2.7
G	14.6	13.5	0.3	31.2	25.7	...	3.5	1.3	1.9	3.0
H	18.1	13.2	0.5	28.2	31.8	4.5	0.1	0.9	0.9	3.0
I	17.0	12.2	0.7	36.8	29.2	0.7	8.9	1.6	0.1	3.5

Fig. 17 Relationship among flux composition, weld-metal oxygen content, and weld-metal Charpy toughness for X-70 linepipe

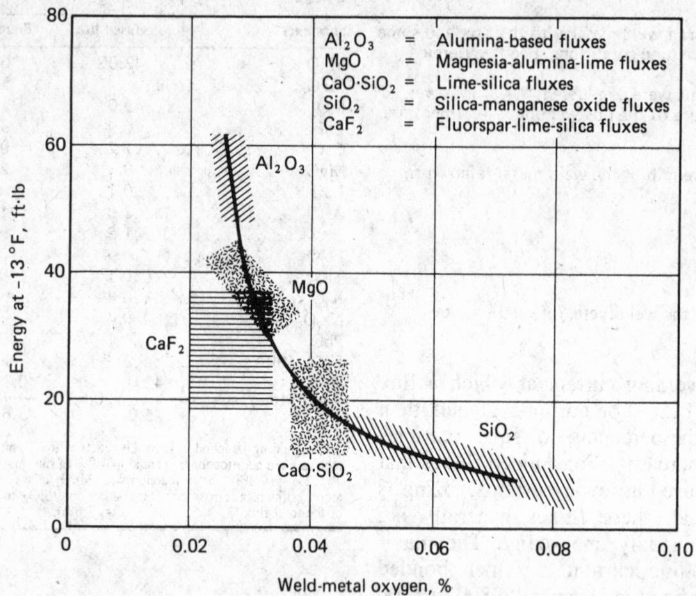

Fluxes recently have been developed for SAW that contain prescribed amounts of titanium and boron compounds. These fluxes deposit welds that have titanium and boron contents, which optimizes weld-metal microstructure (large contents of acicular ferrite) and toughness. Conventional carbon-manganese-molybdenum electrode filler wire can be used with boron- and titanium-bearing submerged arc fluxes to deposit tough weld metals that contain optimum amounts of these elements.

Effect of Flux Composition on Weld-Metal Composition. During SAW, reactions between liquid weld metal and fused flux are similar to those between molten metal and slag in steelmaking. Removal of impurities from the liquid weld metal and transfer of elements, such as manganese and silicon, between flux and metal are similar to molten slag-metal reactions. For example, the manganese content of the weld metal increases rapidly with increase in MnO content of the flux (Fig. 18). Many flux compositions contain about 10% MnO to promote the pickup of manganese by the weld metal. A different relationship exists between the SiO_2 content of the weld metal (Fig. 18). Normally, a rapid pickup of silicon is not encountered until the SiO_2 content of the flux is approximately 40%. As a result, commercial fluxes generally are restricted to a maximum of about 40% SiO_2.

The influence of the various specific flux types on weld-metal oxygen content is shown in Fig. 19. Flux formulations producing the lower deposit oxygen contents can be derived from a consideration of the

Table 8 Fluxes for SAW

Flux type	Chemical constituents	Advantages	Limitations	Basicity	Flux form	Comments
Manganese silicate	$MnO + SiO_2 > 50\%$	Moderate strength, tolerant to rust, fast welding speeds, high heat input, good storage	Limited use for multipass welding, use where no toughness requirement, high weld-metal oxygen, increase in silicon on welding, loss in carbon	Acid	Fused	Associated manganese gain; maximum current, ≅1100 A; higher welding speeds
Calcium-high silica	$CaO + MgO + SiO_2 > 60\%$	High welding current, tolerant to rust	Poor weld toughness, use where no toughness requirement, high-weld metal oxygen	Acid	Agglomerated fused	Differ in silicon gain, some capable of 2500 A, wires with high manganese
Calcium silicate-neutral	$CaO + MgO + SiO_2 > 60\%$	Moderate strength and toughness, all current types, tolerant to rust, single- or multiple-pass weld		Neutral	Agglomerated fused	
Calcium silicate-low silica	$CaO + MgO + SiO_2 > 60\%$	Good toughness with medium strength; fast welding speeds, less change in composition and lower oxygen	Not tolerant to rust, not used for multiwire welding	Basic	Agglomerated fused	
Aluminate basic	$Al_2O_3 + CaO + MgO > 45\%$ $Al_2O_3 > 20\%$	Good strength and toughness in multipass welds No change in carbon; loss of sulfur and silicon	Not tolerant to rust, limited to DCEP welding, poor slag detachability	Basic	Agglomerated	Usually manganese gain; maximum current, ≅1200 A; good mechanical properties
Alumina	Bauxite base	Less change in weld composition and lower oxygen than for acid type, moderate to fast welding speeds		Neutral	Agglomerated fused	
Basic fluoride	$CaO + MgO + MnO + CaF_2 > 50\%$ $SiO_2 \leq 22\%$, $CaF_2 \leq 15\%$	Very low oxygen, moderate to good low-temperature toughness	May present problems of slag detachability May present problem of moisture pickup	Basic	Agglomerated fused	Can be used with all wires, preferably direct current welding, very good weld properties

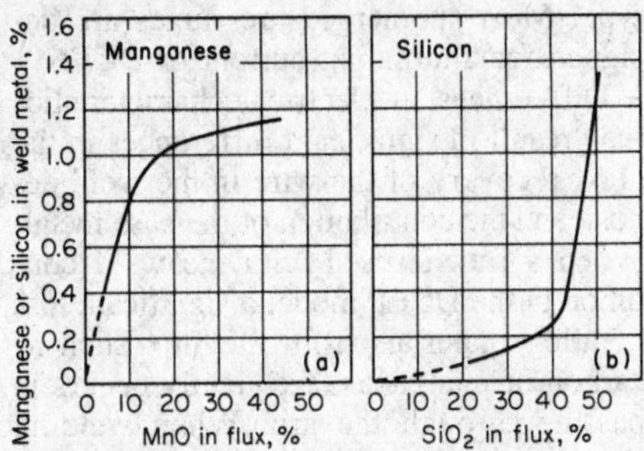

Fig. 18 Effect of MnO and SiO_2 contents of SAW fluxes on manganese and silicon contents of weld metal

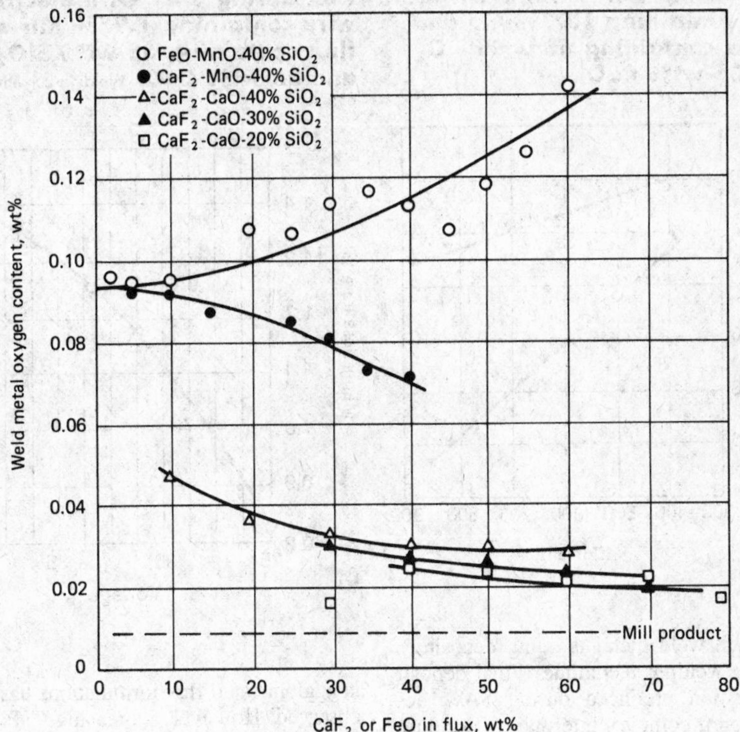

Fig. 19 Weld-metal oxygen content as a function of CaF_2 and FeO additions in the flux

oxygen potential of flux constituents, which can be assessed from the Ellingham diagram. This diagram relates the free energy of oxide formation with temperature and indicates relative oxide stabilities. For example, oxides such as Al_2O_3 are very stable, while oxides such as FeO and MnO are not so stable. Fluxes high in FeO or MnO would then produce welds with high oxygen contents. SiO_2 and TiO_2 have oxygen potentials intermediate between FeO and Al_2O_3.

Oxygen potential is not the only concept used in flux formulation. Basicity, though conceptually imprecise, is more frequently used. Fluxes prepared from predominantly acidic oxides, such as SiO_2 and Al_2O_3, are described as acidic. Welds made with this kind of flux usually have high oxygen concentrations. Fluxes with predominantly basic oxides, such as CaO and MgO, are described as basic and produce low weld-metal oxygen contents. Empirically, weld-metal mechanical properties can be related to flux basicity. High weld-metal toughness usually is observed in welds made with basic flux. Coarse microstructure and low resistance to cleavage usually are reported in welds made with acidic flux.

The basicity index (*BI*) is commonly used to describe different flux types. Fluxes with a basicity index greater than 1.5 are basic in nature; those below 1.0 are considered acidic. Fluxes between 1.0 to 1.5 are considered neutral. Use of the basicity index with fluxes of high transition-metal oxide contents is questionable. The basicity index (see below) is given as the ratio of weight percents of specific compounds.

The relationship between weld-metal oxygen content and flux basicity is shown in Fig. 20. During SAW, slag-metal interaction principally occurs at the droplet formation stage on the electrode tip. The final weld deposit is actually an accumulation of metal that is quenched from high temperatures (approximately 4350 °F). Because droplet transfer frequencies are on the order of 10 droplets per second, chemical reactions occur extremely rapidly. Any factor that affects the residence time of droplets on the electrode tip has a large effect on weld-metal composition. Droplet transfer frequencies increase as arc current is increased and as the voltage is decreased during welding. The importance of welding parameters on weld-metal composition is emphasized in Fig. 21 and 22, which show the influence of increasing current and voltage on manganese transferred to the flux during direct current electrode positive welding using a high-manganese-content electrode wire and a flux comprised of 65 wt% SiO_2 and 35 wt% CaO. The principal role of welding speed in controlling weld-metal composition depends on its effect on dilution occurring during welding. Dilution increases with an increasing ratio of current to travel speed.

Some submerged arc fluxes are formulated to provide control over the transfer of metallic elements. Ferroalloys bonded into the flux can supply alloying elements to the weld metal to obtain desired properties of the deposit. Oxides added to the flux, such as Cr_2O_3 and NiO, help distribute metallic elements between the molten flux and weld metal. The silicon content of the weld metal is influenced by the Cr_2O_3 content of the flux, the composition of the electrode, the composition of the base metal

Fig. 20 Relationship between weld-metal oxygen content and flux basicity index

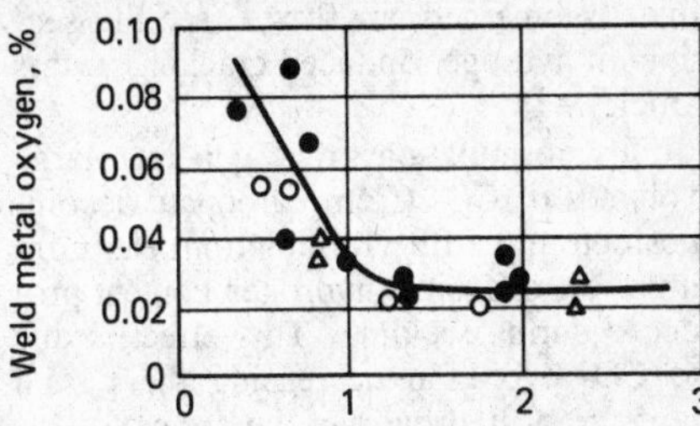

$$BI = \frac{CaO + CaF_2 + MgO + K_2O + Na_2O + Li_2O + 1/2(MnO + FeO)}{SiO_2 + 1/2(Al_2O_3 + TiO_2 + ZrO_2)}$$

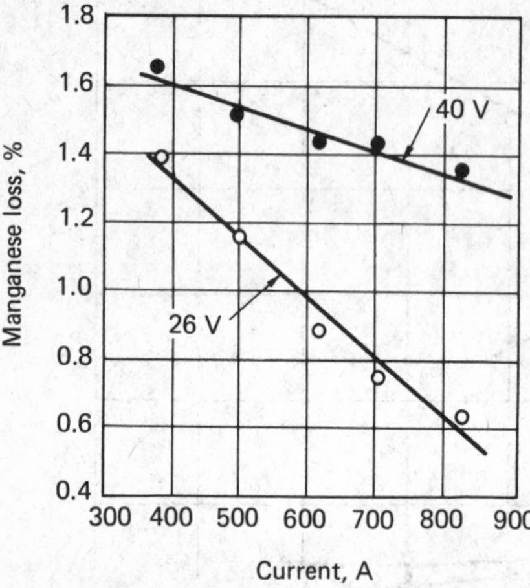

Fig. 21 Effect of current increase on the loss of manganese to the flux during SAW with electrode wire containing 1.93% Mn and a flux containing 65 wt% SiO_2 and 35 wt% CaO

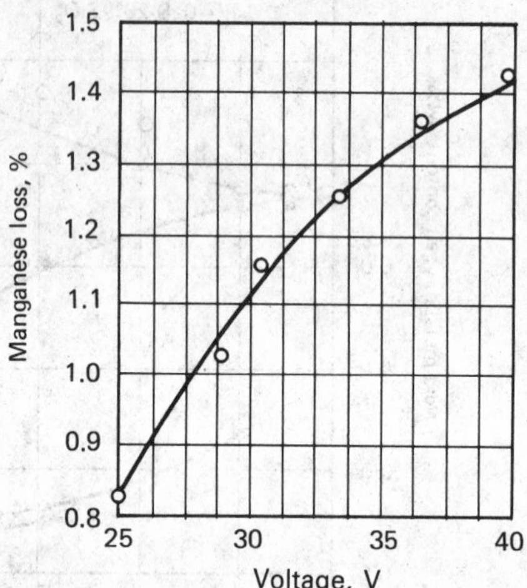

Fig. 22 Effect of voltage increase on the loss of manganese to the flux during SAW with electrode wire containing 1.93% Mn and flux containing 65 wt% SiO_2 and 35 wt% CaO. Welding current, 615 A

on which weld metal is being deposited, and the welding technique. Final deposit composition produced during SAW depends on a complex interplay of welding parameters (current, voltage, and travel speed) and external factors such as flux composition, electrode wire, and base-metal composition.

Flux Moisture and Weld Hydrogen Content. Bonded and agglomerated fluxes can pick up moisture when exposed to high atmospheric temperatures and high humidity for long periods. Flux moisture content produces weld-metal hydrogen during welding operations and can promote cracking. The diffusible hydrogen content produced when welding with a flux containing a given moisture content depends markedly on the flux composition.

Although fused submerged arc fluxes do not absorb moisture during long exposure to high atmospheric temperature and high humidity, they do have an as-received moisture content. Hydrogen-induced cracking in weld metal (chevron cracking) also has been detected when welding with fused submerged arc fluxes (see the section on hydrogen-induced cracking in this article).

Flux formulations that generate large volumes of CO_2 (from carbonate decomposition in the flux formulation) can minimize the diffusible hydrogen content produced during welding. This effect is due to CO_2 evolution decreasing the partial pressure of hydrogen in the arc cavity. A similar effect of increased CaF_2 contents in agglomerated flux formulations has been observed. However, increasing CaF_2 contents may also decrease weld-metal oxygen content (see Fig. 23).

Type of Welding Current. Fluxes usually are designed to operate with either alternating current or direct current. The type of welding current used has an effect on all weld properties. Joint penetration (and resulting dilution) are greatly affected by the type of welding current used. Direct current can be used in the electrode positive (DCEP), also known as reverse polarity, or the electrode negative (DCEN), also known as straight polarity, mode. The joint penetration will change with type of current used; however, that change will be influenced by the current range and flux type. Most submerged arc fluxes are designed for alternating current or DCEN.

Differences in electrical characteristics may result in significant differences in the alloy recovery of the wire in the weld deposit and the contribution of the base metal. When a flux is used with the weld condition in the DCEP mode, a significant loss of alloy materials from the flux such as carbon or manganese frequently occurs in passing through the arc. When welding equipment is switched to the alternating current mode, nearly 100% of the elements may be recovered. Typically, more flux is melted per pound of weld-metal deposit with DCEP electrical connection than with the alternating current connection—an important consideration in applications requiring clean deposits, as fluxes are a potential source of contamination. (See the section on power supplies in this article for more information on welding currents.)

Melting Rate. The amount of flux fused per minute in SAW depends on the welding current and voltage. For a given current, the amount of flux fused per minute increases with voltage, as shown in Fig. 24. Under normal welding conditions, the weight of flux melted is about the same as the weight of electrode melted.

Depth of flux layer affects the shape and penetration of welds, as shown in Fig. 25. When the flux layer is too shallow (Fig. 25a), the arc is exposed and a cracked or porous weld results. When the flux is too

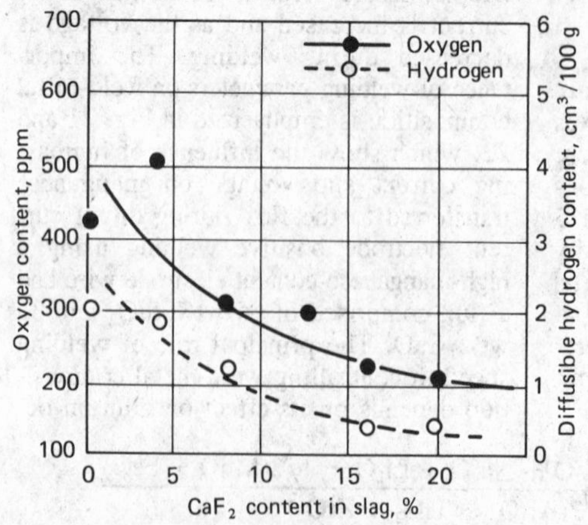

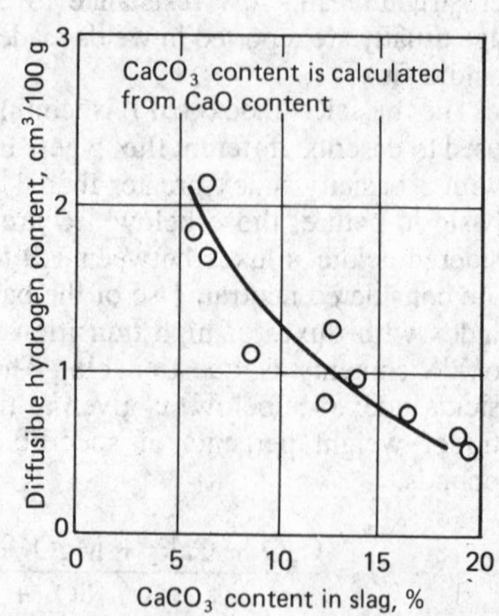

Fig. 23 Effect of CaF_2 and $CaCO_3$ contents in flux on weld-metal oxygen and hydrogen contents during SAW with a highly basic bonded flux

Fig. 24 Effect of increasing voltage on the fusion rate of a calcium silicate

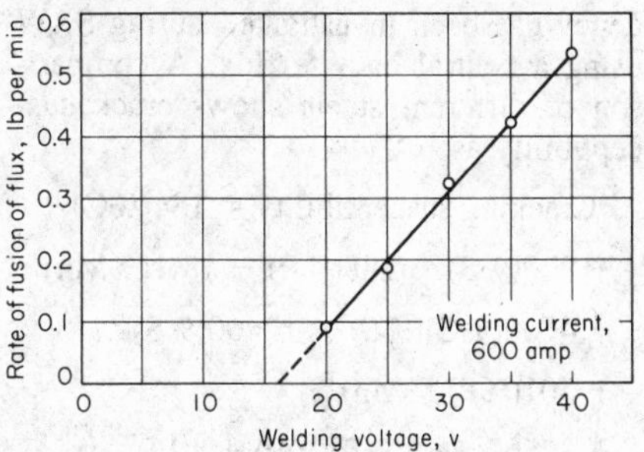

Fig. 25 Effect of depth of flux layer on shape and penetration of submerged arc surface welds made at 800 A. (a) Flux layer too shallow, resulting in arc breakthrough (from loss of shielding), shallow penetration, and weld porosity or cracking. (b) Flux layer at correct depth for good weld-bead shape and penetration. (c) Flux layer too deep, resulting in peaked weld bead with above-average penetration

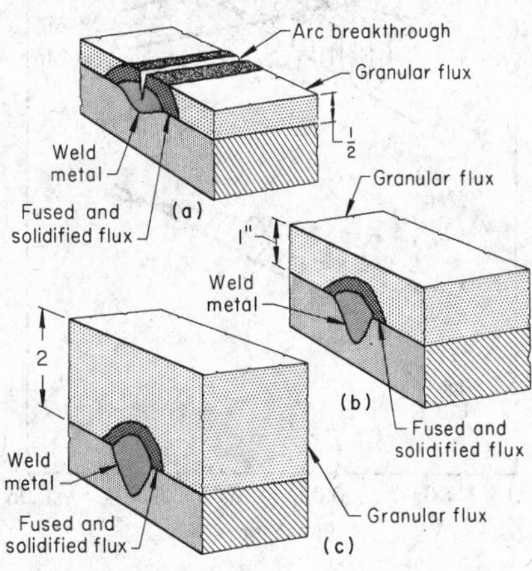

deep (Fig. 25c), the result is peaked weld beads with above-average joint penetration. When the flux is neither too shallow nor too deep, very faint flashes appear around the interface between the electrode wire and the flux, and the weld bead appears as shown in Fig. 25(b).

Viscosity and Conductivity. The viscosity of a welding flux at the weld zone must be high enough to make it impermeable to atmospheric gases and to prevent it from running off the molten weld metal or from flowing in front of the arc, which could lead to molten metal being deposited on top of the flux. On the other hand, during welding, flux must be fluid enough to permit rapid solution of nonmetallic constituents, such as oxides, and the release of gases from molten metal.

To be effective, a welding flux should have a viscosity at high temperatures that provides a fluid blanket over the metal and protects it from oxidation; it should be brittle at room temperature to facilitate the removal of slag. The viscosities of welding fluxes are typically between 2 to 7 poise at 2550 °F. Both the melting temperature of the flux and the density of the molten flux must be lower than that of the weld metal. Consequently, the upper limit for the melting range of SAW fluxes is about 2370 °F.

At room temperature, the granules of SAW fluxes are electrical insulators; electrical resistivity decreases with increasing temperature, and the fluxes are highly conductive at the temperatures prevailing in the welding zone. The effect of temperature on the viscosity and electrical resistance of a typical manganese silicate SAW flux is shown in Fig. 26(a). This flux has a narrow melting range, typical of a single compound. The effect of temperature on viscosity and electrical resistance of the more complex calcium silicate flux modified with alumina is shown in Fig. 26(b).

Weld Porosity

Contaminants in the flux, notably moisture, dirt, and mill scale, promote weld porosity. The presence of these contaminants is more likely in flux that is recycled. Moisture can be removed by heating the flux; however, overheating of fluxes can be detrimental, and the precise temperature and time for baking should be obtained from the flux manufacturer for each specific flux. Such drying operations may not be practical when handling large volumes of flux. Prevention, rather than baking, is the key approach when addressing the problem of flux moisture absorption.

Insufficient Flux Coverage. When flux coverage is insufficient, the arc flashes through and may cause porosity. Insufficient flux cover is more likely to occur on circumferential welds than on flat welds. On small circumferential welds, it is often necessary to provide mechanical support to the flux around the arc. Similarly, if slag spills off a weld that has not solidified, the weld may exhibit surface porosity. Corner welds and multiple-pass horizontal fillet welds are especially susceptible to surface porosity caused by flux spillage.

Excessively viscous fluxes can be the cause of porosity in SAW welds. Gases, which are dissolved in the high-temperature liquid weld metal, evolve to the surface during cooling prior to and on solidification. These gases form bubbles in the flux which must be able to move through the flux, away from the weld-metal surface. If the flux is overly viscous, these bubbles may be trapped in the surface region of the weld metal, resulting in porosity. Adjustment of flux viscosity through composition control can alleviate this concern.

Arc blow is a cause of porosity that is most often encountered in the high-speed welding of thin sheet, using direct current, although thicker work metal can be similarly affected. Although arc blow can occur in any direction, backward arc blow is a frequent cause of porosity. Backward arc blow can result from a variety of causes, including improper placement of the electrical ground, poor electrical contact at the ground due to mill scale, and the development of a secondary magnetic field caused by the welding cable being partially wrapped around the joint being welded and thereby diverting the arc. In sheet metal welds, porosity from back-

Fig. 26 Effect of temperature on the viscosity and electrical resistance of two types of SAW flux

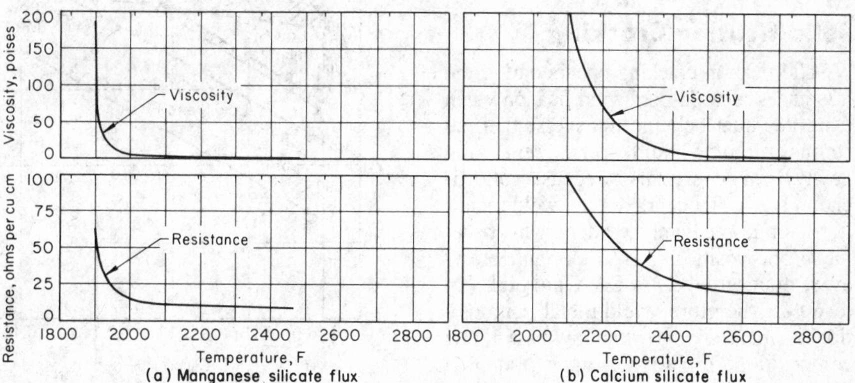

ward arc blow usually occurs in the last few inches of a weld, becoming progressively more severe as the end of the weld is approached. Backward arc blow can be prevented by (1) welding away from the location of the ground, (2) depositing a heavy tack weld at the finished end of a joint, (3) angling the electrode forward, (4) using alternating current welding power, or (5) clamping the ground firmly to the workpiece and welding toward the closed end of the fixture. Fixtures for welding thin metal preferably should be made either of copper or of some other suitable nonmagnetic material.

Prevention. Special procedures can be adopted to overcome the problem of porosity formation. Lower welding speeds permit entrapped gases to escape from the molten weld pool. Multipass welding procedures may alleviate the problem, as weld deposits remelt some of the prior weld beads and may remove some of the porosity. These recommendations may not always be possible to implement, for example, in situations where productivity requirements demand high welding speeds and where multipass welding is impractical.

Press-fitted joints are sometimes coated with white lead (basic lead carbonate) before parts are pressed together. White lead (like most other lubricants) becomes a gas-producing contaminant that may cause porosity, generally in the form of large holes at or near the end of the weld, or weld cracking. It is best to avoid press fits. A gap of 10% of the plate thickness, not to exceed 1/32 in., should be allowed. The edge of one workpiece also can be knurled, thereby providing a path for escape of gas.

Cracking

Two types of cracking generally are encountered in submerged arc welds—solidification and hydrogen-induced cracking. Solidification cracking is limited to the weld deposit, while hydrogen-induced cracking can occur in weld metal or in the HAZ.

Solidification Cracking

Solidification cracking occurs during the last stages of weld-pool solidification when dendrites interlock and transverse thermal strains rupture liquid films, separating newly formed crystals. Because solidification cracking occurs only in weld metal, it is not immediately evident why base-metal composition is so important. However, dilution levels in SAW are high (up to 60%); therefore, weld-metal composition is markedly affected by base-metal composition. Impurity elements, such as sulfur and phosphorous, in the base metal have an important influence on solidification cracking, because they promote formation of low-melting-point films during weld-pool solidification. In SAW, a tendency toward solidification cracking is not generally counteracted by a change in flux or electrode filler-wire composition, as the choice of flux and electrode wire may be restricted by special weld-metal mechanical property requirements, such as high toughness at low design temperatures.

Dilution by elements from the base metal has a marked influence on weld-metal composition and on solidification cracking tendency. Increased amounts of carbon, phosphorous, and sulfur increase solidification cracking tendency, while resistance to cracking is promoted by high weld-metal manganese-to-sulfur ratios. The effect of increasing carbon, manganese, sulfur, and phosphorous contents on cracking tendency has been investigated during SAW using a neutral bauxite flux. A comparison of different steels shows crack susceptibility as:

$$\begin{aligned}\text{Cracking susceptibility} &= 184(\%C) \\ &+ 970(\%S) - 188(\%P) - 18.1(\%Mn) \\ &- 4760(\%C)(\%S) - 12400(\%S)(\%P) \\ &+ 501(\%P)(\%Mn) \\ &+ 32600(\%C)(\%S)(\%P) + 12.9\end{aligned}$$

This equation, with these particular constants, was derived for a single flux type and a given set of welding conditions

Fig. 27 Effect of weld-metal composition on solidification cracking. 0.60-in.-thick transvarestraint specimen, 70° notch 0.18 in. deep, 0.16-in. root radius. Submerged arc welded with neutral bauxite flux, 0.13-in.-diam wire, 500 A, 31 V (DCEP) at 20 in./min

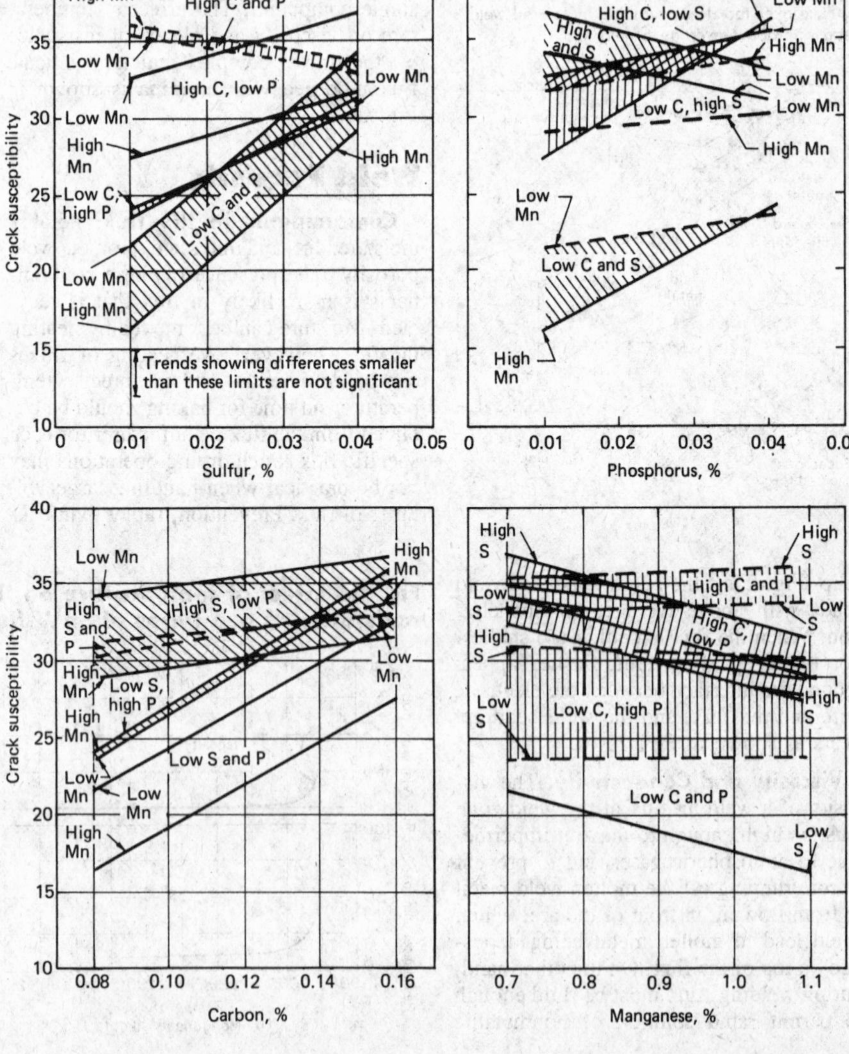

(welding of 0.60-in.-thick plate at 500 A, 31 V, and 20 in./min when using DCEP); however, the general equation is applicable to many other welding conditions. The influence of any element on cracking susceptiblity depends on the levels of the other elements present (Fig. 27). Only a limited amount of work has been conducted on the effects of other elements, such as niobium, vanadium, aluminum, and the rare earth elements, on cracking susceptibility. Increasing niobium contents up to about 0.13% has been reported to decrease cracking resistance, but at high levels (up to about 1%), a beneficial effect has been experienced. Increasing vanadium content has been reported to reduce cracking tendency when welding 0.16%C-1.4%Mn pipe. Aluminum has been found to be beneficial or detrimental, depending on the welding situation. Rare earth elements are known to reduce cracking tendency in a 0.04%C-1.4%Mn weld deposit.

The detrimental role of sulfur in increasing solidification cracking tendency can be partially overcome. When welding steels containing high sulfur contents or steels nominally low in sulfur that exhibit sulfur-rich segregation, special welding procedures can be adopted. Dilution of base metal can be restricted by using low DCEN current and large electrode-wire diameters. The use of high-manganese-content electrode wire promotes the formation of manganese sulfides and reduces cracking susceptibility. Butt joints that normally would have squared edges should be prepared as beveled joints; on joints that are normally beveled, the bevel angle should be increased. These techniques help minimize the dilution produced during welding operations.

Fillet Welds. To minimize internal shrinkage stresses that could result in solidification cracking, the fillet-weld bead should be at least 1.25 times as wide as it is deep. This is especially important for steels susceptible to cracking. Beads deeper than they are wide may crack internally, as shown in Fig. 28(a), and may present difficulties in slag removal. Bead contours and cross sections such as those shown in Fig. 28(b) are satisfactory. These beads are 1.25 times as wide as they are deep, and the bead surface ranges from flat to slightly convex.

Butt Welds. Solidification cracking in butt welds is promoted by making weld deposits that are deep and narrow (Fig. 29). In general, the width-to-depth ratio of welds should not be less than one. Joint preparation prior to welding has a large influence on cracking tendency. For example, use of too narrow a groove angle during welding increases the likelihood of cracking. For applications in which 100% joint penetration welds are required for T-joints between thick plates, a low-hydrogen, shielded metal arc electrode can be employed for the first pass on each side to avoid penetration that will produce weld beads that are deeper than they are wide.

Fig. 28 Effect of the shape of the weld bead on its susceptibility to cracking. (a) Fillet weld beads that have cracked internally because they are deeper than they are wide. (b) Fillet welds that are wider than they are deep, and hence less susceptible to cracking

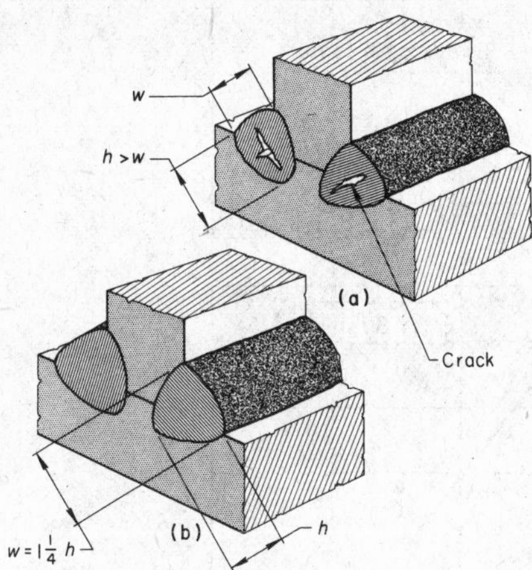

Another type of butt-joint weld bead that may crack is the weld bead made during the first pass on the second side of a double-bevel-groove or double-V-groove joint. When the trough produced by backgouging the root-pass weld bead on the first side of the joint is deep and narrow, the first-pass weld bead on the second side of the joint will also be deep and narrow and is likely to crack. This can be prevented by making the backgouged trough wide enough so that the first bead on the second side can be wider than it is deep. The depth of any one weld pass should not exceed its width. When necessary, a split-weave layer may be used.

Hydrogen-Induced Cracking

Hydrogen-induced cracking is caused by a complex interaction of a diffusible hydrogen supply, tensile residual stresses, and a susceptible microstructure. This form of cracking generally is not encountered when welding plate sections with thicknesses less than 0.40 in. However, when thicker sections are welded, particularly those 2 in. or more in thickness, welds are subjected to more rapid cooling, accompanied by more severe cooling stresses. Hydrogen-induced cracking is promoted by (1) failure to preheat, or preheating at too low a temperature; (2) heavy sections, which rapidly draw heat from the weld zone and thereby intensify thermal stress; (3) workpiece designs that impose a high degree of mechanical restraint on the weld area; (4) the presence of hydrogen-producing contaminants, such as moisture, in the weld zone; (5) failure to adequately postweld heat treat or stress relieve the weldment; and (6) improper bead shape and size.

Chevron Cracking. Transverse cracking in submerged arc welds has been observed when welding a range of steels, including 2.25%Cr-1%Mo steel for pressure vessels, carbon-manganese-niobium steel, and 1%Ni-0.5%Cr-0.5%Mo steel. This form of cracking lies 45° to the weld surface in multipass welds in thick-walled plate. Chevron cracking has been observed when welding 1-in.-thick plate under conditions of severe restraint, and in the final passes of welds in 4-in.-thick plate. Due to the inclination of these cracks, they are called "chevron cracks." They are not easily detected by ultrasonic testing and are usually found only with detailed metallographic examination.

Chevron cracking is associated with the moisture content of submerged arc fluxes. The problem is emphasized when welding is performed with high basicity (low-oxygen-potential) agglomerated submerged arc fluxes. However, if sufficient moisture is present in as-received fused fluxes, cracking can also be promoted.

A detailed investigation of chevron cracking in carbon-manganese-nickel weld deposits (containing 0.1%C-1.5%Mn-1%Ni) has indicated that the likelihood of cracking is extremely low when the diffusible hydrogen content in the weld is less

Fig. 29 Internal cracking occurring when joint penetration is greater than width

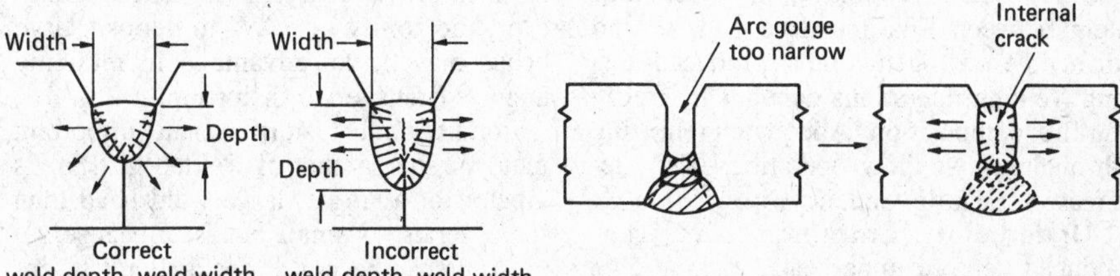

Fig. 30 Relationship between weld-bead maximum hardness and the residual hydrogen content just below the surface of the weld bead to avoid transverse cracking in weld metal

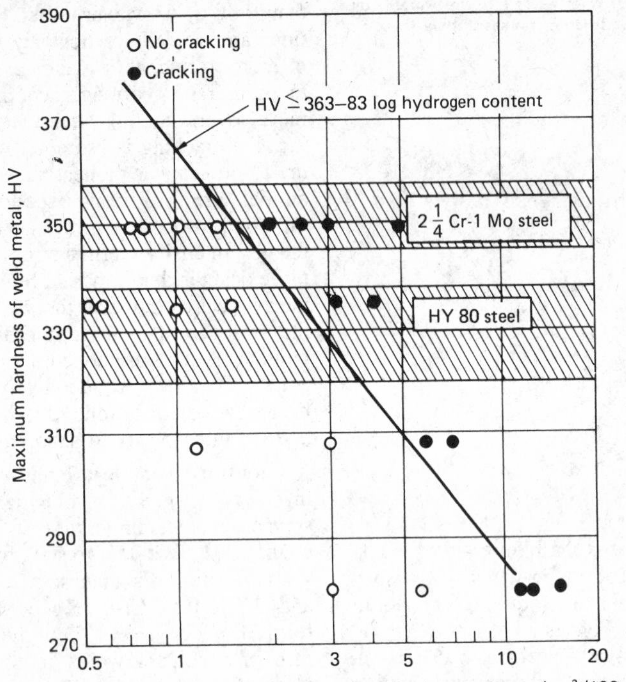

than 10 mL/100 g. This critical hydrogen content depends on weld-metal composition, weld hardness, plate thickness, and the welding parameters employed.

Figure 30 relates the residual hydrogen content just below the final bead and the maximum hardness of the weld bead during multipass welding of 2- to 6-in.-thick 2.25%Cr-1%Mo and HY80 steel. The maximum hardness that can be tolerated without cracking decreases as the diffusible hydrogen content of the weld metal increases. Prevention of transverse cracking during SAW can be achieved by careful control of flux moisture content during storage and in use; operational procedures maintaining fluxes in their as-received condition should be emphasized, such as redrying (or disposal) of flux spillages in high-temperature and high-humidity environments. Some types of flux may be inherently susceptible to cracking even in the as-received condition. The relationship between flux moisture content and diffusible hydrogen content produced during welding operations depends markedly on flux composition. Adequate preheating throughout welding operations also decreases cracking tendency.

Underbead Cracking. Hydrogen-induced delayed underbead cracking also can occur in submerged arc welds produced in higher strength steels. This hydrogen-induced cracking is found at the toes of weld deposits and is promoted by increased base-metal hardenability (high carbon and alloy element contents), by improper welding conditions (depositing small welds so that cooling rates after welding favor the formation of hard, brittle HAZ regions), and by weld designs that impose a high degree of mechanical restraint on the weld area.

Distortion

Because SAW is characteristically well adapted to the joining of heavy sections at high current densities and correspondingly high heat inputs, it is more likely to cause distortion than arc welding methods that employ lower heat input. This generalization, however, depends in part on the number of weld layers deposited. However, the ability of SAW to deposit large beads may be an advantage in reducing angular (transverse) distortion.

Longitudinal distortion, an important factor in the camber of welded girders, is greater for a single, large weld bead than for a series of small beads. Because reverse precambering is not difficult to apply, longitudinal distortion resulting from large weld beads can be readily controlled. In addition, torch straightening can be used after welding.

Angular distortion of a fillet welded member increases with the number of weld layers. Thus, a $^1/_2$-in. fillet weld deposited as a single layer by SAW may exhibit only 1° distortion, whereas a four-layer deposit may be distorted by as much as 3°. For more information, see the article "Residual Stresses and Distortion" in this Volume.

Slag Inclusions

Poor joint fit-up can promote first-pass porosity and slag inclusions by allowing flux to become trapped between the bottom of the weld bead being deposited and the side of the joint opposite the bead. These defects may be subsurface, occurring in the root of the weld, or they may come through to the lower surface.

Slag inclusions due to poor joint fit-up frequently occur in butt welds. In a butt weld, the joint may be backed by a backing strip, by backing flux, or by another weld produced using SMAW or GTAW on the back of the weld or bottom of the root of the weld. If the gap between the plate edges is $^1/_{32}$ in. or more, flux spills into the gap ahead of the arc. Therefore, to avoid inclusions, either the weld bead must penetrate the backing, or penetration must be reduced to allow the weld bead to clear the backing by at least $^5/_{32}$ in.

Power Supplies

Power supplies for SAW include motor-generators and transformer-rectifiers for direct current output and transformers for alternating current output. Both alternating current and direct current produce acceptable results in SAW, although each has distinct advantages in specific applications, depending on the amperage range, diameter of the electrode wire, and travel speed.

In semiautomatic welding with electrode wire $^5/_{64}$ or $^3/_{32}$ in. in diameter, use of direct current predominates at input current of 300 to 550 A. Direct current is also frequently used in automatic welding with a single electrode at low current (300 to 500 A) and high travel speed (40 to 200 in./min). Direct current, as well as alternating current, finds favorable use in automatic welding with a single electrode using medium current (600 to 900 A) and travel speeds of 15 to 30 in./min.

Use of alternating current predominates in automatic welding, with a single electrode, using high current (1200 to 2500 A),

Fig. 31 Oscillograms (superimposed) for welding current and voltage for a motor-generator and a transformer

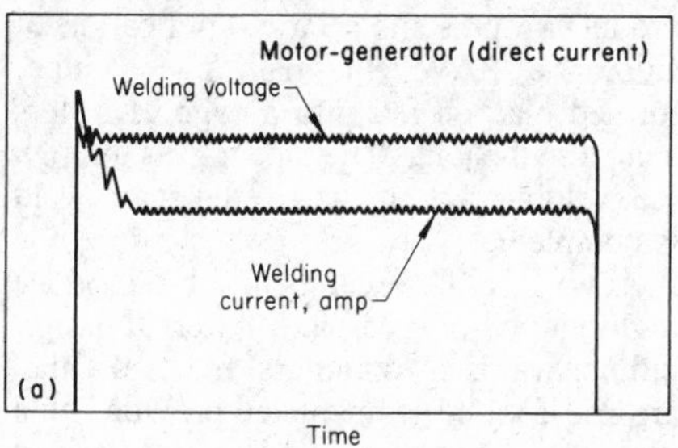

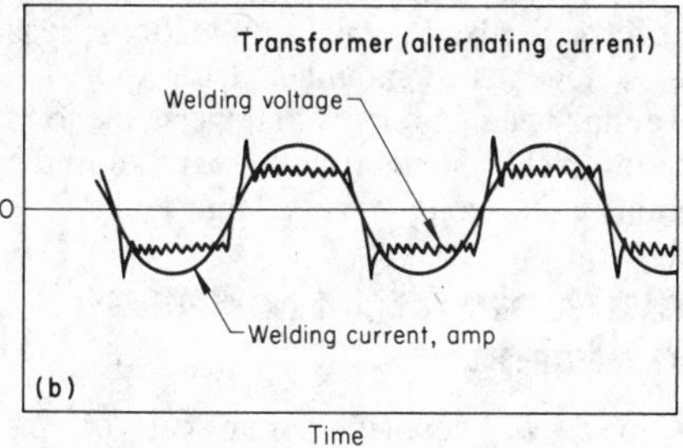

and travel speeds of 5 to 15 in./min. Automatic welding with two or more electrodes in tandem at 500 to 1000 A per electrode frequently uses alternating current on all electrodes (or direct current on the lead electrode). Automatic welding with two electrodes in the transverse position may be done with either alternating or direct current.

Direct Current. The characteristics of the arc circuit for SAW with direct current are essentially continuous current and continuous voltage, uninterrupted by short circuits, as shown in the superimposed oscillograms in Fig. 31(a). At the start of welding, there is a surge of current as the arc is initiated. The current should drop to the desired value within a fraction of a second.

Direct current may be supplied by a motor-generator or a transformer-rectifier. The important features of generators and transformer-rectifiers are the rated and maximum current output (Table 9) and the volt-ampere characteristic curves.

Alternating Current. The power demand for SAW with alternating current from a transformer approximates a sine wave of amperage and a rectangular wave of voltage (Fig. 31b). There is a reversal of polarity at every half cycle, with an associated peak voltage to reignite the arc. The power is supplied by a transformer with a drooping characteristic, that is, a constant-current transformer.

Most transformers are provided with reactance to stabilize the arc. High-current-capacity transformers for SAW have an open-circuit voltage of 80 to 100 V to accommodate the high arc voltage needed to start and stabilize the alternating current arc. For transformers with lower current capacities, the open-circuit voltage usually is 65 to 75 V. Transformers with open-circuit voltage of less than 65 V are not suitable for SAW. There are several satisfactory systems for changing the transformer setting; one of these, the saturable reactor control, is most convenient for remote control.

As previously noted, alternating current is preferred in both automatic welding with a single electrode at a high current level and in automatic, multiple-electrode tandem-arc welding. For high-current applications, transformers are more economical than motor-generators. Furthermore, the high current facilitates arc starting, and alternating current reduces the severe arc blow encountered with direct current at high amperage. For lower current applications,

Table 9 Output ratings of power supplies used in SAW

Volt-ampere characteristic	Rated amperage, A	Rated voltage, V	Duty cycle
Motor-generators			
Drooping dc	300	32	60(a)
(constant current)	400	36	60(a)
	500	40	60(a)
	600	44	60(a)
Flat dc	300	25	100(b)
(constant voltage)	500	40	100(b)
	750	40	100(b)
	900	40	100(b)
Transformer-rectifiers			
Drooping dc	400	36	60(a)
(constant current)	500	40	60(a)
	600	44	60(a)
	800	44	60(a)
Flat dc	300	30	100(b)
(constant voltage)	500	37	100(b)
	750	50	100(b)
	1000	55	100(b)
Transformers			
Drooping ac	400	36	60(a)
(constant current)	500	40	60(a)
	600	44	60(a)
	750	44	75(c)
	1000	44	75(c)
	1500	44	75(c)

(a) Rated current can be delivered continuously for 6 min out of every 10 min. (b) Rated current can be delivered continuously as long as required. (c) Rated current can be delivered continuously for 1 h, and then for 45 min out of every hour for the next 3 h.

direct current is better than alternating current for sharp wire and scratch starts. At high travel speed, constant-voltage direct current provides more uniform weld beads than does alternating current.

Volt-ampere characteristic curves show the voltage at which every value of current is provided under steady-load conditions. The terms "welding current" and "values of current" are interpreted in this discussion as root-mean-square (rms) values of current. A curve is obtained by loading a power supply with different (variable) resistances and plotting the voltage between the electrode and work terminals for each output current.

Two types of characteristic curves are available for SAW—drooping and flat; these are illustrated by curves D and F, respectively, in Fig. 32. Curve D is called a "drooping-characteristic curve" because the voltage between the terminals of the power supply decreases sharply as the current increases. Power supplies having this characteristic are often called "constant-current" power supplies. Such supplies may provide alternating and/or direct current outputs. Curve F is called a "flat-characteristic curve" because the voltage remains almost constant as the current is increased. Power supplies having this characteristic are often called "constant-voltage" power supplies, and their outputs are only direct current.

The difference in welding action between flat-characteristic and drooping-characteristic power supplies is shown by plotting the volt-ampere characteristics of two submerged welding arcs (arcs 1 and 2) on the same graph with curves D and F (see Fig. 32). Arc 2 has higher voltage than arc 1 and therefore is a longer arc. It is assumed that the electrode wire is advancing at constant speed. If, because of a sudden change in level of the workpiece, the arc voltage increases from that

Fig. 32 Static volt-ampere curves of motor-generators or transformer-rectifiers with drooping (curve D) and flat (curve F) characteristics

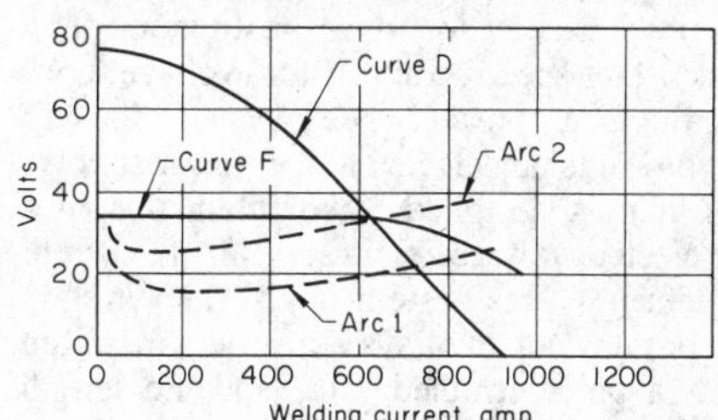

of arc 1 to that of arc 2, the current will shift to a slightly lower value on the drooping-characteristic power supply. However, on the flat-characteristic supply, there is a larger decrease in current. The flat-characteristic power supply returns the arc to arc 1 voltage more rapidly than the drooping-characteristic supply. Thus, a power supply with a flat characteristic has an advantage when used with a constant-speed wire feeder. Such a wire feeder/power supply combination is preferred when welding with small-diameter wires (such as $^1/_{16}$ in.) which are fed at high rates. For fine wires at high feed rates, the fast current response of a flat-characteristic curve is essential for good welding action. With larger diameter wire ($^5/_{32}$ in. and over), the wire feed is so slow that the fast current response of the flat-characteristic curve is of little or no consequence. Indeed, a drooping-characteristic power supply with a voltage-controlled wire feeder (i.e., one that varies the wire-feed speed directly in proportion to the arc length) has superior welding action when using large-diameter wire.

Wire-Feed Systems

Equipment for feeding the electrode wire in SAW employs either of two types of systems for control of wire-feed rate—(1) voltage-sensitive systems with constant-current power supplies, and (2) constant-speed systems with constant-voltage power supplies.

Voltage-sensitive systems are those in which the power supply maintains an essentially constant welding current, and the electrode feeder varies the feed rate of the electrode wire to maintain a constant-arc voltage. This is accomplished by monitoring the arc voltage and having the voltage control momentarily increase the wire-feed rate when the voltage increases beyond a preset level, or decrease the feed rate when the voltage drops. A voltage-sensitive control system is preferred for large-diameter wire welding, wherein the current density is less than 38 000 A/in.^{2} of the cross-sectional area of the electrode wire.

Constant-speed systems are essentially the same as voltage-sensitive systems, except for the substitution of a constant-speed control for voltage control. In a constant-speed system, the arc voltage is preselected at the power supply, which is designed to maintain this preselected voltage regardless of the current demand. This system is also known as a constant-potential system. The wire is fed at a predetermined rate, and arc length is held constant through automatic adjustment of the current by the power supply to maintain fixed resistance across the arc. Compared with a voltage-sensitive control system, a constant-speed control system provides increased voltage stability, consistent scratch starting, and more convenient adjustment of voltage and current. Constant-speed systems are particularly well suited for high-speed welding. Best results are obtained when the current density is in excess of 38 000 A/in.^{2} of the cross-sectional area of the electrode wire.

Automatic Welding

In automatic SAW, the welding equipment performs the entire operation (except for loading and unloading, which can be done manually or automatically) with little or no observation or adjustment of the controls by an operator. The electrode wire is fed automatically by means of an electric motor located in the welding head. The welding head also contains the necessary mounting and adjusting equipment for directing the arc at the work and maintaining proper torch-to-work distance. The welding head is advanced by means of an automatic drive mechanism. It may also be mounted on the articulating arm of a welding robotics system. It can travel over or alongside the workpiece, or can be held stationary while the workpiece travels beneath or alongside it. Arc striking is also automatic. In gravity-feed systems, a flux-dispensing tube and hopper usually are attached to the welding head; however, pressure-feed systems have been used for several years to facilitate welding on inclined surfaces. Contactors, usually copper, provide a current source for the welding electrode as close to the arc as possible to prevent excessive resistance heating of the electrode.

Advantages. In automatic SAW, precise control can be maintained on high-volume production, and more than one machine can be controlled by a single operator. Only periodic spot checking is required. The process is well suited to multiple-electrode welding and also to the completely automatic handling of workpieces during loading and unloading. It also can provide higher welding speed than is possible by a welder using semiautomatic equipment.

Limitations. Automatic SAW requires use of expensive equipment. In common with many automatic processes, justification for capital outlay for tooling and fixturing depends greatly on the production quantities and continuing demand. Automatic SAW may be relatively cumbersome and inefficient when applied to small assemblies.

Applications. In terms of tonnage, steel linepipe used for transporting petroleum products represents a large application of automatic SAW. In making the pipe, formed plate is fed into a cage of rollers that close the joint. The pipe travels through the welding station, where the seam weld is completed.

Automatic SAW can be used to produce high-quality internal and external longitudinal welds in cylinders and tanks that are roll-formed from plate, provided that the internal and external clearances permit the necessary manipulation of the welding head or heads. In some applications, it is advantageous to mount and transport the welding head on booms to traverse the joint or on track systems running parallel to the joint, as in welding very large tanks.

Multiple-Electrode Welding

In SAW, deposition rate can be increased by the use of a multiple-electrode welding system. These systems can be differentiated on the basis of three variables: number of electrodes, type of power connection used, and the position of the electrodes (tandem or transverse) with respect to direction of travel.

Number of Electrodes. Depending on the system selected, the number of electrodes in simultaneous operation is usually from 2 to 4, with as many as 14 being used in special heavy-section applications. Power connection may consist of a multipower connection, a parallel connection, or a series connection.

Multipower Connections. In a multipower connection, each electrode has its own power supply, welding head, voltage-control mechanism, and wire-contact assembly. Single-phase or three-phase power, and many possible combinations of alternating and direct current, can be supplied to the electrodes. The welding ground is attached to the workpiece, and a control unit governs each welding head. With multipower connections, extra-high-speed welds can be made. With the use of alternating current power, magnetic arc blow between electrode is minimized.

Parallel Connections. In parallel connection, electrodes are connected in parallel to the power supply. The wires may be run through a single welding head (twin-arc SAW) from separate spools (see Fig. 3) or twisted into a single wire rope. Attachment of the welding ground to the workpiece is conventional, and one control unit governs the operation. Current densities are reduced, and joint penetration is lower than with a multipower connection. A parallel connection frequently is used with electrodes in the transverse position to make extra-wide welds for poorly fitted joints.

Series Connections. In a series connection, two electrodes are connected in series, and two welding heads are used. Each electrode may have separate wire-feed control units and operates independently. Electrodes may be synchronized to operate from the same wire-feed control. The electrode cable is connected to one welding head. The work cable is connected to the second welding head, instead of to the workpiece as in conventional welding. Welding current travels from one electrode to the other through an arc over the weld pool. There is no connection between the power supply and the workpiece. Almost all of the power is used to melt the electrode, with little power entering the workpiece. With the electrodes in the transverse position, series connection produces a broad, shallow weld bead and thus is well suited to the deposition of a surfacing overlay on a base metal with very little dilution. Deposition rates of 25 lb/h are easily obtained.

Electrode Position. With any of the three types of power connections, different effects can be produced by altering the position of the electrodes with respect to the direction of travel. Electrodes may be arranged in either the tandem or the transverse position; with three or more electrodes, other patterns are also used with series, parallel, or multipower connections.

Usually, electrodes are arranged in the tandem position—with one electrode following behind the other as they move in the direction of travel. Welding with electrodes in the tandem position is sometimes referred to as tandem-arc welding. For welding with electrodes in the transverse position, one electrode is positioned alongside the other as they move in the direction of travel.

Strip Electrode Welding

The use of a metal strip electrode rather than multiple electrodes reduces the overall equipment and consumable cost, while retaining the efficiency and high deposition rates of several wires. For weld overlay and hardfacing applications, wide strip electrode SAW offers considerable advantages (see the articles "Hardfacing" and "Weld Overlays" in this Volume).

In strip welding, alternating or direct current power may be used. The arc is small, similar in size to a wire electrode arc, and moves rapidly across the strip end with erratic direction changes. The advantage of better control of magnetic disturbance fields in alternating current welding is offset by the disadvantages of less stable arc movement and greater susceptibility to weld-metal porosity.

In strip electrode SAW, the equipment setup is nearly identical to wire electrode welding. A coil of metal strip is substituted for the coil of wire, and the electrode guides, straightener, feed rolls, and electrical contacts are modified to accept a metal strip. Commercial strip thickness varies from 0.02 to 0.04 in., and strip width is chosen for the joint size and configuration. For weld overlay and hardfacing, widths of up to 10 in. have been used, with 1.2, 2.4, and 3.6 in. being the most widely used commercial sizes. For butt and fillet welding, commonly used strip is 0.4 to 1.0 in. wide.

All commercially available SAW equipment and fluxes are suitable for modification to strip electrode welding by the addition of a new welding head. For high-deposition-rate welding, strip electrode welding is an ideal application of the submerged arc process. In weld overlay and hardfacing, the process produces excellent surface finish and substantially reduces the potential for defects by decreasing the number of bead-to-bead tie-ins required for a given surface.

Powder Metal Joint Fill

Increased deposition rates can be obtained in SAW by the addition of iron or alloy metal powder to the flux, or by the use of a separate layer of powder under the granulated flux. Typical deposition rate increases of 60 to 100% have been reported with the use of powdered metal fill. Additionally, when coupled with multiple electrodes or strip electrodes, powder metal fill can produce extremely high deposition rates. Ordinary SAW equipment and fluxes can be modified for use with metal powder fill by adding a metal-powder delivery system. Metal powder usually is delivered to the joint before flux delivery, and an accurate powder-metering system is required to ensure correct joint fill and metal deposition rate and to prevent overfill that may result in inadequate fusion or insufficient penetration.

Alloy metal powders can be used to produce specific weld-metal compositions. Stainless steel weld deposits can be produced using low-carbon steel wire with a high chromium and nickel alloy metal powder. In addition, weld overlay or hardfacing deposits can be produced using specially designed powder compositions for which electrode wire or strip compositions are not available.

Semiautomatic Welding

Semiautomatic SAW consists of a hand-held welding gun that feeds a solid wire and flux to the arc area. The flux may be supplied by gravity feed through a cone, or by a pressurized tank system. Nearly all SAW equipment and consumables can be used in the semiautomatic mode. However, the process requires a change in the wire delivery system to accommodate the cables and welding gun. Typically, wires larger than $^3/_{32}$ in. are not used for semiautomatic welding.

This process offers the advantages typically associated with SAW—high deposition rate and decreased exposure of the operator to arc flash, spatter, and fumes. In addition, semiautomatic welding enables use of the SAW process without the expenses, setup requirements, and space requirements of fully automatic welding. However, semiautomatic SAW has two distinct disadvantages. First, the welding operator may have difficulty in precisely locating the arc in the weld groove, because the arc is not visible under the flux blanket. Second, the semiautomatic SAW torch with associated flux-delivery system weighs substantially more and is more cumbersome for the welding operator to manipulate than a semiautomatic flux cored arc welding (FCAW) or GMAW system.

Joint Design

Recommended designs of joints used in SAW are given in the article "Joint Design and Preparation" in this Volume. These joint designs take into consideration the deep penetration that is characteristic of SAW and the difficulty of slag removal that attends deep-groove, multiple-pass welding. The square joint design with a maximum root opening of $^1/_{32}$ in. may be utilized for component thicknesses up to $^5/_8$ in. Beveled joint designs should be used for component thicknesses greater than $^5/_8$ in. Backing rings or bars are recommended to prevent melt-through, loss of flux, and to ensure full-penetration weldments when only one side of the joint is accessible. Backing rings or bars may be made of steel, copper, or ceramic tape. Steel backing rings or bars should be thick enough to prevent melt-through by the molten weld metal. Copper is one of the best nonfusible weld backings. Copper backing rings or bars should be water cooled or sufficiently thick to dissipate the heat developed by welding.

The use of a solid backing ring or strip to obtain complete joint penetration while avoiding melt-through can sometimes be avoided by redesigning the joint, or by the use of a flux bed as backing for the first pass. This is desirable in applications that require removal of the backing strip after welding is complete.

In SAW of pipe and other tubular workpieces, it is advantageous to use a joint design that incorporates a backing lip, thereby eliminating the need for a backing ring or bar, which ordinarily would be required for preventing melt-through.

Effect of Welding Position on Joint Design and Welding Conditions

Welding position during SAW affects current, voltage, speed, joint design, and flux retention. As compared to the preferred flat-position groove welding, horizontal (or 3 o'clock) position groove welding is done with lower current, higher voltage, and higher welding speed, as well as a different groove design. A flux trough is needed to retain the loose flux. To compensate for the greater effect of gravity on the molten weld pool and the molten flux, smaller beads are deposited in a weld groove that has its lower surface beveled at only a few degrees from the horizontal. Narrow stringer beads are used. For the same thickness of work metal, at least twice as many passes are needed as for flat-position groove welding.

When two or more electrodes are used to deposit filler metal, the flat welding position is usually preferred because the bead is least distorted by the unsymmetrical pull of gravity. Loss of flux is also minimized so that there is less need for dams and restraints to keep the flux from spilling off the weld area.

When heavy assemblies that require fillet welds are difficult to manipulate into the flat welding position, the horizontal welding position (for fillet welds) may be used. To compensate for the lopsided flow of the filler metal in horizontal welding, electrodes are spaced far enough apart to make separate weld pools. The leading electrode deposits a layer that sags down to cover more of the flat surface than the vertical surface. Trailing electrodes are angled to deposit most of the metal on the vertical wall of the joint. The composite of all the beads makes a relatively symmetrical fillet.

Submerged arc welded bridge girders and other welded bridge structural components are designed and fabricated in accordance with AWS specification D1.1 to attain requalified status. Although welding in the flat position is generally preferred, the code states that fillet welds may be made in either the flat or horizontal position, except that the size of single-pass multiple-arc fillet welds made in the horizontal position shall not exceed $^1/_2$ in. Single-pass single-electrode or parallel-electrode fillet welds made in the horizontal position shall not exceed $^5/_{16}$ in. in size.

Initial and Interpass Cleaning

The surfaces to be welded should be clean and free of grease, oil, marking paints, scale, moisture, oxides, or other material that is detrimental to welding. When weld metal is to be deposited over a previously welded surface, all slag, surface flaws, arc craters, and welding residue should be removed by brushing, grinding, or filing prior to depositing the next bead.

Although blue mill scale does not adversely affect weld quality, red mill scale, which is reddish brown in color, has the same detrimental effect as rust. For best results, red mill scale should be removed from joint surfaces by a combination of power wire brushing, gritblasting, and torch heating. The brush should be used for preliminary removal of scale, and during welding, the torch should be directed at the joint ahead of the arc to provide a reducing flame and to drive off residual moisture. To be effective, the torch should heat the surface of the base metal to between 400 to 600 °F.

Jigs, Fixtures, and Booms

Because requirements vary with the shape, size, and weight of individual workpieces, jigs and fixtures are subject to few firm rules of design and construction. However, the following design principles are generally applicable:

- Fixtures should position the weld in an acceptable position, usually flat.
- Designs should be simple to minimize cost.
- Clamping must be firm and secure.
- Fast-acting clamps (hand, air, or hydraulic) are recommended.
- Clamps should be adequately spaced from the joint being welded.
- The ground connection, or connections, for the fixture should be located for most effective grounding and best control of arc blow.
- The fixture should minimize or eliminate the need for tack welds in the weldment.
- With rotating work, an electrically efficient rotating ground must be used.
- Fixtures may also be required to firmly hold the flux in place during welding.

In addition to their primary function of holding and positioning the workpiece during welding, fixtures also may serve to control or minimize distortion due to thermal stress. Thus, fixtures should be rigid enough to ensure retention of alignment during welding. Although the movement of critical areas may be restricted, provision must be made for thermal expansion and contraction. Fixtures may be stationary or mobile in relation to the welding head. Lathe fixturing commonly is used in welding plugs or caps to the ends of cylinders and in joining cylindrical sections, because it combines clamping action with centering and controlled rotation. Booms and extensions of various types are used to support and guide the welding head to enable it to reach locations that are not readily accessible, such as the inside surfaces of tubes and cavities.

Tack Welds. For shielded metal arc tack welds that will be covered with a submerged arc weld bead, electrodes of class E7018 should be used because they provide adequate joint penetration and produce porosity-free deposits. With these electrodes, slag removal is easy. Defective tack welds must be removed before proceeding with SAW. Care should be taken to ensure that submerged arc welds completely melt out the tack welds. For example, weld porosity during seam welding of linepipe has been associated with insufficient melting out of tack welds.

Arc Starting

The most commonly used power sources for SAW are constant-voltage or drooping-voltage sources. With constant-voltage direct current power supplies, the welding arc is started by running the wire into the work. If a drooping alternating or direct current power source is used, special consideration must be given to arc starting and wire-feed rate. To ensure that voltage is adequate to maintain the arc with a drooping source, the wire-feed system must be designed to detect arc voltage and vary wire-feed rate to prevent short circuiting. To initiate an arc with a drooping power source, the following methods are used to prevent short circuiting of the wire on the work:

- *High-frequency RF signal* can be added to generate a low-power spark that provides an ionized path for arc initiation.
- *Wire retraction* can be used as the welding voltage is applied to the electrode by a reversing wire-feed system.
- *Fuse ball* of steel wool can be used between the electrode and the work to provide an initial ionized path for initiating the arc.
- *Scratch, pointed wire, and carbon arc starting* depend on light contact between an electrode and the workpiece to develop an arc.

Welding Tabs

Undesirable weld stopping and starting defects, such as weld craters, may be eliminated by using metal starting and runoff tabs. These tabs enable welding variables to stabilize and maintain uniformity at the start and end of the weld seam.

Slag Removal

The size and shape of the weld bead can facilitate or hinder slag removal. Small beads cool more rapidly than large beads, thereby reducing the likelihood that slag will adhere to them. First-pass weld beads that are slightly concave (Fig. 33a) are much easier to remove slag from than are convex beads (Fig. 33b), which provide crevices for slag entrapment.

Because a circumferential weld on a small-diameter workpiece develops a large amount of local heat, slag removal can be made easier by training an air jet on the completed weld. When welding pipe in the horizontal-rolled position, the jet should strike the weld between the 2 and 3 o'clock positions.

Fig. 33 Shapes of weld beads that facilitate and hinder slag removal between passes in multiple-pass SAW

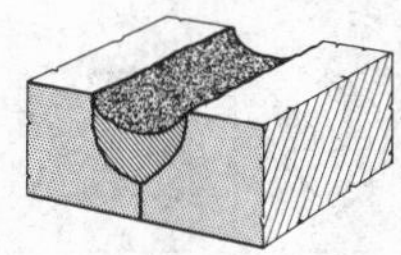

(a) Concave bead (easy slag removal)

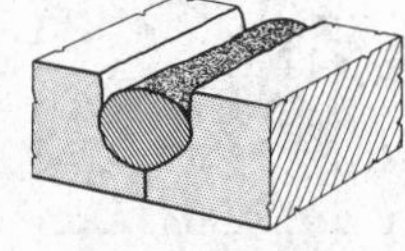

(b) Convex bead (difficult slag removal)

Preheating and Postweld Heat Treating

The principles of preheating and postweld heat treating are the same for SAW as for other arc welding processes. Preheating and postheating are applied to hardenable steels, particularly steels whose carbon content exceeds about 0.30%, in thicknesses greater than $^3/_4$ in. The slower cooling rate that results from preheating increases the time during which the HAZ is in the temperature range above the martensite start temperature and therefore promotes transformation of austenite to soft pearlite instead of to hard martensite. A preheated weld is less likely to have harder zones than a weld that was not preheated. Postweld heat treating is used on hardenable grade steels to reduce the risk of cold cracking resulting from hydrogen in the weld deposit and HAZ. Holding the steel at a temperature above room temperature (200 to 600 °F) allows hydrogen to diffuse from the weld. This prevents or reduces the risk of cold cracking.

Electrical Relations

Typical oscillographic traces of current, voltage, and power during SAW are shown in Fig. 34. Oscillographic, spectrographic, and radiographic studies indicate the presence of a normal arc in the welding zone during SAW. In oscillographic studies, the voltage trace is the most significant record in determining electrical relationships.

Fig. 34 Typical oscillographic traces for SAW with manganese silicate flux. Traces indicate that arc heating is occurring under the flux.

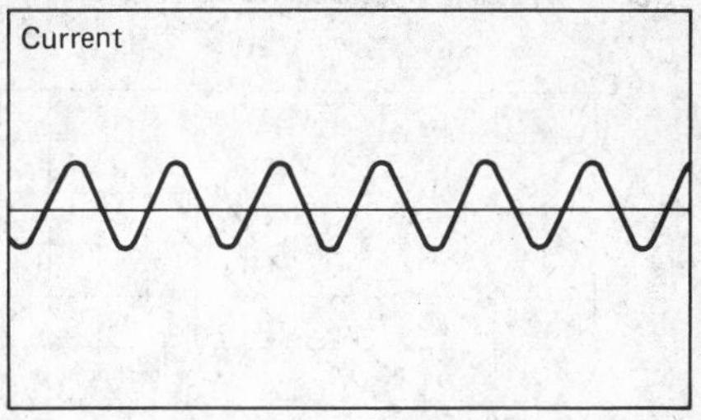

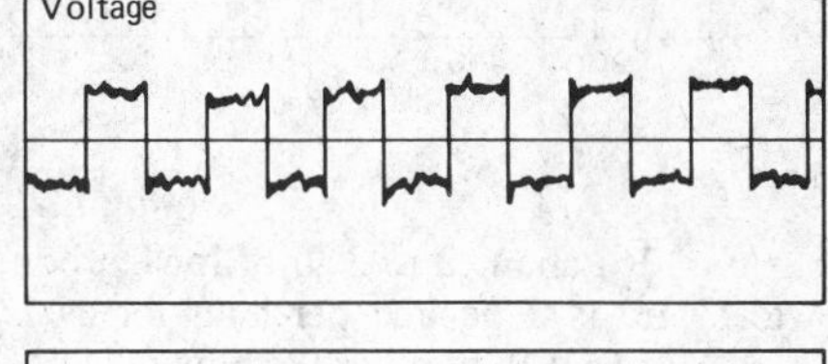

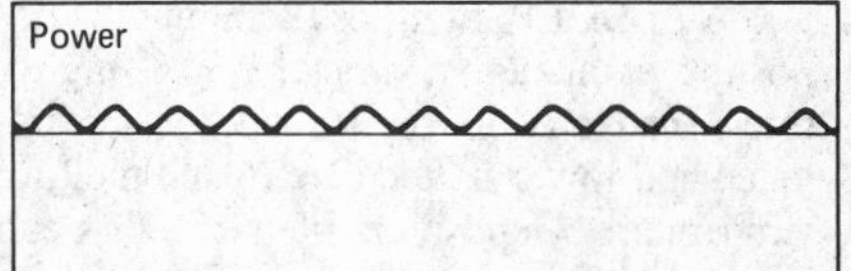

Variables That Affect Weld Size and Shape

Many of the operating variables of SAW have a direct and/or interaction effect on weld size and shape. Therefore, an understanding of these variables and their effects is crucial to planning a properly controlled application of the process. This understanding is also useful for the analyses of quality problems within a given application.

Welding current controls the rate at which the electrode is melted, the depth of fusion, and the amount of base metal melted. If the current is too high for a given travel speed, the depth of fusion or joint penetration will be too great, the weld may melt through the joint, and the weld HAZ will be large. Costs will also be high because excessive power and filler metal will be consumed. However, too low a current will lead to inadequate penetration and inadequate reinforcement. The effect of current (amperage) on penetration and shape of butt welds in carbon steel plates $^1/_2$ and $1^1/_2$ in. thick is shown in Fig. 35.

Fig. 35 Effect of welding current on joint penetration and configuration of submerged arc butt welds. Welds were made in carbon steel plates $^1/_2$ and $1^1/_2$ in. thick, with $^5/_{32}$-in.-diam electrode wire. For the $^1/_2$-in. plate, voltage was 29 V and travel speed was 30 in./min. $1^1/_2$-in. plate was welded at 40 V and a travel speed of 12.5 in./min.

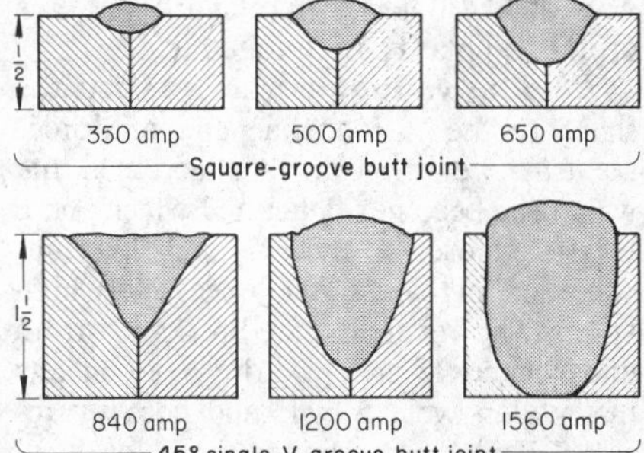

Deposition Rate. The amount of weld metal deposited per unit of time is almost directly proportional to amperage. This is illustrated in Fig. 36, which shows the effect of amperage on deposition rates of $^1/_8$-in.-diam low-carbon steel and type 308 stainless steel electrode wires. As shown

Fig. 36 Effect of current on deposition rates of stainless steel and low-carbon steel electrode wires with DCEP. With DCEN, deposition rates increase about 30 to 50%.

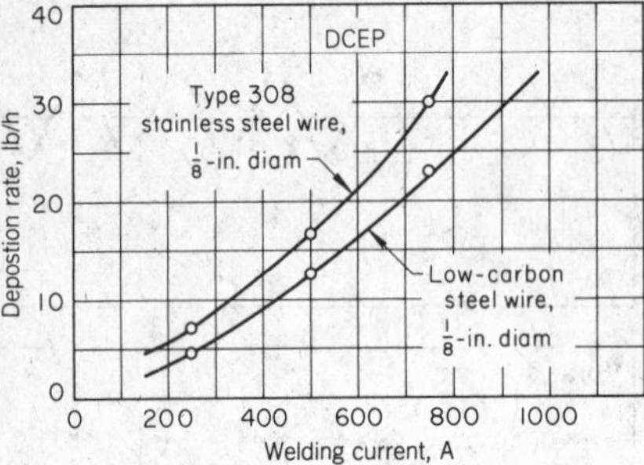

in Fig. 36, about 2 to 3 lb of low-carbon steel wire is deposited per hour for each 100 A of DCEP. Stainless steel wire is deposited at an approximately 30% higher rate because of its higher electrical resistance and lower heat capacity and melting temperature. Deposition rate per 100 A can be increased by increasing the electrode extension beyond the contact tip, but this decreases joint penetration.

In SAW, the current density in the electrode wire (Fig. 12) is several times greater than in SMAW, and the melting rate of the electrode wire and the speed of welding are greater than for SMAW.

In multiple-wire, multiple-power welding, two arcs in close proximity are affected by their respective magnetic fields. Arcs of like polarity flare apart, and arcs of unlike polarity flare together.

Welding voltage primarily affects the shape of the fusion zone and reinforcement. As welding voltage is increased, the weld bead becomes flatter and wider, more flux is exposed to the arc, and flux consumption increases. With excessively high voltage, however, the arc breaks out from under its cover of liquid flux, air contacts the molten weld metal, and porosity results. The effects of increasing levels of voltage on the configurations of submerged arc butt welds are shown in Fig. 37.

If the arc current is held constant and the arc voltage is low in relation to current, the base metal does not melt enough to give a good weld. The molten globules of metal passing from the electrode to the workpiece cause continual short circuiting, which results in spattering and a high bead. As the arc voltage is increased, an optimum point is reached at which the arc no longer sputters but has a steady, sharp cracking sound. At this stage, good joint penetration can be obtained. If the arc voltage is further increased, the arc length also increases. The arc is unstable at this stage and makes a wide, flat bead, usually with an undercut. Increasing electrode extension has essentially the same effect on bead shape as decreasing voltage.

Current-voltage relations typical of industrial applications are shown in Fig. 38. These data suggest that there is a range of about 10 V for each setting of welding current; the lower voltages give a narrow weld bead, and the higher voltages give a wide bead. Outside the 10-V working range, soundness of the weld metal deteriorates.

Welding speed (travel speed) is an important variable governing the production rate and metallurgical quality of welds. Increasing the welding speed decreases the production time per unit on fillet welds, but has little influence on production time per unit length of groove welds for a given current and wire size. Welding speed also affects the rate of heat input. Increasing welding speed and decreasing current are two practical ways to lower heat-input rate.

Welding speed helps to determine the width and depth of the weld, as shown in Fig. 39. A weld bead consists partly of molten filler metal and partly of base metal melted by the arc. Base metal can constitute 10 to 60% of a submerged arc weld, the percentage decreasing with welding speed. Excessively high welding speed decreases wetting action and increases the probability of undercutting, arc blow, weld porosity, and uneven bead shapes. Excessively low speed produces hat-shaped beads that are subject to cracking, causes excessive melt-through, and produces a large weld pool that flows around the arc, resulting in a rough bead, spatter, and slag inclusions.

The maximum speed of welding, beyond which erratic behavior and undercutting occur, also is controlled by the welding current. In Fig. 40, which is a survey of published data, any combination of current and speed of travel to the right of the diagonal line leads to undercutting.

The variation of fillet size with travel speed for various submerged arc welding processes is shown in Fig. 41. Techniques have been developed for welding 14-gage (0.078-in.-thick) steel at speeds greater than 100 in./min. Single-pass welds in plate 1 in. thick can be made with 1500 A at 10 in./min. Multiple-pass techniques have been developed for joints in heavy plate

Fig. 37 Effect of welding voltage on joint penetration and configuration of submerged arc butt welds. Welds were made in carbon steel plates 1/2 and 1 1/2 in. thick, with 5/32-in.-diam electrode wire. For the 1/2-in. plate, welding current was 500 A and travel speed was 30 in./min. 1 1/2-in. plate was welded at 1200 A and a travel speed of 12.5 in./min.

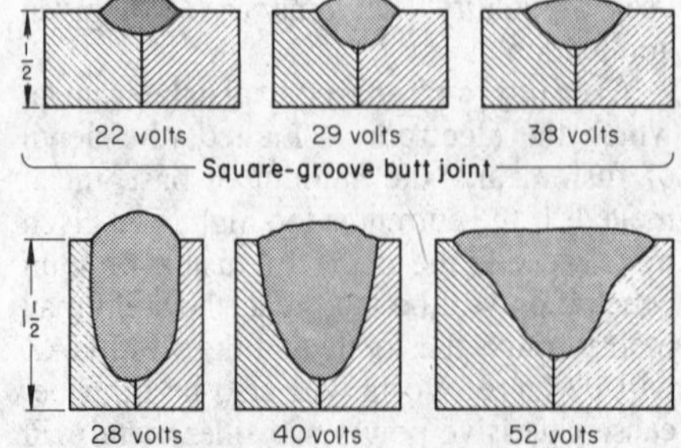

Fig. 38 Current-voltage relations for practical conditions of SAW using a variety of commercial fluxes

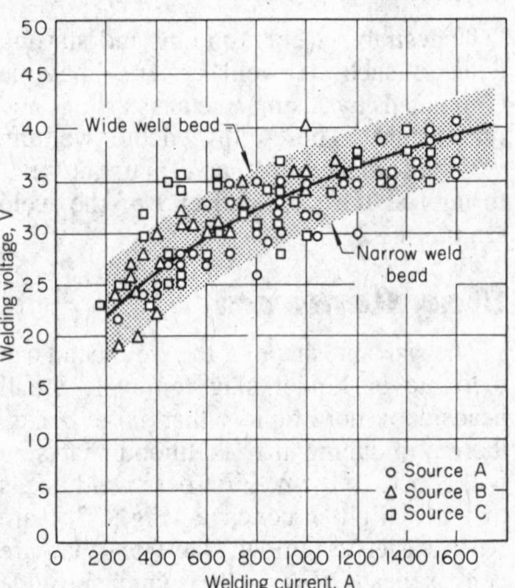

Fig. 39 Effect of welding speed (travel speed) on width and configuration of submerged arc butt welds. Welds were made in carbon steel plates 1/2 and 1 1/2 in. thick, with 5/32-in.-diam electrode wire. For the 1/2-in. plate, welding current was 500 A and voltage was 29 V; the 1 1/2-in. plate was welded at 1200 A and 40 V.

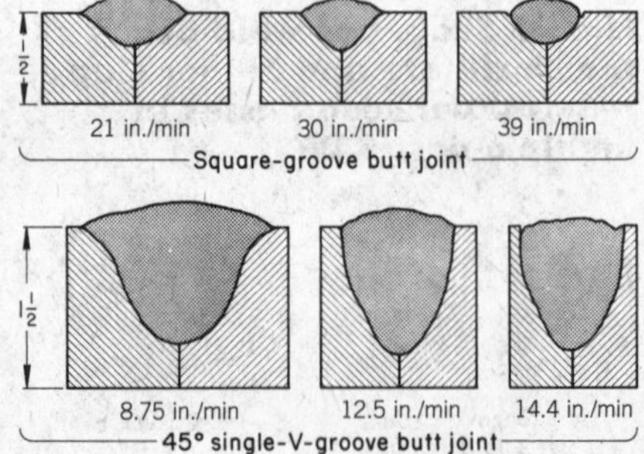

Fig. 40 Current-speed relations for undercut-free welding of steel with a single electrode and various commercial fluxes

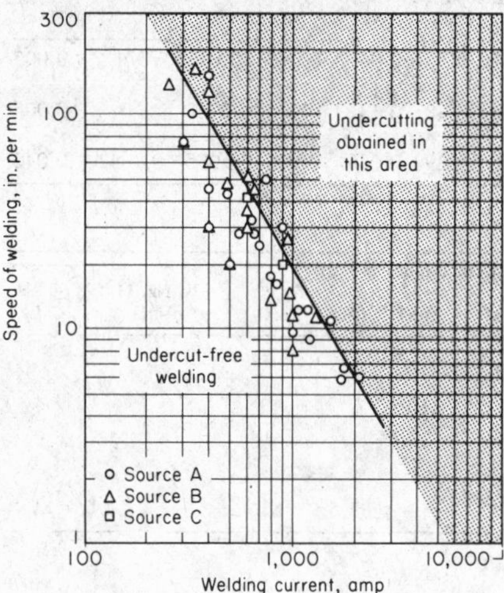

Fig. 41 Deposition efficiencies for various multiple-wire SAW processes. VV, variable voltage with arc voltage control; CV, constant voltage with constant wire-feed speed

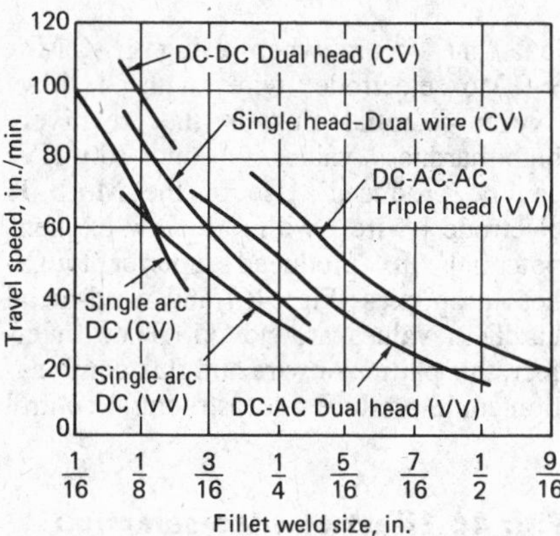

and joints requiring optimum mechanical properties. Two, three, or more electrodes can be fed into a single weld zone in some applications.

Electrode-Wire Diameter. Decreasing the diameter of the electrode wire increases the pressure of the arc, thus increasing joint penetration and decreasing the width-to-depth ratio of the weld bead (see Fig. 42).

Weld-Bead Fusion Area. For any given welding technique, the efficiency of the process is inherent in the measurement of the weld-bead area. Weld-bead area increases with increasing welding current and decreases with increasing travel speed. It is affected only slightly by normal changes in arc voltage. The alignment chart of Fig. 43 and the weld-bead area formula of Table 10 are based on data for SAW, which generally provides for 100% transfer of filler metal.

The effect of the current-to-speed ratio on the cross-sectional weld-metal area is shown in Fig. 44. Penetration of a weld deposited in a groove or on the surface of the base metal is usually defined as the distance below the original surface to which metal has been melted. Welding current is the most significant variable in determining penetration; travel speed and voltage are less important. Figure 45 shows the combined effects of current, voltage, and travel speed on penetration for a series of experiments with SAW. An equation for arc penetration is also given in Table 10.

Dilution of the weld metal by the base metal can be expressed as the percentage of the bead cross-sectional area that is base metal. Dilution of weld metal by base metal increases with an increasing ratio of current to speed of travel. An increase in voltage may increase the amount of dilution, as a consequence of the slightly lower electrode melting rate that results with the higher voltage condition. Dilution for a weld bead can be estimated with the equations in Table 10 if the melting rate of the electrode and the process variables are known.

Fig. 42 Effect of electrode-wire diameter on width and penetration of surface welds. Welds were made on carbon steel plate by fully automatic SAW at 30 V, 600 A, and 30 in./min.

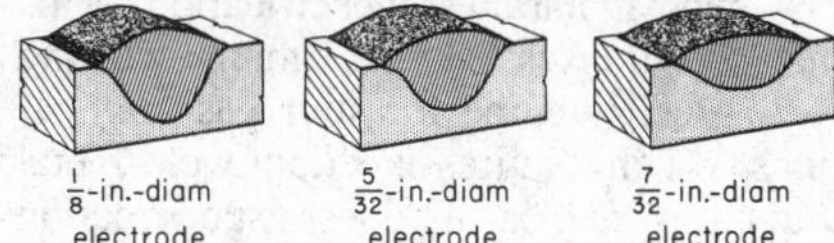

Fig. 43 Alignment chart for determining the area of a weld bead from welding current and travel speed for a given welding technique

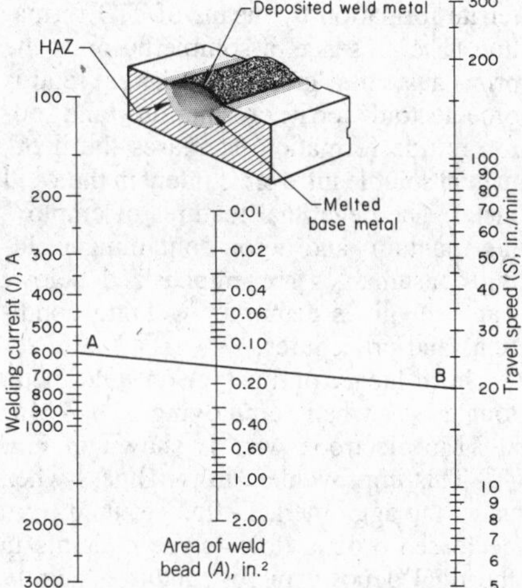

Table 10 Empirical formulas for SAW

Electrode melting rate, lb/min

$$MR = \frac{I}{1000}\left[0.35 + d^2 + 2.08 \times 10^{-7}\left(\frac{IL}{d^2}\right)^{1.22}\right]$$

Area of weld bead, in.²

$$\text{Log } A = 0.903 \log\left(\frac{I^{1.716}}{S}\right) - 3.95$$

$$A = \frac{I^{1.55}}{10^{3.95}\, S^{0.903}}$$

Arc penetration, in.

$$P = K\sqrt[3]{I^4/SE^2}$$

Weld dilution, %

$$\%\text{ dilution} = \left(\frac{A - \dfrac{MR}{0.283S}}{A}\right)100$$

$$= 100 - \frac{353MR}{AS}$$

MR = electrode melting rate, lb/min
I = welding current, A
L = electrode extension, in.
d = electrode diameter, in.
A = area of weld-bead section, in.²
P = arc penetration, in.
K = process penetration constant (0.0012 for calcium silicate flux)
E = welding voltage
S = speed of travel in./min

Safety

Submerged arc welding presents hazards that differ from those found in other arc welding processes. Detailed information on hazards and safe practices in SAW can be found in ANSI Z49.1, "Safety in Welding and Cutting," and in "Safe Practices in Welding and Cutting," *Welding Handbook*, 6th edition, Chapter 9, Section 1, American Welding Society.

Production Examples

The following examples illustrate various SAW procedures. Different approaches to the selection of welding consumables, welding parameters, fixturing, and weld design are described. This information should be used as guidelines and not as specifications.

Fig. 44 Effect of ratio of current to speed on the cross-sectional area of weld metal deposited for four rates of electrode melting

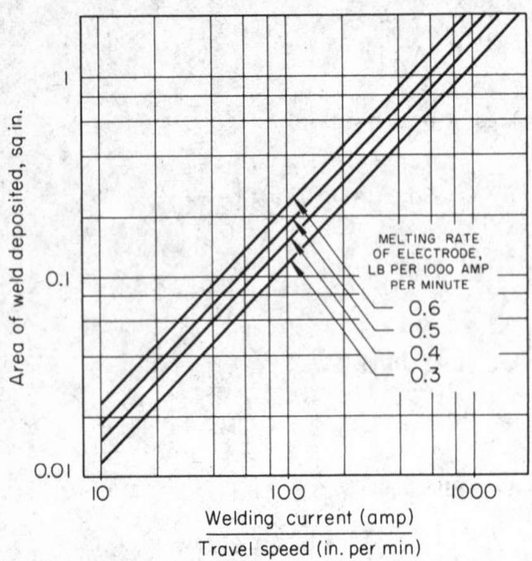

Fig. 45 Effect of welding technique on arc penetration

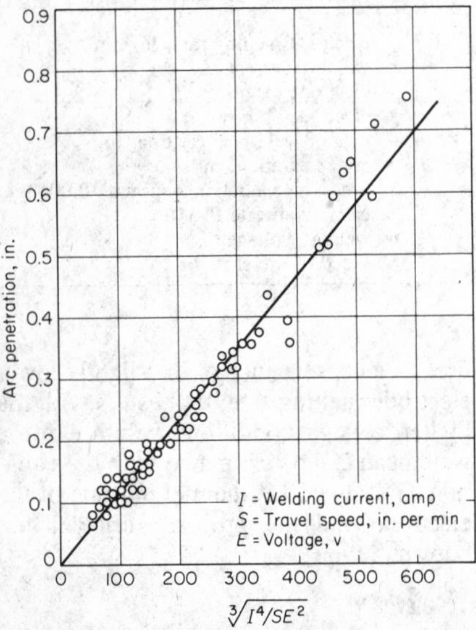

Example 1. Use of Titanium- and Boron-Containing Welding Consumables to Meet Toughness and Hardness Specifications. Weld-metal microstructure and toughness properties depend on a complex interplay of flux type, electrode and base-material compositions, and heat input during welding. In high-heat-input applications, obtaining a desirable deposit microstructure (large contents of acicular ferrite) for any given base-material composition may be further complicated by the requirement of a weld-metal hardness limit for the final deposit. In effect, the alloy elements required to promote a high acicular ferrite microstructure (molybdenum, for example) may have a solid-solution hardening effect that makes it difficult to meet a low deposit hardness specification. This may be the case when manufacturing X-70 steel linepipe for sour gas applications, which requires excellent weld-metal toughness and a hardness specification of 250 HV maximum.

The development of titanium- and boron-containing SAW consumables has provided a means of satisfying high weld-metal toughness and hardness specifications at the same time. Titanium is required to prevent oxidation of boron during welding and to promote acicular ferrite formation by means of TiO formation. The presence of soluble boron at the prior austenite grain boundaries retards proeutectoid ferrite formation, and boron nitride formation decreases the detrimental soluble nitrogen content in the weld metal. The beneficial features of employing titanium- and boron-containing welding consumables are emphasized below. Table 11 gives details of welding conditions and procedures.

The influence of flux type on weld-metal toughness when employing 1½%Mn-½%Mo electrode wire is shown in Fig. 46. This improvement in toughness when using the agglomerated flux resulted from decreased proeutectoid ferrite contents in the final deposit microstructure. A problem that occurs when using 1½%Mn-½%Mo electrode wire with a low-oxygen-potential flux is that relatively high hardness values (around 300 HV) can be produced. Use of the Mo-B-Ti electrode wire with a low-oxygen-potential flux produced superior toughness properties (Fig. 46) and weld-metal hardness values around 250 HV. In effect, this particular wire and flux combination could produce microstructural control

Table 11 Conditions for welding X-70 steel linepipe with titanium- and boron-containing welding consumables

Plate and electrode composition

	Composition, %										
	C	Mn	Si	P	S	Mo	Ni	Nb	V	B	Ti
X-70 plate	0.08	1.38	0.32	0.013	0.007	0.35	0.04	0.046	0.011	Nil	<0.005
1½%Mn-½%Mo filler wire	0.11	1.60	0.27	0.014	0.018	0.48	0.13	Nil	Nil	...	0.005
Mo-Ti-B filler wire	0.084	1.20	0.035	0.01	0.007	0.29	0.085	0.005	0.005	0.0042	0.042

Flux composition

	Composition, %									
	SiO_2	Al_2O_3	MgO	CaO	FeO	MnO	TiO_2	ZrO_2	K, Na, Li oxides	CaF_2
High-oxygen-potential fused flux	15	39	4	1	0.6	10	7	13	0.3	7
Low-oxygen-potential agglomerated flux	13	14	40	11	2	1	0.5	Nil	2	14

Welding parameters

		Current, A	Voltage, V	Speed, in./min	Heat input, kJ/in.
Side 1	Lead arc (dc)	750	30	45	60.9
	Trail arc (ac)	650	38		
Side 2	Lead arc (dc)	850	30	45	63.5
	Trail arc (ac)	650	38		

Fig. 46 Effect of wire selection on Charpy V-notch toughness of tandem submerged arc welds

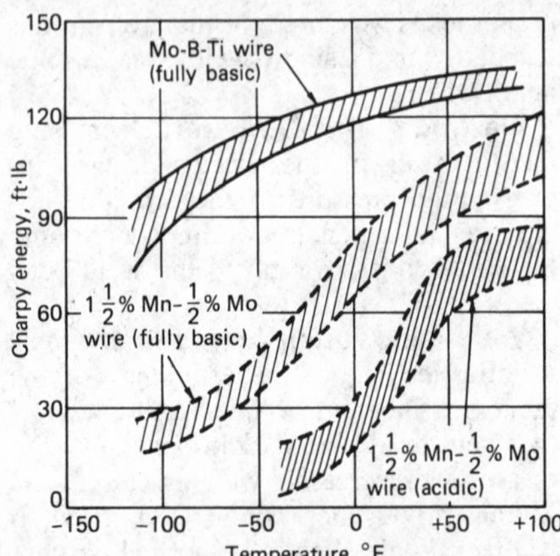

(a high acicular ferrite content microstructure) without excessive solid-solution hardening.

Example 2. Steam-Boiler Header (Fig. 47). In welding radial stub tubes to a steam-boiler header (Fig. 47), production rate using automatic SAW was 2$^3/_4$ times that obtained by SMAW. For both methods, stub tubes were held in place for welding by two tack welds 180° apart, made manually. When SMAW was used, the operator welded 180° on all stub tubes while standing on one side of the header; after removing the slag, he completed the welds from the other side of the header.

Submerged arc welding was done by a rotating welding head that was supported by an overhead-trolley crane. The welding head was located by a pin that was inserted into the stub tube before welding was begun. The arc was started, and the four beads shown in Fig. 47 were deposited without stopping. Unfused flux was removed continuously by a vacuum hose attached to the welding head, and slag was removed manually by the operator.

Example 3. Submerged Arc Welding of a Large Piston. The large hydraulic-jack piston shown in Fig. 48 was

Fig. 47 Submerged arc welding setup for steam-boiler header

Low-carbon steel; low-carbon steel filler metal

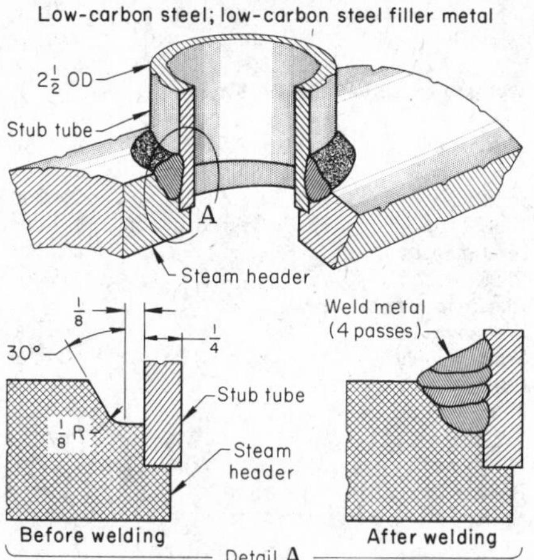

Automatic SAW

Joint type	Corner
Weld type	Single-J-groove
Electrode wire	0.045-in.-diam low-carbon steel(a)
Current	200 to 250 A (DCEP)(b)
Voltage	30 V
Welding speed	6 in./min
Welding position	Horizontal
Number of passes	Four

(a) A $^5/_{32}$-in. E7018 electrode was used for SMAW. (b) Supplied by a 300-A constant-voltage rectifier. Current for SMAW was 195 A, supplied by a 300-A constant-current rectifier.

Fig. 48 Large piston assembled by SAW

Low-carbon steel; low-carbon steel filler metal (EL12)

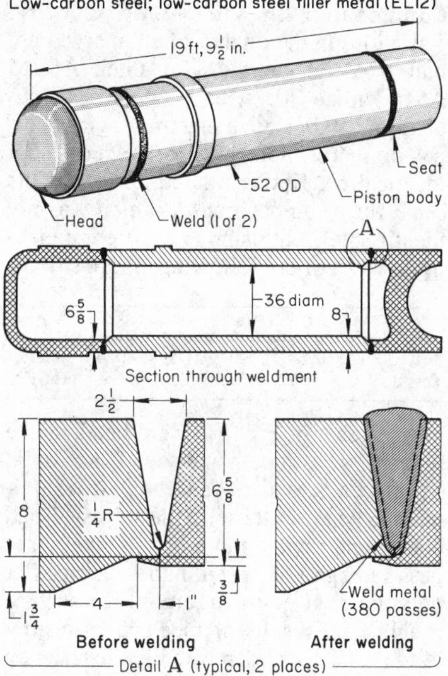

Conditions for SAW

Joint type	Circumferential butt
Weld type	Single-U-groove, integral backing
Joint preparation	Machining
Power supply	1000-A transformer
Wire feed	Fully automatic, constant speed
Welding head	Machine held, air cooled
Fixture	50-ton variable-speed roll
Auxiliary equipment	Exhaust fan, vacuum flux remover, positioning arm
Electrode wire	$^5/_{32}$-in.-diam EL12(a)
Flux	F71(a)
Welding position	Flat (horizontal-rolled pipe position)
Number of passes	380
Current and voltage:	
Passes 1 through 3	700 A, ac; 38 V
Remaining passes	750 A, ac; 40 V
Preheat	400 °F (by torch)
Postheat (stress relief)	7 h at 1115 °F; furnace cool to 600 °F
Welding speed	9$^1/_2$ in./min

(a) Electrode and flux yielded a weld deposit containing 0.12 C, 0.84 Mn, 0.72 Si, 0.018 S.

assembled by welding three low-carbon steel castings (head, piston body, and seat) at girth joints. When similar smaller pistons with wall thicknesses of 3 to 5 in. had been assembled by SMAW, about one welded joint in eight was found to be defective and had to be reworked.

Because of the experience with the pistons with 3- to 5-in. wall, it was decided to use SAW to assemble four large pistons, in which an 8-in. wall was to be joined to a 6$^5/_8$-in. wall, using the joint design shown in detail A. The outside surfaces of the three castings to be welded were rough machined and the joints were prepared by machining. The joints were of the interlocking type (see Fig. 48, detail A) and provided support for the unwelded components during positioning on variable-speed welding rolls. Joint areas were preheated to 400 °F with gas torches as the piston was rotated. The welds were made in 380 passes and were produced over-size and machined to size after magnetic-particle inspection and stress relief. The welded pistons were stress relieved at 1115 °F for 7 h and furnace cooled to 600 °F.

Each welded joint was ultrasonically inspected for a distance of 3 in. on each side of the weld. After inspection, the pistons were hydrostatically tested at 3000 psi. There were no rejections. Production time for welding the large piston was 101 h, which was a considerable improvement over the production time of 212 h for the smaller pistons assembled by SMAW.

Example 4. Submerged Arc Welding of a Press-Gear Blank. The blank for a large press gear was made by welding a forged medium-carbon steel rim to a web and crank arm made of low-carbon steel plate, as shown in Fig. 49. Semiautomatic SAW was used, because the production lot for any one size was too small to justify the cost of more expensive equipment for automatic welding.

Before welding, the web and crank arm were cut to shape with an oxynatural gas torch, guided by a template, and the crank arm was beveled (by gas cutting) to make the groove shown in Fig. 49, detail B. The rim, web, and crank arm were positioned for welding, and the rim was preheated by torch to 150 °F. With the exception of the one groove weld shown in Fig. 49 (detail B), the welds were fillet welds. Depending on the size of the blank, the production time varied from 5 to 8 h per piece. This was 50 to 70% faster than the rate obtainable using SMAW (with low-hydrogen electrodes).

Example 5. Circumferential welds (girth welds) present different problems in flux retention, depending on the orientation of the axis of the cylinder or other shape of workpiece being welded.

When the workpiece axis is horizontal, several techniques are effective in welding on small diameters. The normal electrode position is a few degrees from the vertical to ensure proper solidification of flux and weld metal. This position helps to prevent spillage of loose flux. A nozzle assembly can be arranged to pour the flux directly on the arc, affording the flux less chance to spill. Also, a flexible strip of metallic

Fig. 49 Submerged arc welding setup for typical press-gear blank

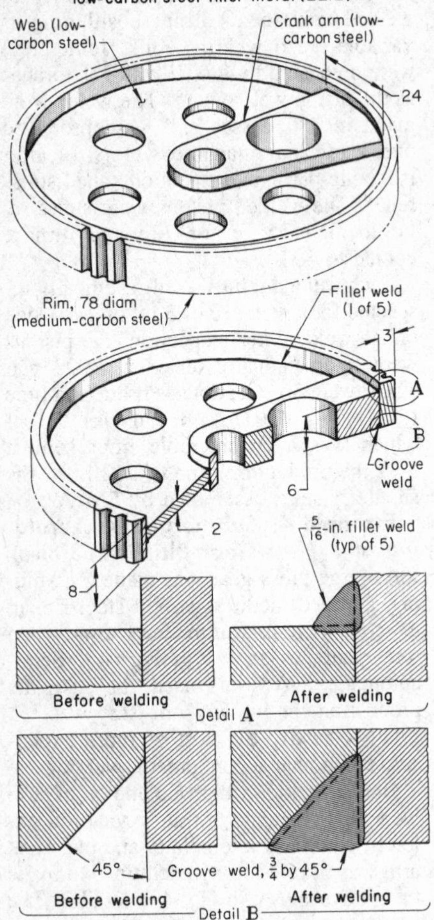

Semiautomatic SAW

Joint types	T; corner
Weld types	Fillet; single-bevel-groove
Power supply	28 to 40-V, 400 to 600-A, constant-voltage motor-generator
Electrode wire	$^5/_{64}$-in.-diam EL12
Flux	F71
Current	400 A (DCEP)
Voltage	28 V
Number of passes	One
Deposition rate	15 to 20 lb/h
Preheat	150 °F(a)
Postheat (stress relief)	3 h at 1175 °F
Electrode consumption	15 lb/h
Flux consumption	15 lb/h
Welding speed	14 to 16 in./min
Production time	5 to 8 h per blank(b)

(a) The rim only was preheated by torch. (b) Depending on size of gear blank

or nonmetallic heat-resistant material may be attached to the nozzle assembly and positioned to ride over the workpiece ahead of the arc so as to retain the flux. Such an arrangement, incorporating a flexible dam, is shown in Fig. 50.

Example 6. Influence of Weld-Metal Composition on Weld-Bead Quality. Variations in the quality of submerged arc butt welds made in $^1/_2$-in.-thick ASTM A515, grade 70, steel plate were attributed to differences in electrode-wire composition. Test joints were welded under identical conditions, including use of the same flux (composition of which was not identifiable), but using two different electrode wires of the following compositions:

	C	Mn	Si	S	P
Test 1	0.14	2.00	0.03	0.024	0.017
Test 2	0.09	0.50	0.03	...	0.017

As shown in Fig. 51, weld beads made in test 1 were smooth and sound and would be expected to pass radiographic examination, whereas those made in test 2 had craters and ripples and probably would not pass radiographic examination even after dressing or grinding to remove the surface roughness. Results of mechanical testing of the weld metal deposited in the tests are shown in Fig. 51.

Analysis of the weld metal in the two beads showed:

	C	Mn	Si
Test 1	0.14	0.74	0.25
Test 2	0.11	0.50	0.22

Because manganese serves both as a deoxidizing agent and as a strengthening element, the cratered surface, irregular outline of the weld, and lower strength of the weld metal in test 2 were attributed to the lower manganese content. If this deficiency had been known in advance of welding, it might have been overcome by using a flux capable of adding an adequate amount of manganese to the weld metal.

Fig. 50 Use of a flexible dam, attached to a nozzle assembly, for retention of flux in circumferential welding

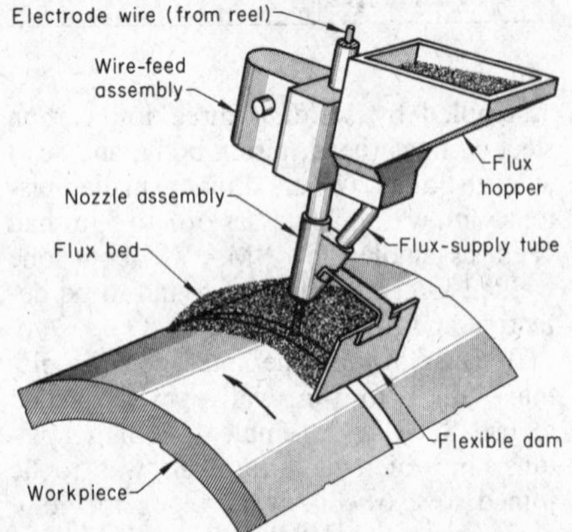

Fig. 51 Weld beads made by SAW under similar conditions but using electrode wires of different compositions. The test 2 weld bead, made with electrode wire containing 0.50% Mn, is deeply cratered, contrasting with the smooth, sound weld bead made in test 1 with electrode wire containing 2% Mn.

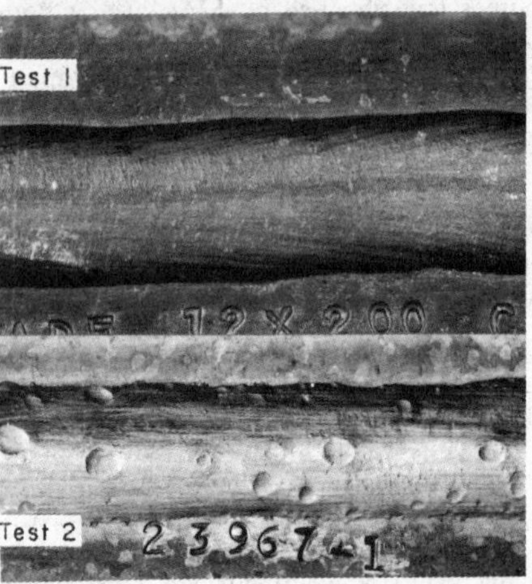

	Test 1	Test 2
Mechanical properties of weld metal		
Yield strength, ksi	65	58
Tensile strength, ksi	85	77
Elongation in 2 in., %	28	26
Reduction in area, %	56	59
Charpy impact, ft·lb:		
Room temperature	25	50
−4 °F	10	18

Welding conditions for both tests

Base metal	ASTM A515, grade 70, steel, $^1/_2$ in. thick
Joint type	Butt
Weld type	Single-V-groove; copper backing
Joint preparation	Machined
Flux	F61
Electrode-wire diameter	$^5/_{32}$ in.
Welding position	Flat
Power supply	1200-A transformer
Current	500 A
Voltage	30 V
Welding speed	30 in./min
Number of passes	One
Preheat and postheat	None

Example 7. Use of Low-Silicon Flux in Welding Joints for Service in a Corrosive Environment. The plates of a pulp-digester tank were butt welded together by SAW in 18 passes. The plates were made of $1^1/_2$-in.-thick ASTM A285, grade C, low-carbon steel. It was important that the silicon content of the metal exposed to the digester fluids be below about 0.015% to minimize corrosion.

Table 12 Conditions and results of tests to determine effect of silicon content of flux on silicon content of weld metal

Flux type, test 1	Special low-silicon
Flux consumption, test 1	0.15 lb/min
Flux type, test 2	General purpose(a)
Flux consumption, test 2	0.20 lb/min

Welding conditions for both tests

Joint type	Butt
Weld type	U-groove (side 1), V-groove (side 2)
Joint preparation	Machining
Welding position	Flat
Power supply	40-V, 600-A transformer-rectifier
Preheat and postheat	None
Electrode	$^3/_{32}$-in.-diam
Electrode extension	1 in.
Deposition rate	0.5 lb/min per 1000 A
Number of passes	18

Sequence and conditions of the 18 passes:

Side	Pass No.	DCEP, A	Voltage, V	Travel speed, in./min
1	1	320	27	20
	2	320	28	18
	3	320	28	18
2	4	350	26	15
	5	330	28	8
	6	330	30	8
	7-10	320	30	8
1	11-18	320	30	8

Chemical compositions

	C	Mn	Si
Base metal	0.20 C,	0.60 Mn,	0.008 Si
Electrode	0.14 C,	2.00 Mn,	0.024 Si
Weld metal, passes 6 and 7:			
Test 1	0.09 C,	1.00 Mn,	0.014 Si
Test 2	0.09 C,	0.95 Mn,	0.400 Si

Low-carbon steel (ASTM A285, grade C); 2% Mn steel filler metal (EH14)

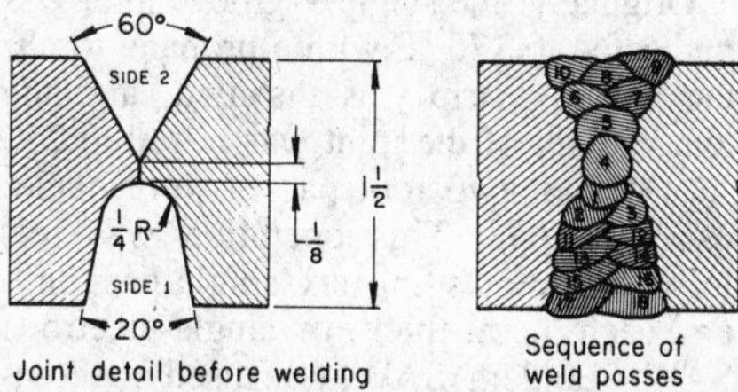

Note: Test welds were made in butt joints as illustrated above. (a) Contained a significant amount of silicon dioxide (silica) and no ferrosilicon or other deoxidizer

To determine the effect of the silicon content of the flux on the silicon content of the weld metal, test plates were welded using a special low-silicon flux (test 1), and other test plates were welded under the same conditions but using a general-purpose flux containing a significant amount of silicon dioxide (silica) and no ferrosilicon or other deoxidizer (test 2).

The carbon, manganese, and silicon contents of weld beads made with the two fluxes are compared in Table 12. The weld bead that was made with the special low-silicon flux (test 1) contained only 0.014% Si, in contrast to the 0.40% Si contained in the bead made with the general-purpose flux (test 2).

Example 8. Elimination of Backing Bars to Avoid Entrapment of Slag. A 12-ft-long header assembly for a large high-pressure heat exchanger was manufactured to Section VIII, Division 1, of the ASME Boiler and Pressure Vessel Code (Fig. 52).

As originally designed (see view at upper left in Fig. 52), for manual FCAW from the outside, the assembly consisted of four steel components and was welded at corner joints that incorporated backing bars, as shown in Fig. 52, section A-A. It was difficult to ensure a uniformly tight fit of the backing in the joint. Under radiographic inspection, slag, which ran between the backing bars and the adjacent $1^1/_2$-in.-thick components, was revealed.

The problem was eliminated by redesigning the header assembly for automatic SAW without backing bars. The redesigned assembly, shown in the upper right view, consisted of two $1^1/_2$-in.-thick channels formed in a press brake. The two components were welded at two longitudinal butt joints of the double-V-groove design shown in Fig. 52, section B-B.

For this improved design, the welding

Fig. 52 Submerged arc welding setup for heat-exchanger header

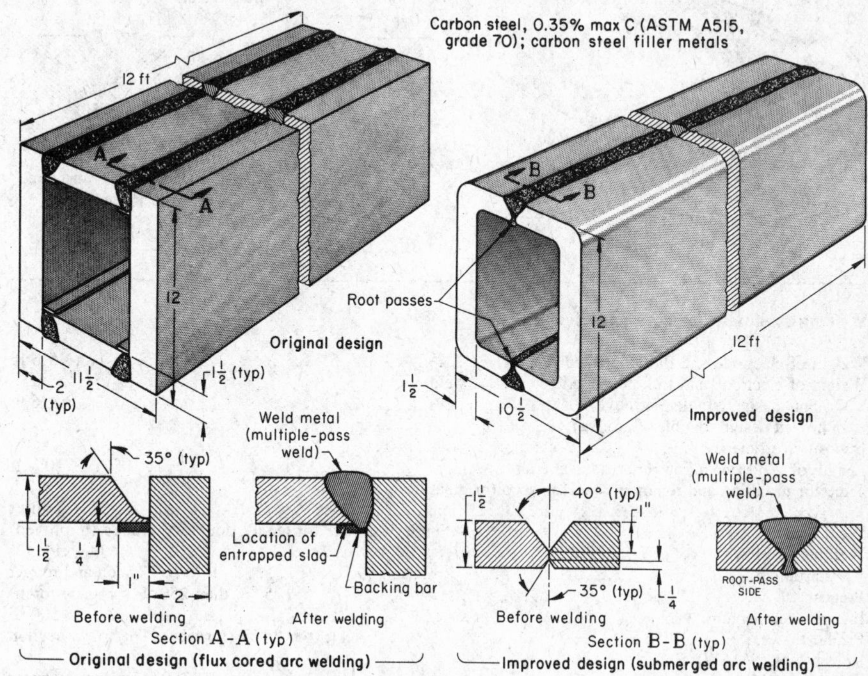

	Original design	Improved design
Welding process	Manual FCAW	Automatic SAW
Electrode	$^3/_{32}$-in.-diam flux cored wire	$^3/_{32}$-in.-diam solid wire
Flux	...	F72
Welding position	Flat	Flat
Root-pass welding conditions		
Welding current(a), A	375-425	460-480
Number of passes per joint	1	1
Welding speed, in./min	6	8
Filler-pass welding conditions		
Welding current(a), A	375-425	400-600
Deposition rate, lb/h	6	18
Number of passes per joint	7-8	5
Welding speed, in./min	10	22(b)

(a) Power supply for welding of both designs was an 80-V (open-circuit) transformer-rectifier. (b) Welding speed for the first filler pass was 28 in./min.

was done by use of a boom-mounted automatic welding head. The formed channels were held stationary while the welding head was advanced along the joints. First, root passes were made along the inside grooves of the two joints, then filler passes were made along the outside grooves. After the root passes, the outside grooves were machined-out to sound metal before the filler passes were begun.

A major benefit of the change to two-piece design was that only about one third as many filler passes were required for the entire weldment (10 passes, as compared with 28 to 32 passes for the original four-piece design).

Example 9. Change From a Single-V-Groove Weld With a Backing Strip to a Double-V-Groove Weld With a Welded-in Center Spacer Rod, To Reduce Costs and Distortion. Figure 53 shows a 120-in.-long steam-drum shell course, roll formed with a welded longitudinal seam.

Originally, the butt joint for this seam was of single-V-groove design, and was welded with the use of a backing strip (see "Original design" in section A-A in Fig. 53). Fit-up and removal of the backing strip were time-consuming operations, and welding from one side distorted the weldment.

The joint was changed to the double-V-groove design shown as "Improved design" in section A-A of Fig. 53. This change resulted in the need for much less weld metal; the need for a backing strip was eliminated; and distortion was reduced by sequential deposition of weld beads on the inside and outside of the joint. The amount of back gouging needed was less than that required to remove the backing strip from the single-V-groove weld. As a result of these improvements, electrode, flux, and labor costs were reduced by 46%, and total cost of welding was reduced by 62%. Welding procedures and postweld operations for the two designs are described below. For both designs, the shell courses were hot roll formed into a cylinder and descaled, and the joint grooves flame cut.

Originally, the single-V-groove joint was preheated to 175 °F with a propane torch, the backing strip was installed, and the temperature of the joint was raised to 250 °F. At least two root passes were made, using SMAW. This operation was followed by depositing six single-pass layers, each $^1/_8$ in. thick, by single-electrode SAW. Tandem SAW was used to complete the weld, single-pass layers $^1/_8$ in. thick being deposited to a weld level of $1^1/_2$ in., followed by two-pass (split) layers $^1/_8$ in. thick. Then the backing strip was removed by air carbon arc gouging and grinding, and back welding was done, as required, to provide a flush joint.

In the improved design, the double-V-groove joint was also preheated in two stages (175 and 250 °F) with a propane torch, except that instead of a backing strip being installed between stages, a spacer rod of $^1/_4$-in.-diam 0.5% Mo steel electrode material was tacked in place and seal welded by SMAW. Shielded metal arc welding was used also for root passes. The first increment of single-electrode sub-

Fig. 53 Submerged arc welding setup for steam-drum shell course

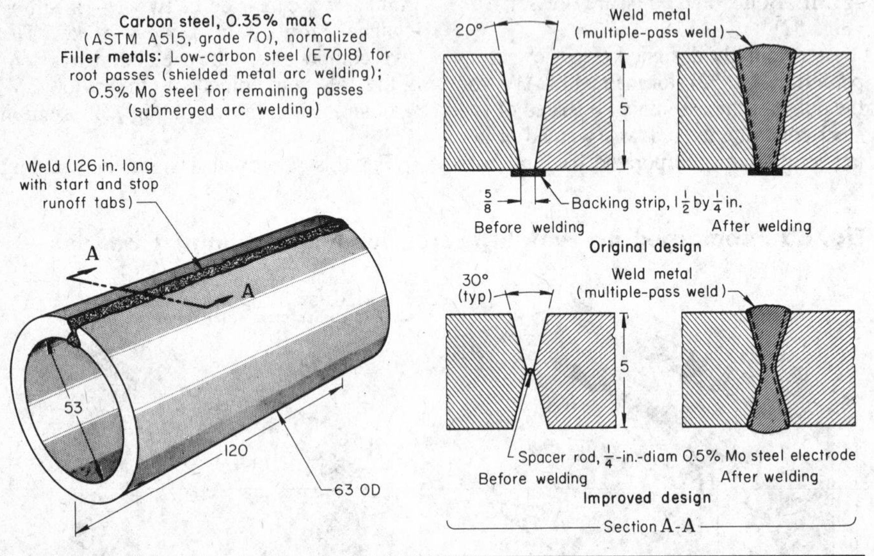

Welding conditions for both joint designs

Weight of electrode and flux deposited per hour	15.6 lb
Weight of electrode and flux deposited per foot of weld:	
Original design (single-V-groove)	26 lb
Improved design (double-V-groove)	14 lb
Deposition efficiency	98%
Length of weld (including runoff tabs at ends)	$10^1/_2$ ft
Time for installing and removing backing strip (original design)	12 h
Joint type	Butt
Weld types	Single-V-groove (original); double-V-groove (improved)
Welding position	Flat(c)
Arc starting	Touch and retract
Preheat	175 °F, then 250 °F (propane torch)
Interpass temperature	500 °F
Postheat	1150 ± 25 °F (furnace), 1 h/in. of section
Root passes (SMAW):	
Power supply	300-A motor-generator
Electrode	$^3/_{16}$-in. E7018
Current and voltage	250 A (DCEP); 24 V
Intermediate passes (single-electrode SAW):	
Power supply	900-A motor-generator
Electrode wire	$^7/_{32}$-in.-diam 0.5% Mo steel(d)
Current and voltage	700 A (DCEP); 30 V
Travel speed	20 in./min
Final passes (tandem SAW)(e):	
Leading head:	
Power supply	900-A motor-generator
Electrode wire	$^7/_{32}$-in.-diam 0.5% Mo steel(d)
Current and voltage	800 A; 30 V
Trailing head:	
Power supply	1000-A transformer
Electrode wire	$^5/_{32}$-in.-diam 0.5% Mo steel(d)
Current and voltage	700 A, ac; 35 V
Travel speed	30 in./min

(a) Pounds of electrode and flux deposited per foot, multiplied by cost per pound of electrode and flux, divided by deposition efficiency. (b) Pounds deposited per foot, divided by pounds deposited per hour, multiplied by labor and overhead cost per hour. (c) Workpiece supported on one power roll and one idler roll. (d) Electrode wire contained 0.11% C, 0.50% Mo, 0.85% Mn, and was used at a 1-to-1 ratio of wire to flux. (e) Tandem welding head was mounted on a boom-type manipulator.

Fig. 54 Use of a lathe fixture for support and rotation during simultaneous welding of two plugs to the ends of a pipe, in the production of a cargo cushion

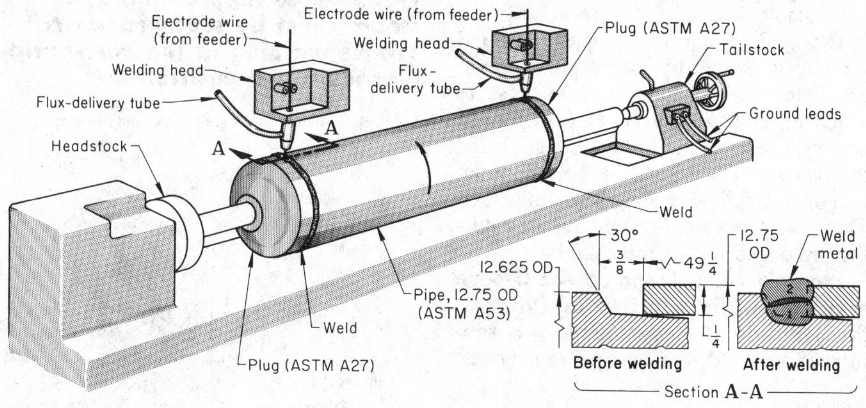

Process ..... Automatic SAW	Welding current ..... 350 A (ac)
Joint type ..... Butt	Welding speed ..... 14 in./min
Joint preparation ..... Machining	Flux consumption per assembly ..... 4.1 lb
Weld type ..... Single-bevel-groove	Electrode consumption per assembly ..... 2.88 lb
Electrodes (two) ..... $^3/_{32}$ in. diam	Number of passes per weld ..... Two
Flux ..... F70; size 12 × 65	Power supplies ..... Two 48-V, 500-A transformers
Electrode ..... $^3/_4$ in.	Preheat and postheat ..... None
Welding position ..... Flat (horizontal-rolled pipe position)	Auxiliary equipment ..... Rework lathe
Welding voltage ..... 29 V	Production rate ..... Six assemblies per hour

merged arc welds consisted of eight $^1/_8$-in.-thick single-pass welds on the outside of the weldment. The workpiece was rotated 180°, and the joint was backgouged by air carbon arc gouging and ground to a radius of $^1/_4$ to $^3/_8$ in. The first increment of welding on the inside of the joint consisted of $^1/_8$-in.-thick single-pass welds to a $1^1/_2$-in. level, using single-electrode SAW. Then the workpiece was again rotated 180°, and the remainder of the outside welding was completed using tandem SAW to deposit two-pass (split) layers of $^1/_8$-in. thickness.

Example 10. Use of Lathe Fixturing in Welding Plug Ends of Cargo Cushions. Cargo cushions (Fig. 54), used to control shock when coupling railroad freight cars, were assembled by welding a plug into each end of a 12.75-in.-OD pipe ($^1/_4$-in. wall thickness), using automatic SAW. The beveled end plugs were tack welded into place before being submerged arc welded to the tube. The welded joints had to withstand a hydrostatic test pressure of 430 psi. The two plugs were welded simultaneously, using two welding heads mounted on separate carriages. The assembly was supported and rotated in a rework lathe. Each weld was made in two passes, as shown in section A-A of Fig. 54. No postweld finishing was done. Sample welds were macroetch-inspected.

Example 11. Use of an Offset in a Pipe To Eliminate Backing Rings. A component of a heat-exchanger shell assembly (Fig. 55) was initially made by submerged arc welding a $^1/_4$-in.-wall cap drawn from medium-carbon steel to a low-carbon steel pipe of the same wall thickness, by means of a circumferential butt joint supported and aligned by a backing ring, as shown in the "Original design" in Fig. 55.

When it became apparent that the wall thickness of the pipe cap could be less than that of the pipe without adversely affecting service performance, the joint was redesigned as a joggled lap joint (see "Improved design," Fig. 55). The offset incorporated in the pipe for the redesigned joint took the place of the backing ring previously used and furnished a locating surface for the cap. The redesigned joint was submerged arc welded under the same conditions as those for the original joint, except that only two passes were required, rather than three.

Cost reduction was realized from elimination of the backing ring, from the saving in material resulting from the use of thinner pipe caps, and from the elimination of one circumferential welding pass. The change in joint design led to a saving of approximately 35% in total factory cost.

All joints were inspected visually and radiographically to check for full penetration and absence of slag inclusions. The rejection rate was less than 1%.

Example 12. Use of a Special Clamping Fixture With an Integral Flux Trough for Making a Horizontal Edge-Flange Weld. The two AISI 1008 steel components of a warm-air furnace combustion chamber were joined by a submerged arc peripheral edge-flange weld while being held in the horizontal position between upper and lower rings of a special clamping fixture. As shown in Fig. 56, the lower clamping ring incorporated a flux-retaining trough to ensure flux coverage of the arc and the weld metal while the joint

Fig. 55 Cap-to-pipe weldment

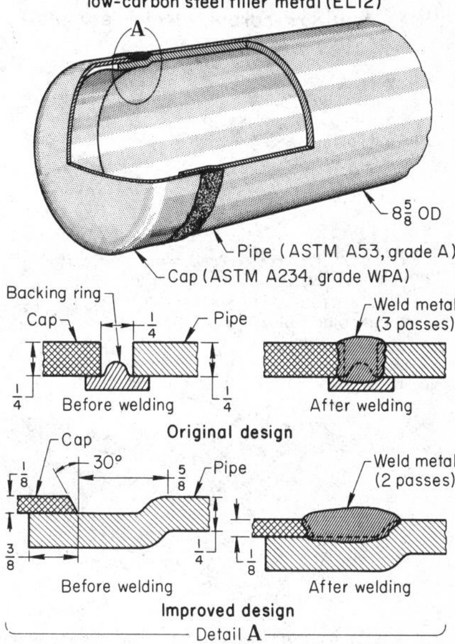

Joint type	Joggled lap
Weld type, original design	Square-groove, with backing ring
Weld type, improved design	Modified single-V-groove, with integral backing
Joint preparation:	
Original design	Backing ring machined
Improved design	Cap end machined, pipe end reduced
Electrode wire	$^1/_8$-in.-diam EL12
Flux	F62
Welding position	Flat (horizontal-rolled pipe)
Welding voltage	25 to 26 V
Welding current	350 to 410 A (DCEN)
Welding speed	18 to 20 in./min
Number of passes, original design	Three
Number of passes, improved design	Two
Power supply	40-V, 600-A transformer-rectifier (constant-voltage)
Fixturing	Chuck-type turning rolls; alignment clamps for tack welding

was being welded. The sheet metal components of the chamber were aligned and located between the clamping rings, which were made of copper.

The welding-head carrier moved along the outside of a geared rack mounted on the underside of the table supporting the lower clamping ring. The gear that engaged the rack was driven by a motor mounted on the welding-head support arm, and the geared rack served as a template to guide the welding head.

In addition to salvaging the unfused flux for reuse, the vacuum pickup cleaned the lower clamping ring well enough to receive the next chamber to be welded. The flux-delivery tube, the welding head, and the flux-recovery tube were guided from the same geared rack and were part of a single welding-head assembly.

An approximately round weld bead was produced along the flange edges, as shown in the joint detail for SAW in Fig. 56. A 3/32-in.-diam electrode wire was used to make alignment of the electrode tip with the flange edges less critical for maintaining a stable arc.

The welded joint was scraped clean and visually inspected by the welding operator. Later, the assembly was air-pressure tested under a waterhead of 6 in. When heated in service, the welded flange flexed to open slightly. Expansion and contraction in heating and cooling were noiseless. There were no weld-related failures during prolonged use of the chambers.

Example 13. Submerged Arc Groove Welding in the Horizontal (Vertical Pipe) Position. A pipe nipple was submerged arc welded to a forged steam chest oriented with the work axis vertical, so that the welding had to be done in the horizontal (vertical pipe) position (see Fig. 57). The two members were held in position by

Fig. 57 Pipe nipple joined to steam chest by submerged arc groove welding in the horizontal (vertical-pipe) position

Low-alloy Cr-Mo steel; low-alloy Cr-Mo filler metal

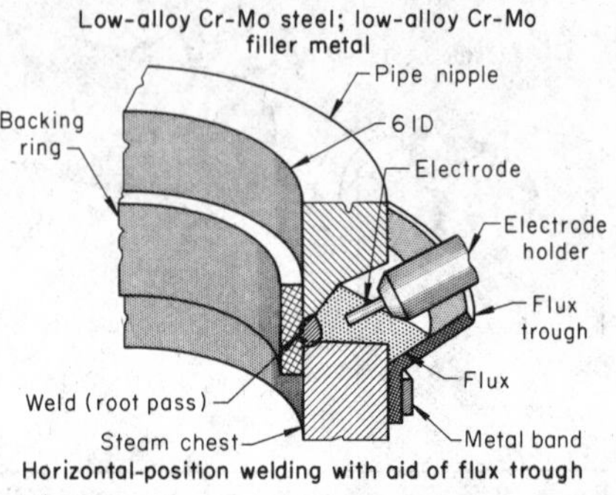

Horizontal-position welding with aid of flux trough

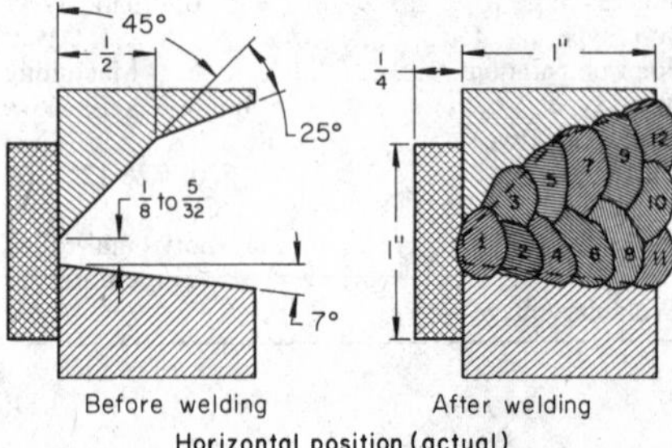

Horizontal position (actual)

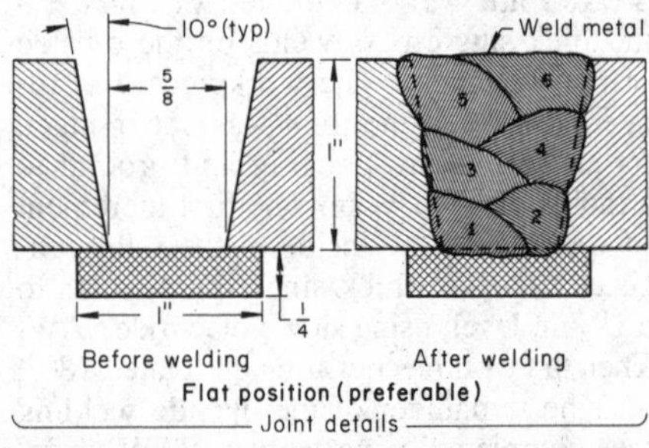

Flat position (preferable)

Joint details

Fig. 56 Use of a clamping fixture with an integral flux trough for SAW of furnace combustion chamber

1008 steel; low-carbon steel filler metal (EL12)

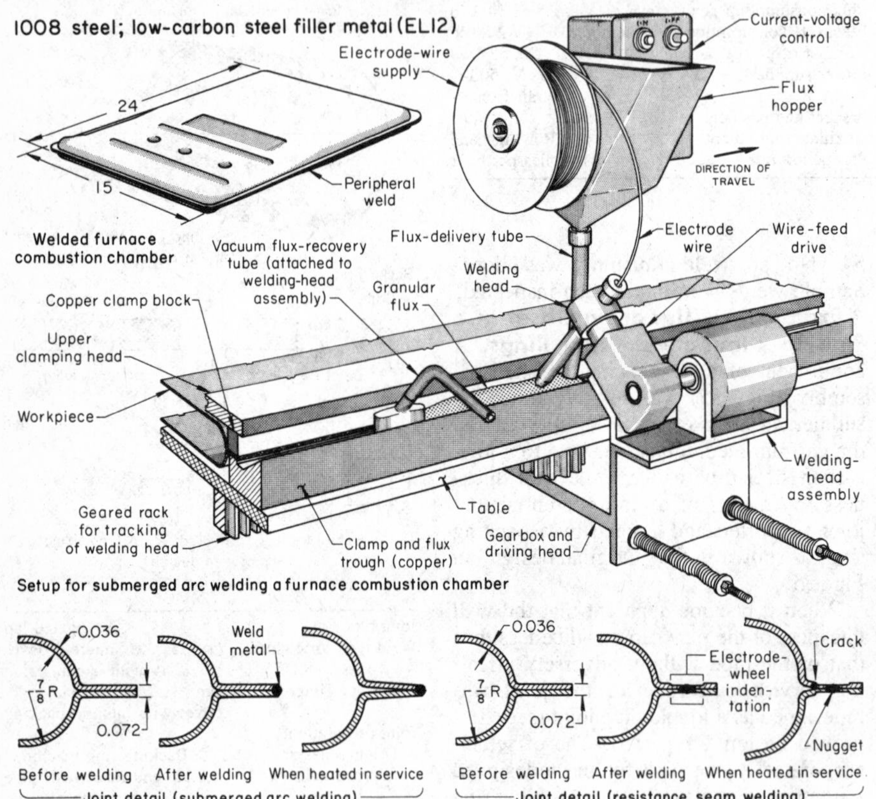

Conditions for SAW

Joint type	Edge
Weld type	Edge-flange
Joint preparation	Die trimmed to square edge
Electrode wire	3/32-in.-diam EL12(a)
Flux	F62
Welding position	Horizontal
Welding voltage	22 V
Welding current	400 A (DCEP)
Wire feed	Automatic
Welding speed	180 in./min(b)
Number of passes	One

(a) Copper plated. (b) In SMAW, welding speed had been 36 in./min.

Conditions for horizontal-position welding(a)

Joint type	Circumferential butt
Weld type	Modified J-groove
Joint preparation	Machining
Electrode wire	1/8 in. diam(b)
Flux	Neutral; size, 20 × 200
Electrode extension	3/4 to 1 in.
Welding position	Horizontal (vertical pipe)
Voltage	28 to 30 V
Current	325 to 350 A (DCEP)
Travel speed	20 to 25 in./min
Flux consumption	10 to 15 lb/h
Deposition rate	10 to 12 lb/h
Number of passes	12 to 14
Power supply	45-V, 600-A transformer-rectifier (constant-voltage type)
Preheat	400 to 450 °F
Postheat(c)	600 °F, 2 h; 1250 °F, 2 h

(a) See text for details of conditions that would have been different for flat-position welding. (b) Same composition as base metal (low-alloy Cr-Mo steel). (c) Stress relief

tack welding them to a machined backing ring, and a sheet metal trough was clamped in place with a metal band to retain the loose flux, as shown. The flux trough was wide enough to retain flux for the entire welding operation, so repositioning was not necessary. The backing ring was removed in a later operation. During welding, the workpiece remained stationary and the welding head was advanced.

Although many of the welding conditions were the same as they would have been for welding in the flat (horizontal-rolled pipe) position, there were differences:

- Electrode wire of smaller diameter (1/8 in. instead of 5/32 in.) was used.
- Current was about 30% lower.

Fig. 58 Dam-gate trunnion pin submerged arc surface welded with stainless steel for corrosion resistance

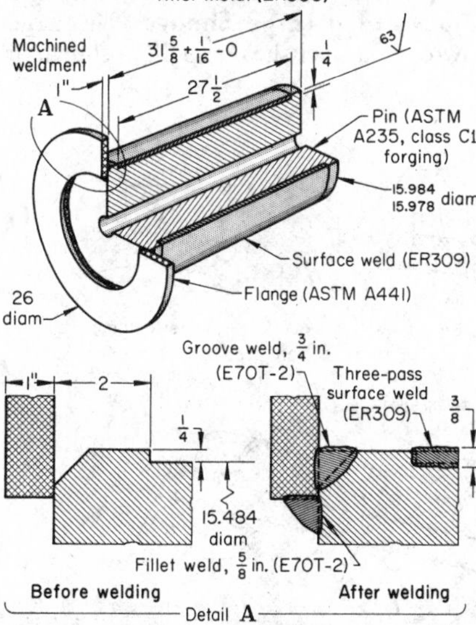

Conditions for surfacing

Weld type	Surfacing
Surface preparation	Machining 1/4-in. undercut
Electrode wire	1/8-in.-diam ER309
Flux	Neutral
Power supply	Two 35-V, 350-A transformers
Current	250 A (ac)
Voltage	25 V
Welding position	Flat(a)
Welding speed	17 in./min
Number of passes	Three
Preheat and interpass temperature	250 to 300 °F
Arc time per pin	23 1/4 h
Flux consumption per pin	40 lb
Weight of filler metal deposited per pin	40 lb

(a) Horizontal-rolled pipe position

- Voltage was about 1 to 2 V higher, to flatten the bead.
- Travel speed was about twice as high.
- Flux depth was increased to maintain 1-in. minimum flux coverage as each bead was deposited.
- A special joint design with a very small bevel angle at the bottom and a generous bevel angle at the top was used.
- The position of the electrode in relation to previously deposited weld beads was particularly important.

Fig. 59 Submerged arc welding setup. See text for details.

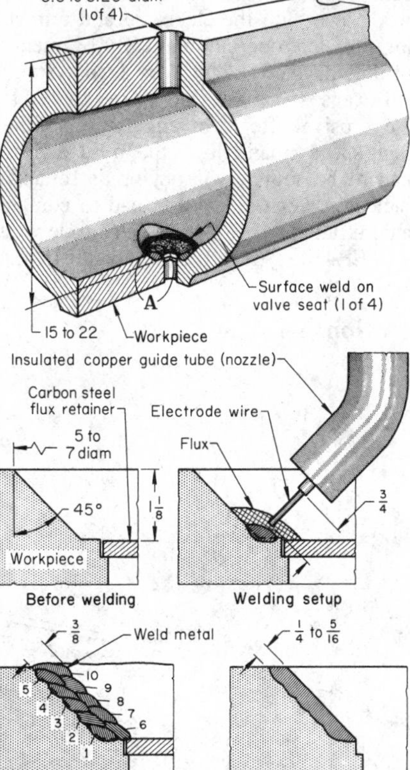

Process	Automatic SAW
Weld type	Surfacing
Power supply	45-V, 600-A constant-voltage rectifier
Voltage	34 to 35 V
Current	320 to 340 A (DCEN)
Electrode wire	5/64-in.-diam ER410
Electrode extension	3/4 in.
Flux	Neutral
Welding speed	15 in./min
Number of passes	10 to 14
Preheat and interpass temperature	250 °F min
Postheat (stress relief)	10 h at 1250 °F
Deposition rate	12 to 14 lb/h
Flux consumption	12 to 15 lb/h
Production time per valve seat	2 h

- Closer control over current and welding speed was required.

Example 14. Surfacing a Low-Carbon Steel Trunnion Pin With Stainless Steel. The dam-gate trunnion pin shown in Fig. 58 forged from ASTM A235, class C1, steel provided adequate strength for the application, but inadequate resistance to atmospheric corrosion. To provide protection from corrosion, a type 309 stainless steel overlay was deposited in an undercut section of the pin by three-pass SAW, after a flange had been joined to the pin by FCAW.

To make the overlay, the pin was rotated in a 1000-lb welding positioner. The pin was supported by a shaft through its cored hole and by a mounting plate (not shown) that was tack welded to the end of the pin opposite the flange. A bridge over the positioner supported a carriage for the welding head. An improvised heater was used for preheating the pin and for interpass heating. In each pass, a bead 1/8 in.

Fig. 60 Submerged arc welding of sheave with a rim groove. See text for details.

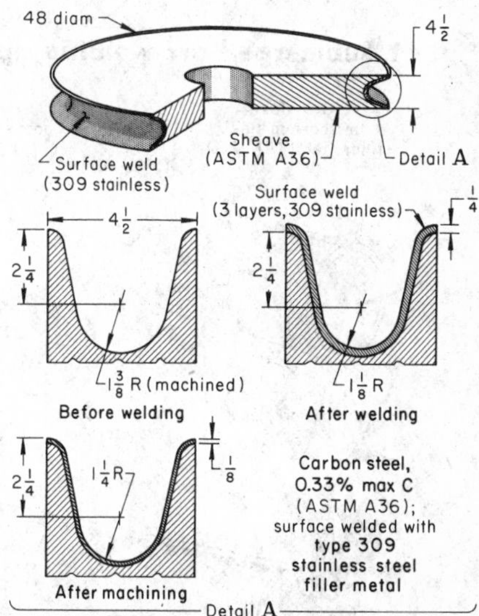

Process	Automatic SAW
Weld type	Surfacing
Surface preparation	Machining
Power supply	750-A, constant-voltage rectifier
Electrode wire	1/8-in.-diam cored 309 stainless
Flux	Proprietary
Welding position	Flat (within 30°)
Current	380 A (DCEP)
Voltage	30 V
Wire feed	Automatic
Welding speed	20 to 22 in./min
Number of weld layers	Three
Number of passes per layer	36
Total arc time per sheave	12 h
Finishing	Machining to dimensions

thick was deposited. After the last pass, the pin was wrapped in 3-in.-thick fiberglass and allowed to cool slowly for 12 to 14 h. It was then machined to final dimensions, leaving a $^1/_4$-in.-thick coat of stainless steel on its shank.

Example 15. Automatic Submerged Arc Surface Welding of Internal Valve Seats. Four internal valve seats in a steam chest were surface welded with stainless steel by automatic SAW (Fig. 59). Manual welding was not feasible, because the only access to the valve seats was through holes $3^1/_2$ to $5^1/_4$ in. in diameter, 15 to 22 in. above the seats (hole diameter and distance from the seats varied with the size of the chests, which varied in diameter and ranged from 8 to 12 ft in length).

The electrode wire was brought to the weld area through an electrically insulated copper guide tube (nozzle) with an angled head (see Fig. 59). The guide tube, lowered through and centered in the hole above each valve seat, was rotated at a fixed speed. During welding, flux, which was supplied manually or by gravity feed through the same hole in the steam chest, was retained by a carbon steel plug, which was machined out after welding. On each valve seat, a $^3/_8$-in.-thick, two-layer surface weld was deposited in 10 to 14 passes.

After each pass, a limit switch automatically stopped the welding and the rotation of the guide tube; then loose flux and slag were removed by a suction system, assisted by a pneumatic hammer that loosened the slag.

Gas burners were used to preheat the valve seats to at least 250 °F before welding and to maintain this minimum temperature throughout the 2-h welding time required for completing each valve seat.

Electrode extension was held constant at about $^3/_4$ in., and the guide tube was relocated upward $^1/_4$ in. and outward (horizontally) $^1/_8$ in. for each pass. The arc for each pass was initiated by contacting the workpiece with the electrode and retracting the electrode automatically by means of the constant-voltage wire-feed system.

Because the valve seats were exposed to the erosive effects of high-temperature, high-speed steam flow, the weld deposit had to be sound, without voids, cracks, inclusions, or oxides, and had to contain a minimum of 11% Cr. The electrode wire was ER410 (type 410 stainless steel), $^5/_{64}$ in. in diameter—the largest wire that could be fed through the guide tube.

By welding with DCEN of 320 to 340 A and an arc voltage of 34 V, supplied by a constant-voltage transformer-rectifier, the small-diameter electrode wire provided a flat bead with good flow and clean overlaps.

The weld-surfaced parts were stress relieved by holding at 1250 °F for about 10 h. Then the flux retainers were machined out, and the weld deposit on each valve seat was machined to final dimensions (see "After machining" in detail A of Fig. 59). Inspection was carried out by visual examination and liquid-penetrant testing. The rejection rate was less than 1%.

Two other automatic processes, GTAW and GMAW, had been evaluated for this application. Submerged arc welding was chosen over the other processes because it required the least setup and cycle time, was easiest to apply through the limited-access hole, permitted greatest control of variables, and entailed the lowest initial and maintenance costs.

Example 16. Surface Welding the Groove of a Large Sheave. The large low-carbon steel sheave (pulley) shown in

Fig. 61 Submerged arc welding application. See text for details.

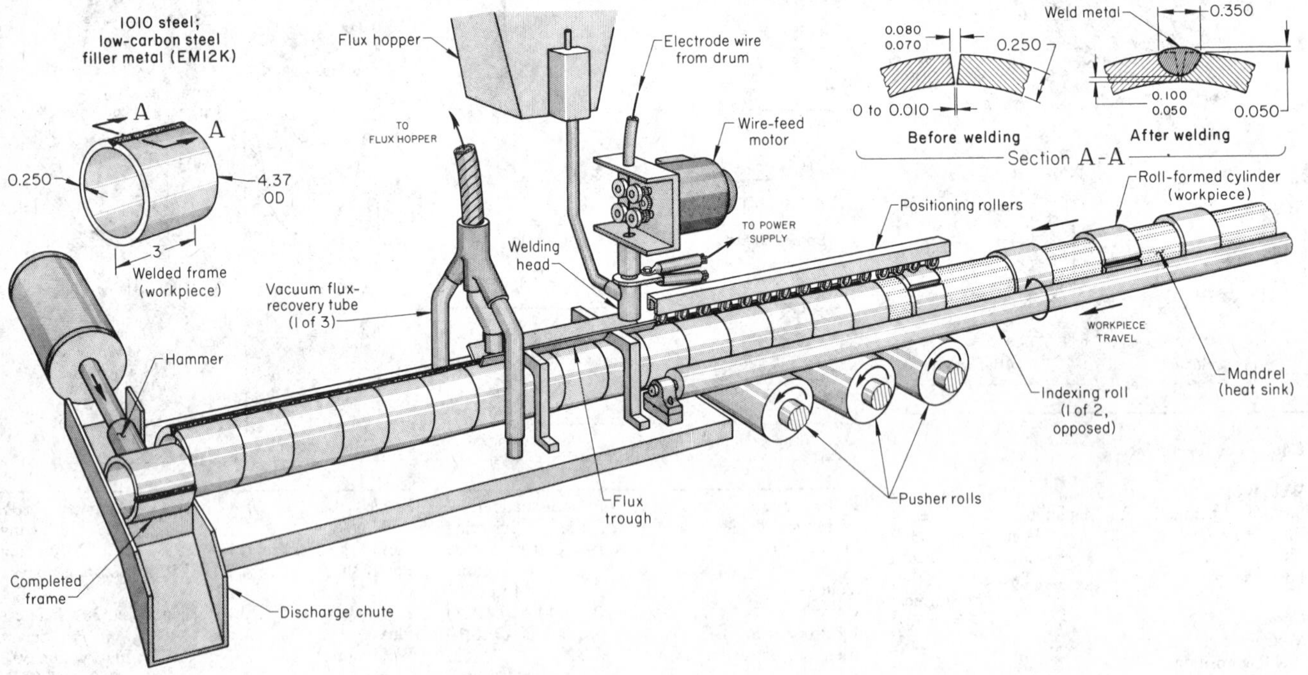

Joint type	Butt	Flux	F72(c)	Welding position	Flat
Weld type	Single-V-groove	Number of passes	One	Arc length	$^1/_4$ in.
Power supply	Motor-generator(a)	Current	420 A (DCEP)	Welding speed	51 in./min
Electrode wire	$^1/_8$-in.-diam EM12K(b)	Voltage	27 V	Preheat or postheat	None

(a) Output rating at 100% duty, 35 V and 1000 A. (b) This medium-manganese wire (0.80 to 1.25 Mn) was used to obtain weld metal with slightly higher manganese content than that of the base metal. (c) Containing manganese silicates

Fig. 60 was used infrequently, and when not in use was exposed to the weather, which led to surface deterioration that caused the grooved rim to seize during operation of the sheave. To prevent this, the rim was surface welded with a three-layer deposit of stainless steel. Automatic SAW was chosen to obtain the best combination of deposition rate and soundness of deposit.

The sheave was gas cut from ASTM A36 steel plate $4^1/_2$ in. thick, and the center hole was rough bored. The rim groove was then machined to the dimensions shown in the "Before welding" view in detail A of Fig. 60, and the sheave was mounted on a welding positioner.

So that the same welding conditions could be applied during the entire surfacing operation, the welding head was mounted on the frame of the positioner with fixturing that held the head in a fixed position with reference to the grooved rim. The sheave was rotated automatically on its axis to give a welding speed of 20 to 22 in./min. The overlay was deposited in three layers, each consisting of 36 passes; total thickness of the overlay was $^1/_4$ in. For the initial passes covering the central portion of the sheave groove, the sheave was in an upright position (with its axis horizontal). As the welding progressed toward the outside of the groove, the angle of the positioner was adjusted in increments to keep the work surface at the welding point within 30° of horizontal for each pass. The same sequence of positioning was repeated to deposit the second and third layers of weld metal.

Initially, solid stainless steel electrode wire was used, but the overlay cracked severely on cooling. The cracking was caused by shrinkage stress in the first layer of the overlay, which was brittle because of dilution of its alloy content and increase in carbon content by diffusion from the carbon steel base metal. To increase the ductility of the first layer deposited, cored tubular electrode wire was used instead of solid wire. The cored wire consisted of a low-carbon steel tube filled with alloying and fluxing ingredients needed to deposit weld metal similar in composition to type 309 stainless steel. The cored wire gave a less penetrating arc, so that bead dilution and carbon diffusion were reduced, thereby eliminating the cracking during cooling.

The completed overlay was visually inspected, and then was machined to final dimensions. The finished surface was dye-penetrant inspected; a few pinholes were the only flaws detected in the 40 sheaves processed. The pinholes were repaired by manual GTAW and hand dressed.

Example 17. Automated Continuous Welding of Roll-Formed Cylinders. Figure 61 shows essentials of the welding station in a continuous line for the production of automobile-starter frames from $^1/_4$-in.-thick AISI 1010 steel flat stock by shearing, three-roll forming to cylindrical shape, and automatic SAW.

The V-groove of the longitudinal joint to be welded (see section A-A in Fig. 61) was that resulting from the roll forming of the sheared stock. After being washed, the roll-formed frames were transferred to a storage tower in the line to ensure continuity of supply for welding in the event of shear or forming-roll failure. From the storage tower, the frames were pushed, one after the other, onto a mandrel to form a tubelike arrangement for welding.

The frames were oriented under the welding head by means of two parallel indexing rolls that rotated the frames until the joints were at the top, where a row of sharp-edge positioning rollers engaged and held the joints in alignment for continuous welding. Pushers advanced the aligned frames along the mandrel.

Through a single tube in the welding head, flux was fed to the joint from an overhead hopper while the electrode wire was motor-fed to the joint from a large cardboard drum. Just beyond the welding head were three vacuum tubes that picked up unfused flux for return to the overhead hopper. One tube was positioned directly above the weld; the other two were positioned on either side of the work to pick up flux that spilled off the frames onto the machinery bed.

The mandrel that supported the frames

Fig. 62 Use of boom-mounted welding heads for two-operation automatic SAW

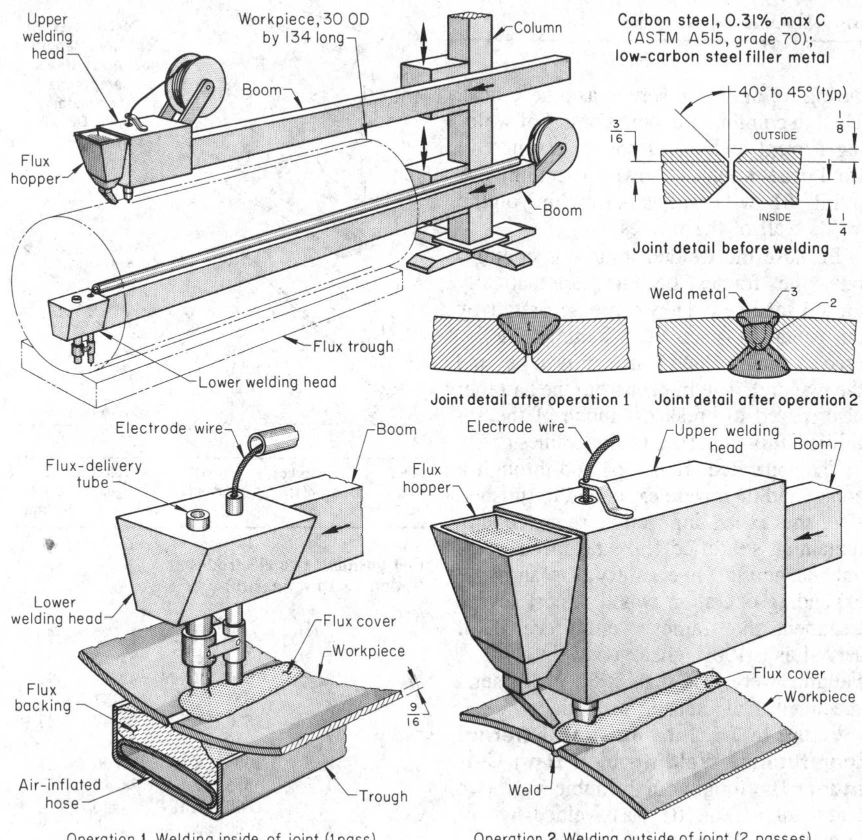

Joint type	Butt
Weld type	Double-V-groove
Joint preparation	Machining
Welding position	Flat
Power supply	Two 900-A motor-generators, connected in parallel
Number of passes	Three

	Inside weld	Outside weld
Electrode-wire diam, in.	$^5/_{64}$	$^7/_{32}$
Welding voltage	37	38
Welding current (DCEP), A	525-575	900
Welding speed, in./min	10	19
Welding time per bead, min	13.4(a)	7

(a) Exclusive of runoff tabs at the two ends of the cylinder

Fig. 63 Longitudinal butt joint in a thick-walled pressure vessel

Carbon or low-alloy steel base metal and filler metal

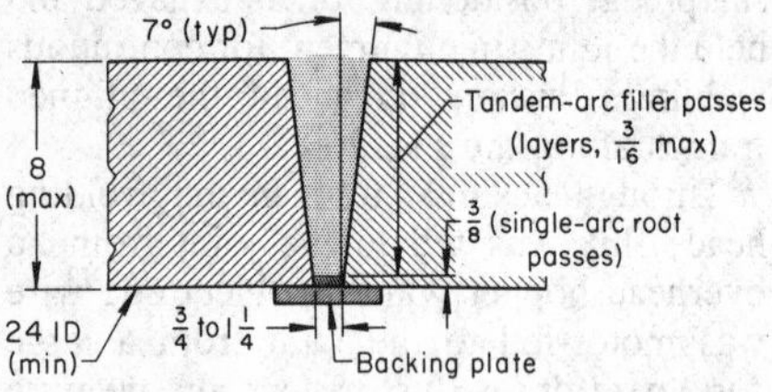

Process	Automatic SAW
Joint type	Butt
Weld type	Single-V-groove
Electrode wire	3/16-in.-diam carbon or low-alloy steel
Power supply	Transformers
Preheat	250 °F

	Single-arc root passes	Tandem-arc filler passes(a)
Current (ac), A	500	750
Voltage, V	33	35
Welding speed, in./min	12	20

(a) Values shown are for each electrode.

during welding also served as a heat sink. This, in combination with control of welding current and travel speed, resulted in weld penetration as shown in section A-A of Fig. 61, with no flash or melting on the inside wall of the frames.

Because the welded joint was continuous, the frames became automatically tacked together. They were separated by horizontal blows from a hammer, delivered as each welded frame moved beyond the mandrel. The impact from the hammer also served to break off much of the solidified flux adhering to the frames.

The separated frames passed through a cooler and then were expanded to finished size; the expanding action removed any remaining solidified flux and made internal machining unnecessary. Because the expanding operation was the most severe treatment the frames would receive, it served as a 100% check on weld quality. Failures were less than 2%. Weld-metal specimens had a tensile strength of 75 ksi.

Example 18. Internal and External Longitudinal Welding of a Long Cylinder. The longitudinal double-V-groove butt joint in Fig. 62 was welded by automatic SAW in two operations. The welded cylinder had to undergo a fairly severe final forming operation, and for this reason it was made from ASTM A515, grade 70, steel (0.31% C maximum), which has good formability and weldability.

Submerged arc welding was used because of its ability to yield sound welds with good penetration. An additional benefit of SAW was that a flux-bed backing could be used in depositing the inside weld bead (the first operation).

First, the plate was beveled on the edges to be joined and roll formed to a cylinder, thus producing the double-V-groove butt joint shown in detail in Fig. 62. The outside of the joint was tack welded, and runout tabs (not shown) were added to each end.

Two welding heads, mounted and transported on booms supported by the same column, were used for successively welding the inside and outside of the joint. The welding setup included a trough in which an air-inflated hose was used to force granular flux against the outside joint groove, thus providing backing to contain the weld metal deposited on the inside of the joint in the first operation. The flux trough also served as a workholding fixture.

The roll-formed cylinder was set on the flux bed with the joint at bottom center, the runout tabs were clamped down, and the air hose in the flux trough was inflated, forcing flux against the joint. Then the inside weld was deposited by the lower welding head in one pass. After the inside of the joint had been welded, the cylinder was removed from the flux trough and the outside of the joint was ground to sound weld metal. Then the cylinder was returned to the flux trough, with the joint at top center, for deposition of the external bead by the upper welding head in two passes. The inside bead acted as the backing. Both operations required some ob-

Fig 64 Tandem-electrode submerged arc fillet welding of bridge girders in the flat and horizontal positions

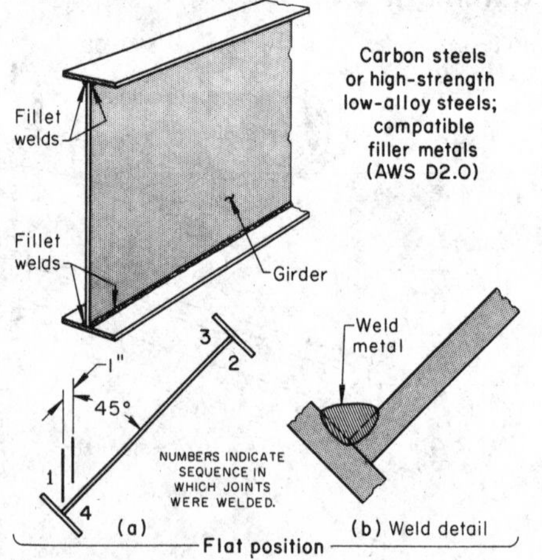

Fillet size, in.	Electrode diam(a), in.	Current, A	Voltage, V	Travel speed, in./min
Flat position, two electrodes (L and T) 1 in. apart(b)				
1/4	L, 3/16	715, DCEN	31-35	60
	T, 5/32	540, ac	31-35	
5/16	L, 3/16	800, DCEN	33-37	47
	T, 3/16	655, ac	33-37	
3/8	L, 3/16	800, DCEN	34-40	37
	T, 3/16	720, ac	38-44	
1/2	L, 3/16	1000, DCEN	34-40	27
	T, 3/16	855, ac	38-44	
5/8	L, 3/16	1000, DCEN	34-40	18
	T, 3/16	900, ac	38-44	
3/4	L, 3/16	1000, DCEN	34-40	13
	T, 3/16	900, ac	38-44	
Horizontal position, two electrodes (L and T) 4 1/2 in. apart(b)				
1/4	L, 5/32	500, DCEP	28-30	40
	T, 1/8	400, ac	32-34	
5/16	L, 5/32	650, DCEP	32-34	32
	T, 1/8	500, ac	32-34	
3/8	L, 5/32	650, DCEP	32-34	26
	T, 1/8	500, ac	32-34	
Horizontal position, three electrodes (L, T, and T) 3 in. and 7/8 in. apart(c)				
5/16	L, 5/32	600, DCEP	28-30	40
	T, 1/8	550, ac	32-34	
	T, 3/32	350, ac	27-29	
3/8	L, 5/32	650, DCEP	28-30	34
	T, 1/8	550, ac	32-34	
	T, 3/32	350, ac	27-29	
Horizontal position, four electrodes (L, T, T, and T) 3 in., 7/8 in., and 4 in. apart(c)				
1/2	L, 5/32	750, DCEP	28-30	28
	T, 1/8	600, ac	32-34	
	T, 1/8	500, ac	30-32	
	T, 3/32	350, DCEN	27-29	

(a) L, leading electrode; T, trailing electrode. (b) As shown in illustration. (c) Not shown in illustration

servation by the welding operator to ensure alignment and to control welding speed and other operating variables.

The welding was done automatically with single electrode wires, using DCEP to control the bead shape and appearance and to produce consistent full penetration. Two 900-A motor-generator sets connected in parallel supplied the welding current. The generators had a steep arc-voltage current characteristic. The electrode wire was low-carbon steel with copper coating to reduce electrical resistance at the contacting surfaces and to resist corrosion. The electrode wire and the contact jaws were kept clean and close fitting to ensure the maintenance of good electrical contact.

For each operation, flux was fed by gravity from a hopper at the welding head. Unmelted flux was returned by vacuum to storage after passing through a screen to remove slag. The flux was selected for good bead shape with reasonable penetration. It produced easily removed slag without significantly altering the composition of the weld metal.

The inside bead was ground flush with the cylinder wall, as required for subsequent operations. The welds were given a 100% radiographic inspection, and all production conformed to Section VIII, Division 1, of the ASME Boiler and Pressure Vessel Code.

Example 19. Tandem-Arc Welding of Pressure-Vessel Plates Up to 8 in. Thick. Figure 63 shows the design of a single-V-groove longitudinal butt joint used in SAW of thick-walled pressure vessels made of carbon or low-alloy steel. The minimum inside diameter of these vessels was 24 in., and the wall or plate thickness did not exceed 8 in.—an upper limit imposed by the need to accommodate the welding head in the joint.

The edges that formed the joint groove were beveled by gas cutting to provide the 7° angle shown. Mismatch allowance after roll forming the plate was $^1/_{16}$ in. maximum. Prior to welding, a backing plate was located at the base of the groove. The root opening varied from $^3/_4$ to $1^1/_4$ in., depending primarily on wall thickness and accessibility required by the welding head. Workpieces were preheated to a minimum of 250 °F to minimize distortion and prevent cracking.

For root passes to a depth of $^3/_8$ in., a single electrode was used, and power input was reduced to a level that would avoid burning through the backing plate. The remainder of the joint was filled by two-electrode (tandem-arc) welding, using alternating current in both electrodes. The current input was increased to provide maximum speed of metal deposition.

Example 20. Change in Welding Position To Increase Welding Efficiency and Reduce Handling of Tandem-Welded Bridge Girders. Originally, I-beam girders (Fig. 64) were welded in the flat position with the web tilted at 45° after assembly by tack welding. The welding head was held in a vertical position during travel along each joint, and the girder was progressively repositioned to make the four welds in the numerical sequence shown in Fig. 64(a). The wires in the two-electrode welding head, each fed by a separate wire-feed motor, were spaced 1 in. apart, so that both fed the same weld pool, developing a single bead (Fig. 64b). Although sound welds were made with this procedure, girders had to be repositioned and aligned for each weld, and only one weld could be made at a time.

To increase the production rate, the gir-

Fig. 65 Submerged arc welding application. See text for details.

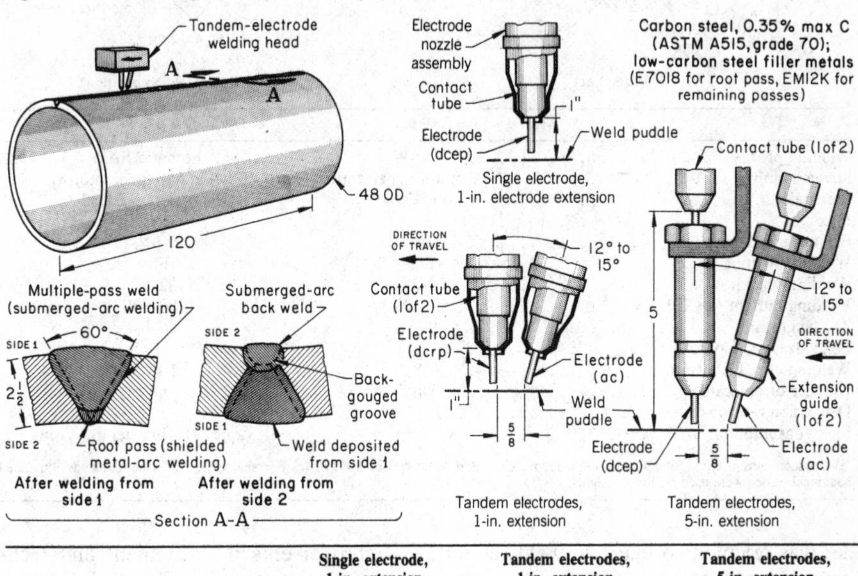

	Single electrode, 1-in. extension	Tandem electrodes, 1-in. extension	Tandem electrodes, 5-in. extension
Arc time, h	8.6	4.6	2.8
Deposition rate, lb/h	14	26	42

Side(a)		Current, A	Voltage, V	Welding speed, in./min
SAW with single (1-in. extension) electrode(a)				
1	First filler	400-450 (DCEP)	32-33	14-15
	Remaining	575-600 (DCEP)	33-34	13-14
2	1 to 3(b)	650-700 (DCEP)	35-38	12-14
SAW with single (1-in. extension) and tandem (1-in. extension) electrodes(a)				
1	First filler (single electrode)	450-500 (DCEP)	32-34	17
	Remaining (leading electrode)	500-600 (DCEP)	34-36	15-16
	(trailing electrode)	475-550 (ac)	...	(tandem)
2	1 to 3 (single electrode)(b)	450-500 (DCEP)	32-34	10-12
SAW with single (1-in. extension) and tandem (5-in. extension) electrodes(a)				
1	First filler (single electrode)	400-450 (DCEP)	33-35	20-21
	Second filler (leading electrode)	550 (DCEP)	35-38	19
	(trailing electrode)	500 (ac)	38-42	(tandem)
	Remaining (leading electrode)	650 (DCEP)	35-38	22
	(trailing electrode)	550 (ac)	38-42	(tandem)
2	1 to 2 (single electrode)(b)	700-725 (DCEP)	35-37	10-12

(a) Joint was sealed from side 1 by a root pass made by SMAW, using a $^1/_8$-in.-diam E7018 electrode at 110 to 140 A, 18 to 20 V. Filler passes were made by SAW, using $^5/_{32}$-in.-diam EM12K electrode wire and F61 flux. Welding was done in the flat position. Power supplies were an 800-A motor-generator for the single electrodes and for the leading tandem electrode, and 1000-A transformer for the trailing tandem electrode. Preheating consisted of heating the vessel to 175 °F minimum. Postweld heating consisted of holding the vessel at 1150 °F minimum for $2^1/_2$ h, then furnace cooling it to 60 °F. (b) After air carbon arc backgouging approximately $^3/_8$ in. deep to reach sound metal

Fig. 66 Submerged arc welding application. See text for details.

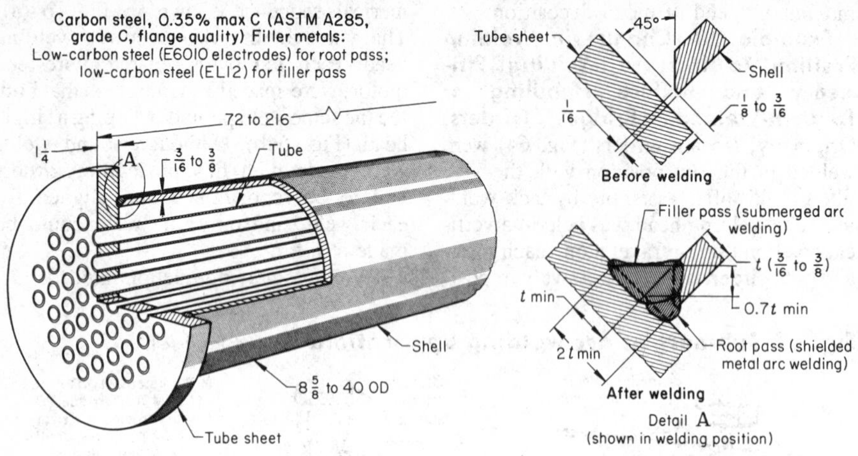

	Root pass	Filler pass
Welding process	SMAW	Automatic SAW
Power supply	400-A motor-generator(a)	600-A motor-generator(a)
Electrode	$^1/_8$-in. E6010	$^5/_{32}$-in.-diam EL12
Flux	...	F72
Electrode extension, in.	...	$^1/_2$-$^3/_4$
Welding position	Flat	Flat
Welding voltage	24-28	32-34
Welding current (DCEP), A	80-120	300-350
Arc length, in.	$^1/_8$-$^3/_{16}$	$^3/_8$-$^1/_2$
Bead size, in. (approx)	$^1/_4$	See figure
Welding speed, in./min	7-8	10-15
Number of passes(b)	One	One(c)
Deposition rate, lb/ft	...	1.2
Flux consumption, lb/ft	...	$^1/_2$ (approx); flux depth, $^3/_4$ in.

(a) Constant-current type. (b) Interpass temperature, estimated to be 400 to 500 °F. (c) For shells $^3/_{16}$ to $^3/_8$ in. thick, as illustrated; additional passes were used on thicker shells.

der was oriented so that two welds could be made simultaneously in the two horizontal positions shown in Fig. 64(c) and (d). For horizontal-position welding, the positions of the electrodes were changed so that the fillet welds would be of proper shape. First, the two electrode wires were spaced 4$^1/_2$ in. apart, so that the bead from the leading wire solidified before the second bead was deposited; welding thus became a multiple-pass operation. This was necessary because gravity pulled the molten metal toward the horizontal member of the girder and the proper shape for the weld bead could not be developed from a single weld pool. In addition, the trailing electrodes were positioned as shown in Fig. 64(c) and (d) to develop the upper portion of the fillet welds. The two-layer weld deposits built up in these horizontal positions are shown in Fig. 64(e).

With the reduction in handling time from that required when one weld at a time was deposited in flat-position welding, and with the increase in deposition rate obtained with the "two-pass" welding effected by the use of spaced electrodes, the time for welding each girder was reduced 65% by welding in the horizontal position.

Improvements in equipment and technique have made it possible to use three, and sometimes four, electrodes in the horizontal position, depending on fillet size and code requirements.

Example 21. Deposition Rates in Welding With a Single Electrode, and With Tandem Electrodes Using 1-In. and 5-In. Electrode Extensions. A cylindrical, high-pressure steam vessel, shown at upper left in Fig. 65, was roll formed. The joint edges were then machined or torch cut to make a 60° groove (see section A-A in Fig. 65), the joint was fitted up, and the sealing root pass was welded using SMAW.

Automatic SAW was then used to complete the joint from the outside. The joint was turned over and the root-pass bead was air carbon arc gouged from the inside to expose sound submerged arc weld metal, and the resulting groove was then submerged arc welded to produce a smooth inside joint.

When single-electrode SAW was used, total arc time required per seam was 8.6 h. To reduce this arc time, two other SAW techniques were evaluated. In both techniques, following a single-electrode first filler pass, tandem-arc welding was employed for the first side of the joint, using a lead electrode with DCEP and a trailing electrode with alternating current. The difference lay in the length of the electrode extension: in one method it was 1 in. on both electrodes; in the other it was 5 in. on both electrodes, using extension guides.

For the tandem-arc arrangement with 1-in. electrode extension, the total arc time for the joint was 4.6 h, or about one half the time consumed by single-electrode welding. For tandem-arc welding with 5-in. electrode extension, arc time per joint was reduced to 2.8 h, because of the increased resistance heating of the wire that resulted from the greater electrode extension.

Example 22. Joint Design and Welding Procedure Without Backing Ring. Fabrication of heat exchangers for condensing or evaporating refrigerant involved welding the heat-transfer chamber directly to the tube sheet (Fig. 66). The chamber consisted of a cylindrical open-top shell and, when welded to the tube sheet, formed a pressure vessel completely enclosing the tube bundle.

Because the inside of the vessel was inaccessible both for welding and for weld inspection, a major problem was that positive assurance could not be obtained that complete joint penetration had been effected, as required by Section VIII of the ASME Boiler and Pressure Vessel Code. When the entire weld was made by SAW, complete penetration without melt-through was not obtained consistently when the joint was prepared by grooving the tube sheet for a single-V-groove and fillet weld. The use of a backing ring was considered, but was ruled out because it would necessitate a reduction in the number of tubes in the bundle.

After experiments with several joint designs, a two-step welding procedure, using the joint shown in detail A of Fig. 66, was adopted. This design called for a single-bevel-groove weld with a fillet-weld reinforcement equal to the shell thickness.

The two-step procedure consisted of the deposition of a root bead by SMAW, followed by a single filler pass by automatic SAW. The welding procedure, as well as the manual welder and the SAW operator, had to be qualified in accordance with the rules of Sections VIII and IX of the ASME Boiler and Pressure Vessel Code. Melt-through and effective penetration were brought under control only by carefully following the welding procedures. Weld-

Fig. 67 Bull-gear blank welded by automatic SAW, SMAW, and semiautomatic FCAW to improve quality and reduce costs

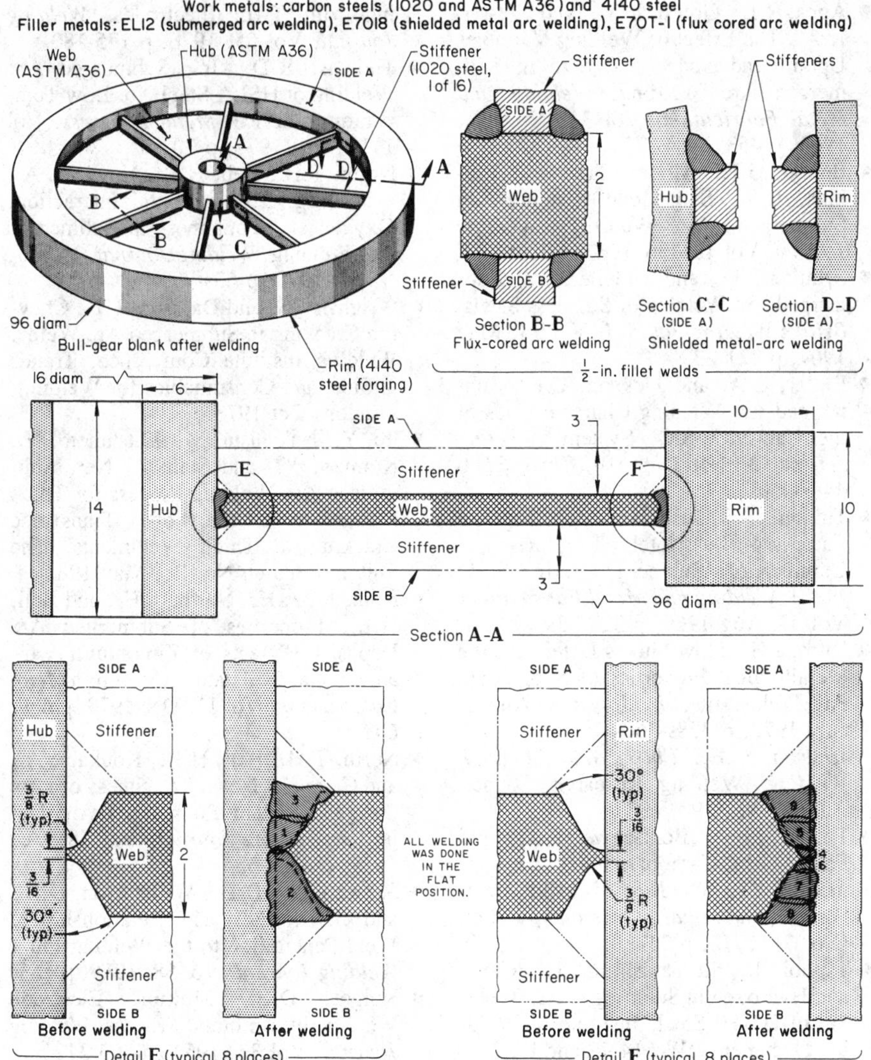

Welding condition	SAW(a)	SMAW(b)	FCAW(c)
Power supply	40-V, 1200-A rectifier(d)	400-A rectifier	40-V, 500-A rectifier(d)
Electrode wire	1/8-in.-diam EL12	3/16-in.-diam E7018	1/8-in.-diam E70T-1
Shielding gas	...	...	Carbon dioxide
Flux	F72	...	...
Current (DCEP), A	380-420	160	450
Voltage, V	30-32	25	30-32
Welding speed, in./min	16-24	...	12-14
Preheat temperature, rim	450 °F	400 °F	450 °F
Preheat temperature, hub	300 °F	...	...

Note: After welding, the gear blank was stress relieved at 1050 °F for 6 h.
(a) Used for all welding operations in joining the web to the hub (operations 1, 2, and 3; see detail E), and for filler passes in joining the web to the rim (operations 5, 7, 8, and 9; see detail F). (b) Used for root passes in joining the web to the rim (operations 4 and 6; see detail F), and for tacking and welding stiffeners to hub and rim (sections C-C and D-D). (c) Used for joining stiffeners to web (section B-B). (d) Constant-voltage type

ers had to be specially trained to make the root pass satisfactorily. During SAW, the electrode had to be carefully positioned to obtain proper sidewall fusion and to avoid undercutting.

Before welding, the joint areas were cleaned of scale and oil. Shell sheets had been sheared to size and beveled by oxyacetylene cutting before forming. In assembling the shell on the tube sheet, spacers were used to obtain the desired root opening and were removed after tack welding. No preheat or postheat treatment was used, normal room temperature (70 °F) being the only requirement. The assembly was mounted on a conventional welding positioner for flat-position welding. After the root pass had been deposited, slag was removed by a needle gun, which consisted of a bundle of impact-driven steel quills capable of conforming to and abrading uneven surfaces. Surfaces were ground where necessary.

Automatic SAW was performed from the carriage of a conventional manipulator on the same positioner and in the same position as the manual welding. Interpass temperature for this weld was 400 to 500 °F.

After welding, the weld surface was cleaned by a needle gun, and the vessel was then tested for leaks by pressurizing it with dry air at 300 psig while submerged in water. Dry air was used because the interior of the vessel had to be dry for operation with refrigerants.

Example 23. Submerged Arc Welding of Large Bull-Gear Blanks (Fig. 67). A large bull gear for a steel-mill drive was machined from the welded blank shown in Fig. 67. The welded blank was made up of the following components: an AISI 4140 steel rim, which had been produced as a roll forging and rough machined; a hub, which had been gas cut from a 14-in.-thick slab of carbon steel (ASTM A36) and then rough machined on the outside diameter; 16 stiffeners, which had been sheared or gas cut from 1-by-3-in. 1020 steel bar stock, with one edge ground to facilitate fit-up; and a web, which had been gas cut from 2-in.-thick carbon steel (ASTM A36) plate. Double-J-grooves were machined into the edges of the web where it was to be joined to the hub and the rim. The sequence of operations for welding the bull-gear blanks was as follows:

Tack weld

- Fit and tack web to hub.
- Fit and tack web to rim (rim was preheated before welding).

Join hub to web by automatic SAW (Fig. 67)

- Preheat hub to 300 °F.
- Weld joint on side A half full (Fig. 67).
- Turn workpiece over and backgouge to sound metal.
- Weld joint on side B full (operation 2, detail E, Fig. 67).
- Turn workpiece over; complete weld on side A (operation 3, detail E, Fig. 67).

Join rim to web by SMAW for root passes and SAW for filler passes (Fig. 67)

- Preheat the rim locally to 400 °F.
- Weld root pass on side A by SMAW (operation 4, detail F, Fig. 67).
- Inspect root pass on side A (magnetic-particle inspection).
- Weld joint on side A half full (operation 5, detail F, Fig. 67); follow with magnetic-particle inspection.
- Turn the workpiece over and back-gouge to sound metal.
- Weld root pass on side B by SMAW (operation 6, detail F, Fig. 67).
- Inspect root pass on side B (magnetic-particle inspection).
- Weld side B half full (operation 7, detail F, Fig. 67); magnetic-particle inspect.
- Weld side B full (operation 8, detail F, Fig. 67); magnetic-particle inspect.
- Turn workpiece over; complete weld on side A (operation 9, detail F, Fig. 67); magnetic-particle inspect.

Fit, tack, and weld stiffeners to web by FCAW
Fit, tack, and weld stiffeners to hub and rim by SMAW

The welded blank was covered with insulation blankets, allowed to cool slowly, and then stress relieved at 1050 °F for 6 h. After cooling, the hub and rim joints were 100% ultrasonically inspected.

SELECTED REFERENCES

- U.S. Patents, 2,043,960; 2,200,737; 2,360,716; 2,474,787; 2,681,875; 2,755,211; 2,895,863; 3,078,193; 3,340,105
- Christensen, N. and Chipman, J., *Slag-Metal Interaction in Arc Welding,* Welding Research Council, Bulletin No. 15, Jan 1953
- Jackson, C.E. and Shrubsall, A.E., Submerged Arc Welding of Chromium-Bearing Steels, *Welding Journal,* Vol 33, 1954, p 752-758
- Jackson, C.E., The Science of Arc Welding, *Welding Journal,* Vol 39, 1960, p 129s-140s, 177s-190s, 225s-230s
- Robinson, M.H., Observations of Electrode Melting Rates During Submerged Arc Welding, *Welding Journal,* Vol 40, Nov 1961, p 503s
- Lewis, W.T., Faulkner, G.E., and Rieppel, P.J., Flux and Filler-wire Developments for Submerged Arc Welding HY80 Steel, *Welding Journal,* Vol 40, Aug 1961, p 337s-345s
- Kubli, R.A. and Sharav, W.B., Advancement in Submerged Arc Welding of High-Impact Steels, *Welding Journal,* Vol 40, Nov 1961, p 497s-502s
- Apps, R.L., Gourd, L.M., and Nelson, K.A., The Effect of Welding Variables Upon Bead Shape and Size in Submerged Arc Welding, *Welding and Metal Fabrication,* Vol 31 (No. 10), 1963, p 453
- Belton, G.R., Moore, T.J., and Tankins, E.S., Slag-Metal Reactions in Submerged Arc Welding, *Welding Journal,* Vol 42, July 1963, p 289s-297s
- Schulten, D. and Mueller, P., Submerged Arc Welding of Stainless Steels, *British Welding Journal,* Vol 14, May 1967, p 221-232
- Butler, C.A. and Jackson, C.E., Submerged Arc Welding Characteristics of the $CaO\text{-}TiO_2\text{-}SiO_2$ System, *Welding Journal,* Vol 46 (No. 10), Oct 1967, p 448s-456s
- Tuliani, S.S., Boniszewski, T., and Eaton, N.F., Notch Toughness of Commercial Submerged Arc Weld Metal, *Welding and Metal Fabrication,* Vol 37, Aug 1969, p 327-339
- Palm, J.H., How Fluxes Determine the Metallurgical Properties of Submerged Arc Welds, *Welding Journal,* Vol 52, July 1972, p 358s-360s
- Jackson, C.E., *Fluxes and Slags in Welding,* Welding Research Council Bulletin No. 190, 1977
- Tuliani, S.S., Boniszewski, T., and Eaton, N.F., Carbonate Fluxes for Submerged Arc Welding of Mild Steel, *Welding and Metal Fabrication,* Vol 40 (No. 7), 1972, p 247
- Tsuboi, J. and Terashima, H., Crack and Hydrogen in Submerged Arc Welding of HT-80 Steel, Institute of Welding Document IIW-614-72 and IX-794-72
- Garland, J.G. and Kirkwood, P.R., Toward Improved Submerged Arc Weld Metal, *Metal Construction,* Vol 7, 1975, p 275-283 (May), 320-330 (June)
- Ferrera, K.P. and Olson, D.L., $MnO\text{-}SiO_2\text{-}CaO$ Welding Flux System, *Welding Journal,* Vol 54, 1975, p 211
- Wittstock, G.G., Selecting Submerged Arc Fluxes for Carbon and Low Alloy Steels, *Welding Journal,* Vol 55, Sept 1976, p 733-741
- Garland, J.G. and Kirkwood, P.R., A Reappraisal of the Relationship Between Flux Basicity and Mechanical Properties in Submerged Arc Welding, *Welding and Metal Fabrication,* Vol 44, April 1976, p 217-224
- Renwick, B.G. and Patchett, B.M., Operating Characteristics of the Submerged Arc Process, *Welding Journal,* Vol 55, March 1976, p 69s-76s
- Hickel, J.E. and Forsthoefel, F.W., High Current Density Submerged Arc Welding with Twin Electrodes, *Welding Journal,* Vol 55, 1976, p 175-180
- Thomas, R.D., Jr., Submerged Arc Welding of HSLA Steels for Low Temperature Service, *Metal Progress,* Vol 111, April 1977, p 30-63
- North, T.H., Bell, H.B., Nowicki, A., and Craig, I., Slag/Metal Interaction, Oxygen and Toughness in Submerged Arc Welding, *Welding Journal,* Vol 57, March 1978, p 63s-75s
- Wright, V.S. and Davison, I.T., Chevron Cracking in Submerged Arc Welds, Welding Institute Conference, Trends in Steel and Consumables for Welding, London, Oct 1978
- Ito, Y., Nakanishi, M., Katsumoto, N., Komizo, Y., and Seta, I., New Submerged Arc Welding Process for Thick Plates—Prevention of Transverse Cracking in Thick Weldments, The Suitomo Search No. 25, May 1981
- Koukabi, A.H., North, T.H., and Bell, H.B., Properties of Submerged Arc Deposits—Effects of Zirconium, Vanadium, and Titanium/Boron, *Metal Construction,* Vol 11, Dec 1979, p 639-642
- North, T.H., Bell, H.B., Koukabi, A., and Craig, I., Notch Toughness of Low Oxygen Content Submerged Arc Deposits, *Welding Journal,* Vol 58, Dec 1979, p 343s-354s
- Schwemmer, D.D., Williamson, D.L., and Olson, D.L. The Relationship of Weld Penetration to the Welding Flux, *Welding Journal,* Vol 58, 1979, p 153s
- Knight, D.E., Multiple Electrode Welding by Uniomelt Process, *Welding Journal,* Vol 33, 1954, p 303-312
- Kubli, R.A. and Shrubsall, H.I., Multipower Submerged-Arc Welding of Pressure Vessels and Pipe, *Welding Journal,* Vol 35, Nov 1956, p 112s-113s
- Hinkel, J.E. and Forsthoefel, F.W., High Current Density Submerged-Arc Welding with Twin Electrodes, *Welding Journal,* Vol 55, March 1976, p 175-180
- Davis, M.L.E. and Bailey, N., International Conference, Trends in Steel and Consumables for Welding, Paper No. 19, Welding Institute, London, 1978
- Davis, M.L.E. and Coe, F.R., Welding Institute Research Report No. 39/1977/M, 1977
- Billy, J., Johansson, T., Loberg, B., and Easterling, K.E., Stress-Relief Heat Treatment of Submerged Arc Welded Microalloyed Steels, *Metal Technology,* Vol 7, Feb 1980, p 67-78

Gas Metal Arc Welding (MIG Welding)

By the ASM Committee on Gas Metal Arc Welding*

GAS METAL ARC WELDING (GMAW), which often is called MIG (metal inert gas) welding, is an arc welding process in which the heat for welding is generated by an arc between a consumable electrode and the work metal. The electrode, a bare solid wire that is continuously fed to the weld area, becomes the filler metal as it is consumed. The electrode, weld pool, arc, and adjacent areas of the base metal are protected from atmospheric contamination by a gaseous shield provided by a stream of gas, or mixture of gases, fed through the welding gun. The gas shield must provide full protection, because even a small amount of entrained air can contaminate the weld deposit.

Gas metal arc welding overcomes the restriction of using an electrode of limited length, as in shielded metal arc welding (SMAW), and overcomes the inability to weld in various positions, which is a limitation of submerged arc welding (SAW).

Gas metal arc welding is widely used in semiautomatic, machine, and automatic modes. In semiautomatic welding, the most popular method of applying this process, the welder guides the gun along the joint and adjusts the welding conditions. The wire feeder continuously feeds the filler-wire electrode, and the arc length is maintained by the power source. In automatic GMAW, the machinery controls the welding parameters, arc length, joint guidance, and wire feed, observed by the operator. Machine GMAW has only limited popularity and is characterized by machine control of the arc length, wire feed, and joint guidance. The operator adjusts the welding parameters.

Metals Welded. The GMAW process was first applied to the welding of magnesium and aluminum alloys and stainless steels, because it was often the only method by which satisfactory welds could be produced at an economical rate. The nature of the process permits its use for welding most metals and alloys. However, some metals are more adaptable than others, and there are a few that cannot be welded. Those metals most easily welded by GMAW include carbon and low-alloy steels, stainless steels, heat-resistant alloys, aluminum and aluminum alloys of the 3000, 5000, and 6000 series, copper and copper alloys other than the high-zinc alloys, and magnesium alloys.

Metals that can be welded by GMAW but that may require special procedures are high-strength steels, aluminum alloys of the 2000 and 7000 series, copper alloys that contain high percentages of zinc (such as manganese bronze), cast iron, austenitic manganese steel, titanium and titanium alloys, and refractory metals. Gas metal arc welding of these metals may require preheating or postheating of the base metal, use of special filler metals, closer than normal control of shielding gas, and use of backing gas.

Metals that have a low melting temperature or a low boiling temperature are not amenable to GMAW (or any other arc welding process). Lead, tin, and zinc are typical of this group. Zinc, for instance, boils at 1665 °F, which is far below the arc temperature, and it produces toxic fumes, as does lead.

High-melting-point metals that are coated with a low-melting-point metal (such as lead, tin, cadmium, or zinc) are difficult or impossible to weld satisfactorily, because the welding heat causes fuming of the coating or alloying with the base metal, or both—the result of which is welds with poor mechanical properties. When coated metals are to be welded, the coating first should be thoroughly removed from the joint areas. Postweld repair of the coating over the weld area is essential if the protection afforded by the coating is required at this location.

Although aluminum and its alloys are readily welded by GMAW, aluminum-coated steel may cause difficulty in welding, because aluminum vaporizes at a low temperature on aluminum-coated steel. For further information and examples of practice that deal with GMAW of metals other than low-carbon steel, see the articles in this Volume on welding of specific metals.

Welding Position. Gas metal arc welding is an all-position welding technique. As in other welding processes, welding is most efficient in the flat and horizontal positions, but GMAW is at least equal to SMAW in the other positions.

Polarity. Most GMAW applications require the use of direct current electrode positive (DCEP), which is also referred to as reverse polarity. This type of electrical connection provides a stable arc, smooth metal transfer, relatively low spatter loss,

*Daniel Hauser, *Chairman*, Senior Research Metallurgist, Battelle Columbus Laboratories; Charles W. Bosworth, Senior Specialist Engineer, Boeing Engineering & Construction Co.; Kevin A. Lyttle, Laboratory Division Head, Linde Division, Union Carbide Corp.; Thomas A. Nevitt, II, Division Welding Engineer, McDermott International, Fabricators Division; Abe Pollack, Head, Joining Branch, David Taylor Naval Ship Research & Development Center; John Springer, Principal Engineer, Boeing Commercial Airplane Co.

and good weld-bead characteristics for the entire range of welding currents used.

Direct current electrode negative (DCEN), which is also referred to as straight polarity, is seldom used, because the arc can become very unstable and erratic even though the electrode melting rate is higher than that achieved with DCEP. Penetration is lower with DCEN than with DCEP.

Alternating current is not normally used with GMAW for two reasons: (1) the arc is extinguished during each half cycle as the current reduces to zero, and it may not re-ignite if the cathode cools sufficiently; and (2) rectification of the reverse polarity cycle promotes erratic arc operation.

Advantages. High productivity, reliability, all-position capability, wide area of application, ease of use, and low cost are all qualities inherent in GMAW. Higher productivity is achieved through the greater deposition efficiency of GMAW—92 to 98% of each pound of wire becomes deposited weld metal, compared with 60 to 70% for SMAW; the small quantity of slag on the bead surface of a gas metal arc weld means less time is spent in interpass cleaning; and the duty cycle of semiautomatic and automatic GMAW is 35 to 50%, compared with 20 to 30% for SMAW.

The GMAW process reliably produces weld metal with excellent mechanical properties—particularly good low-temperature impact strength and elongation are developed using a variety of shielding gases. Resistance to hydrogen-induced cracking is inherent in the process due to the nonhygroscopic nature of the wire and other components of the welding system. Diffusible weld-metal hydrogen levels of 1 to 2 mL of hydrogen per 100 g of deposited metal makes GMAW a reliable choice for joining high-strength steels where hydrogen control is of great concern.

All ferrous and most nonferrous materials can be joined using GMAW. A wide range of material thicknesses can be joined in all positions by selecting the right mode of metal transfer from among those possible with GMAW. The potential for deep weld joint penetration and more consistent root penetration with GMAW can sometimes allow the use of a smaller fillet weld size than when SMAW is used. This can make GMAW a better choice for welding thinner materials, because lower currents and higher travel speeds can be used more effectively.

A visible arc and a minimum amount of slag make it easy for an operator to place the weld metal where it is needed. Operator training is generally easy with GMAW. Minimal smoke levels with all GMAW process variations enhance operator comfort and reduce ventilation and air-processing equipment needs. High deposition efficiency and the ability to achieve high operator duty cycle make GMAW the welding process of choice where minimum cost is a major requirement. These factors, in addition to a generally low cost per pound for a wire electrode, make substantial savings over SMAW easily achievable.

Disadvantages or limitations of GMAW compared with SMAW include:

- Equipment for GMAW is more complex and consequently is more costly and less transportable.
- In GMAW, the welding gun must be close to the work, which makes the process less adaptable than SMAW for welding in difficult-to-reach areas.
- Gas metal arc welding requires positive protection from strong drafts that blow the stream of shielding gas away from the weld; for this reason, GMAW may be less practical than SMAW for welding outdoors.

Principles of Operation

Gas metal arc welding uses the heat of an electric arc produced between a bare electrode and the part to be welded. The electric arc is produced by electric current passing through an ionized gas. Gas atoms and molecules are broken up and ionized by losing electrons and leaving a positive charge. The positive gas ions then flow from the positive pole to the negative pole, and the electrons flow from the negative pole to the positive pole. About 95% of the heat is carried by the electrons, and the rest is carried by the positive ions. The heat of the arc melts the surface of the base metal and the electrode. The molten weld metal, heated weld zone, and electrode are shielded from the atmosphere by a shielding gas that is supplied through the welding gun. The molten electrode filler metal transfers across the arc and into the weld pool, producing an arc with more intense heat than most of the arc welding processes.

The arc is struck by starting the wire feed, which causes the electrode wire to touch the workpiece and initiate the arc. Normally, arc travel along the work is not started until a weld pool is formed. The gun then moves along the weld joint manually or mechanically so that the adjoining edges are welded. The weld metal solidifies behind the arc in the joint and completes the welding process.

Essential requirements for GMAW are: (1) a power supply that provides sufficient voltage to push the current across the gap to make the arc, and sufficient current to melt the electrode to make the weld deposit; (2) a wire feeder that continuously advances the electrode as it melts; (3) a smooth flow of shielding gas; and (4) a welding gun that carries the current, electrode wire, shielding gas and, depending on the gun design, cooling water.

These essentials are illustrated schematically in Fig. 1. The wire-feed system is of the constant-speed, push type, by means of which a specific rate of wire feed can be obtained. The power supply is a constant-voltage type that will maintain any desired voltage output within its capability, regardless of current flow. Under these conditions, only enough current will flow to melt the electrode wire at a rate equal to that of the wire feed.

Metal Transfer

The type of arc obtainable in GMAW is identified by the mode of metal transfer. These modes of transfer are commonly referred to as spray, globular, short circuiting, and pulsed.

Spray Transfer. In this mode, metal is transferred from the end of the electrode wire to the pool in an axial stream of fine droplets. This condition is illustrated in Fig. 2(a). These small droplets emanate from the tapered end of the electrode; one droplet follows another, but they are not connected. The size of the droplets may vary, but in a spray arc the maximum diameter is less than that of the electrode wire.

The spray arc occurs at high current density, generally with argon or an argon-rich shielding gas. A true spray arc cannot be obtained with a shielding gas composed of more than 10 to 15% carbon dioxide. The spray transfer mode gives high heat input, maximum penetration, and a high deposition rate. In welding steel, it generally is limited to welding in the flat position and the horizontal fillet position. This mode also produces the least amount of spatter.

Globular transfer occurs at lower current densities and is characterized by the formation of a relatively large drop of molten metal at the end of the tapered electrode wire (Fig. 2b). The drop forms at the end of the electrode wire until the force of gravity overcomes the surface tension of the molten drop, at which time the drop falls into the weld pool. Globular transfer occurs with all types of shielding gases, but it cannot be used for out-of-position welding. If overhead welding were attempted using globular transfer, the

Fig. 1 Equipment for semiautomatic GMAW

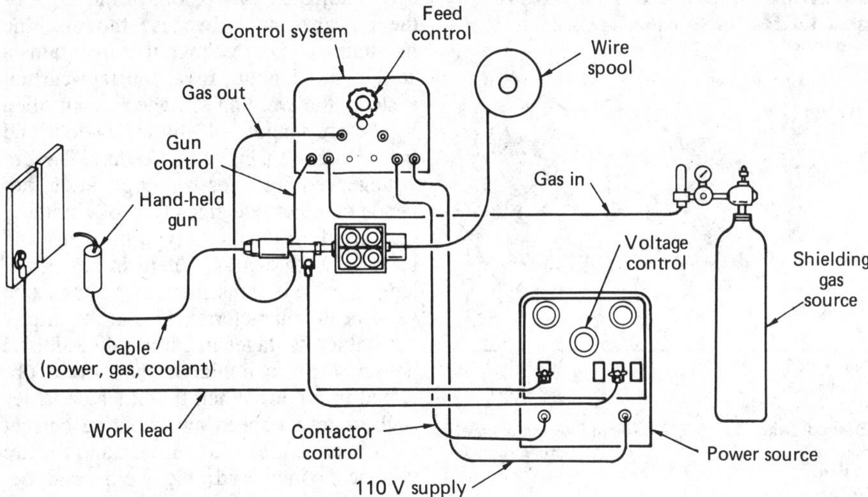

molten electrode metal would fall downward into the welding gun nozzle.

Short circuiting transfer is used in many applications of GMAW. It is especially well adapted to welding thin sections, because heat input is low; it is less often used for welding thick sections. This mode of transfer permits welding in any position and occurs with carbon dioxide, argon-carbon dioxide mixtures, and helium-based shielding gases.

Steps that occur with the short circuiting mode of transfer are shown in Fig. 2. At the start of the short circuiting arc cycle, the end of the electrode wire melts into a small globule of liquid metal (Fig. 2c). Next, the molten metal moves toward the workpiece, taking the form shown in Fig. 2(d). Then the molten metal makes contact with the workpiece, creating a short circuit. At this stage of the cycle, metal transfer is by gravity and surface tension, and the arc is extinguished (Fig. 2e). Finally, the molten metal bridge is broken by pinch force, the squeezing action common to current carriers. The amount and suddenness of pinch is controlled by the power supply. At this stage, the electrical contact is broken, and the arc is re-ignited (Fig. 2f). With the arc renewed, the cycle begins again. Frequency of arc extinction and re-ignition varies from 20 to 200 times per second, in accordance with preset electrical conditions.

Figure 2(c) to (f) shows the low-current short circuiting arc. As the current density is increased, the molten metal transfer moves from a high-frequency rate of short circuits to a lower rate of short circuits, detaching much larger molten metal drops. At the higher current densities and normal arc voltages, metal transfer is much more violent. Short circuiting transfer is usually associated with lower current density and a rather exact arc-voltage setting. If, for example, a 0.035-in.-diam steel electrode wire is fed at a rate requiring 120 A of current, and carbon dioxide shielding gas is used, an excellent short circuiting mode of transfer requires arc voltage of about 19 to 20 V. If the arc voltage is increased to 25 to 26 V, the short circuiting mode of transfer reverts to a mild type of globular transfer. Depth of penetration decreases as current density decreases, except for a minor effect on penetration from arc voltage. A decrease in arc voltage can result in a slight increase in penetration, because the decreased arc voltage with carbon dioxide shielding is accompanied by a shorter arc length.

Pulsed current transfer is a spray-type transfer that occurs in pulses at regularly spaced intervals rather than at random intervals. In the time interval between pulses, the welding current is reduced and no metal transfer occurs.

The pulsing action is obtained by combining the outputs of two power supplies working at two current levels. One acts as a background current to preheat and precondition the advancing continuously fed electrode; the other power supply furnishes a peak current for forcing the drop from the electrode to the joint being welded. The peaking current is a half-wave direct current; because it is tied into line frequency, drops will be transferred from the electrode to the joint 60 to 120 times per second.

Electrode-wire diameters of 0.045 to $^{1}/_{16}$ in. are most commonly used. The pulsed arc mode is capable of welding thinner sections than are practical with conventional spray transfer, because heat input is less. Also, because of the low heat input, distortion is less. The pulsed arc mode of transfer is suited to all welding positions.

Rotating arc transfer is a spray-type transfer where the current, voltage, and electrode extension are increased beyond the levels used to obtain conventional spray transfer. Under these conditions, the lower part of the electrode becomes molten over a considerable length and rotates in a helical spiral under the influence of the magnetic field surrounding the arc. As it rotates, a controlled stream of droplets is transferred from the electrode tip to the

Fig. 2 Modes of metal transfer in GMAW. (a) Spray transfer. (b) Globular transfer. (c), (d), (e), and (f) Steps in short circuiting transfer

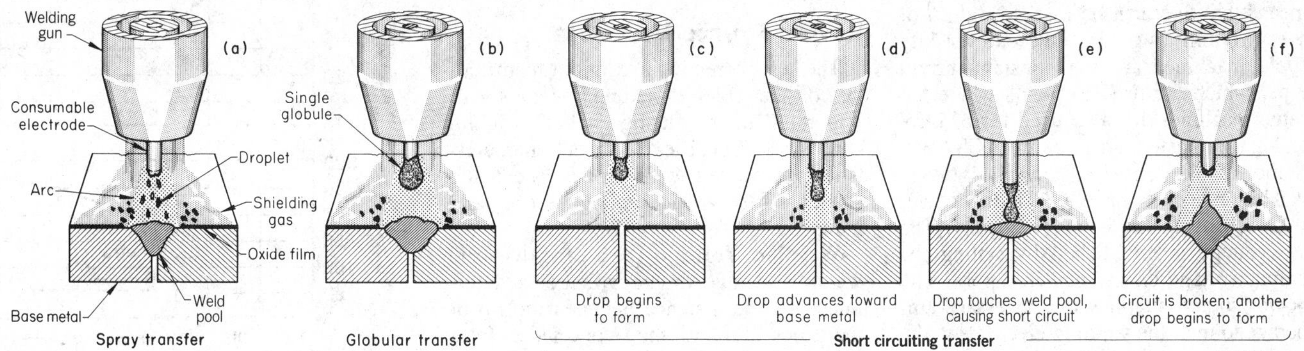

weld pool over a relatively wide area. The arc energy is spread over a wider area, minimizing the fingerlike pattern associated with spray arcs.

The current necessary to induce the rotational mode of spray transfer increases rapidly with wire diameter and decreases with longer electrode extension. Generally, smaller diameter wires (0.035 to 0.045 in. diam) with $^3/_4$- to $1^1/_4$-in. extension are used to keep currents at easily usable levels (320 to 380 A). Deposition rates of 20 to 25 lb/h are achieved with this process in the flat position. One- and two-pass heavy fillet welds are a major area of application for this process variation.

Equipment

The equipment for GMAW consists of a power source, controls, wire feeder, welding gun, welding cables, and a gas shielding system. Accessory equipment such as traversing mechanisms and seam followers can be added to automatic applications.

Power Sources

Alternating current is seldom used in GMAW. Direct current electrode positive is used for most applications, although DCEN is sometimes used when penetration must be minimal. Figure 3 compares the depths of penetration obtained in welding with DCEN and DCEP under otherwise identical conditions.

The GMAW process uses power sources similar to those used with other continuous electrode-feed welding processes, such as flux cored arc welding (FCAW) and SAW. Many types of direct current power sources may be used, including rotating (generator) or static (single- or three-phase transformer-rectifier) welding machines. Any of these types of machines are available to produce constant current, constant voltage output, or both.

Most power sources used for GMAW have a duty cycle of 100%, which indicates that they can be used to weld continuously. Some machines used for this process have duty cycles of 60%, which means that they can be used to weld 6 of every 10 min. In general, these lower duty cycle machines are the constant-current type, which are used in plants where the same machines are also used for SMAW and gas tungsten arc welding (GTAW). Some of the smaller constant voltage welding machines have a 60% duty cycle.

Constant-Voltage Power Supply. The arc voltage is established by setting the output voltage on the power supply. The power source supplies the necessary amperage to melt the welding electrode at the

Fig. 3 Depths of penetration obtained in GMAW with DCEP and DCEN. Under otherwise identical conditions

Low-carbon steel; low-carbon steel filler metal

0.09 — Weld metal — DCEN — DCEP

Joint type	T
Weld type	Fillet
Welding position	Horizontal
Electrode wire	$^1/_{32}$-in.-diam low-carbon steel
Shielding gas	Carbon dioxide
Current	120 A (dc)
Voltage	19 V
Melting rate	280 in./min

rate required to maintain the preset voltage (or relative arc length). The speed of the electrode drive is used to control the average welding current. This characteristic generally is preferred for welding all metals. The use of a constant-voltage power supply along with a constant wire-feed rate results in a self-correcting arc length system.

Constant-Current Power Supply. With this type of power supply, the welding current is established by the appropriate setting. Arc length (voltage) is controlled by the automatic adjustment of the wire-feed rate. This type of welding is best suited to large-diameter electrodes and machine or automatic welding, where very rapid change of wire-feed rate is not required. Most constant-current power sources have a drooping volt-ampere output characteristic. However, true constant-current machines are available. Constant-current power sources are not normally selected for GMAW because of the greater control needed for wire-feed speed. The systems are not self-regulating.

Power Source Variables

The self-correcting arc length feature of the constant-voltage welding system is very important in producing stable welding conditions. Specific electrical characteristics are needed to control arc heat and spatter. These include voltage, slope, and inductance.

Welding Voltage (Arc Length). The welding voltage or arc voltage is determined by the distance between the tip of the electrode and the workpiece. In a constant-voltage system, the welding voltage is adjusted by a knob on the front of the power source, because the machine maintains a given voltage that maintains a certain arc length. In a constant-current system, the welding voltage is controlled by the arc length held by the welder and the voltage-sensing wire feeder. The arc voltage required for an application depends on electrode size, type of shielding gas, welding position, type of joint, and base-metal thickness. There is no set arc length that will consistently give the same weld-bead characteristics. For example, normal arc voltages in carbon dioxide and helium are much higher than those obtained in argon. When the other variables such as travel speed and welding current (in amperes) are held constant and the arc voltage is increased, the weld bead becomes flatter and wider. Penetration will increase up to an optimum voltage level and then begin to decrease, as shown in Fig. 4. A higher voltage is often used to bridge a gap because of the decreased penetration obtained. An excessively high arc voltage causes excessive spatter, porosity, and undercutting. A decrease in arc length produces a narrower weld bead with a greater convexity and, down to the optimum voltage level, deeper penetration. An excessively low arc voltage may cause porosity and overlapping at the edges of the weld bead.

For any power source, the voltage drop across the welding arc is directly dependent on arc length. An increase in arc length

Fig. 4 Effect of travel speed, arc volts, and welding current on penetration

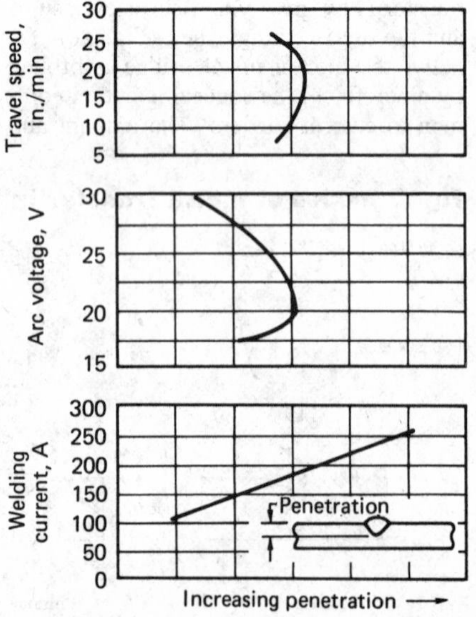

results in a corresponding increase in arc voltage, and a decrease in arc length results in a corresponding decrease in arc voltage. Another important relationship exists between welding current and the melt-off rate of the electrode. With low current, the electrode melts off more slowly and the metal is deposited more slowly. This relationship between welding current and wire-feed speed is definite, based on wire size, shielding gas, and type of filler metal used. A faster wire-feed speed will give a higher welding current. In the constant-voltage system, instead of regulating the wire to maintain a constant arc length, the wire is fed into the arc at a fixed speed, and the power source is designed to melt off the wire at the same speed. The self-regulating characteristic of a constant-voltage power source is shown by the ability of this type of power source to adjust its welding current to maintain a fixed voltage across the arc. With the constant-current arc system with a voltage-sensing wire feeder, the welder changes the wire-feed speed as the gun moves toward or away from the weld pool. Because the welding current remains the same, the melt-off rate of the wire is unable to compensate for the variations in the wire-feed speed, which allows stubbing or burning back of the wire into the contact tip to occur. To lessen this problem, a special voltage-sensing wire feeder is used that regulates the wire-feed speed to maintain a constant voltage across the arc. The constant-voltage system is preferred for most applications, particularly for small-diameter wire. With smaller diameter electrodes, the voltage-sensing system often is not able to react fast enough to feed at the required melt-off rate, resulting in a higher instance of meltback into the contact tip of the gun.

Figure 5 shows a comparison of the volt-ampere curves for various arc lengths. For these particular curves, when a normal arc length is used, the current and voltage level is the same for both the constant-current

Fig. 5 Volt-ampere curves

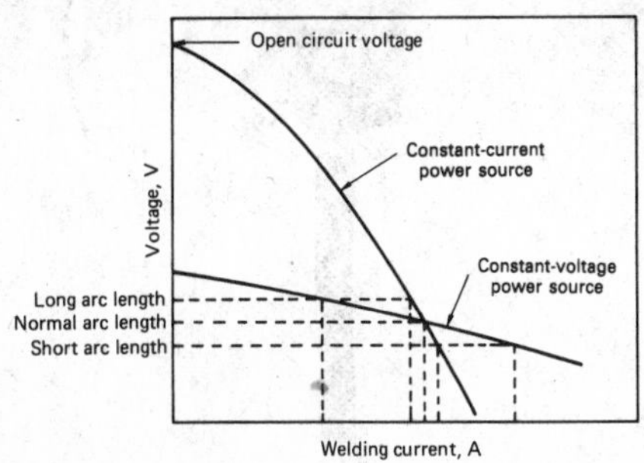

Fig. 6 Slope calculation of a welding system

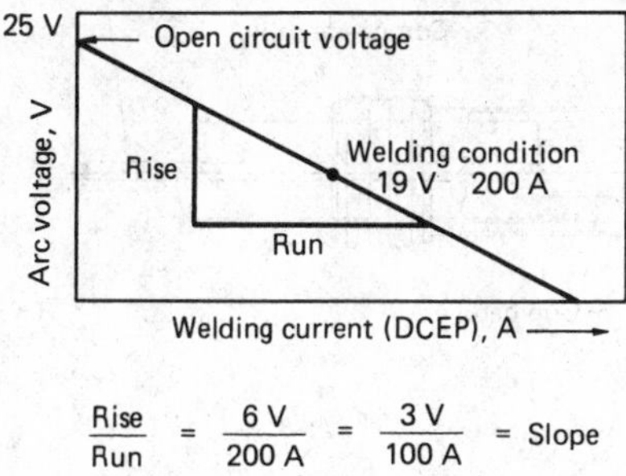

$$\frac{\text{Rise}}{\text{Run}} = \frac{6\text{ V}}{200\text{ A}} = \frac{3\text{ V}}{100\text{ A}} = \text{Slope}$$

and constant-voltage systems. For a long arc length, there is a slight drop in the welding current for the constant-current machine and a large drop in the current for a constant-voltage machine. For constant-voltage power sources, the volt-ampere curve shows that when the arc length shortens slightly, a large increase in welding current occurs. This results in an increased melt-off rate that brings the arc length back to the desired level. Under this system, changes in the wire-feed speed caused by the welder are compensated for electrically by the power source. The constant-current system is sometimes used because the welder can vary the current slightly by changing the arc length. This varies the depth of penetration and the amount of heat input.

Slope. Figure 6 illustrates volt-ampere characteristics for a GMAW power supply. The slant from horizontal of the curve is referred to as the "slope" of the power supply. Slope refers to the reduction in output voltage with increasing current.

As an example of slope, suppose the open circuit voltage is set at 25 V and the welding condition is 19 V and 200 A, as shown in Fig. 6. The voltage decreases from 25 to 19 V in 200 A; the slope is 3 V/100 A.

The slope of the power supply itself is not the total slope of the arc system. Anything that adds resistance to the welding system adds slope and increases the voltage drop at a given welding current. Power cables, connections, loose terminals, and dirty contacts all add to the slope. Therefore, in a welding system, slope should be measured at the arc.

Slope in a GMAW system is used during short arc welding to limit the short circuit current so that spatter is reduced when short circuits between the wire electrode and workpiece are cleared. The greater the slope, the lower the short circuit currents and, within limits, the lower the spatter. The amount of short circuit current must be high enough to detach the molten drops from the wire (but not too high). When little or no slope is present in the welding circuit, the short circuit current rises to a very high level, and a violent reaction takes place.

When a short circuit current is limited to excessively low values by use of too much slope, the wire electrode can carry the full current, and the short circuit will not clear itself. In that case, the wire either piles up on the workpiece or may stub to the pool occasionally and flash off. This causes spatter. When the short circuit current is at the correct value, the parting of the molten drop from the wire is smooth, with very little spatter.

Inductance. Power sources do not respond instantly to load changes. The current takes a finite time to attain a new level. Inductance in the circuit is responsible for this time lag. The effect of inductance can be illustrated by analyzing the curve appearing in Fig. 7. Curve A shows a typical current-time curve with inductance present as the current rises from zero to a final value. Curve B shows the path that the current would have taken if there were no inductance in the circuit. The maximum amount of current attainable during a short is determined by the slope of the power supply. Inductance controls the rate of rise of short circuit current. The rate can be slowed so that the short may clear with minimum spatter. The inductance also stores energy and supplies this energy to the arc after the short has cleared, causing a longer arc.

In "short arc" welding, an increase in inductance increases the "arc on" time. This, in turn, makes the weld pool more fluid, resulting in a flatter, smoother weld bead. The opposite is true when the inductance is decreased. In spray arc welding, the addition of some inductance to the power supply will produce a better arc start. Too much inductance will result in erratic starting.

When conditions of both correct shorting current and correct rate of current rise

Fig. 7 Change in current rise due to inductance

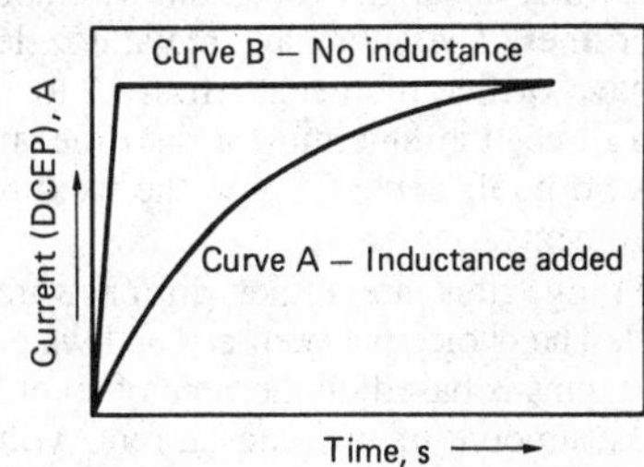

Fig. 8 Typical semiautomatic gas-cooled, curved-neck GMAW gun

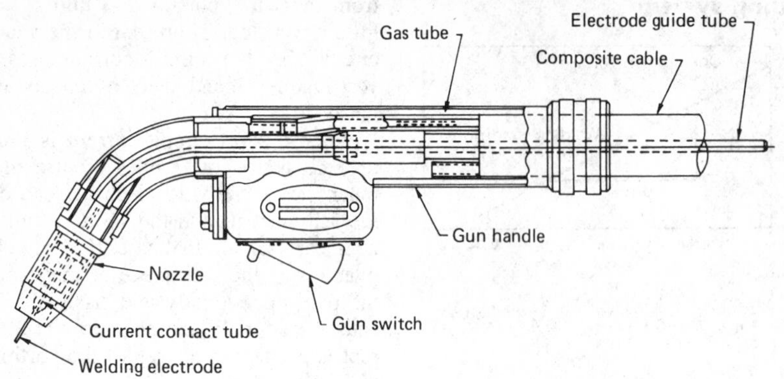

exist, spatter is minimal. The power supply adjustments required for minimum spatter conditions vary with the electrode material and size. As a general rule, both the amount of short circuit current and the amount of inductance needed for ideal operation are increased as the electrode diameter is increased.

Welding Guns

The welding gun used in GMAW transmits welding current to the electrode. Because the wire is fed continuously, a sliding electrical contact is used. The welding current is passed to the electrode through a copper alloy contact tube. The contact tubes have various hole sizes, corresponding to the diameter of the electrode wire. The gun also has a gas supply connection and a nozzle to direct the shielding gas around the arc and weld pool. To prevent overheating of the welding gun, cooling is required to remove the heat generated. Shielding gas or water circulating in the gun, or both, are used for cooling. Some guns are also air cooled. An electrical switch is used to start and stop the electrode feeding, welding current, and shielding gas flow. This is located on the gun in semiautomatic welding and separately on machine welding heads.

Semiautomatic Guns. Handheld semiautomatic guns usually have a curved neck, which makes them applicable to all welding positions. A semiautomatic gun is shown in Fig. 8. The gun is attached to the service lines, which include the power cable, water hose, gas hose, and wire conduit or liner. The guns have metal nozzles that have orifice diameters from $^3/_8$ to $^7/_8$ in. to direct the shielding gas to the arc and weld pool, depending on the welding requirements.

Welding guns are either air or water cooled. The choice between air- and water-cooled guns is based on the type of shielding gas, amount of welding current, voltage, joint design, and the shop practice. A water-cooled gun is similar to an air-cooled gun except that ducts that permit the cooling water to circulate around the contact tube and the nozzle have been added. Water-cooled guns provide more efficient cooling of the gun. Air-cooled guns are employed for applications where water is not readily available, and they are actually cooled by the shielding gas. The guns are available for service up to 600 A used intermittently with carbon dioxide shielding gas. These guns are usually limited to 50% of the carbon dioxide rating when argon or helium is used. Carbon dioxide cools the welding gun, while argon or helium does not. Water cooling permits the gun to operate continuously at the rated capacity with lower heat buildup. Water-cooled guns generally are used for applications requiring between 200 and 750 A. Air-cooled guns of the same capacity as water-cooled guns are heavier, but they are easier to manipulate in confined spaces or for out-of-position applications because there are fewer cables.

There are three general types of guns available. The most common types are shown in Fig. 8 and 9. These have the electrode wire fed through a flexible conduit from a remote wire feeder. The conduit generally is limited to approximately 12 ft due to the wire-feeding limitations of a push-type wire-feed system. Figure 10 shows another type of welding gun that has a self-contained wire-feeding mechanism and electrode wire supply. The wire supply is generally in the form of a 1- to $2^1/_2$-lb spool. This gun employs a pull-type wire-feed system and is particularly good for feeding aluminum and other softer electrode wires that tend to jam in long conduits. Another type of gun has the wire-feed motor on the gun, and the wire is fed through a conduit from a remote wire-feed supply. This system has a pull-type wire feeder and can use longer length conduits.

Machine welding guns use the same basic design principles and features as semiautomatic welding guns. These guns have capacities up to 1200 A and generally are water cooled because of the higher amperages and duty cycles required. The gun is mounted directly below the wire feeder. Large-diameter wires up to $^1/_4$ in. are often used. Figure 11 shows a machine welding gun mounted on a welding head which has the control box wire-feed motor and wire dispenser all attached to the welding torch.

Fig. 9 Various types of semiautomatic guns for GMAW. Courtesy of Hobart Brothers Co.

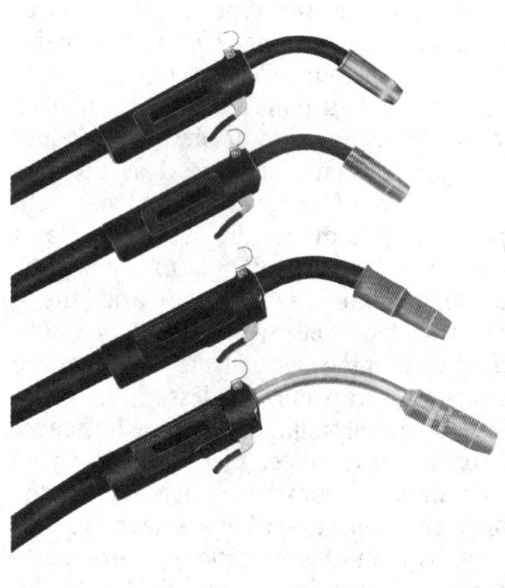

Fig. 10 Semiautomatic welding gun. With self-contained, pull-type wire feeder and spool of electrode wire. Courtesy of Hobart Brothers Co.

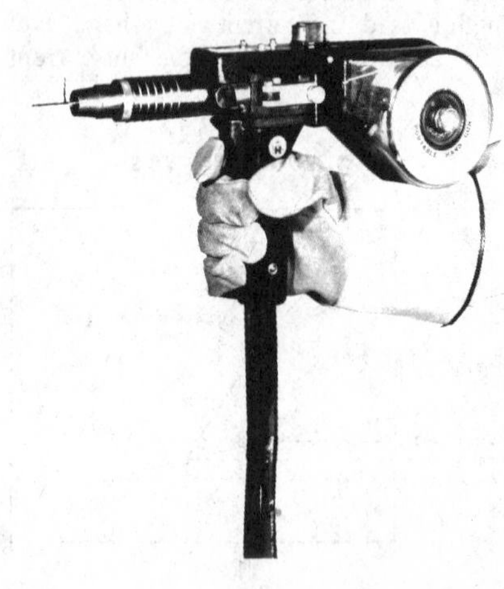

Manipulation of the Welding Gun

For best results in flat-position welding, the welding gun should be held nearly vertical. The angle is not critical and may be as much as 30° from vertical in any direction without any noticeable effect on results; however, if the welding gun is positioned far from vertical, the effectiveness of the gas shielding may be impaired.

The distance from the end of the contact tube to the end of the nozzle should be $^1/_4$ to $^3/_8$ in., although when using the short circuiting mode of metal transfer, the contact tube can be extended beyond the end of the nozzle as much as $^1/_8$ in. The amount of electrode extension from the contact tube to the weld pool is not critical, although it is related to the mode of metal transfer and does affect deposition rate. As electrode extension is increased, deposition rate increases, penetration decreases, and the amount of weld spatter increases. Also, as the electrode extension is increased, so is the distance between the end of the nozzle and the weld pool, and the effect of the gas shield is decreased. Generally, the end of the nozzle should be $^5/_8$ to $^3/_4$ in. from the work (electrode extension of $^1/_4$ to $^1/_2$ in.).

Bead Type. Specific types of beads can be produced by different techniques. Stringer beads generally have deeper penetration, and beads made by weaving have less penetration. In many applications of multiple-pass welding, the first pass is a stringer, and the following one or more passes are made with the weaving technique.

Fig. 11 Machine-type GMAW gun. Courtesy of Hobart Brothers Co.

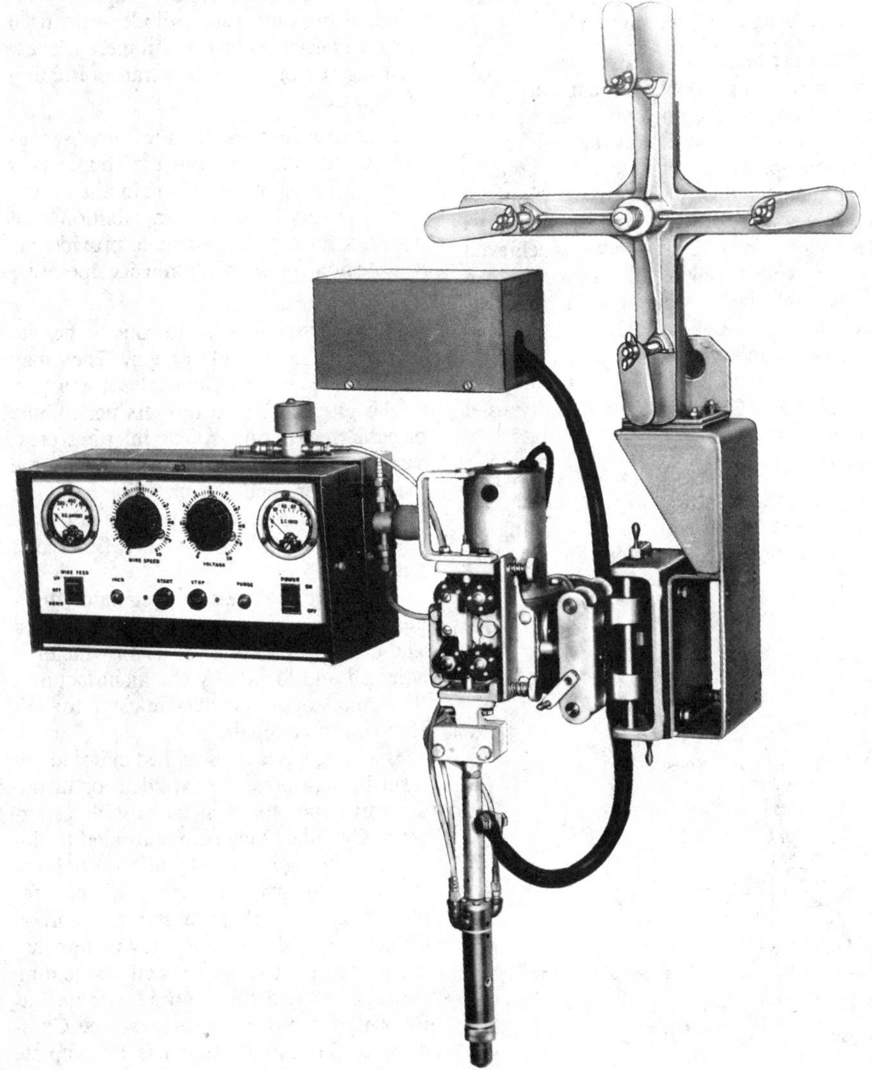

Direction. Both forehand and backhand techniques are used. The choice in any application depends on which technique seems to be most appropriate. The backhand technique allows the welder a better view of the weld pool. Penetration is influenced by technique, and it is greater with the forehand technique than with the backhand technique.

Wire-Feed Systems

The wire-feed motor provides the power for driving the electrode through the cable and gun to the workpiece. Numerous types of wire-feed systems are available, of which most are sufficiently flexible to handle a range of wire compositions and sizes. Most wire-feed systems are of the constant-feed type; that is, the rate of feed (in inches per minute) is selected before welding begins. The feed is started and stopped by a switch on the welding gun for semiautomatic welding, or by other start and stop controls on automatic equipment.

Variable-speed wire-feed systems are adaptable only to constant-current power supplies, and therefore are used less than the constant-feed types. Wire-feed systems may be of the push type, the pull type, or the push-pull type. The type used depends mainly on the size and composition of electrode wire used and sometimes on the distance between the wire reel or spool and the welding gun.

Push-Type Wire-Feed Systems. Most wire-feed systems are of the push type (Fig. 12), in which the wire is pulled from a coil, spool, reel, or drum by means of feed rolls and is pushed through a flexible conduit (wire-feed cable) to which the welding gun is attached. Lengths of conduit are commonly 10 to 15 ft for steel electrode wire and up to 10 ft for aluminum electrode wire, depending on the column strength of the wire.

If spools or coils are used, some type of friction braking device is usually incorporated in the spindle of the spool to prevent overrun of the wire when the feed motor is stopped. Push-type wire-feed systems are available that can handle hard solid wire from 0.023 to 0.125 in. in diameter and soft solid wire from 0.035 to 0.093 in. in diameter. The terms "hard" and "soft" generally refer to ferrous and nonferrous wires, respectively.

The system illustrated in Fig. 12 is typical, although there are many variations among systems from different manufacturers. For instance, the feed-roll mechanism may have up to four driven rolls. Many utilize only two rolls, one of which (usually the lower) is driven. Ordinarily, the lower roll (or each of the lower two rolls) has a circumferential V-groove. The upper rolls do not have grooves; knurled rolls are not recommended as they damage the wire surface. Regardless of type, a feed-roll mechanism must be designed so that roll pressure on the wire can be increased or decreased as required.

Pull-Type Wire-Feed Systems. Welding guns that incorporate a wire-feed driving mechanism are also available. The most popular type has a small drive motor in the handle and an attached 4-in.-OD spool of wire. The unit is compact and can be easily manipulated manually. The equipment is relatively delicate and is best suited to use with electrode wire having a diameter less than 0.045 in. It is particularly useful when welding thin sections, when the total weight of weld deposit is low, and when the work must be done in a confined space.

Push-pull wire-feed systems are particularly well suited to use with soft wires. The welding gun is fitted with a motor and drive rolls and is used as the master unit for controlling the speed of wire feed. It receives wire through a flexible conduit, the other end of which is attached to a remote wire-drive mechanism. The speed of the remote mechanism is adjusted to keep the wire in tension. Small-diameter wires (even soft aluminum alloy wires) can be moved 50 ft or more from the source to the welding gun by a push-pull system. The conduit may have a plastic liner to reduce drag.

Variable-speed wire-feed systems are used only with constant-current power sources, because they depend on deviations in arc voltage to increase or decrease the rate of wire feed. These systems usually employ variable-speed series-wound direct current motors. To start the arc, it is necessary only to ground the electrode wire. When the arc starts, short circuit current closes a control relay, and the drive motor also starts. Variable-speed wire-feed systems are self-regulating, because once the wire-feed rate is set, any deviation in arc voltage results in a corresponding change in motor speed. For instance, if there is a reduction in voltage due to surface irregularity, the drive motor will slow down, allowing the arc length to re-establish itself. Welding is stopped by breaking the arc, thus allowing the current relay to drop out and stop the wire feed.

Controls. For semiautomatic welding, a wire-feed speed control is normally part of the wire feeder assembly or is close by. The wire-feed speed sets the welding current level on a constant-voltage machine. For machine or automatic welding, a separate control box is often used to control the wire-feed speed. If a constant-current power source is used, the control unit varies the wire-feed rate so that a preset arc voltage (length) is maintained.

Maintenance. A regular program of preventive maintenance is mandatory for prevention of feeding difficulties. The electrode-wire conduit must be kept clean. Common practice is to clean the conduit after each spool of wire is used. If spools are large, more frequent cleaning may be necessary. Cleaning the conduit is achieved by removing the wire and blowing clean air through the conduit in the reverse direction from usual flow. In many shops, because clean air is not available, the conduit is cleaned by blowing it out with shielding gas. Oxygen should never be used for this purpose.

Alignment between the feed rolls and the inlet guide to the wire conduit should be checked to minimize any scraping of the wire surface as it enters the conduit. Feed rolls should be checked for wear and for damage to the grooved surface; they should be replaced when a problem is detected to minimize the buildup of any debris in the feeding system.

Fig. 12 Push-type wire feed assembly

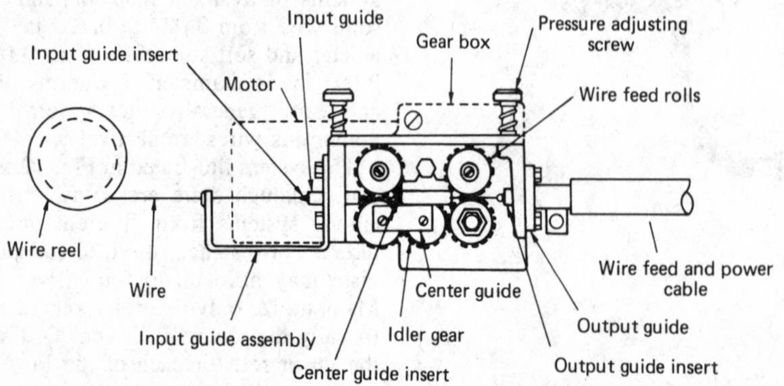

Shielding Gas Equipment

The shielding gas system used in GMAW consists of a gas supply source, a gas regulator, a flowmeter, control valves, and supply hoses to the welding gun. The shielding gases are supplied in liquid form when they are in storage tanks with vaporizers or in a gas form in high-pressure cylinders. Carbon dioxide, however, exists in both the liquid and gas forms when put in high-pressure cylinders. The bulk storage tank system is used when there are large numbers of welding stations using the same type of shielding gas in large quantities. Cylinders are sometimes connected to a manifold which then feeds a single line to the welding stations, which results in increased storage capacity. Individual high-pressure cylinders are used when gas usage is low, when there are few welding stations, or when transportability is required.

Gas pressure regulators are used to reduce the pressure from the source to a working pressure and to maintain a constant delivery pressure. In addition, the regulator must be adjustable to provide gas at a desired pressure within its operating range.

Flowmeters are used to control the rate of gas flow to the welding gun. They may have single or dual flow rates for a particular gas, calibrated in liters per minute or cubic feet per hour. The inlet gas pressure to the flowmeter is specified by the manufacturer, and the pressure regulator must be set accordingly. Gas flow is adjusted with a valve at the flowmeter outlet.

Combination pressure regulator-flowmeter units are available for some gases. The delivery pressure to the flowmeter is adjusted and locked by the manufacturer. It should not be changed except for adjustment after repair.

Shielding gases are supplied either in gas form in high-pressure cylinders or in liquid form and stored in tanks with vaporizers. Cylinders may be manifolded for increased storage capacity at the welding station. Proportioners are available for mixing gases such as argon and carbon dioxide. Regulators and flowmeters are designated for use with specific shielding gases and should only be used with the gas for which they were designed. The Compressed Gas Association has a complete set of specifications for cylinder valves and regulator inlet connections that are not interchangeable.

Hoses are normally connected to solenoid valves on the wire feeder to turn the gas flow on and off with the welding current. A hose is used to connect the flowmeter to the welding gun. The hose is often part of the welding gun assembly.

Welding Cables

Welding cables and connectors are used to connect the power source to the welding gun and to the work. These cables normally are made of copper or aluminum, although copper is more frequently used. The cable consists of hundreds of wires that are enclosed in an insulated casing of natural or synthetic rubber. The cable that connects the power source to the welding gun is called the electrode lead. In semiautomatic welding, this cable is often part of the cable assembly, which also includes the shielding gas hose and the conduit through which the electrode wire is fed. For machine or automatic welding, the electrode lead normally is separate. The cable that connects the work to the power source is the work lead. Work leads are usually connected to the work by pincher clamps or a bolt.

The size of welding cables used depends on the output capacity of the welding machine, the duty cycle of the machine, and the distance between the welding machine and the work. Cable sizes range from the smallest at American Wire Gage (AWG) No. 8 to AWG No. 4/0, with amperage ratings of 75 A and up.

Water-Cooling Systems

Apart from permitting the use of lighter weight welding guns, considerably less copper in the power lead, and decreasing the difficulty of removing weld spatter, water cooling adds to the service life of the welding equipment. Proper design, installation, and preventive maintenance have largely eliminated the corrosion and plugging at one time associated with gas metal arc equipment. Sealed recirculating water-cooling systems are preferred, because they enable corrosion to be stopped by the use of inhibitors, and because dirt particles that normally plug small water passages settle to the bottom of the tank. Tap water can be filtered, but this does not remove materials that cause corrosion, and water conditioning usually is too expensive to be considered. Another advantage of sealed systems is that they do not cause the condensation that results from the use of low-temperature tap water.

Motion Devices

Motion devices, or traversing mechanisms, are used for machine and automatic welding. These motion devices can be used to move the welding head, workpiece, or gun, depending on the type and size of the work and the preference of the user. For more information on the design and use of these devices, see the section on motion devices in the article "Gas Tungsten Arc Welding" in this Volume.

Accessories

Accessory equipment used for GMAW consists of items used for cleaning the weld bead and cutting the electrode wire. In many cases, cleaning is not required, but when some slag is created by the welding, a chipping hammer or grinder is used to remove it. Wire brushes and grinders sometimes are used for cleaning the weld bead. A pair of wire cutters and pliers are used to cut the end of the electrode wire between stops and starts. This ensures a clean, nonoxidized wire end and proper electrode extension for easy starts.

Electrodes

The electrodes for GMAW are usually quite similar or identical in composition to those used for welding with most other bare electrode processes. In many cases, the electrode wires are chosen to match the chemical composition of the base metal as closely as possible. In some cases, electrodes with a somewhat different chemical composition are used to obtain maximum mechanical properties or better weldability.

For example, the electrodes that are most satisfactory for welding manganese bronze, a copper-zinc alloy, are either aluminum bronze or copper-manganese-nickel-aluminum alloys. Somewhat similar, although not to the same degree, are electrodes used for welding some stainless steels. In the 200 series austenitic stainless steels, for example, 300 series austenitic filler metal is usually used, due to a lack of availability of 200 series filler metal. This weld joint generally is weaker than the surrounding base metal. Type 410 and 420 electrodes are the only martensitic stainless steel types recognized by the American Welding Society (AWS). This limitation is the reason why austenitic stainless steel filler metal is often used. Austenitic filler metal provides a weld with lower strength but higher toughness. It also eliminates the need for preheating and postheating. For welding ferritic stainless steels, both ferritic and austenitic filler metal may be used. Ferritic filler metal is used when higher strength and an annealing postheat are required. Austenitic filler metal is employed when high ductility is required.

Almost all electrodes used for GMAW of steels have deoxidizing or other scavenging elements added to minimize porosity and improve mechanical properties. The use of electrode wires with the right amount of deoxidizers is most important when using oxygen- or carbon dioxide-bearing shielding gases. The deoxidizers most frequently used in steel electrodes are manganese, silicon, and aluminum. Titanium and silicon are the principal deoxidizers used in nickel alloys. Copper alloys may be deoxidized with titanium, silicon, or phosphorus, depending on the type and desired end results.

The electrodes used for GMAW are quite small in diameter compared to those used for other types of welding. Wire diameters of 0.035 to $^{1}/_{16}$ in. are about average. However, electrode diameters as small as 0.020 in. and as large as $^{1}/_{8}$ in. are sometimes used. Because of the small sizes of the electrode and the comparatively high currents used for GMAW, the melting rates of the electrodes are very high. The rates range from approximately 100 to 800 in./min for all metals except magnesium, which reaches speeds up to 1400 in./min. Because of this rapid melting, the electrodes always must be provided as long, continuous strands of suitably tempered wire that can be fed smoothly and continuously through the welding equipment. The wires are normally provided on spools or in coils that range in weight from 2 to 100 lb.

Because of the relatively small size of the electrode wire, which gives a high surface-to-volume ratio, cleanliness of the wire is very important. Drawing compounds, rust, oil, or other foreign matter on the surface of the electrode wire tend to be in high proportion relative to the amount of metal present. These impurities can cause weld-metal defects such as porosity and cracking.

Composition of electrode wire has a significant effect on results. Because of the importance of electrode composition, many large users of electrode wire have established their own specifications, which they have developed from their own experience. Table 1 lists AWS classifications and composition limits for different types of electrodes that are used for GMAW of low-carbon steels. For compositions of electrode wire used for welding other metals, see the articles in this Volume that deal with the welding of specific metals. Filler-metal specifications for GMAW of these metals are:

AWS specification	Metal
A5.7	Copper and copper alloys
A5.9	Stainless steel
A5.10	Aluminum and aluminum alloys
A5.14	Nickel and nickel alloys
A5.16	Titanium and titanium alloys
A5.18	Carbon steel
A5.19	Magnesium alloys
A5.24	Zirconium and zirconium alloys
A5.28	Low-alloy steel

Table 1 AWS classifications and composition limits for carbon steel electrode wires for GMAW (AWS A5.18)

AWS classification(a)	Composition, %						
	Carbon	Manganese	Silicon	Phosphor	Sulfur	Copper	Other
ER70S-2	0.07	0.90-1.40	0.40-0.70	0.025	0.035	0.50	Ti, Zr, Al
ER70S-3	0.06-0.15	0.90-1.40	0.45-0.70	0.025	0.035	0.50	...
ER70S-4	0.07-0.15	1.00-1.50	0.65-0.85	0.025	0.035	0.50	...
ER70S-5	0.07-0.19	0.90-1.40	0.30-0.60	0.025	0.035	0.50	Al
ER70S-6	0.07-0.15	1.40-1.85	0.80-1.15	0.025	0.035	0.50	...
ER70S-7	0.07-0.15	1.50-2.00	0.50-0.80	0.025	0.035	0.50	...

(a) ER70S-G, which is not shown in this table, has no chemical requirements.

Mechanical Properties. Table 2 shows the minimum mechanical properties specified by AWS for the classes of electrode listed in Table 1, and indicates the shielding gases and the polarity of direct current suitable for use with the various electrodes.

Classification of Electrodes

The classification system for bare, solid-wire electrodes that is used throughout industry in the United States was devised by AWS. Because of the wide variety of metals that can be welded by this process, there are a wide variety of classifications covering GMAW electrodes. Many of the classifications used are the same as those used to classify filler rods for GTAW.

Most classifications of GMAW electrodes are based on the chemical composition, mechanical properties, or both, of the solid filler metal. A typical carbon steel classification for ER70S-2 is:

- The "E" indicates the filler wire is an electrode that may be used for GMAW. The "R" indicates it may also be used as a filler rod for GTAW or plasma arc welding (PAW).
- The next two (or three) digits indicate the nominal tensile strength of the filler wire.
- The letter to the right of the digits indicates the type of filler metal. An "S" indicates a solid wire and a "C" indicates a metal cored wire that consists of a metal powder core in a metal sheath.
- The digit or letters and digit in the suffix indicates the special chemical composition of the filler metal.

Carbon Steel Electrodes. Characteristics of specific types of carbon steel bare wire electrodes covered by AWS A5.18 are summarized below. Each number in the suffix (for example, S-2 or S-3) indicates the special chemical composition of the filler metal.

ER70S-2 classification covers multiple deoxidized steel filler metals that contain a nominal combined total of 0.20% zirconium, titanium, and aluminum, in addition to silicon and manganese contents. These filler metals are capable of producing sound welds in semikilled and rimmed steels as well as in killed steels of various carbon levels. Because of the added deoxidants, these filler metals can be used for welding steels that have a rusty or dirty surface, with a possible sacrifice of weld quality, depending on the degree of surface contamination. They can be used with a shielding gas of argon-oxygen mixtures, carbon dioxide, or argon-carbon dioxide mixtures and are preferred for out-of-position welding with the short circuiting type of transfer because of ease of operation.

ER70S-3 classification filler metals meet the requirements of this classification with either carbon dioxide or argon-oxygen as a shielding gas. They are used primarily on single-pass welds, but can be used on multiple-pass welds, especially when welding killed or semikilled steel. Small-diameter electrodes can be used for out-of-position welding and for short circuiting type transfer with argon-carbon dioxide mixtures or carbon dioxide shielding gases. However, the use of carbon dioxide shielding gas in conjunction with excessively high heat inputs may result in failure to meet the minimum specified tensile and yield strength.

ER70S-4 classification filler metals contain slightly higher manganese and silicon contents than those of the ER70S-3 classification and produce a weld deposit of higher tensile strength. These filler metals are used in welding applications requiring carbon dioxide or argon-carbon dioxide shielding where a slightly longer arc or other conditions require more deoxidation than provided by the ER70S-3 filler metals. These filler metals are not required to demonstrate impact properties.

ER70S-5 classification covers filler metals that contain aluminum in addition to manganese and silicon as deoxidizers. These filler metals can be used when welding rimmed, killed, or semikilled steels with carbon dioxide shielding gas and high welding currents. The relatively large amount of aluminum ensures the deposition of well-deoxidized and sound weld metal. Because of the aluminum, they are not used for the short circuiting type of transfer, but can be used for welding steels that have a rusty or dirty surface, with a possible sacrifice of weld quality, depending on the degree of surface contamination. These filler metals are not required to demonstrate impact properties.

ER70S-6 classification filler metals have the highest combination of manganese and silicon, permitting high-current welding with carbon dioxide gas shielding even in rimmed steels. These filler metals may also be used to weld sheet metal where smooth weld beads are desired and to weld steels that have moderate amounts of rust and mill scale. The quality of the weld depends on the amount of surface impurities present. This filler metal also can be used in out-of-position welding with short circuiting transfer.

ER70S-7 classification filler metals have a substantially greater manganese

Table 2 Minimum mechanical properties of weld-metal deposits of carbon steel electrodes

AWS classification	Shielding gas(a)	Tensile strength, ksi	Yield strength, ksi	Elongation in 2 in., %	Impact strength, ft·lb at °F
ER70S-2	CO_2	72	60	22	20 at −20
ER70S-3	CO_2	72	60	22	20 at 0
ER70S-4	CO_2	72	60	22	...
ER70S-5	CO_2	72	60	22	...
ER70S-6	CO_2	72	60	22	20 at −20
ER70S-7	CO_2	72	60	22	20 at −20
ER70S-G	(b)	72	60	22	(b)

(a) Used with DCEP. (b) As agreed between the supplier and purchaser

content (essentially equal to that of ER70S-6) than those of the ER70S-3 classification. This provides slightly better wetting and weld appearance with slightly higher tensile and yield strengths and may permit increased speeds compared with ER70S-3 filler metals. These filler metals are generally recommended for use with argon-oxygen shielding gas mixtures, but they can be used with argon-carbon dioxide mixtures and carbon dioxide alone under the same general conditions as for the ER70S-3 classification. Under equivalent welding conditions, weld hardness will be lower than ER70S-6 weld metal but higher than ER70S-3 deposits.

ER70S-G classification includes those solid filler metals that are not included in the preceding classes. The filler metal supplier should be consulted for the characteristics and intended use. This classification does not list specific chemical composition or impact requirements, because these are subject to agreement between supplier and purchaser. However, any filler metal classified ER70S-G must meet all other requirements of this classification.

Shielding Gases

The primary purpose of shielding gas is to protect the molten weld metal and the heat-affected zone (HAZ) from oxidation and other contamination. Reactive metals such as titanium require protection over a much greater area in the weld vicinity.

Originally, only the inert gases (argon and helium) were used for shielding, but carbon dioxide is now used extensively, and oxygen and carbon dioxide are often mixed with the inert gases. Chemical behavior and welding applications of the gases and mixtures of shielding gases commonly used in GMAW are presented in Table 3.

Table 3 Uses of different shielding gases for GMAW

Type of gas	Typical mixtures, %	Primary uses
Argon	...	Nonferrous metals
Helium	...	Aluminum, magnesium, and copper alloys
Carbon dioxide	...	Low-carbon and low-alloy steels
Argon-helium	20-50A-50-80He	Aluminum, magnesium, copper and nickel alloys
Argon-oxygen	1-$2O_2$	Stainless steels
	3-$5O_2$	Low-carbon and low-alloy steels
Argon-carbon dioxide	20-$50CO_2$	Low-carbon and low-alloy steels
Helium-argon-carbon dioxide	90He-$7^1/_2$A-$2^1/_2CO_2$	Stainless steels
	60-70He-25-35Ar-$5CO_2$	Low-alloy steels
Nitrogen	...	Copper alloys

Argon. Welding-grade argon is 99.995% pure. It is a monatomic gas (one atom per molecule), it is inert, and it is insoluble in molten metal. Argon is 38% heavier than air, which is advantageous for welding in the flat position and the horizontal fillet position. As shown in Table 3, pure argon can be used as a shielding gas for virtually all metals, but it is not ordinarily used in welding steels, for which argon-based mixtures are preferred (see the typical mixtures column in Table 3). The use of argon with DCEP when welding plain carbon steel often results in undercutting along the edges of the weld.

Argon shielding results in a bead shape that is different from that obtained with helium or carbon dioxide shielding. Also, under otherwise identical welding conditions, argon shielding results in a different pattern of penetration than carbon dioxide shielding (see Fig. 13).

Argon has a lower ionization potential than helium, which results in lower arc voltage for a given arc length. Consequently, less heat is produced at a given amperage with argon than with helium, which makes argon preferable to helium for the welding of thin sections. Argon is about ten times as heavy as helium, so less argon is required to retain a protective blanket over the weld area in the flat position and the horizontal fillet position.

Argon costs less than helium (per unit of volume purchased) and is more plentiful, which partly accounts for its far more extensive use, compared with helium.

Helium shielding gas is chemically inert and is used primarily on aluminum, magnesium, and copper alloys. Helium is a light gas that is obtained by separation from natural gas. It may be distributed as a liquid, but it is more often used as compressed gas in cylinders.

Helium shielding gas is lighter than air, and because of this, high gas flow rates must be used to maintain adequate shielding. Typically, the gas flow rate is two to three times that used for argon when welding in the flat position. Helium is often preferred in the overhead position because it floats up and maintains good shielding, while argon tends to float down. Globular metal transfer is usually obtained with helium, but spray transfer may be obtained at the highest current levels. Because of this, more spatter and a poorer weld-bead appearance will be produced, as compared to argon. For any given arc length and current level, helium will produce a hotter arc, which makes helium good for welding thick sections and metals like copper, aluminum, and magnesium that have high thermal conductivity. Helium generally gives wider weld beads and better penetration than argon.

Carbon Dioxide. Most reactive gases cannot be used alone for shielding. Carbon dioxide is the outstanding exception; it is extensively used both by itself and as a component of gas mixtures, to which it imparts improved arc action and metal transfer.

Carbon dioxide is widely used in the welding of steel by the short circuiting mode of metal transfer. A true spray arc is not obtained with a shielding gas composed entirely of carbon dioxide. In many applications, carbon dioxide has provided good welding speed and good penetration, and in many applications the welds so shielded have proved to be less expensive than argon-shielded welds.

Carbon dioxide decomposes to carbon monoxide and oxygen at arc temperatures, producing an oxidizing effect approximately equal to that obtained by the use of an inert gas with 8 to 10% oxygen. In spite of this oxidizing effect, sound weld deposits, free of porosity, can be consistently obtained with carbon dioxide shielding when a deoxidizing electrode wire

Fig. 13 Patterns of penetration. Obtained using argon shielding and carbon dioxide shielding in flat-position fillet welding under otherwise identical conditions

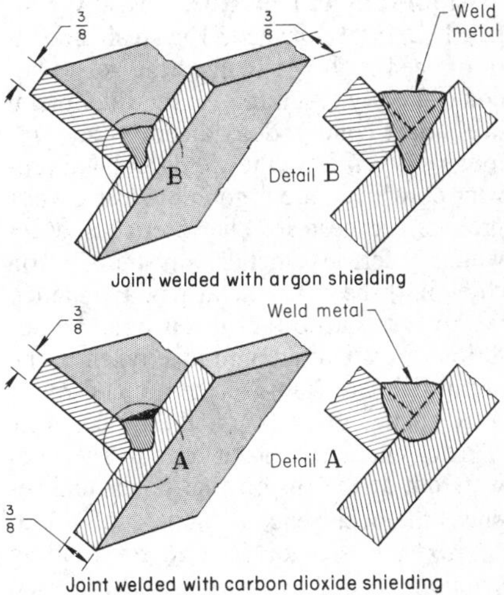

is used with the globular mode of metal transfer.

The major disadvantage of carbon dioxide shielding is the rather "harsh" arc and excessive spatter that it produces. Weld spatter can be minimized by maintaining a short, uniform arc length.

Argon-carbon dioxide mixtures that consist of 3 to 10% carbon dioxide added to the argon are used for the same purposes that argon-oxygen mixtures are used. This mixture maintains a spray-type transfer and improves the weld pool wetting and arc characteristics when welding ferrous metals.

Argon-carbon dioxide mixtures that contain 20 to 50% carbon dioxide are used to weld steel with short circuiting transfer. At higher current levels this mixture tends to promote globular transfer. Because there is a high amount of an active gas in the mixture, argon-carbon dioxide mixtures can cause harmful effects on some steels. These mixtures may be used to weld stainless steels, but they may increase the carbon content in the weld metal and reduce the corrosion resistance of the weld. These mixtures are not used on nonferrous metals because the carbon dioxide would cause contamination.

Argon-helium mixtures often are used to get the best characteristics of both argon and helium. Common mixtures of these gases by volume range from 80% helium-20% argon to 50% helium-50% argon, but there is a wide variety of mixtures available. Argon-helium mixtures give the good penetrating characteristics of helium and the spray metal transfer characteristic of argon. Argon-helium mixtures may be used for all metals, but are primarily used on aluminum, copper, magnesium, and their alloys. A mixture of 75% helium-25% argon is popular for welding aluminum because it also helps to reduce porosity.

Argon-oxygen mixtures usually contain 1, 2, or 5% oxygen. The small amount of oxygen in the gas causes the gas to become slightly oxidizing, so the filler metal used must contain deoxidizers to help remove oxygen from the weld pool and prevent porosity. Pure argon does not always provide the best arc characteristics when welding ferrous metals. In pure argon shielding, the filler metal has a tendency not to flow out to the fusion line. An addition of a small amount of oxygen to the argon helps to stabilize the arc and minimize spatter, while maintaining the spray transfer characteristic of argon. The pool wets out at the fusion line better and reduces the occurrence of undercut. A 1 or 2% oxygen addition is used for welding stainless steel. Up to 5% oxygen is used for welding low-carbon and low-alloy steels and deoxidized copper.

Helium-argon-carbon dioxide mixtures are used for the same reasons argon-carbon dioxide mixtures are used: to promote better weld pool wetting and short circuiting metal transfer. Mixtures of 90% helium, 7$^1/_2$% argon, and 2$^1/_2$% carbon dioxide are used on stainless steel to promote the short circuiting transfer and a less active atmosphere that will not decrease corrosion resistance. Mixtures of 60 to 70% helium, 25 to 35% argon, and 5% carbon dioxide are used for welding low-alloy steels when maximum toughness is desired. This mixture promotes short circuiting transfer and keeps the amount of carbon dioxide to the minimum so that carbon, which reduces toughness, is not added to the weld.

Nitrogen is occasionally used as a shielding gas when welding copper and copper alloys. Nitrogen has characteristics similar to helium because it gives better penetration than argon and tends to promote globular metal transfer. Nitrogen is used where the availability of helium is limited, such as in Europe.

Selection of a shielding gas for a given application depends on the type and thickness of the base metal, cost and effectiveness of the different gases, joint design, position of welding, technique to be employed, fixturing, speed, and required quality. Information on the selection of shielding gases for GMAW of carbon steels, using the short circuiting and spray arc modes of transfer, is given in Table 4.

Holding and Handling Workpieces

Requirements for jigging or fixturing for GMAW are usually based on the same considerations as those for other arc welding processes. The major functions of a weld fixture are to:

- Locate and maintain parts in their position relative to the assembly
- Maintain alignment during welding without excessive restraint, which would induce weld cracking
- Control distortion in the weldment
- Increase the efficiency of the welder

For additional information on jigging, fixturing, and positioning, see the articles "Shielded Metal Arc Welding" and "Gas Tungsten Arc Welding" in this Volume.

Full automation can be achieved in GMAW by the use of appropriate fixturing. The continuity of operation that can result from the use of an electrode of virtually unlimited length enables the use of turning rolls and the mounting of electrodes on carriages, and the high welding speeds attainable result in much less heat transfer to the base metal than occurs in SMAW and SAW. This, in turn, calls for a comparatively small amount of clamping to prevent distortion.

Fixturing for welding in the flat position is generally the simplest and most economical for relatively small weldments. Fillet welds made in the flat position usually are more uniform, less likely to have unequal legs and convex profiles, and less susceptible to undercutting than fillet welds made in the horizontal position (Fig. 14). Fixtures for GMAW should be designed and built so that locating, clamping, and gaging permit good joint accessibility. Fixtures with few protruding clamp arms and gage points allow easy manipulation of the electrode holder.

Table 4 Selection of shielding gases for GMAW of carbon steels

Metal	Shielding gas	Advantages
Short circuiting transfer		
Carbon steel	Argon-20-25% carbon dioxide	Less than $^1/_8$ in. thick: high welding speeds without melt-through; minimum distortion and spatter; good penetration
	Argon-50% carbon dioxide	Greater than $^1/_8$ in. thick: minimum spatter; clean weld appearances; good weld pool control in vertical and overhead positions
	Carbon dioxide	Deeper penetration; faster welding speeds; minimum cost
Spray transfer		
Carbon steel	Argon-3-5% oxygen	Good arc stability; produces a more fluid and controllable weld pool; good coalescence and bead contour; minimizes undercutting; permits higher speeds, compared with argon
	Carbon dioxide	High-speed mechanized welding; low-cost manual welding

Fig. 14 Profiles and leg lengths of fillet welds. Made in the flat position and the horizontal position. Note undercut above horizontal-position weld.

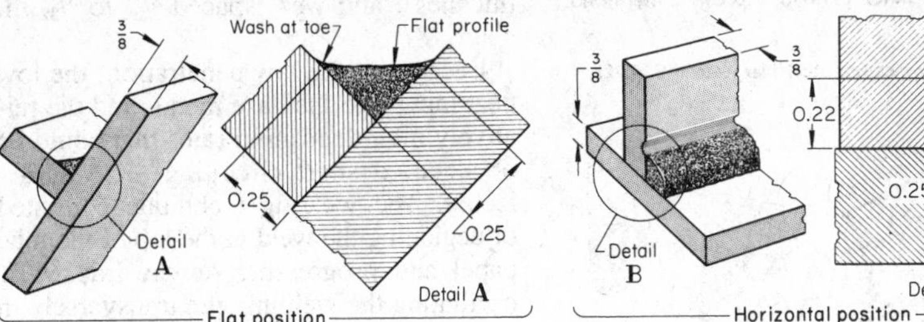

Joint Design

Gas metal arc welding is applicable to all five basic types of joints—butt, lap, T, corner, and edge—and to most modifications of these basic types. The rules of joint design for GMAW are not necessarily different from those for SMAW, but some design considerations are unique to GMAW, mainly because of the differences in size and complexity of equipment. If these differences are not taken into consideration in locating a particular weld joint, the welder may be prevented from properly manipulating the welding gun. Joints must not be located so as to cause an excessive gap between the nozzle of the welding gun and the root of the joint, because such a gap reduces the effectiveness of the shielding gas and may prevent root penetration. Inadequate shielding gas is evidenced by weld discoloration, excessive weld spatter, and porosity.

Recommended proportions of grooves for GMAW are shown in the Appendix to the article "Joint Design and Preparation" in this Volume. With some exceptions, the proportions of joints for GMAW often are the same as those for SMAW.

Selection of Joints for Economy. Careful selection of joint type often can result in manufacturing economy. This economy is not always confined to the welding operation, but may apply in other manufacturing operations, as shown in the following example.

Example 1. Use of a Modified Butt Joint Instead of an Offset Lap Joint To Reduce Tooling and Labor Costs. For welding the spherical refrigerant container shown in Fig. 15, an offset lap joint (middle view in Fig. 15), although frequently used in welding the components of a variety of pressure cylinders and spheres, would have resulted in high labor and tooling costs. The lip of the offset hemisphere would have caused interference in assembly, and additional tooling would have been needed to offset the lip.

The modified butt joint shown at bottom in Fig. 15 allowed use of identical forming tools for both halves of the sphere and was the best compromise among weldability, tooling costs, and labor costs. To reduce labor costs still further, each welder operated two girth welding machines and thus was not able to observe the welding operation; therefore, automatic seam tracking was necessary.

The automatic tracking system consisted of two recirculating ball-screw cross slides mounted at right angles and driven by reversible alternating current motors. A probe, which was mounted to move with the welding gun (see view at upper left in Fig. 15), sensed the location of the joint in relation to the gun. A movement of the probe tip caused the appropriate slide to bring the probe and gun back to the neutral position. The probe was mounted on a small screw-adjusted slide to provide quick and accurate adjustment of both the horizontal position of the welding gun and the distance between it and the work. At the end of the welding cycle, the probe and the welding gun were raised by the vertical slide. After the next assembly was in place, the probe and gun were lowered by the same means and welding was started automatically.

The hemispheres were held in a special welding machine (lathe) consisting of one fixture rotated by a continuously variable drive and a second fixture mounted on the tailstock. Both fixtures were mounted on air-operated slides. Thus, the parts were held together and rotated under the welding gun. When the gun was retracted after completing the weld, the air cylinders separated, releasing the welded workpiece.

The operator loaded the hemispheres in the machine and pushed a button to close the fixtures. The operator then had an option of using automatic start, by which the weld started as soon as the welding gun was in position, or manual start, whereby the position of the gun could be observed, corrections could be made (if required), and the weld could be started by pushing a button. While one container was being welded, the operator loaded and started a second machine. After each container had been welded, the operator checked it for visible defects. Those requiring repair welding were set aside, and those with no visible defects were transferred on a conveyor to the testing area.

The workpiece was a disposable refrigerant container with a water capacity of 25.5 lb produced under a special permit that specified two types of pressure tests. Each welded container was tested by subjecting it to 300-psi internal air pressure while within a heavy steel safety chamber. The pressure in the container then was reduced to 100 psi, the chamber was opened, and the sphere was forced under water to check for leaks. If repairs were required,

Fig. 15 Girth welded refrigerant container. Labor and tooling costs were reduced by use of a modified butt joint instead of an offset lap joint.

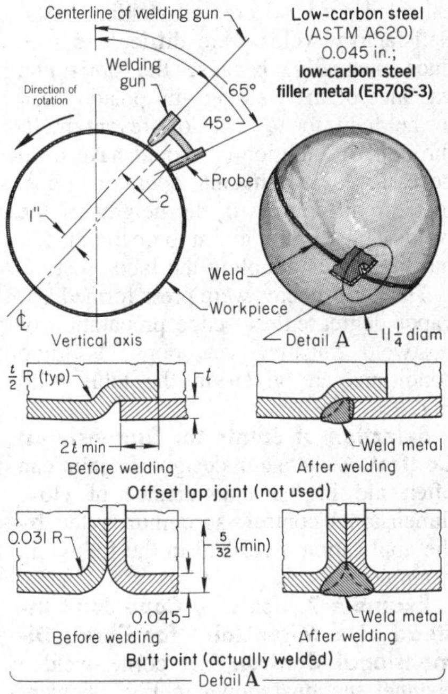

Joint type	Circumferential modified butt
Weld type	Single-flare V-groove
Power supply	300-A transformer-rectifier
Electrode wire(a)	0.030-in.-diam ER70S-3
Welding gun	Mechanized, fixed, water cooled
Wire feed	Push-type motor, on welding gun
Current	170-190 A (DCEP)
Voltage	22-23 V
Shielding gas(b)	98% argon-2% oxygen, at 35 ft³/h
Number of passes	1
Wire-feed rate	340-380 in./min
Electrode extension	1/4-3/8 in.
Welding speed	46.6 in./min
Weld time per container	42 s

(a) Selection of wire wound to a large diameter eliminated need for wire straighteners and reduced leakage rate. (b) Argon of 99.999% purity from bulk-liquid holder

the spheres were retested after repair. A destructive test was required on one container out of each lot of 1000, with a minimum of one per day (although, in practice, at least one container was tested from each machine during each shift). This test consisted of filling the container with water, connecting it to a high-pressure pump, and increasing the pressure until the container burst. The minimum bursting strength was 800 psi. Fewer than 5% of the spheres required weld repairs.

The guidance system caused some problems, primarily because of the maintenance required. Repair and adjustment of the probe switch were difficult, and improperly adjusted probes could cause misplaced welds. Spare systems were available for replacement of defective probe units.

Although the guidance system added to the machine cost and caused maintenance problems, these disadvantages were soon canceled out by decreased welding costs. Satisfactory welds were difficult to produce manually, because the horizontal variance of the welding gun position had to be held to $^1/_{32}$ in. to prevent melt-through. In addition, it would have been necessary for the manual operator to correct for differences in the heights of the weld seams, limiting him to operating one machine, thus doubling the labor cost.

The hemispheres were press formed and vapor degreased. No edge preparation or postweld finishing was done. Welding conditions are given in the table with Fig. 15.

Selection of Joints for Dimensional Control. A change in design of a joint can often aid in the maintenance of close dimensional control, as demonstrated by the application described in the following example.

Example 2. Use of a Cope Joint Instead of a Miter Joint for Closer Dimensional Control. The corner-welded channel sections shown in Fig. 16 were parts of rectangular frames for data-processing machines. Originally, 45° miter joints were used (Fig. 16a), but dimensions after welding were unsatisfactory because of joint location and weld restraint.

To provide a more positive joint location with less weld restraint, cope joints (Fig. 16b) were substituted for the miter joints. Tolerances of ±0.010 in. on length and width, and ±0.032 in. on squareness, were met on channel sections welded with the improved joint design.

The channel sections were contour roll formed from 0.120-in.-thick low-carbon steel strip. Pieces were cut to length by a cutoff die in a press. The cut lengths were also coped by a die in a press, and all parts were inspected. Tolerances on individual pieces were held to ±0.005 in.

Fig. 16 Corner section of a rectangular frame. A cope joint was substituted for a miter joint to improve dimensional control.

Low-carbon steel; low-carbon steel filler metal (ER70S-2)

(a) Original design (miter joint)

Details of cope joint before welding

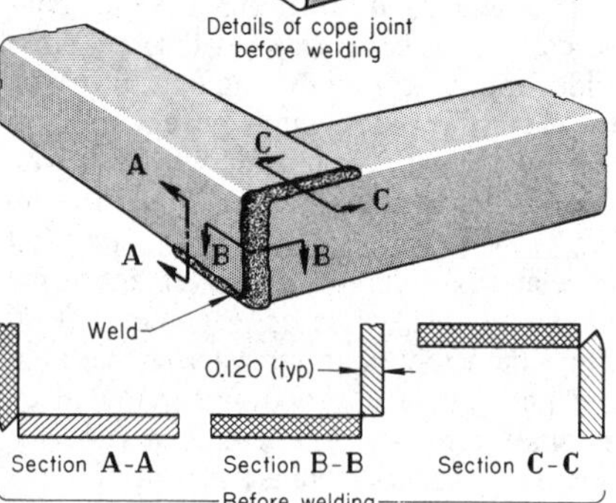

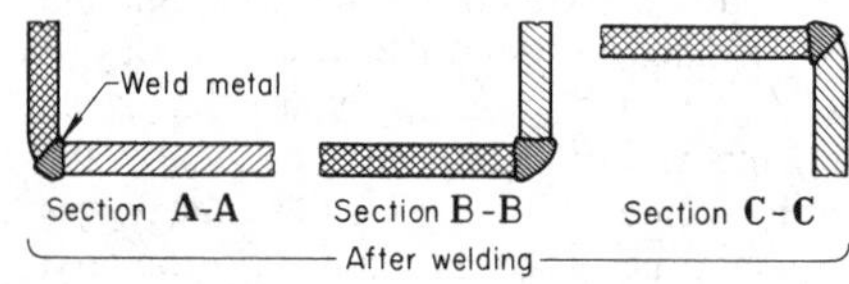

(b) Improved design (cope joint)

Joint type	Corner
Weld type	Fillet and V
Power supply	200-A, constant-voltage rectifier
Electrode wire(a)	0.030-in.-diam ER70S-2
Wire feed	Constant feed
Current	80-85 A (DCEN)
Voltage	26 V
Shielding gas	75% argon-25% carbon dioxide, at 40 ft³/h
Wire-feed rate	30-100 in./min
Welding speed	10 in./min

(a) 0.04% C; triple deoxidized, with flash copper plating

Example 3. Use of Short Circuiting Arc for Welding Between 0.120-in.-Wall Tubes. A package boiler assembly was produced by welding between adjacent low-carbon steel tubes, in the vertical position. As Fig. 17 shows, the tubes were 3 in. in outside diameter, 0.120 in. in wall thickness, and were spaced $^3/_{32}$ to $^5/_{32}$ in. apart.

Because of the low penetration, the low heat input into the base metal, and the relatively high deposition rates that could be obtained, a short circuiting arc was selected. The welding technique consisted of beginning the weld at the top of the tube panel and progressing downward, while oscillating the welding gun transversely in an arc approximately $^3/_{16}$ in. wide. This technique resulted in an effective seal weld. Additional welding details are given in the table that accompanies Fig. 17.

Thin Sections

The minimum metal thickness appropriate to GMAW is about 0.020 in., which is thin enough to present problems. The most common approach to welding of thin sections ($^1/_8$ in. thick or less) is to use a short circuiting arc, which permits metal transfer with a lower heat input than that of a spray transfer. By this technique, adequate penetration is obtained and the possibility of melt-through is minimized.

Example 4. Change From Shielded Metal Arc to Gas Metal Arc Welding With a Short Circuiting Arc To Simplify Welder Training and Increase Welding Speed. Housings for spray-type, conveyorized metal-treatment equipment were made from hot rolled low-carbon steel

Fig. 17 Section of a package boiler assembly. Made by welding between adjacent tubes, using a short circuiting arc

Low-carbon steel (ASTM A192); low-alloy low-carbon steel filler metal

Power supply	300-A, constant-voltage motor-generator
Electrode wire(a)	0.045-in.-diam low-alloy, low-carbon steel
Arc type	Short circuiting
Shielding gas	Carbon dioxide
Deposition rate at 100% arc time	4.1 lb/h

(a) 0.12% C, 1.90% Mn, 0.80% Si, 0.50% Mo, and 0.10% Ni

sheet 0.120 in. thick and ranged in length from 30 to 300 ft. The height and width were 6 to 12 ft and 4 to 12 ft, respectively. Components for housings were made in 4-ft-long sections, partly fabricated in the shop, and shipped "knocked down" for field erection. Installations often involved several hundred feet of welded joints.

Conventionally, welding was done by SMAW, using small-diameter covered electrodes at low amperage. The training of workers to use this process on 0.120-in.-thick sheet proved difficult and expensive. By changing to GMAW with a short circuiting arc, welder training was greatly simplified, and welding speed was more than doubled.

A typical housing, together with details of the welded joints, is shown in Fig. 18. To prevent melt-through, the joints were designed for single-pass welding and heat input was kept low by using the short circuiting type of metal transfer. Conditions that made this type of weld deposition possible are given in the table with Fig. 18, and consisted principally of (1) a constant-voltage power supply with variable inductance, (2) small-diameter electrode wire, (3) low welding current, and (4) a shielding-gas mixture that gave good arc stability and minimized weld spatter.

Because of variations in groove size and because of the fit-up conditions that existed in field assembly, specific welding rates were not determined. However, a comparison of overall job times showed that deposition rate in GMAW was $2^1/_2$ times the rate obtained in SMAW. Furthermore, slag cleaning was eliminated, and weld spatter was considerably reduced. As a result of these improvements in the field, the manufacturer adopted GMAW for welding of sheet metal in the shop.

Intermediate and Thick Sections

In a majority of applications, the most significant difference in technique for GMAW of intermediate and thick sections (as opposed to thin sections) is the type of arc used. For welding thin sections ($^1/_8$ in. or less), a short circuiting arc generally is used to minimize heat input (see Examples 3 and 4). Sections of intermediate thickness (about $^1/_8$ to $^1/_2$ in.) and thick sections (more than about $^1/_2$ in.) require greater heat input, and therefore are more often welded with the use of a spray-type transfer.

For groove welds, the joint design requires additional consideration as section thickness increases (see the Appendix to the article "Joint Design and Preparation" in this Volume). The examples that follow describe typical practice for welding sections $^1/_2$ to $1^1/_2$ in. thick.

Example 5. Gas Metal Arc Welding of Plates 1 in. Thick. The GMAW process and the single-V-groove joint design shown in Fig. 19(a) were selected for butt welding of 1-in.-thick low-carbon steel plates. This joint required 100% penetration and high quality. The gas shield permitted the welder to see the arc action without obstruction, which was helpful in obtaining the required quality. The weld was made in nine passes, using a backing strip, as shown in Fig. 19(a). A short cooling period followed passes 3, 5, and 7. The filler metal was $^1/_{16}$-in.-diam low-carbon steel wire fed at 220 in./min. Direct current electrode positive was used at 29 V. A mixture of argon (at 27 ft³/h) and carbon dioxide (at 18 ft³/h) was used as the shielding gas.

Fig. 18 Field-erected housing. Changing from SMAW to GMAW simplified training of welders and more than doubled the welding speed.

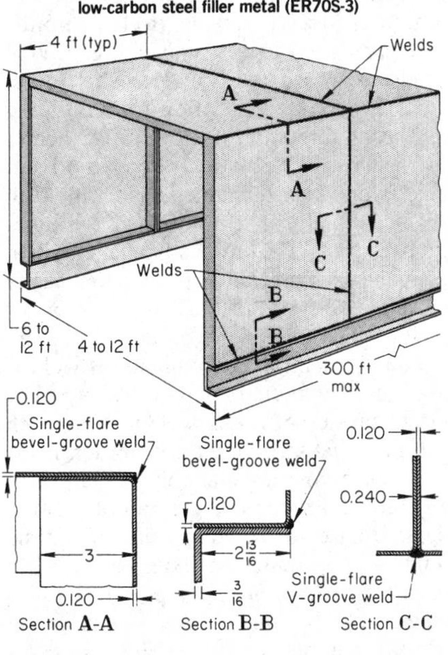

Conditions for GMAW

Power supply	200 A, constant-voltage rectifier, with variable inductance
Wire feed	Push type
Electrode wire	0.030-in.-diam ER70S-3
Welding gun	200 A, air cooled
Current	140 A (DCEP)
Voltage	20 V
Arc type	Short circuiting
Shielding gas	75% argon-25% carbon dioxide, at 15 ft³/h
Welding positions	Horizontal, vertical
Number of passes	1
Wire-feed rate	200 in./min
Electrode extension	$^1/_2$-$^3/_4$ in.

Example 6. Gas Metal Arc Welding of Plates $1^1/_2$ in. Thick. A double-bevel groove was selected for the butt joint in welding the $1^1/_2$-in.-thick plates shown in Fig. 19(b). This type of joint was selected not only to minimize the amount of weld metal used, but also to permit full accessibility of the welding gun and shielding gas to the root of the joint. The GMAW process was selected mainly to avoid the risk of entrapping slag, which could have occurred with SMAW.

The weld was completed in 14 passes (seven on each side) using $^1/_{16}$-in.-diam low-carbon steel electrode wire fed at 210 in./min. Direct current electrode positive was used at 29 V. A mixture of argon and carbon dioxide was used as the shielding gas; argon flow was 27 ft³/h; carbon dioxide, 18 ft³/h.

Sections of Unequal Thickness

The difference in thickness that can be tolerated between sections being welded is generally the same for GMAW as for other arc welding processes. Often the amount of penetration required in the thicker section is the governing factor, because the current used and the resulting heat input must be low enough to avoid overheating the thinner section.

Two approaches are possible to minimize heat-sink differential when there are wide differences in thickness between two sections. One approach is to machine a groove in the thick section close to the joint to be welded; then, in effect, two sections of more nearly the same thickness are welded. This approach is commonly used in making boilers and heat exchangers, where relatively thin-walled tubes must be welded to thick tube sheets.

Another approach is to hold a copper backing bar or strip against the thin section during welding, using the bar or strip as a heat sink. Additional information on backing strips can be found later in this article. In many applications, if requirements are not stringent, a considerable difference in thickness between parts being welded can be tolerated without the use of special techniques. One such application is described in the example that follows.

Example 7. Welding 0.060-in. Sheet to a Thick Section. A dust shield made from 0.060-in.-thick low-carbon steel was welded to a relatively thick yoke forged from low-carbon steel, to make the universal-joint assembly shown in Fig. 20. The dust shield was supported on a

Fig. 19 Joint designs and buildup sequences for butt welding of plates

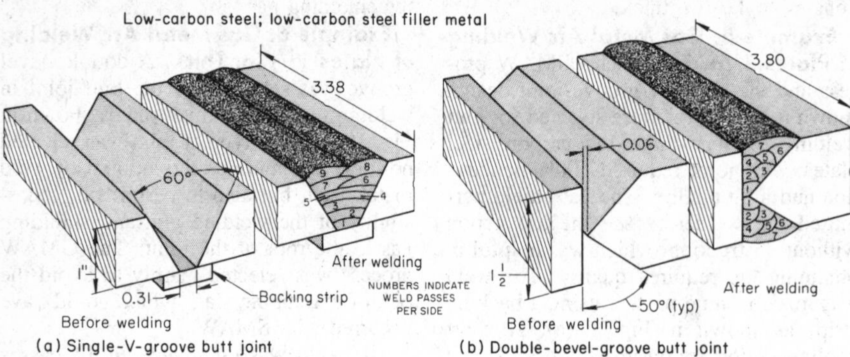

cut-out ring, and clamping pressure was applied on the end of the yoke to hold the parts in position for welding. Welding was done by GMAW, using a hand-manipulated water-cooled welding gun at a 45° work angle and a 0° lead angle. Additional welding details are given in the table that accompanies Fig. 20. Thousands of these assemblies were produced with no melt-through of the 0.060-in.-thick sheet.

Weld Design for Improved Accessibility

The accessibility of the weld joint is an important factor in determining weld joint design. Welds can be made from either one or both sides of the joint. Single-V, J, U, bevel, and combination grooves are used when accessibility is from one side only and on thinner metal. Double-V, J, U, bevel, and combination grooves are used on thicker metal where the joint can be welded from both sides. Double-groove welds have three major advantages over single-groove welds where accessibility is only from one side. The first is that distortion is more easily controlled through alternate weld-bead sequencing. Weld beads are alternated from one side to the other to keep the distortion from building up in one direction. The roots are nearer the center of the plate. A second advantage is that less filler metal is required to fill a double-groove joint than a single-groove joint. The third advantage is that complete penetration can be more easily ensured. The root of the first pass on the plate can be gouged or chipped out before the root pass on the second side is welded to ensure complete fusion at the root. The disadvantages of joints welded from both sides are that more joint preparation is required and gouging and chipping are usually required to remove the root of the first pass. Both of these add to the labor time required. Welding on both sides of a square-groove weld joint provides fuller penetration in thicker metal than metal welded from one side only. This also saves joint preparation time.

The complexity of the welding guns used in GMAW may make the process less versatile than SMAW for applications in which welding is done on intricate assemblies that have joints in difficult-to-reach locations.

Sometimes GMAW can be simplified, or its efficiency can be increased, by minor changes in weld design—particularly in placement of the welds for greater accessibility. Two weldments that serve as examples are shown in Fig. 21 and 22.

Figure 21 shows a three-piece weldment that requires a square-groove butt weld at one end and a fillet weld for securing a section of tubing in place at the other. If the fillet weld were to be made as a $^1/_8$-in. continuous circumferential bead (Fig. 21a), this would present two difficulties: (1) the two welds could not be made in one setup, because the wire-feed rate required for the butt weld would not be appropriate for the fillet weld; and (2) welding completely around a small circumference would be difficult unless the workpiece were rotated. If the weldment were to be redesigned so that a larger ($^3/_{16}$-in.), two-segment fillet weld (Fig. 21b) replaced the complete circumferential fillet weld, the same wire-feed rate could be used for both the butt and the fillet welds. The change from a circumferential to a two-segment weld would also greatly increase accessibility. Also note in Fig. 21 that the square-groove butt weld would require two $^1/_8$-in. welds (one on each side) and that there would be no root opening. In the more efficient design (Fig. 21b), a $^3/_{32}$-in. root opening could be used, which would permit the weld to be made with a single-pass $^1/_4$-in. weld (one side only).

Figure 22(a) shows a weldment designed with four fillet welds. With the fillet placement shown, GMAW would be impractical, because the $1^1/_4$-in. space between the two components would not permit access by the welding gun for the full $1^1/_2$-in. weld length. Changing weld placement as shown in Fig. 22(b) would provide essentially the same fillet length, and would reduce the need for accessibility so that GMAW could be used.

Fig. 20 Universal-joint assembly. A dust shield of formed thin sheet was welded to a heavy-section forging.

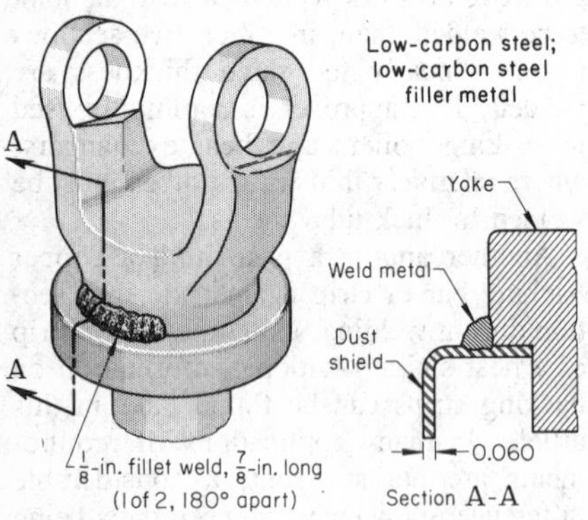

Electrode wire(a)	0.035-in.-diam low-carbon steel
Current	160 A
Voltage(b)	23.5 V
Shielding gas	Carbon dioxide, at 20 ft^3/h
Wire-feed rate	276 in./min
Electrode extension	$^7/_{16}$-$^1/_2$ in.
Welding speed	17.5 in./min
Welding time per piece (two welds)	6 s

(a) 0.17% C, 1.44% Mn, 0.017%, P, 0.023% S, 0.56% Si. (b) Direct meter reading across arc

Fig. 21 Three-piece weldment. A change in size and placement of fillet weld allows use of same wire-feed rate for both welds and improves joint accessibility.

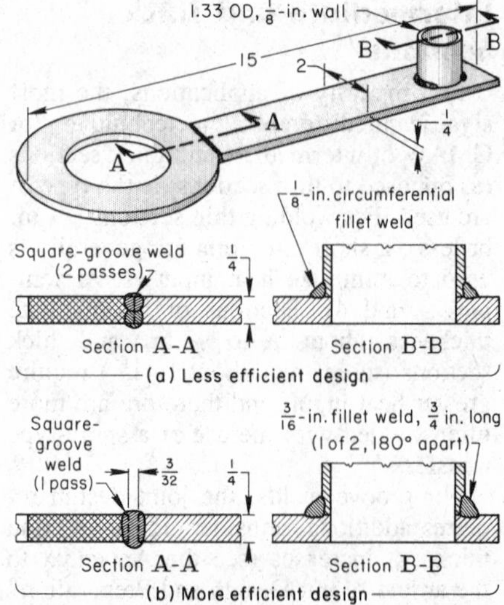

Backing Strips. When backing strips are used, joints are accessible from one side only. Backing strips allow better access to the root of the joint and support the molten weld metal. These strips are available in two forms: fusible or nonfusible. Fusible backing strips are made of the metal being welded and remain as part of the weldment, but may be cut or machined off. Nonfusible backing strips are made of copper, carbon, flux, or ceramic backing in tape or composite form. These forms of backing do not become part of the weld. Backing strips on square-groove joints make a full-penetration weld from one side easier. For this application, using a backing is more expensive because of the cost of a backing strip and the larger amount of filler metal required. However, on V-groove joints, the backing strip allows wider root openings and removes the need for a root face, which reduces the groove preparation costs. Another advantage is that because the root may be opened up, the groove angle may be reduced, which will reduce the amount of filler metal required in thicker metal. These effects are shown in Fig. 23, where single-V-groove joints are shown with and without a backing strip.

Welding Variables

Welding variables affect the characteristics of the weld, such as penetration of the weld, bead height, bead width, and deposition rate. Some of the variables that govern weld characteristics include:

- Welding current
- Arc voltage
- Travel speed
- Electrode extension

Fig. 22 Relocation of fillet welds to improve accessibility and allow use of GMAW

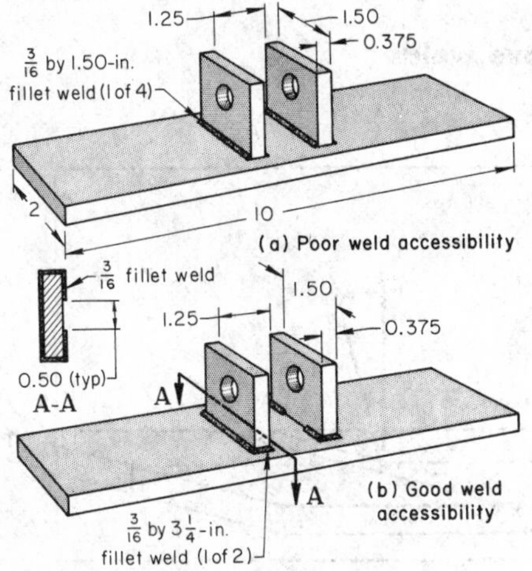

Fig. 23 Single-V-groove joints. With or without backing strip in the same thickness of metal

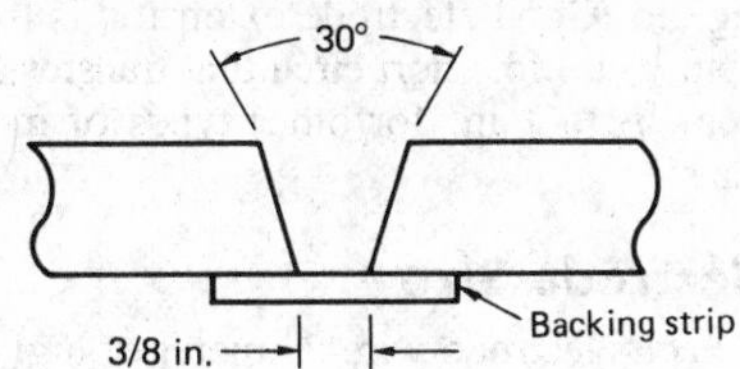

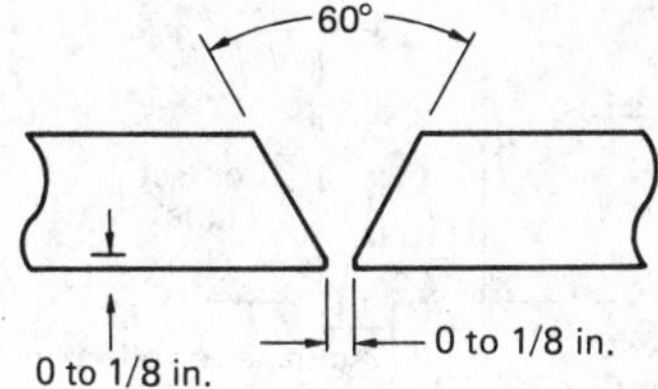

- Electrode size
- Electrode position

Proper control of these variables is essential to produce weld properties that meet or exceed the service requirements for a given joint. All of the above variables are discussed in this article. Weld characteristics are also affected by shielding gas selection. See Table 4 for information on shielding gas selection for carbon steels.

Welding Current

The amount of welding current used has the greatest effect on deposition rate, weld-bead size and shape, and the penetration of the weld. In a constant-voltage system, as the wire-feed speed is increased, the welding current increases. Figure 24 shows typical deposition rates for four sizes of carbon steel electrodes using carbon dioxide shielding gas.

As shown in Fig. 24, the deposition rate of the process increases as the welding current increases. The lower part of the curve is flatter than the upper part, because at higher current levels, the melting rate of the electrode increases at a faster rate as the current increases. This can be attributed to resistance heating of the electrode extension beyond the contact tube. When other welding variables are held constant, increasing the welding current will increase the depth and width of weld penetration and the size of the weld bead. An excessive current level will create a large, deep-penetrating weld bead that wastes filler metal and can burn through the bottom of the joint. An excessively low welding current produces poor penetration and a piling up of weld metal on the surface of the base metal. The deposition rates of GMAW are higher for the same welding currents that are obtained with SMAW.

Fig. 24 Effect of welding current on deposition rates of four carbon steel electrode wires using carbon dioxide shielding gas

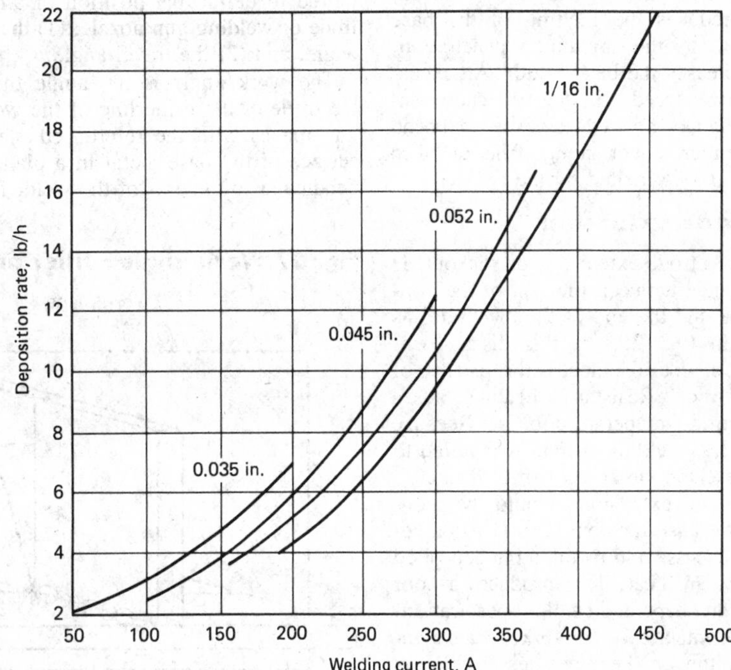

Fig. 25 Electrode extension (or stickout) for GMAW

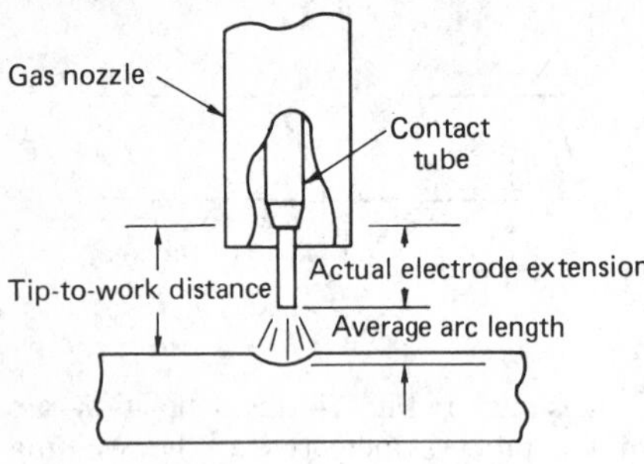

These higher rates occur because there is no electrode coating that must be melted.

Travel Speed

The travel speed is the linear rate at which the arc travels along the workpiece. Travel speed is controlled by the welder in semiautomatic welding. In machine and automatic welding it is controlled by the machine. When the travel speed is decreased, the amount of filler metal deposited per unit length increases, which creates a large, shallow weld pool. The welding arc impinges on this pool rather than the base metal as the arc advances. This limits penetration but produces a wide bead. An excessively slow travel speed can cause excessive piling up of the weld pool, overlapping at the edges, and excessive heat input to the plate, which creates a larger HAZ.

As the travel speed is increased, the heat transmitted to the base metal is reduced, which reduces the melting of the base metal, limits the amount of penetration, and decreases the bead width. An excessive travel speed will tend to cause undercutting along the edges of the weld bead because there is not enough filler metal to fill the groove melted by the arc.

Electrode Extension

The electrode extension, or stickout, is the distance between the end of the contact tube and the end of the electrode, as shown in Fig. 25. As this distance increases, so does the electrical resistance of the electrode. Resistance heating causes the electrode temperature to rise. Because of this, less welding current is required to melt the electrode at a given feed rate.

Electrode extension should be controlled because too long an extension results in excess weld metal being deposited with low arc heat. This produces a poor weld-bead shape and shallow penetration. As the contact tube-to-work distance increases, the arc becomes less stable. A longer electrode extension will also produce a higher deposition rate, as shown in Fig. 26. Good electrode extension is from $^1/_4$ to $^1/_2$ in. for short circuiting transfer and from $^1/_2$ to 1 in. for other types of metal transfer.

Electrode Size

Each electrode wire diameter of a given chemical composition has a usable welding current range. The electrode melting rate is a function of current density. If two wires of different diameters are operated at the same amperage, the smaller will have the higher melting rate.

Penetration is also a function of current density. A smaller electrode wire (0.045 in.) will produce deeper penetration than a larger diameter wire ($^1/_{16}$ in.) at the same current setting. The weld bead, however, will be wider when using the larger electrode wire. The choice of the size of the electrode wire to be used is dependent on the thickness of the metal being welded, the amount of penetration required, the deposition rate desired, the bead profile desired, the position of welding, and the cost of the different electrode wires. A smaller electrode wire is more costly on a weight basis, but for each application there is a wire size that will produce minimum welding costs.

Electrode Position

The position of the welding electrode with respect to the weld joint affects the shape of the weld bead and the amount of penetration obtained. Two angles are required to define the position of an electrode or welding gun nozzle: (1) the work angle, and (2) the travel angle.

The work angle is the angle that the electrode or the centerline of the welding gun makes with the referenced plane or surface of the base metal in a plane perpendicular to the axis of the weld. Figure 27 shows the work angle for a fillet weld and a groove weld.

Fig. 26 Effect of electrode extension on deposition rate. Welding current is constant at 175 A, and arc length also is constant.

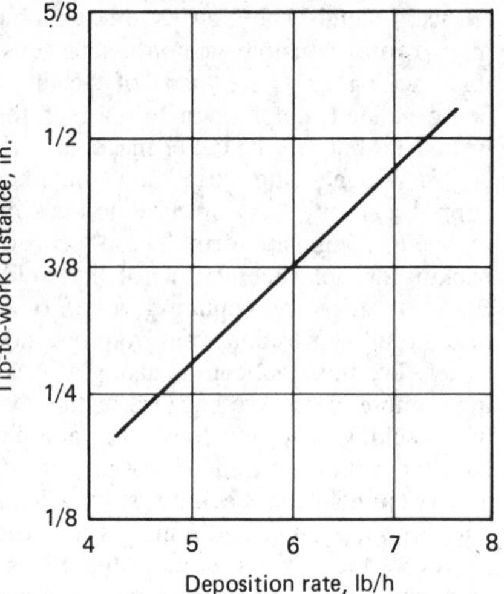

The travel angle is the angle that the electrode, or the centerline of the welding gun, makes with a reference line perpendicular to the axis of the weld in the plane of the weld axis (Fig. 28). The travel angle is further described as either a drag angle or a push angle. Figure 28 also shows the drag and push travel angles. The drag angle, which points backward from the direction of travel, is also known as backhand welding. The push angle, which points forward in the direction of travel, is also known as forehand welding.

Maximum penetration is obtained when a drag angle of 15 to 20° is used. If the gun travel angle is changed from this optimum condition, penetration decreases. From a drag angle of 15° to a push angle of 30° the relationship between penetration and travel angle is almost a straight

Fig. 27 Work angle—fillet and groove welds

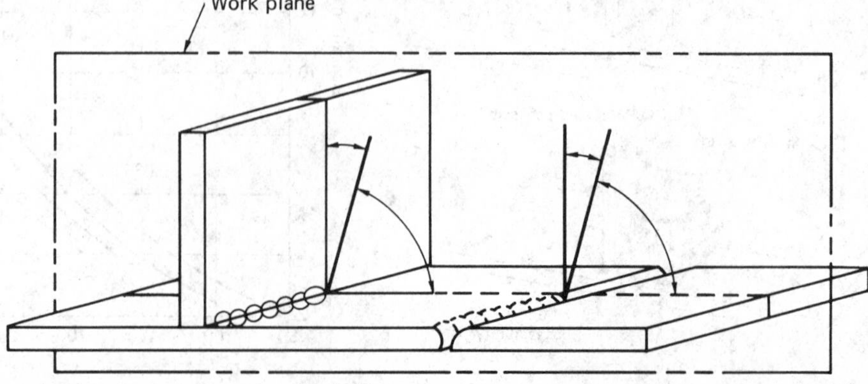

Fig. 28 Travel angle. (a) Fillet weld. (b) Groove weld

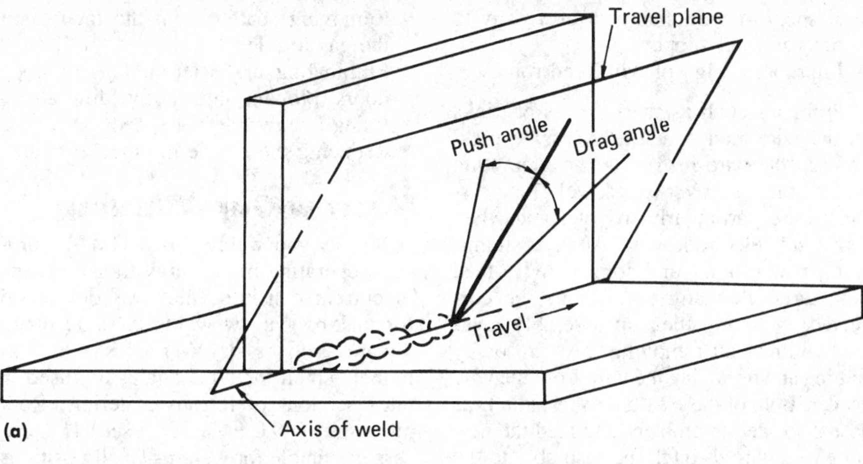

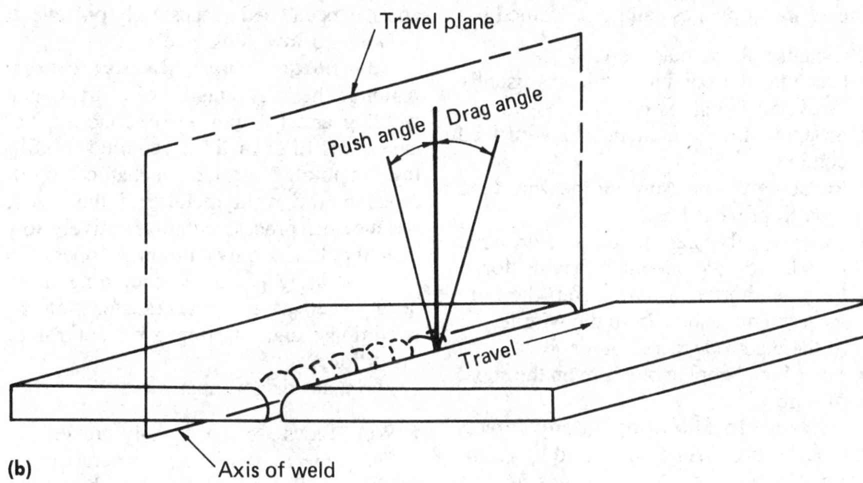

line. Therefore, good control of penetration can be obtained in this range. A drag angle of greater than 25° should not be used. The gun travel angle variable can also be used to change bead height and width, because the gun travel angle does affect bead contour. A drag travel angle, which is commonly used on steel, tends to produce a high, narrow bead. As the drag angle is reduced, the bead height decreases and the width increases. The push travel angle is used on aluminum to avoid contamination and give good penetration but minimize the heat input to the base metal.

Weld Quality

Gas metal arc welding, like the other welding processes, can produce discontinuities or defects such as weld-metal cracks, inclusions, porosity, undercutting, incomplete fusion, excessive melt-through, and lack of penetration. Poor welding technique and improper choice of welding parameters are major causes of weld flaws. Contaminants and the use of improper base metal, filler metal, or shielding gas can also cause flaws. Various weld discontinuities in gas metal arc welds and their possible causes are:

Weld-metal cracks

- Weld depth-to-width ratio too high
- Weld bead (particularly fillet and root beads) too small
- Rapid cooling of the crater at the end of a weld
- Excessive travel speed
- Poor fit-up
- High joint restraint

Porosity

- Inadequate shielding gas flow rate
- Drafts that deflect the shielding gas coverage
- Blockage of the shielding gas flow by spatter build-up on the nozzle
- Contaminated or wet shielding gas
- Excessive welding current
- Excessive welding voltage
- Excessive electrode extension
- Excessive travel speed, causing rapid freezing of the weld pool before gases can escape
- Electrode contamination
- Rust, grease, oil, moisture, or dirt on the surface of the base metal or filler wire, including moisture trapped in aluminum oxide
- Impurities in the base metal, such as sulfur and phosphorus in steel

Inclusions

- Excessively high travel speeds when welding metals with heavy oxide coatings (for example, aluminum)
- Inadequate preweld cleaning of base metal
- Use of multiple-pass, short circuiting arc welding (slag inclusions)

Incomplete fusion

- Excessive travel speed, causing an excessively convex weld bead or inadequate penetration
- Welding current too low
- Poor joint preparation
- Letting the weld metal get ahead of the arc or letting the weld layer get too thick, which keeps the arc away from the base metal

Lack of penetration

- Improper joint design that does not provide adequate access to the bottom of the groove
- Inadequate root opening in butt joints
- Failure to keep the arc on the leading edge of the weld pool
- Inadequate heat input

Undercutting

- Excessive welding current
- Arc voltage too low
- Excessive travel speed, causing an inadequate amount of filler metal to be added
- Erratic feeding of the electrode wire
- Excessive weaving speed
- Incorrect electrode angles, especially on vertical and horizontal welds

Excessive melt-through

- Excessive welding current
- Travel speed too slow
- Root opening too wide or root face too small

Common Causes of Welding Difficulty

In GMAW with either a hand-held or a mechanically held welding gun, production rate or weld quality, or both, may be impaired as a result of defective contact at the electrode tip (in the contact tube), wire-feed stoppages, and inadequate shielding. These common causes of difficulty, together with methods of avoiding or correcting them, are discussed in the paragraphs that follow.

Contact-Tube Problems. Gas metal arc welding depends on a transfer of current from a tip of copper or some low-resistance alloy to the electrode wire by means of a sliding contact within the contact tube. Changes in resistance at this point will change arc voltage and, in turn, arc characteristics and may result in weld flaws. The current density that must be passed across this point is higher than that used in other welding processes such as SAW. Thus, small irregularities on the electrode wire, such as microscopic laps, seams, and slivers, which would have little effect in SAW, may significantly affect GMAW. This effect increases in importance as electrode size decreases and is more significant when welding alloys having higher resistance than carbon steel or low-alloy steel.

Residual drawing lubricants that have been allowed to remain on the electrode wire can result in a glazed condition within the contact tube and can increase the resistance to current passage. Electrode wire that has insufficient temper is a potential source of trouble, because the resulting low bearing pressure of the electrode within the contact tube can cause arc instability.

These problems can best be prevented by making sure that the electrode wire is clean and that the wire-feed rolls are tight enough to feed the wire without creating chips. A wire wipe made of cloth often is attached to the wire feeder to clean the electrode wire as it is fed.

Wire-Feed Stoppages. Gas metal arc welding has a greater problem with wire-feed stoppages compared to other continuous wire-feed welding processes, because of the relatively small-diameter electrode wires used. In addition to the loss of welding time, wire-feed stoppages cause the arc to be extinguished and can create an irregular weld bead because of stops and starts. Wire-feed stoppages can be caused by:

- A clogged contact tube
- A clogged conduit in the welding gun assembly
- Sharp bends or kinks in the wire-feed conduit
- Excessive pressure on the wire-feed rolls that can cause breakage of the wire
- Inadequate pressure on the wire-feed rolls
- Attempting to feed the wire over excessively long distances
- A spool of wire clamped too tightly to the wire reel support
- Improper design of wire-feed rolls

Problems such as sharp bends or kinks in the wire-feed conduit, excessive pressure on the wire-feed rolls, or attempting to feed the wire over excessively long distances are particularly troublesome when using soft electrode wires such as aluminum, magnesium, and copper. Wire-feed stoppages, in many cases, must be corrected by taking the gun assembly apart and cutting and removing the wire or by cutting and removing the wire from the wire feeder. Both of these solutions result in time lost to locate the problem and feed the new length of wire through the assembly to the gun. Wire stoppages can be prevented by:

- Cleaning the contact tube
- Cleaning the conduit, which is usually done with compressed air
- Straightening or replacing the wire-feed conduit
- Reducing the pressure on the wire-feed rolls to prevent breakage
- Increasing the pressure on the wire-feed rolls to provide adequate driving force
- Using a shorter distance from the wire feeder to the gun or from the wire feeder to the electrode wire source
- Reducing clamping pressure on the spool of wire

Inadequate Shielding. Many defects that occur in GMAW are caused by an inadequate flow of shielding gas to the welding area or blockage of the shielding gas flow. An inadequate gas supply can cause oxidation of the weld pool and porosity in the weld bead. This will usually appear as surface porosity. The most common causes of inadequate shielding are:

- Blockage of gas flow in the torch or hoses or freezing of the regulator with carbon dioxide
- A leak in the gas system
- Blockage of the welding-gun nozzle by spatter
- Very high travel speed
- Improper flow rate
- Wind or drafts
- Excessive distance between nozzle and work

The problem can be corrected by:

- Checking the torch and hoses before welding to make sure the shielding gas can flow freely and is not leaking
- Increasing the shielding gas flow to displace all air from the weld area
- Decreasing the shielding gas flow to avoid turbulence and entrapment of air in the gas
- Removing spatter from the interior of the gas nozzle
- Eliminating drafts (from fans or open doors) blowing into the welding arc
- Using a slower travel speed
- Reducing the nozzle-to-work distance

Narrow-Gap Welding

Narrow-gap welding is a GMAW process, operating in the spray-transfer range of current densities, that was developed for making narrow welds in thick plate. Square grooves or V-grooves with extremely small included angles are used in thick sections of ferrous materials. Root openings of $^1/_4$ to $^3/_8$ in. are used. The process is suitable for welding in all positions and has been used successfully on several carbon and low-alloy steels.

Advantages and Disadvantages. Among the advantages of narrow-gap welding are: (1) improved economy, because less filler metal is required for filling the joint; (2) good mechanical properties in the weld metal and the HAZ, because the process entails relatively low heat input; (3) fully automatic operation in all welding positions, including overhead, when using spray-transfer welding conditions; and (4) improved control of distortion.

The main disadvantages are:

- Weld flaws are more easily created.
- Flaws are more difficult to remove.
- Fit-up of the joint must be more precise.
- Placement of the welding gun must be more precise.

Operating Principles. Narrow-gap welds are deposited by using special water-cooled contact tubes that are inserted into the joint. In the preferred configuration, two contact tubes are used in tandem, and the electrodes are oriented so that one weld bead is directed toward one sidewall, and the other weld bead is directed toward the opposite sidewall. Each contact tube is also guided so that it remains a fixed distance from its respective sidewall to ensure proper sidewall fusion regardless of variations in gap width. Narrow-gap welding can also be done using a single contact tube centered in the joint. With this method, shielding gas is introduced in the narrow-gap joint from the plate surface by means of a special gas shield, and the weld is completed from one side of the plate.

Narrow-gap welds have been made with electrode wires ranging from 0.035 to 0.060 in. in diameter. Out-of-position narrow-gap welds preferably are deposited with the use of 0.035-in.-diam electrode wire.

Relatively high travel speeds are used because of the confinement imposed by the narrow-gap joint. If slower welding speeds were used, the weld pool would become too large to control in the narrow gap. The weld beads obtained with high travel speeds are thin and are deposited one on top of the other to fill the joint.

The major problem encountered in narrow-gap welding is incomplete side-wall fusion because of the low heat input in thick metal. Careful placement of the electrode wires and removal of slag islands between passes to prevent slag inclusions can avoid problems. When used for welding metal thicknesses over 2 in., narrow-gap welding is competitive with the other automatic arc welding processes.

Welding Applications. Narrow-gap welding systems have been proposed for both shop and field erection of large pressure vessels. One potential application, the narrow-gap welding of horizontal girth joints in large vertical steam generators, prompted a feasibility study, a part of which is described in the example that follows.

Example 8. Narrow-Gap Flat and Horizontal Welding of Steel Plate. As part of a feasibility study, narrow-gap welds were made in steel plates $6^3/_4$ and 8 in. thick, using a laboratory model narrow-gap welding unit. The $6^3/_4$-in.-thick plates were of ASTM A302, grade B steel (a manganese-molybdenum steel containing a maximum carbon content of 0.20%, 1.15 to 1.50% Mn, and 0.45 to 0.60% Mo); the 8-in.-thick plates were of low-carbon steel. Welds in the A302 plate were made in the horizontal position; those in the low-carbon steel plate, in both the flat and the horizontal positions.

All welds were made using 0.035-in.-diam ER70S-G electrode wire and a mixture of 75% argon and 25% carbon dioxide as the shielding gas. Welding current at each of two electrodes was 260 to 270 A at 25 to 26 V. Welding speed, using special long contact tubes and standard gas shields, was 40 in./min.

Difference in plate thickness had no apparent detrimental effect on the quality of the welds. The weld beads had a bright silvery appearance, and macrosections indicated that weld soundness was satisfactory. Estimated joint completion rate at 100% arc operating factor was $2^3/_4$ ft/h.

Joint Completion Rates.* Figure 29 shows joint completion rates for narrow-gap welding of various thicknesses of low-

*Length of joint welded to full groove depth in 1 h (also called joint finishing rate)

Fig. 29 Joint completion rates. For narrow-gap welding of various thicknesses of low-carbon steel plates at two deposition rates

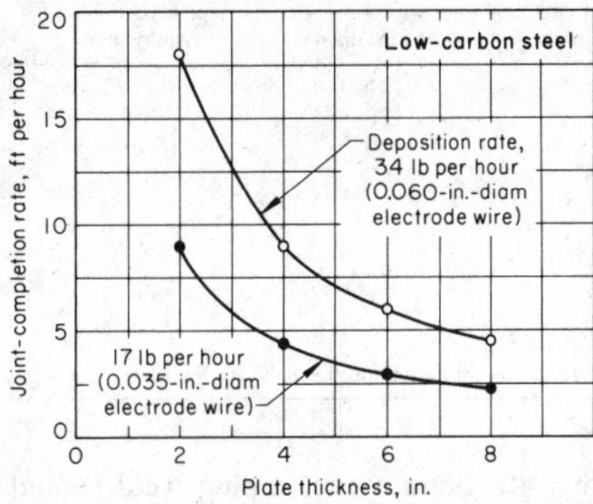

carbon steel plates at deposition rates of 17 and 34 lb/h. The deposition rate of 17 lb/h is obtained in all welding positions with the 0.035-in.-diam electrode wire normally used in narrow-gap welding of low-carbon steels or high-strength low-alloy (HSLA) structural steels. The 34-lb/h deposition rate is obtained with larger wire (0.060-in.-diam) and can be achieved only in flat-position welding.

Figure 30 compares joint completion rates for narrow-gap welding of low-carbon steel plates with those obtained with two typical SAW procedures. These comparisons are for flat-position welding, using a deposition rate of 17 lb/h for narrow-gap and two deposition rates for SAW—48 lb/h, for applications where heat-input control is not important, and 15 lb/h, for applications where heat-input control is important. All of the curves are based on a 100% duty cycle. The SAW data are for double-V-grooves with a 45° included angle and a $^1/_4$-in. root face welded from both sides. The narrow-gap data are for welding from one side only. The data show that, for thicknesses greater than 2 in., narrow-gap welding compares favorably with SAW.

Figure 31 compares joint completion rates for vertical-position welding of low-carbon steel plates by the narrow-gap, SAW, electrogas welding (EGW), and dual-electrode electroslag welding (ESW) processes. Submerged arc welding, EGW, and ESW were done with $^1/_8$-in.-diam electrodes. Spacing between plates was $^3/_4$ in. for EGW, and 1 in. for SAW and ESW. All data in Fig. 31 are for 100% duty cycle. As these data show, for vertical-position welding, the joint completion rates for narrow-gap welding are greater than for any of the other processes except dual-electrode ESW (on 5-to-8-in. plate).

Welding parameters for flat-position narrow-gap welding of 2-in.-thick HY-80 steel plate using 0.045- and 0.062-in. Airco AX-90 filler wire are:

Welding condition	AX-90 filler wire	
Diameter of electrode, in.	0.045	0.062
Arc voltage, V	26-28	26-28
Current, A	280-320	320-360
Wire-feed speed, in./min	535-565	210-270
Carriage speed, in./min	9	11
Preheat and interpass temperature, °F	200-300	200-300
Gas flow rate, ft^3/h	90	90
Type of gas	Argon-2% O_2	Argon-2% O_2
Gap opening, in.	0.25-0.375	0.25-0.375

The mechanical properties of these welds are given in Table 5. For a complete re-

Fig. 30 Joint completion rates. For narrow-gap welding and SAW of various thicknesses of low-carbon steel plates in the flat position

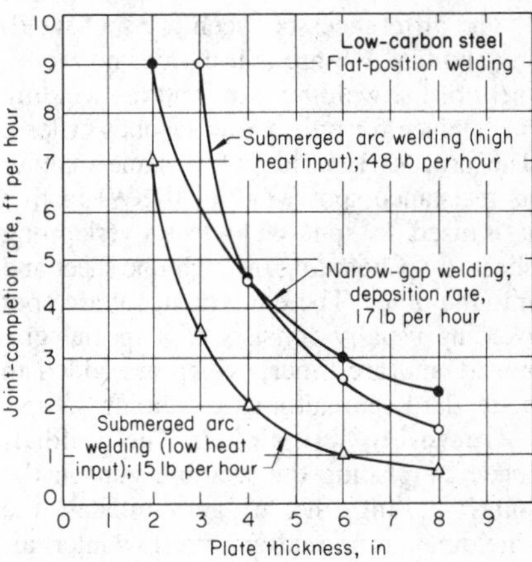

Fig. 31 Joint completion rates. For vertical-position welding of various thicknesses of low-carbon steel plates by narrow-gap welding, dual-electrode ESW, EGW, and SAW

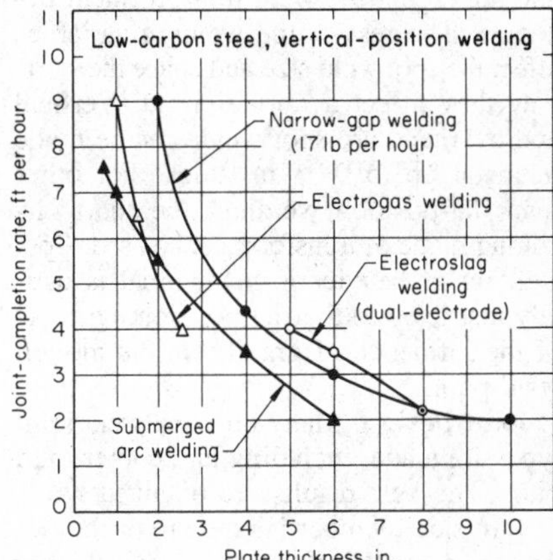

Table 5 Mechanical properties of 2-in.-thick HY-80 steel narrow-gap weldments

Property(a)	Airco AX-90 filler wire 0.045 in. diam	Airco AX-90 filler wire 0.062 in. diam	MIL-E-23765/2 type MIL-100-S-1 requirement
Yield strength at 0.2% offset, ksi	106	105	88-102
Tensile strength, ksi	113	120	(b)
Elongation in 4D, %	23	20	16 min
Reduction in area, %	66	67	(b)
CVN value, ft · lb at:			
60 °F	63	(c)	50 min
0 °F	92	(c)	...
30 °F	98	(c)	...
Room temperature	160	96	...

(a) Values are average except for dynamic tear where individual values are shown. (b) For information only. (c) Not tested

view of developments in narrow-gap welding, the reader should consult:

- Malin, V.Y., The State-of-the-Art of Narrow Gap Welding, *Welding Journal,* Vol 62 (No. 4), p 22-32
- Malin, V.Y., The State-of-the-Art of Narrow Gap Welding, *Welding Journal,* Vol 62 (No. 6), p 37-46

Arc Spot Welding

The differences between arc spot welding and GMAW are that there is no movement of the welding gun, and the welding takes place for only a few seconds or less. The process is used in the same manner as resistance spot welding (RSW) in that it is used to spot weld two overlapping sheets by penetrating through one sheet and into the other. The equipment for arc spot welding usually consists of a special gun nozzle and arc timer, which are added to a standard semiautomatic welding setup.

Operating Principles. The weld is made by placing the welding gun on the joint. Pulling the trigger initiates the shielding gas and, after a preflow interval, starts the arc and the wire feed. When the preset weld time is finished, the arc and wire feed are stopped, followed by the shielding gas flow. The longer the weld time, the greater the penetration obtained and the higher the weld reinforcement becomes. The rest of the welding variables affect the spot weld size and shape the same way they affect a normal weld. Vertical and overhead arc spot welds can be made in metal up to 0.05 in. thick. For other than flat-position welding, the short circuiting mode of transfer must be used. Spot welding of heavier gage material is usually restricted to the flat position, because of the influence of gravity on the molten weld pool.

Joint Design. Many different weld joint types are made, including lap, corner, and plug. The best results are obtained when the arc side member is equal to or thinner than the other. When the top plate is thicker than the bottom one, a plug weld should be made. In plug welding, care must be taken to ensure complete penetration into the base plate and also into the side walls of the top member. Lack of fusion is a common problem in this type of joint. A copper backing block may be required on the back side of an arc spot weld to prevent excessive penetration when the bottom sheet is thinner than the top.

Applicability. Gas metal arc spot welding is commonly applied to low-carbon steel, stainless steel, and aluminum, but can be used on all the metals welded by GMAW. The shielding gases for arc spot welding are selected by using the same criteria as those used for conventional GMAW. The shielding gas performs the same functions in arc spot welding, but a flow rate of 25 to 50% of that normally used for continuous GMAW is adequate.

Advantages and Disadvantages. The advantages of this process over RSW are that the gun is light and portable and can be taken to the weldment, spot welding can be done in all positions more easily, spot welds can be made when there is accessibility to only one side of the joint, spot weld production is faster for many applications, and joint fit-up is not as critical.

The major disadvantage of this process is that the consistency of weld strength and size is not as good as with RSW. This is especially true for aluminum arc spot welds. A typical application of arc spot welding is described in the next example.

Example 9. Arc Spot Welding of Motor-End Panels. Figure 32 shows a motor-end panel to which support bars and channel sections were arc spot fillet welded by timed intermittent GMAW. As shown at the right in Fig. 32, the welding gun had a nozzle-guide tube with a 45° bevel on opposite sides, so that when the tube was seated against two right-angled surfaces, the welding gun was automatically positioned at a 45° angle from each of the two surfaces. For welding in the vertical position, as in this application, the gun was lowered 15° from horizontal (Fig. 32). A ceramic bushing was inserted over the guide tube. The end of the guide tube was preset for a $^5/_8$-in. nozzle standoff. Welding conditions are given in the table that accompanies Fig. 32. In comparison with shielded metal arc intermittent fillet welding, the method previously used for joining of these panels, timed intermittent GMAW resulted in a 30% decrease in welding time and required less welder skill.

Fig. 32 Spot welded motor-end panel. Spot fillet welds were made by timed intermittent GMAW using a portable welding gun with a nozzle-guide tube, positioned as shown.

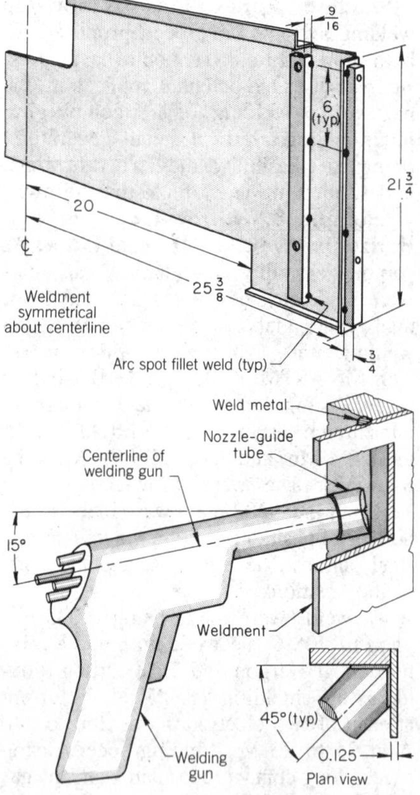

Power supply	Constant-voltage rectifier
Electrode wire	$^1/_{16}$-in.-diam low-carbon steel
Welding gun	Portable automatic, with constant wire feed rate
Current	380-400 A (DCEN)
Voltage	30 V
Arc time(a)	25 cycles
Shielding gas	Argon plus 27% oxygen, at 14 ft^3/h (measured by flowmeter)
Welding position(a)	Vertical
Bead size(a)	$^1/_4$-in. fillet leg and $^1/_4$ in. long

(a) Welding gun was stationary during arc time. On panels of other designs, welds were made in the horizontal position (with the welding gun at 45° to joint), at an arc time of 35 cycles. These horizontal welds also had $^1/_4$-in. fillet legs, but were $^5/_{16}$ in. long.

Automatic Welding

Use of automatic GMAW is considered when one or more of the following conditions prevail:

- Quantity of similar parts to be welded is large enough to warrant the cost of using specially designed fixtures and mechanisms.
- Skilled welders (needed for semiautomatic welding with hand-held welding guns) are not available.
- Workpiece configuration or the type of welding (surfacing, for example) makes the use of a hand-held welding guns impractical.

Tooling for automatic welding is likely to be prohibitively expensive, or to be impractical where:

- The required weld is such that the arc must operate successively in more than one plane.
- Several changes of welding direction or location are required.
- Welding conditions must be changed during the cycle.

The configuration of the workpiece, especially if it requires a more difficult-to-deposit weld bead, is a major factor to be considered when deciding whether or not to use automatic welding. Although automatic welding can be applied to almost any joint or series of joints, the cost of tooling may be prohibitive even if production quantities are large.

In most applications, tooling can be simplified by moving the components to be welded while the welding gun remains stationary, but if the workpiece is large or of complex shape, moving it may not be practical, and the more costly alternative of moving the welding gun will have to be adopted.

Circumferential welding is usually easy to automate, because round workpieces of all sizes can almost always be held and rotated without difficulty. In practice, the welding gun remains stationary and makes the weld as the workpiece slowly rotates. Circular welding is also readily mechanized, using a rotating worktable. The examples that follow describe production applications in which automatic GMAW proved advantageous.

Example 9. Butt Welding of Tube Sections. Sections of steel tubing of a variety of compositions (including ASTM A192) were welded together for use in pressure piping systems, using special automatic welding equipment and a special setup. As shown in Fig. 33, the tubes were held in alignment between hollow universal chucks, with the weld joint (see section A-A in Fig. 33) centered beneath the welding head. At the start of the welding cycle, the tubes were rotated synchronously as the welding gun initiated the arc. The welding head oscillated in a plane transverse to the direction of welding, thereby distributing the weld pool evenly between the two tubes. Although the welding operation was fully automatic up to completion of the weld, loading and unloading of tubes were performed manually.

Fig. 33 Setup and design of joint. Used in automatic GMAW of tubes for pressure piping systems

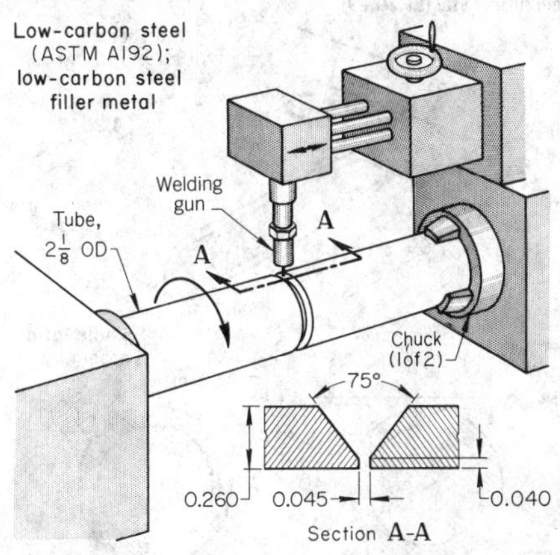

Power supply	300-A rectifier, with variable slope and inductance
Current	DCEP
Electrode wire	0.030-in.-diam low-carbon steel
Shielding gas	80% carbon dioxide-20% argon
Deposition rate	$3^1/_4$ lb/h
Welding time per tube	2 min, 21 s

Example 10. Automatic Welding of Universal-Joint Assemblies for Uniform Strength and Dimensional Accuracy. Figure 34 shows two designs of joints used in welding 1020 steel rods to yokes made of HSLA steel (with a maximum of 0.22% C, 1.25% Mn, 0.04% P, 0.05% S, and 0.30% Si and also 0.02 to 0.07% V and a minimum of 0.20% Cu) in the production of single-yoke and double-yoke universal-joint assemblies. These weldments were used in steering mechanisms for trucks and agricultural machines and were required to withstand a testing torque of 4000 in. · lb. In addition, the welds made in the assemblies of bevel-joint design (Fig. 34b) had to be held to close dimensional tolerances to avoid subsequent machining in the weld area. These requirements were met by the use of automatic GMAW, under the conditions listed in the table with Fig. 34.

On some single-yoke assemblies, the rod was not beveled at the joint with the yoke (Fig. 34a). On other assemblies, both single-yoke and double-yoke, the rod was beveled (Fig. 34b)—not for weld strength, but to avoid a machining operation for maintaining a maximum weld diameter of $1^5/_{32}$ in. This maximum diameter was required for fit of a $1^5/_{32}$-in.-ID washer on the rod against the yoke. Positions of the welding guns in welding joints of both designs were the same, except that a smaller work angle was required for the beveled joint (see views at right in Fig. 34a and b).

Fig. 34 Universal-joint assembly. Automatic GMAW was used on joints of two different designs.

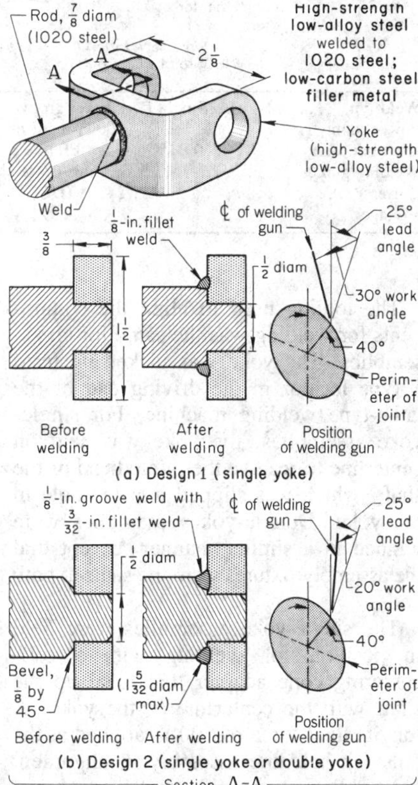

Electrode wire(a)	0.035-in.-diam low-carbon steel
Welding gun	Air cooled; side-delivery gas tube
Current(b)	195 A (DCEP)
Voltage(c)	21 V
Shielding gas(d)	Carbon dioxide, at 50 ft^3/h
Wire-feed rate	350 in./min
Electrode extension	$^1/_2$ in.
Welding speed	18 in./min
Welding time	9 s
Electrode wire consumed per piece	$52^1/_4$ in.

(a) 0.17% C, 1.44% Mn, 0.017% P, 0.023% S, 0.56% Si. (b) From constant-voltage power supply with secondary reactance. (c) Direct meter reading across arc. (d) Flow rate was high to maintain proper shielding; with the air-cooled welding gun and side-delivery gas tube, lower flow rates resulted in poor shielding due to drafts, and reworks were necessary.

Fig. 35 Machine used for automatic GMAW of a spindle-arm assembly with welds on opposite sides

Low-carbon steel; low-carbon steel filler metal (ER70S-3)

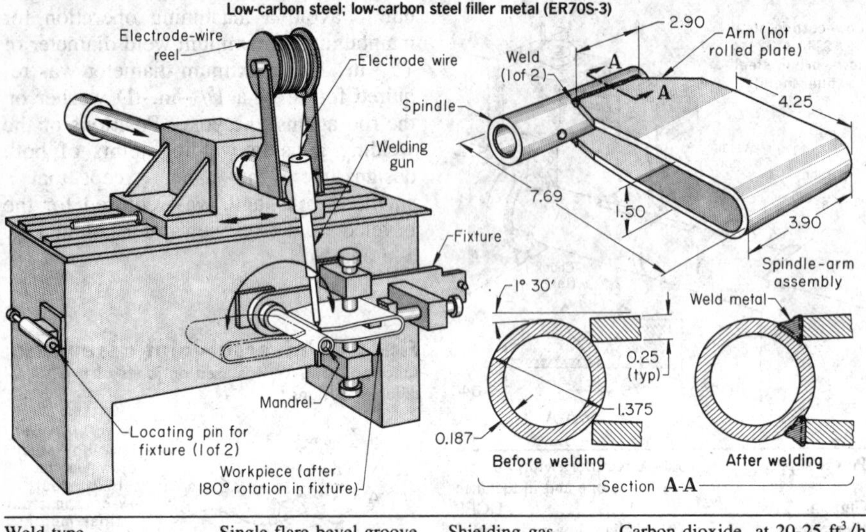

Weld type	Single-flare bevel groove	Shielding gas	Carbon dioxide, at 20-25 ft³/h
Power supply	500-A rectifier	Welding position	Flat
Electrode wire	0.045-in.-diam ER70S-3	Wire-feed rate	400 in./min
Welding gun	500 A, air cooled	Welding speed	30 in./min
Current	240 A (DCEP)	Arc time	6.5 s per side
Voltage	21 V	Floor-to-floor time	76 s per part

Fig. 36 Compressor casing, produced in two halves. Change from SMAW to automatic GMAW decreased welding time and weld repair.

Low-alloy steel; low-carbon low-alloy steel filler metal

Flange; Inner rib, 0.520 by 0.400 deep; 30 ID; Outer rib, 0.110 by 0.670 deep; Section A-A (typ): 0.600, 0.600, 0.270, 0.380, 0.950, 1.150, Weld metal, Before welding, After welding; Section B-B (typ): 0.380, 0.070, 0.060, 0.050, Outer rib, Weld metal, Before welding, After welding

Conditions for GMAW

Joint type	Butt
Weld type	Groove
Power supply	300-A, 40-V rectifier, with variable inductance
Electrode wire(a)	0.030 in. diam
Welding gun	Mechanically held, water cooled
Wire-feed system	Automatically controlled
Current:	
Root pass	180 A (DCEP)
Fill pass	175 A (DCEP)
Voltage	32 V
Shielding gas:	
At electrode	75% argon-25% carbon dioxide, at 30-40 ft³/h
Backing gas	Argon, at 40-50 ft³/h
Welding position	Flat
Number of passes	2
Preheat and postheat	None

(a) 0.20% C, 1.0% Cr, 1.0% Mo, 0.10% V

The tooling used to align the components for welding was the same for all assemblies. The yoke was picked up by a V-type locator in the driving end of the lathe-type welding machine. For single-yoke assemblies, the yoke was held on centerline to receive the piloted end of the shaft, which was slipped into the hole in the yoke. Double-yoke assemblies were aligned in a similar manner, except that the assembly fixtures were the same at both ends.

The single-yoke assemblies were held in position for welding with a self-centering cone adapter that lined up the shaft with the centerline of the yoke. A longitudinal force of 500 lb applied at the tailstock held the assembly in alignment for welding. A variable-speed direct current motor was employed to control the rotational speed of the welding lathe, and a timer controlled the duration of the welding cycle. The two welds required on assemblies with yokes at both ends of the rod were deposited simultaneously.

Example 11. Automatic Welding of a Spindle-Arm Assembly. The automatic welding machine shown at the left in Fig. 35 was used for making two welds 2.90 in. long in a spindle-arm assembly (shown, together with joint details, at right in Fig. 35). This machine was one unit in a partly manual, partly automated production line that saved 3 lb of material and 2 h total manufacturing time per piece in comparison with former methods. The savings of material and time resulted in a 25% reduction in overall cost.

Arc welding equipment included a 500-A rectifier with an automatic sequencer, a heavy-duty automatic wire feed, a heavy-duty air-cooled welding gun, and a flowmeter for control of carbon dioxide shielding gas supplied from cylinders. The mechanical equipment provided for advance-retract motion of the welding gun in the horizontal plane, rotation of the welding gun in the vertical plane, and 180° rotation of the fixtured workpiece to present each joint in turn for welding, without reclamping (Fig. 35).

A control panel provided for regulation of welding current, wire-feed rate, and travel speed of the welding gun, and had push buttons for starting and stopping gas flow, starting the automatic welding sequence, and controlling rotation and lateral motion of the welding gun.

The machine required setting up of travel limits only for the first operation. Subsequent travel was initiated by a push button and terminated by interlock switches. The only manual operations after setup consisted of loading parts, swinging the clamping fixture 180°, pinning the fixture, and unloading the welded assembly.

Welding conditions are given in the table that accompanies Fig. 35. The sequence of operations was as follows:

1 With the welding gun retracted, the two components were manually loaded in the fixture—the spindle placed on the mandrel and the U-shaped arm clamped in position against the spindle.

Fig. 37 Electrical mounting base. Change from SMAW to GMAW improved weld quality and increased productivity.

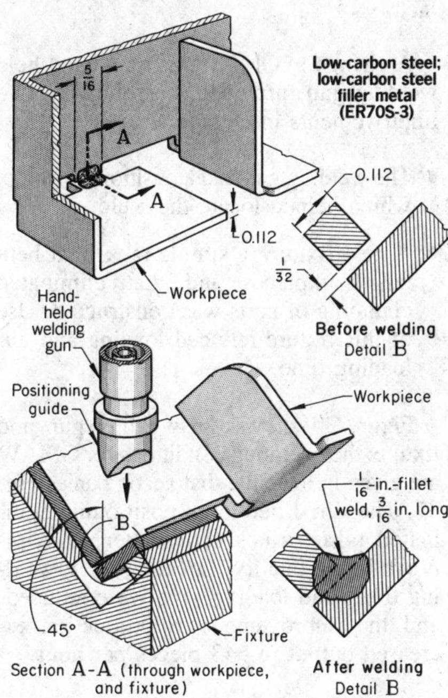

Conditions for GMAW

Joint type	Corner, with 1/32-in. root opening
Weld type	1/16-in. fillet, 3/16 in. long
Power supply	200-A, three-phase, constant-voltage rectifier
Timer	Variable, 0-60 cycles
Electrode wire	0.045-in.-diam ER70S-3
Welding gun(a)	Hand-held, air cooled
Wire feed	Fully automatic, variable speed
Fixture	Rack type
Current	200 A (DCEP)
Voltage	24 V
Shielding gas(b)	Carbon dioxide, at 35 ft^3/h
Welding position	Flat
Wire feed rate	110 in./min
Welding time	22 cycles
Production rate	643 pieces per hour

(a) Timer controlled. Nozzle had a positioning guide (see drawing) to locate it in corner joint and against adjacent flange. (b) From cylinder

2 The welding gun was rotated to a work angle 15° from the vertical.
3 The start button was pushed, feeding the welding gun forward and simultaneously starting gas flow, welding current, and wire feed. At the end of the weld traverse, current downsloped for crater fill; then wire feed and gas flow stopped. The welding gun was rapidly retracted.
4 The fixture was manually pivoted 180°. The welding gun was rotated to an opposite 15° work angle, and the lateral slide was indexed to the new position. Step 3 was then repeated.
5 The welded assembly was manually unloaded; the welds were snag ground, and the spindle was finish machined to size.

Total time cycle per side was 10.5 s, and index time (for 180° rotation) was 10 s, making total machine time per assembly 31 s. Loading and unloading time was 45 s, thus resulting in a floor-to-floor time of 76 s per weldment.

Example 12. Change From Shielded Metal Arc to Automatic Gas Metal Arc Welding of Axial-Flow Compressor Casings. A 30-in.-ID axial-flow compressor casing, 23 in. long, was built in two sections to simplify maintenance and reduce weight and cost. Each section consisted of a ribbed semicylindrical low-alloy steel forging (half of the casing) and two cast low-alloy steel flanges that were welded to the axial edges of the casing half (see Fig. 36). Nominal composition of the castings and of the forging was 0.20% C, 1.0% Cr, 1.0% Mo, and 0.10% V.

When the components were welded by SMAW, 20 h (including 5 h for weld repair) were required for completing one casing (welding four flanges to two casing halves). Changing to automatic GMAW reduced welding time per casing to 2 1/2 h (this also included time for machine maintenance), and less time was needed for weld repair.

For automatic GMAW, fixtures were designed so that both flanges were clamped to each casing half, thus ensuring good fit-up and reproducible final dimensions. Electromechanical relays controlled height and location of the welding gun. Each flange was welded in two passes (see Fig. 36). For the root pass, the welding gun followed the joint line (without being oscillated) and provided an average root penetration of 1/4 in. The second pass was a fill pass, in which the welding gun was oscillated to make a weld of two different widths, for outer and inner ribs; the welding current controlled the feed rate of the electrode wire. Additional welding conditions are given in the table that accompanies Fig. 36.

Weldments were inspected by visual and magnetic-particle methods. Despite the reduction in welding time and in the amount of weld repair needed, automatic welding presented three continuous problems: (1) weld porosity, resulting from nitrogen in the work metal; (2) maintenance of the automatic equipment; and (3) incomplete fu-

Fig. 38 Winch base welded by GMAW. See text for details.

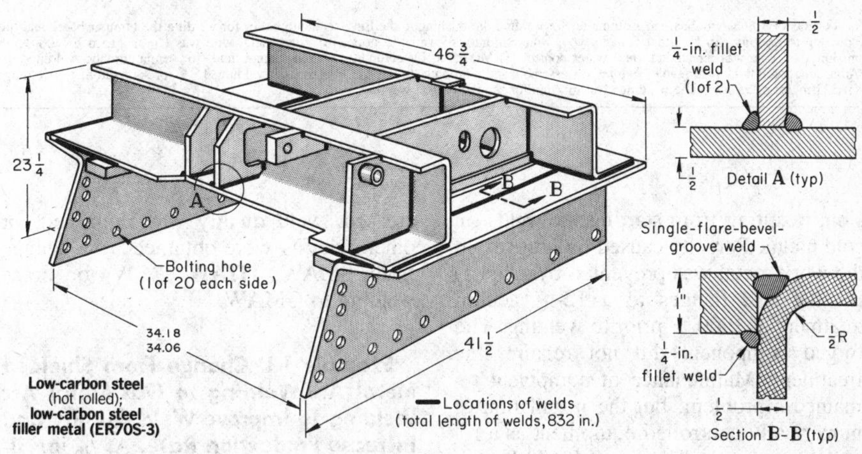

Conditions for semiautomatic GMAW

Joint types	Butt, corner, T, lap
Weld types	Groove, fillet
Power supply	300-A, constant-voltage transformer-rectifier
Electrode wire(a)	0.045-in.-diam ER70S-3
Welding gun	Hand-held, water cooled
Wire feed	Pull type, with dial control
Current	275 A (DCEP)
Voltage	28 V
Shielding gas(b)	75% argon-25% carbon dioxide, at 25 ft^3/h
Welding positions	Flat, horizontal
Number of passes	1
Welding speed	7 in./min
Welding time per assembly(c)	10 h
Preheat and postheat	None

(a) Copper coated. (b) From cylinders; flowmeter controlled. (c) With SMAW, time per assembly had been 13 h.

Fig. 39 Aircraft-wheel link welded by GMAW. See text for details.

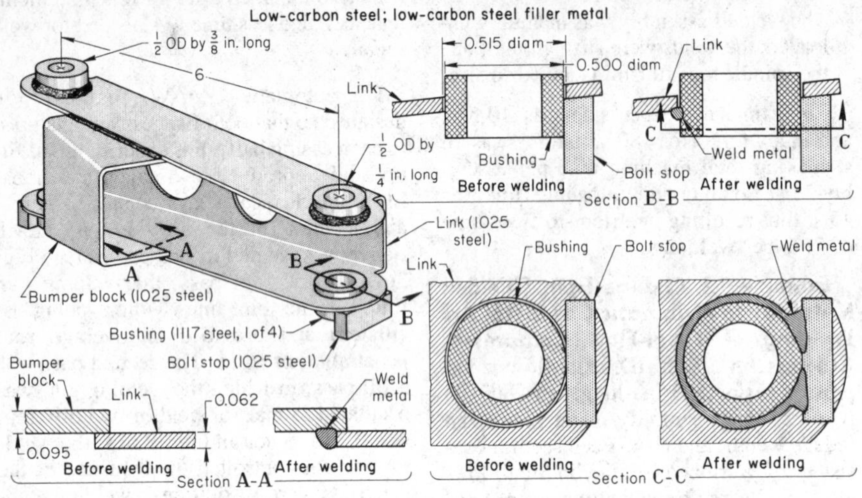

	GTAW	GMAW(b)
Production, weldments per 8-h day	30	~100
Rejections requiring weld repair, %	50	0-6(a)
Welding conditions		
Power supply	250-A motor-generator	Transformer-rectifier(c)
Electrode wire	(d)	0.030-in.-diam ER70S-2
Welding gun	Manual, two types(e)	Air cooled(f)
Wire feed	Manual	Automatic, push type
Special equipment	Tacking jig	(g)
Current, A	40-70, DCEN	300, DCEP
Voltage, V	10	20; 15(h)
Shielding gas	Argon, at 15 ft^3/h	75% A-25% CO_2, at 15 ft^3/h
Welding position	...	Flat
Number of passes per weld	1	1
Wire-feed rate, in./min	...	223; 160(j)

(a) For lots of 100 assemblies. (b) Automatic, for welding the bushing to the link; semiautomatic, for welding the bumper block and the bolt stop to the link. (c) Constant-voltage type, with continuously variable slope. (d) Filler-metal wire was 1/16-in.-diam ER70S-3, 36 in. long. (e) One was air cooled; one, water cooled. (f) Mounted, for automatic welding; hand held, for semiautomatic welding. (g) Adjustable torch holder; turntable; fixtures for assembly and clamping. (h) 20 V for automatic welding; 15 V for semiautomatic welding. (j) 223 in./min for automatic welding; 160 in./min for semiautomatic welding

sion, resulting from starting the welds on cold metal. Porosity caused by nitrogen in the work metal was prevented by subjecting the cast flanges to a 12-h vacuum treatment at 1900 °F prior to welding. The forged component did not require this treatment. Maintenance of equipment remained a problem, but the installation of numerically controlled equipment as a replacement for the electromechanical setup was expected to reduce maintenance costs. Incomplete fusion was minimized through careful selection of welding conditions, but it was not eliminated.

Gas Metal Arc Welding Versus Other Welding Processes

The examples that follow describe applications in which production rate was increased, weld quality was improved, or other benefits were obtained by a change from SMAW, GTAW, RSW, or braze welding to GMAW.

Example 13. Change From Shielded Metal Arc Welding to Gas Metal Arc Welding To Improve Weld Quality and Increase Production Rate. A 1/16-in. fillet weld 3/16 in. long was made in a corner joint on an electrical mounting base (Fig. 37), which was subsequently zinc-plated and chromate-treated. Originally, the joint, which had a 1/32-in. root opening to simplify forming of the part, was shielded metal arc welded, with four workpieces clamped onto a fixture for each setup. This arrangement made accurate location of the weld difficult, and as a result, weld quality was poor. In addition, there was excessive weld spatter, which complicated the zinc-plating operation. Production rate, about 250 pieces per hour, was unsatisfactory.

A change to GMAW, using a hand-held welding gun, provided the following two improvements in technique:

- The welding gun had a positioning guide, which helped locate the weld.
- A new fixture (a simple rack) that held ten workpieces and that eliminated clamping of parts was constructed. Use of this fixture reduced loading and unloading time.

Figure 37 shows the welding gun and fixture; the operating conditions for GMAW are given in the table that accompanies Fig. 37. The simplified torch positioning eliminated false strikes and spatter, and improved weld quality. Locating time, welding time, and loading time were reduced, and the improvement in production increased output to 643 pieces per hour.

Before welding, workpieces were degreased and loaded, ten at a time, into the fixture. As shown in Fig. 37, they were tilted 45°, which permitted the joint to be welded in the flat position and the welding gun to be held vertical. The accessibility of the welding gun was limited by the adjacent flanges, and it was positioned on the first workpiece with the guide against the appropriate flange; it was then triggered and held during timed fillet welding. The procedure was repeated for the nine other workpieces in the fixture. Each weld was checked visually for loose weld spatter before the parts were removed from the fixture, and one part in each batch was checked with a pry bar to proof test the weld. If the weld was good, the test was nondestructive.

To maintain weld quality, the nozzle and the contact tube were cleaned at regular specified intervals. In addition, the settings of the welding machine and timer, the speed of the electrode wire, and the flow rate of the shielding gas were checked at least once every day.

Example 14. Change From Shielded Metal Arc Welding to Gas Metal Arc Welding To Reduce Welding Time, Distortion, Weld Repairs, and Cleanup Time. A 430-lb main base for a truck-mounted 15-ton winch was fabricated by welding components made from hot rolled low-carbon steel $^1/_2$ in. and 1 in. thick, as shown in Fig. 38. The base was subsequently bolted to the truck frame through matching holes, which had been punched in the plates before forming. During forming and welding, opposing pairs of holes had to be kept aligned within a total tolerance of about 0.010 in. to facilitate subsequent use of a bridge reamer and insertion of bolts.

Originally, the base was shielded metal arc welded with a $^3/_{16}$-in. E7024 electrode, using 260-A direct current from a rectifier. The components were in the mill-finish condition and had no edge preparation. Welding time per assembly was 13 h. Two types of distortion occurred: (1) warping of mating surfaces, and (2) misalignment of bolting holes on assembly.

When semiautomatic GMAW with a spray transfer arc mode replaced SMAW, welding time per assembly was reduced to 10 h, and faster welding resulted in less distortion. Also, weld repairs were minimized, and because of the absence of slag, cleanup time (chipping) per assembly was reduced by 30 min. Electrode-drying ovens were not needed. Existing welding machines were used, although machine maintenance increased. Welding conditions are listed in the table with Fig. 38.

As before, the parts were welded in the mill-finish condition, without surface preparation. Subassemblies were welded on locating fixtures to maintain alignment of the bolting holes. The fixture used for final assembly utilized the same holes for locating as were used in subassembly to maintain proper hole locations. All assemblies were shot-blast cleaned after welding; chipping was sometimes needed. The welders were certified according to applicable military specifications, and there was constant visual inspection of work in progress by qualified specialists.

Example 15. Change From Gas Tungsten Arc to Gas Metal Arc Welding. A link for an aircraft nose wheel had four bushings, a bolt stop, and a bumper block welded to it, as shown in Fig. 39, and the welded assembly was subsequently fitted, aligned, reamed, and straightened before being magnetic-particle inspected.

Originally, the components were gas tungsten arc welded after being tack welded, under the conditions listed in the table with Fig. 39. Production rate was 30 weldments per 8-h day. In magnetic-particle inspection, 50% of the weldments were rejected and required weld repairs and subsequent realignment.

To increase production rate and reduce rejection rate, processing was changed to GMAW, which did not require the tack welding operation necessary with GTAW. Automatic tooling was used for the bolt stop and the bumper block. Conditions for GMAW are given in the table with Fig. 39.

The production rate per day by GMAW was 100 assemblies. Magnetic-particle tests of the first production run of 100 assemblies resulted in 6 rejected assemblies, but the defects were minor and no realignment was needed after weld repair. Also, fitting

Fig. 40 Sheave welded by automatic GMAW. See text for details.

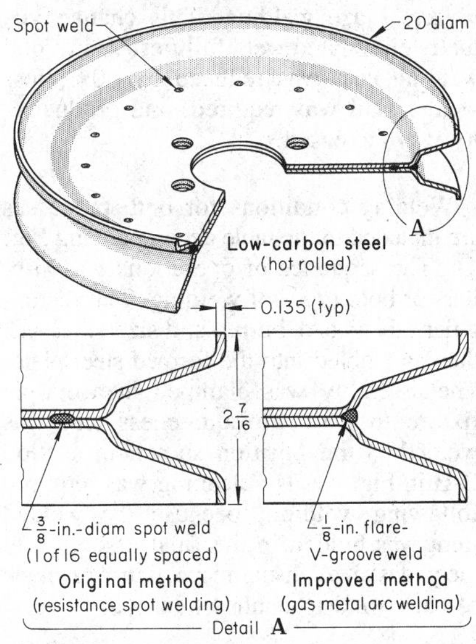

Automatic GMAW

Joint type	Edge
Weld type	Single-flare V-groove
Power supply	500-A, constant-voltage type
Electrode wire	0.045-in.-diam ER70S-3
Current	350-390 A (DCEP)
Voltage	31-33 V
Shielding gas	Carbon dioxide, at 20 ft^3/h
Welding position	Workpiece axis 45° from horizontal; electrode holder at 2 o'clock
Wire-feed rate	600 in./min
Work-rotation speed	$1^1/_2$ rpm
Welding time per sheave(a)	40 s
Production rate(b)	51 sheaves per hour

(a) Compared with 60 s per sheave for spot welding. (b) Compared with 32 sheaves per hour for spot welding

Fig. 41 Overbed table and details of a continuous-weld T-joint at its base

Low-carbon steel; filler metal, copper alloy (ERCuZn-C) (braze welding) or low-carbon steel (gas metal arc welding)

Conditions for braze welding

Torch(a)	Oxyacetylene, No. 53 tip
Filler metal	ERCuZn-C
Flux(b)	Liquid proprietary mixture

Conditions for GMAW

Weld type	Continuous fillet
Power supply	300-A, three-phase, constant-voltage rectifier with slope control
Electrode wire	0.035-in.-diam low-carbon steel
Welding gun	Hand-held, 150 A, air cooled, light duty
Wire feed	Push type
Current	90 A (DCEN)
Voltage	30 V
Shielding gas	Carbon dioxide, at 22 psi (from cylinders)
Number of passes	1
Wire-feed rate	320 in./min

(a) Hand-held; gas mixture adjusted to produce a slightly oxidizing flame. (b) Introduced to the flame by a gas fluxer. The mixture consisted of methyl alcohol, acetone, and boric acid.

required was negligible, and reaming of the bushings was easier than after GTAW.

With GMAW, the 1/4-in.-long bushings for the complete production run were welded first. For this step, a jig was bolted to a turntable. The welding gun was held in place by a clamp on the arm, and handwheels were used to adjust it vertically, horizontally front to back, and horizontally left to right. Before welding, the bushing was nested in the jig through the link, and the link was fastened to the jig with a spring clamp. After the 1/4-in.-long bushings had been welded, the jig was relocated to weld the 3/8-in.-long bushings.

The bumper block also was located by a jig. A pin was inserted through the end of the link containing the 3/8-in.-long bushings and a hole in the end of the welding jig. The bumper block was located by a hole in the face of the jig and the corresponding hole in the bumper block. The bolt stops were located by placing a bolt in the bushing and placing the bolt stop against the bolt face. The bolt stop was then welded to the link and bushing.

Example 16. Change From Resistance Spot Welding to Gas Metal Arc Welding for 60% More Production. Figure 40 shows a sheave consisting of two matching components (of hot rolled, pickled, and oiled low-carbon steel sheet, 0.135 in. thick) welded at an edge joint. The primary requirement of the welded joint was the provision of a solid backing for a belt pulley in fairly severe service.

Originally, the sheave components were joined by 16 resistance spot welds 3/8 in. in diameter, equally spaced around the periphery of the sheave (see left view in detail A, Fig. 40). Welds were satisfactory, but welding time per sheave was 1 min, and production rate was 32 sheaves per hour.

By changing to automatic GMAW, under the conditions listed in the table with Fig. 40, welding time per sheave was reduced to 40 s, and production rate was increased to 51 sheaves per hour. The gas metal arc weld was continuous along the periphery of the edge joint between the two components (see view at right in detail A in Fig. 40). During the GMAW cycle, the operator was able to stack parts and to remove the small amount of weld spatter.

Example 17. Change From Braze Welding to Gas Metal Arc Welding. The welded joint between an upright support tube and a formed plate at the base of a single-pedestal overbed hospital table (Fig. 41) had to resist overloads that resulted when the outboard edge of the table was improperly used as a seat. This eccentric loading subjected the weld to a force several times greater than the weight of the seated person.

Originally, the joint was manually braze welded, but weld failures resulted because the base metal was not always properly heated. Because cold joints were difficult to detect by visual inspection, an alternative welding process was needed. Flash welding was considered, but was rejected because the production volume of only 150 tables per day was too small to warrant the cost of equipment and tooling; also, a flash-removal operation would have been required.

Semiautomatic GMAW was selected to replace braze welding. This change virtually eliminated weld failures. Also, total welding cost was reduced by 50%, less welder skill was required, and productivity was increased.

Welding conditions for both processes are included in the table accompanying Fig. 41. The sequence of operations was similar for both types of welding. The rectangular tube was deburred and sized and was then assembled into the formed steel plate. The assembly was clamped in a simple fixture to maintain squareness and was welded in the position shown in section B-B in Fig. 41. No cleaning was required following welding, because the welded joint was hidden in the final assembly to the pedestal. A visual inspection was made of each weld to control quality.

Safety

The major hazards of concern during GMAW are: the fumes and gases, which can harm health; the high-voltage electricity, which can injure and kill; the arc rays, which can injure eyes and burn skin; and the noise which may be present that can damage hearing.

The type and amount of fumes and gas present during welding depend on the electrode being used, the alloy being welded, and the presence of any coatings on the base metal. To guard against potential hazards, a welder should keep his head out of the fume plume and avoid breathing the fumes and gases caused by the arc. Ventilation is always required.

Electrode shock can result from exposure to the high open-circuit voltages associated with welding power supplies. All electrical equipment and the workpiece must be connected to an approved electrical ground. Cables should be of sufficient size to carry the maximum current required. Insulation should be protected from cuts and abrasion, and the cable should not come into contact with oils, paints, or other fluids which may cause deterioration. Work areas, equipment, and clothing must be kept dry at all times. The welder should be well insulated, wearing dry gloves, rubber-soled shoes, and standing on a dry board or platform while welding. See the National Electrical Code NFPA No. 70 for further information.

Radiant energy, especially in the ultraviolet range, is intense during GMAW. To protect the eyes from injury, the proper filter shade for the welding-current level selected should be used. These greater intensities of ultraviolet radiation can cause rapid disintegration of cotton clothing. Leather, wool, and aluminum-coated cloth will better withstand exposure to arc radiation and better protect exposed skin surfaces.

When noise has been determined to be excessive in the work area, ear protection should be used. This can also be used to prevent spatter from entering the ear.

Conventional fire prevention requirements, such as removal of combustibles from the work area, should be followed. Sparks, slag, and spatter can travel long distances, so care must be taken to minimize the start of a fire at locations removed from the welding operation. For further information, see the guidelines set forth in the National Fire Protection Association Standard NFPA No. 51B, "Fire Protection in Use of Cutting and Welding Processes."

Care should be exercised in the handling, storage, and use of cylinders containing high-pressure and liquefied gases. Cylinders should be secured by chains or straps during handling or use. Approved pressure-reducing regulators should be used to provide a constant, controllable working pressure for the equipment in use. Lubricants or pipe fitting compounds should not be used for making any connections, as they can interfere with the regulating equipment, and in the case of oxygen service, they can contribute to a catastrophic fire.

For further safety information, see American National Standards Institute (ANSI) Z49.1, "Safety in Welding and Cutting."

Gas Tungsten Arc Welding (TIG Welding)

By the ASM Committee on Gas Tungsten Arc Welding*

GAS TUNGSTEN ARC WELDING (GTAW), often called TIG (tungsten inert gas) welding, is an arc welding process in which the heat is produced between a nonconsumable electrode and the work metal. The electrode, weld pool, arc, and adjacent heated areas of the workpiece are protected from atmospheric contamination by a gaseous shield. This shield is provided by a stream of gas (usually an inert gas), or a mixture of gases. The gas shield must provide full protection; even a small amount of entrained air can contaminate the weld.

The arc and weld pool are visible to the welder in GTAW. Slag that may be entrapped in the weld is not produced, and filler wire is not transferred across the arc, thus eliminating weld spatter. Because the electrode is nonconsumable, a weld can be made by fusion of the base metal without the addition of filler metal. A filler metal may be used, however, depending on the requirements that have been established for the particular joint.

Gas tungsten arc welding is an all-position welding process and is especially well adapted to the welding of thin metal—often as thin as 0.005 in. The process can be applied by the manual, semiautomatic, machine or automatic methods. Manual GTAW uses a hand-held torch. Filler metal, if used, is added by hand. Figure 1 illustrates the equipment used for manual welding. Manual welding is used for most GTAW applications. A foot-operated control regulates the amount of welding current, and switches the current on and off. Semiautomatic welding is

Fig. 1 Manual GTAW equipment

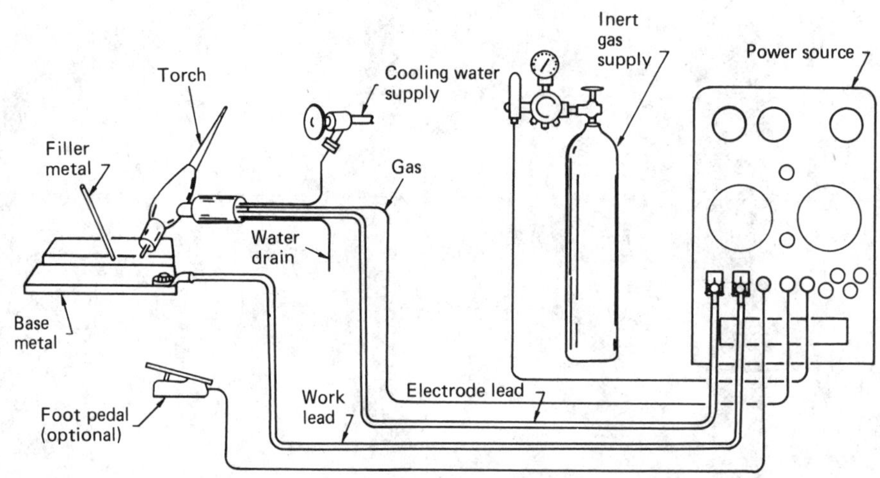

also possible, using a hand-held torch. Filler metal is added automatically by a wire feeder. Semiautomatic GTAW is not widely used. Automatic welding is the most widely used mode for many applications. In machine welding, equipment is controlled by the welding operator. In automatic welding, the unit performs the welding operation, without operator adjustment or control. The amount of automation or mechanization applied to this process depends on the accessibility of the joint, quality control requirements, the degree of accuracy required, and the number of identical welds to be made.

In mechanized applications, filler metal is added to the weld pool in wire form, and may be either cold or hot, depending on the process variation in use. Cold filler wire addition refers to the automatic feed of unheated, or electrically neutral, filler wire into the weld pool, whereas hot wire refers to the feeding of resistively heated filler wire. Resistive heating is accomplished by use of a second power source connected between the wire guide tube and the workpiece.

Metals That Can Be Welded. The nature of GTAW permits its use for welding of most metals and alloys. Metals that are gas tungsten arc welded include carbon and alloy steels, stainless steels, heat-resistant alloys, refractory metals, aluminum alloys, beryllium alloys, copper alloys, magnesium alloys, nickel alloys, titanium alloys, and zirconium alloys.

*Daniel Hauser, *Chairman*, Senior Research Metallurgist, Battelle Columbus Laboratories; George E. Cook, Manager, Research & Development, CRC Welding Systems, Inc.; Emet C. Dunn, Jr., Senior Principal Engineer, Rockwell International; James R. Hannahs, President, Midwest Testing Laboratories, Inc.; Peter C. Levick, Manager, Process & Testing, CRC Welding Systems, Inc.; Daniel O'Hara, Vice President—Engineering, Thermal Dynamics Corp.; Abe Pollack, Head, Joining Branch, David Taylor Naval Ship Research & Development Center; John Springer, Principal Engineer, Boeing Commercial Airplane Co.

Lead and zinc are difficult to weld by GTAW. The low melting temperatures of these metals make control of the process extremely difficult. Zinc boils at 1663 °F, which is far below the arc temperature, and poor welds result from vaporization of the zinc. Steels and other metals that melt at higher temperatures, but that are coated with lead, tin, zinc, cadmium, or aluminum, are weldable but require special procedures.

Welds in coated metals are likely to have low mechanical properties, as a result of interalloying. To prevent interalloying in welding coated metals, the coating should be removed in the area to be welded, then replaced after welding.

Base-Metal Thickness. Gas tungsten arc welding is applicable to a wide range of base-metal thicknesses. The process is well adapted to welding sections $^1/_8$ in. thick or less, because of the intense, concentrated heat produced by the arc, which results in high welding speeds. Multiple-pass welding with the addition of filler metal can be done.

For base metal more than $^1/_4$ in. thick, other welding processes are generally used, although multiple-pass GTAW is used for thick sections in applications where high quality is mandatory (as in aerospace work or in nuclear facilities for welding of piping and pressure vessels). For example, in the fabrication of a 26-ft-diam rocket-motor case with a 0.600-in. wall, longitudinal and girth welds were made by GTAW, using multiple passes and filler metal. Although a slow process for metal of this thickness, GTAW was used because high-quality welds were mandatory.

In the fabrication of pressure vessels, mechanized hot wire GTAW has been found to provide exceptional weld quality along with high completion rates. By using specially designed gas shielding nozzles, the GTAW process may be adapted to welding heavy wall (4-in.-thick) steam generator girth seams with a narrow groove joint preparation. Joint preparations for this thickness are typically single-bevel J configurations with an included angle of 6 to 8°. The root area of the bevel is generally a close butt radiused land of 0.060 to 0.080 in. thickness, which may be penetrated in a single pass. The overall joint opening at the top of the weld groove is 0.9 in.

The GTAW process has been successful in welding various alloys in foil thickness. Thin sheet requires accurate fixturing, and for metal of foil thickness, machine or automated welding is necessary. Plasma arc welding (PAW), often considered as a variation of GTAW, has advantages for welding thin metal (see the article "Plasma Arc Welding" in this Volume).

Workpiece Shape. Manual welding is required where complex shapes preclude the use of machine or automatic methods. Manual torch manipulation is generally used for irregularly shaped parts that require short welds, or for welding in difficult-to-reach areas. Welds can be made manually in the flat, horizontal, vertical, and overhead positions.

Special-purpose automatic equipment lends itself to curvilinear and rectilinear surfaces. Computerized numerically controlled (CNC) machines and robotic welders offer much more flexibility and can be programmed to weld virtually any contour.

Advantages and Limitations

Gas tungsten arc welding enjoys numerous advantages over other welding processes, including:

- The ability to weld most metals and alloys; however, generally not used for very low melting metals such as tin and lead
- The ability to join metals that form refractory oxides, such as aluminum and magnesium, and also reactive metals, which because of their affinity for oxygen and nitrogen can become embrittled if exposed to air while melting
- No slag, which eliminates postweld cleaning
- No weld spatter
- Filler metal not always required
- All-position welding
- Reduced heat input through pulsing
- The ability to join thin base metals because of excellent control of heat input; heat source and filler metal can be controlled separately
- Arc and weld pool visible
- Filler metal does not cross the arc; the amount added does not depend on weld current level.

Some limitations of this process are:

- Slow welding speed
- Electrode contamination
- Low deposition rates; welding thick sections is thus time consuming and expensive
- Requires arc protection from drafts that blow the shielding gas from the arc
- Particles from the tungsten electrode can enter the weld pool

Fundamentals of the Process

Gas tungsten arc welding uses heat produced by the arc between a nonconsumable electrode (tungsten or tungsten alloy composition) and the workpiece. The molten weld metal, heat-affected zone (HAZ), and nonconsumable electrode are shielded from the atmosphere by an inert shielding gas that is supplied through the torch. A weld is made by applying the arc so that the workpieces and filler metal (if used) are melted and joined together as the weld metal solidifies.

The electric arc is produced by the passage of current through the ionized inert shielding gas. The ionized atoms lose electrons and are left with a positive charge. The positive gas ions flow from the positive to the negative pole of the arc. The electrons travel from the negative to the positive pole. The power expended in an arc is the product of the current passing through the arc and the voltage drop across the arc.

To obtain welds of good quality by GTAW, it is essential that all surfaces to be welded and adjacent areas be clean. Filler metal, if used, must also be clean.

Further, it is essential that the components of the assembly being welded be held firmly in the correct position relative to one another. Fixturing is necessary when fit-up is marginal, work metals are thin, shapes are complex, welding is done without filler metal, or mechanized or automatic welding is used.

Arc Initiation. Some preliminary means of initiating the emission of electrons and ionization of the gas is generally used for initiating (striking) the arc. Energy for this emission and ionization can be obtained by (1) touching the energized electrode to the work and quickly withdrawing it to the desired arc length, (2) using a pilot arc that provides an ionized path for the main arc, or (3) using auxiliary apparatus that produces a high-frequency spark between electrode and work. Touching and mechanical retraction of the electrode from the work and pilot arc starting are limited to manual and mechanized welding with a direct-current power supply. High-frequency spark starting is applicable to manual and mechanical welding with either alternating or direct current power supplies. Many power supplies incorporate an apparatus that produces a high-frequency spark to initiate and stabilize the arc.

High-voltage direct current arc initiation is also used in some machines designed for automated welding applications. Such a system requires careful design to ensure a relatively safe output. The distinct advantage of the direct current arc initiation system is less radiated electrical interference.

When beginning to weld, if the electrode is started (or "warmed") on a piece of copper, the arc may be started on the work metal with greater ease. Prewarming also reduces the amount of tungsten that may be lost as the result of forcing the tip of a cold electrode to support maximum current.

Electrode and filler-metal positions in manual GTAW are shown in Fig. 2. Once the arc is started, the torch is held so that the electrode is positioned at an angle of about 75° to the surface of the workpiece and points in the direction of welding, as shown in all views in Fig. 2. To start welding, the arc usually is moved in a circular fashion until enough base metal melts to produce a weld pool of suitable size (Fig. 2a). As adequate fusion is achieved, a weld is made by gradually moving the electrode along the adjoining edges of the parts to be welded, so as to progressively fuse the parts together. Filler metal, when added manually, is often held at an angle of about 15° to the surface of the work and is slowly fed into the weld pool (Fig. 2c). Filler metal must be fed carefully to avoid disturbing the gas shield or touching the electrode and thereby causing oxidation at the end of the filler rod or contamination of the electrode. The filler metal may be added continuously from a rod, or the rod may be dipped in and out.

Filler metal can be added continuously by holding the filler rod in line with the weld (as is often done in multiple-pass welding of V-joints), or by oscillating the rod and the torch from side to side, with the filler rod feeding into the weld pool (a technique often used in surfacing).

To stop welding, first the filler metal is withdrawn from the pool (Fig. 2d), but is momentarily kept under the gas shield to prevent oxidation of the filler metal; then the torch is moved to the leading edge of the pool (Fig. 2e) before the arc is extinguished. The arc can be extinguished by raising the torch just enough to extinguish the arc, but not enough to cause contamination of the weld crater and the electrode. Preferred practice is to decrease the current gradually with a foot control without raising the torch.

Arc Systems

Gas tungsten arc welding uses a constant-current type of power source, like shielded metal arc welding (SMAW), to produce direct and alternating current. A constant-current welding machine provides a nearly constant current during welding. Both SMAW and GTAW can be operated with the same power supply. A high-frequency arc-starting circuit can be added for GTAW. This is not needed for scratch starting the arc.

Constant-current output is obtained with a drooping volt-ampere characteristic. The changing arc length causes the arc voltage to increase or decrease slightly, thus changing the welding current. The steeper the slope of the volt-ampere curve, the smaller the current change for a given change in the arc voltage.

The variations in power sources are caused by differences in power source engineering. Higher short circuit current enhances starting. Steep volt-ampere characteristics are needed to achieve maximum welding speeds. Steeper slopes have less current variation with changing arc length. These machines are used for welding at high current levels. A less steep volt-ampere characteristic is needed for some applications, such as for all-position pipe welding where better arc control with high penetration capability is desired. Machines with a less steep volt-ampere curve also are used to deposit the root passes on joints with uneven fit-up. This type of machine also produces a more driving arc.

Arc length equal to about $1^1/_2$ times the electrode diameter is used in many applications of GTAW, but it may vary for specific applications and, particularly, according to the welder's preference. The greater the arc length, the higher is the heat dissipation into the surrounding atmosphere. Also, long arcs ordinarily interfere, to an extent, with steady progression of the weld. One exception is the bell-and-spigot joint in piping or tubing welded with the pipe axis in the vertical position; here, a long arc will produce a fillet weld of smoother contour than will a short arc.

Welder Skill. To produce high-quality manual welds, a high degree of operator skill is needed. The welder must have his hands free to add filler metal. Selection and training of operating personnel depend greatly on the degree of mechanization used. Because GTAW is most frequently used for joining sheet metal parts in applications where one welder can readily manage the relatively light, small components to be welded, the welder often spends part of his time in cleaning, assembling, fixturing, and tacking. Aside from the high degree of manual dexterity, patience, and training required for producing quality welds, the welder sometimes must have the mechanical skills needed for properly assembling and fixturing the components to be welded.

The specific welding skills required vary from one process to another; for instance, a welder skilled in manual SMAW with a stick electrode will need additional training to qualify for GTAW. In addition, special skills are involved in some applications, such as the placement and welding of consumable backing rings and repair welding.

Welding Current

Current is one of the most important operating conditions to control in any welding operation, because it is related to the depth of penetration, welding speed, deposition rate, and quality of the weld.

Fundamentally, there are but three choices of welding current: (1) direct current electrode negative (DCEN), (2) direct current electrode positive (DCEP), and (3) alternating current. Certain desirable effects can be obtained by superimposing high-frequency current on all three. A guide to selection of the type of current for

Fig. 2 Positions of torch and filler metal in manual GTAW. See text for discussion.

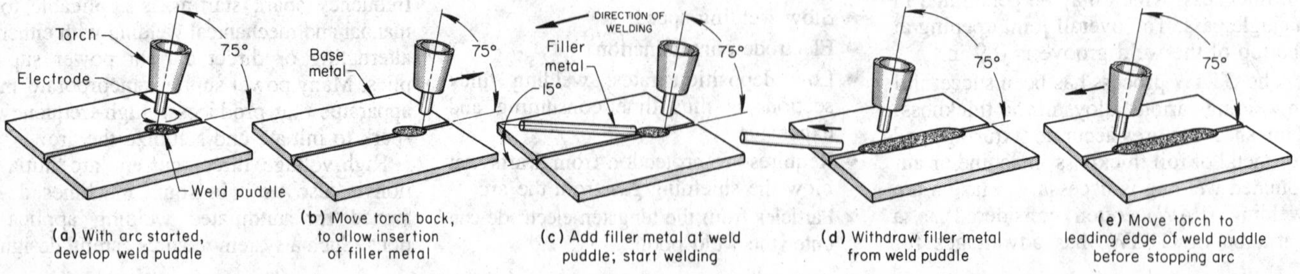

Table 1 Suitability of types of current for GTAW of various metals

Metal welded	Alternating current(a)	DCEN	DCEP
Low-carbon steel:			
0.015 to 0.030 in.(a)	G(b)	E	NR
0.030 to 0.125 in.	NR	E	NR
High-carbon steel	G(b)	E	NR
Cast iron	G(b)	E	NR
Stainless steel	G(b)	E	NR
Heat-resistant alloys	G(b)	E	NR
Refractory metals	NR	E	NR
Aluminum alloys:			
Up to 0.025 in.	E	NR(c)	G
Over 0.025 in.	E	NR(c)	NR
Castings	E	NR(c)	NR
Beryllium	G(b)	E	NR
Copper and alloys:			
Brass	G(b)	E	NR
Deoxidized copper	NR	E	NR
Silicon bronze	NR	E	NR
Magnesium alloys:			
Up to 1/8 in.	E	NR(c)	G
Over 3/16 in.	E	NR(c)	NR
Castings	E	NR(c)	NR
Silver	G(b)	E	NR
Titanium alloys	NR	E	NR

Note: E, excellent; G, good; NR, not recommended
(a) Stabilized. Do not use alternating current on tightly jigged assemblies. (b) Amperage should be about 25% higher than when DCEN is used. (c) Unless work is mechanically or chemically cleaned in the areas to be welded

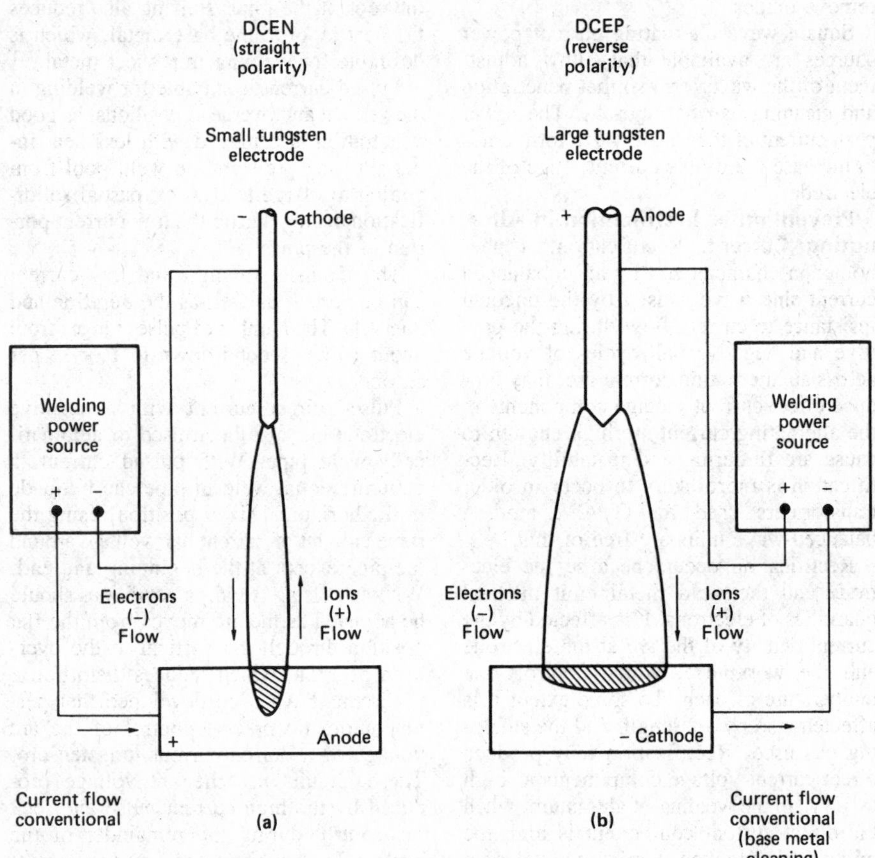

Fig. 3 Gas tungsten arc welding setup. (a) DCEN. Deep penetration, narrow melted area, approximate 30% heat in electrode and 70% heat in base metal. (b) DCEP. Shallow penetration, wide melted area, approximate 70% heat in electrode and 30% heat in base metal

welding various metals is presented in Table 1.

Direct current electrode negative is the type of current most widely used for GTAW. It can produce satisfactory welds in almost all the common weldable metals and alloys. In welding with DCEN, the electrode is negative and the work metal is positive, so that electrons flow from the electrode to the work metal. Because in all direct current arcs 70% of the heat is generated at the positive, or anode, end of the arc, an electrode of a given size will support more negative-polarity than positive-polarity current (see Fig. 3). Consequently, DCEN is the type of current to use if the hottest arc for a specific size of electrode is desired.

Direct current electrode negative produces a deep, narrow weld bead, with penetration superior to that provided by the two other types of current. The narrow bead and deeper penetration, however, may cause difficulty when thin metal is welded with DCEN. Unlike DCEP, or alternating current, DCEN will not remove surface oxides on aluminum, magnesium, or beryllium copper, but aluminum may be welded with DCEN by the use of specialized welding techniques, plus mechanical or chemical cleaning prior to welding (see Table 1).

Direct Current Electrode Positive. In welding with DCEP, the electrode is connected to the positive terminal of the power supply, and the work metal to the negative terminal (see Fig. 3). Thus, electrons flow from the work to the electrode, generating high heat in the electrode and low heat in the workpiece. At the same amperage and arc length, the arc voltage of a DCEP arc is somewhat higher than that of a DCEN arc, so that the DCEP arc has more total energy.

Direct current electrode positive is the least used of the three types of current, because it produces a flat, wide bead with shallow penetration. Welding with DCEP requires great skill and, because of the large size of electrode that must be used with a comparatively low level of welding current, is not generally recommended. Direct current electrode positive has the coldest effective arc of the three types of current, but it does provide superior removal of oxides from the surface of the work metal.

Aluminum is particularly difficult to weld with DCEP because the weld pool is readily attracted to the tip of the electrode, which becomes contaminated when touched by the aluminum. However, DCEP may be effectively used for joining thin aluminum sheet (up to about 0.025 in. thick). Magnesium, on the other hand, appears to be repelled by the arc action inherent to DCEP, and thus contamination is not a problem; DCEP may be used for welding magnesium in thicknesses up to $^1/_8$ in.

Oxide Removal by Direct Current Electrode Positive. When the electrode is positive, argon or helium ions travel to the surface of the base metal. Positively charged gas ions are produced through action of the arc on the surrounding inert-gas atmosphere. The gas ions have considerable mass and hence acquire large amounts of kinetic energy while speeding to the surface of the base metal. When these ions collide with the surface, they clean it by tearing away particles of oxide in a manner somewhat analogous to grit blasting. The ions produce little heating of the base metal, in comparison with the heating that occurs at the anode end of the arc; as a result, the amount of penetration is slight. If the electrode is negative and the work positive (DCEN), the ions travel to the electrode and exert no cleaning action on the work metal, and electrons bombard the metal being welded, thus producing considerable heat and penetration of the work metal.

Metals such as stainless steel, carbon steel, and copper do not form oxide coatings that interfere appreciably with GTAW.

Alternating current may be represented as a series of alternate pulses of DCEN and DCEP, and it reverses direction 120 times per second. With alternating current, the voltage varies from a maximum positive value to a maximum negative value during each cycle, and the arc is extinguished each time this occurs. A conventional arc welding transformer does not produce voltage high enough for positively reestablishing the arc after it has been extinguished when welding in an inert atmosphere; consequently, high-frequency current must be superimposed on the arc at each half cycle, unless the transformer used has a sufficiently high inherent voltage characteristic to render it unnecessary.

Alternating current gives good penetration and surface-oxide reduction, and the form of the gas tungsten arc weld bead it produces when filler metal is added is more nearly that of a satisfactory shielded metal arc deposit but with slightly less reinforcement. The bead produced in GTAW with alternating current is wider and shallower than a DCEN bead, but narrower and deeper than a DCEP bead, and it has more reinforcement than either a DCEN or a DCEP bead. Figure 4 illustrates the effects of the three different types of welding current on weld-bead geometry. Alternating current is therefore preferred for the welding of aluminum, magnesium, and beryllium copper because of its ability to remove oxides.

Fig. 4 Effects of welding current on weld shape

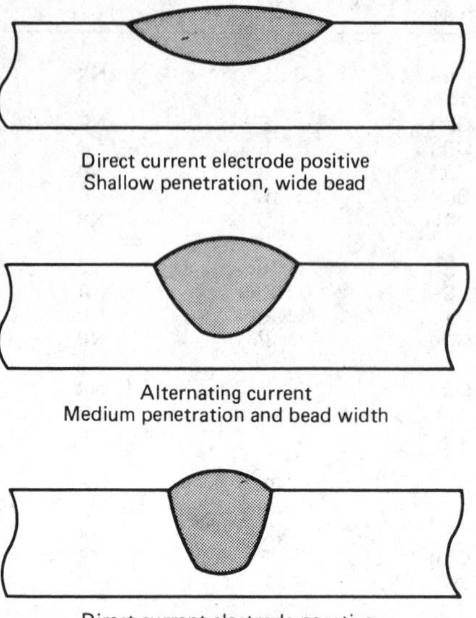

Square wave alternating current power sources are available that allow adjustment of the wave form so that penetration and cleaning can be adjusted. The lower peak current of the square wave form tends to increase the usable current range of the electrode.

Prevention of Rectification in Alternating Current. Rectification, a phenomenon characterized by an unbalanced current sine wave caused by the unequal resistance to current flow during the positive and negative half-cycles of voltage across an alternating current arc, may produce direct current voltage components in the alternating current arc high enough to cause arc fluttering and instability. Rectification is more likely to occur in older transformers used for GTAW; modern balanced-wave units are free of this.

Rectification occurs because the electrode and the weld metal emit unequal quantities of electrons. It is affected by the current density of the arc at the electrode and the workpiece, which controls the temperature of each. To some extent it is affected also by arc length and the shielding gas used. Rectification may produce direct current voltage components as high as 12 V. In the welding of aluminum, when the direct current component is high the bright pool of molten aluminum will darken and film over with oxide—the extent being proportional to the magnitude of the direct current component.

Rectification, and its adverse effects, can be eliminated by the use of a balanced-wave transformer. This unit incorporates into the welding-current circuit a series capacitor (condenser) of a capacity that will permit the alternating welding current to flow efficiently but will block the direct current component. These units are usually designed for open-circuit voltages in the range of 100 to 150 V, require high-frequency current for arc starting only, and are used extensively in welding aluminum alloys and magnesium alloys.

Pulsed-Current Welding. Pulsed-current GTAW employing a high rate of current rise and decay and a high pulse-repetition rate is widely used in the joining of precision parts.

Pulsed-current GTAW has several advantages over constant-current welding of thin materials. This mode of welding is best suited to poor joint fit-up. Also, improved control over distortion is possible. Conventional fixturing can be used with thinner materials. High pulse provides the high current levels needed to complete penetration in open root welding. Low pulse cools the pool, thus preventing melt-through at the root of the joint. Pulsing also reduces the heat input to the base metal, which is desirable for welding thin sheet metal.

Pulsed current is suitable for welding in the vertical and overhead positions, as good penetration is achieved with less heat input. Pulsing prevents the weld pool from getting too large to control; partial solidification occurs during the low-current portion of the pulse.

The intensity of high and low current can be varied, as well as the duration and intervals. The number of pulses ranges from about 10 per second down to 1 or $^1/_2$ per second.

Pulsed direct current with a negative electrode has also been used to automatically weld pipe. With pulsed current, a circumferential weld in pipe can be made in the horizontal fixed position, using the same amount of current and voltage around the pipe except at the beginning and end. Without pulsing, welding conditions should be adjusted as the arc moves from the flat position through the vertical to the overhead position, which reduces distortion.

Circuits have been developed that permit automatic precise control of the arc voltage of a pulsed-current tungsten arc. These circuits use the arc voltage produced by the high current pulse and lock the control during the remainder of the cycle. In sophisticated pulsed-current

Fig. 5 Representative shapes of current pulses with high and low rates of rise

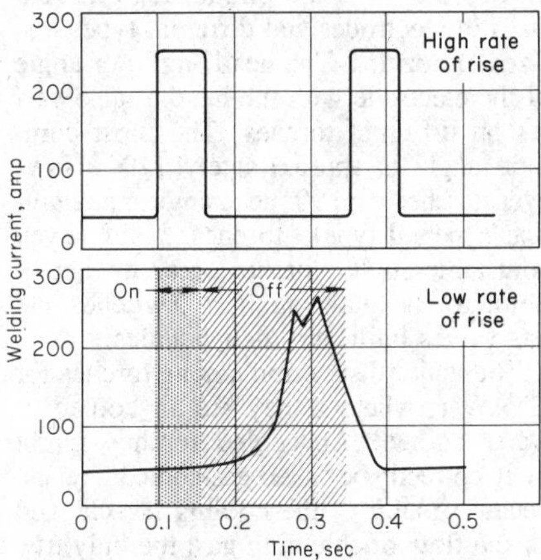

welding power supplies, the following may be set independently:

- Pulse length at low current
- Pulse length at high current
- Amplitude of high-current pulse
- Beginning of pulses

Figure 5 illustrates representative shapes of current pulses with high and low rates of rise, correlated with a scale of time in seconds. The advantages of pulsed-current GTAW are:

- *Increased depth-to-width ratios of weld beads*: By using a short-duration, high-current welding pulse with a small, blunt thoriated tungsten electrode, the arc force generated produces weld beads in stainless steel that have depth-to-width ratios approaching 2 to 1.
- *Elimination of drop-through*: High-current, short-duration pulses melt through root passes or thin work metal and solidify before the weld pool becomes large enough to sag.
- *Minimal HAZ*: The HAZ can be minimized by proper proportioning of the high pulse height and pulse-on time and the low pulse height and pulse-on time. It is sometimes desirable to set the low pulse height at zero while holding a finite space between high-current pulses.
- *Stirring in the weld pool*: The high pulse of current develops arc and electromagnetic forces much greater than those developed with constant-current welding. These high forces produce agitation of the weld pool that reduces the porosity and incomplete fusion that may occur at the bottom of the joint. The pulsing also produces substantial arc stiffness when used for low-current welding, eliminating the arc wander associated with a low constant-current arc.

High-frequency current is produced by a separate power source and is used to maintain a pilot arc and to initiate an arc (see the section of this article on arc initiation). A pilot arc is used to start the welding arc without touching the electrode to the workpiece. With alternating current, a high-frequency current maintains the arc during cycle changes from positive to negative, or vice versa. High-frequency current is used to start the arc when direct current is used. It may be terminated when the arc has started. The use of a high-frequency current is the preferred method of arc initiation. Contamination of the tungsten electrode is eliminated, which may occur when the electrode tip touches the workpiece.

When using high-frequency current for GTAW with alternating current, interference may be generated that disturbs radio and television transmissions. Consequently, such operations are monitored by the Federal Communication Commission (FCC). Special grounding and shielding precautions must be taken, as per manufacturers' instructions. Usually, all metal conductors must be grounded to minimize high-frequency radiation.

Equipment

The equipment used for GTAW includes a power supply, a welding torch, a nonconsumable electrode, and a gas shielding system. Optional equipment also may include a water circulator, foot control pedal, voltage control devices, motion devices, oscillators, and wire feeders.

Power Supplies

Power-supply units for GTAW include (1) power-driven generators, either electric-motor driven or engine-driven; (2) transformer-rectifier welding machines; (3) three-phase rectifier welding machines; or (4) transformer welding machines.

Generator welding machines are operated by electric motor for shop use or by internal combustion engines for on-site use. Generator welding machines used for SMAW can be modified for use in GTAW, by adding gas and high-frequency auxiliary equipment.

Engine-driven welding machines also can be adapted to GTAW, by providing auxiliary modifications. Generator welding machines usually provide direct current, or direct and alternating current, to the arc.

Transformer-rectifier welding machines are widely used for GTAW. These welding machines operate on single-phase input power. Motor-generator welding machines are used less frequently. Transformer-rectifier machines are capable of providing alternating and direct current to the arc. A single-phase transformer with alternating current may be added to the rectifier, thus producing direct current for the arc. Rectifiers change alternating current into direct current.

In recent years, solid-state devices, either power transistors or thyristors, have been employed in many modern power source designs. The thyristor-silicon controlled rectifier (SCR) system should not be confused with the direct transistor approach. The SCR is in effect a gated rectifier by which the maximum output level can be reduced, although at the expense of ripple (which is appreciable even at full output). Furthermore, as its controls are essentially linked to the main power supply, the thyristor-SCR system cannot operate at high frequencies. On the other hand, the transistor regulator is a solid-state-type power supply, but with the capabilities of providing a wide range of output levels over a range of frequencies extending from steady direct current up to several kiloHertz with inherently smooth waveform.

The modern solid-state power supplies lend themselves to very precise control even in the presence of marked variations in input line voltage. These power supplies are frequently programmed and controlled by a microcomputer. The programmed sloping of the current and/or voltage waveforms may be either time- or position-based and may be set and controlled with great accuracy and repeatability.

Transformer-rectifier welding machines are capable of handling applications involving many types of base metals. Programmable transformer-rectifier power sources are available commercially, which enable the welding operator to easily select the transformer or rectifier mode. Depending on the application, alternating or direct current can be indicated.

Transformer-rectifier welding machines enjoy several advantages over rotating power sources:

- Lower operating cost
- Quiet operation
- Lower maintenance cost
- Lower power consumption
- No rotating parts

Three-phase rectifier welding machines furnish direct current to the arc and operate on three-phase input power. Line imbalance that is a problem in single-phase transformer-rectifier welding machines is avoided by three-phase input power. Transformers interact with the rectifier, which produces direct current. Programmable power sources are available commercially.

Transformer welding machines usually are not used for GTAW. These machines produce only alternating current power and operate on single-phase input power. Power comes directly from the line and is transformed into the type of power needed for each given welding application. Low initial cost is a major advantage of transformer welding machines.

Welding Controls

Welding units for GTAW come equipped with the following:

- On-off power switch
- Polarity selection switch (direct current welding)
- Current-control switch that regulates the amount of current delivered to the arc
- High-frequency control to regulate and select type of high-frequency current
- Hot start devices to furnish a surge of current substantially above the normal welding current and to initiate the weld with less time lag. This is particularly beneficial in mechanized or automatic welding. A hot start device incorporated into the circuitry provides this initial surge of current. Ordinarily, the device can be preadjusted to furnish the degree of additional current desired and for the time span required.
- Pulsation controls to begin pulsation and regulate intensity and duration
- Shielding gas controls to start and maintain the flow of shielding gas. Prevents oxidation of the tungsten electrode and contamination of the weld pool
- Up-slope and down-slope controls (optional). Upslope control allows the welding current to build up gradually at a set rate at the beginning of the welding; down-slope control allows welding current to decay gradually to prevent cracking. With modern solid-state machines, the current sloping is accomplished electronically. The operator programs a current slope by specifying the starting current level, the end current level, and the time between. Some programming systems are position-based, in which case the operator programs the start and end positions as well as the corresponding current levels.
- Foot control (optional) to start and vary current flow. Also starts high-frequency current

Welding Torches

A torch for manual GTAW should be compact, lightweight, and fully insulated. It must provide a handle for holding it, a means for conveying the shielding gas to the arc area, and a collet, chuck, or other means for securing the tungsten electrode and conducting welding current to it. The torch assembly normally includes various cables, hoses, and adapters for connecting the torch to sources of power, gas, and water (if water is used for cooling). A cross-sectional view of a water-cooled manual torch is shown in Fig. 6. Figure 7 is an enlarged view of the components for a manual torch. The shielding-gas passage through the entire system must be leakproof. Leaks in the hose or connections result in loss of valuable shielding gas and insufficient shielding at the weld pool. Aspiration of air into the shielding-gas system can be a major problem. Careful maintenance is required to ensure a leakproof gas system.

Torches for GTAW are available in a variety of sizes and types weighing from as little as three ounces to almost one pound. The different sizes of torches are rated in accordance with the maximum welding current that can be used. In addition, they will accommodate different sizes of electrodes and different types and sizes of nozzles. The head angle, or angle of the electrode with the handle, also varies on different torches. The most common angle is approximately 120°. However, torches with 90° head angles, straight-line (pencil-type) torches, and even adjustable-angle torches are available. Some torches have auxiliary switches and gas valves built into their handles.

The major distinction among torches for GTAW is whether they are air cooled or water cooled. Air-cooled torches might more correctly be called gas-cooled torches, because much of the cooling is achieved by the flow of shielding gas; the only true air cooling is by radiation into the surrounding air. On the other hand, for water-cooled torches, some cooling is provided by the flow of shielding gas, but it is supplemented by water that is circulated through the torch (Fig. 6a).

Air-cooled torches are usually lightweight, small, compact, and less expensive than water-cooled torches. They generally have a maximum welding current of about 200 A. They normally are used for welding thin metal and for limited duty cycles. The tungsten electrode operates at higher temperatures in air-cooled torches than in water-cooled torches, and thus may cause tungsten particles to slough off into the weld pool when pure tungsten electrodes are being used at or near current-carrying capacity.

Water-cooled torches are designed for continuous high-current welding. They operate continuously with welding currents up to 200 A, and some are designed for welding currents up to 1000 A. Generally, water-cooled torches are heavier and

Fig. 6 Sectional views of a typical water-cooled torch for manual GTAW

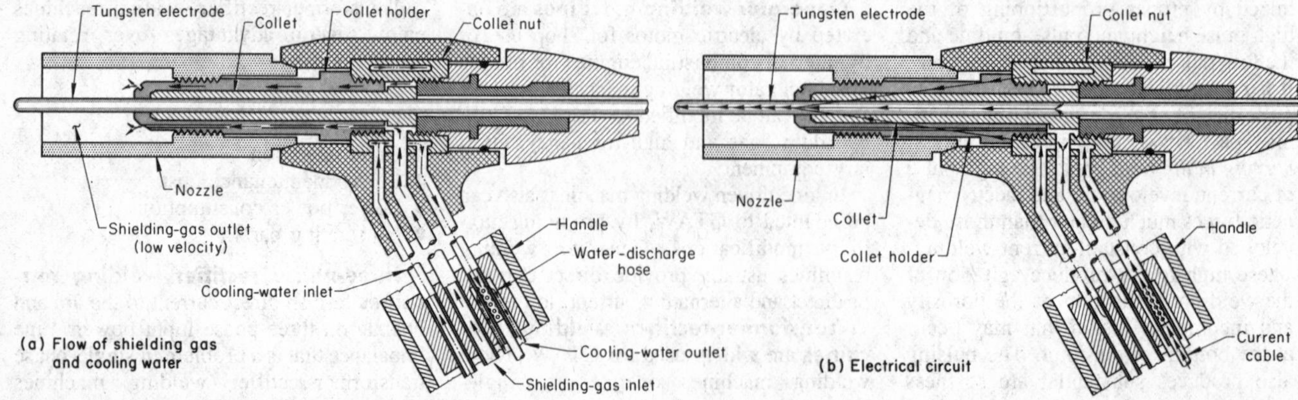

Fig. 7 Manual torch

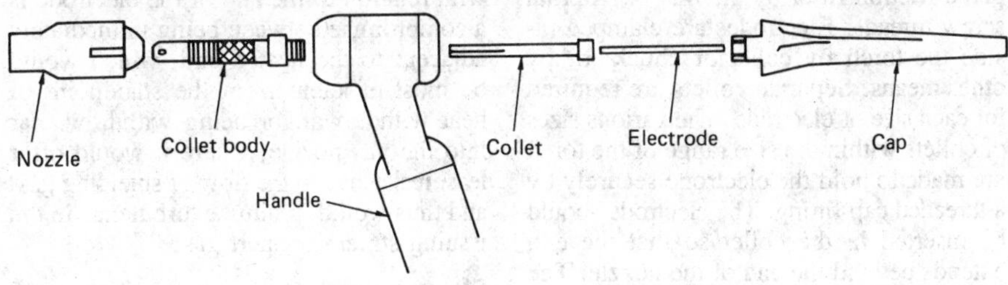

more expensive than air-cooled torches.

The water hose and associated connectors normally are supplied with the torch. Generally, the power cable bringing the welding current from the power supply to the electrode is enclosed in the water-discharge hose (Fig. 6). This cools the cable and allows the use of a small-diameter, lightweight, flexible conductor. Adaptor blocks, and sometimes flow switches and fuses, also are included. Water leaks in the torch or moisture contamination of the gas system will contaminate the weld and render the process inoperable.

General characteristics of air-cooled and water-cooled torches are presented in Table 2. A gas tungsten arc torch for mechanized operations typically is water-cooled and may have a continuous rating as high as 1200 A. It normally is not required to be lightweight; however, its other characteristics are the same as the manual gas tungsten arc torch.

Types of Nozzles. Several types of nozzles or gas cups are used in GTAW: ceramic, metal (cooled by radiation and the shielding gas, or water-cooled), fused quartz, and dual-shield nozzles. Ceramic nozzles cost the least and are the most popular, although water-cooled metal nozzles have longer service lives if properly used.

Ceramic nozzles become brittle after continuous use and must be replaced when the lip of the nozzle becomes rough and uneven. A rough, uneven lip interferes with the flow of shielding gas and causes nonuniform gas coverage of the weld area.

When welding with high-frequency current, ceramic nozzles must be used whenever possible, to prevent arcing to the gas nozzle, which often occurs when metal nozzles are used. However, arcing to the gas nozzle can be offset to a degree by using the largest size nozzle that is practical for the application.

Sleeve-type, or slip-on, metal nozzles have very limited current-carrying capacity. Because they are delicate, they are easily damaged and misused. Also, under the stress of continuous use, they run so hot that arcing destroys them immediately when it occurs.

Water cooling of metal nozzles makes it possible to use welding current as high as 500 A, which is about the maximum practical for manual GTAW, although mechanized operations have used current well above 1000 A.

Fused-quartz nozzles are transparent and are preferred by some welders, who maintain that superior vision of the weld operation is achieved by their use. At the slightest contamination of the electrode, however, violent ejection of metallic vapors away from its surface dulls the inside of the quartz nozzle, and visibility of the weld is thereafter severely impaired. Fused-quartz nozzles are also brittle and will break if not handled carefully.

The substantial difference in cost between ceramic nozzles and water-cooled metal nozzles sometimes is offset by the rapidity with which ceramic nozzles are discarded by welders who fail to give them reasonable care in the course of operation. Several ceramic nozzles may be used in a single shift by one welder if they are subjected to careless use.

The dual-shield nozzle permits a relatively small flow of argon or helium around the electrode to shield the immediate weld pool, while an annular grooved section around the central nozzle sends down an atmosphere of nitrogen or carbon dioxide to exclude air from contact with the central inert-gas column. This type of nozzle, however, is rarely used in industry.

All gas nozzles, of whatever material, should be kept clean at all times, because an accumulation of foreign material on the inner surface—and, particularly, on the lip of the orifice—will eventually interfere with, or set up turbulence in, the gas column or may begin to pass vaporized metal down into the weld pool.

Size of Nozzle. When choosing a ceramic nozzle for a specific job, an attempt should be made to use the smallest nozzle whose lip will not melt under the concentrated heat of the arc. A small nozzle will assist in maintaining a more stable and positive arc, permit welding in more restricted areas, and give better vision of the weld. Larger nozzles give better shielding-gas blankets at a slower gas-discharge rate than is provided by smaller nozzles. For metals such as titanium that are sensitive at elevated temperature to contamination from the ambient atmosphere, larger nozzles are safer.

Shape of Nozzle. The common form of gas nozzle is either cylindrical or tapered in the inner surface. Because commercially available nozzles are almost invariably round, however, does not necessarily mean that they are the most economical for all applications. Greater economy can ordinarily be achieved by using a nozzle designed for a specific production application.

Nozzles to which elongated trailing sections have been added have proved helpful for shielding welds made in metals with high susceptibility to gas contamination at elevated temperature. Nozzles have been made that have a section flared out or shaped in such a manner that the gas flow is altered to achieve a special effect, such as maintaining protection over the finished weld area for a longer period than will a conventionally shaped nozzle. This permits reducing the gas flow to the welding nozzle by a factor of two or three from the flow that would be required without a trailing shield, in order to achieve only a part of the protection offered by the special shield.

Although additional gas is directed into the trailing shield, the gas flows slowly and is ordinarily liberated through a porous metal baffle onto the heated metal surface that is to be protected. The purpose of the porous baffle is to avoid turbulence and to economize on gas. Some metals and al-

Table 2 Typical characteristics of manual torches used for GTAW

Torch characteristic	Torch size: Small	Medium	Large
Maximum current usable (continuous duty), A	200	200-300	500
Cooling method	Air(a)	Water	Water
Diameters of electrodes accommodated, in.	0.010-3/32	0.040-5/32	3/32-1/4
Nozzle-orifice diameters accommodated, in.	1/4, 5/16	1/4, 5/16, 3/8	3/8, 1/2, 3/4

(a) Air-cooled nozzles actually are cooled by the flow of shielding gas and by radiation.

loys susceptible to gas contamination at elevated temperature could not be welded without such devices. The trailing nozzles should be of a form that follows well behind the arc location itself. Other nozzles have been designed so as to direct the gas flow forward to offset a draft effect from the ambient atmosphere caused by a high speed of welding-head traverse over the work surface.

When it is necessary to weld in extremely restricted locations, special torches should be fabricated, if practical. When welds must be made where there is considerable interference, long, narrow nozzles are available or can be made. Round nozzles made from very thin copper brought down to a knife edge have been effective at exceptionally low gas flow rates.

Gas Lenses. Laminar flow has also been achieved by the introduction of a special screen inside the gas nozzle. This device, known as a gas lens, permits projection of an uncontaminated column of inert gas to a considerable distance beyond the nozzle orifice.

Size of Gas Orifices in Torch Collets. In some older torches, many of which are still in wide use in industry, the gas orifices in the electrode-holding collet are so small that considerable turbulence is created in the gas column issuing from the torch nozzle. This distorts the coverage pattern of the gas over the weld. Enlarging the orifices by drilling and reaming overcomes this difficulty, and is worth considering should turbulence be encountered while an older torch is being used. When enlarging the orifices, it is important not to remove so much material that the collet will be weakened to the extent that it might fail structurally to hold the electrode securely.

Less turbulence and mixing with surrounding air will occur if the gas column issues from the nose of the gas nozzle with a minimum of deflection.

Electrode sizes (diameters) for torches with various sizes of nozzle-orifice diameters are typically as follows:

No. 4 (1/4-in. orifice)	Electrode, 0.020 in.
No. 5 (5/16 in.)	0.040 in.
No. 6 (3/8 in.)	1/16 or 3/32 in.
No. 7 (7/16 in.)	1/8 in.

In general, a metal nozzle should have a slightly larger orifice than a ceramic nozzle for the same electrode size. When the diameter of a metal nozzle is too small, overheating occurs, and the nozzle deteriorates rapidly.

Nozzles are attached to the torch by tapered friction fit or by internal or external screw threads. Electrodes are clamped inside the torch by collet or chuck, or by other means. Separate collets are required for each size of electrode. The various sizes of collets within the size range of the torch are made to hold the electrode securely by a threaded cap fitting. The electrode should be inserted in the collet so that the end extends beyond the end of the nozzle. The permissible amount of extension depends on the type of weld joint, the electrode size employed, the size and shape of the nozzle, and the type of torch used (see the section of this article on electrode extension). The collet assembly also carries the current from the torch body to the electrode.

The design of the cap on the torch governs the length of electrode that can be used. A typical cap is shown in Fig. 7. Cap extensions allow the use of electrodes as long as 7 in.

Centering of the electrode in the torch nozzle orifice is important, because a deflection will cause the arc to be offset from the center of the issuing stream of shielding gas. If the electrode is slightly bent, it may be straightened by pressing against it laterally after it has been heated and is red hot. Axial alignment of the electrode is easily checked by looking straight into the gas nozzle; it is unnecessary to apply a gage or similar device to attain precision in centering.

Collets are of either the split type or the draw type. Both types are effective in making positive electrode contact if properly adjusted. The bodies of most collets are made of a copper alloy, to obtain the efficiency of heat transfer and electrical conductivity offered by copper, although some collet bodies are made of heat-resistant alloys, such as 80Ni-20Cr.

The inner surface of a collet should be smooth. Before a collet is placed in service, it should be thoroughly checked for burrs, which might decrease the efficiency of electrode contact. Poor contact contributes to poor current-carrying capacity.

The position of most electrode collets with relation to the end of the electrode is a compromise between being immediately adjacent to the heated end, which would be most efficient from the standpoint of heat withdrawal, or being withdrawn far into the gas nozzle, where it would offer least resistance to the flow of shielding gas, and thus would minimize turbulence in the issuing stream of inert gas.

Electrodes

The use of a nonconsumable electrode—an electrode that does not supply filler metal—constitutes the major difference between GTAW and other metal arc welding processes. Tungsten, which has the highest melting temperature of all metals (6170 °F), has proved to be the best material for nonconsumable electrodes. In addition to having an extremely high melting point, tungsten is a strong emitter of electrons, which stream across the arc path, ionize it, and thus facilitate the maintenance of a stable arc.

Tungsten of commercial purity (99.5% W) and tungsten alloyed with either thoria or zirconia are the electrode materials used in virtually all applications of GTAW. Pure tungsten electrodes cost about 25 to 35% less than the thoriated types, depending on finish.

Table 3 gives AWS classifications and composition for tungsten and tungsten alloy electrodes. Pure tungsten electrodes (EWP) are the least costly type of electrode, and they are used on less critical applications with alternating current. Pure tungsten electrodes have low current-carrying capacities and low contamination resistance. These electrodes are designated with green markings.

Tungsten electrodes with 1 to 2% thoria (EWTh-1 and EWTh-2) are superior to pure tungsten electrodes. They have improved electron emissivity, current-carrying capacity, longer life, and greater contamination resistance. Arc starting is easier, and the arc is more stable.

When striped electrodes (EWTh-3) are used for alternating current welding, the

Table 3 AWS classifications and composition limits for GTAW electrodes (AWS A5.12)

AWS classification	Tungsten (min)(a), %	Thoria, %	Zirconia, %	Other (max)(b), %
EWP	99.5	...	...	0.5
EWTh-1	98.5	0.8-1.2	...	0.5
EWTh-2	97.5	1.7-2.2	...	0.5
EWTh-3(c)	98.95	0.35-0.55	...	0.5
EWZr	99.2	...	0.15-0.40	0.5

(a) By difference. (b) Total. (c) EWTh-3 is a tungsten electrode with an integral lateral segment throughout its length that contains 1.0 to 2.0% thoria; average thoria content of the electrode is as shown in this table.

Table 4 Recommended tungsten electrodes and shielding gases for welding different metals

Type of metal	Thickness	Type of current	Electrode	Shielding gas
Aluminum	All	Alternating current	Pure or zirconium	Argon or argon-helium
	Thick only	DCEN	Thoriated	Argon-helium or argon
	Thin only	DCEP	Thoriated or zirconium	Argon
Copper, copper alloys	All	DCEN	Thoriated	Argon or argon-helium
	Thin only	Alternating current	Pure or zirconium	Argon
Magnesium alloys	All	Alternating current	Pure or zirconium	Argon
	Thin only	DCEP	Zirconium or thoriated	Argon
Nickel, nickel alloys	All	DCEN	Thoriated	Argon
Plain carbon, low-alloy steels	All	DCEN	Thoriated	Argon or argon-helium
	Thin only	Alternating current	Pure or zirconium	Argon
Stainless steel	All	DCEN	Thoriated	Argon or argon-helium
	Thin only	Alternating current	Pure or zirconium	Argon
Titanium	All	DCEN	Thoriated	Argon

starting-capacity characteristics of the thoriated electrodes are combined with the desirable stability characteristics of the 99.5% purity EWP electrodes. The 1% thoriated tungsten electrodes are identified by a red marking, and the striped thoriated tungsten electrodes are designated by a blue marking.

Tungsten electrodes that contain some zirconia (EWZr) generally have welding properties similar to pure tungsten electrodes and thoria-containing tungsten electrodes. Zirconiated electrodes, when used for alternating current welding, combine the stability of pure tungsten electrodes with the capacity and starting ability of thoriated tungsten electrodes. This type of electrode is identified by a brown marking. Table 4 shows the types of tungsten electrodes and shielding gases used for welding different metals.

Finish and Surface Condition. The two main types of finish with which tungsten arc welding electrodes are commercially available are a ground finish and a chemically cleaned finish.

The arc is more stable when electrodes with a ground finish are used. The relative smoothness of the surface of a ground electrode has considerable effect at high current levels; the resistance offered by the rougher surface of a chemically cleaned electrode may reduce the maximum current-carrying capacity. Ground electrodes maintain maximum contact with torch collets. The smooth finish of a ground electrode ensures that a uniform surface is exposed to the arc, the gas stream, and the inner holding surface of the collet.

Electrodes made by drawing with the use of graphite as a lubricant may retain a slight coating of graphite and thus present a black or blue-black appearance. The graphite coating does not affect the arc or welding characteristics. Any other discoloration on the surface of a tungsten electrode, however, indicates the presence of tungsten oxide or some type of contamination. This will result in dirty welds and rapid consumption of the electrode, and may cause an unstable arc.

If the electrode is discolored after use but has not been contaminated with filler metal or base metal, the discoloration is caused by oxidation. Ground or chemically cleaned electrodes will remain bright and shiny if properly protected from air contamination. The one exception is the graphite-drawn electrode, which will always be dark before and after use.

Electrodes having seams, cracks, pipes, slivers, or nonmetallic segregated inclusions should not be used. Any of these structural imperfections will substantially reduce the maximum current density that the electrode will tolerate. Also, irregularity on the surface of the electrode may cause the arc stream to "backfire" and attach itself to the electrode at some distance back from the tip; this results in a welding difficulty that need not have occurred.

Electrodes should be stored in a clean container until they are needed. Grease or dirt on the surface of tungsten electrodes will interfere with good electrical contact when they are inserted into the torch, and may cause damage to the torch through arcing to the collet. Also, dirty electrodes can contaminate the weld metal while welding is in progress.

Size. Standard commercial diameters and lengths of tungsten electrodes are given in Table 5. Table 6 lists typical ranges of current within which the various diameters of tungsten or tungsten alloy electrodes are used when welding with argon as the shielding gas.

Generally, the electrode size should be chosen so that the electrode will operate

Table 5 Standard diameters and lengths of GTAW electrodes

Standard diameters, in.	Tolerance on diameter, in.	Standard lengths, all diameters(a), in.	Tolerance on length, in.
Diameters		**Lengths**	
0.010	±0.001	3, 6, 7	±1/16
0.020	±0.002	12, 18, 24	±1/8
0.040, 1/16, 3/32, 1/8, 5/32, 3/16, 1/4	±0.003		

(a) 0.010-in. electrodes also are available as coils.

Table 6 Typical ranges of current used in GTAW with tungsten electrodes of various diameters

Electrode diameter, in.	DCEN, A EWP, EWTh-1, EWTh-2, EWTh-3	DCEP, A EWP, EWTh-1, EWTh-2, EWTh-3	Alternating current (high frequency), A					
			Unbalanced wave			Balanced wave		
			EWP	EWTh-1, EWTh-2, EWZr	EWTh-3	EWP	EWTh-1, EWTh-2, EWZr	EWTh-3
0.010	Up to15	(b)	Up to 15	Up to 15	(b)	Up to 15	Up to 15	(b)
0.020	5-20	(b)	5-15	5-20	(b)	10-20	5-20	10-20
0.040	15-80	(b)	10-60	15-80	10-80	20-30	20-60	20-60
1/16	70-150	10-20	50-100	70-150	50-150	30-80	60-120	30-120
3/32	150-250	15-30	100-160	140-235	100-235	60-130	100-180	60-180
1/8	250-400	25-40	150-210	225-325	150-325	100-180	160-250	100-250
5/32	400-500	40-55	200-275	300-400	200-400	160-240	200-320	160-320
3/16	500-750	55-80	250-350	400-500	250-500	190-300	290-390	190-390
1/4	750-1000	80-125	325-450	500-630	325-630	250-400	340-525	250-525

(a) Ranges are based on the use of argon as the shielding gas. Other current values may be employed, depending on the shielding gas (lower values would be used with helium as the shielding gas), type of equipment, and application. (b) These combinations are not commonly used.

at near-maximum current-carrying capacity. At such a current level, the heat of the arc is more concentrated. This ensures maximum penetration, a stable arc, high welding speed, and minimum width and convexity of weld bead. The electrode size must be chosen for the current required and the speed desired for the specific application.

Authorities differ as to the electrode size that should be used for a specific current. Most agree that the smallest electrode that will maintain the arc without losing metal from the electrode tip in the form of molten drops or solid pieces should be selected for the unit current. Many users of pure tungsten electrodes grind a needle point on a much larger electrode than would ordinarily be used and permit the electrode tip to find its normal current-density diameter upon the first striking of the arc. This practice, which is applicable only to pure tungsten electrodes, automatically provides an ideal electrode tip diameter.

The most stable arc for any particular electrode is that which supports the maximum amperage without failing. Because of the high current-carrying capacity of tungsten electrodes and the large end-area differential existing between any electrode and the next larger or smaller size, it is not possible to attain an ideal current density merely by changing electrodes.

End Profile. Tungsten electrodes may have an end profile that is pointed, partly or completely hemispherical, or a bulbous mass of greater diameter than the electrode. A pointed end is ideal for welding in restricted locations, such as narrow joints in stainless steel piping, and it permits the current density to be maintained at an extremely high level. An electrode with a hemispherical end profile has the greatest current density, as less of the electrode end is in contact with the arc. The thoriated and zirconiated tungsten electrodes maintain their prepared end section over an extensive heat range, but pure tungsten electrodes change their end profile according to the current density at which they are operating.

The electrode taper angle is the angle that is ground on the end of the tungsten electrode, as shown in Fig. 8. Thoriated tungsten electrodes are ground to a tip to provide improved arc starting with high-frequency ignition. Higher current-carrying capacity enables these electrodes to be tapered. Taper angles usually range from 30 to 120°. Smaller taper angles tend to wear more quickly, especially when the tip of the electrode touches the work. To reduce number of times an electrode tip may be ground, a large taper angle is recommended.

Fig. 8 Electrode taper angle

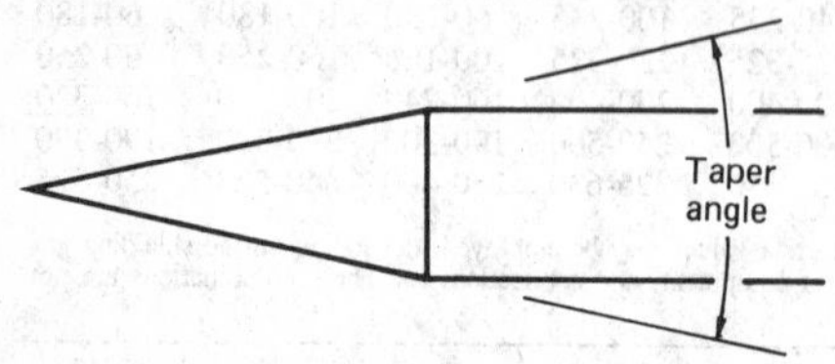

Because tungsten electrodes are not available in a large assortment of diameters (0.04, 1/16, 3/32, and 1/8 in. are the most common), tapering electrodes to a small diameter is recommended. Table 7 gives recommended electrode tip sizes and current ranges.

There is a method, previously mentioned, by which the end diameter of pure tungsten electrodes can automatically be adjusted to the specific job. In this method, after the heat required for the specific joint, metal, and welding position has been determined, an electrode is chosen that is in the general current range for the job. The electrode is then ground to a sharp point, with a taper three to six diameters long. When the pointed electrode is applied to the work, the point will melt back on itself until the tendency of the electrode mass to cool the molten tip balances the heat from the arc tending to melt it. Thus, the molten section on the end of the electrode will have the ideal diameter for the specific job, and the diameter of this end section will rarely be identical with any available electrode size.

This method of preparing an electrode adds greatly to the ease of operation in executing any specific weld. It also avoids a difficulty that attends the use of a blunt-end electrode (particularly when welding in a deep groove)—that of the arc running up the side of the electrode. Several pure tungsten electrodes should be pointed before a job is begun. Then, if the electrode strikes the workpiece or the filler rod, it will be necessary only to change electrodes, instead of waiting for time-consuming regrinding of the point of the electrode being used.

When extremely thin and delicate sections are to be joined, it may be necessary to grind the smallest available electrode to a needle point in order to help stabilize the arc at extremely low current. If welding is to be done automatically, the grinding should be done by machine, so that the end profile is duplicated from electrode to electrode for repetitive use under closely controlled welding conditions.

Current-Carrying Capacity. Polarity, electrode holder, torch design, and the skill

Table 7 Tungsten electrode tip shapes and current ranges

Electrode diameter, in.	Diameter at tip, in.	Included angle, °	DCEN: Constant current range, A	DCEN: Pulsed current range, A
0.040	0.005	12	2-15	2-25
0.040	0.010	20	5-30	5-60
0.062	0.020	25	8-50	8-100
0.062	0.030	30	10-70	10-140
0.093	0.030	35	12-90	12-180
0.093	0.045	45	15-150	15-250
0.125	0.045	60	20-200	20-300
0.125	0.060	90	25-250	25-350

and ability of the welder all have great influence on the maximum current that an electrode will withstand. The surface finish of the workpiece also has some effect; the radiation from a mirrorlike workpiece surface will reduce the current-carrying capability of the electrode because of heat reflection, whereas a dull surface will permit the current to be increased. When the work metal has been preheated, the radiation of heat from the work will reduce considerably the current-carrying capacity of the electrode—although less current will be required for obtaining a given depth of penetration.

If the current range is too low for the electrode being used, the arc will wander over the end of the electrode. Excessive current values will cause the tip of a pure tungsten electrode to "ball up" and vibrate at high frequency, at which point the tungsten begins to transfer across the arc in the form of small particles and metallic vapor. Excessive current may also cause arc instability. When this occurs, there is a strong possibility that the large molten globule suspended from a pure tungsten electrode will drop into the weld pool, contaminate it, and require mechanical removal before welding can proceed. Ideal arc conditions result when the molten tip of a pure tungsten electrode assumes the shape of a hemisphere. Stability of the arc will be promoted by using the smallest electrode at the maximum current value that will form a molten hemisphere at the tip.

Figure 9(a) shows the ends of pure tungsten electrodes after being used for welding at 300, 250, and 150 A with argon shielding. Note the erosion on the end of the electrode used at 150 A and the small projection on that electrode where the arc was concentrated but wandered over the surface of the electrode end. This wandering sometimes leads to erratic following of the weld seam if mechanized welding is being used.

Fig. 9 Ends of pure tungsten electrodes after welding at three levels of current with argon or helium shielding

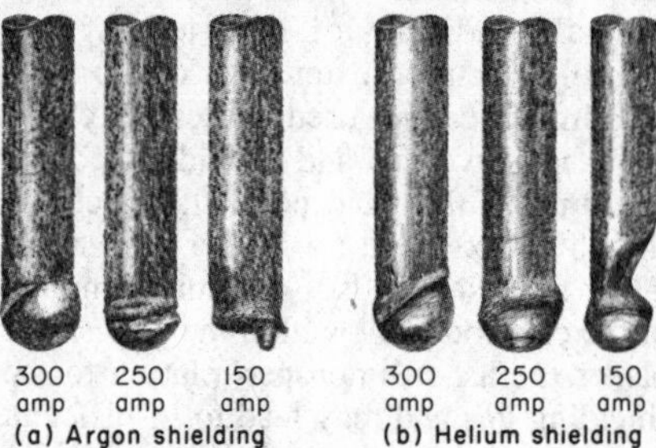

Figure 9(b) shows the ends of pure tungsten electrodes after being used for welding at 300, 250, and 150 A with helium shielding. The undercut appearance of the electrode used at 150 A may result from both low amperage and low gas-flow rate.

Under extended periods of use in automated applications where only the molten hemisphere was permitted to form, no measurable usage of tungsten could be detected after 200 h of continuous welding. When this form of electrode tip is presented to the work, not only is the current density the greatest, but also the arc gap may be the longest for any specific application.

Heating of the Electrode Tip. More heat is liberated on the positive side of a welding arc than on the negative side, because the impact of the electrons on the positive terminal heats it more than the negative. Consequently, when the polarity is reverse (DCEP), the electrode becomes hotter than the workpiece, whereas with straight polarity (DCEN), the workpiece becomes hotter. For example, to withstand the heat of a 125-A current, a tungsten electrode must be $^1/_4$ in. in diameter if DCEP polarity is used, but only $^1/_{16}$ in. in diameter with DCEN.

Greater penetration into the work metal can be obtained using DCEN, because about 70% of the heat of the arc is concentrated on the positive end of the arc stream, and only about 30% on the negative. Thus DCEN provides the most satisfactory heat transfer. However, when the arc is changed to DCEP, the current-carrying capacity of the electrode is reduced so much that continuous welding is maintained only with great manual skill and dexterity.

Excessive heating of an electrode can be avoided by changing to a larger electrode. Increasing the contact surface of the collet may also prove of some benefit. A non-water-cooled pure copper gas nozzle may be substituted for a ceramic nozzle; water-cooled nozzles further assist.

The flow of inert gas has a cooling effect on the electrode, although if the welder attempts to use higher amperage by employing an abnormally high gas flow, the cost of doing so will be excessive.

Consumption of tungsten electrodes while welding is in progress is so small that it can be detected only by elaborate methods. A tungsten electrode loses material by inclusion into the weld metal, by condensation on the surface of the work, and by vaporization into the atmosphere.

The greatest consumption of electrodes results from striking the workpiece or the filler rod. Another cause of electrode consumption is improper shielding of the electrode after the arc has been extinguished. Also, the introduction of a contaminating atmosphere into the gas nozzle along with the inert gas causes tungsten oxide to be passed down from the oxidizing tungsten electrode.

If a tungsten electrode has been oxidized because the shielding gas was shut off too soon after welding was completed, when the arc is restruck tungsten oxide will be thrown onto the work surface in the form of brilliant white particles, which pass through the arc with great speed. This throwdown, or spitting, of tungsten oxide across the arc, coincidental with missed cycles, may result also from instability of the welding transformer. To avoid tungsten spitting, it is necessary, after extinguishing the arc, to permit argon or helium to flow until the tungsten electrode becomes bright, silvery, and shiny. Any discoloration on the surface of the electode is caused by penetration of oxygen into the surface of the metal, and this oxidation may increase tungsten consumption by as much as 20 to 30 times.

If a large rate of consumption of electrodes is anticipated, it is best to use holders that will accommodate long electrodes, in order that the unusable electrode stub will be short in relation to the original length. For example, the unusable stub on a 3-in.-long electrode represents a much greater percentage of the original electrode length than the stub on a 7-in. electrode.

Tungsten Contamination of Weld Pool. When a pure tungsten electrode strikes the weld pool, tungsten is deposited in the weld metal. Because the contaminating tungsten may impair the strength and corrosion resistance of the welded joint, if the weldment is critical the welding operation should be stopped when such a strike has been made, and the tungsten should be mechanically removed from the weld.

To remove the tungsten inclusions, carbide burrs must be used; an ordinary grinding wheel is likely to push the embedded particles even farther into the work as the wheel goes down into the material. For critical weldments, it may be necessary to radiograph the area to ascertain that all of the tungsten has been removed before resuming welding. Tungsten particles show up as white flecks in weld radiographs.

Contamination of Tungsten Electrodes. Contamination of the tungsten electrode can cause discontinuities in the weld, in addition to making it difficult to control the arc. The electrode can become contaminated by (1) contact of the molten weld pool by the electrode, (2) contacting the electrode with the filler metal, (3) inadequate shielding-gas flow, or (4) postweld gas flow time that is too short. During welding, the tip of a pure tungsten electrode reaches a temperature of at least 6200 °F, which is above the melting point of tungsten. Any filler metal or base metal that contacts the molten tip is vaporized to a degree. It also chills the molten tip enough so that the tungsten combines with the filler or base metal to form an alloy. If welding is continued, the arc will be unstable. When this occurs, there are two alternatives: (1) to break off the contaminated section of electrode and start welding with a clean section, or (2) to hold the arc on a section of copper or other material until the electrode has been cleared of contaminating metal through its vaporization. It can be seen through the filter lens that at the instant of clearing, the arc stabilizes itself and regains a normal welding condition.

The contamination of tungsten electrodes results not only in defacing of the workpiece, but also in loss of several minutes of welding time while the electrode is being cleaned of the base metal plated on its surface or alloyed with it. When the electrode is being used at or near its maximum current capacity, if the torch is given a quick flip instantly after the electrode strikes the base metal or filler metal, most of the contaminating metal will slide down and off the end of the electrode. However, it may still be necessary to hold the arc for a short time on a piece of copper or other material, in order to clean the electrode completely. A considerable amount of electrode can be saved by this technique.

When welding stainless steel, if a pure tungsten electrode is contaminated with the base metal or filler metal, it is possible to

Fig. 10 Typical shapes of pure tungsten electrodes, $^1/_{16}$ in. in diameter. After contamination with stainless steel base metal or filler metal during welding

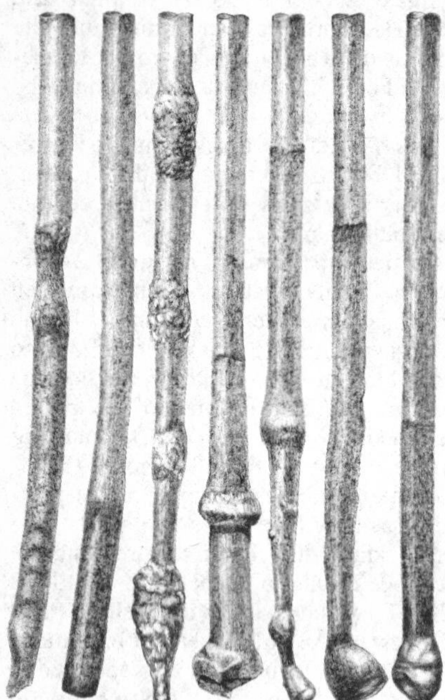

maintain the welding arc without cleaning the electrode, despite the instability of the arc. This is not good practice, however, because contamination of the weld metal is almost certain to result. The arc instability results from the unusual shape of the contaminated electrode tip (see Fig. 10), and also from the passing of metallic vapors across the arc as the base metal or filler metal adhering to the electrode is vaporized. The electrode will not regain its proper form until the contaminated section is removed and readjusted to its proper extension beyond the nozzle orifice.

Once put into use and heated, a tungsten electrode is brittle a short distance back from the point. If the end of an electrode is to be removed, it should be gripped tightly with a pair of pliers and snapped off suddenly in the brittle area, generally within $^3/_8$ in. from the end. If an attempt is made to break the electrode farther away from the point, the electrode bends rather than breaks, thus causing subsequent difficulty in realigning it with the nozzle orifice.

Electrode Extension. Ordinarily, the distance an electrode extends beyond the orifice of the gas nozzle should equal one or two electrode diameters, as shown in Fig. 11. The longer the electrode extension, the more danger there is of striking the work metal or filler rod and contaminating the electrode. The farther the electrode tip is withdrawn into the gas nozzle, the less current the electrode accommodates, probably because more arc heat is reflected from the gas nozzle onto the electrode.

Ordinarily, the amount of electrode extension is dictated mainly by the profile of the joint to be welded. An inside fillet weld requires the greatest electrode extension, so that the electrode may approach the root of the joint and yet permit some vision of the weld pool. At the opposite extreme is an upstanding edge-flange weld, for which only a very slight extension is required, or no extension at all.

In some applications of welding with high-frequency stabilized current, it is practical to withdraw the end of the electrode entirely within the nozzle nose. This has been done, for example, in welding an edge-flange joint between two 0.081-in.-thick sheets of aluminum alloy, wherein high-frequency-stabilized alternating current was used, and the electrode was retracted $^1/_{32}$ in. within the gas-nozzle nose. This practice makes it impossible for the electrode to become contaminated by touching the work surface. It does, however, interfere with weld visibility and demands a high degree of welder skill. Also, when the electrode is retracted inside the gas nozzle and there is no possibility of contaminating atmosphere reaching the weld pool, a black or gray material may appear in the center of the pool if the arc is too long. This phenomenon has occurred while using unbalanced alternating current transformers and may have been caused by direct current component passing across the arc.

For metals that require a very short welding arc, electrode extension should be greater than normal to provide the welder with better vision and to assist in control of the arc. However, the farther the electrode is extended, the greater must be the volume of gas flow to compensate for the greater distance of the gas nozzle from the workpiece surface.

Fig. 11 Electrode extension

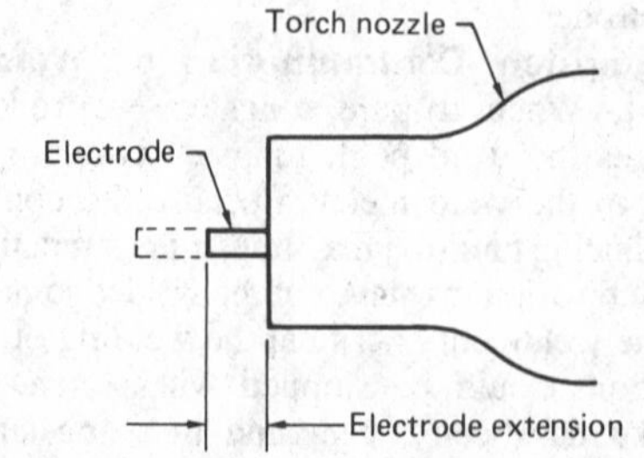

As the inclination of the torch varies from one joint to another, some variation in electrode extension must accompany it. It is seldom necessary to extend the electrode farther than $^1/_2$ in., although under special conditions of inert-atmosphere control, electrode extensions of as much as 3 in. have been used successfully.

To conserve gas and provide adequate shielding of the weld pool, the electrode should be extended no farther than absolutely necessary. Also, any misalignment of the electrode axis within the gas nozzle places the arc in improper relation to the shielding gas and may lead to insufficient coverage of the weld pool. When welding with high current and a ceramic gas nozzle, if the electrode is slightly off center, the nozzle lip nearest the electrode may begin to melt and vaporize and thus contaminate the shielding gas.

Thoriated tungsten electrodes contain up to 2.2% thoria (see Table 3). Unlike pure tungsten electrodes, thoriated tungsten electrodes do not melt at the usual current levels. For example, one $^1/_8$-in.-diam thoriated tungsten electrode ground back $^3/_4$ in. from a needle point held a 450-A DCEN arc for 12 min, at which time the grinding-wheel marks had disappeared from the surface of the ground portion only in the first $^1/_{32}$ in. Failure of thoriated tungsten electrodes from excessive current is characterized by the sudden dropping off of a large length, rather than by the melting off of droplets as occurs with pure tungsten electrodes.

If high frequency is used in striking the arc, the arc starts instantly at lower current values than are required for touch starting. Also, the arc may be started at much greater arc lengths with complete arc stability, which is of prime importance when welding expensive materials that may have to be surface-finished to a high degree.

Touch starting with thoriated tungsten is free of the sputtering and flashing that may occur when pure tungsten is used. The best conditions for easy touch starting are high open-circuit voltage, thoriated tungsten electrodes, and argon shielding. The type of work metal also has a considerable influence on the comparative ease with which touch starting is accomplished; stainless steel is one of the easier alloys from this standpoint, whereas aluminum requires high open-circuit voltage for successful touch starting.

Climbing of the arc up the side of the electrode at reduced current values, which often occurs when pure tungsten electrodes are used, is not encountered with thoriated tungsten electrodes except after long periods of use.

The degree to which thoriated tungsten electrodes are superior to pure tungsten in current-carrying capacity when welding with alternating current may range from 0 to 50% (Table 6), and apparently depends on the amount of direct current component and the amount of rectification in the arc. Current-carrying capacities of the large thoriated tungsten electrodes are considerably higher with helium shielding than when pure argon is used. Compared with pure tungsten electrodes, thoriated tungsten electrodes provide somewhat greater arc stability.

The higher current density that can be tolerated by thoriated tungsten electrodes, and the ability of these electrodes to hold a point without melting, contribute to an extremely stable and "stiff" arc stream. The consumption of thoriated electrodes in manual welding is only 10 to 20% as great as that of pure tungsten electrodes, and in mechanized applications the advantage is many times greater.

At a given current level and arc length, less penetration is obtained when welding with thoriated than with pure tungsten electrodes, because arc voltage is from 3 to 5 V less when thoriated electrodes are used. However, the much greater current-carrying capacity of thoriated tungsten electrodes more than compensates for the decreased arc voltage.

When a thoriated tungsten electrode is dipped into a molten pool of stainless steel weld metal, ordinarily neither the weld metal nor the electrode will be contaminated, particularly if the welding is being done in the flat position. In the vertical, horizontal and overhead positions, however, a portion of the molten metal in the pool often will adhere to the electrode and create a deformed outer profile. It is common practice to break off the section so contaminated, rather than to grind off the contaminating material, because of the difficulty of determining whether all the adhering material has been ground off before welding is resumed.

The pointing of thoriated tungsten electrodes is particularly beneficial for the welding of piping or of joints with deep grooves. With a pointed electrode, there is extreme concentration of the arc force, and the welding technique is assisted by the arc stiffness.

It has been reported that there are certain low-current mechanized welding applications for which pure tungsten electrodes are more satisfactory than the thoriated type. After a period of use, some thoriated tungsten electrodes permit the arc to attach itself to a part of the electrode other than the extreme tip. With continued use, the thoria inclusions may become segregated, and the arc may direct itself to these areas on the surface of the electrode, thus becoming unstable.

Zirconiated tungsten electrodes, which contain 0.15 to 0.40% zirconia (see Table 3), are well suited for applications in which contamination of the weld with tungsten must be minimized. They retain a balled end during welding and are highly resistant to contamination; thus, they perform well when used with alternating current. Zirconiated tungsten electrodes are in common use, but are less widely used than thoriated tungsten electrodes. Both types are comparable in performance and applicability.

Effect of Tip Angle. In GTAW with DCEN, the included angle at the conical tip of the tungsten electrode has a significant influence on (1) the voltage-current characteristics of the arc, (2) the penetration characteristics and the width of the weld pool, and (3) the distribution of temperature along the length of the electrode.

For 2% thoria tungsten electrodes, centerless ground, the width of bead-on-plate welds decreased by a factor of nearly two when the electrode taper angle at the conical tip was increased from 30 to 120°. For the same increase in taper angle, the depth of penetration increased by as much as 45%, while the cross-sectional area of the weld remained essentially unchanged.

The taper angle at the conical tip of a thoriated tungsten electrode should be specified and controlled if reproducible results are to be obtained with automatic GTAW.

Filler-Wire Feeders

In semiautomatic, machine, and automatic welding, filler-wire feeders are required to supply filler metal. For manual welding, filler metal is added by the operator. Hot or cold filler metal can be fed into the molten weld pool.

A cold wire-feed system includes a wire drive, wire speed control, and wire guide to feed the wire into the leading edge of the molten weld pool. The wire drive is controlled by a motor and gears to propel the filler wire. A constant speed can be achieved by electronic or mechanical control. Filler wire usually is directed to the weld pool through a flexible conduit. The wire guide also can be water cooled. Filler-wire sizes range from $^{1}/_{32}$ to $^{3}/_{32}$ in. in diameter.

Hot Wire-Feed System. Equipment for hot wire systems is similar to cold wire equipment. The wire, however, is electrically resistance heated to the desired temperature before it reaches the weld pool. Alternating current from a constant-current power source is used to heat the wire.

Shielding gas may be used to prevent oxidation of the filler wire. Higher deposition rates, comparable to those obtained by GMAW, are possible, as shown in Fig. 12. Hot wire-feed systems are used to weld stainless steel, carbon and low-alloy steels, copper alloys, and nickel alloys. Filler wire should not be preheated for welding aluminum and aluminum alloys due to arc blow. Hot wire usually is fed into the trailing edge of the weld pool.

Welding Cables

Welding cables and fixturing connectors, or work leads, link the power source to the torch and workpiece. Cables normally are made of copper or aluminum and consist of fine stranded wires enclosed in insulated casings of natural or synthetic rubber. A protective wrapping is placed between the stranded conductor wires and insulating jacket to permit some movement between them and provide maximum flexibility.

Cables must be kept to a minimum length, and must be of ample size to carry the welding current without overheating. Excessively long cables or undersize cables will cause loss of current and voltage. Long cables should never be coiled up; rather, the proper length of cable should be used. Welding cables must not be operated at currents in excess of their rated capacity, or overheating and rapid deterioration of the insulation will result. Frayed or worn cables should be replaced or repaired immediately to prevent short circuiting.

For welding with high-frequency arc stabilization, cables should be as short as

Fig. 12 Deposition rates in steel for GTAW with cold and hot filler wire

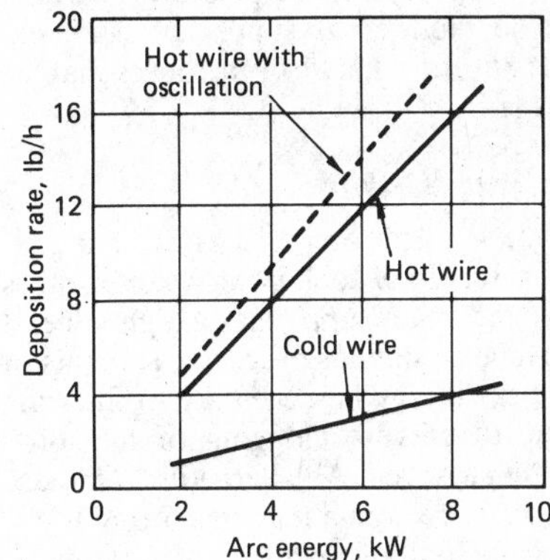

practical to prevent high-frequency attenuation. The cable between the high-frequency unit and the torch should be suspended from insulated hangers and not be taped or looped with other cables. In mechanized welding operations where a jig or holder is used, all metallic parts containing the workpiece must be grounded, to prevent arc-over or burning.

Interfering radiation can escape in four distinct ways from an installation for machine welding by GTAW in which high-frequency stabilization is used:

- *Direct radiation from the welding machine*: This may be minimized if the cover case of the welding machine is properly grounded.
- *Direct feedback to the primary power line*: This may be prevented by enclosing the line in a solid metallic conduit for at least 50 ft away from the welding machine in an unbroken line.
- *Direct radiation from the welding cables*: High-frequency attenuation may be minimized by keeping the welding cable as short as possible.
- *Pickup and reradiation from power lines and unshielded wiring*: Telephone wires, electric wires, and ungrounded metallic objects up to 50 ft away from the welding machine may pick up high-frequency radiation and conduct it for some distance before reradiating a strong interference field in another area as a radiating antenna.

Water Hoses

Water hoses should have a large diameter and be short, to minimize pressure drop in the flow of cooling water. Water flowing at about one quart per minute will provide adequate cooling of the torch. To maintain proper flow, a hose $12^1/_2$ ft long must have a line pressure of at least 25 psi, and a hose 25 ft long should have a pressure of 35 psi. A pressure regulator may be required if supply pressure exceeds 35 psi; otherwise, the hoses may be damaged.

Cooling Water

Water-cooled torches must have a continuous supply of cool, clean water. In areas where the water supply has a high mineral content and contains excessive amounts of particles, it may be necessary to filter the water, to prevent clogging of the torch cooling passages. An alternative is the use of a water-recirculating system, as is required for portable equipment systems.

For portable installations, a water tank with a 10- to 40-gal capacity, equipped with an electrically driven pump, is usually adequate. The water tank should be filled with distilled, demineralized, or deionized water. A bactericide or fungicide may also be added to further ensure its quality.

Gas Shielding Equipment

Shielding gas is supplied from many sources—single cylinders, portable or stationary manifold systems, or bulk storage. Gas flow is controlled by flowmeters and regulators. Different flowmeters are used for different shielding gases, each calibrated separately. The shielding gas is brought to the welding torch through hoses, which may be connected straight to the torch or through the power supply or to an inert gas attachment to the torch.

Hoses. Plastic hoses should be used for transporting helium. At 20 psi, helium diffuses through the walls of rubber or rubber-fabric hoses. The safer procedure is to use plastic hoses for both helium and argon to guard against possible gas diffusion.

When setting up a GTAW operation, it is best to use only new hoses and fittings, because old hoses may have been used for purposes that would have contaminated them. When old hoses are used for argon or helium, they may continue for a long time to contaminate the inert gas transmitted through them.

When argon or helium lines are disconnected, both male and female connections should be closed with pressure-sensitive tape or with plastic plugs to prevent water, moisture, or other contaminants from entering the gas passages. This is an important precaution; even the slightest deposit of water in the line will contaminate the inert atmosphere to the extent that the equipment cannot be used until considerable time and effort have been expended to remove the moisture.

Mixers. When a mixture of two or more gases is used, the mixture is made by metering the separate gases into a mixing chamber. Here the gases are combined and mixed, after which the mixture exits through a single port. Gas mixtures also are available in cylinders.

Motion Devices

Motion devices, also referred to as traversing mechanisms, are used for machine and automatic welding. These devices can be used to move either the welding head, workpiece, or torch, depending on the size and type of workpiece and user preference. Motion devices maintain the position of the electrode within close joint tolerances. Travel proceeds at a uniform rate, without excessive vibration.

Motor-driven mechanisms on tracks or on the workpiece are available in various sizes. They are used for straight line and contour welding. Motion devices also are available for vertical and horizontal welding.

Side beam carriages are available for transporting welding heads and allied equipment in straight line operation. The carriage is supported on the vertical face of a flat track. The track can be mounted to the side of a beam.

Track sections are available in various lengths. The various types of carriages are fitted with mechanical clutches so that the drive motor can be disengaged while the welding head is being positioned.

Welding head manipulators, also referred to as pedestal boom manipulators, can be used for rapid positioning and welding in any direction to produce internal or external longitudinal or circumferential (girth) welds. This equipment is suitable for use with power rolls or headstock-tailstock welding positioners on tanks, vessels, pipes, and similar weldments. Rotating positioners, to which fixtures or other workholding tooling devices are mounted, are readily adapted to use with pedestal boom manipulators for small production runs or special fabrications.

Power supplies, filler-metal supplies, portable coolant tanks, automatic welding heads, and remote controls can be mounted on a pedestal boom manipulator. A typical pedestal boom manipulator that includes some of these features is shown in Fig. 13. Pedestal boom manipulators can be moved manually, on integral wheeled

Fig. 13 Use of a pedestal boom manipulator for making a circumferential (girth) weld

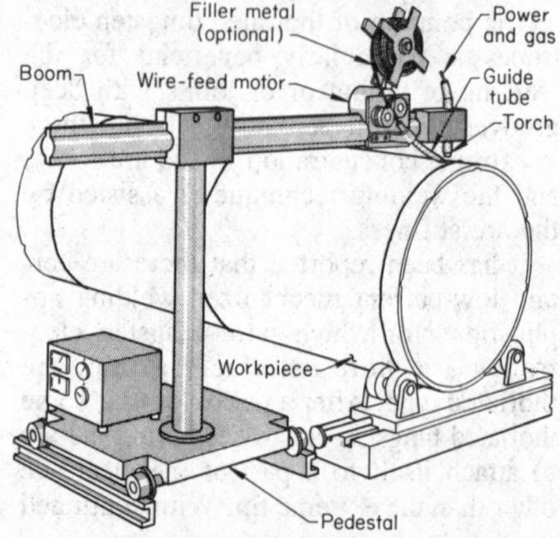

Fig. 14 Positioner that rotates and tilts workpieces

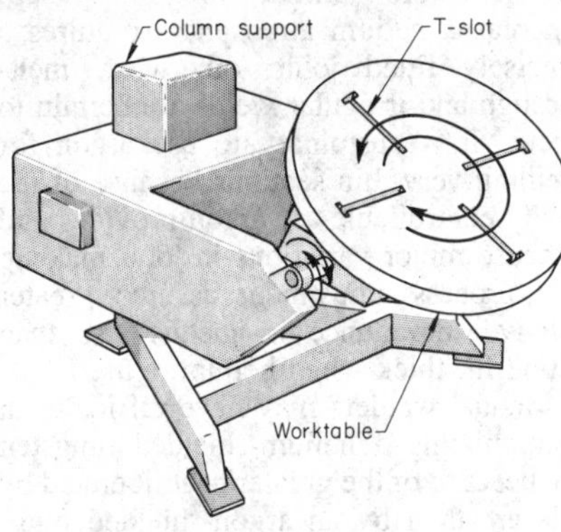

bases, to work locations, or they can be powered to travel at welding speeds on rails.

Rotating positioners, as shown in Fig. 14, are adaptable to manual, machine, or automatic welding. The positioner worktable is a circular or square plate that has been machined to a flat surface and is fitted with T-slots for bolting down various jigs or workholding tools. The worktable can be rotated for 360° circumferential welds in any plane from flat to 10° past vertical. Variable rotational speeds and provision for smooth starting and stopping are mandatory for automatic welding operations. Pedestal boom manipulators are used to position the automatic welding head over the joint, and the positioner worktable may be started manually or sequenced to start automatically.

Oscillators are used to oscillate the torch for surfacing, out-of-position welding, and other welding operations that require a wide bead. They are optional attachments that are mechanically or electromagnetically operated.

Fixtures

Jigs or restraining fixtures should be used whenever possible, to hold workpieces in alignment and to exert pressure at the proper point so that the work cannot move during heating and cooling. If alignment, or joint fit-up, is not maintained, expansion and contraction may cause distortion or mismatch of the welded joint. Principles of fixturing for GTAW are generally the same as for other welding processes. However, fixturing is frequently more critical for GTAW, because the workpieces are often made of thin metal and thus are more susceptible to deformation than heavy structural parts.

Tack welding may be used in manual welding operations in which the shape of the workpieces prevents the use of fixtures, or in which production quantity is too small to warrant the cost of fixtures. However, in mechanized or automatic welding, tack welds may impair control of arc voltage and of filler-wire feeding.

Weld backing improves the uniformity, appearance, and contour of a weld. Thin work metal is usually backed to protect the underside of the weld from atmospheric contamination, which may result in weld porosity or poor surface appearance. Backing bars also absorb some of the heat generated by the arc and can lend support to the assembly and the weld pool. A weld can also be backed up with inert gas, which serves to control penetration and maintain a bright, clean undersurface, but does not function as a support. Sometimes, metal backing bars and an inert gas are used in combination.

Nitrogen has been used effectively to back stainless steel welds, but caution is necessary to prevent contamination of the arc atmosphere with nitrogen, which will cause arc instability. The use of nitrogen for backing can contribute substantially to the economy of the operation.

In applications where the final weld composition must conform to extremely rigid specifications, particular care must be taken to exclude all atmospheric gases from the underside of the weld. This is accomplished by introducing an atmosphere of shielding gas into the relief groove of a backing bar, as shown in Fig. 15.

Shielding Gases

The main requirement of a shielding gas is that it exclude air from the weld pool, the electrode, and the heated end of the filler rod (if used), to avoid contamination of the weld deposit. Shielding gas does not directly add heat to the weld, but it does affect heat input. The gases ordinarily used in GTAW are argon, helium, argon-helium mixtures, and argon-hydrogen mixtures.

The choice of shielding gas can significantly affect weld quality as well as welding speed. Argon, helium, and argon-helium mixtures do not react with tungsten or tungsten alloy electrodes and have no adverse effect on the quality of the weld metal.

Argon Versus Helium. There is considerable difference of opinion about the relative merits of argon and helium for welding purposes. Each gas possesses characteristics that make it more suitable than the other for certain applications (see Tables 8 and 9).

Argon is more widely preferred, because, in addition to it being less expensive, it provides a softer arc, which is smooth and stable. Helium-shielded weld pools are very hot and fluid, the weld metal is difficult to handle when pipe must be

Fig. 15 Use of a backing bar with a relief groove for shielding gas. Excludes atmosphere from underside of weld

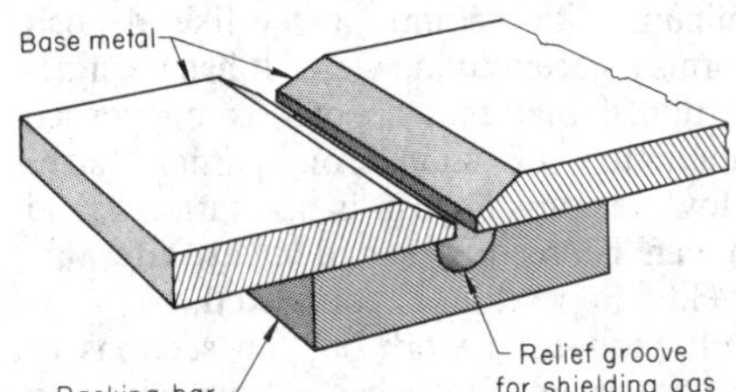

Table 8 Characteristics and comparative performance of argon and helium as shielding gases

Argon

Low arc voltage: Results in less heat; thus, argon is used almost exclusively for manual welding of metals less than $^1/_{16}$ in. thick.

Good cleaning action: Preferred for metals with refractory oxide skins, such as aluminum alloys or ferrous alloys containing a high percentage of aluminum.

Easy arc starting: Particularly important in welding of thin metal.

Arc stability: Greater than with helium.

Low gas volume: Being heavier than air, argon provides good coverage with low gas flows, and it is less affected by air drafts than helium.

Vertical and overhead welding: Sometimes preferred because of better weld-pool control, but gives less coverage than helium.

Automatic welding: May cause porosity and undercutting with welding speeds of more than 25 in./min. Problem varies with different metals and thicknesses, and can be corrected by changing to helium or a mixture of argon and helium.

Thick work metal: For welding metal thicker than $^3/_{16}$ in., a mixture of argon and helium may be beneficial.

Welding dissimilar metals: Argon is normally superior to helium.

Helium

High arc voltage: Results in a hotter arc, which is more favorable for welding thick metal (over $^3/_{16}$ in.) and metals with high heat conductivity.

Small HAZ: With high heat input and greater speeds, the HAZ can be kept narrow. This results in less distortion and often in higher mechanical properties.

High gas volume: Helium being lighter than air, gas flow is normally $1^1/_2$ to 3 times greater than with argon. Being lighter, helium is more sensitive to small air drafts, but it gives better coverage for overhead welding, and often for vertical-position welding.

Automatic welding: With welding speeds of more than 25 in./min, welds with less porosity and undercutting may be attained (depending on work metal and thickness).

Table 9 Suitability of argon and helium for use as shielding gases in GTAW of various metals
Welding with DCEN unless noted otherwise

Aluminum alloys: Argon (with alternating current) preferred; offers arc stability and good cleaning action. Argon plus helium (with alternating current) gives less stable arc than argon, but good cleaning action, higher speed, and greater penetration. Helium (with DCEN) gives a stable arc and high welding speed on chemically clean material.
Aluminum bronze: Argon reduces penetration of base metal in surfacing (for which aluminum bronze is used).
Brass: Argon provides stable arc, little fuming.
Cobalt-based alloys: Argon provides good arc stability, is easy to control.
Cupronickels: Argon provides good arc stability, is easy to control. Used also in welding cupronickels to steel.
Deoxidized copper: Helium preferred; gives high heat input to counteract thermal conductivity. A mixture of 75% helium and 25% argon provides a stable arc, gives lower heat input than helium alone, and is preferred for thin work metal ($^1/_{16}$ in. or less).
Inconel: Argon provides good arc stability, is easy to control. Helium is preferable for high-speed automatic welding.
Low-carbon steel: Argon preferred for manual welding; success depends on welder skill. Helium preferred for high-speed automatic welding; gives more penetration than argon.
Magnesium alloys: Argon (with alternating current) preferred; offers arc stability and good cleaning action.
Maraging steels: Argon provides good arc stability, is easy to control.
Molybdenum-0.5Ti alloy: Purified argon or helium equally suitable; welding in chamber preferred, but not necessary if shielding is adequate. For good ductility of the weld, the nitrogen content of the welding atmosphere must be kept below 0.1%, and the oxygen content below 0.005%.
Monel: Argon provides good arc stability, is easy to control.
Nickel alloys: Argon provides good arc stability, is easy to control. Helium is preferred for high-speed automatic welding.
PH stainless steels: Helium preferred; provides more uniform root penetration than argon. Argon and argon-helium mixtures also have been used successfully.
Silicon bronze: Argon minimizes hot shortness in the base metal and weld deposit.
Silicon steels: Argon provides good arc stability, is easy to control.
Stainless steel: Helium preferred. Provides greater penetration than argon, with fair arc stability.
Titanium alloys: Argon provides good arc stability, is easy to control. Helium is preferred for high-speed automatic welding.

welded in the horizontal fixed position, and welding becomes more susceptible to manipulative errors. Argon is better for welding aluminum alloys, magnesium alloys, and beryllium copper; helium is less efficient in the surface-oxide cleaning characteristic that is essential when these metals are gas tungsten arc welded. Greater penetration is achieved when using helium, and helium is used to good effect when fast, hot welding is desired for stainless steel.

Argon is used more extensively throughout industry than is helium. Argon is 1.4 times as heavy as air, but it will mix with air in confined spaces. Also, it is about ten times as heavy as helium and thus is better suited to welding in certain positions than helium, which rises rapidly from a weld after release from the torch nozzle.

Helium has only about one seventh the weight of air and will mix with air only slowly. For welding of a given joint, a greater amount of helium than of argon must be used for shielding, because it is not feasible to confine helium to the weld area with dams and baffles, as must be done when welding gas-sensitive metals and alloys. The one exception is when welding is being done in the overhead position; then, baffles and dams can be used effectively to confine helium to the weld area and thus achieve a beneficial shielding effect as well as conserve gas.

Helium is ideal for welding in the overhead position, and for shielding the bottom side of joints welded in the flat position. In a vessel, helium will rise to the top. This constitutes a health hazard if the welder is working inside the vessel. In contrast, argon will sink to the bottom, and this must be considered if personnel are to enter and work near the bottom of a vessel.

Schlieren shadowgraphs show that the weight of the gas (whether argon or helium) has little effect on the immediate shield of the arc and weld pool, and that adequate coverage is offered by either argon or helium. There is less entrainment of air into the issuing stream of helium than occurs with argon. Increasing the flow of argon after the necessary coverage has been established brings about an increase of turbulence without increased effectiveness of coverage. With helium, however, which induces almost no turbulence, additional coverage of area is attained by increasing the amount of flow. When welding aluminum with helium, a sootlike deposit forms adjacent to the weld. It has been determined that this deposit is condensed aluminum of submicron particle size. However, this deposit is not harmful, and it can be readily wiped away after the welding operation is completed.

Experimental work on thin sections of heat-resistant alloys has shown argon to be superior to helium for most applications of manual welding. The use of argon permits more latitude in joint fit-up, whereas a helium atmosphere requires a precisely fitted joint; otherwise, melt-through and irregular welds are certain to occur. It is customary to use argon for welding very thin sections because of the easily controllable arc argon provides and because minor variations in joint makeup and process adjustment assume greater relative importance in metals less than 0.062 in. thick when helium is used.

Manual welders may have difficulty in manipulating a helium-shielded tungsten arc because of the greater heat liberated by this arc than by an argon-shielded tungsten arc. The use of helium requires considerably more skill than the use of argon, because argon permits a wide variation in arc length with a relatively small difference in heat input.

A helium-shielded arc emits about one third more heat than an argon-shielded arc at the same current setting, because of the higher arc voltage of the helium arc. Some authorities state that about 10 to 20% more helium than argon is required for effective shielding of the weld pool. However, 30 to 40% greater welding speed will be attained with helium because of the hotter and more intense arc it provides.

With DCEN high-quality welds cannot be made in aluminum when argon shielding is used, but when high-purity helium is used, thoroughly acceptable weave beads and stringer beads can be made.

Argon-helium mixtures are used when the greater penetration of helium is desired but the arc-softening action of argon is helpful from the standpoint of control. A mixture of 80 parts by volume of helium and 20 parts of argon has been effective in such applications. If a hotter arc is desired when welding aluminum with argon, helium can be added until the desired penetration is obtained. Combinations of argon and helium are widely used for automatic welding and are obtainable in various percentages in cylinders.

Argon provides greater coverage of the weld pool at low flow rates, whereas helium gives maximum coverage at high flow rates. Common mixtures of these gases by volume are 75% helium and 25% argon, and 80% helium and 20% argon, but a wide variety of mixtures are available.

Argon-hydrogen mixtures frequently are used to weld stainless steel, Inconel, and Monel alloys, and in applications where porosity is a problem. The purpose of argon-hydrogen mixtures is to increase the welding speed and help control the weld-bead profile. Argon-hydrogen mixtures enhance weld-pool wettability, thus promoting uniform weld-bead geometry. Argon-hydrogen mixtures are not suitable for plain carbon or low-alloy steel applications. However, stainless steel can be welded with mixtures containing up to 15% hydrogen. This mixture is used most often for welding tight butt joints in stainless steel up to $^1/_{16}$ in. thick. The most common argon-hydrogen mixtures are 95% argon and 5% hydrogen, or 85% argon and 15% hydrogen.

Oxygen-Bearing Argon Mixtures. Argon mixed with 1 to 5% oxygen is for use only with GMAW, which uses a consumable electrode. If oxygen-bearing argon is used with GTAW, the oxygen in the gas will cause very rapid deterioration of the electrode.

Nitrogen in a shielding-gas mixture is always detrimental to arc stability, because of the inherent steady attack on, and deterioration of, the electrodes. However, nitrogen markedly increases the voltage and attendant heat in the welding of copper and copper alloys, for which an extremely hot arc is essential.

If high-purity nitrogen is used as the shielding atmosphere for welding deoxidized copper, the higher arc voltage obtainable permits higher current to be used with standard equipment. The efficiency of heat transfer is much higher than when argon or helium is used, and this results in substantial economy.

Nitrogen is used in Great Britain for welding deoxidized copper, because of the shortage of argon and helium there; but this usage is unknown in the United States.

Purity of Shielding Gases

Only welding-grade gases should be used in GTAW. If commercial-grade gases are used, difficulties occur. For welding reactive metals such as titanium, tantalum, and zirconium, which are penetrated at elevated temperature by detrimental gases present in the ambient atmosphere, gas of absolutely predictable purity must be used to avoid difficulty during welding. Assurance that gas purity is being maintained can be obtained by periodically testing cylinders before use on production work. Relative coloration of the surfaces of welds made in inert-gas chambers has been used as a test criterion for purity.

The dew point of shielding gas should be −75 °F or lower (11.4 ppm water vapor by volume). Water vapor in impure helium dissociates in the arc and yields hydrogen and oxygen. The presence of hydrogen in the inert atmosphere causes porosity, and the oxygen forms a film over the weld pool, which may impair the ease of welding and result in poor fusion, together with inclusions. The presence of nitrogen as a contaminant, in any concentration, impairs the speed of weld progression. Helium of less than 99.8% purity contains excessive water. If welding is attempted with helium of such purity, difficulty may be encountered when the cylinder pressure drops below 50 psi. It may be necessary to discontinue use of the cylinder at this point unless the water is removed by use of a drying agent, such as magnesium perchlorate, in the venting system.

Cylinder Gas Contamination. To prevent contamination of inert-gas cylinders with other gases, an arrangement is generally entered into between the customer and the supplier, whereby the user is requested to leave a certain residual pressure in the cylinder at all times. This assures the supplier that he will not need to purge any contaminating gases from the returned cylinder. The customer is permitted a price adjustment on the basis of the volumes of gas returned to the supplier. Usually, a residual pressure of 25 psi is sufficient to ensure against contamination of the cylinder prior to recharging.

Welding difficulty is often attributed to impurity of the shielding gas, but experience has shown that of all possible sources of trouble, the gas in the cylinder itself is least likely to contribute to the difficulty.

Hoses and connections must be thoroughly checked against leakage, which can readily cause contamination of the inert atmosphere flowing to the torch. Every possible care should be taken to ensure that the pure gas from the cylinder is conveyed to the weld pool through a system that is completely leakproof.

If commercial-grade argon or helium is used for welding, the residual gases in the cylinders contaminate the weld pool, resulting in weld porosity. Purity of gas must be considered for critical applications, although the purity of welding-grade argon and helium as now produced (99.995% purity) is generally satisfactory for almost all applications. Passing the gas through a small chamber containing incandescent "getter" metals is one way of reducing impurities in inert gases to only a few parts per million.

Removal of Air From Argon. Figure 16 shows that four volume changes of argon in a 70-ft^3 atmosphere chamber can reduce air content to about 0.1% (which is adequate for welding most reactive metals), if the argon is introduced slowly (at 80 ft^3/h) through a porous bronze or similar diffusing plate. If, on the other hand, the argon is introduced directly (without diffusion through a porous plate) at a higher flow rate (350 ft^3/h, as shown in Fig. 16), and perfect mixing of the gases is assumed, a proportion of 0.1% air in argon is reached after ten volume changes.

Gas Flow

Only enough shielding gas to exclude air from the weld location (and the heated area, for reactive metals and alloys) should be used. Excessive gas flow not only unnecessarily increases cost, but also may cause undercutting and arc instability.

The minimum flow of gas required for maintaining adequate and effective coverage of the welding area is influenced by the following variables:

- Shielding gas
- Distance of gas-nozzle orifice from the work surface
- Design of the weld joint
- Size of gas nozzle
- Shape of gas nozzle
- Size of weld pool
- Amount of welding current
- Presence of drafts or wandering air currents
- Inclination of the torch
- Arc length
- Welding speed
- Position of the workpiece
- Metal or alloy being welded

This list of variables is arranged in the general order of relative importance for most jobs, although specific applications may alter the relationship of these factors. In still atmospheres, effective shielding has been obtained with gas flows as low as 6 ft^3/h for helium and 4 ft^3/h for argon. These flow rates are only about half the rates normally used for average welding conditions, but they indicate the economy that

Fig. 16 Relationship of purging times and gas-flow rates to number of volume changes required for removal of air from argon in a 70-ft^3 chamber

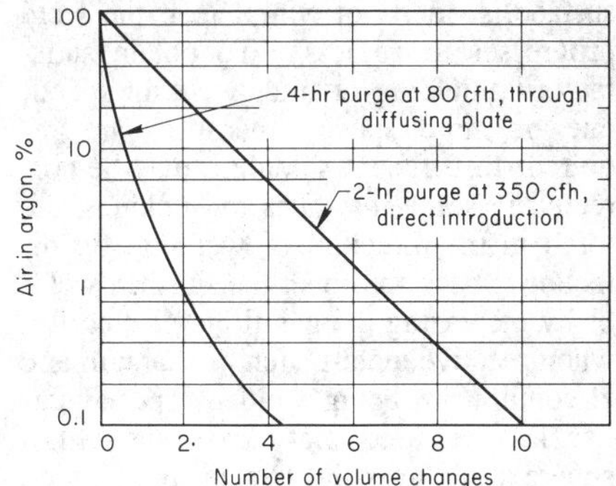

Table 10 Conversion equivalents of gas flow rates

L/min	L/h	ft³/h	Life of 224-ft³ cylinder, h
1	60	2.1	116.2
2	120	4.2	58.1
3	180	6.4	38.1
4	240	8.5	28.7
5	300	10.6	23.0
6	360	12.7	19.2
7	420	14.8	16.5
8	480	16.9	14.4
9	540	19.1	12.8
10	600	21.2	11.5
11	660	23.3	10.5
12	720	25.4	9.6
13	780	27.5	8.9
14	840	29.7	8.2
15	900	31.8	7.7

is possible when all factors are under good control.

Shielding gas issues from the torch nozzle at relatively low velocity—about 10 ft/s—and thus the gas column is comparatively easy to disturb with drafts and air currents. The use of excessive gas flow to prevent the disturbance, however, not only is wasteful but also may be detrimental to the weld metal and the welding operation. Excessive flow also may cause an unstable arc at low welding-current values, and can result in undercutting of the work surface adjacent to the weld bead.

Only enough gas flow is required to exclude the surrounding air from the weld location; for argon, a flow rate of 8 to 12 ft³/h usually should be ample. If substantially greater flow than this is required to provide ample coverage, consideration should be given to the possibility of interference from drafts in the weld area, incorrect nozzle size for the specific job, or improper jig design.

Some flowmeters are calibrated in cubic feet per hour; others, in liters per minute. To facilitate correlation between gas flows, the conversion equivalents in Table 10 may be used.

Maintenance of Adequate Shielding. Protection of the heated and welded surface is often not complete if the weld progresses too quickly. To obtain satisfactory shielding at high welding speed, one or more special measures are required. Insufficient shielding may be corrected in some instances by inclining the torch in the direction of welding, by directing it backward over the finished weld, or by increasing the gas flow. Ordinarily, when reactive metals such as titanium and zirconium are being welded, special gas nozzles are used that direct the atmosphere back over the heated and welded surface. For these metals and alloys, which are subject to gaseous contamination in the solid state at elevated temperature, separate means of protection by inert atmosphere furnished from a separate source must often be provided. Sometimes even this is not wholly effective, and the entire operation may have to be done within an inert-atmosphere chamber.

Drafts and Air Currents. To avoid contamination of the shielding atmosphere brought to the weld area by the torch, the ideal location for GTAW is a draftless area. If conditions of disturbed air exist, suitable baffle screens should be set up around the operation. If work is to be done in the open, a portable framework of convenient size may be erected around the work area. This may be covered by canvas, leaving an opening for entry. When welding is done with any gas shielded process in the field, much time, effort, and difficulty will be avoided by providing some method of proper protection against drafts.

Conditions of disturbed air can be overcome to a degree by delivery of a greater volume of inert gas to the work, but ordinarily this is inadvisable because of the much greater expense involved per unit weld, as well as the fluctuations in shielding obtained under such conditions. Inclining the torch toward the direction from which the draft is coming will help somewhat, but this will hinder the welding operation and make it more difficult to perform.

Filler Metals

The selection of the proper filler metal is based primarily on the composition of the base metal being welded. Filler metals usually are matched as closely as possible to the base-metal composition.

Closer control of composition, purity, and quality is exercised for filler metals than for base metals. Choice of a filler metal depends on the proposed application. Tensile properties and impact toughness, as well as electrical conductivity, thermal conductivity, corrosion resistance, and weld appearance, are important considerations in the choice of a filler metal.

Deoxidizers may be added to improve weld soundness. In GTAW, loss of deoxidizers is minimal; the filler metal is not transferred across the arc. Generally, a solid bare wire (electrode) manufactured for GMAW is suitable for GTAW. Further modifications may be made to some filler-metal compositions to improve postweld heat treatment response.

Table 11 lists AWS filler-metal specifications that are applicable to GTAW. These specifications establish filler-metal classifications based on mechanical properties and chemical compositions. They also set forth the conditions under which filler metals must be tested. For some materials, radiographic standards of acceptability are given. Usability tests also are required by some of these specifications. For more information on the classification system used by AWS, see the section on electrodes in the article "Gas Metal Arc Welding" in this Volume.

Forms, Sizes, and Use. Filler metals are available in the form of rods, spooled wire, and consumable inserts. For straight rods, which are used for manual welding, standard diameters range from 0.030 to 0.25 in.; nominal length is 36 ± 0.375 in. For mechanized welding, filler metals come in continuous spooled wire. The diameter of the wire ranges from 0.020 in. for delicate work to 1/4 in. for high-current welding and surfacing. Consumable inserts are designed to meet size and shape requirements of specific joints; some types of consumable inserts have been standardized.

Filler metals are added either manually or by a mechanized wire feeder during

Table 11 AWS specifications for filler metals suitable for GTAW

Specification	Title
A5.2	Iron and Steel Gas-Welding Rods
A5.7	Copper and Copper Alloy Bare Welding Rods and Electrodes.
A5.9	Corrosion-Resisting Chromium and Chromium-Nickel Steel Bare and Composite Metal Cored and Stranded Arc Welding Electrodes and Welding Rods
A5.10	Aluminum and Aluminum Alloy Welding Rods and Bare Electrodes
A5.13	Surfacing Welding Rods and Electrodes
A5.14	Nickel and Nickel Alloy Bare Welding Rods and Electrodes
A5.16	Titanium and Titanium Alloy Bare Welding Rods and Electrodes
A5.18	Mild Steel Electrodes for Gas Metal Arc Welding
A5.19	Magnesium Alloy Welding Rods and Bare Electrodes
A5.21	Composite Surfacing Welding Rods and Electrodes
A5.24	Zirconium and Zirconium Alloy Bare Welding Rods and Electrodes
A5.28	Low Alloy Steel Electrodes for Gas Shielded Arc Welding
A5.30	Consumable Inserts

welding, or are added as preplaced consumable inserts. Manual additions are made by hand-feeding a filler rod to the weld pool. Mechanized and automatic feed systems supply filler metal, normally as spooled wire, to the weld pool at a predetermined rate. Consumable inserts are placed in the weld zone before the arc is started, and are consumed in the weld pool during welding.

Storage and Preparation. Manufacturers of filler metals employ elaborate techniques to ensure that rod or wire is clean prior to packaging. Storage under improper conditions results in contamination of the filler metal. Any oil or other organic material on filler metal, or a heavy oxide film, interferes with the deposition of sound weld metal.

Filler metal should be stored where condensation will not contaminate it. Normally, an area that has reasonably constant temperature and low humidity, and that is protected from oily shop atmospheres, is adequate. More elaborate facilities may include temperature and humidity controls or inert-gas chambers designed to remove air and moisture.

For most welding applications, filler metal that has been properly stored requires no specific preparation prior to welding. When necessary, straight rods may be degreased, cleaned chemically by etching, or rubbed with steel wool or abrasive to remove traces of oxide. These techniques may also be used to clean coiled filler metals, but uncoiling and recoiling is a time-consuming chore.

Joint Design

The basic types of joints—butt, lap, corner, edge, and T—are all used in GTAW. Variations of these joints can be made to meet special requirements. Selection of the proper design for a particular application depends primarily on:

- Mechanical properties desired
- Cost of preparing the joint and making the weld
- Type of metal being welded
- Size and configuration of the components to be welded

Edge and joint preparation are critical to the obtaining of sound welds, because proper joint fit-up is necessary—particularly for square-groove butt joints and any other joint to be made without adding filler metal.

Butt Joints. A square-groove butt joint is the easiest to prepare, and can be welded with or without filler metal, depending on the composition and thickness of the pieces being welded. A single-V-groove butt joint is used where complete penetration is required on work metal more than $^3/_{16}$ to $^3/_8$ in. thick. The maximum thickness of square-groove butt joint that can be penetrated from one side depends greatly on the composition of the metal being welded. For instance, the maximum practical thickness for stainless steel is approximately $^3/_{16}$ in., whereas for aluminum alloys this limit is about $^3/_8$ in. Filler metal must be used for a V-groove weld. The included angle of the V-groove should be approximately 60°. The root face will be $^1/_8$ to $^1/_4$ in. high, depending on the composition and thickness of the pieces being welded.

A double-V-groove butt joint is generally used on stock thicker than $^1/_2$ in., when design of the weldment permits access to the back of the joint. Butt joints of special design are sometimes used for very thick work metal, which may require many welding passes (often by other processes in addition to GTAW). However, the shape and dimensions of the portion of these joints where welding begins are usually close to those described above. Butt joints of special design are used also in welding with consumable inserts (see Fig. 17).

Fig. 17 Butt joints using consumable inserts. (a) Shape of one type of consumable insert placed between two pipe sections prior to welding. (b), (c), and (d) Joint designs for consumable inserts. Nominal dimensions *d*, *w*, and *h* as shown in (a), for two sizes of inserts, are as follows: $^1/_8$-in. inserts: *d*, 0.125 in.; *w*, 0.047 in.; *h*, 0.055 in.; $^5/_{32}$-in. inserts: *d*, 0.156 in.; *w* and *h*, 0.063 in.

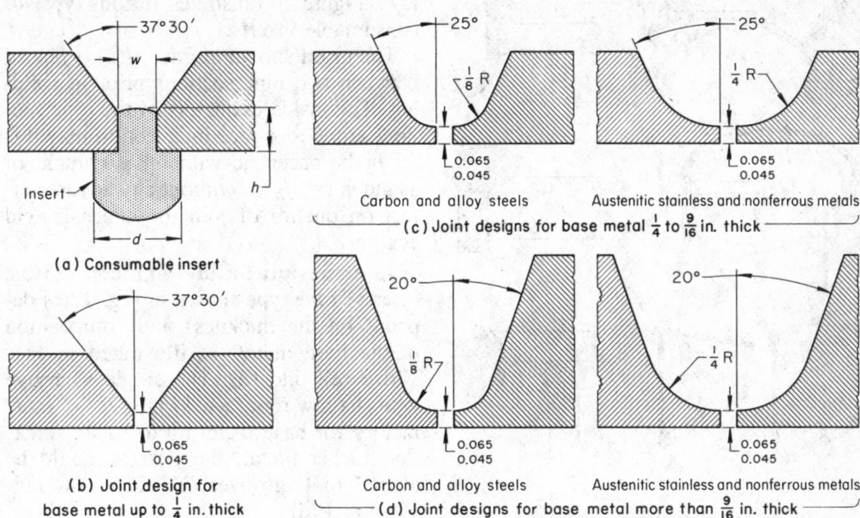

Lap Joints. A lap joint has the advantage of eliminating the need for edge preparation. The only requirement for making a good lap weld is that the plates be in close contact along the entire length of the joint to be welded. This requirement is difficult to meet for thin base metal unless jigs and fixtures of appropriate design to provide adequate clamping are available. Otherwise, closely spaced tack welds may be necessary. Lap joints are difficult to repair, finish, or clean, and they frequently have root defects.

Corner joints are used for fabricating boxlike structures. Two designs of corner joints are illustrated in Fig. 18. Assuming that the work metal is stainless steel, if the metal is no thicker than $^1/_8$ in., the corners can be butted as shown in Fig. 18(a), and a satisfactory weld can be produced without filler metal; if the metal is thicker than $^1/_8$ in., one of the corner members must be beveled or shaped as shown in Fig. 18(b).

Edge joints usually do not require the addition of filler metal. Three common types of edge joints are illustrated in Fig. 19; a feature common to all three types is that the thickness of the two components must be approximately the same at the joint. For joining thin turned-up edges, such

Fig. 18 Corner joints for thin and thick metals. Thickness limits apply to stainless steel or to metals with thermal conductivity similar to that of stainless steel.

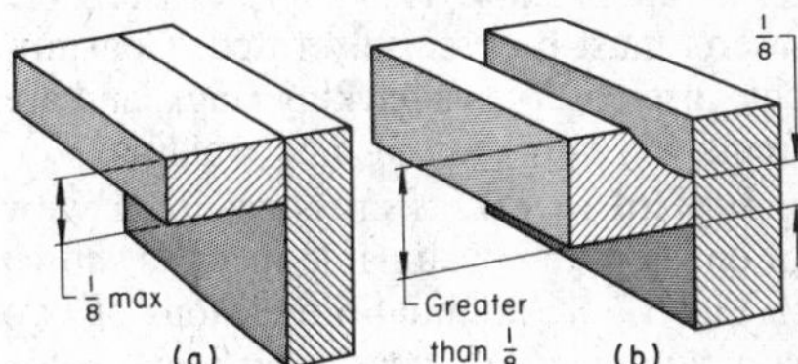

Fig. 19 Typical designs of edge joints

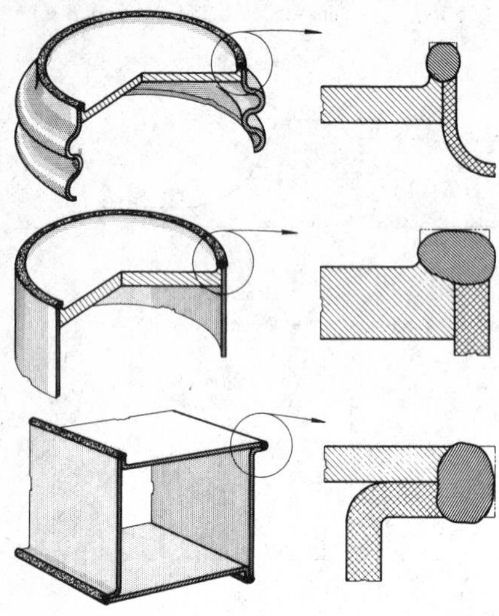

as illustrated in Fig. 19, some sort of backing for a heat sink usually is required to obtain uniformly fused edges.

T-joints require the addition of filler metal to provide the necessary buildup of fillet-weld size. The number of passes that must be made on each side of the joint depends on work-metal thickness and required weld size.

Consumable Inserts

The consumable-insert method of root-pass welding originally was developed for use in the fabrication of nuclear-powered submarines, for which the highest quality joints are essential. This method is intended primarily for applications in which (1) accessibility is limited to one side of the joint; (2) smooth, uniform, crevice-free inner weld surface contours are essential; and (3) the highest quality attainable in the root pass is mandatory.

This method involves the use of an insert that is completely fused by a gas tungsten arc. The insert permits the deposition of a root-pass bead that is smooth and uniform even though welding is done from one side only. It provides full penetration to the root of the joint from the top side of the joint. The insert method is especially useful in butt welding of pipe, although there are no particular restrictions on its application. However, consumable inserts must be precisely fitted. Consumable inserts serve as backing rings, and are frequently specified in pipe welding.

Typical Insert. A cross-sectional view of one type of consumable insert is shown in Fig. 17(a); nominal dimensions of two common sizes of these inserts are noted below Fig. 17. Inserts of this type are available in a variety of compositions, including carbon, alloy, heat-resistant and stainless steels, and nickel and copper alloys. Figure 20 illustrates various types of consumable inserts.

The insert shown in Fig. 17(a) is placed between two pipe sections prior to welding. The inside diameters of the pipe sections at the root-face intersection may differ or be eccentric with misalignment of as much as $^1/_{32}$ in. without any adverse effect on quality of joint or internal weld contour.

Joint design for use with consumable inserts of the type shown in Fig. 17(a) depends on the thickness and composition of the base metal, as illustrated in Fig. 17(b), (c), and (d). The single-V-groove with shallow root faces (Fig. 17b) is satisfactory for base metal up to $^1/_4$ in. thick. For thicker metal, the design should be altered to U-grooves, as shown in Fig. 17(c) and (d).

Welding Procedure. In most applications of joining with consumable inserts, GTAW is used only for tacking and for the root pass. Other welding processes that are capable of depositing metal faster than GTAW ordinarily are used for completing the weld. A typical procedure for using a consumable insert for making a high-quality weld that joins two pipe sections with the axis of the pipe in the horizontal fixed position consists of the following:

- Place an insert (ring form), with its overlap, on one pipe end that has been prepared (usually by machining).
- Using a gas tungsten arc torch, make small tack welds appropriately spaced to obtain a close fit, starting at one end of the insert and continuing halfway around the circumference.
- Using a hacksaw or hand shears, cut off the overlapping ends so that the ends of the insert are butted together and the gap between the ends of the insert does not exceed $^1/_{32}$ in.
- Tack weld the remainder of the insert to the pipe end. One of the tack welds should be at the splice where the insert ends are butted together.
- Butt the second pipe against the insert that has been tacked to the first pipe end. The fit-up tolerance should not exceed $^1/_{32}$ in.
- Midway between the original tacks, tack weld the second pipe to the insert and continue the tack across the insert to include the first pipe section.
- From the top of the joint, weld upward first on one side, and then on the other, fusing the insert with the pipe ends to complete the root pass.
- Complete the joint by any welding process, using filler metal.

Fig. 20 Typical consumable inserts

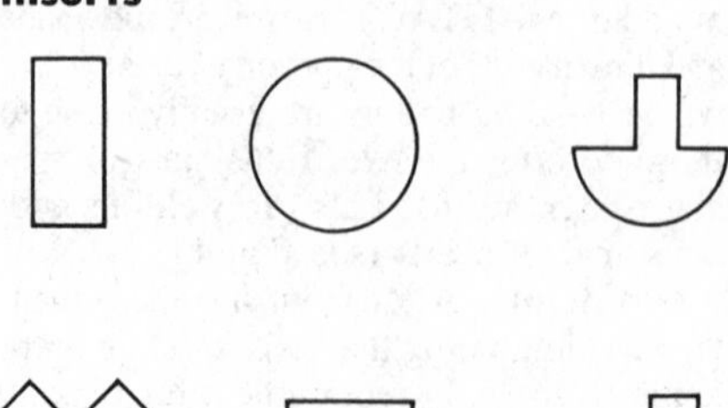

For highest integrity welds, it may be necessary to use a backing of inert gas while welding the root pass. This is done by placing a cap over the end of the pipe and purging with an inert gas.

Although the procedure described above was developed primarily to produce high-integrity welds, it may also prove to be less costly than other methods, as in the following example of an improved joint design.

Example 1. Joint Redesign To Reduce Cost by Permitting a Root Pass To Be Made by Gas Tungsten Arc Welding With a Consumable Insert. The longitudinal butt joints in 20-ft-long sections of SA-106, grade B, carbon steel pipe used for power-boiler headers were originally designed as shown at lower left in Fig. 21. With this design, the root pass and the second pass were made by SMAW using a backing bar, and then the weld was completed by SAW.

To reduce cost, the joint design was changed to that shown at lower right in Fig. 21. This permitted making the root pass by GTAW, using a consumable insert, instead of by SMAW with a backing bar. Then, as with the original joint design, the second pass was made by SMAW, and the weld was completed by SAW. The SMAW process was used for the second pass to provide a deposit thick enough to ensure against melt-through by SAW. The improved joint design and change in welding procedure resulted in a 25% saving in cost (material, labor, and overhead) per foot of seam welded.

Base-Metal Preparation

Welds made by GTAW are extremely susceptible to contamination during

Fig. 21 Revision of joint design. Use of a consumable insert permitted change to a lower cost method of welding boiler-header pipes.

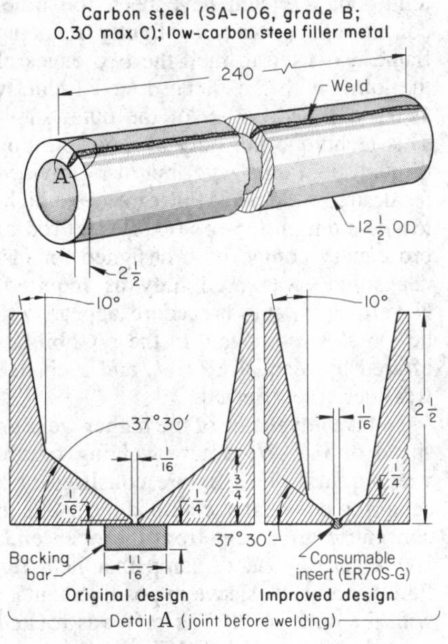

Welding conditions for improved design

Welding process:	
Root pass	GTAW
Second pass	SMAW
Remainder	SAW
Power supply	
GTAW, SMAW	200-A transformer-rectifier(c)
SAW	600-A motor-generator(d)
Edge preparation	Machined
Preheat	250 °F min
Filler metal (low-carbon steel)	
GTAW (argon shielding)	ER70S-G consumable insert
SMAW	E7018
SAW	EL12
Power setting:	
GTAW	90 A (DCEN); 12 V
SMAW	120 A (DCEP); 23 V
SAW	450 A (DCEP); 30 V
Interpass temperature	500 °F max
Postheat	Stress relieve at 1150 ± 25 °F(e)

(a) With the original joint design, the root and second passes were made by SMAW using a backing bar. Weld was then completed by SAW. (b) With the improved joint design, the root pass was made by GTAW using a consumable insert (no backing), the second pass by SMAW, and the remainder by SAW. (c) With high-frequency start, and slope control. (d) Welding head on boom-type manipulator; workpiece supported on power and idler rolls for turning. (e) In furnace: 1 h per inch of section

the welding process. Therefore, the base metal must be free of grease, oil, paint, marking-pencil inscriptions, cutting lubricants, plating, dirt and oxides, or any other foreign material. It is possible, for instance, to pick up enough sulfur from soil contaminants to introduce brittleness in some joints. Heavily soiled workpieces (except those made of titanium alloys) are usually cleaned by immersion in emulsion or solvent cleaners followed by vapor degreasing with trichlorethylene. Phosphorus, lead, zinc, cadmium, and low-melting alloys, and iron contamination from metal stamping dies, also must be removed prior to welding.

Simple degreasing is enough preparation for metals that have oxide-free surfaces. For metals with a light oxide coating, acid pickling treatments are generally used. Heavy oxide scales are removed by mechanical cleaning operations, such as grinding and abrasive blasting. When a plate having an oxide coating is sheared, the oxides may become embedded in the sheared edge. These oxides cannot be removed by chemical means, but must be removed by grinding or filing.

Aluminum Alloys. Exposure to the atmosphere quickly causes the formation of a self-protective oxide coating on aluminum surfaces. This oxide coating, which is highly refractory and has considerable electrical resistance, must be removed by deoxidation with a hot alkaline cleaning solution, followed by rinsing in demineralized or deionized water.

Nickel alloys and stainless steel may be chemically cleaned by pickling to remove sand-blast residue or iron and other contaminants. The pickling solutions usually are composed of 5 to 20% nitric acid plus 0.5 to 2% hydrofluoric acid in water, and are used at 130 to 160 °F; treatment time is 5 to 30 min, depending on the thickness of the contamination or oxide on the surface.

Carbon and low-alloy steels may be chemically cleaned in solutions of 50% hydrochloric acid in water; the acid treatment is followed by desmutting with an electrolytic chromic acid dip and by a cold water rinse.

Titanium alloys may be descaled in molten salt baths or by abrasive blasting. Chlorinated solvents, such as trichlorethylene, should not be used for degreasing titanium alloys, because the chlorine residues cause intergranular attack in subsequent heating operations. Chemical cleaning may be performed by pickling for 1 to 20 min in solutions containing 20 to 47% nitric acid plus 2 to 4% hydrofluoric acid in water, or about a 10-to-1 ratio of nitric acid in hydrofluoric acid; bath temperature is 80 to 160 °F.

Welding of Carbon and Alloy Steels

Carbon Steel. The weldability of carbon steel, for a given thickness and joint design, depends mainly on carbon content, and it is essentially the same for GTAW as for other arc welding processes. As carbon content increases, difficulty in welding also increases—generally, in the form of cracking in the weld or in adjacent base metal. Thus, the requirements for preheating and postheating increase as carbon content increases, as in any other welding process. Other elements in carbon steel, however, such as sulfur, phosphorus, and oxygen, generally have greater influence on weld quality in GTAW than in other arc welding processes, such as SMAW or SAW.

When GTAW was first applied to carbon steels, the filler-metal compositions were selected on the basis of weld-deposit compositions obtained with the older, well-established processes. The result was that high-quality welds could not be produced consistently. Base-metal composition proved to be a major variable. For instance, rimmed steels were more susceptible to porosity in the weld than similar steels that were fully killed. Further, it was found that shielding gas, current density, and other operating variables also had a marked influence on weld porosity in welding of carbon steel and must be kept under close control to obtain successful welds consistently. For further information on welding carbon steels having higher carbon content (generally, more than 0.25% C), see the article "Arc Welding of Hardenable Carbon and Alloy Steels" in this Volume.

Alloy steels of all types are successfully welded by GTAW. Alloy steels, however, particularly those that contain more than about 0.20% C, are far more susceptible to cracking than plain low-carbon steels. For this reason, all aspects of the welding practice must be under more careful control for alloy steels than is usually necessary for plain low-carbon steels. For instance, it is even more important that welds be located as far as possible from points of maximum stress. Joint design should be such that full penetration can be obtained and thorough inspection of the weld is possible. For further information on welding these steels, see the article "Arc Welding of Hardenable Carbon and Alloy Steels" in this Volume.

High-strength structural steels, also known as high-strength low-alloy (HSLA) steels, are welded by virtually all welding processes, including GTAW. As with other alloy steels, susceptibility to cracking in the HAZ is the most common problem in welding of these steels. However, some of these steels have specific properties or contain elements in amounts that intensify cracking difficulties during or after welding. For instance, some HSLA steels contain relatively high percentages of phosphorus. Since phosphorus and sulfur increase susceptibility to cracking, it is desirable to keep the total content of phosphorus and sulfur well below 0.03%. Silicon contributes to cracking in some grades.

In welding these steels, inert gases with leading and trailing shields are often used to prevent weld contamination from the atmosphere. Preheating or postheating, or both, are often used to prevent cracking and to assist in attaining specified mechanical properties. In most applications, the use of GTAW on these steels is restricted to the root pass; another process, such as SMAW or SAW, is used for the remaining passes. For further information on welding these steels, see the article "Arc Welding of Hardenable Carbon and Alloy Steels" in this Volume.

Production Examples of Welding Low-Carbon Steel

Partly because of the difficulty in obtaining consistent results, and partly because GTAW was generally considered a high-cost process compared with gas welding, SMAW, or SAW, there has been some hesitation in adapting GTAW to the welding of low-carbon steels. However, consistent results can be obtained by close control of the steel being welded and the welding operation. Further, the cost of using GTAW for carbon steels is not necessarily higher than the cost of competitive processes; in many applications, the finished weldment has cost less because production rate was higher, because control was better, or because the better appearance made possible by GTAW eliminated, or at least reduced, postweld operations.

Oxyacetylene Welding Versus Gas Tungsten Arc Welding. Although manual GTAW differs greatly from oxyacetylene welding (OAW), the two processes are similar in manipulation techniques. Both processes can be used with or without the addition of filler metal. When filler metal is used, the welder has both hands occupied. However, the tungsten arc produces a higher temperature, resulting in deeper, narrower welds made at higher welding speeds, generally with lower total heat input and narrower HAZ. Both processes are used effectively on thin base metals. These similarities and differences suggest that manual GTAW can be substituted for OAW for selected applications, as described in the following example.

Fig. 22 Double-walled bearing mold welded by GTAW

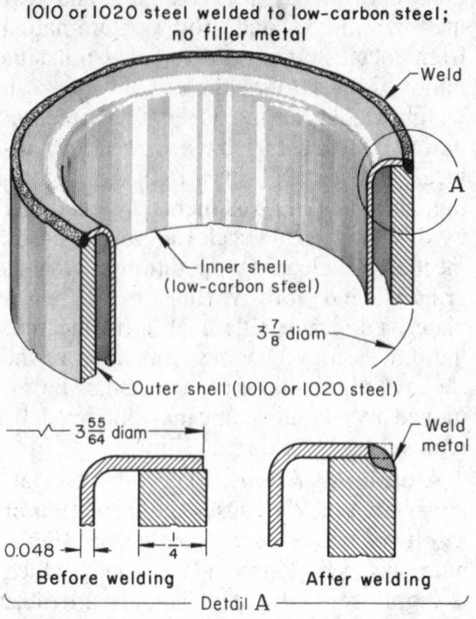

1010 or 1020 steel welded to low-carbon steel; no filler metal

Joint type	Corner
Weld type	Corner fillet
Process	Manual GTAW
Electrode	$^1/_{16}$-in.-diam EWTh-2
Welding torch	250-A, water cooled
Filler metal	None
Shielding gas	Argon, 8 to 10 ft^3/h
Welding position	Horizontal or vertical
Current	120 A (DCEN)
Voltage	14 V
Arc starting and stabilization	High open-circuit voltage
Power supply	250-A transformer-rectifier
Welding speed	25 to 30 in./min
Fixturing	Centering and hold-down clamps; variable-speed drive for rotation

Example 2. Change From Oxyacetylene Welding to Gas Tungsten Arc Welding To Increase Productivity. Double-walled steel cylinders of the design shown in Fig. 22 were used as molds for sleeve and journal bearings. The inner shell was a 0.048-in.-thick sleeve flared over at one end to a diameter $^1/_{64}$ in. less than the outside diameter of the outer shell. When properly centered, the sleeve could be slipped into the outer shell so that the flared edge could be welded to form a concentric annular closure (Fig. 22). The major welding problem was to obtain a leakproof weld without burning through the thin sleeve.

Oxyacetylene welding, which was originally used for this application, produced satisfactory welds. For OAW, the inner and outer shells were positioned in a plug-type fixture that centered the sleeve and permitted the flare to rest concentrically on the outer shell. With the assembly in the vertical position (Fig. 22), welding was done in the horizontal position. Because the flare edge was set back from the outside of the outer shell, a conventional fillet weld could have been used to make the joint, but it would have been too time-consuming. Instead, a slightly reducing flame was used to melt the two edges of the joint, with the flame directed chiefly at the exposed corner of the outer shell. This technique not only avoided burning through the thin sleeve, but also provided the desired weld pool. Filler metal—a high-tensile-strength (75 to 90 ksi) steel rod of proprietary composition designed for fast deposition—was used only as required. Nevertheless, this procedure appeared to be too slow compared to the possibilities offered by manual GTAW, and a change was therefore effected.

To take advantage of the higher welding speed of GTAW, a new welding fixture was required. This fixture consisted of a centering device that clamped the shells in concentric alignment from the open end, and a spring-loaded clamp that held the flared end of the sleeve in positive contact with the outer shell. The unit was rotated by a variable-speed (0.10 rpm) drive actuated by a foot switch, and was additionally gimbal mounted to permit welding in either the vertical or the horizontal position.

Gas tungsten arc welding was done in a single pass, entirely without filler metal. No change in the design of the shells was required, since the welding technique was essentially the same as that described for OAW. To attain high speed with the new process, however, welders required special practice sessions. Welding speed was three to four times greater than that obtained in OAW; the assemblies were gas tungsten arc welded at the rate of 40 per hour.

Gas Tungsten Arc Versus Other Arc Welding Processes. In speed and cost, GTAW is not competitive with other arc welding processes for most joining of carbon and low-alloy steels, provided acceptable results can be obtained by SMAW, GMAW, or SAW. There are, however, some applications in which the degree of control and weld quality afforded by GTAW makes it the most efficient process. The example that follows describes an application in which SMAW was changed to GTAW to obtain the desired results.

Example 3. Change From Shielded Metal Arc to Gas Tungsten Arc Welding to Control Weld Deposit. Short lengths of cold rolled 1010 steel strip $^3/_{32}$ in. thick formed the connections between the turns of a helical resistor and seven $^3/_8$-in.-diam cold finished 1018 steel terminal studs of varying length. Because of the heat and stresses developed during operation of the resistor, the joint between connector strips and studs was welded (see Fig. 23), rather than being soldered or brazed. Originally, these assemblies were shielded metal arc welded in a simple fixture. However, difficulty in maintaining control over the small weld deposit and the necessity for slag removal prompted a change to GTAW.

The new method employed a special holding fixture mounted on a foot-operated turntable. Studs were inserted vertically through holes in a horizontal plate so that their lower ends rested on locating pins, which were adjustable to the several stud lengths. When properly located, each stud projected above the plate surface sufficiently to allow the connector strip to be placed over the stud, leaving a $^1/_{16}$-in. projection for a fillet weld (Fig. 23).

Welding was done manually, without added filler metal. The arc was started at the center of the stud and moved to the edge, at which point the welder rotated the fixture by foot, while holding the torch 30° from vertical. After fusing the edge of the stud to form a $^1/_{16}$-in. fillet through one complete revolution, the arc was again brought to the center of the stud and extinguished.

Five of the seven studs required for each assembly were welded in this manner, under the conditions listed with Fig. 23. The two other studs, which were connected to the extreme ends of the resistor, had a joint design different from that shown in Fig. 23(b). For these studs, the connector strips were larger and were formed with a semicylindrical shape at one end, so that the stud nested in the shaped portion. These joints were also made with a $^1/_{16}$-in. fillet weld, but a 0.040-in.-diam low-carbon steel rod was used as filler metal. One of these studs was welded on the special fixture described above; the other was welded manually without a fixture, after all other components had been assembled and welded.

Fig. 23 Terminal stud gas tungsten arc welded to a strip for connection to a helical resistor

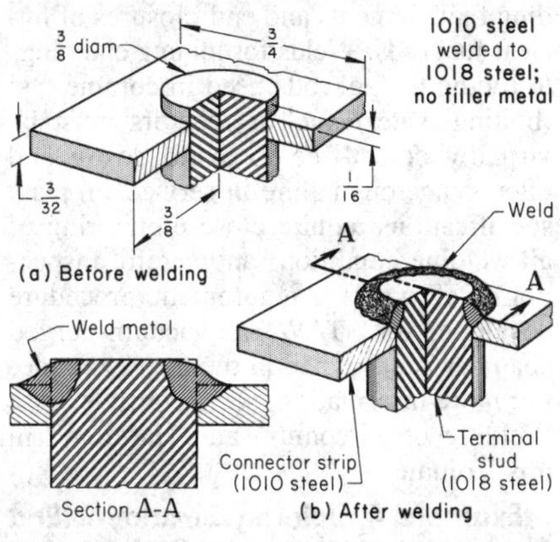

Joint type	T
Weld type	Circumferential fillet
Process	Manual GTAW
Electrode	0.060-in.-diam EWTh-2
Welding torch	150-A, air cooled
Filler metal	None
Shielding gas	Argon, 10 ft^3/h
Welding position	Horizontal
Fixtures	Locating jig and foot-operated turntable
Current	70 A (DCEN)
Voltage	10 to 12 V
Arc starting	High frequency
Arc length	0.06 in., approx
Power supply	200-A transformer-rectifier (constant-current type)
Welding speed	4 to 6 in./min approx

Welding of Stainless Steel

Gas tungsten arc welding is used extensively for the various grades of stainless steel, because the filler metal does not cross the arc; therefore, the composition of the base metal is unchanged. The GTAW process provides an inert atmosphere. Also, there is no slag produced that could react with the base metal.

Lower current levels should be used to weld stainless steels than are used to weld low-carbon steels. The stainless steel has higher thermal expansion, a lower thermal conductivity, and a lower melting point. Lower values of thermal conductivity and higher values of thermal expansion tend to create more distortion and warpage.

Direct current electrode negative is used for most GTAW applications. Tungsten electrodes (2% thoriated) are widely used. Argon, argon-helium mixtures, and helium shielding gases are used; pure argon is preferred. Argon-hydrogen gas combinations may be used to improve the bead shape and wettability.

Gas tungsten arc welding can be used to weld very thin sheet or strip, with or without the addition of filler metal, as well as for critical applications on material 1 in. or more in thickness. Welding equipment, gas shielding, tooling, and other processing conditions must be closely controlled for sound weldments. A filler metal that is compatible with the base metal must be used. For further details on the application of GTAW to stainless steel, see the article "Arc Welding of Stainless Steels" in this Volume.

Welding of Heat-Resistant Alloys

Gas tungsten arc welding is used for most grades of iron-, nickel-, and cobalt-based heat-resistant superalloys. Section thicknesses varying from thin sheet to about 1 in. are readily welded by manual or mechanized methods. Direct current electrode negative is recommended for use when welding these alloys. Alternating current can be used for automatic welding where close control of arc length is possible. Appropriate selection of electrodes and filler-metal composition is especially important. Filler metal differing from base-metal composition may sometimes be used to control porosity or hot cracking. Close control of current, voltage, travel speed, shielding gas, and tooling is mandatory for producing sound welds. For further details on the application of GTAW to heat-resistant alloys, see the article "Arc Welding of Heat-Resistant Alloys" in this Volume.

Refractory Metals. Metallurgical characteristics of the refractory metals (molybdenum, tungsten, niobium, and tantalum) preclude the use of any welding process that affords less protection to the hot metal than does GTAW with the inert gas shield. These metals must be protected from the oxidizing atmosphere. Under certain conditions it is possible to weld in the open with provisions for careful control by inert-gas shielding. Torch gas flow and backup shielding must be closely controlled. Usually, purged chambers containing high-purity inert gas are used. Close control of equipment and process, coupled with careful material handling for preparation and cleanness, is essential for welding the refractory metals. For further details on GTAW of these metals, see the article "Arc Welding of Reactive Metals and Refractory Metals" in this Volume.

Welding of Nonferrous Metals

Aluminum alloys, beryllium, copper alloys, magnesium alloys, nickel alloys, titanium alloys, and zirconium alloys are all welded by GTAW. The application of GTAW to these alloys is discussed briefly in the paragraphs that follow.

Aluminum Alloys. Gas tungsten arc welding can be used to weld thinner aluminum products. Best results are achieved by manual welding on thicknesses ranging from 0.030 to $^3/_8$ in. Mechanized welding usually is done on thicknesses ranging from 0.010 to 1 in. Alternating or direct current power sources may be used, but alternating current is sometimes preferred for manual and mechanized applications.

Filler metal may not be required, depending on the joint geometry and application. Thin materials are frequently welded without filler metal. Choice of filler metal is based on ease of welding, corrosion resistance, strength, ductility, elevated temperature service, and color match with the base metal after welding.

Direct current electrode negative is used for high-current automatic welding applications. Direct current electrode positive is only used for thin metal applications. Pure or zirconium tungsten electrodes usually are used for aluminum. Thoriated tungsten electrodes tend to spatter and cause inclusions with alternating current, and thus are not widely used.

Argon shielding gas usually is used, but argon-helium mixtures sometimes are used to achieve good penetration and faster travel speeds. Argon and helium mixtures are preferred with DCEN.

Good joint fit-up, cleaness of joint and filler metal, and close control of process variables are requisites for producing metallurgically sound welds. For more detailed information, see the article "Arc Welding of Aluminum Alloys" in this Volume.

Beryllium can be welded by GTAW. However, because of inherent characteristics of beryllium, the welds are brittle. Special precautions in handling beryllium are required because of its toxicity. Beryllium should be welded in a hooded enclosure, preferably a glove box. Techniques such as described for titanium alloys (see below) also are applicable to welding of beryllium. For more detailed information, see the article "Arc Welding of Beryllium" in this Volume.

Copper Alloys. Copper, copper-zinc alloys, copper-silicon alloys, copper-tin alloys, copper-aluminum alloys, and copper-nickel alloys are all weldable by GTAW. Each group of copper alloys requires special consideration.

Direct current electrode negative is used because of the higher current-carrying capacity of copper and its alloys. Welding beryllium coppers and aluminum bronzes uses alternating current to prevent buildup of oxides. When welding beryllium coppers, care must be taken to avoid the toxic fumes.

Thoriated or zirconiated tungsten electrodes are recommended; 2% thoriated are preferred. Argon shielding gas generally is used on thinner sections, while helium and mixtures of argon and helium are used on thicker sections.

When welding material thicker than thin sheet, joints must have wide groove angles (60 to 90°) to permit proper manipulation of the torch and to obtain adequate penetration at the root of the joint. Root openings should be used to facilitate complete penetration. For more information, see the article "Arc Welding of Copper and Copper Alloys" in this Volume.

Magnesium alloys are readily welded by GTAW. Magnesium forms an oxide similar to aluminum oxide, which gives these metals similar welding characteristics. Alternating current is used for most magnesium and magnesium alloy welding applications due to its superior oxide cleaning action, which allows higher welding speeds. Direct current electrode positive may be used for welding thicknesses less than $^3/_{16}$ in. Gas tungsten arc welding is preferred for welding metal thicknesses up to $^3/_8$ in. Above this thickness, GMAW usually is used. Argon, argon-helium mixtures, and helium are preferred shielding gases, as magnesium reacts chemically with active gases. All types of tungsten electrodes can be used. For more information, see the article "Arc Welding of Magnesium Alloys" in this Volume.

Nickel alloys are readily welded by GTAW. Direct current electrode negative usually is recommended for manual and mechanized welding. Argon, argon-helium mixtures, and helium are used as shielding gases. Helium is preferred for welding applications that do not require filler metals. Argon-hydrogen mixtures are used frequently for single-pass welding of nickel alloys and for applications in which porosity may be a problem. All types of tungsten electrodes may be used, but alloyed tungsten electrodes are preferred.

Filler metal usually is required to weld nickel and its alloys. These filler metals are similar in composition to the base metal. For more information, see the article "Arc Welding of Nickel Alloys" in this Volume.

Titanium Alloys. Titanium is highly reactive and in welding must be carefully shielded to prevent the absorption of oxygen, nitrogen, hydrogen, or carbon. The absorption of these impurities will adversely affect toughness and ductility of the weld and the base metal. In addition to the gas flow at the torch, shielding-cup leading and trailing gas shields are used. The trailing shield must provide protection for the weld area until the weld area cools below 1000 °F. Leading shields should be used when the metal ahead of the weld is above 1000 °F. Flow-purged chambers and vacuum-purged chambers are used to provide the highest purity inert-gas atmosphere. For quality welding, the base metal and filler metal must be clean and shielding gas must be of high purity. Welding procedures must be specific for each individual setup.

Helium or argon shielding gases are used almost exclusively for welding titanium. Argon-helium mixtures are used infrequently. Thoriated tungsten electrodes are preferred for welding titanium and titanium alloys; 2% thoriated electrodes are preferred. For more information, see the article "Arc Welding of Titanium and Titanium Alloys" in this Volume.

Zirconium Alloys. The principal difficulty encountered in the welding of zirconium alloys is the contamination of the heated metal surface by oxygen and nitrogen. These impurities increase the hardness of the metal and decrease the ductility and resistance to corrosion. Production of ductile, corrosion-resistant welds in zirconium alloys requires protection of the welded surface by a rigidly controlled inert-gas atmosphere, often necessitating welding inside a chamber filled with inert gas. Zirconium alloys are also sensitive to traces of chemical contaminants, and to obtain high-quality welds, all joint surfaces must be carefully cleaned before welding.

Zircaloy-2 is commonly used for cladding fuel elements and end-closures of nuclear fuel rods. Welds for joining end plugs to tubing of fuel rods used in commercial (boiling water) nuclear reactors must be virtually defect-free to avoid costly and often dangerous failure in service. In-plant specifications require close monitoring of all welding conditions and careful inspection of all welds. An automatic procedure developed for GTAW in a vacuum-purged chamber is described in the next example. For more information, see the article "Arc Welding of Zirconium and Hafnium" in this Volume.

Example 4. Joining Zircaloy-2 End Closures in Nuclear Fuel Rods by Automatic Gas Tungsten Arc Welding. Figure 24 shows a nuclear fuel rod consisting of two end plugs and a tube, all of Zircaloy-2, which were joined by automatic GTAW.

Equipment for welding included a cylindrical vacuum-purged welding chamber, about 15 ft long and 4 ft in diameter, specially designed to accommodate a batch of 100 fuel-rod assemblies, a rotatable chill-block fixture to clamp the tube in the horizontal position, and a 110-A air-cooled torch rigidly mounted to hold the electrode in the vertical position for welding.

Fig. 24 Zircaloy-2 nuclear fuel-rod assembly joined by automatic GTAW in a vacuum-purged chamber

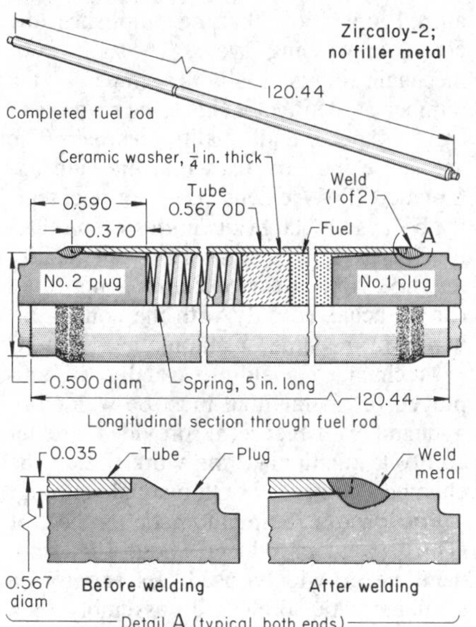

Before welding, a batch of tubes and end plugs was degreased, cleaned with detergent, fired, and loaded into the welding chamber. The tubes were placed in the horizontal position and handled by means of supporting trays. After being evacuated, the welding chamber was backfilled with a mixture of 92% helium and 8% argon; then each tube, manually fixtured, was inserted with a No. 1 end plug through glove ports in the chamber. The tube-and-plug assembly was rotated at a speed of 8 rpm for welding. Current at high frequency was used to start the arc, and the square-groove butt joint between the end plug and tubing (see Fig. 24) was welded in a single pass without added filler metal. Welding was done at 50 to 60 A DCEN, with no inert-gas flow at the torch; welding speed was 14.25 in./min. A 300-A transformer-rectifier was the power supply; the welding electrodes were $^{1}/_{16}$-in.-diam EWTh-2.

After welding of the No. 1 end plugs, each joint was subjected to visual, liquid-penetrant, and radiographic inspection, and to corrosion tests. After inspection, acceptable tubes were loaded with uranium oxide pellets, a ceramic (Al_2O_3) washer $^{1}/_{4}$ in. thick, and a 5-in.-long spring. Then the tubes were assembled to No. 2 end plugs, fixtured, inserted through glove ports into the chamber, which had been evacuated and backfilled as before, and the No. 2 end plugs were welded to the tube, using the same procedure as for the No. 1 end plugs.

The completed fuel rods were subjected to a final leak test. Because helium had been sealed in the rods during final welding, a helium mass spectrometer was used for leak detection. The rods were loaded in a vacuum chamber equipped with a "snifter" probe. The chamber was evacuated, and a leak-rate measurement was taken. If a failure was detected, the rods were removed and tested in smaller lots until the defective rod was sorted.

Welding Dissimilar Metals

Various combinations of dissimilar metals are welded by GTAW. Combinations commonly welded are carbon steel to stainless steel, and carbon steel to copper alloys.

Selection of filler metal generally is more critical for welding of dissimilar metals than for welding of similar metals. In welding steel to a copper-based alloy, the welding is started on the steel, which then serves to preheat the copper alloy.

The two examples that follow describe applications in which SMAW was replaced by GTAW to obtain better control of weld deposit and less distortion in joining of dissimilar metals.

Example 5. Better Control of Weld Deposit. Resistor assemblies of the type shown in Fig. 25 required dependable welds between the resistance coil and seven terminal connector strips for stepped resistances. Example 3 describes welding of these strips to terminal studs. The coil consisted of 24$^{5}/_{6}$ turns of edge-wound flat wire 0.067 in. thick by $^{3}/_{4}$ in. wide, made of a 13Cr-4Al-0.15C iron-based alloy; connector strips were of cold rolled 1010 steel $^{3}/_{32}$ in. thick by $^{3}/_{4}$ in. wide. Welding, rather than soldering or brazing, was used to make the welds, because of the heat and stresses developed in operation. At first, SMAW was used, but the procedure was changed to manual GTAW because of difficulty in controlling the unequal-leg ($^{1}/_{16}$ by $^{1}/_{8}$ in.) fillet weld (see Fig. 25).

Gas tungsten arc welding was done with ER308L stainless steel filler metal, which was selected on the basis of laboratory tests to determine the most economical filler metal for the dissimilar-metal combination, and with the aid of a specially built fixture. This consisted of an L-shaped bracket that permitted the coil to be held in the vertical position by a spring-loaded leather strap. The base of the bracket was drilled in seven locations to accept the stud end of respective connector strips. With the stud located in its proper hole, the end of the connector strip lapped under the edge of the resistor coil at the designated number of turns. The lap was then made snug by adjusting an elevating screw. To tighten the lap joint for welding, the two members were clamped by a vise-grip pliers modified with small tabs at the grips.

Fig. 25 Resistor assembly. Weld joining coil of iron-based resistance alloy and terminal connector strip (1 of 7) of low-carbon steel

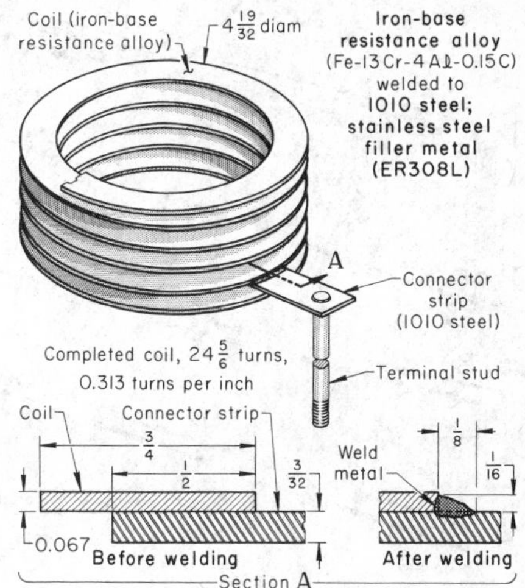

Joint type	Lap
Weld type	Single fillet
Process	Manual GTAW
Electrode	0.060-in.-diam EWTh-2
Welding torch	150-A, air cooled
Filler metal	0.045-in.-diam ER308L wire
Shielding gas	Argon, 10 ft^3/h
Welding position	Horizontal
Fixtures	Coil-supporting jig with stud-locating template
Current	50 A (DCEN)
Voltage	10 to 12 V
Arc starting	High frequency
Power supply	200-A transformer-rectifier

Welding was done by starting the arc approximately $^{1}/_{16}$ in. from the edge of the strip, moving back to the edge and then carrying the pool across the joint to the opposite edge. The torch was held at a 45° angle. After a momentary break in the arc, the weld was carried around the edge of the strip to prevent cracking at this point. This procedure was followed in sequence for the seven joints.

Example 6. Improved Results. The boxlike weldment shown in Fig. 26 consisted of two naval brass bars to which were welded side plates of 1018 or 1020 steel. Used as a sliding ram, the assembly was installed in the front end of a lift truck to push the load forward off the truck. Tolerances on width and height (see Fig. 26)

Fig. 26 Boxlike weldment of steel and brass. Change from continuous welds by SMAW to intermittent welds by GTAW eliminated a distortion problem and saved time and cost.

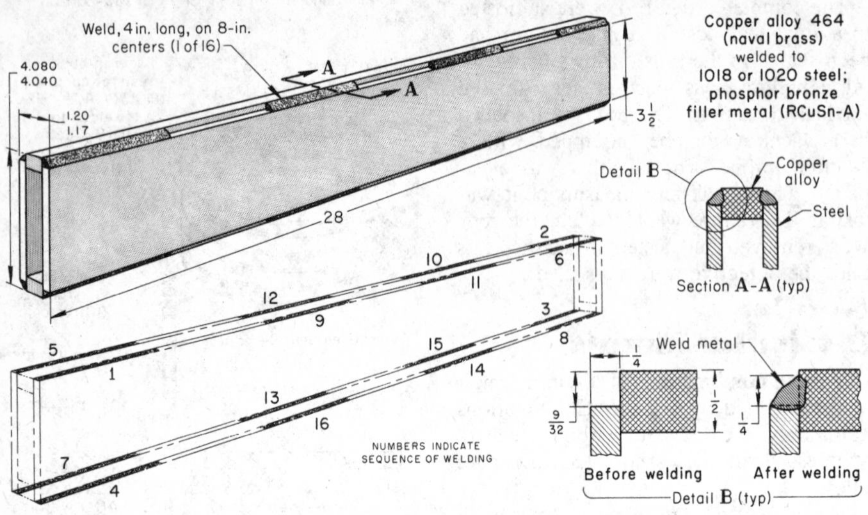

Conditions for GTAW

Joint type	Corner	Welding position	Flat(c)
Weld type	Fillet	Current	Medium A (DCEN)(d)
Process	Manual GTAW	Voltage	30 to 33 V
Electrode	$^1/_{16}$-in.-diam EWTh-2(a)	Arc starting	High frequency
Welding torch	300-A, water cooled	Arc length	$^1/_8$ in., approx
Filler metal	$^1/_8$-in.-diam RCuSn-A(b)	Power supply	300-A transformer-rectifier
Shielding gas	Argon, 20 ft^3/h		

(a) Taper ground. (b) Phosphor bronze. (c) With assembly in a jig and mounted on a positioner. (d) Equipment had a range selector for low, medium, and high amperage.

were important, because the assembly was not machined after welding.

Originally, the four corner joints were continuously fillet welded using SMAW with a phosphor bronze electrode. This procedure came under review because of the amount of time spent in straightening (to correct distortion), in grinding and cleaning the welds, and in removing spatter.

To eliminate these problems, the process was changed to GTAW, using a revised procedure. Before welding, joint surfaces were cleaned by sanding and wiping with clean rags. The components were assembled in a fixture (jig) and mounted on a positioner to orient the assembly for flat position welding. Joints were then tack welded, using GTAW with argon shielding and phosphor bronze filler metal. Although tack welding had not been used in the original procedure, it was specified here to help maintain alignment of the pieces during welding. Then the joints were welded, using the same process and filler metal as for tack welding. Because of the large difference in melting temperature between the steel and the brass, the welding was started on the steel. This preheated the brass. Filler metal was then fed in at 6 to 8 in./min.

In GTAW, instead of continuous fillet welds (as in the previous method), 16 intermittent fillet welds, 4 in. long on 8-in. centers, were deposited in a staggered sequence. The use of intermittent welds did not impair the strength of the assembly, because the service load on the welds was quite low. On the other hand, the intermittent welds greatly reduced heat input and distributed it favorably, both of which helped eliminate distortion. Also, the use of GTAW eliminated the need for grinding and cleaning welds and removing spatter.

Details of the improved welding procedure, which resulted in considerable time and iabor costs, are given in Fig. 26. Although part of the time and cost saving with GTAW resulted because of the change from continuous to intermittent welds, the major portion was due to the elimination of the postweld cleaning and straightening operations.

Mechanized Welding

Mechanized GTAW, which includes the semiautomatic, machine, and automatic modes, is used extensively. The degree of mechanization varies from simple mounting of the welding torch in a bracket that moves over the workpiece (or in a stationary bracket, with the workpiece being moved under the torch), to a fully mechanized operation that accomplishes the complete welding cycle. The degree of mechanization is usually determined by the number of identical welds to be made, and by the speed and quality desired. The aerospace industry uses machine and automatic GTAW extensively, not necessarily because of large quantities of production parts, but because the weld quality required for aerospace components often can be achieved only with the control inherent in machine or automatic welding.

Mechanized welding usually is employed for metals that must be welded in a chamber, because it provides greater ability to manipulate the work within the chamber, compared with manual welding.

Equipment. Semiautomatic torches for GTAW were introduced about 1952, but were never widely used. Essentially, a semiautomatic torch is an assembly of a hand-held water-cooled torch with an attachment that brings the filler metal (wire) into the arc area. The filler metal is fed to the torch through a flexible conduit by means of a motor-driven wire feeder. The wire feeder is controlled by a trigger switch on the torch. In theory, the filler wire fed from the torch helps propel the torch and establish travel speed. The wire is then melted by the arc and deposited in the joint. The introducton of GMAW interrupted the full development of the semiautomatic GTAW system, and it is infrequently used.

For mechanized welding systems, the torch is mounted in a mechanism for moving the work or the torch relative to one another. Specially designed torches are used; two types are available. One, the heavy-duty barrel type, is of metal construction and water cooled. It is equipped with a rack and is long enough to be adjusted and carried in a holder similar to that used for machine gas cutting. Barrel-type holders are fully insulated, can be used with various types of nozzles, and accommodate electrodes up to 18 to 24 in. long, with diameters of 0.040 to $^1/_4$ in. Welding current may be up to 600 A.

The second type of torch is also water cooled, but does not fit mechanized gas-cutting equipment. These torches usually are made to accommodate shorter (7-in.-long) electrodes in a slightly smaller size range (0.020 to $^3/_{16}$ in.). Welding current for these torches may be up to 500 A. In addition, metal or ceramic nozzles can be used. These shorter holders require a special offset bracket for mounting.

Mechanized or gas-cutting carriages are often used for moving the torches. Special hardware and adapters are readily available; torch head controllers are also available. They will electronically maintain a preset arc voltage even though the surface of the work being welded varies substantially.

For completely automatic production-line welding of joints requiring the addition of filler metal, a wire feeder is added. See the section of this article on filler wire feeders for information on cold wire and hot wire feeding systems. Mechanized GTAW may be controlled through various programmable devices, such as punched tapes or cards. In some machines, input data are stored in memory devices. Cam-actuated pressure or mechanical switches also are available. Automatic or machine welding can be controlled by commercially available programmed arc welding power sources that automatically initiate the arc, control the current, control travel speed, and terminate the arc.

Changing from manual to mechanized welding usually requires a good knowledge of welding, a high degree of skill in machine design, and a high initial investment. These requirements necessarily increase with the degree of mechanization desired. Because of the availability of a wide variety of mechanical and electrical controls, operations of seemingly great complexity can often be automated. Because of its inherent arc stability, GTAW is readily adapted to this type of control. Usually, if a repetitive operation can be manually fixtured and manually welded, it can be mechanized. And if the welding machine is used frequently, the initial investment often can be recovered in a short time.

Example 7. Change From Manual to Mechanized Welding To Increase Productivity. The four legs that constituted the supporting members of swivel-chair bases were made by blanking, forming, and welding 0.057-in.-thick low-carbon steel sheet. Figure 27 shows the configuration of a typical leg, with its underside facing up to reveal the weld at the seam.

Originally, the seam was welded by manually clamping the piece in a fixture, to draw the edges closely together, and then fusing the seam with a hand-operated gas tungsten arc torch. Although this method welded 39 legs per hour, it was too slow to satisfy increasing demands. Because the operation was simple, it was possible to develop a machine to solve the production problem without changing part design.

Fig. 27 Swivel-chair leg that was welded automatically at 5½ times the production rate obtained in manual welding

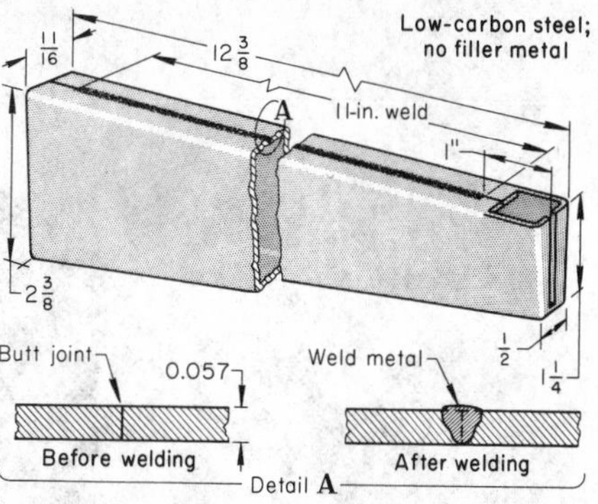

Automatic GTAW

Joint type	Butt
Weld type	Square groove
Electrode	⅛-in.-diam EWTh-2
Filler metal	None
Shielding gas	Argon, 15 ft^3/h
Welding torch	500-A, water cooled(a)
Current	150 A (DCEN)
Voltage	14 V
Arc length	3/32 in.
Power supply	550-A transformer-rectifier with high frequency
Production rate	210 legs per hour(b)

(a) Ceramic cap. (b) Based on welding with two machines operated simultaneously by one operator

The operating cycle of the machine was initiated by inserting the workpiece into the clamping fixture. From this point on, the operator simply monitored the operation. The fixture automatically clamped the workpiece so that the sheared edges (burrs facing out) formed a tight square-groove butt joint. Next, the gas tungsten arc torch started its traverse with a preflow of argon gas for shielding. A limit switch initiated the arc by closing the direct current welding circuit, together with a superimposed high-frequency current circuit used for arc starting only.

At the end of the 11-in. traverse, a second limit switch tripped the welding circuit. After a short argon postflow, the part was automatically ejected. Because neither the mechanical strength of the joint nor the appearance of the weld was critical, visual inspection was made for uniformly good appearance; rejection rate was less than 1%. Smooth-appearing welds with no need for cleanup were obtained more consistently than by manual welding. With the change to automatic welding, not only did production rate increase, but it was soon found that one operator could easily manage two machines simultaneously. Operating two machines on this basis, production rate increased to 210 legs per hour.

Example 8. Simultaneous High-Speed Welding of Four Box Corners. Figure 28 shows a low-carbon steel enclosure for electrical controls, in the form of a rectangular open-top box, that was produced in sizes ranging from 2½ to 6 in. in depth, 5 to 36 in. in width and 8 to 36 in. in length. Thicknesses ranged from 0.030 to 0.078 in. These enclosures were produced by blanking, bending the four sides to form the box shape, and welding the four corner joints in the vertical position. Joint requirements called for

Fig. 28 Electrical-control enclosure. Automatically welded at all four corners simultaneously using water-cooled copper chill bars

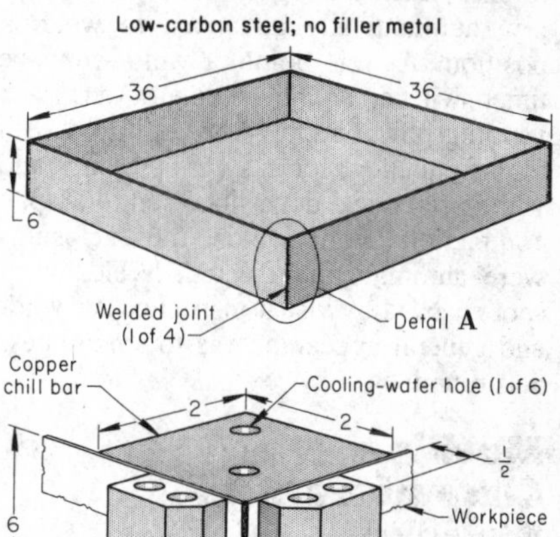

Detail A (workpiece with chill bars; typical, 4 places)

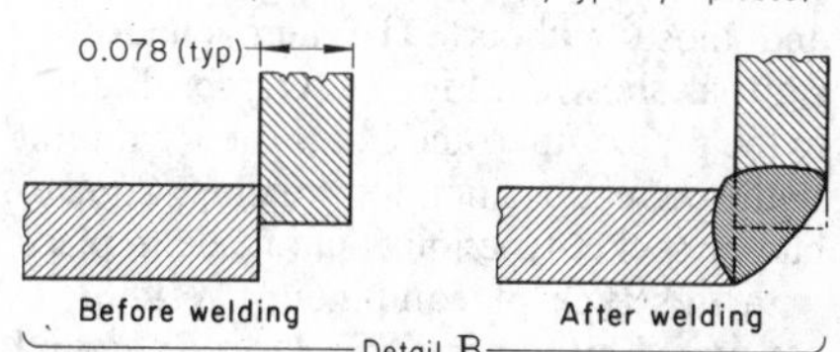

Joint type	Corner
Weld type	Fillet
Process	Automatic GTAW
Electrodes (4)	⅛-in.-diam EWTh-2(a)
Filler metal	None
Welding torches (4)	300-A, water cooled(b)
Shielding gas	Argon, 10 ft^3/h
Current (each circuit)	100-200 A (DCEN)
Voltage (each circuit)	30-34 V
Power supplies (4)	250-A transformer-rectifier (constant current, high frequency)
Power control	Sequencer with slope control
Welding speed	200 in./min

(a) Taper ground to point. (b) With gas cup

a full-penetration seal weld having a smooth appearance suitable for painting. Aluminum-killed steel of drawing quality was selected to avoid weld porosity, as no filler metal was used.

To meet competitive prices on these products, a completely mechanized setup, using high-speed welding, was constructed. The setup involved automatic sequencing of the production cycle, including initial fixturing of parts, synchronizing the simultaneous operation of four welding torches, and the final ejection of the welded product.

After forming, the enclosures were positioned and automatically clamped in a fixture consisting of twelve water-cooled copper chill bars, three bars being located at each of the four corners (see detail A in Fig. 28). Four water-cooled GTAW torches, each having its own power supply, automatically advanced to welding position. At this point, a weld-sequence timer switched on the welding power. Each welding power supply was equipped with high frequency, for easy arc starting, and with upslope and downslope control. Upon completion of the welds, the enclosures were automatically ejected. Welds were spot checked by visual inspection for voids and general appearance as to smoothness and roundness.

Machine Circumferential Welding

Specialized circumferential in-place pipe and tube welding heads have been developed with the need for higher weld quality and fewer joints in nuclear and aerospace applications. These welding heads generally are divided into two groups—those with the capability of feeding filler metal, and those without. The autogenous design, as shown in Fig. 29, clamps directly to the pipe or tube and holds the weld joint in the correct position for welding. Square-butt or socket-type joints are made in place with this type of equipment. Welding is performed by rotating the electrode around the joint on a ring gear. This welding head is limited by the penetration characteristics of GTAW to a maximum of approximately 0.125 in. in butt joints.

An automatic circumferential pipe welding head that incorporates such features as a filler-wire feeder, arc voltage control, and electrode holder oscillation is shown in Fig. 30. As the pipe or tubing remains in place, the welding head is secured in place by spring-type inserts, which accurately accommodate outside-diameter tolerances and position the welding head.

Fig. 29 Automatic gas tungsten arc tube welding head

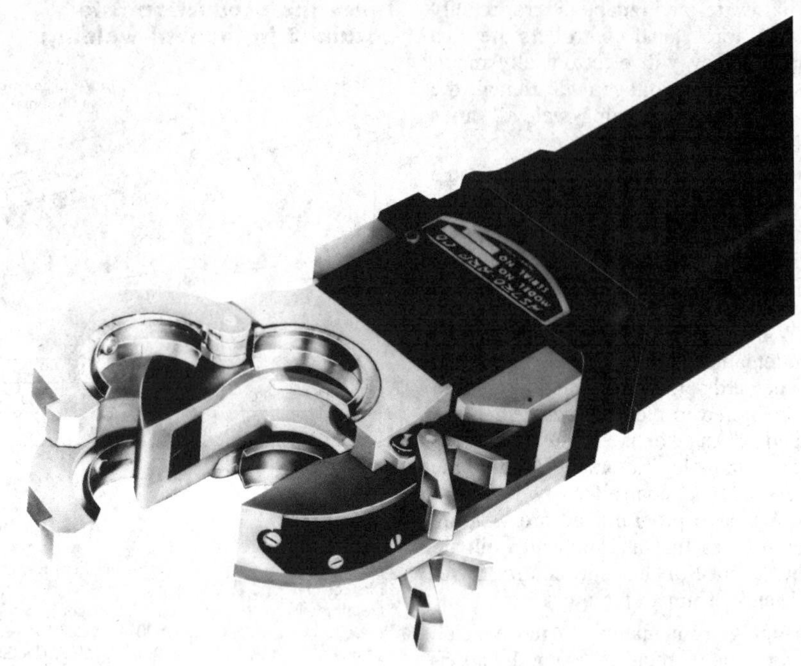

Pipes from 2 to 48 in. in diameter, with walls up to 5 in. thick, can be welded with such equipment.

The electrode holder relies on automatic voltage control to accurately maintain constant arc length during welding. The control mechanism is driven around the joint by a high-torque gear motor with a tachometer mounted in the handle of the welding head. The head is capable of feeding filler wire to fill the groove joint.

Welding current, inert gas, and coolant are transferred by cable to the electrode holder. The path of the current is fully insulated, thus ensuring arc ignition. Due to the use of arc voltage control and microprocessor-controlled programmers on the welding power supply, this type of welding head can be used for irregular as well as regular cross sections.

Automatic circumferential welding systems are used for tube and pipe welding in nuclear power plants, electrical generating stations, and heat-exchanger installations, for rocket engines and in space applications, for hydraulic equipment on jet aircraft, for air cushion and hydrofoil vehicles, as well as in pharmaceutical and chemical production lines.

Spot Welding

Adaptation of GTAW to spot welding permits mechanized arc spot welding of sheet metal assemblies where access to only one side of the joint is possible. The capabilities of this method are especially useful for welding of corrugated structures for aerospace applications. Spot welding may be done with either alternating current or DCEN.

Gas tungsten arc spot welding may be automated to the extent determined necessary to produce quantity and quality welds. The process is used for spot welding of automobile bodies, double-walled structures, aerospace fuel ducts, brackets to thin-walled skins, and foil-thin skins to thicker materials. Gas tungsten arc spot welding is capable not only of making spot welds on overlapped sheets (considered as the conventional procedure), but also of producing short welds that join two abutting edges.

Equipment. Automatic sequencing controls usually are used for arc spot welding because of the relatively complex cycles. These controls establish preweld gas and water flow, initiate the arc, regulate arc duration, and provide the required postweld gas and water flow.

The touch start system, which advances and retracts the tungsten electrode, may be used to start the arc, or it may be started by the use of high frequency or a pilot arc.

The welding gun is usually of pistol-grip design with a finger switch. Metal nozzles are used, because the nozzle is in contact with the work and is used to apply enough pressure to ensure close contact of the parts being spot welded together. Arc-spot guns

Fig. 30 Automatic circumferential pipe welding head with arc voltage control and filler-wire feed capability

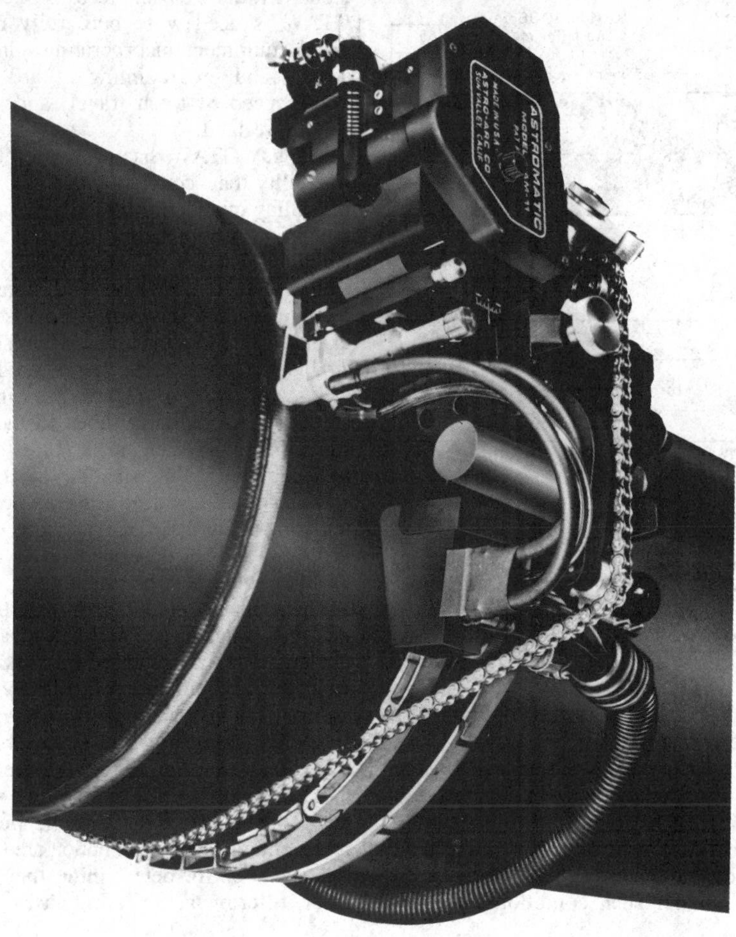

are either air cooled or water cooled, depending on capacity and duty cycle.

Penetration is controlled by adjustment of the amount and duration of current. A reduction in the duration or amount reduces penetration and the diameter of the spot weld. An increase produces the opposite effect. In some applications, multiple pulses of current are preferred over one long, sustained pulse. Variations in shear strength, nugget diameter, and penetration can be minimized with an accurate timer, an ammeter, a flowmeter, and tungsten electrodes.

Use of Filler-Metal Wire. Two major problems encountered in gas tungsten arc spot welding are crater cracking and excessively concave weld surface. A modification of the process to overcome these problems is the addition of filler metal to the arc spot weld. Equipment has been developed to feed filler-metal wire into the arc areas during the arcing period. The major requirement of the equipment is the sequencing of operations. A programmed controller provides for control of gas flow, welding current, tapering (decay) of welding current, and wire feed. The nozzle of the torch is more complicated than the conventional type. Provisions are made for properly feeding the filler wire into the arc without fouling the electrode or sticking in the weld pool. This process has not been widely adopted; it shows the most promise for spot welding aluminum in thicknesses greater than can normally be spot welded by resistance welding.

High production rates and low cost are possible with gas tungsten arc spot welding. The cost of the equipment is low compared to resistance welding equipment. Also, when the proper settings are used, visual inspection of the gas tungsten arc spot welds is more reliable than in resistance spot welding. The example that follows describes the use of gas tungsten arc spot welding equipment and techniques for joining abutting edges by intermittent square-groove welds, in an application for which quality was critical.

Example 9. Rimmed Versus Killed Steels for Spot Welding Application. Steel quality proved critical in spot welding the longitudinal seams of collapsible steering columns made of 0.065-in.-thick, cold rolled 1006 or 1008 steel, perforated and roll-formed into tubes 2 in. in diameter by 30 in. long. As shown in Fig. 31, the perforated (collapsible) section of the steering column was joined with fourteen 0.08-in.-diam butt welds, and the two unperforated sections were joined with seven 0.25-in.-diam butt welds.

Performance specifications required that all welds remain intact when the column collapsed. In view of the collapsibility of the column, this requirement was not considered severe.

Because of the high production rate desired for this part, weld time had to be held to an absolute minimum—which was 15 cycles ($^{1}/_{4}$ s). With such a short weld time, any gas generated during welding was trapped in the weld, resulting in porosity. Because the welds were made by an arc discharge without added filler metal, cleanness of the base metal was an important consideration. Hot rolled steel was not considered, because of the presence of a thin layer of oxide.

Preliminary work on a rimmed cold rolled steel indicated that too much gas was generated. Therefore, an aluminum-killed cold rolled steel, known to be relatively gas-free, was selected for this application. Welds were made with no porosity. As production experience was gained, it was found that weld strength was more than adequate. To lower cost, a trial run was made using semikilled steel. Although some porosity appeared, the welds were strong enough to meet the performance specification.

An automated five-station setup was used for welding. The steering columns were clamped in a jawlike fixture that held the abutting seam edges in correct alignment and position for welding. The fixture-and-part assembly was then shuttled through the five stations, where selected welds were made by energizing prepositioned, gas-shielded tungsten electrodes. Welding was the same as conventional gas tungsten arc spot welding, except that instead of being produced between lapped sheets, the spot welds were centered on the abutting seam edges; thus, they were not true spot welds, but were short intermittent tack-type groove welds. The need for five stations was determined by the proximity, size, and distribution of the welds along the seam and, to some extent, by distortion effects.

Fig. 31 Collapsible steering column that was joined at 21 spots by automatic GTAW

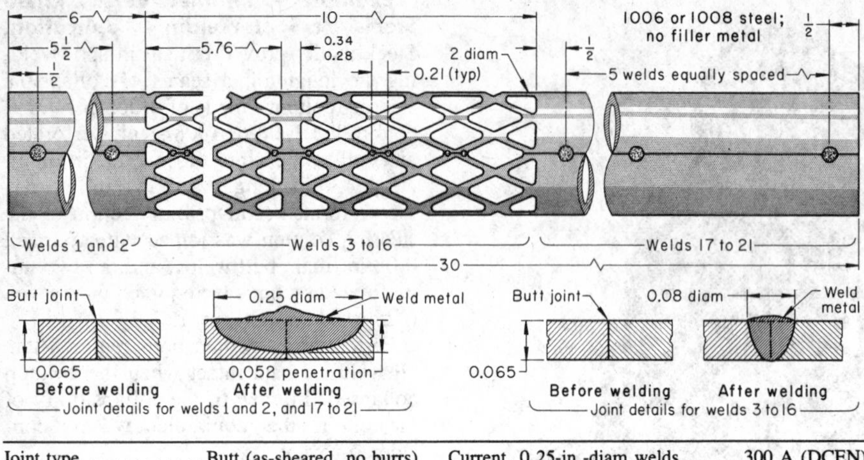

Joint type	Butt (as-sheared, no burrs)	Current, 0.25-in.-diam welds	300 A (DCEN)
Process	Automatic GTAW(a)	Voltage	22 to 23 V
Electrode	$\frac{1}{8}$-in.-diam EWTh-2(b)	Arc starting	High-frequency pilot arc
Filler metal	None	Weld time	15 cycles(d)
Welding torch	Water cooled, gas cup	Welding position	Flat
Shielding gas	Argon(c)	Power supply	300-A transformer-rectifier(e)
Current, 0.08-in.-diam welds	200 A (DCEN)	Production rate	1200 pieces per hour

(a) Five-station automatic sequencing. (b) Tapered to 0.040 in. diam, for welds of both diameters. (c) Bulk; continuous flow. (d) For welds of both diameters; timer controlled. (e) Three-phase, with high frequency and timing controls

After welding was completed, the columns were given a 100% visual inspection. The high production rate of 1200 pieces per hour (see table of welding conditions with Fig. 31) was not obtained without occasional problems. One problem was arc blow, which caused the weld to deviate from true center on the seam. Arc blow was minimized by machine design. In addition, because seam edges were given no special edge preparation, burrs on sheared edges sometimes interfered with weld depositon. This difficulty was overcome by tighter control over the shearing operation.

Cost

For some applications, GTAW may be the only technically acceptable welding process for the work metal, or for the conditions under which welding must be accomplished. Such restrictive considerations are more important than cost in the selection of the welding process.

On the basis of cost alone, the use of GTAW is limited primarily by the cost of the inert gases used for shielding and by generally rather low production rates, compared with those obtained in other arc welding processes. On the other hand, this process in its simplest form requires only a manually operated torch and a gas regulator and flowmeter, in addition to the welding power supply. Accordingly, the initial capital outlay for this equipment is relatively modest.

Gas tungsten arc welding can be employed for production welding most economically in the joining of thin sections, particularly where square-groove welds (no bevel) can be made, and no filler metal is needed. Under such conditions, welding speeds of 150 in./min are easily achieved using relatively simple mechanized equipment. In more specialized applications, GTAW is used with more fully mechanized equipment incorporating automatic control and programming of arc conditions, speed of torch travel, and rate of filler-wire feed.

Manual GTAW has a versatility and flexibility that, coupled with the low capital outlay involved, provide a most useful tool of job-shop maintenance and repair welding for which the work metal or other technical factors rule out the more commonly used and less costly processes. For example, GTAW may be particularly desirable for rework or repair of production welds, for altering of parts machined incorrectly, for welding tubes to the tubesheet in heat exchangers, and for the root pass in pipe welding. In addition, this process is sometimes used for complete piping welds, for the repair of worn or broken dies, and for retrofit modifications of existing equipment or products.

The primary step in determining the cost of a gas tungsten arc weld is to analyze the length of time required for making the weld and the amount of filler wire required. This may be done by observing the actual job, by estimation from tables, or by estimation from past experience. Once these elements are determined, the cost of the weld can be figured if the purchase prices and rates of consumption are known. The methods for determining the cost of labor, filler metal, shielding gas, tungsten

Table 12 Procedure for determining cost of GTAW

$$\text{Labor cost} = \frac{\text{labor + overhead cost/hour} \times \text{pounds of weld deposit/weld}}{\text{deposition rate} \times \text{operating factor}}$$

or

Labor cost = total welding time × labor + overhead cost/hour

Filler metal cost(a) = length of wire used × deposition efficiency × wire weight per unit of length × wire cost/pound

or

$$\text{Filler metal cost(a)} = \frac{\text{weight of deposit} \times \text{filler metal cost}}{\text{deposition efficiency}}$$

Filler metal cost(b) = arc time × wire feed rate × wire weight per unit of length × wire cost/pound

Shielding gas cost = arc time × gas flow rate × cost of gas per cubic foot

$$\text{Electric power cost} = \frac{\text{welding current} \times \text{welding voltage} \times \text{power cost} \times \text{arc time}}{\text{power source efficiency}}$$

Tungsten electrode cost = 0.04 × shielding gas cost

(a) For manual welding. (b) For automatic welding

electrodes, and electricity are outlined in Table 12.

Safety

Gas tungsten arc welding is no more hazardous than other welding processes if adequate precautions are observed with regard to eye protection, protective clothing, and ventilation.

Eye Protection. The filter-plate lenses used in helmets and face shields should be of the deepest shade that permits adequate visibility of the welding operation. (As a guide, AWS recommends shade No. 11 for GTAW of nonferrous metals, and No. 12 for ferrous.) The use of medium-shade (No. 2) flash goggles is recommended in addition to the welding helmet for both the welder and other personnel present in the welding area.

Protective clothing is needed to shield the welder from the intense arc radiation. The tungsten arc is fully exposed, and the ultraviolet and infrared radiations may produce an arc burn that resembles sunburn, except that it is more severe. Dark-colored clothing is preferred to light, because the rays more readily penetrate light-colored fabrics. Light colors are also more reflective, and may cause eye burns even when a helmet is worn. Cotton fabrics should be avoided. Clothing should be made flame resistant by immersing it in sodium tetraborate or sodium stannate and ammonium sulfate solutions. Cuffless trousers should be worn to prevent entrapment of hot slag or spatter. Gauntlet gloves should be worn to protect the hands and wrists from arc burns and possible weld spatter.

Ventilation. Adequate ventilation that does not disturb the gas shield can be obtained by placing a low-velocity suction duct several inches away from the welding operation. Fans or drafts of over 50 sfm may blow the protective gas envelope away from the weld zone, causing oxidation of the heated metal.

During welding operations, ozone and noxious fumes are generated, which may become toxic in large concentrations. The fumes from some chlorinated solvents (three examples are carbon tetrachloride, trichlorethylene, and tetrachlorethylene), when exposed to a tungsten arc, form a toxic gas, phosgene, even at a long distance (hundreds of feet away) unless walls or baffles keep these fumes from the welding area.

Care should be taken when working with toxic metals. Air-supplied face shields or respirators may be required in confined areas that are not adequately ventilated. Argon or helium may displace the air that the welder requires for breathing.

Adequate ventilation is mandatory for all GTAW. Detailed information concerning ventilation is contained in American National Standards Institute (ANSI) Z49.1, "Safety in Welding and Cutting."

Plasma Arc Welding

By the ASM Committee on Plasma Arc Welding*

PLASMA ARC WELDING (PAW) is an arc welding process in which heat is produced by a constricted arc between an electrode and a workpiece (transferred arc), or between a nonconsumable tungsten electrode and a constricting orifice (nontransferred arc). Shielding is generally obtained from the hot, ionized gas issuing from the orifice of the constricting nozzle, which may be supplemented by an auxiliary source of shielding gas. Shielding gas may be an inert gas or a mixture of gases.

Plasma arc welding is closely related to gas tungsten arc welding (GTAW). Plasma is present in all arcs. If a constricting orifice (nozzle) is placed around the arc, the amount of ionization, or plasma, is increased. This results in a higher arc temperature and a more concentrated heat pattern than exists in GTAW. For more information, see the article "Gas Tungsten Arc Welding" in this Volume.

For plasma arc welding, constriction of the arc is produced by the design of the welding torch. Figure 1 shows the heat patterns and arc temperatures for a nonconstricted arc, used in GTAW, and a constricted arc, used in PAW.

Fig. 1 Comparison of a nonconstricted arc used for GTAW and a constricted arc used for PAW. Shows the effect of constriction on temperature and heat pattern

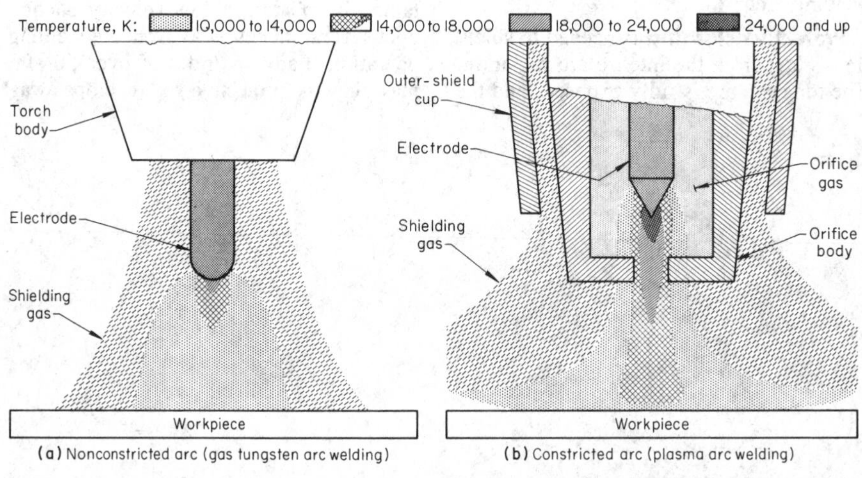

Nonconstricted arc		Constricted arc (3/16-in.-diam orifice)	
Shielding gas	Argon, at 40 ft^3/h	Shielding gas	Argon, at 40 ft^3/h
Current	200 A	Current	200 A
Voltage	15 V	Voltage	30 V

Applicability

Plasma arc welding is adaptable to both manual and mechanized operation, and can be used to produce either continuous or intermittent welds. Welds may be made with or without the addition of filler metal. Plasma arc welding most often is an alternative to GTAW and sometimes is competitive with oxyacetylene welding (OAW), electron beam welding (EBW), and occasionally with resistance seam welding (RSW). For more information, see the section on plasma arc welding versus alternative processes in this article.

Metals Welded. Plasma arc welding is used to join most of the metals commonly welded by GTAW. These metals include carbon and low-alloy steels, stainless steels, copper alloys, nickel- and cobalt-based alloys, and titanium alloys. Thicknesses ranging from 0.001 to 0.25 in. can be welded in one pass. They are welded with a transferred arc using direct current electrode negative (DCEN). Plasma arc welding of aluminum alloys can be done with direct current electrode positive (DCEP) and alternating current with continuous high-frequency stabilization.

Welding Positions. Manual PAW generally is considered to be an all-position process. Mechanized PAW usually is done in the flat and horizontal positions.

Work-Metal Thickness. Plasma arc welding is well adapted to welding thin sections. Because of the stability and dimensional control provided by the constricted arc, foils as thin as 0.001 in. have been welded in production, with current

*Daniel Hauser, *Chairman,* Senior Research Metallurgist, Battelle Columbus Laboratories; George E. Cook, Manager, Research & Development, CRC Welding Systems, Inc.; Emet C. Dunn, Jr., Senior Principal Engineer, Rockwell International; James R. Hannahs, President, Midwest Testing Laboratories, Inc.; Peter C. Levick, Manager, Process & Testing, CRC Welding Systems, Inc.; Daniel O'Hara, Vice President—Engineering, Thermal Dynamics Corp.; Abe Pollack, Head, Joining Branch, David Taylor Naval Ship Research & Development Center; John Springer, Principal Engineer, Boeing Commercial Airplane Co.

as low as 0.3 A. Welding of such thin metal requires extremely precise fixturing and close control of welding parameters. Using the keyhole mode of operation, metals from $^1/_{16}$ to $^1/_4$ in. can be welded. Thickness ranges vary somewhat with different metals. For more information, see the section in this article on keyhole welding.

Advantages. The constricted arc used in PAW offers several advantages over the nonconstricted arc used in GTAW:

- Concentration of energy is greater.
- Arc stability is improved, particularly at low current levels.
- Heat content is higher.
- Less current is required.
- There is less sensitivity to variations in arc length.
- Tungsten contamination is eliminated.
- Less welder dexterity is required for manual welding.
- Solid backing is not required for obtaining complete penetration, because the keyhole technique can be used (Fig. 3).
- Less heat input to the workpiece means less distortion.

Disadvantages. The main disadvantages of PAW, compared with GTAW, are:

- Higher cost of equipment (generally two to five times as much)
- Need for greater welder knowledge, although not necessarily greater dexterity on the part of the welder
- Larger torch diameter

Fundamentals of the Process

In PAW, a tungsten electrode is used, as in GTAW. Two separate streams of gas are supplied to the welding torch. One stream surrounds the electrode within the orifice body and passes through the orifice, constricting the arc to form a jet of intensely hot plasma. This gas must be inert and is usually argon.

The other stream of gas, the shielding gas, passes between the orifice body and the outer shield cup. It prevents the molten weld metal and the arc from becoming contaminated by the surrounding atmosphere. An inert gas, such as argon, also can be used for shielding, but nonoxidizing gas mixtures, such as argon with 5% hydrogen, have often proved advantageous (see the section in this article on orifice and shielding gases).

Distance from orifice to work is commonly maintained at about $^3/_{16}$ in. This distance is less critical than the distance from the end of the electrode to the work in GTAW. Varying the distance from $^1/_8$ to $^1/_4$ in. does not significantly affect welding results.

Arc Modes. Two modes of operation are the nontransferred arc and the transferred arc. In the nontransferred mode, the current flow is from the electrode inside the torch to the nozzle containing the orifice and back to the power supply. The nontransferred mode normally is used for plasma spraying or for generating heat in nonmetals. In the transferred arc, current is transferred from the tungsten electrode inside the welding torch, through the orifice to the workpiece, and back to the power supply. The difference between these two modes of operation is shown in Fig. 2. The transferred arc mode is used for welding metals and is the subject for the remainder of this article.

Plasma Generation. Plasma is generated by constricting the electric arc and hot ionized gases passing through the orifice of the nozzle. Plasma has a stiff columnar form and generally has parallel sides, so it does not flare out in the same manner as the gas tungsten arc. The high-temperature, stiff plasma arc, when directed toward the work, melts the base-metal surface and the filler metal, if used. The plasma acts as an extremely high-temperature heat source to form a molten weld pool in the same manner as a gas tungsten arc. The higher temperature plasma, however, causes this to happen faster. Plasma used in this way is referred to as the "melt-in" mode of operation. The high temperature of the plasma or constricted arc and the high-velocity plasma jet provide an increased heat transfer rate over GTAW when using the same current. This results in faster welding speeds and deeper weld penetration. This method of operation is used for welding extremely thin material and for welding multiple-pass groove welds and fillet welds. Another method of welding with plasma is known as "keyhole" welding. In this method, the plasma jet penetrates through the workpiece and forms a keyhole-shaped opening. Surface tension forces the molten base metal to flow around the keyhole to form the weld. For a detailed description of this technique, see the section on keyhole welding in this article.

Fig. 2 Transferred and nontransferred plasma arc modes

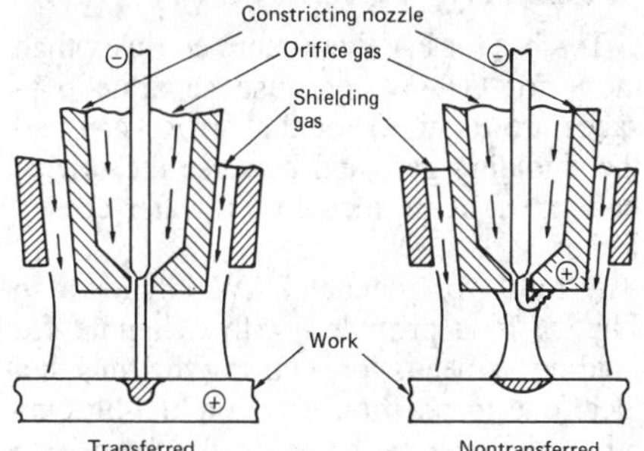

Fig. 3 Fusion zone widths for EBW, PAW (keyhole technique), and GTAW

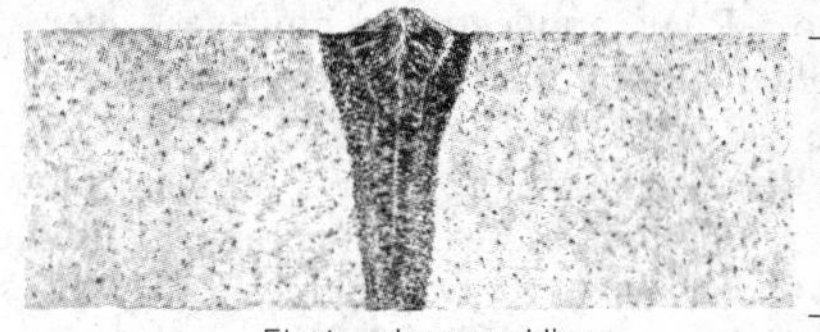

Electron beam welding

Plasma arc welding (keyhole technique)

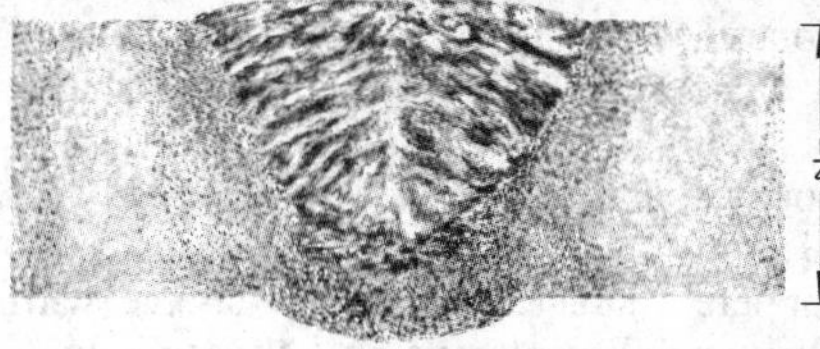

Gas tungsten arc welding or plasma arc welding (melt-in technique)

Heat-Energy Concentration. In concentration of heat energy, the plasma arc falls between the unconstricted arc used for GTAW and the electron beam in EBW. Heat-energy concentrations for the three processes, based on width of fusion zone in $^1/_4$-in.-thick 410 stainless steel, are shown in Fig. 3. The PAW and GTAW processes yield similar welds if the melt-in mode of welding is used. However, by using the keyhole welding technique, deeper, narrower penetration can be achieved, as shown in Fig. 3 (see the description of this technique later in this article).

Current. Plasma arc welding can be done at considerably lower current than GTAW, often at a current below 1 A. Generally, PAW is done at $^1/_2$ to $^2/_3$ of the current required by GTAW for the same metal thickness. High current also is used. The line of demarcation between high-current and low-current operation is arbitrary, although usually low-current welding is considered to be that done in the range of 0.1 to 100 A, and high-current welding to be that done with current above 100 A—usually 100 to 500 A. Welding current never exceeds 500 A and is usually less than 300 A.

Power Sources

Direct current electrode negative from a constant-current power source is used for most PAW applications; however, alternating current or DCEP power sources can be used.

Rectifiers having an open-circuit voltage of 65 to 80 V are most commonly used as the basic unit. Power sources for GTAW can be used for PAW. Also, power sources that have appropriate adjustments are made especially for PAW. The nature of the application usually dictates the adjustments required. For instance, power sources with current slope control are required for welding circumferential joints where a keyhole must be initiated and closed out gradually. Programming equipment that controls current slope and gas flow is available. Pulsed direct current can also be used for PAW. Pulsed-current power sources are similar to those used with GTAW. A conventional drooping volt-ampere characteristic power source with the capability of pulsing between a low-level current, referred to as a "background current," and a high-level current called "peak current" is employed. Pulsed-current power sources used for PAW have variable pulse frequencies and sometimes variable pulse width ratios.

For conventional constant-current power sources, add-on packages are available that provide pulsed current within a limited range of pulse frequencies. The same features of upslope of weld current, taper of weld current, and downslope of weld current can be obtained on pulsed-current power sources similar to constant-current power sources. An essential piece of equipment for PAW is a power source for starting the arc.

Arc Initiation. Because the electrode tip is within the orifice body (Fig. 2 and 4), the arc cannot be started by touching the electrode to the workpiece, as in GTAW. Other means must be employed for arc initiation.

Most PAW torches are started by means of a high-frequency generator, which is used to provide a high-voltage spark in the torch. This spark usually is used to start a pilot arc, which is an arc between the electrode and the constricting orifice. Power from the pilot arc may be drawn from the main power source through a current-limiting resistor or, more commonly, from a separate direct current power source. In either case, the pilot arc remains inside the torch, but gas flowing through the torch is heated (ionized) by the pilot arc and flows out in an incandescent stream.

The plasma that flows out of the torch while it is piloting looks like a flame, but because it consists of argon, an inert gas, no combustion is involved. The incandescent plasma is electrically conductive and, when held close to the workpiece, provides a path between the electrode and workpiece. The welding arc follows this conductive path to establish itself between the electrode and the workpiece.

The pilot arc, which stays inside the torch between the electrode and orifice, is called a nontransferred arc. The welding arc, which is between the workpiece and the electrode, is called a transferred arc. The pilot arc can be maintained independently of the main arc, and it is often left on while the torch or workpiece is repositioned. The main arc is not on during these times. In situations requiring very low welding current, the pilot arc often is kept running even when the main arc is on. This provides a more stable low-current arc. For most welding operations requiring 10 A or more, the pilot arc is turned off during welding.

Some PAW torches for low-current welding use a mechanically adjustable electrode that can be advanced until it contacts the tip and starts current flowing, then retracted gradually, drawing an arc back as it goes. Other systems, usually for high-current operation, use a high frequency injected between the tip and workpiece to start the welding arc directly, without using a pilot arc.

Fig. 4 Low-current PAW system, using a transferred arc

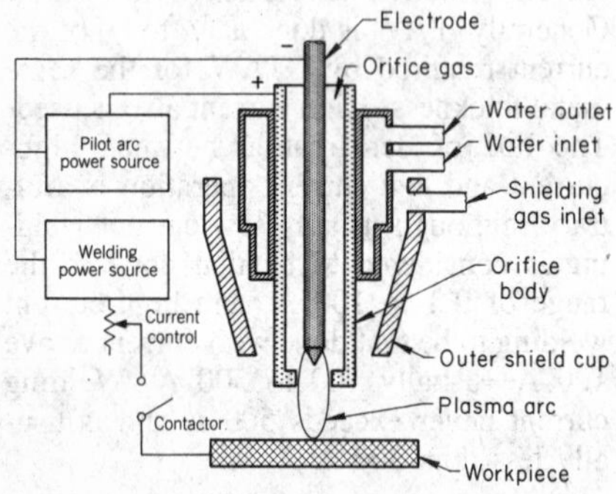

Welding Torches

Torches for PAW are more complex than those for GTAW, because separate passages are required for the orifice gas and the shielding gas, and because the orifice body must be protected by a water-cooled jacket.

A torch for manual PAW is shown in Fig. 5. It is provided with a handle for holding, a means for securing the tungsten electrode in position and conducting current to it, separate passages for the orifice and shielding gases, a water-cooled orifice body (copper), and an outer shield cup (usually of ceramic material). Manual PAW torches are available for operation on DCEN at currents up to 225 A. Controls for gas and welding current usually are separate from the torch and are operated either by a foot control or automatically.

Fig. 5 Torch for manual PAW

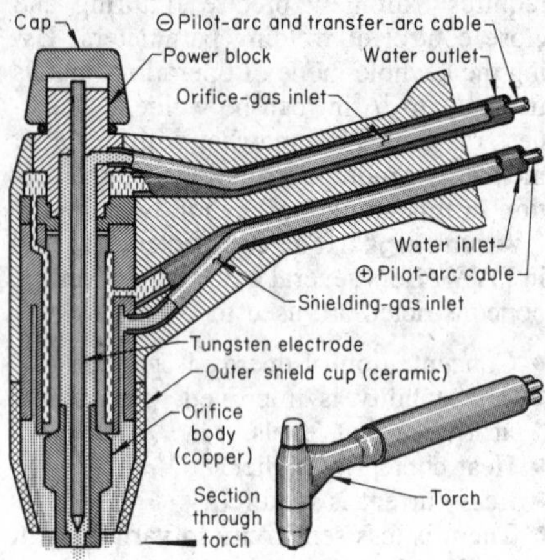

Machine Torches. Torches for mechanized PAW are similar to manual torches, except some are straight-line, offset torches. Mechanized PAW torches are available commercially for operation on either straight or reverse polarity. Direct current electrode negative power is used with a tungsten electrode for most welding applications. Direct current electrode positive is used to a limited extent with tungsten or water-cooled copper electrodes for welding aluminum and also is used with specially designed torches and copper electrodes for joining titanium and zirconium sponge compacts where freedom from tungsten or copper contamination is a prime consideration. Machine torches are available commercially for operation on either DCEN or DCEP at currents up to 500 A.

Torch Position. For manual welding, the torch head is positioned with the plasma jet at a travel angle of 25 to 35° from the vertical and pointing in the direction of welding (forehand technique). The torch (and filler rod, if used) is manipulated in the same manner as a gas tungsten arc torch to control weld-bead shape, size, and penetration.

For mechanized welding, the travel angle is 10 to 15°, with the plasma jet pointed in the direction of travel. For keyhole welding of butt joints, the torch is placed perpendicular to the adjacent work surfaces in a plane transverse to the joint. The torch standoff distance is normally about $^3/_{16}$ in. However, the distance may vary

Fig. 6 Single- and multiple-port orifice bodies

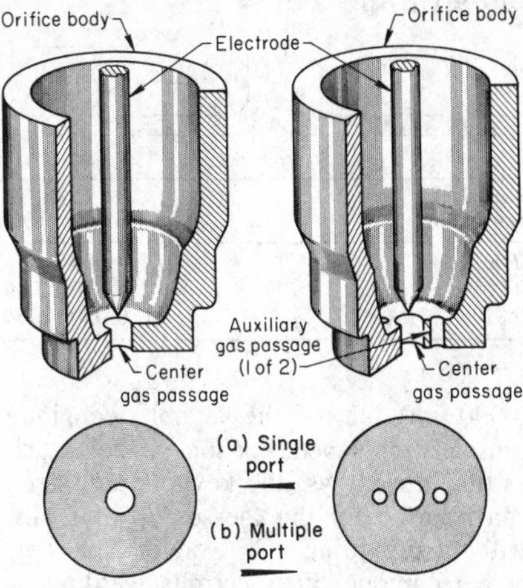

from $^1/_8$ to $^1/_4$ in. without significantly affecting the welding operation.

Torch Cooling Systems. The water-cooling unit for PAW can be either a tank-type water circulator or a water circulator-radiator. The primary function of the water-cooling unit is to remove a specific quantity of heat per hour to properly cool the torch. The water-cooling unit capacity should be defined by the welding equipment manufacturer. The unit must filter out all particles that can possibly block the small torch passages. The torch manufacturer usually specifies the type of water or coolant recommended.

Arc-Constricting Nozzles. A wide variety of nozzles are used in PAW. These include single-port and multiple-port nozzles with holes arranged in circles, rows, and other geometric patterns. Single-port nozzles are most widely used. Among the multiple-port nozzles, the most widely used design has the center orifice bracketed by two smaller ports, with a common centerline for all three openings. These two common nozzle types are shown in Fig. 6.

Orifice Diameter. An orifice diameter as small as 0.030 in. may be used with very low current. The relationship of orifice diameter to current and the increase in orifice-gas flow with an increase in orifice diameter is as follows:

Orifice diameter, in.	Current, A	Flow rate of orifice gas, ft^3/h
0.031	1-25	$^1/_2$
0.062	20-75	1
0.081	40-100	2
0.099	100-200	4
0.125	150-300	5
0.187	200-500	6

Electrodes

The nonconsumable tungsten electrodes used in PAW are the same as those used for GTAW. The most commonly used electrode material for PAW is AWS EWTh-2, which contains 1.7 to 2.2% thoria. EWTh-1 and EWTh-3 also are used. Because the electrode is upstream of the orifice, it is always surrounded by pure, uncontaminated gas, so electrode life is much longer than it is in GTAW, where the electrode is exposed. For additional information on classification, composition, standard lengths and diameters, and typical ranges of current for tungsten electrodes, see the article "Gas Tungsten Arc Welding" in this Volume or AWS A5.12, "Specification for Tungsten Arc Welding Electrodes."

Electrode Shape. The arcing end of an electrode used for PAW is ground to an included angle of between 20 and 60°, and the tip is either sharp or flattened slightly. The diameter of the flat tip is not critical; it generally is about $^1/_{32}$ in. maximum for $^1/_8$- or $^5/_{32}$-in.-diam electrodes and proportionately less for smaller diameter electrodes.

Orifice and Shielding Gases

The orifice gas is always pure argon. Any sources of contamination, including leaks in the gas system that let air diffuse into the plasma gas, must be avoided. If the plasma gas is contaminated, the electrode will quickly become discolored and will require reconditioning.

Argon also is a suitable shielding gas, but it does not necessarily produce optimum results for all metals. Argon is used for welding carbon steel, high-strength steel, and reactive metals such as titanium alloys and zirconium alloys.

Addition of hydrogen to the argon shielding gas produces a hotter arc and more efficient heat transfer to the workpiece, and higher welding speeds are obtained for a given arc current. The amount of hydrogen that can be used is limited, because excessive hydrogen may cause porosity in welds. The ability to use a higher percentage of hydrogen without inducing porosity depends on the techniques used and on the thickness of the metal being welded. Hydrogen is extremely detrimental in welding reactive metals such as titanium and zirconium alloys, because it results in embrittlement.

Argon-hydrogen mixtures commonly are used as the shielding gas for welding austenitic stainless steels, nickel alloys, and copper alloys. Permissible percentages of hydrogen usually vary from 5% used on $^1/_4$-in.-thick stainless steel to 15% used for the highest welding speeds on 0.150-in. and thinner wall stainless tubing in tube mills. In general, the thinner the workpiece, the higher the permissible percentage of hydrogen in the gas mixture, up to 15% maximum.

Addition of helium to argon also produces a hotter arc for a given arc current, but at least 50% helium is needed in a gas mixture before a significant change can be detected. Mixtures containing more than 75% helium behave about the same as pure helium and have the same limitations. Mixtures containing 50 to 75% helium have been employed in making keyhole welds in titanium, where the use of this gas mixture permits a higher travel speed than when pure argon is used and prevents concave root surface, which is sometimes obtained with pure argon. Guidelines for shielding gas selection for aluminum killed carbon steel, low-alloy steel, and stainless steel are given in Table 1.

The use of pure helium as an orifice gas increases the heat load on the torch and reduces its service life and current-carrying capacity. Because helium has a low mass, it is difficult to obtain a keyholing action at reasonable flow rates. Therefore, pure helium is not used for making keyhole welds. For further information on gases used in welding, see the articles "Gas Tungsten Arc Welding" and "Gas Metal Arc Welding" in this Volume.

Table 1 Shielding gas selection guide for low-current PAW

Metal	Thickness, in.	Welding technique: Keyhole	Welding technique: Melt-in
Carbon steel (aluminum killed)	Under $^1/_{16}$	Not recommended	Ar, 25% He-75% Ar
	Over $^1/_{16}$	Ar, 75% He-25% Ar	Ar, 75% He-25% Ar
Low-alloy steel	Under $^1/_{16}$	Not recommended	Ar, He, Ar-H_2 (1-5% H_2)
	Over $^1/_{16}$	75% He-25% Ar, Ar-H_2 (1-5% H_2)	Ar, He, Ar-H_2 (1-5% H_2)
Stainless steel	All	Ar, 75% He-25% Ar, Ar-H_2 (1-5% H_2)	Ar, He, Ar-H_2 (1-5% H_2)

Filler Metals

The use of filler metal in PAW depends on whether or not additional metal is needed. Filler metal is available in wire form in virtually any composition needed. For all but special applications, filler metals are the same as those used for GTAW (see the section on filler metals in the article "Gas Tungsten Arc Welding" in this Volume). Filler metals for PAW covered by American Welding Society (AWS) specifications are:

AWS specification	Filler metals
A5.7	Copper and copper alloy welding rods
A5.9	Corrosion-resistant chromium and chromium-nickel-steel bare electrodes and welding rods
A5.10	Aluminum and aluminum alloy welding rods and bare electrodes
A5.14	Nickel and nickel alloy bare welding rods and electrodes
A5.16	Titanium and titanium alloy bare welding rods and electrodes
A5.18	Low-carbon steel electrodes for GMAW
A5.19	Magnesium alloy welding rods and bare electrodes
A5.24	Zirconium and zirconium alloy bare welding rods and electrodes
A5.28	Low-alloy steel electrodes for gas shielded arc welding

For the melt-in technique of PAW, the filler metal is fed into the leading edge of the weld pool. The height at which the filler metal is fed is not critical in PAW, because the design of the torch ensures that there is no danger of the filler wire touching, and thereby contaminating the electrode. For the keyhole method, filler metal is added to the leading edge of the weld pool formed by the keyhole. The molten weld metal will flow around the keyhole to form a reinforced weld bead.

Hot-Wire Systems. Welding speed can be increased by preheating the filler-metal wire before it enters the weld pool. The usual method is to pass electrical current through the wire for a short distance before the wire enters the weld pool. Figure 7 shows a hot-wire-feeding system commonly used for automatic welding. For more information on wire-feed systems, see the articles "Gas Metal Arc Welding" and "Gas Tungsten Arc Welding" in this Volume.

Accessory Equipment

Accessory equipment required for PAW, including jigs, fixtures, and the various types of positioners and handling devices, is generally the same as for other shielded arc welding processes. For more information, see the sections dealing with these subjects in the articles "Shielded Metal Arc Welding," "Gas Tungsten Arc Welding," and "Gas Metal Arc Welding" in this Volume.

Fig. 7 System for automatically feeding hot filler-metal wire to the weld pool

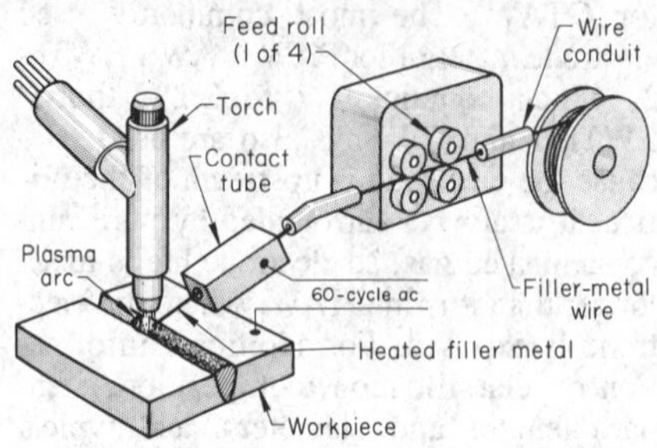

Joint Design

Joint designs for workpieces up to about 0.060 in. thick generally are the same for PAW as for GTAW. Square-groove butt joints are commonly used for most workpieces in the above thickness range. Also, the melt-in (not keyholing) method of welding is usually employed, and fusion is accomplished in the same manner as with GTAW. Root opening and misalignment are less critical in PAW, because of the "stiffness" of the plasma arc and its insensitivity to voltage changes.

Edge-Flange Welds. For joining workpieces 0.002 to 0.010 in. thick, melting down of an edge-flange is commonly used. In effect, this technique at least doubles the thickness at the weld joint of the metal being welded. The pieces to be welded can be flanged on a roll flanger. A typical correlation between metal thickness and flange height is presented in Fig. 8.

Fig. 8 Correlation of metal thickness and flange heights for edge-flange welds

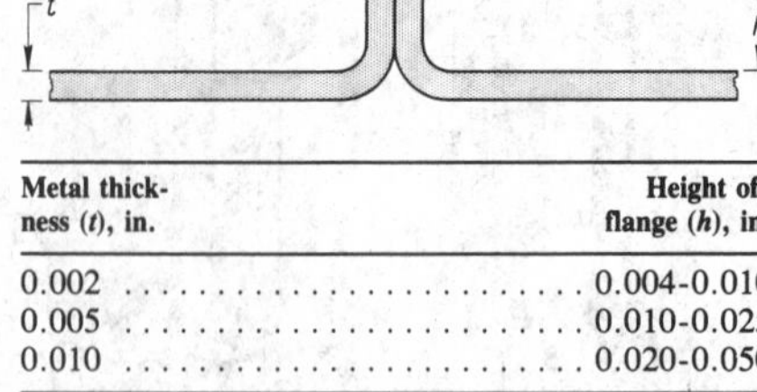

Metal thickness (t), in.	Height of flange (h), in.
0.002	0.004-0.010
0.005	0.010-0.025
0.010	0.020-0.050

Butt Joints in Thin Metal. Low-current PAW has been successfully applied to stainless steels. Typical conditions for making low-current butt welds in thin austenitic stainless steel, either manually or with mechanized equipment, are:

Section thickness(a), in.	Current (DCEN), A	Welding speed, in./min
0.001	0.3	5
0.003	1.6	6
0.005	2.0	5
0.010	6.0	8
0.030	10.0	5

(a) Orifice gas, argon at 0.5 ft^3/h; orifice diameter, 0.030 in. Shielding gas, 99% argon-1% hydrogen

Square-Groove Butt Joints. For joining workpieces in the thickness range of 0.090 to 0.325 in., the keyhole technique is usually employed. For some metals, such as titanium alloys, the keyhole technique can be used for thicknesses up to $^1/_2$ in. without providing a V-groove. The keyhole technique often permits welding of metals up to $^1/_4$ in. thick in one pass, with or without filler metal. However, for welding thicker metal (and sometimes for metal less than $^1/_4$ in. thick), a more uniform weld is obtained by using at least two passes, often the first without filler metal and the second with filler metal.

Machined-groove joints are usually required for PAW of workpieces $^1/_4$ to 1 in. thick, but because of the greater penetration of the plasma arc, compared with the gas tungsten arc, the groove depth can be much less. Consequently, the weld can be completed in fewer passes and with less filler metal. Joint preparation required for PAW and GTAW of a V-groove is compared in Fig. 9.

Figure 10 compares PAW and GTAW for joining $^1/_4$-in.-thick type 410 stainless steel. For GTAW, it was necessary to make a deep V-groove to leave a small root face (about $^1/_{16}$ in. thick) that the arc could penetrate. The weld was then completed by making one root pass (without filler metal) and two filler-metal passes. For PAW, using the keyhole technique, no joint preparation was required, and the weld was completed with one root pass and one filler-metal pass. The arc time per 100 in. of weld in the table that accompanies Fig. 10 shows PAW to be approximately four times as fast as GTAW. An additional saving is obtained in reduced joint-preparation time. Weld procedure schedules for various joint designs in stainless steel, low-carbon steel, copper, and aluminum are given in Table 2.

Keyhole Welding

Because of the intense heat and mechanical force of the plasma arc, the keyhole technique can be used with PAW. A hole is produced at the leading edge of the weld pool by the force of the plasma arc

Fig. 9 Joints for a V-groove weld. Made in $^{3}/_{8}$-in. metal by GTAW and PAW

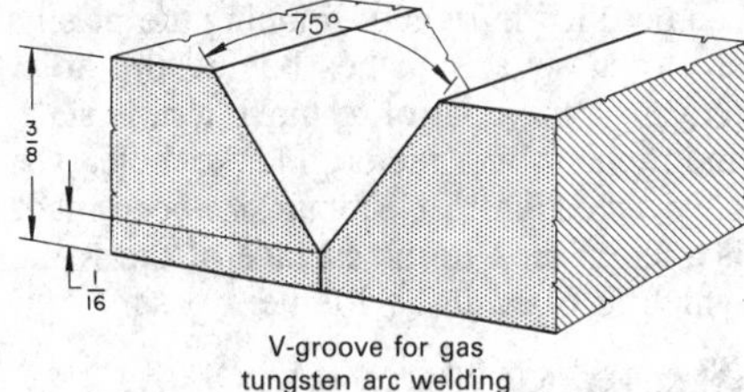

V-groove for gas tungsten arc welding

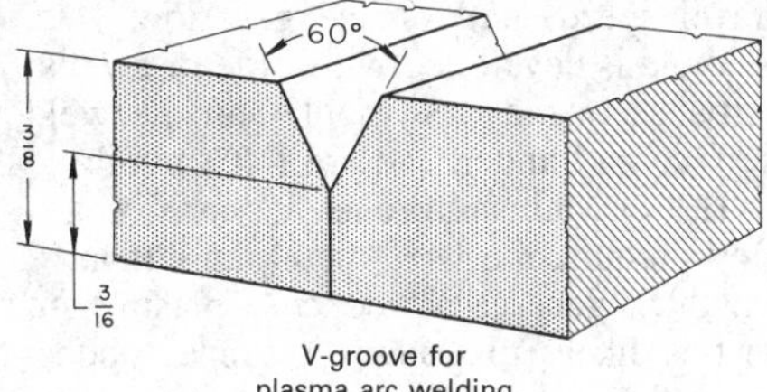

V-groove for plasma arc welding

displacing the molten metal, allowing the arc to pass completely through the workpiece. As welding progresses, surface tension causes the molten metal to flow in behind the hole to form the weld bead. A keyhole weld produced in butt welding $^{1}/_{4}$-in.-thick stainless steel, as revealed photographically from above the weld and schematically at a cross section through the keyhole of the workpiece, is shown in Fig. 11.

The major advantages of the keyhole technique are the ability to penetrate rapidly through relatively thick root sections and to produce a uniform underbead without mechanical backing. Also, the ratio of the depth of penetration to the width of the weld is much higher, resulting in a narrower weld and heat-affected zone (see Fig. 3). Presence of the underbead is proof of complete joint penetration and simplifies weld inspection. Because of the precise control that is required, keyhole welding usually is done automatically. However, a successful keyhole weld can be produced manually by a skilled welder.

The keyhole technique can be applied to carbon and alloy steel and stainless steel in the thickness range of 0.090 to 0.325 in.—total thickness of plate or thickness of root face for thicker plates with prepared joint edges. For metals that have lower density or greater surface tension in the molten state, such as titanium alloys, the keyhole technique can be employed for thicker sections, often up to 0.600 in.

Aluminum and aluminum alloys can also be welded using the keyhole technique using a variable-polarity squarewave direct current power supply. With this power

Fig. 10 Joints welded in $^{1}/_{4}$-in.-thick type 410 stainless steel

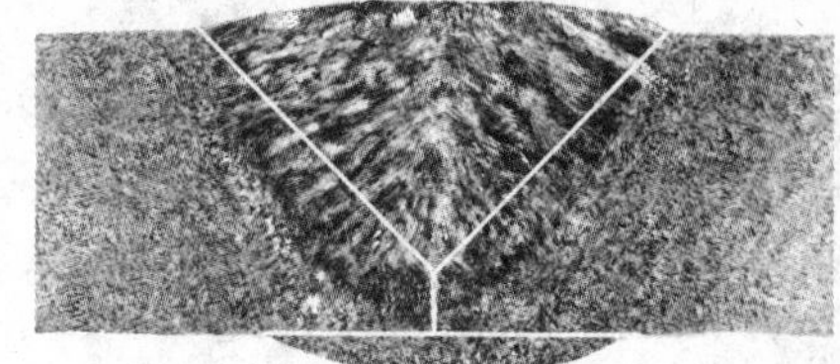

Gas tungsten arc welding

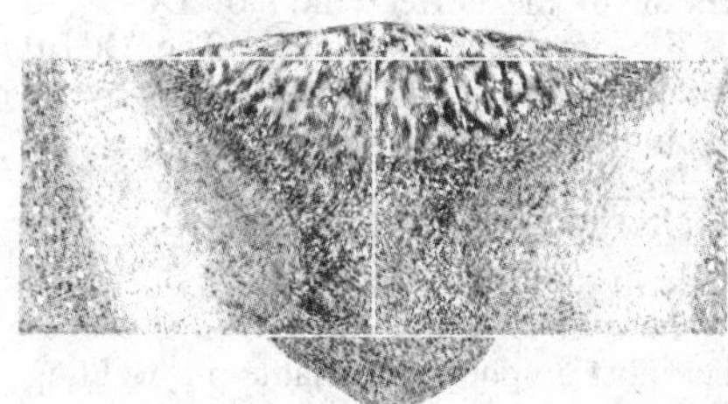

Plasma arc welding using the keyhole technique

Item	GTAW	PAW
Joint preparation	Scarf V-groove	None
Number of passes:		
Fusion	1	1
Filler metal	2	1
Travel speed, in./min	4	10
of weld, min	75	20

supply, the current polarity reversals can be controlled from 2 to 999 ms in 1-ms

Table 2 Manual PAW procedure schedules for various metals

Material thickness, in.	Type of weld	Orifice diameter, in.	Filler diameter, in.	Shielding gas at 20 ft^3/h, %	Argon plasma gas flow, ft^3/h	Welding current (DCEN), A	No. of passes	Travel speed, in./min
Stainless steels(a)								
0.008	Edge butt	0.093	...	Argon	0.5	12	1	7
0.008	Edge butt	0.093	...	Argon-5H_2	0.5	10	1	13
0.020	Square groove	0.046	...	Argon-5H_2	0.5	12	1	21
0.030	Square groove	0.046	...	Argon-5H_2	0.5	34	1	17
0.062	Square groove	0.081	...	Argon-5H_2	0.7	65	1	14
0.093	Square groove	0.081	...	Argon	2.0	85	1	12
0.093	Square groove	0.081	...	Argon-5H_2	2.0	85	1	16
0.125	Square groove	0.081	...	Argon	2.5	100	1	10
0.125	Square groove	0.081	...	Argon-5H_2	2.5	100	1	16
0.187	Square groove	0.081	...	Argon-5H_2	3.5	100	1	7
0.250	V-groove	0.081	...	Argon-5H_2	3.0	100	First	5
0.250	V-groove	0.081	$^{3}/_{32}$	Argon-5H_2	1.4	100	Second	2
Low-carbon steel								
0.030	Square groove	0.081	...	Argon	0.5	45	1	26
0.080	Square groove	0.081	...	Argon	1.0	55	1	17
Copper(a)								
0.016	Edge butt	0.093	...	Helium	0.5	18	1	24
Aluminum								
0.036	Square groove	0.081	$^{1}/_{16}$	Helium	0.05	47(b)	1	24
0.050	Edge joint	0.081	...	Helium	0.5	48(b)	1	22
0.090	Fillet	0.081	$^{3}/_{32}$	Helium	1.4	34(b)	1	4

(a) Backing gas 5 to 10 ft^3/h (argon). (b) Direct current electrode positive used for PAW of aluminum.

Fig. 11 Keyhole weld in $^1/_4$-in.-thick stainless steel plate

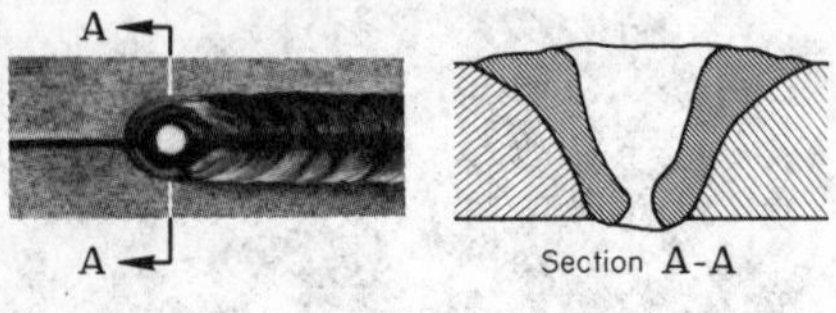

increments. These controls allow the selection of the correct ratio of DCEN to DCEP time and amplitude. The DCEP cycle keeps the base metal clean by its etching action, and the DCEN cycle concentrates the heat in the parts being welded and not on the orifice or the tungsten electrode in the torch.

The ratio of DCEN to DCEP cycles is the most important variable in welding aluminum with the keyhole plasma arc process. The ratio of 20 ms of DCEN to 3 ms of DCEP is correct for welding aluminum. Changes in either cycle may be necessary, depending on other welding variables such as alloy selection, thickness, position, and travel speed.

Backing Requirements. The weld pool of a plasma arc keyhole weld is supported by the surface tension of the molten metal. Close-fitting backing bars, which affect chill and ability to hold assembly tolerances, are not necessary. However, as in GTAW, shielding gas generally is required on the back side of the weld to protect the molten underbead from atmospheric contamination. A backing such as that shown in Fig. 12 provides a duct for flow of the shielding gas at the weld root. This type of backing supports and aligns the workpiece, contains backing gas, and provides a vent space for the plasma jet. The groove generally is about $^3/_4$ in. wide and 1 in. deep.

Starting a Keyhole Weld. When welding metal less than $^1/_8$ in. thick, longitudinal and circumferential keyhole welds can be started at full operating current, travel speed, and orifice gas flow. The keyhole is developed with little or no disturbance in the weld pool, and the weld surface and underbead are left smooth.

The operating currents and orifice gas flows required to weld metal more than $^1/_8$ in. thick generally produce a plasma arc that is likely to gouge or tunnel underneath the molten metal just prior to piercing through the joint. Because this gouging action may cause porosity and surface irregularities, starting tabs are used whenever possible to establish the keyhole off the workpiece. When welding heavy circumferential joints, a smooth transition from shallow penetration to a keyhole can be accomplished with a programmed increase in welding current and orifice gas flow. A typical slope-up control cycle for current and gas flow for starting a keyhole weld is shown on the left-hand side in Fig. 13. These functions are perfomed automatically by commercial welding controls.

Terminating a Keyhole Weld. If the welding current is turned off abruptly at the end of the weld, the keyhole will not close. This is no drawback if the weld can be stopped on an end tab, but if end tabs cannot be used, the keyhole can be closed by use of a slope-down control to reduce the welding current and orifice gas flow gradually, as shown on the right side of Fig. 13. The net effect of a programmed reduction in orifice gas flow and welding current is to reduce arc force and heat input and allow the molten metal to flow gradually into the keyhole and solidify.

In the event of a power failure or other malfunction that would cause welding to stop before the weld is completed, the keyhole is left in the weldment. This hole can be filled in by repositioning the plasma arc torch several inches behind the hole, recycling the control to make a new start, and running over the original keyhole. The only indication that a repair has been made is a slight sink in the surface of the weld, which is filled in by the next pass.

Manufacture of Stainless Steel Tubing

Continuously formed stainless steel pressure tubing conforming to American Society for Testing and Materials (ASTM) A312 is made from strip rolled into tube form and butt welded, using PAW, generally without filler metal. Installation of plasma arc equipment in tube mills has resulted in substantially increased welding speed, compared with GTAW. A comparison of average welding speeds for fabricating stainless steel tubing by GTAW and PAW are:

Wall thickness, in.	Welding speed, in./min GTAW	Welding speed, in./min PAW	Increase with PAW, %
0.109	26	36	38
0.125	22	36	64
0.154	20	36	80
0.216	8	15	88
0.237	6	14	134

Plasma arc welding shows the greatest speed advantage on thicker walled tubing. In addition, the rejection (or repair) rate when using PAW is lower than when using GTAW on thick-walled tubing, because of the uniform penetrating power of the plasma arc.

Fig. 12 Backing used with the keyhole technique

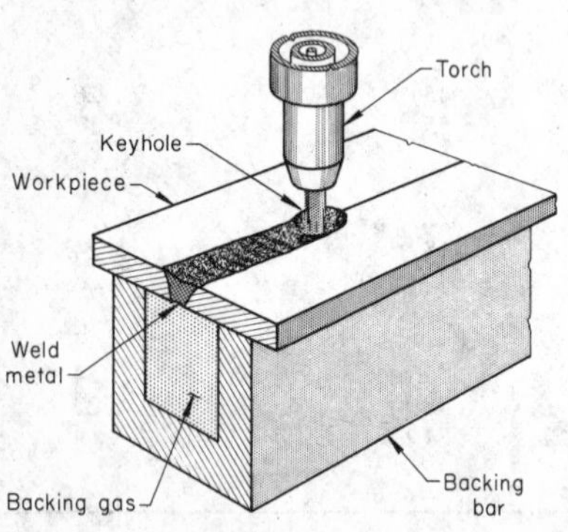

Fig. 13 Slope-control pattern. For welding current and orifice-gas flow for starting and terminating a keyhole weld in $^3/_8$-in. D-6 AC steel. Pattern is typical for welding other thicknesses of other metals.

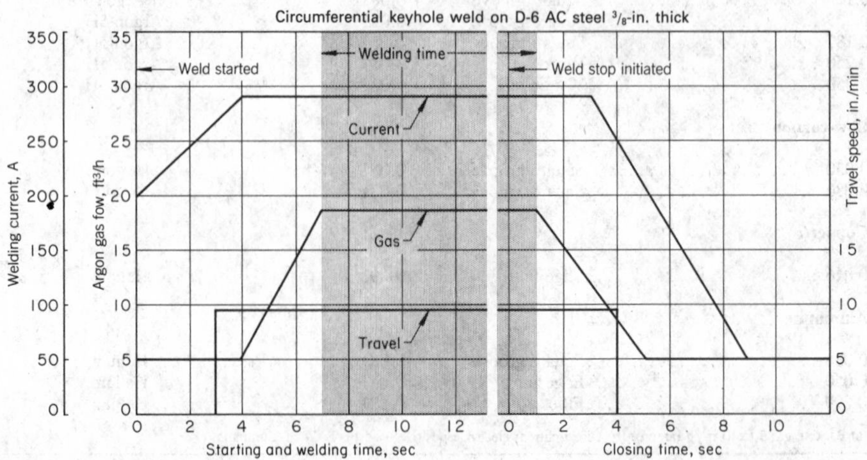

Fig. 14 Setup for welding longitudinal seams in stainless steel tubes. Using internal backing with inert gas

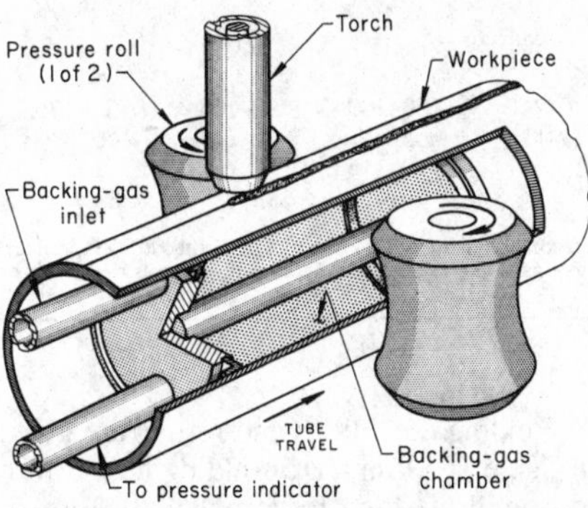

A schematic view of plasma arc tube welding is shown in Fig. 14. Weld-bead shape and reinforcement are controlled by adjusting four variables: the welding current, the location of the arc relative to the centerline of the pressure rolls on the tube mill, the force exerted by the pressure rolls, and the backing-gas pressure inside the tube.

In beginning a production run, the arc is started, and a keyhole is established approximately 1 in. ahead of the centerline of the pressure rolls. Excessive weld metal (reinforcement) at the top and bottom of the tube joint (Fig. 15a) indicates that the weld bead is too hot when it reaches the pressure point between the rolls and is, therefore, soft enough to be upset. This condition can be corrected by moving the torch farther ahead of the pressure rolls, allowing more time for the weld pool to cool before it comes under forging pressure. If there is not enough reinforcement at the top of the joint (Fig. 15b), the torch should be moved closer to the pressure rolls. Pressure-roll force also can be increased or decreased to change the amount of weld reinforcement.

Fig. 15 Three undesirable conditions encountered in plasma arc tube welding. See text for methods of correcting these conditions.

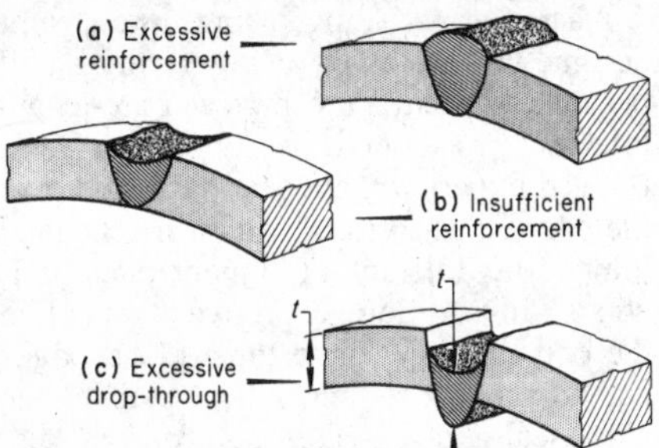

If excessive drop-through occurs (Fig. 15c), it can be corrected by increasing the backing-gas pressure. The backing gas is confined between plugs or diaphragms mounted on a pipe or lance inside the tube. Backing-gas pressure can be controlled by connecting a hose to the backing-gas chamber and exhausting the gas into a beaker or tube of water. Gas pressure is increased until bubbles appear in the water. Thereafter, pressure in the chamber will depend on how deeply the hose is immersed in the water.

Plasma arc tube welding, unlike gas tungsten arc tube welding, does not require arc-voltage control or automatic adjustment of orifice-to-work distance. This may greatly reduce the cost differential for equipment between PAW and GTAW. Constant travel speed for the tube and good joint fit-up are essential for good-quality welds.

When tube diameter is 1 in. or less, the speed obtainable using PAW is about the same as for GTAW. For making this size of tube, a V-groove is formed as the edges of the strip are brought together. Keyholing, with its accompanying speed advantage, becomes impractical, because for the maximum wall thickness that would prevail for a 1-in.-diam tube, there is not enough metal at the joint to support the molten weld pool.

Welding Stainless Steel Foil

At very low currents, PAW can be used for joining stainless steel foils (0.001 to 0.005 in. thick). For this work, a system such as that shown in Fig. 4 is used.

The following current settings (DCEN) and welding speeds have been satisfactory for welding three different thicknesses of stainless steel foil, using argon plus 1% hydrogen at 0.5 ft^3/h through a 0.030-in.-diam orifice and at 20 ft^3/h for shielding:

0.001-in. foil	0.3 A, 5 in./min
0.003-in. foil	1.6 A, 6 in./min
0.005-in. foil	2.4 A, 5 in./min

Only edge-flange welds for butt joints are recommended for welding foil, and precise fixturing is a primary requirement. Figure 16 shows a typical fixturing arrangement for making the edge-flange weld. An essential part of any fixture used is a backing groove at least $^1/_4$ in. deep, because the underside of the joint must be shielded with inert gas during welding. The width of the backing groove is not critical; a range of 10*t* to 24*t* (*t* being the foil thickness) is satisfactory. Also, considerable tolerance is allowed on spacing between hold-down clamps: it can range from 15*t* to 30*t* (Fig. 16).

Circumferential Pipe Welding

Mechanized PAW has been used to make circumferential joints in stainless steel pipe and in pipe of some other materials, in the horizontal and vertical pipe positions. This type of weld normally would be made by multiple-pass GTAW, using a backing ring and filler metal on a prepared joint. The use of PAW permits keyhole welding of square-groove butt joints in one pass in pipe of 0.090- to 0.250-in. wall thickness. On pipe with wall thickness from 0.250 to 0.375 in., a V-groove with a 60° included angle and a root face half the thickness of the wall is used. This type of prepared joint requires two passes—a root pass using the keyhole technique, and a second pass using filler-metal wire. Slope control for welding current and orifice gas flow is employed to start and terminate the keyhole weld (Fig. 13).

Multiple-Pass Welding. As stated previously, a multiple-pass plasma arc weld involves a keyhole root pass and one or more melt-in filler passes, with or without filler-metal addition. The melt-in plasma arc uses a plasma jet with lower force than that used for keyhole welding. Total orifice gas flow rate is reduced; torch standoff is increased. Varying percentages of helium or hydrogen may be mixed with argon in both the orifice and shielding gas circuits to dissipate the arc heat over a larger surface area of the weld joint. Helium can be used, and is preferred for some applications, because it provides a wider heat input pattern and produces a flatter weld bead. The wide range of weld penetration characteristics attainable with high-current plasma arc techniques provides substantially greater process flexibility than is available with other gas shielded arc welding processes.

Plasma Arc Welding Versus Alternative Processes

Plasma arc welding is most often an alternative to GTAW, of which it is a modification. In other applications, PAW competes with such widely different processes as EBW, OAW, or RSW. The examples that follow compare PAW with these processes for specific applications.

Example 1. Plasma Arc Welding Versus Gas Tungsten Arc Welding for Manual Fusion of Thin Joints. Enclosures for small electronic and electromagnetic devices were made of a thin stainless

Fig. 16 Edge-flange joint. Also shows design and dimensions of fixturing for PAW of stainless steel foil

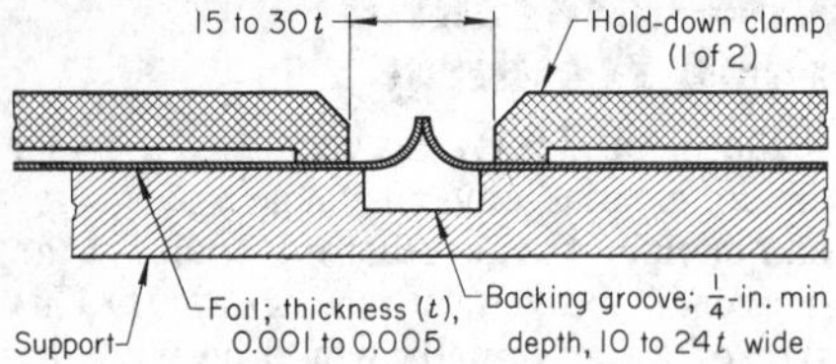

Table 3 Plasma arc welding versus gas tungsten arc welding for making 100-in. welds in three aerospace metals

Condition	410 stainless steel PAW	410 stainless steel GTAW	Maraging steel, 18% Ni PAW	Maraging steel, 18% Ni GTAW	Titanium alloy Ti-6Al-4V PAW	Titanium alloy Ti-6Al-4V GTAW
Thickness of metal, in.	0.250	0.250	0.250	0.250	0.095	0.095
Number of passes	1	2	1	3	2(a)	5(b)
Current, A	240	170-200	240-260	180-200	90-175	120-175
Travel speed, in./min	12	4	12-13	4	15	6
Time per 100 in. of weld, min	8.3	50	7.4	75	13.4	83.5

(a) One keyhole welding pass and one filler-metal pass on the outside of a rocket-motor case. (b) One root pass without filler metal and two passes with filler metal on the outside of a rocket-motor case. The root pass was back gouged, and two filler passes were made on the inside.

steel case into which was fitted a reverse-flanged cover containing glass-insulated connector pins. The enclosures were hermetically sealed by fusing the outer edge to form a corner weld (Fig. 17). The procedure was set up for manual welding by clamping the fitted stainless steel cases in the copper-faced jaws of a small, movable bench vise.

Originally, the weld was made by using a small gas tungsten arc torch at low amperage. Manual seam tracking was difficult and tedious because the low amperage, needed to avoid overheating, required holding the arc length to within about 1/32 in. Small variations resulted either in stubbing the electrode tip or breaking the arc. As a result, arc restrikes, tip dressing, and leaky welds were frequent. When amperage was increased to permit a longer arc, overheating often damaged the glass seals.

These problems were overcome to a large extent by changing over to a small plasma arc unit designed for operation at approximately 1 to 10 A. The lightweight torch (about 3 oz heavier than, and roughly the same size as, the gas tungsten arc torch) had a 0.030-in.-diam orifice. This torch produced a shielded plasma column with an effective operating length of 1/8 to about 3/8 in., when used in the transferred arc mode at 5 A. The arc was initiated in the nontransferred mode—that is, as a pilot arc in a high-frequency circuit between the tungsten electrode and a copper anode within the torch body. When the torch nozzle was brought within operating distance, the direct current welding circuit was closed by transfer of the arc to the grounded workpiece. After welding, the arc returned to the nontransferred mode.

Because of arc-length tolerance, manual welding was easier to control. Low heat input avoided damage to the glass seals, while the absence of frequent starts and stops resulted in better weld quality. Productivity increased, largely due to fewer rejects. The plasma torch electrode required occasional dressing because of arc erosion, but dressing because of base metal contamination was eliminated. Welding details are given in the table with Fig. 17.

Fig. 17 Electronic package that was plasma arc welded to fuse a thin corner joint

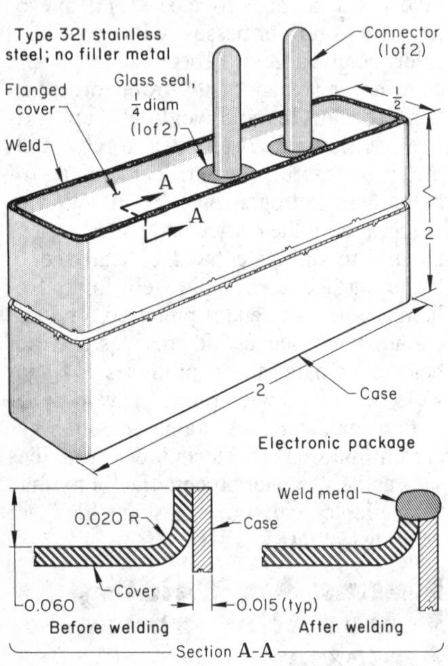

Conditions for PAW

Joint type	Edge
Weld type	Corner flange
Filler metal	None
Orifice gas	Argon, at 0.6 ft^3/h
Shielding gas	95% Ar-5% H_2, at 25 ft^3/h
Orifice diameter	0.030 in.
Power supply	1-10 A (dc) (with continuous high-frequency pilot circuit)
Current	5 A (DCEN)
Welding speed	5 in./min

Example 2. Welding Speed—Plasma Arc Welding Versus Gas Tungsten Arc Welding. Three metals used for aerospace components were welded by PAW and GTAW to compare times required for making welds. Samples welded were: 1/4-in.-thick type 410 stainless steel; 1/4-in.-thick maraging steel (18% Ni); and 0.095-in.-thick titanium alloy Ti-6Al-4V.

Welding conditions and results are given in Table 3. Plasma arc welding took a fifth to a tenth as long to complete 100 in. of weld as GTAW, primarily because fewer passes were required.

Example 3. Change to Plasma Arc Welding From Gas Tungsten Arc Welding for Attaching Tube End Caps. A delicate electrical component required hermetic seals between two end caps and a tube section (Fig. 18). Each end cap contained a connector pin insulated from the cap by a glass seal and joined to an internal circuit by wrapped and brazed wires. Originally, the part was designed for a single-pass weld by GTAW at each end. For this process, end caps did not have the heat-barrier groove and rolled-over edge shown in Fig. 18.

The part was fixtured for rotation about its vertical axis under a stationary electrode holder, and the edge between the tube and cap was gas tungsten arc welded. Several serious problems arose. Because a low heat input arc was essential to avoid overheating the glass seal, very precise electrode positioning was needed to maintain stability of the arc operating on a low amperage. Also, to ensure proper electrical grounding of the two parts during welding, machining tolerances on the end cap and roundness tolerances on the tube were extremely close. When the welds made by this procedure were leak tested, about 80% were rejected; 30% could be salvaged by rewelding. Because the rejection rate was too high, this welding procedure was abandoned.

Plasma arc welding, using a new joint design, was tried. By using a very small plasma arc torch, a much greater arc-length tolerance was obtained. End caps were beveled to permit the tube to be rolled over the edge, thus forming a tight mechanical joint more tolerant of dimensional variations. In addition, a groove was cut in the end caps to isolate the welded edges and to provide a longer heat path to the glass seal and a greater area for heat dis-

Fig. 18 End cap in place on small electrical component. Shows cap and joint design used for PAW

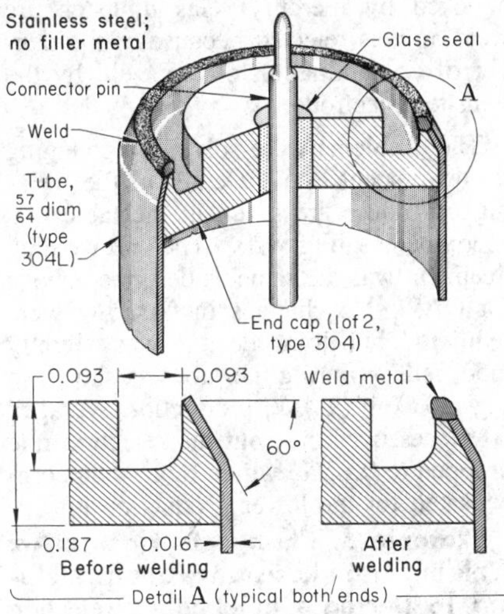

Conditions for automatic PAW

Joint type	Edge
Weld type	Edge-flange
Number of passes	1
Welding position	Flat
Fixtures	Clamping jig; rotating positioner; fixed torch mount
Torch	10 A, water cooled
Electrode	0.020-in.-diam EWTh-2
Filler metal	None
Orifice gas	Argon, at 0.68 ft^3/h
Shielding gas	95% Ar-5% H_2, at 20 ft^3/h
Orifice-to-work distance	0.090-0.120 in.
Power supply	10-A, 110-V (max) control unit
Current	5 A (DCEN)
Voltage	50 V
Welding speed	3 in./min
Preheat, postheat	None

sipation by radiation and convection. The groove also reduced restraint to shrinkage of the weld metal when the assembled relay housings were cooled.

The same clamping fixture, turntable, and torch mount were used as with GTAW. New equipment comprised a low-current PAW unit, separate gas cylinders for orifice and shielding gases, and a plasma arc torch.

The end caps were preplaced in the tube, the tube assembly was loaded in the fixture, the torch was positioned at an angle of 45° from vertical, and the weld was made. After one revolution of the work, the arc was continued for an overlap of about 1/8 in., after which the current was shut off by a foot control. The part was turned over, and the operation was repeated on the other end. Additional details for PAW are given in the table that accompanies Fig. 18.

After assembly with other components, each unit was tested with air at 300 psi while immersed in water and tested with a halide leak detector, using a mixture of 75 lb freon and 300 lb air. Use of PAW and the new joint design decreased rejections to 3%.

Example 4. Selection of Plasma Arc Welding for Assembling Relay Housings. Housings for power relays used in aircraft and missiles were plasma arc welded after consideration of soldering, brazing, GTAW, and EBW. Soldering and brazing were eliminated because welding was more convenient, less expensive, and produced better joints. Electron beam welding was ruled out because of the high cost of equipment, the welder skill required, and the effect of the vacuum on various electrical components. Equipment was available for PAW, and its operation required less welder skill than GTAW.

Housings were required to be hermetically sealed. Each housing was made of four components—a cover, a two-piece shell, and a base, as shown in Fig. 19. The base was a 1% silicon steel casting; the other components were press formed 1018 steel. The formed parts were 0.032 in. thick, and the joint areas of the cast base were also 0.032 in. thick.

Joints (Fig. 19) were designed for ease of assembly and to permit deposition of small edge-seal welds. In addition, this type of joint simplified repair welding.

After joint edges were degreased and wire brushed, the components were first assembled in a special tack welding fixture, with a reasonable allowance made for some mismatch. Tack welds were placed on the sides, at the four corners of the cover, and on the base. The workpiece was then removed from the tack welding fixture and placed in a manually rotated welding fixture, with the vertical axis turned to the horizontal. The piece was held by power-arm-actuated holding clamps. Next, a two-piece band of 1-by-3/8-in. copper was fitted around the shell and fastened with a garter spring, to serve as a heat sink to protect glass seals on terminal studs and connector pins. The straight sides and then the cover were manually welded in a single pass, using the settings and conditions given in the table that accompanies Fig. 19. Mismatch was adjusted (where needed) by melting down the protruding edge. The unit was allowed to cool to room temperature, and then the shell was welded to the cast base.

The plasma arc torch operated continuously—in the pilot arc (nontransferred) mode when not welding and in the transferred arc mode when welding. Filler metal was required only for touch-up or repair, and then a bare low-carbon steel rod was used. Because the power arm attached to the fixture permitted the workpiece to be turned as needed, all welding could be done in the flat position. During welding, the assembly was purged with helium at a rate of 3 to 5 ft^3/h.

Completed welds were inspected visually for pinholes, which were ground out and repaired. The units were given a preliminary leak test by pressurizing them with helium at 15 psi and observing bubble

Fig. 19 Plasma arc welded relay housing and joints used

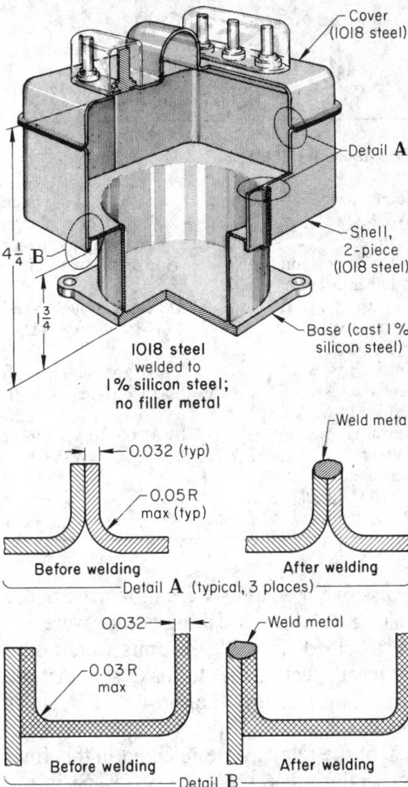

Joint type	Edge
Weld type	Edge-flange; corner-flange
Fixtures	Clamping jigs for tack welding and final welding; copper chill strip
Welding position	Flat
Electrode	0.040-in.-diam EWTh-2
Filler metal(a)	None
Orifice gas	Argon, at 0.6 ft^3/h
Shielding gas	Argon, at 5 ft^3/h; helium, at 20 ft^3/h
Orifice diameter	5/16 in.
Orifice-to-work distance	1/8 to 1/4 in.
Power supply	Special rectifier and control unit for 0.1-to-17-A output, 100% duty cycle
Welding current	17 A (DCEN)
Welding speed	3-4 in./min
Production per 8-h shift	12-16 housings

(a) For touch-up and repair, a bare low-carbon steel rod was used.

Fig. 20 Relay component. Plasma arc welded, requiring fusion of type 304 stainless steel to Kovar. Joint design and weld are shown at right.

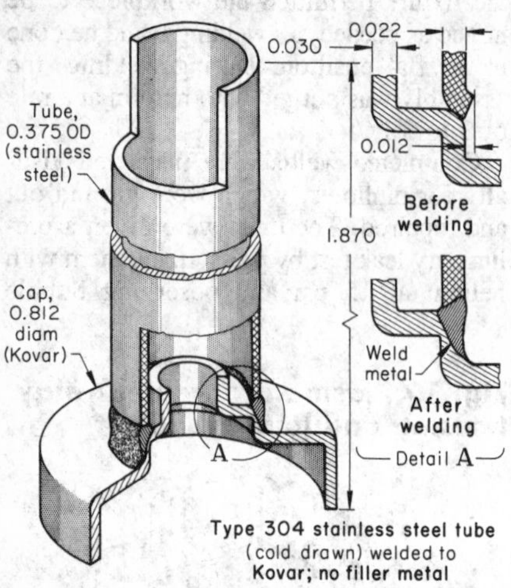

Conditions for PAW

Joint type	Edge
Weld type	Corner-flange
Fixtures	Clamping jig; rotating positioner
Welding position	Horizontal (see figure)
Number of passes	1
Torch	10 A, water cooled
Electrode	3/64-in.-diam EWTh-2
Filler metal	None
Orifice gas	Argon, at 0.6 ft^3/h
Shielding gas	85% Ar-15% H_2, at 18 ft^3/h
Orifice-to-work distance	1/4 in.
Power supply	0.1-10 A, 60-80 V rectifier
Current	9 A (DCEN)
Voltage	70 V
Welding speed	6 rpm
Production rate	60 assemblies per hour

Fig. 21 Double-walled cup. Sealed by PAW to reduce the rejection rate

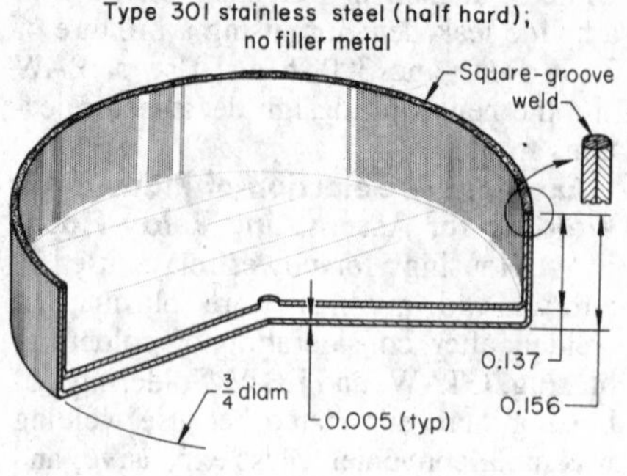

Conditions for automatic PAW

Joint type	Edge
Weld type	Square-groove
Fixtures	Clamping jigs; rotating table; torch mount
Welding position	Flat
Torch	Machine held, water cooled
Electrode	0.020-in.-diam EWTh-2
Filler metal	None
Orifice gas	Argon, at 0.6 ft^3/h
Shielding gas	97.5% Ar-2.5% H_2, at 14 ft^3/h
Orifice-to-work distance	0.050 in.
Power supply	10-A, single-phase rectifier
Current	3.5 A (DCEN)
Welding speed	28 in./min
Annual production	5 million cups

emission when units were immersed in the testing solution. Leaking joints were repair welded. Finally, the units were tested by mass spectrometer to make sure that the leak rate was less than 1.4×10^{-5} cm^3/s per cubic inch of sealed volume. Less than 1% of the relays were rejected in this final inspection. The presence of 1% Si in the cast steel presented no problem in maintaining weld quality.

Example 5. Change From Oxyacetylene Welding to Plasma Arc Welding for Joining Stainless Steel to Kovar. A circumferential weld was required to join a type 304 stainless steel tube to a Kovar electrode cap of a relay component, as shown in Fig. 20. Kovar, an alloy of 29% Ni, 17% Co, 0.2% Mn, remainder iron, is used for making glass-to-metal seals because its coefficient of expansion is the same as that of certain types of glass. The chief requirements of the welded joint were (1) medium joint penetration, (2) smooth appearance, (3) carbon-free surface, (4) strength equal to that of the tube wall, and (5) resistance to penetration of mercury at 80 psi.

Originally, OAW (without filler metal) was used for making the weldment. Prior to welding, both components were annealed in a hydrogen atmosphere. The tube and cap were clamped in a special jig to maintain alignment. The assembly was mounted, axis vertical, on a turntable, and welding was done in a single pass, using a hand-held torch. From 80 to 90 pieces were produced per hour, but welds were of poor quality, and over 50% were rejected for carbon deposits and poor bonding and spreading of metal at the interface between the two components. When OAW was used, the tube ends were not flared, as they were when PAW was adopted (shown in the "before welding" view in detail A, Fig. 20).

Other joining processes, including brazing and GTAW, were investigated. Brazing was unsatisfactory because it introduced metals into the joint that were attacked by mercury. Gas tungsten arc welding was rejected because of the danger of contamination of the weld by the tungsten electrode.

The problems were solved by changing to low-current PAW. Details of the welding procedure are given in the table that accompanies Fig. 20. The prewelding treatment was the same as described above for OAW. No changes in fixturing were required, but joint design was slightly modified by flaring the tubes as shown in Fig. 20. With PAW, production averaged 60 pieces per hour, but the rejection rate dropped to 0.05%, which more than compensated for the lower production rate.

Example 6. Change to Plasma Arc Welding From Resistance Seam Welding To Decrease Rejections. Rejection rate for the small thin-walled double cup shown in Fig. 21 was about 4.5% when RSW was used to make the seal weld. The major requirement was that the press-fit-and-welded joint between the two 0.005-in.-wall cups be pressure-tight under a test applied at a later stage of assembly. Rejections were a significant cost factor because about five million parts were produced per year. By changing to automatic low-current PAW, welding time was increased threefold, but rejections were reduced to less than 1%.

Automation for PAW was accomplished by installing two welding stations, each provided with clamping jigs for holding parts in position and rotating them under the torch. The torch was rigidly mounted on sliding ways so that it could be positioned accurately over the edge joint of an assembly at either station. A cup assembly was loaded on the fixture at one welding station, and rotation and the arc were initiated by a push button. While this assembly was being welded, another was being loaded on the second fixture. After the first cup had been welded, the torch was moved to the other welding station, and the second cup was welded.

Welding details are given in the table that accompanies Fig. 21. The completed welds were examined at a magnification of eight diameters. After further assembly, the finished unit was pressure tested with low-pressure, clean, dry air while immersed in water for 10 s.

Electroslag Welding

By the ASM Committee on Electroslag Welding*

ELECTROSLAG WELDING (ESW) is a process that uses the heat generated by passing an electrical current through a pool of molten slag (flux) to melt the edges of the joint (base metal) and a filler wire (electrode). The electrical resistivity of the molten slag continuously produces the heat necessary to continue the welding process. The molten slag pool also acts as a protective cover over the liquid metal weld pool. The slag pool and weld pool are contained in a cavity formed between the edges of the parts being joined and copper shoes used as dams. The copper shoes are normally water cooled, but this is not mandatory if the necessary amount of heat can be rejected. Typical electroslag weld joint cross sections are shown in Fig. 1.

Fig. 1 Seven joint-weld combinations made by ESW. (a) Butt; square-groove weld. (b) Butt; square-groove weld, transition between two plates of different thicknesses. (c) Corner; square-groove weld. (d) T; square-groove weld. (e) Corner; fillet weld. (f) Double-T; two square-groove welds. (g) Modified butt; square-groove weld

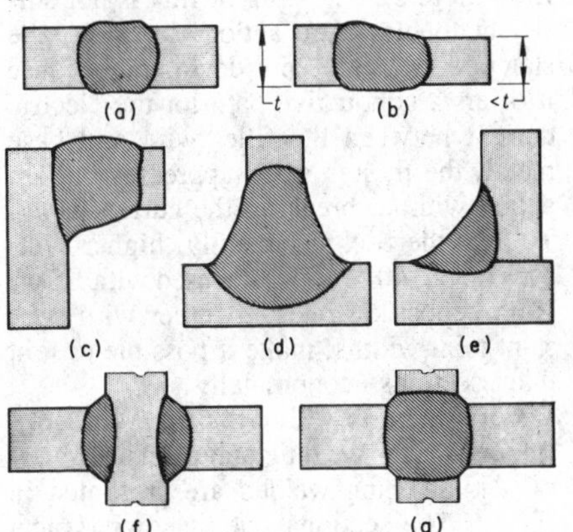

Fig. 2 Setup for conventional ESW. The mechanism for providing vertical movement of the welding machine is not shown.

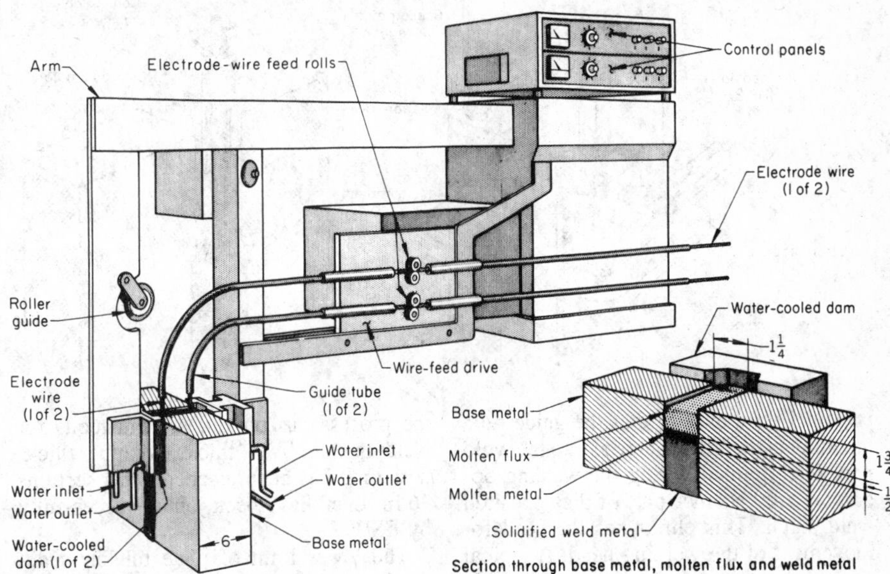

The axis of the weld joint must be approximately vertical. A consumable electrode is fed downward, into the cavity containing the slag pool, through a guide tube.

In the mechanical aspects of equipment and application of the processes, ESW and electrogas welding (EGW) are quite similar (see the article "Electrogas Welding" in this Volume). However, two important differences in the electrical characteristics of the two processes are: (1) ESW may be performed using either direct or alternating current, but EGW normally uses only direct current; and (2) ESW uses the resistivity of a molten slag pool to produce heat for welding, while EGW uses the energy of an electric arc to produce the heat for welding.

Equipment used for ESW falls into two distinct types. The first uses a nonconsumable guide in the form of a tube to direct the electrode into the slag pool. When the nonconsumable guide is used, the welding head must be motorized so it can travel vertically upward. Travel speed of the head is slow so that it can match the fill rate of the joint. The equipment shown in Fig. 2 is of the nonconsumable guide type.

The second type of equipment used for ESW is the consumable guide system. The

*R. David Thomas, Jr., *Chairman*, President, R.D. Thomas & Co., Inc.; Dennis R. Amos, Welding Engineer, Westinghouse Turbine Plant; Glen Edwards, Professor, Metallurgical Engineering, Colorado School of Mines; G.A. Gix, Chief Welding Engineer, American Bridge Division, U.S. Steel Corp.; James R. Hannahs, President, Midwest Testing Laboratories, Inc.; Franklin W. Henderson, Manager—Engineering, Richlin Equipment, Inc.; Guy A. Leclair, Manager, Welding Development Laboratory, Foster Wheeler Energy Corp.; Robert C. Shutt, Vice President, Lincoln Electric Co.; James E. Sims, Manager, Welding Research and Development, Chicago Bridge & Iron Co.

Fig. 3 Setup for consumable guide ESW. The ground cable is normally connected to the sump, not the top of the plate as shown above. It can also be connected to a clamp-on starting sump made from copper.

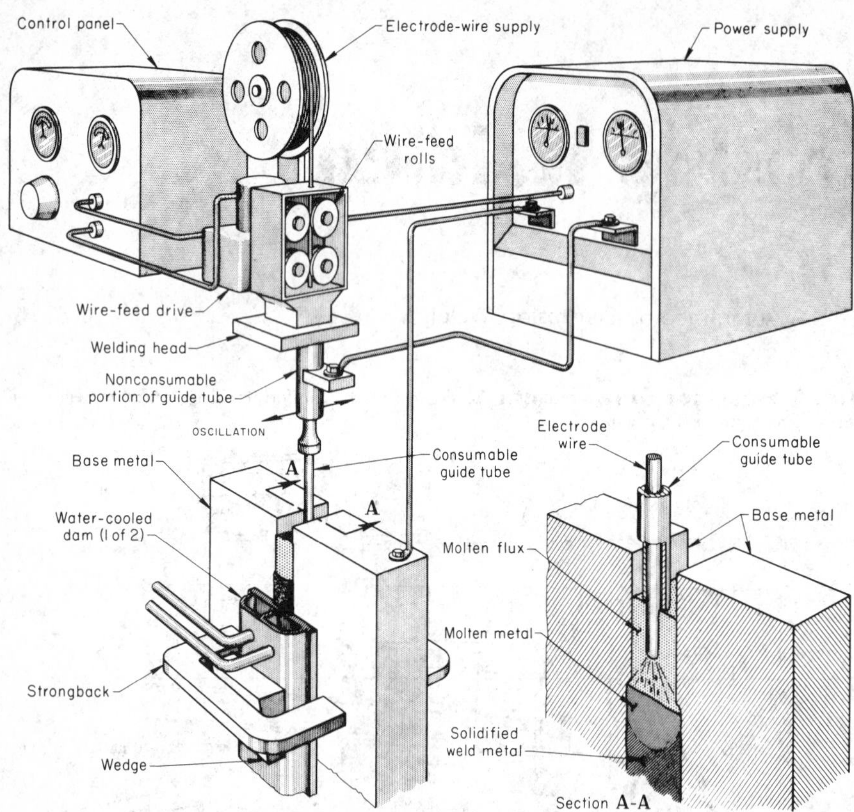

principal difference is that the guide tube that directs the filler material into the weld pool is consumed during the welding operation and becomes part of the deposited weld metal. This eliminates the need for movement of the welding head. A typical equipment setup for consumable guide ESW is shown in Fig. 3. For the most part, the two systems are capable of producing similar results, although there are applications for which one is better suited than the other (see the section on conventional system versus consumable guide system in this article).

Applicability

Materials. Steel- or iron-based alloys are the primary application for ESW. Low- and medium-carbon steels are easily welded using ESW, and it has been widely applied to these grades of material. Low-alloy and structural steels have also enjoyed successful application of the process when proper precautions are taken, as have high-strength steels such as alloy D-6 AC, stainless steels, and nickel-based alloys.

Thickness Range. Electroslag welding is most often used for welding plate from $1^1/_4$ to 12 in. thick. In a few applications, the process has proved advantageous for welding plate $^1/_2$ in. thick. No upper thickness limit has been reached; steel sections 36 in. thick have been joined successfully by ESW.

The lower limit of plate thickness depends mainly on costs compared with other processes. Submerged arc welding (SAW) and EGW are alternative processes that usually compete economically with ESW on thinner sections. Assuming each process can be applied with equal ease, the economic break-even point among the three processes is usually around 2 in. in thickness. Because circumstances vary from plant to plant, this limit varies, and a careful cost analysis must be made to determine the most economical process for each application.

Length of Joint. The maximum length of joint that can be welded by the conventional system is limited only by the auxiliary equipment available. For the consumable guide system, there are other limitations (see the section on conventional system versus consumable guide system in this article).

Types of Joints Welded. Butt joints between plates of equal thickness are those most readily, and hence most frequently, welded by ESW. When other types or shapes of joints are welded, special dams or techniques, or both, are usually required. Figure 1 shows several joints that have been successfully welded by ESW. Circumferential welding can be done by ESW with the use of special techniques (see the section on circumferential welding in this article).

Principles of Operation

Electroslag welding depends upon heat generated by the passage of current through a layer of molten flux (slag), melting the filler metal and surfaces of the joint being welded. Welding current maintains the molten slag at a temperature above the fusion point of the base metal. Fluxes formulated to provide the desired melting temperature, viscosity, and electrical resistivity are essential for satisfactory operation. The principle of operation is the same for both methods of ESW—metal transfer is arcless.

Electroslag welding is started by initiating an arc beneath a layer of granular flux, similar to SAW. As soon as a sufficient layer of flux is melted to form the slag, all arc action stops and current passes from the electrode to the base metal through the conductive molten slag. The process then becomes ESW. Welding progresses in a vertical direction using retaining dams to mold the molten metal. The resulting weld solidification pattern is highly directional, and nonmetallics are rejected upward into the slag pool.

In SAW, by comparison, the arc under granular flux creates the heat necessary for welding. The voltage across the arc ranges from 25 to 32 V. Depth of flux is not sufficient to block arc action. In ESW, the slag pool is $1^1/_2$ to 2 in. deep, and because it offers a conductive path for the electric current between the filler wire and base metal, the molten slag suppresses the arc action without breaking the current flow. To heat the flux sufficiently, higher voltages (about 40 to 55 V) are used with ESW. High deposition rates, together with constant joint widths, make it possible to join thick sections economically.

Conventional Electroslag Welding. Equipment for welding and a detailed view of a joint being welded are presented in Fig. 2. The sectional view shows typical depths of molten weld metal and slag. In Fig. 2, two guide tubes and two sets of electrode wire-feed rolls are shown. One, two, or three guide tubes and sets of feed rolls are commonly used, depending on base-metal thickness, although more may sometimes be required. The electrode wire guides are held a short distance of $^1/_2$ to 2 in. above the molten slag. This distance is controlled by moving the entire welding machine upward at a pre-established rate that is consistent with the deposition rate. Usually, the welding machine is moved on a vertical track or rail. The mechanism for achieving vertical movement of the machine is not shown in Fig. 2.

The rate of vertical travel is usually adjusted by the operator, who increases or decreases the drive motor speed so that the entire welding unit moves upward to maintain the proper distance between wire guides and the surface of the molten slag pool. Adjustments are usually required during the early stages of the welding operation to establish the proper travel speed. Optical sensors focused on the molten slag pool are often used to control the vertical motion automatically.

As shown in Fig. 2, the electrode wires are fed through the guide tubes by feed rolls. Current is carried to the electrode wire by the guide tubes, which may be oscillated (moved horizontally back and forth across the joint) during welding, depending on the thickness of the work and number of electrodes used. The water-cooled dams (mold shoes) move vertically as the machine moves. The dams serve three purposes: (1) to contain the molten weld metal and slag until they solidify, (2) to speed up solidification of the weld metal, and (3) to mold the outside contours of the weld to give the desired reinforcement.

Consumable Guide Electroslag Welding. With the consumable guide system, the mode of metal transfer is also arcless. The principal difference from the conventional system is that none of the welding machine components moves upward during the welding operation; the only vertical movement is that of the electrode wire feeding down through the feed rolls and guide tube. As shown in Fig. 3, the main components of the consumable guide system are the power supply, control panel, wire-feed drive, consumable guide, and water-cooled dams. The wire is fed vertically down through the consumable guide to the molten slag pool. Prior to welding, the consumable guide is lowered to approximately $1^1/_2$ in. of the bottom of the joint; it melts off as the level of the weld rises. Welding begins by arcing between the electrode wire and the base metal, or steel wool, as in the conventional system. As soon as the slag pool is established, it extinguishes the arc and welding proceeds vertically. Most of the filler metal is supplied by the electrode wire, but about 5 to 10% is supplied by the consumable guide.

In this system of ESW, the water-cooled dams (mold shoes) are separate from the other equipment. They are held in place by wedges between the dams and strongbacks, as shown in Fig. 3. For welding a short joint, one set of dams may be enough, but two or more sets may be used. The bottom set is removed from the joint when the weld metal has solidified and is placed above the other set. As welding proceeds, the dams are moved (hopped or leapfrogged) as required. Although a single-wire feed is shown in Fig. 3, multiple-wire feeds are sometimes used. Wire and guide can be oscillated if required, as in the conventional system.

Equipment

Power Source. Direct current, constant-voltage power supplies are recommended for ESW. For continuous-duty welding, power supplies with adequate current and voltage ratings should be used. Usually, the power rating required is 750 A, 50 V at 100% duty cycle. A power source of this rating is necessary for each wire in multiwire applications. Constant-current power supplies may be used for ESW; however, they are not recommended for most applications. Controlling this type of power source makes it difficult to maintain the welding conditions necessary for good welds. Alternating current power supplies can be used and, for some applications, may even be desirable. Generally, alternating current welds run hotter and require a few more volts than direct current welds.

Guide Tube (Nonconsumable Wire Guide). This type of guide tube should be fabricated from a beryllium copper tube supported by two soft copper bars (top and bottom of tube). The inside diameter of the beryllium copper tube should provide clearance (0.020 to 0.040 in.) for the size (diameter) of wire to be used, usually $^3/_{32}$ or $^1/_8$ in. in diameter. The beryllium copper tube is hard and tough to resist wear from the feeding weld wire, but in fabrication can be bent and shaped for most applications. Water jackets can be attached (brazed) along the length of the guide tube to circulate cooling water and extend its usable life.

Guide Tube (Consumable Wire Guide). Consumable wire guides are round, seamless steel tubes manufactured in three sizes—$^3/_8$, $^1/_2$, and $^5/_8$ in. OD with an inside diameter suitable for $^3/_{32}$- and $^1/_8$-in.-diam wire—and should be of material compatible with the base metal being welded. They can be either bare or flux coated. The flux coating insulates the guide tube from the weld joint and replenishes the slag in the pool during welding. For bare tubes, insulator rings can be attached. Consumable guides are best suited for relatively short joints: 8 to 12 ft for stationary guides, and up to 3 ft for oscillating guides.

Holders and Guide Tubes. Holder design for nonconsumable guides varies with the application. Some attach directly to the wire feeder, while others mount in holders remote from the wire feeder. The remotely mounted holders offer more flexibility in adjusting the guide tube in the weld plate seam.

Holder designs for consumable guides also vary with the application. They must be mounted above the weld plate seam in a vertical position. If oscillation is to be used, the holders are independent of the plates; if stationary, they are mounted directly to the top of the plate, in which case the clamping mechanism must be electrically insulated from the plate. Adjustment of the guide tube is very important. Adjustments for centering and angular movement (bidirectional) are a necessity in most cases. Power-line connections are made directly to the holder, and a flexible wire guide is used for the wire feeder connection.

Dams, shoes, backing bars, and molds are required to retain the molten metal and flux between the plates or parts being joined, which become the weld pool. With a dam on both sides of the weld plate seam, they guide and retain the weld pool upwards in the vertical position. All are water cooled. Shoes are made of heavy copper plate, generally 1 to $1^1/_4$ in. thick and measuring approximately 4 in. wide and 6 in. long. They are usually attached to the welding machine, which rises with the weld pool.

Water circulation can be achieved by cross drilling (through the thickness) the length and width of the perimeter of the shoe. Hose connections are made on the back side. Two-shoe applications require one shoe (inboard) to be mounted directly to the welder, while one shoe (outboard) is mounted to a through-the-seam adjustable arm, also attached to the welder and capable of clamping the outboard shoe to the weld plates. Both shoes rise with the welder.

One-shoe applications require a fixed backing bar on the outboard side and a means for applying pressure to the inboard movable shoe. Backing bars are often made of heavy copper plate, usually $1^1/_4$ in. thick and approximately 4 to 6 in. wide. The length varies with the application. Long backing bars are provided with copper tubing brazed to the outside surface for water circulation. When the backing bar is shorter than the joint being welded, two bars may be stacked, with the lower bar removed after the weld has solidified and replaced above the bar where the weld is being made. This technique is referred to as leap-frogging. Because the bars are stationary, they must be forced strongly against the plates by means of wedges held in place by strongbacks. For consumable wire guides, backing bars are used on both sides. Backing bars and shoes may also be machined to various shapes, such as those used to make vertical corner welds.

Molds are made of copper and usually designed to join relatively small, odd-shaped pieces such as car rails, sprocket teeth, bosses on heavy-walled pipe, and simple square or rectangular billet ends. Most molds are made in two pieces with a detachable sump for ease of cleaning. The two pieces are clamped together or may be hinged for automation. Hinge-type molds may be clamped with pneumatic or hydraulic cylinders that have manual or solenoid-operated control valves. All molds must be water cooled.

Wire-Feed Drive Systems. Generally, the wire-feed drive is a gear motor consisting of a gear reducer with an integral or flange-mounted motor. The permanent magnet/direct current motor is most popular, although the shunt wound/direct current motor has been used for years and is still supplied by some manufacturers. Direct current motors are used because of their small size-to-horsepower rating, excellent speed control, and constant torque characteristics. The electrode wire is driven by means of drive rolls mounted on the output of the shaft of the feeder. A similar spring-loaded roller is mounted above the stationary roller to squeeze the wire between them, which drives (pushes) the wire through the conduit or wire guide while pulling the wire from the pay-off reel or pay-off pack.

Oscillators. Oscillation of the wire or wires is often required for ESW, usually when the component thickness being joined exceeds about $2^1/_4$ in. per electrode. Various methods of oscillating the wire have been used, such as motor-driven lead screws or ball-nut screws, as well as rack-and-pinion mechanisms. Recently, pneumatic oscillators incorporating a pneumatic cylinder and controls have been designed. Regardless of the type of oscillator mechanism used, the following adjustments should be possible during operation:

- *Stroke* should be from about $^1/_2$ to 8 in.
- *Velocity of oscillation* should be from 0 to 5 in./s.
- *Dwell time* at each end of the stroke is often necessary, usually 0 to 5 s.

All of these functions may be accomplished using state-of-the-art electronic controls with visual analog or digital readout.

Mounting Systems. Most electroslag welds are performed in the vertical or almost-vertical position with the work stationary. The conventional or nonconsumable guide method requires that the equipment rise as the weld progresses. This is done in a number of ways. Relatively small machines are provided with geared motors fitted to knurled drive-wheels that operate by friction with the workpiece, or cogs that engage tracks or chains fastened to the workpiece. The vertical drive speed is usually adjusted by the operator, who watches the speed of welding. Such arrangements are usually limited to single-wire machines.

For large machines capable of providing three-wire operation, a separate column is used with a drive mechanism capable of not only lifting the welding machine, but also the operator and the controls. Large welding manipulators are also used for mounting the welding head of the large machines. Because the rate of rise may vary during the welding operation due to changes in the welding conditions (particularly amperage) or variations in the width of the gap, the rate of rise must be monitored. In single-wire applications, this is generally done visually by the operator. In large machines, automatic control of the rise is sometimes provided by sensors that view the height of the weld slag surface relative to the top of the shoe.

Controls. Because the power sources are at a distance from the machine and the operators, adjustments of current, voltage, and travel speed are located near the operator. Control panels also include meters for monitoring conditions. Safety devices include water flow alarms. Fast rise and lower switches enable convenient machine setup. Digital read-out meters and chart recorders are sometimes used for more advanced equipment.

Electrode Wires

Electrode wires for ESW are available in two types: solid wire and metal cored wire. Solid wire electrodes similar to those used for gas metal arc welding (GMAW) or SAW are most frequently used. Solid wires used for ESW of low-carbon steel and high-strength low-alloy (HSLA) steel conform to American Welding Society (AWS) A5.25 (Table 1). For welding medium-carbon and alloy steels, an electrode wire with an alloy content that produces a weld matching the base-metal composition is most often used. Metal cored electroslag wire classifications and compositions conforming to AWS A5.25 are given in Table 2.

Electrode wire diameter is most often either $^3/_{32}$ or $^1/_8$ in. The $^3/_{32}$-in. wire is used with the consumable guide system because suitable ($^1/_8$-in. ID) guide tube is readily available, and the smaller wire is more easily fed. The $^1/_8$-in. wire is usually used with the nonconsumable guide system; however, consumable guides can be obtained that permit feeding of $^1/_8$-in.-diam wire. In all cases, sufficient wire

Table 1 Chemical compositions for solid wire electrodes

AWS classification	Composition, %						
	C	Mn	P	S	Si	Cu	Other
Medium manganese classes							
EM5K-EW	0.06	0.90-1.40	0.03	0.035	0.40-0.70	0.30	0.50
EM12-EW	0.07-0.15	0.85-1.25	0.03	0.035	0.05	0.30	0.50
EM12K-EW	0.07-0.15	0.85-1.25	0.03	0.035	0.15-0.35	0.30	0.50
EM13K-EW	0.07-0.19	0.90-1.40	0.03	0.035	0.45-0.70	0.30	0.50
EM15K-EW	0.12-0.20	0.85-1.15	0.03	0.035	0.15-0.35	0.30	0.50
High manganese classes							
EH10Mo-EW(a)	0.07-0.12	1.60-2.10	0.03	0.035	0.50-0.80	0.15	0.50
EH10K-EW	0.06-0.14	1.40-2.0	0.025	0.030	0.15-0.30	...	0.50
EH11K-EW	0.07-0.15	1.40-1.85	0.025	0.035	0.80-1.15	...	...
EH14-EW	0.10-0.18	1.75-2.25	0.03	0.035	0.05	0.30	0.50
Special classes(b)							
EWS-EW(c)	0.07-0.12	0.35-0.65	0.03	0.040	0.22-0.37	0.25-0.55	0.50

Note: Single values indicate maximum percentage.
(a) EH10Mo-EW also contains 0.40 to 0.60% Mo. (b) An additional special class electrode, ES-G-EW, has no chemical requirements. (c) EWS-EW also contains 0.40 to 0.75% Ni and 0.50 to 0.80% Cr.

Table 2 Chemical compositions for deposited weld metal from metal cored electroslag electrode wire(a)

AWS classification	Composition, % C	Mn	P	S	Si	Ni	Cr	Mo	Cu
EWT1	0.13	2.00	.03	.03	0.60	...	...	...	...
EWT2	0.12	0.50-1.60	.03	.04	0.25-0.80	0.40-0.80	0.40-0.70	...	0.25-0.75
EWT3(b)	0.12	1.00-2.00	.02	.03	0.15-0.50	1.50-2.50	0.20	0.40-0.65	...
EWT4	0.12	0.50-1.30	.03	.03	0.30-0.80	0.40-0.80	0.45-0.70	...	0.30-0.75
EWT5(c)	...	...	...	...	...	...	...	...	...

(a) Single values indicate maximum percentage. (b) EWT3 also contains 0.05% V. (c) EWT5 has no chemical requirements.

must be provided on the reel or in the drum so that the weld is completed without interruption.

Fluxes

The molten flux is the slag that gives the electroslag process its name. Fluxes are marketed as proprietary materials; there are no standard specifications other than the classification in AWS A5.25 (Table 3), where fluxes are classified on the basis of the mechanical properties of the weld deposit made with a particular electrode.

Physical Properties. A flux for ESW must possess several specific characteristics regarding such physical properties as electrical resistivity, viscosity, melting range, and volatility. Electrical resistivity is an important characteristic of electroslag fluxes; a relatively high resistivity is required for the flux to generate the ohmic heating that causes fusion. There is an optimal resistivity, however. At a given voltage, a flux of excessively high resistivity draws too little current, allowing the slag (flux) to cool and the wire to drive too deeply into the molten slag. A flux of excessively low resistivity draws too much current, raising the temperature of the pool (which further reduces the resistance). This allows the wire to run with too short a length immersed in the flux. In extreme circumstances, this condition can lead to arcing between the electrode wire and the top of the slag (flux) pool.

Flux viscosity must also be adjusted to an optimal value. An excessively viscous flux prevents rapid settling of the small molten metal droplets down through the slag (flux) pool, does not stir sufficiently to maintain good heat distribution, and promotes slag entrapment. An extremely fluid flux is difficult to contain and tends to leak through small crevices between the dam and the work. The electroslag flux must have a melting range lower than the weld-metal melting temperature. Furthermore, no component of the flux should be volatile at the welding temperature. Preferential loss of any one flux component during welding leads to variations in weld-metal composition along the weld length. Slag removal is seldom a problem in ESW because of the small volume of slag that solidifies on the weld surface.

Chemical Properties. Fluxes for ESW are usually combinations of oxides of silicon, manganese, aluminum, calcium, magnesium, and titanium, with some calcium fluoride always present. Fluxes that are high in calcium, magnesium, and manganese oxides (basic fluxes) produce welds of low residual oxygen and are used for welding high-quality steels. However, these fluxes easily absorb moisture from the atmosphere and require drying prior to use to avoid hydrogen pickup. Fluxes that are high in silica (acid fluxes) are generally less expensive and are not susceptible to moisture pickup, but produce welds higher in oxygen. Compositions of typical proprietary fluxes are:

Proprietary flux	Acid	Neutral	Basic
Calcium fluoride (CaF_2)	9	15	48
Calcium oxide (CaO)	12	20	18
Aluminum oxide (Al_2O_3)	8	15	21
Magnesium oxide (MgO)	2	5	3
Silicon dioxide (SiO_2)	33	35	4
Manganese oxide (MnO)	22	7	1
Iron oxide (FeO)	2	<1	<1
Titanium dioxide (TiO_2)	8	<1	4
Other	4	2	<1

Two components that adjust electrical resistivity of the slag are aluminum oxide and silica; aluminum oxide increases resistivity, while silica decreases resistivity. Titanium oxide also reduces resistivity and is used particularly in starting fluxes. Calcium fluoride reduces the resistivity, viscosity, and melting range of electroslag fluxes and is the most commonly used flux addition for adjusting both resistivity and viscosity.

Starting Flux. Specially compounded fluxes that have lower-than-normal resistivity and that are used with higher-than-normal voltage and current can be used for start-up. These fluxes are specifically formulated to rapidly produce a molten slag pool and weld-metal pool. After start-up, the regular welding flux is added. Many fabricators find the use of special start-up fluxes unnecessary.

Consumable guides (nozzles) are seamless low-carbon steel tubes measuring $^{3}/_{8}$, $^{1}/_{2}$, or $^{5}/_{8}$ in. OD. The inside diameter accommodates $^{3}/_{32}$-in.-diam wire in the two smaller tubes and $^{1}/_{8}$-in.-diam wire in the larger tube. Insulating the tubes to prevent a short circuit to the side wall of the joint is accomplished in one of two ways. Usually, insulating spacer washers made of fiberglass or a tough, high-silica glass are placed on the guide during the setup. Flux-covered tubes are the other method of ensuring insulation. The flux is generally of the same composition as the granular flux used to form the slag. Because the guide tube contributes relatively little to the weld metal, compared to that of the filler wire, low-carbon steel tubes may be used for welding low-alloy steels by using wires of slightly higher alloy content.

For unusual applications, guide tubes are provided in a variety of cross sections (Fig. 4). Clusters of four rods may be taped together, the cavity providing the hole through which the filler wire is fed. Tubes

Table 3 As-welded mechanical property requirements for flux-electrode classifications

AWS flux classification(a)	Tensile strength, ksi	Yield strength at 0.2% offset (min), ksi	Elongation in 2 in. (min), %	Charpy V-notch impact strength (min), ft·lb
Using ASTM A36 plate				
FES6Z-xxxx	60-80	36	24	Not required
FES60-xxxx	60-80	36	24	15(b)
FES62-xxxx	60-80	36	24	15(c)
Using ASTM A242, A441, A572 grade 50, or A588 plate				
FES7Z-xxxx	70-95	50	22	Not required
FES70-xxxx	70-95	50	22	15(b)
FES72-xxxx	70-95	50	22	15(c)

(a) The letters "xxxx" as used in this table stand for the electrode designation EM12K-EW, EM13K-EW, etc. (see Table 1). (b) At 0 °F. (c) At −20 °F

Fig. 4 A variety of consumable guides (nozzles) for ESW. (a) Single flux-covered tube. (b) Cluster of rods taped together. (c) Flux-covered wing nozzle. (d) Flux-covered wing or web nozzle with two tubes. (e) Flux-covered, tapered, and curved nozzle for joining a turbine blade runner

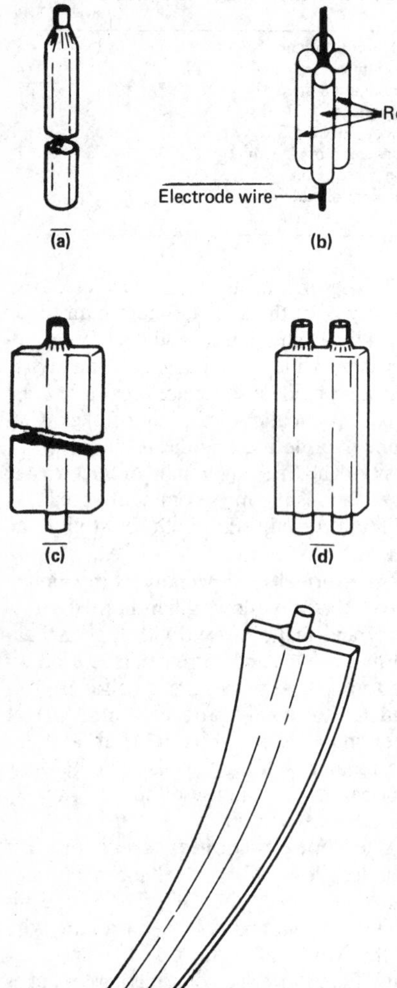

may be attached longitudinally to spacer plates of varying distances between the tubes and subsequently shaped to fit a curved joint of varying cross section, as in a turbine runner blade.

Workpiece Preparation

Joint edges are usually flame cut and ready to weld without further preparation. This simple procedure saves considerable time and money compared with preparation by beveling.

Starting Cavity. Components to be welded are assembled in a vertical position, over a starting trough (or sump) of full plate thickness and about 3 in. deep (Fig. 5). The starting trough can be made by tack welding together three pieces of scrap steel plate, or it can be a clamp-on starting sump made of copper, which can be reused. The faces of the starting trough must be flush with the outside surfaces of the workpiece for proper fit-up. The runoff tabs (Fig. 5) are also full thickness and about 3 in. deep, to bring the weld crater beyond the top of the seam.

Strongbacks. Spacing of the joint opening, usually $1^1/_8$ to $1^3/_8$ in., and alignment of the workpieces must be maintained. A common way of doing this is by use of U-shaped restraining plates (strongbacks) welded across the outer face of the joint, as shown in Fig. 4. For welding by the conventional system, the opening of the strongbacks must be large enough to permit the water-cooled dams to move through. A space of 5 by 3 in. is usually sufficient. With the consumable guide system, dimensions of the strongback are not critical. In this case, the strongbacks, with the help of wedges, hold the stationary dams in place (see Fig. 3).

The number of strongbacks along a joint depends mostly on the degree of difficulty encountered in maintaining joint alignment. Fit-up requirements also dictate whether strongbacks are needed on both sides of the joint. Strongbacks are usually located 12 to 18 in. apart, starting about 6 in. from the bottom of the joint and ending no closer than 6 in. from the top. In some setups, the workpieces are easily positioned, and fewer than the normal number of strongbacks are adequate. When welding with the consumable guide system, placement of the water-cooled dams

Fig. 5 Two 6-in.-thick plates ready for ESW. Shown without wedges and dams

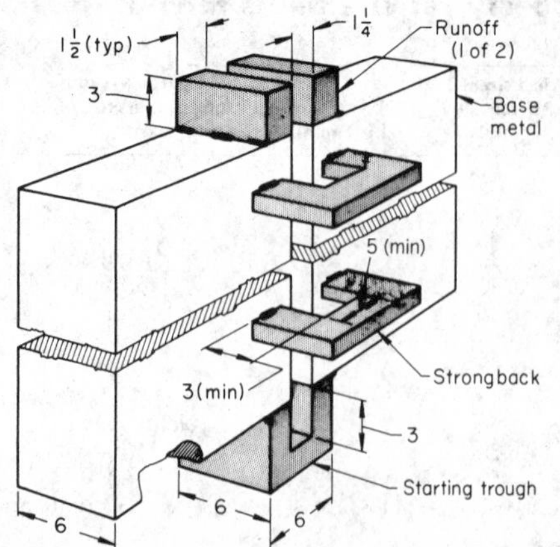

may influence the number and location of strongbacks, because the water-cooled dams depend on the strongbacks to hold them in position. With the conventional system, strongbacks should be used only on the side opposite from where the electrode wires are fed. Strongbacks are removed after the joint is welded, unless equipment clearance requirements during welding make it necessary to remove them sooner. Methods of removal are similar to those used for removing other temporary clips or attachments. Strongbacks often are only tack welded along one side and can be broken off with a few sledgehammer blows.

Operating Procedure

In setting up for ESW, the entire welding machine is positioned over or to one side of the weld joint. With the conventional system, the machine is lowered and raised to ensure that the guide tube (or tubes) enters and leaves the joint opening smoothly and remains centered in it. The roller guide shown in Fig. 2 keeps the machine aligned after the initial adjustments are made. The water-cooled retaining dams are checked to make sure that they slide freely. Welding conditions, including wire-feed rate, voltage, and current should be adjusted for the electrode wire diameter being used, preferably in accordance with a prequalified welding procedure specification. Setup for the consumable guide system is similar to the above procedure. When the machine is ready, the guide tube is fitted with insulating spacers (providing it is not of the flux-covered type) and is adjusted so that it extends nearly to the bottom of the joint. The retaining dams are then wedged in place.

Steel wool may be placed in the starting trough to initiate start-up. As the electrode is lowered and touches the steel wool, violent arcing occurs. Flux is added immediately; a large handful is usually sufficient. As a matter of preference, the initial flux is sometimes placed on top of the steel wool before the current is switched on. As soon as the flux melts and the violent arcing ceases, more flux is added. The arc is soon extinguished and welding begins. The objective is to have the welding action proceeding smoothly by the time the flux level reaches the top of the sump and the bottom of the joint. To obtain a smoother start, especially if a special starting flux is used, the voltage can be increased and the amperage decreased during start-up. After the arc has been extinguished, welding proceeds continuously until the joint is completed. During welding, some adjustments in current or wire feed may be needed.

Flux is added manually as required to maintain a molten slag depth of about $1^1/_2$ to 2 in. On the average, about 1 lb of flux is used per 20 lb of weld metal deposited. The amount of flux consumed, normally very little, depends to large extent on the amount lost as leakage between the retaining dams and the workpiece. Although the depth of the pool is not critical, it should be kept reasonably uniform and generally should not exceed 2 in. If the pool starts to bubble vigorously, it is too shallow and flux should be added; if the surface of the pool is entirely quiet, it is too deep. When welding with the consumable guide system, flux-covered guide tubes are sometimes preferred, because they automatically help to replenish the flux pool as welding proceeds. Although the operation of both conventional and consumable guide systems is largely automatic, additions of flux, minor adjustments of the machine, and relocating of dams demand fairly close surveillance in the consumable guide system.

Process Conditions

Process conditions have a marked effect on the shape of the molten weld pool. High currents and low voltages produce a deep pool (Fig. 6b). Conversely, low currents and high voltages have the opposite effect (Fig. 6a). The ratio of the width-to-depth of the pool, known as the form factor, often governs the tendency of the weld to form centerline cracks when this ratio is small. The size of the spacing, or root opening, and the depth of the slag pool also affect the form factor. The influence of welding conditions on the form factor is indicated in Table 4. As in other welding processes using electrode filler wires, the deposition rate, and therefore the speed of welding, is determined by the current, or wire-feed speed. Figure 7 shows the deposition rate for two sizes of wire used in conventional welding and in consumable guide welding. Typical welding conditions for single-wire conventional ESW are given in Table 5 for plate thicknesses up to 3 in.

Actual operating conditions for electroslag welding a particular component are best determined by beginning with those conditions successfully used in a previous weld of similar composition and design and making adjustments according to the concepts outlined in the preceding paragraphs. Typical process conditions for electroslag welding various thicknesses of steel plate by the consumable guide method are given in Table 6.

Oscillation. Thickness of the weldment is the major factor in determining whether the guide tubes (and electrodes) should be oscillated during welding. When the weldment thickness is less than about 2 in., oscillation is not usually required, although it is sometimes used for plate as thin as $1^1/_2$ in. Better uniformity of heat penetration is obtained when the electrode is oscillated, even in welding thin plates. Because the water-cooled copper dams are efficient heat sinks when the electrode re-

Table 4 Influence of welding conditions on weld-metal pool shape

Weld pool variable	Current or wire-feed speed: Low values	Current or wire-feed speed: High values	Voltage	Slag pool depth	Root opening
Increasing the weld pool variable causes the weld pool form factor to:					
Pool width (w)	Increase	Decrease	Increase	Decrease	Increase
Pool depth (h)	Increase	Increase	Slightly increase	Slightly increase	No change
Form factor (F = w/h)	Slightly increase	Decrease	Increase	Decrease	Increase

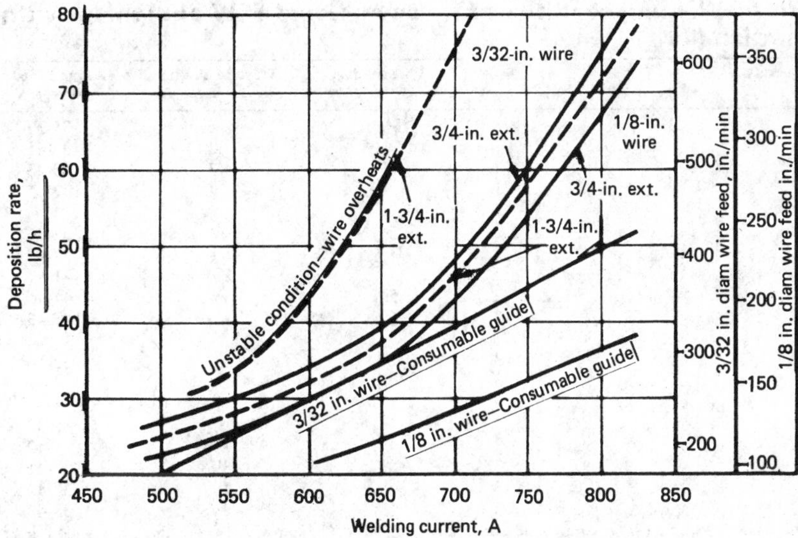

Fig. 7 Deposition rates for consumable and nonconsumable guide tube welding (Ref 1)

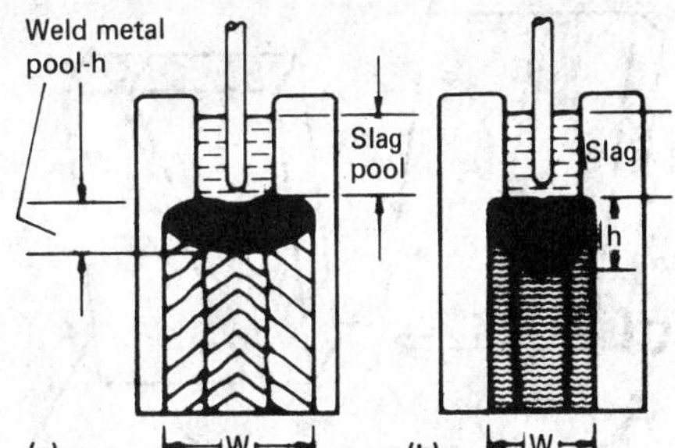

Fig. 6 Influence of welding conditions on weld pool shape. (a) Desirable: shallow weld pool, high form factor, and acute angle between grains. (b) Undesirable: deep weld pool, low form factor, and obtuse angle between grains

Table 5 Welding conditions for conventional single-wire equipment using $^3/_{32}$-in. wire

Plate thickness, in.	Joint: Type	Joint: Dimensions, in.	Current, A	Voltage, V
$^3/_4$	Square butt	$^7/_8$ opening	500	32
1	Square butt	$^7/_8$ opening	525	33
1	Single-V	$^3/_{16}$ root opening, 1 face, no land	450	32
$1^1/_2$	Square butt	$^7/_8$ opening	525	34
$2^1/_2$	Square butt	1 opening	575	40
3	Square butt	$1^1/_4$ opening	575/600	44

Table 6 Typical conditions for consumable guide ESW

Plate thickness, in.	No. of electrodes	Root opening, in.	Total current, A	Voltage, V	Oscillation distance, in.	Oscillation speed, in./min
5/8	1	1 1/4	300	34-36	0	0
1	1	1 1/4	350	37-42	0	0
2	1	1 1/4	450	38-47	0	0
2 1/2	1	1 1/4	575	39-49	1.6	24
3	1	1 1/4	675	42-52	2.4	24
5	1	1 1/4	750	50-55	4.3	47
5	2	1 1/4	1500	40-42	1.2	9
6	2	1 1/4	1500	41-43	2.0	24
8	2	1 1/4	1500	44-46	4.0	47
10	2	1 1/4	1500	47-49	6.0	65
12	2	1 1/4	1500	50-52	8.0	85

Table 7 Welding conditions for conventional ESW equipment with oscillation (Ref 1)

Plate thickness, in.	No. of wires	Wire spacing, in.	Traverse direction, in.	Approximate voltage, V
2	1	...	1	35-37
3	1	...	2	38-41
4	1	...	3	40-43
5	1	...	4	42-45
4	2	2 3/8	5/8	32-36
6	2	3 3/8	1 5/8	37-40
8	2	4 3/8	2 5/8	41-43
10	2	5 3/8	3 5/8	44-46
12	2	6 3/8	4 5/8	47-49
6	3	2 1/4	1/2	32-36
8	3	2 15/16	1 3/16	33-37
10	3	3 9/16	1 13/16	34-38
12	3	4 1/4	2 1/2	35-39
14	3	4 15/16	3 3/16	36-40
16	3	5 9/16	3 13/16	37-41
18	3	6 1/4	4 1/2	37-42
20	3	6 15/16	5 3/16	38-43

Note: All values of voltage for 1/8-in.-diam wire, operating at 600 to 700 A (DCEP) per wire. Traverse time can be set at 1 to 2 s for the traverse distance given. Dwell time at each end is 2 to 3 s. For most one- and two-wire applications, 2 s is adequate. Wire extension (wire guide-to-slag pool distance) can be set at about 1 1/2 in. Wire is kept at about 1/2 in. from each shoe or back-up bar at the extremities of oscillation. For three wires, the center wire is usually set a few volts lower than the outside ones.

mains stationary in the center of the joint opening and midway between the two sides, heat penetration is considerably less at the edges. Oscillation greatly lessens this nonuniform distribution of heat; a dwell time of 1 s or more at each end of the stroke compensates for the heat loss because of the water-cooled dams. Figure 8 illustrates the effect of oscillation on depth of fusion in welding plates. A base-metal thickness of 3 in. generally is considered excessive for welding with one electrode wire without oscillation. As shown at the lower right in Fig. 8, there is little depth of fusion at the four corners adjacent to the dams if oscillation is not used.

Electrode Wires. With consumable guides, one wire with no oscillation is commonly used for plate thicknesses of 5/8 to 2 in., one wire with oscillation for thicknesses of 2 to 5 in., and two wires with oscillation for thicknesses of 5 to 12 in. (see Table 6). Using three electrode wires with consumable guides and oscillation, plate thicknesses above 12 in. can be welded. Conventional electroslag equipment with the capability of up to three wires and oscillation allow for welding a wide range of plate thicknesses. Operating conditions for such equipment are shown in Table 7.

Control of Vertical Travel. Best results are obtained by maintaining close control of vertical travel when the conventional electroslag system is used, to keep the length of wire in the slag pool constant. Even though the equipment may provide for automatic control of vertical travel, the rate of travel requires periodic adjustment during welding, particularly when long welds are being made. With the consumable guide system, there is no vertical travel of the equipment, but electrode feed rate (which controls current) must be closely controlled.

Verticality Tolerances. Electroslag welding in a vertically inclined direction can be done with careful setup and control of process conditions. Two different designs requiring a nonvertical weld are: (1) the weld axis is horizontally inclined but faces of the workpiece are in the vertical plane (Fig. 9a), and (2) the planes of the work faces are inclined but the workpieces are in the vertical plane (Fig. 9b).

Paton (Ref 2) states that inclinations of the first kind (Fig. 9a) can be as much as 45° for plate thicknesses up to 4 in., but because of difficulties in fusing the lower edges, recommends that inclinations be kept to less than 30°. Jones *et al.* (Ref 3) have shown that successful welds of the second kind (Fig. 9b) can occur at inclinations of up to 45° from the vertical when the weld conditions are carefully controlled. In these nonvertical welds, both

Fig. 8 Setup for consumable guide ESW. Note different depths of fusion obtained with and without guide tube oscillation.

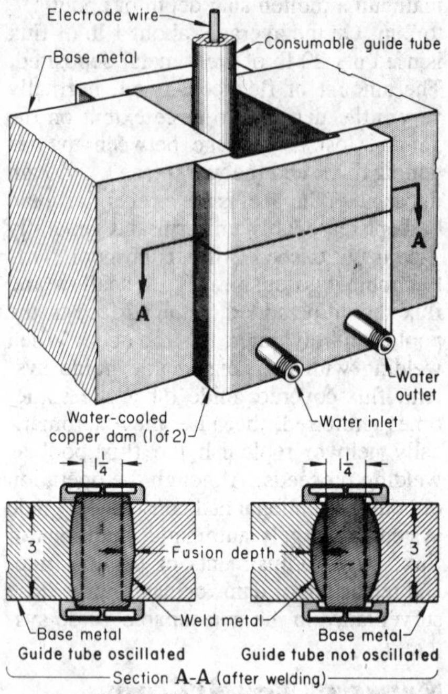

Fig. 9 Two possible orientations in which nonvertical electroslag welds may be made (Ref 2)

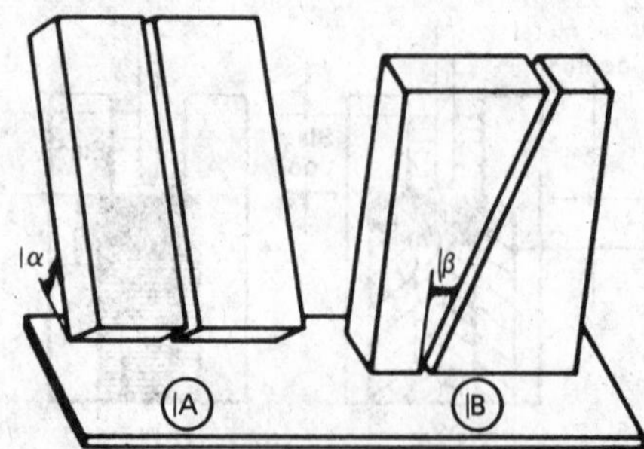

Fig. 10 Circumferential ESW

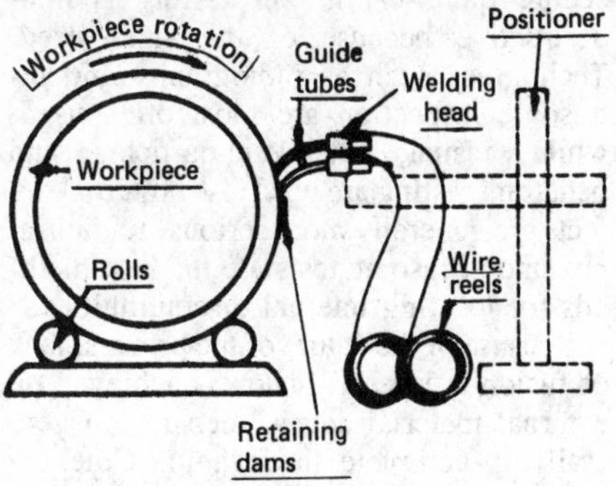

the fusion and heat-affected zones (HAZ) are not symmetrical with respect to the axis of the weld. The observed anisotropy is such that both penetration and HAZ width are greater on the top side of the weld, a fact that has been attributed to radiative preheating of the parent plate on the top side. The more inclined the weld, the more asymmetric is the resulting weld microstructure.

Circumferential Welding

Circumferential (girth) welds for joining sections of thick-walled pipe or pressure vessels can be made by ESW. This method consists of using a stationary welding head and rotating the parts, which are joined in the vertical position by a square-groove butt joint. The welding is done from the 3 o'clock position as shown in Fig. 10. The circular weldment is placed on a set of rolls that are rotated clockwise during the welding procedure. After all but about 18 in. of the circumference has been welded, the electroslag equipment has to be removed and the weld completed by another process.

To start and finish the weld completely by the electroslag process, the simplest procedure is to provide a fixed steel backing bar on the inside of the cylinder (not shown in Fig. 10). The weld is started on a steel bar in the weld cavity that is removed by arc gouging after the weld has proceeded approximately 180° of its circumference. During the gouging, enough of the starting weld metal is removed to provide a surface tangential to the inside surface of the cylinder. During the start and the closure, rotation of the workpiece is stopped, and the elevating mechanism of the equipment allows the weld to proceed upward. Dams are inserted during the closure to contain the slag and weld metal as it rises outside the exterior surface of the cylinder. The inside backing bar and the finish weld metal are then removed.

Metallurgical Considerations

While the temperature of the slag that does the melting of both the electrode and the base metal is less than the arc temperature of arc welding processes, the total heat input per mass is usually much more. This results in the following metallurgical advantages and limitations for ESW, as compared to the arc welding processes:

- In the as-deposited condition, a favorable residual stress pattern is developed with the weld surface and HAZ in compression and the center of the weld in tension.
- Because of the symmetry of most vertical butt welds (square-groove welded in a single pass), there is no angular distortion in the horizontal plane. There is slight distortion in the vertical plane, for which an allowance can be made.
- The weld metal stays molten long enough to permit some slag-refining action, and the progressive solidification allows gases to escape and nonmetallic inclusions to float to the slag above the weld pool. High-quality, sound weld deposits are generally produced.
- The weld deposit contains up to 50% admixed base metal depending on welding procedures; therefore, the steel being welded significantly affects the chemical composition of the deposit and the resultant mechanical properties.
- The protracted thermal cycle results in a weld-metal structure that consists of large prior-austenite grains that generally follow the columnar solidification structure. The grains are oriented horizontally at the weld edges and turn to a vertical orientation at the weld center. The microstructure for electroslag welds in carbon structural steels generally consists of acicular ferrite and pearlite grains with proeutectoid ferrite outlining the prior-austenite grains. Coarse prior-austenite grains can be seen at the periphery of the weld and a much finer grained region can be observed near the center of the weld. This fine-grained region appears equiaxed in the horizontal cross section; however, vertical sections reveal its columnar nature. Changes in weld-metal composition, and to a lesser extent welding procedure, can markedly change the relative proportions of the coarse- and fine-grained regions. Better impact properties are gen-

Table 8 Electroslag welding defects, causes, and remedies

Location	Defect	Causes	Remedies
Weld	Porosity	Insufficient slag depth	Increase flux additions
		Moisture, oil, or rust	Dry plate surfaces and shoes
		Contaminated or wet flux	Dry or replace flux
	Cracking	Excessive weld rate	Slow wire feed rate
		Poor form factor	
		Excessive center-to-center distance between electrodes or guide tubes	Decrease spacing between electrodes or guide tubes
		Insufficient number of electrodes or guide tubes	Increase number of electrodes or guide tubes
		Improper filler materials	Re-evaluate welding procedure
	Nonmetallic inclusions	Copper contamination	Improve shoe cooling and/or design
		Rough plate surface	Grind plate surfaces
		Unfused nonmetallics from plate laminations	
		Foreign material in weld	Exercise care
Fusion line	Lack of fusion	Low voltage	Increase voltage
		Excessive weld rate	Decrease wire feed rate
		Excessive slag depth	Decrease slag additions
		Misaligned electrodes or guide tubes	Realign electrodes or guide tubes
		Inadequate dwell time	Increase dwell time
		Excessive oscillation speed	Slow oscillation speed
		Excessive electrode to shoe distance	Increase oscillation width
		Excessive center-to-center distance between electrodes	Decrease spacing between electrodes
	Undercutting	Excessively slow weld rate	Increase wire feed speed
		Excessive voltage	Decrease voltage
		Excessive dwell time	Decrease dwell time
		Inadequate cooling of shoes	Increase cooling flow to shoes or use larger shoes
		Poor shoe design	Redesign groove in shoe
		Poor shoe fit-up	Improve fit-up or seal gap with refractory cement or asbestos dam

erally observed in a coarse-grained region than in a fine-grained region.

- Grain boundary separation or microfissures have occasionally been observed in the proeutectoid ferrite. They do not propagate to the weld surface or the HAZ, and while at low levels the grain boundary separations still meet or exceed base-metal properties for many uses. These separations should be eliminated now that there is evidence (Ref 4) to show that hydrogen dissolved in the liquid weld metal is the principal cause. Control comes from eliminating moisture and hydrocarbons as much as possible from the electrode, consumable guide (if used), flux, and the joint. More highly restrained joints and more hardenable alloys increase the risk. For critical applications subject to dynamic load, an immediate postweld heat of 570 °F for 3 h can be used to eliminate the grain boundary separations.
- The high heat input and slow cooling rate result in a large HAZ, with much of it coarse grained. This can adversely affect notch toughness. Also, the cooling is slow enough that only high-temperature transformation products of low hardness form. For most steels this is an advantage, particularly if stress-corrosion cracking is a consideration.

Preheating and Postheating. Preheating, other than that necessary to ensure that the materials being welded are dry, is not required or generally used in ESW. By its nature, the process is self-preheating in that the heat generated is conducted into the workpiece and thus preheats the base plate ahead of the weld. Also, unless grain boundary separations are known to be a problem (see the discussion in the preceding list), postheating is usually unnecessary because of the very slow cooling rate after welding.

Most applications of ESW, particularly welding structural steel, do not require postweld heat treatment. As discussed previously, as-deposited electroslag welds have a favorable residual stress pattern that may be lost after postweld heat treatment. Grain-refining postweld heat treatments (austenitizing) are beneficial with respect to notch toughness of the HAZ and are usually beneficial to the weld metal. Such postweld heat treatment is required by the American Society of Mechanical Engineers (ASME) Boiler and Pressure Vessel Code for construction of pressure vessels of ferritic steels over $1^1/_2$ in. thick. The austenitizing treatment is followed by air cooling (normalizing), by normalizing and tempering, or by quenching and tempering. Annealing (furnace cooling after austenitizing) is generally not used because of the inherent loss of strength. Quenched and tempered steels must again be quenched and tempered after ESW to restore HAZ properties. The AWS Structural Welding Code does not permit ESW of quenched and tempered structural steels, because most work covered is too large or irregular to permit quench and temper.

Weld Soundness. Welds made by ESW under normal conditions are generally high-quality welds free from defects. In any welding process, however, abnormal conditions that may exist or occur during welding can result in defective welds. Table 8 offers guidelines for correcting defects.

Inspection

Ultimately, the code or standard used for welding, the governing agency, or the customer specifies the type of inspection required. Generally, electroslag welds are inspected with the same nondestructive examination methods as other heavy-section welds. With the exception of procedure qualification, all testing is nondestructive because of the sizes used. Techniques such as radiography and ultrasonic inspection are most often used, while visual, magnetic-particle, and penetrant testing are used also. Internal defects are generally more serious. Radiography and ultrasonic tests are the best methods for locating internal discontinuities.

Because of the nature of the process, lack of fusion is rare. If fusion is achieved on external material edges, then fusion generally is complete throughout. Cracking may occur either in the weld or the HAZ. Porosity may either take the form of a rounded or piped shape, the latter often called wormhole porosity. Ultrasonic inspection is probably the quickest single method for inspecting any large weldment. If defects should occur, they appear as porosity or centerline cracking. Ultrasonic inspection is effective for locating either type of defect; however, only well-qualified personnel should set up the

Fig. 11 Cylindrical pressure vessel fabricated by hot press forming two halves and welding them by conventional ESW

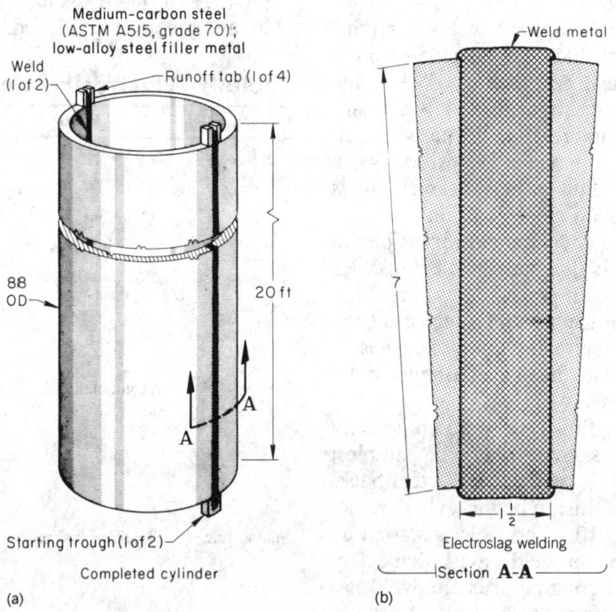

Joint type	Butt	Voltage	45-50 V
Weld type	Square groove	Deposition rate	70 lb/h
Joint preparation	Gas-cut edges	Electrode stickout	2-3 in.
Fixtures	Strongbacks; wedges	Electrode separation	2-3 in.
Power supply	Two 1000-A transformers(a)	Oscillation speed	1 in./s
Electrode wire	Two; $^1/_8$-in.-diam low-carbon steel (Mn-Mo-Si)	Dwell time	5-7 s
		Preheat	None
Flux	Proprietary(b)	Postheat	(c)
Welding position	Flat (work axis vertical)	Welding speed, in./min	...
Number of passes	1	Welding time (per seam)	8 h
Current	500-600 A ac, per wire		

(a) Variable output; drooping to constant voltage characteristics. (b) Composed of silicon oxide, manganese oxide, calcium oxide, and calcium fluoride; quantity used was $^1/_{20}$ of the weight of metal deposited. (c) Normalized at 1650 °F for 1 h per inch of cross section, for grain refinement; stress relieved at 1150 °F for 1 h per inch of cross section

Fig. 12 T-section of a rock-crusher jaw

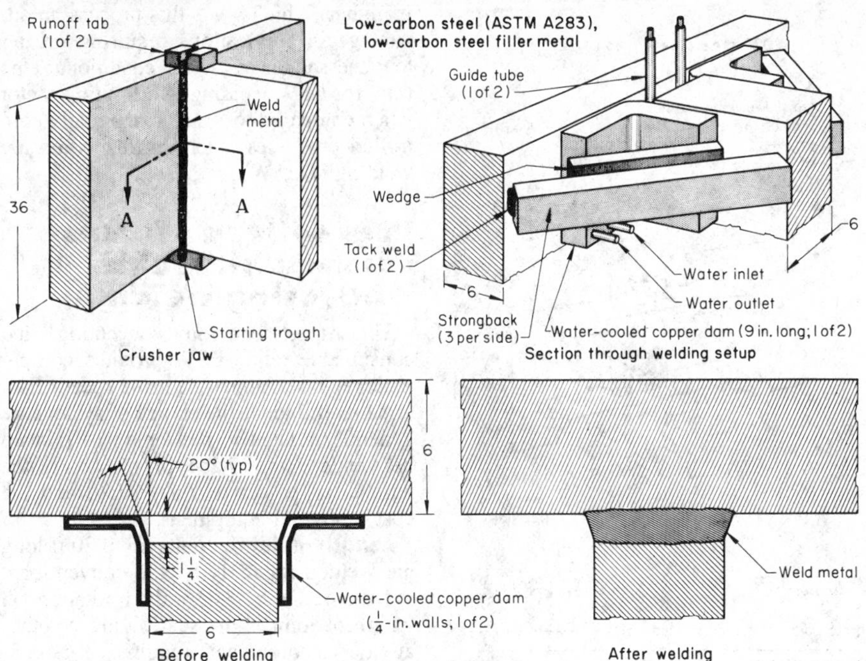

Joint type	T, with square-groove weld	Guide tubes	Two; 5/8 in. OD, 1/8 in. ID(c)
Joint preparation	Gas-cut edges	Number of passes	1
Fixtures	Strongbacks; wedges	Current (DCEP)	750 A per wire; 1500 A total
Power supply	Two 750-A transformer-rectifiers(a)	Voltage	41-43 V
		Preheat	150 °F
Electrode wires	Two; 3/32-in.-diam low-carbon steel(b)	Stress relief	1150 ± 25 °F, 6 h
		Welding speed	0.72 in./min

(a) Constant-voltage type; 100% duty cycle. (b) Wires supplied from 750-lb reels. (c) Low-carbon steel

equipment and interpret the test results.

Electroslag welding results in large dendritic grain sizes because of the slow cooling rate. Inexperienced personnel often use high sensitivity and actually pick up the large coarse grains; when such welds are sectioned, usually no defects are present. Inspectors must learn to use low sensitivity to obtain good results when inspecting electroslag welds. Magnetic-particle inspection is not a particularly good inspection method, because the areas examined by this technique are primarily surface or near surface. This is only a small percentage of the total weld; the only useful information is either checking the ends for craters, cracks, or centerline cracking, or possibly lack of edge fusion on the weld faces. Usually, a visual examination gives the same result unless the defect is subsurface. Visual examinations are only effective for surface defects, which are not common in this process.

Production Examples

The following examples describe ESW by the conventional system and by the consumable guide system. Because ESW is capable of extremely high deposition rates, it is generally less costly than other welding processes for joints in massive parts in which hundreds of pounds of filler metal must be deposited. In some of the applications that follow, weldments were comprised of several castings or of various combinations of castings, wrought shapes, and forgings that were impractical to fabricate by any other process.

Example 1. Conventional Electroslag Welding for Pressure Vessel Wall Fabrication. The 88-in.-OD pressure vessel shell course shown in Fig. 11 was fabricated by welding two halves of cylinders hot press formed from 20-ft-long, 7-in.-thick American Society for Testing and Materials (ASTM) A515, grade 70 plates. Welding of the longitudinal seams was accomplished by conventional ESW with considerable saving in welding time and improvement in the quality of weld over the conventional submerged arc process.

For conventional ESW, joints were prepared by gas cutting the plate edges to form a square groove spaced with a 1 1/2-in. gap (Fig. 11b), which was maintained by strongbacks tack welded to the walls of the vessel on both sides of the joints. Spacer blocks were also tack welded in the seam as a precaution to keep the joint from closing up in the event of a welding interruption. Starting troughs and runoff tabs were attached to the joints, and the vessel was mounted vertically on a flat table and leveled. Welding was done in a single pass with two 1/8-in.-diam electrode wires that were oscillated across the joint between two sliding water-cooled copper dams. The copper dams, welding head, and controls were mounted on an elevator platform that moved upward along the joint as welding progressed. The strongbacks were progressively removed during welding (along with the spacer blocks) to allow the dams to move vertically with the level of the rising weld and slag pool. Welding time per joint was around 8 h at a deposition rate of 70 lb/h. This compares favorably with the tandem-wire SAW time of around 44 h per seam.

The electroslag process produced welds that under radiographic and ultrasonic inspection were nearly free of voids and slag inclusions and required no grinding or repair. For submerged arc welded vessels, by comparison, the rework rate was generally 4 to 6%, whereas for the electroslag welded vessels it never exceeded 2%. This ESW procedure has also been applied successfully to similar pressure vessels made from HSLA steel (ASTM A387, grades 11 and 22).

Example 2. Replacement of a Casting by an Electroslag Weldment. Jaws for rock crushers were originally made as ductile iron castings; to reduce cost and lead time, the castings were replaced by weldments comprised of two 6-in.-thick plates of ASTM A283 low-carbon steel. Electroslag welding, using the consumable guide system and the conditions listed in the table with Fig. 12, was employed. The welded T-section of the crusher jaw is shown at upper left in Fig. 12. At first, because of the joint design, solid copper dams were used (no water cooling). However, these dams overheated and were replaced with shaped, water-cooled dams.

Two consumable guides, inserted at the top of the joint (Fig. 12, upper right) and extending down its entire length, carried the electrode wires into the starting trough. The water-cooled dams were leap-frogged as required during welding. A cross section of the welded joint is shown at the lower right in Fig. 12. After welding, the strongbacks, starting trough, and runoff tabs were removed, and the weldment was stress relieved at 1125 to 1175 °F for 6 h.

Fig. 13 Setup for consumable guide ESW for fabrication of a press frame

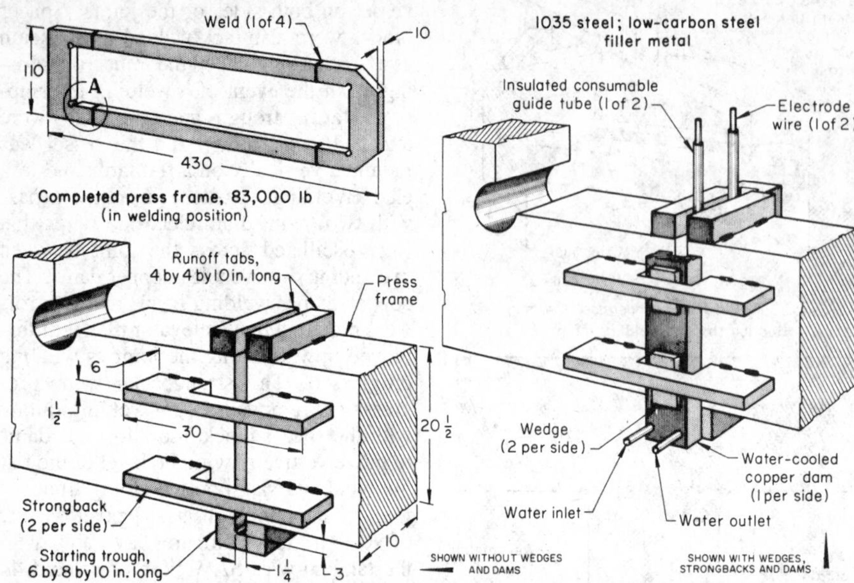

Joint type	Butt	Current (DCEP)	700 A per wire; 1400 A total
Weld type	Square groove	Voltage	52 V
Joint preparation	Gas-cut edges	Oscillation:	
Fixtures	Strongbacks; wedges	Speed	122 in./min
Power supply	Two 750-A transformer-rectifiers	Travel	6 in.
		Dwell time	2 s(c)
Electrode wires	Two; $^3/_{32}$ in. diam(a)	Frequency	20 cycles/min
Guide tubes	Two; $^5/_8$ in. diam(b)	Postheat	Stress relief
Flux depth	$1^1/_2$ in.		

(a) Composition: 0.11% C, 1.20% Mn, 0.45% Si, 0.020% S, 0.020% P. (b) Insulated with ceramic ferrules or glass fiber. (c) Within $^1/_2$ in. from each dam

After stress relief, the weldment was inspected ultrasonically. This joint is typical of those that are better suited to welding by the consumable guide system than by the conventional system.

Example 3. Consumable Guide Electroslag Welding of a Press Frame. Weighing 83 000 lb, the press frame (Fig. 13, upper left) was too large for production as a single piece by casting or cutting from a large plate. Consequently, it was fabricated by joining four sections of 10-in.-thick medium-carbon steel (1035) plate by consumable guide ESW. A typical setup is shown at the lower left in Fig. 13. The guide tubes were insulated with ceramic ferrules or glass fiber to prevent electrical contact with the sidewalls of the joint. During welding, the guide tubes were oscillated along the width of the joint to within about $^1/_2$ in. of the copper dams. The dwell time at each end was 2 s. After the first joint had been completed, the guide tubes were replaced, and the welding head was moved to the second joint, which was welded in the same manner. After completion of the first two welds, the workpiece was inverted, and the remaining two welds were made by the same procedure. The completed weldment was stress relieved. Quality was verified by ultrasonic inspection.

Electroslag Welding Versus Other Welding Processes

Deposition rates are higher for ESW than for any other welding process—at least 25 to 50% higher than for SAW, which is usually the closest to ESW in deposition rate. For circumferential or girth welds in large cylindrical vessels, however, multiwire SAW is generally more economical because of setup difficulties associated with ESW and because of code requirements requiring an austenitizing postweld heat treatment. In some applications, ESW is closely competitive with EGW, particularly when only one wire is to be used during the welding process. Equipment used and principles of operation are similar. The range of base-metal thickness is limited for EGW ($^3/_8$ to 4 in.). As a result, ESW is generally preferred for plate thicknesses requiring two or more electrode wires. In EGW, heat input is less and cooling rate is higher, which gives this process an advantage in applications requiring greater toughness in the as-welded condition. Setup time for ESW welding is a limiting factor when compared to simpler, manually controlled processes, such as flux cored arc welding (FCAW).

Conventional Versus Consumable Guide Electroslag Welding

Under most conditions, conventional and consumable guide ESW produce comparable results at about the same speed. Factors that influence the choice of system are generally related to the length of joint to be welded, the shape of the joint, the number of pieces to be welded, and the cost of capital equipment.

Length of Joint. Joints up to 20 ft long are welded regularly by the conventional system of ESW. The only limitation on length of joint is the scaffolding or other auxiliary equipment required for controlled vertical travel of the welding machine. Although joints over 20 ft long have been successfully welded by the consumable guide system, more care is required to ensure that the guide tube remains properly located throughout the length of the weld. If the thickness is no more than about $2^3/_8$ in., long joints can be welded successfully, using one electrode wire and one guide tube without oscillation. Plate thicknesses which require oscillation are more difficult when long guide tubes are employed, making the conventional system preferable.

Joint Design. For welding square-groove butt joints, the two ESW systems are equally suitable. For other types of joints, however, one system is often preferred. For example, if the joint to be welded is curved in the vertical plane, or if it contains some other type of irregularity that prevents a straight guide tube from extending the full length of the joint, ESW may be restricted to the conventional system, although consumable guides can be bent to conform to some shapes. Welding of a T-joint such as that described in Example 2 is difficult, if not impossible, by the conventional system.

Production Requirements. Both ESW systems are adaptable to repetitive and one-of-a-kind jobs, but the conventional system is better suited to repetitive work and the consumable guide system to a variety of short-run work.

Equipment Cost. Equipment required for the conventional system costs from four to five times as much as that needed to do the same job by the consumable guide system. To a great extent, this difference accounts for the more extensive use of the consumable guide system. Use of the conventional system is restricted mainly to applications in which the practical limits of the consumable guide system are exceeded. Equipment utilization is an important consideration in choosing between the two systems, because the cost of conventional system equipment cannot be justified for part-time use.

REFERENCES

1. *Electroslag Electrogas Tips and Techniques,* Electrotherm Corporation Bulletin, 1973
2. Paton, B.E, *Electroslag Welding,* American Welding Society, New York, 1959, 467 pages
3. Jones, J.E., Olson, D.L., and Martin, G.P., Metallurgical and thermal aspects of non-vertical electroslag welds, *Weld. J.,* Vol 59 (No. 9), 1980, p 245s-254s
4. Konkol, P.J. and Domis, W.F., Causes for grain boundary separations in ESW metals, *Weld. J.,* Vol 58 (No. 6), 1979, p 161-S to 167-S

SELECTED REFERENCE

- Frost, H.R., Edwards, G.R., and Rheinlander, M.D., A constitutive equation for the critical energy input during electroslag welding, *Weld. J.,* No. 1, 1981, p 1s-6s

Electrogas Welding

By the ASM Committee on Electrogas Welding*

ELECTROGAS WELDING (EGW) is an automatic method of gas metal arc vertical butt welding using a solid wire consumable electrode or a flux cored tubular wire consumable electrode. With solid wire electrodes, an external shielding gas is required; with flux cored tubular wire electrodes, the composition of the electrode core supplies all or part of the shielding. An essential feature of the process is the use of copper dams to confine the molten weld metal. The dams are usually water cooled, although air cooling is also used. Often referred to as "molding shoes," the dams shape the weld. Although the axis of the weld is vertical, the process is actually flat-position welding with vertical travel. The consumable electrode is fed down into a cavity formed by the opposing faces of the components to be welded and the two water-cooled dams (Fig. 1).

In its mechanical aspects, and its application to welding practice, EGW resembles conventional electroslag welding (ESW), from which it was developed (see the article "Electroslag Welding" in this Volume). Electrically, EGW differs from ESW primarily in that the heat is produced by an electric arc and not by electrical resistance of a slag.

Fig. 1 Electrogas welding unit

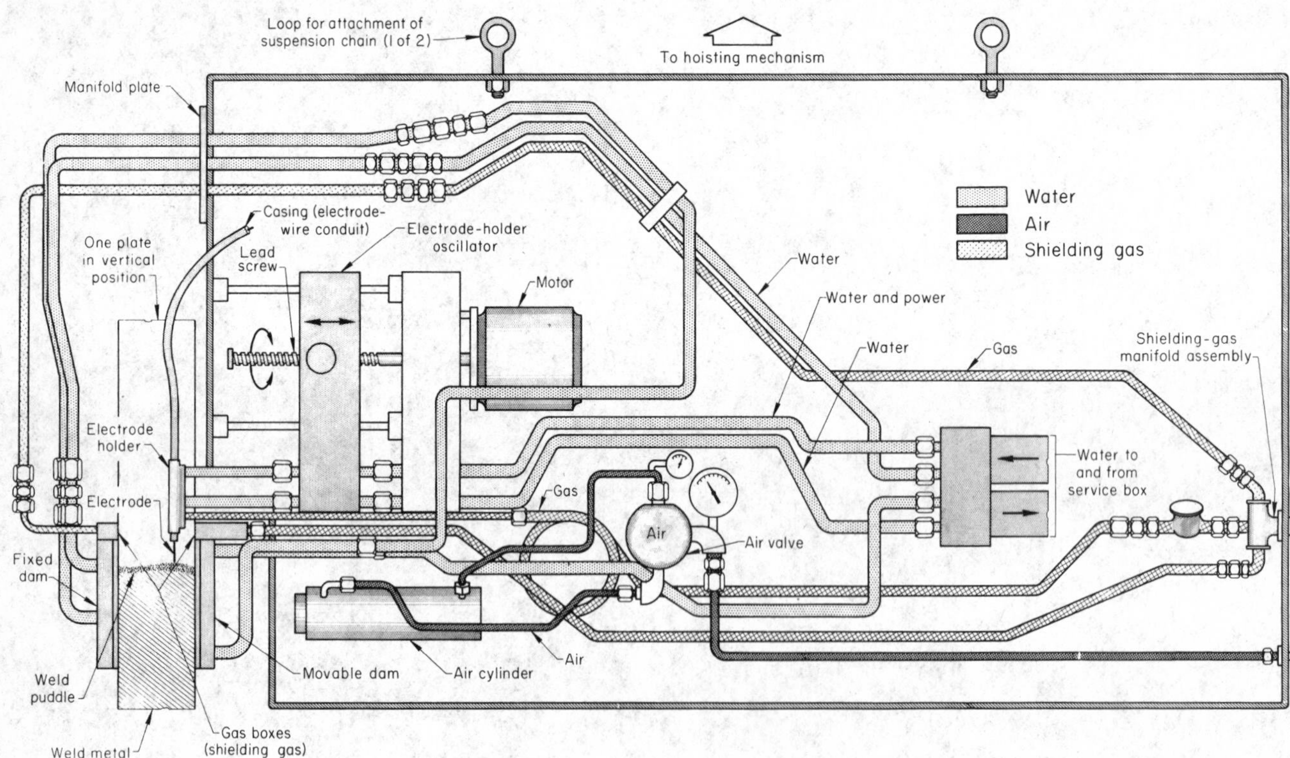

*R. David Thomas, Jr., *Chairman*, President, R.D. Thomas & Co., Inc.; Dennis R. Amos, Welding Engineer, Westinghouse Turbine Plant; Glen Edwards, Professor, Metallurgical Engineering, Colorado School of Mines; G.A. Gix, Chief Welding Engineer, American Bridge Division, U.S. Steel Corp.; James R. Hannahs, President, Midwest Testing Laboratories, Inc.; Franklin W. Henderson, Manager—Engineering, Richlin Equipment, Inc.; Guy A. Leclair, Manager, Welding Development Laboratory, Foster Wheeler Energy Corp.; Robert C. Shutt, Vice President, Lincoln Electric Co.; James E. Sims, Manager, Welding Research and Development, Chicago Bridge & Iron Co.

Applicability

Electrogas welding is most often used for joining relatively thick plates, such as those required in the construction of ships, bridges, storage tanks, and pressure vessels. Large-diameter thick-walled pipes and longitudinal seams of pressure vessels also can be butt welded by EGW.

Metals Welded. The electrogas process has been generally restricted to the welding of low-carbon and medium-carbon steels. These groups include structural steels such as American Society for Testing and Materials (ASTM) A36 and carbon-manganese-silicon steels. It has also been successfully used for the welding of alloy steels and the austenitic grades of stainless steel. Experimental work has also been carried out on aluminum.

Base-Metal Thickness. Electrogas welding is most suitable for the joining of plates within the thickness range of $^1/_2$ to 3 in., although plates $^3/_8$ in. thick and 4 in. thick have been welded successfully by the process. For the welding of plates thinner than $^1/_2$ in. in the vertical position, shielded metal arc (SMAW), gas metal arc (GMAW), and flux cored arc welding (FCAW) usually are more economical. For plates thicker than 3 in., ESW is usually more practical, because of the difficulty in obtaining adequate shielding-gas coverage. Inadequate coverage results in weld porosity and irregular or inadequate penetration of the base metal.

With the use of specially shaped dams it is possible to weld together two plates that differ in thickness by as much as 50%, provided that both plates are within the thickness range of $^3/_8$ to 4 in. prescribed for EGW.

Length of Joint. There is no established limit on the length of joint that can be welded by EGW. In shipyards, joints 80 ft long have been welded in one pass, with the joint in the vertical position. Single-pass welding is the method most frequently employed, but a two-pass method can also be used.

Equipment

The essential components of equipment for EGW are a power supply, an electrode wire guide, water-cooled dams, a system for feeding the electrode wire, a mechanism for oscillating the electrode wire guide, and methods for supplying shielding gas to the area immediately above the weld pool. Except for the power supply, the major components of the equipment are incorporated in an assembly that moves as an integral unit as welding proceeds.

A typical unit for EGW is shown in Fig. 1. This unit is suspended on two chains that raise and lower it by means of a motor-driven hoisting mechanism (not shown in Fig. 1). Units may be suspended by a single chain or by cables, or a track may be employed. In addition to the components shown in Fig. 1, the unit incorporates control devices for maintaining the required welding voltage and current, electrode-wire feed rate, shielding-gas flow rate, welding speed, and cooling-water flow rate.

The arrangement of the joint assembly (plates in a vertical position), water-cooled dams, gas boxes, and the electrode wire guide for one system of EGW is shown in Fig. 2(a). Details of the electrode wire guide, a gas box, and the dam are shown in Fig. 2(b), (c), and (d), respectively. The workpieces are held in position during welding by conventional strongbacks. Shielding gas is supplied through ports of two separate gas boxes (Fig. 2c), one on each side of the joint being welded. In this system, the electrode wire guide (also called the contact-tube guide) has a duct through which a part of the shielding gas is supplied.

Shielding-gas ports are often found unnecessary when self-shielding electrode wires are used as filler metal. Nevertheless, in field sites where it is difficult to shield the weld zone from wind, externally provided gas shielding is often employed.

Figure 3 shows the arrangement and details of essential components for another system of EGW. In this system, shielding gas is supplied through ports incorporated in the water-cooled dams (see Fig. 3c); therefore, no separate gas boxes are needed.

Power Supplies. Electrogas welding is done with direct current electrode posi-

Fig. 2 Electrogas welding unit. Employs water-cooled electrode wire guide and separate gas box

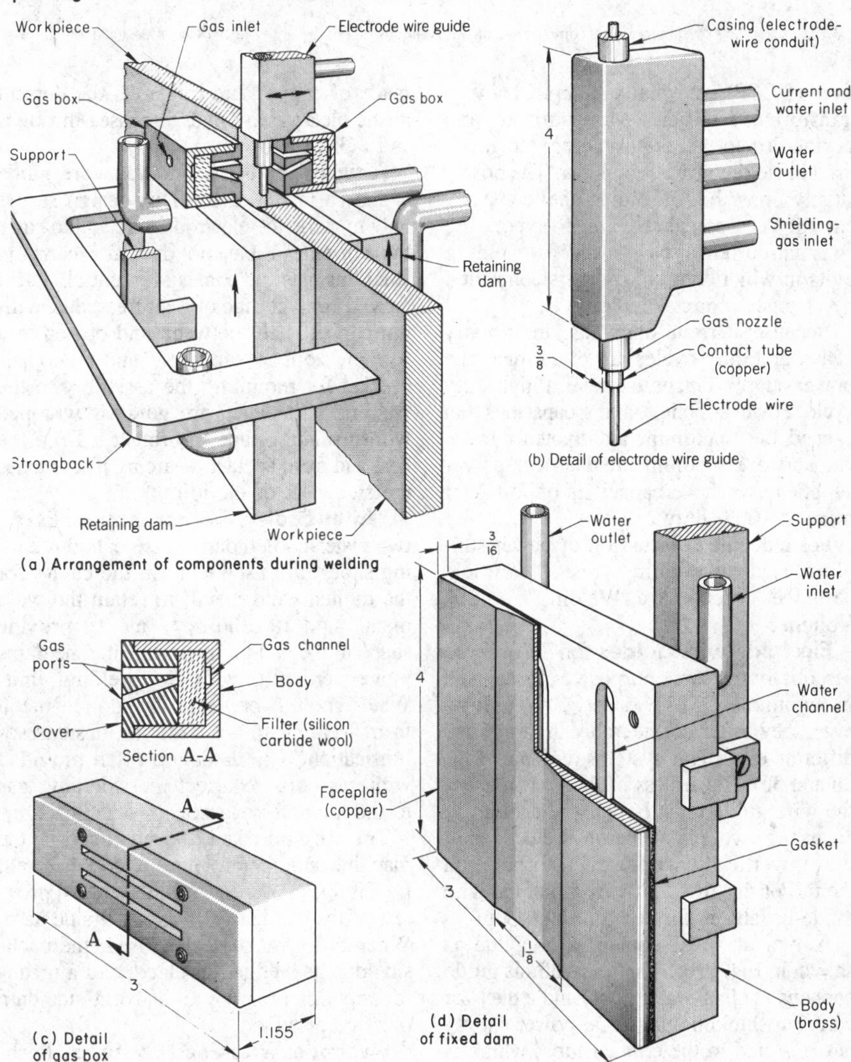

Fig. 3 Electrogas welding unit with insulated electrode wire guide

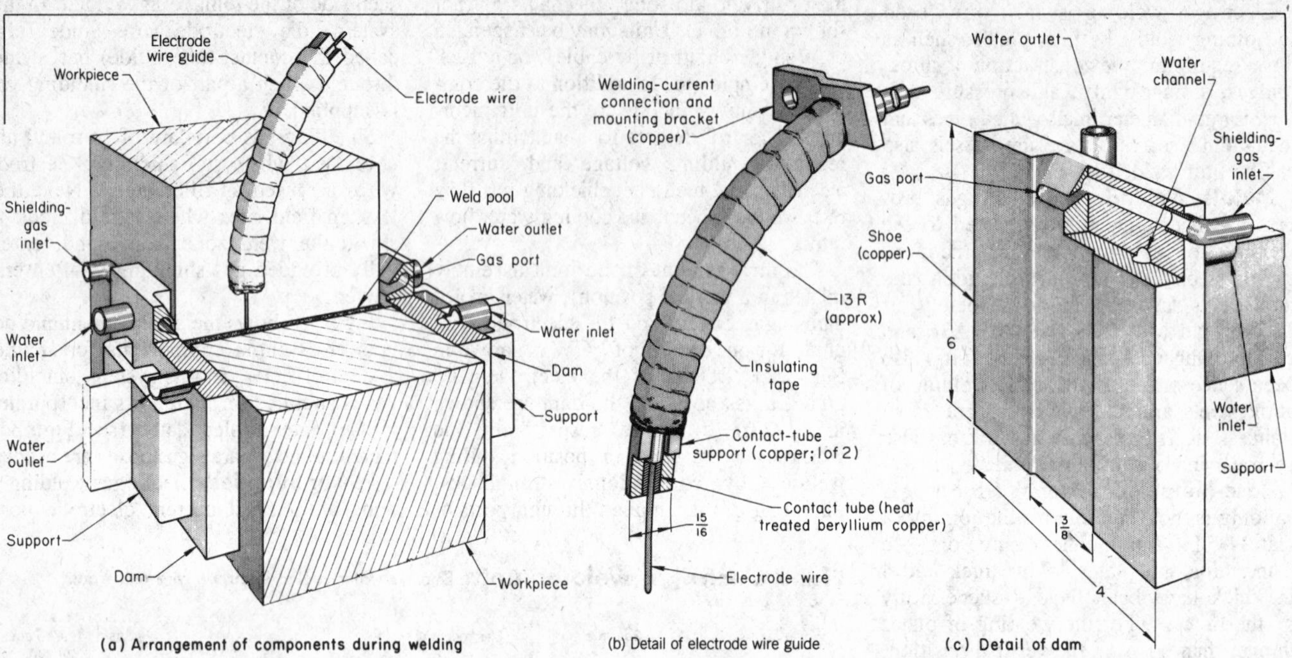

(a) Arrangement of components during welding (b) Detail of electrode wire guide (c) Detail of dam

tive (DCEP), normally supplied by a transformer-rectifier. Motor-driven and engine-driven generators are sometimes used in field construction sites. The power supply may be of either the constant-current or constant-voltage type; the constant-current type is used for welding units in which vertical travel is controlled by changes in arc voltage.

Because current demands are usually heavy and duty cycles are often long, the power-supply unit must have a high duty cycle. Power supplies having capacities that exceed the maximum anticipated current demand are recommended. Most power supplies used have capacities of 750 A or more at 100% duty cycle.

For a detailed discussion of power supplies used in welding, see the article "Shielded Metal Arc Welding" in this Volume.

Electrode wire guides for EGW serve essentially the same purpose as those used in automatic GMAW and FCAW. However, they differ considerably in design for different electrogas systems (compare Fig. 2b and 3b). Regardless of the system used, the wire guide must be narrow enough to clear the two plates being welded as it moves vertically within the gap between them. For this reason, the width of the wire guide usually is limited to about $^3/_8$ in.

A typical water-cooled wire guide is shown in Fig. 2(b). The body of this guide contains cooling-water ducts and a duct for carrying shielding gas. The power supply is connected to the contact tube, which is made of copper and transmits the current to the electrode wire as it passes through, as in GMAW.

A typical non-water-cooled wire guide is shown in Fig. 3(b). This design serves only to guide the electrode wire and to carry the current; it has no duct for carrying shielding gas. It consists essentially of a curved contact tube of heat treated beryllium copper, held between and brazed to a pair of copper supports, and a copper bracket for mounting the assembly to the welding unit. The wire guide is wrapped with insulating tape to protect it from the heat and from accidental arcing if it touches the sidewalls of the joint.

Water-Cooled Dams. In most EGW, two water-cooled dams (also called retaining shoes) are used to form the cavity for the molten weld metal, to retain the weld metal until it solidifies, and to provide shape to the weld reinforcement. The dams move vertically with the welding unit. Water cooling prevents the molten metal from welding to the dam and hastens solidification. The dams are often provided with gas ports to direct the shielding gas to the arc and weld zones (see Fig. 3).

The back-side dam (the side opposite the machine and operator) is sometimes fixed; i.e., it does not rise with the upward progress of the weld as does the front-side dam. When this arrangement is used, the back-side dam is wedged in place, and a means for applying pressure to the front-side dam must be provided.

Electrode Wire-Feed Systems. Push-type systems are commonly used and are similar to those used for GMAW (see the article "Gas Metal Arc Welding" in this Volume for more information on wire-feed systems). In addition to reels of electrode wire, feed rolls, and the casing (wire conduit) through which the electrode wire is pushed, the system may include a wire straightener, located between the reel and the feed rolls. Wire-feed speed depends on the size and type of wire used. A typical $^3/_{32}$-in. flux cored wire may allow speeds up to 400 in./min. With $^1/_8$-in. flux cored wire, speeds up to 550 in./min are obtained. The wire supply must be adequate to complete the weld without stopping.

Electrode Wire Oscillators. Most units for EGW incorporate an oscillator mechanism for the electrode wire. Oscillation is generally required when welding sections thicker than 1 in., but it is sometimes used for thinner sections. There are several types of mechanisms for achieving oscillation of the electrode wire. One common type (Fig. 1) consists of a motor-driven lead screw that moves an oscillator plate connected to the electrode wire guide.

Gas Ports. When shielding gas is required, it must be supplied, uniformly and without turbulence, above the weld pool. In one system, part of the shielding gas is supplied through two separate components, called gas boxes, and supplementary shielding gas is supplied through the electrode holder. One gas box is mounted on the top of each dam, as shown in Fig.

Table 1 Chemical composition for solid carbon steel electrodes

AWS classification	C	Mn	P	S	Si	Ni	Cr	Mo	V	Ti	Zr	Al
					Composition, %							
EGXXS-1	0.07-0.19	0.90-1.40	0.025	0.035	0.30-0.50	...	...	...	...	...	...	...
EGXXS-1B	0.07-0.12	1.60-2.10	0.025	0.035	0.50-0.80	0.15	...	0.40-0.60	...	...	...	...
EGXXS-GB	No chemical requirements											
EGXXS-2	0.06	0.90-1.40	0.025	0.035	0.40-0.70	...	...	...	...	0.05-0.15	0.02-0.12	0.05-0.15
EGXXS-3	0.06-0.15	0.90-1.40	0.025	0.035	0.45-0.70	...	...	...	...	...	...	...
EGXXS-5	0.07-0.19	0.90-1.40	0.025	0.035	0.30-0.60	...	...	...	...	...	...	0.50-0.90
EGXXS-6	0.07-0.15	1.40-1.85	0.025	0.035	0.80-1.15	...	...	...	...	...	...	...

2(a). Each gas box is constructed with a gas inlet (Fig. 2a), a gas channel, and slot-shaped gas ports (Fig. 2c). Common practice is to drill the inlet hole to a diameter that will permit an adequate supply of shielding gas to be fed into the box under some preselected line pressure. In another system, all of the shielding gas is supplied through ports in the dams (see Fig. 3c), and no supplementary shielding is required.

Electrode Wires

Either solid or flux cored electrode wire (filler metal) can be used in EGW. American Welding Society (AWS) Specification A5.26 covers the requirements of both types of these electrodes for welding carbon and high-strength low-alloy (HSLA) steels (non-heat-treatable types).

Solid electrode wires are similar to those used for GMAW, but fall into a different AWS classification. There are six classifications based on the mechanical properties of the deposited weld metal, separated into two levels. One level has a minimum tensile strength of 60 ksi; the other group, 70 ksi. Table 1 gives the chemical composition for the six classifications of solid electrodes used for EGW. Sizes most commonly used are $^1/_{16}$, $^5/_{64}$, and $^3/_{32}$ in. diameter. Typical relations between reference voltage and welding current for these three sizes of electrode wire are shown in Fig. 4.

Fig. 4 Relationship between reference voltage and welding current

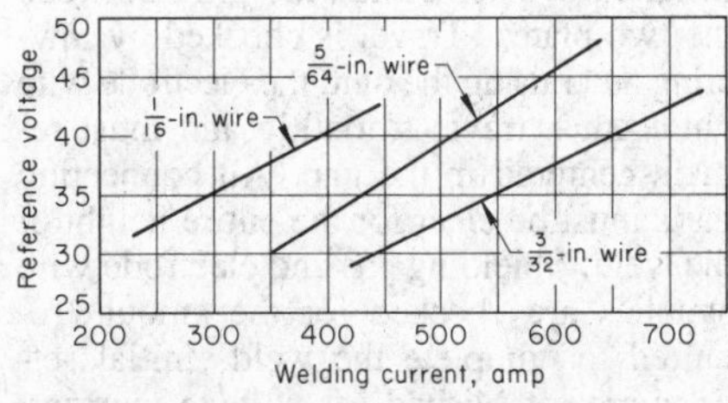

Flux cored electrode wires used for EGW differ from those used in FCAW. The classifications and compositions of flux cored electrode wires used for EGW of carbon and HSLA steels are given in Table 2. Sizes of flux cored electrode wire range from $^1/_{16}$ to $^1/_8$ in. in diameter.

Flux cored electrode wires for EGW contain a lower percentage of slagging-type compounds in their flux fill than standard flux cored electrodes. Electrode wires designed for FCAW provide excessive slag on the molten weld pool, resulting in the choking of the arc when used for EGW of long joints.

In welding of low-carbon steel, the higher cost of flux cored wire is often justified because flux cored electrode wire has abut 20% greater deposition rate than solid electrode wire. Deposition rates typical of two sizes of solid and flux cored electrode wire are:

Diameter, in.	Deposition rate, lb/h Solid	Flux cored
$^1/_{16}$	18	22(a)
$^5/_{64}$	24	29(a)
$^3/_{32}$	32	36(b)
$^1/_8$	...	65(b)

(a) Gas-shielded type. (b) Self-shielded type

Stickout. In most EGW, electrode stickout of about $1^1/_2$ in. is used. This relatively long stickout increases electrode melting efficiency by providing preheating of the wire. Flux cored electrodes often use electrode stickouts of 2 to 3 in.

Shielding Gases

A mixture of approximately 80% argon and 20% carbon dioxide is widely used, and is generally preferred, as a shielding gas for most EGW applications. This mixture is well suited for use with both solid and flux cored electrode wire. Carbon dioxide alone is also used and is particularly satisfactory when employed with flux cored wire. Self-shielded flux cored electrodes contain core materials which generate gases that shield the molten metal from atmospheric contamination.

Workpiece Assembly

Figure 5 shows two steel plates that have been assembled in a typical joint for EGW. Both a starting trough and runoff tabs are used to permit deposition of weld metal beyond the ends of the joint. The need for starting troughs and runoff tabs depends mainly on plate thickness and on the extent to which repair welding (usually by SMAW) can be tolerated. Plates less than 1 in. thick can usually be welded satisfactorily without the use of runoff tabs and by using only a starting block instead of a starting trough. Plates thicker than 1 in. can be welded in the same way, but as plate thickness increases, the amount of repair welding required will be greater if a starting trough and runoff tabs are not used.

As Fig. 5 shows, the starting trough and runoff tabs have the same thickness and width of gap as the plates to be welded and are tack welded in place. The depth of the starting trough and the height of the

Table 2 Chemical composition for deposited weld metal from flux cored electrodes

AWS classification	Shielding gas	C	Mn	P	S	Si	Ni	Cr	Mo	Cu	V
						Composition, %					
EGXXT1	None	...	1.50	0.03	0.03	0.50	0.30	0.20	0.35	...	0.08
EGXXT2	CO_2	...	2.00	0.03	0.04	0.90	0.30	0.20	0.35	...	0.08
EGXXT3	CO_2	0.10	1.0-1.8	0.03	0.030	0.50	0.7-1.1	...	0.30		...
EGXXT4	A-CO_2, or CO_2	0.12	1.00-2.00	0.02	0.03	0.15-0.50	1.50-2.00	0.20	0.40-0.65	...	0.05
EGXXT5	CO_2	0.12	0.50-1.30	0.03	0.03	0.30-0.80	0.40-0.80	0.45-0.70	...	0.30-0.75	...
EGXXTG	Not specified, no chemical requirements										

Note: Single values shown are maximum percentages.

Fig. 5 Setup for EGW of two steel plates

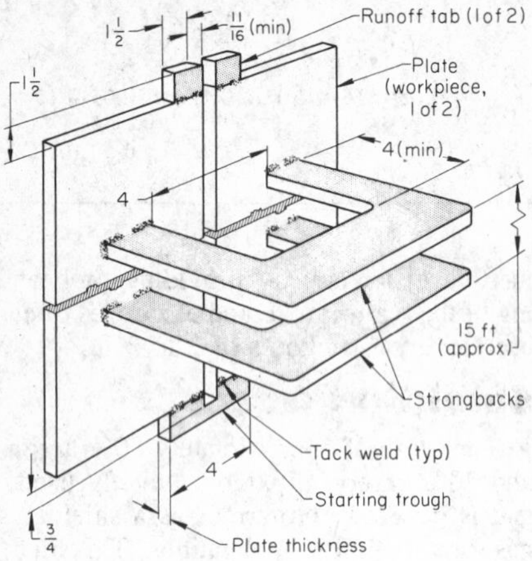

weld metal deposited in the runoff tabs vary, depending on the specific equipment and the application requirements. Ordinarily, a 1/2-in. starting-trough depth and runoff-tab height are sufficient, but these may be increased to as much as 1 1/2 in. for thicker sections. In one plant, 2 in. is the standard depth of starting troughs and height of runoff tabs.

When many identical joints are to be welded, the starting trough and the runoff tabs are often made of copper and are usually air cooled. Instead of being tack welded, they are clamped in place, thereby shortening setup time, and can be reused many times. When copper starting troughs are used, a piece of steel should be placed in the bottom to prevent copper dilution (and possible cracking) of the weld.

Usually, the plates to be welded are temporarily held in position by as many strongbacks as are needed to obtain rigidity. The cut-out portion of strongbacks must accommodate the back-side dam and usually must not be smaller than 4 by 4 in., as shown in Fig. 5, so that the dam can pass through as it travels up the joint.

Most EGW is done in a single pass, using square-groove butt joints, as shown in Fig. 6(a). The gap between plates for the square butt joint does not vary with thickness and is set by clearances for the welding nozzle. Typically, this gap is between 5/8 and 3/4 in. Less effort is required in seam preparation of square butt joints, and plate edges are less likely to damage during handling. One disadvantage of square butt joints, however, is that they cannot be welded conveniently by another method if the EGW equipment is out of commission. For this reason, conventional single- or double-V-grooves, similar in design to the ones commonly used for SMAW or GMAW, are sometimes preferred.

Single-V-grooves (Fig. 6b) may be welded using a fixed copper dam on the back side of the weld that has a smaller reinforcing groove cut into it than those used for square butt welding. With a fixed dam on the back side, it is no longer necessary to provide a gap between the plates being welded, thus reducing the volume of filler metal and consequently increasing the vertical travel speed. A back-up bar may also be used to replace the copper dam on the back side of the weld. When back-up bars are used (Fig. 6c), they must be thick enough to prevent melt-through of the weld pool. A minimum thickness of 1/2 in. is recommended. When the back-up bar can remain in position, some saving of time can be realized. However, when back-up bars must be removed, clean-up of the back side is time consuming. For two-pass welding, specially shaped dams are required. For instance, in welding a double-V-groove joint, the fixed dam must be contoured to fit the V for the first pass (Fig. 6d). The welding unit is then moved to the opposite side and the second pass is made with only the movable dam; the first-pass weld performs the function of a fixed dam. Back gouging or grinding of the root of the first pass may be required before the second pass can be deposited.

Fig. 6 Joint design for EGW

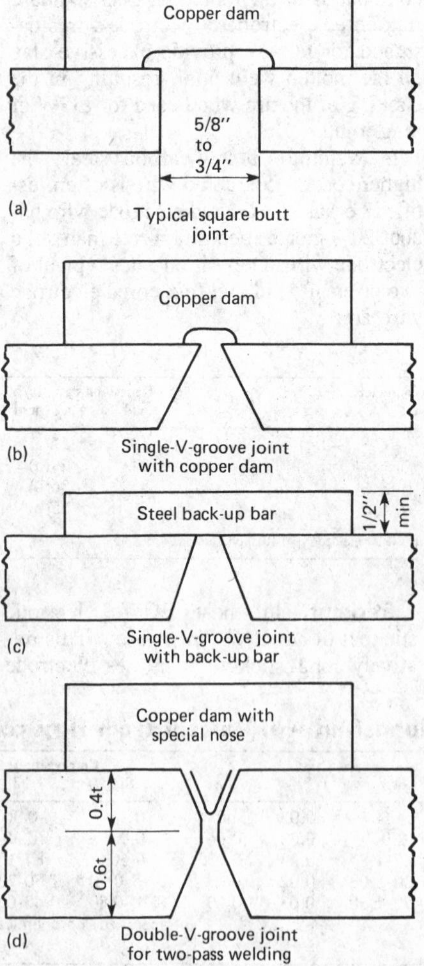

Two-pass welding has two metallurgical advantages over single-pass welding: (1) the smaller weld cross section requires lower heat input, and thus grain growth is minimized; and (2) the second pass generates enough heat to provide a grain-refining heat treatment for the weld made on the first side.

Operating Procedures

Welding Parameters. Selection of operating parameters is a compromise to achieve the most economical joint for the intended service. Factors such as small joint openings, fast fill (high current), low voltage, and fast setups lead to lower costs. Extremely high welding speeds can result in centerline cracking, inadequate penetration, hot metal leakage, and loss of gas shielding. Low welding speeds, however, can result in excessive coarse grain size, reduced toughness, overmelting, and increased cost. Welding procedures should be selected from proven procedures where possible. Manufacturers' recommendations are a reliable source of information, as is AWS C5.7-81, "Recommended Practices for Electrogas Welding." Procedures should be verified against the appropriate specification for the intended service to evaluate equipment, welding operators, and the required mechanical properties.

Operation Setup. The welding unit is installed over or beside the joint between the two plates. Travel is checked by lowering and raising the unit; the electrode wire guide must travel smoothly and must remain centered in the gap. All connecting parts must be clear for the entire height of the weld. Shielding gas and electrode wire supplies are checked for the amount required to complete the weld. Initial settings are established for voltage, current, and travel speed. After prepurging where external gas shielding is used, the weld is started; voltage and current are corrected to the procedure values.

Control of Vertical Travel. Close control of vertical travel is necessary to maintain constant arc length and uniform arc voltage. The method of control depends on the type of power supply used. A constant-current (drooping-voltage) power supply enables travel to be controlled by a change in arc voltage. If the reference voltage (set on the machine) is 35 V, movement will not begin until the arc voltage drops below 35 V. At this point, the travel mechanism is automatically actuated, and the unit moves upward until the arc voltage is restored to its former value. The equipment is sensitive to a reduction as small as $^1/_4$ V.

When a constant-voltage power supply is used, a photoelectric cell that controls the rate of travel by observing the height of the rising weld pool is employed. Another system uses a current pickup device to stop and start the travel at high- and low-current preset limits. Both systems use an adjustable midrange travel speed and make corrections by stopping and starting the travel motors. To obtain the required control of travel speed, minor adjustments by the operator are commonly required.

Oscillation. Thickness of the workpieces determines whether or not the electrode wire guide will be oscillated during welding. When the workpieces are less than 1 in. thick, oscillation usually is not required, although it may be used in certain applications. If oscillation is used, it is necessary to make the adjustments during setup.

The oscillator mechanism moves the electrode wire guide back and forth within the gap. The length of travel can be varied, but is usually set so that the electrode wire guide stops at a distance of $^3/_8$ in. from each dam. The speed of traverse during oscillation is commonly 16 to 18 in./min. The oscillation cycle is normally set to incorporate a dwell time of up to 3 s at each side before the direction is reversed. Dwell time near each water-cooled dam is used to achieve satisfactory penetration into the workpiece sidewall near the outer face that abuts the dam. If the dwell time is insufficient, the cooling effect of the copper dam causes lack of fusion of the base-metal surface.

Restarts and Repairs. Electrogas welding produces the best results when the weld is continuous from the starting-point sump to the completion. All planning and setup checks should be scheduled accordingly. In spite of all planning, restarts may be needed. The manner of repairs and restarts depends on the economics of joint thickness and the related cost and serviceability of removing a partial weld and rewelding or in welding over the top of a previous weld, thus necessitating a manual repair weld. If the welding process is interrupted before the entire weld is completed and the weld is allowed to cool, shrinkage could trap the weld equipment. Wedges should be placed in the joint while the weld is still hot, and the stop area of the joint should be reheated as needed to free the wedges before restarting. Arc gouging a U-groove in the stopped weld with a 45° upward slope from the front to the back side, starting the arc on the lower front side, and recentering the arc after the weld has penetrated the gouged-out area provides a restart technique requiring a minimum of repair.

Welding of Tanks

Self-shielded flux cored electrode wires are widely used for vertical welds on field oil and water storage tanks because of their ability to ensure adequate gas shielding in the presence of drafts. Large electrode extensions of up to 3 in. result in high preheating of the electrode and up to 70 lb/h deposition rates, so the joint fill rate is very fast. In EGW, high fill rates can mean lower heat input, less coarse grain structure, and improved heat-affected zone (HAZ) toughness properties. The range of welding variables that results in an acceptable joint is less than some of the slower adaptions; therefore, qualified welding procedures should be closely followed.

Surge Tanks. Fourteen tanks 70 ft in diameter and 145 ft high were constructed for use as surge tanks and were located between the penstocks and the powerhouse near a large dam. These tanks were fabricated from ASTM A515 or A516 carbon steel plates ranging in thickness from $1^3/_4$ in. for the bottom course to $^3/_8$ in. for the top course. The tanks were built in accordance with Section VIII of the American Society of Mechanical Engineers (ASME) Boiler and Pressure Vessel Code and in accordance with the Corps of Engineers specifications. Welded seams were subjected to x-ray examination.

Cement slurry tanks 150 ft in diameter by 37 ft high were constructed from ASTM 515 or 518 steel to conform with API-12C (American Petroleum Institute) specifications. Plate thickness in these tanks ranged from $1^3/_{16}$ in. for the bottom course to $^3/_8$ in. for the top course.

Storage Tanks. A group of three storage tanks was built from ASTM A285 steel in accordance with Section VIII of the ASME Boiler and Pressure Vessel Code. The 2 bottom courses were $1^1/_2$ in. thick, and the remaining 13 courses were made of plates $1^3/_{16}$ in. thick.

Water Reservoirs. Three tanks for water reservoirs were constructed in accordance with American Water Works Association (AWWA) D-100 and AWS D5.2 specifications. Each of these tanks was 338 ft in diameter and 16 ft high and was constructed with two courses. Plates for the lower course were $1^1/_8$ in. thick and for the upper course were $^9/_{16}$ in. thick.

Fig. 7 Two-pass EGW

Medium-carbon steel (ASTM A515, grade 70); low-carbon steel filler metal

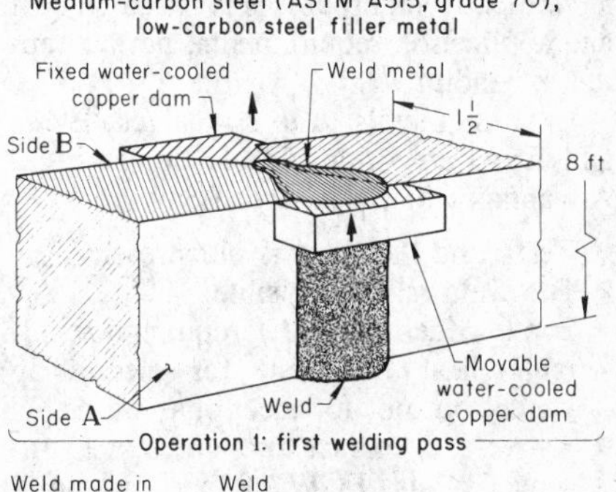

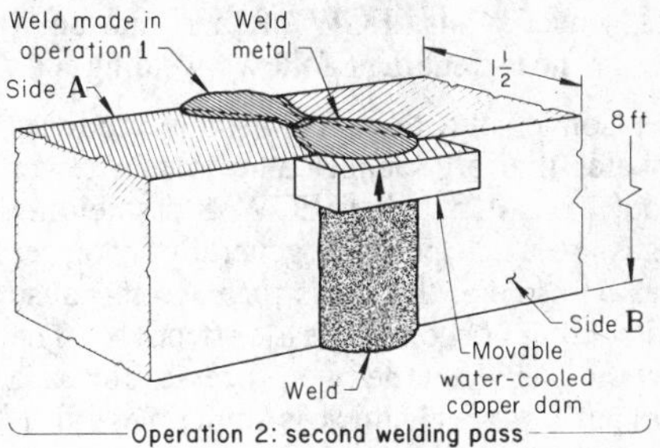

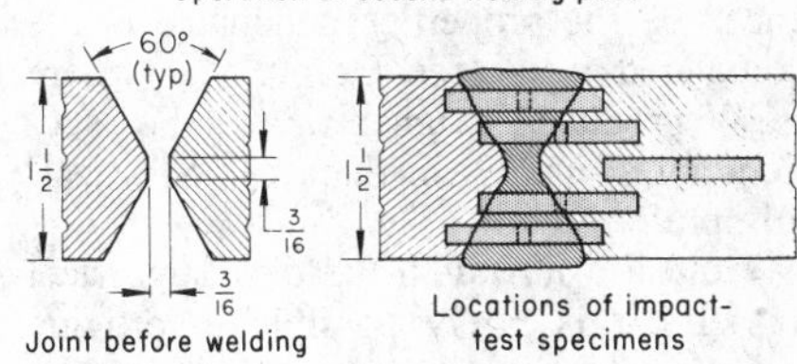

Location of impact test specimen bottom	Testing temperature, °F	Energy absorbed, ft·lb
Charpy V-notch impact strength		
Plate as received	70	17
	50	7
Weld metal, side A	70	80
	30	60
Weld metal, side B	70	65
	30	45
HAZ:		
Side A	70	24
Side B	70	18
One-pass electroslag weld	70	10

Conditions for EGW

Joint type	Butt
Weld type	Double-V-groove
Electrode wire	$^7/_{64}$ in. diam, flux cored
Current (both passes)	475A (DCEP)
Voltage	34 V, first pass; 36 V, second pass
Arc length	$^1/_4$ in. approx
Shielding gas	Carbon dioxide, 30 ft^3/h
Welding position	Vertical
Welding speed, both passes	3 in./min
Preheat	None
Postheat	Stress relief (if required for service)

Electrogas Versus Electroslag Welding

The equipment for EGW closely resembles that for conventional ESW. Therefore, a change from one process to the other entails simply a change from shielding gas to flux, or from flux to shielding gas. Selection between processes is based on cost and application requirements, not on capital expenditure.

For base metals $^3/_4$ to 3 in. thick, EGW and ESW are often closely competitive. Advantages of EGW over ESW are:

- Restarting the weld is much easier.
- The weld is more visible.
- Some codes (ASME) require normalizing heat treatment for electroslag welds, but not for electrogas welds.
- As-welded impact properties are improved; thus, EGW often is chosen if no heat treatment follows welding.

Conversely, the ESW usually produces welds that are cleaner and contain fewer defects. Also, when EGW and shielding gas are used, porosity generally increases as base-metal thickness increases because shielding becomes less effective. The choice between the two processes for base metal $^3/_4$ to 3 in. thick is sometimes influenced by the properties obtainable in the completed weld (see Example 1, where EGW produced tougher welds). For sections more than 3 in. thick, ESW is usually preferred.

Weld Properties. In welding large steel vessels, it is rarely possible to provide postweld heat treatment to improve the mechanical properties of the weld metal and of the HAZ, so the welding process must ensure adequate toughness in the as-welded condition. In some applications, low heat input and fast cooling of the weld metal in EGW produce specified mechanical properties that could not be obtained by ESW.

Example 1. Two-Pass Electrogas Welding Versus One-Pass Electroslag Welding. A vertical weld 8 ft high was made by EGW in welding a vessel fabricated from ASTM A515, grade 70, steel plate. Welding was done in two passes, one on each side of the plate, as shown in Fig. 7. A water-cooled copper dam with a shaped nose was used as a fixed dam for the first pass; the weld metal of the first pass served as the fixed dam for the second pass. For startup, a starting trough was used at the bottom of the joint (not shown in Fig. 7). Joint preparation consisted of beveling the plates as shown in the view at bottom left in Fig. 7.

Impact tests were made using full-sized Charpy V-notch specimens cut from the weld metal and HAZ for each weld pass on a specimen taken from the as-received plate for comparison. A further comparison was made by impact testing the HAZ of a one-pass weld made by ESW. The notch toughness of all specimens from the electrogas welded joint exceeded that of the as-received plate, whereas the notch toughness of the specimen from the HAZ of the electroslag weld was considerably less.

Four side-bend tests and two tension tests of specimens from the electrogas welded vessel were conducted in accordance with Section IX (Welding Qualifications) of the ASME Boiler and Pressure Vessel Code. Code requirements were met in all tests. In the first tension test, tensile strength was 82.5 ksi and elongation in 2 in. was 24%; in the second test, tensile strength was 81.9 ksi, and elongation 30%.

Metallurgical Considerations

As in most high heat input processes such as submerged arc welding and ESW, a tendency for coarse grains to form in the weld and the HAZ exists. Nonmetallic inclusions tend to accumulate at larger grain boundaries, with a resultant loss in toughness. Control of filler- and base-metal chemistry is necessary to achieve the cost advantages of EGW, while achieving the required mechanical properties in the as-welded condition. Because the base metal contributes up to 35% of the resultant weld nugget and greatly affects the HAZ properties, the base metal must be suitable for the process and intended service.

Preheat. The large amount of heat generated in EGW travels upward, providing all the preheat necessary for most low- and medium-carbon steels. However, preheat may be required, where specified by code, for welding high-strength steels, higher alloy steels, for material over 3 in. thick, for high restraint, or where base-metal temperatures are below 32 °F. Excessive preheat affects mechanical properties, causing weld problems such as excessive melting and melt-through.

Postweld Heat Treatment. The majority of EGW applications does not require or facilitate postweld heat treatment. Stress relief, where required by code, normally results in a slight decrease in tensile strength and a slight improvement in toughness. When the weld is subject to extreme in-service temperatures, corrosion, or cyclic loading, normalizing may be required. Quenched and tempered steels require postweld heat treatment to restore the base metal to its original properties.

Inspection

Defects most commonly encountered in EGW are porosity, lack of fusion, and surface undercut. Internal defects are generally detectable by radiographic techniques. Surface defects are readily seen by magnetic-particle inspection, which also uncovers defects below the surface. Centerline cracks are also detectable by radiography. If the cracks, however, are very fine and oriented in the plane of the x-ray beam, they may be overlooked. For such defects, ultrasonic inspection is preferred. The codes to which an electrogas vessel is constructed usually specify the methods of inspection. When defects are detected that warrant repair welding, SMAW is usually employed after removal of the defect.

Arc Welding of Low-Alloy Steels and Other Ferrous Metals

Arc Welding of Hardenable Carbon and Alloy Steels

By the ASM Committee on Arc Welding of Hardenable Carbon and Alloy Steels*

HARDENABLE CARBON AND ALLOY STEELS share certain metallurgical characteristics that govern some of the guidelines used in arc welding. These steels may form martensite and/or bainite as a result of welding. These constituents can cause cracking as a result of welding and have a marked effect on the mechanical properties of the weldment. The presence of martensite and/or bainite and the resulting mechanical properties depend on the chemical composition of the steel, the cooling rate following welding, and any postweld heat treatment. This article discusses plain carbon steels that contain enough carbon (approximately 0.2%) to be hardenable during rapid cooling and certain alloy steels. The alloy steels are divided into the following categories: (1) high-strength low-alloy (HSLA) steels, (2) high-strength low-carbon quenched and tempered steels, (3) high-strength medium-carbon quenched and tempered AISI-SAE and AMS steels, (4) heat-resistant low-alloy steels, and (5) tool steels.

Steel Classifications

The HSLA steels typically contain less than 0.2% C and less than 2% alloy content. These steels usually are produced in the hot rolled condition. Because these steels have the lowest hardenability of the alloy steels considered in this article, they are the easiest to weld from the standpoint of avoiding cracking and meeting service requirements. The high-strength low-carbon low-alloy quenched and tempered steels typically contain less than 0.25% C and less than about 5% alloy addition for hardenability. These steels primarily are strengthened by heat treatments where the steel is quenched to form martensite and then tempered to produce tensile strengths in the range of 65 to 185 ksi.

High-strength medium-carbon quenched and tempered steels typically have more than 0.2% C and less than about 5% alloy content for hardenability. As with low-carbon quenched and tempered steels, these steels are strengthened by transformation to martensite. The combination of both high hardenability and high carbon contents results in susceptibility to weld cracking.

Heat-resistant low-alloy steels generally contain less than about 0.25% C plus alloy additions such as chromium, molybdenum, and vanadium. Tool steels range in composition from plain carbon types to high-alloy grades that may have a total alloy content of over 25%; thus, weldability varies over a wide range.

Hardness and Hardenability

Weldability of the hardenable carbon and alloy steels differs from that of the plain carbon steels because of their greater tendency to form harder regions in the heat-affected zone (HAZ). Hardenability determines the depth and distribution of hardness for a given cooling rate. The thermal cycle of the weld metal and HAZ in a typical arc weld involves a short-time, high-temperature austenitizing treatment, followed by rapid cooling. Hardenability data that have been used in the development of steels can be used as a guide to the hardness levels expected in the HAZ of a weld. Generally, a higher carbon and/or alloy content leads to the production of more crack-sensitive microstructure in the HAZ. Higher carbon content not only produces more martensite, but also increases the crack sensitivity of the martensite. In a carbon steel without any significant alloy content, only very rapid cooling produces 100% martensite and the maximum obtainable hardness. The effect of carbon content and of the amount of martensite formed on the hardness of carbon steel is shown in Fig. 1.

In most arc welding applications involving unalloyed carbon steel, the cooling rate of the weld metal and the HAZ is too low to develop the maximum hardness that the steel of a specific carbon content can attain, because the hardenability of the steel is low. Nevertheless, an undesirable amount of hardening can occur. In welding alloy steels, maximum hardness often is developed in the heat-affected base metal even when the cooling rate is low, because of the high hardenability of alloy steels.

As the carbon content of plain carbon steel is increased, the hardenability (as well

*James M. Sawhill, Jr., *Chairman*, Engineer, Product Metallurgy, Homer Research Laboratories, Bethlehem Steel Corp.; Anthony Di Nardo, Vice-President—Marketing, Materials Development Corp.; Harry W. Ebert, Engineering Associate, Exxon Research & Engineering Co.; Robert B. Lazor, Group Leader, Materials Technology, Welding Institute of Canada; William S. Ricci, Welding Engineer/Metallurgist, U.S. Army Materials and Mechanics Research Center; C.E. Spaeder, Jr., Associate Research Consultant, U.S. Steel Corp.; R. Vasudevan, Project Engineer, Union Carbide Corp.; James T. Worman, Development Engineer, Union Carbide Corp.

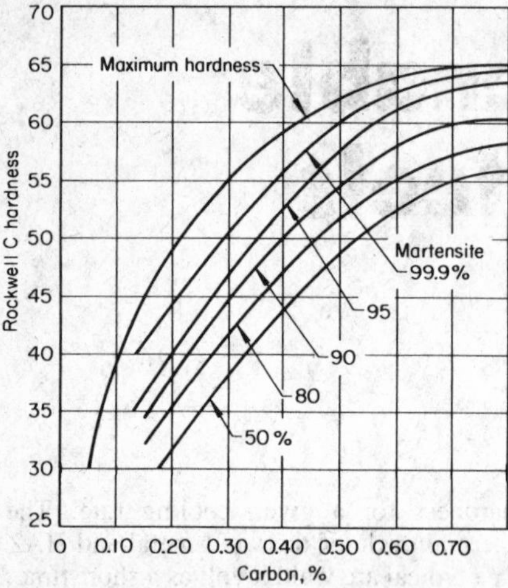

Fig. 1 Effect of carbon content on the hardness of carbon steel cooled rapidly. Cooling produced specific percentages of martensite and (top line) the maximum hardness obtainable in severe water quenching of small specimens of carbon steel.

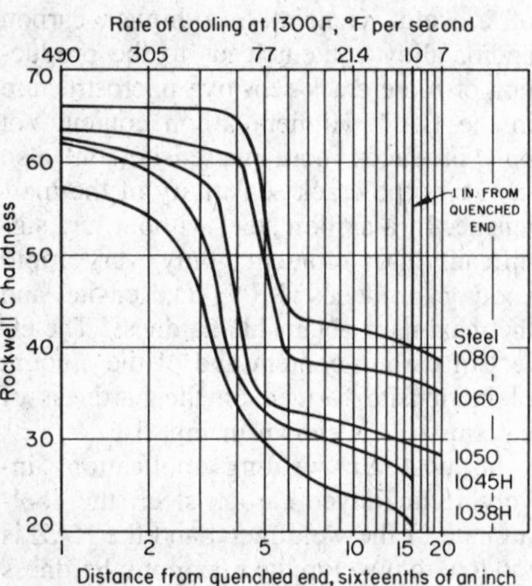

Fig. 2 Maximum hardenability of carbon steels, determined by the standard end-quench test

as the hardness) increases. This is shown by the end-quench maximum hardenability curves for five plain carbon steels plotted in Fig. 2; comparing the five steels at 1 in. from the quenched end of the hardenability specimen (which corresponds to a cooling rate of 10 °F/s at 1300 °F), the hardness increases from 21 HRC for 1038H steel to 40 HRC for 1080 steel. Cooling rates of 10 °F/s are not uncommon in arc welding. If the hardness shown for a particular steel at the 1-in. position in Fig. 2 is unacceptable (or is associated with cracking), measures must be taken to avoid the development of this hardness, or to decrease it.

Carbon and low-alloy steels that have the same carbon content will have the same maximum hardness when cooled rapidly enough to achieve maximum martensite in the microstructure. In Fig. 3(a), the maximum hardness is shown at the $^1/_{16}$-in. end-quench distance for five alloy steels that have the same nominal carbon content of 0.40% (maximum of the specification range is 0.44% for each steel). Despite the major differences in alloy content and hardenability among these five steels, each steel has the same maximum hardness, 60 HRC. The two steels in Fig. 3(a) with slightly lower maximum carbon contents (4037H and 1038H) have slightly lower maximum hardnesses, as shown.

Ten 41*xx*H steels are compared in Fig. 3(b). Each of these steels has essentially the same alloy content (nominally 1% Cr and 0.20% Mo, with 0.80% Mn; there are slight variations through the series), but the steels range in maximum carbon content from 0.23% for 4118H to 0.65% for 4161H. As shown in Fig. 3(b), in this series of ten chromium-molybdenum steels, the maximum hardness (at the $^1/_{16}$-in. end-quench distance) increases from 48 to 65 HRC as maximum carbon content increases from 0.23 to 0.65%.

In the welding of high-strength quenched and tempered steels that contain not more than 0.25% C, welding procedures are deliberately chosen so that martensitic structures are obtained. These steels are intended to be welded in the quenched and tempered condition. The low-carbon martensite formed during postweld cooling has desirable strength, and the as-welded joints have adequate toughness. However, as carbon content is increased, the higher carbon martensites that are formed are harder and less ductile in the untempered condition, and are a major contributor to the cold cracking of welds.

As the alloy content or the carbon content of a steel is increased, the hardenability is increased. Figure 3(a) shows the maximum hardenability of 1038H carbon steel and six widely used alloy steels of 0.40% nominal carbon content. The maximum hardenability of carbon steel 1038H (bottom curve) is low; that is, hardness decreases rapidly as cooling rate (top scale) decreases. As various amounts and combinations of alloying elements (chromium, nickel, molybdenum, manganese) are added to steel containing 0.40% C, the maximum hardenability increases to that shown for 4340H. Figure 3(b) shows the large effect of increasing carbon content on maximum hardenability in 41*xx*H steel. The effect on hardenability of increasing carbon from 0.23% (the maximum in 4118H) to 0.65% (the maximum in 4161H) in these alloy steels is somewhat greater than the effect, shown in Fig. 3(a), of increasing total alloy content (manganese, chromium, molybdenum, nickel) from 1% (maximum in 1038H) to 4.15% (maximum in 4340H).

In general, the weldability of steel decreases as its hardenability increases, because higher hardenability promotes the formation of microstructures which are more sensitive to cold cracking.

In welding, it is seldom possible to achieve cooling rates slower than 5 or 6 °F/s. This cooling rate corresponds to positions of 20 to 24 sixteenths of an inch from the quenched end of the standard end-quench hardenability specimen (Fig. 3a and 3b). Thus, if the hardness corresponding to this cooling rate (end-quench distance) for a particular steel is too high to be acceptable in the HAZ of a weldment, the weldment, or at least the zone of excessive hardness, must receive a tempering treatment after welding.

Weldability of Hardenable Carbon and Alloy Steels

When arc welding hardenable carbon and alloy steels, the primary objective is to obtain a sound weld and a weldment that exhibits adequate mechanical strength and corrosion resistance. Thus, discussion of weldability includes factors that lead to the production of unacceptable discontinuities during fabrication (sometimes called fabricability), as well as the properties of the weld after it is placed in service (sometimes called serviceability). The causes and characteristics of the various forms of flaws that are of primary concern when welding the hardenable carbon and alloy steels are discussed below.

By far the most common form of cold cracking in hardenable carbon and alloy steels is hydrogen-induced cracking. Occasionally, weldability of a steel is taken to mean simply the resistance of steel to this form of cracking. Plain carbon steels also experience hydrogen-induced cracking, but higher hardenability and HAZ hardness associated with the higher carbon and/or alloy steels frequently requires more precautions during welding to avoid hydrogen-induced cracking. Steels of high carbon content can produce martensite of very poor ductility that cannot withstand the shrinkage strains in an arc weld, even in the absence of appreciable hydrogen interaction. In these cases, steels are susceptible to cracking from inadequate ductility. Lamellar tearing, another form of

Fig. 3 Maximum hardenability of alloy steels and 1038H carbon steel. (a) Effect of various amounts and combinations of alloying elements in steel with a nominal carbon content of 0.40%, compared with carbon steel 1038H. (b) Effect of carbon content in 41*xx*H alloy steel (nominal 1% Cr and 0.2% Mo, except for 4118H). All data from SAE Handbook, 1970. See text for discussion.

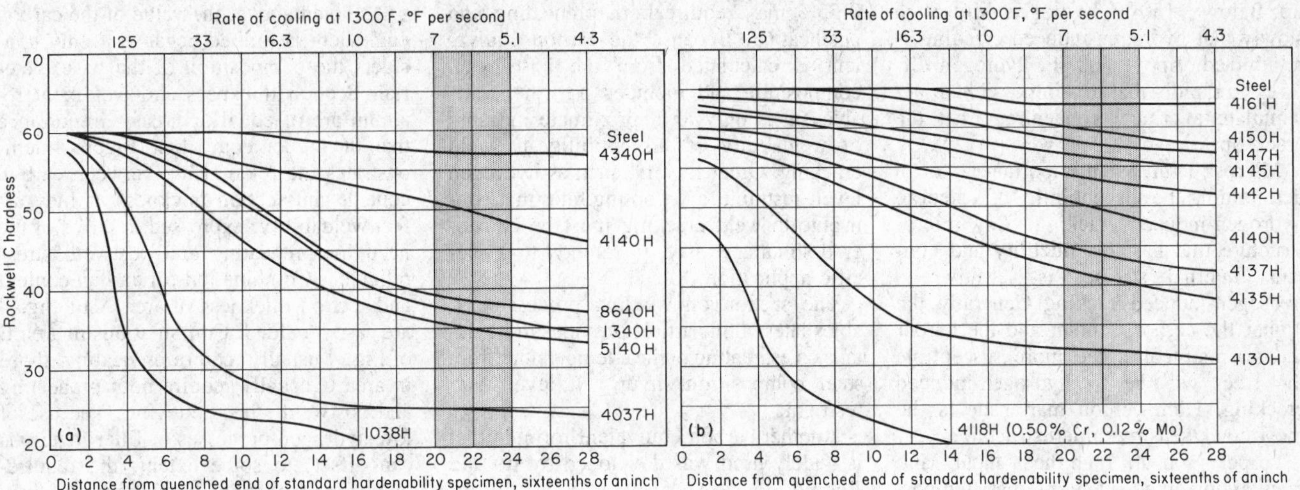

cold cracking, is not unique to these steels and is also common in plain carbon steels. However, the high-strength grades are slightly more susceptible.

Solidification cracking, or hot cracking, is usually not as serious a problem with these steels as with the higher alloy austenitic steels. Hot cracking can occur with the right combination of high sulfur, carbon, and nickel together with high restraint. This form of cracking occurs in hardenable carbon and alloy steels, but the higher strength and/or high-nickel grades are considered more susceptible.

Finally, stress-relief cracking is of concern when welding quenched and tempered grades and heat-resistant steels that contain significant levels of carbide formers such as chromium and vanadium.

Hydrogen-Induced Cracking

Hydrogen-induced cracking is usually more prevalent when welding hardenable carbon and alloy steels than when welding plain carbon steels. When other factors such as hydrogen, restraint, and thermal cycle are equal, a steel with higher carbon and/or alloy contents has a greater tendency to form a harder microstructure, which is more susceptible to hydrogen-induced cracking.

Failures associated with hydrogen-induced cracking usually occur in the HAZ, but may also occur in the weld metal if four conditions are present simultaneously. These conditions are (1) a critical concentration of hydrogen, (2) a stress intensity of significant magnitude, (3) a susceptible microstructure, and (4) a temperature between −150 to 400 °F. Hydrogen-induced cracks are generally transgranular and initiate either immediately after welding or after a delayed period.

Hydrogen in the welding arc atmosphere is converted to the atomic state and readily dissolves in the weld pool. Because the solubility of hydrogen in steel decreases with decreasing temperature, hydrogen is strongly driven out of solution in the HAZ and weld metal during cooling. To escape, atomic hydrogen must diffuse to some interface, collect, and re-form as molecular hydrogen. Atomic hydrogen may, however, interact with dislocations and diffuse to triaxially stressed regions where it acts as an embrittling agent. Because it is difficult for atomic hydrogen to escape from lattice imperfections by diffusion, extremely high internal stresses may develop and cracking may occur.

The amount of hydrogen absorbed by a weldment depends on several factors, including the cooling rate, size of the weld bead, and initial concentration of hydrogen in the arc atmosphere. Generally, the risk of cracking increases with increasing hydrogen concentration. Control of the hydrogen level may be achieved by minimizing the available hydrogen and providing sufficient time for hydrogen to diffuse from the weldment. The major sources of hydrogen are hydrogenous compounds and moisture in fluxes and electrode coatings, contamination of shielding gas, contamination of bare filler wires, and surface contamination of the workpiece.

The potential hydrogen level of a process is determined by measuring the moisture or hydrogen content of welding consumables and the resultant weld deposit. The higher the potential hydrogen level is, the higher the hydrogen content of the weld will be. Control of hydrogen potential of most consumables, particularly shielded metal arc welding (SMAW) electrodes and submerged arc welding (SAW) fluxes, depends on the conditions under which they are stored. Generally, the potential hydrogen level of welding consumables can be decreased by drying or baking.

Welding stresses arise from three factors: (1) external restraint of the welded sections, (2) unequal thermal expansion and contraction of the base metal and weld metal, and (3) volumetric expansion resulting from microstructural changes in a weldment. The stresses acting on a weld are a function of weld size, joint shape, fixturing, welding sequence, and yield strength of the base metal and weld metal. Pre-existing cracks and other discontinuities increase stress concentrations and the risk of failure.

Hydrogen in a weldment lowers the stress at which cracking will occur by reducing the cohesive strength of the lattice and by adding to the localized stresses at discontinuities. Stresses generated interact with hydrogen to enlarge discontinuities to the observed crack size. To prevent cracking, the stresses developed must be accommodated by strain in the weld metal. Selecting the lowest strength weld metal allowable by the design, in conjunction with good welding practice, reduces weld stress and, therefore, the probability of hydrogen-induced cracking.

After application of stress, a specific time must pass for hydrogen-induced cracks to appear. This delay time decreases with increasing hydrogen concentration and stress magnitude. The delay time for hydrogen-induced cracking depends on the time required for hydrogen to diffuse and produce a critical concentration at the crack tip. Below −150 °F, hydrogen diffuses so slowly that hydrogen-induced cracking is minimized. Above 400 °F, hydrogen diffuses so rapidly that it is impossible to accumulate a critical concentration at the crack tip.

Microstructure is an important factor in determining the susceptibility of a steel to hydrogen-induced cracking. Any microstructure that has low ductility and contains internal stresses is sensitive to hydrogen-induced cracking. Generally, the higher the carbon content and the harder the microstructure, the more susceptible the steel will be to hydrogen-induced cracking. High carbon martensite is the most crack-sensitive microstructure.

Proper preheat, high heat input, and maintaining an adequate interpass temperature reduces the quenching rate in the HAZ and provides a softer, less sensitive microstructure. The HAZ also may be softened either by postweld heat treatment or by the tempering effect of subsequent weld passes. Where the procedure allows some flexibility in altering these variables, it is important to recognize these effects and to use caution when conditions increase the tendency for cracking.

Austenitic filler materials sometimes are used to reduce the amount of hydrogen available to sensitive HAZ microstructures and thus reduce the tendency for hydrogen-induced cracking. Austenitic weld metals have a higher solubility for hydrogen than the ferritic weld metal or HAZ; hydrogen in the HAZ escapes at a faster rate than it is replenished by hydrogen in the austenitic weld deposit.

Carbon Equivalent

Several formulas have been developed to assist in evaluating the weldability of hardenable carbon and alloy steels. These formulas reduce the significant composition variables to a single number, known as the carbon equivalent (*CE*). For example:

$$CE = \%C + \frac{\%Mn}{6} + \frac{\%Cr + \%Mo + \%V}{5} + \frac{\%Si + \%Ni + \%Cu}{15}$$

Steels having carbon equivalents of less than 0.35% (using this formula) usually require no preheating or postheating. Steels with carbon equivalents between 0.35 and 0.55% usually require preheating, and steels with carbon equivalents greater than 0.55% may require both preheating and postheating. Because the carbon equivalent is calculated from the base-metal composition and includes no other variables, it is only an approximate measure of weldability or susceptibility to weld cracking. Other factors, such as hydrogen level, restraint, and cooling rate, that contribute to weld cracking must be considered simultaneously, in relation to a specific application.

The section on welding process variables later in this article provides more details on estimating preheat temperature from steel composition, hydrogen level, and restraint.

Another carbon equivalent formula that is widely used was developed by Ito and Bessyo:

$$CE = \%C + \frac{\%Si}{30} + \frac{\%Mn}{20} + \frac{\%Ni}{60} + \frac{\%Cu + \%Cr}{20} + \frac{\%Mo}{15} + \frac{\%V}{10} + 5(\%B)$$

This formula usually is combined with terms that also give an index of the hydrogen content and level of restraint in a weld. Extensive calculations have been developed using these two formulas to estimate the preheat, postheat, heat input, and hydrogen control necessary to avoid cracking when welding steels of known composition and joint geometry. Unfortunately, calculations have not been developed to define procedures for a wide range of steel compositions and welding conditions, and it is unrealistic to expect that a simple carbon equivalent formula based on average steel composition can allow for variations in steel macro- and microsegregation, inclusion content, and the wide range of filler metals and construction variables that affect the residual and applied stresses at the weld. Consequently, most welding procedures are based on practices developed from experience and on data published by the American Welding Society (AWS), including Welding Research Council Bulletin No. 191 and the Structural Welding Code D1.1, manufacturers' literature, and other professional societies, including the American Society of Mechanical Engineers (ASME) and the American Petroleum Institute (API). Occasionally, a carbon equivalent formula will be included in a specification or code, and the level of carbon equivalent will determine certain welding conditions, such as preheat.

As noted above, the value of the carbon equivalent is limited because it only considers the composition of the base material. Section thickness and weldment restraint are of equal or greater importance than the carbon equivalent. Figure 4 demonstrates the relationship between carbon content and section thickness as they affect weldability, expressed in terms of the need for preheating and postweld stress relieving. Combinations of carbon content and section thickness in area A of Fig. 4 are easily welded. Combinations in area B of Fig. 4 usually require preheating; those in area C usually require both preheating and postweld stress relieving. The use of low-hydrogen processes or filler materials can offset, to some extent, the requirement for preheating, by shifting the lines that separate the three zones upward.

Stress-Related Factors

Local cracking or fissuring in the deposited weld metal or HAZ occurs either because the strength or ductility of the steel is locally insufficient to sustain localized stresses without fracturing under the conditions of welding. Under conditions of extreme stress, complete failure of the joint may occur during or shortly after welding. However, it is also possible that stresses of a lesser magnitude, insufficient to cause instantaneous failure, will generate minute cracks in the weld zone that will propagate

Fig. 4 Combined influence of base-metal thickness and carbon content on weldability. Expressed in terms of the need for preheating and postweld stress relieving. (A) Neither preheating nor postweld stress relieving is usually required. (B) Preheating is usually required; postweld stress relieving is not usually required. (C) Both preheating and postweld stress relieving are usually required.

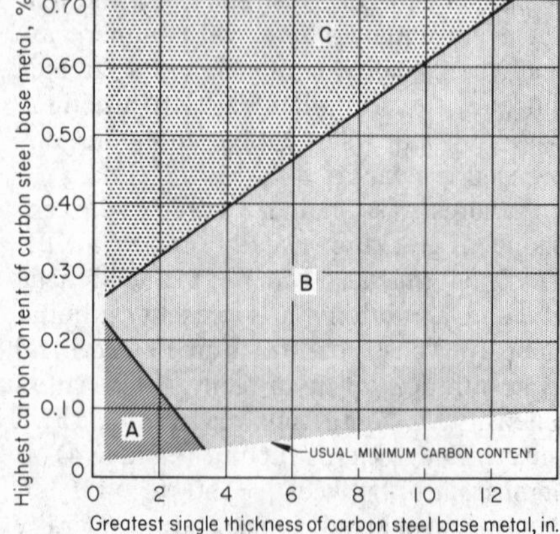

Fig. 5 Semitrailer upper fifth wheel. Design change provided less rigid construction preventing weld failures.

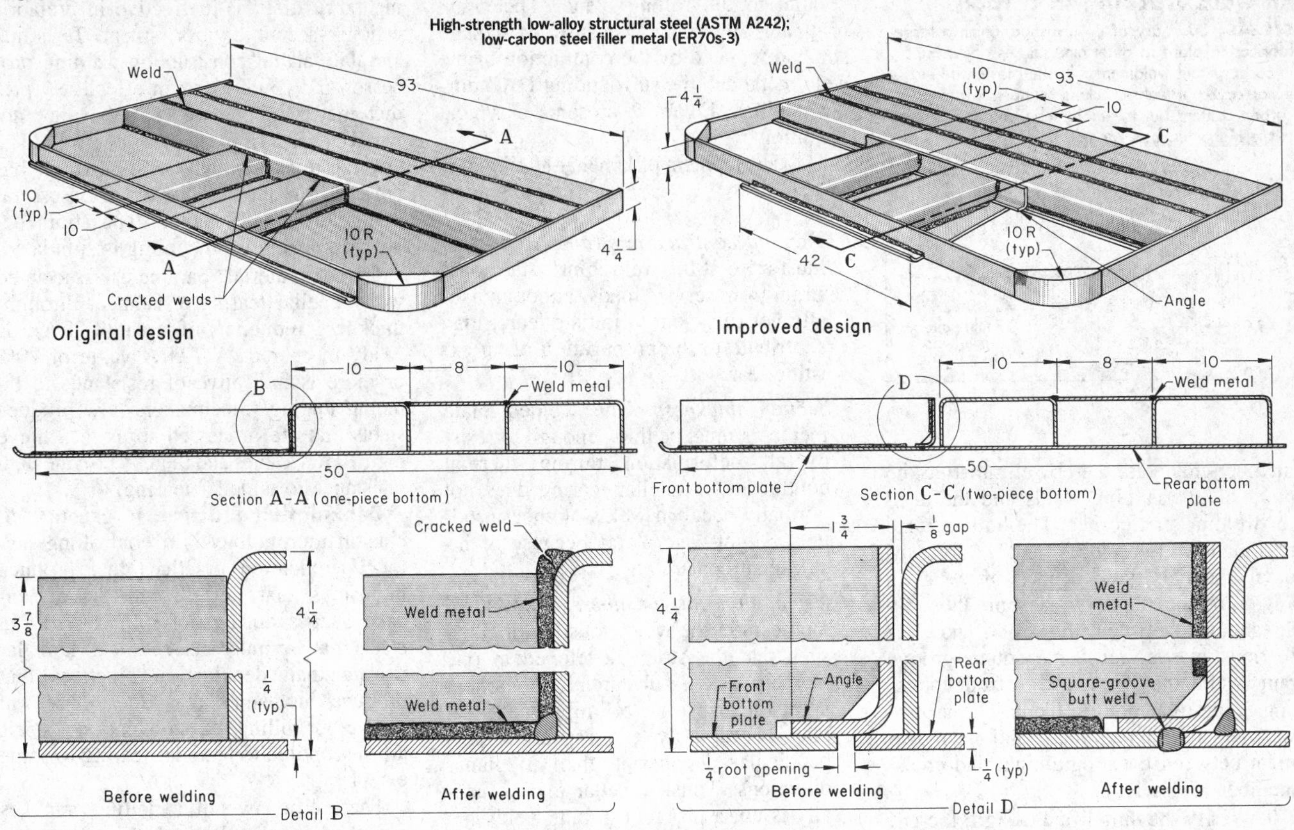

Conditions for GMAW

Joint type	Butt and T
Weld type	Square groove and fillet
Welding position	Flat and horizontal
Number of passes	One
Preheat and postheat	None
Fixtures	All components pneumatically clamped in a fixture before welding
Shielding gas	Argon, at 40 ft^3/h
Electrode wire	0.045-in.-diam ER70S-3
Electrode holder	Hand-held, air cooled
Wire feed	Automatic constant-feed type
Wire-feed rate	300 in./min
Power supply	500-A motor-generator
Current	300 A (DCEP)
Voltage	26
Travel speed	12 in./min

and eventually cause the welded structure to fail in subsequent service.

By far the most significant sources of stress during welding are thermal gradients, which may generate stresses that exceed the yield strength, particularly in light of the fact that yield strength is lower at elevated temperatures than it is at room temperature.

In practice, stresses are accommodated by movement in the parts of the assembly being welded. This movement may consist of plastic or elastic deformation, or gross movement of the parts. Thus, conditions that prevent movement of the parts increase the probability of cracking.

Therefore, thicker and more massive plates or parts being welded have a more rigid design; consequently, with higher hold-down or fixturing loads, the probability of cracking is greater. The root pass in massive weldments is particularly susceptible to cracking, because the relatively small cross section of weld metal is not strong enough to force the movement of the large mass of base metal.

Restraint in a welded joint usually is imposed by component design, although fixtures that hold the assembly for welding can also contribute to restraint. Susceptibility to cracking caused by restraint increases as the hardenability of the base metal or filler metal increases. Thus, cracking caused by restraint is closely related to preheating, postheating, and interpass temperature. The following example describes an application in which fatigue cracking failures in service were traced to rigidity in weldment design and excessive restraint during welding.

Example 1. Prevention of Fatigue Cracks in Welds by Redesigning Weldments for Less Rigidity. An upper fifth wheel for a highway semitrailer, when designed as shown at left in Fig. 5, proved unsatisfactory in service because fatigue cracks developed in the welds, made by semiautomatic gas metal arc welding (GMAW), at the joints between the longitudinal and cross-member channels. Under a 35 000-lb vertical load on the fifth wheel, the vertical displacement of these joints was $^3/_8$ in. relative to the ends of the cross members. Closer control of weld quality at the junction did not eliminate the cracking, nor did the addition of gussets between the two longitudinal channels and the front cross member. Although deflection under load could have been reduced by increasing the section thicknesses, this was undesirable because of the weight penalty.

The assembly was redesigned to omit the welds that were subject to failure. As shown at right in Fig. 5, a gap was left between the longitudinal and cross-

Fig. 6 Effect of carbon content on weld cracking in carbon steels. Summary of an investigation in a large fabricating plant to determine causes of cracking in carbon steel weldments. Brittle martensite was a minor cause of weld cracking in steels of lower carbon content but the major contributor for steels in the higher carbon range.

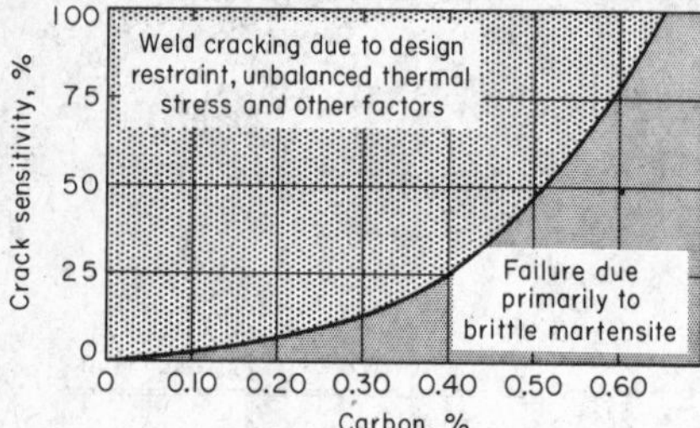

member channels, thereby eliminating the point of overlapping welds and the need for welding in a corner. The bottom plate was made in two pieces, and the longitudinal channels were capped off with an angle. In the improved design, the only attachment between the front and rear sections was a transverse square-groove butt weld that joined the front plate, angle, and rear bottom plate. The groove weld and the gap permitted rotational movement between the longitudinal and cross-member channels.

To verify the simplified design, the entire upper fifth wheel assembly was fatigue tested under a cyclic load of 0 to 35 000 lb at a rate of 19 cycles/min. The revised assembly design was accepted after withstanding 500 000 cycles without failure. Strain-gage measurements of the stress developed in the metal adjacent to the transverse groove weld by a 35 000-lb load showed it to be 24 ksi max. This was well within the 43 ksi fatigue strength limit (rotating-beam method) generally ascribed to high-strength structural steels with 50-ksi minimum yield strength.

Cracking Due to Design Restraint. Weld cracking is correlated with carbon content in Fig. 6, which summarizes a study made in one plant of a variety of welded components. Cracks occurring in weldments made of steels of lower carbon content were attributed chiefly to design restraint and unbalanced thermal stresses. In weldments made of steels with high carbon contents, hardening was the major cause of weld cracking.

Lamellar Tearing*

Lamellar tearing, a condition that may occur in base metal beneath a weld in steel plate, is characterized by a step-like crack parallel to the rolling plane. The crack originates internally because of tensile strains produced by the contraction of the weld metal and the surrounding HAZ during cooling. Figure 7 presents a typical condition.

The development of lamellar tearing may be caused by:

- *Thermal contraction strain*: This strain must arise at the weld joint. Additional strain from service loads, although usually not sufficient to initiate tears, may contribute to the propagation of an existing tear.
- *Section thickness*: The welded plate members must be thick enough to resist overall deformation during thermal contraction. Lamellar tearing does not normally occur in welds of sheet product or light-gage plates because of insufficient constraint.
- *Orientation of members*: Individual members in the welds must be oriented such that the strain developed in one member has a substantial component perpendicular to the rolling plane. Steel plate is ordinarily less resistant to perpendicular decohesion than in planar directions. Thus, lamellar tearing usually is not a problem in butt welds.
- *Susceptible material*: A local concentration of inclusions, particularly those extended in planar directions, must exist in the base metal. Such conditions usually are present in conventionally melted steels.

Many procedures have been used to assess the resistance or susceptibility of steel plate to lamellar tearing. Ultrasonic testing, which is effective in detecting the presence of lamellar tears developed during fabrication, is ineffective in prefabrication susceptibility assessment. Tests that simulate actual construction-welding processes and conditions can effectively predict material behavior, but they are not widely used for quality control. The through-thickness tension test (Ref 4) has emerged as a popular way to assess lamellar tearing resistance. The strain developed during weld cooling is simulated, and the material parameter associated with lamellar tearing resistance, through-thickness reduction of area (TTRA), is easily measured. A TTRA value of 20% or more is indicative of resistance to lamellar tearing. Steelmakers now produce steels that are processed so as to achieve high TTRA values and thereby provide high resistance to lamellar tearing.

Fig. 7 Lamellar tear caused by thermal contraction strain

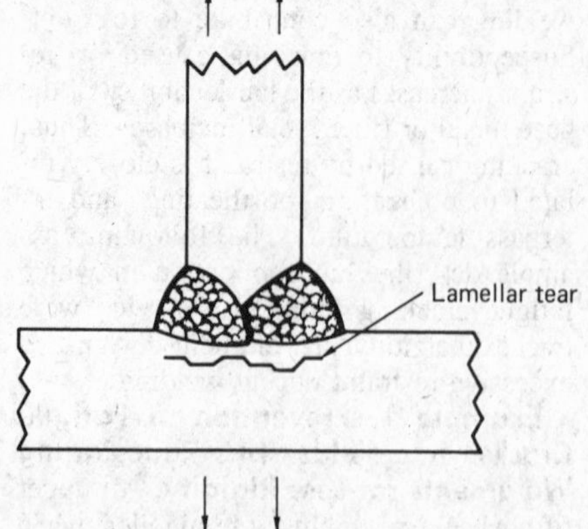

Deoxidation Practice. In general, inclusion morphology is a controlling factor. Deoxidation practices that produce inclusions that deform into large pancake shapes during hot rolling result in steels that are more susceptible to lamellar tearing than deoxidation practices that produce inclusions that do not deform during hot rolling. All conventional steels are susceptible to lamellar tearing to some extent.

The use of low-sulfur melting practices and the use of sulfide shape-control addition agents to prevent the formation of sulfide stringers is an effective way of improving the resistance of a steel to lamellar tearing. Many of these addition agents are also strong deoxidizers. To prevent loss by oxidation, they must be added to the steel only after the oxygen content of the molten metal has been reduced to a very low level by use of strong deoxidizers such as aluminum. Accordingly, the use of many of the most effective shape-control agents is restricted to fully killed steels, and special equipment generally is required to protect the shape control agent from oxidation during addition and during pouring. Examples of shape control agents include magnesium, calcium, rare earth elements, and zirconium.

The relative resistance to lamellar tearing of steels produced by the same deoxidation practice or by using the same sulfide shape-control agents may vary over a wide range. Obviously, an extremely clean steel that is free of pancake-type inclusions should exhibit the optimum resistance to lamellar tearing; a steel containing minimal small pancake-type inclusions usually is more resistant to tearing than a steel containing large numbers of refractory or globular inclusions (particularly if the refractory or globular inclusions tend

*The information contained in this section is based on Ref 1, 2, and 3. Additional sources of information are presented in these references.

to lie on planes or in clustered colonies within the plate).

Practices that combine (1) low sulfur content, (2) sulfide shape control, and (3) use of aluminum or other strong deoxidizers to protect the shape-control addition prevent silicate formation and improve through-thickness ductility. In addition, vacuum melting to reduce oxygen content and incremental rapid solidification (vacuum-arc remelting or electroslag remelting) to reduce segregation can result in still further improvement in through-thickness ductility by producing a higher level of cleanliness and less anisotropy in ductility.

Matrix Properties. The intrinsic fracture toughness of the matrix can also modify the lamellar tearing characteristics of a steel. A highly ductile, low-strength matrix can yield at the sharp edges of inclusions and provide the required plastic deformation beneath the restrained weld zone without propagating cracks that open at the inclusion-matrix interface. In higher strength steels, the overall elastic strain occurring across an inclusion before the matrix yields will be higher, and the ductility of the matrix will be lower. Under these conditions, stress concentrations and restraint existing at the edges of inclusions within the steel may lead to easy crack propagation between inclusions in the presence of high strains due to welding.

For a given morphology and distribution of inclusions, a critical matrix strength level exists where the crack-like interfaces between the inclusions and the matrix propagate through the matrix from inclusion to inclusion by unstable crack growth when the strain level in the base plate beneath a weld reaches a critical level. If the matrix is stronger than the weld metal, the plastic strains resulting from welding could be forced to occur in the lower strength weld metal, thus avoiding lamellar tearing in the base plate.

Steels with low notch toughness often exhibit high susceptibility to lamellar tearing. Because many of the material characteristics that contribute to good resistance to lamellar tearing also improve notch toughness, steels that are designed for resistance to lamellar tearing exhibit high shelf energies and may exhibit low transition temperatures, provided the steel is properly heat treated.

Because lamellar tears generally occur just below the weld HAZ, embrittlement of the matrix by strain aging is one of the factors that may reduce lamellar-tearing resistance of the steel. Hydrogen embrittlement and strain aging are contributory causes to lamellar tearing. Low-hydrogen practices are desirable to minimize lamellar tearing. It has been shown that hydrogen has a deleterious effect on the resistance of a steel to lamellar tearing (Ref 5). Banding in plate steel also has been suggested as a contributing matrix variable influencing lamellar tearing. Because inclusions tend to lie in or along the bands, it is difficult to separate the influence of inclusions from the influence of banding.

Preheat and Heat Input. A high transition temperature and a high dissolved hydrogen content increase the susceptibility of a steel to lamellar tearing. The harmful effects of both of these metallurgical factors can be counteracted by maintaining a high preheat and interpass temperature. Fabricator experience with preheat has been mixed. In some instances, increased preheat has resulted in increased tearing. If transition temperature is not a problem and low-hydrogen consumables are used, high preheat probably does more harm than good.

The effect of heat input is also unclear. High-heat-input welding processes, such as SAW, are believed to diminish the risk of tearing because of (1) deeper weld penetration, (2) greater width of the HAZ, and (3) generally lower strength of the weld metal. However, raising the heat input for a given welding process has not proved beneficial. In rigid structures, high heat input results in higher final shrinkage strains. In some instances, tearing can be avoided by using very low heat input with no preheating. The latter technique has proved successful for repairing cracks in joints of highly restrained structures.

Modified Weld-Pass Procedures. One method of making a modified weld pass is to continue to lay weld beads on the plate subjected to through-thickness stresses after completing the root pass of a T-joint. Weld passes bridging the leg of the joint with the base are made as late as possible in the welding sequence, thus tending to develop the highest weld restraint strain in the underlying weld metal rather than in the base plate. With T-joints, high strains at the weld root caused by rotation of the leg of the joint should be avoided. By making both fillet welds in the joint concurrently, the strains imposed on the plate subject to through-thickness stress are balanced. As a result, the tendency for a lamellar tear to initiate from the root of the weld is reduced because the leg is not pulled to one side by the one-sided shrinkage that results if one of the fillet welds is completed before the second fillet weld is started.

Although certain welding modifications have been found to be helpful under the appropriate conditions, no one procedure or combination of procedures can ensure against lamellar tearing. Accordingly, each situation must be judged separately, and the best method of avoiding lamellar tearing must be selected after careful consideration and a complete understanding of all the factors involved.

Weld Joint Design. Reference 3 details weld joint design considerations that can be used to minimize lamellar tearing in shipbuilding applications. These considerations, which are applicable to other situations as well, indicate that it is advisable to (1) avoid excessive through-thickness strains, (2) reduce joint restraint, and (3) reduce component restraint.

Excessive Through-Thickness Strains. Weld shrinkage strains can be avoided by welding between the ends of plates rather than on the surface of the susceptible material (Fig. 8). This welding technique minimizes the shrinkage strains in the critical through-thickness direction but may require the use of electroslag welding (Fig. 9).

Orienting the weld fusion interface at an angle to the surface of the susceptible plate (Fig. 10) is another method of reducing weld shrinkage strain. Large bevel angles offer less risk of tearing, but edge preparation cost and the volume of weld metal required are also higher than for smaller edge angles. Selection of a cost-effective angle must take into consideration the susceptibility of the plate, the importance of the connection, and the relative cost of fabrication.

Replacing double-sided, full-penetration welds (Fig. 11a) with symmetrical fillet or partial-penetration welds (Fig. 11b) minimizes the volume of weld material and reduces the strain in the through-thickness direction. The total shrinkage of the fillet welds occurs at an oblique angle to the plate surface, thereby further reducing the strain component in the through-thickness direction.

The use of castings or forgings in T- and cruciform joints to eliminate any risk of

Fig. 8 Joint design to reduce weld shrinkage strain.
(a) Susceptible. (b) Improved

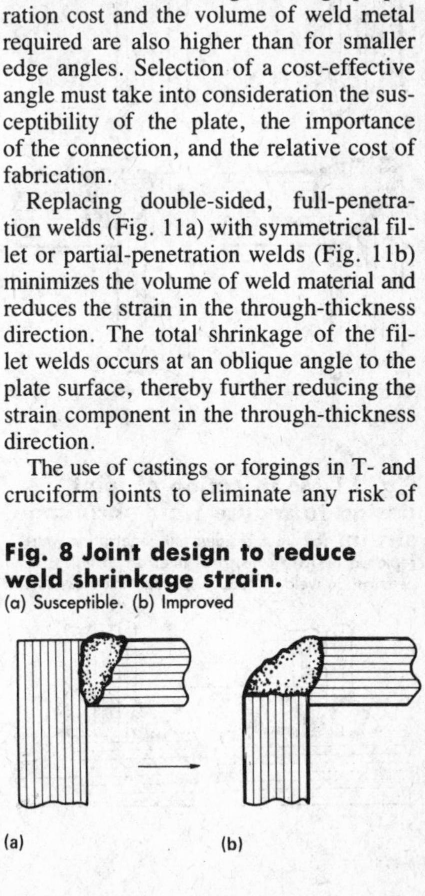

(a) (b)

Fig. 9 Use of electroslag welding to reduce through-thickness strains. (a) and (c) Susceptible design. (b) and (d) Improved design to eliminate weld shrinkage strain

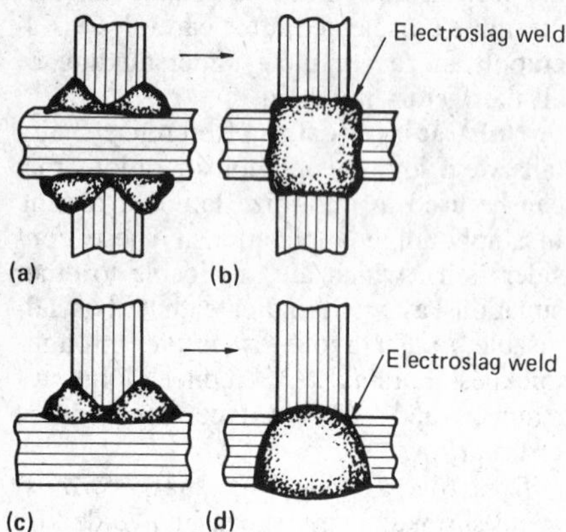

Fig. 10 Angling of weld fusion interface to minimize weld shrinkage strain. (a) Susceptible joint design. (b) Improved, but not as desirable as (c). (d) Joint design offers little improvement.

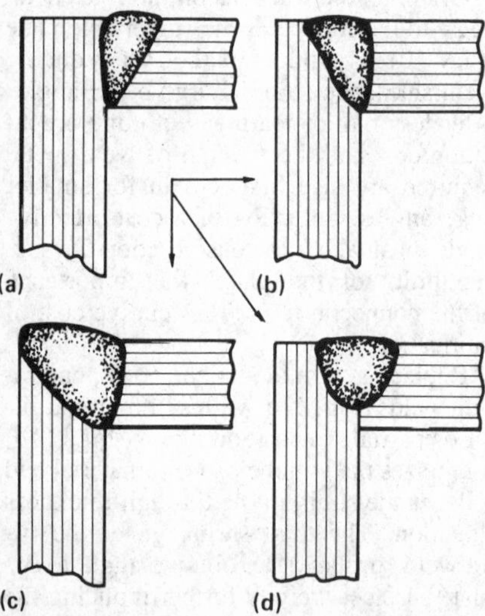

Fig. 11 Modification of joint design to reduce weld shrinkage strain. (a) Double-sided full-penetration weld replaced by (b) symmetrical fillet or partial-penetration weld to reduce through-thickness strain

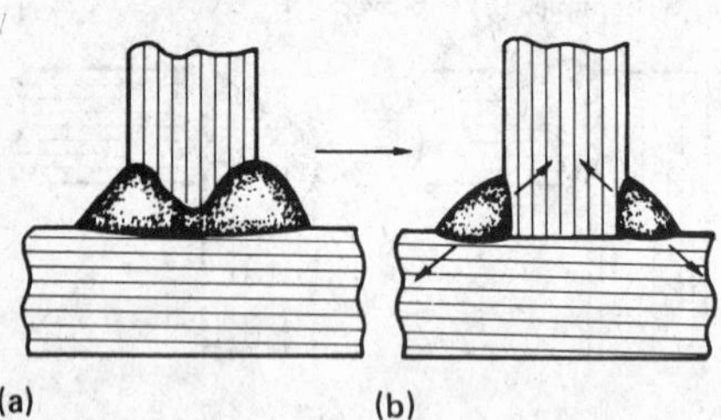

lamellar tearing (Fig. 12) in critical applications, e.g., pressure vessels, is also advised.

Reduction of joint restraint can be accomplished by the following methods:

- Reducing the size of the weld by not using welds larger than necessary to transfer the calculated design loads. For example, full-penetration welds at the deck stringer plate/sheer strake connection (Fig. 13a) often can be replaced by smaller partial-penetration weld (Fig. 13b) or a combination of a partial-penetration weld and fillet weld (Fig. 13c).
- Joining plates of different thicknesses may be used so that weld size may be reduced by placing it in the thinner plate (Fig. 14).
- Replacing large single-sided welds (Fig. 15a and c) with balanced double-sided welds (Fig. 15b and d) is suggested to eliminate the unsymmetrical concentration of strain.
- Selection of weld configurations that distribute the weld metal over more of the surface of the susceptible plate is also recommended (Fig. 16). The use of smaller weld sizes of longer length or double fillets in place of full-penetration welds reduces the volume of weld metal and distributes the shrinkage strains over a larger area of the susceptible plate.

Reducing component restraint can sometimes be accomplished by modifying the structural configuration or cross sections. To decrease the level of restraint:

- Avoid complex, multi-member connections. This is not always practical in structures such as offshore drilling units.
- Minimize member stiffness by using sections of minimum thickness.

Fig. 12 Use of forging or casting to reduce lamellar tearing in critical welds

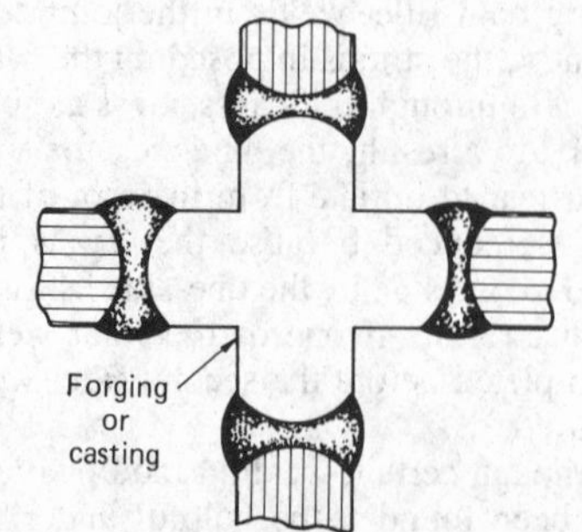

- Use flat plate instead of curved members wherever possible.
- Do not use stiffeners, brackets, or gussets not specifically required by the design. Welded sections and all auxiliary stiffening should be the smallest required to suit the design loads.
- In cruciform joints, stagger the members on opposite sides of the susceptible plate (Fig. 17). This method is not always desirable in highly loaded joints.
- If possible, use lower strength material for the member causing the strain in the through-thickness direction (Fig. 18).

Solidification Hot Cracking

Hot cracks occur at elevated temperatures (1000 °F and above) and usually are located in the weld metal, but may also be found in the HAZ. The fracture path of a hot crack is intergranular. If a hot crack propagates to the surface of the weld bead

Fig. 13 Joint design to reduce joint restaint. (a) Susceptible design. (b) Use of smaller partial-penetration weld. (c) Combination of partial-penetration weld and fillet weld

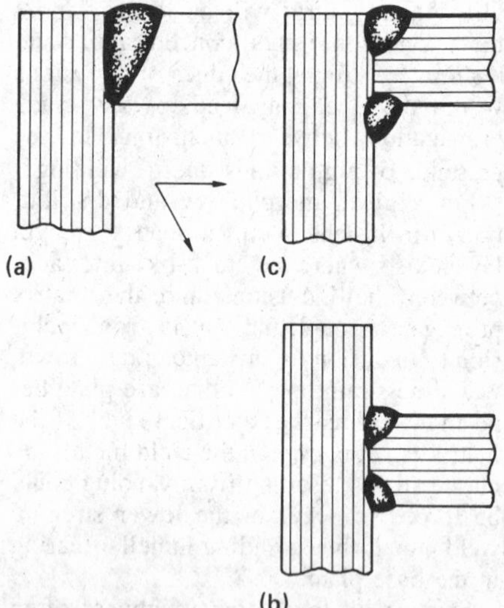

Fig. 14 Weldment design to eliminate joint restraint. (a) Susceptible design. (b) Improved design

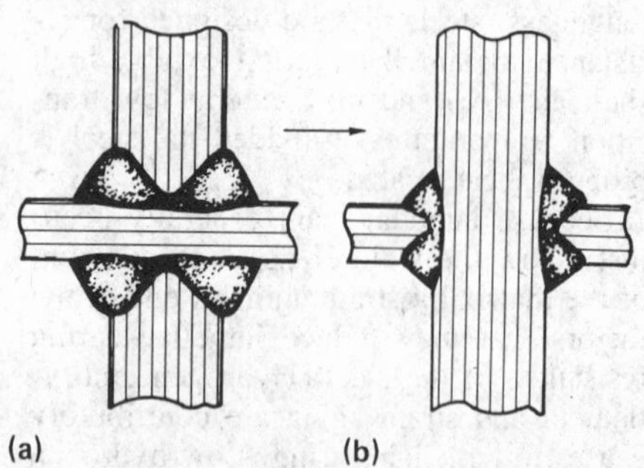

Fig. 15 Elimination of unsymmetrical concentration of strain. Large single-sided welds (a) and (c) are replaced by balanced double-sided welds (b) and (d).

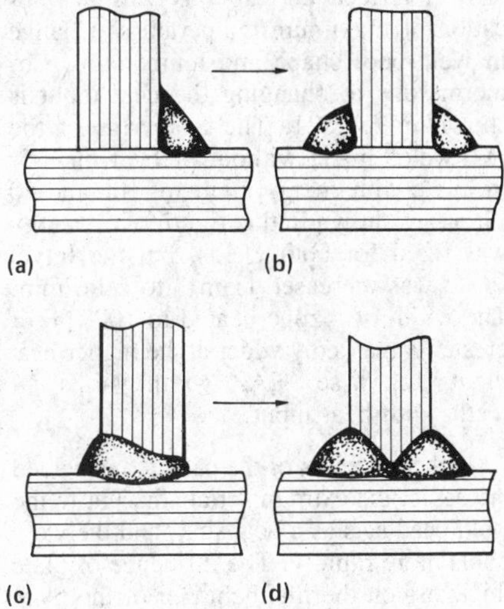

Fig. 16 Weld configurations to distribute weld metal evenly and eliminate joint restraint. (a) Susceptible design. (b) Improved design

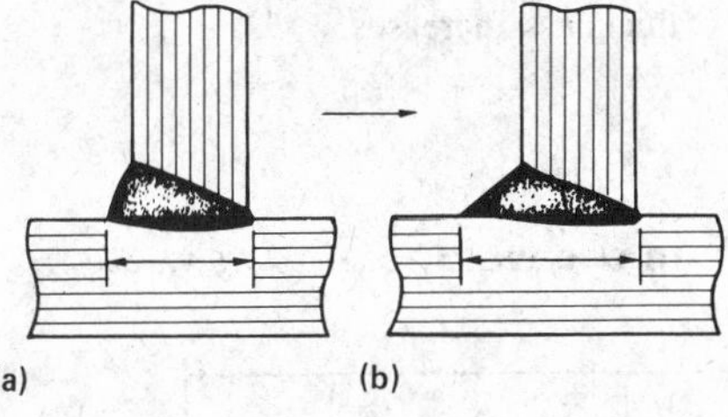

and is exposed to the atmosphere, the fracture surface may be coated with oxides. The color of this oxide, whether temper colored or scaled, provides a rough indication of the temperature at which cracking occurred.

Solid crystals begin to form when the weld metal is cooled below the liquidus temperature. As the temperature of the weld is reduced to near the lower end of the weld-metal solidification temperature range, the weld has little if any tensile strength, although it appears to be completely solidified. If the weld experiences stress of a sufficient magnitude (either from weldment restraint or shrinkage) under these conditions, and deformation of the solid crystals or readjustment of the remaining liquid metal is not possible, the weld may crack through that portion of the remaining liquid metal that would have formed a grain boundary.

Wide weld-metal freezing ranges and the presence of high levels of impurities (particularly sulfur and, to a lesser extent, phosphorus) in the filler metal and base metal increase the tendency for hot cracking. Both of these factors contribute to the formation of continuous liquid films at grain boundaries that are present at lower temperatures and for longer periods of time. By controlling the concentration and shape of these intergranular constituents, hot cracking can be eliminated.

In steel, for example, iron sulfides may form and wet grain boundaries. Manganese, in sufficient concentrations, usually ties up most of the available sulfur as globular manganese sulfide. Manganese sulfide has a much higher melting point than iron sulfide and does not wet grain boundaries, thereby inhibiting hot cracking. Usually, a 20-to-1 ratio of manganese to sulfur inhibits the formation of low-melting intergranular sulfide films. Resistance to hot cracking may be obtained if the individual concentrations of sulfur and phosphorous are reduced. Increased carbon content, however, adversely affects the tolerance for these elements.

When constrained by specific material compositions susceptible to hot cracking, the degree of restraint should be reduced and low-heat-input processes and procedures should be used. Susceptible solidification patterns, such as those with high weld pool depth-to-width ratios and those that include long columnar crystals, also should be avoided.

Fig. 17 Reduction of component restraint. See text for discussion.

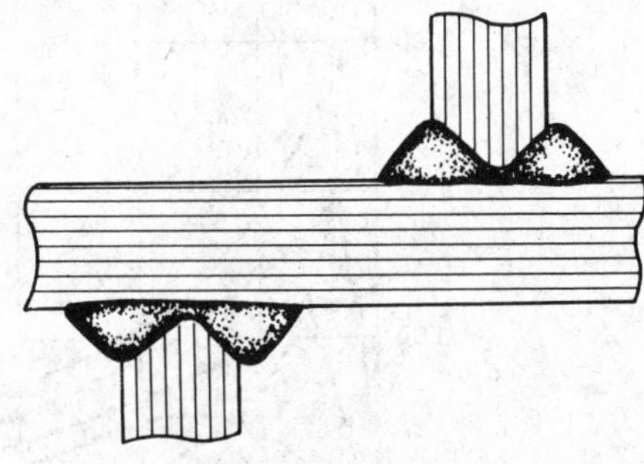

Fig. 18 Reduction of component restraint. See text for details.

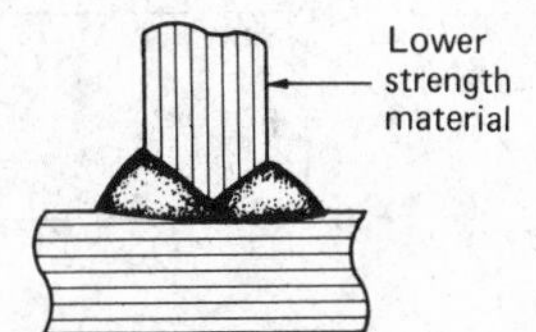

Hot cracking is not usually as frequent a problem with the steels covered in this article and is a more important consideration when arc welding austenitic stainless steels. The effects of travel speed, weld-bead geometry, and restraint are similar in all alloy systems and can be used to design crack-free welding procedures for any steel grade. For example, slower travel speeds reduce the length of the weld pool and also reduce the tendency for hot cracking.

Other Forms of Cracking

Reheat, stress-relief, or stress-rupture cracking are equivalent forms of cracking that occur in quenched and tempered steels. This type of cracking is covered in detail in the section of this article on these steels. Sulfide-stress cracking, or hydrogen sulfide cracking, occurs in welds during service when subjected to high hydrogen sulfide environments. When welding HSLA steels, the hardness of the welded area may have to be held to a low level to avoid this type of in-service hydrogen-induced cracking.

Serviceability

Serviceability of welds in hardenable carbon and alloy steels is frequently of more concern than with plain carbon steel welds, because these steels usually are incorporated into weldments designed for strength, toughness, or corrosion properties. These steels should be welded with greater precautions than when welding plain carbon steels to preserve these properties in the weldment.

Literature on the properties of weldments required to meet specific service levels is extensive; only a summary of general guidelines can be discussed in this article. See the article "Weld Discontinuities" in this Volume for more information on the combined effects of weldment properties, service conditions, and discontinuities and how they relate to fracture-safe design.

In designing any weldment, base-metal properties, anticipated loading and/or corrosive environment, welding procedures, and inspection methods for discontinuities must be considered simultaneously. The steels discussed in this article are usually higher strength than plain carbon steels; therefore, fracture toughness and fatigue properties are more critical when considering weldment design. With higher strength, the restraint resulting in the presence of a notch is also higher. Consequently, high-strength welds may require a higher degree of fracture toughness to meet equivalent service requirements. In the sections of this article on the various steel classifications, weldment toughness properties are discussed, as well as the welding procedures (particularly heat input) for producing required properties. Stress risers and discontinuities that result from arc welds also play a crucial role in determining the fatigue properties of high-strength steels. Most arc welds in high-strength steels, if left in the as-welded condition, exhibit long-life fatigue properties that are only marginally superior to welds in plain carbon steels, if at all. Care must be exercised in using high-strength steels in applications involving fatigue so that the welds do not reduce the serviceability of the structure.

Welding Process Variables

Welding process variables affect the temperature distribution and metallurgical changes of the weld metal and the adjoining base plate. Energy input from the welding arc, preheat and interpass temperatures, chemical composition and thermal properties of the plate, joint design, and physical dimensions of the part being welded all influence the properties of the completed joint. Distribution of peak temperatures in the HAZ and the cooling rate of the weld metal and HAZ are the most important considerations when welding hardenable carbon and alloy steels.

Arc energy is expressed as heat input to the base material (*H*) and is defined in terms of the arc voltage (*E*) in volts, the welding current (*I*) in amperes, and arc travel speed (*S*) in inches per minute as:

$$H = \frac{E \times I \times 60}{S \times 1000} \quad \text{(Eq 1)}$$

Heat input (given in kilojoules per inch), as calculated using Eq 1, is nominal heat input, as arc efficiency is not considered in the calculation. Arc efficiencies range from about 0.7 for shielded metal arc electrodes to 0.95 for SAW. Efficiency should be included in the above calculation to determine true heat input when the temperature distributions and/or cooling rates are calculated. However, in a welding procedure specification, nominal heat input is adequate and provides a useful comparison of energy inputs. Some welding specifications limit heat input, especially for welding quenched and tempered steels. Heat input also is related to other welding conditions. For example, allowable arc heat input usually varies directly with increasing section thickness and inversely with increasing preheat temperature.

Temperature gradients calculated for various locations in the HAZ of an arc weld are shown in Fig. 19. The following conclusions drawn from the curves in this figure are applicable to all arc welds:

- Peak temperature decreases rapidly with increasing distance from the weld centerline.
- Time required to reach peak temperature increases with increasing distance from the weld centerline.
- Rates of both heating and cooling decrease with increasing distance from the weld centerline.

Changes in the HAZ thermal cycles with various energy inputs and preheat temperatures of welds made on 0.5-in. plate using stick electrodes are shown in Fig. 20. The four thermal cycles compare welds made at 100 and 50 kJ/in. with preheat temperatures of 80 and 500 °F. Separation of the curves for the effect of preheat is reduced at the high heat input compared to the 50-kJ/in. welds. At 100 kJ/in., the cooling curves are essentially identical for the first 30 s. In contrast, an increase in preheat temperature from 80 to 500 °F at an energy input of 50 kJ/in. results in the temperature being 340 °F greater after 30 s.

Peak temperatures in Fig. 20 represent the melting temperature. The weld metal-HAZ interface corresponds with the location of maximum temperature. Change in weld-pool shape and temperature isotherms due to changing the heat input is shown in Fig. 21. The isotherms of the two welds made without preheat on 0.5-in. plate with energy inputs of 100 and 50 kJ/in. are shown in this figure. Arc energy was fixed for both welds, but the travel speed was increased from 3 to 6 in./min. The width of a zone heated to 1000 °F or greater is markedly wider at the higher heat input. Likewise, HAZ width varies directly with heat input.

As the thickness of the parts to be welded increases, the extra material adjacent to the weld bead acts as a heat sink, and the weld cools more rapidly. The influence of plate thickness on thermal behavior of the plate can be summarized as:

- Cooling rate of the HAZ increases with plate thickness.
- Time at elevated temperature and the width of the HAZ decreases as the plate thickness increases.

Fig. 19 Thermal cycles in the HAZ of an arc weld. Source: *AWS Welding Handbook*, Vol 1, 7th ed.

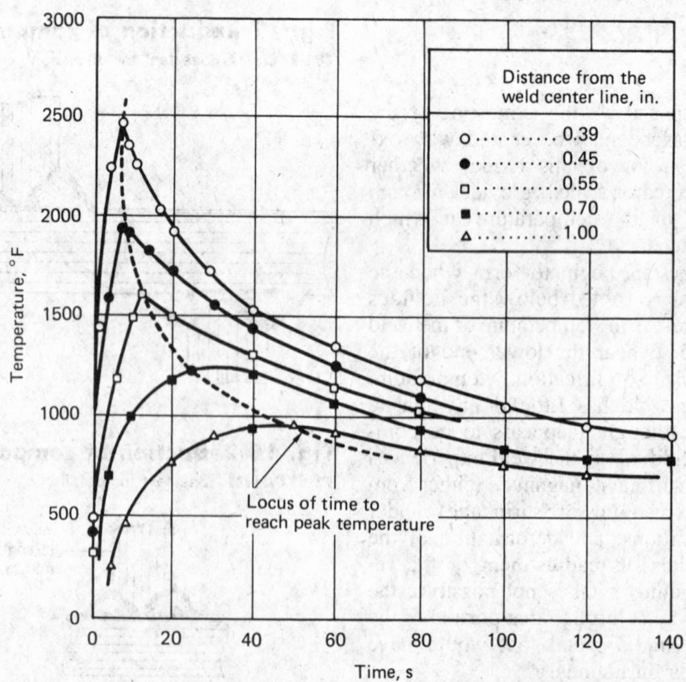

Fig. 20 Effect of initial plate temperature on thermal cycles in the HAZ of arc welds. Source: *AWS Welding Handbook*, Vol 1, 7th ed.

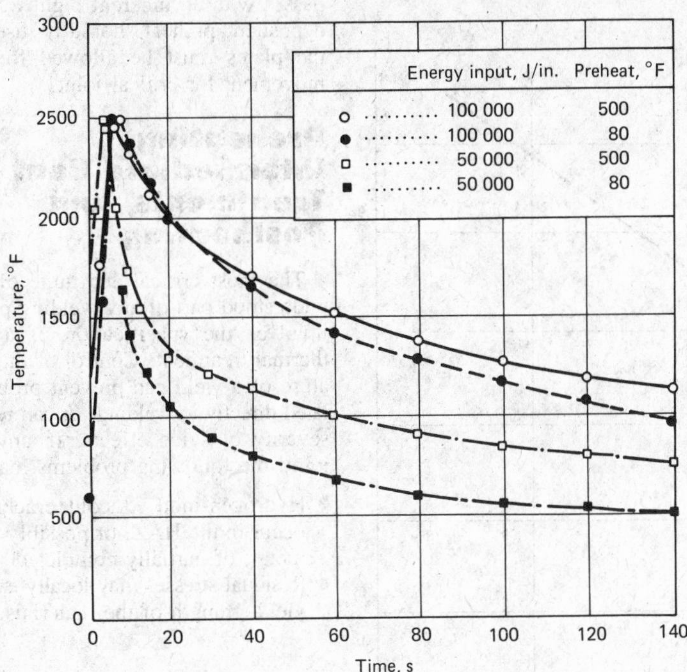

Fig. 21 Effect of energy input on temperature isotherms at the surface of a steel plate. Upper portion 100 kJ/in., 24 V, 208 A, 3 in./min. Lower portion: 50 kJ/in., 24 V, 208 A, 6 in./min. (Source: *AWS Welding Handbook*, Vol 1, 7th ed.)

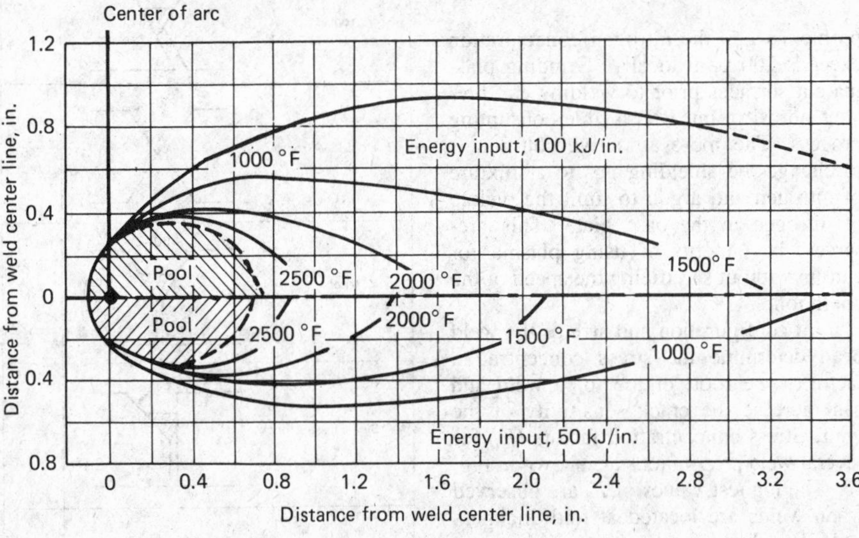

Although weld thermal cycles are useful to illustrate the effect of various process variables, they are difficult to use in a practical manner. A more significant parameter that affects metallurgical transformations, mechanical properties, and weldability of steels is the cooling rate of the weld. Cooling rate either is specified at one particular temperature (in degrees Fahrenheit per second) or is taken as the time required to cool between two arbitrary temperatures (in seconds).

Cooling rate of the HAZ at 1000 °F (R_{1000}) is frequently used to determine welding parameters. This temperature represents the temperature at which austenite transforms to martensite during continuous cooling. It is also referred to as the critical cooling rate—cooling rates faster than a given value may develop hard martensitic structures that are extremely susceptible to cracking, while slower cooling rates yield transformation products that are not prone to cracking.

The average cooling rate between 570 and 120 °F ($t_{570-120}$ in seconds) is used to determine the reduction of hydrogen after welding. Hydrogen content in the weld area is directly related to crack susceptibility of the joint. As with the cooling rate, there exists a critical time to cool ($t_{570-120}$), which defines the crack/no crack threshold. Longer cooling times reduce local hydrogen concentration below the critical level, whereas short cooling times do not remove sufficient hydrogen to prevent cracking. Cooling time is controlled primarily by preheat and interpass temperatures and, to a lesser extent, by the heat input. The increase in cooling time for a butt weld made using a heat input of 20 kJ/in. on 0.56-in.-thick plate for a range of preheat temperatures is shown in Fig. 22.

Joint Preparation

Preparation of a joint for welding and fit-up affect joint integrity and the time required to complete the weld. In butt welding, the widths of the root face and the root gap affect susceptibility to cracking, as they affect the size of the root bead and the resultant stress. If an insufficient amount of weld metal is deposited in the root pass, the bead will lack strength; unless the interpass temperature is carefully maintained, cracking in the root bead is likely to result. Angles between components being welded also influence weld-crack susceptibility, as these angles govern the amount of weld deposit and therefore the heat input.

Joint geometry and location of the root pass, with respect to plate thickness, can produce different stress concentrations and influence crack susceptibility of the root.

Preparation of the joint for welding is chosen so as to provide adequate room for electrode manipulation to ensure root bead penetration and subsequent ease of slag removal, while the amount of filler metal required is kept to a minimum. The recommended proportions of grooves for SMAW, GMAW, gas tungsten arc welding (GTAW), and flux cored arc welding (FCAW) are illustrated in the Appendix of the article "Joint Design and Preparation" in this Volume.

Joint edges can be prepared by shearing, machining, grinding, gas cutting, or

Fig. 22 Increase in cooling time for a butt weld made using a heat input of 20 kJ/in. on a 0.56-in.-thick plate for a range of preheat temperatures

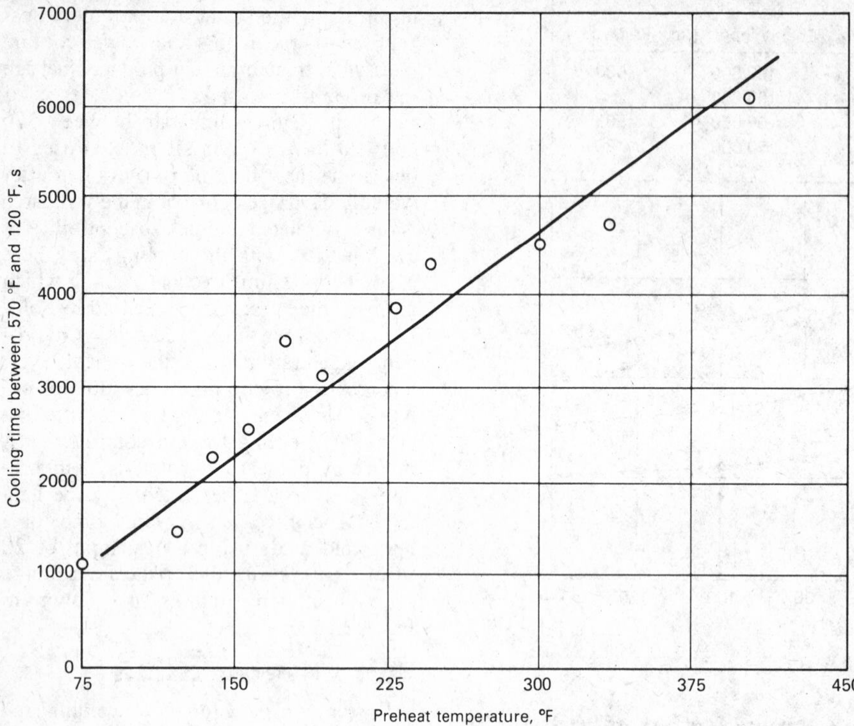

arc cutting. Gas or arc cutting can produce zones of hard, brittle metal, the hardness and depth of which increase with increasing alloy content. The effects of gas cutting on medium-carbon steels and the use of preheating and postheating to obtain an edge with the desired properties are described in the article "Thermal Cutting" in this Volume. Annealing or normalizing may be necessary following gas cutting to restore the original structure of the steel.

Before welding, surface irregularities, such as nicks, cracks, and gouges, that act as stress risers must be ground out. Grease, paint, surface scale, oxide, water, and other foreign material also must be removed. Because some carbon steels are sensitive to hydrogen-induced cracking, water in any form (moisture in scale, humid atmosphere, rain, snow) should be removed, and a low-hydrogen electrode may be recommended.

Welding of joints that have been plasma cut can sometimes lead to porosity. Nitrogen commonly is specified as the shielding gas because it is relatively inexpensive, but it is readily absorbed by steel at elevated temperatures, such as those along the cut surface. Porosity becomes a problem when the nitrogen concentration exceeds 200 ppm locally. Grinding plasma cut surfaces prior to welding can prevent porosity, but it is a time-consuming process. The more attractive solution is to change the shielding gas to a mixture of nitrogen and argon to limit the pickup of nitrogen in the base plate. This preserves the benefits of using plasma for cutting without sacrificing the speed of the operation.

Joint configuration and size of the weld bead determine the stress concentration factor at the root, or toe, of a weld and can increase the crack sensitivity of the joint. Stress concentration factors (K_t) for several weld preparations are shown in Fig. 23. The highest values of K_t are observed when welds are located at mid-thickness and when there is an acute angle between the weld metal and base plate. The lowest K_t values occur when the weld is deposited at the outside edge of the joint. At these locations, the plate bends sufficiently to reduce the sharpness of the notch and hence the stress concentration. A large K_t value alone does not necessarily mean that the weld will crack; a stress applied across the weld when the plates are prevented from contracting freely increases crack susceptibility. Therefore, when root passes with an inherent high K_t value are deposited, preheat should be used and/or the plates must be allowed freedom of movement for critical joints.

Preheating, Intermediate Heat Treatments, and Postheating

The most critical and quite often least understood part of any welding procedure involves the specification of a suitable thermal treatment. Control of the temperature of a weld can prevent problems related directly to welding, or can reduce the severity of some effects. In any welded joint, the following problems may arise:

- Hydrogen-induced cold cracking may occur in the HAZ, or possibly the weld metal, of partially completed welds.
- Residual stresses may locally exceed the yield strength of the materials.

Fig. 23 Stress concentration factors at root and toe weld positions

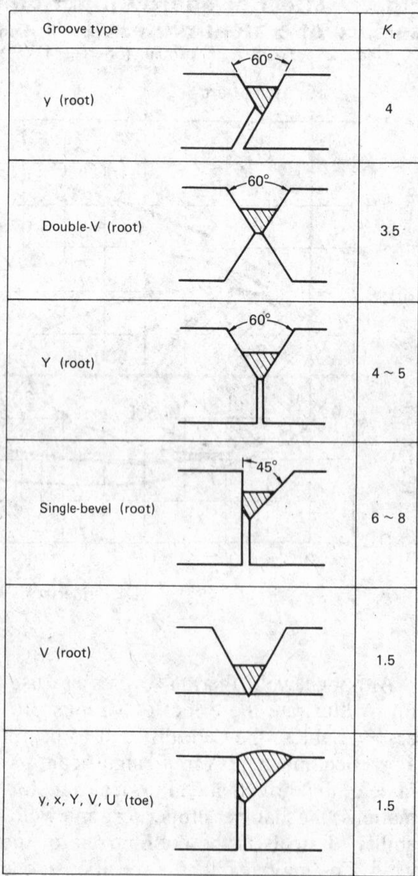

Groove type		K_t
y (root)	60°	4
Double-V (root)	60°	3.5
Y (root)	60°	4 ~ 5
Single-bevel (root)	45°	6 ~ 8
V (root)		1.5
y, x, Y, V, U (toe)		1.5

- Restraint, or rigidity, of the overall structure may contribute to the generation of reaction stresses transverse to the direction of welding.
- Heat of welding may produce unacceptably low toughness or mechanical properties in the weld area compared to the base plate.

To ensure the integrity of a welded structure, the items above should be controlled by welding procedure.

Preheating of a joint is commonly done with torches or electrical strip heaters. If the part is small enough, it can be placed in a furnace. Preheating is specified primarily to prevent hydrogen-induced cracking. Preheat reduces the cooling rate of the HAZ of each weld bead. Slower cooling rates reduce hardness, hydrogen content in the vicinity of the weld, and stresses. Preheat temperature is maintained throughout the welding cycle and is known as the interpass temperature between weld beads. Hydrogen cracking can occur if the interpass temperature becomes too low. An upper limit on the interpass temperature also must be maintained. If the temperature becomes too great, excessive grain growth in the HAZ may result, with a corresponding reduction in mechanical properties.

In welding thick sections, which are prone to hydrogen-induced cracking, an intermediate heat treatment sometimes is used to reduce peak hydrogen concentration. The temperature of the weld zone is increased to 1160 °F for approximately 1 h in a typical heat treatment cycle. An extreme case involves an intermediate heat treatment after each weld bead. Besides being time consuming, this practice requires an undue amount of energy. A low-temperature postweld heat treatment, which decreases the maximum hydrogen content and likewise prevents cold cracking, is preferred.

Increasing the temperature of a completed weld to just below the critical temperature for a specified length of time, followed by uniform cooling, is the usual stress-relief heat treatment. Thermal treatment alleviates stresses related to welding, which results in secondary improvements, such as:

- Improved notch toughness and resistance to fatigue conditions
- Greater resistance to stress corrosion cracking
- Improvement in dimensional stability of a welded structure, especially when welding is followed by machining

Table 1 Recommended minimum preheat and interpass temperatures for arc welding of typical ASTM quenched and tempered steels

Thickness range, in.	Minimum preheat and interpass temperatures, °F(a) A514/A517	A533	A537	A543	A678
Up to 0.50	50	50	50	100	50
0.56 to 0.75	50	100	50	125	100
0.81 to 1.00	125	100	50	150	100
1.1 to 1.5	125	200	100	200	150
1.6 to 2.0	175	200	150	200	150
2.1 to 2.5	175	300	150	300	150
Over 2.5	225	300	225	300	...

(a) With low-hydrogen welding practices. Maximum temperature should not exceed the given value by more than 150 °F.

Stress-relief heat treatments are done at temperatures between 1100 and 1250 °F. The exact temperature is material dependent, with the time at temperature nominally specified as 30 min/cm (0.4 in.) of thickness. Lower stress-relief temperatures require slightly longer lengths of time than higher temperatures for comparable results.

Steels susceptible to hydrogen-induced cracking, thick sections in particular, should not be allowed to cool below the preheat temperature before they are stress relieved. Stresses generated during welding reach a maximum at this time, and hydrogen concentration gradually builds to a peak value also. These maximum values are located in the HAZ of the final weld passes, and cracking is likely to occur when joint temperature decreases.

Preheating

The minimum preheat and interpass temperatures are chosen primarily so as to prevent hydrogen-induced cracking. They are chosen with considerations of the chemical compositions of the base material and electrode, the hydrogen content of the deposited weld metal, and the stresses imposed across the weld. Local stress concentrations at the weld root and toe increase the possibility of cracking (Fig. 23).

The structural welding codes in Canada (CSA W59) and the United States (AWS D1.1) outline the minimum preheat and interpass requirements. Tables 1, 2, and 3 give recommended minimum preheat and interpass temperatures for various alloy steel classifications. The temperatures depend upon the welding process used and are chosen for the more susceptible steel or plate thickness. Welding is not permitted when the ambient temperature surrounding the weld area is lower than 0 °F. When no preheat is specified and the plate temperature is below 32 °F, the base metal should be preheated to at least 50 °F and maintained during welding. When the base-metal temperature is below the specified minimums in Tables 1, 2, and 3, it should be preheated such that the surfaces of the parts being welded are at or above the specified minimum temperature for a distance equal to the thickness of the part being welded, but not less than 3 in., both laterally and in advance of welding.

The temperatures in Tables 1, 2, and 3 are guidelines based on average conditions. The codes recommend higher temperatures for (1) highly restrained welds, (2) certain combinations of steel thickness and energy input, (3) high-strength weld metal, and (4) joints where transfer of ten-

Table 2 Recommended minimum preheat and interpass temperatures for several AISI low-alloy steels

AISI steel	Thickness range, in.	Minimum preheat and interpass temperatures, °F(a)
4027	Up to 0.5	50
	0.6-1.0	150
	1.1-2.0	250
4037	Up to 0.5	100
	0.6-1.0	200
	1.1-2.0	300
4130, 5140	Up to 0.5	300
	0.6-1.0	400
	1.1-2.0	450
4135, 4140	Up to 0.5	350
	0.6-1.0	450
	1.1-2.0	500
4320, 5130	Up to 0.5	200
	0.6-1.0	300
	1.1-2.0	400
4340	Up to 2.0	550
8630	Up to 0.5	200
	0.6-1.0	250
	1.1-2.0	300
8640	Up to 0.5	200
	0.6-1.0	300
	1.1-2.0	350
8740	Up to 1.0	300
	1.1-2.0	400

(a) Low-hydrogen welding processes only

Table 3 Recommended minimum preheat and interpass temperatures for ASTM high-strength low-alloy structural steels using low-hydrogen welding procedures

ASTM steel	Thickness, in.(a)	Minimum temperature, °F
A242; A441; A572, grades 42, 50; A588; A633, grades A, B, C, D	Up to 0.75	32
	0.81 to 1.50	50
	1.56 to 2.50	150
	Over 2.50	225
A572, grades 60, 65; A633, grade E	Up to 0.75	50
	0.81 to 1.50	150
	1.56 to 2.50	225
	Over 2.50	300

(a) Thickness of thicker section at the joint

sile stress occurs in the through-thickness direction of the weld metal. Although the need for higher temperatures is recognized, the increase is left entirely up to the discretion of the fabricator.

The preheat requirements considered in CSA W59 can likewise be reduced for specific single-pass fillet welds made by the submerged arc process with good fit-up when the minimum fillet sizes are in accordance with Table 4. The requirements are based on the material with the higher carbon equivalent when fillet welds are made between materials of different compositions. Preheat temperatures can also be lowered with approval of the engineer if it can be established that the HAZ hardness will not exceed 280 DPH. This reduction of preheat is not intended to be utilized for tack welding, temporary welds, or partial-penetration groove welds. In reducing the preheat levels, it is important to pay attention to the condition of the welding consumables to avoid weld-metal cracking.

Revisions to the preheat requirements in Canada and the United States which will allow individual assessments of minimum temperatures are being incorporated into the various codes. An example of a proposal which includes the effects of hydrogen, steel composition, restraint, and thickness, but which does not incorporate the hardness directly, is outlined below.

Research has shown that predicting preheat temperatures should be based on control of hydrogen. Empirical relationships among the cooling rate between 572 and 212 °F, the composition, and the hydrogen content allow one to predict preheat temperatures which are not overly conservative.

The susceptibility index for cracking is determined from both the composition and the hydrogen content in a susceptibility parameter. This parameter incorporates the Ito and Bessyo carbon equivalent formula, P_{cm}, and the diffusible hydrogen content in mL/100 g of deposited metal measured over mercury. The parameter is as follows:

$$12\,P_{cm} + \log_{10}\mathrm{H}$$

where

$$P_{cm} = \mathrm{C} + \frac{\mathrm{Mn} + \mathrm{Si} + \mathrm{Cu} + \mathrm{Cr}}{20} + \frac{\mathrm{Ni}}{60} + \frac{\mathrm{Mo}}{15} + \frac{\mathrm{V}}{10} + 5\mathrm{B}$$

Three hydrogen levels are selected, corresponding to 5, 10, and 30 mL/100 g. A letter designation can be applied to various combinations of consumables and baseplates, as shown in Table 5.

Analysis used to determine P_{cm} may be obtained from:

- Analysis of the plate material
- Mill test certificate
- In the absence of chemical analysis, P_{cm} can be calculated from the maximum limits specified in material specifications.

Elements present only at residual levels need not be included in P_{cm}.

The relation between hydrogen content and condition of the electrodes is based on both published and unpublished work. The intent of such work is to provide an incentive for lower hydrogen by the adoption of careful handling and drying procedures and use of consumables with lower potential hydrogen. The hydrogen levels discussed in this section are defined as follows.

H1 extra-low hydrogen electrodes are consumables giving a diffusible hydrogen content of less than 5 mL/100 g of deposited metal when measured using International Standards Organization (ISO) standard ISO3690-1976(E). This may be established by a test on each type of electrode and wire-flux combination used. The following electrodes meet this requirement:

- Basic electrodes baked (or rebaked once) at 750 °F and used within 2 h
- Basic electrodes baked at 750 °F and maintained in holding ovens, containers, or dispensers that have been shown to maintain the hydrogen below 5 mL/100 g and used within 2 h of removal
- Gas metal arc welding with clean solid wires and gas shielding using argon, carbon dioxide, or a combination of the two

H2 low-hydrogen electrodes are consumables giving a diffusible hydrogen content of less than 10 mL/100 g of deposited metal. This may be established by

Table 4 Minimum single-pass fillet weld sizes to eliminate preheat for SAW

Plate thickness (t), in.	Carbon equivalents(a) 0.35	0.40	0.45	0.50	0.55	0.60
≤0.5 welded to 1.5(b)	8	8	8	9.5	9.5	13
>0.5 welded to >1.5(b)	8	8	9.5	9.5	13	16

Note: Does not apply to quenched and tempered steels.
(a) $CE = \mathrm{C} + \frac{\mathrm{Mn} + \mathrm{Si}}{6} + \frac{\mathrm{Cr} + \mathrm{Mo} + \mathrm{V}}{5} + \frac{\mathrm{Ni} + \mathrm{Cu}}{15}$. The analysis used in determining the carbon equivalent may be obtained from: (1) analysis of the plate material; (2) the mill test certificates; and (3) the material specification, using the maximum composition allowed in the specification. (b) For ASTM A572, grades 60 and 65, and G40.21, grades 60W, 60WT, 60AT, 70W, and 70WT, this thickness is 3/4 in.

Table 5 Values of the susceptibility index for cracking ($12P_{cm} + \log_{10}H$)

Hydrogen content of filler metal, mL/100 g	P_{cm} ≤0.18	≤0.23	≤0.28	≤0.33	≤0.38	≤0.43
5 (H1)	A	B	C	D	E	F
10 (H2)	B	C	D	E	F	NR
30 (H3)	C	D	E	F	NR	NR

Note: A represents the lowest susceptibility parameter; F represents the highest; NR, not recommended

Table 6 Minimum preheat and interpass temperatures for low-restraint fillet welds

Thickness, in.	Minimum preheat and interpass temperatures, °F					
	Susceptibility					
	A	B	C	D	E	F
<0.4	<68	<68	<68	<68	140	285
0.4-0.8	<68	<68	<68	68	220	285
0.8-1.6	<68	<68	<68	150	240	285
1.6-3.1	<68	<68	115	195	250	285
>3.1	<68	<68	195	212	255	285

Note: Thickness corresponds to the thicker part being welded. For susceptibility values, see Table 5.

a test on each type of consumable, and each wire-flux combination used. The following electrodes meet this requirement:

- Basic electrodes taken from hermetically sealed containers or rebaked once at 750 °F and used within 4 h
- Basic electrodes from hermetically sealed containers or rebaked once at 750 °F and maintained in holding ovens, containers, or dispensers that have been shown to maintain the hydrogen below 10 mL/100 g and used within 4 h of removal from such containers
- Submerged arc welding with dry flux

H3—Hydrogen Uncontrolled. This classification includes all other consumables that do not meet the requirements of H1 or H2, such as some flux cored electrodes and cellulosic coated electrodes, e.g., E6010.

Minimum Preheat and Interpass Levels. Tables 6, 7, and 8 give recommended minimum preheat and interpass levels for three levels of restraint. In using these tables, a qualitative assessment of the restraint of a joint must be made. The restraint is the resistance to shrinkage that is imposed on the plates being welded. Welding personnel learn from experience what types of joints are high restraint and what types are low restraint, and in a few cases calculations have been made. Restraint is higher in thicker plates and is usually higher in groove welds than fillet welds.

Energy input affects cooling time and is inherent in the preheat/cooling time results used to calculate the preheat temperatures. In Tables 6, 7, and 8, an energy input of 40 kJ/in. was assumed for the preheat calculations. Higher energy inputs reduce the preheat requirements, as the HAZ susceptibility is also lowered.

In applications that do not require preheat, good practice includes warming the surfaces of the work (including any joints or crevices where moisture can be entrapped) to drive off moisture, especially in humid atmospheres. Warming slightly with a torch before welding often prevents weld porosity. Preheating also may be necessary if a SAW flux contains moisture or a shielding gas has an unusually high dew point. In summary, preheating may be necessary even if the plate temperature exceeds the minimum listed in the welding procedure.

Table 7 Minimum preheat and interpass temperatures for normal restraint butt welds

Thickness, in.	Minimum preheat and interpass temperature, °F					
	Susceptibility					
	A	B	C	D	E	F
<0.4	<68	<68	<68	<68	140	285
0.4-0.8	<68	<68	68	175	285	300
0.8-1.6	<68	85	212	285	300	300
1.6-3.1	160	230	285	300	300	300
>3.1	212	300	300	300	300	300

Note: For susceptibility values, see Table 5.

Table 8 Minimum preheat and interpass temperatures for high restraint butt welds

Thickness, in.	Minimum preheat and interpass temperatures, °F					
	Susceptibility					
	A	B	C	D	E	F
<0.4	<68	<68	<68	212	300	300
0.4-0.8	<68	140	250	300	300	300
0.8-1.6	195	300	300	300	300	300
1.6-3.1	250	300	300	300	300	300
>3.1	255	300	300	300	300	300

Note: For susceptibility values, see Table 5.

Arc Welding of Hardenable Carbon Steels

Carbon steels are defined as steels that contain up to 2% C, 1.65% Mn, 0.6% Si, and 0.6% Cu. Phosphorus and sulfur may also be present in these steels in concentrations up to 0.04 and 0.05%, respectively. Carbon content and the steelmaking process used (acid or basic) are frequently used as guidelines for the classification of carbon steels. Carbon content, however, because of its well-defined effect on the properties of steel, is considered the most common criterion for classification. Chemical compositions of several common grades of plain carbon steels are given in Table 9.

Low-carbon steels, those containing less than 0.25% C, are generally easy to join by any arc welding process. Arc welding of these steels is discussed in detail in the articles in this Volume that deal with the various arc welding processes. Welds of acceptable quality usually can be produced without the need for preheating, postheating, or special welding techniques, provided the sections being welded are less than 1 in. thick and that severe joint restraint does not exist. Filler-metal selection is seldom critical for welding low-carbon steel and is based mainly on tensile-strength requirements.

Medium-carbon steels, those containing 0.25 to 0.50% C, can also be satisfactorily welded by all of the arc welding processes. Because of the formation of greater amounts of martensite in the weld zone and the higher hardness of the martensite, preheating or postheating, or both, are often necessary.

For joint designs and welding processes and procedures that induce high weld cooling rates, preheat is used to inhibit martensite formation and allow upper transformation products to form. Postweld heat treatment also is used to temper martensite and restore toughness in the HAZ. Modifications in welding procedure—for example, the use of large V-grooves or multiple passes—also decrease the cooling rate and the probability of weld cracking. In multiple-pass welding, the final weld bead should be deposited in such a manner that it is surrounded on both sides by weld metal from previous passes. By so doing,

Table 9 Typical compositions of plain carbon steels

AISI or SAE designation	Composition, % C	Mn	P max	S max	Si max
1006	0.08 max	0.25-0.40	0.040	0.050	0.60
1008	0.18 max	0.30-0.50	0.040	0.050	0.60
1010	0.08-0.13	0.30-0.60	0.040	0.050	0.60
1012	0.10-0.15	0.30-0.60	0.040	0.050	0.60
1015	0.13-0.18	0.30-0.60	0.040	0.050	0.60
1016	0.13-0.18	0.60-0.90	0.040	0.050	0.60
1017	0.15-0.20	0.30-0.60	0.040	0.050	0.60
1018	0.15-0.20	0.60-0.90	0.040	0.050	0.60
1019	0.15-0.20	0.70-1.00	0.040	0.050	0.60
1020	0.18-0.23	0.30-0.60	0.040	0.050	0.60
1021	0.18-0.23	0.60-0.90	0.040	0.050	0.60
1022	0.18-0.23	0.70-1.00	0.040	0.050	0.60
1023	0.20-0.25	0.30-0.60	0.040	0.050	0.60
1025	0.22-0.28	0.30-0.60	0.040	0.050	0.60
1026	0.22-0.28	0.60-0.90	0.040	0.050	0.60
1030	0.28-0.34	0.60-0.90	0.040	0.050	0.60
1035	0.32-0.38	0.60-0.90	0.040	0.050	0.60
1037	0.32-0.38	0.70-1.00	0.040	0.050	0.60
1038	0.35-0.42	0.60-0.90	0.040	0.050	0.60
1039	0.37-0.44	0.70-1.00	0.040	0.050	0.60
1040	0.37-0.44	0.60-0.90	0.040	0.050	0.60
1042	0.40-0.47	0.60-0.90	0.040	0.050	0.60
1043	0.40-0.47	0.70-1.00	0.040	0.050	0.60
1045	0.43-0.50	0.60-0.90	0.040	0.050	0.60
1046	0.43-0.50	0.70-0.90	0.040	0.050	0.60
1049	0.46-0.53	0.60-0.90	0.040	0.050	0.60
1050	0.48-0.55	0.60-0.90	0.040	0.050	0.60
1055	0.50-0.60	0.60-0.90	0.040	0.050	0.60
1060	0.55-0.65	0.60-0.90	0.040	0.050	0.60
1064	0.60-0.70	0.50-0.80	0.040	0.050	0.60
1065	0.60-0.70	0.60-0.90	0.040	0.050	0.60
1070	0.65-0.75	0.60-0.90	0.040	0.050	0.60
1074	0.70-0.80	0.50-0.80	0.040	0.050	0.60
1078	0.72-0.85	0.30-0.60	0.040	0.050	0.60
1080	0.75-0.88	0.60-0.90	0.040	0.050	0.60
1084	0.80-0.93	0.60-0.90	0.040	0.050	0.60
1086	0.80-0.93	0.30-0.50	0.040	0.050	0.60
1090	0.85-0.98	0.60-0.90	0.040	0.050	0.60
1095	0.90-1.03	0.30-0.50	0.040	0.050	0.60

the HAZ that results from the deposition of the previous-pass weld beads is tempered by the heat from the final-pass bead.

Selection of filler materials for arc welding becomes more critical as the carbon content of the steel increases. As discussed earlier in this article, steels with higher carbon contents are more susceptible to hydrogen-induced cracking; therefore, low-hydrogen electrodes and processes ordinarily are used. As the carbon content of the steel being welded approaches 0.50%, low-hydrogen conditions become mandatory.

High-carbon steels, with more than 0.50% C, are difficult to weld because of their susceptibility to cracking. Excessive hardness and brittleness often occur.

For best results in SMAW, the use of low-hydrogen electrodes is recommended. Similarly, for other arc welding processes, low-hydrogen practice is mandatory. Both preheating and postweld stress relieving or tempering usually are required. Austenitic stainless steel electrodes sometimes are used for welding high-carbon steel to obtain greater notch toughness in the joint. However, the HAZ may still be hard and brittle, and preheating and postweld stress relieving may be necessary.

Successful welding of high-carbon steel requires the development of specific welding procedures for each application. Composition, thickness and configuration of the component parts, and service requirements must be considered. For each application, the welding procedure should be qualified by tests before it is adopted. High-carbon steels frequently are welded using procedures suggested for tool steels, which are discussed later in this article.

Weldability

For carbon steels, carbon and manganese contents largely determine the extent of hardening under given heating and cooling conditions. The higher carbon, higher manganese grades of carbon steel can be welded satisfactorily if they are preheated and postheated. Sometimes, peening of the weld bead is helpful, but many codes restrict the use of peening because it is difficult to control.

Carbon equivalent formulas are sometimes used as a rough guide to evaluate weldability. However, the section size being welded has an important influence on the combined effect of carbon and manganese contents on weldability because of its relation to heat input and cooling rate. Figure 24 shows the effect of section thickness and carbon equivalent on weldability, as expressed in terms of notch bend test results.

Several different equations for expressing carbon equivalent are commonly used. The one shown in the horizontal legend of Fig. 24, which considers only carbon, manganese, and silicon, is appropriate for most carbon steels. A steel with a carbon equivalent less than 0.40% is usually considered safe for welding without special precautions, unless very high restraint is present. Another well-known equation for expressing carbon equivalent that is sometimes used for carbon steel considers residual (or higher) amounts of chromium and molybdenum in addition to manganese. Chromium and molybdenum are the two elements most likely to affect hardenability and weldability when present in small amounts. The carbon equivalent equation is:

$$CE = \mathrm{C} + \frac{\mathrm{Mn}}{6} + \frac{\mathrm{Cr}}{5} + \frac{\mathrm{Mo}}{4}$$

This equation yields considerably lower carbon equivalents than the equation used in Fig. 24. For example, for a steel containing 0.35% C, 0.80% Mn, 0.20% Si, and trace amounts of chromium and molybdenum, the carbon equivalent is 0.60% according to the equation in Fig. 24 and 0.48% according to the equation that ignores silicon, but includes chromium and molybdenum.

Susceptibility to weld cracking increases as the number of unfavorable welding conditions increases. For example, 1065 steel welded with an E7024 electrode (not classified as a low-hydrogen electrode), without preheat or postheat, performed satisfactorily when joint restraint was low and cycle loading of the weldment in service was limited. The HAZ exhibited a region of martensite about $^{1}/_{16}$ in. thick, but no cracking. However, when the section thickness was doubled, the thickness of the martensite zone increased, and cracking resulted. Changing to a low-hydrogen electrode (E7018) did not prevent cracking of the thicker section. The use of a 600 °F preheat prevented cracking by retarding the rate of cooling and thereby reducing the amount of martensite formed.

Preheating and Interpass Temperatures

Preheating an assembly for welding is intended primarily to prevent cracking in the weld metal or in the HAZ. With preheating, the hardness in the HAZ is usually lower, because the cooling rate is slower, and residual stress and distortion will be minimized. Many code-governed weldments have mandatory requirements, and specific conditions, for preheating.

Because the need for preheating of carbon steel is based not only on carbon content, but also on the combination of carbon, manganese, silicon, and residual alloy contents along with various aspects of joint configuration, chiefly section thickness, the selection of a preheating temperature is determined largely by this combination. Figures 25 and 26 relate the selection of preheating and interpass temperatures to carbon content and section thickness. The data and experience from which these graphs were developed were obtained with SMAW. Similar relations would apply to GMAW and FCAW. Less restrictive relations would sometimes be applicable for SAW, which characteristically involves lower cooling rates for a given joint.

Welding at low ambient temperature (especially below room temperature) can cause cracking, and preheating to a safe-to-weld temperature usually is the easiest and most effective preventive. If there is doubt concerning the composition of the steel being welded, preheating should be at a temperature conservatively on the high side. The use of low-hydrogen electrodes frequently eliminates the need for preheating of thinner sections.

Fig. 24 Effect of carbon equivalent and plate thickness on the weldability of carbon steels as indicated by notch-bend tests. Ratio (welded to unwelded) of bend angle for normalized steel plates. A high value of the ratio indicates high weldability.

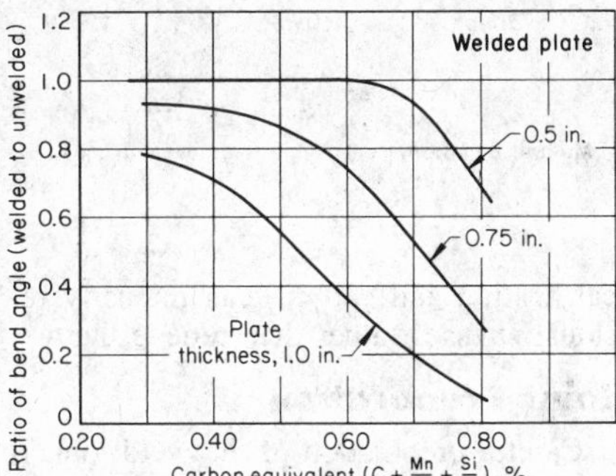

Fig. 25 Effect of base-metal carbon content and thickness on preheating and interpass temperatures for SMAW of carbon steel

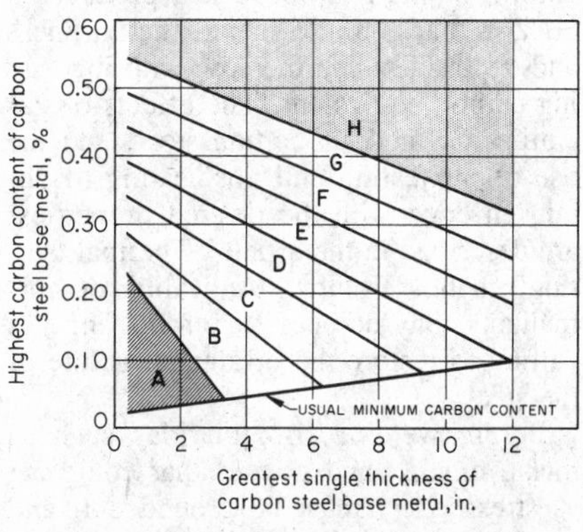

Area	Range of preheating and interpass temperatures, °F: Welding with high-hydrogen electrodes(a)	Welding with low-hydrogen electrodes
A	50-100	50-100
B	100-200	50-100
C	200-300	100-200
D	250-400	150-300
E	300-500	200-400
F	350-600	250-500
G	400-700	300-600
H	450-800	400-800

(a) In practice, it is usually considered mandatory to use low-hydrogen electrodes for welding steels containing 0.50% C or more.

Fig. 26 Effect of base-metal carbon content on preheating and interpass temperatures for six section thicknesses of carbon steel welded by SMAW

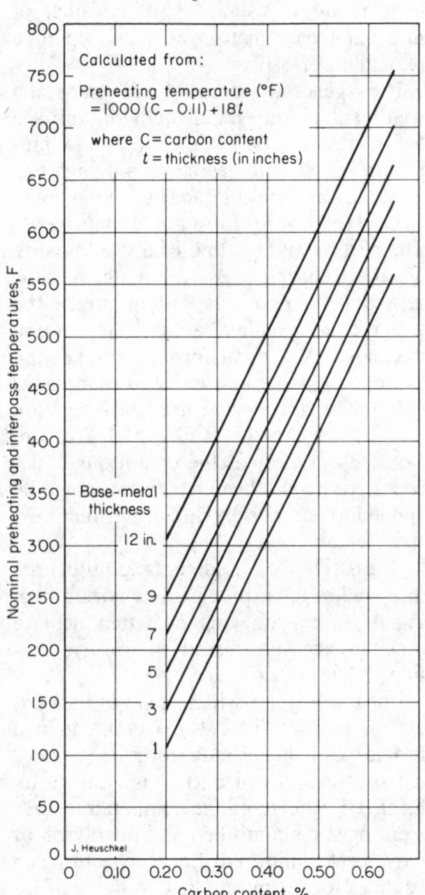

Bead size is also related to preheating. Small beads cool faster than large beads. Increasing the size of the bead by reducing travel speed, for example, increases heat input for a given length of weld, which results in a slower cooling rate, thus producing less martensite and lower hardness in the HAZ. In some applications, the need for preheating has been eliminated by control of bead size.

Multiple-pass welding can sometimes reduce or eliminate the need for preheating. In certain applications where cracking has occurred when welding was done with a single pass and with no preheating, changing to multiple-pass welding has prevented cracking (without preheating). Under these conditions, each pass serves as a preheat for the subsequent pass, while providing a postheating effect for earlier passes.

Methods of Preheating. Ideally, a weldment is preheated in a furnace, but sometimes no furnace is available or the weldment is of such size or shape that furnace heating is not feasible. Under these conditions, preheating is usually done with gas torches, and temperature-sensitive crayons (or a surface pyrometer) are used for monitoring temperature.

When preheating is done with torches, regardless of the temperature selected, the heat should be applied at one face of the joint and measured at the opposite face to ensure that the joint is heated through, or the preheating temperature should extend for a distance of 3 in. in each direction from the joint to be welded.

Interpass Temperature. In multiple-pass welding, it is usually necessary to maintain a minimum interpass temperature equal to the minimum preheating temperature if preheating is required. However, if welding conditions are changed, such as a change in process or electrode, after deposition of the first bead, the interpass temperature should be adjusted to suit the new conditions. Suggested preheating and interpass temperatures, based on carbon content and thickness of the base metal, are given in Fig. 25 and 26.

In many automatic welding processes (especially SAW), enough heat is supplied to the joint so that additional heating between passes may not be required.

Postweld Stress Relieving

Stress relieving or tempering after welding of hardenable carbon steel is often desirable, and in some applications it is mandatory. Many codes define the conditions under which stress relieving is required and the time and temperature of the stress-relieving cycle. The times and temperatures prescribed in these codes can be used for noncode applications of the same steels.

Although other variables, such as joint restraint, influence the need for postweld stress relieving, the principal factors are the same as for preheating—that is, base-metal thickness and carbon content. In Fig. 4, area C defines the combinations of base-metal thickness and carbon content for which postweld stress relieving is usually required. A more detailed chart for postweld stress relieving is presented in Fig. 27. The six areas into which Fig. 27 is divided take into account various conditions of welding and service that relate to the need for stress relieving, as explained below the graph.

Fig. 27 Effect of base-metal thickness and carbon content on requirements for postweld stress relieving or tempering of carbon steels. (A) Postweld heating is seldom required. (B) Postweld heating is required only for dimensional stability, as when parts are to be finish machined after welding. (C) Postweld stress relief is highly desirable for repetitively loaded or shock-loaded structures and for restrained joints having a thickness greater than 1 in. (D) Postweld stress relief is required for all repetitively loaded or shock-loaded structures, for all restrained joints, and for all thicknesses over 2 in. It is also desirable for statically loaded structures.
(E) Postweld stress relief is recommended for all applications. No intermediate cooldown should be permitted for restrained structures or for base metal having a thickness greater than 2 in.
(F) Same as (E), except hazards are greater.

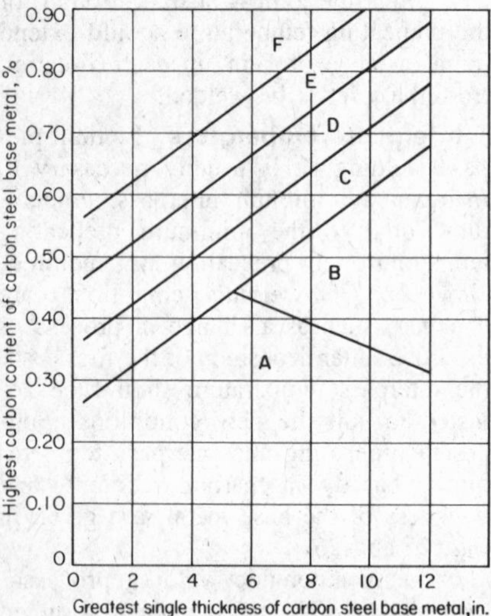

The temperatures generally used for stress relieving of medium-carbon steels range from 1100 to 1200 °F. The time at temperature is usually 1 h per inch of maximum base-metal thickness, up to a maximum of 8 h.

It is generally desirable to place the weldment in the stress-relieving furnace before it cools below the minimum preheat or interpass temperature, although this is not always feasible and is seldom considered mandatory for plain carbon steels. The heating and cooling of a weldment in a furnace during stress relieving must be at a gradual rate that will minimize the temperature gradient across the section thickness. A general formula for heating and cooling rates has not been found suitable for all shapes and materials. Section VIII of the ASME Boiler and Pressure Vessel Code defines the conditions under which postweld heat treatment must be applied for stress relieving of unfired pressure vessels.

Stress relief of weldments requires that the weldment be properly supported during the stress-relieving heat treatment to prevent excessive distortion from relaxation of stresses.

Local reheating will soften hard zones, just as placing the entire weldment in a furnace will, but it may not reduce residual stresses. In order to stress relieve by localized reheating, the temperature gradient must be controlled and restriction of movement minimized. Local reheating can be dangerous if misapplied. A cycle of local heating and cooling can intensify residual stresses rather than relieve them.

Fig. 28 Joint preparation for SMAW of heavy channel sections back-to-back

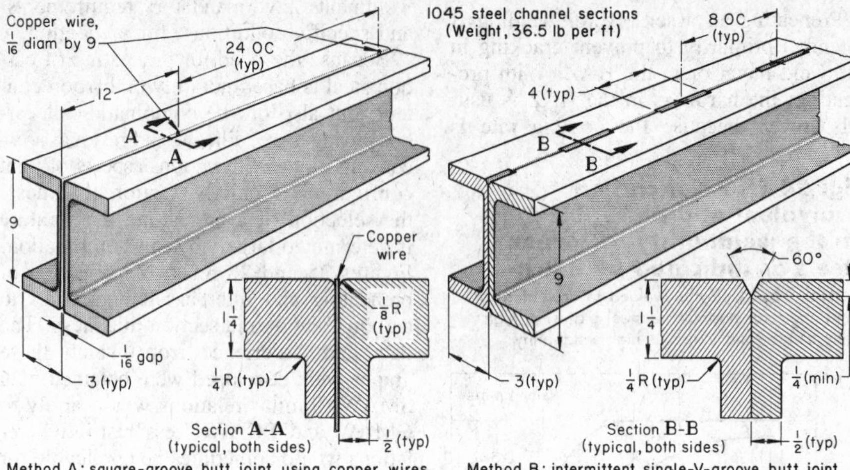

Joint Preparation

Careful preparation of the weld joint is important. The joint edges can be sheared, machined, ground, chipped, gas cut, or arc cut to the desired shape. Cold working of the metal adjacent to a joint, either in preparing the joint or in cold forming after edge preparation, can cause cracks that can propagate and become large fractures. Gas cutting of high-carbon steels produces a HAZ of hard, brittle metal, the hardness and depth of which increase with increasing carbon equivalent. The effects of gas cutting on medium-carbon steel, and the use of preheating and postheating to obtain an edge with the desired properties, are described in the article "Thermal Cutting" in this Volume. Annealing or normalizing may be necessary following gas cutting to restore the original structure of the steel.

Before welding, irregularities such as nicks, cracks, and gouges that could act as stress risers must be ground out, and grease, paint, surface scale, oxide, water, and other foreign material must be removed. Because some carbon steels are sensitive to underbead cracking from hydrogen, water in any form—moisture in scale, humid atmosphere, rain, snow—should be removed, and a low-hydrogen electrode is recommended.

In the example that follows, cracking of weld metal was prevented by two methods of joint preparation. Because large, heavy-section pieces were being welded, preheating and postweld stress relief were not feasible, but cracking was avoided either by reducing joint restraint or by increasing the size and strength of weld metal.

Example 2. Methods of Joint Preparation That Prevented Weld Cracking Due to Joint Restraint and Loss of Ductility. Weld cracking occurred when heavy 1045 steel channel sections 10 to 20 ft long (see Fig. 28) were joined back-to-back with single-pass welds made by SMAW, using low-carbon steel filler metal (E6010). The cracking was ascribed to: (1) the restraint of the heavy sections (36.5 lb/ft) on the welds during cooling and contraction; (2) loss of weld ductility due to carbon enrichment of the weld metal by the 1045 steel base metal; (3) rapid chilling of the weld metal by the large mass of surrounding base metal; and (4) small cross-sectional area of the bead.

In an attempt to prevent cracking, low-hydrogen electrodes, including E7016 and the higher-strength E12018, and 35Cr-15Ni electrodes were tried, but results were little better than those obtained with E6010. Cracking always occurred through the centerline of the weld, which indicated lack of strength or ductility, or both, rather than hydrogen embrittlement.

Next, a revision in joint preparation for welding was made. Two methods were equally effective in preventing weld cracking. In one (method A in Fig. 28), the channel sections were separated by copper wires $^1/_6$ in. in diameter and 9 in. long, which were placed perpendicular to the channel flanges at 2-ft intervals. To hold the wires and the channels in proper position for welding, strong tack welds were placed near each wire. Both sides of the joint were shielded metal arc welded with E6010 or E6011 electrodes, working from the center toward the ends. Because of the length of the channels, welding was alternated from side to side every few feet to avoid cambering. The weld beads were deposited in a single pass, with deep penetration of the base metal. As the welds cooled, the soft copper yielded sufficiently to reduce shrinkage stress.

The other method used to avoid weld cracking (shown as method B in Fig. 28) involved the use of intermittent V-groove welds 4 in. long, spaced on 8-in. centers; joint edges were chamfered to form a 60° V-groove of $^1/_4$-in. minimum depth. After the channels had been placed back-to-back, clamped, and tack welded, they were shielded metal arc welded with low-carbon steel electrodes of the E60*xx* or E70*xx* class. The welds were single-pass welds, deposited from the center toward the ends. Each 4-in. weld was allowed to cool before the next one was begun. To avoid cambering, welding was alternated from side to side after every two welds. Because the V-grooves increased the cross-sectional area of the welds, weld strength and ductility were increased sufficiently to enable the weld to withstand stresses developed by joint restraint and to prevent cracking. Although some of the ductility gained by depositing a greater amount of weld metal undoubtedly resulted from a lesser amount of carbon pickup from the base metal, it was mostly because a greater amount of hot metal was available for elongation or compression.

Shielded Metal Arc Welding

Shielded metal arc welding is used extensively for joining hardenable carbon steel. In general, the equipment, techniques, and difficulties discussed in the article "Shielded Metal Arc Welding" in this Volume apply to the welding of hardenable carbon steel.

The use of low-hydrogen electrodes is important in welding hardenable carbon steel. Electrodes with flux coverings that contain cellulose will release hydrogen, which, when absorbed in the hardened zone of a weld, can cause cracks, especially in steel with a high carbon equivalent. Cracking is minimized by the use of low-hydrogen electrodes, by proper storing and drying procedures, by minimizing carbon pickup from the base metal, by selection of electrodes and welding techniques that minimize melting of the base metal, by preheating, and by maintaining an appropriate interpass temperature (see Fig. 25 and 26). Transferring the weldment to a furnace for postweld stress relieving while it is at or above the interpass temperature is desirable for steels of higher carbon content or higher hardenability.

Large tonnages of hardenable carbon steels are used for applications where weld integrity is imperative and various welding codes and standards have been developed for these applications. The heating requirements set forth in these codes can be used for applications that are not governed by codes.

The type and size of electrode selected depend on the composition and thickness of the steel, joint preparation, welding position, and available welding current. Not all of the electrodes covered by AWS specifications A5.1 and A5.5, which apply to low-carbon steel and low-alloy steel covered electrodes, respectively, are suited for welding hardenable carbon steel. Although hardenable carbon steels are sometimes welded with cellulose-type (high-hydrogen) covered electrodes such as E6010, low-hydrogen electrodes are generally preferred. Among the electrodes most often used are E7015, E7016, E7018, E7028, E7048, E8016-C1, E8016-C3, E8018-C1, E8018-C3, E10016-D2, and E10018-D2. Of this group, E7018 electrodes are the ones most often used. Ranges of welding current for use with various sizes of the above electrodes are given in Table 10.

E7015 and E7016 electrodes are low-hydrogen low-carbon steel electrodes. E7015 electrodes are used with direct current electrode positive (DCEP), and E7016 electrodes are used with alternating current and DCEP. These electrodes are applicable to the welding of hardenable carbon steels that are subject to hydrogen-induced cracking; they can be used without preheating of the metal to be welded or with preheating to a lower temperature than would be necessary with other than low-hydrogen electrodes. However, if E7015 or E7016 electrodes are used without preheating, the notch toughness of the weld will generally be lower than if preheating is employed. The E7015 or E7016 electrodes are also used in the welding of high-sulfur steels; electrodes that are not of the low-hydrogen type are likely to deposit porous weld metal.

With E7015 and E7016 electrodes, the arc, which is moderately penetrating, should be kept as short as possible. Slag is moderately difficult to remove, and bead shape is slightly convex.

E7018 electrodes are low-hydrogen low-carbon iron-powder steel electrodes for use with alternating current and DCEP. These electrodes deposit essentially the same composition of weld metal as E7016 electrodes. The electrode covering contains essentially the same low-hydrogen materials as E7016, but the 25 to 45% iron powder gives E7018 electrodes a higher deposition rate. With E7018 electrodes, the arc, which is moderately penetrating, should be kept as short as possible, and the resulting bead shape is slightly convex. E7018 electrodes can be used in any welding position, and thus are preferred for a wide range of applications.

E8016-C1 and E8018-C1 electrodes are low-hydrogen low-alloy electrodes for use with alternating current or DCEP. Weld deposits from these electrodes contain about 2.5% Ni. The electrodes are widely used when low-temperature notch toughness is required. The operating characteristics of these electrodes are similar to those of E7016 electrodes.

E8016-C3 and E8018-C3 Electrodes. E8016-C3 electrodes are low-hydrogen low-alloy electrodes for use with alternating current and DCEP. The weld deposit contains about 1% Ni. When these

Table 10 Welding current ranges for various sizes of electrodes used in SMAW of hardenable carbon steels(a)

Electrode diameter, in.	Current, A, for electrode of class: E7015, E7016	E7018	E8016-C1, E8018-C1	E8016-C3	E8018-C3	E10016-D2, E10018-D2
3/32	65-110	70-100	80-120	70-90	70-95	60-100
1/8	100-150	115-165	100-150	100-130	110-140	80-120
5/32	140-200	150-220	150-185	130-180	130-200	140-190
3/16(b)	180-255	200-275	200-250	165-230	165-290	180-250
7/32(b)	240-320	260-340	240-325	270-300	280-315	...
1/4(b)	300-390	315-400	300-425	290-330	320-400	300-400
5/16(b)	375-475	375-470	...	...	...	...

(a) Lower sides of current ranges are used for vertical or overhead welding; higher sides of current ranges, for welding in the flat position. (b) Not used in all welding positions

electrodes are supplied as iron-powder types, they are designated E8018-C3. The operating characteristics of these electrodes are similar to those of E7016 and E7018 electrodes, respectively.

E10016-D2 and E10018-D2 electrodes are low-hydrogen low-alloy electrodes for use with alternating current and DCEP. The weld deposit contains about 1.75% Mn and 0.35% Mo. The electrodes were designed for welding high-strength steels of high hardenability, but they have been used for welding carbon steels containing 0.30% C or more, when high strength of the weld deposit is required.

Control of Moisture Content of Electrodes. Because hardenable carbon steels are susceptible to cracking when welded with an arc that contains hydrogen, the moisture content of electrode coverings should be kept at a very low level, which can be done by careful storage and use of electrodes. For information on the care and reconditioning of low-hydrogen electrodes, see the article "Shielded Metal Arc Welding" in this Volume.

Flux Cored Arc Welding

For the most part, the process, equipment, and techniques discussed in the article "Flux Cored Arc Welding" in this Volume apply to the welding of hardenable carbon steel. The procedures for preheating and postheating presented earlier in this article, although generally oriented toward SMAW, apply to a large extent to FCAW. However, in welding steels of undetermined weldability, heating procedures must be established with caution. The process should be qualified for the specific application.

Examples 3, 4, and 5. Flux Cored Arc Welding Versus Shielded Metal Arc Welding. For each of the three applications listed in Table 11, SMAW had been used, and the process was changed to FCAW with auxiliary gas shielding, at a substantial saving in welding time, generally from 35 to 50%. Continuous electrode feed, higher deposition rate, and higher travel speed with FCAW were mainly responsible for the decreased welding time. No preheating or postheating was required in these applications.

Details about the base metal, the type of weld, and welding conditions for FCAW in the three applications are presented in Table 11. The example that follows describes an application in which automatic FCAW was used to produce welds suitable for service under severe loading and extremes of temperature.

Example 6. Welding Bearing-Race Assemblies by the Automatic Flux Cored Arc Process. Ball-bearing inner-race assemblies 2 to 10 ft in diameter, each consisting of a 1050 steel inner race and a 1017 steel flange, were joined by two circumferential welds, using automatic FCAW with auxiliary gas shielding. As indicated in Fig. 29, which shows a typical bearing race, the welds were a 1/2-in. fillet weld and a 3/4-in. single-bevel groove weld.

Because the bearings were the main supports for mobile cranes and shovels, and were subjected to high loads and extreme weather conditions in service, sound welds with good mechanical properties were essential. The welds were specified to have no underbead cracks or other defects visible under liquid-penetrant inspection. With the use of the welding conditions listed in the table with Fig. 29, crack-free welds with no evidence of slag inclusions or porosity were obtained. The weld metal had the following typical tensile properties:

Tensile strength	115 ksi
Yield strength	103 ksi
Elongation in 2 in.	20%
Reduction of area	42%

Although all processing conditions contributed to weld soundness, the most important were the use of preheating, the selection of a suitable low-carbon alloy steel electrode wire (see footnote in table with Fig. 29), and the close control that is characteristic of automatic welding.

Submerged Arc Welding

Submerged arc welding is used both as a fully automatic and as a semiautomatic process for joining and for surfacing of hardenable carbon steels. The article "Submerged Arc Welding" in this Volume discusses in detail the equipment, electrodes, fluxes, and procedures. Although that article primarily concerns low-carbon steels, most of the information in it is pertinent to the welding of hardenable carbon steels.

Procedures for preheating and postheating suggested in the earlier part of this article were oriented toward SMAW and do not necessarily apply to the same degree to SAW. Preheating and postheating are less likely to be required for SAW than for SMAW for two reasons: (1) because of the high heat input and high deposition rate of SAW, a larger area is heated, and consequently, the cooling rate is lower and the probability of hardening is lessened; and (2) the flux blanket helps to retard cooling. However, in many applications, pre-

Table 11 Operating conditions for three applications of automatic FCAW of hardenable carbon steels

Item	Example 3 (House-trailer axle)	Example 4 (Punch-press gear)	Example 5 (Sinter-pot flange)
Steel welded	1045	1020 and 1045	ASTM A285
Thickness	3/16 to 1/4 in.	...	2 in.
Joint type	Butt	T-joint	Edge
Weld type	...	1/2-in. fillet	(a)
Electrode wire	3/32 in. diam E70T-1	3/32 in. diam E70T-1	3/32 in. diam E70T-1
Shielding gas (carbon dioxide) flow rate	...	45 ft^3/h	55 ft^3/h
Welding current (DCEP)	300 to 400 A	600 A	450 A
Arc voltage	25 V	32 V	28 to 30 V
Power supply(b)	500 A	750 A	500 A
Travel speed	50 in./min	10 in./min	18 to 24 in./min

(a) Single-bevel groove weld, 45° by 19/32 in. deep, inside; 3/4-in. flat fillet, outside. (b) Constant-voltage motor-generator

Fig. 29 Ball-bearing inner-race assembly joined by automatic FCAW

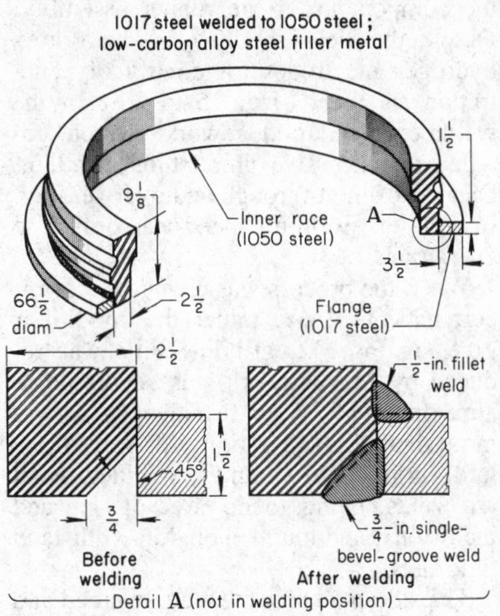

Automatic FCAW(a)

Joint types	Corner and butt
Weld types	Fillet and single-bevel groove
Power supply	750-A rectifier(b)
Electrode holder	Mechanically held, air cooled
Electrode	$^7/_{64}$-in.-diam flux cored wire(c)
Shielding gas	Carbon dioxide, at 40 ft^3/h
Welding position	Flat
Number of passes:	
Single-bevel groove weld	Three
Fillet weld	Two
Current (DCEP) and voltage:	
Root pass	550 A; 28 to 30 V
Filler pass	550 to 600 A; 32 to 34 V
Welding speed:	
Root pass	25 in./min
Filler pass	13 in./min
Preheat temperature	600 °F
Interpass temperature (minimum)	600 °F
Postheat	None(d)

(a) Equipment included an automatic, constant-speed wire feeder; a variable-speed rotating positioner with a universal clamping fixture; a ram-type manipulator for supporting the welding head; and two natural-gas preheating torches on pedestals. (b) Constant-voltage. (c) Composition, 0.06% C, 1.60% Mn, 0.40% Si, 2.30% Ni, 0.40% Mo. (d) Air cooled to room temperature; later the race area only was induction hardened.

heating and postheating are needed in SAW of hardenable carbon steels. Each application must be evaluated separately.

The SAW process is ordinarily capable of depositing low-hydrogen weld metal, provided the flux is properly dried and other precautions, including keeping the area surrounding and including the joint free of moisture, oil, grease, paint, scale, or other material that could be a source of hydrogen, are observed. The flux, which may be hygroscopic, must be kept dry. Proper storage and handling are important. A widely followed practice for maintaining low-hydrogen conditions is to keep the flux in an oven at about 250 °F for several hours before it is needed.

The use of preheat, controlled interpass temperature, and postweld stress relieving in joining low-carbon steel arms to a medium-carbon steel shaft is illustrated by the following example.

Example 7. Submerged Arc Welding of 1015 Steel Arms to a 1038 Steel Shaft. The rotor shaft for a motor (Fig. 30) consisted of four longitudinal arms of 1015 steel joined to a premachined shaft of 1038 steel by two-pass submerged arc welds. The arms were beveled by either machining or gas cutting, and were shot-blast cleaned before welding. The arms were aligned on the shaft by means of two retaining rings, and the entire assembly was preheated to 400 to 500 °F before any welding was done.

After preheating, one side of each arm was tack welded to the shaft, as shown in section A-A in Fig. 30, by SMAW with E7016 low-hydrogen electrodes. The shaft was then placed on the turning rolls of a positioner table, and the four arms were checked for alignment. Next, the tack welds were cleaned of slag, and the first passes by SAW were made (see position 1 in section B-B in Fig. 30). First-pass weld beads were deposited on side 1 of each of the four arms (the side that had not been tack welded); then the shaft was turned end-for-end, and first-pass beads were deposited on side 2.

With the completion of all first passes, the welding electrode and shaft were repositioned for making the second passes (see position 2 in section B-B in Fig. 30). Second passes were completed on one side of each arm. At this point, the assembly was checked for alignment, so that any distortion could be corrected by changing the sequence of the remaining weld passes. The shaft was then turned end-for-end, and the second passes were made on the other side of the joints.

The placement of weld beads was specified to ensure a smooth transition from the arm to the shaft. The weld surface remained as-deposited, receiving no further finishing except for cleaning and slag re-

Fig. 30 Rotor-shaft assembly in retaining ring for SAW. See text for details.

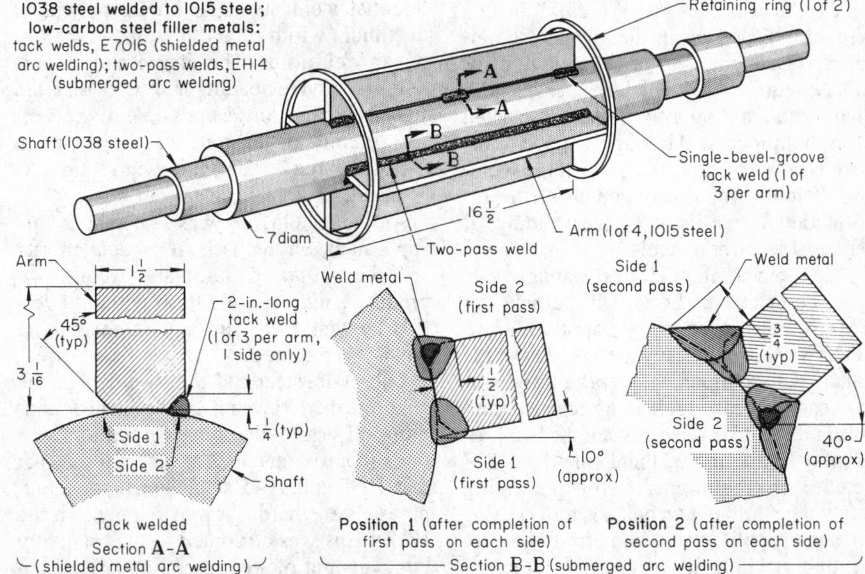

Conditions for SAW

Joint type	T	Current	540 A, ac
Joint preparation	Machining or gas cutting	Number of passes per side	Two
Weld type, each side	Single-bevel groove	Welding speed, first pass	10 in./min
Electrode wire	$^5/_{32}$-in.-diam EH14(a)	Welding speed, second pass	7 in./min
Flux	F72	Preheat (before tack welding)	400 to 500 °F
Welding position	Flat	Interpass temperature (min)	400 °F
Voltage	28 to 29 V	Postheat	1175 °F(b)

(a) Automatic feed. (b) For 1 h per inch of maximum section thickness. Weldment was heated to 1175 °F (and later cooled to 300 °F) at 150 °F/h.

moval. Subsequently, attachments were made to the arms by SMAW with low-hydrogen electrodes.

During the entire sequence of welding operations, the interpass temperature was held at a minimum of 400 °F, and after completion of welding, the assembly was stress relieved at 1175 °F for 1 h per inch of maximum section thickness. The heating rate was 150 °F/h, and the controlled cooling rate was 150 °F/h to 300 °F.

Previously, this rotor shaft had been machined from a fluted open-die forging at a cost that was competitive with the cost of the welded assembly. However, because the rotor shafts were produced in a variety of shapes, and with major shaft diameters ranging from $4^1/_2$ to 10 in., lead times for the forgings gave rise to inventory and handling charges that were considered unnecessary. The change to a weldment eliminated these costs, because the lead time for the components for the welded assembly was less than for the forging. Shafting stock and plate for the arms were readily available, and the arms were gas cut from plate as needed.

Gas Metal Arc Welding

Gas metal arc welding is widely used for joining hardenable carbon steels. The article "Gas Metal Arc Welding" in this Volume discusses in detail the advantages, disadvantages, applicability, principles, equipment, shielding gases, design, costs, and safety problems associated with this process. Although the above article is concerned primarily with the welding of low-carbon steel, much of the information is pertinent to the welding of hardenable carbon steels.

The preheating and postheating practices described earlier in this article for SMAW are frequently applicable to GMAW. However, they are not always the same. For example, the depth of penetration can influence preheating practice, and this is likely to differ between the two processes. Furthermore, there may be a difference in requirements for preheating among the different techniques for GMAW. In some applications, the need for preheating is eliminated by using argon instead of carbon dioxide as a shielding gas, because of the difference in penetration patterns obtained with the two gases. The mode of metal transfer can also affect the need for preheating—the spray arc penetrates considerably deeper than the short circuiting arc. Because of the many variables with GMAW, it is more important to qualify the procedure for a given application than it is for other arc welding processes.

Because a shielding gas is used instead of a flux, the possibility of introducing hydrogen into the weld metal is minimal, but the usual precautions, including keeping the weld joint free of moisture, oil, grease, paint, scale, or other material that could adversely affect weld quality, should be observed to ensure sound welds. The electrode wire is sometimes a source of hydrogen, particularly if the surface is poor. Laps or folds are likely to contain drawing lubricants.

Welding-grade argon is generally low in hydrogen. A minimum dew point temperature of −60 °F should be specified and certified by the vendor. Sometimes gas suppliers fail to drain and dry cylinders before refilling; this results in dew point temperatures of −10 to −30 °F, which are likely to cause cracking. The moisture content of welding-grade carbon dioxide used for GMAW is usually at a dew point of −40 °F or less. In the examples that follow, GMAW was used because of the higher production rate and better quality of welds, and because less operator skill was needed.

Example 8. Automatic Circumferential Welding of a Yoke to a Rod Section. The steel yoke-and-rod assembly shown in Fig. 31 required a circumferential weld about $^3/_4$ in. deep, with a maximum width of about $^3/_8$ in. at the top of the weld joint. The nature and magnitude of service loading required that the weld begin as close as possible to the center, or core, of the rod, which accounted for the design of the joint (see section A-A in Fig. 31).

Automatic GMAW was selected for this application, on the basis of production rate and the ability of automatic welding to produce a narrow weld bead without overfill, so that little subsequent machining would be required.

The 1045 steel rod was welded in the cold finished, ground and polished condition. Despite the relatively thick sections (approximately $2^1/_4$ in.) and the fact that 1035 and 1045 steels are hardenable, no excessive hardness was obtained in the HAZ. This was attributed to the relatively large amount of weld metal deposited and the use of multiple-pass welding, which functioned with about the same effect as preheating and postheating.

Example 9. Change From Shielded Metal Arc to Semiautomatic Gas Metal Arc Welding to Reduce Repair and Finishing Time. A runner for a seed-planter unit, shown in Fig. 32, was subjected to severe abrasion during its use in cleaving soil, usually to a depth of 3 to 4 in. In addition, the runner had to sustain the occasional impact that resulted from contact with stones.

Originally, SMAW was used for joining the components of the runner assembly. Despite the use of small ($^1/_8$-in.-diam) low-hydrogen electrodes for control of penetration and bead size, 15 to 20% of the weldments required rework, which entailed an excessive amount of grinding, because of melt-through and overwelding. In addition, welding speed was only 5 to 7 in./min.

When the process was changed to semiautomatic GMAW, under the conditions given in Fig. 32, grinding time was reduced by 75%, rework was decreased to a maximum of 2%, and welding speed was more than doubled. The welds were more uniform in quality than the shielded metal arc welds, being sound, free of slag and inclusions, and tough enough to withstand field use.

The sides and crosstie were blanked and formed in presses from hot rolled 0.45 to 0.55% carbon steel. The blade was blanked cold from hot rolled 0.65 to 0.80% carbon steel, and hot worked to shape. All components were abrasive blast cleaned

Fig. 31 Yoke-and-rod assembly welded by automatic GMAW

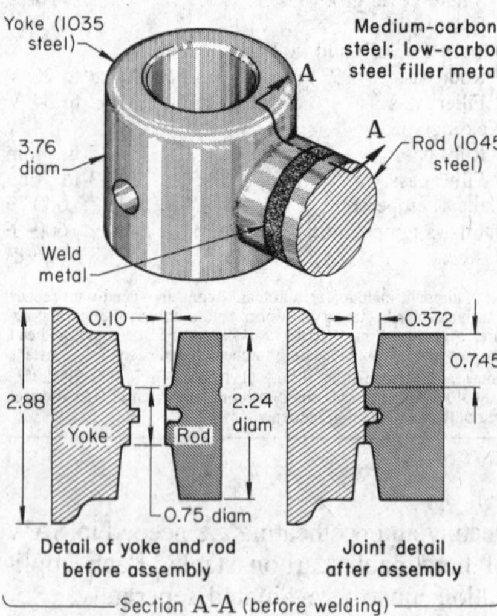

Automatic GMAW

Power supply	300-A rectifier, with controlled slope, voltage, and inductance
Electrode wire	0.045-in.-diam low-carbon steel
Current	DCEP
Arc time	2.56 min
Shielding gas	75%A-25%CO_2, at 28 ft^3/h
Wire-feed rate	295 in./min
Preheat and postheat	None
Electrode consumption per weld	755 in.
Floor-to-floor time	3.43 min

Fig. 32 Changing from SMAW to semiautomatic GMAW to reduce welding and finishing times. See text for details.

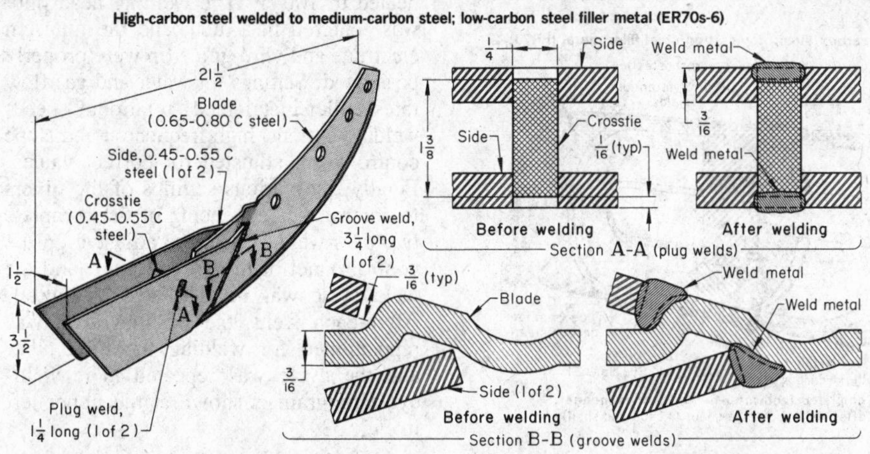

Conditions for semiautomatic GMAW

Joint types	Butt and T	Shielding gas	95%A-5%O_2, at 29 ft^3/h(b)
Weld types	Groove and plug	Current	260 A (DCEP)
Welding position	Flat (15° inclination)	Voltage	29 V
Power supply	300-A rectifier(a)	Stickout	1/2 in.
Wire feed	Attached to power supply	Electrode-wire speed	325 in./min
Electrode holder	Hand held, 300 A	Welding speed	14 in./min(c)
Electrode wire	0.045-in.-diam ER70S-6	Preheat and postheat	None

(a) Constant-voltage type. (b) Supplied from bulk storage. (c) Welding speed by SMAW had been 5 to 7 in./min.

before being delivered to the welding department.

After the components had been located over pins and clamped, the assembly was inclined 15° from the flat position for welding. The welder used a slightly oscillating movement to control penetration and bead shape. The end of the nozzle of the electrode holder was held at a distance of about 1/2 in. from the workpiece; this provided sufficient gas for a porosity-free weld and eliminated excessive spatter. No preheat or postweld heat treatment was used.

Before production was started, three sample weldments were sectioned and etched to check for porosity and cracks. If the samples were satisfactory, the power supply and wire drive were locked and production began. During production, an inspection was made every 30 min, and all defective parts were reworked before production continued.

The change to GMAW resulted in a 25% reduction in cost per unit. For every hundred units, welding time was reduced by 2.013 h, and grinding time was reduced by 2.632 h. With this total saving of 4.645 h per 100 units, and an annual production of 9000 units, the change to GMAW saved 418 h of production time per year.

Although neither preheating nor postweld heating was used in producing the weldment described in the preceding example (which had one component made of steel containing 0.65 to 0.80% C), preheating and postweld heating are ordinarily used. As-welded joints in a steel containing 0.80% C should be employed with great caution, because of low notch toughness in the HAZ.

Gas Tungsten Arc Welding

Gas tungsten arc welding is less frequently used for joining carbon steel than for high-alloy steels and nonferrous metals. With GTAW, as with other welding processes, the use of automatic methods generally is desirable (provided the production volume is great enough to justify the equipment cost), because automatic welding usually increases the uniformity and controllability of weld quality and decreases manufacturing time. The example that follows describes an application in which a change from manual to automatic GTAW resulted in improved weld quality and increased production rate.

Example 10. Automatic Gas Tungsten Arc Welding of Medium-Carbon Steel Tubes to Tube Sheet Overlaid With Low-Carbon Steel. Fillet welds joining tubes to tube sheet for use in heat exchangers in feedwater heaters (see Fig. 33) were formerly made by manually rotating a piloted gas tungsten arc torch centered in the tube hole. Two passes were made around the 3/16-in. tube projection to form a fillet weld, each pass fusing a preplaced filler-wire ring. Although the welds were acceptable, they lacked uniformity; weld size and contour depended largely on the skill of the operator in rotating the torch at a constant speed for proper fusion of the filler-wire ring. In addition, production rate seldom exceeded 80 welded joints per 8-h shift.

The substitution of a new procedure, using fully automatic GTAW equipment under the conditions in Fig. 33, made it possible to obtain uniform, high-quality welds with good reproducibility. Defects that required rework were few, and production rate increased to an average of 190 welded joints per 8-h shift.

The automatic equipment included a custom-built GTAW head that was motor driven to rotate around a water-cooled stationary pilot inserted in the tube hole. Cold filler-metal wire was supplied to the arc from a spool, through a wire-feed drive and an adjustable wire-guide tip, all of which rotated with the welding head. This equipment, together with the slide mounting for horizontal and vertical positioning of the welding head to the tubes, is shown in Fig. 33.

Other equipment features essential to the procedure included: (1) an electric gage for accurate positioning of the tip of the tungsten electrode before the first pass was started; (2) automatic control of the flow of cooling water for the torch pilot, and of the argon shielding gas; (3) an indexing system that automatically retracted the tungsten electrode between passes; (4) a 300-A constant-current rectifier power supply with high frequency and slope control; and (5) an electronic programmer with sequence timers, to coordinate the weld cycles for preweld gas (prepurge), arc initiation, weld time per pass, arc decay, postweld gas (postpurge), and cooling-water flow.

The tube sheet had been surfaced with a low-carbon steel overlay of 1/8-in. minimum thickness after machining. The carbon content of the overlay was held to 0.15% maximum to provide a more ductile HAZ on the tube sheet, since the tube-sheet forging before cladding contained as much as 0.35% C.

Before assembly, tubes and tube sheet were solvent cleaned, and rigid cleanli-

Fig. 33 Automatic GTAW of tubes to tube sheet. Use of orbital welding setup, in which the welding head was automatically rotated, replaced an earlier procedure employing manual rotation. The change to automatic welding increased reproducibility and more than doubled production rate.

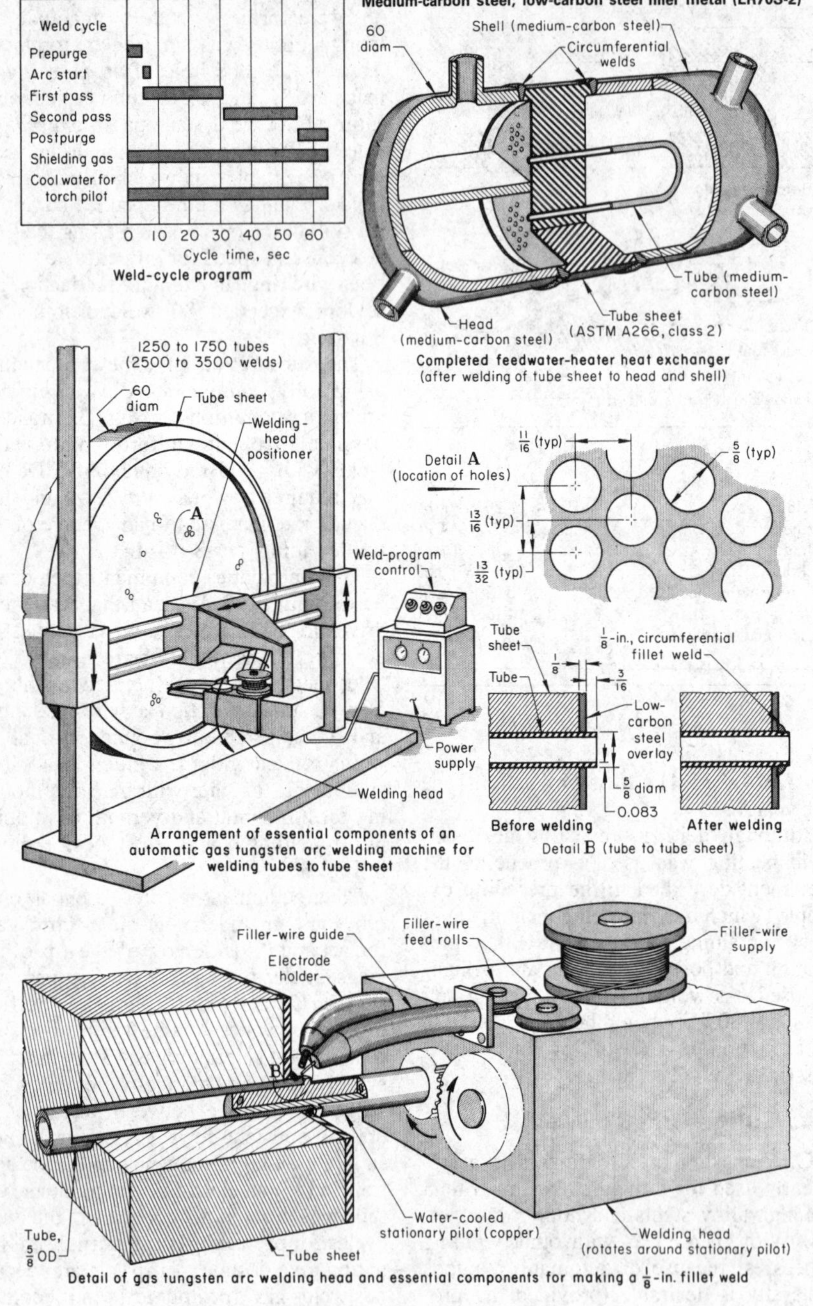

Conditions for automatic GTAW

Joint type	Circumferential T	Filler-wire feed	20 in./min
Weld type	Fillet	Shielding gas	Argon, at 15 ft^3/h
Welding position	Horizontal-fixed pipe	Current	140 to 150 A (DCEN)
Preheat temperature	100 °F	Voltage	9 to 11 V
Power supply	300-A rectifier(a)	Arc starting	High frequency
Electrode holder	Orbiting; argon-gas cooled	Number of passes	Two
Electrode	3/32-in.-diam EWTh-2	Welding speed	5 in./min
Filler metal	0.030-in.-diam ER70S-2 wire	Total time per two-pass weld	1 1/4 min

(a) Constant-current type, with high-frequency generator, slope control, and sequence timers

ness was maintained during welding. With tubes in place, the tube sheet was preheated to 100 °F. The welding-head pilot was centered in a tube, and the tungsten electrode and wire-guide tip were properly positioned. Settings for water and gas flow rates, filler-metal feed, rotational speed, welding current, high frequency, and slope control were adjusted for correct values. Finally, the various settings of the timers to control the sequencing of the complete two-pass weld cycle were adjusted on the control panel of the programmer, and the weld cycle was initiated by pushbutton. After each weld, the welding head was repositioned for welding the next tube, and the cycle was repeated. The weld-cycle program is shown at the upper left in Fig. 33.

After assembly and welding of the completed tube bundle in the vessel chamber (see view at upper right in Fig. 33), welds were tested hydrostatically, at a pressure of 7300 psi (chamber design pressure was 4500 psi). Welds were also inspected visually for apparent defects and mass-spectrometer freon tested for leaks.

Arc Welding of High-Strength Structural Steels and HSLA Steels

The steels included in these groups have higher strengths and better toughness than plain carbon steels; some grades also have improved corrosion resistance compared to carbon steels. This classification includes as-rolled pearlitic structural steels with minimum specified yield strengths of 40 to 50 ksi; normalized carbon steels with minimum yield strengths of 42 to 100 ksi and, if required, specified toughness and impact strengths; and microalloyed HSLA steels with properties that result from a combination of alloy additions and controlled rolling procedures. These steels are frequently used in the as-rolled condition and do not require subsequent heat treatments after forming or welding. They are widely used for structural applications and components for transportation, materials handling, agriculture, and construction.

Three major organizations responsible for the standardization of steels are the American Society for Testing and Materials (ASTM), the Society of Automotive Engineers (SAE), and the Canadian Standards Association (CSA). The specifications applicable to high-strength and HSLA steels are listed in Table 12. These specifications standardize chemical compositions and mechanical requirements and also

Table 12 Specifications for high-strength structural and HSLA steels

Alloy group	Specifications ASTM	SAE	CSA
Niobium or vanadium	A572 A607 A656 A709 A715	J410c, grades ending in "X"	G4021, type W
Low manganese-vanadium-titanium	A606 A633 A715	J410c	...
Manganese or manganese-copper	A537 A606 A633 A662 A678 A690 A737	J410c, grade 950C	G40.21, grade 350G
Manganese-vanadium-copper	A441 A607 A618	J410c, grade 950B	G40.21, grade 350T
Multiple alloy plus copper	A242 A588 A606	J410c, grades 950A, 950B	G40.21, grade 350A
Multiple alloy plus copper and phosphorous	A242 A606	J410c, grade 950D	G40.21, grade 350R
Constructional	A225 A299 A302	J410c	...

indicate the intended uses of each grade.

The ASTM specifications cover chemical compositions and minimum mechanical properties for various hot and cold rolled products. They also describe the intended end uses of each grade, with emphasis on civil engineering applications. The ASTM classifications provide sufficient flexibility in the compositional limits that one particular designation may be found in more than one alloy group. Supplementary requirements for the delivery of the steel by the customer often determine the designation of the alloy.

The SAE specifications are similar to those of ASTM in that they list requirements for chemical compositions and minimum mechanical properties; SAE grades are designated by a three-digit numbering system, suffixed by one or more letters. The last two digits designate the minimum yield strength (in ksi). Suffix letters are used to distinguish different degrees of toughness, formability, and/or weldability among steels of the same strength level. The suffix letter X is used to designate the addition of strengthening elements, such as niobium and vanadium, and the suffix K can be used to specify fully killed steels made to fine grain sizes.

The CSA classification lists the steel grades by a numbering system similar to SAE specifications. The CSA system lists the minimum yield strength by a three-digit number (in MPa) followed by a letter which identifies specific types. The letters provide details on the uses as follows:

- *Type G*: General constructional steel
- *Type W,T*: Weldable steels
- *Type R*: Atmospheric corrosion-resistant structural steel
- *Type A*: Atmospheric corrosion-resistant weldable structural steel
- *Type Q*: Quenched and tempered low-alloy steel plate

All grades are not available in each of the types listed above, and the appropriate standard should be consulted.

High-strength and HSLA steel grades are available in virtually all mill forms, including sheet, strip, plate, bar, rod, wire, and structural shapes. The availability of a particular grade, thickness, or shape is limited by the steelmaking capabilities and demand. The composition limits, available forms, special characteristics, and intended uses for steels listed in the ASTM, SAE, and CSA standards are given in Tables 13, 14, and 15, respectively.

High-strength structural and HSLA steels cover a wide range of applications and steel grades. The information presented in this article is intended only to provide a guideline of the properties of these steels. In summary, the main characteristics of these steel classifications are described below.

As-rolled pearlitic structural steels are a specific group of steels in which enhanced mechanical properties and, in some cases, resistance to atmospheric corrosion are obtained by addition of moderate amounts of one or more alloying elements other than carbon. Some of these steels are carbon-manganese steels and differ from ordinary carbon steels only because they have a greater manganese content. Other pearlitic structural steels contain small amounts of other alloying elements, which are added to enhance weldability, formability, toughness, and/or strength. These steels are characterized by as-rolled yield strengths in the range of 42 to 50 ksi. They are not intended for quenching and tempering and should not be subjected to such treatment. For certain applications, they may be annealed, normalized, or stress relieved, with corresponding changes in mechanical properties.

Microalloyed high-strength low-alloy steels use small additions of alloying elements such as niobium and vanadium or boron to increase the strength of standard carbon-manganese steels. Niobium and vanadium increase the yield strength by precipitation hardening and improve toughness by reducing ferrite grain size. Microalloying additives also include titanium and rare-earth elements. Heat treatment is not required because the properties develop on cooling from hot rolling. The introduction of controlled rolling results in additional benefits to strength and toughness.

High-strength structural carbon steels are basically carbon-manganese-silicon steels, with some grades containing small alloying additions of other elements. Such steels are either normalized or quenched and tempered to produce minimum yield strengths of 42 to 100 ksi and, when specified, minimum toughness or impact values.

Welding Processes

The HSLA steels can be welded with any of the arc welding processes using suitable welding procedures. The selection of a process is dictated by the thickness, position of the joint, and the physical location of the part to be welded. Most of the applications listed in Table 13 are structural uses such as bridges and buildings; in these cases, GTAW and plasma arc welding (PAW) are uneconomical and thus are not used. Electroslag welding (ESW) and electrogas welding (EGW) have been used to a limited extent because the high heat inputs may cause excessive grain growth and precipitation of microalloying elements, both of which lead to reduced mechanical properties. The most commonly used processes are SMAW, FCAW, GMAW, and SAW.

Welding Consumables

To avoid confusion, AWS classifications for consumables will be the only

Table 13 Characteristics and intended uses for high-strength structural and HSLA steels

ASTM specification	Title	Alloying elements	Available mill forms	Special characteristics	Intended use
A225	Pressure vessel plate, alloy steel, manganese-vanadium-nickel	V, Ni, Si	Plate, bar, and shapes up to 6 in.	Yield strength to 70 ksi in two steel grades	Primarily for welded layered pressure vessels
A242	HSLA structural steel	Cr, Cu, N, Ni, Si, Ti, V, Zr	Plate, bar, and shapes up to 4 in. thick	Atmospheric corrosion resistance four times that of carbon steel	Structural members in bolted, riveted, or welded construction
A302	Pressure vessel plate, manganese-molybdenum, and manganese-molybdenum-nickel	Mo, Ni, Si	Plate, bar, and shapes over 0.25 in. thick	Yield strength to 100 ksi in four steel grades	Welded boilers and other pressure vessels
A441	HSLA structural manganese-vanadium steel	V, Cu, Si	Plate, bar, and shapes up to 8 in. thick	Atmospheric corrosion resistance twice that of carbon steel	Welded, riveted, or bolted structures, but primarily welded bridges and buildings
A537	Pressure vessel plate heat treated, carbon-manganese-silicon steel	Cu, Ni, Cr, Mo, Si	Plate, bar, and shapes up to 4 in. thick	Yield strength to 50 ksi in one grade	Welded pressure vessels and structures
A572	HSLA niobium-vanadium steels of structural quality	Nb, V, N	Plate, bar, shapes and sheet to 6 in. thick	Yield strengths of 42 to 65 ksi in six grades	Welded, bolted, or riveted structures, but mainly bolted or riveted bridges and buildings
A588	HSLA structural steel with 50 ksi minimum yield point to 4 in. thick	Nb, V, Cr, Ni, Mo, Cu, Si, Ti, Zr	Plate, bar, and shapes up to 8 in. thick	Atmospheric corrosion resistance four times that of carbon steel; nine grades of similar strength	Welded, bolted, or riveted structures, but primarily welded bridges and buildings where weight savings or added durability is important
A606	Steel sheet and strip, hot rolled and cold rolled, HSLA with improved corrosion resistance	Not specified	Hot rolled and cold rolled sheet and strip	Atmospheric corrosion resistance twice that of carbon steel (type 2) or four times that of carbon steel (type 4)	Structural and miscellaneous purposes where weight savings or added durability is important
A607	Steel sheet and strip hot rolled and cold rolled, HSLA niobium and/or vanadium	Nb, V, N, Cu	Hot rolled and cold rolled sheet and strip	Atmospheric corrosion resistance twice that of carbon steel, but only when copper content is specified; yield strengths of 45 to 70 ksi in six grades	Structural and miscellaneous purposes where greater strength or weight savings is important
A618	Hot formed welded and seamless HSLA structural tubing	Nb, V, Si, Cu	Square, rectangular, round, and special shape structural tubing, welded or seamless	Three grades of similar yield strength; may be purchased with atmospheric corrosion resistance twice that of carbon steel	General structural purposes, including welded, bolted, or riveted bridges and buildings
A633	Normalized HSLA structural steel	Nb, V, Cr, Ni, Mo, Cu, N, Si	Plate, bar, and shapes up to 6 in. thick	Enhanced notch toughness; yield strengths of 42 to 60 ksi in five grades	Welded, bolted, or riveted structures for service at temperatures down to −50 °F
A656	HSLA hot rolled structural vanadium-aluminum-nitrogen and titanium-aluminum steels	V, Al, N, Ti, Si	Plate normally up to $^5/_8$ in. thick	Yield strength of 80 ksi	Truck frames, brackets, crane booms, rail cars, and other applications where weight savings is important
A662	Pressure vessel plate carbon-manganese, for moderate and lower	Not specified	Plate, bar, and shapes up to 2 in. thick	Yield strengths to 90 ksi in three steel grades	Welded pressure vessels with improved low-temperature notch toughness
A690	HSLA steel H-piles and sheet piling for use in marine environments	Ni, Cu, Si	Structural quality H-piles and sheet piling	Corrosion resistance two to three times greater than carbon steel in splash zone of marine structures	Dock walls, sea walls, bulkheads, excavations and similar structures exposed to sea water
A709	Structural steel for bridges	Nb, V, Ti, Al	Plate, bar, and shapes to 4 in. thick	Atmospheric corrosion resistance four times that of carbon steel	General structural purposes, including welded bridges
A715	Steel sheet and strip, hot rolled, HSLA with improved formability	Nb, V, Cr, Mo, N, Si, Ti, Zr, B	Hot rolled sheet and strip	Improved formability compared to A606 and A607; yield strengths of 50 to 80 ksi in four grades	Structural and miscellaneous applications, high strength, weight savings, improved formability
A737	Pressure vessel plate, steel	V, Nb, Si, N	Maximum thickness limited by mechanical property requirements	Yield strengths to 60 ksi in two grades	Welded pressure vessel and piping components

Table 14 Composition, properties, mill forms, and characteristics of HSLA steels described in SAE J410c

Grade(a)	Heat composition limits(b), % C max	Mn max	P max	Other elements(c)	Tensile strength, ksi	Yield strength (0.2% offset), ksi	Elongation in 2 in. min, %	Available mill forms	Special characteristics
942X	0.21	1.35	0.04	Nb, V	60	42	24	Plate, bar, and shapes up to 4 in. thick	Similar to 945X and 945C except to better weldability and formability
945A	0.15	1.00	0.04	...	65	45	22-24	Sheet, strip, plate, bar, and shapes up to 3 in. thick	Excellent weldability, formability, and notch toughness
945C	0.23	1.40	0.04	...	65	45	22-24	Sheet, strip, plate, bar, and shapes up to 3 in. thick	Similar to 950C except that lower carbon and manganese improve weldability, formability, and notch toughness
945X	0.22	1.35	0.04	Nb, V	60	45	22-25	Sheet, strip, plate, bar, and shapes up to 3 in. thick	Similar to 945C except for better weldability and formability
950A	0.15	1.30	0.04	...	70	50	22-24	Sheet, strip, plate, bar, and shapes up to 3 in. thick	Good weldability, notch toughness, and formability
950B	0.22	1.30	0.04	...	70	50	22-24	Sheet, strip, plate, bar, and shapes to 3 in. thick	Fairly good notch toughness and formability
950C	0.25	1.60	0.04	...	70	50	22-24	Sheet, strip, plate, bar, and shapes to 3 in. thick	Fair formability and toughness
950D	0.15	1.00	0.15	...	70	50	22-24	Sheet, strip, plate, bar, and shapes to 3 in. thick	Good weldability and formability; phosphorus should be considered in conjuction with other elements
950X	0.23	1.35	0.04	Nb, V	65	50	22	Sheet, strip, plate, bar, and shapes to 1.5 in. thick	Similar to 950C except for better weldability and formability
955X	0.25	1.35	0.04	Nb, V, N	70	55	20	Sheet, strip, plate, bar, and shapes to 1.5 in. thick	Similar to 945X and 950X except that progressively higher strengths are obtained by increasing the carbon and manganese contents, or by adding nitrogen up to 0.015%; formability and weldability generally decrease with increased strength; toughness varies with composition and mill practice
960X	0.26	1.45	0.04	Nb, V, N	75	60	18	Sheet, strip, plate, bar, and shapes to 1.5 in. thick	
965X	0.26	1.45	0.04	Nb, V, N	80	65	16	Sheet, strip, plate, bar, and shapes to 0.75 in. thick	
970X	0.26	1.65	0.04	Nb, V, N	85	70	14	Sheet, strip, plate, bar, and shapes to 0.75 in. thick	
980X	0.26	1.65	0.04	Nb, V, N	95	80	12	Sheet, strip, and plate to 0.38 in. thick	
980X	0.26	1.65	0.04	Nb, V, N	80	95	12	Sheet, strip, and plate to 0.38 in. thick	

(a) Fully killed steel made to fine grain practice may be specified by adding a second suffix "K"; for instance, 945XK. Steels made to "K" practice are normally specified only for applications requiring better toughness at low temperatures than steels made to normal semikilled practice. (b) 0.05 max P and 0.90 max Si, all grades. (c) Elements normally added singly or in combination to produce the specified mechanical properties and other characteristics. Other alloying elements such as copper, chromium, and nickel may be added to enhance atmospheric corrosion resistance.

designation used in this section. Consumables for welding high-strength structural and HSLA steels are chosen primarily to match the strength level of the base plate, while ensuring that weldability is adequate.

Susceptibility to cracking often requires that a particular filler metal be used. If the weldment is such that optimum preheating and postweld heat treatment practices can be employed, the selection of a filler metal is less critical than if welding conditions do not permit optimum preheating and postweld heat treatment.

Filler metals for arc welding alloy steels can be classified as:

- Low-carbon steels, often used when joint strength requirements are not stringent, or for joining alloy steel to low-carbon steel
- Alloy steels, used when joint strength must equal or approach that of the base metal
- High-alloy (stainless steel or nickel-based) filler metals for special applications, such as welding dissimilar steels

Carbon content of the filler metal should not be higher than that of the base metal unless the base steel has a very low carbon content. This situation would unnecessarily increase susceptibility to cracking. More often, the carbon content of the filler metal is lower than that of the base plate. A sufficiently close match of strength between base metal and weld metal usually can be achieved when the carbon content of the filler metal is no more than half that of the base metal.

Dilution of microalloying elements into the weld metal, and especially for the root pass, can make the weld metal more susceptible to cracking than the HAZ for HSLA steels. The smaller cross-sectional area of the weld bead, compared to the base metal, increases the applied stress across the section; this may increase crack susceptibility. Bead cracking can be avoided by depositing subsequent weld beads without excessive delay.

Filler metals designated for SMAW of high-strength structural and HSLA steels are listed in Table 16. The electrode is selected primarily by strength level and to a lesser extent by chemical composition. Alloying elements are added to the electrodes to achieve the necessary mechanical properties (Table 17) or chemical compositions (Table 18).

Low-hydrogen potassium electrodes or low-hydrogen iron-powder electrodes are recommended for these grades. These steels can also be welded with cellulosic electrodes, provided necessary precautions are taken to prevent hydrogen-induced cracking. Low-hydrogen electrodes often can be used without preheat. Preheating is advised with cellulosic electrodes. The use of adequate preheat and/or postheat can

Table 15 Composition, properties, mill forms, and characteristics of steels described in CSA G40.21M

Grade(a)	Heat composition limits, % C max	Mn	Si	Grain-refining elements(b)	Nominal maximum thickness, in.	Available mill forms	Tensile strength, ksi	Minimum yield point, ksi Up to 1.6 in.	1.6 to 2.5 in.	>2.5 in.	Elongation, minimum in 2 in., %
350G	0.28	1.65 max	0.4 max	0.10 max	1.2	Plate, bar, shapes, and pilings	70-100	51	...	...	19
350W	0.23	0.5-1.50	0.4 max	0.10 max	1.6	Plate, bar, shapes, pilings, and HSS	65-90	44	...	...	22
400W	0.23	0.5-150	0.4 max	0.10 max	0.8	Plate, bar, shapes, and pilings	75-100	58	...	...	...
480W	0.26	0.5-150	0.4 max	0.10 max	0.6	Plate, bar, shapes, and pilings	86-115	70	...	...	...
350T	0.22	0.8-1.50	0.15-0.40	0.10 max	4.0	Plate, bar, shapes, pilings, and HSS	70-94	51	48	46	22
350R	0.16	0.75 max	0.75 max	0.10 max	0.6	Plate, bar, shapes, and pilings	70-94	51	...	...	...
350A	0.20	0.75-1.35	0.15-0.40	0.10 max	4.0	Plate, bar, shapes, pilings and HSS	70-94	51	51	51	21

(a) Suffix designates the steel characteristics according to the following: **Type G**, general construction steel. Steels of this type meet minimum strength requirements; however, the chemical control is not such that all of these steels may be welded satisfactorily under normal field conditions. They are primarily designed for applications involving bolted connections or for welding under carefully controlled shop conditions. **Type W**, weldable steels. Steels of this type meet minimum strength requirements and are suitable for general welded construction where notch toughness at low temperatures is not a prime consideration. Applications may include buildings, or compression members of bridges. **Type T**, weldable steels. Steels of this type are suitable for welded construction and may be specified when low-temperature notch toughness is a prime consideration in design. In such cases Charpy V-notch testing should be specified by the purchaser to ensure that the impact requirements are met. **Type R**, atmospheric corrosion-resistant structural steel. Steels of this type display an atmospheric corrosion resistance approximately four times that of plain carbon steels. Steels whose copper contents do not exceed 0.02% may be readily welded up to the maximum thickness covered by this Standard. Applications include unpainted siding, unpainted light structural members, where notch toughness at low temperatures is not a consideration. **Type A**, atmospheric corrosion-resistant weldable structural steel. Steels of this type display an atmospheric corrosion resistance of approximately four times that of plain carbon steels. Steels whose copper content does not exceed 0.02% may be welded readily under normal conditions and are often used in structures in the unpainted condition. These steels may be specified when low-temperature notch toughness is a prime consideration in design. In such cases, Charpy V-notch testing should be specified by the purchaser to ensure that the impact requirements are met. (b) Niobium, vanadium, or aluminum either singly or in combination

prevent hydrogen cracking in the HAZ. Necessary preheat or postheat temperatures can be calculated when chemical composition of the base plate, weld cooling rate, hydrogen content of the weld metal, and the stresses imposed across the weld are known. A recommended method to calculate preheat temperatures is discussed in the previous section of this article on preheating. Usually, however, preheats are established for standard steel grades by codes or AWS recommended practices (Welding Research Council Bulletin No. 191). Hydrogen in the weld pool is absorbed from the arc atmosphere during welding. Sources of hydrogen include moisture contained in electrode coatings and fluxes, environmental humidity, or oil and grease on the plate surfaces. Moisture absorbed by consumables is a major contributor. Permissible moisture contents of low-hydrogen electodes are given in Table 19.

Measurement of hydrogen usually is accomplished by collecting the diffusible hydrogen of a welded test piece under a liquid. The traditional method for moisture determination of the flux coating of SMAW electrodes is not as useful because the weld "sees" the hydrogen and not the moisture, and the moisture in the coating can be either chemically combined or simply resting on the surface. Moisture on the surface can be evaporated before it enters the weld arc and thus does not enter the weld pool. This procedure for measuring actual hydrogen content of a deposited weld is the most widely accepted.

"Hydrogen-controlled" electrodes are available packaged in hermetically sealed containers to prevent moisture absorption. Once opened, containers should be placed in electrode ovens at temperatures between 200 and 300 °F. Low-hydrogen electrodes exposed to the atmosphere absorb moisture, which ultimately results in greater weld-metal hydrogen contents. The results of hydrogen measurements using a standard test for E7018 electrodes exposed at 90% relative humidity for various lengths of time are shown in Fig. 34. Electrodes used from new undamaged packages or from electrode ovens result in weld-metal hydrogen contents of about 4 mL/100 g of deposited metal. Exposure to humid conditions can increase this value up to a maximum of 30 mL/100 g of deposited metal. Hydrogen values less than 10 mL/100 g are considered hydrogen controlled. Figure 34 shows that this value is exceeded after about 3 h of exposure in very humid conditions; therefore, electrodes should be rebaked or discarded after this time. The increase in hydrogen content is directly related to the relative humidity, but the worst possible conditions should be used as a guideline. Recommended rebaking temperatures for basic electrodes range from 660 to 800 °F; temperatures for storing and rebaking filler metals are listed in Table 20.

Filler metals designated for GMAW

of high-strength structural and HSLA steels are classified on the basis of chemical composition and mechanical properties of the weld metal. Filler metals available with a minimum ultimate tensile strength of 70 ksi usually are not required to undergo mechanical testing in other than the as-welded condition. Some electrodes of 80-ksi strength or greater are required to exhibit minimum mechanical properties in the stress-relieved condition as well.

Designations and chemical compositions of the filler metals suitable for high-strength and HSLA steels are listed in Table 21. Mechanical properties of weld metals conform to the minimum requirements given in Table 22. It should be noted that weld properties may vary appreciably, depending on filler-metal size and current used, plate thickness, joint geometry, preheat and interpass temperatures, surface conditions, base-metal composition and extent of alloying with the filler metal, and shielding gas. For example, when filler metals with composition within the ranges given in Table 21 are deposited, weld-metal chemical composition does not vary greatly from the as-manufactured composition of the filler metal when used with argon-oxygen shielding gas, but does show considerable reduction in manganese, silicon, and other deoxidizers when used with carbon dioxide as the shielding gas. This reduction lowers the tensile and yield strengths of the welds made using carbon dioxide shielding gas, but these values are

Table 16 Covered electrodes commonly used for SMAW of high-strength structural and HSLA steels

Steel grade	SMAW electrode
A225	
Grade C	E11018-M
Grade D	E9018-M
A242	E7016, E7018
A299	E8016-C3, C8018-C3
A302	
Grade A	E7016-A1, E7018-A1
Grade B	E8016-B2, E8018-B2
Grades C, D	E10016-D2, E10018-D2
A441	E7016, E8016-C3, E8018-C3
A537	
Class 1	E7018-A1
Class 2	E8016-C1, C11018-M
A572	
Grades 42, 50	E7016, E7018
Grades 60, 65	E8016-C3, E8018-C3
A588	
Grades A, B, C, D, E	E7016, E7018
Grade F	E8016-B1, E8018-B1
Grade G	E8016-C1, E8018-C1
Grade H	E8015-G, E8018-G
A606	E7016, E7018, E7028
A607	
Grades 45, 50, 55	E7016, E7018, E7028
Grades 60, 65, 70	E9018-M
A618	E7018
A633	E7016, E7018, E7028
A656	E10018-D2
A662	E7016, E7018, E7028
A678	
Grade A	E7016, E7018, E7028
Grade B	E9018-M
Grade C	E10018-M
A709	
Grade 36T	E6012, E6013, E7014, E7016, E7018, E7028
Grade 50T, 50WT	E7016, E7018, E7028
Grade 100T, 100WT	E11018-M
A737	
Grade B	E7016, E7018, E7028
Grade C	E9018-M

Table 17 Minimum mechanical-property requirements for weld-metal deposits of SMAW electrodes

AWS classification	Yield strength (0.2% offset), ksi	Tensile strength (min), ksi	Elongation, minimum in 2 in., %	Charpy V-notch strength (min), ft·lb
E7016-X, E7018-X	57	70	25	...
E8015-X, E8016-X, E8018-X	67	80	19	...
E8016-C1, E8018-C1	67	80	19	20 at 75 °F
E8016-C3, E8018-C3	68-80	80	24	20 at 40 °F
E10016-D2, E10018-D2	87	100	16	20 at 60 °F
E11018-M	98-110	110	20	20 at 60 °F

not less than the minimum values specified in Table 22. Filler metals of the ER70S-X classifications have slightly different chemical compositions, which make them suitable for particular applications. The characteristics of each are given below.

ER70S-2. This classification covers multiple deoxidized steel filler metals that contain a nominal combined total of 0.20% zirconium, titanium, and aluminum, in addition to the silicon and manganese contents. These filler metals are capable of producing sound welds in semi-killed and rimmed steels, as well as in killed steels of various carbon levels. Because of the added deoxidants, these filler metals can be used for welding steels that have a rusty or dirty surface, with a possible sacrifice of weld quality depending on the degree of surface contamination. They can be used with a shielding gas of argon-oxygen mixtures, carbon dioxide, or argon-carbon dioxide mixtures and are preferred for out-of-position welding with short circuiting transfer because of their ease of operation.

ER70S-3. These filler metals meet the requirements of AWS A5.18 with either carbon dioxide or argon-oxygen as a shielding gas. They are used primarily on single-pass welds, but can be used on multiple-pass welds, especially when welding killed or semi-killed steel. Small-diameter electrodes can be used for out-of-position welding and for short circuiting transfer with argon-carbon dioxide mixtures or carbon dioxide shielding gases. However, the use of carbon dioxide shielding gas in conjunction with excessively high heat inputs may result in failure to meet the minimum specified tensile and yield strengths.

ER70S-4. These filler metals contain slightly higher manganese and silicon contents than those of the ER70S-3 classification and produce a weld deposit of higher tensile strength. The primary use of these filler metals is for carbon dioxide shielded welding applications where a slightly longer arc or other conditions require more deoxidation than provided by the ER70S-3 filler metals. These filler metals are not required to demonstrate impact properties.

ER70S-5. This classification covers filler metals that contain aluminum in addition

Table 18 Maximum chemical composition limits for SMAW electrodes

AWS classification(a)	Composition(b), % C	Mn	P	S	Si	Ni	Cr	Mo	V
E7016-A1	0.12	0.90	0.03	0.04	0.60	...	...	0.40-0.65	...
E7018-A1	0.12	0.90	0.03	0.04	0.80	...	...	0.40-0.65	...
E8016-B1	0.05-0.12	0.90	0.03	0.04	0.60	...	0.40-0.65	0.40-0.65	...
E8018-B1	0.05-0.12	0.90	0.03	0.04	0.80	...	0.40-0.65	0.40-0.65	...
E8016-B2	0.05-0.12	0.90	0.03	0.04	0.60	...	1.0-1.50	0.40-0.65	...
E8018-B2	0.05-0.12	0.90	0.03	0.04	0.80	...	1.0-1.50	0.40-0.65	...
E8016-C1	0.12	1.25	0.03	0.04	0.60	2.0-2.75	...	...	...
E8018-C1	0.12	1.25	0.03	0.04	0.80	2.0-2.75	...	...	...
E8016-C3(c)	0.12	0.40-1.25	0.03	0.03	0.80	0.8-1.10	0.15	0.35	0.05
E8018-C3(c)	0.12	0.40-1.25	0.03	0.03	0.80	0.8-1.10	0.15	0.35	0.05
E10016-D2	0.15	1.65-2.0	0.03	0.04	0.60	...	...	0.25-0.45	...
E10018-D2	0.15	1.65-2.0	0.03	0.04	0.08	...	...	0.25-0.45	...
EXX15-G(d), EXX16-G, EXX18-G	...	1.0 min(e)	...	...	0.80 min(e)	0.50 min(e)	0.30 min(e)	0.20 min(e)	0.10 min(e)
E11018-M	0.10	1.30-1.80	0.03	0.03	0.60	1.25-2.50	0.40	0.25-0.50	0.50

Note: Single values shown are maximum percentages, except where otherwise specified.
(a) Suffixes A1, B3, C2, etc., designate the chemical composition of electrode classification. (b) For determining the chemical composition, DCEN may be used where direct current, both polarities, is specified. (c) These classifications are intended to conform to classifications covered by military specifications for similar compositions. (d) The letters "XX" used in the classification designations indicate the various strength levels (70, 80, 90, 100, 110, and 120) of electrodes. (e) To meet the alloy requirements of the G group, the weld deposit must have the minimum, as specified above, of only one of these elements.

Table 19 Permissible moisture content of coverings on low-hydrogen electrodes packaged in sealed containers

AWS classification(a)	Maximum moisture content, wt%
E7015, E7016, E7018	0.6
E7015-X, E7016-X, E7018-X	0.4
E8015-X, E8016-X, E8018-X	0.2
E9015-X, E9016X, E9018-X, E10015-X, E10016X, E10018-X, E11015-X, E11016X, E11018-X, E12015-X, E12016X, E12018-X	0.15
E12018-M1	0.10

(a) Suffix "X" stands for all the suffixes (A1, B2, C3, M, etc.) except for the 12018-M1 classification.

to manganese and silicon as deoxidizers. These filler metals can be used when welding rimmed, killed, or semi-killed steels with carbon dioxide shielding gas and high welding currents. The relatively large amount of aluminum ensures the deposition of a well-deoxidized and sound weld metal. Because of the aluminum, these filler metals are not used for short circuiting transfer, but can be used for welding steels that have a rusty or dirty surface, with a possible sacrifice in weld quality, depending on the degree of surface contamination. These filler metals are not required to demonstrate impact properties.

ER70S-6. Filler metals of this classification have the highest combination of manganese and silicon, permitting high-current welding with carbon dioxide gas shielding even in rimmed steels. They also may be used to weld sheet metal in applications requiring smooth weld beads and in steels having moderate amounts of rust and mill scale. The quality of the weld depends on the degree of surface impurities. This filler metal is also suitable for out-of-position welding with short circuiting transfer.

ER70S-7. These filler metals have a substantially greater manganese content (essentially equal to that of ER70S-6) than ER70S-3 filler metals. This provides slightly better wetting and weld appearance with slightly higher tensile and yield strengths and may permit increased welding speeds compared to ER70S-3 filler metals. These metals generally are recommended for use with argon-oxygen shielding gas mixtures, but they are usable with argon-carbon dioxide mixtures and carbon dioxide under the same general conditions as for ER70S-3 filler metals. Under equivalent welding conditions, weld hardness is lower than ER70S-6 weld metal but higher than ER70S-3 deposits.

ER70S-G. This classification includes solid filler metals that are not included in the preceding classes. The supplier should be consulted to provide required characteristics for the intended use. This specification does not list specific chemical composition or impact requirements. These are subject to agreement between supplier and purchaser. However, any filler metal classified ER70S-G must meet all other requirements of this specification.

The other filler metals listed in Tables 21 and 22 have different chemical compositions to achieve different strength levels in the weld-metal deposits; they also meet minimum impact requirements. Electrodes are comparable to SMAW electrodes with respect to their applicability, as described below.

ER80S-Ni1. These filler metals deposit weld metal similar to 8018-C3 covered electrodes and are used for welding HSLA steels requiring good toughness at temperatures as low as −40 °F.

ER80S-Ni2. These filler metals deposit weld metal similar to 8018-C1 electrodes. Typically, they are used for welding $3^1/_2$% Ni steels and other materials requiring a tensile strength of 80 ksi and good toughness at temperatures as low as −75 °F.

ER80S-Ni3. These filler metals deposit weld metal similar to 8018-C2 electrodes. Typically, they are used for welding $3^1/_2$% nickel steels for low-temperature service where a tensile strength of 90 ksi is required.

ER80S-D2. This classification is the same as E70S-1B of AWS specification A5.18. Filler metals of this classification contain a high level of deoxidizers (manganese and silicon) to control porosity when welding with carbon dioxide as the shielding gas, and molybdenum for increased strength. They produce radiographic-quality welds with excellent bead appearance in readily weldable and difficult-to-weld carbon and low-alloy steels. They exhibit excellent out-of-position welding characteristics with short circuiting and pulsed arc processes. The combination of weld soundness and strength makes filler metal of this classification suitable for

Fig. 34 Effect of humidity on weld-metal hydrogen content

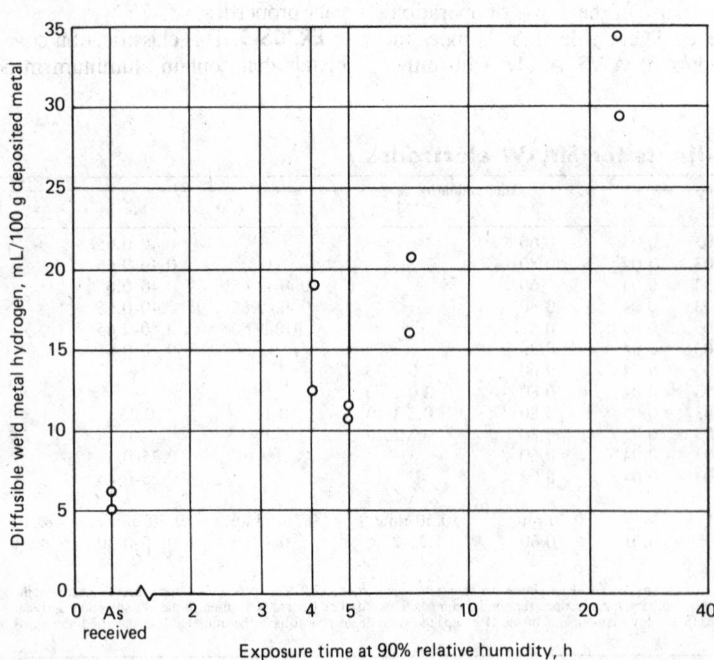

Table 20 Storage and rebaking of filler metals

Filler metal	Storage temperature, °F	Rebaking temperature, °F
Cellulosic electrodes	75-100	Not recommended
Basic electrodes	200-300	650-800
Other electrodes	75-100	250-300
Flux cored wires	70-300	Not recommended
Solid wires	70-100	Not required
Submerged arc flux	200-300	750-850

Table 21 Composition requirements of GMAW electrodes for high-strength structural and HSLA steels

AWS classification(a)	Composition, wt%													
	C	Mn	Si	P	S	Ni	Cr	Mo	V	Cu(b)	Ti	Zr	Al	Total other elements(c)
ER70S-2	0.07	0.90-1.40	0.40-0.70	...	...	...	...	...	...	...	0.05-0.15	0.02-0.12	0.05-0.15	...
ER70S-3	0.06-0.15	0.90-1.40	0.45-0.70	...	...	...	...	...	...	...	...	...	...	...
ER70S-4	0.07-0.15	1.00-1.50	0.65-0.85	0.025	0.035	(d)	(d)	(d)	(d)	0.50	...	...	...	...
ER70S-5	0.07-0.19	0.90-1.40	0.30-0.60	...	...	...	...	...	...	...	...	...	0.50-0.90	...
ER70S-6	0.07-0.15	1.40-1.85	0.80-1.15	...	...	...	...	...	...	...	...	...	...	...
ER70S-7	0.07-0.15	1.50-2.00(e)	0.50-0.80	...	...	...	...	...	...	...	...	...	...	...
ER70S-G	No chemical requirements(f)													
ER80S-Ni1	...	...	...	...	...	0.80-1.10	0.15	0.35	0.05	...	...	...	...	...
ER80S-Ni2	0.12	1.24	0.40-0.80	0.025	0.025	2.0-2.75	...	...	...	0.35	...	...	...	0.50
ER80S-Ni3	...	...	...	...	...	3.0-3.75	...	...	...	...	...	...	...	...
ER80S-D2	0.07-0.12	1.60-2.10	0.50-0.80	0.025	0.025	0.15	...	0.40-0.60	...	0.50	...	...	...	0.50
ER100S-1	0.08	1.25-1.30	0.20-0.50	0.010	0.010	1.40-2.10	0.30	0.25-0.55	0.05	0.25	...	...	...	...
ER100S-2	0.12	1.25-1.80	0.20-0.60	0.010	0.010	0.80-1.25	0.30	0.20-0.55	0.05	0.35-0.65	0.10	0.10	0.10	0.50
ER110S-1	0.09	1.40-1.80	0.20-0.55	0.010	0.020	1.90-2.60	0.50	0.25-0.55	0.04	0.25	...	...	...	...
ER120S-1	0.10	1.40-1.80	0.25-0.60	0.010	0.010	2.0-2.80	0.60	0.30-0.65	0.03	0.25	...	...	...	...
ERXXS-G	Subject to agreement between supplier and purchaser													

Note: Single values shown are maximums. Analysis shall be made for the elements for which specific values are shown in this table. If, however, the presence of other elements is indicated in the course of routine analysis, further analysis shall be made to determine that the total of these other elements, except iron, is not present in excess of the limits specified for "Total other elements" in the last column of this table.

(a) Electrodes classified E70S-1B previously are now classified ER80S-D2. The suffixes B2, Ni1, etc., designate the chemical composition of the electrode and rod classification. (b) The maximum weight percent of copper in the rod or electrode due to any coating plus the residual copper content in the steel shall comply with the stated value. (c) Other elements, if intentionally added, shall be reported. (d) These elements may be present but are not intentionally added. (e) In this classification, the maximum manganese may exceed 2.0%. If it does, the maximum carbon must be reduced 0.01% for each 0.05% increase in manganese or part thereof. (f) For this classification, there are no chemical requirements for the elements listed, with the exception that there shall be no intentional addition of nickel, chromium, molybdenum, or vanadium. (g) To meet the requirements of the G classification, the electrode must have as a minimum one of either 0.50% Ni, 0.30% Cr, or 0.20% Mo.

Table 22 Mechanical-property requirements for weld-metal deposits of GMAW electrodes

AWS classification	Shielding gas	Current and polarity	Minimum tensile strength, ksi	Minimum yield strength at 0.2% offset, ksi	Elongation in 2 in., min, %	Condition	Minimum required impact properties, ft · lb
ER70S-2	CO_2(a)	DCEP	72(b)	60(b)	22	As welded	20 at −20 °F
ER70X-3	CO_2(a)	DCEP	72(b)	60(b)	22	As welded	20 at 0 °F
ER70S-4	CO_2(a)	DCEP	72(b)	60(b)	22	As welded	Not required
ER70S-5	CO_2(a)	DCEP	72(b)	60(b)	22	As welded	Not required
ER70S-6	CO_2(a)	DCEP	72(b)	60(b)	22	As welded	20 at −20 °F
ER70S-7	CO_2(a)	DCEP	72(b)	60(b)	22	As welded	20 at −20 °F
ER70S-G	(c)	DCEP	72(b)	60(b)	22	As welded	(c)
ER80S-Ni1	Argon + 1 to 5% oxygen	DCEP	80	68	24	As welded	20 at −50 °F(f, g)
ER80S-Ni2	Argon + 1 to 5% oxygen	DCEP	80	68	24	PWHT(d)	20 at −80 °F(f, h)
ER80S-Ni3	Argon + 1 to 5% oxygen	DCEP	80	68	24	PWHT(d)	20 at −100 °F(f, g)
ER80S-D2	CO_2	DCEP	80	68	17	PWHT(d)	20 at −20 °F
ER100S-1	Argon + 2% oxygen	DCEP	100	88-102	16	As welded	20 at −20 °F(f, g)
ER100S-2	Argon + 2% oxygen	DCEP	100	88-102	16	As welded	50 at −60 °F(j)
ER110S-1	Argon + 2% oxygen	DCEP	110	95-107	15	As welded	50 at −60 °F(j)
ER120S-1	Argon + 2% oxygen	DCEP	120	105-122	14	As welded	50 at −60 °F(j)
ERXXS-G	(c)	DCEP	(e)	(c)	(c)	As welded	(c)
EXXC-G	(c)	DCEP	(e)	(c)	(c)	As welded	(c)

(a) Carbon dioxide shielding gas. The use of CO_2 for classification purposes shall not be construed to restrict the use of argon-CO_2 or argon-O_2 shielding gas mixtures. A filler metal classified with CO_2 will also meet the requirements of this specification when used with argon-CO_2 or argon-O_2 mixtures. (b) For each increase of one percentage point in elongation over the minimum, the yield strength, or tensile strength, or both may decrease 1 ksi to a minimum of 70 ksi for the tensile strength and 58 ksi for the yield strength. (c) Subject to agreement between supplier and purchaser. (d) PWHT, postweld heat treated in accordance with AWS A5.28-79. (e) Tensile strength shall be consistent with the level placed after the "ER" or "E" prefix; e.g., ER90S-G shall have 90 ksi minimum ultimate tensile strength. (f) The lowest and the highest values obtained shall be disregarded for this test. Two of the three remaining values shall be greater than the specified 20-ft · lb energy level; one of the three may be lower but shall not be less than 15 ft · lb. The computed average value of the three values shall be equal to or greater than the 20-ft · lb energy level. (g) As-welded impact properties. (h) Required impact properties after postweld heat treatment (see AWS A5.28-79). (j) Impact properties for the ER100S-1, ER100S-2, ER110S-1, and ER120S-1 shall be obtained at test temperature specified in the table, ±3 °F. The lowest and the highest impact values thus obtained shall be disregarded for this test. Two of the three remaining values shall be greater than the specified 50-ft · lb energy level; one of the three may be lower but shall not be less than 40 ft · lb; the computed average values of these three shall be greater than the specified 50-ft · lb level.

single- and multiple-pass welding of a variety of carbon and low-alloy steels.

ER100S-1, ER100S-2, ER110S-1, and ER120S-1. These filler metals deposit high-strength, very tough weld metal for critical applications. Originally developed for welding HY80 and HY100 steels for military applications, they also are used for a variety of structural applications where tensile strength requirements exceed 100 ksi and excellent toughness is required to temperatures as low as −60 °F.

ERXXS-G. This classification includes solid electrodes and rods and composite metal cored and stranded electrodes that are not included in the preceding classes. The supplier should be consulted for the characteristics and intended use of these filler metals. This specification (AWS A5.18) does not list specific chemical compositions or impact requirements, which are subject to agreement between supplier and purchaser. However, any filler metal classified ERXXS-G must meet all other requirements of this specification.

Electrodes of the ER70S-3 and ER70S-6 groups are suitable for most of the lower strength grades of high-strength structural and HSLA steels. The higher alloyed wires can be selected to optimize the properties of the weld and obtain a good match with the base plate.

Gas shielding of the arc gives much lower weld-metal hydrogen levels than most processes. The only possible sources of hydrogen in GMAW are from drawing compounds in the wire, moisture in the shielding gas, or contaminants on the plate surface. When normal precautions are observed for welding, hydrogen levels do not exceed the low hydrogen limits; therefore, they are suitable for welding the higher strength grades.

Filler metals designated for FCAW of high-strength structural and HSLA steels are classified on the basis of:

- Whether carbon dioxide is used as a separate shielding gas
- Type of current
- Position of welding
- Chemical composition of the weld metal
- Mechanical properties of the weld metal

Weld-metal properties may vary, depending on electrode size, amperage and voltage used, type and amount of shielding gas, electrical extension, plate thickness and composition, joint geometry, preheat and interpass temperatures, surface condition, and admixture of the base metal with the deposited metal. Electrode manufacturers and suppliers can supply pertinent welding procedure information.

Toughness requirements can be used to select electrodes for applications demanding low-temperature notch toughness. Because properties vary considerably with welding procedures, particular attention must be paid to the preparation of test welds.

Flux cored electrodes are designed primarily for use in the flat and horizontal positions, if designated EX0TX-X. These electrodes may be used in other positions if the proper welding current and electrode diameter are used. Electrode diameters less than $^{3}/_{32}$ in. and currents on the low side of the range recommended by the manufacturer may be used for out-of-position welding. EX1TX-X electrodes are designed for use in all positions.

There are two AWS specifications for flux cored electrodes that are applicable to the welding of high-strength structural and HSLA steels: AWS A5.20, "Specification for Carbon Steel Electrodes for Flux-Cored Arc Welding"; and AWS A5.29, "Specification for Low Alloy Steel Electrodes for Flux-Cored Arc Welding." More information on the classification system for both carbon steel and low-alloy steel flux cored electrodes can be found in the article "Flux Cored Arc Welding" in this Volume.

The most commonly used carbon steel filler metals for FCAW of HSLA steels are those of the E70T1-X, E70T4-X, and E70TX-6 classifications. The low-alloy steel flux cored electrodes have a wide variety of compositions that may or may not match the base-metal composition (Table 23). As can be seen from Table 23, an "A" classification indicates the filler material is a carbon-molybdenum steel electrode, "B" refers to chromium-molybdenum steel electrodes, "C" refers to nickel steel electrodes, "D" refers to manganese-molybdenum steel electrodes, and "K" refers to other low-alloy steel electrodes. Mechanical-property requirements for low-alloy steel flux cored electrodes are given in Table 24.

Filler Metals Designated for SAW. Selection of an electrode and flux combination for welding high-strength structural and HSLA steels requires the consideration of the composition and condition of the base plate, whether single- or multiple-pass welding is used, and the arc voltage and current. Multiple-pass deposits made at moderate welding currents have, as a general rule, better mechanical properties than one- or two-layer deposits made at high currents in similar plate thicknesses.

Chemical composition and mechanical properties of the deposit are influenced by both the wire and flux. Electrodes are classified only on the basis of their chemical composition, as shown in Table 25. Fluxes are classified on the basis of mechanical properties of a weld deposit, using the flux in combination with any of the electrodes classified in Table 25. A specific flux may have many designations, the number of which is limited only by the number of different electrode types with which it can be used. The flux classification system is given in Table 26.

Weld-metal properties available with several flux and wire combinations are given in Table 27. The same AWS classification can be obtained with many consumables; as such, the values listed in Table 27 should be considered as typical values. Additional information on chemical composition and mechanical properties of flux cored deposits can be found in AWS A5.23, "Specification for Bare Low-Alloy Steel Electrodes and Fluxes for Submerged Arc Welding."

Submerged arc consumables suppliers should be contacted for proper recommendations for welding procedures. General guidelines for process and flux selection available from the manufacturers include:

- Low-carbon and low-alloy steel welding
- Sheet metal welding
- Single-pass welding on plate
- Flat fillets
- Horizontal fillets (0.2 in. and under)
- Horizontal fillets (over 0.2 in.)
- Minimum melt-through (high currents)
- Minimum melt-through (low currents)
- Lowest flux consumption
- Welding over rust or mill scale
- Circumferential welds
- Alternate current welding
- Slag removal, deep groove welds
- Resistance to multiple-pass high-voltage cracking

A high-current one-pass SAW procedure can be used to join beams and plates up to 60 ft long with thicknesses ranging from 0.4 to 2 in. Greater thicknesses may require more than one pass. Longitudinal joint welding in box columns for high-rise buildings, machine bases, heavy equipment, shipbuilding, barges, and heavy plate joining are suitable applications. The flux is designed to tolerate the high currents used in the process and provides the proper slag coverage to prevent melt-through, which normally is a problem with high-current welding. Because of the high dilution rate and the nature of the process, high-current welding does not have an AWS classification. Weld-metal properties depend on the base plate used.

Welding of Linepipe Steels

The largest single application of HSLA steels is in linepipe steels. Economics favor the use of higher strength steels for

Table 23 Composition requirements of FCAW electrodes for high-strength structural and HSLA steels

AWS classification	Chemical composition, %(a) C	Mn	P	S	Si	Ni	Cr	Mo	V	Al(b)	Cu
Carbon-molybdenum steel electrodes											
E70T5-A1, E80T1-A1, E81T1-A1	0.12	1.25	0.03	0.03	0.80	···	···	0.40-0.65	···	···	···
Chromium-molybdenum steel electrodes											
E81T1-B1	0.12	1.25	0.03	0.03	0.80	···	0.40-0.65	0.40-0.65	···	···	···
E80T5-B2L	0.05	1.25	0.03	0.03	0.80	···	1.00-1.50	0.40-0.65	···	···	···
E80T1-B2, E81T1-B2, E80T5-B2	0.12	1.25	0.03	0.03	0.80	···	1.00-1.50	0.40-0.65	···	···	···
E80T1-B2H	0.10-0.15	1.25	0.03	0.03	0.80	···	1.00-1.50	0.40-0.65	···	···	···
E90T1-B3L	0.05	1.25	0.03	0.03	0.80	···	2.00-2.50	0.90-1.20	···	···	···
E90T1-B3, E91T1-B3, E90T5-B3, E100T1-B3	0.12	1.25	0.03	0.03	0.80	···	2.00-2.50	0.90-1.20	···	···	···
E90T1-B3H	0.10-0.15	1.25	0.03	0.03	0.80	···	2.00-2.50	0.90-1.20	···	···	···
Nickel-steel electrodes											
E71T8-Ni1, E80T1-Ni1, E81T1-Ni1, E80T5-Ni1	0.12	1.50	0.03	0.03	0.80	0.80-1.10	0.15	0.35	0.05	1.8	···
E71T8-Ni2, E80T1-Ni2, E81T1-Ni2, E80T5-Ni2, E90T1-Ni2, E91T1-Ni2	0.12	1.50	0.03	0.03	0.80	1.75-2.75	···	···	···	1.8	···
E80T5-Ni3, E90T5-Ni3	0.12	1.50	0.03	0.03	0.80	2.75-3.75	···	···	···	···	···
Manganese-molybdenum steel electrodes											
E91T1-D1	0.12	1.25-2.00	0.03	0.03	0.80	···	···	0.25-0.55	···	···	···
E90T5-D2, E100T5-D2	0.15	1.65-2.25	0.03	0.03	0.80	···	···	0.25-0.55	···	···	···
E90T1-D3	0.12	1.00-1.75	0.03	0.03	0.80	···	···	0.40-0.65	···	···	···
All other low-alloy steel electrodes											
E80T5-K1	0.15	0.80-1.40	0.03	0.03	0.80	0.80-1.10	0.15	0.20-0.65	0.05	···	···
E70T4-K2, E71T8-K2, E80T1-K2, E90T1-K2, E91T1-K2, E80T5-K2, E90T5-K2	0.15	0.50-1.75	0.03	0.03	0.80	1.00-2.00	0.15	0.35	0.05	1.8	···
E100T1-K3, E110T1-K3, E100T5-K3, E110T5-K3	0.15	0.75-2.25	0.03	0.03	0.80	1.25-2.60	0.15	0.25-0.65	0.05	···	···
E110T5-K4, E111T1-K4, E120T5-K4	0.15	1.20-2.25	0.03	0.03	0.80	1.75-2.60	0.20-0.60	0.30-0.65	0.05	···	···
E120T1-K5	0.10-0.25	0.60-1.60	0.03	0.03	0.80	0.75-2.00	0.20-0.70	0.15-0.55	0.05	···	···
E61T8-K6, E71T8-K6	0.15	0.50-1.50	0.03	0.03	0.80	0.40-1.10	0.15	0.15	0.05	1.8	···
E101T1-K7	0.15	1.00-1.75	0.03	0.03	0.80	2.00-2.75	···	···	···	···	···
EXXXTX-G	···	1.00 min(c)	0.03	0.03	0.80 min(c)	0.50 min(c)	0.30 min(c)	0.20 min(c)	0.10 min(c)	1.8	···
E80T1-W	0.12	0.50-1.30	0.03	0.03	0.35-0.80	0.40-0.80	0.45-0.70	···	···	···	0.30-0.75

(a) Single values are maximums unless otherwise noted. (b) For self-shielded electrodes only. (c) In order to meet the alloy requirements of the G group, the weld deposit need have the minimum, as specified in the table, of only one of the elements listed.

large-diameter pipelines for oil and natural gas. Linepipe frequently is welded in the field using cellulosic (high-hydrogen) electrodes, requiring steels of low-carbon equivalents. Linepipe is often used in cold-weather regions, making weld-zone toughness a critical requirement, and petroleum products may contain hydrogen-sulfide, which can cause sulfide stress cracking in high-hardness regions of welds.

Field welding of HSLA linepipe steels can be carried out using a variety of processes, including SMAW, semiautomatic and automatic GMAW, and FCAW. Generally, electrodes or consumables whose strengths match the base metal are used. Techniques are slightly different from normal welding procedures, and separate welder qualifications are required.

Standards normally referred to in linepipe welding applications are issued by API and include procedures for welding pipe, qualification of operators, joint design, testing, and inspection. Additional API

Table 24 Mechanical-property requirements for carbon steel and low-alloy steel electrodes used for FCAW of high-strength structural and HSLA steels

AWS classification	Tensile strength range, ksi	Yield strength at 0.2% offset, min, ksi	Elongation in 2 in., min, %
E6XTX-X	60-80	50	22
E7XTX-X	70-90	58	20
E8XTX-X	80-100	68	19
E9XTX-X	90-110	78	17
E10XT-X	100-120	88	16
E11XT-X	110-130	98	15
E12XT-X	120-140	108	14
EXXXTX-G	Properties as agreed upon between supplier and purchaser		

Note: Properties of electrodes that use external gas shielding (EXXT1-X and EXXT5-X) vary with gas mixtures.
Source: AWS A5.29

Table 25 Composition requirements for electrodes for SAW

AWS classification	Composition, %						
	C	Mn	Si	S	P	Cu	Total other elements
Low-manganese electrodes							
EL8	0.10	0.30-0.55	0.05	...	...	...	...
EL8K	0.10	0.30-0.55	0.10-0.20	...	...	...	...
EL12	0.07-0.15	0.35-0.60	0.05	...	...	...	...
Medium-manganese electrodes							
EM5K(a, b)	0.06	0.90-1.40	0.40-0.70	...	...	...	...
EM12	0.07-0.15	0.85-1.25	0.05	0.035	0.03	0.30	0.50
EM12K	0.07-0.15	0.85-1.25	0.15-0.35	0.035	0.03	0.30	0.50
EM13K(c)	0.07-0.19	0.90-1.40	0.45-0.70	0.035	0.03	0.30	0.50
EM15K	0.12-0.20	0.85-1.25	0.15-0.35	0.035	0.03	0.30	0.50
High-manganese electrodes							
EH14	0.10-0.18	1.75-2.25	0.05	0.035	0.03	0.30	0.50

Note: Analysis shall be made for the elements for which specific values are shown in this table. If however, the presence of other elements is indicated in the course of routine analysis, further analysis shall be made to determine that the total of these other elements is not present in excess of the limits specified for "Total other elements" in the last column of the table. Single values shown are maximum. The copper limit includes any copper coating that may be applied to the electrode. The suffix letter "N," when added to classification designations in this table, stands for nuclear grade. Grade "N" restricts sulfur content to 0.013% max, phosphorus content to 0.010% max, copper content to 0.08% max, and vanadium content to 0.05% max. Grade N electrodes shall not be copper coated or coated with a copper-containing material. Analysis shall be made for the elements for which specific values are shown in this table. In addition to the elements listed, other elements intentionally added in a classification shall be reported. The total of both these other elements, and other elements not intentionally added (as indicated above), shall not exceed 0.50%. AISI check tolerances pertaining to carbon steel wire and rod are applicable.
(a) The chemical composition requirements for EM5K classification are similar to the chemical composition requirements for ER70S-2 classification in AWS A5.18, "Specification for Mild Steel Electrodes for Gas Metal Arc Welding." (b) This electrode contains 0.05 to 0.15% Ti, 0.02 to 0.12% Zr, and 0.05 to 0.15% Al, which is exclusive of the "Total other elements" requirement. (c) The chemical composition requirements for the EM13K classification are similar to the chemical composition requirements for ER70S-3 classification in AWS A5.18, "Specification for Mild Steel Electrodes for Gas Metal Arc Welding."

standards include specifications for the types of linepipe used for gas and oil transmission lines. In many cases, these standards are supplemented by more stringent requirements from the user.

The most common steels used in linepipe are covered by API 5LX series alloys—"5L" indicates linepipe, "X" indicates high-test linepipe, and additional numbers indicate the grade (or strength level). Typical mechanical and chemical requirements are given in Tables 28 and 29.

Arc Welding Processes. Of all of the arc welding processes, only two are used to any extent in the field welding of linepipe. Shielded metal arc welding traditionally has accounted for most of the girth welding performed. However, GMAW (mostly carbon dioxide shielded) has been rapidly growing in the past few years. Additionally, a considerable amount of development work is being carried out on FCAW.

In SMAW, two basic techniques are used—vertical down and vertical up. The majority of field welding uses the vertical-down technique, or "stovepipe" welding with cellulosic electrodes. The vertical-up technique is used for specific applications or with low-hydrogen electrodes. Recent developments in electrode manufacturing have led to the increased application of vertical-down low-hydrogen electrodes. The SMAW process can be manipulated to adjust to variations in joint dimensions to achieve high construction rates. Specialized electrodes permit high-quality weld metal to be deposited under difficult field conditions, particularly if careful control of pipe preparation, alignment, interpass cleaning, and inspection is observed. Preheat may be required, depending on ambient temperature, pipe composition, wall thickness, and joint procedures. A typical procedure for a vertical-down pipe field weld is shown in Fig. 35.

Most field welding of linepipe is performed in the vertical-down position with cellulosic electrodes, using specialized procedures to prevent hydrogen damage. These procedures include control of interpass time and use of temper weld beads. Low-hydrogen electrodes for vertical-up welding have been used in some applications. Recently, low-hydrogen electrodes that facilitate vertical-down welding have been developed. Manufacturers should be consulted concerning these developments. The advantages and disadvantages of these processes are compared in Table 30.

Increasingly, field welding is being carried out by semiautomatic or automatic procedures, typically using gas metal arc solid-wire processes. Flux cored arc consumables (either self shielded or gas shielded) have been developed and are being introduced slowly into field practice.

Semiautomatic GMAW uses a procedure that deposits the root pass, followed by a fill-up procedure by SMAW. The automatic method uses GMAW throughout.

The deoxidation process and low heat input used in automatic GMAW produces higher hardnesses than other processes. This consequence must be remembered if pipeline is to be used for sour service (high hydrogen-sulfide applications). Maximum levels such as 260 HV or 22 HRC are typical for many such applications. Developments continue in the consumable and arc welding procedures used for field welding of HSLA steels, and close contact with manufacturers should be maintained.

Longitudinal Seam Welding of HSLA Steels by Submerged Arc Welding

Virtually all longitudinal seam welding of linepipe steels of all grades is carried out in the shop using multiple-electrode SAW. It provides the highest welding speeds and high weld deposition. Multiple-electrode, multiple-power systems (dc-ac, ac-ac Scott, dc-ac-ac, ac-ac-

Table 26 Mechanical-property requirements for fluxes with SAW electrodes

AWS flux classification(a)	Tensile strength, ksi	Minimum yield strength at 0.2% offset, ksi	Elongation (min) in 2 in., %	Minimum Charpy V-notch impact strength, ft · lb(b)
F6Z-EXXX	62-80	50	22(d)	No impact requirement
F60-EXXX(c)	62-80	50	22(d)	No impact requirement
F62-EXXX(c)	62-80	50	22(d)	No impact requirement
F64-EXXX(c)	62-80	50	22(d)	20 at 0 °F
F66-EXXX(c)	62-80	50	22(d)	20 at −20 °F
F7Z-EXXX	72-95	60	22(e)	20 at −40 °F
F70-EXXX(c)	72-95	60	22(e)	No impact requirement
F72-EXXX(c)	72-95	60	22(e)	20 at −60 °F
F74-EXXX(c)	72-95	60	22(e)	20 at −60 °F
F76-EXXX(c)	72-95	60	22(e)	20 at −60 °F

(a) The letters "EXXX" as used in this table stand for the electrode designations EL8, EL8K, etc. (b) The lowest value obtained, together with the highest value obtained, shall be disregarded for the test. Two of the three remaining values shall be greater than the specified 20-ft · lb energy level; one of the three may be lower but shall not be less than 15 ft · lb. The computed average value of the three values shall be equal to or greater than the 20-ft · lb energy level. (c) Note that if a specific flux-electrode combination meets the requirements of a given F6X-XXXX classification, this classification also meets the requirements of all lower numbered classifications in the F6X-XXXX series. For instance, a flux-electrode combination meeting requirements of the F64-XXX classification also meets the requirements of the F62-XXXX, F60-XXXX and F6Z-XXXX classifications. This applies to the F7X-XXXX series also. (d) For each increase of one percentage point elongation over the minimum, the yield strength or tensile strength, or both, may decrease 1 ksi to a minimum of 60 ksi for the tensile strength, and 48 ksi for the yield strength. (e) For each increase of one percentage point elongation over the minimum, the yield strength or tensile strength, or both, may decrease 1 ksi to a minimum of 70 ksi for the tensile strength and 58 ksi for the yield strength.

Table 27 Weld-metal properties available with various flux and wire combinations

AWS flux/wire classification	Composition, wt% C	Mn	Si	S	P	Ultimate tensile strength, ksi	Yield strength, ksi	Elongation in 2 in., %	Charpy V-notch impact, ft · lb
F66-EM12K	0.085	0.67	0.35	0.022	0.020	66	53	30.5	21 at −60 °F
F66-EH14	0.089	1.11	0.22	0.013	0.022	67	57	30.0	33 at −60 °F
F70-EL12	0.098	0.098	0.34	0.022	0.15	72	58	31.5	34 at 0 °F
F70-EM12K	0.10	1.23	0.44	0.023	0.020	77	68	30.5	29 at 0 °F
F72-EM12K	0.074	1.39	0.62	0.019	0.029	80	62	26.5	21 at −20 °F
F72-EM13K	0.073	1.31	0.63	0.016	0.022	93	81	24.0	34 at −20 °F
F72-EH14	0.074	1.70	0.57	0.024	0.036	80	64	22.0	24 at −20 °F
E74-EM12K	0.077	1.33	0.55	0.025	0.025	86	72	26.5	23 at −40 °F
F74-EH14	0.077	1.69	0.46	0.019	0.024	89	61	25.0	52 at −40 °F
E76-EM13K	0.092	0.77	0.51	0.013	0.021	72	62	27.0	35 at −60 °F

required weld-metal impact strength specifications.

The only specifications on weld-metal impact strength are those requested by the customer. Because preheating or postweld heat treatments are not employed during linepipe production, the weld metal should be able to meet the specifications in the as-welded condition. This is in contrast to girth welding where preheating is sometimes necessary for certain steel compositions to avoid hydrogen-induced cracking. In linepipe production by SAW, dilution from the base plate is of the order of 60 to 70%; consequently, the interaction among base material, electrode, and flux in determining the final weld-metal composition, microstructure, and mechanical properties should be considered. Furthermore, the flux should be capable of producing quality welds at high welding speeds. Several fluxes are available that are especially designed to handle high welding speeds.

For lower grade linepipe steels (X-42 to ac) are most commonly used in linepipe fabrication.

Process selection of a specific SAW process for welding linepipe entails evaluation of many combinations of electrodes and types of power. To achieve the highest welding speeds with a minimum of weld heat input, a combination multi-electrode system using an alternating current power source should be used.

Alternating current welding power ensures uniform penetration without the deleterious effects of arc blow, which can be encountered with direct current welding power. With alternating current power, high penetration can be achieved by the use of high welding currents (not limited in value because of arc blow problems) and smaller diameter electrodes to achieve high arc current densities.

Weld Joint Design and Tracking. To achieve high-quality welds, weld joint preparation must be consistent, with joint gap controlled to close tolerances. Weld joint preparation frequently is achieved through mechanical beveling. Weld fixturing is designed so that a joint gap is held to an absolute minimum. The joint design most often used is the double-V-groove. Welding usually is performed in two passes (one pass each side).

Because high welding speeds (function of plate thickness) are employed, it is essential to have a good seam tracking system to provide consistent welds. For this purpose, a bevel frequently is used on thinner plates to provide a means of automatically guiding the welding electrodes in the joint by the use of a V-wheel, which can ride in the groove, or through the use of other tracking mechanisms.

Welding Consumables. Generally, welding consumables for linepipe welding are selected not only for matching base-metal strength, but also to achieve the

Table 28 Mechanical-property requirements of 5LX series high-test linepipe

API grade	Minimum yield strength, ksi	Minimum tensile strength, ksi
X42	42	60
X46	46	63
X52	52	66(a)
X56	56	75(a)
X60	60	77(a)
X64	65	77(a)
X70	70	82(a)

(a) Tensile strength requirements for pipe of 20 in. OD and larger, with 0.375-in. or less wall, are 3 to 6 ksi higher.

Table 29 Chemical requirements (ladle analysis) for welded 5LX series high-test linepipe

API grade	Manufacturing process	Composition, max, % C	Mn	P	S
X42	Nonexpanded	0.28	1.25	0.04	0.05
X46	Nonexpanded	0.30	1.35	0.04	0.05
X52	Nonexpanded	0.30	1.35	0.04	0.05
X42	Cold expanded	0.28	1.25	0.04	0.05
X46	Cold expanded	0.28	1.25	0.04	0.05
X52	Cold expanded	0.28	1.25	0.04	0.05
X56(a)	Nonexpanded, cold expanded	0.26	1.35	0.04	0.05
X60(a)	Nonexpanded, cold expanded	0.26	1.35	0.04	0.05
X65(a)	Nonexpanded, cold expanded	0.26	1.40	0.04	0.05
X70(a)	Nonexpanded, cold expanded	0.26	1.40	0.04	0.05

(a) These steels may also contain small amounts of niobium, vanadium, or titanium. Other analyses may be furnished by agreement between purchaser and producer. Normally the carbon content is minimized as much as possible.

Fig. 35 Circumferential butt weld in API 5LX grade 70 pipe. Vertical-down position was used in making the weld.

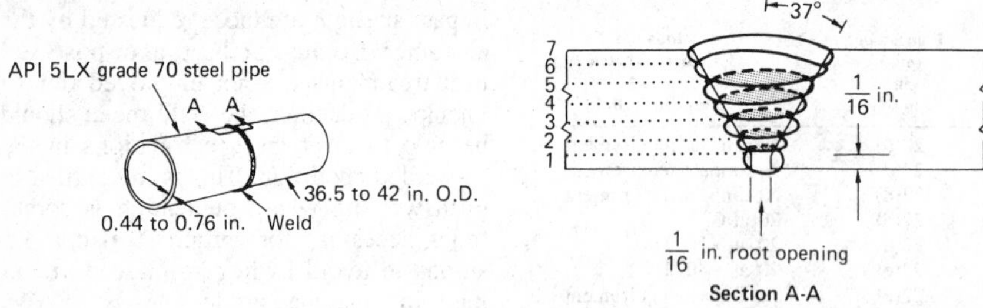

Pass No.	Welding process	Size, in.	Filler-metal designation	Welding parameters: Current (DCEN), A	Voltage, V	Travel speed, in./min
1, root pass	SMAW	5/32	E8010 (6010)	140-170	24-26	9-15
2, hot pass	SMAW	5/32	E8010	160-190	26-32	11-16
3, hot pass	SMAW	3/16	E8010	160-190	26-32	11-16
4, fill pass	SMAW	3/16	E8010	150-175	26-32	7-11
5, fill pass	SMAW	3/16	E8010	150-175	26-32	7-11
6, fill pass	SMAW	3/16	E8010	150-175	26-32	7-11
7, cap pass	SMAW	3/16	E8010	150-175	26-32	5-8

Note: Single-V-groove, 75° included angle; preheat 212 °F, maintained throughout; interpass temperature, 350 °F maximum; time lapse between root and hot pass, 5 min maximum; strip passes used at 3 and 9 o'clock position prior to capping; pipe material, API 5LX grade 70

X-60 grades), weld-metal impact requirements are generally low and the more acidic fluxes may be used. For the higher grades (X-65 and X-70, which are generally lower carbon microalloyed steels) of arctic gas and oil transmission pipe, higher weld-metal impact levels are required. Higher basicity fluxes are required to reduce the oxygen content in the weld metal. To further improve toughness, alloyed electrodes should be used. Unfortunately, the two criteria—high welding speed and high weld-metal impacts—are not easily combined in flux design. To achieve maximum welding speeds, minimum undercuts, fewer internal defects, and reduced hydrogen porosity and cracking, a fused flux usually is used. Fused manganese silicate or calcium silicate fluxes handle high currents, provide freedom from moisture absorption, and by the nature of their manufacture, are consistent in composition. These fluxes, combined with alloyed welding wires, provide weld-metal toughness that meets most established requirements.

A number of pre-established wire flux combinations are available for all grades of linepipe steels from consumable manufacturers. Because the base plate composition is crucial (because of high dilution) in determining the final weld-metal properties, electrode and flux combinations to be used for each grade of base plate should be evaluated separately. This is particularly true for high grade (X-65 and X-70) arctic linepipe steels.

Arc Welding of High-Strength Quenched and Tempered Alloy Steels

The steels discussed in this section are the quenched and tempered weldable alloy steels containing not more than 0.25% C, and with a total content of alloying elements (not including manganese and silicon) ranging from 0.85% to about 16% (see Table 31). These steels are welded in the quenched and tempered condition and have yield strengths of 50 to 180 ksi, depending on alloy content, section thickness, and heat treatment. They have high strength in combination with good ductility. Various combinations of notch toughness, fatigue strength, and corrosion resistance can be developed to meet the requirements of different applications, such as structures and pressure vessels for use at atmospheric, cryogenic, or elevated temperatures.

Although high-strength alloy steels with up to 0.25% C cannot be successfully welded with simple procedures and minimal control, they are less difficult to weld than are the higher carbon alloy steels such as 4140. The quenched and tempered alloy steels with up to 0.25% C were designed to be welded with moderate or no preheat, and to be used in most applications in the as-welded condition. Knowing the correct procedures that should be used for welding these steels, and rigorously following them, are fundamental to welding them successfully.

Composition, Properties, and Microstructure. The composition ranges (ladle analysis) for representative high-strength quenched and tempered alloy steels are given in Table 31. The mechanical properties of pressure vessel-quality plates are given in Table 32. The A517 steels referred to in these tables are also produced, in accordance with ASTM A514, as structural-quality plate steels, and some are available as abrasion-resistant steel plate with minimum hardness of 321, 340, or 360 HB. The A543 steels, classes 1 and 2, are modifications of HY-80 and HY-100 steels, respectively. Some steels, such as A514 and A517 (grades B, F, and H), A543, HY-130, and A553, are available as heat treated shapes and seamless pipe.

A517 is a multiple-alloy steel of 100 ksi minimum yield strength in the heat treated

Table 30 Advantages and disadvantages of various welding positions and consumables

Position and consumable	Advantages	Disadvantages
Vertical-up welding with basic electrodes	Higher crack resistance and lower preheating temperature; slightly improved impact values at low temperatures	Very low welding efficiency, especially during stringer-bead welding; redrying of electrodes on site
Vertical-down welding with cellulosic electrodes	High welding efficiency; no need to redry the electrodes on site	Elevated preheating temperature; slightly lower impact values compared to basic electrodes
Vertical-down welding with special basic electrodes	High welding efficiency; lower preheating temperature; slightly improved impact values compared to cellulosic electrodes	Increased defect sensitivity, especially concerning pore formation at starting points; high fit-up accuracy when setting up joints; more difficult to achieve root fusion, especially in the overhead position (concavities); moving ahead of slag during cap welding; redrying of electrodes on site

Table 31 Compositions of representative high-strength alloy steels (quenched and tempered)

ASTM designation	Composition type	C	Mn	Si	Ni or Cu	Cr	Mo	Other
		Composition, %						
A533, grade B(a)	Mn-Mo-Ni	0.25 max	1.15-1.50	0.15-0.30	0.40-0.70 Ni	...	0.45-0.60	...
A517(b):								
Grade A	Mn-Si-Cr-Mo-Zr-B	0.15-0.21	0.80-1.10	0.40-0.80	...	0.50-0.80	0.18-0.28	0.05-0.15 Zr; 0.0025 max B
Grade B	Mn-Cr-Mo-V-B	0.15-0.21	0.70-1.00	0.20-0.35	...	0.40-0.65	0.15-0.25	0.01-0.03 Ti; 0.0005-0.005 B(c)
Grade C	Mn-Mo-B	0.10-0.20	1.10-1.50	0.15-0.30	...	...	0.20-0.30	0.001-0.005 B
Grade D	Cr-Mo-Cu-Ti-B	0.13-0.20	0.40-0.70	0.20-0.35	0.20-0.40 Cu	0.85-1.20	0.15-0.25	0.04-0.10 Ti; 0.0015-0.005 B
Grade E	Cr-Mo-Cu-Ti-B	0.12-0.20	0.40-0.70	0.20-0.35	0.20-0.40 Cu	1.40-2.00	0.40-0.60	0.04-0.10 Ti; 0.0015-0.005 B
Grade F	Mn-Ni-Cr-Mo-Cu-V-B	0.10-0.20	0.60-1.00	0.15-0.35	0.70-1.00 Ni(d)	0.40-0.65	0.40-0.60	0.03-0.08 V; 0.002-0.006 B
Grade G	Mn-Si-Cr-Mo-Zr-B	0.15-0.21	0.80-1.10	0.50-0.90	...	0.50-0.90	0.40-0.60	0.05-0.15 Zr; 0.0025 max B
Grade H	Mn-Ni-Cr-Mo-V-B	0.12-0.21	0.95-1.30	0.20-0.35	0.30-0.70 Ni	0.40-0.65	0.20-0.30	0.03-0.08 V; 0.0005 min B
Grade J	Mn-Mo-B	0.12-0.21	0.45-0.70	0.20-0.35	...	...	0.50-0.65	0.001-0.005 B
Grade K	Mn-Mo-B	0.10-0.20	1.10-1.50	0.15-0.30	...	...	0.45-0.55	0.001-0.005 B
Grade L	Cr-Mo-Cu-Ti-B	0.13-0.20	0.40-0.70	0.20-0.35	0.20-0.40 Cu	1.15-1.65	0.25-0.40	0.04-0.10 Ti; 0.0015-0.005 B
Grade M	Mn-Ni-Mo-B	0.12-0.21	0.45-0.70	0.20-0.35	1.20-1.50 Ni	...	0.45-0.60	0.001-0.005 B
Grade P	Mn-Ni-Cr-Mo-B	0.12-0.21	0.45-0.70	0.20-0.35	1.20-1.50 Ni	0.85-1.20	0.45-0.60	0.001-0.005 B
Grade Q	Mn-Ni-Cr-Mo-V	0.14-0.21	0.95-1.30	0.15-0.35	1.20-1.50 Ni	1.00-1.50	0.40-0.60	0.03-0.08 V
A542(e)	2¼Cr-1Mo	0.15 max	0.30-0.60	0.15-0.30	...	2.00-2.50	0.90-1.10	...
A543(f)	3Ni-Cr-Mo	0.23 max	0.40 max	0.20-0.35	2.60-4.00 Ni	1.50-2.00	0.45-0.60	0.03 max V
HY-130(g)	5Ni-Cr-Mo-V	0.12 max	0.60-0.90	0.20-0.35	4.75-5.25 Ni	0.40-0.70	0.30-0.65	0.05-0.10 V
HP 9-4-20(h)	9Ni-4Co-Cr-Mo-V	0.17-0.23	0.20-0.30	0.10 max	8.50-9.50 Ni	0.65-0.85	0.90-1.10	4.25-4.75 Co; 0.06-0.10 V
Mod A203, grade D	3½Ni	0.17 max	0.70 max	0.15-0.30	3.25-3.75 Ni	...	...	...
A553, grade A	9Ni	0.13 max	0.90 max	0.15-0.30	8.50-9.50 Ni	...	...	...
A553, grade B	8Ni	0.13 max	0.90 max	0.15-0.30	7.50-8.50 Ni	...	...	...
A333, grade 3	3½Ni	0.19 max	0.31-0.64	0.18-0.37	3.18-3.82 Ni	...	...	...
A514, type F	Mn-Ni-Cr-Mo-Cu-V-B	0.10-0.20	0.60-1.00	0.15-0.35	0.70-1.00 Ni(d)	0.40-0.65	0.40-0.60	0.03-0.08 V; 0.002-0.006 B

Note: Phosphorus content is 0.035 max for all steels except HY-130 and HP 9-4-20, which contain 0.010 max P; and A333, grade 3, which contains 0.05 max P. Sulfur content is 0.040 max for all steels except HY-130 and HP 9-4-20, which contain 0.010 max S; A542, which contains 0.035 max S; and A333, grade 3, which contains 0.05 max S.
(a) See A541, class 3, and A508, class 3, for forging steels. (b) Pressure-vessel quality; see A514 for structural quality. (c) Vanadium, 0.03 to 0.08. (d) Copper, 0.15 to 0.50. (e) See A541, class 6, for forging steel. (f) See A541, class 7, and A508, class 4, for forging steels. (g) See A579, grade 12, for forging steel. (h) See A579, grade 81, for forging steel.

Table 32 Tensile properties and Charpy V-notch impact properties of representative high-strength alloy steels (quenched and tempered)

Steel	Composition type	Thickness range, in.	Yield strength (min or range), ksi	Tensile strength (min or range), ksi	Elongation in 2 in. (min), %	Charpy V-notch impact energy, ft · lb(a) Longitudinal	Transverse
A533, grade B, class 1	MnMoNi	¼ min	50	80-100	18	30 at 10 °F	...
A533, grade B, class 2	MnMoNi	¼ min	70	90-115	16	30 at 10 °F	...
A533, grade B, class 3	MnMoNi	2 max	82.5	100-125	16	25 at −85 °F(b)	...
A517, grades A, B, C, D, J, and K	(See Table 31)	1¼ max	100	115-135	16	15 at −50 °F	15 at −50 °F
A517, grades G, H, L, and M	(See Table 31)	2 max	100	115-135	16	20 at 10 °F 15 at −50 °F	15 at +10 °F
A517, grades E, F, P, and Q	(See Table 31)	2½ max	100	115-135	16	30 at 0 °F 20 at −50 °F	20 at 0 °F 15 at −50 °F
A542, class 1	2¼Cr1Mo	$^3/_{16}$ min	85	105-125	14	35 at 10 °F	...
A542, class 2	2¼Cr1Mo	$^3/_{16}$ min	100	115-135	13	35 at 10 °F	...
A543, class 1	3NiCrMo	$^3/_{16}$ min	85	105-125	14	35 at 10 °F	...
A543, class 2	3NiCrMo	$^3/_{16}$ min	100	115-135	14	35 at 10 °F	...
HY-130	5NiCrMoV	Less than ⅝	130-150	...	14	50 at 0 °F	50 at 0 °F
		⅝ to 4	130-145	...	15	50 at 0 °F	50 at 0 °F
HP 9-4-20	9Ni4CoCrMoV	¼ min	180	190	14	40 at room temperature	...
Mod A203, grade D	3½Ni	2 max(c)	65(c)	75(c)	22(c)	25 at −155 °F	...
A553, grade A	9Ni	2 max	85	100-120	22	25 at −320 °F	...
A553, grade B	8Ni	2 max	85	100-120	22	25 at −275 °F	...

(a) The minimum required value for Charpy V-notch impact energy is subject to agreement between the steel producer and the user; the values shown here for the various steels are typical required minimums. (b) For ½-in.-thick steel. (c) Tentative

condition that is now used extensively in construction equipment, bridges, buildings, pressure vessels, storage tanks, penstocks, spiral cases, and ships.

A533, grade B, steel (manganese-molybdenum-nickel steel) has been used extensively for nuclear pressure vessels of heavy sections. The same steel at a higher strength (class 3) has been used for thin-walled pressure vessels and layered pressure vessels. A542, a 2¼Cr-1Mo steel, has been used for large-diameter hydrocracking pressure vessels that operate at elevated temperature.

The need for tough steels of high yield strength led to the development of HY-80 steel as ASTM A543. Further development resulted in HY-130, a 5Ni-Cr-Mo-V steel, and in HP 9-4-20, a 9Ni-4Co-Cr-Mo-V steel, mostly for hydrospace and aerospace requirements.

Increasing cryogenic applications led to the development of A553, grade A, a 9% Ni steel, for use at temperatures down to −320 °F, and of A553, grade B, a steel of somewhat lower nickel content (8% Ni), for use at temperatures down to −275 °F.

Table 33 Heat treatments for representative high-strength alloy steels (quenched and tempered)

Steel	Treatment	Microstructure
A533, grade B	Water quenched from 1550 °F (min) and tempered at 1100 °F (min)	Tempered bainite and tempered martensite (thin plates); ferrite and tempered bainite (thick plates)
A517	Water quenched from 1650 °F (min) and tempered at 1150 °F (min)	Tempered bainite and tempered martensite
A542	Water quenched from about 1750 °F and tempered at 1050 °F (min)	Tempered bainite and tempered martensite
A543	Water quenched from 1650 °F (min) and tempered at 1100 °F (min	Tempered bainite and tempered martensite
HY-130	Water quenched from 1475 to 1525 °F and tempered at 1000 °F (min)	Tempered bainite and tempered martensite
HP 9-4-20	Normalized from 1675 °F, water quenched from 1550 °F and tempered at 1000 °F	Tempered martensite
Mod A203, grade D	Water quenched from about 1600 °F and tempered at about 1100 °F	Proeutectoid ferrite and fine pearlite
A553	Water quenched from 1450 to 1500 °F and tempered at 1050 to 1125 °F	Ferrite and austenite

The mechanical properties given in Table 32 result from heat treatments described in Table 33.

The purpose of quenching and tempering is to produce low-temperature transformation products, such as bainite and tempered martensite, which have excellent strength combined with toughness. For example, although A533, grade B, steel has limited hardenability, the quenched and tempered microstructure in 1/2-in.-thick plate consists entirely of tempered martensite, which produces a tensile strength of at least 110 ksi and a Charpy V-notch energy absorption of at least 25 ft · lb at −85 °F. In a 7-in.-thick plate, the microstructure consists of ferrite and tempered bainite, and results in a tensile strength of about 90 ksi and a Charpy V-notch energy absorption (1/4-thickness location) of at least 30 ft · lb at 10 °F. If the nickel, chromium, and molybdenum contents are high (as in A543 and HY-130 steels), the microstructure is tempered bainite and tempered martensite, even in very thick steel sections, and results in a desirable combination of strength and toughness. Furthermore, the HY-130 and HP 9-4-20 steels display markedly better toughness when residual or impurity elements such as phosphorus, sulfur, nitrogen, oxygen, and hydrogen are restricted to the lowest possible levels.

Within carefully chosen limits on thickness, the A517 steels provide high strength and toughness with the least amount of alloying elements. The effect of quenching and tempering on the longitudinal mechanical properties of 1/2-in.-thick A517, grade F, steel is shown in Fig. 36. In the hot rolled condition the steel has a microstructure of proeutectoid ferrite and high-carbon martensite, which results in a yield strength of only about 80 ksi and very poor Charpy V-notch energy absorption at −50 °F. In comparison, the as-quenched microstructure contains high percentages of martensite and bainite; consequently, the yield and tensile strengths are much higher than those of the same steel in the hot rolled condition. The toughness of the as-quenched plate, although considerably better than that of the hot rolled plate, is greatly increased when the steel is tempered, particularly at temperatures above 1100 °F, as shown in Fig. 36. Also, as a result of tempering, the yield and tensile strengths are significantly lowered and the tensile elongation is almost doubled.

Fig. 36 Effect of quenching and tempering on the mechanical properties of 1/2-in.-thick ASTM A517, grade F, steel in the longitudinal direction

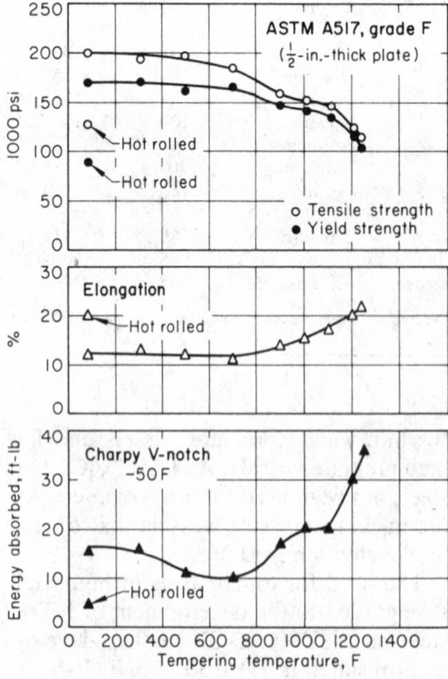

Many studies have been made using underbead-cracking tests, fillet-weld and butt-weld restraint-cracking tests, and hot ductility tests to define the composition limits for susceptibility of steels to hot and cold cracking. Hot cracking normally occurs at a high temperature, usually just below the solidus temperature, whereas cold cracking, or delayed cracking, occurs below the M_s temperature.

Many of the quenched and tempered steels in Table 31 usually are produced with a sulfur content of less than 0.025%, and some, such as HY-130 and HP 9-4-20, are produced with a sulfur content not exceeding 0.010%. The manganese-to-sulfur ratio is generally greater than 30 to 1, so that with a carbon content of about 0.20% or less, the susceptibility to hot cracking is negligible.

The A533 steel, with a somewhat higher carbon content, has negligible susceptibility to hot cracking because it has a high manganese-to-sulfur ratio, usually about 50 to 1. The A543 steel, with a low manganese content, is susceptible to cracking when the carbon content is at the maximum unless sulfur content is extremely low.

Susceptibility to cold cracking under conditions of high restraint decreases with increased M_s temperature. This effect has been attributed to the self-tempering of the martensite that forms at high temperatures within the martensite transformation range. Furthermore, cold cracking is directly proportional to the hydrogen content in the welding atmosphere.

All of the steels in Table 31 have only a low susceptibility to cold cracking, if suitable care is taken to limit the amount of hydrogen in the welding atmosphere.

Another feature of the steel compositions listed in Table 31 is the frequent use of molybdenum to increase resistance to tempering as well as to provide greater hardenability. Molybdenum, like the strong carbide-forming elements such as titanium, vanadium, and zirconium, serves two functions: to retard softening of the steel during tempering at high temperature for increased toughness and ductility, and to provide that portion of the HAZ just below the lower transformation temperature with needed resistance to excessive softening. The retarded softening of A517, grade F, steel at tempering temperatures above 900 °F is illustrated in Fig. 36 by the decreased slope of the strength curves between 900 and 1100 °F. Thus, in a quenched and tempered steel with adequate resistance to softening, 100% joint efficiency at the desired high strength can be readily obtained if uncommonly high heat inputs in welding are avoided. Furthermore, the resistance to tempering contributes to more uniform bendability in welded joints subjected to forming operations after welding.

Steels having excellent toughness in the unaffected base metal may have inadequate toughness in the HAZ. To ensure that this does not occur, much care has been given to the composition limits and recommended welding procedures for several of the quenched and tempered alloy steels in Table 31.

The development of good notch toughness in the HAZ of quenched and tempered alloy steels depends on the rapid dissipation of welding heat to permit the formation of martensite and bainite on cooling. This requirement may increase the susceptibility of the steel to hydrogen-induced cracking. With increasing section thickness of the steel as a plate, structural section, or forging, a higher hardenability is required to secure the desired hardened structure throughout the section. Sections 1 in. thick and greater, however, produce much higher cooling rates in the HAZ during welding than do $^1/_4$- or $^1/_2$-in. sections. Therefore, thicker sections require compositions of higher hardenability principally so that heat treatment will provide the required mechanical properties before welding, whereas thinner sections require compositions of adequate hardenability both to provide the required mechanical properties before welding and to maintain the desired properties after welding.

A microstructure produced by rapid cooling so as to have a high percentage of low-temperature transformation products (martensite and bainite) responds favorably to tempering above 1000 °F, whereas the structure that results from slow cooling is only moderately improved by tempering. The structure will be comprised of upper bainite and high-carbon martensite and will have poor impact properties. These same principles apply to the microstructure produced in the HAZ of a weldment.

Results of fatigue tests on butt-welded and fillet-welded joints in the weldable quenched and tempered alloy steels have shown no special sensitivity of the HAZ or weld metal to fatigue failure, provided that the steel has adequate resistance to softening in the HAZ, as previously discussed, and that the weld metal has adequate tensile strength. However, because fatigue-crack initiation is associated with geometric discontinuities and stress risers, the effect of cutting and welding on fatigue must be carefully considered in relation to shape and soundness.

Location and Design of Joint. To obtain maximum advantage from the weldable quenched and tempered alloy steels, it is necessary that the increase in the yield strength of the steels be accompanied by refinements in design, workmanship, and inspection. Even more thought must be given to joint design and location than with steels of which the properties are less dependent on heat treatment. A design that incorporates abrupt changes in section in a region of high stress cannot be tolerated, particularly if the steel has a yield strength of 80 ksi or higher. Locating welds incorrectly can cause or contribute to abrupt changes in section. The welds must be located at sites where there is sufficient access for the welder and where proper examination can be made by the inspector. For these reasons, butt welds are preferable to fillet welds.

Suitable joint preparation can usually be achieved by gas or arc cutting without preheating, although some codes require preheating. For all the steels in Table 31, the steel temperature should be not lower than 50 °F during cutting, but when cutting A533 and A543 steels the preheat temperature recommended for welding (see Table 35) is suggested.

Arc cutting is often preferred for steels such as A553 and HP 9-4-20. All slag or loose scale from cutting should be removed by light grinding. Prior to welding, joint surfaces should be cleaned to remove organic and hygroscopic materials, and dried. Backing bars, when required, should be made continuous, by butt welding, so as to avoid notches, which can be sites for the start of transverse cracking of weld metal. Preferably, backing bars are removed after welding in applications entailing fatigue loading. If fatigue loading is involved, the root must be inspected or back welded.

Butt joints should be welded so as to avoid creating severe stress risers at the root. Stress risers at this location, like those at the toe of butt or fillet welds with excessive weld reinforcement or inadequate fairing into the adjacent plates, are of greater concern when the steel has high yield strength.

Selection of Welding Process. Shielded metal arc, submerged arc, flux cored arc and gas metal arc welding are most commonly used for joining the quenched and tempered alloy steels of carbon content up to 0.25%. These four processes can be used effectively for welding steels having yield strengths up to approximately 150 ksi. Gas tungsten arc or electron beam welding (EBW) must be used for steels of yield strength over 150 ksi, including the HP 9-4-20 steel.

Weld cooling rates for the arc welding processes are usually so high that the mechanical properties of the HAZ in the high-strength quenched and tempered steels approach those of the steel in the quench-hardened condition. Thus, postweld heat treatment, such as quenching and tempering, is unnecessary unless stress corrosion is a factor. Electroslag welding, which subjects the base metal to prolonged heating and consequently slower cooling rates, generally requires quenching and tempering of these steels after welding.

Welding of structures in the field often requires welding conditions similar to those used in the shop. To minimize the problem of duplicating shop practice in the field, large sections are often fabricated in the shop and transported to the field for final assembly by welding. The example that follows describes an application in which both FCAW and SMAW were used for joining subassemblies of 158-ton dragline buckets.

Example 11. Flux Cored Arc Welding and Shielded Metal Arc Welding for Assembly of Dragline Buckets. Because two dragline buckets, each weighing 158 tons and having a capacity of 220 yd^3, were too large to ship in one piece to the site of use, they were shop fabricated in five sections, as shown at the left in Fig. 37. Field welding of the assembled sections presented the problem of duplicating the favorable shop welding conditions.

The plate material, which was a high-strength alloy steel equivalent to ASTM A514, type F, was quenched and tempered to a tensile strength of 105 to 135 ksi. Castings were of a proprietary alloy steel that had a tensile strength of 120 to 130 ksi. The thickness range of the metal was from $1^1/_2$ to 6 in., and the average thickness of the steel plate was about 2 in.

Groove and fillet welds were used for the various butt, corner, lap, and T-joints. Welding these joints and steels required careful control of preheat and weld heat input to avoid cracking. Shop operations were as follows:

1 Plate edges were prepared by gas cutting square edges, and single and double bevels of 30 and 45°, as required.
2 Each section was assembled and tack welded.
3 The joints that were to be shop welded were preheated to 400 °F by propane

Fig. 37 Dragline bucket that, because of size and weight (158 tons), was produced as five shop-welded subassemblies that were shipped to location for field assembly and final welding. In both shop and field welding, FCAW was used for flat and horizontal positions, and SMAW for out-of-position, short, or difficult-access welds.

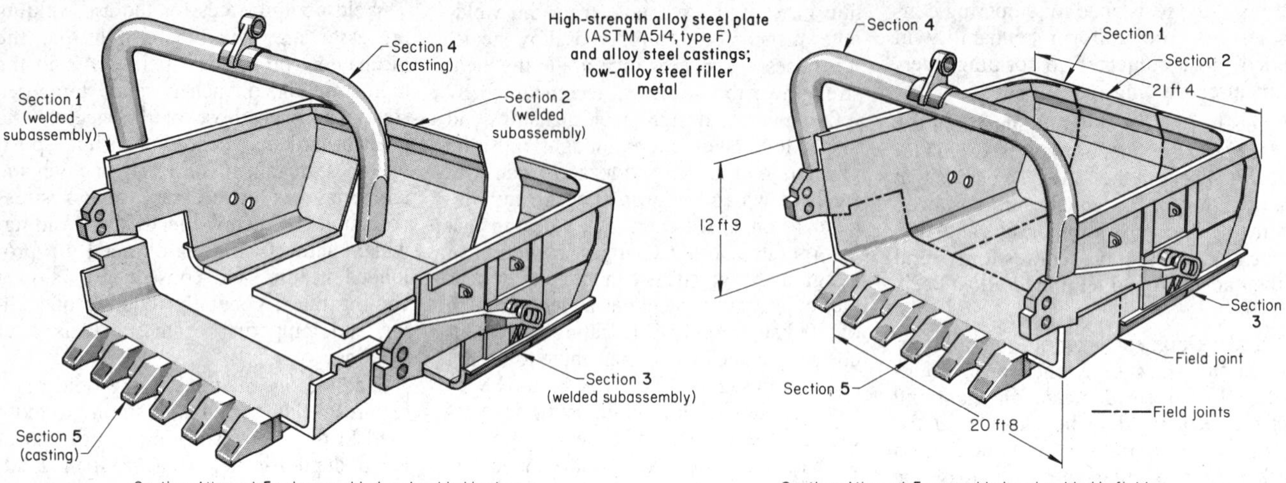

torches. Heat was maintained until all shop welding was completed.

4 All joints except the field joints between the five sections shown in Fig. 37 were welded using FCAW and SMAW. Flux cored arc welding was used for welds that could be made in the flat and horizontal positions; SMAW, for out-of-position welds, for welds of short length, and for those that were difficult to reach. For FCAW, $^{3}/_{32}$-in.-diam electrodes were used; with larger diameter electrode wires, welds cracked, probably because of the higher heat input and dilution effects. The flux cored electrode yielded a deposit of 0.15% C, 0.93% Mn, 0.22% Si, 0.66% Ni, 0.75% Cr, and 0.50% Mo. Flux cored arc welding was done at 400 to 500 A DCEP and 30 to 32 V, using carbon dioxide (−45 °F dew point, from bulk storage) at a flow rate of 30 to 35 ft^3/h for shielding. The electrode holder was a 600-A water-cooled type; a 750-A constant-voltage transformer-rectifier was used. Shielded metal arc welding was done with E11018 electrodes, $^{5}/_{32}$ and $^{3}/_{16}$ in. in diameter, at 150 to 225 A DCEP and at 20 to 30 V.

5 To minimize distortion, long welds were made using a block sequence, in which full-size welds were deposited in the middle and at the ends of the joint and then weld beads were back-stepped to the blocks. Weld penetration was as near 100% as possible without back gouging the root bead.

6 After shop welding, the sections were stress relieved at 1115 °F, furnace cooled, and shipped to the site.

For field welding, the five sections were set up and cribbed in such a manner that all joints were accessible to permit welding in all positions. Because the welding had to be done in cold weather, a shelter was provided to help maintain a uniform temperature during welding. Heating was accomplished by the use of electric resistance strip heaters and large (2500-Btu) electric space heaters. Heating was maintained at all times until the job was completed, even on weekends when work stopped.

Field welding was done by FCAW and SMAW, under essentially the same conditions as those used in shop welding. On long welds, the block sequence described in item 5 above was used. All weld runs were completed before the next weld was started. The body plate was welded first, by joining sections 1, 2, and 3. Section 5, the lip section, was then welded to the body plate. The arch, section 4, was welded last.

Table 34 Filler metals for joining representative high-strength alloy steels (quenched and tempered) by three welding processes

Steel	Filler metal (and suitable flux and shielding gas, as applicable)		
	SMAW	SAW	GMAW(a)
A533, grade B, class 1 and 2	E9018-D1	Mn-Mo wire and neutral flux	Mn-Mo wire and argon-O_2 gas
A533, grade B, class 3	E11018-G(b) (Mn-Ni-Cr-Mo)	Mn-Ni-Cr-Mo wire and neutral flux	Mn-Ni-Cr-Mo wire and argon-O_2 gas
A517	E11018-G(b)(c) (Mn-Ni-Cr-Mo)	Mn-Ni-Cr-Mo wire and neutral flux, or carbon steel wire and alloy flux	Mn-Ni-Cr-Mo wire and argon-O_2 gas
A542, class 1 and 2	E9015-B3	2$^1/_4$Cr-1Mo wire and neutral flux	2$^1/_4$Cr-1Mo wire and argon-O_2 gas
A543, class 1 and 2	E11018-M(b)	Mn-Ni-Cr-Mo wire and neutral flux	Mn-Ni-Cr-Mo wire and argon-O_2 gas
HY-130	Special E14018 (Mn-Ni-Cr-Mo)	(d)	Special Mn-Ni-Cr-Mo wire and argon-O_2 gas
HP 9-4-20	Not recommended	Not recommended	Special Ni-Co-Cr-Mo-V filler wire with tungsten arc and argon gas
Mod A203, grade D	(e)	(e)	(e)
A553, grades A and B	68Ni-15Cr-3Ti-9Fe	65Ni-27Mo wire and neutral flux	70Ni-15Cr-3Ti-9Fe wire; helium-argon gas

(a) Consumable-electrode GMAW, except as noted for HP 9-4-20 steel, which was welded by GTAW. (b) Lower strength low-hydrogen electrodes (depending on design stress) may also be suitable. (c) A higher strength electrode, such as E12018-G (Mn-Ni-Cr-Mo), may be necessary for thin plates of A517, grades A, B, C, D, and J, steels. (d) Suitable filler metal-flux combination being developed. (e) At present, same as for A553 steel, but more economical filler metals are being developed.

Selection and Care of Electrodes. Suitable filler metals for SMAW, SAW, and GMAW of the quenched and tempered alloy steels are described in Table 34. A joint efficiency of 100% can be obtained with most steels.

The covered electrodes listed in Table 34 for joining A533, A517, A542, and A543 steels are those commonly used, but similar low-hydrogen electrodes such as E*xx*15 or E*xx*16, with a different covering, are also suitable. Regardless of the strength of the heat treated steels listed in Table 34, electrodes that deposit weld metal having a lower strength than that of the steel are often adequate, as is true of fillet welds in longitudinal shear; in fact, lower strength weld metal is often desirable to prevent cracking in highly restrained fillet welds.

When a carbon steel electrode and alloy-flux combination is used for SAW, rather than an alloy steel electrode and neutral-flux combination, the welding conditions must be closely controlled to prevent a wide variation in weld-metal composition and consequent wide variation in the cracking susceptibility and mechanical properties of the deposited metal. For example, a fabricator of components for a long-span bridge encountered cracking in submerged arc butt welds made with a normally satisfactory combination of a carbon steel electrode wire and an alloy flux. The cracking was traced to variations in chemical composition of the weld metal, caused by fluctuations in arc voltage. Cracking was eliminated by controlling the voltage to ±1 V, and by ensuring adequate supervision.

Hydrogen must be kept to a tolerable amount. Sources of hydrogen include: (1) organic materials, chemically bonded water, and absorbed water in the electrode covering or welding flux; (2) hydrogen in the filler metal or in contaminants on the surface of the electrode wire, or moisture in the shielding gas or flux; and (3) moisture on the steel surface at the location of welding.

In the welding of quenched and tempered alloy steels by the three metal arc processes, hydrogen is kept to a tolerable amount in SMAW by using properly dried low-hydrogen electrodes; in SAW, by using clean, dry flux and clean low-hydrogen electrode wire; and in GMAW, by using moisture-free shielding gas and clean low-hydrogen electrode wire. For each of these processes, the steel must be dry at the location of welding. Warming with a torch is often used to ensure dry surfaces.

As discussed in the section of this article on HSLA steels, the low-hydrogen characteristic of covered electrodes must not be taken for granted. Normally, low-hydrogen alloy steel covered electrodes are packaged in sealed containers, and the moisture content of the covering is as shown in Table 19. The coverings on special E12018-M electrodes have a moisture content not greater than 0.1%. However, experience has shown that such limits on moisture content are not always sufficiently restrictive. For example, the E9015-B3 electrodes listed in Table 34 for joining A542 steels are suitable for these steels only if the moisture content of the covering is less than 0.15%, rather than the normal 0.4%. If electrodes of a lower strength class than that suggested for any particular steel in Table 34 are used, they should be baked to reduce the moisture content of the covering to that of the higher strength electrode suggested in Table 34. For example, if E7018 electrodes, normally supplied with a maximum moisture limit of 0.6%, are to be used for welding A517 steel, the moisture content of the electrode covering should be reduced to less than 0.2%. It should be noted that the feasibility of, and procedure for, baking depend on the composition of the electrode coating; baking should not be done without the advice of the electrode manufacturer.

The importance of care in the storage of low-hydrogen electrodes is illustrated in Fig. 38, which shows the effect on the moisture content of the covering on E11018-G electrodes of exposure for up to 24 h at 75 °F to atmospheres of 60 and 90% relative humidity. At both humidity levels, the permissible moisture content of 0.2% is exceeded by some of the electrodes after 30 min of exposure, and at 90% humidity the increase in moisture content with time of exposure is markedly accelerated.

To satisfy these rigid moisture requirements, electrodes for the high-strength quenched and tempered steels should not be exposed to the atmosphere for more than ½ h. Electrodes that have been exposed for longer periods must be redried before use. Failure to observe this practice is a common cause for underbead cracking in quenched and tempered alloy steel welds.

For the bare-wire electrodes used in SAW, FCAW, or GMAW, low hydrogen content is as important as for the covered electrodes used in SMAW. The total hydrogen content of bare-wire electrodes, including the hydrogen from contaminants on the wire surface, should not exceed 5 ppm for electrodes used to weld A517, A542, and A543 steels, and should not exceed 3 ppm for electrodes used to weld HY-130 steel.

Preheating. The minimum preheat and interpass temperatures for satisfactory shielded metal arc welds in quenched and tempered alloy steels increase with plate thickness, as shown in Table 35. Preheating is required to prevent weld-metal cracking in restrained welds, and temperature above the minimums shown in Table 35 may be required for highly restrained welds. The use of a preheating temperature of less than 100 °F requires that any moisture on the plate or introduced to the arc atmosphere by the electrode or welding flux be kept to a minimum—by driving it off by mild heating, if necessary.

The maximum preheat and interpass temperatures should, in general, not exceed the minimum values given in Table 35 by more than 150 °F. Otherwise, the upper limit on welding-heat input, as subsequently discussed, is very restrictive. For some grades the maximum preheat and interpass temperatures are more restrictive than for others. Thus, maximum preheat

Fig. 38 Moisture content of covering on E11018-G electrodes as a function of time of exposure to atmosphere of moderate (60%) and high (90%) relative humidity

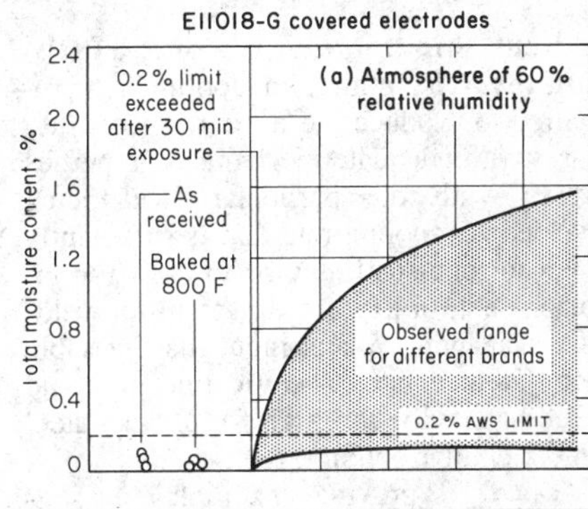

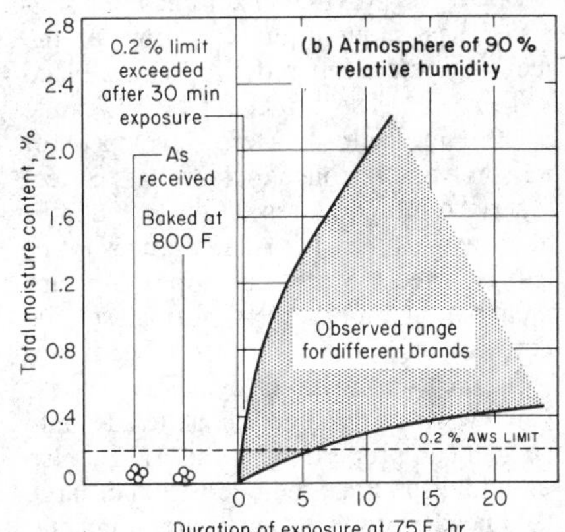

Table 35 Suggested minimum preheat and interpass temperatures for SMAW of representative high-strength alloy steels (quenched and tempered)

Plate thickness, in.	Minimum preheat and interpass temperatures for welding with low-hydrogen electrodes, °F						
	A533, B	A517	A542	A543	HY-130	Mod A203, D	A553
Up to 1/2	50	50	150	100	75	50	50
Over 1/2 to 5/8	100	50	200	125	75	50	50
Over 5/8 to 3/4	100	50	200	125	125	50	50
Over 3/4 to 7/8	100	50	200	150	125	50	50
Over 7/8 to 1	100	50	200	150	200	50	50
Over 1 to 1 3/8	200	150	250	200	200	150	150
Over 1 3/8 to 2	200	150	250	200	225	150	150
Over 2	300	200	300	200	225	200	200

Note: The minimum temperatures shown are suitable also for GMAW and for SAW with a neutral flux. A preheat temperature above the minimum may be required for highly restrained welds. No welding should be done when ambient temperature is below 0 °F. Steel that is at an initial temperature below 100 °F may require preheating to remove moisture from its surface. Maximum preheat and interpass temperatures should not exceed the minimums shown here by more than 150 °F.

and interpass temperatures for various thicknesses of HY-130 steel plate are:

Plate 5/8 in. thick or less	150 °F
5/8 to 7/8 in. thick	200 °F
7/8 to 1 3/8 in. thick	275 °F
Over 1 3/8 in. thick	300 °F

The preheat and interpass temperatures suggested in Table 35 for SMAW are also suitable for GMAW, SAW with a neutral flux, and FCAW with electrodes of a basic slag system. For SAW with an alloy flux, somewhat higher temperatures are required to prevent weld-metal cracking, as shown by the preheat and interpass temperatures recommended for A517 steels in Table 36.

Heat Input and Welding Techniques. The minimum cooling rate required to produce the desired microstructures with adequate mechanical properties varies with the particular steel being welded. A cooling rate that is sufficiently high for one steel may be too low for another steel, or expressed in terms of welding heat input, a heat input that is usable for one steel at a specific thickness and preheat may be too high for another steel under similar conditions.

Table 37 provides examples of suggested heat-input limits for some of the quenched and tempered alloy steels listed in Table 31. The limits for A533, grade B, steel in Table 37 were developed for pressure-vessel applications of class 3 steel at service temperatures down to −85 °F, whereas the limits for A517, grades B, F, and H, steels were developed for applications at more nearly normal service temperatures.

The heat-input limits discussed above, which were established to ensure adequate mechanical properties in the HAZ, also serve to discourage the deposition of large weld beads, produced with large-diameter electrodes and high heat input, and having characteristically poor notch toughness. In welding quenched and tempered alloy steels, it is good practice, whenever possible, to deposit many small beads; this technique improves the notch toughness of the weld metal by the grain-refining and tempering action of successive passes. Such a practice is especially necessary if the welded steel is to endure severe plastic deformation.

Because heat input determines the solidification characteristics and transformation structures (and consequently the properties) of the weld metal, heat input is especially important in the welding of steels of minimum yield strength greater than 100 ksi. In Table 38, specific values of heat input are suggested for welding of HY-130 steel. Such specific conditions ensure that the HAZ and the weld metal meet high strength and toughness requirements.

The stringer-bead technique, rather than the weave-bead technique, is preferred in the welding of quenched and tempered alloy steels, because the heat input per inch of forward travel of the arc is lower, distortion of the joint during welding is less, and notch toughness is better in both the weld metal and the base metal. The partial weaving and uphill motion that is unavoidable when welding in the vertical position should be kept to a minimum. Air carbon arc gouging, at a controlled heat input, followed by cleanup of the cut surface by grinding, is the preferred method for metal removal to facilitate the making of a sound weld at the root of a joint in quenched and tempered alloy steels of yield strength up to about 150 ksi. Machining and grinding only are used for metal removal at the root of a joint in HP 9-4-20 steel.

Postweld Heat Treatment. Of the quenched and tempered alloy steels in Table 31, only A533 and A542 are usually given a postweld stress-relief heat treatment. However, these two steels have been used almost exclusively in thicknesses greater than 2 in.

Stress relieving is necessary for some applications in which the steel: (1) has inadequate notch toughness after cold forming or welding; (2) must retain extremely close dimensional stability during close-tolerance machining after cold forming or welding; and (3) has high residual stresses, from cold forming or welding, that might increase susceptibility to stress corrosion. However, the need for such postweld heat treatment should be thoroughly established for each steel and application, not only because many modern steels for welded construction are designed to be used in the as-welded condition, but also because the properties of the welded steel or the weld metal joining the steel may be adversely affected by a postweld stress-relief heat treatment if the stress-relieving temperature exceeds the tempering temperature. If a weldment must be stress relieved, the temperature should not exceed that used for tempering the steel. Stress-relief heat treatment at a temperature at least 50 °F lower than the tempering temperature is desirable, to avoid the possibility of overtempering.

The alloying elements that contribute most significantly to the attainment of high

Table 36 Suggested minimum preheat and interpass temperatures for ASTM A517, grades B, F, and H, steels(a)

Plate thickness, in.	Min preheat, interpass temperature, °F	SAW	
	SMAW or GMAW	Alloy or carbon steel filler metal, neutral flux	Carbon steel filler metal, alloy flux
Up to 1/2	50(b)	50(b)	50(b)
Over 1/2 to 1	50(b)	50(b)	200
Over 1 to 2	150	150	300
Over 2	200	200	400

(a) Preheat temperatures above the minimums shown may be required for highly restrained welds. Maximum preheat and interpass temperatures should not exceed the minimums shown here by more than 150 °F. (b) Welding of plate that is at any initial temperature below 100 °F will require extreme care to minimize the formation of moisture on the steel being welded.

Table 37 Maximum welding heat input for butt welds in ASTM A533, grade B, steel and ASTM A517, grades B, F, and H, steels

Plate thickness, in.	Maximum heat input, kJ/in., with preheat and interpass temperature of: 70 °F	150 °F	200 °F	300 °F	400 °F
ASTM A533, grade B, steel					
$^1/_4$	23.7	20.9	19.2	15.8	12.3
$^3/_8$	35.6	31.4	28.8	23.8	19.1
$^1/_2$	47.4	41.9	38.5	31.9	25.9
$^5/_8$	64.5	57.4	53.0	42.5	33.5
$^3/_4$	88.6	77.4	69.9	55.7	41.9
ASTM A517, grades B and H, steels					
$^3/_{16}$	17.5	15.3	14.0	11.5	9.0
$^1/_4$	23.7	20.9	19.2	15.8	12.3
$^3/_8$	35.0	30.7	28.0	23.5	18.5
$^1/_2$	47.4	41.9	35.5	31.9	25.9
$^5/_8$	64.5	57.4	53.0	42.5	33.5
$^3/_4$	88.6	77.4	69.9	55.7	41.9
1	Any	120.0	110.3	86.0	65.6
$1^1/_4$ up	Any	Any	154.0	120.0	94.0
ASTM A517, grade F, steel					
$^3/_{16}$	27.0	...	21.0	17.0	13.0
$^1/_4$	36.0	...	29.0	24.0	19.0
$^1/_2$	70.0	...	56.0	47.0	40.0
$^3/_4$	121.0	...	99.0	82.0	65.0
1	Any	...	173.0	126.0	93.0
$1^1/_4$	Any	...	Any	175.0	127.0
$1^1/_2$	Any	...	Any	Any	165.0
2	Any	...	Any	Any	Any

$$\text{kJ/in. of weld} = \frac{\text{Amperes} \times \text{volts} \times 60}{\text{Speed, in./min} \times 1000}$$

Note: heat-input limits for temperatures and plate thicknesses included, but not shown, in this table may be obtained by interpolation; 25% higher heat inputs are allowable for fillet welds.

strength and notch toughness in quenched and tempered alloy steels are usually those that have an adverse effect when such weldments are heat treated after welding.

Postweld heat treatment in the temperature range of 950 to 1200 °F may impair the toughness of the weld metal and of the HAZ. This type of cracking is referred to as reheat, stress-relief, or stress-rupture cracking. The term "stress-rupture cracking" is based on intergranular appearance of the cracks already noted that are similar to those observed in stress-rupture tests. Also, steels that exhibit low ductility (elongation and reduction of area) in the stress-rupture test often are susceptible to reheat cracking. Reheat cracking is not limited to the quenched and tempered steels discussed in this section but may occur in a wide variety of steels.

The extent of impairment depends on composition, heat treatment temperature, and time at temperature and, for some steels, is greater with slow cooling, as in stress relieving. Furthermore, when welds in high-strength alloy steel are given a postweld heat treatment above about 950 °F, intergranular cracking may occur in the grain-coarsened region of the HAZ, usually in the early stage of the stress-relief treatment. Chromium, molybdenum, and vanadium contribute to this crack susceptibility. Nakamura, Naiki, and Okabayaski (Ref 6) observed that susceptibility to stress-rupture cracking (ΔG) in the HAZ of welds is related to composition as follows:

$$\Delta G = \%\text{Cr} + 3.3\,(\%\text{Mo}) + 8.1\,(\%\text{V}) - 2$$

According to this equation, cracking is possible if ΔG is greater than zero. However, experience has shown that this relationship (Table 39) does not always give a reliable estimate of crack susceptibility. Furthermore, factors other than chemical composition are known to affect crack susceptibility.

Table 38 Suggested welding heat input (kJ/in.) for joints in HY-130 steel

Plate thickness, in.	SMAW	GMAW
$^3/_8$ to $^5/_8$	40	35
Over $^5/_8$ to $^7/_8$	45	40
Over $^7/_8$ to $1^3/_8$	45	45
Over $1^3/_8$ to 4	50	50

In A517, A542, and A543 steels, such stress-rupture cracking at the toes of welds has been prevented by properly contouring the welds to minimize points of abrupt change and stress concentration, by light peening at the toes of the welds, by buttering the toes of a fillet weld, or by depositing weld metal having elevated-temperature strength significantly lower than that of the HAZ of the steel during the stress-relief treatment. As shown in Table 39, the Nakamura equation does not always predict the behavior of steels. Nevertheless, it can be used as a guide to determine which steels are likely to be sensitive to this cracking.

Suitable and serviceable welded joints in quenched and tempered steels can be ensured only by the use of welding procedures that have been tested to prove that they can consistently provide sound welds possessing the required service properties. The usual procedure-qualification tests (as in the AWS Structural Welding Code, or the ASME Boiler and Pressure Vessel Code) are generally sufficient for the majority of applications.

Arc Welding of AISI-SAE Alloy Bar Steels

The steels discussed in this section are the widely used hot rolled and cold finished bar steels, ranging in carbon content from 0.09 to 0.64% and in alloy content (excluding manganese and silicon) from zero (the 13*xx* series) to approximately 4.0% (the 48*xx* series). These steels are capable of developing maximum as-quenched hardness from approximately 36 to 65 HRC. The hardenability ranges from only slightly higher than that of plain carbon steels to that characterized by steels such as 4140H and 4340H (see Fig. 3). The weldability of these steels generally decreases as hardenability increases.

Table 40 lists 18 representative AISI alloy steels within the range of 0.18 to 0.53% C content. Alloy steels containing more than 0.53% C are seldom welded. Table 35 includes the alloy steels that are most frequently selected for weldments, with 4130 being the most widely used, because it combines high strength in the heat treated condition with acceptable weldability. AISI steels 4140 and 4340 are also used extensively for weldments, but are more difficult to weld without cracking than 4130 steel.

Filler Metals. Table 40 lists covered electrodes commonly used for producing

Table 39 Steels sensitive to stress-relief cracking

Steel	ΔG cracking	Characteristics	Observed behavior
A212BQ	−2	Safe	No cracking
A533A	−0.5	Safe	No cracking
A533B	−0.25	Safe	Susceptible
A387BQ	0.6	Crack	Cracked
A542	3.4	Crack	Cracked
A543	1	Crack	No cracking
A517A	−1	Safe	No cracking
A517B	−0.6	Safe	Some cracking
A517E	1.4	Crack	Cracked
A517F	0.5 to 0.9	Crack	Bad cracking
A517J	−0.06	Borderline	Borderline
1/2Mo-B-0.01V	−0.04	Borderline	Not susceptible
1/2Mo-B-0.04V	0.2	Crack	Susceptible
1/2Mo-B-0.3V	2.5	Crack	Susceptible
C-Mn	−2	Safe	Not susceptible
C-Mn-0.002B	−2	Safe	Not susceptible
C-Mn steels	−2	Safe	Not susceptible
1/2Mo-B	−0.5	Safe	Susceptible
1/2Cr-Mo-V	2.5	Crack	Susceptible
1/2Cr-Mo	0.5	Crack	Not susceptible
2 1/4Cr-Mo	3.4	Crack	Borderline
5Cr-Mo	4.5	Crack	Not susceptible
9Cr-Mo	9.7	Crack	Not susceptible
C-Mn 0.2Mo-0.04V	−1	Safe	Cracked
C-Mn 0.25Mo-0.15V	0.1	Borderline	Cracked
1/2Cr-1/2Mo-V	>0.5	Crack	Cracked
1Cr-1Mo-0.35V	5.2	Crack	Cracked
1.7Cr-0.2Mo-0.28V	2.6	Crack	Cracked

Source: Nichols, R.W., Reheat Cracking in Welded Structures, *Weld World*, Vol 7 (No 4), 1969, p 36-44

high-strength welds in 18 alloy steels by SMAW. In Table 40, E7018 (or E7018-A1, carbon-molybdenum) is shown as the choice for six steels. This electrode, a low-hydrogen iron-powder type, provides an as-welded joint having minimum tensile strength of 70 ksi. The minimum tensile strength of deposited weld metal is denoted by the first two of four, or the first three of five, digits in each electrode designation. For single-pass welds, the strength specified for the deposited weld metal usually increases as the carbon content of the base metal increases.

All of the electrodes listed in Table 40 are low-hydrogen types. Low-hydrogen electrodes almost always are recommended for welding alloy steels to help prevent cracking.

Unless a joint of maximum strength is required, a lower strength filler metal can often be used and the susceptibility to cracking thereby decreased. For instance, the electrode shown in Table 40 as commonly used for welding 4130 steel is E10016-D2, but if a lower as-welded strength is acceptable, E7018-A1 can be used.

Table 40 Covered electrodes commonly used for SMAW of alloy bar steels

Steel	Electrode	Steel	Electrode
1330	E7018	4320	E7018-A1
1340	E10016-D2	4340	E12018-M
4023	E7018-A1	4620	E8016-C1
4028	E7018-A1	4640	E12018-M
4047	E10016-D2	5120	E8016-B2
4118	E7018-A1	5145	E9016-B3
4130	E10016-D2	8620	E7018-A1
4140	E12018-M	8630	E11018-M
4150	E12018-M	8640	E12018-M

Flux cored electrodes are selected largely on the basis of the mechanical properties of the deposited weld metal. Because susceptibility to cracking increases as strength increases, the lower strength filler metals should be used if the lower strength is acceptable. For joining 4130 to 4130 in the following example, a weld of acceptable strength was made with an electrode wire that resulted in an as-welded composition of 0.15% C, 1.60% Mn, 0.55% Si, 0.70% Cr, and 0.75% Mo. One advantage of FCAW of alloy steels is that cored electrode wires of special composition are more readily obtained than are solid wires of special composition, as described in the following example.

Example 12. Use of Flux Cored Wire of Special Composition To Meet Mechanical-Property Requirements in Welding 4130 Steel. In order to use FCAW to achieve needed production rates, a special composition of flux cored wire was produced to meet the mechanical properties specified for the weldment shown in Fig. 39. The weldment, part of a military tank, consisted of two guides mounted saddle-fashion on and fillet welded to a grouser plate.

The flux cored electrode wire could be obtained more quickly and economically than a solid wire of special composition for GMAW. The flux cored wire, which was supplied in 50-lb coils, was formulated to deposit weld metal having the composition given in Fig. 39. Smooth weld beads were obtained, the welds were sound, with satisfactory strength, ductility, and toughness, and weld spatter was minimal.

As shown in Fig. 39, each guide required two fillet welds, which were made by the automatic method, after the assembly had been tack welded. A special fixture was used to tilt the assembly at 45° to permit welding the backs of the guides

Fig. 39 Use of flux cored electrode wire composition to meet military strength requirements

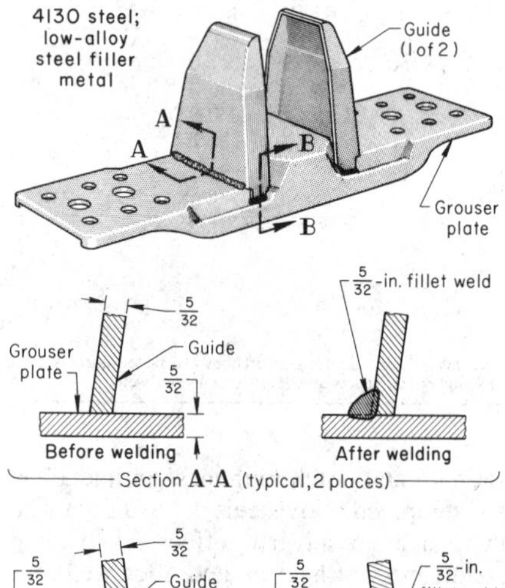

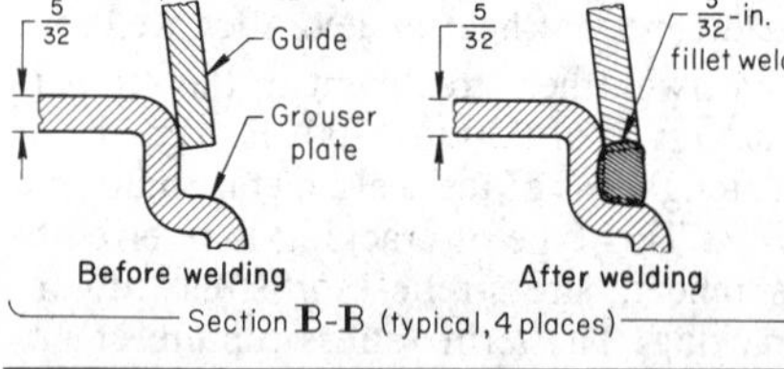

Conditions for FCAW

Joint types	T, corner
Weld type	Fillet
Welding position	Flat
Number of passes	One
Preheat	None
Postweld heat treatment	Quench and temper to 35 and 40 HRC
Fixture	Special, for 45° tilting
Shielding gas	Carbon dioxide, at 35 to 40 ft^3/h
Electrode wire	5/64 in. diam(a)
Electrode holder	500 A
Power supply	500-A constant-voltage rectifier
Current	250 to 300 A (DCEP)
Voltage	23 to 26 V

(a) Specially formulated to produce a deposit with a composition of 0.15% C, 1.50 to 1.70% Mn, 0.50 to 0.60% Si, 0.60 to 0.80% Cr, 0.60 to 0.90% Mo

in the flat position—first one guide, then the other. The fixture was then rotated to deposit the short welds at the feet of the guides. Weld starts and stops were controlled by limit switches. After welding, the assembly was quenched and tempered to a hardness of 35 to 40 HRC.

Electrode Selection. When only moderate strength is required, low-carbon steel electrodes are used. When strength requirements are higher, alloy steel electrode wires that provide weld metal to match or nearly match the alloy composition of the base metal are most commonly used. High-quality electrode wires of virtually any composition are obtainable for GMAW. The carbon content of the electrode wire should generally be lower than that of the base metal to minimize the formation of hard, brittle transformation products that may cause cracking. For instance, in welding 4130 steel, an electrode containing no more than about 0.20% C should be used. If the completed weldment is to be heat treated (for example, quenched and tempered), a filler metal of similar composition should be used. Low-alloy steel filler metals used for GMAW and FCAW of AISI-SAE bar steels are covered in AWS A5.28 and AWS A5.29, respectively. Table 23 in this article lists low-alloy filler metals for FCAW, some of which are applicable to AISI steels.

Table 25 lists designations and compositions of low-carbon steel electrodes for SAW. These electrodes are intended primarily for welding low-carbon steel, but they are also used for alloy steel weldments.

Filler metal that matches the base metal in alloy content is usually recommended when a submerged arc weld of maximum strength is required, but the carbon content of the filler metal should be lower than (often no more than half) that of the base metal, as previously described for GMAW.

In many applications, alloy steels are welded by GTAW without filler metal. Selection of filler metal (if used) for GTAW generally presents fewer problems than for the other arc welding processes. In GTAW, the filler metal is fed into the weld pool, not into the arc, which is an advantage because it results in more efficient alloy transfer and thus enables more accurate control over the composition of the weld metal. It is generally recommended that filler metals match the base-metal composition, but higher alloys are selected whenever special properties are needed.

Preheating. Suggested preheating and interpass temperatures for 18 alloy bar steels are given in Table 41. These suggested temperatures are based on the use of low-hydrogen electrodes. The preheating temperature increases with the carbon content or hardenability of the steel and with the section thickness. If local preheating is used, the full thickness of the joint and about 3 in. (or a distance equal to base-metal thickness, whichever is smaller) on either side of the joint, should be preheated. The preheating temperatures suggested in Table 41 are intended as a guide.

Table 41 Suggested preheat and interpass temperatures for various alloy bar steels

Steel	Preheat and interpass temperature, °F, for section thickness of:		
	To 1/2 in.	1/2 to 1 in.	1 to 2 in.
1330	350-450	400-500	450-550
1340	400-500	500-600	600-700
4023	100 min	200-300	250-350
4028	200-300	250-350	400-500
4047	400-500	450-550	500-600
4118	200-300	350-450	400-500
4130	300-400	400-500	450-550
4140	400-500	600-700	600-700
4150	600-700	600-700	600-700
4320	200-300	350-450	400-500
4340	600-700	600-700	600-700
4620	100 min	200-300	250-350
4640	350-450	400-500	450-550
5120	100 min	200-300	250-350
5145	400-500	450-550	500-600
8620	100 min	200-300	250-350
8630	200-300	250-350	400-500
8640	350-450	400-500	450-550

Under certain circumstances, such as very low restraint, alloy steels can be welded successfully without preheating. Often, preheating of the entire weldment is impractical, so that if preheating is needed, it must be done locally.

Postheating. Generally, postheating is specified for the steels in Table 41 that contain more than about 0.35% C, although there are many exceptions. One practice for welding these higher carbon steels is to use a preheating and interpass temperature that is just above the M_s temperature for the particular steel, and then to hold at this temperature for at least 1 h. The objective is to permit the weld metal to transform from austenite to softer microconstituents, rather than to martensite, and thereby to minimize the possibility of cracking without undue sacrifice in mechanical properties.

Stress relieving usually is required and may be mandatory for weldments of all of the steels listed in Table 41, if the weldment is to be put into service without being quenched and tempered. If a weldment is to be quenched and tempered, stress relieving can usually be omitted. Dimensional stability and notch toughness usually determine the need for stress relief.

In preferred practice, the heating for stress relieving, or for the austenitizing that precedes quenching, should begin before the weldment cools to a temperature below the interpass temperature. However, this procedure is not always practical, and in some applications the weldment remains at room temperature for an indefinite time before being stress relieved. Drafts of air impinging on the weldment while it is cooling to room temperature should be avoided.

For complete, or almost complete, stress relief, the weldment should be heated to 1100 to 1250 °F and held for one hour per inch of maximum base-metal thickness. If heating in this range is impractical, partial stress relief can be attained by heating at a lower temperature (for instance, 900 °F) for several hours.

Arc Welding of High-Strength AMS Alloy Steels

The steels of this group have carbon contents ranging from 0.30 to 0.50%, and total content of chromium, nickel, molybdenum, and vanadium between approximately 2.0 and 7.0%. These alloy steels have high hardenability and, depending primarily on their carbon content, may be capable of developing high hardness and may prove difficult to weld without cracking.

Representative compositions are given in Table 42. Although some compositions are closely similar to those of the AISI-SAE alloy bar steels (4340, for example), they are always purchased to AMS (Aerospace Material Specifications) requirements, and the principal field of application for these steels is in the aerospace industry. They are sometimes welded in the quenched and tempered condition; generally, however, welding is done before final heat treatment. In nearly all applications, a postweld heat treatment is required.

Welding Processes. Shielded metal arc and gas metal arc processes have been used successfully for joining these steels, but GTAW is generally preferred and more often used, because it provides more precise control. Plasma arc welding and EBW have also been used successfully.

Susceptibility to Cracking. With the exception of some high-carbon tool steels, these AMS alloy steels, as a group, are the most susceptible of all steels to the various types of weld cracking. Of the steels for which compositions are given in Table 42, AMS 6431 (well-known as D-6AC) is the most susceptible to crack-

Table 42 Nominal compositions of some high-strength AMS alloy steels containing 0.30 to 0.50% C

AMS No.	Similar to	Composition, % C	Mn	Si	Cr	Ni	Mo	V
6302, 6385	17-22A (S)	0.30	0.55	0.65	1.25	...	0.50	0.25
6415	4340	0.40	0.75	0.27	0.80	1.80	0.25	...
6428, 6434	4335 mod	0.35	0.72	0.27	0.80	1.80	0.35	0.20
6431	D-6AC	0.47	0.75	0.22	1.05	0.55	1.00	0.11

ing, because it has the highest carbon content. However, the demand for AMS 6431 for components such as motor cases, missile tanks, and other aerospace hardware has resulted in the development of welding techniques that produce acceptable results.

Base-Metal and Preweld Heat Treatment. The earliest high-strength rocket-motor cases and missile tanks were fabricated from alloy steels such as 4130 and 4340. Because of low fracture toughness, resulting from attempting to achieve yield strengths above 200 ksi, these steels proved unsatisfactory. To obtain the required high yield strength, the quenched steels had to be tempered in the range of 500 to 700 °F. In addition to the low fracture toughness, these low tempering temperatures did not allow temper straightening of large pressure vessels, which distort during welding and heat treating.

These difficulties led to the development and use of secondary-hardening, low-alloy martensitic steels, such as D-6AC, that can be tempered in the range of 1000 to 1125 °F to provide a combination of high yield strength and good fracture toughness.

D-6AC steel (vacuum melted from consumable electrodes) is used for pressure vessel applications utilizing GTAW. Most of these pressure vessels are tempered at 1000 °F. Large thick-walled vessels, however, require slightly higher tempering temperatures (in the range of 1000 to 1125 °F).

Filler Metal. Criticality in the choice of filler metal for good fracture toughness in D-6AC weldments can be minimized when the base metal is tempered at temperatures higher than 1000 °F. Modified low-alloy filler metal with carbon content somewhat lower than that of the base metal is used. Correct joint preparation and control of heat input in welding are important in maintaining consistent dilution of weld metal by base metal.

For joint thicknesses of about 0.200 in., a low-carbon, $2^1/_4$% Cr alloy steel filler metal can be used to produce a weld in which considerable base-metal dilution is achieved. For joint thicknesses of about 0.900 in., a higher carbon, lower alloy filler metal is used, because there is less dilution in the larger welds. Carbon and alloy contents of the filler metal are adjusted to obtain satisfactory response to heat treatment. The compositions of the two filler metals are given in Table 43.

Welding Procedure. When welding a medium-carbon low-alloy steel of this type, good results are obtained when preweld and postweld heat treatments are employed to control weld cooling rates so as to produce mixed structures of bainite and tempered martensite in the weld zone and to prevent the formation of primary martensite. Such structures can better withstand subsequent processing and final heat treatment to produce high weld joint efficiency and high notch toughness.

Arc Welding of Heat-Resistant Low-Alloy Steels

Heat-resistant low-alloy steels, also referred to as chromium-molybdenum steels, find frequent application in power, refinery, and chemical industries for pressure vessels, piping systems, furnace components, and rotating equipment. The addition of up to 10% Cr and up to 1% Mo enhances high-temperature strength, creep resistance, and corrosion resistance to permit satisfactory operation up to about 1100 °F. These materials exist in many product forms for which separate ASTM specifications have been issued (Table 44). However, with the exceptions of castings and plate, common nominal grade designations are used.

Table 43 Compositions of filler metals used for GTAW of D-6AC alloy steel(a)

Element	2.25% Cr(b)	17-22A(S)(c) (AMS 6458)
Carbon	0.08-0.14	0.28-0.33
Manganese	0.40-0.70	0.45-0.65
Silicon	0.30-0.55	0.55-0.75
Phosphorus	0.008 max	0.008 max
Sulfur	0.008 max	0.008 max
Chromium	2.25-2.75	1.15-1.35
Molybdenum	0.85-1.10	0.40-0.60
Vanadium	...	0.20-0.40
Oxygen	0.0025 max	0.0025 max
Hydrogen	0.0025 max	0.0025 max
Nitrogen	0.005 max	0.005 max

(a) Filler-metal wires were drawn from vacuum-melted stock and had a mirror-finish surface. (b) For joint thickness of about 0.200 in. (c) For joint thickness of about 0.900 in.

Weldability

All materials listed in Table 44 are weldable, but they require a higher degree of quality planning and control than low-carbon steel. The primary difference is the air hardenability of alloy steels. A weld without preheat or even a small arc strike will cause localized hard spots that can initiate solidification or delayed cracking. The risk of such cracking can be increased by the presence of hydrogen. A possible source of such hydrogen includes moisture in electrode coatings and fluxes, cutting oils on machined surfaces such as bevels, and drawing compound on electrode wires. Other potential problems include tenacious chromium oxides that form during thermal cutting and heating and that melt at higher temperatures than the base metal.

Precautions to prevent contamination include thorough cleaning and degreasing of areas to be welded, obtaining welding consumables with low moisture levels (typically no more than 0.2 wt%), and protecting bare metal and electrodes from dirt and moisture through clean, elevated-temperature storage. Because the rate of moisture pickup varies extensively with type of electrode, portable heating or drying ovens should be used, and the number of electrodes issued at one time should be limited. Recent developments in electrode technology include low-moisture absorption types that can retain moisture levels below 0.2% for up to 10 h while exposed to a humidity of 80% at a temperature of 80 °F.

Preheating and postweld heat treating are extremely important when welding air-hardening heat-resistant alloy steels. These treatments reduce stresses, limit or temper martensitic areas, and reduce the amount of hydrogen retained in the weld. The need for these thermal treatments increases with alloy content, air hardenability, wall thickness, and mechanical restraint. Various codes and specifications differ in detailed requirements, and it is beyond the scope of this article to list all conditions and exceptions. While the preheating guidelines shown on Table 45 do not follow any one specification or code, they represent typical practices used by industry. The adequacy of such thermal treatments can be verified by hardness testing of the base metal, HAZ, and weld metal using a portable Brinell tester. Typical hardness limits are 225 HB for the first five grades and 241 HB for the second five grades listed in Table 45.

Table 44 ASTM specifications and compositions for heat-resistant chromium-molybdenum steels

Type	Forgings	Tubes	Pipe	Castings	Plate	Composition(a), % C	Mn	S	P	Si	Cr	Mo
½Cr-½Mo	A182-F2	A213-T2	A335-P2 A369-FP2 A426-CP2	...	A387-Gr 2	0.10-0.20	0.30-0.60	0.045	0.045	0.10-0.30	0.50-0.80	0.45-0.65
1Cr-½Mo	A182-F12 A336-F12	A213-T12	A335-P12 A369-FP12 A426-CP12	...	A387-Gr 12	0.15	0.30-0.60	0.045	0.045	0.50	0.80-1.25	0.45-0.65
1¼Cr-½Mo	A182-F11 A336-F11/F11A A541-C15	A199-T11 A200-T11 A213-T11	A335-P11 A369-FP11 A426-CP11	A217-WC6 A356-Gr6 A389-C23	A387-Gr 11	0.15	0.30-0.60	0.030	0.045	0.50-1.00	1.00-1.50	0.45-0.65
2Cr-½Mo	...	A199-T3b A200-T3b A213-T3b	A369-FP3b	...	...	0.15	0.30-0.60	0.030	0.030	0.50	1.65-2.35	0.45-0.65
2¼Cr-1Mo	A182-F22/F22a A336-F22/F22A A541-C16/6A	A199-T22 A200-T22 A213-T22	A335-P22 A369-FP22 A426-CP22	A217-WC9 A356-Gr10 A643-GrC	A387-Gr 22 A542	0.15	0.30-0.60	0.030	0.030	0.50	1.90-2.60	0.87-1.13
3Cr-1Mo	A182-F21 A336-F21/F21A	A199-T21 A200-T21 A213-T21	A335-P21 A369-FP21 A426-CP21	...	A387-Gr 21	0.15	0.30-0.60	0.030	0.030	0.50	2.65-3.35	0.80-1.06
5Cr-½Mo	A182-F5/F5a A336-F5/F5A A473-501/502	A199-T5 A200-T5 A213-T5	A335-P5 A369-FP5 A426-CP5	A217-C5	A387-Gr 5	0.15	0.30-0.60	0.030	0.030	0.50	4.00-6.00	0.45-0.65
5Cr-½MoSi	...	A213-T5b	A335-P5b A426-CP5b	...	...	0.12	0.30-0.60	0.030	0.030	0.50	4.00-6.00	0.45-0.65
5Cr-½MoTi	...	A213-T5c	A335-P5c	...	...	0.15	0.30-0.60	0.030	0.030	0.50-1.00	6.00-8.00	0.45-0.65
7Cr-½Mo	A182-F7 A473-501A	A199-T7 A200-T7 A213-T7	A335-P7 A369-FP7 A426-CP7	...	A387-Gr 7	0.15	0.30-0.60	0.030	0.030	0.50-1.00	6.00-8.00	0.45-0.65
9Cr-1Mo	A182-F9 A336-F9 A473-501B	A199-T9 A200-T9 A213-T9	A335-P9 A369-FP9 A426-CP9	A217-C12	A387-Gr 9	0.15	0.30-0.60	0.030	0.030	0.25-1.00	8.00-10.00	0.90-1.10

(a) Single values are maximums.

Several special considerations apply to the welding of heat-resistant steels that have been in service. Extra cleaning operations are required to remove hydrogen-containing compounds from the surface and from pitted areas. In addition, there may have been a loss of ductility or strength that can limit the ability of the metal to withstand welding stresses, which include:

- Creep damage
- Hydrogen attack
- Carburization or decarburization
- Temper embrittlement

Depending on severity, these conditions can be overcome or reversed by welding techniques and/or preweld thermal treatments.

Welding Processes

Most fusion welding processes can be used for joining heat-resistant alloy steels. Selection or preference depends on the component to be welded, thickness of the joints, accessibility of the weld, and availability of equipment. Typical welding processes and their application for welding heat-resistant alloy steel components include the following.

Shielded metal arc welding is used widely because of process flexibility, accessibility, welding position, and portability. However, it has two important limitations: (1) because nearly all SMAW electrodes are of the low-hydrogen type, it is not possible to achieve the root pass quality and uniformity that can be obtained when welding carbon steel with cellulose-coated electrodes; (2) because SMAW is not automated easily, it does not provide the productivity or deposition rate that can be achieved by other processes for joining thick plate or long seams.

Gas tungsten arc welding and plasma arc welding provide a high degree of control and quality, but low productivity. Primary applications include root-pass welding of piping and tubing assemblies requiring full penetration,

Table 45 Recommended minimum preheat temperatures for welding heat-resistant chromium-molybdenum steels with low-hydrogen covered electrodes

Steel(a)	ASTM plate specification	Minimum preheat temperature, °F, for thickness, in., of: Up to 0.5	0.5 to 1.0	Over 1.0
½Cr-½Mo	A387-Gr2	100	200	300
1Cr-½Mo	A387-Gr12	100	200	300
1¼Cr-½Mo	A387-Gr11	100	200	300
2Cr-½Mo	...	150	200	300
2¼Cr-1Mo	A387-Gr22	150	200	300
3Cr-1Mo	A387-Gr21	250	300	400
5Cr-½Mo	A387-Gr5	250	300	400
7Cr-½Mo	A387-Gr7	400	400	500
9Cr-1Mo	A387-Gr9	400	400	500

(a) Maximum carbon content of 0.15%. For higher carbon content, the preheat temperature should be increased 100 to 200 °F. Lower preheat temperatures may be used with GTAW.

Table 46 Suggested welding consumables for joining heat-resistant chromium-molybdenum steels

ASTM plate specification	Steel	SMAW(a)	GTAW(b), GMAW(b)	FCAW(c)	SAW(d)
A387-Gr2	$^1/_2$Cr-$^1/_2$Mo	E80XX-B1	ER80X-B2L	E81T1-B1	F8XX-EXXX-F4
A387-Gr12, Gr11	1Cr-$^1/_2$Mo, 1$^1/_4$Cr-$^1/_2$Mo	E80XX-B2 or B2L	ER80X-B2 or B2L	E8XTX-B2 or B2X	F8XX-EXXX-B2 or B2H
A387-Gr22	2Cr-$^1/_2$Mo, 2$^1/_4$Cr-1Mo	E90XX-B3 or B3L	ER90X-B3 or B3L	E9XTX-B3 or B3X	F9XX-EXXX-B3 or B4
A387-Gr21	3Cr-1Mo	E90XX-B3 E502-XX	ER90X-B3 ER502	E9XTXS-B3 E502T-1 or 2	F9XX-EXXX-B3
A387-Gr5	5Cr-$^1/_2$Mo	E502-XX	ER502	E502T-1 or 2	F9XX-EXXX-B6 or B6H
A387-Gr7	7Cr-$^1/_2$Mo	E7Cr-XX	ER502 ER505	E502T-1 or 2 E505T-1 or 2	
A387-Gr9	9Cr-1Mo	E505-XX	ER505	E505T-1 or 2	

(a) Shielded metal arc welding electrodes from AWS specification A5.4-78 or A5.5-81. (b) Welding rod or electrode from AWS specification A5.9-77 or A5.28-79. Argon or argon plus 1 to 5% oxygen is used for shielding with GMAW. (c) Flux cored electrodes from AWS specification A5.22-80 or A5.29-80 using CO_2 or argon plus 2% oxygen shielding. (d) Electrode-flux combination to produce the desired weld metal composition and strength from AWS specification A5.23-80.

a controlled internal reinforcement, and radiographic quality. These processes can be used in an autogenous mode (without the addition of filler metal). This may present some metallurgical limitations if the steel is not sufficiently deoxidized to prevent cracking. With an open root gap and filler metal these processes are also suitable. Filler metal can be selected to metallurgically upgrade the purity of the molten metal and reduce the risk of root-pass solidification cracking. Consumable inserts can be used successfully. This method has the advantage of filler-metal addition and, if properly employed, optimizes uniformity of penetration. However, it requires accurate joint preparation and fit-up plus thorough back purging with argon or nitrogen.

Except when selecting consumable inserts, the use of back purging is optional. Because it prevents the formation of tenacious chromium oxides, its desirability or need increases with an increase in chromium content. Thus, few 1Cr-Mo and many 9Cr-Mo butt joints are back purged.

Submerged arc welding is the primary tool for automated welds using one or more arcs. It is capable of high deposition rates and high service factors or percentages arc time. While most submerged arc welds are deposited in the flat position, recent developments permit SAW of alloy steels in the horizontal position, such as pressure vessel closure joints.

Electroslag welding is limited to joining heavy-walled seams that can be positioned in the vertical position. Typical applications include 2-in. and heavier longitudinal seams in pressure vessels.

Gas metal arc welding and flux cored arc welding have limited applications for joining low-alloy heat-resistant steels. Some of the main applications include semiautomatic welding for piping assembling and thin-walled pressure vessels. Recent developments have adapted the GMAW process to narrow-gap welding of heavy-walled (12-in.) components.

Filler Metals

Two rules apply to the selection of filler metals for heat-resistant low-alloy steels: (1) the filler metal must match the chemical composition and mechanical properties of the base metal, and (2) hydrogen must be kept from the weld. Optimum results can be achieved by matching the chemical and mechanical properties of the base metal. This ensures uniform resistance to stress, heat, and corrosion. Overmatching chemical and mechanical properties can be as undesirable as undermatching, because this can prevent the welded assembly from performing as a uniform structure. An exact match cannot be expected due to differences in product form—a weld is basically a casting, while other components may have been influenced by working and/or special thermal treatments.

Priorities of various properties may have to be established. For example, one must decide whether it is more important to match the minimum percentages of chromium and molybdenum or the maximum tensile strength of the steel to be welded. Typical filler metals selected for the materials and the welding processes discussed in this section are listed in Table 46.

In evaluating final chemical and mechanical properties of a weld deposit, the effect of dilution must be considered. Depending on the specific welding processes, the base metal or any previously deposited beads contribute between 20 and 65% to the weld pool and consequently to final weld-bead composition. This can result in a composition whose mechanical properties differ from the base metal and from the undiluted filler metal.

Mechanical and chemical properties also can be influenced by variations or inconsistencies in the weld deposit. While most filler metals offer a predetermined ratio of wire and flux that melts into the pool, the SAW operator can modify this ratio by changing the welding conditions. This is of concern whenever all or some of the alloying elements are added through active or alloyed fluxes. By using a neutral or near-neutral flux and adding all alloying and most deoxidizing elements through the wire, the influence of SAW variables can be minimized.

Arc Welding of Tool Steels

The compositions of tool steels range from that of the plain low-carbon mold steel P1 to that of the high-alloy high-speed steels, some of which have a total alloy content that exceeds 25%. It follows that their weldability also varies over a broad range. Commercial tool steels are classified into seven major AISI groups. These groups are further classified as to type (hot work, high speed, etc.). Hardening mediums for these seven groups of tool steels are listed in Table 47. Virtually all of these steels are weldable with varying degrees of difficulty. Generally, weldability varies with hardenability.

Steels such as P1, like other plain low-carbon steels, can be welded without special procedures such as preheating and postheating. However, most tool steels have a high carbon content (some as high as 2.50%) and a relatively high content of alloying elements such as manganese, silicon, chromium, molybdenum, tungsten, vanadium, and cobalt. Therefore, most tool steels require the use of carefully controlled preheating and postheating, and in most applications a considerable amount of welder skill is required.

Welding operations for tool steels are generally divided into five classes of work:

1 Assembly of components into a single tool

Table 47 Typical hardening mediums for alloy steel bars

Group	Type	Hardening medium	Typical grades
Water hardening	Plain carbon	W, B	W1
Shock resisting	Medium carbon, low-alloy	O	S1, S5, S6
Cold work	High-carbon, low-alloy	O	O1, O2, O6, O7
	High-carbon, medium-alloy	A	A2, A6, A7
	High-carbon, high-chromium	A	D2, D4, D7
Hot work	Chromium	A	H11, H12, H13
	Tungsten	A	H21
	Molybdenum	A	H42
High speed	Tungsten	A	T1, T4, T15
	Molybdenum	A	M1, M2, M3
Mold steels	Low-carbon, low-alloy	O	P1, P20
Special-purpose	Low-alloy	O	L2, L6

Note: A, air; B, brine; O, oil; W, water

2 Fabrication of a composite tool by depositing an overlay of tool steel weld metal on specific areas of a carbon steel or a less highly alloyed steel (a shear blade, for example)
3 Rebuilding of worn surfaces and edges
4 Alteration of a tool or die to meet a change in design of the product being manufactured by use of the tool
5 Repair of cracked or otherwise damaged tools

For assembly of components (the first item above), the use of a low-cost steel as part of a tool assembly often permits the construction of tooling that is far more economical than if it were made entirely of tool steel. Moreover, this practice often permits use of a tougher steel to provide a backing for the tool steel portion. Purposes covered by items 2 and 3 in the above list represent most of the welding that is done on tool steels by hardfacing. The welding of tool steel described in this article deals mainly with the repair of metalworking tools (item 5 of the above list).

Welding Processes

The repair of tools by welding may be accomplished by virtually all of the arc welding processes. The selection of the process depends largely on the size and complexity of the work, the practicality of observing the recommended precautions relative to preheat, interpass, and postheat temperatures, and the form (rod, continuous wire, etc.) in which filler metals are available.

The SMAW process usually is used. The principal reasons for the use of this process are flexibility in welding position and location, simple equipment that can be quickly moved about, and the wide variety of filler-metal compositions available as covered electrodes.

Small- and medium-sized production sizes frequently are handled by GTAW. This process is particularly well suited to small areas and sharp edges where low heat input and minimum dilution are of particular advantage.

Massive weldments, particularly those involving the repair of deep cracks or the rebuilding of large surface areas, are performed most economically by automatic or semiautomatic continuous wire processes. Submerged arc and plasma hot wire processes are two examples. Filler metals may consist of drawn solid wires or fabricated tube wires and also may include the use of metal powder additions made directly into the weld pool. Enriched (metal-bearing) submerged arc fluxes are a further source of alloy content.

Conditions for Welding. Tool steels are preferably welded in the annealed condition, but often this is impractical as well as costly to the extent that any gain made possible by the use of weld repair would be nullified by the cost of annealing.

Tools and dies in the quenched and tempered condition can often be repaired by welding, but in such a case the preheating and postheating treatments for the base metal must not exceed the original tempering temperature. In some applications, therefore, the microstructure of the base metal, in addition to composition, influences the choice of filler metal.

Most of the rules that apply to the welding of alloy steels apply also the the welding of tool steels, but as the alloy content or carbon content, or both, increase, the importance of adhering to these rules also increases. The following guidelines apply to the arc welding of all grades of tool steel:

- Always use the smallest diameter electrode that will do the job.
- Prepare the surface by machining or grinding. A crack should be gouged out to a U-shape, never to a V-shape, because sharp angles will promote cracking.
- Grind or machine away all of the existing cracks, and provide at least 1/16 in. of excess weld metal for finish grinding.
- Make sure that all electrodes and base-metal surfaces are clean and dry before welding. Absolute freedom from oil and grease is essential.
- Insofar as possible, position the work so that weld beads are laid slightly uphill. This aids joint penetration.
- Never weld any tool steel that is at room temperature; the recommended preheating temperature must be adhered to.
- Keep heat input to the minimum; use minimum arc voltage and amperage, and reduce amperage for secondary and finishing passes.

Selection of Filler Metal. The factors that must be considered in the selection of an electrode for arc welding of tool steel include (1) composition of the base metal, (2) heat treated condition of the base metal (annealed or hardened), and (3) service requirements of the welded area—whether the weld is located at a critical working surface such as a cutting edge or in an area of high wear. Typical electrode compositions (deposit analyses) are presented in the footnotes under Table 48. The selections listed in this table are based on the assumption that the welded area is a working surface of the tool and, therefore, must have approximately the same hardness, or be capable of achieving the same hardness, as the base metal. Selection of an electrode is often simplified when this requirement does not have to be met.

When welding annealed tools, the composition of the deposited weld metal should approximate that of the base metal, so that the weld and base metal will respond similarly to heat treatment and will develop the same hardness after heat treatment.

Selection of electrodes for welding hardened tools requires more consideration than for welding annealed tools. If the weld metal is to be deposited at a functional area of the tool (such as a cutting edge), it is necessary to select a filler metal that will harden as it cools. Under these conditions, the composition of the deposited weld metal may be entirely different from that of the base metal.

Another technique that is often used in repair welding of hardened tools when the weld will be in a working area is to begin the weld with an electrode that deposits weld metal that hardens only slightly or not at all on cooling—for example, a stainless steel or a low-hydrogen low-alloy steel electrode. A major portion of the weld is made with one of these electrodes, but approximately 3/16 in. of the

Table 48 Conditions for arc welding of tool steels

Steel (AISI type)	Type of tool steel electrode(a)	Annealed base metal: Annealing temperature, °F	Annealed base metal: Preheat and postheat temperature, °F	Annealed base metal: Hardening temperature, °F	Annealed base metal: Quenching medium	Annealed base metal: Tempering temperature, °F	Annealed base metal: Resulting hardness, HRC	Hardened base metal: Preheat and postheat temperature, °F	Hardened base metal: C Hardness(b), HRC
W1, W2	Water hardening(c)	1375-1425	250-450	1375-1475	Water	300-650	54-65	250-450	56-62
S1	Hot work(d)	1475	300-500	1750	Oil	300-500	54-57	300-500	52-56
S5	Hot work(d)	1450	300-500	1625	Oil	500 min	55-59	300-500	52-56
S7	Hot work(d)	1500-1550	300-500	1725	Air(e)	400-425(f)	56-58	300-500	52-56
O1	Oil hardening(g)	1450	300-400	1475	Oil	300-450	61-63	300-400	56-62
O6	Oil hardening(g)	1425-1450	300-400	1450-1500	Oil	300-450	61-63	300-400	56-62
A2	Air hardening(h)	1650	300-500	1775	Air	350-400(f)	60-61	300-400	56-58
A4	Air hardening(h)	1425	300-500	1550	Air	350-400(f)	60-61	300-400	60-62
D2	Air hardening(h)	1650	700-900	1850	Air	900-925(f)	58-60	700-900	58-60
H11, H12, H13	Hot work(d)	1600(j)	900-1200	1850	Air	1000-1150(f)	40-50	700-1000	46-54
M1, M2, M10	High speed(k)	1550(j)	950-1100	(m)	Salt(e)	1000-1050(f)	65-66	950-1050	60-63

(a) Nominal compositions of weld deposits are footnoted by type. The compositions of proprietary electrodes vary. (b) As-deposited after postheat. (c) Deposit: 0.95% C, 0.20% Si, 0.30% Mn, 0.20% V. (d) Deposit: 0.33% C, 1.00% Si, 0.40% Mn, 5.00% Cr, 1.35% Mo, 1.25% W. (e) Oil may also be used. (f) Double temper recommended. (g) Deposit: 0.92% C, 0.30% Si, 1.28% Mn, 0.50% Cr, 0.50 W. (h) Deposit: 0.95% C, 0.30% Si, 0.40% Mn, 5.25% Cr, 1.10% Mo, 0.25% V. (j) For H12 and M2 steels, anneal at 1625 °F. (k) Proprietary compositions. (m) 2240 °F for M1, 2260 °F for M2, 2215 °F for M10

weld depth is left for completing the weld with an electrode that deposits approximately the same composition as the base metal. This final layer may need to be thicker than $^3/_{16}$ in., depending on the stresses to which it will be subjected. However, this procedure should not be used if the welded tool will be quenched and tempered, because cracking may occur.

When high hardness is not required in the welded area, the most common practice is to make the entire weld with a low-alloy, stainless steel, or nickel-based alloy electrode.

Preheat and Interpass Temperatures. Regardless of composition or condition (annealed or hardened), tools should never be welded without preheating. Preheating temperature varies with tool steel composition. Sometimes, a steel of a given composition is preheated at different temperatures depending on whether it is in the annealed or the hardened condition. Preheating temperatures commonly used for various grades of tool steel in both annealed and hardened conditions are given in Table 48. As shown in Table 48, preheating temperatures range from 250 °F for the water-hardening tool steels (W grades) to 1200 °F for annealed hot work steels (H grades). When a temperature range, as opposed to a specific temperature, is shown (for instance, 250 to 450 °F for W1 and W2), the lower temperature of the range is used for thin sections and the higher temperature for thick sections and massive tools. The minimum preheating temperature should always be maintained as the interpass temperature during welding.

Postweld Heat Treating. When welding annealed material, best practice is to allow the weldment to cool to about 150 °F, and then to heat it within the appropriate temperature range (see Table 48). When dealing with hardened base metal, the postheating temperature must not exceed the original tempering temperature if the hardness of the tool is to be retained. Common practice is to heat at 25 to 50 °F below the original tempering temperature.

The two examples that follow describe techniques for repair welding. The first example deals with repairing of a cracked die by SMAW. The second example describes how defects in a large roll casting were repaired by SAW.

Example 13. Repair Welding of a D2 Tool Steel Die. The punch-and-draw die shown in Fig. 40 was made from D2 tool steel and was used for blanking and drawing steel shells from low-carbon steel sheet 0.075 in. thick. Improper operating conditions caused the crack, which propagated completely through the sidewall of the die, as shown in Fig. 40. The cracked sidewall was successfully repaired by SMAW, in spite of the difficulty generally encountered in obtaining sound welds in D2 tool steel. Because of its composition (1.5% C, 12% Cr, 1% Mo), D2 tool steel is air-hardening. In addition to its extremely high hardenability, the microstructure of the steel is characterized by massive alloy carbides. Both of these conditions contribute to its high susceptibility to weld cracking.

As shown in Fig. 40, the crack was repaired by a partial-penetration weld, rather than by welding the entire crack. The groove for the weld was ground to less than the full depth and length of the crack. During preheating to 800 ± 100 °F, a slight excess of heat was applied to the cracked side with an oxyacetylene torch to open the crack so that any particles lodged in the crack could be removed with a piece of music wire. A temperature-indicating crayon for 800 °F was used to check preheat temperature, which was maintained during welding.

The weld was made with E312-16 stainless steel electrodes. Welding current was kept as low as possible. Short stringer beads, approximately 1 in. long, were de-

Fig. 40 Repair welding application using SMAW

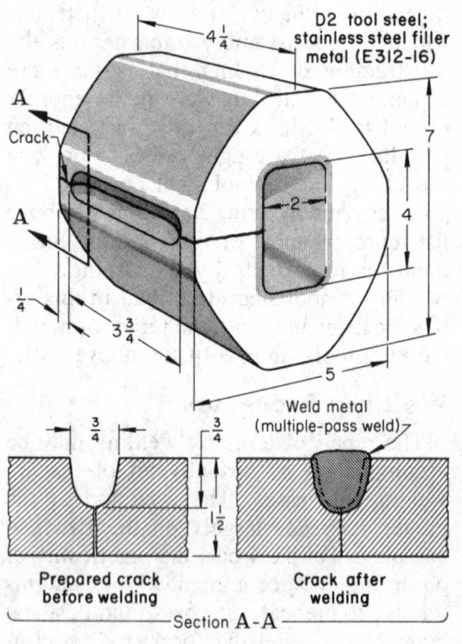

Conditions for SMAW

Joint type	Butt
Weld type	Partial-penetration U-groove
Welding position	Flat
Preheat	800 ± 1.00 °F
Postheat	800 °F(a)
Electrode wire	$^1/_8$-in.-diam E312-16
Power supply	300-A rectifier
Current	150 to 200 A (DCEP)
Hot peening	Required; applied to each bead

(a) For furnace heating, 800 °F for 2 h. For torch heating, after heating to 800 °F, bury in lime for 12 h.

posited with minimum penetration; weaving was carefuly avoided.

Success of the operation depended greatly on the effective use of peening. Immediately after it was deposited, each weld bead was vigorously peened while it was still hot, using an air-operated or electrically operated gun, with a rounded thin-edge tool equivalent to a dull chisel. The hot weld metal was peened to produce a more favorable stress pattern.

The crack closed after deposition of each of the first few beads, but reopened after each of the first few peening operations. Peening did not appear to open the crack when the groove was about one fourth filled, but peening was just as essential from this point on and was done with equal thoroughness for the remaining beads. After welding, the die was allowed to cool slowly to room temperature out of drafts. The die was postheated in a furnace at 800 °F and air cooled.

Example 14. Weld Repair of Casting Tears in a High-Carbon Alloy Steel Roll. The large cast alloy steel mill roll illustrated in Fig. 41 was produced to the following composition: 1.0% C, 0.9% Mn, 0.5% Si, 0.038% S, 0.038% P, 1.0% Cr, 0.5% Ni, and 0.5% Mo. During rough machining of the roll neck to within $^1/_8$ in. of finished size, a number of short discontinuous casting tears or cracks appeared. The cracks were completely removed by cutting a circumferential groove of the size shown in section A-A in Fig. 41; absence of the defects was confirmed by liquid-penetrant inspection. The defective area was then repaired by automatic SAW.

Previous repair welding experience had shown that local preheating was ineffective because of the heat-sink effect of the large mass. Therefore, after machining and degreasing of the groove, the roll casting was heated slowly to 750 °F in a furnace over a period of 27 h (1 h per inch of body radius). When the casting was removed from the furnace, it was completely covered with insulating cloth, except for the area to be welded, and was positioned in a lathe for rotation in the horizontal position. The lathe was equipped with an automatic stepping carriage on which the SAW head was mounted. A gas burner was placed under the roll for use if the interpass temperature dropped below 700 °F, but this did not occur, and welding was completed without the need for additional heat.

The electrode was selected for good color match and toughness, and to provide a deposit that would be equal in hardness to that of the base metal. A tubular electrode was used, consisting of a low-carbon steel sheath and a core of neutral flux and alloying elements. Typical composition of as-deposited weld metal was 0.10% C, 1.35% Mn, 0.75% Si, 2.0% Cr, and 1.0% Mo.

The roll was submerged arc welded. These conditions had been shown by earlier tests to provide the most desirable bead shape with maximum deposit thickness and minimum deposit dilution. There was some carbon pickup that caused increased hardness at the bond line, but the condition was acceptable. The weld metal was built up to approximately $^1/_4$ in. above the surface of the roll neck (see section A-A in Fig. 41) by stringer beads deposited about $^1/_2$ in. wide with a buildup of about $^3/_{16}$ in. A bead overlap of about 40% was obtained by setting the stepping carriage for a $^5/_{16}$-in. advance after each pass.

After being welded, the roll was placed in a furnace and cooled slowly to 600 °F, at which temperature the metal in the weld zone that had been above the austenitizing temperature was transformed to bainite. The roll was then heated to 750 °F (the specified tempering temperature) at 15 °F/h and cooled to room temperature in 50 h.

After heat treatment, the weld buildup was machined flush with the roll-neck surface and liquid-penetrant inspected. The deposit was found to be free from cracks and porosity and was therefore checked for hardness, ground to finish dimensions, and polished in a roll grinder.

Fig. 41 Large cast alloy steel mill roll that was salvaged by automatic SAW

1% carbon alloy steel; Cr-Mo steel filler metal (from low-carbon steel tubular electrode with alloy-flux core)

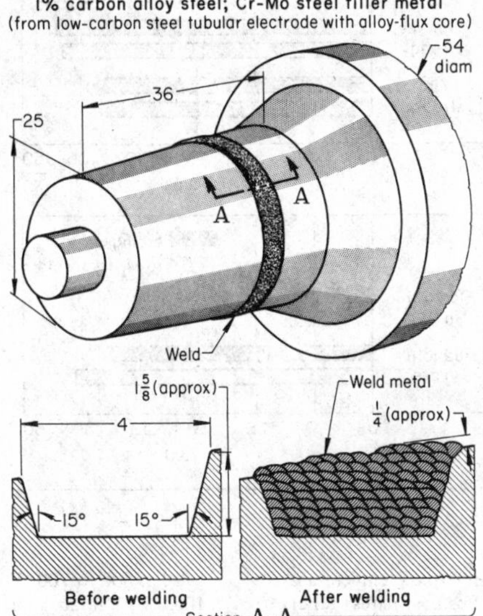

Conditions for SAW

Weld type	Circumferential single-V-groove
Preparation	Machined to remove cracks
Welding position	Flat(a)
Number of passes	About 102 (10 layers)
Preheat	750 °F, in a furnace
Postheat	See text
Electrode wire	$^5/_{32}$-in.-diam low-carbon steel(b)
Flux	Neutral, with alloy content
Flux consumption	15 lb/h
Bead width	$^1/_2$ in., 40% stepover
Power supply	Constant-voltage motor-generator
Current	500 to 550 A (DCEP)
Voltage	25 to 27 V
Electrode extension	$1^1/_2$ in.
Travel speed	20 in./min
Deposition rate	10 lb/h
Finishing	Machined and ground
Auxiliary equipment	Roll lathe, flux-recirculating system, automatic stepping carriage

(a) Horizontal-rolled pipe position. (b) An unclassified low-carbon steel tubular electrode with a core containing flux and alloying elements. Typical weld deposit was 0.10% C, 1.35% Mn, 0.75% Si, 2.0% Cr, 1.0% Mo.

Applicable Welding Processes

This section deals with the arc welding processes that are applicable to hardenable carbon and low-alloy steels and the advantages and disadvantages associated with each process. Process flexibility, portability, cost, and other factors that influence process selection are addressed. In addition, production examples pertinent to each process are provided. Additional information for each process is available in this Volume in articles covering arc welding processes and their application to low-carbon steels.

Shielded Metal Arc Welding

Most alloy steels discussed in this article can be successfully joined by SMAW. This process offers the same advantages for welding alloy steels as it does for plain low-carbon steels.

Shielded metal arc welding is more versatile than the other arc welding processes for work to be done in the field or in drafty areas, because less equipment is required and because under adverse conditions the shielding obtained from decomposition of the electrode covering during welding is more dependable than gas shielding.

An important advantage of SMAW of alloy steel is that flux-covered electrodes (filler metals) are available in a wide range of standardized compositions (see Tables 16, 17, and 18). Because the slag covering the weld metal retards cooling, it is

Fig. 42 Gas-turbine compressor wheel welded by SMAW

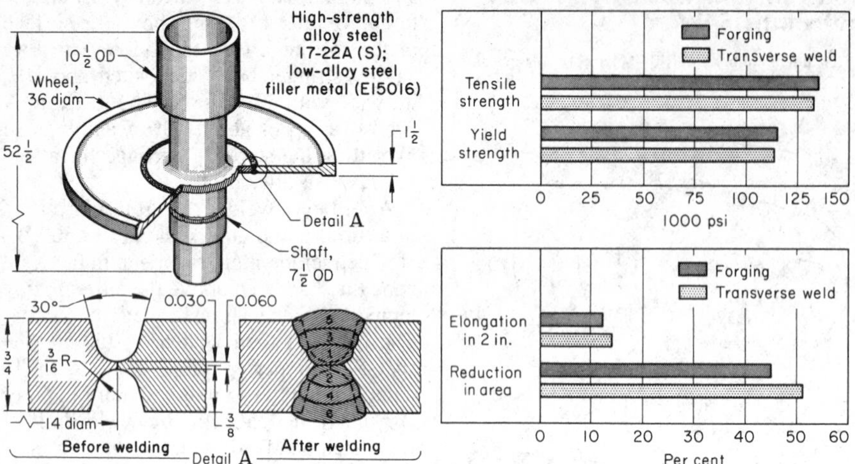

Conditions for SMAW

Joint type	Circumferential butt	Interpass temperature	500 to 700 °F
Weld type	Double-U-groove	Postheat (stress relief)	1065 °F, 2 h; air cool
Welding position	Flat	Electrode wire	5/32-in. diam, low hydrogen(b)
Number of passes	Six	Current:	
Power supply	Constant-current motor-generator	Passes 1 and 2	120 to 130 A (DCEP)
		Passes 3 to 6	170 to 180 A (DCEP)
Preheat	600 °F (a)	Voltage	23 to 26 V

(a) See text for details. (b) Tensile strength, 150 ksi; formerly designated E15016

sometimes an advantage in welding alloy steel.

The major disadvantages of SMAW of alloy steels are those encountered in the welding of other metals—the variation in weld quality, and the loss of time that results from frequent interruptions to replace spent electrodes. This loss of time can be greater for alloy steels than for low-carbon steels because alloy steels generally are welded with smaller diameter electrodes, and thus electrodes have to be replaced more often. Also, the frequent stops make it more difficult to control interpass temperature—a critical factor in some alloy steel applications.

The use of low-hydrogen electrodes and close control of moisture content in the electrode covering are essential if the danger of underbead and toe cracking is to be minimized.

Example 15. Turbine Wheels of High-Strength Alloy Steel Welded by the Shielded Metal Arc Process. Because only six of the assemblies were needed, and because of the simplicity of developing procedures for the process, SMAW was selected for joining a shaft to a compressor wheel for a gas-turbine engine (see Fig. 42).

The wheel and shaft were of 17-22A(S) steel (see Table 42 for composition) in the annealed condition at a maximum hardness of 179 HB. The weld groove shown in detail A in Fig. 42 was machined in the wheel and shaft, and the parts were degreased and wiped with acetone before being preheated in a furnace. Initially, the shaft was preheated to 600 °F and, because the shaft hole in the wheel was 0.002 in. undersize to give a press fit, the wheel was preheated to 800 °F to produce the additional expansion needed for clearance. The temperature of the assembly was equalized at 600 °F after the assembly was placed in a welding positioner. The preheat temperature of 600 °F reduced the cooling rate and was close enough to the M_s temperature (630 °F) of the steel to prevent the formation of martensite in austenitized areas.

The preheat temperature, which was maintained throughout the welding operation by means of a circular gas burner located beneath the joint, was held in the range of 500 to 700 °F, as measured by a surface pyrometer.

The welding sequence, shown in Fig. 42, consisted of six passes—the first two at a current of 120 to 130 A to deposit stringer beads, and the remaining passes at 170 to 180 A to fill the joint. Slag was removed after the first pass, and the bead was wire brushed, before the assembly was turned over in the fixture for the second pass. The current range selected for the first two passes gave 100% penetration of the 0.060-in. root face. The remaining four passes were made at the higher current and with a weave technique to obtain the deposition needed to fill the joint. Immediately after being welded, the assemblies were stress relieved at 1065 °F for 2 h, followed by air cooling, and the welds were then machined flush and inspected by radiographic and magnetic-particle techniques. Final treatment consisted of normalizing for 1 h at 1750 °F, air cooling, tempering for 6 h at 1200 °F, and air cooling.

The weld metal had the following composition: 0.11% C, 0.36% Mn, 0.62% Si, 0.45% Cr, 0.14% V, 0.64% Mo, 1.68% Ni, 0.007% P, and 0.015% S. The weld metal responded favorably to heat treatment, resulting in a hardness of 23 to 26 HRC, compared with 27 to 35 HRC for the base metal. Mechanical properties of the weld in the transverse direction (Fig. 42) were comparable to those of the base metal and within specified requirements.

Flux Cored Arc Welding

Flux cored arc welding can be used to join most hardenable carbon steels and alloy steels. Most FCAW is done in the semiautomatic mode, although some automatic welding is done. Two types of electrodes exist—self-shielded and gas-shielded. They both have high deposition rates and are economical in high-volume welding applications. High heat inputs generally are associated with the high deposition rates of FCAW. This can cause deterioration of properties in hardenable carbon steels and alloy steels. Multipass welds with preheating, regulated interpass temperatures, and postweld heat treatments are helpful with difficult-to-weld steels. The continuous electrode supplies a constant heat input, which makes controlling the interpass temperature easier.

Small diameter electrodes (0.062, 0.052, and 0.045 in.) are used widely, because they provide high out-of-position deposition rates (4 to 8 lb/h) and are easy to manipulate. Porosity from rust and mill scale and cold lap are less likely to occur in FCAW compared to GMAW.

Fume generation rates for FCAW are higher than any other process. Adequate ventilation is necessary to prevent welders from breathing welding fumes. Exhaust fans and various fume extraction systems may be necessary if welding is done indoors because self-shield wires produce their own shielding from decomposed core ingredients.

Electrodes for FCAW are made of a low-carbon steel sheath that surrounds a core of powdered flux material. The flux is a proprietary mixture of alloying elements, deoxidizers, arc stabilizers, slag formers, fluxing agents, and in the case of self-shielded wires, gas formers. Flux cored electrodes for hardenable carbon and alloy steels are covered in two specifications from the American Welding Society: AWS A5.20, "Specification for Carbon Steel Electrodes for Flux Cored Arc Welding," and AWS A5.29, "Specification for Low Alloy Steel Electrodes for Flux Cored Arc Welding." These specifications include gas-shielded and self-shielded electrodes.

Self-shielded flux cored electrodes derive some gas shielding from vaporized core ingredients, but rely on deoxidizers and denitriders to remove atmospheric contaminants from the liquid weld metal. Consequently, self-shielded wires may perform better than gas-shielded wires in drafty areas. Gas-shielded flux cored electrodes rely on an auxiliary shielding gas to protect the arc and weld pool, usually carbon dioxide or a mixture of carbon dioxide and argon. Use of the gas shield makes it unnecessary to add the gas formers and denitriders required for self-shielded wires.

The type of shielding gas used has a significant effect on the mechanical properties of the weld. A change from carbon dioxide to 75% argon and 25% carbon dioxide shielding generally increases tensile strength about 10%, because manganese and silicon strengthen the weld metal. These elements also work as deoxidizers. When a more oxidizing shielding gas is used, the manganese and silicon form oxides that evolve from the weld metal, thus lowering strength. Depending on the original carbon content of the base metal and electrode, a carbon dioxide atmosphere may behave as either a carburizing or decarburizing medium. If the carbon content of the weld metal is below 0.05%, the molten weld pool tends to pick up carbon. If the carbon content of weld metal is above 0.10%, it tends to reject carbon.

Only welding-grade shielding gases with a dew point below −40 °F should be used to avoid moisture problems such as porosity and hydrogen cracking. Flux cored arc welding generally is considered a low-hydrogen welding process because cellulose is not used as a flux ingredient, as with some shielded metal arc electrodes. Most flux cored electrodes produce between 0.03 to 0.08 mL of diffusible hydrogen per gram of deposited weld metal, measured by the glycerin method. Some basic slag system electrodes produce less than 0.01 mL/g of hydrogen.

Example 16. Welding Frigate Plate Using All-Position Flux Cored Arc Welding.* Flux cored welding with all-position continuous electrodes was used to fabricate guided missile frigates for the U.S. Navy. Structures were made of low-carbon steel plate from 0.10 to $1^1/_2$ in. thick. Plate up to 4 in. thick was used in some sections. Joint designs and tolerances were mandated by MIL-STD-0022B, which generally calls for root openings of $^1/_4$ in. or less. A small-diameter FCAW electrode wire that could be used by semi-automatic equipment in any position was originally used. Studies showed that welders could achieve high deposition rates with 0.045- or $^1/_{16}$-in. electrodes, using high current densities that produced a spray-like transfer.

With this opportunity to raise productivity came a problem—these wires did not meet the only existing military specification for a flux cored wire: the MIL-70T-5 class of electrode, which is analogous to E70T-5. Military specification MIL-E-24403/1A, which covers the MIL-70T-5 electrode, gives only flat and horizontal fillet requirements of a low-carbon steel class of electrode.

A need developed to prove that property requirements could be met in the vertical-up and overhead positions as well as horizontal position. To satisfy notch toughness requirements in all positions, a high-tensile $2^1/_2$% Ni electrode was chosen.

A modified weld-metal analysis reduced the maximum carbon and silicon contents (from 0.12 to 0.10% C and 0.90 to 0.50% Si), narrowed the manganese range (from 0.6 to 1.5% Mn to 0.6 to 1.25% Mn), and set a range of 2.00 to 2.75% for nickel content. Typical analysis for the wire was 0.05% C, 0.95% Mn, 0.35% Si, and 2.60% Ni.

Because the nickel alloy electrode displays better wetting action and out-of-position ease of handling with a 75% argon and 25% carbon dioxide mixture, the requirement that test plates be welded only with straight carbon dioxide was altered.

Plates $^5/_8$ and $^3/_4$ in. thick were welded, which qualified the electrode for plates $^1/_8$ to $1^1/_2$ in. thick. Eleven passes in the vertical-up position and ten passes in the horizontal and overhead positions were made using the $^1/_{16}$-in.-diam electrode.

Nondestructive testing included radiographs of welds made in horizontal, vertical, and overhead positions. To further ensure weld-metal integrity, dry magnetic-particle tests with a magnetic field set up in two directions, parallel and perpendicular to the weld, including 1 in. of the HAZ, were performed.

Weld coupons made with the Ni-2 electrode were tested to determine mechanical properties. Required tensile strength of the base metal was 58 ksi minimum; all-weld metal from each welding position was 70 ksi minimum. Specifications also called for transverse face and root bends, and Charpy V-notch impacts of 20 ft · lb at −20 °F; 45 ft · lb at 0 °F; 50 ft · lb at 30 °F; and 60 ft · lb at 75 °F.

The modified electrode has been used extensively for fabricating T-bars, collars, bulk-heads, and other structures. The 0.045- and the $^1/_{16}$-in. electrodes are used. The 0.045-in. type was used in many out-of-position applications.

Welding parameters for the 0.045-in. electrodes were: 125 to 200 A and 24 to 26 V in the vertical-up and overhead positions; 125 to 250 A and 24 to 28 V in the flat and horizontal positions.

The $^1/_{16}$-in. electrode operates at 220 to 240 A and 23 to 25 V in the overhead position; 210 to 230 A and 22 to 24 V in the vertical-up position; and 270 to 290 A and 26 to 28 V in the downhand and horizontal positions.

The use of the all-position wire triples productivity over that achieved by using conventional coated electrodes.

*Adapted from Gas Metal Arc Welds Keep Frigates Sailing, *Welding Design & Fabrication*, May 1980, p 73

Gas Metal Arc Welding

Almost all hardenable carbon and alloy steels can be welded by GMAW in all positions by choosing the appropriate shielding gas, electrode type and size, and welding conditions. Thin-gage materials can be welded in all positions using short circuiting type transfer where metal is transferred as the wire touches the workpiece. Thick materials can be joined in the flat position using standard spray transfer or can be joined out-of-position using a variation of the process called "pulsed spray transfer." With spray transfer, metal is propelled from the wire to the workpiece by the arc forces present. Large, flat, or horizontal single-pass deposits in steel can be made using a second spray transfer variation called "rotating spray transfer." Metal is propelled across the arc gap while the arc, due to the magnetic forces present, moves in a circular manner around the wire axis. Using these modes of metal transfer, heat input and penetration can be controlled closely to obtain optimum weld-metal mechanical properties.

There are numerous advantages to GMAW of hardenable carbon and alloy steels, including: (1) the availability of standardized electrode wire alloys for joining a variety of base materials; (2) the existence of a wide range of wire diameters, which enables the user to select low- to high-current operating ranges to control heat input and provide optimum weldment microstructure; (3) the low weld-metal hydrogen levels (0.5 to 2.5 mL/100 g of deposited metal measured using the glycerin method), which minimize the chance of hydrogen-induced cracking; and (4) excellent weld-metal mechanical properties obtained with a variety of filler metal and shielding gas combinations.

The limitations of GMAW compared with SMAW are essentially the same for joining hardenable carbon and alloy steels as for welding plain carbon steels—namely, the equipment is less portable, is generally more expensive, and process use is limited to areas where drafts do not interfere with shielding.

Shielding Gas Selection. Factors influencing the selection of shielding gas for welding hardenable carbon and alloy steels include base-metal composition, filler-metal composition, mode of metal transfer, welding position, and most importantly, the required weld-metal quality and mechanical-property levels. There is no uniform practice for selection of a shielding gas, although argon-based mixtures are used most frequently for welding hardenable carbon and alloy steels. Carbon dioxide is used widely for welding low-carbon steels—particularly single- and multiple-pass welds where mechanical properties are not a major consideration. Because of the combination globular- and short circuiting type transfer developed with carbon dioxide shielding, its performance features, particularly spatter levels, make it a less desirable gas for joining materials where weld quality is important.

The high oxidizing nature of the gas also creates a detrimental elemental transfer situation, which results in both lower strength and impact properties when compared to similarly made argon-shielded welds. Argon mixtures that include high percentages of argon with 1 to 4% oxygen, or 5 to 8% carbon dioxide, or small additions of both, are used extensively for conventional, rotational, and pulsed spray welding of a variety of alloy steels. Some tri-mix gases, which include a high percentage of helium (with argon and oxygen or carbon dioxide), are used for out-of-position short circuiting welding of alloy steels, particularly high-nickel-bearing alloys. Argon with 25 to 50% carbon dioxide, a commonly used shielding gas for short circuiting welding of plain carbon steel, sometimes is used for alloy steels where mechanical-property requirements are not as stringent.

Electrode Wire Selection. Most electrode filler wires used for joining hardenable carbon and alloy steels are classified under AWS specification A5.28, "Specification for Low Alloy Steel Filler Metals for Gas Shielded Arc Welding." Selection generally is based on compatibility with the base metal and on the desired weld-metal mechanical properties. Alloys not covered by this specification may be included in the appropriate military specifications.

Example 17. Use of Short Circuiting Transfer for Close Control of Deposition in Welding 4130 Steel. A brake frame, half of which is shown in Fig. 43, was a welded assembly of two 4130 steel frame pieces (stampings) and sixteen 4130 steel torque bars (extrusions), joined by GMAW. Forty mounting holes (20 opposing pairs) had to be held in accurate alignment, and each weld had to be closely controlled for bead size and shape, weld-metal flow, and joint penetration. It was required that surfaces and edges as designated in Fig. 43 be free from weld metal and that undercutting along two sides of each weld (see detail C in Fig. 43) not exceed 25% of the stamping thickness. Porosity was limited to a maximum of three pinholes in any one continuous weld, with a maximum pinhole diameter of $^1/_{32}$ in. The close control necessary for compliance with these requirements was achieved with the short circuiting mode of metal transfer.

The stampings and torque bars were assembled in a welding fixture, with a tension bar (not shown in Fig. 43) tack welded to each frame half to maintain alignment of the mounting holes. The joints were then welded by semiautomatic GMAW. Each of the 16 joints on one side of the assembly was welded on three sides in the flat position (detail A in Fig. 43). Then the assembly was inverted, and the process was repeated on the reverse side.

After welding, the assembly was removed from the fixture and stress relieved at 1250 °F for 15 min. On cooling to room temperature, the tension bar was removed, and the tack weld was ground off. Finally, the assembly was shot peened.

Example 18. Change From Semiautomatic to Automatic GMAW of Axle Assemblies. The rear-wheel axle assembly shown in Fig. 44 was originally welded by semiautomatic GMAW at a rate of 59 assemblies per hour. A change to automatic GMAW increased production to 115 assemblies per hour; also, less skill was

Fig. 43 Brake-frame assembly joined by GMAW. Short circuiting mode of metal transfer was used for close control of deposition.

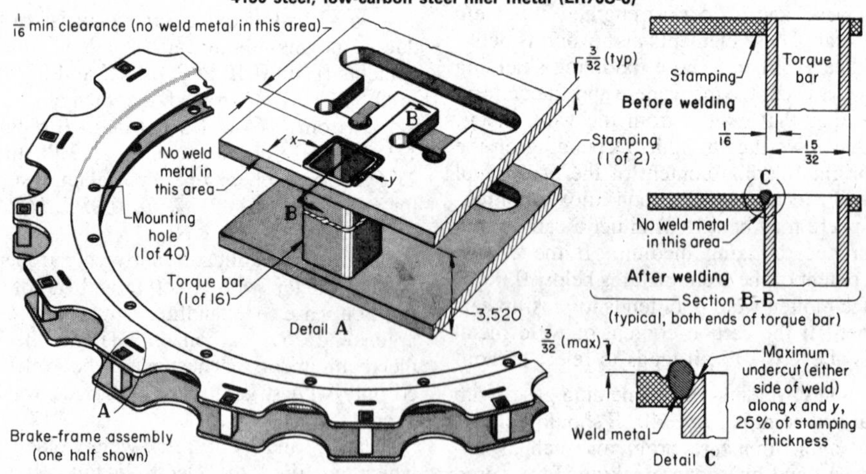

Conditions for GMAW

Joint type	Corner	Electrode	0.035-in.-diam ER70S-6
Weld type	Square groove	Electrode holder	Hand held, air cooled
Welding position	Flat	Shielding gas	Carbon dioxide, at 20 ft^3/h
Number of passes	One	Power supply	300-A constant-voltage motor-generator
Preheat	None	Current and voltage	90 A (DCEP); 18 V
Postheat (stress relief)	1250 °F for 15 min	Travel speed	4.1 in./min
Fixture	Assembly jig		

required for the welding operation, and all welds had a more uniform appearance.

The axle was made of 4140 steel, strain hardened to approximately 125 ksi yield strength. The nut was made of a killed 1146 free-cutting steel. Although this steel is not recommended for welding, the free-cutting additives produced no noticeable detrimental effects.

Before being welded, the axle and nut were degreased and assembled. The assembly was manually loaded in a special chucking fixture, and the axle, at 45° to the horizontal, was rotated clockwise, at 16.8 in./min at the joint. Welding was fully automatic, and the electrode holder was held in a retractable air-operated mechanism. The holder was air cooled. No preheating or postheating was needed, and the weld was made in one pass. Because the purpose of the weld was to fix the nut permanently to the axle, penetration through at least one thread engagement (see Fig. 44) was required.

After welding, the axle assembly was washed, zinc plated, and heated for 3 h at 375 ± 25 °F. This heat treatment served to lessen danger of hydrogen embrittlement resulting from the plating operation. The welds were inspected visually.

Submerged Arc Welding

Submerged arc welding is used as a fully automatic and as a semiautomatic process for joining and for surfacing of hardenable carbon and alloy steels. The article "Submerged Arc Welding" in this Volume discusses in detail the equipment, electrodes, fluxes, and procedures. Electrodes and fluxes with particular reference to hardenable carbon and alloy steels have been discussed earlier in this article.

Procedures for preheating and postheating suggested earlier in this article were oriented toward SMAW and do not necessarily apply to the same degree to SAW. Preheating and postheating are less likely to be required for SAW than for SMAW. This is because the high heat input and high deposition rate in SAW give greater temperature rises to a larger area. Consequently, the cooling rate is lowered, and the probability of hardening is lessened. Further, the flux blanket also helps to retard cooling. However, in many applications, preheating and postheating are needed in SAW of hardenable carbon and alloy steels. Each application must be evaluated separately.

The SAW process ordinarily is capable of depositing low-hydrogen weld metal, provided the flux is properly dried and other precautions—keeping the area surrounding and including the joint free of moisture, oil, grease, paint, scale, or other material that could be a source of hydrogen—are observed. This provides additional ensurance against hydrogen cracking. The flux, which may be hygroscopic, must be kept dry. Fused fluxes that are nonhygroscopic generally are preferred to bonded fluxes that have a tendency to absorb moisture. Proper storage and handling of fluxes are important. A widely followed practice for maintaining low-hydrogen conditions is to keep the flux in an oven at about 250 °F for several hours before it is needed.

The only disadvantage of SAW of alloy steels is that because of the higher heat inputs involved, there is a possible loss of toughness in the HAZ in certain grades of steels. For such grades of steel, the heat inputs employed may have to be restricted to a certain maximum level. Each material and application must be evaluated separately.

Fig. 44 Axle assembly for which change from semiautomatic to automatic GMAW increased production rate and improved weld appearance

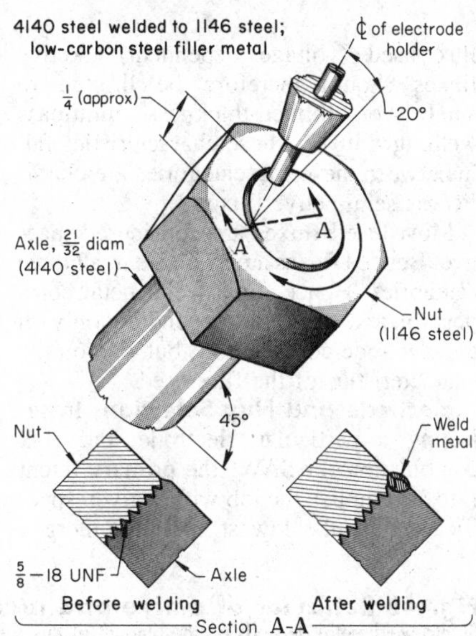

Conditions for automatic GMAW

Joint type	Circumferential butt
Weld type	Square groove
Number of passes	One
Preheat and postheat	None
Shielding gas	Carbon dioxide(a)
Electrode wire	0.035-in.-diam low-carbon steel
Electrode holder	Machine held, air cooled
Wire-feed rate	264 in./min
Power supply	250-A constant voltage transformer-rectifier
Current	160 A (DCEP)
Voltage	19 V
Travel speed	16.8 in./min
Production rate	115 assemblies per hour(b)

(a) Supplied from cylinder, at 5 to 10 psi. (b) In semiautomatic method, production rate was 59 assemblies per hour.

Filler Metals. Submerged arc welding electrodes are selected primarily for their influence on the mechanical properties and/or required weld-metal composition. Carbon and manganese are the most common alloying elements, with additions of silicon, molybdenum, nickel, chromium, copper, and others used to raise weld-metal strength and control low- or high-temperature mechanical properties. Sometimes, alloying elements are added to enhance special physical and chemical properties such as corrosion resistance. For example, copper is added to the electrode to weld weathering steels. Manganese and silicon additions provide deoxidation and assist in eliminating carbon monoxide porosity. In some wires, other elements such as aluminum, vanadium, titanium, and zirconium also are added to provide additional deoxidation.

Most of the commonly used carbon and low-alloy steel electrode compositions are defined by AWS specifications such as A5.17, "Specification For Carbon Steel Electrodes and Fluxes For Submerged Arc Welding," and A5.23, "Specification For Low Alloy Steel Electrodes and Fluxes For Submerged Arc Welding." Table 49 shows several typical electrode classifications based on these specifications.

In these specifications, electrodes are not related to weld-deposit mechanical properties without the associated flux. This is in recognition of the role flux plays in controlling the final mechanical properties of the weld deposit. A summary of electrode alloy types and application areas is shown in Table 50. In the first three groups, the manganese content varies up to 2.4% and the silicon content up to 0.75% to provide flexibility in meeting varying requirements for strength and toughness, welding performance, and deoxidation to ensure weld soundness. The influence of flux and base plate must also be considered, as the molten flux, electrode, and base plate interact to determine the resulting weld-metal composition and properties.

Published literature and manufacturers' product catalogs define many suitable flux and electrode combinations for various materials and applications. Alternative combinations not listed require careful consideration of the interactions among flux, electrode, and base material and require appropriate testing to determine compatibility.

Flux and Slag Systems. The real key to the versatility of SAW is the flux. Fluxes

Table 49 Typical submerged arc electrode classifications for carbon and low-alloy steels

AWS class	Classification	Electrode composition, % C	Mn	P	S	Si	Mo	Ni
AWS A 5.17-80	EL12	0.05-0.15	0.25-0.60	0.035	0.035	0.07	...	...
	EM12K	0.05-0.15	0.80-1.25	0.035	0.035	0.10-0.35	...	...
	EH14	0.10-0.20	1.70-2.20	0.035	0.035	0.10	...	...
AWS A 5.23-80	EA2	0.07-0.17	0.95-1.35	0.025	0.035	0.20	0.45-0.65	...
	EA3	0.10-0.18	1.65-2.15	0.025	0.035	0.20	0.45-0.65	...
	EF2	0.10-0.18	1.70-2.40	0.025	0.025	0.20	0.40-0.65	0.40-0.80
	EM2	0.10	1.25-1.80	0.010	0.010	0.20-0.60	0.25-0.55	1.40-2.10

Table 50 Applications for various SAW electrode groups

Major elements	Electrode classification	Characteristics and uses
C-Mn	All AWS A5.17	General purpose, low cost, suitable for structural steels and many vessel and ship applications. Example: ASTM A515, A516, A36, A287, and A441
C-Mn-Mo	AWS A5.23, classes A1-A3	Similar to above, but Mo addition retards strength reductions from stress relieving. Example: A533, A537, A516
Mn-Mo-Ni	AWS A5.23, classes 1Ni, 3Ni, EF1-EF3	High weld strength and/or toughness with selected fluxes. Example: A302, A537
All other	AWS A5.23, classes EF4, EF5, EM1-EM4, EW	Special high-strength steels, A517, A543, A387, corrosion-resistant steels, A588

perform many functions in addition to providing atmospheric protection for the molten metal and influencing the mechanical properties of the weld deposit. Their actions in controlling arc stability, bead shape, pool fluidity, rust and scale tolerance, slag peeling characteristics, and welding speed capabilities give SAW its inherent application versatility.

Submerged arc welding fluxes are produced in three forms: prefused, bonded, and agglomerated. Method of production, relative advantages and disadvantages, typical compositions, and their characteristics are covered in detail in the article "Submerged Arc Welding" in this Volume.

Fluxes are classified as inactive (neutral) or active. These terms describe flux behavior and refer to the amount of manganese and/or silicon that is transferred to the weld metal (Fig. 45). They are relative terms, depending on the flux composition, wire and flux composition, and slag-to-wire melted ratio. Inactive fluxes are those in which the manganese and/or silicon transfer is not altered significantly by the amount of flux used. The amount of flux used depends on welding voltage—the higher the voltage, the greater the amount of flux melted. Active fluxes, on the other hand, transfer significant amounts of manganese and/or silicon to the weld deposit and thereby affect weld mechanical and chemical properties. The amount of alloying elements transferred to weld metal is proportional to the amount of flux used (voltage dependent). Active fluxes should therefore be limited to single- or limited-thickness multipass welding. Fluxes whose characteristics fall in between these two categories are classified as semi-active fluxes.

Most fused fluxes are considered inactive. Bonded fluxes are available in all three categories. Hence, final weld-metal composition and properties depend not only on the electrode composition, but also on the characteristics of the flux used.

Electrode and Flux Selection. In selecting a particular electrode and flux combination for SAW, the primary intent is to accomplish the job within given specifications at the lowest cost. Submerged arc welding consumables selection must consider primarily two key functional factors: (1) fulfilling mechanical-property requirements and (2) satisfying performance features desired, such as usability, maximum travel speed, rust and scale tolerance, current capability, slag peeling, stability with multiple electrodes, or fillet weld performance.

Present AWS flux and electrode classification standards can be helpful, as they define mechanical properties obtained (strength and toughness), but even these specifications have limitations. AWS A5.17 and A5.23 consider only undiluted welds, deposited under very specific conditions: 550 A, 28 V, 16 in./min provides a heat input of 58 kJ/in.

The AWS test assembly used for defining flux classifications requires a number of small passes (usually 14 to 16), which minimizes dilution to obtain weld metal most representative of the flux and electrode combination tested. It does not resemble most production welds and is not intended to define mechanical-property ranges obtainable in production. The American Bureau of Shipping and Lloyd's specify standards for weld-metal testing in specific diluted two-pass welds; these results are very dependent on the base metal. None of these specifications quantitatively tests many of the flux performance features that influence usability and application economics.

To obtain additional information for materials selection, manufacturers' literature describing the different flux types and flux classification data must be obtained. It is also necessary to consider past usability experiences and ultimately evaluate possible flux and electrode combinations, duplicating job conditions as closely as possible.

Fig. 45 Behavior of active and inactive SAW fluxes. (a) Transfer of Mn and Si into weld metal depends on the amount of flux used. (b) Transfer of Mn and Si is minimal and not significantly affected by the amount of flux used.

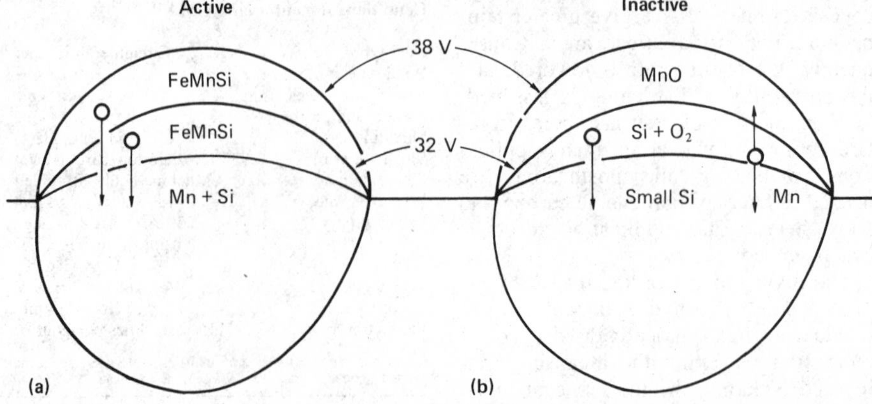

The first priority in materials selection is to define in detail specific application requirements. This summary must include mechanical-property requirements, exact postweld heat treatment, plate condition, thickness, and composition. The number of weld passes, acceptable travel speed, number of electrodes, and required current are also key considerations. Although manufacturers' recommendations are a valuable asset and should be consulted for guidance, ultimately it is necessary to assess material performance using the welding condition in service or the condition in a procedure qualification test made in advance. Each application must be evaluated separately.

Example 19. Welding 3.5% Ni Steel Pipe Sections for Low-Temperature Service. The pipe shown in Fig. 46 was nickel steel (3.2 to 3.8% Ni) and was welded with stainless steel filler metal to meet the mechanical-property requirements of ASTM A333, grade 3, which included a minimum tensile strength of 65 ksi and a minimum notched-bar impact value of 15 ft·lb at −150 °F (0.4-in.-by-0.4-in. specimens).

Fig. 46 Circumferential welding of pipe for low-temperature service

3.5% Ni steel (ASTM A333, grade 3); stainless steel filler metals: E308-16, shielded metal arc welding (root pass) ER308, submerged arc welding (filler passes)

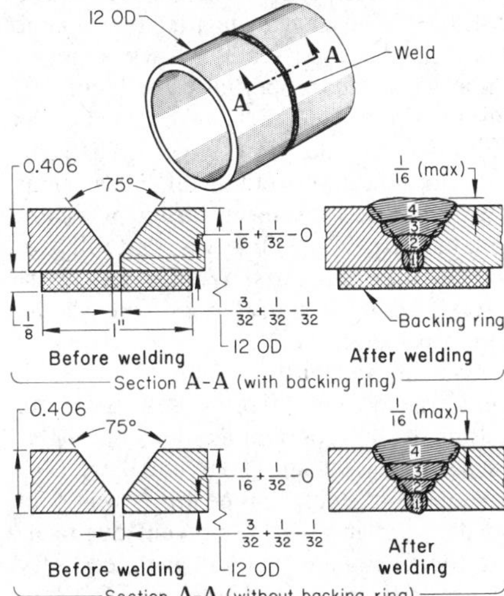

Item	SMAW (root pass)	Automatic SAW (filler passes)
Number of passes	One	Three
Electrode wire	E308-16	ER308
Electrode diam, in.	$^1/_8$	$^1/_8$
Wire feed, in./min	···	15
Current, A	95-125 (DCEP)	250 (DCEN)
Voltage, V	20-24	25
Interpass temp, °F	150-350	150-350
Travel speed, in./min	···	12-15

The joint design shown in Fig. 46 was used for the 0.406-in. wall thickness, as well as for similar pipes having wall thicknesses up to $^3/_4$ in. The mechanical-property requirements could be met by welding either with or without backing rings.

Where service conditions and specifications permitted, backing rings were used (which were allowed to remain in place) to reduce the time and welder skill needed to make the root pass. The root pass was made without a backing ring when internal obstructions were not allowed. Making the root pass required skill to obtain complete fusion of the root with a smooth inside root condition. Care was used with both procedures to ensure full root penetration.

Because it was less costly, automatic SAW was used for all passes except the root pass. It was not suitable for the root pass because drop-through would occur if no backing were used, or the welding flux could be trapped between the backing ring and the pipe, which could cause incomplete fusion and other defects. The root pass was made by SMAW with $^1/_8$-in.-diam E308-16 electrodes. This metal was sluggish, requiring the weld pool to be "pushed" with the arc. A skilled welder could produce a smooth root pass approximately $^3/_{32}$ in. deep. To guard against cracking, weld starts were made by backstepping over previous deposits. Although no preheat was used, interpass temperature was held in the range of 150 to 350 °F to avoid overheating and subsequent cracking. Data for root-pass welding with and without backing were not separated because welding current and speed varied with welders and were not accurately predictable.

Edge preparation and preweld cleaning, as well as quality-control methods, inspection, and testing, were in accordance with the ASME code. Weld surfaces had to be clean of solid matter, oil, and grease. All slag and flux had to be removed before successive beads were deposited. Cracks and blowholes on bead surfaces had to be removed by chipping or grinding. Undercutting was not permitted. Impact values for welds made with and without backing did not differ, provided that the root pass was smooth and uniform. Qualification welds were inspected visually, by radiography, by magnetic-particle or liquid-penetrant methods, and metallographically. Radiographic interpretation was more difficult when backing rings were used.

Essentially the same procedure as that described above was used for welding pipe sections of the following steels: A203, grades D and E; A334, grade 3; A350, grade LF3; and A352, grade LC3.

Gas Tungsten Arc Welding

In GTAW, the composition of the weld metal can be accurately controlled, because in GTAW any filler metal used is fed into the weld pool and not into the arc (see the article "Gas Tungsten Arc Welding" in this Volume). Control of weld-metal composition is generally more important in the welding of alloy steel than in the welding of plain low-carbon steel; in most applications, composition control becomes more important as the alloy content of the base metal increases.

The close control of heat input that is characteristic of GTAW is frequently advantageous when welding alloy steel in the heat treated condition.

The main disadvantage of GTAW of alloy steel is the same as for other steels: the process is slower and consequently somewhat more expensive than the other metal arc processes. However, for many critical applications, the need for the high-quality welds that can be obtained by GTAW does not permit substitution of another metal arc process.

Example 20. Welding of 4140 Steel Stampings to a Heat Treated 4340 Forging. When undesirable vibrations were traced to the 4340 forged steel torque plates of an aircraft brake assembly, the design was revised to incorporate an integral ring of metal that changed the resonance frequency of the part. In order to make use of 1500 existing torque plates, they were altered by gas tungsten arc welding a torque ring stamped from annealed 4140 steel to the inside of each heat treated 4340 steel torque plate, as shown in Fig. 47. This weldment served the same purpose as the redesigned torque plate. As shown in Fig. 47, the torque ring, which was stamped in three segments, had a corrugated configuration to avoid interference with brake flange mounting bolts. The three segments of the ring were joined by square-groove butt welds, and the ring was attached to the torque plate with two fillet welds at each of the 12 corrugations (24 fillet welds), as shown in Fig. 47.

Although the welds were required only to hold the ring in solid contact with the torque plate, no cracking could be tolerated. In addition, the heat treated hardness of the 4340 steel torque plate (35 to 39 HRC) had to be retained. Therefore, it was

Fig. 47 Heat treated torque plate to which a three-piece torque-ring assembly was welded to eliminate vibrations

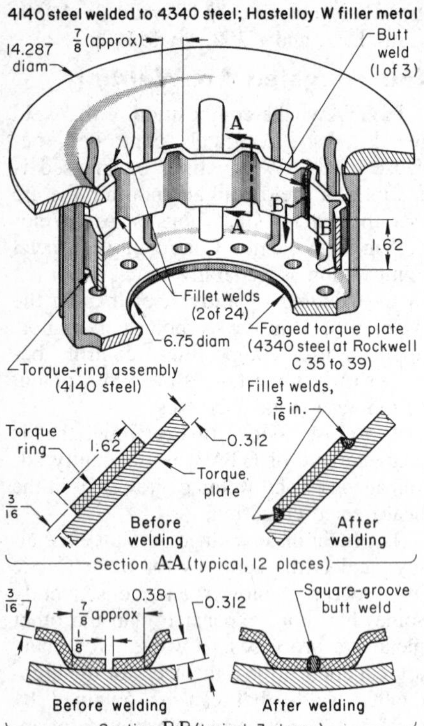

Conditions for GTAW

Joint types	Butt and lap
Weld types	Square-groove and fillet
Number of passes	One
Preheat	400 °F
Interpass temperature	350 to 500 °F
Postheat	Maintained at 400 to 500 °F until stress relieved at 980 to 1000 °F for 3 h; furnace cooled
Fixtures	Locating jig
Shielding gas	Argon
Filler metal	Hastelloy W (AMS 5786)

necessary to establish precise procedures for preheating, maintaining temperature during welding, and postheating.

Before welding, the parts were cleaned in an alkaline cleaning solution and rinsed. The ring segments were positioned and clamped in a locating jig and tack welded to the torque plate. The assembly was then furnace preheated to 400 °F in an argon atmosphere. From this time until the welded assembly was ready for stress relieving, the temperature was not permitted to drop below 350 °F, nor to rise above 500 °F. Temperature-sensitive crayons were used. When temperature dropped to 350 °F, the weldment was heated with torches; when temperature rose to 500 °F, welding was temporarily stopped.

Argon shielding and a Hastelloy W filler rod were used. Heat input was controlled solely by manipulating the arc; no remote current-control device was employed. To minimize residual shrinkage stress, the butt joints were welded last. Each weld was made in one pass, which was interrupted three or four times, depending on heat buildup. Welding conditions were the same for the fillet and butt welds.

After being welded, weldments were maintained at the welding temperature in an argon-atmosphere furnace until a furnace retort load was ready for stress relieving. Weldments were stress relieved at 980 to 1000 °F for 3 h and furnace cooled, after which the joints were sandblasted and inspected for cracks by visual and fluorescent liquid-penetrant examination. Following an alkaline cleaning cycle similar to preweld cleaning, the joints were shot peened, using S-330 shot, to produce a surface compressive stress (150% coverage; 0.010- to 0.014-in. intensity). On the first ten pieces only, radiographs were taken to ensure that the weldments were acceptable.

Example 21. Use of Preliminary Buttering to Permit Welding of 4340 Steel Without Preheating. Conventional GTAW could not be used for joining two shell sections of a naval torpedo, because of difficulties in preheating. The shell sections, as shown at left in Fig. 48, were machined from normalized 4340 steel forgings, which required preheating at 500 to 700 °F to prevent cracking after welding. However, the forward shell contained a shrink-fitted tank forged from aluminum alloy 5254, which could not be heated above 250 °F without loss of strength. This restriction was overcome by a two-stage procedure that permitted the final closure to be made by GTAW without the need for preheating. The procedure involved the use of a submerged arc buttering deposit, as described below and illustrated in Fig. 48.

Both shells were received as oversize pierced forgings, requiring considerable machining. During initial machining, a large circumferential groove was cut in each shell in the location that subsequently would become the joint area (see top middle view in Fig. 48). These grooves were filled by SAW, after a 700 °F preheat, using a low-carbon steel electrode. Immediately after welding, the shells were given a stress-relieving treatment at 1250 °F for 2 h, and then were heat treated to a hardness of 311 to 375 HB.

Machining was resumed to obtain the configurations shown in the bottom middle view in Fig. 48. Next, the aluminum tank was shrink-fitted to the forward shell. The two shells were then aligned in a fixture to form the weld groove shown at upper right in Fig. 48. Finally, the assembly was mounted on a rotating positioner, and joined by automatic GTAW with a low-carbon steel filler metal (Fig. 48, lower right). Because the buttering welds formed the base and sides of a continuous groove of low-carbon steel, the weld that joined the shells could be made without preheating and without the risk of cracking.

Both welding operations were carefully planned and executed, and the welds were thoroughly inspected. Submerged arc welding was done with a $^3/_{32}$-in.-diam electrode and electrode and flux combinations having the AWS classifications EM12K-F71 and EM12K-F72.

Up to eight passes were required for buttering by SAW. Interpass temperature was maintained within the range of 500 to 800 °F. The first two passes were made using the EM12K-F71 combination at 300 A, 27 to 28 V, with straight polarity. Passes 3 to 8 were made using the EM12K-F72 combination at 420 A, 28 to 29 V, with reverse polarity. It was important to avoid excessive alloying of the filler metal with the base metal, which could have caused cracking. The change in flux type (same filler metal) for passes 3 to 8 was made to improve impact properties by reducing dilution of the weld metal. The joint areas (Fig. 48, middle views) were machined from low-dilution weld metal. These areas were required to meet stringent radiographic inspection. Slag was carefully removed after each pass. Weld starts, crossovers, and any doubtful areas were ground to clean metal.

Gas tungsten arc welding of the girth joint was done in five passes, using an automatically programmed procedure. Because both joint and filler metal were of low-carbon steel, preheating was not needed, but in order to prevent overheating the aluminum tank, it was necessary to limit interpass temperature to 200 °F. Temperature at the joint was checked with temperature-indicating crayons. At the end of each pass, current automatically tapered off over a 10-s trailing time, and the weld was wire brushed. After welding, the assembly was finish machined and ground to a 32-μin. finish.

Plasma Arc Welding

Plasma arc welding can be used to join steel sections whose thicknesses range fom 0.004 up to 0.5 in. Plasma arc welding is

Fig. 48 Alloy steel forward and aft shells of a naval torpedo that were joined by GTAW without preheating

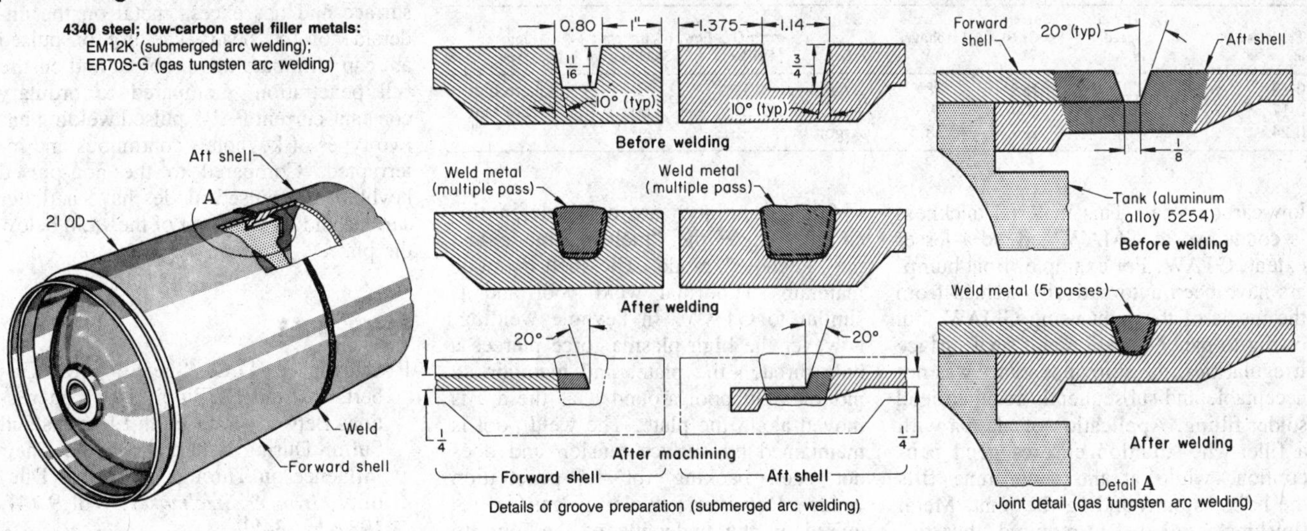

Automatic GTAW

Joint type	Butt, with integral backing
Weld type	V-groove
Welding position	Flat
Fixtures	Clamping jig, rotating positioner
Number of passes	Five
Preheat and postheat	None
Interpass temperature	200 °F max
Shielding gas	Argon, at 24 ft^3/h
Torch	Machine type with gas lens
Electrode	$^3/_{32}$-in.-diam EWTh-2
Filler metal	$^1/_{16}$-in.-diam ER70S-G
Welding speed, all passes	35 in./min
Power supply	300-A, constant-current transformer-rectifier(a)
Current (DCEP) and voltage:	
Pass 1	260 A, $11^7/_8$ V
Pass 2	285 A, 12 V
Pass 3	290 A, $12^1/_4$ V
Pass 4	295 A, $12^3/_8$ V
Pass 5	300 A, $12^1/_2$ V

(a) With slope control and programmer

an all-position process, but may be limited in some instances when filler metal is added. It often is used as an alternative to GTAW, over which it has inherent advantages. The main advantages are due to the narrow, constricted arc, which enhances good penetration and reduces the amount of distortion in the base plate (see the article "Plasma Arc Welding" in this Volume for more information).

Metals are welded using direct current electrode negative (DCEN), with a constant-current direct current power supply. The process often is classified according to the welding current:

- Microplasma welding: 0.1 to 15 A
- Intermediate current plasma welding: 15 to 100 A
- Keyhole plasma welding: Over 100 A

Microplasma welding can be used readily to weld thin steel sheet with thicknesses from 0.004 in. It also can be used for welding of wire and mesh sections. The constricted nature of the plasma column prevents arc wandering and minimizes distortion of thin material. The process is suitable for manual welding because of a wide tolerance of the torch-to-workpiece distance.

Joint Preparation. As with most welding processes, high-quality welds require proper preparation. The most important consideration in thin-gage welding is fit-up and mismatch, which should not be greater than 10% of the sheet thickness. For example, a gap of 0.0004 in. should not be exceeded when welding 0.004-in. sheet. Lap joints and fillet welds are more tolerant to fit-up than butt welds.

As sheet thickness decreases, the joint becomes less tolerant to surface contamination. Microplasma welding should be done in clean conditions, and all dust and other debris must be removed from the surface. To maintain a consistent arc, the tungsten should be machine ground and properly positioned in the nozzle.

Tolerance to Arc Length Variations. The wide conical arc used in GTAW requires that the electrode-to-workpiece distance be maintained to obtain consistent penetration and bead shape. The almost cylindrical plasma arc, however, reduces this problem to a minimum and enables high-quality welds to be made with greater ease. Insensitivity to arc length variations allows manual welding of thin sheet to be carried out by average operators rather than skilled personnel. The fatigue element is also reduced. Plasma welding in the out-of-position mode in areas of limited access can be used successfully, as arc length does not change the resultant bead.

Orifice and Shielding Gas. Orifice gas flow alone is not adequate to protect the weld pool. Shielding gas, usually the same as the orifice gas, provides the necessary protection of the weld during solidification and shortly after to prevent oxidation. Using the same gas prevents variations in the welding arc, which can occur when two different gases are used.

For most welding applications, argon plasma gas and argon shielding gas are used, but a further increase in arc intensity can be obtained by adding 2 to 8% hydrogen to the shielding gas. However, high-strength structural and HSLA steels are susceptible to hydrogen-induced cracking and should be welded with 100% argon shielding gas.

Applications. At intermediate currents, PAW has been used to weld sheet metal in the auto industry (Ref 7), which fabricates automobile bodies in 0.04-in.

Table 51 Typical operating conditions for keyhole PAW of high-strength structural and HSLA steels

Thickness, in.	Travel speed, in./min	Current (DCEN), A	Arc voltage, V	Gas	Gas flow, ft³/h Orifice	Gas flow, ft³/h Shielding	Joint type
0.13	12	185	28	Argon	13	60	Square butt
0.17	10	200	29	Argon	12	60	Square butt
0.24	...	120	28.5	Argon	...	...	...

low-carbon steel. This material thickness is conducive to GMAW and, to a lesser extent, GTAW. For example, front bumpers have been automatically welded from the inside of the joint using GMAW, but owing to panel variation and final surface irregularities, the skin appearance was not acceptable and subsequently required lead solder filling. Application of plasma with a filler-wire addition ensures good penetration, while at the same time fills the hollows surrounding the joint. Metal finishing is all that is required, thus bypassing solder fill requirement.

Production time is reduced in welding operations and in finishing operations using PAW. Lead solder filling is not required prior to painting, thereby reducing environmental hazards. Additional advantages of PAW include: (1) sufficient illumination from the pilot arc to enable the operator to position the torch and filler wire more readily; (2) instantaneous ignition of the main arc, thus eliminating starting problems; (3) lack of contamination of the tungsten electrode; and (4) increased tolerance to standoff variations. Four production jigs for manual PAW are now in use, and the production target aimed at is 2800 automobile bodies per week. With such high production rates, a further benefit from the use of PAW is the ease with which operators can be trained, as welding conditions are far less critical.

Plasma welding between 15 and 100 A is comparable to GTAW. Its advantages over GTAW include a higher depth-to-width ratio, immediate weld starting, and reduced electrode wear.

Keyhole Plasma Welding. Above 100 A, the plasma arc is very intense. Two welding techniques are available in this current range—the "melt-in" mode and the "keyhole" mode. The melt-in mode maintains a normal weld pool and is similar to GTAW. In keyhole welding, however the high plasma force pierces a hole through the plate, and maintains a molten weld pool around it as the arc is moved along the plate. The weld pool is maintained by surface tension and does not need backing for full-penetration welds. However, shielding gas is required on the underside of the joint to prevent atmospheric contamination.

Plates up to 1/4 in. thick can be butt welded from one side in a single pass, using only a square butt preparation. Full penetration can be obtained consistently with mechanized torch travel. The welding speed is usually about 12 in./min for keyhole welding of steels.

Keyhole welding is well suited to welding long lengths of plate or to seam welding of tubes. Run-on and run-off tabs are used whenever possible to ensure that the keyhole is established at the start of the weld for full penetration. Run-off tabs are used at the end of the weld so that an open keyhole does not remain. The keyhole does not close when the arc is extinguished abruptly. Where tabs cannot be used, current and orifice gas flow can be sloped down to close the keyhole. Because of the nature of starting and stopping the keyhole, this mode is used infrequently for circumferential welding of pipe. Some typical operating conditions for keyhole PAW of high-strength structural and HSLA steels are given in Table 51.

Pulsed Plasma Arc Welding. The keyhole formed in PAW sometimes becomes unstable. When this happens, the solidified pool appears concave on the surface and has excess metal on the underside of the bead. The use of pulsed arc can eliminate this problem and ensure full penetration. Compared to ordinary constant-current PAW, pulsed welding has two types of keyhole—continuous and interrupted. Compared to the non-pulsed keyhole, the pulsed modes have a flatter surface and less sagging of the weld below the plate.

REFERENCES

1. Ludwigson, D.C., Speich, G.R., Gilbert, S., and Defilippi, J.D., Interactions Between Rare-Earth Elements and Sulfur During Steelmaking and Their Influence on Through-Thickness Ductility, *Iron & Steelmaker,* Vol 9 (4), 1982, p 38-43
2. Porter, L.F. "Lamellar Tearing in Plate Steels," United States Steel Corporation, ADUSS 16-6919-02, Aug 1976
3. Sommella, J., "Significance and Control of Lamellar Tearing of Steel Plate in the Shipbuilding Industry," SSC-290, Ship Structure Committee, 1979
4. ASTM A770-80, "Specification for Through-Thickness Tension Testing of Steel Plates for Special Applications," American Society for Testing and Materials, Philadelphia, 1980
5. Kaufmann, E.J., Pense, A.W., and Stout, R.D., An Evaluation of Factors Significant to Lamellar Tearing, *Welding Journal,* March 1981, p 43s-49s
6. Nakamura, H., Naiki, T., and Okabayaski, H., Fracture in the Process of Stress-Relaxation Under Constant Strain, Proceedings of the First International Conference on Fracture, Vol 2, Sendai, Japan, 12-17 Sept 1965, p 863-878; and Stress-Relief Cracking in HAZ, IIW-IX-648-69, and IIW-X-531-69
7. *TIG and Plasma Welding,* The Welding Institute, Cambridge, 1978

Arc Welding of Cast Irons

By the ASM Committee on Arc Welding of Cast Irons*

CAST IRONS include a large family of ferrous alloys covering a wide range of chemical compositions and metallurgical microstructures. Some of these materials are readily welded, while others require great care to produce a sound weldment. Some cast irons are considered nonweldable. Consequently, welding procedures must be suited to the type of cast iron to be welded. Various processes for joining iron castings are described in this article, as well as the advantages and disadvantages of each. Iron castings are welded to:

- Repair defects to salvage or upgrade a casting before service
- Repair damaged or worn castings
- Fabricate castings into welded assemblies

Repair of defects in new iron castings represents the largest single application of welding cast irons. Defects such as porosity, sand inclusions, cold shuts, washouts, and shifts are commonly repaired. Fabrication errors, such as mismachining and misaligned holes, are also weld repaired. Restoration of corrosion resistance, structural integrity, or wear resistance may also be achieved by weld overlaying, repair, and/or hardfacing.

Classification of cast iron weldments by design requirements aids in the selection of welding processes and consumables. Three categories of design requirements are (1) highly stressed welds; (2) non-stress-bearing welds; (3) and weld overlay cladding for corrosion, abrasion, and wear resistance. Highly stressed welds should have mechanical properties that are compatible with the cast iron base metal. Non-stress-bearing welds are common where restoration of proper casting size and shape is the primary concern. Both of these categories of welds may also require good machinability and color match between the weld deposit and the iron casting. Weld overlay cladding for specific surface requirements necessitates careful selection of welding consumables and the deposition process.

Welding of simple castings to form assemblies is often more economical than casting a complex shape. Iron castings can be welded to other iron castings as well as to steel, nickel alloys, and many wrought shapes.

Welding Metallurgy of Cast Irons

Cast irons have carbon contents in excess of 2% and silicon in excess of $^1/_2$%. Carbon can be present in the form of (1) eutectic graphite flakes, (2) graphite nodules caused by modifications of eutectic graphite during solidification, (3) pearlitic iron carbide, (4) eutectic iron carbide, or (5) carbon retained in a solid phase such as martensite. Carbon in the form of iron carbide is in a metastable state and can transform to graphite if the kinetics are suitable. Depending on alloy content, melting practice, and thermal treatment, cast irons represent a wide range of alloys, including the following categories: white, malleable, gray, ductile, and compacted graphite irons. Mechanical properties and weldability are dependent on microstructure, which is directly related to partitioning of carbon during solidification and subsequent cooling. White iron is generally considered to be unweldable because of its extreme hardness and brittleness. Ductile iron is easier to weld than gray iron, partially due to the lower levels of sulfur and phosphorus in ductile iron. Ductile iron offers superior base-metal properties; however, weaknesses that are not critical in a gray iron weldment are unacceptable in a ductile iron weldment.

White iron is cast iron in which the majority of carbon occurs as the intermetallic compound Fe_3C (cementite), a very hard constituent. White iron usually consists of a pearlitic matrix with a network of eutectic carbide. Hypereutectic irons have the greatest hardness and are more brittle. Most white irons are of hypoeutectic compositions. The white iron

Fig. 1 Unalloyed chilled cast white iron consisting of coarse lamellar pearlite and ferrite in a matrix of M_3C carbides. Magnification: 100×

*David LeRoy Olson, *Chairman,* Professor of Metallurgical Engineering, Colorado School of Mines; Robert J. Dybas, Manager, Mechanical Processes, Apparatus & Engineering Services Division, General Electric Co.; Richard L.Hellner, Welding Engineer, Public Service of Colorado; Leroy W. Myers, Jr., Welding Engineer, Dresser Clark Division, Dresser Industries; Lewis E. Shoemaker, Metallurgist—Advanced, Huntington Alloys, Inc.; Thomas A. Siewert, Manager, Development, Alloy Rods

Fig. 2 ASTM A602 grade M3210, ferritic malleable iron. Two-stage annealed by holding 4 h at 1750 °F, cooling to 1300 °F in 6 h, air cooling. Type III graphite (temper carbon) nodules in a matrix of granular ferrite; small gray particles are MnS. Magnification: 100×

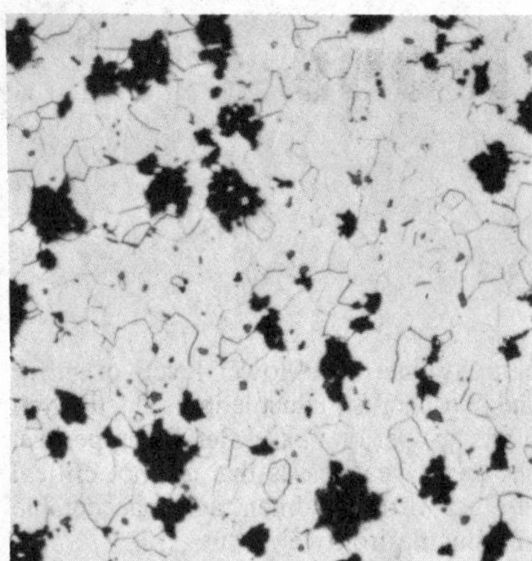

microstructure, as shown in Fig. 1, is a function of both alloy composition and cooling rate during solidification. Unalloyed white iron castings have a hardness of about 400 to 450 HB and must be ground to achieve required size and shape.

Malleable iron is heat treated white iron in which the iron carbides are dissociated when annealed at a temperature above 1600 °F for periods greater than 6 h. Irregularly shaped graphite nodules are precipitated and grow in the solid iron. Figure 2 shows a typical malleable iron microstructure.

Gray cast irons are alloys of iron, carbon, and silicon in which the separation of excess carbon from the liquid during solidification takes the form of graphite flakes, as shown in Fig. 3. Silicon and slow cooling rates promote graphitization in the iron. The low toughness and ductility of graphitic cast iron is neither a direct result of the brittleness of the graphite flakes nor a result of the large proportion of graphite. The continuity of the ductile matrix is interrupted by the shape and distribution of these graphite flakes, resulting in low toughness and ductility. Some castings have insufficient silicon content or cool too quickly to achieve a completely graphitic structure. This results in a white iron microstructure at the surface and a gray iron microstructure at the center; such castings are called mottled iron.

Ductile cast iron, also known as nodular iron or spheroidal graphite iron, has graphite present as small spheroids instead of flakes (Fig. 4). The spheroidal graphite structure is produced by ladle or mold additions (i.e., magnesium and/or rare earths such as cerium) to certain molten gray iron compositions. To achieve favorable mechanical properties, graphite in spheroidal form can be dispersed, causing minimal interruption of the continuity of the metallic matrix. Ductile cast irons combine the principal advantages of gray iron, such as excellent machinability, with the engineering advantages of steel, such as high strength, toughness, ductility, hot workability, and hardenability.

Fig. 3 Class 30 as-cast gray iron in a sand mold. Structure: type A graphite flakes in a matrix of 20% free ferrite (light constituent) and 80% pearlite (dark constituent). Magnification: 100×

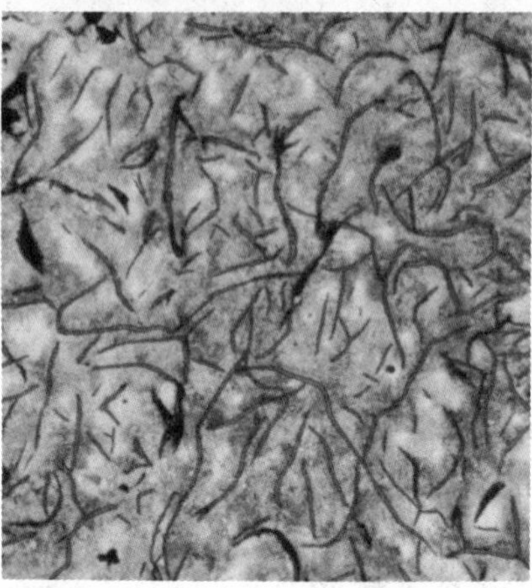

Compacted graphite or vermicular graphite irons have structures and properties that are between those of gray and ductile irons, as shown in Fig. 5. The graphite forms interconnected flakes as in gray iron, but the flakes are thicker and their edges are blunted. This graphite flake shape provides for higher strength levels than gray irons and greater machinability than ductile irons. Graphitic irons are classified according to their graphite shape (ASTM A247). The matrix and graphite shape must be defined to identify the specific type of cast iron. For example, flake graphite in a pearlitic matrix is termed pearlitic gray iron.

Fig. 4 Grade 80-55-06 as-cast ductile iron. Graphite nodules in envelopes of free ferrite in a matrix of pearlite. Magnification: 100×

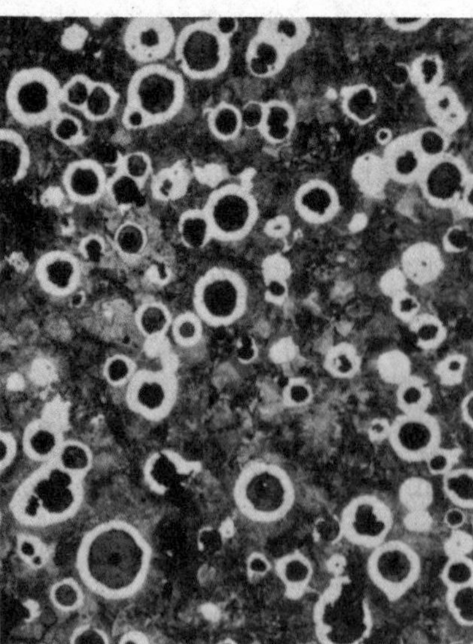

Chemical composition limits for some cast irons are given in Table 1. The influence of alloy additions and contaminants on cast iron microstructure and properties is important in the selection of the proper welding process, consumables, variables, and procedures. For additional information on cast iron materials and their properties, see Volume 1 of the 9th edition of *Metals Handbook*.

Microstructure

Cast irons have varied microstructures and physical properties, resulting in marked differences in weldability. Variations in thermal gradients across the weldment result in varying microstructures and properties. The various microstructures are classified into different zones and regions, as shown in Fig. 6. The nature and size of these zones in cast iron weldments are determined by the thermal weld cycle, composition of the base metal, and welding consumables. To develop welding procedures that minimize the deleterious effects of these zones, the influence of

Fig. 5 Compacted graphite cast iron microstructure. Type I graphite (spherulites) and type III graphite (vermicular) in matrix of ferrite and pearlite. Magnification: 100×

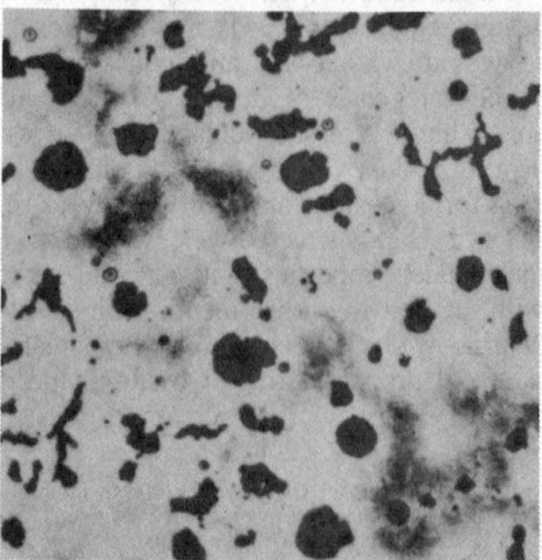

Table 1 Chemical composition limits for cast irons

Type	ASTM specification	C	Si	Mn	S	P	Other
		Composition limits, %					
White	...	1.8-3.6	0.5-1.9	0.25-0.8	0.06-0.20	0.06-0.18	...
Gray	A48, A159, A278, A319	2.5-4.0	1.0-3.0	0.25-1.00	0.02-0.25	0.05-1.0	...
Malleable	A47, A197, A220, A338, A602	2.0-2.6	1.1-1.6	0.20-1.0	0.04-0.18	0.18 max	...
Ductile	A536, A395, A476	3.2-4.0	1.8-2.8	0.10-0.8	0.03 max	0.10 max	0.030 Mg
Compacted graphite	...	3.5	2.0-3.0	0.20-0.40	...	0.05 max	0.015 Mg, 0-1.5 Cu
High chromium white	A532	2.3-3.6	0.8-1.0	0.5-1.5	0.06	0.10	1.4-28.0 Cr, 1.5-7.0 Ni, 0.5-3.5 Mo
Austenitic gray	A436	3.0	1.00-2.8	0.5-1.5	0.12 max	...	28-32 Ni, 1.5-6.0 Cr, 0.50-7.5 Cu
High silicon	A518	0.7-1.1	14.2-14.8	1.5	...	...	0.5 Cr, 0.5 Mo, 0.5 Cu
Austenitic ductile	A439, A571	2.4-3.0	1.0-6.00	0.70-2.5	...	0.08	18.0-36.0 Ni, 0.20-5.5 Cr

welding variables on mechanical properties must be considered.

Heat-Affected Zone

Welding of cast irons is characterized by rapid cooling as compared to cooling rates during casting. Consequently, properties of the weld and the sections of the casting exposed to elevated temperatures (the heat-affected zone, HAZ) differ from the remainder of the casting. Portions of the cast iron HAZ reach temperatures during welding which cause the carbon to diffuse into the austenite. On cooling, this austenite transforms into hard eutectoid decomposition products such as martensite. The amount of martensite formed depends on the cast iron composition and thermal treatment. Ferritic cast irons contain most of their carbon in the form of graphite, which dissolves slowly, thus producing less martensite. The highest percent of carbon in pearlitic cast irons is finely divided into the pearlitic structure. This carbon dissolves readily, producing a large amount of martensite. The brittle martensite may be tempered to a lower strength, more ductile structure through (1) preheating and interpass temperature control, (2) multiple-pass welding, or (3) postweld heat treatments such as stress-relief annealing. Hardness in the HAZ ranges from 250 to 650 HB, with most of the area below 450 HB. Some welding procedures are designed to reduce the size of the HAZ and thus minimize cracking. Methods of accomplishing this are (1) reduction of heat input, (2) use of small-diameter electrodes, (3) use of low-melting-point welding rods and wires, and (4) use of lower preheat temperatures. Typical HAZ widths are in the range of 0.03 to 0.10 in.

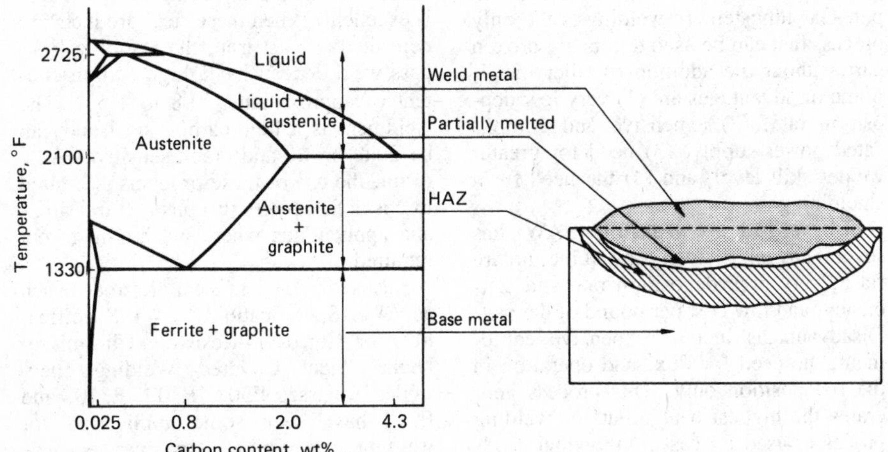

Fig. 6 Schematic representation of the zones in a typical cast iron welding

Partially Melted Region

The partially melted region of a weldment is an extension of the HAZ that occurs where a high peak temperature has caused partial melting of the base metal near the fusion line. The effects of this region on the mechanical properties of ductile iron welds must be taken into account to produce successful welds. The liquid in the partially melted region is similar to liquid eutectic cast iron, which freezes as white iron because of the high cooling rates during the weld heat cycle. The microstructure of the partially melted region is complex and consists of a mixture of martensite, austenite, primary carbide, and ledeburite which surrounds partially dissolved nodules or flakes of graphite. If the amount of dissolved graphite is great enough to form an almost continuous molten matrix, the carbide network is also continuous; less extensive melting produces a discontinuous carbide structure. The partially melted region, which contains a great proportion of hard products, is the hardest zone in the weld.

The high hardness and consequent low toughness of the partially melted region adjacent to the fusion line create mechanical problems in the welding of cast iron. The formation of hard structures results from the transport of carbon from graphite particles into the adjacent matrix. Reduction of peak temperatures and time at high temperature is the most effective method of reducing cracking. This can be done by controlling the heat input, preheat, and interpass temperatures.

Low-melting-point filler materials can also alleviate fusion-line cracking by reducing peak temperatures. If a weldment is sufficiently small, a large heat input may raise base-metal temperatures enough to cause severe fusion-line problems even though no preheat is used. Conversely, a high preheat temperature used to prevent martensite in the HAZ and to reduce the thermal expansion stresses may result in fusion-line cracking even with low heat input. Therefore, trial-and-error procedure development may be necessary. A welding procedure that successfully incorporates the partially melted region and HAZ can produce joints with good mechanical properties.

Fusion Zone

The fusion-zone microstructure and properties are influenced primarily by the selection of the welding consumables. The fusion-zone composition consists of melted welding electrode or wire with some dilution from the iron casting. The weld pool is mixed during the welding process to produce a relatively uniform composition within each weld bead. Dilution should be kept to a minimum. The fusion zone should be machinable, below 300 HB in hardness. If color match between the weld and the base metal is important, filler-metal compositions that match the cast iron must be used.

Welding Procedures and Processes

More than 90% of all industrial welding of cast iron is done by arc welding processes. Arc welding has a lower heat input than oxyfuel gas welding because of a faster welding speed and higher deposition rate. The welding operation can be automated to varying degrees, and distortion due to welding heat is more readily controlled. Arc welding achieves temperatures (approximately 5430 °F) in excess of those required for fusion of the base metal. The intensity of the heat source allows the necessary fusion while heating only a small portion of the weldment. This may cause high cooling rates and may result in large thermal expansion and contraction stresses. However, arc welding processes can produce welds of good quality with proper selection of the welding process, consumables, and procedures. Detailed descriptions are given in this Volume in the articles which deal with each of the various processes.

Shielded metal arc welding (SMAW), the most widely used and best known process, has the advantages of (1) large selection of consumables, (2) good availability, (3) low-cost power supplies, and (4) all-position welding. The disadvantages include a higher cost per pound of weld metal deposited and an associated low rate of deposition. These disadvantages are not important in a low-volume or maintenance application.

Gas metal arc welding (GMAW) has limited use, and its advantages include a high deposition rate and efficiency. Several disadvantages are (1) limited selection of consumables, (2) shielding gas, (3) higher power source cost, and (4) need for wire-feeding equipment. Gas metal arc welding can be operated in various metal transfer modes, including globular, pulsing, spray, and short circuiting. The spray transfer mode is characterized by a relatively high heat input, high deposition rate, and use in high-volume operations, but can only be used in the flat or horizontal position. The short circuiting transfer mode is characterized by lower heat input, moderate deposition rate, and the capability of operating in all positions.

Flux cored arc welding (FCAW), which is growing in usage, has advantages and disadvantages similar to GMAW in the spray transfer mode. The slag coverage provides advantages of (1) better protection during cooling, (2) improved wetting of the bead at the corners, (3) an out-of-position weld capability, and (4) chemical refinement of the weld deposit. In many applications, FCAW has higher deposition rates than GMAW, particularly for out-of-position welding. Disadvantages include the added step of removing the slag and a lower deposition efficiency.

Gas tungsten arc welding (GTAW) generally is used to repair smaller parts. Advantages include the capacity to operate at low power levels, all-position capability, and superior control of heat input. Gas tungsten arc welding is the only process that can be used to repair a broken part without the addition of filler metal. Some disadvantages are (1) very low deposition rate, (2) expensive and complicated power supply, (3) need for greater welder skill level, and (4) the need for a shielding gas.

Submerged arc welding (SAW) has several advantages: very low fume, no arc flash, very high deposition rate and efficiency, and low cost per pound of deposit. Disadvantages include expensive equipment, the need for flux, and operation in the flat position only. This process generates the highest heat input for welding processes used for cast irons, which leads to the greatest dilution and widest HAZ. However, when comparing the same size wire, it has a lower heat input than GMAW in the spray transfer mode. Submerged arc welding has limited application to iron castings and is usually used for weld overlay cladding of cast iron.

Consumables

Filler Metals for Shielded Metal Arc Welding of Cast Irons. A variety of covered electrodes are used for SMAW of iron castings. Economic considerations and weld requirements determine the appropriate product for each application. Electrodes designed specifically for welding iron castings are described in American Welding Society (AWS) Specification A5.15, "Welding Rods and Covered Electrodes for Welding Cast Iron."

Covered electrodes utilizing a cast iron rod as the core are classified by AWS as ECI. They are of comparative low cost and produce a weldment with chemical and mechanical properties similar to the base metal. Color match is excellent; however, ease of operation of these products is marginal, and appropriate procedures must be followed closely to ensure weld quality. If a cast iron electrode is used, graphite flakes or spheroids form in the weld during cooling. The ductility of a repaired malleable or ductile iron casting may be less than that of an unrepaired part.

Electrodes using steel as a core wire but depositing a cast iron type of deposit are similar in usage to the ECI electrodes. These products, classified as ESt, are generally of higher quality because they are produced by extrusion techniques. Welding procedures must be followed closely to avoid brittle or cracked weldments. Because of their ease of operation, widespread availability, and low cost, low-hydrogen types of covered electrodes, e.g., E7015, E7016, E7018, and E7028, are used for welding iron castings. Color match is excellent. When these steel products are deposited on cast iron, the resulting first-pass weld deposit has a high carbon content of approximately 0.8 to 1.5%. The weldment is a high-carbon steel and can be quite brittle and crack sensitive. As a result, the use of these products is limited to cosmetic repairs in nonstructural areas and applications where machining is not required.

Austenitic stainless steel electrodes such as AWS Specification A5.4 ("Specification for Corrosion-Resisting Chromium-Nickel Steel Covered Welding Electrodes") classes E308, E309, E310, and E312 have seen some application for welding iron castings. However, because of chromium carbide formation in the weld as well as tearing due to strength and coefficient of expansion differences, use of these products is marginal and extreme care should be exercised with their use.

Nickel-based electrodes have been widely accepted for welding iron castings. Unlike iron, nickel has a low solubility for carbon in the solid state. Thus, as the weld pool solidifies and cools, carbon is rejected from the solution and precipitates as graphite. This reaction increases the volume of the weld deposit, thereby offsetting shrinkage stresses and lessening the likelihood of fusion-zone cracking.

Electrodes classified as ENi-CI utilize a commercially pure nickel-core wire and thus produce a deposit of high nickel content. Even when highly diluted by the base metal, the deposit remains ductile and machinable. ENiFe-CI electrodes produce a nickel-iron deposit and, as a result, have four distinct advantages over the ENi-CI electrodes:

- The deposits are stronger and more ductile; this increased strength level makes the product suitable for welding the higher strength gray and ductile irons, as well as for many dissimilar metal joint applications.
- Nickel-iron deposits are more tolerant of phosphorus than nickel deposits; thus, ENiFe-CI electrodes are preferred for welding gray iron castings with higher phosphorus contents.
- The coefficient of expansion of the nickel-iron deposit is somewhat less than the nickel deposit; therefore, nickel-iron products may be used to weld heavier sections while still avoiding fusion-line cracking due to expansion differences.
- ENiFe-CI electrodes are generally lower in cost than ENi-CI electrodes.

Welding current ranges for SMAW of cast iron with nickel-iron and nickel electrodes are given in Table 2. For nickel-iron electrodes, the welding current should be reduced 5 to 15 A for overhead welding and 10 to 20 A for vertical welding from the values listed in Table 2.

Electrodes with a nickel-copper deposit classified as ENiCu-A and ENiCu-B have limited use for welding iron castings. Because of the sensitivity of their deposit to iron dilution, they have been replaced for the most part by the ENi-CI and ENiFe-CI products.

Copper-Based Electrodes. Bronze covered electrodes, which are of the copper-tin or copper-aluminum type, are used for welding iron castings. A low melting point allows their use at low current level, thus minimizing dilution and the effect of heat on the base metal. Although deposits are soft and machinable, color match is poor. Care must be exercised when using these electrodes because excessive dilution and contamination readily cause cracking.

Filler Metals for Gas Metal Arc, Gas Tungsten Arc, and Submerged Arc Welding of Cast Irons. In general, bare wires of similar chemical composition to the covered electrodes are used for cast iron welding. The same cautions hold for the solid wire as those listed for the covered electrodes.

Carbon steel filler metals are used for welding iron castings because they are economical and readily available. Because the diluted weld deposit is brittle, use of these products is limited to cosmetic, nonstress-bearing and nonmachined welds. AWS Specification A5.18 ("Mild Steel Electrodes for Gas Metal Arc Welding") class ER70S-6 materials are the most frequently used products. The high silicon content of the wire (0.80 to 1.15% Si) helps deoxidize the weld pool and improves weld-metal flow and wetting of the base metal.

Austenitic stainless steel filler metals are sometimes applied to the welding of iron castings. However, their use should be limited because of the same problems that occur with stainless steel electrodes. These products are classified in AWS Specification A5.9, "Corrosion-Resisting Chromium and Chromium-Nickel Steel Welding Rods and Bare Electrodes."

AWS Specification A5.14 ("Nickel and Nickel Alloy Bare Welding Rods and Electrodes") class ENi-1 nickel wire is commonly used for welding iron castings. As discussed previously, the deposits are relatively strong, machinable, and free of fissuring. Nickel-iron and nickel-iron-manganese filler metals are sold under proprietary labels. Because these products are not classified by specification, they should be judged on their own merits. Welding condition ranges for GMAW of cast irons with nickel-iron and nickel wire electrodes are given in Table 3.

Filler Metals for Flux Cored Arc Welding of Cast Irons. Flux cored filler-metal products that produce deposits with compositions paralleling covered electrodes are available. Because they combine the best features of coated electrodes and solid wire processes, they are increasing in popularity and variety. Welding condition ranges for welding of cast iron with nickel-iron flux cored electrodes are given in Table 3.

A proprietary nickel-iron-manganese flux cored filler metal is gaining widespread use for welding iron castings. This filler metal produces a deposit that is similar in composition to ENiFe-CI except for a higher level of manganese. It is extremely crack resistant and capable of high deposition rates. Applications are similar to those of ENiFe-CI.

Steel flux cored electrodes, specified in AWS Specification A5.20 ("Mild Steel Electrodes for Flux Cored Arc Welding") and A5.22 ("Flux Cored Corrosion-Resisting Chromium and Chromium-Nickel Steel Electrodes") are also used to weld iron castings. Use of these electrodes is subject to the same precautions as described for covered electrodes with similar composition.

Preparation for Welding

Proper preparation of the casting prior to welding is extremely important. When a defect is being repair welded, the defect must first be removed, usually by grinding, gouging, or machining. Attempting to weld over a defect, instead of removing it completely, usually results in poor weld quality. When the defect is a crack contained within the material, a 1/8-in.-diam hole should be drilled at each end of the crack prior to preparation to prevent crack propagation. Air carbon arc gouging is used successfully on iron castings for removing defects; however, at least 1/16 in. of additional material should be removed from

Table 2 Welding-current ranges for SMAW of cast iron with nickel-iron and nickel electrodes(a)

Electrode diameter, in.	ENiFe-CI electrodes(b) Current (dc)(d), A	ENiFe-CI electrodes(b) Current (ac), A	ENi-CI electrodes(c) Current (dc)(d), A	ENi-CI electrodes(c) Current (ac), A
3/32	40-70	40-70	40-80	40-80
1/3	70-100	70-100	80-110	70-110
5/32	100-140	110-140	110-140	110-150
3/16	120-180	130-180	120-160	120-170

(a) For welding in the flat position; current for overhead welding should be 5 to 15 A less than shown; for vertical welding, 10 to 20 A. (b) Percentage composition, based on analysis of deposit on standard test pad: 2.00 C, 4.00 Si, 1.00 Mn, 0.03 S, 45 to 60 Ni, 2.50 Cu, 1.00 other, remainder Fe. (c) Percentage composition, based on analysis of deposit on standard test pad: 2.00 C, 4.00 Si, 1.00 Mn, 0.03 S, 8.00 Fe, 85 min Ni, 2.50 Cu. (d) Either polarity

Table 3 Welding conditions for GMAW and FCAW of cast iron

Electrode	Wire size (diam), in.	Process	Voltage	Current
Spray transfer				
Nickel-iron wire	0.045	GMAW(a)	30-32	170-260 DCEP
	0.078	FCAW(b)	28-30	275-325
	0.093	FCAW(b)	28-32	300-375
Nickel wire	0.035	GMAW(a)	30-32	260-280
	0.045	GMAW(a)	30-32	260-280
	0.065	GMAW(a)	30-34	260-300
Short arc transfer				
Nickel-iron wire	0.035	GMAW(a)	15-19	100-120
Nickel wire	0.062	GMAW(a)	18-19	290-300 DCEP

(a) Pure argon shielding gas. (b) Self-shielding, but CO_2 shielding is optional

the arc-gouged surface by grinding to eliminate the resultant high hardness layer.

The second step consists of inspecting the area to ensure that the defect has been completely removed prior to welding. The liquid-penetrant inspection process is an excellent method of detecting small defects (cracks, pores, shrinkage) that are not readily visible to the naked eye. Low-phosphorus-containing penetrants should be used with nickel consumables.

When the defect is completely removed, the preparation of the joint should be completed. A minimum of 60° included angle generally is required for virtually all of the filler metals and processes listed in this article, and 70° is recommended for most of them. Ends of narrow grooves completely contained within a surface should be beveled to facilitate good fusion in those areas. The size of the weld joint must be kept to a minimum to reduce stress levels resulting from mismatch of thermal contraction rates between weld metal and base metal. The load-carrying ability of a cast iron weldment can be improved through proper joint design. Joints with incomplete penetration reduce load-carrying cross-sectional thicknesses and create potential stress concentrations beneath the weld bead. If possible, the weld joint should be located away from a highly stressed area, especially where there is a change in section thickness. If possible, welded joints should be loaded in compression, torsion, or shear, rather than in tension or bending.

Prior to welding, the casting should be thoroughly cleaned in the area of the weld. If the casting was exposed to dirt, grease, or paint in service, these contaminants must be completely removed from the weld joint, including the back or root side of any full-penetration welds. Repeated degreasing with a good solvent is necessary for castings exposed to grease and oil. Heating the casting to 700 to 900 °F for 15 to 30 min aids removal of oil and grease impregnated in the casting, followed by mechanical cleaning and additional degreasing with solvent. The surface (skin) of the casting also should be removed in the immediate area of the weld, because of contaminants (burned-in sand, dirt, paint) that cause poor-quality welds. All liquid penetrant and developer applied during the inspection operations should be completely removed before welding.

Prior to repair welding, all closed chambers within the casting should be vented before heating to prevent pressure buildup, and distortion and stresses caused by heating and cooling should be minimized. Drilling small holes into cored areas and forming internal passages for venting relieves excessive pressure buildup. Welding a groove from both sides minimizes distortion. If this is not possible, the part must be restrained during welding to minimize distortion. Peening or preheating can be used to prevent cracking of the weld during cooling.

Preheat and Interpass Temperature

In many cases, preheating is not needed to produce an acceptable weld, but can be used to reduce the thermal gradient and conductivity and to decrease the rate of cooling. Preheating is also useful in reducing differential mass effects, such as those encountered when welds are made between light and heavy sections. A thorough analysis and judicious use of preheating for each shape enable the welder to obtain a uniform cooling rate and thus reduce cooling stresses (Fig. 7). A fundamental preheating rule is illustrated in Fig. 7; the preheat is applied in a manner such that on cooling, the weld is under compressive stress. Preheating in the areas

Fig. 7 Preheating castings. Specific part should be heated in such a manner that the weld is under compressive stress on cooling.

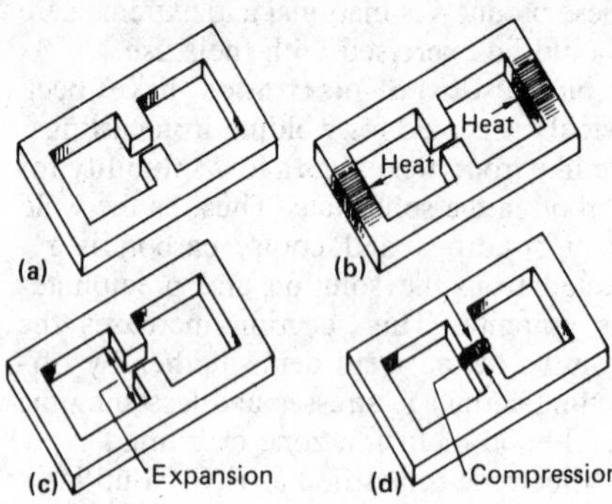

indicated in Fig. 7(b) expands the metal. Thus, the crack is expanded. After welding, the weld is in compression as the part cools and the metal shrinks (Fig. 7d).

The major concerns with use of excessive preheat temperature are the resulting amount of unnecessary base metal melted, a larger HAZ, and the formation of massive continuous carbides along the fusion line. Cast iron welding requires careful selection of a preheat temperature if the weldment is expected to have the same mechanical properties as the unaffected base metal. Advantages of preheating are most evident in welding of gray iron. Continuous carbides are not a major concern, as gray iron is already brittle because of its discontinuous flake graphite microstructure. The influence of preheating on the hardness of gray iron weldments made by SMAW using ENiFe-CI (nickel-iron) type of electrodes is listed in Table 4. Preheating of gray iron is effective in reducing excessive thermal contraction mismatch, thus avoiding fracture during cooling.

The actual preheating temperature range that is best suited to a specific application depends on the hardenability of the base metal, the size and complexity of the weldment, and the type of electrodes to be used. Table 5 illustrates typical preheat temperatures necessary to achieve a spe-

Table 4 Effect of preheat on hardness of shielded metal arc welds in class 20 gray iron welded with $^3/_{16}$-in. nickel-iron (ENiFe-CI) electrodes

Preheat temperature, °F	Hardness, HB: Weld	HAZ	Base metal
No preheat	342-362	426-480	165-169
225	297-362	404-426	165-169
450	305-340	362-404	169
600	185-228	255-322	169-176

Table 5 Influence of preheat on HAZ of cast irons

Preheat temperature, °F	Resulting microstructure
72	Martensite
212	Pearlite transformation occurs
392	A greater proportion of martensite and carbides are replaced by pearlite
572	Almost all martensite is replaced by pearlite
752	All martensite is prevented

cific type of microstructure in the HAZ of cast iron.

Although preheat temperatures for both gray and ductile iron are usually above the M_s temperature, to prevent martensite formation in the HAZ, temperatures for gray iron are typically higher than those for ductile iron. If preheat is required, a temperature of 700 °F typically is used for welding gray iron. Temperatures in the 1100 to 1200 °F range may be necessary when welding very heavy sections because of the heat sinks caused by the large mass of metal. Temperatures as high as 1600 °F may be used in some cases, as in the welding of high-alloy cast irons. The amount of area preheated and the time of preheating become more significant with increasing size of the part to be welded.

When welding ductile iron, choosing an optimum preheating temperature is more complex. Preheating is not always beneficial or necessary. Preheat temperatures as low as 300 to 350 °F are preferred when high thermal stresses are expected. At temperatures above 600 °F, continuous carbides often form along the fusion line. Preheat temperatures above 600 °F may be necessary, however, for production welding where higher deposition rates are required and for welding of heavy sections. Also, temperatures as high as 900 to 1300 °F may be used to offset the chill effect caused by decreasing carbon content in the liquid metal when producing a spheroidal-graphite cast iron fusion zone. In such cases, heat input must be reduced to minimize the formation of continuous carbides. The choice of a preheat temperature depends on several factors, and no single temperature is satisfactory for all base-metal shapes and compositions under all welding conditions.

The martensite formed in the weld deposit and the HAZ behaves in the same way as the martensite in a quenched steel. Martensite can be transformed into softer products by tempering. In multiple-pass weldments, some tempering of the previous layers occurs as each subsequent layer is deposited, causing a fluctuation in hardness measured over the height of the joint. Heat of welding in single-pass arc welding does not temper the martensite enough to provide reasonable machinability or reduce the danger of postweld cracking. Welds should be made with at least two passes to produce good machinability. The interpass temperature should exceed the preheat temperature by more than 100 °F to prevent additional martensite formation and to capitalize on the tempering effect of later deposits. The alternative to tempering is to decrease the amount of martensite by reducing heat input to produce a narrower HAZ.

Fig. 8 Use of short weld beads to avoid cumulative stresses. Source: Procedure Handbook of Arc Welding, The Lincoln Electric Co.

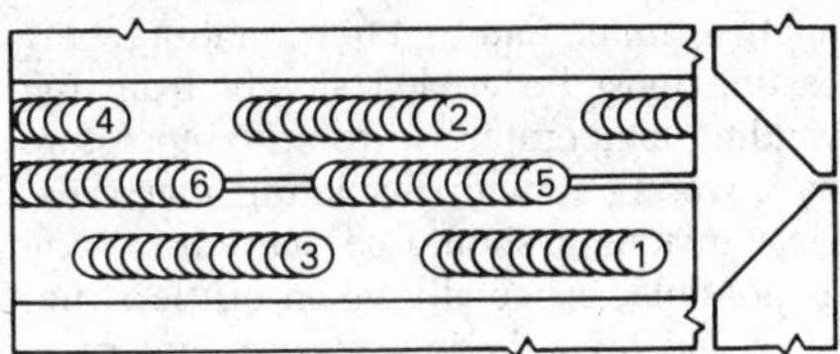

Welding Techniques

Heat Input. One method to control or minimize heat input is by the selection of the type of weld bead deposited. Of the two predominant types of weld beads, stringer and weave, the stringer bead is preferred because heat input is minimized. If weave beads are required, the weave should not exceed two to three times the electrode diameter.

Short Weld Sequence. Another method for controlling weld heat buildup is to deposit small weld beads in various portions of the weld joint and to allow each weld bead to cool to approximately 100 °F (Fig. 8). A sequence of short welds helps avoid the cumulative stresses that occur with long weld beads. Peening these short welds before they cool deforms or works the metal and reduces stress. This technique allows the welder to deposit a weld bead in another location while a previous weld bead cools.

The backstep technique reduces transverse and longitudinal weld shrinkage stress. It utilizes short weld beads which are deposited as shown in Fig. 9. The first weld bead is deposited with a short length of about 2 to 3 in. and 2 to 3 in. inward from the edge of the weld joint. Subsequent weld beads are deposited inward from the previously deposited weld bead. All weld beads should overlap and should be deposited in the same direction.

Fig. 9 Backstep sequence. Source: AWS A3.0-1978

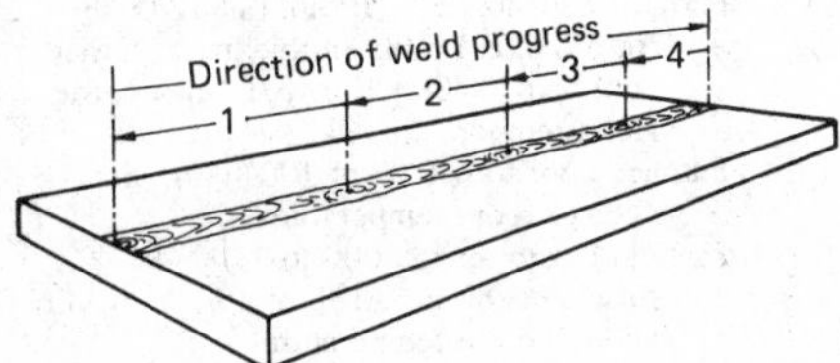

The block and cascade techniques can also be used to minimize shrinkage stresses in large weld grooves. The block sequence utilizes a longitudinal buildup of intermittent weld blocks joined by subsequent weld blocks (Fig. 10). It reduces longitudinal weld stresses, but does not minimize transverse weld stresses. Both longitudinal and transverse shrinkage stresses are reduced with the cascade sequence (Fig. 11). The cascade technique uses a combined longitudinal and overlapping weld-bead buildup sequence, as each succeeding bead tempers the one beneath or adjacent to it. The cumulative heating effect of subsequent passes provides additional tempering.

Fig. 10 Block sequence to minimize weld stress. Source: AWS A3.0-1978

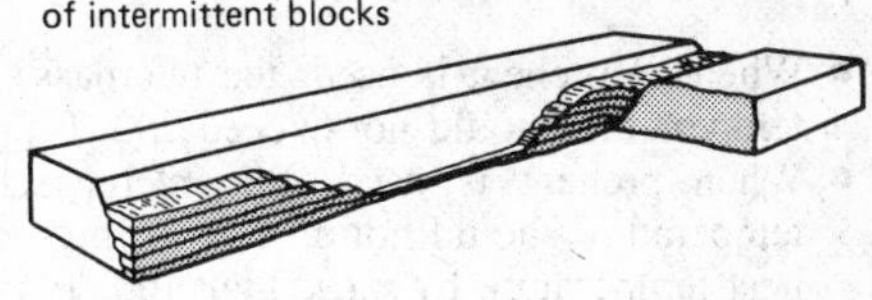

Peening. The principal effects of peening are relief of shrinkage stress, minimizing of distortion, and refinement of microstructure. Peening is best performed at temperatures that turn the metal a dull red. Peening performed below a dull red heat cold works the metal and increases susceptibility to cracking. Peening should be performed utilizing a $^1/_2$- or $^3/_4$-in.-diam round-nosed hammer. Repeated moderate blows should be used instead of a few heavy blows. After peening, all slag should be removed by wire brushing before starting subsequent passes.

Fig. 11 Cascade sequence of reducing weld stress. Source: AWS A.30-1978

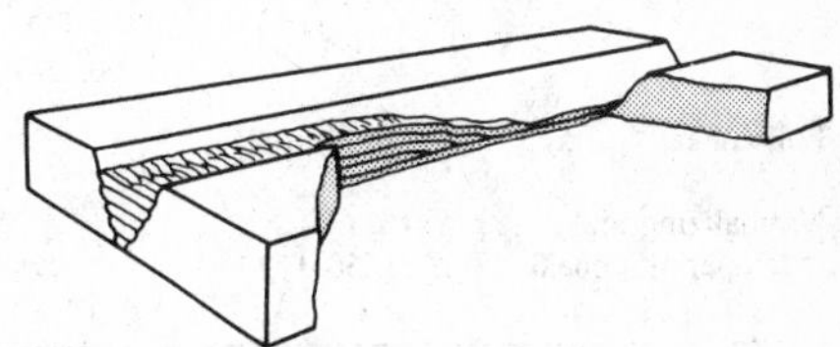

Welding Practices

The following general practices have been found to be useful in the welding process:

- When no preheat is used, the interpass temperature should not exceed 200 °F.
- When preheat is used, the interpass temperature should not exceed the preheat temperature by more than 100 °F.
- To minimize welding stresses, a back-step sequence should be used with stringer beads no more than 2 to 3 in. in length; allow each deposit to cool to approximately 100 °F before making subsequent deposits.
- Avoid melting more of the casting than is necessary.
- Whenever possible, deposit two or more layers for best machinability.
- Always strike the arc in the weld groove, never on the casting.
- The arc length should be kept as short as possible, typically 1/8 to 3/16 in.

Postweld Heat Treatment

Postweld heat treatment may be done for several reasons and can take many forms. The most common treatments are intended to eliminate the hard structures and thermal stresses formed during welding. Practical considerations may require the finished weldment to cool to room temperature before the postheat treatment can be started. When heat treatment is not applied immediately after welding, the casting must be cooled slowly from the welding temperature to room temperature by covering it with insulating materials. Heat treatment should be started as soon as possible, especially when brittle structures and high thermal stresses are present. In such cases, the chance of cracking increases as temperature decreases, and welds may fail catastrophically.

The simplest postweld heat treatments for gray, compacted graphite and ductile irons are stress relief and/or tempering, for which 1 or 2 h at 1100 to 1200 °F are usually sufficient. A heating and cooling rate of 100 °F/h above 600 °F is recommended. Holding times and postweld heat treatment temperatures are specified in Table 6. Furnace cooling rates specified in Table 6 to at least 600 °F are recommended before air cooling. This treatment also tempers the martensite and bainite transformation products and reduces hardness of the products by a considerable degree. When carbides are present, a full anneal may be necessary. Longer times may be needed for dissolving massive continuous carbides formed at high peak temperatures.

The weldment may be fully softened by a ferritize annealing process as specified in Table 6 for gray and ductile iron. This treatment results in a ferritic matrix and tends to dissociate iron carbide and transports the carbon to the graphite. This structure has maximum machinability and ductility. This practice is common for the 60-43-18 and 65-45-12 grades of ductile iron.

Other recommended postweld heat treatment annealing practices are listed in Table 6 for the various types of graphitic cast iron. In critical applications that require radiographic inspection after heat treatment, castings are inspected before heat treatment to save unnecessary costs if an internal defect is present. Additional heat treatment information on specific cast irons can be found in Volume 4 of the 9th edition of *Metals Handbook*.

Table 6 Recommended postweld heat treatment practice for various types of graphitic cast irons

Treatment	Temperature, °F	Holding time, h/in. of thickness	Cooling rate
Gray iron			
Stress relief	1100-1200	1.5	Furnace cool to 600 °F at 100 °F/h, air cool to room temperature
Ferritize anneal	1300-1400	1	Furnace cool to 600 °F at 100 °F/h, air cool to room temperature
Full anneal	1450-1650	1	Furnace cool to 600 °F at 100 °F/h, air cool to room temperature
Graphitizing anneal	1650-1750	1-3 h, plus 1 h/in.	Furnace cool to 600 °F at 100 °F/h, air cool to room temperature
Normalizing anneal	1600-1750	1-3 h, plus 1 h/in.	Air cool from annealing temperature to below 900 °F; may require stress relief
Ductile iron			
Stress relief:			
Unalloyed	950-1050	1.5	Furnace cool to 600 °F, at 100 °F/h, in air from 600 °F to room temperature
Low alloy	1050-1100		
High alloy	1000-1200		
Austenitic	1150-1250		
Ferritize anneal	1650-1750	1 h plus 1 h/in.	Furnace cool to 1275 °F; hold at 1275 °F for 5 h plus 1 h/in. of thickness, furnace cool to 650 °F at 100 °F/h; air cool to room temperature
Full anneal	1600-1650	1	Furnace cool to 650 °F at 100 °F/h, air cool to room temperature
Normalizing and tempering anneal	1650-1725	2 h (2 h min)	Fast cool with air to 1000 to 1200 °F, furnace cool to 650 °F at 100 °F/h, air cool to room temperature

Gray Irons

Because gray iron has graphite in flake form, carbon can be easily introduced into the weld pool, thus causing weld-metal embrittlement. Consequently, techniques that minimize base-metal dilution are recommended. Care must be taken to compensate for shrinkage stresses, and the use of low-strength filler metals helps reduce cracking without sacrificing overall joint strength.

Gray iron weldments are susceptible to formation of porosity. This can be controlled by lowering the amount of dilution of the weld metal, or by slowing the cooling rate so the gas has time to escape. Preheat helps reduce porosity and reduces the cracking tendency. A minimum preheat of 400 °F is recommended, but 600 °F is generally used.

The most common arc welding electrodes for gray iron are nickel and nickel-iron types (AWS A5.15 class ENi-CI and ENiFe-CI). These electrodes have been used with or without preheating and/or postweld heat treatment. The cast iron (ECI) and steel (ESt) types must be used with high preheats (1000 °F) to avoid cracking and hard deposits. The copper-aluminum (ECuAl-Al) and copper-tin (ECuSn) types are used, but color match is poor. Copper-based welds have comparable quality to a brazed joint, and a preheat temperature of at least 300 °F must be employed. The strength of these joints is often poor, partially due to the difference in thermal expansion between the copper alloy and the gray iron. Typical tensile properties for weldments of various grades of gray iron are given in Table 7 as a function of welding process and consumable.

Ductile Irons

Ductile cast irons have greater weldability than gray irons, but require specialized welding procedures and filler materials. Pearlitic ductile iron produces a larger amount of martensite in the HAZ than ferritic ductile iron and is generally more susceptible to cracking.

Shielded metal arc welding using an ENiFe-CI (nickel-iron) electrode is the most common welding technique for welding ductile cast irons. Shielded metal arc welding with an ENiFe-CI electrode sometimes employs a 300 to 400 °F preheat; preheats up to 600 °F are used on large castings. Most ductile iron castings, however, do not require preheating. Electrodes should be baked per manufacturer's recommendation to minimize hydrogen damage and porosity. Direct current electrode positive (DCEP) is usually used at a current sufficient to produce stringer beads with a moderate traveling speed. If machinability or optimum joint properties are desired, castings should be annealed immediately after welding. Transverse joint properties of typical ductile iron weldments are listed in Table 8.

Gas metal arc welding utilizing short arc transfer with nickel (AWS ERNi-1) filler metal and pure argon shielding has been successfully used for welding ductile cast irons. Recommended welding current is 130 to 160 A at 18 to 24 V. Based on the wire-feed rate, these controls produce a short arc transfer mode, which reduces heat input and reduces the amount of HAZ carbides and martensite, but requires a preheat of 400 °F on heavier (over 1/2 in. thick) wall castings. Ferritic ductile iron, however, can usually be welded without preheat. Typical transverse tensile strength properties obtained in 1-in.-thick pearlitic and ferritic ductile iron test plates using this process are listed in Table 9.

Gas tungsten arc welding produces reasonably strong weld joints utilizing either nickel-iron or ductile iron filler metals. The transverse joint properties for ductile iron weldments made with GTAW are given in Table 10.

Table 7 Approximate tensile strength and color match obtainable in welded joints in gray iron using different processes(a)

Process	Filler metal	Base-metal class	Tensile strength of joint(b), ksi	Color match
SMAW	ENi-CI	30	25-30	Fair
	ENiFe-CI	30	25-30	Fair
	ENiFe-CI	40	36	Fair
	ESt	30	30	Poor to fair
	ESt	40	40	Poor to fair
GMAW	Ni or Ni-Fe wire(c)	30	40	Fair
	Ni or Ni-Fe wire(c)	40	40	Fair
FCAW	Ni-Fe wire	30	25-30	Fair
		40	30	Fair
GTAW	Ni or Ni-Fe wire(c)	30	40	Fair
	Ni or Ni-Fe wire(c)	40	40	Fair

(a) Based on results reported from various tests and on production experience. (b) Approximate strength expected if good welding procedures and skilled operators are used. Wide variations in strength may occur as a result of variation in welding practice or in base-metal condition or size. (c) These filler metals are not classified, but wire coils and rods of compositions equivalent to ENi-CI and ENiFe-CI are obtainable.

Table 8 Transverse joint properties of typical SMAW ductile iron weldments using ENiFe electrodes

Plate condition	Tensile strength, ksi	Yield strength 0.2% offset, ksi	Elongation, %	Reduction in area, %	Weld hardness, HB	Maximum hardness in HAZ, HV	Bend angle, °	Unnotched Charpy impact strength, ft·lb
Ferritic (60-45-10)								
Unwelded	68.5-72.5	47.5-57	10-18	...	...	...	78-80	101
As-welded	51-68.5	40-57	1-9	11-18	205-250	665	14-28	14-20
Annealed(a)	54-72	37-57	5-14	5-28	185-235	420	18-38	28-36
Ferritize annealed	61-68	39-52	6-12	3-5	175-180	175	40-45	33-36
Pearlitic (80-60-03)								
Unwelded	98.5	86	3	nil	...	...	5-10	10-11
As-welded	66-82	71-76	1-2	3-4	190-225	535	14-15	13
Annealed(a)	63-71	68	1-2	2-6	175-205	555	...	11
Ferritize annealed	57-65	49-51	6-12	7-15	175-180	185	33-36	36-40

(a) 1100 to 1200 °F for various times
Source: Pease, G. R., The Welding of Ductile Iron, *Welding Journal*, Vol 39, No. 1, 1960, p 1s

Table 9 Transverse tensile properties(a) of gas metal arc butt welds on ductile iron plates
Short arc process used on 1-in.-thick plates

Filler metal	0.2% proof stress, ksi	Maximum stress, ksi	Elongation over weld in 2 in., %	Location of fracture
Pearlitic plate				
Unwelded	56-67.2	89.6-112	1-3	...
Nickel 61	52	75.7	3	Weld fusion zone
	52	83.8	4	Heat-affected zone
Ferritic plate				
Unwelded	36.6-44.8	56-78.4	15-25	...
Nickel 61	43.9	61.8	11.5	Away from weld
	44.4	60.5	11.5	Away from weld

(a) Test pieces 0.564 in. diam over 3.15 in. parallel position
Source: *Iron Castings Handbook*, Iron Casting Society, 1981

Malleable Irons

During welding, the ductility of the HAZ of malleable iron is severely reduced because graphite dissolves and precipitates as carbide. Although postweld annealing softens the hardened zone, minimal ductility is regained. Despite these limitations, for certain applications, malleable iron castings can be welded satisfactorily and economically if precautions are taken. Malleable iron castings should not be repaired by welding to correct a failure caused by overstressing of the part.

Because most malleable iron castings are small, preheating is seldom used. If desired, small welded parts can be stress relieved at temperatures up to 1000 °F. For heavy sections and highly restrained joints, preheating at temperatures up to 300 to 400

Table 10 Transverse joint properties of GTAW ductile iron weldments

Filler metal	Postweld heat treatment	Tensile strength, ksi	Yield strength 0.2% offset, ksi	Elongation, %
Ductile iron	None	55	...	8
	1100 °F/2 h; furnace cool	54	...	5
	1650 °F/4 h; furnace cool	58-61	45-56	4-6
	Full ferritize anneal	45-49	...	...
Nickel-iron (60-40)	None	62-67	...	1-7
	Full ferritize anneal	61	...	8

°F and postweld malleablizing annealing are recommended. However, this costly practice is not always followed, especially when the design of the assembly is based on reduced strength properties of the welded joint. Because no welding procedure can satisfy all types of service conditions, each application for welding malleable iron assemblies should be carefully reviewed and tested before production is begun.

Ferritic malleable grades 32510 and 35018 have the highest weldability of the malleable irons, even though impact strength is reduced. Pearlitic malleable irons, because of their higher combined carbon content, have lower impact strength and higher cracking susceptibility when welded. Small leaks, gas holes, and other small casting defects can be repaired in grades 32510 and 35018 by arc welding. If a repaired area must be machined, arc welding should be done with a nickel-based electrode. Shielded metal arc welding using low-carbon steel and low-hydrogen electrodes at low amperage produces satisfactory welds in malleable iron. If low-carbon steel electrodes are used, the part should be annealed to reduce any increased hardness in the weld (due to carbon pickup) and in the HAZ.

White and Alloy Cast Irons

Chilled and white cast irons are abrasion-resistant cast irons having structures free of graphitic carbon. Because of their extreme hardness and brittleness, they are generally considered unweldable. Alloy irons are used in applications requiring good abrasion, corrosion, or heat resistance properties. The most important consideration of welding alloy cast iron is to achieve equivalent service properties.

Because abrasion-resistant cast irons (such as ASTM A532) have limited resistance to thermal shock, welding is generally not recommended. Welding sometimes is employed in repair operations or to attach parts to other machine components. Gas welding is preferred over arc welding because of the lessened tendency toward cracking.

Arc welding may be done with a type 310 stainless steel electrode (25%Cr-20%Ni) if the welded area is not subject to abrasion. This electrode minimizes cracking in the parent metal. If the welded area is subject to abrasion, an electrode that produces weld metal of the same abrasion resistance as the base metal should be used. In either case, the casting should be preheated from 600 to 900 °F and slowly cooled after welding by covering with insulating material. Stress relieving at 400 °F should follow. Unless welding is done very carefully, the weld metal may exhibit very fine cracks that may cause failure if the casting is subjected to heavy loads in service. Under most operating conditions, however, the repaired casting should perform satisfactorily.

Corrosion- and Heat-Resistant Cast Irons

Corrosion-resistant cast irons are high-silicon, high-chromium, or high-nickel irons. Specifications for many of these irons permit welding for repair of minor casting defects. For more extensive welding, the effect of welding on the service properties of the casting should first be determined; applications of this group of cast irons are highly specialized. Because weld deposits usually need to duplicate base-metal compositions, filler metals may not be generally available.

Heat-resistant cast irons provide high strength at elevated temperatures, as well as resistance to scaling. They are normally produced as flake-graphite irons, but may also be produced with the free carbon in the form of nodules or spheroids. Heat-resistant cast irons may be welded for repair of minor casting defects, but much like corrosion-resistant irons, the effects of welding on service properties must be determined before extensive welding is undertaken.

Practical Applications

Example 1. Repair Welding of Cracks in Gray Iron. Shielded metal arc welding was selected for repairing two cracks in a 25-ton, 20-ft-diam gray iron casting (Fig. 12) because it was simpler, faster, and less expensive than gas welding. Gas welding would have entailed high preheat in a large furnace, difficult welding conditions, and care in maintaining high interpass temperature.

Both cracks were dressed out to form 60° V-grooves by chipping and grinding. Crack depth varied from 1 to 1³/₄ in.; length was about 30 in. Magnetic-particle and liquid-penetrant inspection methods were used to determine completeness of crack removal. The casting was then locally preheated, using gas torches; preheat temperature was checked by temperature-indicating crayons and a surface pyrometer. Stringer beads were deposited the full length of the cracks. After welding, the casting was allowed to cool slowly, with insulating material packed around the weld area. The casting was stress relieved at 1100 °F. Welds were machined.

Example 2. Submerged Arc Welding of Ferritic Malleable Iron to Forged Steel. The snap coupler shown in Fig. 13

Fig. 12 Repair welding of gray iron casting by SMAW

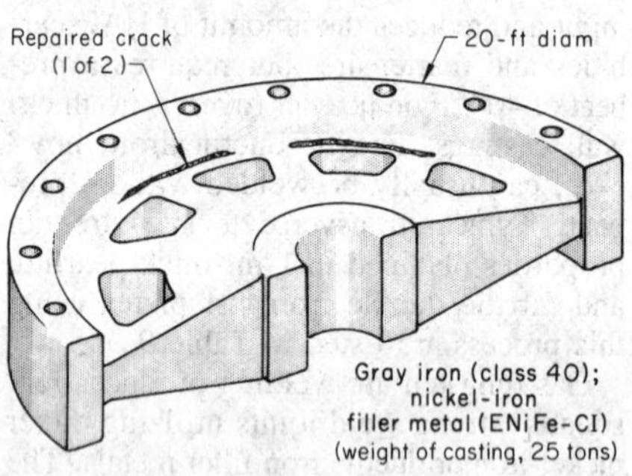

Welding conditions

Process	SMAW
Electrode	³/₁₆-in. ENiFe-CI(a)
Current	160 A (DCEP)
Voltage	24 V
Preheat (local)	400-450 °F
Postheat	Cooled slowly to room temperature, then stress relieved at 1100 °F(b)

Hardness after stress relief

Base metal	178-187 HB
Weld metal	169-176 HB
HAZ	210-241 HB

(a) Covered electrode with 55% Ni-45% Fe. (b) Weld area was packed in asbestos; furnace was not immediately available.

Fig. 13 Use of SAW to join malleable iron to forged steel

Ferritic malleable iron welded to low-carbon steel; mild steel filler metal (EL12)

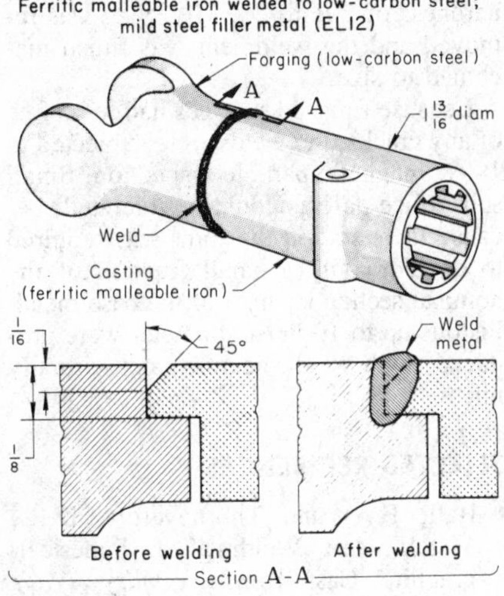

Conditions for SAW

Filler metal	3/32-in.-diam EL12
Flux	F62
Current	275 A (DCEP)
Voltage	24-25 V
Preheat and postheat	None
Production rate, pieces	120/h

was used on various types of farm machinery as a power take-off. Because this part was impractical to make in one piece, it was produced by welding a ferritic malleable iron casting to the stub end of a forged steel universal joint. Ferritic malleable iron was used because it was less costly than steel and offered adequate strength. A 1/8-in. weld was sufficient to carry the service load.

Both mating surfaces were machined to form a stepped joint that provided good alignment and proper joint depth. The assembly was fitted over a spline that rotated on a horizontal axis. The welding head was positioned near 12 o'clock, and SAW was done in a single pass. These snap couplers had a service-failure record of less than 0.1%.

Example 3. Use of FCAW To Repair Weld Gray Cast Iron Molds. Gray cast iron molds are used in the casting of both ferrous and nonferrous metals, as well as nonmetallics such as glass. Tears, cracks, and run-outs may cause defects that necessitate cropping the end off the mold or in many cases scrapping the unit. Repair welding can be accomplished with a nickel-iron flux core filler material by the FCAW process. Repairing the molds increases life an average of 35% in addition to reducing the number of new back-up molds normally in inventory. The following conditions were used for mold repairs with NiFe electrode:

- *Filler metal*: Nickel-iron flux cored filler material similar to AWS ENiFe-CI
- *Filler metal size*: 0.093 in. diam
- *Current*: 350 to 400 A, DCEP
- *Voltage*: 30 to 32 V
- *Shielding gas*: None
- *Weld preparation*: Removal of extraneous materials only; no grinding or gouging repaired in most cases
- *Manipulation*: Manual
- *Preheat/postheat*: None

Example 4. Fatigue Strength of Gas Metal Arc Welded Joints Between Ferritic Malleable Iron and Steel Tubing. Figure 14 shows a weldment on which fatigue tests were made to obtain data for use in production of welded malleable iron assemblies. The test piece, which consisted of two ferritic malleable iron eyes welded to a length of steel tubing, closely simulated an eye-bar or link commonly used to carry axial tension-compression loads.

Fig. 14 Preproduction fatigue test specimen

Ferritic malleable iron (grade 32510) welded to low-carbon steel; 1% silicon mild steel filler metal

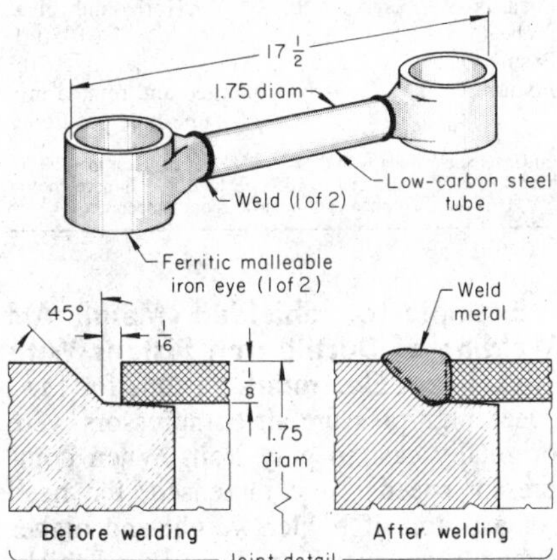

Conditions for GMAW

Electrode wire	0.035-in.-diam E70S-6 type(a)
Wire-feed rate	298 in./min
Shielding gas	Carbon dioxide, at 15 ft³/h
Current	140 A (DCEP)
Voltage	23 V
Welding speed	13 in./min

Fatigue-test results

Max load, ksi	Cycles to failure(b)
25	22 993(c)
25	66 352(c)
25	36 098(c)
25	46 892
20	72 403
20	139 394
20	244 347

(a) Composition: 0.11% C, 1.65% Mn, 1.12% Si, 0.021% P, and 0.024% S. (b) Each cycle consisted of one axial load in tension and compression: 120 cycles/min. (c) Failure occurred in eye of casting, not in weld.

Before welding, the parts were cleaned with a solvent and assembled in a fixture rotating on a horizontal axis. As shown in Fig. 14, the weld groove was formed by the square end of the tubing and the 45° bevel machined on the plug end of the eye. The tapered plug end of the eye was inserted in the tube and served as backing for the weld. Seven test pieces were gas metal arc welded using an E70S-6 type electrode. The arc was held between 1 and 2 o'clock, and rotation was counterclockwise at 2.4 rpm.

An eighth specimen was welded under the same conditions except that electrode composition was 0.12% C, 1.90% Mn, 0.80% Si, 0.020% P, 0.020% S, 0.50% Mo, and 0.10% Ni. This specimen failed after 8692 cycles at 18 ksi tension and compression. Although the single test could not be conclusive, it suggested sensitivity to different steel electrode compositions.

Example 5. Use of Gas Metal Arc Welding of Ductile Iron. Sections of ductile iron pipe, to one end of which ductile iron slip-on flanges had been fillet welded (see Fig. 15), which were flared at the opposite end, were coupled by means of a lock joint. The slip-on flanges were gas metal arc welded to the pipe sections in the manufacturer's plant. Each assembly was annealed after welding. The flanged sections were then assembled in the field and locked together by bolting. When necessary, some flanges were welded in the field, apparently with satisfactory results; but, since postweld annealing in the field was impractical, the heat-affected zones had lower corrosion resistance and toughness.

The shop welding procedure was automatic. First, the joint surfaces were cleaned with detergent and ground to remove the casting skin. Next, the flange and pipe were mounted together on turning rolls under a stationary GMAW head. To reduce the possibility of cracking in the ductile iron flange during cooling after welding, the flange (but not the pipe) was preheated to 600 °F. Both the flange and the pipe were in the annealed condition.

After preheating, the flange was tack welded to the pipe in three places, using a nickel-iron electrode under helium shielding. The electrode holder was then positioned at 45° from the vertical face of the flange, with the electrode aimed at a point about 1/16 in. from the flange, to ensure an even weld on both pipe and flange surfaces. Welding was done in a single

Fig. 15 Use of automatic GMAW

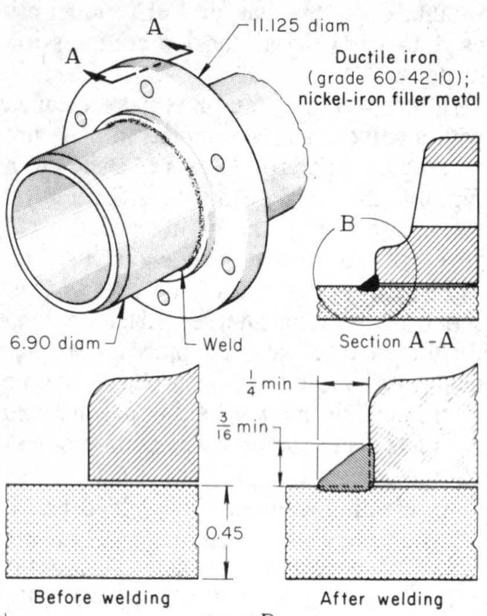

Process	GMAW
Joint type	Lap
Weld type	Fillet
Preheat (flange only)	600 °F
Electrode wire	0.045-in.-diam nickel-iron alloy(a)
Wire feed	432 in./min
Electrode holder	Machine type, water cooled
Shielding gas	Helium, at 30 ft³/h (or argon, at 20 ft³/h)
Welding position	Horizontal rolled
Current	200 A (DCEP)
Voltage	26 V
Number of passes	One
Power supply	200-A rectifier
Welding speed	22 in./min (1 rpm)
Postheat	Ferritizing anneal(b)

(a) Coiled wire, nominally 55% Ni-45% Fe. (b) Heat to 1700 °F in 15 min, hold at 1700 °F for 15 min, cool to 1300 °F in 15 min, air cool

pass, using the same electrode and shielding gas as for tack welding. During welding, gas cup and contact tip were positioned as close to the work as possible, to increase arc efficiency and shielding coverage. After welding, the entire assembly was annealed. The procedure produced joint efficiencies approaching 100%.

Tests made on two-pass welded butt joints in the same pipe gave the following results:

Tensile properties(a)

Tensile strength, ksi	69.1
Yield strength, ksi	48.700
Elongation in 1 in., %	19

Impact properties(b)

	At 70 °F	At −40 °F
Center of weld, ft·lb	14.9	13.3
Fusion zone, ft·lb	17.8	15.8

(a) Specimens measured 0.250 in. in diameter by 1 in. long. Tensile results are average values for ten specimens per weld. (b) Standard V-notch Charpy impact specimens

Fig. 16 Ductile iron piston produced as a two-piece weldment

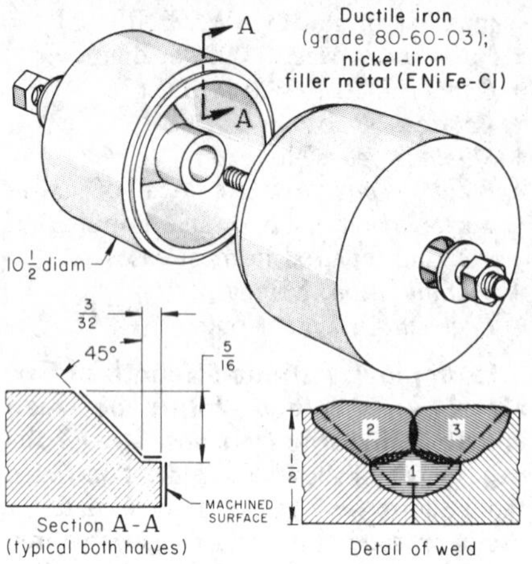

Process	SMAW
Electrode	ENiFe-CI(a), 1/8 in. for root pass; 5/32 in. for remaining passes
Number of passes	Three
Current	100 A (DCEN)
Power supply	300 A (dc)
Welding time	1st pass, 5 min; 2nd and 3rd passes, 6 min each
Position	Horizontal rolled
Preheat	1050 °F
Postheat	(b)
Fixturing	Arbor-mounted and rotated in a lathe-type positioner

(a) Covered electrode with 55% Ni-45% Fe. (b) Heat to 1700 °F, hold 4 h, cool in air; heat to 1050 °F, hold 5 h, furnace cool to 550 °F at 100 °F/h; then cool in air to room temperature

Example 6. Shielded Metal Arc Welding of Ductile Iron Pistons With Nickel-Iron Electrodes. Pistons for low-speed high-pressure air compressors were originally cast in gray iron. When compressor speeds were increased, the need for a piston of lighter weight and higher strength was met by reducing wall thicknesses and casting the pistons in ductile iron. Changing to the two-piece design shown in Fig. 16 simplified casting procedures and permitted further weight reduction. Although aluminum pistons of this design were joined by epoxy bonding, reliable bonding of ductile iron was difficult to obtain, probably because of graphite on the joint surfaces. Shielded metal arc welding with nickel-iron electrodes was finally selected, together with the following procedure.

Mating surfaces and weld bevels (Fig. 16) were machined without cutting fluid, and the piston halves were bolted together on an arbor and furnace heated to 1050 °F. The preheated assembly was placed in an operator-controlled variable-speed positioning lathe, with the arbor horizontal. The partial-penetration groove weld was made in three passes (Fig. 16) and the assembly was then heat treated in a furnace to the desired strength range, corresponding to a hardness of 201 to 269 HB. Slag was removed and the weldment was finish machined to size.

Because finished surfaces had to be free of any cracks, the welds were subjected to 100% magnetic-particle inspection. Small subsurface defects could be tolerated, because in operation the joint was required to develop only a small fraction of the nominal section strength of the base metal. Pistons up to 16 in. in diameter were produced, with weld rejection rates virtually nil.

SELECTED REFERENCES

- Ball, F.A. and Thorneycroft, D.R., Metallic-Arc Welding of Spherical-graphite Cast Iron, *Foundry Trade Journal,* Vol 97, Oct, 1954, p 499
- Bates, R.C. and Morley, F.J., Welding Nodular Iron Without Postweld Annealing, *Welding Journal,* Vol 40 (No. 9), 1961, p 417s
- Bishel, R.A., Flux-cored Electrode for Cast Iron Welding, *Welding Journal,* Vol 52 (No. 6), 1973, p 372
- Bishel, R.A. and Conway, H.R., Fluxed Cored Arc Welding for High-Quality Joints in Ductile Iron, *Mod. Cast,* Vol 67 (No. 1), 1977, p 59-60
- Cookson, C., The Metal Arc Welding of Cast Iron for Maintenance and Repair Welding, *Metal Construction and British Welding Journal,* Vol 3 (No. 5), 1971, p 179
- Cookson, C., Metal-Arc Welding of White Cast Iron, *Metal Construction and British Welding Journal,* Vol 5 (No. 10), 1973, p 370
- Davila, A.M., Olson, D.L., and Freese, T.A., Submerged-arc Welding of Ductile Iron, *AFS Transaction,* Vol 85 (No. 77), 1977
- Davila, A.M. and Olson, D.L., Welding Consumable Research for Ductile Iron, *Modern Casting,* Vol 70 (No. 11), 1980, p 70
- Devletian, J.H., Weldability of Gray Iron Using Fluxless Gray Iron Electrodes for SMAW, *Welding Journal,* Vol 57, 1978, p 183s
- Dixon, R.H.T. and Thorneycroft, D.R., Filler Rod for the Gas Welding of S.g. Iron, *Foundry Trade Journal,* Vol 108, May, 1960, p 583
- Forberg, S.A., Short-Arc Welding of S.g. Iron in the SKF Katrineholm Works, *Sweden Foundry Trade Journal,* Vol 124 (No. 2685), 1968, p 833

- Hogaboom, A.G., Welding of Gray Iron, *Welding Journal,* Vol 56 (No. 2), 1977, p 17
- Hucke, E.E. and Udin, H., Welding Metallurgy on Nodular Cast Iron, *Welding Journal,* Vol 32 (No. 8), 1953, p 378s
- Kiser, S.D., Production Welding of Cast Irons, *Trans AFS,* Vol 85 (No. 37), 1977
- Klimek, J. and Morrison, A.V., Gray Cast Iron Welding, *Welding Journal,* Vol 56 (No. 3), 1977, p 29
- Kotecki, D.J., Braton, N.R., and Loper, C.R., Jr., Preheat Effects on Gas Metal-Arc Welded Ductile Cast Iron, *Welding Journal,* Vol 48 (No. 4), 1969, p 161s
- Kumar, R.L., Welding Gray Iron with Mild Steel Electrodes, *Foundry,* Vol 96 (No. 1), 1968, p 64
- Mohler, R., Repairing Cast Iron by Welding, *Plant Engineering,* Vol 31 (No. 23), 1977, p 171-174
- Nippes, E.F., Savage, W.F., and Owczarski, W.A., The Heat-affected Zone of Arc Welded Ductile Iron, *Welding Journal,* Vol 39 (No. 11), 1960, p 465s
- Pease, G.R., The Welding of Ductile Iron, *Welding Journal,* Vol 39 (No. 1), 1960, p 1s
- Riley, R.V. and Dodd, J., Ferrous Rod for Welding Nodular Graphite Cast Iron, *Foundry Trade Journal,* Vol 93 (No. 1887), 1952, p 555
- Walton, C.F. and Ojar, T.J., Ed., *Iron Castings Handbook,* Iron Casting Society, Cleveland, 1981, p 599-665

Arc Welding of Stainless Steels

By the ASM Committee on Arc Welding of Stainless Steels*

STAINLESS STEELS are defined in Volume 3 of the 9th edition of *Metals Handbook* as iron-based alloys that contain at least 10% Cr. Even though iron-based alloys with as little as 4% Cr are sometimes referred to as stainless steels, this article deals only with alloys containing at least 10% Cr. These alloys generally contain elements in addition to chromium, such as nickel, molybdenum, copper, and manganese, which are added to improve a desirable mechanical property. Generally, these elements do not adversely affect weldability. Other elements, such as carbon, silicon, niobium, and titanium, can enhance properties in carefully controlled amounts or can be detrimental if their presence is uncontrolled. The third group of alloying elements, including sulfur, phosphorus, and selenium, are considered to be detrimental to the weldability of stainless steel even though they may improve some other characteristic of the alloy, such as machinability. Most stainless steels that do not contain amounts above 0.03% sulfur, phosphorus, and selenium are considered to be weldable. All weldable stainless steels can be joined by the various arc welding processes. However, variations in composition and physical and mechanical properties affect their relative weldability, as do fabricating conditions and service requirements.

Shielded metal arc (SMAW), submerged arc (SAW), gas metal arc (GMAW), gas tungsten arc (GTAW), and plasma arc welding (PAW) are used extensively for joining stainless steels. Flux cored arc welding (FCAW) is also used, but to a lesser extent. This article discusses the weldability of the various grades and the suitability of arc welding processes for specific conditions and requirements.

Austenitic Stainless Steels

Differences in composition among the standard austenitic stainless steels affect weldability and performance in service. For example, types 302, 304, and 304L differ primarily in carbon content, and consequently, there is a difference in the amount of carbide precipitation that can occur in the heat-affected zone (HAZ) after the heating and cooling cycle encountered in welding. Types 303 and 303Se contain 0.20% P maximum plus 0.15% Se or S for free machining. These elements are detrimental to weldability and can cause severe hot cracking in the weld metal. Types 316 and 317 contain molybdenum for increased corrosion resistance and higher creep strength at elevated temperatures. However, unless controlled by extra-low carbon content, as in type 316L, carbide precipitation occurs in the HAZ during welding. Types 318, 321, 347, and 348 are stabilized with titanium, or niobium-plus-tantalum, to prevent intergranular precipitation of chromium carbides when the steels are heated to a temperature in the sensitizing range, as during welding.

Welding Characteristics. The austenitic stainless steels, except for the free-machining grades, are the easiest to weld and produce welded joints that are characterized by a high degree of toughness, even in the as-welded condition. Serviceable joints can readily be produced if the composition and the physical and mechanical properties are well suited to the welding process and condition. Heat of welding, contamination, carbide precipitation, cracking, and porosity must be considered before, during, and after welding stainless steels.

Heat of Welding. Excessive heat input may result in weld cracking, loss of corrosion resistance, warping, and undesirable changes in mechanical properties. Welds in stainless steels generally require 20 to 30% less heat input than welds in carbon grades because stainless steels have lower thermal conductivity and higher electrical resistance. Because of low thermal conductivity, heat remains near the weld, so that more heat is available to melt the material, which may produce detrimental results. Too much extra heat produces large thermal gradients across the joint, which can cause distortion. Because heat dissipates slowly in stainless steel, it may lower corrosion resistance and change strength. These effects can be minimized with chill bars and less heat input. Weld-metal cracking, another problem resulting from excessive welding heat, is discussed later in this article.

The high electrical resistivity of stainless steel makes it suitable for welding with low heat inputs. With reduced heat, good penetration and fusion result because low thermal conductivity retains heat in the weld area. Comparative electrical resistivities are 25 to 50 $\mu\Omega \cdot$ in. for carbon steel and 175 to 200 $\mu\Omega \cdot$ in. for austenitic grades. Heat input can be reduced by using low amperages, low voltages (short arc lengths), high travel speeds, and stringer beads. With GMAW and GTAW processes, heat input can be affected by the type of shielding gas. Argon produces a cool, stable arc, while helium produces a hot arc that is somewhat unstable. For manual processes, pure argon is generally

*Robert S. Brown, *Chairman*, Senior Metallurgist, Stainless Alloy Metallurgy, Carpenter Technology Corp.; Daniel A. DeAntonio, Associate Metallurgist, Corrosion and Welding Research, Carpenter Technology Corp.; Michael J. Shields, Associate Metallurgist, Corrosion and Welding Research, Carpenter Technology Corp.; Thomas A. Siewert, Manager, Development, Alloy Rods

best. When working with automatic welding equipment that offers good control of amperage, voltage, and travel speed, mixed gases can be used without risking damage from high heat. Finally, pulsed arc welding techniques can be used to lower heat input.

Weld Contamination. Contaminants not only hinder successful welding, but may also prevent an apparently sound weld from functioning satisfactorily. A contaminated weld has inferior corrosion resistance and strength, and the weldment may fail prematurely. The stainless steel itself may also contain the contaminant. Free-machining stainless steels frequently contain sulfur or selenium. Both elements can make the steel unweldable. Similarly, high concentrations of carbon in high-strength stainless steel can inhibit weld serviceability.

External sources of contamination include carbon, nitrogen, oxygen, iron, and water. Carbon is often picked up from shop dirt, grease, forming lubricants, paint, marking materials, and tools; consequently, steel parts should be cleaned before welding and during welding. Otherwise, carbon contamination can cause welds to crack, change the mechanical properties, and lower the corrosion resistance in weld areas. Although iron contamination generally does not affect weldability, it can lower serviceability. Flakes of iron on surfaces rust, thus speeding localized corrosion. The welder may unknowingly cause the contamination by grinding stainless steel with a wheel previously used on carbon steel. Clean Al_2O_3 grinding wheels, preferably those not used for grinding other alloys, should be used.

One of the most troublesome types of contamination is stainless steel surface contamination by copper, bronze, lead, or zinc from hammers, hold-down fingers on seamers, or tools used in fabrication. Small amounts of these materials on the surface of the stainless steel can lead to cracking in the high-temperature HAZ of the weld. This type of cracking generally occurs in the HAZ, where the contaminant attacks the grain boundaries.

Effect of Carbide Precipitation on Corrosion Resistance of Welded Joints. The precipitation of intergranular chromium carbides is accelerated by an increase in temperature within the sensitizing range and by an increase in time at temperature. When intergranular chromium carbides are precipitated at welded joints, resistance to intergranular corrosion and stress corrosion markedly decreases. The decrease in corrosion resistance is attributed to the presence of the chromium-rich carbides at the grain boundaries and the depletion of chromium in the adjacent matrix material. Although intergranular carbide precipitation generally occurs between 800 to 1600 °F, sensitization is restricted to a narrower range by the fairly rapid heating and cooling that usually occur in welding. The narrower range varies with time at temperature and steel composition, but is approximately 1200 to 1600 °F.

The base metal immediately adjacent to the weld is annealed or solution treated by the heat of welding, and because it generally is cooled rapidly enough to hold the dissolved carbides in solution, this zone usually exhibits normal resistance to corrosive attack. A short distance from the weld, about $^1/_8$ in. (the distance depending on the thermal cycle and material thickness), there is a narrow zone in which lower heating and cooling rates prevail. In this HAZ, intergranular precipitation of chromium carbides is most likely to take place. Harmful carbide precipitation can be overcome or prevented by the use of:

- Postweld solution annealing
- Extra-low-carbon (0.03% C maximum) alloy
- Stabilized alloy containing preferential carbide-forming elements, such as niobium-plus-tantalum or titanium

Solution annealing puts carbides back into solution and restores normal corrosion resistance, but is generally inconvenient. The solution annealing temperature range is very high, 1900 °F minimum, and unless stainless steel is protected from air at these temperatures, it oxidizes rapidly, forming adherent oxide scale. Thin sections, unless adequately supported, may sag or be severely distorted at these temperatures or during rapid cooling from them. Rapid cooling in solution annealing may present other problems. Water quenching, although effective, is seldom feasible except for small workpieces of simple shape. Unless adequate safeguards are available, water quenching of large workpieces from the solution annealing temperature is hazardous. Often solution annealing is impractical because the workpiece is too large for available furnace and cooling facilities.

Extra-low-carbon stainless steels (types 304L and 316L) are resistant to carbide precipitation in the 800 to 1600 °F range and can thus undergo normal welding without reduction in corrosion resistance. Carbides precipitate in significant quantities when extra-low-carbon steels are heated and held in the sensitizing temperature range for an extended period, as in service. These steels are generally recommended for use below 800 °F.

Stabilized Steels. Compared with the extra-low-carbon steels, the stabilized steels exhibit higher strength at elevated temperature. For service in a corrosive environment in the sensitizing temperature range of 800 to 1600 °F, an austenitic steel stabilized with niobium-plus-tantalum or titanium is needed. The filler metal used for welding should also be of a stabilized composition. Because an inert shielding gas is used, GTAW and GMAW are suitable for titanium-stabilized steel without oxidizing the titanium. Under certain conditions, stabilized stainless steel weldments are susceptible to sensitization, which occurs in narrow zones of the base metal immediately adjacent to the line of weld fusion. During welding, stabilized carbides are dissolved and, as a result of rapid cooling, are retained in solution. Subsequent reheating to about 1200 °F results in preferential precipitation of chromium carbides in a narrow zone that exhibits less than normal corrosion resistance.

Microfissuring in Welded Joints. Interdendritic cracking in the weld area that occurs before the weld cools to room temperature is known as hot cracking or microfissuring. The occurrence of microfissuring is related to the:

- Microstructure of the weld metal as solidified
- Composition of the weld metal, especially the content of certain residual or trace elements
- Amount of stress developed in the weld as it cools
- Ductility of the weld metal at high temperatures
- Presence of notches

Susceptibility to microfissuring is highly dependent on the microstructure of the weld metal. Weld metal with a wholly austenitic microstructure is considerably more susceptible to microfissuring than weld metal with a duplex structure of delta ferrite in austenite. The content of alloying elements and residual elements strongly influences the susceptibility of fully austenitic stainless steel weld metal to microfissuring. Susceptibility can be reduced by a small increase in carbon or nitrogen content or by a substantial increase in manganese content. Residual or trace elements that contribute to microfissuring are boron, phosphorus, sulfur, selenium, silicon, niobium, and tantalum.

The amount of stress imposed on austenitic stainless weld metal as it cools from the solidus down to about 1800 °F should be minimized. In this temperature range,

Fig. 1 Schaeffler diagram. See text for explanation. (Source: Anton L. Schaeffler)

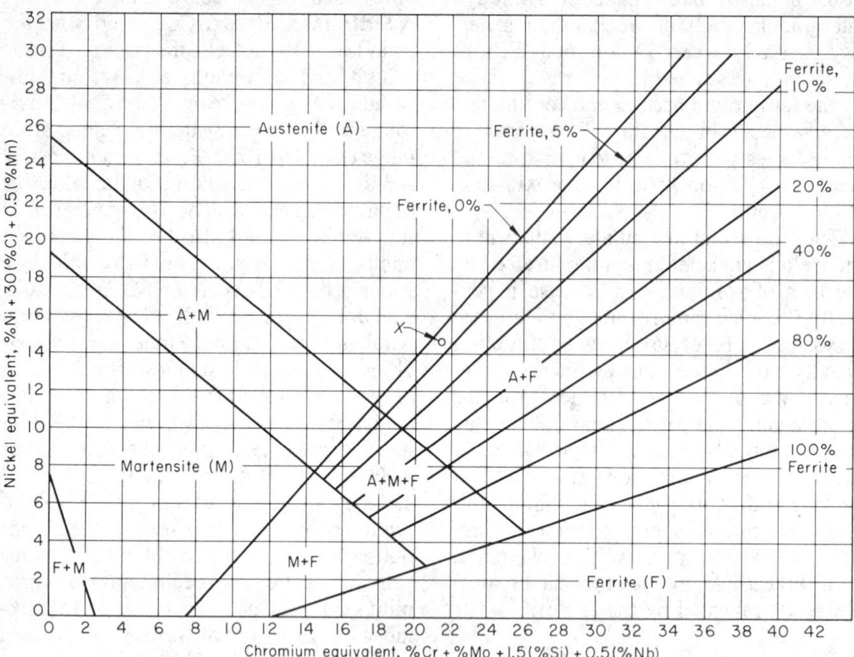

the weld metal is most susceptible to microfissuring, and if the level of stress is high, the fissures propagate to form visible cracks. Peening is not an effective method of preventing this type of cracking, because it can seldom be applied early enough to reduce stress buildup.

Prevention of Microfissuring. To obtain duplex-structured weld metal that has a controlled ferrite content of at least 3 to 5 ferrite number (FN), a filler metal of suitable composition is selected. The ferrite number is a magnetically determined scale of ferrite measurement. The Welding Research Council Advisory Subcommittee on Welding Stainless Steels has determined that the ferrite number of a weld metal, at least from 0 to 6 FN, approximates the average value of "percent ferrite" assigned by laboratories applying metallographic measurements of ferrite to a given weld metal.

Microfissuring can be prevented or minimized by proper control of ferrite in the weld metal. Wide use has been made of the Schaeffler diagram (Fig. 1) to determine the approximate amount of ferrite that will be obtained in the austenitic weld metal of a given composition. Point *X* indicates the equivalent composition of a type 318 (316Cb) weld deposit containing 0.07%C-1.55%Mn-0.57%Si-18.02%Cr-11.87%Ni-2.16%Mo-0.80%Nb. To determine the chromium and nickel equivalents, each percentage was multiplied by the potency factor indicated for the respective element along the axes of the diagram. When these were plotted as point *X*, the constitution of the weld was indicated as austenite plus 0 to 5% ferrite. Magnetic analysis of an actual sample revealed an average 2 FN. For austenite-plus-ferrite structures, the diagram predicts ferrite within 4% for stainless steel types 308, 309, 309Cb, 310, 312, 316, 317, and 318. Actual measurements of ferrite content can be conveniently made with the aid of a magnetic analysis device. American Welding Society (AWS) A4.2-74, "Standard Procedures for Calibrating Magnetic Instruments to Measure the Delta Ferrite Content of Austenitic Stainless Steel Weld Metal," and A5.4-81, "Specification for Covered Corrosion-Resisting Chromium and Chromium-Nickel Steel Welding Electrodes," discuss this measurement.

Because many heats of austenitic stainless steel contain appreciable amounts of nitrogen (a very strong austenitizer), a revised constitution diagram for austenitic stainless steel weld metal has been developed to include nitrogen in the nickel equivalent (Fig. 2). Compared with Fig. 1, Fig. 2 is modified in shape and slope to improve the accuracy of ferrite estimation for types 309, 309Cb, 316, 316L, 317, 317L, and 318. In addition, ferrite calculation for types 308 and 347 weld metal is improved on samples with either high or low nitrogen content. For use in the diagram, actual nitrogen content is preferred. If it is not available, the following nitrogen value shall be used: 0.12% for self-shielding flux cored electrode GMAW welds, 0.08% for other GMAW welds, and 0.06% for welds of other processes.

When the weld metal must be wholly austenitic (i.e., when the metal must be nonmagnetic or when specific corrosive environments which selectively attack delta ferrite will be encountered), the content of crack-promoting residual elements must be stringently controlled, and the composition of the weld metal must be adjusted to increase crack resistance. Crack resistance can be increased by modifying the carbon, manganese, sulfur, phosphorus, silicon, and nitrogen contents of the weld metal. However, even with optimum composi-

Fig. 2 Revised constitution diagram. Source: ASME code, Section III

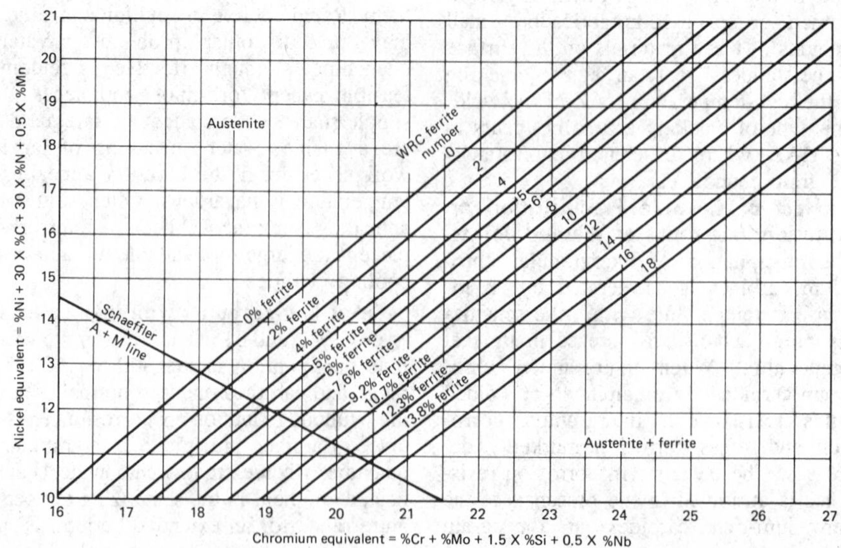

tions and the most favorable welding procedures, wholly austenitic weld deposits are more crack sensitive than those of a duplex structure.

Underbead cracking can occur in the HAZ of the austenitic stainless steel base metal immediately adjacent to the weld metal, especially in sections more than 3/4 in. thick. Serious failures due to underbead cracking have occurred in heavy-walled pressure vessels used in nuclear power units. The vessels were made of type 347 stainless steel, which is stabilized with niobium-plus-tantalum. Similar failures have occurred in thick-walled type 347 pipe used in high-temperature central steam stations. These failures occurred after seemingly sound welded joints had been subjected to high-temperature service for an extended period of time. The underbead cracking that caused the failures was attributed to strain-induced precipitation of niobium carbide, arising from shrinkage stresses in the heavy weldments.

Underbead cracking is not limited to niobium-stabilized steels. Tests have shown that all of the common austenitic stainless steels, with the possible exception of type 316L, are susceptible to this type of defect to some extent. Therefore, if underbead cracking is to be avoided, weld restraint must be kept to a minimum, especially during welding of heavy sections.

Selection of Filler Metals. Electrodes and welding rods suitable for use as filler metal in the welding of austenitic stainless steels are shown in Table 1. These filler metals, with AWS standard composition specifications, are for GMAW, SAW, and SMAW. The notes in this table should be carefully studied because selection of filler metals for welding austenitic stainless steels requires consideration of the microstructural constituents of the as-deposited weld metal. Ultimately, these microstructural constituents determine the mechanical properties, crack sensitivity, and corrosion resistance of the weld. The constituents of principal concern are austenite, delta ferrite, and precipitated carbides.

Some filler metals, such as types 310, 310Cb, 310Mo, and 330, invariably deposit a fully austenitic weld metal. In these alloys, the ratio of ferrite formers to austenite formers cannot, within permissible limits, be raised high enough to produce any delta ferrite in the austenite. Consequently, when these filler metals are applied to restrained joints or to base metals containing additions of elements such as phosphorus, sulfur, selenium, and silicon, only procedures proved suitable by experience should be used.

The compositions of most filler metals are adjusted by the manufacturers to produce weld deposits that have ferrite-containing microstructures. Thus, ferrite-forming elements, such as chromium and molybdenum, are maintained at the high side of their allowable ranges, and austenite-forming elements, such as nickel, are kept low. The amount of ferrite in the structure of the weld metal depends on the ratio or balance of these elements. At least 3 or 4 FN delta ferrite is needed in the as-deposited weld metal for effective suppression of hot cracking. With the proper techniques, however, types 316 and 316L can be welded with as little as 0.5 FN. Ferrite-containing weld metal may have certain disadvantages in a welded austenitic stainless steel. Ferrite is ferromagnetic, and the increased magnetic permeability of the weld metal may be objectionable in applications that require nonmagnetic properties. When exposed to service at elevated temperature, the ferrite in some weld metals may transform to sigma phase and adversely affect mechanical properties and corrosion resistance, a problem that has been encountered in power plant applications.

When a joint is arc welded without the addition of filler metal, the structure of the weld metal is determined by the composition of the base metal. Sometimes this leads to unfavorable results, because wrought base metals may not have the composition limits required for good weld metal.

Preheating. In general, no benefit is derived from preheating austenitic stainless steels. In some applications, preheating can increase carbide precipitation, cause shape distortion of the workpiece, or increase hot-cracking tendencies.

Postweld Stress Relieving. Although the effects of residual stress from welding on the properties of austenitic stainless steels are limited in comparison to the effects of cold working, residual stress may significantly affect the mechanical properties. Because the effective yield strength varies from point to point, the application of further stresses at later stages of fabrication can cause excessive distortion and even premature failure. Nonuniform heating, which relieves some local residual stress, may also contribute to distortion.

Table 1 Electrodes or welding rods as filler metals for use in arc welding of stainless steels

Type of steel welded	Condition of weldment for service(a)	Electrode or welding rod(b)	Type of steel welded	Condition of weldment for service(a)	Electrode or welding rod(b)
Austenitic steels			**Martensitic steels**		
301, 302, 304, 305, 308(c)	1 or 2	308	403, 410, 416,.. 416Se(k)	2 or 3	410
302B(d)	1	309	403, 410(m)	1	308, 309, 310
304L	1 or 4	347, 308L	416, 416Se(m)	1	308, 309, 312
303, 303Se(e)	1 or 2	312	420(n)	2 or 3	420
309, 309S	1	309	431(n)	2 or 3	410
310, 310S	1	310	431(p)	1	308, 309, 310
316(f)	1 or 2	316			
316L(f)	1 or 4	318, 316L	**Ferritic steels**		
317(f)	1 or 2	317			
317L(f)	1 or 4	317Cb	405(q)	2	405Cb, 430
318, 316Cb(f)	1 or 5	318	405, 430(m)	1	308, 309, 310
321(g)	1 or 5	347	430F, 430FSe(m)	1	308, 309, 312
347(h)	1 or 5	347	430, 430F, 430FSe(r)	2	430
348(j)	1 or 5	347	446	2	446
			446(s)	1	308, 309, 310

(a) 1, as welded; 2, annealed; 3, hardened and stress relieved; 4, stress relieved; 5, stabilized and stress relieved. (b) Prefix E or ER omitted. (c) Type 308 weld metal is also referred to as 18-8 and 19-9 composition. Actual weld analysis requirements are: 0.08% C max, 19.0% Cr min, and 9.0% Ni min. (d) Type 310 (1.50% Si max) may be used as filler metal, but the pickup of silicon from the base metal may result in weld hot cracking. (e) Free-machining base metal will increase the probability of hot cracking of the weld metal. Type 312 filler metal provides weld deposits that contain a large amount of ferrite to prevent hot cracking. (f) Welds made with types 316, 316L, 317, and 317Cb electrodes or welding rods may occasionally display poor corrosion resistance in the as-welded condition. In such cases, corrosion resistance of the weld metal may be restored by the following heat treatments: for types 316 and 317 base metal, full anneal at 1950 to 2050 °F; for types 316L and 317L base metal, 1600 °F stress relief; for type 318 base metal, 1600 to 1650 °F stabilizing treatment. Where postweld heat treatment is not possible, other filler metals may be specially selected to meet the requirements of the application for corrosion resistance. (g) Type 321 covered electrodes are not regularly manufactured, because titanium is not readily recovered during deposition. (h) Caution is needed in welding thick sections, because of cracking problems in heat-affected zones. (j) In base metal and weld metal, for nuclear service, tantalum is restricted to 0.10% max and cobalt to 0.20% max. (k) Annealing softens and imparts ductility to heat-affected zones and weld. Weld metal responds to heat treatment in a manner similar to the base metal. (m) These austenitic weld metals are soft and ductile in as-welded condition, but the heat-affected zone will have limited ductility. (n) Requires careful preheating and postweld heat treatment to avoid cracking. (p) Requires careful preheating. Service in as-welded condition requires consideration of hardened heat-affected zones. (q) Annealing increases ductility of heat-affected zones and weld metal. Type 405 weld metal contains niobium, rather than aluminum, to reduce hardening. (r) Annealing is employed to increase ductility of the welded joint. (s) Type 308 filler metal will not display scaling resistance equal to that of the base metal. Consideration must be given to differences in the coefficients of thermal expansion of the base metal and the weld metal.
Source: George E. Linnert, Welding Characteristics of Stainless Steels, *Metals Engineering Quarterly*, Nov 1967

For these reasons, stress relieving may be required to ensure dimensional stability.

Stress relieving can be performed over a wide range of temperatures, depending on the amount of relaxation required. Time at temperature ranges from about 1 h per inch of section thickness at temperatures above 1200 °F to 4 h per inch of section thickness at temperatures below 1200 °F. Because of the high coefficient of expansion and the low thermal conductivity of austenitic stainless steels, cooling from the stress-relieving temperature must be slow. The stress-relieving temperature selected must be compatible with the extent of carbide precipitation acceptable and with the corrosion resistance desired. Nonstabilized stainless steels cannot be stress relieved in the sensitizing temperature range without sacrifice of corrosion resistance. Extra-low-carbon stainless steels are affected much less, because carbide precipitation in these steels is sluggish. Stabilized stainless steels exhibit minimal chromium carbide precipitation tendencies.

For austenitic stainless steels, the estimated percentages of residual stress relieved at various temperatures, for the times previously noted, are:

1550 to 1650 °F	85% stress relief
1000 to 1200 °F	35% stress relief

Porosity can be caused by shop dirt, grease, marking materials, and scale, all of which must be removed before welding. Another cause of porosity is poor gas shielding, which allows nitrogen, oxygen, and moisture to enter the weld pool. It is important that both base metal and filler metal are dry and that proper shielding techniques are used. Excessive gas flow rates must also be avoided. Excessive gas can cause turbulence around the arc, pulling air into the molten metal.

Shielded Metal Arc Welding

The SMAW process (stick electrode manual welding) is widely used for welding stainless steel. For a description of the process, see the article "Shielded Metal Arc Welding" in this Volume. The advantages of this process include: (1) large selection of consumables, (2) readily available and low-cost power supplies, (3) little specialized training required, (4) capability of making repairs in all positions, and (5) no shielding gas requirements.

The disadvantages of SMAW are: (1) the slag blanket constitutes a potential source of inclusions, (2) visibility during welding is impaired by slag, (3) slag removal between passes is necessary, (4) the electrodes may be sensitive to moisture pickup, and (5) the low deposition efficiency means a higher cost per pound of deposited weld metal. This process is often chosen for field erection, repairs, less common materials, and small production runs.

Constant-current (drooping-voltage) power sources, conventionally used for SMAW, are applicable for welding stainless steel. Either alternating current or direct current electrode positive (DCEP) is suitable in most applications. Direct current power supply can be of the motor-generator or transformer-rectifier type.

Electrodes. Table 2 lists austenitic stainless steel electrodes used for SMAW, together with their composition. A suffix number, −15 or −16, relates to the polarity to be used. The suffix −15 indicates

Table 2 Chemical compositions for all-weld-metal deposits of austenitic stainless steel electrodes for use in SMAW

All weight percentages are maximum, unless otherwise noted; total of other elements, except iron, not to exceed 0.50%

AWS classification(a)	C(b)	Cr	Ni	Mo	Nb + Ta	Mn	Si	P	S	Cu
E307	0.04-0.14	18.0-21.5	9.0-10.7	0.5-1.5	...	3.3-4.75	0.90	0.04	0.03	0.75
E308	0.08	18.0-21.0	9.0-11.0	0.75	...	0.5-2.5	0.90	0.04	0.03	0.75
E308H	0.04-0.08	18.0-21.0	9.0-11.0	0.75	...	0.5-2.5	0.90	0.04	0.03	0.75
E308L	0.04	18.0-21.0	9.0-11.0	0.75	...	0.5-2.5	0.90	0.04	0.03	0.75
E308Mo	0.08	18.0-21.0	9.0-12.0	2.0-3.0	...	0.5-2.5	0.90	0.04	0.03	0.75
E308MoL	0.04	18.0-21.0	9.0-12.0	2.0-3.0	...	0.5-2.5	0.90	0.04	0.03	0.75
E309	0.15	22.0-25.0	12.0-14.0	0.75	...	0.5-2.5	0.90	0.04	0.03	0.75
E309L	0.04	22.0-25.0	12.0-14.0	0.75	...	0.5-2.5	0.90	0.04	0.03	0.75
E309Cb	0.12	22.0-25.0	12.0-14.0	0.75	0.70-1.00	0.5-2.5	0.90	0.04	0.03	0.75
E309Mo	0.12	22.0-25.0	12.0-14.0	2.0-3.0	...	0.5-2.5	0.90	0.04	0.03	0.75
E310	0.08-0.20	25.0-28.0	20.0-22.5	0.75	...	1.0-2.5	0.75	0.03	0.03	0.75
E310H	0.35-0.45	25.0-28.0	20.0-22.5	0.75	...	1.0-2.5	0.75	0.03	0.03	0.75
E310Cb	0.12	25.0-28.0	20.0-22.0	0.75	0.70-1.00	1.0-2.5	0.75	0.03	0.03	0.75
E310Mo	0.12	25.0-28.0	20.0-22.0	2.0-3.0	...	1.0-2.5	0.75	0.03	0.03	0.75
E312	0.15	28.0-32.0	8.0-10.5	0.75	...	0.5-2.5	0.90	0.04	0.03	0.75
E316	0.08	17.0-20.0	11.0-14.0	2.0-3.0	...	0.5-2.5	0.90	0.04	0.03	0.75
E316H	0.04-0.08	17.0-20.0	11.0-14.0	2.0-3.0	...	0.5-2.5	0.90	0.04	0.03	0.75
E316L	0.04	17.0-20.0	11.0-14.0	2.0-3.0	...	0.5-2.5	0.90	0.04	0.03	0.75
E317	0.08	18.0-21.0	12.0-14.0	3.0-4.0	...	0.5-2.5	0.90	0.04	0.03	0.75
E317L	0.04	18.0-21.0	12.0-14.0	3.0-4.0	...	0.5-2.5	0.90	0.04	0.03	0.75
E318	0.08	17.0-20.0	11.0-14.0	2.0-2.5	6 × C min to 1.00 max	0.5-2.5	0.90	0.04	0.03	0.75
E320	0.07	19.0-21.0	32.0-36.0	2.0-3.0	8 × C min to 1.00 max	0.5-2.5	0.60	0.04	0.03	3.0-4.0
E320LR(b)	0.035	19.0-21.0	32.0-36.0	2.0-3.0	8 × C min to 0.40 max	1.50-2.50	0.30	0.020	0.015	3.0-4.0
E330	0.18-0.25	14.0-17.0	33.0-37.0	0.75	...	1.0-2.5	0.90	0.04	0.03	0.75
E330H	0.35-0.45	14.0-17.0	33.0-37.0	0.75	...	1.0-2.5	0.90	0.04	0.03	0.75
E347	0.08	18.0-21.0	9.0-11.0	0.75	8 × C min to 1.00 max	0.5-2.5	0.90	0.04	0.03	0.75
E349(c,d)	0.13	18.0-21.0	8.0-10.0	0.35-0.65	0.75-1.2	0.5-2.5	0.90	0.04	0.03	0.75
E16-8-2	0.10	14.5-16.5	7.5-9.5	1.0-2.0	...	0.5-2.5	0.60	0.03	0.03	0.75

(a) Suffix −15 electrodes are classified with DCEP. Suffix −16 electrodes are classified with alternating current and DCEP. Electrodes up to and including $^5/_{32}$ in. diam are usable in all positions. Electrodes $^3/_{16}$ in. and larger are usable only in the flat- and horizontal-fillet positions. (b) Carbon shall be reported to the nearest 0.01%; for E320LR, to the nearest 0.005%. (c) Titanium, 0.15% max. (d) Tungsten, 1.25 to 1.75%

Source: AWS A5.4-81

Table 3 Comparison of the commercially important coating types for austenitic SMAW electrodes

Coating designation	−15	−16	High titania −16
Metal transfer mode	Globular	Globular	Spray
Bead contour	Convex/coarse ripple	Flat/medium ripple	Concave/fine ripple
Low-temperature toughness	Superior	Good	Fair
Slag removal	Fair	Good	Superior
Welding position	All	All	Flat/horizontal, only

that the coating is primarily of the lime type, containing a fair amount of calcium or other alkaline elements and that the electrodes are suitable for use with DCEP. For welding with alternating current, an electrode with the suffix −16 should be selected. Such electrodes can also be used with DCEP and generally have a mixture of lime and titanium in their coating. For welding with alternating current, the coating, besides having alkaline elements, contains readily ionized elements to stabilize the arc. A third type is also available in the United States and Europe. It contains a much higher percentage of titanium in the coating than most −16 coatings, but does not yet have a separate designation and is commonly classified as a −16 type. It is characterized by a very smooth arc and smooth concave bead. These three coating types are compared in Table 3.

Electrodes are available in diameters ranging from 1/16 to 1/4 in. If operating amperages are not given by the manufacturer, the suggested values in Table 4 should be used. Electrodes of the −15 and −16 types up to 5/32 in. in diameter can be used in all welding positions; electrodes 3/16 in. in diameter and larger should be used in the flat and horizontal fillet positions only. The electrodes listed in Table 2 cover most requirements for welding the standard austenitic grades.

For welding special grades of stainless steel, composite electrodes of special composition may be necessary. For SMAW, a composite electrode consists of a flux-covered carbon steel tube containing the alloying elements, in powder form, in the core. These electrodes also are available for depositing the standard compositions of stainless steel.

Table 4 Suggested operating parameters when electrode guidelines are not available
Electrode designations, all classifications; suffix −5 and −16

Electrode size	Average arc current, A	Maximum arc voltage, V
1/16	35-45	24
5/64	45-55	24
3/32	65-80	24
1/8	90-110	25
5/32	120-140	26
3/16	160-180	27
1/4	220-240	28

Source: AWS A5.4-81

Example 1. Use of a Composite Electrode. In Fig. 3, air separators for self-priming pumps were cast in the form of hemispheres and were joined at the equator by SMAW. The composition of the austenitic-ferritic (CD-4MCu) stainless steel castings was 0.04% C maximum, 1.0% Mn maximum, 1.0% Si maximum, 24.5 to 26.5% Cr, 4.75 to 6.0% Ni, and 1.75 to 2.25% Cu. Minimum mechanical properties were tensile strength, 100 ksi; yield strength, 70 ksi; and elongation in 2 in., 20%.

Fig. 3 Fabricated air separator for a self-priming pump used in corrosion service

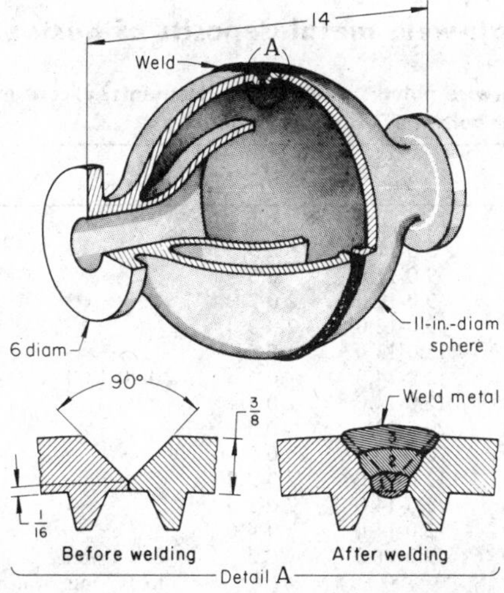

Process	SMAW
Joint type	Circumferential butt
Weld type	90° V-groove
Fixture	Rotating positioner
Power supply	300-A, dc motor-generator
Electrode	5/32-in.-diam composite(a)
Welding position	Flat
No. of passes	3
Current	130 A (DCEP)
Voltage	24 V
Preheat	None
Postheat	(b)
Travel speed	5-8 in./min

(a) A composite electrode having a flux covering and an alloy-powder core formulated to deposit a weld of the same composition as the base metal. (b) Heated to 2050 °F in a furnace; furnace cooled to within the range of 1750 to 1900 °F; held for 1/2 h; quenched in water

Because no standard filler metal was available to match the base-metal composition, a composite electrode was formulated. Low-carbon steel tubes were filled with alloy powder in proportions that would produce a deposit equivalent in composition to the base metal. The electrodes were manufactured as follows:

- Carbon steel strip was partially roll formed and filled with alloy powder.
- Roll forming was continued until the tube was tightly closed and crimped.
- Tube was pulled through dies and cut to length.
- Tube was processed through a press that extruded a flux covering concentric with the tube.
- Hemispheres were cast with 45° bevels at joint edges.
- Casting skin at joint edges was removed by grinding.
- Parts were assembled in lathe-type positioner for flat welding.

Welding was accomplished in three passes. Welds were carefully cleaned between passes, using a stainless steel wire brush. The width of weave was not permitted to exceed four times the electrode diameter, or 5/8 in. maximum. After welding, the assembly was solution heat treated to obtain maximum corrosion resistance. Following heat treatment, weldments were sand blasted, machined, and pressure tested with air at 100 psi.

Welding Procedure. The most significant difference in the procedure used for welding stainless steel as opposed to plain low-carbon steel is that less welding heat is required. The workpieces should be carefully prepared and fitted. Thin-gage stainless steel, in particular, should be properly clamped and held in alignment to reduce buckling. Large electrodes of more than 1/4 in. diam and excessive arc length contribute to loss of chromium in the weld deposit. Excessive weaving of electrodes of any size should be avoided. Maximum width of weave should be limited to four times the core wire diameter. The stringer-bead technique is generally recommended for depositing weld metal.

Special care is required between weld passes to remove all slag from the deposited bead. Only stainless steel wire brushes and tools should be used for this purpose. If grinding wheels are used, they should not be contaminated by use on metals other than stainless steel. Discoloration on either side of the weld metal by a thin oxide layer is a harmless surface condition, but it is often removed by chemical (pickling) or mechanical (sandblasting) means.

Storage of electrodes is important. Electrode coatings pick up moisture, which may lead to weld porosity. Electrode manufacturers generally supply stainless steel electrodes in moistureproof packages. Electrodes exposed to air and humidity can sometimes be restored to their initial condition by baking. Restored electrodes should not be used for critical welding applications. No attempt should be made to restore electrodes that are wet; they should be discarded.

Austenitic stainless steels should be welded in the annealed condition. The stabilized grades, such as types 321 or 347, should be in the stabilized annealed condition, while the nonstabilized grades should be in the solution annealed condition. Because austenitic steels are not hardenable by heat treatment, preheating is not used. Their mechanical properties are not substantially changed by welding, although stainless steel that has been purposely work hardened to increase its strength softens in the HAZ.

For welding in the flat position, the stringer-bead technique should be used for the first pass; when the fit-up is poor, a slight weave may be required. For succeeding passes, stringer beads or a slight weave should be used. The arc should be no longer than $^{1}/_{8}$ in. For vertical-down welding, the same technique is used, but with the electrode tilted so that the force of the arc pushes the molten metal back up the joint. Vertical-up welding calls for a triangular weave with a $^{5}/_{32}$-in.-diam or smaller electrode. A shelf is made at the bottom of the joint (lower end), and succeeding passes are made using a slight weaving motion with the electrode pointed slightly upward. Overhead welding requires a whipping technique and a slightly circular motion. Weaving should be avoided. The arc should be as short as possible, and the movement should be rapid.

Flux Cored Arc Welding

As more electrodes become available, FCAW is gaining wide acceptance for the joining of stainless steel. For a description of the process, see the article "Flux Cored Arc Welding" in this Volume. Advantages of this process include: (1) higher efficiency than SMAW and associated high deposition rate, (2) gas shielding not required for certain consumables, (3) all-position capability for certain consumables, and (4) production in compositions and ferrite ranges that cannot be easily produced in wrought products. Disadvantages include: (1) fewer sources of FCAW electrodes than SMAW electrodes, (2) higher equipment cost, (3) gas shielding required for certain consumables, (4) certain consumables available in flat position only, and (5) slag removal required. This process is well suited to larger volume production applications.

Constant potential power sources used for FCAW are applicable for welding stainless steel. A 300- to 400-A machine with a 1- to 2-V slope per 100 A is recommended for smoother operation with the smaller diameter ($^{1}/_{16}$-in.) electrodes.

Electrodes. Table 5 lists austenitic stainless steel electrodes and compositions used for FCAW. Electrodes are classified on the basis of an all-weld-metal deposit like the SMAW electrodes, and they are

Table 5 Chemical compositions for all-weld-metal deposits of austenitic stainless steel electrodes for use in FCAW

All weight percentages are maximum, unless otherwise noted; total of other elements, except iron, not to exceed 0.50%; X indicates a classification covering the shielding designation for both the 1 and 2 categories

AWS classification	C	Cr	Ni	Mo	Nb + Ta	Mn	Si	P	S	Fe	Cu
E307T-X	0.13	18.0-20.5	9.0-10.5	0.5-1.5	...	3.3-4.75	1.0	0.04	0.03	rem	0.5
E308T-X	0.08	18.0-21.0	9.0-11.0	0.5	...	0.5-2.5	1.0	0.04	0.03	rem	0.5
E308LT-X(a)	...	18.0-21.0	9.0-11.0	0.5	...	0.5-2.5	1.0	0.04	0.03	rem	0.5
E308MoT-X	0.08	18.0-21.0	9.0-12.0	2.0-3.0	...	0.5-2.5	1.0	0.04	0.03	rem	0.5
E308MoLT-X(a)	...	18.0-21.0	9.0-12.0	2.0-3.0	...	0.5-2.5	1.0	0.04	0.03	rem	0.5
E309T-X	0.10	22.0-25.0	12.0-14.0	0.5	...	0.5-2.5	1.0	0.04	0.03	rem	0.5
E309CbLT-X(a)	...	22.0-25.0	12.0-14.0	0.5	0.70-1.00	0.5-2.5	1.0	0.04	0.03	rem	0.5
E309LT-X(a)	...	22.0-25.0	12.0-14.0	0.5	...	0.5-2.5	1.0	0.04	0.03	rem	0.5
E310T-X	0.20	25.0-28.0	20.0-22.5	0.5	...	1.0-2.5	1.0	0.03	0.03	rem	0.5
E312T-X	0.15	28.0-32.0	8.0-10.5	0.5	...	0.5-2.5	1.0	0.04	0.03	rem	0.5
E316T-X	0.08	17.0-20.0	11.0-14.0	2.0-3.0	...	0.5-2.5	1.0	0.04	0.03	rem	0.5
E316LT-X(a)	...	17.0-20.0	11.0-14.0	2.0-3.0	...	0.5-2.5	1.0	0.04	0.03	rem	0.5
E317LT-X(a)	...	18.0-21.0	12.0-14.0	3.0-4.0	...	0.5-2.5	1.0	0.04	0.03	rem	0.5
E347T-X	0.08	18.0-21.0	9.0-11.0	0.5	8 × C min to 1.0 min	0.5-2.5	1.0	0.04	0.03	rem	0.5
E307T-3	0.13	19.5-22.0	9.0-10.5	0.5-1.5	...	3.3-4.75	1.0	0.04	0.03	rem	0.5
E308T-3	0.08	19.5-22.0	9.0-11.0	0.5	...	0.5-2.5	1.0	0.04	0.03	rem	0.5
E308LT-3	0.03	19.5-22.0	9.0-11.0	0.5	...	0.5-2.5	1.0	0.04	0.03	rem	0.5
E308MoT-3	0.08	18.0-21.0	9.0-12.0	2.0-3.0	...	0.5-2.5	1.0	0.04	0.03	rem	0.5
E308MoLT-3	0.03	18.0-21.0	9.0-12.0	2.0-3.0	...	0.5-2.5	1.0	0.04	0.03	rem	0.5
E309T-3	0.10	23.0-25.5	12.0-14.0	0.5	...	0.5-2.5	1.0	0.04	0.03	rem	0.5
E309LT-3	0.03	23.0-25.5	12.0-14.0	0.5	...	0.5-2.5	1.0	0.04	0.03	rem	0.5
E309CbLT-3	0.03	23.0-25.5	12.0-14.0	0.5	0.70-1.00	0.5-2.5	1.0	0.04	0.03	rem	0.5
E310T-3	0.20	25.0-28.0	20.0-22.5	0.5	...	1.0-2.5	1.0	0.03	0.03	rem	0.5
E312T-3	0.15	28.0-32.0	8.0-10.5	0.5	...	0.5-2.5	1.0	0.04	0.03	rem	0.5
E316T-3	0.08	18.0-20.5	11.0-14.0	2.0-3.0	...	0.5-2.5	1.0	0.04	0.03	rem	0.5
E316LT-3	0.03	18.0-20.5	11.0-14.0	2.0-3.0	...	0.5-2.5	1.0	0.04	0.03	rem	0.5
E317LT-3	0.03	18.5-21.0	13.0-15.0	3.0-4.0	...	0.5-2.5	1.0	0.04	0.03	rem	0.5
E347T-3	0.08	19.0-21.5	9.0-11.0	0.5	8 × C min to 1.0 max	0.5-2.5	1.0	0.04	0.03	rem	0.5
EXXXT-G(b)	...	...	...	...	...	...	...	...	...	...	...

(a) Carbon, 0.04% max when suffix X is 1; 0.03% max when suffix X is 2. (b) Composition as agreed upon between supplier and purchaser
Source: AWS A5.22-80

Table 6 Classification system and shielding medium for flux-cored electrodes

AWS designations (all classifications)(a)	External shielding medium(b)	Current and polarity
EXXXT-1	CO_2	DCEP
EXXXT-2	Ar + 2% O	DCEP
EXXXT-3	None	DCEP
EXXXT-G	Not specified	Not specified

(a) The letters XXX stand for the chemical composition. (b) The requirement for the use of specified external shielding media for classification purposes shall not be construed to restrict the use of other media for industrial use as recommended by the producer. Source: AWS A5.22-80

also classified on the basis of shielding medium as shown in Table 6.

Welding Procedure. The same procedures for SMAW apply to FCAW. The higher deposition rate may increase the distortion tendency, which can be minimized by control of the travel speed and weld parameters. One of the advantages of the flux cored process is greater arc force, resulting in deeper penetration. This is a real asset in a fillet weld, where the deeper penetration gives a greater effective weld throat with the same fillet size. This deeper penetration becomes a disadvantage on thinner plate, where even the smallest diameter electrode may have too much penetration.

When ferrite contol is critical, the welding parameters must be tightly controlled when using FCAW, especially with the self-shielded type. Compared to SMAW, where a single knob controls amperage and voltage, FCAW allows separate control of the amperage and voltage. An electrode may tolerate several volts more or less than the recommended value at a given amperage. This voltage change is reflected in a significant change in arc length and in the reaction of the constituents with the atmosphere. A 3-V change could result in more than a 10 FN change in the deposit. The consumable manufacturer can give recommended ranges and estimate the effects of voltage changes.

The joints used with FCAW electrodes can be narrower and contain a smaller included angle because the small electrode diameter provides easy access and better visibility than SMAW electrodes. Again, the greater penetration enables their use on thicker sections.

Submerged Arc Welding

Among the economic advantages of SAW are high deposition rate and high welding speed. The principal disadvantages of the process are higher equipment cost, higher heat input, and use in the flat and horizontal positions only. Most installations are mechanized, although semiautomatic welding is possible. The use of flux to shield the weld requires flux handling and slag removal operations. For a detailed description of the process, see the article "Submerged Arc Welding" in this Volume.

Weld Quality. The quality of welds made by SAW is generally high, but some conditions restrict use of the process. The composition of the weld metal is generally more difficult to control with SAW than it is with GTAW, PAW, or electron beam welding (EBW). The heat input can be much greater than in other arc welding processes, and solidification of the weld metal is generally slower. These conditions can raise problems in welding stainless steel that are not encountered when welding carbon steel. For instance, the silicon content of the weld metal may be much higher in SAW than in other processes, and increased silicon content may cause hot cracking or fissuring. Basic type flux minimizes this effect. When using SAW with some of the more highly alloyed stainless steels (precipitation-hardening and 300 series stainless steels), the effect of the flux on the deposit composition must be considered. In some cases, pickup from the flux can change the deposit composition enough to affect the weld hot-cracking tendencies or the corrosion resistance of the weld.

When the weld deposit must be fully austenitic or when the ferrite content of the deposit must be controlled to less than 4 FN, the risk of microfissuring is considerable, particularly in welding heavy sections that are generally well suited for welding by SAW. Welding of relatively thick sections of type 347 stainless steel also presents a problem. Service failures have occurred because of strain-induced precipitation of niobium carbides in the HAZ during postweld heat treatment or service at elevated temperature.

Welding Current. Direct current is normally used for welding thin sections. For thicker sections, either alternating or direct current can be used. Current for welding stainless steel is usually about 80% of that used for welding carbon steel, section thickness and other conditions being the same. Some typical values of current, specific welding conditions, and a typical joint design are shown in Fig. 4.

Electrode Wires as Filler Metals. Table 7 lists designations of electrode wires applicable to SAW. Selection of filler metal for a particular grade of stainless steel is complicated by the differences in service requirements. If experience with a similar application is not available, use of a filler

Fig. 4 Submerged arc butt welding of 1/2-in.-thick plates

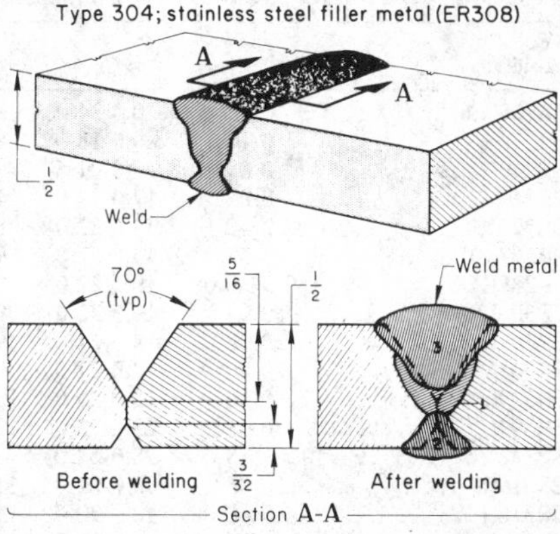

Pass No.(a)	Current (DCEP), A	Voltage, V	Welding speed, in./min
1	400	26	20
2	420	28	20
3	450	32	18

(a) For electrode, 3/32 in. diam; neutral flux 12 × 150

metal similar in composition to the base metal should be considered. Adjustments, based on experience, may be required in the composition of the filler metal, flux, or both to arrive at a suitable composition of weld deposit. In some applications, a filler metal that is higher in alloy content than the base metal is used. Typical electrode wire size and current combinations are given in Table 8. Sizes larger than 3/16 in. diam are seldom used and may result in heavy deposits with undesirable metallurgical and mechanical properties.

Flux. Submerged arc fluxes, either prefused or bonded, are available as proprietary materials for welding stainless steel. Alloying elements, such as chromium, nickel, molybdenum, and niobium, are frequently added to the weld deposit through the flux. If the flux contains no compensating alloying additions, it is neutral.

Weld Backing. Submerged arc welding deposits relatively large volumes of metal that remain fluid for longer periods than are possible with the other arc methods discussed in this article. Thus, for many types of joints, backing is required. The two common types of weld backing are nonfusible and fusible. Copper is one nonfusible type used in the welding of stainless steel. Ceramic backing types are also sometimes used. In effect, the copper bar serves more as a chill than as a backing, helping to distribute heat from the weld metal to adjacent areas. Recommended dimensions for grooves in copper backing

Table 7 Chemical compositions for all-weld-metal deposits of austenitic stainless steel electrodes for use in SAW, GTAW, and GMAW

All weight percentages are maximum, unless otherwise noted; total of other elements, except iron, not to exceed 0.50%

AWS classification	C	Cr	Ni	Mo	Nb + Ta	Mn	Si	P	S	N	Cu
ER209(a,b)	0.05	20.5-24.0	9.5-12.0	1.5-3.0	...	4.0-7.0	0.90	0.03	0.03	0.10-0.30	0.75
ER218	0.10	16.0-18.0	8.0-9.0	0.75	...	7.0-9.0	3.5-4.5	0.03	0.03	0.08-0.18	0.75
ER219	0.05	19.0-21.5	5.5-7.0	0.75	...	8.0-10.0	1.00	0.03	0.03	0.10-0.30	0.75
ER240	0.05	17.0-19.0	4.0-6.0	0.75	...	10.5-13.5	1.00	0.03	0.03	0.10-0.20	0.75
ER307	0.04-0.14	19.5-22.0	8.0-10.7	0.5-1.5	...	3.3-4.75	0.30-0.65	0.03	0.03	...	0.75
ER308(b)	0.08	19.5-22.0	9.0-11.0	0.75	...	1.0-2.5	0.30-0.65	0.03	0.03	...	0.75
ER308H	0.04-0.08	19.5-22.0	9.0-11.0	0.75	...	1.0-2.5	0.30-0.65	0.03	0.03	...	0.75
ER308L(b)	0.03	19.5-22.0	9.0-11.0	0.75	...	1.0-2.5	0.30-0.65	0.03	0.03	...	0.75
ER308Mo	0.08	18.0-21.0	9.0-12.0	2.0-3.0	...	1.0-2.5	0.30-0.65	0.03	0.03	...	0.75
ER308MoL	0.04	18.0-21.0	9.0-12.0	2.0-3.0	...	1.0-2.5	0.30-0.65	0.03	0.03	...	0.75
ER309(c)	0.12	23.0-25.0	12.0-14.0	0.75	...	1.0-2.5	0.30-0.65	0.03	0.03	...	0.75
ER309L	0.03	23.0-25.0	12.0-14.0	0.75	...	1.0-2.5	0.30-0.65	0.03	0.03	...	0.75
ER312	0.15	28.0-32.0	8.0-10.5	0.75	...	1.0-2.5	0.30-0.65	0.03	0.03	...	0.75
ER316(b)	0.08	18.0-20.0	11.0-14.0	2.0-3.0	...	1.0-2.5	0.30-0.65	0.03	0.03	...	0.75
ER316H	0.04-0.08	18.0-20.0	11.0-14.0	2.0-3.0	...	1.0-2.5	0.30-0.65	0.03	0.03	...	0.75
ER316L(b)	0.03	18.0-20.0	11.0-14.0	2.0-3.0	...	1.0-2.5	0.30-0.65	0.03	0.03	...	0.75
ER317	0.08	18.5-20.5	13.0-15.0	3.0-4.0	...	1.0-2.5	0.30-0.65	0.03	0.03	...	0.75
ER317L	0.03	18.5-20.5	13.0-15.0	3.0-4.0	...	1.0-2.5	0.30-0.65	0.03	0.03	...	0.75
ER318	0.08	18.0-20.0	11.0-14.0	2.0-3.0	8 × C min to 1.0 max	1.0-2.5	0.30-0.65	0.03	0.03	...	0.75
ER320LR(c)	0.025	19.0-21.0	32.0-36.0	2.0-3.0	8 × C min to 0.40 max	1.5-2.0	0.15	0.015	0.020	...	3.0-4.0
ER321(d)	0.08	18.5-20.5	9.0-10.5	0.75	...	1.0-2.5	0.30-0.65	0.03	0.03	...	0.75
ER330	0.18-0.25	15.0-17.0	34.0-37.0	0.75	...	1.0-2.5	0.30-0.65	0.03	0.03	...	0.75
ER347(b)	0.08	19.0-21.5	9.0-11.0	0.75	10 × C min to 1.0 max	1.0-2.5	0.30-0.65	0.03	0.03	...	0.75
ER349(e)	0.07-0.13	19.0-21.5	8.0-9.5	0.35-0.65	1.0-1.4	1.0-2.5	0.30-0.65	0.03	0.03	...	0.75
ER16-8-2	0.10	14.5-16.5	7.5-9.5	1.0-2.0	...	1.0-2.5	0.30-0.65	0.03	0.03	...	0.75

(a) Vanadium, 0.10 to 0.30%. (b) Available with 0.65 to 1.00% Si, with Si added to classification designation. The fabricator should consider carefully the use of high silicon filler metals in highly restrained fully austenitic welds. (c) Carbon, nearest 0.01%; E320LR, nearest 0.005%. (d) Titanium, 9 × C min to 1.0 max. (e) Titanium, 0.10 to 0.30%; tungsten, 1.25 to 1.75%
Source: AWS A5.9-81

Table 8 Relation of wire diameter to current for SAW of stainless steel

Wire diameter, in.	Current range, A	Wire diameter, in.	Current range, A
$^3/_{32}$	120-700	$^3/_{16}$	400-1300
$^1/_8$	220-1100	$^1/_4$	600-1600
$^5/_{32}$	340-1200	$^5/_{16}$	1000-2500

bars are given in Fig. 5. Many designs other than that shown in Fig. 5 are also used. With a fusible backing made of stainless steel, the weld penetrates and fuses with the backing, which becomes, temporarily or permanently, an integral part of the welded joint.

Joint Design. Sound single-pass welds can be made in square-groove butt joints up to $^5/_{16}$ in. thick, without a root opening and with suitable nonfusible backing. Two-pass welds (one pass on each side) are made without a root opening on metal up to $^5/_8$ in. thick. Edges of the joint should be closely butted to avoid melt-through if weld backing is not used. A square-groove butt joint requires a minimum of edge preparation and produces sound welds that have adequate penetration.

Single-V-groove welds with a root face are used with nonfusible backing for single-pass butt joints $^5/_{16}$ in. thick or more. For most applications, the maximum thickness of single-V-groove welds is $1^1/_4$ to $1^1/_2$ in., with root-face dimensions $^1/_8$ to $^3/_{16}$ in. This joint design can be used for multiple-pass welds, without backing, where plate thickness exceeds $^5/_8$ in. The first pass is made in the larger V. The work is then turned over and welded on the reverse side. The root face is about $^3/_8$ in. for multiple-pass welds.

The double-V-groove butt joint is the basic design for SAW of material having a thickness of $^1/_2$ in. or greater. A large root face is generally used. Figure 4 shows a typical double-V-groove weld in $^1/_2$-in. type 304 plate. A single-U-groove butt joint

Fig. 5 Copper backing bars used in SAW

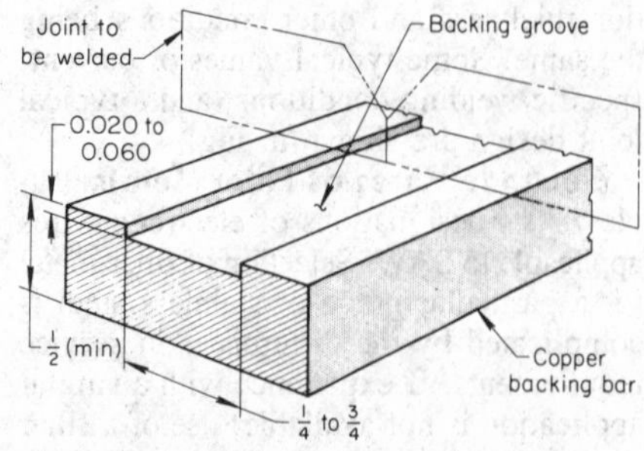

is also commonly used. Shielded metal arc welding is often used to make a root pass, or backing pass, on the reverse side of the joint. For more information on joint design and preparation, see the article "Joint Design and Preparation" in this Volume.

Welding Position and Technique. Submerged arc welding is done in the flat or horizontal position. Operation difficulties increase and preferred shape of beads is sacrificed when the work is inclined.

When welding stainless steel, the most common method of starting the arc is the fuse-ball start. A tightly rolled ball of stainless steel wool, about $^3/_8$ in. in diameter, is placed between the end of the electrode and the work. The electrode wire is inched forward to compress the ball. Flux is added, and the arc is initiated and flashes through the stainless steel wool. For details on other methods of starting the arc, see the article "Submerged Arc Welding" in this Volume.

Once the arc is started, welding conditions must be carefully monitored. The most important conditions are current, voltage, and travel speed. If current is too high, likely results are melt-through, pear-shaped fusion zones, excessive penetration, and overlap or excessively high reinforcement. If current is too low, inadequate penetration and low reinforcement result. When voltage is too high, shallow penetration and wide reinforcement are common. Low voltage results in pear-shaped fusion zones and narrow and steep-sided reinforcement, or overlapped reinforcement. When welding speed is too high, undercutting and low reinforcement are likely results; too low a speed leads to overlap.

Circumferential Welding. Submerged arc welding is used extensively for making girth welds for joining pipe or other cylindrical sections. Equipment required is generally the same as for welding cylindrical workpieces of carbon steel and is described in the article "Submerged Arc Welding" in this Volume. When submerged arc welding stainless steel pipe joints, some difficulty may be encountered in cleaning the root of the joint and in obtaining a sound weld root inside the pipe. These problems can be overcome by (1) using a backing ring and making the root pass by SMAW, followed by SAW for the filler passes; or (2) making the root pass by GTAW, without backing, and completing the joint by SAW.

Example 2. Use of a stainless steel backing ring in welding type 347 stainless steel pipe joints is shown in Fig. 6. Service conditions required the joint between pipe sections to be made with full penetration, but without weld metal on the inside of the pipe. The groove was prepared by gas tungsten arc cutting, with the pipe section in position on turning rolls. Oxide on the surface was removed mechanically, and the pipe sections were assembled with a stainless steel backing ring to form the joint. The root pass was made by SMAW, and the weld was cleaned with a stainless steel brush and inspected visually. Visible flaws were removed by grinding. Welding was completed in five additional passes by SAW, under conditions selected to avoid undercutting on the sidewalls of the weld groove.

After each bead had been deposited, slag was removed by wire brushing, and any cracks and porosity were removed by chipping and grinding. Because of the susceptibility of the bead to cracking, the arc could not be struck on the weld bead already deposited. It was struck on the side of the weld groove, and the stops and starts were subsequently ground out. After completion of the joint, the fused flux was removed by brushing, and the joint was prepared for radiographic inspection. Rejection rate was less than 5%.

Gas Metal Arc Welding

The major advantages of GMAW for welding stainless steel are: (1) the continuously fed electrode wire permits long, uninterrupted periods of welding; (2) the process can be readily automated; (3) the use of shielding gas instead of a flux eliminates the need for slag removal and enables the operator to watch the welding operation; and (4) low-hydrogen deposits are obtained without the need for baking the electrodes, as may be required in SMAW. Also, GMAW offers a variety of means by which transfer of weld metal can occur, such as spray arc, globular, short circuiting arc, and pulse arc.

Limitations of the process are essentially the same for welding stainless steel as for welding low-carbon steel. Compared with SMAW, the equipment costs more and is less portable, and operations must be shielded from drafts. For details of the process, see the article "Gas Metal Arc Welding" in this Volume.

Welding Current. The polarity of the welding current used in GMAW depends primarily on the penetration desired. The greatest penetration is obtained with reverse polarity (DCEP). Regardless of whether the electrons impinge on the electrode wire (as in reverse polarity) or on the work (as in straight polarity), heat builds up in the base metal through the metal drops. Reverse polarity results in the metal drops being subjected to considerable force by the positive ions of gas, resulting in deep penetration. Conversely, when straight polarity (direct current electrode negative, DCEN) is used, the force exerted on the metal drops by the gas ions tends to support them, resulting in shallow penetration. More than 95% of all GMAW of stainless steel is done using DCEP, with DCEN being restricted to applications requiring shallow penetration.

Because the electrical resistivity of stainless steels is greater than that of carbon steels, the electrode melting rate is higher for a given electrode stickout. Allowing for some variation among different grades, the same deposition rate can be achieved for stainless steels at about 80% of the current required for plain carbon steels, other conditions being the same. Some typical current values for specific conditions are given in Tables 9 and 10.

Spray Transfer. High current density is the principal requirement for spray transfer of metal from the electrode to the base metal. At a certain minimum current density, which varies with electrode size and material, metal transfer through the arc changes from very large globular drops, which fall off the end of the electrode, to a stream of extremely fine droplets axially projected from the end of the electrode. The metal transfer changes from a fluttering, erratic discharge with a wandering cathode spot to a steady, quiet arc column. This arc column is a well-defined, narrow, incandescent, cone-shaped core within which metal transfer occurs. Because the method utilizes high current and voltage, the weld pool is quite fluid, limiting welding to the flat or horizontal position. Deposition rates are high, penetration is deep when DCEP is used, and the arc is exceptionally stable. The minimum thickness of stainless steel that can be satisfactorily welded using spray transfer is about 1/8 in. Generally, it is used on materials ranging from 1/4 to 1 in. in thickness. The diameter of electrode wire for spray transfer is usually 0.035 to 1/16 in.

On square-butt welds, a backing strip should be used to prevent weld drop-through. When fit-up is poor or copper backing cannot be used, a short circuiting transfer for the root pass minimizes drop-through.

Short Circuiting Transfer. Welding with a short circuiting arc employs lower current, generally ranging from 50 to 225 A, low voltages of 17 to 24 V, and small-diameter wires, with 0.030-, 0.035-, and

Fig. 6 Stainless steel backing ring for welding pipe sections

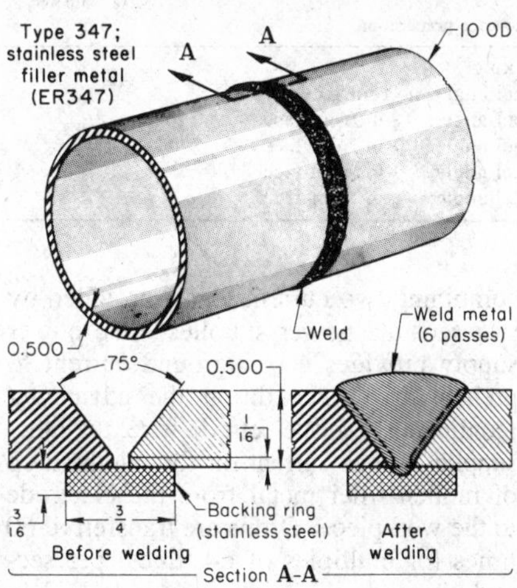

Process	SAW(a)
Joint type	Butt
Weld type	Single-V-groove
Welding position	Horizontal-rolled pipe
Welding head	Fixed position, single wire
No. of passes	6(a)
Preheat	None
Interpass temperature	400 °F max
Postheat	None
Electrode wire	1/8-in.-diam ER347
Flux	Neutral
Current	240-260 A (DCEP)(b)
Voltage	25 V
Welding speed	12-15 in./min

(a) First pass made by SMAW. (b) Actual reading at welding head

Table 9 Nominal conditions for GMAW of austenitic stainless steel with a spray arc

Plate thickness, in.	Joint and edge preparation	Electrode wire diameter, in.	Current (DCEP), A	Wire-feed speed, in./min	Welding speed, in./min	No. of passes
0.125	Square butt with backing	1/16	200-250	110-150	20	1
0.250	Single-V butt, 60° incl angle, no root face	1/16	250-300	150-200	15	2
0.375	Single-V butt, 60° incl angle, 1/16-in. root face	1/16	275-325	225-250	20	2
0.500	Single-V butt, 60° incl angle, 1/16-in. root face	3/32	300-350	75-85	5	3-4
0.750	Single-V butt, 90° incl angle, 1/16-in. root face	3/32	350-375	85-95	4	5-6
1.000	Single-V butt, 90° incl angle, 1/16-in. root face	3/32	350-375	85-95	2	7-8

0.045-in. diameters being the popular sizes. The distinctive feature of the short circuiting arc is the frequent shorting of the electrode wire to the work. All metal transfer occurs at arc outages, which occur at a steady rate and can vary from 20 to more than 200/s. The net result is a stable arc of low energy and heat input, ideally suited to the welding of thin sections in all positions. The low heat input minimizes work-metal distortion. The short circuiting arc is better than the spray arc for welding joints that have poor fit-up.

Short circuiting characteristics cannot be obtained for the full range of operation with ordinary or conventional power-supply units. Machines are available that contain slope, voltage, and inductance adjustments suitable for producing the controlled current surges needed to implement short circuit transfer on stainless steel. Inductance, in particular, plays an important part in obtaining proper weld pool fluidity when short circuiting transfer is used on stainless steel.

Electrode wire extension (stickout) should be kept as short as possible. Backhand welding is normally easier on fillet welds and results in cleaner welds. Forehand welding is used for butt welds, while outside corner welds are made with a straight motion. Table 10 gives typical welding conditions.

Pulsed arc transfer is a spray-type transfer occurring in pulses at regularly spaced intervals rather than at random intervals. The welding current is reduced and no metal transfer occurs in the time interval between pulses. Pulsing is obtained by combining two current levels provided by two separate power supplies. One power supply provides a background current to preheat and precondition the advancing (continuously fed) electrode wire; the other supplies a peak current for forcing the drop of molten filler metal from the electrode to the workpiece. Drops are transferred 60 times (or multiples of 60 times) per second, because the peak current is tied in with the line frequency.

Electrode wire diameters of 0.045 to 1/16 in. are most commonly used for pulsed arc transfer welding of stainless steels. Because of the lower heat input, the process is capable of welding sections thinner than are practical to weld by conventional spray transfer. Work-metal distortion is also lower.

Example 3. Use of Spray Transfer Versus Pulsed Arc Transfer. Excessive distortion and leakage were encountered in welding the supporting flanges of electrical bushings to the cover plate of a switching device (Fig. 7, upper left). The supporting flanges, consisting of stainless steel disks molded to the epoxy body of the electrical bushings, were joined to the cover plates by a single-pass circumferential fillet weld. The principal requirements were that the weld be leaktight under an air pressure of 5 psi and that the cover remain flat, as any distortion of the subassembly would cause misalignment and malfunctioning of parts to be installed later. Distortion was further increased by the number and arrangement of bushings in the cover and by differences in thermal conductivity and expansion of the two metals being welded.

Originally, 1/8-in. fillet welds were deposited at the joint between the flange and cover plate by automatic GMAW with spray transfer. The procedure had to be abandoned because even the smallest welds deposited were oversize in relation to the nominal 1/16-in. size of the joint and resulted in excessive heat buildup and distortion of the welded components. To reduce heat buildup in the subassembly, copper heat sinks were placed in either side of the flange. The upper heat sink, consisting of a heavy thimble-shaped cap, was held firmly against the flange by a clamping screw. The lower heat sink was placed inside a 4-in.-diam schedule 80 pipe and held in place by an adjustable clamping nut. The copper heat sinks were used in pairs and alternated with other pairs, the working pair being removed and cooled in water after two successive runs.

Before welding, the pipe was clamped vertically in a rotatable three-jaw chuck. The cover plate was positioned over the pipe and the bushing hole was centered. The bushing was then inserted in the cover, and the assembly was drawn tight by the clamping nut at the bottom. The upper heat sink was fastened in place, and the bottom (adjustable) heat sink was run up with a special wrench through circumferential slots in the pipe. The electrode holder and wire drive, which were mounted on a manipulator boom, were adjusted so that the electrode holder maintained a 60° angle with the horizontal during welding (Fig. 7, section A-A). Heat dissipation was assisted by welding the nine joints in the numerical sequence shown in Fig. 7, upper left.

Although the use of heat sinks and sequential welding reduced distortion to an acceptable level, about 80% of the weldments still showed small leaks during air testing. Inadequate joint penetration at the overlap of the start and finish of the weld and fast cooling of the oversize bead were suspected of being responsible for the leaks. The difficulties were overcome by changing to pulsed arc transfer using a 250-A dual rectifier and the same setup of equip-

Table 10 Nominal conditions for GMAW of austenitic stainless steel with a short circuiting arc

Electrode wire diameter, 0.030 in.; one pass each; shielding gas containing 90% helium, 7.5% argon, and 2.5% carbon dioxide

Plate thickness, in.	Joint and edge preparation	Current (DCEP), A	Voltage, V	Wire-feed speed, in./min	Welding speed, in./min
0.063	Nonpositioned fillet or lap	85	21	184	18
0.063	Butt (square edge)	85	22	184	20
0.078	Nonpositioned fillet or lap	90	22	192	14
0.078	Butt (square edge)	90	22	192	12
0.093	Nonpositioned fillet or lap	105	23	232	15
0.125	Nonpositioned fillet or lap	125	23	280	16

Fig. 7 Fixtures for welding joints between electrical bushings and cover plate

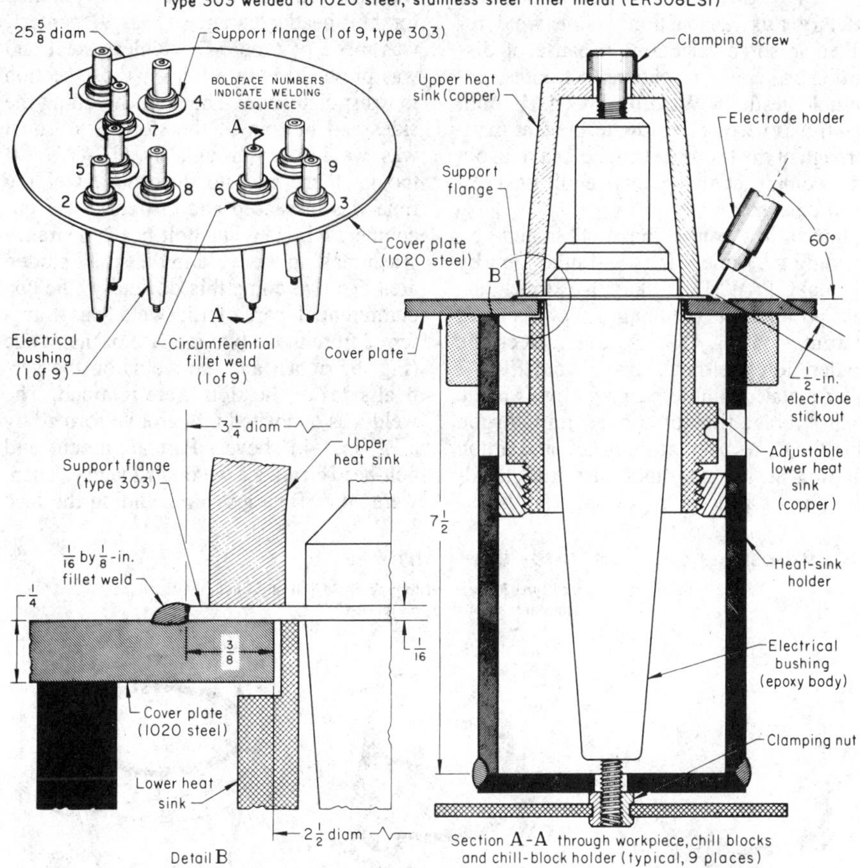

Item	Spray arc transfer	Pulsed arc transfer
Process	Automatic GMAW	Automatic GMAW
Joint type	Lap	Lap
Weld type	Fillet	Fillet
Fixtures	Copper chills(a)	Copper chills(a)
Power supply	500-A transformer-rectifier(b)	250-A dual transformer-rectifier
Electrode holder	···	200 A, air cooled
Electrode	0.035-in.-diam ER308L(Si)(c)	0.035-in.-diam ER308L(Si)(c)
Shielding gas	98% A, 2% O_2	98% A, 2% O_2, at 30 to 40 ft^3/h
No. of passes	1	1
Arc length	···	$^3/_{32}$ in.
Current (DCEN)	200-225 A	110-115 A
Voltage	···	24 V (background); 38 V (pulse)(d)
Electrode-feed rate	Dial adjustable	360 in./min
Electrode stickout	···	$^1/_2$ in.
Pulse frequency	···	60 Hz
Welding speed or time	22 in./min	30 s per pass

(a) Rotatable three-jaw chuck; manipulator boom. (b) With slope control. (c) Low-carbon, high-silicon ER308L-Si electrode was selected to minimize base-metal dilution and for ease of deposition. (d) Maximum setting of pulse voltage is 38 V; peak near 50 V.

ment for heat dissipation from the weld zone as described earlier. With pulsed transfer, penetration was satisfactory, and a smaller bead with a lower heat input was obtained. Distortion was further reduced, and rejections for leaks were decreased to about 20%.

Electrode Wires. The electrode wire diameters for GMAW are generally between 0.030 and $^3/_{32}$ in. For each wire diameter, a certain minimum welding current must be exceeded to achieve spray transfer. For example, when welding stainless steel in an argon-oxygen atmosphere with a $^3/_{64}$-in.-diam stainless steel electrode wire, spray transfer may be obtained at a welding current of about 225 A (DCEP). Figure 8 illustrates typical deposition rates and welding current values for various sizes of ER347 electrode wires, using a mixture of argon with 1% oxygen as the shielding gas. With the minimum current, a minimum arc voltage must also be used. This is generally between 25 and 30 V.

Table 11 lists austenitic stainless steel base metals and the electrode wires recommended for welding. Electrode wires made of high-silicon austenitic stainless steel and wires for welding straight-chromium stainless steel are also available. The austenitic wires with higher silicon content have particularly good wetting characteristics when they are used with short circuiting transfer.

Shielding Gases. Although the range of choice of shielding gas for welding stainless steels is considerably narrower than for carbon and low-alloy steels, several gas mixtures have proved satisfactory. The mode of metal transfer influ-

Fig. 8 Relation of welding current to deposition rate for various sizes of ER347 electrode wires

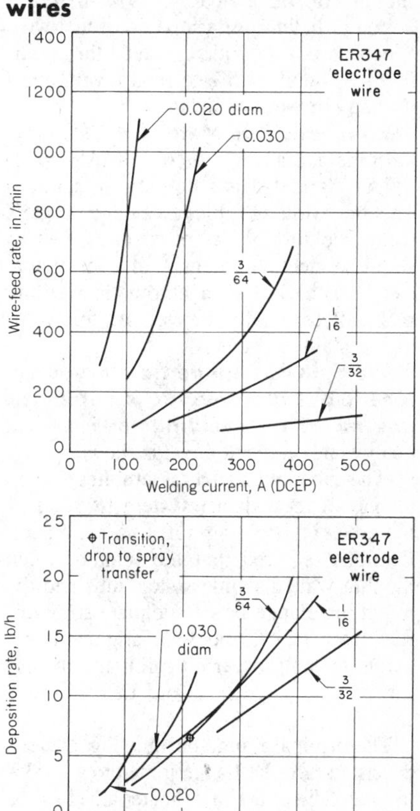

Table 11 Recommended electrode wires for GMAW and SAW of austenitic stainless steels

Base metal	Filler metal	Base metal	Filler metal
301, 302	ER308	316	ER316
304	ER308	316L	ER316L
308	ER308	317	ER317
304L	ER308L	330	330
309	ER309	321	ER321
310	ER310	347	ER347

ences the choice of shielding gas. For instance, with spray arc or pulsed arc transfer, a shielding-gas mixture containing 99% argon and 1% oxygen has been widely used and generally recommended. In some plants, 98% argon and 2% oxygen is used with success. For short circuiting transfer, a mixture of 90% helium, 7.5% argon, and 2.5% carbon dioxide has been extensively used for shielding, but helium is gradually losing favor because of its high cost. However, a mixture in which the above proportions of helium and argon have been reversed (90% argon, 7.5% helium, 2.5% carbon dioxide) has proved successful for short circuiting transfer. Regardless of other variations, the shielding gas for GMAW of stainless steel should contain at least 97.5% inert gas (argon or helium, or a mixture of the two). When carbon dioxide is used, the maximum is usually 2.5% to retain weld quality and corrosion resistance.

Automatic Gas Metal Arc Welding. A principal advantage of GMAW over SMAW is that it is automatic or semiautomatic. Although this advantage prevails in the welding of carbon steel as well as stainless steel, the degree of control that can be achieved with automatic welding makes it especially appropriate for welding of stainless steel.

Example 4. Change from oxyacetylene to gas tungsten arc to automatic gas metal arc welding for increased quality and production rate is shown in Fig. 9. The tube and flange assemblies served as part of the exhaust system for a small reciprocating-type aircraft engine. Specifications required that the joint between the thin-walled stainless steel tube and the mounting flange be of adequate strength, free from leaks, and with alignment accurate enough to permit minimal machining of the flange face and the bolt-head seat.

The original flange design (Fig. 9, left) consisted of a flat flange joined to a 0.035-in.-wall tube of the same material. The tube was expanded from 1.625-in. OD, with a 0.035-in. wall, to 1.740-in. OD for a distance of 1 in. at the joint. A single-pass circumferential fillet weld was made by oxyacetylene welding with a $^{1}/_{16}$-in.-diam rod of type 347 stainless steel and a proprietary flux. Inspection of the weld resulted in some rejections because of distortion and crack indications in fluorescent-particle testing. Welding speed (7 min/weld) was necessarily low to prevent melt-through of the thin-walled tube and to avoid overwelding at the points requiring bolt-head clearance.

Improved results were obtained by welding the same flange and tube joint by manual GTAW. Using a 250-A direct current power supply having a foot-operated current control, a light-duty electrode holder, low welding current, and ER347 filler metal, a single-pass weld was made in an average time of about 5 min. Argon shielding was used at the torch and argon purging inside the tube assembly. Although this process produced less distortion and fewer weld defects, results were still not acceptable.

To eliminate distortion, flange design was changed. The hub-type flange, slotted for bolt-head clearance (Fig. 9, center), was made of type 347 stainless steel and was provided with a thicker cross section to resist distortion. The hub, including the sides and bottom of the slotted portions, was welded all around by GTAW. Although there was no distortion, welding time was increased and difficulty was encountered in leaving bolt-head clearance when making the weld in the small slotted area. To overcome this difficulty, the circumferential part of the weld was transferred from the hub to the face of the flange (Fig. 9, right), and the welds on the vertical sides of the slots were retained. The weld was deposited in a groove formed by a $^{3}/_{32}$-in., 45° bevel. Part alignment and bolt-head clearance were satisfactory. There were no difficulties in machining the face

Fig. 9 Designs for flange-to-tube joint

Type 321 (original design) or type 321 welded to type 347 (first and final modifications); stainless steel filler metal (ER347)

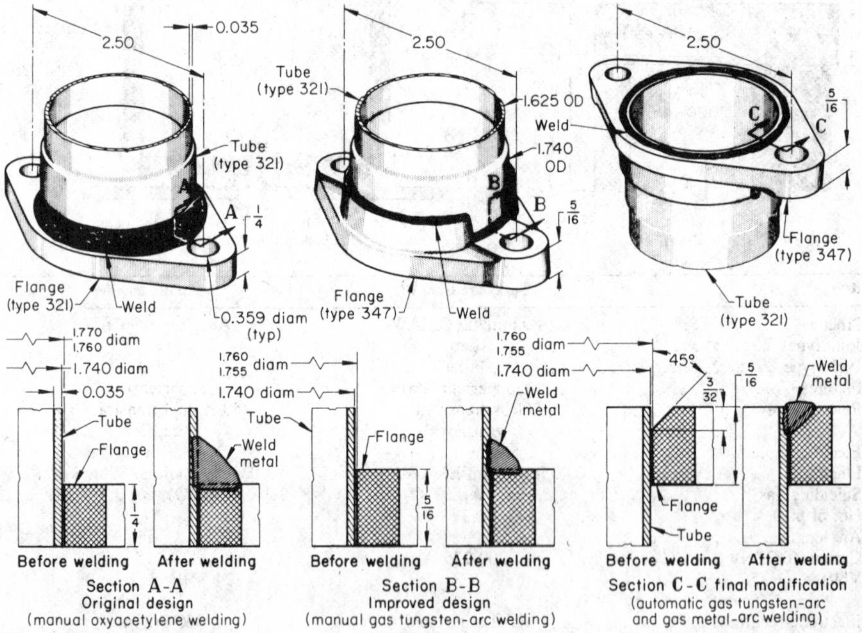

Section A-A
Original design
(manual oxyacetylene welding)

Section B-B
Improved design
(manual gas tungsten-arc welding)

Section C-C final modification
(automatic gas tungsten-arc and gas metal-arc welding)

Conditions for automatic GMAW

Joint type	Corner
Weld type	45° single-bevel groove
Fixtures	Adjustable torch holder, turntable, assembly fixture
Power supply	300-A, constant-potential transformer-rectifier(a)
Electrode holder	Fixed machine type
Electrode	0.030-in.-diam ER347
Shielding gas	90% helium, 7.5% argon, 2.5% CO_2; at 15 ft^3/h
Welding position	Flat
No. of passes	1
Voltage	25 V
Electrode-feed rate	216 in./min
Mode of transfer	Short circuiting
Electrode stickout	$^{3}/_{8}$ in.
Preheat	None

Welding speeds for all processes (different designs):

Gas welding (section A-A)	$^{3}/_{4}$ in./min
GTAW (section B-B)	1.1 in./min
GMAW (section C-C)	30 in./min

(a) With variable slope control

of the flange, and liquid-penetrant inspection indicated very few flaws. Because the weld was circular and easily accessible, the gas tungsten arc process was readily automated, using an adjustable torch mount, wire-feed drive, and a rotatable turntable. To obtain a consistently small weld bead without melt-through or overlapping in the thin-walled tube, voltage was automatically controlled. Despite the automation of equipment, the bead contour was insufficiently uniform.

Finally, automatic GMAW with short circuiting transfer was tried. The low heat input of the process allowed a small weld bead of the desired contour to be consistently deposited at a rate of 30 in./min. Argon purging of the inside of the tube was no longer needed. Once the machine settings were established, parts were welded with repetitive accuracy, meeting all specification requirements. A simple welding fixture was constructed to align the flange with the tube and to hold the assembly in position during welding. Supplementary tests showed that the welds on the vertical sides of the hub slots could be eliminated, further reducing production time. Fluctuations in line voltage had no serious effect, because at the low voltages used, only minor adjustments in the power supply were required.

Example 5. Full-penetration weld in 0.045-in.-wall tubing by automatic GMAW is shown in Fig. 10. Obround tubing for use in low-pressure heat exchangers was welded along the longitudinal seam. By automatic GMAW with short circuiting transfer, full-penetration welds were obtained at a welding speed of 30 in./min with relatively simple equipment and tooling. The as-sheared square-butt joints of the tubing were brushed with a stainless steel wire brush and wiped with a solvent cleaner. The 8-ft length of tubing was placed between the jaws of a hydraulically operated clamping fixture, which forced the joint edges together along the full length of the seam. Starting and runoff tabs of the same material and thickness as the base metal were tack welded in place to ensure sound weld metal at the tube ends.

The electrode holder, wire-feed drive, and filler-metal coil were mounted on a motor-driven carriage that traveled on a horizontal track above the joint. Welding was done in a single continuous pass. The significant variables in the process were joint fit-up, wire feed, and travel speed. Some melt-through occurred, which required repair welding, but rejections resulting from this amounted to less than 2%. After welding, completed tubes were leak tested at an air pressure of 5 psi under

Fig. 10 Obround tubing for heat exchanger

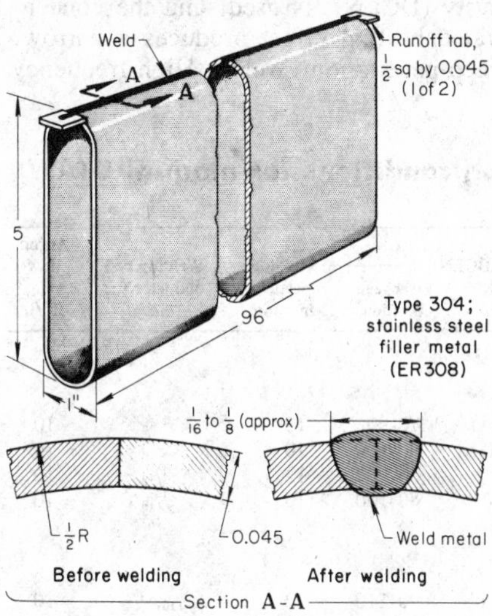

Process	Automatic GMAW(a)
Joint type	Butt
Weld type	Square groove
Fixtures	Seam-alignment clamp, torch carriage(b)
Power supply	300-A constant-potential dc transformer-rectifier(c)
Electrode holder	Machine type, water cooled
Electrode	0.035-in.-diam ER308
Shielding gas	90% argon, 7.5% helium, 2.5% CO_2; at 15 ft^3/h
Welding position	Flat
No. of passes	1
Current	85-95 A (DCEP)
Voltage	14 to 15 V
Electrode-feed rate	200-240 in./min
Electrode stickout	½ in.
Preheat and postheat	None
Welding speed	30 in./min

(a) With short circuiting transfer. (b) Arc starting and runoff tabs were used. (c) Slope and reactance controls

water. Qualification tests for the welding procedure and for operators were conducted according to section IX of the American Society of Mechanical Engineers (ASME) code.

Gas Tungsten Arc Welding

Gas tungsten arc welding is adaptable to both manual and automatic operation in all welding positions. The process can be used to produce continuous welds, tack welds, and spot welds with or without filler metal. The fundamentals of GTAW are covered in the article "Gas Tungsten Arc Welding" in this Volume. Because this basic information applies to all metals that can be welded by this process, only those modifications that apply specifically to stainless steel are considered in this article.

Applicability to Stainless Steel. Gas tungsten arc welding is particularly well suited to the welding of virtually all types and compositions of stainless steel, because (1) the filler metal, when used, does not pass through the arc and, as a result, does not undergo significant alteration in composition; (2) the shielding atmosphere enveloping the arc is chemically inert, which eliminates the hazard of gas-metal reactions; and (3) no flux is used, so there are no slag-metal reactions and no resultant nonmetallic inclusions. Transfer of elements from the filler-metal welding rod to the weld deposit is high, and the pickup of contaminants is extremely low. Because the properties and characteristics of stainless steels depend greatly on the maintenance of a preferred composition within close limits and on the avoidance of pickup of contaminants during the welding cycle, GTAW is well suited for these steels, whether or not a filler metal is used.

Alloys Welded. Gas tungsten arc welding can be applied to all weldable stainless steels in both the wrought and cast form, to clad products (such as stainless-clad carbon steel), and to dissimilar stainless steel alloys. The problems that may occur in welding certain alloys are not unique to GTAW, but rather reflect the specific weldability of the alloys.

Work-Metal Thickness. For stainless steels, GTAW is best suited, although not limited, to welding metal thicknesses of ¼ in. and less. Foil 0.002 in. thick has been welded successfully using automatic equipment. Thicknesses from about 5/16 to ½ in. generally can be satisfactorily welded using multiple passes. Sections thicker than ½ in. are usually not welded by GTAW because of high cost compared to other welding processes.

Workpiece Shape. Stainless steel imposes no limitations on workpiece shape that would not apply equally to the welding of carbon and low-alloy steels. General limitations are considered in the article "Gas Tungsten Arc Welding" in this Volume.

Welding Positions. Gas tungsten arc welding of stainless steel can be performed manually in the flat, horizontal, vertical, and overhead positions. Automatic vertical and overhead welding is possible with the use of special fixtures.

Current Characteristics. Although the general procedures for GTAW of stainless steel are similar to those used for low-carbon ferritic steel, certain of the physical properties of stainless steel necessitate some alteration of current characteristics in welding. Specifically, the stainless alloys have higher thermal expansion and lower thermal conductivity, which increases the likelihood of distortion during welding. They also have generally lower melting points, which results in a slightly higher rate of weld-metal deposition for the same welding current.

Direct current of the constant-current (drooping-voltage) type, with straight polarity (DCEN), is used, and the concentrated heating effect produces a narrow, deep-penetration weld. High-frequency starting, which necessitates the use of a high-frequency generator to provide an ionized path for the welding current, is most commonly used. Although scratch starting may be acceptable in some applications, it can result in electrode contamination of the weld. For this reason and because of the possibility of carbon pickup, the arc must not be struck on a carbon block.

Recommended ranges of welding current for several metal thicknesses and types of joints welded in the flat, vertical, and overhead positions are given in Table 12. Table 13 shows welding current values from actual conditions reported for manual GTAW.

Table 12 Joint designs and operating conditions for manual GTAW of austenitic stainless steel

Type of joint and weld(a)	Electrode diameter, in.	Welding current (DCEN), A: Flat position	Vertical position(b)	Overhead position	Welding speed(c), in./min	Welding rod diameter(d), in.	Argon flow rate, ft^3/h
1/16-in.-thick base metal							
Butt, A or B	1/16	80-100	70-90	70-90	12	1/16	10
Lap, F or G	1/16	100-120	80-100	80-100	10	1/16	10
Corner, H	1/16	80-100	70-90	70-90	12	1/16	10
T or corner, E or J	1/16	90-100	80-100	80-100	10	1/16	10
3/32-in.-thick base metal							
Butt, A or B	1/16	100-120	90-110	90-110	12	1/16, 3/32	10
Lap, F or G	1/16	110-130	100-120	100-120	10	1/16, 3/32	10
Corner, H	1/16	100-120	90-110	90-110	12	1/16, 3/32	10
T or corner, E or J	1/16	110-130	100-120	100-120	10	1/16, 3/32	10
1/8-in.-thick base metal							
Butt, A or B	1/16	120-140	110-130	105-125	12	3/32	10
Lap, F or G	1/16	130-150	120-140	120-140	10	3/32	10
Corner, H	1/16	120-140	110-130	115-135	12	3/32	10
T or corner, E or J	1/16	130-150	115-135	120-140	10	3/32	10
3/16-in.-thick base metal							
Butt, A or B	3/32	200-250	150-200	150-200	10	3/32, 1/8	15
Lap, G	3/32, 1/8	225-275	175-225	175-225	8	3/32, 1/8	15
Corner, H	3/32	200-250	150-200	150-200	10	3/32, 1/8	15
T or corner, E or J	3/32, 1/8	225-275	175-225	175-225	8	3/32, 1/8	15
1/4-in.-thick base metal							
Butt, B or C	1/8	275-350	200-250	200-250	5(e)	3/32-3/16	15
Lap, G	1/8	300-375	225-275	225-275	5(e)	3/32-3/16	15
Corner, H	1/8	275-350	200-250	200-250	5	3/32-3/16	15
T or corner, E or K	1/8	300-375	225-275	225-275	5	3/32-3/16	15
1/2-in.-thick base metal							
Butt, C or D	1/8, 3/16	350-450	225-275	225-275	3(f)	1/8-1/4	15
Lap, G	1/8, 3/16	375-475	230-280	230-280	3(g)	1/8-1/4	15
T or corner, E or K	1/8, 3/16	375-475	230-280	230-280	3(g)	1/8-1/4	15

A - Square groove; B - Square groove; C - Single-V groove; D - Double-V groove (Butt joints); E - Double fillet (T-joint)

F - Double concave fillet; G - Double convex fillet (Lap joints); Square groove (H, J); K - Single-bevel groove (Corner joints)

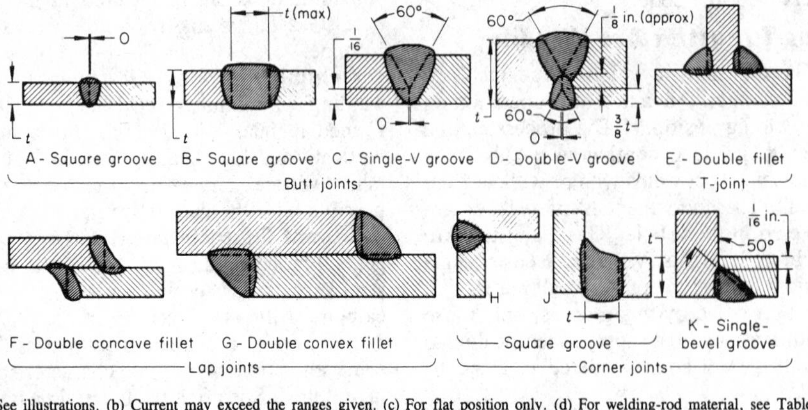

(a) See illustrations. (b) Current may exceed the ranges given. (c) For flat position only. (d) For welding-rod material, see Table 1. (e) Two passes may be needed. (f) Two or three passes may be needed. (g) For three passes
Source: R. P. Sullivan, Fusion Welding of Stainless Steel, *Metals Engineering Quarterly*, Nov 1967

Tungsten Electrodes. Of the five types of tungsten electrodes available for GTAW, the most widely used for stainless steel is EWTh-2 (97.5% W and 1.7 to 2.2% Th). The thoriated electrodes are recommended because of their excellent emissive qualities. They may be used at higher current than pure tungsten electrodes, and they provide exceptional arc stability. The EWTh-2 electrodes are obtainable in standard diameters of 0.010 to 1/4 in. Standard lengths generally used are 3, 6, and 7 in. Contact between the end of the electrode and the molten stainless steel pool must be avoided, as it may result in contamination of the weld metal. This contamination may affect the properties of the weld metal.

Shielding Gases. Depending on the application, argon, helium, or a mixture of the two is used in the welding of stainless steel. The same gases may also be used for backing the weld. Argon is usually preferred because it provides excellent protection at considerably lower flow rates. In welding thicknesses up to 1/16 in., melt-through is less likely when argon is used. Helium produces the higher heat input and deeper penetration required when welding thicknesses of more than about 1/16 in. However, welding is usually easier with argon, especially in manual applications, because heat input to the weld pool is less affected by variation in arc length. Argon is also less expensive than helium and is more readily available. Helium is preferred as a protective gas or backing gas in some applications and may be used with argon.

Filler Metals. Bare stainless steel filler metals for use in GTAW are covered by AWS specification A5.9-81. Designations and compositions of the filler metals are given in Table 7. Table 14 gives the standard sizes (diameters) of filler metals in the three forms used in GTAW of stainless steel. These include rod in straight lengths,

Table 13 Summary of operating conditions for GTAW of stainless steel from actual reported conditions

Base metal: Type of steel	Base metal: Thickness, in.	Electrode	Current (DCEN), A	Voltage, V	Filler metal	Argon flow rate, ft^3/h	Welding speed, in./min
Manual welding							
347	1/4	EWTh-2	85, 100	12-15	ER347(a)	20	...
347(b)	0.50	EWTh-2	70	12	None	15-20	...
304(c)	...	EWTh-1	34	15	None	15	2
304L	0.045, 0.090	EWTh-2	80-100	12-14	ER308L	15	2-4
304	0.109	EWP	70	14	ER308L	15	4
430	0.063	EWTh-2	31	...	ER430	5	...
316L	0.049	EWP	78	...	None	20	...
321 to carbon steel	0.014	EWTh-2	10, 55(d)	15, 9(d)	ER308L	10	(e)
321 to 321	0.032, 0.063	EWTh-2	80	10	ER347	10	3.25
321 to A-286	0.093, 0.063	EWTh-2	130	10	ERNiMo-6	10	2.75
321 to Inconel 718	0.188	EWTh-2	140	11-13	ERNiMo-6	20	3
304 to 348(f)	...	...	110-130	12-14	ER308	10-15	12-18
Automatic welding							
304	3/8	EWTh-2	190-220(g)	11	ER308L(h)	20	7
PH 15-7 Mo	0.040	EWTh-2	26-30	...	PH 13-8 Mo, ELC	(j)	...
PH 15-7 Mo	0.050	EWTh-2	65-70	...	(k)	24	...
446	0.016	EWTh-2	60	8	None	12	(m)
316	0.093, 0.095	EWTh-2	105	15	ER316	(n)	12.75
321	0.034	EWTh-2	40	8	ER347	30	5
347	0.050	EWTh-2	73	7.5	ER347	60	4.5
17-4 PH to AM-350	0.020	...	14	16	None	(p)	5
430	0.024	EWTh-2	80	...	None	25	60
17-7 PH to 316L	0.002	EWP	30	7	None	15	7
301 to 301	0.028	EWTh-2	18.5	15	None	(q)	10
301 to AM-350	0.028, 0.032	EWTh-2	50	15	ER308	(r)	10

(a) Type 347 consumable insert used for root pass. (b) Gas tungsten arc welding used for root pass only. (c) 1/16-in. fillet weld through 1/32-in.-thick shoulder. (d) For first and second passes, respectively. (e) Ten joints per hour. (f) Type 304 baseplates ranging from 1 3/4 to 3 1/2 in. thick were welded to 7 1/16-in. manifolds. (g) Different amperages between 190 and 220 for different passes. (h) Consumable insert ring of type 308L used for root pass. (j) Helium, at 120 ft^3/h. (k) Consumable insert ring of PH 15-7 Mo, ELC. PH 13-8 Mo, ELC, wire added when needed. (m) 1000 pieces per hour. (n) Helium, at 45 ft^3/h. (p) 75% helium and 25% argon, at 20 ft^3/h. (q) Helium, at 30 ft^3/h. (r) Helium, at 35 ft^3/h

Table 14 Standard sizes of filler-metal rod and wire used in GTAW of stainless steel

Form of filler metal	Diameter(a), in.
Rod in straight lengths(b)	0.045, 1/16, 5/64, 3/32, 1/8, 5/32, 3/16
Wire in coils(c)	0.045, 1/16, 5/64, 3/32, 7/64, 1/8, 5/32, 3/16, 1/4
Wire on 12-in.-OD spools	0.030, 0.035, 0.045, 1/16, 5/64, 3/32, 7/64

(a) Tolerances are ±0.001 in. on diameters of 0.045 in. or less, and ±0.002 in. on diameters of 1/16 in. and larger. (b) Standard length is 36 +0, −1/2 in. (c) With or without support
Source: AWS 5.9

wire in coils, and wire on 12-in.-OD spools. Straight lengths are normally used in manual GTAW, and coils and spools are used in automatic welding.

Filler metals are available that deposit weld metal of compositions comparable with those of the standard wrought stainless steels. However, selection of a filler metal is not solved solely by matching alloy designations or compositions. Other factors must be considered, such as the interalloying of filler metal and base metal, filler-metal selection for welding dissimilar metals, and the metallurgical structure and properties of the weld deposit. The filler metals suitable for welding similar stainless steels are given in Table 1, which also lists the metallurgical condition in which the resulting weldment can be put into service, a major consideration in the selection of filler metal. Table 15 shows filler metals suitable for welding joints between dissimilar austenitic stainless steels.

Use of Consumable Inserts. Consumable inserts offer an alternative method of supplying filler metal (1) when making the

Table 15 Filler metals suitable for welding joints between dissimilar austenitic stainless steels

Base metal A(c)	Suitable filler metals(a) — Base metal B(b): 304L	308	309	309S	310	310S	316, 316H	316L	317	321, 321H	347, 347H 348, 348H
304, 304H, 305	308L	308L	308, 309	308, 309	308, 309, 310	308, 309, 310	308, 316	308, 316	308, 316, 317	308	308
304L	...	308L	308, 309	308, 309	308, 309, 310	308, 309, 310	308, 316	308L, 316L	308, 316, 317	308L, 347	308L, 347
308	...	...	308, 309	308, 309	308, 309, 310	308, 309, 310	308, 316	308, 316	308, 316, 317	308	308, 347
309	...	...	...	309	309, 310	309, 310	309, 316	309, 316	309, 316	309, 347	309, 347
309S	...	...	...	...	309, 310	309S, 310S	309, 316	309S, 316L	309, 316	309, 347	309, 347
310	...	...	...	...	...	...	310, 316	310, 316	310, 317	308, 310	308, 310
310S	...	...	...	...	...	...	316	316	317	308, 310	308, 310
316, 316H	...	...	...	...	...	...	...	316	316, 317	308, 316	308, 316, 347
316L	...	...	...	...	...	...	...	...	317	316L	316L, 347
317	...	...	...	...	...	...	...	...	...	308, 317	308, 317, 347
321, 321H	...	...	...	...	...	...	...	...	...	...	308L, 347

(a) Listed in no preferred order; prefix ER omitted. (b) Type of steel being welded to base metal listed in first column. (c) The H suffix indicates a grade used for tubing for high-temperature service.

root pass in butt joints where accessibility is limited to one side of the joint; (2) where smooth, uniform, crevice-free root side contours are mandatory; or (3) where a root pass of highest quality is essential to the integrity of the completed weld. For pipe and tube joints, inserts are available in the form of rings of suitable diameter. Gas tungsten arc welding results in the consumable insert being completely fused with the base metal.

Stainless steel inserts are of compositions formulated to promote weldability and to reduce the possibility of weld cracking. With inserts, smooth, uniform melting on pipe interiors is relatively easy to obtain, and weld contour appears to be relatively insensitive to welding variables. In joints that must satisfy high acceptance standards under visual, liquid-penetrant, and x-ray radiographic examination, the use of insert rings provides a root pass of highest quality. This benefit must be weighed against the cost of using insert rings.

The production of nuclear reactor components required the welding of two circumferential butt joints in stainless steel pipes that, when assembled, served as a housing for drives. This nuclear application imposed rigid acceptance standards on weld quality, pipe alignment, and overall dimensional tolerances. These standards were met by using consumable inserts and automatic welding.

Example 6. Consumable insert rings were used for root-pass welding of circumferential butt joints in piping. A typical joint (Fig. 11) consisted of a single-V-groove with an insert ring fitted to the abutting ends of 5-in. schedule 40S 0.258-in.-wall piping. The piping was used in either the primary or secondary system of a nuclear reactor and was required to withstand pressure of 2 ksi at an operating temperature of 600 °F under water. The welded joints had to meet rigid specifications, which included:

- Complete joint penetration, with no notches at the root
- Smooth inner surface, free from scale and oxides
- Freedom from slag inclusions and cracks
- Minimum porosity, as determined by radiographic inspection

Manual GTAW was selected for welding the root pass. It was also used for the filler passes because the wall thickness of the pipe was not great enough to warrant changing to another process. The single-V-groove joint design was adopted because it was simple to prepare, afforded

Fig. 11 Nuclear reactor pipe with consumable insert ring

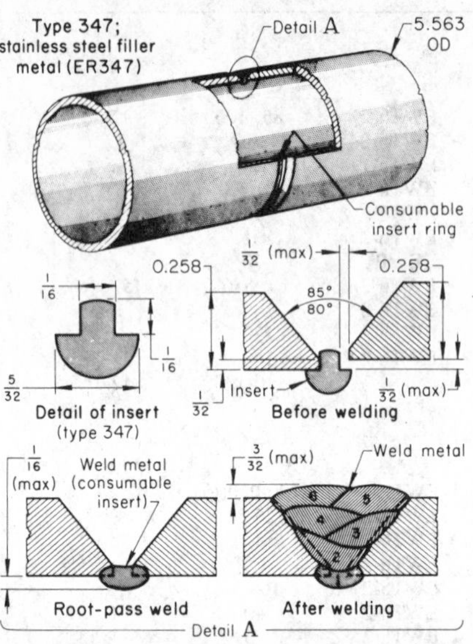

Process	Manual GTAW
Joint type	Circumferential butt
Weld type	Single-V-groove
Fixtures	Turning rolls (shop); none (field)
Power supply	300-A motor-generator or transformer-rectifier
Electrode holder	200 A, water cooled
Electrode	3/32-in.-diam EWTh-2(a)
Filler metal:	
Root pass	Type 347 consumable insert ring(b)
Filler passes	3/32- and 1/8-in.-diam ER347
Shielding gas	Argon, at 20 ft^3/h(c)
Welding position	Horizontal-rolled pipe (in shop)
Arc starting	High frequency
Arc length	3/32-1/8 in.(a)
Current (DCEN)	Root pass, 85 A; filler passes, 100 A
Voltage	12-15 V(a)
Preheat and postheat	None

(a) For root pass and filler passes. (b) Containing 5 to 12% ferrite. (c) Argon, at 20 ft^3/h, was also used for purging and backing.

adequate access to the root of the joint, and did not require an excessive quantity of filler metal to make the weld.

The insert rings were modified to obtain 5 to 12 FN in the weld to control microfissuring and cracking. The melting characteristic of the inserts and the manner in which the weld pool rose in the joint because of the surface tension enabled the welder to see exactly when the root pass had been properly made and the insert ring fused. The insert ring, in wire form, was carefully fitted and tack welded to the inside surface of one of the pipe sections. Any overlap was trimmed off, and the ends were properly matched. The second pipe section was placed in position and aligned, and the complete assembly was held together by regularly spaced tack welds.

Before making the root pass, the inside of the pipe was purged with argon, and the flow of argon was continued during welding to provide backing gas and internal shielding. The general appearance of the root-pass weld is shown in Fig. 11, lower left, and a cross section through the completed joint, showing the number of filler passes, is shown in the lower right. After each operation (root pass and filler passes), welds were closely inspected by visual and liquid-penetrant techniques. The completed weld was then subjected to radiographic examination.

Root-Pass Welds Without Inserts. When the use of backing rings is prohibited, it is not essential to use consumable inserts to obtain smooth internal joint surfaces on pipe welds. An alternative method is to provide a groove weld with a root face of a width that permits complete root fusion in a single pass, using manual GTAW and a gas backing. Filler metal is not used on the initial root pass. Subsequent passes to fill the joint may employ a different process, such as SMAW, because deposition of filler metal by manual GTAW is relatively slow and expensive.

Example 7. Pipe Welding Without Filler Metal for Root Pass. The 10-in. schedule 80S pipe shown in Fig. 12 was used in oil refinery equipment. Specifications called for pipe joints to have sound, complete-penetration welds conforming to the radiographic standards and other rules of the ASME code for unfired pressure vessels. The pipe joints also were required to have smooth, well-rounded inside surfaces; therefore, the use of solid backing strips was prohibited. Because there was no access to the inside of the joints, the success of the operation depended on the skillful execution of the root pass from the outside.

Manual GTAW with argon shielding and backing, but without the use of filler metal, was selected for the root pass. For the filler passes, SMAW was chosen. Before root-pass welding, the pipe ends were beveled to the configuration shown in Fig. 12 (before welding). The joint area was cleaned of all oil, water, and foreign matter. The pipe ends were then aligned, with root faces tightly butted and tack welded. The tack welded pipes were supported in position for welding and purged with argon. This was accomplished by inserting dams as close to the joint as possible on either side and feeding argon through one dam and venting it (under back pressure) through the other. After purging, a low flow of ar-

Fig. 12 Root pass without filler metal used on oil refinery pipe

Type 347 base metal. Filler metals: none for root pass; filler passes, stainless steel (E347-16)

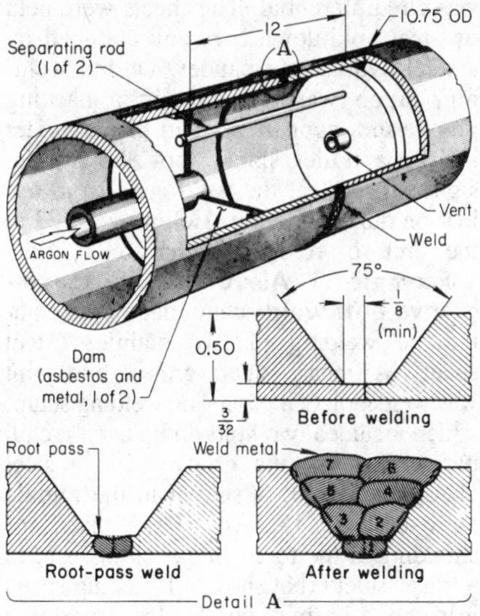

Root pass (GTAW)

Power supply	300-A transformer-rectifier(a)
Electrode holder	200 A, water cooled
Electrode	3/32-in.-diam EWTh-2
Filler metal	None(b)
Shielding gas	Argon, at 15-20 ft^3/h
Backing gas(c)	Argon, at 8 ft^3/h
Welding position	Horizontal-fixed pipe
Arc length	1/8 in.
Current	70 A (DCEN)
Voltage	12 V

Filler passes (SMAW)(d)

Power supply	300-A transformer-rectifier(a)
Electrode	E347-16 (1/8 in., first filler pass; 5/32 in., others)
Welding position	Horizontal-fixed pipe
Current (DCEP)	100 A, first filler pass; 130 A, others
Voltage	18 V, first filler pass; 20 V, others
Interpass temperature	350 °F max

(a) With high-frequency generator, to facilitate arc starting. (b) In repair welding of root pass, 3/32-in.-diam ER347 welding rod was used. (c) Gas backing was used because the welded joint was required to have a smooth, well-rounded inner surface. Dams inserted in the pipe confined argon for purge backing and ensured minimum waste. (d) Two-process welding, resulted in 98% weld acceptability.

gon was maintained to provide backing during welding.

Welding of the root pass was made in two semicircular halves beginning at the 6 o'clock position. The completed root-pass weld is shown in cross section (Fig. 12, lower left). In the horizontal fixed position, fusion of the root faces to form a uniformly smooth inside surface called for a high degree of welder skill because the forces of gravity and surface tension acting on the molten weld pool were in opposite directions at the top of the pipe and co-directional at the bottom, resulting in either convexity or concavity at the weld face. Thus, to ensure complete penetration, the formation of a large superheated weld pool was avoided by control of heat input and welding speed. Both visual and liquid-penetrant examinations were made on the completed root-pass weld. All defects were completely removed by grinding, and the defective areas were rewelded by GTAW. The repair welds were given visual and liquid-penetrant inspection, and if no further defects were found, the joint was cleaned and prepared for making the filler passes.

Filler-pass welding was done by SMAW, using the same starting position and welding directions for each pass as in root-pass welding, but with a weaving technique. During the first filler pass, care was exercised to avoid melt-through of the relatively thin root-pass weld bead. After each of the six filler passes (Fig. 12, lower right), slag was carefully removed by wire brushing or, if necessary, chipped out. Each pass was inspected visually for undercutting, slag inclusions, incomplete fusion, and porosity. Areas with any of these defects were repaired by grinding and rewelding.

Interpass temperature, measured by a contact pyrometer, was maintained at 350 °F maximum, but no preheat, postheat, or postweld heat treatment was used. The completed weld was inspected by either x-ray or gamma-ray radiography. Because of the care exercised in welding the root pass, defects averaged only 2% of the weld length.

Joint Design. Some typical joint designs for GTAW of stainless steel, together with welding conditions for base metal in thicknesses ranging from 1/16 to 1/2 in. inclusive, are given in Table 12. The basic reasons for an altered joint design are:

- *Improved efficiency*: Contributes to basic engineering requirements; promotes welding efficiency; and conserves cost, production time, and labor
- *Structural soundness*: Improves strength or rigidity and avoids welding defects that would reduce structural soundness
- *Accessibility*: Provides physical ability to make a weld
- *Alignment*: Produces accurate alignment where fixtures are unable to be feasibly or fully effective
- *Quality*: Provides thoroughness of quick cleaning and sterilizing by ensuring absence of pits, sharp pockets, or deep corners, where process materials can be entrapped; provides smoothly polished, corrosion-resistant interiors with generously filleted corners and large transition radii often prescribed by sanitary codes; and produces stainless steel weldments used extensively in fabrication of equipment for food processing and for processing organic materials subject to microbial attack or deterioration through decomposition

Manual GTAW is frequently used for these weldments because it provides good deposition control and clean welds, and varying contours can be followed with comparative ease. Wherever possible, butt welds are used in preference to fillet welds because they are easier to grind and finish and are less likely to have crevices that cannot be cleaned.

Example 8. Joint to Eliminate Use of Filler Rod. Production of stainless steel molds for processing meat products required the welding of 19 600 positioning buttons to the surface of the mold tubes. A portion of a meat processing mold with a positioning button welded to its surface is shown in Fig. 13. The original joint design called for placing the flat end of a button on the curved surface of the mold and depositing a 1/16-in. fillet weld around the base of the button, using manual GTAW. A clamping fixture positioned the button and held it on the tube during welding. The welder could have difficulty in maneuvering the electrode holder and filler rod around the button and also in maintaining a uniform fillet size. This was avoided by changing the joint design, so that the buttons could be welded to the mold surface without the use of filler metal by machining the buttons from 3/4-in.-diam bar stock, to provide them with a shoulder 1/16 in. wide and 1/32 in. thick at the joint end. The 3/4-in.-diam shoulder end of the button was welded without filler metal to the curved surface of the mold tube, with the shoulder forming a 1/16-in. circumferential fillet (improved design, Fig. 13). With the elimination of the filler rod, the welder had only the electrode holder to manipulate, resulting in increased welding speed, improved penetration control, and more uniform fillet size.

The completed welds were wire brushed to remove any discoloration and visually inspected for smoothness and fillet uniformity. Production rate averaged 38 buttons per hour. The improved welding speed and elimination of the filler rod more than offset the additional cost of machining the modified buttons.

Example 9. Redesign of tube-to-tube-sheet joint resulted in improved

Fig. 13 Redesigned positioning button for welding without filler metal

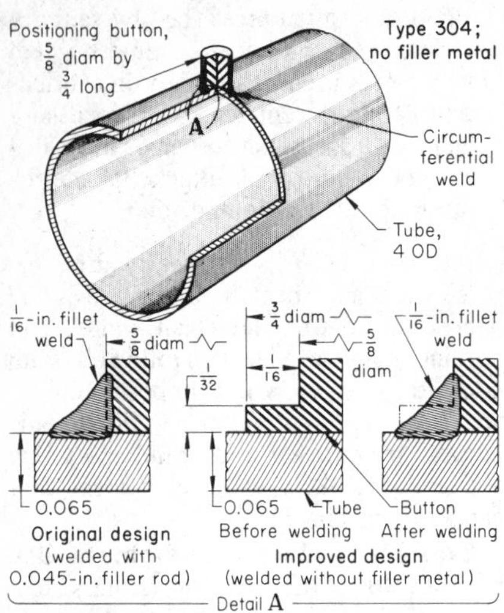

Welding conditions for improved design

Process	Manual GTAW
Joint type	T, with shoulder
Weld type	Circumferential fillet
Fixtures	Clamping jig
Power supply	250-A, three-phase transformer-rectifier
Electrode holder	300 A, water cooled
Electrode	1/16-in.-diam EWTh-1
Filler metal	None (shoulder was fused)
Shielding gas	Argon, at 15 ft^3/h
Welding position	Horizontal
Arc starting	Scratch
Arc length	1/8 in. (approx)
Current	34 A (DCEN)
Voltage	15 V
Welding speed	2 in./min (avg)
Production rate	38 buttons per hour

weld quality, strength, and structural soundness (Fig. 14). Heat exchangers, designed for heating or cooling gases of various types, were made of parallel rows of oval-section tubes seal-welded to a tube sheet. In the original joint design (Fig. 14), the tube sheet was flared inward, and each seal weld was deposited between the rounded corner of the tube sheet and tube wall. The welds, although smooth and pleasing in appearance, were not of uniform quality, and many were structurally unsound. In attempting to obtain penetration to the point of tangency between tube and tube sheet, the welder sometimes melted through the tube sheet. Furthermore, with this joint design, the weld was in the area where high stress concentration was developed in service.

These problems were eliminated by an improved joint design (Fig. 14b). The thickness of the tube sheet was increased from 0.045 to 0.090 in. for greater strength. The tube sheet was flared outward, enabling the welder to deposit a bead on the solid, exposed edges of the tube and tube sheet to avoid melt-through. The improved joint resulted in more uniform welds and a minimum of weld repairs.

After welding, the joints were inspected visually for continuity, size, and general quality. If required, the welds were liquid penetrant inspected for cracks. Each welder was required to pass a qualification test as set forth in section IX of the ASME code on welding qualifications.

Automatic gas tungsten arc welding can be mechanized to perform some or all of the functions and controls of a welder. The degree of mechanization generally is determined by the number of identical welds to be made and by the speed and quality desired (see the article "Gas Tungsten Arc Welding" in this Volume).

Welding of Sheet. Automatic GTAW can be adapted readily to the welding of stainless steel sheet and products formed from sheet. It is widely used in the aircraft industry for this purpose because of its reliability and high joint efficiency.

Example 10. Aircraft-quality butt welds were produced by joining two stainless steel sheets by automatic welding. The joint design and work-holding setup for welding the sheets of slightly different thicknesses in a single pass are shown in the top view of Fig. 15. A sheet, 0.095 in. thick, had been formed to the shape of a tank head, and the butt joint was circumferential. The sheets were held by steel hold-down bars and clamped by a steel hydraulic expander. During welding, argon was fed through a backing groove in a copper insert in the expander and to a trailer shield (not shown) designed to protect the weld zone from oxidation during cooling. Helium was fed to the torch to provide a hotter arc.

Example 11. Aircraft-quality square-groove butt welds were made by the automatic welding of two stainless steel sheets. A linear, square-groove butt joint was welded in one pass. The welding setup, which included two hold-down bars, a chill bar with a backing groove, and a steel clamping fixture, is shown in the middle view of Fig. 15. Argon was used for shielding at the torch, for backing, and in a trailer shield (not shown). Welding speed was only 5 in./min, but this low speed was justified by the high quality of the welds in an application in which welding speed was not a major criterion.

Example 12. Automatic welding was used to produce an aircraft-quality square-groove butt joint in one pass between two stainless steel sheets. The welding setup

Fig. 14 Redesigned joint in a tube-to-tube-sheet weldment

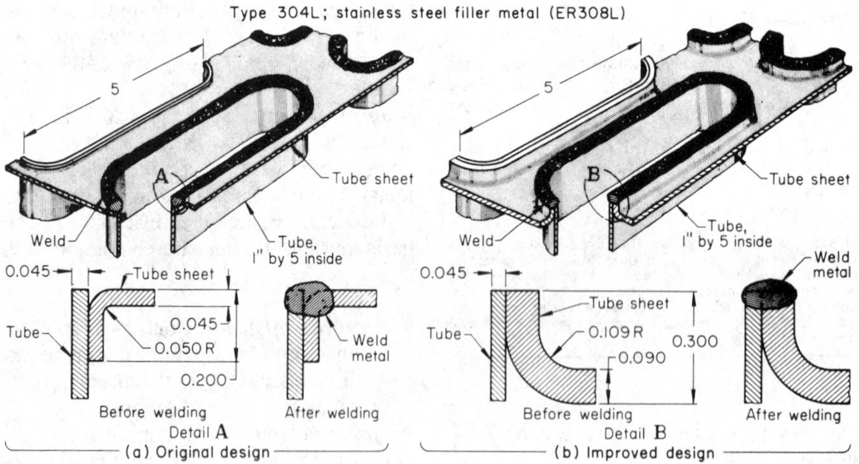

Welding conditions for improved design

Process	Manual GTAW
Joint type	Edge
Weld type	Corner flange
Fixtures	Fit-up and welding table
Preweld cleaning	Solvent; stainless wire brush
Power supply	300-A, ac/dc with high frequency
Electrode holder	300 A, water cooled
Electrode	3/32-in.-diam EWTh-2
Filler metal	3/32-in.-diam ER308L
Shielding gas	Argon, at 15 ft^3/h
Welding position	Flat
No. of passes	1
Arc starting	High frequency
Current	80-100 A (DCEN)
Voltage	12 to 14 V
Welding speed	2-4 in./min (approx)

Fig. 15 Aircraft-quality sheet-to-sheet butt joint by automatic welding

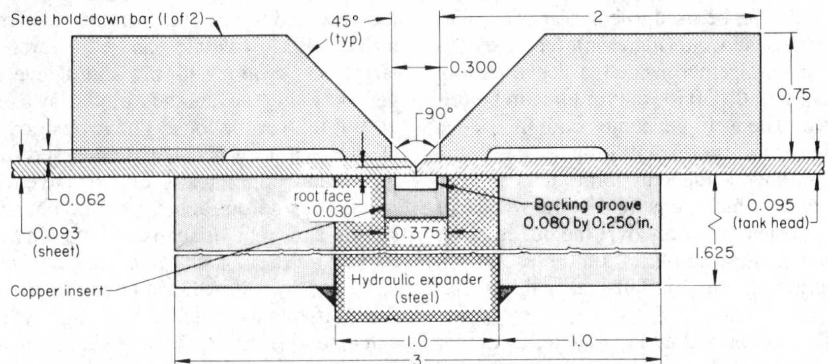

Type 316; stainless steel filler metal (ER316)

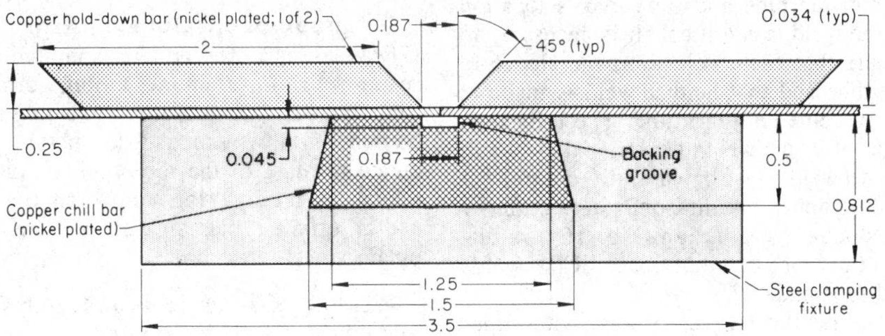

Type 321; stainless steel filler metal (ER347)

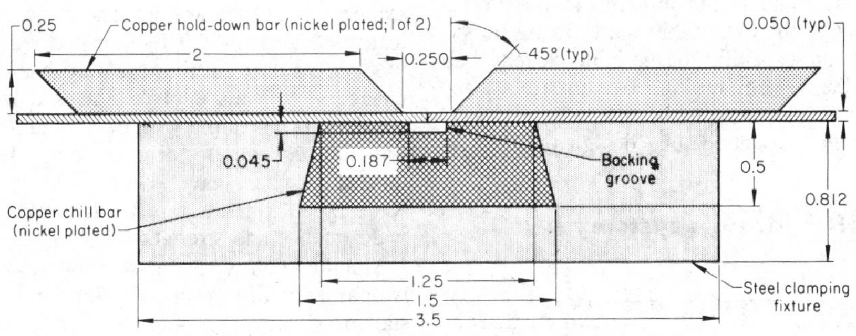

Type 347; stainless steel filler metal (ER347)

Welding condition(a)	Example 10	Example 11	Example 12
Electrode holder(b)	250 A	250 A	250 A
Electrode(c)	$^3/_{32}$-in. EWTh-2	$^1/_{16}$-in. EWTh-2	$^1/_{16}$-in. EWTh-2
Filler metal(c)	$^3/_{64}$-in. ER316	$^1/_{32}$-in. ER347	$^1/_{32}$-in. ER347
Filler-metal feed, in./min	48.5	6	9
Shielding gas:			
At torch	Helium; 45 ft^3/h	Argon; 30 ft^3/h	Argon; 60 ft^3/h
Backing	Argon; 30 ft^3/h	Argon; 30 ft^3/h	Argon; 80 ft^3/h
Trailer shield	Argon; 20 ft^3/h	Argon; 30 ft^3/h	Argon; 60 ft^3/h
Current (DCEN), A(d)	105	40	73
Voltage, V	15	8	7.5
Welding speed, in./min	12.75	5	4.5

(a) For all three welds, joint preparation consisted of either machining or drawfiling the joint edge, followed by cleaning with silicon carbide abrasive and wiping with a solvent. (b) Water-cooled type. (c) Size given is diameter. (d) Power supply for all three welds was a 300-A transformer-rectifier.

is shown in Fig. 15 (bottom view). Welding speed was only 4.5 in./min.

Gas tungsten arc spot welding is well suited to the welding of austenitic stainless steel, particularly in thicknesses of 0.020 to 0.090 in. Manual equipment and completely mechanized equipment are available. The essential components are a torch (usually equipped with a vented nozzle), a power supply, a tungsten electrode, a trigger switch, and a timing unit. In operation, the torch is placed against the work, the trigger is depressed, and a spot weld is made at a predetermined time. Ordinarily, filler metal is not used. Direct current electrode negative is most commonly used.

Arc starting is accomplished by the following methods:

- *High-frequency starting* is most commonly used, although unreliable below about 30 A.
- *Retract starting* consists of advancing the electrode to the work and then retracting it to establish an arc.
- *Pilot arc starting* makes use of a low-intensity arc to establish a high-intensity arc. The low-intensity arc is maintained from the tungsten electrode to an auxiliary electrode, usually the nozzle. It ionizes the shielding gas, thus facilitating arc starting.

Because the electrode is usually recessed and because arc length must be closely controlled, starting of the arc may pose a problem. Manual scratch starting is never used.

Advantages Over Resistance Spot Welding. The advantages of gas tungsten arc spot welding are:

- Welds can be made when only one side of the workpiece is accessible.
- Thickness ratio of pieces being joined is less important.
- Only cleaning typically used for stainless steel is necessary to obtain arc spot welds of good quality.
- On equal cross sections, shear strengths are equivalent to those obtained with resistance spot welds.
- Capital outlay required for equipment is considerably less, particularly for short production runs.

Welding of Tubing. Because of the wide variety of industrial and engineering uses of stainless steel tubing, welding of tube has widespread application both in the manufacture of tubes and in the assembly of tubular components. Manual and automatic gas tungsten arc processes are well suited to the welding of stainless steel tubing, especially when tube-wall thickness does not exceed about $^1/_4$ in. For greater wall thicknesses, GTAW may be used to make the root pass and one or two subsequent passes. To reduce cost, the bulk of the weld metal is more likely to be deposited using a process that has a higher deposition rate, such as GMAW.

Circumferential Welding. Several stainless steels, including types 304L, 321, 347, and AM-350, are widely used in high-reliability hydraulic systems, such as those installed in aircraft. Choice of tube fittings and connections for these systems is often difficult because of the exacting requirements imposed. Mechanical threaded connections have limited ability to seal and to consistently maintain maximum reliability. Welded joints may be unsatisfactory or incapable of meeting all requirements, depending on joint design and welding procedure. One specification requires that the weld penetration be 100%, that the strength of the weld at the service temperature be equal to that of the tubing, and that the weld be capable of meeting all performance requirements, including repeated flexure at 35-ksi stress levels.

Designs of welded tube fittings for hydraulic systems are shown in Fig. 16. The straight butt weld (Fig. 16a) is seldom used for thin-walled hydraulic tubing because fixturing is generally needed for control of alignment and because the tubing is unsupported in the annealed HAZ adjacent to the weld. The wedding-band design (Fig. 16b) constitutes a slight improvement over the butt weld. The band is completely fused during welding, but the HAZ still are unsupported.

The two designs most widely used are the sleeve butt weld (Fig. 16c) and the sleeve double weld (Fig. 16d). In these designs, a loose tubular sleeve supports the ends of the tubes and also serves to provide weld metal. Either a single or a double weld can be used to complete the joint. With a single weld (Fig. 16c), the tube ends are butted together inside the sleeve. A root opening of 0.030 in. maximum can be tolerated. The weld penetrates both the sleeve and tube to fuse the butt joint. Because the sleeve covers the butt joint, it is not visually possible to ensure that the weld will be deposited precisely over the butted ends. Reliance must be placed on the use of appropriate guide marking or gaging fixtures.

The use of a double weld on a tubular sleeve fitting (Fig. 16d) or of an expanded tube lap weld (Fig. 16e) eliminates the need to butt the tube ends and makes location of the weld less critical. It is necessary to ensure that each weld joins the sleeve to the tube and that neither weld is made at the extreme end of a tube. These two designs place the weld in shear, and the width of the weld must be carefully controlled. Radiographic or ultrasonic inspection of the double-weld design (Fig. 16d) is difficult because of interference of the welds and tube ends.

The tubular sleeve designs, with either a single or a double weld, and the expanded tube design support the tube in the HAZ to the extent that the ultimate burst-pressure capability of the joint exceeds that of the system tubing. Because of clearances, however, neither the tubular sleeve nor the expanded tube design provides a snug fit to withstand vibration or flexing. Calculations indicate that the deflection required to take up the clearance exceeds the flex endurance limit of the tube material. This weakness can be overcome largely by swaging tubing and sleeve together. Other advantages of the swaged-on fitting over the loose-sleeve design are: (1) no clearance between sleeve and tube to entrap contaminants or corrosive elements, (2) no entrapment of air between sleeve and tube to cause contamination during welding, and (3) no need to purchase controlled-tolerance tubing or to size the ends of tubes before welding. A patented modification of the swaged-on design is discussed in the section on PH grades in this article.

Fig. 16 Welded tube fittings for aircraft hydraulic systems

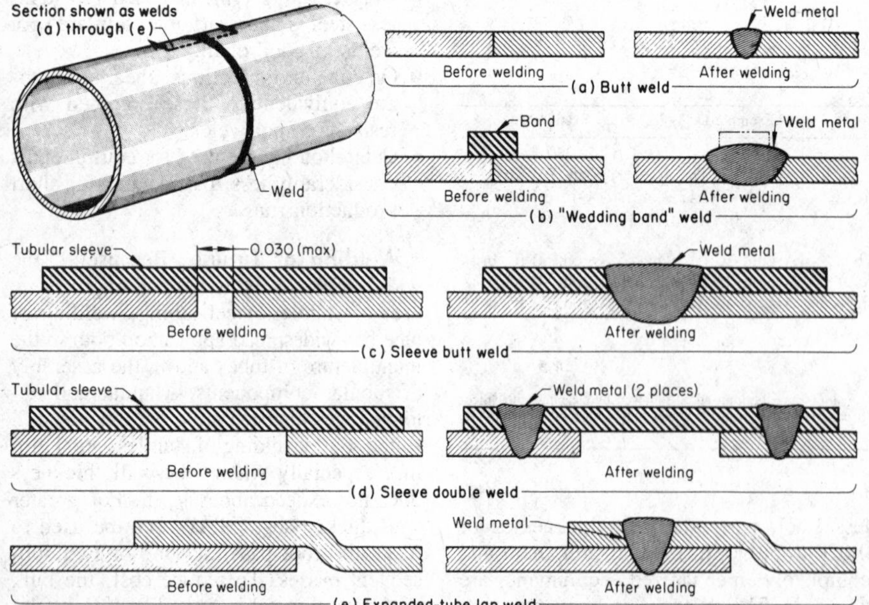

Longitudinal Welding. Automatic GTAW is used for making longitudinal seam welds in structural tubing formed from thin stainless steel sheet. Welding can be done with or without filler metal. Accurate forming of the tube section, close tolerances on fixturing and torch travel, and an efficient heat sink are required to obtain a high production rate.

Welding of Tubes to Tube Sheet. Gas tungsten arc welding is used in joining many types of assemblies that embody stainless steel tubular components. A typical application is the joining of tubes to tube sheets used in high-pressure heat exchangers. Various joint designs and weld configurations, along with manual and automatic procedures for producing these critical welds, have been developed. Redesigned joints can result in improved weld quality and strength.

Much of this type of assembly welding is done manually, especially on small installations, although automatic equipment has been developed for high production. Manual welding requires accurate control over torch movement because the tubes used for these weldments have thin walls and usually are of small diameter. The simplest joint design consists of a tube inserted in a straight drilled and reamed tube hole having no special weld-groove preparation and requiring no filler-metal addition.

Example 13. Manual gas tungsten arc welding of tubes to tube sheet was done without filler metal (Fig. 17). Heat exchangers for use with corrosive gases contained 110 tubes that were seal welded in a tube sheet. The tubes and tube sheet were designed for a maximum working tube pressure of 300 psi at 800 °F. The tube-to-tube-sheet joints were welded by manual GTAW in accordance with the re-

Fig. 17 Tube-to-tube-sheet weldment with simple joints

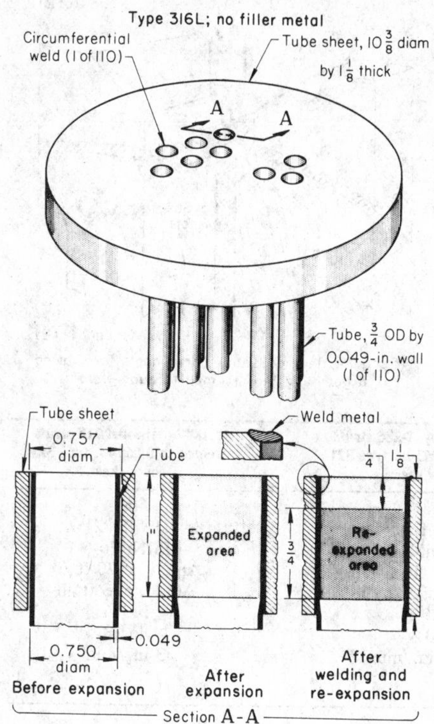

Process	Manual GTAW
Joint type	Square edge
Weld type	Seal
Power supply	400-A motor-generator
Electrode holder	300 A, water cooled
Electrode	1/16-in.-diam EWP
Filler metal	None
Shielding gas	Argon, at 20 ft^3/h
Welding position	Flat
Arc starting	High frequency
Arc length	3/16 in. (approx)
Current	78 A (DCEN)

quirements of the ASME code for unfired pressure vessels.

Before welding, the tubes and tube sheet were deburred and solvent cleaned, using suitable precautions. No special edge preparation was required. The tubes were inserted in the 0.757-in.-diam holes in the tube sheet, with the ends of the tubes flush with the tube-sheet face. The tubes were lightly expanded for a depth of 1 in. to hold them in place. Each was welded in a single pass without filler metal. After welding, the tubes were re-expanded for a distance of 3/4 in., starting 1/4 in. below the welded joint. After assembly in the heat exchanger shell, the tube bundle was subjected to a 600-psi hydrostatic test. Rejections because of tube-weld leakage averaged less than 0.01%.

Welding of thin sections of stainless steel requires the use of low arc currents. The gas tungsten arc process, widely used for joining thin sections, is generally satisfactory for welding at currents down to about 10 A. Below this level, considerably more skill is required when welding manually because a small change in torch standoff and arc length produces a change in arc voltage and results in a fairly large change in current. Automatic welding may therefore be mandatory to prevent significant fluctuation in arc current.

The welding of stainless steel in foil thicknesses can be simplified by converting a butt joint into some type of edge joint. Such a joint is easier to weld because it permits the widest latitude in fixturing tolerance and welding conditions.

Restricting the Heat of Welding.

Small electrical assemblies are often protected by enclosure in sheet metal housings which may be sealed mechanically or by soldering, brazing, or welding, depending on the specific application and on cost. The metal selected for the housing, as well as the method for sealing, usually depends on service requirements. When a small, delicate electrical switch assembly is enclosed in a stainless steel housing sealed by welding, extreme care must be taken to keep the assembled components clean, dry, and relatively cool. Heat from welding must be minimized. Ordinarily, only automatic welding can provide the degree of control required to minimize the heat input and produce a smooth, leakproof joint.

Example 14. Automatic Precision Welding of Microswitch Housing Assembly. Enclosures for protection of microswitch assemblies consisted of rectangular housings with rounded corners welded on one side to a cover and hermetically sealed on the other, after backfilling with nitrogen, with a 0.10-in.-thick layer of dielectric material. Welding of the recessed corner joint between the cover and housing (Fig. 18, before welding) was accomplished in a single pass by automatic GTAW with argon shielding and backing, but without addition of filler metal. The welded joint was required to be leakproof, with the leakage rate for each switch, after sealing, not exceeding 0.2 cm^3/year. Extreme care was required to minimize heat input during welding to prevent damage to the brazed flexible diaphragm and other delicate components of the microswitch.

Several modifications in the process were adopted to minimize heat input. The tungsten electrode was taper ground to a point to obtain a fine, high current density arc at low amperage. Straight polarity was selected over reverse polarity to produce a finer arc and to avoid overheating the electrode tip. The combined effect of these modifications was to increase welding speed while limiting heat input to a narrow weld zone and HAZ. Argon was selected for shielding, rather than helium, to obtain smoother arc action at lower voltage. Arc starting was also easier and heat input more stable because small changes in arc length resulted in less change in arc voltage.

Fig. 18 Microswitch housing assembly

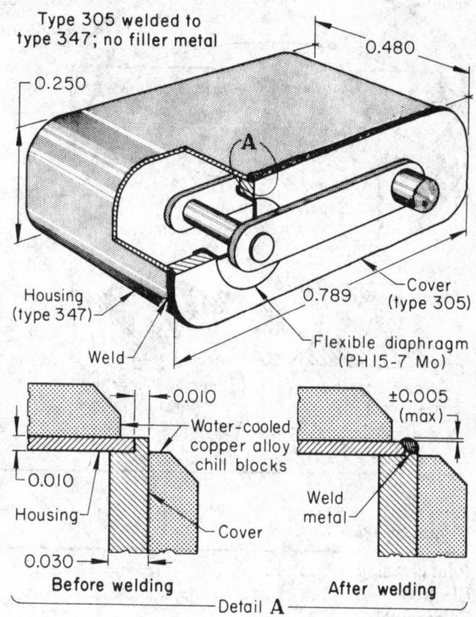

To facilitate rapid withdrawal of heat from the weld zone, the housing and cover were clamped in a water-cooled chill-block fixture that exposed only the corner to be welded (Fig. 18, before welding). The fixture was attached to a sliding mechanism designed for easy loading and unloading of the workpiece and for positioning the joint directly under the electrode. Motion of the electrode holder was controlled by a magnetic tracing unit, which was driven by rotating a knurled magnetic pin around the edge of a steel template. An ultraviolet lamp aimed at the electrode tip provided an ionized path for easy arc starting. Welding current was supplied by a 200-A transformer-rectifier. The entire welding cycle was regulated by a sequence timer equipped with controls for argon preflow and postflow, arc starting and stopping, torch travel, and current downslope. Welding was done in two steps. First, one half of the joint was welded from end to end by melting the corner edge of the cover to form a smooth radius. The assembly was then unclamped, rotated 180°, and reclamped. The other half was

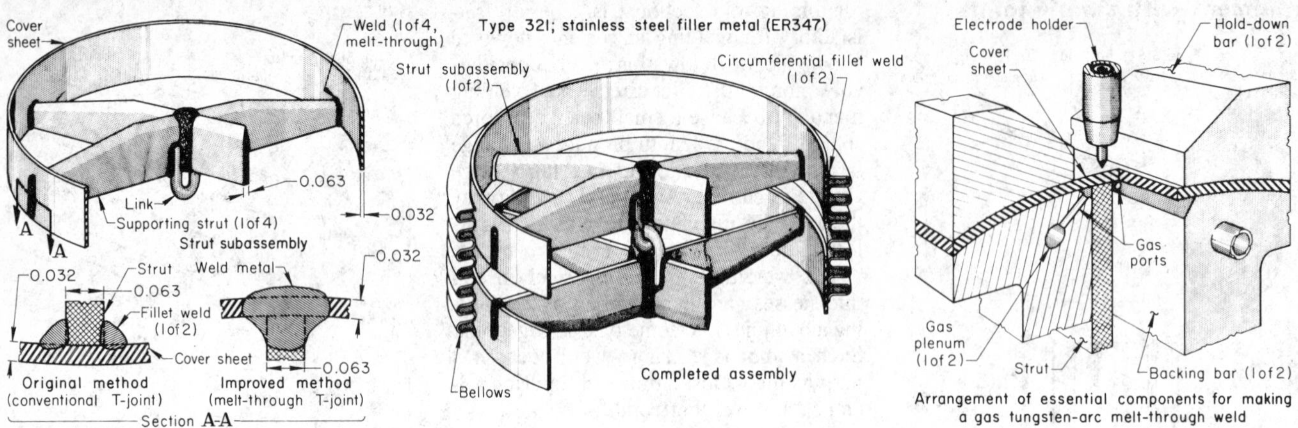

Fig. 19 Single-pass melt-through welding of T-joints on a strut-cover sheet assembly

Welding condition	0.063-in. type 321 struts welded to 0.032-in. type 321 cover sheet, as shown	0.093-in. alloy A-286 struts welded to 0.063-in. type 321 cover sheet	0.188-in. Inconel 718 struts welded to 0.060-in. type 321 cover sheet
Joint cleaning	(a)	(a)	(a)
Electrode	EWTh-2	EWTh-2	EWTh-2
Filler metal	ER347	ERNiMo-6	ERNiMo-6
Shielding gas at nozzle	Argon, at 10 ft^3/h	Argon, at 10 ft^3/in.	Argon, at 20 ft^3/h
Backing gas	Argon, at 3 ft^3/h	Argon, at 3 ft^3/in.	Argon, at 3 ft^3/h
Current (DCEN)	80 A	130 A	140 A
Voltage	10 V	10 V	11 to 13 V
Welding speed	3.25 in./min	2.75 in./min	3 in./min

(a) All joints were degreased, pickled, rinsed, and dried before welding.

then welded in the same manner. Production was 59 switches per hour.

After welding, the switch body was tested for leaks, using a leak tester of the mass spectrometer type. If acceptable, the switch was backfilled with dry nitrogen and sealed off. If a leak was detected, a bubble test was made to locate the leak. If the leak was in the weld, the switch was rewelded and again backfilled with dry nitrogen and resealed. Rejections for defects in welding were generally below 5%.

Melt-through welding refers to the welding of a joint from only one side in a manner that provides complete joint penetration of weld metal with visible root reinforcement. The melt-through technique is limited to the joining of relatively thin sections. Gas tungsten arc welding is a preferred process for such applications, especially when the elements to be joined are made of stainless steel or of a nickel-based alloy. Obtaining a satisfactory weld from only one side is most advantageous when accessibility is largely limited to one side. In such applications, melt-through welding may also be more economical than conventional techniques because less welding is required.

Example 15. Single-Pass Melt-Through Welding. The left view of Fig. 19 shows a welded assembly consisting of a cylindrical cover sheet enclosing a central link and four supporting struts. The cover sheet of two of these assemblies was joined by circumferential fillet welds to the opposite ends of a bellows (Fig. 19, center). The completed weldment was used in an aerospace application to provide a strong, flexible connection to rigid segments of a ducting system. This example deals only with the welding of the supporting struts to the cover sheet of the assembly.

Originally, the T-joint between the ends of the struts and the inside surface of the cover sheet was made by two fillet welds deposited by GTAW. Success of the welding procedure depended largely on the inside diameter of the cover sheet, which, depending on the size of the assembly, ranged from $1^1/_4$ to 8 in. As the diameter decreased, joint accessibility for manipulation of the torch and filler metal also decreased. As a result, weld quality was adversely affected, and meeting radiographic standards became difficult. By joining the cover sheets to the struts from the outside by melt-through welding, the problem was solved. Complete joint penetration was obtained in a single pass. Filler metal was added to supply the volume needed to form a double-fillet-welded T-joint (section A-A, Fig. 19). A minimum fillet-weld size of 0.040 in. was adequate for strength.

To establish welding conditions for various combinations of alloys and work-metal thicknesses (ranging from 0.032 to 0.250 in.), about 250 strut-cover sheet assemblies were joined by melt-through welding and inspected. All welds met the minimum requirements for fillet size and surface appearance. Radiographic examination, which was simplified because of the absence of discontinuities at the root of the weld, showed complete fusion and adequate joint penetration. Porosity was encountered only when preweld cleaning had been inadequate. No significant indications were noted under liquid-penetrant inspection. Macrographs of weld sections showed no significant defects. Failures in peel tests occurred wholly in the base metal. Mechanical tests indicated significant improvement in fatigue life.

Plasma Arc Welding

Plasma arc welding is suitable for both manual and automatic operation. The process is used for continuous welds made with or without filler metal, as well as for spot welds. The fundamentals of the process are covered in the article "Plasma Arc Welding" in this Volume. The PAW method is ideally suited for welding stainless steels because of its cleanliness.

Typical conditions for PAW of various thicknesses of austenitic stainless steel are

Table 16 Conditions for low-current PAW of butt joints in thin-section austenitic stainless steel(a)

Section thickness, in.	Current (DCEN), A	Welding speed, in./min
0.001	0.3	5
0.003	1.6	6
0.005	2.0	5
0.010	6.0	8
0.030	10.0	5

(a) Orifice gas, argon at 0.5 ft^3/h; orifice diameter, 0.030 in. Shielding gas, 99% argon and 1% hydrogen

shown in Tables 16 and 17. For all thicknesses, the electrode tip is 1/8 in. from the nearest point on the inner surface of the orifice body; orifice-to-work distance is 3/16 in.

The use of the keyhole technique on square-groove butt joints in stainless steel is usually limited to metal less than 0.375 in. thick. If all welding conditions are closely controlled, however, 0.375-in.-thick stainless steel can be fused by this technique. Suitable conditions for a typical application include: travel speed, 8 in./min; current, 250 A at 38 V; orifice gas, 99.5% argon and 0.5% oxygen at 15 ft^3/h; and shielding gas, 99.5% argon and 0.5% oxygen at 40 ft^3/h. Even slight variations in these conditions can produce unsatisfactory welds, so keyhole welding of square-groove butt joints in 0.375-in.-thick stainless steel is not recommended.

Manufacture of Stainless Steel Tubing. Continuously formed stainless steel pressure tubing conforming to American Society for Testing and Materials (ASTM) A312 is made from strip rolled into tube form and butt welded using the plasma arc process, generally without filler metal. Installation of PAW equipment in tube mills has resulted in substantially increased welding speed, compared with GTAW. A comparison of average welding speeds for GTAW and PAW on tubing of various wall thicknesses is shown in Table 18. Plasma arc welding shows the greatest speed advantage on thick-walled tubing, as well as a lower rejection or repair rate because of the uniform penetrating power of the plasma arc.

A schematic view of plasma arc tube welding is shown in Fig. 20. Weld-bead shape and reinforcement are controlled by adjusting four variables:

- Welding current
- Location of the arc relative to the centerline of the pressure rolls on the tube mill
- Force exerted by the pressure rolls
- Backing-gas pressure

In beginning a production run, the arc is started and a keyhole is established approximately 1 in. ahead of the centerline of the pressure rolls. Excessive weld metal (reinforcement) at the top and bottom of the tube joint (Fig. 21a) indicates that the weld bead is too hot when it reaches the pressure point between the rolls and is, therefore, soft enough to be upset. This condition can be corrected by moving the torch farther ahead of the pressure rolls, allowing more time for the weld pool to cool before it comes under forging pressure. If there is not enough reinforcement at the top of the joint (Fig. 21b), the torch should be moved closer to the pressure rolls. Pressure-roll force can also be increased or decreased to change the amount of weld reinforcement.

If excessive drop-through occurs (Fig. 21c), it can be corrected by increasing the backing-gas pressure. The backing gas is confined between plugs or diaphragms mounted on a pipe or lance inside the tube. Backing-gas pressure can be controlled by connecting a hose to the backing-gas chamber and exhausting the gas into a beaker or tube of water. Gas pressure is increased until bubbles appear in the water. Thereafter, pressure in the chamber will depend on how deeply the hose is immersed in the water.

Table 18 Average speed for fabricating stainless steel tubing by GTAW and by PAW

Wall thickness, in.	Welding speed, in./min GTAW	PAW	Increase with PAW, %
0.109	26	36	38
0.125	22	36	64
0.154	20	36	80
0.216	8	15	88
0.237	6	14	134

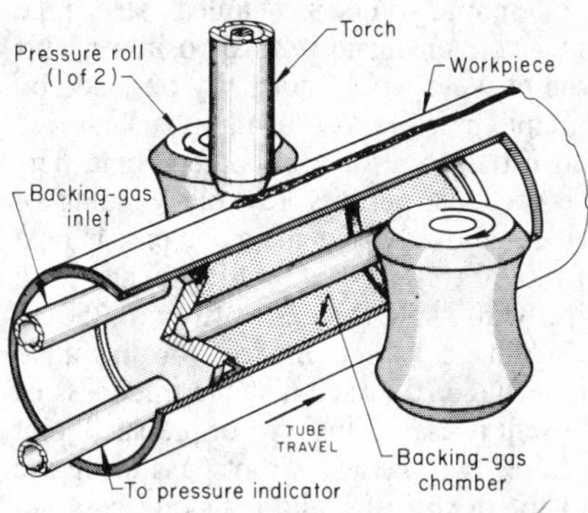

Fig. 20 Welding longitudinal seams in stainless steel tubes. Internal backing with inert gas

Plasma arc tube welding, unlike gas tungsten arc tube welding, does not require arc-voltage control or automatic adjustment of orifice-to-work distance. This may greatly reduce the cost differential for equipment between the plasma arc and gas tungsten arc processes. Constant travel speed for the tube and good joint fit-up are essential for good-quality welds. When tube diameter is 1 in. or less, the speed obtainable using PAW is about the same as for GTAW. For making this size of tube, a V-groove is formed as the edges of the strip are brought together. Keyholing, with its accompanying speed advantage, becomes impractical. Because of the maximum wall thickness that would prevail for a 1-in.-diam tube, there is not enough metal

Table 17 Typical operating conditions for PAW of stainless steel(a)
Square butt joint and keyhole technique(b), except for root and filler passes

Thickness, in.	Travel speed, in./min	Current (DCEN), A	Arc voltage, V	Gas	Gas flow, ft^3/h Orifice gas	Shielding gas
0.093	24	115	30	95% A, 5% H_2	6	35
0.125	30	145	32	95% A, 5% H_2	10	35
0.187	16	165	36	95% A, 5% H_2	13	45
0.250	14	240	38	95% A, 5% H_2	18	50
0.375:						
Root pass(c)	9	230	36	95% A, 5% H_2	12	45
Filler pass(d)	7	220	40	Helium	25	175

(a) Backing gas required. (b) Orifice-to-work distance is 3/16 in. Multiple-orifice torch is used. (c) V-groove joint with 60° included angle; 3/16-in. root face; keyhole technique. (d) Filler technique; 0.045-in.-diam filler-metal wire fed at 60 in./min; joint type not given

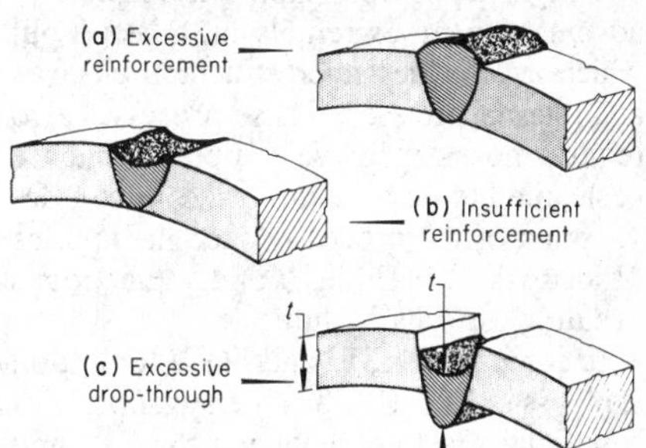

Fig. 21 Three undesirable conditions in plasma arc tube welding

at the joint to support the molten weld pool.

Circumferential Pipe Welding. Plasma arc welding has been used to make circumferential joints in stainless steel pipe in the horizontal and vertical positions. This type of weld would normally be made by multiple-pass GTAW, using a backing ring and filler metal on a prepared joint. The use of PAW permits keyhole welding of square-groove butt joints in one pass in pipe of 0.090- to 0.250-in. wall thickness. On pipe with wall thickness from 0.250 to 0.375 in., a V-groove with a 60° included angle and a root face half the thickness of the wall is used. This type of prepared joint requires two passes, a root pass using the keyhole technique, and a second pass using filler-metal wire. Slope control for welding current and orifice gas flow is used to start and terminate the keyhole weld.

Welding of Vessels. Welding of circumferential joints in tanks or vessels is an extension of the technique used to weld circumferential pipe joints. The vessel must be positioned and rotated so that the joint is in the horizontal position. Whether or not PAW will be more expensive than another welding process will be determined only by careful evaluation of the application and the welding processes.

Equipment. Allowing for the same capacity, PAW equipment is more expensive than GTAW equipment, but is far less costly than EBW equipment. Maintenance cost is also greater for PAW processes, mainly because of frequent replacement of orifices. Before it can be determined whether or not the high cost of equipment would be justified, the high productivity of PAW must be evaluated. In applications where PAW is up to four times as fast as alternative processes, equipment costs are rapidly compensated for if production quantities are high.

Electroslag and Electrogas Welding

Electroslag welding (ESW) and electrogas welding (EGW) are the least commonly used processes for joining austenitic stainless steels. Their primary advantages are nearly automatic operation and high deposition rate. The major disadvantage is an extremely high heat input which accentuates microstructural changes and thermal stresses. These processes also require more expensive equipment and are best suited to large section sizes. For a description of the processes, see the articles "Electroslag Welding" and "Electrogas Welding" in this Volume.

In comparing ESW and EGW for use on stainless steel, ESW has the advantages of beneficial reactions between the slag and molten weld metal that may remove some impurities, improved wetting action due to the slag, and no need for a shielding gas. The electrogas process has the advantages of easier restart and improved view of the weld for operator adjustments. The austenitic stainless steel electrodes for use with ESW and EGW are the same as those used with SAW and are listed in Table 8.

Nitrogen-Strengthened Austenitic Stainless Steels

This family of stainless steels offers two specific advantages over the conventional austenitic stainless steels: (1) increased strength at all temperatures, cryogenic through elevated; and (2) improved resistance to pitting corrosion. They differ from the conventional austenitic stainless steels in that manganese has been substituted for all or part of the nickel, thus allowing greater amounts of nitrogen to be dissolved in the matrix of the alloy. The nitrogen acts as a solid solution strengthener and increases the annealed yield strength to approximately twice that of the conventional austenitic stainless steels.

Welding characteristics of the conventional austenitic stainless steels also apply to the nitrogen-strengthened austenitic stainless steels. The measures taken to prevent unwanted changes in composition in the conventional austenitic stainless steels should be taken with the nitrogen-strengthened steels. Specific attention should be paid to the control of nitrogen when welding these grades. A significant loss of nitrogen often results in a loss of strength or corrosion resistance. A significant increase in the nitrogen content is likely to result in weld porosity or weld hot cracking. To achieve this necessary balance or composition, specific electrode compositions have been developed. These are discussed in the sections on welding processes used for the nitrogen-strengthened austenitic stainless steels in this article.

The most significant difference in the metallurgy of welding the conventional and nitrogen-strengthened stainless steels is in the calculation of the delta ferrite potential. The constitution diagram for the high-nitrogen grades has been modified from that shown in Fig. 1 and 2. The modified diagram (Fig. 22) was developed to correspond to the actual delta ferrite contents observed in weld deposits. The constant factor for manganese has been increased, and the constant factor for nitrogen varies with nitrogen content.

The composition of the weld deposit must be maintained through proper welding procedures to ensure that the delta ferrite content is constant and weld hot cracking is avoided.

In Fig. 22, for average stainless steel welds (5/32-in.-diam electrode, SMAW process), including the modifications (up to 15% Mn and 0.35% N), the percent of ferrite (up to ≈30%) in a matrix of austenite, or austenite and martensite or martensite, can be predicted within ≈4%. As explained in the austenitic stainless steel section, the percent of ferrite is considered

Fig. 22 Schaeffler constitution diagram for stainless steel weld metal modified for manganese with nitrogen. Vanadium, copper, and aluminum added. (Source: *Welding Journal,* May 1982, p 152s. Adapted by R. Harry Espy)

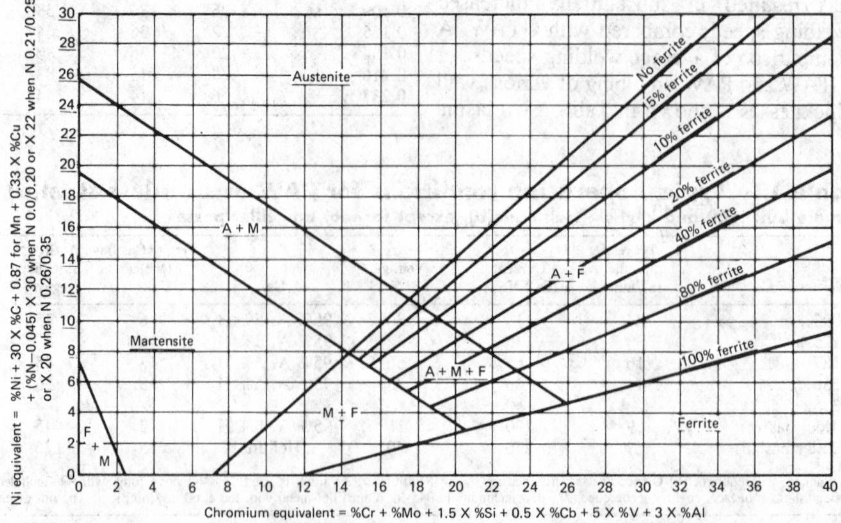

Table 19 Chemical compositions for all-weld-metal deposits of nitrogen-strengthened austenitic electrodes for SMAW
All weight percentages are maximum, unless otherwise noted

AWS classification(a)	C	Cr	Ni	Mo	Mn	Si	P	S	N	Cu	Base metal(c)
E209(b)	0.06	20.5-24.0	9.5-12.0	1.5-3.0	4.0-7.0	0.90	0.03	0.03	0.10-0.30	0.75	S20 910
E219	0.06	19.0-21.5	5.5-7.0	0.75	8.0-10.0	1.00	0.03	0.03	0.10-0.30	0.75	S21 900
E240	0.06	17.0-19.0	4.0-6.0	0.75	10.5-13.5	1.00	0.03	0.03	0.10-0.20	0.75	S24 000, S24 100

(a) Suffix −15 electrodes are classified with DCEP. Suffix −16 electrodes are classified with alternating current and DCEP. Electrodes up to and including $^{5}/_{32}$ in. in size are usable in all positions. Electrodes $^{3}/_{16}$ in. and larger are usable only in the flat- and horizontal-fillet positions. (b) Vanadium shall be 0.10 to 0.30%. (c) UNS designation
Source: AWS A5.4-81

equivalent to the Welding Research Council ferrite number (WRC-FN). The faster freezing rate of a smaller than average size weld gives a lower ferrite content than predicted by the diagram, and the slower freezing rate of a larger than average size weld results in a higher ferrite content. With greater than 2.5% Mn and 0.5% Cu, the austenite resistance to martensite transformation increases, expanding the stable austenite region to a broader area than shown in the diagram.

Shielded Metal Arc Welding

This process is widely used to join the nitrogen-strengthened stainless steels. The various advantages and disadvantages of SMAW are described in the section on austenitic stainless steels in this article. Constant-current (drooping-voltage) power sources, conventionally used for SMAW, are applicable for welding the nitrogen-strengthened stainless steels. Either alternating current or DCEP is suitable in most applications. Direct current power supply can be of the motor-generator or transformer-rectifier type.

Electrodes. Table 19 lists the nitrogen-strengthened stainless steel electrodes used for SMAW and their deposit analyses. They are most often used to join matching composition plate, although they have sufficient alloy to join dissimilar alloys such as low-carbon steel and stainless steel. The common corresponding base-metal designations are also included in Table 19. The suffix number is related to the polarity to be used. The suffix −15 indicates that the coating is primarily of the lime type, containing a fair amount of calcium or other alkaline elements, and that the electrodes are suitable for use with DCEP. For welding with alternating current, an electrode with the suffix −16 should be selected. Such electrodes can also be used with DCEP and can have either a lime-type or titanium-type covering. For welding with alternating current, the coating, besides having alkaline elements, contains readily ionized elements to stabilize the arc.

Electrodes of the −15 and −16 types up to $^{5}/_{32}$ in. in diameter can be used in all welding positions. Electrodes $^{3}/_{16}$ in. in diameter and larger are used only in the flat- and horizontal-fillet positions. Welding procedures discussed for austenitic stainless steels are applicable to the nitrogen-strengthened grades. The arc length must be kept short to control the transfer of the various elements, including nitrogen, into the weld metal.

Submerged Arc Welding

The submerged arc process is described in detail in this Volume. Its application to the austenitic stainless steels is discussed earlier in this article. Its application to the nitrogen-strengthened stainless steels is very similar. A particular advantage is better control of nitrogen transfer into the weld pool. The various filler-metal compositions are listed in Table 20, as are the base-metal designations. See also the section on submerged arc welding of austenitic stainless steel in this article for joint design and procedures.

Gas Metal Arc Welding

This welding method can be used for the nitrogen-strengthened austenitic stainless steels. The procedures are similar to those previously covered in the section on austenitic stainless steels in this article. The major difference is the increased tendency toward porosity due to the increased nitrogen level.

Gas Tungsten Arc Welding

Nitrogen-strengthened austenitic stainless steels can be readily welded autogenously or with a matching filler metal. The filler-metal designations are covered in AWS specification A5.9-81 and are listed in Table 20. The major difference in weldability is due to the nitrogen level. In some cases, this can increase molten weld-metal turbulence and porosity. This turbulence can cause the electrode to deteriorate. To minimize this condition, the filler metals have lower nitrogen content than the base-metal counterparts.

Plasma Arc Welding

Plasma arc welding of the nitrogen-strengthened austenitic stainless steels has some advantages over GTAW. The favorable depth-to-width ratio minimizes the molten-metal volume and thus reduces the amount of outgassing and porosity. The torch configuration also reduces the possibility for electrode erosion. The materials behave generally the same as the austenitic stainless steels.

Electroslag and Electrogas Welding

Descriptions of these processes can be found in the articles "Electroslag Welding" and "Electrogas Welding" in this Volume. Detailed descriptions of their ap-

Table 20 Chemical compositions for nitrogen-strengthened austenitic stainless steel electrodes for SAW
Maximum percentages, unless otherwise noted

AWS classification	C	Cr	Ni	Mo	Mn	Si	P	S	N	Cu	Base metal(a)
ER209(b)	0.05	20.5-24.0	9.5-12.0	1.5-3.0	4.0-7.0	0.90	0.03	0.03	0.10-0.30	0.75	S20 910
ER218	0.10	16.0-18.0	8.0-9.0	0.75	7.0-9.0	3.5-4.5	0.03	0.03	0.08-0.18	0.75	S21 800
ER219	0.05	19.0-21.5	5.5-7.0	0.75	8.0-10.0	1.00	0.03	0.03	0.10-0.30	0.75	S21 900
ER240	0.05	17.0-19.0	4.0-6.0	0.75	10.5-13.5	1.00	0.03	0.03	0.10-0.20	0.75	S24 000, S24 100

(a) UNS designation. (b) Vanadium, 0.10 to 0.30%
Source: AWS A5.9-81

plication to the austenitic stainless steels have also been included and are very similar to those for the nitrogen-strengthened stainless steels. The electroslag process is preferred over EGW due to better control of the nitrogen. Electrodes for use with these processes are listed in Table 20.

Ferritic Stainless Steels

The ferritic stainless steels are generally less weldable than the austenitic stainless steels and produce welded joints having lower toughness because of grain coarsening that occurs at the high welding temperatures. The standard ferritic stainless steels are: (1) type 446 (25% Cr); (2) types 430, 430F, and 430F-Se (17% Cr); and (3) types 405 and 409 (13% Cr). Type 409 is ferritic because it has a low carbon content (0.08% maximum) and a minimum titanium content equal to six times the carbon content. Type 405, which also contains only 0.08% maximum C, contains an average of 0.20% Al, which promotes ferrite formation.

Effect of Welding Heat on Ductility and Grain Size. Although most ferritic stainless steels have compositions that ensure a ductile ferritic structure at room temperature, variations in composition within the standard composition limits can result in the formation of small amounts of austenite during heating to elevated temperature. On cooling, the austenite transforms to martensite, resulting in a duplex structure of ferrite and a small amount of martensite. The martensite reduces both ductility and toughness of the steel. Annealing transforms the martensite and restores normal ferritic properties, but annealing increases cost and can result in an excessive amount of distortion, particularly in parts that were previously formed by a cold working process.

All ferritic stainless steel mill products are normally annealed at the mill to transform any martensite that may be present to a softer structure of ferrite and carbides. In this condition, the steel can be readily cold formed. It is only when the steel is heated near or above the transformation temperature (approximately 1600 °F), as during welding, that the risk of austenite formation and subsequent transformation to martensite arises. In addition, heating to temperatures above 1750 °F results in enlargement of the ferrite grain size, which also reduces the ductility and toughness of the steel. Although martensite can be eliminated by annealing, coarsened ferrite grains remain unaffected. Because martensite responds to annealing and inhibits ferrite grain growth, some applications may benefit from martensite formation, provided that the workpiece can be annealed after welding. In a 17% Cr steel, martensite formation is promoted by lowering the chromium content to 15 or 16%. When this practice is adopted, it is usually necessary to preheat before welding or to select a steel of lower carbon content to guard against cracking in the HAZ.

When postweld annealing is not feasible, the ductility of the welded joint can be controlled by the selection of a stainless steel base metal containing a substantial amount of a strong ferrite-former, such as aluminum, niobium, or titanium. One such steel, which has the commercial designation of type 430Ti, is a 17% Cr steel containing 0.12% C maximum and a minimum titanium content equal to six times the carbon content. The metallurgical functions of the titanium in this steel are to form stable titanium carbides and to promote the formation of ferrite. When a completely ferritic steel is welded, no martensite is formed in the HAZ, although some grain coarsening may occur. Grain coarsening can be controlled, to some extent, by minimizing heat input during welding and by avoiding slow cooling from the welding temperature.

There are a number of ferritic stainless steels that contain very low amounts of carbon and nitrogen. These low-interstitial ferritic stainless steels (including the 26%Cr-1%Mo and 29%Cr-4%Mo alloys) can be welded so that they do not lose any of the base-metal ductility in the weld area, while retaining grain growth. The key to successful welding of these steels is to prevent any carbon, nitrogen, or oxygen contamination during welding. Thus, the part and filler material must be clean before welding, and both the molten weld metal and hot weld-area metal must be fully shielded from the atmosphere. All moisture must also be excluded from the weld area before and during welding.

Effect of Temperature on Notch Toughness. For the 17% Cr steels, the temperature range is just above room temperature for transition from a tough shear-type fracture at the higher temperature to a brittle, cleavage-type fracture at the lower temperature, upon impact at a notch. Under impact loading at room temperature and below, these steels are notch sensitive, and impact test values of less than 15 ft · lb are usual. At 200 to 250 °F, the impact test values increase to approximately 30 to 50 ft · lb.

This relation between notch toughness and temperature is important in the selection of joint design and welding conditions for ferritic stainless steel. In service, a weldment designed primarily to withstand static load may be subjected to accidental or unforeseen impact loading. Furthermore, a weldment with low notch toughness may not withstand appreciable cyclical stresses even under a low rate of loading. Because multiaxial residual stresses are often developed during welding (especially when welding heavy sections), notches and points of stress concentration that might cause failure in service must be avoided whenever possible. Preheating before welding is often useful in preventing cracking during welding.

Preheating. The recommended preheating temperature range for ferritic stainless steels is 300 to 450 °F. The need for preheating is determined largely by the composition, mechanical properties, and thickness of the steel being welded. Steels less than $^1/_4$ in. thick are much less likely to crack during welding than those more than $^1/_4$ in. thick. The type of joint, joint location, restraints imposed by clamping and jigging, the welding process, and the rate of cooling from the welding temperature can also affect weld cracking.

Postweld Annealing. The temperature range for postheating or postweld annealing of ferritic stainless steels is 1450 to 1550 °F, which is safely below the temperatures for austenite formation and grain coarsening. Annealing transforms a mixed structure to a wholly ferritic structure and restores the mechanical properties and corrosion resistance that may have been adversely encountered in welding. Thus, except for its inability to refine coarsened ferrite grains, annealing is generally beneficial. Annealing has two major disadvantages: (1) the time and cost of the treatment, and (2) the need to prevent the formation of the oxide scale or the removal of it. Annealing may also require the use of elaborate fixturing to prevent sagging or distortion of the weldment.

Cooling ferritic stainless steel from the annealing temperature may be done by air or water quenching. To minimize distortion from handling, weldments often are allowed to cool to about 1100 °F before they are removed from the furnace. Slow cooling through the temperature range of 1050 to 750 °F must be avoided because it produces brittleness in the steel. Susceptibility to this type of embrittlement, known as 885 °F embrittlement, normally increases as chromium content increases. Heavy sections may require forced cooling or a spray quench to bring them safely through this embrittlement range.

Selection of Filler Metal. As shown in Table 1, both ferritic and austenitic stain-

Table 21 All-weld-metal chemical compositions for ferritic stainless steel electrodes for FCAW

All are maximum percentages, unless otherwise noted; other elements, except iron, not to exceed 0.50%

AWS classification	C	Cr	Ni	Mo	Mn	Si	P	S	Fe	Cu
E430T-X(a)	0.10	15.0-18.0	0.60	0.5	1.2	1.0	0.04	0.03	rem	0.5
E430T-3	0.10	15.0-18.0	0.60	0.5	1.0	1.0	0.04	0.03	rem	0.5
EXXXT-G(b)	...	...	...	...	...	...	...	...	...	...

(a) X indicates classification covering the shielding designation for both 1 and 2 categories. (b) As agreed upon between supplier and manufacturer
Source: AWS A5.22-80

less steel filler metals are used in the arc welding of ferritic stainless steel. Ferritic stainless steel filler metals offer the advantages of having the same color and appearance, the same coefficient of thermal expansion, and essentially the same corrosion resistance as the base metal. However, austenitic stainless steel filler metals are often used to obtain more ductile weld metal in the as-welded condition.

Although austenitic stainless steel weld metal does not prevent grain growth and martensite formation in the HAZ, the ductility of austenitic weld metal improves the ductility of the welded joint. The selection of austenitic stainless steel filler metal, however, should be carefully related to the specific application to determine whether differences in color or in the physical corrosion and mechanical properties of the weld metal and the base metal cause difficulty.

For weldments that are to be annealed after welding, the use of austenitic filler metal can introduce several problems. The normal range of annealing temperature for ferritic stainless steels falls within the sensitizing temperature range for austenitic steels. Consequently, unless the austenitic weld metal is of extra-low-carbon content or is stabilized with niobium or titanium, its corrosion resistance may be seriously impaired. If the annealing treatment is intended to relieve residual stress in the weldment, it cannot be fully effective because of the difference in the coefficients of thermal expansion of the weld metal and the base metal.

Corrosion Resistance. Ferritic stainless steels usually exhibit less corrosion resistance than austenitic stainless steels. Any condition of the welded joint that might therefore impair corrosion resistance must be avoided. The presence of martensite or the precipitation of sigma phase at the grain boundaries can cause severe intergranular corrosion in the HAZ. Completely ferritic steels, such as types 430Ti and 446, display little or no susceptibility to intergranular attack at the weld joint. Annealing of any welded ferritic steel eliminates the unfavorable structural conditions that promote corrosive attack.

Shielded Metal Arc Welding

The application of SMAW to austenitic stainless steels is discussed in depth earlier in this article. Most of the principles that apply to the austenitic steels also apply to the ferritic stainless steels, once they are modified to the metallurgical considerations detailed above. This process is discussed further in the article "Shielded Metal Arc Welding" in this Volume. Depending on the service conditions, either ferritic or austenitic steel electrodes are suitable for SMAW of ferritic base metals.

Flux Cored Arc Welding

This process is described in the article "Flux Cored Arc Welding" in this Volume, and the general principles of its application to austenitic stainless steels are covered earlier in this article. Flux cored arc welding of ferritic stainless steels is characterized by the potential for higher heat input. If coarse grain size is a concern, heat input must be controlled through higher travel speed, lower interpass temperature, and the use of heat sinks. Table 21 lists the ferritic stainless steel compositions in the AWS filler-metal specification. Other analyses are available by agreement between consumer and manufacturer.

Submerged Arc Welding

This process is described in detail in the article "Submerged Arc Welding" in this Volume, and its application to stainless steels is discussed in the section on austenitic stainless steels in this article. The heat input with this process is generally higher than with FCAW. If coarse grain size is a concern, the heat input must be controlled through higher travel speed, lower interpass temperature, and the use of heat sinks. Table 22 lists the ferritic stainless steel compositions in the AWS filler-metal specification. The ER26-1 grade depends on very high purity to achieve a combination of toughness and corrosion resistance. The interaction with submerged arc flux tends to degrade these properties. Qualification testing is necessary to ensure adequate mechanical properties. The austenitic stainless steel filler metals that may be suitable are shown in Table 8. Other analyses are available by agreement between consumer and manufacturer.

Gas Tungsten Arc Welding

Gas tungsten arc welding is adaptable to both automatic and manual operation in all welding positions. The fundamentals of the process are similar to those covered in the section on austenitic stainless steels in this article. Designations for ferritic stainless steel filler metals are covered in AWS specification A5.9-81 and are listed in Table 22. Austenitic stainless steel filler metals are shown in Table 7.

When welding some ferritic stainless steels, preheating may be essential to prevent cracking in the weld or the heat-affected zone. For low-interstitial ferritics, preheating is not recommended. In these materials, heat input must be minimized to prevent grain growth and embrittlement of the weld.

Closure of Small-Diameter Tubes. Various types of stainless steel are used in the manufacture of small instrument components that embody a welded joint or closure. Accurate placement of small welds often demands machine-guided rather than manual welding, and a welding procedure that does not necessitate postweld cleaning. Gas tungsten arc welding is often selected when automatic welding is required.

Example 16. Automatic Welding Without Filler Metal. A temperature-

Table 22 Chemical compositions for ferritic stainless steel electrodes for SAW and GMAW

All are maximum percentages, unless otherwise noted

AWS classification	C	Cr	Ni	Mo	Mn	Si	P	S	N	Cu
ER430	0.10	15.5-17.0	0.6	0.75	0.6	0.50	0.03	0.03	...	0.75
ER26-1	0.01	25.0-27.5	(a)	0.75-1.50	0.40	0.40	0.02	0.02	0.015	0.20(a)

(a) Nickel, max = 0.5% − Cu
Source: AWS A5.9-81

Fig. 23 Automatic GTAW of temperature-sensing tube

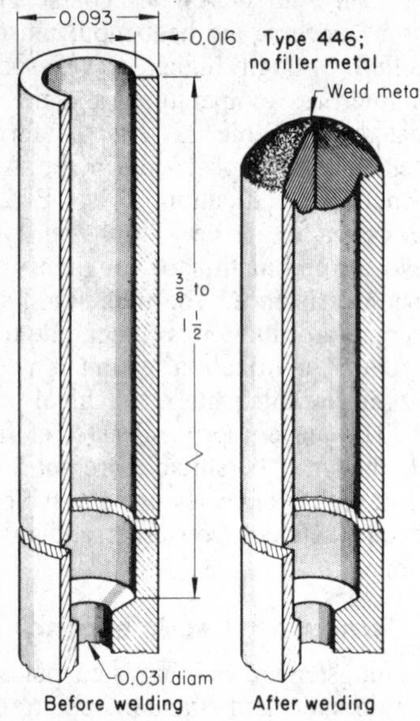

Automation	Rotary indexing table(a)
Power supply	200-A transformer-rectifier
Electrode holder	Machine held, air cooled
Electrode	0.040-in.-diam EWTh-2(b)
Filler metal	None
Shielding gas	Argon, at 12 ft^3/h(c)
Arc starting	High frequency
Arc length	0.030 in.
Current	60 A (DCEN)
Voltage	8 V
Weld time	0.5 s
Production rate	1000 tubes per hour

(a) Equipped with ten workholding chucks and automatic unloading mechanism; indexing speed, 3.6 s per piece. (b) Ground to a pencil-point tip configuration. (c) Argon was also used for purging, immediately before welding.

sensing tube (Fig. 23) was bored at one end to increase the inside diameter from 0.031 to 0.061 in., for a depth of $^3/_8$ to $1^1/_2$ in., depending on application. A leak-proof seal at the bored end of the tube was obtained by automatic GTAW without filler metal. Before being welded, the tubes were vapor degreased and then freed from surface oxides by being heated at 1750 °F in a conveyor furnace containing a hydrogen atmosphere. The tubes were then furnace cooled to room temperature and manually loaded into the ten holding chucks of a rotary indexing table. The chucks accurately positioned each tube under the electrode tip. The tubes were purged with argon immediately before welding to prevent oxidation of the bore during welding. Preweld argon flow, weld time, and postweld argon flow were automatically controlled by a sequence timer. Using an arc time of 0.5 s, the bored ends of the tubes were welded at the rate of 1000/h. After the tubes had cooled sufficiently under argon, they were automatically ejected from the holding chucks. Quality control was maintained by visually inspecting the finished temperature-sensing tubes on a sampling basis. Rejection rate was less than 1%.

Plasma Arc Welding

Plasma arc welding of ferritic steels is similar to austenitic stainless steels. The techniques used are basically the same, with slight modifications because of the metallurgical differences between the two alloy families. These differences are discussed in the beginning of the section on ferritic stainless steels and have been applied in the discussion of gas tungsten arc welding in this article.

Gas Metal Arc Welding

This process is described in detail in the article "Gas Metal Arc Welding" in this Volume, and its application to stainless steels is described in the section on austenitic stainless steels in this article. Because heat input is generally higher than with many other welding processes, its effects must be carefully considered before the process is selected. If coarse grain size is a concern, heat input must be controlled through higher travel speed, lower interpass temperature, and the use of heat sinks. Both ferritic and austenitic stainless steel filler metals are used to weld the ferritic steels. Table 22 lists the ferritic stainless steel compositions in the AWS filler-metal specification. The compositions of the austenitic stainless steel filler metals are shown in Table 7.

Electroslag and Electrogas Welding

These processes are described in detail in their respective articles in this Volume. Electroslag and electrogas welding have the highest heat input and should not be used in situations requiring fine grain size in ferritic stainless steels.

Martensitic Stainless Steels

The standard martensitic stainless steels are types 403, 410, 414, 416, 416Se, 420, 431, 440A, 440B, and 440C. These steels derive their corrosion resistance from chromium, which they contain in proportions ranging from 11.5 to 18%. Martensitic stainless steels are the most difficult stainless steels to weld because they are chemically balanced to become harder, stronger, and less ductile through thermal treatment. These same metallurgical changes occur from the heat of welding. As a result of welding, these changes are restricted to the weld area and are not uniform over the entire section. The nonuniform metallurgical condition of the part is susceptible to cracking when subjected to the high stresses from welding. Increasing carbon content in martensitic stainless steels generally results in increased hardness and reduced ductility. Thus, the three type 440 stainless steels are seldom considered for applications that require welding, and filler metals of the type 440 compositions are not readily available.

Modifications of the standard martensitic steels contain additions of elements such as nickel, molybdenum, vanadium, and tungsten, primarily to raise the allowable service temperature above the 1100 °F limit for the standard steels. When these elements are added, carbon content is increased and the problem of avoiding cracking in the hardened HAZ of weldments becomes more serious.

Martensitic stainless steels can be welded in the annealed, hardened, and hardened and tempered conditions. Regardless of the prior condition of the steel, welding produces a hardened martensitic zone adjacent to the weld. The hardness of the HAZ depends primarily on the carbon content of the base metal. As hardness increases, toughness decreases, and the zone becomes more susceptible to cracking. Preheating and control of interpass temperature are the most effective means of avoiding cracking. Postweld heat treatment is required to obtain optimum properties.

Preheating and Postweld Heat Treating. The usual preheating temperature range of martensitic steels is 400 to 600 °F. Carbon content of the steel is the most important factor in determining whether or not preheating is necessary. On the basis of carbon content alone, a steel containing not more than 0.10% C seldom requires preheating, and one with more than 0.10% C requires preheating to prevent cracking. Other factors that determine the need for preheating are the mass of the joint, degree of restraint, presence of a notch effect, and the composition of the filler metal. The following can be used to correlate preheating and postweld heat treating practice with carbon contents and welding characteristics of martensitic stainless steels:

- *Carbon below 0.10%*: Neither preheating nor postweld annealing generally is required; steels with carbon contents this low are not standard.
- *Carbon 0.10 to 0.20%*: Preheat to 500 °F; weld at this temperature; cool slowly.
- *Carbon 0.20 to 0.50%*: Preheat to 500 °F; weld at this temperature; anneal.

Table 23 All-weld-metal chemical compositions for martensitic stainless steel electrodes(a) for SMAW

All are maximum weight percents, unless otherwise noted

AWS classification	C	Cr	Ni	Mo	Mn	Si	P	S	Cu
E410	0.12	11.0-13.5	0.60	0.75	1.0	0.90	0.04	0.03	0.75
E410NiMo	0.06	11.0-12.5	4.0-5.0	0.40-7.0	1.0	0.90	0.04	0.03	0.75
E502	0.10	4.0-6.0	0.40	0.45-0.65	1.0	0.90	0.04	0.03	0.75
E505	0.10	8.0-10.5	0.40	0.85-1.20	1.0	0.90	0.04	0.03	0.75
E7Cr	0.10	6.0-8.0	0.40	0.45-0.65	1.0	0.90	0.04	0.03	0.75

(a) Suffix −15 electrodes are classified with DCEP. Suffix −16 electrodes are classified with alternating current and DCEP. Electrodes up to and including 5/32 in. in size are usable in all positions. Electrodes 3/16 in. and larger are usable only in the flat- and horizontal-fillet positions.
Source: AWS A5.4-81

Table 24 All-weld-metal chemical compositions for martensitic stainless steel electrodes for FCAW

All are maximum percentages, unless otherwise noted; total of other elements, except iron, not to exceed 0.50%

AWS classification	C	Cr	Ni	Mo	Mn	Si	P	S	Fe	Cu
E410T-X	0.12	11.0-13.5	0.60	0.5	1.2	1.0	0.04	0.03	rem	0.5
E410NiMoT-X	0.06	11.0-12.5	4.0-5.0	0.40-0.70	1.0	1.0	0.04	0.03	rem	0.5
E410NiTiT-X(a)	(b)	11.0-12.0	3.6-4.5	0.05	0.70	0.50	0.03	0.03	rem	0.5
E502T-X	0.10	4.0-6.0	0.40	0.45-0.65	1.2	1.0	0.04	0.03	rem	0.5
E505T-X	0.10	8.0-10.5	0.40	0.85-1.20	1.2	1.0	0.04	0.03	rem	0.5
E410T-3	0.12	11.0-13.5	0.60	0.5	1.0	1.0	0.04	0.03	rem	0.5
E410NiMoT-3	0.06	11.0-12.5	4.0-5.0	0.40-0.70	1.0	1.0	0.04	0.03	rem	0.5
E410NiTiT-3(b)	0.04	11.0-12.0	3.6-4.5	0.5	0.70	0.50	0.03	0.03	rem	0.5
EXXXT-G(c)	...	...	...	...	...	...	...	...	...	...

(a) Titanium = 10 × C min to 1.5% max. (b) When suffix X is 1, 0.04% C max; when X is 2, 0.03% C max. (c) As agreed upon between supplier and purchaser
Source: AWS A5.22-80

- *Carbon over 0.50%*: Preheat to 500 °F; weld with high heat input; anneal.

If the weldment is to be hardened and tempered immediately after welding, annealing may be omitted. Otherwise, the weldment should be annealed immediately after welding, without cooling to room temperature.

Shielded Metal Arc Welding

Table 23 lists martensitic stainless steel electrodes for SMAW. Other compositions are available by agreement between consumer and manufacturer. Austenitic stainless steel filler metals are often used to weld martensitic stainless steels. These electrodes or welding rods provide an as-welded deposit of somewhat lower strength, but of greater toughness, than the martensitic types. Filler metals suitable for welding some martensitic stainless steels are given in Table 1.

Flux Cored Arc Welding

The FCAW process is described in the article "Flux Cored Arc Welding" in this Volume. The particular concerns with martensitic stainless steels are outlined above. The possibility of high heat input with FCAW makes preheat and interpass temperature control even more important in reducing cracking tendency. Matching composition electrodes for welding the martensitic stainless steels are included in Table 24. Other compositions may be available from the electrode manufacturer.

Submerged Arc Welding

This process is described in detail in the article "Submerged Arc Welding" in this Volume. The particular concerns with welding the martensitic stainless steels are covered above. The possibility of greater heat input than with FCAW make preheat and interpass control more important in reducing cracking tendency. Table 25 gives some of the compositions commercially available. Other compositions are available through special order.

Gas Tungsten Arc Welding

This process is described in the article "Gas Tungsten Arc Welding" in this Volume. The specific application of it to austenitic stainless steels is discussed earlier in this article. When GTAW is to be used on the martensitic stainless steels, all of the considerations for austenitic stainless steels must be considered, as well as the potential difficulties that may occur because of the hardening response of the material on heating. Specific attention must be paid to preheating and postheating. The composition of martensitic stainless steel filler metals used with this process are given in Table 25. When it is not necessary to match the base-metal properties, the low-carbon and stabilized austenitic filler metals shown in Table 7 may be used. Use of the austenitic filler metals does not eliminate the need for preheating or postheating.

Precipitation-Hardening Stainless Steels

Designations and nominal compositions of typical precipitation-hardening (PH) stainless steels are given in Table 26. The steels are divided into three groups on the dual basis of characteristic alloying additions, particularly the elements added to promote precipitation hardening, and the matrix structures of the steels in the solution-annealed and aged condition. Because differences among the steels have a direct bearing on the behavior of the steels in heat treatment and welding, the metallurgical characteristics of each are considered separately.

Martensitic Precipitation-Hardening Steels

These steels have a predominantly austenitic structure at the solution-annealing temperature of approximately 1900 to 1950 °F, but they undergo an austenite-to-martensite transformation when cooled to room temperature. The M_s temperature is usually in the range of 200 to 300 °F. When

Table 25 Chemical compositions for martensitic stainless steel electrodes for GMAW, SAW, and GTAW

All are maximum percentages, unless otherwise noted

AWS classification	C	Cr	Ni	Mo	Mn	Si	P	S	Cu
ER410	0.12	11.5-13.5	0.6	0.75	0.6	0.50	0.03	0.03	0.75
ER410NiMo	0.06	11.0-12.5	4.0-5.0	0.4-0.7	0.6	0.50	0.03	0.03	0.75
ER420	0.25-0.40	12.0-14.0	0.6	0.75	0.6	0.50	0.03	0.03	0.75
ER502	0.10	4.6-6.0	0.6	0.45-0.65	0.6	0.50	0.03	0.03	0.75
ER505	0.10	8.0-10.5	0.5	0.8-1.2	0.6	0.50	0.04	0.03	0.75

Source: AWS A5.9-81

Table 26 Nominal composition of typical precipitation-hardening stainless steels

Common designation	UNS designation	C	Cr	Ni	Mo	Cu	Al	Nb	Other
Martensitic steels									
17-4 PH	S17 400	0.04	17.0	4.0	...	4.0	...	...	...
Custom 450	S45 000	0.03	15.0	6.0	0.75	1.5	...	0.3	...
15-5 PH	S15 500	0.04	15.0	5.0	...	4.0	...	0.3	...
PH 13-8 Mo	S13 800	0.04	13.0	8.0	2.0	...	1.0	...	...
Custom 455	S45 500	0.03	12.0	8.5	...	2.0	...	0.3	...
Semiaustenitic steels									
17-7 PH	S17 700	0.07	17.0	7.0	...	...	1.0	...	...
PH 15-7 Mo	S15 700	0.07	15.0	7.0	2.0	...	1.0	...	...
PH 14-8 Mo	S14 800	0.04	14.0	8.0	2.0	...	1.0	...	...
AM 350	S35 000	0.08	17.0	4.0	3.0	...	...	...	...
AM 355	S35 500	0.12	16.0	5.0	3.0	...	...	...	...
Austenitic steels									
17-10 P	...	0.12	17.0	10.0	...	...	...	...	0.25 P

martensite is reheated to 900 to 1100 °F, precipitation hardening and strengthening occur, promoted by the presence of one or more alloying additions. Molybdenum, copper, titanium, niobium, and aluminum (and their compounds) are dissolved during annealing and retained in solid solution by rapid cooling, producing precipitate (usually submicroscopic particles) that increases both the strength and the hardness of the martensitic matrix.

The compositional balance in the martenistic PH steels is critical, because relatively slight variations can lead to the formation of excessive amounts of delta ferrite during solution annealing. If the austenite is too stable, large amounts of austenite can also be retained at room temperature after solution annealing. Either of these two conditions prevents full hardening during aging. Carbon and nitrogen contents can significantly affect this balance. Increased carbon or nitrogen may result in contamination. Typical sources of these contaminants are shop dirt and the atmosphere.

These steels can be readily welded. The welding procedures resemble those ordinarily used for the 300 series stainless steels, despite differences in composition and structure between the two classes. The formation of martensite, which occurs during cooling from elevated temperatures as in welding, does not result in full hardening. These steels are not sensitive to cracking and do not require preheating.

Selection of filler metal depends on the properties required for the welded joint. If strength comparable to that of the base metal is not required at the welded joint, a tough 300 series stainless steel filler metal may be adequate. When a weld having mechanical properties comparable to those of the hardened base metal is desired, the filler metal must be of comparable composition, although slight modifications are permissible to obtain better weldability.

When welds are deposited in a single pass, the weld metal and the HAZ usually respond uniformly to a postweld precipitation-hardening heat treatment. There is seldom any significant variation in hardness across the joint. Multiple-pass welds, however, exhibit less uniform response to the same heat treatment because successive applications of heat during welding result in marked variations in the structure of weld metal, HAZ, and base metal. Annealing eliminates these variations and provides a more uniform microstructure capable of responding uniformly to precipitation hardening.

Semiaustenitic Precipitation-Hardening Steels

Unlike martensitic PH steels, semiaustenitic PH steels are soft enough in the annealed condition to permit cold working. When cooled rapidly from the annealing temperature to room temperature, they retain their austenitic structure, which displays good toughness and ductility in cold forming operations. The M_s temperatures for these steels are well below room temperature, but they vary depending on composition and annealing temperature.

To obtain hardening and strengthening, the austenitic structure must be transformed to an essentially martensitic one. This can be accomplished by treating the steel before subjecting it to the precipitation-hardening heat treatment by (1) heating the steel in the range of 1200 to 1600 °F to precipitate carbides and other compounds, thus depleting the matrix of enough austenite-stabilizing elements to allow transformation of austenite to martensite when the steel is cooled to room temperature; (2) refrigerating the steel to a temperature well below the M_s point (−100 °F, for example); or (3) cold working the steel enough so that the austenite transforms to martensite. After transformation to martensite, the semiaustenitic PH steels respond to precipitation hardening in the temperature range of 850 to 1100 °F like the martensitic PH steels. Whether a precipitate forms or a tempering reaction takes place depends on the steel composition.

The M_s temperature of the semiaustenitic PH steels is controlled by the solution-annealing temperature, as well as by composition. For example, when AM-350 steel is solution annealed at temperatures below 1700 °F, incomplete carbide solution raises the M_s temperature above room temperature. On the other hand, when the solution-annealing temperature is raised above 1700 °F, the M_s temperature drops precipitously. In practice, the solution-annealing temperature is not permitted to exceed about 1925 °F, because higher temperatures promote the formation of delta ferrite.

The semiaustenitic PH steels are normally welded in the annealed condition. The tough austenitic structure imparts welding characteristics similar to those of 300 series stainless steels. The semiaustenitic PH steels are not susceptible to cracking when welded, even when welded after transformation to martensite, because the low-carbon martensite developed is not of high hardness or low ductility. Also, cold cracking does not occur in the base metal adjacent to the weld because the HAZ is austenitized during welding and remains substantially austenitic as the joint cools to room temperature.

The choice of filler metal depends largely on the weld properties desired. The filler metal can be an alloy of precipitation-hardening composition capable of developing mechanical properties comparable to those of the base metal. If high strength is not a requisite, the filler metal can be a 300 series austenitic stainless steel. When these steels are welded in the annealed condition, the following microstructural relations are generally obtained as a result of relatively rapid heating and cooling at the joint:

- Weld metal contains small amount of ferrite in an essentially austenitic matrix; hardness, approximately 90 HRB.
- Base metal immediately adjacent to weld

Table 27 Chemical compositions of welding consumables for precipitation-hardening stainless steels
All are maximum percentages

AWS classification	Composition, % C	Cr	Ni	Mo	Nb + Ta	Mn	Si	P	S	Cu
E630(a), ER630(b)	0.05	16.0-16.75	4.5-5.0	0.75	0.15-0.30	0.25-0.75	0.75	0.04	0.03	3.25-4.00

(a) Undiluted weld-metal composition. (b) Consumable composition
Source: AWS A5.4-81, A5.9-81

displays high-temperature annealed (austenitic) structure; hardness, approximately 90 HRB.

- Base metal in narrow zone just beyond annealed zone next to weld is hardened slightly; hardness, approximately 90 to 98 HRB.

If the welded assembly is given the customary double-aging heat treatment, the three areas identified above, as well as the unaffected base metal, transform and precipitation harden uniformly to a hardness range commensurate with the precipitation-hardening temperature. The weld metal may be somewhat less tough than the wrought base metal, as measured by the results of tensile-elongation and bend tests, depending on the type of joint, the welding process, and the hardening temperature.

Higher precipitation-hardening temperatures ensure good weld toughness with little sacrifice of strength. Maximum toughness requires annealing of the weldment prior to the transformation and hardening treatments. Although other variations in the welding and heat treating sequence are possible and may be desirable at times, the choice of the sequence should ensure that, after welding, the weld metal and the HAZ are in the annealed (austenitic) condition. To harden these areas, both the transformation and the precipitation-hardening heat treatments must be applied. If the components are given the transformation treatment before welding, the precipitation-hardening treatment alone, after welding, produces no significant hardening in either the weld metal or the HAZ.

Austenitic Precipitation-Hardening Steels

The alloy content of these steels is high enough to maintain an austenitic structure after annealing and after any aging or hardening treatment. The precipitation-hardening phase is soluble at the annealing temperature of 2000 to 2050 °F, and it remains in solution during rapid cooling from the annealing temperature. When these steels are reheated to about 1200 to 1400 °F, precipitation occurs, and the hardness and strength of the austenitic structure increase. The hardness attained

Fig. 24 Helium-storage vessel with self-aligning joint

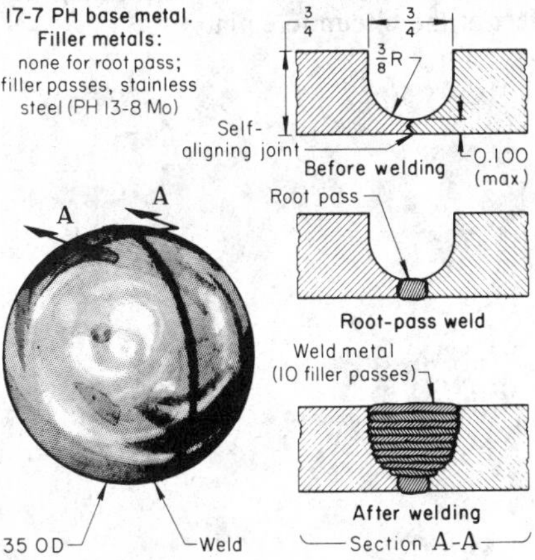

Fig. 25 Welding of thin-skin missile body segments

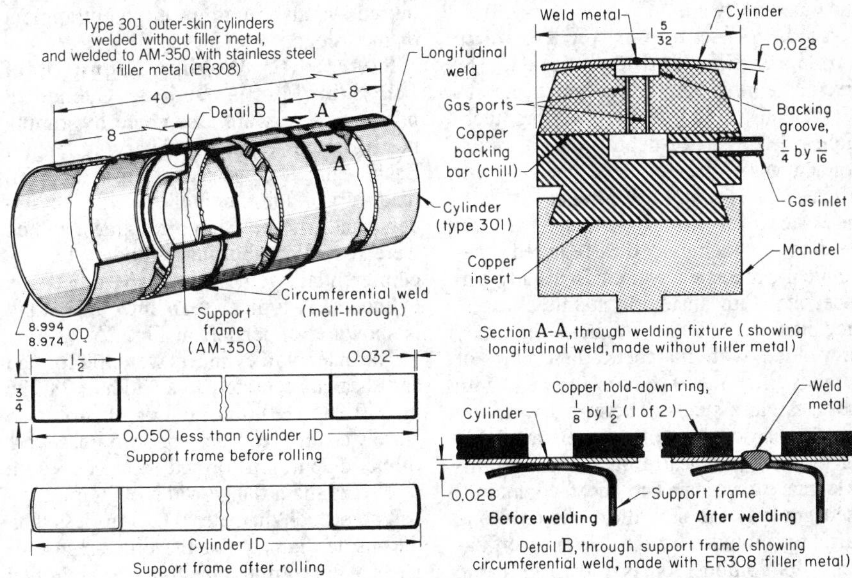

Item	Longitudinal weld	Circumferential weld
Process	Automatic GTAW	Automatic GTAW
Joint type	Butt	Edge
Joint location	Cylinder joint	Frame to cylinder
Weld type	Square groove	Square groove
Fixtures	Backing bar(a), hold-down bars	Copper hold-down rings
Power supply	200-A transformer-rectifier	200-A transformer-rectifier
Electrode	1/16-in.-diam EWTh-2	3/32-in.-diam EWTh-2
Electrode taper	1/4 in. long to point	1/8 in. long to point
Nozzle diameter	3/8 in.	5/8 in.
Filler metal	None(b)	0.030-in.-diam ER308(c)
Filler-wire angle	...	30° from horizontal(c)
Filler-wire tip to electrode	...	3/8 in.(c)
Shielding gas	Helium, at 30 ft^3/h	Helium, at 35 ft^3/h
Backup gas	Helium, at 25 ft^3/h	Helium, at 25 ft^3/h
Starting-tab size	0.028 by 1.0 by 3.0 in.	None
Runoff-tab size	0.028 by 1.0 by 3.0 in.	None
Welding position	Flat	Horizontal-rolled pipe
Current (DCEN) and voltage	18.5 A; 15 V	50 A; 15 V
Filler-wire feed rate	...	28 in./min
Electrode extension	13/32 in.	7/16 in.
Travel speed	10 in./min	10 in./min

(a) Made of deoxidized copper. (b) Except for weld repair. (c) Applies only to circumferential welding of joint between two cylinders and supporting joint providing internal link

is lower than that of the martensitic or semiaustenitic PH steels, but the nonmagnetic properties are retained. Although the austenitic PH steels remain austenitic during all phases of forming, welding, and heat treatment, some contain alloying elements (for precipitation-hardening purposes) that greatly affect behavior in welding.

The one austenitic PH steel (17-10P) listed in Table 26 has very limited weldability. When heated above about 2150 °F, it exhibits hot shortness because of the formation of phosphorus-rich compounds at the grain boundaries. When 17-10P is arc welded, hot shortness causes underbead cracking in the HAZ. However, 17-10P steel has been successfully flash welded, apparently because the upsetting force was adequate to extrude the hot short material in the form of flash and thus produce a sound weld.

Welding Techniques for Precipitation-Hardening Stainless Steels. The precipitation-hardening stainless steels can be welded using the arc welding techniques described in depth under the section on austenitic stainless steels in this article. The major difference is that these steels are usually heat treated after welding to achieve the required mechanical properties, which is usually unnecessary with austenitic stainless steels. Precipitation-hardening stainless steels may be welded with matched, dissimilar, or without filler metals, as is the case with most stainless steels. There is a wide variety of hardenable filler metals available for these PH grades through the consumable manufacturer. The most commonly used grade is the 630 alloy, the only one currently included in the AWS specifications. Its composition is shown in Table 27.

Example 17. Self-Aligning Joint for Welded Hemispheres. Two hemispheres, forged from 17-7PH stainless steel, were welded by automatic GTAW to produce a pressure vessel (Fig. 24) for containment of helium under high pressure. To ensure self-alignment of the hemispheres and to avoid mismatch during welding, a specially designed joint containing a notch (Fig. 24, upper right) was used.

Equipment for welding included a special oscillating head with automatic arc-voltage control. The assembled hemispheres were rotated in a circumferential fixture mounted on turning rolls. Helium shielding was used at the arc, and free-flowing argon was fed to the interior of the sphere through a port (not shown). The root pass was made without the use of filler metal and without oscillation of the welding head. Ten subsequent passes were made with oscillation of the welding head and with filler-metal wire continuously fed to the joint. The oscillating technique minimized the amount of interpass cleaning required and also minimized the entrapment of inclusions.

Example 18. Welding Segments of Thin-Skin Missile Bodies. Cylindrical missile bodies were assembled by joining prefabricated segments of varying length. Each segment consisted of an outer skin, made up of longitudinally welded cylinders, and internal support frames, which were joined to abutting cylinders by circumferential welds that also joined the cylinders. A typical segment 40 in. long is shown at upper left in Fig. 25.

The outer-skin cylinders were roll formed and sheared accurately to a width of 28.236 in. +0.002, −0.000 in. Length was also closely controlled. The support frames were rubber-diaphragm formed to a web width of $1^1/_2$ in. and a flange width of $^3/_4$ in. They had an outside diameter 0.050 in. less than the inside diameter of the outer-skin cylinder. After forming, the frames were rolled to produce a convex radius at the circumference, with the outermost diameter exactly matching the inside diameter of the outer-skin cylinder. A cross section of the frame before and after rolling is shown at lower left in Fig. 25. Before welding, the cylinders and frames were cleaned and prepared as follows:

- Clean in hot (180 to 200 °F) aqueous solution of heavy-duty alkaline cleaner (8 to 12 oz/gal) for 15 to 20 min.
- Rinse in clean water at room temperature.
- Descale in aqueous solution of nitric acid (15 to 20%) and hydrofluoric acid (3 to 5%) at 120 to 140 °F for not more than 30 min.
- Rinse in clean water at room temperature.
- Drawfile and deburr faying surfaces.
- Wipe all surfaces with acetone and remove residues with a lint-free cloth.

The upper right view in Fig. 25 shows the setup used for welding the longitudinal seam of each outer-skin cylinder. Conventional hold-down bars (not shown in Fig. 25) with 1-in.-wide fingers were used to clamp the cylinder at a pressure of 80 psi. The inner surface of the cylinder was supported by a deoxidized copper backing bar, which also served as a chill. Except for occasional repair welds, the longitudinal joints were welded without filler metal, because filler metal increased shrinkage and was not required to ensure x-ray soundness. Helium was used as backup and shielding gas. Starting and runoff tabs were used at the beginning and ends of the weld.

After fabrication of the outer skin, the supporting frames were fitted inside the cylinders supported on a wooden block to prevent belling of the cylinder walls at a distance of 8 in. from the ends (Fig. 25, upper left). The frames were joined to the outer skin of the cylinders by a single-pass circumferential melt-through weld. The frames provided backup for welding, and two copper rings ($^1/_8$ in. thick by $1^1/_2$ in. wide) slipped over the cylinder served as hold-downs (Fig. 25, detail B). This arrangement prevented any drawstring effect at the circumferential joint.

Arc Welding of Heat-Resistant Alloys

By the ASM Committee on Arc Welding of Heat-Resistant Alloys*

HEAT-RESISTANT ALLOYS considered in this article include nickel-based alloys, iron-based alloys, and cobalt-based alloys. The procedures used in welding heat-resistant alloys depend to some extent on the mechanism by which they are strengthened for high-temperature service—primarily solid-solution strengthening or precipitation hardening.

Many heat-resistant alloys are susceptible to cracking during welding or during subsequent heat treatment. The severe service conditions to which heat-resistant alloys are exposed make changes in microstructure and properties, as produced by arc welding, of special importance and make avoidance of cracking imperative. Nominal compositions of commonly arc welded heat-resistant alloys are given in Table 1.

Welding Processes

Heat-resistant alloys can be welded by all the arc welding processes. Gas tungsten arc welding (GTAW) is widely used, especially for joining thin sections. In general, shielded metal arc welding (SMAW) and gas metal arc welding (GMAW) are used in joining sections more than $^1/_4$ in. thick, where the heat input does not adversely affect the weld metal or the base metal. Submerged arc welding (SAW) generally is used in high-volume production welding of sections more than 1 in. thick.

The data in Table 2 are intended to serve as starting points for the establishment of machine settings for GTAW. These conditions are, in general, suitable for welding nickel, iron-nickel-chromium, iron-chromium-nickel, and cobalt-based heat-resistant alloys, when making butt, corner, or T-joints, using an appropriate groove design based on stock thickness and application. An increase in welding current of 10 to 20 A may be needed for melt-through T-joints. Generally, the interpass temperature should range from 200 to 350 °F, depending on the alloy. Oscillation of the welding torch may help to prevent cracking by changing the solidification pattern. This may also improve the appearance of the weld.

Cleaning of Workpieces

The weldability of heat-resistant alloys is markedly affected by the cleanliness of the base metal and the filler metal. Shop dirt, paint, grease, oil, machine lubricants, processing chemicals, temperature-indicating sticks, marking crayons, oxide films, and scale are the main surface contaminants. Sulfur and lead in foreign material on the workpiece surface can diffuse into the base metal when it is heated and result in severe cracking.

Processes for the removal of surface dirt are described in Volume 5 of the 9th edition of *Metals Handbook*. Grease and oil are removed by commercial solvents and by vapor degreasing. Soaps can be removed by rinsing in hot water. Soluble oils, tallow fats, and fatty acids require a more complex cleaning procedure. Techniques for removal of tarnish, oxide films, and heavy scale are described in the articles "Cleaning and Finishing of Heat-Resistant Alloys" and "Cleaning and Finishing of Nickel and Nickel Alloys" in Volume 5 of the 9th edition of *Metals Handbook*.

Dust that has settled on the workpieces after cleaning can be removed by carefully wiping the joint area with clean, lint-free cloths dampened with a solvent such as methyl ethyl ketone. To avoid surface contamination of cleaned workpieces, white gloves are sometimes used during handling. After welding, all slag must be removed. If any surface slag left on a weldment should become molten during service, severe attack on the metal may occur.

Welding Fixtures

Fixtures used in arc welding heat-resistant alloys are generally similar to those used on other metals. Chill bars are often used to cool the weld area rapidly. Backing bars, inserts, and facing plates that are in contact with the workpieces on either the root or the face side of the welds should be located so as not to contaminate the weld metal or base metal or cause gas or flux entrapment. These accessories are usually made of copper.

Hold-down bars and backing bars extend the full length of the weld. The backing bars usually contain passages to facilitate inert gas shielding. When grooved backing bars are used, the grooves should be shallow to minimize melt-through and to limit the height of the root reinforcement. Grooves in backing bars should have rounded corners to be elliptical in shape to prevent entrapment of slag.

Example 1. Fixturing in Gas Tungsten Arc Welding. Two formed rings were

*Charles J. Sponaugle, *Chairman*, Fabrication Manager, Technical and Product Services, Cabot Corp.; Jon R. Bryant, Manager, Metallurgy and Welding Engineering, Otis Engineering Corp.; Charles H. Cadden, Metallurgist, High Technology Materials Division, Cabot Corp.; Harold R. Conaway, Vice President, Major Tool & Machine, Inc.; Samuel D. Kiser, Product Manager, Huntington Alloys, Inc.; Donald E. Wenschhof, Jr., General Manager, High Technology Products Division, A.M. Castle Co.

Table 1 Nominal composition of heat-resistant alloys

	Composition, %										
	Ni	Cr	Co	Fe	Mo	Ti	W	Nb	Al	C	Other
Solid-solution nickel-based alloys											
Hastelloy C-4	63.0	16.0	2.0 max	3.0 max	15.5	0.7 max	...	...	...	0.015	...
Hastelloy C-276	59.0	15.5	...	5.0	16.0	...	3.7	...	...	0.02	...
Hastelloy N	72.0	7.0	...	5.0 max	16.0	0.5 max	...	...	...	0.06	...
Hastelloy S	67.0	15.5	...	1.0	15.5	...	...	...	0.2	0.02 max	0.02 La
Hastelloy X	49.0	22.0	1.5 max	18.5	9.0	...	0.6	...	2.0	0.15	...
Inconel 600	76.0	15.5	...	8.0	...	...	...	...	...	0.08	0.25 max Cu
Inconel 601	60.5	23.0	...	14.1	...	...	...	...	1.35	0.05	0.5 max Cu
Inconel 617	55.0	22.0	12.5	...	9.0	...	...	...	1.0	0.07	...
Inconel 625	61.0	21.5	...	2.5	9.0	0.2	...	3.6	0.2	0.05	...
Precipitation-hardenable nickel-based alloys											
GMR-235	63.0	15.5	...	10.0	5.25	2.0	...	...	3.0	0.15	0.06 B
Inconel 702	79.5	15.5	...	0.4	...	0.7	...	...	3.4	0.04	...
Inconel 706	41.5	16.0	...	37.5	...	1.75	...	2.9(a)	0.2	0.03	...
IN-713 C	74.0	12.5	...	...	4.2	0.8	...	2.0	6.0	0.12	0.012 B; 0.10 Zr
Inconel 718	52.5	19.0	...	18.5	3.0	0.9	...	5.0	0.5	0.08 max	0.15 max Cu
Inconel 722	75.0	15.5	...	7.0	...	2.5	...	...	0.7	0.04	...
Inconel X-750	73.0	15.5	...	7.0	...	2.5	...	1.0	0.7	0.04	0.25 max Cu
Incoloy 901	42.5	12.5	...	36.2	6.0	2.7	...	...	...	0.10 max	...
M-252	56.5	19.0	10.0	<0.75	10.0	2.6	...	...	1.0	0.15	0.005
René 41	55.0	19.0	11.0	<0.3	10.0	3.1	...	...	1.5	0.09	0.01 B
Udimet 700	53.0	15.0	18.5	<1.0	5.0	3.4	...	...	4.3	0.07	0.03 B
Waspaloy	57.0	19.5	13.5	2.0 max	4.3	3.0	...	...	1.4	0.07	0.006 B; 0.09 Zr
Solid-solution iron-based alloys											
16-25-6	25.0	16.0	...	50.7	6.0	...	...	...	...	0.06	1.35 Mn; 0.70 Si
17-14 CuMo	14.0	16.0	...	62.4	2.5	0.3	...	0.4	...	0.12	0.75 Mn; 0.50 Si; 3.0 Cu
19-9 DL	9.0	19.0	...	66.8	1.25	0.3	1.25	0.4	...	0.30	1.10 Mn; 0.60 Si
Incoloy 800	32.5	21.0	...	45.7	...	0.38	...	...	0.38	0.05	...
Incoloy 800H	33.0	21.0	...	45.8	...	...	...	...	...	0.08	...
Incoloy 801	32.0	20.5	...	46.3	...	1.13	...	...	...	0.05	...
Incoloy 802	32.5	21.0	...	44.8	...	0.75	...	...	0.58	0.35	...
Multimet (N-155)	20.0	21.0	20.0	32.2	3.0	...	2.5	1.0	...	0.15	0.15 N; 0.02 La; 0.02 Zr
RA 330	36.0	19.0	...	45.1	...	...	...	...	...	0.05	...
Precipitation-hardenable iron-based alloys											
A-286	26.0	15.0	...	55.2	1.25	2.0	...	...	0.2	0.04	0.005; 0.3 V
Discoloy	26.0	14.0	...	55.0	3.0	1.7	...	...	0.25	0.06	...
Incoloy 903	0.1 max	38.0	15.0	41.0	0.1	1.4	...	3.0	0.7	0.04	...
Haynes 556	21.0	22.0	20.0	29.0	3.0	...	2.5	0.1	0.3	0.10	0.50 Ta; 0.02 La
Solid-solution cobalt-based alloys											
Haynes 25 (L-605)	10.0	20.0	50.0	3.0	...	...	15.0	...	...	0.10	1.5 Mn
Haynes 188	22.0	22.0	37.0	3.0 max	...	...	14.5	...	...	0.10	0.90 La
S-816	20.0	20.0	42.0	4.0	4.0	...	4.0	4.0	...	0.38	...
Stellite 6B	1.0	30.0	61.5	1.0	...	...	4.5	...	...	1.0	...
UMCo-50	...	28.0	49.0	21.0	...	...	...	...	...	0.12 max	...
Precipitation-hardenable cobalt-based alloys											
AR-213	0.5 max	19.0	65.0	0.5 max	...	...	4.5	...	3.5	0.17	6.5 Ta; 0.15 Zr; 0.1 Y
MP-35N	35.0	20.0	35.0	...	10.0	...	...	...	...	...	...
MP-159	25.0	19.0	36.0	9.0	7.0	3.0	...	0.6	0.2	...	...

(a) Nb + Ta

circumferentially butt welded to form an air seal for a compressor stage, as shown in Fig. 1. The thickness of each ring varied, but it was 0.040 in. on each ring at the weld joint. The welding procedure described below was that used for completing Waspaloy rings; rings made of Inconel alloys were also assembled in a similar manner.

The fixture and method of clamping the rings for welding are shown in Fig. 1, detail A. The groove in the backing fixture provided a passageway for the backing gas, which was fed into the groove through a series of small holes to ensure even distribution.

The rings were in the solution-treated condition when welded. Surface oxides were removed for a minimum of $^1/_2$ in. from the joint edges by a 240-grit aluminum oxide wheel mounted on an air grinder, followed by wiping with a solvent. The cleaned rings were assembled in the fixture (Fig. 1, detail A), making certain that the joint line was centered over the backing groove and that joint alignment was within 0.005 in., as determined by dial-

Table 2 Conditions for GTAW of heat-resistant alloys(a)

Base-metal thickness, in.	Diameter of filler metal, in.(b)	Electrode diameter, in.(c)	Shielding gas: Gas	Shielding gas: Flow rate, ft³/h	Welding current, A(d)
0.010	0.020	0.040-0.060	Ar	12-15	10-15
0.020	0.030	0.060	Ar	12-15	15-25
0.030	0.030; 0.045	0.060	Ar	12-15	25-35
0.045	0.045	0.060	Ar	12-15	40-50
0.050	0.045	0.060	Ar	12-15	45-55
0.060	0.045	0.060	Ar	12-15	55-65
0.080	0.060	0.060	Ar	12-15	75-85
0.100	0.060; 0.090	0.093	Ar or He	12-20	95-105
0.125	0.060; 0.090	0.093	Ar or He	12-20	110-135
0.250	0.060; 0.090	0.093	Ar or He	12-20	130-200

(a) The data in this table are intended to serve as starting points for the establishment of optimum machine settings for welding workpieces on which previous experience is lacking. The data are subject to adjustment as necessary to meet the special requirements of individual applications. Torch nozzle diameter was 7/16 in.; nozzle had a gas lens. (b) Minimum wire diameters were applicable. (c) EWTh-2 electrodes. (d) DCEN with high-frequency arc starting. An increase of 10 to 20 A may be needed for melt-through T-joints.

indicator measurement. The backing groove was purged with shielding gas before welding began.

The rings were tack welded by manual GTAW, using 24 welds 3/8 in. long, equally spaced along the joint. A penetration of 0.010 to 0.015 in. was required. After tack welding, the rings were welded in one pass by automatic GTAW, using high-frequency arc starting and Waspaloy filler metal. Shielding gas was dispersed through the torch nozzle and an auxiliary trailing shield. The backing fixture also provided shielding gas to the root area of the weld. After the weld was made, wire feed was stopped, and the current was decreased slowly. Shielding-gas flow was maintained until after the weld metal solidified.

The welded assembly was heat treated at 1600 °F for 1 h and at 1400 °F for 10 h. Before and after heat treatment, welds were inspected for surface cracks by fluorescent-penetrant techniques and for internal defects by x-ray radiography.

Nickel-Based Heat-Resistant Alloys

Some nickel-based heat-resistant alloys (Table 1) are solid-solution strengthened, and others are strengthened by the precipitation of a second phase from a supersaturated solid solution. Many of the solid-solution nickel-based alloys, including Inconel 625 and Hastelloy C-276, are widely used in applications requiring corrosion resistance at temperatures ranging from cryogenic to 800 °F or higher; these alloys, therefore, are classified as being both heat resistant and corrosion resistant. Welding of other corrosion-resistant nickel-based alloys, including nickel-copper alloys, is discussed in the article "Arc Welding of Nickel Alloys" in this Volume. Powder metallurgy (P/M) nickel-based alloys can also be strengthened by the uniform dispersion of refractory oxide particles throughout the matrix. This process is referred to as oxide dispersion strengthening (ODS). When these P/M materials are fusion welded, the oxide particles will agglomerate during solidification. This action negates the original strengthening mechanism provided by dispersion within the matrix. As a result, the weld metal will be significantly weaker than the base metal. Joining processes that do not involve melting of the base metal will retain the high strength of ODS alloys.

The nickel-based alloys are commonly welded by GTAW, SMAW, and GMAW. Frequently, a root pass is made by GTAW and the subsequent passes by GMAW. Submerged arc welding can be used on certain alloys, but the welding flux must be carefully selected to obtain adequate protection and provide correct elemental additions to the weld pool. The welding conditions chosen must avoid excessive heat input. When welding metal more than 3 in. thick, shrinkage stresses decrease ductility slightly, and a postweld stress-relieving treatment may be necessary. The manufacturer of the alloy should be consulted for specific details.

Precipitation-hardenable alloys are susceptible to cracking in the weld metal or in the heat-affected zone (HAZ) unless they are properly heat treated before and after welding. These alloys are normally welded in the solution-annealed condition. If they are welded in the precipitation-hardened condition, a solution anneal is required before returning to high-temperature service. When solution annealing, a rapid heating rate should be used to avoid parent metal strain-age cracking. This usually can be accomplished by charging into a hot furnace. The alloys containing niobium and/or tantalum, such as Inconel 718, have a relatively slow hardening response and can be welded without undergoing spon-

Fig. 1 Waspaloy air-seal ring. Method of holding components for GTAW

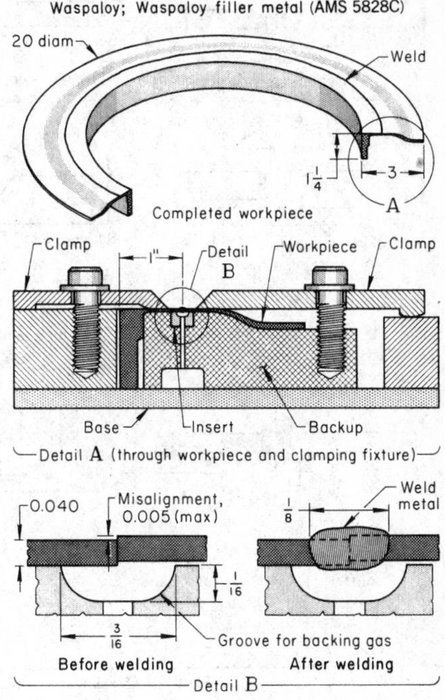

Joint type	Butt
Weld type	Square groove
Power supply	200- to 300-A transformer-rectifier, constant current

Tack welding

Welding process	Manual GTAW
Torch	Manual, water cooled
Electrode	1/16-in.-diam EWTh-2
Filler metal	None
Current	40 A (DCEN)
Voltage	9-11 V
Shielding gas (argon):	
At torch	15 ft³/h
Backing gas	5 ft³/h
Welding position	Flat

Continuous welding

Welding process	Automatic GTAW
Torch	Mechanical, water cooled
Electrode	1/16-in.-diam EWTh-2
Electrode extension	5/16 in. max
Arc starting	High frequency
Current	135 A (DCEN)
Voltage	11 V
Welding speed	16 in./min
Filler metal	0.032-in.-diam Waspaloy (AMS 5828C)
Filler-metal feed	Automatic at constant speed
Filler-metal speed	30 in./min
Shielding gas:	
At torch	Argon at 30 ft³/h; helium at 2 ft³/h
Backing gas	Argon at 5 ft³/h
Trailing shield	Argon at 25 ft³/h
Welding position	Flat (electrode holder, fixed; workpiece revolved)
Number of passes	1

Fig. 2 Joint designs and dimensions for GTAW, GMAW, and SMAW

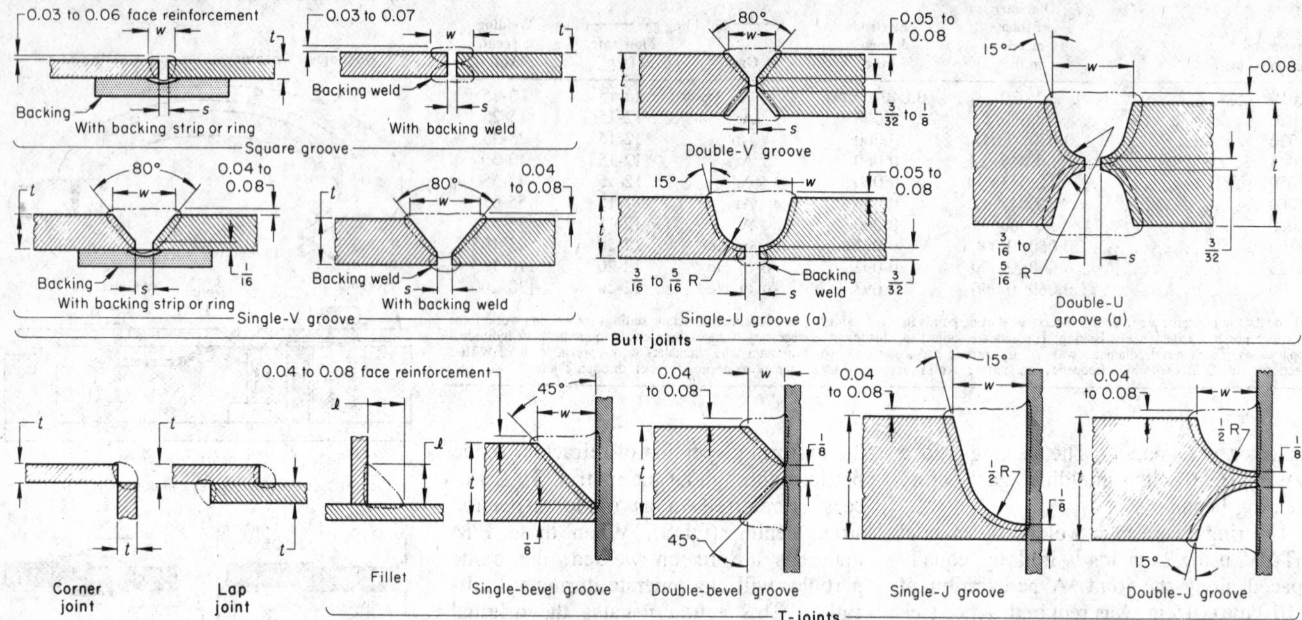

Base-metal thickness (*t*), in.	Width of groove or bead (*w*), in.	Maximum root opening (*s*), in.	Approximate amount of metal deposited lb/ft	Approximate weight of electrode, lb/ft(a)
Square-groove butt joint with backing strip or ring				
0.037	1/8	0	0.02	0.025
0.050	5/32	0	0.04	0.05
0.062	3/16	0	0.04	0.06
0.093	3/16-1/4	1/32	0.06	0.08
0.125	1/4	1/16	0.07	0.09
Square-groove butt joint with backing weld				
1/8	1/4	1/32	0.11	0.15
3/16	3/8	1/16	0.24	0.32
1/4	7/80	3/32	0.31	0.42
Single-V-groove butt joint with backing strip or ring				
3/16	0.35	1/8	0.227	0.31
1/4	0.51	3/16	0.443	0.61
5/16	0.61	3/16	0.582	0.80
3/8	0.71	3/16	0.745	1.02
1/2	0.91	3/16	1.16	1.59
5/8	1.16	3/16	1.61	2.21
Single-V-groove butt joint with backing weld				
1/4	0.41	3/32	0.42	0.58
5/16	0.51	3/32	0.54	0.74
3/8	0.65	1/8	0.73	1.00
1/2	0.85	1/8	1.21	1.67
5/8	1.06	1/8	1.46	2.00
Double-V-groove butt joint				
1/2	0.40	1/8	0.89	1.16
5/8	0.49	1/8	1.08	1.48
3/4	0.62	1/8	1.46	2.00
1	0.81	1/8	2.42	3.34
1 1/4	1.03	1/8	2.92	4.00
Single-U-groove butt joint(b)				
1/2	0.679	1/8	1.03	1.41
5/8	0.745	1/8	1.38	1.90
3/4	0.813	1/8	1.68	2.30
1	0.957	1/8	2.63	3.60
1 1/4	1.073	1/8	3.62	4.96
1 1/2	1.215	1/8	4.79	6.55
1 3/4	1.349	1/8	5.98	8.19
2	1.485	1/8	7.40	10.12
Double-U-groove butt joint(b)				
1	0.679	1/8	2.06	2.82
1 1/4	0.745	1/8	2.76	3.80
1 1/2	0.813	1/8	3.36	4.60
2	0.957	1/8	5.26	7.20
2 1/2	1.073	1/8	7.24	9.92
Corner and lap joint				
1/16	...	...	0.02	0.04
1/8	...	...	0.05	0.07
3/16	...	...	0.10	0.14
1/4	...	...	0.19	0.26
3/8	...	...	0.42	0.57
1/2	...	...	0.74	1.02
T-joint with fillet				
...	...	1/8	0.03	0.04
...	...	3/16	0.07	0.10
...	...	1/4	0.12	0.16
...	...	5/16	0.19	0.26
...	...	3/8	0.27	0.37
...	...	1/2	0.47	0.64
...	...	5/8	0.74	1.01
...	...	3/4	1.07	1.46
...	...	1	1.90	2.60

(continued)

Base-metal thickness (t), in.	Width of groove or bead (w), in.	Maximum root opening (s), in.	Approximate amount of metal deposited lb/ft	Approximate weight of electrode, lb/ft(a)
Single-bevel-groove T-joint				
1/4	0.125	...	0.07	0.09
5/16	0.188	...	0.13	0.17
3/8	0.250	...	0.19	0.26
1/2	0.375	...	0.38	0.52
5/8	0.500	...	0.63	0.86
3/4	0.625	...	0.93	1.28
1	0.875	...	1.77	2.42
Double-bevel-groove T-joint				
1/2	0.188	...	0.25	0.34
5/8	0.250	...	0.39	0.54
3/4	0.313	...	0.56	0.77
1	0.438	...	0.99	1.36
1 1/4	0.563	...	1.54	2.15
1 1/2	0.688	...	2.21	3.03
1 3/4	0.813	...	3.00	4.09
2	0.938	...	3.90	5.35

Base-metal thickness (t), in.	Width of groove or bead (w), in.	Maximum root opening (s), in.	Approximate amount of metal deposited lb/ft	Approximate weight of electrode, lb/ft(a)
Single-J-groove T-joint				
1	0.625	...	1.78	2.4
1 1/4	0.719	...	2.50	3.4
1 1/2	0.781	...	3.23	4.4
1 3/4	0.875	...	4.09	5.6
2	0.969	...	4.93	6.8
2 1/4	1.031	...	5.80	8.0
2 1/2	1.094	...	6.94	9.5
Double-J-groove T-joint				
1	0.500	...	1.48	2.0
1 1/4	0.563	...	1.90	2.6
1 1/2	0.594	...	2.56	3.5
1 3/4	0.625	...	3.11	4.3
2	0.656	...	3.81	5.2
2 1/4	0.688	...	4.51	6.2
2 1/2	0.750	...	5.27	7.2

(a) To obtain linear feet of weld per pound of consumable electrode, take the reciprocal of pounds per linear foot. If the underside of the first bead is chipped out and welded, add 0.21 lb/ft of metal deposited (equivalent to 0.29 lb/ft of consumable electrode). (b) For GMAW (except with the short circuiting arc), root radius should be one half the value shown and bevel angle should be twice as great.

taneous hardening during heating and cooling.

The nickel-chromium and nickel-molybdenum solid-solution alloys are readily welded in the annealed condition. No heat treatment is needed after welding to improve corrosion resistance, and generally the alloys do not become embrittled after long exposure at temperatures up to about 1500 °F. Table 2 lists general conditions for GTAW of heat-resistant nickel-based alloys.

Joint Design

The same joint designs are used for both GTAW and SMAW. Joint design for GMAW requires special consideration, as discussed in the section on gas metal arc welding of nickel-based alloys in this article. Unless it has been proven satisfactory by experience, a joint design that has been developed for another metal should not be used for nickel-based alloys.

Figure 2 shows joint designs used in welding nickel-based heat-resistant alloys and gives the sizes of grooves or welds in different metal thicknesses and the approximate amount of metal deposited. Nickel-based alloy weld metal does not flow as readily or penetrate as deeply as steel weld metal does. Therefore, the joints must be more open to allow placement of weld metal, and lands should be thinner to accommodate lower penetration. Excessive puddling and heat input have a detrimental effect, because loss of residual deoxidizers may result. Use of the single- and double-bevel T-joint may not be suitable in some cases because of lack of joint accessibility.

Preweld and Postweld Heat and Mechanical Treatments

The solid-solution (non-age-hardenable) alloys are welded in both the annealed and moderately cold worked condition. The precipitation-hardenable alloys usually are welded in the solution-treated condition, although test data indicate that welding René 41 in the overaged condition can help prevent strain-age cracking. If a high degree of deformation should occur during preweld forming, or if the alloy has a high work-hardening rate, process annealing on the formed workpieces, before welding, may be required. Usually, preheating nickel-based heat-resistant alloys is neither needed nor recommended. A postweld thermal or mechanical treatment is sometimes needed, especially for the precipitation-hardenable alloys, to redistribute and relieve residual stresses resulting from weld-shrinkage strains. Often, both treatments are used on the same weldment.

Weldments made of solid-solution alloys can be used as welded or after stress relieving, depending on the alloy and application. Stress relieving in the range from 800 to 1600 °F, depending on the alloy and its condition, can be used to reduce or remove stresses in work-hardened solid-solution alloys without producing a recrystallized grain structure. A low-temperature stress-equalizing heat treatment of 600 to 800 °F can be used to redistribute stresses without appreciably decreasing the mechanical strength produced by the previous cold working.

Precipitation-hardenable alloys are given a solution treatment after welding to relieve residual stresses, and then they are hardened by an aging heat treatment. If the normal aging time or temperature is exceeded, overaging occurs; loss of strength and increase in ductility can result.

Example 2. Modification of Postweld Heat Treatment to Minimize Distortion. The gas-turbine shroud shown in Fig. 3 required two circumferential welds to join an outer case, a front case, and a flange, all made of Waspaloy. Single-pass welds were made by automatic GTAW.

Waspaloy is susceptible to strain-age cracking during service in the temperature range of 1200 to 1500 °F. Waspaloy components, therefore, are usually solution treated prior to welding and, after welding, are again solution treated at about 1975 °F for 1 h and then aged at 1400 °F for 16 h. In this instance, excessive distortion was encountered during solution treatment, and therefore a modified heat treatment was developed that consisted of solution treating the components between rough and

Fig. 3 Waspaloy gas turbine shroud. Made by GTAW three ring-shaped parts in two single-pass circumferential welds

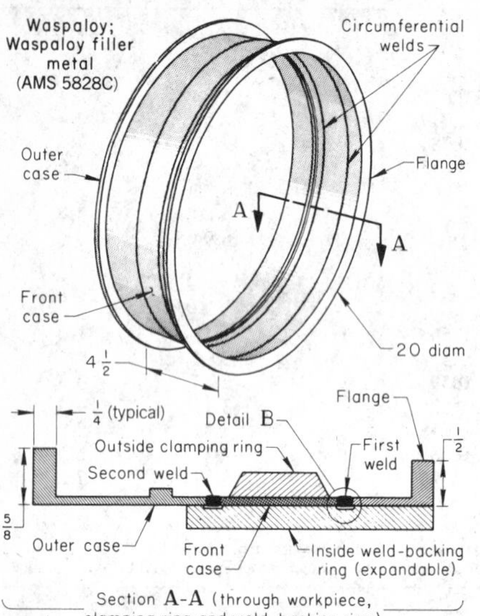

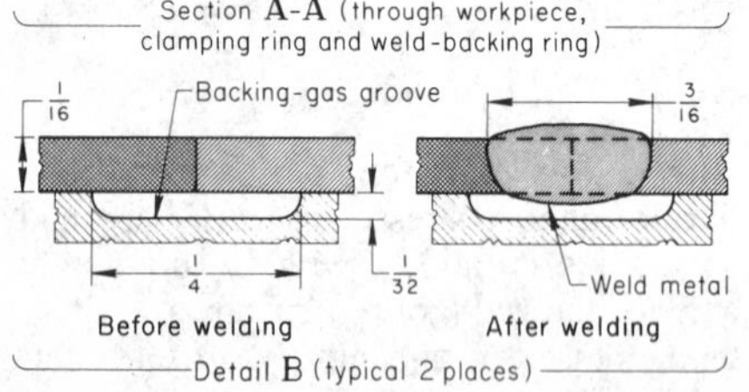

Joint type	Butt
Weld type	Square groove
Welding process	Automatic GTAW
Power supply	200- to 300-A transformer-rectifier, constant current
Torch	Mechanical, water cooled
Electrode	3/32-in.-diam EWTh-2, tapered to 0.025 in. diam
Electrode extension	1/4 in.
Arc starting	High frequency
Current (DCEN):	
First weld	65-70 A
Second weld	70 A
Voltage (both welds)	9-9 1/2 V
Welding speed:	
First weld	11 in./min
Second weld	13 1/2 in./min
Filler metal	0.032-in.-diam Waspaloy
Filler-metal feed	Constant speed, with feedback control
Filler-metal speed	20 in./min
Shielding gas (argon):	
At torch	30-35 ft^3/h
Backing gas	8-10 ft^3/h
Welding position	Flat
Number of passes	1

finish machining (prior to welding), stress relieving the weldments at 1600 °F for 1 h, then aging at 1400 °F for 10 h.

Joint areas were cleaned by wiping with methyl ethyl ketone, and parts were handled with white gloves. The assembly was held together in a fixture with an expandable inside weld-backing ring that forced the assembly against an outside ring, as shown in Fig. 3. The inside ring was relieved at the roots of the joints for backing-gas flow, and the assembly was purged with argon gas for 5 min before welding was started. Welds were inspected for surface cracks before and after heat treatment by fluorescent-penetrant techniques and for internal defects by x-ray radiography.

Overaging René 41. Tests conducted with circular patch test specimens made of René 41 plate (0.60 in. thick) have indicated that this alloy is more resistant to strain-age cracking when preweld solution heat treated at 1975 °F and overaged than when solution heat treated at 2150 °F, or when mill annealed at 1950 °F and then aged at 1400 °F for 16 h. The preweld overaging treatment consisted of holding at progressively lower temperatures as follows:

- Solution treat at 1975 °F for 1/2 h, cool at 3 to 8 °F/min to 1800 °F
- Hold at 1800 °F for 4 h, cool at 3 to 8 °F/min to 1600 °F
- Hold at 1600 °F for 4 h, cool at 3 to 8 °F/min to 1400 °F
- Hold at 1400 °F for 16 h, air cool to room temperature

An alternative overaging treatment was simply to cool slowly at 50 to 100 °F/h from the 1975 °F solution-treating temperature. Although ductility was decreased slightly, the tensile and rupture strengths of René 41 were completely recovered by the conventional solution and aging heat treatment after welding.

Effect of Cold Work. As-deposited weld metal has a cast dendritic structure that is usually lower in ductility than wrought material. Cold working and subsequent annealing of a weldment result in a more uniform structure.

Gas Tungsten Arc Welding

Nickel-based heat-resistant alloys are readily weldable by GTAW. This process is widely used for welding thin sections and for applications where a flux residue would be undesirable. Thin sections of aluminum-containing, precipitation-hardening alloys are frequently joined without filler metal. The addition of filler metal is usually recommended for solid-solution alloys. Direct current electrode negative (DCEN) is recommended for both manual and automatic GTAW. Alternating current can be used for automatic current if the arc length can be closely controlled. The welding arc is started by a high-frequency current. Extensions on the workpiece (start-up and run-off pads) that are machined off before the weldment is put into service are frequently used to ensure full-penetration welds and to minimize cracks in the weld metal caused by starts and stops. Heat input is kept as low as possible to minimize annealing and grain growth in the HAZ. General conditions for GTAW of nickel-based alloys are summarized in Table 2.

Shielding Gas. Argon, helium, or a mixture of argon and helium is used as shielding gas. The arc characteristics and heat pattern are affected by the choice of shielding gas. This choice should be based on welding trials for the particular production operation. Argon is normally used for manual welding; helium has shown some advantages over argon for machine welding thin sections without the addition of filler metal. Settings for welding 0.045-in.-thick nickel-based alloy 718 using argon and helium shielding are:

Operating condition	Shielding gas — Argon	Shielding gas — Helium
Current, A	80	40
Voltage, V	8-16	16-18
Welding speed, in./min	8	6-8
Filler-wire diam, in.	0.030-0.035	0.030-0.035
Wire-feed rate, in./min	12-15	8-9
Torch gas flow, ft^3/h	20-24	20
Backing-gas flow ft^3/h	4	4

Welding-grade argon and helium should be used; oxygen, carbon dioxide, or nitrogen in the shielding gas are not used because they reduce the service life of the tungsten electrode and can cause porosity in certain alloys. An addition of about 5% hydrogen to argon acts as a reducing agent and is sometimes beneficial when the work metal has not been thoroughly cleaned. However, argon with 5% hydrogen should be used only for first-pass or single-pass welding, because porosity can result if this mixture is used for subsequent passes in multiple-pass welding.

Filler metals used with nickel-based heat-resistant alloys usually have the same general composition as the alloy being welded. Because of high arc currents and high welding temperatures, compositions of filler metals are often modified to resist porosity and hot cracking of the weld metal. Tack welding and root-pass welding without filler metal are permissible for some alloys. However, care must be taken to avoid centerline splitting and crater cracking when no filler metal is used. To minimize cracking, concave welds should be avoided. Table 3 gives the compositions of filler metals commonly used in GTAW; several of these filler metals are used for welding metals other than nickel-based alloys.

For welding the precipitation-hardenable nickel-based alloys, either a precipitation-hardenable or a solid-solution filler metal may be used, depending on service requirements. Maximum mechanical properties, particularly in thick metal, are obtained when precipitation-hardenable filler metals are used, because most of the weld deposit is composed of filler metal. The solid-solution filler metals produce welds with lower mechanical properties, but they can be used where maximum strength is not needed. For example, when welding Inconel 718 using filler metal René 41, Inconel 718, GMR-235, Hastelloy S (AMS 5838), or Inconel 82, weld specimens using the first three filler metals (precipitation-hardenable) give tensile properties similar to those of the base metal, but Hastelloy S and Inconel 82 filler metals (solid-solution) give tensile properties about one third lower than those of the base metal.

Filler metal of the ERNiCr-3 classification (Table 3) is used for welding nickel-chromium-iron alloys to each other and to dissimilar metals, for high-temperature service, and for nuclear applications. Filler metal of the ERNiCrFe-5 classification is used to weld nickel-chromium-iron alloys and Inconel 600. The niobium-plus-tantalum content of these filler metals minimizes hot cracking in the weld when high stress is developed, as when welding thick metal.

Filler metal of the ERNiCrFe-6 classification is used for welding some combinations of dissimilar metals. The deposited weld metal responds to age-hardening treatments. The age-hardening response of this filler material is slight and does not exclude its use in the temperature range of 1000 to 1500 °F.

Filler metal of the ERNiCrFe-7 classification contains aluminum, titanium, niobium, and tantalum and is used for welding the precipitation-hardenable alloys. The deposited filler metal responds to aging treatments. The weldment must

Table 3 Compositions of filler metals and electrode wires for arc welding of heat-resistant alloys

AWS classification or trade name	Composition, %												
	C	Mn	Fe	S	Si	Cu	Ni(a)	Co	Al	Ti	Cr	Nb + Ta	Mo	Other
Nickel-based bare electrodes for GTAW and GMAW														
ERNiCr-3	0.10	2.5-3.5	3.0	0.015	0.50	0.50	67 min	(b)	...	0.75	18.0-22.0	2.0-3.0(c)	...	0.50
ERNiCrFe-5	0.08	1.0	6.0-10.0	0.015	0.35	0.50	70 min	...	...	...	14.0-17.0	1.5-3.0	...	1.0
ERNiCrFe-6	0.08	2.0-2.7	10.0	0.015	0.35	0.50	67 min	...	...	2.5-3.5	14.0-17.0	...	...	0.50
ERNiCrFe-7	0.08	1.0	5.0-9.0	0.01	0.50	0.50	70 min	...	0.40-1.00	2.00-2.75	14.0-17.0	0.70-1.20	...	0.50
ERNiCrMo-3	0.10	0.5	5.0	0.015	0.5	...	rem	1.0	0.4	0.4	20.0-23.0	3.15-4.15	8.0-10.0	...
GMR-235	0.16	0.25	9.0-11.0	0.03	0.6	...	rem	2.5	1.75-2.25	2.25-2.75	14.0-17.0	...	4.5-6.5	0.009 B
ERNiCrMo-2	0.05-0.15	1.0	17.0-20.0	0.03	1.0	...	rem	0.5-2.5	...	...	20.5-23.0	...	8.0-10.0	0.2-1.0 W
Hastelloy S	0.01	0.2	1.0	0.005	0.20	...	67	...	0.2	...	15.5	...	15.5	0.009 B, 0.02 La
ERNiCrMo-7	0.007	0.50	1.5	0.005	0.04	...	65	1.0	...	...	16	...	15.5	...
Haynes 556	0.10	1.5	...	0.005	0.40	...	20	20	0.3	...	22	0.1	3	0.9 Ta, 0.2 N, 2.5 W
ERNiCrMo-4	0.01	0.5	5.5	0.005	0.04	...	62	1.2	...	...	16	...	16	3.5 W, 0.35 V
Inconel 601	0.05	0.5	14.1	0.007	0.25	0.25	60.5	...	1.35	...	23.0	...	...	...
Inconel 617	0.07	0.02	0.4	0.005	0.14	...	54	12.5	1.0	0.24	22	...	9	...
Inconel 718	0.08	0.35	rem	0.015	0.35	0.3	50-55	1.0	0.2-0.8	0.65-1.15	17.0-21.0	4.75-5.5	2.8-5.5	(d)
René 41 (AMS 5800)	0.12	0.1	5.0	0.015	0.5	...	rem	10.0-12.0	1.4-1.6	3.0-3.3	18.0-20.0	...	9.0-10.5	(e)
Waspaloy (AMS 5828C)	0.07	0.10	0.75	...	0.1	...	rem	13.5	1.4	3.0	19.75	...	4.45	(f)
Nickel-based covered electrodes for SMAW														
ENiCrFe-1	0.08	1.5	11.0	0.015	0.75	0.50	68 min(a)	...	...	...	13.0-17.0	1.5-4.0	...	0.50
ENiCrFe-2	0.10	1.0-3.5	6.0-12.0	0.020	0.75	0.50	rem	...	...	...	13.0-17.0	0.5-3.0	0.5-2.5	0.50
ENiCrFe-3	0.10	5.0-9.5	6.0-10.0	0.015	1.0	0.50	rem	(g)	...	1.0	13.0-17.0	1.0-2.5(h)	...	0.50
EniMo-1	0.12	1.0	4.0-7.0	0.030	1.0	...	rem	2.5	...	...	1.0	...	26.0-30.0	(j)
ENiMo-3	0.12	1.0	4.0-7.0	0.030	1.0	...	rem	2.5	...	...	2.5-5.5	...	23.0-27.0	(j)
ENiCrMo-3	0.10	0.5	5.0	0.015	0.50	...	rem	1.0(a)	0.40	0.40	20.0-23.0	3.15-4.15	8.0-10.0	...
ENiCrMo-2	0.10	0.5	18.5	0.005	0.5	...	47	1.5	...	...	22	...	9	0.005 B
ENiCrMo-7	0.007	0.5	1.5	0.005	0.10	...	65	1.0	...	...	16	...	15.5	...
ENiCrMo-4	0.01	0.5	5.5	0.005	0.04	...	62	1.2	...	...	16	...	16	3.5 W, 0.35 V
Inconel 117	0.01	0.6-1.4	1.7	0.008	0.50	0.20	52	12.0	0.2	...	23.5	0-0.5	9.0	...
Iron-nickel-chromium, iron-chromium-nickel, and cobalt-based heat-resistant alloy filler metals														
19-9 W (AMS 5782)	0.07-0.13	1.00-2.00	rem	0.030	1.00	0.50	8.00-9.50	...	...	0.10-0.30	19.0-22.0	1.00-1.30	0.35-0.65	(k)
Multimet (N-155) (AMS 5794)	0.1	1.00-2.00	rem	0.030	1.00	...	19.00-21.00	18.5-21.0	...	...	20.0-22.5	0.75-1.25	2.5-3.5	(m)
A-286 (AMS 5804)	0.04-0.05	1.25-1.35	rem	0.008	0.70	...	25	...	0.24-0.32	2.2	15	0.10-0.12	1.25	(n)
HS-25 or L-605 (AMS 5796)	0.10	1.5	3 max	...	10 max	...	10	rem	...	...	20	...	...	15 W
Haynes 188	0.10	0.6	1.5	0.005	0.35	...	22	39	...	...	22	...	...	14.5 W, 0.04 La

(a) Contains incidental cobalt. (b) Cobalt, 0.10% max, when specified. (c) Tantalum, 0.30% max, when specified. (d) Phosphorus, 0.015%; boron, 0.006%. (e) Boron, 0.01%; total of other elements, 0.003%. (f) Boron, 0.005%; zinc, 0.04%. (g) Cobalt, 0.12% max, when specified. (h) Tantalum, 0.30% max, when specified. (j) Vanadium, 0.60%; phosphorus, 0.04%; total of other elements, 0.50%. (k) Phosphorus, 0.04% max; tungsten, 1.25 to 1.75%. (m) Phosphorus, 0.040% max; tungsten, 2.00 to 3.00%. (n) Phosphorus, 0.02% max; boron, 0.0015 to 0.0022%

be stress relieved prior to aging. Filler metals of the ERNiCrMo-3, ERNiCrMo-4, and ERNiCrMo-7 classifications are intended for welding the nickel-chromium-molybdenum alloys. Aerospace Material Specification (AMS) 5838 (Hastelloy S) is used for welding a variety of nickel-chromium, nickel-chromium-molybdenum, and cobalt- and iron-based alloys. It is well suited for dissimilar welding and exhibits excellent high-temperature stability.

The filler metals listed in Table 3 by trade name have no applicable American Welding Society (AWS) classifications, but most have AMS designations that are given in the table. These filler metals are primarily used for welding alloys of the same composition, although they are sometimes used for welding alloys of a different composition. For instance, René 41 and GMR-235 filler metals have been used to weld Inconel 718.

Joint Design. All of the joints shown in Fig. 2 are weldable by GTAW. When no filler metal is used, the sections to be joined must be held tightly together (zero root opening) to promote proper fusion. Figure 2 gives the recommended groove dimensions for joints of each type and the amount of filler metal needed. Because nickel-based alloys are more viscous in the molten condition than steel, when a V-, U-, or J-groove design is used, a slightly larger bevel angle than is needed for steel is used to ensure complete penetration. When welding nickel-chromium-iron alloys, V-grooves should be beveled to a 75 to 80° groove angle; U-grooves are beveled to a 30° groove angle, with a $^3/_{16}$- to $^5/_{16}$-in. radius. A J-groove should have a 15° bevel angle with a radius of at least $^3/_8$ in.; a $^1/_2$-in. radius is preferred. T-joints between members of different thicknesses should have bevel or J-grooves. When welding nickel-chromium-molybdenum alloys, V-grooves are beveled to a groove angle of 60°; the root opening is $^1/_{16}$ to $^3/_{32}$ in. for $^1/_4$-in.-thick metal and $^3/_{32}$ to $^1/_8$ in. for $^1/_2$-in.-thick metal.

A square-groove butt joint in metal up to $^1/_8$ in. thick can be made by welding on one side only, using the proper root opening to provide full penetration. A backing strip usually is needed to produce good back reinforcement. Although a square-groove butt joint can be made in metal up to $^1/_4$ in. thick, provided that a backing weld is used, metal thicker than 0.125 in. should preferably be beveled and welded from both sides. When this is not practical, the root opening should be increased and a backing strip should be used to ensure full penetration. When butt welding two pieces of different thicknesses, the heavier section should be machined to the thickness of the thinner section at the joint for ease of welding and for better stress distribution. Nonuniform penetration can result in undesirable crevices and voids in the underside of the joint and can create stress raisers that act as focal points for mechanical failure in service.

When pipe or tubing is used to carry corrosive materials, backing rings should be avoided if they cannot be removed after welding. Crevices between the backing ring and the tube are highly susceptible to localized corrosion.

When a product made of a precipitation-hardenable nickel-based alloy cannot be heat treated after welding (because of size or shape), the following technique can be used. Connecting pieces made of a solid-solution alloy, or one that is unaffected by the welding heat, are welded on the joint side of each component of the product, and the components plus connecting pieces are given an aging heat treatment. Welding of the final product is done at the connecting pieces. The composition and location of the connecting pieces must be carefully selected so that welds are made in noncritical locations and so that service performance of the weldment is not adversely affected. This approach is often used for vessels that cannot be stress relieved after welding and is often referred to as "safe-ending."

Corner and lap joints should be avoided if service temperatures are high or if service conditions involve thermal or mechanical cycling. When corner joints are used, a full-thickness weld must be made, such as that shown in Fig. 2. Usually, a fillet weld on the root side is also required.

Joint design often affects selection of the welding process and procedure. For example, when joining thin-walled tubes to tube sheets and tubes to flanged connections and when welding bellows joints of various types, differential melting, caused by the varying heat transfer capability of different base-metal thicknesses, may require special welding techniques. Sometimes, differential melting can be prevented by machining the thicker member to the same thickness as that of the thinner member, or by suitable preheating of the thicker member. Directing the heat of welding to the thicker member is also beneficial. When these methods cannot be applied, a combination welding procedure may be successful.

Example 3. Resistance Seam Welding Used with Gas Tungsten Arc Welding. A 10-in.-ID bellows assembly for a fuel duct for a rocket motor (Fig. 4)

Fig. 4 Bellows joint of a rocket-motor fuel duct. Details show how leaks occurred after GTAW and how resistance seam welding together with GTAW resulted in a leaktight joint.

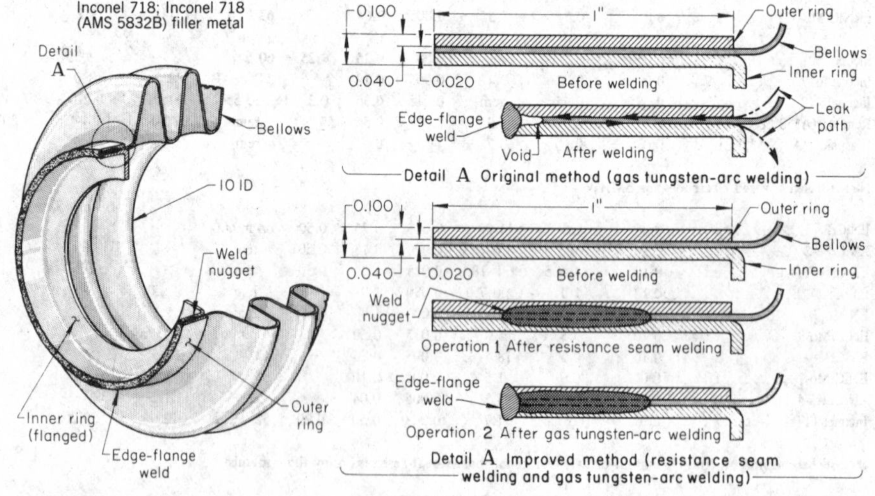

Conditions for GTAW

Joint type	Edge	Current	50-55 A (DCEN)
Weld type	Three-member edge flange	Voltage	10-12 V
Fixture	Rotating positioner	Arc starting	High frequency
Power supply	300-A transformer-rectifier	Arc length	0.040 in. (approx)
Electrode	0.040-in.-diam EWTh-2	Welding speed	60 in./min
Torch	300 A, water cooled	Preweld cleaning	Immerse for 15 min in a solution of 30-40% HNO_3 and 2-5% HF
Filler metal	0.035-in.-diam Inconel 718		
Shielding gas	Argon at 15-18 ft^3/h		

consisted of a straight section of a 0.020-in.-thick bellows wall sandwiched between two 0.040-in.-thick rings, all made from Inconel 718. A high-reliability seal weld was required. Originally, the joint was gas tungsten arc welded. The acid-cleaned assembly was mounted on a rotating turntable and welded with 0.035-in.-diam Inconel 718 (AMS 5832B) filler metal under argon shielding gas. When the joint was tested hydrostatically under an internal pressure of 160 psi, the joint did not leak, but when it was vacuum tested with a helium mass spectrometer, significant leakage was detected.

After considerable investigation, it was determined that during welding the bellows wall melted back faster than the rings, so that weld metal was deposited at the edges of the rings only, resulting in a void that provided a leak path between the rings and the unintegrated surfaces of the bellows (see after welding view in detail A, original method, Fig. 4). One small void was enough to cause a leak.

The problem was solved by resistance seam welding and gas tungsten arc welding the joint. The joint was first resistance seam welded (see operation 1, improved method, Fig. 4) and then machined back to the edge of the seam weld. A gas tungsten arc edge-flange weld was deposited as before (see operation 2, improved method, Fig. 4). The completed combination weld was gastight under both methods of testing. General conditions for resistance seam welding are given in the article "Resistance Seam Welding" in this Volume.

Welding Techniques. When filler metal is used, the hot end of the wire must be kept under the shielding gas, and wire diameter should be no larger than work-metal thickness. Excessive turbulence in the molten weld pool must be avoided; otherwise, the deoxidizing elements will burn out.

To ensure a sound weld, the arc must be maintained at the shortest possible length. When no filler metal is added, arc length should not exceed 0.05 in. and preferably should be 0.02 to 0.03 in. When filler metal is added, the arc is longer, but it should be as short as possible, consistent with filler-metal diameter. Filler metals often contain elements specifically added to improve resistance to cracking and porosity. To obtain the full benefit of these elements, the finished weld should consist of about 50% filler metal.

A greater-than-normal electrode extension is needed for fillet welds and for the first few passes on heavy sections. Small-diameter filler-metal wires and more passes may be used on welds made in other than the flat position, for adequate control of weld metal. When the back sides of butt welds do not show adequate penetration, they should be ground back to sound metal and back beads should be deposited. When possible, backing gas should be provided when welding the first side.

Example 4. Full-Penetration Circumferential Weld in Hastelloy X Pipe. A 6-in. elbow was manual gas tungsten arc welded to the swaged end of a pressure vessel as shown in Fig. 5. The assembly, which was made entirely of Hastelloy X, was part of a helium filter for the gas-cooled loop of a nuclear reactor. Fabrication was done in accordance with the American Society of Mechanical Engineers (ASME) code for nuclear vessels. Carefully selected practices were incorporated in the welding procedure to ensure high joint reliability for preweld cleaning, shielding, weld deposition, heat input control, and quality control as follows:

- *Preweld cleaning*: Parts were received free of scale and with a bright finish. After the weld grooves had been machined (Fig. 5), the parts were immersed and scrubbed in unused, agitated acetone, and then wiped dry with a clean, lint-free white cloth. No wire brushing or sanding was needed.
- *Shielding*: In addition to torch shielding, low positive-pressure argon flow was required for internal backing to avoid oxidation at the weld root face and to limit root-face concavity to 3/32 in. Before welding, the assembly was evacuated to 230 torr, and then purged with argon gas to bring the system up to atmospheric pressure. This was repeated five times to guarantee high purity of the argon during welding. Argon flows of 5 ft^3/h at the root of the weld and 15 ft^3/h at the torch were maintained during welding.
- *Positive internal pressure*: This was ensured by placing a plate with a 1/8-in.-diam hole over the open end of the elbow and sealing the joint with tape. As welding advanced, the tape was removed about 3 in. ahead of the weld. Argon of certified welding quality (99.99% purity) was used. Any length of weld made during a cessation of gas flow was removed and rewelded. As an indication of adequate shielding, beads had to appear bright to dull gray after cooling.
- *Heat control*: To avoid any change in alloy phase distribution, and thus a change in alloy properties, heat input, interpass temperature, and postweld cooling were controlled. Heat input was controlled by depositing thin beads at low amperage. About 100 welding passes were needed to fill the joint. In-

Fig. 5 Welding of an elbow to a pressure vessel for a nuclear reactor application. Shows positioning of vessel for welding, joint design, and maximum root reinforcement and concavity

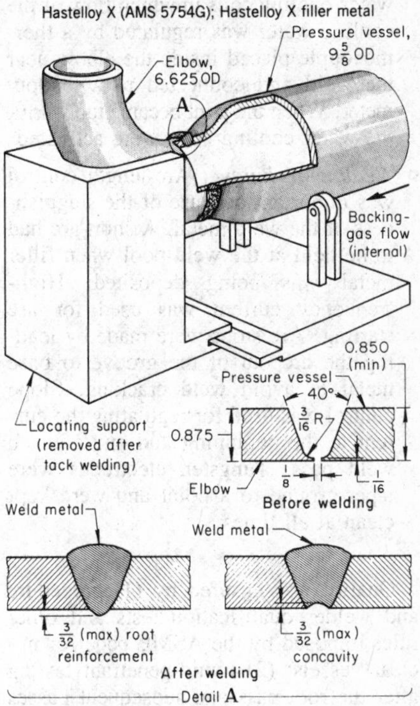

Joint type	Butt
Weld type	Single-V-groove
Welding process	Manual GTAW
Power supply	500-A constant-voltage transformer-rectifier, with slope control and gas preflow
Electrode	1/8-in.-diam EWTh-2, ground to a point
Torch	300 A, water cooled; 7/16-in.-diam cup
Electrode extension	3/16 in. (approx)
Filler metal:	
Root pass	1/16-in.-diam Hastelloy X
Filler passes	3/32-in.-diam Hastelloy X
Shielding gas (argon):	
At torch	15 ft^3/h(a)
Backing gas	5 ft^3/h
Welding position	Horizontal-rolled pipe
Fixtures	Fit-up jig; turning rolls
Current	125-190 A (DCEN) (all passes)
Voltage	30 V
Interpass temperature	150 °F max
Arc length	1/32 in. (approx)
Arc starting	High frequency(b)
Number of passes	100 (approx)
Travel speed	1 in./min
Preheat and postheat	None

(a) Welding grade, 99.997% pure. (b) Welding was terminated by leading the arc out of the groove to base metal. Slope control was used at the beginning and end of each pass.

terpass temperature was not allowed to exceed 150 °F and was measured by temperature-indicating crayons used on the base metal about $^3/_8$ in. from the weld edge. Postweld cooling was assisted by water-cooling coils in which flow of the cooling water was regulated by a thermocouple placed inside the elbow near the weld and connected to a readout meter. When the joint became too warm, the water-cooling coils were activated.

- *Welding technique*: Arc-length control was important because of the sluggishness of the weld metal. A short arc had to be held at the weld pool when filler metal was being deposited. High-frequency current was used for arc starting. Arc stops were made by leading the arc out of the groove to base metal to avoid weld cracking. Slope control was used for regulating the current at the beginning and end of each weld pass. Tungsten electrodes were taper-ground to a point and were kept clean at all times.

Quality was ensured by (1) procedure and welder qualification tests and other rules imposed by the ASME code for nuclear vessels; (2) liquid-penetrant testing after the root pass and subsequent passes throughout the welding operation; (3) 100% radiographic inspection after the root pass and the final weld pass; and (4) a leak testing of the completed weld, using a helium lead rate detector. The five joints welded by this procedure were acceptable.

Gas Metal Arc Welding

Solid-solution-strengthened nickel-based alloys and, with suitable welding procedures, many precipitation-hardenable alloys can be joined by GMAW. Gas metal arc welding is best suited to the joining of thick sections of more than about $^1/_4$ in. thick, where high filler-metal deposition rates are desirable. For a description of this process, see the article "Gas Metal Arc Welding" in this Volume.

Spray, pulsed arc, globular, and short circuiting metal transfer can be used. Optimum metal transfer is obtained when operating slightly above the transition from globular to spray transfer. All of these methods use electrode wire of comparatively small diameter. Incomplete fusion and oxide inclusions can occur when the short circuiting arc is used. Multiple-pass welds should be made by highly skilled welders only. Direct current electrode positive (DCEP) should be used because the greater heating effect of reverse polarity assists in obtaining the required high melting rate.

Shielding gas for nickel-based heat-resistant alloys is argon or an argon-helium mixture. Gas flow rates range from 25 to 60 ft^3/h, depending on joint design, type of metal transfer, and welding position. Generally, flow rate is about 50 ft^3/h for spray transfer. As the percentage of helium is increased, gas flow rate must be increased to give adequate protection. Pure argon is normally used for spray transfer. Other types of metal transfer commonly use argon with 25 to 30% helium added.

Joint designs recommended for GMAW are shown in Fig. 2. For U-groove designs using globular or spray metal transfer, the root radius should be decreased by about 50% and the bevel angle should be doubled, compared with those shown in Fig. 2. When using a short circuiting arc, the U-groove designs shown in Fig. 2 can be used without change.

Welding Techniques. Best results are obtained when the electrode holder is positioned at about 90° to the joint. Some inclination (up to about 15°) is permissible to permit a better view of the work, but excessive inclination can draw the surrounding atmosphere into the shielding gas, resulting in porous or heavily oxidized welds. Arc length is important. Weld spatter occurs if the arc is too short, and loss of control occurs if the arc is too long.

The manipulation and electrode holder angle used with pulsed arc welding are similar to those used with SMAW. A slight pause at the limit of the weave is required to avoid an undercut. Electrode wire compositions for GMAW are the same as those recommended for filler metals for GTAW (Table 3). With globular and spray transfer, wire with 0.035-, 0.045-, or 0.062-in. diameters are used. The short circuiting arc generally requires wire 0.045 in. or less in diameter.

Example 5. Manual Gas Metal Arc Welding for Joining Inconel 600 Plates. Because only a few welds were to be made and the available automatic equipment could not deposit a weaving bead, manual GMAW was used to join 10-ft-long, $^3/_8$-in.-thick plates of Inconel 600 (Fig. 6). A backing strip of Inconel 600 was temporarily tack welded by manual GTAW to the back of the groove. Tack welds 1 in. long on 6-in. centers helped maintain the $^1/_8$-in. root opening.

Fig. 6 Plate of $^3/_8$-in.-thick Inconel 600. Welded by manual GMAW, showing joint design, backing strip, and completed weld. The backing strip, tack welded in place by manual GTAW, served as a fixture to maintain the root opening.

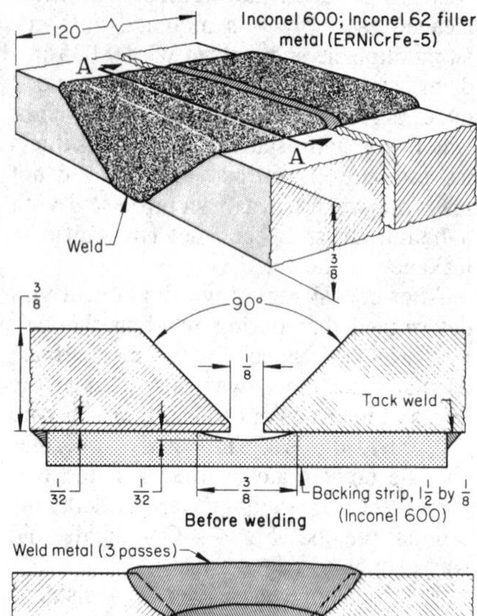

Joint type	Butt
Weld type	Single-V-groove
Welding process:	
Tack weld	Manual GTAW
Groove weld	Manual GMAW

Welding conditions for tack welding by manual GTAW

Power supply	Transformer-rectifier, constant current
Electrode	$^3/_{32}$-in.-diam EWTh-2
Filler metal	0.045-in.-diam Inconel 62
Shielding gas	Argon at 35 ft^3/h
Welding position	Flat
Current	275 A (DCEN)
Voltage	27-29 V
Tack-weld spacing	1 in., on 6-in. centers

Welding conditions for manual GMAW

Power supply	Constant-voltage transformer-rectifier
Electrode holder	Manual, water cooled
Electrode wire	0.045-in.-diam Inconel 62
Shielding gas	Argon at 35 ft^3/h
Current (DCEP)	275 A
Voltage	27-29 V
Welding position	Flat
Number of passes	3
Setup and welding time per plate	1 h, 50 min

Shielded Metal Arc Welding

Shielded metal arc welding is widely used for joining solid-solution nickel-based alloys, but is rarely used for joining precipitation-hardenable alloys. The process is described in more detail in the article "Shielded Metal Arc Welding" in this Volume. Direct current electrode positive is generally used to obtain optimum mechanical properties.

Weaving is sometimes desirable, but the amount should not exceed three times the electrode diameter. Overheating can cause hot short cracking in the weld metal or the base metal and excessive carbide precipitation at grain boundaries in the HAZ. This can be avoided by using the recommended amperage ranges, maintaining a short arc length, and not weaving the electrode excessively.

Electrodes for SMAW are listed in Table 3. Electrode composition should be similar to that of the base metal with which the electrode is to be used.

Welding Conditions. Figure 7 gives welding conditions for making butt, corner, and T-joints in solid-solution-strengthened nickel-based alloys by SMAW.

The weld metal of most nickel-based alloys does not flow readily and must be manipulated or correctly positioned. This often requires a slight weave and a short pause at the sides to allow the undercut to be filled in. When the arc is broken, it should be shortened and the travel speed increased slightly. This reduces the weld pool size. When restarting, a reverse or T restrike should be used. The arc is struck at the leading edge of the weld crater and carried back to the rear of the crater. The travel direction is then reversed and normal weaving started. All welding slag should be removed before placing a weld in service. Catastrophic high-temperature corrosion occurs if this is not done. Adhering slag can also enhance crevice corrosion at lower temperatures. Slag should also be removed between passes to ensure high-quality, metallurgically sound welds.

Causes and Prevention of Weld Defects

Weld defects such as cracks, porosity, inclusions, and incomplete fusion usually are unacceptable in weldments made of heat-resistant alloys. Nondestructive inspection is used on almost all completed weldments; destructive inspection generally is limited to test samples. Various types of leak tests are used on weldments that are to be subjected to pressure in service. Thirteen types of defects that occur in arc welds are shown in Fig. 8.

Porosity and Inclusions. Porosity is the term used to describe gas pockets or voids in the weld metal. Typical causes include improper shielding, moisture, incorrect amperage, and excessive arc length. Dry electrodes are essential. When high amperage or a long arc length is used, deoxidizers that help prevent porosity can be totally consumed when transferring across the arc. Inclusions are usually slag, oxides, or other nonmetallic solids entrapped in the weld metal between adjacent beads or between the weld and parent metal. Excessive weld pool agitation, downhill welding, and undercutting can lead to slag entrapment. These conditions can usually be prevented by good weld practice and proper weld design.

Fig. 7 Conditions for SMAW of solid-solution-strengthened, heat-resistant, nickel-based alloys

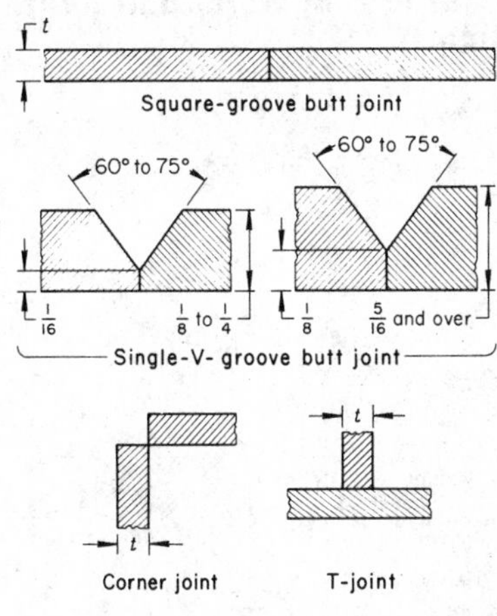

Metal thickness, in.	No. of passes	Current (DCEP), A(a)	Electrode diameter, in.(b)
Square-groove butt joints			
1/16	1	40-70	3/32
5/64	1	40-70	3/32
3/32	2	45-75	3/32
Single-V-groove butt joints			
1/8	2	40-70	3/32
5/32	2	40-70	3/32
3/16	2-3	40-70	3/32
1/4	3-4	40-130	3/32-5/32
3/8	5-6	40-130	3/32-5/16
1/2	8-10	40-130	3/32-5/16
Corner joints and T-joints(c)			
1/16	1	40-70	3/32
5/64	1	40-70	3/32
3/32	1	40-70	3/32
1/8	1	40-70	3/32
5/32	1	40-100	3/32-1/8
3/16	1	40-100	3/32-1/8
1/4	2	40-130	3/32-5/32
3/8	3	40-130	3/32-5/32
1/2	6	40-130	3/32-5/32

(a) Current should be within the range recommended by the electrode manufacturer. (b) Where a range is shown, the smaller diameters are used for the first pass in the bottom of the groove, and the larger diameters are used for the final passes. (c) Fillet welds

Fig. 8 Typical arc welding defects. Heat-resistant alloys (a-d). A weld with no defects and good reinforcement is shown in (e).

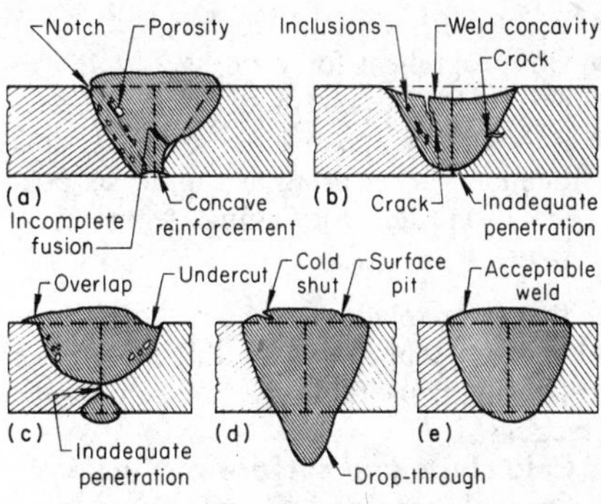

Cracks and Fissures. The six general types of cracks are: (1) transverse cracks in the base metal perpendicular to the weld, (2) longitudinal cracks in the base metal parallel to the weld, (3) microcracks and macrocracks in the weld metal, (4) centerline longitudinal weld-metal cracks, (5) crater cracks, and (6) start cracks or bridging cracks. Transverse cracks usually are the result of external contamination or a base metal with poor weldability. Base-metal cracking parallel to the weld typically is caused by the combination of a strong weld metal and a weak, low-ductility base metal. An example of this is seen when welding medium-carbon, cast stainless steel furnace tubing after service exposure without re-solution annealing.

Weld-metal microfissuring can be caused by contamination or impurities in the metal that lower weldability. Centerline longitudinal cracking is caused by concave beads or a very deep, narrow weld bead. Crater cracking occurs when the arc is extinguished over a relatively large weld pool. The resulting concave crater is prone to shrinkage cracking.

Bridging cracks occur in highly stressed joints where good penetration is not achieved at the arc initiation point. Cracks of any type and size cannot usually be tolerated. If a material proves to be crack sensitive, base-metal cracking can be minimized by reducing heat input and depositing small beads, which results in lowered residual stresses.

Strain-age cracking can occur in precipitation-hardenable nickel-based alloys during the initial postweld heat treatment if the base metal is in the aged condition and at least one area, such as the as-deposited weld metal, is not in the precipitation-treated condition. Cracks that form under these conditions are relatively large, and most of them are in the base metal. The metal is more likely to crack when an aged part is being repair welded. To minimize strain-age cracking:

- Weld in the solution-treated (annealed) condition.
- Weld with minimum restraint.
- Do not preheat for welding.
- Use as low a heat input as possible.
- Accomplish postweld solution heat treatment by heating as rapidly as possible through the aging temperature range.

Test data relating exclusively to René 41 indicate that overaging this alloy prior to welding is useful in avoiding strain-age cracking.

Cold shuts and surface pits in a weld are possible evidence of a subsurface crack or of porosity that has emerged at the surface. Also, there may be lack of fusion between successive layers of weld metal or between adjacent weld beads. Cold shuts can occur between the weld metal and the base metal when hot metal runs ahead onto a cold metal surface and does not fuse properly. Cold shuts should be removed by grinding because they are linear discontinuities that may prevent successful weld qualification bend tests, compromise strength, and contribute to crevice corrosion problems. Both visual and liquid-penetrant inspection are used to locate cold shuts and surface pits.

Voids in welds can occur when gases have been trapped in a groove or between two pieces of metal being welded. When welding a lining in a thick-walled vessel or when welding one tube inside another, entrapped gas can cause a void if it does not have an easy escape path.

Example 6. Elimination of a Void at the Root of a Joint. The elimination of a small tear caused by a small void at the root of a welded joint presented a problem in making the laminated resistance heating element shown in Fig. 9. A thin-walled Hastelloy X tube of high electrical resistance was welded to a heavier, pure nickel conductor tube. An inner tube, also of pure nickel, was separated from the outer tube by a layer of ceramic insulation and completed an electrical circuit when joined to the Hastelloy X tube at the heater end. After welding, the weld metal was machined flush and the tube assembly was swaged to shape.

A small tear occurred in the weld metal after swaging, at the point on the circumference where the final joint closure was made. A small void found at the root of the joint was assumed to have weakened the weld sufficiently to cause the small rupture. The void was caused by a small amount of gas (under pressure) trapped at

Fig. 9 Joint between resistance and conduction segments of a laminated, tubular resistance-heating element, location of void at root of weld, and joint design

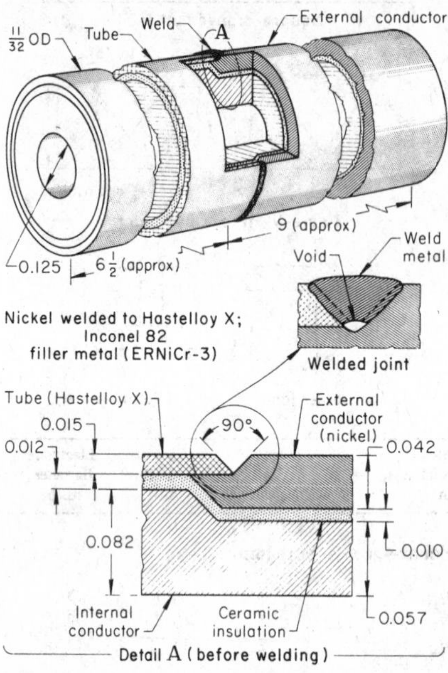

Joint type	Butt
Weld type	Single-V-groove with backing
Welding process	Manual GTAW
Power supply	300-A transformer-rectifier
Electrode holder	250 A, water cooled
Electrode	1/16-in.-diam EWTh-2
Filler metal	1/32-in.-diam Inconel 82 (ERNiCr-3)
Shielding gas	Argon
Welding position	Horizontal-rolled pipe
Current	15 A (DCEN)
Voltage	15 V
Arc starting	High frequency
Number of passes	1
Edge preparation	Machining (see text)
Special equipment	Vacuum pump

the root of the joint. During welding, the gas expanded and found an easy escape path through the unwelded portion of the joint, until the closure was made and the escape path was sealed off. There was no alternate route along the tube interlayers because of the close fit of components.

To ensure evacuation of the tube, a small vacuum pump was connected to one end of the tube and the other end of the tube was sealed with a rubber stopper, leaving a path for escape of gas through the interlayers. Pumping to obtain a low vacuum was carried out before and during welding. The tube was rotated manually during welding. The vacuum hose was long enough to be twisted through 360° of rotation without significant loss of vacuum. Using this procedure, no more tears occurred, and the problem was solved. After swaging, the welded joints were tested by a helium mass spectrometer. The vacuum was applied to the end of the tube (as in welding) and helium test gas to the outside of the tube at the weld.

Inconel 82 (ERNiCr-3) filler metal was selected for this dissimilar metal joint because of its strength and excellent ductility in the as-welded condition. Joint design consisted of a 90° single-V-groove formed by beveling the 0.015-in.-wall Hastelloy X tube 45° and the 0.042-in.-wall nickel tube 45° to a depth of 0.015 in., which ensured a nickel backing for the joint (Fig. 9). The tubes were machined and washed in acetone before assembly for welding. Manual GTAW with argon shielding gas was used to deposit the weld metal in a single pass. No preheat was needed.

Notches are formed when the edge of the weld reinforcement is thick and does not blend smoothly into the surface of the base metal (see Fig. 8). Frequently, the edge of the reinforcement can be ground or machined to obtain suitable blending, but preferably welding practice should be changed to prevent the condition.

Iron-Nickel-Chromium and Iron-Chromium-Nickel Heat-Resistant Alloys

Iron-based heat-resistant alloys include strain-hardenable, solid-solution-strengthened, and precipitation-hardenable types. All contain appreciable amounts of nickel and chromium, with either one or the other of these elements constituting the principal alloying addition. Other alloying elements generally are added to increase high-temperature strength (molybdenum, tungsten, and cobalt), to act as stabilizers (niobium and tantalum), or to promote strengthening (aluminum, titanium, copper, and boron).

The usual range of service temperature for these alloys, 1200 to 1400 °F, limits the selection of filler metals and preheat and postheat treatments for welding. Nominal compositions of the common iron-based alloys are given in Table 1.

The 16-25-6 and 19-9 DL alloys are easily joined by arc welding. Weld deposits can be made with an austenitic stainless steel filler metal, with a nickel-based alloy filler metal, or with a filler metal of the same composition as the base metal. Generally, preheating and postheating are used.

The solid-solution-strengthened alloys, such as N-155, are weldable by SMAW, GMAW, and GTAW; however, the heat input should be kept low, and welds should be cooled rapidly to maintain ductility.

Some precipitation-hardenable alloys, such as A-286, are considerably more difficult to weld. These alloys are extremely sensitive to hot cracking in the weld metal and in the HAZ. Cracking is most likely to occur when aged metal or highly restrained parts are joined. Cracks in root passes or crater cracks can be minimized by using suitable welding procedures and techniques to control heat input during welding. Microcracking (microfissuring) can occur in the weld metal and in the HAZ. It must be controlled by proper preweld and postweld heat treatment and by selection of the most suitable filler metal.

Nevertheless, high joint efficiency can be obtained in arc welding A-286 sheet up to 0.094 in. thick. Joints in thicker sections are more difficult to weld and require special techniques such as automatic welding, a single pass, or the use of weld back cooling.

In all the precipitation-hardenable alloys, aluminum, niobium, or titanium may combine with either iron or the major solid-solution elements to form low-melting eutectic phases in the grain boundaries. Melting of these grain-boundary phases, often called incipient melting, occurs near the fusion line during welding, and thermal stresses incidental to weld cooling may cause the grains to separate. Grain separation produces gross subsurface cracks or microfissures that are difficult to detect by available nondestructive inspection methods.

Joint design for iron-based heat-resistant alloys depends on the welding process, the application, and the filler metal used. For some applications, joint designs similar to those used for stainless steels are appropriate (see the article "Arc Welding of Stainless Steels" in this Volume). For other applications, joint designs similar to those shown for nickel-based alloys are used (Fig. 2), except that the included angle of V-grooves is usually 60°. For SMAW, the included angle should be from 75 to 90°. Although square-groove butt joints usually are used for metal up to $^1/_8$ in. thick, in Example 7 a zero-root-face V-groove was ground in 0.080-in.-thick A-286 sections. Because nickel-based alloys have low fluidity, wider joint bevels and openings are sometimes required when welding with nickel-based filler metal.

Preweld and Postweld Heat Treatment. Strain-hardenable alloys usually are welded in the combined hot-cold worked condition. The metal in the HAZ is essentially solution treated by the welding heat, resulting in a decrease in hardness and strength. Although strain-hardenable alloys frequently are preheated for welding, the preweld heat treatments must be done in a limited temperature range to avoid annealing the metal. This same restriction on heat treating temperature applies to postweld treatments of these alloys.

Solid-solution-strengthened alloys generally are welded in the solution-treated condition and are used without postweld heat treatment. These alloys have a small zone of grain growth adjacent to the weld, but this does not appreciably reduce weld strength.

Precipitation-hardenable alloys are welded in the solution-treated condition because greater ductility of the base metal in this condition permits some relaxation of the shrinkage stresses associated with welding. Postweld heat treatment generally includes a re-solution treatment and an aging treatment.

Gas Tungsten Arc Welding

Gas tungsten arc welding is the most widely used process for joining iron-nickel-chromium and iron-chromium-nickel heat-resistant alloys, especially in thin sections, because the welding operation can be observed and controlled easily and because of the high percentage of alloy transfer that is characteristic of the process. The minimum heat input that is appropriate to work-metal thickness and to filler-wire diameter should be used to reduce the temperature gradient in the HAZ. The amount of heat input is particularly important in welding precipitation-hardenable alloys, because these alloys are subject to grain boundary melting. Direct current electrode negative is used for welding iron-nickel-chromium and iron-chromium-nickel alloys. Thoriated tungsten electrodes (EWTh-2) are most often used.

Argon shielding gas at a flow rate of 15 to 20 ft^3/h is used at the torch. Argon also is used for purging before welding and as a backing gas. If shielding is inadequate, aluminum and titanium, which are present in the precipitation-hardenable alloys, are likely to combine with oxygen in the air to form refractory oxides on the surface of the weld bead being deposited. These oxides may result in oxide inclusions or otherwise interfere in the production of a sound weld. In multiple-pass welding, surface oxides must be removed after each welding pass, preferably by grinding.

Filler metals that produce weld metal with a composition the same as, or similar to, that of the base metal are usually selected. Wire of the same composition as the base metal, austenitic stainless steel wire, or nickel-based alloy wire is used (Table 3). Because heat-resistant alloys have relatively high strength at elevated temperature, the solidifying weld metal frequently is severely stressed. Hastelloy S (AMS 5838) and many other nickel-based filler metals, such as Inconel 82 and Inconel 625, are recommended for joining many of the iron-nickel-chromium and iron-chromium-nickel alloys. These filler metals are not age hardenable, but their as-welded mechanical properties are satisfactory for many applications. For high-stress applications of the precipitation-hardenable iron-based alloys, a filler metal that responds to age hardening and thus develops the highest joint efficiency is recommended.

Procedures for GTAW of thin workpieces of a precipitation-hardenable alloy are described in the following example. Filler metal, joint preparation, method of purging the weld, and welding technique were carefully selected to prevent weld- or base-metal cracking.

Example 7. Prevention of Weld Cracking in A-286 Fuel Ducts. Special techniques were needed to weld sections of A-286 aerospace fuel ducts to meet radiographic inspection requirements. The chief concern in welding this alloy was to prevent cracking, which was likely to occur in the weld metal, in the underbead region, and in the HAZ. The reducing section (Fig. 10) was welded to the two straight-side tube sections in two passes for each weld, using manual GTAW, Hastelloy W (ERNiMo-3) filler metal, and argon gas for external and internal shielding.

Material Condition. The components were welded in the solution-treated condition. If a part was cold worked after solution treatment, it was re-solution treated before welding.

Joint Preparation and Fit-Up. Section ends were beveled 45° to a feather edge (zero root face) by grinding and disk sanding (Fig. 10). Extreme care was needed to prevent burning or overheating the feather edges. Sections to be joined were vapor degreased. After fit-up, and just before welding, joints were again cleaned, using a stainless steel wire brush. Accuracy of fit-up was important: overlaps, misalignment, and root openings were not permitted.

Filler Metal. Hastelloy W filler-metal wire was used for tack and groove welding, in preference to A-286 filler-metal wire, because deposition proceeded more smoothly, with less drop-through and bet-

Fig. 10 Gas tungsten arc welded A-286 fuel duct for which special welding techniques were used to prevent cracking

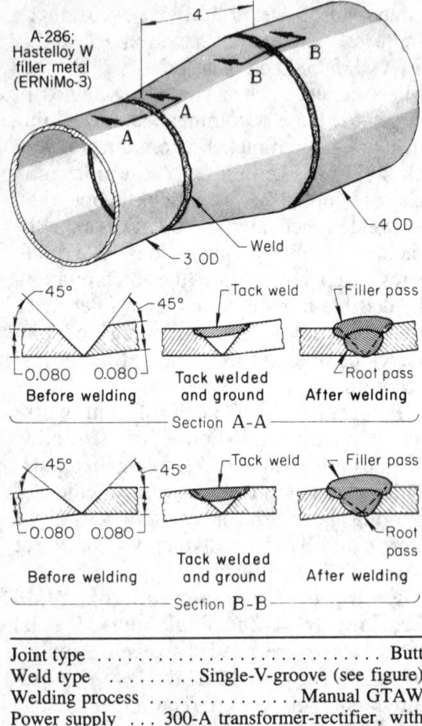

Joint type	Butt
Weld type	Single-V-groove (see figure)
Welding process	Manual GTAW
Power supply	300-A transformer-rectifier, with high-frequency arc starting and automatic time delay for gas and water
Electrode	1/16-in.-diam EWTh-2, tapered to a sharp point
Torch	Water cooled
Filler metal	1/16-in.-diam Hastelloy W (ERNiMo-3)
Shielding gas (argon):	
At torch	10-15 ft^3/h
Backing gas	10 ft^3/h
Welding position	Horizontal-rolled pipe position
Fixtures	V-block; turning rolls
Current	70 A max (DCEN)(a)
Voltage	Not controlled
Arc starting	High frequency

(a) Foot-operated remote control was connected with low-range circuit of power supply.

ter wetting of the groove faces. The incidence of cracking of the weld metal (as detected by liquid-penetrant inspection) was lower, and the Hastelloy W wire was easier to handle.

Tack Welding. During fit-up, tack welds were placed at 1/2-in. intervals around the joint. Each tack was carefully deposited across the top of the groove, instead of at the root. This procedure (contrary to normal practice) was necessary to avoid oxidizing or overheating the root of the joint; internal gas purging was not used during tack welding, except when fit-up at the feather edge was poor, leaving a small gap. The tack welds were ground flush with the tube wall, leaving a thin web of tack-weld metal that was consumed during the intermittent root pass. For tack welding and for the initial welding pass, parts were supported in a V-block.

Argon Purging. After tack welding, purging dams fitted with small-diameter pipe fittings were placed at both ends of the assembly. A hose attached to one dam provided an argon flow of 10 ft^3/h. In the dam at the opposite end, an exit orifice was made by placing adhesive masking tape over the fitting and puncturing the tape with a 3/32-in.-diam electrode.

Welding. The root pass was made by depositing intermittent welds 1 to 1 1/4 in. long, separated by intervals of about the same length. Each end of the weld bead was then ground back approximately 1/8 in. to a feather edge, using a small grinder fitted with a 1-in.-diam by 1/32-in.-thick wheel. When the joint was cool enough to handle with bare hands, the root pass was completed by filling the intervals. Intermittent welding was used during the root pass to reduce heat input and cumulative shrinkage stress. The ends of the weld beads were ground to remove possible cracks at these stress points and to reduce heat buildup on making the tie-ins.

During welding, current was regulated by a foot-operated remote control switch that was connected with the low-range circuit of the power supply. A low welding current (maximum of 70 A) was used.

After it was cooled to room temperature, the root bead was cleaned with a power-operated stainless steel wire brush and inspected visually and by a liquid-penetrant method for cracks, especially at tie-in points. Cracks were removed by grinding, and the areas to be repaired were blended in before being carefully rewelded.

The second pass was made while the duct was being rotated on turning rolls. This prevented heat buildup, which was likely to occur if the operator had to stop to turn the workpiece during welding. Stops and starts caused root cracking that was not detectable by liquid-penetrant inspection, but that was evident in radiographs. If the duct could not be automatically rotated, the second pass was made by intermittent welding.

Gas Metal Arc Welding

Gas metal arc welding is sometimes used for joining iron-nickel-chromium and iron-chromium-nickel heat-resistant alloys when sections are more than about 1/4 in. thick, where joint design and workpiece size can compensate for the high heat input. Generally, the spray transfer mode of GMAW is not recommended for coarse-grained heat-resistant alloys (both wrought and cast), because of the tendency of grain boundary liquation. However, controlled heat input modes of GMAW such as pulsed arc and short arc welding usually can be used with little difficulty provided that the technique used is sufficient to overcome cold lapping tendencies. Electrode material is selected to match the composition of the base metal and to suit service conditions.

Example 8. Use of Gas Metal Arc Welding for Filler Passes on a Circumferential Joint. The 5/8-in.-wall cast tubing shown in Fig. 11 was originally joined by GTAW for the root pass and SMAW for the filler passes. This practice was changed so that filler passes were welded

Fig. 11 Cast HK-40 alloy tubing. Gas tungsten arc welded for the root pass and automatic gas metal arc welded for the filler passes. Joint design and sequence of passes are also shown.

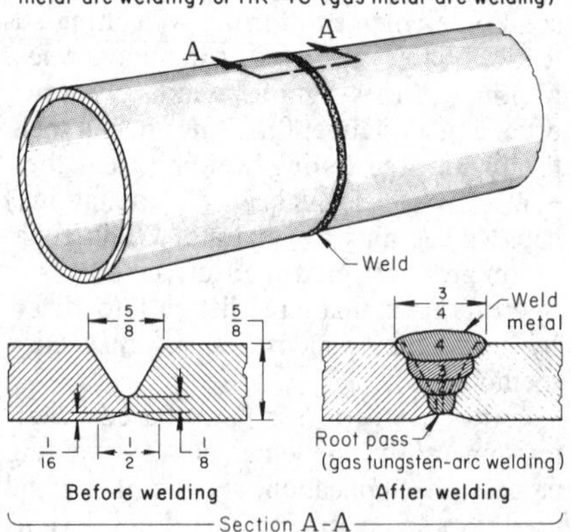

Joint type	Circumferential butt
Weld type	Modified single-U-groove
Welding position	Horizontal-rolled pipe

Root pass

Welding process	Manual GTAW
Electrode	3/32-in.-diam EWTh-2
Filler wire	None
Shielding gas	Argon, at 20 ft^3/h
Current (DCEN)	120 A
Voltage	26 V
Welding speed	2 in./min

Buildup passes	Original method	Improved method
Welding process	SMAW	GMAW(a)
Electrode	Mod E310(b)	HK-40 wire
Electrode	5/32 in. diam	0.045 in. diam
Shielding gas	...	Argon-CO_2
Current	130 A (DCEP)	...
Voltage	28 V	...
Welding speed	0.5 in./min	...

(a) Automatic. (b) 0.35 to 0.45% carbon

by automatic GMAW, because it was faster and offered improved control. In addition, wire brushing and grinding after each pass were eliminated, and less skill was needed.

The HK-40 tubing (25% Cr, 20% Ni, and 0.35 to 0.45% C) was beveled as shown in Fig. 11, section A-A. A gas tungsten arc root pass without filler metal gave a smooth, continuous weld bead on the inaccessible interior of the circumferential joint and provided a sound surface for the weld that was made from the other side. As a result, visual inspection using a boroscope was easily accomplished, and interpretation of radiographs was simplified, compared with a weld in which a backing strip was used.

The filler passes were made with automatic GMAW. Weld metal was deposited by a modified weaving technique to ensure uniform deposition across the groove, low residual stresses in the weld metal, and rapid weld-metal deposition. The final bead extended $^{1}/_{16}$ in. beyond the edge of the weld groove. The root bead and final bead were inspected for surface discontinuities by the liquid-penetrant method. Welded joints were examined radiographically in accordance with paragraph UW-51, Section VIII, ASME Boiler and Pressure Vessel Code, except that 10% random sampling was used.

Shielded Metal Arc Welding

Shielded metal arc welding can be used for joining iron-nickel-chromium and iron-chromium-nickel heat-resistant alloys when GTAW equipment is not available, or when it is not practical to use GTAW. The SMAW process is often preferred because of its versatility, simplicity, and the inexpensive equipment required.

Direct current electrode positive produces the best mechanical properties in the welds. When joint design permits, rapid travel with as little weaving as possible is preferred to minimize heat input. To avoid overheating when starting and stopping a weld, welding currents that are consistent with the metal thickness or the size of the components should be used. Striking the arc on a starting tab adjacent to the joint helps prevent cracking at the beginning of the weld bead. The arc can be broken on a similar tab at the end of the weld, but doubling back on the bead with a slant arc is also acceptable. Whenever possible, welding should be done in the flat position.

Electrodes for SMAW usually are selected so that the composition of the deposited weld metal is close to that of the base metal. Service requirements, such as matching stress rupture properties at operating temperature or the need for crack-free weld metal, often influence selection of electrode metal. For example, for applications in which service temperature is above 1450 °F, Inconel Welding Electrode 112 is recommended for joining Incoloy alloys 800 and 800 H. For temperatures below 1450 °F, Incoweld A is satisfactory. The compositions of electrodes used for welding iron-nickel-chromium and iron-chromium-nickel heat-resistant alloys appear in Table 3.

Example 9. Change in Welding Procedure to Avoid Intergranular Corrosion. X-ray inspection showed scattered gas pockets in cast tubes of Thermalloy 40B, and repair welding was required. The tubes, used in a calciner that operated at 1400 °F to convert uranyl nitrate to uranium oxide, were 8 in. in outside diameter, with a $^{3}/_{8}$-in. wall. The tubes shown in Fig. 12 were 48 in. long.

Because of a misunderstanding as to alloy composition, the first repair welding procedure resulted in carbide precipitation in the grain boundaries and failure of the welded areas by intergranular corrosion within 3 months. Originally, the defective tubes were cleaned in hot trichlorethylene vapor and rinsed in hot water, and the defects were ground out, using a hand grinder. Before welding, the tubes were furnace heated to 600 °F, and around each defect an area about 4 in. in diameter was heated to 1100 °F with a torch. The defects were repaired by SMAW, using a $^{1}/_{8}$-in.-diam austenitic stainless steel electrode, E309-16, and a welding current of 100 A (DCEP). The tubes were then wrapped in an asbestos blanket to obtain slow cooling.

Because of poor performance resulting from the first welding procedure, a new procedure was tried. Tubes were cleaned for welding in the same manner as before, preheating was eliminated, a niobium-tantalum-stabilized stainless steel electrode, E347-16, was used, and the tubes were cooled in air after welding. Elimination of the high preheat and the slow cooling reduced carbide precipitation in the HAZ of the base metal, although the higher rate of cooling did cause some cracking. The niobium-tantalum-stabilized electrode also inhibited carbide precipitation in the weld. Although tubes failed from intergranular corrosion after approximately 3 years of service, the failures were not associated with the weld repairs. Details of the revised welding procedure are given in the table with Fig. 12.

Fig. 12 Cast Thermalloy 40B calcining tube and preparations for weld repair of gas pocket

Thermalloy 40B; stainless steel filler metal (E347)

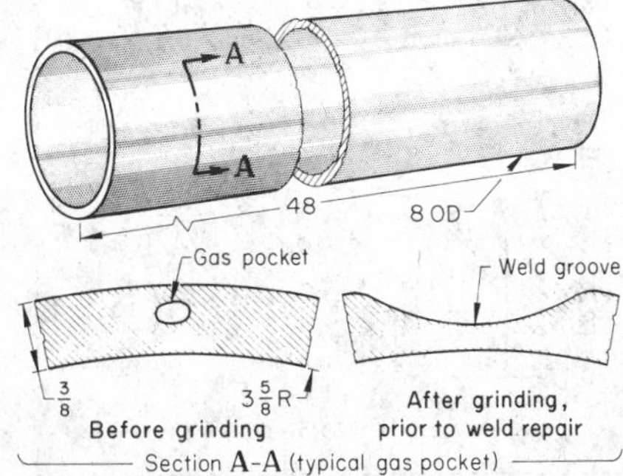

Revised welding conditions

Weld type	Repair
Welding process	SMAW
Power supply	400-A transformer-rectifier
Electrode	$^{1}/_{8}$-in.-diam E347-16
Current	100 A (DCEP)
Voltage	32 V
Preheat and postheat	None

Submerged Arc Welding

Submerged arc welding of iron-nickel-chromium and iron-chromium-nickel alloys normally is limited to applications involving thick sections, where a high deposition rate is desired. For some alloys, the lack of a suitable flux prevents the use of this process.

The types of joints and groove designs are similar to those used for SAW of low-carbon steel. The composition of the weld metal is often critical when joining iron-nickel-chromium and iron-chromium-nickel heat-resistant alloys. Careful selection of the electrode and flux is needed. The electrode generally is of a composition that produces weld metal of the same composition as the base metal. Modified stainless steel electrodes are used for welding castings.

Cobalt-Based Heat-Resistant Alloys

Cobalt-based heat-resistant alloys (Table 1) are available in both cast and wrought forms. Generally, the cast alloys are more difficult to weld than are the wrought alloys. Where the application requires very high reliability of welds, only GTAW and GMAW are recommended; otherwise, SMAW is used.

Joint design and weld grooves for cobalt-based alloys are essentially the same as for nickel-based alloys (Fig. 2). A square-groove butt joint is used for sheet metal up to about $^{7}/_{64}$ or $^{1}/_{8}$ in. thick, a V-groove for plate up to $^{3}/_{8}$ in. thick, a double-V-groove or a double-U-groove for thicknesses of $^3/_8$ to $^5/_8$ in., and a double-U-groove for thicknesses over $^5/_8$ in. Where T-joints are used, the same groove limitations apply as for butt joints. Corner joint welds should be backed by a fillet weld if possible. This type of joint should be avoided where high stresses are likely to occur. V-grooves should have a 60° groove angle for GTAW.

The weld grooves should be machined to ensure proper fit-up. The edges of a sheared plate should be ground or machined back $^1/_{16}$ in. to remove stressed metal. Gas and arc cutting and beveling are not recommended. All joints should be designed to ensure full penetration.

Example 10. Changes in Joint Design, Shielding, and Heat Extraction to Eliminate Weld Defects. The end closures for a sheathed heating element assembly required welding a disk-shaped Haynes alloy No. 25 end-cap stamping to the open end of a 0.2825-in.-diam Haynes alloy No. 25 tubular sheath (Fig. 13). Earlier, the end cap had been welded to a wire coil that served as the heating element of the assembly. After the end cap was welded on, the tube was packed with ceramic insulation material and swaged to reduce the diameter approximately 16%, thus compacting the insulation. Tests were made for leaktightness and for electrical properties.

The original welding procedure consisted of fitting the end cap with its welded-on heating coil into the beveled end of the tube as shown in Fig. 13 (original design and detail A). Then the tube was mounted vertically on a turntable and rotated under a fixed gas tungsten arc torch. The shielding gas was argon. Weld metal was obtained by melting down the protruding tube wall.

The customer's specification of 100% joint penetration (with a joint design not conducive to complete penetration) required the use of excessive heat conducted to the heating coil attached to the end cap, causing the joint between the heating coil and the end cap to soften and fail during the swaging and assembly operations. A second defect was the formation of oxides (because of inadequate shielding and trapped air) on the interior of the tube and at the root of the joint. The oxides that formed on the inside of the tube caused short circuiting during final testing and those at the root of the joint caused incomplete fusion. To eliminate these conditions, a new welding procedure was used.

First, the depth of the bevel on the tube was reduced to provide a root face of 0.015 in. (Fig. 13, detail B), and the diameter of the end cap was decreased to provide a press fit into the tube. Two flats were ground on the end cap to allow the escape of air and permit the use of argon gas for shielding the back of the weld. An O-ring that held the tube in the rotating fixture also served as a seal for the argon backing gas, which flowed through the rotating fixture and into the tube. This procedure was successful in preventing oxide formation on the inside of the tube and at the weld root, but new difficulties developed: the heat produced by welding caused the O-ring to break down and the end cap to become overheated. Placing a copper ring on the tube to act as a heat sink rectified this. The sequence of welding operations was:

- Degrease parts; solution heat treat the tube under vacuum. (This also cleaned the tube.)
- Press fit end-cap assembly into tube.
- Mount tube in rotatable fixture.
- Place copper heat sink around tube.
- Purge air from tube interior with continuous flow of argon.
- Move swing-mounted torch into position ($^1/_{16}$ in. above, and centered on, tube wall).
- Start gas flow, arc, and turntable rotation. After two weld passes, terminate arc, rotation, and gas flow in half a rotation.
- Retract torch, unclamp heat sink, and remove assembly with a special hand tool.
- Test assembly for leaks by holding under water for 10 s while pressurized with air at 100 psi.

This procedure produced 100% joint penetration when used with welding conditions given in Fig. 13. Rejects after swaging amounted to less than 1%, compared with more than 20% previously. Electrodes were changed two or three times during an 8-h shift, because the points became misshapen. After each electrode change, the first piece welded was sectioned and polished to check for penetration and defects.

Fig. 13 End closure detail of sheathed heating element assembly and original and revised joint design

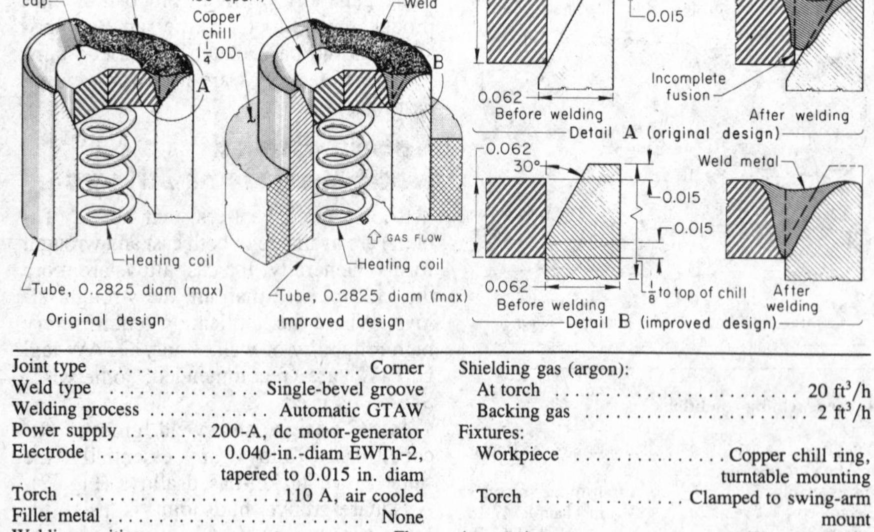

Joint type	Corner
Weld type	Single-bevel groove
Welding process	Automatic GTAW
Power supply	200-A, dc motor-generator
Electrode	0.040-in.-diam EWTh-2, tapered to 0.015 in. diam
Torch	110 A, air cooled
Filler metal	None
Welding position	Flat
Current	45 A (DCEN)
Shielding gas (argon):	
At torch	20 ft^3/h
Backing gas	2 ft^3/h
Fixtures:	
Workpiece	Copper chill ring, turntable mounting
Torch	Clamped to swing-arm mount
Arc starting	High frequency
Arc length	$^1/_{16}$ in.
Production rate	65 pieces per h

Cleaning. The weld joint and adjacent area must be thoroughly cleaned before welding. All foreign matter should be removed by grinding, machining, or scrubbing with a suitable solvent. Shot or sand blasting should not be used because iron and sand particles embedded in the workmetal surface can cause serious contamination. Wire brushing should be done with stainless steel. Copper brushes or carbon steel brushes can contaminate the base metal. If the alloys are heat treated in an oil-fired furnace, fuels of low sulfur content must be used to avoid sulfur contamination and subsequent detrimental effect on mechanical properties and corrosion resistance.

Gas Tungsten Arc Welding

Direct current electrode negative is preferred for GTAW of cobalt-based alloys.

A thoriated tungsten electrode (EWTh-2) of the smallest diameter that can carry adequate current is recommended. Use of electrodes too small in diameter, at high current densities, can cause excessive electrode erosion and result in tungsten inclusions in the weld metal. To ensure sound welds, argon gas at a flow rate of 15 to 25 ft^3/h for weld pool shielding and 5 to 10 ft^3/h for backing is recommended.

Example 11. Elimination of Microcracks in the Heat-Affected Zone of a Haynes Alloy No. 25 Weldment. A bellows 25 in. in diameter and 30 in. long was formed from a cylinder 72 in. long made of 0.012-in.-thick Haynes alloy No. 25 sheet. Two longitudinal welds were required in fabricating the cylinder (Fig. 14). The welds had to be ductile and crack free so that the bellows convolutions could be formed without cracking and without adversely affecting the mechanical properties of the welds.

Originally, the cylinders were welded by GTAW, without filler metal, using a welding positioner with copper hold-down fingers and a copper backing bar. Transverse microcracks, up to about 0.10 in. long, through the work metal, appeared in the HAZ. All attempts to eliminate the cracks, including increasing the welding speed, decreasing the clamping force, changing the shielding gas from argon to 25% argon-75% helium and then to 95% argon-5% hydrogen, and decreasing arc voltage, were unsuccessful.

When 0.030-in.-diam Hastelloy W filler-metal wire was used (instead of no filler metal), cracking was reduced to one or two repairable cracks in four or five welds, provided that the weld was less than 0.065 in. wide. However, at higher welding speeds, such as 60 in./min, it was difficult to feed the limited amount of wire needed to make the narrow weld. In addition, the occasional cracks that did occur were difficult to repair.

Because satisfactory welds had been made in tests on short lengths of Haynes alloy No. 25 using no fixturing, methods of reducing tooling effect while maintaining alignment were considered, and the backing bar was plated with 0.001 in. of hard chromium. Welds were made, with and without filler metal and with different welding currents, and all of the welds were crack free. It was not known whether the cracks were initially caused by copper contamination of the weld metal or by heat dissipation because of the higher thermal conductivity of the copper, but the cracking was eliminated by chromium plating the backing bar. Forming of the convolutions was done with no cracking, indicating no significant adverse effect on the mechanical properties of the welds.

Fig. 14 Haynes alloy No. 25 cylinder that was welded without microcracking in the HAZ when the backing bar was chromium plated

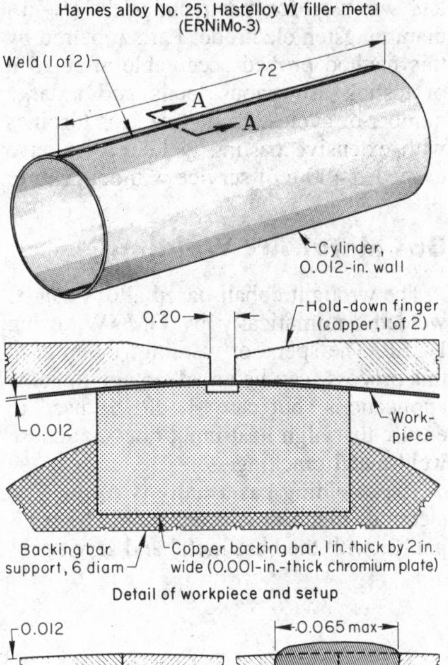

Welding conditions for improved method

Joint type	Butt
Weld type	Square groove
Welding process	Automatic GTAW, automatic arc voltage with 0.25 sensing range
Power supply	300-A transformer-rectifier
Electrode	0.045-in.-diam EWTh-2
Filler metal	0.030-in.-diam Hastelloy W
Torch	Water cooled
Shielding gas:	
At torch	25Ar-75He or 95Ar-5H, at 15-25 ft^3/h
Backing gas	25Ar-75He or 95Ar-5H, at 3-5 ft^3/h
Welding position	Flat
Current	10-30 A (DCEN)
Voltage	11-19 V
Arc starting	High frequency
Travel speed	15-90 in./min
Wire feed	Automatically controlled
Fixture	Hold-down fingers and backing bar

Example 12. Repair of Cracks in Cast HS-21 Gas Turbine Nozzles. During gas turbine overhaul, cracks were often detected in the vanes and shrouds of nozzles, most often appearing on the trailing edges of vanes. A section of a typical 9-in.-diam second-stage nozzle is shown in Fig. 15. Most of the defective nozzles were returned to service after repair by manual GTAW. The repair procedure was:

- Remove heavy carbon coating and other surface contaminants by wire brushing and vapor degreasing.
- Chemically remove residual carbon by using strong alkaline cleaners and oxidizers, followed by light wet blasting of tough scaly areas or porous surfaces.
- Heat treat castings at 2150 °F for 15 min in a vacuum furnace, for final cleaning and solution treating.
- Inspect castings by the fluorescent-

Fig. 15 Section of a gas turbine, second-stage nozzle showing steps in the repair welding of a crack in the trailing edge of one of the vanes

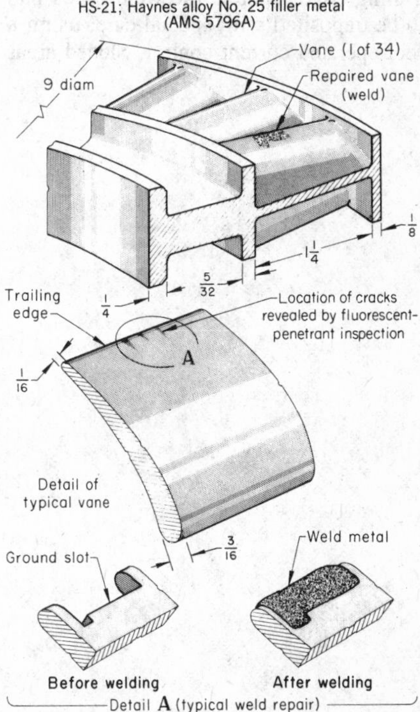

Weld type	Repair
Welding process	Manual GTAW
Power supply	300-A transformer-rectifier, with high-frequency current and time delay for gas and water
Electrode	1/16-in.-diam EWTh-2
Torch	Water cooled, with gas lens
Filler metal	1/16-in.-diam Haynes alloy No. 25 (AMS 5796A)
Shielding gas	Argon at 10-15 ft^3/h
Welding position	Flat
Current	12-70 A (DCEN)
Voltage	Not controlled
Arc starting	High frequency

penetrant method and mark cracks with solvent-soluble ink.

- Grind out all cracks, using a small grinding wheel to cut slots (weld grooves) as shown in detail A of Fig. 15.
- Fill the weld grooves by depositing Haynes alloy No. 25 filler metal (AMS 5796A) with low welding current and argon shielding.
- Grind the weld surfaces to the proper contour.
- Visually inspect ground surfaces.
- Re-solution treat castings at 2150 °F in a vacuum furnace.
- Inspect welds, using fluorescent-penetrant method.
- Wet blast all ground and blended areas for final finish.

The thickness of the vanes varied from $^3/_{16}$ in. at the leading edge to $^1/_{16}$ in. at the trailing edge. Hence, the weld beads had to be deposited with special care, using a foot-operated current control. Slotted areas were gradually built up by welding around the groove edges until the area was filled sufficiently to finish to size. It was not possible to weld the vanes from both sides. A small water-cooled torch with a $^3/_8$- or $^7/_{16}$-in.-diam cup and a $^1/_{16}$-in.-diam gas lens was used for manipulating the $^1/_{16}$-in.-diam tungsten electrode. Parts repaired by this method proved acceptable after 25 h of testing at various loads and a large number of cycles. In service, gas turbines with extensive casting weld repairs have exceeded 400 h of service without failure.

Gas Metal Arc Welding

The wrought cobalt-based alloys can be welded automatically by GMAW, using DCEP. The speed of welding, inherent to this process, can be an advantage in some applications, but care should be used to ensure that high heat input does not cause weld-metal cracking.

The shielding gas usually is argon. The electrode material selected should be compatible with the base metal and service requirements. Typical conditions for GMAW of cobalt-based heat-resistant alloys are:

Base-metal thickness	$^1/_8$ to $^3/_8$ in.
Filler-metal thickness	0.035 in.
Welding current	130 to 160 A (DCEP)
Voltage	22 to 25 V
Shielding gas (argon)	40 ft^3/h
Travel speed	30 in./min

Shielded Metal Arc Welding

Shielded metal arc welding can be used for joining cobalt-based alloys when service conditions permit and when it is not feasible to weld by a gas shielded method. Direct current electrode positive results in the best mechanical properties. Welding should be done in the flat position, with rapid travel and as little weaving as possible.

Types of joint proportions of weld grooves are the same as those used for nickel-based alloys. Cobalt-based alloys are welded in the solution-treated condition with electrodes that have a composition similar to that of the base metal being welded.

Arc Welding of Nonferrous Metals and Alloys

Arc Welding of Aluminum Alloys*

GAS METAL ARC WELDING (GMAW) and gas tungsten arc welding (GTAW) have almost entirely replaced other arc welding processes for aluminum alloys. These gas shielded arc welding processes result in optimum weld quality and minimum distortion, and they require no flux. As a result, difficult-to-reach places and completely inaccessible interiors of welded assemblies are left free from flux residues that could be a potential source of corrosion. Furthermore, welding can be done in all positions, because there is no slag to be worked out of the weld by gravity or by puddling. Visibility is good because the gas envelope around the arc is transparent, and the weld pool is clean.

Base Metals

Most aluminum alloys can be joined by either GMAW or GTAW, and the weldabilities of aluminum alloys are essentially the same for both processes. The most common alloys are grouped by weldability rating as follows:

Readily weldable

- *Wrought alloys*: Pure aluminum, 1350, 1060, 1100, 2219, 3003, 3004, 5005, 5050, 5052, 5083, 5086, 5154, 5254, 5454, 5456, 5652, 6010, 6061, 6063, 6101, 6151, 7005, 7039
- *Casting alloys*: 356.0, 443.0, 413.0, 514.0, A514.0

Weldable in most applications

- *Wrought alloys*: 2014, 2036, 2038, 4032
- *Casting alloys*: 208.0, 308.0, 319.0, 333.0, 355.0, C355.0, 511.0, 512.0, 710.0, 711.0, 712.0

Limited weldability

- *Wrought alloys*: 2024
- *Casting alloys*: 222.0, 238.0, 295.0, 296.0, 520.0

Welding not recommended

- *Wrought alloys*: 7021, 7029, 7050, 7075, 7079, 7129, 7150, 7178, 7475
- *Casting alloys*: 242.0

Wrought and casting alloys are listed above by Aluminum Association designations. Wrought and casting alloys that are weldable in most applications may require special techniques for some applications. Those aluminum alloys with limited weldability always require special techniques.

Wrought alloys most easily welded by gas shielded arc processes are those of the non-heat-treatable 1*xxx*, 3*xxx*, and 5*xxx* series; the alloys in the heat treatable 6*xxx* series are also easily welded. Alloys of the 4*xxx* series and of the high-strength, heat treatable 2*xxx* series can also be arc welded, but special techniques may be required and somewhat lower ductility may be obtained. Of the high-strength heat treatable 7*xxx* series, alloys 7050, 7075, 7079, 7178, and 7475 are not recommended for arc welding.

Alloys 7005 and 7039, however, were developed specifically for welding and have good weldability. Alloys 7005 and 7039 are of special interest for large structures in which the welds must be of high strength, because welds age naturally to 70 to 90% of the strength of the heat treated base metal (depending on the chemical composition of the weld deposit) within 30 to 90 days after welding. The major alloying elements found in aluminum wrought alloys are:

Major alloying element	Designation
99.0% min aluminum and over	1*xxx*
Copper	2*xxx*
Manganese	3*xxx*
Silicon	4*xxx*
Magnesium	5*xxx*
Magnesium and silicon	6*xxx*
Zinc	7*xxx*
Other elements	8*xxx*

The heat of welding removes part or all of the effects of strain hardening; consequently, the strength of the heat-affected zone (HAZ) of a weld in a non-heat-treatable alloy may not exceed that of the annealed alloy. The size of the low-strength zone depends primarily on the speed of welding. On the whole, these weldments exhibit good joint efficiency and ductility.

When a heat treated alloy (T4 or T6 condition) is arc welded, its strength in the as-welded condition is slightly less than that of the unwelded alloy in the T4 condition. This decrease in strength is attributed to the comparative weakness of the HAZ. The zone normally consists of an area of solution-annealed material adjacent to the weld, an area where partial annealing has occurred, and an overaged area. Because of the high strength of the base metal and the low strength of the HAZ, weldments of alloys in the T6 condition have a low as-welded joint efficiency and often lack ductility. Solution heat treatment and aging after welding may restore much of the strength, but ductility loss usually occurs.

Casting Alloys. Most casting alloys can be gas shielded arc welded if they are given the correct edge preparation. Aluminum sand and permanent mold castings are welded to repair foundry defects, to repair items broken in service, or to join cast fittings to wrought members. Formerly, die-cast fittings were seldom used where welded construction was required because they often contained porosity, but recent advances in casting technique, such as vacuum die casting, have resulted in improved quality with satisfactory welding characteristics.

Filler Metals

Classifications and compositions of filler metals for GMAW and GTAW of alumi-

*Revised by the Aluminum Association Technical Committee on Welding and Joining: I.B. Robinson, *Chairman*, Head, Joining Section, Kaiser Aluminum & Chemical Corp.; Paul B. Dickerson, Technical Specialist, Product Engineering Division, Alcoa Laboratories, Aluminum Company of America; Arthur H. Lentz, Welding Engineer, Reynolds Metals Co.; H.L. Saunders, Welding Engineer, Alcan International Ltd.

Table 1 Chemical composition requirements of filler metals for GMAW and GTAW of aluminum alloys

	Composition(a), %										
									Other elements(b)		
AWS classification	Silicon	Iron	Copper	Manganese	Magnesium	Chromium	Zinc	Titanium	Each(c)	Total	Aluminum
ER1100	(d)	(d)	0.05-0.20	0.05	...	...	0.10	...	0.05	0.15	99.00 min(c)
ER2319(e)	0.20	0.30	5.6-6.8	0.20-0.40	0.02	...	0.10	0.10-0.20	0.05	0.15	rem
ER4043	4.5-6.0	0.8	0.30	0.05	0.05	...	0.10	0.20	0.05	0.15	rem
ER4047	11.0-13.0	0.8	0.30	0.15	0.10	...	0.20	...	0.05	0.15	rem
ER4145	9.3-10.7	0.8	3.3-4.7	0.15	0.15	0.15	0.20	...	0.05	0.15	rem
ER5183	0.40	0.40	0.10	0.50-1.0	4.3-5.2	0.05-0.25	0.25	0.15	0.05	0.15	rem
ER5356	0.25	0.40	0.10	0.05-0.20	4.5-5.5	0.05-0.20	0.10	0.06-0.20	0.05	0.15	rem
ER5554	0.25	0.40	0.10	0.50-1.0	2.4-3.0	0.05-0.20	0.25	0.05-0.20	0.05	0.15	rem
ER5556	0.25	0.40	0.10	0.50-1.0	4.7-5.5	0.05-0.20	0.25	0.05-0.20	0.05	0.15	rem
ER5654	(f)	(f)	0.05	0.01	3.1-3.9	0.15-0.35	0.20	0.05-0.15	0.05	0.15	rem
R242.0(g)(h)	0.7	1.0	3.5-4.5	0.35	1.2-1.8	0.25	0.35	0.25	0.05	0.15	rem
R295.0(g)	0.7-1.5	1.0	4.0-5.0	0.35	0.03		0.35	0.25	0.05	0.15	rem
R355.0(g)	4.5-5.5	0.6(j)	1.0-1.5	0.50(j)	0.40-0.6	0.25	0.35	0.25	0.05	0.15	rem
R356.0(g)	6.5-7.5	0.6	0.25	0.35	0.20-0.40	...	0.35	0.25	0.05	0.15	rem

(a) Single values shown are maximum percentages, except where a minimum is specified. Analysis shall be made for the elements for which specific limits are shown. If, however, the presence of other elements is suspected or indicated in the course of routine analysis, further analysis shall be made to determine that these other elements are not in excess of the limits specified for "other elements." (b) Beryllium shall not exceed 0.0008%. (c) The aluminum content is the difference between 100.00% and the sum of all other metallic elements present in amounts of 0.010% or more each, expressed to the second decimal before determining the sum. (d) Silicon plus iron shall not exceed 0.95%. (e) Vanadium content shall be 0.05 to 0.15%. Zirconium content shall be 0.10-0.25%. (f) Silicon plus iron shall not exceed 0.45%. (g) For repair of castings. (h) Nickel content shall be 1.7 to 2.3%. (j) If iron exceeds 0.45%, manganese content shall not be less than half the iron content.
Source: AWS A5.10

num alloys are given in Table 1. In addition, filler metals having the same composition as the base-metal alloy are often used for repairing casting defects.

Selection of Filler Metal. Common criteria to be considered in selecting a filler metal are ease of welding, strength, ductility, corrosion resistance of the filler metal/base metal combination, color match with the base metal after anodizing, and service at elevated temperature. The filler metals listed in Table 1 have been developed to satisfy these requirements. A guide for selection of the filler metal that gives the optimum combination of these criteria for general welding of a selection of alloy combinations is shown in Table 2. Tables 3 and 4 rate filler metals for specific welding criteria—ease of welding, as-welded joint strength and ductility, corrosion and heat resistance, and color match after anodizing.

Joint Design and Edge Preparation

In general, joint design for aluminum alloys is similar to that for steel. Some recommended butt-joint designs for GMAW by direct current electrode positive (DCEP; reverse polarity) and GTAW with alternating current are shown in Fig. 1. When using direct current electrode negative (DCEN; straight polarity) GTAW, the root face can be thicker and the grooves narrower.

Lap joints are used more often for aluminum alloys than for most other metals. The efficiency of lap joints is 60 to 80%, depending on the alloy and temper. Lap joints offer the advantages of no edge preparation being required and ease of fit-up, but have the disadvantage that inspec-

Table 2 Guide to the choice of filler metal for gas shielded arc welding of aluminum

Base metal	319.0, 333.0, 355.0, C355.0	356.0, 413.0, 443.0	511.0, 512.0, 514.0, A514.0,	7005(a), 7039, 710.0, 711.0, 712.0	6070	6061, 6063, 6101, 6151, 6201, 6951	5456
1060, 1350	4145(c)(d)	4043(c)(e)	4043(c)(f)	4043(c)	4043(c)	4043(c)	5356(d)
1100, 3003, alclad 3003	4145(c)(d)	4043(c)(e)	4043(c)(f)	4043(c)	4043(c)	4043(c)	5356(d)
2014, 2024	4145(g)	4145	...	...	4145	4145	...
2219	4145(c)(d)(g)	4145(c)(d)	4043(c)	4043(c)	4043(c)(e)	4043(c)(e)	4043
3004, alclad 3004	4043(d)	4043(c)	5654(h)	5356(f)	4043(f)	4043(h)	5356(f)
5005, 5050	4043(c)	4043(c)	5654(h)	5356(f)	4043(f)	4043(h)	5356(f)
5052, 5652(b)	4043(c)	4043(c)(h)	5654(h)	5356(f)	5356(d)(h)	5356(d)(f)	5356(h)
5083	...	5356(c)(d)(f)	5356(f)	5183(f)	5356(f)	5356(f)	5183(f)
5086	...	5356(c)(d)(f)	5356(f)	5356(f)	5356(f)	5356(f)	5356(f)
5154, 5254(b)	...	4043(c)(h)	5654(h)	5356(h)	5356(c)(h)	5356(d)(h)	5356(h)
5454	4043(c)	4043(c)(h)	5654(h)	5356(h)	5356(c)(h)	5356(d)(h)	5356(h)
5456	...	5356(c)(d)(f)	5356(f)	5556(f)	5356(f)	5356(f)	5556(f)
6061, 6063, 6101, 6201, 6151, 6951	4145(c)(d)	4043(c)(h)	5356(d)(h)	5356(c)(d)(h)	4043(c)(h)	4043(c)(h)	...
6070	4145(c)(d)	4043(c)(f)	5356(d)(f)	5356(c)(d)(f)	4043(c)(f)	...	...
7039, 7005(a), 710.0, 711.0, 712.0	4043(c)	4043(c)(h)	5356(h)	5356(f)	...	...	...
511.0, 512.0, 514.0, A514.0	...	4043(c)(h)	5654(h)(j)	...	...	...	...
356.0, 413.0, 443.0,	4145(c)(d)	4043(c)(j)	...	...	...	...	...
319.0, 333.0, 355.0, C355.0	4145(c)(d)(j)	...	...	...	...	...	...

(continued)

Notes: (1) Alloys listed by Aluminum Association designations. (2) Service conditions such as immersion in fresh or salt water, exposure to specific chemicals or a sustained high temperature (over 150 °F) may limit the choice of filler metals. (3) Recommendations in this table apply to gas shielded arc welding processes. (4) Where no filler metal is listed, the parent alloy combination is not recommended for welding.
(a) 7005 extrusions only. (b) Base-metal alloys 5652 and 5254 are used for hydrogen peroxide service. 5654 filler metal is used for welding both alloys for low-temperature service (150 °F and below). (c) 4047 may be used for some applications.

tion of the weld is difficult. Underbead cracking and crazing in the HAZ has been observed in 6061. Preferred types of lap joints are shown in Fig. 2. T-joints are also widely used. Beveling is seldom required, but it is used on thick material to reduce welding costs and to minimize distortion. Welding a lap joint or a T-joint on one side only is not recommended. A small continuous fillet weld on each side of the joint is preferable.

Edge Preparation and Assembly. Materials up to about $^3/_8$ in. thick can be sheared to a reasonably square edge that can be cleaned readily. Dull or improperly designed tools result in lapping of material on prepared edges that can trap lubricant, which can cause weld porosity.

The extra time needed to ensure a close fit is often less than the extra time required in welding an improperly prepared assembly. Better fit and uniformity of the joint are required for automatic and out-of-position welding than for semiautomatic and flat-position welding. Automatic GTAW of aluminum less than $^1/_8$ in. thick requires that joint fit-up should be held within 0.003 to 0.010 in., depending on metal thickness. A very close fit of the edges is also essential when GTAW without the addition of filler metal.

Aluminum alloy extrusions are sometimes produced with edge designs that facilitate welding. Besides edge preparation, the design may include (1) self-aligning mechanical fitting; (2) integral weld backing; or (3) an increase in section thickness at the joint area to make welding easier, or to compensate for the lower unit strength of the weld area than of the base metal. This is especially valuable in butt welds in heat treatable alloys used in structures too large for most furnaces, used for postweld heat treating, or where heat treating is impractical due to quenching distortions.

Preweld Cleaning

Preweld cleaning of aluminum is essential for optimum weld quality. Precleaning requirements are especially stringent prior to direct current electrode negative GTAW, because under such conditions, the arc exerts no cleaning action.

Surface contaminants that should be removed from the base metal include dirt, metal particles, oil and grease, paint, moisture, and thick oxide coatings. Another source of contamination is oxide film on the filler metal. Base metals such as 1100 and 3003 have a relatively thin oxide coating as-fabricated, while the 5*xxx* and 6*xxx* series alloys generally have a thick, dark oxide coating. The thicker the oxide, the greater its adverse effect on weld-metal flow and solidification and the greater the risk of porosity. Any foreign material that remains on the surfaces to be welded is a potential source of unsound welds. For best results, all cleaning and oxide removal should be done immediately before welding.

First, the work-metal surface should be cleaned of contaminants. The following manual cleaning methods can be used for small production runs. Dirt can be removed easily by washing and scrubbing with a detergent solution; drying is necessary to ensure that no moisture is present on the surfaces to be welded. Removal of grease and oil can be accomplished by swabbing with cloths soaked in an approved nontoxic solvent.

Next, thick oxide layers should be mechanically removed with a wire brush, steel wool, mill file, portable milling tool, or a scraper. The use of abrasive paper or grinding disks alone is not recommended, because particles of the abrasive may become embedded in the aluminum and, unless subsequently removed, can cause inclusions in the weld. Wire brush bristles preferably should be 0.012 to 0.016 in. in diameter and made of stainless steel to minimize iron oxide pickup.

Motor-driven wire brushes should be used carefully. If excessive pressure is applied, a burnishing action in which the oxide is rolled into the freshly exposed surface results, and the weld may be of poorer quality than one made without wire brushing. However, enough pressure has to be used to cause the sharp bristles to break the oxide from the surface of the aluminum alloy.

Table 2 (continued)

Base metal	5454	5154, 5254(b)	5086	5083	5052, 5652(b)	5005, 5050	3004, alclad 3004	2219	2014, 2024	1100, 3003, alclad 3003	1060, 1350
1060, 1350	4043(c)(f)	4043(c)(f)	5356(d)	5356(d)	4043(c)	1100(d)	4043	4145	4145	1100(d)	1100(c)(d)(j)
1100, 3003, alclad 3003	4043(c)(f)	4043(c)(f)	5356(d)	5356(d)	4043(c)(f)	4043(f)	4043(f)	4145	4145	1100(d)	...
2014, 2024	...	...	...	...	...	...	...	4145(g)	4145(g)	...	...
2219	4043(c)	4043(c)	4043	4043	4043(c)	4043	4043	2319(c)(d)	...	...	...
3004, alclad 3004	5654(h)	5654(h)	5356(f)	5356(f)	4043(c)(f)	4043(f)	4043(f)	...	...	...	...
5005, 5050	5654(h)	5654(h)	5356(f)	5356(f)	4043(c)(f)	4043(f)(j)	...	...	...	...	...
5052, 5652(b)	5654(h)	5654(h)	5356(f)	5356(f)	5654(b)(d)(h)	...	...	...	...	...	...
5083	5356(f)	5356(f)	5356(f)	5183(f)	...	...	...	...	...	...	...
5086	5356(h)	5356(h)	5356(f)	...	...	...	...	...	...	...	...
5154, 5254(b)	5654(h)	5654(b)(h)	...	...	...	...	...	...	...	...	...
5454	5554(d)(f)	...	...	...	...	...	...	...	...	...	...
5456	...	...	...	...	...	...	...	...	...	...	...
6061, 6063, 6101, 6201, 6151, 6951	...	...	...	...	...	...	...	...	...	...	...
6070	...	...	...	...	...	...	...	...	...	...	...
7039, 7005(a), 710.0, 711.0, 712.0	...	...	...	...	...	...	...	...	...	...	...
511.0, 512.0, 514.0, A514.0	...	...	...	...	...	...	...	...	...	...	...
356.0, 413.0, 443.0,	...	...	...	...	...	...	...	...	...	...	...
319.0, 333.0, 355.0, C355.0	...	...	...	...	...	...	...	...	...	...	...

(d) 4043 may be used for some applications. (e) 4145 may be used for some applications. (f) 5183, 5356, or 5556 may be used. (g) 2319 or 4043 may be used for some applications. (h) 5183, 5356, 5554, 5556, and 5654 may be used. In some cases they provide improved color match after anodizing treatment, highest weld ductility, and higher weld strength. 5554 is suitable for elevated temperature service. (j) Filler metal with the same analysis as the base metal is sometimes used

Table 3 Filler metals commonly used in arc welding combinations of aluminum alloys

Ratings are relative, in decreasing order of merit, and apply only within a given block; combinations having no rating are not recommended.

Alloys to be welded	Ease of welding						Strength of welded joint (as-welded)(a)						Corrosion resistance(b)						Service at sustained temperature above 150 °F						Color match after anodizing						Ductility(c)					
Filler alloy(d)	1100	4043	5654	5356	5554	5556	1100	4043	5654	5356	5554	5556	1100	4043	5654	5356	5554	5556	1100	4043	5654	5356	5554	5556	1100	4043	5654	5356	5554	5556	1100	4043	5654	5356	5554	5556
To weld alloy 1100 to:																																				
1100	B	A	...	C	...	C	B	A	...	A	...	A	A	A	...	...	...	...	A	A	...	...	...	...	A	...	...	B	...	B	A	D	...	B	...	C
3003, alclad 3003	A	A	...	B	...	B	B	A	...	A	...	A	A	A	...	(e)	...	(e)	A	A	...	...	...	...	A	...	...	B	...	B	A	D	...	B	...	C
3004, alclad 3004	C	A	...	B	...	B	B	A	...	A	...	A	A	A	...	(e)	...	(e)	A	A	...	...	...	...	A	...	...	B	...	B	A	D	...	B	...	C
5005, 5050	B	A	...	B	...	B	B	A	...	A	...	A	A	A	...	...	...	...	A	A	...	...	...	...	A	...	...	B	...	B	A	D	...	B	...	C
5052, 5154, 5454	...	...	...	...	...	...	...	...	...	...	...	...	...	...	...	...	...	...	...	...	...	...	...	...	...	...	...	...	...	...	...	...	...	...	...	...
5083, 5086, 5456	...	...	...	...	...	...	...	...	...	...	...	...	...	...	...	...	...	...	...	...	...	...	...	...	...	...	...	...	...	...	...	...	...	...	...	...
6063(f), 6101(f)	...	A	...	B	...	B	...	A	...	A	...	A	...	A	...	...	...	...	...	A	...	...	...	...	...	...	...	A	...	A	...	C	...	A	...	B
6061(f)	...	A	...	B	...	B	...	A	...	A	...	A	...	A	...	...	...	...	...	A	...	...	...	...	...	...	...	A	...	A	...	C	...	A	...	B
To weld alloy 3003 to:																																				
3003, alclad 3003	A	A	...	B	...	B	C	B	...	A	...	A	A	A	...	(e)	...	(e)	A	A	...	...	...	...	A	...	...	B	...	B	A	D	...	B	...	C
3004, alclad 3004	...	A	...	B	...	B	...	B	...	A	...	A	...	A	...	(e)	...	(e)	...	A	...	...	...	...	...	...	...	A	...	A	...	C	...	A	...	B
5005, 5050	B	A	...	B	...	B	C	B	...	A	...	A	A	A	...	...	...	...	A	A	...	...	...	...	A	...	...	B	...	B	A	D	...	B	...	C
5052	...	A	...	B	...	B	...	B	...	A	...	A	...	A	...	...	...	...	...	A	...	...	...	...	...	...	...	A	...	A	...	C	...	A	...	B
5154	...	A	C	B	C	B	...	B	A	A	A	A	...	C	A	B	A	B	...	...	...	...	...	...	...	...	B	A	A	A	...	C	A	A	A	B
5454	...	A	...	B	C	B	...	B	...	A	A	A	...	C	...	B	A	B	...	A	...	...	A	...	...	...	...	A	A	A	...	C	...	A	A	B
5083, 5086, 5456	...	A	...	A	...	A	...	B	...	A	...	A	...	B	...	A	...	A	...	...	...	...	...	...	...	...	...	A	...	A	...	C	...	A	...	B
6063(f), 6101(f)	...	A	...	B	...	B	...	B	...	A	...	A	...	A	...	...	...	...	...	A	...	...	...	...	...	...	...	A	...	A	...	C	...	A	...	B
6061(e)	...	A	...	B	...	B	...	B	...	A	...	A	...	A	...	...	...	...	...	A	...	...	...	...	...	...	...	A	...	A	...	C	...	A	...	B
To weld alclad 3003 to:																																				
Alclad 3003	A	A	...	B	...	B	C	B	...	A	...	A	A	A	...	B	...	B	A	A	...	...	...	...	A	...	...	B	...	B	A	D	...	B	...	C
3004, alclad 3004	...	A	...	B	...	B	...	B	...	A	...	A	...	A	...	B	...	B	...	A	...	...	...	...	...	...	...	A	...	A	...	C	...	A	...	B
5005, 5050	B	A	...	B	...	B	C	B	...	A	...	A	A	A	...	B	...	B	A	A	...	...	...	...	A	...	...	B	...	B	A	D	...	B	...	C
5052	...	A	...	B	...	B	...	B	...	A	...	A	...	A	...	B	...	B	...	A	...	...	...	...	...	...	...	A	...	A	...	C	...	A	...	B
5154	...	A	C	B	C	B	...	B	A	A	A	A	...	C	A	B	A	B	...	...	...	...	...	...	...	...	B	A	A	A	...	C	A	A	A	B
5454	...	A	...	B	C	B	...	B	...	A	A	A	...	C	...	B	A	B	...	A	...	...	A	...	...	...	...	A	A	A	...	C	...	A	A	B
5083, 5086, 5456	...	A	...	A	...	A	...	B	...	A	...	A	...	B	...	A	...	A	...	...	...	...	...	...	...	...	...	A	...	A	...	C	...	A	...	B
6063(f), 6101(f)	...	A	...	B	...	B	...	B	...	A	...	A	...	A	...	B	...	B	...	A	...	...	...	...	...	...	...	A	...	A	...	C	...	A	...	B
6061(f)	...	A	...	B	...	B	...	B	...	A	...	A	...	A	...	B	...	B	...	A	...	...	...	...	...	...	...	A	...	A	...	C	...	A	...	B
To weld alloy 3004 to:																																				
3004, alclad 3004	...	A	C	B	C	B	...	D	C	B	C	A	...	A	B	(e)	B	(e)	...	A	...	...	A	...	...	...	B	A	A	A	...	C	A	A	A	B
5005, 5050	...	A	...	B	...	B	...	B	...	A	...	A	...	A	...	...	...	...	...	A	...	...	...	...	...	...	...	A	...	A	...	C	...	A	...	B
5052	...	A	...	B	...	B	...	C	...	B	...	A	...	A	...	...	...	...	...	A	...	...	...	...	...	...	...	A	...	A	...	C	...	A	...	B
5154	...	A	C	B	C	B	...	D	C	B	C	A	...	C	A	B	A	B	...	...	...	...	...	...	...	...	B	A	A	A	...	C	A	A	A	B
5454	...	A	...	B	C	B	...	D	...	B	C	A	...	C	...	B	A	B	...	A	...	...	A	...	...	...	...	A	A	A	...	C	...	A	A	B
5083, 5086, 5456	...	A	...	A	...	A	...	C	...	B	...	A	...	B	...	A	...	A	...	...	...	...	...	...	...	...	...	A	...	A	...	C	...	A	...	B
6063(f), 6101(f)	...	A	...	B	...	B	...	C	...	B	...	A	...	A	...	...	...	...	...	A	...	...	...	...	...	...	...	A	...	A	...	C	...	A	...	B
6061(f)	...	A	...	B	...	B	...	C	...	B	...	A	...	A	...	...	...	...	...	A	...	...	...	...	...	...	...	A	...	A	...	C	...	A	...	B
To weld alclad 3004 to:																																				
Alclad 3004	...	A	C	B	C	B	...	D	C	B	C	A	...	A	B	C	B	C	...	A	...	...	A	...	...	...	B	A	A	A	...	C	A	A	A	B
5005, 5050	...	A	...	B	...	B	...	B	...	A	...	A	...	A	...	B	...	B	...	A	...	...	...	...	...	...	...	A	...	A	...	C	...	A	...	B
5052	...	A	...	B	...	B	...	C	...	B	...	A	...	A	...	B	...	B	...	A	...	...	...	...	...	...	...	A	...	A	...	C	...	A	...	B
5154	...	A	C	B	C	B	...	D	C	B	C	A	...	C	A	B	A	B	...	...	...	...	...	...	...	...	B	A	A	A	...	C	A	A	A	B
5454	...	A	...	B	C	B	...	D	...	B	C	A	...	C	...	B	A	B	...	A	...	...	A	...	...	...	...	A	A	A	...	C	...	A	A	B
5083, 5086, 5456	...	A	...	A	...	A	...	C	...	B	...	A	...	B	...	A	...	A	...	...	...	...	...	...	...	...	...	A	...	A	...	C	...	A	...	B
6063(f), 6101(f)	...	A	...	B	...	B	...	C	...	B	...	A	...	A	...	B	...	B	...	A	...	...	...	...	...	...	...	A	...	A	...	C	...	A	...	B
6061(f)	...	A	...	B	...	B	...	C	...	B	...	A	...	A	...	B	...	B	...	A	...	...	...	...	...	...	...	A	...	A	...	C	...	A	...	B

(continued)

Table 3 (continued)

Alloys to be welded	Ease of welding						Strength of welded joint (as-welded)(a)						Corrosion resistance(b)						Service at sustained temperature above 150 °F						Color match after anodizing						Ductility(c)					
Filler alloy(d)	1100	4043	5654	5356	5554	5556	1100	4043	5654	5356	5554	5556	1100	4043	5654	5356	5554	5556	1100	4043	5654	5356	5554	5556	1100	4043	5654	5356	5554	5556	1100	4043	5654	5356	5554	5556
To weld alloy 5005 or 5050 to:																																				
5005, 5050	C	A	...	B	...	B	...	B	...	A	...	A	A	A	...	...	...	...	A	A	...	...	...	...	A	...	...	B	...	B	A	D	...	B	...	C
5052	...	A	...	B	...	B	...	B	...	A	...	A	...	A	...	...	...	...	...	A	...	...	...	...	...	...	...	A	...	A	...	C	...	A	...	B
5154	...	A	C	B	C	B	...	B	A	A	A	A	...	C	A	B	A	B	...	...	...	...	...	...	...	...	B	A	A	A	...	C	A	A	A	B
5454	...	A	...	B	C	B	...	B	...	A	A	A	...	C	...	B	A	B	...	A	...	...	A	...	...	...	...	A	A	A	...	C	...	A	A	B
5083, 5086, 5456	...	A	...	A	...	A	...	B	...	A	...	A	...	B	...	A	...	A	...	...	...	...	...	...	...	...	...	A	...	A	...	C	...	A	...	B
6063(f), 6101(f)	...	A	...	B	...	B	...	B	...	A	...	A	...	A	...	...	...	...	...	A	...	...	...	...	...	...	...	A	...	A	...	C	...	A	...	B
6061(f)	...	A	...	B	...	B	...	B	...	A	...	A	...	A	...	...	...	...	...	A	...	...	...	...	...	...	...	A	...	A	...	C	...	A	...	B
To weld alloy 5052 to:																																				
5052	...	A	B	A	C	A	...	D	C	B	C	A	...	C	B	...	B	...	...	A	...	...	B	...	...	...	A	A	B	B	...	C	A	A	A	B
5154	...	A	B	A	C	A	...	D	C	B	C	A	...	C	A	B	A	B	...	...	...	...	...	...	...	...	A	A	B	B	...	C	A	A	A	B
5454	...	A	B	A	C	A	...	D	C	B	C	A	...	C	B	B	A	B	...	A	...	...	A	...	...	...	B	A	A	A	...	C	A	A	A	B
5083, 5086, 5456	...	...	...	A	...	A	...	...	...	B	...	A	...	...	...	A	...	A	...	...	...	...	...	...	...	...	...	A	...	A	...	...	...	A	...	B
6063(f), 6101(f)	...	A	C	B	C	B	...	B	A	A	A	A	...	A	B	...	B	...	...	A	...	...	A	...	...	...	B	A	A	A	...	C	A	A	A	B
6061(f)	...	A	C	B	C	B	...	D	C	B	C	A	...	A	B	...	B	...	...	A	...	...	A	...	...	...	A	A	B	B	...	C	A	A	A	B
To weld alloy 5083 or 5456 to:																																				
5154	...	...	B	A	B	A	...	...	C	B	C	A	...	...	A	A	A	A	...	...	...	...	...	...	...	...	B	A	A	A	...	...	A	A	A	B
5454	...	...	...	A	B	A	...	...	...	B	C	A	...	...	...	B	A	B	...	...	...	...	...	...	...	...	...	A	A	A	...	...	...	A	A	B
5083, 5086, 5456	...	...	...	A	...	A	...	...	...	B	...	A	...	...	...	A	...	A	...	...	...	...	...	...	...	...	...	A	...	A	...	...	...	A	...	B
6063(f), 6101(f)	...	A	B	A	B	A	...	B	A	A	A	A	...	A	A	A	A	A	...	...	...	...	...	...	...	...	B	A	A	A	...	C	A	A	A	B
6061(f)	...	A	B	A	B	A	...	D	C	B	C	A	...	A	A	A	A	A	...	...	...	...	...	...	...	...	B	A	A	A	...	C	A	A	A	B
To weld alloy 5086 to:																																				
5154	...	...	B	A	B	A	...	...	C	B	C	A	...	...	A	A	A	A	...	...	...	...	...	...	...	...	B	A	A	A	...	...	A	A	A	B
5454	...	...	...	A	B	A	...	...	...	B	C	A	...	...	...	B	A	B	...	...	...	...	...	...	...	...	...	A	A	A	...	...	...	A	A	B
5086	...	...	...	A	...	A	...	...	...	B	...	A	...	...	...	A	...	A	...	...	...	...	...	...	...	...	...	A	...	A	...	...	...	A	...	B
6063(f), 6101(f)	...	A	B	A	B	A	...	B	A	A	A	A	...	A	A	A	A	A	...	...	...	...	...	...	...	...	B	A	A	A	...	C	A	A	A	B
6061(f)	...	A	B	A	B	A	...	D	C	B	C	A	...	A	A	A	A	A	...	...	...	...	...	...	...	...	B	A	A	A	...	C	A	A	A	B
To weld alloy 5154 to:																																				
5154	...	...	B	A	B	A	...	...	C	B	C	A	...	...	A	...	A	...	...	...	...	...	...	...	...	...	A	A	B	B	...	...	A	A	A	B
5454	...	...	B	B	B	A	...	...	C	B	C	A	...	...	A	B	A	B	...	...	...	...	...	...	...	...	B	A	A	A	...	...	A	A	A	B
6063(f), 6101(f)	...	A	C	B	C	B	...	B	A	A	A	A	...	A	B	...	B	...	...	...	...	...	...	...	...	...	B	A	A	A	...	C	A	A	A	B
6061(f)	...	A	C	B	C	B	...	D	C	B	C	A	...	A	B	...	B	...	...	...	...	...	...	...	...	...	A	A	B	B	...	C	A	A	A	B
To weld alloy 5454 to:																																				
5454	...	...	B	A	B	A	...	...	C	B	C	A	...	...	B	B	A	B	...	...	...	...	A	...	...	...	B	A	A	A	...	...	A	A	A	B
6063(f), 6101(f)	...	A	C	B	C	B	...	B	A	A	A	A	...	B	B	...	A	...	...	A	...	...	A	...	...	...	B	A	A	A	...	C	A	A	A	B
6061(f)	...	A	C	B	C	B	...	D	C	B	C	A	...	B	B	...	A	...	...	A	...	...	A	...	...	...	B	A	A	A	...	C	A	A	A	B
To weld alloy 6061 to:																																				
6063(f), 6101(f)	...	A	C	B	C	B	...	B	A	A	A	A	...	A	B	C	B	C	...	A	...	...	B	...	...	...	B	A	A	A	...	C	A	A	A	B
6061(f)	...	A	C	B	C	B	...	D	C	B	C	A	...	A	B	C	B	C	...	A	...	...	B	...	...	...	B	A	B	B	...	C	A	A	A	B
To weld alloy 6063 or 6101 to:																																				
6063(f), 6101(f)	...	A	C	B	C	B	...	B	A	A	A	A	...	A	B	C	B	C	...	A	...	...	B	...	...	...	B	A	A	A	...	C	A	A	A	B

(a) Rating applies particularly to fillet welds. All filler alloys rated develop presently specified minimum strengths in butt welds. (b) Rating based on continuous or alternate immersion in fresh or salt water. (c) Rating based on free-bend elongation of weld. (d) Filler alloy 5183 has the same ratings as 5556, except that welds made with 5183 are slightly more ductile and, in cases where the filler metal controls the weld strength, slightly less strong than welds made with 5556. Because of its lower strength, 5183 filler metal is not recommended for welding 5456. (e) Filler-metal alloys 5356 and 5556 are not recommended for corrosion resistance in welding alloy 1100, 3003, or 3004 to bare alloy 3003 or 3004, but are rated B for corrosion resistance in welding alloy 1100 or 3003 to alclad 3003 or 3004, and rated C for corrosion resistance in welding alloy 3004 to alclad 3004. (f) Ratings do not apply when heat treated after welding.

Table 4 Guide to selection of filler-metal alloys for arc welding various combinations of heat treatable aluminum alloys

Ratings are relative, in decreasing order of merit, and apply to a given base-metal combination and postweld condition; the use of base metals as filler metals, or of combinations indicated here by dots as having no ratings, is not recommended.

		Filler alloys																			
		Ease of welding					Strength(c)					Ductility(d)					Corrosion resistance(e)				
Alloys to be welded(a)	Postweld condition(b)	2319	4043	4145	5556(f)	5554(g)	2319	4043	4145	5556(f)	5554(g)	2319	4043	4145	5556(f)	5554(g)	2319	4043	4145	5556(f)	5554(g)
To weld 2014 or 2024 to:																					
2014, 2024 or 2219	X	C	B	A	…	…	A	B	A	…	…	A	A	B	…	…	A	B	B	…	…
	Y	C	B	A	…	…	A	C	B	…	…	A	B	B	…	…	A	B	B	…	…
To weld 2219 to:																					
2219	X	A	A	A	…	…	A	B	B	…	…	A	B	B	…	…	A	B	B	…	…
	Y or Z	A	A	A	…	…	A	C	B	…	…	A	B	B	…	…	A	B	B	…	…
To weld 6061, 6063 or 6101 to:																					
1100	X	…	A	…	B	…	…	A	…	A	…	…	B	…	A	…	…	A	…	B	…
2014 or 2024	X	…	B	A	…	…	…	A	A	…	…	…	A	B	…	…	…	A	A	…	…
2219	X	…	A	A	…	…	…	A	A	…	…	…	A	B	…	…	…	A	A	…	…
3003, 3004, 5005, or 5050	X	…	A	…	B	…	…	B	…	A	…	…	B	…	A	…	…	A	…	B	…
5052, 5154, or 5454	X	…	A	…	B	C	…	C	…	A	B	…	B	…	A	A	…	A	…	B	A
5083, 5086, or 5456	X	…	…	…	A	B	…	…	…	A	B	…	…	…	B	A	…	…	…	A	A
6061, 6063, or 6101	X	…	A	…	B	C	…	C	…	A	B	…	B	…	A	A	…	A	…	C	B
	Y or Z	…	A	…	(h)	B	…	A	…	(h)	B	…	B	…	(h)	A	…	A	…	(h)	B
To weld 7005 or 7039 to:																					
5052, 5154, or 5454	X	…	A	…	A	B	…	D	…	B	C	…	B	…	A	A	…	B	…	A	A
5083,5086, or 5456	X	…	…	…	A	…	…	…	…	B	…	…	…	…	A	…	…	…	…	A	…
6061 or 6063	X	…	A	…	A	B	…	D	…	B	C	…	B	…	A	A	…	A	…	A	A
	Y or Z	…	A	…	(h)	B	…	C	…	(h)	B	…	B	…	(h)	A	…	A	…	(h)	A
7005 or 7039	X	…	…	…	A	…	…	…	…	B	…	…	…	…	A	…	…	…	…	A	…
	Y or Z	…	…	…	(h)	…	…	…	…	(h)	…	…	…	…	(h)	…	…	…	…	(h)	…
To weld 7075 or 7178 to:																					
7075 or 7178	X	…	A	A	B	…	…	C	C	B	…	…	B	B	A	…	…	B	B	A	…
	Y or Z	…	A	A	(h)	…	…	B	B	(h)	…	…	B	B	(h)	…	…	A	A	(h)	…

(a) Ratings for both bare and alclad materials are the same. (b) X = naturally aged for 30 days or longer; Y = postweld solution heat treated and artificially aged; Z = postweld artificially aged. (c) Ultimate strength from cross-weld tensile test. (d) Ratings based on free-bend elongation of weld. (e) Ratings based on continuous or alternate immersion in fresh or salt water. (f) 5183 and 5356 have the same ratings as 5556. (g) Filler alloy 5554 is suitable for welding 6061, 6063, and 7005 prior to brazing (h) Filler alloy not recommended because of possible susceptibility to stress-corrosion cracking when postweld heat treated.
Source: Aluminum Co. of America, "Welding Alcoa Aluminum," 1972

Chemical removal of oxides can be accomplished by immersion in solutions of the butyl alcohol-phosphoric acid type. After the chemical treatment, the parts should be washed thoroughly with water and dried with hot air. For thick and persistent oxide coatings, immersion in a 5% sodium hydroxide solution at 150 °F for about 30 s is recommended. This treatment may leave a dark smut on the surface. To remove the smut, the treatment should be followed by a cold water rinse, immersion in a solution containing equal parts of commercial nitric acid and water at room temperature, a final water rinse (preferably hot), and hot air drying. Heavy-etching caustic solutions are not recommended, because the rough surface that they produce is likely to collect hydrocarbons and foreign material.

Degreasing and chemical cleaning should be done before the parts are assembled for welding. Cleaning after assembly can result in retention of foreign material and solutions between abutting edges and lapped areas of the joint, and porosity and dross entrapment in the weld are likely to result. Freshly machined and freshly filed surfaces are the cleanest and are often specified when the ultimate in weld quality is demanded. For additional information on cleaning aluminum surfaces, see the article "Cleaning and Finishing of Aluminum and Aluminum Alloys" in Volume 5 of the 9th edition of *Metals Handbook*.

Preheating

In gas shielded arc welding of aluminum alloys, preheating parts to be welded is normally done only when the temperature of the parts is below 32 °F or when the mass of the parts is such that the heat is conducted away from the joint faster than the welding process can supply it. Preheating may be advantageous for GTAW with alternating current of parts thicker than about $^{3}/_{16}$ in. and GMAW of parts thicker

Fig. 1 Recommended butt-joint designs for direct current electrode positive GMAW and alternating current GTAW of aluminum alloys. Joints 2, 3, 4, 5, and 6 should be back gouged to solid weld metal before applying a pass on the root side.

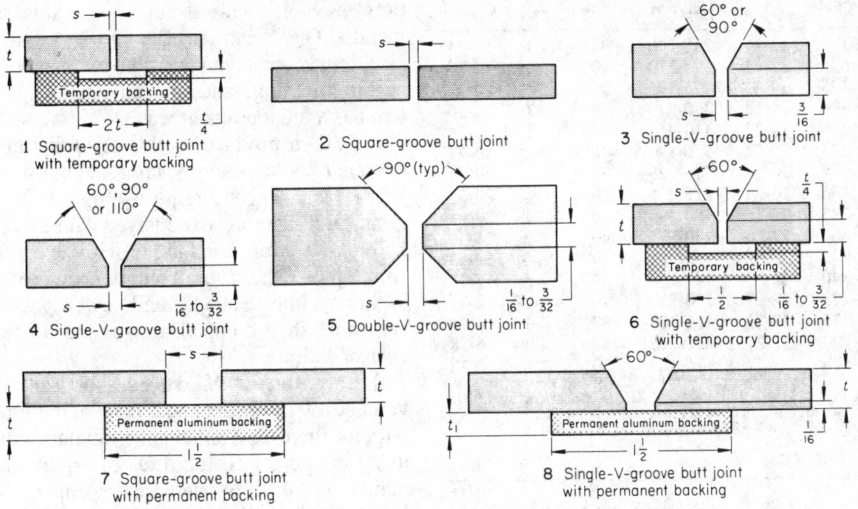

Metal thickness, t, in.	Semiautomatic GMAW Welding position(a)	Joint design(b)	Root opening, s, in.	Manual GTAW Welding position(a)	Joint design	Root opening, s, in.
1/16	F	1, 7	0-3/32	F,H,V,O	2	0
3/32	F	1	0	F,H,V	2	0
	F,H,V,O	7	1/8	O	2	0
1/8	F,H,V	1	0-3/32	F	2	0
	F,H,V,O	7	3/16	H,V,O	2	0
3/16	F,H,V	2	0	F	4-60°	0-1/8
	F,H,V,O	6	0-1/16	H	4-90°	0-3/32
	F,V	8	3/32-3/16	V	4-60°	0-3/32
	H,O	8	3/16	O	4-110°	0-3/32
1/4	F	2	0	F	4-60°	0-1/8
	F,H,V,O	6	0-3/32	H	4-90°	0-3/32
	F,V	8	1/8-1/4	V	4-60°	0-3/32
	H,O	8	1/4	O	4-110°	0-3/32
3/8	F	3-90°	0-3/32	F	4-60°	0-1/8
	F	6	0-3/32	F	5	0-3/32
	H,V,O	6	0-3/32	V	4-60°	0-3/32
	F,V	8	1/4-3/8	H,V,O	5	0-3/32
	H	8	3/8	H	4-90°	0-3/32
	O	8	3/8	O	4-110°	0-3/32
3/4	F	3-60°	0-3/32	...	...	...
	F	8	0-1/8	...	...	...
	H,V,O	8	0-1/16	...	...	...
	F,H,V,O	8	0-1/16	...	...	...

(a) F = flat; H = horizontal; V = vertical; O = overhead. (b) For design 8, $t_1 = t$ for t less than 3/8 in., and t_1 = 3/8 in. for t greater than 3/8 in.

than about 1 in. Gas tungsten arc welding with DCEP is limited to thin material, and preheating is not necessary with this process. Thick parts also should not be preheated when GTAW using DCEN, because of the high heat input provided to the work. Preheating can also reduce production costs because the joint area reaches welding temperature faster, thus permitting higher welding speeds.

Various methods can be used to preheat the entire part or assembly to be welded, or only the area adjacent to the weld can be heated by use of a gas torch. In mechanized welding, local preheating and drying can be done by gas or tungsten arc torches installed ahead of the welding electrode.

The preheating temperature depends on the job. Often 200 °F is sufficient to ensure adequate penetration on weld starts, without readjustment of the current as welding progresses. Preheating temperature for wrought aluminum alloys seldom exceeds 300 °F, because the desirable properties of certain aluminum alloys and tempers may be adversely affected at higher temperatures. Aluminum-magnesium alloys containing 4.0 to 5.5% Mg (5083, 5086, and 5456) should not be preheated to more than 200 °F, because their resistance to stress corrosion cracking is reduced.

Large or intricate castings may be preheated to minimize thermal stresses and to facilitate attainment of the welding temperature. After welding, such castings should be cooled slowly to minimize the danger of cracking. Castings that are to be used in the heat treated condition should be welded before heat treatment or should be re-heat treated after welding. Preheating and the heat of welding may affect the corrosion resistance of some alloys, such as alloy 520, unless welding is followed by heat treatment.

Fixtures

Design of fixtures is based on the expectation that dimensional changes in welding aluminum alloys are twice as great as in the welding of steel. The coefficient of expansion of aluminum is about twice that of steel, and its melting point is about half that of steel. Thus, the change in dimensions from welding heat is in the same range as that for steel. However, the thermal conductivity of aluminum is greater. In general, the amount of expansion is inversely proportional to the speed of welding.

When butt welding aluminum sheet, the fixture should enable the sheet to be clamped with a uniform force of approximately 200 lb per linear inch of seam. This amount of force usually ensures against movement of the sheet during welding. To guard against deflection of the arc, nonmagnetic materials such as austenitic stainless steel, copper, and aluminum should be used for those parts of a fixture within 4 in. of the arc and must be used for fixture parts within 2 in. of the arc.

Although rigid clamping reduces distortion, the inability of the weldment to con-

Fig. 2 Preferred types of lap joints for arc welding of aluminum alloys

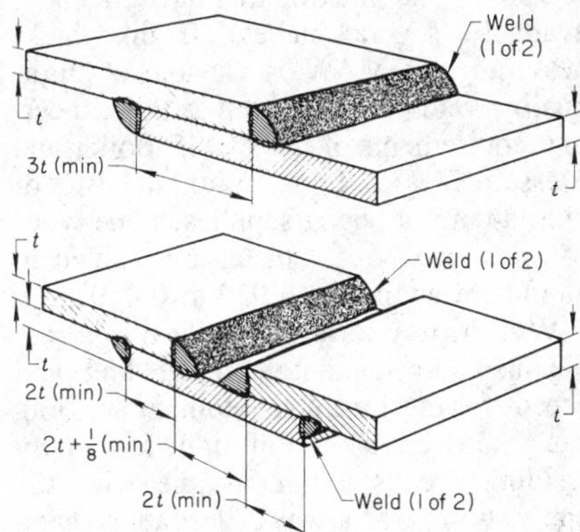

Table 5 Welding procedure schedules for GMAW of aluminum

Material thickness or fillet size, in.	Type of weld fillet or groove	Electrode diameter, in.	Welding power(a) Current, A (DCEP)	Voltage, V	Wire-feed speed, in./min	Shielding gas flow, ft³/h	No. of passes	Travel speed (per pass), in./min
3/64	Square groove and fillet	0.030	50	12-14	268-308	30	1	17-25
5/64	Square groove and fillet	0.030	55-60	12-14	295-320	30	1	17-25
5/64	Square groove and fillet	3/64	110-125	19-21	175-185	30	1	20-27
3/32	Square groove and fillet	0.030	90-100	14-18	330-370	30	1	24-36
1/8	Fillet	0.030	110-125	19-22	410-460	30	1	20-24
1/8	Square grove	3/64	110-125	20-24	175-190	40	1	20-24
3/16	Square groove and fillet	3/64	160-195	20-24	215-225	40	1	20-25
1/4	Fillet	3/64	160-195	20-24	215-225	40	1	20-25
1/4	V-groove	1/16	175-225	22-26	150-195	40	3	20-25
3/8	V-groove and fillet	1/16	200-300	22-26	170-275	40	2-5	25-30
1/2	V-groove and fillet	1/16	220-230	22-27	195-205	40	3-8	12-18
1/2	Double-V-groove	3/32	320-340	22-29	140-150	45	2-5	15-17
3/4	Double-V-groove	1/16	255-275	22-27	230-250	50	4-10	8-18
3/4	Double-V-groove	3/32	355-375	22-29	155-160	50	4-10	4-16
1	Double-V-groove	1/16	255-290	22-27	230-265	50	4-14	6-18
1	Double-V-groove	3/32	405-425	22-27	175-180	50	4-8	8-12

Notes: (1) For groove and fillet welds, material thickness also indicates fillet weld size. Use V-groove for 3/16 in. and thicker. (2) Use argon for thin and medium material; use 50% argon and 50% helium for thick material. Increase gas flow rate 10% for overhead position. (3) Increase amperage 10 to 20% when backup is used. (4) Decrease amperage 10 to 20% when welding out of position.
(a) Direct current electrode positive (DCEP)
Source: Cary, H. B., *Modern Welding Technology*, Prentice-Hall, New York, 1979, p 440

tract, caused by the restraint, may induce residual stress as high as the yield strength of the base metal and may also result in cracking. Hold-down fingers should be designed to permit the joints to accommodate expansion and contraction, yet still maintain proper position for welding. To keep distortion to a minimum, the joint should be designed with minimum separation between members, and welding should be done in the minimum number of passes. A fixture used in welding long pipelines is described in Example 1.

Gas Metal Arc Welding

The ability of GMAW to deposit large quantities of weld metal in a short period of time has played a large part in the increased use of aluminum since the late 1940's. Typical welding schedules for GMAW of aluminum alloys are given in Table 5.

Thicknesses of aluminum alloys commonly joined by GMAW range from 1/8 in. up to the maximum plate thickness available (several inches). In this thickness range, GMAW is capable of high-quality weld deposits, such as those meeting requirements of the ASME Boiler and Pressure Vessel Code. With the use of pulsed-current power supplies, some types of joints can be gas metal arc welded in aluminum as thin as 0.030 to 0.040 in.

Welding speeds up to 55 in./min are obtained with semiautomatic welding, and speeds for machine and automatic welding can be as high as 180 in./min. Maximum welding speeds commensurate with the application are always desirable when welding aluminum alloys. Rapid cooling after welding, which results from high welding speeds, produces fine-grain weld deposits and retards the formation of low-melting constituents at the grain boundaries.

Power Supply and Equipment. Only direct current electrode positive, which gives good penetration and a cathodic cleaning action at the work surface, is used in GMAW of aluminum alloys. The constant and pulsed direct current power supplies, and the wire-feed systems, electrode holders, and control systems used for GMAW of aluminum alloys, are the same as those used for GMAW of other metals. Push-type wire-feed systems can handle aluminum wire down to 0.045 in. diam, but for smaller wires, a pull-type or push-pull system must be used. Grooved drive rolls are preferred; knurled rolls and serrated rolls are likely to chip off small particles of metal that can enter the wire conduit and slow down or stop wire feed. Wire conduits, inlet guides, guide liners, and bushings for aluminum electrode wire should be of all-nylon or all-Teflon construction.

Shielding Gases for Gas Metal Arc Welding

Argon, helium, and mixtures of the two are used as shielding gases in GMAW of aluminum alloys.

Argon is generally preferred when welding thinner metal, mainly because of its lower arc heat. In addition, argon results in a smoother and more stable arc than helium, and thus much less weld spatter is obtained.

Helium, because of its greater arc heat, is capable of producing the deep penetration desirable in weld deposits in thicker metal. The bead profile with helium shielding is wider and less convex than with argon shielding, and the penetration pattern has a broader underbead. Welding with pure helium produces welds of darker appearance with some spatter. Helium is lighter than argon, requires higher flow rates, and is more expensive. Therefore, helium is seldom used alone. However, in some jobs, the use of higher currents, higher welding speeds, and fewer passes can more than compensate for the higher cost of helium.

Argon-Helium Mixtures. To take advantage of the higher arc heat of helium without the disadvantages associated with using the pure gas, mixtures of argon and helium are usually used for joining thick metal. Although users have individual preferences, mixtures ranging between 50 and 75% helium are used. A helium-rich mixture, such as 75% helium and 25% argon, is frequently used when welding workpieces more than 2 in. thick. For workpieces more than 3 in. thick, helium-rich mixtures maximize weld penetration and minimize porosity. When welding workpieces 1 to 3 in. thick in the flat position, increasing the current or voltage, or both, allows the helium content to be decreased.

Flow Rates. Typical shielding gas flow rates for GMAW of aluminum and aluminum alloys using 1/16-in.-diam electrode wire are:

Shielding gas	Flow rate(a), ft³/h
100% argon	30-70
75% helium, 25% argon	50-110
100% helium	60-140

(a) The lower rates are more suitable for indoor work and moderate welding current. The higher rates are more suitable for high current, maximum speed, and outdoor welding.

Note that helium requires about twice the flow of argon. The rate should not be greater than that which will provide laminar flow.

Arc Characteristics in Gas Metal Arc Welding

Increasing the welding current changes the arc from one producing short circuiting transfer to one producing globular transfer and then to one producing spray transfer. Spray transfer produced by either a constant-current arc or by a pulsed-current arc is used for almost all GMAW of aluminum alloys. In some special applications, globular transfer and constant current may be used instead.

To obtain spray transfer from constant-current arcs requires extremely high current densities when welding aluminum alloys. Current densities ranging from 50 000 to 300 000 A/in.[2] of electrode cross section have been used. In contrast, current densities for GTAW of aluminum alloys and for GMAW of steel are about 10 000 A/in.[2]

The constant-current and current-density ranges in which the transition from globular to spray transfer takes place depend on the electrode size and the arc voltage used. For a $^3/_{64}$-in.-diam electrode and 22 to 31 V, the change in type of transfer occurs at about 120 A or about 70 000 A/in.[2] Increasing the electrode diameter to $^3/_{32}$ in. increases the transition current to about 220 A, but decreases the current density to about 30 000 A/in.[2] When electrode diameters are larger than standard, the current density for the transfer transition is further reduced.

Spray Transfer

The notable characteristics of the spray transfer arc are its stiffness and its narrowness. These advantages are described below.

Arc Stiffness for Deep Penetration. There is no lack of weld penetration when using spray transfer. Even in the low range of welding currents, the use of high current density and small-diameter electrode wire establishes a stable arc column with a well-defined pattern on the workpiece. To ensure fusion at the root of a butt joint, root reinforcement is required, usually $^1/_{32}$ to $^3/_{16}$ in., depending on metal thickness and joint design.

Arc Stiffness for Out-of-Position Welding. When using spray transfer, the transfer follows the direction in which the electrode wire is pointed, which makes this type of transfer suitable for out-of-position welding.

Arc Narrowness for Small Fillet Welds. The spray arc has a narrow stable core that concentrates the heat. This property enables fully fused small fillet welds to be made in relatively thick material.

Arc Narrowness for Square-Groove and Narrow-Groove Butt Joints (High Current Density Welding). The concentrated heat of the spray arc can also be used to weld butt joints with square or narrow grooves, thus reducing the amount and cost of the electrode wire required to make the joint. Techniques have been developed to extend the usable current densities into the high range (to 300 000 A/in.^{2}), to take advantage of the very narrow penetrating arc. At these high current densities, the characteristic hissing noise of the arc is replaced by a crackling noise. These techniques are especially suitable when making square-groove butt joints in base metal from $^1/_4$ to $^5/_8$ in. thick. Welding is often accomplished in two passes, one from each side, at a much greater speed than is possible at lower current densities. Back gouging is rarely required, and welding in one pass instead of several stringer-bead passes greatly reduces the total heat input. The reduction in heat input results in less distortion and, in heat treatable alloys, produces better as-welded properties. Most welds made with the square-butt and high-current techniques are of good quality.

Thicknesses greater than $^5/_8$ in. can also be welded with a square-groove butt joint, but the amount of reinforcement may be excessive. Where reinforcement must be minimized, V-grooves can be machined in both sides of the joint to the amount required. Figure 3 shows how two $1^3/_4$-in.-thick plates of alloy 5083 were joined, using only one pass from each side, with 450-A welding current, 28 V, 100 ft^3/h of argon for shielding, and $^1/_{16}$-in.-diam ER5356 electrode wire.

The degree of bevel required with high current density welding is considerably less than with conventional welding. The root face is quite thick—usually about half the thickness of the plate. Best results are achieved with a constant-current power supply.

Pulsed arc transfer is a type of spray transfer that occurs in pulses at regularly spaced intervals. In the time interval between pulses, the welding current is reduced and no metal transfer occurs. The low average current and low heat input associated with pulsed arc welding have allowed the advantages of spray transfer to be extended to the welding of sections thinner than can be spray transfer welded using conventional constant-current power supplies.

Fig. 3 Edge preparation for high current density welding of aluminum plate. Plate is $1^3/_4$ in. thick, with one pass from each side.

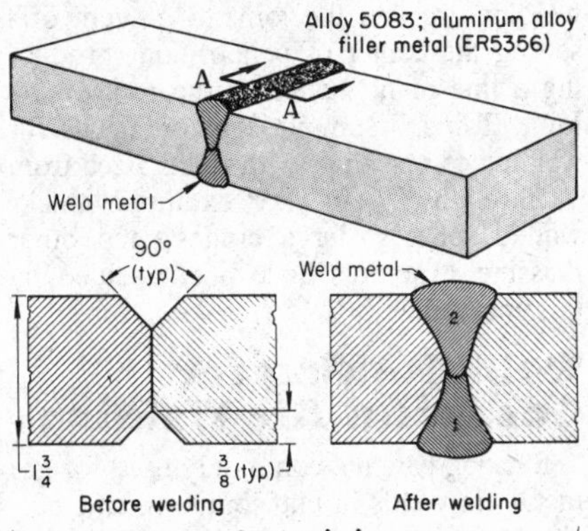

In addition to enabling the spray transfer welding of aluminum 0.030 to 0.125 in. thick, pulsed arc transfer offers other advantages. One is the option of using larger diameter electrode wires, which cost less per pound, are easier to feed, have fewer current-transfer problems in the contact tube, and have a lower probability of weld porosity from surface contamination on the wire because of the lower surface-to-volume ratio of the larger wire. Another advantage is that sheet can be welded to thicker plate, even when the joint has a poor fit. A layer of metal is progressively built up on the thicker section until the gap is bridged.

Well-formed root beads are easily made on thin aluminum with pulsed arc welding, whereas the beads made by short circuiting transfer have high crowns, which consume more filler metal, can cause distortion because of the unbalanced cross section of single-pass welds, and have poor appearance. Use of the short circuiting transfer has been largely replaced by pulsed arc transfer.

Globular Transfer. The type of arc that produces metal transfer by a large drop of molten metal is seldom used when welding aluminum alloys, because the transfer is erratic. Because the arc penetration is shallower and the heat input is lower at the current densities that produce this type of transfer, globular transfer has occasionally been used when welding metal thinner than that normally welded with spray transfer and constant current ($^1/_8$ in. and less).

Electrode Wires for Gas Metal Arc Welding

Classifications and compositions of electrode wires are given in Table 1. Standard wire sizes available on 1- and 5-lb spools are 0.030 to $^1/_{16}$ in. diam, on 10- and $12^1/_2$-lb spools are 0.030 to $^1/_8$ in. diam, and on 30-lb spools are $^1/_{16}$ to $^1/_8$ in. diam. Deposition and wire-feed rates obtained with two common electrodes in the standard sizes are shown in Fig. 4 for various welding conditions.

Electrode-wire feed should be selected so that the wire is consumed as fast as it emerges from the welding torch without extending more than $^3/_8$ in. beyond the shielding gas nozzle. The torch is tilted not more than 10° forehand. The arc length that should be used is governed by the metal thickness, the type of filler-metal alloy,

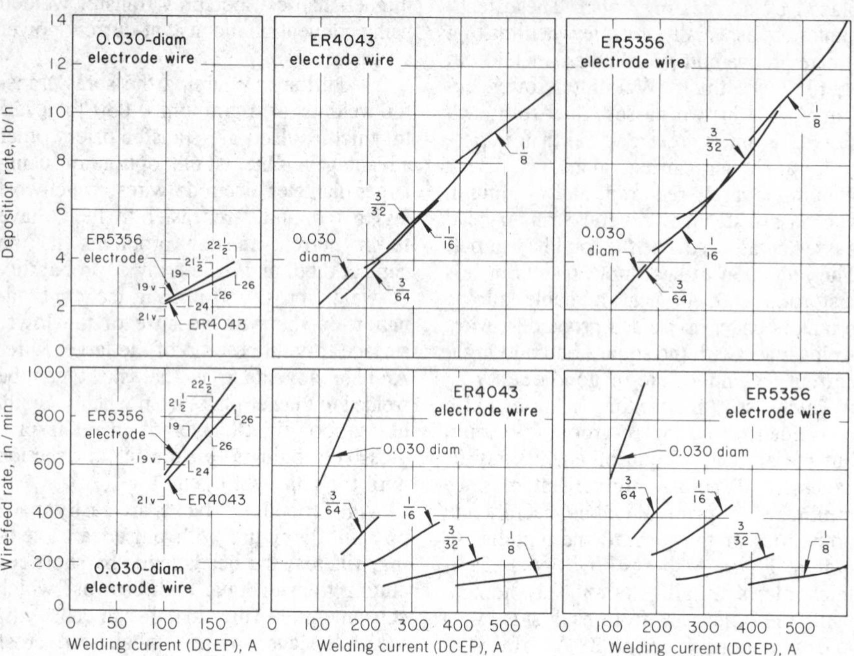

Fig. 4 Typical deposition rates and wire-feed rates for GMAW with ER4043 and ER5356 electrode wire, under argon shielding

and the welding current. When making small fillet welds and welding narrow-groove butt joints, a short arc is preferred. Arc length is usually $1/8$ to $3/8$ in. The wire sizes chosen for each application depend on the requirements and welding conditions for that application.

Welding With Large-Diameter Electrode Wires. Increased use of thick aluminum plate (over 1 in. thick) has created a need for more economical welding procedures. Continued development of the large-wire welding process has resulted in techniques for high-current two-pass welding of aluminum plate up to 3 in. thick using electrodes of $5/32$ through $1/4$ in. diam. Welding power supplies having drooping-type volt/ampere output curves are required for high-current, large-wire welding. They may be either high ampere rating drooping rectifiers or electronically slope-controlled constant-current power sources of sufficient output capacity for the welding current required.

Wire feeders are of the constant-speed type with sufficient output torque to accurately maintain wire feed speed. Special high-capacity GMAW welding barrels are required to weld with the high currents and large wires used. They must be able to withstand the high welding heat input of the process and provide satisfactory shielding over a large molten pool of weld metal. These welding barrels may either have a large-diameter, single gas shielding nozzle, or they may be constructed as a dual flow shielding system of a nozzle within a nozzle. The dual shielding nozzles provide greater flexibility in the selection of inert shielding gas mixtures and flow rates required for welding. The high-current, large-wire welding process is essentially a deep-penetrating welding technique. Therefore, the joint designs use a small bevel on each side of the plate to improve penetration depth and reduce weld bead reinforcement buildup. Approximate welding conditions are given in Table 6.

The high-current, large-wire technique is applicable only to flat-position welding because of the large molten pool characteristic of the process. Welding uphill (4 to 8°) is often helpful in controlling position and flow of molten weld metal. Accurate alignment is required between the welding arc and the joint to prevent offsetting the center of penetration, producing a line of no fusion at the root of the joint. Table 7 summarizes conditions for welding corner joints with fillets sized from $1/2$ through $1\,1/2$ in. For example, thick-walled spheres, large cranes, and other massive structures have been welded by this method.

Weld Backing for Gas Metal Arc Welding

Backing bars are commonly used for gas metal arc welds in butt joints, as this permits welding to be accomplished at higher speed, with less operator skill, and with less control of welding conditions, especially when using spray transfer for joining thin sections. Steel is the material most often used for temporary backing when welding aluminum alloys. Carbon steel is often used, but stainless steel is used when lower thermal conductivity is required in the backing or to avoid iron oxide contamination of the weld. Copper and aluminum may be used when higher thermal conductivity is needed. Backing made of magnetic material sometimes deflects the arc and interferes with welding. When this occurs, nonmagnetic materials such as austenitic stainless steel, copper, and aluminum should be used instead.

Austenitic stainless steel backing bars have reasonable life against arc damage and do not produce arc blow; their use minimizes the possibility of iron or rust pickup in the root bead. When copper backing is used, copper pickup must be prevented. Local deposition of copper or copper-aluminum alloy can result in corrosion in service. The life of copper backing is somewhat less than that of stainless steel, especially under direct arc impingement. Chromium plating of copper backing has been used to reduce copper pickup and increase backing life. Aluminum backing with a hard anodic coating provides adequate chilling; an added advantage is that the arc cannot strike the aluminum backing and cause damage, because the anodic coating is an excellent dielectric.

Backing bars may be temporary, permanent, or integral. Temporary aluminum backing, which is not anodically treated, is removed by chipping after welding. A butt weld need not be completely fused to the temporary aluminum backing, provided that the root pass is back gouged to sound metal after the backing bar has been removed. Temporary backing should be grooved to allow the root surface of the weld to protrude beyond the plane of the back surface of the workpiece, thereby ensuring adequate penetration. This groove should be shallow (0.010 to 0.030 in.) and wider than the width of the root surface of the weld. Too wide a groove provides insufficient support for the metal under the hold-down clamps.

When permanent aluminum backing is used, it is necessary to obtain complete fusion between the backing, the root faces, and the root layer of the weld. This is facilitated by using a greater root opening than is used with temporary backing. Mechanical and magnetic oscillation can be used to help achieve fusion to both root faces in a single pass. Permanent backing

Table 6 Approximate welding procedures for two-bead GMAW aluminum butt joints with a large-diameter electrode

Plate thickness, in.	Electrode diameter, in.	Shielding, ft³/h	Current, A	Voltage, V	Travel speed, in./min	Joint design
1½	3/16	220 (argon)	640-670	27-32	12	70°; 0.4 in.; 0.7 in.; 0.4 in.; 70°
2	3/16	240 (75% helium, 25% argon)	720-760	32-34	9-10	70°; 0.6 in.; 0.8 in.; 0.6 in.; 70°
2¼	3/16	240 (75% helium, 25% argon)	720-760	36-38	8-9	70°; 0.6 in.; 1.05 in.; 0.6 in.; 70°
2½	3/16	250-300 (75% helium, 25% argon)	750-800	36-38	8-9	70°; 0.7 in.; 1.1 in.; 0.7 in.; 70°
3	3/16	250-300 (75% helium, 25% argon)	800-900	37-39	5-7	70°; 0.8 in.; 1.4 in.; 0.8 in.; 70°

Source: Kaiser Aluminum and Chemical Co.

bars are shown with five welds in Fig. 5. Service conditions do not always permit the use of permanent backing, but by eliminating the back gouging required when temporary backing is used, the use of permanent backing (when permitted) can reduce costs. Butt joints in extrusions can be designed so that the weld backing is an integral part of the extrusion. Integral backing is not recommended in environments where the nonwelded portion of the backing can promote crevice-type corrosion.

Multiple-Pass Gas Metal Arc Welding

When welding joints in thin-walled vessels, two passes are sometimes used rather than one to avoid leakage because of porosity. Gases can escape more easily from the smaller weld pools, and the second pass can fill some of the porosity that may exist in the first-pass weld bead. There is little probability that any remaining pore in the first-pass bead can line up with a pore in the second-pass bead to produce a through hole.

For butt welding sections $^3/_8$ in. thick or thicker, welding should be done from both sides when possible. Whether or not the first weld penetrates to the underside, the bead should be back gouged into sound fused weld metal before depositing the backing bead. Back gouging removes entrapped oxide film at the base of the weld bead. Proper depth is attained by gouging to the point where the chip no longer splits along the oxide film entrapped at the original face of the joint. Back gouging can be accomplished by using a pneumatic hammer and knife-edge chisels of proper design, or by machining. Oil should not be used when back gouging, or it should be removed prior to welding. Disk grinders should not be used. Portable back-gouging equipment that uses shaped milling cutters is available. Regardless of method, the back-gouging groove should be of uniform shape, with no torn metal, sharp corners, or crevices from which oil or foreign material cannot be removed.

It is advisable to check that all defects extending to the gouged surface have been removed before depositing the backing bead. This check may be done by liquid-penetrant inspection or by radiographic inspection. Residual oil from penetrant or ultrasonic inspection must be thoroughly removed prior to welding to prevent weld porosity.

The shells of most aluminum vessels are thick enough to require at least two welding passes. In the feed-gas cooler shown in Fig. 6, 45 passes, made in the horizontal welding position, were required to complete the joining of a 3-in.-thick tube sheet to a $1^1/_2$-in.-thick by 18-in.-OD pipe, both made from alloy 5083-O plate. To facilitate welding, the cooler was placed on turning rolls and rotated. Two welders worked simultaneously on the job, one at the 12 o'clock position and the other at the 6 o'clock position. The welding guns were hand held, but were provided with supports to minimize operator fatigue. A constant-voltage power supply, a shielding gas mixture of 75% argon and 25% helium, and ER5183 electrode wire were selected. A $^1/_{16}$-in.-diam wire was used, allowing maneuverability of the welding gun. None of the welded coolers failed the mass spectrometer leak test. Conditions for making multiple-pass welds in butt, corner, and T-joints in $^1/_4$- and $^3/_8$-in.-thick alloy 6061-T6 are given in Fig. 5.

Automatic Gas Metal Arc Welding

When the size of the production run warrants the installation cost and setup time, the use of automatic GMAW equipment offers several advantages, among which are better quality on a more consistent basis and higher welding speed than can be obtained with manual manipulation of the electrode gun. An application where the higher welding speed greatly reduced welding time is described in the following example.

Table 7 Typical corner joints and conditions for GMAW with large-diameter electrode wires(a)

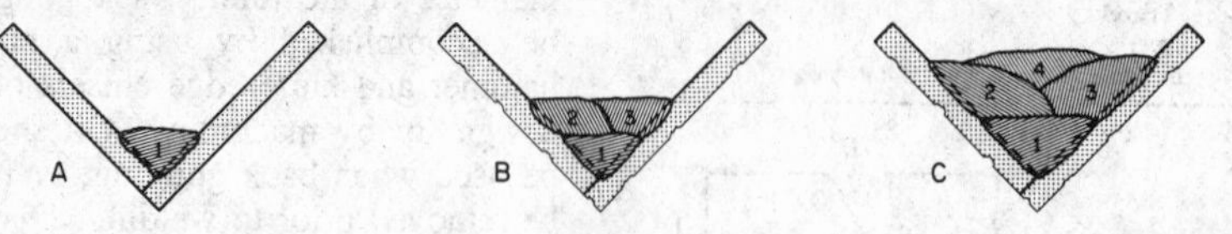

Fillet size, in.	Pass type (see above)	Pass number	Electrode wire diameter, in.	Arc current (DCEP), A	Arc voltage(b), V	Welding speed, in./min
1/2	A	1	5/32	525	22	12
1/2	A	1	3/16	550	25	12
5/8	A	1	5/32	525	22	10
3/4	A	1	5/32	600	25	10
3/4	A	1	3/16	625	27	8
3/4	A	1	7/32	625	22	8
1	B	1	5/32	600	25	12
		2, 3	5/32	555	24	10
1	B	1	3/16	625	27	8
		2, 3	3/16	550	28	12
1	A	1	7/32	675	23	6
1 1/4	B	1	5/32	600	25	10
		2, 3	5/32	600	25	10
1 1/4	B	1	3/16	625	27	8
		2, 3	3/16	600	28	10
1 1/4	B	1	7/32	625	22	8
		2, 3	7/32	625	22	10
1 1/2	C	1	7/32	650	23	6
		2 to 4	7/32	650	23	10

(a) Argon shielding gas, at 100 ft^3/h. (b) Measured from contact tube to test plate

Fig. 5 Conditions of multiple-pass welding by manual GMAW

Alloy 6061-T6; aluminum alloy filler metal (ER5356)

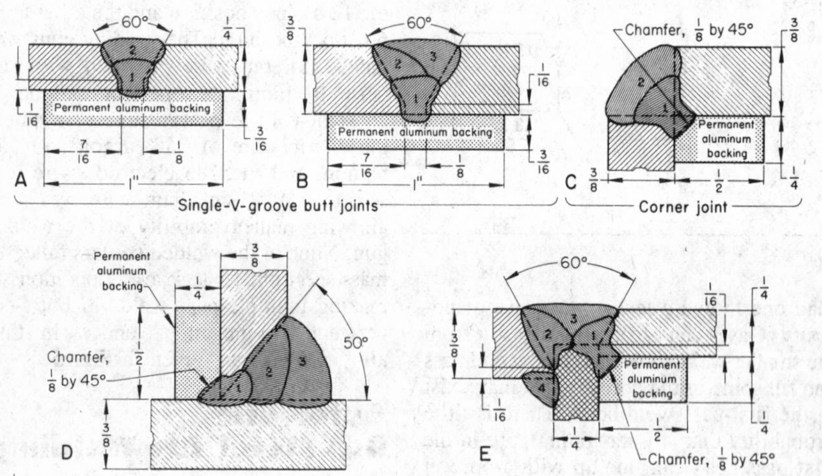

Base metal	Alloy 6061-T6	Welding positions	Flat and horizontal
Electrode wire	3/64-in.-diam ER5356	Precleaning	Chemical
Shielding gas	75% helium, 25% argon; 50 ft^3/h	Interpass cleaning	None

Item	Welding conditions for joints illustrated above: A	B	C	D	E
Joint type	Butt	Butt	Corner	T	T
Current (DCEP), A	180-200	180-200	180-200	190-220	180-210
Voltage, V	22-24	22-24	22-24	22-25	22-24
Welding speed, in./min	25-30	25-30	25-30	20-28	25-30
Number of passes	2	3	2	3	4

Example 1. Use of Automatic Gas Metal Arc Welding of Pipe to Reduce Welding Time. Joining lengths of pipe in the field by automatic GMAW was found to be economical for lengths of 5 miles or more for a 6-in.-diam pipe (Fig. 7). The U-groove joint with ends butted tightly, as shown in Fig. 7, was essential for the process. After cleaning the joint by wiping with solvent, alignment and backing for the root bead were accomplished by an internal tool that could be expanded and collapsed manually by means of an extension handle. Use of the alignment tool made tack welding unnecessary. The alignment tool was withdrawn as soon as the root bead was completed; it was used to align the next joint while welding of the previous joint was continued.

Welding was accomplished with an air-cooled welding torch (with electrode-wire spool attached) mounted on a machine that rotated it around the pipe. In this way, a six-pass weld was finished without stopping. Welding began near the top of the pipe and continued until the six passes were completed (Fig. 7). The welding head was adjusted for bead location and depth, and the direction of rotation was reversed after passes 2 and 5 without extinguishing the arc to ensure adequate penetration and fusion and correct weld-bead shape. Welding speed was fairly high (100 in./min); welding current was high (200 A), and an electrode wire of small diameter (0.035 in.) was used. The estimated production rate, allowing for normal downtime, was 12 welds per hour per machine. The estimated production rate for welding manually was five welds per hour per opera-

Fig. 6 Leak-free joint between tube sheet and pipe in an aluminum feed-gas cooler

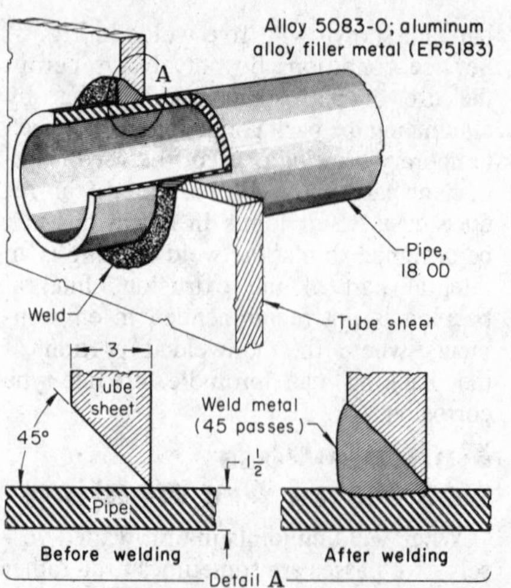

Fig. 7 Large pipe that was automatic gas metal arc welded. Passes 3, 4, and 5 were made in the opposite direction of the other passes.

Alloy 6351-T4; aluminum alloy filler metal (ER5254)

Automatic GMAW

Joint type	Butt
Weld type	Single-U-groove
Root opening	None
Welding position	Horizontal-fixed pipe
Power supply	300-A engine-driven generator
Fixture	Expandable mandrel
Electrode holder	Mechanized, air cooled
Electrode wire	0.035-in.-diam ER5254(a)
Shielding gas	Argon, at 60 ft^3/h
Current	200 A, (DCEP)
Voltage	20-24 V
Welding speed	100 in./min
Arc time per joint	80 s
Number of passes	6
Production rate per machine	12 welds per hour

(a) Chosen because it is compatible with all common aluminum alloys used for pipe and pipe fittings, ER5254 is a former AWS classification that has been replaced by ER5654.

tor. In the example that follows, the reduced welding time and reduced distortion that resulted from the high welding speed obtained by automatic welding were important advantages.

Example 2. Reduction in Welding Time and Distortion by Changing From Manual to Automatic Gas Metal Arc Welding. Originally, the four continuous welds along the full 96-ft length of the alloy 5083-H 112 overhead-crane girder shown in Fig. 8 were made as fillet welds in square-groove joints by manual GMAW. By changing to the single-bevel-groove weld with fillet weld reinforcement shown in section A-A in Fig. 8 and to fully automatic GMAW, welding time was reduced by 80%. The 60° groove provided by beveling the web plates contributed to the reduction in welding time by allowing a 33% decrease in weld metal, at no sacrifice in joint strength. Welding current was increased from 260 to 375 A.

The girders were tack welded at 15-in. centers on the inside to hold them in place for fillet welding. The continuous machine fillet welds were made in the horizontal position. Other benefits of machine welding were a significant reduction in distortion, better joint appearance, and elimination of the defects associated with weld starts and stops. A comparison of manual and automatic welding is given in Fig. 8.

Fig. 8 Overhead-crane girder. Welded by automatic GMAW rather than the manual process to reduce welding time and distortion.

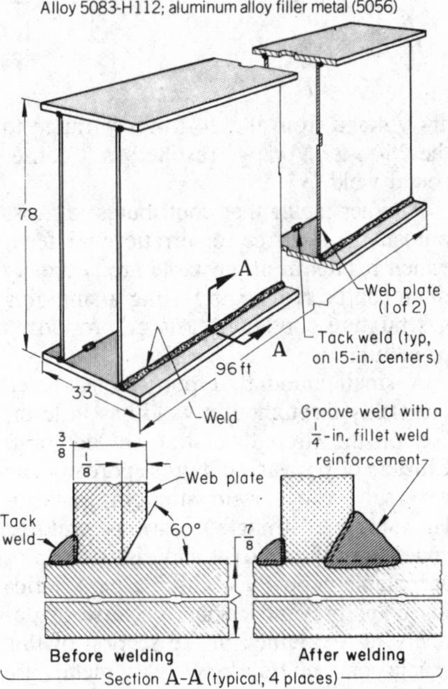

Automatic GMAW

Joint type	T
Weld type	Single-bevel-groove fillet
Root opening	None
Welding position	Horizontal
Power supply	500-A, constant-current transformer-rectifier
Wire-feed system	Constant speed
Electrode holder	Mechanized, water cooled
Electrode wire	1/16-in.-diam alloy 5056(a)
Shielding gas	Argon, at 60 ft^3/h
Current	See below
Voltage	26 V
Wire-feed rate	410 in./min
Welding speed	See below
Number of passes	1

Comparison of manual and automatic GMAW

	Manual	Automatic
Current (DCEP), A	260	375
Welding speed, in./min	12	20
Duty cycle per joint, %	33 1/3	100
Welding time per joint, min	288	58
Shielding-gas consumption per joint, ft^3	64	60
Electrode consumption per joint, lb	12	8

(a) This electrode wire has been replaced by ER5356.

Soundness of Welds Made by Gas Metal Arc Welding

The five principal flaws encountered in GMAW of aluminum alloys are transverse weld-metal cracking, longitudinal weld-metal cracking, crater cracking, porosity, and inadequate penetration. In addition, inclusions and cold laps are occasionally encountered. Transverse and longitudinal weld-metal cracking are usually associated with the higher strength aluminum alloys of the 2*xxx* and 7*xxx* series. In particular, cracks are likely to originate at the starting and stopping points of the arc, because craters caused by normal shrinkage occur in these areas. Therefore, it is good practice to use starting and runoff tabs and not to start or break the arc on the workpiece.

Weld cracking problems can often be attributed to restrained shrinkage of the weld metal. Process variables that produce coarse-grained weld-metal deposits and a large HAZ are contributors. For example, longitudinal weld-metal cracking in a square-groove butt joint in 1/8-in. alloy 5083 sheet was eliminated in one application by sufficiently relieving hold-down clamp pressure to allow slight transverse movement of the workpieces during welding. In addition, weld reinforcement was increased by reducing the welding speed, thereby creating a larger weld cross section that withstood the higher shrinkage stresses without cracking.

In another application, transverse weld-metal cracking in a square-groove butt joint in 3/16-in. alloy 5083 sheet was eliminated by increasing the solidification rate of the weld pool. This was accomplished by reducing the arc voltage from 28 to 24 V and the welding current from 280 to 185 A, and increasing the welding speed from 18-to-25 to 27-to-30 in./min. The lower arc voltage resulted in a narrow, deeply penetrating weld and allowed the welding speed to be increased. In addition, the power supply was changed from the constant-voltage type to the conventional drooping volt-ampere characteristic to stabilize heat input. The electrode diameter remained at 1/16 in. and the number of passes at two.

Hydrogen contamination is the cause of virtually all weld porosity in aluminum alloys. Solidification shrinkage is of minor significance as a contributor to weld porosity. Hydrogen has high solubility in molten aluminum but very low solubility in solid aluminum (a small fraction of the solubility in solid steel and titanium). Hydrogen dissolved in the molten weld pool during welding is released during solidification. The high freezing rate associated with GMAW can prevent the evolved hydrogen from rising to the surface of the weld pool, causing porosity. An extremely small amount of hydrogen source can cause significant amounts of porosity.

The major sources of hydrogen are hydrated oxide, oil, and other hydrocarbon contaminants on the surface of the electrode wire or work being welded. Other sources are moisture, oil, grease, and hydrated oxide on the work-metal surface and moisture in the shielding gas. Factors that can contribute to porosity from contamination are insufficient gas flow, excessive distance from the gas nozzle to the work metal, leakage in the shielding gas hose or fittings, improper fit-up, and erratic electrode-wire feed.

When weld porosity is encountered, the electrode wire should be checked, first to determine whether it is clean, and then to determine whether it is capable of depositing sound metal. In some instances, a radiograph of a weldment made in the overhead position with the wire in question can be obtained from the producer of the wire. Also, a check weld (butt or fillet) can be made, broken open, and examined. If the wire produces sound welds on one machine but not another, the source of the porosity is elsewhere than the wire. Proper wire storage is important. Storage in a cabinet heated to above room temperature prevents moisture condensation. Discarding the outer wrap ensures freedom from contamination when a weld is started. Wire should not be left in machines for extended periods of time.

Next, conditions for GMAW that bring about a slower solidification rate should be selected. Welding in the flat or vertical position also aids escape of gases. Correct voltage and arc length can reduce porosity in a weld, such as the one in the subcooler shown in Fig. 9. This vessel was fabricated from $^{1}/_{2}$-in.-thick alloy 5083 by GMAW in the vertical-pipe position. A constant-voltage power supply and $^{1}/_{16}$-in.-diam ER5183 electrode wire were used with a welding current of 280 to 290 A, a welding speed of 18 to 20 in./min, and 75% argon, 25% helium shielding gas. Eight passes were required. Severe porosity was encountered. After a thorough re-evaluation of the procedure, it was concluded that welding position had little effect on the porosity but that raising the arc voltage from the 24-to-26 V range to the 28-to-29 V range resulted in a dense, sound weld.

Fig. 9 Subcooler shell. Porosity was eliminated by increasing the arc voltage in GMAW.

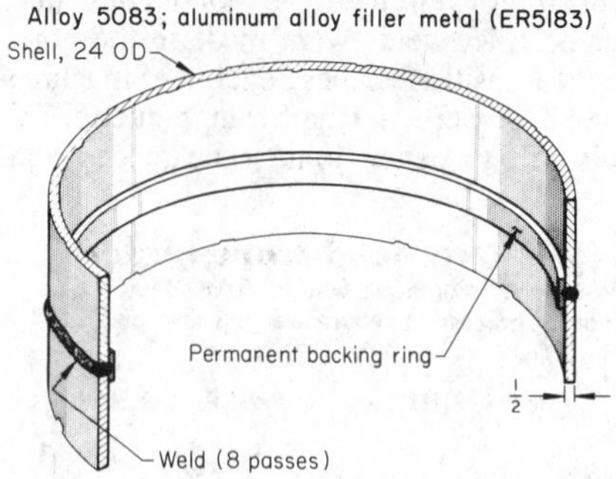

Another factor that contributes to gross porosity is incorrect or erratic wire feed, which results in an unstable arc. Porosity often can be reduced by using a shielding gas mixture containing a large proportion of helium.

A small amount of porosity scattered uniformly throughout a weld has little effect on the strength of the welded joint. Clusters of porosity and gross porosity can adversely affect weld strength, particularly fatigue strength. Various welding codes limit the amount and distribution of acceptable porosity. Multiple-pass welding helps prevent leakage because of porosity, as explained in the section of this article on multiple-pass gas metal arc welding. In the following example, leaktight welds were obtained in joints designed so that complete penetration was obtained; 4 passes were used for one type of joint and about 26 passes for the other type.

Example 3. Producing Leaktight Welds in a Pressure Vessel. The major components of the pressure vessel shown in Fig. 10 were two closed-die-forged domes, a central partition, and an axial conduit—all made of alloy 5254. The domes were joined by a single circumferential weld that also incorporated the partition as backing (Fig. 10, detail B). The conduit was welded to the vessel at both ends. Internal baffles and fittings (not shown in Fig. 10) completed the vessel structure, which later was shrink fitted into the 4340 steel shell of a naval torpedo, the vessel forming part of the fuel system. Joint designs and welds used to join the major components of the pressure vessel are shown in Fig. 10, details A and B.

The centrally located partition and the two dome sections were welded simultaneously (Fig. 10, detail B), in four passes. To ensure weld soundness, joints between the conduit and the $1^{3}/_{8}$-in.-thick bosses were designed for full-penetration single-J-groove welds and were completed in about 26 passes. These joints were economical to make and had a high degree of reproducibility. Radiographically, the welds were of excellent quality.

After edge preparation (machining) and cleaning of joint areas, internal baffles and fittings (not shown in Fig. 10) were installed in the dome sections. These sections, together with the center partition, were then assembled in an alignment fixture, clamped, mounted (axis horizontal) under a fixed machine torch, and welded. The welding area was carefully shielded from drafts. During welding, the workpiece was rotated without stopping, except to change electrode wire, and the welds were wire brushed during rotation. The conduit was then inserted and similarly welded, with the axis of the assembly in a vertical position. After welding, the vessel and welds were machined. Welds were inspected by liquid-penetrant methods, and the vessel was subjected to a mass spectrometer leak rate test.

Mechanical problems, such as meltback of the electrode wire into the contact tube, snagging of the electrode wire either on the wire spool or in the flexible cable joining the spool to the welding torch, and inability of the wire to maintain a constant arc length during welding (shunting), also may occur. These are almost always associated with malfunctions of equipment (especially with wire-feeding equipment), rather than with process or material limitations. These and other difficulties and their usual causes include:

Porosity

- Gas entrapment from poor shielding, shielding gas, air
- Hydrogen from moisture, unclean wire surface, oil on base metal
- Excessive cooling rate of weld pool, erratic electrode-wire feed (see causes under the section on arc length fluctuations below)
- Erratic arc transfer, caused by incorrect current

Transverse weld-metal cracking

- Excessive longitudinal restraint
- Slow solidification of weld pool
- Incorrect combination of base metal and filler metal

Longitudinal weld-metal cracking

- Excessive transverse restraint
- Concave, instead of convex, root-pass weld bead

Fig. 10 Pressure vessel forming part of the fuel system of a naval torpedo

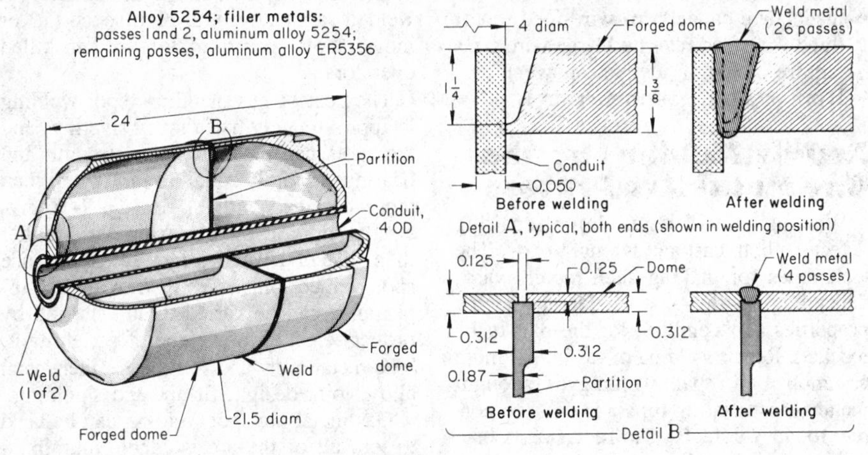

Item	Circumferential joint	Conduit-to-boss joint
Conditions for automatic GMAW		
Joint type	T	Corner
Weld type	See detail B	Single-J-groove
Position	Flat	Flat
Power supply	500-A, constant-voltage transformer-rectifier	500-A, constant-voltage transformer-rectifier
Fixtures	Clamping jig, rotating positioner	Clamping jig, rotating positioner
Electrode holder	Water cooled	Water cooled
Gas-nozzle diameter	3/4 in., all passes	Passes 1-4: 5/8 in.; passes 5-26: 3/4 in.
Electrode wire(a)	Passes 1 and 2: 1/16-in.-diam 5254 Passes 3 and 4: 1/16-in.-diam ER5356	Passes 1 and 2: 1/16-in.-diam 5254 Passes 3 to 26: 1/16-in.-diam ER5356
Shielding gas	75% helium; 25% argon, at 30 ft^3/h	75% helium, 25% argon, at 30 ft^3/h
Current	1st pass: 180-190 A (DCEP) Passes 2 to 4: 190 A (DCEP)	200-210 A (DCEP) (all passes)
Voltage	27 to 27 1/2 V, all passes	27-27 1/2 V, all passes
Welding speed	29 in./min	29 in./min
Number of passes	4	26 (approx)

(a) Electrode wires conformed to AWS 5.10. The 5254 electrode wires (no longer an AWS classification; replaced with ER5654) were selected for corrosion resistance, and the ER5356 electrode wires for better strength.

- Insufficient cross section of root-pass bead
- Excessive current for electrode-wire size

Inadequate penetration

- Insufficient back gouging
- Improper edge preparation for arc characteristics (groove too narrow)
- Insufficient current
- Excessive voltage (arc length)
- Welding speed too high

Incomplete fusion

- Excessive base-metal oxide
- Improper edge preparation for arc characteristics
- Arc too long or improper torch angle
- Dirty workpieces or electrode wire
- Insufficient current

Undercutting

- Excessive arc length
- Excessive current
- Welding speed too low
- Improper torch angle
- Inadequate edge delay for transverse oscillation

Arc-starting difficulty

- Wrong polarity (electrode should be positive)
- Incomplete welding circuit
- Inadequate flow of shielding gas
- Excessive electrode-wire feed rate or insufficient welding current

Arc length fluctuations

- Poor condition of contact tube (rough inner walls, sharp shoulders, contamination by weld spatter)
- Erratic electrode-wire feed, caused by:
  Kinked wire (electrode wire should be evenly wound and free of kinks)
  Excessive or erratic friction in electrode conduit or electrode holder (conduit should be in good condition and of the correct size and length)
  Clogged contact tubes
  Unbalance of wire spool
  Maladjustment of wire-spool brake
  Poor operation of wire-feed motor or wire straightener
  Sharp bends in electrode conduit (suspend equipment overhead)
  Fluctuations in line voltage to wire-feed unit
  Poor ground connection or burned governor control in wire-feed motor
  Slippage or insufficient pressure of drive rolls in wire-feed unit

Meltbacks

- Erratic electrode-wire feed (see causes under the section on arc length fluctuations above)
- Poor condition of contact tube (see the section on arc length fluctuations above)
- Incorrect power-supply or wire-feed rate settings
- Poor functioning of cooling equipment
- Poor contact between voltage pickup lead and work

A meltback occurs when the electrode wire fuses to the copper contact tube and wire feed is stopped. It happens when the wire feed is insufficient for the current being used, which causes the arc to lengthen until it overheats the end of the contact tube. An inexperienced welder may get meltbacks while establishing the correct welding conditions.

Inadequate cleaning action by the arc

- Wrong polarity (electrode should be positive)
- Inadequate gas shielding because of:
  Insufficient gas flow
  Spatter on inside of gas nozzle
  Contact tube off-center in relation to gas nozzle
  Wrong nozzle-to-work distance
  Incorrect torch angle
  Drafty environment (work should be shielded)
- Incorrect torch angle (should be 7 to 15° forehand)

Dirty weld bead

- Dirty workpieces or electrode wire
- Impurities in shielding gas (because of air or water leakage)
- Wrong forehand angle
- Damaged or dirty gas nozzle

- Wrong gas-nozzle size (should be smallest possible)
- Insufficient shielding gas flow
- Drafty environment (work should be shielded)
- Incorrect arc length
- Contact tube recessed too far (should not be more than 1/8 in. inside gas nozzle)

The appearance of small amounts of black smut in welding with aluminum-magnesium alloys is not a fault.

Rough weld bead

- Unstable arc
- Improper arc manipulation
- Improper current
- Welding speed too low

Too narrow weld bead

- Arc length too short
- Insufficient current or voltage
- Welding speed too high

Too wide weld bead

- Excessive current
- Welding speed too low
- Arc too long

Poor visibility of arc and weld pool

- Wrong position of work
- Wrong work angle or forehand angle
- Small, dirty, or wrong lens in helmet (No. 10 or 12 lens should be used)
- Wrong gas-nozzle size (should be smallest size)

The welder must be able to see the arc and weld pool at all times.

Overheating of power supply

- Excessive power demand (two similar welding machines can be used in parallel if the capacity of one is insufficient)
- Poor functioning of cooling fan
- Dirty rectifier stacks (regular maintenance)

High demand can damage the power supply by overheating. Overheating is particularly serious with rectifiers.

Overheating of cables

- Loose or faulty connections
- Cables too small
- Insufficient water supply

Overheating of electrode-wire-feed motor

- Excessive friction between electrode wire and conduit
- Wrong gear ratio of wire-feed unit
- Poorly adjusted wire-spool brake
- Incorrect alignment of gears and rolls in wire-feed unit
- Worn brushes on wire-feed motor
- Inadequate capacity of wire-feed motor (high wire-feed rates and large wire sizes require motor of adequate power)
- Worn or arced governor controls

Repair Welding by the Gas Metal Arc Process

Welding is widely used for repairing defects in both cast and wrought parts. The techniques for making such repairs vary, depending on the type of defect and the properties and condition of the part to be repaired. Repair welding of castings is more common than repair welding of wrought metal. Most casting repair is done to correct foundry defects or to repair parts broken in service. For example, an alloy 355.0-T6 piston, 42 in. in diameter and 31 in. high with a thickness that varied from 1 to 3 in., had a large broken area on the top surface and broken areas on the skirt near the compression ring. For repair, the defective areas were chipped to sound unbroken metal, and all surfaces were washed with a solvent cleaner. The casting was then placed in an insulated box and preheated to 300 °F by wrapping it with a bead-type resistance heater. For the welding operation, a constant-voltage power supply, a shielding gas mixture of 75% argon and 25% helium, and 1/16-in.-diam ER4043 electrode wire were used. Each area of repair was welded without interruption. Adequate weld reinforcement was provided so that the repaired area could be ground flush after cooling. The repairs, five in all, were made with no evidence of cracking, and radiographic inspection showed that they were satisfactory.

Spot Welding by the Gas Metal Arc Process

Spot welding by the gas metal arc process is a quick and reliable method of joining aluminum alloy sheet and extrusions where maximum strength is not required. Normally, only enough force to hold parts in intimate contact is required and only one side of the joint need be accessible. Aluminum alloys can be arc spot welded satisfactorily at high speed (12 or more spots per minute) by unskilled operators.

The cost of gas metal arc spot welding is approximately half that of riveting, and the spots can develop shear strength and bearing strength equal to or better than those obtained with 5/32-in. 2117-T6 rivets. In addition, fatigue strength is higher than that of riveted joints and resistance spot welded joints. Weld strength and appearance can be varied to suit the job by proper selection of degree of penetration, base-metal alloy and thickness, filler-metal alloy, joint design, fit-up, and shielding.

Gas metal arc spot welding can be used to join all of the arc weldable aluminum alloys and some (for example, 7075 and 7079) whose welding by other arc processes is not recommended. A guide to filler-metal selection is given in Table 8. The process is also useful for welding aluminum alloys to steel or copper.

For spot welding, the nozzle assembly on the conventional GMAW torch is changed and circuitry is added to provide automatic control of feed, arc, and gas time. After adjusting the current, wire feed, and other operating conditions in keeping with the thickness of metal and penetration required, the nozzle is brought into contact with the first member and sufficient pressure is applied to hold the two members in contact. The welding process is started and the electrode wire feeds forward, initiating the arc, which melts through the first member into the second member and forms the spot weld. Crater filling is automatically controlled to obtain a shrink-free spot of the correct shape. The weld usually is made in 1 s or less.

Direct current electrode positive is used with argon, helium, or argon-helium gas shielding. A pull-type wire feed with a slow run-in control is recommended to ensure reliable arc starting. A time-lag control between wire-feed shutoff and breaking the welding current prevents the electrode wire

Table 8 Guide to the choice of filler-metal alloys for gas metal arc spot welding(a)

Aluminum base-metal alloys	Preferable filler-metal alloys(b)	Usable filler-metal alloys(b)
1060, 1100, 1350	ER1100, ER4043, ER5556	...
2014, 2024, 2219	ER2319, ER4043, ER4145	...
3003	ER1100, ER4043, ER5556	...
5005, 5050	ER4043, ER5356, ER5556	...
5052, 5083, 5086, 5154, 5456	ER5356, ER5556	ER5554
5454	ER5356, ER5554, ER5556	...
6061, 6063, 6101	ER4043, ER5356, ER5556	ER5554
7075, 7178	ER4043, ER4145	ER5554, ER5556

(a) Based on weld cracking resistance. (b) ER5183 can be used in place of ER5556.

from freezing in the weld pool when the power is cut off.

Constant-voltage power supplies, of either the motor-generator or rectifier type with adjustable slope control, give the best results. They are capable of delivering the high current surge required for arc initiation. Constant-current power supplies are not recommended because aluminum electrode wire is likely to ball up on the end when the arc is extinguished.

Gas metal arc spot welding necessitates the minimum amount of preweld cleaning. Removal of normal oxides and mill finishes is not necessary unless the ultimate in weld quality is demanded, but anodic and chemical-dip oxide finishes, lacquers, and other insulating coatings must be removed. Solvent wiping is recommended for the removal of grease and lubricating compounds. Alcohol and toluol are the preferred cleaning agents. Chlorinated solvents such as carbon tetrachloride and trichlorethylene should be used only with thorough ventilation and in an area remote from the welding area, because they decompose and yield toxic fumes when in the vicinity of an electric arc. Governmental health and safety rules relating to the use of solvents must be followed.

Helium shielding is preferred when welding thin sheet (less than 0.040 in. thick), because its use results in a less sharply pointed weld cone than the use of argon shielding, thus permitting a larger fused area to be made at the faying surfaces. Drawbacks to the use of helium include its limited availability compared with argon, its greater cost, and its higher level of spatter. Helium also results in a rougher weld surface. Low gas-flow rates—approximately 20 ft^3/h—are possible because the spot welding nozzle provides an almost complete enclosure. Gas hoses should be purged with shielding gas before making the first weld in a series. A 5-s purge is usually sufficient, but a longer purge may be needed at the beginning of the day's work. Once the gas nozzle is purged, successive welds can be made as quickly as the welding torch can be repositioned.

Aluminum in thicknesses from 0.020 to $^1/_4$ in. can be joined by gas metal arc spot welding. When the first member (usually, the top member) is $^1/_8$ in. thick or less, no special preparation is needed, but if the first member is $^1/_8$ to $^1/_4$ in. thick, a pilot hole is required. The second member should preferably be at least as thick as the first member, and ideally two to three times as thick. Minimum thickness for the second member is generally 0.030 in. using solid backing; otherwise, it is 0.050 in. Use of chilled or massive backing allows complete penetration through the second member, to achieve maximum nugget diameter and strength.

Spot welds look like slightly crowned buttons on the surface closest to the electrode holder. In cross section, they should be cone-shaped and penetrate through, or almost through, the second member. For sheets of nearly equal thickness, weld strength is highest when the weld penetrates through the second member. When made against a flat, clean metal backing plate, the spots are flush and uniform, leaving a mark resembling a resistance spot weld. For consistent results, clamping may be necessary to ensure uniform contact at the faying surfaces and between the work and backing. Partial-penetration spot welds show no mark at all on the back surface, and stiffeners and brackets can be welded to the back of architectural panels and other products where appearance is of prime importance, without any marring of the opposite surface. Typical welding conditions for gas metal arc spot welding of aluminum alloys with $^3/_{64}$-in.-diam ER5554 electrode are covered in Table 9.

Table 9 Typical conditions for gas metal arc spot welding of various thicknesses of aluminum alloy sheet(a)

Sheet thickness, in.		Partial-penetration welds			Complete-penetration welds		
Top	Bottom	Open circuit voltage, V	Wire feed(b), in./min	Welding time, s	Open circuit voltage, V	Wire feed, in./min	Welding time, s
0.020	0.020	...	...	...	27	250	0.3
0.020	0.030	...	...	...	28	300	0.3
0.030	0.030	25.5	285	0.3	28	330	0.3
0.030	0.050	25.5	330	0.3	31	430	0.3
0.030	0.064	30	360	0.3	31	450	0.3
0.050	0.050	31	385	0.4	32	450	0.4
0.050	0.064	32	400	0.4	32	500	0.4
0.064	0.064	32	420	0.4	32	550	0.5
0.064	0.125	32.5	650	0.5	34.5	675	0.5
0.064	0.187	35	700	0.5	39	700	0.5
0.064	0.250	39	775	0.5	41	800	0.5
0.125	0.125	39.5	800	0.5	41	850	0.6
0.125	0.187	41	850	0.75	41	900	0.75
0.125	0.250	41	900	1.0	...	...	...

(a) Overlap joints; electrode, 0.047-in.-diam ER5554. (b) Welding current (DCEP), in amperes, is approximately equal to one-half wire feed, in inches per minute, of 0.047-in.-diam electrode wire.

Fig. 11 Relative diameter of fused zones for GMAW spot welds at sheet interface for three combinatons of sheet thicknesses

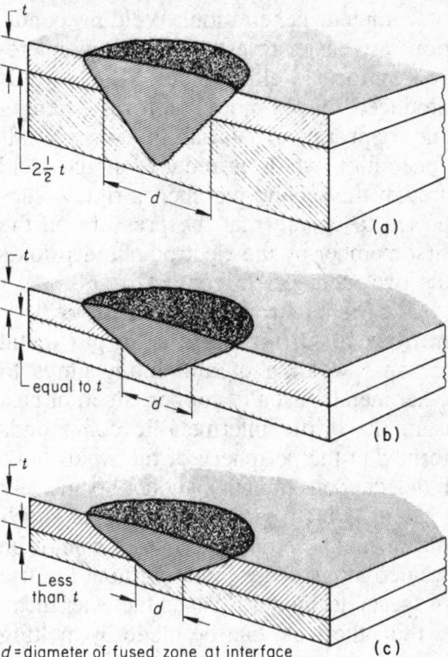

The strength of a GMAW spot weld is a function of the alloy and the area of the nugget at the interface. Welds in alloys of the 5*xxx* series have the highest strength. The higher base-metal strength of the heat treatable alloys of series 2*xxx*, 6*xxx*, and 7*xxx* is not reflected in the as-spot-welded strength. The combination of alloy 5456 sheet and ER5556 filler metal produces the highest tensile-shear strength per spot weld at a given thickness.

Maximum strength is obtained when the bottom sheet is at least twice the thickness of the top sheet, because the interface is nearer to the face of the weld nugget and therefore larger (Fig. 11a). When the bottom sheet is about $2^1/_2$ times thicker than the top sheet, that area of the weld nugget does not vary greatly with small variations in penetration. If the two sheets are of equal thickness (Fig. 11b), or if the top sheet is thicker (Fig. 11c), and if complete penetration cannot be tolerated, the same percentage variation in depth of penetration produces much greater variation in the weld-nugget interface area. If the surface of the bottom sheet is completely penetrated, less variation results in the interface area.

Thicknesses being joined, and strength and appearance requirements, should be considered in determining whether control should be set to produce welds with partial or complete penetration. Welding conditions are easier to establish for complete-penetration welds, and the welds produced have more uniform strength and appearance. Variability in partial-penetration welds is reduced if the weld area of the second member is rigidly supported to ensure that the pressure on the first member by the electrode holder holds the two members in contact.

Gas Metal Arc Spot Welding of Aluminum to Other Metals. In gas metal arc spot welding of aluminum alloys to other metals, usually copper, steel, or cast iron, the brittle intermetallic compounds formed at the periphery of the welds have little effect on strength and ductility in shear loading, which permits fusion spot welding of metal combinations that cannot be welded along seam joints without the use of special techniques. When the other metal is thin, the weld can be made by melting through it with the arc. The in-rushing aluminum alloy filler metal and the force of the arc push the other metal away from the center of the weld so that the core and crown are composed of relatively ductile aluminum. The maximum thickness of the other metal that can be handled in this way is about 0.030 in. A pilot hole (about $^1/_4$ in. in diameter) through the nonaluminum member improves performance by providing a path for the filler metal, so that it remains undiluted and ductile. Pilot holes are essential when the other metal is more than 0.030 in. thick.

When the filler metal is suitable for fusion welding the other metal, joints can be made in which the bottom member is the other metal. However, these filler metals usually have higher melting points than aluminum. A superior joint is obtained if the aluminum is sandwiched between two pieces of the other metal, as in the case of copper-aluminum-copper three-layer joints used in making permanent connections in electrical applications. Two or more pieces of aluminized steel can be joined by gas metal arc spot welding, using an aluminum alloy filler metal and a backing strip of aluminum beneath the bottom steel sheet. In this joint, only the aluminum is exposed to the environment.

Gas Tungsten Arc Welding

The advantages of welding aluminum alloys by GTAW are the same as for other metals, as discussed in the article "Gas Tungsten Arc Welding" in this Volume. Welding can be done with or without filler-metal additions, depending on the aluminum alloys to be alclad and on the joint configuration.

Thicknesses of aluminum alloys commonly welded by GTAW range from 0.040 to $^3/_8$ in. for manual welding and from 0.010 to 1 in. for automatic welding. Gas tungsten arc welding is especially suitable for automatic welding of thin workpieces that require the utmost in quality of finish, because of the precise heat control possible and the ability to weld with or without filler metal. Metal thicker than $^3/_8$ in. can be manually welded; however, either GMAW or automatic GTAW is preferred.

Power Supply and Equipment. For joining aluminum alloys, GTAW uses either alternating or direct current. Both negative and positive electrode polarities are used with direct current welding. The same power supplies, arc-stabilization accessories, torches, and control systems that are used for other metals are used for the GTAW process on aluminum. Single-phase alternating current welding transformers should have an open-circuit voltage of 80 to 100 V. Equipment is available with programming, pre- and postflow of shielding gas, and pulsing. Power supplies may include square wave output and a choice of each polarity that are ideal for aluminum.

The oxide layer on the surface of aluminum alloys gives rise to some arc rectification during the positive electrode half of the alternating current cycle. This arc rectification, either partial or complete, is undesirable, because it results in poor arc stability and possible overheating of the transformer; it can be overcome by the use of condensers in the welding circuit or by placing a battery bias in series with the welding circuit. Battery power of 6 to 8 V is usually sufficient to balance the current. The positive terminal should be in the direction of the electrode. About 100 A/h of storage-battery capacity should be used with every 100 A of welding current. If two or more batteries are needed to obtain the required ampere-hour electrical capacity, they should be connected in parallel.

Electrodes for Gas Tungsten Arc Welding

For alternating current GTAW, unalloyed tungsten and tungsten zirconia electrodes are recommended. Zirconiated electrodes are less likely to be contaminated by aluminum and have a slightly higher current rating. Unalloyed tungsten electrodes minimize inclusions in the weld bead and current imbalance. Electrodes of 1% thoriated tungsten, which have higher current capacity than unalloyed tungsten and cost less than zirconiated electrodes, are also used; their main disadvantage for alternating current welding is a slight tendency to drip. This drip results in some tungsten inclusions in the weld metal. With a skilled operator and good equipment, the inclusions are small and well dispersed. For best results with 1% thoriated electrodes, the electrode should have a ground surface. Electrodes of 2% thoriated tungsten are not generally used for alternating current welding of aluminum alloys.

When welding aluminum by GTAW with alternating current, the tip of the electrode should be hemispherical. The tip is prepared by using an electrode one size larger than required for the welding current, taper grinding the tip, and forming the hemispherical end by welding for a few seconds with a current 20 A higher than needed, holding the electrode vertically up.

Thoriated tungsten electrodes are normally used for direct current GTAW of aluminum. The tip of the electrode should be ground to a blunt conical point, having an included angle between 60 and 120° to attain maximum penetration.

When an electrode becomes contaminated with aluminum, it must be replaced or cleaned. Minor contamination can be burned off by increasing the current while holding the arc on a piece of scrap metal. Severe contamination can be removed by grinding or by breaking off the contaminated portion of the electrode and re-forming the correct electrode contour on a piece of scrap aluminum.

Shielding Gases for Gas Tungsten Arc Welding

Argon, helium, and mixtures of argon and helium are used as shielding gases in GTAW of aluminum alloys. The selection of shielding gas is somewhat dependent on the type of current used. Welds made with alternating current show little difference in soundness or strength whether made with argon or helium shielding.

Helium. With helium shielding, penetration is deeper, but a higher flow rate is required; hence, helium is sometimes used with higher speeds and for thicker sections.

Argon is used as a shielding gas more often than helium, because it:

- Is more readily available and costs less than helium
- Affords better control of the weld pool

• Gives a smoother, quieter arc, greater arc cleaning action, and easier arc starting
• Requires less gas for specific applications
• Has better cross-draft resistance than helium
• Causes less clouding, and the metal stays brighter; the operator can thus see the weld pool more easily

With argon shielding, the arc voltage is lower for a given current value and arc length. The lower arc-voltage characteristic of argon is essential to the successful manual alternating current welding of extremely thin material, because it decreases the probability of melt-through. This same characteristic is advantageous in vertical and overhead welding, because the molten metal is less likely to sag or run. In special applications, where a balance of characteristics is desired, a mixture of argon and helium is used with alternating current welding.

Argon is preferred for direct current electrode positive welding because it establishes an arc more easily and provides better arc control. Helium or an argon-helium mixture is always used with direct current electrode negative welding of aluminum. It assists in providing the deep, narrow penetration essential for the best properties and a minimum HAZ. Helium shielding also prevents development of the rippled surface typical of argon shielding and DCEN welding.

When aluminum is welded with alternating current or DCEP, a white band of varying width appears alongside the weld bead. This white band, disclosed by one analysis to be aluminum oxide, never occurs when welding is done with DCEN, and it is believed to be caused by the emission of electrons from the surface of the aluminum when it is on the cathode side of the arc (DCEP). The electrons leave the aluminum through the aluminum oxide, thus serving to detach the oxide from the surface.

When the white band alongside the weld is of hairline width immediately adjacent to the weld bead itself, the flow of shielding gas is adequate for proper shielding of the tungsten arc. Any increase in gas flow causes widening of the band, indicating that gas is being wasted.

Filler Metals for Gas Tungsten Arc Welding

Gas tungsten arc welding can be done with or without filler metal. Sometimes the joint design permits the base metal to provide the weld metal. In some square-groove butt joints, the weld metal comes from the straight sides of the groove, or extra metal may be provided on a corner or flange that is melted to form the weld. If restraint is likely to cause cracking, best results are achieved by the addition of separate filler metal.

Usually, filler metal is added in the form of a bare rod for manual welding or as a coil of wire for automatic feeding. The filler-metal alloys used for gas shielded arc

Table 10 Welding procedure schedules for alternating current GTAW of aluminum

Material thickness or fillet size, in.	Type of weld fillet or groove	Tungsten electrode diameter, in.	Filler rod diameter, in.	Nozzle size (inside diameter), in.	Shielding gas flow, ft^3/h	Welding current (ac), A	No. of passes	Travel speed (per pass), in./min
3/64	Square groove and fillet	1/16	1/16	1/4-3/8	20	40-60	1	14-18
1/16	Square groove and fillet	3/32	3/32	5/16-3/8	20	70-90	1	8-12
3/32	Square groove and fillet	3/32	3/32	5/16-3/8	20	95-115	1	10-12
1/8	Square groove and fillet	1/8	1/8	3/8	20	120-140	1	9-12
3/16	Fillet	5/32	5/32	7/16-1/2	25	160-200	1	9-12
3/16	V-groove	5/32	5/32	7/16-1/2	25	160-180	2	10-12
1/4	Fillet	3/16	3/16	7/16-1/2	30	230-250	1	8-11
1/4	V-groove	3/16	3/16	7/16-1/2	30	200-220	2	8-11
3/8	V-groove	3/16	3/16	1/2	35	250-310	2-3	9-11
1/2	V- or U-groove	1/4	1/4	5/8	35	400-470	3-4	6

Notes: (1) Increase amperage when backup is used. (2) Data are for all welding positions. Use low side of range for out-of-position welding. (3) For tungsten electrodes: 1st choice—pure tungsten EWP; 2nd choice—zirconiated EWZr. (4) Normally, argon is used for shielding; however, mixtures of 10% or more helium with argon are sometimes used for increased penetration in aluminum 1/4 in. thick or more. The gas flow should be increased when helium is added. A mixture of 75% helium plus 25% argon is popular. When 100% helium is used, gas flow rates are about twice those used for argon.
Source: Cary, H.B., *Modern Welding Technology*, Prentice-Hall, New York, 1979, p. 339

Table 11 Typical conditions for fixed-position pipe welding by manual alternating current GTAW, using joint designs shown in Fig. 12

Pipe: Nominal pipe size, in.	Pipe: Wall thickness, in.	Thickness of backing ring (t in Fig. 12), in.	Electrode diameter, in.	Filler-metal diameter, in.	Gas nozzle diameter, in.	Argon flow, ft^3/h	Current (ac), A	No. of passes(a)
Horizontal-fixed position								
1	0.133	0.072	1/8	3/32	1/2	30-80	90-110	1-2
1 1/4	0.140	0.072	1/8	1/8	1/2	30-80	100-120	1-2
1 1/2	0.145	0.072	1/8	1/8	1/2	30-80	110-130	1-2
2	0.154	0.093	1/8	1/8	1/2	30-80	120-140	1-2
2 1/2	0.203	0.093	1/8	1/8	1/2	30-80	130-150	2
3	0.216	0.093	1/8	1/8	1/2	30-80	145-165	2
3 1/2	0.226	0.093	1/8	1/8	1/2	30-80	150-170	2
4	0.237	0.125	3/16	1/8-3/16	1/2	35-80	160-180	2
5	0.258	0.125	3/16	5/32-3/16	1/2	35-80	180-190	2
6	0.280	0.187	3/16	5/32-3/16	1/2	50-80	195-205	2
8	0.322	0.187	3/16	5/32-3/16	1/2	50-80	210-220	2-3
10	0.365	0.187	3/16	5/32-3/16	1/2	50-80	230-240	2-3
12	0.406	0.187	3/16	5/32-3/16	1/2	50-80	245-255	2-3
Vertical-fixed position								
1	0.133	0.072	1/8	3/32	7/16	25-50	95-115	1-2
1 1/4	0.140	0.072	1/8	1/8	7/16	25-50	105-125	1-2
1 1/2	0.143	0.072	1/8	1/8	7/16	25-50	115-135	1-2
2	0.154	0.093	1/8	1/8	7/16	30-60	125-145	2-3
2 1/2	0.203	0.093	1/8	1/8	7/16	30-60	135-155	3-5
3	0.216	0.093	1/8	1/8	1/2	40-60	150-170	3-5
3 1/2	0.226	0.093	1/8	1/8	1/2	40-60	155-175	3-5
4	0.237	0.125	3/16	1/8-5/32	1/2	40-60	165-185	3-5
5	0.258	0.125	3/16	5/32-3/16	1/2	50-60	185-205	3-5
6	0.280	0.187	3/16	5/32-3/16	1/2	50-60	200-220	3-5
8	0.322	0.187	3/16	5/32-3/16	1/2	60-80	215-235	5-8
10	0.365	0.187	3/16	5/32-3/16	1/2	60-80	235-255	5-8
12	0.406	0.187	3/16	5/32-3/16	1/2	70-80	250-270	6-8

(a) For horizontal-fixed position, more passes are required for the bottom quadrant of the joint.

welding of aluminum are discussed earlier in this article and are listed in Table 1. Suitability of the various filler metals for use with different aluminum alloys, and for different properties, is given in Tables 2, 3, and 4.

Filler-metal rod and wire comes in a wide range of sizes. Straight lengths of rod (36 in. long), packaged in 5- and 10-lb containers of filler metal, are available in diameters of 1/16, 3/32, 1/8, 5/32, 3/16, and 1/4 in. from regular stock, and rod in diameters of 0.030, 0.035, 0.040, and 3/64 in. is available on special order. Wire is available in 1-, 5-, and 12 1/2-lb spools in sizes from 0.030 through 1/16 in. diam and also in 3/32 and 1/8 in. diam in 12 1/2-lb spools.

Weld Backing in Gas Tungsten Arc Welding

Backing bars are commonly used when butt welds are made from one side only, but they are usually not necessary in automatic DCEN welding of square-groove butt joints from one side only. Permanent and temporary backing bars are shown with recommended butt-joint designs for GTAW of aluminum in Fig. 1. For additional discussion of backing bars, see the section on weld backing for gas metal arc welding in this article.

Alternating Current Gas Tungsten Arc Welding

Usually, alternating current welding provides the optimum combination of current-carrying capacity, arc controllability, and arc cleaning action for the welding of aluminum alloys. A short arc length must be maintained to obtain sufficient penetration and to prevent undercutting, excessive width of weld bead, and consequent loss of control of penetration and weld contour. Arc length should be about equal to the diameter of the tungsten electrode. On fillet welds, a short arc and adequate current are needed to prevent bridging the root. A short arc also ensures that the inert gas completely surrounds the weld as it forms.

Manual Welding. Typical welding schedules for manual alternating current GTAW are given in Table 10. Metal that is more than 1/8 in. thick is grooved to ensure complete penetration. A single-V-groove with an included angle of 60 to 90° is used most often. Under certain conditions, a double-V-groove may be advantageous for metal more than 1/4 in. thick. If the operator has difficulty in maintaining a very short arc, a larger included angle or joint spacing may be required to allow a longer arc to be used.

The arc should not be started by touching the tungsten electrode to the aluminum workpiece, because this may mark the work or result in aluminum being picked up on the electrode, which is likely to cause a wild, uncontrollable arc and a dirty weld. The initial arc should be struck on a starting block to heat the electrode to its operating temperature. The arc should never be struck on a piece of carbon, because this contaminates the electrode. The arc is then broken and reignited at the joint. This technique reduces the likelihood of forming tungsten inclusions at the start of the weld, which can happen when a cold electrode is used to start the weld. The arc is struck like striking a match by swinging the electrode holder in a pendulum-like motion toward the starting place. With superimposed high-frequency current for arc stabilization, there is no need to touch the work with the electrode because the arc starts when the electrode tip is brought close to the work surface. Some machines apply the high frequency only when starting; others have it on continuously, or have a switch that permits the operator to cut it on or off at will. The arc is held at the starting point until the metal liquefies and a weld pool is established. Establishment

Table 12 Typical conditions for pipe welding by manual alternating current GTAW, using joint designs shown in Fig. 13

Pipe: Nominal pipe size, in.	Pipe: Wall thickness, in.	Root-face width (*w* in Fig. 13), in.	Electrode diameter, in.	Filler-metal diameter, in.	Gas-nozzle diameter, in.	Argon flow, ft^3/h	Current (ac), A	No. of passes
Horizontal-fixed position								
1	0.133	1/16	1/8	3/32	1/2	30-80	90	3-4
1 1/4	0.140	1/16	1/8	1/8	1/2	30-80	100	3-4
1 1/2	0.145	1/16	1/8	1/8	1/2	30-80	110	3-4
2	0.154	1/16	1/8	1/8	1/2	30-80	120	3-4
2 1/2	0.203	1/16	1/8	1/8	1/2	30-80	130	3-4
3	0.216	3/32	1/8	1/8	1/2	30-80	145	3-4
3 1/2	0.226	3/32	1/8	1/8	1/2	30-80	150	3-4
4	0.237	3/32	3/16	1/8-5/32	1/2	35-80	160	3-4
5	0.258	3/32	3/16	5/32-3/16	1/2	35-80	180	3-4
6	0.280	3/32	3/16	5/32-3/16	1/2	50-80	195	3-4
8	0.322	3/32	3/16	5/32-3/16	1/2	50-80	210	3-4
10	0.365	3/32	3/16	5/32-3/16	1/2	50-80	230	3-4
12	0.406	3/32	3/16	5/32-3/16	1/2	50-80	245	3-4
Vertical-fixed position								
1	0.133	1/16	1/8	3/32	1/2	25-50	90	3-4
1 1/4	0.140	1/16	1/8	1/8	1/2	25-50	100	3-4
1 1/2	0.145	1/16	1/8	1/8	1/2	25-50	110	3-4
2	0.154	1/16	1/8	1/8	1/2	30-60	120	4-5
2 1/2	0.203	1/16	1/8	1/8	1/2	30-60	130	4-5
3	0.216	3/32	1/8	1/8	1/2	40-60	145	4-5
3 1/2	0.226	3/32	1/8	1/8	1/2	40-60	150	4-5
4	0.237	3/32	3/16	1/8	1/2	40-60	160	4-5
5	0.258	3/32	3/16	5/32-3/16	1/2	50-60	180	4-5
6	0.280	3/32	3/16	5/32-3/16	1/2	50-60	195	5-6
8	0.322	3/32	3/16	5/32-3/16	1/2	60-80	210	5-6
10	0.365	3/32	3/16	5/32-3/16	1/2	60-80	230	5-6
12	0.406	3/32	3/16	5/32-3/16	1/2	70-80	240	5-6
Horizontal-rolled position								
1	0.133	1/16	1/8	3/32	7/16	25-40	90	1-2
1 1/4	0.140	1/16	1/8	1/8	7/16	25-40	100	1-2
1 1/2	0.145	1/16	1/8	1/8	7/16	25-40	110	1-2
2	0.154	1/16	1/8	1/8	7/16	25-40	120	3-4
2 1/2	0.203	1/16	1/8	1/8	7/16	30-40	130	3-4
3	0.216	3/32	1/8	1/8	1/2	30-40	145	3-4
3 1/2	0.226	3/32	1/8	1/8	1/2	30-40	150	3-4
4	0.237	3/32	3/16	1/8	1/2	30-40	160	3-4
5	0.258	3/32	3/16	5/32	1/2	30-40	180	3-4
6	0.280	3/32	3/16	5/32-3/16	1/2	35-40	195	3-5
8	0.322	3/32	3/16	5/32-3/16	1/2	35-40	210	3-5
10	0.365	3/32	3/16	5/32-3/16	1/2	35-40	230	3-5
12	0.406	3/32	3/16	5/32-3/16	1/2	35-40	245	3-5

and maintenance of a suitable weld pool is important, and welding must not proceed ahead of the pool. A separate foot-operated heat control, available with certain power supplies, is highly advantageous in preventing uneven penetration by permittting the current to be adjusted as the work becomes hotter. Breaking the arc also requires special care to prevent the formation of shrinkage cracks in the weld crater. Several techniques are used. The arc can be quickly broken and restruck several times while adding filler metal to the crater; or a foot control can be used to reduce current at the end of the weld; or travel speed can be accelerated to "tail out" the weld before breaking the arc. Crater-filling devices may be used if properly adjusted and timed.

By using these techniques and adequate fixturing, weld joints can be made manually in aluminum alloys down to 0.040 in. thick, without excessive distortion. Tack welding before final welding is helpful in controlling distortion. Tack welds should be of ample size and strength and preferably should be chipped out or tapered at the ends before welding over them.

Fig. 12 Edge preparation for fixed-position pipe welding using the conditions given in Table 11. Root opening (*s*) is zero with no backing ring or with a removable backing ring. With an integral backing ring, root opening is 1/4 in. maximum. For thickness of backing ring (*t*) for fixed-position welding, see Table 11.

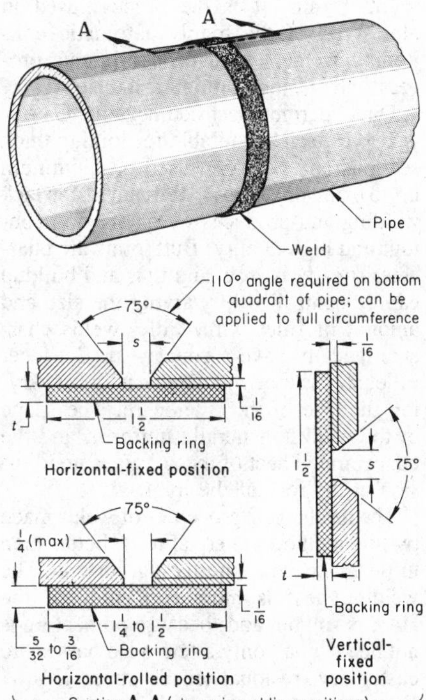

Fig. 13 Edge preparation for pipe welding using the conditions and root-face widths (*w*) given in Table 12

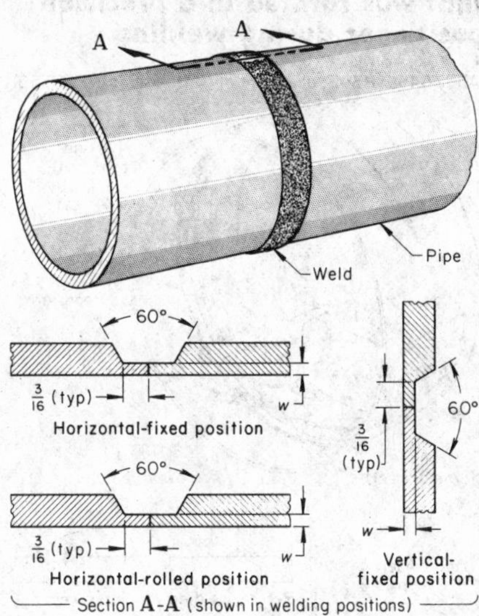

Because GTAW with alternating current affords close control of the weld pool in all positions, it is often used for welding pipe joints. Typical welding conditions are given in Tables 11 and 12, and joint edge preparations are shown in Fig. 12 and 13. Manual alternating current GTAW is an economical process for joining pipe. The equipment required is less expensive than that for GMAW, and it needs less maintenance, a desirable feature when field welding.

Preheating parts thicker than about 3/16 in. may be advantageous when using GTAW with alternating current. The preheating temperature depends on the thickness of the base metal and the particular problems associated with the job. Often 200 °F is a sufficient preheat temperature for material up to 1/2 in. thick, as in the following example, in which the use of preheat and careful selection of electrode material helped in obtaining the necessary mechanical properties.

Example 4. Use of Preheat in Welding 1/2-in.-Thick Alloy 5083 Plate. Figure 14 shows the joint designs and the buildup sequences used in manual GTAW of 3/16-in.-thick sheet and 1/2-in.-thick plate, both of alloy 5083. In welding the 1/2-in. plate, preheating to 200 °F was helpful in producing a weld pool fluid enough for welding in the horizontal position; the 3/16-in. sheet could be welded without preheat. Both the sheet and the plate were welded with EWZr electrodes, which resulted in clean weld metal, with minimum tungsten contamination.

The edges to be welded were grooved by machining. All grease, oil, and oxide were removed by solvent cleaning and brushing with a stainless steel wire brush before welding. The welding current and method of depositing the weld metal were selected to avoid undercutting.

Automatic Welding. The gas tungsten arc process can be mechanized either with or without the addition of filler metal. If filler metal is used, it is normally added mechanically by cold wire-feed units. Automatic GTAW usually employs a shorter arc length and higher welding current and welding speed, and results in deeper penetration than manual welding. Typical conditions for automatic alternating current GTAW for butt joints and metal thicknesses from 1/16 to 1/2 in. are:

Metal thickness, in.	Electrode diameter(a), in.	Argon flow, ft^3/h	Welding current(b), A	No. of passes
1/16	1/16	15	60-80	1
1/8	3/32	20	125-145	1
3/16	1/8	20	190-220	1
1/4	3/16	25	260-300	2
3/8	1/4	30	330-380	2
1/2	1/4	30	400-450	4

(a) Unalloyed tungsten electrodes are recommended. (b) Power supply: alternating current with superimposed high-frequency current, balanced wave. Current ranges are averages for butt welds with square edges. For metal 1/4 in. thick or more, edges are usually beveled.

An application in which automatic alternating current GTAW was used to join relatively thin aluminum was the fillet welding of a 0.087-in.-thick alloy 5456-H343 extruded T-ring stiffener to the inside surface of a 0.071-in.-thick alloy 5456-H343 sheet cylinder, as shown in Fig. 15. The fillet welds were made using an oscillating torch, which was mounted on a boom that reached inside the cylinder. Welding was done in the horizontally rolled position, with the cylinder rotated by a precision positioner. The torch had arc-voltage control. A 300-A balanced-wave power supply and 0.030-in.-diam ER5556 filler-metal wire were used.

The assembly was part of a fuel tank. Manual welding of the stiffeners had been tried, but it resulted in excessive distortion, defects requiring repair, and excessive production time. Automatic welding reduced the number of out-of-tolerance parts by 90%, weld defects by 30%, and welding time per tank by 200 man-hours.

As the thickness of the metal to be welded increases, power requirements increase. When the power requirement exceeds the capacity of the power supply, maximum welding speed cannot be attained and one of the advantages of mechanization is not realized. For single-pass square-groove butt welding, the advantage of higher welding speed is lost when the thickness of the material reaches $^3/_{16}$ to $^1/_4$ in. For example, when welding the circumferential seam of a manifold sump in 0.190-in.-thick alloy 5052-H32, the current was raised from 210 A, which is normally used for automatic GTAW of this thickness of aluminum, to 310 A (the upper limit of the power supply) and the welding speed was raised to only 8 in./min, about the same as for manual welding. The weld joint and the conditions for welding are shown in Fig. 16.

For butt welds in material thicker than $^3/_{16}$ to $^1/_4$ in., the edges of the joint are usually beveled to a single or double-V-groove for multiple-pass welding. An alternative process is GTAW with direct current electrode negative, which can make

Fig. 14 Welds in $^3/_{16}$-in. sheet and $^1/_2$-in. plate, made in the horizontal position by GTAW

Alloy 5083; aluminum alloy filler metal (ER5356)

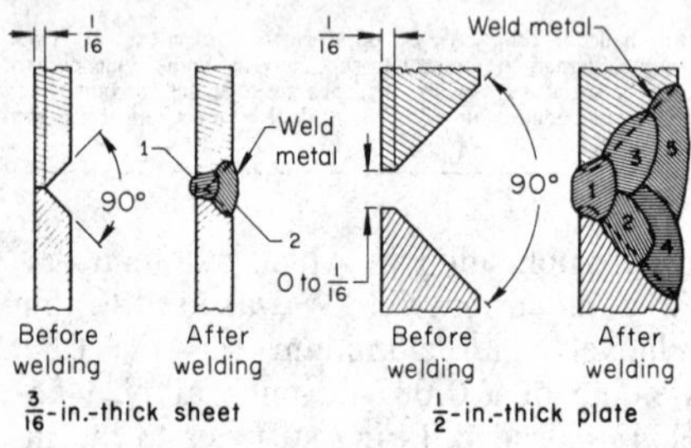

Manual ac GTAW

Joint type	Butt
Weld type	Single-V-groove
Welding position	Horizontal
Power supply	Stabilized
Weld backing	None
Torch	Manual, water cooled
Electrode:	
$^3/_{16}$-in. sheet	$^1/_8$-in.-diam EWZr
$^1/_2$-in. plate	$^1/_4$-in.-diam EWZr
Filler metal:	
$^3/_{16}$-in. sheet	$^5/_{32}$-in.-diam ER5356
$^1/_2$-in. plate	$^3/_{16}$-in.-diam ER5356
Shielding gas	Argon, at 35 ft³/h
Current:	
$^3/_{16}$-in. sheet	160 A, ac
$^1/_2$-in. plate	300 A, ac
Preheat:	
$^3/_{16}$-in. sheet	None
$^1/_2$-in. plate	200 °F
Interpass temperature	200-400 °F
Number of passes:	
$^3/_{16}$-in. sheet	2
$^1/_2$-in. plate	5

Fig. 15 Automatic alternating current GTAW, with an oscillating torch, of an extruded stiffener to a fuel tank cylinder that was rotated in a precision positioner during welding

Alloy 5456-H343; aluminum alloy filler metal (ER5556)

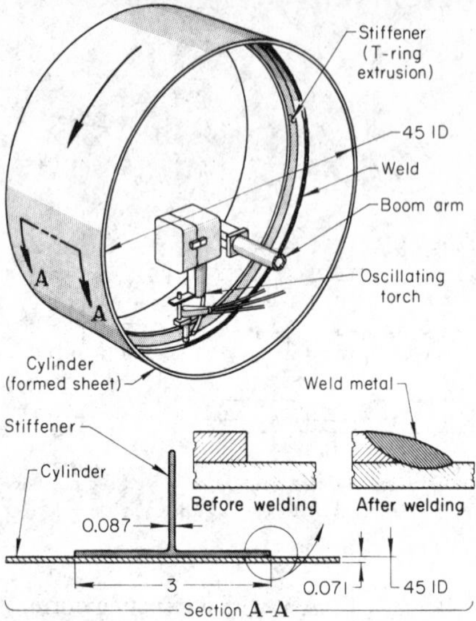

square-groove welds in materials up to $^3/_4$ in. thick without difficulty.

Direct Current Electrode Positive Gas Tungsten Arc Welding

Although GTAW of aluminum with DCEP is seldom used, this process offers certain advantages in the joining or repairing of thin-walled heat exchangers, tubing, and similar assemblies with sections up to about $^3/_{32}$ in. thick. The process is characterized by shallow penetration, ease of arc control, and good arc cleaning action. However, DCEP causes most of the heat of the arc to be generated at the electrode, which necessitates the use of large-diameter electrodes and decreases arc efficiency. If practical electrode sizes are to be used, work must be thin. For example, a $^1/_4$-in.-diam electrode is needed to carry a 125-A current. This would weld aluminum up to about $^1/_8$ in. thick.

Direct current electrode positive welding is useful for small shops because it can be used with almost any general-purpose power supply. Thoriated tungsten electrodes are normally used, and argon shielding is preferred, because it facilitates arc starting and arc control. Typical conditions for manual GTAW of square-groove butt joints in metal up to 0.050 in. thick with DCEP are:

Metal thickness, in.	Electrode diameter(a), in.	Filler-metal diameter(b), in.	Argon flow rate, ft³/h	Current (DCEP)(c), A
0.020	$^1/_8$-$^5/_{32}$	0.030	15-20	40-55
0.030	$^3/_{16}$	0.030 or $^3/_{64}$	15-20	50-65
0.040	$^3/_{16}$	$^3/_{64}$	25-30	60-80
0.050	$^3/_{16}$	$^3/_{64}$ or $^1/_{16}$	25-30	70-90

(a) Thoriated tungsten electrodes. (b) Single-pass welds made in the flat position. (c) Higher currents with larger electrodes may be used for automatic welding.

Direct Current Electrode Negative Gas Tungsten Arc Welding

With direct current electrode negative, the heat is generated at the workpiece surface, producing deep penetration and permitting higher welding currents for a given electrode size than can be used with DCEP. As a result, smaller electrodes can be used with a given welding current, which helps to keep the weld bead narrow. Because of the narrow and deep penetration obtained, less edge preparation and less filler metal are needed, and welding is faster than when using the DCEP process. Because of the high heat generated on the workpiece surface, melting is rapid, no preheat is required, even of thick sections, and little distortion of the base metal occurs. The process has been used for years on as-received material to make irrigation tubing in high-speed tube mills, at speeds up to 50 ft/min. It is the process used in almost all of the highly automatic coil-joining welders used on continuous process lines in the aluminum industry.

Direct current electrode negative GTAW is also especially suitable for joining thick sections and has been used on aluminum up to 1 in. thick. It is well suited for tack welding and produces welds of good contour and high quality. Butt joints are characteristically narrow and flat, and buildup can be controlled by varying the size and amount of filler wire. Fillet welds characteristically have a concave or flat face. Fillet size can be regulated easily by varying the size of the filler wire. The shape of the weld is generally uniform, and the concentrated heat of the arc gives good fusion at the root of the joint.

The mechanical properties of welds made by this method are equal to or better than those made with alternating current. The welding heat is more concentrated, the HAZ is smaller and, because preheating is not needed and only a few weld passes are customary, residual tensile stress is low. These are important advantages when

Fig. 16 Square-groove butt weld that represents practical thickness limit for automatic alternating current GTAW in one pass

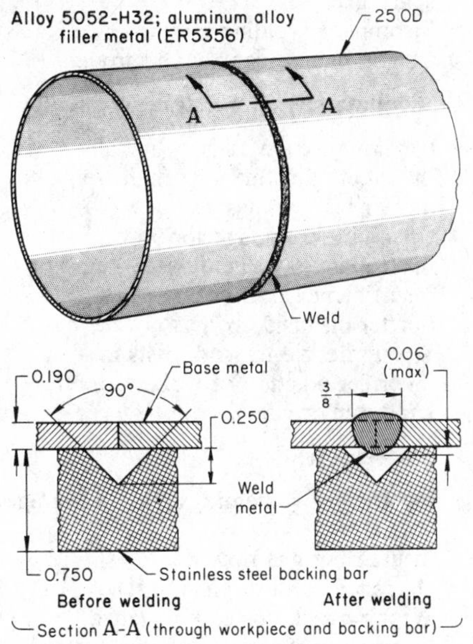

Automatic ac GTAW

Joint type	Butt
Weld type	Square groove
Root opening	None
Welding position	Horizontal-rolled pipe
Power supply	300 A, balanced wave
Fixtures	Clamps, stainless steel backing bar(a), rotating positioner
Torch	Water cooled
Electrode	$^3/_{16}$-in.-diam EWP
Filler metal	0.045-in.-diam ER5356
Shielding gas	Argon, at 20 ft^3/h
Current	310 A, ac
Voltage	11-12 V
Filler-metal feed	88 in./min
Welding speed	8 in./min
Number of passes	1

(a) The segmented backing bar (25 in. in diameter, with a 90° V-groove 0.250 in. deep) was hydraulically expanded to align the joint edges against an external clamping fixture.

joining heavy sections of alloys such as the 5*xxx* series.

Because surface oxides on aluminum are not removed during DCEN welding, thorough preweld cleaning is necessary to ensure that oxide is not trapped in the molten weld pool. Normal practice in the aerospace industry is to clean chemically and to scrape or file the joint area. However, many commercially acceptable welds have been made with no preweld oxide removal treatment.

The surfaces of the welds are not as bright as those made with alternating current because they are coated with a thicker oxide film. The film does not indicate any lack of fusion or the presence of porosity or inclusions and is easily removed by a light wire brushing. Because of the highly penetrating nature of the DCEN arc, melting occurs the instant the arc is struck. Care should be taken to prevent undesirable marking of the workpiece. Although a high-frequency sparking current is not required to stabilize the arc during direct current welding, it is useful for starting the arc without marking the workpiece or causing tungsten inclusions. A starting tab of scrap aluminum can be used for touch starting if a high-frequency circuit is not available. Because continuous high-frequency and wave-balancing circuits are not used in direct current welding, performance can be duplicated easily by using standardized arc voltages and amperages, even on different machines. Normally, thoriated tungsten electrodes are used, with helium shielding.

Manual Welding. Typical conditions for welding by the manual DCEN gas tungsten arc process are given in Tables 13 and 14. Metal thicker than $^1/_4$ in. is grooved (single-V or single-U) to ensure complete joint penetration. A double-V or double-U-groove may be advantageous for metal more than $^1/_2$ in. thick.

The welder must use care in maintaining a suitably short arc length. In addition to the standard techniques (runoff tabs and striking plates) for preventing and filling craters, foot-operated heat controls are used. These controls are also advantageous for adjusting the current as the workpiece heats up and as section thickness changes. The arc is moved steadily forward and filler metal is fed into the leading edge of the weld pool and melted by the arc. Bead size can be controlled by varying filler-metal size.

Automatic Welding. Direct current electrode negative GTAW is readily adaptable to mechanization, which is desirable in order to maintain the required short arc lengths. Mechanization is useful in preventing crater cracking by eliminating all but one start and stop. For best results, a fully mechanized automatic setup should be used. With a fully automatic setup, welds are made in aluminum alloys from 0.010 in. thick to more than 1 in. thick. Because this process allows precise control of weld penetration, it is often selected for joining aluminum in the thickness range from 0.010 to $^1/_8$ in.

Typical conditions for making square-groove butt welds with automatic direct current electrode negative GTAW in aluminum up to $^3/_8$ in. thick are given in Table 15. Automatic welding has also been used to weld square-groove butt joints in aluminum up to 1 in. thick, but V-groove and U-groove edge preparations are often used on thick sections. Mechanical and magnetic oscillation can be used to spread the filler metal and to aid fusion when groove welding with this process.

Because square-groove butt welding using the automatic DCEN gas tungsten arc process results in narrow weld beads, low dilution with filler metal, and excellent weld strength, this process is used for joining rather thick sections of high-strength aluminum alloys. Conditions for welding three such alloys (5083, 2219, and 7039) are given in Table 16. Although the welding of some of the high-strength heat treatable alloys, such as 7075 and 7079, is not recommended, they can be gas tungsten arc welded with DCEN by skilled welders using special techniques. A welding technique that ensures retention of the high strength of the base metal, prevents cracking of the weld metal, and allows development of maximum weld-metal strength is imperative.

Problems in Gas Tungsten Arc Welding of Aluminum Alloys

Many of the problems that are encountered in GMAW of aluminum alloys are

Table 13 Typical conditions for manual GTAW of butt joints, using DCEN, thoriated tungsten electrodes, and helium shielding

Metal thickness, in.	Groove design	Electrode diameter, in.	Filler-metal diameter, in.	Helium flow rate, ft^3/h	Current (DCEN), A	Voltage, V	Welding speed, in./min	No. of passes
0.030	Square	0.040	$^3/_{64}$	20	20	21	17	1
0.040	Square	0.040	$^1/_{16}$	20	26	20	16	1
0.060	Square	0.040	$^1/_{16}$	20	44	20	20	1
0.90	Square	$^1/_{16}$	$^3/_{32}$	30	80	17	11	1
0.125	Square	$^1/_{16}$	$^1/_8$	20	118	15	16	1
0.250	Square	$^1/_8$	$^5/_{32}$	30	250	14	7	1
0.500	90° single-V, $^1/_4$-in. root face	$^1/_8$	$^5/_{32}$	40	310	14	5$^1/_2$	2
0.750	90° double-V, $^3/_{16}$-in. root face	$^1/_8$	$^5/_{32}$	50	300	17	4	2
1.000	90° double-V	$^1/_8$	$^1/_4$	50	360	19	1$^1/_2$	5

Table 14 Typical conditions for manual GTAW of T- and lap joints, using DCEN, thoriated tungsten electrodes, and helium shielding

Metal thickness, in.	Welding position	Fillet size, in.	Electrode diameter, in.	Filler-metal diameter, in.	Helium flow rate (1/2-in. nozzle), ft³/h	Current (DCEN), A	Voltage, V	Welding speed, in./min
0.090	Horizontal	1/8	3/32	3/32	40	130	14	21
0.125	Horizontal	1/8	3/32	3/32	40	180	14	18
0.250	Horizontal	3/16	1/8	5/32	40	255	14	15
	Vertical	3/16	1/8	5/32	40	230	14	10
0.375	Horizontal	5/16	1/8	1/4	50	290	14	7
	Horizontal	3/16	1/8	5/32	50	335	14	14
0.500	Horizontal	5/16	1/8	1/4	50	315	16	7
	Vertical	5/16	1/8	1/4	50	315	16	6

Table 15 Typical conditions for automatic GTAW of square-groove butt joints, using DCEN(a)

Metal thickness, in.	Electrode diameter, in.(b)	Filler metal: Diameter, in.	Filler metal: Feed, in./min	Helium flow rate, ft³/h	Arc: Current (DCEN), A	Arc: Voltage, V	Welding speed, in./min
0.025	3/64	3/64	60	60	100	10	60
0.031	3/64	3/64	76	60	110	10	60
0.040	3/64	3/64	68	60	125	10	60
0.051	3/64	3/64	64	60	150	12	60
0.062	3/64	3/64	99	60	145	13	60
0.080	3/64	3/64	100	60	290	10	60
0.125	1/16	1/16	55	30	240	11	40
0.250	1/16	1/16	40	30	350	11	15
0.375	1/16	1/16	30	40	430	11	8

(a) Single-pass welds made in the flat position. (b) Thoriated tungsten electrodes

Table 16 Typical conditions for automatic GTAW of square-groove butt joints, using DCEN(a)

Metal thickness, in.	Welding position(b)	Filler-metal (1/16-in. diam) feed in./min	Helium flow rate, ft³/h	Arc: Current (DCEN), A	Arc: Voltage, V	Welding: Speed per pass, in./min	Welding: No. of passes
Alloy 5083							
1/4	F	30	100	250	11	25	2, 1 each side
1/4	V	None	50	260	10	20	2, 1 each side
3/8	F, V	None	80	300	12	14	2, 1 each side
3/8	F	12	100	360	10	10	2, 1 each side
1/2	F, V	None	100	400	10	15	2, 1 each side
1/2	F	12	100	390	10	8	2, 1 each side
3/4	F, V	None	100	500	9	5	2, 1 each side
Alloy 2219							
1/4	F, V	36	100	145	12	8	2, 1 side
1/4	H	36	100	135	12	10	2, 1 side
3/8	F, V	32	120	220	12	8	2, 1 side
3/8	H	32	120	180	12	10	2, 1 side
1/2	H, V	10	100	250	12	8	2, 1 each side
5/8	H, V	5-7	120	300	12	7	2, 1 each side
3/4	H, V	5-7	125	340	12	6	2, 1 each side
7/8	H, V	4-6	125	385	12	5	2, 1 each side
1	H, V	3-5	120	425	12	4	2, 1 each side
Alloy 7039							
1/4	F, V, H	None	100	265	10	18	2, 1 each side
1/4	F	40	120	250	14	20	2, 1 each side
3/8	F, V	None	50	300	10	12	2, 1 each side
1/2	F, V	None	100	390	10	15	2, 1 each side
3/4	F, V	None	100	450	9	6	2, 1 each side
3/4	F	48	100	390	10.5	4	2, 1 each side

(a) Thoriated tungsten electrodes: 1/8-in.-diam electrodes with 0.100-in.-diam tip for metal 1/4 to 3/4 in. thick; 5/32-in.-diam tip for metal 7/8 in. thick; 3/16-in.-diam electrodes with 0.156-in.-diam tip for metal 1 in. thick. (b) F = flat; H = horizontal; V = vertical

also present in GTAW. Common problems and their causes are:

Arc-starting difficulty

- Incorrect adjustment of high-frequency spark gap
- Incomplete welding circuit
- Contaminated tungsten electrode

Inadequate cleaning action by the arc

- Excessive oxide on base metal
- Incorrect adjustment of high-frequency unit or battery bias
- Open-circuit voltage too low
- Inadequate gas shielding caused by:
  Insufficient gas flow
  Spatter on inside of gas nozzle
  Wrong nozzle-to-work distance
  Incorrect position of welding torch
  Drafty environment

Dirty weld bead

- Insufficient shielding gas coverage caused by:
  Insufficient gas flow
  Damaged or dirty gas nozzle
  Wrong nozzle-to-work distance
  Incorrect position of welding torch
  Wrong nozzle size (use smallest possible)
  Tungsten electrode not centered in gas nozzle
  Drafty environment
- Impurities in shielding gas because of air or water leakage
- Poor cleaning action by the arc (see causes listed under the section on inadequate cleaning action by the arc)
- Unstable arc
- Electrode contamination
- Dirty workpieces or filler metal

Electrode contamination by aluminum

- Improper filler addition angle or position
- Improper manipulation of torch and filler
- Excessive electrode extension
- Touching of electrode to workpiece

Incorrect electrode contour

- Incorrect electrode size for current
- Incorrect contouring of electrode end before welding
- Wrong electrode material (use pure or zirconiated tungsten electrode with alternating current)

Weld-bead contamination by electrode

- Electrode size too small for current
- Improper manipulation of torch
- Aluminum-contaminated electrode
- Wrong electrode material (use pure or zirconiated tungsten electrode with alternating current)

- Insufficient post flow of shielding gas (oxidized tungsten)
- Touching of electrode with filler or to the workpiece

Rough weld bead

- Excessive filler size or nonuniform additions
- Unstable arc
- Improper manipulation of torch
- Incorrect current

Weld bead too wide

- Excessive current
- Welding speed too low
- Arc too long
- Electrode extension too short
- Incorrect position of welding torch

Inadequate penetration

- Wrong edge preparation for the arc characteristics (groove too narrow or shallow)
- Excessive filler metal in weld pool
- Insufficient current
- Arc too long
- Welding speed too high
- Inadequate back gouging

Difficulty in adding filler metal

- Improper feeding angle or position
- Improper manipulation of welding torch or filler metal, or both
- Unstable arc
- Excessive time or too low background current with current pulsation

Poor visibility of arc and weld pool

- Wrong position of work
- Incorrect position of welding torch
- Small or dirty helmet lens
- Wrong size of gas nozzle (use smallest possible)

Overheating of power supply

- Excessive power demand (two similar welding machines can be operated in parallel if the capacity of one is insufficient)
- Poor functioning of cooling fan
- Poor grounding of high-frequency unit
- Poor functioning of bypass capacitor
- Poor functioning of battery bias
- Dirty rectifier stacks (regular maintenance required)

Overheating of welding torch, leads, and cables

- Loose or faulty connections
- Welding torch, leads, or cables too small
- Inadequate cooling-water flow

Gas Metal Arc Welding Versus Gas Tungsten Arc Welding

The uses of GMAW and GTAW overlap to some extent. The following is a summary of the general merits of the two processes.

Advantages of Gas Metal Arc Welding

Most of the advantages of GMAW over GTAW stem from the fact that in GMAW direct current electrode positive is used at a high current density. This is possible because the electrode is consumable and is melted during the welding, whereas in GTAW, the current is limited by the melting temperature of the electrode. Heat transfer by the gas metal arc is very efficient.

High Welding Rate. Welding speeds two to three times those obtainable by manual GTAW are possible by GMAW, particularly when welding metal more than $^3/_8$ in. thick. When machine welding thinner metal, the welding speed is about the same for the two processes. In manual GTAW, the length of welding is somewhat limited by the length of filler-metal rod that the welder can conveniently handle; a weld longer than 12 in. usually cannot be made without breaking the arc.

In GMAW and automatic GTAW, the filler metal is added mechanically, and the operator is usually able to weld at least 24 in. without having to break the arc and change position. With these processes, less time is lost in starting and stopping, which results in fewer weld craters and more feet of weld per hour. They are also less fatiguing than manual GTAW because the operator does not have to coordinate the movement of both hands and can continue welding for a greater length of time, thus facilitating higher welding rates.

Lower Welding Cost on Metal More Than $^1/_4$ in. Thick. The equipment used for GMAW is more expensive than that used for GTAW, but on thick sections, requiring multiple-pass welding (generally more than $^1/_4$ in. thick), the higher welding rates of the gas metal arc process generally result in lower welding cost.

Low Distortion. Because of the high welding speed, which results in rapid chilling of the weld area, distortion using GMAW is generally low. The distortion produced on aluminum due to heat input with GMAW is not more, and is usually less, than that produced on steel of the same thickness when it is welded with flux cored electrodes. As soon as the arc is established, filler metal is added to the joint, which aids in preventing distortion when sheet is welded to thicker framing members.

Good Weld Quality. The quality of welds produced by GMAW using spray transfer is very high.

Good Out-of-Position Welding. Because the appreciable arc force projects the weld metal across the arc at a high velocity and is not affected by gravity, welding can be done in any position.

High Deposition Rate. Where a high rate of filler-metal deposition is required, as in welding heavy sections or in building up a surface, GMAW has a considerable advantage. High rates of metal deposition are easy to obtain with the large-diameter filler wires (up to $^7/_{32}$ in. diam) when they are used with high welding currents.

Readily Adapted to Machine Welding. Because of its semiautomatic nature, GMAW can be readily adapted to automatic welding for metal from 0.030 in. thick to the thickest commercially available. Automatic GTAW requires good control of joint fit-up, usually within 0.003 to 0.010 in., depending on material thickness, but GMAW is less sensitive to variations in fit-up.

Freedom From Radio Interference. Gas metal arc welding uses direct current, and so it is not necessary to use high-frequency current for arc stabilization. Consequently, there is no radio interference as there may be when using high-frequency current with GTAW.

Advantages of Gas Tungsten Arc Welding

Most of the advantages of GTAW stem from the fact that the filler metal is introduced separately into the arc. This allows the welding current and wire speed to be independently adjusted.

Lower Welding Costs on Metal Less Than $^1/_8$ in. Thick. In automatic welding, where attainable welding speed is nearly equal to that used in GMAW, greater economy is usually realized with GTAW, because welding current and wire speed can be independently adjusted to reduce the consumption of filler metal to the minimum. However, in semiautomatic welding, greater economy is generally realized with GMAW because welding speed is two to three times faster than that obtainable with GTAW. The equipment used for GTAW is less expensive than that used for GMAW and needs less maintenance.

Very Thin Material. Using a pulsed arc, GMAW can be used to weld metal as thin as 0.040 in., but metal as thin as 0.010 in. can be welded by GTAW, provided the workpieces are correctly aligned and held.

Fig. 17 Comparison of three arc welding processes for welding butt joints in $^1/_4$-in.-thick aluminum alloy plate

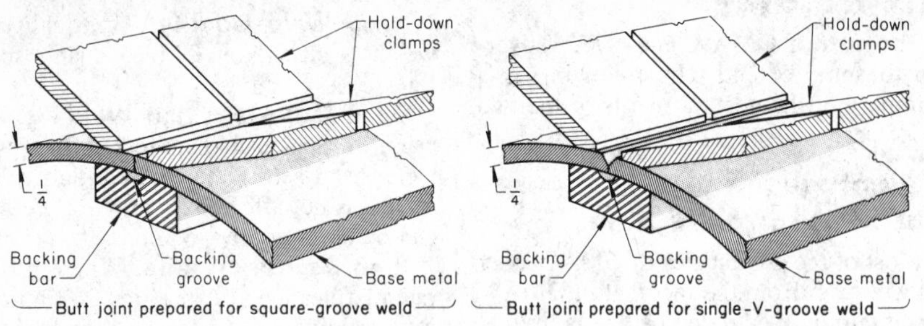

Welding process	Edge preparation (weld type)	Shielding gas	No. of passes	Typical welding speed, in./min	Suitability of process
GTAW (ac)	Single-V-groove	Argon	2	4-8	(a)
GTAW (DCEN)	Square groove	Helium	1	12-20	(b)
GMAW (DCEP)	Square or single-V-groove	Argon(c)	1 or 2	10-50	(d)

(a) Suitable for nuclear and aerospace applications, but too slow. (b) Suitable for nuclear and aerospace applications. (c) Helium additions up to 80% improve radiographic quality. (d) Suitable for commercial applications—both code (pressure vessel, piping) and noncode applications

Making Butt Welds in Small Shapes. Because filler metal is added manually to the weld in manual GTAW, the welder has complete control of the weld pool at all times. This is a definite advantage in butt welding small and medium angles and other shapes in which the heat requirement at the toe of the angle is less than at the heel.

Welds Made Without Filler Metal. This is important in fusing together the edges of a butt joint or the edge of a lap joint to make smooth welds that require little grinding or cleaning.

Excellent Weld Quality. The quality of welds made by GTAW is very high, and the process offers excellent reliability. When welding thick material by GTAW, the filler metal need not be added to the weld until the base metal has been well penetrated; with GMAW, filler metal is added as soon as the arc is struck, which sometimes prevents penetration and causes cold starts. Preheating thick material to 200 to 250 °F ensures satisfactory starts when gas metal arc welding; otherwise, cold starts must be chipped out and rewelded.

Versatility. Gas tungsten arc welds are easily made on edge and corner joints in materal thicknesses impractical to weld by GMAW. Gas tungsten arc welding is also more applicable to fixed-position welding of small pipe because it is slow enough to permit maneuvering around small components.

Selection of Process

Availability of equipment may determine the process to be used for a given application, but generally selection is based on the capability of a process to meet joint requirements and cost. When the thickness of aluminum sheet reaches about $^3/_{16}$ in., the use of automatic GTAW with alternating current for single-pass butt welding becomes prohibitively slow, although joint quality is excellent. The same joints in this thickness and above can be single-pass welded at a much higher speed using GTAW with DCEN, because of the narrower and deeper penetration obtained, but more careful edge cleaning is needed to ensure high joint quality. Both processes are capable of producing welds to high quality standards, and joint thickness becomes a consideration when deciding which to use.

Aluminum sheet $^3/_{16}$ in. or more in thickness can also be welded by GMAW using spray transfer and at much higher speeds. Using DCEP, edge cleaning is less critical than for GTAW. Joint quality is good and easily meets commercial requirements. In Fig. 17, these three processes are roughly compared for mechanized welding of butt joints in $^1/_4$-in.-thick aluminum alloy plate, using hold-down clamps and a grooved backing bar. Figure 17 shows the two most commonly used types of edge preparation for these joints—square grooves and single-V-grooves.

The use of GTAW with alternating current, as indicated in Fig. 17, necessitates edge preparation in the form of a single-V-groove to obtain complete penetration. The welding speed of this process is relatively low, and the effective welding speed for the joint is only half the actual welding speed, because two passes are needed to fill the joint. Sound weld metal is produced.

Gas tungsten arc welding with DCEN can be done at a relatively high welding speed, and deep penetration is obtained. Although the square-groove edge preparation shown in Fig. 17 does not entail beveling to obtain optimum weld-metal soundness, joint edges must be cleaned chemically and mechanically (by filing or scraping) before welding. Starts and stops must be regulated by current-control devices, or must be eliminated completely by means of starting and runoff tabs.

Gas metal arc welding of the joints shown in Fig. 17 can be accomplished at high welding speed, depending on the quality and reliability demanded of the weld, but it is more difficult to establish and maintain precise control over welding variables in GMAW than in GTAW. For example, arc length, melting rate, wire-feed speed, and current and voltage are directly related in GMAW. The butt joints in Fig. 17 could be welded in a single pass with high reinforcement if square-groove edge preparation were used, or in two passes with limited reinforcement if single-V-groove edge preparation were used. Although argon is normally used in GMAW of metal up to $^3/_4$ in. thick, radiographic quality of the weld metal can be improved by the addition of up to 80% helium.

For welding the joints shown in Fig. 17, the first process, GTAW with alternating current, is too costly; it would be used only if no other process were available. In a choice between GTAW with direct current electrode negative and GMAW, weld-quality requirements are the deciding factor. Gas tungsten arc welding is capable of meeting the highest standards of nuclear and aerospace applications; GMAW is capable of meeting the requirements of pressure-vessel and piping codes, as well as other commercial standards, where high deposition rates are more important than in nuclear and aerospace applications.

Other Arc Welding Processes

In addition to being welded by GMAW and GTAW, aluminum alloys are sometimes joined by shielded metal arc, stud, and percussion welding.

Shielded metal arc welding is used primarily in small shops for miscellaneous repair work in noncritical applications. A flux-covered aluminum alloy electrode is used. The flux combines with aluminum oxide to form a slag. The slag must be removed after each weld pass. Weld soundness and surface smoothness are poor. The process is limited to butt welds in $^1/_8$-in. and thicker aluminum. See the article "Shielded Metal Arc Welding" in this Volume for more information.

Stud welding of aluminum alloys is generally accomplished by the capacitor-discharge and the drawn arc methods. See the article "Stud Welding" in this Volume for more information.

Percussion welding of aluminum alloys is used principally for joining wire to wire. Numerous dissimilar metal joints, including aluminum to copper and steel, can be made. See the article "Percussion Welding" in this Volume for more information.

Arc Welding of Copper and Copper Alloys*

COPPER AND COPPER ALLOYS offer a unique combination of properties, among which the most important are conductivity, strength, and corrosion resistance. Other useful attributes include spark resistance, wear resistance, nonmagnetic or low-permeability properties, and color. In manufacturing, copper is often joined by welding, and arc welding is the most important of the processes employed. Arc welding can be achieved by shielded metal arc welding (SMAW); gas tungsten arc welding (GTAW), including the pulsed-current mode; gas metal arc welding (GMAW), including the pulsed-current and fine-wire modes; plasma arc welding (PAW); and submerged arc welding (SAW). In all processes, the dominant factors in establishing weldability are thermal conductivity and the alloy type with reference to solidification range and low-melting-point constituents.

The copper and copper alloys that are most frequently arc welded are listed in Table 1, which gives the alloy number, using the Unified Numbering System (UNS), the nominal composition, its melting point, relative thermal conductivity, and relative weldability employing GTAW, GMAW, or SMAW. Gas tungsten arc welding is the most broadly used arc welding process for joining copper and copper alloys. It features intense localized heat input, which is particularly needed in the higher conductivity alloys, good control, and ready automation. Its variant, PAW, is used where tip life is an important consideration, as in automated processes. It can be used with or without filler metal, but autogenous welding is limited to thinner sections. Sections up to $^1/_2$ in. can be welded by GTAW, but a more common limit is $^1/_8$ in. When welding thin sections where heat input is critical, the pulsed-current mode may be employed. For sections greater than $^1/_2$ in., GMAW is the preferred process. It features high heat input and a high deposition rate. When used in out-of-position welding, pulsed GMAW is preferred.

Shielded metal arc welding is the oldest of the arc processes and is still employed a great deal because of its simplicity, versatility, and ability to reach relatively inaccessible joints; the mobility of the equipment; and the availability of alloy rod. Generally, labor costs are higher than for GMAW, where speeds are typically faster by a factor of four. When welding high-conductivity coppers, even when preheat is used, GMAW is preferred.

The limitations and precautions necessary with each process are discussed later in this article. Because of the reduction in strength, formation of oxides, or volatilization of elements that often accompanies arc welding, brazing is frequently selected for joining copper alloys (see the articles on brazing in this Volume).

Effects of Alloying Elements on Welding

Several alloying elements have pronounced effects on the welding behavior of copper and copper alloys. Small amounts of volatile, toxic alloying elements are often present in copper and its alloys and, as a result, the use of an effective ventilation system to protect the welder or welding machine operator and to recover dusts, fumes, and mists is more critical than for ferrous metals.

Free-Machining Additives. Low-melting elements such as lead, tellurium, and sulfur, which are sometimes added to copper alloys to improve machinability, make them susceptible to hot cracking in welding. The adverse effect on weldability begins to be evident at about 0.05% of the additive and is more severe with larger additions. Lead is the most harmful of these alloying elements with respect to weldability of copper-based alloys. Free-machining copper alloys (which usually contain from 0.5 to 3 or 4% Pb) are not ordinarily welded, and are not listed in Table 1.

Zinc reduces the weldability of all the brasses and nickel silvers shown in Table 1, approximately in proportion to the amount present. Toxic vapors are emitted when copper-zinc alloys are welded. Efficient forced ventilation is mandatory, and a recovery system should be used to condense the fumes.

Tin increases susceptibility to hot cracking during welding in amounts of about 1 to 10%, as in the phosphor bronzes and tin brasses, and thus reduces weldability. Compared with zinc, tin is less volatile and much less toxic. During welding, the tin may oxidize preferentially to the copper, and the strength of the weld may be reduced because of oxide entrapment. Tin oxide, however, is readily fluxed.

Beryllium, aluminum, and nickel form tightly adherent oxides that must be removed by cleaning before welding. Formation of these oxides on the heated copper alloy must be prevented by gas shielding or by fluxes, in conjunction with selection of the appropriate type of welding current. Nickel oxides interfere with arc welding less than oxides of beryllium or aluminum. Thus, the nickel silvers and copper nickels are less sensitive to the type of welding current used.

Oxygen as gas or in cuprous oxide may cause porosity and reduce the strength of welds made in alloys that do not contain enough phosphorus or other deoxidizer. Most copper alloys that are welded contain deoxidizing elements (usually phosphorus, silicon, aluminum, iron, or manganese) that combine with oxygen. The

* Revised by L. McDonald Schetky, Technical Director—Metallurgy, International Copper Research Association, Inc.

Table 1 Nominal compositions, melting points, relative thermal conductivities, and weldabilities of wrought coppers and copper alloys that are commonly arc welded

UNS No.	Alloy name	Nominal composition, %	Melting point (liquidus), °F	Relative thermal conductivity(a)	Weldability(b) GTAW	Weldability(b) GMAW	Weldability(b) SMAW
OFC and ETP coppers							
C10200	Oxygen-free copper (OFC)	99.95 Cu	1981	100	G	G	NR
C11000	Electrolytic tough pitch copper (ETP)	99.90 Cu, 0.04 O_2	1981	100	F	F	NR
Deoxidized coppers							
C12000	Phosphorus-deoxidized copper, low-P (DLP)	99.9 Cu, 0.008 P	1981	99	E	E	NR
C12000	Phosphorus-deoxidized copper, high-P (DHP)	99.9 Cu, 0.02 P	1981	87	E	E	NR
Beryllium coppers							
C17500	High-conductivity beryllium copper, 0.6%	96.9 Cu, 0.6 Be, 2.5 Co	1955	53-66(c)	F	F	F
C17000	High-strength beryllium copper, 1.7%	98.3 Cu, 1.7 Be	1800	27-33(c)	G	G	G
C17200	High-strength beryllium copper, 1.9%	98.1 Cu, 1.9 Be	1800	27-33(c)	G	G	G
Low-zinc brasses							
C21000	Gilding, 95%	95 Cu, 5 Zn	1950	60	G	G	NR
C22000	Commercial bronze, 90%	90 Cu, 10 Zn	1910	48	G	G	NR
C23000	Red brass, 85%	85 Cu, 15 Zn	1880	41	G	G	NR
C24000	Low brass, 80%	80 Cu, 20 Zn	1830	36	G	G	NR
High-zinc brasses							
C26000	Cartridge brass, 70%	70 Cu, 30 Zn	1750	31	F	F	NR
C26800	Yellow brass, 66%	65 Cu, 35 Zn	1710	30	F	F	NR
C28000	Muntz metal, 60%	60 Cu, 40 Zn	1660	31	F	F	NR
Tin brasses							
C44300	Admiralty brass	71 Cu, 28 Zn, 1 Sn(d)	1720	28	F	F	NR
C46400	Naval brass	60 Cu, 39.25 Zn, 0.75 Sn(d)	1650	30	F	F	NR
Special brasses							
C67500	Manganese bronze A	58.5 Cu, 39 Zn, 1.4 Fe, 1 Sn, 0.1 Mn	1630	27	F	F	NR
C68700	Aluminum brass, arsenical	77.5 Cu, 20.5 Zn, 2 Al (0.06 As)	1780	26	F	F	NR
Nickel silvers							
C74500	Nickel silver, 65-10	65 Cu, 25 Zn, 10 Ni	1870	12	F	F	NR
C75200	Nickel silver, 65-18	65 Cu, 17 Zn, 18 Ni	2030	8	F	F	NR
C75400	Nickel silver, 65-15	65 Cu, 20 Zn, 15 Ni	1970	9	F	F	NR
C75700	Nickel silver, 65-12	65 Cu, 23 Zn, 12 Ni	1900	10	F	F	NR
C77000	Nickel silver, 55-18	55 Cu, 27 Zn, 18 Ni	1930	8	F	F	NR
Phosphor bronzes							
C50500	Phosphor bronze, 1.25% E	98.7 Cu, 1.3 Sn (0.2 P)	1970	53	G	G	F
C51000	Phosphor bronze, 5% A	95 Cu, 5 Sn (0.2 P)	1920	18	G	G	F
C52100	Phosphor bronze, 8% C	92 Cu, 8 Sn (0.2 P)	1880	16	G	G	F
C52400	Phosphor bronze, 10% D	90 Cu, 10 Sn (0.2 P)	1830	13	G	G	F
Aluminum bronzes							
C61300	Aluminum bronze D, Sn-stabilized	89 Cu, 7 Al, 3.5 Fe (0.35 Sn)	1915	14	G	E	G
C61400	Aluminum bronze D	91 Cu, 6-8 Al, 1.5-3.5 Fe, 1 max Mn	1915	17	G	E	G
C63000	Aluminum bronze E	82 Cu, 10 Al, 5 Ni, 3 Fe	1930	10	G	G	G
Silicon bronzes							
C65100	Low-silicon bronze B	98.5 Cu, 1.5 Si	1940	15	E	E	F
C65500	High-silicon bronze A	97 Cu, 3 Si	1880	9	E	E	F
Copper nickels							
C70600	Copper nickel, 10%	88.6 Cu, 9-11 Ni, 1.4 Fe, 1.0 Mn	2100	12	E	E	G
C71500	Copper nickel, 30%	70 Cu, 30 Ni	2260	8	E	E	E

(a) Based on the thermal conductivity of alloy C10200 (226 Btu/ft^2 in feet per hour at 68 °F) as 100. For comparison, carbon steel has a thermal conductivity of 30 Btu/ft^2 in feet per hour at 68 °F, which is 13 on this scale. (b) E = excellent, G = good, F = fair, NR = not recommended. (c) In the precipitation-hardened condition. (d) Alloys C44300 and C46500 contain a nominal 0.06% As; alloys C44400 and C46600, a nominal 0.06% Sb; alloys C44500 and C46700, a nominal 0.06% P.

same deoxidizers also are included in filler metals.

The soundness and strength of arc welds made in commercial coppers depend on the cuprous oxide content. Soundness increases as oxide content decreases. Best results are obtained for deoxidized coppers, because they are free from cuprous oxide and contain residual phosphorus. See the section on the effect of cuprous oxide later in this article.

Silicon has a beneficial effect on weldability of copper-silicon alloys, because of its deoxidizing and fluxing action. This effect, combined with low thermal conductivity, makes silicon bronzes the most weldable of the copper alloys by arc processes.

Phosphorus does not adversely affect weldability in the amounts normally present in copper alloys. It is beneficial to certain coppers and copper alloys as a strengthener and deoxidizer. When added to brass, phosphorus inhibits dezincification.

Cadmium in copper (up to 1.25%) poses no serious problem during arc welding. However, it evaporates from copper rather easily at the welding temperature, thereby creating a potential health hazard. A small amount of cadmium oxide may form in the molten weld metal but can be easily fluxed away.

Chromium, like beryllium and aluminum, can form a refractory oxide on the molten weld pool. Arc welding should be done using a protective atmosphere over the weld pool.

Iron and Manganese. Iron, which is present in some special brasses, aluminum bronzes, and copper nickels in amounts of 1.4 to 3.5%, does not significantly affect the weldability of these alloys. Manganese, which is present in some of these alloys in lower concentrations than iron, has no measurable effect on welding.

Factors That Affect Weldability

Other than the elements that comprise a specific alloy, the principal factors that influence weldability are thermal conductivity of the alloy being welded, shielding gas, type of current, joint design, welding position, and surface condition (cleanness). Effect of type of current is discussed under the sections on individual processes and alloys in this article.

Effect of Thermal Conductivity. The welding behavior of copper and copper alloys is strongly influenced by thermal conductivity, which varies greatly among these alloys. Table 1 shows relative thermal conductivities that are based on the conductivity of alloy C10200 (oxygen-free copper), which is 226 Btu/ft^2 in feet per hour at 68 °F, as 100. The range shown in Table 1 is from 100, for alloys C10200 and C11000, to lows of 8 to 12 for nickel silvers and copper nickels, and 9 for alloy C65500 (high-silicon bronze A). For comparison, the thermal conductivity of carbon steel, 30 Btu/ft^2 in feet per hour at 68 °F, is 13 on this scale.

In welding commercial coppers and lightly alloyed copper materials having high thermal conductivity, the type of current and shielding gas must be selected for maximum heat input to counteract the rapid heat dissipation from the weld region.

Even the less conductive copper alloys may require preheating (depending on section thickness), in spite of the concentrated heat input of arc welding processes. The interpass temperature should be the same as that for preheating. Copper alloys are not postweld heat treated as frequently as alloy steels are, but they may require controlled cooling to minimize residual stress and hot shortness.

Shielding gas for the gas shielded arc welding processes is usually argon, or argon plus 25 to 75% helium. Because helium is more expensive than argon, it is advantageous to develop welding procedures compatible with the use of argon or high-argon mixtures. Argon or argon-helium mixtures produce more uniform welds than helium on copper alloys, give a more stable arc, and cause less weld spatter.

Helium alone or in mixtures with argon is preferred where high heat input is needed, as in welding of highly conductive coppers or copper alloys and aluminum bronze. Helium gives about one third greater heat input than does argon at equal welding current.

Joint design for arc welding of copper and copper alloys does not differ greatly from that used in the arc welding of steel. Sections up to $^1/_8$ in. thick can be joined by use of square-groove welds without root openings. Thicker sections ordinarily are joined using either single-V-groove or double-V-groove welds with root faces not more than $^1/_8$ in. wide.

Joint design and fixturing should minimize constraint and contraction stresses and make allowance for the high coefficient of thermal expansion of copper and copper alloys to prevent cracking, which may result from hot shortness of the base metal at temperatures close to and above the solidus.

Backing strips or rings are used more extensively than on steel to avoid loss of molten metal, particularly on the highly fluid coppers and high-copper alloys. Backing strips and rings are usually made either of the alloy that is being welded, or of copper, carbon, graphite, or ceramic tape.

Joint and weld types used for arc welding of copper and copper alloys are illustrated in Table 2. Typical dimensions, root openings, and groove angles for welding the various types of alloys are given in the tables of nominal welding conditions in this article.

Welding Position. The flat position is used whenever possible, because of the high fluidity of copper and most copper alloys. The horizontal position is used in some fillet welding of corner joints and T-joints.

Vertical and overhead positions, and the horizontal position in welding butt joints, are less frequently used. They are ordinarily restricted to GTAW and GMAW of aluminum bronzes, silicon bronzes, and copper nickels. Small electrodes and filler-metal wire and low welding currents are used for out-of-position welding. This can be achieved by using pulsed current to control fluidity, because it combines low average heat input with spray transfer.

When using SMAW, out-of-position welding is usually limited to the joining of aluminum bronzes and copper nickels, but can also be done on phosphor bronzes and silicon bronzes.

Precipitation-Hardenable Alloys. The most important precipitation-hardening reactions in copper alloys are obtained with beryllium, chromium, boron, nickel-silicon, and zirconium. Care must be taken when welding precipitation-hardenable copper alloys to avoid oxidation and incomplete fusion. Wherever possible, the components should be welded in the annealed condition, and then the weldment should be given a precipitation-hardening heat treatment.

Hot Cracking. Copper alloys with wide liquidus-to-solidus temperature ranges, such as copper-tin and copper-nickel, are susceptible to hot cracking at solidification temperatures. Low-melting interdendritic liquid solidifies at a lower temperature than the bulk dendrite. Shrinkage stresses produce interdendritic separation during solidification. Hot cracking can be minimized by reducing restraint during welding, preheating to slow the cooling rate and reduce the magnitude of the welding stresses, and reducing the size of the root opening and increasing the size of the root pass.

Porosity. Certain elements such as zinc, cadmium, and phosphorus have low boiling points. Vaporization of these elements during welding may result in porosity.

Table 2 Nominal conditions for GTAW of commercial coppers(a) using EWTh-2 electrodes, ERCu welding rod, and DCEN

Work-metal thickness, in.	Root opening(b), in.	Electrode diameter, in.	Diameter of welding rod, in.	Shielding gas(c)	Gas-flow rate, ft^3/h	Current, A	Travel speed, in./min.	Number of passes	Preheat temperature, °F
Butt joints—square groove									
1/16	0	1/16	None used	Argon	15	110-140	12	1	None
1/8	0	3/32	None used	Argon	15	175-225	11	1	None
1/8	1/8	3/32	3/32, 1/8	Argon	15	175-225	11	1	None
3/16	3/16	1/8	1/8	Helium	30	190-225	10	1	200
Butt joints—60° single-V-groove, 1/16-in. root face									
1/4	1/16 max	1/8	1/8	Helium	30	225-260	9	1	300
3/8	1/16 max	3/16	3/16	Helium	40	280-320	...	2	500
Butt joints—60° double-V-groove, 1/8-in. root face(d)									
1/2	1/16 max	3/16, 1/4	1/4	Helium	40	375-525	...	3	500
Lap joints—fillet welded(e)									
1/16	0	1/16	1/16	Argon	15	130-150	10	1	None
1/8	0	3/32	3/32, 1/8	Argon	15	200-250	9	1	None
3/16	0	1/8	1/8	Helium	30	205-250	8	1	200
1/4	0	1/8	1/8	Helium	30	250-280	7	1	300
3/8	0	3/16	3/16	Helium	40	300-340	...	3	500
Outside corner joints—square groove									
1/8	1/8 max	3/32	3/32, 1/8	Argon	15	175-225	11	1	None
3/16	3/16 max	1/8	1/8	Helium	30	190-225	10	1	200
1/4	3/16 max	1/8	1/8	Helium	30	225-260	9	1	300
3/8	1/4 max	3/16	3/16	Helium	40	280-320	...	2	500
Outside corner joints—50° single-bevel-groove, 1/16-in. root face									
3/16	1/16 max	1/8	1/8	Helium	30	205-250	8	1	200
1/4	1/16 max	1/8	1/8	Helium	30	250-280	7	1	300
3/8	1/16 max	3/16	3/16	Helium	40	300-340	...	3	500
Inside corner joints—square groove, fillet welded									
1/8	1/8 max	3/32	3/32, 1/8	Argon	15	200-250	9	1	None
T-joints—fillet welded									
1/8	1/16 max	3/32	3/32, 1/8	Argon	15	200-250	9	1	None
3/16	1/16 max	1/8	1/8	Helium	30	205-250	8	1	200
1/4	1/16 max	1/8	1/8	Helium	30	250-280	7	1	300
3/8	1/16 max	3/16	3/16	Helium	40	300-340	...	3	500

Root face
Root face
Square-groove weld
Single-V-groove weld with no root face
Single-V-groove weld with root face
Double-V-groove weld
Butt joints

Root face
Fillet welds with filler metal added
Fillet welds with no filler metal
Lap joints
Square-groove weld
Single-bevel-groove weld
Outside corner joints
Square-groove weld with fillet
Inside corner joint
Fillet welds
T-joint

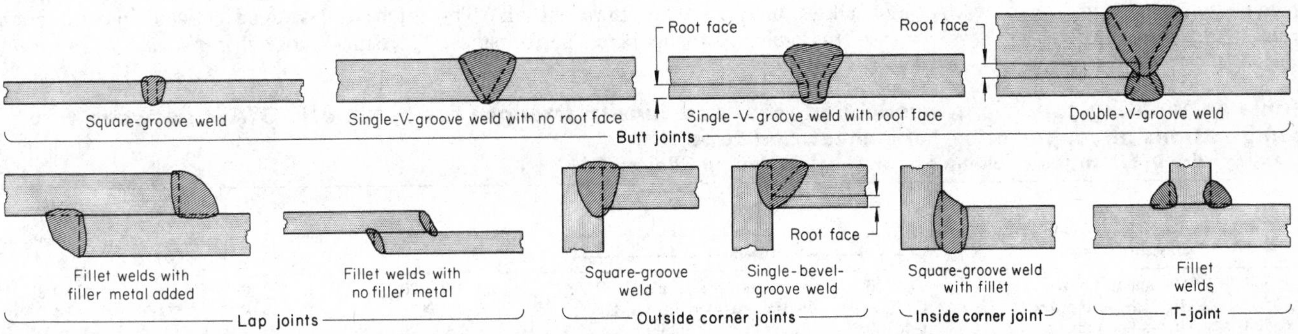

(a) The data in this table are intended to serve as starting points for the establishment of optimum joint design and conditions for welding of parts on which previous experience is lacking; they are subject to adjustments as necessary to meet the special requirements of individual applications. (b) Copper, carbon, graphite, or ceramic tape backing strips or rings may be used (see text). (c) Mixtures of argon and helium are also used (see text). (d) Depth of back V is 3/8 of stock thickness. (e) Use of filler metal optional for thicknesses of 1/4 in. or less

Table 3 Filler metals most frequently used in GTAW of copper and copper alloys(a)

Filler metal	AWS classification	Principal constituents(b)
Copper	ERCu	98.0 min Cu + Ag, 1.0 Sn, 0.5 Mn, 0.50 Si, 0.15 P
Phosphor bronze	ERCuSn-A	93.5 min Cu + Ag, 4.0-6.0 Sn, 0.10-0.35 P
Aluminum bronze	RCuAl-A2	1.5 Fe, 9.0-11.0 Al, rem Cu + Ag
Aluminum bronze	ERCuAl-B	3.0-4.25 Fe, 11.0-12.0 Al, rem Cu + Ag
Silicon bronze	ERCuSi-A	94.0 min Cu + Ag, 2.8-4.0 Si, 1.5 Zn, 1.5 Sn, 1.5 Mn, 0.5 Fe
Copper nickel	ERCuNi	1.00 Mn, 0.40-0.70 Fe, 29.0-32.0 Ni + Co, 0.20-0.50 Ti, rem Cu + Ag

(a) Based on AWS A5.27, A5.7, and A5.6; see current editions of those specifications for complete compositions and qualifications. (b) Single percentages are maximums unless otherwise stated. Optional elements and impurities have been omitted.

When welding copper alloys containing these elements, porosity can be minimized by the use of fast weld speeds and a filler metal low in these elements.

Surface Condition. Grease and oxide on work surfaces should be removed before welding. Wire brushing or bright dipping can be used (see the article "Cleaning and Finishing of Copper and Copper Alloys" in Volume 5 of the 9th edition of *Metals Handbook*).

Mill scale on the surfaces of aluminum bronzes and silicon bronzes is removed for a distance from the weld region of at least 1/2 in., usually by mechanical means. Grease, paint, crayon marks, shop dirt, and similar contaminants on copper nickels may cause embrittlement and should be removed before welding. Mill scale on copper nickels must be removed by grinding or pickling; wire brushing is not effective.

Gas Tungsten Arc Welding

Gas tungsten arc welding is well suited for copper and copper alloys because of its intense arc, which produces an extremely high temperature at the joint and a narrow heat-affected zone (HAZ). In welding copper and the more heat-conductive copper alloys, the intensity of the arc is important in completing fusion with minimum heating of the surrounding, highly conductive base metal. In welding copper alloys that have been precipitation hardened, a narrow HAZ is particularly desirable.

Gas tungsten arc welding of copper and copper alloys is most frequently used for sections up to about 1/8 in. thick that have been prepared with a square edge. Often, no filler metal is used in joining these thicknesses. When sections over 1/8 in. thick are gas tungsten arc welded, filler metal is usually required. Sections more than 1/2 in. thick are gas tungsten arc welded only if GMAW equipment is not available or if special conditions such as hot shortness of the base metal or adjacent heat-sensitive features made it necessary to limit the heat input. Under these conditions, pulsed SMAW would be the preferred process.

Type of Current. Gas tungsten arc welding is done on most copper and copper alloys with direct current electrode negative (DCEN) to permit use of an electrode of minimum size for a given welding current and to provide maximum penetration. Alternating current stabilized by high frequency is used on beryllium coppers and aluminum bronzes to prevent the buildup of tenacious oxide films on these base metals.

Electrodes. Any of the standard tungsten or alloyed tungsten electrodes described in the article "Gas Tungsten Arc Welding" in this Volume can be used in GTAW of copper and copper alloys, and the selection factors discussed in that article apply in general to these metals. Except as noted for specific classes of copper alloys, thoriated tungsten (usually EWTh-2) is preferred for its better performance, longer life, and greater resistance to contamination.

Filler metals most frequently used in GTAW of copper and copper alloys are listed in Table 3. Frequently, the filler-metal composition is matched closely to the base-metal composition, but a filler metal of composition different from that of the base metal may be selected. The reasons are dealt with in the sections on GTAW of various copper alloys in this article.

Mechanized Applications. One application of GTAW of copper and copper alloys that is frequently mechanized to some degree is welding the ends of tubes into tube sheets. In the following example, a change from manual GTAW to automatic welding, with mechanized equipment, increased production rate two to six times and improved quality for such welds.

Example 1. Change from Manual to Automatic Gas Tungsten Arc Welding of Tubes to Tube Sheet. In the fabrication of heat exchangers and condensers, 0.048-in.-wall tubes 3/4 in. in outside diameter of various copper alloys were automatically gas tungsten arc welded to typically 1-in.-thick tube sheet of various other copper alloys (Table 4).

Originally, the welds were made manually, but the operation was difficult and resulted in low productivity and questionable weld quality. Upon changing to automatic welding, joint strength equal to the minimum tensile strength of the base metals was usually attained, and production rates were two to six times as fast as in manual welding. Welding conditions, production rates, and weld tensile strengths for automatic welding are shown in Table 4.

The joint was similar to that shown in Example 8, consisting of a hole through the sheet for insertion of the tube, with an 82° (included angle) countersink 0.040 in. deep on the face of the sheet, and with the tube end swaged outward into the countersink to lock it in place.

Table 4 Welding conditions, production rate, and tensile strength for automatic GTAW between five combinations of copper alloy tube sheet and tube
Welding with DCEN, argon shielding gas at 15 ft³/h, and no filler metal

Alloy numbers (and common names): Tube sheet(a)		Tube(b)		Current, A	Welding time, s	Production rate, welds/h	Average tensile strength, ksi
C61400	(Aluminum bronze D)	C68700	(Aluminum brass, arsenical)	160-170	16	80-85	50.8
C65500	(High-silicon bronze A)	C12200	(Phosphorus-deoxidized, DHP)	170-180	12	90-100	34.3
C65500	(High-silicon bronze A)	C44200	(Admiralty, not inhibited)	160-170	12	90-110	46.9
C65500	(High-silicon bronze A)	C68700	(Aluminum brass, arsenical)	160-170	16	80-85	51.5
C71500	(Copper nickel, 30%)	C71500	(Copper nickel, 30%)	120-130	24	90-110	47.8

(a) Typically 1 in. thick, with holes (for the tubes) having 82° countersinks 0.040 in. deep. (b) 3/4-in. OD by 0.048-in. wall

Equipment used in welding included a mandrel, which was inserted into the end of the tube, a welding torch, and rotary equipment that caused the electrode to trace a circle around the mandrel, and thus around the tube, when the control circuit was closed. The diameter of the welded circle could be adjusted to suit the workpiece.

In welding, control circuitry initiated a timed gas prepurge, and arc starting was by means of a high-frequency pulse. Welding was at full current around the 360° joint and an additional 20° overlap. A current decay was used to avoid crater cracks or pipes when the arc was broken.

Butt Welding. Another application of mechanized GTAW is butt joining the ends of small coils of metal strip to make larger coils. The welding head traverses the length of the joint automatically. The use of this procedure for six alloys is described in the following example.

Example 2. Use of Gas Tungsten Arc Welding to Join Alloy Strips. Strips 0.030 to 0.190 in. thick and up to 25 in. wide of alloy C11000, C12200, C22000, C23000, or C26000 were butt welded by semiautomatic GTAW, under the conditions shown in Table 5, to make longer strips for cold reduction by rolling. Alloy C26000 was soft when welded; the others were hard.

The setup was similar to those shown in Example 9, except that no shielding gas was fed through a backing groove. Filler metal was used only in welding the 0.190-in. strips of alloy C11000 or C12200.

Table 5 Conditions for semiautomatic gas tungsten arc butt welding strips of copper alloys(a)

Strip thickness, in.	Current (DCEN), A	Voltage, V	Travel speed, in./min
Alloy C11000 or C12200			
0.030-0.035	80-95	20	50
0.080	230	21	40
0.190(b)	420-490	13-14	2
Alloy C21000, C22000, or C23000			
0.030-0.035	80-95	20	50
Alloy C26000			
0.090	180	17	35

(a) Strips of each alloy up to 25 in. wide, with ends cut with a shear set at 7° from the normal to the longitudinal direction, were welded with a square-edge butt joint, using an $^1/_8$-in.-diam EWTh-2 electrode, a $^1/_2$-in.-diam torch nozzle, helium shielding gas, and no filler metal (except for 0.190-in.-thick strips of alloy C11000 or C12200). (b) A preheating pass with the torch was made at 4 in./min, by use of 200 A and 18 to 20 V; a lithium-deoxidized copper filler wire was used in making the welding pass.

The ends to be joined were cleaned of oxides and lubricants, sheared square simultaneously, butted together, and held flat with clamps. Tabs of the composition of the metal to be welded were placed under each end of the joint to control melt-through at the ends of the weld and were used for arc starting and stopping. All joints were backed with a carbon block that had a groove $^1/_4$ in. wide by $^1/_{64}$ in. deep cut into it directly beneath the joint. The electrode was $^1/_8$-in.-diam EWTh-2. The torch nozzle was $^1/_2$ in. in diameter. Helium shielding gas was used.

After welding, all of the alloys except alloy C26000 were cold rolled to 25% reduction in cross section, annealed, and cold rolled again. Alloy C26000 was cold rolled to 70 to 80% reduction before annealing.

None of the welds broke in rolling, although all of the brasses (alloys C21000, C22000, C23000, and C26000) fumed during welding and, as determined by examination of samples taken through the weld area, had some weld porosity.

Coppers

Although the weld quality of commercial coppers joined by GTAW differs depending on the cuprous oxide content of the copper, the nominal welding conditions for coppers of a given thickness and joint design are approximately the same (Table 2).

Effect of Cuprous Oxide. Cuprous oxide present in the base metal or introduced through oxidation of the molten metal during the arc welding process migrates to the grain boundaries, thereby lowering the strength and ductility of the weld, adversely affecting fatigue properties.

Best results in the arc welding of copper are obtained on deoxidized coppers because: (1) they are free from cuprous oxide, and (2) they contain residual phosphorus, which combines with oxygen absorbed during heating or welding and thus prevents the formation of cuprous oxide. Strength, ductility, and porosity of welds in alloy C10200 (oxygen-free copper) are intermediate between the corresponding properties for welds in deoxidized coppers (alloys C12000 and C12200) and in alloy C11000 (electrolytic tough pitch copper), which contains 0.02 to 0.05% oxygen.

The decrease in properties obtained by arc welding copper containing cuprous oxide is less than the decrease caused by gassing and embrittlement experienced in oxyacetylene welding of oxygen-bearing copper.

Shielding Gases. Argon shielding is preferred for welding sections up to 0.06 in. thick. With thicker sections, low travel speeds and high preheat are required with argon shielding. Helium is preferred for welding sections over 0.06 in. thick. The weld pool is more fluid and cleaner, and the risk of oxide entrapment is considerably reduced. Compared to argon, helium permits deeper penetration or higher travel speeds at the same welding current.

Mixtures of these gases result in intermediate welding characteristics. For welding positions other than flat, a mixture of 65 to 75% helium-argon produces a good balance between the penetrating quality with helium and the ease of control with argon. Gas-flow rate usually ranges from 15 to 40 ft^3/h, being higher for the higher currents used in welding thicker sections (Table 2).

Type of Current. As indicated in Table 2, DCEN is preferred for GTAW of commercial coppers.

Electrodes. A thoriated tungsten electrode containing 2% thoria (EWTh-2) has the best electrode life and requires minimum tip maintenance in welding copper with DCEN. Electrodes for use on copper are usually pointed, with a cone angle of 60°. The point on the tip is usually broken to a 0.005- to 0.020-in. flat.

Welding Without Filler Metal. Square-groove butt joints in copper up to $^1/_8$ in. thick usually can be gas tungsten arc welded without the use of filler metal, although filler metal is sometimes used in welding thicknesses close to $^1/_8$ in.

Copper in sections thicker than $^1/_8$ in. can be gas tungsten arc welded without the use of filler metal by making two passes, one from each side.

Filler metal is generally used in GTAW of copper thicker than about $^1/_8$ in. (Table 2). The selection of a filler metal containing residual deoxidizer is important because of the adverse effects of oxygen on the ductility, strength, and soundness of the weld metal. The adverse effects of oxygen are even more severe on the filler metal than on the base metal, because filler metal has greater exposure to welding heat.

Generally, copper filler metal that contains a maximum each of 0.15% P and 0.50% Si as deoxidizers (AWS ERCu, see Table 3) is selected. Deoxidized coppers (alloys C12000 and C12200) as filler metal do not contain enough residual phosphorus to ensure sound welds. Other advantages of ERCu filler metal are relatively high electrical conductivity (30 to 40% IACS) and good color match with the copper base metal.

Any of the other filler metals listed in Table 3 can be used in GTAW of commercial coppers (because they contain adequate amounts of deoxidizing elements such as phosphorus, silicon, iron, aluminum, or titanium), but electrical conductivity is limited and color match is poor. However, these filler metals provide greater joint strength.

Welding Technique. Either forehand or backhand welding may be used. Forehand welding is preferred for all welding positions. It can provide a more uniform, smaller bead than backhand welding; however, a larger number of beads may be required to fill the joint.

The joint should be filled with one or more stringer beads or narrow weave beads. Wide oscillation of the arc should be avoided because it intermittently exposes each edge of the bead to the atmosphere and consequent oxidation. The first bead should penetrate to the root of the joint and be fairly thick to provide time for deoxidation of the weld metal and to avoid cracking of the bead.

Joint designs used in GTAW of pure coppers as well as copper alloys are shown in Table 2. A root opening is required for welding butt joints in thicker sections with filler metal because of the high thermal conductivity of copper. The clearance is needed to prevent the base metal from conducting heat away so rapidly that molten filler metal solidifies too quickly and chokes the joint before it is filled.

Backing strips or rings (usually made of copper, carbon, graphite, or ceramic tape) are ordinarily used in GTAW of butt joints in copper, because of its high fluidity. Backing is needed for both tightly fitted butt joints in thin metal and loosely fitted butt joints in thick metal to prevent loss of molten metal. Backing also may be used where needed for this purpose in making other types of joints.

Preheating. In GTAW of sections thicker than $^1/_8$ in., preheating is usually required to maintain the base metal at welding temperature without excessive loss of heat to the surrounding area. The small arc used in GTAW cannot maintain welding heat in thick sections of copper even under the most favorable conditions. Preheat temperatures are given in Table 2.

Deoxidized Coppers. Copper weldments that require the strength of the base metal are usually made from deoxidized copper. Even with the localized heat input of GTAW, special fixturing and other welding conditions may be needed to minimize distortion, as in the following example.

Example 3. Redesign of Joint and Fixture to Minimize Distortion in Gas Tungsten Arc Welding of Alloy C12000. A wafer of nuclear fuel was encapsulated in an alloy C12000 can-and-cover assembly by automatic GTAW. The completed weldment is shown in Fig. 1(a). As shown in detail A in Fig. 1, the can had a machined recess for the wafer and a 0.005-in.-thick bottom. The can-to-cover joint had raised lips, which eliminated the need for filler metal.

The weld had to provide hermetic sealing without distorting the assembly, including the thin can bottom, which was only $^3/_{16}$ in. from the welded area. Leak testing by a mass spectrometer was specified.

The welds were made in a controlled-atmosphere chamber, with the can mounted in a rotating copper chill-block clamping fixture under a stationary electrode. Welding conditions are given in the table accompanying Fig. 1.

Originally, the fixture and joint shown in Fig. 1(b) and detail A (for original method) were used, but at currents high enough for fusion, the welds were porous, the welded joint leaked, and the can warped unacceptably.

The porosity that caused leaking resulted because, in machining the can, lubricant was wiped onto the surface of the metal by the cutting tool and subsequent cleaning failed to remove it. Machining without a lubricant eliminated the difficulty. To minimize distortion, the joint and fixture were improved as follows. The square lips of the joint were chamfered (40°) to a sharp edge, as shown in detail B (improved method) in Fig. 1, to reduce the heat input necessary to make the weld. With the chamfered lips, the welding current could be lower because a smaller mass

Fig. 1 Nuclear fuel container. Distortion in welding was minimized by increasing the mass of the fixture to increase heat withdrawal and by chamfering joint edges to permit reduction of welding current.

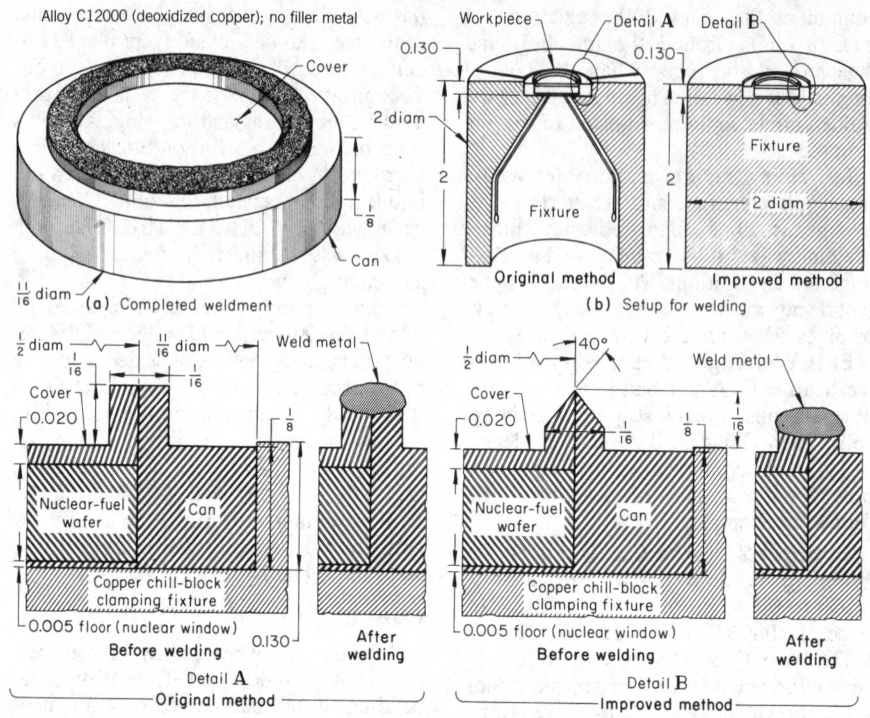

Automatic GTAW

Joint type	Circumferential edge	Current	16 A (DCEN)(c)
Weld type	Edge flange	Arc starting	(d)
Power supply	250-A rectifier	Arc length	0.015 in.
Electrode	0.040-in.-diam EWTh-2	Shielding gas	Argon(e)
Torch	110-A, air cooled(a)	Welding position	Flat
Filler metal	None	Travel speed	5.5 in./min
Fixture	(b)		

(a) Modified by elimination of ceramic cup and by use of bare connecting cable to avoid the presence of organic material in the chamber. (b) Copper chill-block clamping fixture held by a small chuck that was rotated by a variable-speed drive. (c) With continuous superimposed high-frequency current, for arc stability. (d) Superimposed high-frequency current, which was continuous as described in footnote (c). (e) In a vacuum-purged welding chamber; under slight positive pressure

of metal was being melted at any instant. The mass of the fixture was increased, as shown in Fig. 1(b) for improved method, to increase heat withdrawal. The redesigned fixture also provided complete support for the thin can bottom against pressure during leak testing.

Before welding, the assembly was retained in the fixture, which was held by a small chuck with all of the 0.005-in.-thick can bottom resting on the thick copper backing, as shown in the "improved method" view in Fig. 1(b) and in detail B. The fixtured assembly was welded in a chamber that was first vacuum-purged and then filled with 100% argon under slightly positive pressure.

The torch was rigidly mounted on a sliding base with locating stops by which the electrode tip was positioned over the seam and clamped in position. In the argon atmosphere it was not necessary to provide a shielding gas to the electrode tip or to use an electrode cup. The electrode was ground to a sharp point to reduce current flow and to pinpoint the arc.

During welding, the chuck holding the fixture was rotated by a variable-speed drive. The arc was started, and arc stability was maintained by a superimposed high-frequency current that was continuous. Current upslope was not controlled. Two revolutions at $3^1/_2$ rpm were used, because slight variations in pressure between the cover and the can caused variations in the thickness of the weld bead after the first revolution. The second revolution smoothed out these irregularities and ensured adequate penetration of the base metal along the circumference of the joint. After the second revolution, the current was tapered to zero in 5 s.

Distortion was minimized because of the low welding current, the high heat-sink efficiency, and rigid support by the fixture. Control of heat input was also helpful in preventing excessive internal gas expansion after the joint was closed.

After welding, the welding chamber was evacuated to 10^{-6} torr. The interior of the can remained at approximately 15 psi. Under this pressure differential, the can, still in the fixture, was inspected for leaks by use of a mass spectrometer. The fixture supported the thin can bottom against internal pressure.

Oxygen-Free and Tough Pitch Coppers. Gas tungsten arc welding is preferred to GMAW, or other processes generating less localized heat input, for joining either oxygen-free copper (alloy C10200) or electrolytic tough pitch copper (alloy C11000) in thicknesses up to about $^1/_2$ in.

Fig. 2 Alloy C11000 tubing and bar. For inductor coils, by automatic GTAW. Shows position of assembly, guide rolls, and torch during welding

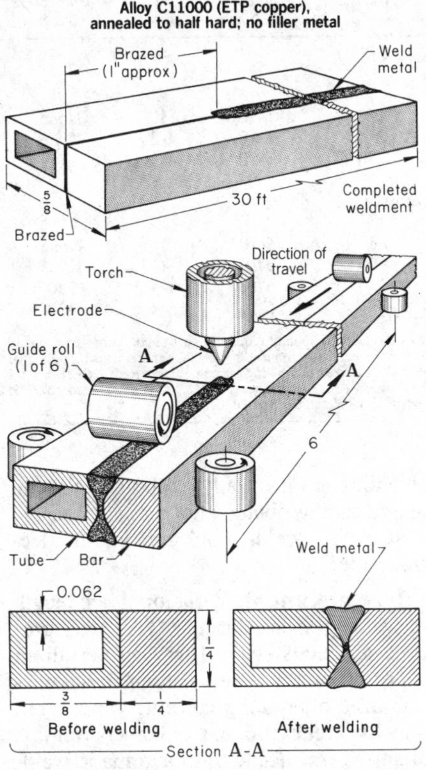

Automatic GTAW

Joint type	Butt
Weld type	Square groove, zero root opening
Power supply	500-A rectifier(a)
Electrode	$^1/_8$-in.-diam EWTh-2(b)
Torch	350-A, water cooled(c)
Filler metal	None
Shielding gas	Helium, at 20 ft^3/h
Fixture	6 guide rolls(d)
Current	220 A (DCEN)
Arc starting	Torch start
Arc length	$^1/_8$ in.
Cup-to-work distance	$^5/_{16}$ in.
Welding position	Flat
Number of passes	2 (one per side)
Welding speed	13 in./min.
Preheat and postheat	None

(a) Constant-voltage. (b) Taper ground. (c) Fixed, on an adjustable mount. (d) Variable-speed drive and chain used to pull assembly through rolls.

Gas tungsten arc welds in alloy C10200 (oxygen-free copper), because of the absence of a deoxidizer, have slightly lower strength and ductility, and are slightly more porous, than those in deoxidized coppers, unless all traces of oxygen are excluded.

Gas tungsten arc welds in alloy C11000 (electrolytic tough pitch copper) have somewhat lower tensile strength than that of welds in deoxidized coppers, and are more porous (see the section on the effect of cuprous oxide in this article). However, the properties of gas tungsten arc welds in electrolytic tough pitch copper (alloy C11000) are acceptable in many applications involving welded electrical conductors—particularly when tensile strength is relatively unimportant, as for the induction-coil weldment described in the following example.

Example 4. Automatic Gas Tungsten Arc Welding of Joints Between Alloy C11000 Bar and Tubing. Alloy C11000 solid bar 30 ft long was gas tungsten arc welded automatically to a 30-ft length of alloy C11000 tubing. The weldment was used in making water-cooled induction coils for low-frequency (60 to 180 Hz) induction heating that required more cross-sectional area for electrical conduction than was provided by the tubular section alone. Figure 2 shows a weldment with a cross section of typical size. The 30-ft length was standard for inductor stock. The weldment was bent into 4-in.-ID coils with the bar on the inside. The joint had to maintain good electrical conductivity.

Originally, the bar and tubing were torch brazed along the full length with a silver alloy filler metal. Not only was the filler metal expensive, but brazing was slow, needing two men to handle the 30-ft lengths of tubing and bar. To increase production rate, a faster and less costly method was developed by mechanizing a GTAW procedure in which no filler metal was used.

Welding conditions are given in the table accompanying Fig. 2. The fixture, which could accommodate the sizes of inductor bar and tubing regularly used, consisted of six guide rolls, as shown in Fig. 2, that properly aligned the assembly during welding. A chain and variable-speed drive were used to pull the assembly through the rolls during welding. The torch was on an adjustable mount that permitted it to be positioned over the joint to suit the size of components being welded.

Before welding, the leading end of the joint was brazed for about 1 in. to prepare the assembly for feeding through the guide rolls. The position of the assembly, guide rolls, and electrode holder during welding is shown in Fig. 2. To make the weld, the assembly was pulled through the rolls at 13 in./min. One pass was used, without coolant, for each side, and the ends were trimmed after welding.

The high thermal conductivity of alloy C11000 caused rapid dissipation of the heat energy. However, this difficulty was overcome without preheating by the use of helium as the shielding gas. Helium provided a hotter arc than could be obtained

Table 6 Nominal conditions for GTAW of beryllium coppers(a)

For butt joints having zero root opening; welding with a zirconiated tungsten electrode, filler metal of the same composition as the base metal, argon-helium shielding gas at 25 ft^3/h

Work-metal thickness, in.	Butt-joint groove(b)	Electrode diameter, in.	Current(c), A	Travel speed, in./min	Number of passes	Preheat temperature, °F
Alloy C17500 (high-conductivity beryllium copper)(d)						
0-0.090	Square	3/32	150	5-10	1	None
0.090-1/8	90° single-V(e)	3/16	250	5-10	1-2	None
1/4	90° single-V(e)	3/16	250	5-10	4-5	800
Alloys C17000 and C17200 (high-strength beryllium coppers)(d)						
0-0.090	Square	3/32	150	5-10	1	None
0.090-1/8	90° single-V(e)	3/32	180	5-10	1	None
1/4-1/2(f)	90° single-V(e)	3/16	250	5-10	3-4	300
Over 1/2(f)	90° single-V(e)	3/16	250	5-10	5-8	400

(a) The data in this table are intended to serve as starting points for the establishment of optimum joint design and conditions for welding of parts for which previous experience is lacking; they are subject to adjustment as necessary to meet the special requirements of individual applications. (b) See Table 2 for illustrations of joints. (c) High-frequency, stabilized alternating current is preferred; DCEN, with a thoriated tungsten electrode, is suitable under some conditions (see text). (d) For composition, see Table 1. (e) Maximum root face is 1/16 in. (f) Gas tungsten arc welding is used on these thicknesses only when GMAW cannot be used.

with argon. Thus, it was possible to produce relatively deep, narrow welds at 13 in./min. Oxygen from the oxide in alloy C11000 caused some porosity, but this was minimized by the fast freezing promoted by the fast travel speed. Thus, required joint properties were obtained.

High-Conductivity Beryllium Copper

Gas tungsten arc welding is preferred to other arc welding processes for joining precipitation-hardened alloy C17500 (high-conductivity beryllium copper, 0.6% Be; composition as shown in Table 1) in thicknesses up to about 1/4 in., because of the narrow HAZ. The maximum thickness that can be gas tungsten arc welded without substantial decrease in strength is about 1/2 in.; thicker workpieces are gas metal arc welded. When heat treatment is required after welding, GTAW generally is used only for thicknesses up to 0.090 in., and GMAW is used for thicker sections.

Alloy C17500 is more difficult to weld than alloys C17000 and C17200 (high-strength beryllium coppers), because of its higher thermal conductivity—which, for alloy C17500 in the precipitation-hardened condition, is 53 to 66% that of tough pitch copper, or about twice that of high-strength beryllium coppers. A difficulty common to the beryllium coppers is the formation of surface oxides (beryllium oxide and cuprous oxide). Beryllium will form a tenacious oxide that inhibits wetting and fusion during welding. Cleanness of the joint faces and surrounding surfaces before and during welding is necessary to ensure sound welds.

Nominal conditions for GTAW of alloy C17500 are given in Table 6. The shielding gas is usually a mixture of argon and helium to obtain a hot arc, smooth and spatter-free welds, and maximum electrode life.

Type of Current. Variation in arc length or welding speed during GTAW can produce tenacious oxide films on beryllium copper. For this reason, high-frequency-stabilized alternating current, which continually breaks up the oxide coating, is preferred (see Table 6) in automatic welding and must be used in manual welding. In automatic welding, advantage can sometimes be taken of the high heat input to the work and the deep penetration of DCEN, if close control is maintained over arc length and welding speed to minimize oxide formation.

Susceptibility to porosity and cracking is greater for welds made in high-conductivity than in high-strength beryllium copper, especially in multiple-pass welding. However, successfully welded joints show less effect on mechanical properties in the HAZ than similar joints in high-strength beryllium copper.

Electrodes. The preferred electrode metal when using alternating current is zirconiated tungsten (EWZr); for economy, unalloyed tungsten (EWP) can be used in noncritical applications. Thoriated tungsten (such as EWTh-2) is preferred when DCEN is used.

Filler metal of the same composition as the base metal (alloy C17500) generally is used, because high electrical conductivity is usually desired in the welds. If maximum electrical conductivity across the joint is not a requirement, silicon bronze filler metal is satisfactory; aluminum bronze filler metals are usable, but less satisfactory.

Joint Design. As Table 6 indicates, the usual joint designs for GTAW of high-conductivity beryllium copper are square-groove or 90° single-V-groove butt joints with 1/16-in. maximum root face and no root opening. See illustrations of joints in Table 2. All joints should be backed with grooved copper or graphite backing strips or rings.

Preheating is not ordinarily needed for welding alloy C17500 up to about 1/8 in. thick, but thicker stock (on which multiple-pass welds are used) is usually preheated to 800 °F.

High-Strength Beryllium Coppers

High-strength beryllium coppers, alloys C17000 (1.7% Be) and C17200 (1.9% Be), are more easily welded than the higher melting and less fluid high-conductivity beryllium coppers. Factors governing the suitability of GTAW for high-strength beryllium copper are as described for high-conductivity beryllium copper. However, contrary to practice with the high-conductivity alloys, GTAW can be used on thicknesses greater than 1/2 in. when it is not practical to weld by the preferred GMAW process.

Nominal conditions, or suggested starting points, for GTAW of high-strength beryllium coppers are given in Table 6. Shielding gas, type of current, and electrode type are the same as for alloy C17500.

Filler metal is almost always used to fill the joint or to provide joint reinforcement in V-groove welds, because joint strength is of primary concern in welding these alloys. Rods or strips of the same composition as the base metal generally are used as filler metal; the standard filler metals of other copper alloys are weaker and offer no advantages.

Joint design, which is indicated in Table 6, is the same as for high-conductivity beryllium copper. The 90° single-V-groove butt welds can be used for thicknesses greater than 1/2 in.

Table 7 Typical mechanical properties of welded joints in beryllium copper

Alloy and condition	Tensile strength, ksi	Yield strength, ksi
C17200 (Cu-2Be)(a):		
As-welded	60-70	30-33
Aged only	130-155	125-150
Solutioned and aged	150-175	145-170
C17500 (Cu-2.5 Co-0.5 Be)(a):		
As-welded	50-55	30-45
Aged only	80-95	65-85
Solutioned and aged	100-110	75-85

(a) Welded in the solution heat-treated condition

Table 8 Three applications of GTAW of high-strength beryllium coppers

	Pressure vessel (Example 5)	Generator liner (Example 6)	Cover to housing (Example 7)
Alloy	C17000	C17000 to C17200	Cast Be-Cu(1.7 Be) to C17000
Size of workpiece, in.	19 diam by 40	19½ ID by 48	11.6 diam
Thickness of work metal, in.	¾	(a)	0.060
Filler metal	0.062-in.-diam alloy C17200	None	None
Current, A	100-150, DCEN	122, ac(b)	275 ac(b)
Shielding gas	Argon and helium	Argon and helium	Argon
Gas flow rate, ft^3/h	25	25	20
Travel speed, in./min	20	10	12
Number of passes	...	1	1
Postheat	650 °F, 3 h	None	None
Testing method	X-ray(c)	...	(d)

(a) Reinforcing rings, 1½ by 1¼ in., welded to ¼-in.[2] longitudinal ribs. (b) High-frequency-stabilized. (c) Weld was tested by dye-penetration techniques, leak testing by helium mass spectrometry and x-ray radiography. (d) Weld depth (0.060 in., minimum) was determined ultrasonically; weld was leak tested by helium mass spectrometry at 11 ksi of helium.

Preheating and Postweld Heat Treatment. Preheating to 300 to 400 °F is recommended for welding of metal thicker than ⅛ in. Maximum weld strength is obtained by solution annealing and aging after welding. Aging treatments are 3 h at 600 °F for alloy C17000 and 3 h at 650 °F for alloy C17200. However, even for welds made under optimum conditions, this postweld treatment does not consistently provide the full strength of solution annealed and aged base metal. Higher strength can be obtained by cold working the annealed metal to a higher temper and modifying the aging treatment.

For some applications, the intermediate weld-metal strength obtained by aging after welding, without solution annealing, is adequate; omission of solution annealing avoids the expense and distortion associated with that high-temperature operation. Table 7 gives mechanical properties of beryllium copper welded joints in the as-welded, aged, and solutioned and aged condition. Three applications of GTAW of high-strength beryllium coppers are in the examples that follow.

Examples 5, 6, and 7. Gas Tungsten Arc Welding of High-Strength Beryllium Coppers. Table 8 gives operating conditions for GTAW of an alloy C17000 pressure vessel (Example 5); a cylindrical generator liner for which alloy C17000 was welded to alloy C17200 (Example 6); and a cover, cast from a beryllium-copper (1.7% Be) alloy that was welded to an alloy C17000 housing (Example 7).

The pressure vessel (Example 5) was used to scavenge propane from a freon bubble chamber. Gas tungsten arc welding was used because the vessel could not be solution annealed after welding because of danger of warping. The service conditions for this vessel included pressure of 6 ksi and rapid thermal cycling.

The cylindrical generator liner (Example 6) was back-extruded from a cast billet, and longitudinal ribs about ¼ by ¼ in. were machined on the outer surface. Nine reinforcing rings, 1½ by 1¼ in., rolled from extruded alloy C17000, were equally spaced along the ribs and welded to them. The liner, used in a magnetohydrodynamic electrical power generator, contained plasma at 3000 to 3500 °F. The cover-to-housing weldment (Example 7) was for ocean-cable use. The cast cover was welded to an extruded housing to provide a joint that was watertight at 12 ksi.

Cadmium and Chromium Coppers

Generally, the procedures recommended for GTAW of copper are good bases for determining welding parameters for cadmium and chromium coppers. These alloys have lower thermal and electrical conductivities than copper and can be welded with lower preheats and heat inputs than those required for copper. In addition to GTAW, cadmium and chromium coppers can be welded by the other gas shielded processes.

Nickel Silvers

Nickel silvers, which are alloys composed of copper (65%), zinc (17 to 27%), and nickel (10 to 18%), can be joined by the GTAW process, although welding of these alloys is not widely practiced. From a welding standpoint, nickel silvers are similar to brasses having comparable zinc content. These alloys are frequently used for decorative purposes where color match is important. However, there are no zinc-free filler metals that are suitable for arc welding that give good color match. Therefore, GTAW is usually restricted to thicknesses of ³⁄₃₂ in. or less without the addition of filler metal. Square-groove butt, lap, or edge joints are used. The joint faces must be in contact before and during welding.

Copper-Zinc (Brass) Alloys

Of the copper-zinc alloys rated as to weldability in Table 1, the low-zinc brasses are shown to have good weldability by GTAW. High-zinc brasses, tin brasses, special brasses, and nickel silvers are shown to have only fair weldability, either because of high zinc content or because of moderate zinc content in combination with other elements, such as oxide-forming aluminum or nickel.

Gas tungsten arc welding, because of its ability to weld rapidly with a highly localized heat input, is sometimes used for welding copper-zinc alloys (20% Zn or less) that contain up to 1% Pb, even though leaded copper alloys are generally not recommended for arc welding.

Maximum thickness of copper-zinc alloys ordinarily gas tungsten arc welded is about ⅜ in., although thick sections of cast alloys such as manganese bronze are sometimes repair welded in small local areas. Preheat is not ordinarily used in joining applications with these alloys.

Shielding gas selection is influenced by the heat requirements, which are related to the thermal conductivity of the base metal. Argon is commonly used for welding those alloys that are least conductive, whereas helium or helium mixtures are preferred for the alloys having greater conductivity. However, helium and helium-rich mixtures with argon are sometimes used on even the less conductive alloys (high-zinc brasses, tin brasses, special brasses, and nickel silvers) to reduce zinc fumes.

Filler metals used in arc welding copper-zinc alloys should not contain zinc. The addition of filler metal is recommended when welding sections over 0.062 in. thick. The arc is struck and held on the filler metal, rather than on the base metal, to help reduce zinc loss and fuming. ERCuSn-A is recommended for the low-zinc brasses, and ERCuSi-A and ERCuAl-2 for the high-zinc alloys.

The silicon in ERCuSi-A helps to decrease zinc fumes. For this reason, and to provide joint reinforcement, ERCuSi-A is sometimes used, with alternating current, on copper-zinc alloys 0.050 in. thick or less. ERCuAl-A2 is sometimes used in welding the high-zinc alloys. It makes sound welds, but does not decrease zinc vaporization.

Fig. 3 Joint preparation for welding a tube into a tube sheet by semiautomatic GTAW

Alloy C44200 (admiralty) welded to alloy C46400 (naval brass); no filler metal

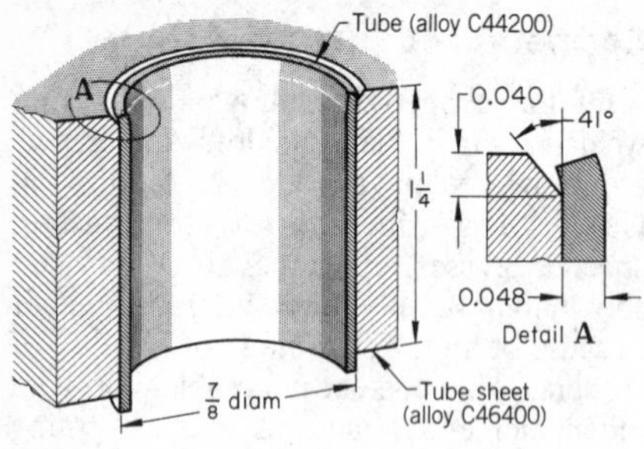

Semiautomatic GTAW

Joint type	Cylindrical edge
Weld type	Bevel
Electrode	1/8-in.-diam EWTh-2
Filler metal	None
Current	185 A (DCEN)
Voltage	(a)
Arc time	6 s
Shielding gas	Argon, at 15 ft³/h
Welding speed	27.5 in./min

(a) Controlled by fixed arc length of 0.040 in.; no readings were taken.

Welding Without Filler Metal. Thin brass sheets less than 0.062 in. thick can be welded without filler metal addition. Use of high electrode travel speed will help to limit the amount of fuming by shortening total arc time. Tube-sheet joints can be made with brass members using the same kind of joint preparation, if high electrode travel speed is maintained, as in the following example.

Example 8. Welding of Copper-Zinc Alloy Tube-Sheet Assemblies. Semiautomatic GTAW was used to weld alloy C44200 (admiralty brass) tubes to alloy C46400 (naval brass) tube sheets, as shown in Fig. 3. The tubes were 7/8 in. in outside diameter with 0.048-in. walls. Linear speed of the welding electrode tip was 27.5 in./min, and each weld was completed in 6 s.

Although helium is the preferred shielding gas for this type of operation, argon was used successfully at a flow of 15 ft³/h. Some zinc fuming occurred when the welds were made, but the joints were uniformly sound and leakproof. Welding details are given in the table with Fig. 3.

Phosphor Bronzes

Gas tungsten arc welding is used to join sheet and other forms of wrought phosphor bronze up to about 1/2 in. thick. This process is also used to join or repair phosphor bronze castings. Copper-tin alloys solidify with large, weak dendritic grain structures. Such structures in the weld metal have a tendency to crack. Hot peening at each layer of multiple-pass welds reduces cracking stresses and, therefore, the likelihood of weld-metal cracking.

Nominal conditions for GTAW of square-groove butt joints in nonleaded phosphor bronzes are given in Table 9. The type of current used is DCEN or stabilized arc.

The nominal conditions shown in Table 9 are based primarily on the three phosphor bronzes shown in Table 1 to have low thermal conductivity (alloys C51000, C52100, and C52400). Higher welding current or lower welding speed is needed for alloy C50500 (phosphor bronze, 1.25% E), which contains 98.7% Cu and has three to four times the thermal conductivity of the three other alloys in this group.

Shielding Gas. Argon is most often used because of the desirability of restricting the size of the HAZ, thus minimizing the area where mechanical properties are decreased. Helium shielding gas, along with an argon backup flow, can also be used.

Filler Metal. ERCuSn-A or wire of approximately the same composition as the base metal is ordinarily used as the filler metal. When matching the composition of the base metal is not essential, ERCuSi-A filler metal can be used instead, producing stronger welds.

Welding Without Filler Metal. If not thicker than about 1/8 in., phosphor bronze strip can be butt welded by GTAW without the addition of filler metal, as in the application described in the next example.

Example 9. Gas Tungsten Arc Welding of Phosphor Bronze Strip Without Filler Metal. Phosphor bronze strip of an alloy similar to C51000 was reduced in a rolling mill from 0.128-in. thickness to 0.002-in. foil. The strip was delivered in coils that weighed up to 500 lb. For efficient handling, four coils were joined into one 2000-lb coil by butt welding the ends of the strip together as shown in Fig. 4 (detail C). The joints had to pass through the mill rolls with a minimum of interference and had to withstand reduction down to 0.002 in.

The joints were welded by automatic GTAW in a machine that combined an uncoiler, a leveler, a shear, hydraulic clamps, electrode holder, and travel carriage, and a recoiler. A 400-A, three-phase rectifier and a 500-A high-frequency oscillator were used. The electrode (EWTh-1) was 1/8 or 5/32 in. in diameter and 18 in. long, and was ground to a point. No filler metal was used.

Some experimentation was necessary on the shape of the backing groove in the backing bar (compare "original method" and "improved method" views in Fig. 4). At first, a groove of rectangular cross section was used to lie beneath the joint (section A-A in Fig. 4). Weld beads made with this backing groove were of generally rectangular shape and did not pass through the rolling mill easily. The backing bar was modified by changing the groove to a radiused cross section, as shown in section B-B in Fig. 4. The change was made by mortising an insert into the original backing bar. The rounded weld bead produced when using this groove passed through the rolling mills without difficulty and withstood reduction to 0.002-in. thickness and elongation to 4 in.

Another improvement was to add gas holes 0.029 in. in diameter and spaced at 1 1/4-in. intervals along the groove to provide shielding gas for the back side of the weld. The groove was supplied with argon at 3 ft³/h from a manifold cut in the backing bar (see improved method, Fig. 4).

At the same time, graphite runout blocks were added at each end of the joint zone to prevent burnout at the edges of the strip (Fig. 4, improved method). This allowed subsequent use of the full width of the strip, without trimming, and thus increased productivity.

The coil-handling equipment, shear, and clamps were operated manually. Arc start, arc length, and torch travel were operated automatically. Approximately 120 welds were made in 8 h. Additional welding

Table 9 Nominal conditions for GTAW of square-groove butt joints in phosphor bronzes(a)

Work-metal thickness, in.	Root opening(b), in.	Electrode (EWTh-2) diameter, in.	Diameter of welding rod(c), in.	Current (DCEN), A	Flow rate of argon, ft³/h	Travel speed, in./min	Number of passes	Preheat temperature, °F
0.012-1/16	0	1/8	None used	90-150	15-35	60-80	1	None
1/16-1/8	0-1/16	1/8	1/16-5/32	100-220	15-35	45-60	1	None

(a) Data here can serve as starting points in establishing optimum joint design and conditions for welding parts for which prior experience is lacking; they may be adjusted as necessary to meet the special requirements of individual applications. Higher current or lower welding speed is used in welding alloy C50500 (see text). (b) See illustration of joint in Table 2. (c) ERCuSn-A, or rod of composition close to that of the base metal

Fig. 4 Setup for welding coil ends of phosphor bronze strip. Lengths are fed into a rolling mill. Shows details of design changes in backing bar

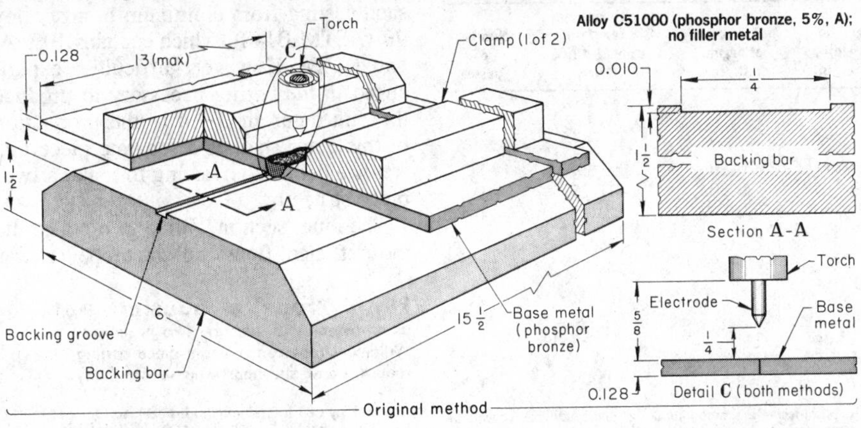

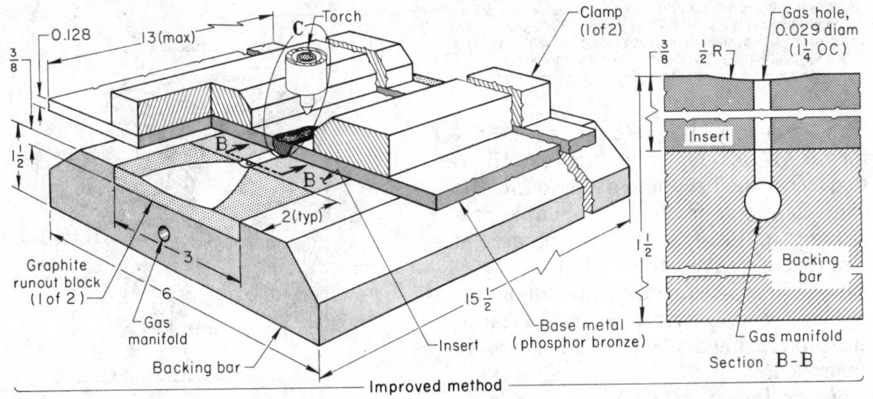

Automatic GTAW

Joint type	Butt	Filler metal	None
Weld type	Square groove	Voltage	15-18 V
Electrode	$^1/_8$ or $^5/_{32}$-in.-diam EWTh-1: 18 in. long, pointed	Current	250-275 A (DCEN)
Shielding gas	Helium, at 45-50 ft^3/h	Power supply	400-A three-phase rectifier
Backing gas	Argon at 3 ft^3/h	Arc start	Nontouch, high-frequency ac
		Travel speed	30-40 in./min

conditions are given in the table with Fig. 4.

Preheating is not normally used in GTAW of thin sections of the phosphor bronzes. Thick sections are usually preheated at 350 to 400 °F and are cooled slowly after welding. Because phosphor bronze is hot short, weld layers should be thin, and interpass temperature should not exceed 400 °F. The weld can be hot peened to refine grain and minimize distortion.

In the example that follows, in which only partial penetration was necessary and high strength was not needed, $^5/_{16}$-in.-thick phosphor bronze was gas tungsten arc welded to $^3/_4$-in.-thick phosphor bronze without preheating and with peening of the weld after cooling.

Example 10. Gas Tungsten Arc Welding of End-Cap Rings to Drilled Phosphor Bronze Bushings. A design change in heavy-duty bushings centrifugally cast in phosphor bronze (89% Cu, 11% Sn) called for gun drilling longitudinal holes that had to be end capped. End capping was done by gas tungsten arc welding a flat ring, also cast from phosphor bronze, to each end of the bushing, as shown in Fig. 5.

Because the function of the weld was to maintain a close fit of the rings during the service life of the bushing, joint design called for only a partial-penetration, single-V-groove weld. This design caused distortion of the rings, because of unbalanced thermal stresses during welding and constriction of the bushing diameter from shrinkage stresses after cooling.

Both difficulties were overcome by designing a welding fixture consisting of a pair of flanged end plugs that could be drawn up tightly by means of a common threaded rod and nuts (Fig. 5). Machining the plugs for a close fit ensured accurate alignment of the parts during fit-up, welding, and cooling, and during subsequent peening. The plugs also served as a heat sink.

Gas tungsten arc welding was done manually on the small quantities of bushings by rolling the bushing and its fixture on a steel-top table. A silicon bronze rod, ERCuSi-A, was selected for filler metal for its weldability, strength, and color match. The shielding gas was argon. The weld was made in two passes at approximately $^1/_2$ rpm. Cracking from hot shortness was avoided by making relatively small weld deposits.

Before assembly, the semifinished machined parts were beveled by machining and detergent cleaned. After cooling from welding, the welds were peened to relieve stresses and examined visually and by liq-

Fig. 5 Heavy-duty bushing and setup. For end-capping holes drilled longitudinally through the bushing wall

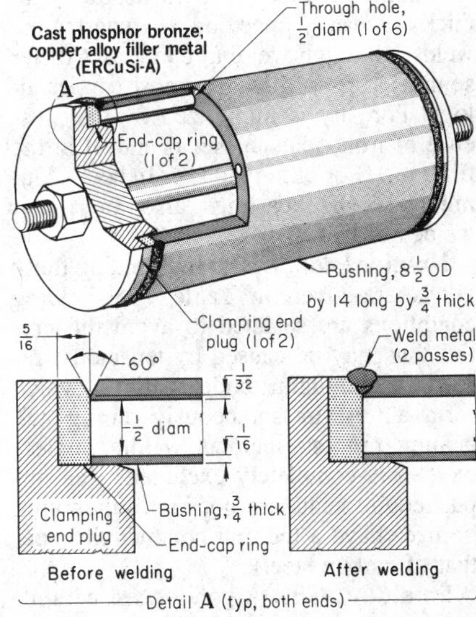

Manual GTAW

Joint type	Circumferential edge
Weld type	60° single-V-groove (partial penetration)
Number of passes	2
Power supply	250-A rectifier
Electrode	$^1/_{16}$-in.-diam EWTh-2
Torch	250-A, water cooled
Filler metal	$^1/_{16}$ or $^3/_{32}$-in.-diam ERCuSi-A
Shielding gas	Argon, at 10-12 ft^3/h
Welding position	Horizontal rolled
Fixtures	Clamping end plugs
Current	130 A (DCEN)
Voltage	12-15 V
Arc starting	High open-circuit voltage (70 V)

Table 10 Nominal conditions for GTAW of aluminum bronzes(a)
For joint configurations, see Table 2

Work-metal thickness, in	Root opening, in.	Electrode diameter(b), in.	Diameter of welding rod(c), in.	Flow rate of argon, ft^3/h	Current (ac, hf-stabilized)(d), A	Number of passes
Square-groove butt joints						
Up to 1/16	0	1/16	1/16(e)	20-30	25-80	1
1/16-1/8	1/16 max	3/32	1/8	20-30	60-175	1
1/8	1/8 max	5/32-3/16	5/32	30	210	1
70° single-V-groove butt joints						
3/8	0	5/32-3/16	5/32	30	210-330	4
Fillet-welded T-joints or square-groove inside corner joints						
3/8	(f)	5/32-3/16	3/32	30	225	3

(a) The data in this table are intended to serve as starting points for the establishment of optimum joint design and conditions for welding parts on which previous experience is lacking; they are subject to adjustment as necessary to meet the special requirements of individual applications. Preheating is not ordinarily used in welding the thicknesses shown. (b) Zirconiated or unalloyed tungsten electrodes are recommended with high-frequency-stabilized alternating current. (c) Preferred welding rod is ERCuAl-A2; otherwise, ERCuAl-A3 or rod of the same composition as the base metal. (d) Direct current electrode negative can also be used in making single-pass welds (see text). (e) Use of welding rod is optional for thicknesses up to 1/16 in. (f) Zero root opening for T-joints; 3/8 in. max for corner joints.

uid penetrants for porosity and cracks. Other details of the welding operation are given in the table with Fig. 5.

Aluminum Bronzes

Aluminum bronzes up to about 3/8 in. thick are readily joined by gas tungsten arc welds, although welding conditions differ somewhat from those for most copper alloys. Porosity is minimized by the presence of iron, manganese, or nickel in the filler metal or base metal, or in both. Aluminum bronze castings also are repair welded by GTAW.

Nominal conditions for welding these alloys are given in Table 10. Welding conditions are selected to avoid difficulties that may be caused by tenacious, refractory aluminum oxide coatings, which form almost instantaneously during any heating process such as welding unless oxygen is completely excluded. Heat input requirements are not high (aluminum bronzes have a thermal conductivity near that of carbon steel).

Shielding Gas. Argon is used with alternating current for shielding. For better penetration or faster travel speed, direct current can be used with argon, helium, or a mixture of the two. The shielding effect of the gas is augmented, where necessary, by the use of a special flux applied to the edge of the joint to increase fluidity and to help protect the base metal from oxide formation. Aluminum oxide forms even at room temperature, and the flux prevents access of air to the prepared edges until the protective argon atmosphere becomes effective.

Type of Current. To prevent oxide buildup, alternating current stabilized by high frequency is preferred to DCEN for GTAW of aluminum bronzes. High-frequency-stabilized alternating current is particularly desirable in multiple-pass welding of these alloys. Direct current electrode negative can be used in single-pass welding, particularly for automatic welding, when surfaces are well cleaned and protected and when the arc is closely controlled.

Electrodes for GTAW of aluminum bronzes are usually zirconiated tungsten (EWZr) or unalloyed tungsten (EWP); thoriated tungsten electrodes cause the arc to wander when they are used with alternating current, although they may be used with direct current. Thoriated tungsten gives a longer tip life and is easier to start. When using this type of electrode, common practice is to use a tapered, cone-shaped tip.

Filler metal ERCuAl-A2 is ordinarily used. When a close match in composition and color with the base metal is needed, ERCuAl-A3 or other aluminum bronze wire of a suitable composition should be used.

Preheating. Aluminum bronzes, which have relatively low thermal conductivity, do not usually need preheating for GTAW of sections thinner than about 1/4 in. Preheating may be necessary in welding thicker sections. Generally, the preheat and interpass temperatures should not exceed 300 °F for alloys with less than 10% Al. A preheat was used in GTAW of 3/8-in.-thick aluminum bronze in the following example.

Example 11. Use of Gas Tungsten Arc Welding Versus Gas Metal Arc Welding To Avoid Melt-Through of Neighboring Section. The propeller housing (Kort nozzle) shown in Fig. 6 had originally been made as a one-piece sand casting from aluminum bronze alloy 9B (ASTM B148), which contains 10% Al and 1% Fe. However, difficulties experienced in the coring necessary to produce the thin outer shell led to the production of this outer shell as a separate piece that was welded to the casting to make a two-piece structure.

The inner section, faired to produce the most efficient flow from the propeller, was

Fig. 6 Propeller housing. Produced as an assembly welded by two processes. Weldment replaced a single-piece casting, reducing cost and improving serviceability.

Alloy C61300 (aluminum bronze), soft temper; aluminum bronze filler metal (ERCuAl-A2)

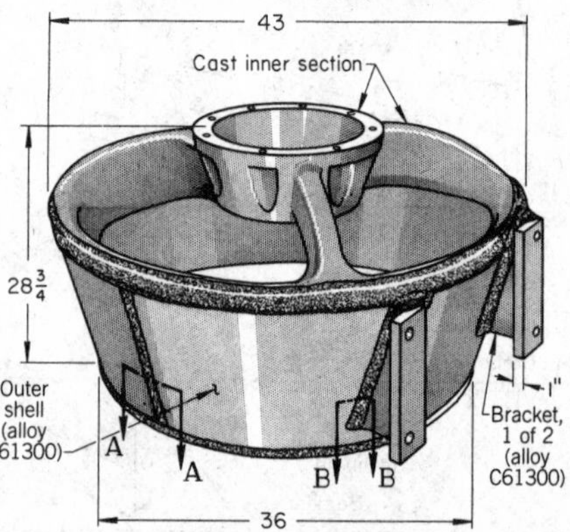

Circumferential welds joining outer shell to cast inner section were made by gas metal arc welding

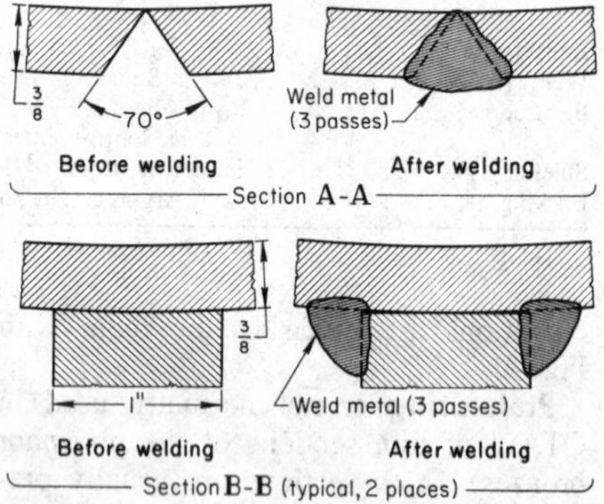

Conditions for GTAW

Joint types	Butt and T
Weld types	Single-V-groove and fillet
Welding position	Flat
Number of passes	3
Power supply	300-A rectifier, constant-current type
Electrode	3/32-in.-diam EWTh-2
Torch	Water cooled
Filler metal	3/32-in.-diam ERCu-Al-A2
Shielding gas	Argon, at 15-20 ft^3/h
Preheat and interpass temperature	300 °F
Voltage	35 V
Current	225 A (DCEN)
Arc starting	Contact and high frequency

cast in propeller bronze in one intricate piece that included an integral hub and shaft-mounting flange connected to the main body of the nozzle by four arms. The outer shell was cut from 3/8-in.-thick alloy C61300 (aluminum bronze), soft temper, and was wrapped around the cast inner section after being cold formed into a truncated cone. Before cold forming, all edges of the shell were machined to a 35° bevel to provide a groove for subsequent welding.

The assembly was tack welded together, and then circumferential seams at the top and bottom were gas metal arc welded. Then, to avoid melt-through of the faired cast inner section (and resultant poor effect on the propeller stream), GTAW, under the conditions in the table with Fig. 6, was used for the longitudinal seam in the shell and for joining two alloy C61300 brackets to the shell for mounting the housing to the hull of the vessel (Fig. 6).

Not only was the cost of production reduced by eliminating the complex and expensive coring in the casting, but the weight of the welded housing was less than that of the completely cast housing. Furthermore, the cast housing had been made of an alloy that (although easy to cast) did not resist erosion, corrosion, and cavitation as well as did the alloys used in the welded assembly.

Silicon Bronzes

Gas tungsten arc welding is used on thin to moderately thick nonleaded silicon bronzes, which are the most weldable of the copper alloys. Characteristics of these bronzes that contribute to weldability are their low thermal conductivity, good deoxidation of the weld metal by silicon, and the protection offered by the resulting slag. Silicon bronzes have a relatively narrow hot short range just below the solidus, and they must be rapidly cooled through this range to avoid weld cracking. The welding of silicon bronze to other copper alloys and to copper is detailed in Table 4.

Nominal conditions for welding of silicon bronzes are given in Table 11. The data for manual welding include welding conditions for various positions. Because they have low fluidity, the silicon bronzes are the only group of copper alloys on which this process is applied extensively in the vertical and overhead positions. Welding is usually done with DCEN and argon or helium shielding, but high-frequency-stabilized alternating current with argon shielding can be used for stock thinner than 0.062 in. Zirconiated (EWZr)

Table 11 Nominal conditions for GTAW of silicon bronzes(a)

Using zero root opening, no preheat, EWTh-2 electrodes, ERCuSi-A welding rod, argon shielding gas, and DCEN; for joint configurations, see Table 2

Work-metal thickness, in.	Current, A	Electrode diameter, in.	Travel speed, in./min	Diameter of welding rod, in.	Shielding-gas flow rate, ft^3/h	Number of passes
Automatic welding						
Square-groove butt joints, flat position						
0.012-0.050	80-140	1/8	60-80	None used	15-35	1
1/16-1/8	90-210	1/8	45-60	None used	15-35	1
1/8	250	1/8	18-20	1/16(b)	15-35	1
Manual welding						
Square-groove butt joints, flat position						
1/16	100-120	1/16	12	1/16	15	1
1/8	130-150	1/16	12	3/32	15	1
3/16	150-200	3/32	...	1/8	20	1
1/4	250-300	1/8	...	1/8, 3/16	20	1
1/4	150-200	3/32	...	1/8, 3/16	20	3
Square-groove butt joints, vertical and overhead positions						
1/16	90-110	1/16	...	1/16	15	1
1/8	120-140	1/16	...	3/32	15	1
60° single-V-groove butt joints, flat position						
3/8	230-280	1/8	...	1/8, 3/16	20	3-4
1/2	250-300	1/8	...	1/8, 3/16	20	4-5
3/4(c)	300-350	1/8	...	3/16	20	9-10
1(c)	300-350	1/8	...	3/16, 1/4	20	13
Fillet-welded lap joints, flat position						
1/16	110-130	1/16	10	1/16	15	1
1/8	140-160	1/16, 3/32	10	3/32	15	1
3/16	175-225	3/32	...	1/8	20	1
1/4	175-225	3/32	...	1/8, 3/16	20	3
3/8	250-300	1/8	...	1/8, 3/16	20	3
1/2	275-325	1/8	...	1/8, 3/16	20	6
3/4(c)	300-350	1/8	...	3/16	20	12
1(c)	325-350	1/8	...	1/4	20	16
Fillet-welded lap joints, vertical and overhead positions						
1/16	100-120	1/16	...	1/16	15	1
1/8	130-150	1/16, 3/32	...	3/32	15	1
Square-groove outside corner joints, flat position						
1/16	100-130	1/16	12	1/16	15	1
1/8	130-150	1/16	12	3/32	15	1
3/16	150-200	3/32	...	1/8	20	1
Square-groove outside corner joints, vertical and overhead positions						
1/16	90-110	1/16	...	1/16	15	1
1/8	120-140	1/16	...	3/32	15	1
50° single-bevel-groove outside corner joints, flat position(d)						
1/4	175-225	3/32	...	1/8, 3/16	20	3
3/8	230-280	1/8	...	1/8, 3/16	20	3
1/2	275-325	1/8	...	1/8, 3/16	20	7
3/4(c)	300-350	1/8	...	3/16	20	14
1(c)	325-350	1/8	...	3/16, 1/4	20	20
Fillet-welded square-groove inside corner joints, flat position(e)						
1/16	110-130	1/16	10	1/16	15	1
1/8	140-150	1/16, 3/32	10	3/32	15	1
3/16	175-225	3/32	...	1/8	20	1

(continued)

Table 11 Continued

Work-metal thickness, in.	Current, A	Electrode diameter, in.	Travel speed, in./min	Diameter of welding rod, in.	Shielding-gas flow rate, ft³/h	Number of passes
Fillet-welded T-joints, flat position						
1/16	110-130	1/16	10	1/16	15	1
1/8	140-160	1/16, 3/32	10	3/32	15	1
3/16	175-225	3/32	...	1/8	20	1
1/4	175-225	3/32	...	1/8, 3/16	20	3
3/8	230-280	1/8	...	1/8, 3/16	20	3
1/2	275-325	1/8	...	1/8, 3/16	20	7
3/4(c)	300-350	1/8	...	3/16	20	14
1(c)	325-350	1/8	...	3/16, 1/4	20	20

(a) The data in this table are intended to serve as starting points for the establishment of optimum joint design and conditions for welding parts on which previous experience is lacking; they are subject to adjustment as necessary to meet the special requirements of individual applications. (b) Wire-feed rate, 115 to 125 in./min. (c) Thicknesses greater than about 1/2 in. are gas tungsten arc welded only when it is not practicable to use GMAW. (d) Root face is 1/16 in. for thicknesses of 1/2 in. or less, and 1/8 in. for thicknesses greater than 1/2 in. (e) Maximum root opening = t (work-metal thickness)

or unalloyed tungsten (EWP) electrodes are recommended.

Filler Metals. The conventional silicon bronze filler metal, ERCuSi-A, which is similar in composition to alloy C65500 (high-silicon bronze A), the most commonly used silicon bronze, can be used to weld any of the silicon bronzes. Thin sections of silicon bronze can be gas tungsten arc welded without the addition of filler metal.

Joint Design. On metal thicker than about 1/4 in., a V-groove with 60° included angle is used. Butt joints in thin stock can be welded without special preparation.

Preheating is not needed on silicon bronzes, some of which have lower thermal conductivity than do carbon steels. Further, because of the hot shortness of these alloys, preheating can be harmful, and interpass temperature on multiple-pass welds should not exceed 200 °F.

Copper Nickels

Gas tungsten arc welding is the preferred process for joining copper nickels in thicknesses up to about 1/16 in. and may be used for greater thicknesses. Data on the automatic welding of alloy C71500 (copper nickel, 30%) tubes to tube sheets of the same alloy are given in Table 4. Manual welding is normally used for sheet and plate up to 1/4 in. thick.

Nominal conditions for automatic and manual GTAW of the most common copper-nickel alloys are given in Table 12 for butt joints with square and single-V-grooves.

Preferred conditions include the use of argon shielding gas, DCEN, and thoriated tungsten electrodes, although these variables are not critical for copper nickels, because these alloys have low heat conductivity. Alternating current can be used for automatic welding, provided the arc length is accurately controlled. Helium can also be used as the shielding gas, but argon provides better arc control.

Slightly higher current or a slower speed can be used for alloy C70600 (as shown in Table 12), because of its higher thermal conductivity. Preheating is not needed. Backing strips or rings for use in welding copper nickels should not be made from carbon, graphite, or steel; instead, copper or copper-nickel backing should be used.

Filler Metal. The only filler-metal composition ordinarily used in GTAW of copper nickels is ERCuNi. This filler metal contains 0.20 to 0.50% Ti to minimize porosity and the possibility of oxygen embrittlement, either in the weld metal or in the HAZ, by acting as a deoxidizer. Because the standard copper-nickel alloys do not contain titanium or a comparable deoxidizer, filler metal should be used even in welding thin sheet of copper nickels, to avoid porosity. However, special compositions of alloys C70600 and C71500 that contain titanium are available, and thin sheet of these special alloys can be welded without the use of filler metal. In multiple-pass welding, the welding-rod size and current may be increased with successive passes, as shown in Table 12, for the most efficient deposition rate.

Autogenous welds can sometimes be made in sheet up to 0.06 in. thick, but porosity may be a problem because of the absence of deoxidizers provided by filler metal.

Dissimilar Metals

Copper and many copper alloys can be gas tungsten arc welded to other copper alloys, steels, stainless steels, nickel, and nickel alloys—usually with the aid of a filler metal. Because the use of GTAW for this purpose usually is restricted to thin metal, the welding is ordinarily done without "buttering" with a preliminary surfacing weld. Combinations of dissimilar metals in thicknesses greater than about 1/8 in. are preferably joined by GMAW or, where GMAW is not applicable, by SMAW. Usually, the arc is directed at the more conductive metal of the combination being welded.

Welding With Filler Metal. Table 13 shows combinations of dissimilar metals that are joined by GTAW with the aid of copper alloy or nickel alloy filler metals and gives the recommended filler metals

Table 12 Nominal conditions for gas tungsten arc butt welding of copper nickels(a)

Work-metal thickness, in.	Butt-joint groove(b)	Current (DCEN), A	Electrode diameter(c), in.	Travel speed, in./min	Diameter of ERCuNi welding rod(d), in.	Flow rate of argon, ft³/h	Number of passes
Automatic welding of alloy C70600 (copper nickel, 10%)							
1/8	Square	310-320	3/16	15-18	1/16	25-30	1
Manual welding of alloy C70600 (copper nickel, 10%)							
0-1/8	Square	300-310	3/16	5	1/8	25-30	1
1/8-3/8	70-80° single-V	300-310(e)	3/16	6	1/8, 3/16	25-30	2-4
Manual welding of alloy C71500 (copper nickel, 30%)							
0-1/8	Square	270-290	3/16	5	1/8	25-30	1
1/8-3/8	70-80° single-V	270-290(e)	3/16	6	5/32	25-30	4

(a) The data in this table are intended to serve as starting points for the establishment of optimum joint design and conditions for welding parts on which previous experience is lacking; they are subject to adjustment as necessary to meet the special requirements of individual applications. Root opening is zero. Preheating is not needed. (b) See Table 2 for illustrations of joints. (c) Preferred electrode material is EWTh-2. (d) Filler metal (ERCuNi) must be used on all welded joints (see text). (e) Current should be increased in equal increments with each pass, up to a maximum of about 375 A, with larger welding rods.

and preheat and interpass temperatures for each combination.

Nearly all of the weldable copper alloys, as listed in Table 1, can be joined to dissimilar metals in this way. Copper-zinc alloys are not gas tungsten arc welded to dissimilar metals, except for the welding of copper to low-zinc brass, which is done with phosphor bronze filler metal.

Ordinarily, joints between a ferrous metal and copper or a copper alloy by this method are welds on the copper side of the joint and braze welds on the iron side of the joint, because the melting point of the filler metal is lower than that of the base metal. Exceptions are gas tungsten arc welds joining a copper or a copper nickel to a ferrous metal, using the nickel filler metal ERNi-3, as listed in Table 13. These are true welds on both sides of the joint.

The aluminum bronze filler rod ERCuAl-A2 is compatible with most of the metals in Table 13, with the notable exception of the phosphor bronzes, and is by far the most widely used filler metal for gas tungsten arc welding the dissimilar metals shown. Unlike copper, phosphor bronze, and silicon bronze filler metals, ERCuAl-A2 can tolerate substantial amounts of dilution with iron, and thus can be used for joining ferrous metals to copper alloys with a minimal danger of cracking.

Preheating. Except where one or both members of a joint have high thermal conductivity, preheating is not needed for welding the dissimilar metals listed in Table 13 if both members are less than about 1/8 in. thick.

Referring to Table 13, in most cases the lower preheat temperature or an intermediate temperature applies when two alloys that have different preheat requirements are welded together. An exception is the joining of copper to other metals, in which the need to provide enough heat in this highly conductive metal overrides other considerations, and the preheat temperature of 1000 °F shown for copper must be used to ensure successful welds. Interpass temperatures should not ordinarily be allowed to rise above the prescribed preheat temperature, because many of the copper alloys are hot short at higher temperatures.

Table 13 Filler metals and preheat and interpass temperatures used in GTAW of coppers and copper alloys to dissimilar metals(a)

Metal to be welded	Filler metals (and preheat and interpass temperatures) for welding metal in column 1 to:				
	Coppers	Phosphor bronzes	Aluminum bronzes	Silicon bronzes	Copper nickels
Copper alloys					
Low-zinc brasses	ERCuSn-A or ERCu (1000 °F)	...	...	...	...
Phosphor bronzes	ERCuSn-A or ERCu (1000 °F)	...	...	...	...
Aluminum bronzes	ERCuAl-A2 (1000 °F)	ERCuAl-A2 or ERCuSn-A (400 °F)	...	...	...
Silicon bronzes	ERCuSn-A or ERCu (1000 °F)	ERCuSi-A (150 °F max)	ERCuAl-A2 (150 °F max)	...	...
Copper nickels	ERCuAl-A2 or ERCu or ERCuNi (1000 °F)	ERCuSn-A (150 °F max)	ERCuAl-A2 (150 °F max)	ERCuAl-A2 (150 °F max)	...
Nickel alloys					
Nickel and Ni-Cu alloys	ERCuNi or ERNiCu-7 (1000 °F)	(b)	(b)	(b)	ERCuNi or ERNiCu-7 (150 °F max)
Ni-Cr, Ni-Fe and Ni-Cr-Fe alloys	ERNi-3 (1000 °F)	(b)	(b)	(b)	ERNi-3 (150 °F max)
Steels					
Low-carbon steel	ERCuAl-A2 or ERCu or ERNi-3 (1000 °F)	ERCuSn-A (400 °F)	ERCuAl-A2 (300 °F)	ERCuAl-A2 (150 °F max)	ERCuAl-A2 or ERNi-3 (150 °F max)
Medium-carbon steel	ERCuAl-A2 or ERCu or ERNi-3 (1000 °F)	ERCuSn-A (400 °F)	ERCuAl-A2 (400 °F)	ERCuAl-A2 (150 °F max)	ERCuAl-A2 or ERNi-3 (150 °F max)
High-carbon steel	ERCuAl-A2 or ERCu or ERNi-3 (1000 °F)	ERCuSn-A (500 °F)	ERCuAl-A2 (500 °F)	ERCuAl-A2 (400 °F)	ERCuAl-A2 or ERNi-3 (150 °F max)
Low-alloy steel	ERCuAl-A2 or ERCu or ERNi-3 (1000 °F)	ERCuSn-A (500 °F)	ERCuAl-A2 (500 °F)	ERCuAl-A2 (400 °F)	ERCuAl-A2 or ERNi-3 (150 °F max)
Stainless steel	ERCuAl-A2 or ERCu or ERNi-3 (1000 °F)	ERCuSn-A (400 °F)	ERCuAl-A2 (150 °F max)	ERCuAl-A2 (150 °F max)	ERCuAl-A2 or ERNi-3 (150 °F max)
Cast irons					
Gray and malleable irons	ERCuAl-A2 or ERCu (1000 °F)	ERCuSn-A (400 °F)	ERCuAl-A2 (400 °F)	ERCuAl-A2 or ERCuSi-A (300 °F)	ERCuAl-A2 (150 °F max)
Ductile iron	ERCuAl-A2 or ERCu (1000 °F)	ERCuSn-A (400 °F)	ERCuAl-A2 (150 °F max)	ERCuAl-A2 or ERCuSi-A (150 °F max)	ERCuAl-A2 (150 °F max)

(a) Filler-metal selections shown in table are based on weldability, except where mechanical properties are usually more important. Preheating is ordinarily used only when at least on member is thicker than about 1/8 in. or is highly conductive (see text). Preheat and interpass temperatures are subject to adjustment on the basis of the size and shape of the weldment. (b) These combinations of work metals are only infrequently joined by welding; as a starting point in developing welding procedures for joining them, the use of ERCuAl-A2 filler metal is recommended, except for welding of combinations that include phosphor bronzes.

Gas Metal Arc Welding

Gas metal arc welding is used to join all of the coppers and copper alloys listed in Table 1. It is preferred for joining the aluminum bronzes, silicon bronzes, and copper nickels in section thicknesses greater than about 1/8 in. Gas tungsten arc welding is preferred for thicknesses less than about 1/8 in.

The major application of GMAW to copper alloys is in joining material from 1/8 to 1/2 in. thick, and the process is almost invariably selected for arc welding sections of copper alloys thicker than about 1/2 in., where its high deposition rate is a major advantage over GTAW or SMAW. The greater rate of heat input to the weld,

Table 14 Nominal conditions for gas metal arc butt welding of commercial coppers and copper alloys(a)

Weld types for butt joints (see Table 2 for illustrations)	Work-metal thickness, in.	Root face, in.	Root opening, in.	Electrode	Electrode-wire diameter, in.	Shielding gas	Gas-flow rate, ft^3/h	Current (DCEP), A	Voltage, V	Travel speed, in./min	Number of passes	Preheat temperature, °F
Commercial coppers												
Square groove(b)	1/8	1/8	0	ERCu	1/16	Argon	30	310	27	30	1	None
Square groove(c)	1/8	1/8	0-1/16	ERCu	1/16	Argon(d)	30-35	325-350	28-33	···	1	None
Square groove	1/4	1/4	0	ERCu	3/32	Argon	30	460	26	20	2	200
	1/4	1/4	0	ERCu	3/32	Argon	30	500	27	20	1	200
75-90° single-V-groove(c)	1/4	1/8	0-1/8	ERCu	1/16	Argon(d)	30-35	400-425	32-36	···	2	400-500
	1/2	0-1/8	0-1/8	ERCu	1/16	Argon(d)	30-35	425-450	35-40	···	4	800-900
90° single-V-groove	3/8	3/16	0	ERCu	3/32	Argon	30	500	27	14	(e)	400
	3/8	3/16	0	ERCu	3/32	Argon	30	550	27	14	(e)	400
	1/2	1/4	0	ERCu	3/32	Argon	30	540	27	12	(e)	400
	1/2	1/4	0	ERCu	3/32	Argon	30	600	27	10	(e)	400
Alloy C17500 (high-conductivity beryllium copper)(f)												
90° single-V-groove	1/4-1/2	1/32	···	Alloy C17500	0.045	A-He	30	200-240	···	···	3-4(g)	600
	3/4	1/32	···	Alloy C17500	0.045	A-He	30	200-240	···	···	6(g)	900
Alloys C17000 and C17200 (high-strength beryllium coppers)(f)												
90° single-V-groove	1/4-1/2	1/32-1/16	···	Alloy C17000, C17200	0.045	A-He	45	175-200	···	···	3-4(h)	300-400
30° double-U-groove(j)	3/4-1 1/2	1/16	···	Alloy C17000, C17200	1/16	A-He	60	325-350	···	···	10-20(k)	300-400
Low-zinc brasses												
Square groove(c)	1/8	1/8	0	ERCuSi-A	1/16	Argon	30	275-285	25-28	···	1	None
	1/8	1/8	0	ERCuSn-A	1/16	Helium	35	275-285	25-28	···	1	None
60° single-V-groove(c)	3/8	0	1/8	ERCuSi-A	1/16	Argon	30	275-285	25-28	···	2	None
	1/2	0	1/8	ERCuSi-A	1/16	Argon	30	275-285	25-28	···	4	None
70° single-V-groove(c)	3/8	0	1/8	ERCuSn-A	1/16	Helium	35	275-285	25-28	···	2	500(m)
	1/2	0	1/8	ERCuSn-A	1/16	Helium	35	275-285	25-28	···	4	500(m)
High-zinc brasses, tin brasses, special brasses, nickel silvers												
Square groove(c)	1/8	1/8	0	ERCuSn-A	1/16	Argon	30	275-285	25-28	···	1	None
70° single-V-groove(c)	3/8	0	1/8	ERCuSn-A	1/16	Argon	30	275-285	25-28	···	2	None
	1/2	0	1/8	ERCuSn-A	1/16	Argon	30	275-285	25-28	···	4	None
Phosphor bronzes(n)												
90° single-V-groove(c)	3/8	0	1/8	ERCuSn-A	1/16	Helium	35	275-285	25-28	···	3-4(p)	200-300
	1/2	0	1/8	ERCuSn-A	1/16	Helium	35	275-285	25-28	···	5-6(p)	350-400
Aluminum bronzes (q)												
Square groove(r)	1/8	1/8	0	ERCuAl-A2	1/16	Argon	30	280-290	27-30	···	1	None
60-70° single-V-groove(c)	3/8	0	1/8	ERCuAl-A2	1/16	Argon	30	280-290	27-30	···	2	None
	1/2	0	1/8	ERCuAl-A2	1/16	Argon	30	280-290	27-30	···	3	Slight
Silicon bronzes(s)												
Square groove(t)	1/8	1/8	0	ERCuSi-A	1/16	Argon	30	260-270	27-30	8 min	1	None
60° single-V-groove(c)	3/8	0	1/8	ERCuSi-A	1/16	Argon	30	260-270	27-30	8 min	2	None
	1/2	0	1/8	ERCuSi-A	1/16	Argon	30	260-270	27-30	8 min	3	None
Copper nickels												
Square groove(c)	1/8	1/8	0	ERCuNi	1/16	Argon	30	280	27-30	···	1	None
60-80° single-V-groove(c)	3/8	0-1/32	1/8-1/4	ERCuNi	1/16	Argon	30	280	27-30	···	2	None
	1/2	0-1/32	1/8-1/4	ERCuNi	1/16	Argon	30	280	27-30	···	4	None
Commercial coppers to steel												
70-80° single-V-groove	3/8	1/16	1/8	ERNi-3	1/16	Argon	60	375	29-31	···	4	800-1000
Copper nickel to steel												
70-80° single-V-groove	3/8	1/16	1/8	ERNi-3	1/16	Argon	60	375	29-31	···	4	150 max

(continued)

Table 14 Continued

Weld types for butt joints (see Table 2 for illustrations)	Work-metal thickness, in.	Root face, in.	Root opening, in.	Electrode	Electrode-wire diameter, in.	Shielding gas	Gas-flow rate, ft³/h	Current (DCEP), A	Voltage, V	Travel speed, in./min	Number of passes	Preheat temperature, °F
Aluminum bronze to steel(u)												
60° single-V-groove	$^3/_8$	0	$^5/_{16}$-$^3/_8$	ERCuAl-A2	$^1/_{16}$	Argon	30	270-280	25-27	...	6	300-500
Silicon bronze to steel(v)												
60° single-V-groove	$^3/_8$	0	$^5/_{16}$-$^3/_8$	ERCuAl-A2	$^1/_{16}$	Argon	30	270-280	28-30	...	6	150 max(w)

(a) The data in this table are intended to serve as starting points for the establishment of optimum joint design and conditions for welding of parts on which previous experience is lacking; they are subject to adjustment necessary to meet the requirements of individual applications. Thicknesses up to about $1^1/_2$ in. are sometimes welded by use of slightly higher current and lower travel speed than shown for a thickness of $^1/_2$ in. (b) Copper backing. (c) Grooved copper backing. (d) Or 75% argon, 25% helium. (e) Special welding sequence is used (see text). (f) See Table 1 for compositions. (g) The final pass is made on the root side after back chipping. Grind after each pass. (h) The final pass is made on the root side after back chipping. Wire brush after each pass. (j) Similar to the double-V-groove weld shown in Table 2, but with a groove radius of $^3/_8$ in. (k) Several passes are made on the face side, then several on the back side, until the weld is completed. Back chip the root pass before making the first pass on the back side. Wire brush after each pass. (m) Should not be overheated; as little preheat as possible should be used. (n) Welding conditions based on alloys C51000, C52100, and C52400; current is increased or speed decreased for alloy C50500. (p) Hot peening between passes is recommended for maximum strength. (q) Slight preheat may be needed on heavy sections; interpass temperature should not exceed 600 °F. (r) With $^1/_8$-by-1-in. aluminum bronze backing. (s) No preheat is used on any thickness; interpass temperature should not exceed 200 °F. (t) With $^1/_8$-by-1-in. silicon bronze backing. (u) Steel should be well penetrated; an overlay is not usually needed. (v) Steel should be well penetrated; an overlay should be applied to avoid excessive dilution of the silicon bronze. (w) Except in welding silicon bronze to high-carbon or low-alloy steel, for which preheat temperature is 400 °F

compared with that for GTAW, is a disadvantage in some applications because of the wider HAZ.

Nominal conditions for GMAW of coppers and copper alloys are shown in Table 14. Direct current electrode positive (DCEP) is used exclusively for GMAW of copper alloys. Argon is normally used for shielding. Helium or mixtures of argon and helium are used where hotter arcs are needed than are possible at given current levels with pure argon.

As shown in Table 14, a square-groove joint is not ordinarily used for welding thicknesses greater than $^1/_8$ in. except for coppers; single-V-grooves are used for thicknesses of $^1/_8$ to $^1/_2$ in. In material thicker than about $^1/_2$ in., joints are usually prepared with double-V or double-U-grooves.

Welding Position. Most GMAW of copper alloys is done in the flat position, with spray transfer; fillet welds acceptable for many applications can be produced in the horizontal position. When it is necessary to weld in positions other than flat, GMAW is preferred to GTAW or SMAW. Gas metal arc welding in the vertical and overhead positions is usually restricted to the less fluid copper alloys such as aluminum bronzes, silicon bronzes, and copper nickels. Small-diameter electrode wire and low currents are preferred for such applications, and a globular or short circuiting mode of transfer is ordinarily used.

Electrode Wires (Filler Metals). Electrode wires used for GMAW and their base-metal applications are given in Table 15.

Coppers

The effects of oxygen in causing porosity and reducing the strength of welds in copper, as described for GTAW of copper, are more pronounced in GMAW, because of the greater heat input associated with this process. Gas metal arc welds of deoxidized coppers compare favorably in density and strength with the welds made using GTAW. However, the greater heat input and the lesser localization of heat obtained with GMAW result in greater porosity and lower strength in the HAZ of welds in coppers that contain insufficient amounts of deoxidizer (especially electrolytic tough pitch copper). Consequently, GMAW has less applicability than GTAW for electrolytic tough pitch copper, oxygen-free copper (alloy C10200) or low-phosphorus deoxidized copper (alloy C12000) up to about $^1/_2$ in. thick.

Nominal conditions for gas metal arc butt welding of coppers are given in Table 14. In welding joints for which the root opening is up to $^1/_{16}$ or $^1/_8$ in., mixtures of 75% helium-25% argon are used instead of pure argon because the preheat requirements are lower, joint penetration is better, and filler-metal deposition rates are higher.

Electrode Wires. The recommended ERCu electrode wire contains phosphorus, tin, silicon, and manganese as deoxidizers to minimize porosity. This electrode produces sound, trouble-free welds that are also a good match to copper in color and have good electrical conductivity. The ERCu electrode was used as a basis for conditions listed in Table 14.

Copper electrode wires of purity higher than ERCu are seldom used in GMAW of copper, because, as a consequence of the absence of deoxidizers, they usually make welds that are porous. Most of the electrodes listed in Table 15 can be used. Any of the electrodes other than ERCu will make dense, strong welds, but the color match will be poor, and the electrical conductivity may be unacceptable. The filler metal should be deposited in stringer beads or narrow weave beads using spray transfer. Wide weaving of the electrode may result in oxidation at the edges of the bead.

The forehand welding technique should be used in the flat position. In the vertical position, the progression of welding should be up. Gas metal arc welding of copper is not recommended for the overhead position because the bead shape will be poor.

Table 15 Filler metals for GMAW of copper and copper alloys

Bare wire(a)	Common name	Base-metal applications
ERCu	Copper	Coppers
ERCuSi-A	Silicon bronze	Silicon bronzes, brasses
ERCuSn-A	Phosphor bronze	Phosphor bronzes, brasses
ERCuNi	Copper nickel	Copper-nickel alloys
ERCuAl-A2	Aluminum bronze	Aluminum bronzes, brasses, silicon bronzes, manganese bronzes
ERCuAl-A3	Aluminum bronze	Aluminum bronzes
ERCuNiAl	...	Nickel-aluminum bronzes
ERCuMnNiAl	...	Manganese-nickel-aluminum bronzes
RBCuZn-A	Naval brass	Brasses, copper
RCuZn-B	Low-fuming brass	Brasses, manganese bronzes
RCuZn-C	Low-fuming brass	Brasses, manganese bronzes

(a) See AWS A5.7-77, "Specification for Copper and Copper Alloy Bare Welding Rods and Electrodes" or AWS A5.27-78, "Specification for Copper and Copper Alloy Gas Welding Rods."

The preferred method for this position is GTAW.

Joint Design. A square-groove joint is used for single-pass welding of coppers up to 1/8 in. thick, in conjunction with a copper backing bar for zero root opening or a grooved copper backing bar where the root opening is 1/16 in. maximum. A square-groove joint is also used for one-pass-per-side welding of coppers up to about 1/4 in. thick.

In welding the single-V-groove joints referred to in Table 14 for thicknesses of 3/8 to about 1/2 in., the metal is deposited on one side in three or more passes, and the root pass is back gouged to sound metal before the last pass is applied to the back of the joint. In some applications, the root pass is applied by GTAW, and subsequent passes are made by GMAW for fast buildup.

Heavy sections (thicker than 1/2 in.) should be prepared with a double-V or a double-U-groove and welded with alternate passes applied to opposite sides of the joint, if readily accessible, to minimize distortion. In a small closed vessel, limited access may prevent the use of this technique, and in addition, heat buildup often prevents welding on both sides.

Preheating. Because of the high thermal conductivity of copper, sections thicker than 1/4 in. are usually preheated. As shown in Table 14, single-V-groove joints in 1/4-in.-thick copper, welded using 1/16-in.-diam electrode wire, are also preheated.

Deoxidized Coppers. Heavy-walled copper pressure vessels (up to about 1 1/2-in. wall), which are usually made from phosphorus-deoxidized copper in which the residual phosphorus helps obtain maximum weld strength and freedom from porosity, are frequently gas metal arc welded. To meet the high heat demand in welding these heavy-walled vessels, welding currents and preheat temperatures may be higher than those shown in Table 14.

Similar operating conditions are employed in welding crucibles used for arc melting of refractory metals in controlled atmospheres. The crucibles are typically heavy-walled deoxidized copper cylinders with wall thicknesses up to 1 1/4 in., diameters of 12 to 48 in., and lengths of 3 to 25 ft. They are made of formed and welded copper plate, because seamless tubes of these sizes are not available. Longitudinal seams are welded, and flanges are welded to each end of the cylinders by GMAW. Such cylinders are heated to 1200 °F before welding. Welding is done automatically to avoid bringing operators into close proximity with heavy metal sections at such high temperature.

Oxygen-Free and Tough Pitch Coppers. The gas metal arc process is used also in welding oxygen-free copper in producing crucibles of the type just described for arc melting of refractory metals, although the weld properties are inferior to those produced with deoxidized copper. Because oxygen-free copper contains no residual deoxidizer, heating and welding cycles must be kept as short as possible, and gas shielding must be completely effective to avoid excessive porosity.

Although the strength and soundness of welds in oxygen-bearing coppers such as electrolytic tough pitch (ETP) copper are substantially less than in oxygen-free copper, the reduced properties are seldom important for welds in electrical conductors. Bus bars made of ETP copper are sometimes joined by GMAW. The adverse effect of the relatively low electrical conductivity of the filler metal, such as ERCu or one of the other standard filler metals listed in Table 15, on the electrical conductivity of the joint can be reduced by providing a large contact area at the joint. For better electrical conductivity, a filler metal with about 0.75% Sn, 0.25% Si, and 0.20% Mn is sometimes used for welding ETP copper bus bars.

High-Conductivity Beryllium Copper

Gas metal arc welding is preferred to GTAW for joining thicknesses of alloy C17500 (high-conductivity beryllium copper) greater than 0.090 in. if the weldment is to be heat treated to obtain maximum weld strength. It also is preferred for joining thicknesses greater than about 1/4 in. if the welding is to be done on precipitation-hardened material. The maximum thickness normally joined by GMAW is about 3/4 in.

As pointed out in the discussion of GTAW of alloy C17500, important factors in arc welding of this alloy are its high thermal and electrical conductivities, oxide-forming characteristics, and response to heat treatment. Although some compromise between weld strength and conductivity may be necessary, high thermal or electrical conductivity is ordinarily the prime objective in welded assemblies of this alloy.

Nominal conditions for gas metal arc butt welding of high-conductivity beryllium copper are given in Table 14. Direct current electrode positive is almost always used. Although argon may be used for shielding gas, greater heat input (usually desirable for welding this alloy) can be obtained with an argon-helium mixture.

Electrode Wires. When high conductivity is desired, as in most welding applications of alloy C17500, electrode wire of the same composition as the base metal is used. When lower conductivity is adequate, electrodes made of high-strength beryllium copper (alloys C17000 and C17200) can be used, because they provide easier welding. Because of the precipitation-hardening characteristics of beryllium coppers, the joint strength is always somewhat lower than that of the base metal, depending on the initial condition of the base metal, the welding conditions, and the selection of filler metal.

Other electrode wires, such as ERCuSi-A or ERCuAl-A2 also are used in GMAW of alloy C17500, but these electrode wires (which contain no beryllium) do not develop the high strength obtained with beryllium copper filler metals.

Preheating and Postweld Aging. Sections of alloy C17500 thicker than 1/8 in. are usually preheated at 600 to 900 °F, depending on section thickness. When beryllium copper filler metal is used, strength can be increased by aging after welding. For alloy C17500 the aging treatment is 900 °F for 3 h.

Properties of Weldments. Because of high thermal conductivity and moderate to high strength, alloy C17500 is used for welded water-cooled assemblies such as tuyeres for blast furnaces, attrition mills for grinding beryllium chips to powder, and molds for the continuous casting of steel. Some strength is lost in the weld metal and in the HAZ when alloy C17500 is welded. The magnitude of this decrease is affected by the condition and thickness of the base metal, the joint design, welding process, and welding conditions. In the following example, the decrease in strength for alloy C17500 was less for GMAW than for GTAW.

Example 12. Comparison of Gas Metal Arc Welding and Gas Tungsten Arc Welding for Joining High-Conductivity Beryllium Copper. Comparison studies were made on joining alloy C17500 (high-conductivity beryllium copper, 0.6%) by GMAW and GTAW for several heat-transfer applications. Test specimens were prepared by welding together pieces 5 by 16 by 9/16 in. thick. Joint preparation consisted of beveling the edges to be joined to make a V-groove 7/16 in. deep with a 90° included angle, leaving a root face of 1/8 in., as shown in the drawing in Table 16.

The gas tungsten arc welds were made by depositing metal in six passes on the face side, back gouging to sound metal, and completing the weld by making two

Table 16 Decrease in strength of alloy C17500 after GMAW and GTAW(a)

	GMAW	GTAW
Properties, using solution-annealed base metal(b)		
Tensile strength, ksi:		
Before welding	47	48.2
After welding	46	37
Decrease	2%	23%
Break:		
Before welding	Normal	Normal
After welding	Outside weld	Outside weld
Properties, using precipitation-hardened base metal(c)		
Tensile strength, ksi:		
Before welding	121	98.9
After welding	110	81.7
Decrease	9%	17%
Break:		
Before welding	Normal	Normal
After welding	In weld	In weld
Welding conditions(d)		
Power supply	900-A motor-generator	500-A alternator(e)
Electrode	1/16-in.-diam alloy C17500	1/4-in.-diam EWTh-2
Filler wire	(Consumable electrode)	1/8-in.-diam alloy C17500
Shielding gas	Argon, at 30 ft^3/h	Helium, at 25 ft^3/h
Current, A	325 (DCEP)	240-260 (ac, hf-stabilized)
Voltage, V	22	...
Wire-feed rate, in./min	200	...
Travel speed, in./min	11	...
Number of passes	4	8
Preheat temperature, °F	1000-1200	800-900

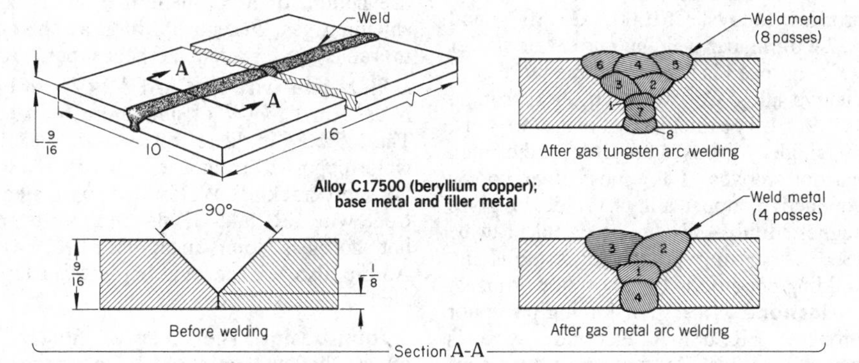

(a) Data are for welding specimens as illustrated above, in the solution-annealed and the precipitation-hardened conditions. Each value for tensile strength is the average of two tests. (b) Solution-annealing treatment consisted of heating to 1700 to 1750 °F and water quenching. (c) Precipitation hardening was done by reheating the solution-annealed specimens for 3 to 4 h at 900 °F and air cooling. (d) For butt joints and single-V-groove welds, as illustrated above; and by use of a manual, water-cooled electrode holder and flat-position welding in both GMAW and GTAW. (e) High-frequency

back passes. Before welding, the joint was wire brushed, and each bead was wire brushed to remove any oxide before the next pass.

The work metal was preheated at 800 to 900 °F and alternating current was used, both to help keep the surface free of oxide, and to avoid longitudinal bead cracking that might have occurred with direct current welding.

The gas metal arc welds were made with three passes on the face side and one back pass, with the root side being ground out to a depth of 1/8 to 3/16 in. before the back pass was made, and with wire brushing of each bead in the same way as for the gas tungsten arc welds. Sound welds free from cracking were produced by each process when using the procedure described above and the welding conditions in Table 16, without using a backing gas.

Gas tungsten arc welding of beryllium copper slightly thicker than 1/2 in. produced specimens showing a 23% decrease in tensile strength for the solution-annealed condition and a 17% decrease in the solution-annealed and precipitation-hardened condition, as shown in Table 16. The usual thickness limit is 1/2 in.

Specimens welded by GMAW, working well within its usual limit of 3/4-in. thickness for this alloy, showed substantially less decrease in strength. Welded joints made in solution-annealed alloys showed only a 2% decrease in tensile strength, while those made in alloys that had been solution annealed and precipitation hardened were 9% lower in strength than the base metal (Table 16). These results may have been affected to a minor extent by the differences in properties of the test specimens used for the two welding processes. As Table 16 shows, in the precipitation-hardened condition, the specimens used for GMAW had a tensile strength of 121 ksi, while those specimens used for GTAW had a tensile strength of only 98.9 ksi. Considerable variation in the results from precipitation hardening is not unusual. In this instance, determining the percentage of decrease for the two welding processes was the primary objective.

Results of bend tests on specimens welded in the solution-annealed condition were acceptable, with no specimen showing more than one crack; the cracks were located near the root edge and were less than 1/8 in. long.

Specimens for all of the tests for qualification of the welding procedure were prepared in the manner described in Section IX of the American Society of Mechanical Engineers (ASME) Boiler and Pressure Vessel Code.

Results by both welding processes were considered acceptable for heat-transfer applications, with gas metal arc welds being preferred where strength requirements were the most critical; both processes were used in welding a limited number of production assemblies.

High-Strength Beryllium Coppers

Alloys C17000 (1.7% Be) and C17200 (1.9% Be), high-strength beryllium coppers, are more easily welded than high-conductivity beryllium coppers, because alloys C17000 and C17200 have lower melting temperatures, greater fluidity, and 50% lower thermal conductivity. As it is for high-conductivity beryllium coppers (see the preceding section), GMAW is generally preferred for welding precipitation-hardened high-strength beryllium coppers in thicknesses of more than 1/4 in., and is preferred for thicknesses down to 0.090 in. if heat treatment is to be done after welding.

Nominal conditions for gas metal arc butt welding of high-strength beryllium coppers are given in Table 14.

Shielding Gas. Mixtures of argon and helium are ordinarily used for shielding gas, in conjunction with DCEP, which helps to prevent oxide buildup during welding. The electrode wire is ordinarily of the same composition as the base metal for joints of maximum strength; electrode wires ERCuSi-A and ERCuAl-A2 can be used where joint strength is less critical. Preheating, postweld heat treating, and joint design are generally the same as described in the section on GTAW of high-strength beryllium coppers in this article.

Brasses

Nonleaded brasses, of both the low-zinc type (red brasses) and the high-zinc type (including yellow brasses, tin brasses, and special brasses), can be gas metal arc welded. Copper-zinc electrodes are not used because of the violent fuming and loss of zinc that accompanies arc welding with zinc-containing electrodes.

Low-Zinc Brasses. Nominal conditions for butt welding the low-zinc brasses (up to 20% Zn) by GMAW are presented in Table 14. Direct current electrode positive is always used.

ERCuSi-A electrode wire provides easy welding because it has good fluidity at low current. A 60° single-V-groove joint is used with ERCuSi-A. When ERCuSn-A electrode wire is used, principally for better color match, its sluggish flow characteristics make a 70° V-groove advisable for the heavier thicknesses shown in Table 14. The wider groove allows more room for manipulation of the molten weld metal. Except for color, weld-metal properties are comparable when using these two types of electrodes. A preheat in the range of 200 to 600 °F is sometimes used for low-zinc brasses because of their relatively high thermal conductivities.

High-Zinc Copper Alloys. Nonleaded copper alloys with zinc contents ranging from about 20 to 40%, or more (see compositions of high-zinc brasses, tin brasses, special brasses, and nickel silvers in Table 1), can be gas metal arc welded, although with greater difficulty than the nonleaded low-zinc brasses. Zinc fumes are more severe, and the welds have greater porosity and lower strength than in the low-zinc brasses. Both wrought and cast alloys are joined by GMAW; massive sections such as manganese bronze ship propellers are regularly repair welded by this process.

Nominal conditions for gas metal arc butt welding of high-zinc copper alloys (high-zinc or yellow brasses, tin brasses, special brasses, and nickel silvers) are given in Table 14. Operating variables are generally the same when using either ERCuAl-A2 (for higher weld strength) or ERCuSn-A (for better color match). Preheating is seldom necessary, because these alloys have relatively low heat conductivity. However, preheating helps to limit fuming in some applications, because it permits use of lower current.

Phosphor Bronzes

Nominal conditions for gas metal arc butt welding of phosphor bronzes are given in Table 14. Lead-bearing or other free-machining types are not welded. As with GTAW, the welding conditions are based on the three poorly conductive phosphor bronzes (alloys C51000, C52100, and C52400); higher welding current or slower welding speed is needed for the more conductive alloy C50500 (phosphor bronze, 1.25% E). For thicknesses of 3/8 to 1/2 in., 90° single-V-grooves are used, rather than narrow grooves as for most other poorly conductive copper alloys. Table 17 gives suggested GMAW conditions that can be used as starting points for establishing welding conditions for phosphor bronzes.

Electrode Wires. For joining phosphor bronzes, ERCuSn-A electrode wire is generally used. Electrode wire that contains about 1/2% Si is sometimes used to minimize porosity in the weld.

Preheating helps in obtaining complete fusion. Also, it minimizes porosity because the freezing rate of the weld pool is decreased, and more gas is permitted to be evolved before solidification. However, preheating increases the susceptibility of the weld to large columnar grain growth and to hot cracking; thus it is common practice to weld with a stringer-bead technique and to peen between layers. A small weld pool and rapid electrode travel are required. Interpass temperature should not exceed the preheating temperature (Table 14), because these alloys are hot short.

Aluminum Bronzes

Gas metal arc welding with aluminum bronze electrode wire is the preferred technique for welding aluminum bronze. Because of the comparatively high surface tension of the molten weld metal and the low thermal conductivity of the base metal, welding can be done in all positions. Welds in the vertical and overhead positions are usually made with either the globular or the short circuiting mode of metal transfer, using electrode wire up to 1/16 in. in diameter.

Nominal conditions for gas metal arc butt welding of aluminum bronze are given in Table 14. The use of aluminum bronze backing strips or rings in welding metal up to 1/8 in. thick may make it necessary to use helium or argon-helium mixtures as shielding gas, instead of argon as shown in Table 14, for adequate heat input.

Electrode wire ERCuAl-A2 ordinarily is used in GMAW of aluminum bronze. The 1.5% Fe in this electrode wire makes welds deposited from it less susceptible to hot short cracking. Welds made with electrode wire ERCuAl-A3 also are free from hot shortness. Joints made with ERCuAl-A3 are also stronger and harder, but less ductile.

Joint Design. The joining of aluminum bronze thicker than about 1/2 in. requires wider root openings and groove angles than are indicated in Table 14, to avoid bridging and incomplete filling of the joint or to improve penetration because of the increased heat sink in the heavier sections. A short arc, less than 1/8 in. long, is preferred for groove welding of aluminum bronze by GMAW.

Preheating is often not needed when welding thin sections of aluminum bronzes. If preheating is needed, the preheat and interpass temperatures should not exceed 300 °F for alloys with less than 10% Al. When the aluminum content is from 10 to 13%, a preheat temperature of about 500 °F is recommended for thick sections. Interpass temperature should not exceed 500 °F, because some aluminum bronzes are hot short.

Table 17 Suggested conditions for GMAW of phosphor bronze

Metal thickness, in.	Joint design: Groove type	Joint design: Root opening, in.	Electrode diam(a), in.	Arc voltage, V	Welding current (DCEP)(a), A
0.06	Square	0.05	0.030	25-26	130-140
0.13	Square	0.09	0.035	26-27	140-160
0.25	V-groove	0.06	0.045	27-28	165-185
0.50	V-groove	0.09	0.062	29-30	315-335
0.75	Double-V or double-U-groove	0-0.09	0.078	31-32	365-385
1.00	Double-V or double-U-groove	0-0.09	0.094	33-34	440-460

(a) ERCuSn-A phosphor bronze electrodes and argon shielding

Silicon Bronzes

The nonleaded silicon bronzes are readily gas metal arc welded.

Nominal conditions for gas metal arc butt welding of silicon bronzes in various thicknesses and joint designs are given in Table 14. Generally, sections over 1/4 in. thick are welded by this process. To prevent excessive heat buildup in these hot short alloys, travel speed should be as fast as possible (minimum of 8 in./min), and current, as shown in Table 14, is slightly lower than for most other copper alloys. The resulting low heat input to the weld is adequate for complete fusion and good penetration, because of the low thermal conductivity of the silicon bronzes (Table 1). Argon shielding is preferred. A thin layer of oxide forms on the weld metal after each pass and must be removed by wire brushing before the next pass.

Electrode Wire. Any silicon bronze alloy can be gas metal arc welded with ERCuSi-A electrodes. These electrodes are similar in composition to alloy C65500 (high-silicon bronze A), the most frequently used silicon bronze, except that they may include up to 1.5% Sn instead of, or in addition to, the 1.5% Mn specified as maximum for alloy C65500.

Joint Preparation. For thicknesses of 1/4 to 3/4 in., single, 60° V-grooves are suitable. For silicon bronze alloys thicker than 3/4 in., U-grooves or 60° double-V-grooves can be used.

Preheating should not be used, because the silicon bronze alloys are hot short; interpass temperature must be held below 200 °F. Because silicon bronzes have low thermal conductivity, there is no difficulty in obtaining enough heat at the joint for satisfactory welding. Stress relieving of silicon bronze weldments is recommended to prevent stress corrosion failure.

Copper Nickels

Gas metal arc welding is the preferred arc welding process for nonleaded copper nickels thicker than about 1/16 in. Table 18 gives welding conditions for GMAW of copper nickel alloy plate. Welding in the flat position is preferred. Argon is the preferred shielding gas, but argon-helium mixtures give better penetration with thick sections. Direct current electrode positive is recommended.

Nominal conditions for gas metal arc butt welding of nonleaded copper nickels in various joint designs and thicknesses are given in Table 14. These alloys can be welded using either spray or short circuiting transfer. Spray transfer is normally used for sections 1/4 in. thick or greater. Spray transfer with pulsed power or short circuiting transfer can be used to weld thin sections. These transfer methods provide better control of the molten weld pool when welding in the vertical and overhead positions. Selection of processing conditions is not critical in GMAW of these poorly heat-conductive alloys. Alloy C70600 is usually welded at slightly higher current or slower speed than alloy C71500.

Table 18 Conditions for GMAW of copper-nickel alloy plate

Thickness, in.	Electrode feed(a), in./min	Arc voltage, V	Welding current (DCEP), A
0.25	180-220	22-28	270-330
0.38	200-240	22-28	300-360
0.50	220-240	22-28	350-400
0.75	220-240	24-28	350-400
1.0	220-240	26-28	350-400
Over 1.0	240-260	26-28	370-420

(a) ERCuNi electrode, 0.062 in. diam

Electrode Wire. Copper nickels are commonly gas metal arc welded with ERCuNi electrode wire, which resembles alloy C71500 in that it has a 70-to-30 ratio of copper to nickel. The titanium content (0.15 to 1.00%) of ERCuNi serves as a deoxidizer to minimize porosity and to prevent oxygen embrittlement. Titanium also improves the fluidity of the weld metal.

Joint Design. For GMAW of butt joints in nonleaded copper nickels 1/8 in. thick, joint design usually includes a square-groove weld and grooved copper backing, as indicated in Table 14. For thicknesses between 1/8 and 1/2 in., single-V-grooves of 60 to 80° and grooved copper backing are usually employed. For thicknesses greater than about 1/2 in., double-V or double-U-grooves are used. Backing should be made of copper (or of copper nickel), rather than of carbon, graphite, or steel, to avoid reaction with copper and nickel.

Preheating and Postheating. The copper nickel alloys have thermal conductivity equal to or lower than that of low-carbon steel, and no preheating or postheating is needed. Interpass temperature should be kept below 150 °F.

Joining of Tubing. Gas metal arc welding is used extensively for joining copper nickel tubing. The automatic method described in the following example is applicable to tubes larger in outside diameter than 5 in. The offset angles for positioning the electrodes (see Fig. 7) were as follows:

5 in. diam	16 ± 2 1/2° offset
6 in. diam	14 ± 2 1/2° offset
8 in. diam or over	12 ± 2 1/2° offset

A semiautomatic method usable for pipes 3 1/2 in. in diameter and larger is compared with the automatic method in the example.

Example 13. Use of Gas Metal Arc Welding to Butt Weld Copper Nickel Tubing. Alloy C71500 copper nickel tubing 6 1/2 in. in outside diameter, as shown in Fig. 7, was successfully gas metal arc butt welded. Wall thickness varied, but was 1/4 in. minimum. The joints were prepared by beveling the ends of the tubes to make a V-groove with an 80° included angle and 0.000- to 0.031-in. root face. A backing ring of copper nickel was provided with a minimum outside diameter 0.015 in. smaller than the inside diameter of the tubes. The backing ring was not removed after welding. Before the tubes and backing ring were assembled, the inner surfaces of the tubes, where they were to be in contact with the backing ring, were cleaned to bright metal, as were the beveled ends of the tubes and the outer surfaces for 1/4 in. to each side of the joint.

The two tubes and the backing ring were assembled, using a 1/4-in. minimum root opening, and tack welded. The tack welds were ground back almost to base metal, and the joint was carefully inspected for starts, stops, crevices, and other irregularities that would need to be ground smooth before welding was continued.

Two different methods were used to weld the tube, a semiautomatic method and an automatic method, as detailed in the table with Fig. 7. In the semiautomatic method, the tube assembly was laid flat on a work surface, and each weld bead was applied in three 120° increments from the 12 o'clock position to the 4 o'clock position, with the assembly stationary during each third of a pass. After each 120° increment, the assembly was rolled counterclockwise to present the next 120° segment of the groove to the welder. The weld was completed in eight passes (including two root passes), for each of which the assembly was in three different attitudes on the table, to weld in the horizontal-fixed pipe position from 12 to 4 o'clock (Fig. 7, section A-A).

In the automatic method, the tube was supported on rolls that turned it continuously while the electrode holder remained fixed for flat-position welding (section B-B).

Development of the automatic process provided data to ensure that satisfactory results were obtained with only one root pass. By carefully controlling clearance of the backing ring, subsequent passes were made without formation of crevices.

Fig. 7 Setups and weld build-up sequences. For gas metal arc butt welding of copper nickel tube by semiautomatic and automatic methods

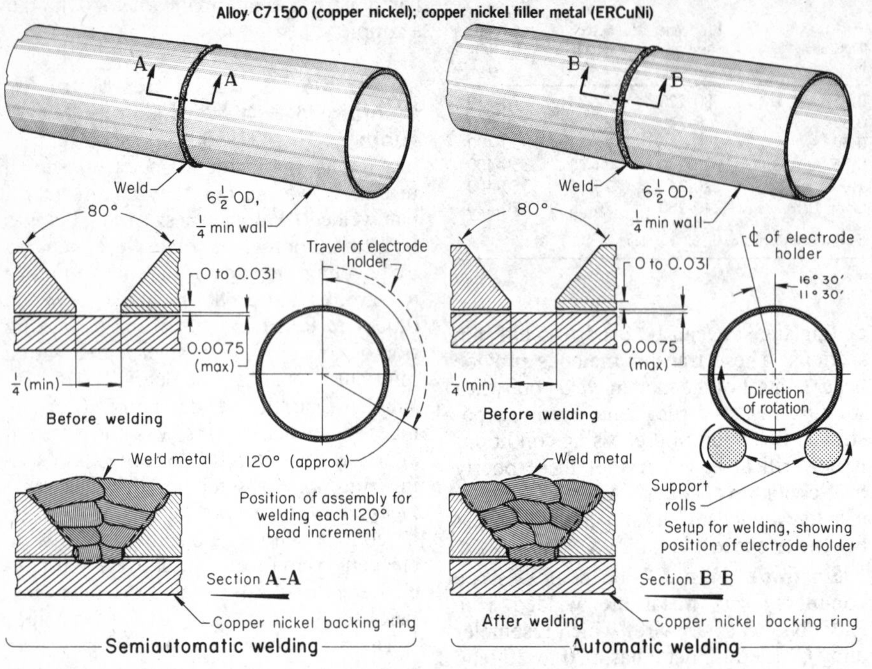

	Semiautomatic welding	Automatic welding
Conditions for GMAW		
Joint type	Butt	Butt
Weld type	Single-V-groove	Single-V-groove
Welding position	Horizontal-fixed pipe(a)	Flat (horizontal-rolled pipe)
Power supply (both methods)	Constant-voltage; variable slope, voltage, and inductance; and compensation for line-voltage variation	
Electrode	0.035-in.-diam ERCuNi	0.062-in.-diam ERCuNi
Shielding gas	50% argon, 50% helium	75% argon, 25% helium(b)
Gas-flow rate, ft^3/h	35	200
Current, A (DCEP)	145-185	230-250
Voltage, V	21-27	28-31
Wire speed, in./min	290-490	...
Travel speed, in./min	15	18
Preheat	None	None
Interpass temperature	150 °F max	150 °F max

(a) Each bead was deposited in three 120° increments, with assembly being turned counterclockwise after each increment. (b) Mixed-gas measurements were based on use of a helium flowmeter.

Dissimilar Copper Alloys

Gas metal arc welding can be used for joining nearly all combinations of the weldable copper alloys, with the major factors in developing the welding procedures being the selection of suitable electrode-wire compositions and preheat temperatures. As in other welding of dissimilar metals, the arc is usually directed at the more conductive metal of the combination.

Electrodes and preheat and interpass temperatures for GMAW of various combinations of unlike copper alloys are given in Table 19. This table includes all of the alloys listed for GTAW in Table 13, plus the high-zinc brasses, tin brasses, and special brasses.

As shown in Table 19, ERCuAl-A2 comes close to being a universal electrode for GMAW of unlike copper alloys, although it is incompatible with phosphor bronzes. Silicon bronze (ERCuSi-A) and phosphor bronze (ERCuSn-A) electrodes are useful for many combinations of copper alloys that do not contain nickel. Copper electrodes (ERCu) are suitable for the final welding passes for welding copper to any of the commonly welded copper alloys, using alloy electrodes to produce the preliminary weld deposits. Copper nickel electrodes (ERCuNi) are generally used to join copper to copper nickel.

Usually, a suitable electrode for welding unlike copper alloys is one that is used for welding one of the metals of the combination to itself. A notable exception to this general rule is in the welding of silicon bronzes to copper nickel. The only electrode that can be used successfully for joining this combination is ERCuAl-A2.

Copper to Copper Nickels. Copper rarely requires direct welding to pure nickel, but joints between copper and the copper nickels are quite common in many heat-exchanger and chemical plant applications. In dissimilar metal joints with copper, it is common practice to use the copper nickel rather than the other copper-based filler metals to minimize the risk of porosity and also to give weld metal of strength compatible with the copper nickel side of the joint. There are no special metallurgical difficulties in producing such a joint by direct weld runs on a conventional weld preparation. The difference in thermal conductivity will, however, mean close attention to preheating the copper side of the joint to ensure that full fusion of the copper is attained. For this reason, it may be helpful to direct the welding arc toward the copper. Dilution of the copper nickel weld metal by copper will occur, and this must be taken into account when selecting the composition of the filler alloy in terms of maintaining matching corrosion resistance. Welding copper nickel alloys to each other presents no problems.

Copper to Aluminum Bronze. Welded joints of this nature are mainly encountered where branches and pump housings utilizing aluminum bronzes for their excellent corrosion resistance to severe turbulence or seawater are joined to the main shell of a copper heat exchanger. In these instances, the aluminum bronze will often be in cast form. Direct welding is suitable using aluminum bronze filler metal.

Aluminum Bronze to Copper Nickel. The increasing use of copper nickel alloys, both 90%Cu-10%Ni and 70%Cu-30%Ni compositions, for heat-exchanger shells can mean welds between the copper nickel and aluminum bronze castings for branches and end plates. Such joints are satisfactorily made directly using an aluminum bronze filler of matching composition.

Preheating. As with GTAW, preheating is not needed in joining metal less than about $^1/_8$ in. thick, unless at least one member has high thermal conductivity. In joining greater thicknesses, the need for

Table 19 Electrodes and preheat and interpass temperatures used in GMAW of coppers and copper alloys to dissimilar metals(a)

Metal to be welded	Electrodes (and preheat and interpass temperatures) for welding metal in column 1 to: Coppers	Low-zinc brasses	High-zinc brasses, tin brasses, special brasses	Phosphor bronzes	Aluminum bronzes	Silicon bronzes	Copper nickels
Copper alloys							
Low-zinc brasses	ERCuSn-A or ERCu (1000 °F)	...	...	...	...	...	...
High-zinc brasses, tin brasses, special brasses	ERCuSi-A or ERCuSn-A or ERCu (1000 °F)	ERCuSn-A (600 °F)	...	...	...	...	...
Phosphor bronzes	ERCuSn-A or ERCu (1000 °F)	ERCuSn-A (500 °F)	ERCuSn-A (600 °F)	...	...	...	...
Aluminum bronzes	ERCuAl-A2 (1000 °F)	ERCuAl-A2 (600 °F)	ERCuAl-A2 (600 °F)	ERCuAl-A2 or ERCuSn-A (400 °F)	...	...	...
Silicon bronzes	ERCuSn-A or ERCu (1000 °F)	ERCuAl-A2 or ERCuSi-A (150 °F max)	ERCuAl-A2 or ERCuSi-A (150 °F max)	ERCuSi-A (150 °F max)	ERCuAl-A2 (150 °F max)	...	...
Copper nickels	ERCuAl-A2 or ERCuNi or ERCu (1000 °F)	ERCuAl-A2 (150 °F max)	ERCuAl-A2 (150 °F max)	ERCuSn-A (150 °F max)	ERCuAl-A2 (150 °F max)	ERCuAl-A2 (150 °F max)	...
Nickel alloys							
Nickel and Ni-Cu alloys	ERCuNi or ERNiCu-7 (1000 °F)	(b)	(b)	(b)	(b)	(b)	ERCuNi or ERNiCu-7 (150 °F max)
Ni-Cr, Ni-Fe, and Ni-Cr-Fe alloys	ERNi-3 (1000 °F)	(b)	(b)	(b)	(b)	(b)	ERNi-3 (150 °F max)
Steels							
Low-carbon steel	ERCuAl-A2 or ERCu or ERNi-3 (1000 °F)	ERCuSn-A (600 °F)	ERCuAl-A2 (500 °F)	ERCuSn-A (400 °F)	ERCuAl-A2 (300 °F)	ERCuAl-A2 (150 °F max)	ERCuAl-A2 or ERNi-3 (150 °F max)
Medium-carbon steel	ERCuAl-A2 or ERCu or ERNi-3 (1000 °F)	ERCuAl-A2 (600 °F)	ERCuAl-A2 (500 °F)	ERCuSn-A (400 °F)	ERCuAl-A2 (400 °F)	ERCuAl-A2 (150 °F max)	ERCuAl-A2 or ERNi-3 (150 °F max)
High-carbon steel	ERCuAl-A2 or ERCu or ERNi-3 (1000 °F)	ERCuAl-A2 (600 °F)	ERCuAl-A2 (500 °F)	ERCuSn-A (500 °F)	ERCuAl-A2 (500 °F)	ERCuAl-A2 (400 °F)	ERCuAl-A2 or ERNi-3 (150 °F max)
Low-alloy steel	ERCuAl-A2 or ERCu or ERNi-3 (1000 °F)	ERCuAl-A2 (600 °F)	ERCuAl-A2 (600 °F)	ERCuSn-A (500 °F)	ERCuAl-A2 (500 °F)	ERCuAl-A2 (400 °F)	ERCuAl-A2 or ERNi-3 (150 °F max)
Stainless steel	ERCuAl-A2 or ERCu or ERNi-3 (1000 °F)	ERCuAl-A2 or ERCuSn-A (600 °F)	ERCuAl-A2 (600 °F)	ERCuSn-A (400 °F)	ERCuAl-A2 (150 °F max)	ERCuAl-A2 (150 °F max)	ERCuAl-A2 or ERNi-3 (150 °F max)
Cast irons							
Gray and malleable irons	ERCuAl-A2 or ERCu (1000 °F)	ERCuAl-A2 or ERCuSn-A (600 °F)	ERCuAl-A2 (600 °F)	ERCuSn-A (400 °F)	ERCuAl-A2 (400 °F)	ERCuAl-A2 or ERCuSi-A (300 °F)	ERCuAl-A2 or ERCuNi (150 °F max)
Ductile iron	ERCuAl-A2 or ERCu (1000 °F)	ERCuAl-A2 (600 °F)	ERCuAl-A2 (600 °F)	ERCuSn-A (400 °F)	ERCuAl-A2 (150 °F max)	ERCuAl-A2 or ERCuSi-A (150 °F max)	ERCuAl-A2 or ERCuNi (150 °F max)

(a) Electrode selections in table are based on weldability, except where mechanical properties are usually more important. Preheating usually is used only when at least one member is thicker than 1/8 in. or is highly conductive (see text). Preheat and interpass temperatures are subject to adjustment based on size and shape of weldment. (b) These combinations are seldom welded; as a starting point in developing welding procedures, use of ERCuAl-A2 electrodes is recommended, except for combinations including phosphor bronzes.

preheating and the choice of preheat temperature are influenced by the joint design and welding conditions. When joining copper to other copper alloys, the need to provide enough heat to melt the highly conductive copper overrides other considerations, and the preheat temperature for copper must be used. For instance, when thick copper is welded to copper nickel, the required preheat temperature is 1000 °F, although copper nickel ordinarily is held below 150 °F between passes. The need for enough heat to melt the copper overrides the concern with hot shortness of the copper nickel. The welded assembly is handled carefully until it has cooled below 150 °F. Table 19 also shows that the heat sensitivity of copper nickel overrides other considerations when it is welded to other copper alloys and to other metals.

Copper Alloys to Dissimilar Metals

The main consideration in GMAW of copper or a copper alloy to a ferrous or nickel alloy is dilution. It is not difficult to make welds as strong as the weaker of the base metals, but it may be difficult to retain the ductility demanded by service requirements. The most serious effect of excessive dilution is unsoundness in the form of shrinkage cracks, which can start in the weld metal and propagate into the base metal.

Either of two methods is frequently used to control dilution when gas metal arc welding combinations involving copper alloys: (1) braze weld one side of the joint, thus promoting a minimum amount of dilution on that side, and weld the other side; and (2) overlay one or both joint surfaces with a buffer metal.

Braze Welding Method. Braze welds can be used to join copper, aluminum bronzes, or copper-zinc alloys to low-carbon or alloy steel, stainless steel, cast iron, or nickel alloys. Other copper alloys are more difficult to join to ferrous and nickel alloys by this method. Silicon bronzes make brittle welds, and phosphor bronzes and copper nickels make porous welds, especially with ferrous alloys, when welded in this way.

Overlay Method. The other method of controlling dilution is the application of an overlay on one or both sides of the joint. The overlay metal can be the same as the electrode (filler metal) chosen to weld the joint, or it can be another type chosen principally for its ability to act as a buffer between two incompatible base metals to minimize or eliminate their mixing.

With one or both joint surfaces overlaid, the electrode for welding the remainder of the joint can be chosen with much more freedom, on the basis of its compatibility with the overlay metal and the opposite base metal, or with the two overlay metals, rather than its compatibility with the two original base metals.

When coppers or copper nickels are welded to nickel alloys by the overlay method, the copper side usually is coated with a filler metal that has a substantial content of nickel (Table 19) to produce a high-nickel surface to weld to the copper or copper nickel.

When silicon bronzes are joined to ferrous or nickel alloys, the silicon bronze side of the joint is ordinarily overlaid with aluminum bronze (ERCuAl-A2). This filler metal is generally suitable for overlaying copper alloys that are being gas metal arc welded to a ferrous or nickel alloy (Table 19).

Copper Alloys to Ferrous Metals. Besides the possibility of dilution of filler metal by copper in welding copper alloys to ferrous metals, there is the possibility of iron pickup from the ferrous side.

Aluminum bronzes can tolerate considerable dilution of this type, and copper nickels can tolerate a somewhat lesser amount. In welding these alloys to ferrous metals, care must be taken to ensure that the iron is well penetrated, so that the joint will be stronger than a braze welded joint, which might separate at the iron interface under heavy stress. Overlaying is not usually necessary in welding copper nickels or aluminum bronzes to steel. Nominal conditions for welding these two combinations are given in Table 14.

Commercial coppers can be joined to ferrous metals, in most cases also without the use of an overlay, when the nickel filler metal ERNi-3 is used. Table 14 shows nominal conditions for gas metal arc welding commercial coppers to steel.

The electrode (filler metal) most often used for GMAW of copper and copper alloys to ferrous metals is ERCuAl-A2 (Table 19). Both overlaying and weld buildup can be done with the same electrode, or a different electrode can be used for buildup. A common practice is to use ERCuAl-A2 for overlaying and to complete the weld with any desired electrode that is shown as compatible in Table 19. This technique cannot be used where one member of the combination to be joined is a phosphor bronze.

Table 14 gives conditions for using ERCuAl-A2 for GMAW of silicon bronze to steel. Excessive dilution of the silicon bronze should be avoided by the use of an overlay. In all welds between copper alloys and ferrous alloys, it is important for maximum weld strength that enough penetration be achieved on the ferrous side of the joint to effect a weld interface rich with an iron alloy.

Copper Alloys to Nickel Alloys. The welding of copper alloys to nickel alloys by the overlay method is generally straightforward, as a result of the mutual solubility of copper and nickel in all proportions. As shown in Table 19, electrodes used in GMAW of nickel alloys to coppers and copper nickels are ERNi-3 (nickel), ERNiCu-7 (nickel copper), and ERCuNi (copper nickel). The copper alloy joint surface should be overlaid with a weld deposit at least $^1/_8$ in. thick before the remainder of the joint is welded.

Aluminum bronze electrodes (ERCuAl-A2) are generally suitable for welding alloy combinations in this category. They are the only electrodes that can be used to weld silicon bronzes to nickel alloys such as Monel without cracking. Even with aluminum bronze electrodes, it is advisable to overlay the nickel with filler metal before completing the weld to minimize the interaction of silicon and nickel that might result in crack-prone weld metal.

Copper may be welded using ECu (copper) covered electrodes, but weld quality is not as good as that obtained with the gas shielded welding processes. Best results are obtained when welding deoxidized copper. The electrodes may also be used to weld oxygen-free and tough pitch coppers, but the welded joints will contain porosity and oxide inclusions associated with these coppers.

Copper also may be welded with one of the copper alloy covered electrodes, such as ECuSi or ECuSn-A electrodes. These electrodes are used for (1) minor repair of relatively thin sections, (2) fillet welded joints with limited access, and (3) welding copper to other metals. Welding should be done using DCEP of sufficient amperage to provide good weld-metal fluidity. Weave or stringer beads may be used to fill the joint groove. Welding should be done in the flat position using a preheat of 500 °F or higher for sections over 0.13 in. thick.

Shielded Metal Arc Welding

Compared with SMAW as applied to low-carbon steel, the process as applied to copper and copper alloys uses larger root openings, wider groove angles, more tack welds, higher preheat and interpass temperatures, and higher currents. For a list of covered electrodes, see Table 20.

Shielded metal arc welding of copper and copper alloys is almost always restricted to flat-position welding; out-of-

Table 20 Filler metals for SMAW of copper and copper alloys

Covered electrode(a)	Common name	Base-metal applications
ECu	Copper	Coppers
ECuSi	Silicon bronze	Silicon bronzes, brasses
ECuSn-A, ECuSn-C	Phosphor bronze	Phosphor bronzes, brasses
ECuNi	Copper nickel	Copper-nickel alloys
ECuAl-A2	Aluminum bronze	Aluminum bronzes, brasses, silicon bronzes, manganese bronzes
ECuAl-B	Aluminum bronze	Aluminum bronzes
ECuNiAl	...	Nickel-aluminum bronzes
ECuMnNiAl	...	Manganese-nickel-aluminum bronzes

(a) See AWS A5.6-76, "Specification for Copper and Copper Alloy Covered Electrodes."

position welding by this process is usually limited to the joining of phosphor bronzes and copper nickels.

Coppers. Problems of porosity and low weld strength due to oxygen content of the base metal and oxygen absorption during welding are more severe in joining coppers by this process than by gas shielded processes.

Copper to Aluminum. A strong, ductile weld between copper and aluminum offers considerable electrical and economic advantages over conventional mechanical joints for electrical bus bars. If a direct fusion weld is attempted, brittle intermetallics, notably $CuAl_2$, are formed which seriously impair the mechanical properties of the joint. A technique has been developed whereby dilution of the copper in the weld pool is inhibited by applying a layer of silver brazing alloy to the weld area of the copper member prior to full fusion welding of the joint with a conventional aluminum-silicon filler metal. The brazed layer is metallurgically compatible with both aluminum and copper. During welding, the weld pool is established toward the aluminum side of the joint by directing the arc away from the silver brazed face. Fusion of the filler metal with the brazed layer is thus effected without excessive melting. For this reason, welding conditions should be as cool as possible and interpass temperatures kept low to avoid overdilution of the brazing alloy.

In satisfactorily made joints of this nature, the electrical and mechanical properties are good, with excellent ductility. They have been produced on a commercial scale with consistent quality and provide satisfactory operation under severe conditions of thermal cycling. However, in service environments where electrochemical corrosion can occur, there is likely to be severe attack on the aluminum and weld metal. Friction welding is increasingly used as the better alternative for joints between copper and aluminum.

Brasses can be welded with phosphor bronze, silicon bronze, or aluminum bronze covered electrodes. Relatively large welding grooves are needed for good joint penetration and avoidance of slag entrapment. Welding should be done using a backing strip of copper or brass.

Phosphor bronze electrodes such as ECuSn-A and ECuSn-C have been used for welding the low-zinc brasses. The base metal is preheated and held at 400 to 500 °F. Weld metal is applied in narrow and shallow stringer beads.

The high-zinc copper alloys can be welded with ECuAl-A2 (aluminum bronze) electrodes. Preheat and interpass temperatures are 500 to 700 °F. The arc is held directly on the molten weld pool rather than toward the base metal and advanced slowly to minimize zinc volatilization.

Phosphor Bronzes. Shielded metal arc welding is done to a limited extent on the phosphor bronzes. Covered electrodes ECuSn-A and ECuSn-C are used interchangeably. These electrodes are designed for use with DCEP.

The phosphor bronzes flow sluggishly and must be preheated to 300 to 400 °F, especially for thick sections. However, because of the hot shortness of these alloys, the interpass temperature must not be permitted to go above the preheat temperature. This is achieved by welding rapidly with light passes. In groove welding, the first two passes are made with a weaving technique. Width of the weave should not exceed two electrode diameters. The remaining passes are made without appreciable transverse weaving and with the use of narrow stringer beads. The development of a coarse dendritic structure of low strength and ductility is minimized by control of preheat and interpass temperature and the use of this method of deposition. Hot peening after welding helps to break up coarse grain structure. For maximum ductility, the welded assembly should be postheated to 900 °F and cooled rapidly. Joint grooves should be wide (80 to 90°) to achieve proper "washing" of the groove walls.

Aluminum Bronzes. Shielded metal arc welding is done readily on aluminum bronzes in both the wrought and cast forms. The aluminum oxides that form on the surface of these alloys are removed by the fluxing action of the electrode coatings.

Except for thin sections, a 70 to 90° V-groove joint is used, usually with a backing strip of the same composition as the base metal. Deposition technique and bead thickness are not critical, because the weld metal has excellent hot strength and ductility.

Aluminum bronze ECuAl-A2 and ECuAl-B electrodes are used for welding the aluminum bronze alloys C61300 and C61400, which contain about 7% Al. Thick sections of these alloys may need preheating, usually to 400 °F, and control of interpass temperatures at 400 °F. Depending on section thickness and overall mass, however, preheat and interpass temperature may vary between about 150 and 800 °F. Weldments of the 7% aluminum bronzes need not be heat treated after welding.

Aluminum bronzes having an aluminum content higher than 7% are usually welded with electrodes that contain more aluminum than do ECuAl-A2 and ECuAl-B. The higher aluminum bronze electrodes, ECuAl-C, ECuAl-D, and ECuAl-E, best known as surfacing electrodes (see AWS A5.13), have nominal aluminum contents of 12.5, 13.5, and 14.5%, respectively, and correspondingly increasing strength. In welding high-aluminum bronzes, thick sections may require preheating up to 1150 °F, and fan cooling may be necessary to avoid cracking. Also, these alloys may require annealing at 1150 °F, followed by fan cooling for stress relief.

Silicon Bronzes. Shielded metal arc welding of silicon bronzes is usually done with ECuAl-A2 aluminum bronze electrodes. Welding temperature is easily attained, because silicon bronzes have low thermal conductivity. However, because the alloys are hot short, the metal should not be preheated and interpass temperature should not exceed 200 °F.

Groove dimensions are similar to those used for welding similar steel joints. Metal thicknesses up to $^5/_{32}$ in. can be welded with square grooves; thicker sections, with a single-V or double-V-groove of 60° included angle.

Properties of welds made in silicon bronzes by SMAW are usually substantially lower than those of welds made by the gas shielded process and may not meet code or design requirements for strength. Peening reduces residual stress and minimizes distortion.

Copper Nickels. Shielded metal arc welding is done on copper nickels in both the wrought and cast forms. Having thermal conductivity close to that of low-carbon steel, these alloys behave like steel in most respects and are as readily welded as steel by SMAW. The 70%Cu-30%Ni copper nickel ECuNi electrodes are used in welding the copper nickel alloys C70600 and C71500, usually with DCEP.

The weld deposits ordinarily have a high center crown, and the slag is viscous when molten and adherent when cold. Therefore, special care is needed to ensure complete slag removal before complete solidification of the weld, to prevent slag entrapment when cleaning between passes.

These alloys can be shielded metal arc welded in the vertical and overhead positions with good results, although best results are obtained in flat-position welding. This process is preferred in some applications where access to the joint is limited, as in the butt welding of copper nickel pipe described in the example that follows.

Example 14. Shielded Metal Arc Butt Welding of Copper Nickel Pipe. Various sizes of pipe made of alloy C71500 (copper nickel, 30%) were butt welded by SMAW, because of limited access to the joint. The pipe ranged in outside diameter from $2^1/_2$ to more than 6 in.

The butt joints most commonly welded were accessible from the outside only, and therefore had V-grooves and backing. For wall thicknesses of $^3/_4$ in. and less, a single-V joint was used with an 80° included angle, $^1/_{16}$-in. root face, and minimum root openings of $^3/_{16}$ in. for pipe of 3-in. OD or less, and $^1/_4$ in. minimum for larger diameters. For thicknesses over $^3/_4$ in., a single-U joint was used with the same root face and root openings. For either single-V or single-U joints, a maximum clearance of 0.015 in. was maintained between the backing ring and the inside diameter of the pipe. Tight fit of the backing in all joints was important to prevent slag from entering between the base metal and backing.

Welds were made in all positions, using ECuNi electrodes with DCEP. Typical electrode diameters and welding currents were:

$^3/_{32}$ in. diam	50-100 A
$^1/_8$ in. diam	70-130 A
$^5/_{32}$ in. diam	90-160 A

No preheat was used, and the interpass temperature did not exceed 150 °F. A stringer-bead technique was used for the root passes and, where necessary, for other passes. At no time did average bead width exceed four electrode diameters. All starts, stops, and crevices were ground or burred to remove porosity or to prevent slag entrapment. Starts and stops were staggered both within and between weld layers. Liquid-penetrant and visual inspection were done on completion of each layer (root, intermediate, and finish). The completed welds were inspected radiographically.

Shielded metal arc welding was also used for butt welding of pipe joints without a backing ring where the inside of the pipe was accessible for welding. Under such conditions, one or more passes were deposited in the root from the back side (inside) of the weld. The opposite side (outside) of the root was ground to sound metal before subsequent passes were deposited on the outside of the joint. From the standpoint of welded accessibility in this type of double-welded joint, the maximum distance, L, of any part of the weld groove from the open end of the pipe varied with the outside diameter of the pipe as follows:

OD, in.	L (max), in.	OD, in.	L (max), in.
$2^1/_2$	4	4	7
3	5	5	8
$3^1/_2$	6	6 and up	10

Recent Developments

Refinements and developments in the arc welding processes are continuously evolving. The increasing application of microelectronics is reflected in a new generation of power sources, process control, and monitoring systems. However, this does not overcome bad practice and lack of attention to design and metallurgical principles fundamental in the achievement of a successful joint. Rather, it has made possible automatic control of various weld parameters that would otherwise require manual intervention. Since the introduction of inert gas shielded arc welding technology, there have been a number of developments applicable to the copper alloy field, as listed below.

Pulsed-Current Gas Tungsten Arc Welding. A series of high-current pulses are superimposed on a continuous, low arc-maintaining background current to give a series of discrete spot fusion welds at controllable frequency to a stage where the spot welds overlap to form, when required, a continuous seam. The main effect is a drastic reduction in overall heat input of considerable advantage in preventing thermal damage and limiting heat spread. The process is normally mechanized. Pulsed-current GTAW has been used in the welding of armature winding and commutator joints in electric motors.

Pulsed-Current Gas Metal Arc Welding. Spray transfer conditions normally required for welding copper alloys are achieved at lower than normal operating currents by superimposing pulses of high, spray-transfer-range current onto a low background current, which premelts the electrode tip. The magnitude, duration, and frequency of the pulses are adjustable to give a low average heat input, less distortion, and a good weld profile free from the spatter often associated with conventional GMAW techniques. The process is ideal for welding out-of-position because of the small, controllable weld pool.

Fine-Wire Gas Metal Arc Welding. Spray transfer conditions at low heat input are achieved also by the use of gas metal arc equipment capable of feeding very fine wire. In this way, spray transfer current densities are achieved at much lower welding currents. Also, heat input problems are reduced.

Plasma Arc Welding. This is a development of the gas tungsten arc process whereby a water-cooled nozzle constricts the arc to give a much higher energy collimated beam of plasma. Current pulsing may also be applied, and the process may be miniaturized (microplasma) for intricate work. The process is used to weld brasses, because the process speed and low overall heat input prevent zinc fuming. Coppers, e.g., C10100, can also be welded with PAW using ERCu welding rods. Argon, helium, or mixtures of the two are used for orifice and shielding gases, depending on the base-metal thickness. As with GTAW, arc energy is higher with helium-rich mixtures.

Submerged Arc Welding. The welding of thick-gage material, such as seamed pipework wrapped from heavy plate, can be achieved by continuous metal arc operation under a powder slag or flux shield. Slag-metal reactions to form the required weld-metal composition, coupled with effective deoxidation, are critical, and the process is still under development for copper-based materials. The process may also be used for overlay weld surfacing.

Arc Welding of Magnesium Alloys

By the ASM Committee on Arc Welding of Magnesium Alloys*

MOST MAGNESIUM ALLOYS can be joined by gas tungsten arc welding (GTAW) and gas metal arc welding (GMAW). Relative weldability ratings, joint design and surface preparation, and the use of filler metals and shielding gases suitable to arc welding are discussed in this article. For more information on arc welding, see the articles "Gas Tungsten Arc Welding" and "Gas Metal Arc Welding" in this Volume.

Weldability

Table 1 lists some magnesium alloys that are weldable, along with their respective weldability ratings based on a scale of A (excellent) to D (limited). This rating is based largely on freedom from susceptibility to cracking, and to some extent on joint efficiency. Under optimum welding conditions, including favorable joint design, joint efficiencies of 60 to 100% can be obtained for virtually all of the magnesium alloys. Alloys rated A in Table 1 are likely to have high joint efficiency ratings.

In the magnesium-aluminum-zinc alloys (AZ31B, AZ61A, AZ63A, AZ80A, AZ81A, AZ91C, and AZ92A), aluminum content up to about 10% aids weldability by helping to refine the grain structure, while zinc content of more than 1% increases hot shortness, which may cause weld cracking. Alloys with high zinc content (ZH62A, ZK51A, ZK60A, and ZK61A) are highly susceptible to cracking and have poorer weldability. Thorium-containing alloys (HK31A, HM21A, and HM31A) have excellent arc weldability and are rated B+ or A in Table 1.

Table 1 Relative arc weldability of magnesium alloys

Alloy	Rating
Casting alloys	
AM100A	B+
AZ63A	C
AZ81A	B+
AZ91C	B+
AZ92A	B
EK30A	B
EK41A	B
EZ33A	A
HK31A	B+
HZ32A	B
K1A	A
QE22A	B
ZE41A	B−
QH21A	B
ZH62A	C−
ZK51A	D
ZK61A	D
Wrought alloys	
AZCOML	A
AZ10A	A
AZ31B,C	A
AZ61A	B
AZ80A	B
HK31A	A
HM21A	A
HM31A	A
ZE10A	A
ZK21A	B
ZK60A	D

Note: A, excellent; B, good; C, fair; D, limited weldability

Welds in magnesium alloys are characterized by a fine grain size averaging less than 0.01 in. Magnesium alloys containing more than 1.5% Al are susceptible to stress corrosion, and residual welding stresses must be relieved.

Filler Metals

Compositions of the four most commonly used electrode wires for GMAW and filler metals (when used) for GTAW are given in Table 2. The choice of electrode wire or filler metal is governed by the composition of the base metal.

Electrode wires or filler metals having compositions conforming to ER AZ61A or ER AZ92A (Mg-Al-Zn) are considered satisfactory for welding wrought alloys AZ10A, AZ31B, AZ31C, AZCOML, AZ61A, AZ80A, ZE10A, and ZK21A to themselves or to each other. ER AZ61A is usually preferred for welding wrought products because of its tendency to resist crack sensitivity. The ER AZ92A filler metal shows less crack sensitivity for welding the cast Mg-Al-Zn and Mg-Al alloys. The same electrode wires or filler metals are used for joining any one of the above alloys to high-temperature alloys HK31A, HM21A, and HM31A. However, when the high-temperature alloys are joined to each other, ER EZ33A is recommended. Joints of wrought or cast alloys welded with ER EZ33A filler metal exhibit good mechanical properties at high temperatures.

The choice of electrode wire or filler metal for welding wrought alloys to cast alloys should be based on the recommendations outlined above, except that ER

*J.H. Waibel, *Chairman*, Group Leader, Dow Chemical U.S.A.; Alan H. Braun, Director of Engineering, Wellman Dynamics Corp.; V.L. Hill, Project Leader, Technical Service and Development, Dow Chemical U.S.A.; Frank Sheara, Executive Vice President, Magnesium Elektron, Inc.

Table 2 Compositions of electrodes and filler metals used in gas shielded arc welding of magnesium alloys (AWS A5.19)

Element	ER AZ61A	ER AZ101A	ER AZ92A	ER EZ33A
Aluminum	5.8-7.2	9.5-10.5	8.3-9.7	...
Beryllium	0.0002-0.0008	0.0002-0.0008	0.0002-0.0008	...
Manganese	0.15 min	0.13 min	0.15 min	...
Zinc	0.40-1.5	0.75-1.25	1.7-2.3	2.0-3.1
Zirconium	...	...	...	0.45-1.0
Rare earth	...	...	...	2.5-4.0
Copper	0.05 max	0.05 max	0.05 max	...
Iron	0.005 max	0.005 max	0.005 max	...
Nickel	0.005 max	0.005 max	0.005 max	...
Silicon	0.05 max	0.05 max	0.05 max	...
Others (total)	0.30 max	0.30 max	0.30 max	0.30 max
Magnesium	rem	rem	rem	rem

AZ101A may be used instead of ER AZ61A or ER AZ92A.

When cast alloys are joined to cast alloys, ER AZ101A or ER AZ92A electrode wire or filler metal is usually recommended. However, for joining HK31A and HZ32A to themselves or to each other, ER EZ33A is preferred; for joining HK31A and HZ32A to any of the other cast alloys, ER AZ101A is used. Filler rod of the same composition as the base metal can be used for most welds.

Shielding Gases

Only the inert gases are used for shielding in arc welding of magnesium alloys. Argon is the most widely used. Helium and various mixtures of argon and helium have also proved satisfactory, but because of the higher cost per unit volume of helium, and because two to three times more helium than argon is required for the same degree of shielding, the use of pure helium has gradually decreased. Pure helium is also undesirable because it raises the current required for spray arc transfer and increases weld spatter. The factors determining shielding-gas selection for magnesium alloys are similar to those for aluminum.

Joint Design

Typical joint designs for gas shielded arc welding of various thicknesses of magnesium alloy sheet and plate are presented in Table 3. The metal thickness ranges vary for GTAW in accordance with the type of current used (alternating current; direct current electrode negative, DCEN; and direct current electrode positive, DCEP) and vary for GMAW in accordance with the mode of metal transfer.

Edges are prepared by milling, sawing, shearing, arc cutting, chipping, planing, routing, sanding, or filing. Sheet up to about 0.080 in. thick is generally single sheared; double shearing is preferred for sheet thicker than 0.080 in. Double-bevel edges are preferred over single-bevel edges because less distortion occurs during welding.

Fit-Up. Components of weldments should fit closely at abutting edges, preferably with no root opening, although a root opening up to 1/16 in. is usually permissible. If tack welds are used, the first tack should be a short distance from the end of a seam. The best practice in tack welding is to use 1/8-in. tacks spaced 1 to 2 in. on centers for sheet up to 0.065 in. thick, and 1/4-in. tacks spaced 4 to 5 in. on centers for sheet or plate from 0.065 to 1/4 in. thick.

Welding fixtures must be rigid enough to resist movement of the workpieces during welding. Hold-down bars must exert enough pressure to keep the edges from overlapping or rising away from the backing bar or plate. A backing bar or strip normally is used when welding sheet metal components to help control joint penetration, root surface contour, and heat removal. The gas is usually supplied through holes in the backing strip. When a backing strip cannot be used because of space limitations, a chemical flux is sometimes painted on the root side of the joint to smooth the root surface and control joint penetration. Backing plates are made of mild steel, stainless steel, magnesium, aluminum, or copper and are cut with a small groove that is positioned directly under the seam. They prevent excessive metal drop-through and minimize distortion. Depth of the grooves in backing plates depends on base-metal thickness, the welding process used, and the absence or presence of a root opening in the joint. Typical groove depths are give in Table

Table 3 Typical joint designs used for gas shielded arc welding of various thicknesses of magnesium alloy sheet and plate

	Applicable range of work-metal thickness(a)					
	GTAW(b)			GMAW(c)		
	ac, in.	DCEN, in.	DCEP, in.	Short circuiting arc, in.	Pulsed arc, in.	Spray arc, in.
A(d)	0.025-1/4	0.025-1/2	0.025-3/16	0.025-3/16	0.090-1/4	3/16-3/8
B(e)	1/4-3/8	1/4-3/8	3/16-3/8	(f)	3/16-1/4	1/4-1/2
C(g)	3/8(h)	3/8(h)	3/8(h)	(f)	(f)	1/2(h)
D(j)	0.04-1/4	0.040-1/4	0.040-1/4	1/16-3/16	1/16-1/4	3/16-1/2
E(k)	3/16(h)	3/16(h)	3/16(h)	(f)	1/8-1/4	1/4(h)
F(m)	0.025-1/4	0.025-1/2	0.025-5/32(h)	1/16-5/32	0.090-3/16	5/32-3/8
G(n)	1/16-3/16	1/16-3/8	1/16-1/8	1/16-5/32	0.090-1/4	5/32-3/4
H(p)	3/16(h)	3/8(h)	1/8(h)	(f)	1/4-3/8	3/8(h)
J(q)	0.040(h)	0.040(h)	0.025(h)	0.040-5/32	0.090-1/4	5/32(h)

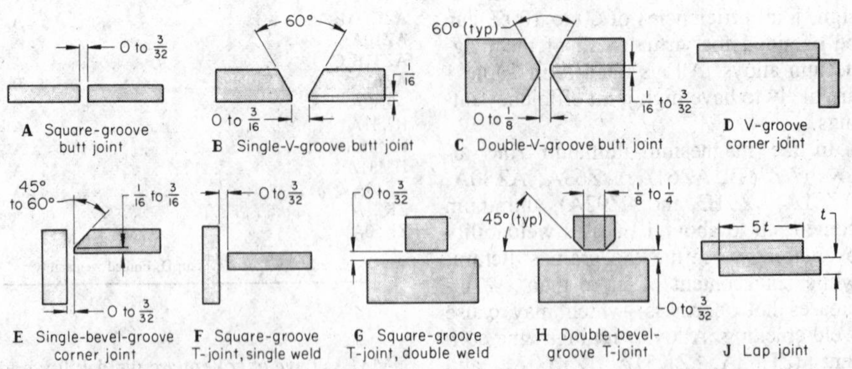

A Square-groove butt joint; B Single-V-groove butt joint; C Double-V-groove butt joint; D V-groove corner joint; E Single-bevel-groove corner joint; F Square-groove T-joint, single weld; G Square-groove T-joint, double weld; H Double-bevel-groove T-joint; J Lap joint

(a) Suggested minimum and maximum thickness limits. (b) Using 300-A ac or DCEN, or 125-A DCEP. (c) Using 400-A DCEP. (d) Single-pass full-penetration weld. Suitable for thin material. (e) Full-penetration weld. Suitable for thick material. On material thicker than suggested maximum, use double-V-groove butt joint to minimize distortion. (f) Not recommended because spray arc welding is more practical or economical, or both. (g) Full-penetration weld. Used on thick material. Minimizes distortion by equalizing shrinkage stress on both sides of joint. (h) No maximum. Thickest material in commercial use could be welded in this type of joint. (j) Single-pass full-penetration especially if a square corner is required. (k) Single-pass or multiple-pass full-penetration weld. Used on thick material to minimize welding. Produces square joint corners. (m) Single weld T-joint. Thickness limits are based on 40% joint penetration. (n) Double weld T-joint. Suggested thickness limits based on 100% joint penetration. (p) Double weld T-joint. Used on thick material requiring 100% joint penetration. (q) Single or double weld joint. Strength depends on size of fillet. Maximum strength in tension on double weld joints is obtained when lap equals five times the thickness of the thinner member.

Table 4 Typical depth of grooves in backing bars or plates used in arc welding of magnesium alloys

Base-metal thickness, in.	Depth of groove GTAW, in.	GMAW No root opening(a), in.	GMAW Root opening, in.
0.025	0.015	0.020	0.020
0.040	0.020	0.030	0.020
0.063	0.025	0.040	0.030
0.090	0.030	0.060	0.040
0.125	0.030	0.070	0.040
0.160	0.040	0.070	0.050
0.190	0.040	0.070	0.050
0.250	0.050	0.070	0.060
0.375	0.060	0.080	0.060

(a) With no root opening, grooves in backing bars or plates are deeper to permit better balance between top and bottom weld reinforcement. Use of a root opening permits balancing reinforcement with a shallower groove depth.

4. When a temporary backing bar or strip is used, the root side of the joint should be shielded with inert gas to prevent oxidation of the root surface.

Surface Preparation

Magnesium alloys are usually supplied with either an oil coating, an acid pickled surface, or a chromate coated surface. Surfaces and edges must be cleaned just before welding to remove oxides and dirt picked up during forming, assembly, and fixturing. Electrode wire or filler metal must be mechanically or chemically cleaned. A partly used spool of electrode wire or filler metal should be properly stored to protect it from dirt.

Mechanical cleaning with aluminum or stainless steel wool, aluminum oxide abrasive cloth, or power wire brushes with stainless steel bristles is preferred for most production jobs. In shops where chemical finishing equipment is available, a cleaning bath of 24 oz/gal chromic acid, $5^1/_3$ oz/gal ferric nitrate, $^1/_{16}$ oz/gal potassium fluoride, and water to make 1 gal can be used. Parts are dipped in this bath, which is maintained at 70 to 90 °F, for about 3 min, then are rinsed thoroughly in hot water and dried in air. Ceramic or stainless steel tanks are preferred; tanks lined with lead, synthetic rubber, or a vinyl-based material are also acceptable. For additional information, see the article "Cleaning and Finishing of Magnesium Alloys" in Volume 5 of the 9th edition of *Metals Handbook*.

Preheating

The need for preheating castings is determined largely by section thickness and amount of restraint. Thick sections, particularly if the magnitude of joint restraint is small, seldom need preheating. Thin sections and highly restrained joints often require preheating to prevent weld cracking, particularly in the high-zinc alloys. Maximum preheat temperatures for all of the common cast magnesium alloys are given in Table 5. In practice, if preheating is used, the temperature is usually less than the maximum shown in Table 5.

Gas Metal Arc Welding

Power supplies used for welding magnesium alloys furnish DCEP. Constant-voltage machines must be used for short circuiting welding and are generally preferred for spray arc welding. Constant-current (drooping volt-ampere output) machines can be used for spray arc welding and are sometimes advantageous for welding that is performed at a current level near the minimum for spray transfer because weld spatter is less. Special constant-voltage machines that are designed to pulse the current output must be used in pulsed arc welding. For detailed information on power supplies and modes of metal transfer, see the article "Gas Metal Arc Welding" in this Volume.

Electrode holders, wire feeders, and related equipment used for GMAW of magnesium alloys are generally the same as those used for welding other metals.

Metal Transfer. Three modes of metal transfer are suitable for welding magnesium alloys: short circuiting, pulsed arc, and spray transfer. Pulsed arc transfer can be achieved only with a power supply designed to produce a pulsing secondary current. Without the pulsing feature, the type of transfer that would be obtained in the specific operating range of current, wire feed, and wire size for this process would be globular, which is not suitable for welding magnesium alloys.

Each of the three modes of transfer is best suited to a specific (sometimes overlapping) range of base-metal thickness and welding current, as shown in Table 6. Curves that show the relationship among current, electrode wire size, and wire-feed rate for each mode are presented in Fig. 1.

Welding positions are restricted to flat, horizontal, and vertical-up by the high deposition rate and fluidity of the weld metal.

Typical operating conditions for gas metal arc butt welding of magnesium alloys in the thickness range from 0.025 to 1 in. are presented in Table 6. These operating conditions are intended as a guide and should be altered to the extent required by experience with specific applications.

Gas Tungsten Arc Welding

Gas tungsten arc welding is used more extensively than GMAW for joining mag-

Table 5 Preheat temperatures and postweld heat treatments for magnesium alloy castings

Alloy	Temper of alloy(a) Before welding	Temper of alloy(a) After treatment	Maximum preheat temperature(b), °F	Postweld heat treatment(c)
AZ63A	T4	T4	720	$^1/_2$ h at 730 °F
	T4 or T6	T6	720	$^1/_2$ h at 730 °F + 5 h at 425 °F
	T5	T5	500(d)	5 h at 425 °F
AZ81A	T4	T4	750	$^1/_2$ h at 780 °F
AZ91C	T4	T4	750	$^1/_2$ h at 780 °F
	T4 or T6	T6	750	$^1/_2$ h at 780 °F + 4 h at 420 °F(e)
AZ92A	T4	T4	750	$^1/_2$ h at 770 °F
	T4 or T6	T6	750	$^1/_2$ h at 770 °F + 4 h at 500 °F
AM100A	T6	T6	750	$^1/_2$ h at 780 °F + 5 h at 425 °F
EK30A	T6	T6	500(d)	16 h at 400 °F
EK41A	T4 or T6	T6	500(d)	16 h at 400 °F
	T5	T5	500(d)	16 h at 400 °F
EZ33A	F or T5	T5	500(d)	2 h at 650 °F(f) + 5 h at 420 °F
HK31A	T4 or T6	T6	500	1 h at 600 °F(f) + 16 h at 400 °F
HZ32A	F or T5	T5	500	16 h at 600 °F
K1A	F	F	None	None
QE22A	T4 or T6	T6	500	8 h at 985 °F(g) + 8 h at 400 °F
ZE41A	F or T5	T5	600	2 h at 625 °F + 16 h at 350 °F(f)
ZH62A	F or T5	T5	600	2 h at 625 °F + 16 h at 350 °F
ZK51A	F or T5	T5	600	2 h at 625 °F + 16 h at 350 °F(f)
ZK61A	F or T5	T5	600	48 h at 300 °F
	T4 or T6	T6	600	2-5 h at 930 °F + 48 h at 265 °F

(a) T4, solution heat treated; T6, solution heat treated and artificially aged; T5, artificially aged; F, as cast. "After treatment" means after postweld heat treatment. (b) Heavy and unrestrained sections usually need no preheat; thin and restrained sections may need preheating, up to maximum temperatures shown, to avoid weld cracking. A sulfur dioxide or carbon dioxide atmosphere is recommended when temperature exceeds 700 °F. (c) Temperatures shown are maximum allowable. Furnace controls should be set so temperature does not exceed indicated maximum. A sulfur dioxide or carbon dioxide atmosphere is recommended when temperature exceeds 700 °F. (d) For $1^1/_2$ h max. (e) 16 h at 335 °F may be used instead of 4 h at 420 °F. (f) This phase of heat treatment is optional and serves to induce greater stress relief. Some loss in high temperature creep stength may occur in EZ33A due to the 650 °F stress relief. (g) Quench in water at 140 to 220 °F before second heat treatment.

Table 6 Typical operating conditions for gas metal arc butt welding of magnesium alloys(a)

Base-metal thickness, in.	Type of groove	No. of passes	Electrode Diameter, in.	Electrode Feed rate, in./min	Electrode Consumption of weld, lb/ft	Current, A	Voltage, V	Argon flow rate, ft³/h
Short circuiting mode of metal transfer								
0.025	Square(b)	1	0.040	140	0.006	25	13	40-60
0.040	Square(b)	1	0.040	230	0.009	40	14	40-60
0.063	Square(b)	1	0.063	185	0.018	70	14	40-60
0.090	Square(b)	1	0.063	245	0.024	95	16	40-60
0.125	Square(c)	1	0.094	135	0.030	115	14	40-60
0.160	Square(c)	1	0.094	165	0.037	135	15	40-60
0.190	Square(c)	1	0.094	205	0.046	175	15	40-60
Pulsed arc mode of metal transfer(d)								
0.063	Square(b)	1	0.040	360	0.014	50	21	40-60
0.125	Square(b)	1	0.063	280	0.028	110	24	40-60
0.190	Square(b)	1	0.063	475	0.047	175	25	40-60
0.250	Single-V, 60°(e)	1	0.094	290	0.065	210	29	40-60
Spray arc mode of metal transfer(f)								
0.250	Single-V, 60°(e)	1	0.063	530	0.042	240	27	50-80
0.375	Single-V, 60°(e)	1	0.094	285-310	0.057	320-350	24-30	50-80
0.500	Single-V, 60°(e)	2	0.094	320-360	0.106	360-400	24-30	50-80
0.625	Double-V, 60°(g)	2	0.094	330-370	0.125	370-420	24-30	50-80
1.000	Double-V, 60°(g)	4	0.094	330-370	0.208	370-420	24-30	50-80

(a) Welding speed, 24 to 36 in./min. (b) Zero root opening. (c) Root opening, 0.090 in. (d) Pulse voltage of 55 V, except for metal 0.190 in. thick, which uses a pulse voltage of 52 V. (e) 1/16-in. root face and zero root opening. (f) Settings also apply to fillet welds in the same thickness of metal. (g) 1/8-in. root face and zero root opening

Table 7 Conditions for manual GTAW of butt joints in magnesium alloys

Base-metal thickness, in.	Joint design(a)	No. of passes	Electrode diameter, in.	ac current(b), A	Argon flow rate, ft³/h	Filler metal Diameter, in.	Filler metal Consumption, lb/ft
0.040	A	1	1/16	35	12	3/32	0.004
0.063	A	1	3/32	50	12	3/32	0.005
0.080	A	1	3/32	75	12	3/32	0.006
0.100	A	1	3/32	100	12	3/32	0.008
0.125	A	1	3/32	125	12	1/8	0.009
0.190	A	1	1/8	160	15	1/8	0.011
0.250	B	2	5/32	175	20	1/8	0.026
0.375	B	3	5/32	175	20	5/32	0.057
0.375	C	2	3/16	200	20	1/8	0.024
0.500	C	2	3/16	250	20	1/8	0.047

(a) A, square groove butt joint, zero root opening; B, 60° bevel, single-V-groove butt joint, 1/16-in. root face, zero root opening; C, 60° bevel, double-V-groove butt joint, 3/32-in. root face, zero root opening. (b) Thorium-containing alloys will require about 20% higher current. With helium shielding, required welding current will be reduced by about 20 to 30 A.

nesium alloys. It is well suited for welding thin sections. Control of heat input and the molten weld pool is better with GTAW than with GMAW. Alternating current machines with a high-frequency current superimposed on the normal welding current for arc stabilization and direct current machines with continuous amperage control are used. For thin sheets, both alternating current and DCEP are used. On material over 3/16 in. thick, alternating current is preferred because it provides deeper penetration. Direct current electrode negative is seldom used on magnesium alloys because the arc lacks cathodic cleaning action. Alternating current machines should be equipped with a primary contactor, actuated by a switch on the torch or by a foot switch, for starting and stopping the arc. Otherwise, the arcing that occurs while the electrode approaches or draws away from the workpiece may result in burned spots on the work. Pure tungsten, zirconiated, and thoriated electrodes, from 0.010 to 0.250 in. in diameter, are used for GTAW of magnesium alloys. Additional information on power supplies, electrodes, and related items for GTAW is given in the

Fig. 1 Effect of wire-feed rate and welding current on mode of metal transfer

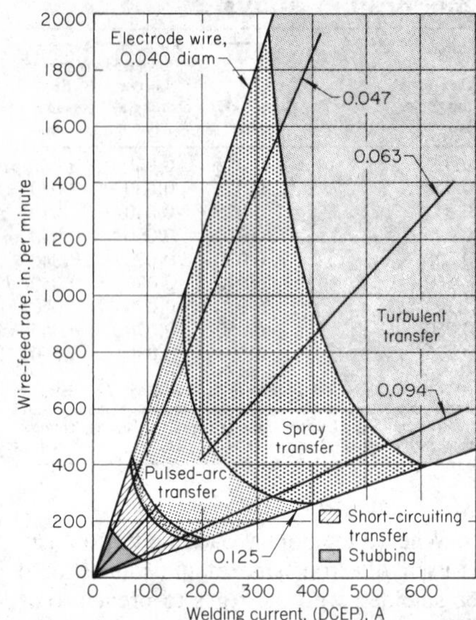

article "Gas Tungsten Arc Welding" in this Volume.

Manual Welding. Data on current settings, electrode diameter, shielding-gas flow, filler-metal diameter, and filler-metal consumption for manual GTAW of butt joints in magnesium alloys 0.040 to 0.500 in. thick are given in Table 7. For best results, the electrode should be held close to the work to produce an arc about 1/32 in. long. The preferred angles of torch, joint, and filler metal are given in Fig. 2. Welding should be done at a uniform speed in a straight line. A weaving or rotary motion should be used only for fillet welds or large corner joints. Forehand welding is preferred; a minimal number of stops produces the best results. If stops are required, the weld should be restarted on weld metal about 1/2 in. from the end of the previous weld.

To minimize the risk of weld cracking, either of the methods shown in Fig. 3

Fig. 2 Gas tungsten arc welding setup

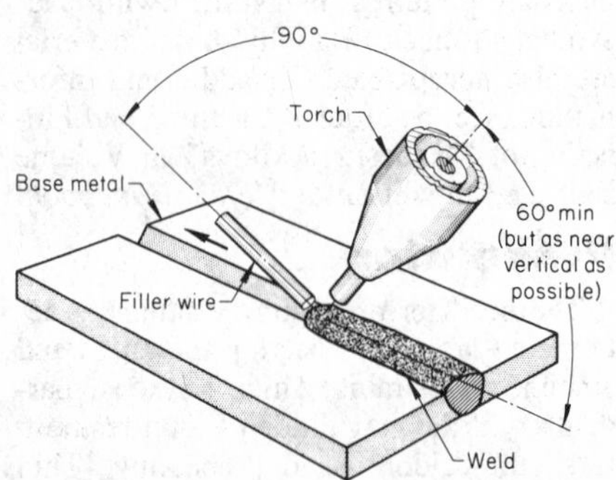

Fig. 3 Welding of magnesium alloys to prevent weld cracking

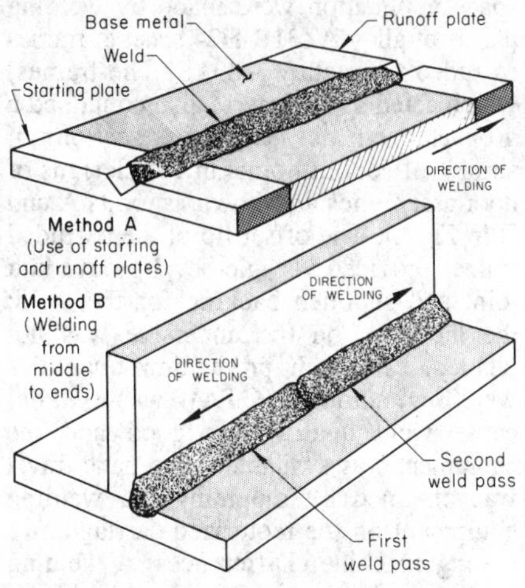

should be used: method A, in which starting and runoff plates (or tabs) are used to start and end the weld; or method B, in which the weld is made in two increments by passes that begin in the middle and progress toward opposite ends. Also, the base metal (and fixture, if used) should be preheated to at least 200 to 300 °F. If the thickness of the two compositions to be welded differs by $^1/_4$ in. or more, the thicker section should be preheated to approximately 300 °F.

Short-Run Production. In short production runs for which the cost of elaborate fixtures cannot be justified, tooling costs can be minimized by fabricating the components so that they can be tack welded into position.

Example 1. Tack Welding for Short-Run Production. To minimize tooling costs on short production runs of electronic deck assemblies, which were essentially two rectangular boxes 2 by 2 by 4 in., as shown in Fig. 4, tack welds were used instead of fixturing to position some of the component pieces. Formed sheet sections of 0.050-in.-thick AZ31B-H24 were tack welded into position by GTAW, using $^1/_{16}$-in.-diam ER AZ61A filler wire. The tack welds were $^1/_8$ in. long and were spaced on 2-in. centers (starting at each corner). A tool plate and toggle clamps held the pieces for tack welding. Tack welds were not used to hold angle pieces.

Welding of the assembled and tack welded components was completed by manual GTAW with $^1/_{16}$-in.-diam ER AZ61A filler wire. The corner joints were welded with continuous beads about 2 in. long, and the flanged bottom of the top part of the assembly was joined to the sides with 1-in.-long fillet welds. Extruded angle sections were fillet welded to the ends of the boxes with welds about 1 in. long (Fig. 4, detail A).

The assembly was repositioned manually so that all welds could be made in either the flat or the horizontal position. A standard alternating current power supply with a high-frequency arc stabilizer was used. Helium was selected as the shielding gas because a hotter and more stable arc was produced than would have been possible with argon shielding gas. Preheating was not used, but after welding, the assemblies were stress relieved at 350 °F for $3^1/_2$ h to prevent stress-corrosion cracking. Welds were inspected visually.

Fig. 4 Manual GTAW of electronic deck assembly

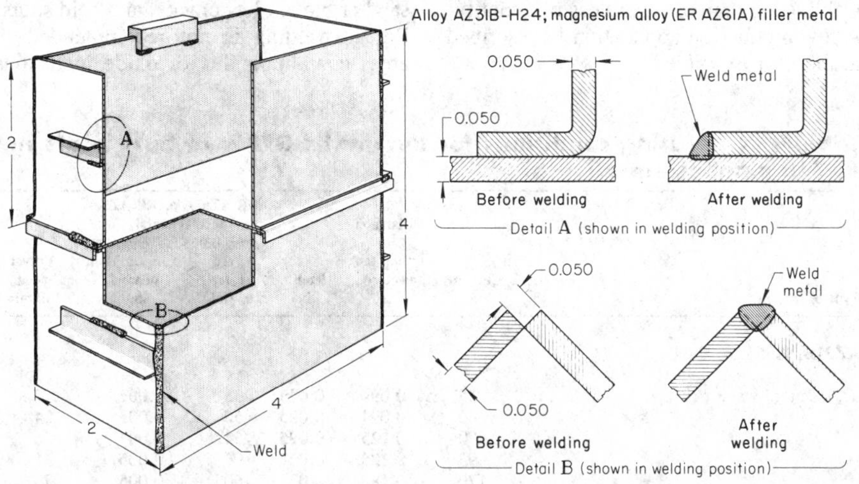

Conditions for manual GTAW

Joint types	Lap and corner	Electrode	0.040-in.-diam EWP
Weld types	Fillet and single-V-groove	Filler metal	$^1/_{16}$-in.-diam ER AZ61A(a)
Welding positions	Horizontal and flat	Torch	350 A, water cooled(b)
Preweld cleaning	Wire brushing	Power supply	300-A transformer(c)
Preheat	None	Current, fillet welds	25 A, ac
Fixtures	Tool plate and toggle clamps	Current, V-groove welds	40 A, ac
Shielding gas	Helium, at 25 ft^3/h	Postweld heat treatment	350° F for $3^1/_2$ h

(a) 36-in.-long rod. (b) Ceramic nozzle. (c) Continuous duty, with high-frequency oscillator

Example 2. Manually Welded T-Joint. The welding of a single $^1/_8$-in. fillet in a 8-in.-long T-joint between sheets of alloy AZ31B, $^1/_{16}$ and $^1/_8$ in. thick (see Table 8), was used as a quick shop test for welder qualification. The welder adjusted the machine settings, gas flow, and welding speed to produce a sound weld with proper contour and adequate penetration, using manual GTAW. After welding, the joint was broken by striking a blow against the unwelded side of the vertical stem to open and expose the root of the weld. The fracture was then examined for depth of penetration, porosity, lack of fusion, and other defects. Welding conditions and a view of the joint are given in Table 8.

Example 3. Manually Welded Single-V-Groove Butt and Corner Joints. Structural frames were fabricated by welding the mitered ends of 1 by 1 by $^3/_{16}$ in. angles extruded from alloy AZ31B. One of the four right angle corners of one of these structural frames is shown in Table 8. Manual GTAW was used because production quantity was less than 50 pieces per month.

The procedure consisted of mitering, edge beveling, cleaning, assembling on a fixture, and welding. Horizontal joint edges were beveled 45° with a 0.040-in.-deep root face; vertical joint edges were beveled to form a similar root face for a 90° corner joint. After acid pickling and rinsing, the angles were assembled on a clamping fixture with flat backing bars for the horizontal joints and corner backing bars for the vertical joints. The backing bars for both the horizontal and vertical joints were provided with backing grooves that were $^1/_8$ in. wide by 0.020 in. deep. The outside corner joint was single-pass welded, vertical-up; the grooved butt joint was single-pass welded in the flat position. No preheat was used, but the frames were postweld stress relieved at 350 °F

Table 8 Manual GTAW of alloy AZ31B

Item	Example 2	Example 3	Example 4
Joint type	T	Corner; butt	Butt
Weld type	Single fillet	Single-V-groove	Square groove
Welding position	Horizontal	Vertical-up; flat	Flat
Shielding gas and flow rate	Argon; 18 ft³/h	Argon; 18 ft³/h	Argon; 18 ft³/h
Electrode (EWP) diameter	3/32 in.	1/8 in.	1/8 in.
Filler metal (ER AZ61A) diameter	1/16 in.	3/32 in.	1/16 in.
Current (ac, HF-stabilized)	110 A	125 A	135 A
Welding speed	10 in./min	5 units/h(b)	10 in./min
Postweld heat treatment	500 °F for 15 min	350 °F for 1 1/2 h	350 °F for 1 1/2 h

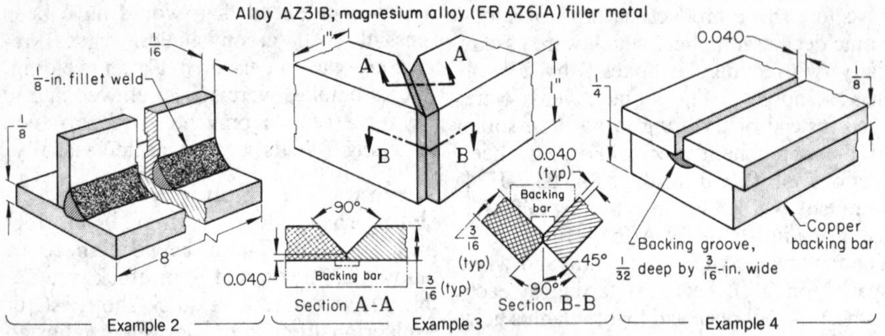

(a) A 300-A ac/dc power supply with continuous high frequency was used for all three applications, with a light-duty water-cooled welding torch. All workpieces were preweld cleaned by chromic-sulfuric pickling. No preheat was used. (b) Includes fixturing, welding, and unloading

for 1 1/2 h. Welding conditions are given in Table 8.

Example 4. Square-Groove-Welded Butt Joints. Gas metal arc welding was to be used on the square-groove-welded butt joints (with 0.040-in. root opening) in 1/8-in.-thick AZ31B-H24 sheet, but contact tube arcing occurred and contamination of the filler metal resulted. Rather than delay production while the arcing problem was being eliminated, manual GTAW was used instead. Good mechanical properties were obtained in single-pass welds, using argon shielding on the torch side and a grooved copper backing bar, but no gas backing, on the underside. Joint alignment was maintained with bolted hold-down bars. Welding conditions are given in Table 8.

Automatic welding of magnesium alloys by GTAW is similar to manual welding, except that higher currents and welding speeds are used. Table 9 can be used as a guide to determine settings for automatic welding. Alternating current is best, although DCEP can be used. A balanced-wave alternating current machine, or a conventional alternating current machine equipped with a battery bias for wave balancing, should be used.

Constant alignment of electrode and filler metal is important in automatic welding. Filler metal is fed into the arc at a low angle to the work so that the filler rod touches the weld surface just ahead of the electrode. The arc length shown in Table 9 should be preset and maintained. Travel speeds up to 95 in./min can be used on thin sheet, but 24 to 36 in./min is more common. Automatic welding is not necessarily restricted to high-production applications; sometimes the high degree of control that can be attained with automatic welding is required regardless of quantity. A low-production application is described in the next example.

Example 5. Automatic Welding of Extrusions. Airtight doors for an aerospace application were made by welding panels of alloy AZ31B-H24 sheet to frames extruded from alloy AZ31B. The frames, which acted as stiffeners, also contained a groove for an air seal. Cross sections of similar offset butt joints in two designs of door assemblies are shown as joints A and B in Fig. 5. The offset lip of the extruded frames provided a single-bevel groove butt joint and supplied backing for the weld; the lap joint on the underside was not welded. Although production quantities were low, automatic GTAW was used because weld quality was good and the equipment was available. Automatic travel was obtained by mounting the welding equipment on the motorized carriage of a cutting machine. Differences in welding conditions for the two joints resulted from operator choice or judgment. Both procedures produced satisfactory welds, but the difference in welding speeds would have been significant had production quantities been large.

Repair Welding of Castings

A significant portion of the total amount of welding that is done on magnesium alloy castings is for repair. Repair welding is usually limited to the repair of defects in clean metal, including broken sections, sand or blowholes, cracks, and cold shuts. Repair welding is not recommended in areas containing flux or oxide inclusions

Table 9 Operating conditions for automatic GTAW of butt joints in AZ31B magnesium alloy(a)

Type of current	Current, A	Diameter of tungsten electrode, in.	Arc length, in.	ER AZ61A or ER AZ92A filler metal, 0.063 in. diam: Feed rate, in./min	Consumption of weld, lb/ft	Travel speed, in./min
AZ31B, 0.063 in. thick						
ac, balanced wave	55	0.094	0.025	35	0.007	12
	60	0.094	0.025	50	0.005	24
	70	0.125	0.025	54	0.004	36
	95	0.125	0.025	96	0.005	45
	170	0.188	0.025	160	0.005	70
	195	0.188	0.025	190	0.006	80
	200	0.188	0.025	203	0.005	95(b)
DCEN	75	0.125	0.025	80	0.004	48
DCEP	120	0.250	0.020	184	0.005	80(b)
AZ31B, 0.190 in. thick						
ac, balanced wave	300	0.250	0.020	159	0.011	34
DCEN	170	0.125	0.030	70	0.008	20
DCEP	120	0.250	0.020	10(c)	0.008(c)	7(b)

(a) With helium shielding. Flow rate, 20 ft³/h. (b) Maximum speed limitation because of undercutting or arc instability. (c) Diameter of filler metal, 0.094 in.

Fig. 5 Automatic GTAW of joints

Alloy AZ31B-H24 welded to alloy AZ31B; magnesium alloy (ER AZ61A) filler metal

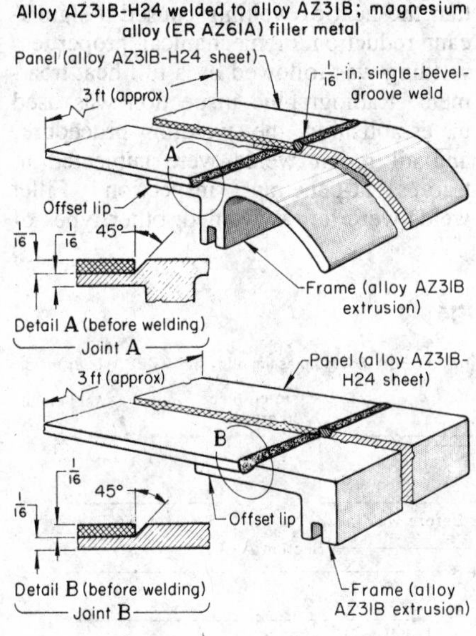

Automatic GTAW

Joint type	Offset butt
Weld type	Single-bevel groove
Preweld cleaning	Chromic-sulfuric pickle
Welding position	Flat
Preheat	None
Shielding gas	Argon, 18 ft^3/h for joint A; argon, 16 ft^3/h for joint B
Electrode	1/8-in.-diam EWP
Filler metal	1/16-in.-diam ER AZ61A
Torch	Water cooled
Power supply	300 A ac (HF-stabilized)
Current (ac)	175 A for joint A; 135 A for joint B
Wire-feed rate	65 in./min
Travel speed	20 in./min for joint A; 15 in./min for joint B
Postweld heat treatment	350 °F for 1½ h

or excessive porosity. Castings that have been organically impregnated for pressure tightness or that may contain oil in pores should not be welded. Many magnesium castings are aircraft parts that are heat treated to meet strength requirements. These castings must be heat treated again after welding. The type, size, and location of defects in castings vary so much that each job presents its own welding problems; procedures cannot be standardized. However, to complete most repairs:

- Castings must be stripped of paint and degreased before welding; conversion coatings should be removed from defective areas with steel wool or with a power wire brush having stainless steel bristles; a rotary deburring tool is recommended for removing defects and preparing the workpiece for welding.
- Castings should be clamped in place, and the surface adjoining the fracture should be beveled for the weld (Table 3).
- Preheating, if necessary, should be done at the temperatures recommended in Table 5. Either local heating with a torch or furnace heating can be used. The use of a protective atmosphere during furnace heating will reduce the possibility of oxidation at temperatures above 700 °F.
- Welding should be done immediately after preheating. Reheating may be necessary if the temperature of the casting drops significantly below the preheat temperature.
- Medium-size weld beads should be used.
- Welding should progress from the center of the break to the outside edges. The arc should not be allowed to dwell too long in any one area because weld cracking may result.
- High welding current, which may cause weld cracking or incipient melting in the heat-affected zone (HAZ), should be avoided; low welding currents, which may cause cold laps, oxide contamination, or porosity, should also be avoided.
- A foot control should be used to fade out the arc gradually to minimize thermal shock from arc stops. Thermal shock may cause cracking.

The examples that follow describe procedures that were used in successful repair welding of specific castings.

Example 6. Repair of Wheel Rim Casting. After machining, aircraft wheel rim castings of AZ91C-T4 were occasionally found to have small sand holes (Fig. 6). Defects, of a size that could be removed by cutting a groove about 1/8 in. deep and 1/2 in. wide, were repaired by welding. The defective area was cleaned by cutting a smoothly blended groove (see Fig. 6, Before welding), using a burring tool. After complete removal of defective material was verified by liquid-penetrant inspection, the groove and surrounding area were wire brushed to prepare for GTAW. The relatively thick section of the casting in the repair area and the wide groove for the deposit ensured that the weld would be under low restraint; therefore, preheating was not necessary. Ordinarily, a standard filler metal, such as ER AZ92A or ER AZ101A, would prove satisfactory for welding the AZ91C base metal; in this application, a filler metal of the same composition as the base metal was specified. The weld was made by striking the arc at the bottom of the groove and proceeding

Fig. 6 Repair welding of aircraft wheel rim casting

Alloy AZ91C-T4; magnesium alloy (AZ91C) filler metal

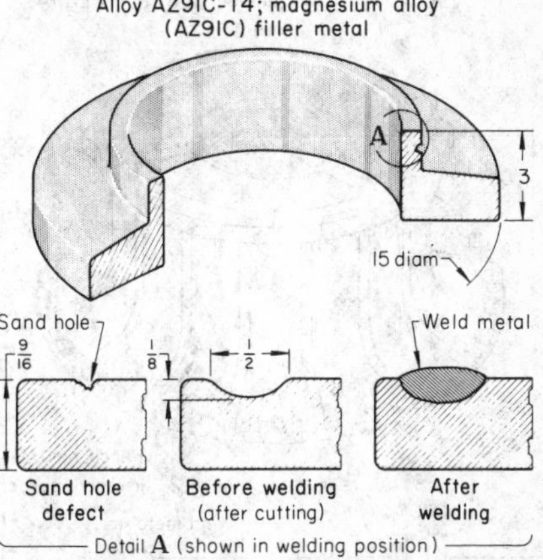

Manual GTAW

Weld type	Surfacing, for repair
Welding position	Flat
Preheat	None
Shielding gas	Helium, 20 ft^3/h
Electrode	3/32-in.-diam EWP
Filler metal	1/8-in.-diam AZ91C
Torch	300 A, water cooled
Power supply	300-A transformer, with high-frequency generator
Current	60 A, ac
Postweld heat treatment	780 °F for 1/2 h plus 420 °F for 4 h

in a circular direction along the groove wall to fill the groove, as shown in the "After welding" view in Fig. 6.

After welding, the weld reinforcement was ground to 1/32 in. above the base-metal surface, and surface weld quality was verified by liquid-penetrant inspection. After passing inspection, the casting was heat treated to the T6 condition. The casting was accepted after radiographic inspection indicated satisfactory internal weld quality.

Example 7. Repair Welding of a Jet Engine Casting. During an aircraft jet engine overhaul, fluorescent-penetrant inspection revealed a 2½-in.-long crack near a rib in the cast AZ92A-T6 compressor housing shown in Fig. 7. The thickness of the section containing the crack ranged from 3/16 to 5/16 in. Repair welding was permissible. The part was vapor degreased to remove surface grease and dirt and soaked in a commercial alkaline paint remover. The crack was then marked with a felt-tip marker, and the part was stress relieved at 400 °F for 2 h. The crack was removed by slotting the flange through to the periphery. Each side of the slot was beveled to approximately 30° from vertical to form a 60° double-V-groove. The

Fig. 7 Repair welding of compressor housing

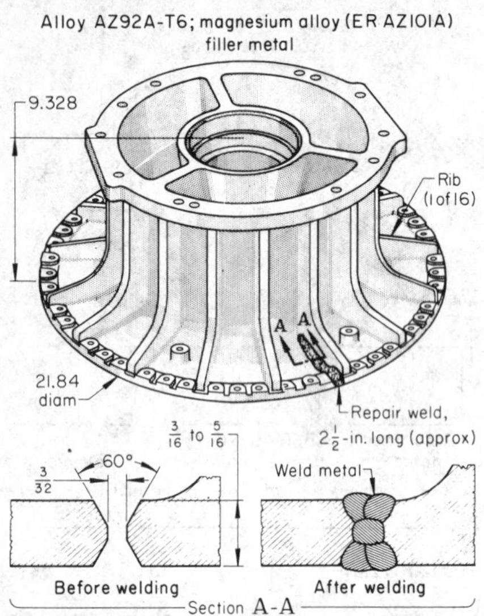

Manual GTAW

Joint type	Butt
Weld type	60° double-V-groove repair
Shielding gas	Argon, 20 ft³/h(a)
Electrode	1/16-in.-diam EWTh-2
Filler metal	1/16-in.-diam ER AZ101A
Torch	Water cooled
Power supply	300-A transformer, with high-frequency starting
Current	Under 70 A, ac(b)
Postweld stress relief	400 °F for 2 h(c)
Inspection	Fluorescent penetrant

(a) Also used for backing. (b) Current was regulated by a foot switch. (c) Also preweld

area to be welded was cleaned with a power wire brush with stainless steel bristles. Welding was done by manual GTAW, without preheating.

The welding technique maintained a low-amperage arc (less than 70 A) directed onto the base metal while filler metal was deposited on the sides of the groove, working from the innermost point outward. After a molten weld pool formed, the arc was weaved slightly while depositing a bead on the sides of the groove. During welding, heat input was adjusted by a foot-operated current-control rheostat to maintain a uniform weld pool. After welding was completed on one side of the slot, the casting was turned over. Excess drop-through and areas of incomplete penetration were removed by grinding. The underside was then welded by the same technique used for the first side. After welding, the casting was stress relieved at 400 °F for 2 h and inspected by the fluorescent-penetrant method.

Sand casting repairs can be made with manual GTAW. Figure 8 shows three repairs for the correction of surface flaws that occur in machining, defects resulting from the casting process, and deficiencies in design. Welding was done with high-frequency stabilized alternating current. The castings were preheated and maintained at temperature during welding to avoid cracking. To avoid distortion, the postweld heat treatments were used as a compromise between stress relieving and an aging treatment. However, when repair-welded areas were large enough so that the heat of welding caused a significant reduction in mechanical properties, welding was followed by a full heat treatment. Radiographic inspection was used in establishing the welding procedure, and all repair welds were subjected to fluorescent-penetrant inspection. Fillet welds were left as welded; other types of

Fig. 8 Repair welding of sand castings

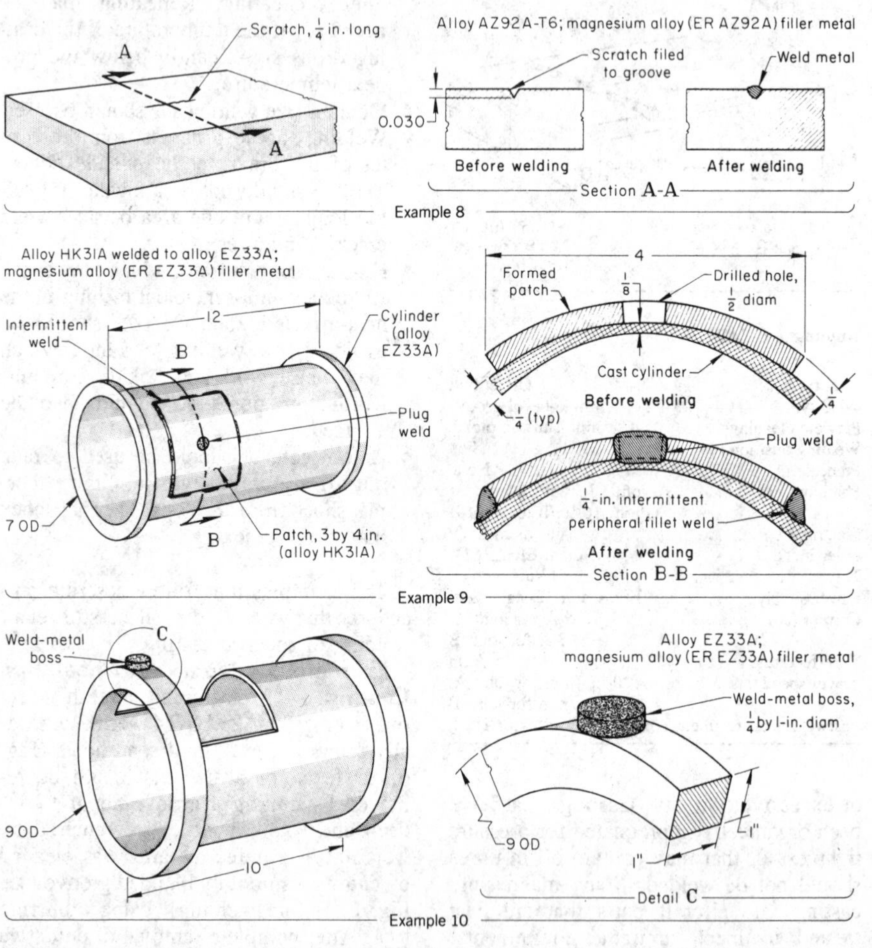

Welding condition	Example 8	Example 9	Example 10
Conditions for manual GTAW(a)			
Fixtures	None	Plates	None
Welding position	Flat	Flat	Flat
Shielding gas	Argon; 20-24 ft³/h	Argon; 24-28 ft³/h(b)	Argon; 24-28 ft³/h
Electrode, 1/8 in. diam	EWP	EWP	EWP
Filler metal, 1/16 in. diam	ER AZ92A	ER EZ33A	ER EZ33A
Current (ac, HF-stabilized)	80-140 A	160-180 A	160-180 A
Preheat and interpass temperature	250 °F	300 °F	250-350 °F
Postweld heat treatment	300 °F for 3 h	400 °F for 2 h	300 °F for 4 h
Total time per piece(c)	1/2 h	2 h	1/2 h

(a) For all three applications, preweld cleaning was done with a wire brush or a rotary file, power supply was a 300-A transformer with high-frequency (balanced wave) stabilizer, and electrode holder was a 300-A water-cooled type. (b) Helium at 8 ft³/h was used for backing. (c) Not including time for postweld heat treatment

Table 10 Postweld stress-relief treatments for magnesium alloys(a)

Alloy	Temperature, °F	Time, min
Sheet		
AZ31B-O(b)	500	15
AZ31B-H24(b)	300	60
HK31A-H24	600	30
HM21A-T8	700	30
HM21A-T81	750	30
ZE10A-O	450	30
ZE10A-H24	275	60
Extrusions		
AZ10A-F	500	15
AZ31B-F(b)	500	15
AZ61A-F(b)	500	15
AZ80A-F(b)	500	15
AZ80A-T5(b)	400	60
HM31A-T5	800	60
AZCOML	500	15
Castings(c)		
AM100A	500	60
AZ63A	500	60
AZ81A	500	60
AZ91C	500	60
AZ92A	500	60

(a) Treatments produce approximately 80 to 95% stress relief in all alloys except HM31A-T5, where 70% stress relief is provided. (b) Requires postweld heat treatment to avoid stress-corrosion cracking. (c) Requires postweld heat treatment for maximum strength (see Table 5 for postweld heat treatments)

welds were finished to conform to the contour of the casting or to the desired shape.

Example 8. Repair Welding of a Machining Scratch. The views at the top in Fig. 8 show a portion of an alloy AZ92A-T6 sand casting with a machining scratch, 1/4 in. long and 0.030 in. deep. This scratch was repaired by gas tungsten arc surface welding. Machining flaws up to 3/8 in. square in castings of various shapes and sizes have been corrected by this method.

Example 9. Repair Welding of Wall Thinning. The middle views in Fig. 8 show a flanged cylindrical alloy EZ33A sand casting that was one of about 200 such castings in which core shift caused the 1/4-in. wall to thin to 1/8 in., beginning about 1 in. away from one of the flanges. To repair these thin areas without distortion of the cylinder, a 3 by 4 in. piece of 1/4-in. alloy HK31A plate was formed to fit the outside wall of the cylinder, drilled in the center with a 1/2-in.-diam hole, and clamped in place over the thin area. The patch was tack welded at the four corners and then gas tungsten arc welded along the periphery with intermittent 1/4-in. fillet welds 1 in. long. During welding, the inside of the cylinder was flushed with helium to prevent oxidation of heated portions. After the patch had been fillet welded to the cylinder, the center hole was plug welded, using successive small beads to ensure fusion to the wall of the casting and to minimize the possibility of cracking, which can occur when large beads are deposited. The part was then finish machined using plates to hold the flanges.

Example 10. Application of an All-Weld-Metal Boss. Because a design change was made after the first run of flanged cylindrical alloy EZ33A castings had been produced, a boss had to be added to one of the flanges on each casting, as shown in the views at the bottom in Fig. 8. Using a buildup welding procedure of successive small beads, a cylindrical boss 1 in. in diameter by 1/4 in. high was produced on the flange by manual GTAW. In later production, the pattern was changed, and the boss was cast integrally with the flange.

Defects in Repair Welds. Oxide inclusions are caused by (1) welding on unsound base metal, (2) inadequate cleaning of base metal or filler metal, (3) maintaining too long an arc, (4) insufficient flow of shielding gas, (5) leaky gas connections, and (6) defective shielding-gas hoses that allow the drawing in of air.

Tungsten inclusions are caused by maintaining too long an arc, using too high a current, and touching the weld pool, base metal, or filler metal with the electrode.

Porosity is usually caused by welding on unsound base metal, poor preweld cleaning of base metal or filler metal, and contamination of the shielding gas.

Microshrinkage, a defect consisting of interdendritic voids detectable on magnification, is caused by too rapid current decay at the end of a weld, which results in too rapid freezing of the weld pool. As a result, the weld pool is unable to serve as a riser to feed the solidifying weld metal below.

Base-metal cracks are usually caused by excessive heat input or by carrying the arc too far over onto the base-metal surface during repair welding.

In repair welds of magnesium alloys that contain rare earth elements or thorium, a light-colored area at the HAZ of the weld is often visible on a radiograph taken with the x-ray beam parallel and the x-ray film perpendicular to the weld edge. This is caused by eutectic-enriched material with high atomic numbers absorbing substantially more radiation. Eutectic enrichment is caused by the melting of eutectic material at a lower temperature and heat expansion of the parent grains, squeezing this material into a layer at the edge of the weld. Below this layer is a darker area on the radiograph where eutectic material has vacated. Above the eutectic-rich layer appears another eutectic-depleted zone due to the first solidification of higher melting constituents. The effect can be minimized by avoiding preheat and by rapid welding.

Heat Treatment After Welding

Heat treated castings are often heat treated again after welding. The heat treatments shown in Table 5 depend on the temper before welding and the temper required after welding. Only the minimum heat treating time (1/2 h) for complete solution is used for welded AZ81A, AZ91C, AZ92A, and QE22A castings to avoid abnormal grain growth in the deposited weld metal. As noted in Table 5, some solution treatments require an SO_2 or CO_2 atmosphere.

If complete solution treatment is not required, magnesium alloys containing more than 1.5% Al should always be stress relieved to prevent corrosion cracking in service. Postweld stress relieving temperatures and times are given in Table 10.

Arc Welding of Nickel Alloys*

NICKEL ALLOYS can be joined by arc welding. The wrought nickel alloys listed in Table 1 can be arc welded under conditions similar to those used in the arc welding of austenitic stainless steel. Cast nickel alloys, particularly those of high silicon content, present difficulties in welding. The arc welding of heat-resistant nickel alloys is described in the article "Arc Welding of Heat-Resistant Alloys" in this Volume.

The most widely employed processes for welding the non-age-hardenable (solid-solution-strengthened) wrought nickel alloys are gas tungsten arc welding (GTAW), gas metal arc welding (GMAW), and shielded metal arc welding (SMAW). Submerged arc welding (SAW) has limited applicability, as does plasma arc welding (PAW). The GTAW process is preferred for welding the precipitation-hardenable alloys, although GMAW and SMAW are also used.

Preweld Heating and Heat Treating. Preweld heating of wrought alloys is not required unless the base metal is below 60 °F, in which case a path 10 to 12 in. wide on both sides of the joint should be warmed to 60 to 70 °F to avoid condensation of moisture that may cause porosity in the weld metal.

Nickel alloys are usually welded in the solution-treated condition. Precipitation-hardenable alloys should be annealed before welding if they have undergone any operations that introduce high residual stresses (see the section on welding of precipitation-hardenable alloys in this article).

Postweld Treatment. No postweld treatment, either thermal or chemical, is needed or recommended to maintain or restore corrosion resistance. Heat treatment may be necessary to meet specification requirements, such as stress relief of a fabricated structure to avoid age hardening or stress-corrosion cracking of the weldment in hydrofluoric acid vapor or caustic soda. If welding induces moderate to high residual stresses, the precipitation-hardenable alloys require a stress-relief anneal after welding and before aging (see the section on welding of precipitation-hardenable alloys in this article).

Table 1 Nominal compositions of weldable wrought nickel and nickel alloys

Alloy designation	Chemical composition, %											
	Ni	C	Mn	Fe	S	Si	Cu	Cr	Al	Ti	Nb	Other
Nickel 200	99.5	0.08	0.18	0.2	0.005	0.18	0.13	...	...	...	...	...
Nickel 201	99.5	0.01	0.18	0.2	0.005	0.18	0.13	...	...	...	...	...
Nickel 205	99.5	0.08	0.18	0.10	0.004	0.08	0.08	...	...	0.03	...	0.05 Mg
Nickel 211	95.0	0.10	4.75	0.38	0.008	0.08	0.13	...	...	...	...	...
Nickel 220	99.5	0.04	0.10	0.05	0.004	0.03	0.05	...	...	0.03	...	0.05 Mg
Nickel 230	99.5	0.05	0.08	0.05	0.004	0.02	0.05	...	...	0.003	...	0.06 Mg
Nickel 270	99.98	0.01	<0.001	0.003	<0.001	<0.001	<0.001	<0.001	...	<0.001	...	Mg <0.001, Co <0.001
Monel 400	66.5	0.15	1.0	1.25	0.012	0.25	31.5	...	...	...	...	...
Monel 401	42.5	0.05	1.6	0.38	0.008	0.13	Bal.	...	...	...	...	...
Monel 404	54.5	0.08	0.05	0.25	0.012	0.05	44.0	...	0.03	...	...	...
Monel R-405	66.5	0.15	1.0	1.25	0.043	0.25	31.5	...	...	...	...	...
Monel K-500	66.5	0.13	0.75	1.00	0.005	0.25	29.5	...	2.73	0.60	...	
Monel 502	66.5	0.05	0.75	1.00	0.005	0.25	28.0	...	3.00	0.25	...	
Inconel 600	76.0	0.08	0.5	8.0	0.008	0.25	0.25	15.5	...	...	...	...
Inconel 601	60.5	0.05	0.5	14.1	0.007	0.25	0.50	23.0	1.35	...	...	...
Inconel 617	54.0	0.07	...	...	...	...	...	22.0	1.0	...	...	12.5 Co, 9.0 Mo
Inconel 625	61.0	0.05	0.25	2.5	0.008	0.25	...	21.5	0.2	0.2	3.65	9.0 Mo
Inconel 671	Bal.	0.05	...	...	...	...	...	48.0	...	0.35	...	...
Inconel 702	79.5	0.05	0.50	1.0	0.005	0.35	0.25	15.5	3.25	0.63	...	...
Inconel 706	41.5	0.03	0.18	40.0	0.008	0.18	0.15	16.0	0.20	1.75	2.9	...
Inconel 718	52.5	0.04	0.18	18.5	0.008	0.18	0.15	19.0	0.50	0.90	5.13	3.05 Mo
Inconel 721	71.0	0.04	2.25	6.5	0.005	0.08	0.10	16.0	...	3.05	...	...
Inconel 722	75.0	0.04	0.50	7.0	0.005	0.35	0.25	15.5	0.70	2.38	...	...
Inconel X-750	73.0	0.04	0.50	7.0	0.005	0.25	0.25	15.5	0.70	2.50	0.95	...
Inconel 751	72.5	0.05	0.5	7.0	0.005	0.25	0.25	15.5	1.20	2.30	0.95	...
Incoloy 800	32.5	0.05	0.75	46.0	0.008	0.50	0.38	21.0	0.38	0.38	...	...
Incoloy 801	32.0	0.05	0.75	44.5	0.008	0.50	0.25	20.5	...	1.13	...	...
Incoloy 802	32.5	0.35	0.75	46.0	0.008	0.38	...	21.0	0.58	0.75	...	...
Incoloy 804	41.0	0.05	0.75	25.4	0.008	0.38	0.25	29.5	0.30	0.60	...	...
Incoloy 825	42.0	0.03	0.50	30.0	0.015	0.25	2.25	21.5	0.10	0.90	...	3.0 Mo
Ni-span-C 902	42.25	0.03	0.40	48.5	0.02	0.50	0.05	5.33	0.55	2.58	...	...

*Revised by S.D. Kiser, Product Manager, Welding Products Co., Huntington Alloys, Inc.

Cleaning of Workpieces

Nickel and nickel alloys are susceptible to embrittlement by lead, sulfur, phosphorus, and other low-melting-point metals and alloys. These materials may be present in grease, oil, paint, marking crayons, marking inks, forming lubricants, cutting fluids, shop dirt, and processing chemicals. Before workpieces are heated or welded, they must be completely free of foreign material. Both sides of the workpiece in the area that will be heated by the welding operations should be cleaned. When no subsequent heating is involved, the cleaned area may be restricted to 2 in. on each side of the joint.

Shop dirt, oil, and grease can be removed by vapor degreasing or by swabbing with acetone or another nontoxic solvent. Paint and other materials not soluble in degreasing solvents may require the use of methylene chloride, alkaline cleaners, or special proprietary compounds. If alkaline cleaners containing sodium carbonate are used, the cleaners themselves must be removed prior to welding; spraying or scrubbing with hot water is recommended. Marking ink can usually be removed with alcohol. Processing material that has become embedded in the work metal can be removed by grinding, abrasive blasting, and swabbing with a 10 vol% hydrochloric acid solution, followed by a thorough water wash. Further information on the cleaning of nickel and nickel alloys is given in Volume 5 of the 9th edition of *Metals Handbook*.

Oxide must also be removed from the area involved in the welding operation, largely because of the difference in melting point between the oxide and the base metal. Oxides are normally removed by grinding, machining, abrasive blasting, or pickling.

Joint Design

Various joint designs are used in the arc welding of nickel alloys. The same design can be used for GTAW and SMAW, but special considerations are required for GMAW and SAW. A joint design developed for other metals is not necessarily suitable for nickel alloys and should not be used unless proved satisfactory by experience or tests.

Beveled Joints. Beveling is usually not required for metal 0.093 in. or less in thickness, although thinner sections of certain alloys are sometimes beveled (Table 2). Metal thicker than 0.093 in. should be beveled to form a V-, U-, or J-groove and should be welded using a backing material, unless welding from both sides. Otherwise, erratic penetration results, leading to crevices and voids that are potential areas for joint weakness and accelerated corrosion in the underside of the joint. For the best underbead contour on joints that cannot be welded from both sides, GTAW should be used for the root pass.

Table 2 Conditions for manual GTAW of 0.062-in.-thick Monel 400

Joint type	Beveled butt, 1/16-in. root opening
Electrode	3/32-in.-diam EWTh-2, tapered to 1/64 in. diam
Filler metal	1/16-in.-diam Monel 60 (ERNiCu-7)
Number of passes	3
Welding current	70-90 A (DCEN)
Voltage	10-12 V
Shielding gas:	
At torch	20 ft^3/h
Backing gas	3-5 ft^3/h
Preheat and postheat	None
Interpass temperature	350 °F max

For metal more than 3/8 in. thick, a double-U-groove or double-V-groove design is preferred. The added cost of preparation is justified, because less welding time and material are required and less residual stress is developed than with a single-groove design.

Beveling is accomplished best by machine, usually a plate planer or other machine tool. Plasma arc and electric arc cutting can be used for joint preparation, but all oxidized metal must be removed from the joint area by grinding or chipping for a depth of 1/32 to 1/16 in.

Corner and Lap Joints. These joints may be welded where service stresses will not be excessive, but should be avoided if service temperatures are high or if service conditions involve thermal or mechanical cycling. When corner joints are used, a full-penetration weld must be made, usually with a fillet weld on the root side. Lap joints should be welded on both sides.

Design Considerations. Unlike steel weld metal, molten nickel alloy weld metal does not flow. This characteristic cannot be compensated for by extra heat input or pooling, because serious loss of residual deoxidizers may result. For joints in metal up to 5/8 in. thick, ample accessibility is provided by V-groove butt joints beveled to an 80° groove angle. For thicker work metal, U-groove butt joints are machined to a 15° bevel angle and a 3/16- to 5/16-in. radius. Single bevels used to form T-joints should have an angle of 45°. The bottom radius of a J-groove for T-joints should be at least 3/8 in. and the bevel angle should be used for J-grooves and U-grooves. For full-penetration welds, the bevel and groove angles should be increased about 40% above those used for carbon steels.

Welding Fixtures

Proper fixturing and clamping hold the workpieces firmly in place, minimize buckling, maintain alignment, and when needed, provide compressive stress in the weld metal. A backing bar or any portion of a fixture that might be contacted by the arc should be made of copper. Copper has such high thermal conductivity that inadvertent welding against the copper will not result in fusion of the bar to the weldment. Also, the grooves in copper bars are often used to contour the underneath side of the weld, without sticking.

Backing bars, when used, should incorporate a groove of suitable contour to permit penetration of weld metal and avoid the possibility of trapping gas or flux at the bottom of the weld. Grooves in backing bars for GMAW usually have a shallow semielliptical shape—0.015 to 0.035 in. deep and 3/16 to 1/4 in. wide. Square-corner grooves are used with GMAW and GTAW to accommodate the backing gas; a machined passageway is connected to the gas supply, and holes 1/16 in. in diameter are drilled 3 in. apart from the bottom of the groove to the passageway, permitting the backing gas to flow along the weld. The gas flows out of the groove at the ends of the bar. Figure 1 shows groove designs for backing bars.

Fig. 1 Groove designs for backing bars. (a) Standard groove for use without a backing gas. (b) Square-corner groove employed with backing gas (groove depth depends on gas flow required and length of bar)

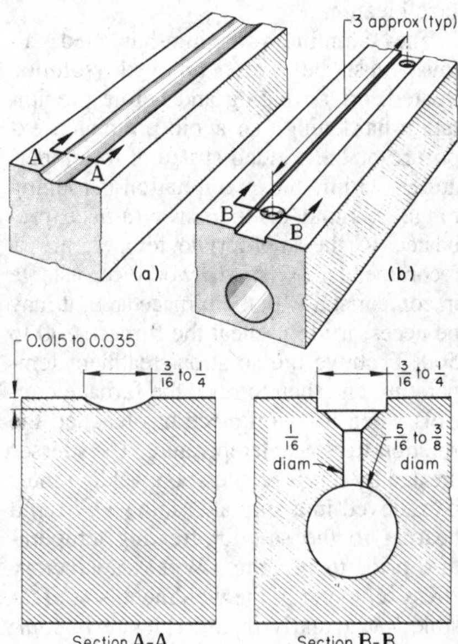

Clamping and restraint required for nickel alloys are about the same as for low-carbon steel. The hold-down bars should be located close enough to the line of weld to maintain alignment and provide the proper degree of heat transfer. Generally, hold-down pressure should be just enough to maintain alignment, but high restraint can be advantageous in welding square-groove joints in thin metal by the GTAW process without a filler metal. When the pieces to be welded are positioned with a zero root opening and hold-down bars are brought very close to the line of welding, and a high hold-down pressure is used, the heat of welding creates an expansive force that results in a compressive force in the line of weld. This compression will have an upsetting effect on the hot weld metal, causing the weld to develop a slight top and bottom crown, or weld reinforcement, without the use of filler metal.

Welding of Precipitation-Hardenable Alloys

The precipitation-hardenable alloys require special welding procedures because of their susceptibility to cracking. Cracks can occur in the base metal heat-affected zone (HAZ) on aging or in service at temperatures above the aging temperature, as a result of residual welding stress and stress induced by precipitation.

Preweld and Postweld Treatments. Any part that has been subjected to severe bending, drawing, or other forming operation should be annealed before welding. If possible, heating should be done in a controlled-atmosphere furnace to limit oxidation and minimize subsequent surface cleaning.

The aluminum-titanium-hardened alloys must be stress relieved (solution treated) after welding and before precipitation hardening. To avoid prolonged exposure of the welded structure to temperatures within the precipitation-hardening range, a rapid heating in a furnace preheated to the appropriate temperature is recommended. When the workpiece is large in comparison with the furnace area, it may be necessary to preheat the furnace 200 to 500 °F above the solution-treatment temperature and then to reset the furnace controls when the workpiece has reached the solution-treatment temperature. The stresses created by repair or alteration welding must be relieved in a similar fashion—by rapid heating to the solution-treating temperature prior to re-aging. When satisfactory stress relieving of the weldment is not feasible, particularly if the structure is complicated, preweld overaging treatments may be helpful. Preheating, however, is not a satisfactory substitute for postweld heat treatment. Precipitation-hardenable alloys can be welded in the aged condition, but if temperatures encountered in service are in the precipitation-hardening range, the weldment must be solution treated and re-aged.

General Welding Procedures. Precipitation-hardenable alloys are usually welded by the GTAW process, but SMAW and GMAW are also applicable. Heat input during the welding operations should be held to a moderately low level to obtain the highest possible joint efficiency. For multiple-bead or multiple-layer welds, many narrow stringer beads should be used rather than a few large, heavy beads.

If oxides form during welding, they should be removed by abrasive blasting or grinding. If such films are not removed as they accumulate on multiple-pass welds, they may become thick enough to inhibit weld fusion and to produce unacceptable laminar-type oxide stringers along the weld axis.

Welding of Cast Nickel Alloys

Cast nickel alloys can be joined by GTAW, GMAW, and SMAW. For optimum results, castings should be solution annealed before welding to relieve some of the casting stresses and provide some homogenization of the cast structure.

Light peening of solidified metal after the first pass will relieve stresses and thus reduce cracking at the junction of the weld metal and the cast metal. Peening of subsequent passes is of little if any benefit. Stress relieving after welding is also desirable.

Gas Tungsten Arc Welding

Nickel alloys, both cast and wrought, either solid-solution-strengthened or precipitation-hardenable, can be welded by the GTAW process. The addition of filler metal is usually recommended. Direct current electrode negative (DCEN) is recommended for both manual and machine welding. If close control of arc length is feasible, alternating current can be used for machine welding, but a superimposed high-frequency current is required.

The welding torch should be set or held at an angle of 90° to the work. A slight deviation is permissible to provide a better view of the work, but an acute angle can lead to aspiration of the surrounding air and contamination of the shielding gas. The largest gas nozzle size applicable to the job should be used, and a minimum practical distance between the nozzle and the work should be maintained. A gas lens is sometimes used to improve shielding. For details of the process and equipment, see the article "Gas Tungsten Arc Welding" in this Volume.

Shielding Gas. Either argon or helium, or a mixture of the two, is used as a shielding gas for welding nickel and nickel alloys. Additions of oxygen, carbon dioxide, or nitrogen to argon gas will usually cause porosity or erosion of the electrode. Argon with small quantities of hydrogen (approximately 5%) can be used for single-pass welding and may help to avoid porosity in pure nickel. The advantages of helium over argon in welding thin metal without the addition of filler metal are:

- *Improved soundness*: Porosity-free welds are more easily obtained in Monel 400, and porosity in welds in Nickel 200 is less.
- *Increased welding speed*: With DCEN, welding speed can be increased as much as 40% over that achieved with argon at the same current setting, because welding speed is a function of heat input, and heat input is considerably greater with helium.

Because purity and dryness of the welding gas are important, welding-grade shielding gas should be used. Furthermore, the welding area should be screened to avoid drafts, because disruption of the shielding atmosphere and drawing in of air can cause porosity in the weld metal.

Electrodes. Pure tungsten electrodes or tungsten alloyed with thoria or zirconia can be used. The alloyed electrodes are more economical because of their lower vaporization loss. Overheating through use of excessive amperage must always be avoided.

The best arc stability and penetration control are achieved by tapering the electrode tip. The taper angle should be approximately 30°, with a flat land of about 0.015 in. diam on the tip end. Larger taper angles are used to produce a narrower bead and deeper penetration. After selection, the taper angle should be maintained, because changes in the angle can affect penetration and bead width.

Electrode extension (stickout) should be short and should be based on joint design. For example, a maximum extension of $^{3}/_{16}$ in. is used for butt welds in thin metal, but a $^{3}/_{8}$- to $^{1}/_{2}$-in. extension may be required

Table 3 Nominal compositions of filler metals for GTAW, GMAW, and SAW of nickel alloys

Electrode	Chemical composition, %												
	C max	Mn	Fe	S	Si	Cu	Ni + Co	Al max	Ti	P	Nb	Mo	Cr
ERNi-1	0.15	1.0 max	1.0 max	0.015 max	0.75 max	0.25 max	93.0 min	1.5	2.0-3.5	0.030 max	...	...	...
ERNiCu-7	0.15	4.0 max	2.5 max	0.015 max	1.25 max	rem	62.0-69.0	1.25	1.5-3.0	0.020 max	...	...	...
ERCuNi	0.02	0.75	0.50	0.005	0.10	67.5	31.0	...	0.30	...	...	...	...
ERNiFeCr-1	0.05	1.0 max	22.0 min	0.03 max	0.50 max	1.5-3.0	38.0-46.0	0.20	0.60-1.20	0.03 max	...	2.50-3.50	19.5-23.5
ERNiCrFe-6	0.03	2.30	6.60	0.007	0.10	0.04	71.0	...	3.2	...	...	...	16.4
ERNiCr-3	0.10	2.5-3.5	3.0 max	0.015 max	0.50 max	0.50 max	67.0 min	...	0.75 max	0.03 max	2.0-3.0	...	18.0-22.0
ERNiCrFe-5	0.02	0.10	7.50	0.005	0.10	0.03	74.0	...	...	...	2.25	...	16.0
ERNiMo-1	0.05	1.0-2.0	18.0-21.0	0.03	1.0	1.5-2.5	rem	...	...	0.04 max	1.75-2.50	5.5-7.5	21.0-23.5
ERNiCrMo-3	0.02	3.00	1.00	0.007	0.20	0.04	72.0	...	0.55	...	2.55	...	20.0
ERNiCrMo-4	0.02	1.0 max	4.0-7.0	0.03	0.08	0.50	rem	...	...	0.04	2.5 max	15.0-17.0	14.5-16.5
ERNiCrMo-9	0.015	1.0 max	18.0-21.0	0.03	1.0	1.5-2.5	rem	...	0.50 max	0.04	...	6.0-8.0	21.0-23.5

for some fillet welds or for reaching the bottom of groove welds.

The electrode should be inclined slightly in the forehand position. When used, filler metal should be added carefully at the leading edge of the weld pool to avoid contact with and contamination of the electrode. If contamination occurs, the electrode should be cleaned and reshaped.

Filler Metals. Compositions of the filler metals used with GTAW are, in general, similar to those of the base metals with which they are used. Because of high arc currents and high welding temperatures, filler metals are alloyed to resist porosity and hot cracking of the weld metal. Filler-metal additions and dilution ratios should be adjusted to ensure that the weld metal contains about 75% filler metal.

Table 3 lists the compositions of filler metals used for GTAW of nickel and nickel alloys. Filler metals of the ERNi-1 classification are used for welding the high-nickel alloys, such as Nickel 200 and 201, and those of the ERNiCu-7 classification are used for welding nickel-copper alloys, such as Monel 400 and 404.

Joint Design. All joint designs shown in the article "Arc Welding of Heat-Resistant Alloys" in this Volume can be used in welding nickel alloys by the GTAW process. Where no filler metal is used, the sections to be joined must be held tightly together to promote satisfactory fusion.

Welding Techniques. When a filler metal is used, the wire diameter should be related to the work-metal thickness. For example, a $^1/_{16}$-in.-diameter wire is generally used to weld $^1/_{16}$-in.-thick sheet. During welding, the hot end of the wire must remain under the shielding gas to avoid oxidation. The molten pool must be maintained as quiet as possible to prevent the deoxidizing elements from burning out.

To ensure a sound weld, the arc should be maintained at the shortest possible length. When no filler metal is added, the arc length should be 0.05 in. maximum and preferably 0.02 to 0.03 in. When filler metal is used, the arc can be longer, but should be kept as short as possible, consistent with filler-metal diameter, to avoid loss of gas coverage. This loss can result in oxidation and reduced penetration. It also reduces weld spatter with filler wire.

The speed of welding affects penetration, width of weld, and weld soundness, especially when no filler metal is added. For a given thickness of metal, there is a range of welding speeds that results in minimum porosity. Speeds outside this range, either faster or slower, result in increased porosity. Also, for a given welding speed, the likelihood of porosity decreases as metal thickness increases.

Complete-penetration welds call for the use of grooved backing bars that permit local gas shielding, such as that shown in Fig. 1(b), or the purging of the inside of the workpiece.

Gas Metal Arc Welding

The high-nickel and nickel-copper alloys can be joined by GMAW. With special procedures, precipitation-hardenable alloys, such as Monel K-500, can be gas metal arc welded as well.

Spray, globular, and short circuiting metal transfer are suitable. Varying the power input produces the different types of metal transfer. The pulsed arc process is also used. In these methods, the filler metal is a current-carrying consumable and is typically 0.035, 0.045, or 0.062 in. in diameter. Table 4 lists typical conditions for GMAW of Nickel 200 and Monel 400 using spray-type metal transfer, short circuiting metal transfer, and pulsed arc transfer.

Welding current is most often provided by constant-voltage power supplies, although all standard direct current power supplies are satisfactory. Direct current electrode positive (DCEP) is used, because it is virtually impossible to transfer metal in a controlled manner with DCEN. Amperage should be well within the maximum rating of the equipment, but sufficient to obtain the desired melting rate.

Shielding gas used for GMAW of nickel alloys by spray or globular transfer may be argon, which yields good results. The addition of 15 to 20% helium is beneficial when welding nickel alloys with short circuiting arc or pulsed arc transfer. As the helium content is increased from 0 to 20%, the weld beads become progressively wider and flatter, and penetration decreases. The addition of oxygen or carbon dioxide to argon stabilizes the arc, but results in heavily oxidized and irregular bead surfaces. Helium alone has been used as a shielding gas, but creates an unsteady arc with excessive weld spatter and is a difficult atmosphere in which to initiate an arc.

Gas flow rates vary from 25 to 100 ft^3/h, depending on joint design and welding

Table 4 Typical conditions for GMAW of Nickel 200 and Monel 400

Item	Nickel 200	Monel 400
With spray transfer(a)		
Electrode wire	ERNi-1	ERNiCu-7
Voltage, V (avg)	29-31	28-30
Current, A (avg)	375	290
Wire feed, in./min	205	200
With short circuiting transfer(b)		
Electrode wire	ERNi-1	ERNiCu-7
Voltage, V (avg)	20-21	16-18
Current, A (avg)	160	130-135
Wire feed, in./min	360	275-290
With pulsed arc transfer(c)		
Electrode wire	ERNi-1	ERNiCu-7
Peak voltage, V	46	40
Voltage, V (avg)	21-22	21-22
Current, A (avg)	150	110
Wire feed, in./min	160	140

(a) Argon shielding gas at 60 ft^3/h; flat welding position; 0.062-in.-diam electrode wire. (b) Argon-helium shielding gas at 50 ft^3/h; vertical welding position; 0.035-in.-diam electrode wire. (c) Argon or argon-helium shielding gas at 25 to 35 ft^3/h; vertical welding position; 0.045-in.-diam electrode wire

position. A flow rate of 50 ft^3/h is most frequently used. The choice of gas for use with short circuiting metal transfer is influenced by the type of equipment available. Argon, without additions, is suitable for use with equipment that has both inductance and slope control. Argon gas produces convex beads, which may cause cold lapping (lack of fusion), but also provides a pronounced pinch effect that can be controlled by inductance. An addition of helium is helpful when induction and slope cannot be varied. Helium imparts a wetting action, and the arc is hotter; these factors greatly decrease the possibility of cold lapping. As the percentage of helium is increased, the gas flow rate must be increased to ensure adequate protection.

The size of the gas nozzle is important when using a short circuiting arc. When using a one-to-one mixture of argon and helium at a flow rate of 40 ft^3/h through a $^3/_8$-in.-diam nozzle, a wire-feed rate of 250 in./min and the current can be increased to 160 to 180 A without affecting weld quality adversely. An argon-helium mixture is recommended as the shielding gas with a pulsed arc. Gas flow rate should be from 25 to 45 ft^3/h. Excessive flow rate can interfere with the arc.

Filler metals used in GMAW are listed in Table 3. The appropriate wire diameter depends on the method used and the thickness of the base metal. With globular or spray transfer, 0.035-, 0.045-, and 0.062-in.-diam wire is used. The short circuiting and pulsed arc processes generally require wire 0.045 in. or less in diameter.

Joint designs recommended for the GMAW process are discussed in the article "Arc Welding of Heat-Resistant Alloys" in this Volume. For U-groove designs that use globular or spray transfer, however, the root radius should be decreased by about 50% and the bevel angle should be doubled. With these types of transfer, the use of high amperage on small-diameter wire produces a high level of arc force. As a result, the arc cannot be deflected from a straight line, which is possible when welding is done with covered electrodes. Because the arc must contact all areas to be fused, the joint design must provide intersection with the arc force line. When a short circuiting arc is used, the U-groove designs are the same as those that are employed with the other arc welding processes.

Welding techniques for best results include positioning the electrode holder vertically along the centerline of the joint. Some slight inclination is permissible to allow a better view of the work, but excessive displacement can cause the surrounding atmosphere to be drawn into the shielding gas, causing porous or heavily oxidized welds. Arc length is important; too short an arc causes spatter, and too long an arc results in loss of control and reduced penetration.

Because arc length is directly proportional to voltage, particular care must be paid to the voltage setting to ensure properly controlled arc length. In addition, once a procedure has been qualified, consistent torch-to-work positioning should be maintained to prevent possible changes in transfer mode caused by loss (or gain) of energy available at the arc. The alteration in energy at the arc is caused by I^2R energy changes corresponding to stickout (wire extension) changes.

Cold lapping (lack of fusion) can occur with the short circuiting, globular, and pulsed arc processes if manipulation is faulty. The torch should be advanced at a rate that will keep the arc in contact with the base metal just ahead of the weld pool. The manipulation and angle used with the pulsed arc torch are similar to those used in SMAW. To avoid undercutting, a slight pause should be made at the limit of the weave.

Plasma Arc Welding

Plasma arc welding (PAW), using the keyholing mode, can produce acceptable welds in nickel alloys up to about 0.3 in. thick. Argon-hydrogen mixtures are used as orifice and shielding gas, 5 to 8% H_2 being optimum. Current needed for keyholing decreases as hydrogen content is increased, up to about 7% H_2, above which (or when helium is used) torch starting is more difficult.

Typical relations of travel speed and current for keyhole welding are:

Travel speed, in./min	Current, A Nickel 200	Monel 400
8	185	155
10	200	175
12	220	195
14	235	215

Typical conditions for welding 0.235-in.-thick Nickel 200, using gas of 95% argon and 5% hydrogen gas, are 245 A; 31.5 V; 14 in./min; and gas flow of 10 ft^3/h (orifice) and 45 ft^3/h (shielding). (References: A.C. Lingenfelter, *Welding Engineer*, Jan 1970, p 42-45; see also A.C. Lingenfelter, *et al.*, *Welding Journal*, May 1966, p 417-422.)

Shielded Metal Arc Welding

The SMAW process can be used for welding nickel and nickel alloys. Although the minimum metal thickness is usually about 0.050 in., thinner metal can be welded when appropriate fixtures are provided. The types of joints used and bead and groove dimensions are given in the article "Arc Welding of Heat-Resistant Alloys" in this Volume.

Electrodes. Composition of electrodes used for SMAW are given in Table 5. Electrode composition is selected to be similar to that of the base metal with which the electrode is to be used. Electrodes of the ENi-1 classification are used for welding wrought and cast forms of nickel and nickel alloys to themselves and to steel. ENiCu-7 electrodes can be used for welding nickel-copper alloys to themselves, for surfacing steel with a nickel-copper alloy, for welding the clad side of a nickel-copper clad steel, and for welding nickel-copper alloys to steel.

ENiCrMo-3 electrodes typically are used for joining a wide range of pitting and crevice corrosion-resistant alloys. The lowest alloyed materials employed with these electrodes are the molybdenum-modified 317 stainless steels. ENiCrMo-3 electrodes are used for richer alloy compositions of the Ni-Cr-Fe-Mo families, up to and including Inconel 625. Intermediate alloys welded with ENiCrMo-3 include Incoloy 825, Cartech 20, and Hastelloy G and G-3. This type of electrode is widely used because its high nickel content allows good dissimilar weldability, while the high molybdenum content matches or exceeds the pitting resistance of the base alloys being welded.

Nickel-chromium-iron electrodes such as ENiCrFe-2 and ENiCrFe-3 are designed for welding the same family of alloys as well as dissimilar metal joints involving carbon steel, stainless steel, nickel, and nickel-based alloys. Nickel-chromium-molybdenum electrodes such as ENiCrMo-4 are used for welding alloys of similar composition to themselves and to steel.

Nickel-molybdenum electrodes such as ENiMo-1 are designed for welding nickel-molybdenum alloys to themselves and to other nickel-, cobalt-, and iron-based metals. They are normally used in the flat position only.

Prior to use, covered electrodes should be kept in sealed, moisture-proof containers in a dry storage area. All opened containers of unused electrodes should be stored in a cabinet equipped with a des-

Table 5 Composition of electrodes used for SMAW of nickel, high-nickel alloys, and nickel-copper alloys

Electrode	Chemical composition, %													
	C	Mn	Fe	S	Si	Cu	Ni + Co	Al	Ti	Nb + Ta	Cr	Mo	P	W
ENi-1	0.03	0.30	0.05	0.005	0.60	0.03	96.0	0.25	2.5	...	...	...	...	...
ENiCu-7	0.15	4.0	2.5	0.015	1.0	rem	62.0-68.0	0.75	1.0	...	...	...	0.02	...
ENiCrMo-3	0.05	0.3	4.0	0.010	0.40	...	61.0	...	...	3.65	21.5	...	...	...
ENiCrFe-1	0.04	0.75	8.5	0.006	0.20	0.04	73.0	...	...	2.1	15.0	...	...	...
ENiMo-1	0.05	1.0-2.0	18.0-21.0	0.03	1.0	1.5-2.5	rem	...	...	1.75-2.50	21.0-23.5	5.5-7.5	0.04	1.0
ENiCrFe-3	0.05	7.75	7.5	0.008	0.50	0.10	67.0	...	0.40	1.75	14.0	...	...	...
ENiCrMo-4	0.02	1.0	4.0-7.0	0.03	0.08	0.50	rem	...	...	...	14.5-16.5	15.0-17.0	0.04	3.0-4.5
ENiCrFe-2	0.03	2.0	9.0	0.008	0.30	0.06	70.0	...	...	2.0	15.0	1.5	...	...
ENiCrMo-9	0.015	1.0	18.0-21.0	0.03	1.0	1.5-2.5	rem	...	...	0.50	21.0-23.5	6.0-8.0	0.04	1.5

Table 6 Recommended electrode diameter and welding current for SMAW of various thicknesses of high-nickel alloys and nickel-copper alloys in the flat position

Base-metal thickness, in.	Electrode diameter(a), in.	Current(b), A
High-nickel alloys		
0.037	3/32	(c)
0.043	3/32	(c)
0.050	3/32	(c)
0.062	3/32	75
0.078	3/32	80
0.093	3/32	85
0.109	1/8	105
0.125	1/8	105
0.125	3/32-5/32	80-150
0.140	5/32	130
0.156	5/32	135
0.187(d)	5/32	150
Nickel-copper alloys		
0.037	3/32	(c)
0.043	3/32	(c)
0.050	3/32	(c)
0.062	3/32	50
0.078	3/32	55
0.093	3/32	60
0.109	3/32	60
0.109	1/8	65
0.125	3/32-5/32	60-140
0.140	3/32-5/32	60-140
0.156	3/32-5/32	60-140
0.250	3/32-5/32	60-140
0.375	3/32-3/16	60-180
0.500(d)	3/32-3/16	60-180

(a) Where a range is shown, the smaller diameter electrodes are used for the first passes at the bottom of the groove, and the joints are completed with the larger diameter electrodes. (b) Current should be in the range recommended by the electrode manufacturer. (c) Use minimum amperage at which arc control can be maintained. (d) And thicker

iccant or heated to 10 to 15 °F above ambient temperature prior to use. If exposed to excessive moisture, the electrodes can be reclaimed by rebaking at 500 °F for 2 h or at 600 °F for 1 h. Choice of electrode diameter should be based on quality requirements and position of welding, rather than on speed of production.

Welding Current. Shielded metal arc welding is done with DCEP. Each electrode size has an optimum amperage range in which it has good arcing characteristics and outside which the arc becomes unstable or the electrode overheats. Suggested electrode diameters and current settings for various metal thicknesses are shown in Table 6. Variables such as type of backing, tightness of clamping, and joint design influence the current density needed. Actual welding currents should be developed by making sample welds in metal of the same composition and thickness as the metal to be welded.

Welding Position. Flat-position welding should be used whenever possible because it is faster, more economical, and produces a weld of good quality. The recommended electrode position for flat-position welding is an inclination of about 20° from the vertical, ahead of the weld pool. This position facilitates control of the molten flux and prevents slag entrapment. A short arc must be maintained to prevent atmospheric contamination of the arc and weld and to prevent porosity.

For vertical welding, the arc should be slightly shorter than for flat-position welding, and the amperage should be 10 to 20% lower than the values in Table 6. The electrode should be at approximately a 90° angle to the joint. For overhead welding, the arc should be slightly shorter than for flat-position welding, and the welding current should be reduced by 5 to 15 A from the values shown in Table 6.

Welding Techniques. Because molten nickel alloy weld metal does not flow, it must be deposited where needed. Therefore, a slight weave is needed. The amount of weave will depend on joint design, welding position, and type of electrode. Although a straight stringer bead laid down without weaving can be used for single-pass work and is satisfactory at the bottom of a deep groove on thicker sections, a weave is generally desirable.

When used, a weave should be no wider than about three times the electrode diameter. Some deviation from this rule may be necessary during vertical welding.

Weld spatter should be avoided. Spatter indicates that the arc is too long, excessive amperage is being used, or current is straight polarity (DCEN). Other causes are wet electrodes or base metal and slag running under the arc during welding.

Arc blow can occur when the arc is deflected from its normal path by a magnetic force in the work metal. One method of overcoming arc blow is to change the location of the ground connection on the workpiece or change the direction of the electrical path to the arc.

When the arc is to be broken, it should be shortened slightly, and the rate of travel should be increased to reduce the size of the weld pool. This practice reduces the probability of crater cracking, ensures that the crater does not develop a rolled leading edge, and prepares the way for restriking the arc.

The manner in which the arc is restruck has a significant effect on the soundness of the weld. A reverse (or T) restrike is recommended. The arc should be struck at the leading edge of the crater and carried back to the extreme rear of the crater at a normal welding speed. The direction is then reversed, weaving is commenced, and the weld continued. Advantages of this procedure are: (1) the welder has an opportunity to establish the correct arc length before actual welding commences; (2) some preheat is applied to the cold crater; and (3) the first drops of rapidly cooled weld metal are deposited where they will be remelted, thus restricting porosity to a minimum.

Another technique is to make the restrike where the weld metal can be readily removed—for example, 1/2 to 1 in. behind the crater on top of the previous pass. Later the restrike area can be ground level with the rest of the bead. This technique is used when the weld must meet rigid radiographic standards and calls for less welder skill than the reverse-restrike method.

Welding procedures should be qualified before production begins. In one plant where this was not done, qualifying tests

Fig. 2 Joint design for welding 1/2-in. Nickel 200 and crack in weld metal caused by excessive heat input

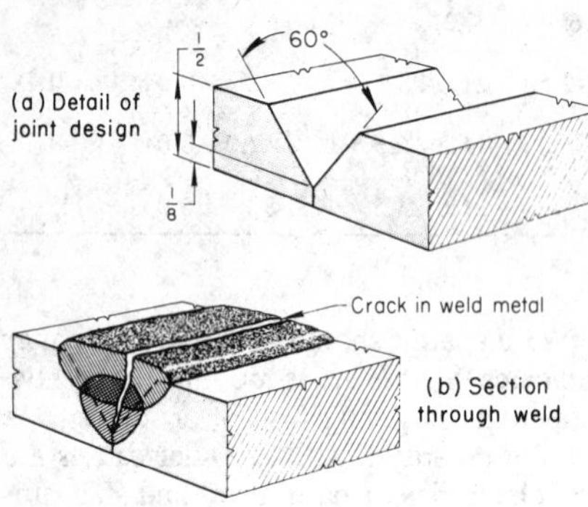

of the welding procedure were conducted before further welds were made when cracks appeared in the weld metal of a joint in a 1/2-in.-thick Nickel 200 cylinder.

The original procedure for this weld specified a V-groove butt joint with a 60° included angle, a 1/8-in. root face, and zero root opening, as shown in Fig. 2(a). An ENi-1 electrode that was 14 in. long by 3/16 in. in diameter was used with a welding current of 235 A. On the first pass, a weld bead 6 1/2 in. long was deposited with each electrode, and on the second pass, a weld bead 3 in. long was deposited with each electrode. A cross section of the completed weld is shown in Fig. 2(b). Investigation showed that cracks in the weld metal were caused by the high welding current and low travel speed, which had resulted in the weld pool being held in a superheated condition for an excessively long time.

A qualified procedure was established, therefore, that specified a welding current of 210 A and an increased travel speed which resulted in a 12-in.-long weld for each electrode on the first pass and 9-in.-long welds per electrode on the the second pass. Electrode type and length, and all other welding conditions, were unchanged. An included angle of 80° and a 1/16- to 3/32-in. root face would have produced a better welded joint, but the workpieces had been machined and roll formed before the problem arose.

Cleaning the Weld Bead. In multiple-pass welding, all flux and slag must be removed before each succeeding bead is deposited. All slag should be completely removed from completed welds, especially if service is to be at high temperature. The slag is easily removed with hand tools or a hand or power wire brush.

Submerged Arc Welding

Submerged arc welding can be used for joining solid-solution nickel alloys. Monel 400 is the alloy most frequently welded by this process. Joints in metal up to 3 in. thick have met American Society of Mechanical Engineers (ASME) codes and other specification requirements. The SAW process cannot be used for welding the precipitation-hardenable nickel alloys.

Joint Design. Some joint designs used in SAW of nickel alloy plate are shown in Fig. 3. Single-compound-angle grooves, single-U-grooves, and double-U-grooves are used for metal 3/4 in. or more thick. The double-U-groove is usually preferred, because it results in a lower level of residual stress and can be completed in less time, with less filler metal. A single-V-groove is used for stock up to 1 in. thick.

Electrodes. The compositions of the electrode wires used in SAW are the same as the filler metals or electrode wires used for GTAW and GMAW (Table 3). Submerged arc welding is possible with the following: ERNi-1, ERCuNi, ERNiCu-7, ERNiCr-3, ERNiCrFe-5, and ERNiCrMo-3.

Wire ranging from 0.045 to 3/32 in. in diameter can be used for all nickel alloys. The 1/16-in.-diam wire is generally preferred. Small-diameter wires are used for welding metal up to 1/2 in. thick, and 3/32-in.-diam wires for heavier sections. The approximate amount of filler metal needed for single-compound-angle, single-U-, and single-V-grooves is given in Table 7.

Fluxes. When welding nickel alloys, the fluxes used in SAW of carbon and stainless steels are not satisfactory. Special proprietary fluxes are available and must be used. Poor weld contour, flux entrapment, weld cracking, and inclusions can result when the wrong flux is used.

Fig. 3 Joint designs for SAW of nickel alloys. Backing strips are optional with zero root openings.

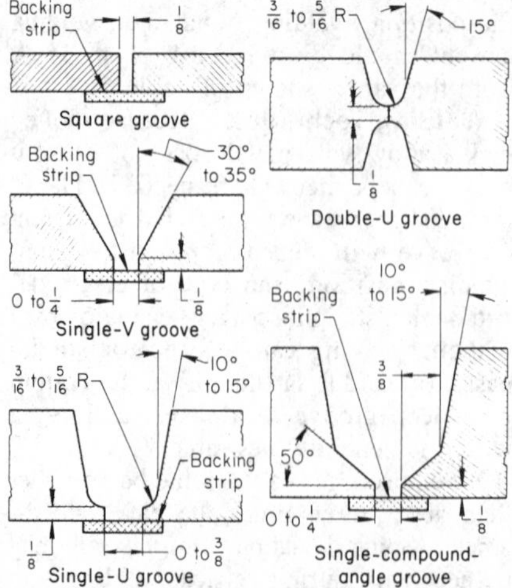

Only enough flux cover to prevent arc breakthrough should be used; excessive flux can cause a deformed bead surface. Slag entrapment can be prevented by appropriate joint design and by correct placement of beads. Slag is easily removed from welds in the bottom of grooves. Fused flux is self-lifting from exposed welds and should be discarded. Unfused flux can be recovered by means of a vacuum system and, if clean, can be reused. Screening to adjust particle size is not required.

Fluxes should be stored in dry storage areas. Opened containers should be resealed to prevent moisture pickup, but flux that has absorbed moisture can be reclaimed by drying at 600 °F for 1 h. Hoppers used for fluxes for welding steel and other metals should be thoroughly cleaned before being filled with flux for welding nickel alloys.

Welding Current. Direct current electrode negative or direct current electrode positive is used. Reverse polarity (DCEP) is preferred for butt welding because it produces a flatter bead with deeper penetration at a rather low arc voltage (30 to 33 V). Straight polarity (DCEN) results in a slightly higher rate of deposition at increased voltage (over 35 V) and requires oscillation. The flux covering, however, must be appreciably deeper. Consequently, flux consumption and the risk of slag entrapment increase. Alternating current and the two-wire series technique for multiple-arc welding are not suitable for use with the available fluxes.

Bead Deposition. Location of a bead in a multiple-pass layer should provide an open, or reasonably wide, root area for the next bead. Slightly convex beads are preferred to concave beads. Bead contour is controlled by voltage, travel speed, and the position of the electrode.

Causes and Prevention of Weld Defects

The defects and metallurgical difficulties encountered in arc welding of nickel alloys are porosity, susceptibility to high-temperature embrittlement by sulfur and other contaminants, cracking in the weld bead because of high heat input, and stress-corrosion cracking in service.

Porosity. Oxygen, carbon dioxide, nitrogen, or hydrogen can cause porosity in welds. In SMAW and SAW, porosity can be minimized by using electrodes that

Table 7 Amount of filler metal required for submerged arc welds in three types of grooves

Joint design: Angle α	Angle β; or radius, r	Root opening, s, in.	Dimension, s_1, in.	Approximate amount of metal required, lb per linear foot, for plate thickness, in., of: 1	$1^1/_4$	$1^1/_2$	$1^3/_4$	2	$2^1/_2$	3	$3^1/_2$	4
Single-compound-angle groove												
10°	40°	0	$^3/_8$	2.19	3.10	4.10	5.17	6.33	8.90	11.77	14.95	18.50
10°	40°	0	$^1/_2$	2.74	3.56	4.84	6.16	7.61	10.45	13.73	17.32	21.00
10°	50°	0	$^3/_8$	1.95	2.83	3.79	4.82	5.93	8.40	11.18	14.00	17.70
15°	50°	0	$^3/_8$	2.03	2.95	4.08	5.29	6.62	9.68	13.20	17.30	21.78
10°	50°	$^1/_4$	$^3/_8$	2.87	3.98	5.17	6.43	7.77	10.70	13.94	17.22	21.38
10°	40°	$^1/_4$	$^3/_8$	3.11	4.25	5.48	6.78	8.17	11.20	14.53	18.17	22.18
Single-U-groove												
10°	$^3/_{16}$ in.	0	...	1.46	2.07	2.76	3.53	4.37	6.33	8.60	11.20	14.10
10°	$^1/_4$ in.	0	...	1.77	2.47	3.25	4.12	5.06	7.20	9.67	12.45	15.60
10°	$^5/_{16}$ in.	0	...	2.03	2.86	3.68	4.71	5.76	8.10	10.70	13.70	17.00
15°	$^3/_{16}$ in.	0	...	1.60	2.36	3.23	4.22	5.34	7.94	11.00	14.60	18.68
15°	$^1/_4$ in.	0	...	1.90	2.73	3.68	4.76	5.96	8.74	12.00	15.78	20.00
15°	$^5/_{16}$ in.	0	...	2.17	3.08	4.14	5.30	6.60	9.55	12.90	16.90	21.40
10°	$^3/_{16}$ in.	$^1/_4$	...	2.38	3.22	4.14	5.14	6.21	8.63	11.36	14.42	17.78
10°	$^1/_4$ in.	$^1/_4$	...	2.69	3.62	4.63	5.73	6.90	9.50	12.43	15.67	19.28
Single-V-groove												
30°	...	0	...	1.63	...	...	...	...	...	...	...	...
35°	...	0	...	1.96	...	...	...	...	...	...	...	...
30°	...	$^1/_4$	...	2.55	...	...	...	...	...	...	...	...

Single-compound-angle groove   Single-U groove   Single-V groove

contain deoxidizing or nitride-forming elements such as aluminum and titanium. These elements have a strong affinity for oxygen and nitrogen and form stable compounds with them.

In GMAW and GTAW, porosity can be avoided by preventing access of air to the molten weld metal. Gas backing on the underside of the weld is sometimes used.

In GTAW, the use of argon with up to 20% hydrogen as a shielding gas helps prevent porosity. The hydrogen acts as a scavenger; bubbles of hydrogen that form in the weld pool gather the diffusing nitrogen. Too much hydrogen in the shielding gas can result in hydrogen porosity.

Cracking. Hot shortness of welds can result from contamination by sulfur, lead, phosphorus, or low-melting-point metals such as bismuth, which form intergranular films that cause severe embrittlement at elevated temperature. Hot cracking of the weld metal usually results from such contamination. Cracking in the HAZ is often caused by intergranular penetration of contaminants from the base-metal surface. Sulfur, which is present in most cutting oils used for machining, is a common cause of cracking in nickel alloys. Removal of foreign material from the surfaces of the work metal is imperative and is described in the previous section of this article on the cleaning of workpieces. Cracking of the weld metal also can be caused by heat input that is too high as a result of high welding current and low welding speed (Fig. 2).

In addition, cracking may result from undue restraint. Particularly when conditions of high restraint are present, as in circumferential welds that are self-restraining, all bead surfaces should be slightly convex. Although convex beads are virtually immune to centerline splitting, concave beads are particularly susceptible to centerline cracking. In addition, excessive width-to-depth or depth-to-width ratios can result in cracking that may be internal.

Stress-Corrosion Cracking. Nickel and nickel alloys generally do not experience any metallurgical changes, either in the weld metal or in the HAZ, that affect normal corrosion resistance. When the alloys are to be used in contact with substances such as concentrated caustic soda, fluosilicates, and some mercury salts, however, the welds may need to be stress relieved to avoid stress-corrosion cracking. Nickel alloys have good resistance to dilute alkali and chloride solutions. Resistance to stress-corrosion cracking increases with nickel content; therefore, stress relieving of welds in high-nickel alloys is not usually needed.

Effect of Slag on Weld Metal. Because fabricated nickel alloys are ordinarily used in high-temperature service and in aqueous corrosive environments, all slag should be removed from finished weldments. If slag is not removed for applications in an aqueous corrosive media, crevices and accelerated corrosion can result. Slag inclusions between weld beads reduce the strength of the weld. If the service temperature approximates the melting point of the slag, severe corrosion can occur when slag is present on the weld surfaces, particularly in oxygen-containing atmospheres.

Slag also acts as an accumulator of sulfur, particularly in reducing atmospheres, which can lead to service failure in atmospheres that would be considered adequately low in sulfur. For example, in one instance, although only 0.01% S was present in the atmosphere, the sulfur content of the slag on the weld surface rose from 0.02% to 2 or 3% after a 1-month exposure. Sulfur pickup also depresses the melting temperature of the weld slag and consequently the maximum safe operating temperature of the weldment.

Joining of Dissimilar Metals

Nickel and nickel alloys have been successfully joined to other nickel alloys, low-carbon steel, stainless steel, and copper alloy 260 (cartridge brass). Filler metal, or consumable electrode, must be selected to ensure a compatible metallurgical relationship between the two base metals. Several factors are involved: differences in thermal expansion of the base metals, the possibility of permanent changes in volume after extended service at elevated temperature, and the effect of weld-metal dilution at the interfaces with the base metal. All of these factors influence the choice of filler metal and welding process.

An example of a good metallurgical combination is the welding of nickel to Monel 400. Because these metals are completely compatible, they can be welded to each other by any welding process, using any compatible filler metal, without difficulty.

Dilution of Weld Metal. The composition of the weld metal can be expected to differ from that of the filler metal or consumable electrode or the base metals being joined, because some elements are transferred from the base metal to the weld metal during welding. This mixing results in the formation of another alloy. The composition of the weld metal is fairly uniform for a given bead, except in a narrow band at the edge of the joint. The amount of dilution varies from bead to bead. The welding process, current density, welding speed, work-metal thickness, and welding technique influence the amount of dilution.

The dilution of a nickel-based alloy by a dissimilar metal can be tolerated only to a limited degree. When a stainless steel filler metal is used in welding Monel 400 to an austenitic stainless steel, any significant amount of copper pickup from the Monel 400 causes the weld metal to become hot short and crack. Thus, stainless steel filler metal should not be used when the welding process can cause a considerable amount of dilution. A Monel alloy type of filler metal cannot be used in this instance because chromium from the stainless steel dilutes the weld metal and causes cracking. Although nickel or an Inconel filler metal is best, these should only be used after the welding procedure has been qualified by tests.

Filler Metal. Selection of the correct filler metal helps produce a weld that meets the service requirements for strength, corrosion resistance, and soundness, and that permits dilution to occur without causing susceptibility to cracking in the weld metal. Crack sensitivity of the weld metal is proportional to the amount of dilution, especially where there is considerable difference in the composition of a base metal and a filler metal.

Some combinations of base metal and filler metal produce undesirable weld-metal compositions:

- A ferritic weld-metal deposit diluted by nickel, chromium, or copper
- An 18-8 stainless steel weld-metal deposit diluted by more than 3% Cu
- An 18-8 type of weld-metal deposit sufficiently diluted by nickel or chromium to result in the crack-sensitive weld-metal composition of 35% Ni and 15% Cr
- Any Monel alloy weld-metal deposit diluted by more than 6 to 8% Cr

Manipulation of the welding arc so that it impinges mainly on the base metal nearest in composition to the filler metal helps reduce dilution. For most combinations of dissimilar metals, suppliers of filler metal should be consulted before a filler metal is selected.

The coefficients of thermal expansion of the base metals must be considered in selecting filler metals for joining dissimilar metal combinations. A large difference in expansion can induce stresses of sufficient magnitude to produce cracking in the weld metal. Some of the Inconel electrodes (ENiCr, ENiCrFe, ERNiCr, and ERNiCrFe) and filler metals can be used for welding a wide range of base-metal combinations. These electrodes and filler metals can tolerate considerable dilution without loss of strength or ductility.

Filler metal ERNiMo-6 (Hastelloy W) can be used for joining nickel-based alloys, cobalt-based alloys, and stainless steels to themselves and to other metals and alloys. The low chromium content of this filler metal, however, should be carefully considered when welding oxidation-resistant materials.

Although GTAW and GMAW can be employed, SMAW is the most widely used process for joining nickel and nickel alloys to dissimilar metals. The welding current should be maintained near the middle of the recommended range for the

Fig. 4 Fluorine-generator cover made by SMAW of Monel 400 to Monel 400 and to low-carbon steel (ASTM A285). Both welds used nickel alloy filler metal and ENiCu-7 electrode.

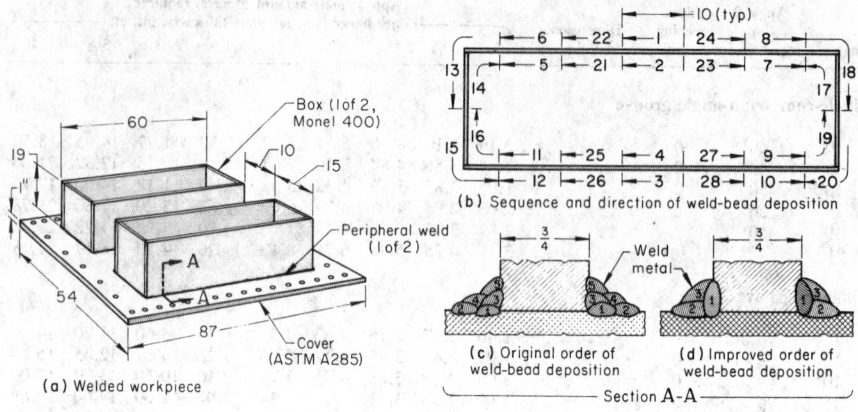

Item	Original method	Improved method
Type of joint and weld:		
Monel 400 to Monel 400	Corner, double fillet	Corner double fillet
Monel 400 to steel	T, double fillet	T, double fillet
Welding passes:		
Monel 400 to Monel 400	1	1
Monel 400 to steel	5	3
Fixtures	None	Clamps
Electrode:		
Monel 400 to Monel 400	5/32-in.-diam ENiCu-4	5/32-in. diam ENiCu-4
Monel 400 to steel	5/32-in.-diam ENiCu-1	3/16-in. diam ENiCu-1
Power supply	400 A, dc	400 A, dc
Welding current:		
Monel 400 to Monel 400	130 A (DCEP)	130 A (DCEP)
Monel 400 to steel	154 A (DCEP)	145 to 150 A (DCEP)
Stress relief	1 h at 1150 ± 25 °F	1 h at 1150 ± 25 °F
Leak test(a)	Welds rejected	Welds accepted
Welding time per cover	40 h	25 h
Weld penetration:		
Bead 1	1/16 in. into steel	1/8 in. into Monel 400, 1/64 in. into steel
Bead 2	1/16 in. into steel	1/32 in. into steel
Bead 3	1/16 in. into Monel 400	1/32 in. into Monel 400
Bead 4	None in base metal	...
Bead 5	1/32 in. into Monel 400	...
Iron in weld metal(b):		
Bead 1	30%	7%
Bead 2	30%	10%
Bead 3	20%	4%
Beads 4 and 5	7%	...

(a) Helium mass spectrometer, internal vacuum of 0.025 mm of mercury. (b) An ENiCu-7 electrode has a nominal iron content of 2.5%.

electrode to control dilution. Dilution usually can be kept below 25% by manipulating the electrode to dissipate the arc force on weld metal already deposited. When depositing the first bead, the arc should be directed toward the member from which dilution of the weld metal will be least detrimental. Welding procedures that reduce the amount of penetration (which can cause dilution of the weld metal by the base metal) increase the probability of securing a good weld.

Example 1. Elimination of Cracking of Welds and Distortion. Welding of a Monel 400 skirt box to a low-carbon steel (ASTM A285) cover for a fluorine generator resulted in cracking of the weld metal and warping of the cover (Fig. 4). Cracks resulting from excessive dilution were eliminated by modifying the way in which the beads were deposited. Weld crater cracks were prevented by care in ending the weld bead and in breaking the arc. The Monel 400 skirt boxes replaced low-carbon steel skirt boxes that had corroded.

The original procedure for removing corroded skirt boxes, making new boxes, and welding the new boxes to the cover plate was:

1 Remove corroded low-carbon steel skirt boxes by air carbon arc cutting.
2 Clean surfaces of low-carbon steel cover by surface grinding.
3 Clean all joint areas on the Monel 400 plates with trichlorethylene and remove mill scale with a surface grinder.
4 Assemble Monel 400 skirt box and weld the four corners in one pass each with a $^5/_{32}$-in.-diam ENiCu-4 (Monel 130) electrode, using a welding current of 130 A.
5 Weld the Monel 400 skirt box to the low-carbon steel cover with intermittent welds, using the sequence of weld-bead deposition shown in Fig. 4(b). The box and cover were first tack welded at the starting place for each of the 10-in.-long welds. Deposit five beads as shown in Fig. 4(c), with a $^5/_{32}$-in.-diam ENiCu-1 (Monel 140) electrode, using a welding current of 154 A.
6 Stress relieve at 1150 ± 25 °F for 1 h in a reducing atmosphere. The assembly was placed in the furnace when furnace temperature was less than 600 °F, and after heating was allowed to cool to 600 °F before it was removed.
7 Test the welds with a helium mass spectrometer under an internal vacuum of 0.025 mm of mercury. Welds that failed the test were rewelded.

During deposition of the first bead, considerable arc blow toward the Monel 400 skirt box was encountered. The ground connection was repositioned several times, but no improvement was obtained. When the welds were tested with a helium mass spectrometer, leakage due to cracks in the weld beads was detected. Investigation revealed that some cracks had resulted from excessive dilution of the weld metal by iron (analysis showed 30% Fe in bead 1 in Fig. 4b), a result of the high welding heat, and that cracks in the weld craters had been caused by ending beads too abruptly. In addition to the cracking, there was excessive distortion of the cover, the cover edges being warped upward toward the skirt boxes $^1/_8$ in. maximum along the sides and about $^1/_4$ in. across the ends. The distortion was not reduced during stress relieving (step 6).

Experimentation resulted in the adoption of a new welding procedure (step 5) that minimized distortion and eliminated weld-metal cracking. The remaining steps in the original procedure were retained unchanged. In this procedure, when welding the box to the cover plate, both parts were securely clamped to a welding surface plate. The sequence of weld-bead deposition was the same as for the original method (Fig. 4b), but only three weld beads were deposited (Fig. 4d). Bead 1 was placed mostly on the Monel 400, which resulted in considerably less dilution by iron (7% compared with 30%). The bead was ended very slowly to avoid the danger of subsequent crater cracking. The welds were made with an electrode slightly larger than the original ($^3/_{16}$ in. diam instead of $^5/_{32}$ in. diam) and using 145- to 150-A welding current instead of 154 A. These modifications reduced the current density and thus the heat input.

When the clamps were released, no significant distortion of the cover plate was observed, and all joints passed the helium leak detection test.

Arc Welding of Reactive Metals and Refractory Metals

Arc Welding of Titanium and Titanium Alloys

By the ASM Committee on Arc Welding of Titanium*

TITANIUM and most titanium alloys can be welded by the gas tungsten arc, plasma arc, and gas metal arc processes. Procedures and equipment are generally similar to those used for welding austenitic stainless steel or aluminum. Because titanium and titanium alloys are extremely reactive above 1000 °F, however, additional precautions, exceeding those required during the welding of austenitic stainless steel or aluminum alloys, must be taken to shield the weld and hot root side of the joint from air. Figure 1 compares welding setups.

Weldability

Unalloyed titanium and all alpha titanium alloys are weldable. Although the alpha-beta alloy Ti-6Al-4V and other weakly beta-stabilized alloys are also weldable, strongly beta-stabilized alpha-beta alloys are embrittled by welding. Most beta alloys can be successfully welded, but because aged welds in beta alloys can be quite brittle, heat treatment to strengthen the weld by age hardening should be used with caution.

Unalloyed titanium is generally available in several grades ranging in purity from 98.5 to 99.5% Ti. These grades are increased in strength by variations in oxygen, nitrogen, carbon, and iron. Strengthening by cold working is possible but is seldom used. All grades are usually welded in the annealed condition. Welding of cold worked alloys anneals the heat-affected zone (HAZ) and negates the strength produced by cold working.

Alpha alloys Ti-5Al-2.5Sn, Ti-6Al-2Sn-4Zr-2Mo, Ti-5Al-5Sn-2Zr-2Mo, Ti-6Al-2Cb-1Ta-1Mo, and Ti-8Al-1Mo-1V are always welded in the annealed condition.

Alpha-beta alloys of Ti-6Al-4V can be welded in the annealed condition or in the solution-treated and partially aged condition, with aging completed during postweld stress relieving. In contrast to unalloyed titanium and the alpha alloys, which can be strengthened only by cold work, the alpha-beta and beta alloys can be strengthened by heat treatment.

The low weld ductility of most alpha-beta alloys is caused by phase transformation in the weld zone or in the HAZ. Alpha-beta alloys can be welded autogenously or with various filler metals. It is common to weld some of the lower-alloyed materials with matching filler metals. Where strength is not critical and more toughness is desired, unalloyed alpha titanium filler metals may be used. Filler metal of matching composition is used to weld the Ti-6Al-4V alloy. This extra-low-interstitial (ELI) grade is employed to improve ductility and toughness. The use of filler metals that improve ductility may not prevent embrittlement of the HAZ in susceptible alloys. In addition, low-alloy welds can be embrittled by hydride precipitation. It should be noted, however, that with proper joint preparation, filler-metal storage, and shielding, hydride precipitation can be avoided.

Sheet thicknesses 0.100 in. and thinner can be welded without filler metal additions by the fused-root welding technique. Filler metal may be added to repair unfused and sunken weld-metal areas. The lack of a joint line on the root face of the weld indicates 100% penetration.

Metastable beta alloys Ti-3Al-13V-11Cr, Ti-11.5Mo-6Zr-4.5Sn, Ti-8Mo-8V-2Fe-3Al, Ti-15V-3Cr-3Al-3Sn, and Ti-3Al-8V-6Cr-4Zr-4Mo are weldable in the annealed or solution heat treated condition. In the as-welded condition, welds are low in strength but ductile. Beta alloy weldments are sometimes used in the as-welded condition. Welds in the Ti-3Al-13V-11Cr alloy embrittle more severely when age hardened. To obtain full strength, the metastable beta alloys are welded in the annealed condition; the weld is cold worked by peening or planishing, and the weldment is then solution treated and aged. This procedure also obtains adequate ductility in the weld.

Welding Processes

Gas tungsten arc welding (GTAW) is the most widely used process for joining titanium and titanium alloys, except in large thicknesses. Square-groove butt joints can be welded without filler metal in base metal up to 0.10 in. thick. For thicker base metal, the joint should be grooved and filler metal is required. Where possible, welding should be done in the flat position. Hot wire GTAW can be used for welding titanium alloys more than $^1/_4$ in. thick.

Gas metal arc welding (GMAW) is employed to join titanium and titanium alloys more than $^1/_8$ in. thick. It is applied using pulsed current or the spray mode and is less costly than GTAW, especially when base-metal thickness is greater than $^1/_2$ in.

Plasma arc welding (PAW) is also applicable to the welding of titanium and titanium alloys. It is faster than GTAW and can be used on thicker sections, such as one-pass welding of titanium alloy plate up to $^1/_2$ in. thick, using square-groove butt joints and the keyhole technique. Titanium and titanium alloys are also welded by the electron beam process, which is de-

*Herbert Nagler, *Chairman*, Research Engineer, Rohr Industries, Inc.; Dennis F. Hasson, Associate Professor, Department of Mechanical Engineering, U.S. Naval Academy; Charles S. Young, Manager, Metallurgical Development, Titanium Industries

Fig. 1 Typical setups for inert gas shielding for GTAW. (a) Conventional setup for welding aluminum alloys and stainless steel. Gas shielding is from the torch. Use of shielding gas in the backing groove is optional. (b) Setup for welding titanium and titanium alloys outside a welding chamber. Gas shielding is from the torch and through parts in hold-down bars, backing bars, and from trailing and backup shields.

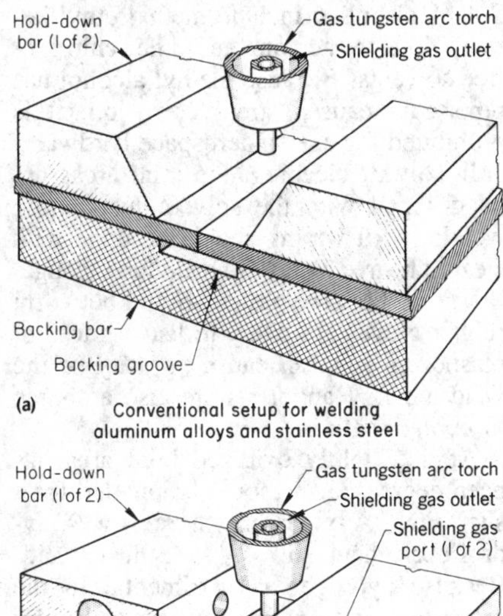

scribed in the article "Electron Beam Welding" in this Volume.

Filler Metals

For welding titanium thicker than about 0.10 in. by the GTAW process, a filler metal must be used. For PAW, a filler metal may or may not be used for welding metal less than 1/2 in. thick.

Composition. Fourteen titanium and titanium alloy filler-metal (or electrode) classifications are given in AWS A5.16. Five of these are essentially unalloyed titanium and the remainder are titanium alloy filler metals. Maximums are set on carbon, oxygen, hydrogen, and nitrogen contents. Compositions for titanium and titanium alloy filler metals are given in Table 1.

Filler-metal composition is usually matched to the grade of titanium being welded. For improved joint ductility in welding the higher strength grades of unalloyed titanium, filler metal of yield strength lower than that of the base metal is occasionally used. Because of the dilution that occurs during welding, the weld deposit acquires the required strength. Unalloyed filler metal is sometimes used to weld Ti-5Al-2.5Sn and Ti-6Al-4V for improved joint ductility. The use of unalloyed filler metals lowers the beta content of the weldment, thereby reducing the extent of the transformation which occurs and improving ductility. Engineering approval, however, is recommended when employing pure filler metal to ensure that the weld meets strength requirements. Another option is filler metal containing lower interstitial content (oxygen, hydrogen, nitrogen, and carbon) or alloying contents that are lower than the base metal being used. The use of filler metals that improve ductility does not preclude embrittlement of the HAZ in susceptible alloys. In addition, low-alloy welds may enhance the possibility of hydrogen embrittlement.

Preparation. The filler metal, as well as the base metal, should be clean at the time of welding. As shown in Fig. 2, wires of the size used for filler metals have a large surface-to-volume ratio. Therefore, if the wire surface is slightly contaminated, the weld may be severely contaminated. Some procedures require that the filler wire be cleaned immediately before use. The use of an acetone-soaked, lint-free cloth serves to assess surface contamination caused by the die lubricant used in the wire drawing operation, in addition to cleaning the filler wire. Pickling in nitric/hydrofluoric acid solution is also used for cleaning.

Shielding Gases

In welding titanium and titanium alloys, only argon and helium, and occasionally a mixture of these two gases, are used for shielding. Because it is more readily available and less costly, argon is more widely used. Argon shielding gas was used in the examples given in this article.

Because of high purity (99.985% min) and low moisture content, liquid argon is often preferred. The argon gas should have a dew point of −75 °F or lower. The hose used for the shielding gas should be clean, nonporous, and flexible, made of Tygon or vinyl plastic. Because rubber hose absorbs air, it should not be used. Excessive gas flow rates that cause turbulence should be avoided, and flowmeters are usually employed for all gas shields. Pressure (psi) gages may be employed for trailing and backup shields.

Characteristics. The type of shielding gas used affects the characteristics of the arc. At a given welding current, the arc voltage is much greater with helium than with argon. Because the heat energy liberated in helium is about twice that in argon, higher welding speeds can be obtained, weld penetration is deeper, and thicker sections can be welded more rapidly using helium shielding. However, when using pure helium for welding,

Table 1 Chemical compositions of titanium and titanium alloy filler metals(a) AWS A5.16

AWS classification	Composition, %													
	C	O	H	N	Al	V	Sn	Cr	Fe	Mo	Nb	Ta	Pd	Ti
ERTi-1(b)	0.03	0.10	0.005	0.012	...	...	...	...	0.10	...	...	...	...	rem
ERTi-2	0.05	0.10	0.008	0.020	...	...	...	...	0.20	...	...	...	...	rem
ERTi-3	0.05	0.10-0.15	0.008	0.020	...	...	...	...	0.20	...	...	...	...	rem
ERTi-4	0.05	0.15-0.25	0.008	0.020	...	...	...	...	0.30	...	...	...	...	rem
ERTi-0.2Pd	0.05	0.15	0.008	0.020	...	...	...	...	0.25	...	...	...	0.15-0.25	rem
ERTi-3Al-2.5V	0.05	0.12	0.008	0.020	2.5-3.5	2.0-3.0	...	...	0.25	...	...	...	...	rem
ERTi-3Al-2.5V-1(b)	0.04	0.10	0.005	0.012	2.5-3.5	2.0-3.0	...	...	0.25	...	...	...	...	rem
ERTi-5Al-2.5Sn	0.05	0.12	0.008	0.030	4.7-5.6	...	2.0-3.0	...	0.40	...	...	...	...	rem
ERTi-5Al-2.5Sn-1(b)	0.04	0.10	0.005	0.012	4.7-5.6	...	2.0-3.0	...	0.25	...	...	...	...	rem
ERTi-6Al-2Nb-1Ta-1Mo	0.04	0.10	0.005	0.012	5.5-6.5	...	...	...	0.15	0.5-1.5	1.5-2.5	0.5-1.5	...	rem
ERTi-6Al-4V	0.05	0.15	0.008	0.020	5.5-6.75	3.5-4.5	...	...	0.25	...	...	...	...	rem
ERTi-6Al-4V-1(b)	0.04	0.10	0.005	0.012	5.5-6.75	3.5-4.5	...	...	0.15	...	...	...	...	rem
ERTi-8Al-1Mo-1V	0.05	0.12	0.008	0.03	7.35-8.35	0.75-1.25	...	...	0.25	0.75-1.25	...	...	...	rem
ERTi-13V-11Cr-3Al	0.05	0.12	0.008	0.03	2.5-3.5	12.5-14.5	...	10.0-12.0	0.25	...	...	...	...	rem

(a) Single values are maximum. (b) Extra-low interstitials for welding similar base metals

Fig. 2 Ratio of filler-metal (electrode) wire surface area to volume for various wire diameters. Shaded portion indicates range of wire diameters most often used for welding titanium.

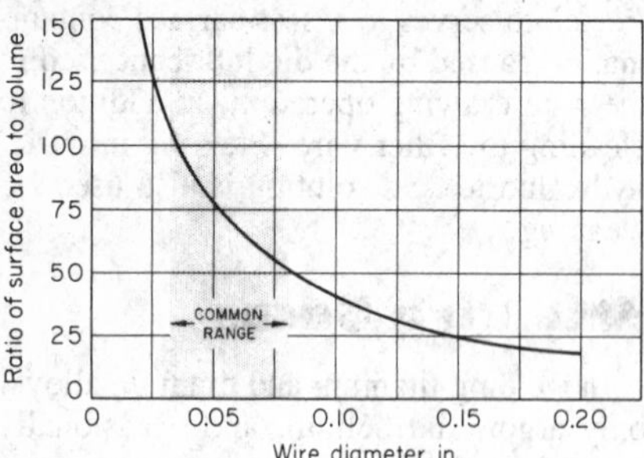

arc stability and weld-metal control are sacrificed.

Argon is used in the welding of thin and thick sections where the arc length can be altered without appreciably changing the heat input. Argon-helium mixtures are also employed; particularly 75% argon, which improves arc stability, and 25% helium, which increases penetration. The 75%Ar-25%He mixture is also frequently utilized

Table 2 Dimensions of typical joints for welding titanium and titanium alloys
t is base-metal thickness

Base-metal thickness, in.	Root opening, in.	Groove angle,°	Weld-bead width, in.
Square-groove butt joint			
0.010-0.090	0	...	...
0.031-0.125	0-0.10*t*	...	...
Single-V-groove butt joint			
0.062-0.125	0-0.10*t*	30-60	0.10-0.25*t*
0.090-0.125	(a)	90	...
0.125-0.250	0-0.10*t*	30-60	0.10-0.25*t*
Double-V-groove butt joint			
0.250-0.500	0-0.20*t*	30-120	0.10-0.25*t*
Single-U-groove butt joint			
0.250-0.750	0-0.10*t*	15-30	0.10-0.25*t*
Double-U-groove butt joint			
0.750-1.500	0-0.10*t*	15-30	0.10-0.25*t*
Fillet weld			
0.031-0.125	0-0.10*t*	0-45	0-0.25*t*
0.125-0.500	0-0.10*t*	30-45	0.10-0.25*t*

(a) Root face, 0.030 in.
Source: J.J. Vagi, *et al.*, "Welding Procedures for Titanium and Titanium Alloys," NASA TMX 53432, 1965

as the shielding gas at the torch in automatic operations. Furthermore, helium is used in shielding for out-of-position welds.

Joint Preparation

If welding is done outside a controlled-atmosphere welding chamber, joints must be carefully designed so that both the top and the underside of the weld can be shielded (Fig. 1). Dimensions of typical joints are given in Table 2. For welding titanium alloys, joint fit-up should be better than for welding other metals, because of the possibility of entrapping air in the joint. The joint should be clamped to prevent separation during welding.

Cleaning

To obtain a good weld, the joint and the surfaces of the workpieces at least 2 in. beyond the width of the gas trailing shield on each side of the weld groove must be meticulously cleaned. As shown in Fig. 3, the cleaning procedure depends on whether the oxide layer in the joint area is light or heavy.

Degreasing. Grease and oil accumulated during forming and machining must be removed before welding to avoid weld contamination. Scale-free metal requires only degreasing. Degreasing precedes descaling for metal with an oxide scale. Methods of degreasing include steam cleaning, alkaline cleaning, vapor degreasing, and solvent cleaning.

For vapor degreasing, toluene rather than a chlorinated solvent should be used, because residues from chlorinated solvents (and also from silicated solvents) may contribute to cracking of titanium weldments. Solvent cleaning is frequently used, especially for large components that cannot conveniently be placed in a vapor degreaser or washer for alkaline cleaning. Solvents applied include methyl ethyl ketone, toluene, acetone and other chlorine-free solvents. Because methyl alcohol has reportedly caused stress corrosion, it is prohibited for use on aerospace hardware.

In solvent cleaning, the joint areas are hand-wiped with the solvent just before welding. All wiping should be done with clean, lint-free cloths or a cellulose sponge. Plastic or lint-free gloves should be worn; rubber gloves are likely to leave traces of plasticizer that can cause porosity in the weld metal. Handprints are also a source of contamination.

After a lightly oxidized joint area has been degreased, it should be pickled for a short time. A typical mixture is 4 wt% hydrofluoric acid and 40 wt% nitric acid. Because hydrogen is detrimental to the properties of titanium, causing embrittlement and sometimes contributing to weld porosity, pickling should be performed cautiously. Industrial practice usually is to maintain the acid bath at a high oxidation potential of 30 wt% or more nitric acid, which simultaneously holds the ratio of nitric acid to hydrofluoric acid at 15 to 1 strictly as a factor of safety. For additional

Fig. 3 Flow chart of procedures for cleaning titanium alloys

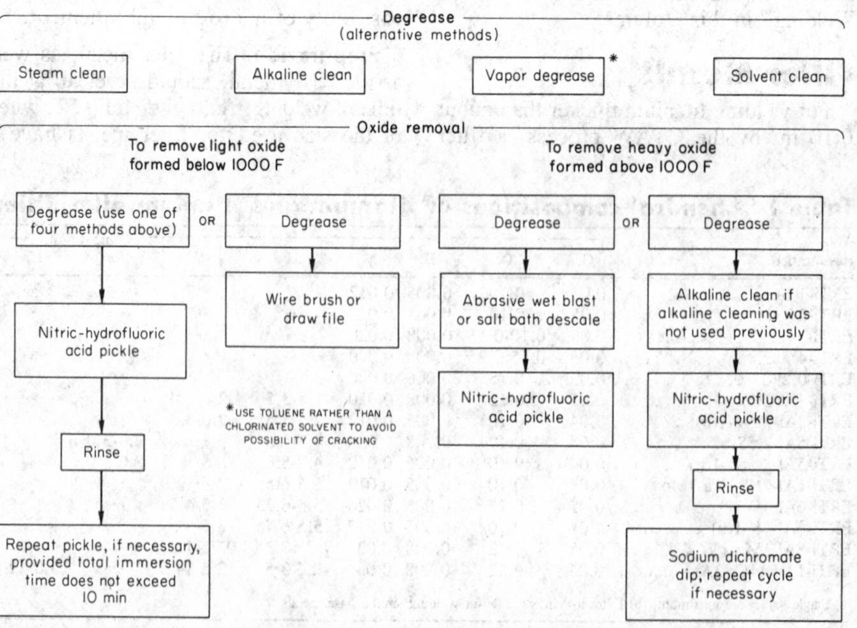

information on acid pickling of titanium, see pages 654 to 655 in Volume 5 of the 9th edition of *Metals Handbook*. If the nitric acid content falls below 30 wt% and the ratio of nitric acid to hydrofluoric acid falls below 10 to 1, excessive hydrogen pickup is possible.

Oxide Removal. Lightly oxidized joint areas may also be cleaned by brushing with a stainless steel wire brush or by draw filing. When weld corrosion resistance is important, however, these cleaning methods should be followed immediately by acid pickling. Steel wool or abrasives should never be used, because of the danger of contamination.

If grinding is required, the use of silicon carbide burrs is preferred, because wheels produce residues of rubber or resin on the surface that contaminate the weld. To prevent burning, excessive heat should be avoided while grinding and low rotating speeds should be used.

If titanium has been exposed to temperatures above 1000 °F, a more complex removal treatment such as chemical, salt bath, or mechanical, or combinations of these treatments, is required. The salt baths are basically sodium hydroxide to which oxidizing agents or hydrogen has been added to form sodium hydride (see the article "Salt Bath Descaling" in Volume 5 of the 9th edition of *Metals Handbook*).

Two alternative procedures for removing heavy scale are shown in Fig. 3. In one, the parts are subjected to liquid abrasive blasting or salt-bath descaling after degreasing. These treatments are usually followed by pickling in nitric-hydrofluoric acid, as for the removal of light scale. When salt bath descaling is used, oxide removal can be hastened by removing the workpieces from the bath, scrubbing them with brushes, and then re-immersing them. To prevent hydrogen pickup during salt-bath descaling, time cycles must be short (preferably no more than 2 min) and bath temperature must be carefully controlled.

In the second method for removing heavy scale described in Fig. 3, parts are alkaline cleaned after grease removal (unless alkaline cleaning was used for grease removal) and then pickled, rinsed, and dipped in a sodium dichromate solution. Selection of cleaning method depends largely on the size and shape of the parts and on the cleaning methods available in a particular plant.

Welding in Chambers

For successful arc welding of titanium and titanium alloys, complete shielding of the weld is necessary, because of the high reactivity of titanium to oxygen and nitrogen at welding temperatures. Excellent welds can be obtained in titanium and its alloys in a welding chamber, where welding is done in a protective gas atmosphere, thus giving adequate shielding. Welding in a chamber, however, is not always practical. For example, in manual welding, the location of the glove ports and the presence of a chamber wall impose limitations on visibility, movement, and accessibility.

For large assemblies, welding in a chamber requires unloading of the chamber after each weldment is completed, accompanied by the loss of purging gas. The chamber must be repurged for welding the next assembly. Such procedures are time consuming and expensive. Various types of modified chambers, such as clamp-on chambers, have been tried to remedy these problems.

Welding in Metal Chambers. Welding of titanium was first done in metal chambers that can be evacuated and then backfilled with argon or helium. Such chambers are equipped with glove ports, so that the welder can handle the torch, separate filler metal (if used), and the weldment without admitting air to the chamber. Viewing ports enable the welder to see the welding operation. Although expensive to operate, especially for large weldments, metal chambers are frequently used in aerospace applications.

Generally, shielding gas is not supplied to the welding torch when welding titanium in a metal chamber, and excellent welds can be made if the chamber atmosphere is maintained properly. In some applications, however, where heavy or long welds are required, gas is supplied to the torch to improve shielding (Example 1).

With flow-purged chambers, the atmosphere is often tested by welding a piece of scrap titanium before making the assembly weld. The color of the solidified weld metal is observed by gradually pulling the torch away from the molten pool. The weld-metal colors, in increasing order of contamination, are bright silver, light straw, dark straw, light blue, dark blue, gray-blue, gray, and white loose powder. A light straw color is generally considered acceptable for all but the most stringent requirements, although a bright silvery color like a newly minted dime is preferred.

To continuously monitor the inert atmosphere, a heated tungsten filament may be placed inside the chamber. Any discoloration or ignition of the filament indicates that the purity of the atmosphere has become degraded.

Example 1. Welding in a Metal Chamber. The center strut subassembly of the complete hydrofoil strut assembly shown at the top of Fig. 4 consisted of two machined pieces of Ti-6Al-4V, approximately 112 in. long by 12 in. wide, which were welded by the manual GTAW process in a metal chamber $3^1/_2$ ft in diameter by 24 ft long equipped with four welding ports. The chamber was flow-purged with argon at a slight positive pressure and was equipped with a fixture-positioner that allowed 360° rotation of the work, if needed, and full conveyance of the workpiece past all of the welding ports. The two components were clamped in a welding fixture, as shown at lower right in Fig. 4, and the fixture was placed in the positioner. The components were clamped at approximately 18-in. intervals along the 112-in. length. Manual welding was used because variations in contour and cross section limited accessibility, and the small quantity of production weldments that were required ruled out automatic welding.

Before welding, the parts were solution heat treated at about 1750 °F, water quenched, and partially aged at 950 °F for 4 h. The joint areas were degreased, alkaline cleaned, and acid pickled in a mixture of concentrated acids (30 volumes of 42° Bé nitric acid and 2 volumes of 70% technical-grade hydrofluoric acid) at room temperature for 5 min. Welding was started within 4 h after pickling. Workpieces were maintained within the welding chamber until all welds were completed.

The root pass was made without filler metal to ensure complete root penetration and to minimize melt-through. To obtain the required notch toughness in the joint, the second pass was made with unalloyed titanium filler metal (ERTi-1), although this decreased the strength of the joint slightly. Subsequent passes were made with ERTi-6Al-4V filler metal.

Direct current electrode negative (DCEN) was supplied from a 300-A transformer-rectifier with a saturable-core reactor and conventional drooping-voltage characteristic. A foot rheostat controlled the current to the electrode. After welding, the assembly was heat treated at 950 °F for 4 h in argon to complete age hardening, partially relieve welding stresses, and increase weld ductility.

The weld-chamber atmosphere was checked periodically by the following tests: weld-bead color, test-coupon weld-bend ductility, dew point, and a heated tungsten filament. All assemblies were examined visually and inspected by liquid-penetrant and radiographic methods. Also, representative sample welds were tested me-

Fig. 4 Hydrofoil center strut subassembly. Welded by manual GTAW in a metal chamber

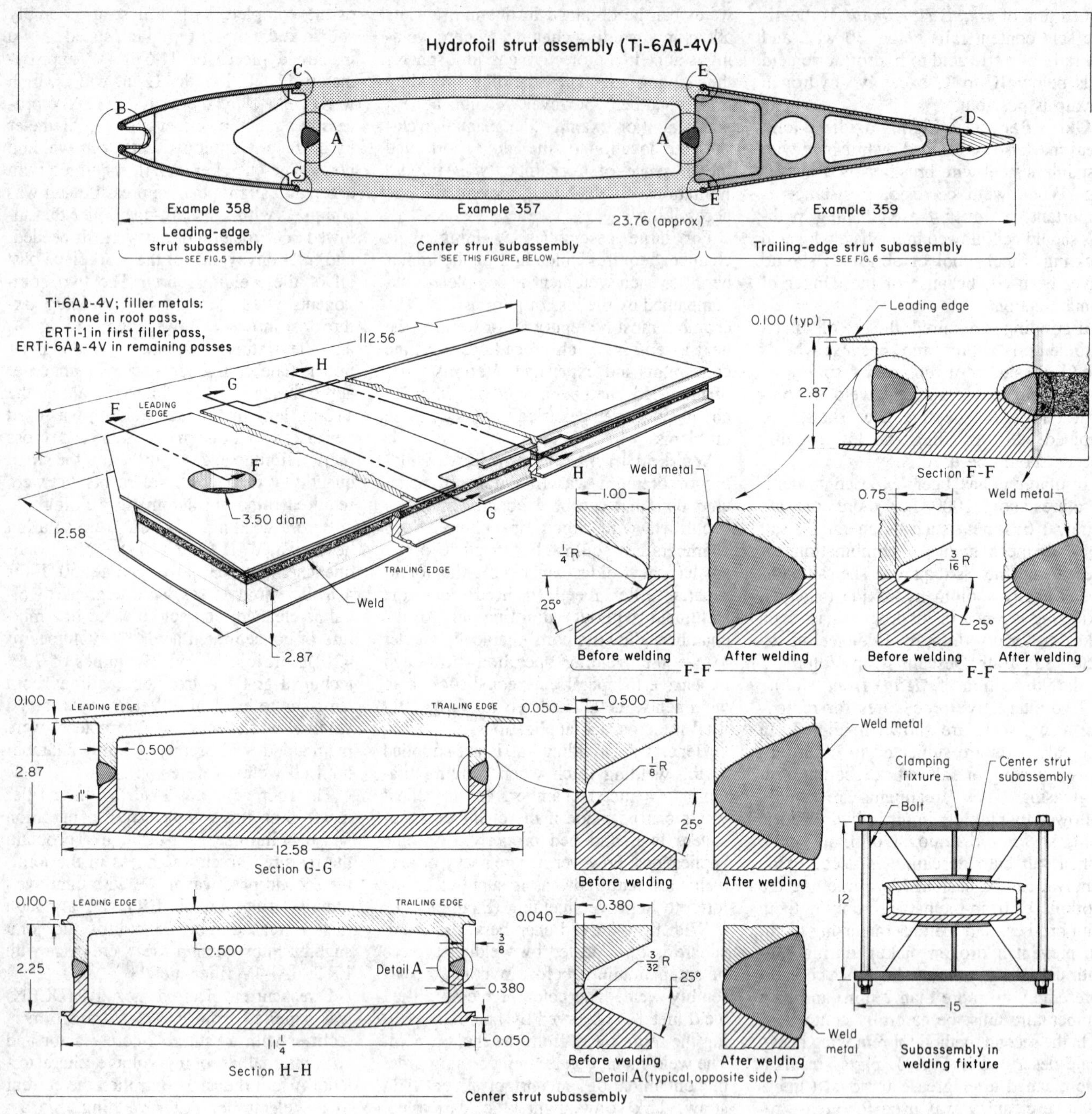

Joint type .... Butt
Weld type .... Single-U-groove
Power supply .... 300-A, 40-V transformer-rectifier
Shielding gas:
At nozzle .... Argon, at 15 ft^3/h
In chamber ... Argon, flowing purge with slight positive pressure
Electrode (EWTh-2)
diameter .... 3/32 in. for sections F-F and G-G; 1/16 in. for section H-H

Torch .... 350 A, water cooled
Filler metal:
Root pass .... None
Second pass .... 3/32- or 1/8-in.-diam ERTi-1
Subsequent passes .... 3/32- or 1/8-in.-diam ERTi-6Al-4V
Nozzles .... 9/16- to 3/4-in. ID
Welding position .... Horizontal
Number of passes .... Various
Type of bead .... Convex; stringer

Current:
Section F-F .... 150 to 160 A (DCEN)
Section G-G .... 130 to 140 A (DCEN)
Section H-H .... 110 to 120 A (DCEN)
Arc initiation .... High frequency
Arc length .... About 1/8 in.
Welding speed .... 2 to 4 in./min
Interpass temperature .... Not controlled
Postweld heat treatment .... 4 h at 950 °F, in an argon atmosphere

chanically and examined metallographically as a further check on weld quality.

Welding in Plastic Chambers. Rigid or collapsible chambers made of transparent plastic can be used where production runs are short, the assembly is large or complicated, and manual welding is required. Rigid plastic chambers are flow-purged with argon or helium, in volumes equal to five to ten times the volume of the chamber, before welding is started. Collapsible plastic chambers are first collapsed and then flow-purged with argon or helium; they require less gas for purging than do rigid chambers.

Advantages of plastic chambers (either rigid or collapsible) are low cost and good visibility of the work. Because there is generally a greater probability of leakage occurring in a plastic chamber than in a metal chamber, the atmosphere must be checked frequently to ensure that it is of proper purity. In addition, torch shielding is usually employed to make certain that the weld zone is adequately protected.

Examples 2 and 3 describe welding procedures used in plastic chamber welding of leading- and trailing-edge strut assemblies, respectively. Setup, preparation, and evaluation were the same for both. A reinforced, collapsible plastic chamber, 3 ft high by 3 ft wide by 20 ft long, having three welding ports, was used for manual GTAW of the strut subassemblies of Ti-6Al-4V alloy and for joining them to a center strut subassembly (Example 1). The chamber was flow-purged with argon at a slight positive pressure. Purity of the chamber atmosphere was checked by weld-bead color and weld-bend tests on Ti-6Al-4V test pieces, by a hot tungsten filament in the chamber, and by dew-point measurements.

Manual welding was used because accessibility difficulties and costs precluded automatic welding. All welding was done on components that had been solution treated, water quenched, and partially age hardened (950 °F for 4 h).

Joints were machined smooth, degreased in an alkaline cleaner, and pickled in a mixture of concentrated acids (30 volumes of 42° Bé nitric acid and 2 volumes of 70% technical-grade hydrofluoric acid) at room temperature for 5 min. Welding was started within 4 h after pickling. Parts remained in the welding chamber until all welds were completed.

Direct current electrode negative was supplied by a 300-A transformer-rectifier with a saturable-core reactor and conventional drooping-voltage characteristic. Welding current was controlled with a foot rheostat. Although Ti-6Al-4V has good weldability, copper chill bars were placed adjacent to all welds to increase the cooling rate and improve weld ductility. Chill bars were held in close contact with the joint by simple toggle clamps. After welding, the parts were heated to 950 °F and held for 4 h to complete age hardening, partially relieve stresses, and increase the ductility of the welds.

Welds were examined visually and inspected by liquid-penetrant and radiographic methods on each weldment. In addition, representative samples welded at the same time as the production assemblies were tested for mechanical properties and examined metallographically.

Mechanical properties of the welds were controlled primarily through filler-metal selection, as shown in the following comparison of minimum weld properties (weld metal only):

	Tensile strength, ksi	Yield strength, ksi	Elongation in 2 in., %
Unalloyed Ti	100	90	5.0
ERTi-6Al-4V	130	120	4.0

Example 2. Leading-Edge Strut Subassembly. First, two curved panels, 83 in. long by 4.62 in. wide and 0.100 in. thick, were joined to a U-shaped insert, 83 in. long and 0.050 in. thick, by square-groove edge welding (Fig. 5, section J-J and detail B). Next, truncated triangular ribs, 0.100 in. thick, were welded to the panels at 15-in. intervals, using a 0.100-in. fillet weld on one side of each rib (section K-K in Fig. 5). Finally, the open side of the subassembly was welded to the center subassembly (Fig. 4), using single-V-groove welds (detail C in Fig. 5).

Copper chill bars were used. Where greater ductility was required, as in the joint shown in section K-K in Fig. 5, unalloyed titanium filler metal (ERTi-1) was used. Where greater strength was required, as in the joints shown in detail C of Fig. 5, ERTi-6Al-4V filler metal was used. Tack welding was employed as needed.

Example 3. Trailing-Edge Strut Subassembly. Two straight panels, 63 in. long by approximately 7 in. wide and 0.050 in. thick, were welded to a solid triangular piece (detail D in Fig. 6) using V-groove butt joints. Stiffeners 0.100 in. thick were joined to the panels at 15-in. intervals by fillet welds (section M-M in Fig. 6). The stiffeners were also plug welded to the panels at 1-in. intervals on one side (section N-N in Fig. 6). The purpose of plug welding was to provide additional stiffener-to-panel attachment in areas where there were no fillet welds because of lack of accessibility. On the side where fillet welds were continuous, no plug welding was needed. Finally, the trailing-edge strut subassembly was welded to the center strut subassembly (detail E in Fig. 6).

Copper chill bars were used adjacent to the butt welds. Where good ductility was required, as in the joints shown in detail D and in sections M-M and N-N of Fig. 6, unalloyed titanium filler metal (ERTi-1) was used. Where greater strength was required, as in the joints shown in detail E of Fig. 6, Ti-6Al-4V filler metal was used. Tack welding was employed as required.

Out-of-Chamber Welding

With proper tooling, joints in titanium can be adequately shielded for welding without using a chamber. Both the weld and the HAZ must be shielded during welding and until the temperature of the metal in the area of the weld is below 1000 °F. If shielding is inadequate, the welds are brittle.

The welding torch (or electrode holder) is usually equipped to supply a trailing shield that provides a diffuse, nonturbulent flow of gas to the solidifying weld. A typical trailing shield is shown in Fig. 7. The length of the trailing shield must be adjusted to the speed of welding. Both straight and curvilinear welding can be shielded. In addition, the welding station must be shielded by curtains to prevent drafts. Most shields are designed and/or handcrafted for the particular weld.

Example 4. Welding Ti-6Al-4V in Air. Titanium alloy (Ti-6Al-4V) fail-safe straps for transport airplanes were welded in air, with adequate shielding provided by a continuous, enveloping flow of argon. Because of their length (36 to 42 ft), the straps were made by welding together two or more pieces, using manual GTAW. A typical 6-in.-wide joint in a 40-ft-long by 0.188-in.-thick fail-safe strap is shown in Fig. 8 (upper left). Because the straps were too long for the available vacuum welding chamber, a special shielding method had to be developed.

The first attempt at providing adequate shielding was to construct a clamp-on displacement chamber, approximately 3 ft long, 2 ft wide, and 2 ft high. This chamber contained glove ports, a viewing port, an argon-flow control system, a flapper valve for venting excess gas pressure, a feed-through plate for torch leads, and slots at both ends for the fail-safe straps to pass

Fig. 5 Hydrofoil leading-edge strut assembly. Welded by manual GTAW in a plastic chamber

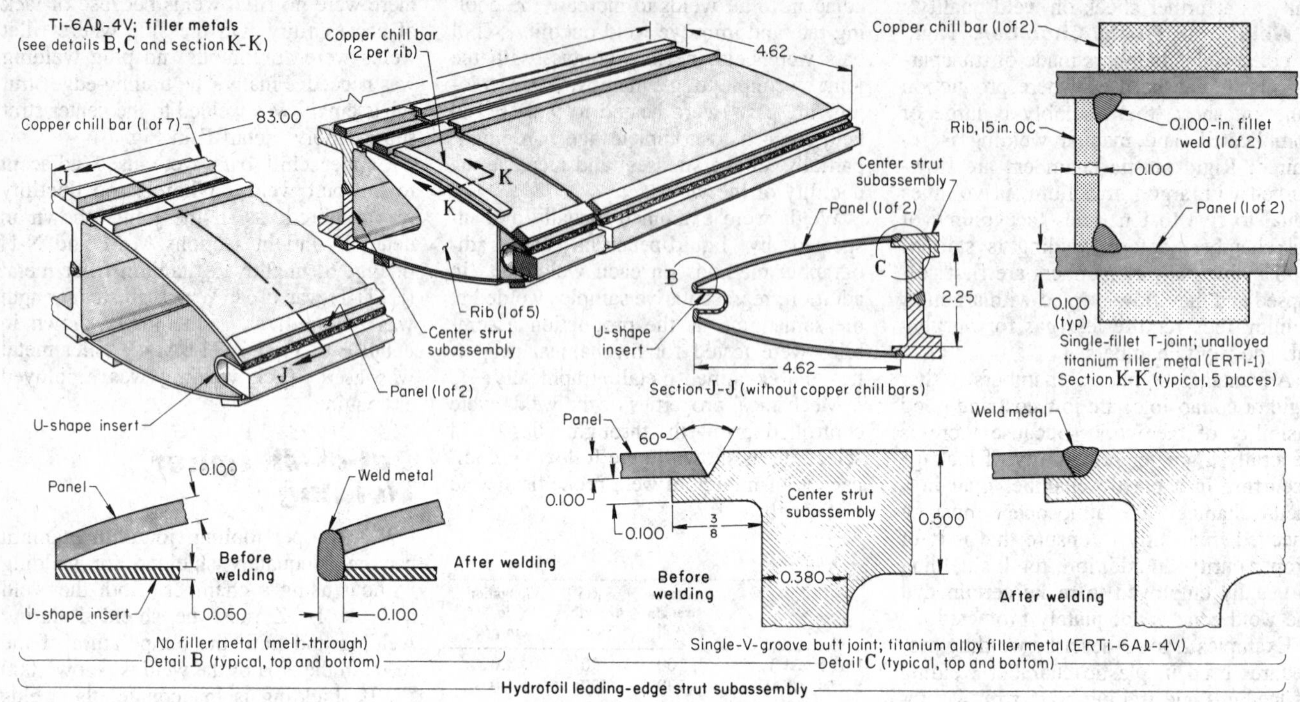

Power supply	300-A 40-V transformer-rectifier, with drooping-voltage characteristic
Chamber	Collapsible plastic, with 3 ports
Electrode	$^3/_{32}$-in.-diam EWTh-2
Torch	350 A, water cooled
Fixture	Inexpensive frame
Shielding gas:	
At nozzle	Argon, at 15 ft^3/h
In chamber	Argon, flow-purge with slight positive pressure
Nozzle(a)	$^9/_{16}$-in. ID
Current	120 to 130 A (DCEN)
Voltage	10 V
Arc initiation	High frequency
Arc length	$^1/_8$ in. (approx)
Type of bead	Convex; stringer

Interpass temperature	Not controlled
Preheat	None
Postweld heat treatment	4 h in argon at 950 °F

	Detail B	Section K-K	Detail C
Joint type	Edge	T	Butt
Weld type	Square groove	Fillet	V-groove
Filler metal	None	ERTi-1	ERTi-6Al-4V
Filler metal diameter, in.	None	0.062	0.062
Welding position	Flat	Horizontal	Flat
Passes	1	1	1
Speed, in./min	4-6	4-6	3-5

(a) Occasionally, nozzle was removed for better accessibility to the joint.

through. Several straps were manually gas tungsten arc welded in the chamber, and although weld quality was satisfactory, it was apparent that this method of shielding was inadequate for production use.

A second method, based on flooding the immediate area of the 6-in.-long weld with a continuous flow of argon, proved successful (Fig. 8, left and section B-B) and was simpler to operate. The joint to be welded was centered over the groove of a copper backing bar that was set flush with the surface of a long aluminum support plate (Fig. 8, section B-B). The entire assembly was placed on a long welding table. Next, two copper hold-down bars were bolted tightly in place, $^1/_2$ in. on either side of the joint. The hold-down bars and the backing bar had been drilled so that shielding gas could be supplied by manifolding arrangements (Fig. 8, bottom left, and section B-B). This tooling provided a heat sink, good joint alignment, effective gas shielding, and a high degree of accessibility. Additional argon shielding gas was supplied through the torch.

Joints were provided with integral starting and runoff tabs. A tack weld was placed on the runoff tab, and welding was started at the opposite end and completed in a single pass, using the forehand technique (filler metal preceded the torch in the direction of welding).

First results, which were used to qualify the procedure, were quite satisfactory. On both sides, the weld color resembled that of newly minted silver. Radiographic and liquid-penetrant examination showed no defects. Specimens bent over a 6t ($1^1/_8$-in.) radius showed no cracks on the outer surface at a magnification of 10×. All tensile-test specimens failed in the base metal at 135 ksi or more.

Specifications called for testing three preproduction specimens and one post-production specimen per lot. The first production lot consisted of 16 straps. The test specimens and the 16 production welds were inspected by visual, radiographic, and liquid-penetrant methods. All welds were acceptable under visual and liquid-penetrant examinations, and only two production welds showed slight porosity (which could be repaired) under radiographic examination. Two bend-test specimens and one tensile-test specimen were machined from the post-production weldment. Bend

Fig. 6 Hydrofoil trailing-edge strut subassembly. Welded by manual GTAW in a plastic chamber

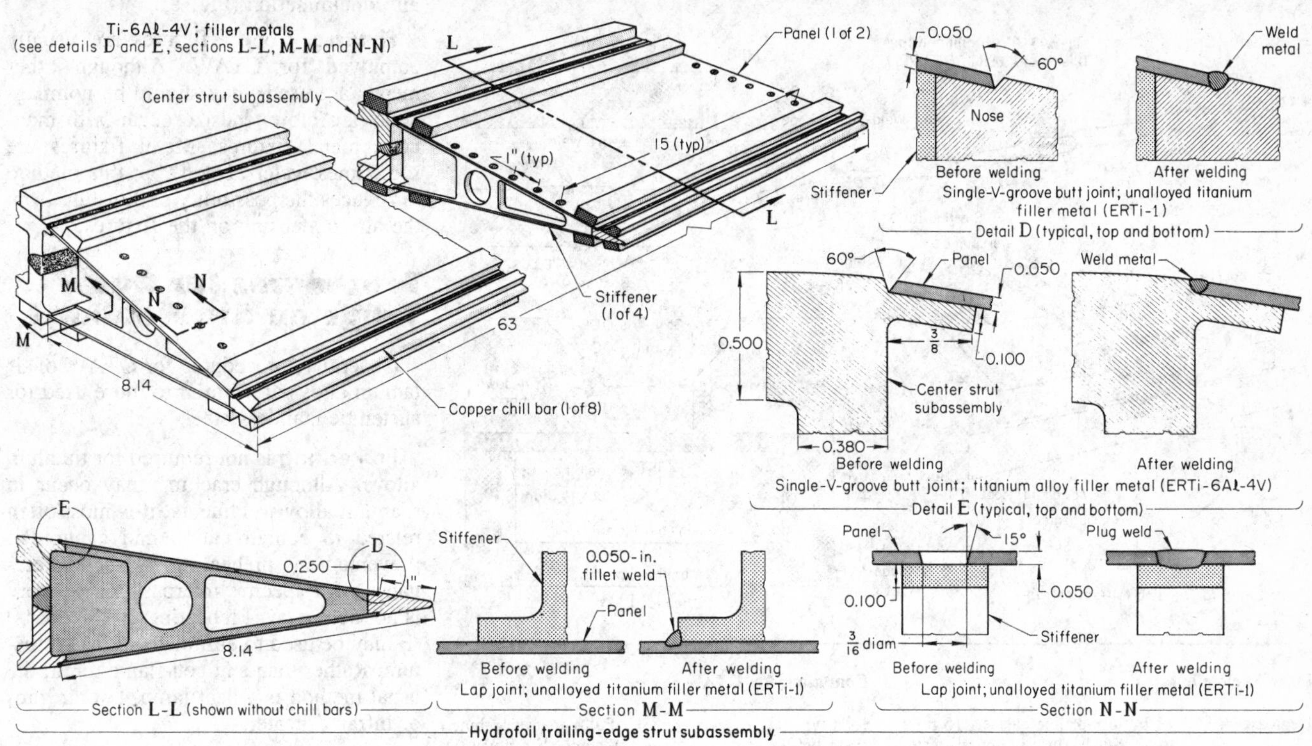

Power supply	300-A 40-V transformer rectifier, with drooping-voltage characteristic
Chamber	Collapsible plastic, with 3 ports
Electrode	1/16-in.-diam EWTh-2
Torch	350 A, water cooled
Fixture	Inexpensive frame
Nozzle(a)	9/16-in. ID
Voltage	10 V
Arc initiation	High frequency
Arc length	About 3/32 in.
Type of bead	Convex; stringer
Interpass temperature	Not controlled
Preheat	None
Postweld heat treatment	4 h in argon at 950 °F

	Detail D	Sections M-M and N-N	Detail E
Joint type	Butt	Lap	Butt
Weld type	V-groove	Fillet; plug	V-groove
Filler metal	ERTi-1	ERTi-1	ERTi-6Al-4V
Filler-metal diameter, in.	0.062	0.062	0.062
Shielding gas	Argon	Argon	Argon
Flow at nozzle ft^3/h	10	15	10
Flow in chamber	Purge(b)	Purge(b)	Purge(b)
Current, A	85-95	120-130	85-95
Welding position	Flat	Horizontal; flat(c)	Flat
Passes	1	1	1
Speed, in./min	4-6	4-6	3-5

(a) Occasionally, nozzle was removed for better accessibility to the joint. (b) Flow-purge with slight positive pressure. (c) Fillet welds were made in the horizontal position; plug welds were made in the flat position.

Fig. 7 Arrangement for shielding in automatic welding of titanium alloys in air. The baffle shown on the leading side of the torch (or electrode holder) is seldom used for GTAW, but is used for GMAW.

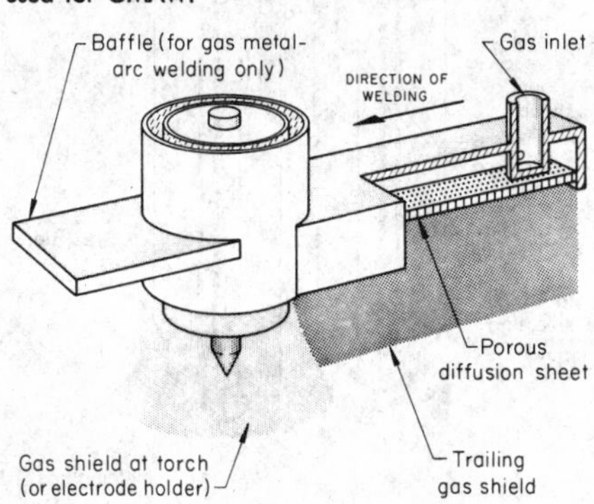

tests showed no cracks at 10× magnification, and the tensile-test specimen failed in the base metal at 140 ksi.

In general, welds made with this shielding technique were comparable in quality to those obtained in a conventional welding chamber and better than those obtained in the clamp-on displacement chamber. Making a weld in the clamp-on displacement chamber required 60 min and 100 ft^3 of argon, compared to 10 min and 15 ft^3 of argon for the out-of-chamber technique.

Equipment for Gas Tungsten Arc Welding

Transformer-rectifiers are the preferred power supply for welding titanium, because the current can be controlled more closely than with motor-generator sets; slight variations in welding current may cause variations in penetration. Direct current electrode negative is always used for GTAW of titanium because deeper weld penetration and a narrower bead can be obtained than with direct current electrode positive (DCEP). Also, in manual welding, DCEN is easier to control.

The power supply should include accessories for arc initiation because of the danger of tungsten contamination of the weld if the arc is struck by torch starting. If welding is to be done in air, controls for extinguishing the arc without pulling the torch away from the workpiece are needed, so that shielding-gas flow contin-

Fig. 8 Tooling setup for supplying shielding gas to weld area in GTAW

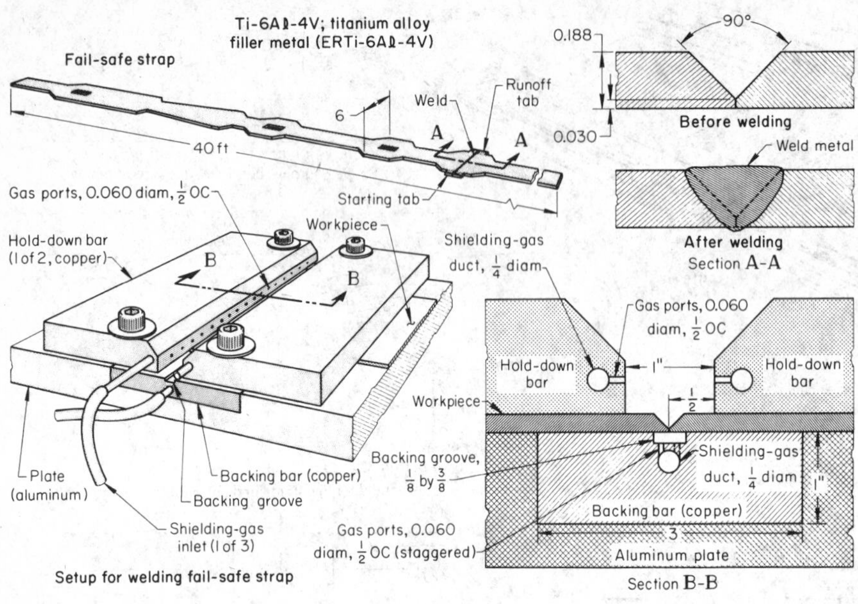

Preweld cleaning

Degrease	In hot alkaline cleaner, 15 min; brush; rinse in hot tap water
Pickle	In solution of $20HNO_3$-3HF, 30 s; rinse in hot, then cold, tap water
Final rinse	In deionized water; dry; weld within 4 h or reclean

(a) With 0.030-in. root face

Conditions for GTAW

Joint type	Full-penetration butt
Weld type	90° single-V groove(a)
Filler metal	$^1/_8$-in.-diam ERTi-6Al-4V
Current	155-195 A (DCEN)
Voltage	12-16 V
Shielding gas:	
At nozzle	Argon, at 20 ft^3/h
Through backing bar	Argon, at 15 ft^3/h
Through hold-down bars	Argon, at 40 ft^3/h
Welding time	About 4.5 min per joint

ues and the hot weld metal is not contaminated by air. For further details on power supplies and other equipment, see the article "Gas Tungsten Arc Welding" in this Volume.

Electrodes. The conventional thoriated tungsten types of electrodes (EWTh-1 or EWTh-2) are used for GTAW of titanium. Electrode size is governed by the smallest diameter able to carry the welding current. To improve arc initiation and control the spread of the arc, the electrode should be ground to a point. The electrode may extend one and a half times the size of the diameter beyond the end of the nozzle.

Shielding. To ensure a diffuse, nonturbulent flow of shielding gas, nozzles of torches for welding titanium are larger than those used for welding other metals. With a $^1/_{32}$-in.-diam electrode, a $^9/_{16}$-in.-ID nozzle is ordinarily used, and with a $^1/_{16}$-in.-diam electrode, a $^3/_4$-in.-ID nozzle is used. Phenolic or other plastic nozzles should not be used to avoid the danger of contaminating the weld with carbon.

Because titanium has low thermal conductivity, the area ahead of the arc does not get heated above 1000 °F; therefore, leading shields are seldom required when welding is done by the GTAW process. For welding operations where a trailing shield (Fig. 7) is not adaptable, the nozzle of the torch is fitted with a concentric outer shroud through which a supplementary supply of shielding gas is fed. In Fig. 9, which shows this type of torch, a shielding gas is diffused through copper shavings contained in the outer shroud that cool the gas substantially, helping to protect the metal near the weld.

Shielding of the underside of a weld is provided by slotted backing bars, usually copper, through which a diffuse flow of argon or helium is maintained (Fig. 1 and 8). Gas channels in the clamping fixtures also provide diffuse flow of inert gas to the weld area. These fixtures are placed close to the weld to avoid the danger of air contamination (Fig. 8).

Fixtures. Copper fixtures are usually employed for GTAW. Although other metals are used, they should be nonmagnetic; arc blow tends to occur with magnetic metal fixtures. Metal fixtures are sometimes water cooled, but this method introduces the possibility of moisture from the air condensing on the fixtures.

Procedures for Gas Tungsten Arc Welding

Generally, procedures for GTAW of titanium alloys are similar to those used for austenitic stainless steel.

Preheating is not required for titanium alloys. Although cracking may occur in titanium alloy weldments, it is most often related to contamination and cannot be prevented by preheating. Also, maintenance of a specific interpass temperature is not necessary. Preheating of 180 to 200 °F may be used to eliminate surface moisture. Rather than an open-flame torch, the usual method is a heat lamp, hot air gun, or infrared heater.

Tack welding is used to pre-position parts or subassemblies for final welding operations. Elaborate fixturing often can be eliminated when tack welds are used to their full advantage. Various tack welding procedures can be used, but in any procedure good cleaning practices and adequate shielding must be provided to prevent contamination of the welds. Contamination or cracks developed in tack welds can be transferred to the finish weld. One procedure is to tack weld in such a

Fig. 9 Sectional view of torch nozzle equipped with an outer shroud. Copper shavings in the outer shroud provide additional gas shielding for manual inert gas welding.

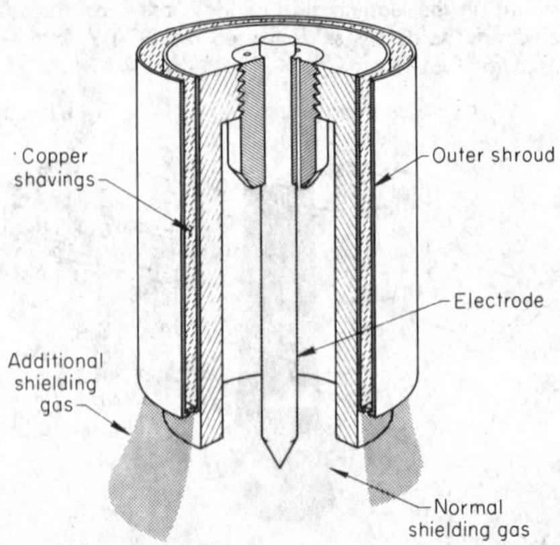

way that the finish weld never crosses over a previous tack weld. To accomplish this, sufficient filler metal is used in tack welding to completely fill the joint at a particular location. The finish weld beads are blended into the ends of the tack welds.

Arc length for welding without filler metal, as with stainless steel and the nickel-based alloys, should have a maximum size about equal to the electrode diameter. With longer arc length, there is danger of turbulence, which may draw air into the weld pool. In addition, increasing the arc length produces wider weld beads. When filler metal is used, the maximum arc length should be about one and a half times the electrode diameter, depending on the thickness of the base metal.

Welding conditions or schedules for GTAW of sheet are given in Table 3. In welding titanium alloys, the best heat input to use is a temperature just above the minimum required to produce the weld. If heat input is greater, the possibility that the weld will become contaminated, distorted, or embrittled increases.

Avoiding porosity in welds is an important consideration in welding titanium alloys. If the joint and filler wire are properly cleaned and the tooling does not chill the weld too rapidly, porosity can be reduced or eliminated by using a slower welding speed, which will retard weld solidification and allow entrained gases to escape.

Hot Wire Gas Tungsten Arc Welding

The hot wire process combines conventional single-wire arc welding with a unique method of filler-metal addition and may be used for welding or surfacing. A 1/16-in.-diam filler wire is preheated to the molten state as it enters the weld pool. The wire is mechanically fed into the weld pool through a holder from which argon gas flows to protect the liquid and solid metal from oxidation. No flux is used when welding titanium.

Welding equipment for automatic hot wire GTAW utilizes two separate power supplies. The power supply for the hot wire addition is connected across the contact tube of the hot wire torch and workpiece, causing a current to flow through the wire. The wire feeder adds a continuous supply of filler wire to the weld deposit, constantly maintaining the electrical circuit that produces the molten metal. Because the hot wire system is independent of the welding arc, adjustments in the wire-feed speed and melting rate of the hot wire can be made to suit the requirement of the weld without changing the arc current.

Power supplies used for melting the hot wire are single-phase, constant-potential transformer-type units with alternating current welding output. The power supply operates from a 230- or 460-V single-phase alternating current primary power source without affecting its rated output of 500 A at 47 V. Open circuit voltage, 0 to 50 V, is provided in three ranges from external output bus taps: 0 to 18, 16 to 34, and 32 to 50 V. Continuous (motorized) voltage adjustment, under load, is possible within each range. Three fixed ranges of volt-ampere slope are also provided from externally accessible output taps.

Hot wire addition is melted by its own alternating current power supply; therefore, the amount of hot wire addition can be controlled independently of the arc wire. The amount of hot wire addition depends on the quantity of metal that can be usefully deposited while maintaining good weld-bead geometry and uniform penetration.

Trailing shields must be capable of providing sufficient gas coverage to produce a sound, bright-silver weld deposit. In addition, heat-resistant tape is used around the trailing shield edges to contain the gas-shielding envelope.

Table 3 Welding procedure schedule for GTAW of titanium

Material thickness(a), in.	Tungsten electrode diameter, in.	Filler rod diameter, in.	Nozzle size ID, in.	Shielding gas flow, ft³/h	Welding current(b), A	Number of passes	Travel speed(c), in./min
Square-groove and fillet welds							
0.024	1/16	...	3/8	18	20-35	1	6
0.063	1/16	...	5/8	18	85-140	1	6
0.093	3/32	1/16	5/8	25	170-215	1	8
0.125	3/32	1/16	5/8	25	190-235	1	8
0.188	3/32	1/8	5/8	25	220-280	2	8
V-groove and fillet welds							
0.25	1/8	1/8	5/8	30	275-320	2	8
0.375	1/8	1/8	3/4	35	300-350	2	6
0.50	1/8	5/32	3/4	40	325-425	3	6

Note: Tungsten used for the electrode; first choice 2% thoriated EWTh2, second choice 1% thoriated EWTh1. Use filler metal one or two grades lower in strength than the base metal. Adequate gas shielding is essential not only for the arc, but for heated material also. Backing gas is recommended at all times. A trailing gas shield is also recommended. Argon is preferred. For higher heat input, on thicker material use argon-helium mixture. Without backing or chill bar, decrease current 20%.
(a) Or fillet size. (b) Direct current electrode negative. (c) Per pass

Table 4 Typical conditions for manual and automatic GMAW of Ti-6Al-4V plate

Using a 0.062-in.-diam electrode

Plate thickness, in.	Current (DCEP), A	Voltage, V	Welding speed, in./min	Argon flow rate, ft³/h: Torch	Trailing	Backing
Manual welding						
0.625	310	38	...	36	(a)	(a)
2.00(b)	310	38	...	36	(a)	(a)
Automatic welding						
0.625(b)	360	45	15	50	60	6
2.00(b)	325	33	25	(c)	(c)	(c)

(a) Not reported. (b) Multiple passes. (c) Argon chamber

Table 5 Typical conditions for GMAW of 1/2-in.-thick Ti-5Al-2.5Sn plate

Electrode wire	1/16-in.-diam ERTi-5Al-2.5Sn
Wire-feed rate	300 in./min
Current	300 A (DCEP)
Voltage	30 V
Nozzle	1-in. ID
Backing bar	Copper; with 1/16-in.-deep by 1/4-in.-wide groove
Shielding gas flow:	
At torch	Argon, at 50 ft³/h
Trailing	Argon, at 50 ft³/h
Backing	Helium, at 20 ft³/h
Welding speed	20 in./min

Gas Metal Arc Welding

Gas metal arc welding is normally used for welding titanium and titanium alloys 1/8 in. or more in thickness. This process is frequently used for welding 1/2-in. plate.

Metal transfer through the arc in GMAW can lead to difficulty in meeting stringent aerospace quality requirements. For example, weld spatter is often associated with inferior weld quality, and arc instability, which can occur in GMAW, is

Table 6 Typical operating conditions for PAW of titanium alloys
Backing gas and trailing shield required; keyhole technique used with orifice-to-work distance of $^3/_{16}$ in.

Thickness, in.	Travel speed, in./min	Current (DCEN), A	Arc voltage, V	Gas	Gas flow, ft³/h Orifice gas	Shielding gas	Joint type
0.125	20	185	21	Argon	8	60	Square butt
0.187	13	175	25	Argon	18	60	Square butt
0.390	10	225	38	75He-25A	32	60	Square butt
0.500	10	270	36	50He-50A	27	60	Square butt
0.600	7	250	39	50He-50A	30	60	V-groove(a)

(a) 30° included angle; $^3/_8$ in. root face.

a potential cause of weld contamination and defect formation. Some users of titanium alloys prefer GTAW over GMAW (even for joining thick plate), because with the gas tungsten arc process more uniform and predictable transverse shrinkage is obtained.

Typical conditions for GMAW of Ti-6Al-4V and Ti-5Al-2.5Sn plate are given in Tables 4 and 5. Electrode wires for GMAW are available in several grades of unalloyed titanium and in titanium alloys that match the composition of the base metal. Joint preparation and cleaning procedures are the same as those described earlier in this article.

Shielding for out-of-chamber welding is provided by inert gas being fed through the nozzle of the electrode holder, through the backing bar or plate, and as a trailing shield, much as in GTAW (Fig. 7). The electrode holder is basically the same as for GMAW of steel (see the article "Gas Metal Arc Welding" in this Volume). To avoid contamination and porosity in GMAW, a leading shield is necessary, as well as a trailing shield and a suitable baffle added on the leading edge of the electrode holder (Fig. 7). A leading shield prevents oxidation of spatter before it is melted in the weld metal.

Plasma Arc Welding

The joining of titanium alloys is one of the major applications of PAW. Because titanium has a lower density, keyhole welds can be made through thicker titanium square butt joints than for steel. As with GTAW, PAW requires backing gas and a trailing gas shield to prevent atmospheric contamination of the weld and adjacent weld metal. Typical operating conditions for PAW of titanium alloys are given in Table 6.

Stress Relieving

Most titanium weldments are stress relieved after welding to prevent weld cracking and susceptibility to stress-corrosion cracking in service. Stress relief also improves fatigue strength. An assembly subjected to a substantial amount of welding and severe fixturing restraint may require intermediate stress relieving of the partially welded structure, which should be done in an inert atmosphere; otherwise, the unwelded joints may have to be recleaned before being welded.

With unalloyed titanium and alpha titanium alloys, time and temperature should be controlled to prevent grain growth. Stress relieving times and temperatures for several weldable titanium alloys are given in Table 7. For alloys not mentioned in Table 7, tests should be conducted to ensure that stress relieving does not reduce fracture toughness, creep strength, or another property of importance.

Stress relieving of Ti-13V-11Cr-3Al weldments causes aging and subsequent embrittlement of the weld and HAZ and, therefore, is not recommended. Resolution heat treatment (re-annealing) may be used to relieve stresses if the welded assembly is amenable to such treatment. All titanium surfaces should be free of dirt, fingerprints, grease, and residues before stress relieving. Contaminated surface metal must be removed from the entire weldment by machining or descaling and pickling to remove 0.001 to 0.002 in. per surface.

Table 7 Stress-relieving times and temperatures for six titanium alloys

Grade	Temperature, °F	Time, h
Unalloyed Ti-0.15Pd	800	8
	900	3/4
	1000	1/2
Ti-5Al-2.5Sn(a)	900	20
	1000	6
	1100	2
	1200	1
Ti-6Al-4V:		
Annealed(a)	900	20
	1000	2
	1100	1
Solution treated	900	15
	1000	4
Aged	900	15
	1000	5
Ti-8Al-1Mo-1V:		
Single and double annealed	1100	2
	1200	1 1/2
	1300	1/2
Triple annealed	1100	5
	1200	2
	1300	1/2
Ti-5Al-5Sn-5Zr	1200	3
	1300	1/2
Ti-7Al-12Zr	1200	3
	1300	3/4

(a) Data apply also to extra-low-interstitial (ELI) modification.

Repair Welding

Repair welds should follow the established specification requirements for the original welds and be made prior to final heat treatment. Manual or automatic GTAW is generally used for repairing butt and fillet welds. Repairs can also employ a combination of welding processes such as GTAW and the initial welding process (GMAW, PAW, or hot wire welding).

Repair welds always must be carefully executed, and all traces of liquid-penetrant inspection material must be removed. Generally, inspection is performed on both faces of the repair weld and several inches beyond the repaired area.

Arc Welding of Zirconium and Hafnium

ZIRCONIUM, ZIRCONIUM ALLOYS, AND HAFNIUM are most commonly welded using the gas tungsten arc welding (GTAW) process. Other arc welding processes can be used, but are uncommon because most zirconium and hafnium alloys are used in applications that require very high weld integrity. Electron beam welding is used for welding thick sections. Because zirconium and hafnium are extremely reactive at high temperatures, the weld and the surrounding area must be carefully shielded from air. Shielding by a standard tungsten arc torch is insufficient to provide adequate protection. Critical nuclear welds are made in controlled atmosphere boxes. For nonnuclear industrial applications, trailing shield torches and gas backup should be used along with a temporary purge chamber or box. Weld backup shielding set up for titanium welding is adequate for most zirconium and hafnium welding for industrial applications.

Weldability

Zirconium, its alloys, and hafnium can be welded without difficulty provided the

Table 1 Chemical compositions of commercial zirconium and zirconium alloys

Element	Composition, wt% R60702	R60704	R60705	R60706
Zirconium + hafnium, min	99.2	97.5	95.5	95.5
Hafnium, max	4.5	4.5	4.5	4.5
Iron + chromium	0.2 max	0.2-0.4	0.2 max	0.2 max
Tin	...	1.0-2.0	...	...
Hydrogen, max	0.005	0.005	0.005	0.005
Nitrogen, max	0.025	0.025	0.025	0.025
Carbon, max	0.05	0.05	0.05	0.05
Niobium	...	...	2.0-3.0	2.0-3.0
Oxygen, max	0.16	0.18	0.18	0.16

Table 2 Chemical compositions of the most common grades of nuclear-grade Zircaloys

Element	Composition, wt% R60802 (Zircaloy-2)	R60804 (Zircaloy-4)	R60901 (Zr, 2.5 Nb)
Tin	1.20-1.70	1.20-1.70	...
Iron	0.07-0.20	0.18-0.24	...
Chromium	0.05-0.15	0.07-0.13	...
Nickel	0.03-0.08	...	...
Niobium	...	...	2.40-2.80
Oxygen	...	...	0.09-0.13
Iron + chromium + nickel	0.18-0.38	...	...
Iron + chromium	...	0.28-0.37	...
Hafnium	0.010	0.010	0.010
Zirconium	rem	rem	rem

weld and surrounding area can be protected from nitrogen, oxygen, or carbon dioxide, which would be absorbed at welding temperatures. Contamination of the weld decreases the corrosion resistance and tends to make the weld brittle. Zirconium has a low-temperature, hexagonal, close-packed structure (alpha zirconium). Pure zirconium transforms to the body-centered cubic structure (beta zirconium) at about 1580 °F. The addition of alloying elements causes the transformation to occur over a range of temperatures. The relatively low thermal expansion value and very small volume change associated with the phase transformation permits the welding of complex structures with minimum distortion. Hafnium has a hexagonal close-packed structure up to 3200 °F, and a body-centered cubic structure above this temperature.

Zirconium and its alloys are available in two general categories: (1) commercial-grade zirconium and (2) reactor-grade zirconium. Commercial-grade zirconium designates zirconium containing hafnium as an impurity. Reactor-grade zirconium designates zirconium from which most of the hafnium has been removed to make it suitable for nuclear reactor applications. Because pure zirconium has low mechanical properties, various alloying elements are added to increase its mechanical properties. Alloys of zirconium for nuclear application are commonly called zircaloys. Zirconium and its alloys are available in plate, sheet, bar, rod, and tubing form to a variety of material specifications. The four most common grades of commercial zirconium alloys as specified in American Society for Testing and Materials (ASTM) B550 are given in Table 1. For reactor grades, the hafnium content is reduced to a minimum and other impurities are closely controlled. Chemical compositions of the most common grades of nuclear-grade Zircaloys as specified in ASTM B351 are given in Table 2.

Table 3 Chemical compositions of zirconium and zirconium-alloy electrodes

Element, %	ERZr2	ERZr3	ERZr4
Carbon	0.05	0.05	0.05
Chromium + iron	0.020	0.020-0.040	0.040
Niobium	...	...	2.00-4.00
Hafnium	4.5	4.5	4.5
Hydrogen	0.005	0.005	0.005
Nitrogen	0.025	0.025	0.025
Tin	...	1.00-2.00	...
Zirconium + hafnium	rem	rem	rem

Hafnium is used mainly in unalloyed form for control rods in nuclear reactors and containers for corrosive media in spent nuclear fuel reprocessing plants. Hafnium is readily weldable by the same processes and procedures described for zirconium.

Both zirconium and hafnium are cleaned and stored for welding in the same manner described for titanium. Postweld annealing of zirconium welds is not usually required because of its good ductility. Zirconium weldments can be stress relieved, however, at about 1250 °F for 15 to 20 min, followed by air cooling. Hafnium weldments can be stress relieved at about 1000 °F.

Zirconium and hafnium can only be fusion welded to themselves or to reactive metals such as titanium, niobium, or tantalum. When welded to nonreactive metals, intermetallics form that embrittle the weld metal.

Filler Metals

Three zirconium and zirconium-alloy electrodes are given in American Welding Society (AWS) A5.24-79; their chemical compositions are given in Table 3. The filler-metal composition is selected to match the base-metal composition. Type ERZr2 is used to weld commercially pure zirconium (Grade R60702), type ERZr3 to weld Zr-1.5Sn alloy (Grade R60704), and type ERZr4 to weld Zr-2.5Nb alloy (grades R60705 and R60706). For critical nuclear applications, the impurities are more closely controlled and hafnium is minimized.

Table 4 Gas tungsten arc welding of zirconium (direct current electrode negative without filler)

Sheet thickness, in.	Tungsten electrode diam, in.	Argon shield nozzle diam, in.	Argon flow, L/s Torch	Backing	Current, A	Welding speed, in./min	Mode
0.008	0.047	0.31	5.7	2.4	45	20	Auto
0.016	0.047	0.31	6.6	2.4	80	25	Auto
0.016	0.047	0.31	6.6	2.4	100	25	Auto
0.024	0.059	0.31	6.6	2.4	50	...	Manual
0.024	0.059	0.47	7.6	2.4	125	20	Auto
0.039	0.079	0.47	7.6	2.4	150	20	Auto
0.059	0.118	0.55	7.6	2.4	160	20	Auto
0.059	0.118	0.59	7.6	2.4	120	...	Manual
0.079	0.118	0.59	7.6	2.4	180	20	Auto

Source: Welding Properties of Non-Ferrous Metals (Data series 2), Sheet 27: Zirconium, *Metal Construction*, Vol 13 (No. 10), Oct 1981, p 632-634

Fig. 1 Exploded view of typical rod assembly

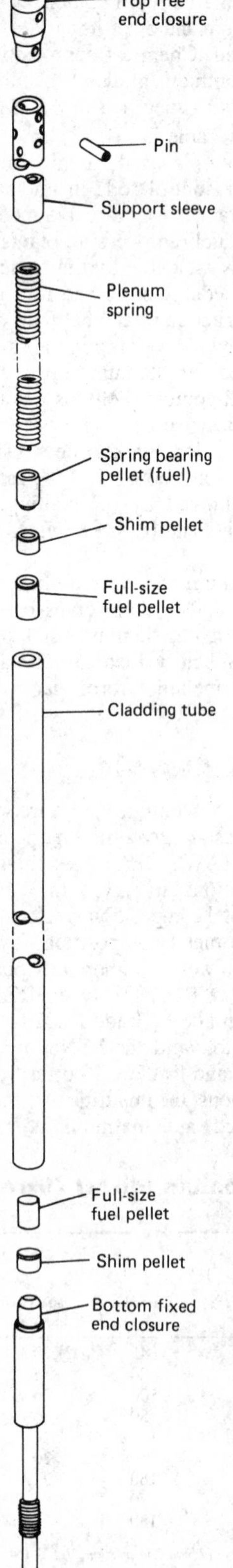

Table 5 Welding parameters for fuel rods

Specific parameters	Diam/wall 0.310/0.024 in.	Diam/wall 0.530/0.028 in.	Diam/wall 0.526/0.030 in.	Diam/wall 0.836/0.043 in.
Speed, rpm	7 ± 3%	4 ± 5%	4 ± 5%	3 ± 5%
s/rev	8.6 ± 0.3	15.0 ± 0.7	15.0 ± 0.7	20.0 ± 1.0
Weld time ($1^1/_4$ rev)	10.7 ± 0.3	18.8 ± 0.7	18.8 ± 0.7	25.0 ± 1.0
Arc delay, s	8.6 ± 0.3	7.5 ± 0.7	7.5 ± 0.7	10.0 ± 1.0
Amperage, A	43 ± 2	66 ± 3	66 ± 3	85 ± 3

Note: General parameters common to all fuel rod assemblies: electrode, 2% thoriated tungsten, $^3/_{32}$ in. diam, ground at a 15° taper to one third the diameter at the tip; weld passes, two $1^1/_4 \pm ^1/_{16}$ revolution passes; arc gap, 0.030 ± 0.002 in.; atmosphere, high-purity helium at 1 atm pressure

Hafnium is welded with bare wire of matching composition.

Shielding Gases

As is the case with all reactive metals, argon, helium, or mixtures of these two gases are the only gases used for shielding. Argon is more widely used than helium because it is more readily available and less costly. Welding-grade argon containing less than 0.2 vol% impurities is desirable to minimize contamination.

Welding Process

Gas tungsten arc welding is the most widely used process for joining zirconium and its alloys. Most of the world's annual output of zirconium is fabricated into components for nuclear reactors. The two examples presented later in this article deal with gas tungsten arc welding of nuclear components.

Standard welding equipment can be used for Zircaloy welding. Direct current electrode negative and thoriated tungsten electrode type EWTh-2 are commonly used. A high-frequency arc starter is recommended to prevent contamination of the weld with tungsten. Table 4 shows typical weld parameters for GTAW of thin sheets.

Because Zircaloys are sensitive to very small amounts of contamination, the weld joint should be clean. Proper preparation is important to remove scale and dirt from the edges to be welded. The edges should be cleaned with a stainless steel brush and degreased with alcohol or acetone. If filler metal is used, it should also be cleaned. During manual welding with filler wire, care should be taken to keep the hot tip within the gas shield.

Example 1. Fuel Rod Welding in the Light Water Breeder Reactor (LWBR) Core.* The LWBR fuel rods are approximately 10 ft long and consist of Zircaloy-4 seamless tube, fuel pellets, a plenum spring, and two welded Zircaloy-4 end closures. Figure 1 shows a typical rod assembly. There are four different sizes of seamless tubes. Dimensions and weld parameters are given in Table 5. The welding is accomplished by the automatic gas tungsten arc process in high-purity helium at 1 atm in a weld chamber. The first operation is to weld the bottom end closure (Fig. 2) to the seamless tube. The weld setup is shown in Fig. 3. After inspecting the bottom end weld, the assembly is loaded with fuel pellets, placed in the weld chamber, and the chamber evacuated and backfilled with helium to 1 atm. The top end closures are then pressed into place and welded. The weld is made in two complete passes rather than a single pass; this substantially reduces weld porosity and the likelihood of repairs. The integrity of the final weld is verified by ultrasonic and radiographic inspection.

*Source: Bickel, W.L., Fuel Rod Welding (LWBR Development Program), WAPD-TM-1235, Feb 1979

Fig. 2 Cylindrical end closure insert design

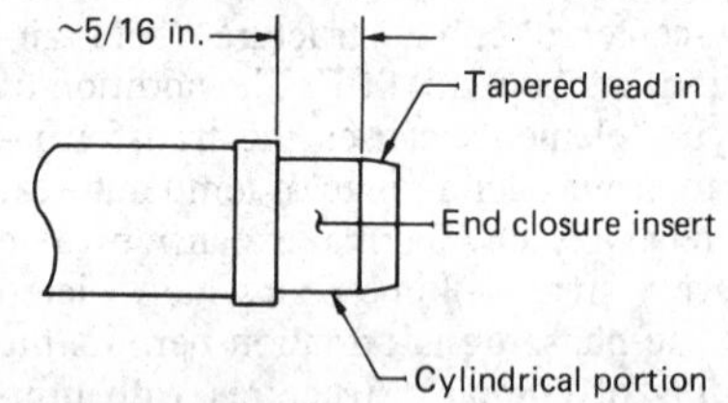

Fig. 3 Weld setup

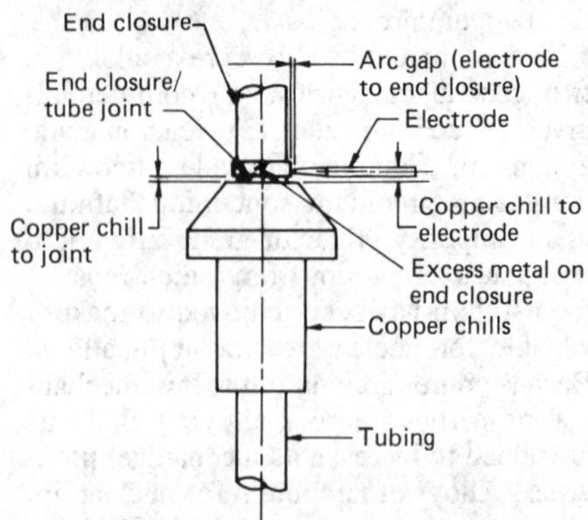

Fig. 4 Typical control rod. Weld tab locations are indicated in (a). (1) Zircaloy-2 adaptor, (2) Zircaloy-2 adaptor plate (0.25 in.), (3) hafnium, (4) Zircaloy-2 extension, (5) Zircaloy-2 weld tab, (6) hafnium weld tab

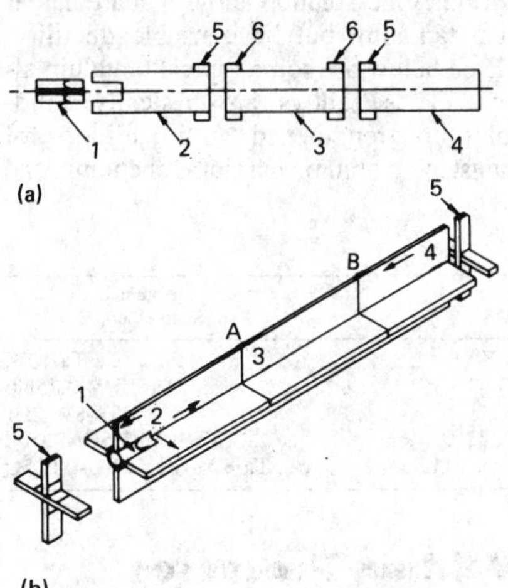

Example 2. Control Rod Welding in a Nuclear Reactor.* A control rod in a nuclear reactor controls the operation of the reactor by absorbing the neutrons produced during the nuclear reaction. Figure 4 illustrates a typical control rod made of hafnium. Because hafnium is expensive, only that part of the control rod which actually serves the neutron-absorbing function is made of hafnium. An upper extension made of Zircaloy-2, identified as the "adapter plate," is used to connect the control rod to the control-rod-actuating mechanism. A Zircaloy-2 extension is sometimes attached at the bottom of the hafnium section. In Fig. 4, both hafnium-to-hafnium and hafnium-to-Zircaloy-2 welds are to be made; the welds required are shown by heavy lines. Prior to fusion welding, the cross sections are held in place by tack welding. The tack welding parameters for $^1/_4$-in.-thick hafnium and Zircaloy are 125 to 135 A at 14 to 18 V.

After tack welding, the control rod assembly is loaded into an inert-atmosphere welding chamber equipped with suitable fixtures. The chamber is evacuated and backfilled with helium or argon. Thoriated tungsten electrodes are used and the weld started and stopped on weld tabs. Welding conditions for butt welding $^1/_4$-in.-thick hafnium or hafnium-Zircaloy plates are approximately 220 A at 16 V, with an arc speed of $6^3/_4$ in./min. Both sides of the butt joint must be fused to ensure complete penetration. The melting point of hafnium is about 1000 °F above that of zirconium. Because of this, a butt-welded joint between these two materials is made by positioning the electrode on the hafnium side $^1/_{32}$ to $^1/_{16}$ in. away from the hafnium-zirconium joint. To further reduce the effect of the difference in melting point, and also to reduce the possibility of undercut of the Zircaloy joint, a thin strip of zirconium is tack welded parallel to the seam on the Zircaloy side to serve as filler metal. The quality of the joint is verified by metallographic evaluation of the weld tabs.

*Source: Sayell, E.H., Joining of Hafnium, in *The Metallurgy of Hafnium*, Thomas, D.E. and Hayes, E.T., Ed., Government Printing Office, 1960

Arc Welding of Niobium

By R. Terrence Webster
Principal Metallurgical Engineer
Teledyne Wah Chang Albany

NIOBIUM AND ITS ALLOYS are easily welded. Niobium is similar to tantalum; the major differences are its density, which is half that of tantalum, and its lower melting point (4474 °F). Niobium is a reactive metal that combines readily with oxygen, nitrogen, hydrogen, and carbon at temperatures above 750 °F. Consequently, precautions should be taken to shield the top and heated root side of the weld from contact with air until cooled below 750 °F.

Weldability

Unalloyed niobium and most niobium alloys, even alloys containing up to 11% W, have good weldability. Most niobium alloys are intended for use in aerospace or nuclear applications. They are solution strengthened and are welded in the fully annealed condition. Niobium alloys are stronger and less ductile than unalloyed niobium and consequently are much more sensitive to contamination from the air during welding and heat treating. Listed below are some typical niobium-based alloys that usually are available in sheet or plate form:

Alloy	Nominal composition, %
Cb1Zr	Nb-1Zr
AS-55	Nb-5W-12k-0.06C + Y
B66	Nb-5Mo-5V-12
C103	Nb-10Hf-1Ti
C129Y	Nb-10W-10Hf + 0.1Y
Cb752	Nb-10W-2.5Zr
D43	Nb-10W-12-0.1C
FS-85	Nb-27Ta-10W-1Zr
SCb291	Nb-10W-10Ta
SU31	Nb-17W-3.5Hf-0.12C-0.03Si

Source: Ref 2

Niobium and its alloys cannot be welded to other structural metals such as steel, copper, and aluminum; the resulting welds are brittle.

Welding Processes

Gas tungsten arc welding (GTAW) is the most commonly used method of welding niobium and its alloys. Unalloyed niobium can be welded with inert gas shielding on both sides of the weld in the same manner as titanium using the same equipment, procedures, and precautions. Fluxes cannot be used because they cause severe weld embrittlement. Niobium can be welded both by manual and automatic procedures provided that adequate inert gas shielding is employed. Niobium alloys are best welded in a vacuum-purged chamber that is backfilled with argon or helium to avoid contamination from the air. Vacuum chamber weld procedures and processes are the same

Table 1 Typical conditions for GTAW of niobium alloy sheet

Sheet thickness, in.	Shielding gas	Welding current, A	Arc voltage, V	Travel speed, in./min
0.020	80He-20Ar	40	12	7.5
		55	12	15.0
		70	11.5	30.0
0.030	He	80	17.5	7.5
	80He-20Ar	90	11.5	15.0
	80He-20Ar	130	11.5	30.0
0.060	He	130	13.5	7.5
		160	15.0	15.0
		200	15.0	30.0

Note: These conditions normally apply to B66, FS-85, C129Y, Cb752, and SCb291 alloys.

as for titanium and zirconium. Plasma arc welding also is used for welding niobium. Gas metal arc welding is used rarely and only in a vacuum-purged chamber. Electron beam welding is used for critical applications, such as gas turbines and rocket motor parts, where maximum strength and ductility are required.

Surface Preparation

The weld joint and the surfaces on both sides at least 1 in. from the joint must be chemically clean when welding niobium, much like when welding titanium. If the surfaces have been oxidized by exposure to temperatures above 500 °F, the oxide scale should be removed by abrasive blasting, grinding, or machining.

Surfaces that are free of oxide scale can be cleaned by first degreasing with alkaline detergent or suitable solvents, followed by pickling in an acid mixture of 10 to 20% hydrofluoric acid, 30 to 40% nitric acid, and the balance water. Pickling should be followed by water rinsing and air drying. For further information on surface cleaning, see Volume 5 of the 9th edition of *Metals Handbook*.

Gas Tungsten Arc Welding

The same process technology and equipment that is used for GTAW of titanium is also suitable for welding niobium alloys. Unalloyed niobium can be welded in or out of chamber using argon or helium for inert gas shielding. Niobium alloys are best welded in vacuum-purged chambers backfilled with argon, helium, or a mixture of both. Direct current electrode negative should be used (Ref 1). High-frequency arc initiation should be used to avoid tungsten electrode contamination. Tungsten electrodes should be the thoriated tungsten type (EWTH-1 or EWTH-2). Neither preheating nor postweld stress relieving is required. Age-hardened alloys should be annealed after welding to restore ductility. Typical conditions for GTAW of niobium are given in Table 1.

REFERENCES

1. Donnelly, R.G. and Slaughter, G.M., *Weldability Evaluation of Advanced Refractory Alloys,* American Welding Society, Welding Research Supplement, June 1966
2. Lessmann, G.G., *The Comparative Weldability of Refractory Metal Alloys,* American Welding Society, Welding Research Supplement, Dec 1966

Arc Welding of Tantalum

By R. Terrence Webster
Principal Metallurgical Engineer
Teledyne Wah Chang Albany

TANTALUM AND ITS ALLOYS can be welded by gas metal arc welding (GMAW), gas tungsten arc welding (GTAW), plasma arc welding, and electron beam welding processes. Resistance spot welding of tantalum is also feasible, but adherence and alloying between copper alloy electrodes and tantalum sheet is a problem. Welding of tantalum is similar to welding of titanium and zirconium with respect to equipment and technology. Tantalum is extremely reactive at temperatures above 600 °F; consequently, great care must be used to prevent contamination from shop oils and dust as well as from gases in air.

Weldability

Tantalum and its alloys have good weldability, provided the welds and the heated base metal are free from contamination. Contamination by oxygen, nitrogen, hydrogen, and carbon should be avoided to prevent embrittlement of the weld. If required, fixturing devices should not come in close contact with the weld joint. Tantalum melts at 5430 °F; consequently, fixture materials may melt and alloy with tantalum to produce brittle welds. Graphite should not be used as a fixturing material, because it reacts with hot tantalum to form carbides, which also cause brittleness.

Tantalum cannot be welded to common structural metals because they form brittle intermetallic compounds, which in turn embrittle the welds. Tantalum can be welded to other reactive metals, with which it forms solid-solution alloys that are harder than tantalum but have usable ductility. Listed below are some typical tantalum alloys. These alloys are basically solid-solution strengthened with additions of tungsten, niobium, hafnium, rhenium, and carbon.

Alloy	Nominal composition, %
Ta10W	Ta-10W
FS 63	Ta-2.5W-0.15Nb
T-111	Ta-8W-2Hf
T-222	Ta-10W-2.5Hf-0.01C
Astar 811C	Ta-8W-1Re-0.7Hf-0.025C

Welding Processes

Tantalum usually is welded by GTAW. Unalloyed tantalum can be welded with inert gas shielding on both sides of the weld using the same techniques that are used to weld titanium and zirconium. Tantalum is the most sensitive of the reactive metals to contamination; therefore, great care must be used in the preparation of the metal and the weld joint. Because of tantalum's sensitivity to contamination, inert gas shielding of the molten pool and the heated part of the metal at temperatures above 600 °F is necessary. None of the tantalum alloys should be welded with localized shielding.

Tantalum and its alloys are welded in vacuum-purged weld chambers that are backfilled with either argon or helium, or a mixture of the two gases. Tantalum can also be welded in a flush chamber when the parts to be welded cannot be shielded properly for open-air welding.

Table 1 Conditions for GTAW of tantalum

Thickness of weld, in.	Welding rate, in./min	Arc voltage, V	Current, A	Argon flow rate at the joint, ft^3/min
0.06	4.0	18	120	0.53
0.06	10.0	9	245-260	...
0.02	16.5	8-10	65-78	0.44
0.04	15.0	8-10	140-150	0.44
0.06	14.0	9-11	200	0.44
0.08	10.5	10-12	235-240	0.44
0.10	9.5	10	250-260	0.44
0.12	16.5	8	350	0.44
0.14	13.0	8	350	0.44
0.15	12.5	8	350	0.44

Source: Adapted from Pelleg, J. and Grill, A., Tantalum: Overcoming the oxidation problems, *Welding and Metal Fabrication,* Vol 47 (No. 7), 1979, p 443-449

Plasma arc welding and GMAW can also be used, but little information on these processes as they relate to tantalum is available. Electron beam welding often is used for tantalum parts used in critical applications.

Surface Preparation

Tantalum weldments must be chemically clean and free of oxide scale prior to welding. Joint preparation should include machining, and rough edges should be filed smooth. The joint should be tight-fitting and restrained to prevent opening during welding.

Heavy oxide should be removed by abrasive blasting or grinding, followed by pickling. Tantalum surfaces free of heavy oxide should be detergent or solvent cleaned and pickled. Tantalum should be pickled in an acid mixture of 10 to 20% hydrofluoric acid, 30 to 40% nitric acid, and the balance water. Detailed cleaning procedures for reactive and refractory metals are discussed in Volume 5 of the 9th edition of *Metals Handbook*.

Gas Tungsten Arc Welding

The procedures and equipment used for GTAW of tantalum are the same as those used for titanium and zirconium. Tantalum can be welded either manually or automatically using standard power supplies and torches. Fluxes cannot be used because they cause weld embrittlement. For welding with localized shielding, torches should be used that provide a lamellar gas flow from the torch to the work. Direct current electrode negative with thoriated tungsten electrodes (EWTH-1 or EWTH-2) should be used. Neither preheating nor postweld heat treatment is required for unalloyed tantalum. Some of the alloys do require annealing to restore ductility. Conditions for GTAW of tantalum are given in Table 1.

Applications

Chemical resistance of tantalum, particularly against acids, makes it attractive to the chemical industry for components such as tube coils, pipelines, valves, and agitators. Its high melting point (5425 °F) and good mechanical properties at elevated temperatures make tantalum suitable for high-temperature applications. The electronics industry uses tantalum in tubes, capacitors, and other applications where gas absorption and high-temperature characteristics are important.

Arc Welding of Beryllium

By Haskell Weiss
Welding Engineer
Lawrence Livermore National Laboratory

BERYLLIUM is a material which is important in the nuclear and space industries because of its unique physical and mechanical properties. It is used in nuclear applications because of its low neutron cross section and in space because of its low density. The absorption cross section of beryllium for thermal neutrons is 0.010 barn; the scattering cross section is 7.0 barns. The melting point and specific heat of beryllium are almost twice those of magnesium and aluminum. Physical properties that affect the weldability of beryllium are:

Crystal structure	Close-packed hexagonal
Melting point, °F	2332
Density, lb/in.^{3} at 68 °F	0.067
Thermal conductivity, Btu/ft · h · °F	87 at 300 °F 104 at 68 °F
Specific heat, Btu/lb · °F at 68 °F	0.45
Thermal expansion, μin./in. · °F	6.4×10^{-6} in./in.

The density of beryllium is less than 70% that of aluminum, but its modulus of elasticity (40 to 44 million psi at 68 °F) is about four times greater. Poisson's ratio of beryllium is almost zero (0.024 to 0.030). A high elastic modulus-to-density ratio makes beryllium useful for lightweight aerospace applications.

Disadvantages of beryllium include a close-packed hexagonal (cph) crystal structure with a very high *c/a* ratio (1.5680 Å), which partially accounts for its limited ductility at room temperature. Like other reactive metals, beryllium is included in the difficult-to-weld category. It has a high affinity for oxygen and cannot be welded by any process that utilizes fluxes, or where heated metal is exposed to the atmosphere. Minor amounts of impurities cause embrittlement, and inert gas shielding or vacuum must be used.

Beryllium is commercially produced as a powder metallurgy product. A cast ingot is first crushed and ground to a controlled sieve size and then hot pressed to form a working billet. The tensile properties of beryllium vary, depending on the type of processing, as shown in Table 1.

Surface Preparation

Like aluminum and magnesium, an adherent oxide film forms on beryllium, which inhibits wetting, flow, and coalescence during welding. As a result, workpieces must be carefully cleaned prior to welding. This is done by degreasing using a combination of acetone and alcohol wipes. If the surfaces appear heavily oxidized or not properly degreased, they should be chemically etched with a 5% solution of hydrofluoric acid and deionized water. An alternative cleaning solution, which reduces the pitting tendencies of hydrofluoric acid, is the use of a 70% phosphoric acid, 25% sulfuric acid, and 5% nitric acid solution.

Beryllium is also highly sensitive to severe twinning damage by heavy tool forces in the course of machining. The consequences can be nucleation sites for cracking during welding. Controlled machining should be used to minimize damage, the three final cuts each being less than 0.003 in. deep. The final cuts are sometimes replaced by a 0.006- to 0.010-in. etch.

Arc Welding

The most common problems associated with welding of beryllium are the susceptibility to hot cracking and the control of grain size in the weld metal and heat-affected zone (HAZ). Cracking in welds is caused by impurities, such as iron and aluminum, in the beryllium. By controlling both the absolute amounts and the ra-

Table 1 Mechanical properties of beryllium

Type	Testing temperature, °F	Tensile yield strength at 0.2%, ksi	Ultimate tensile strength, ksi	Elongation, %
QMV (vacuum hot pressed powder)	Room	27-38	33-51	1-3.5
	390	...	30-43	6-15
	750	...	22-27	19-40
	1110	...	20-22	15-25
	1470	...	7	7-8
Hot extruded powder(a)	Room	45	70-100	10-20
Hot extruded powder(b)	Room	...	50-60	1
Cross-rolled powder(c)	Room	40-60	60-90	10-40

(a) Longitudinal direction. (b) Transverse direction. (c) Values depend on reduction ratio and rolling temperature.

tio of iron to aluminum, hot cracking and lack of ductility in the melted weld zone can be reduced (Ref 1). A low effective heat input to provide a narrow HAZ helps minimize grain size.

The gas metal arc welding process using an aluminum alloy filler wire has been used successfully for welding of beryllium. Although several different alloys of aluminum have been used, the most successful welding has been with alloy 718, which contains 12% Si. Its success lies mainly in its closeness to the eutectic composition, resulting in high fluidity.

Lubricant trapped externally or internally in the filler wire contributes to weld porosity. To avoid lubricants, wire can be vacuum extruded; however, this leads to a rough surface, which in turn affects the design of the weld contact tip.

For 0.03-in. wire, the contact tip opening should be 0.05 in. The wide opening overcomes the rough wire surface. Maintaining electrical contact with this size opening may be difficult. Consequently, the upper portion of the collet has a 5° angle, which causes the wire to bend and continuously contact the wall.

Wire-feed rate is approximately 450 to 900 in./min, depending on groove size. The initial feed rate for arc starting is generally 25% of the regular rate. Weld travel speeds are on the order of 40 to 50 in./min.

At direct current electrode positive, settings of 30 to 32 V and 140 to 150 A are typical. Machines with voltage characteristics of medium slope perform somewhat better than a constant flat potential, because the arc is slightly smaller, thus avoiding arc blast at the start.

Shielding Gases

A combination of 5 parts helium to 1 part argon is used. Typical rates are 49 to 98 ft^3/h of helium and 10 to 20 ft^3/h of argon. Higher flow rates are deciding factors in obtaining contaminant-free welds, depending on the type of weld chamber.

Joint Design

Weld joint designs for plate and tube (Fig. 1) are based on butt joints prepared with a narrow J or U or a narrow V with root radius, each groove following the shape of the welding arc to minimize thermal shock to the base metal. Weld penetrations of up to 0.25 in. may be obtained without preheat; beyond that, preheating to 390 °F by torch or induction heating is necessary. Depths to 0.35 in. have been obtained without cracking the base metal. Minimum weld root depth should not be less than 0.06 in. For purposes of alignment, this can take the form of a step.

Fig. 1 Double-U-groove joint commonly used in GMAW of beryllium plate and tube

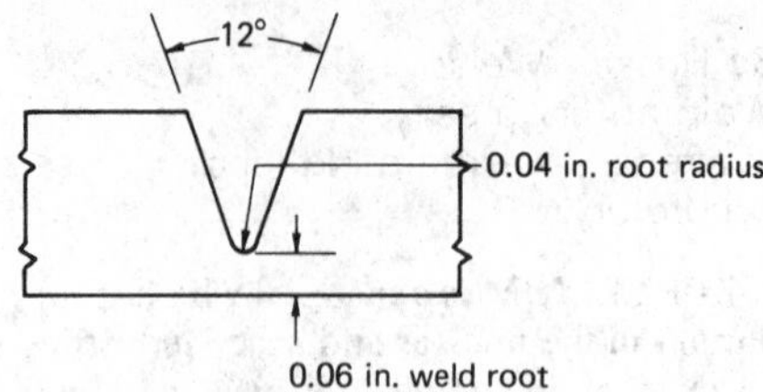

Generally, the amount of beryllium melted down from the side walls and root is approximately 0.02 in.

Weld Repair

Welds containing porosity can be repaired by first machining out the defect and then preheating to 212 °F and rewelding. If a multiple-pass weld must be stopped, the part should be preheated to a minimum of 212 °F before arc restart to reduce the possibility of cracking.

Safety

Beryllium as a fine powder, a pure metal, or an oxide is considered toxic; it can cause beryllosis, a respiratory disease (Ref 2). To minimize health hazards, welding should be done in a glove box with a filtered exhaust system.

REFERENCES

1. Beryllium, Hot Pressed Blocks and Shapes, Lawrence Livermore National Laboratory Specification MEL 71-001162A, Aug 1971
2. Williams, W.J., The Nature of Beryllium Disease, Beryllium 1977, Fourth International Conference on Beryllium, The Metals Society, London, 1977, p 47/1-47/9

Arc Welding of Molybdenum and Tungsten

By the ASM Committee on Arc Welding of Molybdenum and Tungsten*

*Anthony J. Bryhan, *Chairman*, Senior Research Associate, Climax Molybdenum Company of Michigan; C.C. Clark, Manager of Market Development, Tungsten Division, AMAX of Michigan, Inc.; William C. Hagel, Manager, High Temperature Materials Development, Climax Molybdenum Company of Michigan

MOLYBDENUM AND TUNGSTEN metals and alloys may be joined by welding, brazing, or mechanical means. Resistance, diffusion, inertia, friction, explosion, laser, and electron beam welding and brazing are discussed in this Volume. Each of these techniques offers advantages for joining particular items based on the design of the items and the desired mechanical properties of the weldment. Four inherent characteristics of molybdenum and tungsten are of significance to welding because of their effect on the mechanical properties of the resulting weldment. Molybdenum- and tungsten-based metals (1) are arranged in a body-centered cubic (bcc) lattice, (2) melt at very high temperatures, (3) have a tendency to form relatively coarse-grained microstructures during solidification or recrystallization, and (4) are strain-rate sensitive. The exact manner and extent to which these characteristics affect weldment mechanical properties is described briefly below.

Metals with bcc crystal lattice tend to be more sensitive to the presence of interstitial atoms (for example, oxygen, carbon, and nitrogen) than metals with face-centered cubic structures (Ref 1). Molybdenum and tungsten are particularly sensitive because of their low solid solubility for interstitials as compared to iron, niobium, or tantalum (Ref 2). Excessive interstitial contamination results in reduced ductility. The strength, ductility, and toughness of a metal tend to be inversely proportional to the grain size (following the Hall-Petch relationship). Thus, the grain growth that occurs as the result of fusion welding molybdenum and tungsten causes applied tensile strains to be concentrated in the lower strength, narrow, coarse-grained region of the weld rather than distributed uniformly. The actual strain rate in this region may greatly exceed the nominal strain rate, and because of the strain-rate sensitivity of these two metals, the ductile-to-brittle transition temperature (DBTT) increases (Ref 3). These metallurgical characteristics of molybdenum and tungsten preclude the use of any arc welding process that affords less protection to the hot metal than does the gas tungsten arc process with the inert gas shield.

The effects of these characteristics on weldment ductility may be minimized by (1) proper control of the welding atmosphere and chemical cleaning of the base metal and filler metal prior to welding to reduce contamination, (2) use of filler metals alloyed to improve weldment mechanical properties, (3) use of as low a heat input as practical to minimize grain growth in the heat-affected zone (HAZ) and use of appropriate preheating and stress relief treatment, and (4) design of weldments to eliminate notches and to have relatively large cross-sectional areas so that unit stresses are reduced, as discussed in the article "Joint Design and Preparation" in this Volume.

Interstitial Contamination

Weldment mechanical properties can be degraded by the presence of excessive contamination. Three interstitial elements, oxygen, nitrogen, and carbon, are particularly harmful. Oxygen and nitrogen may be introduced into a weldment because of inadequate gas shielding and by surface films on the base metals and filler metals. Carbon may be introduced by hydrocarbon contamination from residual gaseous hydrocarbons in the shielding gas, oils on the components being welded or, if an inert atmosphere welding chamber is being used, from vacuum pump oils or the breakdown of elastomers in the chamber because of the effects of heat, vacuum, or ultraviolet light produced during welding.

Figure 1 shows the effect of varying amounts of oxygen, nitrogen, and carbon on the bend ductility transition temperature in molybdenum. The same trends exist in tungsten; however, the DBTT is elevated by approximately a factor of two (Ref 5). The addition of very small quantities of these elements greatly affects the transition temperature. Oxygen is particularly deleterious to ductility. A few parts per million (ppm) is enough to raise the ductile-to-brittle transition temperature from well below room temperature to well above it. An early study by Platte (Ref 6) reported that oxygen in the base metal or in

Fig. 1 Effect of oxygen, nitrogen, and carbon on the bend ductility transition temperature of molybdenum (Ref 4)

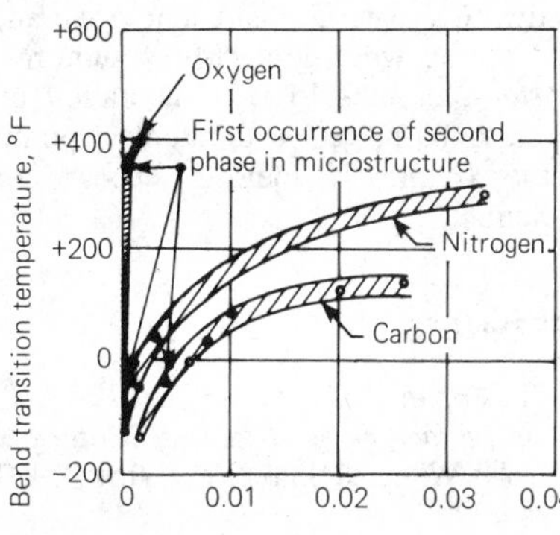

Table 1 The bend transition temperature of alloys welded in helium

Oxygen in helium, ppm	90° bend transition temperature, °F: Molybdenum-zirconium-boron	Molybdenum-aluminum-boron
0.1	10	−30
50	95	54
100	160	86
700	430	212
4000	>440	>440

the welding atmosphere could promote hot cracking and porosity in addition to reducing ductility. Nerodenko *et al.* (Ref 7) reported 90° bend transition temperatures of +10 °F and −30 °F for 0.040-in.-thick molybdenum-zirconium-boron and molybdenum-aluminum-boron alloys, respectively, when gas tungsten arc welded in helium containing 0.1 ppm oxygen. The bend transition temperature steadily increased as the oxygen content was raised, as shown in Table 1. Nerodenko *et al.* recommended that oxygen in the solidified weld metal, which is a function of both the amount transferred from the shielding atmosphere and the concentration in the base and filler metals, should be a maximum of 50 to 70 ppm. The use of specially purified shielding gas, argon, helium, or combinations of the two is necessary.

Surface Cleaning

Prior to welding, the surface of the base metal should be degreased and chemically cleaned. Two cleaning procedures for molybdenum-based metals are recommended. The first, devised by Ryan (Ref 8) and discussed by Thompson (Ref 9), is an alkaline-acid wash. The second, reported by Moorhead *et al.* (Ref 10), is a dual-acid wash. In an evaluation of both cleaning procedures, Moorhead stated that the alkaline-acid wash gave better results based upon the visual appearance of the parts. However, removal of smut deposited during cleaning in the alkaline solution was difficult, especially from internal surfaces. Equal weld quality was reported for both wash procedures. Thus, the choice of cleaning procedures depends on such factors as the amount of contamination of the component or the accessibility for smut removal (Ref 10). The two systems are:

Alkaline-acid procedure (Ref 9)

- Degrease with acetone
- Immerse for 5 to 10 min at 150 to 180 °F in an alkaline solution of 10 wt% sodium hydroxide, 5 wt% potassium permanganate, and 85 wt% distilled water
- Rinse with cold tap water, scrubbing off loose smut
- Immerse for 5 to 10 min at room temperature in a bath of 15 vol% concentrated sulfuric acid, 15 vol% concentrated hydrochloric acid, and 70 vol% distilled water plus 6 to 10 wt% chromic acid
- Rinse in cold tap water and allow to drain dry*

Dual-acid procedure (Ref 10)

- Degrease with acetone
- Immerse for 2 to 3 min at room temperature in a solution of 15 vol% hydrochloric acid (37 to 38% HCl), 15 vol% sulfuric acid (95 to 97% H_2SO_4), 15 vol% nitric acid (90% HNO_3), and 55 vol% distilled water
- Rinse in flowing tap water
- Immerse for 3 to 5 min at room temperature in a solution of 15 vol% sulfuric acid (95 to 97% H_2SO_4), 15 vol% hydrochloric acid (37 to 38% HCl), and 70 vol% distilled water plus 6 to 10 wt% chromium trioxide (CrO_3)
- Rinse in flowing tap water
- Rinse with distilled water
- Dry with forced hot air

Tungsten metals should be degreased and then acid cleaned using a mixture of 90 vol% concentrated nitric acid and 10 vol% hydrofluoric acid (Ref 5 and 11). Following cleaning, the parts should be handled only with clean cloth gloves and should be welded as soon as possible.

Stress Relieving

Molybdenum- and tungsten-based components are normally stress relieved after any processing operations that could contribute residual stresses. The appropriate stress relief temperature and time for the various molybdenum-based alloys can be obtained from the manufacturer of the particular alloy. Typical stress relief times and temperatures for several of the more common commercial materials are:

Material	Stress relief
Arc-cast molybdenum	1 h at 1650 to 1700 °F
Powder metallurgy (P/M) molybdenum	1 h at 1650 °F
TZM (Mo-0.5Ti-0.08Zr)	1 h at 2100 to 2150 °F

*Moorhead (Ref 10) modifies this step by rinsing with distilled water after the flowing tap water rinse and then drying with forced hot air.

A typical stress-relief treatment for P/M tungsten is 1 h at 2550 °F (Ref 11). Stress relieving should be performed in either a vacuum or a hydrogen furnace to prevent atmospheric contamination (Ref 12).

Welding Filler Metals

The need to use filler metals can be prompted by both design and mechanical considerations. In the first case, increasing the base-metal thickness or the distance between the pieces being welded may make filler metals necessary. In the second case, filler metals may be used to assist in filling surface pores that may occur while welding P/M materials, or to improve the weldment ductility or strength by alloying. Examples of the latter are the use of TZM molybdenum or molybdenum-rhenium filler alloys when welding molybdenum, or the use of tungsten-rhenium alloys when welding tungsten.

The TZM molybdenum alloy offers the advantage of titanium and zirconium additions for strengthening and weld-metal grain size control. The various molybdenum or tungsten and rhenium combinations (for example, molybdenum 10 to 50 wt% rhenium, tungsten 5 to 27 wt% rhenium) may be used to take advantage of the effect of rhenium on ductility. The addition of rhenium to molybdenum and tungsten can lower the DBTT of the weld metal to below room temperature (Ref 5 and 13-15).

Preheating and Welding Heat Input

As is the case with many metals, molybdenum- and tungsten-based alloys are susceptible to cracking because of thermal shock and residual stresses produced by welding. Preheating can reduce these tendencies. The specific welding preheat or interpass temperature to be used when welding an assembly depends on both its size and shape and the susceptibility to cracking of the particular alloy. Joining objects of widely differing size can produce uneven heating and cooling because of the ability of a larger object to dissipate heat more rapidly than a smaller object, producing stresses that may cause cracking. Preheating results in a more uniform thermal gradient that reduces the tendency toward cracking.

The current density variations between different welding processes can affect the amount of preheating required. For example, the current density of a gas tungsten arc welded plasma is relatively low. This can result in more preheating prior to obtaining a molten zone than occurs for an electron beam or laser weld in the same material. The benefits of preheating to reduce thermal shock must, however, be balanced with the use of as low a heat input as possible to minimize the extent of recrystallization and grain growth in the HAZ.

Gas Tungsten Arc Welding Procedures

Inert Atmosphere Welding Chamber. To ensure optimum mechanical properties, molybdenum- and tungsten-based metals should be welded in an inert atmosphere chamber. Welding in air, using only gas shielding from the welding torch, results in severe contamination of the weld, with resultant reduced ductility.

To minimize moisture permeation and outgassing from the chamber walls, a metal enclosure is superior to a plastic enclosure. The metal chamber should be capable of being evacuated to a pressure in the range of 10^{-3} to 10^{-6} mbar and should have a helium mass spectrometer leak rate of 1×10^{-8} atm cm^3/s, or better. The chamber should be evacuated and backfilled rather than simply purged with shielding gas to obtain the lowest residual atmospheric contamination.

The use of plastics and elastomeric tubing in evacuated areas should be kept to a minimum. Copper or stainless steel tubing should be used for all-rigid plumbing. Viton may be used for O-ring seals and electrical insulation of wiring inside the chamber. Where flexible water lines are required, an elastomer having low outgassing and moisture permeability, such as butyl rubber, is recommended. All valving should be suitable for vacuum exposure, preferably bellows-sealed valves. Tubing connections to valving or chamber parts may be made using swage-type fittings, although welded or silver soldered (not soft soldered) fittings with copper gaskets are best.

Port gloves should be made from butyl rubber to control moisture permeation. For stringent control of outgassing from these gloves, sulfur-free butyl rubber is recommended. Protective welding gloves, to cover port gloves during manual welding, should be made from aluminized fiberglass rather than leather.

To facilitate the elimination of adsorbed moisture from the chamber walls during evacuation, heat may be applied to the chamber by using heating tapes or infrared lamps. A bake-out temperature of 105 to 170 °F is adequate to reduce the residual moisture in the chamber to less than 1.0 ppm (dew point <-110 °F) in a vacuum of 10^{-4} mbar, or better. All surfaces exposed to the vacuum should be degreased and kept clean and dry to reduce contamination and pumping time. Monitoring the chamber atmosphere for oxygen and moisture is recommended. Oxygen and moisture levels of less than 10 ppm can be obtained from a system such as that described above. Welds produced in this atmosphere are shiny with no discoloration.

Welding Conditions

Gas tungsten arc welding (GTAW) is performed using direct current electrode negative (DCEN). Thoriated tungsten electrodes are used. Grinding the electrode to a sharp point facilitates arc placement and manipulation. A downslope of the welding current eliminates crater cracking. The high melting point and relatively high thermal diffusivity of molybdenum- and tungsten-based metals necessitate the use of slightly more current than is used for metals having a lower melting temperature. The electrode diameter is determined on the basis of the current required. Because welding is performed in an inert atmosphere, no additional torch shielding is required. The welding torch may be either water or air cooled. Water cooling permits the use of higher currents, but can introduce moisture contamination because of permeation through the walls of the elastomeric tubing. For very critical, small-scale welding, a ceramic-handled, noncooled torch can be fabricated. This design minimizes possible sources of contamination.

Automatic GTAW can introduce contaminants into the chamber shielding atmosphere from motor lubricants and insulation. The severity of this contamination should be determined on an individual basis depending on the intended service of the weldment. Because of the high melting temperature and thermal conductivity of molybdenum and tungsten, the welding fixtures used to hold the pieces during welding should be either water cooled or faced with a material that cannot melt and bond to the weldment. Molybdenum-, tungsten-, or nickel-based superalloy inserts may be placed at the contact points in the fixture if melting becomes a problem.

REFERENCES

1. Tetelman, A.S. and McEvily, A.J., Jr., *Fracture of Structural Materials*, John Wiley & Sons, New York, 1967, p 178
2. Semchyshen, M. and Harwood, J.J.,

Ed., *Refractory Metals and Alloys,* Metallurgical Society Conferences, Vol 11, Interscience Publishers, New York, 1961, p 54-63

3. Chewey, P.M., "Molybdenum Joining Techniques," Lockheed Missiles and Space Co. Inc., Project No. MSD 008, 1981
4. Olds, L.E. and Rengstorff, G.W.P., Effect of Oxygen, Nitrogen and Carbon on the Ductility of Cast Molybdenum, *Journal of Metals,* Vol 8 (No.1), Feb 1956, p 150
5. Yih, S.W.H. and Wong, C.T., *Tungsten,* Plenum Press, New York, 1979, p 275-277
6. Platte, W.N., Influence of Oxygen on Soundness and Ductility of Molybdenum Welds, *Welding Research Supplement,* Aug 1956, p 369s-381s
7. Nerodenko, M.M., Polishchuk, E.P., and Kovalenko, R.I., The Optimum Oxygen Content of the Shielding Atmosphere of Helium for Welding Molybdenum Alloys, *Automatic Welding,* Vol 32 (No. 4), 1979, p 35-37
8. "Ryan Devises Molybdenum Cleaner," Climax Molybdenum Co., Technical Bulletin, Oct 1957
9. Thompson, E.G., "Welding of Reactive and Refractory Metals," Welding Research Council Bulletin 85, Feb 1963
10. Moorhead, A.J., DiStefano, J.R., and McDonald, R.E., Fabrication Procedures for Unalloyed Molybdenum, *Nuclear Technology,* Vol 24, Oct 1974, p 50-63
11. Mullendore, J.A., Sylvania Chemical and Metallurgical Div., GTE Products Corp., private communication, June 1982
12. Semchyshen, M. and Barr, R.Q., "Mechanical Properties of Molybdenum and Molybdenum-Base Alloy Sheet," ASTM STP 272, 1960, p 12-35
13. Stephans, J.R. and Witzke, W.R., "Alloy Softening in Group VI A Metals Alloyed with Rhenium," NASA TN D-7000, National Aeronautics and Space Administration, Washington, D.C., Nov 1970
14. Savitskii, E.M., Tylkina, M.A., and Konieva, L.Z., Mechanism of the Rhenium Effect, *Study and Uses of Rhenium Alloys,* Savitskii, E.M. and Tylkina, M.A., Ed., Amerind Publishing Co., Ltd., New Delhi, India, 1978
15. Savitskii, E.M., Povarova, K.B., and Makarov, N.V., Tungsten-Base Alloys, *Study and Uses of Rhenium Alloys,* Savitskii, E.M. and Tylkina, M.A., Ed., Amerind Publishing Co., Ltd., New Delhi, India, 1978

Resistance Welding

Resistance Spot Welding*

RESISTANCE SPOT WELDING (RSW) is a process in which faying surfaces are joined in one or more spots by the heat generated by resistance to the flow of electric current through workpieces that are held together under force by electrodes. The contacting surfaces in the region of current concentration are heated by a short-time pulse of low-voltage, high-amperage current to form a fused nugget of weld metal. When the flow of current ceases, the electrode force is maintained while the weld metal rapidly cools and solidifies. The electrodes are retracted after each weld, which usually is completed in a fraction of a second.

The size and shape of the individually formed welds are limited primarily by the size and contour of the electrode faces. The weld nugget forms at the faying surfaces (Fig. 1), but does not extend completely to the outer surfaces. In a cross section, the nugget in a properly formed spot weld is obround or oval in shape; in plan view, it has the same shape as the electrode face (which usually is round) and approximately the same size. The spots should be at a sufficient distance from the edge of the workpiece (edge distance) so that there is enough base metal to withstand the electrode force and to ensure that the local distortion during welding does not allow expulsion of metal from the weld. Also, spacing between adjacent spot welds or rows of spot welds must be enough to prevent shunting or to limit it to an acceptable amount.

Applications

Spot welded lap joints are widely used in joining sheet steel up to about $^1/_8$ in. thick and are used occasionally in joining steel $^1/_4$ in. or more in thickness. Thicknesses of 1 in. or more have been joined by spot welding, but this requires special equipment and would not ordinarily be economical. Many assemblies of two or more sheet metal stampings that do not require gastight or liquid-tight joints can be more economically joined by high-speed RSW than by mechanical methods. Containers frequently are spot welded. The attachment of braces, brackets, pads, or clips to formed sheet metal parts such as cases, covers, bases, or trays is another common application of spot welding.

Fig. 1 Resistance spot welding setup. Sectioned to show shape of nugget and position of nugget relative to inner and outer surfaces of workpieces

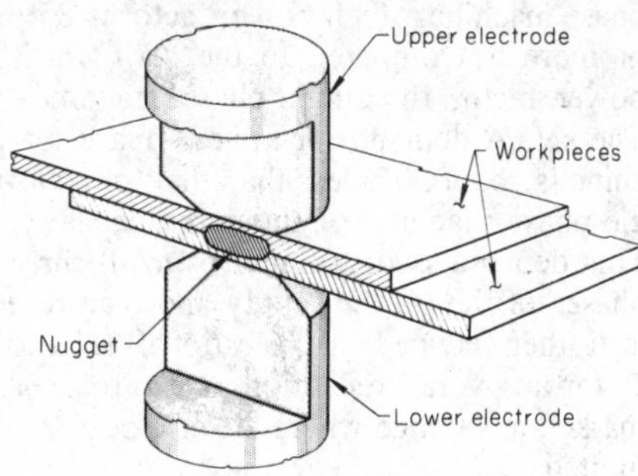

Major advantages of RSW are high speed and suitability for automation and inclusion in high-production assembly lines with other fabricating operations. With automatic control of current, timing, and electrode force, sound spot welds can be produced consistently at high production rates and low unit labor costs by semiskilled operators.

However, a resistance spot weld in steel, which typically draws a current of about 5000 to 20 000 A at about 5 to 20 V, imposes a heavy kilovolt ampere (kV · A) demand. The kV · A demand is even higher for the RSW of more conductive metals, such as many aluminum and copper alloys. Even though the duration of current flow for a single weld is only a few cycles, the corresponding power demand may be an undesirable load, especially if power is drawn from only one phase of a three-phase supply. Also, the initial cost of equipment is generally much higher for resistance welding than for most arc welding processes.

Although the most common application of RSW is the joining of two sheets of metal having the same composition and thickness, the process is also used to join more than two sheets of metal, to join metals that are dissimilar in thickness or composition (or both), and to join steel coated with another metal.

Equipment

Nearly all RSW of low-carbon steel is direct energy welding, in which single- or three-phase 60-cycle alternating current, drawn ordinarily from 220- or 440-V in-plant power lines and stepped down to about 2 to 20 V, is fed directly to the electrodes as each weld is made. The equipment needed for RSW may be simple and inexpensive, or complex and costly, depending on the degree of automation. Machines for direct energy welding generally are composed of these principal elements:

- *Electrical circuit*: Consists of a welding transformer, a tap switch, and a secondary circuit. The secondary circuit includes the electrodes that conduct the welding current to the workpieces.
- *Control equipment*: Initiates and times the duration of current flow and may also be used instead of (or in addition to) the transformer tap switch to regulate the welding current. Controls may also sequence, time, and regulate the overall operation of the machine, including initiation, automatic adjustment, and termination of welding force and current. Feedback circuits monitor the development of a spot weld by sensing one or more of the following: temperature, expansion, contraction, acoustic emissions, ultrasonic signals, voltage, current, power, resistance, and energy. The feedback circuits relay this information into a microprocessor which controls and adjusts weld parameters during each nugget formation.
- *Mechanical system*: Consists of the frame, fixtures, and other devices that

*Revised by David W. Dickinson, Supervisor, Welding and Flat Rolled Research & Development, Republic Steel Corp.

hold and clamp the workpieces and apply the welding force.

Specifications for resistance welding equipment have been standardized by the Resistance Welder Manufacturers Association (RWMA) and for controls by the National Electrical Manufacturers Association (NEMA).

Single-Phase and Three-Phase Direct Energy Welding Machines

The choice between single-phase and three-phase direct energy machines for RSW is based mainly on machine capability and on initial, operating, and maintenance costs. Power factor and load balance among the three phases of the power supply should also be considered, particularly when a high-capacity machine is needed in a plant where the supply of electric power is limited.

Single-phase machines are more widely used than three-phase machines because they are simpler to operate and cost less to buy, install, and maintain. When equipped with suitable controls, however, a single-phase machine has the same performance capabilities as a three-phase machine of the same size and rating.

The two disadvantages of single-phase machines are low power factor, about 40 to 50%, and high kV · A demand (about double that for three-phase machines of the same capacity). Because the demand time for a resistance weld is very short, the amount of electric power consumed in the welding may be insignificant. Nevertheless, if the welding machine load is an appreciable part of the total electrical load in a plant, fixed charges for standby service, connected load, and power factor may be quite high when single-phase machines are used, even though they draw power from only one phase of the three-phase supply. If the welding machine load is a small part of the total plant electrical load or is used in industrial areas where ample electric power is available, these fixed charges usually are minor.

The low power factor and high kV · A demand of single-phase machines are outweighed in most situations by their cost advantages. In plants where many single-phase machines are in use, they can be connected so that the welding current load is distributed among the three phases of the power supply to reduce the kV · A demand.

Three-phase machines, which take electric power from all three phases of the power supply line, are of two general types: frequency converter and rectifier. They differ from single-phase machines only in electrical construction. The transformer of a single-phase machine has only one primary winding. The frequency converter three-phase machine has a transformer with three primary windings. The rectifier type has a three-phase-to-three-phase transformer, the secondary circuit of which supplies current to low-voltage rectifiers. These deliver high-amperage, low-voltage direct current to the electrodes.

Three-phase machines have advantages over single-phase machines in relation to power factor and maximum kV · A demand. Because of the decreased frequency of the secondary current of three-phase machines, their power factor is 85% or more as compared to the 40 to 50% power factor for single-phase machines. The kV · A demand for a three-phase machine is about 50% less than that for a single-phase machine of the same capacity. This demand is distributed over all three phases of the power supply and therefore is further reduced in the ratio of 1.73 to 1, for an overall reduction of $3^1/_2$ to 1—a marked advantage where electric power is limited.

The additional cost of three-phase machines may not be justified where power supply is adequate, or where a large number of machines is involved. Also, a three-phase machine is not generally suitable for welding of thin metal or where weld times of three cycles or less, with 60-cycle current, are required. Three-phase transformers generally have not been applied to multiple-electrode or multiple-transformer machines, because the necessary phase balance can be attained through distribution of single-phase welding transformers over the three phases of the power supply.

Controls for Direct Energy Machines

Electrical controls for direct energy resistance welding machines perform three principal functions: (1) initiating and terminating the flow of current to the welding transformer, (2) controlling the magnitude of the current, and (3) timing and controlling the mechanical operations of the welding machine. The controls fall into three groups: welding contactors, timing and sequencing controls, and other current controls and regulators.

Welding contactors are devices for making and breaking an electric power circuit. On resistance welding machines, contactors and other controls are applied to the primary circuit of the welding transformer. A welding contactor should be large enough to handle the maximum input from the electric power line to the machine with the tap switch at the highest position. Three types of contactors are used on resistance welding machines: mechanical, magnetic, and electronic.

Mechanical contactors are of either the single-pole or the double-pole type and are operated by a foot pedal or a motor-driven cam. In foot-pedal operation, movement of the pedal first applies the squeeze pressure and then closes the electrical contacts. Further movement of the pedal opens the contacts and at the same time increases the welding force.

A magnetic contactor uses an electromagnet for closing the electrical contacts. When the magnet is de-energized, the contacts are opened by gravity and spring pressure. Single-pole, double-pole, and synchronously interrupting types of magnetic contactors are available. The synchronously interrupting type opens the power circuit when the alternating current wave approaches zero and thus reduces arcing, for longer electrode tip life and less marking on the work.

Electronic contactors use ignitron tubes, thyratron tubes, or silicon-controlled rectifiers to control the flow of current to the primary winding of the welding transformer. Ignitron tubes are used for applications requiring an extremely high welding machine current or a very large number of welding operations per minute. Thyratron tubes or silicon-controlled rectifiers are used to control currents that are too low for ignitron tubes to handle (less than 40 A). Electronic contactors, either synchronous or nonsynchronous, open the circuit when the current wave passes through zero.

An ignitron contactor consists of two ignitron tubes connected in inverse parallel so that one tube carries the positive half cycle of the welding current and the other tube carries the negative half cycle. Figure 2 shows the power-supply circuit and control circuit of an electronic contactor with two ignitron tubes. Semiconductor-type rectifiers are used to allow current flow from the ignitor to the mercury pool cathode only. A control circuit fuse provides protection of the ignitrons and isolation of the control circuit from the power circuit. In machines that are provided with electronic heat control, these rectifiers are replaced by thyratron tubes or by silicon-controlled rectifiers.

Most ignitron tubes are water cooled and have a stainless steel water jacket to prevent corrosion. A thermostatic switch (not shown in Fig. 2) is mounted on one of the

Fig. 2 Power-supply circuit and control circuit of an electronic contactor

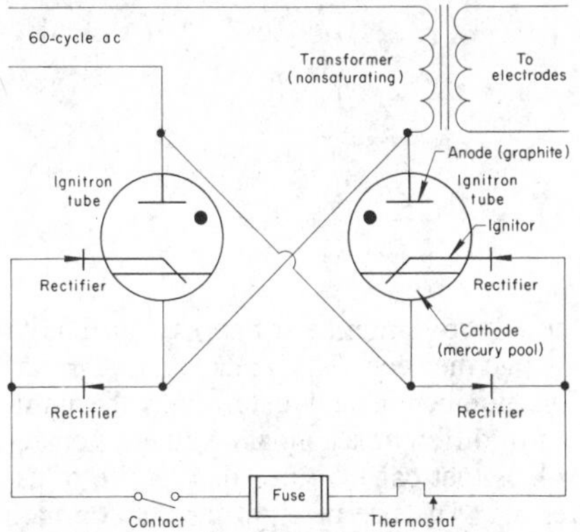

tubes to stop operation if the temperature of the tube becomes too high. The other ignitron tube has a thermostatic switch that controls a solenoid-operated water valve, which conserves cooling water.

The contactor energizes the welding circuit when the control circuit is closed. The voltage for the control circuit is obtained from the power-supply circuit. Each ignitron tube is fired for each conducting half cycle by an ignitor that is energized through the control circuit whenever the control circuit is closed by the timer.

When the duty cycle required of an electronic contactor exceeds the rating of the ignitron tubes, two contactors may be used with an auxiliary control to switch the load alternately from one to the other. When two contactors are operating in this manner, the permissible duty cycle and the averaging time are both doubled. The peak load cannot be exceeded, however, because the entire load at any instant is carried by one contactor only.

Time and Sequence Controls. The duration of current flow to the welding machine is controlled by a welding timer. The overall operating cycle of the welding machine is timed and controlled by a sequence control or timer. Timing elements are of two types: nonsynchronous and synchronous.

A nonsynchronous timer, by NEMA standards, may start and stop the flow of welding current at random points with respect to the line-current wave form. Variations of timing and of current input to the machine result from closing and opening the welding contactor at random points on the wave form. The time variable is at least plus or minus one half cycle and sometimes more. Ordinarily, nonsynchronous timing is sufficiently accurate for weld times of 20 cycles or longer, because the percentage variation is low and can usually be neglected. The nature and quality of the work being done determine the type of timing required for weld times of 10 to 20 cycles. Nonsynchronous timing is not recommended for weld times that are shorter than 10 cycles.

A synchronous timer provides a more accurate timing period and closes the primary circuit of the welding transformer at the same point (electrical angle) with respect to the power circuit voltage in making each weld. Thus, the current wave form is consistent, and the energy delivered to the welding transformer is the same for consecutive operations. Besides providing accuracy and reproducibility in these respects, a synchronous timer also eliminates variation in initial current caused by load transients. Load transients result from failure to initiate the flow of current at the power factor angle in the highly inductive secondary circuit of the welding transformer.

A synchronous timer can be used in any application for which a nonsynchronous timer is suitable, but it is more expensive and, to be effective, must be accompanied by heat control. Sequence timers are used to control four basic steps for spot welding cycles:

1 Close electrodes and apply force
2 Initiate and maintain welding current
3 Turn off welding current and maintain electrode force until weld nugget solidifies
4 Open electrodes

The duration of step 1 is referred to as squeeze time; of step 2, weld time; of step 3, hold time; and of step 4, off time. Off time is provided by means of a two-position selector switch. One position of the selector switch permits the machine to continue cycling as long as the initiating switch is closed; the other position permits one complete sequence only, after which the machine cannot be restarted until the initiating switch is opened and reclosed.

Heat Control. The tap switch on the primary circuit of the welding transformer is used to change the ratio of transformer turns for major adjustment of the welding current. When an intermediate setting is needed, or for fine adjustment of the welding current, electronic heat control is used. Electronic heat control is standard equipment when synchronous timing controls are used, and it can be added to nonsynchronous controls.

For electronic heat control, the semiconductor-type rectifiers in an ignitron contactor (see Fig. 2) are replaced as firing devices by thyratrons or silicon-controlled rectifiers. These tubes or rectifiers control firing of the ignitrons during each half cycle of the welding current. Firing of the ignitron can be delayed by delaying application of a firing signal to the thyratron grid. This delay is usually controlled by a heat adjustment dial. As the firing of the applied voltage wave is retarded, the root mean square (rms) current is reduced. The reduction in heat or energy varies as the square of the applied current, in amperes.

Because ignitron contactors require a certain minimum voltage and current to fire properly, complete control from 100% to zero is not feasible. It is necessary to limit the minimum value to 40% rms current for 220-V equipment and to 20% rms current for equipment operating on 440 V or more. Heat control switches are calibrated as a percentage of the increment in primary voltage between tap settings.

Automatically controlled phase-shift heat control circuitry forms the basis of all accessories that change the level of welding current during a welding sequence. Therefore, use of current and voltage regulators, upslope and downslope controls, and quench and temper controls requires that the basic welding machine control be equipped with a heat control unit. Microprocessor feedback control units have recently been built which can monitor the condition of weld development and adjust phase-shift heat control during the weld cycle to meet a preprogrammed weld nugget growth.

To minimize variations in welding current, heat control should be operated as near full heat as possible. Where a welding current of 30% of the maximum rms current is being used, only a few degrees of delay angle produces a 10% change in welding current. Variations in line voltage can distort the sinusoidal line voltage sufficiently to produce another 10% change at such a low value. Controllers with automatic line voltage compensation are available when this becomes a problem. Therefore, a tap switch should be used to change the ratio of transformer turns for major changes in current magnitude, and heat control should be used only for fine adjustment of current magnitude.

Power demand is always greater when heat control is used to adjust the magnitude of the welding current. If heat control is used, the kV · A demand, in relation to maximum, generally follows a linear relationship with current. If the ratio of transformer turns is changed by a tap switch, the kV · A demand varies as the square of the secondary current value. For example, if the welding current is adjusted to 80% of maximum by heat control, the kV · A demand is also 80% of maximum. If the secondary voltage is reduced to 80% of maximum by changing the ratio of transformer turns through a tap switch, the kV · A demand is about 64% of maximum or 20% less than when using heat control.

Upslope control is used for starting the welding current at a low value and controlling its rate of buildup to full welding current. The use of upslope control in RSW is discussed in the section on the welding cycle in this article, and its application to the single-impulse welding cycle is shown in Fig. 18(b).

Downslope control is used for decreasing the value of the welding current at the end of the welding cycle from the maximum value to a lower value. The use of downslope control in RSW is also discussed in the section on the welding cycle, and its application is shown in Fig. 18(b).

Current and Voltage Regulators. Electronic current regulators are used to maintain a constant welding current under severe conditions by compensating for line voltage variation or impedance changes resulting from insertion of magnetic materials into the throat of the welding machine. Current regulators provide feedback control of the phase-shift network, based on the primary current.

Voltage regulators designed for use in ordinary welding operations can maintain the welding current within ±2% of its preset operating value for any combination of load-circuit changes that would normally cause unregulated current to change ±20%. The response time of these devices is about 3 cycles at best, and it is recommended that the weld time be a minimum of 6 cycles to allow time for regulation of heat. Voltage across the welding electrodes can be maintained at a constant preset value by a device that compares the actual electrode voltage with a preset reference voltage and makes automatic corrections in the phase-shift heat control.

Shunting decreases the current passing through the metal between the electrodes by introducing another path for the welding current, which causes a drop in the welding voltage. A voltage regulator automatically increases the welding voltage when shunting occurs, thus maintaining the correct current level at the weld.

Large changes in the secondary-current requirements cannot be compensated for by voltage or current regulators, and controls must be reset to provide and maintain suitable welding current at the workpieces. Special in-process feedback control units that automatically adjust the current (phase-shift type) or the timing are sometimes used to compensate for (1) different work-metal thicknesses or number of pieces joined, (2) changes in contour or condition of electrode tips, and (3) changes in work, equipment, or ambient conditions that affect welding results.

Equipment for Direct Energy Machines

The electrical system of a direct energy resistance welding machine is shown in Fig. 3. The electrodes are considered here from their electrical standpoint; metallurgical and welding characteristics of electrodes are discussed in subsequent sections in this article. Controls in the primary circuit are not part of the machine; these are discussed in the preceding section on controls for direct energy machines.

Welding Transformer. The transformer used in a direct energy resistance welding machine changes the alternating current from high-voltage, low-amperage current in the primary winding, or coil, to low-voltage, high-amperage current in the secondary winding. The primary winding is connected to the power supply and is made from edge-bent strip copper, insulated between turns with the entire coil also thoroughly insulated.

There are three principal arrangements of transformer windings: multistep, series-parallel, and a combination consisting of multistep and series-parallel. The multistep winding consists of one primary coil with one or more intermediate taps to vary the effective number of turns and thus to change the secondary voltage and current. The number of taps determines the number of different secondary-voltage steps or values that can be furnished by the transformer. The use of eight taps is common and is covered in RWMA standards.

Fig. 3 Electrical system of direct energy resistance welding machine

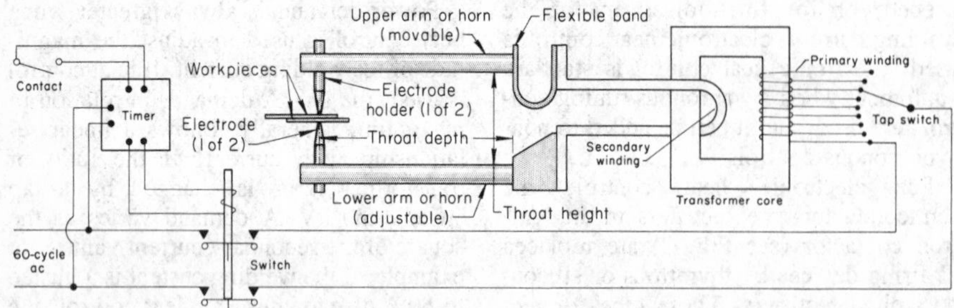

Fig. 4 Connection diagram for primary winding of a series-parallel transformer

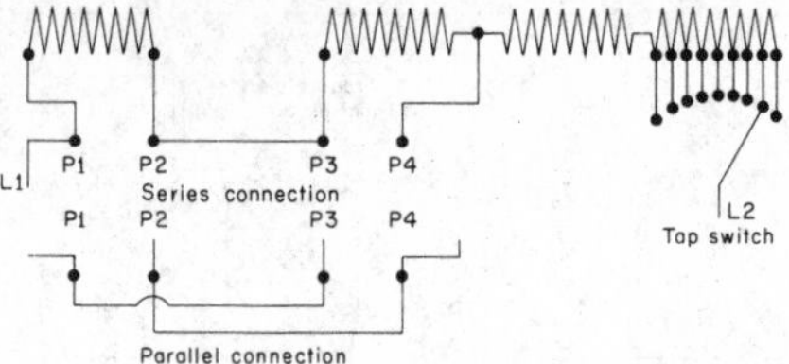

The series-parallel winding consists of two primary coils that can be connected in series or in parallel. There are no intermediate taps on the coils, and thus this winding provides only two values of secondary voltage. A multistep series-parallel winding results when taps are added to a series-parallel winding. This type of winding provides two different secondary voltages for each transformer tap. Thus, a transformer with a series-parallel winding with 8 taps produces 16 secondary-voltage values.

The primary taps are numbered progressively, with the highest open circuit secondary voltage associated with the highest numbered tap. In a resistance welding transformer having a series-parallel primary winding, the terminal identification is P1, P2, P3, and P4, as shown in Fig. 4.

The secondary winding must carry high currents and therefore has a large cross section. Except in low-capacity equipment, the secondary winding is cooled by circulating water. The primary winding is cooled by conduction to the water-cooled secondary winding.

Transformer ratings for resistance welding machines are expressed in kV · A for a specified duty cycle. Standard practice is to rate resistance welding transformers on a 50% duty cycle. This duty cycle rating is a thermal rating and states the amount of power the transformer can deliver for a stated percentage of a time period, usually 1 min, without exceeding a specified temperature rise. Thus, a welding machine rated at 100 kV · A for a 50% duty cycle can deliver 100 kV · A for 30 s of each minute without the transformer components reaching a temperature greater than that for which they were designed. For a repetitive load, one "on" time and one "off" time are considered to be an integrating period, which must not be in excess of 1 min. The kV · A demand rating of a welding transformer is the secondary open circuit voltage multiplied by the secondary (or welding) current, divided by 1000.

The maximum permissible kV · A demand for a transformer used at a duty cycle other than the one for which the transformer is rated can be calculated from the factors listed in Table 1. The maximum kV · A demand is found by multiplying the rated kV · A by the factor for the required duty cycle. For example, a 100-kV · A transformer may be operated at a 25% duty cycle (15 s of each minute) at a maximum demand of 141 kV · A (100 × 1.41). The rated kV · A needed to supply a maximum demand of 90 kV · A for a duty cycle of 15% (rating factor 1.82) is 90/1.82, or 50 kV · A.

Tap switches are switching devices for connecting the various taps on the primary winding of the transformer to the power-supply line. The tap setting determines the number of effective turns in the primary winding and thus the secondary voltage and current. The switches are usually of the rotary dead-front type and are arranged for flush mounting in openings in the machine frame or directly on the transformer. Most of the switch handles have locking buttons so that the contacts are centered for each operating position. Some switches have an "off" position, which acts as a disconnect. On very large transformers, spring contact switches are inadequate, and bolted bar jumpers are used.

Tap switches should not be operated with the current flowing, because the resulting arc-over between contacts eventually causes damage to the contact surfaces. The secondary voltage may also be controlled by electronic phase-shift devices. These are discussed in the preceding section on controls.

Characteristics of the Direct Energy Secondary Circuit

The secondary circuit carries the welding current from the transformer to the electrodes, and thus to the workpiece (Fig. 3). The dimensions of the secondary circuit, or secondary loop, determine to a large extent the output performance of the welding machine. The secondary loop is roughly the area defined by the throat height times the throat depth (see Fig. 3). Resistance and reactance in the secondary circuit at all points except at the weld should be kept as low as practical, to obtain sufficient welding current with minimum secondary voltage at the lowest possible kV · A demand on the alternating current power line.

Table 1 Rating factors for resistance welding machines (RWMA)

Duty cycle, %	Rating factor	Duty cycle, %	Rating factor
1	7.08	30	1.29
2	5.00	35	1.195
3	4.07	40	1.115
5	3.15	50	1.000
7.5	2.57	60	0.912
10	2.23	70	0.843
15	1.82	80	0.787
20	1.58	90	0.745
25	1.41	100	0.707

The materials used in the components of the secondary circuit are selected for low electrical resistivity, and conductors should be large in cross section but no longer than is necessary. Throat depth and height should be no greater than those needed for the work to be handled.

The relation between throat depth and secondary current (welding current) for spot welding small assemblies of low-carbon steel sheet that do not extend into the throat has been studied for commercial spot welding machines. Some typical results are that at:

- 50 kV · A power input to the transformer, the welding current when the throat depth was 42 in. was about half that with a throat depth of 12 in.
- 75 kV · A, the welding current with a throat depth of 42 in. was about 70% of that with a throat depth of 12 in.

Considering the results of the study from the viewpoint of the kV · A demand on the alternating current power line for a given welding operation, one third more power was needed to produce a given secondary current when the throat depth was 42 in. than when it was 18 in. Exact results would vary somewhat, depending on the machine, the work, and the welding parameters.

Magnetic material interposed between the arms or horns of the machine reduces the secondary current in relation to the size and thickness of the material; consequently, the extension of steel work metal or fixtures into the secondary loop should be kept to a minimum. An increase in the frequency of the alternating current also reduces the secondary current for a given kV · A demand.

Machine Construction

On the basis of mechanical construction, there are four basic types of RSW machines: rocker arm, press, portable, and multiple-electrode machines.

Rocker arm machines are the simplest stationary spot welding machines and are made for use with foot lever, air cylinder, or motor operation. The machine, regardless of method of operation, consists of a frame that houses the transformer and tap switch, a vertically adjustable lower horizontal horn, and an upper horn mounted in a rocker arm that is pivoted at the front top edge of the frame.

Rocker arm machines are available in throat depths of 12 to 48 in. and in transformer capacities of 10 to 300 kV · A. Because the upper electrode of a rocker arm machine moves in an arc, the electrodes should be set in the closed (or welding) position so that the upper electrode is perpendicular to the workpiece, and the two horns are parallel. Settings with the horns out of parallel can result in electrode skidding and marking of workpieces.

A foot-operated rocker arm machine consists of two simple levers connected by a rod and a compression spring. Force exerted on a foot lever is transmitted through the spring to the rocker arm lever and then to the welding electrode. This machine is best suited for job shop work with short production runs.

An air-operated machine has an air cylinder that replaces the foot lever, connecting rod, and spring of a foot-operated machine. The stroke of the air cylinder must be proportioned to the needed electrode opening, and the diameter of the cylinder must be proportioned to the needed electrode force and throat depth. For any given cylinder diameter, stroke, and operating air pressure, welding force decreases and electrode opening increases as the throat depth of the machine increases, provided the distance between the rocker arm pivot and cylinder connection is not changed. The welding force provided by any given cylinder is in direct proportion to the air pressure and is controlled by a pressure regulator. Air-operated spot welding machines are best suited for short or medium production runs where minimum setup time is needed.

Motor-operated machines are similar to foot-operated machines except that the rocker arm is operated by a power-driven cam instead of by a foot lever. The machine must not be started with the spring solidly compressed because this can cause the motor to stall. The electrode opening is determined by the rise of the welding cam and by the throat depth. Welding force depends on the amount of spring compression and the ratios of leverage involved.

Motor-operated machines are usually

more difficult to set up and adjust than foot- or air-operated machines. These machines are best suited for long production runs or for use where compressed air is at a premium or unavailable.

Press-type machines have an upper electrode and welding head that move vertically in a straight line. The welding head is guided in bearings or ways of sufficient proportions to withstand the offset loads put on them. These machines have throat depths up to 48 in. and transformer capacities from 5 to 600 kV · A and greater. Some bench-type models used for radio, instrument, dental, and jewelry work have throat depths of only a few inches, and may be rated at less than 5 kV · A. Welding force is provided by air or hydraulic cylinders; manual operation is used on small bench-type machines.

Air cylinders are generally used in machines that have welding transformers with capacities up to 300 kV · A. Hydraulic operation is seldom used in machines with capacities of less than 200 kV · A, but is used in practically all machines with transformer capacities greater than 500 kV · A.

Press-type welding machines generally are designed for both spot and projection welding. Tables or platens are provided for mounting dies for projection welding and fixtures and electrode holders for spot welding. The platens, ram, and air or hydraulic cylinder have a common centerline. On standard machines, the spot welding electrodes are mounted 6 in. in front of this centerline. The knee supporting the lower platen can be vertically adjusted to compensate for reasonable variations in thickness of projection welding dies or in length of spot welding electrodes. For some spot welding applications, the knee can be replaced with an arm or horn to obtain clearance for the workpiece.

Portable machines or guns are used in spot welding when it is impractical or inconvenient to bring the work to the machine and usually consist of:

- Portable welding gun
- Electrical controls (welding contactor and sequence timer)
- Welding transformer
- Secondary cables and hose needed to carry power between the transformer and the welding gun

The portable welding gun consists of water-cooled electrode holders, an air or hydraulic actuating cylinder, hand grips, and an initiating switch. The gun usually is suspended from an adjustable balancing unit. Welding force is supplied by the air or hydraulic cylinder. Hydraulic pressure usually is supplied by an air-hydraulic booster. Because of the high secondary losses of portable machines, transformers used in these machines have secondary voltage two to four times as great as the voltage of transformers used in stationary machines of equal rating.

The transformer, tap switches, electrical controls, and air-line accessories are mounted on one end of a beam balancer above the work area. On the other end of the beam is a spring balancer that counterbalances the welding gun. The gun has a capacity for vertical movement equal to that of the balancer and equal to or greater than the reach of the operator. The beam can be rotated 360° and provides an operating area dependent only on the length of the secondary cables.

Welding current is transmitted between the transformer terminals and gun terminals through a secondary cable, usually of the low-impedance or kickless type. The reactance of this type of cable is near zero, which results in a high power factor and reduced kV · A demand. However, water cooling is necessary. Included with the secondary electric cables are air or hydraulic pressure hoses, cooling-water hoses, and cable to the initiating switch. This switch is usually operated at low voltage, for reasons of safety.

Current density in low-reactance water-cooled cables is as high as 50 000 A, 1 000 000 circular mils. Cable sizes are in American Wire Gage (AWG) or thousand circular mils (MCM). Some types of portable RSW machines combine the gun and transformer in one portable unit, but these have lower capacity than the separate units described.

Many large unassembled sheet metal components without reinforcement are awkward to handle, position, and clamp for welding because of their bulk and lack of rigidity and may require more than one person to manipulate them. When production volume is high enough to justify the capital investment, it may be economical to adapt a portable gun to a special welding machine to allow efficient handling by one person.

Example 1. Use of a Portable Spot Welding Gun to Improve Weld Quality and Reduce Labor Requirements. The tractor fender assembly illustrated in Fig. 5(a) was made by spot welding a 2-by-5-ft fender top to a 3-by-5-ft side panel. The lap joint was approximately 5 ft long and required about 60 spot welds spaced 1 in. apart. A section through the spot welded joint is shown in Fig. 5(b).

Originally, the two parts were pinned together through alignment holes in each part and were welded in a standard press-type spot welding machine. Two people were needed to handle the pinned assembly, and production rate was eight fenders per hour. Weld quality was impaired by warping of the formed sheet metal fender top during welding. The resulting mismatch and opening of the joint caused changes in the faying surface contact area and therefore in the electrical resistance of the workpieces. These changes affected the dimensions of the weld nugget and the strength of the welded joint and could not be tolerated.

To avoid these difficulties and to allow the job to be performed by one person, a special welding machine was built that incorporated a portable spot welding gun (Fig. 5c). A flexible mounting allowed the gun to be moved to any position along the weld line, following a guide (not shown in Fig. 5) attached to the clamping arm.

In operation, insulated pins were passed through the lower electrode and into positioning holes in the side panel and fender top to align the lap joint. The clamping arm was then lowered and held down by an air cylinder, and the welding gun was moved by hand progressively into position for each spot weld along the joint. Once placed in position, the welding gun was triggered, and the spot weld was completed automatically. To minimize changes in impedance due to variations in the length of the lower electrode as the welding gun was moved along the joint, the lower electrode was a casting (Fig. 5d) that provided support for the parts along the weld line and a path of fairly uniform resistance for the electrical return to the transformer.

Using this welding machine, one person placed the side panel and fender top on the clamping fixture, raised the locating pins, activated the clamping arm, and made the 60 spot welds along the joint. After being welded and unclamped, the fender assembly was removed by the same operator.

Nondestructive testing of the spot welds was done by attempting to pull the end welds apart manually, and the die side (outer side) of the welded assembly was inspected visually. In addition, test coupons were made and tested to destruction to ensure that the welding machine settings had not changed. The test coupons used for destructive testing were random samples obtained during production. The aluminum clamping arms had little effect on the reactance of the machine. Also, the amount of steel in the throat of the machine did not vary enough to affect the resistance or the power factor. These conditions helped in maintaining consistent

Fig. 5 Use of a portable spot welding gun to improve weld quality and production rate

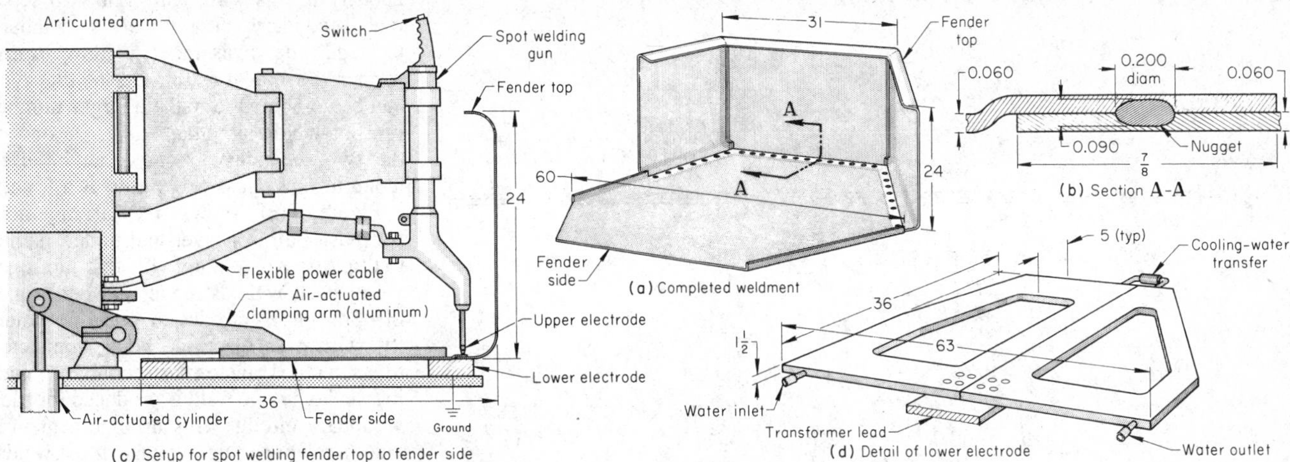

Equipment details		Welding conditions	
Power supply	220-V, 3-phase frequency converter	Heat control setting	70%
Welding gun	Heavy duty, portable, mounted on an articulated arm	Electrode force	500 lb max
Rating at 50% duty cycle	75 kV·A	Squeeze time	15 cycles
Upper electrode	Adaptor shank with No. 5 taper, Type A (pointed) cap, RWMA class 2	Weld time, approx	72 cycles (5 or 6 low-frequency 12-cycle pulses)
Lower electrode	Casting, RWMA class 2	Hold time	15 cycles
Welding controls	Synchronous electronic with phase-shift heat control	Production rate(a)	
		Fenders(b)	15/h
		Welds(b)	900/h

(a) With one operator; previous method required two. (b) Production rate in previous method had been 8 fenders (480 welds) per hour.

weld quality. Production was increased from 8 fenders (480 welds/h) to 15 fenders (900 welds/h). Equipment details and welding conditions are given in the table that accompanies Fig. 5.

Multiple-electrode machines are considered special-purpose machines, usually designed and built for a specific job. These machines are used when there are so many assemblies to be welded or so many welds per assembly that, even though the machines have a high initial cost, it is more economical to use them than to make the spot welds one at a time. Most of these machines are of the multiple-transformer type, in which each welding gun or pair of guns is connected to an individual transformer and all welds can be made simultaneously, as well as sequentially. Transformers built for such service are provided with two secondaries, enabling each transformer to supply two guns. Often two welds can be made in series, allowing four welds for each transformer.

Guns with 2-in.-diam tandem air cylinders are used for spot welding 0.025-in.-thick cold rolled steel. Tandem cylinders 4 in. in diameter can weld steel up to 0.125 in. thick. Spacing of spot welds is governed by the cylinder diameter. Guns with small-diameter, high-pressure hydraulic cylinders can be used for thicker metal.

Roll spot welding machines are essentially the same as resistance seam welding machines, which are described in the article "Resistance Seam Welding" in this Volume.

Direct and Series (Indirect) Welding

The arrangement of the electrodes and workpieces in the secondary circuit in RSW determines whether welding is direct or series (indirect). Welding is direct if all of the secondary current passes through the weld nugget, or nuggets, being formed, so that no effective shunt path bypasses the nugget. Welding is series if there is a shunt path that allows a portion of the secondary current to bypass the weld nugget. Some arrangements of secondary circuits for direct and series welding of two thicknesses of metal are described below; in general, these arrangements also apply for welding of three or more thicknesses.

Direct Single-Spot Welding. Single spot welds are usually made by direct welding. Figure 6 shows schematically three arrangements used for making this type of weld; these arrangements may be modified to meet special requirements. In all of the arrangements shown, one transformer secondary circuit makes one spot weld.

The simplest and most common arrangement, in which two workpieces are sandwiched between opposing upper and lower electrodes, is shown in Fig. 6(a). In

Fig. 6 Setup of work metal and electrodes for making single spot welds

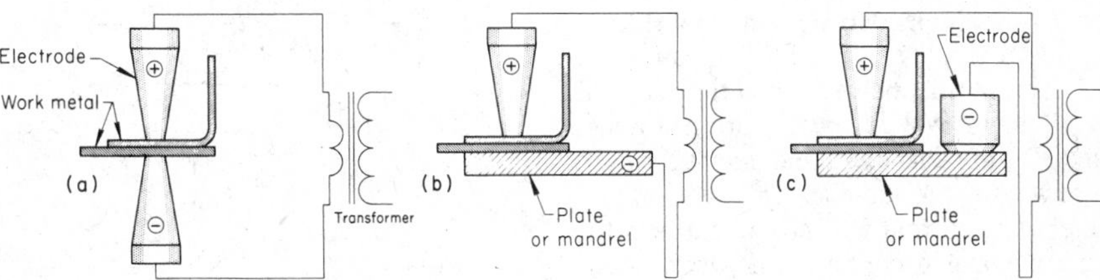

Fig. 7 Setup of work metal and electrodes for making multiple spot welds using direct and series welding

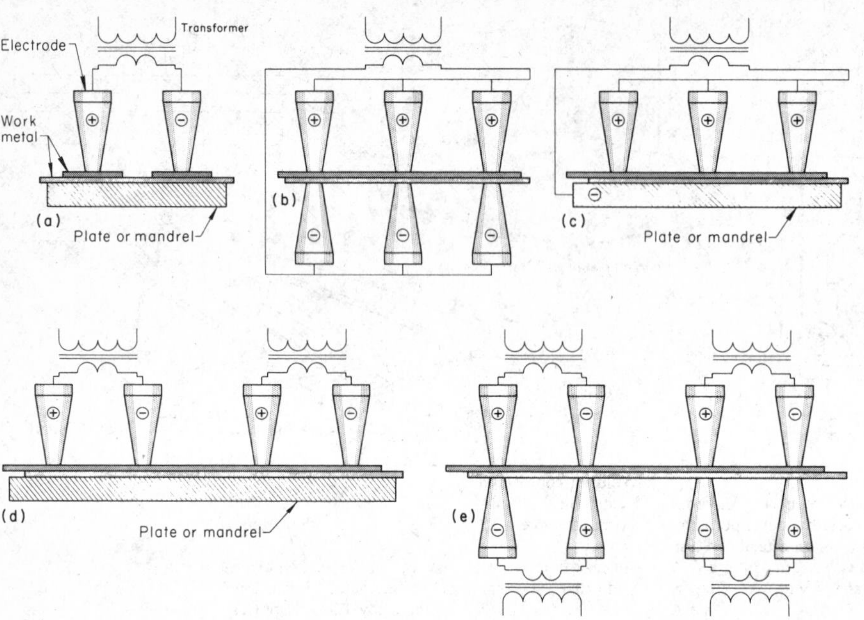

Fig. 6(b), a conductive plate or mandrel having a large contacting surface is used as the lower electrode; this reduces marking on the lower workpiece and conducts heat away from the weld more rapidly and may be necessary because of the shape of the workpiece. In the arrangement in Fig. 6(c), a conductive plate or mandrel beneath the lower workpiece is used for the same purposes but in conjunction with a second upper electrode. Because this second upper electrode is intended to serve as a contact only and not to make a weld, it may be larger (Fig. 6c) to avoid overheating.

Direct Multiple-Spot Welding. Three arrangements of the secondary circuit for making two or more spot welds simultaneously by direct welding are shown in Fig. 7(a), (b), and (c). One transformer secondary circuit can be arranged as shown in Fig. 7(a) to make two spot welds, joining two upper workpieces to one lower workpiece. In this application, the plate or mandrel need not be an electrical conductor.

A single transformer secondary circuit can also be arranged as shown in Fig. 7(b) and (c) to make two, three, or more spot welds simultaneously by direct welding, joining two workpieces. Special care must be taken in this method (sometimes called parallel spot welding) to ensure that resistance, or impedance, in the circuit for each spot weld is the same as for all the others; otherwise, current will not be uniform for each weld. Tip contour and surface condition must be the same for each electrode. Also, the force exerted by all the electrodes on the workpieces must be equal, regardless of inequalities in work-metal thickness. The force can be equalized by using a spring-loaded electrode holder or a hydraulic equalizing system. The use of a conductive plate or mandrel, as in Fig. 7(c), minimizes weld marks on the lower workpiece.

Series Multiple-Spot Welding. Two arrangements for making a number of spot welds simultaneously by series welding are shown in Fig. 7(d) and (e). In Fig. 7(d), each of the two transformer secondary circuits makes two spot welds. A portion of the current bypasses the weld nuggets through the upper workpiece. Four spot welds are conveniently made simultaneously in this way, using the two separate secondary circuits of a standard package-type resistance welding transformer. Several transformers of this type can be combined to make a larger number of spot welds simultaneously, or one of the two secondary circuits can be used alone to make just two welds at a time.

Figure 7(e) shows the arrangement for push-pull, or over-and-under, series multiple-spot welding. In this arrangement, each weld is made between a pair of opposing electrodes of opposite polarity, with each electrode being connected to a separate transformer secondary winding, and only one weld is produced for each secondary circuit. This arrangement provides a relatively high voltage at the welds, because the voltages of the two secondary windings for each pair of spot welds reinforce each other. This arrangement also permits the use of transformers having lower kV · A ratings than can be used with the arrangement shown in Fig. 7(d), but care is needed to limit shunting to an acceptable amount.

Heat for Resistance Welding

The secondary circuit of a RSW machine, including the work being welded, is a series of resistances, the total of which affects the flow of current. The current flow (in amperes) must be the same in all parts of the circuit, regardless of the resistance at any point; however, the heat generated at any point is directly proportional to the resistance at that point. Electrical systems in the secondary circuit are designed to produce heat where it is wanted, leaving the other components of the circuit relatively cool.

Figure 8 shows the composite effects of

Fig. 8 Resistance spot welding setup. Showing major points of heat generation and temperature gradients after 20 and 100% of weld time

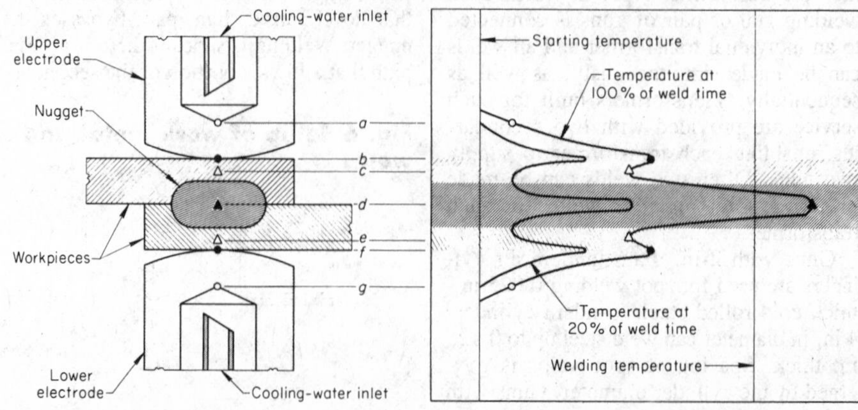

Table 2 Recommended practices for single-impulse resistance spot welding of 1010 steel with class 2 electrodes(a)

Thickness of thinnest outside piece (t), in.(b)	Electrode dimensions: Body diameter, in.	Electrode dimensions: Face diameter, in.(c)	Net electrode force, lb	Weld time, cycles (60 cps)	Welding current (approx), A	Contacting overlap (min), in.	Weld spacing (d)(min), in.	Nugget diameter (approx), in.	Breaking load (min) in shear, lb, for work metal with tensile strength of: Less than 70 ksi	Breaking load (min) in shear, lb, for work metal with tensile strength of: 70 ksi or more
0.010	3/8	1/8	200	4	4 000	3/8	1/4	0.10	130	180
0.021	3/8	3/16	300	6	6 500	7/16	3/8	0.13	320	440
0.031	3/8	3/16	400	8	8 000	7/16	1/2	0.16	570	800
0.040	1/2	1/4	500	10	9 500	1/2	3/4	0.19	920	1200
0.050	1/2	1/4	650	12	10 500	9/16	7/8	0.22	1350	...
0.062	1/2	1/4	800	14	12 000	5/8	1	0.25	1850	...
0.078	5/8	5/16	1100	17	14 000	11/16	1 1/4	0.29	2700	...
0.094	5/8	5/16	1300	20	15 500	3/4	1 1/2	0.31	3450	...
0.109	5/8	3/8	1600	23	17 500	13/16	1 5/8	0.32	4150	...
0.125	7/8	3/8	1800	26	19 000	7/8	1 3/4	0.33	5000	...

(a) Steel should be free from scale, oxides, paint, grease and oil. (b) Thickness of thinnest outside piece (t) determines welding conditions. Data are for total thickness of pile-up not exceeding $4t$; maximum ratio between two thicknesses, 3 to 1. (c) For type A, D and E faces. Body diameters apply also to type F electrodes with 3-in. spherical-radius face. (d) Center-to-center spacing for two pieces for which no special precautions need be taken to compensate for shunted-current effect of adjacent welds; for three pieces, increase spacings here by 30%.
Source: Recommended Practices for Resistance Welding, AWS C1.1.

heat generation and dissipation in the workpieces and electrodes. There are seven resistances connected in series for a two-thickness joint: (*a*) the upper electrode, (*b*) the contact surface between the upper electrode and upper workpiece, (*c*) the upper workpiece, (*d*) the contact surface between the upper and lower workpieces, (*e*) the lower workpiece, (*f*) the contact surface between the lower workpiece and lower electrode, and (*g*) the lower electrode.

Heat is generated at each of these points in proportion to the resistance at that point. The greatest amount of heat is desired at the weld point or interface between the workpieces (point *d* in Fig. 8), and steps must be taken to reduce the heat as much as possible at the other points.

The temperature of all parts at the start of the weld is represented by the line marked "Starting temperature" in Fig. 8. Temperature is rapidly increased at point *d*, the interface between the workpieces. At points *b* and *f*, the temperature rises rapidly, but not as fast as at *d*. The heat generated at points *b* and *f* is rapidly dissipated into the water-cooled electrodes at points *a* and *g*; the heat generated at point *d* is dissipated much more slowly.

After about 20% of the weld time has elapsed, the heat gradient corresponds to the inner curve in Fig. 8. The outer curve represents the heat gradient at the end of the weld time (100% of weld time). When welding conditions are properly controlled, the welding temperature is first reached at sites near *d*, the interface between workpieces. During the heating period, these tiny regions of molten metal enlarge and become continuous to form the weld nugget. The temperature gradients shown in Fig. 8 are also affected by the relative thermal conductivities of work metal and electrodes and by the size, shape, and cooling rate of the electrodes.

Welding Schedules

Low-carbon steel can be satisfactorily resistance spot welded using wide ranges of time, current, and electrode force. Limitations in machine capabilities in any of these variables can be partly compensated for by making suitable adjustments of the others. Practices recommended by the American Welding Society (AWS) for RSW of low-carbon, medium-carbon, and low-alloy steels are listed in Tables 2, 3, and 4.

The data in these tables serve as starting points for the establishment of optimum settings for welding of workpieces on which previous experience is lacking and as an approximate guide to the results that may be expected when good shop practice is followed in RSW. They are subject to adjustment as necessary to meet the requirements of individual applications.

Setting Up Welding Schedules. A typical sequence of steps for determining the most satisfactory conditions for RSW work for which previous experience is lacking is given below. The usual criteria for satisfactory welds in steps 1 to 4 are penetration, nugget diameter and indentation suitable for the application, and the

Table 3 Recommended practices for single-impulse resistance spot welding and postheating of medium-carbon and low-alloy steels with class 2 electrodes(a)

Steel to be welded: Type and condition(b)	Steel to be welded: Thickness (t) of each piece, in.(c)	Dimensions of upper and lower electrodes: Body diameter, in.	Face diameter, in.	Face radius, in.	Net electrode force (weld and temper), lb	Time, cycles (60 cps): Weld	Quench	Temper	Welding current (approx), A	Tempering current, % of welding current	Contacting overlap (min), in.	Weld spacing (min)(d), in.	Nugget diameter (approx), in.	Breaking load (min) of weld, lb: In shear	In tension	Ratio, tensile to shear breaking load, %
1020 HR	0.040	5/8	1/4	6	1475	6	17	6	16 000	90	1/2	1	0.23	1 360	920	68
1035 HR	0.040	5/8	1/4	6	1475	6	20	6	14 200	91	1/2	1	0.22	1 560	520	33
1045 HR	0.040	5/8	1/4	6	1475	6	24	6	13 800	88	1/2	1	0.21	2 000	680	34
4130 HR	0.040	5/8	1/4	6	1475	6	18	6	13 000	90	1/2	1	0.22	2 120	640	30
4340 N&T	0.031	5/8	3/16	6	900	4	12	4	8 250	84	7/16	3/4	0.16	1 084	290	27
	0.062	3/4	5/16	6	2000	10	45	10	13 900	77	5/8	1 1/2	0.27	3 840	1440	37
	0.125	1	5/8	10	5500	45	240	90	21 800	88	7/8	2 1/2	0.55	13 680	4000	29
8630 N&T	0.031	1/2	3/16	6	800	4	12	4	8 650	88	7/16	3/4	0.16	1 220	524	43
	0.062	5/8	5/16	6	1800	10	36	10	12 800	83	5/8	1 1/2	0.27	4 240	2200	52
	0.125	1	5/8	10	4500	45	210	90	21 800	84	7/8	2 1/2	0.55	13 200	4500	34
8715 N&T	0.018	1/2	1/8	6	350	3	4	3	3 900	85	7/16	5/8	0.10	400	200	50
	0.062	5/8	5/16	6	1600	10	28	10	12 250	85	5/8	1 1/2	0.27	3 300	1800	55
	0.125	1	5/8	10	4500	45	180	90	22 700	85	7/8	2 1/2	0.55	12 760	4500	35

(a) Steel to be welded should be pickled, or otherwise cleaned, to obtain a surface contact resistance not exceeding 200 microhms. (b) HR, hot rolled; N&T, normalized and tempered. (c) Welding conditions are for joining two pieces of equal thickness, each of thickness t. (d) Minimum spacing for which no special precautions are needed to compensate for shunted-current effect of adjacent welds
Source: Recommended Practices for Resistance Welding, AWS C1.1.

Table 4 Recommended practices for multiple-impulse resistance spot welding of 1010 steel with class 2 electrodes(a)

Thicknesses of steel to be joined, in.		Electrode dimensions		Net electrode force, lb	Weld time, impulses(c)			Welding current (approx), A	Contacting overlap (min), in.	Nugget diameter (min), in.	Breaking load (min) in shear(d), lb
						Adjacent welds with center-to-center spacing of:					
t-1	t-2	Body diameter, in.	Face diameter, in.(b)		Single welds	1-2 in.	2-4 in.				
1/8	1/8	1	7/16	1800	3	5	4	18 000	7/8	3/8	5 000
1/8	3/16	1	7/16	1800	3	5	4	18 000	7/8	3/8	5 000
1/8	1/4	1	7/16	1800	3	5	4	18 000	7/8	3/8	5 000
3/16	3/16	1 1/4	1/2	1950	6	20	14	19 500	1 1/8	9/16	10 000
3/16	1/4	1 1/4	1/2	1950	6	20	14	19 500	1 1/8	9/16	10 000
3/16	5/16	1 1/4	1/2	1950	6	20	14	19 500	1 1/8	9/16	10 000
1/4	1/4	1 1/4	9/16	2150	12	24	18	21 500	1 3/8	3/4	15 000
1/4	5/16	1 1/4	9/16	2150	12	24	18	21 500	1 3/8	3/4	15 000
5/16	5/16	1 1/2	5/8	2400	15	30	23	24 000	1 1/2	7/8	20 000

(a) Steel should be free from scale, oxides, paint, grease and oil. (b) For type A, D and E faces. Body diameters apply also to type F electrodes with 3-in. spherical-radius face. (c) Each impulse consists of 20 cycles on (heating) and 5 cycles off (cooling), at 60 cycles per second. (d) For steel with a tensile strength of less than 70 ksi
Source: Recommended Practices for Resistance Welding, AWS C1.1

absence of porosity, cracks, excessive expulsion, and gross defects.

1. *Make a preliminary selection of electrode force*: This is completed for the work to be welded and the electrodes to be used. Tables of recommended practices such as Tables 2, 3, 4, and 7 provide starting points for this selection, as well as guidance in choosing preliminary values of current, weld time, and hold time for making trial welds to verify or correct this preliminary selection of electrode force.
2. *Establish the weld time and hold time*: This is done by evaluating trial welds made at several levels of current for each of a number of combinations of weld time and hold time. Squeeze time is not critical in welding trials and is usually set at a convenient value that is long enough to allow for a wide range of test conditions.
3. *Select electrode force*: Using the established combination of weld time and hold time, make welds at several different current levels, using a number of values of electrode force to cover a wide range of force.
4. *Select the welding current*: Using the established weld time, hold time, and electrode force, make test welds at current levels that cover a wide range of amperage.
5. *Verify selection of conditions*: Make trial runs under the welding conditions established by steps 1 to 4, to verify these selections as well as to establish reference data on weld quality and reliability for use in process control. A more complete evaluation than for steps 1 to 4 is performed at this stage (see the discussion in the section on quality control in this article).

If the results of trial runs in step 5 are not satisfactory, the procedure may be repeated, with changes in the welding variables being made as indicated by the test results. Changes in equipment, electrode material, and electrode design may also be made at this time, and the procedure, or a shortened version, repeated until acceptable results are obtained.

These five steps not only determine optimum values for each of the welding variables, but also establish ranges of satisfactory values for them and the criticality of each variable. The sequence of steps followed in setting up a RSW schedule may vary, depending on special aspects of the equipment or the work, the extent of experience with similar types of work, and requirements imposed by the purchaser or applicable specifications. When adjustment of welding conditions is needed at the start of or during production runs, the adjustment is usually made on weld time, hold time, or current, whichever is most convenient and effective.

A recent addition to the family of spot weldable steels is the high-strength low-alloy (HSLA) steels. These steels are supplied in either the hot rolled or cold rolled condition and have yield strength levels ranging from 35 to 140 ksi. The increased strength over that of plain carbon steel is obtained by small additions of niobium, vanadium, and/or titanium together with controlled mill processing. The American Iron and Steel Institute (AISI) has compiled a listing of available materials. Because strength levels are obtained by varying alloying amounts and by processing, weld schedules may also vary. A method of determining the weld schedule compatibility or weldability of HSLA materials is the lobe curve weld schedule analysis. A typical lobe curve for one specific electrode force is presented in Fig. 9.

Figure 9(a) schematically illustrates spot weld nugget diameter as a function of current at weld time (A). As the current increases, the nugget diameter increases up to and, in most cases, beyond the expulsion current level. Specifications generally establish a minimum acceptable nugget size, illustrated in Fig. 9, and indicate that nuggets produced with expulsion are unacceptable regardless of size. It is possible, therefore, to establish a current range over which spot welds having acceptable nugget diameters are obtained for a particular weld time. This acceptable range can be expanded for other weld times to produce a region (or lobe curve) over which acceptable nuggets are produced. Welds made with currents and/or times exceeding the upper curve experience expulsion on welding and, therefore, are considered unacceptable. Welds made with currents or times below the lower curve have nuggets of insufficient size or exhibit interfacial tearing during testing and are likewise considered unacceptable. Only welds made with weld currents and times lying within the lobe area are acceptable. The size of the acceptable current range or the overall size of the lobe curve is usually taken as a measure of the weldability of HSLA steels. Other criteria such as relative position of the lobe curve (especially when compared to the lobe curve for plain carbon steel) and relative electrode wear are also considered as factors in weldability.

Fig. 9 Development of a lobe curve. Source: Ref 1

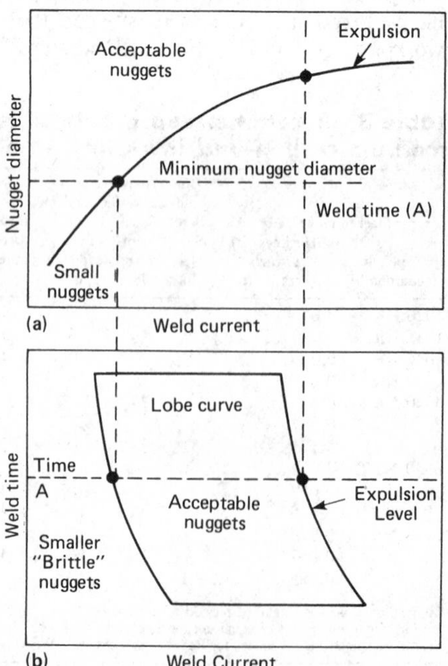

Function of Electrodes

Electrodes in RSW perform three major functions:

- Conducting the welding current to the workpieces
- Transmitting the amount of force needed to the workpieces in the weld area to produce a satisfactory weld
- Rapidly dissipating the heat from the weld zone

During the welding operation, the electrodes are subject to great compressive stresses at elevated temperature and must be frequently dressed and periodically replaced. Because the current conducted to the workpieces must remain localized within a fixed area, the electrodes must resist these stresses without excessive deformation. The electrode force, in addition to forging the heated workpieces together, influences the passage of current to the localized area.

Maintenance. The shape, dimensions, and surface condition of the electrode tips or contact surfaces are important for consistent weld quality in RSW. Shape and dimensions of electrode tips are affected by mechanical wear and deformation, or mushrooming, at a rate depending on tip material and design, operating temperatures, rates of heating and cooling (thermal shock), and welding force.

Alloying between the electrode tip and the work metal can greatly increase the rate of deterioration of the electrode tip. Deterioration is especially rapid when copper alloy electrodes are used in welding work metal coated with tin, zinc, or aluminum, which alloy readily with the electrode metal.

Careful attention to electrode tip condition is needed to avoid such defects as weak or missed welds, irregularly shaped welds, erratic indentation, burning or discoloration of the work surface, surface melting, and electrode deposits on the work surface. Electrode tips should be dressed and replaced at scheduled intervals. Preventive maintenance is particularly important when multiple electrode holders are used to ensure uniformity of resistance of each spot.

Electrode Materials

Materials for spot welding electrodes should have sufficiently high thermal and electrical conductivities, and sufficiently low contact resistance, to prevent burning of the workpiece surface or alloying of the electrode face with it and should have adequate strength to resist deformation at operating pressures and temperatures. Because the part of the electrode that contacts the workpiece becomes heated to high temperatures during welding, hardness and annealing temperature must also be considered.

Electrode materials have been classified by RWMA into two composition groups: copper-based alloys and refractory metal compositions. These classifications cover a wide range of resistance welding electrode materials to meet most applications.

Copper-Based Alloys (RWMA Group A). Table 5 gives the minimum properties for copper-based alloys used as electrode materials for RSW of steel.

- *Class 1 material* is non-heat-treatable and is strengthened and hardened by cold working, which does not affect the high electrical and thermal conductivity. Class 1 material is recommended for spot welding of low-carbon steel coated with tin, terne metal, chromium, or zinc; scaly hot rolled low-carbon steel; and some nonferrous metals, such as aluminum and magnesium alloys. It is available as drawn rod and bar, forgings, strip, and plate.
- *Class 2 material* has higher mechanical properties but lower electrical and thermal conductivities than those of class 1 material. Optimum mechanical and physical properties are developed by heat treatment or by a combination of heat treatment and cold working. Class 2 material is the best general-purpose electrode material and can be used with a wide range of metals and conditions. Class 2 material is used in electrodes for spot welding of cold rolled low-carbon steel, hot rolled pickled low-carbon steel, nickel-plated steel, stainless steel, nickel alloys, and copper-based alloys such as silicon bronze and nickel silver. It is also suitable for shafts, arms, dies, fixtures, platens, gun jaws, and other current-carrying structural members of resistance welding equipment. It is available as drawn rod and bar, strip stock, plate, castings, and forgings. forgings.
- *Class 3 material* is a hardenable alloy with higher mechanical properties but lower electrical and thermal conductivities than those of either class 1 or class

Table 5 Minimum properties for RWMA group A, classes 1, 2, and 3 copper-based electrode materials(a)

Diameter or thickness of electrode material, in.	Proportional limit, ksi			Hardness, HRB			Electrical conductivity(b), % IACS			Tensile strength, ksi			Elongation(c) in 2 in., %		
	Class 1	Class 2	Class 3	Class 1	Class 2	Class 3	Class 1	Class 2	Class 3	Class 1	Class 2	Class 3	Class 1	Class 2	Class 3
Round rod stock															
Up to 1	17.5	35	50	65	75	90	80	75	45	60	65	100	13	13	9
1-2	15	30	50	60	70	90	80	75	45	55	59	100	14	13	9
2-3	15	25	50	55	65	90	80	75	45	50	55	95	15	13	9
Square, rectangular, and hexagonal bar stock															
Up to 1	20	35	50	55	70	90	80	75	45	60	65	100	13	13	9
Over 1	15	25	50	50	65	90	80	75	45	50	55	100	14	13	9
Forgings															
Up to 1	20	22(d)	50	55	65	90	80	75	45	45	55	94	12	13	9
1-2	15	21(d)	50	50	65	90	80	75	45	40	55	94	13	13	9
Over 2	15	20(d)	50	50	65	90	80	75	45	40	55	94	13	13	9
Castings															
All	...	20	45	...	55	90	...	70	45	...	45	85	...	13	5

(a) Nominal compositions: class 1, 1% Cd, rem Cu; class 2, 0.8% Cr, rem Cu; Class 3, 0.5% Be, 1.0% Ni and/or 1.0% Co, rem Cu. (b) International Annealed Copper Standard. (c) Or in a length equal to four times the diameter of the test specimen. (d) For hot worked and heat treated (but not cold worked) electrode material
Source: "Resistance Welding Equipment Standards," American National Standard C-88.2

Table 6 Minimum properties for RWMA group B electrode materials (refractory metal compositions properties are for rod, bar, and inserts

Class(a)	Rockwell hardness	Electrical conductivity(b), % IACS	Compressive strength, ksi
10	72 HRB	35	135
11	94 HRB	28	160
12	98 HRB	27	170
13	69 HRA	30	200
14	85 HRB	30	...

(a) Classes 10, 11, and 12 are copper-tungsten mixtures; class 13 is unalloyed tungsten; class 14 is unalloyed molybdenum. (b) International Annealed Copper Standard.
Source: "Resistance Welding Equipment Standards," American National Standard C-88.2

2 materials. The high hardness, good wear resistance, and high annealing temperature of class 3 material, coupled with medium electrical conductivity (45 to 50% IACS), make it a good material for electrodes used in spot welding applications where pressures and workpiece resistance are high. It is used on thick sections of low-carbon steel and on such materials as stainless steel, Monel, and Inconel. Class 3 material is available as drawn bar and rod, strip castings, and forgings.

Refractory Metal Compositions (RWMA Group B). Table 6 gives minimum properties for refractory metal compositions used as electrode materials for RSW of steel. This group includes classes 10 through 14.

These electrode materials are used where high heat, long weld time, inadequate cooling, or high pressure would cause rapid deterioration of the copper-based alloys. In choosing among them, each application must be considered separately on the basis of design of electrode and workpieces, type of opposing electrode, and type of spot welding equipment. When a copper alloy is being welded to steel, a group B electrode is used to contact the copper alloy, and a group A electrode of class 1 or 2 is used to contact the steel.

- *Class 10 material* is a high melting point copper-tungsten refractory metal recommended for use in facings on spot welding electrodes for applications requiring a compromise between the high electrical and thermal conductivities of the copper-based alloys and the high hardness and strength provided by the other refractory metal compositions.
- *Class 11 material* is a 42% Cu, 58 vol% W refractory metal with a higher hardness but lower electrical conductivity than those of class 10 material. It is specially recommended for spot welding of ferrous metals having high electrical resistance, such as stainless steel.
- *Class 12 material* is a copper-tungsten refractory metal that has higher hardness and lower electrical conductivity than those of class 11 material.
- *Class 13 and 14 materials* consist of unalloyed tungsten and unalloyed molybdenum, respectively. Their properties are not usually needed in RSW of low-carbon steel, except for joining it to such metals as copper alloys.

Special alloys, such as copper-zirconium and copper-cadmium-zirconium, have properties similar to class 1 and class 2 materials and have been used as resistance welding electrodes. They are suitable for spot welding of steel coated or plated with metals such as zinc, aluminum, tin, terne metal, and cadmium and for spot welding of aluminum and magnesium alloys.

In any application where a class 1 material can be used, a copper-zirconium alloy can be used if increased resistance to annealing, or softening, of the electrode face is needed.

In addition to the electrodes listed above, recent work (Ref 2) has been undertaken to develop an aluminum-oxide dispersion-strengthened copper electrode. These electrodes must be supplied in the cold forged condition. They have higher strength than the standard spot welding electrodes, and they exhibit considerably less mushrooming. Results of electrode life tests conducted using the 2.68 vol% Cu Al_2O_3 electrode on galvanized and plain carbon material are presented in Fig. 10. Dispersion-strengthened electrodes change in face length more slowly than conventional class 2 electrodes. Some mill experience with these electrodes in welding various combinations of cold rolled, galvanized, and Zincrometal steels has indicated that the dispersion-strengthened electrodes exhibit from four to ten times the electrode life of conventional class 2 electrodes. A conventional copper electrode with a flame-sprayed Al_2O_3 coating is also being evaluated for its increased electrode life, particularly on coated materials.

Fig. 10 Electrode mushrooming for dispersion-strengthened electrodes. Source: Ref 2

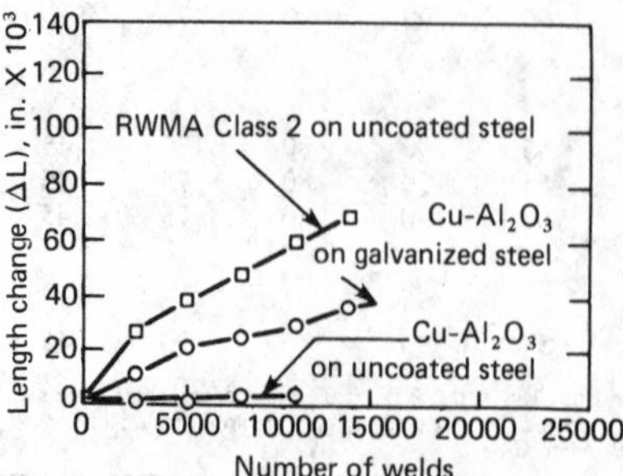

Fig. 11 Comparative hardness of electrode materials. Source: Ref 2 and 3

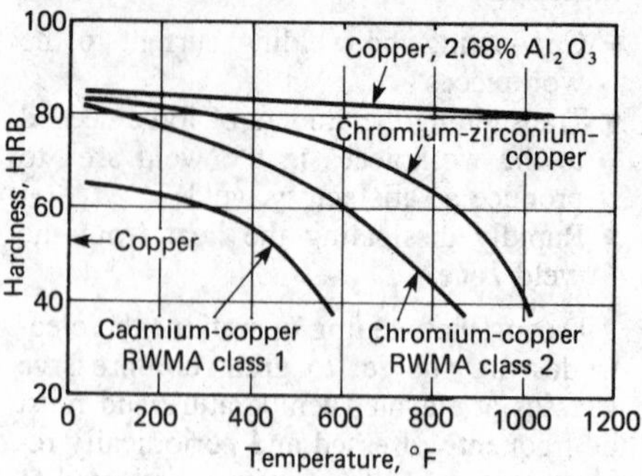

Recent work (Ref 3) has shown that the effect of temperature on hardness is a major factor in electrode wear. Conventional chromium copper alloys (Class 2 materials) exhibit characteristic deterioration of the electrodes at grain boundaries. The copper-zirconium electrodes have increased grain boundary strength. A high temperature hardness comparision of several electrodes is presented in Fig. 11.

Electrode Design

Electrode design involves four structural features of the electrode: face, shank or body, means of attachment to the electrode holder, and provision for cooling (Fig. 12). The face of a spot welding electrode contacts the workpiece directly above or below the point of fusion, and this small area is subject to repeated application of high temperature and pressure in production welding. The probability of pickup, alloying, and deformation is a major consideration in electrode design. In minimizing pickup or alloying of the work metal and electrode material, the affinity of one for the other is important. Resistance to deformation depends on the hardness and strength of the electrode material. In choosing the electrode material that will produce the best results, a compromise between properties is frequently necessary. The size and shape of the electrode tip sometimes can be modified to compensate for compromises in electrode material.

Dimensions of the electrode face are governed by thickness of the work metal, desired size of the weld nugget, and shape and size of the assembly. Electrode face

Fig. 12 Standard electrode face and nose shapes. Type D was formerly called offset.

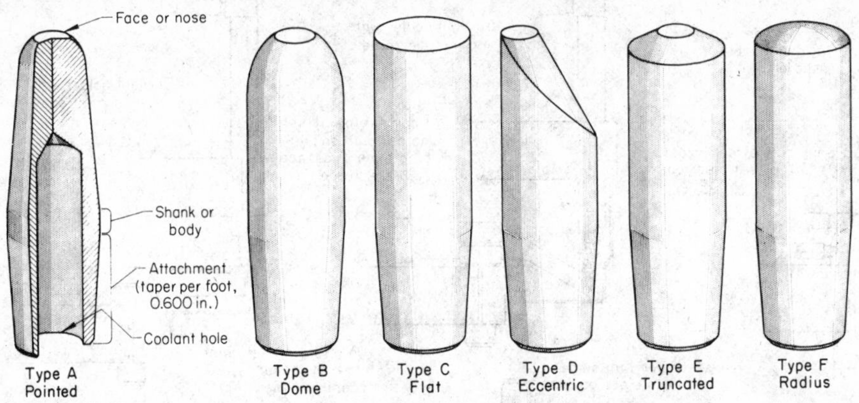

shapes have been standardized by RWMA. Figure 12 shows the six standard face or nose shapes, identified by letters A through F. Electrodes with type A (pointed) tips are used in applications for which full-diameter tips are too wide. Type D (eccentric; formerly called offset) faces are used in corners or close to upturned flanges. Special tools are available for dressing electrode faces, either in or out of the welding machine.

Shank ends or attachments of standard RWMA electrodes are made with Jarno tapers 3, 4, 5, 6, and 7. Body diameters are equal to the Jarno taper number divided by eight, or $^3/_8$, $^1/_2$, $^5/_8$, $^3/_4$, and $^7/_8$ in., respectively. Some manufacturers continue to follow an earlier standard, and use the Morse taper numbers 1, 2, and 3, which are essentially the same as RWMA (Jarno) tapers 4, 5, and 7.

Coolant holes in electrodes are either round or fluted. Fluted holes offer more cooling surface than do round holes. Coolant holes should extend as close to the face of the electrode as possible, without endangering electrode strength.

The electrode designs shown in Fig. 12 cannot be used for all applications. Also available are electrodes having the shape of buttons about $^3/_{16}$ in. high, and electrodes with single-bend and double-bend offsets, irregular offsets, square or rectangular faces, and caps with male or female tapers or threads. Buttons and caps of various nose designs, which are used with special adapter shanks that are placed in standard holders, are interchangeable in their respective adapters.

The diameter of the tip or contact face of the electrode controls the size of the weld nugget. If the diameter of the tip is too small, the resulting spot weld may be weak, even though it is sound. Small-diameter tips also may cause severe heat concentration and surface marking or indentation.

Electrodes with large-diameter faces may overheat because of insufficient electrode pressure, especially at high welding current, and cause voids, blowholes, or a poor surface appearance.

The minimum face diameter for type A, B, D, and E electrodes can be determined by using one of the following formulae:

$$\text{Face diameter, in.} = 0.10 + 2t$$

$$\text{Face diameter, in.} = 0.10 + 1.5t$$

where t is the thickness, in inches, of the base metal contacting the electrode, or by using the diagram presented in Fig. 13. Electrodes with larger face diameters should be used for welding the higher strength steels.

RWMA Designation Code. For identification of straight group A (copper-based alloy) electrodes, RWMA has established four-part code numbers:

Fig. 13 Relationship between electrode diameter and sheet thickness as derived by Leng, AWS, and Ayers. t is the thickness, in inches, of the base metal contacting the electrode. Dotted line represents alternative curve derived by authors. (Source: Ref 4)

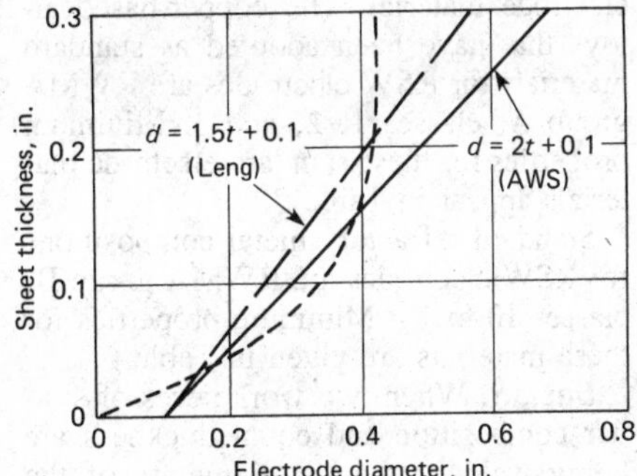

- A letter, A through F, that indicates the type of face (Fig. 12)
- A numeral, 1 through 5, that indicates the RWMA class of electrode material
- A numeral, 3 through 7, that designates the RWMA (Jarno) taper of the shank end and, when divided by 8, indicates nominal body diameter in inches
- A two-digit number that indicates the length of the electrode in increments of $^1/_4$ in.

Thus, an electrode having an RWMA code number A-2516 has a type A (pointed) face, is made of class 2 material, has a No. 5 RWMA taper on the shank end and a $^5/_8$-in.-diam body, and is 4 in. long. No designation code of this type has been established for group B (refractory metal) electrodes, because these electrodes are used as facings, inserts, buttons, or other nonstandard forms.

Electrode Holders

In RSW, electrodes are placed in holders that are mounted in the machine so that the position of the electrode can be adjusted to suit the workpieces. Most holders have an ejector mechanism for removing the electrodes and have hose connections to provide water cooling. Holders are made in straight and offset designs for tapered shank and threaded shank electrodes or adapters, and for button-type and cap-type electrodes.

Holders for tapered shank electrodes may either be straight or have 90 or 30° offsets. Most holders for threaded shank electrodes are straight but have 90 and 30° adapters. Universal and paddle-type holders have a relatively long, thin extension into which button-type electrodes are inserted.

Some combinations of standard holders, adapters, and electrodes that permit the handling of various workpiece configurations and joint arrangements are shown in Fig. 14.

Multiple-electrode holders are used for simultaneously making two or more spot welds in an assembly, using conventional spot welding machines. These holders, which are generally used for the upper electrodes, have springs, mechanical devices, or hydraulic equalizers so that the same force is applied to the workpiece by each electrode. Equal electrode force for each spot weld helps to maintain uniform weld quality. The electrodes can be connected to the transformer in different arrangements for direct or series spot welding (see Fig. 7 for typical arrangements of transformer secondary circuits).

Fig. 14 Standard holders, adapters, and electrodes for RSW

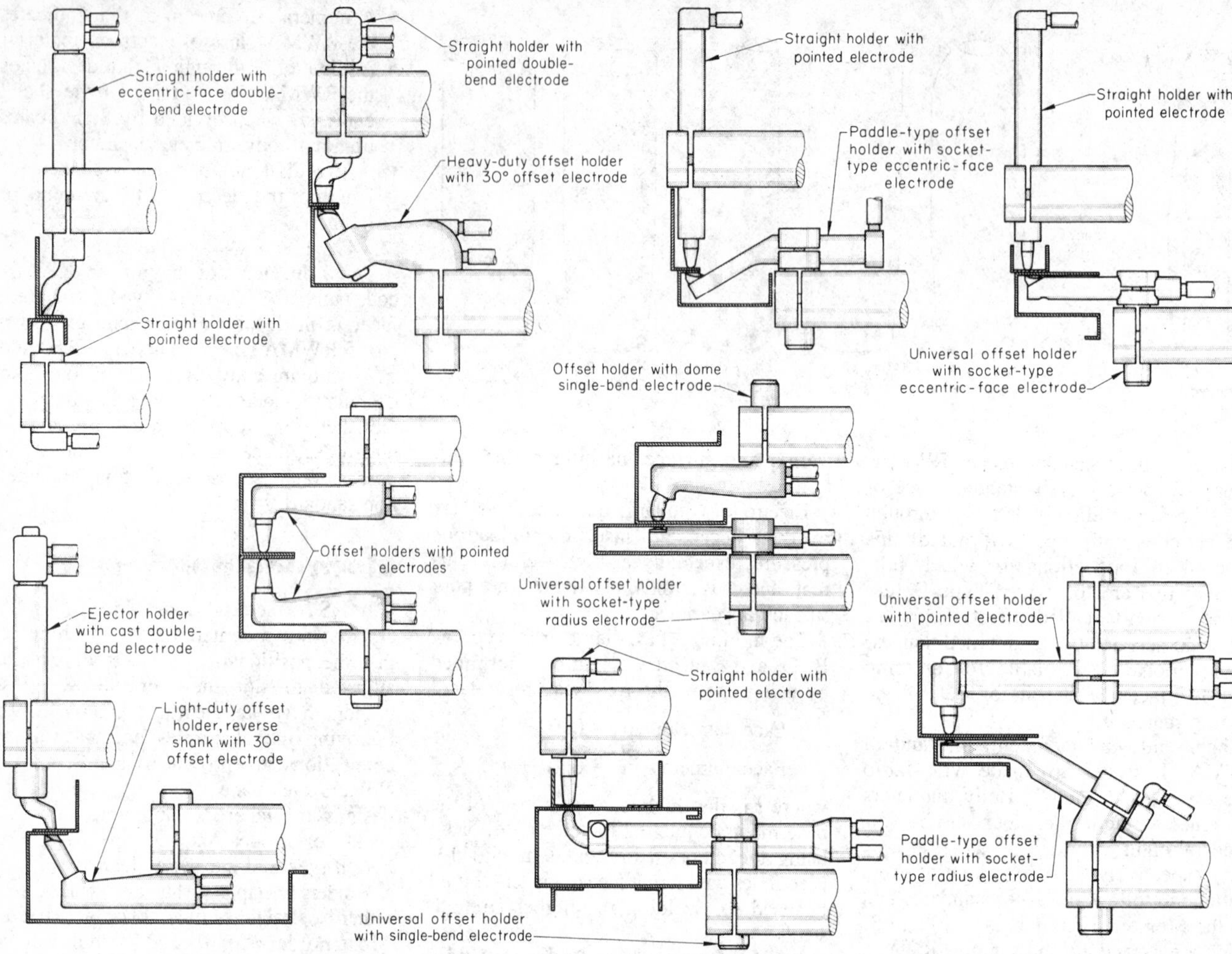

The lower electrode may be a solid piece of metal large enough to oppose all the upper electrodes, a solid piece of metal with inserts positioned opposite to the upper electrodes, or it may consist of individual tips mounted in a fixed holder. Provision must be made for the necessary electrical connection of the electrodes to the machine and for adequate water cooling.

Effect of Electrode Composition and Design on Heating

Electrodes should have high electrical conductivity and low contact resistance to minimize electrode heating, and high thermal conductivity to dissipate the heat from the contact area between the electrode tip and the workpiece (zones *b* and *f* in Fig. 8). Electrodes should also be strong enough to resist deformation caused by repeated applications of high welding force.

Composition. Generally, the harder an alloy is, the greater is its thermal and electrical resistance. In choosing the best alloy for electrodes, a suitable compromise among electrical, thermal, and mechanical properties must be found.

Commercially pure copper has excellent electrical conductivity, but because of its low resistance to compressive forces and low annealing temperature, it has been replaced by other copper-based alloys as an electrode material. The copper-based alloys that have been adopted as standard materials for RSW electrodes are RWMA group A, classes 1, 2, and 3. Minimum properties for these standard electrode materials appear in Table 5.

Standard refractory metal compositions for RSW electrodes are RWMA group B, classes 10 to 14. Minimum properties for these materials are given in Table 6.

Design. When two workpieces of similar composition and equal thickness are being welded, the tip diameters of the electrodes should be the same. However, if the workpieces are of unequal thickness, the tip diameter of the electrode contacting the thicker piece may need to be larger to maintain proper heat balance. In welding dissimilar metals, the same considerations hold true if one piece has higher electrical resistivity than the other. This dissimilarity can be compensated for by increasing the diameter of the electrode tip contacting the higher resistance workpiece, or by using a material of higher resistance for the electrode contacting the lower resistance workpiece.

The diameter of the weld nugget is slightly less than the diameter of the electrode contact area. As the electrode tips are worn or increase in diameter because of mushrooming, the diameter of the weld nugget increases. An increase in tip diameter of more than 5% can affect the weld quality because the current density is decreased and the heat generated may be insufficient to produce a good weld.

Effect of Welding Current on Heating

Welding current flows through a secondary circuit that consists of the transformer secondary coil, the flexible bands connecting the coil to the horns, the horns, the electrodes, and the workpiece. Heat is generated in all portions of the circuit according to the following formula:

$$H = I^2Rt$$

where H is the heat (energy) in watt-seconds or joules; I is the current in amperes; R is the resistance in ohms; and t is the duration of current flow in seconds. Some heat is lost through conduction, convection, and radiation from the electrodes and the workpieces. The magnitudes of these losses are generally not known.

The thermal conductivity of steel is about 12% that of copper; therefore, if sufficient welding current is used in welding steel with copper-based electrodes, the heat generated along the interface of the workpieces (point *d* in Fig. 8) is conducted away from the weld zone more slowly than the heat generated at the electrode faces (points *b* and *f* in Fig. 8) is conducted into the water-cooled electrodes (points *a* and *g* in Fig. 8). Thus, the interface of the workpieces reaches the fusion temperature first, and a weld is produced at this interface.

There is a lower limit for current density below which fusion cannot be obtained. Enough heat must be generated to offset the heat that is lost through conduction into the electrodes, surrounding air, and that part of the work metal not between the electrodes.

As current density is increased, weld time can be decreased sufficiently to produce a weld without heating the electrode contact surfaces to more than a few hundred degrees.

There is also an upper limit for the welding current. If the welding current is too high, the entire thickness of the work metal between the electrodes is heated to the plastic range by the time the weld zone reaches the fusion temperature, and the electrodes embed themselves deeply into the metal. The outer surfaces of the electrodes may also be overheated and burned. This is especially true when the current is high enough to produce expulsion. The result is reduced electrode service life.

For a given electrode force, there is an upper limit of current density above which pitting and expulsion of hot metal occur at one or more workpiece surfaces, causing low-quality welds. Maximum spot strength is obtained by welding at a current density slightly below the value at which expulsion takes place. Setting of the current for production runs is commonly based on this relation.

The effects of welding current on nugget diameter, joint tensile-shear strength, and electrode indentation are shown in Fig. 15. Increasing the current above 13 500 A did not significantly increase the nugget diameter, $^1/_4$ in., but caused a marked increase in electrode indentation. Tensile-shear strength increased rapidly until the optimum current was reached, but decreased slightly when the current was increased to slightly above 14 000 A. Indentation increased from about 2% of the sheet thickness at a welding current of 13 500 A to about 10% at a welding current slightly above 14 000 A.

The use of pulsations of current with varying degrees of "off" time (from 1 to as much as 10 cycles) between pulses has been used on thicker plain carbon and HSLA steels. In parts where material thickness or springback (HSLA steels) prevented attainment of complete intimate contact, pulsations were beneficial. The sheets could be brought into the desired contact as a result of heating during the initial pulse with weld completion during the subsequent pulses. In some HSLA steel parts where springback is a particular problem, dual pulsing is used for materials as thin as 0.08 in.

Fig. 15 Effects of welding current on nugget diameter, joint tensile-shear strength, and electrode indentation during spot welding

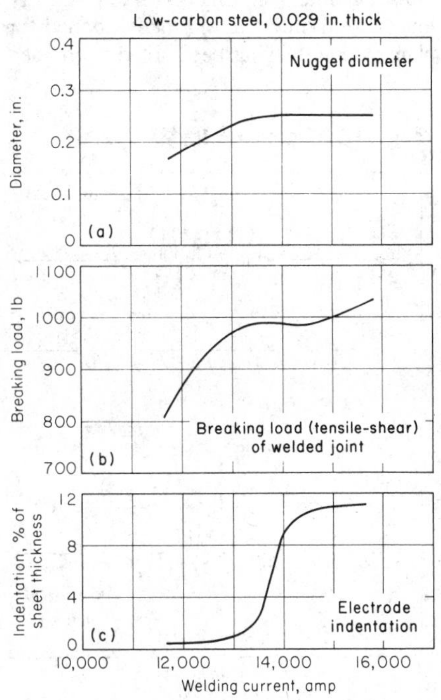

Fig. 16 Variation in total resistance with electrode force and sheet surface condition. First sample was pickled to remove oxide scale which resulted from hot rolling. Second sample was wire brushed. With progressively increased electrode force, oxides remaining on the wire-brushed sample break down, resulting in surface asperity collapse, which in turn will reduce interfacial resistance. (Source: Ref 1)

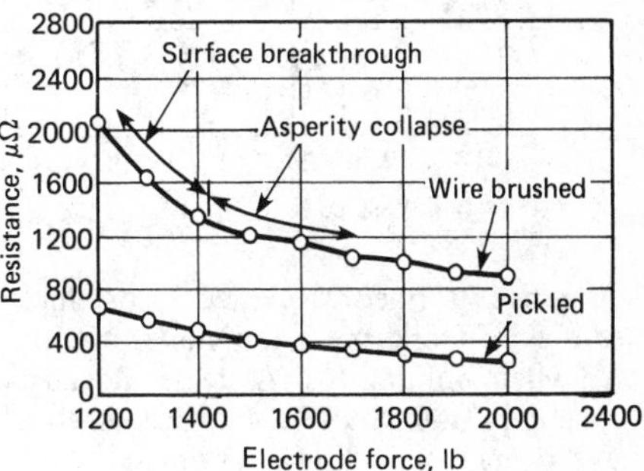

Effect of Welding Force on Heating

Welding force, or electrode force, is the force applied to the workpieces by the electrodes during the welding cycle. Electrode force, usually measured and expressed as a static value, is a dynamic force in operation and is affected by the friction and inertia of the moving parts of the welding machine.

The workpieces to be spot welded must be held tightly together at the indented location of the weld to allow passage of current. Because increasing the electrode force decreases the contact resistance of the work metal (Fig. 16), and therefore decreases the total heat generated between the faying surfaces of the workpieces by the welding current, electrode force should not be excessive (see the section on effect of surface condition on heating in this article).

The electrode force must be compatible with a welding current that is within the capacity of the equipment and must permit the use of a welding time long enough to be reproducible. Also, the workpieces must be in reasonably intimate contact at the weld area without excessive electrode force. If the workpieces are deformed so that contact is not intimate in the weld zone, an excessively high force may be needed to overcome the deformation. Variations in weld strength and quality often result from the variations in electrode force required to bring the workpieces into proper contact, especially in the spot welding of

Fig. 17 Effect of electrode force on the lobe curve of hot rolled HSLA steel. Source: Ref 5

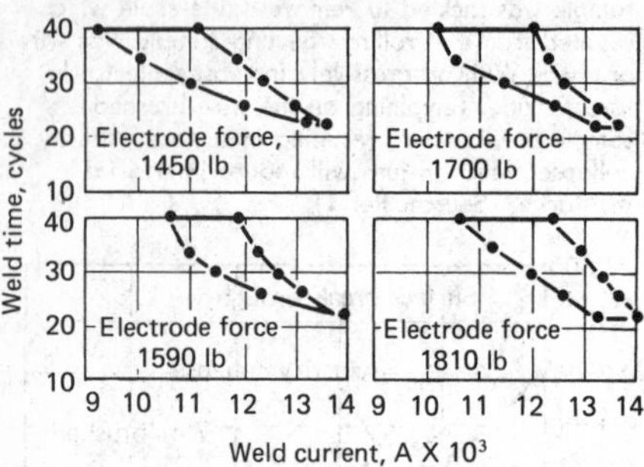

stampings, formed workpieces, or thick sections of work metal. This is especially true when welding HSLA steels. Because of the high strength of these materials, they have considerably higher springback and thus require higher electrode forces than plain carbon steel.

Sometimes, a squeeze time longer than that normally used is needed to force the workpieces together. Also, because of the possibility of springback, hold time must be sufficiently long to permit solidification of the weld metal.

Because variations in electrode force cause changes in resistance and heating, these force variations can cause changes in the size and location of the lobe curves and current level as seen in Fig. 17. Increased electrode force shifts the curves to higher current levels. This shift in the lobe curve indicates that weld current and weld time combinations which produced acceptable weld nuggets at one electrode force may no longer produce acceptable weld nuggets at the new electrode force. Thus, care should be taken when making electrode force changes, especially in alloys that have a limited acceptable weld current range, such as some of the more difficult HSLA steels with yield strengths above 100 ksi.

Effect of Time on Heating

The effect of time on the temperature distribution in workpieces and electrodes during spot welding is shown by the two curves in Fig. 8. The inner curve represents the temperature in each zone after 20% of the weld time has elapsed and shows that the temperature rise at the faying workpiece surfaces (point *d*) during this period is proportionately lower than in the other zones in relation to temperature rise during the remainder of weld time.

The heat (energy) formula $H = I^2Rt$ shows that, with the total resistance held constant, the amount of heat generated in any part of the circuit is proportional to both the weld time (the time the welding current is on) and the square of the welding current. Because heat transfer is a function of time, the time required for development of the proper nugget size can be shortened only to a limited extent, regardless of how much the current is increased. Pitting and spitting (expulsion of metal), especially at the electrode contacting surfaces, results when the heat is generated too rapidly at the three contacting surfaces (zones *b*, *d*, and *f* in Fig. 8).

Some reduction in weld time can be accomplished if the welding current, initial pressure, and follow-up pressure are increased. Higher initial pressure is needed to prevent spitting, because of the increase in current. Higher follow-up pressure and fast follow-up are needed to maintain firm contact and adequate pressure of electrode on the workpiece until the metal of the weld has solidified.

The welding cycle is divided into four major time segments: squeeze, weld, hold, and off. These are shown in Fig. 18.

Squeeze time is an interval of delay between closing of the initiating switch and application of the welding current. It provides time for the solenoid-actuated head cylinder valve to operate and for the welding head to bring the upper electrode in contact with the workpiece and develop full electrode force. This time should be sufficient to ensure that the parts have maintained intimate contact.

Weld time is the interval during which the welding current flows through the circuit. On many HSLA steels, a slightly longer weld time than normally applied for plain carbon steel has been noted to broaden the lobe curve and thus increase acceptable weld current ranges.

Hold time is the interval during which, after the welding current is off, the electrode force is held on the workpiece until the metal of the spot weld has solidified. Some HSLA materials are hold-time sensitive. With hold times as great as 30 to 60 cycles ($^1/_2$ to 1 s), these materials tend to experience interfacial tearing when peel tested.

Off time is the interval from the end of the hold time until the beginning of the squeeze time for the next cycle. In an automatic cycle, off time is the time needed to retract the electrodes and to index, remove, or reposition the work. In manual operation, it is not fixed as a maximum period by the control equipment, but depends on time taken by the operator to start a new cycle. Figure 18(a) shows the relative durations of these four basic segments for a single-impulse RSW cycle. All of the segments are usually expressed in cycles, meaning the number of cycles in a 60-cycle system, where one cycle is $^1/_{60}$ s.

The simplest welding cycle supplies uniform welding current and electrode force throughout the weld interval, but the addition of slope control enables the welding current to be varied. As shown in Fig. 18(b), a welding cycle that incorporates slope control has an increase in current before the weld and a decrease after the weld.

Upslope permits the welding current to be increased over several cycles from a low value to that needed for welding, instead of having the full welding current applied instantly. A low initial or welding current reduces or prevents expulsion of metal, or

Fig. 18 Types of RSW cycles

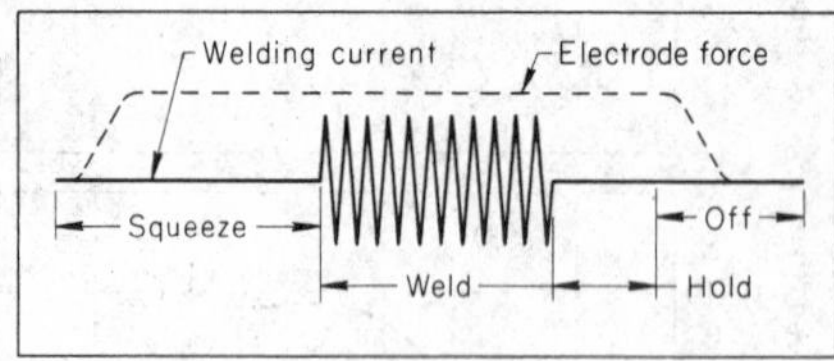

(a) Single-impulse welding cycle

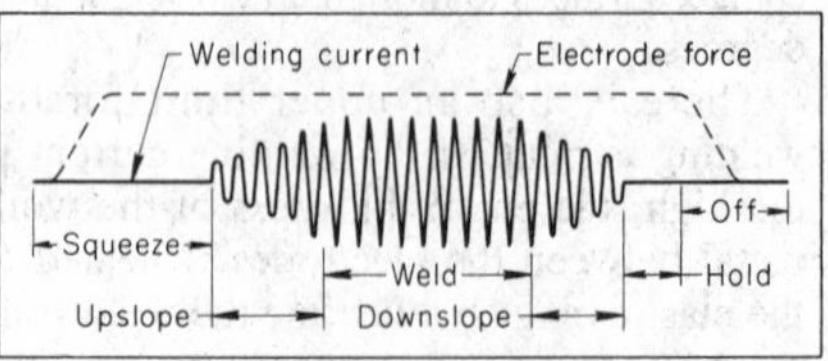

(b) Single-impulse welding cycle with upslope and downslope heat control

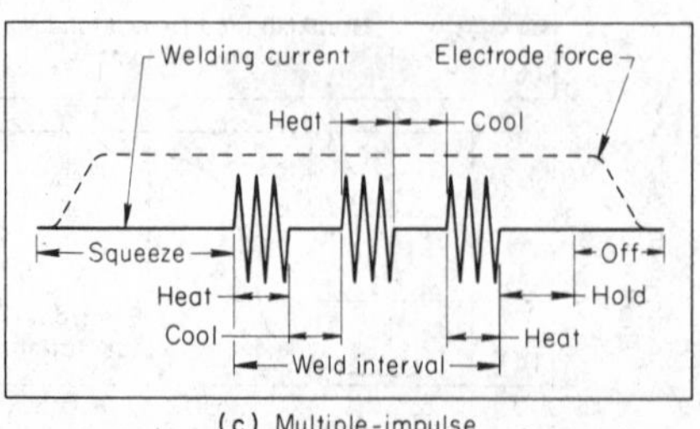

(c) Multiple-impulse welding cycle

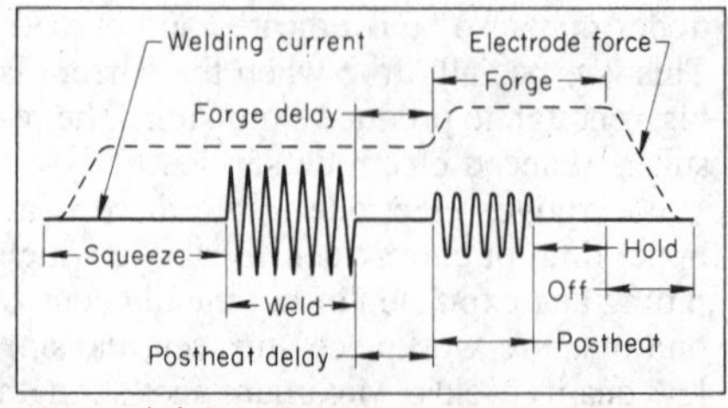

(d) Single-impulse welding cycle with forging force and postheat

spitting, when the current is first applied. Upslope control is used for welding at high current values, and for welding scaly stock as well as most kinds of plated metals.

Downslope permits the welding current to decay gradually to a low value instead of ending suddenly, and it helps to produce good welds in some types of heat treatable metals by lengthening the cooling time gradient. It is rarely needed in welding low-carbon steel, particularly if the carbon content does not exceed 0.15%, but is used when cooling rate must be limited, as in welding hardenable steels.

The high welding currents and long weld times needed for welding sheets greater than $^1/_8$ in. in thickness may cause overheating of the electrodes. This can be minimized by applying the current in pulses during the weld interval, as shown in Fig. 18(c). Heat is dissipated more rapidly from the electrodes than from the workpieces. Therefore, during the cool time, when the welding current is off, the electrodes dissipate most of their heat while the workpieces lose very little. With a series of impulses, each followed by a cooling period, the workpieces are brought to welding temperature while the electrode temperature remains at a safe level. This multiple-impulse technique is also useful to help bring high-strength material with high springback tendencies into intimate contact. The first impulse heats and deforms the metal while the subsequent impulses consummate the weld.

Two other elements that can be added to a welding cycle are forge and postheat, both shown in Fig. 18(d). They are both used chiefly for grain refinement on hardenable carbon and alloy steels and are not used on low-carbon steel. After welding, there is a short delay, or current-off period, to allow the weld to cool before the application of a tempering current. Postheat, during which the current is on at a low value, is followed by a hold time. During postheating, the electrode force may be increased, as shown in Fig. 18(d). This increased force is called a forging force and is applied during the postheat and hold times. The welding force usually is maintained until the postheat current is applied, after which it is increased to the forging force.

Effect of Workpiece Surface Condition on Heating

For consistent production of spot welds of the highest quality, the resistance at the workpiece surfaces that contact the electrodes (zones *b* and *f* in Fig. 8) must be kept to a minimum. This can be done by having smooth, clean work-metal surfaces and by controlling the electrode force. If the workpiece surfaces that contact the electrodes have too high a contact resistance, the temperature rise at these surfaces is almost as fast as at the faying surfaces (zone *d* in Fig. 8). Also, inconsistent results can be obtained because of variations in the contact resistance and corresponding variations in the time it takes for current flow to be established.

The surfaces of metal sheets are not smooth on a microscale, and the actual metal-to-metal contact area may be only a small percentage of the entire contact surface when light electrode pressures are used. As the electrode force is increased, the high spots are depressed, increasing the actual metal-to-metal contact area and thus decreasing the electrical resistance. Increased electrode force also decreases the resistance at the interface of the workpieces. When the electrode material is softer than the work metal, the application of a given electrode force results in better contact at the contacting surfaces between electrodes and workpieces than at the interface of the two workpieces.

Although electrode force does not enter directly into the heat formula $H = I^2Rt$, its effect on electrical resistance has a direct effect on the flow of the welding current. Surface resistance is inversely proportional to electrode force, as illustrated in Fig. 16.

Surface Preparation. Recommended practices for spot welding steel, as given in Tables 2, 3, and 4, call for the work metal to be free from scale, oxides, paint, grease, and oil. The work to be welded, or at a minimum the faying surfaces, should be cleaned to ensure that the welds are free of inclusions. Dirt, scale, rust, and oxide film that may come in contact with the electrodes should be removed or reduced to ensure good surface appearance of the welds. Also, removal of foreign substances from workpiece surfaces reduces electrode pickup and consequently increases electrode life.

A film of dirt or oil can be removed from the surfaces of the workpieces by vapor degreasers and chemical baths; however, careful hand wiping of the surfaces to be spot welded may be sufficient. Oxide films can be removed by mechanical methods; the action must be severe enough to cut through the film, but must not be so severe as to cause the formation of a rough or scratched surface.

Where it is impractical to use pickled or cold rolled steel, the surfaces to be welded can be machined, ground, or wire brushed far enough back from the edge of the workpiece to clear the electrodes, or for a distance equal to the overlap. Annealing in a reducing atmosphere is also an acceptable cleaning method for some metals.

Abrasive blast cleaning methods using sand, coarse grit, or shot usually are not satisfactory because particles of sand or scale are left embedded in the surface. Fine, sharp steel grit is satisfactory in some applications.

Effect of Oil Coatings. Thin coatings of oil on cold rolled or hot rolled pickled and oiled steel have little effect on the quality of spot welds. Tests have shown that the strength of spot welds made on steel having a thin coating of oil is ordinarily about 2 to 3% less than that of spot welds made on the same metal after removing the oil by degreasing.

Excessive amounts of oil should be wiped off or removed in a degreasing operation. The oil itself may not be detrimental to the weld, but the dirt and other contaminants adhering to the oil may cause poor welds.

Effect of Rust, Scale, or Oxide. Steel coated with rust, scale, or thermally produced black or blue oxide finishes can be resistance spot welded in production, but quality and consistency of welds are lower than on steel from which these coatings have been removed. Thin films that have low and uniform electrical resistance have the smallest effect on welding. Steel that is coated with extremely thick and nonuniform mill scale may not be weldable on a practical production basis without first removing the scale.

Steel on which uniform but heavy mill scale or oxide coatings are present can be welded in production by applying low-to-medium current in a series of pulsations at fairly high electrode pressure. The electrical conductivity of mill scale or oxide increases with temperature, and these coatings become fairly good conductors when red hot. Because of the variable time needed to break through the coating and establish current flow, manual timing gives more satisfactory results than automatic timing.

When metal having heavy scale or oxide coatings is welded, much or all of the scale or coating on the faying surfaces remains in the welds, regardless of current, surface resistance, or electrode pressure. These inclusions in the weld metal can cause voids, blowholes, and other internal defects that may sometimes be difficult to detect.

Effect of Surface Finish Requirements on Heating

In spot welding of assemblies that are to be porcelain enameled or painted, or are to receive other decorative surface finishes, the surface condition and close fit of parts after welding are as important as weld strength. Excessive indentation, overheating of the outside surfaces, spatter, and crevices interfere with the finishing operations and must be avoided.

Welding schedules and conditions must be selected that produce a weld of adequate strength with a minimum of indentation and minimum evidence of heating. This requires a uniform welding procedure that is best obtained with automatic control of current, time, and force. Electrode faces should be dressed at regular intervals, before they have worn so much that unsatisfactory spot welds are produced. Workpieces should be cleaned thoroughly before welding.

Effect of Weld Spacing on Heating

Shunting occurs when a second spot weld is made so close to the first one that the welding current can flow either through the metal at the first weld or through the metal between the electrodes at the point of the second weld. The welding current flows in inverse proportion to the resistance of the two paths. Division of current depends chiefly on the ratio of resistance of the base metal to interface resistance at the point of the second weld.

When making a second and following spot welds, the metal between the electrodes becomes a divided circuit; part of the current travels through the metal to the previously made spot weld, while the remainder travels through the metal between the electrode faces at the point of the second weld. If the distance to the first spot weld is great enough, the resistance of the path through the first spot weld, compared to that directly through the metal, is high, and the shunting effect can be neglected. If the distance to the first spot weld is short, a significant fraction of the current is shunted through the first spot weld.

As the temperature of the metal between the electrode tips rises, the resistance at that point increases, thus adding to the shunting effect. Metals having high electrical resistivity are less influenced by the shunting effect than are low-resistivity metals.

The minimum spacing of spot welds in low-carbon steel workpieces depends on stock thickness, diameter of the fused zone, and cleanness of the faying surfaces. The minimum recommended weld spacing for low-carbon, medium-carbon, and low-alloy steels is given in Tables 2, 3, and 4. Welds can be made at less than recommended minimum spacing without significant shunting by using higher current and electrode force, shorter weld time, and fast follow-up.

If spot welds are made too close to the edge of a sheet or flange, there is an insufficient volume of base metal to withstand electrode pressure and heating. This results in a reduction in the effective force along the edge and uneven heating, causing the hot metal to be expelled from the weld. When spot welds are made too close to an upright flange or sidewall, arcing may occur between the electrode and workpiece, or there may be a poor fit at the faying surface because of the bend radius. If large-diameter electrodes are needed because of the metal thickness, eccentric, or offset, faces may be used.

The minimum contacting overlap between two pieces of metal depends somewhat on the metal thickness, which in turn governs the electrode diameter and diameter of the fused zone. The minimum overlaps for low-carbon, medium-carbon, and low-alloy steels are given in Tables 2, 3, and 4.

Effect of Composition on Spot Weldability

Carbon content has the greatest effect on weldability of steels. Weld hardness increases rapidly with small increases in carbon content. This high hardness causes nugget interfacial tears and nugget deterioration. To obtain acceptable weld performance, carbon content should be kept below 0.10% + 0.3t, where t is the sheet thickness in inches. For materials above this range, postweld tempering may be necessary.

Low-carbon steels generally are considered to be spot weldable. These steels can be obtained from a number of suppliers with reasonable certainty of only minor variations in weldability. On the other hand, HSLA steels are sold on the basis of strength, with each steelmaker producing its own composition. Therefore, this variation in composition may cause variations in weldability which the steel user must consider.

Phosphorus and Sulfur. Phosphorus and, to some extent, sulfur are generally considered to promote nugget interface tearing. When the total content of phosphorus, sulfur, and carbon exceeds a critical value, spot weld interfacial failure is observed. This interfacial tearing during peel testing results in reduced current ranges and minimal lobe curve areas.

Titanium. Data on the effect of titanium on both hot and cold rolled steel indicate that the spot weldability of these steels is not as good as that of the HSLA steels containing niobium and/or vanadium. Increasing titanium content generally reduces maximum button diameter, shear and cross-weld tensile strength, and welding current range. Titanium content should not exceed 0.18%, and the use of oversized electrodes and higher force may be required.

Nitrogen in HSLA tests promotes interfacial nugget failure. Nitrogen appears to be more critical in unkilled cold rolled gages. However, the sensitivity can be reduced by decreasing the nitrogen content or by tying up the nitrogen with aluminum as in aluminum-killed steels.

Oxygen. High oxygen levels also promote interfacial failure either by causing weld centerline hot cracks or through a weld embrittlement mechanism. Oxygen level can be controlled by reduction of oxygen in the melt and by avoiding welds on rusty surfaces.

Hydrogen which emanates from surface oils also assists in weld-metal failure. Wiping the oil off the surface before welding or holding the test sample for 1 h after welding before peel testing (allowing diffusion of hydrogen from the weld area) helps prevent weld-metal fracture in borderline rephosphorized steels.

Effect of Material Processing on Spot Weldability

Sheet steels may be processed in different ways to obtain additional strength. One method of strengthening a material is by cold rolling. A heavily cold rolled material is generally not ductile. Normally, annealing heat treatments are performed to restore ductility, but this also lowers strength. A heat treatment designed to promote recovery but not complete recrystallization of the cold worked structure is called a recovery anneal. This treatment restores some ductility while still maintaining fairly high strength levels. Other methods of increasing strength are through a post rolling austenitization treatment followed by various quenching and tempering treatments to produce martensitic or bainitic microstructures.

When welding any of these types of steels, variations in the hardness profiles across the heat-affected zones (HAZ) have been noted. The spot weld hardness traverses of a cold rolled steel which has been

Fig. 19 Microhardness traverses across the HAZ of cold rolled titanium-containing HSLA steels. Source: Ref 6

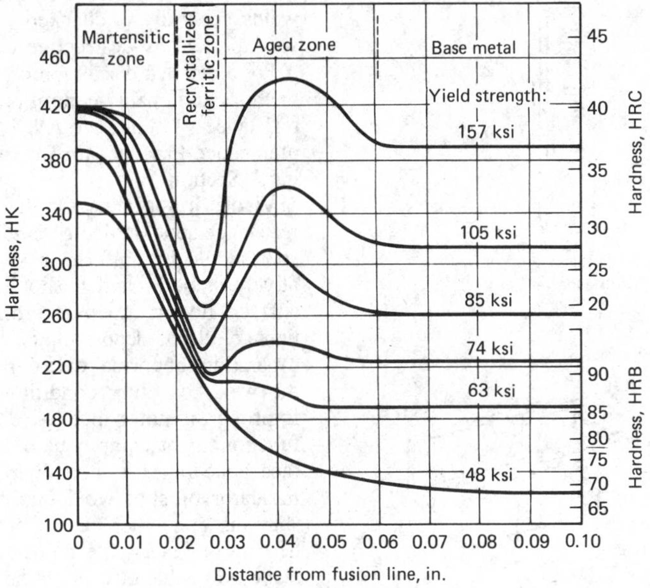

annealed back to various strength levels are presented in Fig. 19. Note that several zones, including an aged zone and a recrystallized zone, can be identified. A drastic softening is noted in the recrystallized zone. The hardness traverses for martensitic and bainitic materials also show softened regions within the HAZ. This softened region may lead to spot weld mechanical property alteration. Longer weld times have been found to promote increased softening (Fig. 20). Thus, spot weld design and weld procedures must be jointly considered when welding these types of steels.

Mechanical Properties of Spot Welds

Tests that are used to obtain some measure of the mechanical properties of spot welds include peel tests, chisel tests, tensile tests, torsion tests, impact tests, and fatigue tests.

Peel Test. This is the most commonly used mechanical test for spot welds. It is the test used for establishing spot weldability in all the automotive specifications. There are several reasons for the generally wide acceptance of this test: (1) ease of performance, (2) low cost, and (3) the ability to use it on the shop floor as a quality-control test. Unfortunately, although the test can identify whether a nugget of acceptable size (without interfacial failure) is being obtained, only in restricted cases are quantitative values relating to the strength of the weld nugget or its performance in service obtained.

The peel test requires a spot weld sample to be peeled, generally by using manual procedures, although some automated procedures have been used. This peeling exposes a weld nugget torn from one sheet which can then be examined for size and material tear location. A full nugget pullout is generally required. Interfacial failures (often called brittle nuggets, even though the fracture may not be of a brittle morphology) are generally not allowed.

Fig. 20 Effect of weld time on microhardness traverses across a recovery annealed, cold rolled HSLA steel. Hardness values converted from Knoop. (Source: Ref 5)

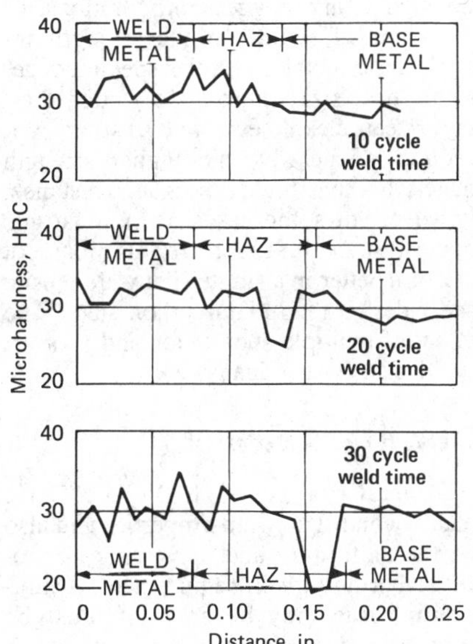

Chisel Test. This is quite similar to the peel test with the exception that the load is applied by driving a chisel or wedge between the sheets near the spot weld location. The presence of the chisel establishes a tearing-type load at the nugget edge. Studies indicate that this tearing-type loading results in the lowest strength value for a weld. Therefore, it is assumed that the chisel test results in the most stringent test condition; welds in service should be more reliable than would be predicted by chisel test results. Additionally, a limited amount of work comparing lobe curves determined using both peel testing and chisel testing has indicated that the nuggets exposed by chisel testing were somewhat smaller (i.e., deformation during tearing exposed the nugget closer to the fusion line), resulting in a smaller lobe curve by about 10%.

Tensile Test. Numerous tensile test specimen designs have been used to determine the tensile properties of spot welds. Some of these designs are illustrated in Fig. 21. In general, the tests in which these specimens are used can be broken down into two types: those designed to measure normal stresses, and those designed to measure shear stresses. Tests designed to measure normal stresses are the cross tension test, the U-tension test, the triangle direct tension test, and the bonded block test. Tests designed to measure shear stresses are the tensile shear test, the reduced section tensile shear test, the double lap tensile shear test, and the Battelle Dynamic Stiffness Method.

It is known that the localized state of stress developed around the spot weld within a test specimen determines, to some measure, the specimen's ability to carry a load. For convenience, the load-carrying capability of a spot weld is defined as the spot weld strength measured in units of force rather than stress. The flange-type specimens, i.e., the peel and chisel test specimens and/or the carriage joint, have the least strength; the specimens designed to carry normal loads are next strongest; and the shear specimens are the strongest.

Besides specimen design, other factors that affect spot weld strength include base-metal strength, sheet size, nugget size, weld time, hold time, spot array, coatings, and testing procedures.

Torsion Test. The torsion test has also been used to evaluate spot weld properties. In general, the angle of twist before

Fig. 21 Tensile specimen designs. Source: Ref 1

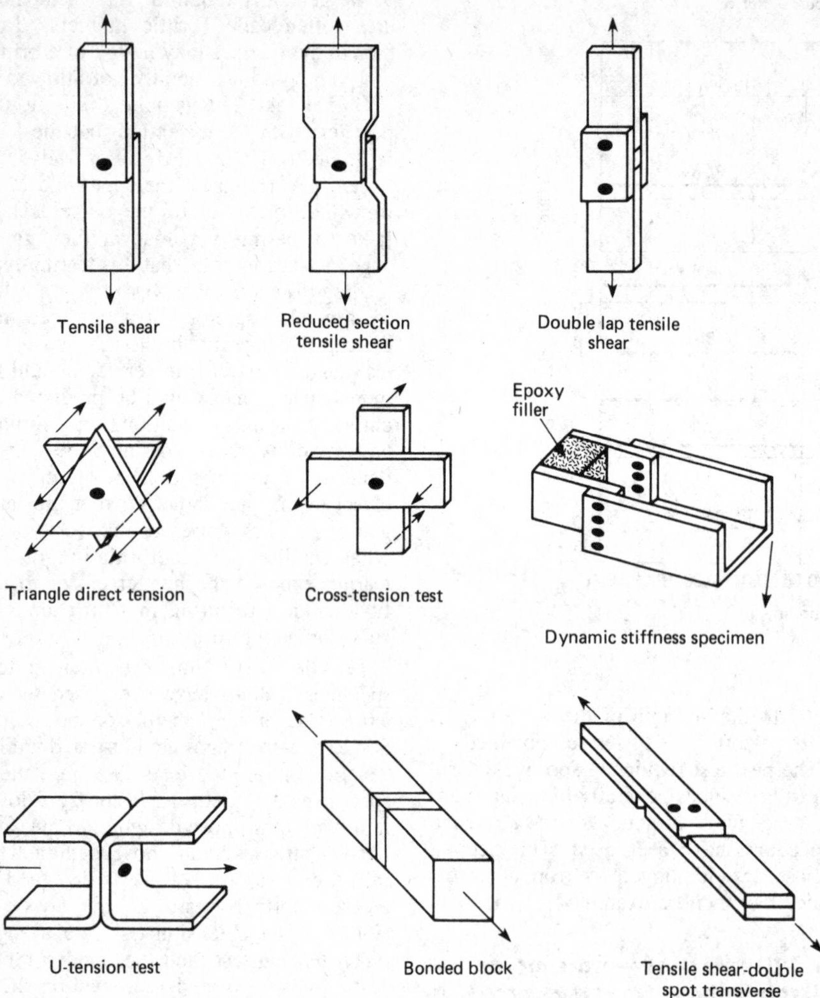

nugget failure has been used as a measure of spot weld ductility, while the maximum torque has been used to evaluate spot weld strength. Several different specimen designs have been used as shown in Fig. 22. In each sample, the twisting occurs with the weld nugget at the center of rotation. The critical data obtained are usually the angle of twist at failure or torque/twist curves. Usually, good forming steels that have high ductility have low torques but large twist angles. In contrast, high-strength steels tend to have higher torques with lower twist angles. When a weld is made with some expulsion, it tends to reduce both the torque and the twist angle. All the factors that affect tensile tests also have been observed to affect torsion testing.

Fatigue Test. Fatigue testing of spot welds has gained considerable interest in recent years. Numerous spot weld fatigue test specimen designs (Fig. 23) have been used, including direct tension, tensile shear, and tack weld designs of varied descriptions, size, and nugget configurations.

Factors affecting fatigue strength include base-metal strength, specimen design, sheet size, nugget size, nugget array, steel cleanliness, and testing procedures. Generally, the higher strength materials have higher fatigue resistance; however, in some cases at high fatigue cycles, some HSLA materials may not perform better in a single spot weld tensile shear design than plain carbon steel. The effect of multiple-spot welds and spot array design is still unanswered.

Quality Control

General practice in RSW is to base quality control on weld properties and also on the uniformity and consistency of results. For any given requirement, a minimum value may be set that must be equaled or exceeded by all test results; it is often also required that all tests in a given sample should fall within ±10% of the mean for that sample. The quality of spot welds generally is checked by visual inspection and by destructive testing.

For a detailed discussion of standard test methods and quality-control procedures for resistance welding, see AWS C1.1, "Recommended Practices for Resistance Welding," Section V.

Visual Inspection. On the surface of a resistance spot welded assembly, the weld spot should be uniform in shape and relatively smooth, and it should be free of surface fusion, deep electrode indentations, electrode deposits, pits, cracks, sheet separation, abnormal discoloration around the weld, or other conditions indicating improper maintenance of electrodes or functioning of equipment. However, surface appearance is not always a reliable indicator of spot weld quality, because shunting and other causes of insufficient heating or inadequate penetration usually leave no visible effects on the workpiece.

Destructive testing can be performed on the actual workpiece or on test specimens. For small, inexpensive parts, actual production samples from each machine are taken at random or at prescribed regular intervals for destructive testing. Test cou-

Fig. 22 Torsion test specimens. In sample B, torque is applied by gripping and twisting the drilled holes (see top view). Sample C is a torsion test for spot welded sheet metal parts. (Source: Ref 1)

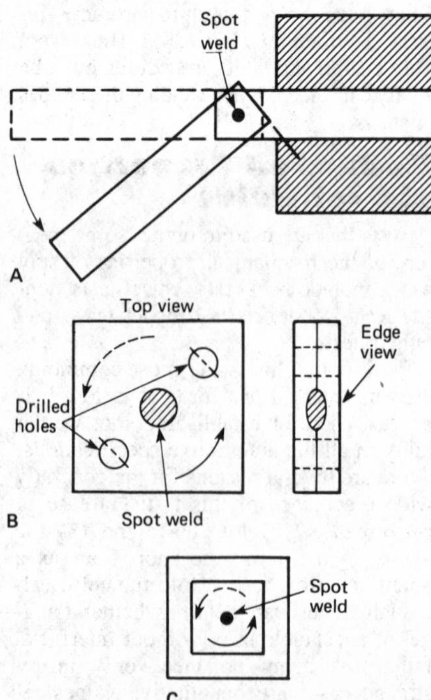

Fig. 23 Fatigue test specimens. Source: Ref 1

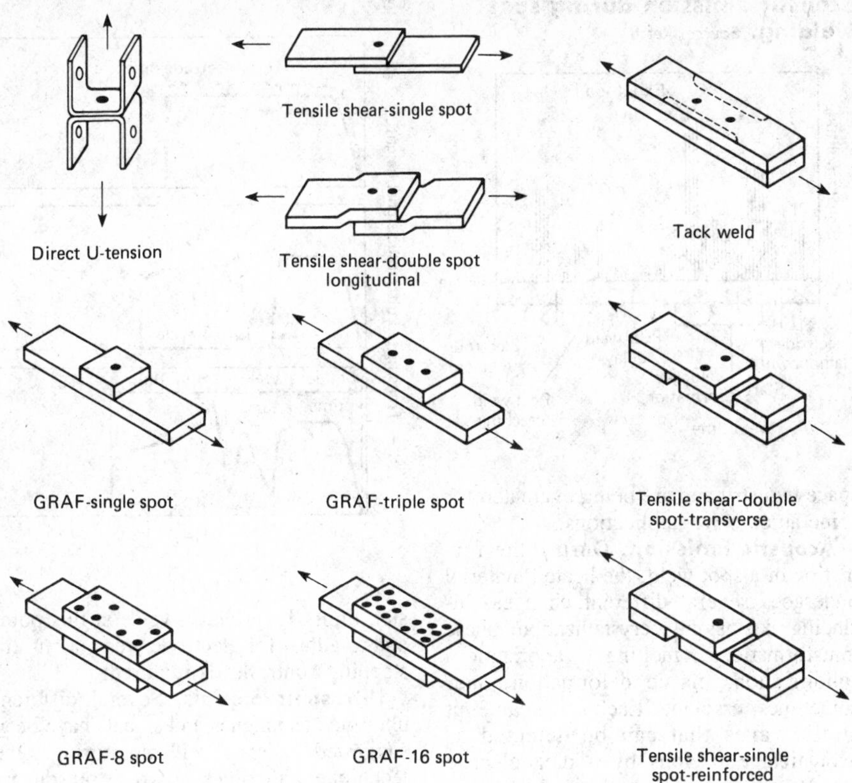

pons are not entirely satisfactory because of the effect of a different amount of magnetic material in the throat of the welding machine on welding current. They are used, however, where production parts are large or costly and, with experience in interpretation of results, can give valuable information on the quality of production welds.

Destructive tests made on low-carbon spot welds are discussed in the section on mechanical properties of spot welds in this article. Strength of the weld is usually determined by the tension-shear and tension tests. Separation of the joint as observed in one of the first five tests can be used to approximate the diameter of the weld nugget. Examination of macroetch specimens can reveal the nugget diameter and penetration and the structure of the weld.

Quality standards on nugget diameter and penetration vary with the requirements of the specific application. For maximum weld strength, a rule of thumb that can be applied as a starting point in welding most metals, in the absence of more specific requirements, is that the minimum nugget diameter should be 0.06 in. plus three times the thickness of the thinnest workpiece. Tables 2, 3, and 4, as developed by AWS, give recommended nugget diameters for spot welding low-carbon, medium-carbon, and low-alloy steels, in relation to shear strength of the welds. These nugget diameters are for welds of the highest average shear strength that can ordinarily be obtained in production with high reliability.

Penetration, or the depth that fusion extends into the outer workpieces, should be 20 to 80% of the workpiece thickness, with welds of maximum strength usually being obtained when penetration is about 70% of stock thickness. When workpieces of unequal thickness are welded, penetration into the thicker piece ordinarily need not exceed the penetration into the thinner piece. Where surface finish is critical, extent of indentation, discoloration, spatter, and sheet separation may be as important as weld strength.

The structure of a spot weld in low-carbon steel ordinarily has:

- A columnar dendritic structure in the fusion zone
- A heat-affected outer zone showing gradual transition from a coarse overheated structure through a normalized region, to the original structure of the unaffected base metal
- A narrow ferritic zone in the interface of the overheated and unaffected zones.

This zone is not always well defined, particularly at the interface of the two workpieces.

Process Variables that Affect Weld Quality. In addition to current, electrode force, and timing, the process variables that affect weld quality in RSW include:

- Ability of the welding equipment to operate consistently at the specified current in continuous production
- Closeness of fit of the components to be welded
- Strength and rigidity of fixtures
- Adequacy of clamping

When only one or a few spot welds are used to join workpieces, the probability of exceeding the duty-cycle rating of the machine is much less than when many spots are made in rapid succession. Exceeding the duty-cycle rating causes the transformer to overheat, resulting in underheating of the workpieces and production of poor-quality welds. In extreme cases, or when the occurrence is of sufficiently long duration, the equipment may shut down.

Poorly fitting workpieces may not permit proper contact. Poor contact can cause insufficient flow of current, small-diameter welds, or welds that have inadequate fusion and pull apart because of springback.

Magnetic materials in or near the throat of the machine affect kV · A demand and power factor, and should not be used in fixtures that are to be placed in the throat of the machine. The fixtures should have sufficient strength and rigidity to prevent warping of the workpieces.

As the section lengths of thick material increase, the four process variables listed above have a critical effect on spot weld quality.

Monitoring and Feedback Control*

Methods of measuring nugget formation include temperature, weld expansion and contraction, acoustic emissions, ultrasonic signals, and electrical properties such as resistance, voltage, and energy. Monitoring provides information about the quality of the product, while the controlling process corrects deviations from previously determined courses of action. To do this, the change in the monitored variable (such factors as nugget temperature, expansion, and contraction, etc.) during the production of an acceptable spot weld is re-

*This section on monitoring and feedback control has been adapted from AISI publication SG-936, Feb 1982.

corded, thus producing a continuous curve. This curve is then used as a baseline to which all subsequent production welds can be compared. When a significant variation between the monitored variable during a production run and the baseline curve occurs, microprocessor computers feed back and correct the welder setting to bring the production weld back into acceptable limits. This ensures the production of acceptable spot welds each time.

Nugget Temperature. As the spot weld nugget grows, it heats the surrounding metal. Based upon fairly rigorous calculations, the surface temperature can be related to the maximum nugget temperature at the nugget center. Several methods are used to measure the temperature of the spot weld. These include thermocouples and detection of infrared radiation emitted from the metal surface near the nugget. However, the latter method is not widely used because dirt and fumes cause spurious results and surface oxides may cause variations in surface infrared emissivity.

Weld Expansion and Contraction. Monitoring systems based on thermal expansion of the weld nugget and surrounding material during heating and melting have been developed. The thermal expansion causes electrode separation during spot weld formation, which can be monitored by a displacement transducer attached to the movable welding electrode. A typical trace obtained is illustrated in Fig. 24. An acceptable nugget can be followed through the time cycle. Welding at a low heat setting produces too small a spot weld size, whereas welding at a higher current level can produce a rapid nugget growth, which may show expulsion. When this occurs, some liquid is pushed out from the nugget itself, culminating in a rapid electrode contraction. Expansion monitoring units have been successfully used in the aerospace industry and are being evaluated for some automotive applications.

Fig. 24 Thermal expansion monitoring system. Source: Ref 7

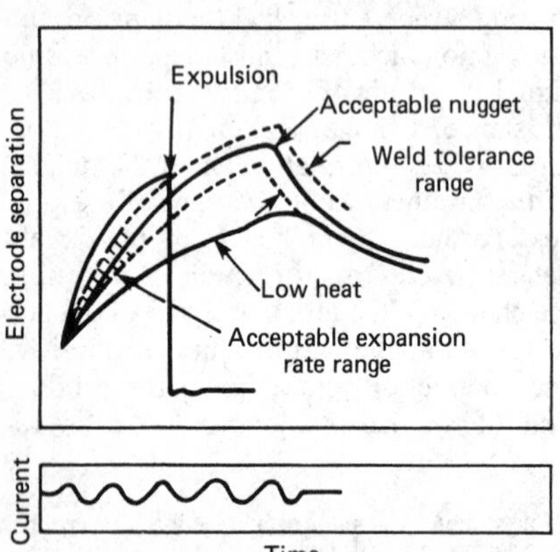

Fig. 25 Schematic signature of acoustic emission during spot welding. Source: Ref 8

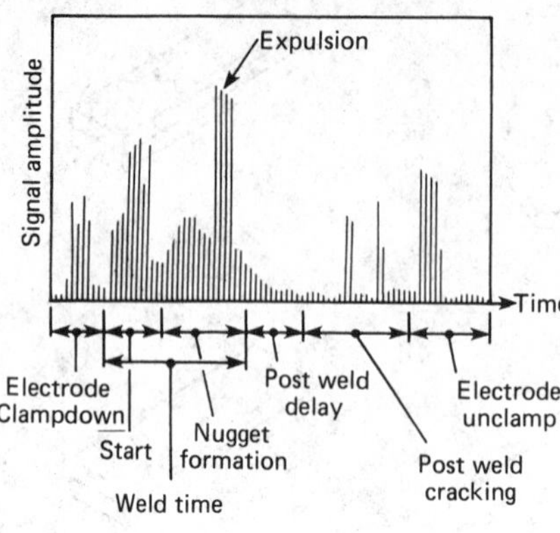

Acoustic Emission. During the formation of a spot weld, the heated material undergoes several different changes, including expansion, recrystallization, phase transformations, melting, vaporization, solidification, plastic deformation, and, sometimes, cracking. Each stage can emit stress waves that can be detected as acoustical emissions by a piezoelectric transducer mounted on the welding machine. A schematic signature of acoustic emission during spot welding is shown in Fig. 25. The first high emissions result from electrode clampdown. Additional high emissions activity occurs during welding. The optimum weld line is generally considered as lying slightly under the expulsion limit. To produce welds at this point and to adjust for electrode wear, a current stepping controller is required.

Fig. 26 Welding electrodes containing ultrasonic crystal transducers. Source: Ref 9

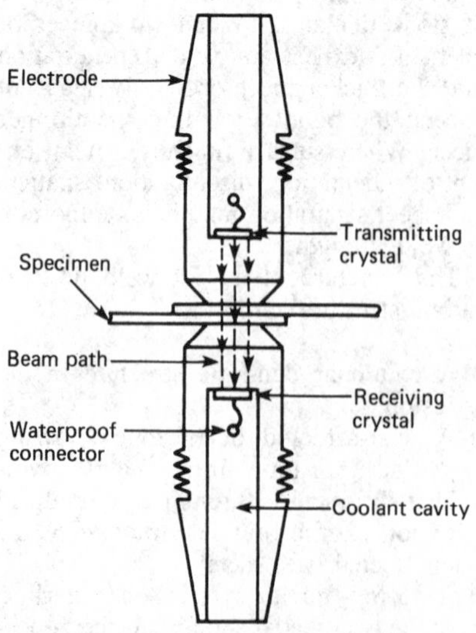

Fig. 27 Ultrasonic patterns. Source: Ref 9

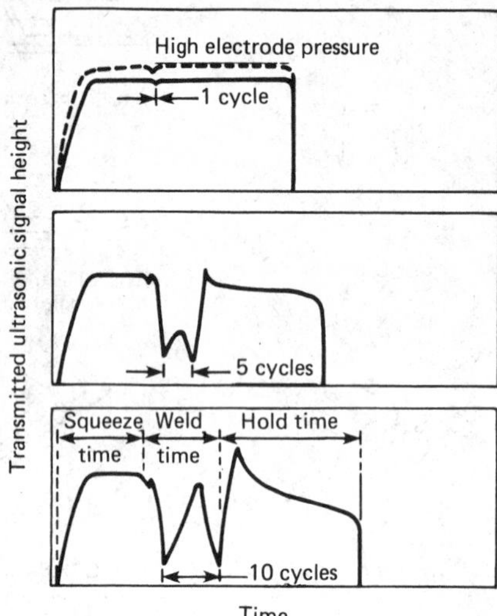

Ultrasonic Signals. Several different ultrasonic techniques and signals have been suggested for spot weld monitoring. One technique involves two transducers mounted as illustrated in Fig. 26. Ultrasonic energy pulses are transmitted from the transmitting crystal and collected at the receiving crystal. Ultrasonic patterns as a function of weld time are shown in Fig. 27. For a very low weld time (1 cycle) there is a slight signal. For a longer period of time (5 or 10 cycles), a larger curve depression is shown. By calibrating the signal, one can establish that a sufficiently large nugget can be produced.

Ultrasonic signals also can be used to test the quality of the weld after it is made. This technique utilizes a pulsed echo, single transducer method to inspect previously welded spot welds. Pulse echo peaks in inspection of a spot weld are shown in Fig. 28.

A number of difficulties have been encountered in the use of the ultrasonic monitoring techniques: a reasonable surface is necessary for good coupling, weld defects are often overlooked, and thin fragile transducers do not lend themselves to heavy-duty usage.

Electrical Characteristics. The utilization of electrical characteristics as monitoring signals has become the most popular technique. Values that are monitored include voltage, current, and time. From these values, dynamic resistance, power, and energy can be calculated. A schematic illustration of the variations in these pa-

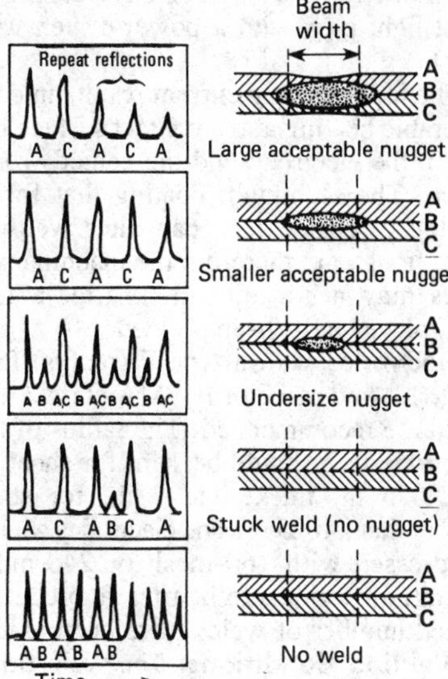

Fig. 28 Pulse echo peaks in ultrasonic inspection of spot welds. Source: Ref 10

rameters as a function of time is shown in Fig. 29. Each of these curves can be used as monitor and control signals. For instance, when using the voltage curve, a critical drop in voltage (ΔV) can be programed. When the weld conditions have met this requirement, the weld cycle is terminated, resulting in controlled nugget growth. Similarly, nugget growth can be monitored by use of the other dynamic electrical parameters as well.

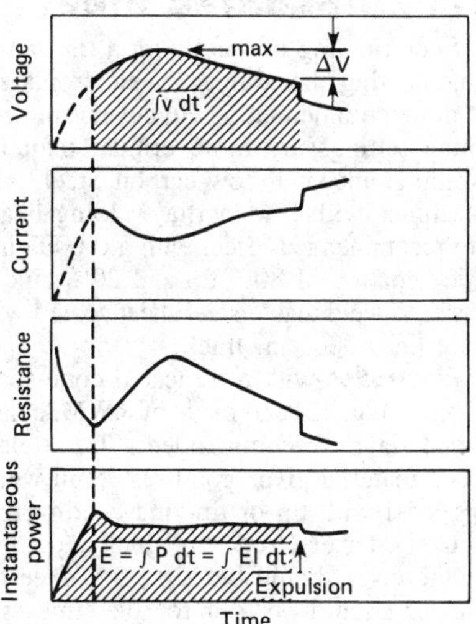

Fig. 29 Integrals of dynamic electrical characteristics during spot welding of uncoated steels. P, power; E, energy; I, current; dt, interval of time. (Source: Ref 1)

Spot Welding of Coated Steel

Low-carbon steel sheet that has been coated by hot dipping, electroplating, or other processes with corrosion-resistant metals or alloys, such as zinc, aluminum, tin, and tin-zinc, is spot welded in large-scale production. Terne-coated steel, steel coated with a lead-tin alloy, is spot welded in large volume in making automobile mufflers, but otherwise not many individual plants use spot welding on this material.

Steel sheet with metallic coatings (such as nickel and chromium plating) that are intended to serve primarily as decorative coatings can be spot welded, sometimes (depending on plating thickness and properties) under the same welding conditions as those used in welding uncoated steel. However, because of the need to avoid marking of decorative surfaces, spot welding of steel that has been plated for this purpose is relatively infrequent.

Steel that has been electroplated with a functional coating of another metal is spot welded in some high-production applications. For instance, nickel-plated steel is spot welded for use in electrical equipment, principally for terminals or contacts.

Steel sheet with phosphate or other conversion coatings is not ordinarily spot welded without removing the coating first. The high electrical resistance of these nonmetallic coatings, which varies with type and thickness of coating, makes it necessary to use high electrode forces and also may cause excessive sparking from between the workpieces when the flow of welding current is initiated. Variation in the thickness of a coating affects its electrical resistance and may interfere with welding. The electrode tips may become badly pitted, requiring frequent cleaning and reconditioning, and particles of coating material may be intermixed with the weld metal.

Steel sheet is not ordinarily resistance spot welded through paint or plastic coatings on a production basis. However, techniques that use specially designed and operated electrodes to heat and penetrate organic coatings have been used with success in volume production. The coating is softened at the weld site and is displaced into the immediately adjacent areas of the work.

Spot Welding of Zinc-Coated Steel

Direct spot welding (see circuits in Fig. 7) is recommended for zinc-coated steel, which may be either hot dip galvanized or electroplated, because the shunting current associated with series welding, when added to the higher-than-normal current needed to weld zinc-coated steel, results in excessive electrode heating and short electrode life. When large panels are welded, the welding machine can be set up as shown in Fig. 7(e) to reduce the throat depth.

Weldability of thin sheets electroplated with zinc decreases as the coating thickness increases, in the range of 0.0002 to 0.001 in. However, as thickness of sheets increases above 0.060 in., weldability increases regardless of coating thickness. The welding behavior of hot dip galvanized steel is affected by the thickness and uniformity of the zinc-iron alloy layer, as well as the thickness of unalloyed zinc.

Electrode Design. Truncated-cone (type E) electrodes (see Fig. 12) are recommended for spot welding zinc-coated steel. The included angle of the cone should be from 120 to 140°, with the smaller angle being used where fit-up problems exist. The face diameter of the electrode should be four to five times the thickness of the thinner sheet for two-thickness welds. A larger face diameter requires greater current and results in shorter electrode life.

Flat-faced (type C) electrodes are not recommended, because the melted zinc forced from the weld area forms an annular deposit on the electrode body surrounding the face. As the thickness of this deposit increases, the weld quality decreases.

When portable welding guns are used, a type F electrode having a radius of 1 to 2 in. is recommended. A smaller radius can result in increased indentation, and a larger radius in reduced electrode life.

Electrode Composition. RWMA class 2 (copper-chromium) electrode material generally provides the best electrode life. A copper-zirconium alloy also produces good electrode life, provided it has been sufficiently cold worked to achieve the hardness needed to resist plastic deformation of the electrode face. The buildup of zinc on and around the electrode face is a problem with both copper-chromium and copper-zirconium alloys, and the electrodes must be cleaned or replaced at regular intervals.

Where weld appearance is important and electrode sticking must be avoided, class 1 material can be used. The high conductivity of this alloy reduces heating at the interface of the electrode and the work. Resistance to deformation is not as great as that of class 2 material. Aluminum oxide dispersion-strengthened copper electrodes are being investigated for increased electrode life, as well as electrodes that have a flame-sprayed coating of aluminum oxide.

Cooling of electrodes is ordinarily done with a minimum water-flow rate of 2 gal/min; cooling requirements vary with electrode material and dimensions, work-metal and zinc thickness, and the welding conditions. If the water temperature exceeds 85 °F, the rate of flow to the electrode should be increased. Overheating of an electrode softens it, and deformation of the electrode face occurs. Also, the higher temperature of the electrode results in an accelerated alloying of the copper electrode material with the zinc coating on the work metal.

Welding Conditions. To produce nuggets of adequate size in zinc-coated steel, longer weld times (25 to 50% longer) and higher currents are required than for uncoated low-carbon steel. Also, the range of welding times is narrower for zinc-coated steel.

The welding current can be as much as 50% greater than that required for uncoated steel, depending on work-metal thickness. Thinner work metals require proportionately higher currents than do thick metals.

The electrode forces used for welding zinc-coated steel are 10 to 25% greater than those used for welding uncoated steel. The increase in electrode force is necessary because the coating reduces the faying surface contact resistance to almost zero; also, the softened zinc must be extruded from between the sheets as quickly as possible, lest the base metal be softened.

To produce true welds of steel to steel, which must be done for strength greater than that of a soldered-type joint, the welding conditions must be such as to melt the zinc completely between the faying surfaces of the workpieces, over an area approximately the size of the electrode face, and to squeeze out this molten zinc in this weld nugget area. Melting of the zinc on the surfaces of work in contact with the electrodes should be avoided if possible, to minimize electrode alloying and sticking, and to retain a uniform layer of zinc on the exposed work surfaces for corrosion resistance. The hold time should be of sufficient duration so that cooling by the electrodes continues long enough to avoid melting of the zinc on the work surfaces in contact with the electrodes, from the residual heat in the nugget and adjacent portions of the steel, and to allow any molten zinc in the vicinity of the weld to resolidify before the clamping force is released. The use of upslope and downslope of the welding current is sometimes helpful in producing the desired weld characteristics. Table 7 compares welding conditions for zinc-coated sheet steel.

For welding heavy-coated steel (1.50 oz of zinc per square foot), weld time must be increased by up to 45% over that used for welding commercial-coated steel (1.25 oz/ft^2), or welding current increased by up to 10%. For welding light-coated steel (0.9 oz/ft^2), weld time is decreased by 5 to 10% from that used for welding commercial-coated steel, or welding current decreased by up to 5%. The nominal coating weights given here are based on sheet size, not on total surface of two sides; for coating weight per unit of surface area, multiply by two.

Spot Welding of Aluminum-Coated Steel

Steel coated on both sides with aluminum in a hot dip process is available in two types. One type is coated with an aluminum-silicon alloy approximately 0.001 in. thick and is used in high-temperature applications up to 1250 °F. The second type is coated with pure aluminum approximately 0.002 in. thick. Spot welds having good strength can be made on both types of aluminum-coated steel.

These coatings require a slightly higher welding current, because they have high electrical and thermal conductivities. No special cleaning of the coating surface is needed. Grease and other foreign material should be removed by solvent cleaning or by a light pass with a power-driven wire brush.

The high welding current results in considerable heating at the contact surface between the electrode and the coated workpiece. The aluminum coating that forms on the electrode face can alter welding conditions, and therefore the machine settings may need adjustment after a few workpieces have been welded.

Electrodes with a type F (radius) face (Fig. 12) and made of RWMA class 2 material are recommended. The radius of the electrode face should be 1 in. for sheet up to 0.025 in. thick, and 2 in. for sheet thicker than 0.025 in. The electrodes should be dressed with 160-mesh or 240-mesh aluminum oxide cloth after a predetermined number of welds have been made.

Welding Conditions. The weld time, welding current, and electrode force are about the same as, or slightly greater than, those recommended for uncoated low-carbon steel. For work metal up to 0.030 in. thick, the welding current may be increased 15 to 25% over that used for uncoated low-carbon steel. Aluminum-zinc alloy coatings with a typical coating chemistry of 50 to 55% Al, 40 to 50% Zn, and 1 to 1.5% Si can also be readily spot welded. The spot welding parameters are close to those for galvanized steels, but welding currents are somewhat higher than for aluminum-coated steels. The electrode life for both aluminum- and aluminum-zinc coated steels is considerably higher than for galvanized steels.

Spot Welding of Tin-Coated and Tin-Zinc-Coated Steel

Spot welding of steel with a tin or tin-zinc coating, hot dipped or electroplated, is done commercially. However, the machine settings are more critical than for welding uncoated low-carbon steel, and machines with low-inertia welding heads are recommended. Steel with a 0.0003-in.-thick coating of 80% tin and 20% zinc is easier to weld than a steel sheet coated with pure tin 0.0001 in. thick.

Electrodes with a truncated-cone (type E) face (Fig. 12) and made of RWMA class 2 material are recommended. Class 1 electrode material gives good results in welding steel with tin or tin-zinc coating, but with shorter electrode life. Included angle of the cone should be 120°, and face diameter should be four to five times the thickness of the thinnest sheet.

Table 7 Conditions for RSW of 1010 steel, with and without zinc coating(a)

	Steel 0.040 in. thick		Steel 0.125 in. thick	
Welding condition(b)	Zinc coated(c)	Uncoated	Zinc coated(c)	Uncoated
Weld times, cycles (60 cps)	13	10	42	26
Welding current, A	14 000	9500	20 000	19 000
Electrode force, lb	650	500	2000	1800
Minimum welding space, in.	0.75	0.75	2	1.75
Nugget diameter, in.	0.21	0.19	0.48	0.33
Minimum contacting overlap, in.	5/8	1/2	1 1/8	7/8

(a) Data for zinc-coated steel are from AWS C1.3, "Recommended Practices for Resistance Welding Coated Low Carbon Steels." Data for uncoated steel are from Table 2 in this article. All data are for welding with class 2 electrodes. (b) For joining two sheets of steel in each of the two thicknesses. (c) Coated with 1.25 oz of zinc per square foot of sheet size

Where marking of one sheet is undesirable, a large flat-face (type C, Fig. 12) electrode can be used in contact with that sheet, and a radius-face (type F) electrode used with the second sheet, if the sheet thicknesses permit. Additional current is needed because the current density is low. Greater indentation occurs on the second sheet, and electrode life is reduced. Initially, the electrodes may adhere to the surface of the sheet, but sticking decreases after a number of welds have been made.

Welding Conditions. Relatively short weld times are used, and the welding current should be adjusted so that there is no expulsion of steel from between the sheets. However, none of the tin or tin-zinc coating should be entrapped at the faying surfaces, and to prevent formation of solder-type joints, electrode force must be sufficient to extrude the melted coating before the base metal begins to soften. Optimum electrode pressure is about 10 ksi; greater pressures cause indentation.

Spot Welding of Dissimilar Metals

Low-carbon steel can be resistance spot welded to most other ferrous metals and to many nonferrous metals, producing a weld nugget that is an alloy of the two metals. In making a spot weld between dissimilar metals, a heat balance must be achieved that compensates for the differing properties of the two metals and results in the production of a weld nugget having approximately the same thickness on each side of the interface. More heat must be provided to the more thermally and electrically conductive metal, which generates less resistive heat and has greater loss of heat by conduction. The higher melting of two dissimilar metals of approximately equal conductivity also needs more heat.

If two metals being welded do not differ greatly in conductivity, a satisfactory heat balance can be obtained by using an electrode of smaller face diameter, or an electrode fitted with a facing of higher resistivity, such as an RWMA group B electrode material, on the more conductive member of the joint. To compensate for a greater difference in conductivity between the two work metals, both techniques can be used simultaneously.

In a variation of facing an electrode with a higher resistivity material, the more conductive workpieces can be provided with a thin layer of a less conductive metal on its surface, either by a coating method such as electroplating, or by inserting a layer of poorly conducting foil between the workpiece and the electrode.

Another technique, which can be used alone or in combination with one or more of the other methods, is to increase the thickness of the more conductive member of the joint.

Roll Spot Welding

Resistance roll spot welding consists of making a series of separate, spaced spot welds in a row by means of resistance seam welding machines and electrode wheels. The roll spot welds are made without retracting the electrode wheels or removing the electrode force between spots.

Current, welding time, and electrode force are essentially the same as for RSW of similar workpieces. An individual roll spot weld has the same appearance as a spot weld made in the usual manner, except that the weld nugget is usually slightly elongated in the direction of electrode travel.

Roll spot welds are made using either continuous or interrupted motion of the electrode wheel, depending on the weld time needed to develop the necessary heat, and the hold time needed to solidify the weld metal (with continuous motion of the electrode wheel, hold time is zero). The desired weld spacing is obtained by adjustment of the electrode wheel speed and of the current on and off times.

Nugget width, transverse to direction of travel, is determined primarily by the contacting width of the periphery of the electrode wheel. When electrode motion is continuous during welding, nugget length is a function of electrode diameter, rotation speed, and heating time; when interrupted electrode motion is used, nugget length depends mainly on wheel diameter and heating time. For some roll spot welded assemblies, nugget width may be governed chiefly by the shape and dimensions of one of the components. Design and arrangement of fixtures and electrodes must be coordinated to ensure accurate matchup and intimate contact of the workpieces at all weld spots for roll spot welding.

REFERENCES

1. Dickinson, D.W., AISI Report SG 81-5, Aug 1981
2. Nadhorni, A.V. and Weber, E.P., *Welding Journal Research Supplement,* Vol 56 (No. 11), Nov 1977
3. Connel, L.D., *Metal Construction,* Vol 9 (No. 1), Jan 1977
4. Williams, N.T. and Jones, T.B., *Metal Construction,* Vol 11 (No. 10), Oct 1979
5. Dickinson, D.W. and Natale, T.V., SAE Paper 810353, Feb 1981
6. Schneider, E.J., *et al.,* Proceedings of the 22nd Mechanical and Seal Processing Conference, Oct 29, 1980
7. Johnson, K.I., *Metal Construction,* Vol 5 (No. 5), May 1973
8. Havens, J.R., SME Paper AD76-279, 1976
9. Burbank, G.E. and Taylor, W.D., *Welding Journal,* Vol 44 (No. 5), May 1965
10. Papadakis, E.P., *Physical Acoustics,* Vol XII, Academic Press, New York, 1976

Resistance Seam Welding*

RESISTANCE SEAM WELDING (RSEW) is a process in which heat caused by resistance to the flow of electric current in the work metal is combined with pressure to produce a welded seam. This seam, consisting of a series of overlapping spot welds, is normally gastight or liquid-tight. Two rotating, circular electrodes (electrode wheels), or one circular and one bar-type electrode, are used for transmitting the current to the work metal. When two electrode wheels are used, one or both wheels are driven either by means of a gear-driven shaft or by a knurl or friction drive that contacts the peripheral surface of the electrode wheel.

The series of spot welds is made without retracting the electrode wheels or releasing the electrode force between spots, although the electrode wheels may advance either continuously or intermittently. The magnitude of the current, the duration of current flow, the electrode force, and the speed of workpiece or electrode travel are all related and must be properly chosen and controlled to produce a satisfactory resistance seam welded joint. The principles described in the article ''Resistance Spot Welding'' (RSW) in this Volume are applicable also to RSEW.

Applications

Resistance seam welding can be applied to a variety of workpiece shapes. Girth welds can be made in round, square, or rectangular parts by using electrode wheels of suitable diameter. Longitudinal welds can be made by using two electrode wheels or by using one wheel and a mandrel or stationary bar over which the wheel travels. When two electrode wheels are used, they can be mounted either on parallel shafts or, if necessary for workpiece clearance, on shafts at an angle to each other.

Most resistance seam welds are lap seam welds. However, by special techniques, discussed in the sections on butt seam welds and foil butt seam welds in this article, butt seam welds can be made by resistance heating the two abutting edges and lightly forging them together or by placing foil strip on one or both surfaces adjacent to the joint before applying heat and force with the electrodes.

Advantages of RSEW, as compared with resistance spot and projection welding, are:

- Gastight or liquid-tight joints can be produced.
- Overlap can be less than for resistance spot welds or projection welds, and seam width can be less than the diameter of spot or projection welds.

Limitations of RSEW, apart from those it shares with spot and projection welding, are:

- The weld ordinarily must proceed in a straight or uniformly curved line.
- Obstructions along the path of the electrode wheel must be avoided or be compensated for in the design of the wheel.
- Sharp corner radii or abrupt changes in contour along the path of the electrode wheel must be avoided.
- Length of joints made in a longitudinal seam welding machine is limited by the throat depth of the machine.
- Fatigue life of resistance seam welds is usually less than that for welds made by other seam welding methods.
- Stock thicknesses greater than $^1/_8$ in. are more difficult to weld than by spot or projection welding.

Metals Welded. Low-carbon, high-carbon, low-alloy, high-strength low-alloy (HSLA), stainless, and many coated steels can be resistance seam welded satisfactorily. Alloys with carbon levels above about 0.15% may tend to form areas of hard martensite upon cooling. In critical applications the welds may require a postweld tempering to reduce the hardness and brittleness. In some instances this can be done in the welding machine (see the section on methods of welding—intermittent motion in this article). Aluminum, aluminum alloys, nickel, nickel alloys, and magnesium alloys can be seam welded, but seam welding is not recommended for copper and high-copper alloys. Compatible combinations of dissimilar metals and alloys also can be seam welded. For information on the seam welding of stainless steel, aluminum alloys, and copper alloys, see the related articles in this Volume.

Preweld Cleaning. Resistance seam welding of surfaces contaminated with grease, paint, scale, or rust results in nonuniform welds, porosity in the weld nugget, and excessive burning and marking of workpiece surfaces in the region of contact with the electrodes. It can also cause rapid deterioration of the electrode wheel face. Methods suitable for cleaning the surfaces of ferrous and nonferrous metals prior to welding are described in Volume 5 of the 9th edition of *Metals Handbook*. See also the section on the effect of the workpiece surface condition on heating in the article ''Resistance Spot Welding'' in this Volume.

Seam Welding Machines

A seam welding machine is similar in construction to a spot welding machine, except that one or two electrode wheels are substituted for the spot welding electrodes. Generally, seam welding is done in a press-type resistance welding machine. Most seam welding machines are powered by alternating current; some are designed for use with three-phase, but a majority operate on single-phase alternating current. Stored-energy seam welding machines have been built, but this type finds little use. Equipment needed for seam welding must be capable of:

- Delivering low-voltage, high-amperage current
- Supporting the electrodes and workpiece and applying the electrode force

*Revised by David W. Dickinson, Supervisor, Welding and Flat Rolled Research & Development, Republic Steel Corp.

- Moving the workpiece or driving the electrodes
- Regulating, timing, and sequencing the application of the welding current and force and controlling the rate of movement of the work between the electrodes

Power supplies used for RSEW are similar to those used for RSW. They consist of a welding transformer, usually with a tap switch in the primary circuit, and a secondary circuit with electrodes for transferring the welding current to the work metal.

Electrode Force and Support. The upper electrode wheel is mounted to, and insulated from, the operating head. The head, which is actuated by a direct-acting air or hydraulic cylinder, applies the electrode force. The lower electrode is either a wheel, a platen, or a mandrel and is mounted on a supporting arm, table, or knee. The supporting element can be adjustable for applications in which the work metal must be maintained at a constant level above the floor.

Electrode or Workpiece Drives. Workpieces are moved by rotating the electrodes with knurl or friction drive, with gear drive, or by clamping the workpieces to a bar electrode and moving the bar electrode.

In knurl or friction drive, one or both of the electrodes are rotated by a knurl or friction wheel on the peripheral surface of the electrode wheel to provide constant linear speed, regardless of the diameter of the electrode wheel. The electrode wheel is continuously trimmed by the drive wheel and thus is prevented from mushrooming. Knurl drive, in which ridges or beads on the roll-shaped steel drive wheel aid in turning the electrode, minimizes slippage and is more positive than friction drive, in which the turning surface of the drive wheel is smooth. Knurl drive is used when welding coated metals, such as galvanized steel and terneplate, in any application where the electrodes are likely to pick up material from the work metal and for scaled stock that cannot be precleaned. The knurled wheel removes much of the pickup from the electrode wheels. Knurl drive usually is not employed where the highest weld quality and appearance are required. Knurl or friction drive can be used only with electrode wheels that have a diameter large enough to allow clearance between the drive wheel and the workpiece.

Gear drive is used with small-diameter electrode wheels that cannot be driven by the use of knurled or friction wheels because of interference with workpiece clearance or where the application cannot tolerate an electrode wheel that has been roughened by a knurled drive roll. With gear drive, in which the mounting shaft of the electrode wheel is rotated, there is no minimum limitation on electrode wheel diameter.

In a standard seam welding machine, there is a minimum distance between electrode wheel centers, and if one wheel must be small to clear the workpiece, the other must be correspondingly large. If the ratio of the electrode wheel diameters is greater than 2 to 1, the smaller wheel should be driven, to minimize slippage. Welding speed can be kept constant in spite of wheel wear and re-dressing by a variable-speed gear-drive mechanism.

When a mandrel or a platen is used as the lower electrode, the operating head carrying the upper electrode may be mounted in a carriage. The carriage is moved by an air or hydraulic cylinder, or by a motor-driven screw, and thus the upper electrode is passed over both the workpiece and the lower electrode. The upper electrode is free-wheeling, but it may be equipped with an idling knurl for dressing. In some machines, the workpiece is clamped to a bar-type lower electrode and moved under an idling fixed-position upper electrode wheel.

Types of Machines. There are four basic types of RSEW machines:

- *Circular*: Axis of rotation of the electrode wheels is at right angles to the front of the machine (Fig. 1a). This type is used for circular work, such as welding the heads on containers and for flat work requiring long seams.
- *Longitudinal*: Axis of rotation of the electrode wheels is parallel to the front of the machine (Fig. 1b), and throat depth is typically 12 to 36 in. This type is used for welding short longitudinal seams in containers, for attaching pieces to containers, and for similar work. Machines with traveling heads or traveling electrodes, in which a mandrel or a platen is used for the lower electrode, are normally of the longitudinal type.
- *Universal*: A swivel-type head and interchangeable lower arms allow the axis of rotation of the electrode wheels to be set either at right angles or parallel to the front of the machine.
- *Portable*: Work is clamped in a fixture, and a portable welding head is moved over the seam. This type of machine is used for workpieces that are too bulky to be handled by regular machines or when it is more efficient to move a portable welding head to the work rather than to move the work to a regular ma-

Fig. 1 Position of electrode wheels on RSEW machines

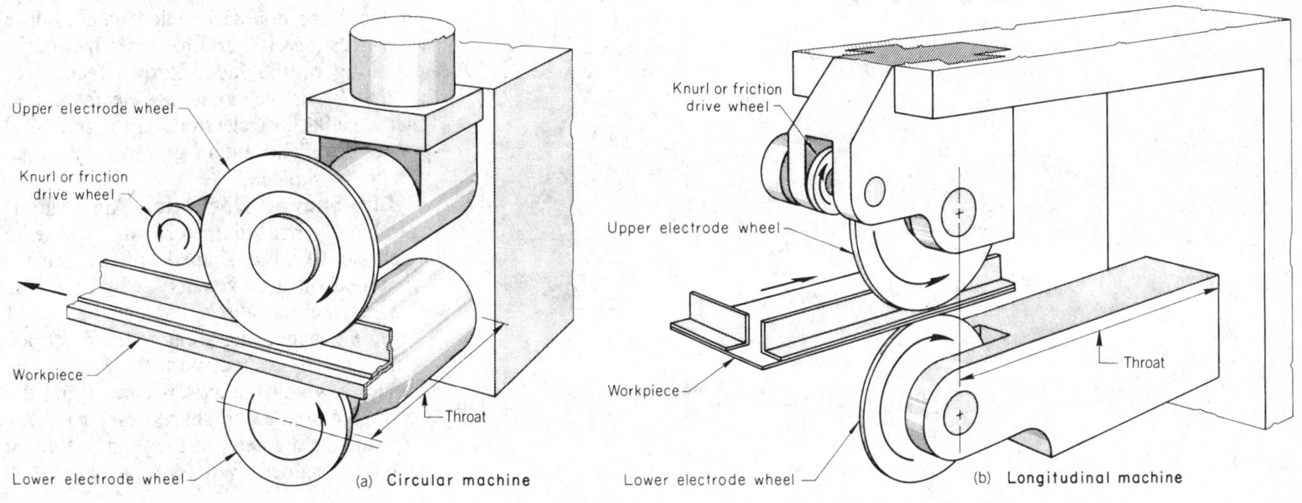

(a) Circular machine

(b) Longitudinal machine

Table 1 Common sizes of electrode wheels used for RSEW

Machine size	Wheel diameter, in.	Wheel width(a), in.
Small	7	$^3/_8$
Medium	8	$^3/_8$-$^1/_2$
Large	10-12	$^3/_8$-$^3/_4$

(a) Data are for body width; see Fig. 4 and footnote (c) in Table 2.

chine. The portable welding head is moved by motor-driven wheels, and an air cylinder mechanism provides the electrode force.

Controls for RSEW are generally similar to those for RSW, except for those differences related to the relative motion of the work and the electrodes. Because of the brevity of the heating and cooling (current on and off) cycles, a synchronous timing control connected to the primary circuit of the transformer is necessary for consistently accurate timing, unless the welding speed is such that the frequency of the welding current itself acts as an interrupter. A phase-shift heat control is also used, as well as tap switches for changing the transformer output.

Electrodes

Electrode wheels made of Resistance Welder Manufacturers Association (RWMA) class 1 material have been used for seam welding aluminum and magnesium alloys, galvanized steel, and tin-plated steel. With adequate cooling, the high electrical and thermal conductivities of this material keep the electrode temperature below that at which work metals such as aluminum readily alloy with copper from the electrode to cause electrode sticking or pickup.

High-production seam welding of low-carbon and low-alloy steels is usually done with electrodes made of class 2 or 3 material. Class 3 materials have higher mechanical properties and lower electrical conductivity and are used in applications where electrode pressure and workpiece resistance are high. In such applications, the resistance of the electrode wheel to deformation is more important than its electrical conductivity.

Size of Electrodes. Electrode wheels range in diameter from 2 to 24 in. Table 1 shows the diameters and body widths of electrodes most often used, in relation to the size of the welding machine. Narrower wheels are used in machines with knurl or friction drive; wider wheels, in gear-drive and idler machines.

A small-diameter electrode wheel (Fig. 2a), or a large-diameter wheel mounted on a canted axis (Fig. 2b), can be used to avoid interference with a sidewall when a narrow flange is being welded on an inside radius or on a re-entrant curve. In both applications, the diameter of the opposing wheel must be selected in accordance with the spacing of the mounting shafts. Canting of electrode wheels usually is limited to 6 to 10° to minimize the effect of this type of loading on the machine components. Larger angles can be used if the machine is suitably designed.

When the sides and flanged ends of square or rectangular containers are being joined, welding of corners requires electrode wheels of different diameters. An electrode wheel with a radius smaller than the corner radius of the workpiece is used to contact the inside of the flange joint at the corners. Because the inner driving wheel has a shorter circumference, it has a shorter life than the outer wheel. Beads, ribs, and other extensions on the surfaces to be seam welded can be passed over by cutting notches in the electrode wheels (Fig. 3).

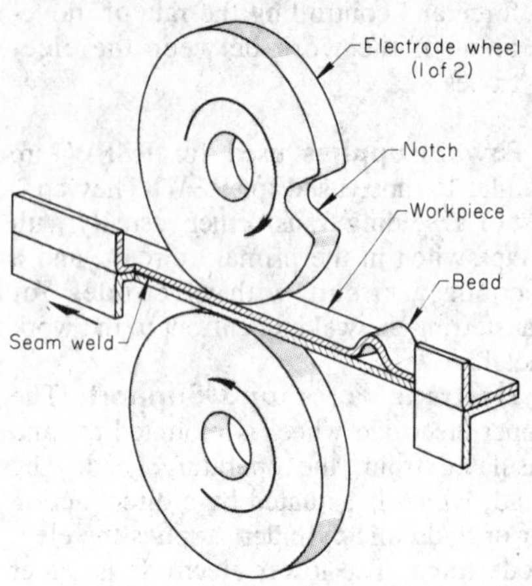

Fig. 3 Notched electrode wheel for seam welding of a workpiece having obstructions in the path of the wheel

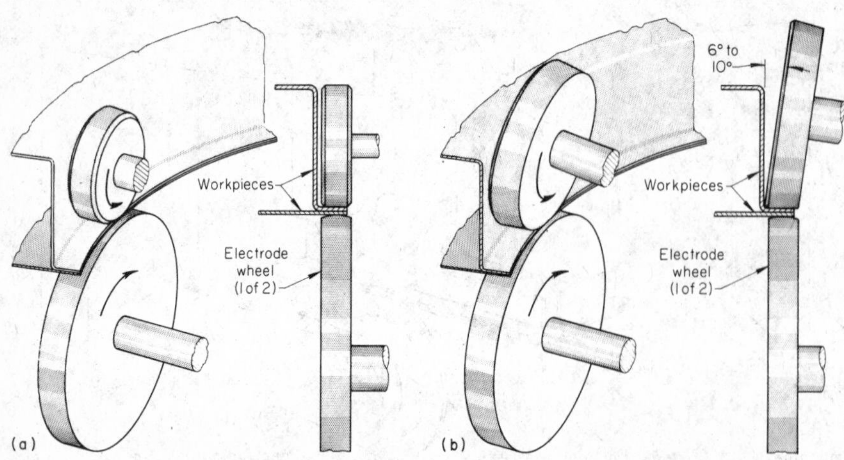

Fig. 2 Upper electrode wheels used to avoid interference with a sidewall. (a) Small-diameter wheel. (b) Canted, large-diameter wheel

When thick work metal is being welded, the higher current and force requirements necessitate increasing the electrode face width (Table 2) to avoid excessive pressure and current density. Typical electrode face widths range from $^3/_{16}$ to $^1/_2$ in. for the wheel body widths listed in Tables 1 and 2. The width of a resistance seam weld is usually about 80% of the width of the electrode face.

Electrode Face Contours. The four basic electrode face contours in common use are straight flat, single-bevel flat, double-bevel flat, and radius (Fig. 4). Flat-face electrodes, although more difficult to set up, control, and maintain than radius-face electrodes, may be necessary for welding certain workpieces. Electrode wheels must be of sufficient body width (W in Fig. 4) to provide stability, but can be single or double beveled to provide a narrower face (w) and to minimize mushrooming of the face. Radius-face electrodes (R), or one radius-face electrode used with one flat-face electrode, give good weld appearance and aid in guiding the travel of the workpiece.

Cup-Shaped Electrodes. Applications producing small-diameter, ring-type seam welds have a cup-shaped upper electrode mounted on a canted axis (Fig. 5). The cup axis gyrates in a cone-shaped path, bringing succeeding areas of the electrode cup rim into contact with the work. The angle at which the axis is canted and the diameter of the cup-shaped electrode determine the diameter of the seam weld that can be obtained. With this arrangement,

Table 2 Recommended practices for seam welding of low-carbon steel(a)

Thickness (t) of thinnest outside piece(b), in.	Electrode dimensions: Wheel width (W)(c), in.	Electrode dimensions: Face width (w)(c), in.	Net electrode force, lb	Heat time, cycles (60 Hz)	Cool time (pressure tight), cycles (60 Hz)	Welding speed, in./min	Welds per in.	Welding current (approx), A	Minimum contacting overlap(d), in.
0.010	3/8	3/16	400	2	1	80	15	8 000	3/8
0.021	3/8	3/16	550	2	2	75	12	11 000	7/16
0.031	1/2	1/4	700	3	2	72	10	13 000	1/2
0.040	1/2	1/4	900	3	3	67	9	15 000	1/2
0.050	1/2	5/16	1050	4	3	65	8	16 500	9/16
0.062	1/2	5/16	1200	4	4	63	7	17 500	5/8
0.078	5/8	3/8	1500	6	5	55	6	19 000	11/16
0.094	5/8	7/16	1700	7	6	50	6	20 000	3/4
0.109	3/4	1/2	1950	9	6	48	5	21 000	13/16
0.125	3/4	1/2	2200	10	7	45	5	22 000	7/8

(a) Data are for 1010 steel; material should be free of scale, oxide, paint, grease, and oil. Welding conditions are determined by thickness of thinnest outside piece. (b) Data for total thickness of pile-up not exceeding $4t$. Maximum ratio between thicknesses, 3 to 1. (c) For RWMA class 2 electrode material (minimum conductivity, 75% IACS; maximum hardness, 75 HRB. Wheel width is overall width of a flat-face wheel or of a wheel with a 3-in.-radius face; face width is that of the contacting surface of flat-face wheels (Fig. 4). (d) For larger assemblies, minimum contacting overlap indicated should be increased 30%.
Source: "Recommended Practices for Resistance Welding," AWS C1.1; also *Welding Handbook*, 6th Ed., Section 2, American Welding Society, 1969. With permission.

Fig. 4 Face contours of electrode wheels used for seam welding. *W is wheel width and w is width of contact surface on flat and beveled wheels*

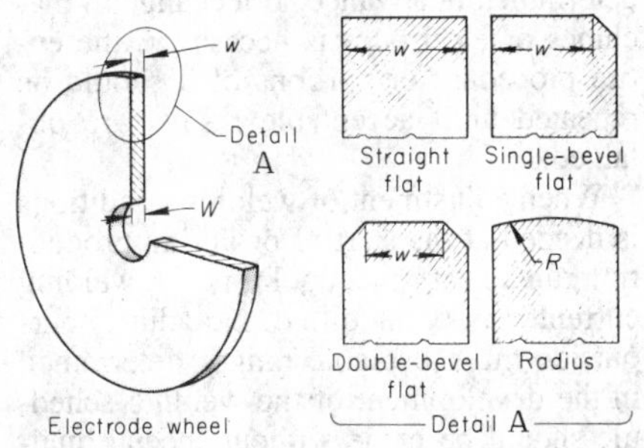

the work is engaged by the radial contact surface of the electrode. The design of the lower electrode depends on the workpiece. Figure 5 shows an annular lower electrode used for welding a flanged workpiece. A disadvantage of a gyrating cup-shaped electrode is that it may skid and cause distortion of the work metal, especially when welding thin material.

Bar electrodes, made of the same materials as electrode wheels, are sometimes used as lower electrodes. The size and shape of a bar electrode depend on the size, shape, and arrangement of the work.

Cooling of electrodes and work during welding is mandatory. It is usually done by flooding or by directing jets of water on the electrodes and the work. In applications where such cooling techniques have adverse effects on workpieces, mounting shafts can be cooled, and internally cooled electrodes can be designed. Inadequate cooling results in overheating of the work metal and shortening of the life of the electrodes. In welding low-carbon steel, a 5% borax solution is sometimes used to minimize rusting.

Maintenance of the wheel face contour within established limits is essential to the control of current density, electrode pressure, and consistency of weld quality. Only light cleaning of the electrode face should be done while the electrode is in the machine. Dressing the electrode in a lathe after a predetermined number of welds or length of weld seam maintains a uniform contour, saves electrode material, and reduces the number of rejected workpieces.

Automatic wheel dressers have been used in large high-production machines, such as those used in seam welding of roof decking. The use of knurl or friction drive or of automatic wheel dressers does not necessarily ensure removal of pickup. The electrode wheels should be periodically inspected to determine the need for supplementary, off-machine dressing.

Methods of Welding

Continuous motion is used for joining workpieces less than about 3/16 in. thick. Optimum electrode wheel speed depends on wheel diameter, cooling methods that are available or permissible, and the thickness and surface condition of the work metal. For a given cycle of heat and cool time, the peripheral speed of the electrode wheel determines the number of welds per inch and, therefore, the tightness of the joint. Table 3 lists the number of spot welds per inch needed to produce a gastight seam in various thicknesses of low-carbon steel.

Fig. 5 Free-rotating, cup-shaped upper electrode mounted on a canted axis for making a flat, circular seam weld

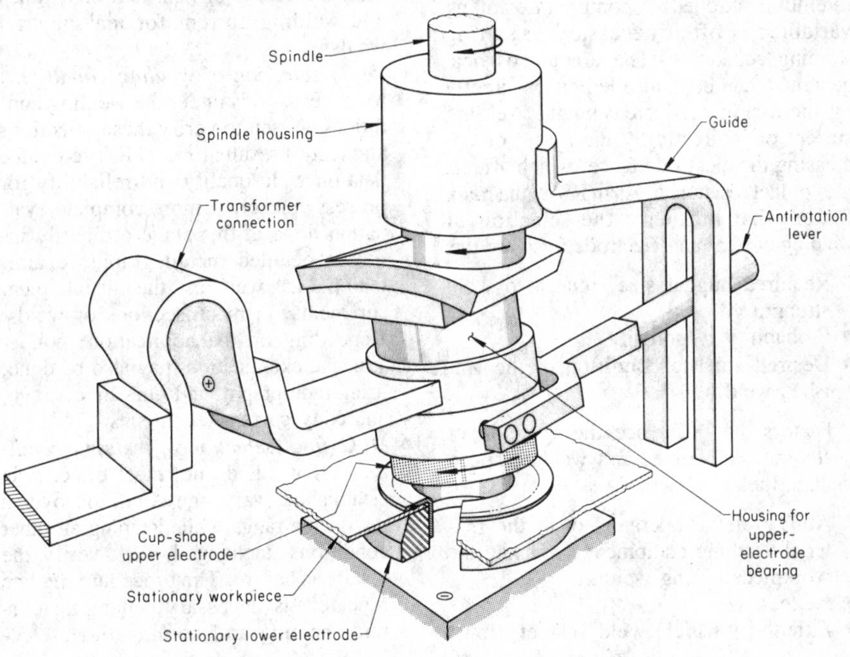

Table 3 Approximate number of spot welds per inch needed to produce a gastight seam in low-carbon steel

Thickness of one sheet, in.	Spot welds per in.
0.010	18-24
0.016	16-20
0.020	15-18
0.025	14-16
0.032	12-14
0.051	12-14
0.064	11-13
0.081	10-12
0.102	9-11
0.125	8-10

The application of the welding current can be either continuous or interrupted as explained in the section on the effect of welding current in this article.

Intermittent motion is used for joining metals more than $^3/_{16}$ in. thick or where postheating or forging cycles are used. The electrode wheels are stopped as each spot weld is made and then are automatically rotated to move the work the proper distance for the next weld. A synchronous precision electronic timer or a solid-state nonsynchronous timer is recommended for controlling the electrode motion and the application of current. Mechanical timers are slow, nonsynchronous, and otherwise inaccurate and are not recommended.

Control of Welding Conditions

Welding current, electrode force, heat time, cool time, and welding speed all directly affect weld quality and must be carefully controlled. A compromise among variables is often necessary in setting welding schedules. The amount of heat generated can be controlled either directly by increasing or decreasing the welding current or indirectly by increasing or decreasing the electrode force, which affects the contact resistance. As in RSW, the main factors that influence the selection of welding current and electrode force are the:

- Required nugget size, related to weld strength
- Probability of porosity
- Desired surface condition of the finished weld

Factors that influence the selection of heat time, cool time, and welding speed include the:

- Ability of the operator, or of the material-handling equipment, to handle the workpiece during welding
- Nugget size
- Extent to which weld nuggets must overlap to ensure pressure-tight joints
- Probability of electrode pickup

Suggested practices for RSEW of low-carbon steel are given in Table 2; values listed are starting conditions and should be adjusted to suit the application. When welding high-strength steel and HSLA steels, additional electrode force is often recommended to overcome increased resistance to deformation and springback.

Welding Schedules. The sequence of steps followed in setting up a RSEW schedule may vary, depending on:

- Special aspects of the equipment or work
- Extent of experience with similar types of work
- Requirements imposed by the purchaser or applicable specifications

When previous experience is lacking, the most satisfactory conditions for RSEW can be achieved with this typical sequence:

- *Establish criteria for satisfactory welds*: Generally, the weld nugget should have no porosity. To yield leaktight seams, nugget overlap should be about 15 to 20% of the nugget length. Nugget penetration should average 45 to 50% of the thickness of the thinnest sheet, but it should be within the limits of 30 to 70%.
- *Make preliminary selection of welding conditions*: Use working values that are not near the upper or lower limits of the welding machine's capability. The suggested practices shown in Table 2 provide starting points for choosing preliminary values of heat and cool times and welding current for making trial welds.
- *Verify selection of welding conditions*: Make test welds under the welding conditions chosen to verify these selections and record results to establish reference data on weld quality and reliability for process control. A more complete evaluation done at this stage ordinarily involves detailed metallographic examination, as well as the usual measurements on macrosections of welds. Depending on the application, nondestructive examination may also be done, using radiography and ultrasonic, sonic, and eddy current techniques.
- *Make final selection of welding conditions*: If welds do not meet the criteria established, vary one condition over a reasonable range while keeping all other conditions unchanged, then verify the results as before. The procedure may be repeated as necessary, changing current, heating and cooling time, electrode force, and welding speed until criteria are met. The information in the following sections on effect of welding current, effect of electrode force, effect of heat time and cool time, and effect of welding speed must be understood in order to estimate which condition should be increased or decreased to produce the desired results. Because each condition affects the results of the remaining conditions, the creation of a welding schedule must be approached with considerable thought.

The steps above not only determine optimum values for each of the welding variables, but also establish ranges of satisfactory values for them and the criticality or noncriticality of each. Changes in welding equipment, electrode material, and electrode design may also be made at this time, although these are ordinarily well determined in advance. If a change in machines or electrodes is necessary, the entire procedure or portions of it should be repeated until acceptable results are obtained.

When adjustment of welding conditions is needed at the start of or during production runs, heat time, cool time, or welding current may be modified. No adjustments outside the acceptable ranges determined in the development of the welding schedule should be made without special qualification by the evaluation of a suitable number of trial welds.

Welding of a corner arc is more difficult than welding of straight sections, especially when the corner radius is small. Different and more closely controlled operating conditions generally are needed for optimum weld quality and usually involve decreased welding speed to permit good control at the workpiece corners by the machine operator. When the corner radius is large and quality requirements are not unduly critical, machine settings based on welding of the corners may also be satisfactory for the straight sections.

Where a machine setting does not produce an acceptable seam weld on corners and straight sections, a two-speed control or automatic changing of heat settings can be used. The two-speed system permits taking advantage of the higher welding speed possible on straight sections.

Quality Control. General practice in quality control is similar in RSEW and RSW. One test method that is not applicable to spot and projection welding, but is especially useful in evaluating seam welds, is pressure testing to determine if seams are liquid-tight or gastight. A standard method for pressure testing of resis-

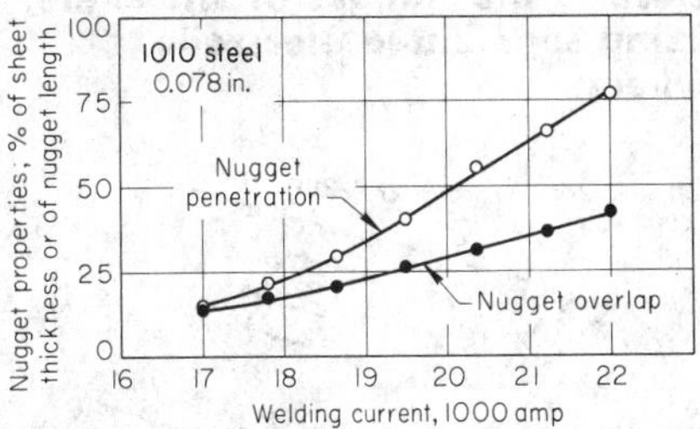

Fig. 6 Effects of welding current on nugget penetration and nugget overlap in seam welding. Welding conditions: heat time, 6 cycles; cool time, 5 cycles; electrode force, 1500 lb; welding speed, 55 in./min. (Source: RWMA Bulletin 23)

tance seam welds is the pillow test. For a comprehensive discussion of the pillow test and other standard methods for testing resistance welds, see Section V in American Welding Society (AWS) C1.1-66, "Recommended Practices for Resistance Welding."

Effect of Welding Current

In RSEW, much as in RSW, heat is generated by resistance to the flow of electric current (1) at the contact interfaces between the electrode wheels and the surfaces of the workpieces, (2) in the work metal itself, (3) at the interface of the workpieces, and (4) in the electrodes.

For a given heating-and-cooling cycle and a given welding speed, the magnitude of the current determines the depth to which the weld metal penetrates the base metal while the cycle and weld speed together control the nugget overlap. The effects of welding current on nugget penetration and nugget overlap in welding 0.078-in.-thick 1010 steel are shown in Fig. 6. See also Fig. 7 for nugget penetration and nugget width obtained at three levels of current.

Welding current values in excess of those giving full joint strength increase nugget penetration and nugget overlap, but do not add to joint strength and are uneconomical. Also, they can cause excessive indentation and, if extreme, burning of the weld. Generally, resistance seam welded joints are sufficiently strong that test samples fail in the heat-affected base metal adjacent to the weld.

Either continuous alternating current or interrupted current are used for RSEW. Continuous current can be used for high-speed operation or where the wave form can be adjusted to produce the proper nugget size and spacing at the available welding speed. Interrupted current is used for most seam welding operations and offers the following advantages:

- Better control of heat
- Provision of time between each spot weld in the seam for cooling of each nugget under pressure
- Less distortion of workpieces resulting from overheating of metal adjacent to the weld
- More uniform welds with fewer surface defects

Welding current is selected by trial and destructive testing of the weld, in relation to the other operating variables, when developing a welding schedule. With shorter heat times or with faster welding speeds, more current is required, and the probability of electrode deterioration is greater. Currents higher than those used for spot welding are required for seam welding, because of the continuous shunting of the current through the preceding welds.

Effect of Electrode Force

The amount of heat generated in the weld can be controlled by varying the welding current and, to a lesser degree, by varying the electrode force which affects contact resistance. The selection of the proper electrode force limits the range of welding current values that will produce satisfactory joints. With low electrode force, small variations in welding current have considerable influence on weld quality; therefore, electrode force should be high enough to permit a wide variation in current values.

Up to a certain point, increasing the electrode force causes a slight increase in nugget penetration and nugget width. Beyond that point, nugget penetration decreases while nugget width increases more rapidly (Fig. 7). With excessive electrode force, welding current is distributed over a wider surface because of workpiece indentation.

The effect of the electrode force on nugget penetration in seam welding 0.078-in.-thick 1010 steel using three different values of welding current is shown in Fig. 7(a). The effect of the electrode force on nugget width under the same welding conditions is shown in Fig. 7(b).

Electrode force must be sufficient to provide good electrical contact and thus permit flow of current through the circuit. To avoid low-quality welds in workpieces that do not fit-up properly, sufficient total force must be applied to bring the workpieces into contact and to maintain the force needed for welding. If the applied force is constant, variations in the force needed to bring the workpieces into contact will cause variations in the welding force, thereby affecting weld quality. Excessive electrode force results in severe indentation of the workpieces, rapid mushrooming of the electrodes, and large reduction of contact resistance, requiring an increase in current.

Fig. 7 Effects of welding current and electrode force on nugget penetration and nugget width in seam welding. Welding conditions: heat time, 6 cycles; cool time, 5 cycles; welding speed, 55 in./min. (a) Nugget penetration. (b) Nugget width. (Source: RWMA Bulletin 23)

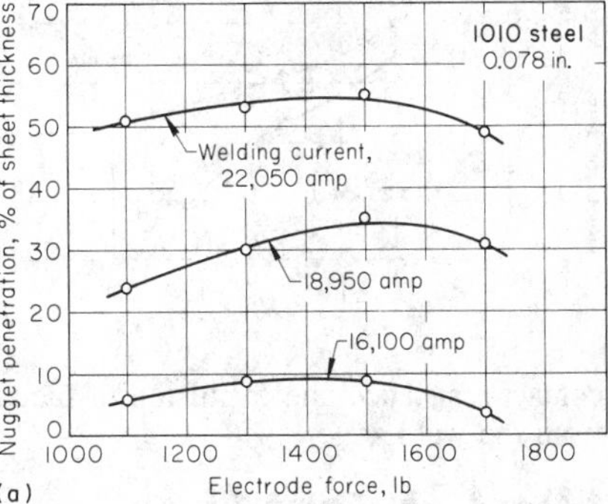

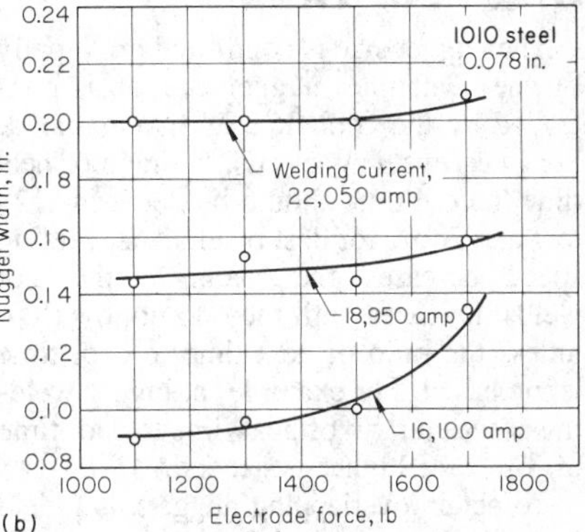

Insufficient electrode force results in improperly forged welds and in excessive contact resistance between the electrodes and the work, causing burning and, therefore, short life of the electrodes. It can also permit the expulsion of molten metal at the interface. Thus, as with excessive electrode force, surface indentation occurs at the welded seam.

The pressure at the weld is constantly changing because of the rapidly changing temperature and strength of the metal in the heated area. For best results, the welding machine should be designed with a low-inertia follow-up system. This system applies reactive force to the welding head to maintain the proper distance between the electrodes and to avoid low-pressure, high-

Fig. 8 Effects of cool time on nugget penetration and nugget overlap in seam welding. Welding conditions: heat time, 6 cycles; electrode force, 1500 lb; welding current, 18 950 A; welding speed, 55 in./min. (Source: RWMA Bulletin 23)

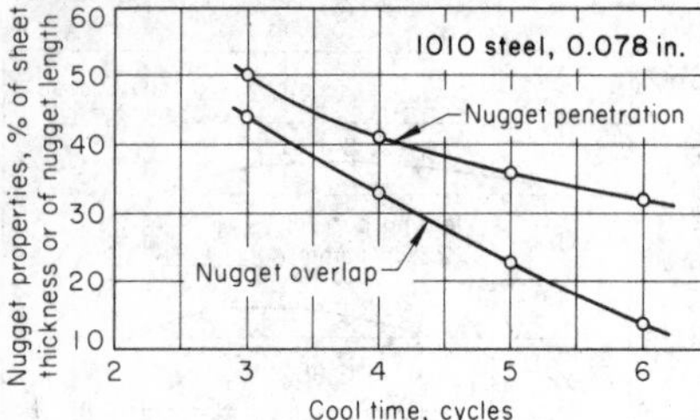

contact resistance and expulsion of metal from the weld zone.

Effect of Heat Time and Cool Time

The nugget size is controlled principally by the heat time; nugget overlap is controlled by the cool time and weld speed. For lower welding speeds, the ratio of heat time to cool time should be between 1.25 to 1 and 2 to 1 for best results. As welding speed increases, the spacing of the spot welds increases until they do not overlap, unless the ratio of heat time to cool time is increased. For example, at higher welding speed, a ratio of heat time to cool time of 3 to 1 or higher is needed.

To obtain overlapping nuggets that yield leaktight joints at a given welding speed, heat time should be adjusted to produce the required nugget properties, and cool time should be short enough to provide the required nugget overlap. The reductions in nugget penetration and nugget overlap that occur with increasing cool time in RSEW of 0.078-in.-thick 1010 steel are shown in Fig. 8.

Effect of Welding Speed

The speed at which low-carbon steel can be seam welded depends on the desired weld quality and on the design, thickness, and surface condition of the workpieces. The use of excessive welding speed causes a rapid reduction in weld strength, even with higher welding currents. As welding speed is increased, the welding current must be increased to develop sufficient heat in the work metal. Increased welding current causes surface burning of the work metal and electrode pickup, and thus, even with flood cooling, welding speed is limited.

The effect of welding speed on joint strength in welding 0.078-in.-thick 1010 steel at three different levels of welding current is shown in Fig. 9. Materials with nonuniform surface contaminants or heavily oxidized surfaces may require slower welding speeds.

Fig. 9 Effect of welding speed on joint strength in seam welding. Welding conditions: heat time, 2 cycles; cool time, 1 cycle; electrode force, 1500 lb. Pillow-type samples were seam welded at three welding current values. The samples were then inflated with air until failure occurred. (Source: RWMA Bulletin 23)

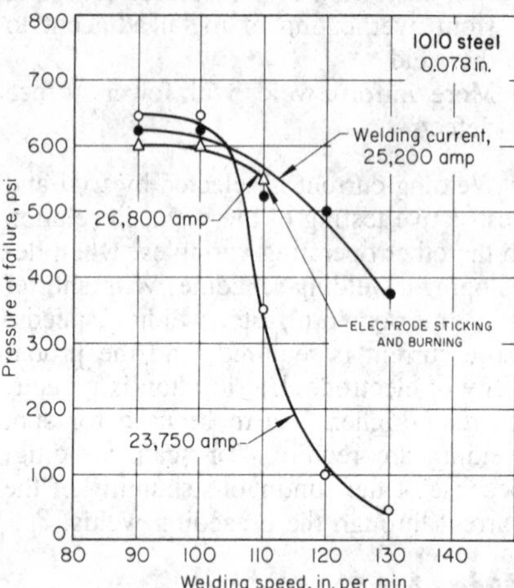

Effect of Workpiece Design on Electrode Shape

Design of workpieces can affect the shape and contour of the electrode face. A small-width seam can be made in a narrow flange by reducing the body width of the wheel at the wheel face. Sheets having stiffeners that cross the path of a seam weld require electrodes with notches, as shown in Fig. 3, if the welds are to be made without lifting the electrode from the sheet. Starting a seam weld close to the flange or sidewall of components that cross the path of a seam weld can be done by using segments of an electrode wheel (Fig. 10).

Width of Joint Overlap

The width of overlap at the joint (contacting overlap) must be sufficient to prevent collapse of the work metal at the edge adjacent to the seam. Table 2 lists recommended minimum widths of contacting overlap for RSEW of low-carbon steel ranging in thickness from 0.010 to 0.125 in.

Excessive width of unwelded overlap between parts of seam welded assemblies can result in entrapment of dirt and moisture, presenting problems in subsequent manufacturing operations or in service. By use of the proper electrode force, welding current, heat and cool time, and electrode shape, the width of the weld nugget can be adjusted to minimize unwelded overlap.

Fig. 10 Resistance seam welding close to the flanges of stiffeners, using segmented electrode wheels

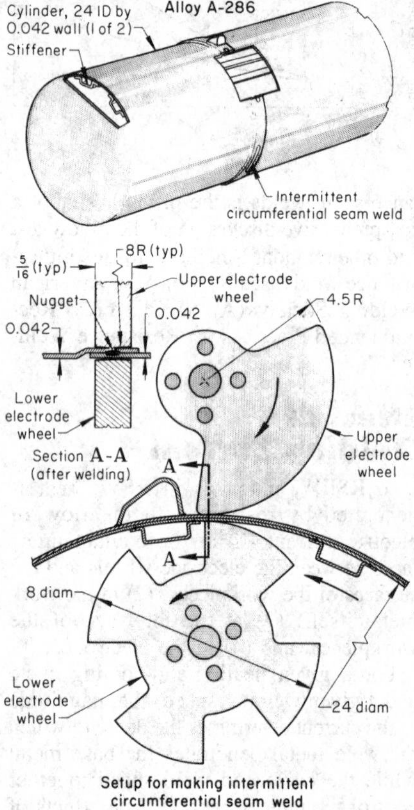

Setup for making intermittent circumferential seam weld

Equipment details

Power supply	440 V, three phase
Welding machine	Semiautomatic, circular
Rating at 50% duty cycle	125 kV·A
Heat control	Phase shift
Upper electrode	5/16 in. wide by 9 in. diam; 8-in. face radius; segmented
Lower electrode	5/16 in. wide by 8 in. diam; 8-in. face radius; notched
Electrode material	RWMA class 3

Welding conditions

Heat control setting	70%
Heat time	5 cycles
Cool time	0.5 cycle
Heating impulses	3
Electrode force:	
Welding	1500 lb
Forging	2000 lb
Spot spacing	13/in.
Welding speed	11 in./min

Fig. 11 Typical arrangements of electrode wheels for various flange-joint lap seam welds

Electrode wheel (1 of 2)
Workpiece
Single flange
Electrode wheel (1 of 2)
Workpiece
Double flange
(a) One member flanged outward
Electrode wheel (1 of 2)
Workpiece
(b) Two members flanged outward
Electrode wheel (1 of 2)
Workpiece
(c) One member flanged inward

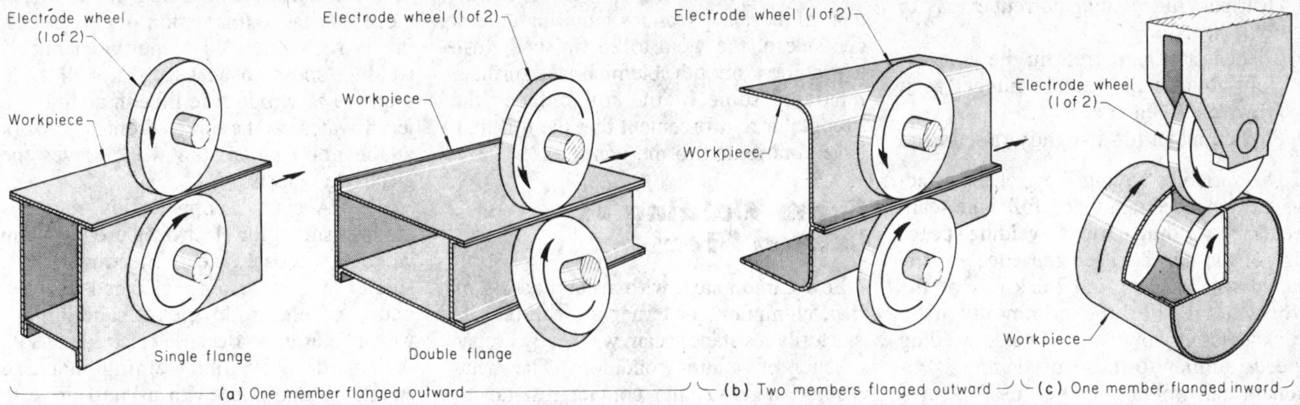

Types of Seam Welds

Several types of resistance seam welds can be made: (1) lap seam welds joining flat sheets, (2) flange-joint lap seam welds with at least one flange overlapping the mating piece, and (3) mash seam welds with work metal compressed at the joint to reduce joint thickness. Two other types of resistance seam welds made with the use of special techniques are butt seam welds and foil butt seam welds. In addition, both low- and high-frequency induction and resistance welding are used for butt welding strip stock into tubes and other shapes.

Lap Seam Welds. The most common type of seam weld is a simple lap seam in which the pieces to be welded are lapped sufficiently to prevent expulsion of weld metal. Applications include sealing of cans, water tanks, and mufflers, in which watertight or gastight joints are needed.

Flange-joint lap seam welds are used in joining assemblies having one straight member and one outward-flanged member at the joint to be welded, such as the duct in Fig. 11(a) or assemblies with two outward-flanged members, such as automotive gasoline tanks in Fig. 11(b). Common applications include containers with outturned bottoms or tops and ducts or structural parts with out-turned sides. Unless the workpiece length is less than the usable throat depth of a longitudinal machine, seam welds of this type are made on a circular machine.

Often container ends are dished for added strength, and one or both electrode wheels must be mounted at an angle to clear the workpiece. Wheel-diameter limitations may also necessitate setting a wheel at an angle (Fig 11a, right view).

Flange-joint seam welds can also be used for welding assemblies in which the flanges face inward at the joint (Fig. 11c). However, to reduce overhang of the arm supporting the lower electrode or the required throat depth of the machine, the length of the workpiece should be kept to a minimum. If the lower electrode support is small, too much overhang can result in excessive deflection of the support and thus cause inconsistent weld quality and unacceptable welds because of reduced electrode pressure.

Mash seam welds are produced by overlapping two sheets by an amount ranging from 1 to $1^1/_2$ times the sheet thickness and applying a high electrode force and a high continuous welding current. The resulting weld thickness is 10 to 25% greater than a single sheet thickness. Flat-face electrodes that are wide enough to control the weld thickness are used. Electrode force, welding current, welding speed, weld thickness, and amount of overlap are interrelated and must be accurately controlled to ensure consistent results. Mash seam welds require higher electrode force, welding current, and welding speeds than are used in conventional seam welding.

The sheet overlap determines the amount of metal to be redistributed and thus affects weld thickness. Therefore, the workpieces must be rigidly and accurately clamped or tack welded to prevent lateral motion during the welding operation. In applications where the mashed surface must be as flat as possible to facilitate porcelain enameling and to present a good appearance, a bar-type electrode is used against the surface to be enameled, and an electrode wheel is used against the other surface. When finished appearance is important, the weld can be ground or roll planished to remove surface defects.

The most satisfactory mash seam welds are made in low-carbon steel. The maximum stock thickness that can be successfully mash seam welded is about 0.060 in. This process is widely used in the manufacture of refrigerator cabinets, stoves, laundry equipment, and other products that receive porcelain enamel coatings. Stainless steel has been mash seam welded in some applications. Because of their narrow plastic range, nonferrous metals generally cannot be mash seam welded.

Butt seam welds are made by a special RSEW technique in which the two abutting edges are heated and lightly forged together. There is a slight depression where the electrode wheels compress the plastic metal. This type of weld usually has low strength.

High-quality butt seam welds can be made in tubing by resistance heating the edges of roll-formed strip stock and forging them together with squeeze rolls. The welded tube has a small upset area on each side of the joint, which is continuously removed by scarfing tools. This technique differs from upset butt welding in that the entire joint is not forged together simultaneously, but rather is forged together progressively as the electrode wheels traverse the seam.

Foil Butt Seam Welds. The strength of a butt seam weld can be increased by adding metal to the surfaces perpendicular and adjacent to the abutting edges of the joint. A thin, narrow strip of metal, usually 0.010 in. thick and $^1/_8$ to $^5/_{32}$ in. wide, is introduced on one or both surfaces adjacent to the joint as the workpiece is passed between conventional seam welding electrodes. This technique produces a smooth, nonoverlapping seam that is high in strength with good appearance. Highest quality foil butt seam welds show nugget penetration to the foil-workpiece interface, complete foil bonding, and smooth, regular surface contours. The only edge preparation required is to ensure that the sheared edges are clean and straight without excessive burr or gap when they are butted together.

The thin strip or foil acts as a bridge to:

- Distribute the welding current evenly to both sheets
- Concentrate the current in the joint
- Help contain the molten nugget as it grows and then cools
- Provide metal for a slightly raised bead

The variables having the greatest effect on the tensile strength of foil butt seam welds are welding current, welding speed, size of the gap between the edges of the butted workpieces, and thickness of the work metal. For the commonly used thicknesses of low-carbon steel, welding speeds similar to those used in conventional seam welding are permissible, provided that the gap is no wider than 0.015 in. Either interrupted current or continuous current is used for foil butt seam welding.

Foil butt seam welded assemblies are ordinarily used in the as-welded condition. In those applications requiring at least one side of the weld to be finished flush to provide a smooth, blemish-free surface, removing some of the foil reduces the strength or reinforcement that the foil may have contributed to the weld.

Seam Welding of Coated Steel

Low-carbon steel with a thin coating of zinc, aluminum, or terneplate can be satisfactorily resistance seam welded by proper selection of welding conditions. The metal coating increases the contact resistance, and thus welding of coated steel requires higher welding current and electrode force than are needed for welding similar thicknesses of uncoated steel. The welding speed must be limited to avoid excessive heating of the electrode and arcing at the workpiece interface. Immersion or flood cooling is recommended to permit a higher welding speed to be used, as well as to improve electrode face life. In addition, a jet of water can be directed onto the work at the point where the work leaves the electrode wheels.

RWMA class 2 copper alloy materials are best suited for electrodes used in seam welding of coated steel. Electrode life is shorter than in welding uncoated steel because of electrode pickup and higher welding currents. Because electrode pickup is considerable, the welding machine should be equipped with a knurl drive to break up and remove the pickup and to maintain the face width of the electrode. A scraper blade may also be needed for proper electrode maintenance.

Projection Welding

PROJECTION WELDING is a resistance welding process in which current flow and heating are localized at a point or points that are predetermined by the design or shape of one or both of two parts to be welded. The process is closely related to resistance spot welding (RSW), in which current flow and heating are localized by one or both electrode contact faces, which determine the location, size, and shape of the weld produced.

This article is concerned primarily with projection welding of low-carbon and low-alloy steels, although much of the information presented is applicable to other work metals as well. Projection welding of stainless steel, aluminum alloys, copper alloys, and nickel-copper alloys can also be performed satisfactorily.

In the most common application of projection welding, a projection, specially designed and formed on one of two conforming workpieces, is used to concentrate current flow and heating at the point where the weld is to be made.

Projections may be of any practical shape that can properly concentrate the welding current. In cross wire projection welding, the curved surfaces of two intersecting wires perform the function of a projection. The shapes of parts may also take the place of conventional projections in projection welding special types of joints.

Electrodes or welding dies are used to conduct current to the workpieces and to apply the welding force. Force is always applied before, during, and after the application of current to ensure a continuous electrical circuit and to forge the heated workpieces together. Welding dies may also hold and clamp the workpieces in proper alignment with each other before and during the welding operation.

Formation of Weld. The formation of a projection weld nugget, which depends on the design of the projection, the selection of welding conditions, and the adequacy of the resistance welding equipment, is shown in Fig. 1. In this application, 0.092-in.-thick low-carbon steel using embossed spherical projections and a weld time of 20 cycles ($^1/_3$ s) is projection welded.

In the first stage of fabrication (Fig. 1a), the workpieces are brought together under pressure without application of welding current, and the projection may be slightly compressed and indented into the surface of the mating workpiece. Figures 1(b), (c), and (d) show the stages of formation of a typical weld at 20, 60 to 70, and 100% of the weld time (or heat time).

At about 20% of weld time (Fig. 1b), collapse of the projection is nearly complete, and a pressure weld is formed. A nugget of fused weld metal does not begin to form until about 50% of the weld time has elapsed. At about 60 to 70% of weld time (Fig. 1c), fusion has progressed a

Fig. 1 Development of weld nugget during projection welding of embossed spherical projections

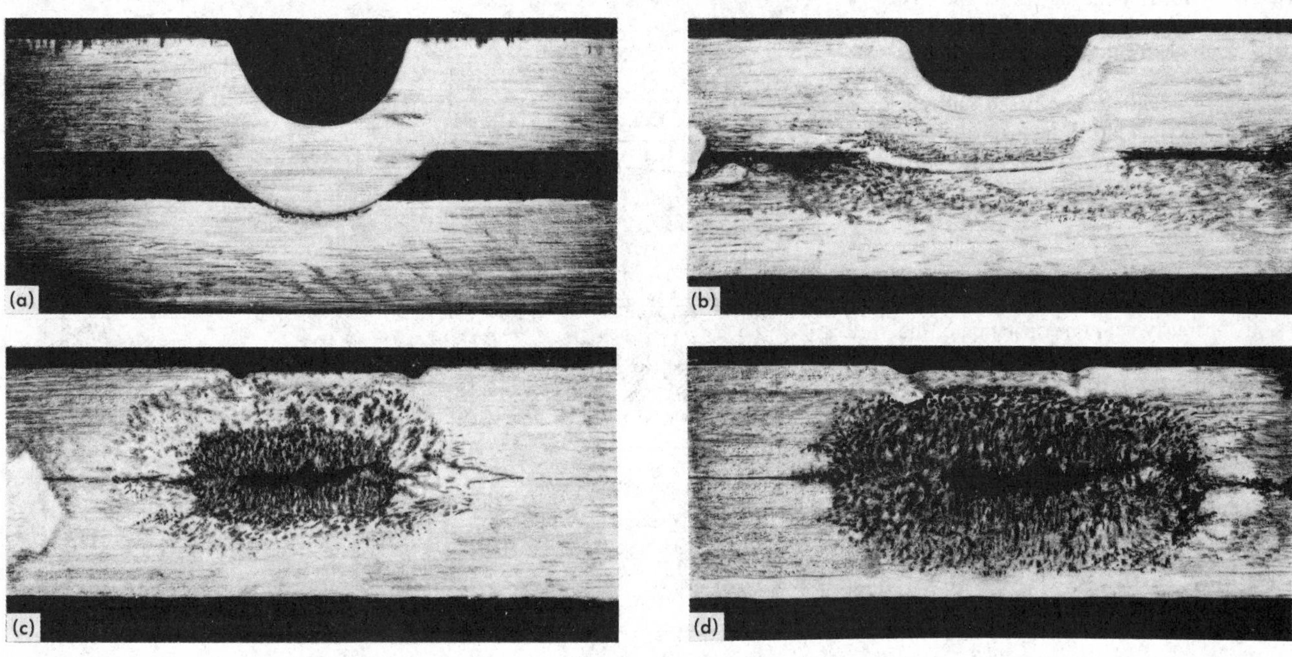

sufficient distance from the interface to produce a well-defined weld nugget that has about half its final thickness (penetration) and diameter. Sheet separation adjacent to the weld nugget has been reduced to zero, and the softened metal above the nugget has been flattened against the face of the upper electrode (compare with Fig. 1b). Nugget diameter, penetration, and shear strength continue to increase as weld time progresses. Figure 1(d) shows the fully developed weld nugget.

Applicability

The principal application of projection welding is the joining of stamped low-carbon, low-alloy, and high-strength low-alloy (HSLA) steel parts. During stamping (punching, drawing, or forming), one of the parts must have a projection which has been formed during the stamping operation to enable projection welding. Projection welding is also used for joining screw-machine parts to stamped parts; the projection is machined or cold formed on the end of the screw-machine part. Fasteners or mounting devices, such as nuts, screws, brackets, pins, bosses, handles, and clips, can be attached to various products by projection welding. This technique is of special value in mounting attachments to surfaces that have back sides that are inaccessible to a welding operator and in applications where the mounting surface must be leakproof at the weld joint.

Projection welding is most successful in workpieces 0.022 to 0.125 in. thick. Stock 0.010 in. thick has been projection welded; projection design is critical, however, and machines with low-inertia heads and fast follow-up are needed. Sections less than 0.010 in. thick are more adaptable to spot welding.

Projection welding of crossed wires is used for making stove and refrigerator racks, soap dishes, lampshade frames, gratings, grills, and electrical connector networks for electronic applications.

Advantages. The principal advantages of projection welding are:

- The number of welds that can be made simultaneously with one operation of the welding machine is limited only by the ability of the controls to regulate current and force.
- Because of greater current concentration at the weld, and thus less chance of shunting, narrower flanges can be welded, and welds can be spaced closer together by projection welding than by spot welding.
- Electrodes used in projection welding have faces larger than the projection or pattern of projections and larger than the faces of electrodes used for making spot welds of comparable nugget diameter. Consequently, because of lower current density, electrodes require less maintenance than do spot welding electrodes.
- Tooling construction for projection welding usually combines welding dies, or electrodes and electrode holders, with workpiece locators in one assembly, which frequently can be designed for welding two or more small workpieces to one larger one.
- Projection welds can be made in metal that is too thick to be joined by RSW.
- Flexibility in the selection of projection size and location allows welding of workpieces in thickness ratios of 6 (or more) to 1. Workpieces in thickness ratios greater than about 3 to 1 sometimes are difficult to spot weld.
- Weldments requiring a minimum of surface marking on one side can be produced by embossing projections on the component with less critical appearance requirements. The slight bump raised in welding (if objectionable) can be removed in a simple mechanical finishing operation.
- Projection welds can be located more accurately and are more consistent in diameter and thickness (penetration) than spot welds (and thus, for a given strength, can be smaller in average size than spot welds).
- Oil, rust, scale, plating, and other work-metal coatings interfere less with projection welding than with spot welding.

In spite of the additional cost of embossing or otherwise providing projections, projection welding often is more economical than spot welding. A greater number of welds can be made simultaneously, surface finish can be better preserved, and handling sometimes can be reduced, as in the application described in the example that follows.

Example 1. Change From Resistance Spot Welding to Projection Welding. Originally, spot welding was used for attaching four brackets to the low-carbon steel door of a clothes dryer (Fig. 2, upper left). The 16 spot welds (4 per bracket) were made one at a time.

Fig. 2 Setup for joining brackets in clothes dryer door assembly

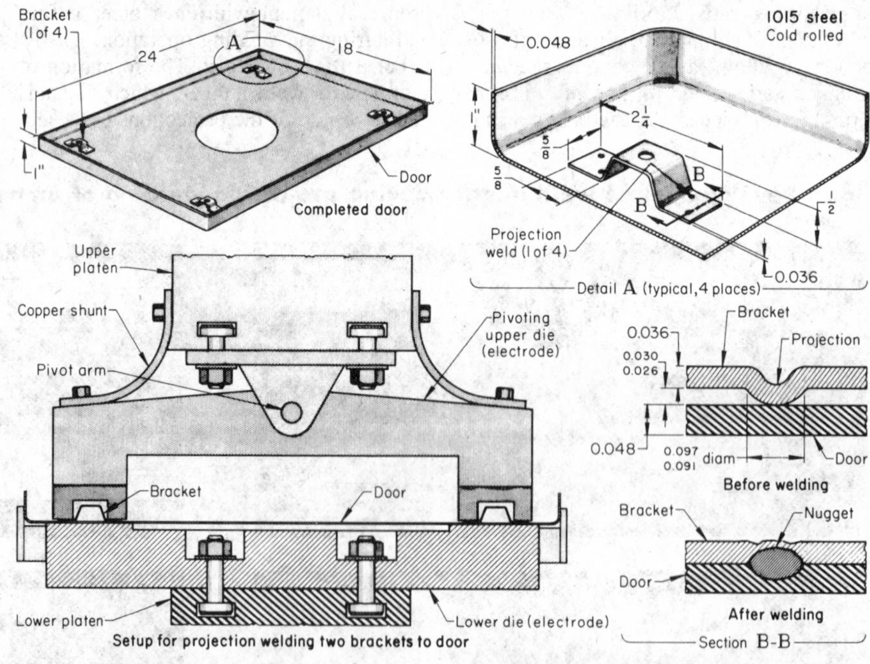

Equipment details for projection welding

Welding machine	Press type, RWMA No. 2
Rating at 50% duty cycle	100 kV·A
Current, max	46 000 A
Electrode material	RWMA class 2
Electrode force, max	1800 lb

Conditions for projection welding

Welding current	32 000 A
Heat control setting	90%
Electrode force	1200 lb
Squeeze and hold times	10 cycles each
Weld time	6 cycles

Fig. 3 Projection welding of mercury-switch contact assembly

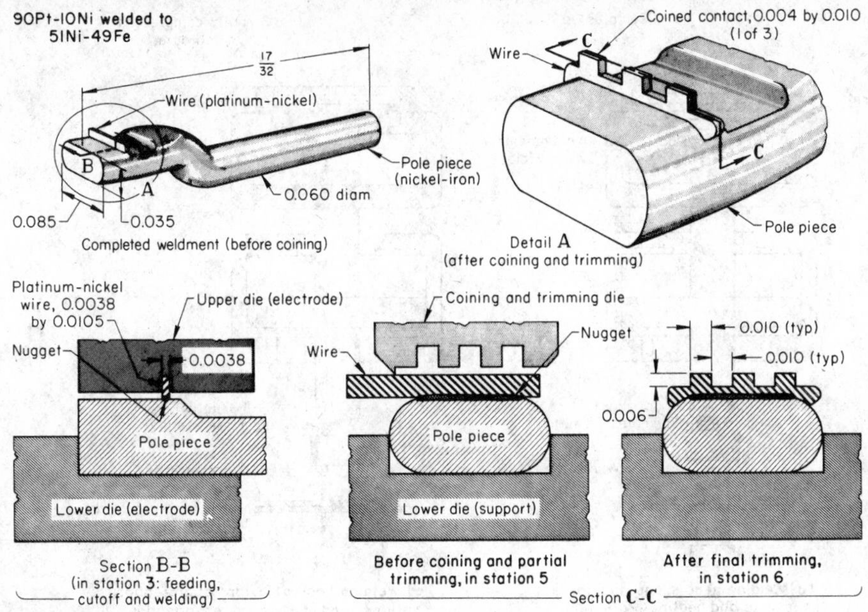

Conditions for automatic projection welding

Transformer rating at 50% duty cycle	5 kV·A	Electrode force	0.6 lb
Electrode material	RWMA class 10 (Cu-W)	Squeeze and hold times	5 cycles each
Welding voltage	1.5 V	Weld time	2 cycles

(a) For cutoff, projection welding, coining, and trimming

Production rate was increased from 75 to 210 doors per hour, and cost was reduced significantly when spot welding was replaced by projection welding for attaching the brackets. With the setup shown at lower left in Fig. 2, projection welds were made eight at a time, attaching two brackets to the door simultaneously. Loading of parts into the welding machine was faster than with the fixture used for spot welding.

The upper die (electrode) was pivoted to equalize the welding force on each of the brackets. A means for locating and holding the brackets for welding was provided in the upper die, and the lower die incorporated gages for positioning the door. Surface finishing was done in less time after projection welding than after spot welding. Yearly production was 750 000 doors.

Because of its speed and adaptability to automatic control, projection welding is readily integrated with high-speed forming, trimming, and other press operations in automatic multiple-station machines, as in the welding of electrical contacts described in the following example.

Example 2. Projection Welding Combined With Cutoff, Coining, and Trimming Processes. The mercury-switch contact assembly shown in Fig. 3, which consisted of a platinum-nickel wire 0.0038 in. thick by 0.0105 in. wide and a nickel-iron pole piece, originally was projection welded manually in a separate operation. To increase production and reduce cost, projection welding was done automatically and was combined with cutoff, coining, and trimming in an eight-station automatic machine. After welding, the wire was coined to form three contact surfaces 0.004 in. wide by 0.006 in. high by 0.010 in. long, as shown in Fig. 3. The operations, performed on a cam-operated eight-station rotary-indexing table, consisted of:

1 Manually loading pole piece into the lower die (the lower welding electrode, which also served as a support during coining and trimming)
2 Automatically aligning pole piece and applying a drop of alcohol for cleaning and to provide cooling during welding
3 Feeding and cutting off platinum-nickel wire, and projection welding to pole piece (Fig. 3, section B-B)
4 Idling
5 Coining three contacts on wire, and partly trimming excess wire (Fig. 3, section C-C)
6 Removing excess wire (Fig. 3, section C-C)
7 Ejecting finished weldment
8 Idling

Prior to assembly, the nickel-iron pole piece was degassed in a hydrogen atmosphere. In station 3, after the pole piece was located in the lower die, the upper die was lowered to within 0.002 in. of the welding position. The rectangular-section wire was fed from a spool and was guided across the pole piece by a slot in the upper die. The wire was cut off, and the upper die was lowered to make the weld.

Previously, the same basic operations had been performed manually. The improved (automated) process required greater care in aligning the workpieces and the dies. Production of 800 pieces per man-hour more than quadrupled the original production (Fig. 3), and there were fewer rejections. The resulting decrease in production cost allowed amortization of the cost of the machine in 147 working days, or after welding of 938 000 assemblies.

Quality Control. Five assemblies were inspected every 15 min. First they were dimensionally inspected in an optical comparator at a magnification of 50 diameters. Then they were tested destructively by applying force to the faces of the three coined contacts. The weld was required to hold until the contacts had been bent 90°.

Limitations of projection welding include:

- Forming of one or more projections on one of the workpieces may require extra operations unless the parts are press formed to design shape.
- When several welds are made at once with the same electrode, alignment of the work and dimensions (particularly height) of the projections must be held to close tolerances to obtain consistent weld quality.
- In any one operation, welding is limited to joining two thicknesses of metal, and the shape of the work and location of the projections must be compatible with application of the electrode force.
- Nugget size is limited by the size of the projection.
- Multiple welds must be made simultaneously, requiring higher capacity equipment than spot welding.

Welding Machines

Press-type machines, with either single-phase or three-phase transformers, usually are used for projection welding. The welding head in these machines is guided by bearings or ways and moves in a straight line. Platens with T-slots or tapped holes

Fig. 4 Low-inertia welding head for projection welding setup

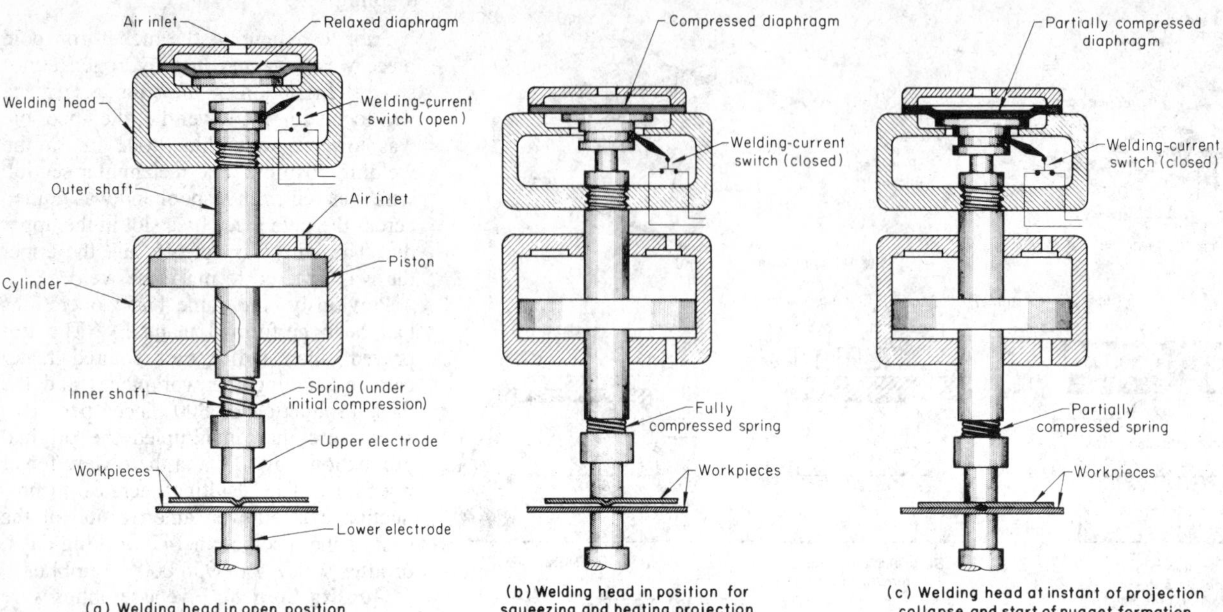

(a) Welding head in open position

(b) Welding head in position for squeezing and heating projection

(c) Welding head at instant of projection collapse and start of nugget formation

are used for mounting the welding dies or electrodes. Rocker-arm machines generally are not used for projection welding because the electrode moves in an arc that can cause slippage between the components as the projection collapses.

The welding head can be hydraulic, spring, magnetic, or air actuated. Many machines have a low-inertia welding head, or a means of uncoupling the electrode holder, to provide fast follow-up as the projection collapses. Figure 4 illustrates a low-inertia welding head.

At the start of the welding stroke (Fig. 4a), equal air pressure is applied to the top of the piston and to the diaphragm. By the time the piston has completed its travel, the diaphragm has been compressed by the retraction of the internal shaft. Retraction of the internal shaft simultaneously compresses the diaphragm and closes the welding-current switch; the control lever for the switch rides in the shaft collar.

When the welding current is initiated (Fig. 4b), the only remaining mass to be moved is the internal shaft and its attached electrode. Air pressure acting on the diaphragm and the force of the compressed spring between the inner and outer shaft easily overcome the low inertia of the system and move the upper electrode downward as the projection collapses. Thus, the workpieces are kept in close contact under partial electrode force until the welding head follows and forges the weld region with the full electrode force (Fig. 4c).

Welding machine controls are usually of the synchronous type. Phase shift and pulsation timing are often included on single-phase or three-phase machines to regulate welding current. Pulsation timing may be helpful when welding thick metal or unequal thicknesses and to help compensate for slow follow-up. On single-phase machines, slope control is sometimes used in special applications, generally for the same purposes as pulsation timing. For additional information on resistance welding machines and controls, see the article "Resistance Spot Welding" in this Volume.

Metals Welded

The most frequently projection welded metal is low-carbon steel (0.20% C maximum). Low-alloy steels, naval brass, Monel (nickel-copper) alloys, and austenitic stainless steels are also projection welded. Coated metals, such as galvanized steel, terneplate, tin plate, and aluminized steel, can be successfully welded, but considerable electrode maintenance is usually needed.

Not all metals can be projection welded, because some are not strong enough to support projections; some brasses cannot be projection welded because the projections collapse too rapidly under pressure. Aluminum has been projection welded only to a limited extent, and best results are obtained with extruded parts using projections similar to those used for low-alloy steel, the exact size being developed by trial. Thin steel (less than 0.010 in. thick) usually is more easily spot welded than projection welded because of the difficulty of forming projections that do not collapse before reaching the welding temperature. Free-machining steel sometimes can be projection welded using specially designed joints and electrodes (see Example 7), but generally it is difficult to projection weld because sulfur and phosphorus segregation causes brittle welds.

Metallurgical Effects

The rate of quenching projection welds is extremely rapid in thin members because of the closeness of the water-cooled electrodes to the weld nugget. Although the total amount of heat is low because of the short weld times and the small amount of metal heated, severe solid-metal quenching is obtained. Under some conditions, quench rates for projection welds may be so severe that the fully hardened condition is achieved even in low-carbon steels. Therefore, recommended practice limits the carbon content of carbon steels to 0.20% maximum, unless special techniques are used.

Cleaning of Workpieces

Optimum welds on all metals are obtained when surfaces are clean and without scale, oxide, excessive oil, grease, or other foreign material. A thin coating of oil normally does not cause poor results, but if optimum quality welds are required, the material should be degreased before welding. Unclean surface conditions result not only in lack of weld uniformity but also in gas and inclusions in the weld nugget, which reduce weld strength.

Surfaces coated with scale, oxide, oil, or other surface films can be more readily projection welded than spot welded. When projection welding on such surfaces, the concentration of the electrode force on the tip of the projection must be sufficient to break through the coating and begin the weld. The heat from this start can then melt and vaporize volatile surface contaminants; other foreign matter is broken up and at least partially expelled during collapse of the projection.

Thin surface films that have low and uniform electrical resistance have the smallest effect on welding. Steel that is coated with extremely thick and nonuniform mill scale may not be weldable on a practical production basis unless the scale is removed prior to welding.

As workpieces are prepared for projection welding, burrs around sheared edges of the components and around pierced holes in the components should be removed. As the projection collapses, these burrs, if not removed, form shunting paths for both current and electrode force. Uncontrolled shunting makes it difficult to obtain consistently high weld quality.

Process Variables

The major variables that affect projection welding are welding current, electrode force, and weld time. Other factors are heat balance, number and placement of projections, and other aspects of projection and joint design.

Variations in weld quality are minimized when electrode force, welding current, and weld time are maintained at constant values, and when the electrodes are kept clean and in good condition. Other equipment-related factors that affect the flow of welding current across the weld interface include: (1) quality of power regulation on the high-voltage lines that supply the welding machines, (2) adequacy of low-impedance substation transformers, (3) capacity of bus lines to the welding machines, (4) the use of suitable current or voltage regulators on the welding machines, and (5) proper allowance for the introduction of magnetic metals into the secondary loop of the machine. In addition, accurate fit-up of the workpieces and uniform height of the projections are necessary.

Schedules for projection welding of low-carbon steel from 0.014 to 0.125 in. thick are given in Table 1. These data are intended to serve as starting points and should be adjusted to suit the specific application and the equipment used.

Welding current required for projection welding, although slightly less per weld than that needed for spot welding, must be high enough to cause fusion before the projection is completely flattened. The recommended current is the highest current that, when used with the correct electrode pressure, does not cause excessive expulsion of metal. For a specific projection size, expulsion of metal increases as welding current increases, because the current must flow through a small contact area. Slope control, a controlled change in current while the welding cycle progresses, sometimes can be used to minimize expulsion.

The rigidity of a projection against collapse, which is a function of work-metal properties and thickness, has an effect on selection of the welding current. For welding thin metal (less than 0.020 in. thick), the ranges of welding current and weld time are narrow.

Progressive heat buildup can also be a problem when a number of projection welds are made in sequence on a single part. Heat buildup can be compensated for by changing the heat control settings, or otherwise adjusting the current, for some of the welds. In addition, shunting effects may vary after the first weld is made, and heat sink behavior may differ for some of the welds, which further affect the heat input from weld to weld. To compensate for such effects in making a number of welds on a single part, suitable changes in heat control settings can be made manually or can be programmed into automatic control equipment for high-production applications.

Electrode force used in projection welding depends on the work metal, the size and design of the projection, the number of projections in the joint, and the welding machine. Excessive force causes the projection to collapse before the weld area has reached the proper temperature, resulting in the formation of ring welds, in which fusion occurs around the periphery of the projection but is incomplete at the center. For best weld appearance, the electrode force should be such that the projection is flattened completely after the metal has reached welding temperature. Also, force must be sufficient to produce a sound weld with minimum separation of the workpieces.

If the projection is high, a welding machine having heavy, slowly moving parts should not be used, or provision must be made for quick follow-up as the projection is melted. Slow follow-up causes expulsion of molten metal before the pieces can be brought together and welded.

Application of an initial force high enough to cause a definite indentation of the lower workpiece by the projection is sometimes used to help prevent expulsion of molten weld metal when follow-up is too slow or follow-up force is insufficient.

Electrode force influences development of porosity in the weld nugget. The force must be low enough to provide properly timed projection collapse and yet high enough to produce sound welds. Sometimes, especially in welding thick workpieces, a weld force is used to make the welds, and a forging force is applied after current flow has stopped to minimize porosity.

Weld time for a given type and thickness of work metal depends on welding current and rigidity of the projection. Weld time is less important than electrode force in projection welding low-carbon, low-alloy, and HSLA steel, provided the time is sufficient to produce a nugget of adequate size at the chosen welding current. A short weld time creates higher production efficiency and less discoloration and distortion of the workpiece. After the proper electrode force and welding current are determined, the weld time is adjusted to make the desired weld.

Projection welds made in the shortest times are not necessarily of the best quality. A short weld time requires a correspondingly high welding current, which initially must pass through a small contact area, increasing the current density and the possibility of metal expulsion.

When several projection welds are made at once using the same electrode (multiple-projection welding), pulsation timing

Table 1 Conditions for projection welding of 1010 steel 0.014 to 0.125 in. thick using RWMA class 2 electrodes

Thickness of thinnest outside piece(a), in.	Electrode face diam (min)(b), in.	Net electrode force, lb	Weld time, cycles (60 cycles/s)	Hold time, cycles (60 cycles/s)	Welding current(c), A
0.014	1/8	175	7	15	5 000
0.021	5/32	300	10	15	6 000
0.031	3/16	400	15	15	7 000
0.044	1/4	400	20	15	7 000
0.062	5/16	700	25	15	9 500
0.078	3/8	1200	30	30	13 000
0.094	7/16	1200	30	30	14 500
0.109	1/2	1700	30	45	16 000
0.125	9/16	1700	30	45	17 000

Note: Steel to be welded should be free from scale, oxides, paint, grease and oil.
(a) Data based on thickness of thinner sheet and for two thicknesses only. Maximum ratio between two thicknesses, 3 to 1. (b) Face diameter equals twice the diameter of the projection. (c) Approximate current at electrodes, using 60-cycle ac
Source: "Recommended Practices for Resistance Welding," AWS C1.1

and somewhat longer weld times than for comparable single welds should be used to ensure high-quality weldments. Variation in height of projections has an effect on timing of initial contact, and the use of slope control provides slow initial heating and allows enough time for all of the projections to make contact with the mating workpiece before the high welding current is applied, thus producing more uniform welds and minimizing expulsion.

Heat Balance. Maintaining the proper heat balance between workpieces sometimes is difficult in projection welding. If heat balance is incorrect, the projection can be melted away before the mating surface is brought to welding temperature.

The factors that affect heat balance are: (1) design of the projection, (2) electrode material, (3) thickness of the workpieces and thermal and electrical conductivity and other properties of the work metals, and (4) heating rate.

Projections must be designed to withstand the initial electrode force needed for the proper flow of current, yet collapse fully at welding temperature to produce a sound weld with minimum or no sheet separation. When multiple projections in groups of four or more are welded, slight variations in projection height can affect heat balance in all the projections and can make it difficult to obtain simultaneous collapse of the projections.

Fig. 5 Projection welding of clutch-mount assembly

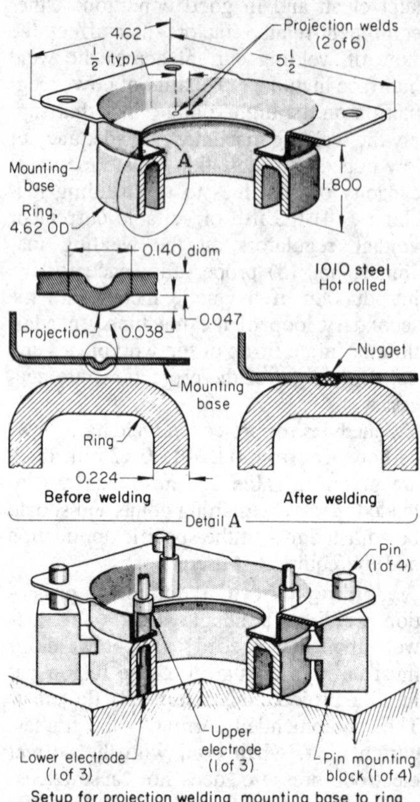

Equipment details

Power supply	440-V 60-cycle ac
Welding machine	Press type, manually operated, spot and projection
Rating at 50% duty cycle	250 kV·A
Heat control	Eight-tap transformer and phase shift
Electrode material	RWMA class 11

Welding conditions

Welding current	42 000 A
Electrode force	2700 lb
Squeeze time	35 cycles
Weld time	20 cycles
Hold time	15 cycles
Assemblies per hour	200 (1200 welds)

A water-cooled electrode made of Resistance Welders Manufacturers Association (RWMA) class 2 or class 3 electrode material may prevent a thin sheet from heating sufficiently if the projection is on a thicker component. This can be avoided by the use of a hard material of low electrical and thermal conductivity, such as RWMA class 10, 11, or 12, in contact with the thinner sheet.

As discussed previously, not all metals can be projection welded, because in some, projections collapse too rapidly under pressure. Also, the thermal conductivity of some metals is such that heat is dissipated away from the projections too quickly for them to reach welding temperature. When dissimilar metals are welded, projections should be formed on the metal of higher electrical and thermal conductivity to more nearly equalize the rate of heating in the two workpieces. Rate of heating is important because the two mating surfaces must be brought to welding temperature at the same time.

When similar metals of equal thickness are welded, projections can be formed on the component that is easier to handle or that has the less critical requirements for surface appearance. In welding workpieces of unequal thickness, projections should preferably be formed on the thicker workpiece to ensure that both of the mating surfaces reach welding temperature at the same time. Projections on the thinner workpiece can melt off before the thicker workpiece reaches welding temperature, unless tooling and welding procedure are specially developed and carefully maintained and controlled to overcome this problem. However, with suitable precautions, good results can be obtained using projections formed on the thinner workpiece, even when the workpiece thickness ratio is as great as 5 to 1, as in the following example.

Example 3. Projection Welding Application in Which the Projections Were Formed on the Thinner Workpiece. The clutch mount for the compressor in an automobile air conditioner was made by projection welding a drawn ring with a 0.224-in.-thick wall to a mounting base 0.047 in. thick (Fig. 5). Both components were made of 1010 steel. In the original design of the assembly, the projections were on the thicker (ring) section, as conventionally recommended. However, these projections were difficult to form, so the assembly was redesigned with the projections on the thinner (base) section. With the new design, satisfactory welds were made without burning the thin mounting base.

The projections were formed in pairs 120° apart on the contact surface of the mounting base. The projections in each pair were 1/2 in. apart. The outside diameter of the ring was required to be concentric with the bolt circle on the mounting base within 0.010 in. TIR. This requirement was met by use of the fixture shown in Fig. 5. Four pins that extended upward from mounting blocks positioned on the base of the fixture passed through locating holes in the mounting base, while machined inner faces on the pin-mounting blocks located the outside diameter of the ring. To maintain the specified tolerances on the assembly, the outside diameter of the ring had to be held within ±0.005 in. in the forming operation. The diameter of the locating holes was held to +0.002, −0.001 in., and the diameter of the bolt circle was held to ±0.001 in. The three upper electrodes were lowered to make the six welds simultaneously.

The finished assemblies were visually examined. Destructive tests were made on two assemblies per each hundred assemblies welded. For acceptance, five welds on each assembly tested had to pull buttons. The rejection rate because of weld failures was negligible. Assemblies were inspected for compliance with the concentricity requirement in a rotating fixture using a dial indicator. Samples were inspected on the basis of a 5% acceptable quality level. The welded assemblies consistently remained within tolerance, provided the components were within their assigned manufacturing tolerances.

Welding Schedules. Practice in setting up welding schedules for projection welding of work for which previous experience is lacking is generally the same as for RSW.

Tables 1 to 7 give recommended practices for projection welding of low-carbon

Table 2 Details of design of projection welds in low-carbon and stainless steels(a)

Thickness (t) of thinnest outside piece(b), in.	Diameter (D) of projection(c), in.	Height (h) of projection(d), in.	Minimum shear strength (single projections only), lb — Tensile strength below 70 ksi	Tensile strength 70 to 150 ksi	Tensile strength 150 ksi and above	Nugget diameter (min) at weld interface, in.	Minimum contacting overlap(e), l, in.
0.010	0.055	0.015	130	180	250	0.112	1/8
0.012	0.055	0.015	170	220	330	0.112	1/8
0.014	0.055	0.015	200	280	380	0.112	1/8
0.016	0.067	0.017	240	330	450	0.112	5/32
0.021	0.067	0.017	320	440	600	0.140	5/32
0.025	0.081	0.020	450	600	820	0.140	3/16
0.031	0.094	0.022	635	850	1100	0.169	7/32
0.034	0.094	0.022	790	1000	1300	0.169	7/32
0.044	0.119	0.028	920	1300	2000	0.169	9/32
0.050	0.119	0.028	1 350	1700	2400	0.225	9/32
0.062	0.156	0.035	1 950	2250	3400	0.225	3/8
0.070	0.156	0.035	2 300	2800	4200	0.281	3/8
0.078	0.187	0.041	2 700	3200	4800	0.281	7/16
0.094	0.218	0.048	3 450	4000	6100	0.281	1/2
0.109	0.250	0.054	4 150	5000	7000	0.338	5/8
0.125	0.281	0.060	4 800	5700	8000	0.338	11/16
0.140	0.312	0.066	6 000	...	...	7/16	3/4
0.156	0.343	0.072	7 500	...	...	1/2	13/16
0.171	0.375	0.078	8 500	...	...	9/16	7/8
0.187	0.406	0.085	10 000	...	...	9/16	15/16
0.203	0.437	0.091	12 000	...	...	5/8	1
0.250	0.531	0.110	15 000	...	...	11/16	1 1/4

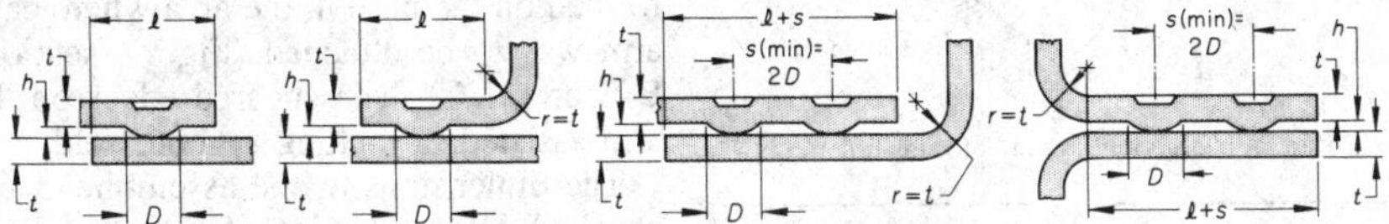

(a) Welding conditions are for 1010 steel, types 309, 310, 316, 317, 321, and 347 stainless steel. Surface of steel to be welded should be free from scale, oxides, paint, grease, and oil. (b) Size of projection is normally determined by thickness of thinner piece, and projection should be on thicker piece where possible. Data are based on thickness of thinner sheet, and for two thicknesses only. (c) Projection should be made on workpiece of higher conductivity when dissimilar metals are welded. For diameter of projection D, a tolerance of ±0.003 in. in material up to and including 0.050 in. in thickness and ±0.007 in. in material over 0.050 in. thick may be allowed. (d) For height of projection h, a tolerance of ±0.002 in. in material up to and including 0.050 in. thick and ±0.005 in. in material over 0.050 in. thick may be allowed. (e) Contacting overlap l does not include any radii from forming. Weld should be located in center of overlap.
Source: "Recommended Practices for Resistance Welding," AWS C1.1

steel, which are intended to serve as starting points. Table 2, which gives details of projection design, also applies to austenitic stainless steel.

Electrodes, Welding Dies, and Fixtures

An electrode designed for RSW can be used for projection welding if the electrode face is large enough to cover the projection being welded, or the pattern of projections being welded simultaneously. To minimize marking and indentation of workpieces, the recommended electrode face diameter for making a single projection weld usually is two or more times the diameter of the projection. In multiple-projection welding, the electrode face should be large enough to extend beyond the boundaries of the pattern of projections by approximately the diameter of one projection.

Simultaneous welding of a number of projections that are far apart requires a rigid electrode large enough to cover the projections. The electrode should be shaped so as not to deform tabs, flanges, or other features of the component being welded. The use of a pivoted electrode that is properly designed and insulated to avoid excessive shunting may be helpful in some applications, as in Example 1. With proper fixturing, the accuracy attainable with projection welding is equal to that of other joining processes.

Electrode Materials. Flat electrodes, or local contact-surface electrodes, usually have acceptable life when made from RWMA class 2, 3, or 4 materials (see the article "Resistance Spot Welding" in this Volume for identification of electrode materials). RWMA class 2 electrode materials are generally preferred because they provide the best compromise among electrical conductivity, strength, hardness, and temperature resistance.

If a harder electrode material or a material with lower electrical conductivity is needed, an economical solution is to use a composite electrode with copper backing and a copper-tungsten alloy facing. The facing material usually is RWMA class 10, 11, or 12. Classes 11 and 12 have lower electrical conductivity and higher hardness and strength than does class 10.

The ideal electrode material is one that is as hard as possible and does not crack or cause surface burning on the weldment. If cracking or surface burning is encountered, a softer alloy with higher conductivity and ductility should be used.

Electrode Design. There are three basic types of electrodes used in projection welding:

- Round flat-faced electrodes of the type used for spot welding
- Large flat-faced electrodes of bar stock
- Bar-type electrodes, in which a series of local-contact surfaces are made by relieving the bar between intended contact locations, or by brazing or clamping contact inserts of hard copper alloy to the bar. (Figures 2, 16, and 17 show electrodes of this type.)

Standard spot welding electrodes are used to weld single projections; electrode diameter is usually at least twice that of the projection. Large flat-faced electrodes are used for welding a few projections in a localized area, but they are not recommended for welding components that may be distorted from the projection-forming operation.

Bar-type electrodes usually can equalize current and force regardless of slight distortion or other variations in the workpieces. They are recommended for all multiple-projection welding applications, especially those involving four or more projections.

Welding Dies and Fixtures. Welding dies hold and clamp the workpieces in correct alignment for welding. They also serve as electrodes or as holders for inserted electrodes or electrode buttons. Fixtures are auxiliary positioning devices that do not conduct current. Welding dies and fixtures should be designed to:

- Accurately position the workpieces
- Permit rapid loading and unloading; air jets, levers, or mechanical loading and unloading devices can be used. A combination of manual loading and automatic unloading is used in many applications.
- Allow initial workpiece contact at the weld, to provide the shortest possible current path between the electrodes, and to avoid energy losses from unnecessary resistive heating of thin sections of work metal in the current path

Fig. 6 Automatic projection welding setup

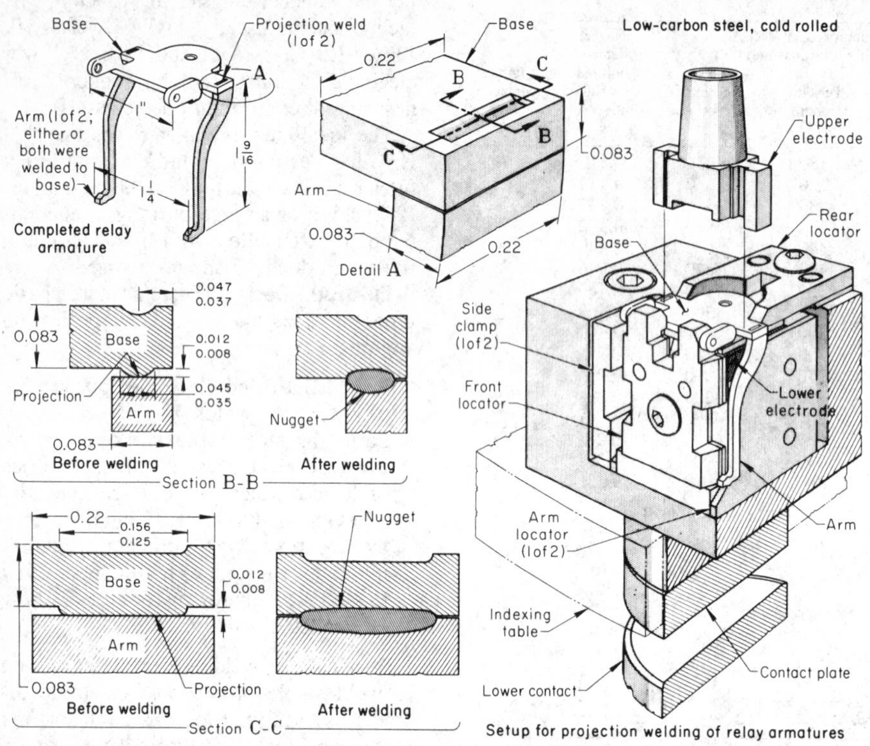

Equipment details	
Welding machine	440 V, single phase, automatic
Rating at 50% duty cycle	30 kV·A
Current, max	30 000 A
Secondary voltage	2.4 to 4.8 V
Electrode material	RWMA class 2
Electrode force, max	550 lb
Heat control	Eight-tap transformer and phase shift
Auxiliary equipment	Rotary-indexing table with loading and unloading devices

Welding conditions	Single arm	Double arm
Welding current, A	3400	4400
Secondary voltage, V	2.7	3.1
Heat control setting, %	80	86
Electrode force, lb	115	160
Clamp force, lb	50	50
Squeeze and hold times, cycles	6 each	6 each
Weld time, cycles	5	6
Production, assemblies per hour	1500	1500
Production, welds per hour	1500	3000

Steel or other magnetic materials should not be used in fixtures for welding with alternating current, because they reduce the electrical capacity of the machine and heat up when within the secondary loop. Nonmetallic materials that have good strength and are electrical insulators, such as filled phenolics, help prevent unintentional shunting of current and are often used in fixturing for small parts.

Small parts can be located by the lower die, and larger parts can be aligned by stops, pins, or other types of locators. When a small part is to be positioned on top of a larger part, a removable gage can be used that aligns the parts while the upper die makes the weld. Parts also can be held in the upper die by spring clips, plungers, or vacuum.

Projection welding of assemblies that have long, thin components may demand special care in the design of fixtures. Heating must be concentrated at the weld interface. To avoid waste of energy and overheating of workpieces, fixtures should be designed to provide electrode contact to the long, thin component as close as possible to the joint, as demonstrated in the following example.

Example 4. High-Production Welding of a Relay Armature. The relay armature shown at upper left in Fig. 6 was made by projection welding the two arms to the formed base. Some of the armatures required only one arm, on either the left or the right side. Fixtures for locating and clamping the components in proper position were mounted on a rotary-indexing table (Fig. 6, right view), and an automatic welding machine with a low-inertia welding head (Fig. 4) was designed that was capable of projection welding either or both arms to the armature base at a rate of 1500 assemblies per hour.

During welding, air-actuated side clamps held the arms against the lower electrode, which was $^1/_4$ in. thick. Below this electrode, the fixture was recessed to restrict the current flow to the joint end of the arms. The cross-sectional area of the lower portions of the arms was not large enough to carry the welding current. This arrangement minimized heat losses.

The upper electrode (shown in Fig. 6) was attached to the lower end of a spring-loaded inner shaft of an air-actuated ram on the welding machine. This low-inertia, fast follow-up welding head provided a precisely timed follow-up to complete the weld, which consequently avoided undesirable arcing.

To concentrate the welding current at the joint, an elongated projection was formed on the base in the area where each arm was to be attached (Fig. 6, sections B-B and C-C). The sheared edges on the arm assisted in making a good weld.

The dimensions of the assemblies were checked after each lot of 500 had been welded. The welds were tested by bending the arms to the side or by twisting them; destructive tests were made to determine tensile strength. Minimum breaking load was specified at 150 lb for each weld. The joints were tested to a load of 300 lb. If a weld failed in the range of 150 to 300 lb, the nugget was required to pull base metal over at least 50% of its area. The electrode force, current, and weld time were established to produce a weld nugget with a strength of 200 lb. The upper electrode was changed after welding each lot of 5000 assemblies.

Electrode Holders. In projection welding, electrodes are placed in holders similar to those used in RSW, and they can accommodate single electrodes, as well as two or more per holder. Holders for single electrodes can be used for both the upper and lower electrodes. Multiple-electrode holders, of either standard or special design, are generally used for the upper electrodes; electrode forces are equalized by springs, mechanical devices, or hydraulic equalizers.

Hydraulic equalizers can transmit high forces equally to electrodes that are spaced closer than is practical with mechanical linkage or spring-type balancers. The amount of electrode force that can be transmitted through mechanical or spring-type equalizers is generally low. Also, it

Fig. 7 Projection welding utilizing multiple-electrode holder. See text for application details.

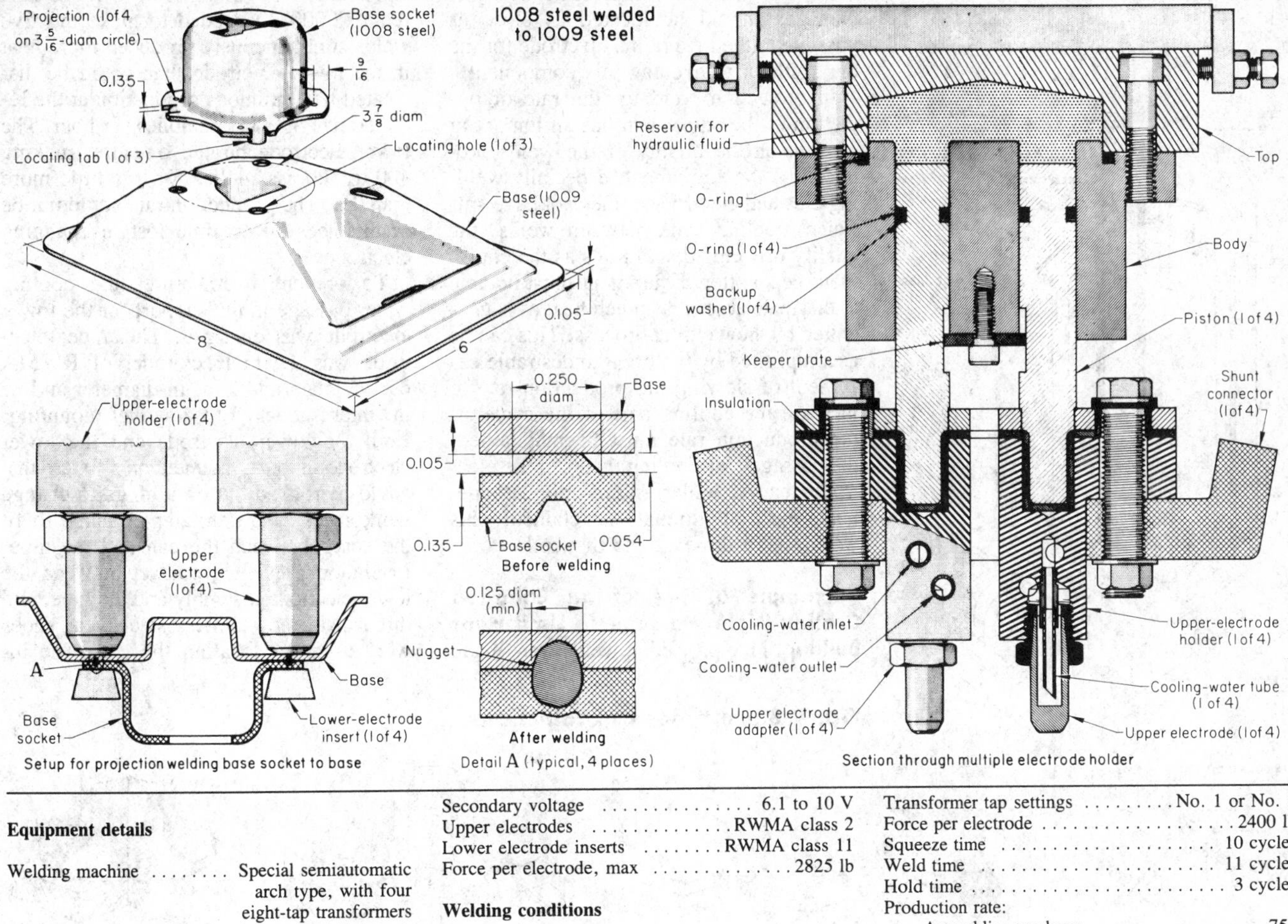

Equipment details

Welding machine	Special semiautomatic arch type, with four eight-tap transformers
Rating at 50% duty cycle	75 kV·A each
Current per electrode, max	12 300 A
Secondary voltage	6.1 to 10 V
Upper electrodes	RWMA class 2
Lower electrode inserts	RWMA class 11
Force per electrode, max	2825 lb

Welding conditions

Current per electrode	8 000 to 10 000 A
Transformer tap settings	No. 1 or No. 2
Force per electrode	2400 lb
Squeeze time	10 cycles
Weld time	11 cycles
Hold time	3 cycles
Production rate:	
Assemblies per hour	750
Welds per hour	3000

is difficult to adjust spring-type equalizers so that the same force is exerted on the workpiece through each electrode. When electrode forces greater than about 1000 lb are used, the size of the spring and holder can limit the minimum center distance between the electrodes.

For the application in the next example, a special hydraulic-actuated multiple-electrode holder was designed that would equalize the high electrode force while making four closely spaced projection welds in thick workpieces.

Example 5. Use of a Custom Designed Multiple-Electrode Holder. A base for an automobile jack was made as a projection weldment of two press-formed components: a base socket of 0.135-in.-thick 1008 steel, and a base of 0.105-in.-thick 1009 steel (Fig. 7). Four projections were formed on a 3 5/16-in.-diam circle on the lower side of the base socket.

Because the spacing of the projections was comparatively close for simultaneous welding with electrodes in separate holders and because the thick work metal required use of high electrode force, a special welding machine with a hydraulic-equalizing upper electrode holder that contained four electrodes was designed. Each of the electrodes was connected to a separate transformer, and all four welds were made simultaneously. The four transformers had a common ground attached to the lower electrode assembly.

Force for all four upper electrodes was provided by a single air cylinder. The hydraulic equalizer consisted of a main chamber with four pistons arranged in a keystone-shaped pattern coinciding with that of the four projections (approximately 2 3/8 in. apart). Each piston was hardened and contained an O-ring and a leather backup washer. A hardened keeper-plate limited piston travel and prevented rotation of the piston from the inductive field created by the adjacent electrode feeder shunts. A flange on the end of each piston was used for attaching a copper alloy electrode holder. Fiber bushings and washers electrically isolated the electrode holder from the piston.

The upper electrodes, made of RWMA class 2 material, were attached to the holders through adapters that provided for water cooling of the electrodes. The lower electrode was a large copper block with replaceable inserts of RWMA class 11 material (copper-tungsten alloy) under each weld location. The lower electrode block had a locator for positioning the base socket. Although the components had matching tabs and holes for final location, the frame for an automatic unloading device was used as an approximate locator when the base was placed on top of the base socket.

The assembly was welded in the inverted position, as shown in Fig. 7. Production rate was 750 pieces per hour. The welds were tested several times each hour by inserting a wedge between the welded

Fig. 8 Projection welding of a can to a tube

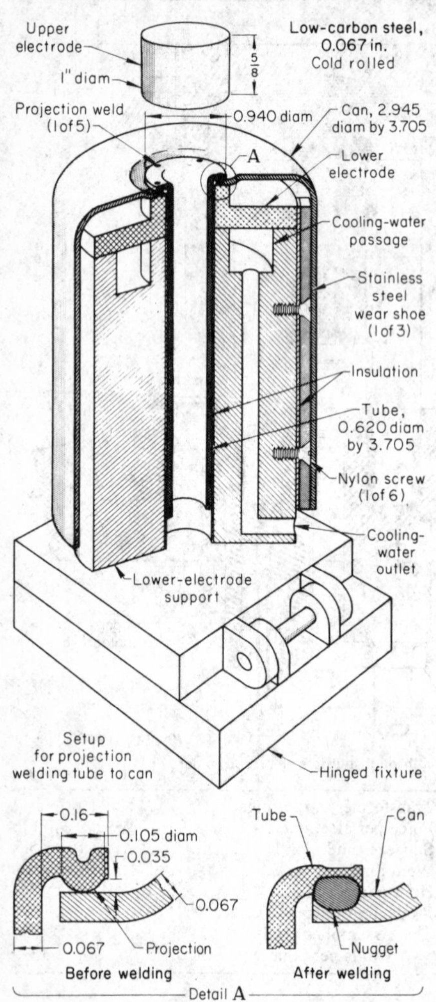

Equipment details

Power supply	440-V 60-cycle ac
Welding machine	Press type, spot and projection, with air-operated ram
Rating at 50% duty cycle	200 kV·A
Heat control	Eight-tap transformer and phase shift
Electrode material	RWMA class 3

Welding conditions

Welding current	32 000 A
Electrode force	2000 lb
Squeeze time	15 cycles
Weld time	26 cycles
Hold time	10 cycles
Off time	10 cycles
Assemblies per hour	200 (1000 welds)

surfaces with a hydraulic jack. The separated nuggets were required to be at least $^1/_8$ in. in diameter.

Cooling of Electrodes and Dies. Water cooling, either directly on or within the electrode or die, is used to avoid overheating of dies and workpieces. The water passages should be located as close as possible to the face of the electrode (or the die) without weakening this component.

In projection welding, the rate of operation of the equipment has an important bearing on the quality of the welds produced. If the equipment, especially welding dies and electrodes, does not have sufficient cooling time between welds, the quality of welds deteriorates as the equipment gets hotter. Equipment must be able to dissipate at least as much heat as is generated by the welding process. This can be accomplished by the often undesirable expedient of slowing the production rate to increase the cooling part of the cycle. If the production rate must be maintained, the passages for cooling water in the electrodes can be enlarged, as long as adequate electrode strength is retained. This was done in the example that follows.

Example 6. Use of an Enlarged Cooling Water Passage To Limit Heat Buildup. Five projection welds were made simultaneously in joining a flanged tube and a deep drawn can (Fig. 8). The current (32 000 A) needed to make the five welds simultaneously produced more heat in the lower electrode than could be dissipated in continuous production at the required rate of 200 assemblies per hour. The lower electrode burned out after making 400 to 500 assemblies, or in a little more than 2 h. The production rate could not be maintained due to time lost in changing electrodes.

To prevent overheating, the cooling water passage in the support for the lower electrode was enlarged. The upper electrode was a simple cylinder of RWMA class 3 material, 1 in. in diameter and $^5/_8$ in. thick, attached to a swivel mounting. Both the lower electrode and the lower electrode support, mounted on a fixture that could be rocked out on a hinge to change workpieces, were annular in shape to fit the space between the can and the tube. Insulation prevented contact between the lower electrode assembly and the tube, and three insulated stainless steel wear shoes were used for locating the can. The in-

Fig. 9 Basic types of projections

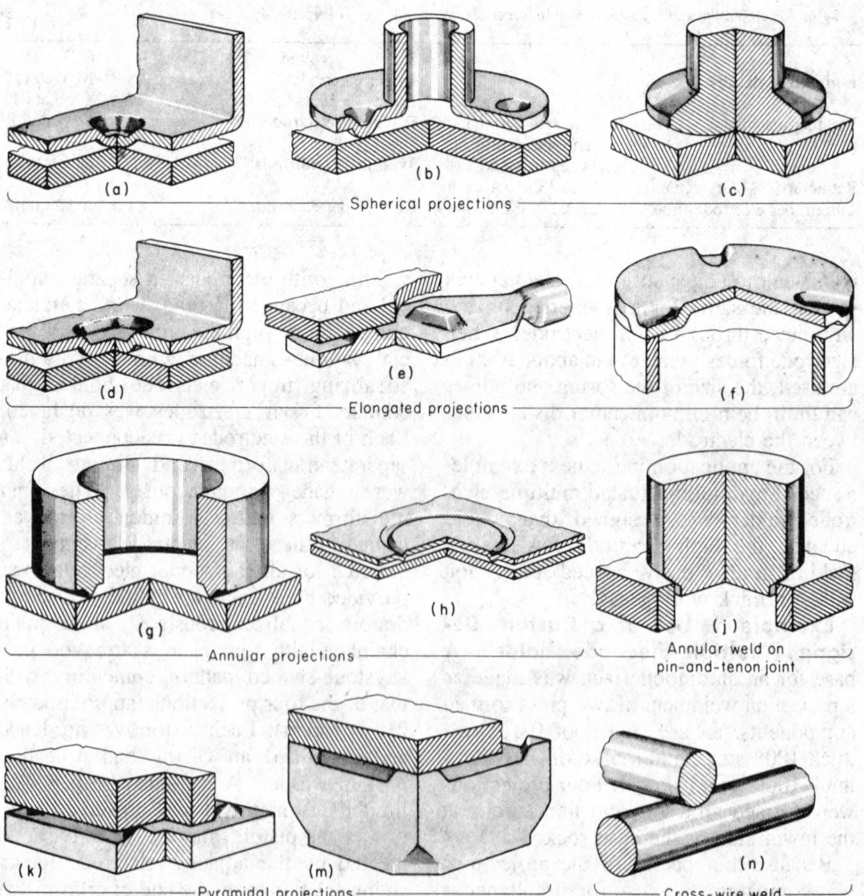

sulated shoes were fastened to the lower electrode support with nylon screws. Electrodes were removed and redressed after every 4000 assemblies, or after every 20 h. Electrodes were redressed two or three times before being replaced.

Each assembly was visually inspected for fusion at each projection. A full weld nugget was required on at least two of the five projections. No definite scrap rate was established because the rate was so low.

Projections

The size and shape of projections depend on the application and the required weld strength. Properly proportioned projections localize or concentrate the welding current, accurately control the location of welds, and ensure uniformity of weld nuggets. Projections often are spaced to give a three-point bearing. Three projections readily equalize themselves and give uniform contact with the mating workpiece if the electrodes are in reasonable alignment. For best results, the three projections usually are equally spaced and, on circular workpieces, are located radially, as far as possible from the axis of the workpiece.

Design. Proper projection design helps to provide consistent results and to reduce breakage of forming tools. Proper design criteria include:

- Projections should be easy to form without distorting the workpiece.
- When projections are formed in sheet metal parts, the metal in and around the projection must not be cracked or made markedly thinner.
- Projections should be strong enough to support the initial electrode force before the current is applied.
- Projections should collapse during welding without excessive expulsion of metal, leaving the two components in intimate contact.

Types of Projections. The five basic types of projections—spherical, elongated, annular, pyramidal, and cross wire—are shown in Fig. 9, along with two special types of projection welds for which conventional projections are not used. All types produce strong welded joints if properly selected and applied.

Spherical projections are recommended for welding assemblies made of steel sheet and plate. They also can be coined or forged on the ends or faces of screws, nuts, and similar fasteners. Several projections can be used, as in Fig. 9(b), or a single large gently radiused projection can cover the entire face of the mating workpiece, as in Fig. 9(c). Table 2 provides information on the size, overlap, and strength of conventional spherical projections. Recessed spherical projections are shown in Table 4, which gives dimensions for recessed projections used in welding steels of different thicknesses.

Table 3 Conditions and joint strength for projection welding thin 1010 steel sheets of equal thickness using annular projections(a)

Item	Sheet thickness (t), in. 0.0105	0.0179
Weld time, cycles	6	6
Electrode force, lb	110	225
Welding current, A	5200	5400
Breaking load of joint in shear test, lb:		
One projection	190	400
Two or more projections(b)	145	280

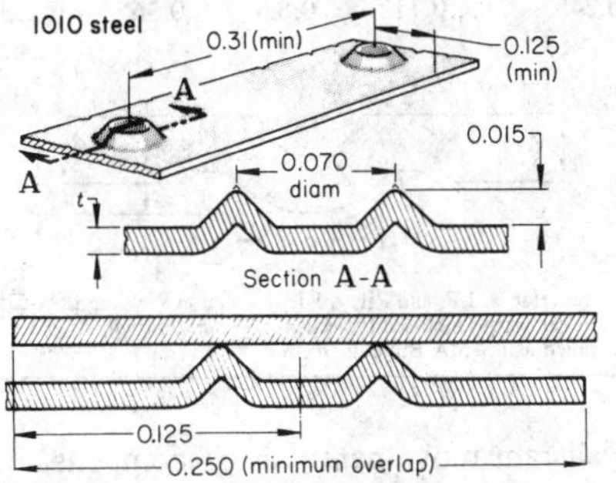

1010 steel
0.31 (min)
0.125 (min)
A
0.015
0.070 diam
t
Section A-A
0.125
0.250 (minimum overlap)

(a) Surface of steel may be oiled slightly, but should be free of grease, scale, and dirt. (b) Approximate strength of joint at each projection
Source: Harris, J. F. and Riley, J. J., Projection Welding Low-Carbon Steel Using Embossed Projections, *Welding Journal*, April 1961; reprinted as RWMA Bulletin No. 31

Elongated projections often are used instead of spherical projections where the shape of the parts makes an elongated weld more suitable and where welds made with spherical projections do not meet strength requirements. For a given work-metal thickness, the height of an elongated projection should be approximately the same as the height of a spherical projection, the width should be approximately equal to the diameter of a spherical projection, and the length should be made to suit the applications, usually two to three times the width.

Figure 9(d) shows an elongated projection that has been embossed in a sheet metal part. This type of projection was used in welding the flat base of a relay armature to the 0.083-by-0.22-in. end of the relay arm in Example 4. The use of elongated projections on sheet metal that is to be welded to the sidewall of a tube or other cylindrical workpiece is described and illustrated in Fig. 12. Elongated projections also can be coined or forged on fasteners, as in Fig. 9(e). Another type of elongated projection, of which the full width and height extend to the edge of a sheet metal part that is to be welded to the edge of a mating part, is shown in Fig. 9(f).

Annular projections are used for welding tubing to sheet metal, as shown in Fig. 9(g). Annular projections are also used for joining sheet metal parts, particularly thin parts, where spherical projections may collapse prematurely during welding (Fig. 9h) and for making liquid-tight or gastight connections, as in attaching mounting studs to housings (Example 14). Accurate alignment of electrode faces is especially critical when annular projections are used.

In welding bar stock or screw-machine parts to sheet or plate in a joint such as the one shown in Fig. 9(j), the shape of the parts in the annular region of contact serves the function of a conventional projection. Pyramidal projections coined or forged on the face of a nut are shown in Fig. 9(k) and (m).

Cross wire welds (Fig. 9n) are used in making various wire products. No separate projection is formed; the shape of the wires produces point contact where the wires intersect and functions as a projection. Round wire is welded to flat stock, using line contact.

In addition to cross wire welds (as shown in Fig. 9n and Examples 9 and 10) and annular welds on pin-and-tenon joints (as shown in Fig. 9j and Examples 7 and 8), other types of joints also can be projection welded without the use of conventional projections. In Example 13, the curved outer surface of a steel tube functioned as a projection in welding the tube to the flat end of a cylindrical slug.

Spacing of Projections. The minimum spacing of projection welds can be somewhat less than that of spot welds; during projection welding, the current is concentrated on a smaller area. The minimum spacing between projections is twice the projection diameter. A spacing substantially greater than the minimum is often used to make embossing of the projections easier.

As shown in Tables 2 and 5, the recommended minimum contacting overlap for projection welding is from 2.3 to 2.8 times the projection diameter.

Projection Welding of Sheet Metal Parts

The principal uses of projection welding are those in which blanked, stamped, and formed sheet metal parts are joined with the aid of projections that have been produced during the stamping or forming operation. Spherical and elongated projections are the types most commonly used for welding sheet metal parts; however, annular projections are used in welding parts less than 0.020 in. thick and in applications where gastight or liquid-tight connections are needed.

Thin workpieces, from 0.010 to 0.021 in. thick, can be projection welded using carefully designed projections and welding machines that have rapid follow-up. The short time required to heat a spherical projection to welding temperature and the relatively small height of spherical projections necessitate extremely fast follow-up of the electrode. However, electrode force need not be as high as for welding thicker work metal. Lack of rigidity in the projection and lack of resistance to collapse before the welding current is applied require careful adjustment of the initial electrode force.

The annular projection shown in Table 3 has greater resistance to cold collapse and can be used in place of a spherical projection for welding thin sheet metal parts. The welding conditions listed in Table 3 are for welding 0.0105- and 0.0179-in.-thick 1010 steel using annular projections. The 6-cycle (0.1-s) weld time allows latitude in selection of electrode force and welding current.

Intermediate thickness workpieces, from 0.022 to 0.135 in. thick, are the most adaptable to projection welding. Workpieces up to 0.135 in. thick are more easily welded and more commonly used than workpieces greater than 0.135 in. in thickness. Table 1 gives recommended conditions for welding low-carbon and low-alloy steel in thicknesses up to 0.125 in. with one projection. For multiple-projection welding, electrode force and welding current are increased in direct proportion to the number of projections.

Thick Workpieces. Projection welding of workpieces 0.136 to 0.250 in. thick is limited because of the large equipment and the high power required. Welding in this thickness range generally results in (1) an increase in weld porosity, (2) an increase in sheet separation because of the difficulty of obtaining complete projection collapse, and (3) an increase in expulsion of metal because the large machines needed usually have sluggish follow-up.

Weld porosity is difficult to eliminate, but can be minimized by using a lower electrode force during welding to reduce the current required and to prevent premature projection collapse. Applying forge force immediately after the weld time also minimizes porosity. The use of pulsation timing may also be helpful.

Expulsion of metal resulting from sluggish ram movement can be compensated for by using upslope control to increase the current gradually during weld time. This procedure decreases the rate of follow-up needed to maintain proper welding force.

Sheet separation is minimized through the use of either forge force or a recessed type of projection. Table 4 shows a recessed projection and gives dimensions for recessed projections in low-carbon steel 0.123 to 0.245 in. thick. The tools needed for making recessed projections have short life. Because this type of projection is coined and not embossed, press capacity for making a projection in 0.250-in.-thick low-carbon steel must be about 30 tons.

Table 4 Dimensions of recessed projections for welding of low-carbon steel 0.123 to 0.245 in. thick

Steel thickness (*t*), in.	Dimensions of projection, in. Height (*h*), ±2%	Diameter (*d*), ±5%	Spherical radius, *R*	Inside radius, *r*
0.123	0.058	0.270	0.196	0.065
0.135	0.062	0.300	0.215	0.072
0.153	0.064	0.330	0.235	0.078
0.164	0.068	0.360	0.248	0.083
0.179	0.080	0.390	0.274	0.091
0.195	0.084	0.410	0.286	0.095
0.210	0.090	0.440	0.305	0.101
0.225	0.100	0.470	0.325	0.108
0.245	0.112	0.530	0.365	0.121

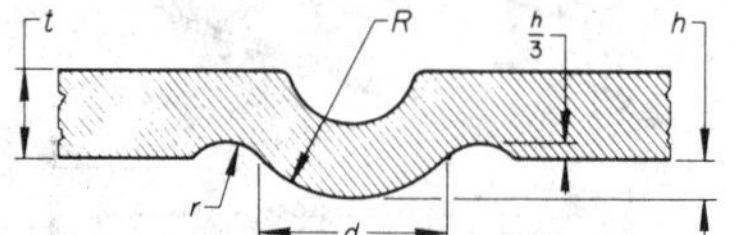

Source: Harris, J.F. and Riley, J.J., Projection Welding Low-Carbon Steel Using Embossed Projections, *Welding Journal*, April 1961; reprinted as RWMA Bulletin No. 31

Recessed projections limit weld strength. The metal that flows into the recess does not contribute appreciably to weld strength, but the recess permits complete collapse of the projection at lower electrode forces than are needed for collapsing conventional spherical projections.

Suggested projection size and welding conditions for projection welding two equal thicknesses of low-carbon steel from 0.153 to 0.245 in. thick using normal-size and small-size projections are given in Table 5. Dimensions given for normal-size projections are slightly different from those given in Table 2. The small-size projections can be used when two or more are needed, or to reduce the welding-machine size and the power requirements.

A special technique that is sometimes used in low-production projection welding of thick work metal is to insert a slug of metal between two flat workpieces to serve as a projection, thus avoiding the difficulties and cost of producing a conventional projection.

Workpieces of unequal thickness are projection welded about as often as are workpieces of equal thickness. Projections can be placed in either workpiece, although weld properties are less affected by variations in welding conditions if the

Table 5 Projection dimensions and welding conditions for projection welding of low-carbon steel 0.153 to 0.245 in. thick(a)

Steel thickness, in.	Projection size: Diam, in.	Projection size: Height, in.	Projection spacing (min), in.	Contacting overlap (min), in.	Electrode force, lb: Weld	Electrode force, lb: Forge	Upslope time, cycles	Weld time, cycles	Welding current, A	Shearing load of joint(b), lb
Normal-size projections										
0.153	0.330	0.062	1.75	0.90	2000	4000	15	60	15 400	7 500
0.164	0.350	0.068	1.80	0.95	2300	4600	15	70	16 100	8 100
0.179	0.390	0.080	1.90	1.00	2630	5260	20	82	17 400	9 500
0.195	0.410	0.084	2.00	1.05	2930	5860	20	98	18 800	11 300
0.210	0.440	0.092	2.10	1.15	3180	6360	25	112	20 200	12 500
0.225	0.470	0.100	2.30	1.20	3610	7220	25	126	21 500	15 000
0.245	0.530	0.112	2.50	1.30	3900	7800	30	145	23 300	17 300
Small-size projections										
0.153	0.270	0.058	1.60	0.75	1400	2800	15	60	11 100	5 100
0.164	0.290	0.062	1.65	0.80	1425	2850	15	70	11 800	5 500
0.179	0.310	0.067	1.70	0.85	1500	3000	20	82	12 800	6 500
0.195	0.330	0.072	1.75	0.90	1600	3200	20	98	13 900	7 700
0.210	0.350	0.077	1.80	0.95	1730	3460	25	112	14 900	8 500
0.225	0.370	0.082	1.90	1.00	1870	3740	25	126	16 000	10 400
0.245	0.390	0.088	2.10	1.10	2100	4200	30	145	17 300	12 000

(a) For welding two equal thicknesses of metal. As thickness ratio increases, welding current and weld time are increased to obtain a sufficiently large nugget diameter and adequate penetration and to minimize sheet separation. (b) Approximate shear strength of joint for each projection weld; shear strength varies with joint design.
Source: Harris, J. F. and Riley, J. J., Projection Welding Low-Carbon Steel Using Embossed Projections, *Welding Journal*, April 1961; reprinted as RWMA Bulletin No. 31

projections are formed in the thicker component. In welding workpieces in thickness ratios greater than 3 to 1, it may be more practical to place the projection in the thinner component.

The conditions listed in Tables 1 and 2 can be used for welding workpieces of unequal thickness. The welding current and weld time may have to be increased to obtain a sufficiently large nugget diameter and adequate penetration and to minimize sheet separation. The added energy increases both the diameter of the nugget and the shear strength of the joint.

Projection Welding of Bar Stock to Sheet Metal

Projection welding of bar stock to sheet metal can be done easily if the parts and projections are correctly designed. Fasteners such as pins, studs, and nuts are welded to sheet metal using spherical (Fig. 9a, b, and c), pyramidal (Fig. 9k and m), and annular (Fig. 9g and h) projections as an alternative to attaching them by stud welding or percussion welding.

Provision must be made for holding the component made of bar stock so that it can be welded at the correct angle with the mating part and so that the current path is as short as possible. Flow of current through a long slender shaft should be avoided, when possible, by applying the current to the shaft near the weld area.

In Example 14, in which a headed stud was welded to a sheet metal housing, an annular projection was formed on the underside of the head, and the shank of the stud was passed through a hole in the housing. Thus, the upper electrode contacted the top of the head and provided a short path for the welding current.

In another type of joint design (Fig. 9j), a tenon on the end of the pin acts as a pilot for inserting the pin into the hole in the sheet metal part, and an adjoining conical tapered section makes annular contact with the sharp edge of the hole, which localizes the current and heating. This joint design, which was used in Examples 7 and 8, facilitates fixturing and assembly and is well suited to high-production welding in semiautomatic and automatic resistance welding machines.

In Example 7, three pins were projection welded at both ends to two platens 1.25 in. apart, making six welds at once, with the electrodes directly above and below each pin. In Example 8, in which a shaft was welded to a plate, a short current path was provided by resting the upper electrode on the plate and clamping the lower electrode to the shaft below the plate.

Besides being well suited to high-production welding, this type of joint permits the use of free-machining steel in the pin or shaft, because it allows close control of heating and a minimum of penetration into the pin or shaft, thus avoiding weakness and porosity that results from segregation of sulfur or phosphorus in the free-machining steel during welding. The use of this type of joint design (Fig. 9j) to allow high-production welding of free-machining steel is described in the next two examples.

Example 7. Use of Pin-and-Tenon Joint for High-Production Projection Welding. The platen-frame assembly shown in Fig. 10 consisted of two end plates made of cold rolled low-carbon steel and three $^{5}/_{16}$-in.-diam separating pins made of free-machining 1213 steel. Although 1213 steel generally is unsuitable for resistance welding, it was used for the pins because most of the molten metal for the welds was provided by the low-carbon steel plates and because machining time was about half that for non-free-machining steel.

A 45° bevel on the pin between the $^{3}/_{16}$-in.-diam tenon and the $^{5}/_{16}$-in.-diam body of the pin was used to contact the sharp edge of the hole in the end plate, and thus the sharp edge acted as a projection and was melted to make the weld (Fig. 10, section A-A). Penetration into the free-machining steel, inclusions in the weld nugget, and dilution of the low-carbon steel, which destroys the bonding properties of the weld metal, were minimized.

To make the six welds simultaneously, the lower electrodes were fixed in position, while the vertical positions of the upper electrodes were hydraulically equalized to compensate for variations in pin length and electrode wear. The welding fixture included locating pins that fitted tightly into the two pairs of locating holes shown in Fig. 10 to maintain alignment of the two platens during welding.

One weldment from every hundred was checked by the operator in a go/no-go gage for dimensional accuracy. One of each 300 weldments was tested to destruction; each of the broken welds was required to show evidence of torn metal.

Previously, these platen-frame assemblies had been joined by riveting over the ends of the pins. The changeover to projection welding resulted in increased production and significant cost saving. Projection welding required the purchase of a new resistance welding machine and welding dies, but these expenditures were outweighed by the saving in time and also by a saving in material. The saving in material was obtained because the added strength and rigidity of the projection welded assembly allowed a reduction in the number of pins per assembly from four to three.

Fig. 10 Projection welding of platen-frame assembly

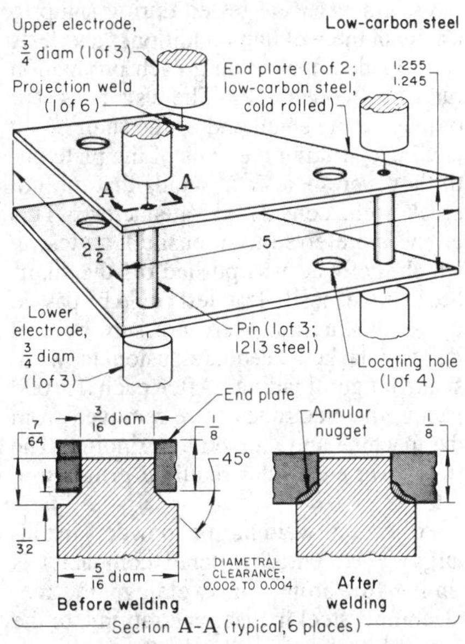

Equipment details

Welding machine	Semiautomatic press type
Rating at 50% duty cycle	200 kV·A
Current, max	75 000 A
Electrodes	$^{3}/_{4}$-in.-diam RWMA class 2
Heat control	Phase shift

Welding conditions

Welding current, approx	70 000 A
Heat control setting	90%
Electrode force, total	3500 lb
Squeeze time	30 cycles
Weld time	20 cycles
Hold time	30 cycles
Production per hour	42 assemblies (252 welds)

Example 8. High-Speed Automatic Projection Welding of a Free-Machining Steel Shaft. The plate-to-shaft assembly shown in Fig. 11 was made in an automatic resistance welding machine equipped with a four-station indexing turntable. Before welding, as shown in detail B in Fig. 11, the edge of the hole in the plate rested on the conical land on the shaft.

In the first station, the large-diameter end of the shaft was placed in the lower electrode; in the second station, the plate was dropped onto the shaft; in the third station, the upper electrode was lowered to contact the plate, and the joint was

Fig. 11 Automatic projection welding of free-machining steel shaft and plate

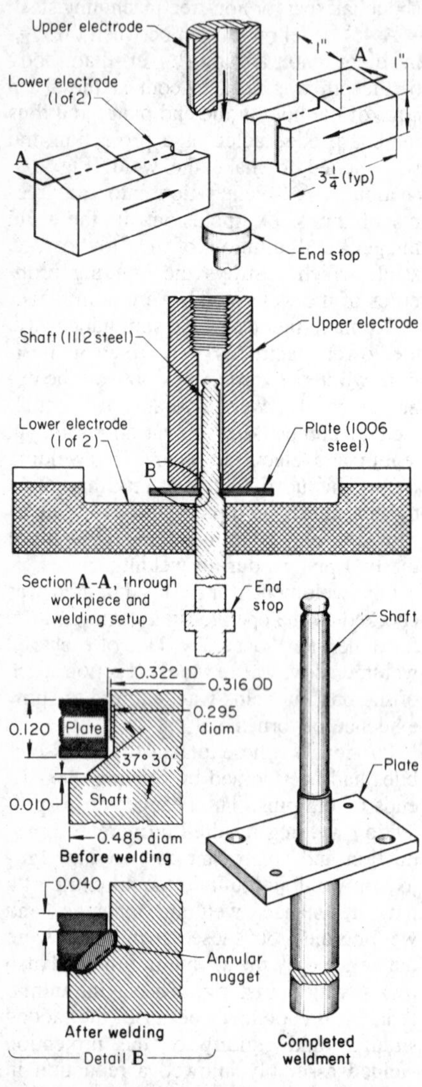

Equipment details

Welding machine	Automatic, with four-station indexing turntable
Rating at 50% duty cycle	100 kV·A
Current, max	65 000 A
Heat control	Eight-tap transformer and phase shift
Electrode material	RWMA class 3
Electrode force, max	2300 lb, approx

Welding conditions

Welding current, approx	45 000 A
Welding voltage	7.8 V
Electrode force	1550 lb
Squeeze time (including machine delay time)	99 cycles
Weld time	6 cycles
Hold time	11 cycles
Production per hour	666 assemblies

welded; and in the fourth station, the weldment was ejected.

As shown in detail B in Fig. 11, an annular weld nugget was formed by fusion of metal from the lower edge of the hole and from the contacting region of the beveled portion of the shaft. Also, plastic metal was forced into the 0.295-in.-diam recess in the shaft, reinforcing the joint.

As shown in Fig. 11, the upper electrode surrounded the small-diameter end of the shaft and bore against the plate when in welding position, and two lower electrodes centered and clamped the shaft. An end stop kept the shaft from moving downward under the force of the upper electrode.

Originally, an angle of 26° was specified for the conical land on the shaft, but with this angle satisfactory welds were not always produced. Use of a 37½° cone angle and the welding conditions given in Fig. 11 produced sound welds. A few faulty welds were caused by a short between the shaft and the upper electrode.

Weldments were tested during setup to establish the welding conditions, and tests were made after welding each production run of 500 assemblies. The tests consisted of placing the small end of the shaft on an anvil and striking the ends of the plate until they were bent 30°. Welds that did not break were considered satisfactory. This test was preferred to a tensile-load test in which the plate was pushed off the shaft, because the tensile-load test could be passed by weldments that were pressure bonded but that lacked adequate fusion to withstand fatigue loading. After each 15 000 welds, the electrodes were removed from the machine and dressed by grinding. The two lower electrodes required grinding in sets.

Projection welding of powder metallurgy (P/M) parts (sintered compacts) is similar to projection welding of free-machining steel in that penetration into the sintered metal must be kept to a minimum. In joining sintered iron to 1010 steel (Example 12), this was accomplished by producing an annular weld nugget at the contact line between the sharp edge of a hole in the steel and the beveled surface of a specially formed truncated conical boss on the sintered iron.

Projection Welding of Tube to Sheet Metal

Standard tube, pipe, and other parts of cylindrical shape can be projection welded to sheet metal at the sidewall or at the end of the tube. The sidewall of a tube can be welded to sheet metal using elongated projections embossed in the sheet metal component, as shown in Fig. 12. The elongated projection provides localized contact with the tube and requires less accurate positioning than does a spherical projection.

The end of a tube can be welded to sheet metal by the use of annular projections, which results in uniform, liquid-tight welds. Annular projections are easily machined on tube ends.

Projection Shape. Strength and final appearance of tube-to-sheet welds depend principally on projection shape and welding conditions. The recommended projection, referred to as a full-width projection, has a 90° included angle machined to intersect both the outside and inside surfaces of the tube and a base width equal to wall thickness (Fig. 13a). A reduced-width projection, which has a base width less than wall thickness (Fig. 13d and Fig. 9g), can be used to minimize deformation of the tube wall.

The appearances of tube-to-sheet welds made with full-width annular projections using short, correct, and excessive weld times are shown in Fig. 13(a), (b), and (c). The appearance of a weld made with the correct weld time and a reduced-width projection is shown in Fig. 13(d).

Figure 13(a) illustrates the use of a short weld time to avoid full collapse of the pro-

Fig. 12 Projection welding of tubes

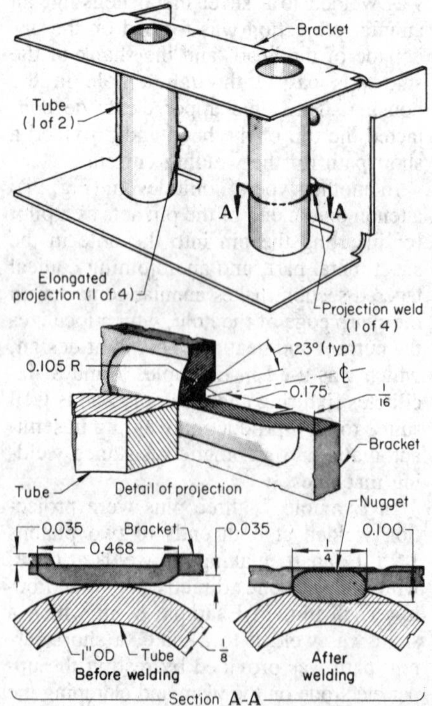

Fig. 13 Effects of weld time and width of annular projection on weld appearance

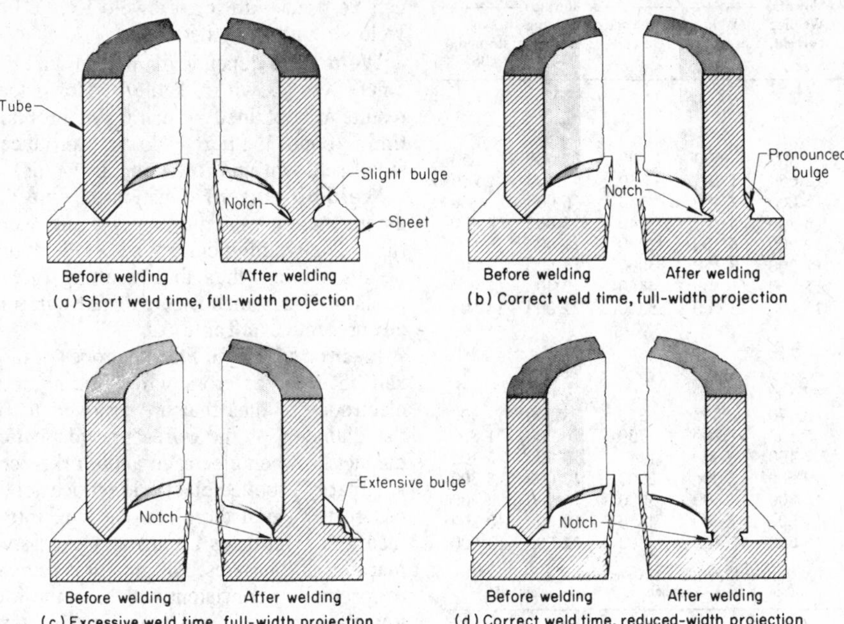

jection. During welding, the outside diameter of the tube is slightly increased at the top of the projection.

With a longer weld time, width of the weld produced is approximately equal to the wall thickness of the tube. As shown in Fig. 13(b), the outside diameter of the tube is increased, and the inside diameter is decreased at the top of the projection.

The appearance of a weld made with excessive weld time is shown in Fig. 13(c). A well-defined horizontal weld-metal displacement is formed around the outside of the tube. Bulging of the weld metal inside the tube is uniform and blends into the tube wall. The weld metal that bulges beyond the original inner and outer tube walls is not subjected to enough electrode force to become properly bonded and, therefore, does not add to weld strength.

The reduced-width projection shown in Fig. 13(d) can minimize deformation of the tube wall. The weld width is less than the thickness of the tube wall, and the shape of the weld metal generally is symmetrical about the initial-contact circle.

Regardless of weld time, initial joint preparation, or width of the projection, a notch is visible at the intersection of the inner tube wall and the surface of the mating workpiece (Fig. 13). Unless weld time is excessive, a notch is visible also at the intersection of the outer tube wall and the mating workpiece. There is always some bulging of the weld metal beyond the original inner and outer tube walls when intended weld width approaches wall thickness.

Heat Balance. The two components of a tube-to-sheet joint in which an annular projection is used can have unbalanced heating due to the different shapes. The welding current flowing through the tube wall near the weld interface heats the tube wall enough to prevent heat loss to this area from the weld interface during welding. However, as the current leaves the projection and enters the sheet, heat starts to flow away from the weld area into the surrounding cold metal. Heat flow in the sheet is limited by its thickness and thermal conductivity. In general, heat loss by thermal conduction is not a major problem in projection welding of low-carbon steel using annular projections.

As sheet thickness increases beyond tube-wall thickness, the assembly becomes increasingly difficult to weld because additional time or welding current is needed to bring the contacting surfaces to welding temperature. The tube wall behind the weld interface collapses more readily because of a large heat-affected zone (HAZ), and metal expulsion increases during welding.

Electrode Design. The electrodes that clamp the tube should be designed to make contact at two areas—one near the weld for electrical contact and the other near the top of the tube for alignment. This arrangement localizes the welding current near the weld and helps to hold the tube perpendicular to the sheet so that the apex of the projection is in uniform contact with the sheet. Clamping force causes the tube to conform to the electrode shape so that current density is uniform at the electrode contact surface.

If the welding current is unevenly distributed around the tube, unequal heating can occur at the weld interface. If localized melting occurs on the tube near the edge of the electrode, the electrode-contact surface must be enlarged or made uniform by remachining the electrode or the tube. Other alternatives are increasing the clamping pressure, or removing scale, varnish, or other foreign material from the surface of the tube to reduce contact resistance.

The initial extension of the tube from the clamping electrode should be about twice the wall thickness. Longer extensions may cause more heating and upsetting during welding, which can decrease the weld pressure because of a larger weld interface.

Weld Strength. Tube-to-sheet weldments, when tested to destruction, can fail in the tube wall, in the weld, or by tearing a slug of metal from the sheet around the weld circumference. Those assemblies in which the sheet thickness is equal to or greater than tube-wall thickness can be expected to fail at the interface, because the sound weld area is smaller than the cross-sectional area of the tube.

The tensile strength of properly welded joints made with reduced-width projections is nearly the same as that of joints made with full-width projections, whether failure in testing is at the interface or in the base metal.

Cross Wire Welding

Resistance welding of crossed wires, generally in a grid arrangement in which a number of parallel wires are welded at right angles to one or more other wires, is a form of projection welding. Cross wire products can be welded in a press-type resistance welding machine using a welding die or special individual electrodes.

If the wires are close together, a seam welding machine can be used. The wires are assembled in a fixture and the current-carrying member of the fixture can be placed on a lower platen or can contact the lower electrode wheel. The upper electrode wheel contacts the upper tier of crossed wires. Welding current flows continuously, and weld time for each cross wire weld is determined by the speed of the electrode wheels.

Resistance between electrodes and work metal should be uniformly low to mini-

Table 6 Conditions for cross wire welding of low-carbon steel wire

Wire diam, in.	Weld time(b), cycles	Conditions for 15% setdown(a): Electrode force, lb	Welding current, A	Weld strength, lb	Conditions for 30% setdown(a): Electrode force, lb	Welding current, A	Weld strength, lb	Conditions for 50% setdown(a): Electrode force, lb	Welding current, A	Weld strength, lb
Cold drawn wire										
$^{1}/_{16}$	5	100	600	450	150	800	500	200	1 000	550
$^{1}/_{8}$	10	125	1 800	975	260	2 650	1 125	350	3 400	1 250
$^{3}/_{16}$	17	360	3 300	2 000	600	5 000	2 400	750	6 000	2 500
$^{1}/_{4}$	23	580	4 500	3 700	850	6 700	4 200	1240	8 600	4 400
$^{5}/_{16}$	30	825	6 200	5 100	1450	9 300	6 100	2000	11 400	6 500
$^{3}/_{8}$	40	1100	7 400	6 700	2060	11 300	8 350	3000	14 400	8 800
$^{7}/_{16}$	50	1400	9 300	9 600	2900	13 800	11 300	4450	17 400	11 900
$^{1}/_{2}$	60	1700	10 300	12 200	3400	15 800	13 600	5300	21 000	14 600
Hot drawn wire										
$^{1}/_{16}$	5	100	600	350	150	800	400	200	1 000	450
$^{1}/_{8}$	10	125	1 850	750	260	2 770	850	350	3 500	900
$^{3}/_{16}$	17	360	3 500	1 500	600	5 100	1 700	750	6 300	1 800
$^{1}/_{4}$	23	580	4 900	2 800	850	7 100	3 000	1240	9 000	3 100
$^{5}/_{16}$	30	825	6 600	4 600	1450	9 600	5 000	2000	12 000	5 300
$^{3}/_{8}$	40	1100	7 700	6 200	2060	11 800	6 800	3000	14 900	7 200
$^{7}/_{16}$	50	1400	10 000	8 800	2900	14 000	9 600	4450	18 000	10 200
$^{1}/_{2}$	60	1700	11 000	11 500	3400	16 500	12 400	5300	22 000	13 000

(a) Setdown, % = (decrease in joint height ÷ diameter of smaller wire) × 100. (b) For 15, 30, and 50% setdown.
Source: Resistance Welding Manual, 3rd ed., Vol 1, p 88, Resistance Welder Manufacturers' Association, 1956; reprinted 1959

mize power loss and surface damage, but resistance between faying surfaces should be uniformly high to concentrate heat for welding. Grooved electrodes provide low contact resistance between electrode and work metal. The point contact resulting from the shape of the crossed wires provides high resistance at the work-metal interface. Bars that are square, rectangular, or hexagonal in cross section can be cross wire welded if positioned so that the joint is formed by edges, rather than by flat surfaces.

Wires or bars from 0.020 to $^{1}/_{2}$ in. in diameter have been cross wire welded. The upper and lower limits of wire diameter are governed mainly by the capability of accurately controlling the welding current and the electrode force for small-diameter wires, or of providing adequate current and force for large-diameter wires.

Metals Welded. Low-carbon steel, stainless steel, copper, and nickel-based alloys such as Monel are among the metals commonly cross wire welded. Table 6 gives recommended conditions for welding low-carbon steel wire.

Type 304 stainless steel requires the same weld time, 60% of the weld current, and $2^{1}/_{2}$ times as much electrode force as is needed for welding low-carbon steel (see the article "Resistance Welding of Stainless Steels" in this Volume), whereas Monel requires the same weld time and current but twice the electrode force.

Cross wire welding of galvanized and cadmium-plated wire can be done, but this destroys the plating surrounding the weld. Also, welding of cadmium-plated metal requires special safety precautions because of the high toxicity of cadmium vapors.

Electrode force needed for a cross wire weld depends on wire diameter, setdown requirements (the amount the wires are to be embedded into each other), desired appearance, and required weld strength. The values for electrode force, weld time, and current given in Table 6 produce strong welds with good appearance. Lower forces can be used with longer weld times, but weld strength may decrease.

Weld time depends mainly on wire diameter, as shown in Table 6. Consistent results are obtained by using synchronous timing control. Short weld time produces small setdown and low-strength welds.

Welding current depends on wire diameter, setdown requirements, and weld time. The welding current selected should be slightly less than that resulting in expulsion of hot metal. Suggested currents are presented in Table 6.

Electrode Design. Flat electrodes or dies can be used for cross wire welding, but electrodes or dies that are grooved to fit the diameter of the wires provide better contact between electrode and workpiece. A suitably grooved electrode reduces contact resistance at the electrode-wire interface and minimizes arcing and excessive marking of the wires. Alignment is critical for producing consistent welds. In the following example, resistance cross wire welding with grooved electrode faces partly replaced shielded metal arc welding (SMAW) in producing cage assemblies.

Example 9. Partial Change From Shielded Metal Arc to Resistance Cross Wire Welding. The cage assembly shown in Fig. 14 was one of three put together to form a protective guard for a twister winder used in the textile industry for making thread. The components, made from $^{5}/_{16}$-in.-diam 1118 steel bar (except for the base ring, which was made from

Fig. 14 Cross wire resistance welding application

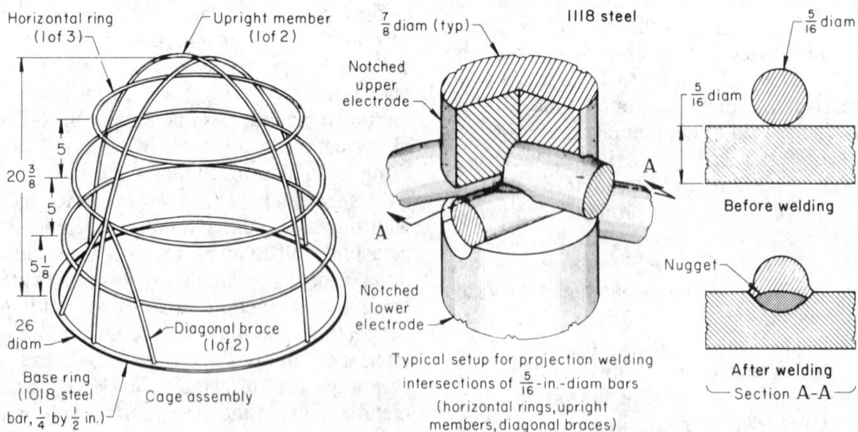

Welding machine	Three-phase manually operated resistance	Electrode force	700 lb
Rating at 50% duty cycle	100 or 150 kV·A	Squeeze time	25 cycles
Electrodes	$^{7}/_{8}$-in.-diam RWMA class 2	Weld time	4 cycles
Welding current	1000 A	Hold time	18 cycles
		Production rate	2 cages per hour(a)

(a) Includes 15 cross wire resistance welds on $^{5}/_{16}$-in.-diam 1118 steel bars and 8 shielded metal arc welds on $^{1}/_{4}$-by-$^{1}/_{2}$-in. 1018 steel base ring for each cage

$^1/_4$-by-$^1/_2$-in. 1018 steel bar), were cut to length, shaped, and fitted.

Originally, the cut, formed, and fitted bars were joined at all contact points by SMAW. The resulting welds were then ground smooth. When arc welding was replaced by resistance cross wire welding for making all welds not involving the base ring, rejects were eliminated, weld strength was adequate, appearance was improved, and both production time and cost were reduced.

In the improved method, the ends of the two upright members, one end of each of the two diagonal braces, and three brackets (not shown in Fig. 14) were joined by SMAW to the base ring while the components were clamped in a welding fixture. The assembly was then placed in an aluminum alloy fixture, where 15 resistance cross wire welds were made, joining the three horizontal rings to the diagonal braces and to the upright members.

An aluminum alloy was used for the fixture material because aluminum is light and easily handled. It is also nonmagnetic and thus has no effect on resistance and reactance, regardless of the amount introduced in the throat of the machine.

The electrodes for resistance welding were notch contoured, as illustrated in Fig. 14. The cage bars were placed in these notches and were forced together to make the weld. The notches were shallow enough to avoid contact and short circuiting between the electrode and the opposing bar. In addition to considerable savings, production time for a three-cage assembly was reduced by $2^1/_2$ h, because resistance welding was faster than SMAW. Postweld grinding was not needed to clean the assembly. All joints were inspected visually for weld soundness.

Techniques have been developed for welding crossed wires that have a plastic coating or are separated by a thin layer of plastic. During the squeeze time, sufficient electrode force is applied to cause the wires to penetrate the plastic material and make electrical contact. Sometimes the penetration can be facilitated by preheating the electrodes by using a special heating circuit. In the example that follows, no preheating was needed.

Example 10. Projection Welding of Crossed Wires. The circuit board shown in Fig. 15 was produced by projection welding crossed 0.020-in.-diam Nickel 200 wires that were separated by a sheet of teflon or similar plastic 0.010 in. thick. A specially designed kV·A spot welding machine equipped with a heat-programmed timer was used for making the welds through the plastic sheet. Optimum welding current and electrode force were determined experimentally. Both upper and lower electrodes were made of RWMA class 3 material.

The electrode force of 23 lb exerted on each pair of crossed wires corresponded to a local pressure of about 7500 psi, which was sufficient to cause the wires to penetrate the layer of plastic and make electrical contact. Another plastic material found to have sufficiently good cold flow characteristics for use in this application is polyethylene.

The transverse and longitudinal wires were assembled in a grid wire-holding fixture. Spacing of parallel wires in both directions was 0.100 ±0.005 in. Some welds were made to maintain parallelism of wires, and others were made to stop off unneeded wires. The portion of a wire not needed in the circuit was removed by piercing a hole through the plastic sheet at the wire-cutoff point (Fig. 15, detail A). The welded circuit board was later encapsulated in plastic to maintain dimensional tolerances and strength.

Parallelism and perpendicularity were measured mechanically; wire positions were held to ±0.005 in. Weld strength was tested by making sample cross wire welds and pulling the welded crossed wires into a straight line. Production rate was 10 000 cross wire welds in 8 h (including testing).

Projection Welding of Dissimilar Metals

Metals differing in thermal and electrical conductivity can be projection welded, provided the surfaces being welded are brought to welding temperature simultaneously. This usually can be done by forming the projection in the higher conductivity metal, by using a low-conduc-

Fig. 15 Application using projection welding of crossed wires

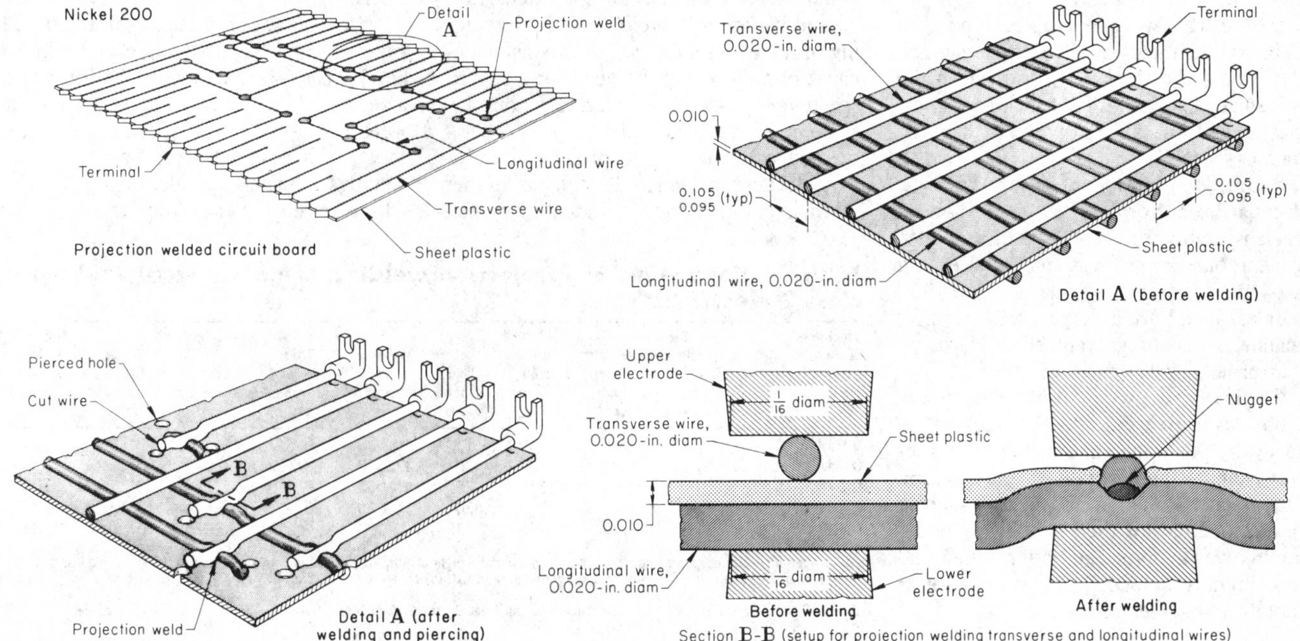

Fig. 16 Projection welding application. See text for details.

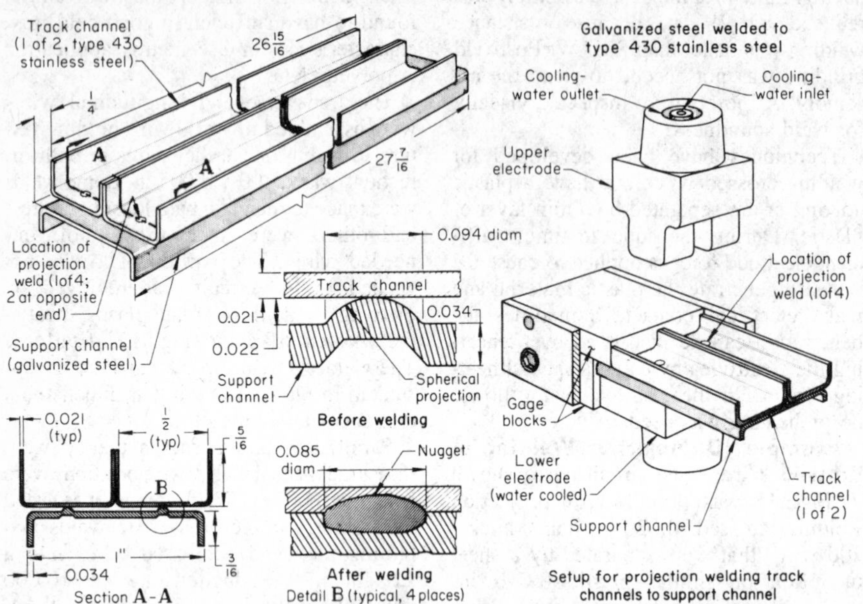

tivity copper-tungsten electrode (class 10, 11, or 12) in contact with the higher conductivity metal, or by combining these two conditions.

If the conductivities of the components are not widely different, a satisfactory heat balance usually can be obtained without using electrodes made of different materials. Differences in thermal and electrical conductivity between two metals ordinarily are closely parallel; the more conductive metal generates less resistive heat and dissipates heat more rapidly because of its higher thermal conductivity.

In the following example, type 430 stainless steel was welded to galvanized steel; the projections were formed on the galvanized steel, which has higher thermal and electrical conductivity than the stainless steel.

Example 11. Use of Spot Welding Versus Projection Welding. The lower-track assembly for a sliding mirror door in a bathroom cabinet was made by joining two track channels made of type 430 stainless steel to a larger track-support channel made of galvanized steel, using four projection welds (Fig. 16).

Originally, the welds were made one at a time by spot welding. The track channels were positioned manually on the support channel, and two spot welds, one in each track channel, were made in succession at one end of the assembly. The assembly was then turned, and the other end was similarly welded. The upper and lower spot welding electrodes had 1/8-in.-diam faces.

Production rate for spot welding was 123 assemblies per hour, and about 10% of the welded assemblies were rejected because of weld breakage. A major factor contributing to weld breakage was pitting of the lower electrode face, which occurred because the copper electrode became alloyed with zinc from the surface of the support channel. This zinc pickup caused changes in the contact resistance and resulted in poor welds. This necessitated frequent dressings of the lower electrode. Dressing was done manually with a file and produced a nonuniform face contour, which also caused variation in weld quality. Rewelding of faulty assemblies was expensive because of rework and the need for close inspection imposed by the high rejection rate.

To improve weld quality, the joining method was changed to projection welding. Because galvanized steel has higher thermal and electrical conductivity than type 430 stainless steel, the spherical projections were formed on the support channel, as shown in Fig. 16. This permitted a reduction in electrode force and current density on the external zinc-coated surface. The change to projection welding reduced zinc pickup by the electrode while allowing the use of sufficient electrode force to ensure consistently sound welds. Frequent removal of zinc from the lower electrode was not needed, and the rejection rate was reduced from 10 to 0.1%.

The upper and lower electrodes used for projection welding had large, flat contact faces and were water cooled to dissipate heat generated at the welds. Gage blocks were attached to the lower electrode for automatic positioning of the three channels for welding. The two projection welds at each end of the track assembly were made simultaneously. Production rate was increased from 123 to 173 assemblies per operator per hour, and welding cost was reduced 27%.

Projection Welding of Coated Steel

Low-carbon steels coated with zinc, lead, tin, or aluminum are projection welded in large quantities. Low-carbon steels with thin coatings of copper, tin, nickel, chromium, or cadmium are easily projection welded using a welding current generally about 10% greater than that used for welding bare low-carbon steel of equal thickness. Cadmium fumes may constitute an occupational hazard; therefore, adequate fume collection and disposal must be provided when welding cadmium-coated steels.

Schedules for projection welding of galvanized low-carbon steel are given in Table 7. These schedules apply for galvanized steel with a commercial coating weight (1.25 oz of zinc per square foot of sheet). Zinc-coated steel having thinner coatings is usually welded using the conditions given for bare low-carbon steel in

Table 7 Conditions for projection welding of galvanized steel using class 2 electrodes(a)

Thickness of each steel piece(b), in.	Electrode dimensions: Body diameter, in.	Electrode dimensions: Face diameter, in.	Net electrode force, lb	Weld time, cycles	Welding current (approx), A	Diameter of projection, in.	Height of projection, in.	Nugget diameter, in.	Minimum tension-shear strength(c), lb
0.039	5/8	3/8	250	15	10 000	0.187	0.041	0.15	925
0.063	5/8	7/16	400	20	11 500	0.218	0.048	0.25	2050
0.078	3/4	1/2	550	25	16 000	0.250	0.054	0.25	2700
0.093	3/4	1/2	750	30	16 000	0.250	0.054	0.30	4300
0.108	7/8	1/2	950	33	22 000	0.250	0.054	0.31	4900

(a) Data are for welding galvanized steel with a commercial-weight coating of 1.25 oz per square foot. Steel to be welded must be free from dirt, grease, and paint, but may have a light coating of oil. (b) Data are for welding two pieces of equal thickness. (c) For single projections only

Source: Recommended Practices for Resistance Welding Coated Low-Carbon Steels, AWS C1.3-70, American Welding Society, 1970. With permission

Table 1, or conditions in between those in Tables 1 and 7.

Electrode pickup is common in projection welding of steel coated with aluminum, tin, or cadmium because these metals readily alloy with copper-based electrodes. Frequent cleaning and redressing of the electrodes are necessary to remove any pickup on the electrode faces. Electrode pickup also occurs in projection welding of zinc-coated steel, but although there is some alloying between the zinc coating and copper-based electrodes, most of this pickup results from "soldering" of the zinc to the electrode face. This occurrence necessitates frequent cleaning. Copper, nickel, and chromium do not diffuse readily into copper-based electrodes, and pickup is less troublesome in projection welding steel coated with these metals.

Generally, less electrode pickup occurs in projection welding than in spot welding of coated steel because the initial heating is concentrated at the tip of the projection and current density is lower on projection welding electrodes than on spot welding electrodes.

For instance, in Example 11, in which galvanized steel was welded to type 430 stainless steel, changing from spot welding to projection welding allowed reduction of electrode force and current density and thus reduced deterioration of the electrode face.

Electrode pickup in projection welding of coated low-carbon steel can be reduced if careful attention is given to joint design, electrode design, and adjustment of hold time.

The joint should be designed so that the metal coating (particularly zinc and tin) can be burned or extruded from between the surfaces to allow formation of a weld nugget in the base metal. Size and shape of the electrode face should be such that heating is adequate but current density at the electrode face is low.

Projection Welding of Powder Metallurgy Parts

In welding P/M parts (sintered compacts), it is essential to restrict the depth of melting (penetration) of the sintered metal at the joint interface to the shallowest depth that produces consistent welds and adequate bond strength. In the following example, projection welding was used for attaching P/M parts to a low-carbon steel rim. The upper electrode was contoured to make contact with the entire upper surface of the P/M part. An annular type of projection, provided by the sharp edge of the hole in the rim, was used to minimize nugget penetration in the sintered iron and thus to produce a strong weld.

Example 12. Projection Welding of a Powder Metallurgy Part. Three sintered P/M parts, consisting of a friction material and a backing that was made from low-density iron powder, were welded to a 1010 steel brake-shoe rim, as shown in Fig. 17, to serve as brake-lining segments. The friction material developed for heavy-duty use in automotive drum-type brakes could not be expected to form acceptable projection welds with the brake-shoe rim because of its composition. The sintered iron-powder backing layer of each brake-lining segment was joined to the rim by four projection welds.

Truncated-cone projections on the sintered backing (Fig. 17, section A-A) were designed so that their conical surfaces contacted the sharp edges of the holes in the rim to produce an annular nugget with shallow penetration into the sintered metal. This design avoided the formation of porous, weak welds that could have resulted from excessive melting of the sintered metal.

The rim was clamped for welding in an indexed position on the fixture that held the lower electrodes (Fig. 17). The first lining segment was positioned with the projections inserted in the holes in the rim, and the welding cycle was initiated. An air-operated ram lowered the upper electrode to make contact with the lining segment. This actuated the weld sequence timer, and the four welds were made. Then the ram and upper electrode were retracted, the assembly was indexed, and the operational sequence was repeated to weld the second and third lining segments to the rim.

The upper electrode, made of RWMA class 2 material, was radiused to fit the

Fig. 17 Projection welding setup. See text for details.

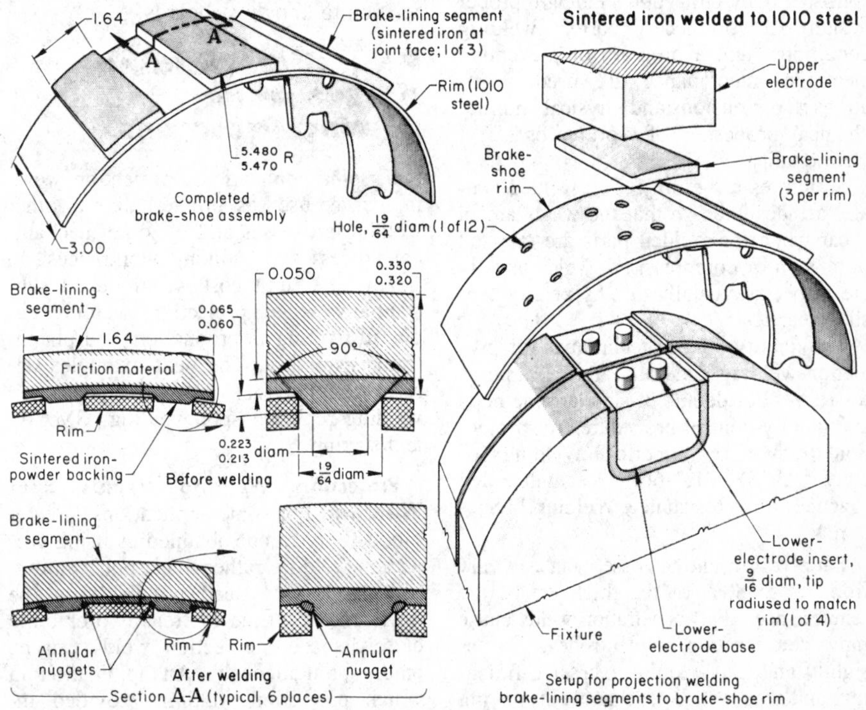

Equipment details	
Welding machine	Manual press type, with 16-tap transformer
Rating at 50% duty cycle	175 kV·A
Electrodes	RWMA classes 2 and 11(a)
Electrode force, max	4500 lb
Heat control	Electronic tube type

Welding conditions	
Welding current	35 000 A
Heat control setting	90%
Electrode force	1000 lb
Squeeze time	50 cycles
Weld time	22 cycles
Hold time	35 cycles

(a) Upper electrode was made of RWMA class 2 material and was radiused to fit the surface of the brake-lining segment; lower electrode consisted of four 9/16-in.-diam RWMA class 11 inserts radiused to conform to the inside surface of the rim and silver brazed to a class 2 base.

lining segment. The lower electrode consisted of a base made of class 2 material with four $^{9}/_{16}$-in.-diam inserts made of class 11 material. The inserts were silver brazed to the base in positions that aligned with the holes in the rim. After about 12 000 assemblies had been welded, the lower electrode was removed from the machine and returned to the toolroom for dressing of the insert tips to the radius matching that of the rim.

Weld quality was spot checked with a destructive push-off test. In this test, the outside diameter of the rim was supported on both sides of a lining segment while force was applied hydraulically to the four projection welds to break them apart. The breaking load for the welds was the essential criterion of acceptance; weld shape and appearance were of minor significance in this application.

Control of Weld Quality

High-quality joints can be produced consistently by projection welding if proper design of workpieces, correct welding conditions, and a program of preventive electrode maintenance are used. Variations in dimensions and physical and mechanical properties of workpieces affect weld quality.

Structures to be projection welded usually are designed so that the welds are in shear when the welded parts are stressed in tension or compression. Welds usually are inspected visually and by peel or tensile-shear tests.

The quality-control techniques for projection welding generally are the same as for RSW. For details of standard methods and quality-control procedures for projection welding, see American Welding Society (AWS) C1.1-66, "Recommended Practices for Resistance Welding," Section V.

Penetration into the base metal can vary from 20 to 80% of the thickness of the thinner sheet. Full-penetration welds cause rapid deterioration of electrodes, are unsightly and may require excessive finishing, and are no stronger than welds having 60 to 80% penetration. Insufficient penetration (less than 30%) results from inadequate or unbalanced heating.

Porous welds usually are caused by overheating, inadequate welding or forging pressures, or late application of adequate forging pressure. Premature collapse of the projection and the presence of scale, oxide, or other foreign material on the surface to be welded also can cause porous welds.

Effect of Work-Metal Properties. The composition, hardness or temper, dimensions, and mechanical properties of the work metal must be controlled within a suitably narrow range to consistently produce welds having uniform and acceptable size, penetration, and strength.

Nondestructive techniques for pretesting weldability of workpieces may be of special value in applications requiring extremely high weld reliability, such as those in which the workpieces are costly, in which many critical welds must be made on a complex assembly, or in which weld repair is difficult, costly, or impossible. One nondestructive method for the testing of weldability that sometimes can be applied to 100% of the work metal (instead of only to samples, as when destructive tests are used) is measurement of a magnetic or electrical property of the workpiece that is related to weldability. Inductance, for instance, can be measured accurately and rapidly on parts of suitable shape and controlled dimensions. It can also be used as an index of weldability in relation to known standards.

Projection Welding Versus Other Joining Processes

In some applications, projection welding can be used as an alternative to other welding processes and to some mechanical processes for joining workpieces. In Example 7, unit cost was reduced substantially when projection welding replaced riveting in production of a platen-frame assembly. The next examples describe applications in which projection welding replaced spot welding, SMAW, and staking.

Projection Welding Versus Spot Welding. For some applications, greater productivity can be obtained by using projection welding rather than spot welding, because closely spaced spot welds must be made one at a time whereas two or more closely spaced projection welds can be made simultaneously. An application in which projection welding provided increased production rate and lower cost, in comparison with spot welding, is described in Example 1.

Projection Welding Versus Arc Welding. Nonweldable components are sometimes joined to weldable components by means of a mechanical fastener that has been arc welded to the weldable component. Where the assembly sequence or the part design and dimensions prevent access for welding in this manner, arc plug welds can be deposited through holes in the nonweldable member.

Projection welding also can be used for attaching fasteners in joining nonweldable and weldable components; it is fast and can be done where there is no access for arc welding. Also, projection welding allows the use of a low, controlled heat input to avoid damage to adjacent heat-sensitive components, as shown in the example that follows.

Fig. 18 Hot upset projection welding application. See text for details.

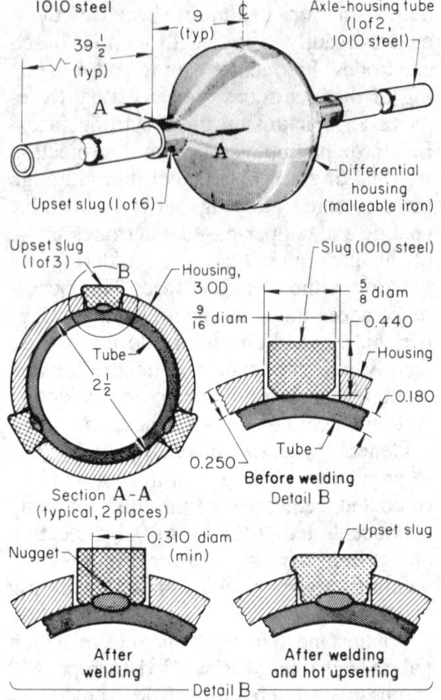

Equipment details

Welding machine .... Three-station projection welding machine with two eight-tap transformers per station and heat control for welding and upsetting
Rating at 50% duty cycle .... 100 kV·A(a)
Electrodes .... RWMA class 2, $^{5}/_{8}$-in. diam, type C (flat) face(b)
Current, max .... 35 000 A ac(a)

Conditions for welding and upsetting

Current .... Welding, 22 000 A upsetting, 15 000 A
Electrode force .... Welding, 2000 lb; upsetting, 3000 lb
Weld time .... 20 cycles
Cool time .... 20 cycles
Upset time .... 30 cycles
Production rate:
Subassemblies per hour 150
Welds per hour 900

(a) For each transformer. (b) Electrodes hydraulically driven

Example 13. Change From Shielded Metal Arc Plug Welds to Upset Projection Welded Slugs. A housing subassembly for a drive train (Fig. 18) consisted of a malleable iron differential housing and two 1010 steel axle-housing tubes, which were press fitted into the neck of the housing. Originally, the subassembly was joined by means of shielded metal arc plug welds deposited in equally spaced holes drilled through the neck of the cast housing; fusion of the deposits was mainly with the steel tubes. The weld-metal plugs acted as drive pins to transmit torque or thrust from one component to the other. This method of joining, however, was unsatisfactory for several reasons: (1) heat from welding often cracked the malleable iron, (2) fusion to the axle-housing tubes was erratic, (3) welds were porous because of an oil-like sealing compound entrapped between the housing and the tubes, and (4) the confined working area made accessibility for arc welding difficult. Other arc welding methods, including automatic gas metal arc welding, were considered for the plug welding, but were rejected because all were slow and involved a high heat input.

In an improved method of joining the three components (Fig. 18), in which a three-station projection welding machine was used, three slugs of 1010 steel were inserted into the holes in each end of the cast housing and were projection welded to the steel axle-housing tubes. The curvature of the tube functioned as a projection for welding.

Two welds were made at once, one on each end of the cast housing. The welds were series welds; a return electrode was clamped to each tube. The subassembly was rotated 120° and transferred to the next station to make the second pair of welds, and these operations were repeated to make the third pair of welds. After each projection weld was completed, the slug was heated by a second impulse and was hot upset by the electrode. The upset slugs partly filled the holes in the housing and thus served as mechanical retaining pins.

The much lower heat input to the welds, as compared to that for shielded metal arc plug welding, greatly reduced the volatilization of sealing compound and resulted in an acceptably low level of weld porosity. Subassemblies joined by the improved method were able to withstand 38 000 lb · in. of torque. One subassembly per shift was torque tested to destruction. Production rate was 150 subassemblies per hour.

Projection Welding Versus Staking. Studs in housings are commonly staked and are thus held with adequate rigidity until the assembly is completed and a nut on the end of each stud holds it secure. But in housings for which exceptionally high integrity is needed (as, for instance, in a housing required to be leakproof under air pressure), projection welding can sometimes produce a superior joint, as in the following example.

Fig. 19 Projection welding application. See text for details.

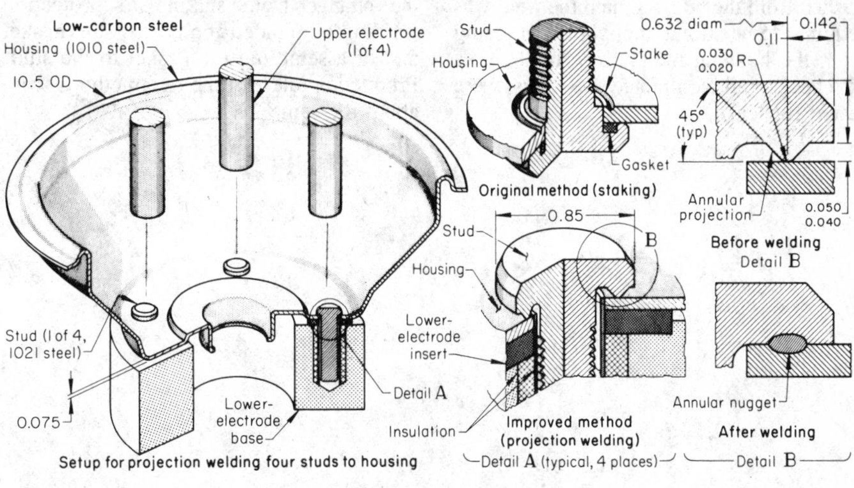

Equipment details

Welding machine	Special four-post, semiautomatic, with two eight-tap transformers
Rating at 50% duty cycle	150 kV·A
Upper electrodes	RWMA class 2
Lower electrodes	RWMA class 11 inserts, silver brazed to a class 2 body
Electrode force, max	6650 lb/electrode
Controls	Individual solid-state control devices for each stud

Welding conditions

Welding current	50 000 A
Heat-control setting	80% and No. 7 tap
Electrode force	4700 lb per electrode
Squeeze time	55 cycles
Weld time	6 cycles
Hold time	10 cycles
Production rate:	
Assemblies per hour	180
Welds per hour	720

Example 14. Change From Staking to Projection Welding. In the brake-booster housing shown in Fig. 19, the joints between the four studs and the housing were required to be leakproof under air pressure of 25 psi. Also, each stud had to be perpendicular to the mounting surface of the housing, within 0.005 in. at the tip of the protruding end of the stud, and had to withstand 30 lb · ft of torque.

Originally, the four studs were assembled to gaskets and were joined to the housing by staking (Fig. 19, detail A). With staking, however, the rejection rate was 15% because of joints that failed leak tests.

The rejection rate was reduced to 0.6% by changing the method of joining to projection welding, using an annular projection formed on the lower surface of the head of the stud (Fig. 19, details A and B). The projection welds were made in a four-electrode fixture that was designed to receive the housing in the open-side-up position (view at left in Fig. 19) and that consisted of four lower electrode inserts, each with a hole in the center to accept the threaded end of the stud. The inserts were mounted on a common base that made it possible to remove, dress, and replace them as a unit.

The welding machine had two transformers for supplying the welding current and four air cylinders for supplying the electrode force. One upper electrode was mounted on each air cylinder, and the welds were made one at a time, using a separate transformer for each weld.

For welding, the housing was placed open side up on the lower electrodes, and studs were inserted into holes in the housing. The welding cycle was started by depressing double palm buttons. The two front upper electrodes were brought in contact with the studs; the left stud was welded using the left transformer, and after a few cycles delay, the right stud was welded with the right transformer. Then the two front electrodes were retracted, the two rear electrodes were brought in contact with the other two studs, and welding was done in the same sequence. After the two rear electrodes were retracted, the assembly was automatically ejected onto a conveyor.

The parts were spot checked for leakage, which was seldom found, and for strength. In the strength test, the stud was pushed from the housing, and the weld was required to be strong enough to tear a slug from the housing metal. In a final series of checks, all completed housings were tested for leaks.

In addition to reducing the rejection rate, the change from staking to projection welding eliminated the cost of the gasket and of assembling the gasket to the stud. Production rates for the two methods were about the same.

The electrodes were removed after each shift, or after welding about 1440 assemblies, and were sent to the toolroom for machine dressing. After being redressed five times, the upper electrodes and the lower electrode inserts were replaced.

Resistance Welding of Stainless Steels

By the ASM Committee on Resistance Welding of Stainless Steels*

STAINLESS STEELS are readily resistance welded by spot, seam, and projection methods. Generally, the weld time and welding current are less than those used for welding carbon steel, but the electrode force is usually greater. Austenitic stainless steels of the 300 series are resistance welded more often than any other metal except low-carbon steel. Ideally, stainless steel to be resistance welded should contain a maximum of 0.08% C (as in types 304, 316, and 347), although steels with a higher carbon content (such as types 301, 302, 309, and 310) can be successfully resistance welded.

The martensitic and ferritic types of stainless steel can be welded satisfactorily. The martensitic types are less frequently resistance welded, because joints made in them are hard and brittle in the as-welded condition. Exact control of welding conditions is required, and for best results, a postweld tempering treatment should be used. Table 1 gives nominal compositions of the types of stainless steel that are usually resistance welded.

Equipment

Resistance welding of stainless steel requires less transformer capacity than for equivalent thicknesses of any other metal, under the same conditions. The welding machines use a single-phase alternating current power supply, a three-phase rectifier, or a frequency-converter power supply. A weld made with a three-phase power supply usually requires slightly higher current and a longer weld time than a weld made with a single-phase supply. The force, material, and shape of the electrodes are similar for both single-phase and three-phase power supplies. Synchronous timing controls are preferred for resistance welding stainless steel, because a variation of one or two cycles in the short weld time results in a large percentage change in weld time. Current and voltage regulators may be desirable, depending on machine size and power-supply capacity.

Factors Affecting Resistance Welding of Stainless Steels

The characteristics of stainless steel that affect resistance welding include electrical resistivity, thermal conductivity, melting temperature, strength at elevated temperatures, thermal coefficient of expansion, and contact resistance. Table 2 lists some of the pertinent physical properties of stainless steels that are usually resistance welded.

Electrical Resistivity. Stainless steels have a much higher electrical resistance than carbon steels. As a result, more heat is generated in a stainless steel with the same current. Therefore, resistance welding of a stainless steel requires lower currents or shorter weld times than welding a carbon steel. Generally, the austenitic, nitrogen-strengthened austenitic, and precipitation-hardening stainless steels have a slightly higher electrical resistance than the ferritic and martensitic (straight chromium) types.

Thermal Conductivity. Stainless steel has a lower thermal conductivity than carbon steel and, therefore, heat is conducted away from the weld zone more slowly. This, in conjunction with the electrical resistivity, means less heat has to be applied to reach the melting temperature. The martensitic and ferritic steels have a slightly higher thermal conductivity than other types; however, all types of stainless steel have a much lower thermal conductivity than carbon steel.

Melting temperatures of stainless steels have an effect on the amount of heat required to produce fusion for welding. The austenitic stainless steels melt in various ranges between 2500 and 2650 °F, and the martensitic and ferritic alloys melt in ranges between 2550 and 2790 °F. Plain low-carbon steels melt at temperatures between 2700 and 2800 °F.

Coefficient of Thermal Expansion. Austenitic and nitrogen-strengthened austenitic stainless steel expands and contracts with changing temperature almost 15% more than does plain carbon steel. These dimensional changes and the slower heat diffusion in austenitic grades result in greater thermal stress, which leads to warping. The ferritic, martensitic, and precipitation-hardening grades of stainless steel have coefficients of thermal expansion from 6 to 11% lower than that of plain carbon steel. The effect of thermal expansion on assemblies that are joined by spot and projection welding usually is minimal, because of the relatively small area that is heated during welding. In seam welding, however, the effect is significant because welding is continuous. The effect

*Robert S. Brown, *Chairman*, Senior Metallurgist, Stainless Alloy Metallurgy, Carpenter Technology Corp.; Daniel A. DeAntonio, Associate Metallurgist, Corrosion and Welding Research, Carpenter Technology Corp.; Michael J. Shields, Associate Metallurgist, Corrosion and Welding Research, Carpenter Technology Corp.; Thomas A. Siewert, Manager, Development, Alloy Rods

Table 1 Nominal compositions of resistance welded stainless steels

Type	Nominal composition, %									
	Carbon(a)	Manganese(a)	Phosphorus(a)	Sulfur(a)	Silicon(a)	Chromium(a)	Nickel	Molybdenum	Copper	Other
Austenitic stainless steels										
301	0.15	2.0	0.045	0.03	1.00	17.0	7.0	...	...	...
302	0.15	2.0	0.045	0.03	1.00	18.0	9.0	...	...	...
304	0.08	2.0	0.045	0.03	1.00	19.0	9.3	...	...	...
304L	0.03	2.0	0.045	0.03	1.00	19.0	10.0	...	...	...
309	0.20	2.0	0.045	0.03	1.00	23.0	13.5	...	...	...
310	0.25	2.0	0.045	0.03	1.50	25.0	20.5	...	...	...
314	0.25	2.0	0.045	0.03	2.00	24.5	20.5	...	...	...
316	0.08	2.0	0.045	0.03	1.00	17.0	12.0	2.5	...	...
316L	0.03	2.0	0.045	0.03	1.00	17.0	12.0	2.5	...	...
317	0.08	2.0	0.045	0.03	1.00	19.0	13.0	3.5	...	...
321	0.08	2.0	0.045	0.03	1.00	18.0	10.5	...	...	Ti(min), 5 × C
347	0.08	2.0	0.045	0.03	1.00	18.0	10.5	...	...	(b)
348	0.08	2.0	0.045	0.03	1.00	18.0	11.0	...	...	(c)
Nitrogen-strengthened austenitic stainless steels										
18Cr-2Ni-12Mn	0.15	12.0(d)	0.06	0.03	1.00	18.0	2.0	...	...	0.30 N
21Cr-6Ni-9Mn	0.08	9.0(d)	0.04	0.03	1.00	21.0	6.0	...	...	0.28 N
18-18 Plus	0.15	18.0(d)	0.04	0.04	1.00	18.0	...	1.0	1.0	0.50 N
22Cr-13Ni-5Mn	0.06	5.0(d)	0.04	0.03	1.00	22.0	13.0	2.0	...	0.30 N, 0.2 Nb, 0.2 V
Martensitic stainless steels										
403	0.15	1.0	0.04	0.03	0.50	12.3	...	...	...	...
410	0.15	1.0	0.04	0.03	1.00	12.5	...	...	...	...
414	0.15	1.0	0.04	0.03	1.00	12.5	2.0	...	...	...
431	0.20	1.0	0.04	0.03	1.00	16.0	2.0	...	...	...
440A	(e)	1.0	0.04	0.03	1.00	17.0	...	0.75(a)	...	...
Ferritic stainless steels										
405	0.08	1.0	0.04	0.03	1.00	13.0	...	...	...	0.1-0.3 Al
430	0.12	1.0	0.04	0.03	1.00	17.0	...	...	...	...
430F	0.12	1.25	0.06	(f)	1.00	17.0	...	0.6(g)	...	...
442	0.20	1.0	0.04	0.03	1.00	20.5	...	...	...	...
446	0.20	1.5	0.04	0.03	1.00	25.0	...	...	...	0.25 N max
Precipitation-hardening stainless steels										
17-7 PH	0.07	0.70	0.02	0.01	0.40	17.0	7.0	...	...	1.15 Al
PH 15-7Mo	0.07	0.70	0.02	0.01	0.40	15.0	7.0	2.25	...	1.15 Al
17-4 PH	0.07	1.00	0.04	0.03	1.00	17.0	4.0	...	4.0	0.03 Nb
Custom 450	0.05	1.00	0.03	0.03	1.00	15.0	6.0	0.75	1.5	Nb, 8 × C min
Custom 455	0.05	0.50	0.04	0.03	0.50	12.0	8.5	0.20	2.0	1.1 Ti, 0.30 Nb
15Cr-5Ni	0.07	1.00	0.04	0.03	1.00	15.0	5.0	...	3.5	0.30 Nb
PH 13-8	0.05	0.20	0.01	0.008	0.10	13.0	8.0	2.25	...	1.1 Al, 0.01 N

(a) Maximum percentage. (b) Minimum niobium-plus-tantalum content, 10 × C. (c) Minimum niobium-plus-tantalum content, 10 × C; cobalt, 0.20% max; tantalum, 0.10% max. (d) Nominal value. (e) Carbon range, 0.60 to 0.75%. (f) Minimum sulfur content, 0.15%. (g) Optional

of heat distortion is discussed later in this article.

High strength at room and elevated temperatures of nitrogen-strengthened austenitic precipitation-hardening stainless steels and, to a lesser extent, of straight-chromium grades makes it necessary to use greater electrode force than is required for carbon steel to bring the work-metal surfaces together for the required intimate contact at points of welding.

Contact resistance of stainless steels is higher than that of carbon steel and, therefore, greater electrode pressure is needed to make good resistance welds. Preweld cleaning is necessary for maintaining uniform contact resistance. Surface preparation for welding is discussed later.

Welding Characteristics of Stainless Steels

Austenitic Stainless Steels. All of the austenitic stainless steels (Table 1) in the solution-annealed condition contain carbon in solution. When the steel is heated to the temperature range of 800 to 1500 °F, as in the weld heat-affected zone (HAZ), carbon combines with chromium, resulting in chromium carbide precipitation at the grain boundaries. Precipitation of chromium carbide (sensitization) makes the material susceptible to intergranular corrosion. Sensitization to intergranular corrosion is influenced in austenitic stainless steel by (1) the carbon content, (2) the solubility of the carbon, (3) the presence of one or more stabilizing elements, (4) the proximity of the actual metal temperature to 1200 °F, and (5) the time period during which the metal is held in that temperature range. The first three factors are controllable only through selection of composition; the latter two are controllable by adjustment of welding conditions such as heat input, size of spot, production rate, and provisions for cooling.

Spot, projection, and seam welds in austenitic stainless steel have high corrosion resistance in most atmospheric environments because appreciable formation

Table 2 Typical physical properties of wrought stainless steels in the annealed condition

Type	Mean coefficient of thermal expansion, μin./in. · °F 212 °F	600 °F	1000 °F	Thermal conductivity, Btu/h · ft · °F 212 °F	930 °F	Electrical resistivity, μΩ · cm	Melting ranges, °F
Austenitic stainless steels							
301	9.4	9.6	10.1	9.4	12.4	72	2550-2590
304	9.6	9.9	10.2	9.4	12.4	72	2550-2650
316	8.8	9.0	9.7	9.4	12.4	74	2500-2550
321	9.2	9.6	10.3	9.3	12.8	72	2550-2600
347	9.2	9.6	10.3	9.3	12.8	73	2550-2600
Martensitic stainless steels							
410	5.5	6.3	6.4	14.4	16.6	57	2700-2790
414	5.8	6.1	6.7	14.4	16.6	70	2600-2700
431	5.7	6.7	···	11.7	···	72	···
440A	5.7	···	···	14.0	···	60	2500-2700
Ferritic stainless steels							
405	6.0	6.4	6.7	15.6	···	60	2700-2790
430	5.8	6.1	6.3	15.1	15.2	60	2600-2750
444	5.6	5.9	6.3	15.5	···	62	···
446	5.8	6.0	6.2	12.1	14.1	67	2600-2750
Precipitation-hardening stainless steels							
17-7 PH	6.1	6.4	···	9.5	12.6	83	2560-2630
17-4 PH	6.0	6.4	···	10.6	13.1	80	2560-2630
15-5 PH	6.0	6.3	···	10.3	13.1	77	2560-2630
PH 13-8Mo	5.9	6.2	6.6	8.1	12.7	102	2560-2630
Carbon steels							
Plain-carbon steel(a)	···	···	7.1	···	31.0	10	2700-2800

(a) For comparison only, average readings

of intergranular chromium carbide is not likely to occur in resistance welds during the relatively short weld times and the low heat input used for most resistance welding applications. Resistance to intergranular corrosion by the solution-treated base metal is less likely to be affected adversely by resistance welding than by conventional arc welding, with its longer weld times and higher heat input.

Variations in composition among standard austenitic stainless steels affect both the behavior of steels in welding and their performance in service. Types 302, 304, and 304L differ chiefly in carbon content, which determines the amount of chromium carbide precipitation that can occur in the HAZ of the base metal. In extra-low-carbon austenitic steels, the carbon content does not exceed 0.03%; thus, chromium carbide precipitation and susceptibility to intergranular corrosion are negligible. Types 316, 316L, and 317 contain additions of molybdenum for improved corrosion resistance; however, they are still susceptible to intergranular corrosion. Stabilized stainless steels, including types 321, 347, and 348, contain additions of titanium or of niobium-plus-tantalum that preferentially form carbides, thus effectively preventing intergranular chromium carbide precipitation during welding.

The effect of welding heat in producing carbide precipitation is discussed more fully in the article "Arc Welding of Stainless Steels" in this Volume. The additional nitrogen in nitrogen-strengthened austenitic stainless steels must be considered in the same manner as carbon. Nitrogen combines with chromium to form nitrides, which have the potential of reducing the corrosion resistance of the material. Thus, if the optimum corrosion resistance is required, postweld solution annealing should be considered.

Martensitic stainless steels (Table 1) most often welded are types 403, 410, 414, and 431. These steels can be resistance welded in the annealed, hardened, or hardened and tempered condition. Regardless of prior condition, welding produces a hardened martensitic zone adjacent to the weld. The hardness of this zone depends mainly on carbon content, although it can be controlled to a degree by the welding procedure. As the hardness of the metal in the HAZ increases, its susceptibility to cracking increases and its toughness decreases. Steels having a maximum carbon content of 0.15%, such as types 403 and 410, often produce satisfactory welds without postweld heat treatment. Steels with higher carbon contents, such as types 420 and 440A, generally require postweld heat treatment.

Ferritic grades of stainless steel (Table 1) most often welded are types 405, 430, 442, and 446. Resistance welds in these steels can exhibit reduced ductility at room temperature because of grain growth in the HAZ. Ferritic stainless steels can also become embrittled if heated to approximately 885 °F. This phenomenon, known as "885 °F embrittlement," may be encountered in the HAZ of a ferritic stainless steel weld. If properties in the as-welded condition are unsuitable, annealing must follow welding. Postweld annealing, in addition to improving ductility, helps to restore normal corrosion resistance.

Precipitation-hardening grades of stainless steel (Table 1) can be resistance welded in the annealed or aged (hardened) condition. If they are welded in the aged condition, the strength of the weld area is significantly reduced because the heat of welding anneals the material subjected to it. Generally, most of the strength may be restored by simply reaging the weldment. When maximum strength is required in the weld area, it is best to resistance weld these grades in the annealed condition and then age the weldment.

Surface Preparation

Surface preparation is critical to the success of resistance welding. Anything that prevents the passage of the welding current from the electrode to the stainless steel or between the pieces of stainless steel prevents a sound weld from being made. Burrs should be removed from the overlapping edges of components being resistance welded. Failure to remove burrs can cause some of the current to shunt through the burr instead of passing through the workpieces at the area of electrode contact.

The chromium oxide that is formed as a result of hot rolling, hot forming, or heat treating the stainless steel must be removed by pickling before the metal is resistance welded. The protective film that forms at ordinary temperatures after pickling is of microscopic thickness and usually does not interfere with resistance welding. Other types of surface contamination may not prevent a weld from being made, but may adversely affect the serviceability of the weldment. For this reason, paint, oil, or grease must be removed from the surfaces to be welded. Otherwise, carbon from the oil or grease can be absorbed into the steel, increasing its susceptibility to intergranular corrosion.

Stainless steel sheet should not be ground or filed with tools used for carbon and low-alloy steel because even very light iron contamination can reduce the effective chromium content in the ground or filed region and cause rejection of the weldment. Wire brushing must be done only with brushes made of stainless steel. The adhesive used for the protective paper used to prevent damage to the surface of polished sheet must be removed before resistance welding. Special etching or chemical cleaning sometimes is needed. For more information on chemical, acid, and alkaline cleaning of stainless steels, see the article "Cleaning and Finishing of Stainless Steels" in Volume 5 of the 9th edition of *Metals Handbook.*

Resistance Spot Welding

Resistance spot welding of stainless steel does not differ greatly from resistance spot welding of carbon steel, but the composition and physical properties of stainless steel make it necessary to control weld time and welding current more closely. Table 3 gives suggested practices for resistance spot welding of 300 series austenitic stainless steels. The values in Table 3 are intended for use as starting points and should be adjusted to suit conditions and job requirements. Procedures for establishing spot welding schedules in the article "Resistance Spot Welding" in this Volume can be followed in making the adjustments.

Welding Current. The electrical conductivity of stainless steel is 15 to 25% that of carbon steel of similar carbon content. Also, the melting temperature of stainless steel is about 90% that of carbon steel. Therefore, the heat required for successful welding, and thus the welding current, should be less than for carbon steel. The values for welding current suggested in Table 3 for welding stainless steel vary from half as much as, to the same as or slightly more than, those for welding of carbon steel as suggested in the article "Resistance Spot Welding" in this Volume. The welding currents are not in strict proportion to the electrical conductivities and melting temperatures, because other variables require that adjustments be made in weld time and electrode force.

Weld Time. The relative thermal conductivities of austenitic, nitrogen-strengthened austenitic, and precipitation-hardening stainless steels are 40 to 50% that of carbon steel (Table 2). Because heat is not conducted away from the weld area as rapidly as in carbon steel, less weld time is needed. The weld time is usually so short that it must be precisely controlled because a variation of ±1 cycle would be ±10% of the weld time for 0.062 in. thickness. For a thickness of 0.006 in., a variation of ±1 cycle would correspond to a variation of ±50% of the weld time.

Electrode Force. The contact resistance and strength of stainless steel remain high at elevated temperatures. Therefore, the electrode force required to make the weld is much higher than that required for welding carbon steel. A large electrode force is needed to forge the weld and to minimize cracks and voids in the weld nugget. The increased electrode force requires that the welding current be decreased, particularly in metal of greater thickness, because of the decrease in contact resistance. With an electrode force that is adequate for a given work-metal thickness, insufficient welding current results in an undersized nugget, whereas excessive welding current causes excessive indentation.

Electrodes. Spot welding electrodes for stainless steel are about the same size as those used for welding carbon steel, but because the electrode force is greater for stainless, the unit pressure at the electrode face is greater. Therefore, the electrode should be made of a harder material than

Table 3 Suggested practices for resistance spot welding of 300 series austenitic stainless steels(a)

Thickness (t) of thinnest outside piece, in.(b)	Electrode dimensions(c): Body diameter (min), in.	Electrode dimensions(c): Face diameter (max), in.	Net electrode force, lb	Weld time (single impulse), cycles (60 Hz)	Welding current (approx) for work metal tensile strength, A: Less than 150 ksi	Welding current (approx) for work metal tensile strength, A: 150 ksi and above	Minimum contacting overlap, in.	Minimum spot spacing (center to center), in.(d)	Nugget diameter (approx), in.	Minimum breaking load of weld in shear for work metal tensile strength, lb: 70-90 ksi	90-150 ksi	150 ksi and above
0.006	3/16	3/32	180	2	2 000	2 000	3/16	3/16	0.045	60	70	85
0.008	3/16	3/32	200	3	2 000	2 000	3/16	3/16	0.055	100	130	145
0.010	3/16	1/8	230	3	2 000	2 000	3/16	3/16	0.065	150	170	210
0.012	1/4	1/8	260	3	2 100	2 000	1/4	1/4	0.076	185	210	250
0.014	1/4	1/8	300	4	2 500	2 200	1/4	1/4	0.082	240	250	320
0.016	1/4	1/8	330	4	3 000	2 500	1/4	5/16	0.088	280	300	380
0.018	1/4	1/8	380	4	3 500	2 800	1/4	5/16	0.093	320	360	470
0.021	1/4	5/32	400	4	4 000	3 200	5/16	5/16	0.100	370	470	500
0.025	3/8	5/32	520	5	5 000	4 100	3/8	7/16	0.120	500	600	680
0.031	3/8	3/16	650	5	6 000	4 800	3/8	1/2	0.130	680	800	930
0.034	3/8	3/16	750	6	7 000	5 500	7/16	9/16	0.150	800	920	1100
0.040	3/8	3/16	900	6	7 800	6 300	7/16	5/8	0.160	1000	1270	1400
0.044	3/8	3/16	1000	8	8 700	7 000	7/16	11/16	0.180	1200	1450	1700
0.050	1/2	1/4	1200	8	9 500	7 500	1/2	3/4	0.190	1450	1700	2000
0.056	1/2	1/4	1350	10	10 300	8 300	9/16	7/8	0.210	1700	2000	2450
0.062	1/2	1/4	1500	10	11 000	9 000	5/8	1	0.220	1950	2400	2900
0.070	5/8	1/4	1700	12	12 300	10 000	5/8	1 1/8	0.250	2400	2800	3550
0.078	5/8	5/16	1900	14	14 000	11 000	11/16	1 1/4	0.275	2700	3400	4000
0.094	5/8	5/16	2400	16	15 700	12 700	3/4	1 3/8	0.285	3550	4200	5300
0.109	3/4	3/8	2800	18	17 700	14 000	13/16	1 1/2	0.290	4200	5000	6400
0.125	3/4	3/8	3300	20	18 000	15 500	7/8	2	0.300	5000	6000	7600

(a) Steel should be free from scale, oxide, paint, grease, and oil. (b) Welding conditions are determined by thickness (t) of thinnest outside piece. Data are for total thickness of pile-up not exceeding $4t$. Maximum ratio between two thicknesses, 3 to 1. (c) Body diameters apply to electrodes with types A, D, and E faces with the face diameters listed, and to type F electrodes with 3-in. spherical-radius faces. Electrode material, RWMA class 2, class 3, or class 11. (d) Minimum spot spacing for two pieces is that spacing for which no special precautions need be taken to compensate for shunting of current through adjacent spot welds. For three pieces, increase spacing 30%.
Source: "Recommended Practices for Resistance Welding," AWS C1.1; also, *Welding Handbook*, 6th ed., Section 2, American Welding Society, 1969

that used for spot welding carbon steel. Current requirements generally are low, and consequently, an electrode material with reduced electrical conductivity is permissible. Resistance Welder Manufacturers Association (RWMA) class 2 and 3 materials (copper alloys) are generally recommended, but for some applications, the refractory metals (classes 10 to 14) can be used. Compositions and properties of electrode materials of classes 2 and 3 and of classes 10 to 14 are given in the article "Resistance Spot Welding" in this Volume, as well as information on electrode design and holders.

Spot Spacing. The shunting effect from one spot to another is somewhat less in welding stainless steel than in welding carbon steel, because of the lower electrical conductivity of stainless steel. Therefore, spots can be more closely spaced, and the contacting overlap can be less than that for carbon steel. Suggested practices for minimum spot spacing and contacting overlap are given in Table 3.

Effect of Heat on Distortion. Because the coefficient of thermal expansion of austenitic and nitrogen-strengthened austenitic stainless steel is considerably higher than that of carbon steel, expansion and contraction of the heated metal should be considered. When metal is heated and cooled rapidly as in resistance welding, an upsetting action results, followed by shrinkage of metal in the weld zone. This can cause waviness in sheet and, when a long series of spot welds are made, an overall shrinkage of the part. In spot welding, weld shrinkage can be reduced by decreasing the size of the individual welds in the joint. The safe upper limit for spot-weld diameter is four to five times the thickness of the thinnest sheet being welded. If more spots of smaller diameter are used in such a way that the total volume of weld metal is the same, distortion is less. Adequate water cooling helps in controlling heat distortion.

Expulsion of Molten Metal. Arcing or flashing during spot welding can be caused by using a high welding current, a low electrode pressure, a small area of contact at the interface of the two workpieces, or by incomplete cleaning. The small area of contact generally is the result of using sharply domed electrodes or a face radius which is too small. A copious supply of water around both the upper and lower electrodes can minimize flashing and allow wider ranges of usable welding current and electrode pressure.

Multiple-Impulse Resistance Spot Welding

Multiple-impulse resistance spot welding consists of transmitting two or more impulses of welding current without removing the electrode force. Multiple-impulse spot welding adds the selected amount of heat to the weld with minimum damage to the electrodes. Additionally, excessive weldment heating is avoided, thus reducing harmful carbide precipitation in heat-sensitive grades, as well as reducing thermal distortion. Multiple-impulse spot welding is useful for joining austenitic stainless steels in thicknesses greater than 1/8 in.

Suggested practices for multiple-impulse spot welding of austenitic stainless steels are given in Table 4. Impulses are shorter and fewer; welding current is slightly greater, and electrode force is much greater than for multiple-impulse spot welding of carbon steels with comparable thicknesses. Note that the pulsations can also be used for postheat cycles.

Resistance Seam Welding and Roll Resistance Spot Welding

Resistance seam welding operates on a time cycle of current-on (heat) and current-off (cool) while the workpieces are traversed and pressure is exerted on them by rotating electrode wheels that are in continuous contact with the work metal to produce a line of overlapping spot welds. The overlapping spot welds provide gas-tight or liquid-tight seams. The spot welds can be spaced apart, which is known as roll resistance spot welding. Seam welds are usually made in lap joints and can be made in a straight line of flat metal, carried around corners of drawn parts by using a small electrode wheel on the inside, or installed as girth joints between cylinder walls and inserted head flanges. Mash seam welding can be used to produce flush joints on thin metal. Roll resistance spot welding is done with the same equipment used for resistance seam welding and is similar to it except that the spots are not overlapping. The spots can be made at any desired spacing, and the seams are not necessarily gastight or liquid-tight.

Mash resistance seam welding can be used to produce flush joints in stainless steel up to about 0.062 in. thick by mashing down the double metal thickness of a lap joint to approximately that of one member of the assembly. The operation is made possible by (1) reducing the overlap below that ordinarily used for seam welding, (2) increasing the electrode pressure, (3) decreasing the welding speed, and (4) increasing the rate of heat generation in the work metal by eliminating cool time from the weld cycle.

The overlap is a maximum of 1 1/2 times the thickness of one sheet of work metal.

Table 4 Suggested practices for multiple-impulse resistance spot welding of 300 series austenitic stainless steels(a)

Thickness (t) of thinnest outside piece, in.(b)	Electrode dimensions(c): Body diameter (min), in.	Electrode dimensions(c): Face diameter (max), in.	Net electrode force, lb	Weld time, pulsations at 15 cycles on, 6 cycles off (60 Hz)	Welding current (approx) for work metal tensile strength, A: Less than 150 ksi	Welding current (approx) for work metal tensile strength, A: 150 ksi and above	Minimum contacting overlap, in.	Minimum spot spacing (center to center), in.(d)	Nugget diameter (min), in.	Minimum breaking load of weld in shear for work metal tensile strength, lb: 90 to 150 ksi	Minimum breaking load of weld in shear for work metal tensile strength, lb: 150 ksi and above
0.156	1	1/2	4000	4	20 700	17 500	1 1/4	1 7/8	0.440	7 600	10 000
0.187	1	1/2	5000	5	21 500	18 500	1 1/2	2	0.500	9 750	12 300
0.203	1	5/8	5500	6	22 000	19 000	1 5/8	2 1/8	0.530	10 600	13 000
0.250	1	5/8	7000	7	22 500	20 000	1 3/4	2 3/8	0.600	13 500	17 000

(a) Steel should be free from scale, oxide, paint, grease, and oil. (b) Welding conditions are determined by thickness (t) of thinnest outside piece. Data are for total thickness of pile-up not exceeding $4t$. Maximum ratio between two thicknesses, 3 to 1. (c) Body diameters apply to electrodes with types A, D, and E faces with the face diameters listed, and to type F electrodes with 3-in. spherical-radius faces. Electrode material, class 2 or class 3. (d) Minimum spot spacing for two pieces is that spacing for which no special precautions need be taken to compensate for shunting of current through adjacent spot welds. For three pieces, increase spacing 30%.

Source: "Recommended Practices for Resistance Welding," AWS C1.1; also, *Welding Handbook*, 6th ed., Section 2, American Welding Society, 1969

The sheets must be assembled carefully before welding and held rigidly during welding. Electrodes are machined to flat faces to straddle the overlap and produce a uniform joint. Electrode pressure has an effect on thickness of the joint and on contact resistance. Square edges are satisfactory on metal thicknesses up to about 0.050 in. On thicker metal, beveling the overlapping edges may be helpful because it reduces the amount of plastic-metal movement needed for thickness control.

Seam Welding Machines. Seam welding of stainless steel is done with the same machines used for seam welding of carbon steel. Welding pressures are greater, but weld times and currents are less than in welding carbon steel and, therefore, more accurate control over welding conditions is needed. The annealed austenitic and nitrogen-strengthened austenitic stainless steels are nonmagnetic and do not influence the reactance of the secondary circuit of the welding machine. Therefore, on longitudinal seam welding machines, welding current does not change as the welded asssembly is fed into the throat of the machine. However, cold working induces some magnetic properties into austenitic stainless steels, but not the nitrogen-strengthened austenitic stainless steels. The martensitic and ferritic precipitation-hardening grades are ferromagnetic and have a marked effect on the characteristics of the secondary circuit.

Heat Distortion. Installation of a closely spaced succession of spot welds generates considerable heat in workpieces—a condition that can cause unwanted metallurgical effects or excessive warpage. The best method of controlling thermal distortion on any grade of stainless steel is to prevent excessive heat buildup during seam welding. Forced cooling of the work metal is usually done with jets or sprays of water directed on the work at the point of contact with the wheels. The water spray should be directed to the top and bottom of the sheet before and behind the electrode wheels. A copious flow of water over the electrode wheel is sometimes used.

Electrodes. Electrode wheels for resistance seam welding of stainless steel are often made of RWMA class 3 material to withstand high electrode forces. With softer electrode materials, mushrooming of the contact face occurs and weld quality is reduced. RWMA electrode specifications are given in the article "Resistance Spot Welding." Electrode face shapes are illustrated in the article "Resistance Seam Welding." Both articles can be found in this Volume.

Welding Schedules. Suggested practices for resistance seam welding of 300 series austenitic stainless steels are given in Table 5. Test welds should be made before starting production to determine optimum conditions, using the conditions listed in Table 5 as starting points. Conditions can be adjusted according to the procedure given for setting up a schedule in the article "Resistance Seam Welding" in this Volume. The wide latitude of conditions under which seam welds can be made between stainless steel sheets of dissimilar thicknesses is illustrated in the examples that follow.

Conditions for Seam Welding Stainless Steel Sheet of Dissimilar Thicknesses

Table 6 lists equipment details and welding conditions for continuous seam welding of type 310 sheet 0.050 and 0.043 in. thick (Example 1), type 316 sheet 0.062 and 0.093 in. thick (Example 2), and types 316 and 321 sheet 0.093 and 0.062 in. thick, respectively (Example 3). For all three applications, the electrode wheels were made of RWMA class 3 material to reduce deformation of the contact surface resulting from high heat, long weld time, inadequate cooling, and highly concentrated electrode force. The welding machine for the three applications was a circular type. Power was supplied by a three-phase half-wave frequency converter. The high electrode forces and low welding speeds used for joining the sheets provided consistency in obtaining high-strength, crack-free seam welded joints.

Projection Welding

Projection welding offers advantages in the joining of stainless steel similar to those obtainable in welding carbon steel. The current and electrode force are concentrated at well-defined spot areas in pro-

Table 5 Suggested practices for resistance seam welding of 300 series austenitic stainless steels(a)

Thickness (t) of thinnest outside piece, in.(b)	Electrode-wheel width (min), in.(c)	Net electrode force, lb	On time, cycles (60 Hz)	Off time for obtaining pressure-tight joints at maximum welding speed for total pile-up thickness, 60 Hz cycles		Maximum welding speed for total pile-up thickness, in./min		Welds per inch for total pile-up thickness of		Welding current (approx), A	Minimum contacting overlap, in.(d)
				2t	4t	2t	4t	2t	4t		
0.006	3/16	300	2	1	1	60	67	20	18	4 000	1/4
0.008	3/16	350	2	1	2	67	56	18	16	4 600	1/4
0.010	3/16	400	3	2	2	45	51	16	14	5 000	1/4
0.012	1/4	450	3	2	2	48	55	15	13	5 600	5/16
0.014	1/4	500	3	2	3	51	46	14	13	6 200	5/16
0.016	1/4	600	3	2	3	51	50	14	12	6 700	5/16
0.018	1/4	650	3	2	3	55	50	13	12	7 300	5/16
0.021	1/4	700	3	2	3	55	55	13	11	7 900	3/8
0.025	3/8	850	3	3	4	50	47	12	11	9 200	7/16
0.031	3/8	1000	3	3	4	50	47	12	11	10 600	7/16
0.040	3/8	1300	3	4	5	47	45	11	10	13 000	1/2
0.050	1/2	1600	4	4	5	45	44	10	9	14 200	5/8
0.062	1/2	1850	4	5	7	40	41	10	8	15 100	5/8
0.070	5/8	2150	4	5	7	44	41	9	8	15 900	11/16
0.078	5/8	2300	4	6	7	40	41	9	8	16 500	11/16
0.094	5/8	2550	5	6	7	36	38	9	8	16 600	3/4
0.109	3/4	2950	5	7	9	38	37	8	7	16 800	13/16
0.125	3/4	3300	6	6	8	38	37	8	7	17 000	7/8

(a) Steel should be free from scale, oxide, paint, grease, and oil. (b) Welding conditions are determined by thickness (*t*) of thinnest outside piece. Data are for total thickness of pile-up not exceeding 4*t*. Maximum ratio between two thicknesses, 3 to 1. (c) Electrode material, RWMA class 3. Face radius, 3 in. (d) For large assemblies, the values listed should be increased 30%.
Source: "Recommended Practices for Resistance Welding," AWS C1.1; also, *Welding Handbook*, 6th ed., Section 2, American Welding Society, 1969

Table 6 Continuous resistance seam welding of three assemblies made from mill-pickled stainless steel sheets of dissimilar thicknesses (Examples 1, 2, and 3)

Item	Example 1	Example 2	Example 3
Details of sheets welded			
Upper sheet (alloy and thickness, in.)	Type 310, 0.050	Type 316, 0.062	Type 316, 0.093
Lower sheet (alloy and thickness, in.)	Type 310, 0.043	Type 316, 0.093	Type 321, 0.062
Overlap of sheets at joint, in.	0.400	1.000	0.508
Equipment details			
Power supply	Three-phase half-wave frequency converter; 100-kV·A rating		
Welding machine	Circular, with synchronous controls and phase-shift heat control		
Upper electrode(a):			
Diameter, in.	10	10	14
Face width, in.	0.280	0.375	0.362
Face radius, in.	3	12	5.5
Lower electrode(a):			
Diameter, in.	6.5	10	10
Face width, in.	0.280	0.375	0.312
Face radius, in.	3	12	6
Welding conditions			
Welding current and voltage	Three-phase half-wave rectified current; 3 to 6 V across weld		
Heat setting, %	30	30	30
Electrode force, lb	1500	2200	2200
Squeeze time, cycles(b)	72	36	36
Hold time, cycles(c)	36	36	36
Heat time, cycles	10	12	12
Cool time, cycles	6	16	16
Impulses per spot	1	1	1
Nugget width and height, in.	0.21, 0.048	0.22, 0.078	0.25, 0.078
Spots per inch	14	16	16
Welding speed, in./min	16	8	8

(a) Electrode material for all three applications was RWMA class 3. (b) Squeeze time at beginning of seam, before welding had started. (c) Hold time at end of seam, after welding had stopped

Table 7 Suggested practices for projection welding of 300 series stainless steels(a)

Thickness (t) of thinner piece (nominal), in.(b)	Electrode face diameter, in.(c)	Net electrode force, lb	Weld time, cycles	Hold time, cycles	Welding current at electrodes (approx), A (60 Hz ac)
0.014	1/8	300	7	15	4 500
0.021	5/32	500	10	15	4 750
0.031	3/16	700	15	15	5 750
0.044	1/4	700	20	15	6 000
0.062	5/16	1200	25	15	7 500
0.078	3/8	1900	30	30	10 000
0.094	7/16	1900	30	30	10 000
0.109	1/2	2800	30	45	13 000
0.125	9/16	2800	30	45	14 000

(a) Steel should be free from scale, oxide, paint, grease, and oil. (b) Welding conditions are based on thickness (t) of thinner piece, and for two thicknesses only. Maximum ratio between two thicknesses, 3 to 1. (c) Electrode face diameter is equal to twice the diameter of the projection. Electrode material, RWMA class 2 or class 12. See the article "Projection Welding" in this Volume for design of standard projections.

jection welding, which minimizes the adverse effects of excessive heat on heat-sensitive grades. Such effects are characteristic of spot welding, in which contact areas are larger. Also, in projection welding, shunting effects of surface irregularities are eliminated and areas of contact at individual points are constant.

The same projection designs and methods of forming are used for stainless steel as for carbon steel. When three or more projections are welded simultaneously, variations in diameter, contour, and height of the projections result in variations in current density, and thus weld quality. When components are of different thicknesses, projections should be in the thicker piece to maintain better heat balance. Projection welding design data are given in the article "Projection Welding" in this Volume.

Welding Schedules. Suggested practices for projection welding of austenitic stainless steels are given in Table 7; the values listed may need adjustment to suit the requirements of individual applications.

Cross Wire Welding

Cross wire welding is a form of projection welding in which the curved surfaces of the wires form the projected or localized contact areas for the welds. Conventional press-type resistance welding machines are used for cross wire welding. Cross wire resistance welds can be made singly or in multiples, as permitted by part design and the availability of suitable welding machines. Flat-face (RWMA type C) spot welding electrodes or electrodes with a V-groove across the face are used for single welds. Flat bar-type electrodes are used for multiple welds. Cut and formed wire sections are generally held in fixtures during welding.

The strengths of cross wire welds in types 304 and 430 stainless steels are about the same. However, the austenitic stainless steels provide a sounder and more dependable welded joint. Weld strength depends on the setdown of the wires at the joint and is not affected by any temper in the wire resulting from cold drawing. Maximum strength is achieved in all wire sizes at about 30% setdown. Best appearance and avoidance of excess flash from the joint are obtained with about 20% setdown, but lower weld strength is obtained with 20% setdown than with 30% setdown. The optimum electrode force and tensile breaking-load values for 20 and 30% setdown for cross wire welding of three sizes of type 304 wire are:

Wire diameter, in.	Optimum electrode force, lb	Tensile breaking load for setdown(a), lb 20%	30%
0.125	350	1350	1800
0.156	600	···	3350
0.250	1700	5800	8100

(a) Percentage of setdown is determined by dividing the setdown (difference between the combined height of the two wires before and after welding) by the diameter of the smaller wire, and multiplying the result by 100.

In cross wire welding, electrode force is related to the diameter of the smaller wire. Welding current and weld time control the amount of setdown, flash, and discoloration. For the application in the following example, strength and appearance were the criteria used for selecting the welding conditions.

Example 4. Cross Wire Welding of Stainless Steel Trays. Food-handling trays measuring 18 by 36 in. were designed for fabrication by cross wire welding. The trays, consisting of a frame made of 1/4-in.-diam type 304 stainless steel wire and 1/8-in.-diam cross wires of the same material, were set up for welding as shown in Fig. 1. Welding was done with the wires held in position by means of a fixture. The 350-lb electrode force was selected on the basis of the smaller (1/8-in.-diam) wire. The short weld time (4 cycles) was selected to avoid discoloration. The 3000-A welding current was selected to produce a setdown of only 0.015 in. (12% setdown), which provided adequate strength for the application. The small setdown resulted in negligible flash and also helped to avoid discoloration. As a result, bright finishing of the as-welded trays was easily accomplished by electropolishing.

Fig. 1 Cross wire welding of a food-handling tray. A low setdown provided adequate strength while minimizing flash and discoloration.

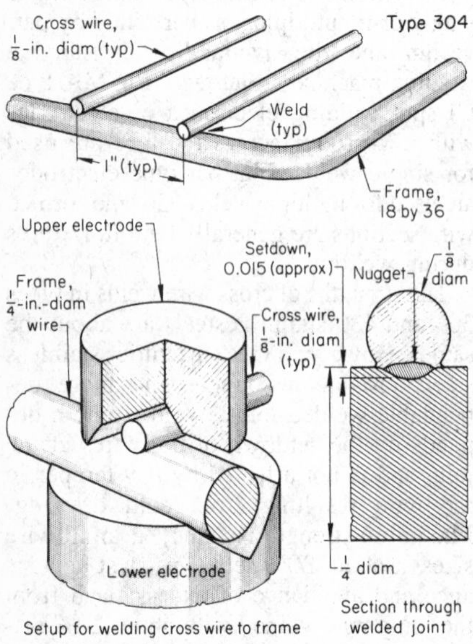

Equipment details

Power supply	440 V, single phase, 60 cycle
Welding machine	Press-type spot, semiautomatic
Rating at 50% duty cycle	60 kV·A
Secondary current, max	70,000 A
Secondary voltage, max	3 V
Transformer taps	8
Upper electrode	5/8 in. diam (grooved face)
Lower electrode	3/4 in. diam (grooved face)
Electrode material	RWMA class 2
Electrode force, max	3000 lb
Welding controls	Synchronous, with phase-shift heat control

Conditions for cross wire welding

Welding current	3000 A, ac
Welding voltage	1 V
Heat-control setting	No. 1 tap and 50%
Electrode force	350 lb
Squeeze time	10 cycles
Weld time	4 cycles
Hold time	10 cycles
Production per hour	20 assemblies

Resistance Welding Dissimilar Metals

In resistance welding of stainless steel to another metal, the maximum electrical resistance may not be at the interface of the two workpieces, but may be within the stainless steel workpiece because of its high electrical resistivity. The low thermal conductivity of the stainless steel can result in a greater heat loss in the weld zone of the other work metal than in the weld zone of the stainless steel workpiece when heat is conducted away rapidly.

The low electrical conductivity of stainless steel reduces the shunting effect from one spot to another. In resistance welding of stainless steel to another metal, the heat and follow-up by the electrode must be carefully controlled. Too little heat does not give enough nugget penetration into the stainless steel to produce adequate weld strength, while too much heat can cause indentation and expulsion of the other metal at spots close to the edge of the joint.

Nickel-based heat-resistant alloys of both solid-solution and precipitation-hardenable types can be resistance spot welded to austenitic stainless steels. In a study of the resistance welding behavior of the solid-solution alloy Hastelloy X and the precipitation-hardening alloy GMR-235, sound spot welds were produced when 0.090-in.-thick type 321 austenitic stainless steel was welded to 0.063-in.-thick Hastelloy X, and when it was welded to 0.090-in.-thick GMR-235. Best results were obtained by multiple-impulse spot welding using high welding current, high electrode force, and a long welding time. Grain-boundary melting in the nickel-based alloys at the periphery of the weld nugget was minimized by directing a stream of cooling water at the electrodes during welding.

The spot welding was done on pickled and degreased sheet. The Hastelloy X was in the solution heat treated condition, and the GMR-235 was in the precipitation-hardened condition. Spot welding conditions and test results are given in Table 8. Evaluation of welds consisted of (1) macroscopic examination for soundness, penetration, and nugget diameter; (2) measurement of shear strength and cross tensile strength; and (3) x-ray examination to detect porosity and cracking. Maximum indentation was 10% and penetration was 50 to 60%—both based on the total thickness of the sheets.

Table 8 Multiple-impulse resistance spot welding of 0.090-in.-thick austenitic stainless steels to 0.063-in.-thick Hastelloy X and GMR-235

Equipment details and welding conditions

Power supply	440-V, three-phase, 60-cycle ac
Welding machine	100-kV·A air-operated press type, with electronic timing and phase-shift heat control
Electrodes	Flat, 5/16 in. diam; RWMA class 3; internal and external water cooling
Welding current	26 600 A
Welding pulsations	8 impulses, each of 10 cycles heat time plus 2 cycles cool time
Electrode force, lb	Weld, 2000; forge, 4000

Property of weld	Type 321 stainless welded to Hastelloy X	Type 316 stainless welded to GMR-235
Test results on spot welds		
Shear strength, lb(a):		
Average	3574	4050
Minimum	3360	3930
Maximum	3750	4160
Variation from	+4% to −6%	+3% to −3%
Min allowable(b)	2530	2530
Tensile strength (cross), lb:		
Average	2250	2675
Minimum	1770	2490
Maximum	3010	3050
Tensile strength/shear strength(c)	0.63	0.66
Nugget diam (avg), in.	9/32	9/32

(a) Beam load. (b) For 0.062-in. sheet thickness, per MIL-W-6858, 90- to 150-ksi class. (c) Based on average values

Example 5. Control of Heat and Electrode Follow-Up in Spot Welding. The spring assembly shown in Fig. 2 (upper left) was used in an electronic memory unit to provide uniform pressure for holding a magnet card against its magnetizing plane. The flat ends of the two halves of the spring, which was made of 0.003-in.-thick phosphor bronze (copper alloy 510) foil, were sandwiched between a U-shaped nosing strip of 0.012-in.-thick type 304 stainless steel. The welding conditions were carefully controlled; adequate electrode force and a low-inertia welding head with fast follow-up were needed to prevent cracking in the stainless steel. The assem-

Fig. 2 Assembly consisting of a foil-thin phosphor bronze spring and stainless steel nosing strip that was resistance spot welded. Results of peel tests after welding are also shown. Close control of heat was needed to obtain sufficient weld strength and to avoid expulsion of phosphor bronze; a low-inertia, fast follow-up head was used to prevent weld cracking.

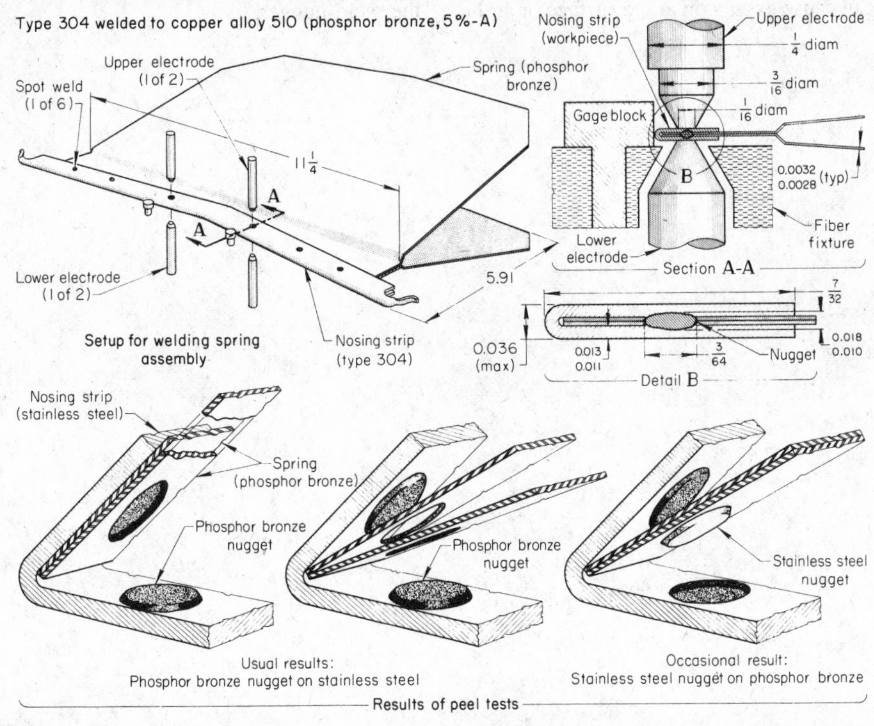

Equipment details

Power supply	460 V, single phase, 60 cycle
Welding machine	Press type, semiautomatic
Transformer rating at 50% duty cycle(a)	3 kV·A
Secondary current and voltage(b)	5400 A, 6 V
Transformer taps	10
Electrodes	RWMA class 2(c)
Electrode force, max	60 lb per electrode
Welding controls	Synchronous, with phase-shift heat control

Conditions for resistance spot welding

Welding current	2000 A, ac
Welding voltage	3.25 V
Heat-control setting	No. 4 tap and 70%
Electrode force	50 lb per electrode
Squeeze time	(d)
Weld time	2 cycles
Machine cycle time	1 s
Hold time	Provided in machine cycle time
Production per hour	200 assemblies(e)

(a) Two transformers, one for each set of electrodes used in making the spot welds two at a time. (b) Maximum. (c) Type E (truncated), 1/4 in. diam, 1 3/4 in. long, 60° chamfer, 1/16-in.-diam face. (d) Not required, because of delay in initiating the welding current. (e) 1200 welds

bly was joined by means of six spot welds. The design of the joint avoided melt-through, which is often encountered in welding such thin metal in the usual sheet-to-sheet arrangement. The welded assembly was required to be of uniform thickness (0.036 in. maximum), except for a slight depression in the nosing strip at each bearing point of an electrode. The presence of waviness and weld spatter was unacceptable.

A press-type semiautomatic spot welding machine that had a head with a low-inertia, inner-shaft assembly capable of providing fast follow-up was used. Holders for the two upper electrodes were mounted to the inner shaft, and holders for the two opposing lower electrodes were mounted on the platen. The six spot welds were made, two at a time, by direct welding. Conditions for welding were adjusted to avoid expulsion of the copper alloy and to obtain adequate weld penetration in the stainless steel. Expulsion of the copper alloy was avoided as long as the spots were made with the specified welding current and the correct edge distance. Because stainless steel has higher electrical resistivity at room temperature than does phosphor bronze, only a small percentage of the current was shunted around the nosing strip.

During welding, the assembly was positioned and supported by a fixture made of a nonconducting material and was held against gage blocks in three locations (one for the position of each pair of spot welds). The upper electrode had a portion of its shank reduced to 3/16 in. diam to clear the gage blocks (Fig. 2, upper right). As the welding-machine head was lowered by an air cylinder, deflection of a leaf spring, incorporated in the head and bearing on an inner shaft that carried the electrode holder, provided the actual electrode force. The leaf spring also served as a decoupling device and allowed the low-inertia inner shaft, electrode holder, and electrode assembly to follow quickly as the metal in the weld zone softened on reaching welding temperature. The resulting electrode indentation was shallow, and the fast follow-up made sound welds and prevented cracks in the stainless steel at the weld. Deflection of the leaf spring also closed a limit switch to start the welding current. When follow-up was slow and force was insufficient, cracks were visible on the outer surface of the stainless steel and extended through to the phosphor bronze. In peel tests that were conducted at regular intervals, the nugget usually was pulled from the phosphor bronze, but occasionally, if the welds were made with higher heat, the nugget was pulled from the stainless steel (Fig. 2, bottom view).

Causes and Prevention of Weld Discontinuities

Indentations and discoloration of resistance welds are common weld discontinuities that detract from the appearance of finished welds. Cracks and voids in the weld nugget and cracks in the base metal are also common flaws. Indentations are the result of plastic yielding of base metal under the force of the electrodes at the welding temperature. Improper maintenance of the electrode face, excessive electrode force, and welding current and shrinkage in the weld nugget are common causes of indentations.

All stainless steels are susceptible to discoloration at temperatures above about 400 °F. The oxygen in the atmosphere causes tinting if the electrode is withdrawn before the metal has cooled sufficiently. To safeguard corrosion resistance in atmospheres that are more than mildly corrosive, and to improve appearance, the work-metal surfaces should be cleaned after welding. Cleaning can be done by light pickling, electropolishing, light buffing, or scouring with mild abrasives.

Cracks and voids in the weld nugget generally are a result of insufficient forging force or a hold time too brief to permit the weld metal to solidify under a given force. The high strength of stainless steel at elevated temperatures necessitates the use of high electrode force to make a sound weld. Flat-face, large-diameter electrodes provide a low unit pressure on the work-metal surface. An electrode having a small tip radius may produce a good weld, but can cause expulsion at the surface or at the interface of the two sheets, or can cause excessive indentation. When the welding current is too high in relation to the electrode force, weld time, and hold time, cracks can occur in the HAZ as well as in the weld nugget.

Resistance Welding of Aluminum Alloys

ALUMINUM ALLOYS, both the non-heat-treatable and heat treatable types, either wrought or cast, can be resistance welded. Some of these alloys are welded more readily than others. Characteristics of aluminum alloys include comparatively high thermal and electrical conductivity, a relatively narrow plastic range (about 200 to 400 °F temperature differential between softening and melting), considerable shrinkage during cooling, a troublesome surface oxide, and an affinity for copper electrode materials.

Resistance spot and seam welding of aluminum alloys are used in manufacturing cooking utensils, tanks (both for seams and for securing baffles), bridge flooring, and many aircraft and automotive components. Resistance welding of aluminum aircraft components, such as wing-skin sections, deck sections, brackets and cowling, usually entails many high-quality welds in one structure and may require elaborate and expensive equipment for cleaning, welding, and controlling weld quality in contrast to many commercial applications done with less cleaning and a lower level of acceptable weld quality.

Weld Strength

The strength of resistance welds in aluminum alloys varies with the alloy composition and thickness. Resistance welds should be located so that the weld is shear loaded. For tensile or combined loading, the tensile strength of the welded joint is only about 25% of the shear strength. Nuggets with diameters equal to two thicknesses of base metal plus 0.06 in. should have shear strengths greater than the values given in Table 1.

The heat of resistance welding decreases the strength and hardness of strain-hardened and of solution-treated and precipitation-hardened aluminum alloys, depending on the temperature attained and the length of time that a temperature of 400 °F or more is maintained during welding.

Table 1 Minimum shear strength of resistance spot welds in aluminum alloys

Thickness of thinnest sheet, in.	Minimum shear strength, lb per spot, for alloy: 1100-H14, 1100-H18	3003-H12, 3003-H18, 5052-O	2036-T4, 6009-T4, 6010-T4, 5182-T4, 5052-H32, 5052-H38, 6061-T4, 6061-T6, 5050-H34	2024-T3, alclad 2024-T3, 7075-T6, alclad 7075-T6
0.016	40	70	98	108
0.020	55	100	132	140
0.025	70	145	175	185
0.032	110	210	235	260
0.040	150	300	310	345
0.051	205	410	442	480
0.064	280	565	625	690
0.081	420	775	865	1050
0.102	520	950	1200	1535
0.125	590	1000	1625	2120

Source: "Welding Alcoa Aluminum," Aluminum Co. of America, Pittsburgh, 1972

Base-Metal Characteristics

Although all aluminum alloys can be resistance spot and seam welded, some alloys or combinations of alloys have higher as-welded properties than others. Table 2 gives melting ranges, electrical and thermal conductivities, and resistance weldability of some wrought alloys and casting alloys.

Alclad Alloys. Resistance welding is also done on alclad products made by roll cladding some of the alloys listed in Table 2 with a thin layer of aluminum or an aluminum alloy. Because this layer is anodic to the core alloy, it provides electrochemical protection for exposed areas of the core. Alclad alloys 2219, 3003, 3004, 6061, and 7075 have a cladding of alloy 7072, which contains 1% Zn; alclad alloy 2014 has a cladding of alloy 6003 or sometimes alloy 6053, both of which contain about 1.2% Mg; and alclad alloy 2024 has a cladding of alloy 1230, which contains a minimum of 99.3% Al.

Effects on Weldability. The hardness of an alloy is one variable influencing weldability. Any alloy in the annealed condition (O temper) is more difficult to weld than the same alloy in a harder temper. In general, alloys in the softer tempers are much more susceptible to excessive indentation and sheet separation and to low or inconsistent weld strength. Greater deformation under the welding force causes an increase in the contact area and variations in the distribution of current and pressure. Therefore, welding of aluminum alloys in the annealed condition or in the softer tempers is not recommended without special electromechanical or electronic controls.

High-strength alloys such as 2024 and 7075 are easy to resistance weld, but may require application of a forge pressure, because they are more susceptible to cracking and porosity than the lower strength alloys. Sheet separation in welding high-strength alloys is low, and weld strength is consistent. Alloys clad with alloy 1230 or 7072 require higher current to resistance weld than bare alloys, because of the low electrical resistance and high melting point of the cladding at the contacting interfaces.

Although the strength of welds made in low-strength alloys such as 1100 and 3003 may vary, these alloys can be resistance welded readily in most applications. High welding current or low electrode force may be needed to compensate for the low electrical resistance of these alloys.

Shrinkage cracks in the weld metal are confined almost exclusively to welds made in the copper- and zinc-bearing alloys such as 2024 and 7075. These high-strength alloys, as well as the chromium-bearing alloys, such as 5052 and 6053, may develop

Table 2 Melting ranges, electrical and thermal conductivities, and resistance weldability of common aluminum alloys

Wrought and casting alloys are identified by Aluminum Association designations.

Alloy and temper	Melting range, °F	Electrical conductivity, % IACS(a)	Relative thermal conductivity(b), %	Resistance weldability(c)
Non-heat-treatable wrought aluminum alloys				
1350-H19	1195-1215	62	60	ST
1060-H18	1195-1215	61	57	ST
1100-H18	1190-1215	57	55	RW
3003-H18	1190-1210	40	39	RW
3004-H38	1165-1205	42	42	RW
5005-H38	1170-1205	52	51	RW
5050-H38	1160-1205	50	49	RW
5052-H38	1100-1200	35	35	RW
5083-H321	1065-1180	29	30	RW
5086-H34	1084-1184	31	32	RW
5154-H38	1100-1190	32	32	RW
5182-O	1065-1185	31	31	RW
5454-H34	1115-1195	34	34	RW
5456-H321	1060-1180	29	30	RW
Heat treatable wrought aluminum alloys				
2014-T6	950-1180	40	39	ST
2024-T361	935-1180	30	31	ST
2036-T4	1030-1200	41	40	RW
2219-T37	1010-1190	28	29	ST
6009-T4	1040-1200	44	43	RW
6010-T4	1085-1200	39	38	RW
6061-T6	1100-1200	43	43	RW
6063-T6	1140-1210	53	51	RW
6101-T6	1140-1205	57	55	RW
7075-T6	890-1180	33	33	ST
Aluminum casting alloys				
413.0-F	1065-1080	31	32	LW
443.0-F	1065-1170	37	37	RW
308.0-F	970-1135	37	37	ST
238.0-F	945-1110	25	26	LW
513.0-F	1075-1180	34	34	ST
520.0-T4	840-1120	21	22	NR
333.0-T6	960-1085	29	30	ST
C355.0-T61	1015-1150	39	38	ST
356.0-T6	1035-1135	39	38	ST
712-F	1120-1190	40	39	RW

(a) International Annealed Copper Standard, volume basis at 68 °F. For comparison, copper alloy 102 (oxygen-free copper) is 101% and low-carbon (1010) steel about 14%. (b) Based on copper alloy 102 as 100%, which has a thermal conductivity of 226 Btu/ft · h · °F at 68 °F. Low-carbon steel has a thermal conductivity of about 13% on this relative scale. (c) RW, readily weldable; ST, weldable in most applications but may require special techniques for specific applications; LW, limited weldability and usually requires special techniques; NR, welding not recommended

some porosity in the weld metal, particularly when they are welded in the hardened condition.

In joining dissimilar aluminum alloys, those with similar electrical conductivities and melting temperatures are easiest to weld together; those pairs with the greatest difference in these properties are the most difficult to weld together.

Wrought aluminum alloys are frequently spot welded to permanent mold, sand, and die castings made from aluminum alloys. Of these types, permanent mold castings are easiest to weld, because the thickness is more nearly uniform. Such castings are sound and have smooth surfaces and a low as-cast surface resistance, if welded within a reasonable time after casting. Die castings also are dimensionally accurate, but may require special cleaning to prepare the surface for welding. Occasionally, special dies and foundry practice are needed to ensure the soundness of the metal in the region to be welded.

Corrosion Resistance. Spot and seam welds in non-heat-treatable alloys, such as 1100, 3003, and 5052, are not selectively attacked by corrosion, but have the same corrosion resistance as the unwelded metal. Welds in heat treatable magnesium-silicon alloys, such as 6061 and 6063, also have good corrosion resistance.

Welds in unclad *2xxx* and *7xxx* series alloys may be attacked preferentially under severely corrosive conditions; therefore, weldments made from these alloys should not be used in a corrosive environment unless properly protected. When these same alloys are resistance welded to aluminum-clad parts of corresponding composition, however, the cladding electrochemically protects the unclad base metal at the interface, thus improving the overall resistance to corrosion at the joint. Maximum resistance to corrosion at the welded joint is achieved when each of the parts being welded is an alclad alloy.

Factors Affecting Resistance Welding of Aluminum Alloys

Because of the inherent characteristics of aluminum alloys, resistance welding of these alloys requires procedures that are different from those used for resistance welding of steel. Included among these characteristics are high electrical and thermal conductivities, low melting-temperature ranges and low strengths at elevated temperatures, narrow plastic ranges, high shrinkage during solidification, and the presence of natural oxide coatings.

Electrical and Thermal Conductivities. Aluminum alloys are much higher in electrical conductivity than most metals that are commonly resistance welded. For example, the electrical conductivity of alloy 2024 (one of the low-conductivity aluminum alloys) is more than twice that of low-carbon steel.

This high electrical conductivity necessitates the use of high-capacity welding machines capable of supplying high welding currents, because high current density is needed to generate enough heat to melt the aluminum alloy and produce the weld. To minimize electrical shunting, minimum spot spacing values are usually increased over those acceptable for ferrous alloys. The high thermal conductivity of aluminum alloys necessitates rapid welding to avoid dissipation of heat into the workpieces. In general, aluminum spot welds are made with about three times the welding current used for steels, but for only one tenth the time.

Rising Temperature. As the temperature is increased during resistance welding, aluminum alloys, when compared to steel, soften more rapidly and at lower temperatures. Low-inertia welding-machine heads are needed to enable the electrodes to make the rapid movements necessary for maintenance of weld force and workpiece contact. Although these movements are small, they must take place during an interval of about 2 to 5 ms.

Plastic Range. For aluminum alloys, the plastic range in which a weld can be made is very narrow. Therefore, the energy input to the weld must be precisely controlled for proper nugget formation.

Shrinkage During Cooling. Aluminum alloys exhibit considerable shrinkage during cooling from the liquidus temperature to solidus temperature. This property is most pronounced in the high-strength heat treatable alloys, such as 2024 and 7075, and can result in cracking. The non-heat-treatable alloys and the *6xxx* series alloys are less likely to crack as a result of shrinkage of the nugget.

Porosity and cracking can result from shrinkage unless the electrodes can maintain proper pressure on the nugget until solidification is completed. Machines for welding aluminum alloys generally have, in addition to low-inertia heads, a means of increasing the electrode force as the nugget solidifies. Because this permits forging of the nugget, it also improves weld soundness.

Surface Oxide. Aluminum combines almost instantaneously with oxygen in the atmosphere to produce an aluminum oxide coating. The high and somewhat erratic electrical resistance of this coating affects the amount of heat produced in the metal during resistance welding. Therefore, for aircraft-quality welds, this oxide film should be removed or changed to a film of uniform electrical resistance before welding.

Commercial spot or seam welding often can be done on aluminum without cleaning or oxide removal, but electrode pickup increases, electrode life decreases, and welds are variable in quality and erratic in shape. The amount of surface preparation needed depends on the strength and quality requirements of the welded product and on the alloy being welded.

Resistance Welding Machines

Aluminum alloys can be resistance welded with single-phase direct-energy, three-phase direct-energy, and stored-energy machines. Best results are obtained by using a machine that has these features:

- Ability to handle high welding currents for short weld times
- Synchronous electronic controls for weld time and welding current
- Low-inertia welding head for rapid follow-up of electrode force
- Slope control (for single-phase welding machines)
- Multiple-electrode-force system to permit proper forging of the weld nugget and re-dressing of electrodes

The article "Resistance Spot Welding" in this Volume contains additional information on each of these subjects.

Single-phase direct-energy machines have high intermittent kV · A demand and low power factor. These machines may disturb other electrical equipment, however. Although a wide variation in line voltage can cause nonuniformity in welding, adequate transformers or substations reduce this variation. The addition of upslope and downslope control to a single-phase machine is recommended for spot welding of aluminum (see the section on slope control in this article).

Three-phase direct-energy machines, the frequency-converter and dry-disk rectifier types, produce excellent resistance welds in aluminum because of the partial control these machines exert on the shape of the welding current wave. The gradual increase in current at the beginning of the weld cycle and the decay at the end of the cycle are similar to the upslope and downslope used with single-phase machines.

Stored-energy machines employ a three-phase full-wave rectifier to charge a capacitor bank to a predetermined voltage. The weld is made by discharging the band of capacitors through a suitable welding transformer. Use of these electrostatic machines has been largely superseded by three-phase direct-energy machines.

Synchronous controls for weld time and welding current are required for resistance welding of aluminum alloys, because they provide precise control of short weld times and high welding currents. Magnetic and mechanical controls are not suitable.

Slope control permits adjustment of the rate of rise and fall of the welding current. Upslope control causes the welding current to increase gradually during the first few cycles of the weld time. The maximum current level is not reached until the electrode face has seated itself into the softened aluminum; therefore, excessively high currents do not normally occur at localized points. Overheating of metal at the interface of the electrode and the workpiece is reduced, which results in increased weld quality and electrode life and better surface appearance.

Downslope control tapers the current off gradually, preventing rapid chilling of the nugget. Downslope also permits better forging, which results in finer grain structure, eliminates cracks and voids in the nugget, and permits wider variations in amount and time of applications of forging force.

Multiple-electrode-force cycles usually consist of three stages. In the first stage, a high precompression force is exerted to seat the electrodes firmly on the work metal and establish good electrical contact. The electrode force is then reduced to increase contact resistance while the weld is being made. After the nugget has formed, the force is increased again to forge the nugget.

A low electrode force for welding permits the use of a lower welding current and minimizes sheet separation. Use of high forging force reduces cracking. Multiple-electrode-force cycles make the accuracy of the machine settings less critical and increase the range in which high-quality welds can be produced. Some welding machines can produce an electrode force of 100 to 200 lb for use in redressing the electrodes with a paddle-type dresser.

Electrodes and Electrode Holders

Selection of electrode material and face shape, maintenance of the face, and cooling of the electrode are important in producing consistent spot and seam welds in aluminum alloys.

Copper alloy electrodes, Resistance Welder Manufacturers Association (RWMA) classes 1, 2, and 3, are used when welding aluminum alloys. These electrode materials have high electrical and thermal conductivities, which combined with adequate cooling help keep the temperature of the electrode below the temperature at which aluminum will alloy with copper and cause electrode pickup.

Design of spot welding electrodes suitable for spot welding of aluminum includes both straight and offset electrodes. Construction details of each are shown in Fig. 1. Straight electrodes should be used whenever possible, because deflection and skidding may occur with offset electrodes under similar welding conditions. If offset electrodes are used, the amount of offset should be the minimum permitted by the shape of the assembly being welded.

Only electrodes that have the cooling water hole within $^3/_8$ in. of the face surface should be used. A design utilizing fluted cooling water holes provides more cooling surface than one specifying round holes.

Face contour of at least one electrode must be shaped so that the current is highly concentrated at the weld. For most spot welding applications, an RWMA type F electrode, with a face having a spherical radius greater than the diameter of the electrode, is used on one or both sides of the workpieces. A radius face provides easy alignment and minimum sheet separation, concentrates the welding current, and is easier to clean and maintain than a flat face. One flat-face (RWMA type C) electrode can be used to minimize indentation on one workpiece, although higher joint strength is usually obtained using electrodes with equal radii in joining workpieces of approximately equal thickness (see the article "Resistance Spot Welding" in this Volume).

Fig. 1 Construction details of straight and offset radius-face electrodes used in resistance spot welding of aluminum alloys

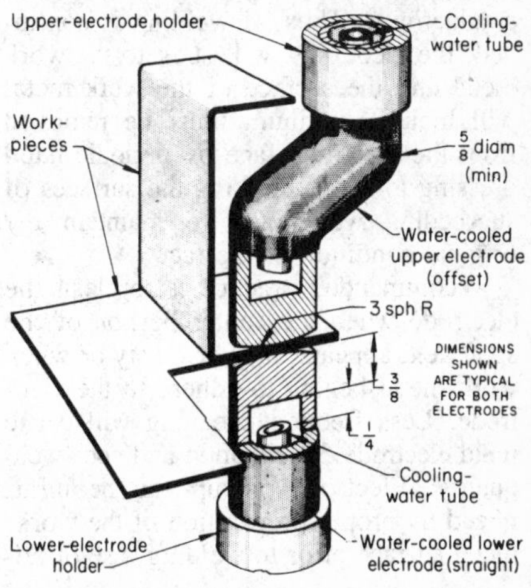

Table 3 Face radii for resistance spot welding electrodes or seam welding electrode wheels for use on aluminum alloys of various thicknesses

Condition of work metal	Radius, in., for work-metal thickness, in., of: Up to 0.020	0.021-0.032	0.033-0.064	0.065-0.094	0.095-0.125
Annealed or as-extruded	2	3	4	4	...
Intermediate tempers of non-heat-treatable alloys	2	3	3	4	6
Heat treated and hard tempers	1	2	3	4	6

Note: Spherical radii used for faces of spot welding electrodes and transverse radii for edges of seam welding electrode wheels. When a flat surface is needed on one side of the workpiece, one electrode is made to the above radius, and the electrode that contacts the surface to be flat is either flat or has a 10-in.-radius face or edge
Source: "Welding Alcoa Aluminum," Aluminum Company of America, Pittsburgh, 1972

Electrode-face radii for a variety of metal thicknesses and tempers are given in Table 3. When workpieces of dissimilar thickness are being welded, a radius-face electrode of correct dimensions should be employed in contact with the thinner member; good-quality welds are more difficult to make if an electrode with too large a face radius or a flat face is used against the thinner workpiece.

Electrode maintenance, or the correct maintenance of electrode faces, is essential if spot welds of uniform size and shape are to be made. The quality of resistance welds in aluminum alloys is more dependent on the contour, surface finish, and cleanness of the electrode face than the quality of welds made in other metals and alloys. When signs of wear or an appreciable change in contour is apparent, the faces should be redressed and periodically replaced or remachined to their original shapes with properly designed tools. Shaping of electrode faces by hand with a file is inaccurate and should be avoided.

In the spot welding of aluminum alloys, electrode life is determined by metal pickup on the electrode face, not by deformation, as in the welding of steel. The copper-aluminum alloy formed on the electrode face by metal pickup has low electrical conductivity. Thus, if welding is continued, the electrodes will stick to the work metal and the surface of the work metal will melt. Aluminum must be removed from the electrode face by periodic hand dressing to avoid marking the surfaces of succeeding welds and to maintain the original condition of the face.

Recommended practice is to clean the electrode when the center portion of the spot weld appears dirty or crusty or when work metal begins to adhere to the electrode. Less frequent cleaning will cause rapid electrode deterioration and poor weld quality. Electrode pickup can be minimized by proper preparation of the work-metal surface prior to welding, use of adequate electrode force, and avoidance of excessive welding current.

The cleaning operation must remove all the aluminum pickup, but must not change the face contour by removing an appreciable amount of electrode material. To maintain the original shape of radius-face electrodes, a dressing tool in the shape of a paddle with two depressions contoured to match the desired face radius should be used. The two depressions are faced with No. 240 or No. 320 abrasive cloth.

The dressing tool is clamped between the electrodes with a force of 100 to 200 lb and is rotated a few times to remove the copper-aluminum alloy. The faces are then cleaned using a cloth dampened with a solvent to remove any clinging abrasive dust and are wiped dry. Abrasive grit or deep scratches in the electrode face caused by coarse grit can be transferred to the workpiece during welding and thus should be avoided. A rubber block and two pieces of abrasive cloth can be used instead of a metal dressing tool.

Electrode cooling, or an adequate means of cooling the electrode face, must be provided. The usual method is a flow of water or refrigerated coolant through the inside of the electrode. Use of electrodes with fluted water holes provides more cooling surface than those with round water holes. For continuous welding, each water-cooled electrode should be operated at a cooling water flow rate of at least 1 gal/min, and preferably $1^1/_2$ to 2 gal/min. The cooling water should be brought to within $^1/_4$ to $^3/_8$ in. of the electrode face, and the inner cooling water tube should extend to within $^1/_4$ in. of the bottom of the cooling water hole to provide good circulation (Fig. 1). The end of the cooling water tube should be cut at an angle; thus if it bottoms in the cooling water hole in the electrode, the water flow is not stopped.

Cooling water temperature should be 60 °F or less and should not vary more than 10 °F. If the water temperature is above 60 °F or varies widely, the use of a refrigerated coolant may be helpful in improving electrode life and weld consistency. By cooling the electrodes with refrigerated coolant at temperatures of 38 to 40 °F, the number of spot welds made between face cleanings can be increased appreciably. When a refrigerated coolant is used, however, condensation of water on the electrodes and electrode holders may present a problem.

Electrode holders that are commercially available are suitable for use in welding aluminum alloys. Offset electrode holders are sometimes required for spot welding assemblies not accessible to straight holders and are preferred for offsetting electrodes because they are more rigid and less likely to cause skidding.

Seam welding electrode wheels are from $^3/_8$ to 1 in. thick and from 6 to 12 in. or more in diameter. An electrode wheel less than 6 in. in diameter is occasionally required, depending on the parts being welded. Electrode wheels usually have two curvatures: the radius of the wheel, which changes only slightly when the wheel is dressed, and the transverse radius of the crown or the edge of the wheel. Table 3 gives transverse radii of electrode wheels used for welding aluminum alloys of various thicknesses and tempers. Electrode wheels with flat faces or with crowns 10 in. or more in radius can be used against work-metal surfaces to avoid marking or indentation.

Maintenance practices for seam welding electrodes are essentially the same as for spot welding electrodes. Electrode pickup can be removed from the face of an electrode wheel by dressing with a suitable grade of abrasive cloth. A moderately coarse grade of cloth produces a rough surface that prevents slippage between the wheel and the work metal. For continuous dressing of electrode wheels, a medium-fine grade of abrasive can be held against the wheel under a load of 5 to 10 lb. Knurl-driven electrode wheels should not be used, because they roughen the surface of the wheel excessively, mark the aluminum work metal, and cause electrode pickup. Electrode wheels are usually cooled by a flow of water directed against the periphery of the wheel at the weld. Occasionally, they are cooled internally.

Preweld Surface Preparation

Although welds for some purposes can be made satisfactorily without any preweld surface preparation, welds that are free of cracks, porosity, and sheet separation, and that have the most uniform strength and symmetry, are obtained only with correct procedures for cleaning and reduction or removal of oxide film. In addition, adequate surface preparation reduces electrode contamination.

Commercial spot and seam welding of aluminum alloys, such as 1100, 3003, and 5052, can be accomplished with only a degreasing operation for surface preparation. Consistency in commercial welding of the more highly alloyed compositions, however, generally requires additional mechanical or chemical cleaning. Aircraft construction, regardless of the alloy used, demands the utmost in cleaning and oxide removal, as well as continual checking of contact resistance.

Oxide can be removed by mechanical or chemical methods. Usually, the length of time that is tolerable between oxide removal and welding varies from 48 h to several days, depending on methods of oxide removal and handling and storage environment.

Cleaning begins with the removal of any stencil identification marks with alcohol, paint thinner, or other suitable solvent. Then, parts heavily soiled with dirt, oil, grease, or lubricants from forming operations are cleaned with commercial solvents by wiping, dipping, washing, spraying, or vapor degreasing, depending on the size and quantity. Vapor degreasing is generally used whenever a large number of parts is involved. If only a few parts are to be cleaned, they can be immersed in or wiped with acceptable chlorinated solvents or acetone, but only if the workplace ventilation system meets health and safety regulations.

Degreasing is often followed by treatment with a nonetching alkaline cleaner specially formulated to produce low, consistent contact resistance on aluminum (see the article "Alkaline Cleaning" in Volume 5 of the 9th edition of *Metals Handbook*). After alkaline cleaning operations are completed, the work-metal surface should be able to support a water film without a break. Lightly soiled workpieces can be alkaline cleaned without degreasing.

Mechanical removal of oxides is primarily a hand operation; therefore, its effectiveness depends on the skill of the operator. The contact resistance of surfaces cleaned by mechanical removal of oxides is sometimes lower than that of surfaces cleaned by a chemical method. The cleaning action must be severe enough to cut through the hard oxide film, yet gentle enough to prevent formation of an excessively rough surface in the comparatively soft metal underneath.

When welding is confined to a small area and the oxide film is thin, mechanical cleaning provides quick and complete removal of the oxide film over that portion of the surface where the welds are to be made, with little investment for equipment, and without danger of subsequent formation of a high-resistance film, as it can be done immediately preceding welding. Mechanical methods are useful for removing oxide from workpieces that are too large to be dipped and rinsed. After the workpieces have been degreased, the oxide film can be removed with a fine grade of abrasive cloth, a fine stainless steel wool, or a motor-driven brush made of fine stainless steel wire. Glass brushes and aluminum wool have also been used. Carbon or low-alloy steel brushes and steel wool contaminate the aluminum surface and are not suitable.

Chemical removal of oxides by an acid dip is used after heavy oxide removal or directly after cleaning to obtain uniform surface resistance. A typical dip solution contains 12 vol% concentrated nitric acid, 0.4 vol% hydrofluoric acid, and 0.2 wt% wetting agent. The typical solution is used at a temperature of 70 to 80 °F with an immersion time of 2 to 6 min, followed by rinsing in clear water and drying.

Forgings, castings, extrusions, or similar parts having thick oxide accumulation may need an alkaline etch before the acid dip. A frequently used alkaline-etch treatment is immersion in an aqueous solution of 5% sodium hydroxide at 150 to 160 °F for 20 to 50 s, followed by rinsing in water. The alkaline-etch solution may contain additives to prevent the formation of scale in the tank.

Aluminum alloys 1060, 1100, and 3003 and aluminum-clad alloys are frequently immersed for $1^1/_2$ to 3 min in a warm (190 °F) solution containing 10 vol% concentrated nitric acid, 6 oz/gal sodium sulfate, and 0.1 wt% wetting agent. The parts are then rinsed in cold running water for 5 min and dried rapidly.

Heat treatable alloys such as 6061 are immersed for $1^1/_2$ to 3 min in a warm (185 °F) solution containing 15 vol% concentrated nitric acid, 13 oz/gal sodium sulfate, and 0.1 wt% wetting agent. The parts are then rinsed in cold running water for 5 min and dried rapidly. For more detailed information, see the article "Cleaning and Finishing of Aluminum and Aluminum Alloys" in Volume 5 of the 9th edition of *Metals Handbook*.

To ensure satisfactory preparation of the surfaces for spot and seam welding, the adequacy of cleaning and deoxidizing procedures and correctness of solution strengths need to be determined. This judgment should be made by measuring the contact resistance after surface treatment of small coupons from each batch of the material to be welded; visual inspection is not a satisfactory control method. Corrections in the procedures or additions to the solutions should be enacted as necessary to maintain the contact resistance of the control samples and of the workpieces at the desired value.

Contact-resistance test is a standard test used to determine the electrical resistance of an interface between aluminum alloy samples or workpieces. The two samples or workpieces are placed between two spot welding electrodes $^5/_8$ in. in diameter with faces 3 in. in spherical radius and are clamped under a static force of 600 lb. A 50-mA direct current is transmitted through the electrodes, and the resistance between the workpieces is determined with a suitable instrument.

The tests must be performed under identical conditions of force, applied current, and electrode size and contour. The measurement is sensitive to small changes in test procedure, and values are usually obtained by averaging at least two or three readings on each of the four possible interface combinations of a pair of specimens. After the electrode pressure is applied, movement of either specimen breaks the oxide coating, causing false low readings. The presence of burrs on the specimens also results in low readings.

The contact resistance of well-cleaned aluminum alloys is from 10 to about 200 μΩ, while that of unclean stock can be 1000 μΩ or more. For noncritical welding, contact resistance of 200 to 500 μΩ is satisfactory. For best results (as required for military applications), contact resistance should be about 50 μΩ. A narrow spread in readings is generally preferable to low average values with an occasional high reading.

Seam sealants, in the form of elastic materials, are used between members of resistance welded joints to make containers fluid-tight and to limit interface corrosion, especially in high-humidity environments. For instance, welds of military quality can be made through gun-grade caulking compounds consisting of finely divided aluminum powder in a special elastic binder with no significant change in machine setting. Some tapes and paints provide equally good results.

Characteristics of Resistance Spot Welding

All commercial aluminum alloys produced as sheet, extrusions, forgings, or

Fig. 2 Typical spot weld nugget in a heat treated alclad aluminum alloy. Made with a three-phase direct-energy resistance welding machine, using forging pressure and postheating. (Source: RWMA Bulletin 18)

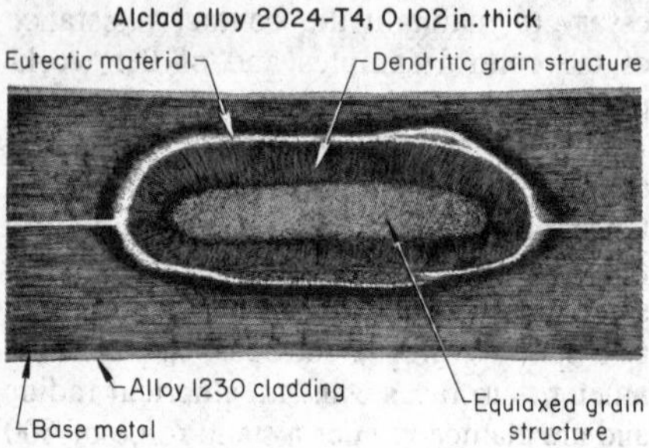

Alclad alloy 2024-T4, 0.102 in. thick
Eutectic material
Dendritic grain structure
Alloy 1230 cladding
Equiaxed grain structure
Base metal

castings can be spot welded; the combined maximum thickness with ordinary equipment is between 1/2 and 3/4 in. Figure 2 shows a typical spot weld between two 0.102-in.-thick sheets of alclad alloy 2024-T4 (alloy 1230 cladding; 99.3% minimum Al). In the center of this spot weld is an oval zone with equiaxed grain structure. Surrounding the central zone is a zone made up of a dendritic (or columnar) type of grain structure. These two zones constitute the spot weld nugget. Each is essentially a cast structure. The large size of the dendritic zone is the result of a welding technique that employed a postweld heat treatment permitting grain growth.

Surrounding the dendritic zone in Fig. 2 is a band of light color consisting of a region of metal that has been heated close to the melting point and in which the temperature has been high enough to put soluble phases into solution. The dark ring surrounding the lighter band is a zone that reached lower temperature than the lighter zone, but in which a high degree of precipitation occurred. The cladding at the interface was completely absorbed with the base metal into the nugget. Indentation was not severe and the outer surfaces of the work metal in contact with the electrodes show no effect of heating.

Welding current, as well as the heat input for welding, must be greater than that used for welding steel of equivalent thickness, because the thermal and electrical conductivities of aluminum alloys are about two to four times those of low-carbon steel. Tables 4, 5, and 6 list typical conditions for spot welding aluminum alloys in single-phase direct-energy, three-phase direct-energy, and capacitor-type stored-energy welding machines. These values are starting points and should be adjusted to suit the particular alloy and job requirements.

The higher the current, the more rapidly

Table 4 Typical conditions for resistance spot welding of aluminum alloy sheets in 60-cycle single-phase direct-energy welding machines

Thickness of thinnest sheet, in.	Electrode diameter, in.	Face radius, in. — Upper electrode	Face radius, in. — Lower electrode	Electrode force, lb.	Weld time, cycles	Welding current, A	Diameter of nugget, in.
0.016	5/8	1	Flat	320	4	15 000	0.110
0.020	5/8	1	Flat	340	5	18 000	0.125
0.025	5/8	2	Flat	390	6	21 800	0.140
0.032	5/8	2	Flat	500	6	26 000	0.160
0.040	5/8	3	Flat	600	8	30 700	0.180
0.051	5/8	3	Flat	660	8	33 000	0.210
0.064	5/8	3	Flat	750	10	35 900	0.250
0.072	5/8	4	4	800	10	38 000	0.275
0.081	7/8	4	4	860	10	41 800	0.300
0.091	7/8	6	6	950	12	46 000	0.330
0.102	7/8	6	6	1050	15	56 000	0.360
0.125	7/8	6	6	1300	15	76 000	0.425

Source: "Welding Alcoa Aluminum," Aluminum Company of America, Pittsburgh, 1972

Table 5 Typical conditions for resistance spot welding of aluminum alloy sheets in 60-cycle three-phase direct-energy welding machines

Thickness of thinnest sheet, in.	Electrode diameter, in.	Electrode face radius, in.	Electrode force, lb — Weld	Electrode force, lb — Forge	Time, cycles — Weld	Time, cycles — Postheat	Current, A — Weld	Current, A — Postheat	Diameter of nugget, in.
Three-phase rectifier-type machines									
0.016	5/8	3	440	1000	1	None	19 000	None	0.110
0.020	5/8	3	520	1150	1	None	22 000	None	0.125
0.032	5/8	3	670	1540	2	None	28 000	None	0.160
0.040	5/8	3	730	1800	3	None	32 000	None	0.180
0.051	5/8	8	900	2250	4	4	37 000	30 000	0.210
0.064	5/8	8	1100	2900	5	5	43 000	36 000	0.250
0.072	5/8	8	1190	3240	6	7	48 000	38 000	0.275
0.081	7/8	8	1460	3800	7	9	52 000	42 000	0.300
0.091	7/8	8	1700	4300	8	11	56 000	45 000	0.330
0.102	7/8	8	1900	5000	9	14	61 000	49 000	0.360
0.125	7/8	8	2500	6500	10	22	69 000	54 000	0.425
Three-phase frequency-converter-type machines									
0.020	5/8	3	500	None	1/2	None	26 000	None	0.125
0.025	5/8	3	500	1500	1	3	34 000	8 500	0.140
0.032	5/8	4	700	1800	1	4	36 000	9 000	0.160
0.040	5/8	4	800	2000	1	4	42 000	12 600	0.180
0.051	5/8	4	900	2300	1	5	46 000	13 800	0.210
0.064	5/8	6	1300	3000	2	5	54 000	18 900	0.250
0.072	5/8	6	1600	3600	2	6	61 000	21 350	0.275
0.081	7/8	6	2000	4300	3	6	65 000	22 750	0.300
0.091	7/8	6	2400	5300	3	8	75 000	30 000	0.330
0.102	7/8	8	2800	6800	3	8	85 000	34 000	0.360
0.125	7/8	8	4000	9000	4	10	100 000	45 000	0.425

Source: "Welding Alcoa Aluminum," Aluminum Company of America, Pittsburgh, 1972

Table 6 Typical conditions for resistance spot welding of aluminum alloy sheets in capacitor-type stored-energy welding machines

Thickness of thinnest sheet, in.	Electrode diameter, in.	Electrode face radius, in.	Electrode force, lb — Weld	Electrode force, lb — Forge	Condenser capacity, μF	Condenser charge, V	Transformer ratio	Total energy, W · s	Diameter of nugget, in.
0.020	5/8	3	376	692	240	2150	300:1	555	0.125
0.032	5/8	3	580	1300	240	2700	300:1	875	0.160
0.040	5/8	3	680	1580	360	2550	300:1	1172	0.180
0.051	5/8	3	890	2100	600	2560	300:1	1952	0.210
0.064	5/8	3	1080	2680	720	2700	300:1	2622	0.250
0.072	5/8	3	1230	3150	960	2750	450:1	3630	0.275
0.081	7/8	3	1550	4000	1440	2700	450:1	5250	0.300
0.091	7/8	3	1830	4660	1920	2650	450:1	6750	0.330
0.102	7/8	3	2025	5100	2520	2700	450:1	9180	0.360

Source: "Welding Alcoa Aluminum," Aluminum Company of America, Pittsburgh, 1972

the nugget is formed. The longer the duration of current flow, the greater the penetration of the nugget into the base metal. If the current is too low, the nugget forms too slowly, and excess heating and warping occur in the surrounding area. If the current is too high, formation of gas pockets and expulsion of metal result, causing sheet separation and weak welds.

Weld time is the actual time that the welding current flows through the workpieces, and it must be sufficient to form the weld without excessively heating the remainder of the weld area. As work-metal thickness increases, longer weld times are required, as shown in Tables 4 and 5.

Within a narrow range of weld time, weld strength is approximately proportional to weld time. Using weld time beyond this range produces no appreciable further increase in weld strength.

Electrode force used for spot welding of aluminum alloys is generally greater than that needed for spot welding of steel of equivalent thickness. Low-strength aluminum alloys usually need lower electrode force than high-strength alloys.

Use of insufficient electrode force can result in expulsion of weld metal, internal defects, surface burning, and excessive electrode pickup. Use of excessive electrode force can result in extreme indentation, sheet separation, work distortion, and asymmetrical welds. Low electrode force with its high contact resistance requires a higher current than needed when high electrode force with its low contact resistance is used.

A multiple-electrode-force cycle, in which the weld is made at a low electrode force followed by application of a higher force during hold time, is used to provide forging action during solidification of the nugget. To minimize internal defects, timing of the application of forging force is important. If the forging force is applied before completion of weld time, the contact resistance is decreased excessively, and welds may be unsound. If forging force is applied too late, after the weld has solidified, forging will be ineffective.

Spacing of spot welds is an important consideration, because when rows of spot welds are made, each successive weld can be affected by shunting of part of the welding current through the preceding welds. The shunting effect increases as spot spacing, work-metal thickness, and electrical resistance of the alloy being welded decrease. The shunting effect also increases as the contact resistance between the workpieces becomes greater.

To eliminate most of the shunting effect, the minimum spacing of spot welds made in aluminum alloy sheet normally should be no less than eight times the sheet thickness ($8t$) and preferably not less than the values given in Table 7. Where it is necessary to space welds at less than $8t$, the current can be increased after making the first spot weld to offset the loss from shunting. An alternative is to use two or more rows of spot welds with the welds in each row spaced apart by at least $8t$. Table 7 gives the suggested minimum distance between rows of staggered spot welds.

If a spot weld is made too close to the edge of a workpiece, the metal between the weld and the edge may bulge out or split, and molten metal may be expelled from the joint. The minimum distance from a spot weld to the edge of a workpiece should be one half the minimum joint overlap given in Table 7. If more than a single row of spot welds is made, the minimum overlap must be increased by the distance between rows.

Resistance Spot Welding Practice

Use of the welding conditions given in Tables 4, 5, and 6 should produce welds with shear strengths exceeding those given in Table 1. Larger welds with proportionately higher shear strengths sometimes can be obtained with stored-energy equipment. Smaller welds with strengths lower than those given in Table 1 should be avoided. Settings that result in small-diameter welds in aluminum are likely to cause a substantial percentage of unacceptable welds under production conditions.

Table 7 Suggested minimum joint overlap, weld spacing, and distance between rows for resistance spot welds in aluminum alloys

Thickness of thinnest sheet, in.	Minimum joint overlap(a), in.	Minimum weld spacing(b), in.	Minimum distance between rows(c), in.
0.016	5/16	3/8	1/4
0.020	3/8	3/8	1/4
0.025	3/8	3/8	5/16
0.032	1/2	1/2	5/16
0.040	9/16	1/2	3/8
0.051	5/8	5/8	3/8
0.064	3/4	5/8	3/8
0.072	13/16	3/4	7/16
0.081	7/8	3/4	1/2
0.091	15/16	7/8	1/2
0.102	1	1	1/2
0.125	1 1/8	1 1/4	5/8

(a) Minimum edge distance is equal to one half of minimum overlap. (b) Measured from center to center. (c) For rows of staggered welds at minimum weld spacing
Source: "Welding Alcoa Aluminum," Aluminum Company of America, Pittsburgh, 1972

Fig. 3 Spot welding of a handle flange to an aluminum alloy frying pan

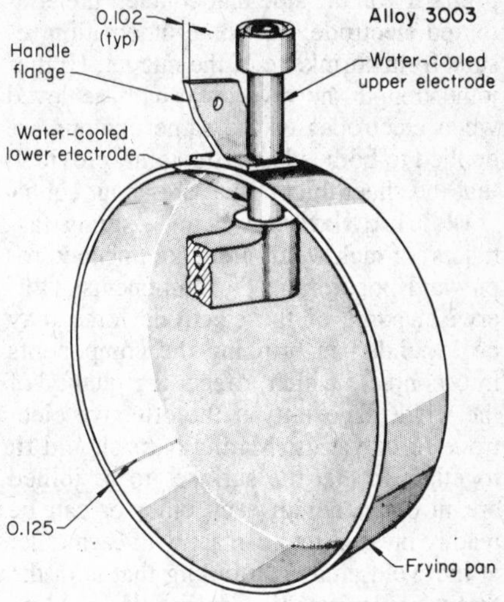

Although the welding conditions given in Tables 4, 5, and 6 are good starting points, optimum conditions must be determined by making and testing welds in sample setups of the job and adjusting the given conditions accordingly. During production, welds must be tested frequently to ensure that optimum conditions are maintained.

The optimum welding conditions for a specific application may vary substantially from those given in Tables 4, 5, and 6, as in welding a 0.102-in.-thick handle flange to a 0.125-in.-thick frying pan (Fig. 3), both made of alloy 3003, using a three-phase frequency-converter welding machine. An electrode force of 1440 lb was used, and a welding current of 70 000 A was applied in two 5-cycle weld pulses separated by a 2-cycle off time for cooling the electrodes. With these conditions, output was 600 assemblies in 8 h.

Workpiece Thickness. In welding workpieces of unequal thickness, the thinner workpiece governs the welding conditions that should be used and the diameter of the resulting nugget. Three or more thicknesses of aluminum alloy can be spot welded simultaneously. Conductivity of the alloy welded and thickness of the outside sheets govern the choice of machine settings. Because of the added resistance of the additional interfaces, electrode pressures are higher than those used for welding two sheets.

When spot welded aluminum assemblies must have smooth outer surfaces for purposes of aerodynamics or appearance, a flat-face electrode can be used on one side of the joint and a radius-face electrode on the other. Indentation will occur primarily on the side that contacts the contoured electrode, but some indentation results from shrinkage of the nugget. Higher joint strength, however, is usually achieved when electrodes of the same contour are applied to both sides of the joint, provided that the sheet thicknesses are about equal.

Welding fixtures, clamps, spring fasteners, or tack welds are recommended for preweld positioning of components. Otherwise, some of the electrode force may be expended in bringing the components into contact, which affects the quality of the weld, especially if the effective electrode force varies. Mating parts should fit together so that the surfaces to be joined are in contact with each other or can be readily pressed into contact with each other at the weld area. All tooling that is in the throat of the machine during the welding operation should be nonmagnetic; aluminum, fiberglass, and various types of plastics are often used.

Order of Welding. The order in which spot welds are made in a multiple-weld assembly is important in controlling warpage of the assembly and maintaining the strength of the spots. Making a spot weld produces slight lateral expansion of the work metal. Therefore, spot welding preferably should be started from the center of the sheet and made in succession at the desired spacing toward the ends. Additional spot welds should not be made between two existing welds, because shunting through adjacent spot welds could result in poor-quality welds. Also, expansion or warpage could prevent good contact of the sheets during welding. If three or more rows of welds are being made, the center row or rows should be made first.

Roll Resistance Spot Welding

Roll resistance spot welds can be made in conventional seam welding machines with continuously rotating or intermittent-motion electrode wheels. In welding aluminum alloys, better surface appearance and weld quality are obtained with intermittent motion. Weld spacing is obtained by proper adjustment of electrode speed, or indexing time, and hold time. The individual roll resistance spot welds are essentially the same as resistance spot welds made in the conventional manner, except that shorter hold times are employed. With continuously rotating electrode wheels, heat times are usually shorter than those normally used for spot welding. The high current employed sometimes requires the use of high electrode force. Because of electrode travel, nuggets made by continuously rotating electrode wheels are usually elongated.

Fig. 4 Resistance seam welding of an aircraft integral wing fuel tank using continuous electrode motion

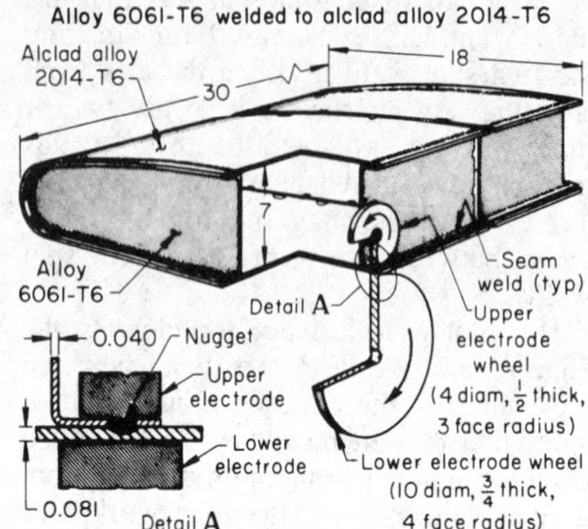

Resistance Seam Welding

The overlapping spot welds produced in resistance seam welding result in a continuous gastight and liquid-tight seam. The welding current can be applied either when the wheels are moving or while they are momentarily stopped. Intermittent motion results in better surface appearance and better weld quality in aluminum alloys and is ordinarily used for work of military quality when the work metal being seam welded is 0.080 in. or more in thickness.

Seam welding of aluminum alloys with continuously rotating electrodes was used in Fig. 4. In this application, 0.040-in.-thick alloy 6061-T6 sheet was joined to 0.081-in.-thick alloy 2014-T6 clad sheet. Nylon rollers were used to guide and support the workpieces and move the assembly more easily between the electrode wheels without marring the work metal surface. The welding force was 2800 lb. The weld current of 58 000 A was supplied from a three-phase frequency-converter welding machine in a single impulse of 2 cycles preheat time and 4 cycles heat time, with 6 cycles cool time. The overlapping spot welds were made 12 to the inch at a rate of 25 in./min.

Typical conditions for seam welding alloy 5052-H34 using a conventional single-phase direct-energy seam welding machine are given in Table 8. Data in this table also can be used as a basis for developing conditions for welding other alloys or tempers. The maximum heat time should be between one fifth and one third the total time. Use of a heat time shorter than one fifth the total time helps reduce electrode pickup, but also reduces nugget size. Electrode force, welding current, heat time, cool time, and welding speed are adjusted to produce the desired spacing and weld width. Width of a seam weld should be twice the thickness of the sheet being welded plus 0.06 in. ($2t + 0.06$ in.). Quality control for seam welding is the same as for spot welding except that an additional test, such as the pillow test, may be required for determining gas or liquid tightness.

Projection Welding

Aluminum alloys are seldom projection welded, because extremely close control

Table 8 Typical conditions for making gastight resistance seam welds in alloy 5052-H34 sheets with 60-cycle single-phase direct-energy welding machines

Electrode wheels and work metal must be cooled with 2 to 3 gal of water per minute.

Sheet thickness, in.	Spots per inch	Total weld time, cycles(a)	Wheel speed(b), ft/min	Heat time, cycles(c) min	Heat time, cycles(c) max	Electrode force(d), lb	Welding current(d), A	Width of weld, in.
0.010	25	3½	3.4	½	1	420	19 500	0.08
0.016	21	3½	4.1	½	1	500	22 000	0.09
0.020	20	4½	3.3	½	1½	540	24 000	0.10
0.025	18	5½	3.0	1	1½	600	26 000	0.11
0.032	16	5½	3.4	1	1½	690	29 000	0.13
0.040	14	7½	2.9	1½	2½	760	32 000	0.14
0.051	12	9½	2.6	1½	3	860	36 000	0.16
0.064	10	11½	2.6	2	3½	960	38 500	0.19
0.081	9	15½	2.1	3	5	1090	41 000	0.22
0.102	8	20½	1.8	4	6½	1230	43 000	0.26
0.125	7	28½	1.5	5½	9½	1350	45 000	0.32

(a) Heat time plus cool time. (b) Wheel speed is adjusted to give desired number of spots per inch. (c) Heat time must be set at full-cycle setting if total time is set at full-cycle setting. (d) Electrode force and welding current are adjusted to give desired width of weld. Values are for 5052-H34 aluminum alloy. Lower forces should be used for 5052-O and 3003-H14 aluminum alloys.
Source: "Welding Alcoa Aluminum," Aluminum Company of America, Pittsburgh, 1972

is needed in order to produce acceptable welds. Embossed projections like those used in low-carbon steel sheet, if made in aluminum alloy sheet, would collapse prematurely, and therefore are not used. Projections that have been coined in aluminum alloy workpieces are stronger and better able to resist collapse than are embossed projections, and are thus preferred.

Because of the narrow plastic range of aluminum alloys, the quality of projection welds is less uniform than that of spot welds of equivalent size.

Cross Wire Welding

Aluminum alloy wires, round and other shapes, can be cross wire welded to make racks, grills, or screens. Press-type resistance spot welding machines with low-inertia heads and fast follow-up are needed to maintain proper welding force on the joint as the wires deform and melt. Each cross wire weld must be made individually; multiple-wire welding with bar electrodes has been unsatisfactory on aluminum alloys.

Selection of alloy and temper is the same as for spot welding. All aluminum alloy wires from about $^1/_{16}$ to $^3/_8$ in. in diameter can be cross wire welded. Wires of equal diameters or of unequal diameters in ratios up to 2 to 1 are readily weldable. Low-strength ring-type welds can result if the ratio of wire diameters is greater than 2 to 1.

Wire can be welded to a tube if the thickness of the tube wall is equal to or greater than the diameter of the wire. Wire can be welded to extruded aluminum alloy angles or other structural shapes that have elongated projections extending at right angles to the wire direction. For optimum results, the cross-section radius of the projection should be approximately equal to the radius of the wire.

Optimum welding conditions are established by twist tests and by setdown measurement. Twist tests are made by twisting the wires to tear the area between wires. Tests of properly made cross wire welds usually result in base-metal fracture outside the weld.

Setdown is the difference in the combined height of both wires before and after welding. For joints between wires of equal diameter, setdown should be 25 to 35% of the combined height before welding. For joints between wires of unequal diameter, setdown should be 25 to 35% of twice the diameter of the smaller wire. The quality of cross wire welds is usually consistent when the variation in setdown does not exceed ±5% of the nominal value during a production run.

Flash Welding

All wrought aluminum alloys can be joined by flash welding. The process is particularly suitable for making butt and miter joints between two workpieces of similar cross-sectional shape. Aluminum alloy bars and tubes can be flash welded to copper bars and tubes.

Machines for flash welding of aluminum alloys require much larger transformer capacity than machines for flash welding of steel. For best results, the welding machine must be capable of supplying a current density of 100 000 A/in.^{2} while the workpieces are held in firm contact, without arcing. A secondary voltage of 2 to 20 V is required, and upsetting pressures are 8 to 40 ksi.

Copper alloy clamping dies, RWMA classes 1 and 2, are used when flash welding aluminum alloys, if a long die life is required. Hard-drawn copper may be used instead, if a limited number of pieces are to be welded. Initial die spacing varies according to work-metal thickness. About $^1/_2$ in. of initial die spacing is suitable for thin-walled tubing and thin sheets or extrusions, and about 1 in. is needed for flash welding a $^3/_4$-in.-diam bar.

Pinch-off dies provide the best weld quality in flash welding of aluminum alloys. These dies are made of hardened steel and are sharpened to a cutting edge. They are used to trim the upset metal or flash from the joint at the end of the upsetting stroke. For additional information, see the article "Flash Welding" in this Volume.

Inspection and Testing

Peel, twist, or tear tests are common methods of determining the quality of resistance welds made in aluminum alloys. To meet commercial requirements, a button of metal with a diameter at least twice the thickness of the thinnest workpiece plus 0.06 in. ($2t$ + 0.06 in.) should be pulled from one of the workpieces at each weld. To meet military standards, the button diameter should be equal to or exceed the nugget diameter given in Tables 4, 5, and 6. Spot welds in aluminum alloys 2014, 2024, 6061, and 7075, either bare or clad, will not always pull buttons completely through the sheet when the work-metal thickness is greater than about 0.080 in. In these alloys and metal thicknesses, the weld is peel tested and the diameter of the fractured area is determined.

When peel testing indicates that welds of the proper size are being produced, test welds can be made to determine shear strength. Welds also can be sectioned for metallographic examination to determine nugget diameter, penetration, and microstructure. Radiographic examination can be used to determine the soundness of test welds.

During production, visual examination can be used to detect electrode pickup, surface burning, cracks, skidding, expulsion of molten metal, and excessive indentation. In some alloys, radiographic inspection can be used to detect internal defects, such as cracks, porosity, and expulsion, and to evaluate the size and shape of the nugget and the structure of the weld metal. In general, radiographic examination is used principally in the establishment of optimum welding conditions and not as a production-control method. Other inspection methods sometimes used are eddy current, ultrasonic, and sonic.

Indentation of the base metal can be determined by measuring the thickness of the weld with ball-tip micrometers or by determining the difference in height between the weld and the surrounding area using a dial-indicator depth gage. Sheet separation can be measured with feeler gages.

Causes and Prevention of Weld Defects

The most frequent resistance weld defects in aluminum alloys include electrode pickup, cracks and porosity, expulsion of molten metal, indentation and sheet separation, and irregular-shaped and unfused welds.

Electrode pickup, or alloying of the work metal with the electrode material, is usually the result of excessive heating at the interface between the electrode and the work metal. In severe cases, actual melting or burning of the work-metal surface occurs.

Electrode pickup can be minimized by using adequate electrode force during welding or avoiding excessive welding current or weld time. Pickup may be caused by improper cleaning of work metal or electrodes; use of improper electrode material, size, or contour; inadequate cooling of the electrodes; or skidding. Electrode pickup can occur with all types of welding equipment, but usually occurs sooner when welding is done with single-phase alternating current equipment.

Cracks and porosity in the weld metal may result from too rapid heating of the weld metal, an excessively high cooling rate, or inadequate or incorrect application of electrode force during or after welding. Spot welds in some high-strength alloys, such as 2024 and 7075, are subject to cracking if welding current is too high or electrode force during welding is too low.

Cracking may also result from too rapid quenching of the weld metal after the welding current has been turned off. Changes in the downslope of the current or the use of postheating may eliminate cracking. On machines with dual electrode force, readjustment of the forge-starting time may also eliminate cracking.

Expulsion of molten metal from the weld area by flashing or arcing usually can be eliminated by better cleaning methods or a slight reduction in welding current. Another cause of expulsion is an initial electrode force that is too low, followed by excessive forging pressure.

Indentation and sheet separation generally occur together. One of the major causes of these defects is work metal with a temper that is too soft. Occasionally, indentation and sheet separation can be minimized by decreasing the electrode force, increasing the face radius of the electrode, or reducing the welding current or time. Sheet separation also can be caused by lack of flatness of the sheets being welded.

Irregular shape and incomplete fusion in welds are caused by incorrect fit-up of workpieces, incorrect electrode alignment, skidding of the electrodes, inadequate surface preparation, or an irregular electrode contour. Inadequately fused welds in clad sheet may have some unfused cladding metal in the nugget. This defect can usually be avoided by a moderate increase in welding current.

Burning of holes through the workpieces can result from (1) insufficient electrode force to squeeze the metal effectively; (2) the presence of foreign material such as paper or steel wool between the workpieces; (3) emery dust or emery cloth adhering to one or both electrodes; and (4) attempts to spot weld at points where there are screws, projections, or drilled holes. If none of these appears to be the cause of the difficulty, the welding machine and its controls should be inspected.

Resistance Welding of Copper and Copper Alloys

COPPERS AND COPPER ALLOYS that are frequently resistance welded are listed in Table 1, which gives nominal compositions, melting points (liquidus temperatures), relative thermal and electrical conductivities, and welding indexes for resistance spot welding (RSW). Leaded and other free-machining copper alloys, which are seldom resistance welded, are not listed in this table.

Resistance spot welding is widely used for joining coppers and copper alloys. Principal applications include welding sections up to about 0.060 in. thick, particularly those alloys with low electrical conductivities. Many copper alloys with low conductivities can be seam welded easily. Coppers are difficult to seam weld. Projection welding is not recommended for copper or for most brasses. Bronzes can be projection welded with satisfactory results in many applications. Flash welding can be used for joining round stock, tubing, sheet, and mill shapes made of copper and copper alloys. Abutting ends, as they become plastic, must be pushed together with minimum upsetting force to produce a satisfactory weld. See the articles "Resistance Spot Welding," "Resistance Seam Welding," "Projection Welding," and "Flash Welding" in this Volume for more detailed information on these processes.

Welding Characteristics

The resistance weldability of any copper or copper alloy is inversely proportional to its electrical and thermal conductivities. Generally, alloys with lower conductivities are easier to weld (see Table 1) and require lower welding currents (see Tables 1 and 2). Compared to steel, most copper alloys require shorter weld time, lower electrode force, higher current, and different electrode materials that are compatible with the alloy being welded. The conditions for spot welding various copper alloys are given in Table 2.

Minimum spot spacing and contacting overlap for RSW of high-zinc brasses are given in Table 3. When workpieces of unequal thickness are welded, spot spacing should be equal to the minimum spacing recommended for the average thickness. The values listed for contacting overlap are designed to prevent bulging of the edge and expulsion of weld metal in welding workpieces that are manually positioned between the electrodes. The contacting overlap can sometimes be less than shown when the workpieces are held in fixtures. Breaking loads in shear of spot welded joints are also listed in Table 3.

Welding Equipment

The moving-force member of a spot, seam, or projection welding machine should be as light as possible to reduce inertia, and relatively free of friction to ensure rapid follow-up during welding. Low-inertia heads provide a fast follow-up of the upper electrode as the weld nugget is formed or as the projection collapses. Because copper alloys have a narrow plastic range, the force member must follow rapidly to maintain pressure on the joint and to prevent expulsion of weld metal.

Single-phase and three-phase direct-energy and electrostatic stored-energy (capacitor-discharge) welding machines are used for resistance welding of copper and copper alloys. The addition of slope control to single-phase direct-energy welding machines is not necessary for spot welding most copper alloys. In welding high-zinc brasses, the use of upslope can result in an increase of as much as 20% in weld strength. Downslope is not recommended for welding any of the copper alloys.

Electrostatic stored-energy machines use for the welding current an energy charge that is stored in a bank of capacitors. A three-phase, full-wave, grid-controlled thyratron rectifier is used for furnishing a direct current power supply to the bank of capacitors. A vacuum-tube leveling circuit is used to predetermine and maintain the capacitor-voltage level.

During the instant of recharging the capacitors, the value of the rectifier current is determined by the impedance of the rectifier transformer and the resistance of a resistor in the circuit. The leveling circuit generally consists of an ordinary radio-tube-type circuit and the necessary direct current positive-bias and negative-bias voltages for the proper on and off conduction time of the rectifier tube.

When the welding contactor is closed, the capacitor energy is discharged into a center-tapped welding transformer, and current flows into one half of the transformer. At the instant of closure, the rectifier tubes are automatically cut off by reversing the bias on the grid of the thyratron tubes to a high negative value. After discharge, the welding current ceases to flow, the rectifier tubes are made conductive by the reversal of the negative bias to a positive value, and the capacitors are automatically recharged. When successive welds are made, two welding contactors alternate the welding current between the two halves of the welding transformer to prevent saturation.

Generally, an electrostatic machine can be recharged rapidly enough to be used for roll-spot and seam welding as well as for spot and projection welding.

Welding Machine Controls. Copper alloys are particularly sensitive to variations in welding conditions, and therefore all direct-energy machines used for welding these alloys should be equipped with synchronous electronic controls, especially in applications requiring short weld times. These devices are capable of controlling weld time and welding current for repeated operations with extreme accuracy, and of eliminating transient cur-

Table 1 Nominal compositions, melting points, relative thermal conductivities, relative electrical conductivities, and RSW indexes for some coppers and copper alloys

UNS No.	Alloy name	Nominal composition, % Cu	Zn	Sn	Ni	Other	Melting point (liquidus), °F	Relative thermai conductivity(a), %	Relative electrical conductivity(b), % IACS	Welding index(c)
OF and ETP coppers										
C10200	Oxygen-free copper (OF)	99.95	...	...	...	...	1981	100	101	(g)
C11000	Electrolytic tough pitch copper (ETP)	99.90	...	...	...	0.04 O	1981	100	101	H-350
Deoxidized coppers										
C12000	Phosphorus-deoxidized copper, low residual phosphorus (DLP)	99.9	...	...	...	0.008 P	1981	99	98	(g)
C12200	Phosphorus-deoxidized copper, high residual phosphorus (DHP)	99.9	...	...	...	0.02 P	1981	87	85	(g)
Beryllium coppers										
C17500	High-conductivity beryllium copper, 0.6%	96.9	...	...	...	0.6 Be, 2.5 Co	1955	53-66(d)	45(d)	(h)
C17000	High-strength beryllium copper, 1.7%	98.3	...	...	...	1.7 Be	1800	27-33(d)	22(d)	C-150
C17200	High-strength beryllium copper, 1.9%	98.1	...	...	...	1.9 Be	1800	27-33(d)	22(d)	C-150
Chromium copper										
C18400	Chromium copper	99.2	...	...	...	0.8 Cr	1967	83(d)	80(d)	(g)
Low-zinc brasses										
C21000	Gilding, 95%	95	5	...	...	...	1950	60	56	H-200
C22000	Commercial bronze, 90%	90	10	...	...	...	1910	48	44	H-200
C23000	Red brass, 85%	85	15	...	...	...	1880	41	37	H-200
C24000	Low brass, 80%	80	20	...	...	...	1830	36	32	G-175
High-zinc brasses										
C26000	Cartridge brass, 70%	70	30	...	...	...	1750	31	28	E-150
C26800, C27000	Yellow brass, 65%	65	35	...	...	...	1710	30	27	F-150
C28000	Muntz metal, 60%	60	40	...	...	...	1660	31	28	F-150
Tin brasses										
C44300 to C44500	Inhibited admiralty brass	71	28	1	...	(e)	1720	28	25	(h)
C46400 to C46700	Naval brass	60	39	0.8	...	(f)	1650	30	26	(h)
Special brasses										
C66700	Manganese brass	70	28	...	...	1.2 Mn	2000	25	17	(h)
C67500	Manganese bronze A	58.5	39	1	...	1.4 Fe, 0.1 Mn	1630	27	24	C-125
C68700	Aluminum brass, arsenical	77.5	20.5	...	...	2 Al, 0.06 As	1780	26	23	(h)
C69400	Silicon red brass	81.5	14.5	...	...	4 Si	1685	7	6.2	(h)
Nickel silvers										
C74500	Nickel silver 65-10	65	25	...	10	...	1870	12	9	C-125
C75200	Nickel silver 65-18	65	17	...	18	...	2030	8	6	C-125
C75400	Nickel silver 65-15	65	20	...	15	...	1970	9	7	C-125
C75700	Nickel silver 65-12	65	23	...	12	...	1900	10	8	(h)
C77000	Nickel silver 55-18	55	27	...	18	...	1930	8	$5^1/_2$	(h)
Phosphor bronzes										
C50500	Phosphor bronze, 1.25% E	98.7	...	1.3	...	0.3 P	1970	53	48	G-200
C51000	Phosphor bronze, 5% A	94.8	...	5	...	0.2 P	1920	18	15	D-125
C52100	Phosphor bronze, 8% C	92	...	8	...	0.3 P	1880	16	13	D-125
C52400	Phosphor bronze, 10% D	90	...	10	...	0.3 P	1830	13	11	D-125
Aluminum bronzes										
C61300	Aluminum bronze D, Sn-stabilized	92.7	...	0.30	...	7 Al, 3.5 Fe	1915	14	12	(h)

(continued)

Table 1 (continued)

UNS No.	Alloy name	Nominal composition, % Cu	Zn	Sn	Ni	Other	Melting point (liquidus), °F	Relative thermal conductivity(a), %	Relative electrical conductivity(b), % IACS	Welding index(c)
C61400	Aluminum bronze D	91	...	...	...	7 Al, 2 Fe, 1 Mn	1915	17	14	(h)
C63000	Aluminum bronze	82	...	...	5	1.5 Mn, 3 Fe, 10 Al	1930	10	7	(h)
Silicon bronzes										
C65100	Low-silicon bronze B	98.5	...	...	...	1.5 Si	1940	15	12	B-125
C65500	High-silicon bronze A	97	...	...	...	3 Si	1880	9	7	B-125
Copper nickels										
C70600	Copper nickel, 10%	88.6	...	...	10	1.4 Fe, 1.0 Mn	2100	12	9	(h)
C71500	Copper nickel, 30%	69.5	...	...	30	0.5 Fe	2260	8	4.6	C-125

(a) Based on alloy C10200, which has a thermal conductivity of 226 Btu · ft/h · °F at 68 °F, as 100%. For comparison, 1010 steel has a thermal conductivity of 30 Btu · ft/h · °F or about 13% on this relative scale. (b) The ratio of the resistivity of the International Annealed Copper Standard at 20 °C to the resistivity of the material at 20 °C, expressed in percent and calculated on a volume basis. (c) Welding index from the Third Edition of Volume I of the "Resistance Welding Manual," published by the Resistance Welder Manufacturers Association. A is the basis of comparison, and is the equivalent of conditions for clean, cold rolled steel. B indicates that the alloy is readily weldable, but not as easily as steel. C, D, E, F, and G represent progressive stages of increasing difficulty. G indicates that the alloy can be successfully welded, but 100% uniform results cannot be expected. H includes those metals which are considered commercially impractical to spot weld. The numbers following the letters indicate the approximate percentage of secondary current required, based on mild steel = 100. (d) In precipitation-hardened condition. (e) Alloys C44300, C44400, and C44500 contain 0.02 to 0.10% As, Sb, or P, respectively. (f) Alloys C465,00, C46600, and C46700 contain 0.02 to 0.10% As, Sb, or P, respectively. (g) Alloys not having a spot welding index by RWMA, but which are listed as not recommended for spot welding by the Copper Development Association. (h) Alloys not having a spot welding index by RWMA, but which are listed as good or excellent under suitability for being joined by spot welding by the Copper Development Association

Table 2 Conditions for RSW of various copper alloys(a)

UNS No.	Alloy name	Weld time, cycles	Electrode force, lb	Welding current, A
C23000	Red brass	6	400	25 000
C24000	Low brass	6	400	24 000
C26000	Cartridge brass	4	400	25 000
C26800-C27000	Yellow brass	4	400	24 000
C28000	Muntz metal	4	400	21 000
C51000-C52400	Phosphor bronze	6	510	19 500
C62800	Aluminum bronze	4	510	21 000
C65100-C65500	Silicon bronze	6	400	16 500
C66700	Manganese brass	6	400	22 000
C68700	Aluminum brass	4	400	24 000
C69200	Silicon brass	6	510	22 000

(a) For spot welding 0.036-in.-thick sheet using RWMA type E electrodes with 3/16-in.-diam face and 30° chamfer and made of RWMA class 1 material

Table 3 Recommended spot spacing, contacting overlap, and approximate shear load of joint for RSW of high-zinc brasses

Thickness of thinnest sheet, in.	Spot spacing, in., min	Minimum contacting overlap(a), in.,	Shear load of joint, lb
0.032	5/8	1/2	330
0.050	5/8	5/8	512
0.064	3/4	3/4	680
0.094	1	1	1168
0.125	1 1/2	1 1/4	1872

(a) Minimum edge distance is equal to one half the contacting overlap.

rents that sometimes result from random switching of power to the welding machine.

The modern resistance welding machine, which often employs tape-wound transformers, is very sensitive to transient currents. This requires that the control portion of the circuit be synchronously coupled to the input voltage waveform. Thus, the firing point for the silicon-controlled rectifiers (SCR), which control the current to the weld, is related to input waveform rather than an arbitrary timing generator signal. Since the electrode force is applied before welding current is initiated, there is usually a 1/2-s force build-up time built into the programmer. This force initiation is not nearly as critical as the SCR firing point. Unlike hardenable steels, copper alloys do not require postweld heating to modify their microstructure. As a result, the resistance welding controller can be simpler.

Electrodes

The current used for resistance welding of copper alloys is much higher than that used for welding low-carbon steel, and therefore, the electrode must have high electrical conductivity to minimize heat buildup. Also, the thermal conductivity of the electrode must be high so that heat generated at the workpiece-electrode interface can be dissipated, preventing these surfaces from fusing. Proper cooling of the electrode is especially important for these alloys. Because of the low electrode force that is used in resistance welding of copper and copper alloys, deformation and wear of the electrode face have less effect on electrode life than does alloying of the face with the work metal.

Electrode Materials. The Resistance Welder Manufacturers Association (RWMA) class 1 electrode materials, containing copper and cadmium, are sometimes used for welding copper and high-conductivity brass and bronze. Class 2 materials, containing copper and chromium, are used on low-conductivity brass and bronze and the copper-nickel alloys. Class 3 materials are used in electrodes for seam welding.

The RWMA group B materials (refractory metal compositions) are used as facings on electrodes for spot welding copper alloys because these materials do not readily alloy with copper alloys. Class 11 (a mixture of copper and tungsten), class 13 (commercially pure tungsten), and class 14 (commercially pure molybdenum) electrodes give good performance and long life when used in the resistance welding of copper and copper alloys that have high electrical conductivity.

Properties of electrode materials are given in the article "Resistance Spot Welding" in this Volume.

Dispersion-strengthened copper is an unclassified material that may be used for electrodes. The high recrystallization temperature of the wrought material provides excellent resistance to softening and mushrooming of electrodes, which contributes to long electrode life. Mechanical properties and electrical conductivity of dispersion-strengthened copper bars meet the requirements of RWMA group A, class 1 and 2 alloys.

Electrodes must be efficiently water cooled to minimize sticking to the work metal and to prolong their life. Face contours must be carefully prepared and the electrodes must be properly aligned for welding. Type F (radius-face) electrodes having a face radius of 3 to 6 in. are often used for spot welding copper alloys. The smaller face radii are used for welding thinner workpieces. A type C (flat-faced) electrode is used where good appearance and minimum identation are required on the surface of one workpiece. Rigid holders are needed to avoid deflection or skidding of the electrodes on the work when force is applied.

Selection of Process

Weldability of the work metal often determines which process should be used for a given application. Some of the coppers and copper alloys can be spot welded, but not seam welded because of high conductivity, and not projection welded because of low compressive strength of the projections at elevated temperature. For further discussion, see the subsequent sections of this article on specific alloys. In welding dissimilar metals, heat balance can also be important in the choice of process.

Spot and seam welds can be made in work metal as thin as 0.001 in. Spot welding of metal as thick as 0.125 in. has been reported for copper alloys. Projection welding is best suited for work thicker than 0.020 in.

The use of projection welding frequently can increase the quality of joints in high-conductivity alloys, because welding current can be concentrated where needed. Distortion and electrode pickup are minimized because the electrode contacts a large area of the work metal. Projection welding may be preferred when the components are self-locating or to simplify fixturing or improve dimensional accuracy.

Lap joints that must be liquid-tight usually are made most efficiently by seam welding. However, if a seam does not require the leaktightness provided by overlapping spots, spot welding is frequently preferred.

Electrode forces lower than those needed for welding low-carbon steel are used, but extremely low forces, which can cause electrode pickup and weak welds, should be avoided. Low electrode force can also cause high-zinc alloys to flash or burn through.

Seam welding is nearly impossible on copper and many of the high-copper alloys, but most low-conductivity copper alloys can be seam welded readily using higher welding current and lower electrode force than those used for welding low-carbon steel. The usual spot spacing is 12 to 18 spots per inch. If fewer than 12 spots per inch are used, the spots sometimes do not overlap. Spots that are too closely spaced can cause excessive hot working of the base metal. Cooling by flooding, immersion, or mist protects the work metal and electrodes from overheating and electrode pickup.

Projection welding is best suited for copper alloys of less than 30% IACS electrical conductivity. The design of the projections in relation to the thickness and type of work metal is important. In general, to prevent collapse of the metal in the projection before welding temperature is reached, coined projections are preferred to formed projections.

Preweld Cleaning

Dirt, scale, oil, drawing compound, or other foreign matter on the surface of the workpiece should be removed before resistance welding. An indication of surface cleanness is the surface contact resistance, which should be uniform for best welding results. High or erratic contact resistance usually causes poor welds rather than reduced electrode life, although increased electrode pickup is an indication of high surface resistance. Surface cleaning can be done by various mechanical or chemical methods.

Coppers

Coppers and copper alloys having electrical conductivity higher than about 30% IACS (see Table 1) are the least well suited for resistance spot, projection, or seam welding, mainly because of severe electrode pickup. Thin copper stock can be welded using electrodes faced with RWMA class 13 (tungsten) or class 14 (molybdenum), but surface appearance is poor and frequent electrode maintenance is required. A tinned coating on wire or sheet is helpful in welding copper.

When welding braided copper wires, best results are obtained if the wires are first bright dipped, then electrotin plated to a maximum thickness of 0.0002 in. Copper wires treated in this manner can be stored for 2 to 3 months, then welded with good results. The groove in the face of the electrode contacting the wire should have a depth not greater than one half the wire diameter. Tapering the electrode tip so that the length of the groove is equal to or slightly greater than the wire diameter will help prevent brittle welds. The sides of the electrode should be tapered so that the face of the electrode is almost square in shape. The use of electrodes made of class 10 to 14 materials (refractory metals) helps heat the workpieces.

An internally water-cooled electrode with a molybdenum face shaped to conform to the top surface of the workpiece and water cooling of the joint were used in the example that follows. Clean joints were made at a higher production rate by spot welding than by resistance brazing.

Example 1. Joining a Braided Copper Conductor to a Chromium Copper Blank by Resistance Spot Welding. For joining a braided copper (alloy C11000)

Fig. 1 Braided copper conductor and chromium copper blank joined by RSW. To avoid use of a filler metal, minimize distortion of the blank, and increase production

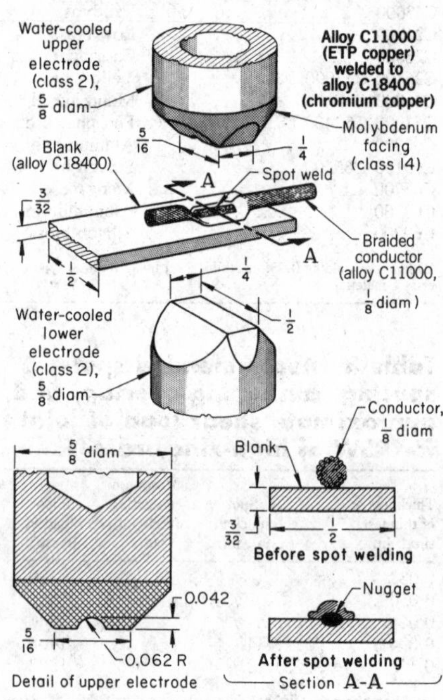

Conditions for RSW

Lower electrode	5/8 in. diam, RWMA class 2, with 1/4-by-1/2-in. face
Upper electrode	5/8 in. diam, RWMA class 2, 1/4-by-5/16-in. class 14 face
Welding current	8500 A, ac
Electrode force	350 lb
Squeeze time	10 cycles
Weld time	12 cycles
Hold time	16 cycles
Production rate	420 welds per hour

conductor to an alloy C18400 (chromium copper) blank, RSW was used instead of resistance brazing to avoid using a filler metal and to minimize annealing and distortion of the blank (see Fig. 1). Spot welding produced a joint of excellent conductivity and an assembly that was inherently clean because no flux or filler metal was used. The production rate for spot welding was more than twice that for resistance brazing, and thus a less costly part was produced by spot welding.

The 5/8-in.-diam water-cooled electrodes were made of RWMA class 2 material. The high conductivity of the copper conductor required that the upper electrode be faced with molybdenum (class 14 material) and be grooved, as shown in Fig. 1, to localize the contact area. The groove in the electrode face also helped to gather and hold together the strands of the conductor end as it was mashed to make a compact and dense weld. The lower electrode had a flat rectangular face 1/4 by 1/2 in. As the weld was made, cooling water was directed on the joint to minimize annealing and to avoid burning of the copper conductor.

The production rate for spot welding was 420 welds per hour. Previously, when resistance brazing was used, production rate was 180 joints per hour. Resistance brazing required preplacement of a filler-metal disk, application of flux, use of a brazing time longer than the weld time used in spot welding, and excessive cleaning of the brazing electrodes.

Welding Plastic-Coated Metals. Often one or both members of a copper joint are coated with a protective or insulating material that interferes with direct welding of the joint. However, other considerations, such as expediency, loss of protection, or cost, may dictate that the coating not be stripped or cleaned away in a separate operation. Some coatings, like paint or insulation, can be penetrated by melting, burning, or volatilizing them away during resistance welding. In one application, a plastic sheet between two nickel wires was penetrated by pressure and heat that were applied to the rounded surfaces of the wires. In the example that follows, a patented electrode-heating circuit was incorporated into the welding machine to burn or melt plastic coatings from copper conductors in welding the conductors to copper pins.

Example 2. Use of a Special Auxiliary Electrode-Heating Circuit To Melt Plastic Insulation From a Copper Conductor in Resistance Spot Welding. Plastic-coated conductors 0.050 in. wide by 0.002 in. thick were spot welded to 0.052-in.-diam pins using electrodes that were heated by an auxiliary alternating current circuit to melt the 0.005-in.-thick insulation at the weld zone (see Fig. 2). Both components were made of copper alloy C11000 (ETP copper). Thirteen conductors were contained in a flat connector board that was about $1^{7}/_{16}$ in. wide, as shown in Fig. 2. The insulation used on the parts described in this example was a polyester, but a higher temperature-resistant plastic was sometimes used. Spot welding produced good joints at a high

Fig. 2 Three types of patented electrode-heating circuits. Setup for melting plastic insulation from copper conductors and resistance spot welding the conductors to copper pins, using a capacitor-discharge welding machine and the heating circuit shown at lower left. (a) Induction circuit using alternating current for heating both upper and lower electrodes, and direct current for welding. When a high-temperature-resistant plastic insulation was used, electrode temperature offered considerable resistance to the flow of welding current. (b) Separate alternating current circuit for heating the upper electrode only, and direct current for welding. (c) Alternating current for both heating and welding; shunt-path heating of upper electrode only

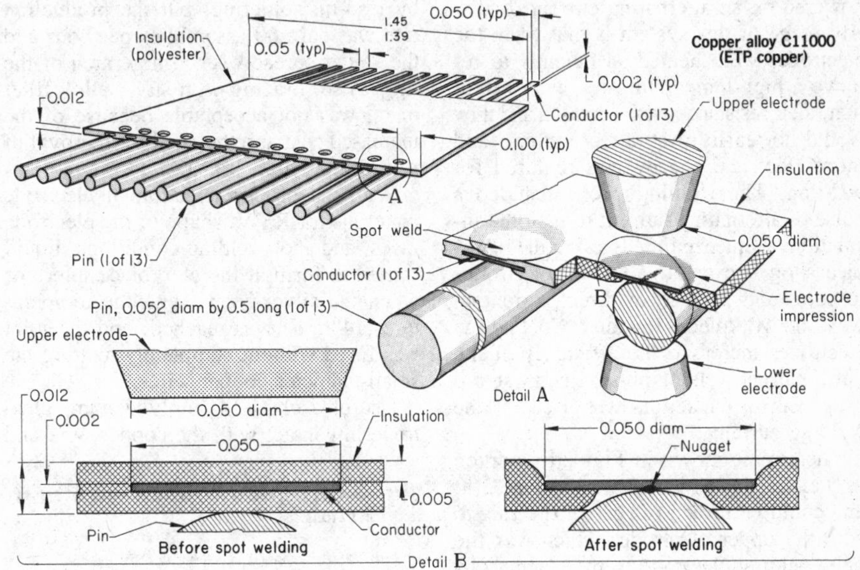

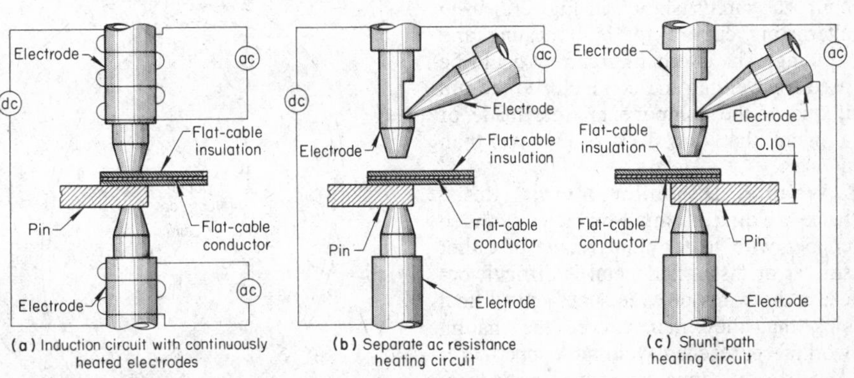

Equipment details

Power supply	110-V single-phase ac(a), 220-V single-phase ac(b)
Welding machine	Press type, stored energy
Rating of capacitor bank	755 μF
Storage voltage	13 to 408 V, dc
Energy storage:	
Low	0.06 to 9 W · s
High	0.1 to 45 W · s
Electrodes	1/6 in. diam by $1^{1}/_{4}$ in. long, tapered to 0.05 in. diam at tip, class 13
Electrode force	0.5 to 15 lb

Conditions for RSW

Welding current	200 A, dc
Welding voltage	0.6 V, dc
Electrode force	3 lb
Melting time	0.3 s
Squeeze time	0.0025 s
Weld time	0.0045 s
Hold time	0.5 to 1 s
Area of weld	40 mil^2 (0.00004 $in.^2$)
Current density	5 A/mil^2
Electrode gap	0.10 in.

(a) For heating the electrode to melt plastic insulation from the conductor. (b) For welding

production rate without the need for stripping the insulation from the conductors.

Any of the patented heating circuits shown in Fig. 2 could be incorporated into a welding machine to melt through the plastic coating on the conductors. The system shown in Fig. 2(a) was selected for this application because it was superior to the other two with respect to all-around capabilities and performance. The direct current welding circuit, because of its speed and dependability, was preferred for welding the copper workpieces. The electrodes were heated by heating elements powered by an alternating current. A disadvantage of this system is that when the electrodes were heated sufficiently to remove a high-temperature-resistant insulation, their resistance was increased and they would not easily pass direct current, and more electrical energy was required for welding. The welding cycle included a pulse of alternating current to melt the insulation, a squeeze time to bring the workpieces together and to allow the electrodes to cool, and a pulse of direct current to weld the workpieces together. Cooling the electrodes increased the efficiency of the direct current welding pulse. Energy stored in a bank of capacitors was used for the welding current.

The circuit shown in Fig. 2(b) worked as well as the circuit shown in Fig. 2(a) and could be turned off between welds to cool the upper electrode, which was the only welding electrode in the heating circuit. The circuit shown in Fig. 2(c) used alternating current for both heating and welding. The alternating current could be used for welding the conductor to the pin if one of the components were made of a metal that was less conductive than copper.

Welding Dissimilar Metals. Resistance welding of small-diameter stranded copper wire to terminals made of either similar or dissimilar metal is difficult because the wires provide such a small heat sink that they melt before the mating workpiece reaches welding temperature. A copper-to-copper joint was made successfully by adding BCuP-5 filler metal and resistance brazing at a temperature below the melting point of the 0.014-in.-diam wires. Making the joint without a filler metal was unsuccessful because the small-diameter copper wires melted before a weld was made. In Example 1 of this article, a braided copper conductor was successfully spot welded to a small chromium copper blank. In the example that follows, a stranded conductor containing 0.016-in.-diam wires was spot welded to a ring made of coin silver (90%Ag-10%Cu). The welding conditions were selected so that the maximum temperature reached in the two workpieces was above the melting point of the silver-copper eutectic (1436 °F), but below the solidus temperature of ETP copper (1949 °F).

Example 3. Selection of Welding Conditions To Control Heating of Workpieces in Spot Welding. A silver-coated seven-strand lead wire of alloy C11000 (ETP copper) was spot welded to a coin silver (90%Ag-10%Cu) ring at a production rate of 190 welds per hour (see Fig. 3). Originally, the assembly was made by lead-tin soldering, but the production rate was only 60 assemblies per hour and the solder spread over a large area of the ring. Also, brazing with silver alloy filler metal was not acceptable because of the increased cost and the need for removal of flux residue after brazing.

With the proper selection of electrode materials for RSW, shape of the electrode faces, and spot welding conditions, liquid started to form at the workpiece interface when the silver-copper eutectic temperature (1436 °F) was reached, and the joint was fused without completely melting the small-diameter copper wires.

Use of a class 14 (molybdenum) electrode in contact with the copper wire and a class 2 (chromium copper) electrode in contact with the coin silver ring helped to maintain heat balance. The molybdenum electrode material promoted heating at the workpiece interface and prevented electrode pickup. A semicircular groove was machined into the upper (molybdenum) electrode to contain the stranded wire. The lower electrode had a flat face that made line contact with the ring and concentrated the welding heat. Both upper and lower electrodes were cooled by water flowing through passages in the electrode holders, and were cleaned by wire brushing after making 200 to 250 welds. Electrode life was about 5000 welds between redressings.

The coin silver ring was degreased with trichlorethylene before welding. Removing the insulation was the only preweld preparation needed for the wire. The ring was held in a vertical position against a stop; the wire was manually located beneath the upper electrode. The welding cycle was initiated with a pedal.

Welding Metal-Coated Base Metals. Spot welds can be made between metal-plated or metal-coated base metals, provided the thickness, composition, and density of the coating are consistent. The ambient atmosphere and time between

Fig. 3 Stranded copper wire and coin silver ring joined by RSW. See text for details.

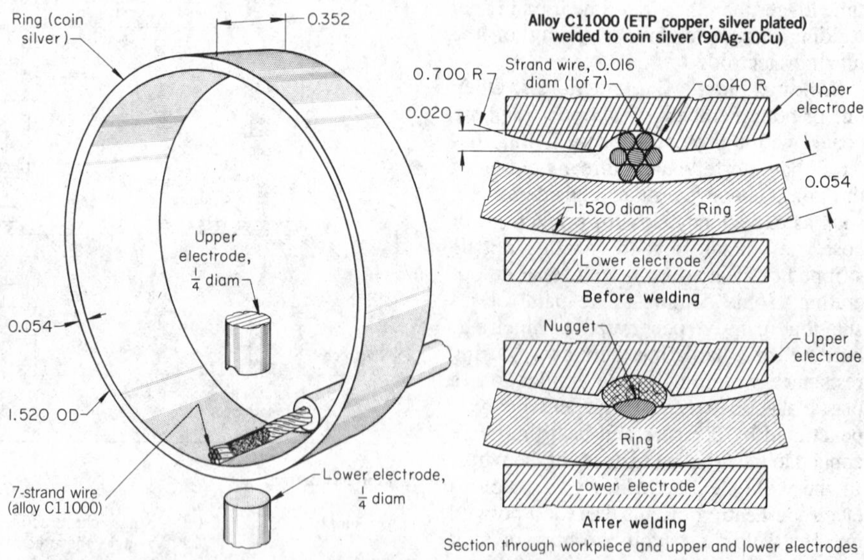

Equipment details	
Power supply	440 V, three phase
Welding machine	Press-type
Rating at 50% duty cycle	100 kV · A
Lower electrode	Type C, 1/4 in. diam, RWMA class 2
Upper electrode	1/4 in. diam, RWMA class 14, with grooved face

Conditions for RSW	
Welding current	70 A, ac
Welding voltage	3.7 V
Heat-control setting	62%
Electrode force	120 lb
Squeeze and weld times	5 cycles each
Hold time	30 cycles
Production rate	190 welds per hour

coating and welding can have an effect on the weldability of coated base metals. The contact resistance and the electrical resistivity of the coating can have a critical effect on spot welding of coated metals, as in the next example.

Example 4. Effect of Plating Composition on Quality of Spot Welds. A gold-plated nickel jumper strip was joined, as shown in Fig. 4, to a copper terminal clad with BCuP-5 (copper alloy brazing filler metal), by capacitor-discharge RSW. The filler-metal coating was used for making other joints. The spot welded joints were welded successfully for several years but, suddenly, joints made under the established welding conditions failed the specified 6.5-lb tensile-shear test. Adjustment of welding conditions did not produce acceptable welds.

The 99.99% gold plating on the Nickel 200 (UNS N02200) (99.5%Ni+Co) jumper strip was 0.00015 to 0.00030 in. thick. The terminal was made of 0.004-in.-thick alloy C11000 with a 0.003-in.-thick coating of BCuP-5 filler metal.

Fig. 4 Gold-plated nickel jumper strip and clad copper terminal joined by RSW. Presence of more than 2% Co in the gold plating affected weld strength adversely.

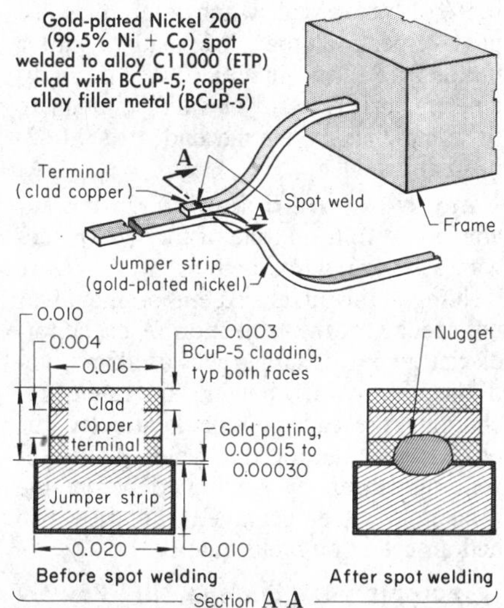

Equipment details

Welding machine	Press type, stored energy
Power supply	110 V, single-phase, 500 W
Rating of capacitor bank	1500
Electrodes	RWMA class 2, 0.071 in. diam(a)

Conditions for RSW

Welding current	2500 A
Electrode force	2.5 lb
Welding heat	20 W·s
Weld time	0.0012 s
Hold time	1.0 s

(a) Ends of electrodes were cut at 30° angle to provide elliptical faces; positive electrode was used in contact with the clad copper terminal.

When tensile-shear breaking loads from 0.9 to 4.5 lb were obtained, instead of the specified 6.5-lb minimum value, it was found that the 99.99% gold plate had been changed to a hard gold plate containing a nominal 2% Co. Tests showed that 2% Co could be tolerated, but that the plating actually contained considerably more than 2% Co. When the plating was changed back to 99.99% Au, no further difficulty was encountered in maintaining the minimum 6.5-lb tensile-shear value.

Although the mechanism by which the cobalt weakened the joint was not known, variation in the thickness of the Au-2%Co plating caused a variation in contact resistance and a nonuniform electrical response to capacitor discharge. The IACS electrical conductivities of the metals in the joint were:

- Copper, 101%
- Gold, 73.4%
- Nickel, 18%
- BCuP-5, 9.9%

Gold containing 1% Co has an electrical conductivity of 9.6%; a larger cobalt content further decreases the conductivity of the plating.

The weld nugget penetrated deeper into the nickel jumper strip than into the copper of the terminal, and there was no electrode identation on either side of the joint.

The electrodes were made of class 2 material and were 0.071 in. in diameter. The end of each electrode was cut at an angle of 30° to the centerline of the electrode, resulting in a face of elliptical shape with a major diameter of about 9/64 in. The positive terminal of the welding machine was connected to the electrode that was in contact with the copper terminal.

The high-conductivity coppers have been projection welded to silver alloy contacts successfully. The copper workpieces were cleaned and plated with tin to a maximum thickness of 0.0002 in. The projections were coined on the silver contacts. Low electrode forces were used to prevent premature collapse of the projections during heating. An increased welding force and a fast electrode follow-up were used to ensure a good weld.

Tin plating kept the workpieces clean, and the workpieces could be stored for 2 to 3 months without affecting weldability. There was a minimum amount of tin vaporization, and the tin melted and alloyed with the base metal with only a trace of flow during welding. Nickel plating on copper can serve a similar function, and plating thickness and composition must be carefully controlled.

Beryllium Copper

Beryllium copper alloys can be resistance welded most successfully in thin gages. Spot welding produces satisfactory welds; seam welding is less successful. Projection welding is satisfactory, provided that the projections can be formed with the work metal in the annealed condition and without cracking the work metal around the projection. Close control of welding conditions is required for consistent weld size and joint strength. Oxide films produced by heat treating must be removed to ensure low and consistent contact resistance. Work metals that have not been heated after rolling frequently need only degreasing before welding.

Low electrical conductivity (22% IACS for alloys C17000 and C17200) contributes to the weldability of beryllium copper alloys. However, they are more difficult to resistance weld than low-carbon steel. Alloy C17500 has an electrical conductivity of 45% IACS and is more difficult to resistance weld than higher strength, lower conductivity beryllium copper.

Resistance welding of beryllium copper requires the rapid heat input that can be provided by the short impulse length of a capacitor-discharge power supply. A low-inertia welding head is needed for fast follow-up to prevent expulsion of molten metal by arcing and to forge the weld nugget. Because capacitor-discharge equipment supplies direct current, polarity can be a factor in weld strength when beryllium copper is welded to another copper alloy or to another metal.

In the following examples, beryllium copper was resistance welded either to itself or to a dissimilar metal. Capacitor-discharge welding was employed in both examples to produce high heat for a short time, which is needed to minimize diffusion of heat.

The size and location of the spot weld and the strength of the welded joint were important in obtaining maximum service life for the switch component described in the following example. Weld quality was maintained by setting up the welding machine carefully and by frequent electrode maintenance.

Example 5. Control of Welding Machine Setup and Electrode Maintenance To Reduce Rejection Rate in Spot Welding. The snap-action switch element shown in Fig. 5 was produced by capacitor-discharge spot welding a blade made of alloy C17200 (beryllium copper)

Fig. 5 Switch element that was made by capacitor-discharge RSW. Beryllium copper blade was joined to a cartridge brass terminal. Rejection rate was reduced by careful control of welding machine setup and electrode maintenance.

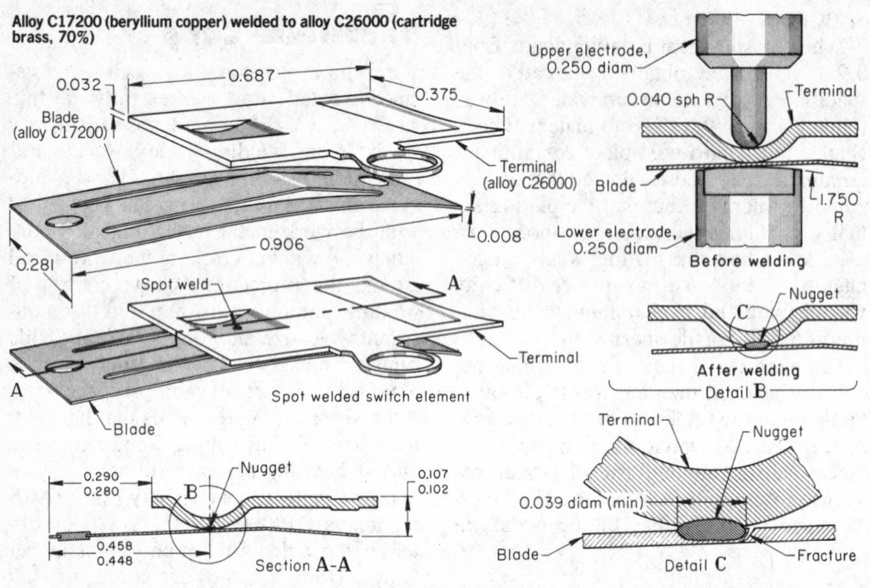

Equipment details

Power supply	110 V, single phase 60 cycle, 20 A
Welding machine	Press type, stored energy
Rating of capacitor bank	400 μF
Storage voltage	100 to 1500 V, dc
Energy storage:	
Low	2 to 50 W · s
High	20 to 500 W · s
Output voltage	1.5 to 25 V
Electrodes	1/4 in. diam, RWMA class 2
Upper-electrode face	0.040-in. spherical radius
Lower-electrode face	1.750-in.-radius crown, 0.125 in. wide

Conditions for RSW

Welding current	8000 A
Welding voltage	12.5 V, dc
Welding energy	250 W · s
Electrode force	22 lb
Pulse time	2.5 ms
Production rate	300 welds per hour

to a terminal made of alloy C26000 (cartridge brass, 70%). The location and size of the spot weld and the strength of the welded joint were critical in the performance and life of the switch.

If the weld joining the blade and terminal was too large, the blade failed prematurely by fatigue because of hardening in the heat-affected zone (HAZ). As shown in Fig. 5 (detail C), the fracture developed in the HAZ.

Mislocation of the weld either caused the blade not to have the snapping action, or changed other operating characteristics of the switch unacceptably. A riveted joint could not be used because the rivet hole in the blade so weakened it that the blade failed to have the snap action.

The rejection rate for the welded assemblies in snap-action testing sometimes was 25%, which could not be tolerated. Variations in the work-metal properties and in the welding equipment were at first suspected of causing poor switch life. Finally, the test failures were reduced to 5% by careful attention to the setup of the welding machine and to the shape and surface condition of the electrodes.

The electrodes were 1/4 in. in diameter and were made of class 2 material. The end of the upper electrode was reduced in diameter and had a 0.040-in. spherical radius; the lower electrode had a 1.750-in.-radius crown 0.125 in. wide.

During machine setup, the upper electrode was carefully clamped in its holder, and the upper electrode holder was lowered and temporarily locked so that a plate cam on the holder contacted the cam follower of a limit switch with the switch in the closed position. This switch actuated the capacitor-discharge welding current. The lower electrode was aligned, raised until it touched the upper electrode, and then clamped in its holder. The spring that provided the electrode force was then compressed so that it would exert a force of 22 lb on the workpieces during welding.

The components were held in a fixture during welding. The blade nest was adjusted to give the blade firm contact with the lower electrode, and the terminal nest was similarly adjusted for proper contact of terminal and blade so that the blade could not flex when force was exerted on the terminal by the upper electrode.

The electrodes were cleaned with crocus cloth after welding every 25 assemblies. When pitting developed on the lower electrode surface, the electrode was removed for refinishing. Refinishing of electrodes usually was a successful remedy when poor welds were produced.

The welded assemblies were subject to 100% inspection for snap action. The weld nugget was to be a minimum of 0.025 in. from the edge of the center leg, and molten metal formed during welding was to be confined between the blade and the protrusion on the terminal. If molten metal was ejected from this area, malfunction of the switch was possible. After welding, the blade surface had to be smooth and free of pitting; pitting would reduce service life. Also, the weld nugget was required to be strong enough to withstand 0.23 lb-in. of torque without permanent set in a torsion-shear test.

Destructive tests, in which a blade was peeled from the terminal, were made on 0.5% of the welded switch assemblies. To pass inspection, the blade had to pull a button 0.0012 in.^{2} in area (0.039 in. in diameter), or greater, yet small enough to fit completely under the end of a 0.062-in.-diam rod.

Improving Weld Quality. In the example that follows, one of the workpieces was a nickel wire, and as in crosswire welding, a manufactured embossment was not needed for the projection. A capacitor-discharge power supply was used and, until the system was changed, the rate of recharging the capacitor bank affected the production rate, or the weld quality, if welding current was initiated before the capacitors were recharged. Polarity also had an effect on weld quality.

Example 6. Change in Power-Supply-and-Control System To Reduce Number of Faulty Welds in Projection Welding. The gate module shown in Fig. 6 was used in a telephone thin-film circuit pack, and consisted, in part, of an electronic circuit deposited on a small glass substrate with 11 beryllium copper (alloy C17200) clips 0.005 in. thick that served as electrical leads. The beryllium copper clips were joined to 0.023-in.-diam flash gold-plated nickel wires that were in a phenolic molded body. The completed module was subsequently installed on a printed circuit board. A stored-energy (capacitor-discharge) welding machine was used for the operation.

Fig. 6 Gate module that was assembled by capacitor-discharge resistance projection welding nickel wires to beryllium copper clips. Change in supplying and controlling power reduced the number of faulty welds.

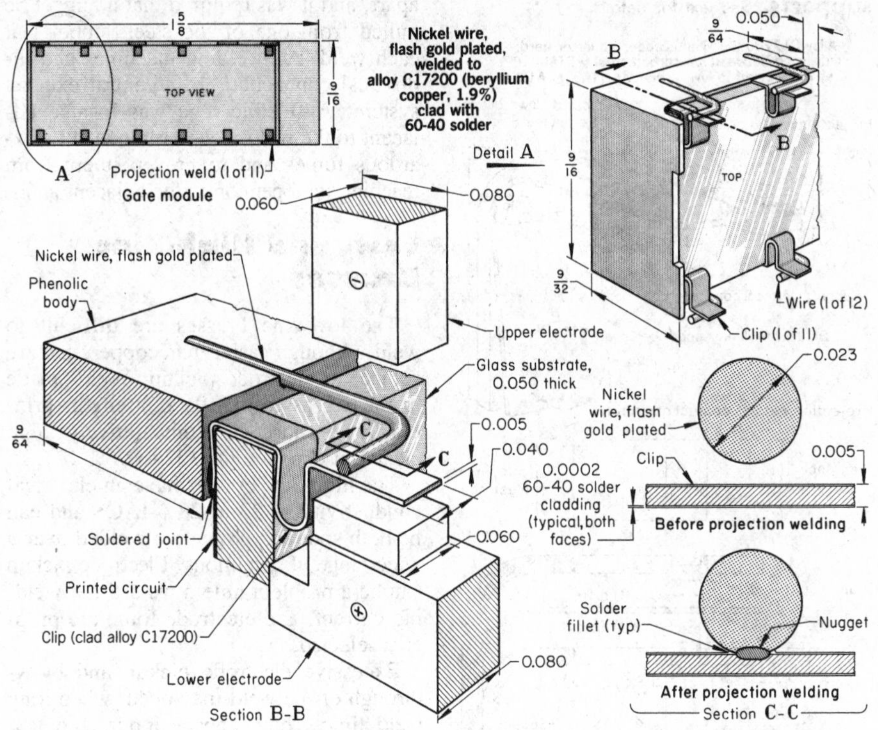

Equipment details	
Power supply	120 V single phase, 60 cycle, 6 A (max)
Welding machine	Bench type, stored energy, with foot-operated low-inertia head
Rating of capacitor bank	14, 28 or 56 μF
Storage voltage	675 to 1500 V, dc
Secondary current, max	3000 A
Energy storage	0-63, 0-32 or 0-16 W · s
Pulse width	0.0018 to 0.0166 s
Electrodes	RWMA class 2(a)
Electrode force, max	15 lb

Conditions for RSW	
Welding current	1900 A
Welding voltage	2.8 V, dc
Welding energy	27 W · s
Electrode force	10 lb
Pulse time	0.005 s
Production rate	1500 welds per hour

(a) Upper electrode was 3/16 in. in diameter by 1 1/2 in. long, with a 0.060-by-0.080-in. face; lower electrode was a bar 0.080 by 3/8 by 1 3/8 in. long, with a 0.060-by-0.080-in. contact surface. Weld strength was significantly greater when the positive electrode was used in contact with the beryllium copper clip.

The beryllium copper clips were formed and mechanically attached to the glass substrate and soldered to the thin-film circuit, so that each clip was aligned and held in position during the clip-to-wire welding operation. The substrate and phenolic subassemblies were positioned by hand with the nickel wires and the beryllium copper clips overlapping for welding as shown in Fig. 6. The wire and clip were then manually positioned between the welding electrodes to make each of the 11 welds needed to complete the assembly.

The beryllium copper used for the clips was coated with 60%Pb-40%Sn solder to a thickness of 0.0002 in. The weld nugget undoubtedly contained some solder from the coating on the beryllium copper clip and some gold from the plating on the nickel wire. However, it would have been difficult to isolate the actual nugget metal because of the solder fillet that surrounded the nugget and joint. If the solder coating was too thick, welding heat was inadequate, but with the coating thicknesses used, the solder and gold did not affect the strength or ductility of the joint. Welding response was determined largely by the thermal conductivity of the beryllium copper, which required that the welding-current pulse be delivered quickly to minimize diffusion of heat.

The short pulse length of the capacitor-discharge welding current necessitated the use of a welding head with a low-inertia, fast follow-up force system. The system used consisted of a decoupling spring between the ram and the electrode. Deflection of the spring actuated the welding switch on the downstroke when the correct welding force was reached. Because a capacitor-discharge pulse is direct current, weld properties are affected by the polarity of the electrodes during welding. The breaking load of the clip-to-wire welded joint was 13 lb or greater when the negative electrode was used against the nickel wire and the positive electrode against the beryllium copper clip, and only 4 lb when the electrodes were reversed.

Originally, the capacitor bank was discharged through a relay, and thus the welding current could be initiated before the capacitors were completely recharged. The pedal-release and relay-reset time was only 0.15 to 0.20 s, while the capacitor-recharge time was 1.0 s, so even an experienced operator, when making closely spaced welds or welds in closely spaced components, could initiate the welding current before the capacitor bank was sufficiently recharged. This led to faulty welds and it was necessary to change the system of supplying and controlling the power.

The system that was finally used was fully electronic with a capacitor-recharge time of 0.30 s. Recharge was started automatically after a 0.15-s shutoff pulse, even with the pedal depressed. In addition, a lockout circuit prevented the welding current from being initiated before the capacitors were 90% recharged. This feature reduced the number of faulty welds.

Electrodes of special shape helped to position the workpieces during welding and to control weld quality. The lower electrode positioned the glass substrate horizontally and supported the beryllium copper clip. Solder pickup was removed from the electrode faces by placing a fine emery cloth between the electrodes and reciprocating and oscillating the cloth while the electrodes were lightly closed.

The welding machine settings were established experimentally by making and testing samples, then adjusting the machine settings until the samples met the minimum strength requirements. The test consisted of holding the wire of a simulated clip-to-wire weldment and pulling the tab at right angles to the wire. A minimum breaking load of 5 lb was required in this test. An average value of about 13 lb was observed on welds made with the correct polarity. Pull tests were made approximately every 4 h. In addition, 10 welds per hour were tested by probing with a force gage that was calibrated at 200 g to verify that there had been no change in the welding conditions that could result in weak welds. Weak welds found by probing were usually repairable by rewelding.

Projection Welding. A beryllium copper blade that was welded to a steel support by projection welding had inadequate service life because of fatigue failure of the beryllium copper blade in the HAZ. A successful redesign in which the beryllium copper blade was sandwiched between two steel supports is described in the following example.

Example 7. Projection Welding Beryllium Copper to Steel. The oscillating-contact assembly shown in Fig. 7 was made by sandwiching a beryllium copper alloy C17200 blade between two supports made of cadmium-plated cold rolled low-carbon steel strip (ASTM A109), and then projection welding the supports together through holes in the beryllium copper blade so that the welds mechanically held the blade. The steel supports were 0.036 in. thick, and the beryllium copper blade was 0.010 in. thick.

Formerly, the blade was projection welded to a single steel support by use of projections in the support. The weld was well formed, but the blade had a life of only 4 to 5 million cycles, compared with the required life of 10 to 15 million cycles. Failure was caused by cracks that originated in the HAZ of the blade.

In the improved design, two projections in one steel support contacted two opposing matched projections in the other support through mating holes in the beryllium copper blade. The projections were 0.060 in. in diameter by 0.012 in. high, and the holes in the blade were 0.070 in. in diameter. There was almost no welding of the beryllium copper blade to the steel supports.

The electrodes were $^5/_8$ in. in diameter with type C (flat) faces, and were made of class 1 material. Two locating pins, made of an electrically nonconductive material and fitted into the lower electrode, passed through matching holes and notches in the workpieces (see Fig. 7) to provide positive location of the assembly. The upper electrode had matching clearance holes for the locating pins.

In operation, the three workpieces were placed on the lower electrode over the two locating pins with the blade between the two steel supports. The welding machine was then actuated to make the weld, and simultaneously a magnetic force system was actuated to forge the weld metal. The forging force closed the gap between the steel supports and expanded the plastic metal against the holes in the blade. The assemblies made in this way proved completely satisfactory, and in service, the silver contact on the blade (see Fig. 7) was worn away before the blade failed in the weld area; failure, if any, occurred elsewhere in the blade.

Fig. 7 Oscillating-contact assembly made by projection welding two low-carbon steel supports. See text for details.

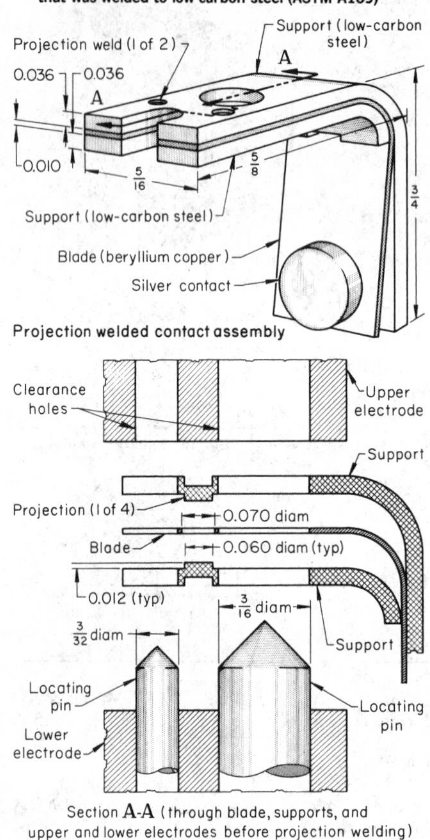

Equipment details

Power supply	480 V, single-phase, 60 cycle
Welding machine	Press type with air and magnetic-actuated electrode force
Rating at 50% duty cycle	75 kV·A
Welding current, max	30 000 A
Electrodes	$^5/_8$-in.-diam type C, RWMA class 1
Electrode force, max	Air actuated, 600 lb; magnetic actuated, not recorded
Welding controls	Synchronous, with phase-shift heat control
Fixturing	Pins fitted into lower electrode

Conditions for projection welding

Welding current	8500 A, ac
Electrode force	Air actuated, 350 lb; magnetic actuated, not recorded
Squeeze time	30 cycles
Weld time	5 cycles
Hold time	5 cycles
Production rate	360 assemblies per hour

Five welded assemblies were peel tested by an inspector during each hour. In this test the two steel supports were peeled apart, and it was required that a nugget be pulled from one of the steel supports at each weld. All welding machines had individual connections to a central exhaust system. A flexible tube was located adjacent to the weld area to remove any hazardous fumes and to prevent them from reaching the operator or the adjacent areas.

Low- and High-Zinc Brasses

The low-zinc brasses are difficult to weld, although easier than copper, and are subject to electrode pickup. Welds made in these brasses may lack strength, principally because of comparatively high electrical conductivity (32 to 56% IACS).

The high-zinc brasses have an electrical conductivity of 27 to 28% IACS and can be both spot and projection welded over a wide range of conditions. Electrode pickup can be a problem unless weld time, welding current, and electrode force are properly selected.

Excessive electrode pickup and blow-through of the weld may occur when long weld times, high energy input, and low electrode forces are used. Yellow brasses (alloys C26800 and C27000) are less susceptible to electrode pickup than cartridge brass, except when long weld times and high energy input are used. Electrode force should be sufficient to prevent arcing or expulsion of molten metal, to which these alloys are subject because of their 30 to 40% content of zinc, which boils at about 1665 °F. As shown in Table 2, the recommended electrode force, when using electrodes having a face diameter of $^3/_{16}$ in., is approximately 400 lb. For weldability ratings, see "Welding index" entries in Table 1.

In projection welding, the projections should be as small as practical, especially the height. A short weld time and a low-inertia welding head with fast electrode follow-up should be used. Tin plating the workpieces generally will improve weld quality, although expulsion of molten tin from the weld interface may occur.

In the example that follows, cartridge brass lead wires were welded to red brass terminals. Heat balance was obtained by flattening the wires, which were square in cross section and thicker than the terminals. Spot welding was replaced by projection welding so that large electrodes, which would not stick to the work metal, could be used. Projections that were carefully designed to trap weld metal and limit

Fig. 8 Cartridge brass lead wires joined to red brass terminals by projection welding

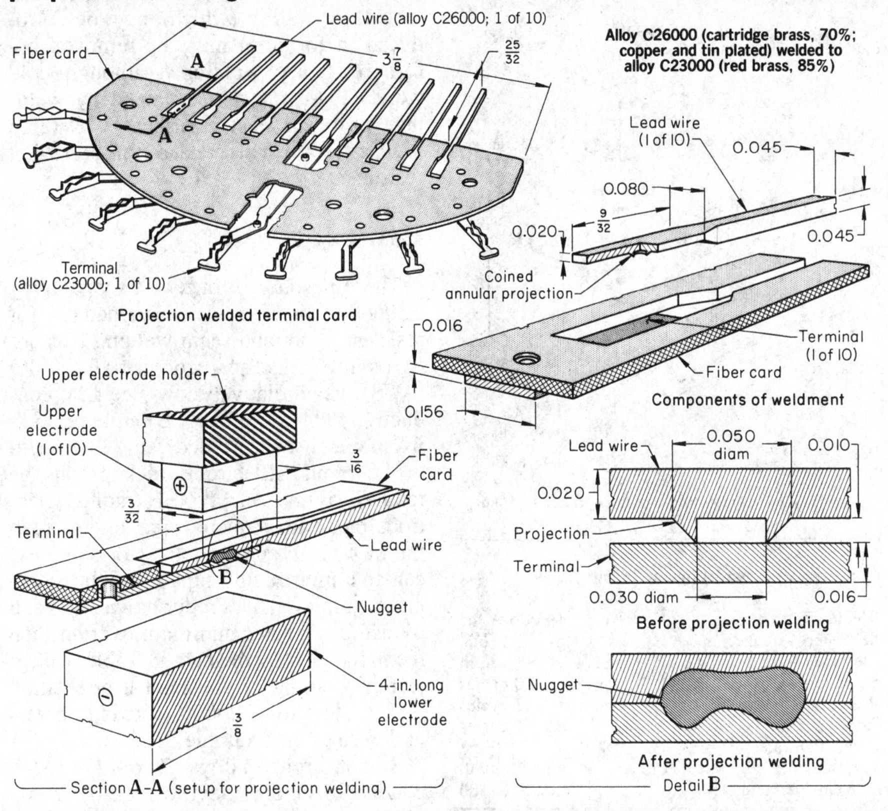

Equipment details	
Welding machine	440-V ac automatic press type, with ten welding heads and ten transformers
Rating at 50% duty cycle	20 kV · A each
Rating of secondary circuit	4.6 to 6.1 V, 2000 to 4000 A
Electrodes	RWMA class 2(a)
Controls	Synchronous, with sequential firing
Auxiliary equipment	Indexing table(b)

Conditions for projection welding	
Welding current	2000 A, ac
Welding voltage	5 V
Electrode force	50 lb
Squeeze time	5 cycles, min(c)
Weld time	1 cycle
Hold time	5 cycles, min(c)
Production rate:	
Assemblies per hour	1100
Welds per hour	11 000

(a) Ten upper electrodes, $^{3}/_{32}$ by $^{3}/_{16}$ in.; one lower electrode, $^{3}/_{8}$ by 4 in. (b) 24-stop rotary table. (c) Squeeze time and hold time varied because the upper-electrode holders were attached to the press ram, and times were determined by mechanical motions of the ram.

expulsion were coined on the flat section of the wire.

Example 8. Use of Projection Welding With Annular Projections To Eliminate Electrode Sticking and To Limit Expulsion of Weld Metal. A cartridge brass (C26000) lead wire was joined to a red brass (C23000) terminal by resistance projection welding, as shown in Fig. 8. The terminal was 0.016 in. thick, and the lead wire was 0.045 in. square and was plated with 0.0002-in.-thick copper and with 0.0002-in.-thick tin over the copper. The end of the cartridge brass lead wire was flattened to 0.020 in. thick by 0.080 in. wide by $^{9}/_{32}$ in. long, and an annular projection was coined on one surface. Coining, rather than forming, produced a projection that was strong enough to withstand the electrode force used without collapsing prematurely. The annular projection was designed to prevent expulsion of molten metal.

Originally, the 0.045-in.^{2} lead wire was resistance spot welded to the terminal without flattening the end of the wire. Spot welding was not successful, and electrode pickup from the cartridge brass wire was excessive because of the small area of contact between the electrode and the work metal.

Projections of several shapes, including a transverse bead and a longitudinal bead, were tried before the annular projection shown in Fig. 8 was developed. Besides allowing excessive expulsion of weld metal, the transverse bead had to be at the center of the electrode to prevent tilting of the workpiece and contact of workpiece surfaces elsewhere than at the projection. Circular, square, and cross-shaped projections provide much more stability than bead-type projections. The longitudinal bead also caused excessive expulsion of work metal because the volume of weld metal was too great and melting was erratic and inconsistent.

The annular projection had a steep internal slope (about 90°) and a 45° external slope so that, as the projection melted, the molten metal was confined by the remaining metal of the projection.

The welds were made in an automatic welding machine that had a 24-stop rotary-indexing table and ten welding heads, each with a 20-kV · A transformer. Electrode material was class 2. The lower electrode was made in one piece and was connected to the negative terminals of the transformer secondary circuits. Each of the ten upper electrodes was mounted to a welding head and was connected to the positive terminal of a transformer.

Ten red brass terminals were permanently mounted on a fiber card and were degreased and cleaned with a rotary wire brush after the card was placed on the indexing table. The cartridge brass wires were cleaned and lubricated by wiping with a felt pad as they were fed into the flattening, coining, and cutoff die. In the welding position, the flattened end of each of ten cartridge brass square wires extended beyond a wire-guide groove in the die and over one of the ten terminals fastened to the fiber card. A notch in the card at each terminal helped to position the lead wire. The ten welding transformers had synchronous controls and were fired in sequence.

Advantages of Projection Welding. The use of projection welding offers several advantages for the joining of high-zinc brasses, which is often done with close weld spacings and edge distances. The projection defines the size and location of the weld. There is no marking of the parts from electrodes that are misaligned, too large in diameter, or have improper face shape. Distortion and sticking of electrodes are minimized because a relatively large area of the workpiece is in contact with the projection welding electrode. Also, localized heating produces a minimum decrease in the hardness and strength of the base metal. Projection welding also permits multiple welds to be made simultaneously at a closer spacing than is possible with spot welds, as illustrated in the following example.

Fig. 9 Switch component made by projection welding a cartridge brass terminal to a phosphor bronze blade

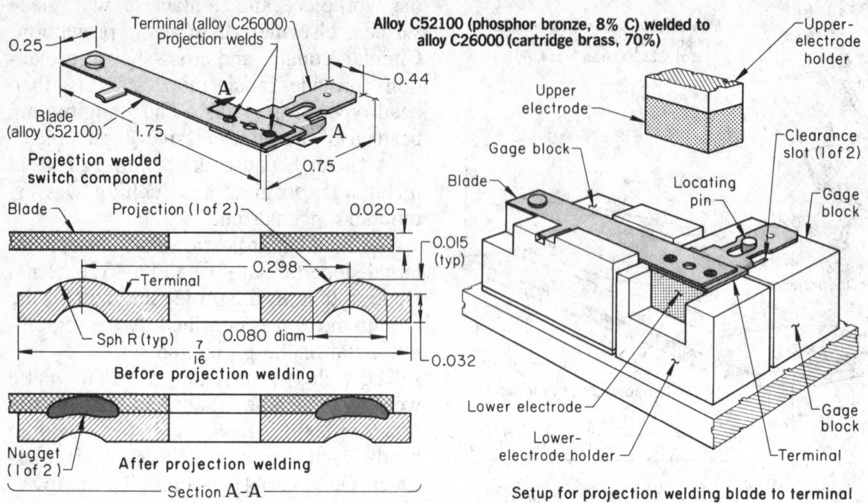

Equipment details	
Welding machine	440-V, single-phase ac press type, with synchronous controls
Rating at 50% duty cycle	30 kV · A
Current, max	22 500 A
Upper electrode	RWMA class 13
Lower electrode	RWMA class 11
Electrode holders	RWMA class 2
Electrode force, max	750 lb

Conditions for projection welding	
Welding current	19 200 A, ac
Secondary voltage	6 V
Electrode force	180 lb
Squeeze time	20 cycles
Weld time	3 cycles
Hold time	5 cycles
Production rate:	
Assemblies per hour	750
Welds per hour	1500

Example 9. Projection Welding a Cartridge Brass Terminal to a Narrow Phosphor Bronze Blade. The switch component shown in Fig. 9 was made by projection welding an alloy C26000 (cartridge brass, half hard) terminal to an alloy C52100 (phosphor bronze, spring temper) blade. The low electrical conductivity of these alloys (28% IACS for alloy C26000 and 13% for alloy C52100) made them relatively easy to join by resistance welding. Two welds were used for joining the terminal and the blade, and the use of projection welding allowed closer spacing of the welds than normally allowed by spot welding, and made both welds simultaneously.

The terminal, with two spherical projections 0.080 in. in diameter by 0.015 in. high and spaced at 0.298 in., was made in a progressive die. The blade was a flat strip with a hole at one end and a contact fastened to the other end. The slot in the terminal was used for locating the part during projection welding. Before welding, both parts were degreased and bright dipped, then carefully rinsed.

The electrodes were made of refractory metal compositions (RWMA group B) to reduce electrode pickup, and were rectangular in shape, which made them easy to clean and replace. The upper electrode was made of class 13 material (tungsten) and the lower electrode was made of class 11 material (a mixture of copper and tungsten). Both electrodes were fastened to holders made of class 2 material. A gage block supported and located the blade, and a pin located the terminal during welding.

Copper Nickels

The copper-nickel alloys have electrical conductivities of 4.6 to 11% IACS, are readily spot and seam welded with relatively low welding current, and generally do not alloy with the electrode material and cause electrode pickup. Proper preweld cleaning is required to ensure low and consistent contact resistance. Short weld times prevent electrode indentation.

Nickel Silvers

Nickel silvers, which have about the same conductivities (6 to 10.9% IACS) as copper nickels, are spot welded as readily as copper nickels (see Table 1), but are more difficult to seam weld. Surface contaminants such as lead and bismuth (which form low-melting eutectics with copper and nickel) or sulfur (which may be introduced in forming) must be removed before resistance welding. Although cold worked base metal is softened by welding, postweld heat treatment is seldom used except when needed for corrosion resistance.

Bronzes

The phosphor bronzes, except alloy C50500, which is not recommended for resistance spot and seam welding because of its high electrical conductivity (48% IACS), have relatively low electrical conductivity (11 to 20% IACS) and are readily spot and seam welded using low welding currents. Electrode pickup can be reduced by use of a type F (radius) electrode face and frequent redressing to keep the face clean and smooth. Hot shortness can be minimized by supporting the workpieces to prevent strain during welding and by using a greater minimum overlap than recommended by the data in Table 3. Projection welding of phosphor bronze alloy C52100 to alloy C26000 (cartridge brass) is described in Example 9.

Silicon bronze alloys (7 to 12% IACS conductivity) are the most easily resistance spot and seam welded of all copper alloys. Low welding current and low electrode force (see Table 2) usually are required. Short weld times should be avoided, to prevent shrinkage voids. The surface oxides that develop during annealing must be removed to ensure low and consistent surface contact resistance.

Resistance welding of aluminum bronze is similar to resistance welding of silicon bronze, but with much more electrode pickup and expulsion of weld metal, which can be controlled by careful adjustment of weld time, welding current, and electrode force.

Safety

In resistance welding of copper alloys, a ventilation system may be needed because many copper alloys contain at least small amounts of toxic alloying elements. However, the need for ventilation is less critical for resistance welding operations than for arc welding operations, because a smaller volume of metal is heated and a smaller volume of fumes is generated.

Detailed information and references on safe practices, cautions, contaminants, and hazards in welding are presented in ANSI Z49.1, "Safety in Welding and Cutting."

Flash Welding

By the ASM Committee on Flash Welding*

FLASH WELDING commonly is used to join sections of metals and alloys in production quantities. It is a resistance/forge welding process in which the items to be welded are securely clamped to electric current-carrying dies, heated by the electric current, and upset. Clamping ensures good electrical contact between the current-carrying dies and the workpiece and it prevents the parts from slipping during the upsetting action. Flash welding equipment must be durable to withstand high clamping force and upset pressures without deflecting. If deflection occurs, misalignment of workpieces may occur during welding. Flash welding is rapid and economical, and when properly executed, welds of uniform high quality are produced. A typical flash welding machine comprises a horizontal press-transformer combination with conducting work-clamping dies mounted on the press platens.

Earlier machines were manually operated; the operator moved the press platen through a long lever, initiating and interrupting current flow with a foot switch. To make a weld, the operator energized the transformer and brought the pieces to be joined into light contact, so that localized melting would occur at a few spots. Molten metal was violently expelled, and heat flowed back into the workpiece from the molten layers on the surfaces. The operator continued to move the platen to sustain the violent action to heat the workpiece. When a sufficiently high temperature had been reached some distance from the abutting surfaces, the operator rapidly moved his lever to extinguish the arcing, push out the molten and overheated metal on the surfaces in the form of flash, and complete the weld by forging the two pieces together while interrupting the current, thus making a solid-state weld.

Early applications included welding joints in wagon wheel tires, joining coil ends in steel mills, and making fifth wheels for wagons. Later, the automotive industry employed the process for pneumatic tire rims and starter ring gears.

World War II provided new applications in the large-diameter bearing races for tank and armored vehicle turrets. These applications, because of the alloys used, required fully automatic machines. Airframe structural members were also produced by flash welding. Jet engines and space vehicles are newer applications that require sophisticated machines and rigid quality control. In its earliest application, flash welding replaced the blacksmith's forge fire welding. It is a preferred method of welding because little or no joint preparation is necessary, and a properly made flash weld does not include any entrapped foreign material or molten metal.

Although the process has remained essentially the same since its inception, major advances include automated control, increased knowledge of the interactions of process variables and their influence on the metals being welded, and improved weld quality. The most significant advancement is the development of flash welding machines with adaptive controls. During the welding process, adaptive controls sense various process parameters, such as voltage, current, weld energy, platen speed, and/or platen force, and compare these parameters against preselected reference signals. After comparison is made between operating and reference signals, error signals are generated, which in turn modify the welding process. Utilization of adaptive controls has not only greatly increased the number of applications and quality of flash welding, but has also resulted in less material being consumed during welding when compared with manual and semiautomatic processes.

Applications

Flash welding can be used for joining many ferrous and nonferrous alloys and combinations of dissimilar metals. In addition to low-carbon steels, metals that are flash welded on a production basis include low-alloy steels, tool steels, stainless steels, aluminum alloys, magnesium alloys, nickel alloys, and copper alloys. Commercial alloys that have been successfully flash welded include:

- *Low-carbon steels*: All types
- *Carbon and alloy steels*: AISI 4130, AISI 4140, AISI 4150, AISI 4340, AISI 4337, AISI 52100, AISI 8740, AISI 8630, Jethete, EN 16, EN 8 series, Chromalloy
- *300 series stainless steels*: 304, 309, 310, 314, 316, 321, 347
- *400 series stainless steels*: 403, 410, 422, 430
- *Other stainless steels*: AMS-5616 Greek Ascoloy, S110, S129, S130, RR517, EN 56, EN 58, Jessops G 88, Jessops G 183, FV 448, FV 535
- *Corrosion- and heat-resistant alloys*: Inconel 600, Inconel 625, Inconel 718, Inconel X-750, Inconel 706, Inconel 700, A286, 17-4 PH, PH 15-7, CG-27, AM 355, Hastelloy X, Hastelloy N, Hastelloy C, Hastelloy S, Hastelloy W, L605, HS-188, Incoloy 801, Waspaloy, René 41, 19-9DL, 19-9DX, U-500, N-155, D6AC, Nimonic 75, Incoloy DS, Nilo K, Nimonic 75, Nimonic 80A, Nimonic 90, Nimonic 105, Nimonic PE

*Lawrence W. Weller, Jr., Engineering Supervisor of Welded Rings, The American Welding & Manufacturing Co.; Robert H. Foxall, Assistant Sales Manager, Wean United, Inc.; I.A. Oehler, retired

7, Nimonic PE 11, Nimonic PE 13, Nimonic PE 16, Nilo K, Nimonic C263, Nimonic C475, EPK 21, EPK 33, Inconel W

- *Titanium alloys*: Ti-commercially pure, Ti-6Al-4V, Ti-6Al-2Sn-4Zr-2Mo, Ti-8Al-1Mo-1V, Ti-5Al-2.5Sn, Ti-6Al-2Sn-4Zr-6Mo, Hylite 50, Titanium 230, Titanium 317, Titanium 318A, Titanium 684
- *Aluminum alloys*: 2017, 2024, 2014, 6061, 7075, 6063, 6033, 5083
- *Other nonferrous alloys*: Copper, yellow brass, magnesium alloys

In general, copper alloys with a high zinc content weld better than those with a low zinc content; the high-leaded copper alloys may give brittle welds. The ductility of welds in some copper alloy rods allows the rods to be flattened under repeated hammering to approximately 15% of the original thickness without indications of fracture at the weld.

Flash welding of lead, zinc, tin, bismuth, and antimony, and alloys of these metals, is impractical. Toxic metals, such as beryllium and beryllium alloys, are not suitable for flash welding without stringent precautionary measures to prevent contamination of air and equipment and danger to personnel by flash particles and dust.

Titanium alloys can also be flash welded. An inert gas shield to displace air from the joint is sometimes used to minimize embrittlement.

Many combinations of dissimilar metals are easily flash welded if the proper welding conditions and workpiece design are used. Table 1 lists combinations of metals that have been joined successfully by flash welding.

A wide variety of cross-sectional shapes can be flash welded. Typical flash welding applications include:

- Sheet in steel mill pickling lines, cold reduction lines, and recoiling lines
- Strip
- Plate
- Rolled or extruded bar
- Tube and pipe
- Chain
- Rings for aircraft engines and missiles
- Automotive wheel rims (steel and aluminum)
- Miter welds for windows and doors
- Unequal sections and joining different alloys, such as carbon steel to tool steel for drills
- Railroad rails
- Landing struts for airplanes
- Wire
- Bearing races (steel and aluminum)

Table 1 Combinations of base metals that have been flash welded

	Metals that have been flash welded to base metals listed in the first column										
Base metal	Al alloys	Cu alloys	Mg alloys	Mo	Ni alloys	Carbon and alloy steels	Stainless steels	Tool steels	Ta	Ti alloys	W
Aluminum alloys	X	X	X	...	X	...	...	...	...	...	...
Copper alloys	X	X	X	...	X	X	X	X	...	X	...
Magnesium alloys	X	X	X	...	...	...	...	...	...	...	...
Molybdenum	...	X	...	X	X	X	X	X	...	...	...
Nickel alloys	X	X	...	X	X	X	X	X	X	...	X
Steels, carbon and alloy	...	X	...	X	X	X	X	X	X	X	X
Steels, stainless	...	X	...	X	X	X	X	X	X	X	X
Steels, tool	...	X	...	X	X	X	X	X	X	...	X
Tantalum	...	X	...	...	X	X	X	X	X	...	...
Titanium alloys	...	X	...	...	...	X	X	...	...	X	...
Tungsten	...	...	...	...	X	X	X	X	...	...	X

Welding Parameters

Fundamentally, flash welding involves heating the ends of the pieces to be welded and subsequently forging them together. During the heating phase, a thermal distribution pattern is established along the axial length of the pieces being joined, which is characterized by a steep temperature gradient. The highest temperatures occur at the abutting surfaces (flashing surfaces) and rapidly decrease in the area adjacent to where the material is clamped in the machine. In this area, the temperature is at, or near, ambient temperature. The major difference between the temperature pattern developed in flash welding and that developed in resistance welding is that flash welding produces a much steeper thermal gradient. This steep thermal gradient, combined with the resulting characteristic upset pattern, enables flash welding to accommodate a much greater variety of materials and shapes than can be welded by resistance welding.

Once the proper temperature distribution pattern has been established, the abutting surfaces are rapidly forced together. The parts must be securely held together during the forging process to prevent slipping. Three distinct peaks are characteristic of flash welds (Fig. 1). The two peaks on either side of the weld line represent the material displaced by the upsetting action; the center peak is the molten metal extruded out of the weld, including oxides or contaminants formed during heating.

The weld is completed in the plastic region, some distance from the original flashing surface. Thus, a properly performed flash weld should be free of cast metal, as well as any oxides or contamination formed during the heating process. The cleaning process which occurs during upset is so complete that reactive metals, such as aluminum and titanium, can be welded without any protective atmosphere. This fact has been verified by microprobe analysis across the weld joint. The actual bond interface in a flash weld is only a few thousandths of an inch wide and, in some cases, can be virtually eliminated by subsequent thermal heat treatment or by mechanical working of the weld.

Weld properties are as good as, or nearly as good as, base-metal properties. Other than the mode in which the heat for welding is generated, the flash weld (which uses electric energy for heating) is analogous to the stored energy or conventional friction weld (which uses friction from a mechanical system for heating).

Heat (Energy) Sources

One of the major considerations for flash welding is the electrical power service. Flashing is a term used to describe the major heating process during flash welding. When the ends of the workpiece are brought together under light pressure, an electrical short circuit is established through the material. Because the abutting surfaces are not perfectly matched, the short circuit current flows across the joint only at a few

Fig. 1 Typical peaks and flow lines in a flash weld

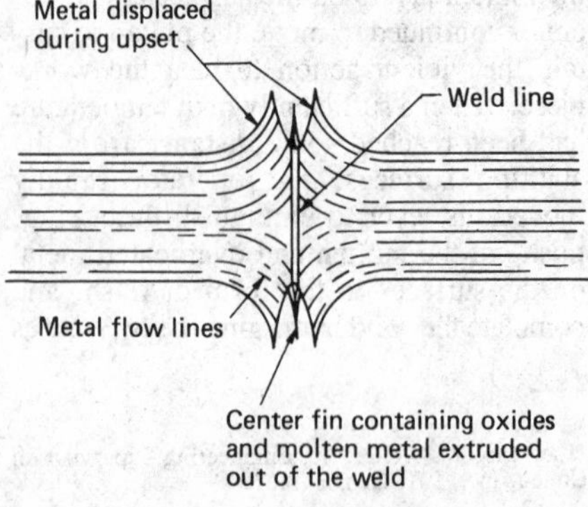

Fig. 2 Effect of energy input on bridges produced during flash welding

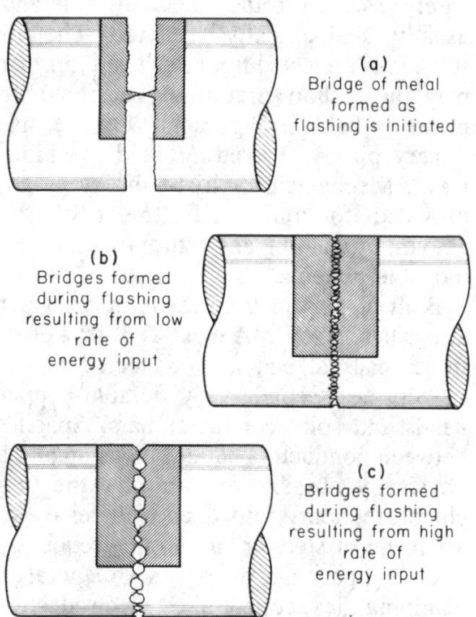

small contact areas. The large amount of current flowing through a relatively small area causes very rapid heating to the melting point. Heating is so rapid and intense that molten metal is expelled explosively from the joint area. Following this explosion, a brief period of arcing occurs. Arcing is not sustained due to the low voltages employed. Studies have shown that stable flashing can be maintained with as low as $2^1/_2$ to 3 V, but typically the flashing voltage for alternating current machines is from 5 to 10 V.

Following the expulsion of molten metal and subsequent arcing, small craters are formed on the ends of the abutting surfaces. The pieces are steadily advanced toward one another, and other short circuits are formed and additional molten metal is expelled. This process continues as random melting, arcing, and expulsion occur over the entire cross-sectional surface. During flashing, many active areas are in various stages of this sequence (Fig. 2). Flashing surfaces act as heat sources, and the steep thermal profile is established primarily from these heat sources. Temperatures of the flashes are at or above the melting point of the material and are progressively lower as distance progresses from the flashing surface toward the clamp (Fig. 3). Some resistive heating effect exists due to the current passing through the material. This I^2R type of heating has only a secondary effect during flashing. If the pieces are advanced too rapidly during flashing, the entire cross section shorts out, and stable flashing stops. This condition is called a "freeze," "stick," or "butt-up." If pieces are advanced too slowly, insufficient flashing occurs, and the required thermal profile is not established.

The higher the voltage used, the easier it is to initiate and maintain flashing. This may cause, however, very coarse flashing, and deep craters may be formed (Fig. 2). These deep craters trap molten metal and oxides that are not completely removed during upset, thus causing a poor-quality weld. Therefore, the lowest voltage that will sustain stable flashing should be used (Fig. 2). Initiating flashing with a low voltage is difficult. This problem can be overcome by the following procedures:

- If the flash welding unit is equipped with a reduced voltage control, flashing can be initiated at a higher voltage and then reduced during the cycle.
- The ends of the parts being welded can be chamfered or tapered, thus producing a smaller cross section and a higher current density to initiate flashing.
- Prior to flashing, the ends of the workpieces can be brought together under constant pressure, thus shorting out the workpiece. Current flows through the workpiece, and resistive heating, or preheating, occurs. This resistive heating is not uniform due to the quality of the abutting surfaces; consequently, localized hot spots may form. To compensate for this, the short circuit is maintained for only a brief time; the workpiece ends then are separated to allow the heat to diffuse through the part. This process may be repeated several times until the desired preheat is established. Preheating can be performed either manually or automatically.

Fig. 3 Temperature gradient of a typical flash weld prior to upset

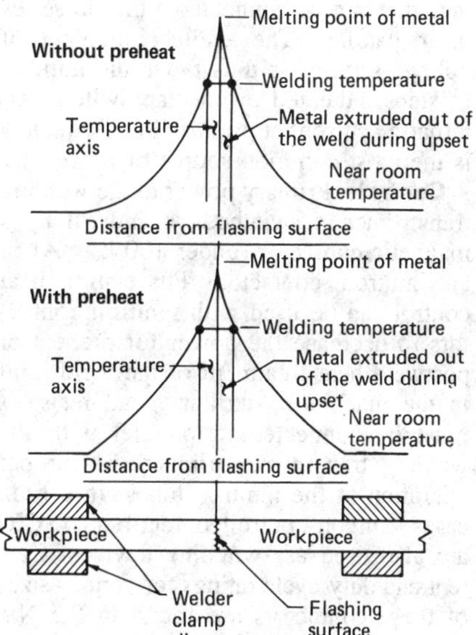

Once flashing action is established, workpieces are advanced toward one another at a rate that maintains flashing action and compensates for the material lost due to the expulsion of molten metal. The nature of this flashing action, i.e., coarse versus fine, has a direct relationship on the thermal profile established and thus the final weld quality. Flashing action can be controlled either manually or automatically. A high-current, low-voltage source of electrical power must be available for successful flash welding. Most flash welding units receive power through a single-phase alternating current welding transformer. Due to the high currents required and the short circuits created during the process, this equipment must be located where sufficient input power is available. With the development of high-current, solid-state devices, other power sources have been developed, each claiming certain advantages over the single-phase alternating current source. Alternative power supplies include:

- Three-phase power supplies
- Direct current supplies
- Frequency converters and/or wave-shaping power supplies

These power supplies can supply power at frequencies other than line frequency or can supply square wave power rather than sine wave.

Force

Parts to be welded must be clamped or fixtured securely to provide good electrical contact with current-carrying dies and to transmit the upset forces. Also, a reliable source of force for the upsetting action is required. Systems for generating these forces vary greatly and are determined by the cross sections to be welded. The simplest welding machines derive their power for clamping and upsetting from the operator, i.e., the parts are clamped to the dies via screw, lever, cam, or toggle force multiplying linkages. Upset forces are generated in a similar manner. In larger machines, these forces are generated by pneumatic systems, oil hydraulic systems, combination air/oil hydraulic systems, or motor-driven cam systems. In selecting a machine for a given application, the method of generating clamping and upset forces, as well as workpiece fixturing, cost,

Fig. 4 Flash welding unit with alligator-type clamps and hydraulic upset cylinder. Courtesy of Wean United, Inc.

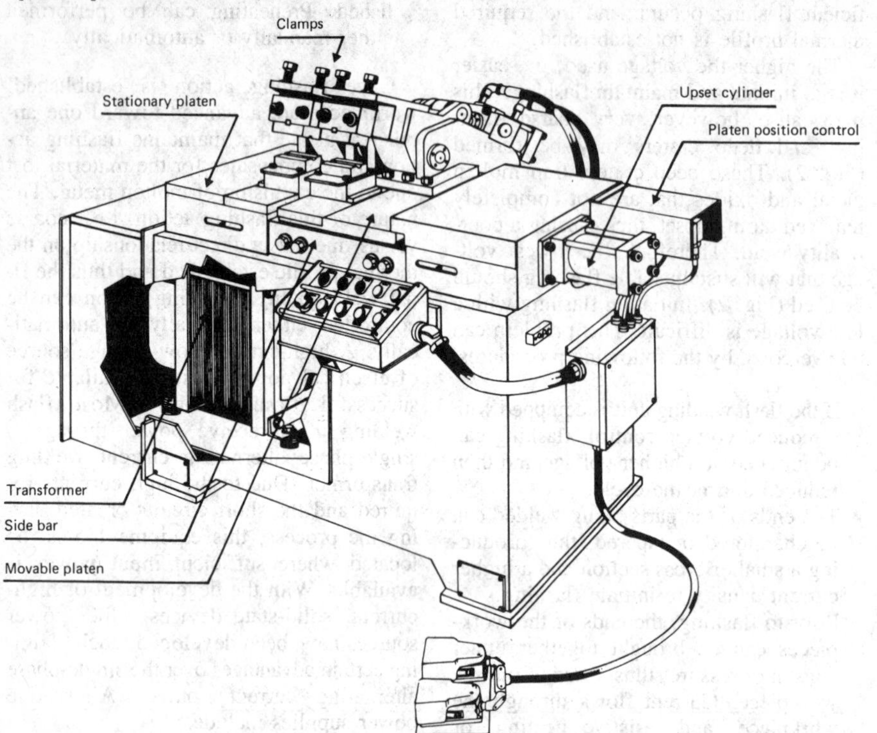

and productivity, must be considered carefully.

Machine Design

Machines used for flash welding generally consist of a mainframe, a low-impedance welding transformer, a stationary platen, a movable platen on which clamping dies, electrodes, and other tools needed to position and hold the workpieces are mounted, flashing and upsetting mechanisms, and the necessary electrical, air, or hydraulic controls (Fig. 4).

Flash welding machines may be manual, semiautomatic, or fully automatic. With manual operation, the operator controls the speed of the platen from the time flashing is initiated until the upset is completed. In semiautomatic operation, the operator manually initiates flashing and then actuates an automatic cycle that completes the weld. In fully automatic operation, after the operator initiates the welding sequence, a fully automatic cycle can be used through the use of position-indicating devices and timers. Values for the various welding parameters are preselected by the operator. Automatic feedback control is used in some applications. These fully adaptive circuits vary from current to voltage feedback circuits, and data are obtained to control the welding sequences. Figure 5 shows various adaptive controls used by different flash welding unit manufacturers.

Transformers are used to supply electrical energy for the welding operation. The transformer tap switch frequently is a rotary eight-step, knife-type, fully enclosed locking switch, which provides convenient adjustment of the welding voltage to suit work requirements. Voltages from 4 to 16 V are common. Welding transformers are available with primary windings that can be connected either in series or in parallel. The available number of voltage settings is thus twice the number of steps indicated on the tap switch. Alternating current at primary line frequency is the most common source of power.

Controls. Primary power to the welding transformer is switched on and off by a magnetic contactor (under 100 kV · A) or an ignitron contactor. Phase-shift heat control can be used with ignitron contactors to decrease the power for preheat or postweld heat treatment of the weld joint in the machine. Auxiliary load resistors must be connected in parallel with the welding transformer primary for proper operation of the ignitron tubes. In certain cases, silicon-controlled rectifiers (SCR) are also used as switching devices. Current and duty-cycle ratings for various sizes of these contactors are found in the National Electrical Manufacturers Association (NEMA) Standards Publication Part ICS 5-005, revised October 1979.

Platens. The stationary platen is securely fastened to the machine frame and usually is electrically insulated. The insulation is protected from flash and dirt that may cause short circuiting. The movable platen is held in alignment with the stationary platen by bearings and machineways. Mechanical or hydraulic stops are provided for quick adjustment of the amount of welding and return travel of the movable platen.

Both platens may be made of a copper alloy such as RWMA class 2, 3, or 4 electrode material and are connected to the transformer secondary by flexible copper bands and solid conductor bars. Spacing between conductors should be as small as physically possible to minimize the machine secondary impedance. Platens also are made of steel or cast iron—frequently with copper alloy inserts. Generally, clamping dies, copper alloy electrodes, or dual-purpose clamping electrodes are mounted on the platens. Steel sometimes is used for welding electrode material in some aluminum welding applications.

Flashing and Upsetting Mechanisms. Flashing and upsetting forces may be applied manually, mechanically through a cam, or hydraulically during flash welding. Manually operated mechanisms consist of a movable platen operated through a hand lever and toggle arrangement. The toggle mechanism is attached to and insulated from the movable platen. One of the toggle links is a turnbuckle that is used for adjusting the final die opening. Initial die opening is adjusted by a screw-type stop. A pointer on the operating lever indicates the amount of material that is flashed off and upset on a stationary graduated dial. Movement of the lever permits preheating (if desired), flashing, and upsetting, in one continuous operation. The operator controls the amount and variation of upset pressure and platen speed used in the forging action (Fig. 6).

A motor-driven cam is used to move the platen when higher production and greater consistency among welds are needed than are possible with manual operation. The cam is operated by the motor through a gear-speed reduction unit with a single-speed or variable-speed drive, or by a variable-speed motor and clutch arrangement. Closing and locking of the clamping dies and electrodes are accomplished either manually or by an air or hydraulic cylinder.

Three rotating cams, arranged as shown in Fig. 7, control the operation of the welding machine. Forward movement of

Fig. 5 Adaptive controls for flash welding

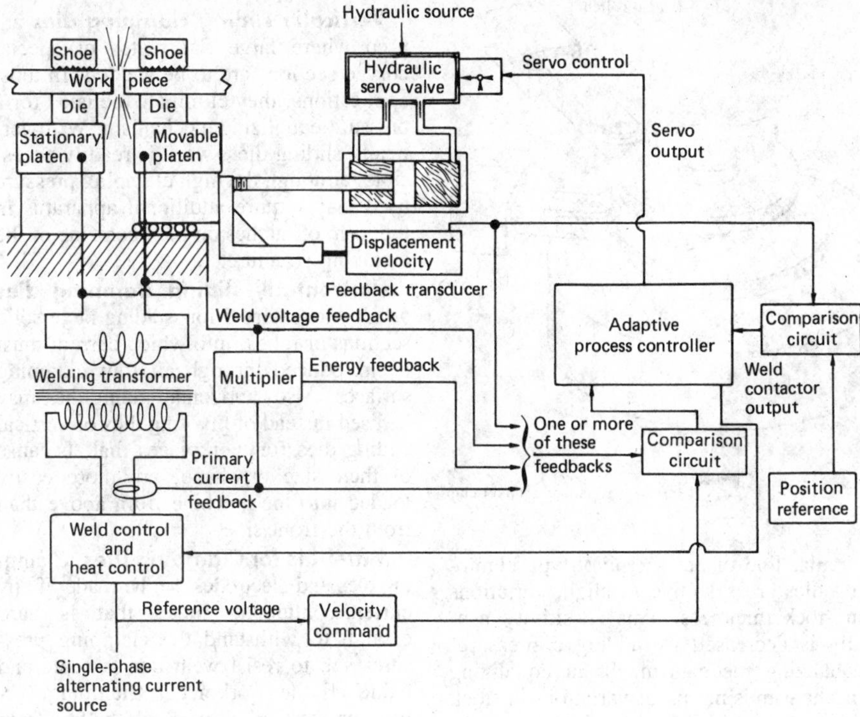

Fig. 6 Manually operated flash welding unit. Courtesy of Berkeley-Davis

the platen and its velocity are controlled by the contour and the rotational speed of the flash and upset cam. The welding current switch cam turns the power on and off at the correct time, relative to the position of the flash and upset cam. The motor-switch cam shuts off the power to the motor or to the clutch at the correct time; this action is relative to the position of the flash and upset cam.

Motor-type flash welding machines are automatic, are the simplest to use, and are predominantly single-purpose, high-production machines. When a general-purpose flexible machine is needed that can weld sections requiring more upset pressure than is obtainable with manual or motor-type machines with a high degree of weld consistency at relatively high production rates, a machine with hydraulic platen actuation usually is selected. These machines generally are equipped with hydraulically actuated clamps. Machines with hydraulically actuated platens can be completely automatic, semiautomatic, or manual in operation. Completely automatic operation requires little operator skill; semiautomatic and manual operation demand increasing degrees of skill.

Machine Selection. Different metals have different strengths at flash welding temperature, and in each application, welding current and upsetting capacity of the machine must suit the properties of the metal being welded. Each machine also has mechanical limitations. Clamping space determines the maximum size of workpiece that can be welded, and clamping area clearance determines the size of part or assembly that can be properly clamped. Size and shape of an assembly to be welded may not be compatible with the design of the machine. For example, if four tubes are to be welded 90° apart to an X-shaped forging, three of the tubes can be welded to the forging with the usual equipment, but a special clamping mechanism with suitable clearance is needed to make the fourth weld.

Clamping Dies and Fixtures

Workpieces must be accurately clamped to maintain alignment, to allow the secondary current to pass into the workpieces, and to apply the upsetting force properly. Generally, the parts of the clamping mechanism that actually grip the workpiece are the electrodes, often called clamping dies. The surface of the material in contact with the dies must be cleaned of scale, rust, paint, or contaminants to ensure proper current flow. Fundamentally, the clamping mechanism allows the

Fig. 7 Rotating cams used in flash welding machine. See text for details.

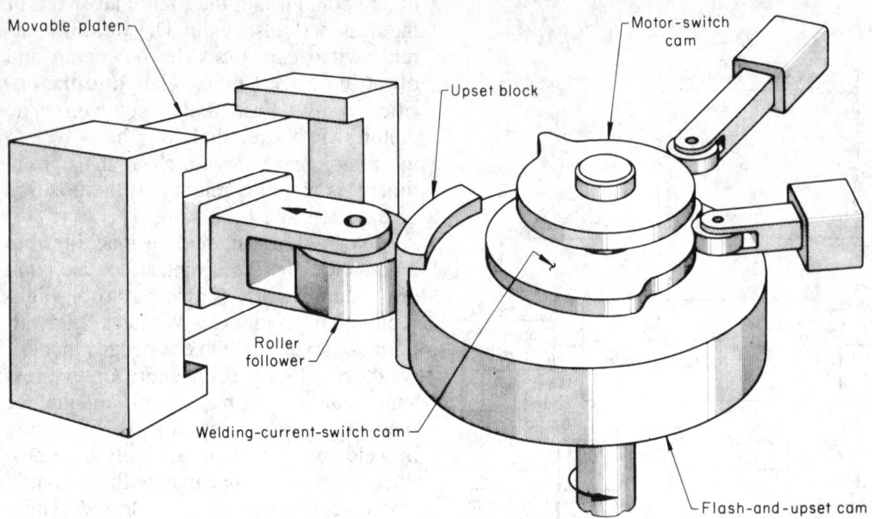

dies to move horizontally or vertically. If the clamping dies move in a straight or curved line nearly parallel to the plane of the platen top, the dies are designated as horizontal. If the dies move in a straight or curved line at right angles to the plane of the platen top, the dies are designated as vertical.

In most machines, the principal moving clamping-die member either slides or pivots. Many workpieces can be held equally well in either type of clamping die. Pivot-type clamping dies, because of their mechanical advantage, exert more force on the workpiece than do sliding clamping dies. Use of fixtures, in addition to or in place of clamping dies, is governed by the size and shape of the workpiece and by the location of areas that must be aligned after welding.

Pivot clamping dies, or alligator clamps, essentially are first-class levers pivoted near one end; the force is applied at one end, and the clamping-die half is attached at the other end. Because of their circular line of action, pivot-type clamping dies are sensitive to slight variations in stock thickness. This sensitivity usually is decreased by adding a pressure-equalizing mechanism. If an equalizing mechanism is not used, variations in stock thickness cause variations in the contact area between work metal and electrode and may damage the contact surfaces because of insufficient area for passage of the secondary current. If one workpiece is clamped incorrectly because it is slightly thicker than the other workpiece, a poor weld can result from the twisting motion that occurs during upset.

When variations in stock thickness are slight, pivot-type clamping dies have an advantage for production purposes in that workpieces are readily accessible and thus can be easily loaded, positioned, and unloaded. Vertically pivoting clamping dies normally are selected for welding joints in sections such as flats and stampings, or for welding small-diameter tubing or solid bars where current distribution in the work metal permits the transformer to be connected to either the upper or the lower electrode.

Vertically sliding clamping dies are used where large flat plates or special curved sections are to be welded. In most applications, they eliminate the need for a pressure-equalizing mechanism. With vertically sliding dies, work is readily accessible, although the high clamping pressure used may require additional apparatus in the form of latches on the open side of the clamping structure.

Horizontally sliding clamping dies usually are selected for welding large solid sections or tubes, into which current must be introduced through all four clamping surfaces. Also, horizontal sliding dies may be used instead of pivoting dies or vertical sliding dies for workpieces that, because of their size or shape, are more easily loaded into the machine from above than from the front side.

Materials for Clamping Dies. Clamping dies and electrodes can be made of any current-conducting metal that is hard enough to withstand the clamping pressures and to resist wear due to scale and oxide on the workpiece. Clamping surfaces on the workpiece should be free from scale, oxide, paint, and oil. When backups are used, clamping dies generally act only as electrodes and do not resist all of the upsetting force.

The materials commonly used for clamping dies are RWMA class 3 and hardened tool steels such as H11, L6, or O1. Bronze and other copper-based electrode materials can be used in some applications. The die half (upper or lower) that conducts the current usually is made of a copper-based material. The other half often is made of the same copper-based material or of hardened steel.

Cooling of Electrodes. The need for water cooling of flash welding electrodes depends on the size of the electrode, die, or fixture, the magnitude of the flashing

Table 2 Minimum lengths of clamping dies, with and without backup, for flash welding various diameters of workpieces made from steels of low or medium forging strength(a)

Workpiece diameter, in.(a)	Minimum length of clamping die, in. With backup	Minimum length of clamping die, in. Without backup(b)	Workpiece diameter, in.(a)	Minimum length of clamping die with backup, in.	Workpiece diameter, in.(a)	Minimum length of clamping die with backup, in.
0.250, 0.312	0.375	1.00	2.00	1.25	6.00	3.25
0.375	0.375	1.50	2.50	1.75	6.50	3.50
0.500	0.375	1.75	3.00	2.00	7.00	3.75
0.750	0.500	2.00	3.50	2.25	7.50	4.00
1.000	0.750	2.50	4.00	2.50	8.00	4.25
1.50	1.000	3.00	4.50	2.75	8.50	4.50
			5.00	2.75	9.00	4.75
			5.50	3.00	9.50	5.00

(a) Diameter of rounds or tubing, or minimum dimension of other sections. (b) Backup is recommended for all workpiece diameters or minimum dimensions over 1.50 in.

Fig. 8 Flash welding setup using backups and locators to position and hold workpiece

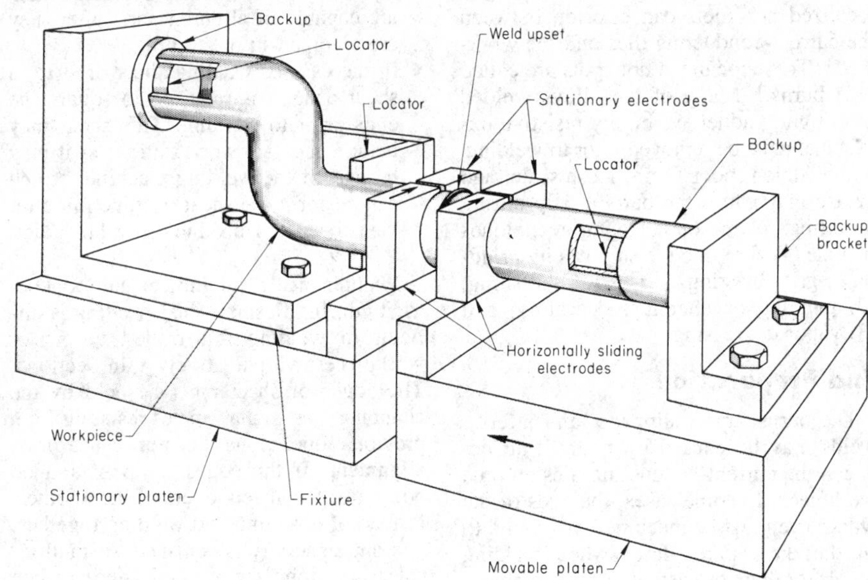

current, and the production rate. When the mass of the electrodes is large compared with that of the workpiece, the heat sink generally is large enough to dissipate the heat generated by the resistance to current flow and the heat absorbed from the workpiece. With high flashing currents and high production rates, electrodes generally must be cooled.

Electrodes, clamping dies, and fixtures used on large flash welding machines usually are water cooled. Those used on smaller machines are less frequently water cooled, unless the flashing current or the production rate is high. Cooling is accomplished by water flowing through internal passages.

Shape and Size of Clamping Dies. Generally, the shape of the clamping-die surface is such that the die encompasses almost the entire workpiece surface. The required area of clamping-die contact depends on the current needed for heating the workpiece and the pressure needed for holding it. Semicircular dies are used where the line contact provided by V-shaped dies gives insufficient surface for the current to flow without burning the workpiece, or for holding the workpiece without marking it.

Table 2 gives the minimum lengths of clamping dies for welding rounds, tubing, and other sections of various diameters or minimum dimensions; these data are for welding steels of low and medium forging strength.

Fixtures are used to support workpieces during welding and to provide orientation and alignment of workpiece surfaces that cannot be obtained by the use of clamping dies alone. Generally, fixtures have backup surfaces that absorb all or part of the upsetting force. When fixtures are used, electrodes are positioned adjacent to the joint in the usual manner for carrying current to the surfaces to be welded. The electrodes usually are attached to the clamping mechanism of the machine and are seldom part of the fixture.

The fixture shown in Fig. 8, fastened to the stationary platen of a flash welding machine, locates and holds a formed tube in alignment during welding. The fixture includes a backup that helps resist the upsetting force. A combination backup and locator also may be fastened to a bracket on the movable platen.

Horizontally sliding electrodes mounted to each platen hold the abutting ends of the tubes in alignment in addition to carrying the welding current. Because the electrodes in this case open in a horizontal plane, workpieces are loaded and unloaded from the top.

Miter fixtures are used for welding miter or corner joints in steel and aluminum sash, in screen and door sections, and in structural frames such as those used for refrigerator bases. The following example describes simultaneous flash welding of two forged steel clevises in precise angular alignment with the tubular midsection of a brace.

Example 1. Use of a Special Fixture for Maintaining Angular Alignment in Simultaneous Welding of Two Joints. Figure 9 shows the use of a special fixture for holding two forged 1020 steel clevises in angular alignment while simultaneously flash welding them to the ends of a length of 1008 steel tubing to produce a brace. Tolerance on center-to-center dimensions of the holes in the clevises was ±0.010 in., and exact angular alignment of the holes was required.

To achieve heat balance between the clevises and the tubing, the clevises were counterbored to approximately the inside diameter of the tubing and to a depth of about twice the amount of material lost

Fig. 9 Flash welding setup. See text for details.

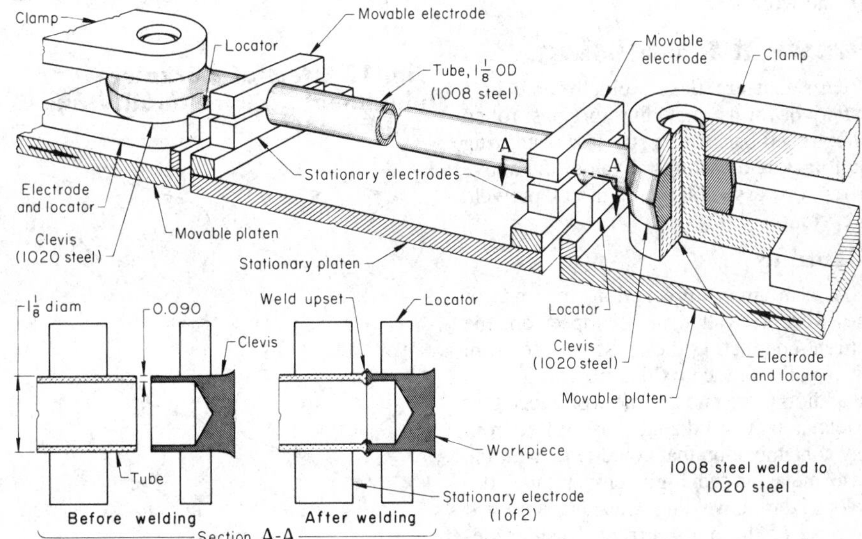

during flashing and upsetting (see section A-A in Fig. 9).

The two joints were welded simultaneously in a synchronized double-end flash welding machine equipped with two 50-kV · A transformers. Each clevis was positioned on a closely fitting pin that formed part of the lower half of an electrode. Proper alignment was established by placing the tubular section of each clevis in a recessed die and clamping it with the upper half of the electrode. The tube was positioned by two electrodes that were mounted on the stationary platen. Vertically moving clamps held the tube in place during welding. Both movable platens were actuated by cams that controlled platen motion during flashing and upsetting and controlled the force applied during upsetting.

Flash welding produced these parts at least cost because of the high production rate and because of the low rate of rejection. Arc welding would have increased the weight of the parts by requiring a tongue-and-socket type of joint. Friction welding would have required a complicated setup to maintain angular alignment. With pinned or riveted construction, the joints would have been weaker and the parts heavier.

Backups are used when clamping dies cannot provide the workpiece with enough resistance to the upsetting force. This usually occurs either when the length of the workpiece adjacent to the joint is too short for effective clamping, or when the workpiece is unable to withstand the clamping force without being collapsed, scored, or otherwise damaged. A backup often consists of a steel bracket that can be bolted to the platen in various positions. Brackets can have either fixed or adjustable stops for the workpiece.

Preweld Processing

Parts that are flash welded come in a variety of forms, including forgings; rolled or extruded bar; sheet, strip, or plate; ring preforms; and castings. Each of the above parts or assemblies requires preweld preparation.

Cleaning

As a minimum precaution, the ends of the workpiece that is clamped on the current-conducting die must be free from dirt, scale, surface oxidation, and grease. In addition, the ends of the workpiece that extend into the weld zone must be free from any contamination that could react with the base metal at the high temperatures developed during welding. Cleaning is needed because of the high current densities developed in the workpiece at the current-conducting dies. If insufficient electrical contact is made, weld quality is poor and localized hot spots can develop between the current-conducting dies and the workpiece. These localized hot spots are called "die burns." Many of the alloys welded have tightly adherent, highly resistive oxides that must be removed prior to welding.

In addition, poor fit-up, loose scale, and grease may cause the parts to slip during upsetting. Common cleaning techniques include (1) abrading the surfaces by grinding, grit blasting, or wire brushing; (2) pickling or chemical descaling; and (3) vapor degreasing.

End Preparation

Die burns, upset slippage, and inferior welds may be caused by poor fit-up between the current-conducting dies and the workpiece. In some cases, the ends of the workpiece must be machined or ground to fit the dies, particularly when welding rough forgings or castings. Also, a chamfer may be required on the ends of the workpiece to initiate flashing action.

Alignment of Workpiece

After the workpieces are clamped in the welding unit, the alignment of the workpieces with respect to each other along the axis of upset must be maintained. After the flashing process and during upsetting, any misalignment may cause the parts to overlap one another or cause a skewed weld. If this occurs, insufficient metal is extruded out of the weld zone during the upset and a poor-quality weld will be formed (Fig. 10). Types of alignment mechanisms include:

- Special fixturing or backup devices that ensure initial alignment that is maintained through upsetting
- Hydraulic or mechanical positioners that are capable of aligning parts after they are clamped in the tooling
- In the case of welding sheet or strip, a shear usually is required to square the ends prior to welding. This shear may be at a separate work station, or it may be part of the welding machine, which can perform the shearing function after the parts are clamped in the welding dies.

Fig. 10 Effect of workpiece alignment on joint quality and weld upset

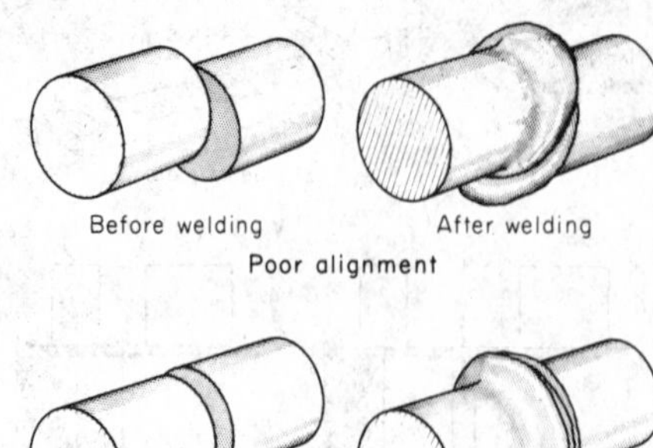

Maintenance of alignment and concentricity during flashing and upsetting is difficult in welding thin-walled tubing or workpieces with relatively thin sections. The edges of sheet must be held by the clamping dies so that ripples resulting from previous operations do not cause misalignment. If the edges are bent and do not mate, the sheets overlap one another, instead of upsetting and welding together. Overlap generally is confined to small local areas along the welded edge, and in these areas, poor fusion occurs. Reducing the thickness of the welded sheet by rolling can cause weld failure. The example that follows describes flash welding of a ring that was roll formed from a stainless steel band. Alignment was maintained by leaving two unformed (flat) ends at the joint to be welded and by using clamping dies that had the same radius as the roll-formed ring.

Example 2. Flash Welding of a Stainless Steel Ring. The ring shown in Fig. 11 was fabricated from 17-7 PH stainless steel coiled strip $1^{21}/_{32}$ in. wide by 0.078 ± 0.002 in. thick. The strip, annealed and pickled and with a 2B finish, was uncoiled, flattened, pierced with 16 holes $^{1}/_{4}$-in. square, and cut to a length of 18.08 in. Each length was then roll formed into a ring with a $^{3}/_{16}$-in.-long flat on each end. These flats helped to keep the ends in alignment during flashing and upsetting. The ends of the band were trimmed to ensure uniform flashing at the start. The band was manually loaded and clamped in the flash welding electrodes (Fig. 11). A clamping force of 4600 lb was exerted on an area of about $1^{1}/_{2}$ in.^{2}

The clamping dies were made of RWMA class 3 electrode material. Power leads from the 50 kV · A (50% duty cycle) transformer were attached directly to the electrodes. The primary voltage input to the transformer was 220 V, and the secondary output voltage was 4.1 to 6.9 V in eight steps. Platen speed and travel were controlled by a cam mechanism driven by a synchronous motor. The operation was semiautomatic in that the workpiece was

Fig. 11 Flash welding setup. See text for details.

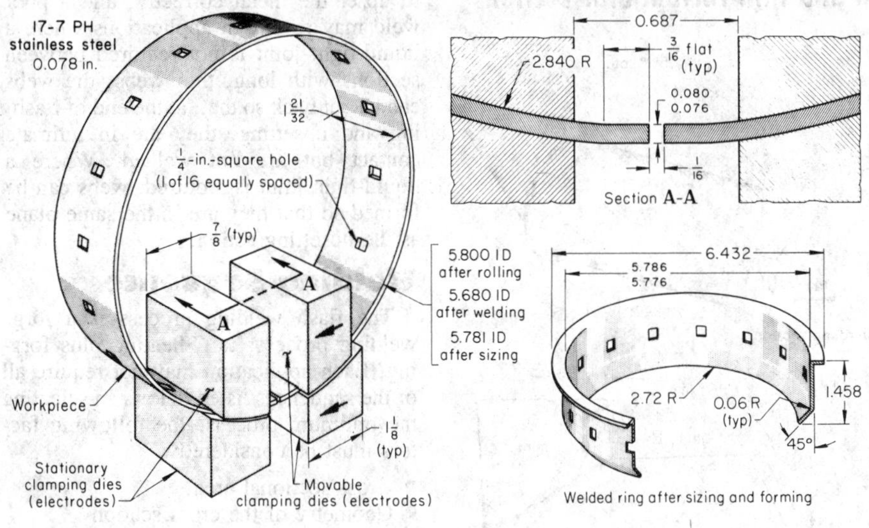

Welding dimensions and conditions

Initial die opening	0.687 in.	Total metal lost	0.474 in.
Initial extension per piece	0.312 in.	Flashing current	10 000 A
Initial gap at joint	0.063 in.	Flashing time	2.50 s
Final die opening	0.150 in.	Upsetting current	27 000 A
Flash-off length, total	0.314 in.	Upsetting time	0.08 s
Upset length, total	0.160 in.	Upsetting velocity	2 in./s

loaded and clamped manually and then the welding cycle was started by the operator. After the automatic flashing and welding cycle was completed, the ring was manually unclamped and transferred to a machine for scarfing the weld upset. Production rate was 1 ring in 24 s, or 150 rings per hour. Flashing and upsetting required 2.58 s.

The flash welded rings were given an intermediate anneal at 1400 °F for $1^1/_2$ h and then were cooled to 60 °F within 1 h. The annealed rings were stretched to an inside diameter of 5.776/5.786 in. and then were formed with a 90° stretch flange and a 45° shrink flange, which proof tested the welded joint. Tension testing of the flash welded joint showed that the metal in the weld zone had a strength equal to that of the base metal. The rings were given a final heat treatment at 1050 °F for $1^1/_2$ h and then were air cooled. This produced a hardness of 40 to 50 HRC. Hardness tests indicated that the metal in the weld zone responded to the heat treatment in the same manner as did the base metal.

Dimensional Tolerances. Some combinations of joint length and stock thickness are difficult or even impossible to weld. Table 3 gives the relationships of joint length (sheet width) to sheet thickness that have proven successful in a large number of production jobs. Relationships of tube diameter to wall thickness that have been successfully welded are listed in Table 4. When the ratio of tube diameter to wall thickness is too large, clamping dies

Table 3 Lengths of joints commonly flash welded in various thicknesses of flat sheet

Sheet thickness, in.	Joint length, in.	Sheet thickness, in.	Joint length, in.
0.010	1	0.060	25
0.020	5	0.080	35
0.030	10	0.100	45
0.040	15	0.125	57
0.050	20	0.187	88

Note: Length of joints in stock thicknesses of 0.050 in. and greater can be up to 100 in., depending on material extension, platen travel, die alignment, and clamping.

Table 4 Maximum flash-weldable diameters of tubing of various wall thicknesses

Wall thickness, in.	Maximum diameter, in.	Wall thickness, in.	Maximum diameter, in.
0.020	$^1/_2$	0.125	4
0.030	$^3/_4$	0.187	6
0.050	$1^1/_4$	0.250	9
0.062	$1^1/_2$	0.375	15
0.080	2	0.500	20
0.100	3		

may deform the tube enough to cause mismatch of the tube ends, producing an incomplete or defective weld in that area.

Diametral tolerances are most critical when extremely thin-walled tubes are being welded because the ends of out-of-tolerance tubes do not mate accurately at all points around the circumferences of the tubes. A difference in tube diameters can cause a telescoping action, which reduces the mechanical properties and quality of the weld.

Standard steel tubing generally is oversize; therefore, dimensions and tolerances of pieces to be welded should be based on actual rather than nominal dimensions. Military specifications require that the ratio of diameter to wall thickness not exceed 30 to 1. Although good welds can sometimes be obtained when this ratio is exceeded, selective assembly or sizing of the tubes may be necessary.

Miter Joints

Flash welded miter joints inherently have less-than-optimum weld quality because the line of application of upsetting force is not parallel to the centerline of the workpiece, and there are variations in the resistance to the upsetting force in the horizontal plane of the abutting surfaces. However, miter joints have been satisfactorily flash welded in bar stock, flat rectangular sections, solid and hollow extruded sections, contour roll formed sections, and structural sections. Solid, compact sections such as bar stock must be joined at an included angle of not less than 150°, as shown in Fig. 12(a). Poor bonding occurs at the outer corner as a result of the absence of backup material.

Satisfactory 90° welds have been made in thin rectangular sections and in similar extruded or roll formed sections, where the width of the joint (w) is greater than 20 times the stock thickness (Fig. 12b). Slippage of the workpiece during upsetting is a normal consequence unless the clamping dies are properly designed and unless workpieces are carefully placed and aligned in the dies.

It is impossible to obtain 100% bonding throughout all parts of a flash welded miter joint because some portions of the weld area are not backed up with cold metal to give support during upset. The lack of backup metal prevents the development of adequate forging pressure and results in poor bonding at outside corners (see details A and B in Fig. 12). Thus, miter joints are inadvisable for assemblies subject to high stresses in tension or torsion. To safeguard against stress concentration at light and medium loads, the poorly bonded

Fig. 12 Typical flash welding setups and dimensional relationships for welding mitered corners in round and thin rectangular sections

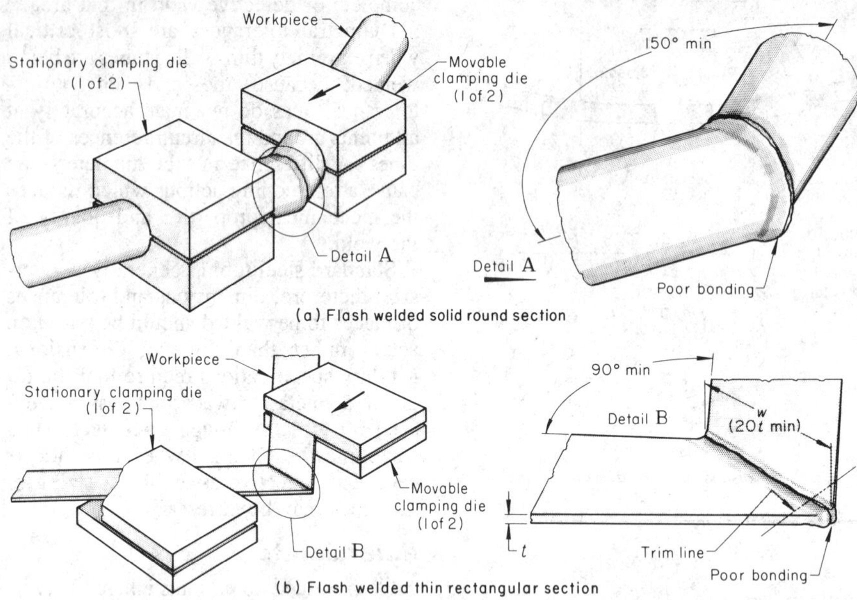

(a) Flash welded solid round section

(b) Flash welded thin rectangular section

metal at the outside corner of the joint should be removed, as indicated by the trim line in detail B in Fig. 12. When thin rectangular sections are joined at an angle, little heat flows beyond the welding surfaces, because welding time is short; in thin sections, however, a deep zone of plasticity is unnecessary.

Accurate cutting of the ends of workpieces for miter joints is not essential. Slight inaccuracies in straightness or angularity of the abutting ends cause initial point contact at the start of flashing and help to initiate a smooth flashing action. The values given in Table 5 can be used as a guide in determining dimensional allowances for bar stock, extrusions, tubing, and sheet stock.

Extrusions having different cross-sectional shapes, such as the header, the jambs, and the sill of a window frame, frequently must be flash welded together in a miter joint (Fig. 13). When the ends are prepared for welding, portions that are not to be welded, or any nonmatching portions of the extrusions, must be cut back for a distance equal to or greater than the length of metal lost during flashing and upsetting. Otherwise, flashing occurs between these portions and other parts of the workpieces or the clamping dies. When properly trimmed, ends of the extrusions terminate in the miter exactly at the weld line.

Adequate support and clamping are important in miter joint welding of extrusions and roll-formed sections that incorporate thin webs, because the plane of the thin web is not in the plane of the upsetting force. Where the length of the thin vertical web is greater than twice the web thickness, upsetting force is more likely to displace metal from the joint rather than to upset the metal correctly, and a poor weld may result. In applications where a liquid-tight joint is not required between sections with long, thin webs, the webs can be cut back so that, at the end of flashing and upsetting, they are in intimate contact but are not welded. Where a liquid-tight joint is required, webs can be formed so that they are in the same plane as the upsetting action.

Welding Sequence

The flash welding process is a forge welding process, i.e., heating plus forging. Each application may not require all of the sequences listed below. In selecting the optimum process, the following factors must be considered:

- Cross-sectional area
- Geometry of the cross section
- Alloy properties, such as high temperature, yield strength, thermal diffusivity, and electrical conductivity
- Welding machine capabilities, such as upset force, clamping force, transformer kV · A, transformer secondary short circuit current capability, and machine control (manual, semiautomatic, automatic)
- Quality of the weld required

Table 5 Typical dimensional allowances for flash welding of low-forging-strength and medium-forging-strength steel tubing, flat stock, and solid bar(a)

Thickness (t) of tube wall or flat stock(b), in.	Dimensional allowance, in. (see illustration for location and definition)						Flashing time, s
	A(c)	B	C	D	E	F	
Tubing or flat stock							
0.010	0.110	0.050	0.020	0.010	0.030	0.055	0.40
0.020	0.215	0.100	0.040	0.018	0.058	0.108	0.80
0.030	0.325	0.150	0.063	0.025	0.088	0.163	1.25
0.040	0.430	0.200	0.083	0.032	0.115	0.215	1.75
0.050	0.530	0.250	0.103	0.037	0.140	0.265	2.25
0.060	0.620	0.290	0.120	0.045	0.165	0.310	2.75
0.070	0.715	0.330	0.140	0.053	0.193	0.358	3.50
0.080	0.805	0.370	0.158	0.060	0.218	0.403	4.00
0.090	0.885	0.410	0.173	0.065	0.238	0.443	4.50
0.100	0.970	0.450	0.188	0.072	0.260	0.485	5.00
0.110	1.060	0.490	0.205	0.080	0.285	0.530	5.75
0.120	1.140	0.530	0.220	0.085	0.305	0.570	6.25
0.140	1.320	0.620	0.255	0.095	0.350	0.660	7.75
0.150	1.390	0.660	0.265	0.100	0.365	0.695	8.50
0.160	1.470	0.700	0.280	0.105	0.385	0.735	9.00
0.170	1.540	0.740	0.290	0.110	0.400	0.770	9.75
0.190	1.690	0.820	0.315	0.120	0.435	0.845	11.25
0.200	1.760	0.860	0.325	0.125	0.450	0.880	12.00
0.250	2.010	1.000	0.365	0.140	0.505	1.005	16.00
0.300	2.245	1.125	0.405	0.155	0.560	1.123	21.00
0.400	2.640	1.350	0.465	0.180	0.645	1.320	33.00
0.500	2.910	1.500	0.510	0.195	0.705	1.455	45.00
0.600	3.135	1.630	0.543	0.210	0.753	1.568	56.00
0.700	3.360	1.750	0.580	0.225	0.805	1.680	70.00
0.800	3.525	1.850	0.605	0.233	0.838	1.763	83.00
0.900	3.660	1.930	0.625	0.240	0.865	1.830	97.00
1.000	3.800	2.000	0.650	0.250	0.900	1.900	110.00

(continued)

Table 5 (continued)

Diameter (or least sectional dimension) of bar(d), in.	Dimensional allowance, in. (see illustration for location and definition) A(c)	B	C	D	E	F	Flashing time, s
Solid bar (round, hexagonal, square, or rectangular)(c)							
0.050	0.100	0.050	0.020	0.005	0.025	0.050	0.40
0.100	0.182	0.100	0.031	0.010	0.041	0.091	0.75
0.150	0.270	0.150	0.045	0.015	0.060	0.135	1.15
0.200	0.350	0.200	0.055	0.020	0.075	0.175	1.50
0.250	0.430	0.250	0.065	0.025	0.090	0.215	1.90
0.300	0.510	0.300	0.075	0.030	0.105	0.255	2.25
0.350	0.600	0.350	0.090	0.035	0.125	0.300	2.75
0.400	0.685	0.400	0.103	0.040	0.143	0.343	3.25
0.450	0.770	0.450	0.115	0.045	0.160	0.385	3.75
0.500	0.850	0.500	0.125	0.050	0.175	0.425	4.25
0.550	0.940	0.550	0.140	0.055	0.195	0.470	5.00
0.600	1.025	0.600	0.153	0.060	0.213	0.513	5.50
0.650	1.100	0.650	0.163	0.062	0.225	0.550	6.75
0.700	1.180	0.700	0.175	0.065	0.240	0.590	7.50
0.750	1.260	0.750	0.188	0.067	0.255	0.630	8.25
0.800	1.340	0.800	0.200	0.070	0.270	0.670	9.00
0.850	1.420	0.850	0.213	0.072	0.285	0.710	9.75
0.900	1.500	0.900	0.225	0.075	0.300	0.750	10.50
1.000	1.660	1.000	0.250	0.080	0.330	0.830	13.00
1.100	1.820	1.100	0.275	0.085	0.360	0.910	16.50
1.200	1.980	1.200	0.300	0.090	0.390	0.990	20.00
1.250	2.060	1.250	0.313	0.092	0.405	1.030	22.50
1.300	2.140	1.300	0.325	0.095	0.420	1.070	25.00
1.500	2.460	1.500	0.375	0.105	0.480	1.230	38.00
1.600	2.620	1.600	0.400	0.110	0.510	1.310	45.00
1.700	2.780	1.700	0.425	0.115	0.540	1.390	54.00
1.800	2.940	1.800	0.450	0.120	0.570	1.470	63.00
1.900	3.100	1.900	0.475	0.125	0.600	1.550	75.00
2.000	3.260	2.000	0.500	0.130	0.630	1.630	90.00

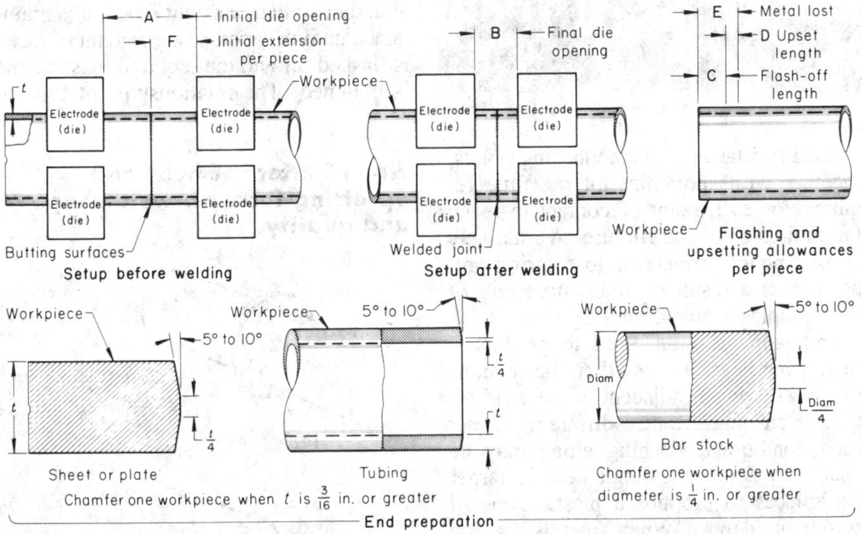

(a) Data are based on welding, without preheating, of two pieces with the same welding characteristics. (b) In the views above, the boldface-letter identifications of dimensions are applicable also to flat stock—for which t would be what is shown in these views as the wall thickness of the tubing. When tubing or flat stock has t of 3/16 in. or more, ends should be chamfered as shown in "End preparation" view above. (c) When an initial gap is used between the workpieces at setup, the initial die opening (Dimension A) is increased by an equal amount. (d) In the views above, the boldface-letter identifications of dimensions are applicable also to bar stock—for which diameter (or least sectional dimension) would be what is shown in these views as the outside diameter of the tubing. When diameter (or least sectional dimension) of bar stock is 1/4 in. or more, ends should be chamfered as shown in "End preparation" view above. (e) Values apply only when ratio of maximum to minimum cross-sectional dimension is less than 1.5.

The heat needed to make a flash weld is generated by flashing or by a combination of preheating and flashing. Enough heat must be generated in each piece to produce a plastic zone deep enough to permit adequate upsetting (Fig. 3).

The depth of the plastic zone and the degree of plasticity affect the slope of the upset metal. An upset with a slope between 45 and 80° (Fig. 14a) usually indicates that the correct amount of heat has been used and that enough upsetting has occurred. If the slope of the upset is steep and longitudinal cracks appear in the upset (Fig. 14b), too great an upsetting force has been applied at too low a metal temperature. These cracks are similar to forging cracks and result from the application of upsetting force before the work metal has become sufficiently plastic. If the slope is much less than 45° (Fig. 14c), the heat and upsetting force have been too low. Heat generated during flashing is concentrated at the flashing surfaces, behind which there is a steep decline in the temperature gradient toward the electrodes or clamping dies (Fig. 3).

The shape of the temperature gradient can be altered by the flash weld processing sequence, as well as the initial clamping distance. Consequently, the characteristics of the upset are changed. For this reason, flash welding is adaptable to a broad range of different alloys and configurations.

Burnoff

After the parts are clamped in the welding machine and the welding power is activated, the abutting surfaces are brought together for a brief period of violent flashing. This phase of the welding sequence is called "burnoff." It serves the function of squaring off the abutting surfaces and compensates for inconsistencies in end preparation. Burnoff is characterized by a substantial loss of material due to the violent flashing, but results in minimal heating of the workpieces; i.e., most of the heat generated is lost with the material that is flashed away. Voltage used for burnoff is usually higher than that required for flashing due to the fact that workpieces are initially cold.

Preheating

Preheating is a resistive heating phase of the welding process in which the heat is generated by the electrical resistance of the workpieces. It is accomplished by bringing the workpieces together under light pressure, creating a short circuit. The pressure must be great enough to prevent flashing, but not so great that workpieces are prematurely welded. Heating occurs over the entire distance between the two clamping dies (between the movable and fixed platens) and is similar to other applications of resistance heating. The short circuit can only be maintained for a brief period because the high current densities and localized irregularities of the workpiece surfaces can cause localized overheating, hot spots, and die burns. Therefore, in practice, the process is performed in a cyclic manner; workpieces are brought

Fig. 13 Use of miter clamps on automobile window frames. Courtesy of Berkeley-Davis

together for a brief period of time, then separated for a brief period of time to allow the heat generated to diffuse into the workpieces. This sequence is repeated several times. In manual and semiautomatic machines, the operator usually performs preheating by oscillating the platen while observing the workpiece, and preheating is terminated when the desired heat is achieved. Fully automatic machines perform this function, thus eliminating operator variables.

Preheating has three advantages: (1) a proper flashing action can be started more easily on heavy sections that demand power levels beyond the electrical capacity of a given welding machine, because the elevated temperature of the preheated contacting surfaces reduces the voltage required to initiate flashing; (2) a weld can be made with less metal loss than when flashing is used alone, because preheated workpieces do not require as long a flashing action; and (3) for large workpieces, platen travel may be inadequate to generate the proper temperature gradient by flashing alone (Fig. 3). Faster heating of the contacting surfaces results from the increased resistance of the contacting points, together with shortening of the time required for each point of contact to reach the molten state and rupture. A relatively small amount of metal is lost during preheating as a result of slight upsetting of the semiplastic metal under pressure.

Preheating is often used to produce a more even heat balance when the sections of the workpieces adjacent to the weld interface are significantly different. Under such conditions, flashing alone may be unable to develop enough heat in larger workpieces to produce a plastic zone of sufficient depth. When metals having widely dissimilar elevated-temperature compressive strengths are welded, only the metal with the higher compressive strength is preheated. For example, a piece of stainless steel may have to be preheated before being welded to a piece of low-carbon steel.

Occasionally, the flashing characteristics of two metals are so different that one heats rapidly, while the other hardly heats at all. When upsetting occurs, the plastic zone is large in the workpiece at the higher temperature and is very small in the cooler workpiece, which results in poor bonding. This problem is likely to occur when high-silicon steel is welded to low-carbon steel and can be corrected by preheating the low-carbon steel only.

Flashing

Flashing derives its name from the rapid expulsion of incandescent particles of molten metal from the bridges or minute points of contact between the two workpieces. The metal in these bridges is heated to the melting point almost instantaneously and subsequently explosively expelled. During flashing, the oxygen content of the atmosphere at the weld region is reduced because of oxygen absorption during heating of the bridges and because of the burning of metal in air. This reduction of oxygen content produces a semiprotective atmosphere, which, if sufficient, allows for highly reactive metals such as aluminum and titanium to be welded without additional inert atmosphere protection.

The primary purpose of flashing is to generate enough heat to produce a plastic zone that permits adequate upsetting. The rate of energy input must be in proper proportion to the travel of the platen or movable die, so that constant flashing is maintained until the appropriate amount of metal is flashed off and the required plastic zone is obtained. The relationships of time to

Fig. 14 Effect of heat and upsetting force on weld shape and quality

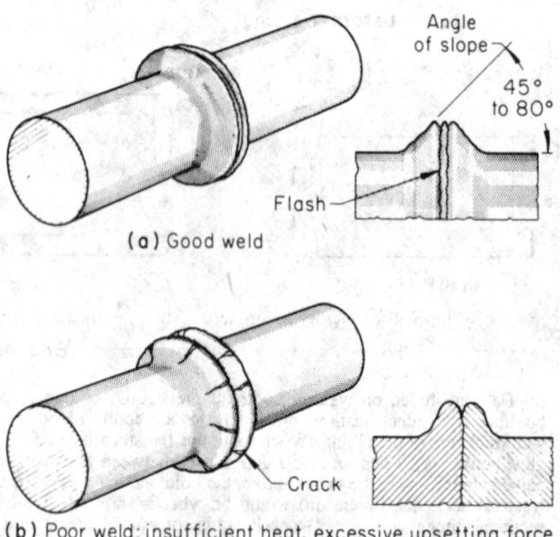

(b) Poor weld: insufficient heat, excessive upsetting force

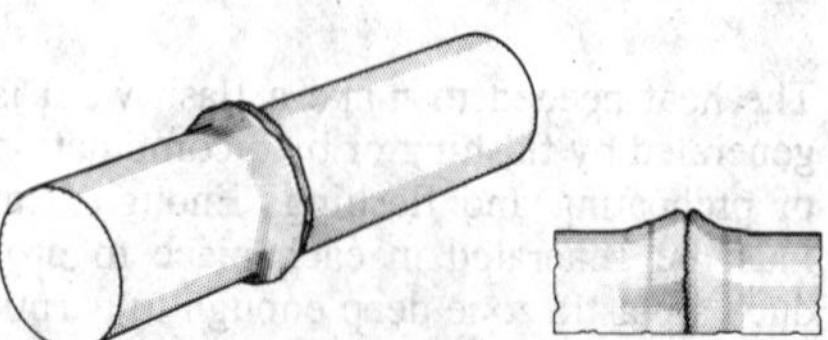

(c) Poor weld: insufficient heat and upsetting force

Fig. 15 Typical flash welding cycle

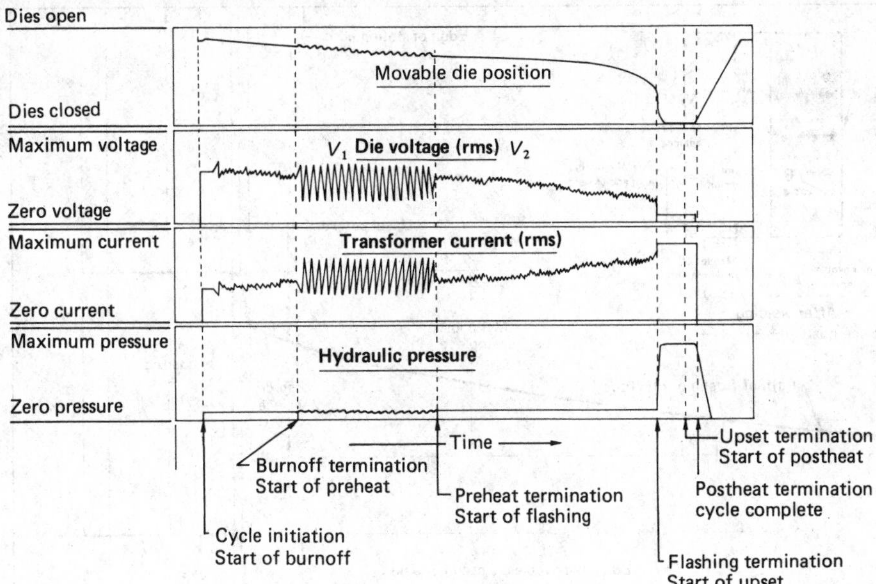

the flashing current and to the travel of the movable die are shown in Fig. 15.

Reduction in the rate of energy input produces a decrease in the violence of the flash expulsions, but with an increase in number and continuity (as shown by the number of small craters in Fig. 2b), and higher rates of energy input increase the size and depth of the craters left as the result of the expulsions (Fig. 2). Large craters, such as those shown in Fig. 2(c), are undesirable because their presence makes it difficult to upset to sound metal.

On manual machines, flashing is controlled by the operator; whereas on semiautomatic and automatic machines, flashing is controlled either by a cam or by a hydraulic servo-valve, which can be open-loop controlled or controlled by closed-loop feedback. On machines using closed-loop servo-control, the feedback signal to the control circuit can be primary voltage or current, secondary voltage or current, or primary or secondary power.

Automatic and semiautomatic machines can be adjusted so that the platen advances at either a constant rate (velocity), or it can be accelerated as the flashing progresses. A stable thermal gradient can be achieved with shorter workpiece travel by using acceleration during flashing. Thus, the volume of material lost during flashing can be reduced by using the acceleration of the platen.

The rate and duration of energy input also affect the depth of the heat-affected zone (HAZ). A low rate of heat input and a long input time result in the greatest depth of HAZ because there is more time for the heat to flow farther back into the workpiece and cause metallurgical changes. A high rate of energy input with a short flashing time causes a large flashing gap and increases the possibility of inclusions and porosity in the weld. Inclusions and porosity can be prevented by increasing the amount of upset, but this increases the amount of metal lost in making the weld.

When a welding machine with only single-voltage control is used for welding heavy, compact sections, ease of starting the flashing action sometimes is sacrificed to prevent excessively high current from causing unsatisfactory flashing action. Flashing action can be started more easily and can be accomplished more satisfactorily by using a dual-voltage system. With this system, a high secondary voltage (V_1 in Fig. 15) is applied for a short time to initiate flashing. Energy input is then reduced (V_2 in Fig. 15) to provide correct flashing conditions.

Upsetting (Forging)

Bonding takes place during the upsetting action, and some metal must be extruded from the weld zone to remove slag and other inclusions not expelled during flashing. The extruded metal must extend beyond the cross-sectional boundaries of the workpiece to ensure that maximum amounts of slag and inclusions are removed when the weld upset is removed during subsequent trimming. When the weld upset is removed, no evidence of the weld should remain. Porosity near the outer surface on an etched section of the weld, or a crevice around the workpiece after the weld upset has been removed, indicates incomplete bonding because of either insufficient upsetting force or insufficient plasticity during upsetting.

A large upset produced by an excessive upsetting force can be detrimental to weld quality, can waste metal, and indicates that most of the plastic metal has been extruded, requiring bonding to take place in metal where plasticity may not have been sufficient to ensure a good weld. Producing a large upset also is equivalent to introducing less total heat into the weld, which can result in poor weld quality. Excessive upsetting force bends the flow lines at the weld in a direction 90° from the flow lines in adjacent unwelded metal. Welds made under these conditions can have low fatigue strength.

The quality of a flash weld is greatly affected by the upsetting force. During the upsetting action, force should be sufficient to close all voids, to expel molten metal, slag, or other impurities, and to allow complete solid-state joining of the two workpieces. The ability of the upsetting force to dispel molten metal, slag, and other impurities depends on several factors, including the actual force, shape of the section, velocity of upset, and the degree of plasticity of the metal at the time of upset. The shorter the expulsion paths of the inclusions, the easier the inclusions are to expel. The larger the ratio of perimeter to area of section in the workpiece, the more easily slag and impurities can be expelled by the upsetting force applied during flash welding.

The optimum velocity of upset is determined by the mass of the moving parts and construction of the welding machine, the manner and rate at which the upsetting force can be applied, and the amount of upsetting force exerted. Some machines have the ability to vary both the upsetting speed and the upsetting force during the upset action. The relation of time to the magnitude of upsetting force is shown in Fig. 16. Pressure on the platen is relatively constant during flashing, but is increased rapidly for upsetting. Upset pressure usually is constant during cooling, postheating, and holding times.

The necessary amount of upsetting force is related to the temperature of the metal in the upset area and to the compressive strength of the metal at welding temperature. Table 6 lists the pressures required for upsetting some ferrous and nonferrous alloys without preheating. With preheating, and/or adaptive controls, the values in Table 6 may be reduced by up to 50%

Fig. 16 Typical flash welding cycle. Illustrates relationship of welding variables and workpiece position before and after welding

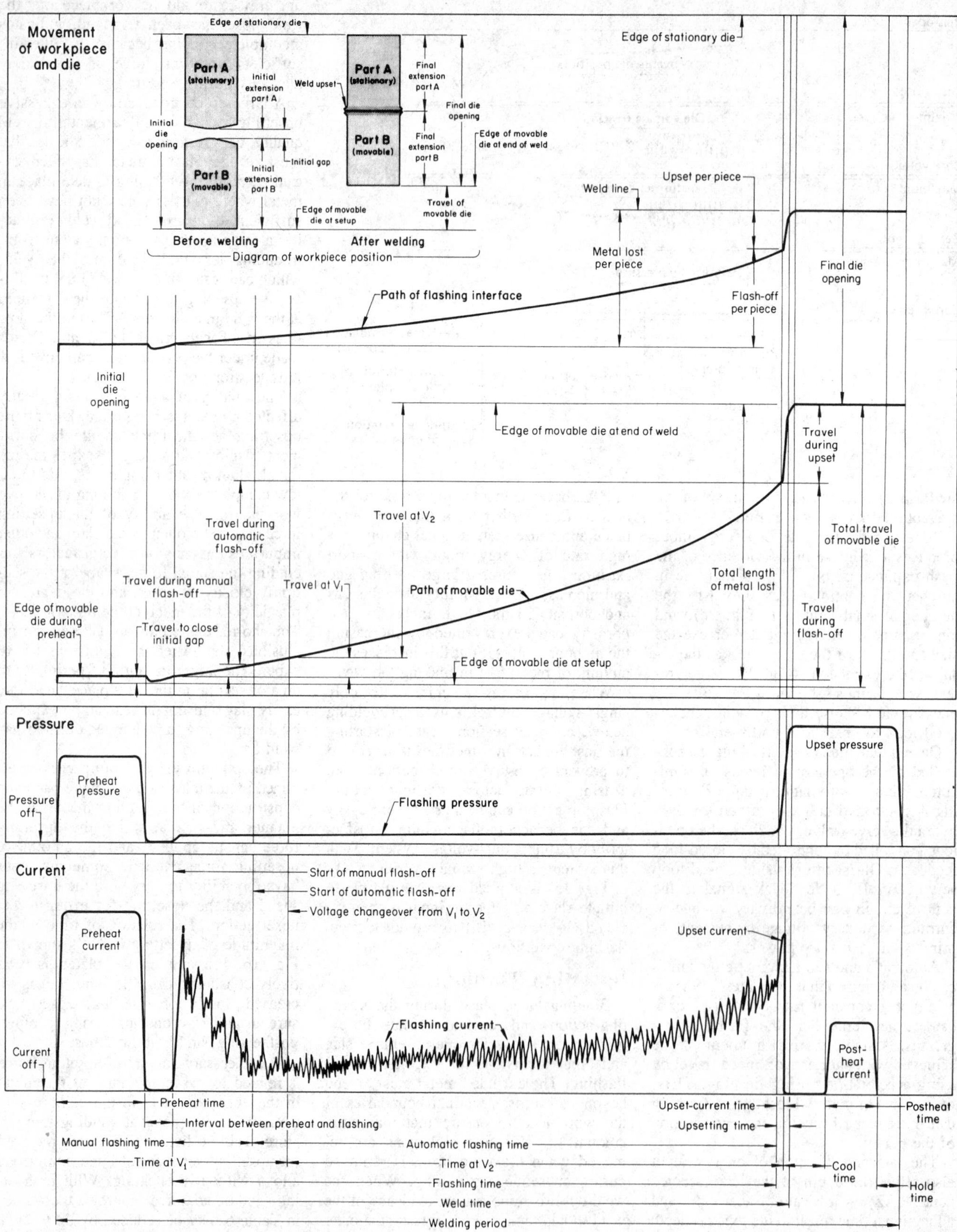

Table 6 Pressures required for upsetting several ferrous and nonferrous metals of various thicknesses(a)

Work metal	Thickness, in.	Pressure(b), ksi
Aluminum alloys:		
Soft	Up to $^5/_8$	8 to 10
High-strength	...	Up to 20
Copper, plain	Up to $^3/_4$	5 to 7.5
Heat-resistant alloys	...	Up to 35
Magnesium alloys	...	6 to 8
Stainless steels	Up to 2	13 to 25
Carbon and low-alloy steels:		
1020	Up to $^3/_8$	6 to 10
	$^3/_8$ to 1	7 to 10
	1 to 2	9 to 11
	2 to 4	10 to 12
2340, 3140	Up to 1	10 to 12
4140, 6145	Up to $1^1/_4$	11 to 18

(a) Based on welding temperature obtained without preheating. Pressures may be as much as 50% less if preheating is used. (b) Pounds per square inch of cross-sectional area of weld

or more. The effect of varying the upset distance is to complete the weld at a higher or lower temperature; for small upset distances, the weld is consummated in higher temperature metal, whereas welds consummated at large upset distances are formed in colder metal. Often, fine tuning of the welding process is accomplished by varying the upset distance.

Upset Current

Most machines are equipped with an adjustable timer that maintains weld current after upset has been initiated, which is called "upset current." It is maintained from 0 to 3 s in extreme cases. Upset heat can improve the plasticity of the metal during upset, provide some homogenization of the weld zone, and broaden the HAZ when required. A patented device, known as a proportional energy regulator, measures flashing energy and automatically controls the amount of upset energy to some preselected proportion of the flashing energy (Ref 1).

Extension of Workpieces From Dies

The amount of initial extension of the workpiece from the clamping die or electrode is determined principally by the ability of the metal to reach the correct temperature in the plastic range during preheating and flashing. Some of the heat generated by flashing is lost by radiation, and some is lost by conduction to the dies. The initial extension must be large enough to minimize conduction loss. However, the extension must not be too large, because a large extension decreases the temperature gradient and thus reduces the effect of the upsetting force and its ability to purge the weld through expulsion of molten metal.

Initial extensions that are larger than required can contribute to misalignment of workpieces after welding. Often, compromise must be made between welding conditions and permissible misalignment, especially in welding sheets or thin workpieces.

The size of the HAZ is partly dependent on the initial and final extensions. It may be desired to harden a workpiece before welding and to retain a specified hardness after welding in a certain portion of the weld zone. By proper manipulation of the initial and final extensions of the workpiece, the desired hardness can be satisfactorily retained.

When workpieces are of the same size, shape, and work metal, the initial extensions are made equal. Unequal extensions are used to assist in producing a better heat balance when equal extensions give unsatisfactory results. Typical dimensional allowances, including lengths of initial and final extensions, for flash welding of steel tubing, flat stock, and solid bars are given in Table 5.

Workpiece Design

In designing parts to be flash welded, an even heat balance should be obtained in the parts, so that the two ends to be welded attain the same degree of plasticity and depth of plastic zone during flashing. The length of metal lost during flashing and upsetting must be added to the initial length of the components if the overall length after welding is important. In designing components to be miter joined, the angle between the axis of the part and the direction of platen travel must be considered in allowing for metal lost.

The parts should be designed so that the reaction to the upsetting force can be resisted in a direction parallel to the line of application of force and so that it is possible to back up and hold the parts securely during welding. The parts should be designed so that they can be held in accurate alignment during upsetting. This can be done by machining locating surfaces on forgings or by making use of the unformed ends inherent in roll forming.

For sound welds in workpieces to be miter joined, any nonmatching sections must be cut back so as not to interfere with flashing, and the webs must be supported or otherwise strengthened to withstand the upsetting forces. The abutting ends should be designed so that the incandescent particles are not trapped in the weld, and on large pieces, flashing should be started at the center of the cross section of the workpiece. In designing hollow members, consideration should be given to the possible disadvantage of upset at the inner surface. If the part must withstand fatigue loads, the stress-raising effect of the upset is important. Under corrosive conditions the metal in the upset can suffer severe attack.

Good joint design generally results in high-quality welds and consistency in production. Poor design does not necessarily mean that the parts are not weldable, but that the weld quality will be impaired and consistency in production will not be obtained.

Heat Balance Between Workpieces

Because the same welding current and upsetting force are applied to both workpieces, the thermal conditions in the two pieces must be approximately equal. To achieve heat balance, the two workpieces should have the same or nearly the same cross-sectional shape, area of contact, and contour. In general, the ratio of the areas of the two abutting surfaces should not be greater than 5 to 4.

The good designs shown in Fig. 17 are predicated on the provision of similar sections and areas in the vicinity of the weld. In the poor designs shown in Fig. 17, the heavier sections would heat more slowly, making it difficult for both pieces to reach equivalent plasticity for welding. If the

Fig. 17 Effect of joint design on heat balance between workpieces in flash welding

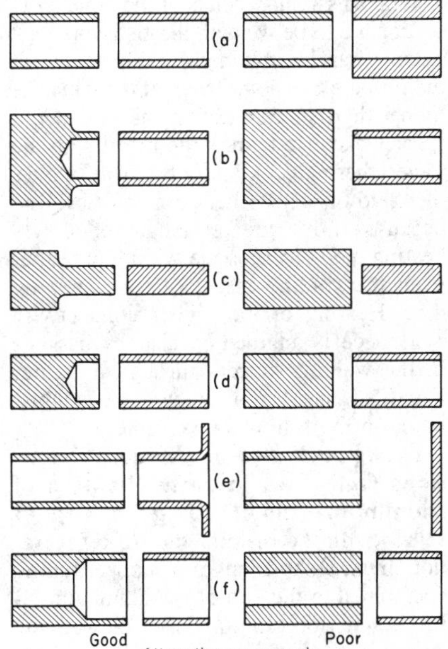

thick-walled tube of the poor design in Fig. 17(a) cannot be replaced with a tube of the same wall thickness as the adjoining piece (as suggested for good design), the thick wall should be thinned for a distance from the joint by machining the outer surface of the tube. The inner surface of the thick-walled tube in Fig. 17(f) should be machined for the same reason. In Fig. 17(d), it would not be possible to obtain a weld with the strength of the base metal if the tube were welded to the solid bar; therefore, the bar should be drilled as shown.

A degree of sectional unbalance is permissible in flash welding. When two tubes or two pieces of solid cylindrical stock are joined, the wall thicknesses of the tubes, or the diameters of the solid stock, should not differ by more than 15%. This allowance compensates for original part tolerances or applicable tolerances on misalignment or concentricity.

In each example of good design shown in Fig. 17, the optimum length of the altered section is equal to the initial extension. The minimum length is about 125% of the metal lost during flashing and upsetting. Initial extensions for sections made of steels of low and medium forging strength are given in Table 5.

Another means of adjusting the heat balance is by special preparation of the ends of the parts to be welded. With solid bar stock, for example, a chamfered end on one of the pieces (or unequal chamfers on both pieces) affects the heat balance by reducing the volume of metal extending beyond the clamping die. The metal with the greater heat conductivity should have the steeper chamfer.

When dissimilar metals of the same cross section are to be welded, a satisfactory heat balance can be obtained by adjustment of the initial extension. The metal having the higher thermal conductivity is extended a greater distance than is the metal with the lower thermal conductivity. This moves the weld off-center between the dies, and because of the greater length of slowly heating metal, places the weld line farther from the cooling effect of the clamping dies. Heating of the higher conductivity workpiece is augmented also by passage of the welding current through a greater length of metal. This technique is illustrated in the following example.

Example 3. Use of Unequal Extensions From Dies in Flash Welding of Aluminum Alloy Tubing to Copper Tubing. In the manufacture of refrigeration units, a short length of copper tubing was joined to the end of the aluminum alloy condenser tubing. This allowed the connection of the condenser to the remainder of the refrigeration system to be made by a copper-to-copper brazed or soldered joint.

Both tubes had outside diameters of $^5/_{16}$ in. Wall thickness was 0.050 in. for the aluminum tube and 0.060 in. for the copper tube. Heat balance would have been better if the wall of the aluminum tube had been thicker than the wall of the copper tube. The unbalance was compensated for by making the initial extension of the aluminum tube from the face of the clamping die ten times that of the copper tube, as shown in Fig. 18. Tool steel pinch-off dies were used to assist in making a better weld and to remove the weld upset from the exterior of the tubes. Flash was removed from the inside by reaming. No special end preparation of the tubing was required.

Fig. 18 Use of unequal extensions from dies to compensate for heat unbalance in flash welding copper tubing to aluminum alloy tubing. See text for details.

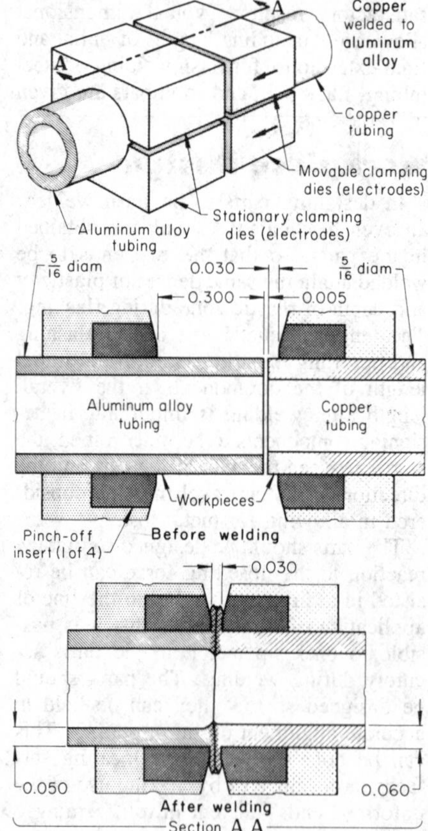

Flashing current	9000 A at 4.5 V
Flashing time	1 s
Flashing travel	0.200 in.
Upsetting current	19 000 A
Upsetting time	2 cycles (0.033 s)
Upsetting travel	0.100 in.
Upsetting force	5000 lb
Clamping force	1000 lb
Total metal lost	0.300 in.

Postweld Processing

In many instances, flash welds are used as welded, with no additional processing. However, in most cases, welds receive some kind of postweld processing.

Second Upset. In some instances after the flash weld is consummated, the weld joint is resistively heated in the welding machine and subsequently upset or forged. Typically, this second heating and forging improves the weld joint properties for certain alloys.

Weld Flash Removal. The weld flash may be removed, if required, by air chisels, machine cutting tools, grinding wheels, high-speed burring wheels, die trimmers, sanders, or machine-mounted flash trimming tools. The choice depends on the shape of the part and production requirements.

It is easier to remove the flash immediately after welding while the metal is still hot. When alloy steels are welded, flash removal by cutting tools is difficult because of the hardness of the flash. Grinding is usually employed. Common methods of flash removal include:

- *Chipping*: This is usually performed immediately after welding, with a manual air-operated chipping chisel, when the welding flash is still hot. Because it is performed manually, chipping usually is not feasible for high-volume operations. Chipping is often followed by a hand grinding operation to blend out the weld.
- *Scarfing*: For low-volume applications, electrical scarfing is used to remove weld flash. As with chipping, scarfing can be followed by a hand grinding operation.
- *Shearing, broaching, and planishing*: For high-volume production, there are many varieties of fully automatic machines that shear and/or broach weld flash. In many cases, the weld does not receive additional blending operations. When an additional operation is required, planishing usually is chosen. During the planishing operation, a set of opposed rollers squeeze the weld to improve surface finish. Cold working due to the rolling improves weld properties.

Pinch-type dies are used in welding many nonferrous metals to contain the HAZ during upsetting and to remove the upset metal. Pinch-type dies have nose inserts shaped to encircle the workpiece and to keep the plastic metal in the weld zone.

Fig. 19 Hydraulic flash welding unit with built-in vertical flash trimmer for transmission bands. Courtesy of Berkeley-Davis

This minimizes the amount of plastic metal that is extruded out of the weld and, immediately at completion of upsetting, trims any upset that is formed. Figure 19 shows a hydraulically operated machine with a built-in vertical flash trimmer for transmission bands (see also Example 3 and Fig. 18).

Heat Treating

To develop the full mechanical properties in the weld and to equalize welding stresses, many flash welding applications require postweld heat treatments. Heat treatment may be as simple as stress relieving or annealing. In the case of high-strength hardenable grades of steel, this heat treatment may be followed by austenitizing, quenching, and tempering. In precipitation-hardening alloys, welding usually is followed by a solution heat treatment with a subsequent full aging cycle.

Sizing, Straightening, and Forming

Due to the superior quality of flash welds, many applications exist where welds are hot or cold worked after welding. These operations may entail minor shape restoration or bulk deformation of the welds. Postweld operations on welded materials include (1) hot rolling, (2) cold rolling, (3) cold roll forming and contour roll forming, (4) hot forging or upsetting, and (5) cold forging or upsetting.

Testing

Due to the nature of a flash weld, nondestructive testing can be difficult. Recent improvements in nondestructive testing techniques, namely in the field of ultrasonics, have increased its usability. One of the most reliable testing techniques is to proof test welds. Most military and aerospace specifications call for a specific amount of deformation across the weld as a final qualification of the weld. Also, postweld forming or forging serves to qualify the welds. Nondestructive techniques used to test welds include:

- Radiographic
- Ultrasonic
- Fluorescent-penetrant inspection
- Magnaflux
- Eddy-current testing
- Acoustic emission
- Proof testing

Destructive testing of flash welds covers the entire range from typical tensile testing, stress rupture, and Charpy to low-cycle fatigue testing.

Weld Quality Variables

Aside from the ability of the flash welding machine to heat and upset a given cross section, many variables that are not related to the process and machine or operator skill affect weld quality.

Section Shape. If the cross section is severely nonuniform (i.e., very heavy areas adjacent to light areas), heating is nonuniform and the subsequent upset is nonuniform. Therefore, it may not be possible to obtain a good weld in both the heavy and light sections simultaneously.

Weldability Ratio. For a given cross-sectional area, differences in weldability exist depending on the weldability ratio. This is a ratio of the perimeter of the cross section, divided by the area. Generally, it indicates the relative ease or difficulty of welding a given section. If the weldability ratio is large, it indicates that there is a short distance for the heated metal to be extruded out of the weld area during upset, thus facilitating welding. For small weldability ratios, the opposite is true—there is a longer distance for the heated metal to be extruded out of the weld area. Thus, there is greater difficulty in welding.

Alloy. One of the distinct advantages of flash welding is the broad range of alloys that can be welded (see the section in this article on applications). Often alloys that are difficult or impossible to weld by other techniques can be joined by flash welding. With few exceptions, alloys that can be hot forged can also be flash welded, although alloys exhibiting hot short behavior are difficult to weld. Another advantage of the process is the capability of producing welds between dissimilar alloys (see Table 1).

The relative difficulty of welding a particular alloy is governed by its high-temperature workability range. Alloys that exhibit a narrow high-temperature workability range also exhibit a narrow welding temperature range or "welding window." In addition, certain alloys—particularly precipitation-hardenable alloys—require special welding setup procedures beyond the scope of this article to obtain uniform quality welds.

Composition. Depending on the alloy being welded, variations of 3 to 5% in chemical composition can cause variations in weld quality. This is particularly true for nickel-based, age-hardenable alloys.

Prior Heat Treatment. A major consideration when selecting material for flash welding is that it be free from stringers or severe banding. Alloys that are prone to forming stringers or banding or that are severely anisotropic must be heat treated to reduce this condition to an absolute minimum. The longitudinal fiber of the metal is turned at an angle during upset. Thus, the strength of the weld is governed by the transverse properties of the metal. In some cases, exposing the end grain at the upset area, as with aluminum alloys, can cause a decrease in the corrosion resistance.

Alloy Depletion. Controversy exists as to the significance of alloy depletion in a flash weld. During the heating process of flash welding, metal at the abutting surfaces is molten, and in some cases, alloying constituents with lower melting points may preferentially dissolve in this molten layer and subsequently are expelled with the flash. This may lead to alloy depletion in the weld area. In addition, elements such as carbon may migrate to the weld area and may be oxidized, thus creating a decarburized zone in the weld. This phenomenon is the probable cause of characteristic white spots in the weld area of etched carbon steel welds.

Suitability of Welds for Subsequent Processing

The upsetting action that occurs in flash welding not only forces molten metal and oxide impurities out of the weld, but also produces a microstructure resembling that of a forging. The grain structure at the weld interface is essentially the same as that of the HAZ on either side of the weld interface, except for size. The high ductility of flash welds in low-hardenability metals permits subsequent cold forming operations of varying degrees of severity without postweld heat treatment. Cold rolling, cold bending, spinning, hot forming, and wiredrawing also can be applied to flash welded products.

Automotive wheel rims are made at a high production rate from low-carbon steel or aluminum by flash welding. The flash welded bands are roll formed cold without prior heat treatment. No fracturing of the weld that could cause air leaks is tolerated (see Example 4).

Example 4. Production of a Rim for an Automotive Wheel. The rim portion of an automotive wheel, shown in Fig. 20, was made from 1012 steel strip stock by automatically flash welding a ring-rolled band, then progressively roll forming the cross-sectional contours. The surface of the rolled ring was shot blasted before welding. Initial die opening was $^7/_8$ in.; final die opening was 0.410 in. minimum. Initial extension per piece was $^{13}/_{32}$ in.

The input rating of the flash welding machine was 600 kV · A, 440 V, single-phase, 60 cycles. The maximum output rating of the secondary was 2000 A at 15 V. The current used for this workpiece was 1800 A at 12 V.

The movable platen was actuated by a cam mechanism. Clamping dies were made of RWMA class 3 electrode material. An air cylinder, using an air pressure of 70 psi, applied the clamping force on electrodes $^5/_8$ in. wide by $8^1/_2$ in. long with knurled or serrated surfaces. The unformed flat areas along the abutting edges of the band, which are common in roll forming, extended on both sides of the jaws and facilitated clamping and ensured alignment of the abutting edges.

A strip $43^1/_8$ in. long was sheared from the coil and was automatically transferred to a ring-rolling station and then to the welding station. The ends of the circular strip were gripped and forced together under high pressure during welding. Metal lost during flashing and upsetting was approximately $^3/_8$ in. Flashing and upsetting time was 90 cycles ($1^1/_2$ s). The coil stock was 0.130 in. thick and $8^7/_{16}$ in. wide.

After the weld upset was scarfed, the band passed through a bath of lubricating oil and into a series of four forming rolls and an expander that progressively formed the band into the cross-sectional shape shown in Fig. 20. After the contour roll forming operation, portions of the fin-

Fig. 20 Flash welding setup. See text for details.

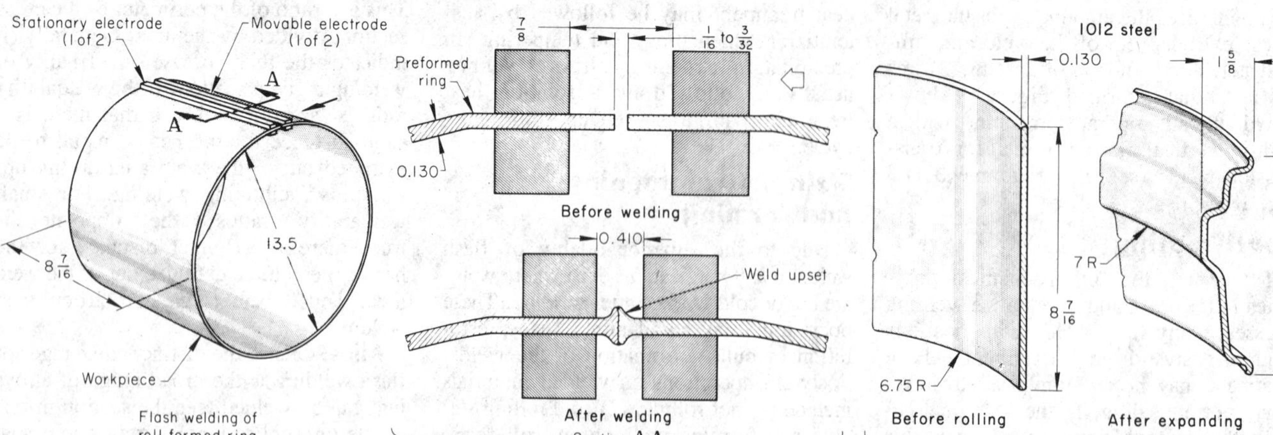

ished rim were as much as 25% greater in diameter than the flash welded band.

The completed rim was then degreased, dried, and conveyed to a station where an operator placed the spider into the rim. The assembly was moved automatically to a press where the spider was press fitted into the rim for resistance spot welding.

The completed wheel assembly was visually and mechanically inspected for weld integrity and dimensional accuracy. To withstand the forming operations, the flash weld had to be free from pinhole defects caused by sulfur, aluminum, or slag in the steel. Pinholes are also caused by a cold weld or by discontinuous flashing. The finished rim had to have an airtight joint to maintain inflation of the tire. Production rate was 720 rims per hour.

Wiredrawing Applications. Increased speeds of cold rolling mills and wiredrawing mills have resulted in the production of a higher percentage of off-gage strip and wire while the mills are accelerating and decelerating. The need for uniformity of thickness and for larger coils has led to the adoption of flash welding for joining the ends of several small coils into one large coil before the stock is reduced in thickness or diameter. In these applications, characteristics of the weld must approach very closely those of the base metal, without varying in uniformity, and must be able to undergo the same deformation by cold reduction.

Example 5. Joining of Aluminum Alloy Rods by Flash Welding. Aluminum alloy 5056-O rods, $^3/_8$ in. in diameter, were joined by flash welding to permit continuous wiredrawing. The high joint efficiency and consistently high quality of flash welds permitted the rods to pass through five drawing operations with a minimum of downtime due to breakage. The die diameter and reduction in area of the rods for each wiredrawing operation were as follows:

Operation No.	Die diameter, in.	Cross-sectional area, in.^{2}	Reduction in area, %
1	0.296	0.0688	37.7
2	0.234	0.0430	37.5
3	0.196	0.0302	29.8
4	0.168	0.0222	26.5
5	0.120	0.0113	49.0

Fig. 21 Pickle line flash welding unit with 1000-kV·A transformers. Courtesy of Wean United, Inc.

Fig. 22 Flash butt welding unit with built-in trimmer between dies. Courtesy of Wean United, Inc.

Pinch-type dies were used to ensure high-quality welds and to assist in removing the weld upset. The welding conditions were as follows:

Initial die opening	0.760 in.
Final die opening	0.010 in.
Flash-off length, total	0.500 in.
Upset length, total	0.250 in.
Flashing velocity	0.25 in./s
Upsetting velocity	5 in./s
Flashing time	2 s
Upsetting time	0.05 s (3 cycles)
Open-circuit voltage	12.6 V
Clamping force	6500 lb
Upsetting force	3200 lb

Flash welding units are also used in steel mill applications for strip and sheet in lines such as pickle lines and coil build-up lines (Fig. 21). Pipe mills employ flash welding units for continuous mill operation. Figure 22 shows a skelp welder with a trimmer between the welder dies.

Typical Welding Schedules

Flash welding schedules are based primarily on the alloy being welded, on the diameter of bar stock, on the thickness of sheet or plate, or on the wall thickness of tubing, and not on the entire cross-sectional area of the workpiece. Because of the distribution of electrical energy and the upset extrusion path, it is easier to weld a plate 1 in. by 12 in. than a bar 3 in. by 4 in., even though both have the same cross-sectional area. The maximum cross section that can be successfully welded depends primarily on the strength of the work metal at elevated temperature. Table 7 gives welding schedules for plates, bars, and tubes made of various metals and alloys.

Heat-Affected Zone

In a flash welded part, there are significant metallurgical differences between the highly heated structure at the center of the weld and the unaffected base metal. These differences occur because there is a large temperature gradient in the work metal between the clamping dies or electrodes. The greatest heat is generated at the faces of the weld because of the flashing action, while at the electrodes the metal is essentially at room temperature. The HAZ is the

Table 7 Typical flash welding schedules for plate, bar, and tube of various metals

Work metal, and section size in inches	Die opening, in. Initial	Die opening, in. Final	Flash-off length, in.	Upset length, in.	Total metal lost, in.	Flashing time, s	Upsetting time, cycles	Heat setting, V
Plate								
1020 steel, 0.25 by 16.75	2.00	0.75	0.75	0.50	1.25	12	8	6.02
Alloy steel(a), 1.165 by 9.95	3.00	1.25	1.00	0.75	1.75	25	50	8.17
Inconel X-750, 0.5 by 6.5	3.44	1.25	0.81	1.38	2.19	18	40	5.8
Inconel X-750, 1.5 by 6.0	4.44	1.75	1.06	1.63	2.69	25	75	8.17
Solid bar								
Low-carbon steel welded to M10 tool steel, 0.875 diam.	2.25(b)	1.50	...	...	0.50	7(c)	60	6.28 max
8620 steel, 0.405 by 0.505(d)	0.94	0.44	0.31	0.19	0.50	3.5	0.5 s	4.5
8620 steel, 2.56 diam	3.75	2.25	1.10	0.40	1.50	57	40	7.2
8640 steel welded to HS-31 alloy, 1.5 diam	3.41(e)	2.31	...	...	0.88	25	...	45 A
Tube								
1008 steel welded to 1010, 1.125 OD by 0.090 wall	0.75	0.25	...	...	0.50	4	...	(f)
AMS 6324 steel (8740 mod), 4.94 OD by 0.52 wall(g)	3.25	2.03	0.86	0.36	1.22	32	40	6.8
AMS 6427 steel(h), 1.94 OD by 0.22 wall	1.76	1.00	0.55	0.21	0.76	20	...	4.8
AMS 6427 steel(h), 8.75 OD by 0.78 wall(g)	5.12	2.02	2.57	0.53	3.10	120	40	10.9
Titanium alloy Ti-6Al-4V, 3.5 OD by 0.5 wall(g)	2.60	1.50	0.55	0.55	1.10	30	9	5.2

(a) 0.20% C, 1.0% Cr, 1.0% Mo, 0.1% V, 0.75% Si, 0.5% Mn. (b) Workpiece extensions: low-carbon steel, 1.62 in.; M10, 0.38 in.; 0.25-in. gap. (c) Includes 3-s preheat. (d) Ring-gear blank. (e) Workpiece extensions: 8640, 1.59 in.; HS-31, 1.82 in. (f) No. 5 tap of eight taps on a 50-kV · A welding transformer. (g) Welds were made in protective atmospheres: 1000-Btu city gas for the AMS 6324 and AMS 6427 steels, argon for the Ti-6Al-4V titanium alloy. (h) 0.28 to 0.33% C, 1.8% Ni, 0.85% Cr, 0.40% Mo, 0.07% V.

region on each side of the bond interface that has been heated during welding to temperatures above or within the transformation-temperature range. The portions of the work metal other than the weld and the HAZ remain relatively cool and conduct heat away from the weld. Consequently, the weld cools at a much higher rate than would be obtained during simple air cooling and, in steel, hardening results, particularly when the welded section is small. There is not only a marked change in microstructure across the weld, but also a variation in hardness—to a small extent in steels of low hardenability, and to a greater extent in steels of high hardenability. Thermomechanically worked, solution quenched, annealed, or other types of structures appear in alloys not subject to martensitic hardening.

In some alloys, the desirable characteristics of flash welds include a narrow hardened zone or HAZ and strength and ductility as nearly equal to those of the base metal as possible. A weld with these characteristics can undergo subsequent cold forming or cold reduction to the same degree as can the base metal.

Hardness of Welds

A hardness survey across a flash weld in cold rolled low-carbon steel shows little change in hardness. Therefore, heat treatment is unnecessary for obtaining uniform hardness. When a hardenable steel is flash welded, considerable hardness differentials are produced across the HAZ. If uniform hardness is desired, heat treatment after welding is necessary.

Flash welds in cold rolled low-carbon steel have a coarse structure near the weld and contain some martensite. During flashing, considerable grain growth occurs in the metal at the joint because the temperature is near or above the melting point, which is far above the transformation-temperature range. Theoretically, all the metal that is melted at the bond line is extruded into the weld upset. Proceeding from the bond line along the workpiece, the temperature decreases, and any martensite present becomes increasingly finer in appearance. Near the edge of the HAZ, where the metal has been heated to a temperature between the lower and upper transformation temperatures, only partial transformation occurs, which results in softening of the steel.

Variations in hardness across the HAZ in cold rolled low-carbon steel are illustrated in Fig. 23(a). The hardness of the metal in the HAZ is lower than that of the unheated base metal because heating has removed the work hardening resulting from cold rolling. The bond interface and points near the edges of the HAZ are slightly softer than the rest of the heated metal. During a tension test, fracture can be expected to occur near the edges of the HAZ because of the slight softening.

After flash welding of hot rolled steel, the metal in the HAZ is harder than the base metal as a result of heat treatment during welding. Therefore, in a tension test of a flash weld in hot rolled steel, fracture usually occurs in the base metal.

A flash weld in an alloy steel containing 0.25 to 0.28% C, 0.60 to 0.80% Cr, 0.80 to 1.00% Ni, and 0.50 to 0.60% Mo has a fully martensitic structure, because of the high hardenability of the steel. The martensite becomes more finely divided as the distance from the bond line increases, because of the decreasing temperature gradient. In a region near the edge of the HAZ where the metal temperature approaches the lower transformation temperature, the original base metal is only partly transformed; after cooling, the microstructure is ferrite and fine-grained martensite.

Variations in hardness across a weld in a Cr-Ni-Mo alloy steel of the composition noted above are illustrated in Fig. 23(b). From the original hardness of the base metal (about 260 HV), the hardness rises sharply at the edge of the HAZ and is higher than 500 HV (48 to 50 HRC) near the weld interface. The dip in hardness at the weld interface is probably caused by decarburization. During welding, the regions on either side of the bond line are heated to a temperature high enough to permit full hardening during cooling, and the hardness decreases with the distance from the bond interface.

The lowest hardness in the welded joint was just beyond the HAZ, where a temperature slightly higher than the tempering temperature (1200 °F) of the original base metal had been reached. Had the original tempering temperature been lower, and the

Fig. 23 Variation in hardness across the HAZ and into the base metal for flash welded joints. (a) Cold rolled low-carbon steel. (b) Cr-Ni-Mo alloy steel. (c) 1040 steel welded to 4150 steel

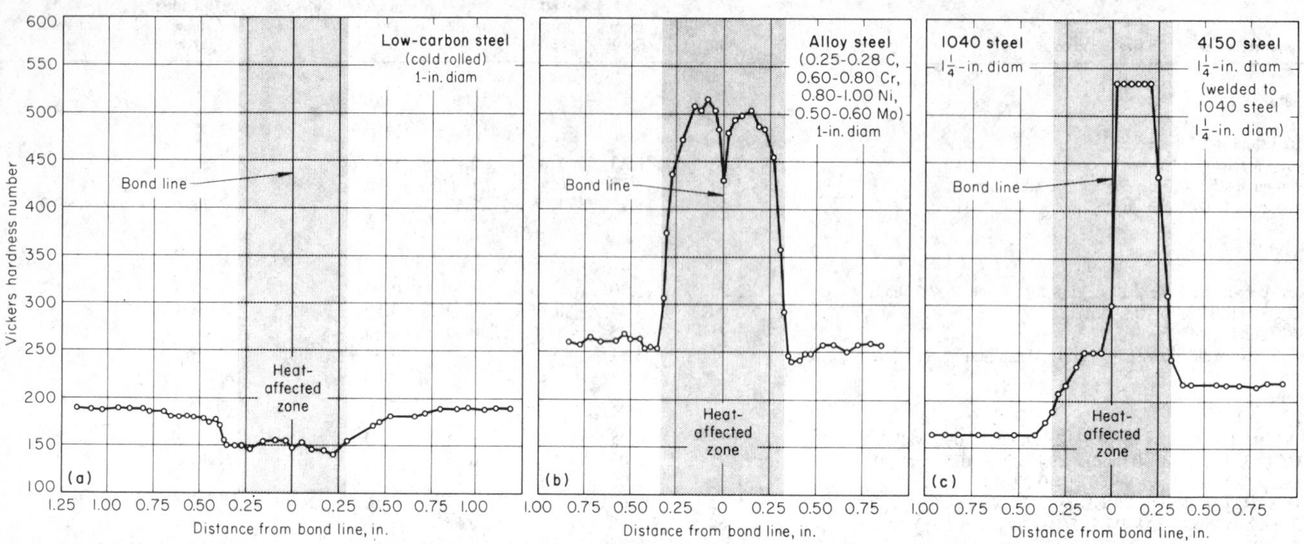

hardness of the metal consequently higher, the dip in hardness would have been more noticeable.

A hardness plot taken across the center of $1^1/_4$-in.-diam bars of 4150 steel and 1040 steel that were flash welded together is shown in Fig. 23(c). Although similar in carbon content, the two steels differ greatly in hardness within the HAZ, because of the greater hardenability of the 4150 steel. During welding, the 4150 steel was hardened to such an extent that it cracked at the outer edge of the HAZ during cooling. Both of the bars were fully annealed before welding; therefore, there was no dip in hardness between the HAZ and the unheated base metal.

Full heat treatment can restore a flash welded steel part to the uniform hardness of the unwelded stock, but merely tempering the hard weld metal to the original hardness of the base metal does not result in full homogeneity, because between the hardened weld metal and the unaffected base metal there is a narrow zone that has been overtempered by the welding heat and that is softer than the original base metal. This softened zone can be brought to uniform hardness only by full heat treatment. The softened zone may or may not be detrimental to the service life of the weldment.

Tempering in a flash welding machine is impractical for two reasons: (1) it is difficult to maintain a uniform temperature in the metal between the electrodes; and (2) proper tempering of the edge of the HAZ normally cannot be done without overtempering and softening of the areas nearest the bond interface.

Strength of Flash Welds

Because flash welds are made from the base metal without any filler metal, the welds can be made to have almost the same properties as the base metal. In welding steel, upsetting diverts the grain flow of the steel from its normal direction, and this may have some effect on strength.

In tension tests, yielding often occurs on both sides of the weld before fracture takes place. Fracture occurs in the base metal if it is the softer structure, or in a region softened by the heat of welding if the base metal is hard.

When workpieces are uniformly heat treated after welding, the strength of the weld is theoretically as high as that of the base metal, except where the base metal has a relatively high tensile strength. If a flash welded carbon or alloy steel workpiece is heat treated so that it has a tensile strength less than about 150 ksi, the workpiece is homogeneous throughout and may fracture either at or away from the bond interface. At a tensile strength higher than about 150 ksi, the fracture is more likely to occur at the bond interface, presumably because of the removal of some carbon from the steel at the bond interface during welding or because of the presence of inclusions. In the following example, in which a stainless steel ring was annealed after flash welding, the weld had only slightly less strength than the base metal.

Example 6. Production of a Stainless Steel Ring by Roll Forming and Flash Welding. A jet-engine rear-mount ring, a section of which is shown in Fig. 24, was made by roll forming a type 410 stainless steel Aerospace Materials Specification (AMS) 5613 extruded section and then flash welding it into a continuous ring. The tensile strength of the welded joint was 98.8 to 99.7% that of the base metal.

The ring was welded in a 750-kV · A, single-phase, 440-V machine with a maximum secondary output rating of 7500 A at 16.9 V. The clamping mechanism and the movable platen were hydraulically actuated. To ensure good conduction of electrical current, the 3-in.-long clamping dies were made of RWMA class 2 copper electrode material, and the clamping surfaces on the workpiece were polished to remove oxide film, scale, and other contaminants that could hinder current flow. The operations for making the ring were:

- Cut extruded section to length.
- Roll form in a horizontal three-roll forming machine equipped with rolls shaped to fit the extrusion.
- Polish end surfaces for electrode contact.
- Flash weld.
- Remove flash and polish HAZ.
- Anneal to 187 HB.
- Size, round up, and proof test.
- Heat treat (if required by customer).

The welded joint in each ring was proof tested in a hydraulically actuated expander that applied uniform pressure on all sections of the ring simultaneously. The proof test consisted of 1% minimum cold expansion of the metal across the weld in a 2-in. gage length. The testing operation also made the ring round and sized it to the specified inside diameter.

Fig. 24 Flash welding application. See text for details.

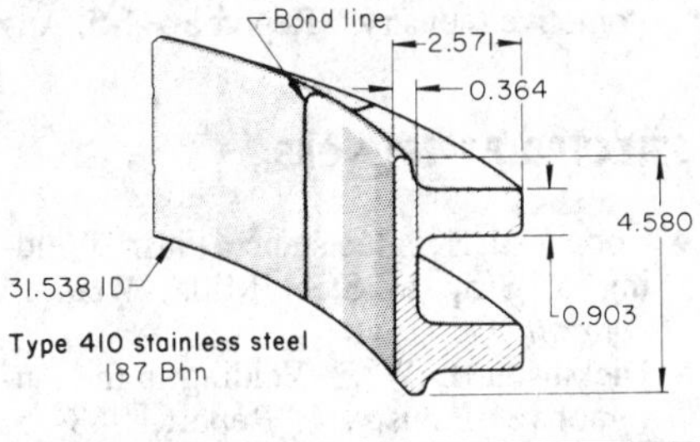

Initial die opening	$3^5/_8$ in.
Final die opening	$1^1/_4$ in.
Metal lost	$2^3/_8$ in.
Preheat time	0.6 s
Flashing time	25 s

One ring from each production lot was tension tested according to AMS 7493, which covers flash welding of martensitic stainless steel. Standard American Society for Testing and Materials (ASTM) tension-test specimens were used. The base metal had a hardness of 187 HB, tensile strength of 95 ksi, and 29% elongation. Two test specimens taken across the weld had the following properties: hardness, 187 HB; tensile strength, 94 and 94.9 ksi; joint efficiency, 98.8 and 99.7%; and elongation, 29 and 26%.

Before proof testing, all rings were inspected visually for alignment and for surface defects. Random samples were checked radiographically for internal defects in the HAZ.

The use of an extruded section for the ring instead of a rectangular bar resulted in material saving and additional saving in machining costs, because there was less metal to remove. Production rate was seven rings per hour, including polishing the clamping surfaces, flash welding, and removing the weld upset.

Alpha-beta titanium alloys can sometimes be strengthened by aging at moderately elevated temperatures. The rapid quenching effect of clamping dies used in flash welding alters the transformation of the elevated-temperature beta phase that occurs with slow cooling. Aging usually decreases the ductility of the work metal with little or no change in the hardness, as in the following example.

Example 7. Production of a Titanium Ring by Flash Welding. The ring shown in Fig. 25 was used, after machining, as a jet-engine inlet-case seal. The workpiece was a titanium alloy Ti-6Al-4V extrusion with a hardness of 34 HRC, tensile strength of 145 ksi, and 14% elongation. The welded joint had the same hardness, but a tensile strength of 152 ksi, and elongation of 12.5%. Thus, the joint efficiency was 105%. The welded area was about 0.5 $in.^2$

The joint was made in a 750-kV · A, single-phase, 440-V flash welding machine with a maximum secondary output rating of 7500 A at 16.9 V. Hydraulically actuated clamps, 3 in. in length and made of high-conductivity copper (RWMA class 2), held the roll formed ring during welding. The movable platen was also hydraulically actuated.

The extruded section was cut to length, heated to 1250 °F, and then roll formed into a circle in a horizontal three-roll forming machine. To ensure maximum current flow, the contact surfaces of the workpiece at the clamping dies were polished before the ring was placed in the welding machine. After welding, the weld upset was removed, the weld area was ground by hand to the same dimensions as the adjoining area, and the joint was inspected for alignment and for defects. Then the welded rings were heat treated in accordance with AMS 4928.

The joint was proof tested by placing the welded ring in a hydraulically actuated expander and applying uniform pressure to all sections of the ring simultaneously. The circumference of the ring was increased about 2 in., or $1^1/_4$%, during this operation, which also made the ring round and sized the inside diameter to specification.

One ring from each production lot was used for destructive tests on the base metal and the welded joint. The tests were conducted according to AMS 7498, which covers flash welding of titanium alloys. Production rate was 35 rings per hour.

Fig. 25 Flash welding application. See text for details.

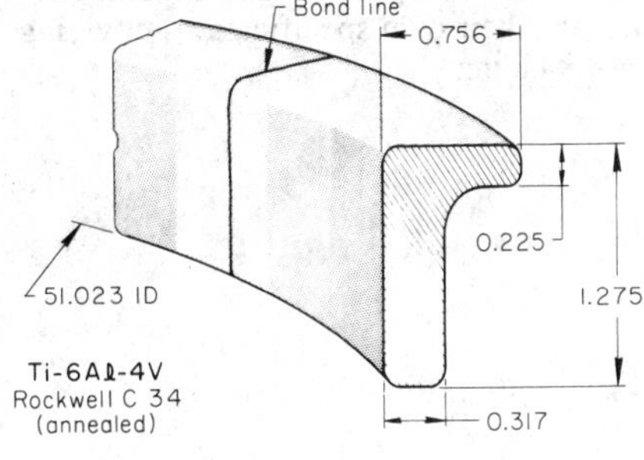

Initial die opening	$2^1/_8$ in.
Final die opening	$^3/_4$ in.
Metal lost	$1^3/_8$ in.
Preheat time	10 s
Flashing time	10 s

Weld Flaws

Common flaws in flash welds include: circumferential crevices at the bond line that appear after removal of the upset; cracking caused by brittleness in the HAZ; cast metal in the weld; voids, oxides, and inclusions in the weld; and intergranular oxidation and decarburization in the HAZ.

Circumferential Crevices. If either heating or upsetting is insufficient, bonding will not extend across the entire workpiece and into the weld upset; removal of the weld upset will reveal the presence of an unwelded area in the workpiece. Such a crevice indicates that a 100% bond has not been obtained and that the strength of the workpiece may be less than minimum requirements. This type of flaw usually can be prevented by the use of greater upsetting pressure or the use of more heat, or both.

Cracking in the heat-affected zone, caused by rapid cooling, can be minimized by releasing the electrodes immediately after upsetting and transferring the workpiece to a furnace for temperature equalization and controlled cooling. After the workpieces have been uniformly heated in the furnace, they can be buried in lime, powdered mica, or other insulating material to ensure slow cooling. If a furnace is not available and the workpieces are of carbon steel, the weld can be reheated in the welding machine by passing current through the finished weld until a dull red color is seen. Then the current is turned off, the electrodes are quickly unclamped, and the workpiece is buried in lime or other insulating material for slow cooling.

In the flash welding of parts made of dissimilar steels, such as welding of high-speed steel drills to medium-carbon steel shanks, slow cooling must be ensured to avoid cracks in the steel of higher hardenability. Drills made by flash welding normally are fully heat treated after welding. When drills are repaired, or are made longer by flash welding a short drill to a shank extension, rehardening after welding may not be feasible, and slow cooling is necessary—not only to prevent cracking, but also to minimize brittleness in the weld zone. Such brittleness could lead to failure of the tool in service. In a few applications, the parts to be welded are furnace preheated, welded, and immediately furnace heat treated to prevent cracking.

Cast Metal in the Weld. A certain amount of molten metal forms at the weld faces during flashing and should be extruded from the weld during upsetting, leaving the weld area in a homogeneous condition. However, pockets of molten metal sometimes are trapped in the weld and appear as cast metal under a microscope. Because cast metal is likely to contain small shrinkage cracks and is of lower ductility than worked metal, its presence should be reduced to a minimum.

Cast metal may be trapped in welds between dissimilar alloys that have widely differing hot compressive strengths. The metal or alloy with the lower hot compressive strength upsets to a greater degree, and so entraps cast metal in the weld. Metals with high electrical resistivity become hotter during welding, increasing the probability of entrapment of cast metal.

Voids frequently are formed when the piece being welded is too large for the welding machine and the upset pressure is insufficient to close all the craters formed during flashing. Sometimes, molten metal formed at the weld interface may be extruded into a void during upsetting, and can be seen under the microscope as a globule of cast metal in the void.

Oxides and other inclusions may be present in joints that have been welded with insufficient upsetting force to expel them, or at too low a temperature to give plasticity to the weld region. Inclusions are very small but they weaken the weld, and heat treatment does not remove them or improve the properties of the weld.

Another type of inclusion is referred to as a penetrator. Penetrators are more prevalent when a high flashing voltage is used and are caused by large temperature gradients, occluded gas, and nonmetallic inclusions that are trapped in craters in the flashing surface during upsetting. The occurrence of penetrators can be minimized by using the correct voltage, platen speed, and flashing gap, and by using a protective atmosphere.

Decarburization. During flash welding, high-carbon steel loses carbon along the flashing surfaces, especially when welding is done in air or in an atmosphere of moist gas. The decarburized metal is not entirely expelled during upsetting and some remains in the weld as a zone of weakness.

Intergranular oxidation ("burning") generally is caused by improper material extension during welding of dissimilar metals. Sufficient heat can be developed to oxidize the metal at a point behind the weld. Sometimes, oxidation can cause the workpieces to fall apart in the oxidized region. Improper contact between the electrode and the workpiece can cause a locally melted or burned area beneath the electrode.

Flash Welding Specifications

Aerospace Material Specifications

- AMS 7488 for aluminum and aluminum alloys
- AMS 7490 for corrosion- and heat-resistant austenitic steels
- AMS 7493 for nonaustenitic corrosion-resistant steels
- AMS 7496 for carbon and low-alloy steels
- AMS 7498 for titanium and titanium alloys

Federal (Military) Specification

- MIL-W-6873-B for carbon and alloy steels

ASME Boiler and Pressure Vessel Code

- Section IX: Welding Qualifications

American Welding Society (AWS)

- C1.1 Recommended Practices for Resistance Welding
- D8.1 Recommended Practices for Automotive Flash Butt Welding

In addition to the above specifications, many specialized industries such as airframe and aerospace engine manufacturers have their own specifications covering flash welding.

REFERENCE

1. Dickinson, D.W., "Welding in the Automotive Industry," Report S681-5, Aug 1981

SELECTED REFERENCES

- Cooper, J.H., Resistance Flash Welding of Strip in Steel Mills, *Welding Journal*, Oct 1940
- Dickinson, D.W., "Welding in the Automotive Industry," Report S681-5, American Iron and Steel Institute, Aug 1981
- Makara, A.M., *et al.*, The Flash Welding of High Tensile Steels, *Automatic Welding*, Vol 21 (No. 1), 1968
- Moore, C.D., Flash Welding Aluminum to Copper, *Machinery*, Vol 99, Oct 1961
- Nippes, E.F., *et al.*, Temperature Distribution During the Flash Welding of Steel, *Welding Journal*, Vol 30, Dec 1951
- Nippes, E.F., *et al.*, A Mathematical Analysis of the Temperature Distribution During Flash Welding, *Welding Journal*, Vol 34, June 1955
- "Recommended Practices for Resistance Welding," AWS C1.1-66, American Welding Society
- *Resistance Welding Manual*, Vol 2, 3rd ed., Resistance Welder Manufacturers Association, 1961
- Savage, W.F., Flash Welding: Process Variables and Weld Properties, *Welding Journal*, Vol 41, March 1962
- Savage, W.F., Flash Welding: The Process and Applications, *Welding Journal*, Vol 41, March 1962
- Stieglitz, H.W., Flash Welding Copper to Steel, *Metal Progress*, Vol 80, Nov 1961
- Sullivan, J.F. and Savage, W.F., Effect of Phase Control During Flashing on Flash Weld Defects, *Welding Journal*, Vol 30, May 1971
- *Welding Handbook*, Vol 3, 7th ed., American Welding Society, 1980

Oxyfuel Gas Welding

Oxyfuel Gas Welding Processes and Their Application to Steel

By the ASM Committee on Oxyfuel Gas Welding*

OXYFUEL GAS WELDING (OFW) is a manual process in which the metal surfaces to be joined are melted progressively by heat from a gas flame, with or without filler metal, and are caused to flow together and solidify without the application of pressure to the parts being joined. The most important source of heat for OFW is the oxyacetylene welding torch.

The simplest and most frequently used OFW system consists of compressed gas cylinders, gas pressure regulators, hoses, and a welding torch. Oxygen and fuel are stored in separate cylinders. The gas regulator attached to each cylinder, whether fuel gas or oxygen, controls the pressure at which the gas flows to the welding torch. At the torch, the gas passes through an inlet control valve and into the torch body, through a tube or tubes within the handle, through the torch head, and into the mixing chamber of the welding nozzle or other device attached to the welding torch. The mixed gases then pass through the welding tip and produce the flame at the exit end of the tip. This equipment can be mounted on and operated from a cylinder cart, or it can be a stationary installation. Filler metal, when needed, is provided by a welding rod that is melted progressively along with the surfaces to be joined.

Capabilities, Advantages, and Disadvantages

In OFW, the welder has considerable control over the temperature of the metal in the weld zone. When the rate of heat input from the flame is properly coordinated with the speed of welding, the size, viscosity, and surface tension of the welding pool can be controlled, permitting the pressure of the flame to be used to aid in positioning and shaping the weld. The welder has control over filler-metal deposition rates, because the sources of heat and filler metal are separate. Heat can be applied preferentially to the base metal or the filler metal without removing either from the flame envelope. With these capabilities, OFW can be used for joining thin sheet metal, thin-walled tube, small pipe, and assemblies with poor fit-up, as well as for smoothing or repairing rough arc welds. Heavy sections can be joined by OFW, but arc welding is more economical.

The equipment is versatile, low-cost, self-sufficient, and usually portable. It can be used for preheating, postheating, welding, braze welding, and torch brazing, and it is readily converted to oxygen cutting. The process can be adapted to short production runs, field work, repairs, and alterations.

Metals That Can Be Oxyfuel Gas Welded. Most ferrous and nonferrous metals can be oxyfuel gas welded. Oxyacetylene supplies the heat intensity and flame atmosphere necessary for welding carbon steel, cast iron, and other ferrous, copper, and nickel alloys. Aluminum and zinc alloys can also be welded by the oxyacetylene process. Oxyfuel gas welding of steel is done almost exclusively with an oxyacetylene flame. Hydrogen, natural gas, propane, and several proprietary gases are used as fuel gases in welding metals with lower melting temperatures, such as aluminum, magnesium, zinc, lead, and some precious metals. Metals unsuited to OFW are the refractory metals, such as niobium, tantalum, molybdenum, and tungsten, as well as the reactive metals, such as titanium and zirconium.

Fluxes. Except for lead, zinc, and some precious metals, OFW of nonferrous metals, cast irons, and stainless steels generally requires a flux. In welding carbon steel, the gas flame shields the weld adequately, and no flux is required. Adjustment for correct flame atmosphere is important, but the absence of flux results in one less variable to control.

Major Applications. Oxyfuel gas welding can be used to join thin carbon steel sheet and carbon steel tube and pipe. The advantages of OFW include: the ability to control heat input, bridge large gaps, avoid melt-through, and clearly view the weld pool. Carbon steel sheet, formed in a variety of shapes, can often be welded more economically by OFW than by other processes. Oxyfuel gas welding is capable of joining small-diameter carbon steel pipe (up to about 3 in. in diameter) with resulting weld quality equal to competitive processes and often with greater economy. Pipe with wall thickness up to $^3/_8$ in. can be welded in a single pass.

Welder Skill. Oxyfuel gas welding requires skill in manipulating the welding rod and the torch flame. In depositing a weld, the welder uses both hands as he works at melting base metal and filler metal, controlling the weld pool, and obtaining progressive solidification of weld metal in the correct bead shape.

Gases

Oxygen and acetylene are the principal gases used in OFW. Oxygen supports

*Gene Meyer, *Chairman*, Manager, Product Distribution and Warranty Service, Victor Equipment; Joseph E. McQuillen, Product Manager, Air Products and Chemicals Inc.; Clarence N. Vaughn, Corporate Burning Manager, Chicago Bridge & Iron Co.

combustion of the fuel gases. Acetylene supplies both the heat intensity and the atmosphere needed to weld steel. Hydrogen, natural gas, propane, and proprietary gases are used only to a limited extent in oxyfuel gas welding or brazing of metals with a low melting temperature.

Oxygen. Only by burning selected fuel gases with high-purity oxygen in a high-velocity flame can the high heat transfer intensity required in OFW be obtained. Oxygen is supplied for oxyfuel gas welding and cutting at a purity of 99.5% and higher, because small percentages of contaminants have a noticeable effect on combustion efficiency.

When the consumption requirement is relatively small, the oxygen is supplied and stored as a compressed gas in a standard steel cylinder under an initial pressure of up to 26 ksi. The most frequently used cylinder (Fig. 1) has a capacity of 244 scf.* The gas is distributed for use under reduced pressure. When consumption of oxygen is somewhat greater, banks of cylinders are joined through a manifold to permanent pipeline systems that terminate at various stations of use.

When oxygen consumption exceeds approximately 244-scf cylinders per week, it may be more economical to obtain and store oxygen in liquid form. Liquid oxygen can be supplied in portable cryogenic cylinders of 4500-scf capacity, or it can be delivered in bulk to an on-site cryogenic storage tank. The gas is then piped to points of use. The distribution and use of oxygen are covered by laws and safety regulations designed to help prevent injury to persons and damage to property.

Acetylene is a hydrocarbon gas with the chemical formula C_2H_2. When under pressure of 29.4 psi and above, acetylene is unstable, and a slight shock can cause it to explode, even in the absence of oxygen or air. Safety rules for the use of acetylene and the handling of acetylene equipment are extremely important.

Acetylene should not be used at pressures greater than 15 psi. Acetylene generators for on-site gas production are constructed so that the gas is not given off at pressures much greater than 15 psi. Commercially supplied portable cylinders are specially constructed (Fig. 1) to store acetylene under high pressure. By dissolving acetylene in liquid acetone, a cylinder such as that shown in Fig. 1 can be used to store about 275 scf of acetylene under a pressure of 250 psi. This pressure must be reduced to 15 psi or less by the regulator (Fig. 1) before the gas enters the hose or distribution line.

*A standard cubic foot (scf) of gas is defined as equivalent to 1 ft^3 of gas at 70 °F and 1 atm (14.7 psi) pressure. This definition, which is used in the gas industry and some engineering practice, differs from the standard temperature and pressure of 32 °F and 1 atm pressure used in scientific work.

Fig. 1 Gas cylinders and regulators used in OFW

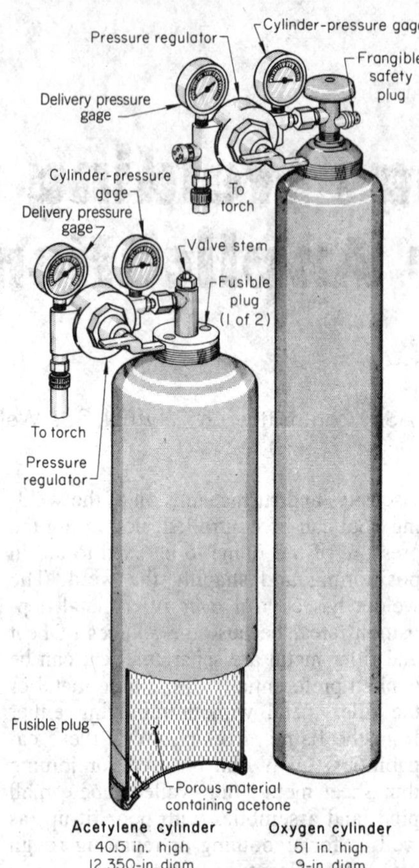

Acetylene cylinders must not be subjected to sudden shock and should be stored well away from any source of heat or sparks. The cylinders must be stored in an upright position to keep the acetone from escaping during use. Under normal sustained use, withdrawal rate from an acetylene cylinder should not exceed one seventh of the cylinder capacity per hour.

Hydrogen is used chiefly for welding lower melting temperature metals, such as aluminum, magnesium, and lead. It cannot be used to weld common thicknesses of steel sheet, because it results in a flame temperature that is too low to produce good fusion. Hydrogen has, however, been used in welding thin sheet, where its lower combustion intensity (about 60% of that of acetylene) can be an advantage. It is used for brazing and, to some extent, for braze welding. Hydrogen is available in compressed-gas cylinders of various sizes.

Natural gas, propane, and proprietary gases can be used with oxygen to weld some lower melting temperature metals. Special welding rods may be required depending on the gas, base metal, and quality of metal desired. These gas mixtures cannot usually be applied to the welding of steel, because when they are burned at temperatures high enough for welding, their flame atmospheres are excessively oxidizing. When ratios of oxygen to fuel gas are reduced to a carburizing condition, flame temperatures are too low. Therefore, these gases are usually limited to brazing, braze welding, heating, and related processes.

Equipment

The principal function of OFW equipment is to supply the oxyfuel gas mixture to the welding tip at the correct rate of flow, exit velocity, and mixture ratio. The rate of gas flow affects the quantity of metal melted; the pressure and velocity affect the manipulation of the weld pool and the rate of heating; and the ratio of oxygen to fuel gas determines the flame temperature and the atmosphere, which must be chemically suited to the metal being welded. Important elements in an OFW system include: (1) gas storage facilities, (2) pressure regulators, (3) hoses, (4) torches, (5) related safety devices, and (6) accessories.

Gas Storage Equipment. Compressed-gas cylinders and storage tanks are used as on-site supply sources for gases. They are constructed under regulations of the Department of Transportation and are regulated to some extent by federal, state, and local laws. Cylinders are designed for specific gases and are not generally interchangeable. Sizes and threading of cylinder connections for oxygen, for example, differ from those for acetylene, hydrogen, and other gases. Only the appropriate fittings can be used for delivery of compressed gas. Users should not tamper with valves or safety devices on cylinders.

When portable equipment is not required and gas consumption is large enough, permanent installations are constructed. Gases can be supplied from a variety of bulk-storage vessels, manifolded gas cylinders, or gas generators. The gases are then separately piped, at suitable pressure, to terminal stations, where they are drawn off through station regulators and used as in a portable setup. See the National Fire Protection Association Bulletin No. 51 ("Oxygen-Fuel Gas Systems for Welding and Cutting") for more details.

Pressure regulators reduce the supply gas pressure to a desired delivery pres-

Fig. 2 Delivery pressure for three pressure regulators

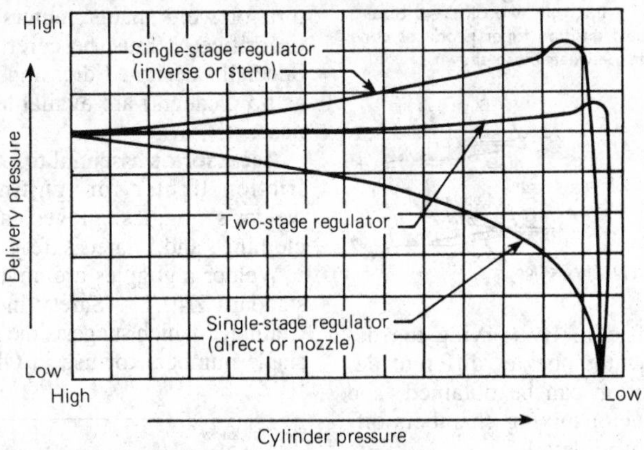

sure. They are designed for specific gases and are not generally interchangeable. In Fig. 1, oxygen and acetylene pressure regulators used in OFW are shown attached to their respective cylinders and hoses. In operation, gas enters the inlet side of the regulator at cylinder pressure and emerges from the outlet side at the desired delivery pressure. Regulators are made for various ranges of inlet and outlet pressure. They can be adjusted within their delivery pressure range by turning an adjusting screw. Although some regulators do not have gages and are preset to deliver at a specific and constant pressure, most are equipped with two pressure gages. The one on the outlet side permits the operator to read the adjusted delivery pressure; the one on the inlet side indicates the pressure in the cylinder.

Two basic types of regulators are used in OFW: single-stage and two-stage. Both are available with either direct or inverse actuation of the valve mechanism. Regulators with direct actuation are known also as positive or nozzle regulators; those with inverse actuation, as negative or stem regulators. All single-stage regulators reduce cylinder pressure in one step. Pipeline regulators, made for lower pressures, must not be used on cylinders (the higher cylinder pressure can blow apart a pipeline regulator which is not designed for the higher pressure).

When adjusted for a desired delivery pressure, regulators continue to deliver gas at the pressure shown on the outlet gage within a fairly narrow range of deviation, as the cylinder pressure drops. The amount and direction of the deviations depend on the design of the regulator. The variation in delivery pressure as cylinder pressure decreases is shown in Fig. 2 for three types of regulators. Both types of single-stage regulators may require occasional adjustment as cylinder pressure drops. The more costly two-stage regulator supplies a more constant working pressure until cylinder pressure nearly equals delivery pressure. On full oxygen cylinders exposed to low outdoor temperatures in winter, two-stage regulators are more reliable than single-stage regulators. The temperature drop of expanding gas is less severe, because pressure is reduced in two steps. Less chance exists for ice to form and clog regulator passages if water vapor is present.

Hoses. Flexible hoses permit gas cylinders and regulators to be kept at a safe distance from the working area and also allow the welder freedom of movement. Hoses usually range from $^1/_8$ to $^1/_2$ in. in inside diameter, with larger sizes available for special applications, and are usually available in 25-ft lengths. If hoses are longer than 25 ft, either the next larger size hose or a length of larger diameter hose with a short length of the usual size hose should be used to connect to the welding torch. This facilitates ease of movement. To ensure that oxygen and fuel gas hoses are used with the proper gas and correct fittings, oxygen hoses are generally green, with right-hand threaded fittings; acetylene and other fuel gas hoses are generally red, with left-hand threaded fittings. Left-hand threaded fittings are grooved on the outside so they can be readily identified.

Welding torches control the operating characteristics of the welding flame and enable the flame to be manipulated during welding. The choice of torch size and style depends on the work to be performed. Aircraft welding torches, for example, are small and light to permit ease of handling. Most torch styles permit one of several sizes of welding tips or a cutting attachment to be added.

The general construction of an OFW torch is shown schematically in Fig. 3. The principal operating parts are inlet valves, rear body, handle, and head. A mixing chamber with a welding tip, heating nozzle, or cutting attachment is added to the

Fig. 3 Oxyfuel gas welding torch

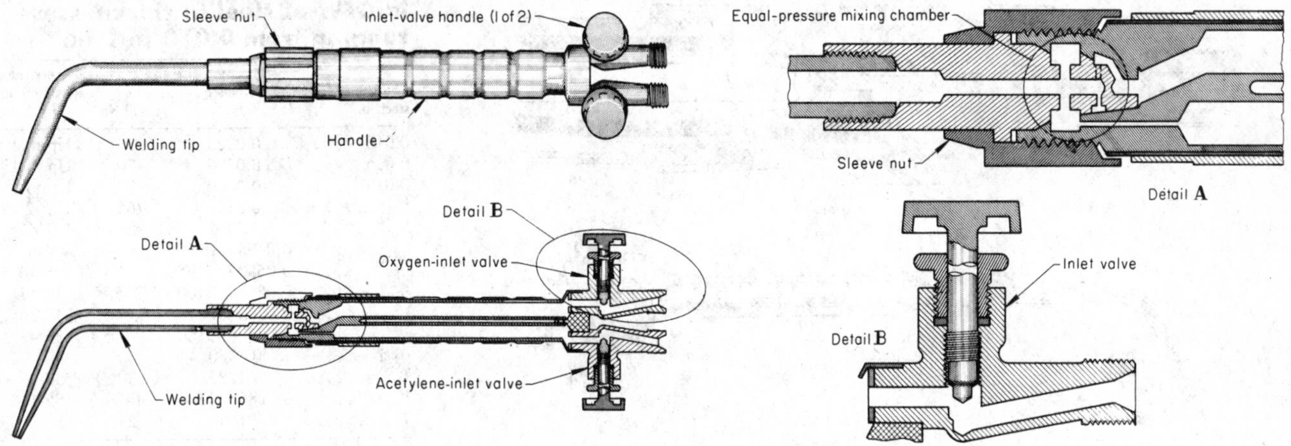

Fig. 5 Inner cones of oxyacetylene welding flames. Produced by welding tips with two different bore shapes. Varying the transition taper produces cone shapes intermediate between those shown.

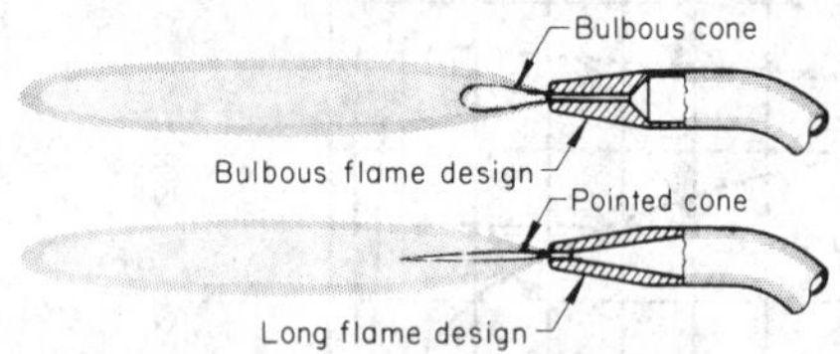

head. By unscrewing the sleeve nut (Fig. 3), the welding tip and mixing-chamber assembly can be removed and replaced by units with different capacity.

Torch inlet valves provide the welder with two important controls. First, the gas pressure, velocity, and flow in cubic feet per hour can be adjusted within the limits set by the pressure regulators and the practical requirements of the welding flame. Second, the ratio of oxygen to fuel gas can be varied. The ability of the oxyfuel flame to produce the combustion intensity needed for welding, while providing a suitable protective atmosphere for the weld metal, is largely a property of the fuel gas, but correct torch settings must be used.

Mixing Chambers. In Fig. 3, the mixing chamber is shown as part of an assembly with the welding tip. Two general mixing chambers are available: equal-pressure (also called positive-pressure and medium-pressure) and injector. In an equal-pressure mixing chamber, gases are at approximately the same pressure and are mixed by directing the fuel gas into the oxygen stream. Detail A in Fig. 3 shows an equal-pressure mixing chamber assembled in position. Figure 4 is a spiral equal-pressure mixer.

In the injector mixing chamber, low-pressure fuel gas is aspirated by directing it into a high-velocity stream of oxygen. A nozzle system based on the flow principles of the venturi tube is used. Injector torches are useful when fuel gases are supplied at pressures too low to produce a flame of adequate combustion intensity. Supplying oxygen at desired pressure is not usually difficult. By varying the design of the injector nozzle, different degrees of aspiration can be obtained, and designs for injector mixing chambers differ considerably in detail.

Welding tips are replaceable nozzles that control gas flow through the diameter of the exit orifice. Tips of various orifice diameter are usually available for any welding torch. Orifice diameters are identified by drill-size number, decimal size in inches, or manufacturer's code number. Because code numbers of different manufacturers do not necessarily correspond, drill sizes or decimals are needed to compare the orifice size of different makes of tips. The performance of tips of equal size at equal pressure-regulator settings may differ, however, because of differences in torch and mixing-chamber designs.

Small-diameter tips produce small flames for welding thin sections; large-diameter tips are required for heavier work. Welding tips are made with a smooth bore at the exit end to ensure laminar flow and a uniform flame. The influence of bore shape on the shape of the flame is shown in Fig. 5. When foreign matter, such as carbon, dirt, or weld spatter, enters the welding-tip orifice, it must be carefully removed. Specially designed dressing tools (known as tip cleaners) are available for this purpose.

Accessories essential to OFW include a friction lighter for igniting the torch, welder's goggles, gloves, and protective clothing, and related safety devices.

Welder's goggles are covered by ANSI standard Z49.1, "Safety in Welding and Cutting," which suggests the following lens shade numbers for use in OFW of steel:

Steel thickness	Shade No.
Up to $1/8$ in.	4 or 5
$1/8$ to $1/2$ in.	5 or 6
Over $1/2$ in.	6 to 8

For general OFW, goggles must have side shields.

Selection of Tip-Orifice Size and Gas Pressure

Torch manufacturers provide charts that give tip-orifice sizes to be used for welding different thicknesses of metal. Table 1 presents the data from a chart showing these relationships. Tip size number and drill size are not precise means of comparing tips, because significant differences exist between tips of the same dimensions when operated under positive pressure and when operated under injector pressure. The only valid basis of comparison is the volume of fuel passing through the tip in a unit of time. Most tips operate within a range, usually recorded as cubic feet of fuel per hour. The manufacturer's recommendations should always be followed for optimum performance and safety.

Table 1 Approximate tip-orifice sizes and acetylene consumption in OFW of steel in thicknesses ranging from 0.010 to 1 in.

Thickness of steel, in.	Tip-orifice size: Diameter, in.	Tip-orifice size: Drill No.	Acetylene consumption, ft^3/h
0.010	0.0225	74	Up to 1
0.016	0.0280	70	Up to 1
0.019	0.0280	70	Up to 1
$1/32$	0.0350	65	$1/2$-2
$1/16$	0.0465	56	1-4
$3/32$	0.0465-0.0550	56-54	4-6
$1/8$	0.0550-0.0595	54-53	6-10
$3/16$	0.0595-0.0700	53-50	10-17
$1/4$	0.0700-0.0810	50-46	17-30
$3/8$	0.0810-0.0860	46-44	30-45
$1/2$	0.0980	40	40-60
$5/8$	0.1285	30	50-75
$3/4$	0.1285-0.1360	30-29	65-100
1	0.1540	23	85-140

Fig. 4 Spiral equal-pressure mixer. (1) Welding torch head. (2) Oxygen tube from torch head. (3) Acetylene (fuel gas) passages from torch head. (4) Nozzle nut. (5) Welding nozzle cone end. (6) Spiral in welding nozzle. (7) Mixer orifice and mixing chamber

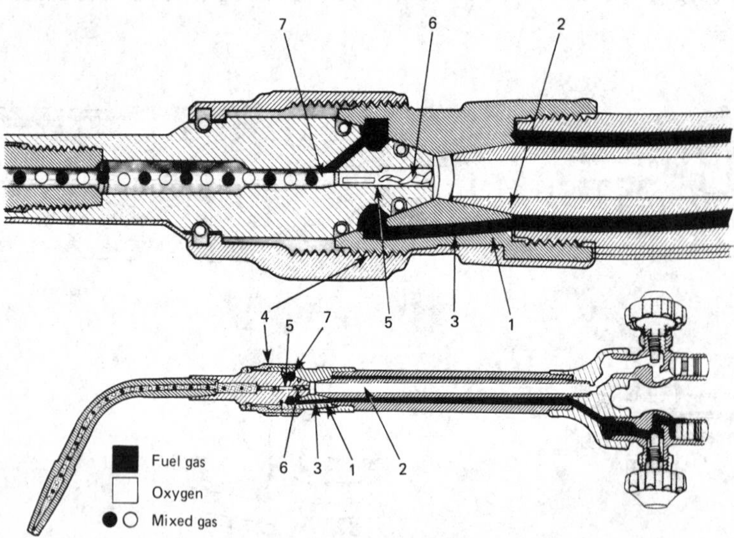

Flame Adjustment

Different welding atmospheres and flame temperatures can be produced by varying the relative amounts of oxygen and fuel gas in the gas flowing to the tip of the torch. Usually, a welder makes the appropriate adjustments in gas flow based on the appearance of the flame. This is not true for oxyhydrogen welding, however. The sequence for setting up a positive-pressure welding outfit is:

- Check all parts of the apparatus, making sure they are free of dirt, oil, or grease and in proper working condition.
- Crack each cylinder valve individually to blow out foreign matter. Make sure vented gases are safely dispersed. Wipe out cylinder-valve outlet with a clean, lint-free cloth.
- Attach the oxygen regulator to the oxygen cylinder or manifold.
- Attach the acetylene (or fuel gas) regulator to the fuel cylinder or source.
- Connect the welding hose to the regulators and the welding torch.
- Open the cylinder valve slowly and carefully. The operator should never stand in front of the regulator when opening the cylinder valve.
- Purge the oxygen line while the acetylene line is closed and the acetylene line while the oxygen line is closed. Vent gases safely.
- Set the oxygen and fuel gas regulators to the recommended working pressure with appropriate torch valve open.
- Open the acetylene (or fuel gas) inlet valve and light the welding torch, using a spark lighter.
- Open the oxygen inlet valve and adjust the flame, using both inlet valves.

For injector equipment, the sequence and method of setting up differ, because high-pressure oxygen aspirates low-pressure acetylene into the torch. The differences among brands make it essential to always follow the manufacturer's directions for setting up.

Oxyacetylene Combustion

As the acetylene/oxygen mixture burns from the tip of the welding torch, it displays several clearly recognizable zones of combustion. The overall chemical equation for the complete combustion of acetylene is:

$$2C_2H_2 + 5O_2 \rightarrow 4CO_2 + 2H_2O \qquad \text{(Eq 1)}$$

Combustion takes place in two stages. The first stage:

$$2C_2H_2 + 2O_2 \rightarrow 4CO + 2H_2 \qquad \text{(Eq 2)}$$

uses the oxygen supplied from the cylinder and available in the oxyacetylene mixture. The reaction can be seen as the small inner cone of the flame. The highest temperature is at the point of this cone. The second stage:

$$4CO + 2H_2 + 3O_2 \rightarrow 4CO_2 + 2H_2O \qquad \text{(Eq 3)}$$

uses the oxygen supplied from the air surrounding the flame. This combustion zone constitutes the outer envelope of the flame.

About two fifths of the oxygen necessary for the complete combustion of acetylene comes from the oxygen cylinder; the remainder comes from the air. Because of the need for supplemental oxygen from the atmosphere, the acetylene/oxygen flame cannot be used inside tubes of structures subject to oxygen depletion from OFW. By varying the relative amounts of acetylene and oxygen in the gas mixture in the torch, a welder can produce different flame atmospheres and temperatures.

The second equation shows that in the first stage, when equal amounts of oxygen and acetylene are burning, neither excess acetylene nor excess oxygen is present at the high-temperature tip of the inner cone. For this reason, the flame is called neutral, and the gas mixture is often described as an acetylene-to-oxygen ratio of 1 to 1, or an equal ratio. The condition of the flame is important, because the inner cone is held close to, but not touching, the work metal. If excess oxygen is present, the molten metal foams and sparks, and brittle oxides may form in the weld metal. If acetylene is present in sufficient excess, indicated by an acetylene feather greater than about one half the length of the inner cone, carbon can enter the metal. Some of the carbon forms carbides and some burns, causing gas and porosity in the solidified weld metal. In austenitic stainless steels, carbides formed in this manner aggravate susceptibility to intergranular corrosion. In welding steel, the flame should be as nearly neutral as possible, but a perfectly neutral flame is difficult to recognize, and a flame containing a slight excess of acetylene is often used.

A neutral flame is obtained by observing the size and color of the combustion zones in the flame as the oxygen-to-acetylene ratio is changed by adjusting the torch inlet valves. Figure 6 shows five typical flame conditions that appear as oxygen flow is increased from zero to excess oxygen, as well as a separated flame condition that results from excessive gas pressure.

Fig. 6 Flame conditions

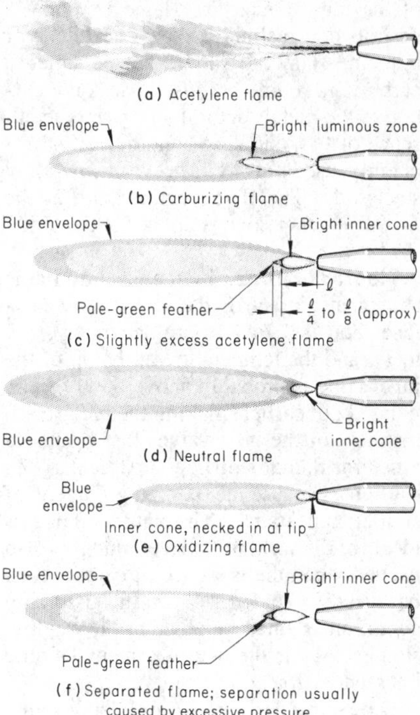

Acetylene Flame. When acetylene alone is burned in air, it produces a flame that varies in color from yellow near the torch tip to orange-red at the outer extremity. If soot particles are present at the end of the flame, the volume of gas should be increased until the flame burns clear (Fig. 6a).

Carburizing Flame. As the oxygen valve in the torch is progressively opened and the ratio of oxygen to acetylene increases, the flame becomes generally luminous. Then, the luminous portion contracts toward the welding tip, forming a distinct bright zone within a blue outer envelope (Fig. 6b). This is a carburizing flame because it has a large excess of acetylene, sometimes described as a soft flame because it has very little force. It has a relatively low temperature and is used in silver brazing and soldering, as well as in the welding of lead.

Reducing Flame. As more oxygen is introduced, the bright zone of the flame contracts farther and is seen to consist of two parts: a bright inner cone and a pale-green streamer, or feather, trailing off its end into the blue envelope. The streamer or feather is caused by a slight excess of acetylene. It disappears as the oxygen-to-acetylene ratio approaches 1 to 1. In Fig. 6(c), the feather is shown adjusted to about one quarter the length of the inner cone. For welding steel, the length of the feather should be about one eighth to one quarter, but never more than one half, the length of the inner cone. The flame is properly described as a slightly excess acetylene or reducing flame. It should not be called a carburizing flame because it does not carburize the metal, but it does ensure the absence of the oxidizing condition. It is frequently used for welding with low-alloy steel rods. The flame temperature at the tip of the inner cone is about 5300 to 5500 °F.

Neutral Flame. In the neutral flame shown in Fig. 6(d), the oxygen-to-acetylene ratio is 1 to 1 (more accurately, 1.1 to 1), and the temperature at the tip of the inner cone is probably above 5500 °F. As pointed out earlier, making the precise adjustment of the inlet valves that results in a neutral flame is difficult, particularly in sunlight, because the oxidizing flame is of similar appearance. The neutral flame is ideal for the welding of steel, and it is also needed when the presence of carbon must be strictly avoided. When the oxidizing condition is unacceptable, as in welding stainless steel, the use of a neutral flame is essential for good results.

Oxidizing Flame. An oxidizing flame is shown in Fig. 6(e). The adjustment of the inlet valves for this flame is difficult, because it cannot be made on the basis of luminosity. When the flame tends to neck in at the juncture of the inner cone and the tip of the torch, this best indicates an oxidizing condition. When the flame is adjusted to be extremely oxidizing, it may also produce a hissing sound. An oxidizing flame should never be used in welding steel. It is used only in welding copper and certain copper-based alloys, and the flame should be just sufficiently rich in oxygen to ensure that a film of oxide slag forms over the weld to provide shielding for the weld pool. With oxygen-to-acetylene ratios of about 1$^3/_4$ to 1, flame temperature can approach 6000 °F.

Separated Flame. When gas pressure is too high for the size of the tip used, the flame can separate from the tip (Fig. 6f) and may even blow out. This is an unusable flame condition and must be avoided. Another cause of this condition is a clogged tip orifice.

Shape of Flame Cone. The design of the welding tip largely determines the shape of the inner cone of the oxyacetylene flame (Fig. 5); however, as the amount of oxygen used in the gas mixture is increased, the shape of the cone becomes more pointed. Both bulbous and pointed cones produce sound welds. The flame with the bulbous inner cone is preferred for welding in deep grooves in heavy sections. The flame with the long, pointed inner cone is softer and is generally preferred for welding thin sheet and aircraft assemblies. Some welders prefer the long, pointed inner cone for joining pipe because that shape facilitates penetration and fusion for the root pass in single-V-grooves.

Oxyhydrogen Combustion

Complete combustion of hydrogen requires an oxygen-to-hydrogen ratio of 1 to 2, as can be seen from the following equation:

$$2H_2 + O_2 \rightarrow 2H_2O \quad \text{(Eq 4)}$$

This gas mixture produces a strongly oxidizing flame having a temperature of about 5000 °F.

It is impossible to obtain a neutral oxyhydrogen flame by the visual methods of flame adjustment described for the oxyacetylene flame. The oxyhydrogen flame itself is scarcely visible, and no combustion zones, typical of the hydrocarbon gases, can be determined. To avoid an oxidizing flame, the pressure regulators must be set to ensure an excess of hydrogen. The flame is then reducing, but not carburizing. It has no carbon, and the temperature is several hundred degrees lower than that of the neutral flame. Metering flow regulators permit establishing the desired ratio of hydrogen to oxygen, usually 4 to 1. The oxyhydrogen flame is useful for welding and brazing aluminum alloys and lead.

Combustion of Natural Gas and Propane

Complete combustion of natural gas (methane) and propane is shown, respectively, by the following equations:

$$CH_4 + 2O_2 \rightarrow CO_2 + 2H_2O \quad \text{(Eq 5)}$$

$$C_3H_8 + 5O_2 \rightarrow 3CO_2 + 4H_2O \quad \text{(Eq 6)}$$

When the flame temperature is high enough to weld steel, the flame atmosphere is excessively oxidizing, but when the ratio of oxygen to fuel gas is decreased to produce a carburizing condition, flame temperature is too low for welding steel.

Welding Rods

Filler metal for OFW of low-carbon steel is available in the form of cold drawn steel rods 36 in. long and $^1/_{16}$ to $^1/_4$ in. in diameter. Welding rods for OFW of other metals are supplied in various lengths, depending on whether they are wrought or cast.

Specifications. Steel welding rods have been standardized in the American Welding Society (AWS) specification A5.2, "Iron and Steel Gas Welding Rods." This specification shows three classifications of welding rods based on the minimum tensile strength of all-weld-metal and transverse weld test specimens:

AWS classification	Minimum tensile strength, ksi	Minimum elongation in 4D(a), %
RG65	67(b)	16
RG60	60(b)	20
RG45	45(b)	...

(a) Elongation in approximately 1 in. (b) All-weld-metal test specimens, 0.252 ± 0.005 in. in diameter; as welded. (c) Tension tests based on transverse weld test specimens, 2 by $^3/_8$ in. in cross section; as welded

The mechanical properties specified for the RG65, RG60, and RG45 rods are obtained by welding in a neutral to slightly excess acetylene atmosphere.

The specification covers the chemical composition of the welding rods only to the extent of limiting sulfur and phosphorus to 0.040% maximum, and aluminum, if present, to 0.02% maximum. Therefore, welding rods of different manufacturers may vary appreciably in chemical composition.

Rods for OFW of steel have no flux covering. In the absence of flux coverings, weld-metal properties depend on chemical composition of the weldiing rod, control of the welding atmosphere, and techniques used to provide for mixing of base metal and filler metal.

Weld-Metal Strengthening. The ability to control the properties of the weld by mixing base metal and filler metal in the weld pool means that the choice of welding rod can influence weld strength to a considerable extent. Fully reinforced welds in thin-walled tubes of 4130 welded with RG45 rod consistently showed tensile strengths of 90 to 100 ksi. When the welds were made with RG60 rods, the strengths increased to 100 to 125 ksi, and with RG65 rods, strengths as high as 145 ksi were attained when the joint was heat treated after welding.

Class RG65 welding rods have a low-alloy steel composition and are used for OFW of carbon and low-alloy steels that have strengths of 65 to 75 ksi. They are used on sheet, plate, tube, and pipe. These rods give the highest strengths in welding 4130, 4340, and 8630 alloy steels when base metal and filler metal are mixed properly. The end use has a marked effect on selection of filler metal. If the base metal was selected to meet a specific corrosion- or heat-resistant application, the filler metal should be of similar composition. However, if a room-temperature mechanical property is the primary requirement, the strength and ductility of the filler metal should be the basis of selection.

Class RG60 welding rods are probably used most widely. They are generally made of low-alloy steel and are preferred for OFW of carbon and low-alloy steels in the tensile-strength range of 50 to 65 ksi. Class RG60 rods are most commonly used for welding carbon steel pipes for power plants.

Class RG45 welding rods have a simple low-carbon steel composition. These rods can be used for OFW of carbon and low-alloy steels.

Techniques

Forehand and backhand techniques are used in OFW, and the usual welding positions (flat, horizontal, vertical, and overhead) are used. Techniques vary to some extent for different joint thicknesses, welding positions, and welder preferences.

Forehand Welding. In forehand welding (Fig. 7a), the flame is pointed away from the completed weld in the direction of welding. The torch is held at about 45° to the workpiece. The welding rod is held at an angle of about 40° to the workpiece, and the flame is between the tip of the rod and the weld. The inner cone of the flame is held close to the work without actually touching it. The torch is moved from side to side or oscillated so that the flame heats the welding rod, the edges to be welded, and the weld metal ahead of the flame. The rod may be oscillated in a direction counter to the side-to-side movement of the torch. Both the welding-rod tip and the weld pool should be kept under the shielding influence of the flame. The forehand technique requires careful manipulation to guard against excessive melting of the base metal. This results in considerable mixing of base metal and filler metal.

Fig. 7 Manipulating welding tip and welding rod in OFW

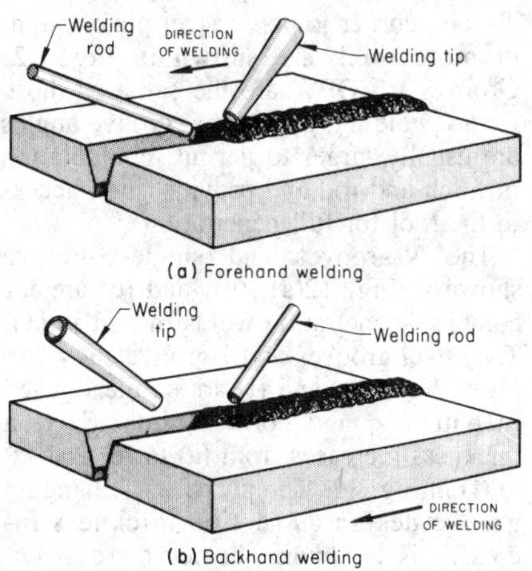

Fig. 8 Preventing cold laps and oxide inclusions. In welding unequal sections, the flame is directed at thicker and continuous sections.

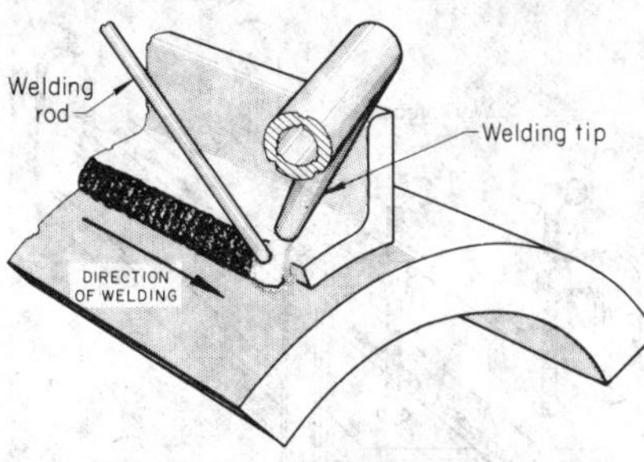

Backhand Welding. In backhand welding (Fig. 7b), the flame is pointed toward the completed weld. The tip of the welding rod is held between the flame and the weld at an angle of about 40° to the workpiece, and the torch is held at an angle of about 50° to the workpiece. Rod and torch are usually oscillated slightly in opposite directions. Because of less side-to-side movement of the flame and less melting of the joint edges, a backhand weld is more likely to retain the properties of filler metal unaltered by the base metal. The weld pool receives less agitation, the weld metal is better protected by the flame, and the weld cools more slowly than in forehand welding.

Backhand welding is often preferred for joints in metal thick enough to require beveled edges. Smaller grooves can be used with this method because the flame does not need to be moved around the rod to melt the edges of the groove. Proponents of this technique believe that it saves time, filler metal, and gas. When working with a weld fillet, the flame should be directed primarily at the heavier and continuous section of base metal to ensure against cold laps and oxide inclusions (Fig. 8).

Flat and Horizontal Positions. Oxyfuel gas welding using flat and horizontal positions presents few problems. However, when root openings must be large because joint thickness is large, the backhand technique is preferred. If the forehand technique is used, the molten weld pool must be worked from side to side and forward with the aid of the flame.

Vertical and Overhead Positions. In vertical and overhead welding, heat input should be carefully controlled to keep the viscosity of the weld pool high enough to prevent the weld metal from dripping. In addition, the force of the flame must be directed to ensure that the weld pool solidifies in the correct position.

Joint Design and Edge Preparation

Joints used in OFW are butt, lap, edge, T-, and corner joints. Either fillet or groove welds are used, depending on workpiece and strength requirements.

Sheet. Five joints used for single-pass OFW of low-carbon steel sheet are shown in Fig. 9. For OFW, beveling of joint edges is not needed in sheet up to $^3/_{16}$ in. thick, and if proper shearing practice has been followed in trimming sheets to size, no special edge preparation is required. However, edges must be free of rust, dirt, oil, and grease.

Square-groove butt joints (Fig. 9a) can be single-pass welded from one side in sheet up to approximately $^3/_{16}$ in. thick. Complete-penetration welds, properly made, develop the same or greater strength in low-carbon steel as the members joined.

Short-Flanged-Edge Butt Joints. For making butt joints in sheet up to about $^3/_{32}$ in. thick, a short flanged edge (Fig. 9b) is beneficial. The weld is called a melt-through weld, because the flanges are melted down to form the butt joint. Filler metal is not required. The oscillating technique for melting down the flanges is shown in Fig. 10. The flanges keep the sheet in flat alignment.

Fig. 9 Joints and corresponding single-pass welds. Used in OFW of thin sheet

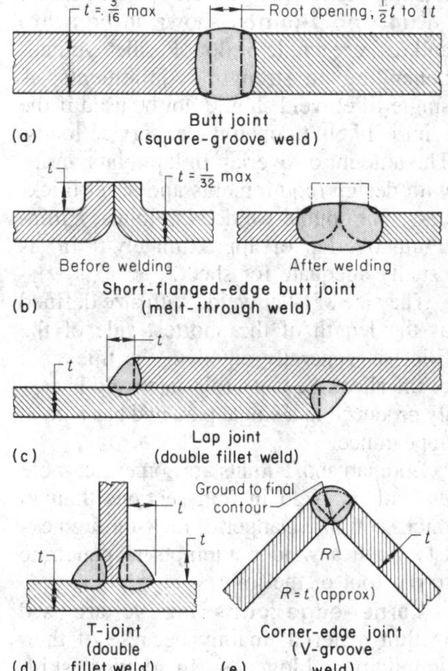

Fig. 10 Oxyfuel gas welding technique for making a short-flanged-edge butt joint. Used in thin sheet (up to $^3/_{32}$ in. thick)

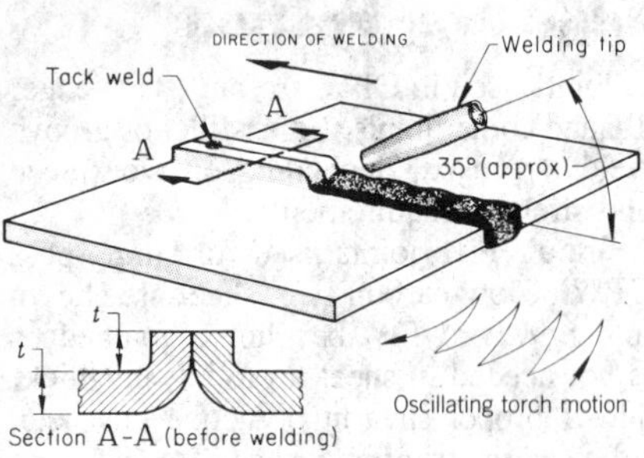

Although the sheet has to undergo a special flanging operation, this joint is practical for long production runs when filler-metal deposition and melt-through are difficult to control. High welding speeds can be attained. Sheet thicknesses are limited to approximately $^1/_{32}$ in., because short flanges are difficult to bend in heavier material. In other flanged-edge joints, the weld is deposited on the edges, and the flanges retain their structural identity.

Without support from adjacent sheet, long sections of thin material cannot be welded without buckling and distortion. The amount of distortion resulting from expansion during welding and contraction during cooling of the welded sheet-metal section or structure is proportional to the amount of heat applied to the metal as well as to its inherent stiffness and rigidity. Jigs and fixtures are essential to counteract the adverse effects of welding heat. Figure 11(a) shows the elements of a jig used for making butt or flange welds in thin sheet.

Lap and T-joints, shown in Fig. 9(c) and (d), are single or double fillet welded, depending on strength requirements. A single fillet weld should not be used if the joint is likely to open under service loads. The amount of overlap in lap joints varies with design requirements and sheet thickness. For double-welded lap joints, a minimum overlap of approximately 1 in. is usually adequate for sheet.

The size of fillet weld, with size defined as the length of the shortest side of the fillet, is generally equal to the thickness of the sheet, because this size weld is easily produced in a single pass and has a good appearance.

Both lap and T-joints are somewhat more difficult to weld in thin sections than in thick sections. Danger of melt-through exists when a welder attempts to penetrate to the root of the joint.

Corner-edge joints (Fig. 9e) are used in thin material, mainly because of their simplicity and low cost. In addition, skillful OFW can produce a weld of good external appearance in a single pass.

Fig. 11 Jigs used with thin sheet. Used to maintain flatness and minimize distortion

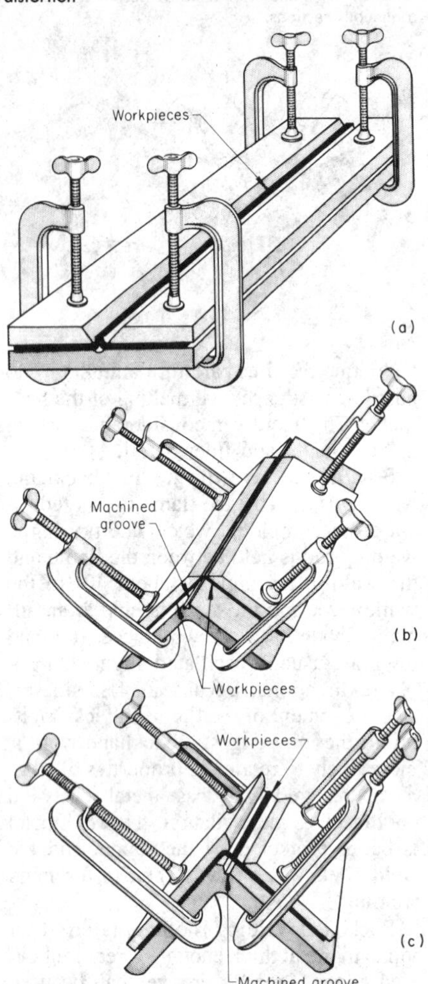

To improve internal appearance and to provide added strength, an internal pass or removable ceramic backing is needed. Square corner joints in thin material can be made with good appearance and strength and with little or no distortion by using a jig as shown in Fig. 11(b). A recess is machined or ground in the corner of the backing plate to provide for root reinforcement of the V-groove weld.

When parts with rounded corner joints are designed in thin sheet, corners should be bent rather than joined. Joints next to corners can be made by butt welding in a jig constructed as shown in Fig. 11(c). Note the use of a recess in the backing member to provide for root reinforcement of the weld.

Plate. Butt, lap, T-, and corner joints are oxyfuel gas welded in steel plate ($^3/_{16}$ in. and thicker). Butt joints welded from one side only require beveled edges in thicknesses $^3/_{16}$ and over. However, complete-penetration welds are made in square-groove butt joints up to $^5/_{16}$ in. thick by single-pass welding from both sides and back gouging the first, or root, pass. Joints are considered only partially penetrated when welded from both sides without back gouging the root pass.

Lap and T-joints can be oxyfuel gas welded in any plate thickness, using single-pass or multiple-pass fillet welds. When these joints are in heavy plate, the size of fillet welds should not exceed the section thickness to avoid costly overwelding. Corner joints can also be oxyfuel gas welded in any thickness, but in heavy sections, this joint design produces an unsatisfactory stress distribution in the weld. One alternative is to extend one of the plates past the other to form a T-joint, which can be double fillet welded; another is to reinforce the corner joint with structural sections. For heavy plate, OFW is slower than arc welding and is not widely used.

Fillet Welds. When compared to groove welds, fillet welds have advantages and disadvantages. No edge preparation such as chamfering or beveling of joint edges is required. If the weld serves mainly to hold structural members together, or if forces transmitted by the weld are low, fillet welds can be relatively small and economical. If welds are required to develop the full strength or a high percentage of the full strength of the members joined, fillet welds become more difficult to achieve and more costly as section thickness increases.

Groove welds are used in butt joints and can be used in lap, T-, and corner joints as an alternative to fillet welds. In low-carbon steel, properly made complete-penetration groove welds develop strength equal to the strength of the members joined.

Typical grooves used for OFW of butt, T-, and corner joints in steel plate $^3/_{16}$ in. or more thick are shown in Fig. 12. Grooves for OFW are the same as those for arc welding, except that groove angles are usually larger to permit manipulation of torch and rod and provide good access to the root for full penetration.

The V-grooves and single-U-groove shown in Fig. 12(a), (b), and (c) are for complete-penetration welds in butt joints. The bevel grooves and J-groove shown in Fig. 12(d), (e), and (f) are similarly used in butt, T-, and corner joints. Section thickness increases from (a) to (c) and (d) to (f) in Fig. 12. The purpose of changing groove design as section thickness increases is to obtain the least cross-sec-

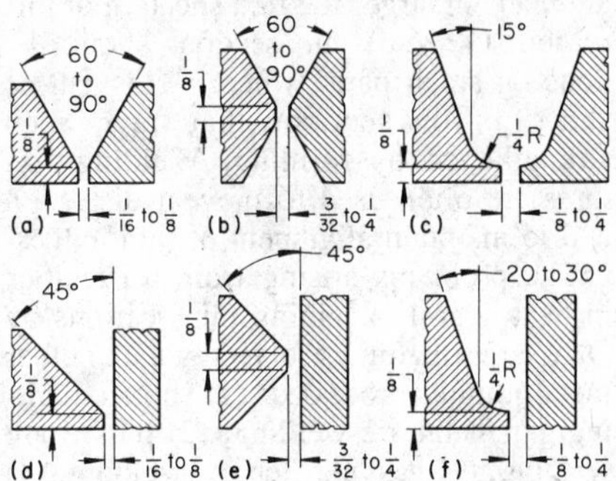

Fig. 12 **Grooves used in OFW of plate.** Plate thickness $^3/_{16}$ in. or more

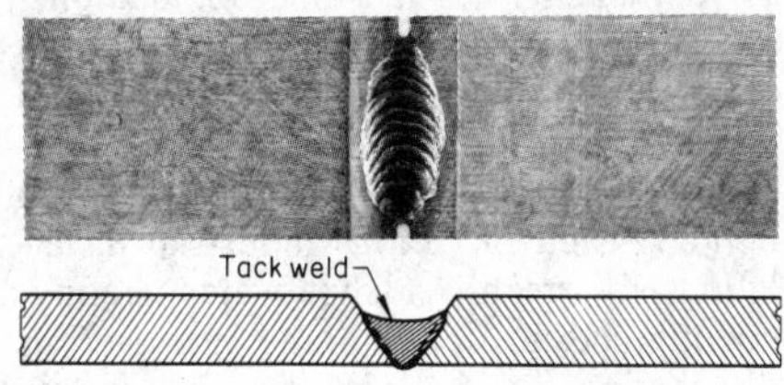

Fig. 13 Preferred contour and relative size of tack weld. Needed to obtain successful fusion with main butt weld in pipe

tional area of weld and reduce welding cost. The dimensions shown in this figure are typical rather than mandatory; reasonable variations are permitted. A double-U-groove and a double-J-groove are not shown because they are used for very thick sections, and OFW is rare. The edges of joints intended for groove welds can be prepared by gas cutting, plasma arc cutting, milling, shaping, planing, or turning, depending on the shape and size of the part as well as other conditions.

Welding of Pipe

Most pipe welding is done on circumferential butt joints. Wall thicknesses up to about $^3/_8$ in. can usually be welded in a single pass. A greater wall thickness requires more than one pass. Grooves must be large enough to permit manipulation of the torch and facilitate adequate root penetration. The root pass must be made carefully to obtain complete penetration and smooth fusion at the inner surface of the pipe. Joints for pipe of up to about $1^1/_8$-in. wall thickness are usually made with single-V-grooves. In pipe with thicker walls, the single-U-groove may be preferred, as it requires fewer passes and less weld metal to fill.

On walls up to and including $^1/_8$ in. thick, a 30° bevel (60° included angle) can be made by grinding the pipe ends. When wall thickness is greater than $^1/_8$ in., the pipe ends are usually beveled to a $37^1/_2$° angle, which provides a 75° groove angle. These limitations are not mandatory and can be varied somewhat. Smaller grooves are sometimes used by experienced welders.

Bevels can be cut by chipping, grinding, turning, or oxyfuel gas cutting (OFC). Beveling machines using OFC torches or cutting tools are used when large quantities of pipe are involved. Manual OFC is used when some form of guided cutting is not available, but manually gas-cut edges are likely to be rough and may have adhering slag. A light grinding operation is often needed to clean the beveled edges and remove gouges before welding.

Elbows, return bends, T's, crosses, Y-branches, reducing sections, end caps, and flanges that are to be welded to small-diameter pipe are usually obtained as forged fittings prepared for circumferential butt welding. These fittings have been forged with their ends beveled to an angle of 30 to $37^1/_2$°, which on fit-up provides welding grooves of 60 to 75° included angle. When joints must be made in intersecting pipes that are not coaxial, or when pipe diameters differ, the angle of bevel must suit the contour at the intersection.

After beveling and subsequent grinding if necessary, the pipe is aligned and tack welded in three or four places to maintain alignment during welding. The pipe ends are usually separated by $^1/_{16}$ to $^1/_8$ in. to provide a root opening.

Tack welds, whether used to join pipe ends or to support the adjacent edges of a seam to be welded, should be small and have the general appearance shown in Fig. 13. Such a tack weld makes it possible to blend the tack with the main weld as it is deposited and produce a smooth finished weld. Also, a tack weld with a smooth contour makes fusion easy to obtain where the main weld must be connected with the tack weld.

Welding positions for pipe are determined by the conditions under which the weld can be made. The three basic positions are horizontal rolled, horizontal fixed, and vertical (Fig. 14).

Horizontal-Rolled Position. In the horizontal-rolled position (Fig. 14a), the axis of the pipe is essentially horizontal, and the pipe is rotated about its axis. For circumferential butt joints, welding is usually done within an angle of about 20 to 45° from the top center of the pipe.

Horizontal-Fixed Position. In the horizontal-fixed position (Fig. 14b), the axis of the pipe is essentially horizontal, and the pipe is not rotated. For circumferential butt joints, welding can start at the center of either the top or the bottom and proceed

Fig. 14 Positions for welding pipe

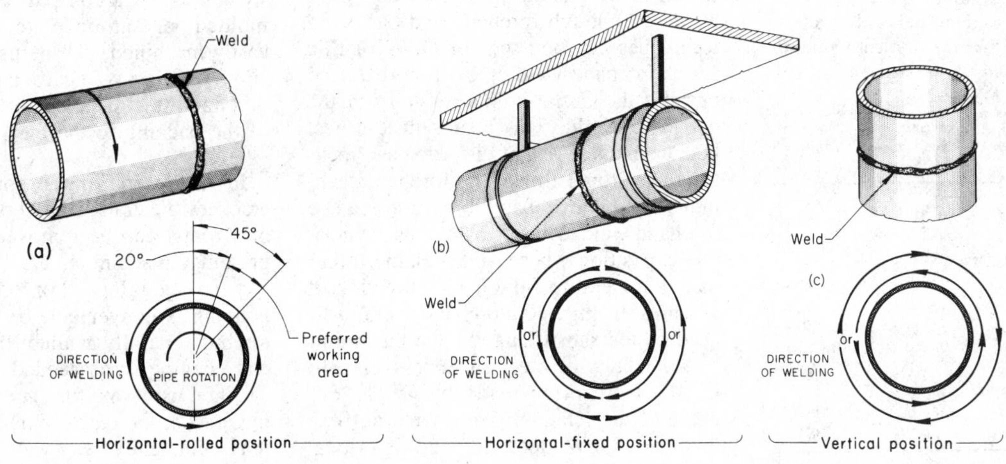

to the opposite point. The procedure is repeated for the other side of the pipe. Welding actually proceeds from flat to vertical to overhead, or vice versa, depending on whether the start is made at the top or bottom of the joint.

Vertical Position. In the vertical position (Fig. 14c), the axis of the pipe is essentially vertical, and the pipe may or may not be rotated. With the pipe vertical, a circumferential butt weld is in a horizontal position regardless of whether the pipe is rotated or not.

Applications. Although OFW can be used to join pipe of any size and of many different materials, the process is most useful for welding small-diameter carbon steel pipe. Joining carbon steel pipe in diameters up to about 4 in. and in wall thicknesses up to 3/8 in. by OFW has often proved more economical than arc welding.

Oxyfuel gas welding offers control over the viscosity of the weld pool, which is useful when pipe is welded in difficult positions. Gas welding equipment can be converted easily to gas cutting equipment, and a pipe welder who is also an experienced pipe fitter should be capable of cutting almost any transition joint that may be required.

Standard-weight (schedule-40) carbon steel pipe up to about 10 in. in diameter can also be oxyfuel gas welded. Although these larger diameters can be welded more efficiently by other methods, OFW is often more expedient than changing to another process during a piping installation. Data for single-pass OFW of standard and schedule-40 low-carbon steel pipe, in sizes from 1/2 to 10 in. and for wall thicknesses from 0.109 to 0.365 in., are given in Table 2.

Table 2 Conditions for single-pass OFW of standard schedule-40 pipe.

Based on pipe ends beveled to 45 ± 2 1/2° (90° included angle), and welded in the horizontal-rolled position; for welds made in the horizontal-fixed or the vertical position, welding time should be increased approximately 15%.

Nominal pipe size, in.	Wall thickness, in.	Welding rod size, in.	Tip-orifice diameter, in.
1/2	0.109	3/32	0.0420
3/4	0.113	3/32	0.0420
1	0.133	1/8	0.0520
1 1/4	0.140	1/8	0.0520
1 1/2	0.145	1/8	0.0520
2	0.154	1/8-5/32	0.0520
3	0.216	1/8-5/32	0.0595
4	0.237	1/8-5/32	0.0700
5	0.258	5/32-3/16	0.0700
6	0.280	5/32-3/16	0.0810
8	0.322	5/32-3/16	0.0810
10	0.365	5/32-3/16	0.0810

In the pipe industry, arc welding plays the dominant role. In some applications, special arc welding techniques have been developed to obtain high joint integrity and less human error. However, OFW applications include pipe systems used in buildings and in various industrial installations for steam and hot water heating, cooling and air conditioning, gas and air distribution, water distribution, certain types of electrical transmission, and transmission of various commercial fluids. The welding of small-diameter low-pressure steel pipe is one of the major applications of OFW.

Welding of Thin-Walled Tube

In many respects, the welding of thin-walled tube (1/8-in. wall or less) is similar to the welding of sheet of the same thickness; in other respects, it is similar to the welding of pipe. Oxyfuel gas welding is often preferred for welding thin-walled tube, especially in small diameters, because the equipment is relatively easy to manipulate in confined spaces. Also, risk of overheating and melting through the tube wall is lessened. Weld spatter is negligible.

Joints in thin-walled tube vary from simple circumferential butt or lap joints to any shape resulting from the intersection of a tube with another shape. Whereas pipe is usually beveled, thin-walled tubing, like thin sheet, requires no particular edge preparation. Standard pipe fittings do not generally fit tubes.

Welding of Thin Sheet

Techniques used in OFW of thin sheet depend on joint design and welding position, as well as on the style used by the welder. Although forehand and backhand techniques can be used for OFW of thin sheet, forehand welding is often preferred because the flame points away from the completed weld. This results in less heat buildup and less risk of distortion and melt-through. Proponents of the forehand technique also believe that thin sections can be forehand welded faster using flat or horizontal positions, because the flame forces the weld metal to flow in the direction of welding. In Fig. 10, forehand welding is used on the short-flanged-edge butt joint.

Assemblies of steel sheet less than 1/8 in. thick are often made by OFW. Because of the relatively long welding time involved, OFW does not permit critical control over total heat input, and distortion can occur when long, continuous welds are made in large, flat thin sheets. For this reason, OFW of thin sections should be done on small parts, where welds can be short, or on sections that have been strengthened by forming. Welding fixtures are often used to prevent distortion and to maintain alignment of joint edges.

The problems arising from difficulties in heat contol in joining thin sections by OFW are sometimes solved by using other metal-joining processes: (1) short circuiting gas metal arc welding, (2) pulsed arc welding, (3) gas tungsten arc welding, (4) plasma arc welding, (5) electron beam welding, (6) ultrasonic welding, (7) resistance welding, (8) brazing, and (9) braze welding.

Products made by OFW of thin sheet use materials ranging in thickness from 0.0239 in. (24 gage) to 0.1196 in. (11 gage). Typical items include: truck bodies, furniture, office equipment, utensils, enclosures, and refrigeration equipment.

Repairs and Alterations

Oxyfuel gas welding is frequently used for repairs and alterations because the equipment is portable, welding can be done in all positions, and acetylene and oxygen are readily available. Also, gas cutting, braze welding, brazing, or flame heating is often needed, and the same personnel, equipment, and gases required for these operations can be used for OFW.

Edge preparation or repair welding generally involves cutting a V-groove to expose clean metal down to the root of the joint. The groove must be large enough to allow room for manipulation of the torch and welding rod.

On many repair welding jobs, sections that must be rebuilt to specific contours or dimensions require backing to support the weld metal. Carbon paste, which can be molded, or carbon plate, which can be easily machined, can be used for backing (Fig. 15). To avoid carbon pickup, copper or a thick steel plate can be used. Welding of the backing to the part should be prevented.

In weld repairing of parts that have cracked, the crack should be fully dressed out. Metal can be removed by chipping, grinding, machining, arc air gouging, or oxyfuel gouging. For OFW, oxyfuel gouging is convenient, because a cutting torch is generally available and only a special gouging tip is needed.

When using oxyfuel gouging for dressing cracks, the crack remains visible until it is completely removed. Grinding may

Fig. 15 Use of carbon backing and support. Used to expedite repair welding of surfaces and contours

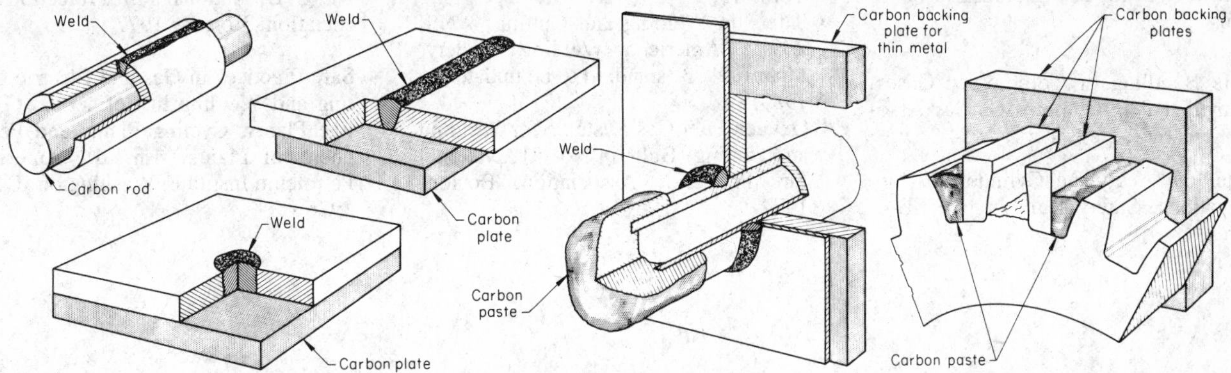

result in a fine crack being smeared over, and chipping is also likely to cover the crack. The disadvantage in using oxyfuel gouging is that the heat input may cause the crack to propagate through differential expansion in the workpiece. The risk is especially great if some of the stresses that caused the crack are still present. Drilling a 1/4- to 1/2-in. hole at the end of the crack may prevent the crack from propagating. Cleanliness of the joint is also a prerequisite for OFW. Rust, scale, dirt, grease, oil, paint, and slag must be removed from the joint areas. The soundness of the weld metal depends in large part on the care used in cleaning. To ensure complete crack removal, liquid-penetrant or magnetic-particle testing can be used.

When OFW is used to correct production parts that would be rejected because of a small error or defect in manufacture, it must be determined if the salvaged part, after welding, is as good as a new part for the required service. Some fabricating codes identify the defects that may be rectified by welding and define the standards for repair. Rectification may include correction of defects, building up of undersize parts, and filling in of drilled holes.

Preheating and Postheating

Preweld and postweld heat treatments are not usually required to reduce the hardness of low-carbon steel, but either or both treatments are beneficial and can be effective in avoiding or reducing distortion. Most steels that are oxyfuel gas welded fall within the low-carbon range.

Oxyfuel gas welding distributes a large amount of heat over a wide area. This results in relatively slow cooling rates and relatively low stress gradients, which, in turn, reduce the degree of hardening and the magnitude of residual stresses that are usually associated with welding heat cycles.

Example 1. Use of Oxyfuel Gas Welding To Eliminate Postweld Tempering. Aircraft landing gear side stay assemblies made of 1330 steel (Fig. 16) were joined in the form of a trapezoid. The assembly was originally welded by the shielded metal arc process; E7016 electrodes were used. Several assemblies cracked in a tube adjacent to a weld after a short time in service. Examination revealed high hardness (50 HRC) in the heat-affected zone (HAZ) was caused by the chilling effect of the large mass of metal in the end fittings. A postweld tempering treatment at 1200 °F to reduce the high hardness could not be used, because the tubes became distorted.

Fig. 16 Oxyfuel gas welded aircraft landing gear suppport. Oxyfuel gas welding eliminated the need for postweld tempering.

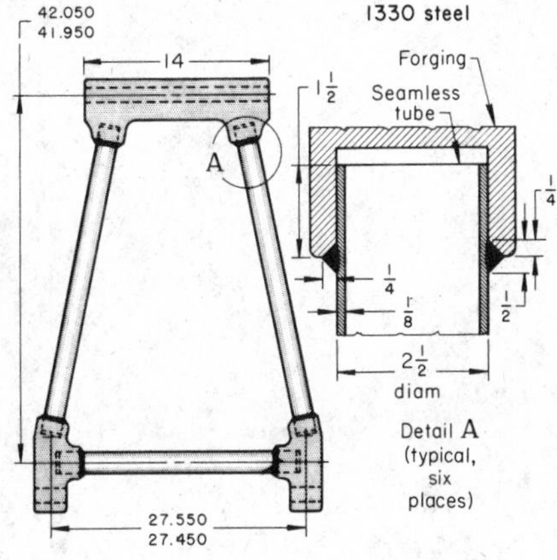

Welding rod	3/32 in., RG60
Torch	Injector
Tip-orifice diameter	0.0595 in.
Acetylene pressure	5 psi
Oxygen pressure	28 psi
Flame adjustment	Neutral

Local stress relief of the welded joint with a gas torch was successful but laborious, and it was finally decided to oxyfuel gas weld the assembly. Hardness tests in the area of the welded joint showed that the maximum hardness in the HAZ was 285 HB (29 HRC). This was only moderately higher than the hardness of the end fittings and tubes, which had a hardness of 223 HB (98 HRB). The improvement was believed to result from the wider area heated during OFW, which produced a lower temperature gradient.

Bridging Gaps in Poor Fit-Ups

The ability to control the flow of molten weld metal in OFW provides a high degree of versatility in bridging large gaps caused by a poor fit-up. By manipulating the gas torch, weld-metal temperature can be held to the minimum, and the pressure of the gas flame can be used to help support the weld pool in any position, including overhead.

Safety

The gases used in OFW can form explosive mixtures if improperly handled, and burning of these gases must be controlled within recommended limits for the safety of the welder and surrounding property. Workers must be familiar with safe welding practices and should be aware that any deviation from recognized safety standards can result in damage to apparatus or personal injury. Workers in charge of welding operations should be instructed and judged competent by their employers.

Rules and instructions covering safety, operation, and maintenance procedures are readily available. Federal, state, and local governments, in addition to equipment manufacturers, gas suppliers, and regulatory agencies, have cooperated in publishing documents on safe welding practices.

This information is included in, but not limited to, the following industrial publications:

- Safe Handling of Compressed Gases, Pamphlet P-1, Compressed Gas Association, Inc., New York, 1974
- Handling Acetylene Cylinders in Fire Situations, Safety Bulletin No. SB-4, Compressed Gas Association, Inc., New York, 1972
- Safety in Welding and Cutting, ANSI Z49.1, American Welding Society, Miami, 1973 (standard to be updated in 1983)
- Oxygen-Fuel Gas Systems for Welding and Cutting, Bulletin No. 51, National Fire Protection Association, Boston, 1977
- Cutting and Welding Practices, Bulletin No. 51B, National Fire Protection Association, Boston, 1977
- Safe Practices in Gas and Electric Cutting and Welding in Refineries, Gasoline Plants, Cycling Plants and Petrochemical Plants, 4th ed., American Petroleum Institute, Washington, D.C., 1976

Oxyacetylene Pressure Welding

OXYACETYLENE PRESSURE WELDING is a process in which heat from oxyacetylene flames is used in conjunction with pressure to produce a solid-phase weld. Applications include joining pipe sections, aircraft landing gear components, and drill bits (to shanks). Most higher melting alloys, including carbon and low-alloy steels, stainless steels, and heat-resistant alloys, can be successfully oxyacetylene pressure welded. The process has not proved successful for welding aluminum and magnesium alloys.

Principles of Operation

Although there is more than one method of oxyacetylene pressure welding, this article deals only with the closed-gap method. Closed-gap welding is accomplished by either of two techniques. In single-pressure welding, workpieces are butted together under a pre-established imposed pressure; the joint is then surrounded with oxyacetylene torches to heat it until a given amount of upsetting has taken place. In double-pressure welding, workpieces are butted together under an initial pressure; the joint is heated to a given temperature, and then pressure is increased to accomplish the upsetting. Heating setup for closed-gap oxyacetylene pressure welding is shown in Fig. 1.

Advantages

Oxyacetylene pressure welding is faster than welding that entails fusion and does not develop a cast structure in the weld zone. Inclusions or porosity is not introduced during welding, and molten metal is not squeezed out as flash (as in flash welding). Metal in the weld zone is capable of the same response to heat treatment as the base metal. Susceptibility to cracking is minimized, because temperature gradients developed during welding are not steep.

Fig. 1 Setup for heating workpieces for closed-gap oxyacetylene pressure welding

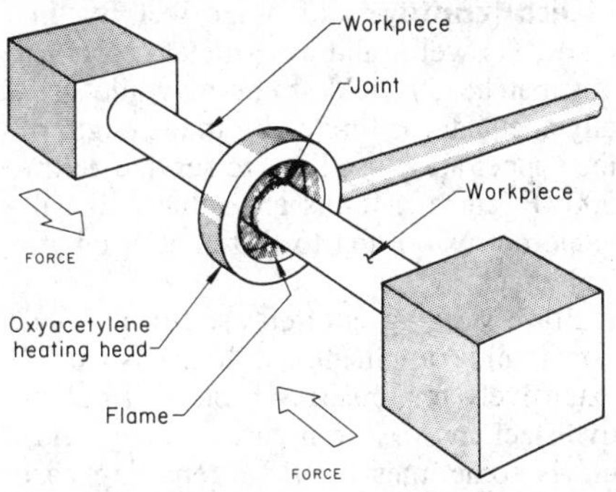

Limitations

Oxyacetylene pressure welding is slow compared with other pressure welding processes. Butting surfaces of the workpieces must be flat, parallel, and clean, thereby increasing the cost of surface preparation.

Bars up to 3 in. in diameter and tubes as large as 25 in. OD have been welded by the closed-gap method. However, in welding thick-walled tubing or large-diameter bars, additional time is required to transmit enough heat inward to metal farthest from the surface. This can be alleviated by the use of a tapered joint design.

Joint Design

The butt ends of thicker sections (tube walls thicker than $^1/_4$ in. or solids thicker than $^1/_2$ in.) are usually tapered. Tapering facilitates heat penetration and helps to control the shape of the upset. Carbon steels are generally beveled to a 6 to 10° included angle, with the open side, or gap, exposed to the flames. This gap is closed during upset. For alloy steels, a much flatter included angle, determined experimentally, may be required.

Typical Procedure

A typical sequence of operations for oxyacetylene pressure welding of hollow 4135 steel landing gear components by the closed-gap method is:

- Butting surfaces of parts are tapered on the outside and inside diameters to control upset contour and are machined to a finish of 100 μin. or smoother.
- Parts are loaded into the welding machine, and accuracy of alignment is checked by inserting a sheet of vellum between butting surfaces and pressing them together. When pressure is released, the vellum is examined for uniformity of contact around the entire circumference. Adjustments are made until uniformity is achieved.
- Weld faces are cleaned with solvent and lint-free cloths, and a pressure of 4 ksi is applied to the weld area.
- Heating head is brought into position and lighted, and the oscillating mechanism is started. The heating head is oscillated longitudinally (and sometimes radially, as well) to prevent localized overheating.
- When the predetermined amount of upset has taken place, thrust and heat are discontinued, and the part is allowed to cool to a black heat in the machine.

Typical conditions for oxyacetylene pressure welding pipes of various wall thicknesses are presented in Table 1.

Table 1 Typical conditions for oxyacetylene pressure welding of carbon steel pipe(a)

Wall thickness, in.	Welding time, s	Shortening, in.	Pipe-end bevel, °
0.187	45	0.25	4
0.203	50	0.265	4
0.238	55	0.300	4
0.250	60	0.312	5
0.375	90	0.412	6

(a) Using the double-pressure method, with initial pressure 1 ksi of pipe cross section and final pressure 4 ksi of pipe cross section

Oxyacetylene Braze Welding of Steel and Cast Irons

OXYACETYLENE BRAZE WELDING is a method of oxyfuel gas welding (OFW) capable of joining many base metals, but it is used primarily on steel, cast iron, and malleable iron with a copper alloy filler metal (rod) and a flux. Braze welding is similar to torch brazing with a filler rod, except that joint openings are wider and distribution of filler metal takes place by deposition rather than by capillary melting action. Equipment and some filler metals used in braze welding are the same as those used in torch brazing (see the article "Torch Brazing of Steels" in this Volume).

In braze welding of ferrous metals, the base metal is not melted. The filler-to-base-metal bond is the same as in torch brazing. Flux is applied to the joint surfaces, which (together with the surrounding area) are preheated to the point where the filler metal wets or "tins" these surfaces. During welding, tinning precedes the weld pool.

Applicability

Braze welding is used for making groove, fillet, plug, or slot welds in metal ranging from thin sheet to heavy castings. Weld layers can be built up, as in OFW. The process is often used as a low-temperature substitute for OFW. Braze welding resembles brazing in that nonferrous filler metals are used, and bonding is achieved without melting the base metal. Braze welding resembles welding because it can be used for filling grooves and for building up fillets as may be required.

Advantages of braze welding include: (1) joints are made at a lower temperature than in OFW or arc welding, thus minimizing thermal stress and distortion, as well as susceptibility to cracking; (2) the weld deposit is relatively soft and ductile, providing machinability and low residual stress; (3) joints with good strength and generous fillets can be produced (fillets often add strength for specific types of loading in service); (4) dissimilar metals, such as steel and cast iron, may be joined; and (5) the equipment is simple and well suited to on-site repair applications.

Disadvantages of braze welding include: (1) weld- and base-metal colors do not match; (2) weld strength, while usually adequate, is limited by the strength of the copper alloy used and to service below 500 °F; and (3) joints are subject to galvanic corrosion and to differential chemical attack.

Braze welding is often used in production joining applications. It is also used extensively for repairing broken or defective steel and cast iron parts. Braze welding is sometimes used for repairing cast iron castings in the foundry, although color mismatch imposes restrictions on its use for this purpose.

Another application of braze welding is in machine shops, for correcting machining errors or modifying in-process parts. Shop maintenance departments and toolrooms use braze welding for repairing tools and pieces of equipment ranging from small hand tools to large cast iron press frames. Mobile repair units, such as those that accompany grain harvesting crews, usually carry equipment for braze welding so that harvesting machinery can be quickly repaired in the field.

Flame Adjustment

Oxyacetylene flame for braze welding is adjusted to the neutral condition for carbon steels. For cast irons, the flame is adjusted to a slightly oxidizing condition by increasing oxygen flow, which reduces the telltale acetylene "feather." When the feather disappears and the inner cone becomes slightly necked, the flame is oxidizing. This type of flame helps remove the graphite from surfaces of cast iron. Air is not used as the combustion agent in braze welding because it results in lower flame temperature.

Filler Metals

Compositions of four common copper-zinc or copper-zinc-nickel filler metals often used for braze welding are given in Table 1. These four compositions are available as welding rods in the standard diameters from $^1/_{16}$ to $^1/_4$ in.; length is usually 36 in. The order in which the four filler metals are listed in Table 1 is roughly that of increasing tensile strength. Weld-metal minimum tensile strengths of the compositions shown range from approximately 40 to 60 ksi. Melting temperatures of these alloys range from about 1600 to 1750 °F.

RBCuZn-A (naval brass) rods contain up to 1% Sn, which improves strength and corrosion resistance. These filler-metal rods are especially suited for use with oxyacetylene and are considered as general-purpose rods for braze welding of steel and the various grades of graphitic cast iron.

RBCuZn-B (low-fuming bronze) rods are similar to RBCuZn-A, but contain additions of iron and nickel, which increase strength. RBCuZn-B rods are used for

Table 1 Approximate compositions of four common filler metals used for braze welding

AWS classification(a)	Chemical composition, % Cu	Zn	Sn	Fe	Ni	Minimum tensile strength, ksi	Liquidus temperature, °F
RBCuZn-A	60	39	1	...	...	40	1650
RBCuZn-B	60	37.5	1	1	0.5	50	1630
RBCuZn-C	60	38	1	1	...	50	1630
RBCuZn-D	50	40	...	...	10	60	1715

(a) See AWS Specifications A5.7 and A5.8 for additional data. Additional filler metals are available. User should follow manufacturer's recommendations.

joining either steel or cast iron, and sometimes these rods are also used for surfacing steel or cast iron for wear resistance.

RBCuZn-C rods are also low-fuming bronze rods with compositions similar to RBCuZn-B, except they do not contain nickel. RBCuZn-C rods provide as-welded mechanical properties that are higher than those obtained with RBCuZn-A and are widely used as general-purpose rods for braze welding of both steel and cast iron.

RBCuZn-D (nickel silver) rods have lower copper and higher nickel contents than those of the three other filler metals listed in Table 1. Because of this difference in composition, the deposit from an RBCuZn-D rod is whiter and is thus used for braze welding when closer color match is important. RBCuZn-D provides the highest as-welded strength of these four braze welding filler-metal alloys.

Joint Properties. Joints braze welded with one of the RBCuZn filler metals have tensile strengths at room temperature that usually range from 40 to 60 ksi, depending on the filler metal used. Strength of the joint decreases at temperatures above 500 °F. Color match with the base metal is not usually obtained; where color is important, RBCuZn-D (nickel silver) rod is used. The bimetallic joint is subject to galvanic corrosion and is less resistant to alkaline solutions than the ferrous base metal.

The following example illustrates the use of braze welding in a conventional production operation on thin low-carbon steel, where difficulties were overcome by a change in filler metal.

Example 1. Filler Metal Use in Braze Welding. Metal furniture parts made of low-carbon tubular steel were braze welded at corner joints. High rejection rates (25%) because of poor appearance were encountered when mitered joints of the type shown in Fig. 1 were nickel-chromium plated after braze welding. Analysis of the fabricating procedure showed that nearly all the rejections could be eliminated by a change in filler metal.

Preparation for braze welding was the same for both the original and revised procedures. Tubing joints were sawed at a 45° angle, and joint edges were then beveled at 45° on all four sides. Without cleaning, the parts were assembled in a fixture to give a $^1/_{32}$-in. maximum clearance at the joint. An oxyacetylene torch with a 0.0595-in.-diam tip was used with both procedures. In both cases, flux was added by gas fluxing through the flame. Gas fluxing is described in the article "Torch Brazing of Steels" in this Volume.

Originally, a rod of copper-zinc alloy RBCuZn-C was used as filler metal. Before being plated, the weld was ground flush, buffed, and chemically cleaned. The following difficulties occurred during cleaning and plating:

Fig. 1 Mitered corner joint in a tubular part. Used in manufacture of metal furniture

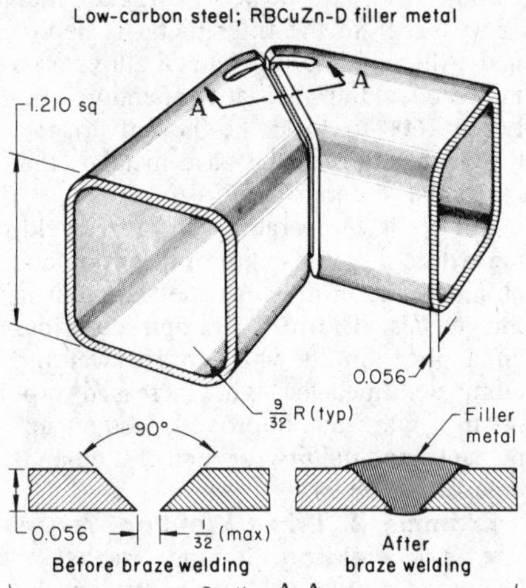

- Conventional alkaline and acid solutions used in the cleaning process attacked the filler metal, causing voids and, in some assemblies, roughness, all of which were apparent after plating.
- Time of exposure during cleaning was critical. Overexposure caused severe etching. Cleaning time was difficult to control because of lack of uniformity in heating during brazing.
- Because of the softness of the filler metal, feather edges were left on the surface at the bond lines after grinding and buffing. The feathers were etched off during cleaning, but a sharp ridge was left that became an area of apparent nickel buildup during plating.
- Color difference between filler and base metal was apparent after plating, outlining the joint.

By changing the filler metal to a rod of a different copper-zinc alloy (RBCuZn-D), difficulties were minimized or overcome. The improvements reduced the rate of rejection after visual inspection to less than 1%. This saving more than offset the higher cost of the RBCuZn-D filler rod. In addition, higher joint strength was obtained. The change did not affect the production rate of 400 welds per 8-h shift.

Fluxes

Fluxes for braze welding are not the same as those used for capillary brazing. Because the temperatures used in braze welding (often higher than 1800 °F) are higher than those used in most capillary brazing and because the time of exposure to elevated temperature is longer in braze welding, the flux used must have a higher melting point and must be able to withstand sustained exposure to higher temperatures.

There are no standard specifications for braze welding fluxes. The few existing government specifications are based on composition, and the fluxes so specified are not considered as effective as the proprietary flux formulations that are available commercially. Three types of flux are in general use:

- A basic type of flux, which facilitates braze welding by cleaning the base metals and aiding in the tinning operation
- A flux, available in paste form, that performs the functions of the basic flux described above and suppresses the formation of zinc oxide fumes from the filler metal
- A flux, sometimes called a tinning flux, that is formulated expressly for use in braze welding of gray or ductile iron

The first two fluxes are generally satisfactory for braze welding of steel and malleable iron. The second type is sometimes used with copper-zinc filler metals in capillary brazing. The third type contains iron oxide or manganese dioxide, either of which combines with surface carbon of the gray and ductile irons; consequently, this type of flux is preferred for braze welding these cast irons.

Application of flux is done in any of four ways: (1) dipping the heated filler-metal rods in the flux, (2) brushing flux on the joint before brazing, (3) using flux-coated filler-metal rods, or (4) fluxing through the gas flame (see the article "Torch Brazing of Steels" in this Volume).

The use of flux-coated filler rods or gas fluxing ensures uniform application. Gas fluxing is used chiefly on steel in production braze welding.

Joint Preparation

In base metals thicker than about $^3/_{32}$ in., the edges of butt joints for braze welding are prepared with a 90 or 120° V-groove to provide a wide bonding area. Fillet, plug, and slot welds present naturally open faces. Edge preparation in base metals less than about $^3/_{32}$ in. thick can be optional; either a square groove with a root opening comparable to the thickness may be used, or a V-groove may be cut. The V-groove makes it possible for the braze welding operator to see whether the joint is properly filled; this is not always possible in torch brazing. In applications involving thin joints with parallel-side joint surfaces and relatively close clearances, it is sometimes difficult to determine whether the joining process qualifies as brazing or braze welding, because there is some capillary action.

Surface Preparation

For satisfactory results in braze welding, joint edges must be as clean as possible before the operation begins. To obtain maximum bond strength, joint surfaces must be bright and free of oil, rust, or other foreign matter. Also, the metal surrounding the joint edges must be cleaned, both on bottom and on top.

The use of a salt bath is best for cleaning any of the cast irons prior to braze welding, just as it is for brazing (see the section on cleaning procedures in the article ''Brazing of Cast Irons'' in this Volume).

If a salt bath is not available and the surface of cast iron has been ground, the graphite smear can be removed by quickly heating the surface until it is dull red in color. After cooling, the surface should be wire brushed. If greasy or oily cast iron is ground, some of the grease or oil may penetrate the surface; the resulting film should be removed by painting the surface with hydrochloric acid. After 15 min, the surface must be scrubbed with a wire brush and cold clean water.

Preheating and Postheating

Although it is not always necessary, iron castings are usually preheated before being braze welded to ensure success of the operation. Preheating may be local or general, depending on size of the casting. Large castings require extensive preheating. A black preheat, or a low red heat visible in darkness (obtained at approximately 900 °F), is generally used. It has been advantageous in some applications to preheat to as high as 1650 °F, but temperatures above 1000 °F may have an adverse effect on the wetting action of the filler metal, because of oxidation of the surfaces before braze welding.

Postheating is not necessary after braze welding of cast iron. However, cooling of the braze welded assemblies is preferably retarded by wrapping or covering with heat-retaining material or by the use of similarly effective methods.

Braze Welding of Steel

Braze welding of steel is faster than OFW, because braze welding requires less heat. Overheating of the base metal must be avoided to prevent the filler metal from failing to tin the joint surfaces. Low-carbon steels are heated no higher than 1350 °F before the filler metal is deposited. Although the filler-metal alloys used in braze welding melt at temperatures between 1600 to 1800 °F, the only rise in the temperature of the base metal is that incidental to deposition.

Low peak temperatures in braze welding reduce the probability of distortion and eliminate the problem of melt-through in thin metals. The next example describes an application in which braze welding eliminated melt-through, increased production rate, and improved product appearance and quality for a steel tube-and-bracket assembly.

Example 2. Braze Welding Versus Arc Tack Welding. Joining brackets to lengths of tubing was originally done by shielded metal arc tack welding. Figure 2 shows a typical tube-and-bracket assembly. The tack welding operation was generally unsatisfactory. When a single-part fixture was used, the operation was too slow. By using a multiple-part fixture, production was speeded up, but inaccurate arc strikes and melt-through resulted in many rejects and some service failures. The welder found it awkward and time-consuming to raise and lower his welding hood for each tack weld, but failure to do so caused erratic arc control. By changing to braze welding, it was possible to use a 25-part fixture and to braze weld each part successively without interruption. A lightly flux-coated, low-fuming bronze rod (RBCuZn-C) was used to deposit a fillet weld that was stronger and of much better appearance than the tack welds. The oxyacetylene torch was adjusted for maximum heating with a neutral flame. In playing the flame on the joints, the torch tip was generally pointed in the direction of welding progress, making a separate preheating operation unnecessary. In addition to achieving a satisfactorily high production rate (125 pieces per hour, including loading and unloading), rejections were negligible. After several years in service, no tube failures were reported.

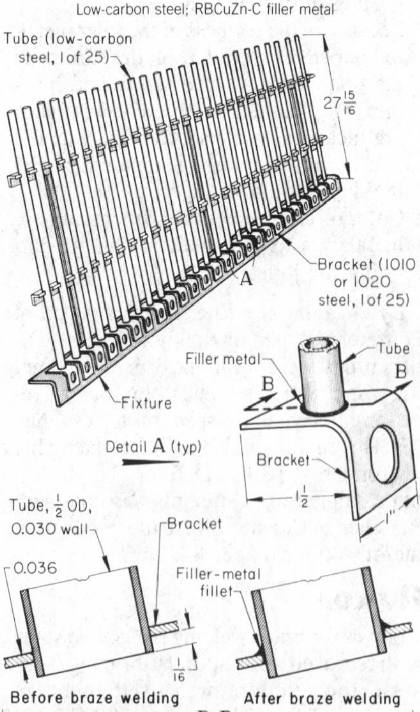

Fig. 2 Tube-and-bracket assemblies. Oxyacetylene braze welded on 25-assembly fixture, at a rate of 5 fixture loads per hour

Production braze welding of metal furniture and similar products that are to be plated requires joints that are free of surface oxidation, overheated flux, and smut film. In addition to pickling, it may be necessary to grind, brush, buff, and polish the joint for satisfactory plating. The cost and delay of the latter operations often can be avoided by using the procedures described in the next example.

Example 3. Copper-Zinc Braze Welding With Gas Fluxing. Braze welding with gas fluxing was the most practical method of joining the various low-carbon steel parts used in the manufacture of chromium-plated wheelchairs. More than 200 joints per chair were involved in this assembly of tubing, wire, formed sheet, and formed or machined bar stock. Two typical structural subassemblies are shown in Fig. 3. The major requirements for both subassemblies were strength, smooth appearance, platability, and speed of production. Strength requirements were not difficult to meet, because the joints were designed so that a smoothly filleted weld deposit would provide more than adequate strength. Meeting the other requirements was a matter of skill in consistently applying optimum technique.

Efficiency of the procedure was based on eliminating operations that would slow production, add cost, or produce questionable results. Thus, to avoid the separate application of a flux and subsequent manual flux cleaning, gas fluxing was used (see the article ''Torch Brazing of Steels'' in this Volume). In addition, gas fluxing eliminated postweld grinding, sanding, and buffing prior to electroplating.

The parts, after solvent cleaning and subsequent drying, were clamped in fix-

Fig. 3 Manually braze welded subassemblies. Used in manufacture of wheelchair tubular supports

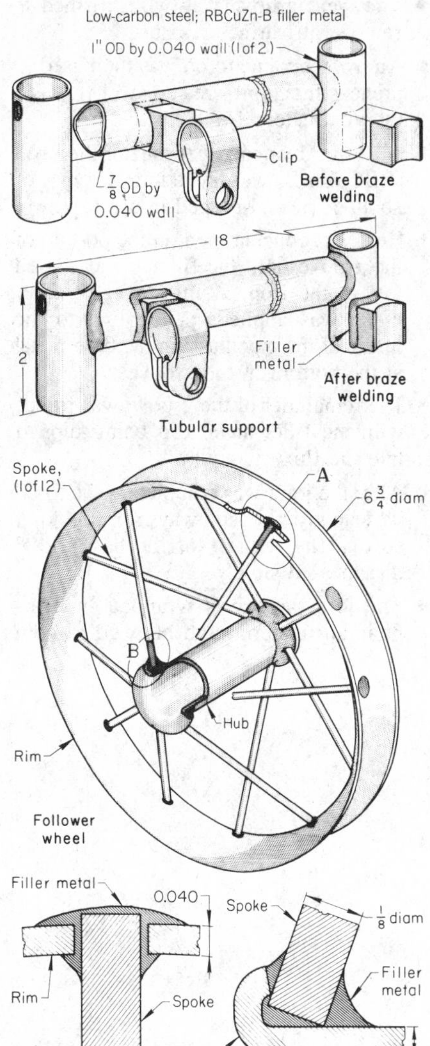

Torch type	Oxyacetylene
Tip-orifice size	No. 57 drill (0.043 in. diam)
Filler metal	$^1/_8$-in.-diam wire, RBCuZn-B
Flux	Gas flux
Preheat and postheat	None
Assembly and brazing time	3 min
Production, tubular supports	120/8-h shift

tures providing joint clearance for the tubular parts that varied from about 0.005 to 0.015 in. before heating, except for the joint at the circular clip, which flared out to about $^1/_8$ in. The oxyacetylene torch flame was adjusted to a very slight acetylene feather. Because of the brilliant green flame resulting from the presence of the flux in the fuel gas, suitably tinted goggles were worn during the adjustments.

The flame was "brushed" around each joint for a few seconds, heating the area and depositing a thin film of flux. A rod of RBCuZn-B was touched to the joint at several places, heated, and melted. Molten filler metal flowed through the remaining joint area and formed well-rounded fillets. Care was exercised to avoid overheating either the base metal or the filler metal. Overheated areas on the assembly, which were readily identified by burned and blackened flux and metal, could not be plated without considerable dressing. Skilled operators were able to avoid overheating; consequently, the few overheated assemblies encountered did not justify the cost of reclaiming and were scrapped. After braze welding, the assemblies were washed, pickled, and nickel-chromium plated. Hot water washing was all that was necessary to remove the flux.

Repair of Iron Castings

Unlike other welding processes, braze welding with copper alloy filler metals is equally effective on any type of cast iron. Peak temperatures of the base metal in braze welding can be low enough to avoid, or at least to cause very little, transformation during the heating cycle. Brittle transformation products in the heat-affected zone of the joint, therefore, can be largely prevented. The weld-metal strength is comparable to the gray irons, the ferritic malleable irons, and the lower strength ferritic ductile irons. Tension tests of braze welded joints have sometimes shown a disturbingly frequent occurrence of parting at the bond line, but this is generally attributed to improper cleaning, fluxing, or tinning, and it emphasizes the need for care and skill.

Low temperature requirements of braze welding make the process particularly suitable for joining malleable iron. Cast iron base metal rarely needs to be preheated to more than 1000 °F, and lower temperatures are often used. In fact, the base metal itself determines the correct preheating temperature; if too hot or too cold, the filler metal will not weld (tin) the joint. Heat added during braze welding is normally of short duration.

Principal drawbacks of braze welding as applied to cast iron are: (1) the color of the copper alloy filler metal does not match that of the iron; (2) the corrosion resistance of the weld metal differs from that of the base metal, being particularly low when the weld metal is exposed to strong alkalis; (3) galvanic corrosion, due to dissimilar metals, may be a problem; and (4) the strength of a braze weld decreases rapidly with increasing temperature, so that the service temperature of the casting is limited to 500 °F maximum.

Braze welding is used to join cast iron to itself or to other metals for production assemblies, but it is more widely used for repairing castings worn or broken in service. In addition to its low cost in comparison with the cost of OFW with cast iron rods, the chief advantages of braze welding are:

- Low thermal stresses minimize the possibility of cracking in the cast iron, which is a brittle material.
- Low peak temperatures avoid the formation of brittle transformation products encountered with other joining processes.
- Copper alloy weld metal is sufficiently ductile to absorb most shrinkage stresses without cracking or parting at the bond.
- Where preheating requirements can be met by local torch preheating, the repair can often be made in position.
- Postheating requirements usually involve only slow cooling from the braze welding temperature.
- Braze welding is better suited to the repair of malleable iron castings than is arc or oxyfuel gas welding.

Fig. 4 Oil-soaked pump casting. Crack repaired by braze welding

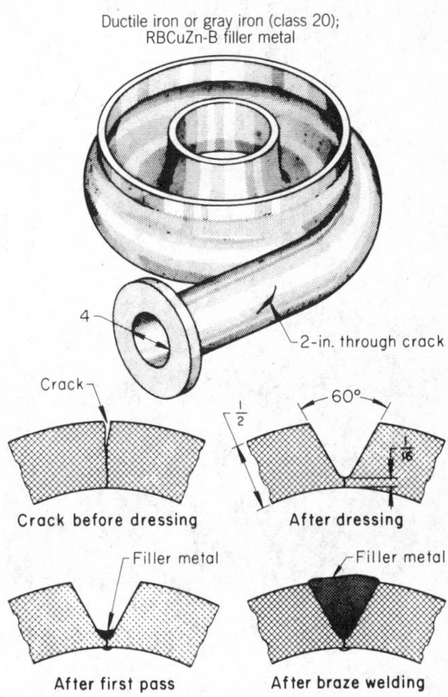

Filler metal	RBCuZn-B
Tip-orifice size	No. 38 drill (0.1065 in. diam)
Flame	Neutral
Preheat	1400 °F, local
Postheat	None (slow cooling under insulation)

Braze welding of cast iron usually requires extra care in joint preparation and cleaning. The two main considerations are the presence of graphitic carbon and the possibility of oil or other impregnation existing as a result of service. The best method of removing cracks or preparing broken edges is by chipping and abrasive blasting. Other methods of metal removal result in smearing of the graphite, which prevents bonding.

A salt bath offers the best method for cleaning cast iron. However, smeared graphite as well as impregnations can be removed by heating the surface of the casting to a dull red heat (approximately 1200 to 1400 °F). Heating the casting surface to this temperature range may cause the formation of surface oxides that prevent wetting by the filler metal; for this reason, the use of a tinning flux is helpful if heating has been used to remove graphite before braze welding. The following example indicates the steps that were necessary for successful repair braze welding of an iron pump casting.

Example 4. Repair of Oil-Soaked Pressure Castings. Centrifugal pump castings of either ductile or gray iron (Fig. 4) were designed to pump transformer oil at 50 psi pressure. Defects, which sometimes became evident after the pump had been in operation, were difficult to repair because oil had penetrated deep into the material. The manufacturer found that even after cutting out the defect, burning out the oil at high temperature, and wire brushing the surface, neither OFW nor arc welding produced a leaktight repair. Braze welding, however, proved successful. Some residue apparently remained below the surface of the base metal after heating; in gas or arc welding, this residue caused excessive porosity during fusion of the base metal. Because the base metal was not melted in braze welding, a leaktight bond was made on the cleaned surface. Figure 4 shows the location of a fine crack that was repaired by braze welding using the following procedure:

- With an abrasive cutoff wheel mounted on a hand grinder, the crack was dressed out to form a 60° V-groove with a small radius at the bottom, leaving approximately $^1/_{16}$-in. thickness.
- The inside of the pipe along the crack area was cleaned with a rotary file.
- The joint area was heated by an oxyacetylene torch to cherry red (approximately 1400 °F) to burn off the oil.
- The weld groove was wire brushed to remove all surface residue.
- An oxyacetylene torch was then used to preheat the joint area to a cherry red (approximately 1400 °F).
- A $^1/_8$-in. manganese bronze filler rod (RBCuZn-B) was heated for a length of about 1 in. and dipped into a flux.
- Heat was concentrated on the bottom of the groove until the filler metal started to wet the iron. A thin layer of filler metal was applied, penetrating to the inside of the pipe through the base metal at the bottom of the groove.
- The remainder of the groove was filled, with the filler-metal rod being dipped into the flux as required.
- After the joint was completed, the torch was slowly worked away from the joint so that the redness of the base metal disappeared slowly.
- The joint area was wrapped in heat-retaining material and allowed to cool slowly.

Oxyacetylene Welding of Cast Irons

By the ASM Committee on Oxyfuel Gas Welding*

OXYACETYLENE WELDING (OAW) is widely used on gray iron, to a smaller extent on ductile iron, and only to a minor extent on malleable iron. Cast iron filler metal is melted together with the base metal to form the joint. An oxyacetylene flame has a maximum temperature of about 6000 °F, which is several thousand degrees less than that of a welding arc. Oxyacetylene welding is therefore slower than arc welding and results in greater total heat input and wider heat-affected zones. This heat results in expansion and localized stress, particularly in large castings. For this reason, high preheats of 1100 to 1200 °F are generally used for OAW of cast irons. However, for local repairs and small unrestrained castings, lower preheat temperatures, often as low as 800 °F, are used. Depending on the mass, composition, and structure of the casting, postheating requirements vary from slow cooling for complete stress relief (1150 °F) to full annealing (1650 °F). Because of higher preheat, and thus lower cooling rates, OAW produces less hardening of the heat-affected zone (HAZ) than does arc welding. Cast irons that are preheated, welded, and slow cooled are readily machinable, which can be important in repair work.

Like the castings themselves, welds in gray iron castings have nil ductility. In a series of tests, joints in class 30 gray iron that were oxyacetylene welded with RCI rods had a somewhat lower average tensile strength than the same joints oxyacetylene welded with RCI-A rods or shielded metal arc welded with ENiFe-CI electrodes. In these tests, the castings were preheated to 400 to 500 °F for arc welding and to 1100 to 1200 °F for OAW, and all welds were cooled slowly under insulation. In spite of the greater total heat input and the high preheat temperatures used, welders skilled in OAW of cast iron produced results comparable to the best arc welded joints.

In the same tests, joints in malleable iron that were oxyacetylene welded using the ductile iron rod RCI-B were not of good quality. The poor results were attributed to the low melting point of the RCI-B rod, which is 200 to 300 °F lower than that of the base metal, and to the hardening of the weld deposit, together with the reversion to white iron that takes place in the HAZ.

Oxyacetylene welding is not recommended for joining of malleable iron (see the section on welding of malleable and white irons in this article). Good results have been obtained in oxyacetylene welding ferritic ductile iron with ductile iron rods. In production welding, however, the speed of OAW cannot compare with that of arc welding. In addition, welds deposited by OAW with cast iron rods usually are less machinable than nickel or nickel-iron welds deposited by arc welding. Porosity, which is a common problem in OAW, can be minimized by using a slightly reducing flame.

As a general practice, OAW has been used on cast irons for:

- Repair of minor casting defects in gray iron (see Example 1) and ductile iron. Minor surface blemishes in malleable irons are sometimes repaired; generally, however, OAW of malleable iron is avoided whenever possible.
- Repair of service-incurred wear and damage (Example 2), mostly on gray and ductile irons
- Production of gray and ductile iron weldments involving either two parts made of cast iron, or one of cast iron and one of another metal, usually steel

Repair of Casting Defects. One of the most common applications of welding cast iron is the repair of rough gray iron castings. Although the majority of this repair work is done by OAW, some is done by arc welding (see the article "Arc Welding of Cast Irons" in this Volume). If repair welding is confined to correction of small defects that affect only the appearance of the casting, inferior mechanical properties and machinability of the weld are of no consequence. The defect must be in an unstressed area that requires no machining and, as a rule, should not extend through the section. Typical defects include sand holes, porosity, washouts, cold shuts, and shift.

Reclaiming defective castings by repair welding is common practice. One foundry has reported that the average cost of repairing leaks in pressure castings was 9% of the selling price of the casting. The repairs were made by OAW, using a cast iron welding rod, to salvage castings that would otherwise have been scrapped. The 9% repair cost was allocated for:

Material	2.5%
Labor	1.5%
Burden, including gas, electricity, and amortization	5.0%

*Gene Meyer, *Chairman*, Manager, Product Distribution and Warranty Service, Victor Equipment; Joseph E. McQuillen, Product Manager, Air Products and Chemicals Inc.; Clarence N. Vaughn, Corporate Burning Manager, Chicago Bridge & Iron Co.

Castings that have defects resulting from machining errors also can be repaired by OAW, provided that the heat of welding does not cause distortion. Usually, arc welding is preferred to OAW for correction of machining errors, because arc welding is faster, has a lower total heat input (and therefore causes less distortion), and produces welds with adequate properties. Good color match of weld and base metal generally is an additional requirement.

Repair of Damaged Castings. Iron castings that have become cracked, broken, or worn in service are regularly repaired by OAW. Braze welding (see the article "Oxyacetylene Braze Welding of Steel and Cast Irons" in this Volume) is used in many applications because of its simplicity and low preheat requirements and because color match is seldom important in such repair. If welding must be done under adverse conditions, extra care and attention to procedural detail are required. Because repair of damaged castings often is a major welding operation, in that a considerable mass of base metal is subjected to high temperatures, preheating is required. A temporary oven around the part or a means of providing localized heating may be needed, depending on size and shape of the part, the required temperature, and duration of heating. See Example 2, which describes an application in which OAW was used in preference to braze welding for repairing a cracked press frame.

Repair of Worn Castings by Hardfacing. Oxyacetylene welding is often used to repair (build up by hardfacing) specific areas of worn gray or ductile iron castings. Malleable iron castings are not well suited to repair by hardfacing.

The choice of a welding process depends largely on service requirements and the equipment available. If the properties obtained are acceptable, braze welding is the logical choice because the casting is far less likely to crack than when arc or oxyacetylene welded. The choice between arc and oxyacetylene welding for repairing worn castings by hardfacing usually depends on the equipment available. Similar results can be obtained with both processes (see the article "Arc Welding of Cast Irons" in this Volume).

The cost of repairing worn castings must be weighed against the cost of replacing them. Consider the cost of building up cast iron, a relatively cheap material, with costly hardfacing alloys. In practice, however, there are considerations other than just the price of the casting, such as delays in getting replacement castings and downtime for replacing a casting in a machine. Repair by hardfacing eliminates procurement delays. A casting surface that has been hardfaced by welding often lasts two to five times as long as the original surface.

Mill-roll journals, rolling-mill guides, wire-spinning rolls, and cast components of mills that process abrasive materials like cement and clay products are typical applications of repair by hardfacing.

The overlay (hardfacing) material applied to castings by OAW usually contains at least 3.0%, and more often 4.0 to 5.0%, carbon. This usually equals or exceeds the carbon content of the base metal. In addition to having high carbon contents, most hardfacing materials also have high alloy contents. Most hardfacing materials are proprietary alloys; three typical ones have nominal (iron-base) compositions:

- 3.9 C, 32.0 Cr, 6.0 Mo
- 4.1 C, 16.0 Cr, 2.0 Ni, 8.0 Mo, 1.0 V
- 4.3 C, 16.0 Cr, 6.0 Ni, 8.0 Mo

Hardfacing alloys such as these are available in rod form for OAW.

Preheat of the casting prior to welding the overlay minimizes cracking. For castings having reasonably uniform sections, a preheat of 650 to 700 °F is sufficient; however, for castings having a wide variation in section thickness, preheating in the range of 1100 to 1200 °F is generally recommended. Time at preheating temperature should be sufficient to ensure that the casting has been uniformly heated.

Postheating is not necessary, but the welded casting must be cooled slowly. In preferred practice, the welded casting is immediately placed in a furnace maintained at or near the preheating temperature used and then is cooled in the furnace. If a furnace is not available, the welded casting should be buried immediately in an insulating material such as lime or spent carburizing compound and should be allowed to remain buried until it has cooled to near room temperature.

Preparation of Castings

If a casting to be repaired has been in service, preparation of the casting for welding requires, in addition to edge preparation, the removal of surface contaminants. Oil, grease, and paint should be removed with solvents, commercial cleaners, or paint removers. Impregnated oil or other volatile matter can be eliminated by heating the casting or weld groove to approximately 900 °F—a dull red heat—for about 15 min and then wire brushing, grinding, or rotary filing to remove the residue. Casting skin on surfaces adjacent to the joint area can be removed by grinding, chipping, shot blasting, or rotary filing. Defects such as porosity, inclusions, and cold shuts should be gouged out, and the bottom of the cavity should be well rounded rather than V-shape.

Completely broken sections should be dressed to form a single-V joint, with a $^1/_{16}$- to $^1/_8$-in. root face to align the parts. Gas welding requires a V-groove with an included angle of 60 to 90° to permit proper manipulation of the torch and welding rod. For heavy sections, a double-V joint should be prepared whenever feasible, with a root face at or near the center of the workpiece.

Preheating

Adequate preheat decreases the rate of cooling of the weld metal and adjacent base metal, thereby minimizing the formation of brittle microstructures in the zone around the weld which might cause cracking immediately after welding or in service. Softer, less brittle microstructures are obtained with high rather than low preheat temperatures. Recommended preheats are discussed in the sections of this article on gray and ductile irons.

To ensure that the preheat temperature is maintained throughout the welding operations, it may be necessary to insulate the heated casting. Heat input from welding must not be permitted to increase the interpass temperature above the maximum preheat temperature, or welding must be stopped until the temperature drops to the preheat range. The temperature should be measured by contact pyrometers or temperature-indicating crayons at or near the weld zone and at one or more other places as required. See the article "Arc Welding of Cast Irons" in this Volume for additional information on preheating.

Postweld Heat Treatment

Postweld heat treatment may be either stress relieving or full annealing. Stress relieving at 1150 °F and then furnace cooling to 700 °F or lower is recommended whenever feasible. Full annealing at 1650 °F produces greater softening of the weld zone and more nearly complete stress relief, but it also lowers the tensile strength of all but the softest irons.

Welding Rods

Cast iron rods used in the welding of cast iron should contain enough carbon and silicon to allow for losses of these elements during welding. The silicon content

Table 1 Chemical compositions of cast iron filler metals used for OAW of cast iron (AWS A5.15)

Classification	C	Si	Mn	Ni	Mo	P	S	Ce
RCI	3.25-3.50	2.75-3.00	0.60-0.75	Trace	Trace	0.50-0.75	0.10 max	...
RCI-A	3.25-3.50	2.00-2.50	0.50-0.70	1.20-1.60	0.25-0.45	0.20-0.40	0.10 max	...
RCI-B	3.25-4.00	3.25-3.75	0.10-0.40	0.50 max	...	0.05 max	0.03 max	0.20 max

of the filler metal must be high enough to permit carbon to precipitate as free carbon during solidification and to promote a soft, machinable matrix as the weld cools to room temperature. Chemical compositions of cast iron filler metals used for OAW of cast iron are given in Table 1.

Rods made of cast iron usually are 24 in. long and 1/4 in. square, although rods 1/8 to 1/2 in. square are available. The RCI filler metals and a number of proprietary rod compositions are used for welding gray irons of classes 20 to 35 (20 to 35 ksi tensile strength). Gray irons that have tensile strengths of 35 to 40 ksi can be welded with RCI-A filler metal, which is similar to RCI but also contains 1.20 to 1.60% Ni and 0.25 to 0.45% Mo. Many proprietary rods contain chromium, nickel, molybdenum, copper, or vanadium, either singly or in combination, to produce high-strength welds; carefully controlled procedure is required with these rods to avoid obtaining a hard weld deposit.

Two basic types of welding rods have been successfully used for welding ductile iron; RCI-B, which is generally higher in carbon and silicon content and lower in manganese than the gray iron rods and contains cerium as a nodularizing agent, and proprietary rods in which magnesium is used as the nodularizing agent.

Fluxes

A flux is required in OAW of cast iron to increase the fluidity of the fusible iron-silicate slag, as well as to aid in the removal of the slag. Fluxes for gray iron rods are usually composed of borates or boric acid, soda ash, and small amounts of other compounds such as sodium chloride, ammonium sulfate, and iron oxide.

Fluxes suitable for welding ductile iron rod are similar to those used for welding gray iron, but are formulated to produce a slag with a lower melting point. Some proprietary fluxes contain inoculants.

Welding of Gray Iron

Oxyacetylene welding of gray iron generally is done for repair, but also can be done for production of simple assemblies. The cavity is first prepared for welding by the usual methods (see the section on preparation of castings in this article) and then ordinarily is tested by liquid-penetrant or magnetic-particle inspection to ensure freedom from defects.

The casting preferably is preheated to 1150 °F in a furnace and then covered with heat-retaining material, exposing only the cavity to be welded. If a furnace is not available, the casting can be covered with heat-retaining material and locally heated by gas flame.

A high-velocity torch that produces a concentrated flame pattern similar to that used for welding low-carbon steel should be used. The cavity surface is dusted with a thin layer of flux; the heated rod is dipped in flux and positioned in the cavity; and both are heated with an oxyacetylene torch adjusted for a neutral or slightly reducing flame. The rod and a small area of the cavity soften under the flame, and as the rod melts off and combines with the base metal, the torch and rod are slowly moved along the cavity. The flame should be directed at the bottom of the cavity with the tip held 1/4 to 1/8 in. from the metal until a molten pool up to 1 in. long begins to form. The torch is moved from side to side until the walls of the cavity start to melt into the pool. This process is continued until the entire cavity is filled with weld metal 1/16 to 1/8 in. thick. Playing the welding flame back over the previously deposited metal to retard its cooling rate will reduce residual stress and permit escape of entrapped gas.

Postweld stress relieving is recommended, particularly for complex castings or where accurate machining will follow. Immediately after completion of welding, the casting should be placed in a furnace heated to the same temperature as the casting, and gradually heated to 1100 to 1200 °F, where it is held for 1 h per inch of thickness. The casting may then be cooled to 500 °F or below at a rate no faster than 50 °F/h. When the welded casting is not stress relieved, it should be cooled slowly either in a furnace or by covering it with heat-retaining material, sand, or some other insulating material.

When class 30 irons are preheated to 1100 to 1150 °F before OAW, some fine pearlite and ferrite are present in the final weld metal and in the HAZ; weld tensile strength approaching 30 ksi and hardness less than 200 HB can be obtained. When class 40 irons are oxyacetylene welded under similar preheat conditions, higher proportions of fine pearlite, less ferrite, and some cementite are present, and weld strength may reach 35 ksi, with hardness as high as 200 HB. Although higher strength irons containing chromium and molybdenum have been welded, welding rods compatible with these irons are not readily available. Welding generally is not attempted in gray iron stronger than class 50 because of the high probability of cracking.

Difficulty in machining is seldom encountered if high preheat and interpass temperatures are used. Class 40 irons usually can be oxyacetylene welded to a hardness of 200 HB or less in weld metal and the HAZ. Cast iron welding rods provide good color match.

More welding of cast iron is done in foundry repair of defects in new castings than in any other type of application. Processes most frequently used are OAW and shielded metal arc welding. The example that follows describes a production line repair setup that is typical of automotive foundries.

Example 1. Repair of Defects in Automotive Engine Blocks. In an automotive engine foundry, defects in gray iron engine block castings were repaired by OAW at a rate of 16.9 castings per hour. The castings weighed about 225 lb each, measured 22 by 18 by 10 in., and had section thicknesses varying from approximately 3/16 to 3/8 in. The welding was done to seal leaks and to repair surface imperfections that had been caused by sand inclusions. Repairable castings were moved by conveyor through six operations.

After shakeout, in order that defective areas would be visible after preheat, castings were inspected and marked for repair before being conveyed to a gas-fired tunnel furnace where they were heated slowly to 1200 °F and soaked for a minimum of 30 min. A cutting torch was then used for thoroughly cleaning out and preparing defective areas for welding. Cavities were filled by OAW, using a 1/4-in. cast iron rod for filler metal. Cleaning and welding had to be done rapidly because welding was not permissible if the casting temperature dropped below 850 °F. Similarly, welding was not done if the temperature of the preheat furnace fell below 1150 °F or if the postheat furnace temperature fell below 1000 °F. After welding, castings were conveyed to a furnace, heated to 1050 °F, and then were furnace cooled to 600 °F over a period of at least 45 min. The final operation consisted of removing the flux deposit and grinding the weld surface flush.

Example 2. Repair of a Cracked Punch Press Frame. When a crack appeared in the frame of an old punch press (Fig. 1), OAW was selected over braze welding as the repair method for improved strength and appearance. The base metal was cast iron; it was assumed to be gray iron, probably of class 30 or 40.

When chipped out, the frame member was laid on its side and firmly braced, and the area near the weld zone was packed in fireclay. The weld area was then preheated to 1200 °F with a heating torch. Welding was done with a carbon plate for backing. During welding, heat had to be applied from time to time to maintain an interpass temperature of about 1200 °F.

Welding of Ductile Iron

Oxyacetylene welding is often used for the repair of defects in ductile iron castings. It has also been used successfully for hardfacing of specific areas on castings to increase abrasion resistance. Oxyacetylene welding of ductile iron has been most successful using a ductile iron rod.

Fig. 1 Repair welding of gray iron punch press frame

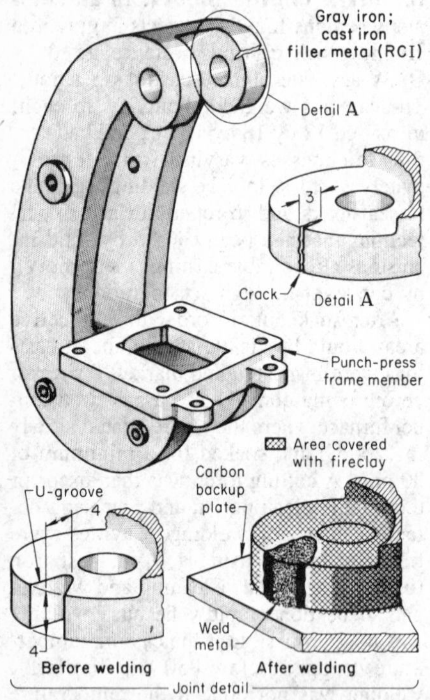

Filler metal, square rod, RCI	1/4 in.
Flux	Proprietary(a)
Oxygen pressure	90 psi
Acetylene pressure	17 psi
Tip size, No. 16 drill, diam	0.177 in.
Preheat	1200 °F
Postheat for 6 h	1200 °F
Welding time (total)	16 h

(a) Mainly sodium borates and carbonates

The repair of ductile iron castings is complicated by the fact that the only way to obtain graphite in nodular form in the weld deposit is to cause it to precipitate from the liquid. Processes have been developed to cause nodularization by introducing magnesium or cerium, or both, into the weld zone by means of a special cast iron filler rod or special flux. Filler metal RCI-B should be used for welding ductile iron.

Magnesium and cerium are carbide formers and consequently must not be present in amounts beyond those required for nodularization. If these elements are present in excessive amounts, a postweld ferritizing anneal is necessary to restore ductility to the weld area; annealing reduces the strength of pearlitic irons and causes distortion of machined surfaces.

Joint preparation and joint cleanness require the same careful attention and the same procedures described earlier for welding gray iron. Preheating practice and interpass temperatures are essentially the same for welding ductile iron as those described for gray iron.

In welding ductile iron, a reducing flame is used to minimize oxidation of the volatile nodularizing elements contained in both the base metal and the welding rod. The welding tip should produce a concentrated flame pattern. The weld area or the complete casting is first preheated to a dull red, and flux is applied to the bottom of the weld groove. Heat is directed at the bottom of the weld groove until a pool begins to form. Walls adjacent to the weld groove are then softened to blend into the pool. The recommended length of the molten pool should be limited to 1 in. A major problem in the welding of ductile iron is the complete loss of ductility in the HAZ.

Welding of Malleable and White Irons

The effect of OAW on malleable iron is to create a wide HAZ of white iron, the material from which the malleable iron was originally produced by applying a malleablizing heat treatment. Because of the hardness and brittleness of the white iron, base-metal properties are lost, and the joint is prone to cracking. The hardened zones in either ferritic or pearlitic malleable iron can be reduced by annealing, but only by processing the casting through a special heat treating procedure for which the original foundry is best equipped.

Oxyacetylene welding is used in the foundry repair of small defects on rough castings while the casting is in the white iron condition, before malleablizing. White iron welding rods are used, which are cast by the foundry to match the base-metal composition, after allowing for constituent losses in deposition. Repair procedures are the same as for gray iron except that, after welding, the casting is given its normal malleablizing heat treatment.

Braze welding, using a copper alloy welding rod, is a more satisfactory method of obtaining relatively strong, machinable joints in malleable iron. Because the base metal is not melted in braze welding, peak temperatures are relatively low, and there is very little hardening in the HAZ. Color match between the weld and the base metal is poor, however, and the service temperature of the casting is limited to about 500 °F. Details of the process are given in the article "Oxyacetylene Braze Welding of Steel and Cast Irons" in this Volume.

Gas metal arc welding with bare wire steel electrodes has been used in joining

Fig. 2 Fabrication of corrosion-resistant high-silicon cast iron pipe

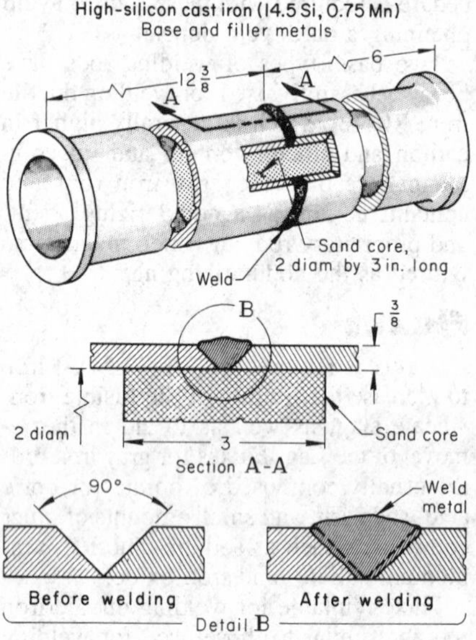

Filler metal	Same as base metal(a)
Flux	50% lime, 50% sodium bisulfate
Backing	Solid cylindrical sand core
Flame adjustment	Oxidizing
Welding position	Horizontal rolled
Number of passes	2
Preheat, furnace	1600 °F
Interpass temperature	1000-1650 °F
Postheat, furnace, for 2 h furnace cool	1600 °F
Fixtures	Clamping jig, turning rolls

(a) Proprietary high-silicon cast iron containing 14.5% Si and 0.75% Mn

malleable iron where end uses do not require full base-metal strength properties. This process produces high peak temperatures, which are of short duration because of high welding speed. Heat-affected zones are therefore narrow. This process is discussed in the article "Arc Welding of Cast Irons" in this Volume.

Welding of Corrosion-Resistant Cast Irons

Corrosion-resistant cast irons usually are identified as high-silicon, high-chromium, or high-nickel irons, specifications for some of which permit welding for repair of minor casting defects. Weld deposits must usually duplicate base-metal compositions, but filler metals may not be generally available.

Example 3. Assembly Welding of High-Silicon Cast Iron Pipe. High-silicon cast iron pipe, used for transporting liquids corrosive to steel, was normally manufactured in diameters ranging from 1 to 8 in. and in standard lengths of 3, 4, or 5 ft, depending on diameter. Standard pipe lengths were cast with beaded ends to permit coupling with split flanges, gaskets, and bolts. Nonstandard lengths were made to order by welding together two pipe ends of suitable length. The manufacturer of this alloy iron found that a special OAW procedure was the only practical method of obtaining a satisfactory weld.

Figure 2 shows the weld joint in an $18^{3}/_{8}$-in. length of 2-in. pipe, normally supplied in 4-ft lengths. The sequence of operations for welding this pipe was:

- Grind both joint edges to 45° bevel.
- Align parts in a clamping fixture and insert a 3-in.-long sand core to straddle the joint to be welded.
- Preheat to 1600 °F in a furnace.
- Adjust oxyacetylene torch for oxidizing flame. (Extensive trials with reducing flames resulted in poor fusion and cracking, apparently because sufficiently high temperature could not be attained.)
- Remove from furnace and tack weld in three places, using a high-silicon cast iron welding rod of the same composition as the base metal, together with a lime-base flux.
- Place on turning rolls and weld in two passes, using the same rod and flux as for tacking. Place sheet insulation over the pipe, leaving opening over weld area for access. Maintain interpass temperature between 1000 and 1650 °F.
- Return to furnace for 2 h at 1600 °F; cool in the furnace with doors shut and burners off.
- Inspect welds visually for cracks and other defects.
- Test hydrostatically at 75 psi.

Welds produced by this method had the same corrosion resistance as the base metal. By welding special pipe lengths to order, inventory requirements were reduced and pattern alterations were avoided.

Special Welding Processes

Electron Beam Welding

By the ASM Committee on Electron Beam Welding*

ELECTRON BEAM WELDING (EBW) is a high-energy density fusion process that is accomplished by bombarding the joint to be welded with an intense (strongly focused) beam of electrons that have been accelerated up to velocities 0.3 to 0.7 times the speed of light at 25 to 200 kV, respectively. The instantaneous conversion of the kinetic energy of these electrons into thermal energy as they impact and penetrate into the workpiece on which they are impinging causes the weld-seam interface surfaces to melt and produces the weld-joint coalescence desired. Electron beam welding is used to weld any metal that can be arc welded; weld quality in most metals is equal to or superior to that produced by gas tungsten arc welding (GTAW).

Because the total kinetic energy of the electrons can be concentrated onto a small area on the workpiece, power densities as high as 10^6 W/cm^2 can be achieved. That is higher than is possible with any other known continuous beam, including laser beams. The high-power density plus the extremely small intrinsic penetration of electrons in a solid workpiece result in almost instantaneous local melting and vaporization of the workpiece material. That characteristic distinguishes EBW from other welding methods in which the rate of melting is limited by thermal conduction.

Principles of Operation

Figure 1 shows an electron beam weld being performed at a weld chamber pressure of approximately 2×10^{-2} torr. The "beam" shown in Fig. 1 is the visible glow that results from the residual (ambient) gas molecules that are excited by the electrons in the actual electron beam.

Basically, the electron beam is formed (under high-vacuum conditions) by employing a triode-style electron gun consisting of a cathode, a heated source (emitter) of electrons that is maintained at some high negative potential; a grid cup, a specially shaped electrode that can be negatively biased with respect to the hot cathode emitter (filament); and an anode, a ground potential electrode through which the electron flow passes in the form of a collimated beam. The hot cathode emitter (filament) is made from a high-emission material, such as tungsten or tantalum. This emitter material, usually available in wire, ribbon, or sheet form, is fabricated into the desired shape for being either directly or indirectly heated to the required emitting temperature of about 4500 °F.

Electrons emitted from the surface of the filament are accelerated to a high velocity and shaped into a collimated beam by the electrostatic field geometry generated from the cathode/grid/anode configuration employed, thus producing a steady stream of electrons that flows through an aperture in the ground plane anode. By varying the negative potential difference between the grid and cathode, this flow of electrons can be altered easily (i.e., gated "on/off" or ramped up/down to different levels) in a precisely controlled manner.

Diode-style electron guns are also employed, but not to the extent that triode-style electron guns are. In a diode gun, the specially shaped electrode (grid cup) is maintained at the same voltage as the emitter, thus making the diode gun a two-element (cathode and anode) device. With this design, the flow of electrons from a diode gun cannot be adjusted by simply varying a grid voltage, as is done with triode guns, and beam current adjustments are usually accomplished by varying the operating temperature of the cathode emitter instead.

Once the electrons exit the anode, they receive the maximum energy input allowable from the operating voltage being applied to the gun. Electrons then pass down through the electron beam column assembly and into the field of an electromagnetic focusing coil (a magnetic lens). This focusing lens reduces the diameter of the electron beam, as it continues in its passage, and focuses the stream of electrons down to a much smaller beam cross section in the plane of the workpiece. This reduction in beam diameter increases the energy density, producing a very small, high-intensity beam spot at the workpiece. In addition, an electromagnetic deflection coil (positioned below the magnetic lens) can be employed to "bend" the beam, thus providing the flexibility to move the focused beam spot. Figure 2 illustrates the main elements of the electron beam welding head. As described above and shown in Fig. 2, electrons are emitted from the cathode and are accelerated to high speed by the voltage between cathode and anode.

The "gun" portion of an electron gun/column assembly generally is isolated from the welding chamber through the use of valves when desired. The gun may be maintained in a vacuum on the order of 1×10^{-4} torr when the welding chamber is vented to atmosphere (for access reasons). This level of vacuum in the gun region is needed to maintain gun component cleanliness, prevent filament oxidation, and impede gun arcing (high-pressure short circuiting between electrodes at different voltages). During welding, this same degree of vacuum is required in both the gun column and welding chamber areas to minimize scattering of beam electrons by excitation collisions with residual air molecules as they traverse the distance from

*Edward A. Metzbower, *Chairman*, Supervisory Metallurgist, Naval Research Laboratory; Robert Bakish, President, Bakish Materials Corp.; Hugh Casey, Associate Group Leader—Materials Technology, Los Alamos National Laboratory; Jeff Flynn, Materials Project Engineer, Pratt & Whitney Aircraft; Stanley E. Knaus, Senior Research Engineer, Rockwell International; Donald E. Powers, Manager—Advanced Product Development, Leybold-Heraeus Vacuum Systems Inc.

Fig. 1 Electron beam welding application. Weld chamber pressure of approximately 2×10^{-2} torr

the gun to the workpiece. This type of interaction tends to produce a broader beam spot and a resulting decrease in energy density. Generally, electron guns are operated with applied voltages that vary from 30 to 200 kV, and they employ beam currents that range from 0.5 to 1500 mA. Electron beam welding equipment with power levels up to 30 kW is common, and several units with power levels of up to 100 kW are commercially available.

Typically, high-vacuum EBW beams can be focused down to spot sizes in the range of 0.010 to 0.050 in. in diameter, with a power density of about 10^6 W/in.^{2} This high level of beam spot intensity is sufficient to vaporize almost any material, forming a vapor hole that penetrates deep into the workpiece. When this vapor hole is advanced along a weld joint, the weld is produced by three effects that occur simultaneously: (1) the material at the leading edge of the vapor hole melts; (2) this molten material flows around the sides of the vapor hole to the trailing edge; and (3) this continuous flow of molten material fills in the trailing edge of the advancing vapor hole and solidifies as the vapor hole moves forward to produce a continuous weld.

Originally, EBW generally was performed only under high-vacuum (1×10^{-4} torr or lower) conditions; because an ambient vacuum environment was required to generate the beam, welding the part within the same clean atmosphere was considered beneficial. However, as the demand for greater part production increased, it was found that the weld chamber vacuum level need not be as high as that needed for the gun region; ultimately, the need for any type of vacuum surrounding the workpiece was totally eliminated for some applications. Currently, three distinct modes of EBW are employed: (1) high-vacuum (EBW-HV), where the workpiece is in an ambient pressure ranging from 10^{-6} to 10^{-3} torr; (2) medium-vacuum (EBW-MV), where the workpiece may be in a "soft" or "partial" vacuum ranging from 10^{-3} to 25 torr; and (3) nonvacuum (EBW-NV), which is also referred to as atmospheric EBW, where the workpiece is at atmospheric pressure in air or protective gas. In all EBW applications, the electron beam gun region is maintained at a pressure of 10^{-4} torr or lower.

Advantages

One of the prime advantages of EBW is the ability to make welds that are deeper and narrower than arc welds, with a total heat input that is much lower than that required in arc welding. This ability to achieve a high weld depth-to-width ratio eliminates the need for multiple-pass welds, as is required in arc welding. The lower heat input results in a narrow workpiece heat-affected zone (HAZ) and noticeably less thermal effects on the workpiece.

In EBW, a high-purity vacuum environment can be used for welding, which results in freedom from impurities such as oxides and nitrides. The ability to employ higher weld speeds, due to the high melting rates associated with the concentrated heat source, reduces the time required to accomplish welding, thereby resulting in an increased productivity and higher energy efficiency for the process. Total energy conversion efficiency of EBW is approximately 65%, which is slightly higher than so-called conventional welding processes and much higher than other types of high-energy-density welding processes, such as laser beam welding (LBW).

These characteristics (1) minimize distortion and shrinkage during welding; (2) facilitate welding of most hardened or work-strengthened metals, frequently without significant deterioration of mechanical properties in the weld joint; (3) facilitate welding in close proximity to heat-sensitive components or attachments; (4) allow hermetic seal welding of evacuated enclosures, while retaining a vacuum inside the component; and (5) permit welding of refractory metals, reactive metals, and combinations of many dissimilar metals that are not joinable by arc welding processes. The ability to project the electron beam a distance of over 20 in. under high-vacuum conditions, as well as the low end of medium-vacuum conditions, al-

Fig. 2 Main elements of the electron beam welding head

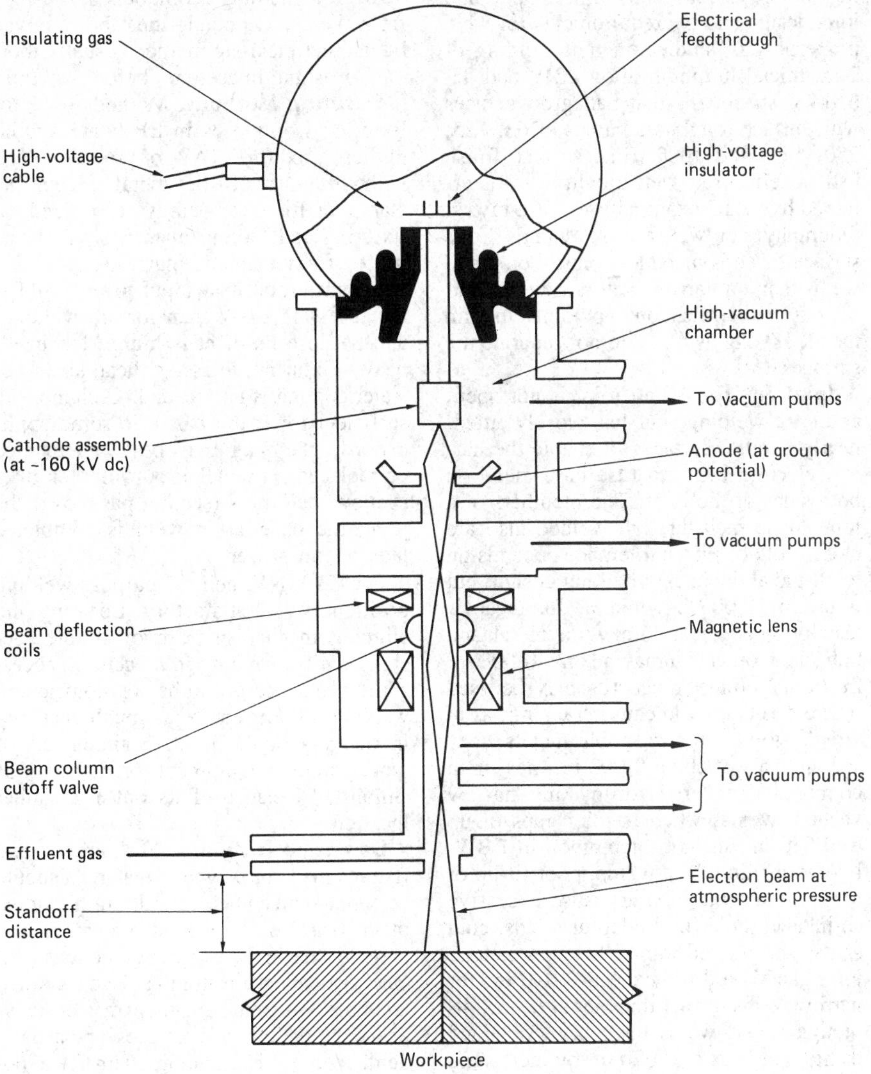

lows otherwise inaccessible welds to be completed.

Disadvantages

Equipment costs for EBW generally are higher than those for conventional welding processes. However, when compared to other types of high-energy density welding (such as LBW), production costs are not as high. The cost of joint preparation and tooling is more than that encountered in arc welding processes, because the relatively small electron beam spot size that is used requires precise joint gap and position.

The available vacuum chamber capacities are limited; workpiece size is limited, to some degree, by the size of the vacuum chamber employed. Consequently, production rate (as well as unit cost) is affected by the need to pump down the chamber for each production load. Because the electron beam is deflected by magnetic fields, nonmagnetic or degaussed metals must be used for tooling and fixturing that are near the beam path.

Although most of the above advantages and disadvantages generally are applicable to all modes of EBW, several do not specifically apply to EBW-NV. Nonvacuum EBW does not offer the advantage of a high-purity environment (unless some form of inert gas shielding is provided), and it is not subject to vacuum chamber limitations. Because welding is not done within the confines of a vacuum environment, the maximum practical "standoff," the working distance between the bottom of the electron beam column and the top of the workpiece, currently used on EBW-NV systems is limited to approximately 1$^{3}/_{8}$ in. The maximum attainable penetration is currently limited to about 1 to 1.25 in. for 60-kW, 165-kV machines. These limitations, however, are offset by higher production rates and a lower rate of vacuum-related equipment problems that are typical of EBW-NV systems.

Process Control

Basic variables employed for controlling the results of an electron beam weld include accelerating (applied gun) voltage, beam current, welding (beam spot travel) speed, focusing current, and standoff (gun column assembly to workpiece) distance. The final beam spot size that is produced in the plane of the workpiece is determined by (1) the characteristics of the electron beam gun/column assembly (gun and electron optics); (2) the focusing current, which controls the focal length of the lens and resulting beam focus location; (3) the gun column assembly-to-workpiece distance; (4) the accelerating voltage; and (5) the beam current. Each of these variables, separately and jointly, affects final beam spot size.

Increasing the accelerating voltage or beam current increases the depth of penetration; the product of these two variables—the beam power—determines the amount of metal melted for a given exposure time. Increasing welding speed—relative travel motion between beam spot and workpiece—without changing any other process variable, reduces depth of penetration and correspondingly reduces weld width. Changing any of the other basic control variables to increase beam spot size, thereby lowering the beam spot intensity (e.g., reducing the power density), reduces depth of penetration and increases weld width, if welding speed is left unchanged. Beam deflection can be used to change the impact angle of the beam (slightly) or to produce controlled patterns of beam oscillation to create greater beam spot size or for other special effects. The beam can also be pulsed to vary the average amount of power inputed per unit of time.

With EBW-NV, the accelerating voltage and focus current are normally preset and held at a fixed value. Beam current may be preset for any application and then turned "on/off" (or programmed to vary in predetermined fashion) with workpiece travel. Workpiece speed and, in certain instances, standoff distance may also be preset to a desired value or varied with workpiece travel. On many of the latest computerized numerical control (CNC) types of EBW-NV systems, beam current, standoff distance, and welding speed are varied simultaneously, as required, throughout the length of the weld. Because of the limited standoff distance involved with EBW-NV, beam deflection normally is not feasible, and beam oscillatory motion generally is not employed.

Operation Sequence and Preparation

Tooling and welding procedures for each EBW application are developed first on experimental workpieces. Details of the welding sequence vary somewhat, depending on equipment differences and application requirements. A typical sequence of operation for welding in vacuum is to:

- Assemble and prepare work and fixtures for welding. This includes cleaning and may include demagnetizing, preheating, and tack welding.
- Load fixtured work onto worktable or work-holding mechanism in welding chamber.
- Start chamber pumpdown.
- After chamber pressure has been reduced to 10^{-4} to 10^{-1} torr, focus on a target block and set beam parameters.
- Align joint to the beam position, using very low power beam spot.
- Begin welding; this usually is performed automatically, but can be performed manually.
- Terminate the welding cycle.
- Allow work to cool sufficiently if made of reactive material, then admit air to the chamber and remove fixtured work.

For nonvacuum welding, the electron gun/column assembly is maintained at established pressure levels; work-handling and welding operations for EBW-NV usually are mechanized for high-speed production, and work and fixturing are prepared before a production run is begun, or they are completed as part of the production-line operation. Beam parameters, alignment of joint with beam position, and transfer and movement of work for welding are established before a production run. Alignment of the joint with beam position is slightly less critical than for vacuum electron beam welding, because beam spot size and weld width are slightly larger in nonvacuum welding.

Joint Preparation. A joint for EBW ordinarily has closely fitted, abutting, square-groove faces, and filler metal usually is not used. Filler metal can be used if desired, as in EBW of dissimilar metals and alloys. Generally, the faces of the joint are machined to a surface roughness of about 125 μin. or less.

Surface finish on the weld groove faces may not be critical, depending on part and joint design and the requirements for weld properties. In studies on butt welding of 2-in.-thick aluminum alloy 2219 and Ti-6Al-4V specimens that had groove faces with surface roughness values of 63, 125, 250, 500, and 1000 μin., surface finish had no effect on weld quality, as determined by visual examination and x-ray radiography; all welds were sound. Edge surface finish is much less critical on broad welds than on narrow welds. Edge roughness is not critical on lap joints in thin metal, as long as burrs do not separate the surfaces.

Joint Fit-Up. A butt joint is not open, as in arc welding, but has closely fitted, nearly parallel surfaces to enable the narrow electron beam to fuse base metal on both sides of the joint. The members of a joint to be melt-through welded also are closely fitted. Fit-up tolerance depends on work-metal thickness and joint design, but is usually 0.005 in. or less. Joint gap is usually smaller for thin work metal and unbacked joints and may be only 0.002 in. maximum. Interference fits may be used where shrinkage can cause cracking, as in circular joints on hardenable metals.

Joint gaps of about 0.003 in. maximum commonly are used for making narrow vacuum welds, whereas joint gaps of up to 0.030 in. are not uncommon in EBW-NV. In making deep welds, poor fit-up or too large a joint gap can cause excessive shrinkage, underfill, undercut, voids, cold shuts, and missed joints. In most metals, joint gap should not exceed 0.010 in. for narrow welds deeper than about 1/2 in., although sound welds have been obtained using joint gaps of 0.020 in. by increasing the weld width to about 0.25 in.

Cleaning. Workpiece surfaces must be properly cleaned for in-vacuum EBW. Inadequate surface cleaning of the weld metal can cause weld flaws and a deterioration of mechanical properties. Inadequate cleaning of weld surfaces also adversely affects pumpdown time and gun operational stability, as well as contributing to the rapid degradation of the oil used in the vacuum pumps.

In EBW-NV, cleanliness of workpiece surfaces is important, but not as critical as with vacuum EBW. Although the effect of workpiece cleanliness on pumpdown time and gun stability is reduced in nonvacuum welding, it still affects final weld quality.

Wire brushing generally is not recommended, because contaminants may become embedded in the metal surface. Acetone is preferred for cleaning electron gun components and workpiece parts. If workpieces are cleaned in chlorine or other halogen-containing compounds, residue from these compounds must be removed by another cleaning method (usually thorough washing in acetone) before welding.

Fixturing Methods. Methods used for fixturing workpieces in EBW are similar to those used for GTAW of precision parts without the use of filler metal, except that clamping force is usually lower and all fixturing and tooling materials should be made of nonmagnetic materials.

Because total heat input to the weld is much less in EBW than for arc welding, and because the heat is highly localized, heavy fixturing, massive heat sinks, or water cooling is not needed. C-clamps are sufficient for many parts. In some applications, clamping may be supplemented or replaced by small tack welds or by a shallow weld pass (sealing pass) over the joint; the penetration weld is completed later at full power.

For EBW-NV, general-purpose welding positioners are satisfactory. Locating and aligning mechanisms are of simpler design than for welding in a vacuum; accessibility and beam widths for nonvacuum welding do not require as much accuracy in tracking of the joint. Maintenance of work-handling equipment for EBW-NV is simplified because of its out-of-chamber location.

Demagnetization. Workpieces and fixtures made of magnetic materials should be demagnetized before welding. Residual magnetism may result from magnetic-particle testing, magnetic chucks, or electrochemical machining. Even a small amount of residual magnetism can cause beam deflection. Workpieces are usually demagnetized by placing them in a 60-cycle inductive field and then slowly removing them. Equipment used for magnetic-particle testing may be used. Before welding, workpieces should be checked with a gauss meter. Acceptable gauss-meter readings vary from 1/2 gauss for very narrow (highly critical) welds up to as much as 2 to 4 gauss for relatively wide welds.

Pumpdown. The time required to pump the work chamber of an in-vacuum EBW unit down to the desired ambient pressure depends on the chamber size used, type of pump employed, and level of vacuum required. Electron beam welding equipment currently employs either computer or program logic controls, normally providing automatic vacuum sequencing.

For EBW-HV, a pressure of 10^{-4} torr or less is produced by a mechanical pump (either a simple piston, a rotary vane type, or a combination pump-and-blower package), operating in conjunction with an oil-diffusion pump. The size of these pumps depends mainly on the size of the weld chamber used, as well as on the final operating vacuum level and total pumpdown time needed.

Generally, a 40-ft^3 chamber employing a 10-in., 8820-ft^3/min diffusion pump and a 600-ft^3/min mechanical pump package can be expected to reach a welding pressure of less than 3×10^{-4} torr in approximately 4 min, while a 400-ft^3 chamber employing a 20-in., 37 800-ft^3/min diffusion pump and a 1300-ft^3/min mechanical pump package can be expected to pump down to the same pressure in approximately 12 min.

If desired, the pumpdown time of the 400-ft^3 chamber could be reduced to under 6 min by increasing the size of the pumping system to a 32-in., 109 200-ft^3/min diffusion pump and a 2750-ft^3/min mechanical pump package; however, the higher cost associated with this increased pumping capability would have to be evaluated against the financial benefit of the saving in pumpdown time to determine whether it would be economically feasible. In the examples above, weld chambers are assumed to be clean, dry, and empty. The cleanliness of the work chamber, the amount of water vapor (humidity) in the ambient air, and the workpiece assembly (material and shape of both weldment and fixturing) thus affect the actual pumpdown time obtained in any production application.

For partial-vacuum EBW, where the pressure level required in the weld chamber is approximately 10^{-1} torr (100 μm) and a pressure level of 10^{-4} torr is provided only in the electron beam gun/column regions, diffusion pumps are not needed for the weld chamber, and weld chamber pumping is accomplished strictly by mechanical pumping. For high-production EBW of small parts in a partial vacuum, pumpdown time ranges from about 20 s for relatively large-volume (general-purpose) partial-vacuum weld chambers of about 5 ft^3 down to less than 5 s for fairly small-volume chambers of around 0.5 ft^3, with the use of a mechanical pump rated at 1300 ft^3/min.

Preheat and Postheat. Most commonly welded metals can be processed with EBW methods, even in thick sections without preheating, because of the extremely narrow width of the HAZ. Hardenable and difficult-to-weld metals may need to be preheated, especially for thick sections and in applications when the weld is restrained.

High-strength alloy steels and tool steels thicker than about $^3/_8$ in. ordinarily must be preheated before EBW to prevent cracking. Deep circular welds, especially partial-penetration welds, in thick sections of carbon steel containing more than about 0.35% C usually require preheating. However, welds subject to less restraint, such as circumferential welds on cylindrical shapes, can be made on $^1/_2$-in.-thick 0.50% C steel without preheating.

Preheating, when required, is usually done before the work is placed in the work chamber. Selection of heating method depends on the size and shape of the work and the preheat temperature; a combination of methods can be used. Torch and furnace heating are widely used; induction and infrared-radiation heating are also used. On small parts or where distortion from localized heating is not a problem and where increased cycle time can be tolerated, heating is sometimes done with a defocused electron beam; this method can also be used to supplement other methods of heating.

When postheating is used on EBW parts, it is accomplished by conventional means (i.e., stress relieving or tempering) after removing the work from the welding chamber.

Operating Conditions. Starting and stopping the weld usually require special consideration to avoid nonuniformity of the weld at these points and possible melt-through and loss of metal.

One technique that is used to avoid these difficulties is to start the weld at full beam power on a starting tab of the work metal that is tightly fitted against one end of the joint and to conclude the weld on a runoff tab at the other end of the joint. The use of starting and runoff tabs prevents underfill at the ends of the joint, which is caused by the introduction or exit of the beam. Workpieces also can be made oversize to provide extra material for starting and stopping. The tabs or the extra material can be machined off after welding. Starting and runoff tabs are used mainly in low-production operations.

Another technique is to start and stop the weld on the work, raising the current gradually (upslope) at the beginning of the weld and reducing it gradually (downslope) at the end of the weld. Upslope and downslope power may be used at controlled rates and time intervals, as established for a specific application, and can be programmed into the welding procedure. The use of upslope and downslope is of special value where the weld is a closed path, as in welding circular and circumferential joints.

In many applications of closed-path welds, the weld can be started at full power, with downslope at the end providing a sufficiently gradual termination and a suitable distance of overlap. Downslope of current after overlapping the beginning of the weld on circular welds in thick sections of high-hardenability steels is critical to avoid porosity and cracking. Other parameters may also be adjusted to avoid these defects. Upslope is usually rapid; downslope is from a few degrees to a major portion of a revolution.

A "cosmetic pass" is made when needed to smooth or flatten the crown of a weld that is irregular or too high. Such a pass is used to correct undercut or underfill. The beam is usually defocused or reduced in power, or both, in making a cosmetic pass. Filler metal may also be added in making a cosmetic pass intended to correct undercut or underfill.

Weld Geometry

Part Shape. The shape of the parts to be welded and the corresponding joint designs are critical to the successful application of EBW in vacuum or at atmospheric pressures (nonvacuum). While minimum heat input and low thermal distortions are important advantages of EBW, the molten metal still shrinks as it solidifies. Shrinkage stresses may lead to microcracks if parts, due to design restrictions, are unable to shrink at corresponding rates and the joint volume is completely constrained. A circular weld in the axial direction, for example, that joins a disk-shaped member (gear or cup) to a shaft (Fig. 3) may experience severe constraint. Such joints are practical in easily welded metals—those that are soft, have low yield points, and exhibit low shrinkage—but not in difficult-to-weld steels that are used for improved strength, hardenability, or machinability.

Fig. 3 Circular weld joining disk-shaped member to a shaft

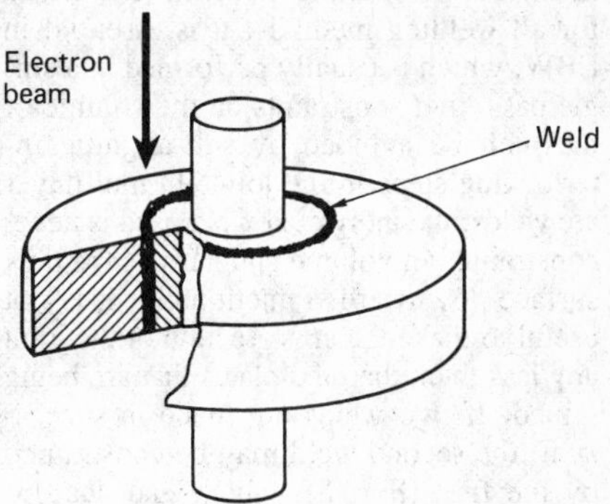

Fig. 4 Shrinkage stress in radial weld

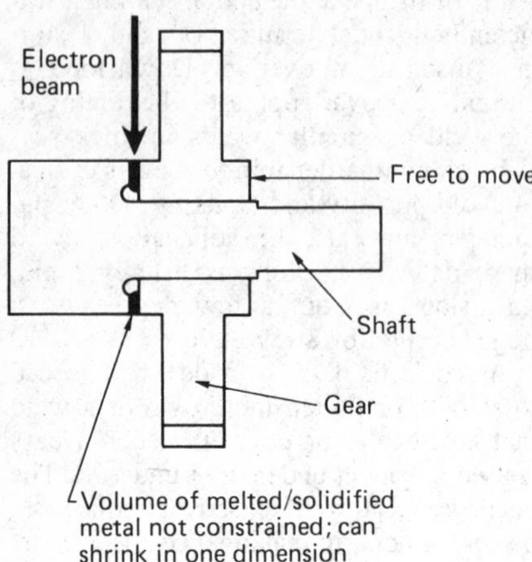

The difficulties encountered in welding shapes such as Fig. 3 are aggravated if the weld zone is broadened by defocusing the electron beam. This is a frequent problem if exact beam position cannot be controlled, either because of beam instability, magnetism of the parts, or excessive tolerances in fixturing on high throughput machines. Beam broadening can be eliminated by the use of closed loop beam position control systems.

Shrinkage stress is minimal in radial welds, as shown in Fig. 4. The volume of molten metal is not constrained by the axial direction if the gear hub is allowed to move by using push fit and no restraints. The weld must also have root clearance—the weld zone must not extend into the shaft. Stresses are lower if the melt zone has parallel side boundaries. Triangular weld zones that are caused by broadened beams, uneven beam power distribution, or incomplete penetration have greater internal stresses.

The same principles of joint design discussed above for joints between shafts and disk-like members apply equally to T-joints and corner joints between plates. Figure 5 shows recommended and nonrecommended weld configurations. Although true for all welding methods, it is essential in EBW, which is usually performed in a single pass, that constraints on the volume of the melt be avoided by self-aligning interlocking steps in the joint. In multilayer arc welds, the individual weld bead is never constrained in volume and shrinks from its surface downward. Functionally, it is not useful to make the cross section of the joint any less thick than the plates that are being joined. If two welds are made in succession, the second weld may be constrained by the first (Fig. 5), which may lead to solidification cracking. It is thus advisable to direct the beam at the joint parallel to the faying surfaces and to cause melting over the whole contact area.

Fig. 5 Preferred and nonrecommended weld configurations. (a) Not recommended—maximum confinement of molten metal; minimum joining cross section (arrows); wastes beam energy for melting nonfunctional metal. (b) Most favorable—volume of melt not confined; maximum joining cross section (arrows). (c) Not recommended—maximum confinement of melt (unless gap is provided); joining cross section less than plate cross section. (d) Most favorable—minimum constraint and confinement of melt; minimum internal stresses; warpage can be offset by bending prior to welding; tilt can be offset by location of T-arm at less than 90° to base prior to welding. Fillet obtained by placing wire in right corner and melting it with the beam. (e) Not recommended—two successive welds; second weld is fully constrained by the first weld and shows strong tendency to crack.

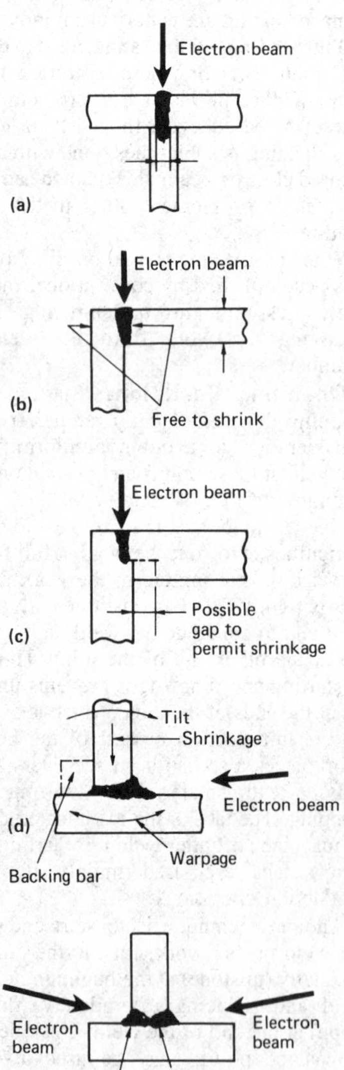

Surface Geometry. Usually, EBW does not use or need filler wire. Therefore, neither V-grooves nor corners can be filled; in fact, they are detrimental. Surface tension pulls some of the molten metal up the V-shaped surfaces so that it is no longer available to fill the space created by the keyhole that is generated by the moving beam. Voids remain in the actual joint plane (Fig. 6).

A step in the surfaces at the joint line is also undesirable. Any small lateral shift of the beam from the low to the high side, and vice versa, changes the penetration to some extent. In the extreme case, this shift may be equal to the height of the step, as shown in Fig. 7. On circular welds, with joint configuration similar to Fig. 7(a), a noncircular weld profile may develop with irregularities that can lead to uneven applied stresses. By contrast, a mere shift in the melt zone, as shown in Fig. 7(b), does not affect weld quality, as long as the entire joint line is covered.

Configurations for Wide Welds Bridging a Gap. The high speed and autogenous nature of an electron beam weld can be advantageous when a wide weld (1/4 to 1/2 in.) is needed without great depth-to-width ratio. The atmospheric electron beam has been used extensively to make such welds in mass-production automotive parts such as catalytic converter cases and die-cast aluminum intake manifolds.

The easiest weld to make with an electron beam is an edge-weld in the vertical position of the beam (weld horizontal and flat), as illustrated by the edge-flange weld in Fig. 8. In vacuum, the beam must be defocused; in air, the standoff must be sufficient to prevent substantial beam width

Fig. 6 Gap in joint plane created by surface tension

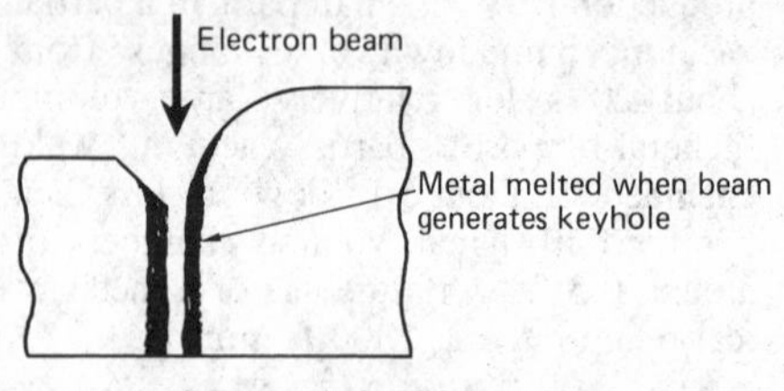

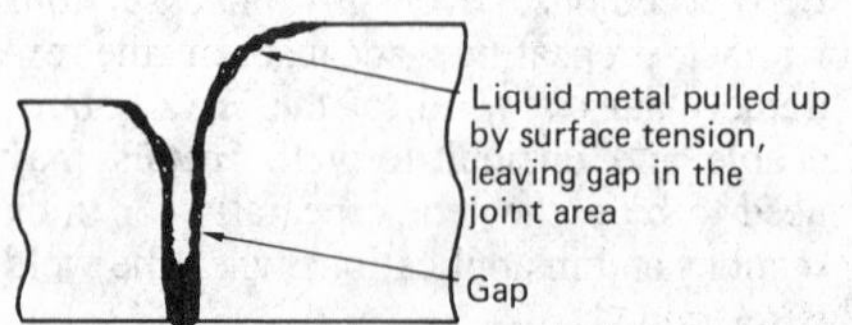

Fig. 7 Effect of beam and melt zone position on weld penetration

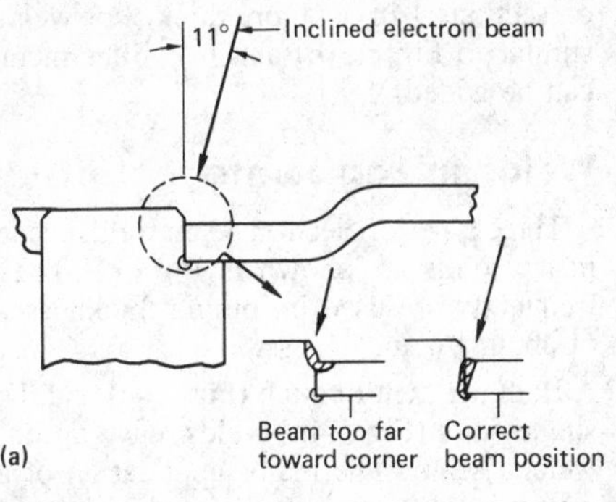

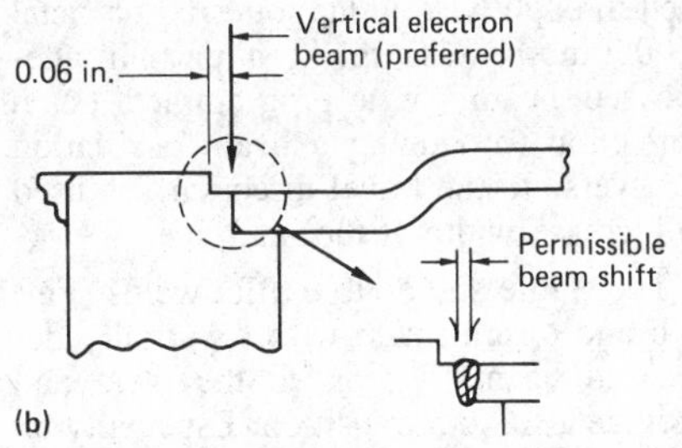

Fig. 8 Edge-flange weld. Preferred geometry for thin panels

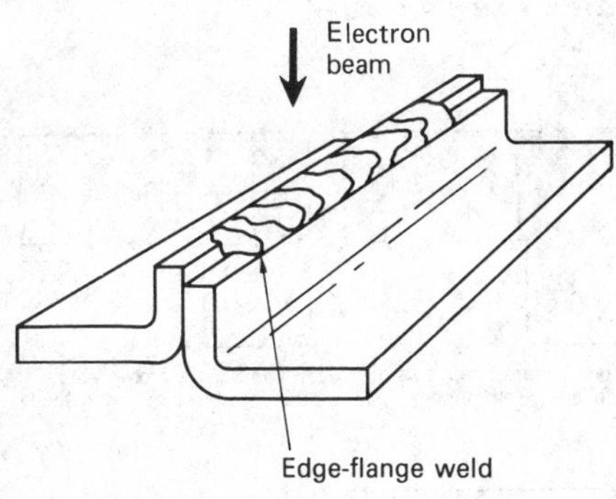

scattering. The standoff is not critical, however; the energy flows into the weld even without its use. Energy flow is not influenced by standoff as it is in electric arc welding.

For edge welds, changes in energy input per unit length and changes in power density affect only the amount of molten metal. Offset and gaps up to the thickness of the sheet to be joined often can be tolerated (Ref 1). In making an edge-flange weld (Fig. 8), the same advantages exist, provided the flange is high enough. For joining thinner sheet that is tightly clamped, edge welds also can be made with the electron beam in the horizontal position.

The EBW-NV process can tolerate and fill large gaps if sufficient sacrificial metal is available to be melted by the beam (Fig. 9). If not, surface tension may pull whatever metal is melted to either side of the weld joint.

Melt-Zone Configuration. Welds with parallel sides of the melt-zone (Fig. 10a) show minimum warpage and only transverse shrinkage when they are unconstrained. Triangular melt zones (Fig. 10b) cause warpage. If there are constraints, the hindered warpage and shrinkage cause internal stresses. A parallel-sided melt zone is preferred in either case; to obtain this configuration, joint design must allow for a full-penetration weld.

Joint Design

Butt, corner, T-, lap, and edge joints can be made by the EBW process using square-groove or seam welds. Fillet welds, which are difficult to make with vacuum EBW, are readily made using EBW-NV. Square-groove welds require fixturing to maintain fit-up and alignment of the joint. They can, however, be self-aligning if a rabbeted joint design is used. Self-alignment is particularly important in batch loading the vacuum chamber for efficient work manipulation. The weld-metal area can be increased using a scarf joint, but fit-up and alignment of the joint are more difficult than with a square-groove weld. Most joints are designed to be welded in a single pass with full penetration or penetration to a specified depth. Figures 11 to 15 show the commonly used joint and weld types for EBW. Different joint preparations, joint designs, and welding positions are used to meet special requirements. For preferred joint configurations and shrinkage stresses encountered in various joint designs, see the section on weld geometry for electron beam welding in this article.

Welds in Butt Joints

Butt joints are the most common of the five basic joint types used in EBW. The basic square-groove weld, which is shown in Fig. 11(a), (b), and (c), needs only the simplest and least expensive joint preparation. It is suitable for either partial- or full-penetration welds. Good fit-up and external fixtures are needed. The flush joint (Fig. 11b) is preferred to the stepped joint (Fig. 11c) for joining unequal thicknesses, chiefly because control of conditions for making sound full-penetration welds is less critical than for the stepped joint. A wider beam is used in welding the stepped joint, and the beam angle must be carefully controlled to avoid scarfing the upper edge of the thicker member (angle too small) or missing the bottom of the joint (angle too large), especially if the thinner member is more than about $^3/_8$ in. thick.

Fig. 9 Use of sacrificial metal to fill a large gap in a square-groove weld

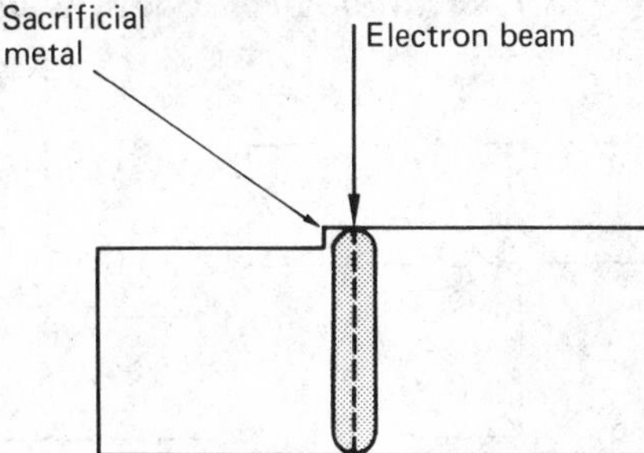

Fig. 10 Difference in warpage for a triangular and a nearly parallel-sided melt zone. Butt welds in 304 stainless steel 0.10 in. thick; made with 12-kW beam in the atmosphere; helium shielding gas; 0.45-in. standoff; speed, 280 in./min (a) and 165 in./min (b)

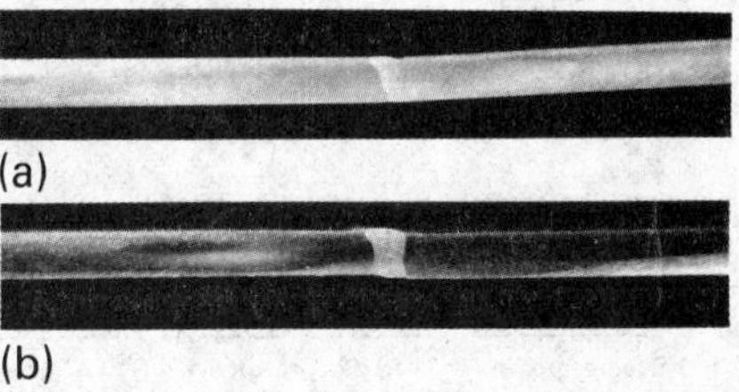

In Fig. 11(d), the joint is rabbeted to make it self-aligning, but the offset is small to avoid leaving an unwelded seam near the root of the weld (compare Fig. 11e). In Fig. 11(d) to (g), the joints are self-aligning and may be self-fixturing in circular, circumferential, and certain other joint arrangements. In Fig. 11(e) and (f), the joints are both self-aligning and self-backing; each, however, leaves an unwelded seam near the root of the weld.

In Fig. 11(g) and (h), two ways are shown of providing integral filler metal. The lip of the joint in Fig. 11(g) provides more filler metal than the shoulder of the joint in Fig. 11(h), but because it conceals the joint, a scribed line or other means for beam placement and for scanning must be provided before the welding operation can proceed.

The slant-groove weld (also called angular or scarf) shown in Fig. 11(j) is used in butt joints to facilitate fixturing. It is also used where limitations on the beam location prevent fusion for the entire groove depth if the basic square-groove weld (Fig. 11a) were used. Greater weld depth, as measured parallel to the beam axis, is needed for full penetration of a given work-

Fig. 11 Types of butt joints and welds used in EBW

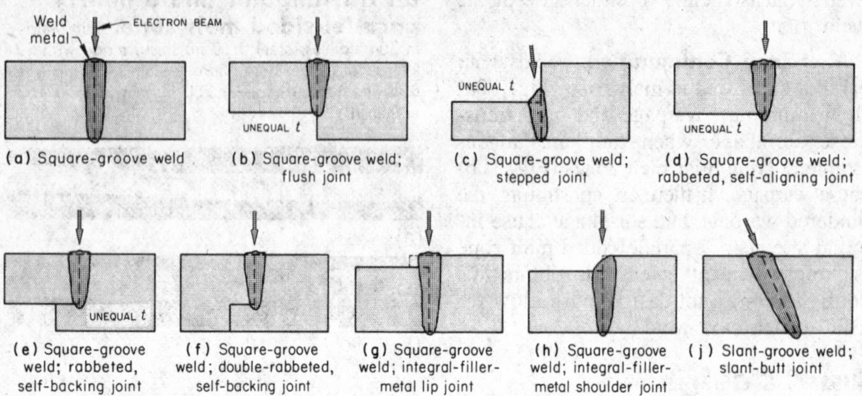

metal thickness than with the square-groove weld. This type of weld and variations of it are used both in butt joints and in corner joints.

A slant-butt joint can also be welded with beam alignment set at 90° to the surface of the work. This beam angle permits the production of defect-free welds where fit-up is poor or where the joint opening is larger than can be tolerated when the beam is aligned with the groove. In a related type of weld, the joint has a square-groove or modified square-groove preparation, and the beam is at an angle to the groove. The main reason for using this type of weld is difficulty of access with other types of welds.

Welds in Corner Joints

Eight types of electron beam welds frequently made in corner joints are shown in Fig. 12. Corner joints are second only to butt joints in frequency of use for EBW.

Important differences between butt and corner joints are notch sensitivity and suitability for nondestructive testing. To compare alignment, self-fixturing, and the occurrence of portions of unwelded seam, see Fig. 11 and 12.

The two simplest and most economical welds in corner joints are the melt-through (Fig. 12a) and basic square-groove (Fig. 12b) welds. Neither is self-aligning or self-fixturing; manipulation of the work for weld (b) in Fig. 12 can be simplified by using a horizontal electron beam and corresponding work orientation (rotated 90° from the orientation shown). The melt-through weld (Fig. 12a), unless made with a fusion zone wide enough to eliminate the unwelded seam completely, is weaker than the square-groove weld and is notch sensitive.

The corner-flange weld (Fig. 12h), usually made only on thin stock, requires precision forming of the 90° bend. Welds in corner joints are subject to high stress concentrations.

Welds in T-Joints

Three types of electron beam welds made in T-joints are shown in Fig. 13. The melt-through or blind weld (Fig. 13a), like melt-through welds in butt or corner joints, leaves an unwelded seam with resulting low strength, notch sensitivity, and corrosion susceptibility. Figure 13(b) is the preferred joint design. The double-square-groove weld (Fig. 13c) is used primarily on sections 1 in. or more thick. For welds similar to Fig. 13(b) and (c), filler metal can be added.

Welds in Lap Joints

Three types of electron beam welds made in lap joints are shown in Fig. 14. They frequently are used in joining thicknesses of about $^1/_{16}$ in. or less.

Both the melt-through (Fig. 14a) and the single fillet (Fig. 14b) welds leave an unwelded seam. The major application of a melt-through weld in lap joints is for metal 0.005 in. or less thick; a partially defocused beam, or a programmed beam deflection (producing a beam oscillation transverse to the travel direction), is used to increase width at the interface.

The single and double fillet welds (Fig. 14b and c) are made with a partially defocused beam to broaden the weld and prov.de a smooth transition. Especially on thick stock, filler metal may be added to increase fillet size.

Fig. 12 Types of corner joints and welds used in EBW

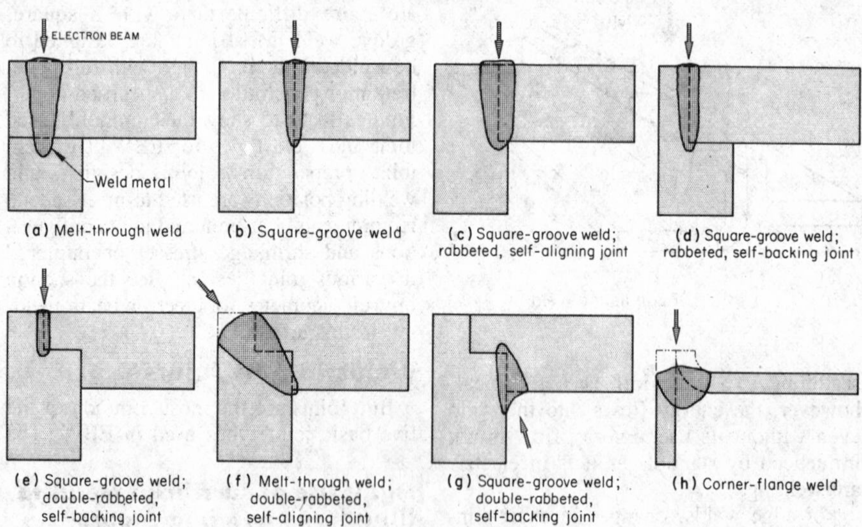

Fig. 13 Typical electron beam welds in T-joints

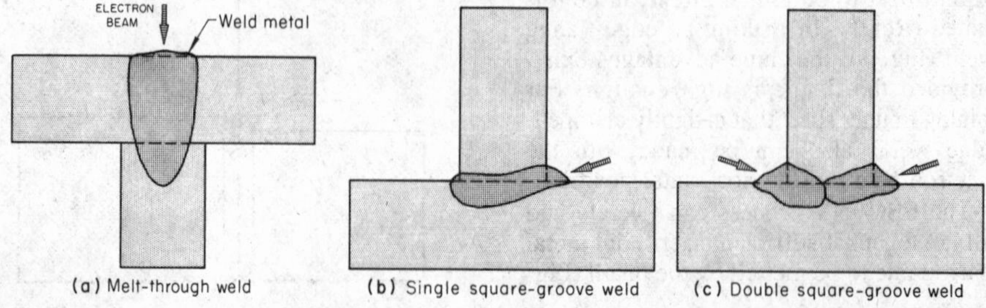

Fig. 14 Typical electron beam welds in lap joints

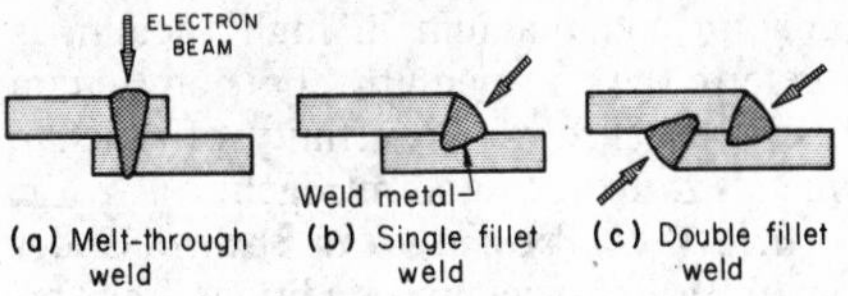

Fig. 15 Typical electron beam welds in edge joints between members of similar and dissimilar section thicknesses

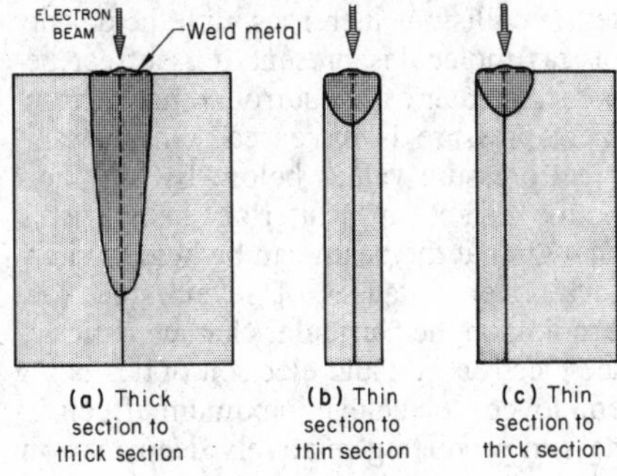

Welds in Edge Joints

Three types of electron beam welds made in edge joints are shown in Fig. 15. Thick sections can be joined by deep, narrow square-groove welds (Fig. 15a). Shallow welds made with a low-power, partially defocused beam are used to join thin sections to each other or to thick sections, as shown in Fig. 15(b) and (c). Shallow edge welds can be made at high speed. Welds of this type are particularly useful on hermetically sealed assemblies, which may be designed for planned salvage by removal of the weld region and subsequent rewelding.

Butt Joints Versus Corner and T-Joints

The concentration of bending stresses at the weld in corner and T-joints is sometimes avoided by designing the work for a butt joint on a straight section near the corner or T (flange neck or welding neck construction). This principle of design is applicable also to butt joints between members having different thicknesses, and it can be applied to straight line, circular, and circumferential joints, as shown in Example 1.

In welding such a joint in magnetic metals, if the beam path is too close to the corner, T, or shoulder, the beam may be deflected away from the metal that it passes near on its way to the joint, causing it to enter the work at an angle and possibly to miss the lower portion of the joint. To avoid this, the beam clearance should be 0.030 in. or more, depending on the height of the magnetic projection (Fig. 16).

Example 1. Use of an Oversize, Self-Aligning, Self-Backing Butt Joint in a High-Strength Flange-to-Tube Weld. Joint design was critical in the carefully planned and executed procedure used to produce a flange-to-tube assembly of 4340 steel for an aircraft engine mount. Basically, the procedure consisted of welding an oversize assembly and then machining to finished dimensions, as indicated in detail A of Fig. 17. Process selection, fixturing, welding, heat treatment, inspection, and final machining were also important to the success of the application.

Electron beam welding was selected primarily to obtain deep, narrow welds at high speed with low total heat input, for high strength. Tests had shown that high joint efficiencies were obtained in static tensile and fatigue loading. In addition, EBW produced low distortion and provided excellent control of penetration.

Fusion under a high vacuum promoted weld cleanness, as determined by x-ray radiography. Very clean welds could be obtained by using a vacuum-melted steel; in this application, however, air-melted 4340 steel was used.

The type of butt joint used is shown in detail A of Fig. 17. The normalized, seamless 4340 steel tubing was finish machined on the inside and rough machined on the outside. The flange was of welding neck design, which located the weld away from major flange bending stresses and simplified nondestructive inspection. The outside of the flange hub was rough machined to the outside diameter of the tube, and the inside diameter of the hub was machined $^1/_4$ in. undersize.

The overlapping lip, made by cutting a recess in the hub, served for alignment and weld backing. The lip was sufficiently thick (0.125 in.) to permit controlled-penetration welding beyond the inside diameter of the tube, so that any root voids would be in the region to be machined away after welding and so that there would be no weld spatter on the interior of the tube. Diametral clearance was zero to 0.005 in., for accurate axial alignment. Joints were cleaned by vapor degreasing and wiping with acetone. The parts were then demagnetized and assembled.

To permit welding in the horizontal-rolled pipe position, a lathe-type posi-

Fig. 16 Joint location and beam clearance for a butt joint near a projecting corner, T, or shoulder in magnetic work metal

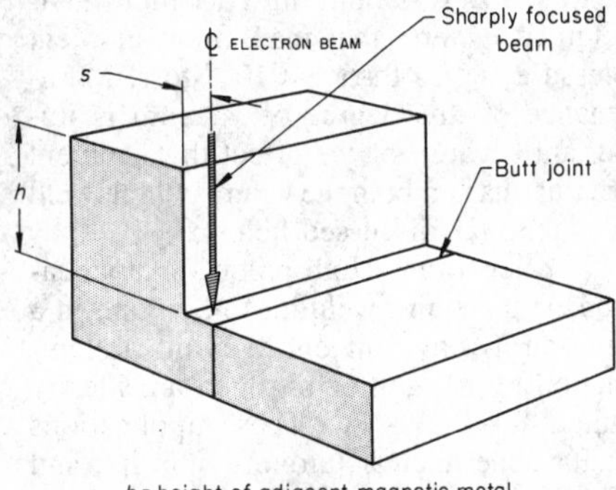

h = height of adjacent magnetic metal above work surface at joint
s = minimum beam clearance; 0.030 in., or 0.030 in. x *h*, whichever is greater

Fig. 17 Precision welding of flange-to-tube assembly

4340 steel

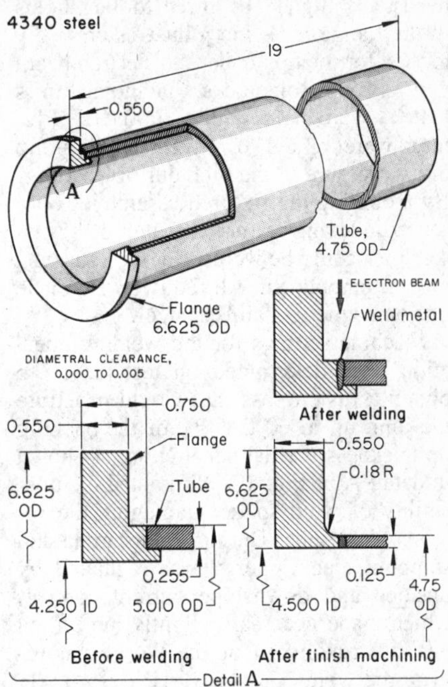

Joint type	Circumferential butt; self-backing
Weld type	Square groove
Weld depth-to-width ratio (typical)	3.7
Machine capacity	150 kV at 40 mA
Gun type	Pierce; fixed
Vacuum chamber size	56 by 56 by 108 in.
Fixtures	Lathe-type positioner with plug and face plate; rotary indexing table
Preheat	None
Welding power	130 kV at 24 mA
Welding vacuum	Less than 10^{-4} torr
Beam focal point	0.250 in. above work surface
Welding speed	31.4 in./min (2 rpm)
Weld time	30 s
Production rate	12 pieces (one load) per hour
Postheat	Stress relieve (950 °F for $^1/_2$ h); quench and temper to a tensile strength of 150 to 172 ksi (see text)

tioner was used. With a plug at the tailstock to fit the tube and a face plate at the headstock aligned by three tooling holes in the flange face, the joint was precisely located for rotation under the electron beam gun. Fit-up tolerances, including mismatch, were held within 0.005 in. Fixtures were vented to prevent any buildup of air pressure in the tube during welding. By mounting the fixtured assemblies on a rotary-indexing fixture, a batch of 12 assemblies could be welded in one hour during one pumpdown. All fixtures were made of nonmagnetic stainless steel.

Machine settings for the welding operation were determined on test pieces to obtain a fusion pass that would penetrate the joint up to 60 to 70% of the backing lip thickness. Partial penetration, as shown in detail A of Fig. 17, eliminated the possibility of weld spatter damaging the interior of the tube. During weld starting and stopping, beam energy was regulated by upslope and downslope current control, which made necessary slightly more than one full revolution of the work to complete the weld.

Immediately after a batch was welded, the parts were stress relieved at 950 °F for $^1/_2$ h and heat treated to a tensile strength of 150 to 172 ksi by heating to 1550 °F for 1 h, quenching, and tempering at 1250 °F for 2 h.

Before final machining, all welds were inspected visually and by magnetic-particle testing. After final machining, the joints were inspected by radiographic and ultrasonic examination. Acceptable defects were limited to 5% of minimum thickness, or approximately 0.006 to 0.007 in. in size.

The flange was rebored to the same inside diameter as the tube. The outside tube surface and the flange hub were machined to finished wall thickness, as shown in detail A of Fig. 17. Roughness was held to 63 μin., and tolerances required concentricity over the tube length within 0.005 in.

Special Joints and Welds

Many variations of the basic joint and weld types described in the preceding section and illustrated in Fig. 11 to 15 are used to meet the needs of individual applications. Some of these special joints and welds are discussed below.

Plug and puddle welds are usually made by manually (or automatically) manipulating the work under a fixed beam at low power. Filler metal used for plug welds is often preplaced at the weld site. Puddle welding is used mainly to fuse shallow defects together locally.

Multiple-Pass Welds. Most electron beam welds produce the desired penetration in a single pass. Tack welds and cosmetic or smoothing passes are not considered penetration passes. Welds several inches deep can be made in most metals in a single pass. Weld depth obtainable in a single pass can be nearly doubled by welding from both the face and the back of the work.

In a variation on straight-through two-pass welding from opposite sides, separate welds that meet or almost meet can be made at an angle of 90° to each other in a rabbeted square-groove joint.

Tangent-tube welds are longitudinal welds joining two parallel tangent tubes (or cylinders). The tubes may differ in size; the beam is perpendicular, or nearly perpendicular, to the common axial plane of the two tubes. Added filler metal may be used for reinforcement. Thinner wall tubes can be joined more easily by EBW than by arc welding methods.

Three-Piece Welds. Welds can be made that join three or more pieces in which there is penetration in a single pass into all of the pieces. Many of the difficulties encountered in welding very thin metal or foil are eliminated by sandwiching it between two thicker sections.

Multiple-tier welds are welds made simultaneously in in-line, separated joints (usually butt joints) in a single pass of the electron beam. A more detailed discussion is given in the section of this article on multiple-tier welding.

Welds using integral filler metal may be of the types shown in Fig. 11(g) and (h), in which an overhanging lip or a shoulder provides filler metal to a butt joint, or both members may be made thicker at the joint than elsewhere.

Welding in High Vacuum

Most EBW is done in a vacuum environment where the maximum ambient pressure is less than 1×10^{-3} torr. Maintenance of this degree of vacuum is important because of the effect that ambient pressure has on both the beam and the weld produced, as discussed below.

Applications. High-precision applications that require welding to be done in a high-purity environment to avoid contamination by oxygen or nitrogen are ideally suited for EBW-HV. These applications include the nuclear, aircraft, missile, and electronics industries. Typical products include nuclear fuel elements, pressure vessels for rocket propulsion systems, special alloy jet engine components, and hermetically sealed vacuum devices. Welding of reactive metals such as zirconium and titanium in high vacuum is preferred over medium- or nonvacuum EBW because of the affinity of reactive metals to oxygen and nitrogen.

Effect of Pressure on Beam. Under ambient high-vacuum conditions, the frequency of collisions between beam electrons and residual gas molecules is extremely low, and any dispersion (i.e., beam broadening effect) that would result from such scattering collisions is minimal. However, because the frequency of scattering collisions increases with the density of gas molecules present, this effect becomes greater as the surrounding environment pressure is increased. Only at ambient pressure values below 1×10^{-3} torr is the effect of scattering insignificant enough that the beam can be held in sharp focus over distances of several feet (depending on the particular characteristics of the electron gun and electron optics being employed) to achieve maximum effectiveness in producing relatively deep, narrow welds.

Beam scattering from collisions with residual gas molecules begins to occur at ambient pressures above 10^{-3} torr and initially results in decreased penetration and increased weld width. This scattering effect reduces the maximum working distance that can be employed for producing a similar electron beam weld as obtained under high-vacuum conditions at ambient pressures greater than 10^{-3} torr.

In EBW-NV, where the ambient pressure is 760 torr and above, the scattering dispersion of the beam increases so that penetration and working distance capacity are appreciably reduced below values obtainable in either high- or medium-vacuum processes. However, penetration and working distance capability of EBW-NV is appreciably greater than conventional arc welding processes. Electron beam dispersion characteristics at various ambient pressures are illustrated in Fig. 18.

Effect of Pressure on Weld and Heat-Affected Zone. The maintenance of a high vacuum in the work chamber protects the weld metal and the HAZ from oxidation and contamination by harmful gases, serving the same function as inert shielding gases do in arc welding, while also degassing the weld. The contamination level as a function of pressure is discussed in the section on welding in medium vacuum in this article.

The nearly complete absence of gaseous impurities eliminates the serious contamination difficulties that usually are en-

Fig. 18 Electron beam dispersion characteristics at various pressures

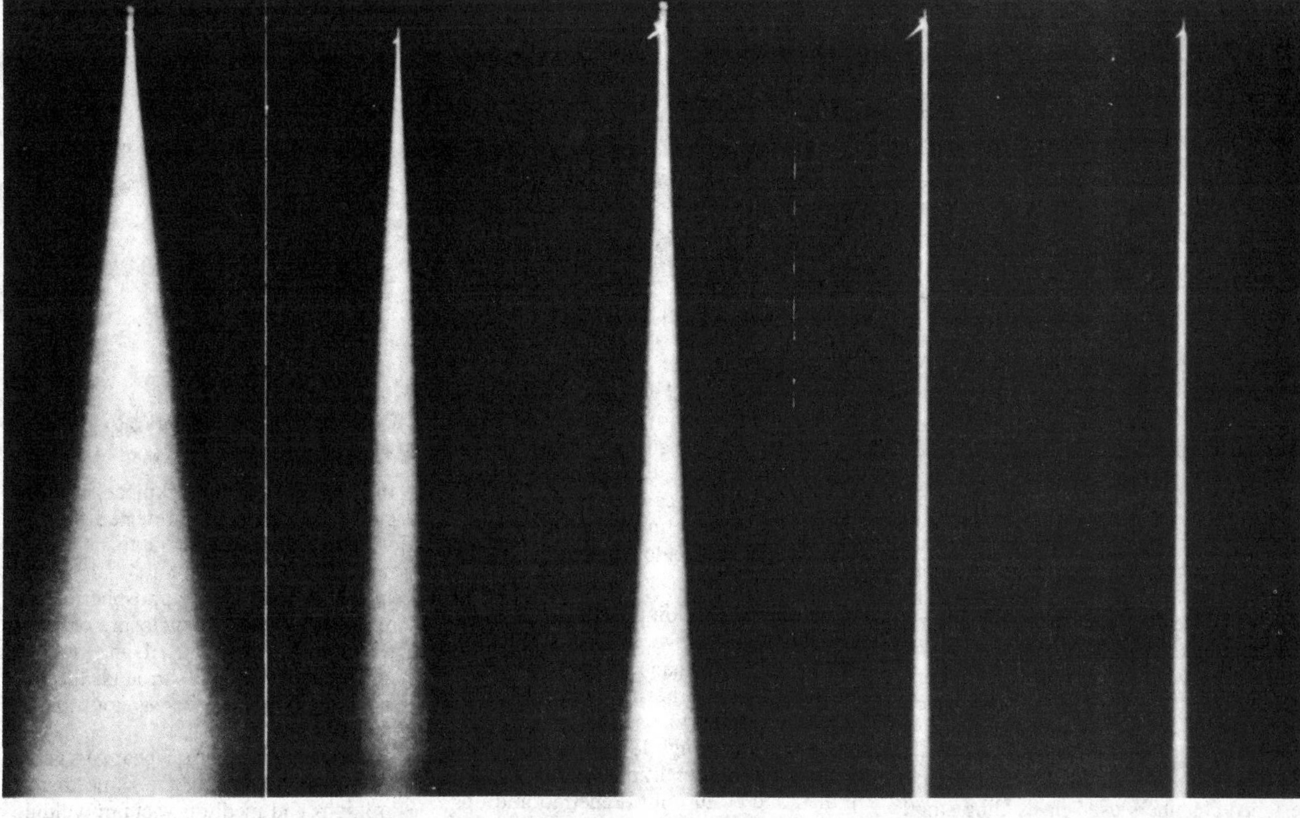

countered when attempting to arc weld reactive metals such as titanium and zirconium.

Width of Weld and Heat-Affected Zone. Electron beam welds made in a high vacuum are narrower and have a narrower HAZ than comparable welds made in medium vacuum or at atmospheric pressure, and they are much narrower than the narrowest welds made in production welding by GTAW. The narrow width of the tempered zone in hardenable steels and other hardenable alloys permits welding of these metals after heat treatment, in many instances without loss of strength (see Fig. 36 and 37).

Limitations. A major limitation in the use of high vacuum in the work chamber is the effect on unit production time, because of the need to pump the chamber down before each load is welded. Pumpdown time is typically about 3 min for a 30-ft^3 chamber and 10 min for a 300-ft^3 chamber. The above pumpdown times are realistic only for a very clean and well-maintained system. In production, pumpdown may be nearly double the above times.

The effect of this limitation on unit production time is reduced by welding a number of assemblies in each load and by keeping chamber size as small as possible. Small chambers specially designed for use with small workpieces have typical pumpdown times of 1 min (225-in.^{3} chamber) and 15 s or less (60-in.^{3} chamber).

A second major limitation in the use of high vacuum is that work size is limited by chamber dimensions. This limitation is sometimes circumvented by the use of a chamber having special openings and seals that permit oversize work to extend outside the chamber, or by the use of a portable clamp-on chamber.

Welding Conditions. Welding in high vacuum is done with low-voltage, as well as high-voltage, equipment. Beam voltage ranges from 15 to 185 kV; beam current, 2 to 500 mA; and welding speed, 14 to 119 in./min. The energy input to the work needed to produce a weld of the required depth of penetration and width is the basis for selecting the welding conditions.

Depth of penetration is increased by increasing the voltage or current for greater beam power, or by decreasing the welding speed. Table 1 lists approximate energy inputs per inch of weld length for making narrow single-pass electron beam welds 0.25 to 3 in. deep in the weldable alloys of copper, iron, nickel, aluminum, and magnesium. These values are intended to serve as guidelines for establishing conditions for welding work for which no previous experience is available. Energy-input requirements for specific applications depend on the alloy composition and special operating conditions such as the use of beam oscillation or a defocused beam.

Welding in Medium Vacuum

Electron beam welding in a medium vacuum usually is performed at a welding (work) chamber pressure of about 75×10^{-3} torr (75 μm), or at a work chamber pressure ranging from 3×10^{-3} to 3×10^{-1} torr (30 to 300 μm)—the total partial-vacuum region. The gun chamber, where the beam is generated, is held at 10^{-5} to 10^{-4} torr, as in high-vacuum welding. A diffusion pump is required to provide this gun chamber vacuum, but only a

Table 1 Energy input at the weld for single-pass EBW in high vacuum for various depths of penetration(a)

Depth of penetration, in.	Weld length, in kJ/in., for welding alloys of: Cu	Fe	Ni	Al	Mg
0.25	7	5	4	2	1
0.50	15	10	8	5	3
0.75	25	18	15	8	5
1.00	37	27	22	13	8
1.50	62	46	39	23	15
2.00	87	68	60	35	22
2.50	112	90	80	47	29
3.00	137	112	100	59	36

(a) Values are approximate and apply to the commonly welded alloys of the metals. They are intended to serve as starting points for establishing conditions for making narrow welds. Energy input may vary substantially from these values in specific applications, depending on the composition of the alloy and special operating conditions. Energy input at weld (kilojoules per inch of weld length) = beam voltage (kV) × beam current (mA) × 0.06 ÷ welding speed (in./min).

mechanical-type pump is required to evacuate the work chamber to the final partial vacuum welding pressure desired.

Comparison With High-Vacuum Welding. The chief advantage of welding in medium vacuum, compared with welding in high vacuum, is the short pumpdown time for the welding chamber, which ordinarily does not exceed 40 s for a general-purpose chamber (of about 4 ft^3) and which is less than 5 s for the specially designed chambers used in welding small parts. Accordingly, EBW-MV permits mass production of parts, using a chamber of minimum volume.

Production rates for welding in medium vacuum depend on part design and other factors; maximum production rate is typically about 60 pieces per hour for general-purpose, manually operated equipment, upwards of 600 pieces per hour for specially tooled semiautomatic machines, and 1500 pieces per hour for fully automatic single-purpose machines. General-purpose chambers and tooling are used for short runs.

Welds on reactive metals and other welds for which the effects of contamination by gases are especially critical are preferably made in high vacuum. The contamination level of air (total concentration of gases present) is proportional to pressure, as follows:

Pressure, torr	Gases, ppm
10^{-5}	0.01
10^{-3}	1.3
10^{-1}	132
4×10^{-1}	500

Lower contamination levels are obtainable than those actually observed for arc welding with inert gas shielding.

Fig. 19 Effect of welding-chamber pressure on penetration and weld shape

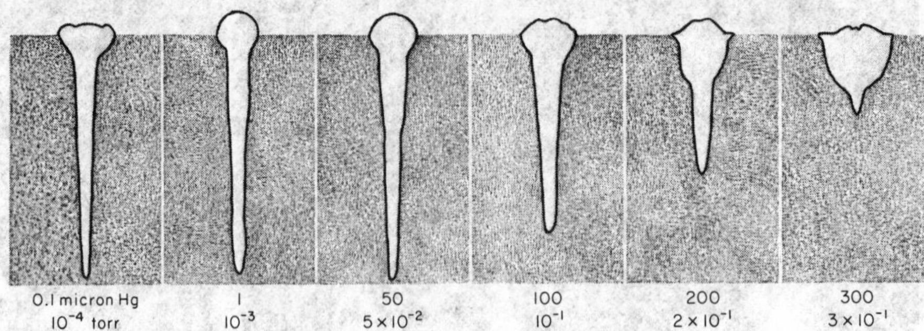

Penetration and Weld Shape. Penetration at a given power level is ordinarily 5 to 10% less than for EBW-HV, and welds are wider and slightly more tapered. These changes are caused by the scattering of the electrons in the beam by collision with gas molecules, which makes the beam broader.

The effect of welding chamber pressure on penetration and weld shape is shown in Fig. 19 for welds made on type 304 stainless steel, using a beam length of 16 in. without changing focus. Maximum penetration shown in Fig. 19 is approximately $^1/_2$ in. Welding conditions were 150 kV, 30 mA, and 60 in./min. Penetration drops off rapidly at pressures of 10^{-1} torr or more at this beam length and at somewhat lower pressures when the beam path is longer.

Comparison With Nonvacuum Welding. Disadvantages of medium-vacuum welding, compared with welding at atmospheric pressure, include limitations on work size imposed by the need to enclose the work in a vacuum chamber and the time needed for chamber pumpdown.

Beam deflection and oscillation, which cannot be used in welding at atmospheric pressure, can be used in EBW-MV.

Applications. Most parts that can be welded in high vacuum can also be welded in medium vacuum. Depth of penetration is less than in high-vacuum welding, and depth-to-width ratio is less than about 10 to 1. Reactive metals are less readily welded in medium vacuum than in high vacuum. Medium-vacuum welding can be acceptable, however, for refractory metal welding when absorption of small amounts of oxygen and nitrogen can be tolerated. Medium-vacuum welding has been used for welding mass-produced automobile parts, such as gears and shafts, that are welded to close tolerances and that are not subsequently finished.

Nonvacuum (Workpiece Out-of-Vacuum) Welding

In EBW-NV (workpiece out-of-vacuum), which is also referred to as atmospheric EBW, the work to be welded is not enclosed in a vacuum chamber. Instead, it is welded at atmospheric pressure. A radiation-tight enclosure or chamber, similar to that used in high- or medium-vacuum EBW, surrounds the weld area. A vacuum, however, is not provided in this welding enclosure.

A nonvacuum electron beam is generated at high vacuum in the same manner as in high- and medium-vacuum welding. It is focused down through a series of individually pumped stages, each connected to the other by concentrically aligned orifices, decreasing in diameter in the direction of increasing pressure; this differential pumping scheme provides the high vacuum (gun region) to atmospheric (workpiece region) pressure gradient necessary to allow the beam to pass down through the column and exit out into the ambient atmosphere.

Figure 20 depicts a typical EBW-NV gun/column assembly. This unit can be mounted in either a vertical or a horizontal position and is capable of motion during welding. The gun/column assembly depicted in Fig. 20 is capable of beam power levels greater than 35 kW and beam transmission efficiencies in excess of 90%.

Production rates are generally much higher for EBW-NV than for EBW-HV and EBW-MV, because no work chamber vacuum pumpdown time is required for each chamber load. This results in a lower per part production cost. Because pumpdown time for high- and medium-vacuum welding varies with chamber cleanliness and ambient humidity, elimination of work chamber pumpdown time in EBW-NV produces appreciable time savings. This saving, however, is gained at the expense

Fig. 20 Typical EBW-NV gun/column assembly. Courtesy of Leybold-Heraeus Vacuum Systems Inc.

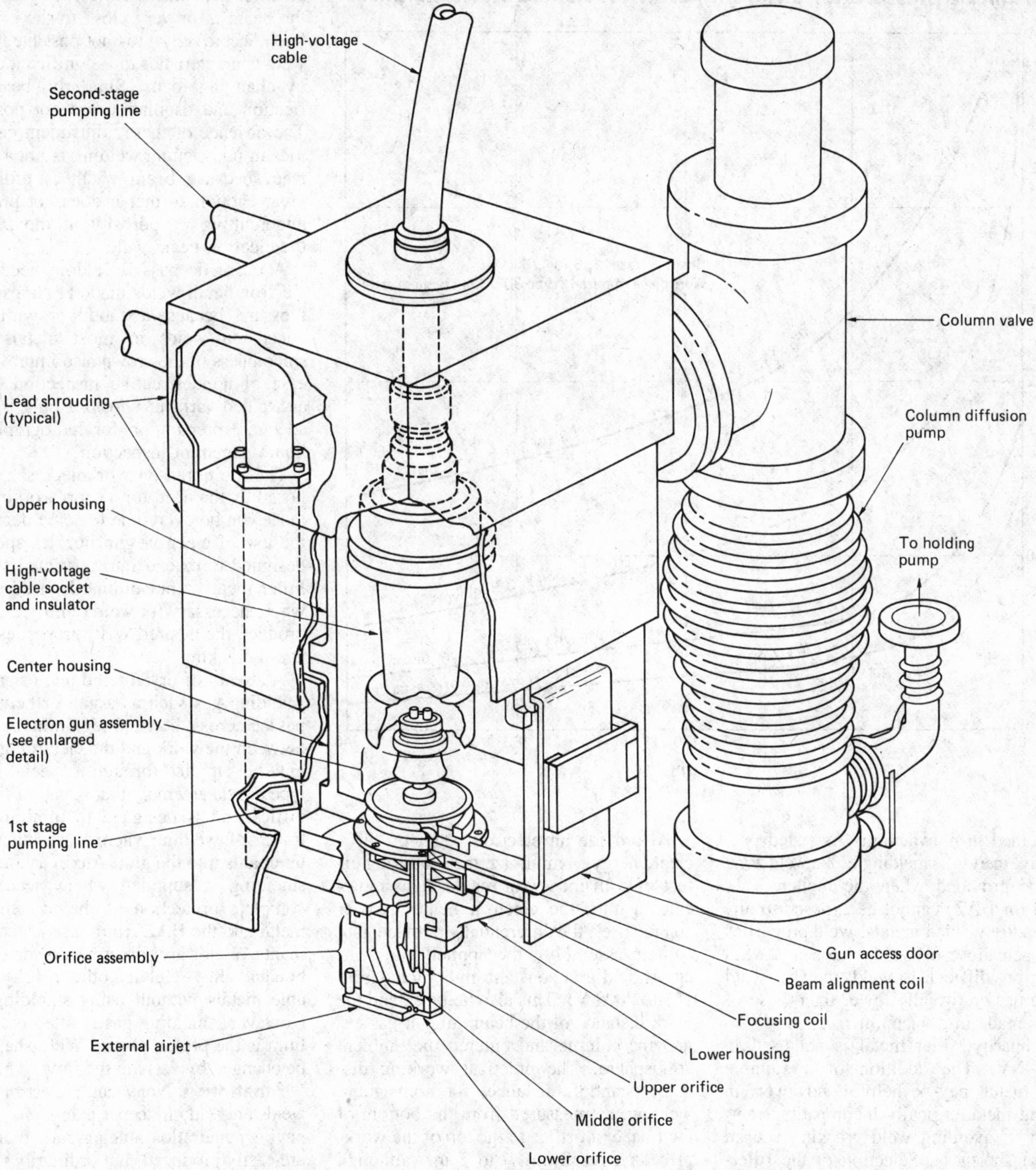

of a reduction in penetration and working distance capability, thus imposing limits on the thickness and shape of the workpiece. Additionally, most (but not all) out-of-vacuum welds cannot be made as narrow and parallel sided as welds made in a vacuum.

When welding in ambient atmospheric air alone, the practical working distance of the nonvacuum process is normally limited to a range of from $^3/_8$ to $^7/_8$ in.; however, when welding in the presence of a lighter ambient gas, such as helium, this practical working distance range is extended to 1 to 2 in. For this reason, most commercial nonvacuum EBW units are equipped to provide a "helium blow-down," which is an effluent of helium that extends out from the nonvacuum column and along the beam travel path. The effect of this helium effluent is that it reduces the degree of beam scattering incurred using only air, thus allowing the working distance capability of the nonvacuum unit to be approximately doubled.

Most metals that can be arc welded or electron beam welded in a vacuum also can

Fig. 21 Effect of welding speed on penetration for air and helium at atmospheric pressure for different power levels and work distances

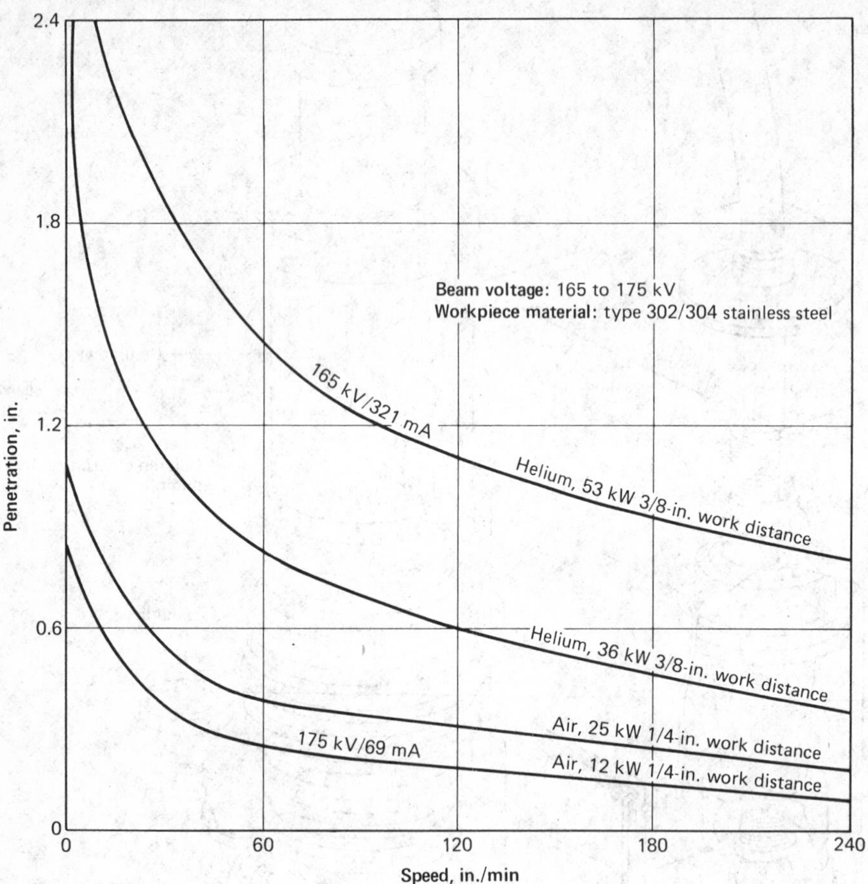

be welded in nonvacuum. As stated previously, inert gas shielding of the weld zone can be provided when the molten weld metal (or HAZ) cannot be exposed to air. On readily welded metals, weld properties approach those of welds made in a vacuum; on difficult-to-weld metals, weld properties are usually lower than those of welds made in a vacuum.

Ordinarily, filler metal is not used in EBW-NV. The addition of a suitable filler metal may be helpful, however, in joining metallurgically incompatible metals or in avoiding weld cracking caused by hot shortness. Selection of the filler-metal composition in such applications is critical.

Operating conditions for nonvacuum welding differ substantially from those for welding in high or medium vacuum. Because beam dispersion increases in proportion to pressure and distance traveled, at pressures above the high-vacuum level, the nonvacuum beam becomes dispersed at travel distances greater than several inches in air at atmospheric pressure.

To provide an electron beam of sufficient energy to minimize the scattering effect of collisions with residual gas molecules within the electron beam transfer column (i.e., the differentially pumped gun/column assembly), the applied gun voltage is held at a constant high value (from 150 to 200 kV). This also helps extend the work distance of the beam after it has exited the column and entered the ambient atmosphere. The practical working distance (standoff distance) for nonvacuum welding, as measured from the bottom of the final exit orifice to the top of the workpiece, ranges from $^1/_2$ to 2 in., although some nonvacuum welding is performed at both shorter and longer standoff distances.

With voltage held constant, beam current, working distance, and welding speed are selected to provide the required penetration and weld shape for a sound weld. Because it is usually desirable to keep working distance as short as possible and welding speed high, beam current is the primary control variable in nonvacuum welding.

The small size of the exit orifice on the electron gun makes it necessary to focus the beam at or very close to the exit orifice. Accordingly, it is not possible to vary weld characteristics to a significant degree by changing focus. Similarly, beam deflection and oscillation are not possible. The absence of these adjustment capabilities in nonvacuum welding is not a handicap, because beam width is ordinarily great enough so that it does not promote undercutting or porosity at the root of the electron beam weld.

A cosmetic pass is seldom needed on electron beam welds made at atmospheric pressure, because a smooth crown is produced on welds in most metals. The smoothness of the crown also improves the ease of nondestructive inspection of the as-welded surface; wire brushing is often adequate preparation for radiographic or liquid-penetrant inspection.

The limitation on workpiece shape imposed by the need for a short working distance can be overcome to some degree by the use of electron gun nozzles specially designed to extend into restricted spaces. Filler metal is not ordinarily used except where necessary for weld reinforcement to produce the desired weld properties or to avoid cracking.

A stream of dry filtered air, or an inert shielding gas such as argon or helium, is passed across the weld region in the space between the work and the electron gun, or may be supplied through a special insert-type nozzle assembly that is part of the exit orifice and is designed to minimize entrance of welding vapors and other contaminants into the gun. Auxiliary inert gas shielding is supplied where needed for complete protection of the molten weld metal and the HAZ from gaseous contaminants. Welds are sometimes made on carbon and alloy steel and other readily weldable metals without using shielding gas. However, shielding gas is often used; helium is the preferred gas. Weld shape can be changed by varying the flow of helium.

Penetration. Nonvacuum electron beam welds are seldom deeper than $^3/_8$ in. Welds having penetration much greater than 1 in. are easily produced, but ordinarily this requires a reduction in welding speed, with a corresponding increase in production cost.

Figure 21 shows the effect of welding speed on penetration of nonvacuum welds made in austenitic stainless steel and the effect of ambient gas and power on the penetration versus speed curve. Curves identified as helium penetration-versus-speed curves were performed using a helium effluent-style lower orifice design, i.e., using a lower (exit) orifice that is ca-

Fig. 22 Effect of welding speed on penetration. Beam power, 9 kW

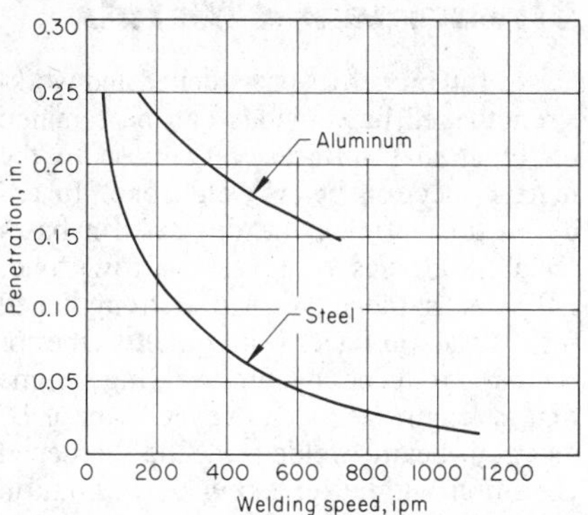

Fig. 23 Effect of varying working distance in EBW-NV

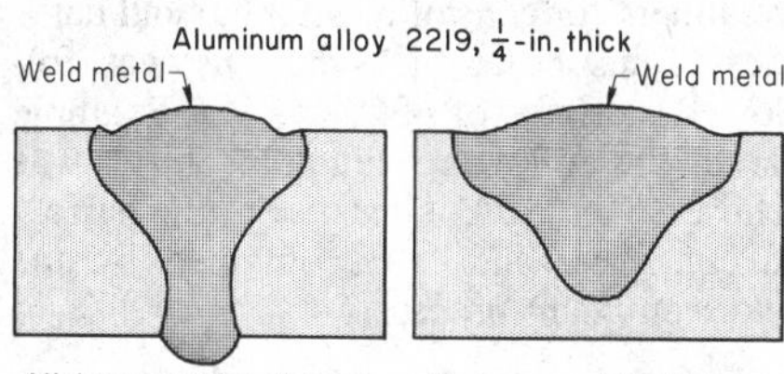

Fig. 24 Effect of shielding gas, working distance, and welding speed on penetration for EBW-NV of 4340 steel. Beam power, 6.4 kW

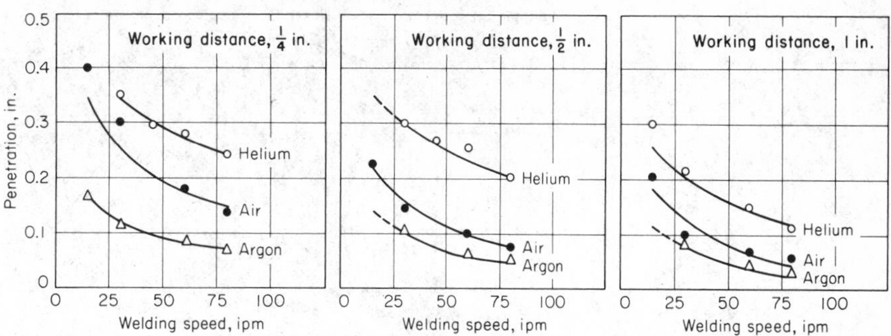

pable of blowing helium into the atmosphere with enough directed flow to provide a helium gas column in which the beam can travel. Figure 22 provides a comparison of penetration, as a function of weld speed, in steel and aluminum, employing a fixed power level and work distance and using a helium shield to improve power density.

The effect on penetration of varying working distance (within practical operating limits and with other variables held constant) is shown in Fig. 23 for $^1/_4$-in.-thick aluminum alloy 2219. When maximum working distance was used, penetration was only about 80%, and maximum weld width was about one third greater than with minimum working distance.

When a shielding gas is substituted for air in nonvacuum welding, assuming the same operating conditions, penetration is greater when a gas is used that has a lower molecular weight (or nominal density) than that of air. Penetration is less when a gas that has a higher molecular weight than that of air is used. In tests on welding 4340 steel, using a fixed beam power of 6.4 kW, working distances of $^1/_4$, $^1/_2$, and 1 in., and welding speeds of about 15 to 80 in./min, penetration was increased when helium was substituted for air and was decreased when argon was substituted for air, as shown in Fig. 24. The relative capacities of helium, air, and argon to scatter the electrons of the beam by collisions of electrons with gas molecules are controlled by their molecular weights and relative densities. On the average, when helium was used as a shielding gas, penetration depth was about twice that obtained when welding was done in air, and when argon was used, penetration depth was about half that when welding was done in air. Thus, with helium shielding, penetration averaged about four times that with argon, especially at higher welding speeds. The speed at which a given thickness of 4340 steel could be welded with full penetration, using helium, was two to four times that with air. Using argon, speed was 25 to 50% that with air.

Studies made of nonvacuum welding of steel pipe showed that, at speeds of 100 to 1000 in./min, full-penetration welds could be made in metal twice as thick if helium shielding gas was used instead of air. Table 2 gives the results of this comparison in terms of welding speed for joining steel pipe 0.050 to 0.150 in. thick in air or in helium. Welding speeds were 2.16 to 2.24 times as great in helium as in air. Beam power used was 12 kW.

Weld Shape and Heat Input. Nonvacuum electron beam welds are generally wider and more tapered than high-vacuum welds (medium-vacuum welds differ only slightly in shape from high-vacuum welds).

Table 2 Speeds for nonvacuum full-penetration welding of carbon and low-alloy steel pipe at a beam power of 12 kW

Wall thickness, in.	Welding speeds, in./min: In air	In helium	Relative welding speed in helium(a)
0.050	417	900	2.16
0.080	241	528	2.20
0.100	187	412	2.21
0.150	118	264	2.24

(a) Based on welding speed in air as 1.00

Typical shapes of full-penetration welds made in 0.224-in.-thick aluminum alloy 2219 by EBW-NV, EBW-HV, and GTAW are shown in Fig. 25.

Comparing the nonvacuum and high-vacuum electron beam welds in Fig. 25(a), heat input for the nonvacuum weld was about 40% greater, producing a midpoint weld width about 33% greater and a maximum crown width 160% greater than those of the high-vacuum weld. Comparing the nonvacuum electron beam weld to the gas tungsten arc weld in Fig. 25(b), heat input for the typical nonvacuum weld was 32% of that for the arc weld, producing a midpoint weld width about 25% and a maximum weld width about 56% of the corresponding widths for the arc weld. Compared to the arc weld, the widest weld produced by EBW-NV under workable operating conditions was about 42% as wide at midpoint and about 70% as wide at the crown.

Weld shrinkage across nonvacuum electron beam welds is generally less than half that of gas tungsten arc welds on the same work metal and thickness and is typically about twice that for high-vacuum electron beam welds.

Studies on welding $^1/_4$-in.-thick aluminum alloy 2219 showed average shrinkage of 0.003 to 0.005 in. for high-vacuum welds and shrinkage of 0.014 in. for high-heat-input nonvacuum welds, compared with a normal shrinkage of 0.030 in. for gas tungsten arc welds.

Tooling generally is designed specifically for welding a particular assembly in mass production. Providing fixtures and equipment that permit welding at high speeds and efficient work handling is the key to obtaining high production rates and low cost.

The work-handling equipment is at atmospheric pressure instead of in a vacuum; hence, general-purpose welding positioners and related fixtures usually are

Fig. 25 Comparison of full-penetration welds made by various processes. (a) EBW-NV versus EBW-HV. (b) EBW-NV versus GTAW

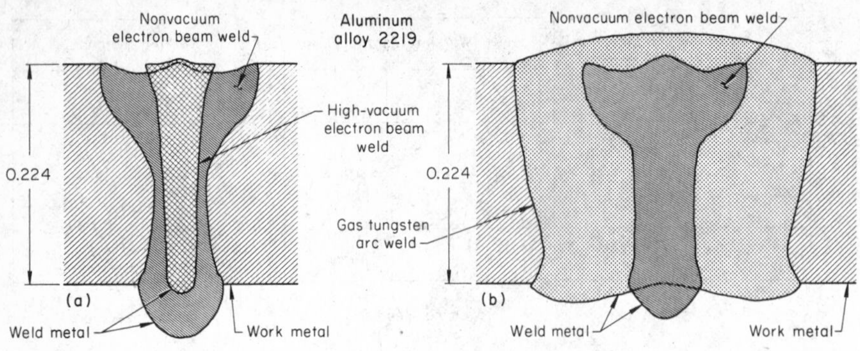

	Fig. 25(a)		Fig. 25(b)	
Welding conditions	High-vacuum EBW	Nonvacuum EBW	GTAW	Nonvacuum EBW
Voltage, kV	150	140	13 V	140
Current, mA	20	30	270 A	20
Welding speed, in./min	100	100	20	50
Welding energy, kJ/in.	1.8	2.5	10.5	3.36

satisfactory. Hold-down clamping devices can be simpler than those often needed for arc welding similar parts, because angular distortion or dihedral warpage is generally low. Locating and alignment mechanisms are simpler than for welding in a vacuum, because of their accessibility and because beams are wider and lower accuracy is required in tracking the joint. Maintenance of work-handling equipment is greatly simplified by its out-of-chamber location.

Applications. Nonvacuum electron beam welding is used in the production of transmission-train components, steering column jackets, die-cast aluminum intake manifolds, and catalytic converters. In addition, several appliance (and other consumer-related) items also are produced using the EBW-NV process. Figure 26 shows a catalytic converter that was nonvacuum electron beam welded with the beam in a horizontal position.

Fig. 26 Catalytic converter assembled by EBW-NV process

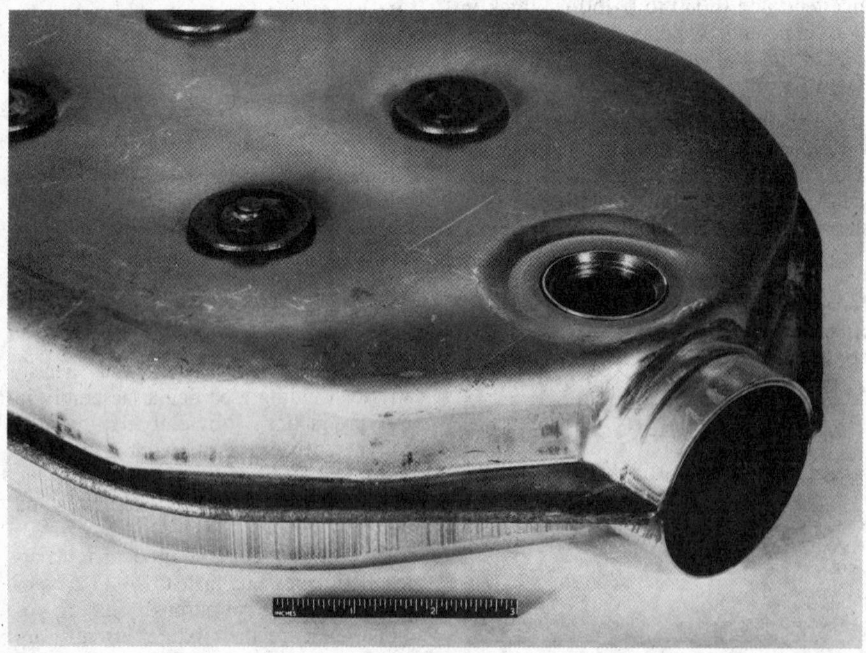

Welding of Hardened and Work-Strengthened Metals

The full properties of hardened and work-strengthened base metals can be retained on functional surfaces very close to the narrow electron beam weld zone. In addition to confining the decrease in base-metal properties to a very narrow zone, EBW of hardened or work-strengthened metals also produces joint quality superior to that produced by arc welding. Some high-strength steels, however, cannot be electron beam welded in the hardened condition without cracking or without an unacceptable decrease in strength and hardness. These steels, such as 52100 bearing steel, are normally annealed before welding and given a postweld heat treatment for controlled strength and hardness. This order of operations was followed in EBW of 4340 steel in Example 1. In the following example, EBW was chosen over GTAW because of the narrow fusion zone and HAZ obtained by the electron beam process.

Example 2. Selection of Electron Beam Welding To Minimize Tempering in the Heat-Affected Zone. Fabrication of armor for military aircraft involved the welding of corner joints in 5/16-in.-thick armor plate. The armor plate was made by roll-bonding H11 tool steel to an alloy steel (AMS 6545) containing 9% Ni, 4% Co, and 0.30% C. Welding was done with the plate in the hardened condition; the H11 steel was hardened to 60 HRC and the alloy steel to 50 HRC. The plate was not difficult to weld. Either GTAW or EBW could be used to obtain sound welds using the joint designs shown in Fig. 27.

The weld and HAZ of the electron beam process are compared with those of GTAW in Fig. 27. Joint designs differed; if the joint for GTAW had been made the same as the joint for EBW, a V-groove would have been required, and about the same amount of weld metal would have been needed. Gas tungsten arc welding was done in one fusion pass and six filler-metal passes, using 0.062-in.-diam Hastelloy W wire.

Electron beam welding was done in a single pass. Because of occasional undercutting, a second (cosmetic) pass was sometimes used. Even when the second pass was used, the weld and the HAZ were much narrower than were those created by arc welding, and they were unaffected below the crown of the weld. Because the general effect of the welding heat was to

Fig. 27 Cross sections through corner joints and welds showing differences in size of welds and HAZ resulting from EBW and GTAW

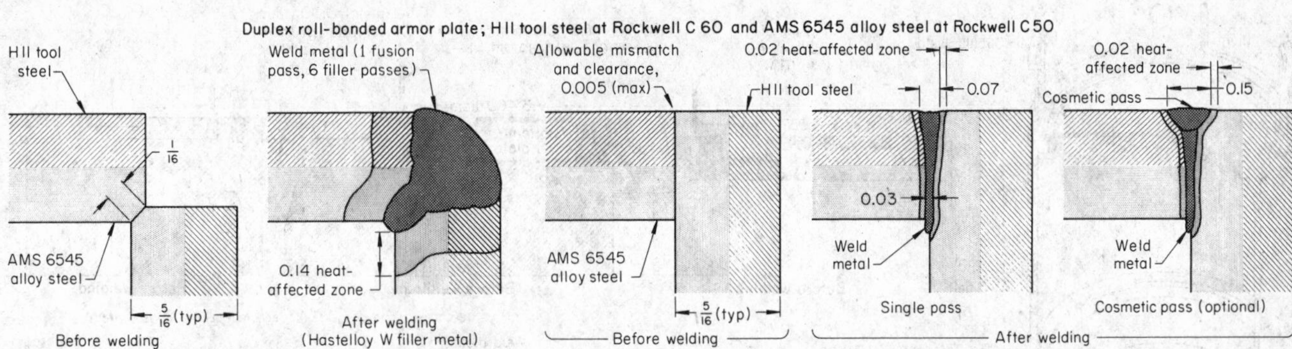

Joint type	Corner	Welding power:		Beam oscillation	None
Weld type	Square groove	Penetration pass	30 kV at 170 mA	Welding position	Flat
Machine capacity	30 kV at 500 mA	Cosmetic pass	19 kV at 95 mA	Welding speed:	
Vacuum chamber size	112 by 30 by 62 in.	Welding vacuum	10^{-5} torr	Penetration pass	60 in./min
Maximum vacuum	10^{-7} torr	Pumpdown time	10 min	Cosmetic pass	32 in./min
Fixtures	Assembly jig; travel carriage	Beam focal point, penetration pass	At surface	Preheat and postheat	None

decrease the hardness and strength of the base metal, EBW, which produced less heat, was selected.

There were no unusual problems associated with the EBW procedure. Plates were machined square and then were degreased. Just before welding, surface films were removed from all surfaces to be welded by brushing with a stainless steel wire brush followed by wiping with alcohol. The plates were then fixtured into an assembly roughly resembling a bathtub having sides about 10 in. high and a floor area of about 20 by 40 in. A maximum tolerance of 0.005 in. was allowed on fit-up for joint mismatch and clearance. Starting and runoff tabs, made of the same duplex plate, were attached to joint ends to contain variations in beam initiation and termination. The fixture was a simple arrangement of toggle clamps and stainless steel blocks that could be set up for flat-position welding on a track-guided carriage in the vacuum chamber.

After beam alignment and vacuum pumpdown, each joint was welded in a single pass without tack welding and without beam oscillation. Because undercutting occurred where fit-up approached the maximum tolerance, a second (cosmetic) pass occasionally was used. The effect of the cosmetic pass on the weld cross section is shown in Fig. 27; the machine settings for the defocused beam used to make the cosmetic pass are in the table with Fig. 27.

Controlling Heat Effects

The extent of heat-affected base metal in EBW is controlled mainly by regulating the width and shape of the weld. In general, the width of the HAZ is proportional to the average width of the weld, which also strongly influences shrinkage and distortion.

Damage to heat-sensitive attachments, inserts, or encapsulated materials close to the weld joint is controlled mainly by regulating the width and shape of the weld bead. Damage to nearby finished functional surfaces on the workpieces or to nearby drilled or tapped holes or other critical features of the workpieces can often be avoided in EBW because of the narrowness of the welds.

Joint designs that provide constricted sections near the weld to retard heat transfer from the weld, or that provide an integral heat sink, and the use of external heat sinks or chill bars are also helpful in controlling heat effects in EBW.

Heat-Sensitive Attachments and Inserts. Narrow, closely controlled electron beam welds, with low heat input, can be made in close proximity to heat-sensitive attachments, inserts, or encapsulated materials without damaging them, often when no other welding process can meet this requirement, as in the next example.

Example 3. Precise, Low-Heat Welds Made in an Aluminum Alloy. Electron beam welding was selected for the delicate final closure of the loaded fuse assembly shown in Fig. 28. The enclosure, which consisted of a shell, top and bottom cover plates, and two identical, oppositely located diaphragms, contained two stab detonators and two high-explosive (RDX) leads. Because the joints between the shell and the diaphragm were less than $^1/_4$ in. from the explosive, the safety of the operator and of the equipment depended on small, accurately placed welds with low total heat input and on careful handling.

Design considerations called for attaching the two cover plates and two diaphragms by controlled-penetration welds of medium strength (18 ksi) and capable of leak rates of less than 10^{-7} cm^3/s of helium at a difference in pressure of 1 atm. Preheating and postheating were not permitted. All of the components were made of aluminum alloy 6061-T6. Filler metal was BA1Si-4 brazing alloy (12% Si), in the form of foil 0.004 to 0.006 in. thick cut to widths of 0.030 and 0.060 in. to fit tightly in the joint openings.

Weld width was about 0.020 in. on the diaphragms and 0.032 in. on the cover plates. Penetration was just beyond joint depth to ensure consistent joint strength. At the joints between the shell and the covers, a circular recess was cut inside the joint (see details A and B of Fig. 28). The recess served to reduce the heat path to the cover, and the small lip allowed movement to help reduce residual stress from weld shrinkage on cooling. The reduced

Fig. 28 Electron beam welding of aluminum alloy 6061-T6 fuse assembly. See text for details.

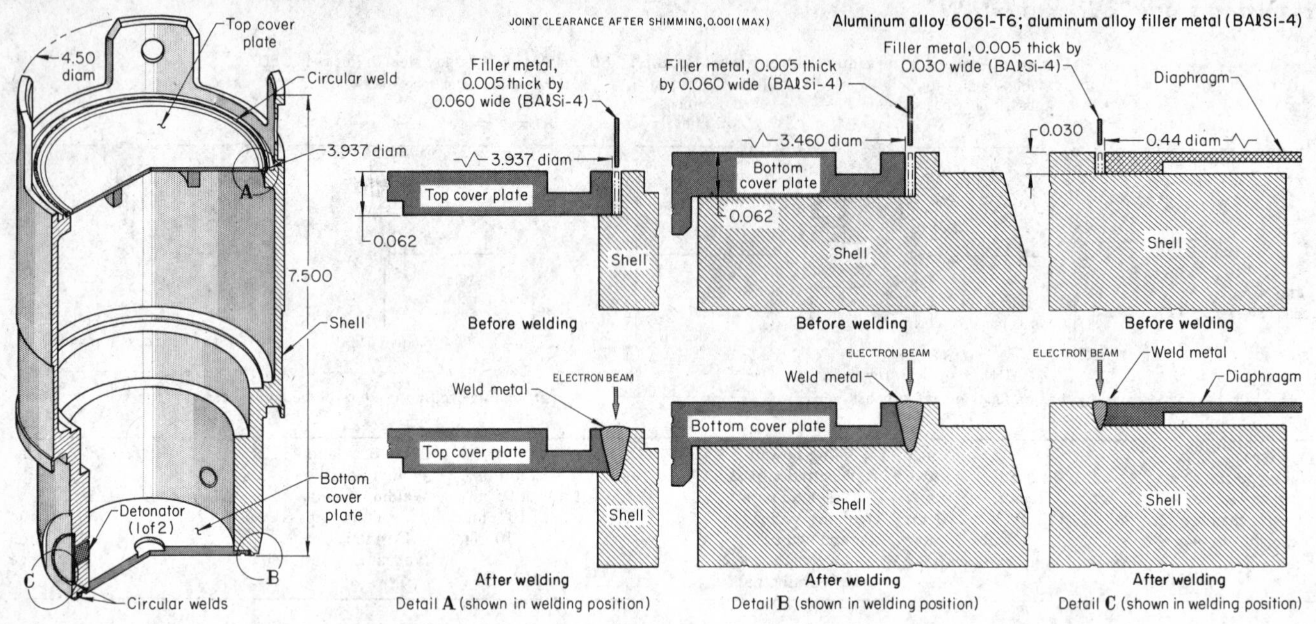

Joint types	Circular corner and butt; rabbeted, self-backing	Filler metal	0.005-in.-thick foil of BAlSi-4	Beam oscillation(a)	0.010-in. diameter, at 1000 Hz(b)
Weld type	Square groove	Welding power(a):		Welding speed(a):	
Machine capacity	150 kV at 40 mA	Top cover (detail A)	130 kV at 6 mA	Top cover (detail A)	62 in./min (5 rpm)
Gun type	Steigerwald; fixed	Bottom cover (detail B)	110 kV at 5 mA	Bottom cover (detail B)	54 in./min (5 rpm)
Vacuum chamber size	36 by 23 by 30 in.	Diaphragm (detail C)	130 kV at 4 mA	Diaphragm (detail C)	58 in./min (42 rpm)
Maximum vacuum	About 10^{-5} torr	Welding vacuum	10^{-4} torr or less (all welds)	Working distance	$7\frac{1}{2}$ in. (all welds)(c)
Fixtures	Universal chuck; turntable	Beam focal point	Sharp at surface (all welds)	Production rate	2.7 parts per hour

(a) Same for tack and final welds; weld time for tack welds was 0.05 s. (b) Circular deflection. (c) Measured from top of chamber to workpiece

section of the diaphragms (about 0.010 in. thick) facilitated detonation.

After machining, the components were cleaned and degreased. Just before assembly for welding, the oxide was removed from welding surfaces by brushing with a small stainless steel wire or fiberglass brush. Dust was removed by wiping with a solvent, and care was taken to leave no fingerprints on the cleaned surfaces.

Each of the covers and diaphragms was assembled, fixtured, tack welded, and welded as a separate operation, the vacuum chamber being loaded and unloaded four times to complete one unit. Tack welding was used (1) to eliminate the need for an additional costly fixture, (2) to prevent displacement of parts from thermal expansion during welding, and (3) to anchor the filler-metal strips to prevent them from curling out of the joint during welding, and (4) to prevent distortion (curling) of the thin diaphragms and covers during welding.

With the top cover plate shimmed tightly in place by filler metal, the subassembly, with its axis vertical, was chucked on a turntable inside the chamber. Using a 20-power telescope, the weld joint was checked for alignment, adjusted, and rechecked. A circular weight was placed on the cover to hold it in place for tack welding. Beam power and beam oscillation settings were made (these were the same as for the final welds), and the automatic weld timer was set for 0.05 s. After vacuum pumpdown ($2\frac{1}{2}$ to $2\frac{3}{4}$ min), four tack welds were placed at 90° intervals by pressing the "beam on" switch and rotating the part.

To make the final circular weld, the machine settings were adjusted as necessary, and the power supply was set for automatic upslope and downslope. Rotation was started and the "beam on" switch was pressed. Upon completing one revolution under welding power, the "beam on" switch was released, thereby actuating the downslope power mode, which carried the beam for a short distance, to taper the weld overlap. The overlap could be distinguished from the remainder of the weld only with difficulty. There was no evidence of the tack welds.

After the top cover was welded, the machine was unloaded. The bottom cover (detail B of Fig. 28) and the two diaphragms (detail C) were then welded similarly.

Testing of the completed units consisted of the leak-rate test described earlier. In addition, destructive tests were made on test pieces that simulated the weld areas of the fuse assembly. Periodically, three such samples were tension tested and examined macroscopically for weld soundness. Similar tests were also made on a regular sampling basis of 1 part per 20 produced.

Welding of Thin Metal

Electron beam welding can be advantageous for joints in which at least one member is made of thin metal (less than $\frac{3}{32}$ in. and as thin as 0.001 in.). Applications include instrument parts, instrument enclosures, pressure or hermetic seals, diaphragms, encapsulations, electrical connectors, and electronic devices, of various metals.

The joint area must be heated to melting temperatures in joining thin sections, just as in joining thick sections, but the rate of heat transfer away from the joint is much

Fig. 29 Comparison of GTAW and EBW applications to rupture-disk assembly

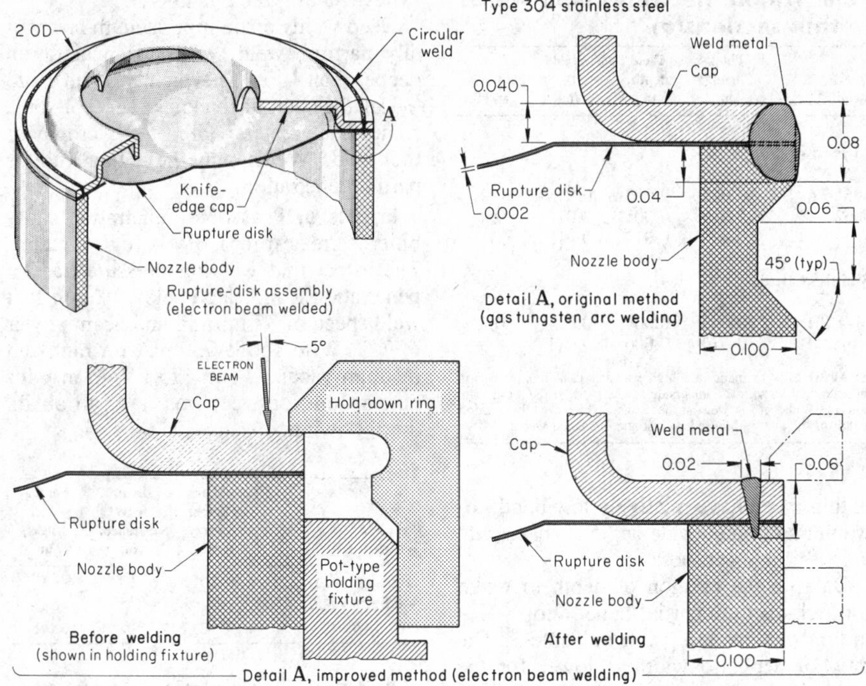

Joint type	Circular, three-piece corner	Welding vacuum	5×10^{-6} torr
Weld type	Melt-through (spike)	Pumpdown time	10 min
Machine capacity	50 kV at 250 mA	Working distance	1.5 in.
Gun type	Diode; fixed during welding	Beam focal point	Sharp at work surface
Vacuum chamber	Steel; 54 by 48 by 48 in., with full-width end doors	Beam spot size	About 0.020 in. diam
Fixtures	56-piece holding tool with individual rotation; table with *x-y* motion	Welding speed	100 in./min
		Number of passes	One, plus 30° downslope
		Setup time	48 min
Welding power	26 kV at 16 mA	Production rate	44½ pieces per hour

lower because of the reduced cross section. Hence, local heat buildup is greater, which increases weld width and decreases depth-to-width ratio. In addition, minimum beam spot size has a much greater effect on depth-to-width ratio in joining thin metal than in joining thick metal. The minimum usable beam spot size obtainable generally ranges from about 0.005 to 0.020 in. in diameter. Because of these conditions, depth-to-width ratio for full-penetration welds joining two sections thinner than about $^3/_{32}$ in. usually does not exceed about 5 to 1 and may be less than 1 to 1.

When voltage, current, welding speed, and other variables have been optimized for a specific joint, use of a heat sink may still be necessary to avoid overheating thin material, depending on base-metal properties (melting point, specific heat, and thermal conductivity) and metal thickness. Best results usually are obtained by the use of copper chill blocks machined to fit the workpiece closely at the joint area. Fixtures made of other metals may be effective as chill blocks if contact area is sufficiently large and close to the heat source.

Joining Thin Sections to Thick Sections. In joining thin sections to thick sections by EBW, the joint is usually designed so that the thick member serves as a heat sink for the thin member, and the point of beam impingement is slightly removed from the thin metal. When these precautions are taken, welding behavior and heat dissipation are generally the same as in welding thick sections under similar conditions. Depth of penetration in the thicker section is usually made only slightly greater than the thickness of the thin section.

Even with the aid of the heat-sink capacity of a thicker member, heat input by arc welding often cannot be localized sufficiently and controlled closely enough to avoid damage to metal of foil thickness. In the example of welding type 304 stainless steel that follows, in which 0.002-in.-thick foil was sandwiched between 0.040-in.-thick sheet and the edge of a 0.100-in.-thick cylinder wall, GTAW was replaced by EBW to eliminate excessive foil melting.

Example 4. Melt-Through Welding of 0.002-in.-Thick Foil. A rupture-disk assembly (Fig. 29) designed for gas-pressure relief consisted of three components: (1) a 2-in.-diam nozzle of 0.100-in.-thick type 304 stainless steel attached to an outer pressure vessel (not shown); (2) a 0.002-in.-thick cupped rupture disk of type 304 stainless steel; and (3) a 0.040-in.-thick cap, also of stainless steel, containing an off-center flued-in hole with downward-pointing knife edges.

The three components were joined by a single weld that was required to have a leak rate of less than 10^{-8} cm^3/s when tested with a helium mass spectrometer. When pressure in the nozzle exceeded the design limit, the cupped rupture disk reversed its shape from concave to convex, puncturing itself against the knife-edge of the cap. Leaktightness, rather than high tensile strength, was the welding objective. The material was selected for long, maintenance-free service, rather than for resistance to a corrosive environment.

Originally, the components were joined in one pass by GTAW, using the design shown in detail A, original method, Fig. 29. A circular groove was cut in the nozzle wall to help localize the weld bead, which was made without filler metal. Light pressure was applied at both ends of the assembly to hold the parts in alignment for welding.

This procedure resulted in numerous rejects because of leaks. Often, difficulty was encountered in getting the disks to lie absolutely flat on the joint surface. The 0.002-in.-thick rupture disks, having been blanked and formed in the work-hardened condition, were tough, springy, and sometimes burred. As a result, they occasionally were burned back, instead of being fused into the weld bead, causing leaks. It was usually impossible to salvage this type of defect by rewelding.

To eliminate these difficulties, the operation was converted to EBW, even though initial equipment and tooling costs were high. The EBW machine used for this application had a relatively large vacuum chamber with full-width doors to facilitate loading and unloading of large, multiple-part fixtures. The holding fixture used had 56 receptacles to position and individually rotate 56 assemblies under a stationary gun. The fixture was mounted on a horizontal table equipped with *x-y* motion control to permit optical alignment of the beam for welding each part in turn.

Each assembly was fitted into a cylindrical receptacle or pot that was equipped with a hold-down ring that forced the periphery of the disks and caps into close alignment, as shown in detail A, improved method, Fig. 29. A melt-through or spike weld was made, rather than an edge-flange weld at the periphery.

With this setup, the blanking and forming operations on the disks could be allowed the more realistic tolerance of ±0.002 in. on flatness. The electron beam was angled slightly and was positioned close to the hold-down ring so that, in the event of a leak, the joint could be rewelded by running a second weld just inside the first weld.

The weld-localizing groove needed for the original GTAW was eliminated, and the electron beam melt-through weld was made at a heat input of 0.25 kJ/in. of weld length, which was much lower than for arc welding. The hold-down ring provided only a slight chill effect, but enabled the nozzle body to function effectively as a heat sink.

Welding results from the electron beam procedure were good. Joint defects of all types dropped to 3%, and most of the rejects could be salvaged by rewelding. A production rate of 44$^1/_2$ assemblies per hour was a significant improvement over the original arc method.

Partial-Penetration Welds. The precise control of heat input obtainable in EBW permits partial-penetration welds having uniform depth to be made in metals as thin as 0.010 in. and less.

Work-metal thickness must be uniform, and workpiece fixture arrangement and dimensions must be selected to provide a constant rate of heat loss adjacent to the path of the weld along its entire length, with special provisions made for weld starting and stopping. Beam power, focus, and workpiece travel speed must be closely regulated to minimize variation in penetration.

For the closest control of penetration, both beam power and focus can be regulated by sensitive, fast-response closed loop servomechanisms or feedback controllers, and workpiece travel speed can be held within extremely narrow limits by precision, mechanical worktable traverse, or rotary-motion systems. To compensate automatically for the effect of any variation in voltage on focus, the servomechanism regulating the beam focus can be coupled to the voltage supply by solid-state devices that have a response time of less than 17 μs.

Partial-penetration welds in thin metal are not subject to root voids, root porosity, or cold shuts, which are flaws frequently encountered in partial-penetration welds in thick metal, because such welds in thin metal have relatively low depth-to-width ratios and a wide-angle V-shape with a gently radiused bottom.

Table 3 Relation of depth to width (D/W) for partial-penetration electron beam welds in thin sections(a)

Work metal	Thickness, in.	Weld depth, in.	Weld width(b), in.	D/W for weld
Aluminum alloys				
6061-T6	0.005	0.003	0.015	0.2
5052	0.010	0.005	0.012	0.4
	0.020	0.017	0.015	1.1
Stainless steels				
17-4 PH	0.005	0.004	0.014	0.3
Type 301	0.010	0.006	0.011	0.5

(a) All welds were made in a high vacuum; chill tooling was in intimate contact with the underside of the work at the weld location. (b) Width of the weld at the crown, which was the greatest width for the gently radiused wide-angle V-shaped welds

Data on the relation of depth to width for high-vacuum partial-penetration welds in thin sections are given in Table 3. The ratio of depth to width is lower for the welds in the thinner sections. Penetration was from 50 to 85% of the work-metal thickness, and all welds were made with chill bars in intimate contact with the underside of the work at the weld location.

Depth of penetration generally varies 10 to 15% from a mean value for a work thickness of about 0.020 in., and the percentage variation is greater for welds in thinner metal. Accordingly, partial-penetration welds are not ordinarily made in metal thinner than about 0.010 in. Beam oscillation is sometimes used to smooth out variations in penetration, but such oscillation must be controlled closely in order to be effective.

Welding of Thick Metal

For joining thick sections, electron beam welding in a vacuum has three principal advantages over other welding processes:

- Much deeper penetration can be obtained in a single pass. Extremely narrow, only slightly tapered welds and a small HAZ can be produced.
- For joints on most weldable metals, no filler metal is needed. Where a filler metal is required, the quantity is usually very small.
- Joints have closely fitted parallel groove faces requiring no edge preparation, instead of V-grooves or U-grooves.

The disadvantages of EBW for joining thick metal include: high equipment cost, size limitations imposed by the dimensions of the vacuum chamber in which the work must be placed, and the time needed for evacuating the chamber.

Deep welds made in a vacuum are usually narrow; weld width for penetration deeper than $^1/_4$ in. is typically about $^1/_5$ to $^1/_{20}$ of the section thickness, except for a somewhat greater width at the crown of the weld. Welds may be either full or partial penetration.

Effects of Pressure. Penetration capability is greatest for EBW in high vacuum. The effect that welding pressure has on penetration is indicated below. A constant weld speed of 35 in./min and beam power of 7.5 kW to 150 kV/50 mA for high and medium vacuum, and 175 kV/43 mA for nonvacuum, on steel were used to obtain the following:

Type of system	Ambient workpiece pressure, torr	Beam travel distance, in.	Pentration (max), in.
High vacuum	1×10^{-4}	18	1
Medium vacuum	1	8	$^5/_8$
Nonvacuum	760	$^1/_2$	$^5/_{32}$

Full-Penetration Welds. In welding carbon steel, single-pass, full-penetration, vacuum (high/medium) electron beam welds are made commercially in thicknesses up to approximately 2 in.; production EBW-HV has been done on aluminum alloy plate 6 in. thick, and 9-in.-thick aluminum alloy sections have been welded experimentally under near-optimum conditions.

Two-Pass Full-Penetration Welding. Penetration can be nearly doubled by making two passes, one on each side of the work, but cold shuts, voids, and porosity are encountered at the root of the second weld unless conditions are adjusted for the pass made on the second, or back, side. A broader second-side weld with a larger radius at the root helps to reduce the incidence and severity of these flaws, but they often remain a problem.

The example that follows describes the full-penetration welding of 5$^3/_8$-in.-thick metal in two passes, one on each side, in welding together large billets of unlike metals for subsequent rolling into thin bimetal strip.

Example 5. Full-Penetration Welding of a 5$^3/_8$-in.-Deep Joint Between Dissimilar Metals in Two Passes. Electron beam welding was used to join massive bars of Invar and a manganese-copper-nickel alloy into bimetal billets that were rolled into thin strip for use in thermostats (Fig. 30). Nominal composition and room-temperature coefficients of thermal expansion for the two alloys, which show a wide difference, were:

Alloy	Nominal composition, %	Expansion, μin./in. · °C
Invar	36Ni-64Fe	0.877
Mn-Cu-Ni	18Cu-10Ni-72Mn	28.5

Normally, bimetal billets were joined by furnace brazing, but an alternative method was sought for joining this combination because of difficulties in brazing and an undesirably low production rate.

A uniform metallurgical bond over the entire interface was required, since after bonding, the billets were hot rolled and then cold rolled to final thicknesses of 0.005 to 0.060 in., depending on the application.

Bars of the two metals were machined to the dimensions shown in Fig. 30, and the surfaces to be welded were ground to a finish of 32 μin. maximum in preparation for joining. Initial thicknesses of the bars to be joined were $1^1/_2$ in. for the manganese alloy and 1 in. for the Invar to allow for lower hot strength and greater reduction of the manganese alloy during hot rolling. Final thickness ratio was 9 to 11 (manganese alloy to Invar).

The first method tried as an alternative to furnace brazing was diffusion welding, because large vacuum hot-pressing furnaces were available in which many pieces could be bonded in a single load. However, good pressure bonds were difficult to obtain, even in small laboratory furnaces at vacuums less than 5×10^{-2} torr, because of the ease with which the manganese alloy oxidized.

Several attempts were made to produce the diffusion bond in a production furnace having a normal operating vacuum of about 3×10^{-1} torr. No filler metal was used. Results obtained after hot and cold rolling the bonded billets were not satisfactory. Temperature seemed extremely critical (1900 °F +10, −20 °F), and control within this narrow range was not feasible on a production basis.

When EBW was tried, excellent bonding results were obtained. The two bars were clamped in heavy-duty vises and C-clamps on a traversing table in the vacuum chamber, with the joints in the vertical position as shown in section A-A in Fig. 30. A 3-in. depth of penetration was achieved in each pass, which provided full penetration with an overlap of $^5/_8$ in.

One of the problems encountered was that, when the billet was removed from the clamping fixture to turn over for making the second weld, the heated piece bowed. Rather than attempt to reclamp the billet in this condition, it was allowed to return

Fig. 30 Stages in two-pass EBW

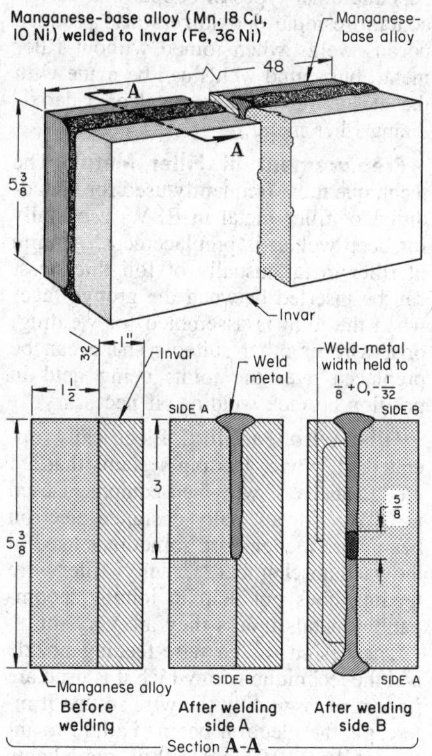

Joint type	Edge
Weld type	Double-square groove
Machine capacity	150 kV at 200 mA
Gun type	Fixed
Vacuum chamber size	56 by 56 by 114 in.
Fixtures	Vises; C-clamps
Welding power	150 kV at 170 mA
Welding vacuum	5×10^{-4} torr
Beam focal point	$^1/_4$ in. below work surface
Welding speed	18 in./min

to room temperature and thereby to resume its original shape.

There was some concern over maintaining uniform weld width through the depth of the joint, because nonuniform weld width might ultimately result in irregular bimetal deflection. However, except for a depth of about $^1/_2$ in. at each surface where the electron beam impinged, the weld width was held to $^1/_8$ in. +0, $-^1/_{32}$ in. After reducing the $2^1/_2$-in.-thick billets to an average strip thickness of 0.020 in., the weld fusion zone could not be identified. After four bimetal billets were welded by this procedure, EBW was accepted as an alternative method to furnace brazing.

Partial-Penetration Welding. Welds that do not penetrate completely through the work metal are satisfactory in many applications in welding thick metal; joint design and product requirements often rule out full penetration. See Example 1 for another partial-penetration weld application.

Problems and Flaws. In making vacuum electron beam welds $^1/_8$ to $^1/_2$ in. deep in weldable metals, weld quality is ordinarily equal to or better than that of arc welds in the same metal. Special precautions are needed to avoid certain types of discontinuities in welds about $^1/_2$ in. deep or deeper. Fusion-zone porosity, gas pockets, and cold shuts are more serious problems than in arc welding, particularly in deep welds.

Reducing the welding speed usually helps to reduce porosity and gas entrapment. Extreme care in cleaning the work metal may be necessary where the problem is severe. When flaws are found near the root of the weld, they can be avoided or minimized by any adjustment of welding conditions that broadens the weld and increases the radius of the weld at the root. Sometimes the joint can be designed so that any root flaws are in a noncritical region, or in integral or separate backup metal that will be machined away after welding.

As in arc welding, cold shuts (incomplete fusion) are sometimes encountered at the root of a deep partial-penetration weld or of a full-penetration weld that has a poorly fitted backing strip. These discontinuities are troublesome because they are difficult to detect. Normally they cannot be detected by radiographic methods. Ultrasonic testing can detect the larger cold shuts. Many cannot be observed on roughly polished macrosections of electron beam welds, but require a metallographic polish and examination at a magnification of at least 100 diameters for detection. Cold shuts can usually be avoided or minimized by reducing the welding speed or by otherwise changing conditions so as to broaden the weld and increase the radius of the weld at the root.

Poor fit-up or excessive joint gap can cause excessive shrinkage, underfill, undercut, voids, and cold shuts. Joint gap should not exceed 0.010 in. for narrow welds deeper than about $^1/_2$ in. in most metals, although sound welds have been obtained using joint gaps of 0.030 in., by use of a procedure that increased weld width to 0.275 in.

Welding with the beam at a slight angle to the joint is helpful in avoiding flaws related to poor fit-up, but may increase weld width excessively and require an unduly low welding speed for deep welds.

Excessive mismatch can cause discontinuities of the same general types as poor fit-up; mismatch limits depend on joint design and dimensions, and on operating conditions. Weld quality in joining thick sections is ordinarily unaffected by changes in surface roughness of the joint faces between about 63 and 1000 μin.

A problem that is to some extent unpredictable and inconsistent in making electron beam welds deeper than about $^1/_2$ in. is arc-outs, or sudden failures of electron emission during welding. However, problems of arc-out have decreased with the newer machines and experience accumulated in operation.

Work-metal composition and quality also influence arc-outs, which are more frequent when welding materials with low vaporization points, or when welding materials susceptible to outgassing in a vacuum or those that contain nonmetallic inclusions.

Use of Filler Metal

Fusion of closely fitted groove faces generally provides sufficient weld metal, or extra metal is provided where necessary by including extra stock thickness or integral shoulders or lips in the joint preparation. Where product requirements or other circumstances prevent the use of a joint preparation and welding conditions that will provide sufficient weld metal, filler metal is added.

Prevention of Cracking. Filler-metal (shim stock) additions can prevent cracking in electron beam welds in a crack-susceptible metal or combination of dissimilar metals. It may be needed even when joint design, fit-up, and operating conditions are selected to achieve minimum joint restraint and residual stress. In other applications, the use of preplaced filler metal can provide weld metal that is less brittle than would be obtained by welding a base metal (or metals) alone.

Among the base metals for which added filler metal is often required are heat treatable aluminum alloys 6061, 6063, and 6066, free-machining steels, and other free-machining alloys.

Many combinations of dissimilar metals that are susceptible to cracking when welded directly, usually because of the formation of brittle intermetallic phases, can be electron beam welded with the aid of filler metal that has a composition compatible with both metals of the combination.

Preventing Porosity. The EBW of rimmed steel without filler metal ordinarily results in severe porosity in the weld, even when welding speed is slow and other conditions are selected to increase the time during which gas can escape from the molten weld metal. Inserting a filler metal that contains a deoxidizer such as aluminum, manganese, or silicon helps to minimize porosity.

Copper alloy C11000 (tough pitch copper) and other types of copper that do not contain residual deoxidizers also produce porous welds when joined without filler metal, but sound welds can be made with the aid of nickel filler metal or a deoxidizing filler metal.

Preplacement of Filler Metal. The technique most frequently used for the addition of filler metal in EBW, especially for deep welds, is preplacement. A shim of filler metal, usually of foil thickness, can be inserted between the groove faces when the joint is assembled for welding, or a wire or other suitable shape can be preplaced over the joint, being held in position by tack welding, if necessary.

Filler-Wire Feeding. Electrically operated filler-wire feeding systems that are modifications of wire-feed equipment used in GTAW, or specially designed electron beam wire feeders, are sometimes used in the vacuum chamber. Usually, filler-wire feeding does not help in joining incompatible metals unless they are very thin.

The design of the wire-feeding nozzle and the technique with which it is used are important in guiding the wire so that it intercepts the electron beam. Failure to intercept the 0.010- to 0.030-in.-diam beam results in irregular application of filler metal. Any filler metal inserted into the beam path absorbs energy and affects penetration.

The wire-feeding nozzle is normally held as close to the weld pool as is possible without damage to the nozzle, and the wire is directed into the forward edge of the pool. The tip of the nozzle should be made of a heat-resistant material and should be coated so as to prevent molten weld metal from adhering to it if inadvertent contact is made. The nozzle is usually pointed in the direction of workpiece travel.

Wire-Feeding Equipment. Several design features are desirable in a wire feeder for EBW. These include means of making positioning adjustments from outside the vacuum chamber during welding, such as: changing the proximity of the nozzle to the weld pool, changing the angle of incidence between the workpiece and nozzle, and controlling the movement of the wire relative to the beam to ensure accurate interception.

A simple, variable-resistance, controlled motor drive is ordinarily satisfactory, as a 10% variation in wire speed causes no adverse effect other than a slight variation in weld width. It is desirable to be able to control the timing so that wire feed can be started and stopped independently of the beam and table. Being able to adjust the timing of the wire feeder eliminates sticking of the wire to the workpiece at the end of the weld. Vibrating the nozzle at subsonic rates promotes flow of molten metal into the weld joint and results in less wire sticking at the end of the weld.

Welding Technique When Feeding Filler Wire. The wire-feed rate is usually set at approximately the same rate as the welding speed. Beam power must be sufficient to melt the wire as fast as it is fed. The wire diameter is selected so as to provide 1.25 times the volume of filler metal needed to fill the joint cavity. Where needed, backing strips or rings are used to prevent loss of molten weld metal.

The filler metal can be deposited during the weld pass, using a low welding speed to allow time for the molten metal to fill the joint gap, but for deep welds it is preferably deposited in a separate pass. If cracking is a problem, another method used on thin metal is to deposit filler metal on the joint, and then make the welding pass to distribute the filler metal more effectively. Additional information on the wire-feed process and related equipment can be found in the section of this article on electron beam welding as a repair method.

Tack Welding

In fixturing work for EBW, clamping often is supplemented or replaced by tack welding, which is usually done with the electron beam. Fixturing time and cost are often substantially reduced by the effective use of tack welds.

In some applications, selection of the tack welding procedure is critical, to avoid unacceptable variation in the shape and dimensions of the final weld bead, and to avoid root porosity or cold shuts. Reduced power and other special operating conditions are often used.

Techniques. Perhaps the most common technique is to make one or more suitably spaced tack welds at full or reduced beam power before making the welding pass. For circular or circumferential welds, one or more tack welds are usually made. Melt-through (spike) tack welds can be used on straight-line lap joints.

A shallow seal pass at reduced beam power can also be made along the full length of the joint to keep the workpieces aligned, to be followed by the full-power penetration pass. For some applications, it is convenient to leave occasional unwelded gaps as an aid for tracking the joint during the penetration pass.

Welding of Poorly Accessible Joints

One of the advantages of EBW is the capability of reaching into areas lying deep within narrow openings. This is accomplished by virtue of the electron beam having a small diameter, long working distance, and frequently, the capability of being projected at an angle. Many joints that are inaccessible for welding by other processes can be electron beam welded. This capability is used both in fabrication and in repair work. It is especially useful in salvaging intricate castings.

Workpiece Requirements. For limited access welding, the workpiece must satisfy three general requirements: (1) the weld area must be on a line-of-sight path from the beam source; (2) there must be sufficient sidewall clearance to avoid beam-fringe interference; and (3) the beam path must be free of magnetic fields.

Beam characteristics that determine applicability of EBW to poorly accessible joints are beam diameter and the working distance or effective beam length available between the exit end of the beam transfer column and the work. These characteristics in turn depend on the beam power used, which is selected to produce the desired penetration, and on beam focal length, which is influenced by the design of the gun, the chamber, and the focusing coil.

Sidewall Clearance. Parts intended for fabrication by EBW can often be designed with sufficient sidewall clearance for the beam power required. Where insufficient sidewall clearance exists in repair welding, lowering the beam power or slightly changing the angle of incidence of the beam often avoids interference. In repair welding, tests should be made on simulated joints, using the same type of work metal and the required beam power, rather than risking damage to difficult-to-replace parts.

Where close sidewall clearances are involved, it is especially important to avoid magnetic fields, which can cause damage to the part by deflecting the beam. A test run on low power is commonly used to detect beam deflection; however, allowance must be made for the increase in deflection at full welding power. Magnetically soft materials with induced magnetism can usually be demagnetized with magnetic-particle inspection equipment or with coils.

Minimum beam clearance for making welds close to magnetic metal that projects above the work surface at the joint is given in Fig. 16.

Multiple-Tier Welding

The deep-penetration properties of the electron beam make it possible to weld two or more tiers of joints simultaneously. The joints can be separated by an air space as great as several inches, provided that the space can be evacuated and the joints aligned in the path of the beam. This type of weld cannot be made by any other welding process.

In multiple-tier welding, the electron beam must pierce at least the upper tiers in such a way that the molten metal flows in behind the advancing beam, as in keyholing. The molten weld metal is held in place by the combined forces of capillarity, surface tension, and viscosity. Welding conditions must be carefully selected and controlled.

Difficulties. In-line welds are progressively different in shape and size, because of the scattering and reduction in beam density and other effects on the beam that take place when it penetrates each layer of metal in turn.

Internal weld spatter may cause difficulty if it interferes with service performance and there is no access for cleaning. Other difficulties such as undercut, underfill, and excessive root bead may be correctable. Hidden welds, which are difficult or impossible to inspect, require some acceptable indirect method of controlling joint reliability.

Applications. Multiple-tier welding has been applied to a variety of joints. Two-tier joints separated by as much as 3 in. have been welded in a single pass in joining honeycomb panels to end frames or structural shapes where only one side was accessible.

The example that follows describes the problems encountered, and the methods used to solve them, in the two-tier welding of a gas-turbine component in which two layers of 0.075-in.-thick René 41, about 1/2 in. apart, were welded simultaneously.

Example 6. Two-Tier Welding. A unique solution to the problem of designing and fabricating a component of an aircraft gas-turbine engine was achieved by electron beam tier welding. As indicated in Fig. 31, the component consisted of a cylinder with an external flange on one end, an internal flange on the other, and a tubular annulus between. The components were assembled by welding the trough-shape ends of two subcomponent cylinders by a single two-tier circumferential weld.

The components were made of René 41, for service at elevated temperature. The chief welding objectives were to obtain sound welds and to avoid distortion of the part, and especially the alignment of the 288 holes located in the annulus less than 1/2 in. from the joints. The holes had to be drilled before welding because they could not be deburred if drilled after welding.

Arc welding was rejected as a joining method because it would have been necessary to use internal chills with gas backing in the annulus to minimize distortion and avoid atmospheric contamination. Electron beam welding not only met the basic requirements but also made both welds simultaneously, even though the two joints were separated by approximately 1/2 in., as shown in detail A of Fig. 31.

Fixturing was relatively simple. The joints were accurately machined square and the components were assembled between two aluminum plates fitted over the flanges. The plates were connected and forced together by bolts located inside the inner flange. This fixture was then mounted on the faceplate of a welding positioner in the vacuum chamber so that the part would rotate with its axis horizontal. The electron beam gun was in a fixed, overhead position.

Success of the two-tier welding operation depended on careful control of part alignment, beam alignment, beam focal point, power adjustment, and travel speed. The joints for the upper-tier and lower-tier welds had to rotate in the same vertical plane, although, by direct viewing, they could not be observed simultaneously. In addition, beam impingement on the joint of the lower-tier weld could be verified only by emergence of the weld bead from the underside, or by sectioning of test pieces.

Alignment of the part for true horizontal-axis rotation was done with the aid of a precision level (sensitivity of 0.0005 in. per foot) and precision spacer blocks placed on the face of the 24.5-in.-OD flange. Beam alignment was done by centering the beam spot on the reticle of the scope and moving the joint to this position. Beam focal point, beam power, and travel speed were adjusted by trial and error on test components until a satisfactory welding procedure was established. The final settings are shown in the table with Fig. 31. By adjusting the beam focus for an indicated setting midway between the two tiers, the weld shapes of Fig. 31, details B and C, were obtained.

The mushroom-head shape of the upper-tier weld was caused by the defocused condition of the beam at that point, while the somewhat oversize root reinforcement resulted from the excess of power needed to penetrate to and through

Fig. 31 Tiered welds made simultaneously by the EBW process

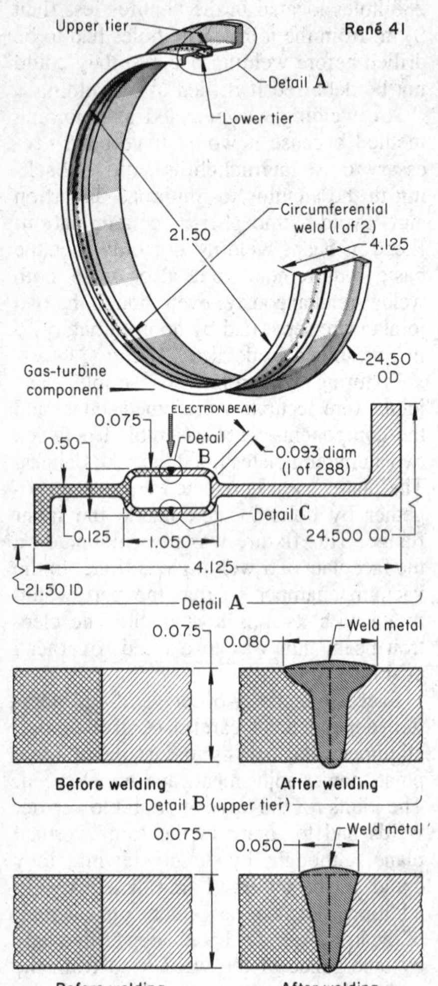

Joint type	Circumferential, two-tier butt
Weld type	Square groove
Machine capacity	150 kV at 40 mA
Gun type	Fixed
Maximum vacuum	10^{-5} torr
Fixtures	Bolted end plates; rotating positioner
Preheat	None
Welding power	125 kV at 9.3 mA
Welding vacuum	10^{-4} torr (min)
Working distance	12 in.
Beam focal point	Midway between tiers(a)
Welding speed (at 0.42 rpm):	
Upper tier	30 in./min
Lower tier	28.7 in./min
Beam oscillation	None
Number of passes	One, plus 30° downslope
Postweld heat treatment	Aged 16 h at 1400 °F

(a) Indicated machine setting. See text of example for explanation of beam focal point.

the lower tier. The relatively narrower weld face of the lower-tier weld, as well as the narrowing of the weld in its progress through the joints, was explained as an effect of a charged plasma that surrounded and refocused the beam on its passage through the material. The plasma, having a net negative charge, repelled the beam electrons, causing the beam to constrict and to change its focal point. The net result on the lower tier was to produce a weld closely approaching the contour of a normal single-thickness weld made with a tight, surface-focused beam. Thus, the indicated focal-point setting was more virtual than real.

Both welds were satisfactory as to soundness and shape. Because of lack of access to the interior of the joint, weld spatter and undercutting were of concern, and spatter associated with penetration of the upper tier was a problem. Most of the particles were loosened with pipe cleaners and were flushed out with solvent at high pressure. The few small particles that remained were judged to be acceptable after radiographic examination. Undercutting was not a problem. The René 41 material was capable of withstanding considerable excess beam power, which was especially important in making the upper-tier weld.

The two-tier welding procedure was used to produce three components, which met all test requirements and were accepted.

Welding With Extreme Accuracy

The development of feedback servomechanisms for controlling variations in beam power and beam spot size, as well as in work travel and speed, permits EBW within tolerances of a few thousandths of an inch in high-vacuum machines.

With the use of closed loop feedback control systems, variations in beam voltage and current can be held to approximately 1% for a 10% variation in the supply line, which is well within the tolerances that can be maintained by the remainder of the system.

Although the diameter of the beam spot cannot be measured directly with great accuracy, it can be estimated by measuring the width of the weld or by calibrating the controls for the focusing and beam-power settings. Depending on machine characteristics, beam spots effective for most welding purposes can be set up in the range of 0.010 to 0.030 in. in diameter without difficulty. Assuming a very small beam variation, an application requiring a beam spot of only 0.010 in. diam would require a means of holding the joint-to-beam coincidence with a runout of less than ±0.005 in.

For welding with a 0.010-in.-diam beam, a pileup of manufacturing tolerances could easily cause the beam to miss the joint. Therefore, for most production purposes, small beam spots are not economically feasible.

Various types of travel mechanisms of rigid construction are available with runout tolerances of approximately 0.001 in. per foot or with total deviations of 0.003 to 0.005 in. for a complex traverse pattern. However, joint runout depends also on the manufacturing tolerances of the part and on accuracy of fit-up and fixturing. In some applications, 20 to 25 possible sources of error, such as magnetic deflection of the beam, and human and systematic errors in beam alignment, can be found.

Use of Scanning

Scanning is a method of checking the runout between the beam spot and the joint to be welded. Its purpose is to indicate the final adjustments needed to align the beam with respect to the joint and to expose possible unsuspected discrepancies in beam behavior or workpiece travel that would interfere with welding. Details of the procedure vary with joint design and the type of equipment available; in one widely used method, a low-power beam is used for visual scanning of the joint during a simulated welding pass.

Accuracy of Beam Alignment. Beam alignment is generally a matter of workpiece alignment, as even guns of the adjustable type are usually fixed during welding. Where the gun does traverse the joint, the principles of alignment are the same. Most joints are designed for simple path shapes—straight lines and circles—that can be adequately traced with precision-made carriages, cross-feeds, turntables, eccentric tables, and spindles.

Scanning Procedure. The workpiece is fixtured, and the joint is aligned for travel direction and is moved to its approximate location under the gun. The chamber is closed and pumped down to a vacuum sufficient for safe operation of the gun. A low-power beam is turned on and focused sharply on the workpiece surface. The specific power settings wil vary with the working distance selected and the power ratings of the machine; however, they should be sufficient only to create a detectable spot without overheating the workpiece.

The joint line is then adjusted to its precise location with respect to the beam spot, the workpiece is traversed through a complete cycle, and the runout is equalized. Maximum allowable tolerance on runout is determined from a consideration of joint design and weld width, which is mainly a

Fig. 32 Electronic scanning setup

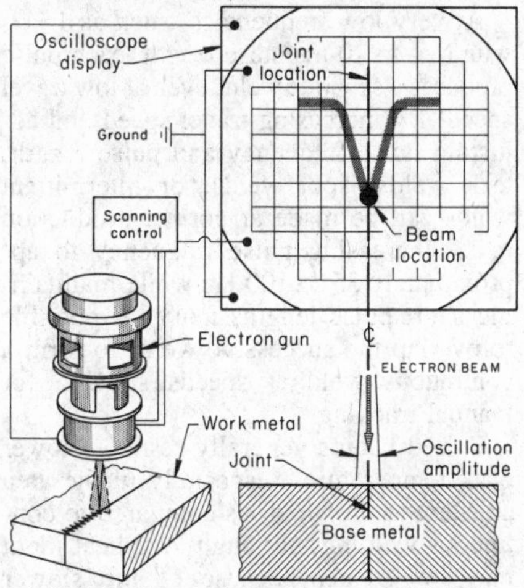

(a) Electronic scanning setup, beam on joint

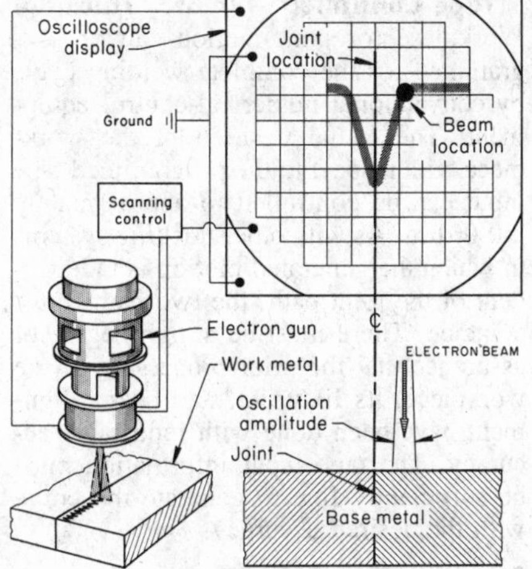

(b) Electronic scanning setup, beam displaced to right of joint

function of the spot size and travel speed.

Considerable latitude in required accuracy of alignment can be gained where beam oscillation can be used. Usually, if the runout approaches one half the effective welding beam spot width (spot size plus oscillation amplitude), the cause must be determined and corrected. If the runout is within tolerances, production parts ordinarily need only periodic scanning as a quality-control measure.

Problems in Scanning. Some of the causes of excessive runout and other difficulties encountered during scanning are as follows:

- Poor joint design, inaccurate joint preparation
- Improper fit-up, inadequate fixturing
- Lack of precision in workpiece traversing mechanism
- Obstructions in the beam path
- Indistinct or undetectable joint line
- Beam deflection by electric or magnetic fields

The corrections required for mechanical discrepancies are self-evident; the effects of electric and magnetic fields may be less obvious.

The negatively charged electrons that comprise the electron beam constitute a negative space charge. When the electrons strike an insulated metal part, negative charges are built up, causing mutual repulsion between the part and the beam. By grounding the part, excess electrons are conducted away. To avoid unintentional deflection of the beam by nearby magnetic fields, workpieces, fixtures, and tooling components made of magnetic materials should be demagnetized before welding.

Scanning techniques vary in detail, depending on the type of equipment used and also on joint design. It is not always necessary to position the beam spot precisely on the joint line. When welding dissimilar metals, dissimilar thicknesses of the same metal, or certain self-locating joints, it is often necessary to position the beam a short distance from the visible joint line. A corner edge or a scribed line can often serve as a reference, or measurement can be made with an optical grid.

Some special-purpose EBW machines are not intended for scanning. Nearly all general-purpose machines, however, make some provision for this operation. On machines with a fixed gun located in an upper section of the vacuum chamber, a built-in internal optical viewing system is generally provided for scanning. Machines having movable guns can be equipped with external telescopes and internal mirror systems.

Optical Scanning. Optical systems usually consist of horizontally mounted monocular or binocular telescopes with internal illumination and reflecting mirrors that provide a line of sight coaxial with the beam path. Magnifying power may be 10 to 40 times, depending partly on working distance. High-precision alignment can be obtained.

There are two common methods of using this equipment. One method makes use of a low-power beam that is sighted telescopically on the workpiece surface. Using a calibrated grid, measurements to 0.001 in. can be made. In the other method, the beam is centered on a crosshair reticle embodied in the optical system. The beam is then turned off and the workpiece joint is moved into coincidence with the crosshair, under internal illumination. The advantage of the latter method is that, after the beam is turned off, the chamber need no longer remain under vacuum, and further alignment work can be done in the open, if necessary.

Before scanning, the mirror surfaces of the internal optical system should be cleaned, because vapor deposition takes place during welding. The frequency of cleaning may vary from a few to many hours of operation, depending on the amount of vapor created and the proximity of the weld. Mirror surfaces are usually protected with glass shields that can be removed for cleaning or replacement.

Scanning Without Optics. Scanning can also be done without visual aids where great precision is not required. On machines that are not equipped with viewing optics, scanning is done by direct observation through sight-glass ports. From one to four such ports may be located in the chamber wall, usually spaced 90° apart, to permit checking beam alignment from several angles. The beam spot is generated the same as for optical scanning.

Electronic Scanning. Checking the location of an electron beam with respect to the joint can be done with an electronic device that displays the relationship on an oscilloscope. The equipment makes use of a low-power electron beam that is made to oscillate at 60 Hz transversely across the joint. The amplitude of the oscillation is sufficient to display the workpiece surface for a short distance on either side of the joint, as well as the joint line itself. In addition, each oscillation pulse is interrupted momentarily to permit the beam to assume its normal welding position, causing a relatively large spot to appear on the display. When the spot coincides with the joint line of the display, the beam is accurately aligned for welding.

Figure 32(a) shows the display condition for a beam centered exactly on the joint. The V represents the joint line, with the beam spot centered accurately; the two horizontal arms (of equal length with a centered beam) at the top of the V represent the adjacent workpiece surface. The broken lines above the work (at left in Fig. 32a) show the plane of oscillation; the wavy line along the joint in the same view shows the approximate amplitude of oscillation.

If the beam location were displaced to the right of the joint, the display would look like Fig. 32(b); an opposite-handed display would indicate displacement to the left. By the relative depth of the V, the display also indicates the amount of joint separation from apparent zero to approximately 0.006 in. A joint mismatch of a few thousandths of an inch is indicated by the relative height of the horizontal arms. In setting up for electronic scanning (under vacuum), the work is provisionally aligned as for optical scanning, making sure that the joint line is parallel to the direction of travel.

Joint Tracking

Joints that have a path that deviates from a straight line or circle usually require an automatic tracking device for EBW of production quantities. For small production lots, manual tracking, using a coordinate drive having separate controls, sometimes can be used. Whether the method is controlled manually or automatically, the motion required to generate the curve is usually imparted to the workpiece, although limited motion can be imparted to some guns.

Manual Joint Tracking. In manual tracking, the welding operator must closely observe the joint at the point where the weld is being made to anticipate any change in direction. The conditions best suited to this type of operation are: (1) a joint path consisting of a smooth curve, (2) slow travel speed, and (3) a low-power beam with oscillation, to form as wide a weld as possible. The welding operator is capable of regulating two motions of the workpiece (or gun) by manually adjusting the remote controls of the travel mechanism.

For a low heat input, to permit the use of the low welding speeds needed when manual tracking is used, beam pulsation can be combined with beam oscillation.

Automatic Joint Tracking. Mechanical, electromechanical, numerical control, and computerized systems are sometimes used for welding joints that deviate from straight lines or circles. The use of computerized numerical control (CNC) has become highly prevalent in EBW equipment. The capability of a CNC for allowing the weld seam path of a part to be quickly "digitized" and "edited" makes it an ideal tool for programming and tracking the actual weld seam path.

Electromechanical joint tracking systems make use of a stylus, or probe, that rides in the joint. Lateral movements of the stylus (which precedes the beam), resulting from a change in joint position, are converted to electrical signals by a transducer. The electrical signals drive a positioning servomotor that maintains preset alignment of the joint-to-gun focus. These electrical signals define a right error, a left error, and the correct gun position. Precision is usually better, response-delay time is shorter, and other problems associated with mechanical tracking with templates are avoided.

Tape-Controlled Joint Tracking. Workpiece (or gun) motion can be programmed for the complete welding cycle by conventional numerical-control equipment. The actual position of the workpiece with respect to the programmed tape input can be controlled to approximately ±0.001 in. As with other tracking systems in which the generated motion is independent of the joint path, the two paths must coincide. There must be a high degree of assurance that the prior processing of the workpiece, its fixturing, and beam alignment have been done with repeatable accuracy. The tape input information must be sufficiently fine to generate the curve with the desired accuracy.

Beam Oscillations

Through the use of a two-axis deflection coil, either a static (fixed) or an oscillatory (periodic) motion can be imparted to the beam spot through a separate deflection control module. Generally, this control consists of a set of amplifiers for driving the x-y axes and a function generator for programming the beam spot motion desired. Beam spot motions employed comprise a variety of shapes and may consist of straight, fixed, or periodic deflection, or a combination of both. Square wave, circular, hyperbolic, or parabolic beam spot motions can easily be programmed, employing angular beam deflections of up to 15° and frequencies greater than 5000 Hz. This beam oscillation capability allows wider welds, slower cooling rates, and more uniform weld shapes to be produced, without necessarily using beam defocusing. Also, beam oscillation capability reduces the need for accurate beam-to-seam alignment and makes precise joint tracking less critical. In addition, programmed beam oscillation can be used to control microshuts and porosity in partial-penetration welds, thus reducing undercut and underfill for better bead control and for maximizing penetration capability.

Use of a Pulsed Beam

Beam operation can be changed from continuous to intermittent on machines equipped with pulsing controls. Although available repetition rates (frequencies) and pulse lengths depend on the control unit, frequencies in the range of 0.1 to 3000 Hz with pulses up to about 60 to 70 ms in duration are representative. Beam pulsation reduces the rate of heat input, but is independent of other beam conditions. Therefore, pulsation can be combined with oscillation and deflection, as well as travel speed, to influence weld behavior.

At very low frequencies, such as 1 Hz, with a 5- to 10-ms pulse length, each pulse produces a separate weld, even at low travel speed. By increasing travel speed and adjusting pulse frequency and pulse length, tack welds, spot welds, or intermittent welds can be made at normal production rates. Increasing pulse frequency to approximately 35 to 100 Hz while maintaining short pulse lengths makes it possible to overlap the successive welds to form a continuous weld at speeds suitable for manual tracking.

Pulsed beams generally result in lower peak temperatures, especially in the area adjacent to the weld, as compared to continuous welding, although total heat input may be greater because of the slower welding speed used. At higher pulse frequencies, pulsation has been used to control the solidification pattern of the weld and the microstructure of the HAZ. Pulsation also has been used in rewelding to fuse cold shuts, fill gas pockets, and smooth irregular root areas.

Electron Beam Welding as a Repair Method

Restoration of worn or damaged components traditionally has been an important application for welding technology. Some repairs are approached by welding a new detail in the area requiring restoration. However, it is frequently more suitable and cost effective to use a weld buildup process to restore the dimensions and structural integrity of the affected area. Welding processes such as manual or automatic GTAW or gas metal arc welding (GMAM) traditionally have been utilized for this type of repair; these processes facilitate the addition of filler wire to accomplish a weld buildup. While these processes enable efficient and relatively fast buildup of filler metal, they have a potential drawback of relatively high heat input, which may cause distortion when section thicknesses are small.

The EBW process has been employed extensively for original equipment manufacture mainly because of its ability to minimize or eliminate distortion through high weld speed, small HAZ, and small bead size. In the past, application of EBW for repair has been relatively limited, confined primarily to butt welding of repair insert details, penetration of parent metal (e.g., to eliminate a crack), or melting of preplaced filler wire. The matching of automatic wire-feed equipment to EBW equipment, however, has led to the development of a system that incorporates the inherent advantages of EBW systems and offers the additional capability of achieving a precise, multiple-pass weld buildup with minimum heat input. The addition of this capability has greatly increased the utilization of EBW as a repair mechanism.

Electron Beam Wire-Feed Process and Equipment for Repairs

Electron beam welding units used for weld buildup repairs are high-voltage, high-vacuum machines (150 kV, 10^{-4} torr). The availability of a highly controlled welding atmosphere is a distinct advantage for the electron beam process, particularly when welding reactive materials such as titanium. High-voltage, high-vacuum EBW machines are equipped with signal generators that oscillate and deflect the beam in the *x-y* directions. They are also capable of making circles and other geometric figures. Beam deflection frequency can be varied, and the beam can be focused above or below the point of impingement. Beam pulsing and modulation of the pulse are also possible. All these tools can be used in weld buildup repairs. Current electron beam vacuum chambers can accommodate parts up to 60 in. in diameter. On the floor of the chamber is an *x-y* table; fixed to this table is a variable-speed rotary table. The component to be repaired, typically a cylindrical ring or case, is fixtured to this rotary table. An automatic filler-wire feeder delivers wire to the beam impingement point, enabling a continuous, multiple-pass weld buildup when the component is rotated under the beam. Typically, a fused spot is less than 0.020 in. in diameter, allowing for an extremely small HAZ.

The wire-feed equipment is an adaption of standard GTAW wire-feed equipment, and the technique is similar to that employed in cold wire circumferential automatic GTAW. The precision wire feeder delivers wire through a nozzle, which can be positioned in the *x*, *y*, and *z* directions and which is easily visible in the coaxial optical system of the electron beam welding machine. Wire entry angle, nozzle position, wire speed, weld speed, accelerating voltage, focal position, deflection, and amperage are critical factors in EBW buildup. Both inside and outside surfaces can be restored, depending on the position of the wire feeder within the chamber.

A relatively small filler-wire diameter (0.035 in. or less) typically is employed to attain a precise low-heat input and a weld buildup typically 0.040 to 0.050 in. wide and 0.100 to 0.300 in. high, although much larger buildups could possibly be attained by multiple applications of the process. Electron beam weld buildup lends itself readily to automation, thereby opening the process to use by less skilled operators. On some machines, all parameters are preset. The operator merely lines the machine up on the surface to be welded and pushes the start button. The wire feeder automatically activates and adjusts itself to produce a smooth, precise buildup.

Application of Electron Beam Wire-Feed Process for Repairs

The electron beam wire-feed process has been applied primarily for restoration of worn or damaged gas turbine aircraft engine components, which are typically cylindrical in shape with diameters ranging from 4 to 60 in., lengths from 1 to 36 in., and section sizes from 0.050 to 0.500 in. Repair of gas turbine engine components has become increasingly attractive to commercial and military operators because of the cost and lead time to procure spare parts. Repairs are typically conducted at 50 to 70% of the cost of a new component, with three to six times improvement in replacement time.

One of the key gas turbine engine component repairs is the restoration of the rotating labyrinth-type air seals used to maximize engine pressure gradients. As seals wear or are damaged, engine efficiency drops and fuel consumption rises. Hence, restoring minimal seal clearances has become increasingly attractive. Seal design is typically of the straight knife edge or tapered knife edge type (Fig. 33). Materials vary from iron-based alloys (AMS 6508 and 5732) to high-nickel alloys (AMS 5660 and 5828) to titanium alloys (AMS 4928, 4973, and 4976).

Currently, virtually all air seal restoration is done by welding. Gas tungsten arc welding, gas metal arc welding, and electron beam welding have all been successfully applied depending on requirements of engine manufacturers, component design, material, and experience and equipment of repair stations. Prior to the establishment of the electron beam wire-feed process, utilization of the standard electron beam weld process for this type of repair was limited to competitive weld buildup processes (GTAW and GMAW) and was primarily applied where distortion and HAZ size were a primary concern.

Initial application of the standard electron beam process involved welding a machining ring onto a premachined pedestal on which the worn air seal was removed, as shown in Fig. 34. The knife edge configuration was then machined into the welded ring. Later, a split ring concept was introduced. In this procedure, an edge-rolled ring was assembled into a shallow groove premachined into the pedestal top (Fig. 34). Circumferential spike welds from each side welded the ring in place and the ends of the split ring were radiused to blend the discontinuity in the ring. The wall thickness of the as-rolled ring was the finished thickness of the knife edge, typically about 0.010 in., and the outside diameter was finish machined after welding.

Fig. 33 Typical air seal designs

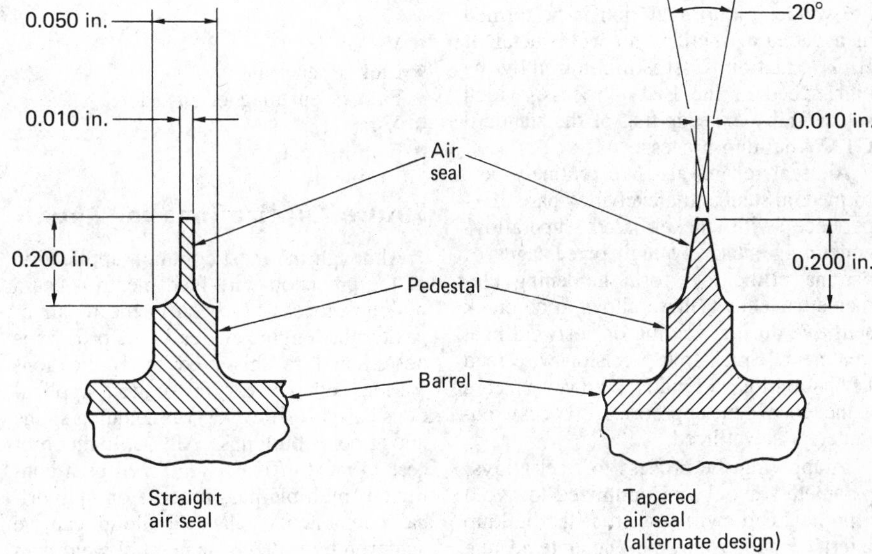

Fig. 34 Electron beam air seal weld repairs

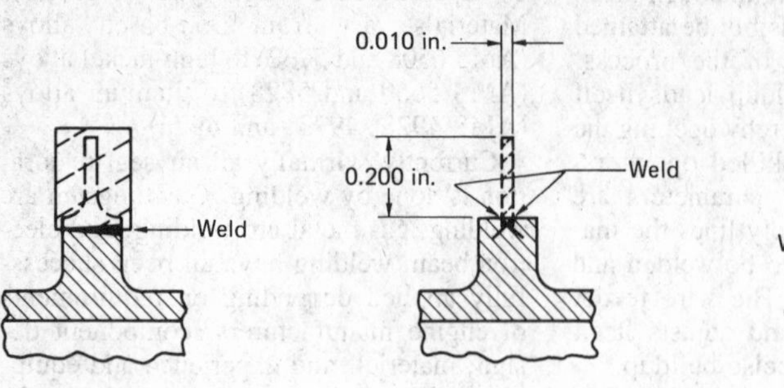

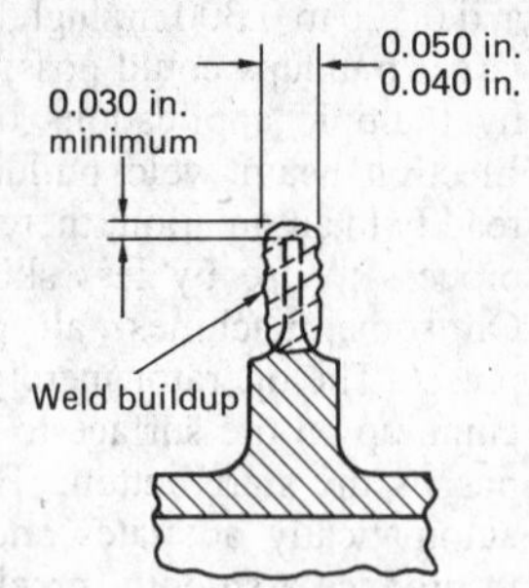

Disadvantages of this weld buildup system include ring detail cost and the amount of post-repair machining required.

The application of the precision wire-feed electron beam process (Fig. 34) to gas turbine seal repairs provides both the material application efficiency and cost advantages of weld buildup systems and the inherent low heat input and controlled atmosphere advantages of the EBW process. A typical automatic wire-feed electron beam air seal repair, applied to a compressor rotor spacer for a gas turbine engine, is shown in Fig. 35. This part is made from a titanium (Ti-6Al-4V) forging. The outside diameter of the knife edges wear in service, and the knife edges must be restored to maintain the proper tip clearance. The knife edge material is machined close to the diameter of the pedestal; a weld buildup is made using Ti-6Al-4V titanium filler wire, and a new knife edge is machined. The multiple-pass technique typically deposits 0.010 to 0.015 in. of radial buildup per pass. Penetration of the HAZ below the pedestal top typically does not exceed 0.015 in. Because the entire weld buildup operation is performed in a vacuum, there is no weld metal or HAZ oxidation. Cost is minimized by the high speed of the EBW process, which substantially exceeds that of the standard GTAW buildup process.

Air seal repairs also are routinely performed on steel and nickel alloy parts. Experience with nickel-based superalloys containing relatively high percentages of gamma prime that form hardening elements has shown these alloys to be crack sensitive during welding or postweld heat treatment. Crack-free precision wire-feed EBW buildup repairs, however, are achieved when appropriate process parameters are utilized.

In applying the process to steel alloys, parameters should be optimized to avoid a pronounced swirl pattern at the buildup interface which may be encountered due to uneven mixing of the weld buildup and substrate. Potential swirling is associated with certain steel alloys and may be a potential source of porosity or cracking.

Quality assurance testing is an integral part of all gas turbine engine repairs. The high degree of control inherent in the precision wire-feed electron beam repair process facilitates in-process control and quality assurance. Because of the high degree of control the electron beam process offers, process variables can be kept to a minimum. This built-in consistency tends to ensure a long-term high level of quality assurance.

Weld parameters are developed by the analysis of microsections of sample welds and by nondestructive testing. Once an acceptable process has been established for a component, a weld schedule is initiated and weld parameters are maintained. Any changes in equipment or process must be qualified by the analysis of new microsections. Quality of finished air seals is ensured by one or more of the following tests, depending on customer requirements:

- Visual
- Liquid penetrant
- Fluorescent magnetic particle
- X-ray
- Etch inspection

Future Applications for Repair

Although the most common application of the precision wire-feed electron beam buildup process has been in the repair of gas turbine engine air seals, the process is not limited to this type of application. Various other wear restoration applications have been developed, including spline and groove buildups. Although the process is most efficiently utilized as a continuous multiple-pass process on cylindrical components, planer buildup can be achieved by utilizing sequential weld pass patterns or oscillation. The process is also potentially applicable for new part manufacture to enable a component to be machined from a smaller, less costly forging with selective electron beam buildup to provide material for specific structural

Fig. 35 Typical automatic wire-feed electron beam air seal repair. Magnification: ×15

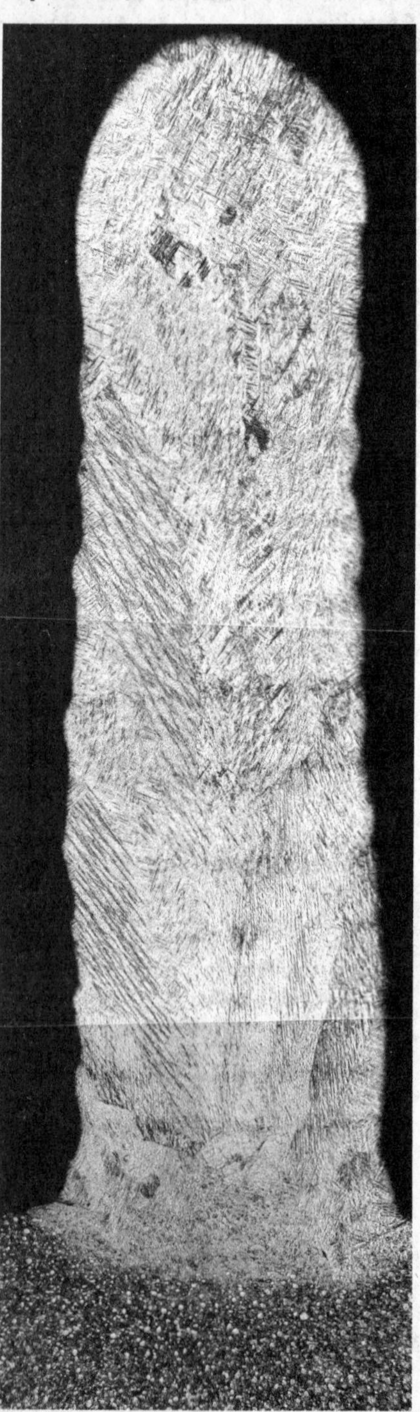

features (e.g., outside diameter bosses on cylindrical cases).

Welding of Low-Carbon Steel

Low-carbon steels are readily electron beam welded. Grain size in both the weld and the HAZ is significantly smaller than in arc welds, because of the lower heat input and extremely rapid heating and cooling. Rimmed steel causes excessive gassing. A technique used to weld rimmed steel is to sandwich an aluminum alloy shim about 0.010 in. thick in the joint to provide deoxidizing action during welding. Welding at a low travel speed or otherwise adjusting welding conditions to produce a more tapered weld and thus to allow more time for the escape of gases from the molten weld metal also helps to reduce porosity. Semikilled steel is easier to weld than rimmed steel, but some oxygen may remain to cause porosity. Fully killed steel is readily weldable.

Welding of Hardenable Steel

The variation in weldability by EBW among hardenable steels follows the same pattern as for arc welding of these steels. Properties of electron beam welds in a given hardenable steel differ from those of arc welds mainly because of the narrowness of the fusion and HAZ and the extremely rapid heat-melt-cool cycles.

Hardness traverses across the midpoint of electron beam (high-vacuum) and gas tungsten arc welds in 0.062-in.-thick heat treated 4340 steel are plotted in Fig. 36. For both welds, maximum hardness was produced uniformly in the weld metal, and hardness dropped off abruptly at the edge of the weld metal.

Similar hardness traverses for electron beam welds in 0.380-in.-thick heat treated 1024 (modified) are shown in Fig. 37. The hardness profile across this weld was similar to that for the electron beam welds in 0.062-in.-thick 4340 steel. Tempering after welding eliminated or minimized the hardness differential in the HAZ (Fig. 37).

Prevention of Cracking. Cracking can be a problem in EBW of hardenable steel, just as in arc welding, if the weld is made in a highly restrained joint, especially in welding parts that have been hardened. Deep circular welds in heavy sections are particularly troublesome. Partial-penetration welds are more likely to crack under restraint than full-penetration welds. It is desirable to place the joint in a location that will allow the part to shrink freely as it cools after welding. Welding through carburized or nitrided cases is not recommended.

Cracking of electron beam welds in hardenable steel can also be minimized by reducing the welding speed to allow more buildup of heat in the base metal, and by preheating, postheating, and allowing the work to cool in the chamber after welding.

Fig. 37 Hardness traverses across EBW-HV butt joints

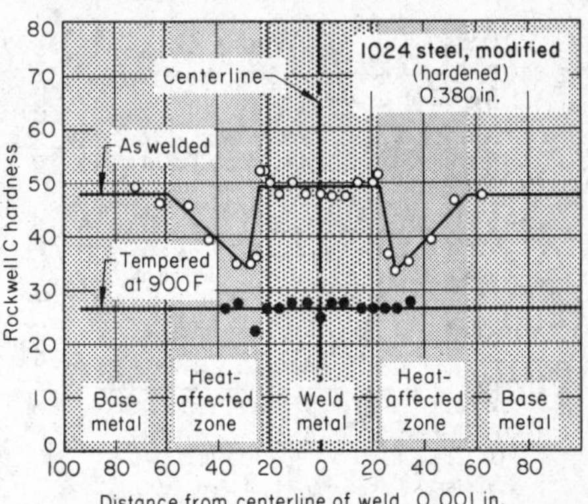

Medium-carbon steels, except for the free-machining types, are readily electron beam welded (for hardness traverses, see Fig. 37). Weldability decreases with increasing carbon content.

Low-alloy steels containing less than 0.30% C are usually electron beam welded without preheating or postheating. When preheating is used—to prevent cracking in highly restrained joints, for example—a temperature of 500 to 600 °F is usually adequate. If the part has been hardened and tempered prior to welding, the postweld tempering must be done at a temperature slightly below that at which the base metal was originally tempered.

Two of the most frequently electron beam welded alloy steels of this type are 8620 and 9310. Components made of these steels often are case hardened, then assembled by EBW, without distortion or heat damage to the case. Before welding, the case should be removed from the immediate vicinity of the joint to prevent microcracks from forming in it.

High-strength alloy steels containing more than 0.30% C are electron beam welded either in the annealed or the normalized condition or in the quenched and tempered condition, although weldability is better for the annealed or normalized condition. The hardness profile for electron beam welds in hardened high-strength alloy steel sections no thicker than about 1/4 in. is ordinarily like that shown for 0.062-in.-thick heat treated 4340 steel in Fig. 36, and joint strength can approach that of the base metal, without preheating or postweld treatment other than stress relief.

In electron beam welding thicker sections of high-strength alloy steels, pre-

Fig. 36 Hardness traverses across EBW-HV and GTAW butt joints

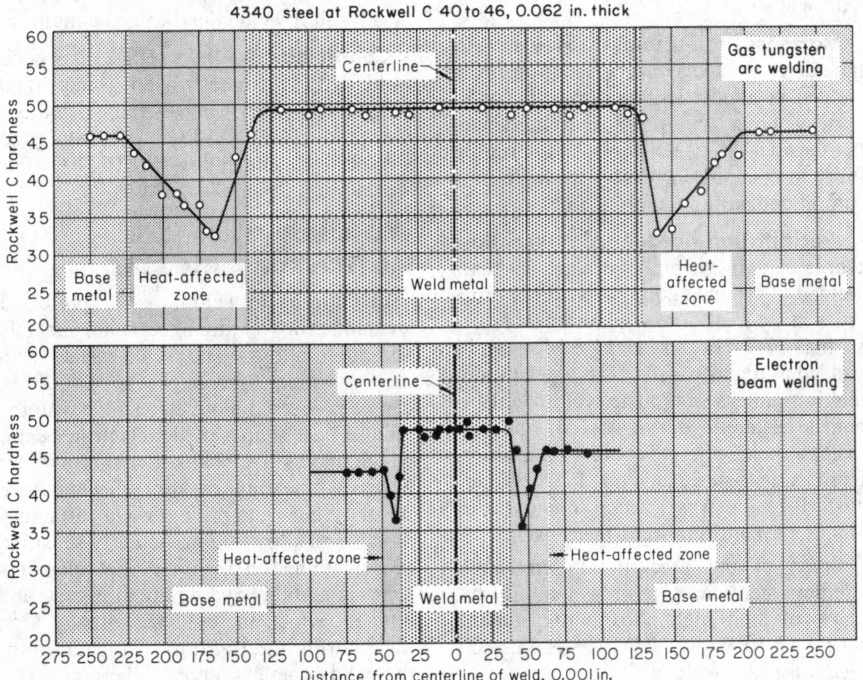

Fig. 38 Effect of preheating on hardness

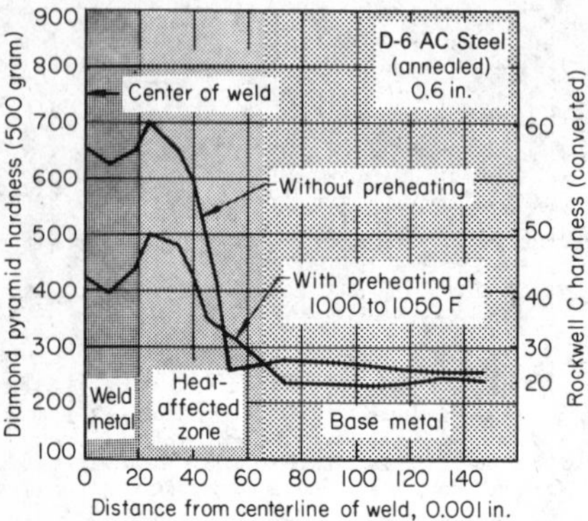

heating or postheating, or both, is usually needed to prevent cracking. In studies on full-penetration, single-pass welding of 0.6-in.-thick annealed D-6 AC steel, which is used frequently for missile casings, cracking was observed when preheating was not used and was eliminated by preheating at 1000 to 1050 °F. Welding was started with the joint at a temperature of 700 °F, and the joint temperature remained above 650 °F for 8 to 10 min after completion of the weld. Hardness profiles for the joints welded with and without preheat are shown in Fig. 38. Sound welds were also obtained in 2.2-in.-thick annealed D-6 AC steel, after preheating to 800 to 850 °F; welding was started when joint temperature was no lower than 500 °F.

High-Carbon Steels. Because of the low total heat input and the rapid thermal cycling that are characteristic of EBW, weld cracking of steels containing more than 0.50% C is less likely than in arc welding. Even bearing steels such as 52100, which are seldom arc welded, have been electron beam welded on a high-production basis. The 52100 steel is welded in the spheroidize-annealed condition in applications where joint performance is not critical to the function of the part and where service stresses in the joint are low. Preheating and postheating are necessary.

Welding of Tool Steels

The chief advantage of EBW over other welding processes in joining tool steels is its ability to produce joints at high speed without annealing or other heat treating operations. Hardness profiles similar to those shown for low-alloy steels in Fig. 36 and 37 were also obtained for H11 tool steel. For full-penetration butt welds on hardened (50 HRC) sections 0.225 in. thick, the hardness of the weld metal and the immediately adjacent base metal was 56 to 57 HRC, and hardness dropped to a mimimum of 43 to 46 HRC in the HAZ, as measured after double tempering at 1025 °F. The overtempered portion of the HAZ was only 0.005 in. wide.

Small dies of D2 tool steel have been electron beam welded in production. A larger operation is the production of bimetal band saws. Specially designed high-vacuum machines, which are referred to as "air-to-air" machines, are used to produce band-saw blades by welding $^{1}/_{16}$-in.-wide M2 high-speed-steel cutting-edge strips to 6150 steel bands. The composite blade is about $^{1}/_{32}$ in. thick and is welded at speeds above 300 in./min in a special vacuum chamber. In "air-to-air" bimetallic strip welding, the backing band and edging strip are continuously fed through slots in a series of differentially pumped chambers and welded in a high vacuum without any interruption of this continuous product flow.

Welding of Stainless Steels

The properties and the welding metallurgy of stainless steels are discussed in the article "Arc Welding of Stainless Steels" in this Volume.

Austenitic Stainless Steels. The high cooling rates typical of EBW help to inhibit carbide precipitation, because of the short time during which the steel is in the sensitizing-temperature range. Filler metal is used in welding joints that cannot be fitted together closely.

Martensitic Stainless Steels. Although these steels can be electron beam welded in almost any heat treated condition, welding will produce a hardened, martensitic HAZ. Hardness and susceptibility to cracking increase with increasing carbon content and cooling rate.

Precipitation-hardenable stainless steels can, in general, be electron beam welded to produce good mechanical properties in the joint. The semiaustenitic types, such as 17-7 PH and PH 14-8 Mo, can be welded as readily as the 18-8 types of austenitic stainless steel. The weld metal becomes austenitic during welding and remains austenitic during cooling. In the more martensitic types, such as 17-4 PH and 15-5 PH, the low carbon content precludes formation of hard martensite.

Some of the precipitation-hardenable stainless steels have poor weldability because of their high phosphorus content. Steels 17-10 P and HNM are not usually electron beam welded.

Welding of Heat-Resistant Alloys

Because of the marked differences in composition and weldability among heat-resistant alloys (nickel-based, iron-based, and cobalt-based), generalizations concerning EBW of these alloys are not useful.

Solid-Solution Nickel-Based Alloys. Hastelloy N, Hastelloy X, and Inconel 625 are readily electron beam welded. Hastelloy B and Inconel 600 can be welded to type 304 stainless steel and to themselves.

Precipitation-hardenable nickel-based alloys that are rated good in weldability by the electron beam process include Inconel 700, alloy 718, Inconel X-750, and René 41. Alloy 718 can be welded in either the annealed or the aged condition. Inconel X-750 should be welded in the annealed condition, and René 41 should be welded in the solution-treated condition. Alloys of this group that have fair weldability include casting alloys 713C and GMR-235 and wrought Udimet 700 and Waspaloy.

Iron-Nickel-Chromium-Based Alloys. This group of alloys is also known as the iron-based heat-resistant alloys. The most readily welded alloy of this group, using EBW, is 19-9 DL, which has excellent weldability and which produces the best results when preheating is used. Alloy N-155 has good weldability, and alloys 16-25-6 and A-286 are rated fair. Alloy A-286 is usually welded in the solution-treated condition; hot cracking may result if welded in the aged condition.

Cobalt-Based Alloys. HS-21 has good weldability in unrestrained joints (and is generally poor in restrained joints). Cast alloy H-31 (X-40) has fair-to-good weldability, and alloy S-816 has fair weldability.

Welding of Refractory Metals

Electron beam welding is the preferred technique for welding of refractory metals and their alloys. These materials, with melting temperatures in excess of 4000 °F, can be processed more efficiently with the intense heat source of the electron beam, which allows for localized fusion without excessive heating of the adjacent base metal. Weld-zone properties in refractory metals and alloys are particularly sensitive to contamination by interstitial elements such as carbon, hydrogen, nitrogen, and oxygen, which increase the already high ductile-to-brittle transition temperature of the weld zone and increase the notch sen-

Table 4 Typical applications and conditions for EBW of refractory metals and alloys

Application	kV	mA	Beam setup in./min	Focus	Comments
Tungsten heat pipe, flange-butt joint, 0.100-in. penetration	100	6	30	Surface	Preheat to 1475 °F
Tungsten crucibles, standing edge joint, 0.080-in. penetration	100	10	60	Above surface	Preheat with beam to 1475 °F
Molybdenum heat pipe, step joint/partial-penetration weld to root of step, 0.062-in. penetration	125	25	30	Surface	High-frequency (~3 kHz) deflection to produce elliptical pool
Mo-TZM alloy straight butt weld in heat shield, 0.060-in. penetration	125	15	30	Above surface	Ductility of weld zone improved by postweld heat treatment
Nb-1Zr reactor tubing, butt weld, 0.060-in. penetration	100	10	30	...	Full penetration weld, excellent as-welded properties
Mo-13 wt% Re test reactor tube, 0.032-in. penetration	100	7	30	Sharp	Excellent weld appearance, no test data available

sitivity beyond the limits required for practical processing of these materials.

The possibility of atmospheric contamination during welding is essentially eliminated by EBW in a high-vacuum environment. Interstitial impurities already present in the base metal, however, have a significant effect on the weldability of the metal and alloy system. The Group V metals (e.g., niobium and tantalum) are softer, less subject to strain hardening, and less prone to the brittle behavior usually associated with the Group VI metals (molybdenum and tungsten). The very low room-temperature ductility of the Group VI metals gives rise to cracking problems due to cyclic thermal stresses during welding. Preheating above 575 °F reduces the susceptibility to this problem.

Excessive grain growth in the weld HAZ and fusion zones can produce cracking during the welding process or during subsequent application of the weldment. Grain coarsening can be controlled to some extent by specifying a fine-grained starting material and minimizing the total heat input during welding. The morphology of the fusion zone is also important. High welding speeds (>24 in./min) selected to minimize heat input can promote pronounced directionality of grain growth in the weld pool, leading to a plane of weakness at the weld centerline. High-frequency (>1 kHz) beam oscillation can be applied to restore an elliptical weld-pool geometry at high travel speeds and eliminate the pronounced delineation of the weld centerline.

As with any low-ductility weldment, the degree of restraint imposed on the joint during the welding operation should be minimized. Most refractory metals and alloys can be welded in thin (<0.04 in.) sheet. The weldability decreases with increasing section thickness in accordance with the increase in restraint. Weld joints should be designed to reduce the restraint by effective use of flange butt and standing edge designs, and by the incorporation of relief grooves to accommodate elastic/plastic strain in the material adjacent to the weld zone.

In contrast to many other alloy systems, the weldability of refractory metals tends to improve with the addition of alloying elements. Specific alloying elements can improve the low-temperature ductility of refractory metals and thus directly affect the weldability and weld mechanical properties. The overall weldability of tungsten is improved by the addition of substitutional alloying elements such as rhenium and molybdenum. As indicated previously, interstitials also raise the transition temperature; this effect can be reduced by alloying with elements that form stable carbides, oxides, and nitrides. For example, titanium and zirconium in molybdenum "scavenge" the interstitial impurities to improve low-temperature ductility and provide a dispersion-strengthening effect. Examples of successful electron beam welds made in refractory metals and alloys are given in Table 4.

Tungsten is the most difficult refractory metal to weld because of its high melting point (6170 °F), sensitivity to thermal shock, and room-temperature brittleness of welds and the HAZ. High-purity tungsten sheet can be electron beam welded if the weld area is preheated and if external restraint is minimal. Successful welds have been made in sections as thick as 0.16 in.; however, the resulting as-welded material is brittle and highly notch sensitive. Tungsten metal, manufactured by a chemical vapor deposition process, has also been proven to be weldable in sections <0.02 in.

Molybdenum can be readily welded by EBW if chemical purity, grain size, and

Fig. 39 Electron beam weld zone in high-purity molybdenum. (a) Molybdenum heat pipe test assembly, end caps joined by EBW. (b) Morphology of weld fusion zone (Magnification: ×25)

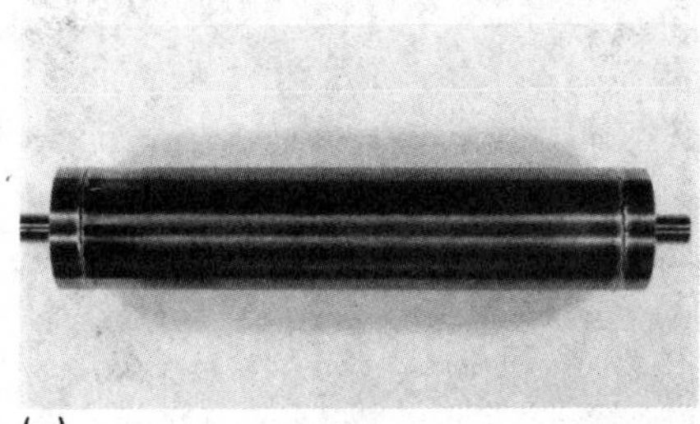

(a)

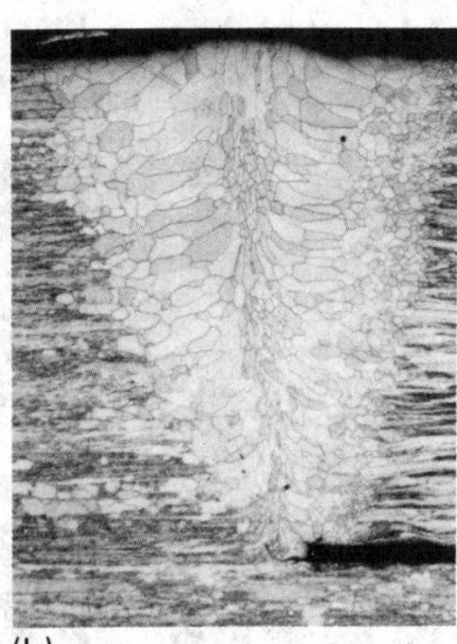

(b)

Table 5 Effect of preheat on the weldability of beryllium and conditions for full-penetration welds

Thickness, in.	Voltage, kV	Current, mA	Travel speed, in./min	Preheat to prevent cracking, °F
0.100	110	7.5	20	500
0.075	100	7.0	20	500
0.055	90	5.0	20	250

texture of the base metal are carefully monitored. Partly because of its lower melting point (4730 °F), molybdenum has better thermal shock resistance than tungsten; thus, narrower welds and higher welding speeds can be used. Preheating may or may not be necessary, depending on the starting material and the degree of restraint on the weld joint. When preheating is used, the workpiece should be heated to 400 °F or higher, followed by a postweld stress relief at 1600 to 1800 °F.

Figure 39 shows an electron beam weld zone in high-purity molybdenum. Penetration is approximately 0.12 in. to the root of a step-butt weld joint. This weld was made without preheat but with a high-frequency oscillation to maintain an elliptical weld pool and influence the direction of grain growth in the solidifying metal. The growth direction at the center of the fusion zone is parallel to the welding direction, and the section is therefore an end view of the columnar grains. This type of solidification structure should not be confused with the grain refinement phenomena sometimes claimed to result from stirring of the weld pool. Because solidification in high-purity metals is essentially a columnar-planar growth, refinement most probably cannot be achieved by stirring the weld pool.

Titanium and zirconium are the common alloying elements added to molybdenum to scavenge interstitial impurities and improve low-temperature ductility. The transition temperatures of welds in these alloys, however, are still above room temperature. Significant increases in the low-temperature ductility of these alloys can be obtained by solution and aging heat treatments. The greatest improvements in ductility are obtained by the addition of rhenium; a 50 wt% addition of rhenium lowers the transition temperature of recrystallized molybdenum from ~125 to ~400 °F.

Niobium, which melts at 4474 °F, is easier to weld than molybdenum or tungsten. Gas tungsten arc welding is often used, but EBW provides better protection, narrower welds, and lower heat input. Preheating is not necessary, but postweld vacuum stress relief is used to restore

Fig. 40 Electron beam braze weld in beryllium. Courtesy of Rockwell International Energy Systems Group

0.04 in.

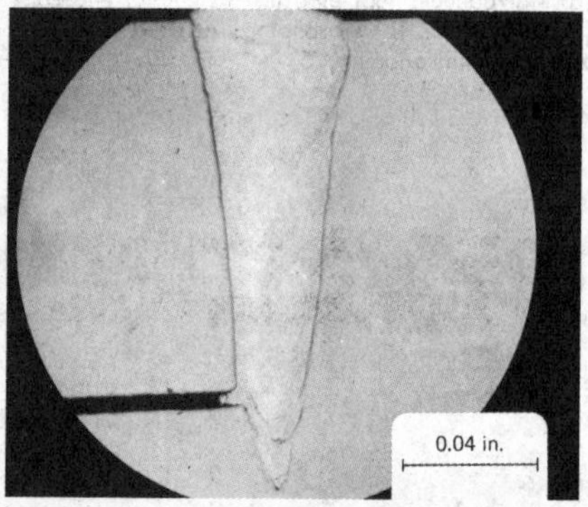

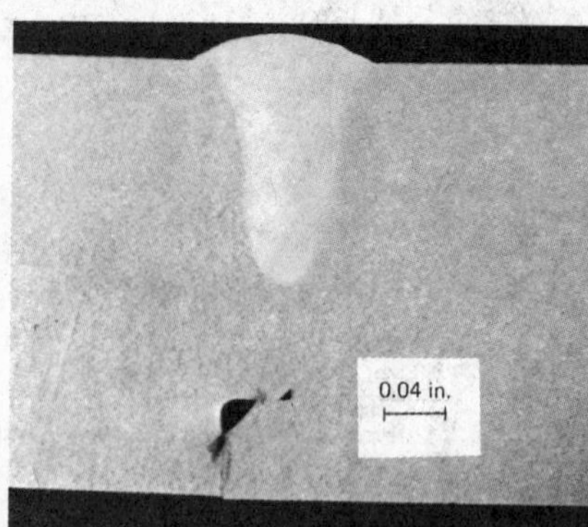

Fig. 41 Electron beam cold wire-feed weld in beryllium. Courtesy of Rockwell International Energy Systems Group

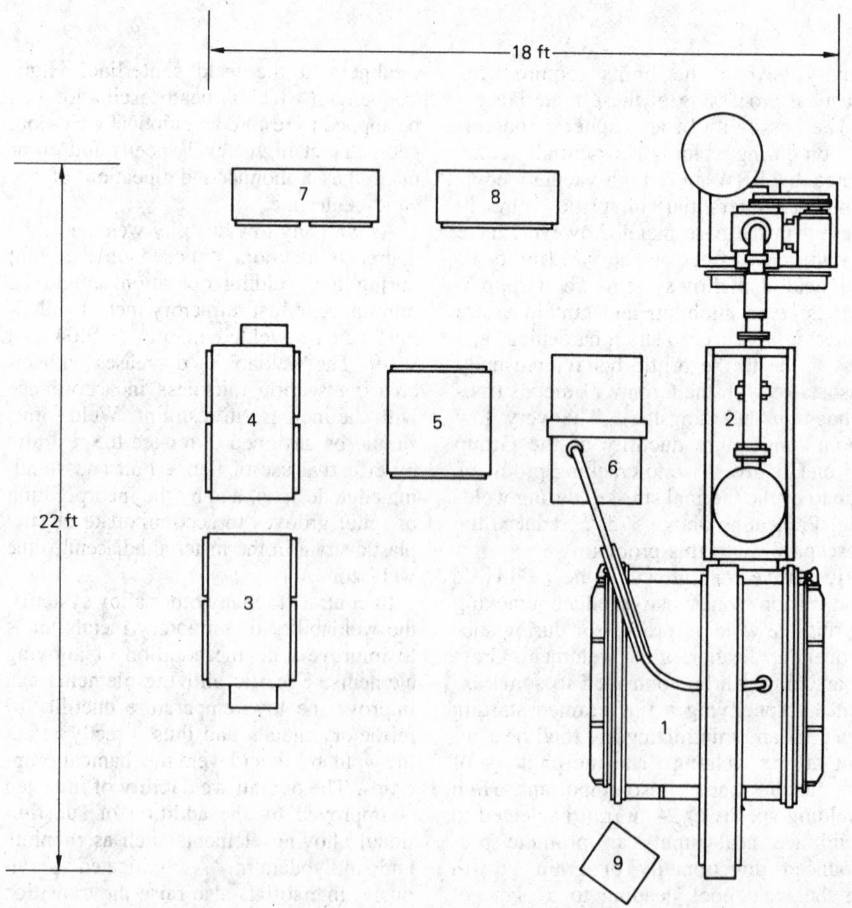

Fig. 42 Block diagram of high-voltage, high-vacuum EBW system. (1) Vacuum chamber and manipulating mechanisms. (2) Vacuum pumping system. (3) Computer control cabinet. (4) Servo cabinet. (5) Automatic voltage regulator. (6) High-voltage power supply. (7) Power distribution cabinet. (8) Pumping control cabinet. (9) Operator console

Fig. 43 Typical EBW-HV unit. Chamber size, 60 by 50 by 71 in. (Courtesy of Leybold-Heraeus Vacuum Systems Inc.)

Fig. 44 Standard low-voltage EBW-HV unit. Chamber size, 138 by 112 by 107 in. (Courtesy of Sciaky Bros., Inc.)

ductility and toughness, especially in the niobium alloys.

Unalloyed niobium is easily electron beam welded, but has relatively low strength. More alloys have been developed, mainly for higher strength, than for the other refractory metals. Alloying usually reduces weldability, but most weld joints in alloys retain about 75% efficiency, and good structural properties can be obtained for service temperatures in the range of about 2000 to 3000 °F, depending on the alloy. Chill tooling is often used, but copper, nickel, and stainless steel should be avoided to prevent contamination of the weld metal.

Tantalum is the most easily weldable of the four refractory metals. Because of its high melting point (5425 °F) and good thermal conductivity, thicknesses of 0.060 in. or more must be rapidly heated to high temperatures for welding. Copper chills are used to avoid distortion and weld sagging and to shorten the time the assembly has to remain in the vacuum chamber, thus limiting grain growth.

Weldability of tantalum alloys containing other refractory elements is somewhat lower than that of unalloyed tantalum; however, successful welds have been made in alloys Ta-10W, T-111 (Ta-8.0W-2.5Hf-0.003C), and T-222 (Ta-9.64W-2.4Hf-0.01C). Because of high vapor pressure, alloys containing vanadium are better welded by the gas tungsten arc process.

Welding of Aluminum Alloys

Single-pass, full-penetration welds with depth-to-width ratios of more than 20 to 1 have been made in 6-in.-thick aluminum alloy plates. The low heat input, narrow HAZ, and short heat cycle result in minimum decrease in mechanical properties, especially in assemblies of some of the heat treatable alloys.

Non-Heat-Treatable Alloys. The mechanical properties of electron beam welds in the non-heat-treatable alloys (1*xxx*, 3*xxx*, and 5*xxx* series) are virtually the same as those obtained by GTAW. If the alloy is in a strain-hardened condition, the weld properties closely approach the annealed properties of the base metal.

Heat treatable alloys of the 2*xxx*, 6*xxx*, and 7*xxxx* series are crack sensitive to varying degrees when welded. Some may also be prone to porosity. The condition of the zone immediately adjacent to the weld is critical in determining weldability.

The 6*xxx* series alloys are only slightly affected by the heat cycles of EBW. Al-

Fig. 45 Twin EBW-MV unit for joining ring gear on flywheel. Courtesy of Leybold-Heraeus Vacuum Systems Inc.

Fig. 46 Custom-designed EBW-MV system for production of automotive parts. Courtesy of Sciaky Bros., Inc.

loys 2219, 7039, and 7005 appear to be the least affected of the alloys of their respective series. The techniques used to prevent cracking in EBW of the heat treatable aluminum alloys are:

- Special care in preweld cleaning to avoid porosity and oxide inclusions in the weld metal
- Prestressing joints in compression by using interference fits when possible
- Use of a ductile filler metal, usually in the form of a thin strip preplaced in the joint, which will yield under shrinkage stresses and fill joint openings
- Selection of beam power, beam spot size, and welding speed to create as narrow welds as practical during short welding heat cycles, thus avoiding excessive grain-boundary melting
- Use of postweld heat treatment to restore the strength and ductility of the weld and HAZ
- Designing or locating joints to be free of externally imposed restraints, reinforcing of joint areas and, when possible, welding in the solution-treated condition, with postweld aging

Some aluminum alloy welded joints (e.g., 1.5-in.-thick welded 7075-T651 plate) exhibit lower mechanical properties than unwelded plate due to overaging in the HAZ. Postweld solution and aging heat treatments improve joint properties. At high travel speeds, porosity can develop in the weld from vaporization of certain elements, such as zinc, in the alloy. Aluminum alloy 7075, which contains 5.6% Zn, is an example of this. Travel speed should be reduced in EBW of high-zinc aluminum alloys so that the vapor has time to escape to the surface before the weld metal solidifies. Zinc-free alloys (the 6*xxx* series alloys, for example) can be welded at higher speeds without the threat of porosity.

Welding of Copper and Copper Alloys

The EBW of copper and its alloys is influenced by the same factors that affect the arc welding of these metals (see the article "Arc Welding of Copper and Copper Alloys" in this Volume). The high thermal conductivity of copper causes less difficulty in electron beam than in arc welding.

Molten metal may be expelled from the weld joint during EBW of nondeoxidized coppers (especially alloy C11000, tough pitch copper), causing spatter and uneven weld surfaces. This can usually be remedied by the use of a cosmetic pass. The vacuum environment eliminates possible hydrogen embrittlement; nevertheless, root voids and porosity still can occur.

The presence of low-melting elements ordinarily makes the welding of free-machining copper alloys impractical, and the volatility of zinc prevents the welding of the brasses and other zinc-containing copper alloys. The remaining zinc-free copper alloys can generally be elec-

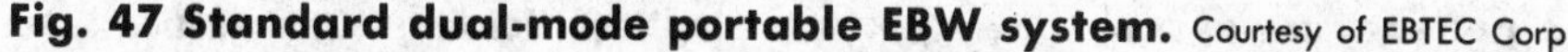

Fig. 47 Standard dual-mode portable EBW system. Courtesy of EBTEC Corp.

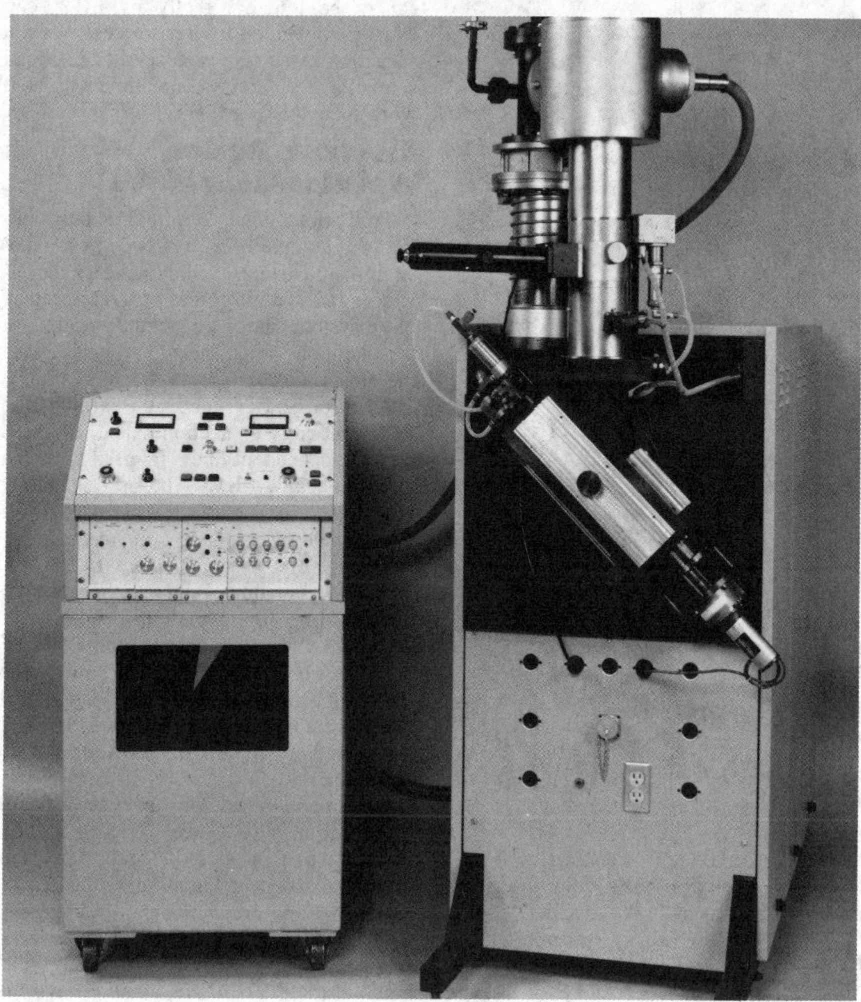

tron beam welded without any unusual problems.

Welding of Magnesium Alloys

Electron beam welding is used to a limited extent, chiefly for repair, on commercial wrought and cast magnesium alloys that contain less than 1% Zn. The relative suitability of alloys for EBW is generally the same as for arc welding, as discussed in the article "Arc Welding of Magnesium Alloys" in this Volume.

Techniques. Special techniques and close control of operating variables are needed to prevent voids and porosity at the root of the weld, because of the high vapor pressure of magnesium, which has the lowest boiling point (2025 °F) of any commonly welded metal. This difficulty is aggravated by the presence of zinc, which has a still lower boiling point (1663 °F). It is ordinarily impractical to electron beam weld magnesium alloys that contain more than 1% Zn.

Circular oscillation of the beam or the use of a slightly defocused beam is helpful in obtaining sound welds. The most satisfactory technique is to use integral or tightly fitted backing of the same alloy and to ensure that welding conditions are held to values that trial welds have shown will either minimize porosity and voids or localize them in an area where they can be tolerated or removed by machining.

Welding of Titanium Alloys

All of the commercial alloys of titanium that can be joined by arc welding can also be joined by EBW. Their relative weldability and response to heat cycling in EBW are generally the same as in arc welding, which is discussed in the article "Arc Welding of Titanium and Titanium Alloys" in this Volume.

Applicability. The vacuum environment of EBW prevents exposure to the atmospheric contaminants that cause embrittlement of titanium alloys, whereas arc welding processes must use elaborate and costly shielding methods to accomplish this. Cost studies show that direct labor costs for EBW of titanium sections more than 1 in. thick are less than for arc welding, provided a suitably large vacuum chamber is available.

Techniques. Filler metal is not ordinarily used, and the work is not preheated. Tack welding, contrary to experience in GTAW, presents no difficulties in EBW. For optimum results, welding is done in a high vacuum, but medium-vacuum welding is satisfactory for many applications.

Alloy Ti-6Al-4V, the alloy most frequently used in assemblies to be welded, can be electron beam welded in either the annealed or the solution-treated-and-aged condition. For weldments that will be used at elevated temperatures, a preferred process sequence is anneal, weld, solution treat, and age. For other service conditions, a process sequence of solution treat, age, and weld gives almost the same strength properties and only slightly lower fracture toughness.

Welding of Beryllium

Beryllium is well suited for EBW. Narrow fusion zone and low heat input help control cracking. Also, the vacuum atmosphere controls toxic beryllium fumes and prevents oxidation. Weldability primarily is affected by the chemical composition of the metal. Beryllium oxide is the major contaminant in powder source material. Lower beryllium oxide content has been found to improve ductility, as well as weldability (Ref 2). Generally, the beryllium oxide content is limited to 1.5 wt% maximum. Lower aluminum content also is reported to improve weldability by decreasing crack susceptibility. The aluminum tends to produce a hot short condition. This hot short condition is reduced if the aluminum is tied up in a compound, such as $AlFeBe_4$, rather than existing as free aluminum in the grain boundaries (Ref 3).

Autogenous Welds. Cracking can be minimized in autogenous welds (no filler metal added) by reducing joint restraint during cooling. This is accomplished by appropriate fixturing, proper joint design, thin members, and good part fit-up. Most welding must be accomplished at sharp focus to minimize the size of the fusion zone. For deep welds, preheat has been found to improve weldability. For example, welds up to 0.050 in. thick have been

Fig. 48 Standard dial index table-type nonvacuum system. Courtesy of Leybold-Heraeus Vacuum Systems Inc.

made without preheat in low-oxide powder source beryllium (Ref 4). The same metal can be welded up to 0.1 in. thick with a preheat of approximately 650 °F. In addition to cracking, high oxide levels can cause porosity and excessive cratering in the weld. Typical parameters for autogenous welds are given in Table 5.

Braze Welds. The greatest success in EBW of beryllium is obtained by using a ductile filler metal in the weld; generally, aluminum is used. Optimum properties are obtained when the fusion-zone microstructure consists of beryllium dendrites surrounded by a ductile aluminum matrix. A fusion zone with less than 30% Al contains large beryllium dendrites which are susceptible to microcracking (Ref 3). Welds can be made with less than 30% Al in the fusion zone, but caution must be used.

Filler metal may be added by several methods. One method is to preplace aluminum shim stock in the joint. Care must be taken to ensure the shim stays in the joint during the weld. This may be accomplished by tack welding the joint prior to the penetration pass. By using a 0.020-in.-thick by 0.100-in.-deep 1100 aluminum shim, welds to 0.120-in. penetration have been made without preheat. A typical weld is shown in Fig. 40.

An alternative method to brazing welds is to use an electron beam cold wire-feed process in a manner similar to stainless steel (Ref 5). In this process, a narrow, deep groove in the beryllium is filled with aluminum wire fed into the weld pool. A typical joint uses 718 aluminum wire, 0.030 in. in diameter, fed into a modified J-groove. The weld joint has a 0.025-in. root radius with a 10° 20′ included angle. Welds to 0.100 in. have been successfully made this way (Fig. 41).

Joining of Dissimilar Metals

The joining of dissimilar metals is an important application of EBW. The weldability of two dissimilar metals or alloys depends primarily on their physical properties, such as melting points, thermal conductivities, and coefficients of thermal expansion. Weldability generally can be determined by examination of the phase diagram of the alloys to be welded. If intermetallic compounds are formed by the metals to be welded, the weld metal will be brittle. However, many combinations that are difficult to weld or unweldable by other processes can be welded by the electron beam process.

Indirect Joining. Combinations that cannot be joined directly by EBW can usually be welded by this process when a transition piece or a preplaced filler-metal shim of a metal that is compatible with both work metals is used.

Electron Beam Welding Machines

This brief review is aimed at introducing the potential user to some of the EBW equipment currently available. Further advances in the control and programming of these systems can be expected because of the likelihood of uninterrupted progress in the electronics and computer industries.

As discussed earlier in this article, the EBW process is accomplished at three pressure-dependent lines, referred to as the three modes of EBW. The original mode is the high or "hard" vacuum mode in which welding is carried out in the pressure range of 10^{-6} to 10^{-3} torr. In the second mode, medium vacuum, the pressure ranges from 10^{-3} to 25 torr. The term "medium vacuum" includes the range of pressure (10^{-3} to 1 torr) referred to as the "soft" or partial vacuum. The third mode is called nonvacuum or atmospheric, with welding carried out at atmospheric pressure.

All three modes employ an electron beam gun/column, a power supply with controls, one or more vacuum pumping systems, and work-handling equipment. Although in nonvacuum welding the workpiece is not placed in an evacuated work chamber, the electron beam gun/column must be in a vacuum environment. The electron gun in all three modes is held at a pressure of 10^{-4} torr or less; otherwise, the high voltage required for the acceleration of the electrons could not be sustained.

Electron beam welding equipment comes in two basic designs: (1) the low-voltage system, which uses accelerating voltages in the 30- to 60-kV range; and (2) the high-voltage system, with accelerating voltages in the 100- to 200-kV range. Beam powers up to 100 kW are available with both high-voltage and low-voltage equipment. Low-voltage equipment is more suitable for high- and medium-vacuum operations, while nonvacuum welding is carried out at higher voltages (130 to 175 kV minimum). All three modes, however, are operational with high-voltage equipment. The nonvacuum, high-voltage systems generally are used for welding materials less than 1 in. thick. With low-voltage systems, the gun may be fixed in position on the chamber or may be mobile inside the chamber. With high-voltage equipment, the gun is

Fig. 49 Electron beam welding computer control system. Courtesy of Sciaky Bros., Inc.

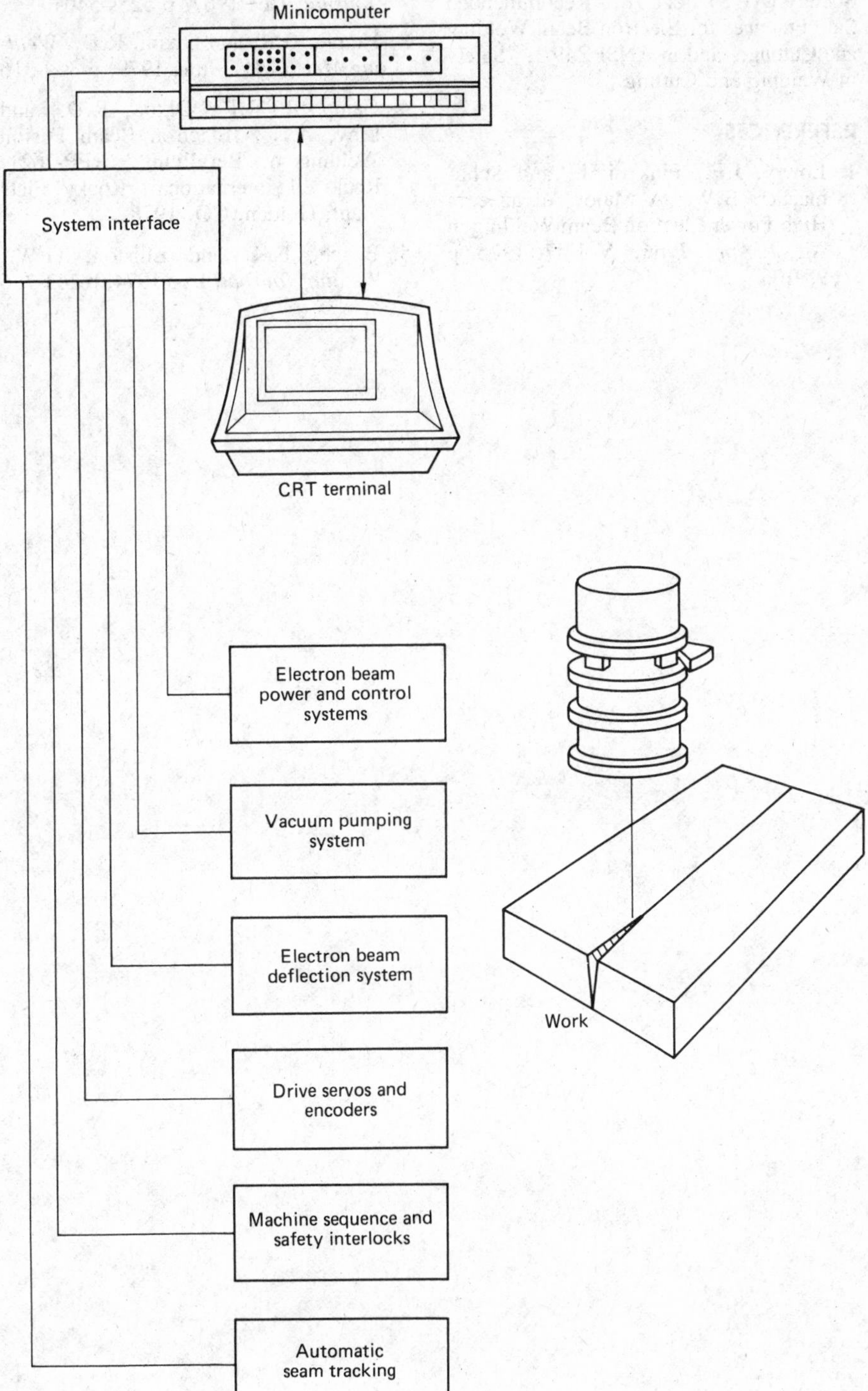

generally fixed in position on the chamber. Figure 42 shows a typical floor plan of a high-voltage, high-vacuum welding facility with a welding chamber size of 68 by 68 by 78 in.

Figure 43 illustrates a standard-type high-vacuum machine with a chamber size of 112 by 60 by 72 in. This type of unit is available with power ratings of 7.5 kW (150 kV, 50 mA) and 25 kW (150 kV, 267 mA), respectively. The gun/column in this unit is fixed, and the work travels under the gun. Work-handling equipment makes processing of workloads of up to 3000 lb at welding speeds of 130 in./min possible. This system can be obtained with various degrees of automation; most utilize CNC operations. Figure 44 shows a high-vacuum unit with a chamber size of 138 by 112 by 106 in. This installation comes with a movable gun. Low-voltage, high-vacuum systems with movable guns are available with power ratings of 7.5 kW (60 kV, 125 mA), 15 kW (60 kV, 250 mA), 30 kW (60 kV, 125 mA), and 42 kW (60 kV, 700 mA).

Medium-vacuum machines are frequently special-purpose units tooled for particular assemblies. Figure 45 shows a dual-system partial-vacuum welding installation for joining ring gear and counterweight onto flywheels. Dual medium-vacuum systems of this type, which are used for high-production applications, can be readily modified to produce alternate parts by simple changes in the work-handling components and weld programming. As with the system shown in Fig. 43, a wide range of automation and computer control is available.

Figure 46 shows a custom partial-vacuum system, typical of the variety used for high-production automotive parts production. Installations of this type currently are available with power ratings of 100 kW (100 kV, 1 A) for welding heavy cross sections from 2 to 8 in. thick.

Some smaller units are also available that feature dual-mode operation; they operate in either the high- or medium-vacuum mode, depending on the application. Figure 47 shows a dual-mode portable machine rated at 6 kW (60 kV, 100 mA) with fully automatic and preset pumping and welding cycles. The small weld chamber can be pumped down to 5×10^{-4} torr in 10 s.

The system shown in Fig. 48 is an example of a nonvacuum unit featuring a standard dial index table. Parts are loaded into a fixture nest in the table and are indexed through a small, radiation-tight enclosure in which parts are welded at 1 atm. This nonvacuum unit has power ratings of 17.5 kW (174 kV, 100 mA) or 35 kW (175 kV, 200 mA) and can be used either as a standard system or as a custom-designed special task system.

Figure 49 shows the schematic of a computer control system that is available in many EBW units. In addition to the controls shown in Fig. 49, control of welding operations can be expanded to include joint locator, five-axis programmable contouring, and several digitizing provisions which can be actuated on the basis of information generated by the seam locator. Many EBW systems can also be adapted for electron beam heat treating applications.

Safety

Protection must be provided by equipment design and arrangement, and by safety precautions in EBW and related operations, against the usual hazards of welding and the special hazards of exposure to (1) the high voltages involved in generating the electron beam, (2) the beam itself, and (3) radiation of x-rays produced by impingement of the beam on the work or other materials. Suitable precautionary measures are described in American Welding Society (AWS) F2.1-78, "Recommended Safe Practices for Electron Beam Welding and Cutting," and in ANSI Z49.1, "Safety in Welding and Cutting."

REFERENCES

1. Lowry, J.F., Fink, J.H., and Schumacher, B.W., A Major Advance in High-Power Electron Beam Welding in Air, *J. Appl. Phys.*, Vol 47, 1976, p 95-106
2. Hauser, D., Mishler, H.W., Monroe, R.E., and Martin, D.C., *Welding Journal*, Dec 1967, p 525s-540s
3. Heiple, C.R. and Dixon, R.D., *Welding Met. Fabr.*, June 1979, p 309-316
4. Campbell, R.P., Dixon, R.D., and Liby, A.L., "Electron Beam Fusion Welding of Beryllium," RFP-2621, Rockwell International, Rocky Flats Plant, Golden, CO, 1978
5. Bench, F.K. and Elliston, G.W., *Welding Journal*, Dec 1974, p 763-766

Laser Beam Welding

By the ASM Committee on Laser Beam Welding*

LASER BEAM WELDING (LBW) is a joining process that produces coalescence of materials with the heat obtained from the application of a concentrated coherent light beam impinging upon the surfaces to be welded. The word laser is an acronym for "light amplification by stimulated emission of radiation." The laser can be considered, for metal joining applications, as a unique source of thermal energy, precisely controllable in intensity and position. For welding, the laser beam must be focused to a small spot size to produce a high-power density. This controlled power density melts the metal and, in the case of deep penetration welds, vaporizes some of it. When solidification occurs, a fusion zone or weld joint results. The laser beam, which consists of a stream of photons, can be focused and directed by optical elements (mirrors or lenses). The laser beam can be transmitted through the air for appreciable distances without serious power attenuation or degradation.

The two kinds of industrial laser welding processes, conduction limited and deep penetration, are normally autogenous; they use only the parent metal with no added filler. In conduction-limited LBW, the metal absorbs the laser beam at the work surface. The subsurface region is heated entirely by thermal conduction. Conduction-limited LBW uses solid-state and moderate power carbon dioxide (CO_2) lasers and is normally performed with low average power ($\lesssim$1kW).

Deep-penetration LBW requires a high-power CO_2 laser. Thermal conduction does not limit penetration; laser beam energy is delivered to the metal through the depth of the weld, not just to the top surface. Deep-penetration LBW is similar to high-power electron beam welding (EBW) in vacuum.

Applications

The automotive, consumer products, aerospace, and electronics industries all use LBW to join a variety of materials. Among the weldable metals are lead, precious metals and alloys, copper and copper alloys, aluminum and aluminum alloys, titanium and titanium alloys, refractory metals, hot and cold rolled carbon and low-alloy steels, high-strength low-alloy (HSLA) steels, stainless steels, and heat-resistant nickel and iron-based alloys. Porosity-free welds can be attained with tensile strengths equal to or exceeding those of the base metal.

In the automotive industry, lasers were first used for surface hardening of parts subjected to wear. There are two major advantages of laser hardening, both of which result from the small size and ease of manipulation characteristic of the laser beam. The first is the ability to localize the hardening at the exact spot where it is needed. This minimizes the total heat input to the material and thus eliminates the distortions that often accompany other hardening techniques. The second advantage is the ability to apply the heat to otherwise inaccessible areas—for example, small bores or complicated shapes not amenable to induction hardening.

In welding applications, lasers have been used to join stamped steel panels that form underbodies. In this computerized process, welding is performed at a rate of 400 to 450 in./min. The laser welds are continuous, which results in high structural integrity and eliminates the need for a sealing operation. The programmability of the LBW system offers the advantage that underbodies for any car model can be welded by calling up the correct program from the computer memory.

Laser spot welding also has become an accepted technique for providing a secure, current-carrying joint on the oxygen sensor used to meet pollution control standards. These devices are screwed into the exhaust manifold and are exposed to continuous high operating temperatures. Because the electrical information signal in this application is very sensitive to small changes in resistivity, the electrical contact must be immune to corrosion. A welded contact provides this quality, and laser welding ensures that a uniform part is achievable at high production rates.

In the electronics industry, LBW is used for sealing electronic devices that are either high-value, low-quantity production items or items that must meet stringent reliability and other special end use requirements. Examples of the latter type include hermetically sealed devices for commercial and military aircraft applications. These devices usually must maintain highly reliable operating performance under extremely severe environments.

The first high-production application of laser sealing of an electronic device was the use of a yttrium aluminum garnet (YAG) laser pulsing at rates up to 10 times per second that produced seam welds in relays fabricated to military specifications at rates up to 6 in./min. Pulse delivery rates in excess of 100 times per second that are capable of welding at more than 60 in./min are currently available.

Laser welding of relay containers has proven to be an effective way of hermetically sealing each package. A fusion seam

*Edward A. Metzbower, Supervisory Metallurgist, Naval Research Laboratory; Charles G. Albright, Assistant Professor, Department of Welding Engineering, Ohio State University; Conrad M. Banas, Chief, Industrial Laser Processing, United Technologies Research Center; Daniel S. Gnanamuthu, Member of Technical Staff, Rockwell International; Matthias J. Kotowski, Safety Engineer, Lawrence Livermore Laboratory; Kim W. Mahin, Welding Metallurgist, Lawrence Livermore Laboratory; John Nurminen, Manager, Metals Joining Research, Westinghouse Research & Development Center; Stanley L. Ream, Consultant, Laser Manufacturing Technologies, Inc.; Lyle B. Spiegel, Manager, Applications Development, Avco Everett Metalworking Lasers

Fig. 1(a) Laser welded relay case

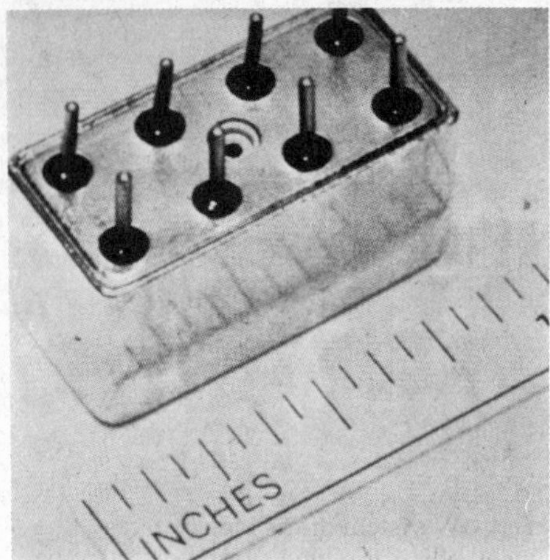

Fig. 1(b) Typical laser setup for hermetic sealing relays

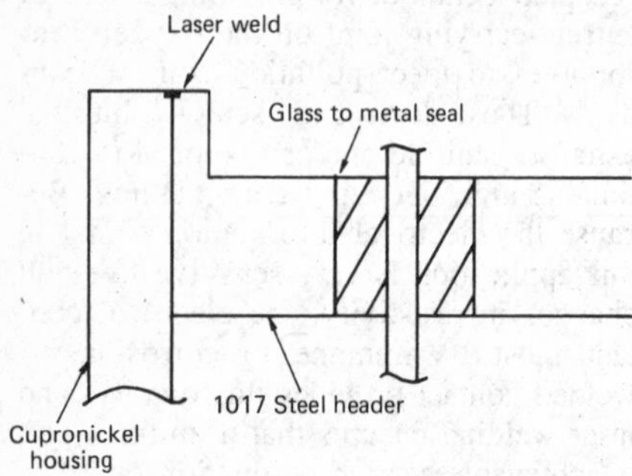

weld is produced with the high-repetition-rate pulsed laser laying down individual spots with 80 to 90% overlap. Laser welding has proven extremely useful in such applications because of its ability to produce welds near heat-sensitive, glass-to-metal seals. Figure 1(a) shows a typical laser welded relay case. The case shell is made of a cupronickel alloy while the header is of Kovar or low-carbon steel. Figure 1(b) illustrates how glass-to-metal feed-throughs are located as close as 0.050 in. from the edge of the fusion weld.

The production of heart pacemaker devices is another application requiring high-quality welded construction. Previously, nonwelded construction resulted in expensive recalls. Today, laser welding has become the most widely accepted technique for producing hermetic welds in titanium and stainless steel pacemaker cases.

The principal power source for pacemakers is the lithium battery. Because of the highly reactive nature of the cell's contents, these units must be hermetically sealed. The very low heat input of lasers facilitates their use in such applications, where welds are made near glass-to-metal seals. More recently, the lithium cells have entered the consumer and industrial marketplace as long-lived power sources for digital watches, calculators, and backup power for computer memories. The characteristic small size, coupled with the requirement for a fusion welded seal in close proximity to the reactive contents of the cell, again make the laser ideal for the job. In addition, the ability of the laser to produce a unique welded product makes the use of LBW suitable for a variety of sheet metal welding tasks in the appliance industry.

Future applications of LBW may include:

- Tube-to-tube welding (up to 0.8-in. wall thickness, with 20-kW beam) using a fixed head, rotating tube configuration. Long lengths of the tube of varying diameter could be fabricated and then fashioned so that only tube-to-tube plate welds would be necessary to complete internal boiler fabrication.
- Platen welding, where long runs of downhand welding in typical wall thicknesses of 0.32 in. are required. At present, autogenous gas tungsten arc welding (GTAW) with a pure argon shield involves high heat losses to the closely butted tubes and a weld-pool penetration of only 75%. Laser beam welding would be faster by an order of magnitude and give full penetration.
- Deep-penetration welding. At present, multipass techniques using GTAW, gas metal arc welding (GMAW), and submerged arc welding (SAW) are used for workpieces thicker than 0.25 in. Narrow-gap GTAW reduces the number of passes required, but still provides a slow welding speed. Laser beam welding has the potential for welding workpieces up to 1.25 in. thick in a single pass using 20-kW beam power. Research is proceeding on increasing the penetration limits of laser welding at a given power, but information is invariably proprietary.
- Difficult or dangerous access welding. It may be possible to mechanically/optically transmit beam power over distances of 20 to 100 ft for in-reactor repair welding and similar hostile environments.

Advantages of LBW include:

- LBW is a noncontact process, which eliminates mechanical distortion of workpieces, particularly when welding thin sheet.
- Because contact with the workpiece is not necessary, joints in restricted areas can be welded, provided that a line of sight to the weld is available.
- Absence of contact makes LBW ideal for use in high-speed, automated welding systems. Parts to be spot welded can move at 60 to 120 in./min and seam welding of thin sheet can be accomplished at speeds up to 200 to 250 ft/min.
- The specific energy input to the workpiece is very low. Distortion from the laser beam process, which produces narrow weld widths and heat-affected zones (HAZ), is minimal.

Fig. 2 Cross-sectional views of 0.5-in. HY-130 weldments etched in 10% ammonium persulphate. (a) Shielded metal arc weld. (b) Electron beam weld. (c) Gas metal arc weld. (d) Laser beam weld

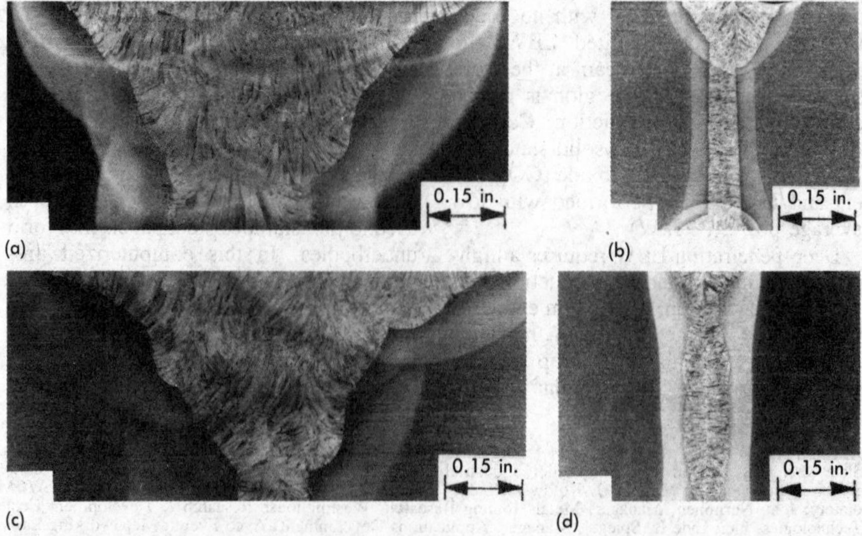

Cross-sectional views of weld zones in HY-130 weldments are compared in Fig. 2.

- Because of the slightly lower energy density of the focused laser beam when compared to an electron beam, the tendencies for spiking, underbead spatter, incomplete fusion, and root porosity are reduced.
- The high-power density of the laser beam facilitates welds between metallurgically compatible dissimilar metals with widely varying physical properties, metals of high electrical resistivity, and parts differing greatly in mass and size.
- The process does not require a vacuum, because the laser beam is easily transmitted through the air.
- Filler metal is not required for most joint preparations.
- Because of the well-defined focused beam, spot weld sizes as small as a few thousandths of an inch can be achieved.
- The process utilizes a simple weld joint geometry.
- The beam produces localized heating and melting.
- Preferential absorption of the beam by nonmetallic inclusions in steels leads to their vaporization and removal from the weld zone.
- Unlike EBW, x-rays are not generated in the laser beam/workpiece interaction.
- The laser beam may be readily focused, aligned, and redirected by optical elements.

Process Fundamentals

Laser beam welding requires a precisely focused, coherent laser beam. When focused to the appropriate spot size, the beam melts the metal, rapidly producing a narrow weld with high joint efficiency and minimal distortion. Accurate control of the power of the laser is necessary during welding to keep the melting zone localized. Usually, control is accomplished by varying pulse width, pulse repetition rate, spot size, and/or power level.

A laser is a device that produces an intense, concentrated, and highly parallel beam of light. Every laser must consist of three fundamental distinct parts: (1) a laser material or medium, (2) a method of excitation, and (3) a resonant cavity. The following description of a laser is paraphrased from *Fundamentals of Optics* (Ref 1).

Consider a gas, enclosed in a vessel, containing free atoms with a number of energy levels, at least one of which is metastable. By shining white light into this gas, many atoms can be raised, through resonance, from the ground state to excited states. As the electrons drop back, many of the atoms are trapped in the metastable state. If the pumping light is intense enough, a population inversion, i.e., more atoms in the metastable state than the ground state, may be obtained.

When an electron in one of these metastable states spontaneously jumps to the ground state, a photon of energy is emitted. As this photon passes by another nearby atom in the same metastable state, it can, by the principle of resonance, immediately stimulate that atom to radiate a photon of the exact same frequency and return to its ground state. This stimulated photon has exactly the same frequency, direction, and polarization as the primary photon (spatial coherence) and exactly the same phase and speed (temporal coherence).

Both of these photons may now be considered primary waves, and when passing close to other atoms in their metastable states, they stimulate them to emission in the same direction and with the same phase. Transitions from the ground state to the excited states can also be stimulated, however, thereby absorbing the primary wave. Therefore, an excess of stimulated emission requires a population inversion, i.e., more atoms in the metastable than the ground state. If the conditions in the gas are right, a chain reaction can be developed, resulting in high-intensity coherent radiation.

To produce a laser, the stimulated emission must be collimated by designing a proper resonant cavity in which the waves can be used over and over again. Suppose two highly reflecting mirrors are placed at the ends of the cylindrical vessel, and the electrons in these atoms are excited to produce a population inversion. If one or more atoms in the metastable state spontaneously radiate, those photons moving at an appreciable angle to the walls of the vessel escape and are lost. Photons emitted parallel to the axis reflect back and forth from end to end. The chance of these photons stimulating emission now depends on a high reflectance at the end mirrors and a high population density of metastable atoms within the vessel or cavity. If both of these conditions are satisfied, the buildup of photons surging back and forth through the cavity can be self-sustaining and the system oscillates, or lases, spontaneously.

Thus, the laser consists of: (1) a material or medium that is capable of maintaining a population inversion of metastable atoms, (2) a method of producing and maintaining the population inversion, and (3) a collimation system that reflects the waves back and forth in the cavity. Every laser system can be broken down into these fundamental components.

A ruby laser, for example, consists primarily of a transparent crystal of corundum (Al_2O_3) doped with approximately 0.05% trivalent chromium atoms in the form of Cr_2O_3, the latter providing its pink color. The aluminum and oxygen atoms are inert; the chromium ions are the active ingredients that provide the population inversion.

When white light (the method of excitation) enters the ruby crystal, strong absorption by the chromium atoms in the blue-green part of the spectrum occurs. Therefore, light from an intense source surrounding the crystal raises many electrons to a wide band of energy levels. These electrons quickly drop back, many returning to the ground state. Some of the electrons drop down to intermediate levels, however, not by the emission of photons, but by the conversion of the vibrational energy of the atoms forming the crystal lattice. Once in the intermediate levels, the electrons remain there for several milliseconds (about 10 000 times longer than in most excited states) and randomly jump back to the ground level, emitting visible

Fig. 3 Ruby laser using a helical flash lamp for optical pumping

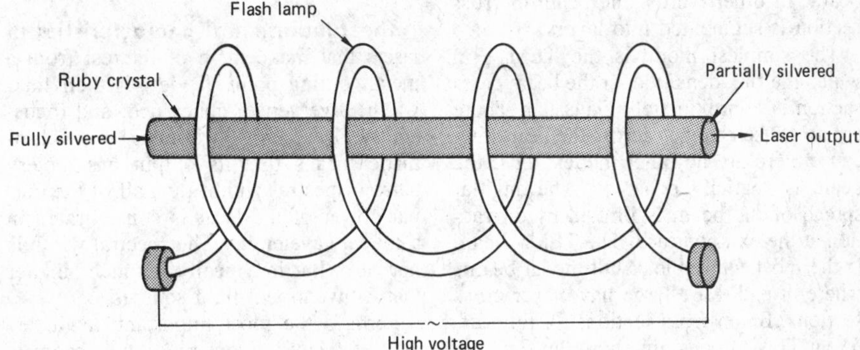

red light. This fluorescent radiation enhances the pink or red color of the ruby laser and gives it its brilliance.

To greatly increase the electron population in the metastable levels, very intense light sources (Fig. 3), as well as light-gathering systems, have been developed. A high-intensity, helical flash lamp can surround the ruby crystal, providing adequate pumping light to produce a population inversion. Another effective arrangement is to place a strong pulsed light source at one focus of a cylindrical reflector of elliptical cross section and the ruby rod at the other focus.

By pumping from a strong, surrounding light source, a large part of the stored energy is converted into a coherent beam. Coherent waves traveling in opposite directions in the ruby crystal set up standing waves comparable to a resonating cavity in microwaves. With one end only partially reflecting, part of the internal light is transmitted as an emerging beam. Thus, in the ruby laser system, the lasing material is the Cr_2O_3-doped corundum crystal (ruby); the method of excitation is the intense white light; and the cavity is the ruby crystal, with one end completely reflecting and the other end partially transmitting.

A laser cavity can be operated in a variety of oscillation modes similar to those of a waveguide. As waves travel back and forth between the mirrors, standing waves can be set up. In a spectrum-rich source, single wavelengths can be selected for oscillation by inserting a silvered prism for one of the mirrors. Owing to the dispersion of the prism, the optical path can be "tuned" to be collinear for the desired wavelength only. In addition to the longitudinal modes of oscillation, transverse modes of oscillation can be sustained simultaneously. Because the fields within a gas are nearly normal to the cavity axis, these are known as transverse electric and magnetic (TEM_{mn}) modes. The subscripts m and n specify the integral number of transverse nodal lines across the emerging beam. In other words, the beam in cross section is segmented into layers.

The simplest mode is the TEM_{00}, in which the flux density over the beam cross section is approximately Gaussian. There are no phase changes across the beam, because there are no other modes; thus, the beam is spatially coherent. The angular spread of the beam is limited by diffraction at the exit aperture. The TEM_{00} beam is the most focusable. Multimodal beams where m and n are large have layer cross sections. Beam cross sections for four different TEM modes are shown in Fig. 4.

Fig. 4 Beam cross sections for four different TEM modes

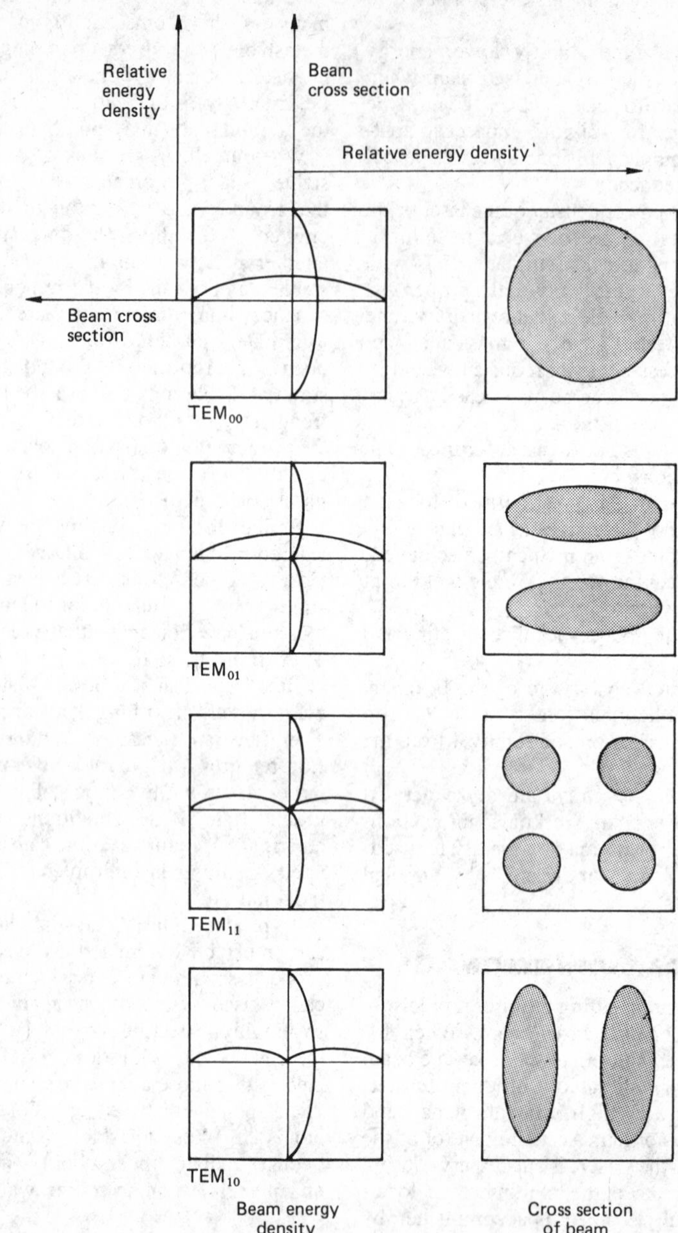

The fundamental characteristics of lasers that make them of interest from a metalworking point of view is their linewidth, divergence, coherence, and focusability. Describing a laser as monochromatic means that its output has a very narrow spectral width; i.e., all of the output power of the laser is concentrated in a single wavelength. The spectral width is not zero, but it typically is much smaller than conventional light sources.

Perhaps the most important characteristic of a laser is that the output beam is highly directional and collimated. Because the laser beam is collimated (its angular spread is small), it can be collected by a lens and focused to a small area. In conventional light sources, the radiation spreads into a solid angle of 4π steradians, and it is virtually impossible to collect the energy in the beam.

Because a laser beam is coherent as the waves travel back and forth between the mirrors, the amplitudes of these waves superimpose to a first approximation. This is a result of the stimulated photon having the same speed and phase as the primary photon. This is temporal coherence. The stimulated photon has the same frequency, direction, and polarization as the primary photon. This is spatial coherence. The results of both spatial and temporal coherence are constructive interference and a very intense laser beam.

The power density of the laser beam at the workpiece is a function of the power of the laser beam and the minimum spot size to which it can be focused. There is a minimum spot size, called the diffraction limit, to which any beam can be focused. The TEM_{00} laser beam is essentially a Gaussian beam; i.e., it has the same intensity distribution in both near and far fields and the same phase across the entire wavefront. A Gaussian beam can always be focused to a minimum spot size of the order of its wavelength. Uniphase Gaussian beams can, in principle, be focused to smaller spots than incoherent beams. A higher order, multimode (TEM_{mn}) beam will always have a larger spot size than a Gaussian (TEM_{00}) beam. As a result of the collimation and coherence of the laser beam, the laser is a source of energy that can be concentrated by a lens to achieve extremely high-power density at the workpiece surface.

Lasers Suitable for Welding

Industrial lasers used for material processing and welding can be divided into two categories: solid state and gas, depending on the lasing medium. Corresponding to this division, interestingly, is the fact that solid-state lasers have a wavelength of 1.06 μm, whereas gas lasers have a wavelength of 10.6 μm. The active element in solid-state lasers is the neodymium (Nd) ion; in gas lasers, the active media is the CO_2 molecule.

Solid-state lasers are characterized by an active medium of an impurity in some host material. For material processing, the dopant is the neodymium ion (Nd^{+++}) in either glass or YAG. The output wavelength is dictated by the Nd^{+++} and is 1.06 μm. The laser material is in the form of a cylindrical rod with ends polished flat and parallel. The excitation is by means of intense optical lamps (krypton or xenon). A simplified schematic arrangement of the rod, the lamps, and the mirrors is shown in Fig. 5.

The selection of a host for the neodymium ion depends on several factors, including the ability to prepare the material in reasonably large samples of good optical quality, the hardness of the material, its ability to be polished easily, the ease and repeatability of production, the thermal conductivity, the fluorescent lifetime and efficiency, and the optical absorption bands. All of these factors affect the ability of the particular system to emit reasonable amounts of energy in a single pulse. The successful materials are those from which large amounts of energy can be extracted.

Fig. 5 Arrangement of a solid-state laser

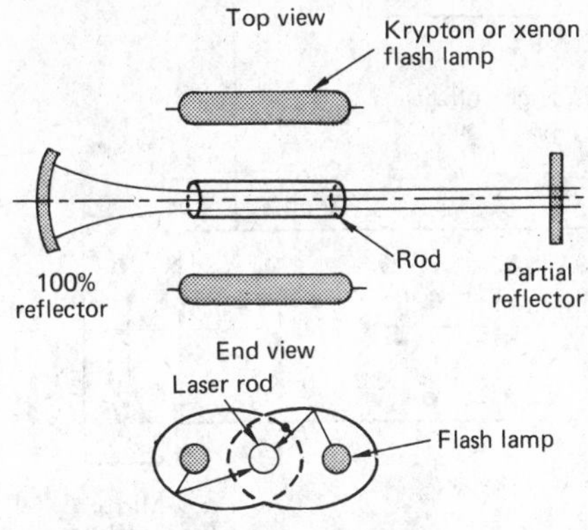

Yttrium aluminum garnet has the chemical composition $Y_3Al_5O_{12}$. It is a hard isotropic crystal that can be grown to yield material of high optical quality and can be polished to a good optical finish. Yttrium aluminum garnet offers low values of threshold and high values of gain. Furthermore, high hardness allows for good optical polishing characteristics, and the thermal conductivity of YAG is over ten times greater than that of glass. The absorption spectrum of Nd in YAG contains many narrow lines.

The output characteristics of Nd:YAG lasers depend on the excitation method and may be continuous or repetitively pulsed. In continuous operation, the laser is excited with either xenon lamps for power levels of the order of 10 W or krypton lamps for power levels of the order of 100 W or more. For repetitively pulsed lasers, the output characteristics depend on the lamp configuration. The most common configuration is shown in Fig. 5 (straight lamps), whereas a useful variation is shown in Fig. 3 (helical lamps). Table 1 gives the characteristics of Nd:YAG lasers and offers some idea of the capability for tradeoffs between the average power, pulse energy, pulse duration, and pulse repetition rates for such lasers.

The relatively narrow line width of 1.06-μm radiation from Nd:YAG lasers facilitates continuous wave operation at room temperature. The result is that the continuous wave Nd:YAG laser is second only to the continuous wave CO_2 laser in terms of continuous wave power generation. However, it does have a considerably lower overall efficiency compared to CO_2 lasers (typically <2%).

Table 1 Output of a Nd: YAG laser

Continuous wave operation	
Average power	<1000 W (multimode) <20 W (TEM_{00})
Divergence	1-20 mrad
Beam diameter	0.04-0.4 in.
Pulse length of 0.1 to 20 ms	
Output energy	<500 J/pulse (multimode) 5 J/pulse (TEM_{00})
Repetition rate	200 H_z
Divergence	10 mrad (multimode) 3 mrad (TEM_{00})
Beam diameter	0.2-0.4 in.
Pulse length of 0.1 to 1μs (repetitive switch)	
Output energy	1 mJ/pulse
Repetition rate	50-100 kHz
Average power	10-100 W
Peak power	10-50 kW

Glass has a number of desirable characteristics as a laser host material. Large pieces of high optical quality can be fabricated into a variety of sizes and shapes, ranging from fibers with diameters of a few microns to rods up to $6^1/_2$ ft long, with diameters of approximately 4 in. The thermal conductivity of glass is lower than that of most crystalline hosts. Therefore, cooling is a problem which limits the maximum repetition rate at a given pulse energy; i.e., average output power is limited. The emission lines of ions in glass are broader than those in crystalline materials. This raises the threshold of the glass for laser action, because a higher population inversion is required to achieve the same gain. The output characteristics of Nd-glass lasers suitable for LBW are given in Table 2.

Gas Lasers. The most efficient laser currently available for material processing applications is the CO_2 laser, which can be utilized in both the high-power continuous wave and pulsed operating modes. Carbon dioxide lasers use an electric discharge as the source for exciting the lasing medium, which is the CO_2 gas molecule.

Table 2 Output of a Nd-glass laser for a pulse length of 1 to 10 ms

Output energy	20 J/pulse (multimode)
Repetition rate	10 Hz
Divergence	5-10 mrad
Beam diameter	0.2-0.4 in.

Fig. 6 Basic CO_2 laser configuration

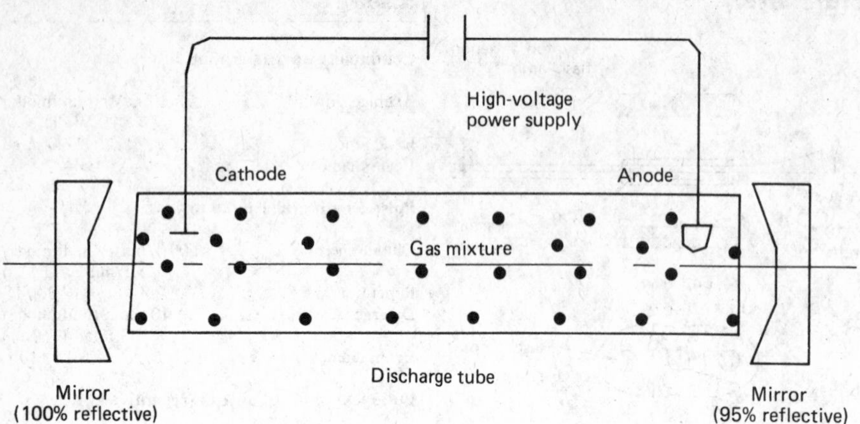

The gas mixture for the laser is a combination of helium, nitrogen, and carbon dioxide. Carbon dioxide lasers can be classified according to their gas flow system.

The simplest CO_2 laser has an axial-flow system; the gas flow is in the same direction as the laser beam and the electric field (Fig. 6). The axial flow of gas is maintained through the tube to replenish molecules depleted by the effects of the multikilovolt discharge of electricity used for excitation. A mirror is located at each end of the discharge tube to complete the resonator cavity. Typically, one mirror is totally reflective and the other is partially transmissive and partially reflective. An axial-flow laser is capable of generating a laser beam with a continuous-power rating in excess of 50 W for every metre of resonator length. A folded tube configuration is used to achieve power levels of 50 to 1000 W in a small volume.

The transverse excited atmospheric CO_2 laser is capable of producing pulsed output laser beams of very high peak power. The gaseous lasing medium is maintained at atmospheric pressure and is excited by an electric discharge from electrodes placed longitudinally along the optical resonator (Fig. 7). Because of the proximity of the electrodes, a relatively low potential is required to maintain high field strength. Very short discharge times facilitate a uniform electrical discharge in the gas at a pressure of 1 atm or more. Transverse excited atmospheric lasers can generate 10 MW or more of power in a single pulse less than 1 μs long. These lasers usually operate at the rate of a few pulses per second.

Fig. 7 Transverse excited atmospheric laser

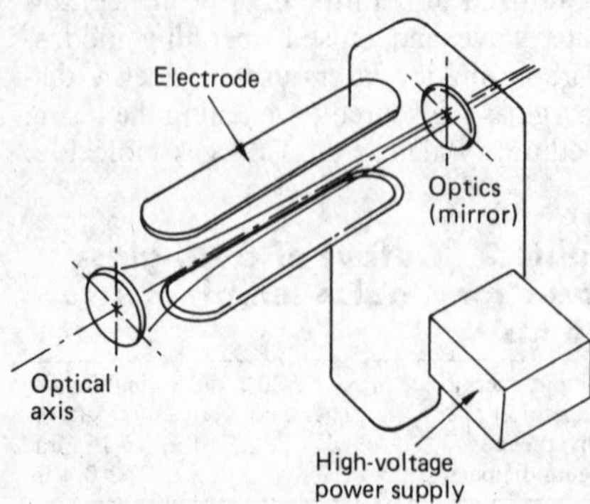

The gas transport laser operates by continuously circulating the gas across the resonator cavity by means of a high-speed blower, while maintaining an electric field perpendicular to both the gas and the laser beam. Because the volume of the resonator is large relative to its length, large mirrors can be placed at each end to reflect the beam through the discharge region several times before it escapes through the output coupler. A gas transport laser is shown in Fig. 8. The ability to achieve a long effective optical path in a short actual distance allows the gas transport laser to be a compact structure that generates high output power. Continuous wave lasers capable of output power between 1 and 25 kW are available.

Fig. 8 Gas transport laser

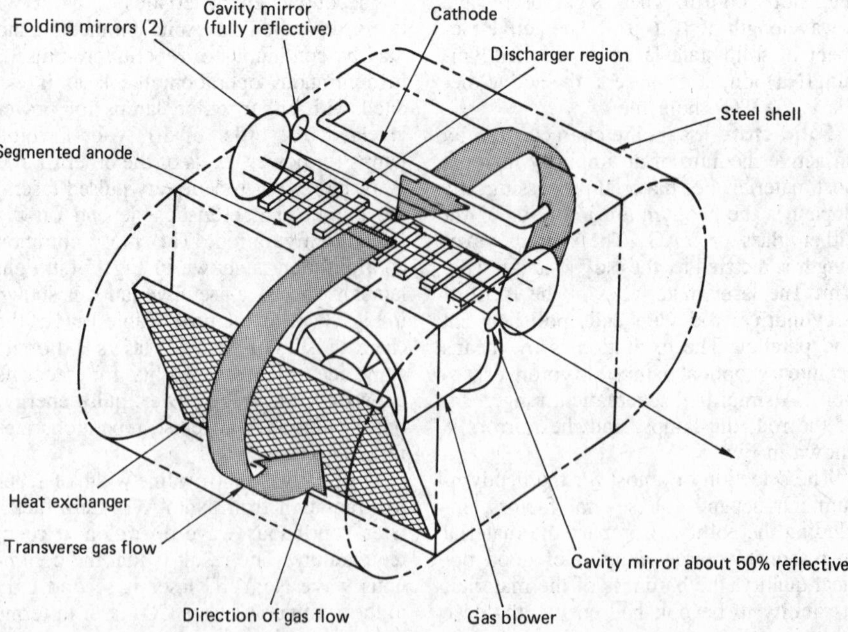

The output characteristics of commercially available continuous wave CO_2 lasers are listed in Table 3. Characteristics of moderate power, high power, and pulsed industrial lasers are compared in Table 4. Additional information on laser optics design, beam transport, and workpiece/beam handling devices is given later in this article.

Laser Selection

Considering the variety of lasers that are available for welding purposes, the selection of a given laser type and/or power level can be expected to encompass many selection criteria. Certainly, the most important consideration in laser selection for a welding application is the required welding penetration.

Penetration. The maximum penetration that can be attained with the laser is a function of beam power, power density,

Table 3 Output of commercial continuous wave CO_2 lasers

Gas flow system	Output power	Beam divergence, mrad	Beam diameter, in.	Available features
Conventional axial flow	10 W	1-5	0.2	Sealed operation, high stability, tunable
	10-100 W	1-3	0.2-0.4	Gas recycling, pulsed operation, tunable
	100-1000 W	1-2	0.2-0.4	Gas recyling, pulsed operation, tunable
Fast axial flow	2 kW	1.2	0.5	Pulsing, gas recycling
Gas transport	6-15 kW	~1	2.7	...

and focusing characteristics. Weld penetration has been shown to increase to a maximum of approximately 0.80 in. at the 20 kW level, as shown in Fig. 9. The shaded area of the curve represents data for several metals, including stainless steel, aluminum, and titanium. Maximum penetration for a given laser power can be altered by changing the relative position of the focal spot with respect to the workpiece. Maximum penetration occurs when the beam is focused slightly below the surface; penetration decreases when the beam is focused on the surface or deeper in the workpiece. The fusion zone appears to have somewhat of an hourglass shape (Fig. 2). The location of the waist of the hourglass is controlled by the position of the focal point relative to the workpiece.

Penetration is not only a function of power, but is also very dependent on welding speed. Figure 10 shows the variation in penetration in type 304 stainless steel as a function of power for several welding speeds.

Many of the early investigations into LBW were directed towards proving that the laser had the capability of penetrating the material at a particular power and welding speed. The soundness of the laser welds was not determined until much later, when nondestructive testing methods were developed and mechanical properties and microstructures of laser beam welds were determined. Some materials are readily weldable using the laser, whereas others, for a variety of different reasons, are less readily weldable.

Although considerable empirical and theoretical information regarding the penetration capability of specific laser beam types does exist, the variations that occur between similar lasers of different manufacturers are of greater practical significance in laser selection. For instance, an axial-flow, 1-kW CO_2 laser may provide greater penetrating capabilities than a transverse-flow, 1-kW CO_2 laser. Such differences may be simply the result of small variations in output beam divergence, or they may be the result of less obvious factors, such as optics vibrations. In any event, the most effective characterization of laser welding penetration capability is likely to be that provided by individual laser vendors, especially as the number and type of lasers capable of performing welding continue to increase.

Productivity. After deciding upon a penetration requirement for the laser welding application under consideration, a welding speed that relates directly to required productivity must be selected. For instance, a 500-W pulsed, axial-flow laser and a 20-kW electron beam, ionized, transverse-flow laser are both capable of welding 0.04-in.-thick stainless steel, but the welding speed obviously varies greatly. Again, as in the case of maximum expected penetration, laser vendors are likely to be the most accurate source of information on the relationship of speed to penetration for their individual lasers.

Table 4 Industrial lasers for welding

Type	Designation	Output, kW	Maximum pulse rate, pps	Characteristics
Moderate power	CO_2	To 1.5	1000	Continuous or pulsed from 0.2 ms up
	YAG (solid-state)	To 2	...	Continuous
High power	CO_2	1.5-20	...	Continuous
Pulsed	YAG (solid-state)	100(a)	200	Pulse 0.1-8 ms, 200 W max, average power
	Nd-glass (solid-state)	100(a)	4 (per min)	Pulse 0.6-8 ms
	Ruby (solid-state)	100(a)	1	Pulse 1-5 ms, 12 W max, average power

(a) Output given in J/pulse, maximum

Fig. 9 Effect of power on maximum weld penetration

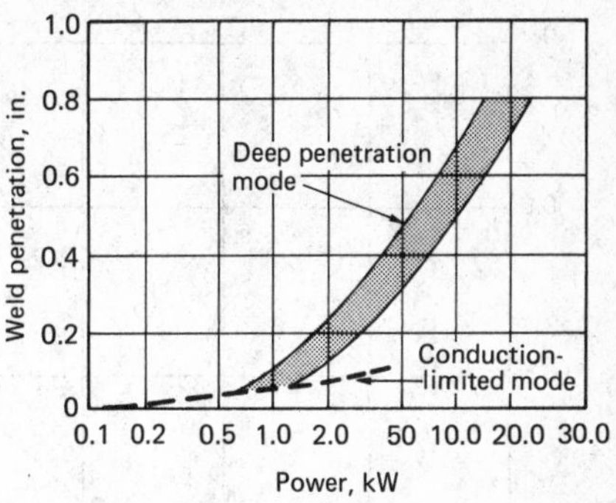

Weld quality is the next major factor in laser selection. Weld quality considerations, such as weld-bead width, aspect ratio, HAZ influence, and distortion, must be addressed. Each of these factors may be affected by the beam quality and power level of the laser system. High-powered, high-quality beams may be attractive from the standpoint of producing minimal distortion; however, such beams may be unattractive from the standpoint of weld joint to focal spot location tolerances.

Issues of weld quality may affect not only the laser welding step in a manufacturing process, but they may also dramatically affect costs encountered in previous and/or subsequent steps in the manufacturing process. A reduction in laser welding distortion resulting from the application of higher powered lasers may result in savings in subsequent machining, straightening, or grinding of the as-welded component. Similarly, the weld joint preparation machining costs may be substantially affected by the selection of high-aspect-ratio versus low-aspect-ratio laser weld nuggets.

Every manufacturing process has tolerances on its many parts. Weld joint preparation, surface finish, joint positioning, travel speed, gas shielding, laser power and quality, and weld joint cleanliness may all vary in the manufacturing setting. Thus, it is important that a laser welding system be designed with sufficient process latitude to accommodate small variations. Such process latitude is usually achieved by providing a laser energy input that is slightly higher than the minimum required for successful welding under ideal conditions.

Expandability relative to possible increased productivity, future needs, or al-

Fig. 10 Effect of beam power on penetration in type 304 stainless steel at various welding speeds

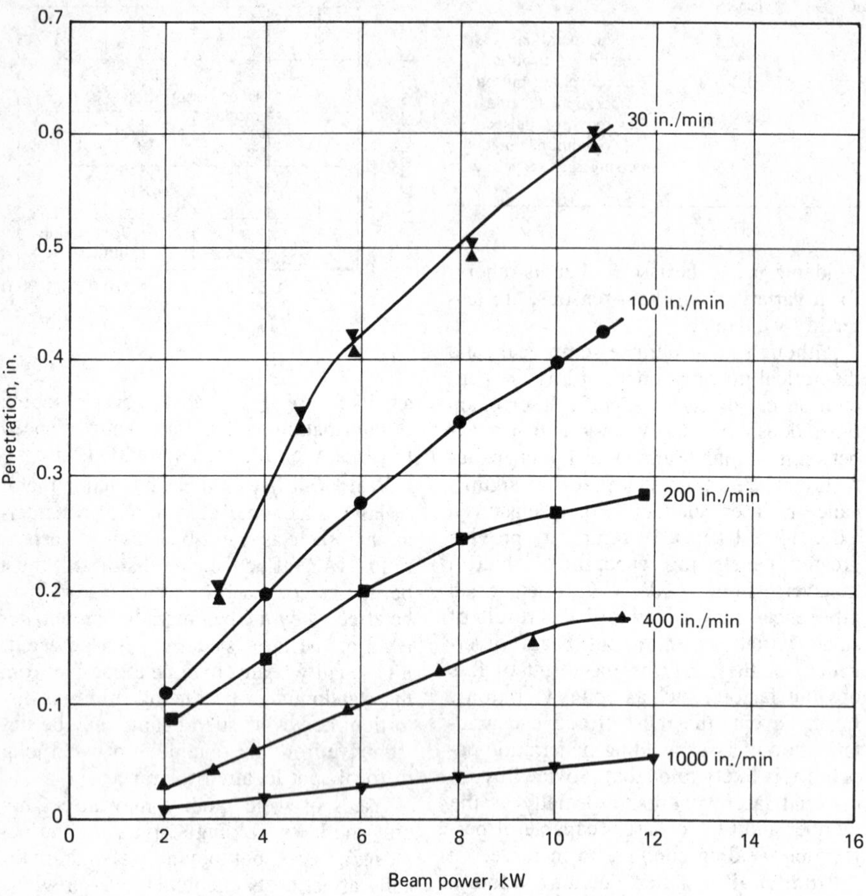

ternative laser materials processing is another consideration in selection of a laser system. In a high-volume laser welding situation, such as that encountered in the automotive industry, this question of expandability and future needs may not exist; however, for a variety of other applications, consideration of optional laser processing capabilities is important.

One obvious example of the need for an expandable or flexible laser welding system is a laser job shop. Such a shop, whether it be a contract job shop or a captive one, initially may be established based on a single application need; future applications may not be identifiable in a detailed sense. If this is the case, it may be advisable to choose a laser power level in excess of that required for the specific application that initially justified the establishment of the laser facility, in the event of future expansion. Considering all of the costs associated with establishing a general-purpose or limited-purpose laser facility, an increase in the size of the selected laser may not impose a substantial increase in the overall facility cost.

Cost. Not only is the initial cost of capital equipment for a laser system a serious concern, but so too are the questions of operating costs and system reliability. Operating costs are considered later in this article.

In a laser welding production application, the capital equipment costs are often significantly lower on an hourly basis than the cost of operation consumables and labor. Thus, the uptime of the laser becomes the primary determining factor when systems of several types and/or manufacturers are considered. Not surprisingly, final laser selection is based largely on the demonstrated performance of a particular laser system in a production setting.

Among all of the relatively intangible factors affecting laser selection, the consideration of system uptime is of particular importance. Even a small percentage of lost production time resulting from laser downtime may easily cost more than the savings of capital equipment dollars that might have been achieved through selection of a lower cost laser system. Of course, an important element of uptime is the availability of spare parts and/or service for the laser system. It is not unrealistic to expect that, in some cases, the availability of service may outweigh other considerations in laser selection.

Laser Material Interactions

When the laser beam interacts with a material, part of the energy of the laser beam is absorbed by the material and the rest of it is reflected. This interaction and partitioning of energy is dependent on characteristics of both the beam and the material. The properties of the laser beam that affect the interaction are: intensity, wavelength, pulse length, divergence, and energy density at the material. The properties of the material that affect this interaction are: reflectivity, absorption coefficient, density, thermal diffusivity, thermal conductivity, heat capacity, and heat content.

The emissivity of a material (E) is the fraction of incident energy absorbed during laser irradiation at normal incidence, when R_o is the reflectivity of the material; therefore:

$$E = 1 - R_o$$

Values of E at room temperature for various metals irradiated at room temperature by Nd:YAG (1.06 μm) lasers are very low and indicate that only a small amount of radiation is absorbed by the metal. However, during the period of time that a pulse interacts with the metal, the reflectivity of the metal changes dramatically, as shown in Fig. 11. This figure shows data on the reflectivity of a 304 stainless steel impinged upon by 200-ns duration pulse from a transverse excited atmospheric (TEA) CO_2 laser that delivered a power density of 1.5×10^9 W/cm^2. The reflectivity dropped rapidly in a few hundred nanoseconds. Thus, the emissivity rose rapidly at the same time. After approximately 1000 ns, the emissivity is about equal to the reflectivity.

The effect of changes in the intensity or in the energy density at the material surface is more difficult to assess quantitatively. However, from LBW experiments in which the laser power and the welding speed are varied, it can be inferred that, above some critical intensity or energy density, increasing either intensity or density effectively compresses the time scale of Fig. 11. Compressing the time scale means that more energy is absorbed by the

Fig. 11 Specular reflectivity at 10 μm as a function of time for a stainless steel surface. Using a transverse excited atmospheric CO_2 laser delivering 1.5 × 10^9 W/cm^2 in a pulse 200 ns wide

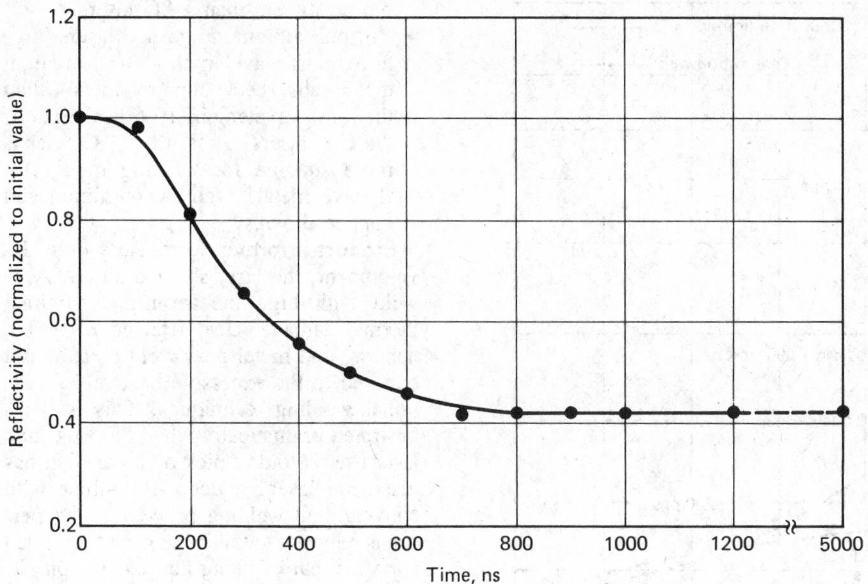

material in the same time, or the same amount of energy is absorbed sooner.

The surface condition of the metal greatly influences the emissivity. In recent experiments, the total energy retained in the metal divided by the incident energy (the thermal coupling coefficient) was measured for different metal surface conditions (polished, grit blasted, and oxidized). The polished surface had a thermal coupling coefficient of 0.083; the grit-blasted metal had a value of 0.25; and the oxidized metal had a value of 0.38. These experiments were carried out in an Fe-4%C alloy using a continuous wave CO_2 laser. Thus, by altering the surface properties, a dramatic change in the thermal coupling coefficient can be achieved and a corresponding increase in the efficiency of the laser/metal interaction occurs.

Once the energy of the laser beam begins to be absorbed by the metal, the amount of energy that is required to either melt or vaporize the metal is a function of the thermophysical properties of the metal, such as density, heat capacity, melting and vaporization temperatures, and latent heats of fusion and vaporization. This can be written as a form of the first law of thermodynamics, in which the total energy required to bring a mass of metal from room temperature to its vaporization temperature can be expressed as:

$$q = \rho V[C_{p_s}\Delta T + H_f + C_{p_l}\Delta T + H_v]$$

where ρ is density, V is volume of metal, $C_{p_s}\Delta T$ is heat required to raise the solid metal to its melting point, H_f is latent heat of fusion, $C_{p_l}\Delta T$ is heat required to raise the liquid metal from its melting point to its boiling point, and H_V is latent heat of vaporization. Thus, by knowing the thermophysical properties of the alloy, a first order approximation of the amount of energy required to melt or to vaporize the alloy can be calculated.

General Welding Considerations

Laser beam welds can be made at a variety of laser powers depending upon whether a conduction-limited or deep-penetration weld is desired. A conduction-limited weld is one in which the laser power melts the materials without vaporization. The delivery of laser power may be continuous or pulsed. If a pulsed laser is used, the pulse rate should be fast enough to allow for the overlap of pulses and to form a weld seam rapidly. Conduction-limited welds are limited to sheet thicknesses less than 0.08 in. Deep-penetration welds, also called keyhole welds, occur when a focused, high-power laser beam vaporizes the material through a thickness. This vapor cavity, or keyhole, is surrounded by molten metal. By translating the beam or the workpiece, a weld is made.

Threshold Effects. In LBW, the maximum laser power is determined by the laser, and the thickness of the material to be welded is determined by the application. The majority of current LBW applications is accomplished in the conduction-limited mode on thin sheets. Figure 12 is a plot of thickness as a function of welding speed for 300 series stainless steel for various low-power lasers, both pulsed and continuous. Several comments should be made about this comparative data:

- For welding speeds of 20 to 60 in./min, the 150-W pulsed Nd:YAG laser can weld thicker sections than the 400-W continuous wave Nd:YAG laser. This is accomplished by changing the pulse rate from 50 to 100 pulses per second (pps).
- Below 120 in./min, the 400-W pulsed Nd:YAG laser welds thicker sections than the 375-W continuous wave CO_2 laser at the same welding speed.
- Above 120 in./min, the 375-W continuous wave CO_2 laser welds the same thicknesses as the 400-W pulsed Nd:YAG laser at the same speed.
- The 400-W pulsed Nd:YAG laser operating at 200 pps makes a penetration weld. The penetration weld approximately doubles the thickness that can be welded at the same welding speed in the conduction-limited mode.

Another example of the relationship between thickness of the material to be welded and the welding speed for various laser powers is shown in Fig. 13. Again, several comments are appropriate:

- Three materials were welded: stainless steel, titanium, and aluminum. For a given power and welding speed, differences in the thickness of the materials are due to the thermophysical properties of the material.
- For a given thickness, increasing the laser power greatly increases the welding speed.
- The maximum thickness that can be welded is less than 0.08 in. For 0.06 in. stainless steel, the 400-W laser can weld at 1.2 in./min, whereas the 100-W laser can weld at only 0.012 in./min.

Laser beam welding with a high-power continuous wave CO_2 laser is most effective in the deep-penetration or keyhole mode, in which the laser melts a small cylindrical volume of material through the thickness of the material. A column of vapor is produced in this hole and is surrounded by a liquid pool (Fig. 14). As the column is moved along the joint, the material on the advancing side of the hole is melted throughout its depth. The molten metal flows along the base of the hole and solidifies along the rear side. The vapor column is stabilized by a balance between the energy density of the laser beam and the welding speed. Thus, the energy den-

Fig. 12 Effect of welding speed on penetration in a 300 series stainless steel

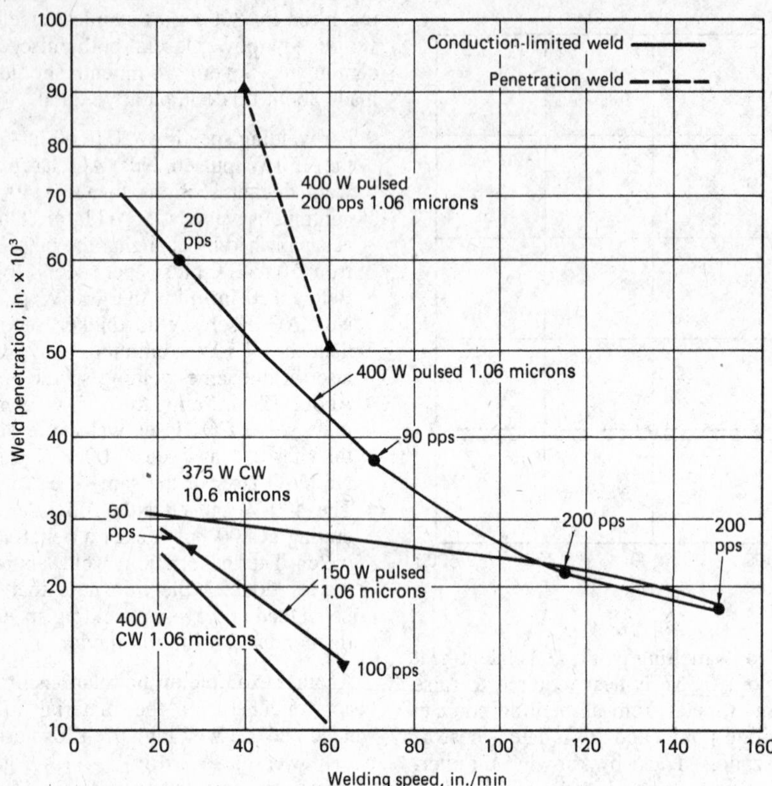

sity of the laser beam at the workpiece and the welding speed must be chosen to complement each other. An energy density that is too high results in an unstable hole that can cause drop-through; an energy density that is too low does not permit vaporization and formation of a keyhole. A welding speed that is too fast results in incomplete penetration; a welding speed that is too slow results in very wide fusion zones and possible drop-through. The depth-to-width ratio of a keyhole laser weld should be greater than 4 to 1.

Pulsed Laser Beam Welding

In selecting the appropriate laser for welding, the following differences between pulsed CO_2 and YAG lasers should be considered:

- Pulsed YAG lasers are considerably smaller than CO_2 lasers of equal average power.
- Carbon dioxide lasers are available at much higher average power levels.
- Yttrium aluminum garnet lasers cannot cut or melt transparent materials such as glass or plastic; CO_2 lasers can.
- Higher peak-to-average-power-output ratios are available from pulsed YAG lasers than from pulsed CO_2 units.
- Yttrium aluminum garnet lasers use glass optics; CO_2 lasers require reflective optics or exotic lens materials.
- Carbon dioxide lasers are electrically more efficient than YAG lasers.
- Yttrium aluminum garnet lasers produce short wavelength (1.06 μm) light that is absorbed better by metals than the long wavelength (10.6 μm) light of the CO_2 lasers. Thus, the YAG laser is more suitable for welding highly reflective metals such as aluminum and copper alloys.

Product Improvement. Early in its development, the laser showed an ability to weld with high precision and minimal thermal damage. Heat-affected zones are narrow, and metal near weld areas is not exposed to the excessive heat of conventional welding techniques. This is demonstrated dramatically when the first-time laser user is told to pick up a part that has just been laser welded. If familiar with conventional welding processes, the person is reluctant to touch the recently welded parts for fear of being burned. In spite of the fact that metal has just been melted, however, a temperature rise often is not even noticeable, because the laser provides intensities an order of magnitude higher than conventional welding techniques. As a result, conventional welding techniques require substantially greater heat input than laser welding for the same joint formation.

Because so little heat is put into the part, virtually no thermal distortion takes place as a result of laser welding. There is no damage to thermally sensitive components such as glass feed-throughs. The lower heat input also reduces fixturing demands, because fixtures are not required to with-

Fig. 13 Power curves for Nd: YAG pulsed lasers with approximate penetrations achievable for high-quality welds

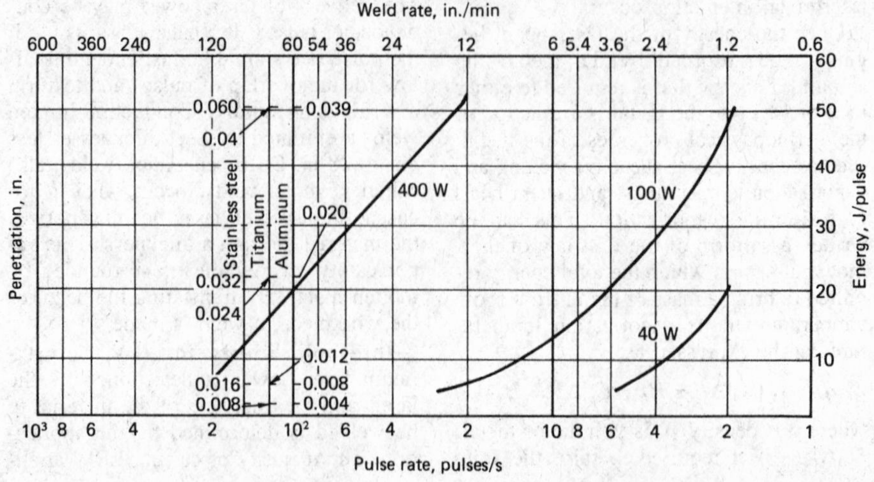

Fig. 14 Deep-penetration weld characteristics

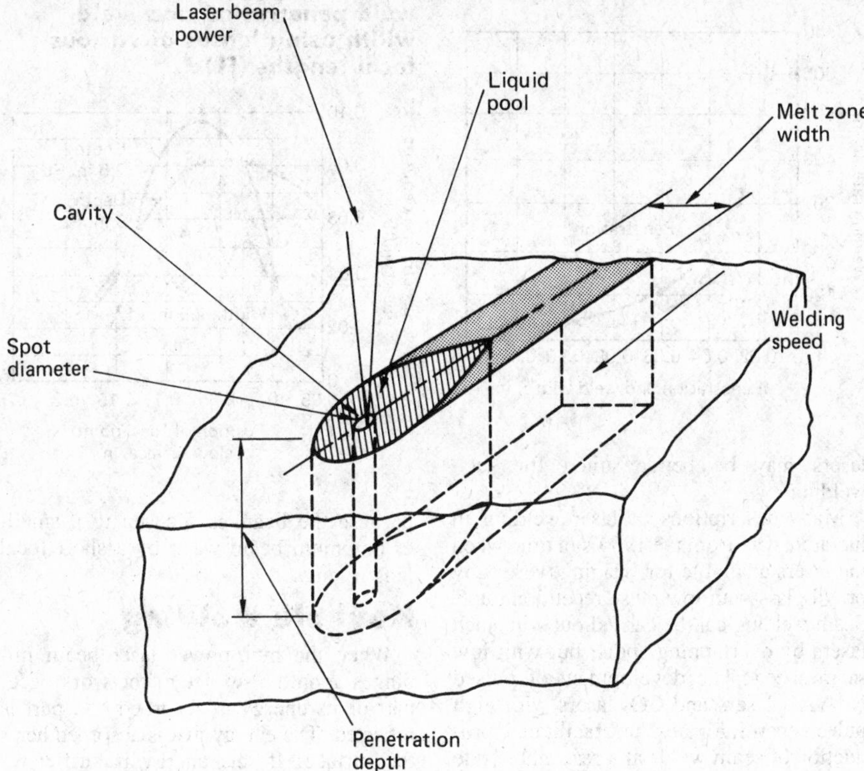

stand large thermal expansion forces or to act as heat sinks.

Lower Cost. Laser welding reduces costs in several areas. First, there is often less pre- and postprocessing of parts that are laser welded. Second, the laser welding process exhibits excellent weld-to-weld repeatability and is particularly amenable to automation. Coupling the precision of the computer controller to the speed of the laser holds down production costs. Third, because the laser welding process is inherently more controlled and repeatable than conventional welding processes, scrap rates are drastically reduced; in some cases, scrap is totally eliminated.

Reflection figures in laser welding vary from about 70% for most metals to as high as 98% for silver or gold. Thus, spot size and peak laser power parameters are chosen so that when reflection is considered, absorbed power concentrations are on the order of 10^5 W/cm^2 or less.

Although various theories relate weld speed and penetration to thermal diffusivity, pulse length, and reflectivity, actual weld energies and pulse lengths are determined empirically for each material based upon penetration and weld nugget size desired.

The welding ability of a pulsed laser is determined by both its energy-per-pulse capacity and its average power output capability. In general, a high energy-per-pulse capacity is necessary to produce welds of a substantial size (penetration and nugget diameter). Desired production rate, such as spot welds per second or seam weld rate, determines the average capability required.

Spot Welding

In principle, production of spot welds at rates up to the maximum pulse rate of the laser welding machine should be possible. Laser welding equipment is now available that produces 1-ms-duration individual pulses at rates of more than 200 times per second. In reality, however, two limitations prevent this. For most applications, it is not possible to position parts under the laser beam at these rates. Second, because the laser pulse requires a finite amount of time, the laser spot may blur when parts continue to move at high rates during the welding pulse. For example, at 200 pps, if laser spots are to be made 0.3 in. apart, the parts pass the laser beam at a rate of 60 in./s. For a 1-ms pulse length, this produces a blur of 0.06 in. at the point to be welded. This amount of blur reduces penetration and creates a larger than acceptable weld spot or weld size. Although scanning techniques have been suggested as a means of eliminating such blurring, to date none has proven operationally feasible.

Presently, several successful spot welding applications are known where welding takes place at 20 spots per second. Such applications use computer-controlled systems with parts moving under a fixed laser beam in a continuous manner. Laser firing is controlled by the computer that monitors the position of the table moving the part.

Seam Welding

Often, seam welding is employed to hermetically seal components. The pulsed YAG laser produces hermetic seams by overlapping spots at a high-repetition rate. Pulsed operation for production of continuous seams is not unique to laser welding; it also is used in advanced resistance and inert gas arc welding equipment. Pulsed operation inherently gives less total heat input and generally provides better control of weld penetration, integrity, and appearance.

With the high pulse rates available from modern laser equipment, production welding rates are quite adequate. Figure 13 shows weld speeds and penetration for three Nd:YAG lasers on 300 stainless steel, aluminum, and titanium.

Continuous Wave Laser Beam Welding

At laser power levels of a few hundred watts of average power, the laser energy is absorbed at the surface of the workpiece, and the penetration of the laser energy into the weldment is by thermal conduction. This fact limits the depth of penetration, so that the weld depths are relatively thin (maximum thickness about 0.080 in.). Most penetrations are of the order of less than 0.040 in. The depth of penetration can be increased by long dwell times (slower welding speeds), which results in a larger HAZ.

At 100-W average power, a repetitively pulsed laser is a welding tool for some metals. A continuous laser must emit several hundred watts to have comparable metalworking capabilities, because the high peak power in the pulse can break down the surface reflectivity, allowing more efficient absorption of the laser energy. At powers of hundreds of watts, continuous lasers (both Nd:YAG and CO_2) offer welding capabilities, and the weld speed and smoothness of the weld bead obtained are satisfactory.

The tradeoff between penetration depth and welding speed for seam welding with a 375-W continuous wave CO_2 laser is shown in Fig. 15. Data are shown for welding of two different types of steel. These results are for butt welding, with full penetration of the weld zone through the material. Many potential applications in welding can be satisfied by combinations of penetration depth and speed, as shown. High speeds are possible only with thin material, however, because of constraints imposed by thermal conduction. Even at very slow rates, penetration is limited to approximately 0.040 in. for stainless steel and approximately 0.060 in. for carbon steel.

Figure 16 shows a similar curve for type 302 stainless steel, welded by a 1500-W continuous wave CO_2 laser. The increased power allows somewhat deeper penetration, but even at slow speeds the maximum penetration is limited to about 0.12 in.

It is worthwhile to compare the results obtained using a CO_2 laser and a Nd:YAG laser for welding. Figure 17 shows the weld rates obtained with a continuous Nd:YAG laser welding 304 stainless steel. These results indicate that the power required to achieve a given welding speed on a given thickness of material is approximately the same for a Nd:YAG laser as for a CO_2 laser at comparable output power. The difference in reflectivity at 1.06 and 10.6 μm is thus not too important for steel. Breakdown of the surface reflectivity by the laser means that the energy is absorbed reasonably well in either case. However, for high-conductivity metals, the difference in reflectivity is more significant, and Nd:YAG lasers may be better suited for such welding.

Fig. 15 Effect of welding speed on penetration depth. Source: Photon Sources, Inc.

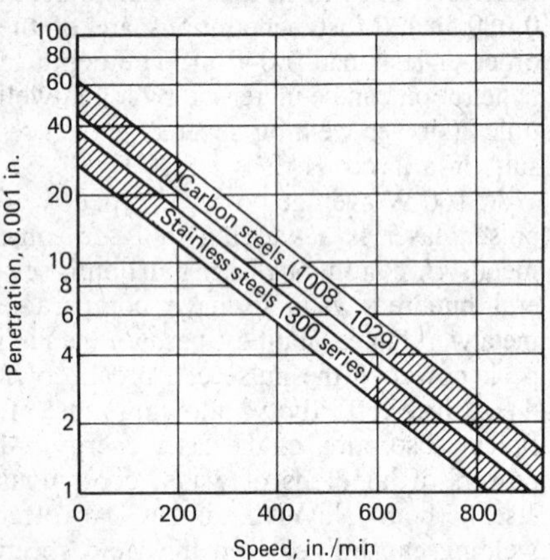

Fig. 16 Effect of welding rate on penetration and weld width

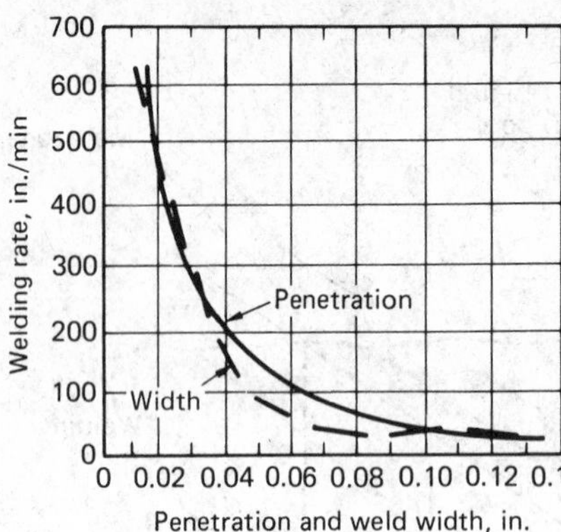

Many descriptions of laser welding in literature date from the 1960's, a time when the lasers available for welding were ruby or Nd-glass with low pulse repetition rates. Seam welding can be carried out with such lasers by overlapping spots, but with low seam speed. The development of pulsed Nd:YAG lasers and CO_2 lasers with high pulse repetition rates has facilitated production of seam welds at a reasonable rate by overlapping spots.

Another consideration is the effect of varying the focus of the beam. Figure 18 shows how the penetration depth and bead width vary for welding of 1018 steel. The welding was done with a 1500-W continuous wave CO_2 laser at a speed of 50 in./min. The effect of changing the focus is apparent, with maximum penetration occurring when the beam is focused at a point slightly below the surface. When a shorter focal length lens is used, the penetration is deeper, but the half-width of the penetration depth curve is greater, because the beam is focused to a smaller minimum beam waist by a short focal length lens.

Fig. 17 Weld rates obtained with a continuous wave Nd:YAG laser on type 304 stainless steel

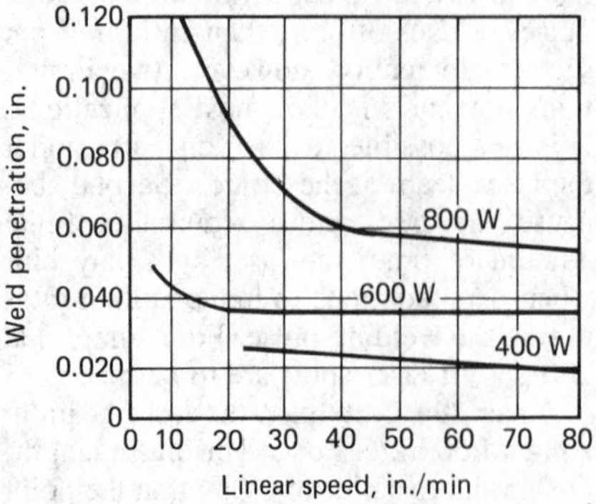

Fig. 18 Effect of focal position for welding of 1018 steel on weld penetration and weld width using lenses of various focal lengths (FL)

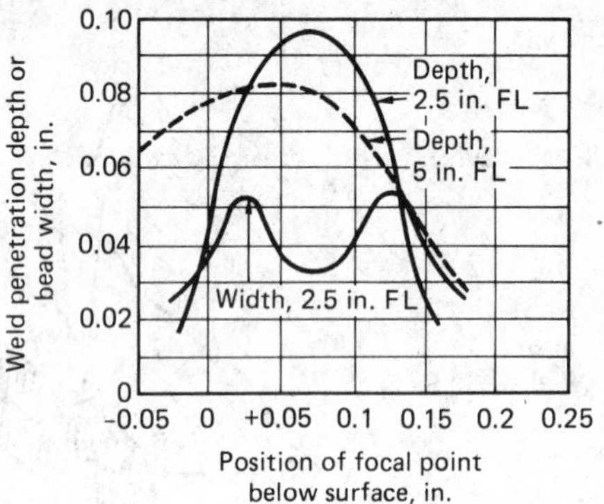

Keyhole Welding

When the high-power laser beam impinges upon the surface of the workpiece, part of its energy is absorbed and part is reflected. The energy that is absorbed heats the surface. If this energy is sufficient, melting occurs. The keyholing that results from deep-penetration welding with high-power lasers was described in the section on threshold effects in this article. The importance of choosing complementary energy density and welding speed was also explained.

When the high-power laser beam interacts with the workpiece, vaporization occurs and a plasma is formed. This plasma, consisting of vaporized metal ions and electrons, is opaque to the laser beam, moves over the surface, and effectively decouples the laser beam from the workpiece. Therefore, to weld with a laser beam, plasma must be minimized. Usually, this is accomplished by directing a high-velocity jet of inert gas into the area of interaction and moving the plasma to one side. A discussion of plasma effects follows this section.

Because the laser beam vaporizes and remelts the material in the fusion zone, this area needs to be protected from the atmosphere. The hot fusion zone can absorb gases that do not readily diffuse when the weld cools, especially hydrogen, oxygen, and nitrogen. Excessive pickup of these gases results in degradation of the mechanical properties of the welds. Absorption of these gases is minimized by protecting the fusion zone by inert gases, as in GMAW.

Gas Shielding

Laser beam welding normally uses an inert gas shield, typically argon or helium. For welding with a pulsed laser or a moderate-power continuous wave laser, the shielding gas feeds to the weld along the laser beam axis. Production welding with a high-power CO_2 laser requires a shielding fixture that completely covers the weld.

To save on operating expense, pulsed solid-state laser spot welding can be performed without shielding gas, resulting in only a slight sacrifice in mechanical properties of the weld. Single-pulse, solid-state laser welding is possible without shielding gas, because the weld pool is molten for only a very short time. The laser pulse itself lasts for 1 to 10 ms, and the pool life is even shorter. Seam welding by using overlapping spot welds is a different situation, and a shielding gas is recommended.

Plasma Effects

An unwelcome consequence of the 10^6 to 10^7 W/cm^2 power densities associated with laser keyhole welding is the formation of a beam-absorbing plasma at the beam/material interaction point. Plasma ignition occurs when the degree of ionization in the local medium increases to a critical breakdown level as the result of interaction with the intense electric field associated with the focused laser radiation. At power densities in the keyhole welding range, the generated plasma propagates at subsonic speeds; such plasmas are often referred to as laser-supported combustion (LSC) waves (Ref 2). Higher intensities lead to supersonic propagation speeds and formation of so-called laser-supported detonation (LSD) waves.

Plasma is undesirable in LBW, because it can absorb a significant fraction of the beam energy and prevent effective radiant energy transfer to the workpiece (Ref 2). Energy absorption within the plasma occurs by the process of inverse Bremsstrahlung (Ref 3) and can readily exceed 50% of the incoming laser energy. Furthermore, in addition to absorption, the plasma can degrade welding efficiency by distorting beam optical characteristics and reducing focusability (Ref 2). A brief review follows on the aspects of plasma formation, its influence on high-power LBW performance, and means for its control.

Fundamental Concepts. A beam of laser light propagates as an electromagnetic wave characterized by electric and magnetic field vectors. The intensity of the electric field associated with the propagating laser beam increases with the optical intensity of the beam and is inversely proportional to the square of the radiated wavelength (Ref 3). If plasma ignition is assumed to be caused by an avalanching ionization process, then it can be compared with the phenomenon of spark breakdown in gases. In this regard, it is noted that an electric field of 30 000 V/cm causes a spark breakdown between parallel plates separated by dry air (Ref 4). At the wavelength of CO_2 laser radiation (10.6 μm or 106 000 Å), the intensity required for spontaneous breakdown is of the order of 10^9 W/cm^2. Such intensities are readily attainable with pulsed lasers, and hence, spontaneous breakdown has been demonstrated in many gases (Ref 3).

Smooth keyhole welding, on the other hand, requires power densities two or three orders of magnitude below those required for spontaneous breakdown, i.e., power densities of 10^6 to 10^7 W/cm^2. Lower power densities do not result in generation of the keyhole mode essential to effective LBW, and higher power densities lead to excessive vaporization. The latter are more suitable for drilling and/or cutting than for welding. Therefore, plasma ignition problems may not be anticipated under normal continuous LBW conditions.

Just as impurities reduce the breakdown potential of a spark gap, however, so do impurities reduce the plasma ignition threshold for a focused laser beam. In the laser welding process, a small quantity of metal is vaporized at the beam/material interaction point. This vaporization is inherent in the formation of the deep-penetration welding keyhole. Thermal ionization of the heated metal vapor produces sufficiently large numbers of free electrons in the interaction zone to promote plasma ignition at beam power intensities substantially below those required for spontaneous breakdown. Some insight into this behavior can be obtained by comparison of the ionization potentials of common gases and metals, which are given in Table 5 (Ref 4). The ionization potentials of the metals are seen to be of the order of one half that of the gases, and because the degree of thermal ionization is an exponential function of the energy level, ionization at comparable temperatures differs by the order of e^2 between metals and gases. An example of the consequence of this difference is provided in Fig. 19, which compares the radiative absorption of aluminum and air as a function of temperature.

Thus, it is apparent from the above discussion that plasma ignition is a function not only of the intensity of the incident beam and the identity of the surrounding medium, but also of the characteristics of the material being welded. Because certain metallic species within an alloy may be preferentially vaporized (Ref 5), ignition can be dominated by trace elements within the workpiece. This behavior, as well as complex interaction processes that include multiphoton absorption, further reduces the initiation threshold.

Table 5 Ionization potentials of common gases and metals

Material	First ionization potential, eV
Argon	15.68
Helium	24.46
Oxygen (O_2)	12.50
Neon (N_2)	15.54
Carbon dioxide (CO_2)	14.41
Water vapor (H_2O)	12.56
Aluminum	5.96
Chromium	6.74
Nickel	7.61
Iron	7.83
Magnesium	7.61
Manganese	7.41

Effects on Welding Performance. The series of bead-on-plate penetrations shown in Fig. 20 illustrates the manner in which plasma ignition influences the laser welding interaction. All weld penetrations were formed at a laser power of 5 kW and a welding speed of 80 in./min. The material was 0.25-in.-thick type 316 stainless steel. The penetrations were formed with the material located in an inert atmosphere within an enclosed chamber. Atmospheric

Fig. 19 Effect of temperature on the plasma absorption coefficient. Pressure = 1 atm

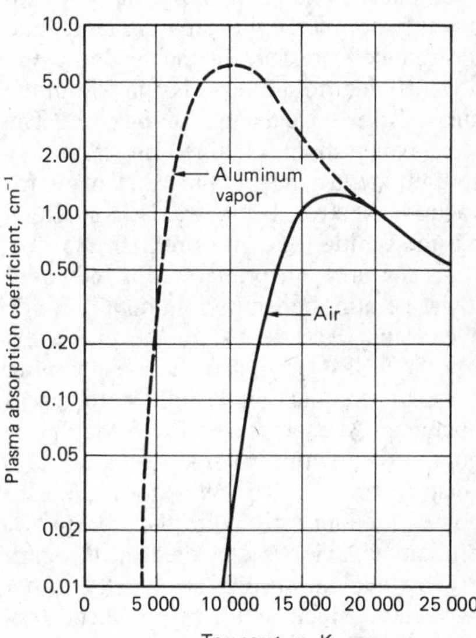

Fig. 20 Effect of plasma on welding performance. Base metal was 316 stainless steel, laser power was 5 kW, and weld speed was 80 in./min.

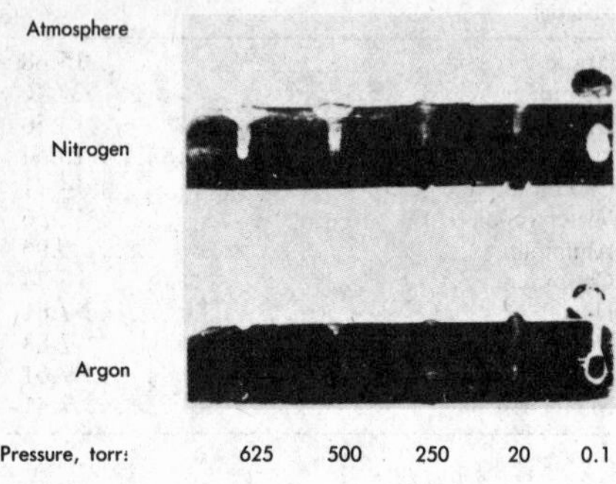

conditions within the chamber were varied from soft vacuum to essentially atmospheric pressure. For the two samples shown, nitrogen and argon were utilized as the background environment.

Inspection of the penetrations formed at 10^{-1} torr ambient pressure indicates a narrow initial penetration with a bulbous lower fusion zone and incomplete penetration. This behavior is assumed to be caused by the ignition of a metal vapor plasma within the deep-penetration cavity itself. Such ignition could have been engendered by the reduction in boiling point of the constituents of 316 stainless steel associated with the reduction in pressure. The inverted fusion zone profile extended along the entire length of the penetration pass and was not just a local phenomenon. This profile persists over a broader range of pressures in argon than in nitrogen.

At a pressure of approximately $^1/_3$ atm, full penetration was obtained in both nitrogen and argon. Conditions for ignition of an internal metal vapor plasma were apparently not met at this pressure level, and the reduced pressure in the medium was not sufficient to ignite a plasma within the atmosphere above the interaction point. The weld penetrations obtained are relatively straight-sided and reminiscent of hard-vacuum electron beam welds, for which plasma ignition is not a problem.

As the ambient pressure was increased toward 1 atm, penetration in both nitrogen and argon decreased, and the characteristic hourglass bead shape associated with plasma absorption and beam distortion was obtained. At a pressure of 625 torr in argon, little melting interaction is seen to occur, even though 5 kW of focused beam power is being directed to the workpiece. Similar behavior is observed in nitrogen, except that the absorption penalty is not as severe, which indicates a reduced tendency for plasma formation. This result is somewhat surprising in view of the nearly identical first ionization potentials for these two gases (Table 5). In similar experiments with other gases, helium exhibited the highest resistance to plasma ignition.

High-speed motion pictures of the interaction zone reveal that the generated plasma is not a steady-state phenomenon. Analysis of the films shows that initial interaction of the beam with a workpiece results in the generation of a thermally ionized metal vapor that triggers plasma ignition. Once ignited, the plasma inhibits further beam interaction with the workpiece and propagates away from the workpiece, expanding into the conical focal volume of the incoming laser beam. When the plasma reaches a point at which the incident beam power intensity is less than the plasma losses, the plasma extinguishes. This again permits the beam to interact with the material, initiating another plasma puff.

Suppression Techniques. The aspects of plasma ignition and its deleterious effect on high-power LBW performance were recognized by early experimenters (Ref 6, 7). Plasma ignition, it was discovered, could be prevented by removal of thermally ionized material from the interaction zone by use of a crossflow of gas over the weld point. The effects of transverse gas flow on the breakdown threshold are noted in Fig. 21. Helium is customarily used in LBW because of its high resistance to breakdown. Effective plasma control has been demonstrated to continuous power levels of 100 kW (Ref 8).

Plasma suppression with gas crossflow is not without its drawbacks, however. The high flow velocities required for plasma suppression can promote aspiration of air into and contamination of the weld. At sufficiently high velocities, the molten pool itself may be disrupted by the flow. Furthermore, shielding of weld material from atmospheric contamination is enhanced by the generation of relatively inert gas curtains—usually involving argon because it is heavier than air. These conditions are clearly at odds with plasma suppression requirements. Therefore, a compromise between shielding and suppression provisions is necessary for optimum welding performance and has led to the development of numerous specific suppression/shielding devices.

Fig. 21 Effect of ignition power threshold on transverse gas velocity for a 0.0008 in. CO_2 laser spot (Ref 2)

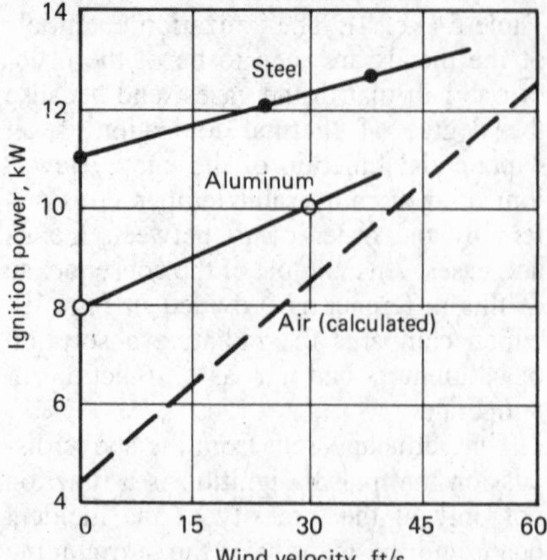

One example of a device developed for plasma control is a small-diameter tube that is accurately positioned at a specified angle above the deep-penetration welding keyhole (Ref 9). Inert gas is directed from the tube into the keyhole. Because of the small quantity of gas involved, the momentum in this tiny jet is not large enough to cause molten pool disruption. Plasma suppression is effectively obtained and some interaction occurs with the liquid within the fusion zone, possibly contributing to increased penetration. Disadvantages of this technique include the high precision to which the jet must be oriented and the thermal frailty of the unit.

Another approach to plasma suppression utilizes the knowledge that ignition is not spontaneous, but instead persists over a time period generally exceeding 500 μs under keyhole welding conditions (Ref 10). Thus, periodic interruption of the laser beam on a time scale shorter than the plasma formation time results in complete elimination of plasma breakdown. Tests using this approach were conducted at interruption frequencies to 24 000 Hz. A typical result is shown in Fig. 22 for an interruption frequency of 6000 Hz (167 μs) and an average beam power of 14 kW. The duty cycle for this test was 75%; i.e., the beam-on time was 128 μs and beam-off time was 39 μs. No plasma ignition was noted in high-speed motion pictures of the interaction zone.

The penetration profile obtained in Fig. 22 is extremely narrow and relatively straight-sided compared with those obtained with a steady 14-kW beam using crossflow suppression techniques. The latter exhibit the characteristic nailhead associated with incomplete plasma extinction. Surprisingly, however, complete extinction of the plasma did not lead to increased maximum penetration. This suggests that improvements in penetration due to "plasma control" devices (Ref 11) may be due more to influence of the fluid

Fig. 22 Laser weld penetration obtained with beam interrupted at 6000 Hz. Material was alloy steel, 0.46 in. thick, pulse welded at a speed of 80 in./min using 14 kW.

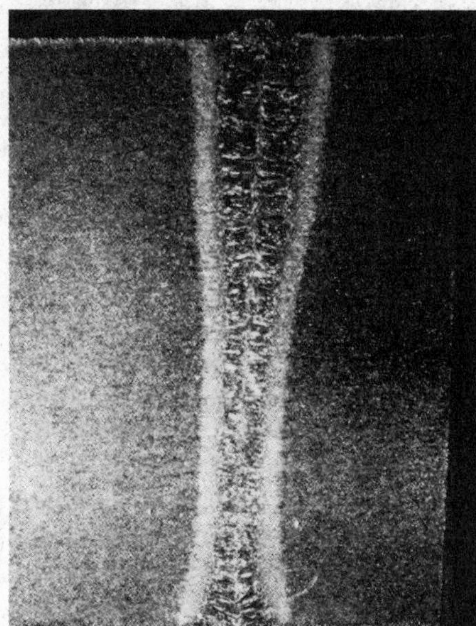

dynamics of the deep-penetration cavity rather than plasma effects. Elimination of the nailhead, however, did lead to an increase in welding speed of about 60%. In effect, the volume of material represented within the nailhead was transferred to the central portion of the weld, creating a longer fused joint for equivalent energy input.

In high-speed welding of thin materials, plasma suppression may be inherent in the process itself. Material vaporized by interaction of the beam with the workpiece moves away from the surface. The initial direction of propagation is normal to the surface, because the workpiece boundary constitutes a constant pressure line. If welding speed is high enough, the vaporized material is left behind within an angular zone with a boundary tangent that is the ratio of the vaporized material propagation speed to the welding speed. For sufficiently high welding speeds, the beam does not propagate through sufficient vaporized material to cause adequate thermal ionization for plasma ignition.

Finally, less severe plasma ignition problems pertain at shorter radiation wavelengths (Ref 3). Although the CO_2 laser is currently the only suitable unit for high-power industrial applications, tests were conducted with a 3.8-μm deuterium fluoride chemical laser to explore wavelength effects in plasma formation (Ref 12). In fact, it was found that plasma suppression problems were substantially less severe than those at 10.6 μm. As in the case of high-frequency beam interruption, however, no increase in maximum attainable penetration at a given power level was obtained.

Weld Properties

Laser beam weldments fabricated by pulsed lasers are rarely tested for conventional mechanical properties. The soundness of the joint is the most important aspect of the process. Often, a hermetic seal is required and is one of the main considerations in selecting the welding process. Shear testing of the seam or spot weld is another concern. Because the strength of the weldment is of secondary importance, mechanical property data are rarely reported in articles describing pulsed LBW.

The mechanical properties of laser beam weldments fabricated by the continuous wave mode are often available. The results of bend tests, hardness values, fracture toughness, and occasionally, fatigue data are also accessible. Several reviews (Ref 13, 14) which have included these results are summarized here by alloy type.

Aluminum and Its Alloys. Although the use of high-power CO_2 lasers for welding titanium, stainless steel, and alloy steels is well researched, their application to aluminum welding did not attract the attention it deserved. To date, very limited data are available in the open literature. One of the reasons that aluminum alloys have proven to be very difficult to weld is the high initial surface reflectivity for 10.6-μm radiation from CO_2 lasers.

A laser welding study of aluminum alloys 5456 and 5086 (Ref 13) revealed that these two aluminum alloys appear to differ in their welding response to a significant degree, both in ability to penetrate under a given set of conditions and in bead appearance. Penetration in 5456 was substantially greater, the difference being primarily attributed to the 1.5% greater magnesium content of this alloy. Porosity was present in unacceptable amounts in all weld specimens. Excessive drop-through of the bead was encountered in all full-penetration welds. This problem is related to liquid-metal viscosity and surface tension. The interaction of shielding gas and/or plasma with the beam and the workpiece is very much a part of this phenomenon. It was concluded that aluminum alloys are very sensitive to the intensity of the input energy and the welding variables.

Some success in LBW of aluminum alloys 2219 and 5456 has been achieved. Both of these materials have been welded in thicknesses of up to 0.4 in. with acceptable microstructure and bead profile, but porosity has been a problem. Tensile tests on these welds resulted in failure by diagonal shear through the welds at an average strength of 49.7 ksi, whereas the ultimate tensile strength value for the parent metal was 50 ksi (Ref 14). Bend test results of laser butt welds were acceptable (Ref 15). Bend test results for laser fillet weld joints of aluminum alloy 5456 0.2-in. thick exhibited sharper bend radii without fracture of the weld when compared with those of similar welds made by GMAW.

The mechanical properties of laser welded 0.5-in.-thick aluminum alloy (5456) have been measured (Ref 16). Tensile tests were performed on a standard ASTM round specimen 0.252 in. in diameter. The weld was transverse to the loading direction in the tests. All of the aluminum laser welds failed in the weld zones. The ductility of these specimens was low, and the amount of porosity visible on the fracture surfaces was high. A nonstandard 0.5-in.-thick dynamic tear test specimen with the notch at the weld was used for fracture toughness tests (the standard size is 0.625 in.). Porosity was also observed in dynamic tear fracture surfaces.

The feasibility of laser welding of three other aluminum alloys (2036, 5182, 6009) has been studied (Ref 17). A 1350-W continues wave CO_2 laser was used for bead on plate and lap welding on 0.04-in.-thick aluminum plates. The composition of the alloy was found to be the determinant of whether the irradiation parameters were critical. Crack-free laser welds with optimum irradiation parameters for aluminum alloy 2036 were produced. Aluminum alloy 5182 was the easiest to weld, compared to 2036 and 6009. Laser irradiation parameters were not very critical for successful welds, but a pronounced loss of magnesium by evaporation was observed. Aluminum alloy 6009 demonstrated very poor weldability. Laser irradiation parameters required for successful welds seemed to be extremely critical. An excessive precipitation of Al-Mg-Si was observed and may have been responsible for the cracking of these welds. However, additional cooling of the irradiated material by a chill plate facilitated successful welding.

A systematic study of laser welding of $^1/_4$-in.-thick Al-Mg alloy (5083) has also been conducted (Ref 18). A 10-kW CO_2 laser and a gas shielding system in which plasma formed during laser materials interaction is pushed into the keyhole was used. This experimental approach successfully produced apparently porosity-free welds.

Steels. Most of the initial parametric laser welding studies were conducted on stainless steel, because of the importance of stainless steel in power plant and chemical industries (Ref 19, 20).

High-power LBW welding experiments with 300 series stainless steel were conducted (Ref 21). Welds were formed in stainless steel with aspect ratios (depth-to-width ratios) as large as 12 to 1. Data were obtained in a series of bead-on-plate penetration tests conducted on samples exposed to the ambient atmosphere using laser powers up to 5.5 kW. It was concluded that the depth-of-penetration at constant laser power is a relatively weak function of welding speed.

Radiographic inspection of selected laser welds in stainless steel has shown that high-density, nonporous welds can be achieved (Ref 22). Tensile tests of stainless steel welds have shown that the joint strength can, with appropriate selection of weld parameters, equal that of the parent material. Similar observations were made for stainless steel 316 (Ref 19), but data obtained for stainless steel 310 were less encouraging.

Tensile tests on laser butt welded stainless steels 316 and 310 and ferritic steel DUCOL W30 were performed (Ref 19). Flat tensile specimens were used with the plane of the weld running transversely across the gage length at its center. The room temperature uniaxial tensile tests were performed at a crosshead velocity of 0.064 in./s, corresponding to a strain rate of 2.8×10^{-2}/s. Laser beam welding of 0.24-in.-thick stainless steel 316 and DUCOL W30 was compared to GTAW, EBW, and plasma arc welding (PAW) methods (Ref 19). The comparison indicated that for high productivity and better weld quality, EBW and LBW techniques are preferable. However, one should keep in mind that electron beams require a vacuum chamber while lasers do not.

Successful autogenous square-butt welds of X-80 Arctic pipeline steel using a high-power continuous wave CO_2 laser were reported (Ref 21, 23). Both single- and dual-pass techniques were used to weld 0.52-in.-thick material. Dual-pass welds exhibited smaller grain structure than single-pass welds. In addition, the upper shelf for dual-pass welds was greater than 264 ft · lb, and the transition temperature was below −60 °F. The mechanical properties of the laser welds appeared better than those of the base metal. Therefore, LBW promises to be suitable for high-quality large-diameter pipe welding applications.

One of the principal commercial areas for the welding of thin material is that of can manufacture. The possibility of high-speed welding of steels used in can making (tin plate and tin-free steels) with a 2-kW continuous wave CO_2 laser was evaluated (Ref 24, 25). A welding speed in excess of 62 ft/min was achieved for bead on plate welding of 0.008-in.-thick tin plate using 1950 W of laser output energy. The laser welding process was compared with other can-making processes and found to be the only method capable of welding tin-free steel (with a 0.01-μm layer of chromium and a 0.04-μm layer of chromic oxide as a corrosion inhibitor) without auxiliary preparation. Although a 2-kW continuous wave CO_2 laser by itself cannot reach the required welding speeds, arc augmentation of the 2-kW laser appears capable of doing so (Ref 25, 26).

Mechanical properties of laser welds in tin plate and tin-free steels seem to be at least as good as those of the base material. All fractures during tensile tests were observed in the base material (Ref 27). Laser welds were also found to be radiographically sound (Ref 25). Simple bending fatigue tests revealed an endurance ratio of 0.45 to 0.5. Data for corrosion rates of the laser welds, using the Tafel extrapolation method, have shown that the weld zone is at least as corrosion resistant as the base metal, if not better. In conclusion, sound welds of good appearance and mechanical properties can be made using a laser in tin plate and tin-free steel. The laser welds have a very narrow HAZ and could be made through painted areas. They are autogenous and, therefore, present no recycling problems such as solder does in lock-seam soldered cans.

The mechanical properties of laser beam weldments of the A36 steels are given in Table 6 (Ref 28). These data are those of the base plate, because all of the specimens fractured in the base plate. The weld itself, therefore, is stronger than the base plate. Figure 23 shows the energy absorbed by the Charpy V-notch (CVN) specimen for A36 base plate and laser beam welds in the as-welded and stress-relief condition as a function of temperature.

Titanium and Its Alloys. The EBW technique has been used more frequently than LBW for the welding of Ti-6Al-4V, an alloy widely used in the aerospace industries for its high strength-to-weight ratio. However, the deep penetration of EBW can be obtained only up to a short distance, under nonvacuum conditions. For optimum efficiency, EBW is carried out in an evacuated chamber (Ref 29). In contrast, CO_2 laser beams can be transmitted for appreciable distances through the atmosphere without serious attenuation or optical degradation. Thus, the laser offers an easily maneuvered, chemically clean, high-intensity, atmospheric welding process that produces deep-penetration welds (aspect ratio greater than 1 to 1) with a narrow HAZ and subsequent low distortion.

The application of the laser technique to a metal such as titanium alloy, which requires extreme cleanliness for attainment of sound welds, is of great interest to the aerospace and chemical industries. More generally, laser techniques are of interest from the standpoint of welding a chemically sensitive metal with a complex, temperature-dependent structure. The importance and the need for better join-

Table 6 Mechanical properties of laser beam weldments for alloy A36 (Ref 28)

Thickness, in.	Yield strength, ksi	Ultimate strength, ksi	Elongation, %
0.25	40	64	23
0.38	39	62	22
0.50	40	65	24
0.63	39	63	26

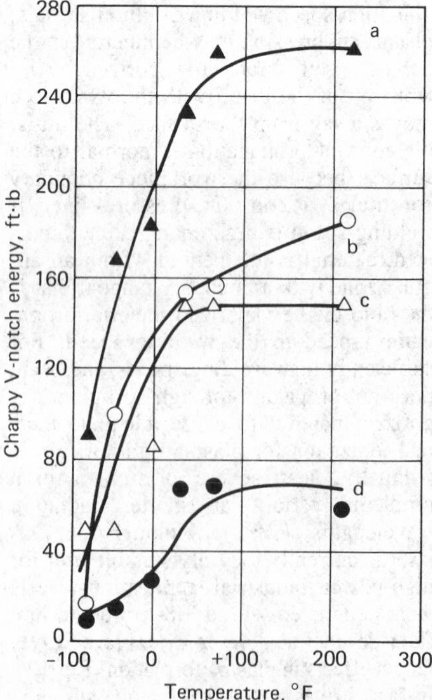

Fig. 23 Effect of temperature on absorption of Charpy V-notch energy. (a) Stress-relieved laser beam welds. (b) As-welded laser beam welds. (c) Longitudinal-transverse loaded A36 base plate. (d) Transverse-longitudinal loaded A36 base plate

ing methods for titanium and its alloys resulted in several investigations of LBW techniques over various power ranges. The relationship between LBW parameters and the metallurgical and mechanical properties of laser welded Ti-6Al-4V and commercially pure titanium was reported (Ref 30). Welding speeds in excess of 50 ft/min were obtained for 0.04-in.-thick Ti-6Al-4V using 4.7 kW of laser power.

X-ray radiographs of successful laser butt welds of Ti-6Al-4V and commercially pure titanium showed no cracks, porosity, or inclusions (Ref 30). Low porosity in laser welded titanium alloy was also observed by Seaman (Ref 31), and radiographically sound welds were produced by Banas (Ref 32). Undercutting was not prominent.

Tensile tests performed on laser butt welds of titanium alloys revealed that laser welds are at least as strong as the base metals (Ref 24, 30). Under simple bending fatigue, the endurance ratio for welded specimens (with a transverse central weld) was found to be 0.40 to 0.47, and for unwelded specimens, 0.50 (Ref 30). Adams (Ref 33) reported that, under proper welding conditions, laser welds can be made in Ti-6Al-4V that exhibit the same fatigue characteristics as the base metal.

The martensitic structure in weld zone is responsible for the good tensile properties of the laser welded titanium alloy. By comparing the microstructure with the prediction of a three-dimensional heat transfer model, the cooling rate of the weld zone was estimated to be 10^4 °C/s. This high cooling rate is responsible for the martensitic structure. The total oxygen analysis of a weld sample indicates that no significant oxygen contamination takes place during laser welding when shielding gas is used (Ref 30).

A comparative study of electron beam, laser beam, and plasma arc welds in Ti-6Al-4V alloy has been conducted by Banas (Ref 32). Radiographically sound welds were produced by all three techniques. The electron beam welds were quite narrow and exhibited a somewhat nonuniform radiographic appearance due to lower surface weld spatter, whereas the arc welds were considerably broader, but also quite uniform in density. Laser welds were narrower than arc welds and were comparable to, but more uniform than, electron beam weld beads. Following stress relieving for two hours at 1000 °F, the welds produced by all three techniques had tensile strengths equivalent to or exceeding those of the base metal.

Laser beam welding of a titanium alloy (Ti-6Al-2Nb-1Ta-0.8Mo) has resulted in satisfactory mechanical properties and fracture toughness (Ref 34). The plates were grit blasted, and the edges to be welded were pickled in an HF-HNO_3 solution, washed with alcohol, and dried with nitrogen. The butt welds had a flow of helium gas as a bottom shield. An area in front of the laser beam/workpiece interaction zone was flooded with helium and a long trailing shield was used. The results of testing the weldments are given in Table 7. The CVN and dynamic tear values were determined at both 32 and 77 °F. These values compare favorably to those of the base plate.

A precracked cantilever beam specimen was fractured in air prior to stress corrosion cracking testing. A K value of 82 ksi$\sqrt{\text{in.}}$ was obtained for the air break. As seen from Table 7, this value falls within the K_{ISCC} range, indicating nonsusceptibility of the Ti-6Al-2Nb-1Ta-0.8Mo laser beam welds to stress corrosion cracking under the test conditions.

Iridium Alloys. High-power continuous wave CO_2 lasers are an attractive tool for joining difficult-to-weld alloys, such as thorium-doped iridium alloys (DOP-14 and DOP-16). These alloys crack severely during GTAW or welding with a highly defocused electron beam. Successful laser welds free from hot cracking have been reported in DOP-14 by David and Liu (Ref 35). This is due to the characteristics of the highly concentrated heat source available with the laser and the refinement in the fusion zone structure. The fusion zone structure is a strong function of the welding speed.

Laser welds with a refined fusion zone structure have also been reported in DOP-26 alloy. The fusion zone structure of the laser welds compares well with the best structures obtained by arc welding using arc oscillation.

Joint Design for Laser Beam Welding

Most weld geometries used in traditional thermal welding processes are suitable for LBW. The most frequently used laser weld geometries are butt and lap type. The fit-up requirements in terms of material thickness are shown in Fig. 24 and 25. Although distortion during the welding process is minimal, some clamping is recommended to retain the pieces in position while being welded.

Table 7 Mechanical properties of Ti-6Al-2Nb-1Ta-0.8Mo laser beam welds

Yield strength	121-128 ksi
Ultimate tensile strength	134-137 ksi
Elongation	10-19%
Reduction in area	15-43%
Charpy V-notch energy	31-39 ft · lb
Dynamic tear energy	181-189 ft · lb
K_{ISCC}	75-86 ksi$\sqrt{\text{in.}}$

Fig. 24 Typical butt joint tolerances

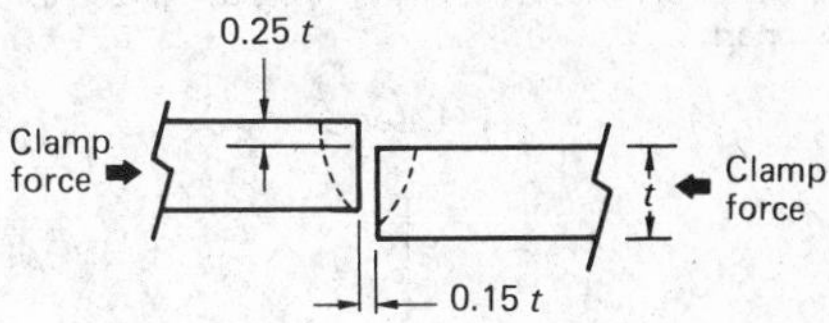

The tolerances shown are applicable to ferrous and nickel alloys. Somewhat tighter tolerances are required for the good conductors, such as copper and aluminum alloys. The amount of gap in butt welds also determines the degree of concavity of the weld, because no filler metals are normally used.

In lap welds (Fig. 25), the 25% material thickness tolerance refers to the thinner of the two metals. It is preferable to weld the thinner metal to the heavier one. Excessive air gap causes the top piece to burn through rather than to weld to the lower piece.

Interest in the welding of sheet metal components has resurfaced both in the automotive and the appliance industries. Figure 26 shows a laser welded corner seam, where the bent edges are melted down to form a rod-like weld bead that acts as a stiffener. With the proper weld parameters, the metal can be made to fill in the gap between the two sheets completely, forming an inside corner that is easy to clean and keep hygienic. When compared with conventional double-lock seams, this joint offers material and time savings, lower die costs, and allows the use of metals that have limited plastic flow. This weld is made possible by the precise

Fig. 25 Typical lap joint tolerances

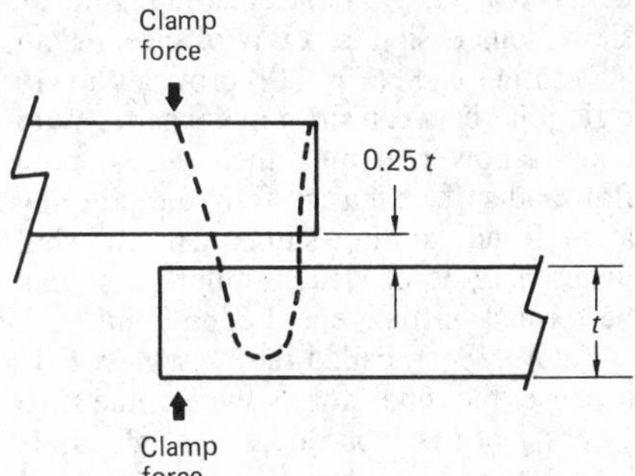

Fig. 26 Laser-formed corner seam. Note the recommended heat sinking and material dimensions to achieve the rod-like corner support.

Fig. 27 Laser welded single-lock seam

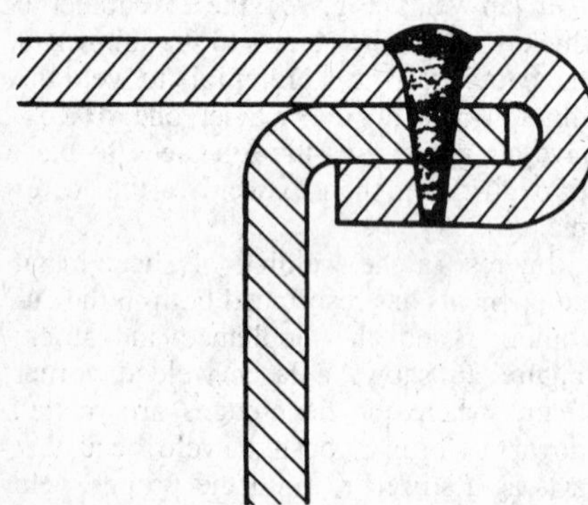

Fig. 28 Joint design for laser beam welds on sheet metal. Arrows show direction of laser beam.

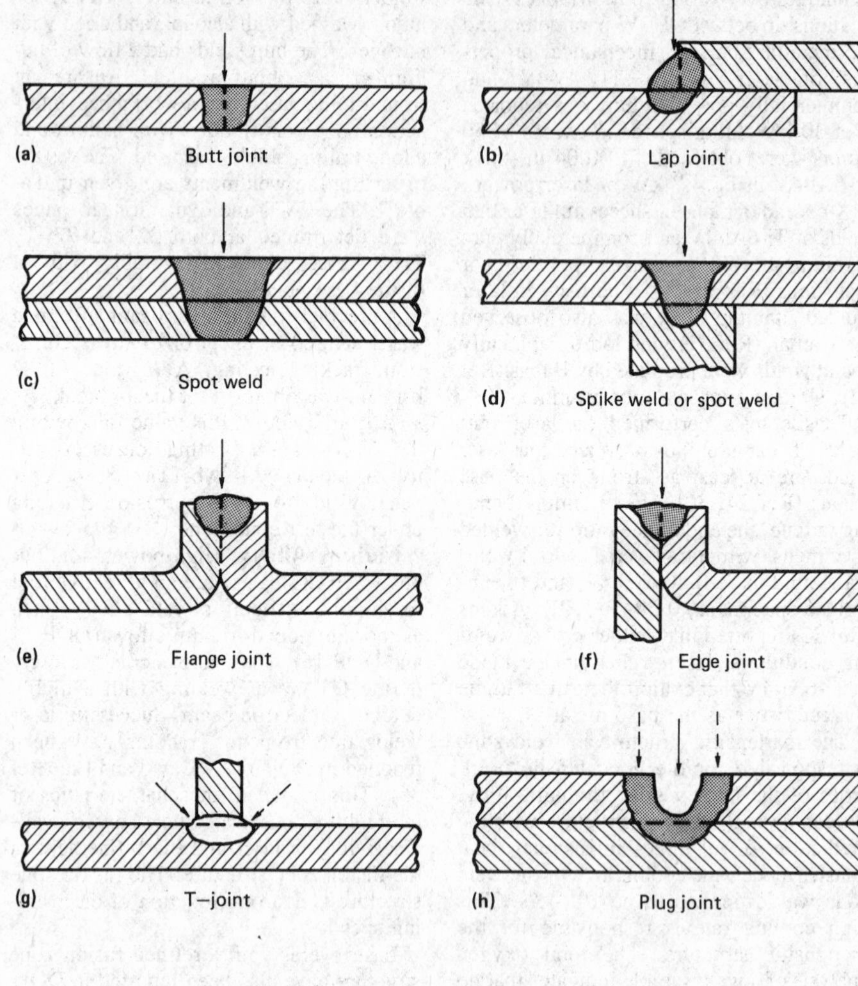

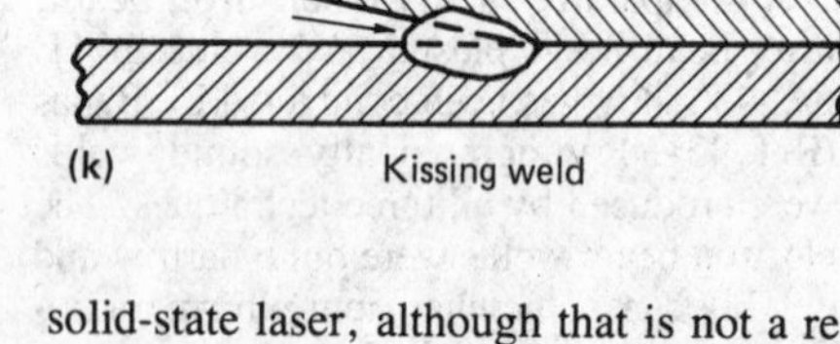

power control and smooth turn-on and turn-off of the laser beam.

Figure 27 shows a laser welded single-lock seam. The welding process provides a hermetic seal; however, the lap joint tolerances of Fig. 25 must be observed, especially between the lowest two of the three plates to be joined. Minimum or no heat sinking is required for this joint, while good heat sinking is required for the joint shown in Fig. 26. Often, joint design is very similar to that for EBW (Ref 36). Figure 28 shows joint design for LBW of sheet metal. The corner weld (Fig. 28j) provides a very rigid joint between sheet metal parts. With enough power to melt the flanges, pool flow and surface tension form a bead (grey area). Rigid fixturing is important, and chill plates may be needed. Similar beads can be formed on flange and edge joints.

Figure 28(k), called the "kissing weld" because the pool forms where the two pieces just kiss, produces a small angle between the two parts that traps most of the energy of the laser beam. Very little pressure, if any, is required for welding, but the faying surfaces must fit well. A gap between the sheets allows radiation to escape. This joint successfully joins thin foils in cases where spot welding burns through. Beam trapping, as in the kissing weld, allows full-penetration welding of a T-joint from one side only, an advantage when the far side of the T is inaccessible. Beam-trapping aids deep penetration of the laser beam for welding.

Joint configurations for wires (Fig. 29) were developed by the electronics industry. These welds most often use a pulsed, solid-state laser, although that is not a requirement. For wire-to-wire joints, the two wires must share the incident laser energy. In the cross joint, Fig. 29(c), for example, the laser beam is directed at the intersection of the wires, so that both wires get the direct energy of the beam.

The terminal or lug weld (Fig. 29e) is best made when the thermal mass of the lug equals the thermal mass of the wrapped wire. Directing the laser beam at the lug helps to avoid melt-through of the wire. Laser welds for electrical connections can be made without stripping of the wire. The laser beam vaporizes the insulation.

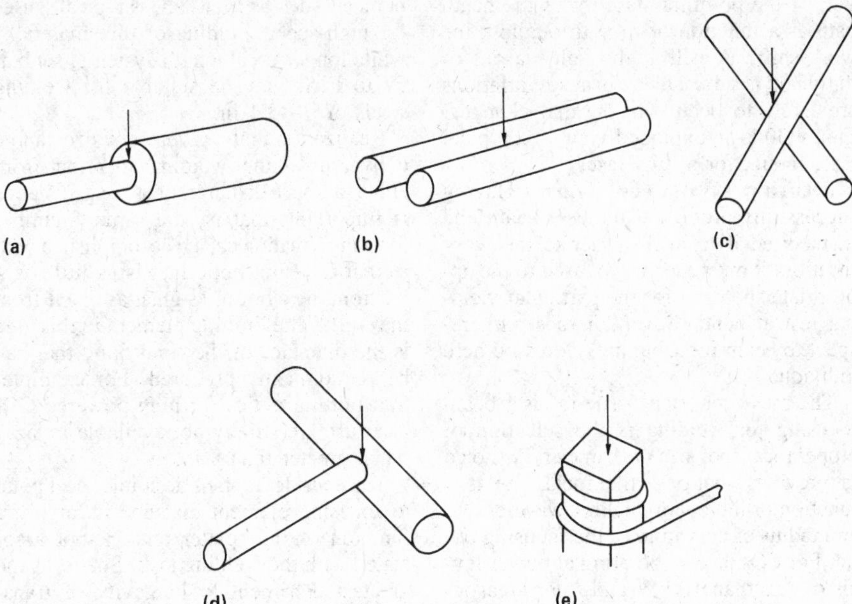

Fig. 29 Joint designs for laser beam welds on wire. Arrows show direction of laser beam. (a) Butt weld. (b) Round-to-round lap weld. (c) Cross-joint weld. (d) Spot weld for T-joint. (e) Terminal or lug weld

Optics Design and Beam Transport

Among the lasers that are appropriate for welding, the nature of the output laser beam varies rather dramatically. Not only are there differences in wavelength, as in the comparison between neodymium and CO_2, but there are differences in beam diameter, mode configuration, divergence, and, certainly, power. These differences in output power configuration require different treatment of the beam transport system and beam-shaping optics. In this section, the several elements of optics system design are considered individually.

Power Measurement. Immediately after the laser, incorporation of a means of measuring the laser beam power is advisable. Power-monitoring devices are usually included as a standard feature in most high-powered laser sources. These devices measure output power in a continuous fashion or intermittently between uses of the beam. Devices that measure power continuously are desirable because they do not require time away from the welding tasks. However, continuous power-monitoring devices may introduce very small oscillations in output beam power, if they are the type of power monitor that samples the output beam with a rapidly spinning blade or wire. In most cases, this method of beam sampling or power measurement is not detrimental; however, in applications requiring exceptionally consistent output power, such small variations may be detrimental. In another convenient technique for power monitoring in relatively low-powered lasers (about 1 kW), the power value is determined through measurement of a partially transmitting rear reflector in the laser resonator. Although this technique does provide a continuous measurement of output power, the measurement is somewhat subjective, because it measures a value that is not truly the output power.

Auxiliary Helium-Neon Pointing Lasers. In addition to power-monitoring devices, many laser sources, particularly those involving higher powered CO_2, utilize low-powered helium-neon (He-Ne) pointing lasers to assist in the alignment of external optics. These prealigned helium-neon pointing lasers are designed to interject their visible beams along a path coaxial with that of the high-powered infrared beam and are extremely useful in the alignment of long or complex optical systems that are not safely aligned by other means because of the relatively high powers involved. In the case of low power (subkilowatt beams of 10.6- or 1.06-μm wavelength), such pointing systems often are not incorporated because of their impact on the laser source cost. In these cases, external optics alignment may be a more tedious procedure.

Beam Transport. Given a laser source with a specific wavelength, divergence, and output power, the task of external optics design can proceed. The first element in this task is the establishment of a beam transport path from the laser source to the workpiece. In a dedicated-production laser system, this path may be short and involve only one or two mirrors or lenses. However, in many other laser system designs, particularly those involving multiple work-station concepts, the beam path and beam transport optics systems may become substantially longer and/or complicated. One of the first questions to be addressed in the beam transport portion of the external optics system is the distance from the laser source to the work. Because all of the lasers under consideration here exhibit some level of divergence or variation from the perfectly collimated beam source, the diameter of mirrors or lenses suitable for dealing with the specific beam, at distance, must be evaluated. To illustrate this point, consider the difference between a 3-mrad divergent, 1-in.-diam CO_2 beam and a 20 mrad-divergent, $^1/_4$-in.-diam Nd:YAG laser beam. Near the laser source, the Nd:YAG beam requires a significantly smaller mirror and/or lens than does the CO_2 beam, but after only a few yards of travel away from the YAG laser, this beam has diverged to a diameter greater than that of the CO_2 beam. Whenever divergence is shown to restrict the ability to transport the collimated laser beam over the required distances, recollimation of the laser beam to a larger diameter is advisable, before transport to the work location. Very simply, if the beam is recollimated to twice its original diameter, the divergence is halved. Thus, beam expansion or recollimation is a convenient means of providing enhanced transportability of the raw output beam.

Optics Materials. The choice of mirror materials for the beam transport system is a function of wavelength, power, and beam diameter. Most importantly, the mirror diameter should be chosen so that the largest anticipated beam diameter falls within the central two thirds of the mirror diameter. This consideration may be viewed as overly conservative by mirror manufacturers; usually, however, it is the case that (1) some variation in beam diameter occurs, (2) variation in beam pointing may occur, and (3) with rare exception, optical elements are most precise near their central portion.

Having selected an appropriate mirror diameter for the beam transport system, the next most important consideration is the ability of the mirror to deal with the power level of the specific laser beam under consideration. Because all reflective surfaces absorb some measurable portion of laser power, this power must be dissipated in a manner that does not induce distortion in the mirror face by virtue of thermal gradients. Most mirror manufacturers specify continuous and/or peak power densities that may be used with their devices. Caution should be exercised in this area, however, because mirror reflectivities degrade as a function of time and use.

The ability to maintain a consistent, high-reflectivity surface on laser mirrors is a function of the mirror material and the quality of the atmosphere surrounding the mirror. Dust particles, moisture, and chemical vapors are examples of conditions that present potential hazards to the maintenance of mirror quality. Among the most detrimental contaminants to high-powered laser optics are those produced by the welding process itself. Although LBW is typically cleaner in operation than conventional welding processes, it still liberates significant amounts of metal vapor and/or spatter.

Atmospheric Effects. At very high power levels in excess of 5 kW, a disruptive phenomenon called "beam blooming" or "thermal blooming" may be encountered. Thermal blooming occurs when absorbers in the beam path, such as dust particles, chemical vapors, or even moisture, are heated by the beam. This heating of the beam path atmosphere causes refraction to occur, which results in distortions to the beam direction and/or quality. Because most high-powered laser beam paths are enclosed in tubes for protection of personnel, often the introduction of clean dry air or nitrogen into these tubular beam paths is convenient. Alternately, the forced flow of air across the beam path has been used effectively as a means of minimizing beam blooming, because the more rapidly moving air does not reside in the beam path for a period of time suitable to produce the blooming effect.

Optics Mounting. Finally, to complete the beam path, the beam transport mirrors must be mounted to some stable surface. The mounting location should be vibration-free and thermally stable, if consistent results are to be expected. In many instances, the beam path optics have been mounted directly to the laser and to the work-station enclosure. Caution should be exercised when attaching optics devices directly to the laser enclosure, because this enclosure may produce vibrations of its own.

In applications involving multiple work stations, it has been consistently demonstrated that the floor upon which the laser and work station are positioned may serve equally well for the positioning of beam path optics. Regardless of the choice of beam path mounting locations, examination of the pointing stability of the beam path as a function of time throughout the work shift, as well as through seasons of the year, is advisable, because variations are likely to occur via thermal changes. Figure 30 is an exploded view of a typical laser mount for a ruby laser.

Focusing Optic Selection. Having successfully produced the laser beam and transported it from the laser to the work location, it must then be focused to the appropriate spot size for the particular welding test at hand. Several issues with respect to beam focusing are addressed here individually.

The most important issue in laser beam focusing for welding is the selection of proper focal spot size or diameter. As noted in previous sections, the focal size is a function of the beam quality, wavelength, and radius of curvature of the focusing optic. For CO_2 lasers operating at power levels of less than 10 kW and, more reasonably, less than 5 kW, focusing may be accomplished with transmission optics such as zinc selenide or gallium arsenide. The selection of focal length for a particular welding application is always a compromise among welding speed, penetration, fit-up, and joint positioning requirements. For example, higher speed welding often can be accomplished with smaller focal spot diameters; however, correct positioning of a weld joint for a presentation to an exceptionally small focal spot may be difficult. As a starting number for consideration in the context of this general discussion, it is recommended that the focal spot size selected for a welding application be from one seventh to one fourth of that of the desired penetration. Especially for good fit-up and joint location situations, the focal spot size may be smaller; situations involving lower quality fit-up and joint positioning may require larger focal points to guarantee involvement of both portions of the intended weld joint.

Determination of the focal spot size of a given optical system is possible by multiplying the focal length of the focusing element by the divergence of the laser beam. For example, a laser beam with divergence of 3 mrad can be focused with a 10-in. focal length lens to a spot size of about 0.03 in. In discussions of focal lengths and spot sizes, the concept of *f*-number frequently arises. The *f*-number of a focusing device is simply the focal length divided by the diameter of the collimated beam at the focusing element. Assume in the example just cited that the input beam diameter was 1.6 in. Thus, the *f*-number of this focusing system would be 10 in. divided by 1.6 in., or F6.25. A low *f*-number, such as F3 to F5, is typically used for high-speed welding of thin materials, while longer focal length systems, such as F7 to F15, may be suitable for welding steels of $^1/_4$ to 1 in.

Realization that a given spot size can be delivered to the welding location from either a short distance or a long distance is important; that is, the same *f*-number focusing situation can be accomplished with a final focusing optic that is as little as 4 in. from the work or as great as $3^1/_4$ ft from the work. The limiting element in this area is the diameter of the final optic that can be economically procured. For example, transmitting optics for high-powered CO_2 laser use are simply not available in sizes much greater than 3 in.

The consideration of desirable focal point to focusing element distance is largely a function of the spatter and/or soot associated with the welding task. Spatter is the greatest detriment to longevity of transmitting focusing optics. When an element of spatter impinges on a transmitting focusing element, the optical coating is destroyed at the spatter impingement location, and this location is then likely to absorb more beam energy than other non-damaged surface areas. As the focusing element becomes less and less transmitting, by virtue of spatter damage or dirt accumulation, the heating of the optical element by the beam energy is increased. If the heating becomes excessive, the resulting distortion can cause catastrophic failure or breakage of the lens. Thus, provision of a maximum cleanliness situation for lens performance longevity is important. This cleanliness can be accomplished with coaxial flowing gases that tend to resist the deposition of spatter and soot, or it may be accomplished with rapidly flowing transverse curtains of air intended to deflect spatter and dirt.

When using transmitting focusing devices for high-powered (1 kW and greater) LBW, the understanding that individual focusing elements exhibit a slightly different focal point position is essential. Because the effective focal point position in a given welding position may be limited to ± 0.02 in. or less, the assumption that replacement focusing optical elements exhibit exactly the same focal point position as an original is not realistic. Furthermore, the heating of the transmitting element in the presence of the laser beam frequently causes a shift in focal point position to occur. This thermally induced focal point position shift usually occurs in a direction

Fig. 30 Laser mount for a ruby or YAG laser rod. Mount includes the rod, about 6 in. long; a helical flash lamp; and the reflectors and mechanical fittings to support the rod and flashlight together in alignment.

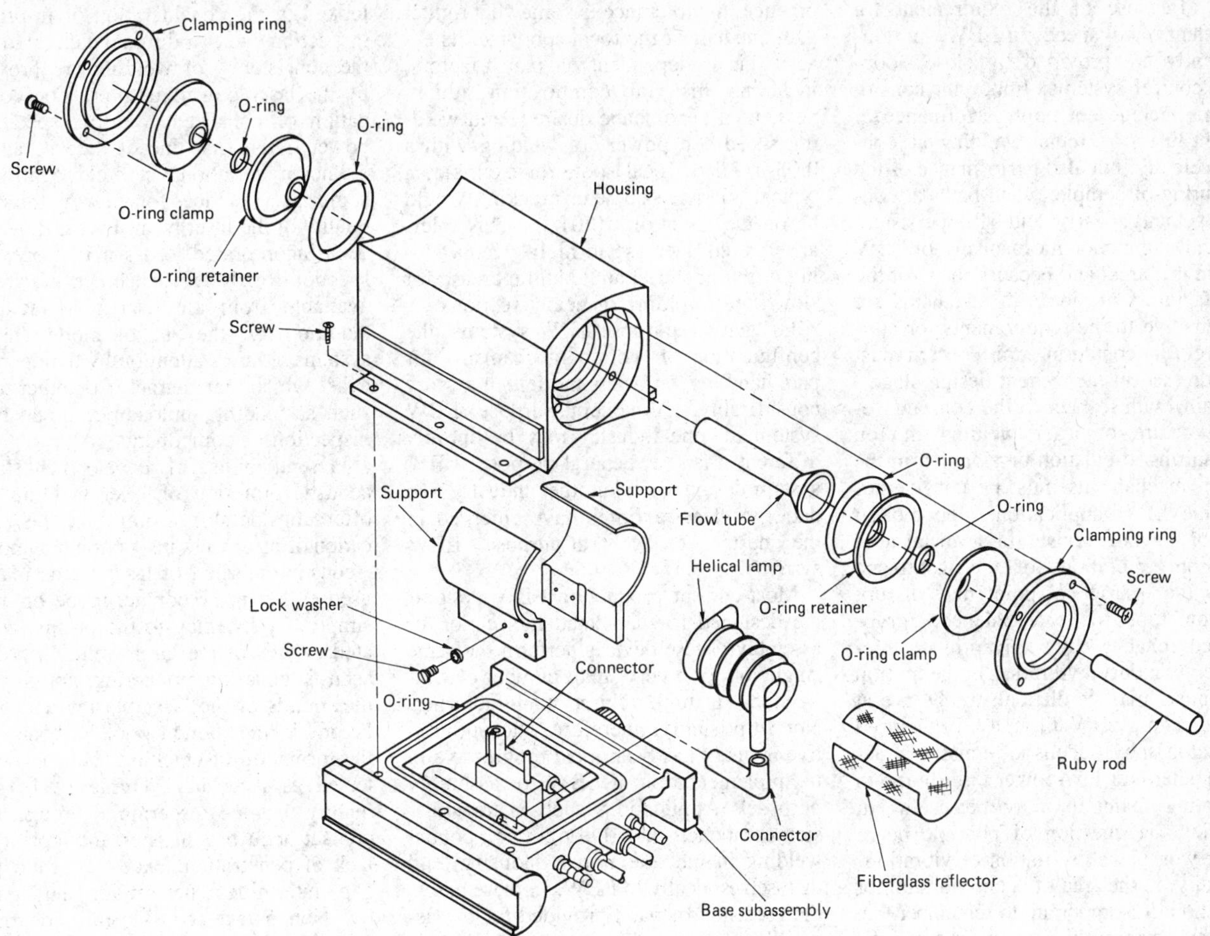

toward the focusing element and usually occurs in less than 1 s. In some cases, programming of a shift in the focal point position to accommodate this short-term, thermally induced focal point shift is necessary. Note that such focal point shifts do not occur with properly cooled metal-substrate reflective optics, only with transmitting ones. A 5-kW CO_2 laser beam, for example, can cause a 2.2-in.-diam, 10-in. focal length zinc selenide lens to shift its focal point as much as 0.04 in. within the first second of beam-on time. Considerable confusion can be saved in the initial setup of focal point position for transmitting laser optics if this consideration is kept in mind.

If long optic-to-work distances are required, either to avoid a high degree of spatter or to accommodate fixturing in complex weldments, larger diameter reflective optics can be utilized. The number of reflecting optic focusing systems available continues to increase with the rapid acceleration in LBW applications.

Focal Point Travel. If the welding motion of the focal point is to be accomplished through motion of the focusing system, as opposed to motion of the work, consideration of the vibration that the motion mechanism may introduce to the focal spot is necessary. Because the focal spot occurs at some distance from the focusing element, small vibrations of the focusing element mounting system can cause larger oscillations in the focused spot itself. Such vibrations cause reduced penetration, increased porosity, and inconsistent welding performance. Vibrations of this sort should be held to less than one quarter of the focal spot size. Even this small degree of vibration may induce inconsistencies in welding performance. It should be noted that longer working distances may imply a greater possibility of detrimental vibration amplitude.

Finally, multidirectional contour welding involving focal point motion in more than one direction requires constant tangential velocity of the focal point. The motion of the focal point around circles or other complex shapes must be smooth, and the tangential velocity should not vary beyond allowable limits (usually several percent of the nominal velocity). This issue is discussed further in the following section on work-handling devices.

Work-Handling and Beam-Handling Devices

As higher power industrial lasers have become available, the laser welding speed for a given thickness has increased. For thin penetration processes, even at moderate laser powers, welding speeds are typically higher than those encountered in conventional welding. Because of the higher speed welding requirements, laser motion mechanisms must be capable of providing more rapid accelerations. The speed control is likely to be a more demanding task in laser welding than in other welding processes. For instance, a butt weld in a laser welding situation may be tolerant only to ±5% variation in travel speed. Because of the requirement for consistent travel speeds in LBW, motions frequently are provided by closed-loop-servo control systems. Important considerations include not only performance of straight-line or circular welding at constant velocity, but also performance of the contouring of complex weld paths at constant velocity. Early and still persistent difficulties in motion mechanisms for LBW exist in this area, and because many of the manufacturers of motion mechanisms are not sensitive to the requirements for constant velocity contouring, this concern must be addressed in the system design stage.

Again, with respect to the constant velocity nature of laser welding motion mechanisms, oscillation or vibration in the motion mechanisms must be considered. In some LBW applications, the molten pool or keyhole exists in a metastable condition. Vibrations of the workpiece and/or the laser focal point may disrupt this sometimes delicate balance. Servo-induced chatter or conventional mechanically induced vibration in the motion mechanism may be difficult to observe at the source of a LBW difficulty. High-speed cinematography, acoustic emission, accelerometers, or low-power laser burns in replicating materials may be useful in resolving the question of possible beam and/or workpiece oscillation or vibration.

Finally, in the area of beam vibration or oscillation, it is important to remember that the distance from the focal point to the final optical element represents a lever arm through which the amplitude of possible optics vibrations may be magnified. This concern for lever-arm-amplified focal point vibration is one reason motion mechanism manufacturers prefer to provide systems with moving workpieces, as opposed to moving optics. Several attempts at providing precision high-speed optics mechanisms failed in the early and mid-1970's, while more recent attempts at optics motion mechanisms have been considerably more successful. In any event, it is reasonable to expect that optics motion mechanisms for LBW will require higher quality and will be more expensive than motion mechanisms that are associated with conventional flame cutting, plasma cutting, or welding torch motion mechanisms.

Seam tracking requirements for LBW are substantially more stringent than those for conventional welding processes. It should be readily apparent that tracking tolerances cannot exceed one half of the beam focal spot size, as misalignment by that distance will result in beam impingement on one side of the joint only. In practice, the tolerance is somewhat tighter than one half of the focal spot size; its exact value is dependent on many factors, including material composition, thickness, final bead shape desired, and welding speed and power. In welding with a 0.03-in.-diam focal spot, for example, a typical tolerance on seam tracking would be on the order of ±0.01 in. This tolerance would decrease in high-speed welding of thin material and could increase for slow-speed welding of heavy sections.

For general-purpose LBW systems, the configuration of motion mechanisms for part handling or beam motion may vary considerably. A general-purpose LBW system for one industry may be entirely different than a general-purpose LBW system designed for another industry. To date, no clear favorites have emerged in the design of general-purpose LBW systems.

Most current production LBW systems are designed for dedicated fabrication or assembly. These devices perform the same single job, or a very small number of similar jobs, in high-volume manufacturing. Not surprisingly, therefore, the automotive industry has been most progressive in its implementation of LBW for assembly purposes. Applications such as transmission component or drive train component welding in the automotive industry lend themselves ideally to LBW. In these cases, the precision of parts provided for the assembly task is well controlled. Fit-up is therefore well controlled and consistency of material, surface finish, and cleanliness is favorable. In these applications, provision of automatic loading, inspection, and unloading features for the LBW system is often desirable. Devices such as dial-feed part loading and handling devices are common, and transfer-line type equipment is also incorporated. In these systems, incorporation of fixturing techniques that are associated in particular with or are appropriate for LBW often is convenient. These fixturing techniques may include magnetic or vacuum chucking. Notably, neither of these latter parts-chucking techniques may be employed conveniently in EBW, which is the nearest competitive process to LBW in the automotive or other high-volume assembly industries.

In-Process Inspection

Because most LBW systems are accompanied by precision motion mechanisms of one sort or another, monitoring the precision of speed control is convenient as a first measure of welding performance. This piece of information is among the most easily acquired and should not be overlooked in the consideration of in-process inspection data. Additional data relative to the consistency of welding are provided by the laser power monitor. The combination of consistent welding speed and power is the first line of defense against unknown variations in weld quality. In some cases, devices for viewing the mode quality of the laser beam in real time have been incorporated to assist in-process inspection. Such devices, however, are not available from all laser manufacturers. Additionally, the use of mode viewing systems requires attention by the operator, a task which may detract from other tasks, such as loading, unloading, or postweld inspection of components.

Presently in its early development stages, acoustic emission of laser welding may offer considerable promise in the application of in-process inspection techniques. Acoustic emission of laser welds may be used either in the contact mode or, more simply, by listening to the plasma-generated sounds of the laser weld. In contact acoustic emission monitoring, not only can the sounds of the laser plasma and oscillations in the molten pool be heard, but the formation of cracking also can be detected as it occurs. Variations in weld quality that are discernible through noncontact acoustic emission monitoring are lack of penetration, excessive weld joint gap, intermittent penetration, and porosity. Noncontact acoustic emission monitoring has the obvious advantage of remote sensing; thus, considerations of contact probe coupling are eliminated and system complexity may be reduced.

Another method of noncontact in-process inspection incorporates either on-axis or off-axis video observation of the laser welding action. A wealth of information exists in the color, intensity, stability, and position of the laser-generated welding plasma with respect to the weld joint location. Additionally, information regarding solidification rate of the molten pool is available in a video presentation of the laser welding operation. While this area of LBW in-process inspection may offer the greatest ultimate quality assurance, considerable hardware and software development of this in-process inspection technique remains to be accomplished.

With few exceptions, in-process inspection of LBW does not take place. The techniques suggested above have all been demonstrated to be effective in research environments; however, they have not been instituted in large part by industry. Industrial practitioners of LBW seem to be more inclined to rely on postweld inspection techniques. In one instance, an automotive manufacturer "overwelded" by as much as 25% to provide a convenient postweld inspection of the underside of a lap weld. Not only does such an inspection technique require an input of labor, but it also reduces the productivity of the welding system. Successful utilization of an in-process inspection technique would, therefore, provide an increase in productivity. As these techniques are developed more fully, they certainly will be utilized, considering their potential impact on productivity.

Operating Costs

Given the rate at which high-power laser equipment is being modified and improved, the presentation of any actual costs for the operation of these devices is inappropriate. Nevertheless, certain operating cost elements are likely to continue to exist, and these are listed here.

Electricity. Regardless of the type of laser, whether it be solid-state or gas, the conversion of electrical energy input to light energy output may be characterized as inefficient. Conversion efficiencies ranging from 2 to 15% are typical. Thus, electric power consumption is a substantial portion of the cost of laser operation.

Cooling. Considering the relative inefficiency of laser light production, considerable waste heat is generated by these devices. The disposal of waste heat introduces another consideration for operating costs. In the case of solid-state Nd:YAG lasers, for example, the disposal of waste heat is accomplished through heat exchangers that incorporate nonconductive deionized water. From time to time, replacement of laser flash lamps (a considerable expense themselves) may be necessary, and some portion of the rather expensive cooling water may be lost. In higher powered CO_2 laser systems, very large amounts of cooling water are required. If in-house chilled water systems are not available, city water must be purchased and then discarded as the cooling medium. In some cases, cooling water costs may be equivalent to electric energy costs.

Medium. In all laser systems, there is a tendency for the lasing medium to degrade. Whether the lasing medium be a glass, crystal, or a gas, the medium must be replaced on some scheduled basis. For solid-state lasers, this schedule may be determined by the number of shots, while gas lasers replace their medium by slowly evacuating and replenishing the gas volume. With more advanced laser discharge systems that are expected to be introduced throughout the 1980's, reduction of the consumption of lasing gases may be possible. This consideration is particularly important for those countries that do not have bountiful supplies of helium, one major element of CO_2 laser gas mixes.

Output Coupler or Window. In addition to the obvious consumables of electricity, cooling water, and lasing media, certain optical elements of the laser may require routine replacement. These routine replacements constitute a consumable cost. In the case of solid-state lasers, the major optical consumable is the flash lamp itself. In high-power CO_2 lasers, the optical consumable is usually associated with extraction of the laser beam from the low-pressure lasing region into the outside world. In the case of lasers of power levels less than 9 kW, an output coupler serves this purpose, and it has a finite lifetime. For lasers of powers higher than 10 kW, aerodynamic windows or curtains that accelerate the use of laser gases when they are opened are incorporated. Furthermore, they are subject to damage by the beam if misalignment of internal optics should occur. In this sense, consideration of the aerodynamic window or curtain as a consumable may be inappropriate; however, experience to date suggests that this must be included in the operating cost summation.

Maintenance. Costs of maintaining high-powered lasers vary dramatically. Electrode cleaning, mirror cleaning, blower bearing replacement, routine optics alignment, and many other factors may contribute to LBW system maintenance costs. Each system must be considered individually to determine expected maintenance costs.

As noted in previous sections, the laser vendor is likely to be the most accurate source of information for laser consumable and maintenance costs. In addition, consultation with users of similar LBW systems may provide a fuller picture of the actual costs to be expected in a production environment than can be gained from distributors. Among all the costs of laser operation, the cost of nonperformance is certainly the highest. System uptime should be considered heavily in the total evaluation of laser operational costs.

Safety

Laser beam welding hazards differ substantively from hazards encountered in other welding techniques. The hazards are not readily apparent, and inexperienced personnel may suffer permanent injury before the existence of hazardous conditions is recognized. For this reason, the American National Standard Institute specification ANSI Z136.1, "Safe Use of Lasers" (latest edition), requires that each facility using lasers designate an individual as "laser safety officer." This individual should be familiar with laser safety and ANSI Z136.1. The officer should monitor the use of lasers to ensure adherence to safe laser practice and the ANSI requirements. Because of the complexity of laser safety, this approach is strongly endorsed. The following brief review of laser safety by itself is not sufficient to ensure personnel safety.

Electrical Hazards. All lasers used for welding employ high voltages capable of lethal electric shocks. Therefore, maintenance should be performed by personnel familiar with high-voltage safety procedures.

Power supplies for high-power lasers contain capacitors capable of lethal shock even after initial discharge due to a phenomenon known as charge buildup. To facilitate safe maintenance access, the following provisions should be included:

- An automatic discharge and grounding circuit that is actuated when the laser is turned off
- Discharge and grounding interlocks on all access panels
- Grounding rods for manual verification of complete discharge. Safety glasses should be used, because explosion-like discharges are possible on partly charged capacitors.
- Grounding straps to short out capacitors to prevent charge build-up

In addition, all capacitors should be discharged and grounded before any work is performed on or near high-voltage components.

Control of laser performance usually requires switching of capacitors. Insulated switches that do not expose personnel to electric conductors are preferred. If switching requires work on bus bars, the procedure listed above should be followed, and the use of insulated tools should be considered. Cooling water leaks are not acceptable, especially when electrical and cooling lines share the same umbilical.

Eye Hazards. Any laser beam capable of welding metals is also capable of causing serious eye damage. Personnel exposure to the beam and any specularly reflected beam must therefore be prevented at all times. Certain lasers, however, are also capable of producing diffuse reflections that can cause permanent eye damage. Hence, viewing of the impact area of the laser beam or reflected beams also must be prevented. The preferred method for this is by complete enclosure.

Enclosures can range from a simple sleeve between the laser optic and the part to be welded to fully automatic operations in enclosed rooms. The following are general guidelines:

- The enclosure must be opaque to the laser wavelength. Metal enclosures are generally suitable, but plastics also are a possible selection. Infrared lasers, for example, can be enclosed in clear poly methylmethacrylate. The required thickness can be calculated using ANSI Z136.1 and the transmittance properties of the material.
- The enclosure must be interlocked to prevent firing of the laser beam when personnel could be exposed. When the piece to be welded is a part of the enclosure, the interlocks should also prevent firing, unless the piece is in place. In the case of pulsed lasers, breaking of interlocks should also discharge the stored energy into a dummy load.
- Signs are required at access points to the enclosures; see ANSI Z136.1.
- Viewing of the weld area can be accomplished in several ways. Most common are viewing ports with filters and television monitors. For microwelding, microscopes should have filtered viewing optics or flip mirrors that permit either welding or viewing.
- Alignment of laser welding systems should be accomplished using low-power lasers.

Laser welding can also be performed with personnel in attendance. When done in this manner, the hazard must be evaluated for each welding process using the ANSI standard. General requirements include:

- The laser welding area must be completely enclosed and access must be restricted.
- In most cases, laser eye protection is required for all personnel in the area. The optical density of the eye protection must be calculated to reduce the potential eye exposure to less than the maximum permissible exposure (MPE) level.
- The welding beam must be carefully controlled and should still be enclosed to the greatest extent possible. Eye protection is likely to fail on exposure to the primary beam.
- Compliance with personnel training requirements is exceedingly important.

Skin Exposure. Skin exposure to the primary beam obviously can result in thermal burns and must be prevented by complete enclosure and/or operator training. Even in attended operations where the operator is permitted to view the welding process, provision of partial enclosures to prevent personnel from placing any part of the body into the beam path is desirable.

The ANSI standard also prescribes MPE values for skin exposure. In the visible and near infrared regions, these are much greater than the MPE values for the eyes; hence, a problem from excessive exposure due to diffuse reflections at these wavelengths rarely occurs. Excessive skin exposures in the ultraviolet and far infrared regions are possible, however. Typically, ultraviolet exposures can be controlled with clothing of tightly woven material and barrier creams applied to exposed skin. Harmful levels of ultraviolet light can also be generated by flash lamps; hence, covers should be kept in place.

Chemical Hazards. Laser welding generates metal fumes similar to other welding processes, and the hazard is largely dependent on the composition of the welded metals. Ventilation is required to meet OSHA standards and American Conference of Governmental and Industrial Hygienists (ACGIH) threshold limit values. In high-power welding applications, fumes can be generated in sufficient quantities to make local exhaust ventilation, in addition to general room ventilation, both necessary and economical.

Harmful fumes or vapors can also be generated when laser energy is deposited in unwanted materials, such as breakdown of plastic enclosure materials due to laser exposure and failure of exotic lens materials due to thermal runaway. These conditions are best controlled by careful material selection and monitoring of welder performance. Finally, laser optics cleaning agents may be toxic and flammable and should be handled accordingly.

Training, Medical Examinations, and Documentation. ANSI Z136.1 requires that training in the potential hazards and control measures be provided to operators, engineers, technicians, maintenance, and service personnel. Special training on subjects such as potential hazards (including biological effects), control measures, and applicable standards is required for the laser safety officer. A model safety and training program is outlined in Appendix D of the standard.

Routine medical surveillance of laser users is no longer mandated by the standard. However, an employer may wish to provide such for medical/legal reasons; i.e., to document what eye damage existed prior to commencement of laser work and that no additional damage has occurred.

Good documentation includes:

- Listing of lasers, duration of use
- Results of hazard surveys and calculations of accessible laser radiation
- Interlock tests
- Laser use procedures
- Employees working in laser area
- Date and extent of employee training
- Dates and results of medical examinations (if any)
- Training and qualifications of laser safety officer

REFERENCES

1. Jenkins, F.A. and White, H.E., *Fundamentals of Optics,* 4th ed., McGraw-Hill, New York, 1980, p 632-651
2. Fowler, M.C. and Smoth, D.C., Ignition and Maintenance of Subsonic Plasma Waves in Atmospheric Pressure Air by CW CO_2 Laser Radiation and Their Effect on Laser Beam Propagation, *Journal of Applied Physics,* Vol 46 (No. 1), Jan 1975, p 138-150
3. Hughes, T.P., *Plasmas and Laser Light,* John Wiley & Sons, New York, 1975
4. *Handbook of Chemistry and Physics,* 46th ed., CRC Press, Cleveland, 1964
5. Block-Bolten, A. and Eagar, T.W., Selective Evaporation from Weld Pools, MIT Report 5-55-82
6. Brown, C.O. and Banas, C.M., Deep-Penetration Welding, AWS Welding Conference, San Francisco, April 1971
7. Locke, E.V., Hoag, E.D., and Hella, R.A., Deep-Penetration Welding with a High-Power CO_2 Laser, *IEEE Journal of Quantum Electronics,* Vol QE-8 (No. 2), Feb 1972
8. Banas, C.M., High-Power Laser Welding—1978, *Optical Engineering,* Vol 17 (No. 3), May/June 1978
9. British Patent GB1591793-D26, U.S. Patent 4127761-A49
10. U.S. Patent 4 152 575, 1 May 1979
11. Minamida, K., Yamaguchi, S., Sakufai, H., and Takafugi, H., CO_2 Laser Welding with Plasma Utilization, First International Congress on Applications of Lasers and Electro-Optics, Boston, Sept 1982, p 21-23
12. Banas, C.M., Alholm, H.A., and Olihan, W.T., Laser Welding at 3.8 Microns, Golden Gate Welding Symposium, San Francisco, 9-11 Feb 1983
13. Snow, D.B. and Breinan, E.M.,

Evaluation of Basic Welding Capabilities, Department of Navy, No. R78-91189-14, July 1978
14. Breinan, E.M., Banas, C.M., and Greenfield, M.A., Laser Welding—The Present State of the Art, DOC IV-181-75, Tel Aviv, 6-12 July 1975, p 1-53
15. Schwartz, M.M., *Metal Joining Manual,* McGraw-Hill, New York, 1979
16. Metzbower, E.A. and Moon, D.W., Mechanical Properties, Fracture Toughness, and Microstructures of Laser Welds of High Strength Alloys, *Proc. Conf. on Applications of Lasers in Materials Processing,* American Society for Metals, 18-20 April 1979, p 83-100
17. Mazumder, J., Laser Welding of Aluminum Alloys, Alcoa Laboratories Joining Division, Alcoa Center, PA
18. Blake, A. and Mazumder, J., Control of Composition During Laser Welding of Aluminum-Magnesium Alloys Using a Plasma Suppression Technique, *Materials Processing*, Vol 31, 1982, p 33-50
19. Willgoss, R.A., Megaw, J.H.P.C., and Clark, J.N., *Welding and Metal Fabrication,* 9-11 May 1978, p 267-278
20. Crafer, R.C., Advances in Welding Processes, *Proc. 4th Int. Conf. Harrogate, Yorks.,* No. 46, June 1981, p 19-25
21. Banas, C.M., Electron Beam, Laser Beam, and Plasma Arc Welding Studies, Contract No. NASA1-12565, NASA, March 1974
22. Banas, C.M., Laser Welding Developments," *Proc. CEGB Int. Conf. on Welding Res. Related to Power Plants,* Southampton, England, 17-21 Sept 1972
23. Breinan, E.M. and Banas, C.M., Preliminary Evaluation of Laser Welding of X-80 Arctic Pipeline Steel, *WRC Bull.,* Vol 201, Dec 1971
24. Mazumder, J., Ph.D. thesis, London University, 1978
25. Mazumder, J. and Steen, W.M., Laser Welding of Steels Used in Can Making, *Welding J.,* Vol 60 (No. 6), June 1981, p 19-25
26. Steen, W.M. and Eboo, M., Arc Augmented Laser Welding, *Metal Construction,* Vol 11 (No. 7), 1979, p 332-335
27. Shewell, J.R., *Welding Design & Fabrication,* June 1977, p 106-110
28. Metzbower, E.A., Moon, D.W., and Fraser, F.W., Mechanical Properties of Laser Beam Weldments, *Trends in Welding Research in the United States,* American Society for Metals, 1982, p 563-580
29. Yessik, M. and Schmatz, D.J., Laser Processing in the Automotive Industry, SME Paper, MR74-962, 1974
30. Mazumder, J. and Steen, W.M., Structure and Properties of Laser Welded Titanium Alloy, TMS-AIME Fall Meeting, No. F79-17, Sept 1979
31. Seaman, F.E., Establishment of a CW CO_2 Laser Welding Process, USAF Tech. Report, AFML-TR-76-158, Sept 1978
32. Banas, C.M., Electron Beam, Laser Beam, and Plasma Arc Welding Studies, Contract No. NASA CR-132386 NASA, March 1975
33. Adams, M.J., CO_2 Laser Welding of Aero-Engine Materials, No. 335/3/73, British Welding Institute, 1973
34. Fraser, F.W. and Metzbower, E.A., Laser Welding of a Titanium Alloy, *Advanced Processing Methods for Titanium,* The Metallurgical Society of AIME, Warrendale, PA, 1982, p 175-188
35. David, S.A. and Liu, C.T., High-Power Laser and Arc Welding of Thorium Doped Iridium Alloys, No. ORNL/TM 7258, Oak Ridge, TN, May 1980
36. Shewell, J.R., Design for Laser Beam Welding, *Welding Design & Fabrication,* June 1977

SELECTED REFERENCES

- Baardsen, E.L., Schmatz, D.J., and Bisaro, R.E., *Welding J.,* Vol 52, April 1973, p 227-229
- Seaman, F.D. and Hella, R.A., Establishment of a Continuous Wave Laser Welding Process, IR-809-3 (1-10), AFML Contract F336 15-73-C5004, Oct 1976
- Banas, C.M. and Peters, G.T., Study of the Feasibility of Laser Welding in Merchant Ship Construction, Contract No. 2-36214, U.S. Dept. of Commerce, Final Report to Bethlehem Steel Corp., Aug 1974
- American National Standard for the Safe Use of Lasers, ANSI Z136.1 (latest edition), American National Standards Institute, Inc., New York
- Threshold Limit Values for Chemical Substances and Physical Agents in the Workroom Environment with Intended Changes for 1980, American Conference of Governmental and Industrial Hygienists, Cincinnati
- Sliney, D. and Wolbarsht, M., *Safety with Lasers and Other Optical Sources,* Plenum Press, New York, 1980

Solid-State Welding

By James L. Jellison
Supervisor, Process Metallurgy
Sandia National Laboratories
and
Frank J. Zanner
Member of the Technical Staff
Sandia National Laboratories

SOLID-STATE WELDING (SSW) processes are those that produce coalescence at temperatures below the melting point of the base metal being joined. These processes involve either the use of deformation, or diffusion and limited deformation, to produce high-quality joints between both similar and dissimilar materials.

One form of SSW, called diffusion welding, is accomplished by bringing the surfaces to be welded together (faying surfaces) under moderate pressure and elevated temperature in a controlled atmosphere so that a coalescence of the interfaces or faying surfaces can occur. The other form, called deformation welding, is accomplished by subjecting the surfaces to be welded to extensive deformation. Melting or fusion is not associated with either process.

Because diffusion welding and deformation welding both may be accomplished by the application of heat and pressure, some specific processes may share characteristics of both methods. In general, diffusion welding involves little bulk deformation and is conducted at temperatures greater than one half the absolute melting point ($>^1/_2\ T_m$). Conversely, deformation welding typically is conducted at temperatures ranging from room temperature to less than one half the absolute melting point. However, some deformation welding processes such as forge welding, roll welding, and extrusion welding may be performed at high temperatures. Another distinction is that deformation welding is usually accomplished within seconds, whereas diffusion welding typically requires weld durations ranging from minutes to hours. Accordingly, diffusion welding is defined as a solid-state welding process wherein coalescence of contacting surfaces is produced with minimum macroscopic deformation by diffusion-controlled processes that are induced by applying heat and pressure for a finite time interval. By contrast, deformation welding includes those SSW processes where mating of weld pairs is accomplished by gross plastic flow, and surface contaminants are primarily disrupted by plastic deformation.

Selecting Solid-State Welding Processes

Solid-state welding processes are selected according to the materials involved and the application. Typical reasons for selecting SSW are listed below.

Near Base-Metal Properties in the Joint. Because solidification microstructures are avoided, the properties of solid-state welds can, in some cases, be equivalent to base-metal properties. In deformation welds, strain hardening can actually result in a joint that is stronger than the base metal.

Weldable Large Surface Areas. Processes such as roll welding and gas pressure diffusion welding are capable of joining large surfaces that would be impossible to join by fusion welding. Also, they eliminate the need for low-melting-point filler metals such as required in brazing.

Welding Complex Assemblies to Near Net Shape. Gas pressure diffusion welding is particularly adaptable to welding complex assemblies of various alloys to produce an assembly requiring little forming or machining. Other SSW processes can be used to attach a stud, bolt, or subassembly to a larger assembly.

Making Multiple Joints in One Operation. Because SSW generally involves the application of heat and pressure over a large area, multiple joints can often be accomplished in one operation. There may be several joints on one assembly or the welding of a large number of assemblies, as in a gas pressure diffusion welding batch process.

Welding Difficult-to-Weld Metals. Metals and alloys that are generally considered unweldable (refractory metals and even cermets) can be solid-state welded. This is frequently accomplished by diffusion welding with foils or coatings as intermediate layers in the weld joint.

Welding Dissimilar Metals Exhibiting Metallurgical Incompatibilities. Metals that normally cannot be welded to one another, typically due to the formation of intermetallic compounds, generally can be solid-state welded. Selection of an interlayer metal may be required to promote compatibility in the solid state. Examples of metallurgical incompatibility are aluminum to austenitic stainless steel. Welds of dissimilar alloys may not perform well in some environments even though as-welded joint properties are acceptable—for example, highly active galvanic corrosion couples in an aqueous environment.

Reduction in Distortion and/or Residual Stresses. Because SSW generally can avoid the application of localized heat, distortion and thermally induced residual stresses can often be minimized. An important exception, however, is diffusion welds involving materials with very different coefficients of thermal expansion.

High Production Rates via Automation. Some SSW processes are particularly adaptable to automation to achieve low-cost production. Examples are multiple upset butt welding, lap welding of sheet metal, and various forms of thermocompression welding and ultrasonic welding employed in microminiature welding of electronic circuits.

Welding Combined With Forming and Heat Treat Processing. Cost advantages can sometimes be realized by combining SSW with another thermal or thermomechanical process required in the production of a part. An outstanding example is the combined superplastic forming and diffusion welding of titanium alloy parts.

Reduction of Material Cost. Cladding of an inexpensive alloy with a more expensive alloy by roll welding can significantly reduce the cost of materials. Also, solid-state welds can be employed to make complex assemblies where expensive alloys are used only in the section required to achieve wear properties, toughness, and strength.

Diffusion Welding

Because diffusion welding requires the application of heat and pressure, the specific equipment utilized is a function of part shape and size and the type of atmosphere and temperature required. In most cases the equipment is custom built by or for the user, and the actual welding is accomplished in a vacuum, inert gas, or reducing atmosphere.

A protective envelope is supplied by either a stationary hard shell furnace or by encapsulating the parts to be welded in a thin sheet metal container or retort made from a suitable alloy, usually low-carbon steel or 304 stainless steel. These thin containers are generally fabricated by fusion welding and subsequently evacuated through a small tube welded to the container. In some instances evacuation continues during the application of temperature and pressure, and in other cases the tube is closed by pinch welding after an initial evacuation is completed.

A number of techniques are used to apply load to the faying surfaces. The method selected is strongly constrained by part shape and temperature requirements. Load is transmitted to the faying surfaces of encapsulated parts by isostatic gas pressure on the evacuated container. This gas pressure can range from atmospheric pressure to more than 30 ksi. Load can also be applied by hydraulic presses with heated platens, rolling mills, or mechanical clamping devices custom built for the application. When diffusion welding is carried out in stationary or hard shell furnaces, the load is applied by press platens contained within the enclosure or by differential thermal constraint. Loading by differential thermal constraint is accomplished with special jigs or fixtures made from materials that have a low coefficient of thermal expansion. Molybdenum is a commonly used low-expansion material.

Heat typically is applied in stationary hard shell furnaces by electrical resistance elements and in some cases by induction heating. If the part to be welded is encapsulated in a retort, the entire assembly can be heated in a conventional air resistance furnace or by resistance heaters surrounding the retort.

All of the above techniques are covered in greater detail in a later section on applications in this article. Terms that sometimes are used synonymously with diffusion welding include diffusion bonding, solid-state bonding, pressure bonding, isostatic bonding, and hot press bonding.

Deformation Welding

Cold Welding

Cold welding is accomplished at or near room temperature. The major divisions of cold welding are lap welding, butt welding, and slide welding. Other deformation welding processes such as roll welding and ultrasonic welding also are often performed at room temperature.

Lap welding involves the welding of ductile sheet metal by overlapping the two pieces and applying pressure by means of suitable tools. Normally this is accomplished by means of tool steel indentors, although end lap joints have sometimes been produced employing flat anvil tools. A typical application of indentor tools to lap welding is illustrated in Fig. 1. The indentation may be from both sides (opposed indentors) or from one side only (indentor/flat anvil). The indentors may be of round or rectangular cross section. The diameter or minimum cross section of the indentors is typically one to three times the thickness of the sheets, with larger indentors tending to reduce the amount of deformation required for welding. Deformations in the 50 to 90% range are usually needed.

As shown in Fig. 1, indentors usually are equipped with shoulders that (1) control the amount of deformation, (2) minimize distortion, and (3) promote welding in the peripheral area surrounding the indentor when deformation exceeds approximately 70%. A similar approach is to employ inner and outer punches where the outer punch clamps the sheets at pressures approximating the yield stress and the inner punch (or indentor) produces the required deformation. Both annealed and strain-hardened sheet metals may be lap welded. Annealed materials promote uniformity of welding from the center to the edge of the weld, whereas strain-hardened materials tend to weld at less total deformations. In dissimilar metal welding, it is often necessary to increase the size of the indentor or to employ a flat anvil on the side of the softer metal.

The most common metals joined by lap welding are aluminum, aluminum alloys,

Fig. 1 Indentor tool used in lap welding. Opposed indentors with shoulders

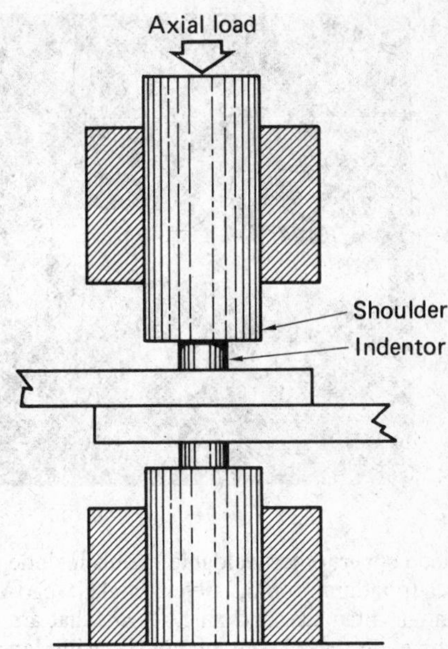

Fig. 2 Butt welding with beveled clamps that facilitate flow of upset metal away from weld zone

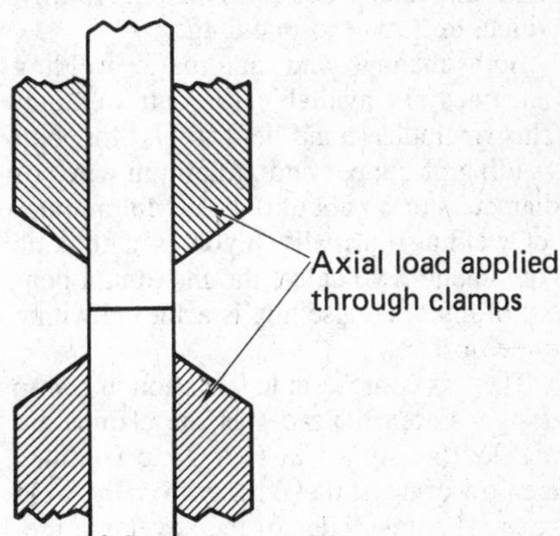

Fig. 3 Tongue-and-groove joint using interference fit with ribbed tongue to create cold weld. (a) Before welding. (b) After welding

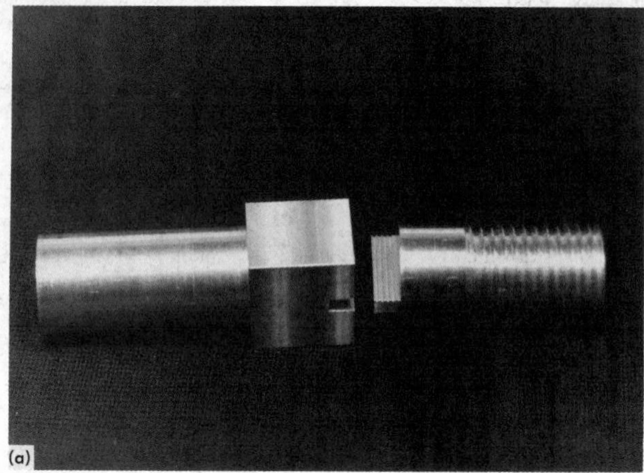

and copper; other weldable metals include lead, indium, gold, silver, nickel, cadmium, titanium, and zinc. Terms that are sometimes used synonymously with lap welding include indentation welding, small tool welding, and cold sheet welding.

Butt Welding. In butt welding, the ends of pairs of wires and rods are pushed against each other with sufficient force to upset the ends. Normally this is accomplished by mechanically clamping both parts at some distance from their ends and applying pressure with the clamps, which results in the upset materials extruding between the clamps (Fig. 2). The extension beyond the clamps must not exceed the length that will result in buckling (usually 1 to 2 diam). However, this does not limit the amount of deformation that can be obtained, because the parts can be reclamped following upsetting and deformed in multiple steps. Although clean, ductile metals can be welded in one upset, multiple upsetting is commonly used to increase tolerance to surface contaminants, to weld less ductile metals and alloys, and to weld small-diameter wires (less than 0.013 in.), which are prone to buckling.

Both manual and automatic welding machines are available for butt welding. This includes hand-held tools for butt welding of copper and aluminum wires of diameters up to about 0.1 in. Multiple upset welding normally involves automatic equipment. Typically, the maximum benefit of multiple upsetting is achieved within three upsets.

There is considerable variation in clamp design. Often the faces of the clamps are beveled (as shown in Fig. 2) to facilitate the flow of upset metal away from the weld zone. The inner edge of the bevel is sometimes designed to cut away the upset material. The inner surfaces of the clamps may be smooth or serrated. The axial pressures required for butt welding are typically several times the yield stress. Clamping forces range from somewhat less than the upset force for small-diameter wires or rods held with serrated clamps to about 150% of the upset force for the welding of rods or tubes with smooth clamps. In the weld-

Fig. 4 Thermocompression weld of gold-plated copper alloy CDA 194 to gold metallized ceramic substrate. Welds made at 300 to 335 °F, ~5 s duration

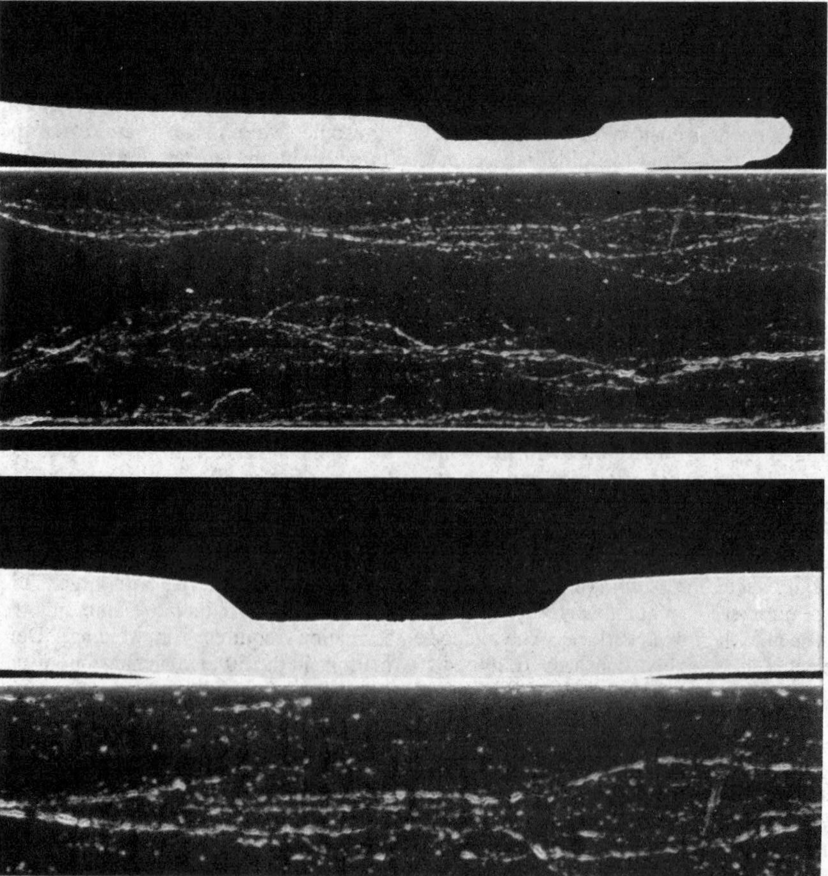

Fig. 5 Design for butt welding of tubes by extrusion

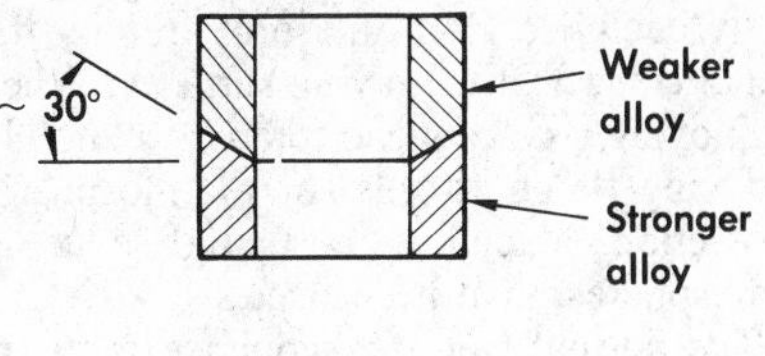

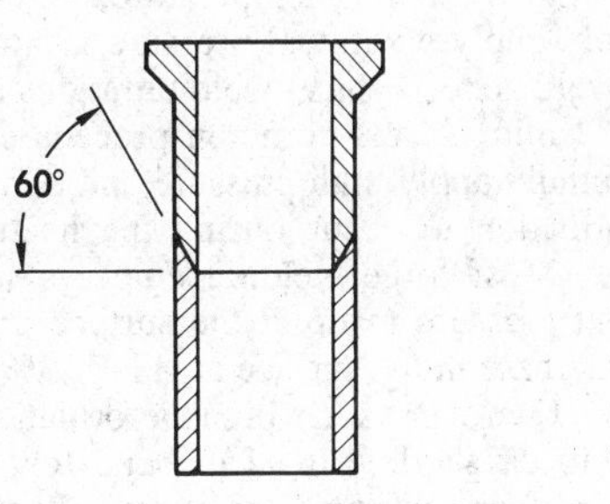

ing of tubes, internal mandrels are sometimes used to minimize internal upset.

The most common applications of solid-state butt welding are joining of electrical cables, particularly aluminum to copper, and joining of tubes. Aluminum and copper are the most common metals welded, but commercially pure iron and nickel can be joined by multiple upsetting. A term that is sometimes used synonymously with butt welding is cold upset welding.

Slide welding is a much less common form of cold welding that involves sliding the surfaces to be welded relative to one another during deformation. The advantage of the technique is that welding can be achieved at lower total deformation than without sliding. The maximum benefit of sliding is generally achieved within sliding distances of 0.3 in. or less. In a typical application involving the joining of tubes, slide welding appears to be a modification of butt welding. Most commonly, one part is twisted relative to another while the upsetting force is applied. Upsetting forces are reduced compared to normal butt welding (to approximate yield stress). Also, tubes have been joined by beveling their ends so that relative movement between faying surfaces is achieved by application of an axial upsetting force.

Another form of slide welding involves the pressing of a ribbed tongue into a tapered groove, where the interference is sufficient to nearly totally deform the ribs at the bottom of the groove. Aluminum tongue-and-groove parts are shown in Fig. 3 before and after pressing. In addition to aluminum and aluminum alloys, copper is commonly joined by slide welding.

Thermocompression Welding

Thermocompression welding includes a group of techniques that produce deformation welds of ductile face-centered cubic (fcc) metals at temperatures ranging up to about half of the melting point ($\sim 1/2$ T_m). Thermocompression welding typically is employed as a microwelding process for joining of interconnections in electronic circuitry. The specific techniques have evolved to accommodate the various conductors and lead geometries used in packaging of integrated circuits and hybrid microcircuits. The most common metals joined by this process are gold or gold-plated conductors welded to components that have been metallized by aluminum. Aluminum and copper conductors are also joined by thermocompression welds.

Thermocompression welds of ribbon conductors and package component leads that are rectangular in cross section resemble indentation welds in cold lap welding. In thermocompression welding, however, the ribbon or lead is normally welded to a metallized electronic device or ceramic package. Consequently, the indentor tool is only applied to the ribbon or lead and little bulk deformation occurs in the metallized substrate. Cold welding normally is not applicable for these microwelds because the high deformations required would greatly weaken the conductors. Increasing welding temperatures to 0.4 to 0.5 T_m significantly reduces the required deformation.

Fig. 6 Minimum pressure-temperature conditions required for hot isostatic pressing welding. Source: Bryant, W.A., *Welding Journal*, Vol 54 (No. 125), Dec 1975, p 433s

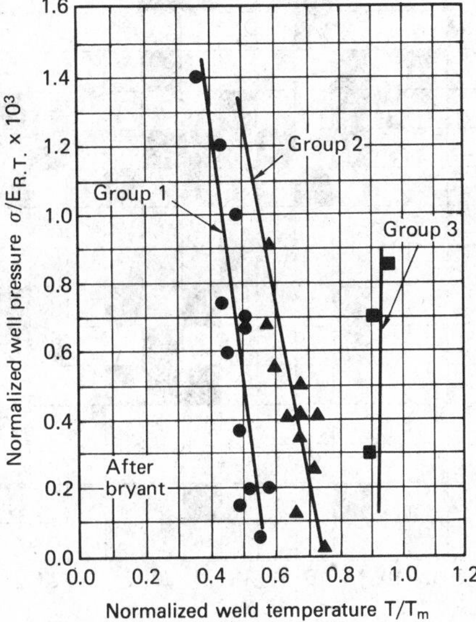

Fig. 7 Effect of organic surface contamination on the thermocompression welding of gold. 1-s pulsed welds. Data are for ball bonds of 25-μm-diam gold wire bonded to gold-chromium metallization

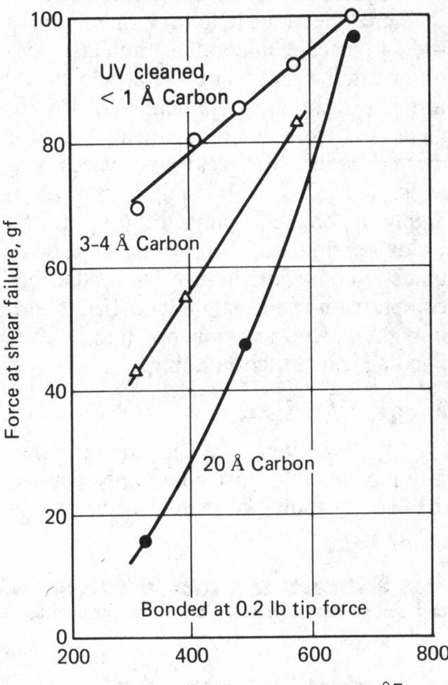

Resistance heating is the common heat source, and heat is transmitted to the weld zone through the indentor tool, which may be made of tungsten, tungsten carbide, ceramic, glass, diamond, or high-temperature nickel-based alloys. The indentor sometimes doubles as the resistance heating element. Resistance heating of the substrate is often used to raise the temperature of the metallization to 300 to 485 °F. Weld durations of $^1/_2$ to 1 s are typical. To increase the rate of heat input and to ensure reaching welding temperatures in these short times, the indentor tool is sometimes pulse heated. This permits longer tool life than if the tools were continuously heated. In addition, because higher tool temperatures can be used in the pulsed mode, acceptable weld interface temperatures can often be achieved without substrate heating. Parallel gap resistance welding machines have also been used in thermocompression welding of gold-plated leads of iron-nickel-cobalt alloys and copper alloys.

Figure 4 illustrates typical lead attachment to hybrid microcircuits by means of thermocompression welding. Also, thermocompression welding of gold wires used as interconnections in hybrid microcircuits and large-scale integrated circuits is described in the section of this article on deformation welding applications. Equipment is commercially available for thermocompression welding of wires, ribbons, and leads. This equipment is often highly automated, permitting up to 60 welds per minute. Terms that are sometimes used synonymously with thermocompression welding include ball bonding, thermocompression bonding, wedge bonding, and stitch bonding.

Forge Welding

Forge welding is a hot deformation welding process most commonly applied to the butt welding of steels. As contrasted with thermocompression welding of ductile fcc metals, which is normally performed at temperatures of $<^1/_2$ T_m, forge welding is typically conducted at temperatures in the 0.8 to 0.9 T_m range. The forge welding temperature is generally selected to be as high as possible with due consideration to avoiding such metallurgical problems as hot shortness, embrittlement, sensitization, and excessive grain coarsening. This implies an understanding of the unique metallurgical problems of the alloy to be welded.

Hydraulic presses are typically employed to apply pressure. Presses are often highly automated, featuring microprocessor control of pressure and temperature cycles. Heat is applied locally to the joint area by multiple-tip oxyacetylene torches, resistance heating, or induction heating. Often the oxyacetylene torches are oscillated to ensure uniformity of heating. In a closely related process, magnetically induced arc butt welding, the surfaces to be welded are heated by a rapidly rotating arc plasma. Generally, the process is conducted in the open air, with oxygen partially occluded from the joint area by the initial contact of the faying surfaces. When employing oxyacetylene torches, a slightly reducing flame affords some atmospheric protection. Vacuum, inert, and reducing atmospheres have been used.

The normal welding sequence is to (1) apply sufficient pressure to firmly seat the faying surfaces against one another, (2) heat the joint to welding temperature, and (3) rapidly apply additional pressure to upset the weld zone. Typical weld durations are 1 to 2 min. A less common procedure is to initially apply high pressure and permit deformation to occur during the heating cycle. Most forge welding employs sufficient pressure to upset the surface until the increase in the surface area is 125% or more. However, such high deformations tend to cause flow lines to bend toward the surface during upsetting. Conse-

Fig. 8 Direct and indirect bonds between various materials. Black squares indicate bonding compatibility; numbers across top and bottom represent materials in the same order as those in the column on the right. (Source: P.M. Bartle, Welding Institute, Great Britain)

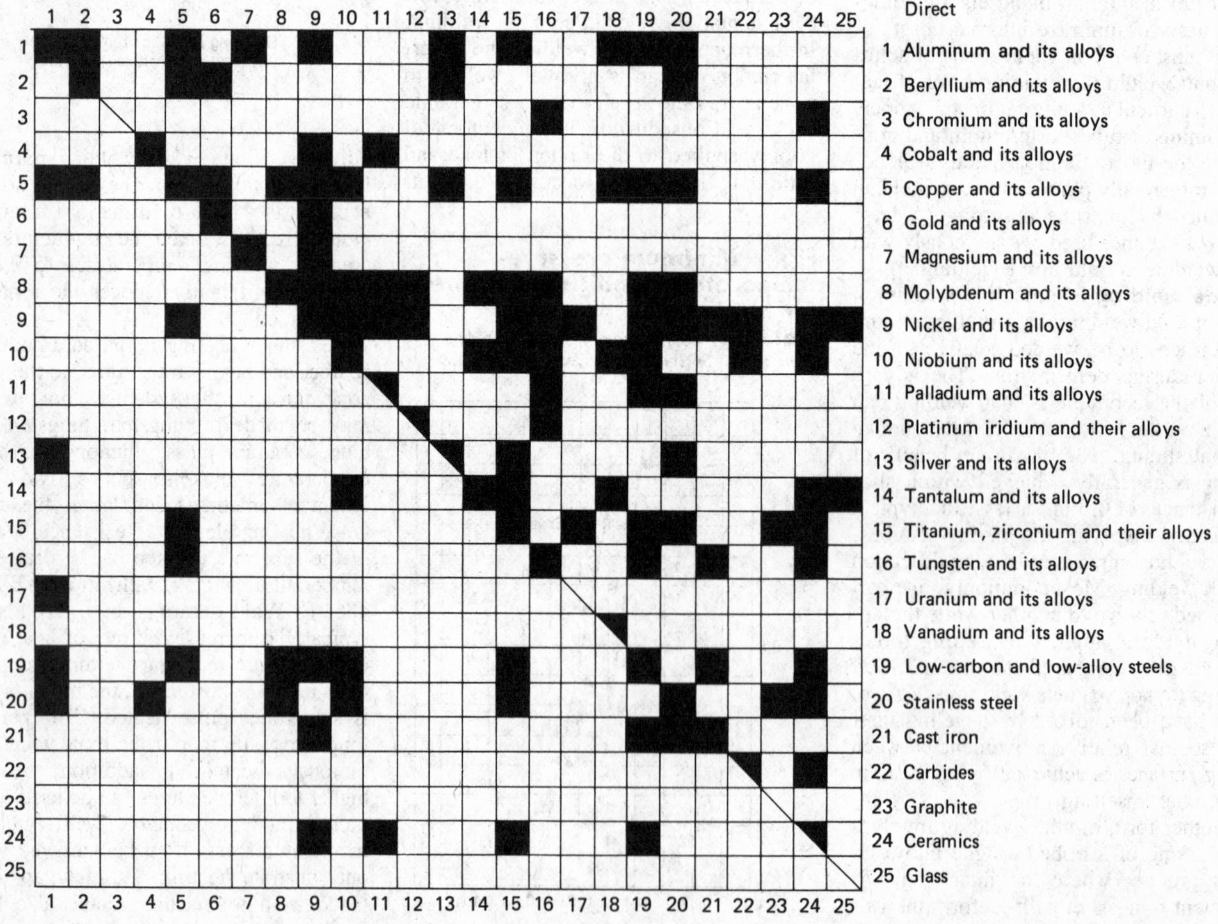

quently, alloys that contain significant stringers and inclusions may exhibit poor impact or fatigue properties when welded with high amounts of upset. This effect may be minimized by reducing the upset, which normally requires increasing welding temperature and/or time to ensure complete elimination of voids and surface oxide.

Forge welding is most commonly applied to carbon and low-alloy steels, with typical welding temperatures of about 2060 °F. Low-carbon steels can be used in the as-welded condition, but medium-carbon steels and low-alloy steels normally are given full heat treatments following welding. In those cases where full heat treatment is impractical, but hardening due to rapid cooling has occurred, induction heating may be used to temper the weld zone. Other metals welded by forge welding include high-alloy steels, nickel-based alloys, cobalt-based alloys, aluminum alloys, titanium alloys, and tungsten. Applications of this process include welding rods, bars, tubes, rails, aircraft landing gear, chains, and cans. The forge welding process is competitive with flash welding and friction welding.

The sealing of nuclear waste canisters by means of solid-state resistance welding is described in the section on deformation welding applications. Weld durations for resistance welding are very short (seconds) compared to those for forge welding where gas torches are employed. This is because heat is generated internally in resistance welding, but externally with gas torches. Terms that are sometimes used synonymously with forge welding include pressure welding, upset welding, and solid-state resistance welding.

Roll Welding

In roll welding, two or more sheets or plates are stacked together and then passed through rolls until sufficient deformation has occurred to produce solid-state welds. Two modes of roll welding are common. In the first, the parts to be welded are merely stacked and passed through the rolls. The second method, usually termed pack rolling, involves sealing the parts to be rolled in a pack or sheath and then roll welding the pack assembly. The first method is more generally employed in the cold welding of ductile metals and alloys. Sometimes the stack to be welded is first tack welded at several locations to ensure alignment during rolling. Also, when using this method, the deformation during the first rolling pass must exceed the threshold for welding (typically greater than 60% for cold rolling) to keep the parts together. The required first pass reduction can be reduced by hot rolling, if the metals to be rolled can tolerate preheating without excessive oxidation. Once the first pass has been accomplished, the reduction per pass can be decreased, as is often desirable because roll-separating forces increase as the parts to be rolled become thinner. However, the nonuniform stress distribution that builds up during a sequence of very light passes can cause the weld to open up or "alligator." Therefore, the reduction for subsequent passes is generally a compromise between applying excessive separating forces and "alligatoring."

In pack roll welding, the parts to be welded are completely enclosed in a pack that is sealed (typically by fusion welding) and often evacuated to provide a vacuum

Table 1 Diffusion welding conditions for various material combinations

Material combination	Intermediate material	Welding conditions: Press, ksi	Temperature, 10^2 °F	Time, min	Atmosphere
Mo to Mo	Ti foil	10	17	120	Argon
Mo to Mo	Ti foil	12.5	16	10	Vacuum
René 41 to René 41	Cu foil	5	16	10	Vacuum
TZM to TZM	Pd film	20	20	1	Air
TD nickel to TD nickel	...	24	16	0.3	Air
TD nickel to TD nickel	Ni	9	20	45	Vacuum
T-111 to T-111	Ta foil	10-40	22-26	1	Argon
Ta to Ta	...	10	24-26	180	Inert
Ta to Ta	Ti foil	10	16	10	Vacuum
Ta-10W to Ta-10W	Ta foil	10-20	26	0.3	Argon
Ti-75A to Ti-75A	...	4-10	13	10	Vacuum
Ti-75A to Ti-75A	Al foil	10	10.5	10	Vacuum
U-80Pd to U-80Pd	...	4	11-12	10	Vacuum
W to W	Ni-Pd	10	18	90	Hydrogen
W to W	Re-Ta	20	18.5	30	...
Zircaloy-2 to 302 SS	...	...	187	80	Vacuum
Zircaloy-2 to 302 SS	...	...	187	3	Helium
Zr to Zr	...	10	15.5	210	Inert
Zr to U	...	22	15.5	2160	Inert
Zr to U-10Mo	...	10	12	360	Inert
Alumina to Ta	...	10	28	120	Inert
Nickel-bonded carbide to nickel	...	...	24.5	10	Inert
UC to Nb	...	10	21-24	180	Inert
UO_2 to Zircaloy	...	10	15.5	240	Inert
UO_2 to Zircaloy	C	10	15.5	240	Inert
Al to Al	Si	1-2	10.8	1	Vacuum
Al to Cu	...	40	8.8	4	Vacuum
Al to Cu	...	22-50	10	15	Vacuum
Al to Ni	...	22-52	9.3	4	Vacuum
Al to U	Ni plate	5	9	...	...
AM-355 to AM-355	...	0.5	20.5	10	Vacuum
Be to Be	...	10	15-16.5	240	Inert
Be to Be	Ag foil	10	13	10	Vacuum
Beryllium copper to Monel	Au-Cu, Au-Ag	1.5-6	6.5	180	Helium
Cast iron to cast iron	...	0.51	14.7	10	Vacuum
Nb to Nb	...	10	21-24	180	Inert
Nb to Nb	Zr	10	15.4	240	Argon
Nb to Nb	In	0.13	20	60	Vacuum
Cb-752 to Cb-752	Ti foil	5	14.5	10	Vacuum
Cu to Cu	...	20.3	6	1	Hydrogen
Cu to Ti	...	0.7	15.6	15	Vacuum
Cu to Al	...	1	9.5	15	Vacuum
Cu to Kovar	...	1	17.4	10	Vacuum
Cu to (Nb-1Zr)	...	...	18	240	Vacuum
Cu to 316 SS	...	...	18	120	Vacuum
Haynes 25 to Haynes 25	...	0.5	21	10	Vacuum
Inconel to Inconel	...	0.5	20-21	10	Vacuum
Low-carbon steel to low-carbon steel	...	0.5	20	10	Vacuum
Mo to Mo	...	10	23-26	180	Inert

atmosphere. This may be accomplished by a frame that surrounds the parts to be welded, which is sandwiched by two lids, or may simply consist of two covers formed to encapsulate the parts to be welded. Semikilled or killed low-carbon steel is a common material for the pack, but is not suitable for all alloy and temperature combinations. Although the preparation costs of pack roll welding are significant, the process has the advantages of (1) providing atmospheric protection, which may be particularly important for reactive alloys such as those of titanium, zirconium, niobium, and tantalum; and (2) permitting welding of complex assemblies involving several layers of parts. A significant limitation of the process is that packs become difficult to process when their length exceeds several feet. Nonpack rolling, of course, can employ continuous strip rolling. A seldom used hybrid version of roll welding is to seal the edges of the parts to be welded by fusion welding prior to rolling. This method does not provide atmospheric protection to the outer surfaces, as in pack roll welding, but does occlude air from the more critical inner surfaces.

In roll welding, successful welding depends not only on the total amount of deformation but also on the pressure during rolling. A high isostatic pressure component during rolling is promoted by large reductions per pass, large roll diameters, and rough roll surfaces. A nonuniform distribution of normal stresses, referred to as pressure hill effects, is caused by frictional and mechanical restraint and results in lower pressure near the outer edges of the parts being rolled. Therefore, rolling conditions must be such to ensure sufficient pressure near the edges to avoid weak welds in these regions.

Copper, aluminum, and their alloys are often cold roll welded. Hot rolling is applicable to a wide spectrum of alloys. Typically, hot rolling is performed above recrystallization temperatures. In practice, weldability is often improved by raising the temperature to the region where oxygen mobility becomes significant to promote oxide assimilation by diffusion. Also, welded assemblies are often postheat treated to cause recrystallization (in cold rolling), oxygen diffusion, and/or sintering. In welding dissimilar metals, selection of temperatures for rolling and postheating must take into consideration the possibility of intermetallic compound formation, eutectic melting, and/or porosity due to Kirkendall diffusion. The most common metals that are roll welded are low-carbon steels, aluminum, aluminum alloys, copper, copper alloys, and nickel. Low-alloy

Fig. 9 Three basic types of superplastically formed/diffusion welded structures. (a) Reinforced sheet. (b) Integrally stiffened. (c) Sandwich

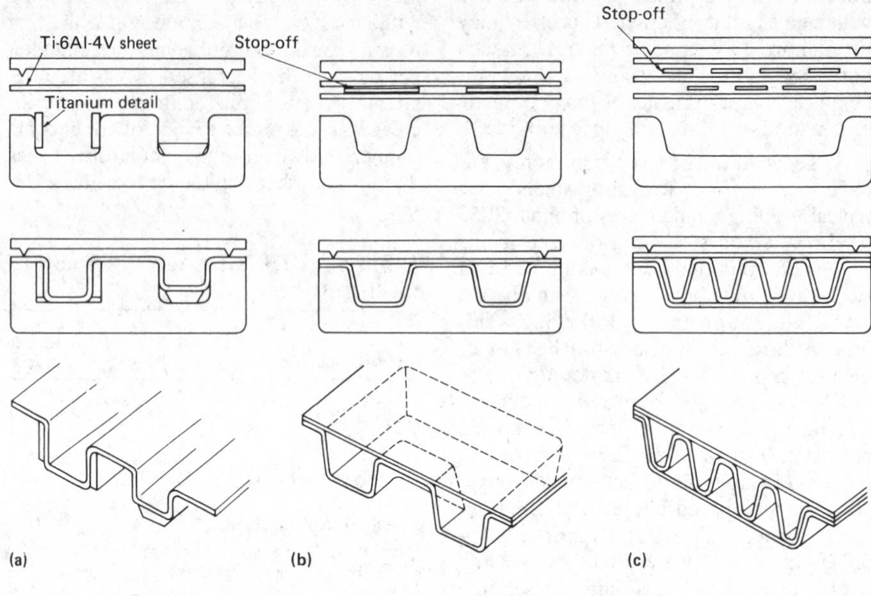

Fig. 10 Press capable of exerting pressure of 2 ksi in all directions on a heated retort. Courtesy of Rockwell International Corp.

Fig. 11 Location of diffusion welds in hollow fan hub for the JT9D-70 turbofan engine

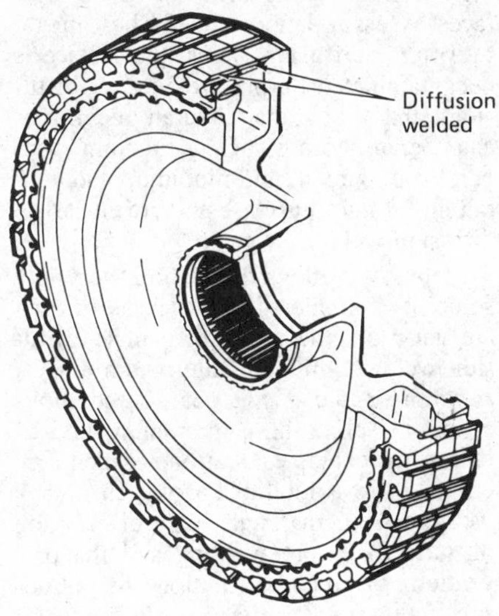

steels, high-alloy steels, and nickel-based alloys have also been roll welded.

One of the more important applications of roll welding is the fabrication of heat exchangers. A patented roll welding process is described in the section on roll welded heat exchangers in this article. This process is now gaining new importance as a method for the fabrication of flat solar collectors. In another method of heat-exchanger fabrication, three plates are stacked, with the middle plate containing the flow channels. The stacked plates are initially held together by tack fusion welds or resistance spot welds. This approach has been employed for both aluminum and steel heat exchangers. A second important application of this process is in the cladding of sheet metal products. Terms that are sometimes used synonymously with roll welding include roll bonding, roll cladding, and metal cladding.

Fig. 12 Diffusion welded hollow-rim fan disc and split spacer ring

Extrusion Welding

Extrusion welding or bonding generally involves the coextrusion of two or more metal parts. The process typically is conducted at elevated temperatures not only to improve welding but also to lower extrusion pressures. Some cold extrusion welding of aluminum and copper has been performed. For hot extrusion, the parts to be welded are often assembled in a can or retort that is designed with the appropriate leading taper and wall thickness to promote initiation of extrusion. For reactive metals, such as zirconium, titanium, and tantalum, the retort may be evacuated and sealed. Both forward and back extrusion have been employed, but forward extrusion is the usual mode. A principal advantage of extrusion welding is that the high isostatic pressures associated with the process are favorable to the deformation welding of low-ductility alloys.

In addition to coextrusion, extrusion welding has been used to butt weld tubes. The ends of the tubes are prepared for extrusion by beveling at a 45 to 60° angle to produce an overlapping joint (Fig. 5). The leading tube contains the female portion of the beveled joint and is the stronger of the two metals in dissimilar metal joints. Extrusion press die angles of 30 to 35° are common. An advantage of extrusion welding over other methods of deformation butt welding of tubes is that there is no flash or upset to remove following extrusion.

The most common metals welded by the extrusion process include low-carbon steel, aluminum, aluminum alloys, copper, and copper alloys. Additional applicable materials include nickel, nickel-based alloys, zirconium, titanium, tantalum, and niobium.

Fig. 13 Vacuum hot press (3000 ton) used for diffusion welding turbofan components. Courtesy of Pratt & Whitney Aircraft Group

Other Solid-State Welding Processes

Ultrasonic welding or bonding, inertia and continuous-drive friction welding, and explosion welding are all classified as solid-state welding processes. Each of these methods is described in a separate article in this Volume.

Fundamentals of Solid-State Welding

To produce solid-state welds, material surfaces must be brought close enough together that short-range interatomic forces operate. In general, this involves both intimate mating of surfaces and removal of surface barriers to atomic bonding. In both diffusion welding and deformation welding, the stress state at the weld interface is controlled by design and the method by which pressure is applied. Geometrical and frictional restraint may give rise to two conditions that influence SSW: (1) a nonuniform distribution of normal stresses (pressure hill); and (2) a nonplastic zone where, although isostatic compressive stresses may be high, the difference in principle stresses is insufficient to cause plastic flow.

In diffusion welding, extensive plastic deformation is not necessarily required. However, the stresses normal to the weld interface must be sufficient to promote mating. The pressure hill effect may result in insufficient mating of surfaces, particularly near specimen edges. A similar phenomenon is often observed in roll welding where, due to the pressure hill effect, insufficient pressure is applied near the edges of the weld. Often a change in the method of pressure application, such as from uniaxial pressure to isostatic pressure, will minimize pressure hill effects.

Deformation welding requires both mating of surfaces and extensive surface deformation. Deformation welds are often incomplete due to the presence of a nonplastic zone within a portion of the weld interface. This occurs at the point of greatest restraint. In lap welding with indentor tools, geometric restraint often creates nonplastic zones near the edges of the indentor. Conversely, in butt welding, a nonplastic zone is often centrally located due to frictional restraint.

Diffusion Welding

A diffusion weld is created when the surfaces of materials are brought sufficiently close together so that short-range interatomic forces operate. If the surfaces are free from contamination, the driving force for the weld is a lowering of surface energy. Because real surfaces can never be perfectly smooth, initial contact on the faying surfaces is made between surface asperities when the load is applied. Further contact is made by plastic yielding and creep deformation. At the same time, diffusion at the clean surfaces eliminates the remaining interfacial boundary. Many mechanisms have been proposed for removal of this interfacial boundary, including:

- Surface and volume diffusion from surface sources to an asperity neck
- Diffusion along the weld interface from interfacial sources to an asperity neck
- Volume diffusion from interfacial sources to an asperity neck
- Evaporation and recondensation

Factors Affecting Weldability. Provided that the surfaces are atomically clean and the materials are compatible, all of the above mechanisms probably operate to some extent simultaneously with creep deformation; the original interfaces disappear, and the weld is completed. The coalescence of these interfaces appears to be analogous to sintering phenomena. In the real world surface contaminants are invariably present, and depending on the materials being welded, mechanisms must exist for dispersion of contaminants away from or into localized areas on the faying surface. Metals that have a high solubility for interstitial contaminants, such as oxygen, can easily accommodate removal of these contaminants from the faying surfaces by assimilation into the base metal by volume diffusion. Thus, the surface is decontaminated during welding by diffusion, and short-range interatomic forces can operate. Metals such as titanium, copper, iron, zirconium, niobium, and tantalum fall into this class and are easiest to diffusion weld.

Metals and alloys that exhibit very low solubility for interstitials such as oxygen are generally the most difficult to diffusion weld. Aluminum, aluminum alloys, and iron-, nickel-, and cobalt-based alloys containing chromium, aluminum, and titanium are examples of strong oxide formers with low solubility for oxygen. When welding these materials, careful cleaning of surfaces before welding and the prevention of recontamination by atmospheric impurities before and during welding is mandatory. This can be accomplished by roughing one or both surfaces (with wire brushing or other machining techniques) prior to welding. When the joint is loaded, high spots deform in shear, resulting in surface extension, and the contaminants are pushed aside.

Welding Conditions. Diffusion welding is usually performed at a welding tem-

Fig. 14 Wing carry-through section for advanced prototype bomber fabricated by diffusion welding 533 individual details. Courtesy of Rockwell International Corp.

Table 2 Equivalent properties of weld and base metal for iron and steel diffusion welded at 1590 °F

Base metal	Tensile strength, ksi	Elongation, %
Iron	58	25.38
Low-carbon steel	99	24.75
0.24% carbon steel	73	28.25
0.45% carbon steel	133	18.38
0.57% carbon steel	133	15.88
Diffusion welding combination		
Iron to iron	71	29.50
Low-carbon steel to low-carbon steel	76	32.50
0.24% carbon steel to 0.24% carbon steel	75	33.75
0.45% carbon steel to 0.45% carbon steel	91	18.75
0.57% carbon steel to 0.57% carbon steel	134	24.00
Iron to low-carbon steel	72	29.25
Iron to 0.24% carbon steel	71	29.13
Iron to 0.45% carbon steel	71	22.00
Iron to 0.57% carbon steel	71	23.25

Note: Times <30 min welding pressures up to 7.25 ksi vacuum $<10^{-4}$ torr

perature equal to or greater than one half the melting temperature of the material being welded. However, the choice of welding temperature is strongly influenced by the time required for surface contaminants to diffuse away the tendency to weld above or below a phase transformation and the amount of load available at the faying surfaces. For example, Fig. 6 shows minimum pressure-temperature combinations required for welding different material combinations as a function of their oxide stability. Materials in group 1 have the lowest oxide stability and group 3 the highest. Diffusion rates are highly temperature-dependent. For example, if welding temperatures in excess of one half the melting point are used for iron, copper, and titanium, then diffusion times required for assimilation of oxides into the base metal would be a few minutes.

To obtain a solid-state weld, the faying surfaces must be in atomically close contact so that short-range interatomic forces can operate. Because the fabrication of perfectly smooth, atomically clean surfaces is not practical, the application of temperature to enhance atomic transport and of moderate deformation at the faying surfaces to enhance contact is necessary to provide the means for establishing this short-range interatomic contact. For materials that have high solubility for interstitial contaminants, probably only diffusion and/or vapor transport are necessary to decontaminate the faying surface and remove the interfacial boundary.

For materials that have a low solubility for interstitial contaminants, additional efforts to decontaminate the faying surfaces are required. These efforts include careful preweld cleaning and handling, possible introduction of reducing atmosphere during welding, and the application of surface extension. Even after taking all of these precautions, a weld joint of less than 100% area fraction could result. Diffusion welding schedules are a strong function of the materials being welded, part design, and part application. Based on the above general principles and some preliminary prototype work, it is usually easy to establish whether diffusion welding is feasible for a particular application.

Deformation Welding

In general, there are five stages involved in deformation welding: (1) mating of surfaces, (2) fracture of surface oxide and/or strain-hardened surface, (3) extension of surface to expose areas of nascent metal, (4) extrusion of the metal through the cracks of the surface layer, and (5) establishment of welds between the nascent metal that has extruded through the surface layer. Because plastic deformation does not occur uniformly along a faying surface, various stages of the deformation welding process may occur simultaneously at different locations. Also, for some processes and some materials, all five stages may not be required. For example, in the deformation welding of gold, which is free of oxide, contact between nascent metal occurs without the necessity of fracture of the surface layer, although surface extension is still useful in overcoming the effects of organic surface contaminants. In slide welding, nascent metal surfaces are exposed by a plowing action that minimizes the need for surface extension and extrusion of the metal through fractured surface layers.

Mating of Surfaces by Plastic Yielding. Intimate mating of surfaces is generally achieved at approximately 10% deformation. However, because plastic deformation is generally inhomogeneous, particularly on the microscale in the region of local inclusions, elimination of all voids may not occur during cold welding, even at deformations as high as 90%. Because alloys employed in deformation welding are generally very ductile, the presence of these microsize defects is usually of little consequence. However, they can significantly reduce the toughness of welds of some alloys, such as precipitation-hardened aluminum alloys.

Breakup of Oxide Films and Surface Contaminants by Surface Extension. Typical minimum deformations for cold welding range from 25 to 90% for indentation welding and from 25 to 45% for roll welding. These deformations are far in excess of that required for either general mating of surfaces or fracture of brittle surface layers (such as oxides). These high deformations generally are required for surface extension and extrusion of metal through gaps in the fractured surface layer. Several mathematical models have been developed on the basis of the amount of overlapping nascent metal produced by lap welding or roll welding. Some of the general observations that have been made which influence the assumption necessary in modeling are: (1) opposing surface layers generally break up independently in lap welding; (2) if surface preparation has eliminated most of the organic contaminants, opposing surface layers generally break up together in roll welding; conse-

Fig. 15 Microstructures of diffusion welds. (a) Low-carbon steel specimen (100×). (b) 0.45% C steel specimen (125×). (Source: Taylor, D.S., and Pollard, G., Proceedings of Advances in Welding Processes Conference, Harrogate, Paper No. 4, The Welding Institute, Abington, 1978)

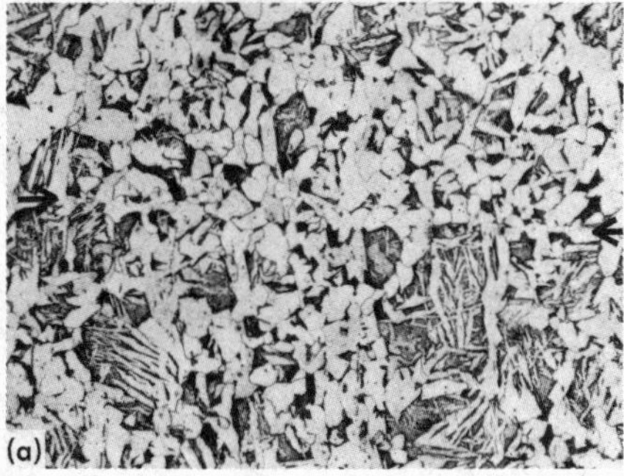

quently, the minimum threshold deformation for welding is often lower for roll welding than lap welding; and (3) interfacial shear displacements reduce the amount of surface extension required for lap or butt welding. Interfacial shear displacements are promoted by dissimilar metal welds or, in lap welds, unequal thicknesses of weld pairs. However, excessive differences in yield strengths in dissimilar metal welds can result in too little surface extension in the stronger alloy.

Thermal Desorption and Dissociation of Surface Contaminants. Although deformation welding is generally conducted at temperatures that are too low for elimination of oxides by diffusion, thermal desorption and/or dissociation of surface contaminants often occurs. Temperatures ranging from 450 to 1000 °F are adequate to reduce organic contaminants to monolayers or less on most metals. Hydrocarbons can be reabsorbed from the atmosphere within minutes or, at the most, hours if surfaces are allowed to cool to room temperature in air prior to welding.

Solid-state sintering is generally considered as a phenomenon important only to diffusion welding. However, in certain instances, solid-state sintering is also an important factor in deformation welding processes. For example, in the case of thermocompression welding of gold, sintering phenomena have been observed at temperatures as low as 315 °F. Sintering has been observed as a normal second stage of deformation welding where the metal-to-metal interface continues to grow at the expense of metal contaminant interfaces after macrodeformation has ceased. A similar phenomenon has been observed during the postheating of gold-to-gold welds. It is not unusual to find that 25 to 50% of the strength of gold-to-gold thermocompression welds is attributable to sintering, rather than to mechanical disruption of surface films. Sintering is also used to increase the strength of welds produced by cold rolling.

Surface Roughness. At the high deformations normally utilized, surface roughness generally is not critical in deformation welding. As-rolled, machined, cut, and scratch brushed surfaces have all been employed successfully. Some evidence suggests that a moderately rough machined surface (120 to 140 rms) actually promotes welding by increasing the interfacial shear displacements. However, at deformations below approximately 20%, surface roughness becomes more critical because of the difficulty of closing voids.

Surface Contaminants. Brittle oxide films, such as on aluminum, readily fragment and permit deformation welding. Inorganic films (such as oxides) generally do not prevent deformation welding because they are mechanically disrupted. The manner in which they break up influences the amount of surface extension required to achieve welding. The more mobile organic contaminants constitute the major barrier to deformation welding. A single fingerprint has been known to prevent deformation welding. Organic contaminants fall into two categories: (1) those that are absorbed from the atmosphere; and (2) those that are incurred by contact, such as oils, greases, and waxes.

Generally, organic atmospheric contaminants are relatively small, volatile molecules and tend to be limited to mono-

Fig. 16 Diffusion welded flueric device. 410 stainless steel. (Source: Spurgeon, W.M., Rhee, S.K., and Kiwak, R.S., Diffusion Bonding of Metals, *Bendix Technical Journal—Materials and Processing*, Vol 2 (No. 1), 1969, p 24-41)

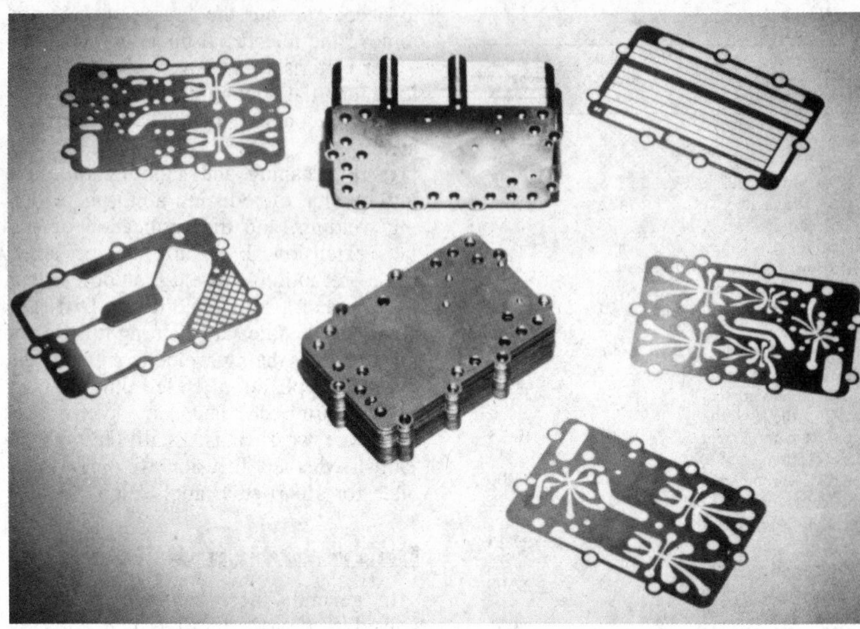

Fig. 17 Burner can for TF30-P-100 jet engine of Hastelloy X Finwall diffusion welded panels. Closeup view of Finwall structure on right. (Courtesy of Pratt & Whitney Aircraft Group)

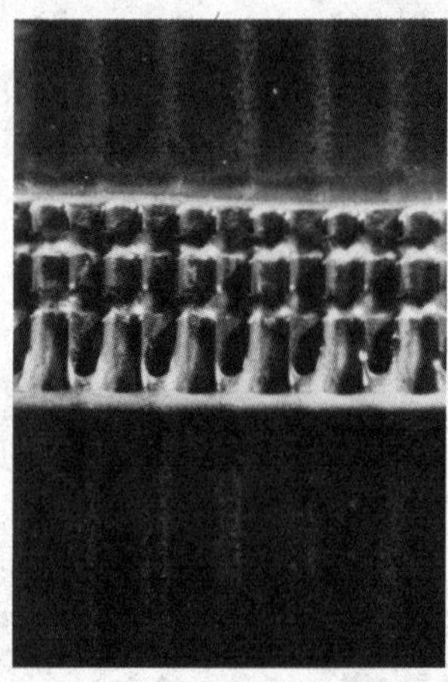

Fig. 18 Fracture surface of a BORSIC reinforced titanium composite axial tensile specimen. Source: Prewo, K.M. and Kreider, K.G., The Deformation and Fracture of Basic Reinforced Titanium Matrix Composites, *Titanium Science and Technology*, Vol 4, Plenum Publishing, New York, 1972, p 2333-2345

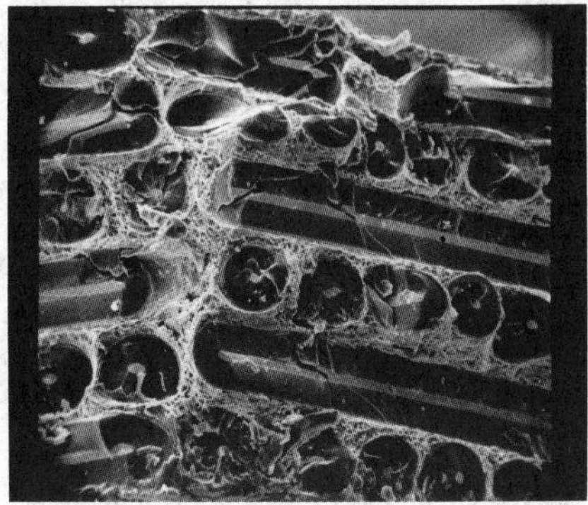

layers on surfaces. Conversely, oils and greases often result in barriers of considerable thickness. Aside from being thicker, nonatmospheric organic contaminants tend to differ from atmospheric contaminants by consisting of larger, more strongly bonded molecules. The detrimental effect of organic contaminants generally increases with increasing chemical activity, molecular size, and film thickness.

For most forms of deformation welding, such as lap welding, butt welding, and roll welding, solvents such as trichlorethylene are used to remove organic contaminants. Vapor degreasing and ultrasonic cleaning are commonly employed. In ultrasonic cleaning, use of detergents to remove the more polar contaminants and a final alcohol rinse is common.

For processes involving high surface extension such as multiple upset butt welding, solvent cleaning adequately removes organic contaminants. However, the residual contaminants left from solvent cleaning are detrimental to many processes. For roll welding and lap welding, the most common procedure to remove residual contaminants is to follow solvent cleaning by scratch brushing with rotary stainless steel wire brushes. Scratch brushing results in thermal desorption of organic contaminants. Surface speeds of 33 to 66 ft/s are used. Scratch brushing also results in a strain-hardened surface layer that helps control the fragmentation of surface oxides. Parts should be welded immediately following scratch brushing to avoid recontamination from the atmosphere, which can occur in less than 10 min.

Other methods of desorbing organic contaminants, such as vacuum baking or hydrogen firing, have been used, but recontamination from air atmospheres generally is a problem unless welding immediately follows.

Microminiature deformation welding processes such as thermocompression welding of gold are highly sensitive to organic contaminants (Fig. 7). Processes such as ultraviolet-ozone cleaning and radiofrequency plasma cleaning have been used to reduce organic contaminants to nondetectable levels, resulting in marked improvements in weldability.

Processes in which parts were surface machined and welded in vacuum (by rolling or lap welding) are characterized by very low required minimum deformations for welding (5 to 10%). The significant benefit of machining in vacuum is believed to be related to the avoidance of both surface oxides and recontamination by organic contaminants.

Pressure. Most deformation processes are controlled by the amount of deformation rather than by pressure. Required pressures are generally two to seven times the uniaxial yield stress of the material being welded. Higher pressures are required when geometric restraint is high. Some evidence suggests that the high isostatic pressure associated with high geometric restraint is beneficial in eliminating microdefects.

Metallurgical Considerations

Whenever a solid-state welding program is contemplated for a particular application, the first steps before implementation should be a search of the literature for previous work on similar alloy systems, followed by inspection of equilibrium phase diagrams for the materials to be welded. If previous work does not exist, then metallurgical decisions whether to apply deformation welding or diffusion welding can be made based on inspection of these phase diagrams. In general, solid-state phase fields can be classified as solid-solution systems and pure elements, two-phase systems, and intermetallic systems.

Pure elements. Joining a pure element to itself or two different metals or alloys together that have unlimited solubility as a single phase (for example, copper-to-nickel) are the easiest systems to handle. For these systems, metallurgical decisions on deformation welding versus diffusion welding could be based on the solubility of the system for interstitials. Some pure metals are subject to allotropic transformations as they are heated, and alloys could be subject to phase transformations, resulting in a volume change. If the volume change is significant (for example, the ferrite-to-austenite transformation in dilute iron-carbon alloys results in a volume change of ~4%), this factor must be evaluated both with respect to making the weld and possible service failure during elevated temperature cycling.

Two-phase systems for engineering alloys usually involve one phase as a metastable precipitate dispersed within a matrix phase. With these systems, options include deformation welding at low temperature with the material in the final aged condition, selecting the diffusion welding temperature so that age hardening and welding occur simultaneously, or selecting a higher temperature where the second phase is in solution. The latter choice requires that the welded part be quenched from the welding temperature or resolution treated and subsequently aged.

Intermetallic systems are by far the most common systems used when couples of dissimilar metals are made. Often it is impossible to fusion weld these systems and obtain joints with significant strength and toughness, making SSW the only welding option. Because of the low temperatures used, deformation welding is a prime candidate for joining these materials, provided that the part design is acceptable. If diffusion welding is required, care must be exercised to keep the welding temperature and time low enough so that a continuous intermetallic film does not form at the weld interface. Another

Fig. 19 Micrograph of an aluminum/boron composite panel made by diffusion welding individual monolayer tapes together

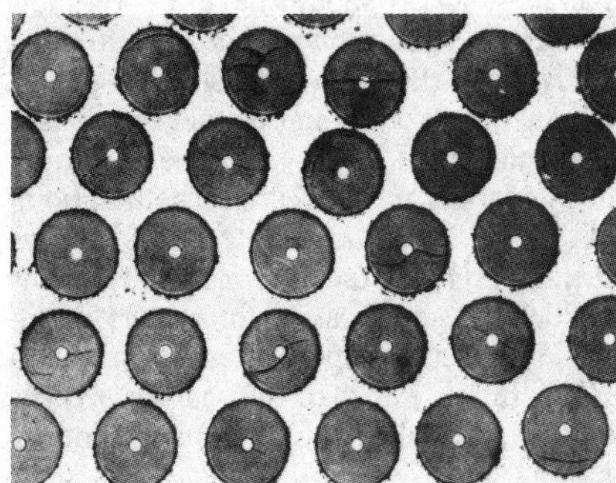

Fig. 20 Stratapax stud assembly

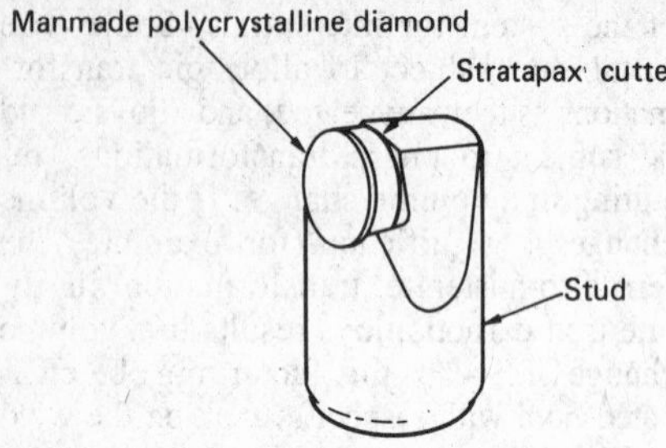

option for diffusion welding of these systems involves the use of an interlayer material that is compatible with both alloy systems.

Interlayers

Interlayers of other metals in the form of foils or coatings may be used to overcome metallurgical problems or simply to facilitate processing. Common reasons for employing interlayers are to (1) reduce welding temperature, (2) restrict high deformation to the interlayer zone, (3) minimize intermetallic formation, (4) improve mating of rough surfaces, (5) minimize porosity resulting from interdiffusion, (6) promote improved sintering by higher diffusion rates, and (7) minimize oxidation.

Selection Criteria. The selection of an interlayer material must be made with a clear idea of its purpose. Interlayers are used more often in diffusion welding than in deformation welding. Therefore, an understanding of diffusion of both the alloying components and interstitial elements as well as the resulting metallurgical phases is essential. Some general guidelines for selection of interlayers are that they: (1) have adequate strength; (2) are usually 0.6 to 2.0 mils thick for diffusion welding, or $^1/_{16}$ to $^1/_8$ in. thick for deformation butt welding; (3) should be metallurgically compatible with base metals; (4) should not form a low-toughness zone due to diffusion of interstitial elements from the base metals; and (5) should have a coefficient of thermal expansion intermediate to those of the base alloys for high-strength alloys. Because the coefficient of thermal expansion generally decreases with melting point, this generally means that the melting point of the interlayer will be intermediate to the melting points of the dissimilar alloys being welded. In some cases, interlayers of silver, gold, and copper, often having melting points lower than the materials being welded, are selected for the purpose of obtaining low welding temperatures.

Problems in Solid-State Welding

Like any other joining process, SSW is subject to limitations that must be understood before selection of a joining process is made. As stated previously, dissimilar metal couples usually have the capability at elevated temperature to nucleate and grow intermetallic phases at the weld interface. If the particular intermetallic phases tend to assume the morphology of a continuous brittle film, the overriding concern in designing the welding schedule should be to limit the growth of these brittle phases by minimizing welding temperature and time. Equilibrium phase diagrams predict the formation of intermetallic phases, but the kinetics of nucleation and growth will have to be determined from a literature search or trial coupons. This is especially important when welded parts are subjected to elevated service temperatures for long periods of time.

Porosity. If the diffusion rate of one metal in a couple is much greater than the other metal, Kirkendall porosity or vacancy condensation will occur on the side of the weld which originally contained the metal with the rapid diffusion rate. Large differences in atomic sizes are frequent contributors to this type of porosity. In many instances, this porosity can be eliminated or minimized by a postweld heat treatment under hot isostatic gas pressure or by limiting the weld temperature and time.

Thermal Expansion. When service conditions for solid-state welded parts involve thermal cycling, possible generation of residual stresses due to thermal expansion differences must be considered. Graded welds involving successive layers of compatible materials that differ by small amounts in thermal expansion are designed to control this problem. If the thermal expansion differences are very large, weld failure can even occur when the part is cooled from the welding temperature.

Corrosion. Economic and engineering criteria that determine weld pairs often dictate metallic combinations that can lead to galvanic corrosion failures. For example, steel-to-aluminum welds are subject to crevice corrosion in the presence of an electrolyte. Corrective measures include removal of the electrolyte (use only in dry conditions), electrical isolation of the couple from the rest of the structure, the use of paint sealers to eliminate the electrolyte path, and the application of sacrificial anodes. Whenever dissimilar metals are joined, the service conditions should be carefully evaluated and, if necessary, steps should be taken to prevent corrosion.

Brittleness. When intermediate foils or interlayers are used, care must be exercised to choose interlayers that do not act as interstitial sinks for surface and bulk metal contaminants and/or alloy elements. Materials such as titanium, zirconium, hafnium, vanadium, and niobium, which have high solubility for interstitials, are commonly used as interlayers. When these materials are used incorrectly, diffusion transforms them to brittle interstitial compounds, resulting in brittle weld joints. For example, when a titanium interlayer is used to diffusion weld TZM (Mo-0.5Ti-0.08Zr-0.015C), the carbon in TZM will diffuse to the titanium interlayer and form brittle carbides.

When high-reliability diffusion or deformation welds are required, all pertinent variables, including metallurgical, electrochemical, and environmental service conditions, must be carefuly evaluated. However, the same evaluations should be carried out when using any high-reliability joining process.

Diffusion Welding Applications

A summary of some material combinations that have been joined by diffusion welding is illustrated in Fig. 8. In this figure, "direct" refers to contact between the faying surfaces of both materials with no interlayer material, and "indirect" means

Fig. 21 Stratapax bit ($4^3/_4$ in. diam)

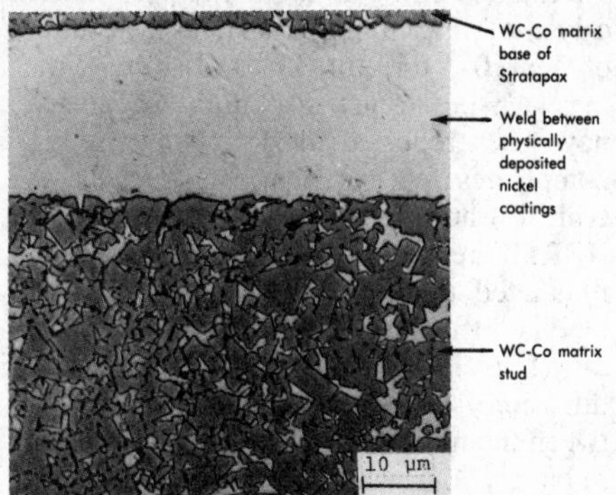

Fig. 22 Cross section of a diffusion welded Stratapax stud assembly. Typical area of weld with no evidence of original interface

that a separate interlayer-layer material was used between the faying surfaces. General welding conditions for a number of the above material combinations are summarized in Table 1.

Titanium and Titanium Alloys. More diffusion welding has been conducted on titanium and its alloys than any other material. Titanium diffusion welding is readily adaptable to production applications because:

- Titanium and its alloys have a high solubility for interstitial oxygen and can assimilate contaminant films into the bulk material during welding.
- Conventional fusion welds in titanium are not as strong as the base metal, whereas titanium diffusion welds possess properties equivalent to the base metal.
- Most titanium structures or components are high-technology items utilized in aerospace applications, and extra costs for diffusion welding can be justified based on improved performance.

Reinforced panels and honeycomb structures usually are produced by diffusion welding. Figure 9 illustrates examples of processing steps for three panel designs fabricated from Ti-6Al-4V alloy. A press used in the process is shown in Fig. 10.

Another example involves the production of Ti-6Al-4V alloy turbofan aircraft engine discs with hollow hubs. Figures 11, 12, and 13 show a schematic illustration of the disc profile, location of the diffusion welds, the welding assembly, and the hot press used for welding. Fabrication of this disc by diffusion welding resulted in weld properties equivalent to the base metal and a weight savings of 105 lb (about 30%) without sacrificing stiffness or durability.

Other titanium alloys that have been successfully diffusion welded include: Ti-8Al-1Mo-1V, Ti-13V-11Cr-3Al, Ti-8Mo-8V-3Al-2Fe, Ti-6Al-6V-2Sn, Ti-6Al-2Sn-4Zr-6Mo, and Ti-11.5Mo-6Zr-4.5Sn.

As many as 66 different titanium components have been fabricated successfully by diffusion welding for use in the B1 bomber. Very large parts have been fabricated in this manner. Figure 14 shows a wing carry-through section that contains 533 individual parts that were diffusion welded together. A 200-ft-diam titanium alloy chamber for the zero gradient synchrotron at Argonne National Laboratory also has been produced by diffusion welding.

Iron-Based Alloys. In general, diffusion welding of steels has not been pursued because of the ease with which they can be fusion welded. However, investigators in the USSR have claimed impressive economic gains for diffusion welding of high-throughput, low-cost steel parts. Research on diffusion welding of iron, low-carbon steel, and various plain carbon steels is summarized in Table 2, which shows that weld properties equivalent to the base metal can be attained. Photomicrographs of as-welded microstructures for diffusion welds made on low-carbon steel and a 0.45% C steel are illustrated in Fig. 15. Other work with low-carbon steel has shown that welds with base-metal properties can be obtained when up to 20% of the faying surfaces remain unwelded.

An interesting application of diffusion welding of 300 and 400 series stainless steels involves the fabrication of flueric integrated circuits (logic circuits using fluids). These devices consist of labyrinths in which impinging jets of gas or liquid perform functions of sensing, compensation, or actuation. Because of the complex internal design and the degree of miniaturization necessary, photochemical machining is used to form openings of the desired configuration in thin foils. The foils are stacked and aligned on a jig and then diffusion welded under loads generated in a differential thermal expansion press while heating in dry hydrogen or vacuum.

This operation converts the multitude of thin laminations into one monolithic block of metal, forming and sealing the flow channels without plugging or distortion. The differential thermal expansion press consists of platens made of a high thermal expansion material, held together with bolts or wires made of a low thermal expansion material. The assembly (platens plus laminations) is preloaded in a hydraulic press so that a uniform pressure is exerted by all fasteners and a larger actual contact area is established. As the assembly is heated, the pressure is relieved by the yielding of asperities and/or fasteners. Advantages of this method include low equipment cost because a vacuum hot press is not re-

Fig. 23 Diffusion welded niobium panel. (a) External surface. (b) Internal surface. (c) Weld microstructure

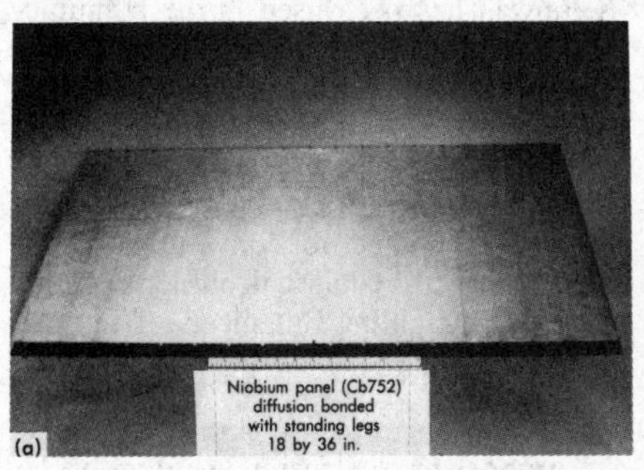

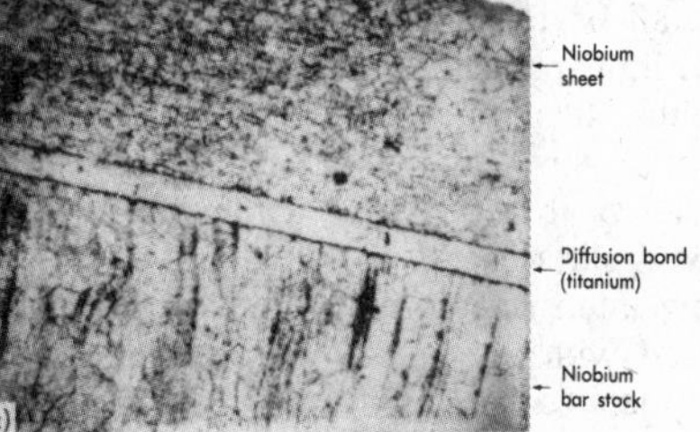

Fig. 24 Apparatus to generate differential thermal constraint for diffusion welding

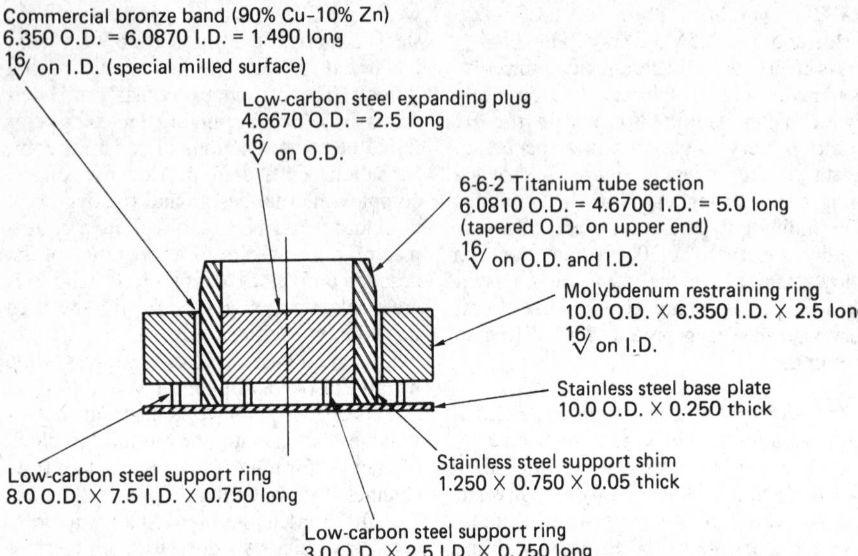

quired. A typical diffusion welded flueric device made of 410 stainless steel is shown in Fig. 16. Stable adherent oxide films are present on stainless steels. Thus, cleaning procedures for the above laminates involve first cleaning the parts anodically and then reversing the current in the same bath to electroplate a thin layer of copper onto the steel. During subsequent handling the copper surface layer is oxidized, but the oxide is easily reduced in hydrogen, and the remaining copper is assimilated into the steel by diffusion during welding. Steels generally are good candidates for diffusion welding. However, applications depend on economics, with the greatest advantages to be gained in parts with difficult designs or in assemblies requiring large weld areas and/or critical performance.

Nickel-Based Alloys. The driving force for the application of diffusion welding to nickel-based alloys stems from their poor fusion weldability and needs of the aerospace industry to produce reliable welds for high-performance hardware. Materials such as Udimet 700, Inconel 718, and Inconel 600 have been successfully diffusion welded. However, all have low carbon contents coupled with the presence of carbide formers (chromium, titanium, and molybdenum). Because of this, organic surface contaminants can be reduced to stable carbides on the faying surfaces. Also, these materials have low interstitial solubility for oxygen coupled with the presence of stable oxide formers (chromium, aluminum, and titanium). This makes the base metal highly susceptible to environmental contamination.

Because of the above reasons, great care must be exercised to thoroughly clean surfaces before welding, prevent recontamination during welding, and provide surface extension so that clean surfaces can come into intimate contact. Finwall is the registered trademark for a hollow sandwich panel structure fabricated by diffusion welding two sheet metal face sheets to a corrugated center section. Hastelloy X Finwall has been used in the manufacture of TF30-P-100 jet engine burner cans. This part, along with a closeup of the Finwall structure, is shown in Fig. 17. This construction results in substantially increased cooling efficiency of the burner liner as compared to conventional construction. The cylindrical burner sections are made by rolling flat sheets of Finwall material into the desired diameter and then butt welding.

Composites. Diffusion welding is an integral part of the fabrication process for many composites. Both titanium and aluminum matrix composites reinforced with BORSIC (silicon-carbide-coated boron fibers) are produced by diffusion welding. Titanium matrix material, for example, is commonly fabricated by winding fiber over titanium or titanium alloy foil and spraying with a polystyrene-xylene mixture that hardens to bond fiber and foil together. The tapes are then diffusion welded together in a two-stage process. First, the organic binder is removed by heating the composite unconsolidated laminate to 840 °F for 30 min in a vacuum. This is followed by diffusion welding at about 1490 °F, under a load of 1.45 to 2.175 ksi, in a vacuum of 5×10^{-6} torr for 30 to 60 min. Ultimate tensile strengths as high as 186 ksi have been reported for Ti-6Al-4V matrix composites fabricated in this manner. A typical transverse fracture surface of a BORSIC reinforced titanium composite is illustrated in Fig. 18. Aluminum matrix tape is diffusion welded in a similar fashion (Fig. 19). In both cases, the use of diffusion welding inhibits reaction between the matrix and fibers because of the relatively low welding temperatures.

Another application of diffusion welding of composites involves the production of rock drill bit cutters by joining a tungsten carbide cermet sandwiched with polycrystalline diamond (manufactured under the registered name of Stratapax) to a cobalt matrix cermet stud (Fig. 20). Diffusion welding was chosen for this application because the stability of the diamond cermet dictated that the welding temperature not exceed 1290 °F. The welding procedure consisted of plating the faying surfaces with nickel by an electron beam ion plating technique, encapsulation of the parts along with graphite powder in a semikilled low-carbon steel evacuated retort, followed by hot isostatic pressing for 4 h at 1190 °F under a gas pressure of 30 ksi. A photograph of the assembled drill bit and a photomicrograph of the weld cross section are shown in Fig. 21 and 22. Weld shear strengths exceeding 82.6 ksi were obtained, and drilling rates as high as 89 ft/h were achieved in Sierra white granite. Other parametric studies involving diffusion welding have been conducted on sys-

Fig. 25 View normal to surface of a typical rough machined surface

Fig. 26 Commercial bronze deposit on titanium alloy after shear failure at the weld interface. Diffusion welded at 900 °F. Typical field for titanium alloy polished with 280 grit SiC paper, commercial bronze rough machined

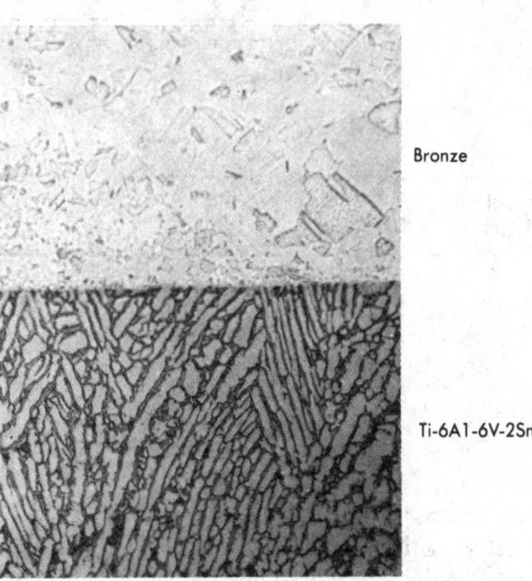

Fig. 27 Weld interface diffusion welded at 900 °F. Typical field for titanium alloy rough machined, commercial bronze polished with 280 grit SiC paper. Note deformation of titanium alloy microstructure at joint interface.

tems such as thoria-dispersed nickel and thoria-dispersed nickel-chromium.

Refractory Metals. The primary impetus for diffusion welding of refractory metals (zirconium, niobium, molybdenum, tungsten, tantalum, rhenium, and their alloys) has resulted from the needs of the aerospace and nuclear industries to procure complex hardware that cannot, in many cases, be fabricated by fusion welding or brazing. For example, when tungsten, molybdenum, and some of their alloys are subjected to the high temperatures created during fusion welding and brazing, areas adjacent to the joint recrystallize, causing embrittlement. In addition, transport of interstitials such as oxygen into the metal via the molten state is very rapid, resulting in embrittlement. Much of the work on refractory metals has involved the use of an interlayer material to lower the welding temperature below the recrystallization temperature of the base material. For example, titanium interlayers are commonly used in diffusion welding T-111 (Ta-8W-2Hf), pure molybdenum, and niobium alloys. Zirconium is also used for molybdenum and niobium alloys. If higher temperatures can be tolerated, refractory metals are also good candidates for self-diffusion welding. When diffusion welding alloys with interlayers, the interlayer must be chosen carefully because of the possibility that it may act as an interstitial sink. For example, the problem with TZM was discussed in a preceding section.

Successful techniques have been developed to produce diffusion welded niobium alloy (Cb-752) panels out of 0.02-in.-thick sheet for potential use as a heat shield on a spacecraft. External and internal surfaces of this panel are illustrated in Fig. 23, along with a typical microstructure of the weld joint. Welds were made utilizing a 0.001-in.-thick titanium interlayer and a 321 stainless steel retort with dynamic evacuation. Welding temperatures in the range of 1950 to 2000 °F with faying surface pressures of 1.5 ksi were used. Welding time was 5 to 7 h. Attempts were made to test the weld joints in shear, and failure always occurred outside the weld joint. In addition, 2500 °F creep tests were conducted on coated diffusion welded specimens, and no detectable creep or preferential oxidation of the weld joint was observed.

Other Metals. One of the most important areas where diffusion welding is applied is the joining of dissimilar materials. Successful welds have been made either in prototypic or parametric studies with the following materials: Ti to low-alloy steel, Ti to Be, Ti to Al_2O_3, Al to Ni, Al to Ti, Al to steel, Cu to steel, Ni alloy to steel, ZrC to ZrB_2, ZrN to ZrB_2, ZrC to ZrN, ZrN to Al_2O_3, ZrB_2 to Al_2O_3, TD Ni to Cr, Cu-10Zn to Ti alloy, Inconel 718 to steel, Ni alloy to TD Ni or TD NiCr, B 1900 to Udimet 700, Al to Cu, Al to U, dispersion-strengthened Al to steel, Cu-2Be to Monel, 80Cu-10Sn-10Pb to steel, Cu to Ag-Cu-In alloys, and Fe to PbTe. The complexity and diversity of the above materials suggest that most materials can be diffusion welded provided that proper attention is given to preparation of the faying surfaces, the metallurgical and chemical compatibility of the materials, and the welding environment.

For example, with an application involving welding of commercial bronze to titanium alloy (Ti-6Al-6V-2Sn), selection of welding parameters is severely constrained by the presence of brittle intermetallics in the copper to titanium system and by the aging temperature of the titanium alloy (900 °F). In this application, bronze bands were successfully diffusion welded to 6-in.-diam titanium alloy tubes at 900 °F in dynamic vacuum while the faying surfaces were loaded by differential thermal constraint. The thermal constraint was generated by a molybdenum ring and a steel plug as illustrated in Fig. 24. Parameter studies showed that surface extension was required to fracture surface oxide films because of the low welding temperature. Extension was achieved by rough machining the bronze faying surface (Fig. 25) so that soft high spots would deform under the welding load. This rough surface resulted in a large unbonded area (Fig. 26), but failure in shear always occurred in the bronze adjacent to the weld. A cross-sectional microstructure of the weld is also presented in Fig. 27.

Deformation Welding Applications

Thermocompression welding is commonly used to interconnect gold wires in microelectronic circuits. The following

Fig. 28 Thermocompression welding of gold wire ball and stitch welds. Process begins by melting the tip of the wire with a hydrogen cut-off torch. (a) Surface tension creates a ball 2$^1/_2$ to 3 times the wire diameter. (b) Capturing the ball within the inside chamber, heat and pressure are applied to the ball weld. (c) The tool is raised, and the wire is fed out. (d) After repositioning over the substrate, the stitch weld is made. (e) The tool is then lifted, breaking the wire from the joint. (Courtesy of E.I. Du Pont de Nemours)

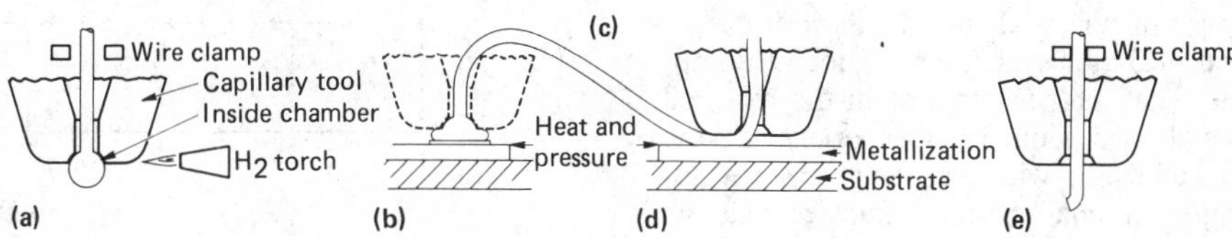

example describes the process for a typical application, in which 0.001-in.-diam gold wire was welded to gold-chromium metallized alumina substrates. The evaporated gold layer was 12 μin. thick.

In the semiconductor industry, the process is known as thermocompression ball and stitch bonding and is schematically illustrated in Fig. 28. The fine gold wire was fed through a capillary, and a ball was formed at the end of the wire by fusion. In this example, the ball was formed by heat from a miniature hydrogen torch. Following formation of the ball (in this case, 3 to 3.5 mil diam), the capillary tool was lowered to push the ball against the metallized surface (0.225 ft · lb applied to tool). To supply heat to the weld interface, the tungsten capillary was pulse resistance heated to raise the weld interface to 480 °F. Weld duration was 1 s.

Following welding of the ball, the capillary tool was raised to feed out more wire and was repositioned to make a stitch bond as shown in Fig. 28(d). The tip pressure was adjusted so that sufficient deformation occurred to ensure that the wire ruptured at the end of the wedge-shaped stitch bond when the tool lifted with the wire clamped, as in Fig. 28(e).

In this process, control of organic contaminants on the surface of the gold metallization is extremely important. For this application, ultraviolet-ozone cleaning was used to remove organic surface contaminants just prior to welding (within 2 h). The substrates were placed in a chamber which exposed the metallized surfaces to radiation from a mercury vapor ultraviolet lamp (radiation peaking at 2537 Å) in air at a spacing of 2.2 in. Filtered air was passed between the lamp and the substrate at 118 ft/s. Radiation intensity 0.3 in. above the substrate was 10 mW/cm^2. Exposure was 16 to 20 h. The combination of this cleaning procedure and welding parameters produced ball bonds with shear strengths in the 0.180 psi range (~0.176 lb-force).

Seal welds on canisters employed in long-term disposal of high-activity nuclear waste must be highly reliable. The fill hole on proposed 24-in.-diam, 118-in.-high 304L stainless steel canisters was 5 in. in diameter. An upset welding process utilizing resistance heating was developed to plug this hole. Fusion welding processes were eliminated from consideration because of potential solidification problems and difficulties of maintenance of equipment and process control in the required remote operation. Prior experience by the manufacturer with resistance upset welding on a smaller scale indicated that the process should afford the necessary reliability without a large amount of equipment maintenance.

In resistance upset welding, the parts are heated by resistance to the flow of an electric current passing through the interface of the surfaces in contact under high pressure. Solid-state deformation welds are produced if the pressure is sufficiently high to cause extensive plastic deformation. As in forge welding with oxyacetylene torches or induction heating, temperatures of

Fig. 29 Design of plug joint for upset welding of canister closure. Courtesy of E.I. Du Pont de Nemours

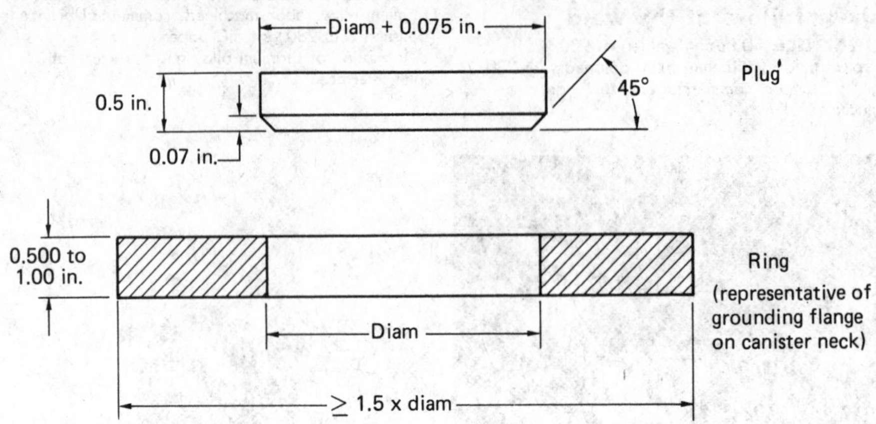

Fig. 30 Welding results (amount of plug inset) as a function of welding parameters. Two-in.-diam upset plug welding study. Force versus current at constant 99-cycle (1.5-s) weld time. (Courtesy of E.I. Du Pont de Nemours)

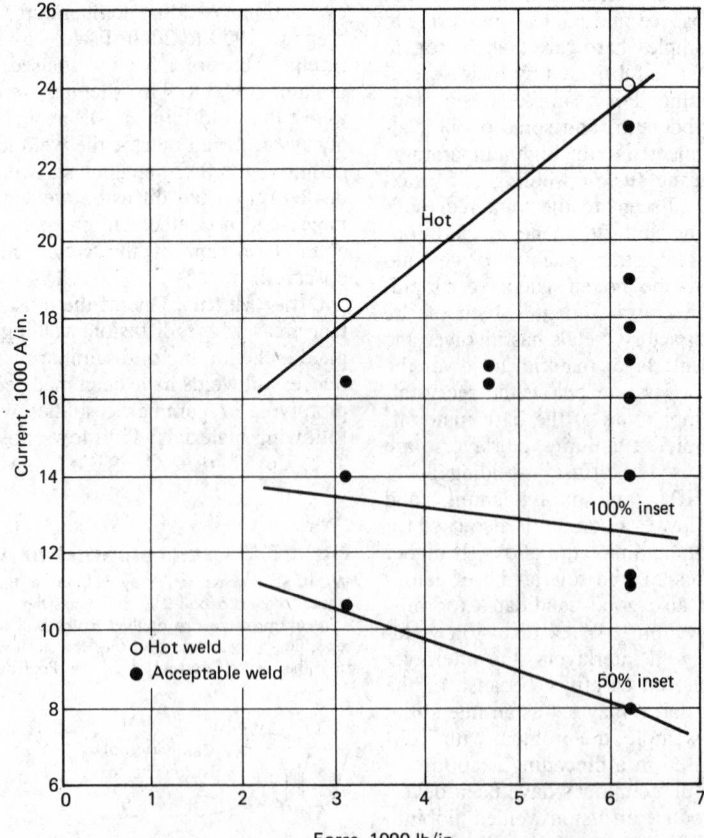

0.8 T_m are achieved in resistance upset welding.

Two-in.-diam plugs were used in the development of this weld. Figure 29 illustrates the joint design. During welding, the upper electrode, which was 0.5 in. larger in diameter than the plug, pneumatically forced the plug into the fill hole, resulting in extensive shear deformation at the weld interface. This shear deformation is characterized by the distance that the plug is inset into the hole. Welds were made using a single-phase alternating current welding machine capable of producing 230 000 A at 75 000 lb electrode force. Typical welding parameters and resulting weld quality are summarized in Fig. 30. The upper limit shown in Fig. 30 indicated by the word "hot" corresponds to the onset of melting. Because welds with only 50% inset were strong and leaktight, Fig. 30 illustrates that the process is intrinsically tolerant to a wide latitude of welding conditions. A metallographic section of a typical 304L stainless steel weld is shown in Fig. 31. The weld zone is wrought metal that is stronger (42 to 44 ksi yield stress) than the annealed 304L stainless steel base metal. To evaluate the tolerance of the process to oxidized surfaces, some of the rings were oxidized for 16 h at 2900 °F and then welded to unoxidized plugs. These welds exhibited up to 50% decrease in strength compared to welds with unoxidized rings, but were nevertheless hermetic (less than 1×10^{-8} standard cm^3/s).

This application illustrates the potential of the high reliability of upset welding. The welding organization responsible for the development of this application has also demonstrated the feasibility of the process for the joining of other alloys, such as Haynes 25, TD nickel, tungsten, 1100 aluminum alloy, 6061 aluminum alloy, and 2024 aluminum alloy. Of these, TD nickel, 2024 aluminum, and tungsten are particularly difficult to join by fusion welding, while amenable to deformation welding.

Fig. 31 Metallographic section of typical solid-state upset weld in type 304L stainless steel

Roll Welded Heat Exchangers. A principal application of roll welding is the production of heat-exchanger panels with flow tubes as an integral part of the composite. Millions of refrigerator evaporator plates have been formed by this process. A cross section of a new application, solar collector panels, is illustrated in Fig. 32. Roll welded heat-exchanger panels are most often made of aluminum, although copper and stainless steel panels have also been produced commercially.

The first major step in the production of roll welded heat exchangers is to thoroughly clean the surfaces of the sheets to be joined by combinations of solvent cleaning, chemical cleaning, and wire brushing. Next, a stop-weld pattern is reproduced on one of the sheets that will later form the passageways in the heat exchanger. This is accomplished by transferring ink onto the surface by silkscreen processing. These stop-off inks contain materials such as graphite or titanium oxide.

The prepared plates are placed together and temporarily joined by spot welding to form a sandwich that is then roll welded, with at least partial welding occurring during the first rolling pass. The sandwich sometimes is heated prior to this first rolling pass. Whether hot rolling or cold rolling is employed, the major reduction takes place in this first pass (50 to 80%). Subsequently, cold rolling may be employed to reduce the panel to final gage. Also, panels are generally given a postweld heat treatment. This heat treatment results in an annealed part and, particularly in the case of cold rolled panels, may improve the weld quality through recrystallization and diffusion. Finally, the passageways are formed by inflating them with air pressure while the sandwich is held between platens. The panel is then ready for trimming, additional forming, and painting, as required for the application.

This process has been used to produce heat exchangers for refrigerators, solar panels, poultry incubators, electronic equipment temperature controllers, industrial heat exchangers, and special applications such as thermal control of spacecraft. Special modifications of the process permit as many as six sheets to be welded at one time and inclusion of passageways oriented along different principal directions at the different layers. Also, it is possible to produce panels where the passageways are expanded on one side only.

Fig. 32 Section of aluminum solar collector panel made by roll welding

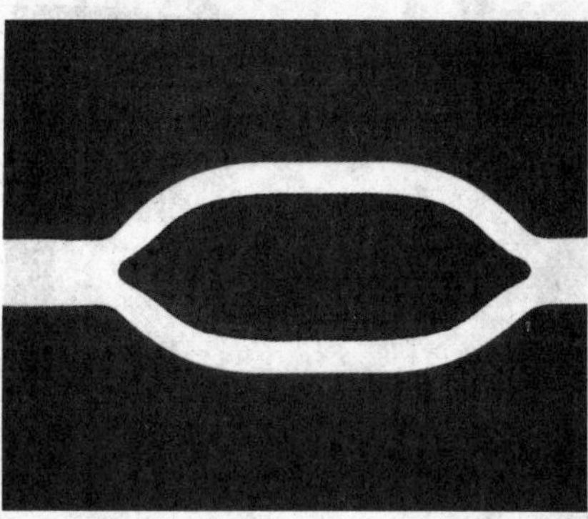

Cladding of Metals by Strip Roll Welding. The common early application of strip roll welding was the fabrication of bimetallic strips for thermostats. This remains an important application, with thermostats taking many complex forms and finding use in furnaces, color televisions, cars, and numerous industrial controls. A more recent application is in the production of the U.S. Mint coins.

Silver shortages resulted in the introduction of coins made from new materials. These coins required a unique set of properties for acceptance by the general public and use in automatic vending machines. Copper cladded with cupronickel was found to meet these requirements (Fig. 33). The cladding process is relatively simple, although stringent in its requirement for surface preparation. Cladded strips are produced by continuous rolling with the surfaces prepared just prior to rolling by processes such as wire brushing. Welding typically is accomplished in a single rolling pass. Subsequent heat treatments may be employed to improve the weld quality by processes such as sintering, diffusion, and recrystallization.

Similar processing is applied to many new products, including electrical contacts, conductive springs, automotive windshield wiper sockets, automotive trim, clad cookware, wrap for underground cable, and electromagnetic shielding.

Essentially all combinations of ductile metals and alloys can be clad by roll welding, although the oxides of some metals may make welding difficult and other combinations may suffer from intermetallic compound formation on heating. Most

Fig. 33 Metallographic cross section of edge of cupronickel cladded copper coin

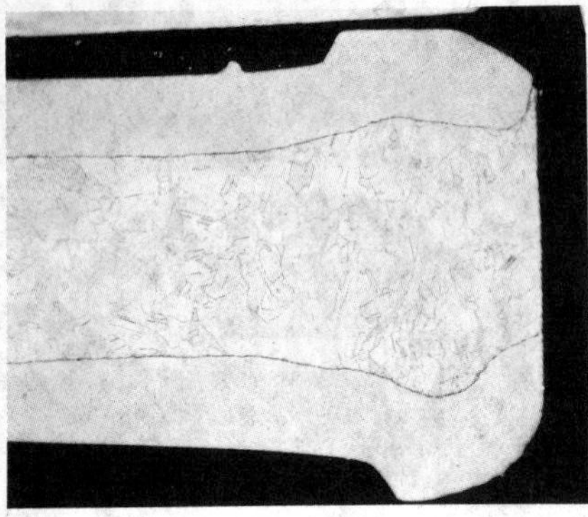

engineering metals and alloys can be processed into clad-laminate form, but some cannot. One characteristic that makes some metals difficult or impossible to combine by deformation is low ductility. Such metals cannot be co-reduced to the extent necessary to achieve a metallurgical bond. Another characteristic that can cause problems is the tendency of some metals to form a tenacious oxide film that inhibits weld formation. Finally, some metals are thermally unstable; they form brittle intermetallic compounds when heated above a certain temperature. Those metals that are readily processed into clad laminates have opposite characteristics. They are ductile and easily cleaned of oxides, and they form no intermetallic compounds. In most cases, when a metal is named, alloys of that metal also apply. Some examples of easy, difficult, and impossible combinations follow:*

Easy to weld

- Copper to steel
- Copper to nickel
- Copper to silver
- Copper to gold
- Aluminum to aluminum alloys
- Tin to copper
- Tin to nickel
- Gold to nickel

Impractical

- Gold to aluminum
- Zirconium to aluminum
- Cobalt to aluminum

Difficult but possible

- Copper to aluminum
- Aluminum to carbon steel
- Stainless steel to aluminum
- Copper to tantalum, niobium, or titanium
- Titanium to carbon steel or stainless steel
- Stainless steel to carbon steel
- Aluminum to nickel
- Nickel to steel
- Copper to stainless steel
- Manganese to nickel to copper
- Copper to manganese
- Silver to manganese
- Silver to steel
- Uranium to zirconium
- Zirconium to copper, steel, or stainless steel
- Platinum to nickel, copper, or steel
- Tantalum to niobium

Impossible

- Alfesil to anything
- Beryllium to anything

*From Designing With Clad Metals, Richard G. Delagi, Metallurgical Materials Division, Texas Instruments, Inc., Attleboro, MA.

Cold Welding of Aluminum Tubing. Aluminum alloy (1100) tubing, 0.187 in. in diameter with a 0.02-in. wall, was cold pinch welded to produce a hermetic seal closure in a pressurized line.

Special tooling was designed that restrained radial deformation of the tube but allowed movement in the longitudinal direction. This tooling is illustrated in Fig. 34. The special radius anvils of high-speed tool steel are housed in the clamshell steel fixtures shown. Approximately 90% deformation (reduction of wall thickness) was achieved by application of hydraulic pressure. Welds were made in tubing in both the O and H-14 tempers. Sound welds were obtained in both cases with no metallurgical evidence of the original faying surfaces in the weld joint. At the high deformations employed, contaminants such as oil, water, and dust were not a problem. No special cleaning procedures were used. Welds of 1100 aluminum alloy tubes in the H-14 temper could withstand internal pressure of up to 2 ksi.

Fig. 34 Tooling used in cold welding of aluminum. Source: Irons, J.L., Cold Welding of Al Tubing, SME Paper, 1975

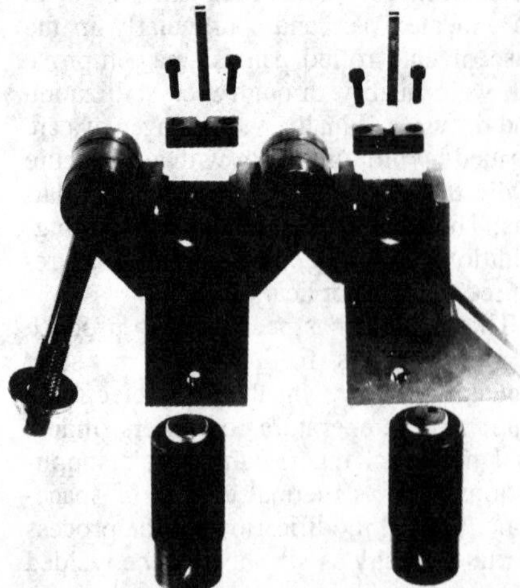

SELECTED REFERENCES

- Alm, G.V., Guide to Diffusion Bonding, *Mater. Eng.*, Vol 70 (No. 3), Sept 1969, p 24-29
- Baboian, R. and Haynes, G.S., Joining Dissimilar Metals With Transition Materials, *Automotive Engineering,* Vol 84 (No. 12), 1976
- Bartle, P.M., Introduction to Diffusion Bonding, *Metal Construction and British Welding Journal,* May 1969, p 255-258
- Billard, D. and Trottier, J.P., Original HC1 Surface Treatment for Diffusion Bonding of Nickel Superalloy Specimens, *Metals Technology,* Vol 504, Sept 1978, p 309-319
- Bryant, W.A., A Method For Specifying Hot Isostatic Pressure Welding Parameters, *Welding Journal,* Vol 54 (No. 12s), Dec 1975, p 433-435
- Derby, B. and Wallach, E.R., Theoretical Model for Diffusion Bonding, *Metal Science,* Vol 16, Jan 1982, p 49-56
- Dunning, J.S. and Metcalfe, A.G., Basic Metallurgy of Diffusion Bonding, AFML-TR-65-34 Tech. Report, Oct 1965
- Duvall, D.S., Owczarski, W.A., Paulonis, D.F., and King, W.H., Methods for Diffusion Welding the Superalloy Udimet 700, *Welding Journal,* Vol 42 (No. 2), Feb 1972, p 41-49s
- Gerken, J.M. and Owczarski, W.A., A Review of Diffusion Welding, *Welding Research Council Bulletin,* No. 109, Oct 1965
- Jellison, J.L., Effect of Surface Contamination on Solid Phase Welding—An Overview, *Surface Contamination,* Vol 2, Plenum Publishing, New York, 1979, p 899-923
- Jellison, J.L., Gas Pressure Bonding of Stratapax, ASME Publication 77-Pet-72, Sept 1977
- Jellison, J.L., IEEE Trans. Parts, Hybrids and Packaging, Vol PHP13, 1977, p 132
- Korb, L.J., Beuyukian, G.S., and Rowe, R.J., Diffusion Bonded Columbium Panels for the Shuttle Heat Shield, *SAMPE Quarterly,* Vol 3, 1972, p 1-11
- Maher, D.M., The Formation of Po-

rosity During Diffusion Processes in Metals, Lawrence Livermore Laboratory, UCRL-10383, Sept 1962

- Milner, D.R. and Rowe, G.W., *Metal Rev.*, Vol 7, 1962, p 433
- Mohamed, H.A. and Washburn, J., Mechanism of Solid State Pressure Welding, *Welding J.*, Sept 1975, p 302s
- Prewo, K.M. and Kreider, K.G., The Deformation and Fracture of Basic Reinforced Titanium Matrix Composites, *Titanium Science and Technology*, Vol 4, Plenum Publishing, New York, 1972, p 2333-2345
- Owczarski, W.A. and Paulonis, D.F., Application of Diffusion Welding in the USA, *Welding Journal*, Feb 1981, p 22-23
- Spurgeon, W.M., Rhee, S.K., and Kiwak, R.S., Diffusion Bonding of Metals, *Bendix Technical Journal—Materials and Processing*, Vol 2 (No. 1), 1969, p 24-41
- Taylor, D.S. and Pollard, G., Proceedings of Advances in Welding Processes Conference, Harrogate, Paper No. 4, The Welding Institute, Abington, 1978
- Tylecote, R.F., *Solid Phase Welding of Metals*, St. Martin's Press, New York, 1968
- Vaidyanath, L.R., Nicholas, M.G, and Milner, D.R., *British Welding Journal*, Vol 6, 1954, p 13
- Weisert, E.D. and Stacker, G.W., Fabricating Titanium Parts With SPF/DB Process, *Metal Progress*, Vol 111 (No. 3), March 1977, p 33-37
- Wright, D.K., Snow, D.A., and Tay, C.K., Interfacial Conditions and Bond Strength in Cold Pressure Welding by Rolling, *Metals Tech.*, Jan 1978, p 24
- Zanner, F.J. and Fisher, R.W., Diffusion Welding of Commercial Bronze to a Titanium Alloy, *Welding Journal*, Vol 54 (No. 4), April 1975, p 105-1125

Thermit Welding

By Hans D. Fricke
President
U.S. Thermit Inc.

THERMIT* WELDING is a process that produces coalescence of metals by heating them with superheated liquid metal from an aluminothermic reaction between a metal oxide and aluminum with or without the application of pressure. Filler metal is obtained from the liquid metal. The aluminothermic reaction associated with the thermit process produces pure carbon-free heavy metals, such as chromium, manganese, or vanadium from ores, oxides, or chlorides. Introduction of this process constituted a major technological breakthrough; when these metals are generated by the electrothermic processes, they have high carbon contents that are unacceptable for many metallurgical applications.

Dr. H. Goldschmidt first introduced this process to the technical community at a convention of the German Electrochemical Society in 1898. The aluminothermic process was described as exothermic, i.e., able to create large amounts of heat without the application of any external heat source. Goldschmidt later suggested utilizing the heat of this exothermic reaction for welding and was instrumental in the development of a powder mixture of aluminum and iron oxide, which is now registered as the tradename Thermit.

Metallurgical Principles

In order for an exothermic reaction to occur, the reductant must have a great affinity for the oxygen associated with the metal to be produced:

$$MeO + R \rightarrow Me + RO$$

where Me is metal, R is reductant, and O is oxygen.

The difference in the free energy of oxide formation is the determining factor as to whether or not a chemical reaction between two components is exothermic or endothermic. Figure 1 shows the free energy of oxide formation in relationship to the temperature for selected metal oxides. For exothermic reactions, those metals characterized by the lowest energy are best suited for use as MeO for reductants with high energy of oxide formation. If the difference in free energy is not high enough between a metal oxide and the reductant, external energy must be supplied to initiate and complete the reaction, which is then endothermic.

Principles of Thermit Welding

The aluminothermic reaction that utilizes aluminum as a reductant takes place according to the following general equation:

$$\text{Metal oxide} + \text{aluminum} \rightarrow \text{aluminum oxide} + \text{metal} + \text{heat}$$

The intense superheat set free during the reaction generates iron and aluminum oxide in liquid form (Fig. 2). Because each component has a different density, they separate automatically within seconds, and the liquid iron can be used for different welding applications or for the production of metals and alloys (Fig. 2). The theoretical temperature achieved by reducing iron oxide with aluminum is about 5600 °F.

The most commonly used oxides for thermit welding and their corresponding reactions are:

$$3Fe_3O_4 + 8Al \rightarrow 9Fe + 4Al_2O_3 \quad (\Delta H = 3350\ kJ)$$

$$3FeO + 2Al \rightarrow 3Fe + Al_2O_3 \quad (\Delta H = 880\ kJ)$$

$$Fe_2O_3 + 2Al \rightarrow 2Fe + Al_2O_3 \quad (\Delta H = 850\ kJ)$$

$$3CuO + 2Al \rightarrow 3Cu + Al_2O_3 \quad (\Delta H = 1210\ kJ)$$

$$3Cu_2O + 2Al \rightarrow 6Cu + Al_2O_3 \quad (\Delta H = 1060\ kJ)$$

Depending on the application, aluminum can be replaced by magnesium, silicon, or calcium as reductant. However, characteristics such as the low boiling point of magnesium and calcium, the hygroscopic nature of calcium, and the high melting points of calcium and magnesium oxides have limited the use of these two elements to applications where they are alloyed with other metals or where special precautions, such as pressurized vessels, are taken. A combination of silicon and aluminum frequently is used in aluminothermic mixtures where a liquid phase is not desired, such as for heat treatment purposes or as a hot top riser compound. In most technical applications, silicon alone does not create sufficient heat to sustain the reaction without external heat sources. Factors affecting the use of various elements for reducing metal oxides are listed in Table 1.

Because most aluminothermic reactions are extremely violent if metal oxide and reductant only are used, portions of non-reacting constituents, such as ferroalloy

*Thermit, which is the term commonly used to identify this welding process, is a worldwide registered trademark of Th. Goldschmidt AG, Essen, West Germany.

Fig. 1 Free energy of oxide formation versus temperature. ΔG° = standard free energy; R = gas constant; T = temperature, °K; lnpo_2 = natural logarithm of the equilibrium oxygen pressure

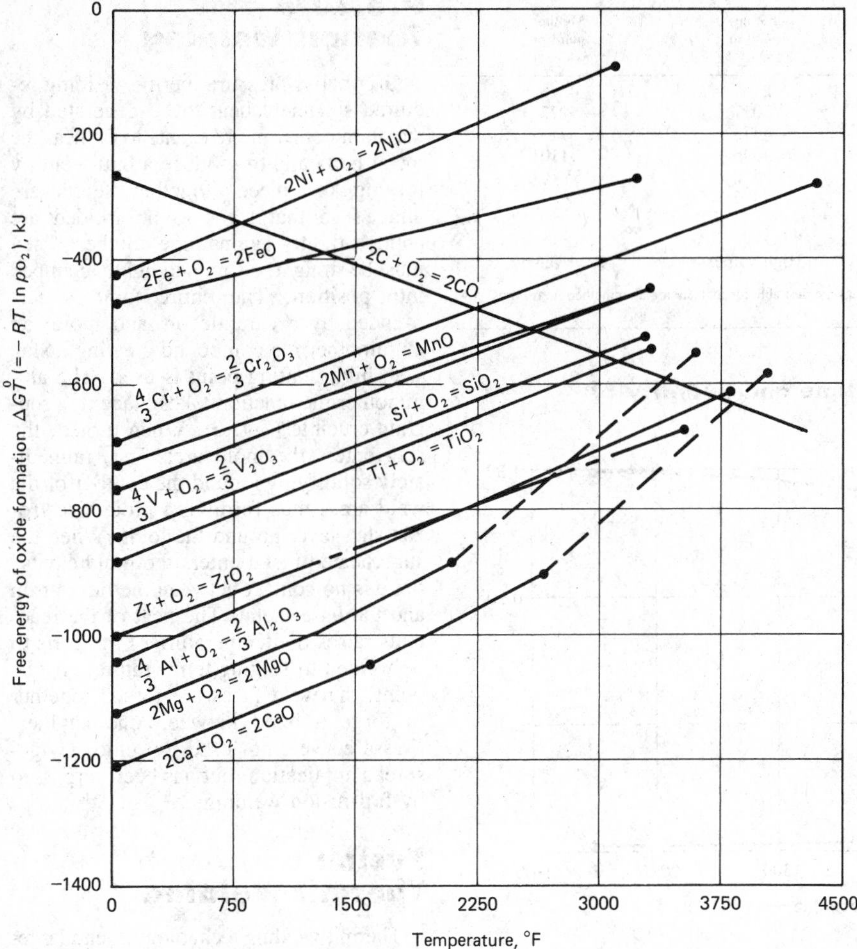

pellets, are added. These additives, as well as radiant heat loss and losses to the reaction vessel, reduce the maximum temperature to about 4500 °F. This is the most efficient temperature range to obtain separation of maximum iron and aluminum oxide. Operating temperatures above 4500 °F will result in aluminum losses due to vaporization (4470 °F), whereas lower temperatures will result in Al_2O_3 slag inclusions due to premature solidification of Al_2O_3 (3700 °F).

The amount of heat loss is highly dependent on the quantity of aluminothermic compound reacted at one time. With larger quantities, the heat loss per weight unit is considerably lower and the reaction more complete compared with smaller quantities of aluminothermic compound, as shown in Fig. 3. An almost linear correlation exists between reaction time and quantity of aluminothermic compound for amounts less than 200 lb; reaction time, related to the weight unit, is ultimately independent of the amount contained in the reaction.

Fig. 2 Aluminothermic reaction. Courtesy of Th. Goldschmidt AG, Essen, West Germany

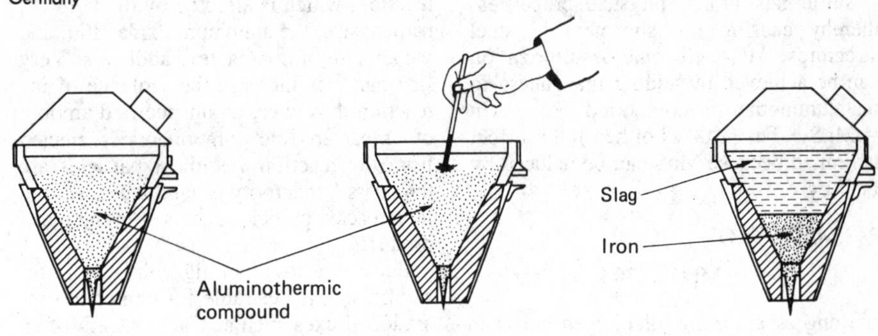

$$Fe_2O_3 + 2\,Al \rightarrow Al_2O_3 + 2Fe + 850\text{ kJ}$$

$$1000\text{ g aluminothermic compound} \rightarrow 476\text{ g slag} + 524\text{ g iron} + \text{heat}$$

Aluminothermic compound in its granular form is not explosive or hazardous. Depending on the grain size of its constituents, as well as the amount of added nonreacting components, the ignition temperature varies; typically, it is set at about 2730 °F. An external heat source, such as special ignition powder or an ignition rod, is required to initiate the aluminothermic reaction. If sufficient heat is generated, the reaction, once started, is self-sustaining. Ignition devices are composed of materials that can be ignited by a match, flint gun, or gas striker.

Metallurgical Parameters

The proper ratio of aluminum and oxide is necessary for completion of the aluminothermic reaction. For the iron aluminothermic reaction, the following types of iron oxide are available: FeO (oxygen content, 22.2%); Fe_2O_3 (oxygen content, 30.0%); and Fe_3O_4 (oxygen content, 27.6%).

Other alloys and metals can be added to match the composition and mechanical properties of the parts to be welded. The proper balance of aluminum and oxygen determines the microstructure of the resulting iron as well as the yield of any alloying additions. If a deficiency of aluminum exists, excess oxygen supplied by the iron oxide continues to react with other elements contained in the aluminothermic compound, or with the reaction vessel itself, in the order of their difference of free energy of oxide formulation with iron oxide (see Fig. 1).

Phosphorus and sulfur are the most detrimental elements in iron or iron-alloy welds and as such are kept to a minimum. Both elements usually are associated with iron oxide when mineral ores are used as

Table 1 Factors affecting the use of various elements for reducing metal oxides

Reductant	Heat of oxide formation at 77 °F, kJ/mol O_2	Boiling point of element, °F	Melting point of oxide, °F
Calcium	−1270	2700	4676
Magnesium	−1205	2017	5072
Aluminum	−1120	4473	3713
Silicon	−905	4496	3110
Carbon	−395 (CO_2)	7592	−135 (CO_2)
Hydrogen	−485	−487	...
Desirable for aluminothermic reaction	High (−) values	High values	Low values

Note: High values (−) in heat of oxide formation and boiling point columns are desirable for maintaining aluminothermic reaction; low values are desirable in melting points for sustaining the reaction.

Fig. 3 Relationship between reaction time and quantity of aluminothermic compound

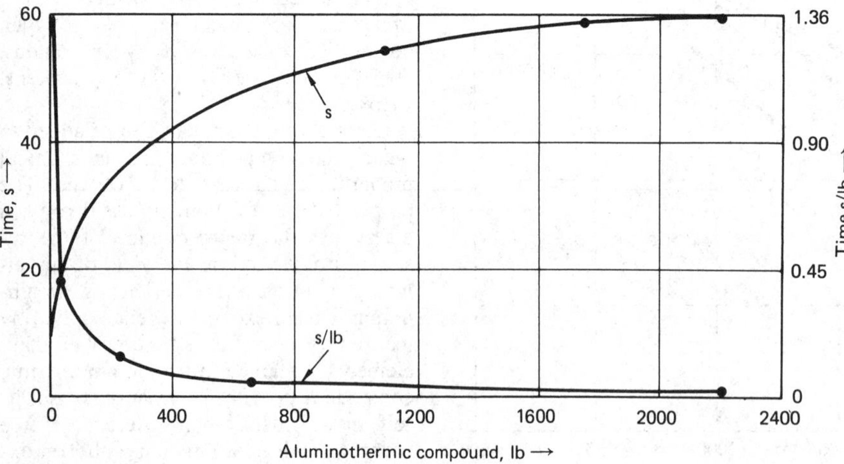

the oxygen-carrying agent. Phosphorus is soluble in liquid iron, but its solubility decreases during solidification of the iron; Fe_2P and Fe_3P tend to segregate and settle between the iron crystals, thereby weakening the microstructure.

Sulfur has similar physical properties, thereby causing hot shortness in steel structures. Although some desulfurization can be achieved by adding manganese to the aluminothermic compound ($FeS + Mn = MnS + Fe$ + 100 kJ of heat), this effect is limited because MnS can be reduced by FeO by:

$$MnS + FeO = FeS + MnO + 38 \text{ kJ of heat}$$

as long as a surplus of oxygen exists in the system. Common steel manufacturing treatments to reduce phosphorus or sulfur in liquid steel are not applicable due to their detrimental effects on quality and reactivity of the aluminothermic compound.

With the exception of special applications where abrasion resistance is important, the metal produced by the aluminothermic reaction must be free of Al_2O_3 slag. Separation of the liquid components, slag and iron, depends on the nature of the reaction, which is affected by the volume, particle size, the aluminum/oxide ratio, and the amount of nonreacting additives. Very fine particles increase the violence of the reaction; however, an unspecified amount of unreacted fine components is ejected from the reaction vessel. If particles are too coarse, reactivity is subdued, resulting in unreacted particles enclosed in the metal product.

Because a low melting point of the resulting slag is desirable for complete separation, fluxes such as CaO or CaF_2 often are added. Due to the unwanted side effects of calcium, the melting point of the Al_2O_3 slag also can be lowered by a surplus of oxygen-carrying components (see Fig. 4). The choice of which material to apply depends on the metallurgical requirements demanded of the end product.

Pressure Thermit Welding

Originally, pressure thermit welding required sufficient heat to be generated by the aluminothermic reactants to preheat the metal parts and to produce a butt weld by forging the pieces together. Parts are aligned so that faces to be welded are abutted tightly against one another. They must be straight, clean, and lightly clamped into position. The entire joint is surrounded by a ceramic or sand mold; an aluminothermic compound creating a slag with high melting point is used. The aluminothermic reaction takes place in a separate crucible (Fig. 5). When turned, the slag enters the mold area first, immediately solidifying around the contour of the weld area, thus forming a protective frozen slag layer around the joint. When the superheated metal enters the mold cavity, there is no contact between the liquid iron and the base metal. The heat of the reactants raises the temperature of the parts to be welded to forging temperature. At this point, the weld faces are forced together to forge a bond between one another. Pressure welding used to enjoy widespread application, but has been surpassed by full-fusion welding.

Fusion Thermit Welding

Thermit welding technology found large-scale application with the introduction of the "centerpour" weld design, where the exothermically produced iron was used as a metallurgical means of joining two parts and not merely as a heating device. When workpieces are properly aligned, with an adequate gap between the parts, a mold that is either built on the parts or premanufactured on a pattern from the parts is placed around the area to be welded.

Depending on the process and the cross section of the parts to be welded, the weld ends must be preheated to provide conditions suitable for complete fusion between the aluminothermic steel and the base metal. Although termed welding, fusion thermit welding, by definition, resembles metal casting. Therefore, various features essential in foundry technology, such as the proper location and dimension of runners, gates, and risers, are instrumental to control and direct the flow of the liquid metal within the mold, to avoid turbulences, and to compensate for shrinkage during the transition from liquid to solid.

Fig. 4 Influence of metal oxides on melting temperature of Al_2O_3. Source: Ref 1

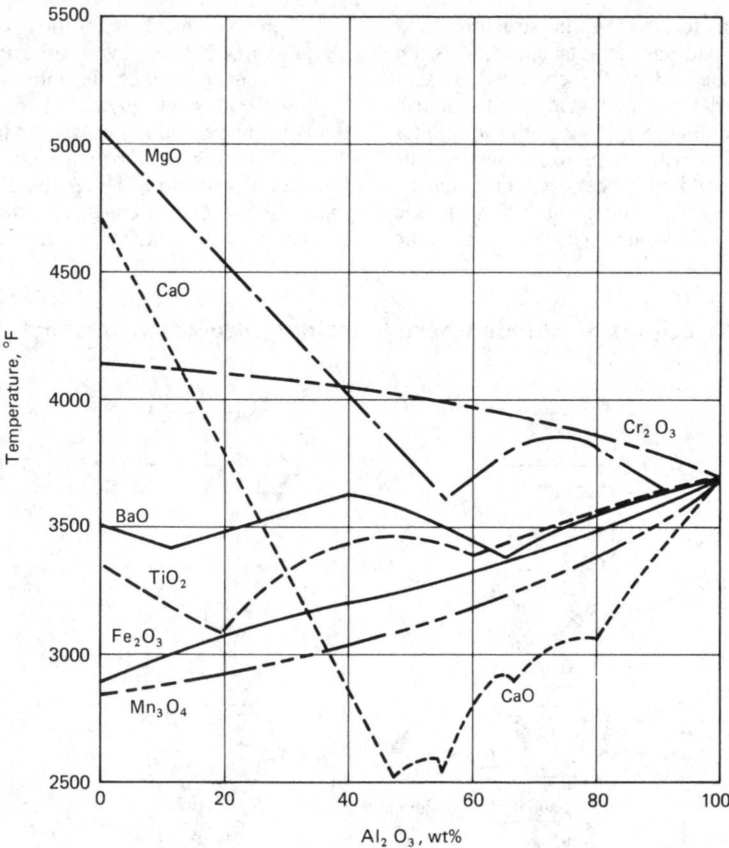

Rail Welding

Heavy power generation equipment is not required in thermit rail site welding. The thermit welding process was used to join trolley track in the United States in the early 1920's. The first long-weld railroad sections were not placed in track until 1933 at the Delaware & Hudson Railroad at Albany.

Continuous welded track currently is recognized by railroad and transit authorities as the most effective means of cutting maintenance cost by eliminating bolted joints. In coal mines, the main haulage track often is welded to reduce maintenance and, more importantly, to reduce excessive coal spillage. The impact of mechanical joints in crane runways on the surrounding equipment and building structures is eliminated by thermit welding crane rail sections. Thermit welding technology has been developed to the extent that proprietary mixtures for all rail composition are available to meet the specifications set by the steel mills, as well as by railroads and other operators of track structures.

Welding With Preheat

Successful full-fusion thermit welding depends not only on the quality of the aluminothermic compound, but also depends to a great extent on the preparation of the rail sections. Besides arranging for proper gap size between the rail ends, the welder must ensure that the welding surface is free of oil, rust, or slag deposits from a preceding torch cut. Premanufactured molds must be aligned so that the centerline of the mold coincides with the centerline of the gap.

Depending on the process, rail ends are preheated from 1100 to 1800 °F, with a gas torch flame as shown in Fig. 6. The refractory-lined crucible, containing the aluminothermic compound, is positioned above the mold halves after the preheating cycle is completed. The reaction is carried out in this crucible, allowing the Al_2O_3 slag to separate before the liquid steel is tapped from a bottom hole of this crucible.

Most welding setups include a self-tapping device; therefore, the welder does not have to choose the proper tapping time. Figure 7 illustrates the flow of material, where the aluminothermic steel enters the chamber between the rail ends first. The slag, floating on top, enters the mold halves only after the gap is completely filled with steel. Part of the slag is discharged into an attached slag pan, but most serves as an insulator against premature heat loss. This insulating layer allows the head riser liquid to act as a feeder basin of liquid steel while the solidification between the rail ends propagates, starting in the lower section of the rail up into the head. After solidification is completed, the excess material is removed manually or by hydraulic shearing devices.

Thermit welding processes with short preheating times require larger amounts of aluminothermic compound than those with long preheating. The lack of preheat, which cannot be provided within the shortened preheating period, must be compensated for by additional liquid superheated aluminothermic steel that provides effective preheating and washes the rail ends.

Figure 8, a cross section of a completed weld with preheat, shows the weld area as well as the various heat-affected zones (HAZ) and their corresponding microstructure. This microstructure is typical of rail steel (132RE) specified by the American Railway Engineering Association

Fig. 5 Principles of aluminothermic pressure welding. Source: Ref 2

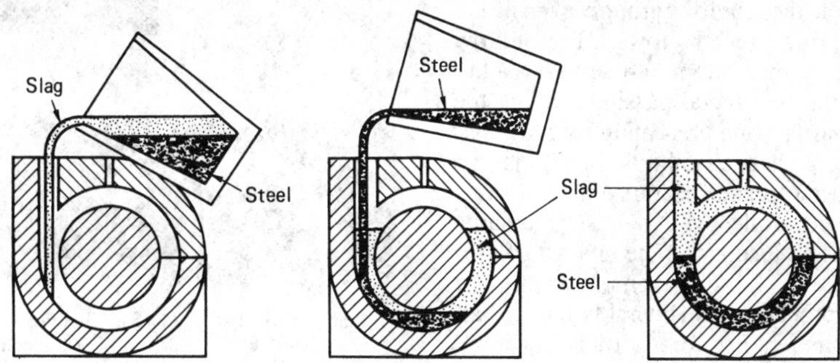

Fig. 6 Assembly of aluminothermic welding process with short preheat. Source: Ref 3

(AREA) with the following chemical composition:

Carbon	0.69-0.82%
Manganese	0.70-1.00%
Phosphorus, max	0.04%
Sulfur, max	0.05%
Silicon	0.10-0.25%

Welding Without Preheat

Thermit welding without preheating is designed to eliminate all equipment associated with torch preheating. Crucible and mold form one solid unit (Fig. 9) and are not reusable. They are manufactured of sand bonded with phenolic resin. Rail end preparation is identical to welding with preheating; preheating is accomplished by using a portion of the molten metal produced by the aluminothermic reaction. After the reaction is completed, the molten steel melts through a stack of disks, separating the crucible area from the joint area of the mold, without passing through the atmosphere. Figure 10 illustrates the shape of the cavity through which the molten filler metal flows. The hollow chamber in the bottom part serves as a basin for the first metal passing through the mold, washing and preheating the rail ends. After the preheat chamber is filled, the gap between the preheated rail ends is gradually filled, while the Al_2O_3 slag remains on top as an insulator. The amount of aluminothermic welding compound needed for this process is considerably larger than the amount required for external preheating.

Metallurgical Aspects of Thermit Rail Welds

The microscopic grain structure of a thermit weld and its adjacent HAZ is not only influenced by the chemical composition of the base and weld metal, but also by the cooling rate of the entire joint after completion of the pour. Because the thermit rail welding process is applied at the tracksite with uncontrolled atmospheric conditions, precautionary steps must be taken to prevent the formation of embrittled microstructures such as martensite and bainite.

The prevention of embrittled microstructures has become even more important in recent years with the introduction of alloyed rail steel. Figure 11 shows a comparison of various microstructures in thermit welds and the adjacent heat-affected rail structure. The ferritic grain is typical for low-carbon conductor rail used, for example, as a third rail in subway

Fig. 7 Principles of aluminothermic welding process with short preheat

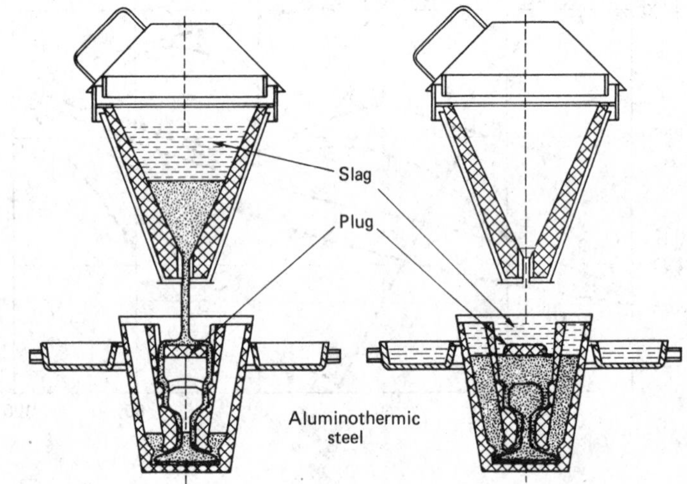

Fig. 8 Cross section and microstructure of an aluminothermic rail weld

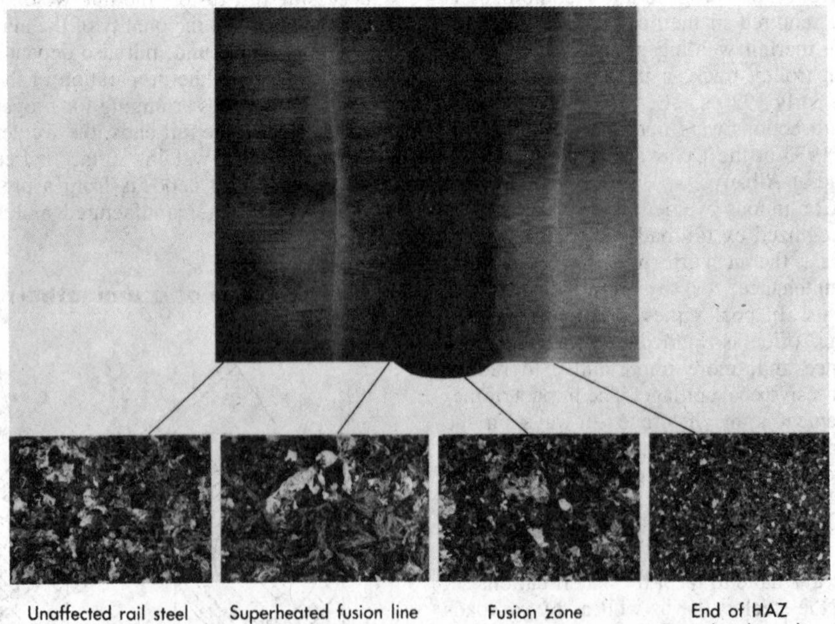

Fig. 9 Aluminothermic welding without preheating

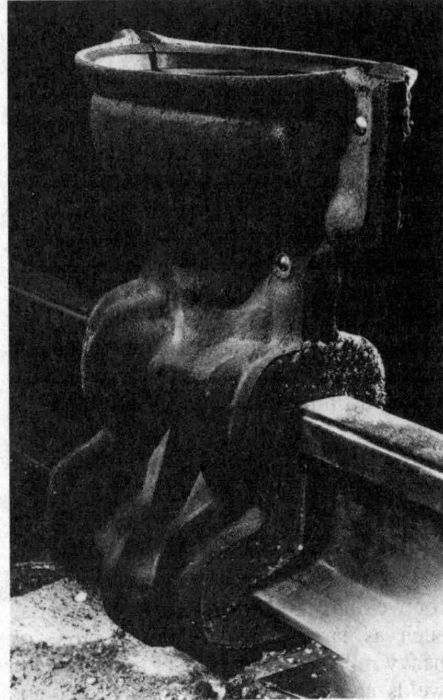

structures. The various pearlitic structures reflect the effect of either carbon or the intensity of heat transformation in the HAZ of the base metal. Martensite and bainite are shown as they develop in or adjacent to welds in alloyed rail steel at extremely high cooling rates. The austenitic microstructure is characteristic for thermit weld with high manganese, e.g., Hadfield steel used for frogs and switches. Each steel and its microstructure is characterized by its transformation behavior while cooling from high to low temperatures. Whereas the steel of a standard United States rail end transforms fully to pearlite after cooling from 1690 to 975 °F in 20 s (Fig. 12), the aluminothermic steel characterized by a higher aluminum content requires about 88 s for the same transformation from austenite to 100% pearlite (Fig. 13). For a chromium-manganese alloyed rail with increased minimum tensile strength of 157 ksi, about 220 s are required for the same drop in temperature to transform completely into pearlite (Fig. 14). The transformation to pearlite is guaranteed for all thermit rail welds because the cooling rate of approximately 15 min in the critical temperature area is slow enough to prevent hardened structures.

Testing of Thermit Rail Welds

The two most common testing procedures for all types of rail welds are the slow bend test (Fig. 15) and the rolling load test (Fig. 16).

Slow bend testing applies a static load to the welded rail joints until the joint either fails or a maximum load equivalent to 100-ksi modulus of rupture with the base of the rail in tension is accomplished. This maximum load, as well as a deflection of 1 in. in the center of the weld, has been adopted by the Association of American Railroads (AAR) as well as the AREA as acceptance criteria for aluminothermic welds performed on controlled-cooled rail steel. No acceptance standards have been developed for welded heat treated and alloyed rail steels, characterized by increased tensile strength and improved wear characteristics.

Rolling load testing is performed on a 12-in.-stroke rolling load machine, intended to simulate the relative movement between the wheel of a railroad car and the rail. The cantilever arrangement shown in Fig. 16 subjects the head of the welded rail to repeated loadings from zero to maximum tension stress. The applied wheel load for these rolling load machine tests depends on the size of the rail section being tested. In this type of testing, two million cycles of repeated loadings without failure is considered a runout.

Pulsating Fatigue Testing. Outside the United States, the rolling load test is rarely used and is, therefore, replaced by the pulsating fatigue test under bending stresses in the bottom of the weld when dynamic

Fig. 10 Principles of aluminothermic welding without preheating. (a) Mold design with small web bracket and double sump. (b) Mold design with web bracket

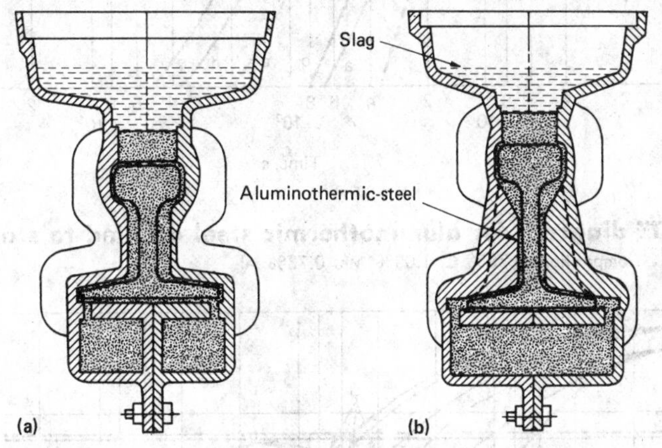

Fig. 11 Typical microstructures obtained by thermit welding

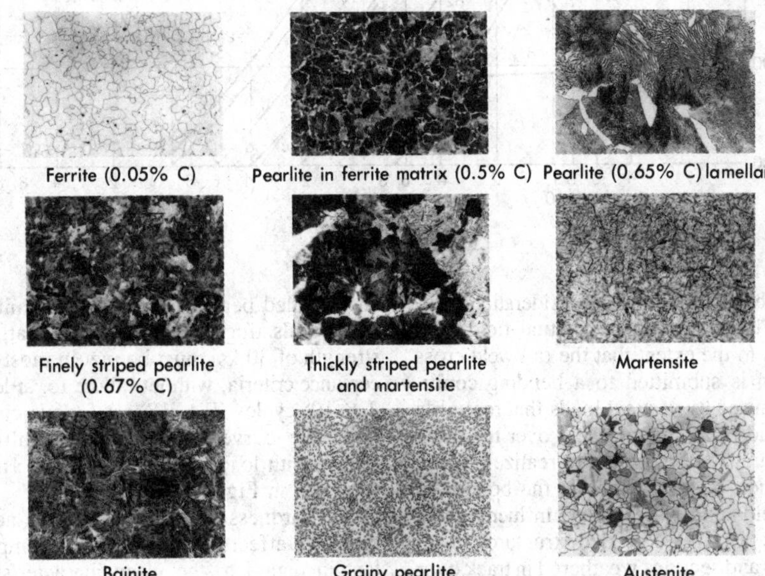

Fig. 12 TTT diagram for standard rail steel. Composition: 0.67 to 0.80% C, 0.70 to 1.00% Mn, 0.10 to 0.23% Si. (Source: Ref 4)

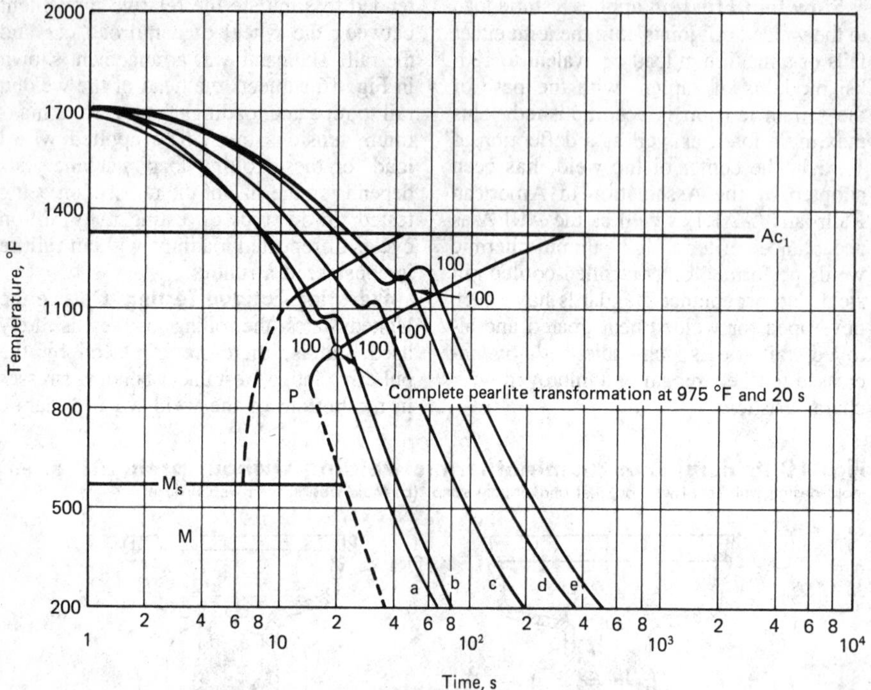

Fig. 13 TTT diagram for aluminothermic steel applied to standard rail steel. Composition: 0.78% C, 1.03% Mn, 0.73% Al

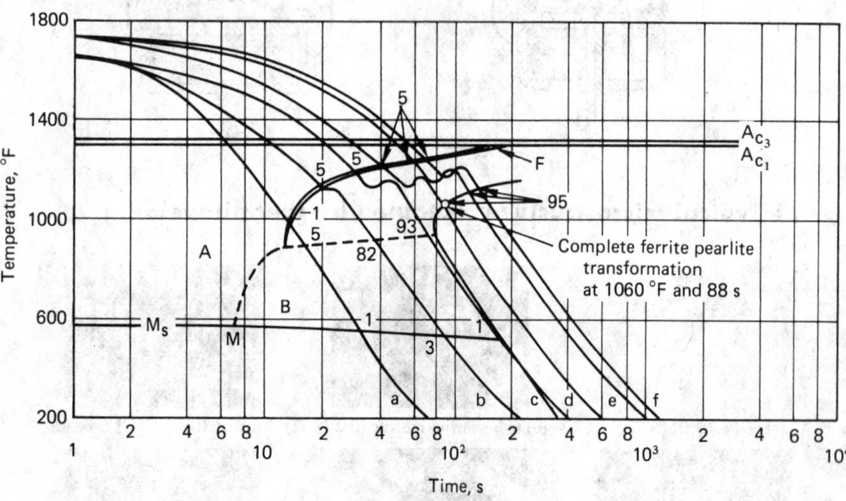

stress behavior is under consideration (Fig. 17). This test simulates actual field conditions to the extent that the rail weld cross section is submitted to a bending condition caused by vertical loads that resemble the load of railcars passing over the track structure. It is important to realize that the condition of the surface at the bottom of the weld and rail decisively influences test results. Even if corroded structures of all grades and sections weathered in track over an extended period of time are submitted to the pulsating test, a minimum fatigue strength of 30 ksi must be met in most acceptance criteria, without failure for at least 2×10^6 cycles (Fig. 18). A fatigue curve (Woehler curve), obtained by multiple testing with loads varied from test to test, is shown in Fig. 18.

The hardness in the weld and the adjacent heat-affected rail steel is very important in regard to the wear characteristics of a welded joint made by the thermit process. Typical hardness values on the running surface of a welded rail joint are plotted in Fig. 19.

The hardness of the weld is slightly higher than that of the unaffected rail steel. Because the rolled steel has a tendency to work harden more than the weld, the difference in hardness values balances out once the rail is used in service. Eventually, the weld area and rail steel exhibit the same hardness, thereby wearing at the same rate. Welded joints that are too soft show increased wear in the weld area. The HAZ of the rail, adjacent to the fusion line, shows the highest peak in hardness, caused by the maximum grain growth of the austenite. Minimal hardness is registered at the interface of heat-affected base metal, where the microstructure consists of a spheroidized coagulated pearlite (Fig. 8).

Repair Welding

Thermit welding is particularly suitable for welding together massive steel parts, such as large-diameter rolls, shafts, or heavy mill housings of steel mills. Repair welds are customized to a particular application; as such, they are nonrepetitive. Contrary to rail welding, where the molds are premanufactured at a central location, a mold for a repair weld is made on site to conform to the shape of the part.

Joint preparation is the same as for rail welding. Proper alignment and removal of foreign particles that may contaminate the weld are essential; the overall shape of fractured parts before and after completion of a weld must be identical. Firm identification marks are made at points outside the area to be covered by the mold box for the purpose of repositioning the pieces after the fractured surface has been removed by torch cutting, as well as after the proper gap size has been established following the guidelines set forth in Tables 2 and 3.

To allow for the contraction of the weld during the cooling process, the parts are initially spaced $^1/_{16}$ to $^1/_4$ in. farther apart than their original position, using the markers as reference points.

Applying the Mold. After aligning, spacing, and cleaning the parts, a wax pattern is built around the weld area to shape the mold cavity. For complete fusion at the peripheral line of the gap, the wax pattern overlaps the base metal, creating a "collar" covering the entire contour of the weld. A sand mold is then built around the wax pattern; a mold box serves as a container for sufficient back-up sand, which is subsequently rammed around the parts.

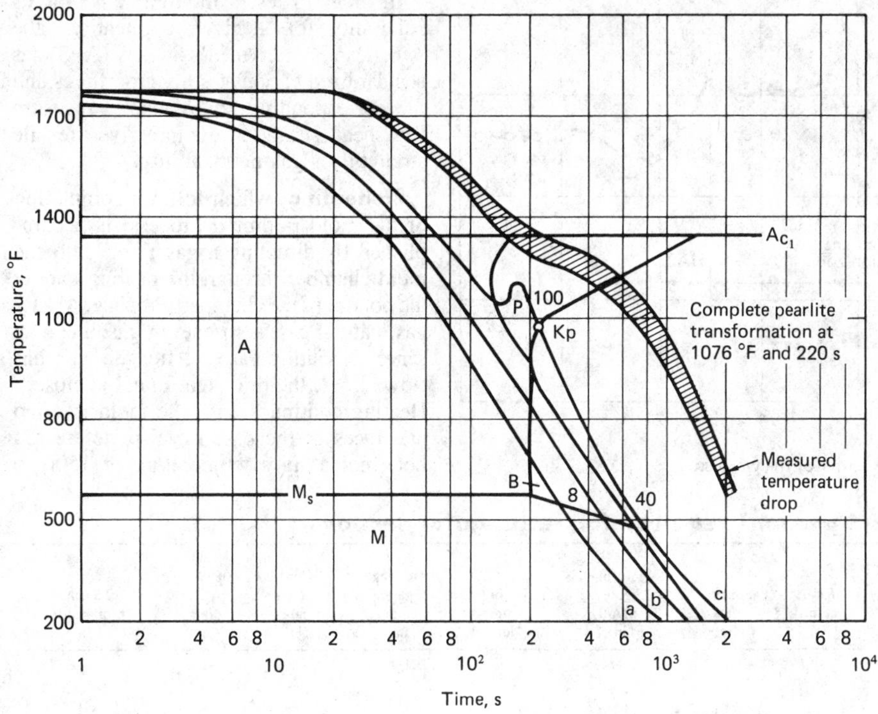

Fig. 14 TTT diagram for a special-alloy highly wear-resistant chromium-manganese rail steel. 157 ksi minimum tensile strength

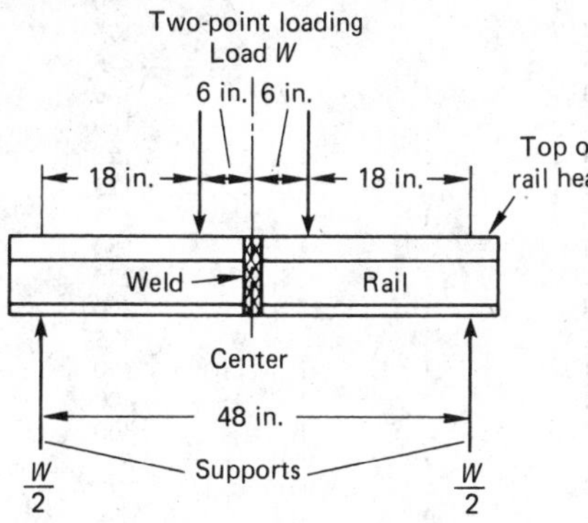

Fig. 15 Slow bend testing arrangement. Source: Ref 5

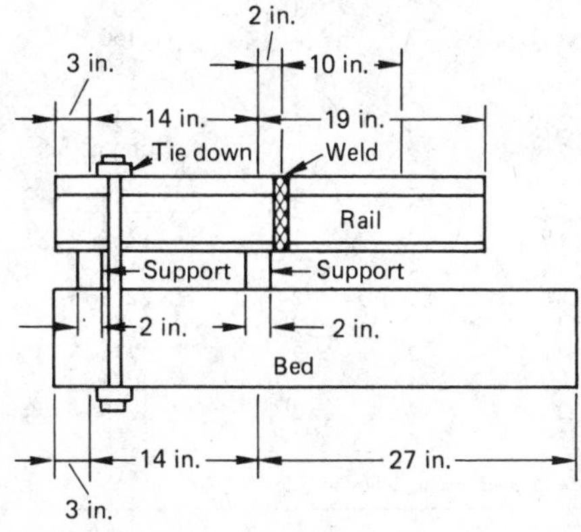

Fig. 16 Rolling load testing arrangement. Source: Ref 5

Pouring and heating gates are used to guide the flow of heating gases during the preheating, as well as to guide the liquid aluminothermic steel as it is poured into the mold (Fig. 20). Heating gates also serve as drainage funnels for the wax, which melts during initial preheating. Gates may be positioned opposite drainage funnels to prevent interference with the burner and the draining liquid wax, as shown in Fig. 20. Risers on top of the weld should be sized and positioned to facilitate gas escape and to channel liquid aluminothermic steel into the weld area after completion of the pour to compensate for shrinkage in volume that occurs during solidification of the weld metal.

Molding Sand Characteristics. After pouring the thermit steel into the mold, the molding material is exposed to severe conditions, such as the high temperature of the aluminothermic steel (about 4000 °F, compared to 3000 °F for normal foundry practice) and the chemical aggressiveness of surplus aluminum. The higher temperature tends to weaken the strength of the binder, causing the aluminum to enter into secondary reactions with the molding sand and binder components. Contamination of the sand by components such as feldspar, lime, or iron oxide will considerably reduce its strength, resulting in sand inclusions inside the weld, as well as rough and rigid surfaces on the weld collar. Poor permeability of the molding sand might also

Fig. 17 Pulsating fatigue testing arrangement. F = applied load; a = distance between points of load initiation, 4 to 6 in.; l = distance between support points, 40 to 75 in. (Source: Ref 3)

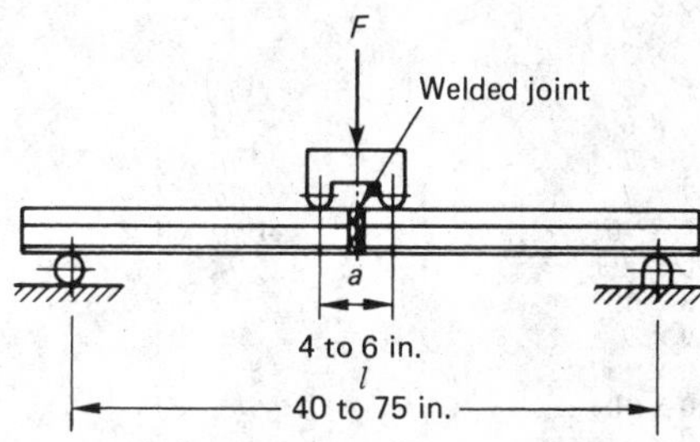

Fig. 18 Woehler curve obtained by applied varied loads. F = applied load; l = distance between load initiation point and support point. (Courtesy of Elektro Thermit GmbH, Essen, West Germany)

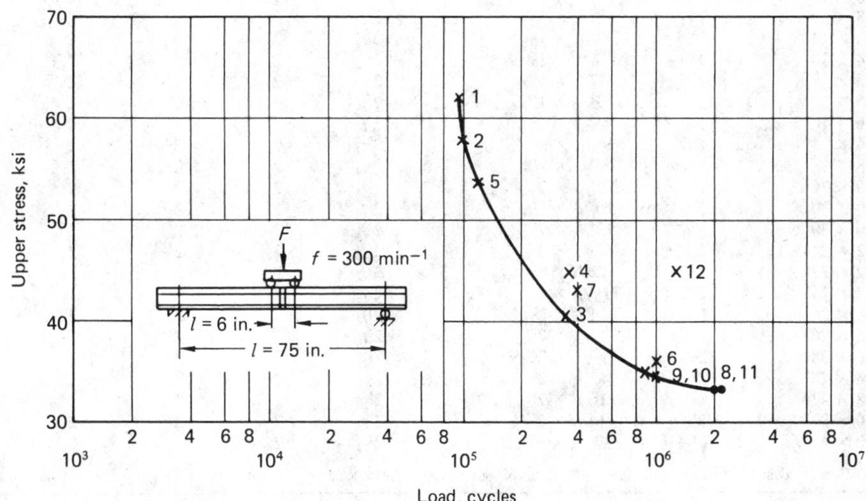

Fig. 19 Vickers hardness on running surface of a welded rail joint

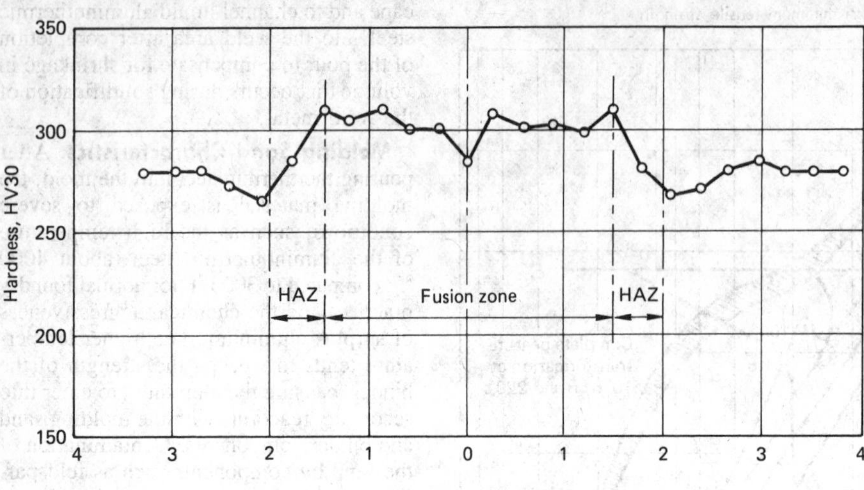

result in gas inclusions (porosity) inside the weld.

In order to test a molding sand for its suitability for a given application, the American Foundrymen's Society (AFS) has standardized various testing procedures and the corresponding equipment to measure for shear strength, compressive, tensile strength, or gas permeability.

Preheating, which follows completion of the mold ramming process, is accomplished by directing a gas flame into the mold chamber through the heating gate at the bottom of the chamber. Propane, MAPP gas, natural gas, kerosene, or gasoline may serve as heating gases. After the wax has flowed out, the drainage funnel is closed. Heating continues until the mold is dried and faces of the base-metal parts are red hot, indicating a temperature of 1500 to

Table 2 Thermit weld dimensions and portions required for rectangular sections

Section size, in.	Gap, in.	Collar, in.	Riser diam, in., No.	Riser diam, in.	Pouring gate No.	Pouring gate diam, in.	Heating gate No.	Heating gate diam, in.	Connecting gate No.	Connecting gate diam, in.	Thermit required(a), lb
2 × 2	$\frac{7}{16}$	$1\frac{1}{2} \times \frac{7}{16}$	1	$\frac{3}{4}$	1	$\frac{3}{4}$	1	$1\frac{1}{4}$	...	...	6
2 × 4	$\frac{9}{16}$	$1\frac{5}{16} \times \frac{9}{16}$	1	$\frac{3}{4}$	1	1	1	$1\frac{1}{4}$	...	...	12
3 × 3	$\frac{9}{16}$	$2\frac{1}{16} \times \frac{9}{16}$	1	1	1	1	1	$1\frac{1}{4}$	...	...	12
3 × 6	$\frac{11}{16}$	$2\frac{3}{4} \times \frac{11}{16}$	1	1	1	1	1	$1\frac{1}{4}$	...	...	25
4 × 4	$\frac{11}{16}$	$2\frac{5}{8} \times \frac{11}{16}$	1	1	1	1	1	$1\frac{1}{4}$	...	...	25
4 × 8	$\frac{7}{8}$	$3\frac{7}{16} \times \frac{7}{8}$	1	1	1	1	2(b)	$1\frac{1}{4}$	...	...	50
5 × 5	$\frac{13}{16}$	$3\frac{1}{8} \times \frac{3}{16}$	1	1	1	1	1	$1\frac{1}{4}$	...	...	50
5 × 8	$\frac{15}{16}$	$3\frac{13}{16} \times \frac{15}{16}$	1	1	1	$1\frac{1}{4}$	2(b)	$1\frac{1}{4}$	...	...	75
6 × 6	$\frac{15}{16}$	$3\frac{5}{8} \times \frac{15}{16}$	1	$1\frac{1}{4}$	1	$1\frac{1}{4}$	2(b)	$1\frac{1}{4}$	...	...	75
6 × 9	$1\frac{1}{16}$	$4\frac{5}{16} \times 1\frac{1}{16}$	1	$1\frac{1}{4}$	1	$1\frac{1}{4}$	1	$1\frac{1}{4}$	1	$1\frac{1}{4}$	100
7 × 7	1	$4\frac{1}{8} \times 1$	1	$1\frac{1}{2}$	1	$1\frac{1}{4}$	2(b)	$1\frac{1}{4}$	...	...	100
7 × 10	$1\frac{1}{8}$	$4\frac{13}{16} \times 1\frac{1}{8}$	1	2	1	$1\frac{1}{4}$	1	$1\frac{1}{4}$	1	$1\frac{1}{4}$	125
8 × 8	$1\frac{1}{8}$	$4\frac{5}{8} \times 1\frac{1}{8}$	1	$1\frac{3}{4}$	1	$1\frac{1}{4}$	2(b)	$1\frac{1}{4}$	...	...	125
8 × 12	$1\frac{1}{4}$	$5\frac{1}{2} \times 1\frac{1}{4}$	1	$1\frac{3}{4}$	1	$1\frac{1}{4}$	1	$1\frac{1}{4}$	1	$1\frac{1}{4}$	175
9 × 9	$1\frac{3}{16}$	$5\frac{1}{8} \times 1\frac{3}{16}$	1	2	1	$1\frac{1}{4}$	2(b)	$1\frac{1}{4}$	...	...	150
9 × 13	$1\frac{3}{8}$	$5\frac{5}{16} \times 1\frac{3}{8}$	1	2	1	$1\frac{1}{4}$	1	$1\frac{1}{4}$	1	$1\frac{1}{4}$	225
10 × 10	$1\frac{5}{16}$	$5\frac{9}{16} \times 1\frac{5}{16}$	1	$2\frac{1}{4}$	1	$1\frac{1}{4}$	1	$1\frac{1}{4}$	1	$1\frac{1}{4}$	200
10 × 15	$1\frac{1}{2}$	$6\frac{5}{8} \times 1\frac{1}{2}$	1	$2\frac{1}{4}$	1	$1\frac{1}{2}$	2(b)	$1\frac{1}{4}$	...	...	325
11 × 11	$1\frac{3}{8}$	$6\frac{1}{16} \times 1\frac{3}{8}$	1	$2\frac{1}{4}$	1	$1\frac{1}{2}$	2(b)	$1\frac{1}{2}$	1	$1\frac{1}{4}$	250
11 × 16	$1\frac{9}{16}$	$7\frac{1}{16} \times 1\frac{9}{16}$	1	$2\frac{1}{4}$	1	$1\frac{1}{2}$	2(b)	$1\frac{1}{2}$	1	$1\frac{1}{2}$	400
12 × 12	$1\frac{7}{16}$	$6\frac{1}{2} \times 1\frac{7}{16}$	1	$2\frac{1}{2}$	1	$1\frac{1}{2}$	2(b)	$1\frac{1}{2}$	1	$1\frac{1}{2}$	300
12 × 18	$1\frac{11}{16}$	$7\frac{3}{4} \times 1\frac{11}{16}$	1	$2\frac{1}{2}$	1	$1\frac{1}{2}$	2(b)	$1\frac{1}{2}$	1	$1\frac{1}{2}$	500
13 × 13	$1\frac{9}{16}$	$7 \times 1\frac{9}{16}$	1	$2\frac{1}{2}$	1	$1\frac{1}{2}$	2(b)	$1\frac{1}{2}$	1	$1\frac{1}{2}$	375
13 × 19	$1\frac{3}{4}$	$8\frac{3}{16} \times 1\frac{3}{4}$	1	$2\frac{1}{2}$	2	$1\frac{1}{2}$	2	$1\frac{1}{2}$	2	$1\frac{1}{2}$	675
14 × 14	$1\frac{5}{8}$	$7\frac{7}{16} \times 1\frac{5}{8}$	1	$2\frac{1}{2}$	2	$1\frac{1}{2}$	2	$1\frac{1}{2}$	2	$1\frac{1}{2}$	500
14 × 20	$1\frac{13}{16}$	$8\frac{5}{8} \times 1\frac{13}{16}$	1	$2\frac{1}{2}$	2	2	2	$1\frac{1}{2}$	2	$1\frac{1}{2}$	800
15 × 15	$1\frac{11}{16}$	$7\frac{7}{8} \times 1\frac{11}{16}$	1	$2\frac{3}{4}$	2	$1\frac{1}{2}$	2	$1\frac{1}{2}$	2	$1\frac{1}{2}$	600
15 × 22	$1\frac{15}{16}$	$9\frac{5}{16} \times 1\frac{15}{16}$	1	$2\frac{3}{4}$	2	2	2	$1\frac{1}{2}$	2	$1\frac{1}{2}$	975
16 × 16	$1\frac{3}{4}$	$8\frac{15}{16} \times 1\frac{3}{4}$	1	$2\frac{3}{4}$	2	2	2	$1\frac{1}{2}$	2	$1\frac{1}{2}$	700
16 × 24	2	$9\frac{15}{16} \times 2$	1	$2\frac{3}{4}$	2	2	2	$1\frac{1}{2}$	2	$1\frac{1}{2}$	1150
18 × 18	$1\frac{15}{16}$	$9\frac{3}{16} \times 1\frac{15}{16}$	2	$2\frac{1}{4}$	2	2	2	2	2	2	950
18 × 26	$2\frac{3}{16}$	$10\frac{13}{16} \times 2\frac{3}{16}$	2	$2\frac{1}{4}$	2	2	2	2	2	2	1525
20 × 20	$2\frac{1}{16}$	$10\frac{1}{16} \times 2\frac{1}{16}$	2	$2\frac{1}{4}$	2	2	2	2	2	2	1225
20 × 30	$2\frac{5}{16}$	$12\frac{1}{16} \times 2\frac{5}{16}$	2	$2\frac{1}{4}$	2	2	2	2	2	2	2000
22 × 22	$2\frac{3}{16}$	$10\frac{15}{16} \times 2\frac{3}{16}$	2	$2\frac{1}{2}$	2	2	2	2	2	2	1550
22 × 32	$2\frac{1}{2}$	$12\frac{7}{8} \times 2\frac{1}{2}$	2	$2\frac{1}{2}$	2	2	2	2	4	2	2475
24 × 24	$2\frac{5}{16}$	$11\frac{13}{16} \times 2\frac{5}{16}$	2	$2\frac{1}{2}$	2	2	2	$1\frac{3}{4}$	2	$1\frac{3}{4}$	1875
24 × 36	$2\frac{5}{8}$	$14\frac{1}{8} \times 2\frac{5}{8}$	2	$2\frac{1}{2}$	2	2	2	2	4	2	3125
26 × 26	$2\frac{7}{16}$	$12\frac{11}{16} \times 2\frac{7}{16}$	2	$2\frac{1}{2}$	2	2	2	2	2	2	2275
26 × 38	$2\frac{3}{4}$	$14\frac{15}{16} \times 2\frac{3}{4}$	2	$2\frac{1}{2}$	2	2	2	$2\frac{1}{4}$	4	$2\frac{1}{4}$	3700
28 × 28	$2\frac{9}{16}$	$13\frac{1}{2} \times 2\frac{9}{16}$	3	$2\frac{1}{2}$	2	2	2	2	2	2	2775
28 × 42	$2\frac{15}{16}$	$16\frac{1}{8} \times 2\frac{15}{16}$	3	$2\frac{1}{2}$	2	2	2	$2\frac{1}{2}$	4	$2\frac{1}{2}$	4675
30 × 30	$2\frac{11}{16}$	$14\frac{3}{8} \times 2\frac{11}{16}$	3	$2\frac{1}{2}$	2	2	2	$2\frac{1}{4}$	2	$2\frac{1}{4}$	3275
30 × 45	$3\frac{1}{16}$	$17\frac{3}{16} \times 3\frac{1}{16}$	3	$2\frac{1}{2}$	2	2	2	$2\frac{1}{2}$	4	$2\frac{1}{2}$	5525

(a) Thermit required includes provision for a 10% excess of steel in slag basin for a single pour and a 20% excess for a double pour. (b) Includes one separate back heating gate

Table 3 Thermit weld dimensions and portions required for round sections

Diameter, in.	Gap, in.	Collar, in.	Riser diam, in.	Pouring gate No.	Pouring gate diam, in.	Heating gate No.	Heating gate diam, in.	Connecting gate No.	Connecting gate diam, in.	Thermit required(a), lb
2	7/16	1 3/8 × 7/16	3/4	1	3/4	1	1 1/4	...	...	5
3	9/16	1 7/8 × 9/16	1	1	1	1	1 1/4	...	...	11
4	5/8	2 3/8 × 5/8	1	1	1	1	1 1/4	...	...	25
5	3/4	2 13/16 × 3/4	1 1/4	1	1	1	1 1/4	...	...	25
6	7/8	3 5/16 × 7/8	1 1/4	1	1 1/4	1	1 1/4	...	...	50
7	15/16	3 3/4 × 15/16	1 1/4	1	1 1/4	1	1 1/4	...	...	75
8	1	4 3/16 × 1	1 1/2	1	1 1/4	1	1 1/4	...	...	75
9	1 1/8	4 5/8 × 1 1/8	1 1/2	1	1 1/2	1	1 1/2	...	...	125
10	1 3/16	5 1/16 × 1 3/16	1 1/2	1	1 1/2	2(b)	1 1/2	...	...	150
11	1 1/4	5 7/16 × 1 1/4	1 3/4	1	1 1/2	2(b)	1 1/2	...	...	175
12	1 5/16	5 7/8 × 1 5/16	1 3/4	1	1 1/2	1	1 1/2	1	1 1/2	200
13	1 7/16	6 5/16 × 1 7/16	1 3/4	1	1 1/2	1	1 1/2	1	1 1/2	250
14	1 1/2	6 11/16 × 1 1/2	1 3/4	1	1 1/2	1	1 1/2	1	1 1/2	300
15	1 9/16	7 1/8 × 1 9/16	2	1	1 1/2	1	1 1/2	1	1 1/2	350
16	1 5/8	7 1/2 × 1 5/8	2	1	1 1/2	1	1 1/2	1	1 1/2	425
18	1 3/4	8 5/16 × 1 3/4	2 1/4	1	2	1	2	1	2	575
20	1 7/8	9 1/8 × 1 7/8	2 1/2	1	2	1	2	1	2	750
22	2	9 7/8 × 2	2 3/4	2	2	2	2	2	2	1050
24	2 1/8	10 5/8 × 2 1/8	3	2	2	2	2	2	2	1325
26	2 1/4	11 7/16 × 2 1/4	3 1/4	2	2	2	2	2	2	1600
28	2 3/8	12 3/16 × 2 3/8	3 1/2	2	2	2	2	2	2	1925
30	2 1/2	12 15/16 × 2 1/2	3 1/2	2	2	2	2	2	2	2275

(a) Thermit required includes provision for a 10% excess of steel in slag basin for a single pour and a 20% excess for a double pour. (b) Includes one separate back heating gate

1800 °F. Upon completion of the preheating cycle, the heating gate is closed to prevent leakage of the weld.

Charging the Crucible. The aluminothermic reaction occurs in a crucible (Fig. 21), positioned above the mold (Fig. 20). When the reaction is complete, molten steel is released through an opening in the bottom of the crucible upon removal of a tapping pin. Metal and Al_2O_3 slag flow is the same as in the rail welding process.

The quantity of aluminothermic welding compound required for a specific application can be ascertained by the following equation:

$$X = \frac{E}{0.5 + S \times 10^{-2}}$$

where X is the quantity of aluminothermic compound required, lb; S is the percent of cold metal added; and E is the quantity of liquid steel required, lb, according to cross section and gap width. An additional 10% of material should be added to compensate for losses. Generally, each pound of wax requires 25 lb of aluminothermic welding compound.

Fig. 20 Mold and crucible arrangement for thermit repair welding

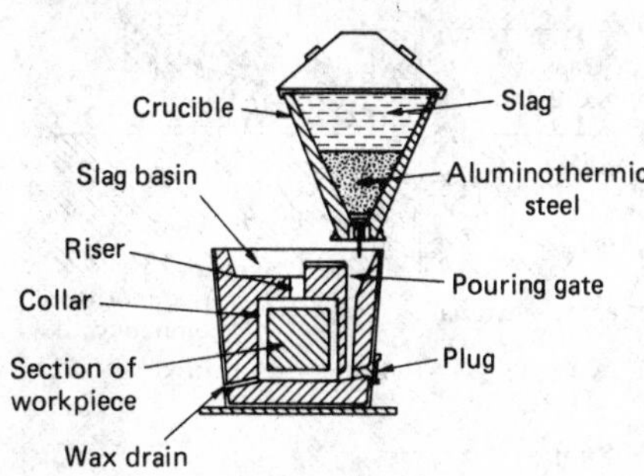

Repair Welding Applications. Besides the applications previously mentioned, the thermit process is used for the repair of ingot molds at significant savings over replacement. Two procedures are commonly applied: an eroded cavity in the bottom of the mold can be filled with aluminothermic steel, or the entire bottom of the ingot mold can be removed and completely rebuilt with aluminothermic steel. The latter process is more sophisticated and requires larger quantities of aluminothermic steel, but the lifetime of the ingot mold is nearly doubled with one repair weld. The built-up weld repair without prior removal of the worn ingot mold bottom must be repeated after every second or third pour.

Thermit welding is regularly used for assembling large parts to make dredge cutters. Several thousand pounds of thermit welding compound are reacted in several crucibles simultaneously and poured into the mold at one time. A typical thermit repair welding application is shown in Fig. 22 and 23, which show a broken and a repair welded roll.

Reinforcement Bar Welding

Reinforced concrete structures generally exhibit monolithic behavior. However, due to practical limitations in normal construction, a structure is built piece by piece, which means that the bars have to be spliced either mechanically or by means of welding. Continuous welded reinforcement bars permit the design of concrete columns and beams that are smaller in section than bars that are not welded together. The thermit process can be used to accomplish reinforcement bar splicing through use of a full-fusion method or by mechanically clamping the bars with a cast filler metal.

The full-fusion method uses two mold halves, manufactured by either the CO_2 or shell mold process, which are positioned at the joint of the aligned bars and sealed

Fig. 21 Cross section of thermit crucible for repair welding. A, magnesite stone; B, magnesite thimble; C, plugging material; D, metal disk; E, tapping pin

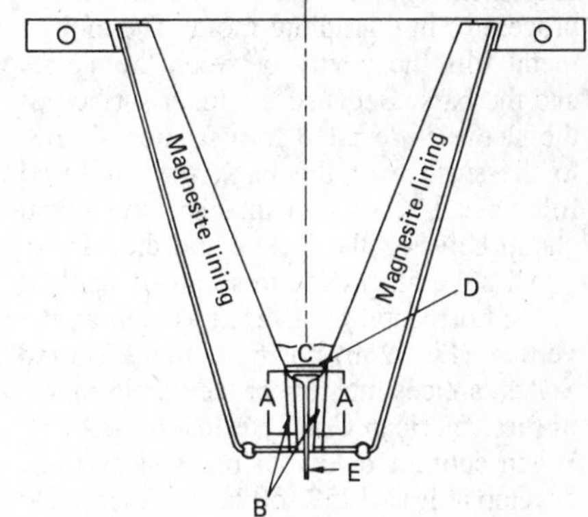

Fig. 22 Broken roll

Fig. 23 Roll welded by thermit repair process

Fig. 24 Full-fusion reinforcement bar welding. (a) Horizontal position. (b) Vertical position. (Source: Ref 6)

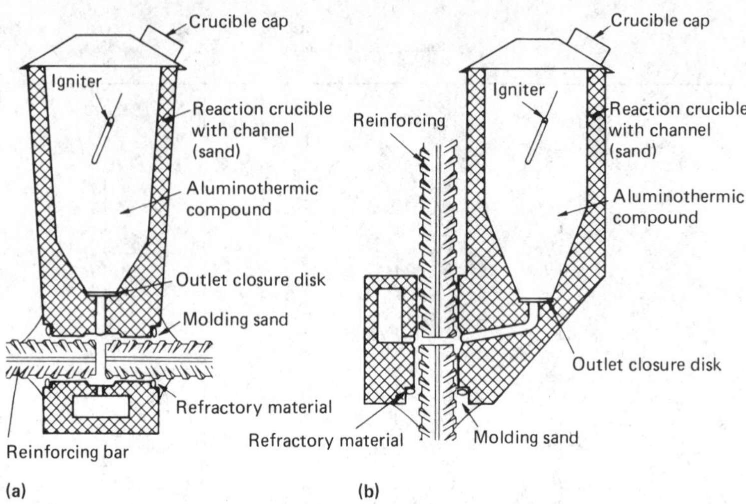

to the bars to avoid loss of molten metal. A tapping disk is placed in a well at the base of the crucible section to prevent premature tapping prior to or during the aluminothermic reaction. The aluminothermic welding compound is placed in the crucible where the reaction takes place and from where the steel automatically pours into the mold chamber to fill the gap between the two bars. The arrangements for horizontal and vertical welding are shown in Fig. 24(a) and (b).

As an alternative to the full-fusion welding process, the bars can be joined by inserting the ends of both bars into a common sleeve and filling the sleeve, which has an inside diameter slightly larger than the diameter of the bars, with an aluminothermic filler metal (Fig. 25). The filler metal is reacted in a mold and crucible arrangement like that used in the full-fusion process or in a graphite mold. The molten metal fills the cavity between the sleeve and the bars. Because the inner surface of the sleeve is serrated in a fashion similar to the surface of the bars, the solidified filler metal acts as an intense mechanical clamp between the sleeve and the bars.

All splicing procedures can be applied in the horizontal (Fig. 25a) as well as the vertical (Fig. 25b) position. In the United States, splices must meet the requirements of the American Concrete Institute (ACI). When submitted to tension, splices must develop at least 125% of the specified yield strength of the plain reinforcement bar (Table 4). Samples of welded reinforcement bars, after being submitted to tensile and bend testing, are shown in Fig. 26; Fig. 27 shows a completed sleeve joint.

Thermit Welding of Electrical Connections

For welding electrical conducting joints, an aluminothermic compound based on cupric oxide and aluminum is applied, which reacts according to the following equation:

$$3Cu_2O + 2Al \rightarrow Al_2O_3 + 6Cu$$

Depending on the application, other metals and alloys can be added to produce alloys to meet individual specifications.

Table 4 American Concrete Institute code requirements

Grade of bar	Specified minimum yield strength for bar, ksi	Tensile strength required for weld, ksi
40	40	50
50	60	75
75	75	93

Fig. 25 Reinforcement bar welding with sleeve. (a) Horizontal position. (b) Vertical position. (Source: Ref 6)

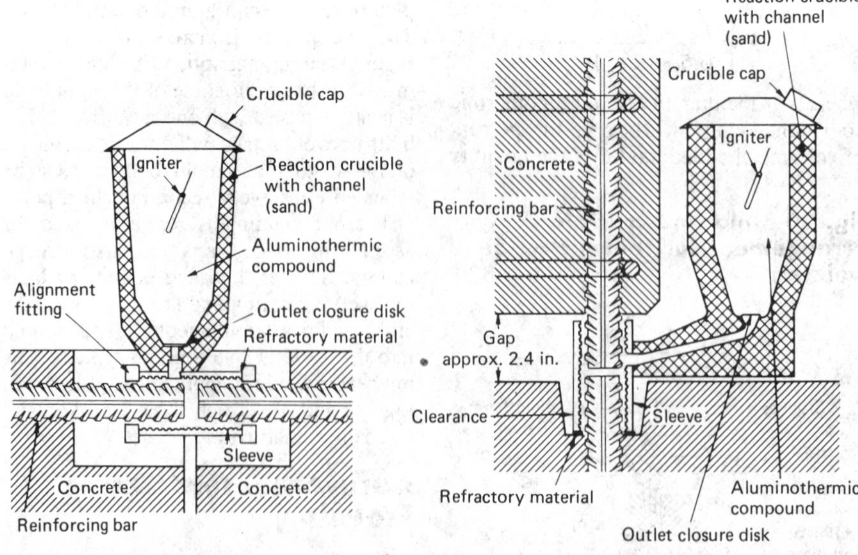

Fig. 26 Full-fusion reinforcement bar welds after tensile and bend testing. Source: Ref 6

Fig. 27 Reinforcement bar sleeve joint. Source: Ref 6

This process preferably is used for welding cables and wires, as well as solid copper or steel conductors, against construction parts, such as steel rails, that serve as grounding devices (Fig. 28). Graphite is the most suitable material to be used as a mold. Small quantities of carbon from the graphite mold participate in the reaction with the oxygen, thereby creating CO_2.

Sufficient quantities of CO_2 gas keep the amount of hydrogen going into solution in the liquid copper melt to a minimum; excess hydrogen in any copper melt segregates during solidification of the copper and leads to porosity in the finished product.

This process has widespread application in railway track structures, where the rail serves as a grounding device for the electrical signal circuits. Due to the small volume of aluminothermic copper welding compound and consequently very rapid cooling rate of the weld, the transition zone between weld and rail steel transforms into a brittle, martensitic microstructure. It is therefore necessary to apply a postheating treatment to the weld area. This can be accomplished with premanufactured exothermic bricks, which generate sufficient heat to transform martensite into pearlite. An exothermic heat treatment block for cable bonds welded to a rail base is shown in Fig. 29. The resulting temperature curve during heat treatment at the fusion line between copper weld and rail base is shown in Fig. 30. Once ignited with a torch, the exothermic brick reacts like a regular aluminothermic compound; however, liquid does not result from the reaction due to the addition of heat absorbers (sand). Consequently, fusion does not occur with the base metal. The microstructure of an as-welded and a heat treated cable bond in the transition zone between copper and steel base metal is shown in Fig. 31.

Fig. 28 Copper cable welding. Courtesy of Elektro Thermit GmbH, Essen, West Germany

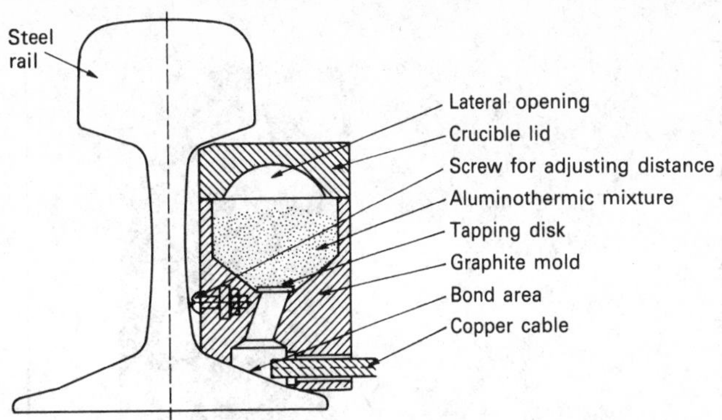

Fig. 29 Exothermic heating block for cable bond welds. Source: Ref 7

Fig. 30 Temperature curve during heat treatment of cable bonds. Source: Ref 7

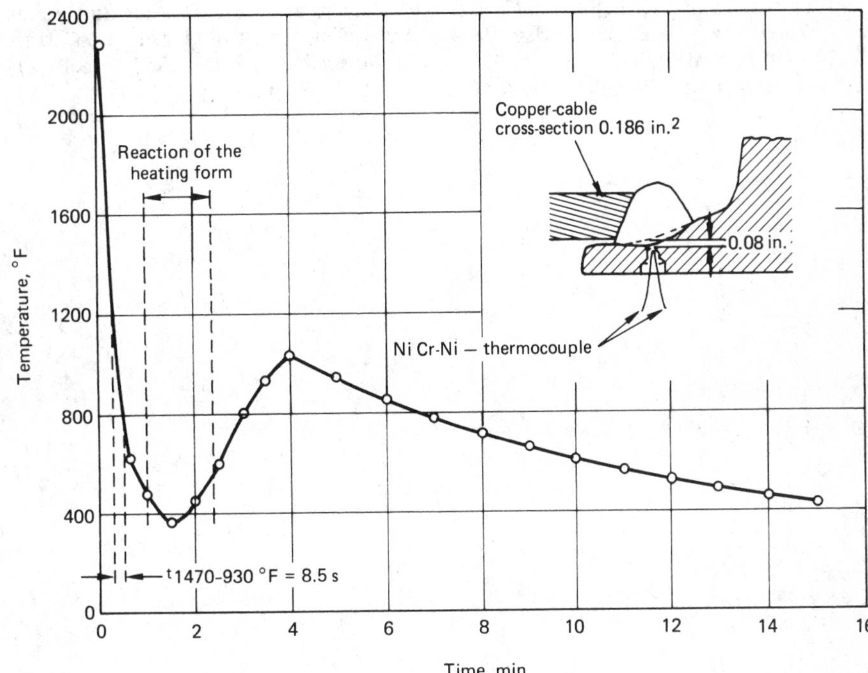

Fig. 31 Microstructure of heat treated and as-welded copper cable bond. Source: Ref 7

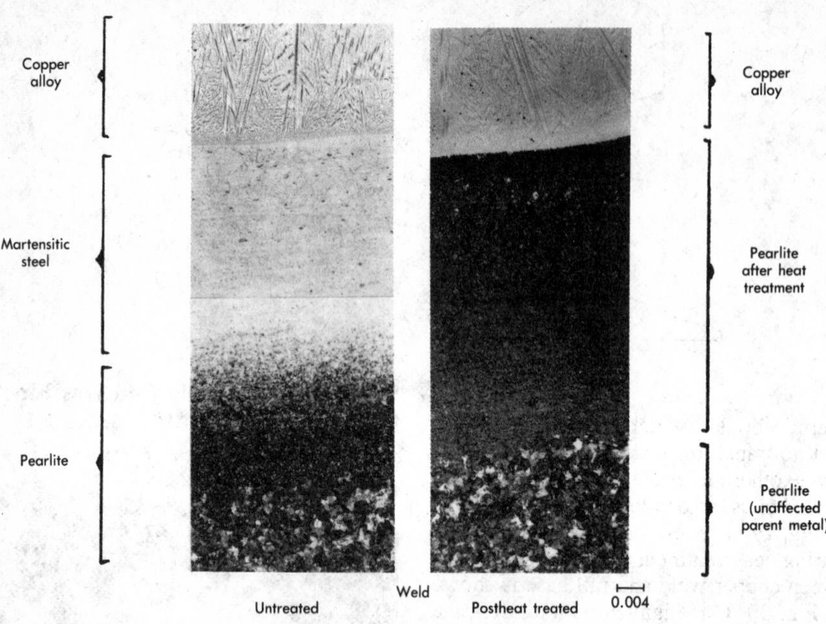

Similar exothermic compounds in solid form have been developed for preheating and postheating of various structures where the aluminothermic mixture, using special binders such as sodium silicate, is formed to the shape of the parts to be heat treated.

REFERENCES

1. *Uhlmann Enzklopaedia der Technischen Chemie,* 4th ed., Vol 7, Verlag Chemie, Weinheim, 1974, p 351-361
2. Erdmann-Jessnitzer, F., *Werkstoffe und Schweissung I,* Akademie Verlag, Berlin, 1951, p 618-680
3. *Aluminothermic Welding of Rails,* Merkblatt 241, 3rd ed., Varein Deutscher Eiseuhutteuleute, Duesseldorf, 1983
4. Heller, W. and Beck, G., Transformation Characteristics of Rail Steels and Consequences for Welding and Flame Cutting, *Archiv fuer das Eisen—Huettenwesen 39,* No. 5, Duesseldorf, 1968, p 375-386
5. *AREA Manual,* Chapter 4, Rail, American Railway Engineering Association
6. Guntermann, H., Thermit-Butt Joints for Concrete Steel Construction, *Maschinenmarkt 75,* No. 75, Vogel Verlag, 1969, p 2-4
7. Jacoby, N., Special Processes of the Thermit Welding Technique, *Der Eisenbahningenieur,* No. 3, 1977, p 91-99

SELECTED REFERENCES

- Anderson, A.H., *et al.,* Underwater Application of Exothermic Welding, Paper Number OTC 1910, American Institute of Mining, Metallurgical and Petroleum Engineers, 1973, p 847-858
- Belitkus, D., Aluminothermic Production of Metals and Alloys, *Journal of Metals,* Vol 24, 1972, p 30-34
- Carter, G.F., *Principles of Physical and Chemical Metallurgy,* American Society for Metals, 1979
- Dautzenberg, W., Metallurgy of the Metallothermic Melting, *Erzmetall,* Vol 3, 1950, p 341-347
- Guntermann, H., The Application of the Thermit Welding Process in Areas Besides Rail Welding, *ZEV—Glaser Annalen,* 1975, p 17-22
- Kubaschewski, E., Evans, L.L., and Alcook, C.B., *Metallurgical Thermo-Chemistry,* 5th ed., Pergamon Press, 1979
- Lowry, D.G., *Reinforcing Bar Splices in Reinforced Concrete Columns,* Division of Solid & Structural Engineering, Faculty of Engineering, Carleton University, Ottawa, Canada, March 1972
- Schroeder, L.C. and Poirer, D.R., Physical Metallurgy and Process Improvement of Thermite Rail Welds, University of Arizona, Department of Metallurgical Engineering, 1982
- *Welding Handbook,* Vol 1 and 3, 7th ed., American Welding Society, Miami, 1980

Explosion Welding

By the ASM Committee on Explosion Welding*

EXPLOSION WELDING (EXW), or explosion bonding and explosion cladding, is a method in which the controlled energy of a detonating explosive is used to create a metallurgical bond between two or more similar or dissimilar metals. No intermediate filler metal, e.g., a brazing compound or soldering alloy, is needed to promote bonding, and no external heat need be applied. Diffusion does not occur during bonding.

Explosion welding was first observed as far back as World War I, when pieces of bombshells stuck to metallic objects in the vicinity of the explosion. The ordnance and explosive specialists who observed this accidental phenomenon did not recognize its industrial potential.

In 1962, a U.S. patent was issued to Philipchuk and Bois covering a method using explosive detonation to weld metals together in spots along a linear path. They were the forerunners of the current explosion welding or cladding industry. This was the first patent issued on EXW. The patent teaches a method that is not practical, and is not used in industry today.

Du Pont explosion welding pioneers developed a phenomenological theory that explained how and why cladding occurred. This theory permitted control of the process. Their intensive research effort culminated in a series of 27 U.S. process patents, the first of which was issued in June 1964. U.S. Patent No. 3,233,312 was issued in Feb 1966 and covered a wide variety of products made by explosion cladding. Subsequent key process patents were U.S. 3,397,444 and U.S. 3,493,353. These patents set the standard for the worldwide explosion cladding industry. They cover successful welding techniques using the variables developed for parallel cladding, by far the most widely used method of EXW.

Process Fundamentals

Explosion bonding is a cold pressure welding process in which the contaminant surface films are plastically jetted off the base metals as a result of the high-pressure collision of the two metals. During the high-velocity collision of metal plates, a jet is formed between the metal plates if the collision angle and the collision velocity are in the range required for bonding. Contaminant surface films that are detrimental to the establishment of a metallurgical bond are swept away in the jet. The metal plates, cleaned of any surface films by the jet action, are joined at an internal point under the influence of the very high pressure that is obtained near the collision point.

Parallel and Angle Cladding. Figures 1 and 2 illustrate the principles of explosion cladding. Figures 1(a) and (b) illustrate the principle of angle cladding, which is limited to cladding for relatively small pieces. Large-area clad plates cannot be made using this arrangement, because the collision of long plates at high standoffs on long runs is so violent that metal cracking, spalling, and fracture occur at downstream edges. The parallel arrangement shown in Fig. 2 is the simplest and most widely used in industry.

Fig. 1 Angle arrangements to produce explosion clads

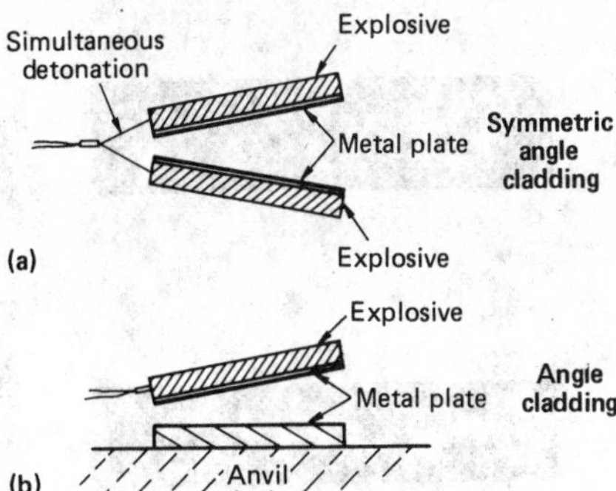

Fig. 2 Parallel arrangement for explosion cladding and subsequent collision between the prime and backer metals that leads to jetting and formation of wavy bond zone

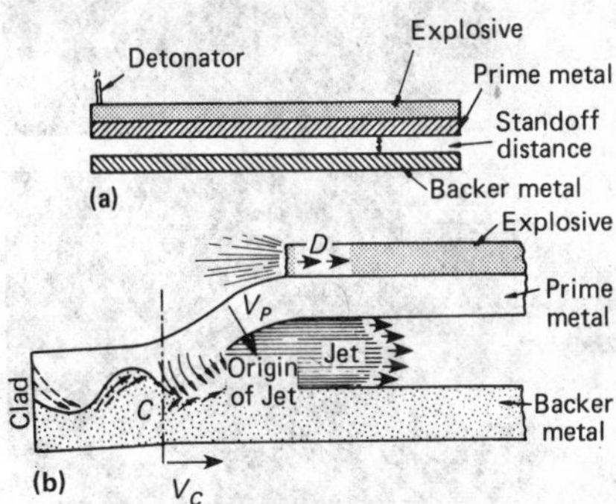

Jetting Phenomenon. A layer of explosive is placed in contact with one surface of the prime metal plate, which is maintained at a constant parallel separation from the backer or base plate, as shown in Fig. 2(a). The explosive is detonated at a point or line, and as the detonation front moves across the plate, the prime metal is deflected and accelerated to plate velocity (V_P), shown in Fig. 2(b), thus establishing an angle between the two plates.

The ensuing collision region then progresses across the plate with a velocity equal to the detonation velocity (D). When the collision velocity (V_C) and angle

*Andrew Pocalyko, *Chairman*, Technical Superintendent, Detaclad Operations, E.I. Du Pont de Nemours & Co., Inc.; Akira Kubota, Manager, Technical Department, Explosives Division, Asahi Chemical Industry Co., Ltd.; Vonne D. Linse, Associate Section Manager, Metals Joining and Nondestructive Testing, Battelle Columbus Laboratories

Fig. 3 Cladding of aluminum to aluminum showing jet formation

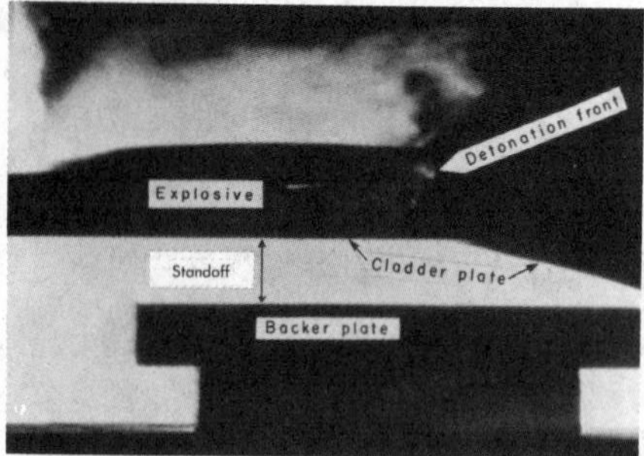

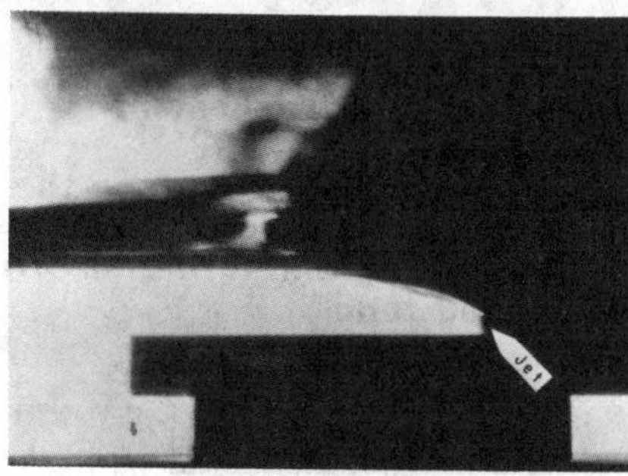

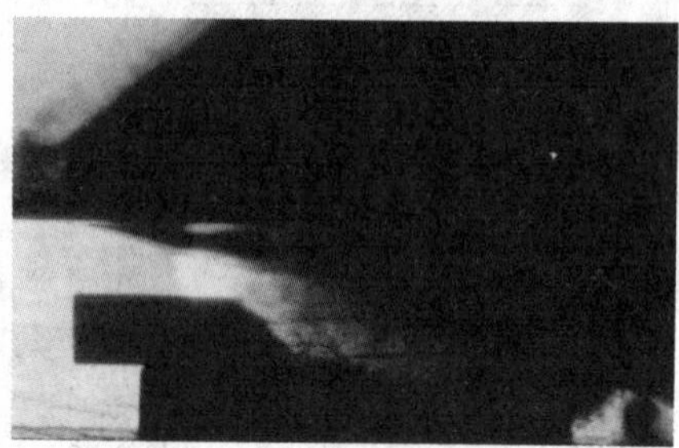

are controlled within certain limits, high-pressure gradients ahead of the collision region in each plate cause the metal surfaces to flow hydrodynamically as a spray of metal from the apex of the angled collision. The flow process and expulsion of the metal surface is known as jetting. The existence of a jet during explosion bonding has been verified by high-speed photography (Fig. 3) and plated surface experiments.

The ability to obtain good explosion bonds is directly related to jet formation. Typically, jet formation is a function of variables such as plate collision angle, collision point velocity, cladding plate velocity, pressure at the collision point, and the physical and mechanical properties of the plates being bonded. Experiments have established that there is a collision velocity below which no welding occurs. For jetting and subsequent cladding to occur, the collision velocity has to be substantially below the sonic velocity of the cladding plates.

There also is a minimum collision angle below which no jetting occurs, regardless of collision velocity. In the parallel plate arrangement shown in Fig. 2, this angle is set by the distance between the plates, called "standoff." For angle cladding (Fig. 1a and b), the preset angle determines the standoff and the attendant collision angle.

Nature of Bonding

Explosively welded metals that are commercially manufactured preferably exhibit a wavy bond zone interface. Aside from its technological importance, the wavy bond is remarkable because of its very regular pattern. Bond zone wave formation is analogous to fluid flowing around an obstacle. When the fluid velocity is low, the fluid flows smoothly around the obstacle; above a certain fluid velocity, the flow pattern becomes turbulent (Fig. 4).

The obstacle in explosion bonding is the point of highest pressure in the collision region. Because the pressures in this region are many times higher than the dynamic yield strength of the metals, they flow plastically in a manner similar to fluids. The microstructure of the metals at the bond zone shows clearly that the metals did not melt but flowed plastically during the process. Electron microprobe analysis across such plastically deformed areas showed that no diffusion occurred due to extremely rapid self-quenching of the metals. The bond represents the "frozen" flow pattern of the plastic metal flow during bond formation.

Under optimum conditions, the metal flow around the collision point is unstabie and oscillates, generating a wavy interface. Figure 5 shows an explosion welding interface between Incoloy 800 and

Fig. 4 Bond zone formation. (a) Fluid flow behind cylinders at increasing flow velocities. (b) Nickel-to-nickel bond zones made at increasing collision velocities: (top to bottom) 4600 ft/s; 5900 ft/s; 9200 ft/s

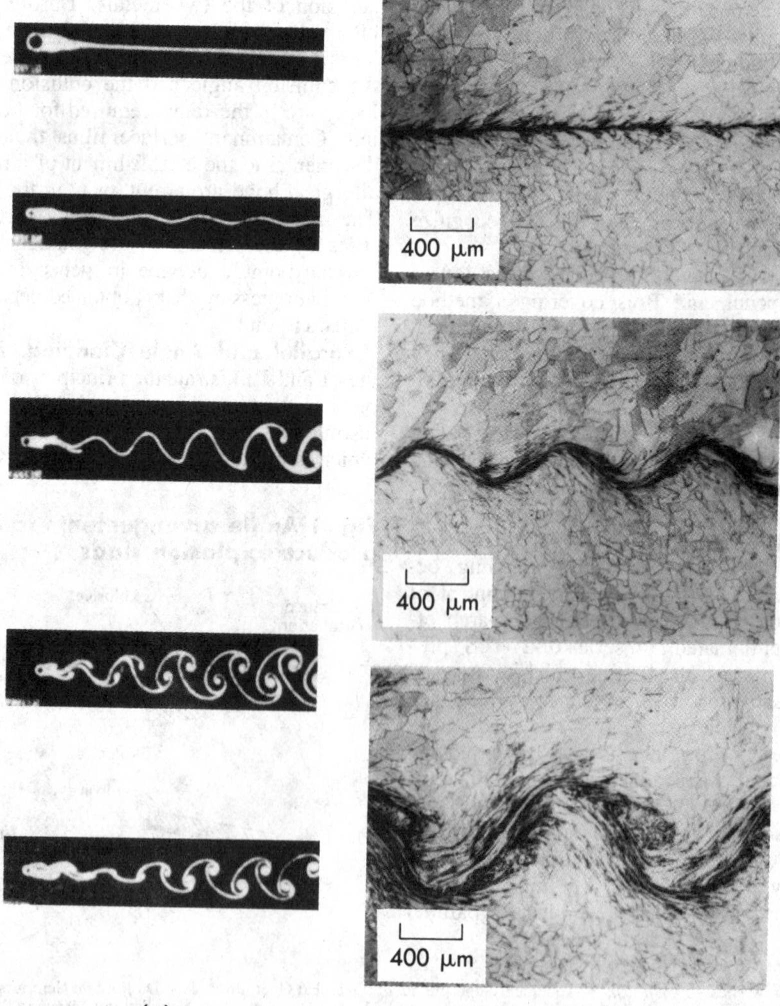

Fig. 5 Incoloy 800 (top) explosively welded to ASME SA-516-70 carbon steel (bottom). Magnification: 100×

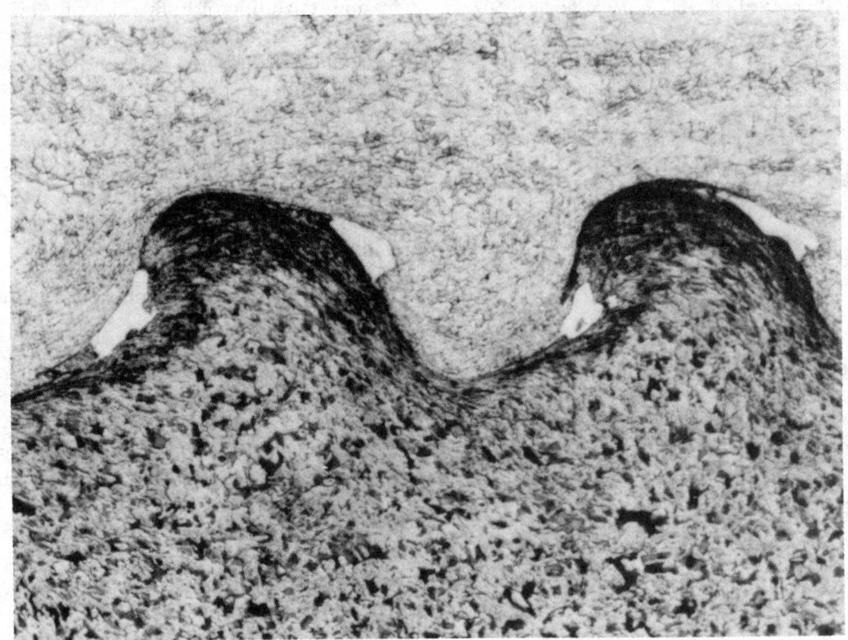

carbon steel. Under the curl of the waves, small pockets of solidified melt may be seen. Some of the kinetic energy of the driven plate was locally converted into heat in these pockets as the system came to rest. These discrete regions are completely encapsulated by the ductile prime and base metals. The direct metal-to-metal bonding between the isolated pockets provides the ductility necessary to support stresses during routine fabrication.

Metallurgical Effects. When the collision velocity is excessive, some cladding metals develop thermal adiabatic shear bands at the interface. Titanium (Fig. 6) and Inconel 718, for example, show this behavior.

Thermal shear bands can lead to shear cracks that penetrate to the cladding surface. However, such cracks also can remain hidden beneath the clad surface, as shown in Fig. 7. This internal defect can be detected and culled by shear wave ultrasonic inspection. When cladding conditions and collision energy are optimum, thermal shear bands are minimized or eliminated, and shear cracks can be avoided.

The quality of welding is related directly to the size and distribution of solidified melt pockets along the interface, especially for dissimilar metal systems that form intermetallic compounds. The pockets of solidified melt in compound-forming systems are brittle and contain localized defects that do not affect composite properties. Explosion bonding parameters for dissimilar metal systems normally are chosen so that pockets of melt associated with the interface are kept to a minimum. When cladding conditions are such that the metallic jet is trapped between the prime metal and the backing metal, the energy of the jet causes surface melting between the colliding plates. In this type of clad, alloying through melting is responsible for the metallurgical bond. As shown in Fig. 8, this type of bond is not desirable.

Applicability

Advantages. The EXW process has a number of characteristics that make it unique compared to other metal bonding processes. Inherent process flexibility provides the following advantages:

- A metallurgical, high-quality bond can be formed not only between similar metals, but also between dissimilar metals classified as incompatible when fusion or diffusion joining methods are used. Brittle intermetallic compounds that are formed in an undesirable continuous layer at the interface during bonding by conventional methods are minimized and, more importantly, are isolated and surrounded by ductile metal in explosion clads. Examples of these systems are titanium to steel, tantalum to steel, aluminum to steel, titanium to aluminum, and copper to aluminum. Even immiscible metal combinations such as tantalum to copper can be clad.
- Explosion cladding can be achieved over areas limited only by the size of the available cladding plate and the size of the explosion that can be tolerated. Areas as small as 0.02 in.^{2} and as large as 300 ft^2 have been bonded; the latter case is illustrated in Fig. 9.

Fig. 6 Titanium (top) explosion clad to carbon steel (bottom). Pockets under the waves that contain brittle intermetallics are surrounded by ductile metal-to-metal bond. Adiabatic shear bands (indicated by arrows) are visible in the titanium. Magnification: 50×

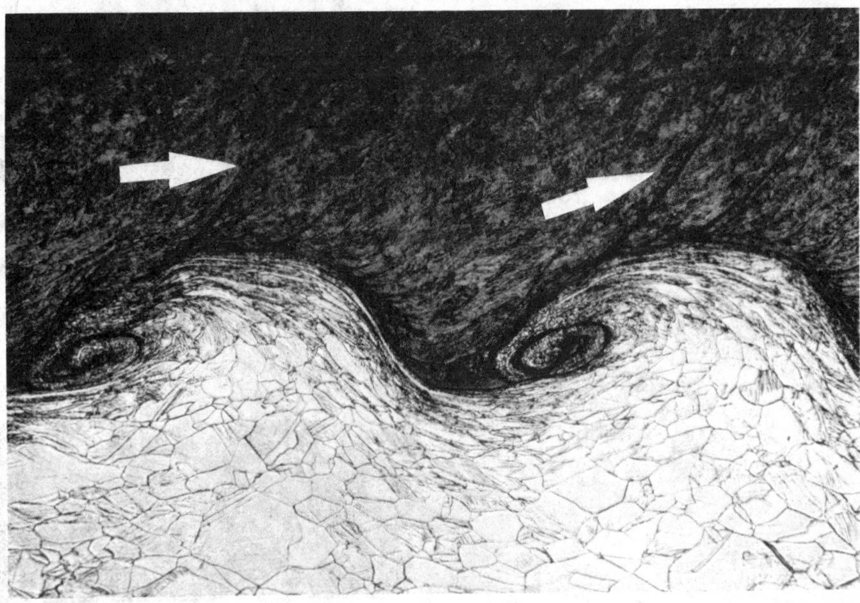

Fig. 7 Internal shear cracks (indicated by arrow) in grade 1 titanium cladding metal (top) explosion bonded to AISI 1008 carbon steel (bottom). After bonding, clad was stress-relief annealed at 1150 °F. Magnification: 50×

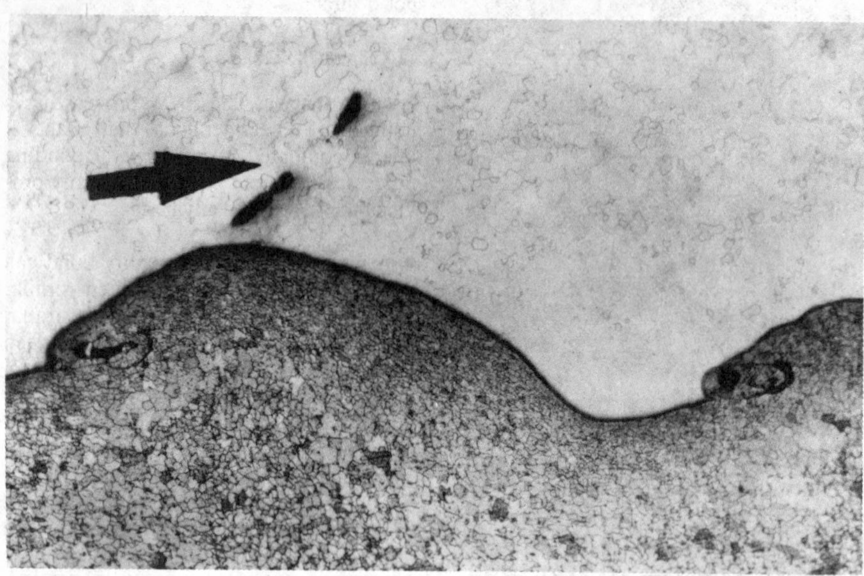

- Metals with tenacious surface oxide films that make roll bonding difficult can be explosion clad.
- Metals having widely different melting points, such as aluminum and tantalum (1220 and 5425 °F, respectively), can be clad.
- Metals with widely different properties, such as copper and maraging steel, can be readily bonded.
- Wide clad-backer ratio limits are available by explosion cladding. Stainless steel-clad components as thin as 0.001 in. and as thick as $1^1/_4$ in. have been explosion clad.
- The thickness of the stationary or backing plate in explosion cladding is essentially unlimited. Backers over 20 in. thick and weighing 50 tons have been commercially clad, as illustrated in Fig. 10.
- High-quality wrought metals are clad without altering chemical composition.
- Different types of backers can be clad. Clads can be bonded to forged members as well as rolled plate.
- Clads can be bonded to plate that is strand cast, annealed, normalized, or quenched and tempered.
- Multiple-layered composite sheet and plate can be bonded in a single explosion. Cladding of both sides of a backing metal can be achieved simultaneously. When two sides are clad, the two prime or clad metals need not be of the same thickness or of the same metal or alloy.
- Nonplanar metal objects can be clad; e.g., the inside of a cylindrical nozzle can be clad with a corrosion-resistant liner.

Limitations of the explosion welding process include:

- The inherent hazards of storing and handling explosives, the difficulty in obtaining explosives with the proper uniform energy, form, and detonation velocity, and the undesirable noise and blast effects.
- Although the nature of the process makes it satisfactory for bonding a wide variety of alloys ranging from plain carbon steel to refractory metals and alloys, metals to be explosively bonded must possess some ductility and impact resistance. Alloys with as little as 5% tensile elongation in a 2-in. gage length and backing steels with as little as 10 ft · lb Charpy V-notch impact resistance at ambient temperature have been bonded. Brittle metals and metal alloys cannot be used in EXW because they fracture during bonding.
- In certain metal systems in which one or more metals to be explosion clad has

Fig. 8 Solidification defects and directional grain growth in explosion-clad copper to copper caused by melting at the interface. Magnification: 100×

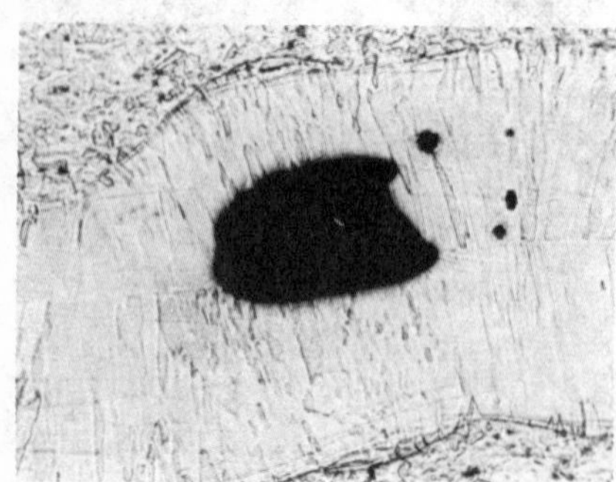

Fig. 9 Type 304L $^3/_{16}$-in.-thick clad explosion bonded to 4-in.-thick SA-516 grade 70 carbon steel. Four 5-by-15-ft stainless plates were fusion welded together (note cruciform outline in center of plate) before explosion bonding. The 10-by-30-ft panel was then clad all at once.

Fig. 10 Explosion-clad plate consisting of ½-in.-thick type 321 stainless steel bonded to 18-in.-thick SA-387 grade D carbon steel. The original 9-by-7-ft clad was conversion hot rolled to a final thickness of about 2.3 in.

a high initial yield strength or a high strain-hardening rate, a high-quality bonded interface may be difficult to achieve. Metal alloys of high strength (>100 ksi yield strength) are difficult to bond. This phenomenon is magnified when there is also a large density difference between the metals. Such combinations are often improved by using a thin, low-yield strength interlayer between the metals.

- Designs lending themselves to explosion bonding are those that allow straight-line egression of the high-velocity jet emanating from between the metals during bonding. In general, the process is best suited to the bonding of flat and cylindrical surfaces.
- Backer thinness rather than thickness is limiting. Thin backers must be supported, thus adding to manufacturing cost.
- The preparation and assembly of clads is not amenable to automated production techniques. Each assembly requires considerable manual labor. The process is thus labor intensive.

Figure 11 illustrates a flow sheet for manufacturing explosively welded flat plate.

Metals Welded

Metals that are difficult or impossible to bond by other welding methods can be explosively welded. Flat plates, solid rods, and the outside and inside surfaces of cylinders and pipes can be explosively welded. Tubes can be machined from thick clads or made by overlap cladding. Explosion welding is employed for the joining of pipes, bonding tubes to tube sheets, and plugging leaking tubes in heat exchangers.

Over 300 dissimilar combinations of metals have been explosion welded, as well as numerous similar combinations. Many of these combinations were only in small sample configurations to demonstrate that a metallurgical bond could be achieved. The industrially useful combinations that are available in commercial sizes are shown in Fig. 12. The chart does not include triclads or combinations that corrosion or materials engineers or equipment designers may yet envision. The combinations that explosion cladding can provide are virtually limitless.

Explosives. The pressure (P) generated by the detonating explosive that propels the prime plate is directly proportional to its density (ρ), and the square of the detonation velocity (V_D^2) and can be reasonably approximated by the following equation:

$$P = \frac{1}{4} \rho V_D^2$$

The detonation velocity is controlled by adjusting the packing thickness, density, and/or the amount of added inert material. Types of explosives that have been used are:

High velocity (14 750 to 25 000 ft/s)

- Trinitrotoluene (TNT)
- Cyclotrimethylenetrinitramine (RDX)
- Pentaerythritol tetranitrate (PETN)
- Composition B
- Composition C4
- Plasticized PETN-based rolled sheet and extruded cord
- Primacord

Low to medium velocity (4900 to 14 750 ft/s)

- Ammonium nitrate
- Ammonium nitrate prills (pellets) sensitized with fuel oil
- Ammonium perchlorate
- Amatol
- Amatol and sodatol diluted with rock salt to 30 to 35%
- Dynamites
- Nitroguanidine
- Diluted PETN

Metal Preparation. Preparation of the metal surfaces to be bonded usually is required, because most metals contain surface imperfections or contaminants that undesirably affect bond properties. The

Fig. 11 Explosion welding process flow sheet

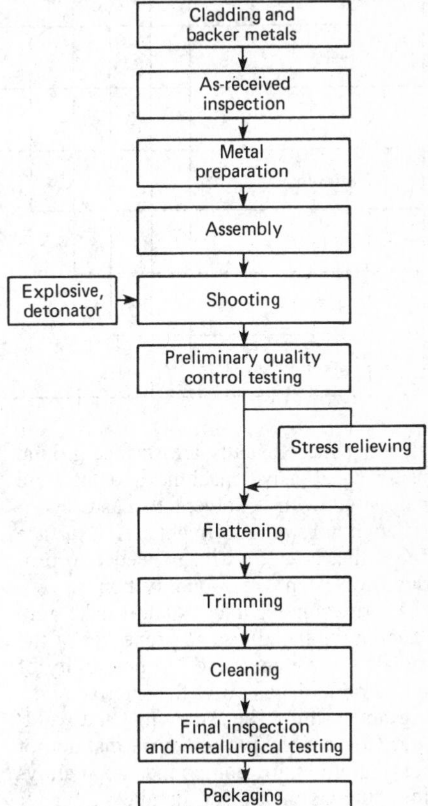

Fig. 12 Commercially available explosion-clad metal combinations

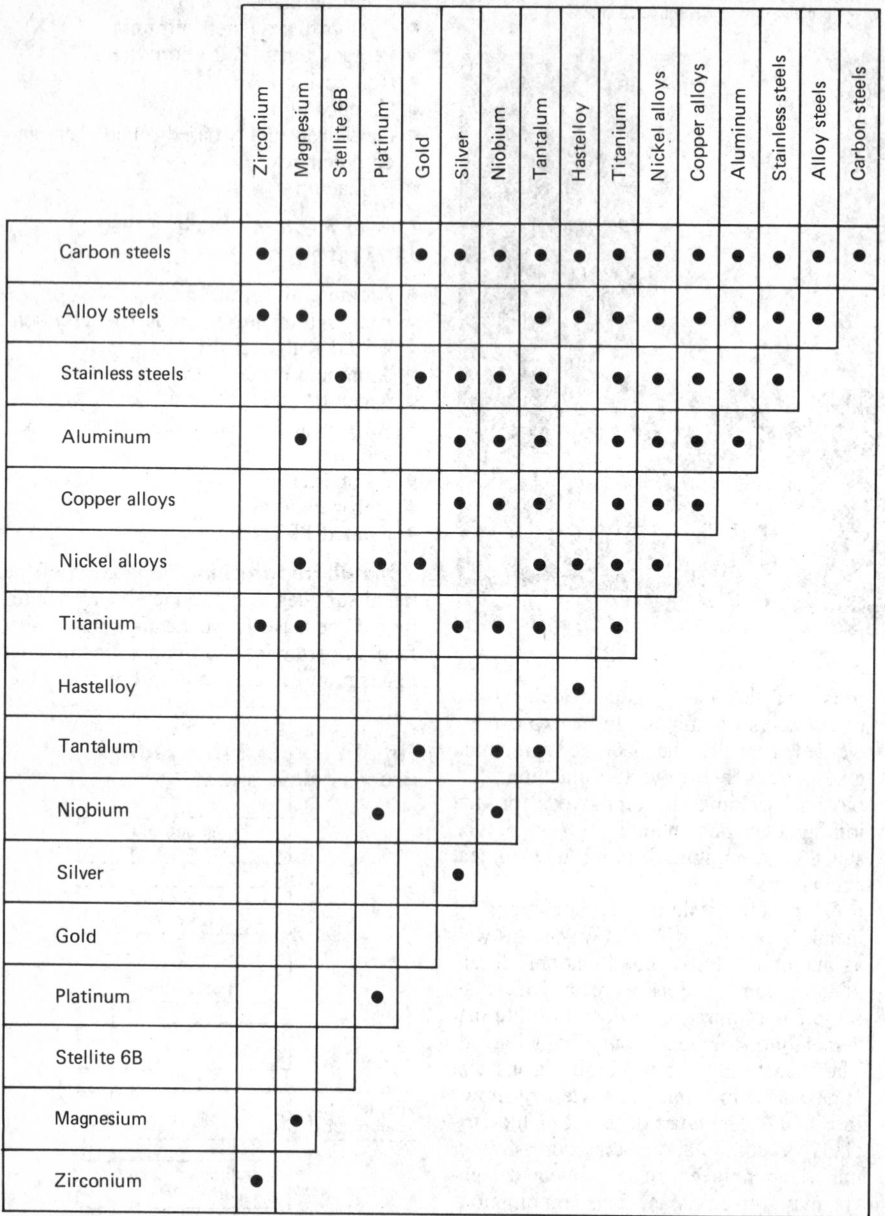

	Zirconium	Magnesium	Stellite 6B	Platinum	Gold	Silver	Niobium	Tantalum	Hastelloy	Titanium	Nickel alloys	Copper alloys	Aluminum	Stainless steels	Alloy steels	Carbon steels
Carbon steels	●	●			●	●	●	●	●	●	●	●	●	●	●	●
Alloy steels	●	●	●					●	●	●	●	●	●	●	●	
Stainless steels			●		●	●	●	●		●	●	●	●	●		
Aluminum		●				●	●	●		●	●	●	●			
Copper alloys						●	●	●		●	●	●				
Nickel alloys		●		●	●			●	●	●	●					
Titanium	●	●				●	●	●		●						
Hastelloy									●							
Tantalum					●		●	●								
Niobium				●			●									
Silver						●										
Gold																
Platinum				●												
Stellite 6B																
Magnesium		●														
Zirconium	●															

cladding faces usually are surface ground using an abrasive machine and then are degreased with a solvent to ensure consistent bond strength. In general, a surface finish that is ≥150 μin. is needed to produce consistent, high-quality bonds.

Fabrication techniques must take into account the metallurgical properties of the metals to be joined and the possibility of undesirable diffusion at the interface during hot forming, heat treating, and welding. Compatible alloys—those that do not form intermetallic compounds upon alloying, such as nickel and its alloys, copper and its alloys, and stainless steel alloys clad to steel—may be treated by the traditional techniques developed for clads produced by other processes. On the other hand, incompatible combinations such as titanium, zirconium, or aluminum to steel require special techniques designed to limit the production of undesirable intermetallics that jeopardize bond ductility at the interface.

Assembly. The air gap present in parallel explosion cladding can be maintained by metallic supports that are tack-welded to the prime metal and backer plates or by metallic inserts that are placed between the prime metal and backer. The inserts usually are made of a metal that is compatible with one of the cladding metals. If the prime metal is so thin that it sags when supported by its edges, other materials—for example, rigid foam—can be placed between the edges to provide additional support; the rigid foam is consumed by the hot egressing jet during bonding. A moderating layer or buffer such as polyethylene sheet, water, rubber, paints, or pressure-sensitive tapes may be placed between the explosive and prime metal surface to attenuate the explosive pressure or to protect the metal surface from explosion effects.

Facilities. The preset assembled composite is placed on an anvil of appropriate thickness to minimize distortion of the clad product. For thick composites, a bed of sand usually is a satisfactory anvil. Thin composites may require a support made of steel, wood, or other appropriate materials. The problems of noise, air blast, and air pollution are inherent in explosion cladding, and clad-composite size is restricted by these problems. Thus, cladding facilities should be located in areas remote from population centers. Using barricades and burying the explosives and components under water or sand lessens the effects of noise and air pollution. An attractive method for making small-area clads using light explosive loads employs a low-vacuum, noiseless chamber. Underground missile silos and mines also have been used as cladding chambers.

Product Quality

When the explosion bonding process distorts the composite so that its flatness does not meet standard flatness specifications, it is reflattened on a press or roller leveller (see American Society of Mechanical Engineers specification ASME SA-20). However, press-flattened plates sometimes contain localized irregularities that do not exceed the specified limits, but which generally do not occur in roll-flattened products.

Nondestructive inspection of an explosion welded composite is almost totally restricted to ultrasonic and visual inspection. Radiographic inspection is applicable only to special types of composites consisting of two metals having a significant mismatch in density and a large wave pattern in the bond interface.

Ultrasonic inspection is the most widely used nondestructive test method for explosion welded composites. Pulse-echo procedures (ASTM A435) are applicable for inspection of explosion welded composites used in pressure applications. The acceptable amount of nonbonded area depends on the application. In clad plates for heat exchangers, >98% area bond usually is required, based on a 4-in. grid. Other applications, such as pressure vessels, may require only 95% of the total area to be bonded, based on a 9-in. grid; 100% area inspection also can be specified.

In some applications, configurations of a nonbond area are specified, such as in heat exchangers where a nonbond area may not be >3 in.^{2} or 3 in. long. The number of allowable nonbond areas generally is specified. Ultrasonic testing can be used on seam welds, tubular transition joints, clad pipe and tubing, and in structural and special applications.

Radiographic Inspection. Radiography is an excellent nondestructive test method for evaluating the bond of aluminum-to-steel electrical and aluminum-to-aluminum-to-steel structural transition joints. It is capable of accurately defining all nonbond and flat-bond areas of the aluminum-to-steel interface, regardless of their size or location.

The clad plate is x-rayed perpendicular from the steel side, and the film contacts the aluminum. Radiography reveals the wavy interface of explosion welded aluminum-clad steel as uniformly spaced light and dark lines with a frequency of one to three lines per centimetre. The waves characterize a strong and ductile transition joint and represent the acceptable condition. The clad is interpreted to be nonbonded when the x-ray shows complete loss of the wavy interface.

Destructive testing is used to determine the strength of the weld and the effect of the EXW process on the base metals. Standard testing techniques can be utilized on many composites; however, nonstandard or specially designed tests often are required to provide meaningful test data for specific applications.

Pressure-Vessel Standards. Explosion-clad plates for pressure vessels are tested according to applicable ASME Boiler and Pressure Vessel Code Specifications. Unfired pressure vessels using clads are covered by American Society for Testing and Materials specifications ASTM A-263, A-264, and A-265; these include tensile, bend, and shear tests. Tensile tests of a composite plate having a thickness of <1½ in. require testing of the joined base metal and clad. Strengthening does occur during cladding, so tensile strengths generally are greater than for the original materials. Some typical shear strength values obtained for explosion-clad composites covered by ASME SA-263, SA-264, SA-265, which specify 20 ksi minimum, and ASTM B-432, which specifies 12 ksi minimum, are:

Cladding metal on carbon steel backers	Typical shear strength, ksi
Stainless steels	65
Nickel and nickel alloys	55
Hastelloy alloys	56
Zirconium	39
Titanium(a)	35
Cupronickel	36
Copper	22
Aluminum (1100-H1)	13

(a) Stress-relief annealed at 1150 °F

Chisel testing is a quick, qualitative technique that is widely used to determine the soundness of explosion welded metal interfaces. A chisel is driven into and along the weld interface, and the ability of the interface to resist the separating force of the chisel provides an excellent qualitative measure of weld ductility and strength.

Ram tensile testing evaluates the bond-zone tensile strength of explosion bonded composites. The specimen is designed to subject the bonded interface to a pure tensile load (Fig. 13). The cross-sectional area of the specimen is the area of the annulus between the internal and overall diameters of the specimen. The specimen typically has a very short tensile gage length and is constructed to cause failure at the bonded interface. The ultimate tensile strength and relative ductility of the explosion bonded interface can be obtained by this technique.

Fig. 13 Machined explosion-clad test sample and fixture for ram tensile testing of bond zone

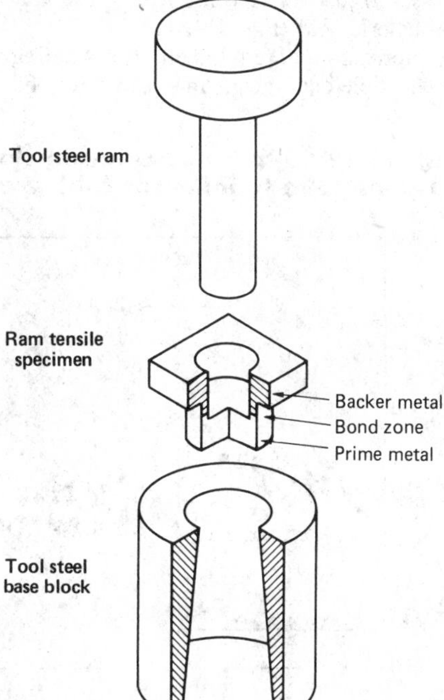

Fig. 14 Tension-shear test specimen

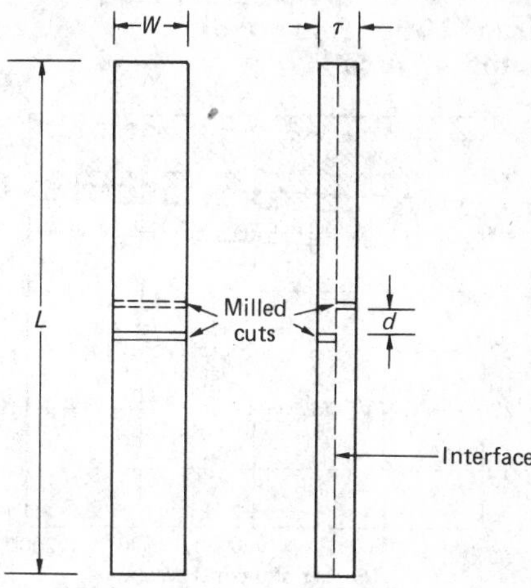

Tension-shear testing is designed to determine the shear strength of the weld. A typical test specimen is shown in Fig. 14. Equal thicknesses of the two components are preferred. The length of the shear zone (d) should be selected so that little or no bending occurs in either component. Failure should occur by shearing parallel to the weld line. If failure occurs through one of the base metals, the shear strength of the weld is obviously greater than the strength of the base metal. Results are useful for comparison purposes only, when using a common test specimen.

Thermal Fatigue and Stability. Explosion welded plates have performed satisfactorily in several types of thermal tests. In thermal-fatigue tests, samples from bonded plate were alternately heated to 850 to 1000 °F at the surface and were quenched in cold water to less than 100 °F. The 3-min cycle consisted of 168 s of heating and 12 s of cooling. Bond-shear tests were performed on samples before and after thermal cycling. Stainless steel clads survived 2000 such thermal cycles without significant loss in strength (Fig. 15). Explosion bonded and tested grade 1 titanium-to-carbon steel samples performed in a similar satisfactory fashion.

Stainless steel clads survived repeated quenching in water from heat treatment temperatures of 1750 to 2000 °F. Water quenching of only the clad side failed to affect the bond of such samples. Long-term air exposure at 600 °F did not alter the strength and ductility of titanium clads. Room temperature bond-shear strength and ductility of test specimens were not altered after such exposure for 10 000 h.

Fig. 15 Effect of thermal cycling type 304L stainless steel and SA-516-70 carbon steel clad from 1000 °F to room temperature

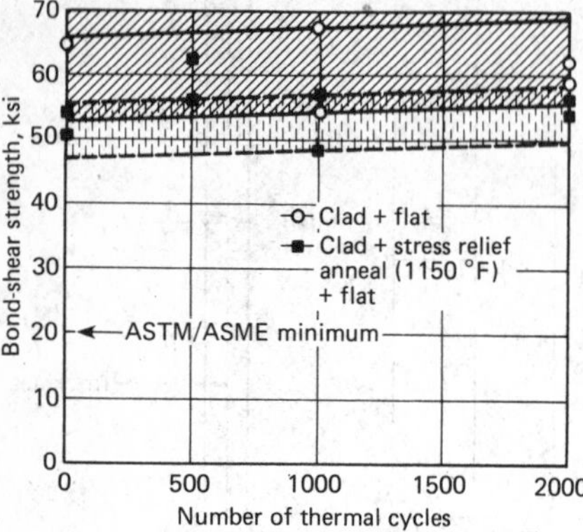

Metallography. The weld interface is inspected on a plane parallel to the detonation front and normal to the surface. A well-formed wave pattern without porosity generally is indicative of a good bond. The amplitude of the wave pattern for a good weld can vary from small to large, without a large influence on the strength. Small pockets of melt can exist without being detrimental to the quality of the bond. Samples should be taken from an area that is representative of the entire weld. Edge effects may result in local nonbonded areas along the edges of a weld. Samples from such locations are not representative of the entire weld.

Hardness and Impact Strength. Microhardness profiles on sections from explosion bonded materials show the effect of strain hardening on the metals in the composite. Figure 16 illustrates the effect of welding a strain-hardening austenitic stainless steel to a carbon steel. The austenitic stainless steel is hardened adjacent to the weld interface by explosion welding, whereas the carbon steel is not hardened to a great extent. Similarly, aluminum does not strain harden significantly.

Impact strengths also can be reduced by the presence of the hardened zone at the interface. A low-temperature stress-relief anneal decreases hardness and restores impact strength. Alloys that are sensitive to low-temperature heat treatments also show differences in hardness traverses that are related to explosion welding parameters. Low welding impact velocities do not develop as much adiabatic heating as higher impact velocities. Adiabatic heating anneals and further ages the alloys. Hardness traverses indicate the degree of hardening during welding and the type of subsequent heat treatment required after explosion bonding. Explosion bonding parameters also can be adjusted to prevent softening at the interface.

Applications

Because explosion bonding is a cold process, thermal damage to the metals that may reduce corrosion resistance or impair other properties is avoided. Cladding and backing metals are purchased in the appropriate heat treated condition, and the desired corrosion resistance is retained through bonding. Composites customarily are supplied in the as-bonded condition, because hardening usually does not affect engineering properties. Occasionally, a postbonding heat treatment is used to meet required properties on specific combinations.

Vessel heads can be made from explosion bonded clads either by conventional cold or hot forming techniques. The latter involves thermal exposure and is equivalent to a heat treatment. Backing-metal properties, bond continuity, and bond strength are guaranteed to equal the same specifications as the composite from which the head is formed.

Chemical Process Vessels. Explosion bonded products are used in the manufacture of process equipment for the chemical, petrochemical, and petroleum industries, where corrosion resistance of an expensive metal is combined with the strength and economy of another. Typical applications include explosively clad titanium to Monel tube sheet (Fig. 17) and a tantalum-to-copper-to-carbon steel-clad vessel to recover chlorine from waste hydrochloric acid (Fig. 18).

Fabrication of equipment for welding compatible clad metals may be performed by traditional techniques. However, special precautions must be taken for incompatible alloy systems. The preferred technique for butt welding involves a batten-strap technique with a silver, copper, or steel inlay (Fig. 19). Precautions must be taken to avoid iron contamination of the weld either from the backer steel or from outside sources. Stress relieving is achieved at normal steel stress-relieving temperatures. Special welding techniques must also be used in joining tantalum-to-copper-to-steel clads.

Conversion-Rolled Billets. Large tonnage of clad plate and strip have been made by hot and cold rolling of explosion bonded slabs and billets. Explosion bonding is economically attractive for conversion rolling, because the capital investment for plating and welding equipment for conventional bonding methods is avoided. Highly alloyed stainless steels and some copper alloys that are difficult to clad by roll bonding are practical candidates for plate made by converting explosion bonded slabs and billets. Conventional hot rolling and heat treating practices are used when stainless steels and nickel and copper alloys are converted. Hot rolling of explosion bonded titainum, however, must be restricted below about 1550 °F to avoid diffusion and attendant formation of undesirable intermetallic compounds at the bond interface. Hot rolling titanium also requires a stiff rolling mill because of the large separation forces required to accomplish reduction. A conversion-rolled titanium-carbon steel assembly is shown in Fig. 20.

The most notable application of conversion-rolled explosion bonded clads was United States coinage in the

Fig. 16 Two types of explosion clads showing widely divergent response due to inherent cold work hardening characteristics

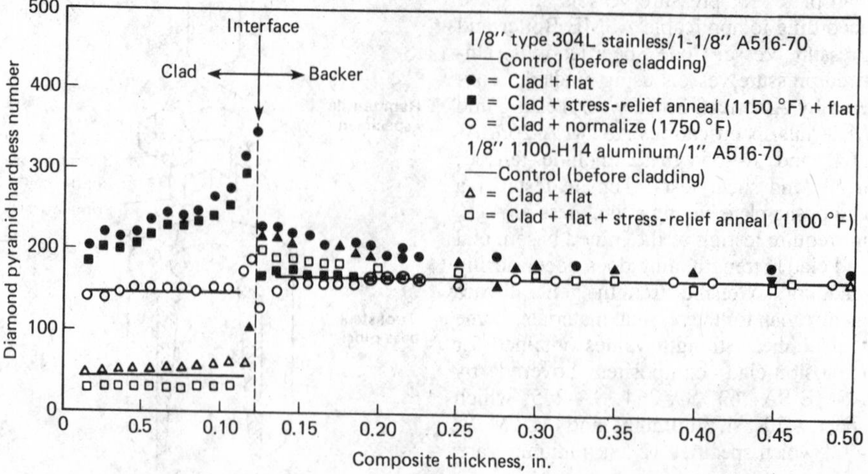

Fig. 17 Titanium tube sheet explosively clad to Monel 400. Courtesy of Nooter Corp.

Fig. 18 Interior of 8- by 60-ft-diam vessel fabricated from explosion-clad tantalum to copper to carbon steel. Courtesy of Nooter Corp.

mid-1960's. After explosion bonding of triclad composites consisting of 70Cu-30Ni-to-copper-to-70Cu-30Ni, conventional hot and cold rolling methods were used to convert the clad billets to strip from which 10- and 25-cent coins were minted (Fig. 21).

Transition Joints. Use of explosion-clad transition joints avoids the limitations involved in joining two incompatible materials by bolting or riveting. Many transition joints can be cut from a single large-area clad flat plate. Conventional fusion welding processes then can be used to attach the members of the transition joint to their respective similar metal components.

Electrical Applications. Aluminum, copper, and steel are the most common metals used in high-current, low-voltage conductor systems. Use of these metals in dissimilar metal systems often maximizes the effects of the special properties of each material. However, joints between incompatible metals must be electrically efficient to minimize power losses. Mechanical connections involving aluminum offer high resistance because of the presence of the self-healing oxide skin on the aluminum member. Because this oxide layer is removed by the jet, the interface of an explosion-clad aluminum assembly essentially offers no resistance to the current. Thus, welded transition joints, which are cut from thick composite plates of aluminum to carbon steel or aluminum to copper, permit highly efficient electrical conduction between dissimilar metal conductors. Sections can be added by fusion welding the aluminum side of the transition joint to the adjoining aluminum member. This concept is routinely employed by the primary aluminum reduction industry in anode rod fabrication, as shown in Fig. 22.

Usually, copper surfaces are mated when joints must be periodically disconnected, because copper offers low resistance and good wear. Joints between copper and

Fig. 19 Double-V inlay batten-strap technique for fusion welding an explosion-clad plate containing titanium

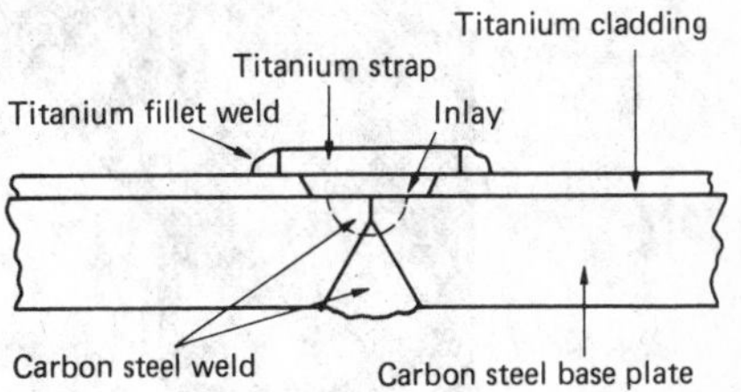

Fig. 20 End plates of rotary dryers for pulp and paper manufacture made from conversion-rolled explosion welded titanium-carbon steel clads. Courtesy of Ingersoll-Rand Co.

Fig. 21 Explosion-clad billet to be converted to strip for minting coins

Fig. 22 Aluminum reduction cell anode with aluminum-to-steel explosion welded transition joint

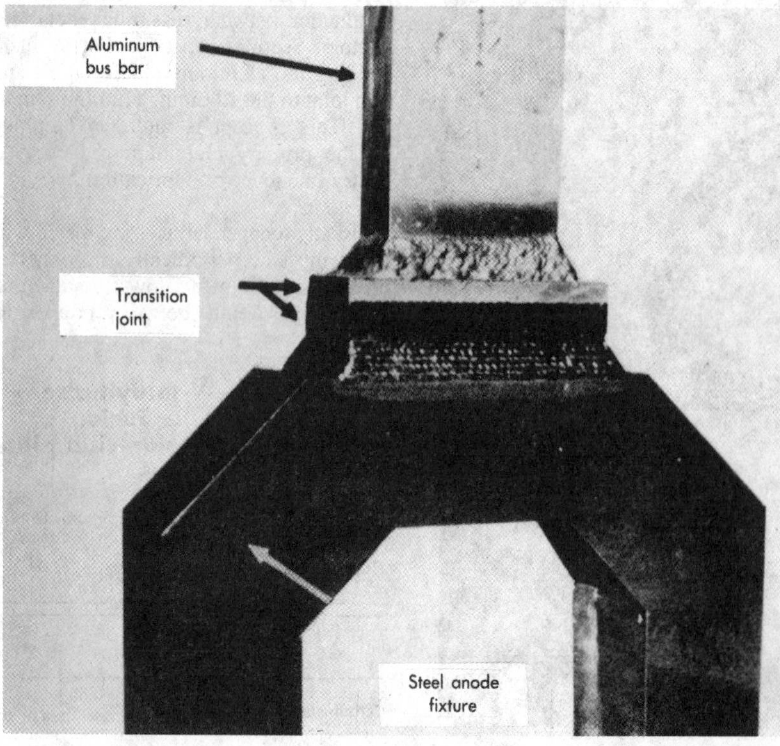

aluminum bus bars are improved by using a copper-to-aluminum transition joint that is welded to the aluminum member. Deterioration of aluminum shunt connections by arcing is eliminated when a transition joint is welded to both the primary bar and the shunting bar.

Marine Applications. The use of aluminum as a superstructure material in the shipbuilding industry has highlighted the shortcomings of bolted or riveted dissimilar metal combinations. In the presence of an electrolyte such as seawater, aluminum and steel form a galvanic cell, and corrosion takes place at the interface. When the aluminum superstructure is bolted to the steel bulkhead in a lap joint, crevice corrosion is masked and may go unnoticed until replacement is required. By using transition joint strips cut from explosion welded clads, corrosion can be eliminated. Because the transition joint is metallurgically bonded, there is no crevice in which the electrolyte can act, and galvanic action cannot take place. Steel corrosion is confined to external surfaces, where it can easily be detected and corrected by simple wire brushing and painting.

Joints between the aluminum superstructure of the ship and the explosively bonded aluminum cladding usually are made by gas metal arc welding. Mechanical testing, such as tensile, shear, impact, fatigue, and explosive bulge, showed that the all-welded construction had equivalent or better properties than the more complicated riveted systems. Peripheral benefits include weight savings and perfect electrical grounding. In addition to lower initial installation costs, welded systems require little or no maintenance and, therefore, minimize life-cycle costs. Applications of structural transition joints include aluminum superstructures welded to decks of sophisticated naval vessels and commercial ships (Fig. 23).

Tubular Applications. Explosion welding is a practical method for joining dissimilar metal pipes, such as aluminum, titanium, or zirconium, to steel or stainless steel, using standard welding equipment and techniques. The process provides a strong metallurgical bond that is maintenance-free throughout years of thermal and pressure-vacuum cycling. Explosion welded tubular transition joints are used in many diverse applications in aerospace, nuclear, and cryogenic industries. Figures 24 and 25 show two such applications. Tubular joints operate reliably through the full range of temperatures, pressures, and stresses normally encoun-

Fig. 23 Aluminum superstructure and deck connection using explosion-clad aluminum-to-carbon steel transition joint

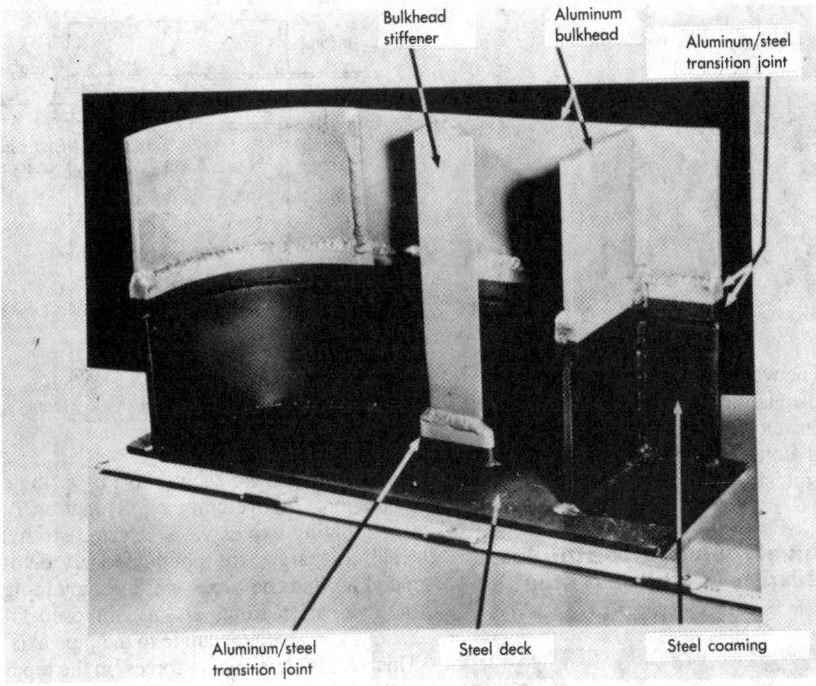

Fig. 24 Aluminum-to-stainless steel transition joints explosion welded in cylindrical form by parallel-overlap technique for cryogenic use

tered in piping systems. Tubular transition joints in various configurations can be cut and machined from explosion welded plate, or they can be made by joining tubes by overlap cladding. Standard welding practices are used to make the final joints.

Nonplanar Specialty Products. The inside walls of hollow forgings that are used for connections to heavy-walled pressure vessels have been metallurgically bonded with stainless steel. These bonded forgings, or nozzles, range from 2 to 24 in. in inside diameter and are up to 3 ft long. Large clad cylinders and internally clad heavy-walled tubes have been extruded using conventional equipment. Figures 26 and 27 show two specialty products that have been explosion bonded.

Tube Welding and Plugging. Explosion bonding is used to bond tubes and tube plugs to tube sheets in heat exchanger fabrication. The commercial process resembles the cladding of internal surfaces of thick-walled cylinders or pressure vessel nozzles, as shown in Fig. 28; angle cladding is used. Countersink machining at the tube entrance provides the angled surface of 10 to 20° at a depth of 0.5 to 0.6 in. The exploding detonator propels the tube or tube plug against the face of the tube sheet to form the proper collision angle, which in turn provides the required jetting and attendant metallurgical bond. Tubes may be welded individually or in groups. Most applications of explosion welding in tube-to-tube-sheet joints involve tube diameters of 0.5 to 1.5 in. Metal combinations that are welded commercially include carbon steel to carbon steel, titanium to stainless steel, and 90Cu-10Ni to carbon steel.

Pipeline Welding. Explosion welding methods have been developed to join sections of large-diameter pipe of the kind used

Fig. 25 Tubular transition joints for cryogenic use. Tubes were made from clad flat plate containing aluminum alloy to aluminum alloy to titanium to nickel to stainless steel.

Fig. 26 Heat exchanger made by explosion welding

Fig. 28 Tube-to-tube-sheet plugging

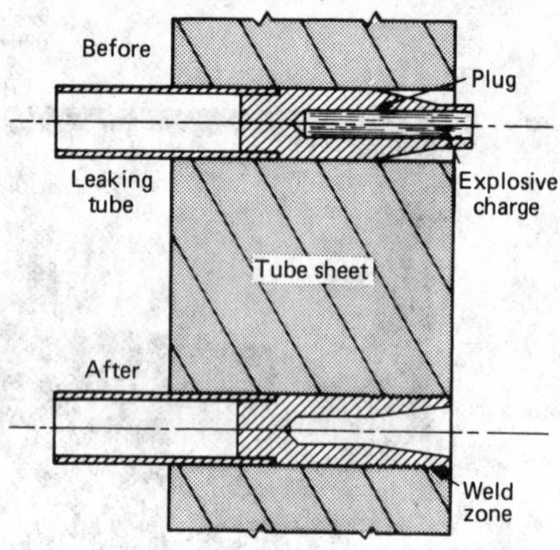

in the construction of high-volume oil and gas pipelines. In one method, internal and external band charges of welding explosive are simultaneously detonated adjacent to overlapped telescoped pipe ends (Fig. 29). The welding charges are detonated simultaneously by means of very high detonation velocity initiation explosive charges placed along a leading edge of each welding explosive charge. The initiating explosive charges are each set off by a single detonator. The velocity of detonation of the initiating explosive is selected so that welding charges are simultaneously detonated around the pipes while maintaining the detonation fronts at superimposed locations and at a large angle to the pipe axis. This avoids unbalancing forces on the pipes and minimizes the collision of circumferential shock waves, thus preventing pipe damage and imperfect welds.

Fig. 27 Inside liner of Hastelloy X explosively welded to stainless steel ribs that were first explosively welded to a stainless steel outer shell. Heat exchanger was welded in cylindrical form for a rocket motor.

Buildup and Repair. Explosion welding also is used for the repair and buildup of worn flat and cylindrical components. The worn area is clad with an appropriate thickness of metal and subsequently machined to the proper dimensions. In some cases, the repair can be made with a material that exhibits superior strength or abrasion and corrosion resistance in comparison with the original material. Figure 30 illustrates the use of explosion bonding to repair a turbine shaft bearing surface.

Safety

All explosive materials should be handled and used in compliance with approved safety procedures either by, or under the direction of, competent, experienced persons in accordance with all applicable federal, state, and local laws. The Bureau of Alcohol, Tobacco, and Firearms, the Hazardous Materials Regulation Board of the Department of Transportation, the Occupational Safety and Health Administration, and the Environmental Protection Agency have federal jurisdiction on the sale, transport, storage, and use of explosives. Many states and local counties have special explosive requirements. The Institute of Makers of Explosives provides educational publications to promote the safe handling, storage, and use of explosives.

Fig. 29 Explosion welding of large-diameter pipe sections

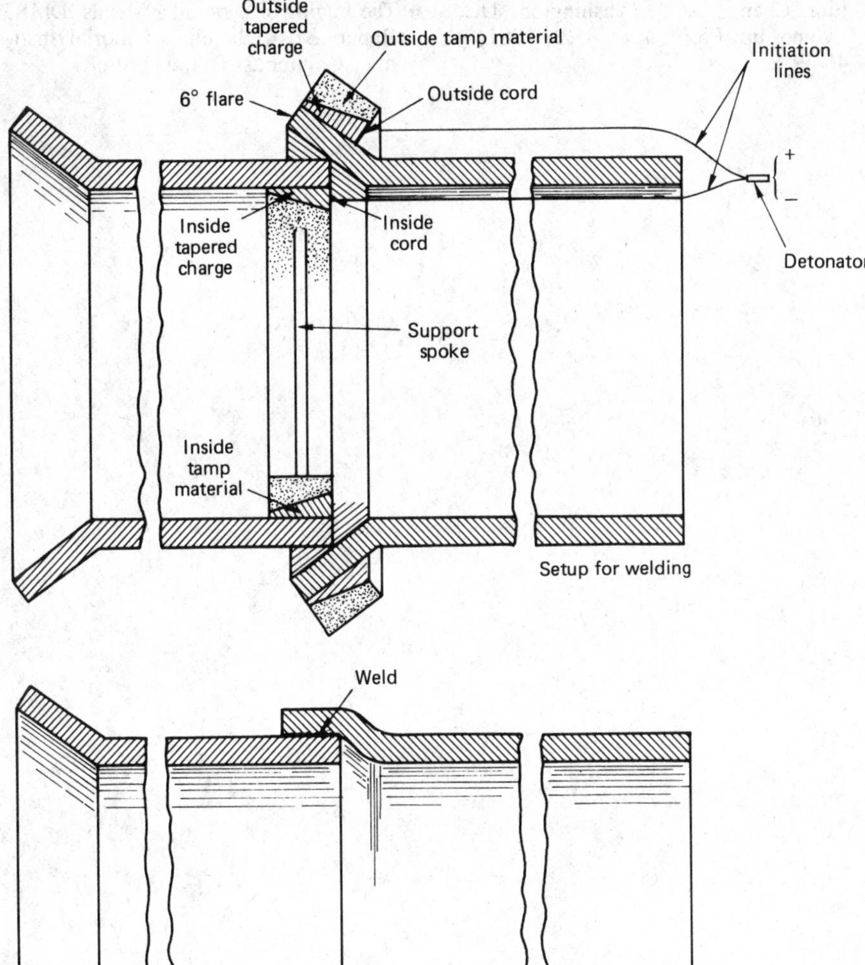

Fig. 30 Bearing of helicopter turbine shaft repaired by explosion welding a steel sleeve over the premachined worn area. Welded sleeve was hardened by induction heat treatment.

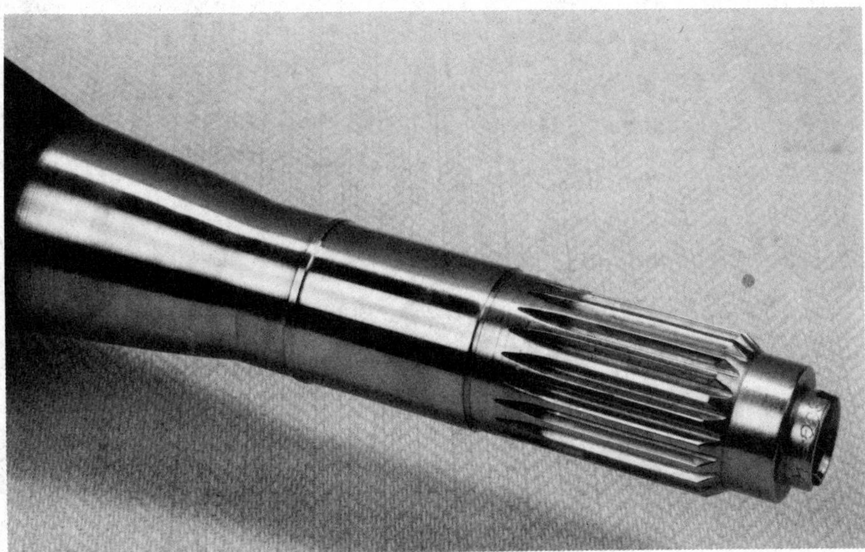

The National Fire Protection Association similarly provides recommendations for the safe manufacture, storage, handling, and use of explosives.

SELECTED REFERENCES

- Carpenter, S.H. and Wittman, R.H., *Annual Rev. Mat. Science*, Vol 5, 1975, p 177-199
- Cook, M.A., *The Science of High Explosives,* Reinhold Publishing, New York, 1966, p 274
- Cowan, G.R. and Holtzman, A.H., *J. Appl. Phys.*, Vol 34 (No. 4), Part 1, 1963, p 928-939
- Cowan, G.R., Bergmann, O.R., and Holtzman, A.H., *Met. Trans.*, Vol 2 (No. 11), 1971, p 3145-3155
- Crossland, B. and Bahrani, A.S., *Contemp. Phys.*, Vol 9 (No. 1), 1968, p 71-87
- Crossland, B., *Explosive Welding of Metals and Its Applications,* Clarendon Press, Oxford, 1982
- Davenport, D.E. and Duvall, G.E., American Soc. Tool and Mfg. Engrs., Dearborn, MI, Technical Paper SP60-161, 1960-1961
- DeMaris, J.L. and Pocalyko, A., Am. Soc. Tool and Mfg. Engrs., Dearborn, MI, Technical Paper AD66-113, 1966
- Deribas, A.A., Kudinov, V.M., and Matveenkov, F.I., *Fizika Goreniya i Vzryva,* Vol 3 (No. 4), 1967, p 561-568
- Enright, T.J., Sharp, W.R., and Bergmann, O.R., *Met. Prog.*, Vol 98 (No. 1), 1970, p 107-114
- Ezra, A.A., *Principles and Practices of Explosives Metal Working,* Industrial Newspapers, Ltd., London, 1973
- Hardwick, R., *Welding J.*, 1975, p 238-244
- Holtzman, A.H. and Cowan, G.R., *Weld Res. Counc. Bull. No. 104,* April 1965, p 1-21
- Kubota, A., *Chemical Economy & Engrg. Rev.*, Vol 7 (No. 12), Dec 1975
- Linse, V.D., Wittmann, R.H., and Carlson, R.J., Defense Metals Information Center, Columbus, OH, Memo 225, Sept 1967
- Linse, V.D., ASME Publication 74-GT-85, Nov 1973
- McKenney, C.R. and Banker, J.G., *Marine Technology,* 1971, p 285-292
- Moro, S., Kubota, A., and Hamada, K., PNC Paper No. A-3, 2nd Joint DOE/PNC LMFBR Steam Generator Seminar, USA, June 1981
- Otto, H.E. and Carpenter, S.H., *Weld-*

ing J., Vol 51 (No. 7), 1972, p 467-473

- Pocalyko, A., *Mat. Protection,* Vol 4 (No. 6), 1965, p 10-15
- Pocalyko, A., High Pressure as a Reagent and an Environment, 17th State-of-the-Art Symposium, Div. of Ind. and Eng. Chem., ACS, Washington, DC, Symposium Proc., June 8-10, 1981, p 46-59
- Popoff, A.A., *Mechanical Engrg.*, Vol 100 (No. 5), 1978, p 28-35
- The Joining of Dissimilar Metals, DMIC Report S-16, Battelle Memorial Institute, Columbus, OH, Jan 1968

Friction Welding

By the ASM Committee on Friction Welding*

FRICTION WELDING (FRW) is a process in which the heat for welding is produced by direct conversion of mechanical energy to thermal energy at the interface of the workpieces without the application of electrical energy, or heat from other sources, to the workpieces. Friction welds are made by holding a nonrotating workpiece in contact with a rotating workpiece under constant or gradually increasing pressure until the interface reaches welding temperature and then stopping rotation to complete the weld. The frictional heat developed at the interface rapidly raises the temperature of the workpieces, over a very short axial distance, to values approaching, but below, the melting range; welding occurs under the influence of a pressure that is applied while the heated zone is in the plastic temperature range.

Friction welding is classified as a solid-state welding process, in which joining occurs at a temperature below the melting point of the work metal. If incipient melting does occur, there is no evidence in the finished weld because the metal is worked during the welding stage. A section through a friction weld joining two dissimilar steels is shown in Fig. 1. The steel with the greater forgeability has the greater amount of weld upset. When similar metals are friction welded, the amount of weld upset is about the same on both sides of the weld interface.

Friction welding has been used in high production of hollow precombustion chambers for diesel engines, in welding trunnions to mounting blocks for air and hydraulic cylinders, in welding connectors to piston rods, and in fabricating track roller hubs and ball shaft linkages.

In the automotive industry, FRW is used in fabricating drive shafts, axles, steering shafts, and bimetal valves, and for joining hubs to gears. Another application is welding bar stock to small forgings or to plate to produce parts that would otherwise be forged. Blanks for cutting tools are made by welding low-carbon or low-alloy steel shanks to tool steel bodies. Jet engine parts are made by welding components made of a heat-resistant alloy to components made of a hardenable or wear-resistant alloy.

Fig. 1 Cross section through a friction weld. The greater amount of weld upset occurred in the 1045 steel due to greater forgeability. When same or closely similar metal workpieces are welded, approximately equal amounts of weld upset form on both sides of bond line.

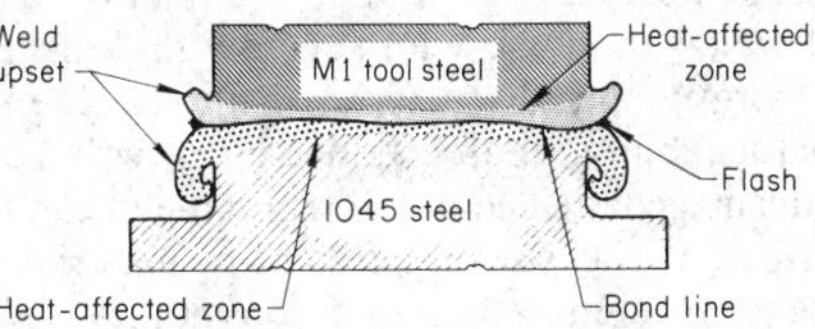

There are two methods of joining workpieces by FRW: continuous drive FRW and inertia drive FRW. In the continuous drive method, which is also referred to as direct drive friction welding, mechanical energy is converted to heat energy by rotating one workpiece while pressing it against a nonrotating workpiece. After a specific period of time or upset distance, rotation is suddenly stopped, and the pressure is increased and held for another specified period of time, thus producing a weld.

In the inertia drive method, the workpiece component that is to be rotated is held in a collet chuck-flywheel assembly. The assembly is then accelerated to a predetermined speed, at which time the flywheel is disconnected from the power supply and the workpieces are brought into contact under a constant force. Flywheel energy is rapidly converted to heat at the interface, and welding occurs as rotation ceases.

Process Capabilities

Many ferrous and nonferrous alloys can be friction welded. Friction welding also can be used to join metals of widely differing thermal and mechanical properties. Often combinations that can be friction welded cannot be joined by other welding processes because of the formation of brittle phases that would make such joints unserviceable. The submelting temperatures and short weld times of FRW allow many combinations of work metals to be joined.

End preparation of workpieces, other than that necessary to ensure reasonably good axial alignment and to produce the required length tolerance for a specific set of welding conditions, is not critical. Frictional wear removes irregularities from the joint surfaces and leaves clean, smooth surfaces heated to welding temperature. In some applications where weld integrity is important, a small projection at the center of one of the weld members is used to ensure proper heating and forging action and to eliminate center defects. This projection is especially helpful in welding large-diameter bars.

Automatic loading and unloading of the welding machines permit high production rates. For instance, bimetal valves are produced two-at-a-time at a rate of 1200/h. Other advantages of FRW include:

- Flux, filler metal, or protective atmospheres are not needed.
- Electric power and total energy requirements are less than for other welding processes.
- The operation is relatively clean; there

*Frank J. Wallace, *Chairman,* Chief, Welding Development Programs, Pratt & Whitney Aircraft Group, United Technologies; Dietmar E. Spindler, Vice President, Manufacturing Technology, Inc.; J. Paul Thorne, Vice President, Research and Development, Bay City Division, Newcor, Inc.

is little spatter, and no arcs, fumes, or slag is developed.

- The heat-affected zone (HAZ) is very narrow and has a grain size that frequently is smaller than that in the base metal.

Limitations of FRW are:

- One workpiece must be round, or nearly round, at the interface and must have a size and shape that can be clamped and rotated. Hexagon-shaped bars have been friction welded to billets.
- Workpieces must be able to withstand the torque and axial pressure imposed during heating and forging.
- Workholding devices must be strong enough to withstand heavy shock and torque loads.
- The process is restricted to flat and angular butt welds that are concentric with the axis of rotation (see Fig. 10).

Example 1. Valve Stem. The use of FRW reduced material and machining costs in the manufacturing of a bronze valve stem (Fig. 2). Originally, the valve stem was machined complete from a $^5/_8$-in.-diam bar. By friction welding two pieces and producing a weld upset large enough to provide material from which the flange could be machined to the required final dimensions, machining time was reduced 95%, and material loss was reduced 98%. The friction weld was made in 7 s, using a spindle speed of 2200 rpm, a heating pressure of 14 ksi, and a welding pressure of 65 ksi. Total axial shortening of the workpieces during upsetting was 0.150 in. Production rate was 180/h.

Fig. 2 Friction welded bronze valve stem. Center flange machined from the upset produced by friction welding

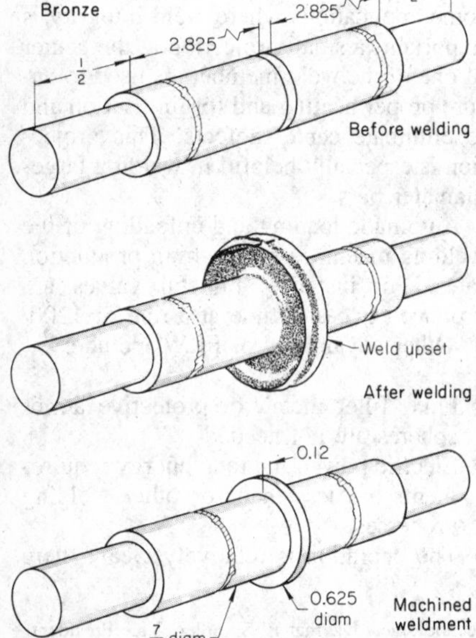

Weld Strength. For most metals, the strength of a friction welded joint is usually as good as or better than that of the base metal. The metal at the weld interface is hot worked, which refines the grain structure. During the final portion of the weld cycle, upsetting and extrusion of flash ensure removal of oxidized metal that may have been produced during heating. This flash usually appears in the valley at the intersection of the two weld upsets (see Fig. 1). The relatively large unheated areas adjacent to the joint extract heat quickly from the small mass of the HAZ, thus keeping the zone shallow in the welded part. It is very difficult, however, to achieve a wide HAZ and a refined grain size. A wide HAZ usually means that more energy was used to make the weld. This slows down the cooling rate and increases grain growth. Irreparable damage in the HAZ can occur if welds are made with severely high heat (energy) input. In materials with very high hardenability, more weld energy will be advantageous since it will retard the quench effect of the surrounding base material, thereby eliminating the possibility of quench cracks during the cooling phase. The HAZ is a small band of material approximately parallel to the weld face. While the flash, or upset, is also heat affected, it is not considered as the HAZ since it does not contribute to the structural integrity of the weld and is usually machined away. In most ferrous alloys it does not make any difference how wide the HAZ is as long as the grain structure is refined, and as long as it is possible to postheat treat the weld in such a way as to eliminate any structural damage due to the HAZ. In some materials such as nickel alloys, the HAZ must be kept as narrow as possible since an effective heat treatment to restore the weld zone would negatively influence the surrounding base metal.

Example 2. Flash Welding Versus Friction Welding of Joints. A 1026 steel shaft 0.75 in. in diameter was joined to the stub shaft of a worm gear made of 5120 steel by continuous drive FRW and by flash welding. Tension tests, rotating-beam fatigue tests, and reversed torsion fatigue tests were made on the weldments. The results of these tests are given in Table 1. Examination of the microstructures of the welded joints showed a much narrower HAZ and a finer grain structure in the friction welded joint than in the flash welded joint. The friction welded part had a higher tensile strength than the flash welded part.

Because there is no spatter of weld flash during FRW, the time required to clean the machines was reduced from 18 h to less than 5 h/week. The increased productivity that resulted from less downtime and higher production rates made it possible to use two friction welding instead of five flash welding machines.

Sections Welded. In FRW, the joint face of at least one of the workpieces must be essentially round. The rotating workpiece should be somewhat concentric in shape because it revolves at a relatively high speed. Workpieces that are not round, such as hexagon-shaped workpieces, have been friction welded successfully, but the resulting weld upset is rough, asymmetrical, and difficult to remove without damaging the welded assembly. For special applications, welding machines have been modified so that the spindle stops at the same place each time, thus making it possible for workpieces to be oriented to each other.

Solid bars of 1018 steel from 0.040 to 7 in. in diameter can be friction welded. Welding of large diameters, although feasible, is limited by machine cost.

Wire and tubing of like and unlike metals 0.040 to 0.10 in. in diameter have been fricton welded in special machines to plates 0.010 to 0.10 in. thick. Wires of unlike metals 0.060 to 0.10 in. in diameter have been joined.

Tubular sections can be much larger in diameter than the rated capacity of the welding machine for solid bars, and the maximum weldable tube diameter depends primarily on wall thickness. For example, a machine capable of welding a 4-in.-diam 1018 steel bar can weld a 1018 steel tube 30 in. in diameter with a $^3/_{16}$-in.-thick wall. The maximum diameter decreases to about $7^1/_2$ in. when the wall thickness is 1 in.

Equipment is currently available for welding up to 75 in.^{2} of medium-carbon

Table 1 Comparison of properties of flash and friction welded joints made between 1026 and 5120 steel bars

Welding type	Tensile strength, ksi	Bending fatigue time, min	Torsional fatigue time, cycles	Metal lost(c), in.
Flash	70	32	1 000 000(a)	0.75
Friction	75	180	4 000 000(b)	0.25

(a) Cycles to failure. (b) Cycles without failure. (c) Axial shortening of bars during welding

steel. The size of section that can be friction welded depends somewhat on the distance the plastic metal must travel to be extruded from the weld interface. Metal in solid bars must travel outward from the center of the bar; metal in tubes can travel both inward and outward from the center of the wall.

Metals Welded

Friction welding can be used to join almost any metal that can be forged and that is also not a good dry-bearing metal. The alloying elements that provide dry lubrication (or do not seize under normal operating conditions when used without grease or other lubricants) prevent the interfaces from being heated to welding temperature by friction. Metals that contain free-machining additives are likely to be hot short or exhibit a lower than expected strength due to reorientation of the inclusions to the short transverse direction at the weld and are generally unsatisfactory for welding.

Many similar and dissimilar metal combinations can be friction welded, and in most combinations, a sound metallurgical bond is formed. In some combinations, the bond is not as strong as the base metal, and postweld heat treatment may be needed to develop full weld zone strength in alloy steels and hardenable stainless steels.

Carbon and alloy steels are relatively easy to friction weld. Low- and medium-carbon steels can be welded under a wide range of welding conditions. High-carbon and alloy steels are easily joined; however, the welding conditions must be controlled within narrower ranges than are permissible for welding low-carbon steels, and the axial pressure must be increased to compensate for lower forgeability.

High-speed tool steel can be welded to carbon and alloy steel shanks for making drills, reamers, and other cutting tools. Steel balls made of 52100 steel, which is normally difficult to weld, are welded to one or both ends of carbon steel rods to make linkage rods. Frequently, the rods are made of 1045 steel, and one end is induction hardened before the 52100 steel ball is welded to the opposite end. The weldments are tempered after welding.

Free-machining steels, except those having a high sulfur and low manganese content, can be welded, but the free-machining elements result in undesirable directional properties in the weld zone. Friction welds in free-machining steels have fatigue strength less than 80% of the base metal and should not be used in applications where high stresses are involved and high fatigue strength is required. Friction welding of free-machining steels should be limited to those with 0.08 to 0.13% S, Pb, or Te. For example, 1141 steel, which has a sulfur content of 0.08 to 0.13%, can be welded satisfactorily.

Heat treated steels can be friction welded with only localized changes in hardness because the heating is confined to a very narrow zone. Also, the rapid self-quenching restores hardness to the weld zone. For instance, as shown in Fig. 3, when 8630 steel bars were hardened to 35 to 50 HRC and then were friction welded, the minimum hardnesses (33 HRC for the 50 HRC steel; 31 HRC for the 35 HRC steel) occurred at about 0.065 in. from the bond line; the HAZ extended about 0.24 in. into the base metal. In welded hardenable steel, the weld upset usually reaches a high hardness because of rapid self-quenching. Therefore, the weld upset can be removed by grinding or machining immediately after welding on equipment with flash removal services. In some applications, the upset is removed by hot shearing before it has cooled from the welding temperature to the M_s temperature.

Sintered metals are further compacted during FRW and have a wrought structure in the weld zone. The joint usually is stronger than the base metal. The results of friction welding steel forgings and steel casting are about the same as those of friction welding steel bars of similar composition.

Stainless steels are comparatively easy to friction weld. The same guidelines apply with respect to hardenability and free-machining additions as have been discussed for plain carbon and alloy steels. Both austenitic and ferritic stainless steels are commonly joined to themselves, each other, other types of steels, and even copper and aluminum.

Fig. 3 Relation of metal hardness in HAZ to distance from the bond line after FRW. Bars for upper curve postweld tempered at 400 °F; those for lower curve at 1000 °F

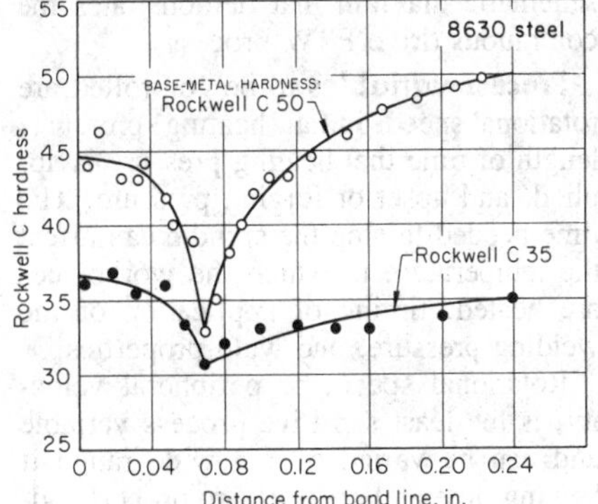

Fig. 4 Pump shaft joined by inertia drive friction welding 1018 steel to stainless steel

1018 steel welded to stainless steel (0.05 C, 0.75 Mn, 1.0 Si, 20 Cr, 29 Ni, 2.2 Mo, 3.2 Cu, rem Fe)

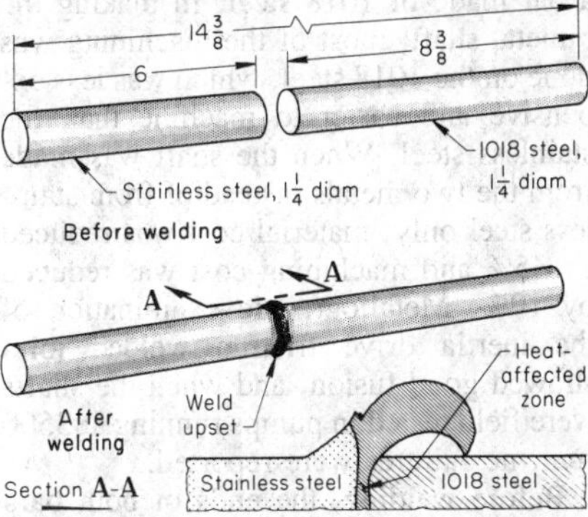

Conditions for inertia drive FRW	
Flywheel moment of inertia	50 lb · ft^2
Spindle speed	3150 rpm
Weld energy(a) ft · lb	84 000 ft · lb
Axial force	60 000 lb
Heat-and-weld time	2-4 s
Weld area	1.23 in.^{2}
Metal lost (max)(b)	5/16 in.
Production rate, welds/h	
Manual	120
Automatic	360

(a) Calculated from flywheel size (moment of inertia) and spindle speed. (b) Total axial shortening of the workpieces during welding

Since the heat treatable stainless steels are sensitive to heat and pressure, a postweld thermal treatment may be required to achieve optimum properties. Precipitation-hardenable stainless steels are often solution-treated and aged after welding, although in some applications heat treated stock is used with only a stress-relief after friction welding.

Care should be taken to locate the weld zone outside a corrosive service environment if practical. It has been suggested that an increased corrosion rate occurs near the weld. The "end grain corrosion" could be promoted by the slow characteristics of the interface material extruded during the welding process.

An important application of FRW of stainless steel is the production of bimetal shafts that are exposed to a corrosive atmosphere or to wear in service. To provide the correct type of resistance where needed, to reduce work-metal cost, or to increase machinability, an alloy that will withstand a corrosive atmosphere can be joined to a metal that is less expensive and easier to machine (Example 3).

Example 3. Manufacturing Pump Shafts by Machining Stainless Steel Inertia Weldments. The production of the pump shaft shown in Fig. 4 from one piece of stainless steel (0.05 C, 0.75 Mn, 1.0 Si, 20 Cr, 28 Ni, 2.2 Mo, 3.2 Cu, rem Fe) was compared with the production of the shaft by inertia drive friction welding a bar made of the same stainless steel to a bar made of 1018 steel. In making the bimetal shaft, most of the machining was done on the 1018 steel, which was less expensive and easier to machine than the stainless steel. When the shaft was made from the two metals instead of from stainless steel only, material cost was reduced by 45% and machining cost was reduced by 10%. Metallographic examination of the inertia drive friction welded joint showed good fusion, and when the shafts were field tested in pumps running at 3500 rpm, no failures were reported.

Before welding, the ends of both bars were ground to remove any mill scale. After welding, weld upset was removed as the shaft was machined to size. To produce good welds, leaded low-carbon steel was not used.

Cast iron in any form—gray, ductile, or malleable—has not been friction welded satisfactorily in production. Joining of ductile iron to steel in laboratories has been reported. Free graphite gathers at the interface and acts as a lubricant, which limits its friction heating. Also, these materials are not forgeable, which is a general requirement for FRW.

Nonferrous Metals and Alloys. Aluminum alloys are friction welded to similar and dissimilar aluminum alloys, copper alloys to similar and dissimilar copper alloys, and aluminum alloys to copper alloys. Most applications of friction welding these metals are in joining aluminum and copper alloys to steel, although problems are presented by high thermal conductivity, large differences in forging temperatures, and the formation of brittle intermetallic compounds.

Joints between aluminum alloy 6061 and copper have a tensile strength near that of the copper. Joints between aluminum alloy 1100 and stainless steel have a strength near that of the aluminum alloy. Friction welding of other aluminum alloys may develop a joint strength of only 60 to 70% that of the weaker base metal. Even though these joints are relatively weak, they are useful for pressure sealing and for joining assemblies that require good electrical and thermal conductivity, rather than high strength.

Titanium, titanium alloys, zirconium alloys, and magnesium alloys can be friction welded to themselves. Most nickel-based and cobalt-based alloys, including the heat-resistant alloys, are easily friction welded to themselves and to alloy steels. The nickel-based alloy GMR-235 can be welded to 1040 steel, Inconel 718 to Inconel 713C, and Inconel 713 to 8630 steel in producing jet engine parts that require high-strength bonds.

The refractory metals—tungsten, molybdenum, niobium, and tantalum—can be welded to themselves. Friction welds between molybdenum rods are ductile enough to withstand substantial reduction by wire drawing.

Continuous Drive Friction Welding

Continuous drive FRW, which is also referred to as direct drive FRW, requires a machine resembling an engine lathe equipped with an efficient spindle braking system, a means of applying and controlling axial pressure, and a weld cycle timer and control. The equipment is simple in principle, but the machines are complex when of a size capable of welding large workpieces.

Principles of Operation. The workpiece to be rotated is clamped in the spindle chuck, and the spindle is brought to a predetermined speed. The nonrotating component is clamped in a chuck or fixture mounted to a hydraulically actuated tailstock slide. To heat the workpieces to welding temperature, the tailstock slide is advanced to bring the workpieces in contact under a constant or gradually increasing axial pressure. When the workpieces are at, or slightly above, the welding temperature, the spindle brake is applied, which suddenly stops the spindle rotation. Simultaneously, the tailstock pressure is increased to complete the weld.

The spindle speed, axial pressure, and length of time the pressure is applied for a given weldment depend on (1) the cross-sectional area of the workpieces to be welded, (2) the melting point and thermal conductivity of the work metal, and (3) the metallurgical changes that occur during the heating cycle, particularly when dissimilar metals are being welded. Figure 5 is a schematic diagram that demonstrates the continuous drive FRW process.

Process variables to be controlled are rotational speed, initial (heating) pressure, length of time that heating pressure is applied, and upset or forging pressure. The time needed to stop the spindle can affect the temperature to which the workpieces are heated, timing of application of the welding pressure, and weld properties.

Rotational speed, or peripheral velocity, is the least sensitive process variable and can be varied over a wide range if heating time and pressure are properly adjusted. However, heating time must be limited to prevent excessive depth of heating. The peripheral velocity recommended for welding most low-, medium-, and high-carbon steels is 250 to 750 sfm.

Fig. 5 Schematic representation of continuous drive FRW. Consists of friction, stopping, and forge phases

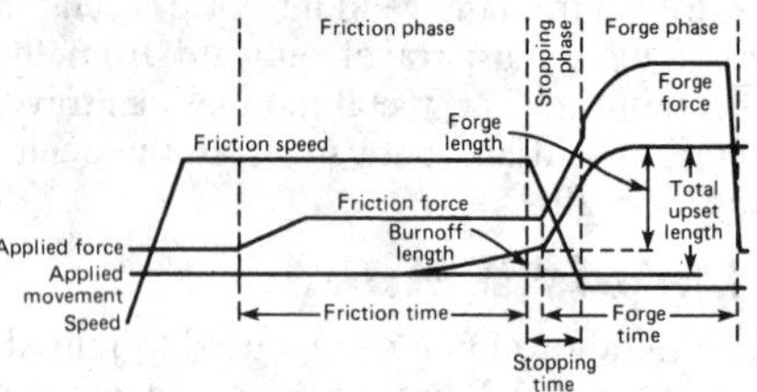

Heating pressures used for welding low-carbon and low-alloy steels are from 6 to 12 ksi. Welding pressures for these steels are from 12 to 25 ksi. Usually the welding pressure is higher than the heating pressure, but sometimes they are nearly the same. For medium- and high-carbon steels, heating pressures are from 6 to 15 ksi, and welding pressures are from 15 to 60 ksi. Preheating pressures are sometimes used prior to the actual heating pressure for large workpieces so that the power requirements do not exceed the capacity of the welding machine.

Heating time varies with the heating pressure, the carbon and alloy content of the steel, and the diameter of the workpiece. Usually heating time is determined by trial.

The spindle should be stopped rapidly to keep the weld from twisting or tearing. For a workpiece less than $1/2$ in. in diameter, stopping time should be within $1\,1/2$ s; a 3-in.-diam bar should be stopped within $1/2$ to 2 s.

Applications. Only minor alterations in design are needed to adapt to FRW a workpiece that previously was butt welded by other processes. Generally, in FRW, less metal is lost during heating and upsetting than in flash welding, and it is not necessary, as it is in flash welding, to machine the interface so that heating will start at the center of the workpiece section. Where allowances are made in the size of a forging for differences in metal lost and for welding in the as-forged condition, forging costs can be reduced.

Workpieces made of dissimilar metals are frequently used to minimize costs while providing a work metal that meets the necessary service requirements. Many corrosion- and heat-resistant alloys are welded to less expensive alloys to reduce material and machining costs, or to provide a wear-

Fig. 6 Diesel engine exhaust valve

Alloy 2112N welded to 4140H steel

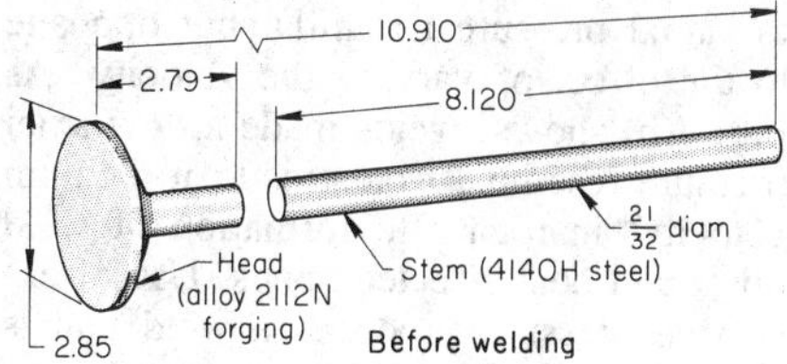

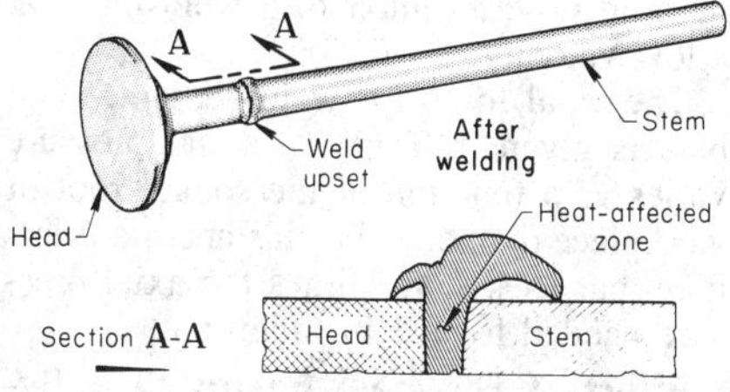

resistant surface (see Examples 3 and 4).

Example 4. Friction Welding Dissimilar Metals. A large exhaust valve for a diesel engine, shown in Fig. 6, was made by continuous drive friction welding a head made of a heat-resistant alloy to a low-alloy steel stem. This permitted using the more expensive (heat-resistant) alloy only where required. The head was forged from alloy 2112N (a 21Cr-12Ni austenitic iron-based alloy used for exhaust valves), and the stem was made of 4140H steel $^{21}/_{32}$ in. in diameter. The initial weld area was 0.34 in.^{2} Overall length of the part after joining by FRW was 10.712 in.

A 25-hp continuous drive FRW machine was used, although the peak requirement was only 5 hp. In operation, the valve head was clamped in an air-actuated fixture mounted in the tailstock and the stem in an air-actuated chuck attached to the machine spindle.

The machine cycle was started, and the spindle was rapidly accelerated to 2700 rpm as the tailstock advanced to bring the workpieces into contact. The pressure of contact initiated a signal in the control circuit that began cycle timing and started a gradual increase in axial pressure by a slope-control unit.

The pressure, applied through the tailstock by two hydraulic cylinders, was smoothly increased to 11 ksi in 4 s. The pressure level was maintained for an additional 4 s, during which time the workpieces were heated to welding temperature. At the end of the 4-s heating period, the braking system quickly stopped spindle rotation, and a welding pressure of 32 ksi was applied. The resulting weld had a good metallurgical bond that was free of oxides, cast metal structure, porosity, inclusions, and other defects. Total axial shortening of the workpieces during welding was 0.195 to 0.200 in.

The weld upset was removed in an automatic lathe prior to heat treating of the valve. After heat treatment, the stem was ground to finished diameter. The production rate, with an unskilled operator, was 190 to 200 pieces per hour. The total input energy per weld was 6900 W · s, with a peak power draw of 5 hp, including all losses.

Inertia Drive Friction Welding

Inertia drive FRW makes use of the kinetic energy of a freely rotating flywheel for all of the heating required to produce a weld.

Equipment. The welding machine is similar in construction to a continuous drive friction welding machine, with the rotating spindle mounted in the headstock. It is capable of moving either the tailstock or headstock relative to each other to apply the necessary welding force. The spindle is generally driven by a hydraulic motor; however, direct current or alternating current drives (for small flywheel masses) can also be used. Flywheel size (moment of inertia of the flywheel or spindle) is adjusted by adding or removing flywheel disks. The spindle speed and axial pressure are adjusted by dials on the control panel.

Principles of Operation. In inertia drive FRW, as in the continuous drive method, one workpiece is clamped in a nonrotating vise or fixture, and the other workpiece is clamped in a chuck mounted to a rotating spindle. The drive motor accelerates the rotation of the flywheel spindle assembly to a predetermined speed (energy level), and then the rotating drive power is shut off. The parts to be welded are brought together, and the kinetic energy of the freely rotating flywheel is rapidly converted to heat at the weld interface as axial pressure is applied. The weld itself is the brake, consuming all the stored energy. No other outside force (clutch brake, etc.) influences the weld. Once the axial pressure, flywheel moment of inertia, and spindle speed have been established for a given workpiece, uniform welds are produced repetitively.

During spindle acceleration the flywheel acts as a capacitor, storing energy from the spindle drive motor. The drive motor and transmissions, therefore, need not be sized for the weld. When the correct speed has been reached, acceleration is discontinued and the stored energy is discharged rapidly into the weld. The amount of power which is consumed during welding can be varied by the axial weld pressure. High-pressure welds consume the energy faster (high power) than low-pressure welds (low power) even if the stored energy is the same. By rapid application of small amounts of energy (high-power welds), inertia drive friction welding produces a narrow HAZ.

In inertia drive FRW, intense hot working of the weld zone in conjunction with rapid cooling immediately after hot working results in a very small grain size in the as-welded condition. Subsequent heat treatment, if necessary, will restore the grains to their normal size.

Since inertia drive friction welds are independent of the drive motor, large tubular welds can be made with the same specific power consumption per welded area as with small tubes. The mathematical models used to determine the weld parameters of the smaller diameter welds also pertain to the larger diameter welds. To weld a 27-in.-OD tube-to-tube mild steel joint, with a wall thickness of 0.5 in., 2 060 000 ft · lb of energy would be used. The drive motor would be rated at 150 hp. The energy consumed (weld time) would be 4.2 s. To have the equivalent input power on a continuous drive friction welder, a motor of about 900 hp would have to be used or the weld time would have to be increased to accommodate a smaller drive motor. This, however, increases the chance for grain growth.

Process Variables. Three variables control the characteristics of an inertia friction drive weld: initial peripheral velocity of the rotating piece (rpm on a machine setting), flywheel size (moment of inertia), and axial pressure. Weld energy is a function of rpm and the flywheel:

$$E = \frac{\text{rpm}^2 \times Wk^2}{C}$$

where E is energy in ft · lb; rpm is rotation per minute of the flywheel/chuck assembly; Wk^2 is moment of inertia in lb · ft^2; and C is the constant for converting to energy (5753).

Since the flywheel is fixed on the machine, the energy becomes directly proportional to the square of the rpm. Because most materials have a wide welding range of peripheral speeds, it is not necessary to change flywheels to accommodate changes for different piece parts. A small change in rpm will automatically accomplish a large change in energy. For most materials and geometries, specific weld energy and pressures have been established. Additional energy may be needed if the surfaces to be welded are rough or out of square with the axis of rotation. Very

Table 2 Conditions for inertia drive friction welding 1-in.-diam bars in combinations of similar and dissimilar metals

Work metal	Welding conditions: Spindle speed, rpm	Axial force, lb	Flywheel size(a), lb · ft²	Resultant weld conditions: Weld energy, ft · lb	Metal lost(b), in.	Total time(c), s
Metals welded to themselves						
1018 steel	4600	12 000	6.7	24 000	0.10	2.0
1045 steel	4600	14 000	7.8	28 000	0.10	2.0
4140 steel	4600	15 000	8.3	30 000	0.10	2.0
Inconel 718	1500	50 000	130.0	50 000	0.15	3.0
Maraging steel	3000	20 000	20.0	30 000	0.10	2.5
Type 410 stainless steel	3000	18 000	20.0	30 000	0.10	2.5
Type 302 stainless steel	3500	18 000	14.0	30 000	0.10	2.5
Copper, commercially pure	8000	5 000	1.0	10 000	0.15	0.5
Copper alloy 260 (cartridge brass, 70%)	7000	5 000	1.2	10 000	0.15	0.7
Titanium alloy, Ti-6Al-4V	6000	8 000	1.7	16 000	0.10	2.0
Aluminum alloy 1100	5700	6 000	2.7	15 000	0.15	1.0
Aluminum alloy 6061	5700	7 000	3.0	17 000	0.15	1.0
Dissimilar-metal combinations						
Copper to 1018 steel	8000	5 000	1.4	15 000	0.15	1.0
M2 tool steel to 1045 steel	3000	40 000	27.0	40 000	0.10	3.0
Nickel alloy 718 to 1045 steel	1500	40 000	130.0	50 000	0.15	2.5
Type 302 stainless to 1020 steel	3000	18 000	20.0	30 000	0.10	2.5
Sintered high-carbon steel to 1018	4600	12 000	8.3	30 000	0.10	2.5
Aluminum 6061 to type 302 stainless	5500	5 000	3.9	20 000	0.20	3.0
		15 000	...	...	...	...
Copper to aluminum alloy 1100	2000	7 500	11.0	7 500	0.20	1.0

(a) Moment of inertia of the flywheel. (b) Total axial shortening of the workpieces during welding. (c) Includes heat time and weld time. (d) The 5000-lb force is applied during the heating stage of the weld; force is increased to 15 000 lb near the end of the weld.

high energy input causes excessive loss of metal, but generally does not affect the strength and quality of the weld. Table 2 gives representative conditions for inertia drive FRW of 1-in.-diam bars of various metals and alloys, in similar and dissimilar combinations.

Peripheral Velocity of Workpiece. For each combination of work metals, there is a range of peripheral velocity that produces the best weld properties. For welding steel to steel, the recommended initial peripheral velocity of the workpiece ranges from 500 to 1500 sfm; however, welds can be made at velocities as low as 275 sfm. As illustrated in Fig. 7(a), low velocities (less than 300 sfm) can reduce center heating and produce rough, uneven weld upset. At medium velocities (300 to 900 sfm), the heating pattern in steel has an hourglass shape at the lower value and gradually flattens as the upper velocity is approached. The heating pattern is essentially flat and uniformly thick across the workpiece at velocities of 900 to 1200 sfm. At high initial velocities (above 1200 sfm), the weld becomes rounded and is thicker at the center than at the periphery.

Spindle speeds in revolutions per minute for inertia drive friction welding 1-in.-diam bars of various metals and alloys are given in Table 2. The relationship of total time and weld upset to bar diameter is approximately linear.

Axial Pressure. The effect of varying the axial pressure is similar but opposite to the effect of varying the velocity. As Fig. 7(b) shows, welds made at low axial pressure resemble welds made at medium velocity, in regard to formation of weld upset and heat-affected zones. Use of excessive pressure produces a weld that is poor at the center and has a large amount of weld upset, similar to a weld made at a low velocity.

The axial force for welding 1-in.-diam bars is given in Table 2. Axial pressure varies as a function of the square root of workpiece diameter. For instance, a 2-in.-diam bar uses 1.414 times the axial pressure needed for a 1-in.-diam bar.

Effect of Flywheel Energy. The flywheel moment of inertia is selected to produce the desired amount of kinetic energy and the desired amount of forging. Forging results from the characteristic increase in torque that occurs at the weld interface as the flywheel slows and comes to rest. This increased torque, in combination with the axial pressure, produces forging. Because forging begins at some critical velocity (about 200 sfm for low-carbon steel),

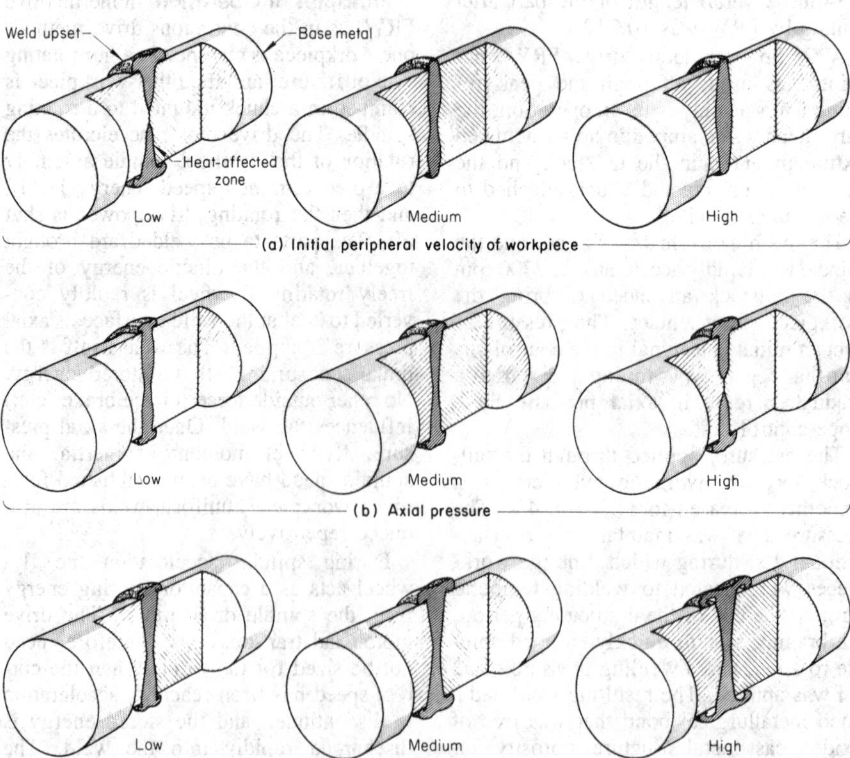

Fig. 7 Effect of welding variables on depth and uniformity of heating and on size and shape of weld upset

the amount of forging depends on the amount of energy remaining in the flywheel, which is a linear function of the flywheel moment of inertia. Large, low-speed flywheels produce greater forging than small, high-speed flywheels, even though they contain the same amount of kinetic energy. Although low, medium, and high amounts of flywheel energy produce similar heating patterns, the amount of energy greatly affects the size and shape of the weld upset, as shown in Fig. 7(c).

Applications. Inertia drive FRW is used in the manufacture of bimetal exhaust valves for internal combustion engines, bimetal shafts for pumps, and cluster ring gears for automotive and aircraft applications, oil drill pipes, aircraft engine rotor assemblies, and turbochargers. The process has been incorporated into the redesign of many parts to reduce costs or improve service life.

Example 5. Inertia Drive Friction Welding Versus Casting or Shrink-Fit Assembly. A high-production part, consisting essentially of a $1^1/_4$-in.-ID cylinder with a bolting flange, served as a suction valve cover. The manufacturing process and design of the part evolved through three successive stages, as shown in Fig. 8. The first stage was a one-piece casting (Fig. 8a); the second, a shrink-fitted assembly (Fig. 8b); and the third, the inertia drive friction welded assembly shown in Fig. 8(c). The slight differences among the three designs are the results of changes in the design of the mating piece. The inertia drive friction welded component was used as welded, without removing weld upset from either the inner or the outer surface of the cylinder.

Originally, the part was machined from a gray iron casting (Fig. 8a). Some difficulty was experienced in obtaining the desired finish on the inner surface of the cylinder, but the most serious problem was rusting of the inside of the cylinder. Attempts were made to improve corrosion resistance by burnishing the cylinder wall, and then plating with cadmium, zinc, or tin, or phosphate coating. Although phosphate coatings were best on the cast iron surface, none of these coatings proved satisfactory in service.

The second method consisted of using a low-carbon steel plate that was bored to accept a shrink-fitted cylinder (Fig. 8b) for the flange. Cylinders were made of anodized aluminum alloy or of stainless steel. When the anodized aluminum alloy was used, the cost per part was slightly higher than that of the gray iron casting. Cylinders made of type 416 stainless steel gave excellent results, but cost considerably more than the castings.

In the final design, shown in Fig. 8(c), inertia drive FRW was used to join a type 416 stainless steel tube to a 1020 steel plate. Valve covers made by this method cost 15% less than the original cast iron covers, and the corrosion and surface finish problems were eliminated.

For inertia drive FRW, flanges were gas cut from 1020 steel plate, and the cylinders were saw cut from type 416 stainless steel tubing. Both components were finish machined, and the joint surfaces were cleaned carefully before welding.

Conditions for FRW are given in the table with Fig. 8. With these settings, an upset of 0.060 ± 0.006 in. was a good indication of an acceptable weld. Part specifications required that a 100-psi pressure test be applied to 5% of the weldments and that 1% of the weldments be examined for weld configuration by sectioning. The test samples were selected on a random basis. Under these conditions, the rejection rate was less than 0.3%, and most rejections were caused by defects in the base metal rather than in the weld.

Example 6. Forging Versus Inertia Drive Friction Welding To Produce a Shaft. The long, slender power-control drive shaft shown in Fig. 9 originally was made from a 1045 steel forging, but was severely warped during forging and heat treating. Nineteen manufacturing operations were needed to complete the forged shaft, including hardening and tempering of the forging to 23 to 30 HRC, cleaning by shot blasting, three straightening operations, and ten machining operations.

The number of operations was reduced to ten when forging was replaced by inertia drive FRW. A spline blank of 1045 steel 1.938 in. in diameter by 1.19 in. long was inertia drive friction welded to a 1045 steel shaft 40.12 in. long (Fig. 9). The shaft had a hardness of 27 HRC and, before welding, was ground and polished to 0.996 in. in diameter. The jaws for holding the shaft were designed so as not to damage the ground and polished surface.

Hardening, tempering, shot blasting, straightening, and three grinding opera-

Fig. 8 Suction valve cover produced by three processes

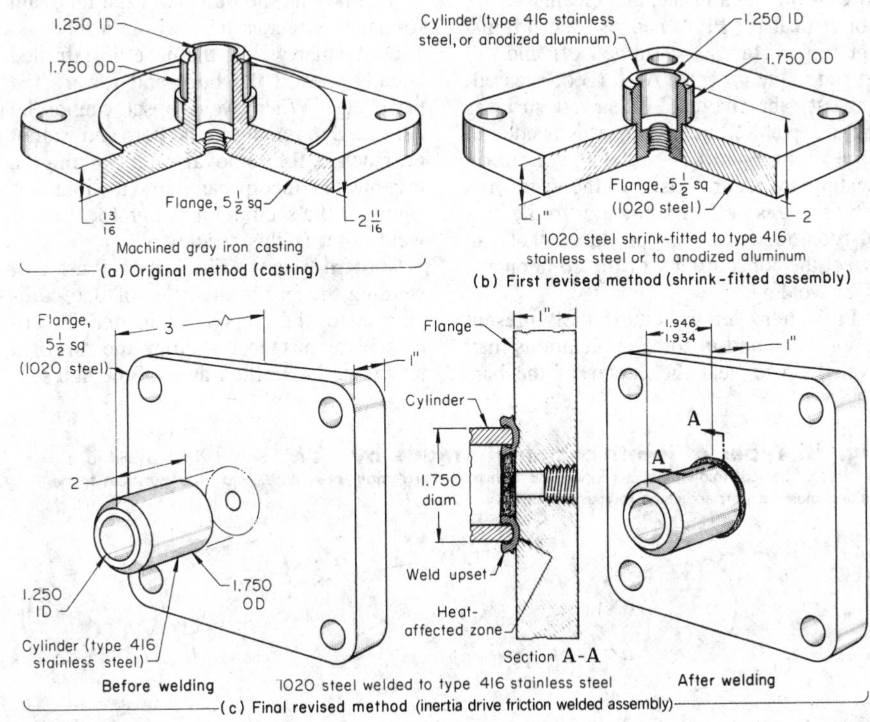

Conditions for inertia drive FRW

Machine capacity		Spindle speed	3 900 rpm
Part diameter	11 in.	Weld energy(a)	51 800 ft · lb
Spindle speed (max)	8 000 rpm	Axial force	35 000 lb
Axial force (max)	45 000 lb	Weld area	1.18 in.^{2}
Flywheel moment of inertia	20 lb · ft^2	Metal lost(b)	0.060 ± 0.006 in.

(a) Calculated from flywheel size (moment of inertia) and spindle speed. (b) Total axial shortening of the workpieces during welding

Fig. 9 Drive shaft manufacturing at reduced costs by inertia drive FRW

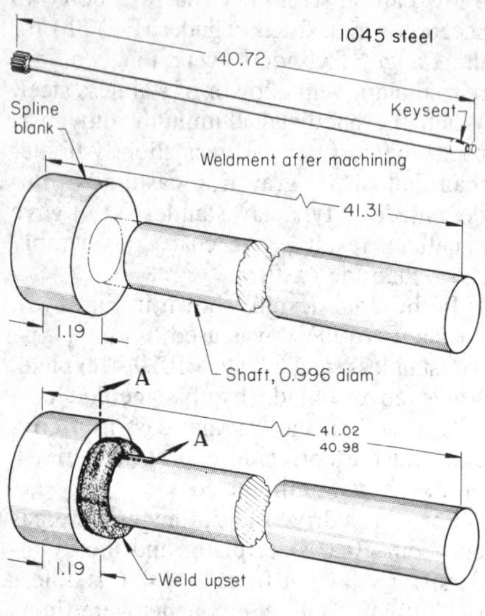

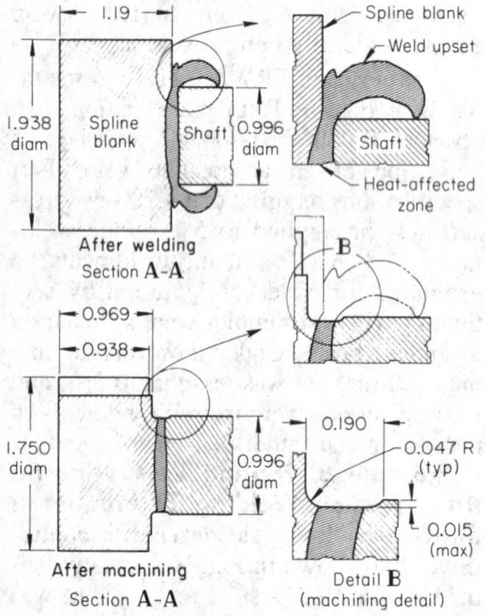

Conditions for inertia drive FRW(a)

Flywheel moment of inertia	150 lb · ft^2
Spindle speed	1850 rpm
Weld energy(b)	87 500 ft · lb
Axial force	21 000 lb
Weld area	0.78 in.^{2}
Metal lost(c)	0.31 ± 0.02 in.
Production rate	70 welds/h

(a) In this application, lower-than-normal spindle speeds and larger-than-normal flywheels were used for convenience in changeover from one workpiece to another. (b) Calculated from flywheel size (moment of inertia) and spindle speed. (c) Total axial shortening of the workpiece during welding

tions were eliminated when forging was replaced by inertia drive FRW. Three operations—hardening a bearing surface, deburring, and washing—were included in both manufacturing sequences. Machining of the spline end of the weldment included removing the weld upset and relieving the area to a minimum diameter of 0.966 in. and to a width of 0.19 in., which eliminated a possible stress raiser.

Conditions for inertia drive FRW are given in the table with Fig. 9. These conditions were selected to produce a weld upset of 0.31 ± 0.02 in., which was greater than usual, so that the upset could be machined in the as-welded condition. Failure of the shaft during fatigue testing occurred at the keyseat, not in the weld.

Joint Design

The mechanics of FRW restrict its use to flat and angular butt welds that are perpendicular to and concentric with the axis of rotation. Flat joints are the most common and can be classified as (1) bar to bar, (2) bar to tube, (3) tube to tube, (4) bar to plate, and (5) tube to plate, as shown in Fig. 10. These classifications refer to the joint itself and not to the shape of the parts. The joint used in making the rotor bodies described in Example 1 is classified as a bar-to-plate joint because a 0.62-in.-diam rod was joined to a 2-in.-diam rod.

Joint surface conditions, such as surface finish, squareness, and cleanness, are not critical for FRW because the original abutting surfaces are rubbed off and extruded in the process. As-forged, sheared, gas-cut, abrasive-cut, or sawed surfaces are acceptable, but extra heat is needed to remove irregularities and allow uniform heating to occur. Also, if the face of a workpiece is not perpendicular to the axis of rotation, forces are produced that can affect the concentricity of the components after welding.

Projections left by cutoff tools present no problem and in some applications may even help to heat the center of the bar. However, center drill holes must be avoided; when the upset metal compresses the air entrapped in a center drill hole during upsetting, a weld defect usually occurs.

Heavy mill scale, thick chromium plating, or a thin carburized or nitrided case acts as a bearing surface and cannot be extruded from the weld area. When deeply carburized workpieces are to be welded, the joint surface and the adjacent surfaces must either (1) be machined before welding (or before hardening, if it precedes welding) to remove the carburized metal; or (2) be copper plated or otherwise covered during the carburizing operation to prevent them from being carburized. In Example 1, the end to be welded was copper plated before carburizing.

Tubular Welds. Tube-to-plate welds are not as strong as tube-to-tube welds, because rounding of the end of the tube during heating and upsetting reduces weld effectiveness. Fatigue life of a tube-to-plate weldment increases with removal of the sharp notch at the tube base before the part is put to use. On some metals, when heating is too slow, or when excessive weld energy is applied, the upset metal flows up and around the outside of the tube and forms, in effect, a tube within a tube.

On tubular welds, the upset is extruded equally toward the bore and toward the periphery. When weld upset cannot be permitted to remain in the bore and cannot be reached for removal after welding, a trap must be incorporated into the joint design (see the section on design for flow of weld upset in this article).

Conical joints (Fig. 10) are used in welding the inside diameter of one component to the outside diameter of another—for instance, welding the rim of a jet engine fan to the flange of the hub sec-

Fig. 10 Types of joints commonly made by FRW. Bars and tubes and joint surfaces of conical joints must be concentric with axis of rotation; bar, tube, and plate surfaces to be welded must be perpendicular to axis of rotation.

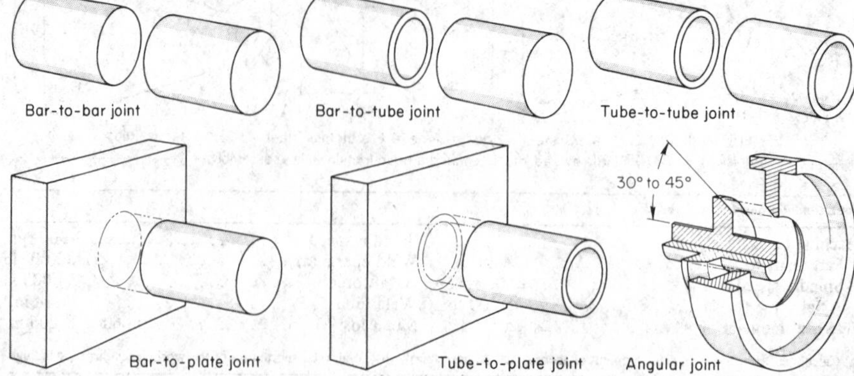

tion or welding a flange to a shaft at some position between the two ends. The joint can be made where a chamfered shoulder can be mated with an equally chamfered bore in the flange. The interfaces of conical (angular) joints must have equal included angles, and the tapered bore in the outer component must have sufficient strength to withstand the axial pressure required to make the weld. Conical joints are usually designed with faces 30 to 45° from the centerline (an included angle of 60 to 90°) to prevent one part from being pushed through the hole in the other part. Some nickel-based alloys have been welded using smaller angles.

Design for Heat Balance. When similar metals or dissimilar metals having about the same forging temperature and thermal conductivity are friction welded, there are no restrictions on the relationship of the cross-sectional area of a bar or tube to the size of plate to which it is welded. Heating rates as high as 100 000 °F/s make the weld cycle very short and keep heat losses to the cold metal in the plate adjacent to the weld very low. Thus, bar-to-plate and tube-to-plate welds are feasible for joining similar and some dissimilar metal combinations.

In welding dissimilar metals with widely different forging temperatures and thermal conductivities, adjustment of size or area adjacent to the weld interface may be necessary in one of the workpieces. Area differentials must be determined experimentally because each metal reacts differently. For instance, in welding a nickel-based alloy shaft 1 in. in diameter to an alloy steel shaft, the alloy steel shaft should be $^1/_{16}$ to $^1/_8$ in. in diameter larger than the nickel-based alloy shaft. If the components are tubes, the inside diameter of the alloy steel tube should be $^1/_{16}$ to $^1/_8$ in. less than that of the nickel alloy tube. Thus, the alloy steel tube would have a smaller inside diameter and a greater wall thickness and outside diameter than the nickel alloy tube. In making a cutting tool blank, a tool steel shank can be readily welded to a tool steel body using a bar-to-plate joint, but if the shank is changed to alloy steel, the tool steel body must be modified to produce a bar-to-bar joint.

When a tube is welded to a thin plate with a hole of the same diameter as the inside of the tube, the hole in the plate frequently is made smaller than the inside diameter of the tube to avoid excessive heating of the plate around the hole. This difficulty does not occur with thick plates because the metal around the hole is not heated through.

Design for Flow of Weld Upset. In applications where the presence of weld upset on one or more work-metal surfaces is undesirable and where the upset cannot be removed after welding, traps can be incorporated into the joint design to provide clearance for the flow of weld metal from the interface. When a plate is welded to a shaft extending through a hole in the plate, or a boss on one workpiece extends into a hole in the other workpiece, enough clearance must be provided to prevent rubbing of the shaft or boss against adjacent surfaces. Rubbing of adjacent surfaces parallel to the axis of rotation uses flywheel energy unpredictably and diminishes reproducibility of the weld.

Typical designs of traps for weld upset are shown in Fig. 11. An assembly of a headed shaft and a plate is shown in Fig. 11(a) and (b). In Fig. 11(a), weld upset could not be tolerated on the surface of the plate opposite the joint and could not be removed after welding. A counterbore in the hole in the plate, extending inward from the joint surface of the plate, served as a trap for weld upset, which was permitted to flow out around the head of the shaft. In Fig. 11(b), weld upset was not permitted on the joint surface of the plate, and the trap was formed in the head of the shaft. A surface on the head touched the plate as rotation ceased and provided a seal.

An end-cap-to-tube weld is shown in Fig. 11(c). A tight internal corner joint free of weld metal was obtained by making the trap in the boss on the end cap. In Fig. 11(d), the step on the joint end of the shaft provided the trap for weld upset, leaving the outside corner formed by the intersection of the plate and the shaft free from upset metal.

Parts that require difficult-to-machine internal recesses or chambers can be made by joining two specially designed and machined parts. Removal of weld upset from the inside surfaces of such parts is difficult; therefore, the joint must be designed with a trap so that weld upset cannot form on the internal surface.

Example 7. Inertia Drive Friction Welding of a Precombustion Chamber. Precombustion chambers for diesel engines (Fig. 12) originally were made by furnace brazing two sections. The brazed joint was subjected to precombustion temperatures as well as a combustion pressure of about 1.7 ksi. Precise machining of the joint surfaces and careful inspection of the parts were needed.

When furnace brazing was replaced by inertia drive FRW, close joint tolerances and meticulous inspection procedures were no longer needed, and product quality was improved. The mating surfaces were designed to contain the internal weld upset so that it could not flow into the chamber.

The two components were made of leaded 5120 steel in a multiple spindle bar machine. The outside diameter of the chamber at the weld interface was 1.875 in., and the width of the weld interface was about 0.22 in. A small enclosed internal cavity, shown in Fig. 12 (section A-A, after fit-up), contained the internal weld

Fig. 11 Recesses used as traps for weld upset in friction welded joints

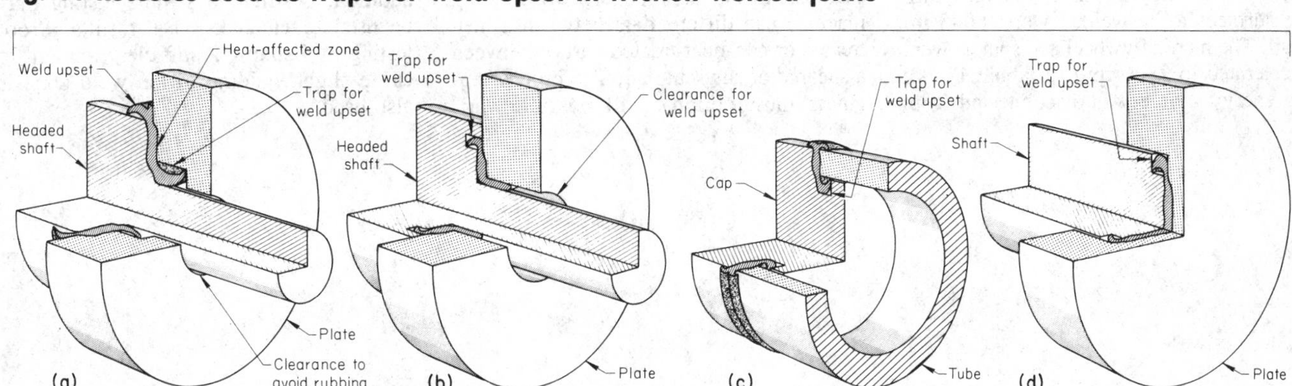

Fig. 12 Components of diesel engine precombustion chamber. Joint was designed with a trap to prevent formation of weld upset on internal surface of chamber.

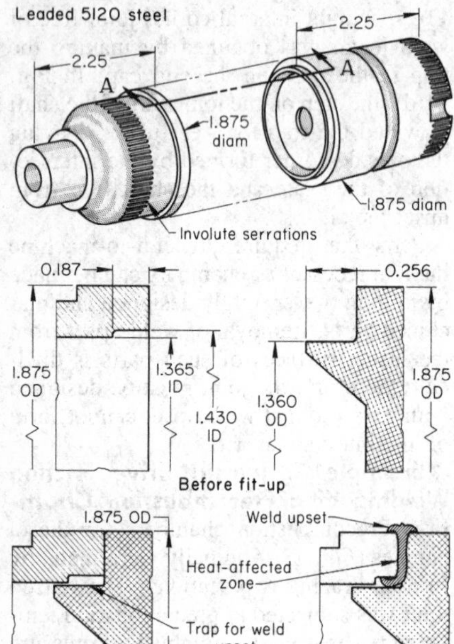

Conditions for inertia drive FRW	
Flywheel moment of inertia	10.75 lb · ft²
Spindle speed	2800 rpm
Weld energy(a)	14 300 ft · lb
Axial force	17 000 lb
Heat-and-weld time	0.7 s
Machine cycle time	4 s
Weld area	1.36 in.²
Metal lost(b)	0.020 ± 0.003 in.
Production rate	280 parts/h

(a) Calculated from flywheel size (moment of inertia) and spindle speed. (b) Total axial shortening of the workpieces during welding

upset and prevented it from flowing into the chamber.

In operation, the two sections were clamped in their respective collet chucks, and the tailstock was moved forward until the surfaces to be welded were 0.045 in. apart. The inertia flywheel and spindle were accelerated to 2800 rpm (in about 1.2 s); the energy source was disconnected from the spindle, and immediately the two sections were brought into contact under a force of 17 000 lb, which remained constant until rotation stopped. Heat-and-weld time was 0.7 s. Strength of the welded joint was equal to that of the base metal. Machine cycle time, exclusive of handling time, was 4 s. Metal lost was 0.020 ± 0.003 in.

Design for Machine Size. The amount of energy needed for FRW depends partly on the distance metal must be extruded from the innermost point of the interface of the workpieces. Thus, more weld energy and pressure are required, per square inch of weld area, for joining two 1-in.-diam bars than for joining equivalent-area tube.

Some joints (usually in torsional applications) can be designed with a center relief in one or both workpieces to eliminate the nonworking but hard-to-weld metal at the center. This also may permit the part to be welded on a smaller capacity machine or the length of the welding cycle to be reduced. The size of the center relief must be large enough so that compression of the entrapped air will not cause a weld defect and to provide adequate space for the weld upset.

Similarly, a bar-to-plate joint can be redesigned as a tube-to-plate or a bar-to-bar joint if desirable to bring it within the capacity of a particular machine. A bar-to-bar joint requires less energy than a bar-to-plate joint because the plate acts as a heat sink and draws heat away from the weld.

Control of Weld Quality

Weld quality depends as much on the quality of the workpiece material as it does on the machine function. Once extensive physical tests have been completed on weld samples and the optimum welding procedure and parameters have been established, the material specifications must be established and adhered to. No welding machine can differentiate between a heat treated or non-heat-treated part or between a sheared or saw-cut end. The process engineer must eliminate all possible variables prior to welding and specify the following:

- Chemical analysis of material with maximum limitations on nonmetallic inclusions
- Heat treatment, hardness, etc.
- Surface condition: pickled and oiled, sand blasted, machined, maximum allowable paint, rust, or oxides
- End preparations: sheared, saw cut, torch cut, etc.
- Inclusions, cracks, laps, etc.

The FRW machines can be monitored during the weld cycle and high and low limits can be set for tolerance bands, shutdown, or expelling a possibly defective weld. Assuming a continuous drive FRW machine works on a pressure time program, the following items may be monitored:

- Spindle speed
- Heat pressure
- Heat time
- Run downtime
- Forge pressure
- Forge time
- Loss of length during heating or forging

An inertia drive FRW machine can be monitored as follows:

- Spindle speed
- Weld pressure
- Loss of length
- Weld time

Most manufacturers today offer additional monitors or even microprocessors which will monitor all of the above welding parameters as well as weld torque or input power and the rate of loss of length. While the monitors are getting more complicated, there is still no guarantee of good weld quality. Postweld inspection may consist of nondestructive testing (NDT) using liquid-penetrant or magnetic-particle inspection for outer-diameter cracks and ultrasonic testing for center cracks. In addition, destructive sample testing may be used to enhance NDT results. Proof testing by bending a sufficient amount to cause slight yielding in the weld zone is also used.

Stud Welding

By the ASM Committee on Stud Welding*

STUD WELDING is an arc welding process in which the contact surfaces of a stud, or similar fastener, and a workpiece are heated and melted by an arc drawn between them. The stud is then plunged rapidly onto the workpiece to form a weld. Arc initiation, arc time, and plunging are controlled automatically.

The two basic methods of stud welding are known as stud arc welding and capacitor discharge stud welding. Both methods involve direct current and arcing. For stud arc welding, a motor-generator, a transformer-rectifier, or a storage battery provides the power supply. The power supply for capacitor discharge stud welding is a low-voltage electrostatic storage system, and the arc is produced by a rapid discharge of stored electrical energy. A welding current controller and a welding tool or gun (stud gun) complete the necessary welding equipment for both methods of stud welding. In the arc method, a ceramic arc shield known as a ferrule is generally used to shield the arc and retain the molten weld metal.

In both methods, the stud, or fastener, serves as the electrode; the gun is the electrode holder. Flux is generally used for stud arc welding of ferrous alloys and is an integral part of the stud. Flux provides cleaning action, arc stability, and a protective atmosphere. The arc time for capacitor discharge welding is so short that flux is not needed. In welding aluminum alloys by the stud arc welding method, a shielding gas is introduced, but is not required with the capacitor discharge method.

In stud arc welding, direct current electrode negative (stud negative and workpiece positive) is generally used for ferrous alloys, and direct current electrode positive is used for nonferrous alloys. Direct current electrode negative is used in capacitor discharge stud welding except when welding to galvanized and plated surfaces.

Process Capabilities and Limitations

Stud welding is a rapid process. Welding time, which depends on the method and on the diameter of the stud, varies from 1 to 6 ms for the capacitor discharge method and from 0.10 s to slightly more than 1 s for stud arc welding. In both methods, melt-through and distortion are minimal.

Studs can be welded in places that are not readily accessible, and the welded area need not be in view of the operator. Stud welding does not require access to the back of the workpiece. Stud welded fasteners can often replace fasteners normally secured by riveting, drilling, and tapping; manual arc welding; resistance welding; or brazing.

The shank of a stud or other weld fastener can be of any size, shape, or type that can be gripped in the stud holder. Usually, the weld base, which is the end of the stud or other fastener that is to be welded, is round for both processes. However, square- and rectangular-shaped weld-base studs can be welded by the arc method.

Suitable stud and base metals are the same as those welded by other arc welding processes. The metals most frequently stud welded are low-carbon steel, high-strength low-alloy (HSLA) steel, austenitic stainless steel, aluminum alloys, and some copper alloys. Alloy steels, magnesium alloys, titanium alloys, zirconium alloys, and zinc alloy die castings have also been stud welded. The capacitor discharge welding method also makes it possible to weld many dissimilar metal combinations.

Fig. 1 Shear connector studs welded to an I-beam through a metal deck

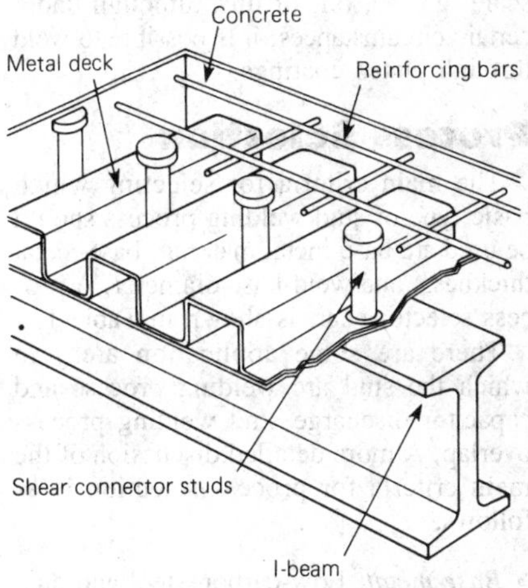

Stud welding is most often used to end-weld a stud or other fastener to a plane surface. The arc method can be used to weld fasteners to curved surfaces such as pipe, as well as inside and outside corners of structural members. Applications for the stud arc welding method include: (1) attaching wood floors to steel decks or framework; (2) securing special linings in tanks, boxcars, and other containers; (3) studding boiler tubes; (4) assembling electric panels; (5) securing inspection covers of various kinds; (6) securing air, water, hydraulic, and electrical lines to buildings, vehicles, and large appliances; (7) welding shear connectors and concrete anchors to structures; and (8) securing handles and feet to small appliances and cookware. Figure 1 illustrates shear connector studs that are stud arc welded

*Thomas E. Shoup, *Chairman*, Manager, Product Development, Nelson Division, TRW, Inc.; Charles C. Pease, Chief Metallurgist, KSM Division, Omark Industries; Mikio Tamiyasu, Director of Engineering, Erico-Jones Co., Inc.

through a metal deck to create a composite structure of metal and concrete. This technique eliminates the need for costly burning, cutting, or prepunching holes in most decks. Applications for the capacitor discharge welding method are name plates, electronic instrumentation panels, trim molding and emblems for automobiles, and applications involving dissimilar metals and thin material where backside marking is critical. Systems with automatic stud feed are capable of welding studs at a rate of 60/min or more. Accuracy of location depends on product design and process equipment.

Stud welding does have limitations. Studs should be of such a size or shape to allow chucking. Only one end of a stud can be welded to the base material. As in all types of welding, areas to be welded must be clean and free from excess paint, scale, grease, oil, or dirt; although under certain circumstances, it is possible to weld through plated coatings.

Process Selection

The main criteria for selecting which basic type of stud welding process should be used are base-metal material, base-metal thickness, and weld-base diameter. A process selector chart is shown in Table 1.

There are some application areas in which the stud arc welding process and capacitor discharge stud welding process overlap. A more detailed discussion of the main criteria for process selection is as follows:

- *Base metal*: Low-carbon steel and austenitic stainless steel are weldable by both arc and capacitor discharge methods. Of the aluminum alloys, the 1100, 3000, and 5000 series are considered the best; the 6000 series considered passable; and the 2000 and 7000 series a poor choice for the stud arc welding process. With capacitor discharge methods, the 1100, 3000, 5000, and 6000 series are considered excellent and the 2000 and 7000 series passable for certain applications. Copper, brass, and galvanized steel sheet usually are welded by the capacitor discharge method only.
- *Base-metal thickness*: If the base metal is less than 0.039 in. thick, the capacitor discharge method should be used. Using the stud arc welding process, the thickness of the base metal should be at least one third the weld-base diameter of the stud in order to ensure development of the full strength of the fastener. The base-metal thickness may be one fifth the weld-base diameter if strength is not the foremost requirement.

Table 1 Applicability of arc and capacitor discharge stud welding

Item	Stud arc welding	Capacitor discharge stud welding: Initial contact; initial gap	Capacitor discharge stud welding: Drawn arc
Base metal			
Carbon steel	A	A	A
Alloy steel	B	A	C
Stainless steel	A	A	A
Aluminum alloys	B	A	A
Copper alloys	C	A	A
Base-metal thickness			
Under 0.015 in.	D	A	B
0.015 to 0.062 in.	C	A	A
0.062 to 0.125 in.	B	A	A
Over 0.125 in.	A	A	A
Stud metal			
Carbon steel	A	A	A
Alloy steel	B	C	C
Stainless steel	A	A	A
Aluminum alloys	B	A	A
Copper alloys	C	A	A
Weld-base shape			
Round	A	A	A
Square	A	C	C
Rectangular	A	C	C
Irregular	A	C	C
Weld-base diameter			
$^1/_{16}$ to $^1/_8$ in. diam	C	A	A
$^1/_8$ to $^3/_8$ in. diam	A	A	B
$^3/_8$ to $^1/_2$ in. diam	A	B	C
$^1/_2$ to $1^1/_4$ in. diam	A	D	D

Note: A: applicable without special techniques and equipment. B: applicable with special techniques or on specific applications that justify preliminary trials or testing to develop welding procedure and technique. C: limited applicability. D: not recommended; welding methods not developed at this time

- *Weld-base diameter*: The diameter range for stud arc welding is $^1/_8$ to $1^1/_4$ in. The diameter range for capacitor discharge welding is $^1/_{16}$ to $^1/_2$ in. For applications in the overlap range, the capacitor discharge process is better suited for welding to thin base materials where lack of backside marking is a foremost requirement and for welding nonferrous materials. Otherwise, the stud arc welding process is generally used.

The stud arc welding method produces a larger fillet and greater base-metal penetration than the capacitor discharge method (Fig. 2). This is because of the basic differences in the weld time/weld current relationship between the two methods. With either process, consideration for the dimensions of the weld fillet must be given in the design of parts which are held by studs (Fig. 3).

Fig. 2 Sections through low-carbon studs arc and capacitor discharge stud welded to low-carbon steel base metals. (a) The stud arc weld exhibits a large amount of weld metal around the stud base and a relatively deep penetration of the weld into the base metal. (b) The capacitor discharge stud weld has a very small amount of weld metal around the stud base and only shallow penetration of the weld into the base metal. Melting of the face of the stud and of the workpiece surface opposite the stud face is greater in stud arc welding than in capacitor discharge stud welding.

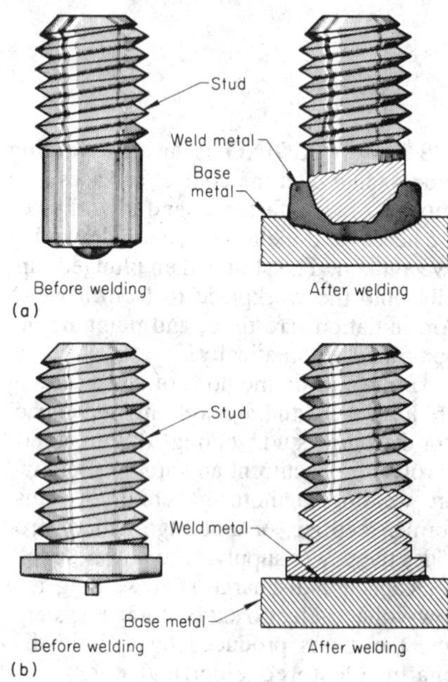

Stud Arc Welding

The stud arc process sequence consists of drawing an arc between the stud and workpiece and then bringing the stud to

Fig. 3 Methods of accommodating weld fillets

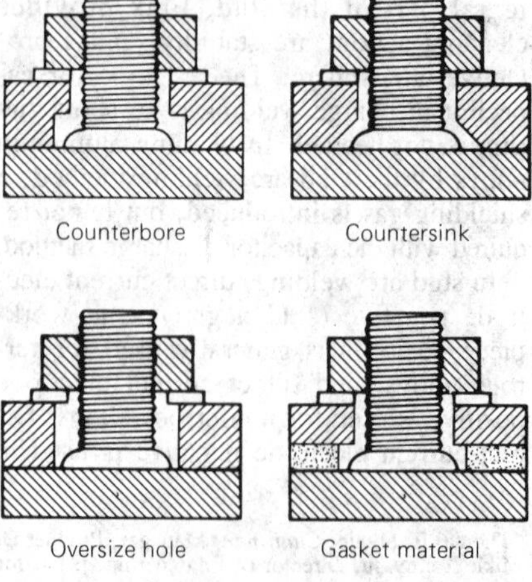

the workpiece when the proper temperature is reached. The equipment consists of a stud gun, a control unit for timing the weld, a direct current power source, and suitable weld cable. Power sources that are specifically designed for stud welding generally combine or make the timing controls integral with the power source. Figure 4 illustrates a stud arc welding system.

The welding sequence begins by loading a stud into the chuck and a ferrule, or arc shield, into the ferrule holder. Figure 5 depicts the stud arc welding sequence. The relationship between the stud and ferrule prior to positioning the stud on the workpiece is shown in Fig. 5(a). The stud protrudes beyond the ferrule to allow for stud burnoff and to enable the stud to plunge fully into the molten metal once the arcing time is completed. The stud and ferrule are then placed on the workpiece as shown in Fig. 5(b). With the stud now flush with the face of the ferrule, the mainspring in the gun is compressed. When the gun button is operated, a solenoid coil within the gun body is energized, causing the stud to lift from the workpiece and create an arc as shown in Fig. 5(c). The heat from the arc causes both stud and plate to melt. When the arc period, as preset and maintained by the control unit, is completed, the solenoid coil is de-energized, and the weld current is automatically shut off. De-energizing the solenoid coil allows the mainspring in the gun to force the stud into the molten pool on the plate to complete the weld (Fig. 5d). The gun is then lifted from the welded stud, and the ceramic ferrule or arc shield is removed (Fig. 5e). Studs may be welded in the flat, vertical, or the overhead position.

Fig. 4 Connection diagram for stud arc welding system

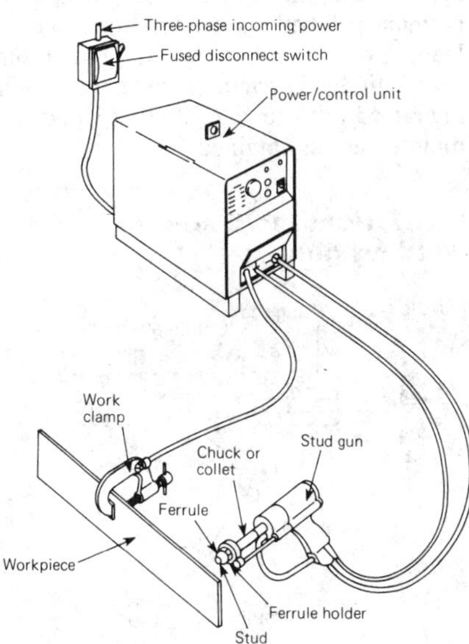

Fig. 5 Stud arc welding sequence. (a) Stud position prior to placing on workpiece. (b) Weld position. (c) Stud lifted and weld arc initiated. (d) Arc period complete and stud plunged into molten pool of metal on plate. (e) Weld complete, ferrule removed

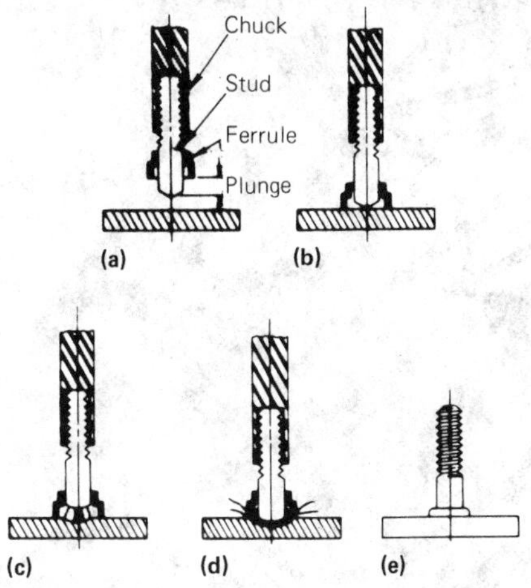

Studs and Ferrules

The most common stud materials being used commercially with the stud arc welding process are low-carbon steel, stainless steel, and aluminum. Other stud materials are being used, but only for special applications. Tensile strength for low-carbon steel studs is 60 ksi minimum; for stainless steel, 85 ksi minimum; and for aluminum, 40 ksi minimum. High-strength studs, meeting Society of Automotive Engineers (SAE) grade 5 tensile strength of 120 ksi minimum, are also available. These studs are made of low-carbon steel heat treated and tempered to a specific hardness to meet the tensile-strength requirement. The weld-base diameter range is $^1/_8$ to $1^1/_4$ in. for low-carbon steel and $^1/_4$ to $^1/_2$ in. for aluminum. The stainless steel stud diameter range is $^1/_8$ to $^3/_4$ in.

Low-carbon steel studs and stainless steel studs require a flux for deoxidizing and arc stabilization. The fluxing agent is permanently affixed to the end of the stud and is generally aluminum.

Aluminum studs do not require any flux on the weld end. However, aluminum studs generally have a small cylindrical or cone-shaped tip to help establish the longer arc length required for aluminum stud welding. Aluminum studs welded by the stud arc welding process require the use of shielding gas, such as argon or helium. For aluminum stud welding, a special foot/ferrule holder is used on the stud welding gun to shield the weld area properly with inert gas.

The stud weld base is generally round, but a number of applications utilize square- or rectangular-shaped weld-base studs. Tests with rectangular weld-base studs show that to obtain satisfactory results, the width-to-thickness ratio should not exceed five.

Stud designs above the weld base have a wide variety of styles, as illustrated in Fig. 6. Stud design is limited by two factors. Studs must be of a size and shape that permit chucking, and the cross-sectional area of the weld base must not be greater than approximately 1 in.^{2} Because a portion of the stud is burned off during welding, the length of a stud before welding is longer than it is after welding. The amount of burnoff depends on the diameter of the stud and to some degree on the application involved.

Ferrules are required for the stud arc welding process except under highly specialized conditions. The ferrule surrounds the weld area and performs several important functions during the weld cycle: (1) concentrating the heat of the arc in the weld area during the weld; (2) reducing oxidation of the molten metal during welding by restricting passage of air into the weld area; and (3) confining the molten metal to the weld area. The ferrule also protects the eyes of the operator from the arc; however, safety glasses with side shields and shade No. 3 filter lenses are recommended. Two types of ferrules are used: expendable and semipermanent.

The expendable ferrule has the broadest commercial use. It is composed of a ceramic material and breaks easily for removal. Because the expendable ferrule is designed for only one weld, it is much smaller, and its design, relative to venting and fillet cavity dimensions, can be optimized. Better fillet control and weld quality can be achieved with the expendable ferrule than with the semipermanent ferrule. Stud shape is not limited, because the ferrule does not have to slip over the stud shank of the welded stud for removal.

The semipermanent ferrule is seldom used and is suitable for special applications involving automatic stud feed systems in which fillet control is not important. The number of welds that can be obtained with a semipermanent ferrule varies considerably, depending on the stud diameter, weld setup, and weld rate, but is generally between 2500 and 7500. The ferrule fails because of the gradual erosion of the ferrule material by the molten metal, causing welds to become unacceptable.

Fig. 6 Stud designs made for stud arc welding

Stud Arc Welding Equipment

Guns. Two basic stud arc welding guns are used, portable and fixed (production type). The principle of operation is the same for both. The portable or hand-held gun resembles a pistol and is designed to be lightweight and durable. Two gun sizes are used. A small gun, weighing approximately $4\frac{1}{2}$ lb, is used for welding $\frac{1}{8}$- to $\frac{1}{2}$-in.-diam studs and a larger gun, weighing approximately 11 lb, is used for welding $\frac{5}{8}$- to $1\frac{1}{4}$-in.-diam studs. Gun bodies are made of high-impact strength plastic. A schematic drawing illustrating a hand-held gun is shown in Fig. 7.

Stud arc welding guns may also incorporate a means for causing the stud to plunge or move slowly as it enters the molten pool of metal at the completion of the weld. This cushioning effect reduces weld splatter considerably and also improves weld integrity.

Power/Control Units. A direct current (dc) power source is required for stud arc welding. Although some work has been done on stud arc welding equipment operating from an alternating current (ac) power source, stud arc welding equipment currently in use is designed for use with direct current. Direct current provides a very stable arc and also allows three-phase incoming power to be used for the power source. The latter is important because the weld current range used for stud arc welding extends above 2500 A at 50 V dc.

Standard direct current power sources designed for manual shielded metal arc welding (SMAW) have excellent characteristics for stud arc welding, because they have drooping volt-ampere characteristics and an open circuit voltage of 70 V dc or higher. However, the welding current range requirement for stud arc weldng extends considerably beyond that used for SMAW. Therefore, standard direct current power sources are not generally used for welding studs larger than $\frac{1}{2}$ in. diam, and a power source specifically designed for stud welding is used. These special-purpose stud welding power sources have approximately two to three times the weld current output capability of a conventional power source of the same size and weight. This is possible because the duty cycle requirements for stud arc welding are one third to one fifth those required for manual SMAW. The types of special power sources that have been developed specifically for stud arc welding include transformer-rectifier, rotating (motor- or engine-driven generator), and battery units. The trend in new equipment has been toward the transformer-rectifier-type power source.

Interfacing the gun with the power source is accomplished either by using a separate control unit or by having gun controls as an integral part of the power unit. The separate control unit is used when the power source is a motor-generator unit, battery unit, or standard manual SMAW power source. Some special transformer-rectifier power sources have been designed for stud arc welding that utilize a separate control unit. However, with the advent of the silicon-controlled rectifier (SCR), the trend has been to combine the power source and stud welding gun controls into a single unit for transformer-rectifier-type equipment designed for stud arc welding.

Separate stud arc welding control units consist of a timer and a contactor. The contactor must be capable of initiating and interrupting relatively high direct currents (600 to 2500 A) and must be designed specifically for stud arc welding. Stud arc welding control units generally utilize the direct current output voltage of the welding power source for their operation.

Combining stud arc welding controls with the power source in transformer-rectifier-type equipment eliminates the mechanical contactor and enables the system to be completely solid-state. Power/control units have been developed for both three-phase and single-phase incoming power. Three-phase units are preferred for studs larger than $\frac{3}{8}$ in. diam because they provide a balanced load on the incoming power line. Some power/contol units provide weld quality assurance features, such as fully regulated current output and a means to alert the operator if the machine is not producing the weld current indicated by the dial setting.

Automatic Feed Systems. Stud arc welding systems with automatic feed are available with both portable and fixed welding guns. Studs are automatically oriented in a parts feeder and transferred through a flexible feed tube into the welding gun chuck. A ferrule or arc shield is hand-loaded for each weld. For special applications, inert gas shielding or a semi-permanent ferrule is used to eliminate the loading of a ferrule for each weld. Using automatic feed systems such as this, welding rates in the range of 20 to 45 studs per minute can be obtained.

Fig. 7 Hand-held stud arc welding gun

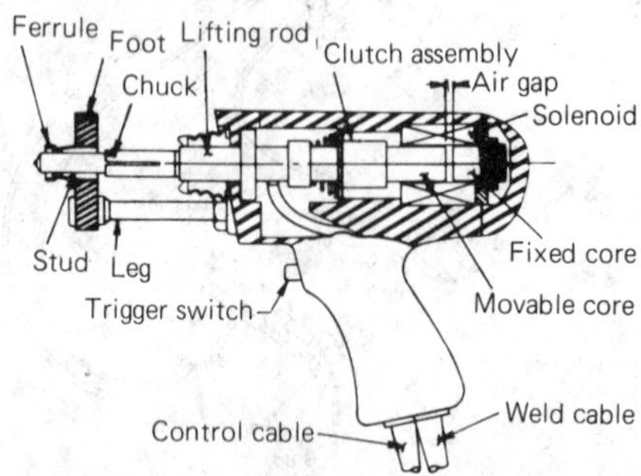

Weld Current/Weld Time Relationships

The weld current and weld time required for a proper stud arc weldment depend on the cross-sectional area of the stud. The total energy input (watt-seconds) to the weld zone is a function of the weld current, weld time, and weld arc voltage. However, the weld arc voltage is determined by the arc length or lift distance set in the gun. Because this is fixed and the variations in arc voltage are relatively minimal, the energy input is basically a function of the weld current and weld time.

The same energy input can be obtained using different combinations of weld current and weld time. In addition, the stud arc welding process tolerates some variations in total energy input. Therefore, the weld current and weld time combinations that result in the production of satisfactory welds are fairly flexible. Although energy input is the basic criterion required for a proper weld, the relationship between weld current and weld time does have limitations. Every stud size has a certain minimum and maximum current level at which satisfactory welds can no longer be obtained, regardless of adjustment of weld time. For example, a long weld time cannot compensate for an inadequate power source, even if the energy input or watt-seconds are within the weldable range. The current and time relationships possible for various diameters of low-carbon steel studs are shown in Fig. 8. A fairly broad weldable range exists for each given stud size. Under some conditions, however, the range is restricted, e.g., welding studs to the vertical, welding studs to thin base material, and welding studs through galvanized sheet steel. Although energy level is the basic criterion required for a satisfactory weld, the converse is not necessarily true; i.e., proper energy level does not always ensure a proper weld. Arc blow, plate surface condition (rust, scale, moisture, or paint), and operator technique can be causes of improper welds, even if the proper weld energy has been used.

Fig. 8 Weld current and weld time range for various diameter low-carbon steel studs. Source: Shoup, T.E., Stud Welding, *WRC Bulletin,* No. 214, April 1976, p 5

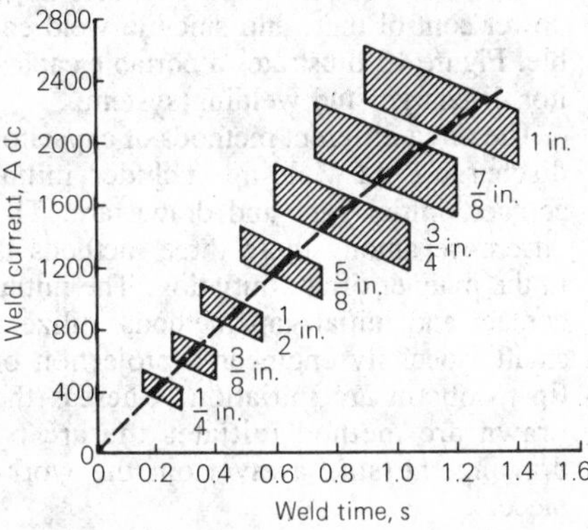

Fig. 9 Macrosection of low-carbon steel stud weld

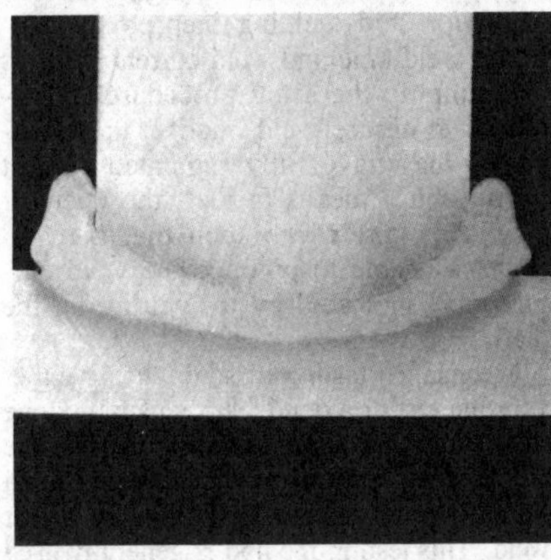

Metals Welded

Metals that are readily welded by other arc welding processes are suitable as base metals for stud welding with proper selection of stud metal. Because stud welding is an arc welding process, the weld metal structure in the various heat-affected zones (HAZ) is similar to that obtained with other arc welding processes (Fig. 9). Combinations of base metal and stud metal that have been welded successfully include:

Base metal	Stud metal
Low-carbon steel, 1006 to 1025	Low-carbon steel(a); stainless steel, 300 series(b)
Stainless steel, 300 series(b), 405, and 430	Low-carbon steel(a); stainless steel, 300 series(b)
Aluminum alloys	Aluminum alloy 5356

(a) 0.23 C max, 0.70 Mn max, 0.04 P max, and 0.05 S max. (b) Except for the free-machining type 303 stainless steel

Carbon steel workpieces with a carbon content of 0.30% or less can be stud arc welded without preheating. Stud arc welding of medium-carbon steels (1030 to 1050) is generally successful, provided that correct preheating and postheating procedures are followed, which are described in the section on high-carbon steels.

High-Carbon Steel. In stud arc welding steel with 0.45% C or more, martensite (a hard, brittle constituent) forms in the steel base metal under the stud. The base metal has been spot hardened, and the hardening is usually accompanied by cracking. In welding plain carbon steel of high carbon content, the base metal must be preheated to prevent formation of cracks. Postheating is less effective because cracks have already formed, but it can prevent the propagation of cracks. The hardenable alloy steels are either preheated, postheated, or both, depending on product requirements. When considering stud arc welding high-carbon steel, sample pieces should be welded experimentally to determine whether the operation is feasible.

High-strength low-alloy steels and most structural steels with a carbon content of more than 0.15% may require moderate preheating to obtain required toughness in the weld area. An oxyfuel gas heating torch can be used because only a localized preheated zone is needed. Some heat treatable structural steels are sufficiently hardenable in the HAZ to be sensitive to underbead cracking. For maximum toughness, preheating at 600 to 700 °F is necessary.

Galvanized Steel. In the construction of buildings and bridges, stud arc welding can be used to weld low-carbon steel shear connectors and concrete anchors directly to steel beams by burning through the galvanized metal decking without prepunching it. Ductwork and other products of galvanized steel can have low-carbon steel studs welded to them. Such studs are used to support insulation, wiring, and piping; the welds have adequate strength for this purpose.

Stainless Steel. The austenitic stainless steels, with the exception of free-machining type 303, can be stud arc welded. Martensitic stainless steels are subject to air hardening and are usually brittle in the weld area and HAZ unless annealed after welding. Stainless steel studs can be welded to stainless steel and to low-carbon steel base metals.

Aluminum alloys of the 1100, 3000, and 5000 series generally are excellent for stud arc welding; alloys of the 4000 and 6000 series are acceptable; alloys of the 2000 and 7000 series are poor.

Studs are made of aluminum magnesium alloys, including 5086, 5356, and 5456. Aluminum magnesium alloys have high strength, good ductility, and metallurgical compatibility with most other aluminum alloys. Aluminum studs range in weld-base diameter from 1/4 to 1/2 in. Figure 10 shows a macrosection of a 3/8-in.-

Fig. 10 Macrosection of $^3/_8$-in.-diam type 5356 aluminum stud welded to $^3/_{16}$-in. type 6061 aluminum plate

diam type 5356 stud arc welded to $^3/_{16}$-in.-thick type 6061 aluminum plate.

Inspection and Quality Control

Weld quality begins with a correct setup and a trained operator. Both weld quality and setup corrections can generally be determined by visual inspection of the weld fillet. Figure 11 illustrates how both weld quality and setup correction can be determined by visual inspection. Proper setup includes a broad range of details: gun settings (arc length and stud protrusion beyond the ferrule), proper energy input into the weld zone, and use of correct accessories and ferrules. The operator should hold the gun perpendicular to the work surface. The gun should not be moved during the weld cycle and cannot be pulled away from the welded stud immediately after the weld is completed.

The work surface should be free of contaminants such as rust, paint, millscale, and water. In some applications, namely concrete anchor and shear connector, limited amounts of rust and millscale can be tolerated. In these applications, contaminants are removed to the extent necessary to obtain satisfactory welds. However, to ensure maximum weld quality, clean work surfaces should be used.

Special instrumentation, designed specifically for measuring the weld current and weld time of a stud weld, is available for setup and monitoring purposes. Generally, power/control units designed specifically for stud welding incorporate calibrated weld time and weld current controls that simplify the setup procedure. In addition, to ensure weld quality, these machines may have fully regulated current output and a means to alert the operator should conditions be such that the machine is unable to produce the weld current level indicated by the setting on the dial.

Mechanical testing of studs, using either a torque test or a bend test, is usually done before starting a production run. For the torque test, a stud may be twisted until failure occurs in the stud, base metal, or weld. This testing method is generally used when studs are welded to thin base material or have no external threads. Studs with external threads may be torque tested in the conventional manner, using a torque wrench to secure the nut until a predetermined load is reached or the stud fails. For the bend test, a stud is either struck with a hammer or bent with a short piece of tube placed over it. Some applications specify that the stud be bent 45 to 90° from its axis without weld failure, whereas a bend of 10 to 15° may be adequate for other applications. Because bend tests can damage studs, bending should be done on samples, not production parts. However, studs can be bent and used in some applications, e.g., concrete anchor and shear connector applications.

Fig. 11 Satisfactory and unsatisfactory stud arc welds. (a) Optimum weld. Fillet is well formed, has a bluish cast, and is wetted to the base plate. (b) Partial fillet or undercut. Condition is caused by insufficient plunge (stud protrusion beyond ferrule) or mechanical hang-up. (c) Cold weld. Fillet has a grayish appearance and is not wetted to base metal. Fillet may have starlike fingers. Correct by increasing weld current. (d) Hot weld. Fillet is very low, washed out, has a bluish cast. Correct by decreasing weld current. (e) Poor alignment. Partial fillet because of tilted gun. (f) Arc blow. Stud burned off on one side. Sometimes corrected by changing grounding procedure. May require technical assistance. Under some conditions, may be caused by insufficient weld current.

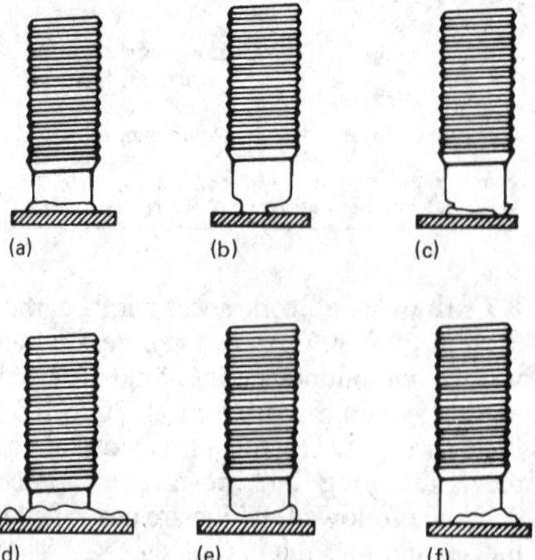

Fig. 12 Bend test. Arrow indicates the bending tool used to bend welded studs.

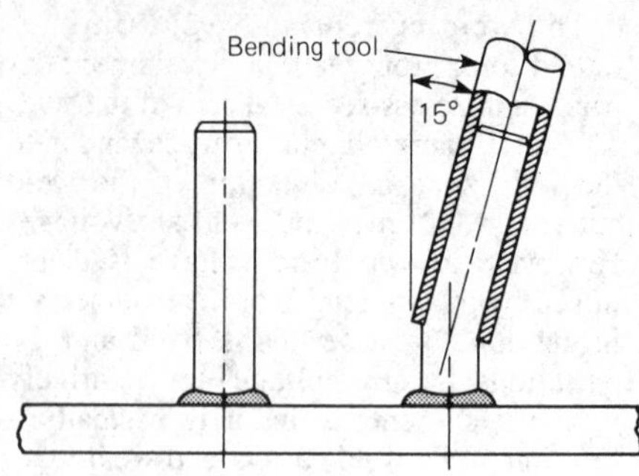

Mechanical testing of aluminum studs, either by torque test or bend test, is required because visual appearance is not completely indicative of a good weld, as is the case with ferrous metals. Visual inspection does, of course, show an obviously bad weld, because of incomplete fillet or undercut. Bend testing of aluminum studs should be done with a tube bending tool and not by striking with a hammer (Fig. 12). A suitable weld is indicated by bending the stud approximately 15° away from its axis before failure of the stud, base metal, or weld.

Capacitor Discharge Stud Welding

The capacitor discharge process uses direct current produced by the rapid discharge of a capacitor bank to create an electric arc and the heat needed for melting both stud and workpiece. The weld time is very short, ranging from 1 to 6 ms. Because the weld time is so short, no ferrule or fluxing is required. The equipment consists of a stud gun, capacitor discharge power/control unit, and suitable weld cable. Figure 13 illustrates a portable capacitor discharge stud welding system.

The three different methods of capacitor discharge stud welding include: initial contact, initial gap, and drawn arc. The difference among these three methods is in the manner of arc initiation. The initial contact and initial gap methods utilize a small, specially engineered projection or tip to obtain arc initiation, whereas the drawn arc method initiates the arc by drawing the stud away from the workpiece.

Fig. 13 Connection diagram for a portable capacitor discharge stud welding system

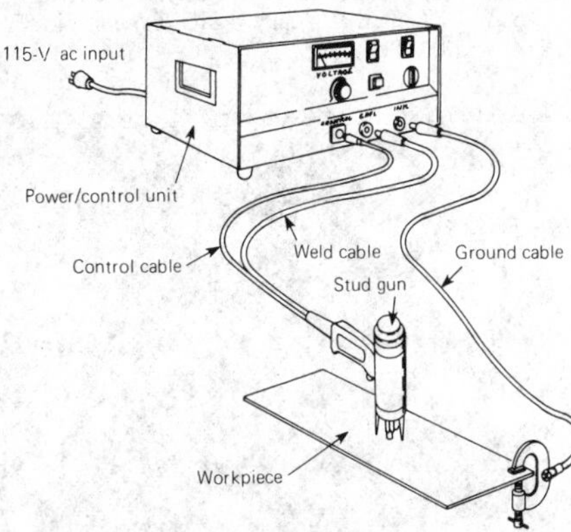

Initial contact capacitor discharge stud welding begins with a stud in contact with the workpiece, shown schematically in Fig. 14(a). Pressure urges the stud toward the workpiece. Capacitors are then discharged, causing a very high current to flow through the small projection on the base of the stud (Fig. 14b). The small projection presents a high resistance to the discharge current, heats up, and rapidly disintegrates, creating an arc that melts the surfaces to be joined (Fig. 14c). Because pressure is urging the stud toward the workpiece, movement of the stud starts when the arc initiates. The welding arc is quenched when the stud contacts the workpiece, and some charge still remains on the capacitors (Fig. 14d). Fusion takes place, and a weld is produced between the stud and workpiece (Fig. 14e).

Initial gap capacitor stud welding starts with a gap between the stud and the workpiece, shown schematically in Fig. 15(a). With the stud positioned above the workpiece, the charged voltage of the capacitors is applied between the stud and the workpiece. The stud is then driven toward the workpiece by spring action, air pressure, or gravity (Fig. 15b). When the tip contacts the workpiece, the capacitors immediately start to discharge, causing a very high current to flow through the tip. This high current flow causes the tip to flash off, creating an arc (Fig. 15c). The arc thus formed heats the surfaces of both stud and work, causing them to become molten as the stud approaches the workpiece (Fig. 15d). The arc is quenched when the stud is seated on the workpiece (Fig. 15e and f). Because the stud is in motion before the tip contacts the workpiece, weld time is shorter with the gap method than with the initial contact method.

Fig. 14 Steps in initial contact method of capacitor discharge stud welding

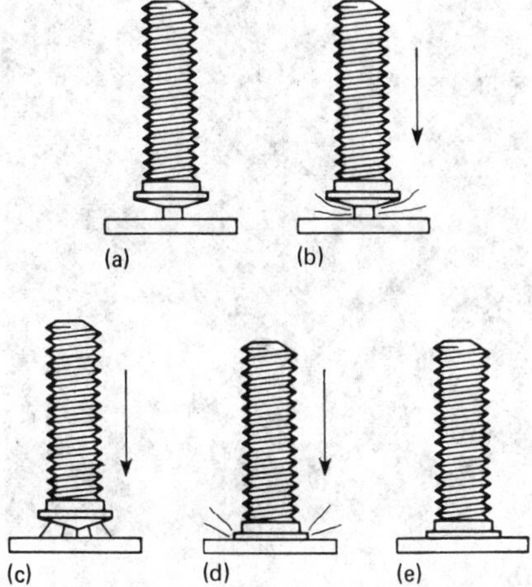

Drawn arc capacitor discharge welding initiates the arc by drawing the stud away from the workpiece to establish the arc. Electronic timing circuitry is used to control the weld arc time. The drawn arc capacitor discharge welding method does not require the projection necessary for both the initial contact methods.

The sequence of operation for the drawn arc process is illustrated in Fig. 16. The stud is positioned against the workpiece (Fig. 16a). The trigger switch on the stud welding gun is actuated, causing the solenoid coil in the gun to be energized. This sets the gun lift mechanism in motion, causing the stud to be lifted from the workpiece and initiating a low-current pilot arc (Fig. 16b). The solenoid coil is then de-energized, allowing the stud to return to the workpiece under spring pressure. While the stud is returning to the workpiece, the welding capacitors are discharged, creating the welding arc and melting the stud and workpiece (Fig. 16c). The welding arc is quenched when the stud contacts the workpiece with some charge still remaining on the capacitors (Fig. 16d). Maintaining the welding arc until the stud contacts the workpiece reduces oxidation and prevents solidification of the weld metal until the stud and workpiece are joined.

Studs

Low-carbon steel, stainless steel, aluminum, and brass are the materials used most often for studs welded commercially with the capacitor discharge method. The preferred weld base of a capacitor discharge welding stud is round, and the shank may be of almost any shape—threaded, unthreaded, round, square, rectangular, tapered, or grooved. The maximum weld-base diameter being welded commercially with the capacitor discharge welding process is $^1/_2$ in.

Fig. 15 Steps in initial gap method of capacitor discharge stud welding

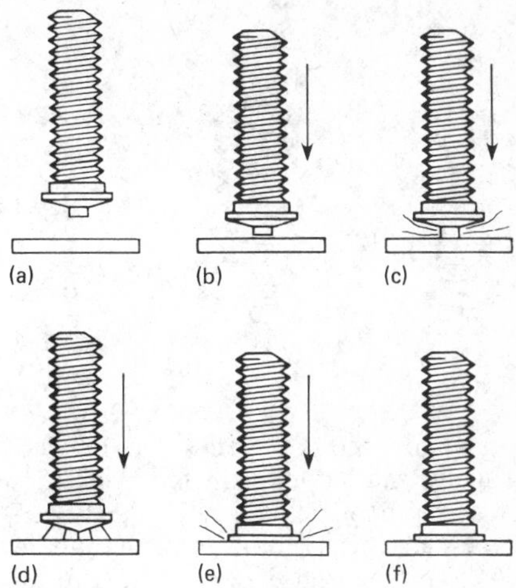

Studs welded with either the initial contact method or initial gap method of capacitor discharge welding require a special tip or projection. The size and shape of this tip is very important, because it is one of the variables that affect weld quality. A cylindrical-shaped tip (Fig. 17a) is the most commonly used shape, because it is suitable for manufacturing on high-speed cold heading equipment. However, a machined conical-shaped tip (Fig. 17b) is frequently used for critical or special applications and is preferred for applications involving nonflanged aluminum studs. Studs welded with the drawn arc method do not require a special tip (Fig. 17c).

Fig. 16 Steps in drawn arc method of capacitor discharge stud welding

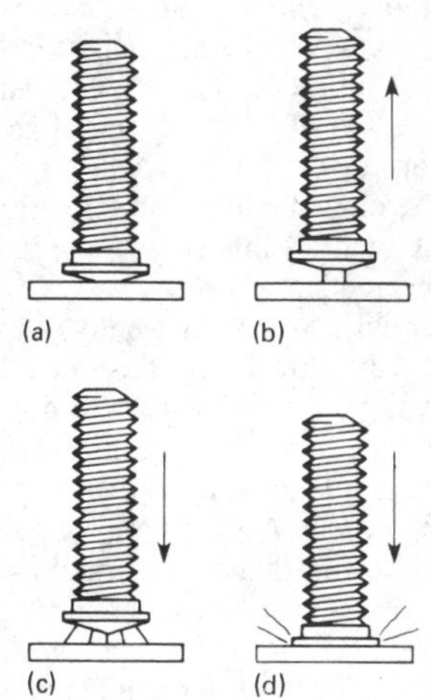

Fig. 17 Weld-end shapes for capacitor discharge studs. Source: Shoup, T.E., Stud Welding, *WRC Bulletin,* No. 214, April 1976, p 14

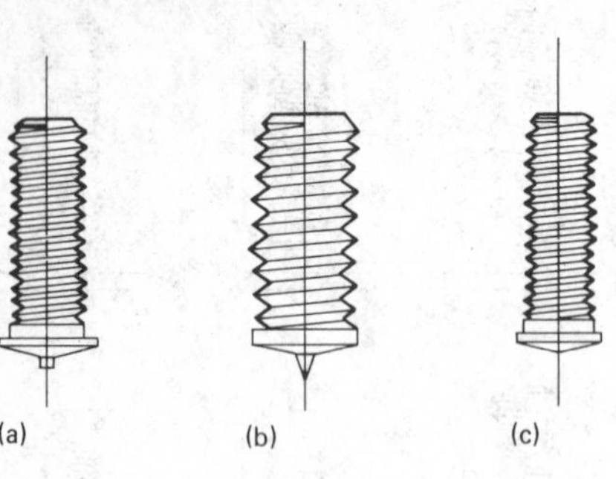

However, the weld end is pointed or slightly rounded to facilitate arc initiation at the center of the weld base. The flanged-type stud, because of its larger weld base, is the recommended standard stud for all three methods. Stud burnoff with the capacitor discharge method is almost negligible when compared to the stud arc welding method. Stud burnoff is generally in the range of 0.008 to 0.015 in.

Capacitor Discharge Stud Welding Equipment

Capacitor discharge stud welding equipment consists of two basic components, a gun and a power/control unit. Portable or stationary production equipment is available. Portable equipment is generally limited to $^5/_{16}$-in.-diam studs maximum, whereas production equipment is capable of welding $^1/_2$-in.-diam studs. Capacitor discharge stud welding equipment is also available which incorporates automatic stud feed for high-production applications.

Guns. Capacitor discharge stud guns are of two basic types: portable (hand-held) and fixed production. Construction and principle of operation between the portable guns and fixed production guns differ considerably. The portable or hand-held tool resembles a pistol and weighs less than 4 lb. The gun mechanism design depends on the method of welding: initial contact, initial gap, or drawn arc. Although a gun can be developed that is suitable for welding studs with all three methods, it is not common practice to do so.

The initial contact hand-held gun is the simplest capacitor discharge stud welding gun. This gun is basically a spring coupled to the stud holding member and mounted in a pistol-shaped enclosure with a trigger switch. A schematic drawing of an initial contact hand-held gun is shown in Fig. 18. The initial gap hand-held gun is similar but incorporates a solenoid coil. The solenoid coil is energized at all times

Fig. 18 Hand-held initial contact capacitor discharge gun

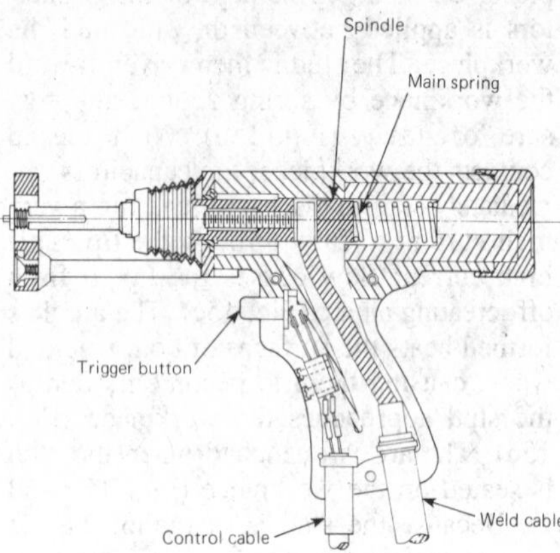

except when making a weld. Thus, it holds the stud above the workpiece until de-energized by depressing the gun trigger switch. The drawn arc hand-held gun is similar to the stud arc welding gun shown in Fig. 7.

The fixed or production gun is usually an initial gap or drawn arc gun. The drawn arc production guns are also similar to stud arc welding guns.

The initial gap production gun may use gravity, air, or a spring to drive the stud to the workpiece. The most consistent means is gravity, and a gun utilizing this method is called a gravity drop gun. This gun consists of a freely mounted ram to which a means for holding studs is attached (Fig. 19). A double-acting air cylinder raises the ram and holds it above the workpiece. By operating the air cylinder rapidly in the flat position, the stop is pulled away, allowing the ram to drop to the workpiece. The ram is designed to allow various weights to be added, and the air cylinder is arranged so that adjustments in drop height or initial gap setting can be made. The gravity drop gun is preferred

Fig. 19 Gravity drop capacitor discharge production gun

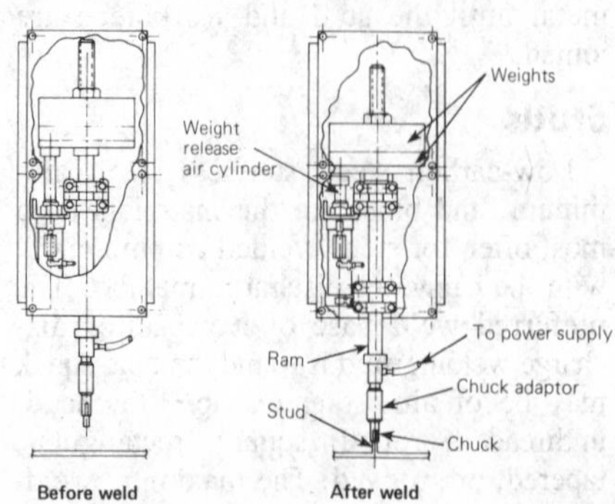

Fig. 20 $^3/_8$-in.-diam 6061-T6 aluminum stud welded to $^1/_4$-in.-thick 6061-T6 plate. (a) With detergent solution. (b) Without detergent solution. Source: Shoup, T.E., Stud Welding, *WRC Bulletin,* No. 214, April 1976, p 18

(a)

(b)

for welding larger capacitor discharge studs, $^1/_4$ to $^1/_2$ in. diam.

Initial gap production equipment incorporates an automatic fluid dispenser for flooding the weld area with a detergent solution. Welding with a detergent solution on the workpiece reduces arc blow, provides a more consistent arc, improves weld quality, and reduces weld splatter. Figure 20 shows the difference in appearance between $^3/_8$-in.-diam aluminum stud welds made with and without the use of a detergent solution.

Power/Control Units. Portable power/control units weigh approximately 60 to 100 lb and operate on 115 V, 60 Hz power. As with guns, power/control units are designed for particular capacitor discharge stud welding methods. Portable power/control units can have a capacitance of 50 000 up to 100 000 μF and are capable of welding up through $^5/_{16}$-in.-diam studs at a rate of 8 to 10/min. The equipment utilizes solid-state components throughout for high reliability.

Power/control equipment for production or fixed equipment is designed to meet the specific requirements of the application for which it is being used, because the part usually requires special fixtures for clamping, indexing, and unloading. To tailor the equipment to a customer's particular welding requirements, the capacitance used in production power/control units may range up to 200 000 μF. Capacitor charge voltage, as with portable control units, does not exceed 200 V. Suitable safety circuitry is incorporated in all capacitor discharge welding systems to prevent the capacitor charge voltage from being at the chuck except during welding. Power input for production equipment is 230 to 460 V single-phase or three-phase to meet recharge time and current requirements for production applications.

Automatic Feed Systems. Capacitor discharge stud welding, because no ceramic ferrule is required, is suited for high-speed automatic stud feed applications. Portable capacitor discharge equipment with automatic stud feed is available for studs ranging from No. 6 through $^1/_4$ in. diam. Studs are automatically oriented in a parts feeder and transferred through a flexible feed tube into the welding gun chuck. The automatic feed attachments add very little weight to the gun and do not encumber its use (Fig. 21). Welding rates with portable equipment range up to 60 studs per minute on applications where stud location tolerances are such that no templating or only a loose-fitting templating is required. Figure 22 shows an initial gap automatic stud feed production unit system.

Fig. 21 Hand-held gun with automatic stud feed attachments

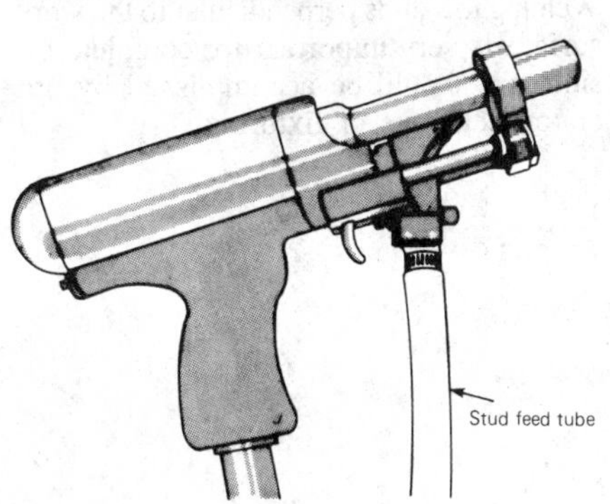

Fig. 22 Initial gap automatic feed production unit

Weld Current/Weld Time Relationships

The weld current used in the capacitor discharge welding process ranges from approximately 600 to 25 000 A, depending on stud size and weld method. The weld current range with the drawn arc method is approximately 600 to 3000 A, whereas with the initial contact or gap method the range is approximately 2500 to 25 000 A. Weld time with the drawn arc method is generally in the range of 4 to 6 ms, whereas with the initial contact and gap method it is generally in the range of 1 to 3 ms. Figure 23 illustrates the differences in the time/current relationship among the three methods when welding a $^1/_4$-in.-diam steel stud to 0.062-in.-thick low-carbon steel plate.

Fig. 23 Time/current relationships among three capacitor discharge stud welding methods. Oscilloscope representations, showing the welding of $^1/_4$-in.-diam steel studs to 0.062-in.-thick low-carbon steel plate. Source: Shoup, T.E., Stud Welding, *WRC Bulletin,* No. 214, April 1976, p 15

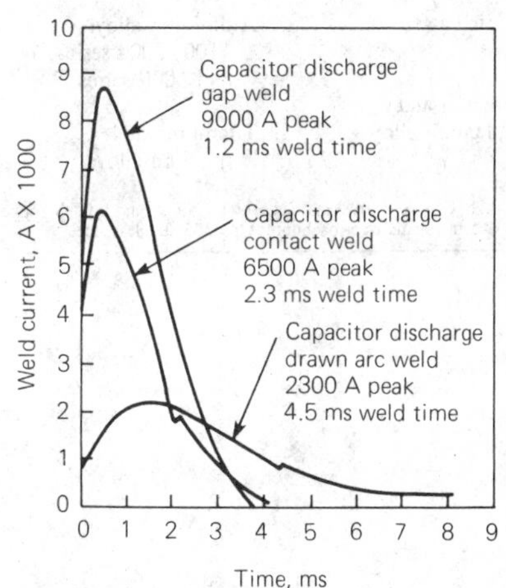

Metals Welded

The capacitor discharge stud welding process is an arc welding process, but differs from other arc welding processes in that the weld is made in a few milliseconds. For this reason, acceptable metallurgical results can be obtained without shielding the arc or using a flux. The welding of aluminum using the drawn arc method of capacitor discharge stud welding does require a shielding gas, however. Because of the speed at which the weld takes place, very little metal is melted. Weld penetration is slight, and very little stud metal intermixes with the plate metal. Therefore, studs may be welded to very thin metals without melt-through, and many dissimilar metals can be welded with acceptable results. Various stud and plate combinations can be welded:

Base metal	Stud metal
Low-carbon steel, 1006 to 1025	Low-carbon steel(a); stainless steel, 300 series(b); lead-free copper alloys
Stainless steel, 300(b) and 400 series	Low-carbon steel(a); stainless steel, 300 series(b); lead-free copper alloys
Aluminum alloys, 1100, 3000 series, 5000 series, 6000 series	Aluminum alloys 1100, 5000 series, and 6000 series
Copper, lead-free brass	Low-carbon steel(a); stainless steel, 300 series(b); lead-free copper alloys
Zinc alloys (die cast)	Aluminum alloys 1100, 5000 series, and 6000 series
Titanium and titanium alloys	Titanium and titanium alloys

(a) 0.23 C max, 0.60 Mn max, 0.04 P max, and 0.05 S max. (b) Except for the free-machining type 303 stainless steel

Fig. 24 Capacitor discharge stud welds. (a) 1/4-in.-diam mild steel stud welded to 1/8-in.-thick mild steel plate. (b) 1/4-in.-diam brass stud welded to 1/8-in.-thick brass plate. (c) 1/4-in.-diam 5356-H32 aluminum stud welded to 1/8-in.-thick 6061-T6 aluminum plate.

(a)

(b)

(c)

Figure 24 shows macrosections which illustrate capacitor discharge welds in mild steel (Fig. 24a), brass (Fig. 24b), and aluminum (Fig. 24c).

The welds obtained with the three methods of capacitor discharge stud welding differ slightly. This is caused not only by the differences in weld time/weld current relationships (Fig. 23), but also by the mechanics of the weld sequence. With the initial gap method, the stud is in motion before the welding arc is initiated. This moving start, in combination with the rapid arc initiation and disintegration of the small projection tip, results in shorter weld time and slightly less weld metal at the interface than with the other two methods. The very short weld time created by the moving start allows less atmospheric contamination and makes the initial gap method generally superior to the drawn arc and the initial contact method for welding dissimilar metals and aluminum. The weld characteristics of the drawn arc method and the initial contact method are quite similar. These two methods are generally used with low-carbon steel applications.

Inspection and Quality Control

Ensuring weld quality of capacitor discharge stud welds is accomplished through the use of proper setup procedures and good operator technique. Weld quality is best determined by running hammer bend tests, tensile tests, or torque tests on studs that have been welded to the production part or material that is similar to that of the production part. Once satisfactory welds are achieved, the production run can begin. Periodic checks should be made to ascertain that the weld quality is being maintained.

Factors influencing weld quality are work surface, cleanliness, and condition. The work surface should be free from contaminants such as excessive oil, grease, rust, millscale, or paint. The work surface should not have imperfections such as porosity, extreme roughness, or other distortions. Welding the studs perpendicular to the work surface is very important for complete fusion and should be accomplished by operator technique or fixturing.

Percussion Welding

PERCUSSION WELDING (PEW) is an arc welding process in which the heat is obtained from an arc produced by a rapid discharge of electrical energy and force is percussively applied during or immediately after the electrical discharge. A shallow layer of metal on the contact surfaces of the workpieces is melted by the heat of the arc produced between them, and one of the workpieces is impacted against the other, extinguishing the arc, expelling molten metal, and completing the weld.

Arc initiation, arc time, and welding force are controlled and synchronized automatically. The power supply usually is a welding transformer or a capacitor (or bank of capacitors). The welding force (forging force) is applied by electromagnetic devices, electromechanical devices, cam-actuated direct drive, springs, or gravity.

The heat input is intense, but extremely brief and localized, and enables the PEW of one small component to another, a small component to a larger one, and dissimilar metals that differ considerably in electrical resistivity and melting temperature. The electrical resistivity of the parts being welded does not noticeably affect the amount of heat generated at the joint. The arc supplies the heat for welding.

The workholding clamp, jaws, or chuck of the welding head does not have to be a good electrical conductor, as in resistance welding, because the amount of current passed is comparatively small and the duration of current flow is extremely brief.

Relation of Percussion Welding to Stud Welding

Percussion welding and stud welding are similar in three important respects:

- Welding heat is obtained from an arc.
- Force is applied percussively.
- Arc-starting methods used in the several variations of the two processes are similar.

Percussion welding and stud welding differ in certain aspects of equipment, technique, and process variables. The similarities and differences among the commonly used methods of PEW and stud welding are compared in Table 1. Although a clear-cut distinction exists between PEW and stud arc welding, there are close similarities between the capacitor-discharge methods of percussion and stud welding.

Percussion welding is used principally for making electrical connections and electrical contact devices; stud welding is used chiefly for joining studs or similar shapes to larger parts for fastening other components.

Applications

Percussion welding can be employed to join like and unlike metals that cannot usually be flash or stud welded. It is used for welding fine wire leads to filaments in lamps and terminals of electrical and electronic components where a reliable joint is needed to withstand shock, vibration, and extended service at elevated temperature. Additionally, PEW is used in making telephone equipment and other electronic and electrical devices and for attaching large-area contacts to switch components. Because the total heat input is small and can be highly localized, these welds can be made a few thousandths of an inch away from glass seals or other heat-sensitive materials without damaging them.

Percussion welding can be used for butt joints between two wires or a wire and a rod and for T-joints between a wire and a workpiece with either a flat or curved surface. The workpiece can be massive or it can be thin sheet metal, as in a chassis or a terminal. Stranded wire, as well as solid wire, can be percussion welded in joints of these types. Flat workpieces of any shape can be percussion welded to mating flat surfaces with the aid of an arc-starting nib or projection on one of the members.

Design and Size of Workpieces. The workpieces must be separate; the ends of a continuous workpiece cannot be joined to make a ring. One of the components of the assembly must be designed so that it can be clamped securely in a welding head and can be impacted against a stationary component without slippage.

Capacitor-discharge PEW can be used to butt weld wires of similar diameters or of greatly different diameters. For some metals, wire diameter can be as small as about 0.005 in. Wire can also be capacitor-discharge percussion welded to a workpiece having a large surface area.

Table 1 Comparison of process variables in percussion and stud welding

Process variable	Percussion welding: Capacitor-discharge method: Low-voltage		Percussion welding: Capacitor-discharge method: High-voltage	Percussion welding: Magnetic-force method	Stud welding: Arc method(a)	Stud welding: Capacitor-discharge method		
Power supply	Capacitor	Capacitor	Capacitor	Transformer(b)	Rectifier(c)	Capacitor	Capacitor	Capacitor
Current supplied to arc	dc	dc	dc	ac	dc	dc	dc	dc
Voltage, V	50-150	12-120	1000-3000	10-35	60-100	100-200(d)	100-200(d)	100-200(d)
Arc-starting method	Nib plus dc voltage(e)	High-frequency ac pulse plus dc voltage(e)	dc voltage(e)	Nib plus first half-cycle of ac (initial contact)	Nib plus dc voltage (draw arc after contact)	Nib plus dc voltage(e)	Nib plus dc voltage (initial contact, no retraction)	dc voltage (draw arc by retracting after contact)
Arc time, ms	0.15-1	1 max	1 max	8 max	100-1000	1-6	3-6	6-12

(a) A ferrule is used to confine the molten metal in this method; flux may be used, depending on workpiece and size. (b) Resistance welding transformer. (c) Arc-welding type of rectifier or motor-generator; no energy storage. (d) Approximate. (e) Initial gap

Magnetic-force PEW can join flat workpieces with a weld area of 0.04 to 0.70 in.^{2} in production. An arc-starting nib or projection must be provided on one of the workpieces.

Workpiece Condition. Heat treated, cold worked, or prefinished metals are virtually unaffected by the heat of PEW because the heat-affected zone (HAZ) is very shallow. As discussed in the section on arc time and heat-affected zone in this article, the heat input is so concentrated at the weld interface and of such short duration that metal more than a few thousandths of an inch away from the weld interface is not affected by the welding heat.

Cleaning is not critical for the production of a sound percussion weld, because at least a thin layer of metal (flash) is melted from each workpiece and expelled from the joint.

Weldable Metals

Almost any like or unlike metals or alloys can be joined by PEW. Work metals of widely dissimilar composition, melting temperature, electrical conductivity, and thermal conductivity can be readily welded together.

Like metals that can be joined include tantalum, copper alloys, aluminum alloys, nickel alloys, low-carbon steels, medium-carbon steels, and stainless steels. Various combinations of these alloys can also be joined.

Gold, silver, copper-tungsten, silver-tungsten, and silver-cadmium oxide have been percussion welded to copper alloys for assemblies for electrical contacts. Copper is routinely percussion welded to molybdenum. Although true welds between these two metals were formerly considered impossible because of mutual insolubility, electron beam microprobe analysis has shown copper penetration of 0.0004 in. into molybdenum at the weld interface.

Thermocouple alloys and low-expansion alloys can be percussion welded to molybdenum and other metals. Welds between molybdenum wires, Nb-1Zr wires, 85Zr-15Nb wires, and tantalum wires can be made in air, with a stream of argon gas directed at the joint during welding.

In some applications, diffusion during prolonged service at high temperature produces weak or brittle structures in direct joints between dissimilar metals (for instance, in joints between copper and stainless steel for continuous service at 1200 °F). Transition joints that include a compatible third metal can be used to prevent this condition.

Percussion welding schedules for joining dissimilar metals are based primarily on the melting temperatures of the metals being joined, although the thermal and electrical conductivities of work metals and the shape and size of workpieces must also be considered. Table 2 gives conditions for and results of capacitor-discharge PEW of several combinations of wire-to-wire and wire-to-stainless plate.

Power Supply

Three types of power-supply units are used for PEW. The first two, low-voltage and high-voltage capacitors, are energy-storage devices that are charged by direct current from a rectifier or motor-generator. The third type, a resistance welding transformer, uses an input of 60-cycle alternating current and does not involve storage of electrical energy.

Low-voltage capacitors that have high capacitance are used as power supplies in capacitor-discharge PEW. The capacitor is charged by direct current from a rectifier or motor-generator. The welding energy is stored at about 50 to 150 V (or occasionally up to 300 V) and later discharged to make the weld.

Low-voltage capacitors are similar to those used to supply power for capacitor-discharge stud welding, but are discharged in about one fourth to one sixth of the time. The low voltage makes this kind of power supply appropriate for use with hand-held PEW guns, for which protection from high voltage would be difficult to provide. Low-voltage capacitors are also used with bench-mounted heads in PEW systems that are less expensive than those using high voltage and are adaptable to mechanized high-speed production.

High-voltage capacitors that have low capacitance are also used to supply power for capacitor-discharge PEW. They function the same as low-voltage capacitors, but store the welding energy at 1000 to 3000 V (or occasionally up to 6000 V).

High-voltage capacitors can produce a more uniform arc discharge, and no arc-starting nib is required. The high voltage also allows more latitude in controlling operating variables. However, for hand-held equipment, it is more difficult and costly to provide operator protection from voltages above 1000 V than those below 150 to 300 V.

Resistance welding transformers used as power-supply units for magnetic-force PEW supply 60-cycle alternating current at low voltage. They are used at a lower impedance and deliver current at a higher voltage (10 to 35 V) than those used in ordinary resistance welding.

Two transformers are used, as explained in the section on magnetic-force percussion welding in this article. The weld is made during the first half-cycle of current flow; thus, this system functions essentially like a low-voltage direct current system without an auxiliary energy-storage device.

Arc Time and Heat-Affected Zone

Arc time is the interval that begins when the arc is initiated and ends when one workpiece strikes the other and the arc is

Table 2 Conditions for and results of PEW of seven combinations of work metals

The welding machine was a low-voltage capacitor-discharge unit; the capacitor bank had a variable output rated at 20 to 400 μF and 600 V max; the series resistor was a 0 to 7.5-Ω, 50-W stepless potentiometer; effective weight of the cantilever was about 6 oz, but additional weight could be added if needed.

Workpieces welded	Welding voltage, V	Resistance(a), Ω	Initial gap(b), in.	Tensile strength, ksi	Location of failure
Wire size of 0.015 in. diam					
Chromel wire to Alumel wire	130	1.0	3/8	77.3	Alumel
Copper wire to Nichrome wire	160	1.5	3/8	39.1	Copper
Copper wire to stainless plate	150	1.0	1 1/4	40.0	Copper
Nichrome wire to stainless plate	350	1.0	2 1/4	145.5	Weld
Wire size of 0.150 in. diam					
Chromel-Alumel wire to stainless plate	350	1.0	2 1/4	65.0	Weld
Wire size of 0.040 in. square					
Thorium wire to thorium wire	350	1.0	2 1/4	(c)	Wire
Thorium wire to Zircaloy-2 wire	350	1.0	2 3/4	(c)	Zircaloy-2

(a) Setting of potentiometer. (b) Distance of fall of wire (workpiece) attached to pivoted arm. (c) Not measured
Source: Owczarski, W.A. and Palmer, A.J., *Metalworking Production*, Aug 9, 1961, p 57-59

quenched. Factors affecting arc time include the work metal or combination of work metals, mass of the moving workpiece and moving parts of the machine, nib dimensions, welding voltage and current, welding force, and synchronization of arc initiation with the application of welding force.

Generally, the shortest arc time that will permit the formation of a sound metallurgical bond with some penetration into both workpieces is used to minimize heating effects on adjacent areas of the workpieces. Typical arc times in PEW are up to 1 ms for capacitor-discharge welding and up to 8 ms when using a transformer, as in magnetic-force PEW.

The heat-affected zone is very shallow, because of the short arc time. For capacitor-discharge welding this zone is often only about 0.0015 to 0.005 in. In percussion welds between metals that have widely different melting temperatures, the HAZ may be only a few millionths of an inch in the higher melting metal and 0.015 to 0.025 in. in the lower melting metals. Because the heat-affected depth is so shallow, heat treated metals can be welded without softening them. The heat input is so concentrated and of such short duration that heat-sensitive components near the weld area are not affected by the welding cycle. Because the capacitor-discharge method permits a shorter arc time, this method can be expected to produce somewhat shallower heat-affected zones in a given joint than the magnetic-force method can.

Welding Energy

The charge on the capacitor (or bank of capacitors) and the voltage give an approximate measure of the welding energy expended at the joint in the arc discharge. The energy can be calculated by the following equation:

$$E = \frac{1}{2} C v^2$$

where E is energy in watt-seconds (joules), C is capacitance in farads, and v is voltage.

The amount of energy used in making a percussion weld depends on the cross-sectional area of the joint, the properties of the work metal or metals, and the depth to which metal is melted on the workpieces.

Welding Current

The welding current pattern (or arc-discharge pattern) in PEW varies with the application and is not usually measured. Figure 2 shows the changes in welding current during capacitor-discharge PEW of wires.

A peak current density of approximately 300 000 A/in.^{2} is obtained when a low voltage of 80 to 90 V is used in capacitor-discharge PEW of steel to steel, steel to copper, aluminum to aluminum, and brass to aluminum.

Polarity is of no consequence in making percussion butt welds between members made of the same metal with the same cross section, but can affect results in other types of percussion welds. When the members have different cross sections or different melting points, and the amount of heat input is a critical factor in obtaining a sound weld, the member requiring a greater amount of heat input (the member having the larger cross section or higher melting point or thermal conductivity) is ordinarily given a positive polarity.

The selection of polarity is of special importance in the PEW of unlike metals that differ greatly in melting temperature, because it is used to minimize the depth of melting, as well as the depth of the HAZ, in the lower melting metal.

The transformer core sometimes must be defluxed during loading time to ensure the correct polarity when alternating current is applied in magnetic-force PEW of work metals for which polarity affects weld quality. Defluxing is done automatically on some machines by passing the second half-cycle at reduced current through the transformer in the direction opposite that of the first (welding) half-cycle.

Welding Force

The force used in PEW is difficult to measure, because it is dynamic rather than static and depends on the velocity and mass of the moving workpiece and moving parts of the machine. Peak loading of 15 to 30 ksi has been observed in dynamic measurements for capacitor-discharge PEW.

To produce good welds, the welding force must be adjusted empirically until the desired weld quality is obtained. Welding force can be supplied by an electromagnet, gravity, a cam-actuated direct drive, or a spring, depending on the PEW method and the size, shape, and arrangement of the parts being welded. The welding force must be great enough to accelerate one of the parts being welded and the moving parts of the machine to a high velocity within the short gap characteristic of PEW

An impact velocity of 80 to 150 in./s has been used with spring drive in low-voltage PEW of wires up to about 0.010 in. in diameter. A range of 10 to 60 in./s has been used in joining wires only a few mils in diameter. The application of the percussive force and the welding current must be precisely coordinated.

Because a force-applying component may rebound and put a tensile load on the weld metal while solidifying, a means of damping the rebound must be provided. Damping is particularly critical in the welding of small parts.

Arc Starting

Three methods of starting the arc are used in PEW. In one method, the arc is started by applying a direct current voltage to the parts being welded that is high enough to overcome resistance of the air in the gap between the parts as one moves toward the other. Thus, the air is ionized, and the flow of welding current is started. This method is used in high-voltage capacitor-discharge PEW.

Another method of arc starting employs the superimposition of an auxiliary high-frequency, high-voltage alternating current on a low-voltage direct current across the gap between the workpieces. The high-frequency alternating current ionizes the air in the gap, and the low-voltage direct current maintains the arc. This method is used in some low-voltage capacitor-discharge PEW. It eliminates the need for preparing a nib on one of the workpieces.

In the third method of arc starting, a starter nib is prepared on one of the workpieces by cutting it at an angle, or in the shape of a chisel tip or other projection, or by attaching or forming a projection on it. Either a low-voltage direct current (in some low-voltage capacitor-discharge PEW) or the first half-cycle of low-voltage alternating current from a transformer (in magnetic-force PEW) will create enough heat to melt the nib. The nib is heated so rapidly when the arc forms that it explodes and molten particles are expelled from between the work surfaces at high velocity. These particles help to form the electric arc, which then spreads progressively over the workpiece surfaces at the interface of the joint being welded.

Capacitor-Discharge Percussion Welding

In capacitor-discharge PEW, a direct current with a low voltage of about 50 to 150 V (or occasionally up to 300 V) or a direct current with a high voltage of 1000 to 3000 V (or occasionally up to 6000 V) is supplied by a capacitor or bank of capacitors (see the section on power supply in this article).

In low-voltage welding, arc starting is accomplished with the aid of a nib or a high-frequency pulse of alternating current (see the sections on nib-starter and high-frequency-start machines in this article). In high-voltage welding, no auxiliary arc starter is needed.

The mechanism of arc starting at voltages below the ionization potential of air (about 450 V) is not clearly understood, but is believed to involve a cold cathode discharge. Arc starting is less consistent below 450 V than at higher voltages, particularly in welding workpieces of large diameter.

Because arc equilibrium conditions are not achieved in the short interval of the arc discharge, the repeatability of percussion welds is affected by variations in arc-starting behavior, and hence is related to the voltage used. Voltages near 450 V are not usually selected, as small variations in the operating conditions could change the mechanism of arc starting.

The use of a nib or a high-frequency current pulse at low voltages greatly improves the consistency of arc starting. With close control of the welding conditions, the quality of welds produced in low-voltage systems can be uniformly high, with the incidence of defective welds often being 0.1% or less.

Arc starting ordinarily takes place just before the workpieces come into contact, at a separation of about 0.0002 to 0.0004 in. in low-voltage systems in which a starter nib is used. The discharge takes place across a typical gap of 0.010 to 0.030 in. in low-voltage systems which use a high-frequency pulse. In studies of PEW of workpieces 0.040 in. in diameter using a voltage of 1300 V, the arc discharge began when the gap between workpieces was 0.004 to 0.006 in.

Sequence of Steps. In operation, first a capacitor (or bank of capacitors) is charged by direct current from a rectifier or motor-generator. Then one of the parts being welded is advanced toward the other and rapidly accelerated. When the two parts are close enough, the capacitor discharges—exploding the nib (arc starter), if one is used. The intense arc formed between the parts by the discharge heats the work surface of each part to the melting temperature in a fraction of a millisecond. Then, as one part is impacted against the other at high velocity, molten metal is expelled from the joint and the parts are forged together to complete a weld. The sequence of steps in capacitor-discharge PEW is shown in Fig. 1.

Control. Close control of voltage, capacitance, impact velocity, and limiting resistance is important for producing a good weld. The voltage and capacitance determine the amount of energy stored in the system, and hence the heating capacity of the arc. Impact velocity (together with the mass of the moving workpiece and clamping members of the machine) determines the amount of forging energy. The limiting resistance is generally adjustable and controls the peak discharge current.

Fig. 1 Sequence of steps in capacitor-discharge PEW. (a) Contactor is momentarily closed to charge the capacitor. (b) One of the parts to be welded is advanced toward the other and rapidly accelerated. (c) The arc forms across the gap just before the parts meet, melting the work surface of each part and exploding the nib (arc starter), if one is used. (d) The arc is extinguished as one part is impacted against the other, expelling molten metal as flash and forging the parts together to complete the weld.

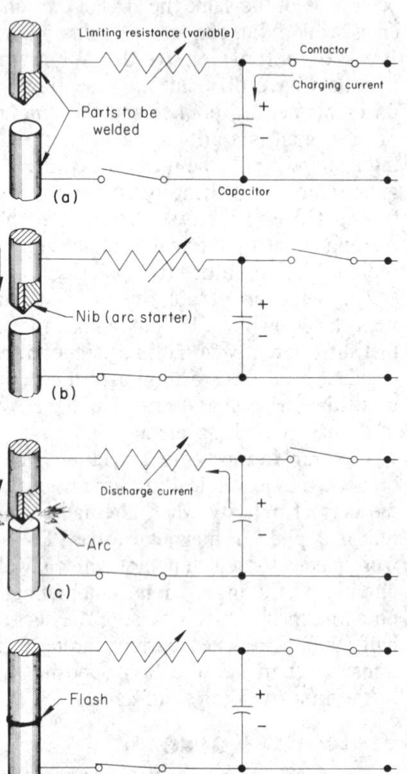

These factors interact to determine the arc duration and timing with respect to arc discharge. The approach of the workpieces serves as a switch to trigger the arc discharge.

Usually, conditions are adjusted to give the shortest arc time that will permit consistent production of welds with desired properties. If the parts being welded are forced together too soon, the arc is extinguished before the work surfaces of both parts are melted. If the impact is delayed too long after arc initiation, the melted interfaces may solidify too soon to permit expulsion of excess molten metal.

Preparation of Workpieces. For capacitor-discharge PEW, the need for preparation of workpieces varies widely, depending on the application. If a nib is needed for arc starting, a shear or cutter usually produces the desired tip configuration.

Displacement, Current, and Voltage. The changes in displacement of the moving workpiece, welding current, and voltage across the weld that take place during capacitor-discharge PEW, based on oscillograph records for the welding of wires, are shown in Fig. 2.

Displacement of the moving workpiece in the direction of travel proceeds at a uniform rate during the arc discharge and while molten metal is being expelled, but is slowed as the forging action takes place. Displacement continues beyond the final displacement on the completed weldment because of deflection of the mechanical system and the workpieces. The maximum displacement because of deflection is reached after about 3 ms. The oscilla-

Fig. 2 Changes in the displacement of the moving workpiece, welding current, and voltage across a weld during capacitor-discharge PEW. D_1 is displacement during arc discharge. D_2 is displacement during expulsion of molten metal. D_3 is displacement during forging and deflection. D_4 is displacement during deflection. Diagrams are based on oscillograph records made in studies on the welding of wires. Source: Bradley, Irving, Use of Percussive Welding in Electronics, *Electro-Technology*, Sept 1968, p 72

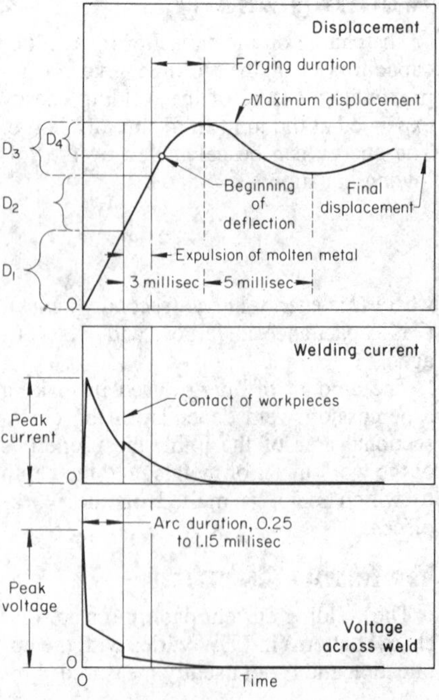

tions are damped rapidly, becoming insignificant after a reverse-oscillation peak that occurs after an additional interval of approximately 5 ms.

Peak welding current is achieved almost immediately on arc initiation and decays rapidly during the arc discharge. Figure 2 shows that the current increases to a secondary peak on contact of the workpieces, because of the sudden drop in electrical resistance, before tapering off to zero in an additional 3 to 5 ms.

Voltage across the weld decreases very rapidly to a fraction of its initial open-circuit value when the arc is initiated by the close approach of the moving workpiece to the stationary workpiece. The voltage decreases less rapidly as the arc discharge continues. When the arc is extinguished on contact of the workpieces, after a typical arc time of 0.25 to 1.15 ms, the voltage drops instantly to nearly zero.

The magnitude and rate of change of displacement, current, and voltage vary with the size and nature of the workpieces and the characteristics of the welding equipment. Figure 2 is based on the low-voltage capacitor-discharge PEW of 0.015- to 0.040-in.-diam wire of various materials, including Dumet, nickel, nickel-plated steel, and copper-plated steel.

Low-Voltage Capacitor-Discharge Percussion Welding

Low-voltage capacitor-discharge PEW is done on many types of equipment. Simple and relatively inexpensive machines are used for low production. For welding in intermediate to large quantities, commercially available and specially designed machines of widely varying capability, complexity, and cost are used. These generally can be adapted for application with specially designed work-handling equipment for semiautomatic or automatic high-volume production.

The amount of energy stored in the capacitor bank to make a weld is usually regulated by selection of the number of capacitors charged and by charging to a controlled voltage. The initial-gap technique is used throughout. Except for machines that use a high-frequency pulse of alternating current for arc starting, the approach of the workpieces initiates the arc.

Machines for low-voltage capacitor-discharge PEW can be classified by the arc-starting technique as nib-starter or high-frequency-start machines.

Nib-Starter Machines. In one type of machine for low-voltage capacitor-discharge PEW, a nib is used for arc starting (see the section on arc starting in this article). This machine usually consists of a portable power supply and either a hand-held gun or a bench-mounted welding head, although these components can also be built into a specially designed integral machine. Voltages used with the bench-mounted machine usually range from 50 to 150 V; for operator safety, the voltage used with the hand-held gun is ordinarily about 50 V.

The bench-mounted and integral machines are readily adaptable to mechanization for high-speed, high-volume welding operations, but the hand-held gun has the advantage of portability. Otherwise, the machines operate in much the same manner.

Nib-starter machines have two sets of jaws—one set movable and the other stationary—with provision for precise alignment of the two sets of jaws and setting the initial gap. One workpiece, usually a wire, is held in the movable jaws; the second workpiece, usually a terminal, is held in the stationary jaws. Wire size is about 0.006 to 0.100 in. in diameter. Terminals are usually at least 0.006 in. thick. Some hand-held guns are made to hold only the movable wire (workpiece). This type of gun, which is held against the stationary workpiece, is suitable for welding to large, relatively flat workpieces.

When the switch is actuated, the wire moves toward the terminal at high velocity. Impact velocity varies with the wire size and the mass of the moving jaws, but is ordinarily between 80 and 150 in./s. Total arc time is about 0.15 to 1.0 ms.

The basic electrical system for nib-starter low-voltage capacitor-discharge machines is shown in Fig. 3. The limiting resistor controls the peak discharge current during welding. Another resistor is used as charging ballast to limit the peak charging rate. Moving the switch to the welding position (Fig. 3) isolates the rectifier and charging circuit from the capacitor bank and discharge circuit. In repetitive welding, the switch remains in this position long enough for the weld to be completed, after which it is returned to the charging position to allow recharging of the capacitor bank. The switching is controlled automatically, often by a system of cams.

Fig. 3 Basic electrical system for nib-starter low-voltage capacitor-discharge PEW machines

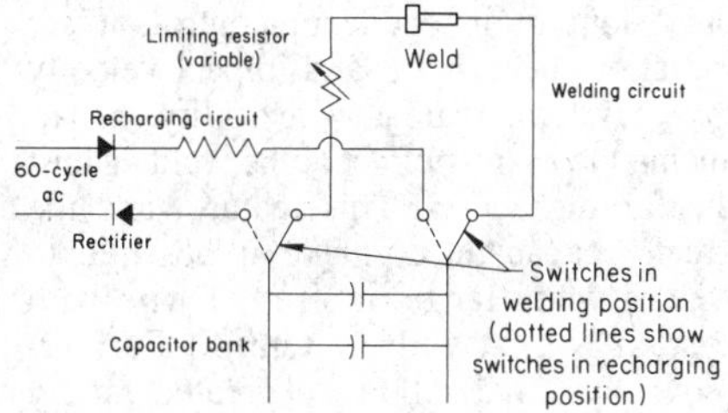

Nib-starter machines vary in details of construction and many are designed for specific welding applications. The driving force for propelling one of the workpieces in nib-starter machines is provided by springs, cams (or cams plus levers), or electromagnets. Timing is automatic for some or all machine functions. Rotating cams on a central shaft are often used to actuate switches, relays, or other devices, as well as to provide machine motions directly. Solid-state timers or other types of timing devices are used in some machines.

Precise alignment is important. For example, when wires of 0.010 in. diam are welded, a misalignment of 0.002 in. can reduce weld size and cause poor welds if wires deflect on impact.

Production rate is usually limited by the time needed for loading and unloading the workpieces or for other work-handling operations. With manual loading and unloading, the production rate for a single welding head is often about 200 to 500 welds per hour.

The simplest type of low-production PEW machine is a nib-starter machine driven by a gravity-operated cantilever arm, which is pivoted on a low-friction bearing to provide the welding force. Figure 4 shows capacitor-discharge PEW of a 304 stainless steel support wire to an Inconel tube.

In fully mechanized welding, the repetition rate of the machine can become a limiting factor for the production rate in welds per head. The charging time for the capacitor bank usually is an approximate measure of the total weld cycle time, which can vary from about 0.2 to 1 s or more, corresponding to production rates of 18 000 and 3600 welds per hour, respectively.

High-Frequency-Start Machines. In the second frequently used type of machine for low-voltage capacitor-discharge PEW, a high-frequency, high-voltage pulse of alternating current serves as an arc starter by ionizing the air in the gap between the workpieces. Similar to the nib-starter machine, this machine usually consists of a portable power supply and either a hand-held gun or a bench-mounted welding head. These components also can be built into a specially designed integral machine. Suitability for mechanization and arrangements for holding and aligning the workpieces in this machine are also similar to the nib-starter machine.

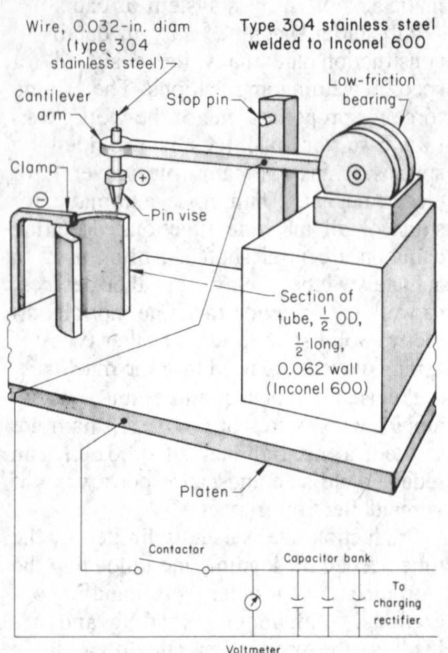

Fig. 4 Setup for capacitor-discharge PEW of a type 304 stainless steel support wire to an Inconel 600 tube on a low-production, nib-starter machine

The main application of high-frequency-start machines is in welding a wire to another wire, pin terminal, or large flat surface. Solid wires about 0.100 in. to less than 0.005 in. in diameter, or stranded wires as small as 0.010 in. in overall diameter (four strands of 0.004-in.-diam wire), can be welded. Special provisions are made for damping the rebound on impact in welding the smaller diameter wires.

The high-frequency-start machine differs from the nib-starter machine because (1) a nib does not have to be prepared on the wire by an angle cut or other means, and the wire can be cut off square or in any convenient shape that gives reproducible dimensions; (2) solid-state devices time the machine function using a switching time of 0.001 to 0.002 ms; (3) a solenoid or electromagnet provides the driving force to propel the moving workpiece; and (4) the approach of the workpieces does not initiate the arc.

A simple form of the high-frequency-start machine (system A) has a basic electrical system similar to the nib-starter machine in Fig. 3. The limiting resistor, however, is replaced by an autotransformer, which is connected to a solenoid and a high-frequency pulse generator (toroid) in series. When the work has been loaded in the preset position with the desired gap, the operator closes a switch to make the weld. The switch (as in Fig. 3) moves to the weld position, closing the discharge circuit from the capacitor bank through the autotransformer and toroid to the weld gap. A high-frequency pulse bridges the weld gap during the first half-cycle of the pulse, ionizing the air in the gap. The arc discharges across the gap, then starts about 0.010 to 0.015 ms after the operating switch is closed. The flow of current through the solenoid actuates it to propel the movable workpiece toward the stationary workpiece at high velocity. Typical total time elapsed at the beginning of the forging action is 1.25 to 2 ms.

Contact of the parts extinguishes the arc. The switch remains in the weld position for a total time of about 6 to 15 ms to allow completion of the weld and almost complete decay of the welding current and voltage. The switch then moves automatically to the charge position. Next, the capacitor bank is recharged by the rectifier to the desired preset voltage, and the machine is ready to repeat the welding operation, after a total time of 1.5 s or less for a complete welding sequence.

The largest percentage of the weld cycle time is charging time, which depends on the characteristics of the charging unit and the capacitor bank, as well as the amount of energy in the charge. Maximum charging time for the type of equipment described is less than 1.5 s.

In system A, the simple form of high-frequency-start machine, two basic adjustments are used to optimize weld properties. Heat input to the weld is adjusted by selecting the number of capacitors to be charged and setting the charging voltage. Impact velocity (and timing) is adjusted by setting the gap. The adjustment of heat input, however, also affects impact velocity (and timing) by changing the amount of current passed through the solenoid. Thus, heat input and impact velocity (and timing) cannot be changed independently by routine adjustments.

A more versatile form of the high-frequency-start machine (system B) has greater capacitance, and the solenoid, heat input, and arc initiation are independently powered. Therefore, the high-frequency pulse can be timed to occur at some definite point during workpiece travel. The provision for independent control of arc initiation, heat input, and impact velocity gives system B greater flexibility in balancing the heating effect at the weld against the forging action. This feature, and the greater capacitance, make it possible to weld a wider range of sizes of wire with system B than with system A. Furthermore, system B allows a greater variety of metal combinations to be welded. Comparison of the characteristics of systems A and B is as follows:

Characteristic	System A	System B
Capacitance, max μF	800	4000
Solenoid connection	Series	Parallel
Voltage, V	12-75	12-120
Gap, in.	0.008-0.010	0.015-0.020
Wire diam(a), in.	0.015-0.070	0.004-0.102

(a) Size that can be welded

High-Voltage Capacitor-Discharge Percussion Welding

High voltage, usually 1000 to 3000 V (but occasionally up to about 6000 V), is used only in bench-mounted machines or integral machines which have been specially designed for capacitor-discharge PEW. The use of hand-held guns in high-voltage systems is precluded by the difficulty and expense of providing protection for the operator against high voltage.

Applications of high-voltage welding are similar to those of low-voltage welding, but the diameter of the smaller workpiece ordinarily does not exceed about 0.060 in., compared to about 0.040 in. diam in low-voltage systems. Maximum diameter of the smaller workpiece is about 0.70 in., compared to about 0.10 in. in low-voltage systems. No nib or other arc-starting aid is needed; the tip of the moving workpiece can have any convenient shape.

The welding operation is generally very close to that described for nib-starter machines in the section on low-voltage capacitor-discharge percussion welding in this article. The basic electrical system is similar to that in Fig. 3.

The amount of energy stored in the capacitor bank to make a weld is usually regulated by selecting the number of capacitors charged and by charging to a controlled voltage. The initial-gap technique is used throughout, and the approach of the workpieces initiates the arc discharge. Impact velocity in high-voltage welding is typically about 40 in./s for wire 0.040 in. in diameter.

The machines are usually designed to suit specific applications and are intended for semiautomatic or fully automatic operation in intermediate- to high-production welding.

In fully mechanized systems, charging time for the capacitor bank may limit the number of welds that can be made per minute by a single welding head. A typical charging time is 300 ms, corresponding to a maximum production rate of 120 welds per minute for a single welding head and capacitance of 75 to 240 μF.

Some machines are fitted with more than one welding head, each with a capacitor bank, to increase production rate. Each capacitor bank is recharged during loading and unloading time.

Magnetic-Force Percussion Welding

In magnetic-force PEW, a resistance welding transformer supplies the welding current at a voltage of about 10 to 35 V. This transformer has lower impedance and higher secondary voltage than the transformers ordinarily used in resistance welding. The system functions like a low-voltage direct current system without an auxiliary energy-storage device, because the weld is made during the first half-cycle of current flow of a 60-cycle alternating current.

Welding force is developed by an electromagnet, and the magnitude of this force, which is controlled by a separate transformer, can be varied without affecting the welding current. The required welding force depends on the size and composition of the parts being welded, which determine the response of the parts to heating.

Arc starters are usually nibs or projections on one of the parts to be welded. For welding elongated parts, two nibs are sometimes necessary. An air cylinder holds the workpieces together at low pressure while a current path is established through the nib. A magnetic-force PEW machine, shown in Fig. 5, is similar to a press-type resistance welding machine.

Fig. 5 Magnetic-force PEW machine and setup for welding

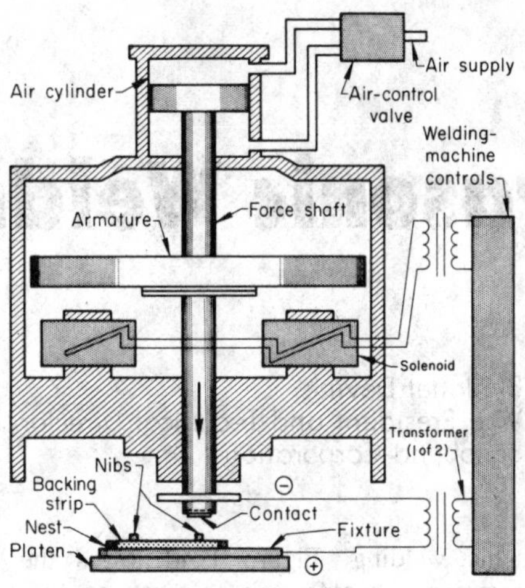

The diameter and height of an arc-starting nib or projection depend on the application. The diameter must be large enough to prevent the nib or projection from collapsing under the initial pressure, but must not be large enough to permit the nib or projection to carry the welding current. The height determines the gap between the workpieces to be welded and the arc voltage that is necessary to explode the nib.

Arc time, up to 8 ms, is governed by the amount of energy discharged, the magnitude of the magnetic force, and the dimensions of the arc-starting nib. The interval between initiation of the arc and application of the magnetic force has an influence on both the arc time and the resulting weld quality.

Weld areas of 0.04 to 0.70 in.^{2} can be welded in production. Some meltback (in addition to the explosive removal of the nib) occurs at the weld interface, and in most applications, some flash must be removed. Weld metal is not always completely expelled from the joint.

Applications of magnetic-force PEW include joining silver-cadmium oxide to brass, cadmium-plated brass, or copper; copper to copper; copper to silver-tungsten; and copper to silver-cadmium oxide.

Safety

An operator or anyone near PEW equipment is exposed to the following hazards. The noise level is high for welding large parts that have nib-type arc starters; for parts about 1/2 in. in diameter, the noise is similar to that created by firing a 12-gage shotgun. Weld flash and molten particles may be expelled from the joint at high velocity. Highly toxic vapors, such as silver-cadmium oxide, are released in the welding of some metals.

In general, the severity of the hazards is approximately in proportion to the area of the weld, and thus is minimal for the welding of fine wires. Hazards from noise and weld flash can be minimized by shielding the weld area. Toxic vapors can be drawn off by an exhaust system. All three hazards can be minimized by enclosing the operation in a well-ventilated cabinet. Furthermore, the operator should be protected from the welding voltage, particularly when high-voltage capacitors are used. Additional information on safe welding practices can be found in the American National Standard Z49.1 on safety in welding and cutting.

Ultrasonic Welding

By Janet Devine
Vice President and Technical Director
Sonobond Corporation

ULTRASONIC WELDING (USW) is used effectively for joining both similar and dissimilar metals with lap joint welds. High-frequency vibrations, introduced into the areas to be joined, disrupt the metal atoms at the interface of the weld components and produce an interlocking of these atoms to achieve a mechanical joint. No significant heating is involved; the maximum temperature at the weld interface is usually in the range of 35 to 50% of the absolute melting point of the metal. The base metal does not melt and subsequently solidify with a brittle cast structure, as with high-temperature joining processes. Moderate pressure is applied during joining to maintain intimate contact between the parts, but the pressure does not cause significant deformation in the weld zone—seldom more than about 10%. Preweld cleaning requirements are minimal, while postweld cleaning or heat treatment is not necessary. Fluxes or filler metals also are not used.

Ultrasonic energy is produced through a transducer, which converts high-frequency electrical vibrations to mechanical vibrations at the same frequency, usually above 15 kHz (above the audible range). Mechanical vibrations are transmitted through a coupling system to the welding tip and into the workpieces. The tip vibrates laterally, essentially parallel to the weld interface, while static force is applied perpendicular to the interface.

Spot welding and continuous seam welding can be accomplished by this method. In spot welding, the duration of the ultrasonic pulse (usually 1 s or less) is selected according to the thickness and hardness of the materials being joined. For seam welding, the tip is a roller disk that rotates across the parts to be joined.

Most industrial metals can be joined to themselves or to other metals by ultrasonic welding. Figure 1 identifies the combinations of metals that can be successfully joined by USW. Ductile metals, aluminum and copper alloys, as well as the precious metals (gold, silver, platinum, and palladium), are among the easiest to weld. Both aluminum alloys and the precious metals can be joined to semiconductor materials such as germanium and silicon. Various types of iron and steel are somewhat more difficult to join. The most troublesome are refractory metals and some of the higher strength metals, including titanium, nickel, and zirconium, and their alloys, which should be welded only in thin gages.

Welding difficulty is related to the ultrasonic energy (power × time in joules or watt-seconds) required to complete a joint. The harder metals, such as various

Fig. 1 Ultrasonically weldable metal combinations. Blank areas in the chart represent combinations that have not been successfully joined or in which welding has not been attempted. From the American Welding Society, with permission.

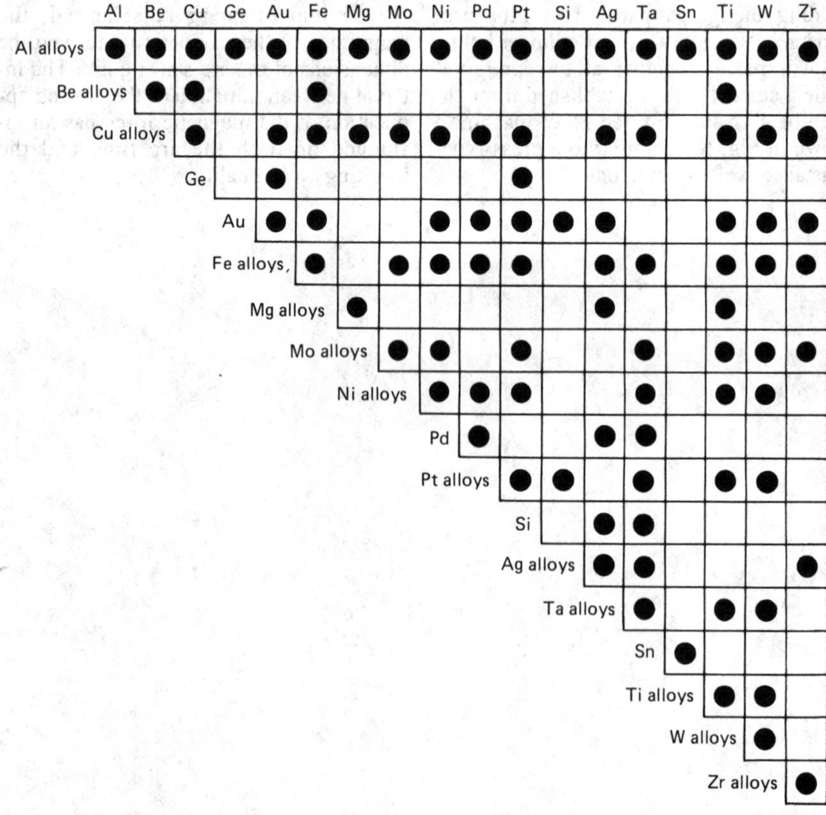

	Al	Be	Cu	Ge	Au	Fe	Mg	Mo	Ni	Pd	Pt	Si	Ag	Ta	Sn	Ti	W	Zr
Al alloys	●	●	●	●	●	●	●	●	●	●	●	●	●	●	●	●	●	●
Be alloys		●	●			●										●		
Cu alloys			●		●	●	●	●	●	●	●		●	●		●	●	●
Ge					●						●							
Au					●	●			●	●	●	●	●			●	●	●
Fe alloys						●		●	●	●	●		●	●		●	●	●
Mg alloys							●						●			●		
Mo alloys								●	●		●			●		●	●	●
Ni alloys									●	●	●			●		●	●	
Pd										●			●	●				
Pt alloys											●	●		●		●	●	
Si													●	●				
Ag alloys													●	●				●
Ta alloys														●		●	●	
Sn															●			
Ti alloys																●	●	
W alloys																	●	
Zr alloys																		●

types of iron and steel, require higher energies than softer materials for a given sheet thickness. Welding-tip wear is also more severe in harder metals. However, the maximum weldable thickness is limited only by the power output limitations of the available welding equipment. The upper limit is currently about 0.125 in. for soft aluminum or copper and generally in the range of 0.010 to 0.040 in. for some of the harder materials. The minimum weldable thickness is represented by thin foils 0.00017 in. thick or fine wires 0.0005 in. in diameter. Thickness limitations apply only to the weldment component adjacent to the welding tip.

Ultrasonic welding is generally restricted to lap joints; butt welds cannot be made because there is no effective means of supporting the workpieces and applying static force. Lap joints can be made between sheet and foil materials. Metal wires and ribbons can be welded to flat surfaces and to each other. Irregularly shaped components can be joined if support is available for the workpieces. Production applications are found in electronic and electrical contact assembly, in the splicing of metal foil and thin sheets, in the fabrication of solar energy systems, in the packaging industry, and in certain structural joining for the automotive and aircraft industries.

Welding Equipment

The flow of energy through a USW system occurs in the following manner. Standard electrical power at 50 to 60 Hz is introduced into a frequency converter, where the frequency is converted, or raised, to the desired high frequency for operation of the equipment. The output from the converter is amplified and fed to the transducer, which converts the electrical energy to mechanical vibratory (ultrasonic) power. An appropriately designed coupling system transmits the vibratory power to the welding tip and then into the metals being joined. An anvil is essential to support the workpieces, and a mechanism must be provided for applying static force. Various electrical, electronic, and mechanical control devices may be incorporated into the welding setup as dictated by the complexity of the equipment and the type of application.

Frequency Converters. The frequency converter, or power supply/generator, is usually separate from the welding head and is connected to it by a lightweight coaxial cable. The converter, like the welding head, is rated according to the operating frequency and the power output. Practical and effective welding frequencies are usually within the range of 15 to 60 kHz, and power outputs may range from a few watts to 16 kW.

Usually the frequency converter is designed to operate at a selected nominal frequency with a narrow range of 1 to 2% above or below this frequency; it is essential to successful welding that the converter be tuned precisely to the operating frequency of the welding head. Automatic frequency control is standard on most units. This provides for automatic tracking of the actual operating frequency of the welding system, as frequency may vary slightly with changes in operating conditions or workpiece variations.

Most welding frequency converters are designed with solid-state circuitry of the transistorized or silicon-controlled rectifier (SCR) type. Such devices are small, lightweight, and compact; they provide high efficiency with minimum maintenance. Frequency converters consist of an oscillator stage, in which the input 60-Hz power is raised to the required high frequency, and an amplifier stage, in which the power output is amplified. Failsafe devices are usually incorporated to protect the circuitry from power overload and overheating.

Transducers accept the high-frequency electrical output from the frequency converter and convert it to mechanical vibrations at the same high frequency. Both magnetostrictive and piezoelectric transducers may be used. The magnetostrictive type is usually a laminated stack of nickel or nickel alloy sheet, which is capable of changing length under the influence of magnetic flux density. Such transducers are rugged and serviceable for continuous-duty operation; however, they have low electromechanical conversion efficiencies, generally in the range of 25 to 35%. Consequently, they are used almost exclusively for low-power welding systems.

Piezoelectric ceramic materials such as lead zirconate titanate change dimensions under the influence of an electrical field. Their conversion efficiencies are about twice those of the magnetostrictive type. Because such ceramic materials have low tensile strengths, they usually are enclosed in rugged assemblies, with the ceramic disks installed under a bias compressive load to ensure that they do not fail in tension during operation. Figure 2 shows typical piezoelectric transducers of several sizes and frequencies. The smaller transducers operate at the higher frequencies, and the larger ones operate at lower frequencies.

There is no critical frequency for welding specific metals or thicknesses. However, the amplitude of vibration is inversely proportional to the frequency. Consequently, where high powers are required, lower frequencies (15 to 20 kHz) are most effective, while higher frequencies (40 to 60 kHz) are used for low-power machines.

Fig. 2 Ultrasonic piezoelectric transducers with frequencies ranging from 15 to 60 kHz

Both magnetostrictive and piezoelectric transducers are equipped with air cooling systems to prevent loss of transduction characteristics with continuous-duty operation.

Coupling Systems. The coupling system must be efficiently designed to conduct the high-frequency mechanical vibrations from the transducer to the welding tip and into the materials being welded. For joining metals, the tip must vibrate in the lateral direction. Two methods are used for achieving this lateral vibration—the lateral-drive system and the wedge-reed system.

All continuous seam welders and low-power spot welders use a lateral-drive system such as that shown in Fig. 3. Here, the welding tip is attached to a lateral coupling member that is driven longitudinally by the transducer, producing tip motion parallel to the weld interface. Clamping force at the tip results when a bending moment is applied to the coupling member as shown by the arrows in Fig. 3. For continuous seam welders, the tip is replaced by a rotating disk, where the entire transducer-coupling-disk assembly is rotated by antifriction bearings and a motor

Fig. 3 Lateral-drive spot and seam welding system. From the American Welding Society, with permission.

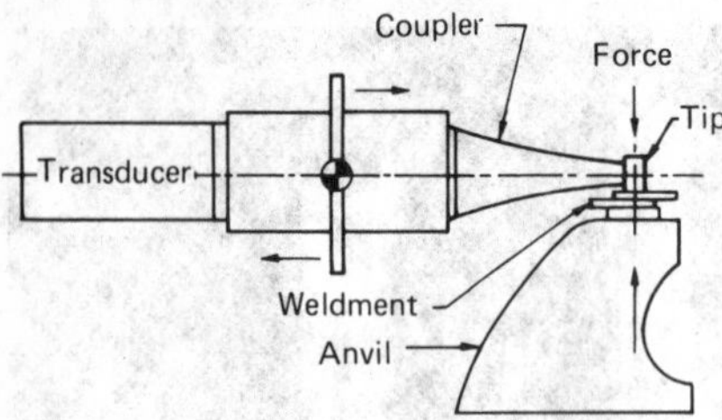

drive so that the disk tip maintains rolling contact with the work.

Intermediate and high-power spot welders incorporate a wedge-reed system, as shown in Fig. 4. The transducer drives a wedge-shaped coupler that is rigidly attached to a reed. Longitudinal vibration of the wedge drives the reed in flexural vibration. With such flexural vibration, there are maximum and minimum displacements (nodes and antinodes) of the sinusoidal wave along the reed length, with spacing dependent on the transducer frequency and the reed material. Reed length, therefore, is selected to provide maximum vibratory amplitude at the tip end, and this vibration occurs parallel to the weld interface. In high-power equipment, two or more transducer-wedge assemblies are used to drive the reed.

Materials for a coupling system are selected to provide low energy losses and high fatigue strength to ensure maximum energy transmission and endurance under the applied static and vibratory stresses. Stainless steel and titanium alloys are among the most effective materials. The coupling system usually includes a tapered or stepped horn (mechanical transformer) to concentrate the vibratory energy into a smaller cross-sectional area and thus increase the amplitude of vibration. The joints between the transducer and coupler and between the various coupling members must have high integrity and good fatigue strength to ensure equipment reliability. Some welding systems use mechanical joints for ease of interchangeability. Brazed or welded joints are also acceptable.

Fig. 4 Wedge-reed USW system. From the American Welding Society, with permission.

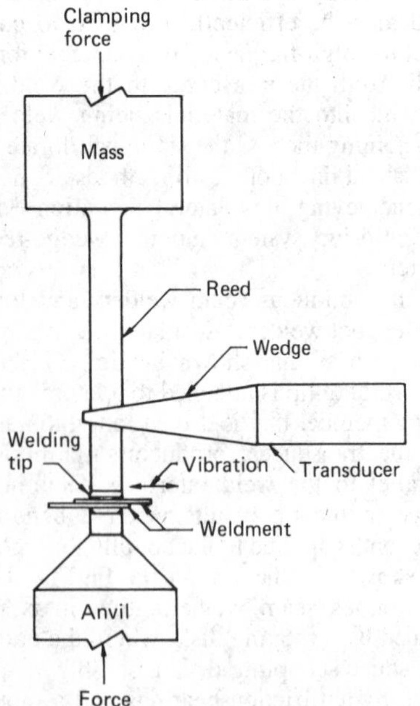

Welding Tips. Coupling members and tips are always designed for resonance at the transducer frequency. Standard spot welding tips, used particularly for joining flat components, are contoured to a hemispherical radius of 50 to 100 times the thickness of the adjacent workpiece. Thus, if the workpiece is 0.040 in. thick, the tip radius would be in the range of 2 to 4 in. When joining two wires or a wire to a flat surface, the tip is usually grooved to the wire diameter so that the wire is not excessively deformed during welding. Large-area tips, elongated tips, multiple-spot tips, and other configurations are possible according to the desired joining area.

Seam welding tips are resonant disks designed to transmit the vibrations from the center coupler attachment to the disk periphery. The disks usually have a crown radius in the range between 1 and 6 in. For specific applications, such as for continuous seam welding of small ribs to tubes, the disk periphery may be slotted or otherwise contoured.

Welding tip materials are selected to provide good fatigue strength and service life for the materials being welded. The material may be high-speed tool steel for welding relatively soft materials such as aluminum, copper, or low-carbon steel, or precipitation-hardenable nickel-based alloys for welding hard, high-strength materials.

Tips are sometimes provided with rough contact surfaces by electrodischarge machining (EDM) or sandblasting. Such abrasive tips or serrated tips tend to prevent gross slippage of the tip against the workpieces and also permit the use of lower powers and clamping forces than are required with smooth tips. For welding copper, hardened tool steel tips are used; tool life in excess of 100 000 welds before redress is routine. In general, weld tip life is dependent upon the weld material and hardness; aluminum and copper offer the longest tip life. The EDM tip treatment produces a shorter tip life.

When tips begin to show wear, they may be reconditioned. With smooth tips, light sanding with 400-grit silicon carbide paper usually is adequate. Worn abrasive tips may be given additional EDM or sandblasting treatment. All spot welding tips are designed for interchangeability so that they may be replaced when wear becomes excessive.

Adhesion of the tip to the weld surface may be a problem, especially if improper machine settings are used. Adjusting the machine settings may alleviate the situation, or a lubricant such as a faint trace of dilute soap solution may be applied to the tip surface. With extreme tip sticking, the welding machine may be designed to provide an afterburst of ultrasonic energy after the weld is completed to effectively release the tip.

Anvils. Whereas welding tips are resonant members of the system, supporting anvils are designed to be contraresonant to prevent transmission of ultrasonic energy through the anvil structure to consequently concentrate the energy within the workpieces. The anvil is designed to support and accommodate the workpieces. Flat surfaces are frequently satisfactory, but the surface may also be contoured, like the welding tips, for special applications. For joining two or more wires, for example, the anvil may be grooved like the tip. It may also consist of a clamp to support a tube or other massive workpiece member.

The anvil, like the tip, is designed to be interchangeable. Anvil mounts are adjustable to accommodate special tooling heights. Material and maintenance requirements for anvils are the same as required for tips.

Force Application Systems. In all welding systems, a method must be provided for applying a static force perpendicular to the weld interface to hold the workpieces in intimate contact. The required force may range from a few ounces to several hundred pounds. Force systems may be hydraulic for high-power welding machines, pneumatic for intermediate-power welding machines, or spring-actuated or deadweight-loaded for low-power systems. Clamping force may be applied through the coupler-tip assembly or from below through the anvil. Generally, hand materials require lower forces and higher power levels; softer materials require higher forces and lower power levels. In addition, the force requirement is directly proportional to the weld contact area.

Welding Machines. The components of a welding system—transducer, coupling system, tip, anvil, and force application system—are designed into a welder frame that is rigid enough to ensure precise alignment and prevent undesirable deflection. Usually, the transducer-coupling assembly is installed with a force-insensitive mounting system to prevent energy losses to the support structure or shift in the resonant frequency when static force is applied. Ultrasonic systems without such force-insensitive mounts have been known to stall under static loads.

With a spot welding machine, after the workpieces are positioned on the anvil, the welding cycle involves lowering the welding tip (or raising the anvil) so that the tip contacts the workpieces, applying the required clamping force, introducing ultrasonic energy for the required time period, then retracting the tip to permit removal of the welded part. For seam welding, the disk tip is lowered, clamping force is applied, ultrasonic power is introduced, and the tip rotates at a preset welding speed for the desired length of seam or to a preset stop position. The entire cycle is completed automatically and may be triggered by dual anti-tiedown, anti-repeat palm buttons in accordance with Occupational Safety and Health Administration (OSHA) regulations, by a foot switch, or by an external relay or microswitch.

Simple controls are provided for setting the primary variables of ultrasonic power, clamping force, and weld time for spot welding or welding rate for seam welding. Appropriate readouts are provided for all controls. Other controls include a radio-frequency switch for introducing power to the unit, means for tuning the frequency of the converter to that of the welding system, and often an emergency stop button to shut off the equipment in case of malfunction. Other controls and adjustments may be provided for a specific application or to provide flexibility of use. For example, a spot welding machine may include a mechanism for adjusting coupler stroke length, speed of welding tip advance and retraction from the workpieces, or anvil height and orientation.

Equipment may also include weld-quality monitors. One such device is a weld power meter that indicates the power delivered into the weld. A substantial change in the load power indicates a faulty weld, perhaps because of variation in part dimensions or surface finish, improper placement of parts, or machine malfunction. On some machines, the high and low limits of acceptable power can be preset, and deviations from this range can trigger a visual or audible signal to alert the operator or to actuate a reject mechanism.

Fig. 5 1400-W ultrasonic spot welding machine

Fig. 6 4-kW ultrasonic spot welding machine

Spot welding machines are available with power capacities ranging from 10 W to 16 kW. The 1400-W machine in Fig. 5 incorporates a wedge-reed coupling system, and clamping force is actuated pneumatically. The 4-kW machine in Fig. 6 has a wedge-reed coupler and a hydraulic clamping force system. This machine demonstrates use of a custom-designed anvil.

Continuous seam welding machines range in power from 100 to 500 W. In one arrangement, used for aluminum foil splicing in rolling mills, the disk tip rotates and traverses across a stationary anvil, as shown in Fig. 7. Other systems draw the work between a rotating tip and a counter-rotating anvil, as in Fig. 8, which provides greater flexibility in manipulating the workpieces.

Automated Production Equipment. Ultrasonic welding systems can be modified to fit into close-packed production lines in which welding is only one of several operations being performed. For example, the transducer-coupling-tip assembly can be removed from the standard frame and

Fig. 7 100-W ultrasonic foil splicer with traversing welding head

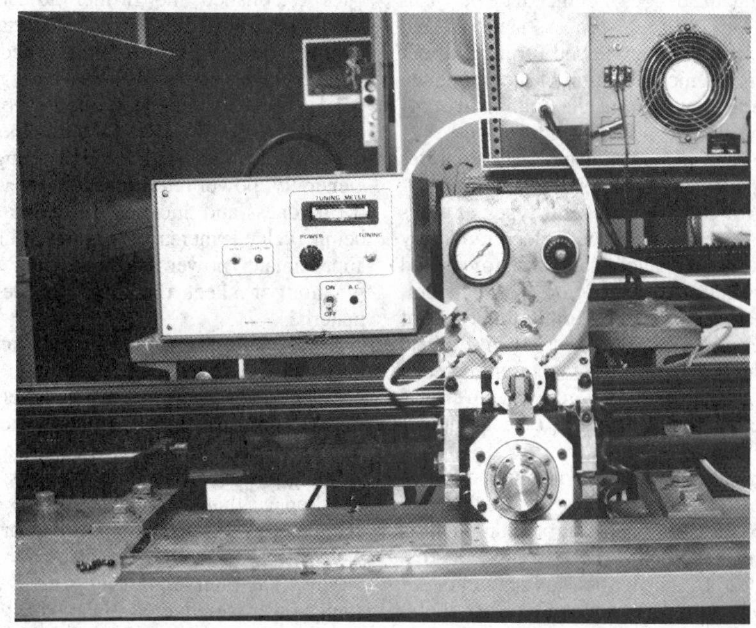

Fig. 8 100-W rotating-tip welding machine for joining foils and thin sheet metals

installed on any rigid structure so that the tip contacts the work from above, below, or laterally. Work-handling equipment can be incorporated for loading the parts into the welding machine and removing them after welding is complete. Spot welding times are usually a fraction of a second, and production rates are limited primarily by the speed of the work-handling equipment. No extensive heating of the workpieces or equipment occurs, so that the welding head does not need to be insulated from adjacent process stations, and the work can be processed further without cooling. Remote location of the frequency converter, away from the continuous processing line, is a further advantage.

One automated machine incorporates multiple channels of power-force-time settings for use when several types or sizes of workpieces are to be welded randomly on the same machine. Power, force, and time values are preset for each channel, and a selector switch permits selection of the desired combination for any given weld without the necessity for resetting each value. Channel selection for a given sequence of welds can also be computer controlled.

Typical installations for automatic processing include systems for high-speed electronic component fabrication such as the attachment of wires to semiconductors or of leads to capacitors, fabrication of solar collectors or photovoltaic devices, aluminum foil splicing in foil mills, bonding of wires to commutators for motor armatures, and attachment of clips to tubes for nuclear fuel elements.

Welding Procedures

Operating Variables. The welding variables requiring control by the operator include ultrasonic power, clamping force, and weld time or rate. These must be adjusted for each specific application, but once established, no further adjustment should be required unless the materials or quality of the workpieces is changed.

The power required for welding depends primarily on the properties of the materials to be joined and the thickness of the component adjacent to the welding tip. Generally, power requirements increase as the hardness and thickness of the material being welded increase. Figure 9 shows approximate power requirements as a function of sheet thickness for several materials.

Static clamping force depends on the material properties and thickness, joint geometry, and ultrasonic power level. Excessive force produces needless deformation of the parts and increases the required welding power. Insufficient force permits tip slippage that may cause surface damage, excessive heating, tip sticking, or poor welds.

The time interval during which ultrasonic power is delivered to the workpieces in spot welding usually ranges from 0.01 s for very fine wires to about 1 s for heavy sections. Longer weld times usually indicate insufficient power. Short weld times and high powers usually produce welds superior to those obtained with longer weld times and lower powers. Excessive weld time may cause poor surface appearance or internal cracks.

For continuous seam welding, the welding rate also is dependent on material properties and thickness, and power and rate should be adjusted to provide the maximum power per unit time. Rates as high as about 500 ft/min may be used with thin metals, while thicker sheet may require rates as slow as 5 ft/min.

Interaction of Welding Variables. A suitable power-force-time combination for a given application must be determined experimentally. Prior experience with similar assemblies can provide approximate guidelines. On this basis, if it appears that good spot welds should be possible with power ranges of 1200 to 1600 W, clamping forces of between 600 and 1000 lbf, and weld times of 0.5 s or less, welds can be made at various combinations within these ranges until satisfactory joints are obtained.

Relatively simple means are available for evaluating weld quality in such experimentation, particularly if thin or ductile workpiece materials are involved. The weld components can be peeled apart, either manually or with a pair of pliers. If the peel leaves the original surfaces intact, the weld has low strength. However, if there is evidence of material transfer from one part to the other, or if a nugget is pulled from one part, the weld is considered

Fig. 9 Ultrasonic power requirements for welding selected materials

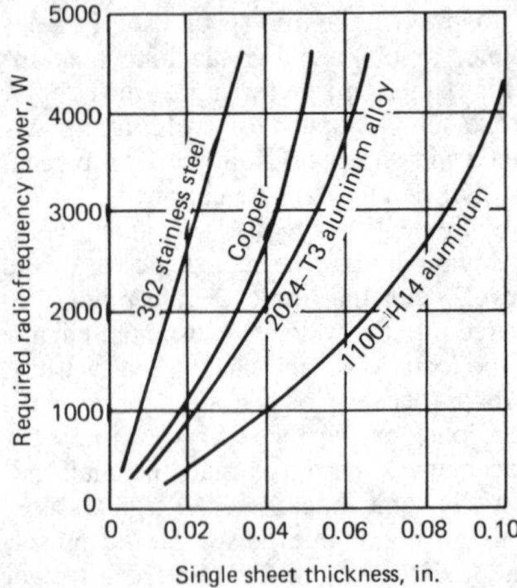

Fig. 10 Threshold curve for establishing welding machine settings. From the American Welding Society, with permission.

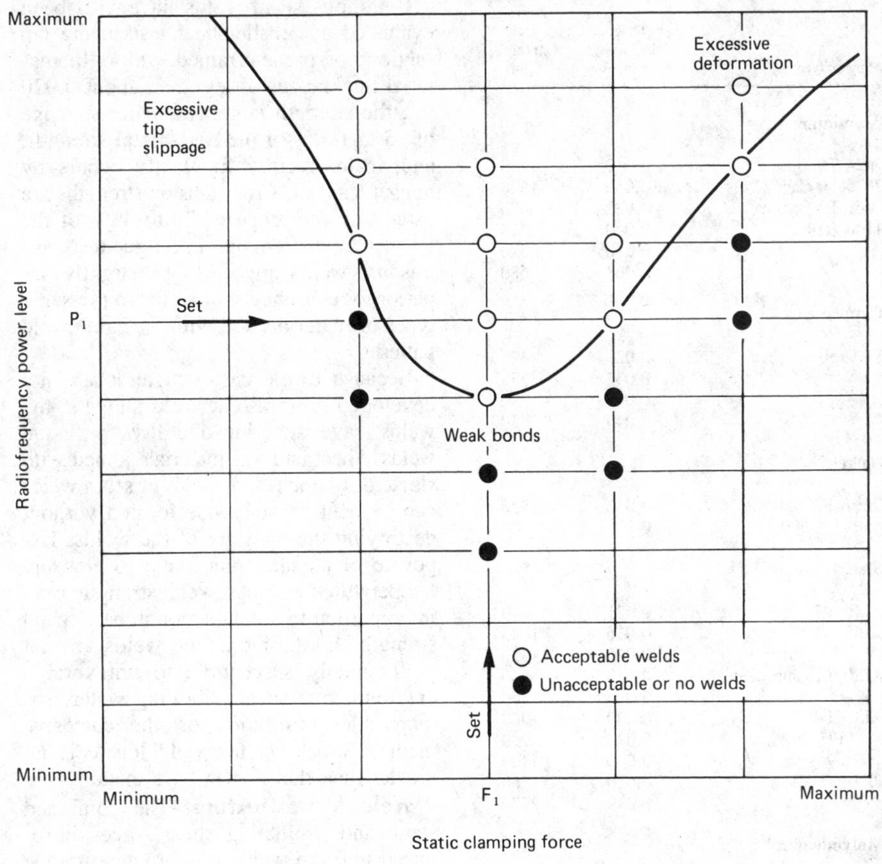

acceptable. With thick or hard materials, a tensile-shear strength test may be required.

In evaluating the various combinations of settings, it is sometimes helpful to construct a threshold curve (Fig. 10), in which the selected power values are plotted as a function of the selected force values at a fixed weld time. This produces a curve separating acceptable from unacceptable welds. Optimum values are selected from within the lower part of this concave curve. Adjustments in time can then be made while retaining the same overall energy level (power × time). The same procedure can be used for seam welding when rate is substituted for time.

For materials that are difficult to weld with fixed settings, as with some of the high-strength or refractory metals, weld quality may be improved by varying the power and force during weld formation. The weld cycle is initiated at low power and high force to establish good coupling into the workpiece. As the cycle progresses, power is increased, and clamping force is reduced. With special logic circuitry, these changes can be accomplished automatically. Such programming may produce welds of higher strength and lower incidence of cracking than are obtainable with fixed settings.

Surface Preparation. Meticulous cleaning of the part surfaces is not essential for USW, although a good surface finish contributes to the ease with which welds are made. Many materials can be welded in the mill-finish condition if not heavily oxidized. Normally, thin oxide films do not inhibit the process; such films are disrupted and dispersed by the ultrasonic energy during welding. The only requirement for most metals is the use of a mild detergent to remove surface lubricant.

It is possible to produce welds through certain coatings or insulations, although somewhat higher ultrasonic power levels may be required. Anodized coatings on aluminum and metal platings on metal surfaces have been successfully penetrated. Welds have been made using wires covered with insulating enamel or plastic films such as polyvinyl chloride or polyethylene. Insulations having a silicone base are more difficult to penetrate and may require removal prior to welding.

Welding of materials with heavy heat-treat scale is best accomplished after mechanical abrasion or descaling in a chemical etching solution. Once the scale is removed, the elapsed time before welding is not important.

Welding of certain materials such as titanium may produce some discoloration of the surface. If such discoloration is undesirable, it can be minimized by welding in an inert argon or helium atmosphere, either by directing gas jets toward the tip contact area or by enclosing the tip/anvil section of the welding unit in a chamber filled with inert gas.

Joint Design. There are no special requirements in the design of joints for USW except the necessity for rigid support for the workpieces. If the lower member of the weldment is a rigid surface, a second component may be welded to it without using an anvil. In effect, this is one-sided welding. It can be used, for example, in welding shims or reinforcements around drilled holes in massive surfaces.

In designing lap joints, material and weight can be saved by providing minimum overlap, because edge distance is not critical. Seam welds can be made with almost no material between the seam and the sheet edge. Spot welds can also be placed near the edge; the only restriction is that the welds must not crush or gouge the sheet edges. Similarly, spot and row spacing are not critical. Spots or seams may be overlapped to provide continuous coverage.

Workpiece Resonance Control. Occasionally, when multiple spots are produced on the same workpiece, the parts may be excited to resonance vibration with the welding system, resulting in cracked welds or fracture of welds previously made. This resonance may be eliminated, or at least alleviated, by changing the workpiece dimensions or altering the orientation of the workpiece in the welding machine. A piece of pressure-sensitive tape applied to one of the parts may eliminate resonance. In the most difficult cases, the clamping of masses to the part or clamping the part in a massive fixture is effective.

Weld Properties

Ultrasonic welds have distinct surface characteristics, high strength, and unique metallographic structures when compared to other metallurgical joints.

Surface Characteristics. The surface of an ultrasonic weld is usually slightly roughened by the combined static and oscillating shear stresses; however, if properly made, there is no warpage, shrinkage, or distortion of the base metal in the weld vicinity. A spot weld is usually slightly elliptical in shape because of the linear displacements of the welding tip. The size of the spot depends on the tip radius and the hardness of the metal being welded. Larger welds are obtained in soft metals such as aluminum than in harder materials.

With flat sheet, a slight indentation of the workpiece occurs in the vicinity of the welding tip, particularly in soft metals. An indentation exceeding about 10% is usually indicative of improper machine settings, especially excessive clamping force. With contoured parts such as wires, deformation may be somewhat greater unless the tip is contoured to mate with the part.

Weld Strength. The strength of ultrasonic welds is usually determined by tensile-shear tests on simple lap joint specimens. Such welds can be made with a wide variety of similar and dissimilar metal combinations. Cross-tension and fatigue tests also are used to characterize welds.

In tensile-shear tests, the nature of the fracture is often indicative of the integrity of the weld. In thinner, more ductile materials, fracture frequently occurs by tear-out of a weld nugget or by failure of the metal outside the weld joint. In such cases, the weld strength approaches or exceeds the base-metal strength of the material. With bimetallic welds, the fracture or nugget invariably occurs in the softer metal of the joint members.

For lack of other criteria, ultrasonic spot weld strengths are frequently compared with military specifications for minimum strength in resistance spot welds. Figure 11 shows such a comparison for bare and clad 2024-T3—typical structural aluminum alloys. Except in very thin gages, ultrasonic weld strengths are more than twice the minimum resistance weld requirements. Typical average strengths for spot welds in a variety of other materials are provided in Table 1.

Fig. 11 Strength of ultrasonic welds in structural aluminum alloys

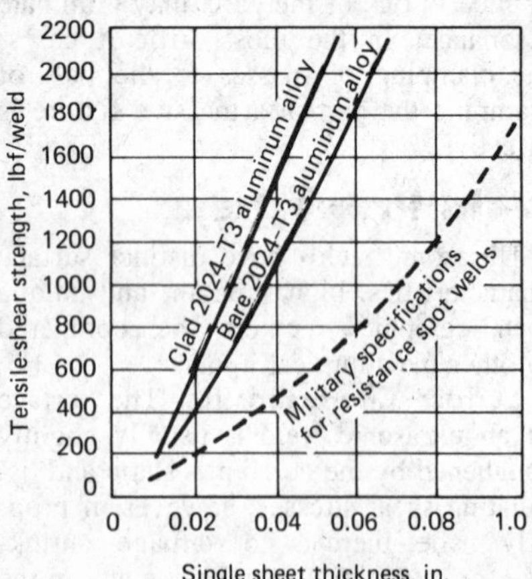

Table 1 Typical shear strengths of ultrasonic welds

Alloy or type	Sheet thickness, in.	Average shear strength, lbf
Aluminum		
2024-T6	0.040	1240
7075-T6 clad	0.050	1540
6061-T6	0.032	870
1100-H14	0.040	470
	0.050	570
	0.063	680
Copper		
Electrolytic	0.020	320
	0.032	530
	0.040	780
	0.045	850
Nickel		
Inconel X-750	0.032	1520
Monel K-500	0.032	900
Steel		
AISI 1020	0.025	910
AISI 301 FH	0.016	1040
	0.025	1350
AISI 302, annealed	0.018	630
	0.032	1300
AISI 304	0.040	1060
17-7 PH	0.020	990
	0.040	1400
PH 15-7	0.020	1360
	0.040	2370
Molybdenum		
Stress relieved	0.010	180
Powder metals	0.020	380
Arc cast	0.025	360
Mo-0.5Ti	0.020	240
	0.032	450
Titanium		
8Mn	0.032	1730
5Al-2.5Sn	0.028	1950
6Al-4V	0.040	2260
Zirconium		
Zircaloy-2	0.020	620

Such welds also show excellent reproducibility. The variability is usually less than ±10% of the average strength. Experience in making hundreds of welds in aluminum alloys, for example, at the same machine settings on different days at different times of the day and by different operators have confirmed such reproducibility.

Continous seam welds have also been evaluated by tensile-shear tests using lap sections of predetermined width. In material thicknesses of less than about 0.020 in., the strength is generally in the range of 85 to 100% of the base-metal strength, and again fracture frequently occurs by nugget tear-out. Cross-tension strengths are usually in the range of 20 to 40% of the tensile-shear strengths. In fatigue tests, ultrasonic welds appear to be slightly superior to resistance spot welds in the same types of materials and with the same weld patterns.

Because brittle cast structures are not developed in ultrasonic weld nuggets, the welds have superior ductility to fusion welds. Sheet and foil materials joined with single or multiple spot welds or seam welds can be bent or otherwise formed without destroying the integrity of the welds. Exposure of an ultrasonic weld to elevated temperatures reduces weld strength only in proportion to the reduction in base-metal strength. Furthermore, the welds are not preferentially susceptible to unfavorable environments, such as boiling water, sodium chloride solution, or other corrosive media. Attack on the weld locale is no greater than that on the base metal.

Weld Microstructure. The combined static and oscillating shear forces introduced into the weld zone produce unique phenomena at the weld interface, depending on the materials being joined and the welding conditions. There may be disruption of surface films, plastic flow, atomic interdiffusion, grain refinement, recrystallization, phase transformation, or other phenomena. With dissimilar metals, there is frequently interpenetration of the two materials; such effects are shown in Fig. 12.

The interfacial disturbance in soft metals is illustrated in the 1100 aluminum weld in Fig. 12(a). The edge of the weld is shown at the left. The dark areas represent oxide surface film which has been broken up and dispersed by the ultrasonic energy, permitting coalescence of the base metal in the two components. Still greater disturbance and swirling effects are evident in the electrolytic copper weld in Fig. 12(b). Again, surface oxides are dispersed, and grain growth occurs across the interface.

The cartridge brass weld of Fig. 12(c) shows recrystallization and grain refinement in the vicinity of the interface, probably the result of heating that occurred during welding. This phenomenon also has been observed in other materials, includ-

Fig. 12 Typical ultrasonic welds. (a) 0.012-in. 1100-H14 aluminum welded to itself. (b) 0.025-in. electrolytic copper welded to itself. (c) 0.010-in. spring-temper cartridge brass welded to itself. (d) 0.024-in. AISI 430 stainless steel (upper portion) welded to 0.028-in. 5Al-2.5Sn titanium alloy. Magnification: (a) 500×, (b) 100×, (c) and (d) 150 ×. Etchants: (a) Kellers, (b) and (c) ammonium hydroxide plus hydrogen peroxide, (d) hydrofluoric plus nitric acid

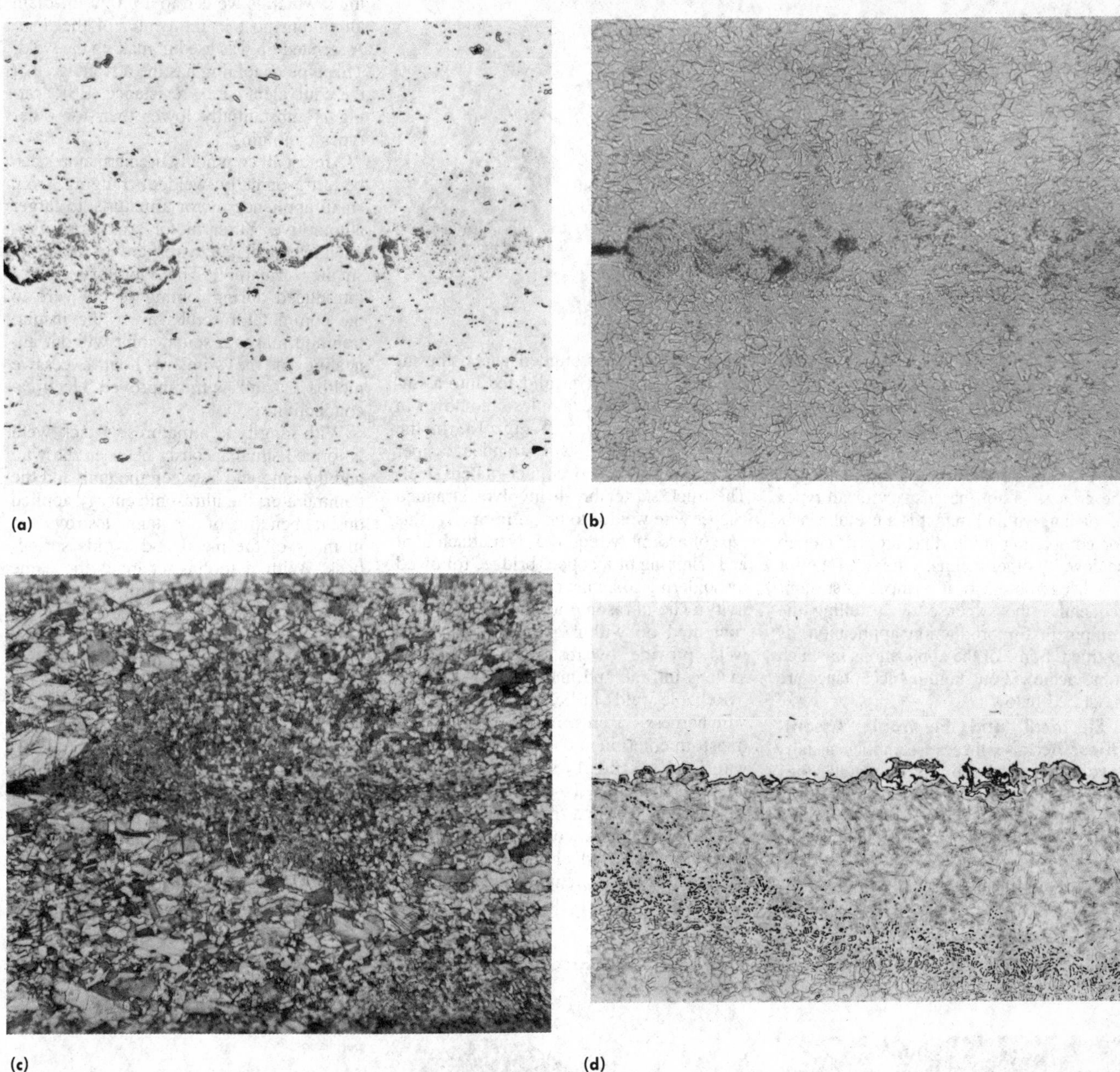

ing beryllium, titanium, and low-carbon steel, even though the metals were not in the cold worked condition prior to welding.

With bimetallic welds, particularly with metals of different hardnesses, there may be interpenetration between the two metals, sometimes in a regular wave-like pattern. For example, in an application in which a Kovar (Fe-29Ni-17Co-0.2Mn) foil was welded to gold-plated nickel foil, the Kovar penetrated as much as 75% into the nickel. The gold plating was dispersed throughout the highly worked region. The steel-to-titanium alloy weld in Fig. 12(d) shows the interpenetration on a lower scale. This specimen was etched with hydrofluoric plus nitric acid solution to reveal the titanium alloy grain structure but not that of the steel. The titanium close to the interface has been recrystallized from the heat effects.

More than one of the interfacial effects noted above may be present in the same weld, and different effects may occur in welds in the same materials made with different machine settings. The various effects are not necessarily related to weld strength and integrity.

Applications

The unique characteristics of ultrasonic welds make USW particularly effective for a wide variety of applications. The capabilities of metallurgically bonding dissimilar as well as similar metals without spe-

Fig. 13 Ultrasonically welded brush assemblies. (a) Starter brush plate. (b) Truck starter brush

cial precleaning, without heat or electrical current, without fluxes or filler metals, without brittle cast structure or alloying, and without excessive part deformation have encouraged industry to adopt USW for difficult bonding problems. Although the process is not the answer for all types of joining, within limits it is a useful means for completing joints that are difficult to achieve by other means. Often, a superior product is obtained at reduced cost; consequently, the expense of installing the equipment for production applications is justified. Some of the applications in which it has achieved outstanding acceptance are described below.

Electrical and Electronic Assemblies. Because these assemblies usually involve relatively thin-gage metals, they offer a range of possibilities for USW. Many such applications involve combinations of aluminum, copper, and brass, which are readily joined.

Both single and stranded wires can be welded to other wires and to terminals. In joining stranded or braided wires, the individual wires are consolidated into a single mass. Typical examples are shown in the brush assemblies of Fig. 13. In the starter brush plate, two stranded copper wires are welded to copper-plated steel. The truck starter brush involves stranded copper wire welded to brass. Formerly, this type of assembly required rivet attachment and crimping of a copper bridge, followed by soldering to achieve electrical conductivity. The ultrasonic joints, made with a contoured tip within about 1 s for each weld, provide low resistance, as well as savings in time and material.

Reliable welds in field coil assemblies, wire harness systems, induction coils, and transformer terminations can also be made with USW. Figure 14 shows one joint of a typical field coil. To assemble the complete device, stranded copper wire is joined to aluminum ribbon, copper ribbon to aluminum ribbon, and aluminum ribbon to itself. These joints are made ultrasonically to achieve high strength and conductivity.

An application gaining increasing acceptance is the attachment of capacitor leads to terminals. The capacitor in Fig. 15 has an aluminum lead that is 0.010 in. thick which is welded to a 0.090-in.-diam aluminum post in the center of the disk. A serrated tip was used in making the weld. This type of joining has the advantage that the equivalent series resistance (ESR) rating is substantially lower than for other types of joining.

Almost all commonly used armatures can be ultrasonically welded, ranging from small appliance motor armatures to larger automotive starter motor armatures. Frequently, insulated wires are used in the smaller armature. The vibratory energy introduced during joining of the wire to the commutator scrubs away the insulation and makes a sound joint without annealing of the wire, without excessive meltback, and with improved electrical conductivity.

With tang-type armatures, a tack weld is formed simultaneously between the wire and the tang and between the tang and the commutator; the ultrasonic energy applied during bending of the tang destroys the memory of the metal and avoids springback. With a tooling change, the same equipment can be used with slotted commutators. The insulated debris from the wire is dispersed, and a sound electrical and mechanical joint is made without risk of shorting between commutator bars.

Completely automated equipment has been developed for welding automotive starter motor armatures such as that in Fig. 16. The armature has a slotted commutator with risers into which pairs of copper wires 0.125 in. thick are inserted. Previously, the wires were soldered into the slots; when the armature was exposed to excessive electrical current, however, as in cold starting conditions, the solder would

Fig. 14 Portion of ultrasonically welded field coil

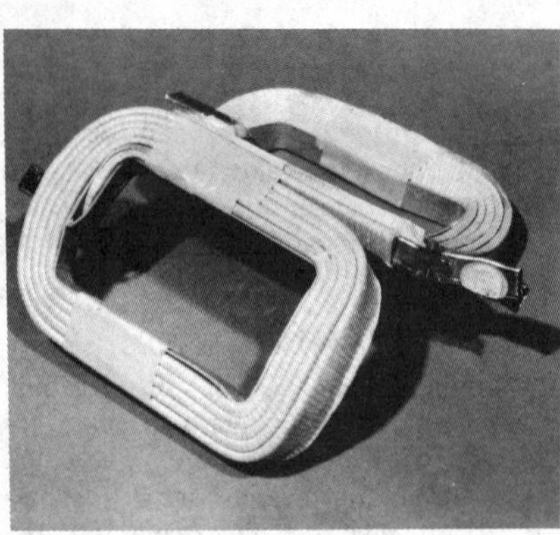

Fig. 15 Ultrasonically welded capacitor lead

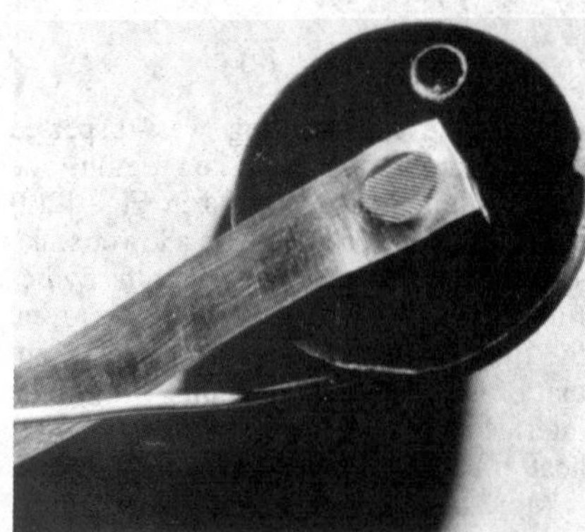

Fig. 16 Automotive starter motor armature with ultrasonically welded wires to commutator. From the American Welding Society, with permission.

Fig. 17 Contact button welded to switch gear arm

soften and melt, the wires would spin free, and the armature failed. With USW, the wires are welded to each other and to the sides of the slot with a single weld pulse. Such joints do not lose strength at temperatures up to at least 300 °F. An automated index-and-weld system permits joining pairs of wires into a 25-slot commutator in a cycle time of 20 s for a production rate of three armatures/min or 180/h.

Dissimilar metals frequently are involved in attaching contact buttons for relays, circuit breakers, and similar components. The assembly in Fig. 17, used in a switch gear, consists of a silver-tungsten contact welded to a copper arm. Other applications include the attachment of thermocouple wires of a variety of materials to various surfaces, the assembly of battery components, including lithium to stainless steel for lithium batteries and lead connections in automotive batteries, and the attachment of fine wires or thin ribbons to semiconductor elements.

Solar Energy Systems. Ultrasonic technology offers the manufacturers of solar photovoltaic and collector systems the opportunity to cut costs and improve quality significantly. The process is reported to operate cleanly and to eliminate the need for auxiliary materials such as solder.

Systems for converting solar heat to electricity frequently involve photovoltaic modules or rows of 0.006- to 0.008-in.-thick silicon cells joined to 0.002- to 0.005-in.-thick aluminum conductors. The cells are aligned on the aluminum sheet, and the connections are made ultrasonically using two ultrasonic seam welding heads, each with a counter-rotating anvil. One head rolls across the connecting points and welds the silicon cells to the top of the aluminum sheet, with the counter-rotating anvil located underneath the sheet. The second welding head is located beneath the sheet, with the anvil above. These staggered systems weld the cells to both sides of the aluminum simultaneously at a rate substantially greater than can be achieved with individual spot welds or soldered joints. Improved conductivity is also obtained with such a streamlined system.

Solar collectors for hot-water systems usually consist of copper or aluminum tubes attached to a 0.008-in.-thick collector plate. Ultrasonic welding offers substantial time and cost savings over joining by resistance welding, soldering, or roll bonding. Automatic systems have been assembled for making successive spot welds between plate and tubing on 1-in. centers as the assembly is passed beneath the welding tip. A mandrel is inserted into the tubing to prevent its collapse under welding loads. With this process, a tube about 36 in. long can be welded to a plate in about 2 min at minimal energy cost.

Foil and Sheet Splicing. Aluminum foil rolling mills routinely use continuous seam welding to repair breaks and join random lengths of thin foil and sheet. Previously, adhesives were used for such joints, but this necessitated expensive downtimes for drying the adhesive, and joint cleanup usually was required. The process currently involves overlapping the ends of the strips to be joined, rolling a USW disk tip such as that shown in Fig. 7 across the width of the foil, and tearing off the tails on both sides of the joint. The joint is made cleanly and effectively in a few seconds and is immediately ready for further processing. The splices are not affected by annealing or other subsequent operations and are almost undetectable. Etched aluminum and spray-coated foils can be processed without preweld treatment.

Another production assembly operation involves the fabrication of aluminum heater ducts. Sheets of 0.0003-in. aluminum foil are wrapped around insulation supported on a mandrel, and the overlapping edges are seam welded with a traversing-head welding machine.

Packaging. Various types of packages can be sealed by seam or overlapping spot welding. Square or rectangular packets are sealed by intersecting seam welds on each of the four edges. Such a process is used, for example, in packaging x-ray film. Other shapes can be accommodated using a roller-roller seam welding machine similar to that illustrated in Fig. 8. The ends of squeeze tubes are sealed by this method. If the packaging material is too thick to be handled by seam welding, overlapping spots can be used. Small pinch tubes are simultaneously pinched and sealed with a single spot weld. Such welds are leaktight and do not lose their hermeticity when exposed to elevated temperatures or other unfavorable environments.

The process is useful for packaging materials that are sensitive to heat or electrical current. Ultrasonic welding can be performed in a protective atmosphere or vacuum and thus permits sterile packaging of hospital supplies, precision instrument parts, ball bearings, and other items that must be protected from dust or contamination.

Structural Applications. Within the limitations of weldable sheet thickness, USW provides joints of high integrity for certain structural parts, such as aircraft or automotive secondary structures. A helicopter access door consisting of inner and outer skins of aluminum alloy was previously assembled with adhesive bonding, which required expensive forms, heating, and lay-up and curing time. This assembly was subsequently made with multiple ultrasonic spot welds, as shown in Fig. 18. Significant time and cost savings were achieved. The individual welds had 2.5 times the minimum average strength requirements for resistance spot welds, and the assembled doors in air load-tests sustained loads of from 5 to 10 times the design load without weld failure.

In another application, small clips were attached to cylindrical reactor fuel tubes with ultrasonic spot welds. Using a contoured welding tip, 8 clips were attached

Fig. 18 Inner and outer skins of helicopter access door assembled by ultrasonic spot welds. From the American Welding Society, with permission.

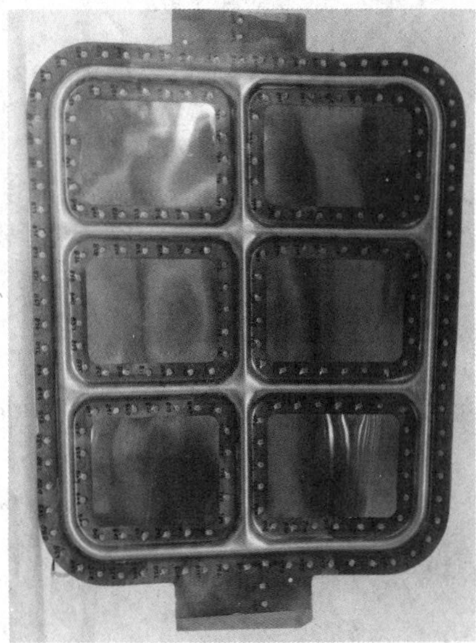

to each tube, and production rates of about 200 assemblies per hour were made with one welding machine in a semiautomated setup.

Ultrasonic welding also has potential for cladding steel with a copper-nickel alloy for protection of steel in a corrosive environment. Using a 8-kW ultrasonic spot welding machine, copper alloy (90Cu-10Ni) 0.01 in. thick can be welded to AISI 1020 steel sheet. High weld strength is obtained, and failure under tensile loading invariably occurs by tear-out of a weld nugget from the copper alloy sheet rather than at the bond line.

SELECTED REFERENCES

- Devine, J., Ultrasonic Bonding for Motor Manufacture, Proceedings of the ICWA Conference, Boston, p 1-5, 1979
- Devine J., Joining Electric Contacts?, *Welding Design and Fabrication,* Vol 53 (No. 3), p 112-115, 1980
- Devine, J., Ultrasonic Welding Helps Lighten Aircraft, *Welding Design and Fabrication,* Vol 51 (No. 8), p 74-76, 1978
- Devine, J. and Vollmer, R.G., Ultrasonic Bonding Arrives, *US Army Man Tech Journal,* Vol 3 (No. 1), p 11-14, 1978
- Devine, J., Dingle, G.K., and Vollmer, R.G., Ultrasonic Bonding, Panacea or Pie in the Sky, Proceedings of the Ultrasonic Industries Association, New York, p 1-15, 1977
- Kelly, T.J., Ultrasonic Welding of Cu-Ni to Steel, *Welding Journal,* Vol 60 (No. 4), p 29-31, 1981
- Meyer, F.R., Ultrasonic Welding Process for Detonable Materials, *National Defense,* Vol 70 (No. 334), p 291-293, 1976
- Meyer, F.R., Ultrasonics Produces Strong Oxide-Free Welds, *Assembly Engineering,* Vol 20 (No. 5), p 26-29, 1977
- Renshaw, T., Curatola, J., and Sarrantonio, A., Developments in Ultrasonic Welding for Aircraft, Proceedings of the 11th National SAMPE Technical Conference, Boston, p 681-693, 1979
- Renshaw, T., Wongwiwat, K., and Sarrantonio, A., A Comparison of Properties of Single Overlap Tension Joints Prepared by Ultrasonic Welding and Other Means, Proceedings of the AIAA/ASME/ASCE/23rd Conference, New Orleans, p 1-8, 1983

High Frequency Welding

By the ASM Committee on High Frequency Welding*

HIGH FREQUENCY WELDING includes a number of processes in which metal-to-metal bonding is accomplished by using heat caused by the flow of high frequency current at the faying surfaces, with upsetting forces perpendicular to the interface added in most cases. Although similar in many respects, two separate high frequency welding processes can be identified: high frequency resistance welding (HFRW) and high frequency induction welding (HFIW), sometimes referred to as induction resistance welding. In HFRW, heating current enters the work through electrical contacts on the surface. In HFIW, heating current is induced in the work through an external induction coil and no physical or electrical contact between the workpiece and the power supply is needed.

In more conventional resistance welding processes, heating is accomplished at the joint interface by the flow of current across the interface as it passes between two electrodes pressed against the work. The current is normally direct current (dc) or low frequency alternating current (ac), from 60 to 360 Hz. In some cases, the current may be the result of a capacitor discharge. Very high currents are required to heat the metal; large, well-cooled electrical contacts must be placed as close as possible to the desired weld, and high contact pressures are normally required.

In high frequency welding, current flows in work surfaces parallel to the joint. For HFRW, the location of the current flow in the work surfaces is determined by the location of the electrical contacts and external electrical conductors, and for HFIW, by the design and location of the induction coil. The shape and relative position of the workpiece surfaces as they are brought together immediately before welding have a major influence in determining the current flow and concentration in continuous seam welding by both processes.

The depth to which current flows depends on the frequency and the resistive and magnetic properties of the workpiece. At higher frequencies, current flow is shallower and more concentrated near the work surface. The range of frequencies used most often in this process is from 300 to 450 kHz, although frequencies as low as 10 kHz may be used in some instances. This concentration of current at the surface permits welding temperatures to be achieved with power consumption at much lower levels than with conventional resistance welding. The efficiency of welding is greatly increased, and relatively small contacts may be used. The placement of contacts is determined by the desired current flow path in the workpiece. This usually uses the shape of the part to produce current flow in the areas to be heated; however, the placement of induction coil parts or special proximity conductors where contacts are used may also be employed to cause or enhance current flow in desired areas.

Several factors must be considered in making successful high frequency welds. Because of the advantages of concentrated heating, the process is inherently fast. Excessive conduction of heat away from the faying surfaces can diminish weld quality; thus, seam welding is performed at relatively high throughput speeds. Materials of high thermal conductivity are ordinarily run at higher speeds than those of low conductivity.

Satisfactory welds are normally produced in air and are possible with water or soluble oil coolant present in the weld area. However, in special situations where inhibiting oxidation or atmospheric pickup of the workpiece materials is desired (as is the case with titanium), inert gas shielding of the weld area can be provided. Flux is not normally required; however, in the case of some copper alloys it has been used to advantage. A number of typical weld shapes are shown in Fig. 1.

Principles of Operation

The phenomenon in which high frequency current flows in the surface of conductors is known as the skin effect. The relationship of depth of current penetration to frequency for a variety of materials at different temperatures is shown in Fig. 2. Typical values vary from about 0.004 in. to about 0.25 in., depending on frequency and metal temperature. The depth of current penetration varies in the workpiece as it heats up.

Control of the current flow path within the workpiece to heat only in desired areas is accomplished by use of the proximity effect. As shown in Fig. 3, high frequency current in the surface of a part flows as close as possible to another conductor carrying current that is 180° out of phase and the flow of current in the workpiece images the shape of a return conductor immediately over the surface. The concentration of current in the surface opposite the return flow conductor increases with frequency and with the nearness of the conductor to the surface (Fig. 4).

The current can be concentrated even more narrowly in the surface of the workpiece by the placement of magnetic cores on either side of the return path conductor (Fig. 4c). At frequencies up to 20 kHz, these magnetic cores are usually laminated iron. At frequencies of 100 kHz and over, ceramic ferrite cores are used.

By far the most common use of high frequency welding is in longitudinal butt

*Wallace C. Rudd, *Chairman*, Consulting Engineer, Thermatool Corp.; Edgar D. Oppenheimer, Consulting Engineer; Humfrey N. Udall, Director of Research & Development, Thermatool Corp.

Fig. 1 High frequency welding applications

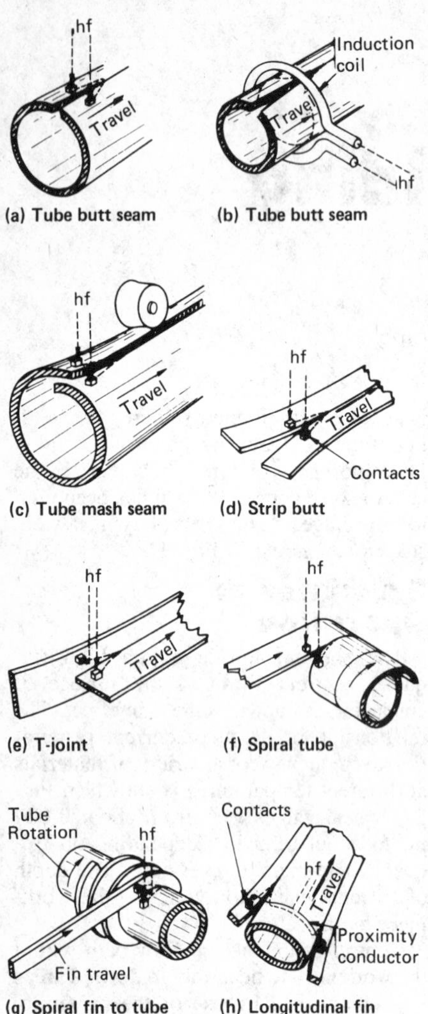

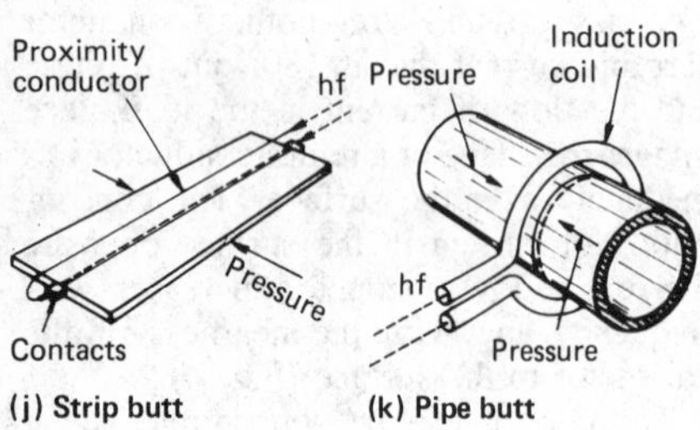

Fig. 2 Frequency versus depth of current penetration into metals at selected temperatures

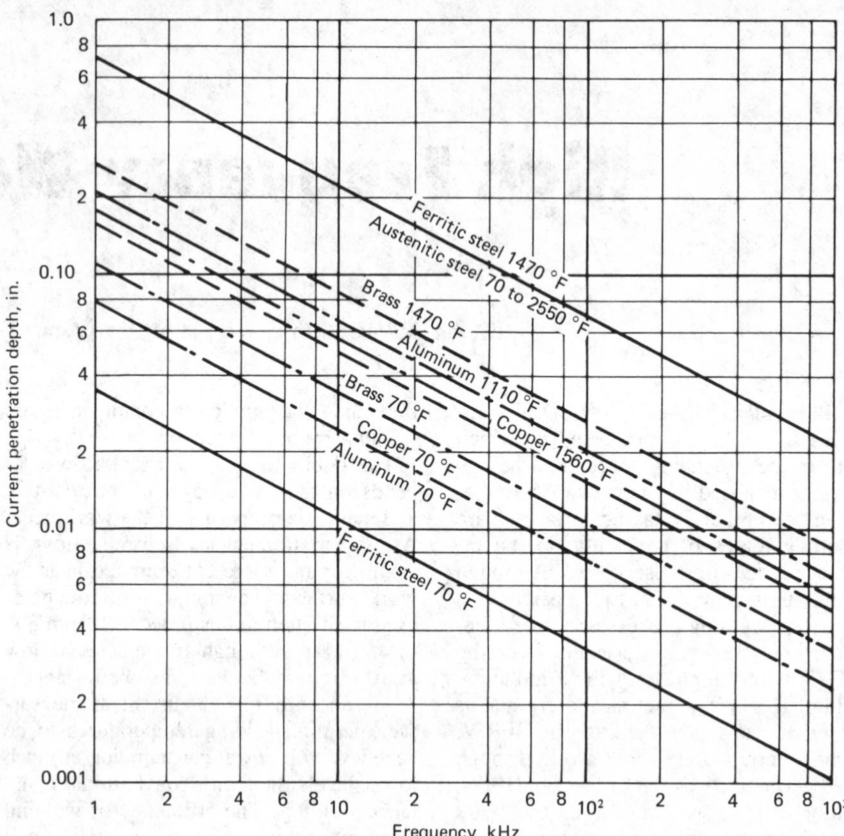

seam joining of the edges of metal pipe and tubing. In this application, opposing edges act as return conductors for each other, causing concentration of the current at the faying surfaces.

In almost all cases of high frequency welding, faying surfaces are heated apart from each other and brought together in a continuous operation. As they come together, they are forged into each other in such a way to extrude foreign matter from the interface and cause a certain amount of upset material to remain on the outside of the joint. This upset material may or may not be removed in subsequent operations, depending on the final use of the product.

High Frequency Resistance Welding

High frequency resistance welding is a resistance welding process that produces coalescence of metals by the heat generated from the resistance of the workpieces to a high frequency current (10 to 500 kHz) and the rapid application of an upsetting force once heating is completed. The current is applied to the base metal by using electrical contacts that are held against the surface of the workpiece. The principal applications of HFRW are for making such structural shapes as pipe and tubing from continuous strips of metal. Two process variations of HFRW are continuous seam welding (Fig. 1a, c, d, e, f, and g) and finite length butt welding (Fig. 1j).

Continuous seam welding can be described by its most common application, longitudinal seam butt welding of pipe and tubing (Fig. 5). A pair of sliding contacts is placed on either side of an open V formed by the converging edges of the strip being made into tubing. The angle of this V typically varies from 4 to 7°. The edges of the strip converge between the weld pressure rolls, where a certain degree of forging upset is accomplished.

As the material moves through the rolls from the contacts to the point of welding, high frequency current introduced at the contacts flows on the opposing faying surfaces of the strip, down one side of the V to the point of welding, and back to the other contact. Because this process is performed at frequencies of about 400 kHz, the depth of heating is quite shallow and heating occurs rapidly. Power is adjusted to bring the edges to a welding temperature at the time they join, and contaminants on the surfaces, as well as most of the molten material, are squeezed out of the seam toward the inside and outside of the tube when the weld is completed. This upset material can be trimmed off flush to the surfaces of the tube if desired. The impeder (Fig. 5) is used to improve welding efficiency by increasing the impedance of

Fig. 3 Control of the flow path of high frequency current by the proximity effect of the return conductor

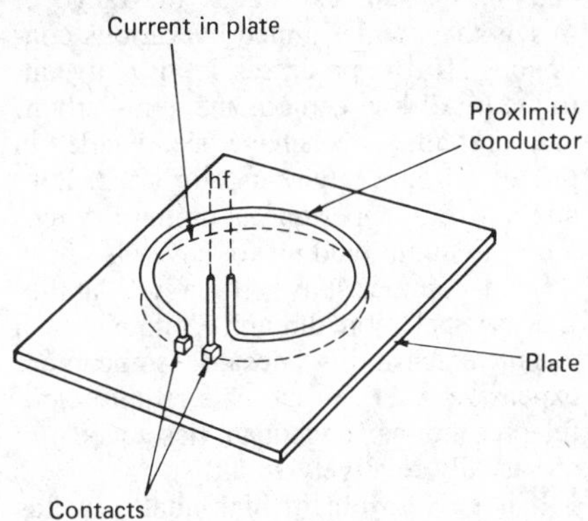

Fig. 4 Depth and distribution of current adjacent to proximity conductors

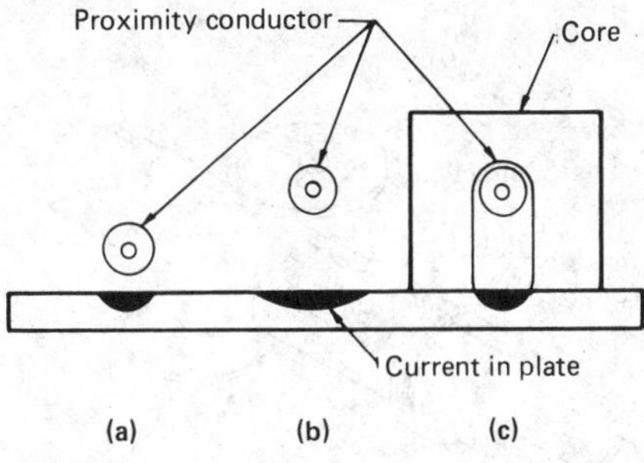

Fig. 5 High frequency longitudinal butt seam welding of tube

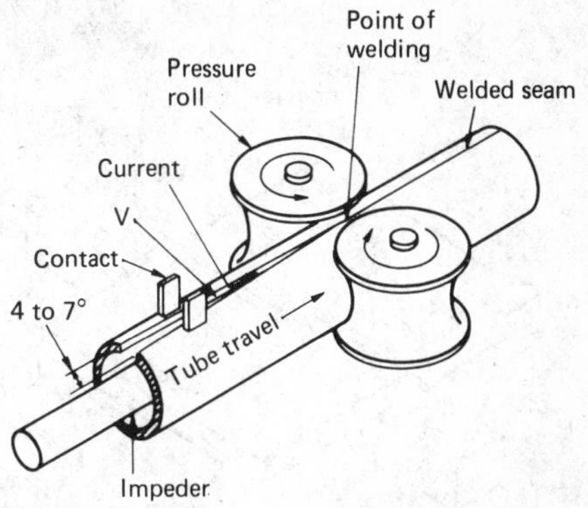

the current path around the inside of the tube. This reduces the waste of energy because of this wasted current that is parallel to the current used in the welding V.

The other applications of continuous high frequency resistance seam welding (Fig. 1) vary from tube welding only in the shape of the workpiece. Contacts introduce current at the open part of a V that is being closed by rolls at the point at which the material has been heated to a high enough temperature to weld.

Finite Length Welding. The process shown in Fig. 1(j) and Fig. 6, in which a butt joint of finite length is welded between two parts, is used for applications in coil end joining as well as in the manufacture of strip and sheet blanks for subsequent press forming. In this process, heating is accomplished along a seam that has already been closed under light pressure. The path of the heating current is controlled by the use of a proximity conductor positioned over the seam. The high frequency heating causes the edges to be brought to welding temperature, after which pressure is increased to weld the joint. The appropriate frequency is selected to achieve heating through the material thickness. For example, 10 kHz can be used up to 0.25 in. thickness. Joints made this way can be produced very quickly, typically in 1 s. Compared to flash welding, the amount of upset is small, no shower of sparks and debris is emitted from the weld, and very little metal is consumed in making the joint.

High Frequency Induction Welding

In HFIW, current for heating is transmitted to the workpiece by means of an induction coil that surrounds but does not touch the work metal. When high frequency current flows through the coil, a highly concentrated magnetic field is established. This magnetic field induces an electric potential, causing a flow of current in the closed circuit provided by the workpiece. Resistance of the work metal to this current flow provides the heat necessary for welding. Two process variations of HFIW are continuous seam tube welding (Fig. 1b) and butt end tube welding (Fig. 1k).

Continuous Seam Tube Welding. Induction welding is used in the manufacture of longitudinal butt seam tubing (Fig. 7). The induction coil surrounds the entire tube at the open portion of the V some distance up from the pressure rolls. Current is induced in the tube surface under the induction coil, except where the V is open. For the induced current to complete its circuit, it must travel down one side of the V and back up the other side, where heating of the faying surfaces occurs. Some power loss occurs in the tube wall underneath the coil; however, in smaller diameter tubing this is not important. An impeder is normally used inside the tube.

Induction welding has the advantage of not placing contacts on the workpiece surface, and coated material or thin-walled tubing can be welded without surface damage. Two common applications of the process are welding of aluminum-coated steel and welding of stock to be chromium plated. Because an aluminum coating melts at a lower temperature than the steel base metal, the molten aluminum often builds up and short circuits the sliding contacts used in HFRW. This buildup, however, does not affect the function of an induction coil. For stock to be chromium plated, HFIW is preferred over HFRW because sliding contacts occasionally arc and leave marks on the workpiece surface near the weld area. These marks are difficult to remove and appear as blemishes in a chromium-plated surface.

High frequency induction welding, however, is inherently less efficient than HFRW, and this becomes an important factor for larger sizes of pipe and tubing. In tube above about 4 in. in diameter, the current flowing around the outside of the tube causes a significant energy loss. This does not occur with HFRW. High frequency induction welding is not practical for such applications as are shown in Fig. 1(d), (e), (f), (g), (h), and (j), because the opportunity to complete the induced current circuit requires a part made from a single piece of metal rather than two parts.

High frequency induction welding of high-conductivity, low-permeability materials, such as aluminum and copper, requires relatively high primary currents in the induction coil, resulting in poor electrical efficiency. Where the product is small and adequate power is available, induction can be used, but the mill operation should not be slowed when adequate power is not available. As a result, HFRW is often a more effective way to weld high-conductivity materials.

Tube Butt End Joining. The butt welding of tube to tube (Fig. 1k) is carried out in much the same way as finite length butt welding with contacts; however, because the entire seam is continuous, it can be heated by an induction coil surrounding the joint. Heating caused by induced current raises the joint temperature very rapidly, causing the axial upset force to produce a solid-phase forge weld. The upset material is usually left in place. This process has been used for welding tube ends 1 to 12.6 in. in diameter and wall thicknesses up to 0.38 in. in times ranging from 10 to 60 s.

Advantages and Limitations

The principal advantages of high frequency welding over low frequency (60

Fig. 6 Joining strips together using HFRW

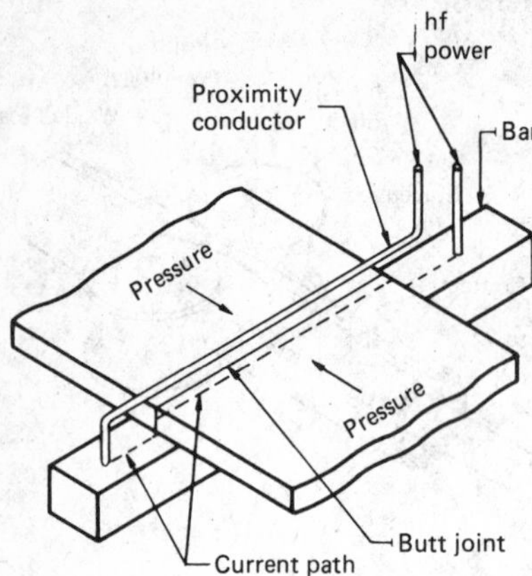

Fig. 7 Joining a tube seam by HFIW

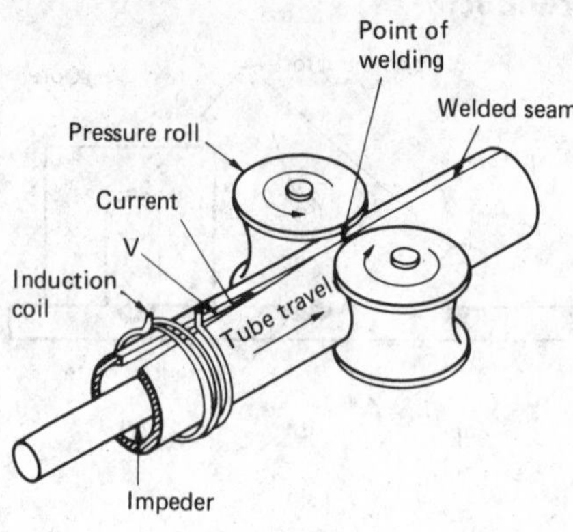

Hz) or direct current resistance welding processes result from the very shallow heating of only the faying surfaces caused by skin effect and proximity effect. Not only does this lead to greater efficiencies and higher speeds of the high frequency welding processes, but it also tends to produce a superior joint metallurgically by creating a very narrow heat-affected zone (HAZ). The hot forging of the weld area removes virtually all of the molten material created during heating and hot works the remainder, producing an autogenous weld with physical properties equal to or better than the parent material in most cases. Usually no postweld heat treatment is required.

The behavior of contacts in HFRW produces fewer problems than in low frequency resistance welding. Contact life is often long enough in HFRW to produce hundreds of thousands of feet of product. Contacts do not have to be dressed or shaped, because they wear into the shape of the part. Worn tips can be quickly replaced by changing the mount and contact tip assemblies. Another advantage of high frequency welding is that preweld removal of oxides from workpiece surfaces is not necessary. With low frequency welding, the contact has to break through this oxide layer and make contact with the metal. With HFRW the voltage is high, and there is continuous puncturing of oxide films (caused by arcing) under the contact. The contact does not have to break through the surface layer, and only enough pressure to ensure continuous contact with the surface is necessary. With these lighter contact forces, very thin material, such as tubing with wall thicknesses of 0.030 in. and less, can be butt welded by high frequency processes.

The extremely localized heating inherent in high frequency welding limits oxidation and distortion of the work and avoids heating of nearby material. In one application, metal sheathing is welded as a tube around electrical conductors without damaging the insulation of the conductors inside. Because of the very shallow heating, workpiece edges must be carefully prepared, aligned, and brought together along well-controlled paths. Variation in the shape of the V or the edges can result in poor weld quality.

The relatively high speed of the process makes it most applicable to products requiring substantial production. Typically, a million feet of tubing might be produced on an average tube mill in a month of single-shift operation; however, small quantities of special products such as titanium I-beams have been produced in cases where weld quality was important and the cost was justified.

High frequency welding has limitations. Equipment usually is built into a mill or line operation and must be fully automated. At the high rates of welding inherent in high frequency welding, an improperly adjusted mill can produce significant quantities of scrap in a short time. The process is also limited to the use of flat, coil, or tubular stock with a constant joint symmetry throughout the length of the part.

Because high frequency equipment operates near the broadcast radio frequency range, special care should be taken during installation, operation, and maintenance to avoid radiation interference in the vicinity of the plant. The manufacturer's recommendations should be followed, and any equipment of this type must meet Federal Communications Commission (FCC) regulations.

Applications

Whereas finite length high frequency welding is usually limited to carbon steels and ferritic stainless steels, the range of metals that can be joined by various continuous HFRW processes is virtually unlimited. All low-carbon, medium-carbon, and most alloy steels have been welded in production in a large range of sizes. Pure and alloyed copper and aluminum, zirconium, titanium, and nickel have also been welded commercially. Often weldability is determined more through economics than technical feasibility, because some more expensive reactive materials require careful precautions, and quantities to be run are usually relatively small.

It is easy to obtain high-quality welds in low-carbon steels, which require relatively less skill and care than most other materials; however, very high-quality welds can be obtained in difficult materials, such as 300 series stainless steels, provided that adequate care is exercised in the production operation. Joint fit-up and alignment are particularly important, and gas shielding may be necessary.

High frequency resistance welding can be used to join dissimilar metals. Examples that have been used in production quantities include: double-walled tubing made of two different types of steel to obtain both corrosion resistance and sound damping in automotive exhaust systems, type 430 stainless steel to galvanized carbon steel strip for automotive trim, tool steel to alloy steel for band saw and hacksaw blades, and copper tube to aluminum strip for solar absorber plates. Commercial-grade copper tube made with UNS C12200 is used for this application, but other grades of copper or copper alloys can be welded also.

Pipe and Tube Welding. The single largest application of high frequency welding is the manufacture of longitudinal butt seam pipe and tubing, particularly in aluminum and low-carbon, medium-carbon, and alloy steels. Both HFRW and HFIW are used. Because of the very high welding speeds that are attained, the process has replaced arc welding and low frequency and direct current resistance welding wherever product size and production quantities are appropriate. Table 1 shows typical values of weld speeds for various wall thicknesses in low-carbon steel and aluminum.

Figure 8 shows a mill that produces large steel pipe up to 20 in. in diameter with a typical wall thickness of 0.38 in. Flat strip enters the mill as shown in the lower right-hand corner of the photograph and travels

Table 1 Typical weld rates for high frequency tube and pipe welding

High frequency resistance welded steel — 600 kW, 12 in. diam		High frequency resistance welded steel — 200 kW, 3.0 in. diam		High frequency induction welded aluminum — 200 kW		
Wall thickness, in.	Speed, ft/min	Wall thickness, in.	Speed, ft/min	Wall thickness, in.	Diameter, in.	Speed, ft/min
0.24	393	0.06	571	0.06	1.0	518
0.32	272	0.08	428	0.08	1.0	389
0.39	191	0.12	285	0.12	1.0	258
0.47	137	0.16	213	0.06	2.0	467
0.55	99	0.24	132	0.08	2.0	351
0.63	69	0.32	81	0.12	2.0	233
...	...	...	...	0.16	2.0	176
...	...	...	...	0.08	3.0	289
...	...	...	...	0.16	3.0	145
...	...	...	...	0.24	3.0	85

through forming rolls to the weld area in the extreme upper left-hand corner. The upset metal is then trimmed off the welded pipe, the weld is locally heat treated, and the pipe is cut to length on the fly. Figure 9 is a close-up of the welding section of a pipe mill.

Pipe and tubing produced by high frequency welding have gained wide acceptance in the petroleum industry as well as for structural purposes and high pressure pipe and tubing in power generation. A number of American Petroleum Institute (API) and American Society for Testing and Materials (ASTM) specifications describe products made by this process. Pipe and tubing ranging in diameter from less than 0.5 in. to over 50 in. and with wall thicknesses from less than 0.005 in. up to 1.0 in. have regularly been run in production. Weld speeds range from less than 25 ft/min to over 1000 ft/min, depending on the material and the product size. High frequency induction welding is usually limited to smaller diameter and thinner walled tubing because of the inherently lower efficiency of the process, particularly at larger diameters. Weld quality of high frequency welded pipe and tubing under properly controlled conditions is so high that numerous products previously produced by seamless practice, such as boiler tubing and hydraulic cylinder tubing, are currently manufactured by high frequency welding.

Structural Sections. High frequency resistance welding is used extensively for the production of structural shapes such as I- and H-beams in sizes from 3 to 20 in. and with web thicknesses up to 0.38 in. Weldable materials range in yield strength from 30 to 80 ksi. Shapes can be produced with mixed materials, such as carbon steel webs and high-strength steel flanges. A particular advantage of this process is the ability to make relatively short runs with inexpensive tooling; usually only an adjustment of rolls is required to change sizes. Thinner section components can be manufactured by welding than by rolling without causing distortion of the product, and special and asymmetrical shapes are easily created without expensive tooling. Major areas of application include highway trailer frames, mobile home framing, and general light structural use. Another important advantage of the process is the very low installed plant cost for shape production when compared to conventional hot rolling facilities. Figure 10 illustrates the setup for HFRW of two pieces of steel into a T-section.

Spiral seam pipe and tube can be manufactured by HFRW using either a butt or a lap joint. Figure 11 illustrates typical setups for spiral butt welding and spiral lap welding of tubing by HFRW. The very high welding speeds obtainable by HFRW, coupled with the extremely simple and economical mechanical forming equipment, make this a useful way for produc-

Fig. 8 Mill that produces large steel pipe

Fig. 9 Welding station of a pipe mill

Fig. 10 Setup for HFRW of a T-section

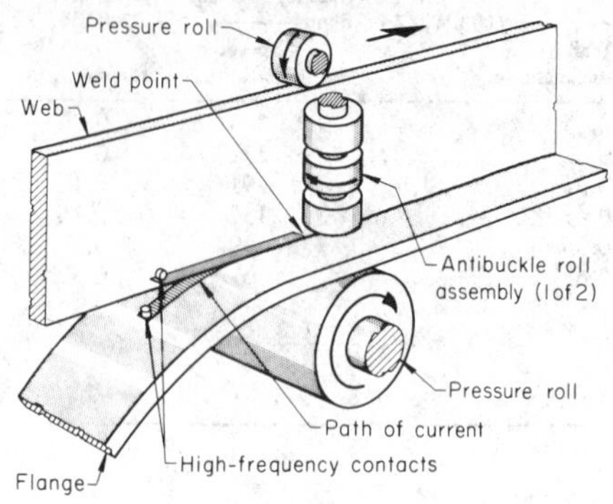

Fig. 11 Setups for production of tubing using HFRW

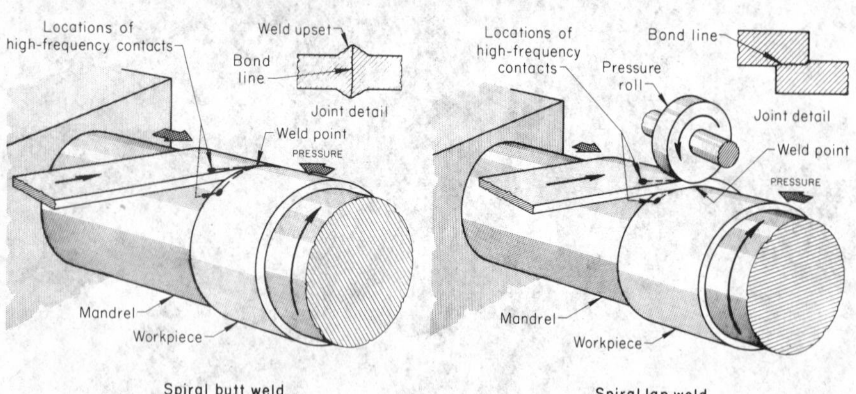

ing larger diameter tubular material from narrow strip. The best attainable weld quality is ordinarily not as good as that produced on longitudinal butt seam HFRW mills, but is quite satisfactory for structural and low-pressure purposes. Pipe produced in this way is used for piling and drainage culvert.

Finned Tube. An important area of application for HFRW is the manufacture of finned heat exchanger tubing. Longitudinal fins can be welded to continuous or cut length tubing for improved heat exchange or for mechanical assembly of boiler water-wall panels. Figure 1(h) shows the schematic arrangement. Spiral fins are readily welded to tubing for extended surface heat exchangers (Fig 1g and 12) and to boiler water-wall tubes (Fig. 13). Tubing diameter ranges from 0.63 to 10 in., with typical fin heights equal to as much as half the tube diameter. Fin thickness is determined both by fin height and the wall thickness of the tube to which it is welded. The tube must support the weld forging force without significant distortion.

A wide range of materials has been used in the production of finned tubing. Low-carbon steel fins are usually put on low-alloy steel tubing; however, stainless steel fins have been welded to carbon and stainless steel tubing in commercial production. In addition, other combinations, such as aluminum fin on cupronickel tube and zircalloy fin on zircalloy tube, have been produced satisfactorily.

Fin-to-tube welding by HFRW has the advantages of producing an excellent mechanical and thermal joint between fin and tube surface without causing damage to the tube itself, as well as an economical use of fin material and very high production speeds. Typical fin welding speeds range from 50 to 150 ft/min.

Fig. 12 Welding of spiral fins to a tube for use in heat exchangers

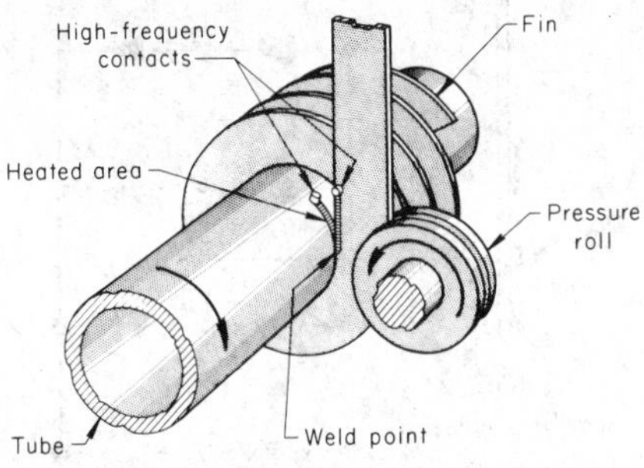

Welding Procedures

The inherent characteristics of HFRW impose certain requirements on the mechanical equipment used to form and support the material being welded. Although cleaning workpiece surfaces for high frequency contacts is not necessary, faying surfaces should be reasonably clean and smooth and parallel to each other during heating and upsetting. Special processes such as solvent cleaning or brushing usually are not used.

Joint Fit-Up and Alignment. The shape of the V during continuous seam butt welding is an important factor in producing satisfactory welds. As the faying surfaces converge, the included angle of the V must be maintained from 4 to 7°. If the angle is less than 4°, the apex (weld point) does not remain fixed, and poor weld quality can result. If the angle is greater than 7°, wrinkling or stretching of the joint edge may occur. In low-carbon steels, variations in V length and width are more tolerable than in other materials. Failure to keep the faying surfaces vertically parallel in the V may result in excessive melting of the edges and arcing near the apex of the V. Foreign material must not enter the V. It can cause short circuiting of the weld current, resulting in variation of heating. These requirements underscore the importance of accuracy and rigidity of the forming machinery as well as control of the dimensions and properties of the welded material. Mill maintenance and accurate tooling are essential to successful HFRW.

Fig. 13 High frequency resistance welding of longitudinal fins to a boiler water-wall tube

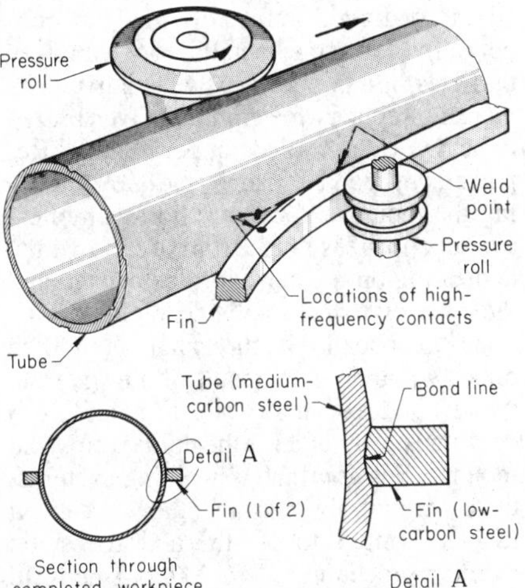

Postweld Procedures. The upset material resulting from the forging of the weld can be left in place on some products, but is frequently removed from external and internal surfaces of pipe and tubing. This process, known as scarfing or bead trimming, is performed by single point cutting tools mounted on the welding mill immediately downstream of the welding rolls. The chips that this produces are normally continuous and must be carefully removed. Postweld heat treatment of the weld zone is commonly performed on large high-strength pipe and tubing by in-line induction equipment on the pipe mill; however, in some cases the entire product is heat treated either in a tunnel furnace or in batches after cutting to length.

Most continuous welding mills have provisions for cooling and straightening and, if necessary, final sizing before the product is cut into lengths. Continuous nondestructive test equipment is usually mounted on the mill over the weld seam before cutoff to perform 100% inspection of the joint. Cutoff of the product is performed on the fly by a variety of means, including shears, abrasive wheels, and saws.

Weld Quality

High frequency welding is characterized by extremely rapid heating of the faying surfaces, usually to the melting point of one or both of the surfaces being joined. Heating can take place in as little as 0.01

Fig. 14 Cross section through high frequency weld. Upset material present. Shows characteristic hourglass-shaped HAZ. Steel: 8.74-in. OD, 0.32-in. wall. Magnification: 7.8×

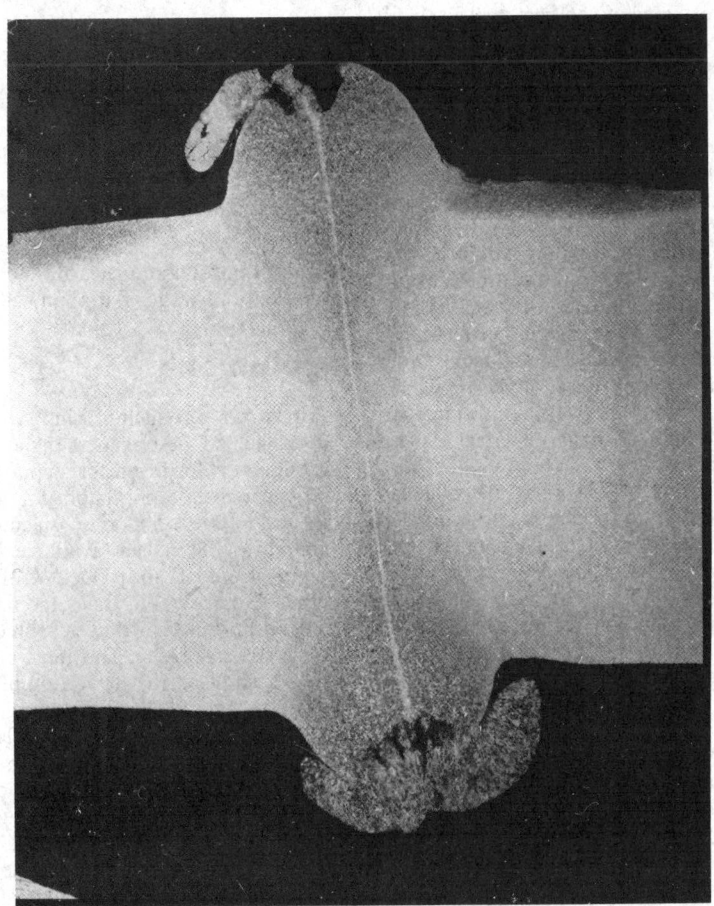

s or as much as 1 s. Immediately after heating, hot forging of the faying surfaces takes place, typically in one third of the heating time. During the hot forging process, most of the hottest material is extruded from the weld interface. This extruded material carries with it contaminants and leaves a clean and well-bonded joint.

As soon as heating stops, the joint starts cooling very rapidly through the conduction of heat to adjacent unheated metal. As a result of this high-speed process, little or no grain growth occurs. In fact, some grain refinement frequently occurs as a result of hot working. Reorientation of directional structures occurs as a result of flow in the upset forging of the joint. In the case of material in which laminations are present, this can create difficulties. Figure 14 shows a section through a high frequency weld with upset material present. The characteristic hourglass-shaped HAZ is visible. The bond interface looks slightly lighter because of the characteristics of the etchant.

Mechanical properties of the weld zone may vary from the surrounding material, depending on the thermomechanical characteristics of the material and its state before welding. Local annealing can occur in the weld zone of non-heat-treatable hard-rolled material, resulting in a weld area that is softer than the surrounding material. Conversely, in heat treatable materials weld zones can be produced that are substantially harder than the adjacent material because of rapid self-quenching. Local heat treatment of the weld is practical in most cases and is carried out where necessary on the welding mill. Very little if any residual cast material is left in the joint interface, and modification of local chemical compositions is insignificant because no filler material is used, and exposure to an oxidizing atmosphere is very brief. As a result, finding the weld joint in the finished products of many materials is often difficult.

The joining of dissimilar materials can be done to create special products. In cases where brittle intermetallic compounds form, joint strength is enhanced by the very thin, uniform layer of the mixture at the joint interface which is inherent to the process.

Weld quality is typically very high; however, failure to perform the process correctly can result in the presence of residual contaminants at the weld interface. The most common causes of poor weld quality are:

- Insufficient or nonuniform power in the V
- Insufficient or nonuniform speed

Fig. 15 Circuit of oscillator power source. Range of 200 to 400 kHz

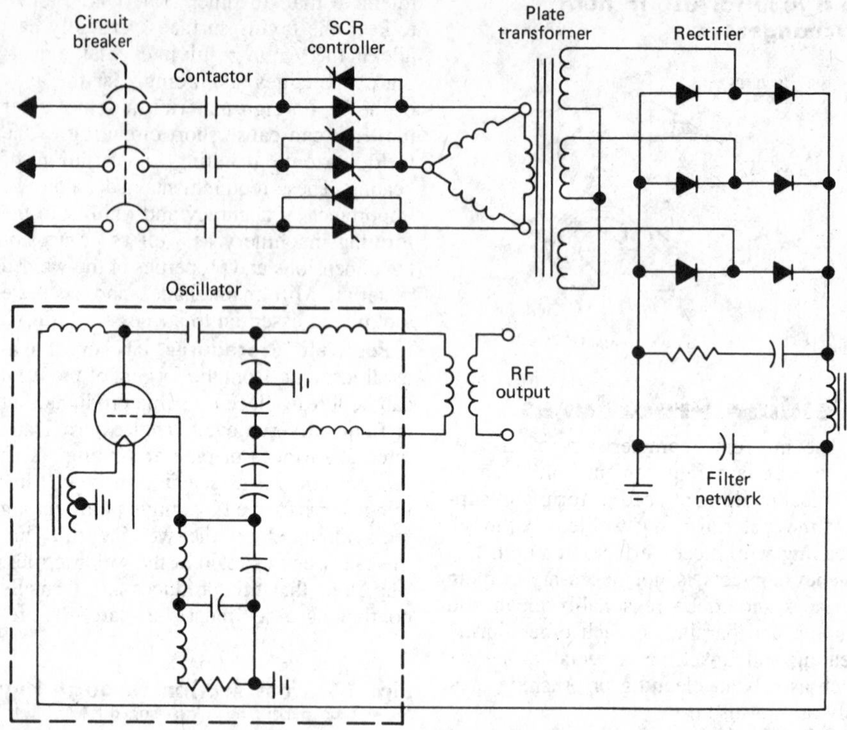

- Insufficient or nonuniform upset
- Poor shape or condition of faying surfaces before welding
- Improper positioning or movement of surfaces during welding
- Presence of excessive contamination in the weld area
- Use of unsuitable material, such as the presence of large, flat nonmetallics

Equipment

Units for producing high frequency power include motor generators and solid-state inverters for frequencies up to 10 kHz and vacuum tube oscillators for frequencies from 100 to 500 kHz. Vacuum tube welding oscillators are available in sizes ranging from 30 to 1000 kW of power output. Figure 15 illustrates the basic circuit for an oscillator power source in the 200 to 400 kHz range. The transformer and rectifier are used for converting plant line voltage to high-voltage direct current power for use in the oscillator. The oscillator circuit converts direct current to high frequency current at high voltage. The output transformer converts the high-voltage, low-amperage power to low-voltage, high-amperage power for welding. The output of the transformer may be fed into a set of contacts for HFRW or into an induction coil for HFIW.

For frequencies in the range of 3 to 10 kHz, power is usually produced by solid-state inverters that operate on direct current produced by a rectifier. In recent years, these inverters have replaced motor generator/motor alternator power sources. The inverter portion of the equipment is a solid-state device, in which silicon-controlled rectifiers are fired by a timing circuit, thus creating high frequency power.

Contacts. High frequency resistance welding current transfer contacts, either sliding or fixed, are usually made of copper alloy or hard metallic or refractory particles in a copper or silver matrix. The blocks of contact material are silver brazed to heavy water-cooled copper mounts. Replacements can be made rapidly by changing the mount and contact tip assemblies. Contact tip sizes range from 0.25 to 1 in.^{2}, depending on the current to be carried and the type of work involved. Welding currents are usually in the range of 500 to 5000 A, and as a result, both internal (water) and external (water or soluble oil) cooling is provided for the contact tips and mounts. For continuous welding systems, the contact tip force against the work metal ranges from 5 to 50 lb, and for static welding procedures, 5 to 100 lb. The weld current as well as workpiece thickness and surface condition affect these forces.

Contacts used for continuous seam welding systems for nonferrous materials may have triple the life of those used for ferrous materials. Welding 300 000 ft of nonferrous tube with one set of contacts is possible. On stationary welding operations, contacts may last several thousands of operations before requiring dressing or replacement.

Induction coils, sometimes called inductors, are generally made of copper tubing, copper bar, or copper sheet and are water cooled. Optimum efficiency is obtained when an induction coil completely surrounds the workpiece. The coil may have one or more turns, depending on the application. It is designed to fit close to the workpiece to be heated. The strength of the magnetic field, which induces the heating current in the workpiece, falls off rapidly as the distance between the coil and work is increased. The sharpness with which the heating pattern in the workpiece mirrors the shape of the coil improves as frequency increases and distance decreases between the coil and the workpiece. Typical spacing between coil and workpiece is from 0.08 to 0.5 in.

Impeders. In tube and pipe welding with either high frequency process, current can flow on the inside surface of the tube as well as on the outside surface, in addition to the desired flow of current in the V. The only current that is accomplishing the desired heating is that flowing on the edges to be welded; therefore, the additional currents flowing inside the tube result in a power loss and are wasted. To minimize this loss, a magnetic core, or impeder, is placed inside the tube in the weld area. The impeder increases the inductive reactance of the current path around the inside surface of the tube. This in turn reduces the undesired inside current and increases the current flowing in the weld zone. The reduction of inside current permits the transformer output to rise in voltage, permitting higher welding speeds for a given power output. Impeders are usually made of one or more ferrite bodies. They are water cooled to keep their operating temperatures below their Curie point, the temperature at which they lose their magnetic properties.

When internal scarfing is required, a mandrel must pass inside the tube through the weld area. This mandrel must be made of a nonmagnetic material, such as austenitic stainless steel, and should have impeders mounted on or around it to prevent it from being induction heated.

Testing

Weld imperfections of very small size can be detected successfully at high production speeds by a variety of commercially available nondestructive testing devices. Usually bond plane nonmetallic inclusions, which are well below the acceptable levels for weld quality, can be detected immediately after welding, thus permitting tight control of the welding operation. In addition to nondestructive testing, a number of small-sample mechanical testing procedures are normally used in the course of production to assess joint strength and ductility. Some weld testing procedures commonly used in arc welding quality control, such as x-ray and magnetic-particle or liquid-penetrant crack detection, are inappropriate for high frequency welding except to detect very large and obvious defects.

Nondestructive Testing

Three methods of nondestructive testing are in general use to inspect high frequency welded joints: ultrasonic, eddy current, and flux leakage. Hydrostatic testing may also be considered as a nondestructive test, although a failure can cause a crack to propagate into good material. In a few cases magnetic-particle testing has also been used.

Ultrasonic Testing. The general requirements for ultrasonic testing of tube and pipe are given in ASTM E213, "Recommended Practice for Ultrasonic Testing of Metal Pipe and Tubing," and ASTM E273, "Ultrasonic Inspection of Longitudinal and Spiral Welds of Welded Pipe and Tubing." Angle projection of a pulsed ultrasonic beam is used. A reference standard is made of tube of the same diameter and wall thickness with one or more notches or holes of specified shape and size. An indication of a discontinuity greater than that of the reference discontinuity is cause to reject the tube.

Eddy Current Testing. The general requirements for eddy current testing of tube and pipe are given in ASTM E309, "Eddy Current Examination of Steel Tubular Products Using Magnetic Saturation," and ASTM E426, "Electromagnetic (Eddy Current) Testing of Seamless and Welded Tubular Products, Austenitic Stainless Steel and Similar Alloys." Eddy currents are induced in the tube by an alternating current in an exciter coil in close proximity to the tube. A sensor coil detects the resulting electromagnetic flux related to these currents. The presence of discontinuities affects the normal flow of current, and this change is detected by the sensor. In the case of ferromagnetic materials, a strong external magnetic field is applied in the region of the exciting and sensor coils to make that region of the tube effectively nonmagnetic. The exciting and sensor coils may completely encircle the tube or may be smaller localized probe coils placed in the area of the weld. In some cases, one or more coils may function concurrently as both exciters and sensors. Reference standards with known discontinuities are used to calibrate the system and to set up the acceptance levels.

Flux Leakage Examination. The general requirements for flux leakage examination are given in ASTM E570, "Flux Leakage Examination of Ferromagnetic Steel Tubular Products." The tube wall is magnetized to near saturation by the application of a strong, adjustable transverse magnetic field. Flux sensors are used to detect the flux leakage directly from a discontinuity in the tube. Reference standards with known discontinuities are used

Fig. 16 High frequency resistance welded tubing for hydraulic cylinders. 11.25-in. OD, 0.65-in. wall. As-welded condition, scarfed inside and outside. Before heat treatment and drawing over mandrel. Magnification: 6×

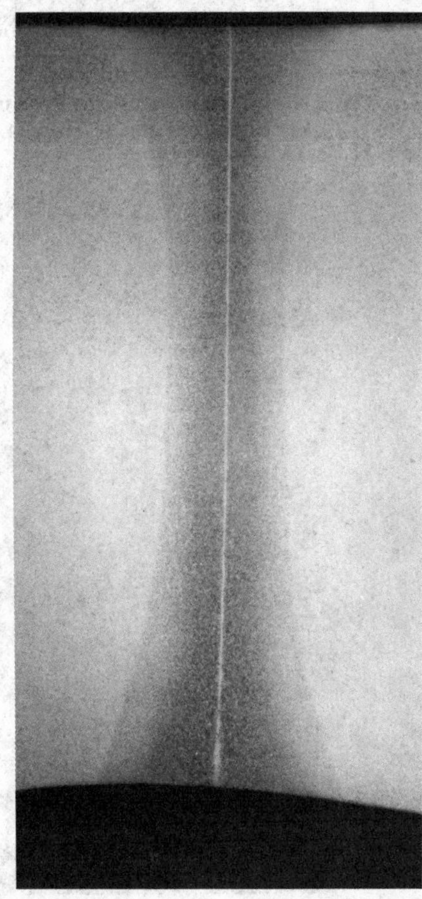

Fig. 17 Cross section of high frequency resistance weld in API K-55 oil well casing. 7.0-in. OD, 0.242-in. wall. As welded, prior to hot stretch reducing to 5.5-in. OD. Magnification: 17×

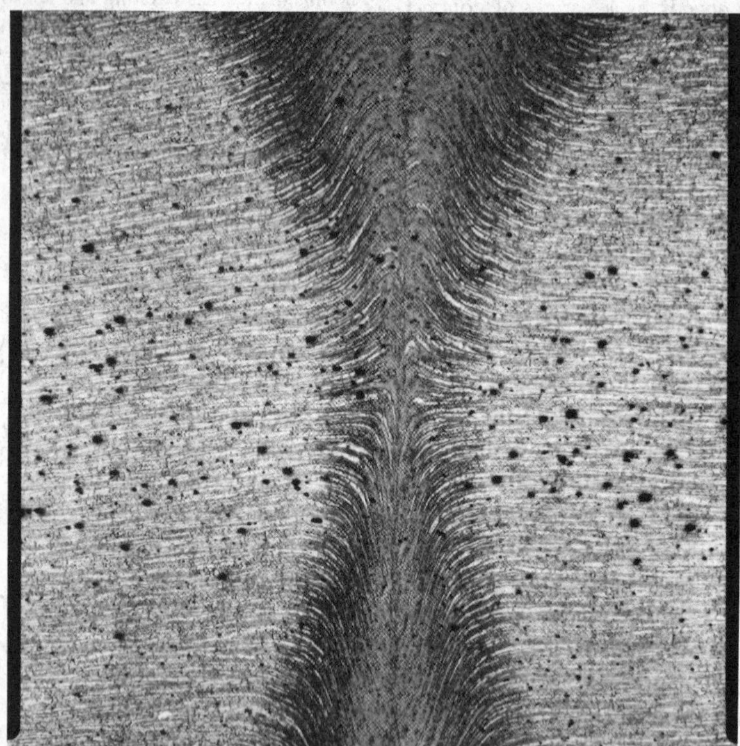

Fig. 18 Cross section of high frequency induction weld in aluminum alloy 3105 for furniture tubing. 1.0-in. OD, 0.065-in. wall. As welded with inside upset left unscarfed. Magnification: 50×

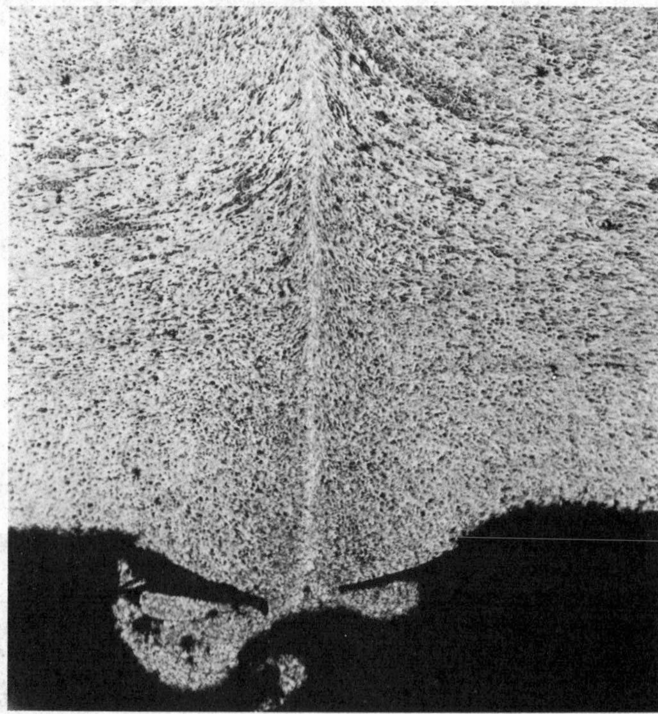

to calibrate the instrument and to set up the acceptance levels.

Hydrostatic Testing. The general requirements for hydrostatic testing are given in ASTM A450, "General Requirements for Carbon Ferritic Alloy and Austenitic Alloy Steel Tubes." The tubes are tested to a minimum hydrostatic test pressure which is given in the applicable standard or specified by the purchaser. This test pressure may be a relatively low pressure which serves only as a leak test or may be high enough to stress the material to close to its yield point. In some cases, hydraulic expansion is used to bring the pipe to its final size and to provide a severe test of the welded joint.

Destructive Testing

A number of destructive tests using small samples of the product are in general use for testing high frequency welds in tube and pipe. The requirements for these tests are given in ASTM A450.

Flattening Test (Crush Test). This test requires that a short sample of the welded product be placed in a press between parallel platens. The sample is then flattened to a specified height with the weld at 90° from the direction of applied force and examined for cracks or breaks in the weld area. The flattening is then continued until the specimen breaks or the opposite walls of the tube meet. This test imposes maximum extension on the outer surface of the weld. The test is also performed with the weld under the platen to stretch the weld at the inside of the pipe. These tests can reveal both imperfections in the weld and a lack of ductility in the parent metal adjacent to the weld. In the case of failure, examination of the fracture surface can often assist in determining the nature of the imperfections and the corrective steps to be taken.

Reverse Flattening Test. When a more severe test of the weld at the inside is required, a section of the tube is split longitudinally 90° on each side of the weld, opened up, and flattened. It is then bent back with the weld at the point of maximum bend. Cracks and other surface defects at the weld may be cause for rejection. Examination of the fracture surface can be valuable in determining the reasons for any failure.

Flaring Test. The end of a short length of tube is pressed over a conical mandrel, causing it to flare to a specified diameter. If cracks or imperfections are revealed, the tube is rejected. Although this test is in general use (particularly for welds in duc-

Fig. 19 Cross section of high frequency resistance weld for solar absorber plate. Between 0.375-in.-OD by 0.020-in.-wall copper tube and 4.5-in.-wide by 0.010-in.-thick copper strip. Magnification: 75×

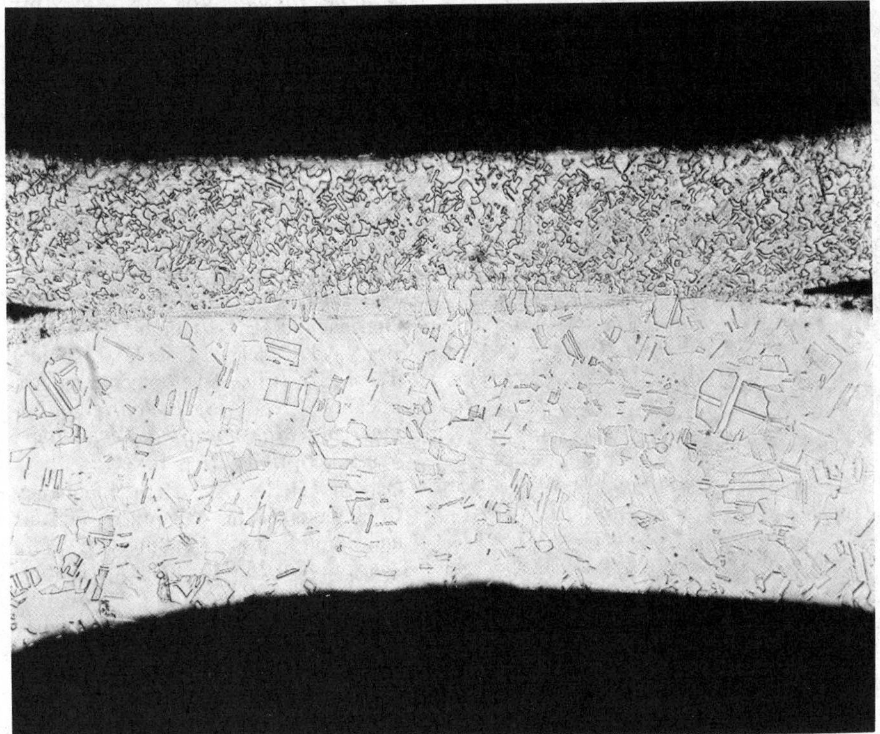

Fig. 20 Cross section of high frequency resistance weld in API J-55 well casing. 4.5-in. OD, 1.89-in. wall. Shows crush test failure near bond plane. Magnification: 19×

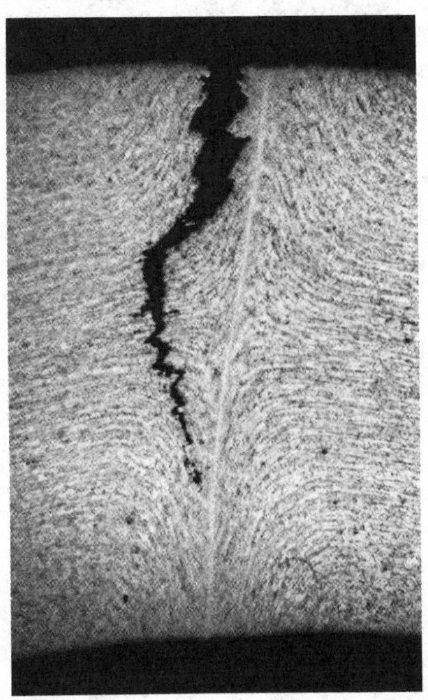

tile materials), it can give misleading results, especially when the weld upset has not been removed from the inside of the tube. The presence of upset tends to act as a stress raiser and can lead to premature failure next to the weld.

Metallographic Examination of High Frequency Welds

Metallographic examination of high frequency welds is frequently undertaken to assist in determining the quality and soundness of a weld as well as the microstructure in the HAZ. Figure 16 is a cross section of a typical high frequency weld in the longitudinal seam of a steel pipe in the as-welded condition prior to heat treatment. It was welded at 11.25-in. OD with a 0.65-in. wall. This material is subsequently heat treated and drawn over a mandrel to make smooth inside diameter drawn-over-mandrel tubing for hydraulic cylinders and similar products. The very narrow HAZ is typical of high frequency welds. The light etching bond plane is also frequently seen.

Figure 17 is a cross section of API grade K-55 oil well casing. It was welded at 7.0-in. OD by 0.242-in. wall and is shown in the as-welded condition. The upsetting pattern is clearly evident in this section, caused by slight banding of the parent metal. Note the hourglass shape of the HAZ. This material was subsequently hot stretch reduced to 5.5-in. OD.

Figure 18 is a weld in 3105 aluminum alloy. It is 1.0-in. OD by 0.065-in. wall, in the as-welded condition, and not subsequently heat treated. The outside upset has been removed, but the inside bead has been left in. This example is typical of structural quality aluminum tubing used for such applications as lawn furniture.

Figure 19 shows a weld between a copper tube 0.375-in. OD by 0.020-in. wall and a copper strip 4.5 in. wide by 0.010 in. thick. This weld is used to transfer heat from the fin to a liquid in the tube in a solar absorber plate. This weld is also characterized by a very narrow HAZ.

Figure 20 is a section of API J-55 casing which failed in a flattening test. The failure is beside the bond plane and has propagated along flow lines that were originally parallel to the surfaces of the strip but have been rotated during the forging process.

Figure 21 is a high magnification photomicrograph of the failed bond plane in Fig. 20. This figure shows the elongated inclusions in the rolling direction of the parent strip that gave rise to the planes of weakness along which the fracture tended to propagate. Double ligament tests of the ductility of this material showed that the elongation of the material in the rolling direction was over 40%, whereas in the through thickness direction, it was less than 12.5%. Calculations indicate that a through thickness elongation of about 20% would be required to pass the flattening test in pipe of this size.

Safety

Serious consideration must be given to the health and safety of welding operators, maintenance personnel, and other personnel in the area of the welding operations. Good engineering practice must be followed in the design, construction, installation, operation, and maintenance of equipment, controls, power supplies, and tooling to conform to federal, state, and local safety regulations, as well as those of the using company.

High frequency welding power sources are electrical devices and require all the usual precautions in handling and repairing such equipment. Voltages are in the range of 400 to 25 000 V and are lethal. These voltages may be low frequency or high frequency. To prevent injury, proper care and safety precautions should be taken while working on high frequency gener-

Fig. 21 Parent metal from pipe of Fig. 20 showing nonmetallic inclusions leading to low ductility in direction of through thickness. Magnification: 630×

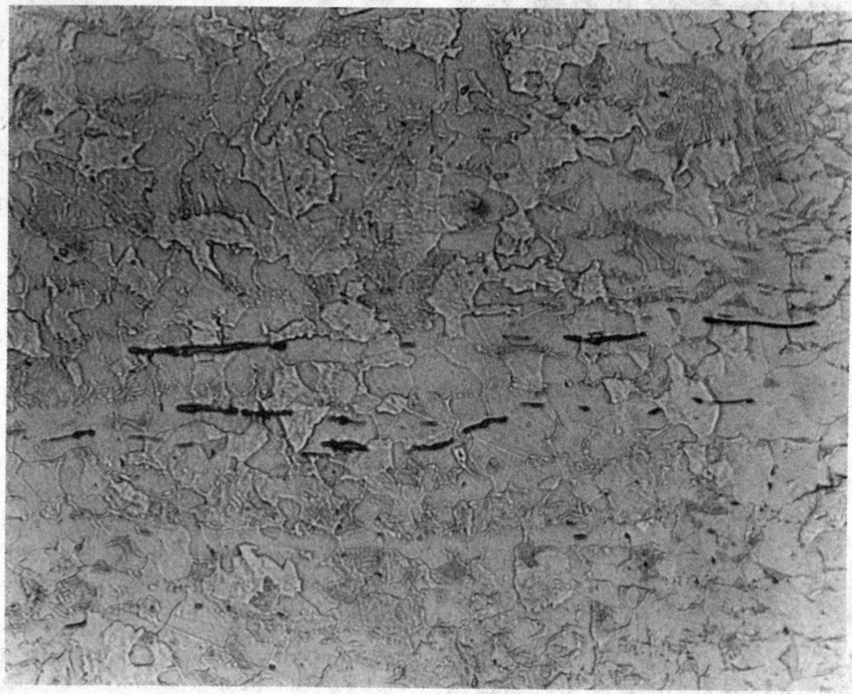

ators and their systems. Modern units are equipped with safety interlocks on access doors and automatic safety grounding devices that prevent operation of the equipment when the access doors are open. The equipment should not be operated with panels or high-voltage covers removed or with interlocks and grounding devices blocked.

The output high frequency primary leads should be encased in metal ducting and should not be operated in the open. The induction coils and contact systems should always be properly grounded for operator protection. High frequency currents are more difficult to ground than low frequency currents, and grounding lines should be kept extremely short and direct to minimize inductive reactance. Care should be taken that the magnetic field from the output system, particularly the output transformer, does not heat adjacent metallic sections by induction. Injuries from high frequency power, especially at the upper range of weld frequencies, tend to produce severe local surface tissue damage. Metal or flux fumes present little or no hazard to the welder, and although safety glasses are recommended for eye protection, arc burns pose no danger.

SELECTED REFERENCES

- Brown, G.H., Hoyler, C.N., and Bierwith, R.A., *Theory and Applications of Radio Frequency Heating*, D Van Nostrand, New York, 1957
- Daily, R.F., Induction Welding of Pipe Using 10,000 Cycles, *Welding Journal*, Vol 44 (No. 4), June 1965, p 475-479
- Harris, S.G., Butt Welding of Steel Pipe Using Induction Heat, *Welding Journal*, Vol 40 (No. 2), Feb 1961, p 57s-65s
- Johnstone, A.A., Trotter, F.J., and a'Brassard, H.F., Performance Record of the Thermatool High Frequency Resistance Welding Process, *British Welding Journal*, Vol 7 (No. 4), April 1960, p 238-249
- Koppenhofer, R.L., *et al.*, Induction-Pressure Welding of Girth Joints in Steel Pipe, *Welding Journal*, Vol 39 (No. 7), July 1960, p 685-691
- Oppenheimer, E.D., Helical and Longitudinally Finned Tubing by High Frequency Resistance Welding, ASTME Tech. Paper AD 67-197, Society of Manufacturing Engineers, Dearborn, MI, 1967
- Oppenheimer, E.D., Kumblé, R.G., and Berry, J.T., The Double Ligament Tensile Test: Its Development and Application, *Trans. ASME, Journal of Engineering Materials and Technology*, Vol 97, April 1975, p 107-112
- Osborn, H.B., Jr., High Frequency Continuous Seam Welding of Ferrous and Non-Ferrous Tubing, *Welding Journal*, Vol 35 (No. 12), Dec 1956, p 1196-1206
- Osborn, H.B., Jr., High Frequency Welding of Pipe and Tubing, *Welding Journal*, Vol 42 (No. 7), July 1963, p 571-577
- Rudd, W.C., High Frequency Resistance Welding, *Welding Journal*, Vol 36 (No. 7), July 1957, p 703-707
- Rudd, W.C., High Frequency Resistance Welding of Cans, *Welding Journal*, Vol 42 (No. 4), April 1963, p 279-284
- Rudd, W.C., High Frequency Resistance Welding, *Metal Progress*, Oct 1965, p 239-240, 244
- Rudd, W.C., Current Penetration Seam Welding—A New High Speed Process, *Welding Journal*, Vol 46 (No. 9), Sept 1967, p 762-766
- Wolcott, C.G., High Frequency Welded Structural Shapes, *Welding Journal*, Vol 44 (No. 11), Nov 1965, p 921-926
- Udall, H.N., Berry, J.T., and Oppenheimer, E.D., A High Speed Welding System for the Production of Custom Designed H.S.L.A. Structural Sections, in *Welding of HSLA (Microalloyed) Structural Steels*, American Society for Metals, 1978, p 706-733

Surfacing

Hardfacing

By the ASM Committee on Hardfacing*

HARDFACING is the application of a hard, wear-resistant material to the surface of a component by welding, spraying, or allied welding processes to reduce wear or loss of material by abrasion, impact, erosion, galling, and cavitation. The stipulation that the surface be modified by welding, spraying, or allied welding processes excludes the use of heat treatment or surface modification processes such as flame hardening, nitriding, or ion implantation as a hardfacing process. The stipulation that the surface be applied for the main purpose of reducing wear excludes the application of materials primarily used for prevention or control of corrosion or high-temperature scaling. Corrosion and/or high-temperature scaling may, however, have a major effect on the wear rate and, hence, may become a significant factor in the selection of proper materials for hardfacing.

Hardfacing applications for wear control vary widely, ranging from very severe abrasive wear service, such as rock crushing and pulverizing, to applications to minimize metal-to-metal wear such as control valves where a few thousandths of an inch of wear is intolerable. Hardfacing is used for controlling abrasive wear, such as encountered by mill hammers, digging tools, extrusion screws, cutting shears, parts of earthmoving equipment, ball mills, and crusher parts. It is also used to control the wear of unlubricated or poorly lubricated metal-to-metal sliding contacts such as control valves, undercarriage parts of tractors and shovels, and high-performance bearings.

Hardfacing also is used to control combinations of wear and corrosion as encountered by mud seals, plows, knives in the food processing industry, valves and pumps handling corrosive liquids, or slurries. In most instances, parts are made of either plain carbon steel or stainless steel, materials that do not provide desirable wear on their own. Hardfacing alloys are applied to critical wear areas of original equipment or during reclamation of worn parts.

Typical applications of hardfacing of parts in original equipment include:

- Seating surfaces of some diesel engine valves are subjected to extremely severe conditions, involving erosion and fatigue combined with hot corrosion. To extend the life of the valve, the seating surfaces are hardfaced with cobalt-based alloys, and the valves themselves are made of inexpensive alloys.
- Chain saw manufacturers typically hardface the nose of saw guides with cobalt-based alloys to resist wear from high-velocity sliding of the blades.
- Steam control valves subjected to high stresses are hardfaced with cobalt-based alloys. In this type of application, most other materials would be subjected to galling and seizure.
- Fluid-control valves are hardfaced with cobalt-based alloys to minimize damage due to cavitation erosion from turbulent flow.
- Food products such as dry pet foods, breakfast cereals, meat extenders, soybean meat substitutes, and party snacks are produced by hot extrusion. The extruder usually consists of a screw and barrel combination. An extrusion die at the end of the extruder barrel forms the material into the desired shape, such as stars, rings, etc. Heat needed for cooking comes from the friction generated by the screw flights. Figure 1 shows hardfacing of the flight area of a hot extrusion screw. Similar applications are found in extrusion of thermoplastic resins for manufacturing plastic components.
- Industrial knives used for pelletizing plastic strands or textile fibers are frequently hardfaced with wear-resistant alloys to retain an edge.
- Horseshoes are subjected to extremely severe abrasion. In normal use, a horseshoe made of carbon steel by a blacksmith shop would last from 4 to 6 weeks.

Fig. 1 Hardfacing a food processing extrusion screw by GMAW

*Kenneth C. Antony, *Chairman*, Manager, Technical Services, Wear Technology Division, Cabot Corp.; Kirit J. Bhansali, Group Leader, Chemical Metallurgy, National Bureau of Standards; Robert W. Messler, Jr., Director of Welding Research & Development, Eutectic Corp.; Albert E. Miller, Professor of Metallurgical Engineering and Materials Science, University of Notre Dame; Marianne O. Price, Senior Project Engineer, Engineering Products Division, Union Carbide Corp.; R.C. Tucker, Jr., Associate Director, Materials Development, Coatings Service Department, Union Carbide Corp.

Fig. 2 Hardfacing a horseshoe by OFW

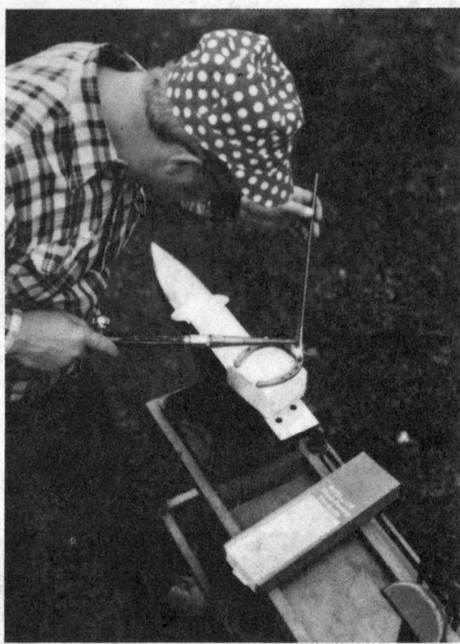

Figure 2 shows hardfacing of a horseshoe with tungsten carbide tube rod containing 60% W. The life of a horseshoe can be extended five to ten times by hardfacing.

- Hammers in large impact mills such as those used to crush limestone rock wear rapidly due to impact and abrasion. Maintenance downtime because of wear is a serious economic problem. Rebuilding with an iron-based buildup alloy and hardfacing by shielded metal arc welding (SMAW) with a high-carbon chromium white iron hardfacing alloy on a regular basis significantly extends the service life of the rotor hammers and reduces maintenance downtime. The hardfacing operation is illustrated in Fig. 3.
- The wheels on mine cars used to transport ore from underground mine sites to ore treatment plants suffer rapid metal-to-metal wear under high impact loading conditions. Mine car wheel replacement is expensive. Hardfacing or, more properly, rebuilding worn mine car wheels with a work-hardening manganese steel by submerged arc welding (SAW), as shown in Fig. 4, provides equivalent improved service life at approximately two thirds the replacement cost.

Fig. 3 Rebuilding and hardfacing a limestone crusher by open arc welding

Generally, hardfacing offers the following economic advantages: (1) increased productivity through less downtime for repair and replacement, (2) increased efficiency by permitting use of higher applied loads, (3) reduced maintenance cost through reclamation of worn parts, and (4) optimum compromise between wear and toughness through the use of less expensive, tougher base metals.

Wear

Economics frequently plays a major role in the selection of a hardfacing material. As a result, there is a temptation to hardface with whichever material is cheapest, applied fastest, or most readily available. Also, the belief is still prevalent that all hard metals resist wear equally well. Such a haphazard approach has only a limited chance for success. If optimum results are to be achieved from hardfacing, an engineering approach must be taken to the problem of wear. Selection of proper materials should be made only after careful consideration of part design, wear mode, and material and environmental interactions.

Fig. 4 Rebuilding and hardfacing a mine car wheel by SAW

There is some debate concerning wear mode classification. However, most tribological investigators agree on the following classifications of wear:

Adhesive wear

- Oxidative wear
- Metallic wear
- Galling

Abrasive wear

- Low-stress scratching abrasion
- High-stress grinding abrasion
- Gouging abrasion

Erosion

- Particle impingement
- Cavitation

Fretting

Adhesive wear generally describes wear due to sliding action between two metallic components where no abrasives are intended to be present—for example, wear due to sliding caused by slippage of roller bearings or sliding between a valve and a seat. When the applied load is sufficiently low, an oxide film usually is generated as a result of frictional heating accompanied by sliding. The oxide film prevents the formation of a metallic bond between the sliding surfaces, resulting in low wear rates. This form of wear is called oxidative or mild wear.

However, if the applied load is high, formation of a metallic bond occurs between the surfaces of mating materials. The resulting wear rates are extremely high. This form of wear is called severe or metallic wear. The load at which the mild to severe wear transition occurs is called the transition load.

Another form of wear, called galling, is a special form of severe adhesive wear. Galling occurs if the wear debris is larger than the clearance and if seizure of the moving component results.

Frequently, only small amounts of sliding result in galling and subsequent failure of a component. For example, in a high-pressure gate valve, high seating stresses may result in the formation of a metallic bond between the surface asperities of the mating materials. When the valve is actuated, material transfer occurs from one point to another due to locking of the surface asperities. The resulting surface damage prevents leak-free reseating of the valve, and valve failure occurs. This type of damage is called galling or scoring.

In situations where lubrication is not possible, hardfacing is recommended to minimize adhesive wear, such as in automotive exhaust valves that experience extreme temperatures at which lubricants are not stable. Generally, most moving components should be designed to resist mild wear. Nickel-based hardfacing alloys usually have very high transition loads compared to cobalt-based hardfacing alloys. However, cobalt-based alloys are extremely resistant to galling compared to nickel- or iron-based alloys.

Abrasive Wear. Low-stress scratching abrasion is defined as wear resulting from a cutting action by sliding abrasives stressed below their crushing strength. Usually, the resulting wear pattern shows scratches, and the amount of subsurface deformation is minimal.

High-stress grinding abrasion describes wear resulting under conditions of stress that is high enough to crush the abrasives. Stresses are usually high enough to also cause plastic deformation of the ductile constituents (matrix) of the materials being abraded. Ball and rod mills are typical examples of high-stress grinding abrasion.

The term "gouging abrasion" is used to describe high-stress abrasion where sizable grooves or gouges are created on the wearing surface. The gouges may be a result of a sliding motion, followed by impact such as that encountered in gyratory crushers. In this type of application, high stresses transferred through large chunks of abrasives are experienced in the crushing regions of the crushers. The interfacial stresses are high even on a macroscale.

Generally, abrasion resistance of hardfacing alloys increases as the carbide or hard phase content increases. As a result, in extremely abrasive situations, alloys with large amounts of carbides are recommended. In addition, the type of hardfacing process used has a significant influence on dilution and on the size of the carbides. For example, depositing certain cobalt-based hardfacing alloys by gas tungsten arc welding (GTAW) creates finer carbides than the oxyacetylene process. As a result, in an abrasive wear test, the wear resistance of oxyacetylene deposits are superior to those of gas tungsten arc deposits. Thus, consideration should be given to the hardfacing process used for optimum performance.

Erosion. Impingement erosion occurs when material loss is caused by the cutting action of a fluid-borne moving particle. Usually, resistance of materials to impingement erosion varies with the angle of impingement, as shown in Fig. 5. At low impingement angles (<15°), hardfacing alloys with large amounts of carbides (such as hypereutectic alloys) are recommended. At high impingement angles (>80°), hardfacing alloys with large amounts of matrix (hypoeutectic alloys) are recommended.

Fig. 5 Effect of impingement angle on erosion rate. (a) Ductile materials. (b) Brittle materials

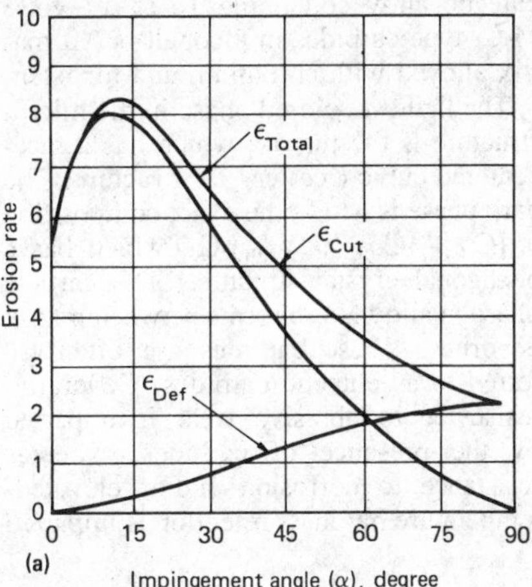

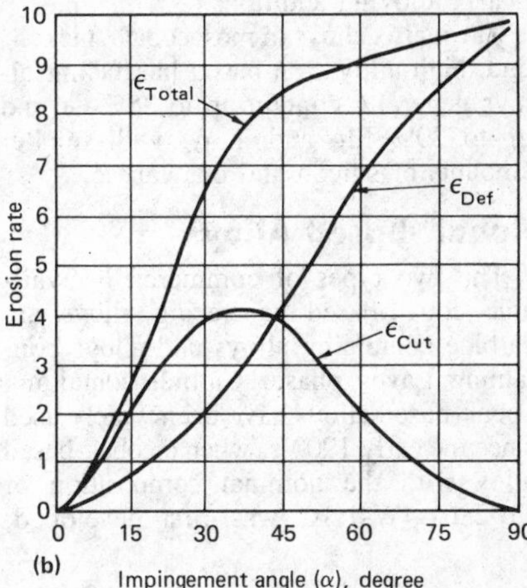

Frequently, in fluid handling valves or pumps, a phenomenon known as "cavitation" is caused by turbulent flow. Cavitation is characterized by collapsing of air bubbles caused by the turbulence. When the air bubbles collapse on the metal surface, wear is caused by the resultant shock waves. This type of wear is called cavitation erosion. Typical examples include fluid flow valves with large pressure differentials, high-velocity pumps, and leading edges of steam turbine blades. Cobalt-based hardfacing alloys are the best suited for resisting this type of wear.

Fretting describes material loss due to very small amplitude vibrations at mechanical connections (such as riveted joints or joints using other fasteners). This type of wear is a combination of oxidative and abrasive wear. Usually, oscillation of two metallic surfaces at a joint results in the formation of an oxide. Wear is subsequently caused by abrasion of this oxide. Cobalt-based hardfacing alloys usually are recommended for resisting this type of wear.

Hardfacing Materials

For simplicity, it would be convenient to have one material that would be able to resist all different types of wear. However, the complexity of the environment along with economic considerations has led to the development of a large number of hardfacing materials. Furthermore, labor and production costs have necessitated the development of a large number of deposition methods and product forms (bare rod, coated electrodes, powder, wire).

Selection of an appropriate hardfacing alloy for a given application is best achieved through careful analysis of all service conditions. A balance among wear properties, environmental resistance, and weldability must be achieved. Balancing properties within each alloy system has led to the development of a great many commercial hardfacing alloys. These are marketed as proprietary alloys. It is not uncommon to find one alloy composition identified by many different commercial designations.

Hardfacing materials include a wide variety of alloys, ceramics, and combinations of these materials. Conventional hardfacing materials are normally classified as steels or low-alloy ferrous materials, chromium white irons or high-alloy ferrous materials, carbides, nickel-based alloys, or cobalt-based alloys. A few copper-based alloys are sometimes used for hardfacing applications, but for the most part, hardfacing alloys are either iron-, nickel-, or cobalt-based. Microstructurally, hardfacing alloys generally consist of hard phase precipitates such as borides, carbides, or intermetallics bound in a softer iron-, nickel-, or cobalt-based alloy matrix.

Carbides are the predominant hard phases in iron- and cobalt-based hardfacing alloys. Carbon contents of iron- and cobalt-based hardfacing alloys generally range up to 4 wt%. Borides, as well as carbides, are the predominant hard phases in nickel-based hardfacing alloys. Combined carbon plus boron contents generally range up to 5 wt%. The specific carbide and/or boride phases that form are determined by matrix alloying additions.

The matrix alloys in most cobalt-, nickel-, and high-alloy iron-based hardfacing alloys generally contain up to 35% Cr and up to 20% Mo and/or W, with smaller amounts of silicon and manganese.

Cobalt-Based Alloys

The two types of commercially available cobalt-based hardfacing alloys are carbide-containing alloys and alloys containing Laves phase. Carbide-containing cobalt-based alloys have been widely used since the early 1900's, when a cobalt-based alloy with the nominal composition of Co-28Cr-4W-1.1C was first developed. Typically, this alloy, referred to in the hardfacing industry as Alloy No. 6, is a eutectic alloy containing 16 to 17 vol% M_7C_3-type carbides in a cobalt-based matrix alloyed with chromium and tungsten.

The lightest colored phase in the microstructure is the matrix, which has a face-centered cubic (fcc) crystal structure. The dark phase is M_7C_3; the exact composition is $(Cr_{0.85}, Co_{0.14}, W_{0.01})_7C_3$, which has a hexagonal crystal structure. The carbide phase in alloy No. 6 forms between matrix dendrites. These carbides are often referred to as eutectic carbides. Generally, resistance to abrasive wear is imparted by the presence of carbides, whereas resistance to corrosion and/or elevated-temperature hardness retention is imparted by the matrix. Galling and metal-to-metal wear properties are similarly determined by the matrix alloy.

There are five main alloys in the Co-Cr-W-C system that are suitable for use in hardfacing applications. The composition, bulk hardnesses, and hardnesses of the microconstituents in these alloys are listed in Table 1. Bulk hardness generally increases as carbon content increases. High-carbon-content alloys are hypereutectic and contain primary M_7C_3 carbides, which are modified hexagonal in shape and are considerably larger than the eutectic carbides. High-carbon-content alloys also contain M_6C-type carbides, the exact composition of which may vary from $(Co, W)_6C$ to $(Co_{0.66}, W_{0.34})_6C$.

The alloys listed in Table 1 that contain 2.5% C have more than 30 vol% total carbides, which results in extremely high abrasion resistance. The microstructure of the Co-30Cr-12W-2.5C alloy, sometimes referred to as Alloy No. 1, reveals a large volume fraction of carbides. As the carbon content is increased, the volume fraction of the matrix is decreased, and the impact resistance, weldability, and machinability also decrease. Thus, the gain in abrasive wear resistance is at the expense of other properties that may be more desirable.

For proper selection of a hardfacing alloy, service conditions should be carefully considered. Table 2 lists the available wear, mechanical, and corrosion properties of the cobalt-based hardfacing alloys listed in Table 1.

Laves phase is a type of topologically closed-packed intermetallic compound. Historically, metallurgists have avoided the presence of Laves phase in most alloys due to its detrimental effect on mechanical properties. However, in the early 1960's, the usefulness of Laves phase in resisting metal-to-metal wear was discovered; subsequently, alloys containing Laves phase have become commercially available.

There are two different Laves phase-containing cobalt-based alloys commercially available for hardfacing applications: Co-28Mo-8Cr-2Si and Co-28Mo-17Cr-3Si. Both of these alloys contain at least 50 vol% of Laves phase, bound in a cobalt-based matrix alloyed with chromium and molybdenum. Laves phase has a hexagonal structure similar to M_7C_3 carbides, but a hardness value between 1000 to 1200 DPH (kg/mm²), which is less than that of carbides. Consequently, the Laves phase-containing alloys are less abrasive to mating materials than carbide-containing alloys in metal-to-metal wear situations.

Table 1 Composition and hardness of selected cobalt-based hardfacing alloys

AWS designation or tradename	Nominal composition	Nominal macrohardness DPH	Nominal macrohardness HRC	Approximate hardness of microconstituents: Matrix, DPH	Hard particles Type	Hard particles DPH
Alloy 21	Co-27Cr-5Mo-2.8Ni-0.2C	255	24-27	250	Eutectic	900
RCoCrA	Co-28Cr-4W-1.1C	424	39-42	370	Eutectic	900(a)
RCoCrB	Co-29Cr-8W-1.35C	471	40-48	420	Eutectic	900(a)
RCoCrC	Co-30Cr-12W-2.5C	577	52-54	510	M_7C_3	900(a)
					M_6C	1540
						1700
Alloy 20	Co-32Cr-17W-2.5C	653	53-55	540	M_7C_3	900
					M_6C	
Tribaloy T-800	Co-28Mo-17Cr-3Si	653	54-64	800(b)	Laves phase	1100

(a)Matrix and M_7C_3 eutectic. (b) Matrix and Laves phase eutectic

Table 2 Properties of selected cobalt-based hardfacing alloys

	Alloy 21	RCoCrA	RCoCrB	RCoCrC	Tribaloy T-800
Density, lb/in.³	0.30	0.30	0.31	0.31	0.31
Ultimate compressive strength, ksi	188	220	256	280	258
Ultimate tensile strength, ksi	103	121	120	90	...
Elongation, %	8	1.2	1	1	
Coefficient of thermal expansion, (°C⁻¹)	14.8×10^{-6}	15.7×10^{-6}	14×10^{-6}	13.1×10^{-6}	12.3×10^{-6}
Hot hardness, DPH at:					
800 °F	150	300	345	510	659
1000 °F	145	275	325	465	622
1200 °F	135	260	285	390	490
1400 °F	115	185	245	230	308
Wear					
Unlubricated sliding wear(a), mm³ at:					
150 lb	5.2	2.6	2.4	0.6	1.7
300 lb	14.5	18.8	18.4	0.8	2.1
Abrasive wear(b), mm³					
OAW	...	29	12	8	...
GTAW	86	64	57	52	24
Unnotched Charpy impact strength, ft·lb	27	17	4	4	1
Corrosion resistance(c):					
65% nitric acid at 150 °F	U	U	U	U	S
5% sulfuric acid at 150 °F	E	E	E	E	...
50% phosphoric acid at 750 °F	E	E	E	E	E

(a) Wear measured from tests conducted on Dow-Corning LFW-1 against 4620 steel ring at 80 rpm for 2000 revolutions varying the applied load. (b) Wear measured from dry sand rubber wheel abrasion tests. Tested for 2000 revolutions at a load of 30 lb using a 9-in.-diam rubber wheel and AFS test sand. (c) E, less than 2 mils/yr; S, over 20 to less than 50 mils/yr; G, less than 20 mils/yr; U, more than 50 mils/yr

Table 3 Composition and hardness of selected nickel-based hardfacing alloys

AWS designation or tradename	Nominal composition	Nominal macrohardness DPH	Nominal macrohardness HRC	Matrix, DPH	Approximate hardness of microconstituents — Hard particles Type	DPH
RNiCr-C	Ni-15Cr-4Si-3.5B-0.75C	633	57	420	Primary boride	2300
					Secondary boride	950
					Eutectic	750
					Carbide (M_7C_3)	1700
RNiCr-B	Ni-12Cr-3.5Si-2.5B-0.35C	530	51	410	Primary boride	2300
					Secondary boride	950
					Eutectic phase	750
Hastelloy C	Ni-17Cr-17Mo-0.12C	200	HRB 95	180	M_6C	1700
Haynes 716	Ni-11Co-26Cr-29Fe-3.5W-3Mo-1.1C-0.5B	315	32	215	M_7C_3	1500
Tribaloy T-700	Ni-32Mo-15Cr-3Si	470	45	800(a)	Laves phase	...

(a) Matrix and Laves phase eutectic

Table 4 Properties of selected nickel-based hardfacing alloys

	RNiCrC	RNiCrB	Hastelloy C	Tribaloy T-700
Density, lb/in.^{3}	0.28	0.29	0.32	0.32
Coefficient of thermal expansion, (°C^{-1})	14.3	14.3	13.7	11.9
Hot hardness, DPH at:				
800 °F	555	...	190	500
1000 °F	440	...	185	485
1200 °F	250	...	170	400
1400 °F	115	...	145	280
Wear				
Unlubricated sliding wear(a), mm^3 at:				
150 lb	0.15	0.3	0.4	0.1
300 lb	0.3	0.4	...	0.3
Abrasive wear(b), mm^3				
OAW	12	18	...	...
GTAW	11	12	105	43
Unnotched Charpy impact strength, ft · lb	2	2	29	1
Corrosion resistance(c):				
65% nitric acid, at 150 °F	U	U	E	E
5% sulfuric acid at 150 °F	U	U	E	E
50% phosphoric acid at 150 °F	U	U	E	E

(a) Wear measured from tests conducted on Dow-Corning LFW-1 against 4620 steel ring at 80 rpm for 2000 revolutions varying the applied loads. (b) Wear measured from dry sand rubber wheel abrasion tests. Tested for 2000 revolutions at a load of 30 lb using a 9-in.-diam rubber wheel and AFS test sand. (c) E, less than 2 mils/yr; S, over 20 to less than 50 mils/yr; G, less than 20 mils/yr; U, more than 50 mils/yr

Nickel-Based Alloys

Most commercially available nickel-based hardfacing alloys can be divided into three groups: boride-containing alloys, carbide-containing alloys, and Laves phase-containing alloys. The compositions of some typical nickel-based hardfacing alloys are listed in Table 3. Selected properties of these alloys are listed in Table 4.

The boride-containing nickel-based alloys were first commercially produced as spray-and-fuse powders. The alloys are currently available from most manufacturers of hardfacing products under various tradenames and in a variety of forms such as bare cast rod, tube wires, and powders for plasma weld and manual torch. This group of alloys is primarily composed of Ni-Cr-B-Si-C. Usually, the boron content ranges from 1.5 to 3.5%, depending on chromium content, which varies from 0 to 15%. The higher chromium alloys generally contain a large amount of boron, which forms very hard chromium borides with hardnesses of approximately 1800 DPH (kg/mm^2). Other borides high in nickel and with lower melting points are also present to facilitate fusing.

The abrasion resistance of these alloys is a function of the amount of hard borides present. Alloys containing large amounts of boron such as Ni-14Cr-4Si-3.4B-0.75C are extremely abrasion resistant but have poor impact resistance. Because most of the boride-containing nickel-based alloys contain only small amounts of solid-solution strengtheners, considerable loss of room-temperature hardness occurs at elevated temperatures.

The use of carbide-containing nickel-based alloys has been extremely limited. However, these alloys are gaining popularity as low-cost alternatives to cobalt-based hardfacing alloys. The most popular and widely used alloys in this group are included in the Ni-Cr-Mo-C system.

Generally, carbide-containing nickel-based alloys have poor oxyacetylene weldability, which has made them unsuitable for hardfacing applications in the past. Carbide-containing nickel-based alloys of the Ni-Cr-Mo-Co-Fe-W-C system are more readily weldable by oxyacetylene methods and are gaining popularity as low-cost alternatives to cobalt-based alloys. These alloys, depending on precise composition, contain M_7C_3- or M_6C-type carbides similar to cobalt-based alloys.

Only one Laves phase-containing nickel-based alloy is commercially available —Ni-32Mo-15Cr-3Si. This alloy, like most nickel-based alloys, is difficult to weld using the oxyacetylene process, but can be readily welded using GTAW or plasma transferred arc process. It can also be applied using the plasma or detonation gun spray technique. Although it has excellent metal-to-metal wear resistance and moderate abrasive wear resistance, it possesses poor impact resistance.

Iron-Based Alloys

Iron-based hardfacing alloys are more widely used than cobalt- and/or nickel-based hardfacing alloys and constitute the largest volume use of hardfacing alloys. Iron-based hardfacing alloys offer low cost and a broad range of desirable properties. Most equipment that undergoes severe wear, such as crushing and grinding equipment and earthmoving equipment, is usually very large, rugged, and often subject to contamination. Parts subjected to wear usually require downtime for repair. For this reason, there is a general temptation to hardface them with the lowest cost and most readily available materials. As a result, literally hundreds of iron-based hardfacing alloys are in use today.

Due to the great number of alloys involved, iron-based hardfacing alloys are best classified by their suitability for different types of wear and their general microstructures rather than by chemical composition. Most iron-based hardfacing alloys can be divided into the following classes:

- Pearlitic steels
- Austenitic steels
- Martensitic steels
- High-alloy irons

Pearlitic steels are essentially low-carbon steels with minor adjustments in composition to achieve weldability. These alloys contain low carbon (0.25%) and low amounts of other alloying elements, resulting in a pearlitic structure. Pearlitic steels are useful as buildup overlays, primarily to rebuild machinery parts back to size. Examples include shafts, rollers, and other parts in heavy machinery subjected to rolling, sliding, or impact loading. Typically, this group of alloys has high impact resistance and low hardness (in the range 25 to 35 HRC), as well as excellent weldability.

Austenitic Steels. Austenite in this group of alloys is usually stabilized by manganese additions. Austenitic iron-based hardfacing alloys essentially are modeled after Hadfield steels. Most commercially available alloys in this category can be broadly subdivided into low-chromium and high-chromium alloys.

Low-chromium alloys usually contain up to 4% Cr and 12 to 15% Mn and some nickel or molybdenum. Low-chromium austenitic steels generally are used to rebuild machinery parts subjected to high impact (impact crusher or shovel lips). Low-chromium austenitic steels are not recommended for joining manganese steel parts due to the possibility of cracking. Welding to plain carbon steels should be carefully executed. Martensite sometimes forms in zones that are low in manganese content, thereby embrittling the interdiffusion zone.

High-chromium austenitic steels, which may normally contain 12 to 17% Cr in addition to about 15% Mn, were developed to preclude diffusion zone embrittlement. In addition, the as-deposited hardness of high-chromium steels is higher (16 to 20 HRC) than that of low-chromium steels (86 to 88 HRB). High-chromium austenitic steels typically are used for rebuilding manganese steel and carbon steel parts subjected to high metal-to-metal pounding, such as railroad frogs and steel mill wobblers, as well as for joining manganese steels.

Martensitic Steels. Alloys in this category are designed to form martensite on normal air cooling of the weld deposit. As a result, these steels are often termed "self-hardening" or "air hardening," and they resemble tool steels with hardnesses in the range of 40 to 45 HRC. The carbon content of the martensitic steels ranges up to 0.5%. Other elements such as molybdenum, tungsten, nickel (up to 3%), and chromium (up to 15%) are added to increase hardenability and strength and to promote martensite formation. Manganese and silicon usually are added to aid weldability.

The major hardfacing applications for martensitic steels include unlubricated metal-to-metal rolling or sliding parts such as undercarriage parts of tractors. The impact resistance of martensitic steels is inferior to that of pearlitic or austenitic alloys, but there is a compensating increase in hardness and resistance to abrasive wear.

High-Alloy Irons. This group of alloys is referred to as irons because of their similarities to cast irons. These alloys contain large amounts of chromium and/or molybdenum carbides in what is essentially a martensitic matrix. The carbon content ranges from 2 to 6%, providing large amounts of carbides. Accordingly, resistance to abrasion is improved considerably over the low-carbon iron-based alloys. Impact resistance and toughness of these alloys, however, are correspondingly lower. High-alloy iron hardfacing deposits usually are limited to one or two passes in thickness due to an extremely high susceptibility to form tension cracks. In some alloys containing 4 to 8% Mn or up to 5% Ni, the austenitic phase is stabilized in the matrix, which reduces the tendency to crack on cooling. Most high-alloy irons have extremely high compressive strength, but very low tensile strength. The hardness range for high-alloy irons is typically 52 to 62 HRC.

Most iron-based alloys lack the corrosion, oxidation, or creep resistance of cobalt- or nickel-based hardfacing alloys and thus are limited in their applications. As mentioned previously, there are literally hundreds of iron-based alloys commercially available today. The properties of several typical iron-based hardfacing alloys of major types are listed in Table 5 as a general guide for selection of hardfacing materials.

Table 5 Composition and hardness of selected iron-based hardfacing alloys

Nominal composition	Nominal hardness DPH	HRC	Unlubricated sliding wear(a), mm^3	Abrasive wear(b), mm^3	Density, lb/in.^{3}
Pearlitic steels					
Fe-2Cr-1Mn-0.2C	318	32	0.5	55	0.28
Fe-1.7Cr-1.8Mn-0.1C	372	38	0.6	67	0.27
Austenitic steels					
Fe-14Mn-2Ni-2.5Cr-0.6C	188 RHB	88 RHB	0.4	86	0.28
Fe-15Cr-15Mn-1.5Ni-0.2C	230	18	0.3	113	0.28
Martensitic steels					
Fe-5.4Cr-3Mn-0.4C	544	52	0.4	54	0.27
Fe-12Cr-2Mn-0.3C	577	54	0.3	60	0.27
High-alloy irons					
Fe-16Cr-4C	595	55	0.3	13	0.27
Fe-30Cr-4.6C	560	53	0.2	15	0.26
Fe-36Cr-5.7C	633	57	0.1	12	0.27

(a) Wear measured from tests conducted on Dow-Corning LFW-1 against 4620 steel ring at 80 rpm for 2000 revolutions varying the applied loads. (b) Wear measured from dry sand rubber wheel abrasion tests. Tested for 2000 revolutions at a load of 30 lb using a 9-in.-diam rubber wheel and AFS test sand.

Carbides

The amount of carbides used for hardfacing applications is small compared to iron-based hardfacing alloys, but carbides are extremely important for severe abrasion and cutting applications. Historically, tungsten-based carbides were used exclusively for hardfacing applications. Recently, however, carbides of other elements such as titanium, molybdenum, tantalum, vanadium, and chromium have proved useful in many hardfacing applications.

Generally, the melting points of carbides are extremely high, and in many cases, they tend to disintegrate prior to forming a molten weld pool. Consequently, carbides themselves cannot be deposited by conventional welding techniques. To overcome this problem, carbides usually are inserted in a steel or alloy tube that is used as the weld consumable. For applications requiring resistance to corrosion in addition to abrasion resistance, weld rods containing sintered carbides in a corrosion-resistant matrix are produced by powder metallurgy (P/M) techniques. Composite powders of carbides with nickel-, cobalt-, or iron-based alloys also are available for plasma transferred arc or spray-and-fuse applications.

The microstructures of tungsten carbide composite weld deposits are essentially mixtures of tungsten carbide particles embedded in an alloy steel matrix. Any partial dissolution of carbides results in a stronger matrix. Depending on the process, the amount of tungsten dissolved in the matrix varies. A matrix containing high amounts of tungsten or molybdenum may undergo martensitic transformation on cooling. The martensitic matrix provides a stronger base to hold carbides than the pearlitic steel matrix and, hence, improved resistance to abrasive wear.

Occasionally, M_6C-type carbides, precipitated on cooling, are also found in the microstructure. Hardfacing products containing vanadium carbides in steel tubes are also found to be very satisfactory in severe abrasion applications. However, little technical information is available on weld products containing vanadium carbides.

Another major approach to carbide hardfacing is by plasma spraying (PSP) or detonation gun. These processes are not truly welding techniques, as only a mechanical bond is obtained between the overlay and the substrate. To maintain good bond strength, overlay thicknesses of no more than 0.010 to 0.020 in. are recommended. A wide variety of carbides are commercially available for deposition by these techniques.

The widespread use of carbides for hardfacing is primarily based on the general belief that all carbides, due to their high hardness, resist abrasion. In reality, wear resistance of carbide composites is a function of the abrasion resistance of the matrix, as well as the resistance of the carbides to fracture and fragmentation, especially under high-stress applications. While the hardness values of various carbides are readily available, crushing strengths unfortunately are not. As a result, a general tendency to select carbides based solely on their hardness is widespread. Table 6 lists the hardnesses of various carbides and selected other materials of comparison.

Table 6 Approximate hardness of selected materials

Material	Hardness DPH	HK	Mohs
Diamond	...	8000	10
SiC	3200	2750	9.2
W_2C	3000	2550	+9
VC	2800		+9
TiC	2800	2750	+9
Cr_3C_2	2700		
Alumina	...	2100	9
WC	2400	1980	+9
Cr_7C_3	2100	...	...
$Cr_{23}C_6$	1650	...	...
Mo_2C	1570	...	8
Zircon	...	1340	...
Fe_3C	1300	Cementite	...
Quartz	1000	800	7
Lime	...	560	...
Glass	...	500-600	...

Copper-Based Alloys

The copper-based hardfacing alloys are similar to bronzes and are used in applications where copper-based bearing materials normally are employed as homogeneous parts. It is often more economical to apply copper-based hardfacing alloys as overlays in less expensive base metals such as low-carbon steels.

The properties of copper-based hardfacing alloys are similar to the properties of corresponding bronzes. Copper-based hardfacing alloys are used for applications where resistance to corrosion, cavitation erosion, and metal-to-metal wear is desired, as in bearing materials. Copper-based hardfacing alloys have poor resistance to corrosion by sulfur compounds, abrasive wear, and elevated-temperature creep, are not as hard as all the classes of alloys previously discussed, and are not easily welded.

Hardfacing Alloy Selection

Hardfacing alloy selection is guided primarily by wear and cost considerations. However, other manufacturing and environmental factors must also be considered, such as base metal, deposition process, and impact, corrosion, oxidation, and thermal requirements. The factors affecting hardfacing process selection are discussed in other sections of this article. Usually, the hardfacing process dictates the hardfacing or filler-metal product form.

Hardfacing alloys usually are available as bare rod, flux-coated rod, long-length solid wires, long-length tube wires (with and without flux), or powders. Table 7 lists various welding processes, type of heat source, and the proper forms of consumable for each process. In general, the impact resistance of hardfacing alloys decreases as the carbide content increases. As a result, in situations where a combination of impact and abrasion resistance is desired, a compromise between the two must be made. In applications where impact resistance is extremely important, austenitic manganese steels are used to build up worn parts.

Frequently, wear is accompanied by aqueous corrosion from acids or alkalis, such as those encountered in the chemical processing or petroleum industries or in flue gas scrubbers. Few of the iron-based hardfacing alloys possess the necessary corrosion resistance in such aqueous media. As a result, nickel- or cobalt-based hardfacing alloys generally are recommended when corrosion resistance combined with wear resistance is required. For example, a knife used to cut tomatoes in a food processing plant can last many times longer than a tool steel knife if the edge is made of a cobalt-based alloy. Oxidation and hot corrosion resistance of iron-based alloys also is generally poor. For the most part, boride-containing nickel-based alloys do not contain sufficient chromium in the matrix to resist oxidation. Hence, Laves phase or carbide-containing nickel- or cobalt-based alloys typically are recommended for applications where wear resistance, along with oxidation or hot corrosion resistance, is required.

The ability of an alloy to retain strength at elevated temperatures is important for wear applications such as hot forging dies or valves for service at 1600 °F, as well as service in coal gasification and liquefaction applications. Iron-based alloys with martensitic structures lose their hardness at elevated temperatures. In general, the high-temperature strength retention of a hardfacing alloy increases with its tung-

Table 7 Hardfacing processing

Process	Heat source	Mode of application	Hardfacing alloy form
Oxyfuel gas welding	Oxyfuel gas	Manual or automatic	Bare cast rods or powder
Shielded metal arc welding	Electric arc	Manual	Flux coated rods
Open arc welding	Electric arc	Semiautomatic	Flux cored tube wire
Gas tungsten arc welding	Inert gas shielded electric arc	Manual or automatic	Bare rods or wire
Submerged arc welding	Flux covered electric arc	Semiautomatic	Bare solid or tubular wire
Plasma transferred welding	Inert gas shielded plasma arc	Automatic	Powder, hot wire
Plasma arc welding	Inert gas shielded plasma arc	Manual or automatic	Same as GTAW
Spray and fuse	Oxyfuel gas	Manual	Powder
Plasma spray	Plasma arc	Manual or automatic	Powder
Detonation gun	Oxyacetylene detonation	Automatic	Powder

Table 8 Hardfacing alloy selection

Hardfacing materials	Service conditions
Hypoeutectic cobalt-based Alloy No. 1, Laves phase alloys	Metal-to-metal sliding, high-contact stresses
Low-alloy hardfacing steels	Metal-to-metal sliding, low-contact stresses
Most cobalt-based alloys or nickel-based alloys depending on corrosive environment	Metal-to-metal sliding in combination with corrosion or oxidation
High-alloy cast irons	Low-stress abrasion Particle impingement Erosion at low angles
Carbides	Low-stress severe abrasion Cutting edge retention
Cobalt-based alloys	Cavitation erosion
High-alloy manganese steels	Heavy impact
Hypoeutectic cobalt-based alloys	Heavy impact with corrosion, oxidation
Hypoeutectic cobalt-based Alloys No. 21 and 6, cobalt-based Laves phase alloys	Galling
Austenitic manganese steels	Gouging abrasion
Cobalt-based alloys and nickel-based carbide-type alloys	Thermal stability Creep resistance at elevated temperatures (>1000 °F)

sten or molybdenum content. In applications requiring elevated-temperature strength and wear resistance, cobalt-based alloys of Laves phase alloys are strongly recommended.

A general guide for hardfacing alloy selection based on service conditions is given in Table 8. The following steps should be taken in selecting a hardfacing alloy:

- Analysis of the service conditions to determine the type of wear and environmental resistance required
- Selection of several hardfacing alloy candidates
- Analysis of the compatibility of the hardfacing alloys with the base metal, taking into consideration thermal stresses and possible cracking
- Field testing of hardfaced parts
- Selection of an optimum hardfacing alloy, considering cost and wear life
- Selection of the hardfacing process for production of wear components, considering deposition rates, the amount of dilution, deposition efficiency, and overall cost, including the cost of consumables and processing

Example 1. Selection of Hardfacing Materials. A manufacturer of vegetable oil uses an auger to extract oil from oil seeds such as soft peanuts, rapeseed, sunflower, and cottonseed. The auger is made of case-hardened steel. When a decision was made to process palm kernels and copra, changes in wear life of the case-hardened augers were not anticipated. However, copra and palm kernels are among the most fibrous of commercial oil seeds. As a result, the wear life of the steel augers was considerably reduced and the capability of the machines decreased about one third.

Design engineers considered various ways of improving service life. Finally, a decision was made to try cobalt-based alloys. Alloys RCoCrA, RCoCrB, and RCoCrC from Table 1 were selected to minimize corrosion from fatty acids and to provide the required abrasion resistance. Initially, oxyacetylene deposition was used. Alloy RCoCrC performed best in field tests, but was extremely crack-sensitive and required excessively high preheat. Alloy RCoCrB performed better than alloy RCoCrA in field tests, only required moderate preheat, and was ultimately selected as the best hardfacing alloy for this particular application.

Deposition rates using oxyacetylene are low, resulting in high labor costs. Another test was made using alloy RCoCrB deposited by GTAW. The test showed slightly inferior performance of the gas tungsten arc deposit compared to the oxyacetylene deposit due to higher dilution and small carbides. However, the decrease in wear life was economically compensated for by the reduced welding costs. Furthermore, there was no need to case harden the steel auger, thus offsetting the cost of hardfacing. Design engineers thus recommended routine use of alloy RCoCrB on steel augers to increase wear life.

Hardfacing Process Selection

Hardfacing process selection may be as important as hardfacing alloy selection, depending on the engineering application. Service performance requirements not only dictate hardfacing alloy selection, but have a strong influence on hardfacing process selection as well. Other technical factors involved in hardfacing process selection include (but are not limited to) hardfacing property and quality requirements, physical characteristics of the workpiece, metallurgical properties of the base metal, form and composition of the hardfacing alloy, and welder skill. Ultimately, economic considerations predominate, and cost is the determining factor in the final process selection.

For years, hardfacing has been limited, by definition, to welding processes. The definition adopted in this article has been expanded to include thermal spraying (THSP) as a hardfacing process. The first consideration in hardfacing process selection is frequently to determine if welding processes or THSP processes are preferred or required. As a rule, welding processes are preferred for hardfacing applications requiring dense, relatively thick coatings with high bond strengths between the hardfacing and workpiece. Thermal spraying processes, on the other hand, are preferred for hardfacing applications requiring thin, hard coatings applied with minimal thermal distortion of the workpiece.

Property and Quality Requirements. Welded hardfacing deposits are, in effect, mini-castings characterized by variable composition (segregation) and solidification kinetics that influence deposit microstructure. It is not surprising, therefore, that the properties and quality of welded hardfacing deposits should depend on welding process and technique, as well as on alloy selection.

Composition variations also derive from base-metal dilution during welding, although carbon pickup in oxyfuel gas welding (OFW) processes and solute volatilization during arc welding processes are additional factors. Dilution is the interalloying of the hardfacing alloy and the base metal and is usually expressed as the percentage of base metal in the hardfacing deposit. A dilution of 10% means that the deposit contains 10% base metal and 90% hardfacing alloy. The wear resistance and other desirable properties of the hardfacing alloy are generally thought to degrade as dilution increases. The maximum amount of allowable dilution depends on specific service requirements. Welding process and technique should, however, be selected so as to control dilution to less than 20%.

Dilution is nil in THSP and related processes and tends to be acceptably low in conventional OFW processes. Dilution is generally more of a problem in arc weld hardfacing processes, ranging from approximately 5% in plasma arc welding (PAW) hardfacing processes to as much as 60% in SAW. High-dilution welding processes can be tolerated in applications requiring relatively thick hardfacings that can be applied in multiple layers. The effects

Fig. 6 Effect of dilution on the microstructure of a Fe-28Cr-4Mo-0.4Mn-4.6C hardfacing alloy. (a) First layer over mild steel showing a hypoeutectic structure of metal dendrites and interdendritic eutectic, 55 HRC. (b) Second layer showing a fine hypoeutectic structure with a large percentage of eutectic matrix, 57 HRC. (c) Third layer showing a coarse hypereutectic structure of primary M_7C_3 carbides in a eutectic matrix, 61 HRC. (d) Fifth layer showing a similar structure to both the third and fourth passes 60 HRC.

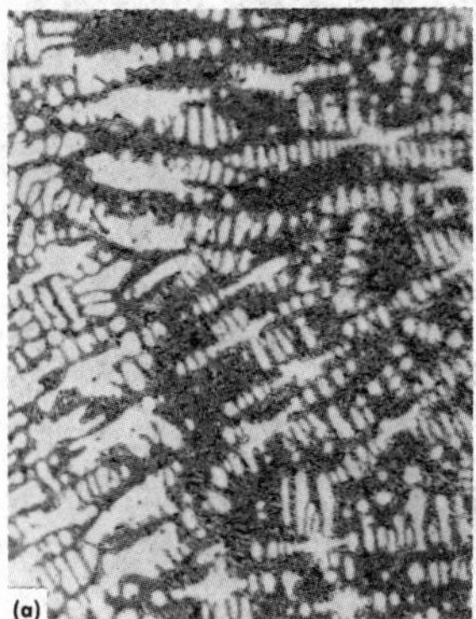

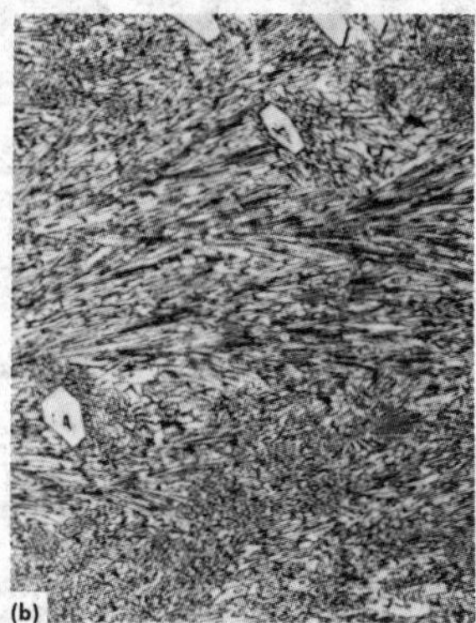

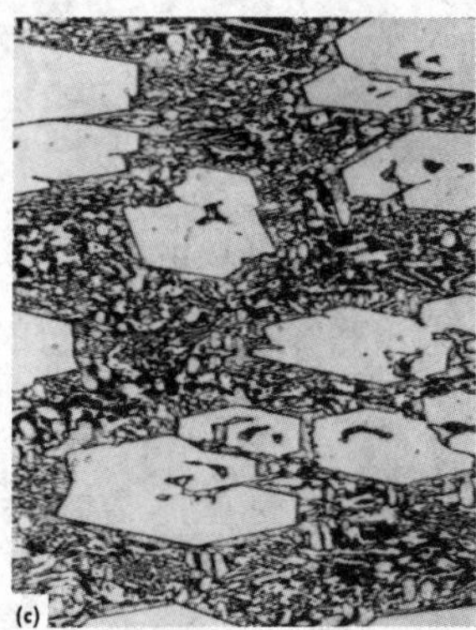

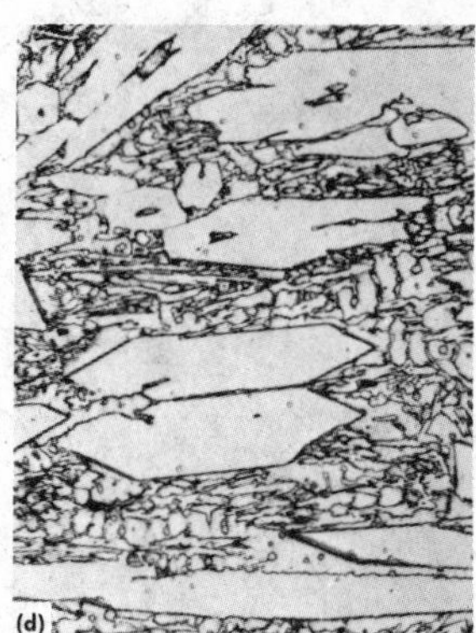

of dilution on the microstructure of a high-chromium, white iron hardfacing alloy deposited in multiple layers on low-carbon steel substrate using SMAW are illustrated in Fig. 6. The first layer (Fig. 6a) microstructure in this particular case was hypoeutectic with interdendritic eutectic and exhibited a hardness of 55 HRC. The second layer (Fig. 6b) consisted of a fine hypereutectic microstructure with a large percentage of eutectic and exhibited a hardness of 57 HRC. The third (Fig. 6c) and fifth (Fig. 6d) layers were microstructurally equivalent, consisting of primary M_7C_3 carbides in eutectic matrices. The hardness in the third and subsequent layers ranged from 60 to 61 HRC.

The microstructures and mechanical properties of hardfacing deposits, like castings, vary depending on solidification kinetics as well as on dilution. Solidification generally is rapid in THSP processes, with near amorphous structures being achieved with some THSP processes. Solidification kinetics tend to be somewhat slower in conventional weld hardfacing processes. As a rule, hardfacing deposits produced by OFW processes tend to solidify slower than hardfacing deposits produced by arc welding processes. These differences in solidification rate produce widely different microstructures and widely different properties regardless of dilution. The effects of welding process on the microstructure of a Co-28Cr-4W-1.1C hardfacing alloy deposited on a low-carbon steel substrate are illustrated in Fig. 7.

Physical Characteristics of the Workpiece. The size, shape, and weight of the workpiece have a strong influence on hardfacing process selection. Large, heavy workpieces that are difficult to transport are most conveniently hardfaced using manual or semiautomatic hardfacing processes in which the hardfacing equipment can be moved to the workpiece. Gas shielded metal arc and open arc welding processes, for which portable equipment is readily available, are well suited to field applications involving inaccessible surfaces on large, difficult-to-transport workpieces. Thermal spraying processes and equipment also are available for in situ field hardfacing applications.

Hardfacing process selection is more involved in applications where the workpiece is sufficiently small so that it can be readily transported to the welding equipment. Thermal spraying processes and OFW, SAW, GTAW, PAW, and gas metal arc welding (GMAW) are all well suited to in-plant hardfacing of small workpieces. Preference is usually given to THSP and/or OFW or GTAW for small workpieces requiring thin, accurately placed hardfacing deposits. Large quantities of small workpieces can usually be hardfaced most economically using dedicated, specially designed, fully automatic hardfacing equipment and processes.

Metallurgical Characteristics of the Base Metal. Base-metal surface preparation is important in THSP hardfacing processes. However, base-metal composition, melting temperature range, and thermal expansion and contraction characteristics have a significant effect on welding process selection.

Steels are generally suitable base metals for hardfacing. Low-alloy and medium-carbon steels with carbon contents up to 0.4% can be easily hardfaced by all welding processes with generally excellent results. Higher carbon steels are similarly readily amenable to hardfacing, but preheating is necessary to minimize deleterious martensitic reactions in the heat-affected zone (HAZ). Austenitic stainless steels, with the exception of the free-machining and titanium-stabilized grades, and most nickel-based alloys can also be easily hardfaced by all welding processes with excellent results. Martensitic stainless steels, tool steels, die steels, and cast irons are also amenable to hardfacing, but greater attention to preheat, interpass temperature, and postheat is required. Age-hardenable base metals present special hardfacing problems and generally require solution and/or overaging heat treatments prior to hardfacing, as well as careful attention to preheat, interpass temperature, and postheat.

Hardfacing welding processes require that the base-metal melting temperature range be greater than or at least equal to that of the hardfacing alloy. Thermal spraying processes, on the other hand, do not require a base-metal melting temperature advantage.

Base-metal thermal expansion and contraction characteristics affect solidification strain. Workpieces frequently have to be uniformly heated to relatively high temperatures during welding to enable the deposition of crack-free, hard, brittle hardfacing alloys. Welding process selection is important in that it directly affects heat input and preheat requirements.

Thermal expansion and contraction differences between the hardfacing alloy and base metal are important in applications involving thermal cyclic service conditions. Large differences can result in shear failures in THSP deposits or thermal fatigue failures in welded deposits. Buffer layers are often deposited between the base metal and hardfacing alloy to counteract large differences in thermal expansion and contraction characteristics. Frequently, dilution, which is generally regarded to be

Fig. 7 Effect of welding process on the microstructure of a Co-28Cr-4W-1.1C hardfacing alloy. (a) Oxyfuel gas welding. (b) Plasma transfer arc welding. (c) Gas tungsten arc welding. (d) Open arc welding. (e) Submerged arc welding. (f) Shielded metal arc welding

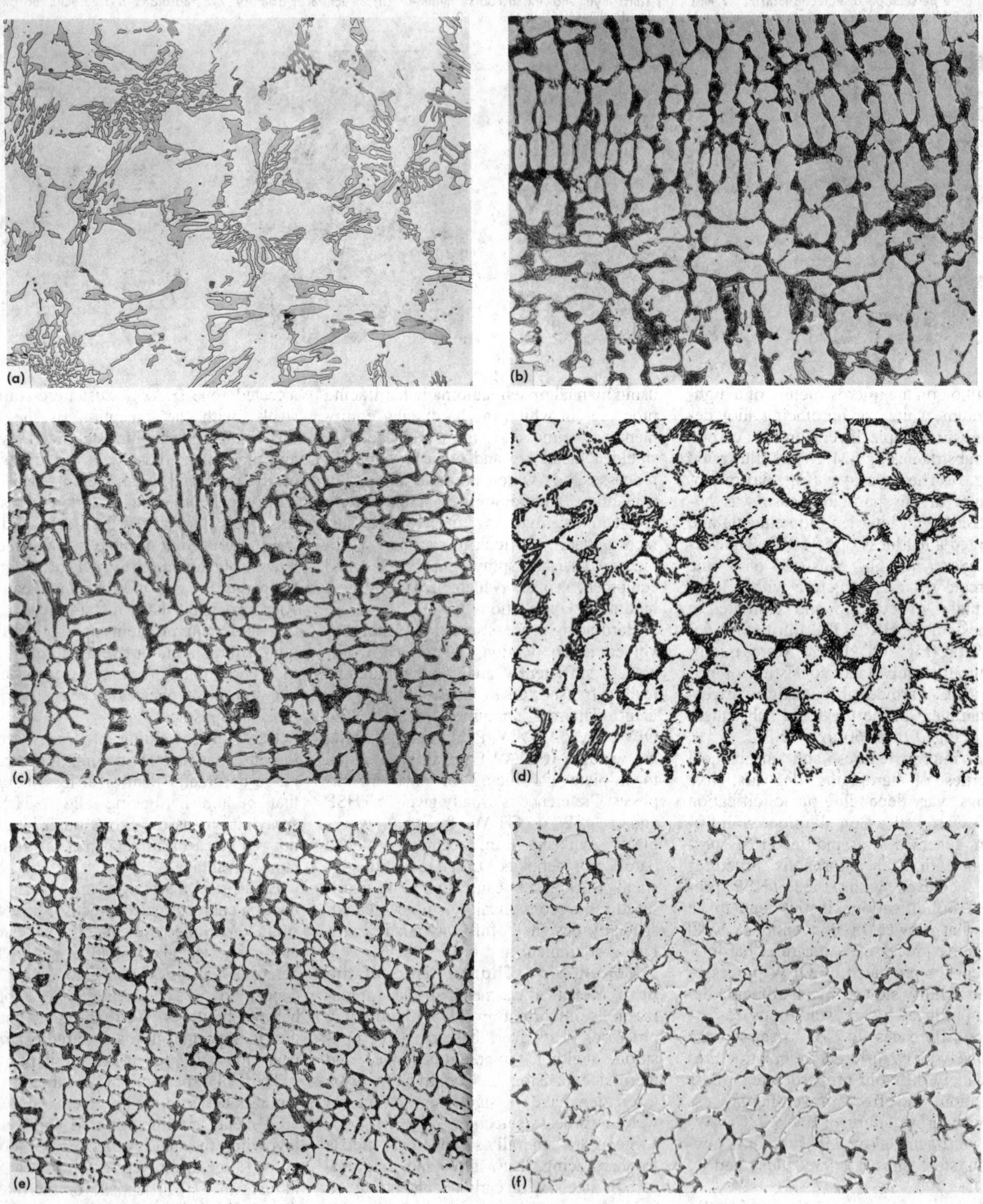

detrimental, can be used to good advantage to form in situ buffer layers and minimize the risk of thermal fatigue failure. Hardfacing process selection is an important consideration in this regard.

Hardfacing Product Forms. Hardfacing process selection is often limited by product form. Hardfacing alloys normally are available as bare cast rod, covered electrodes, wire, and powders. However, not all alloys are available in all product forms.

Virtually all hardfacing alloys can be produced in powder form, and most hardfacing alloys can be produced in rod form.

Rods generally are available in bare cast form or in tubular form. Bare cast rod diameters are limited to a large extent by alloy characteristics and casting processes to a range of $^1/_8$ to $^5/_{16}$ in. Bare cast rod lengths usually range from 10 to 28 in., but some hardfacing alloys can be cast in extra long, continuous lengths with diameters as small as $^1/_8$ in. Tubular rods consist of highly alloyed powder cores encased in low-alloy sheaths. Tubular rods can be produced in discrete lengths or in spooled continuous coils. Discrete-length tube rods generally are available in the same diameter range as bare cast rod. Long-length tube rods, or tube wire, as they are properly named, normally are available in diameters ranging from $^5/_{32}$ to $^1/_{16}$ in. Some of the tougher, more ductile hardfacing alloys can be produced as solid drawn wire. Some of the more brittle hardfacing alloys can be produced, at a premium cost, as solid wire using extrusion techniques.

Welder Skill. It is essential to consider hardfacing quality requirements in relation to welder/operator skill, as well as to the hardfacing process. As a rule, hardfacing by manual welding processes such as OFW or GTAW requires high welder proficiency, whereas hardfacing by automatic welding processes, such as SAW, requires a minimum of operator skill, assuming machine control settings have previously been established. Thermal spraying and related powder welding processes generally require intermediate welder/operator skills.

Manual GTAW can be used to obtain high-quality deposits on relatively small areas, such as the interlock regions of turbine blades. In this type of application, thin layers can be deposited with dilution as low as 10%, but relatively high welder skill and close control of the welding operation are necessary. In contrast, earthmoving and mining equipment can be hardfaced adequately in the field by relatively unskilled welders. Selection of process is usually based on maximum deposition rate: high dilution rates seldom significantly affect the suitability of the coating for service.

Cost. Ultimately, cost is the determining factor in hardfacing process selection. Hardfacing costs basically consist of labor, materials, and in some instances, new equipment cost. Transportation cost also must be considered.

Labor costs depend primarily on the level of welder skill required and process deposition rates, but surface preparation and finishing costs also must be considered. Deposition rates are normally lowest with manual hardfacing processes. Considerably higher deposition rates are possible with automatic arc welding hardfacing processes. Deposition rates as high as 60 lb/h are possible with multiwire or powder and wire submerged arc hardfacing processes, but dilution is also high. An ideal welding process for hardfacing would be one with the ability to limit dilution while achieving a high deposition rate. Typical dilution percentages, deposition rates, and minimum deposit thicknesses for different welding processes and various forms, compositions, and modes of application of hardfacing alloys are given in Table 9.

Hardfacing material costs depend on the prevailing price of raw materials and product form. Raw material cost is predominant in tungsten-, cobalt-, and nickel-based hardfacing alloys, whereas product form is the predominant cost in iron-based hardfacing alloys. As a rule, tube rod and wire are the least expensive hardfacing product forms. In contrast, solid wire is the most expensive hardfacing product form. Cast rod and powder are generally intermediate in cost. Powder, depending on specific powder size requirements, tends to be less expensive than cast rod. However, deposit efficiency, measured as the percentage of hardfacing consumable retained on the workpiece surface, is generally lower for hardfacing processes that utilize powder consumables than for hardfacing processes that utilize rod consumables. Thermal spraying processes are particularly poor in this regard, with typical deposit efficiencies as low as 70%.

Equipment costs vary from simple arc welding equipment to sophisticated, fully automated low-pressure plasma spraying (PSP) systems. The more advanced automatic hardfacing equipment often costs 100 to 1000 times as much as simple arc welding equipment, depending on the level of instrumentation and computer control.

It is relatively easy to estimate or predetermine the cost of hardfacing. It is more difficult to estimate the value of hardfacing. For maintenance applications, it has been suggested that hardfacing is preferred over total replacement when the cost advantage (CA) is positive. The cost advantage can be calculated by:

$$CA = \frac{CN}{PN} - \frac{CR}{PR} \quad \text{(Eq 1)}$$

where CN is the cost of a new component; CR is the cost of hardfacing plus downtime costs; PN is the work output during the sevice life of a new component; and PR is the work output during the service life of a hardfaced component. Most of the data needed for the above calculation usually can be obtained; PR is perhaps the most difficult value to assess, but reasonable

Table 9 Characteristics of welding processes used in hardfacing

Welding process	Mode of application	Form of hardfacing alloy	Weld-metal dilution, %	Deposition, lb/h	Minimum thickness(a), in.	Deposit efficiency, %
OAW	Manual	Bare cast rod, tubular rod	1-10	1-4	$^1/_{32}$	100
	Manual	Powder	1-10	1-4	$^1/_{32}$	85-95
	Automatic	Extra-long bare cast rod, tubular wire	1-10	1-15	$^1/_{32}$	100
SMAW	Manual	Flux covered cast rod, flux covered tubular rod	10-20	1-12	$^1/_8$	65
Open arc	Semiautomatic	Alloy cored tubular wire	15-40	5-25	$^1/_8$	80-85
	Automatic	Alloy cored tubular wire	15-40	5-25	$^1/_8$	80-85
GTAW	Manual	Bare cast rod, tubular rod	10-20	1-6	$^3/_{32}$	98-100
	Automatic	Various forms(b)	10-20	1-10	$^3/_{32}$	98-100
SAW	Automatic, single wire	Bare tubular wire	30-60	10-25	$^1/_8$	95
	Automatic, multiwire	Bare tubular wire	15-25	25-60	$^3/_{16}$	95
	Automatic, series arc	Bare tubular wire	10-25	25-35	$^3/_{16}$	95
PAW	Automatic	Powder(c)	5-15	1-15	$^1/_{32}$	85-95
	Manual	Bare cast rod, tubular rod	5-15	1-8	$^3/_{32}$	98-100
	Automatic	Various forms(b)	5-15	1-8	$^3/_{32}$	98-100
GMAW	Semiautomatic	Alloy cored tubular wire	10-40	2-12	$^1/_{16}$	90-95
	Automatic	Alloy cored tubular wire	10-40	2-12	$^1/_{16}$	90-95

(a) Recommended minimum thickness of deposit. (b) Bare tubular wire; extra-long (8 ft) bare cast rod; tungsten carbide powder with cast rod or bare tubular wire. (c) With or without tungsten carbide granules

estimates can be made based on relative laboratory wear test results. If the workpiece is hardfaced with an alloy of essentially the same chemical composition as the worn material, it is reasonable to assume that $PR = PN$. On the other hand, if the hardfacing material is more wear resistant than the worn material, then $PR > PN$. Therefore, from Eq 1, CA is likely to be positive, which is the primary objective of hardfacing.

Hardfacing by Oxyfuel Gas Welding

Oxyfuel gas welding is a process in which the heat from a gas flame is used to melt hardfacing materials onto workpiece surfaces. During hardfacing, the workpiece surface is only superficially melted. The gas flame is created by the combustion of a hydrocarbon gas and oxygen. Acetylene is the most widely used fuel for OFW. In fact, OFW is commonly referred to as oxyacetylene welding (OAW), even though other fuel gases such as methylacetylene propadiene (MAPP) occasionally are used in certain OFW processes.

Hardfacing by OFW is a versatile process, readily adaptable to all hardfacing product forms, including powder. Oxyfuel gas welding hardfacing processes that use fillers in powder form are commonly referred to as powder welding or manual torch hardfacing. Oxyfuel gas welding equipment is simple, and capital equipment costs are low. Furthermore, OFW equipment is mobile and can be used for fusion welding and cutting operations, as well as for hardfacing. More importantly, OFW processes generally produce hardfacing deposits with optimum microstructures for wear-resistant applications.

Hardfacing by OFW, however, is not without limitations. Hardfacing by manual OAW requires a high degree of welder skill to obtain smooth, high-quality deposits, as both the welding rod and the torch flame have to be separately manipulated. The speed of welding and low deposition rates often preclude the use of OAW for hardfacing applications on large components that require large quantities of facing alloy, unless automation of the operation is warranted by configuration of the surface and a large number of identical parts.

Oxyacetylene welding is best suited to hardfacing applications in which the surface area to be hardfaced is minimal. Thus, for hardfacing steam valves, automotive and diesel engine valves, cutter blades for wood and plastic, chain saw bars, and plowshares and other agricultural implements, it is often the most satisfactory process as it can be easily employed in field applications.

Principles of Operation. Oxyfuel gas welding equipment includes a device in which oxygen and fuel are mixed in controlled ratios and at controlled flow rates. The oxygen and fuel gases are delivered from storage tanks through pressure regulators and hoses to a welding torch comprised of inlet valves, a mixing chamber, and a welding tip. The orifice diameter of the welding tip is varied, depending on the thickness or mass of the workpiece. Tables relating welding-tip orifice diameter and gas pressure to workpiece thickness are readily available for fusion gas welding processes. As a rule, welding tips one size greater than required for fusion gas welding are recommended for use in hardfacing applications. The majority of hardfacing applications using acetylene as the fuel gas are accomplished at oxygen and acetylene pressures corresponding to 20 to 40 psi and 4 to 10 psi, respectively.

The combustion of acetylene with oxygen at the tip of an OFW torch produces a multicolored flame, consisting of an inner cone and an outer envelope. The inner cone, which is generally pale blue in color, is the zone in which acetylene and oxygen from the storage tanks combine to form carbon monoxide and hydrogen:

$$C_2H_2 + O_2 \rightarrow 2CO + H_2$$

The highest flame temperature occurs at the tip of the inner cone. The torch is usually positioned so that the tip of the inner cone is close to, but not touching, the workpiece. The flame temperature and flame atmosphere are controlled by varying the relative amounts of oxygen and acetylene at the torch tip. The inner-cone combustion reaction requires about equal volumes of oxygen and acetylene. However, an excess of acetylene is normally used in hardfacing applications, producing a reducing flame with a maximum flame temperature ranging from 5300 to 5500 °F. When equal volumes of oxygen and acetylene are combusted, the flame is neutral, with a maximum temperature greater than 5500 °F.

The outer envelope, which is generally light blue in color, describes the complete combustion of the inner-cone combustion products with oxygen from the surrounding air to form carbon dioxide and water vapor:

$$4CO + 2H_2 + 3O_2 \rightarrow 4CO_2 + 2H_2O$$

An intermediate zone occurs between the inner cone and the outer envelope when there is an excess of acetylene. This intermediate zone, which is referred to as the oxyacetylene feather, contains white-hot carbon particles and is whitish in color. The carbon particles in the intermediate zone partially dissolve in the molten hardfacing deposit, thereby increasing deposit hardness and wear resistance. The carbon particles in the intermediate zone also serve to reduce some of the oxides that form during hardfacing. An excess acetylene flame is thus both carburizing and reducing.

The amount of excess acetylene required for hardfacing varies depending on alloy composition. The amount of excess acetylene is conveniently measured in terms of the flame geometry as the ratio between the length of the oxyacetylene feather and the inner cone. The length of the inner cone, indicated by the value X, is the distance from the torch tip to the extreme end of the inner cone. The length of the oxyacetylene feather is measured from the torch tip to the extreme end of the intermediate zone, generally expressed in multiples of the length of the inner cone. A neutral flame is thus classified as 1X. A flame in which the oxyacetylene feather extends from the torch tip outward a distance twice the length of the inner cone is classed 2X; three times the length of the inner cone is classed as 3X, and so on. Nickel-based hardfacing alloys normally are deposited with a 1X flame, but occasionally a slight excess of acetylene, not exceeding 2X, is used to promote flow of the hardfacing alloy on the workpiece. Iron-based hardfacing alloys generally are deposited with 2X flames. Cobalt-based alloys normally are deposited with 3X flames. The amount of excess acetylene for tungsten carbide hardfacing materials is normally dictated by the matrix base alloy.

Most steel workpieces can be hardfaced by OFW processes. High-manganese and high-sulfur steels are exceptions. High-speed steel workpieces are difficult to hardface and should be fully annealed before hardfacing. Cast iron workpieces also can be hardfaced by OFW processes, but cast iron substrates generally require the use of fluxes and special welding precautions.

Dilution is generally low in oxyfuel gas welded hardfacing deposits, as only the surface of the workpiece in the immediate area being hardfaced is brought to the melting temperature. The hardfacing rod is melted and spread over the surface of the workpiece. The melted rod does not mix or alloy with the base metal, but only with the melted surface. Superficial workpiece melting of steel workpieces produces a wet appearance, called "sweating."

Operating Procedures. Workpiece surfaces must be thoroughly cleaned before hardfacing. If impurities are present on surfaces of the workpiece, or if an excessive amount of oxide is produced in heating (usually, a neutral or reducing flame is used), it may be necessary to dislodge the solid contaminants by prodding or rubbing with a hardfacing rod tip. Oxide trapped beneath the hardfacing overlay can react with carbon in the hardfacing alloy to produce a gas and thus cause porosity. Hardfacing alloys usually contain a deoxidizer that controls a moderate amount of oxidation.

It is often necessary to preheat workpieces prior to hardfacing, regardless of the welding process, to prevent cracks in the hardfacing deposit. Preheating not only reduces tensile strain associated with thermal contraction differences between the hardfacing alloy and the workpiece, but in the case of OFW, also permits the use of a softer flame and results in less dilution. Small low-carbon steel workpieces can be preheated locally with a torch, but large workpieces are best preheated with gas burners in a brick muffle furnace, allowing sufficient soak time to ensure even heating.

With practice, one-layer hardfacing deposits up to $^1/_{16}$ in. thick can be deposited, particularly if the workpiece is inclined so that the deposition is slightly uphill. Alternatively, thick deposits can be applied in several layers. Hardfacing deposits by OFW can be remelted as necessary to improve deposit quality or surface appearance.

Powder Welding. Hardfacing also can be accomplished by OFW techniques using powder filler instead of rod filler. This method of hardfacing is frequently referred to as powder welding or manual powder torch welding. Hardfacing by powder welding is similar to hardfacing by conventional OFW methods, except for the obvious difference in filler product form and a difference in flame adjustment and torch-to-workpiece distance.

The powder filler in powder welding usually is fed from a small hopper mounted on a gas welding torch into the fuel gas supply and conveyed through the flame to the workpiece surface. Manufacturer's recommendations for nozzle size and gas pressures should be followed. Experience has shown that the flame should be adjusted to neutral or very slightly carburizing to avoid porosity in the deposit. Hardfacing alloys deposited by powder welding techniques invariably contain silicon and boron, which slag off any oxide skin that may form on the molten powder particles, thereby enabling them to coalesce and form a clean weld pool on the workpiece surface despite the absence of a reducing flame.

The workpiece surface should be clean, as in conventional hardfacing by OFW techniques, but preheat should be limited below the oxidation range of the workpiece substrate material. The torch tip should be held vertical to the workpiece and approximately $^1/_2$ in. from the workpiece surface. The gas flame should be concentrated on the workpiece surface area to be hardfaced until the surface area reaches a dull red heat, after which the powder feed is opened briefly to spray a thin layer on the workpiece. After the initial powder layer melts and wets the workpiece surface and forms a molten metal pool, the powder feed is opened to allow the weld pool to build up. Deposits ranging from $^1/_{16}$ to $^3/_{16}$ in. in thickness can be applied by the powder welding method at a rate slightly less than $^1/_{16}$ in. per pass.

Common Flaws. Hardfacing deposits applied by OFW generally are characterized by low dilution and high deposit quality. Discontinuities are possible, however, depending on the skill of the welder and the quality of the filler material.

Porosity is probably the most common flaw observed in hardfacing deposits applied by OFW. Porosity in hardfacing deposits, like porosity in castings, can result from inadequate feeding or gas evolution during solidification. Porosity resulting from inadequate feeding is sometimes referred to as microshrinkage. Microshrinkage is most common at the termination of the hardfacing deposit and usually results from the inability of the hardfacing deposit to accommodate solidification shrinkage after the flame is withdrawn. This type of flaw can be minimized or prevented by withdrawing the flame very slowly and allowing solidification to progressively occur from the bond line to the surface.

Porosity can also occur by gas evolution during solidification if the total gas content in the molten deposit is greater than the solid solubility limit. Gas-induced porosity generally derives from slag or nonmetallic inclusions trapped in or on the surface of the base metal. However, the initial gas content of the filler materials, coupled with possible gas pickup during OFW, are also contributing factors.

Hardfacing by Arc Welding

In hardfacing by arc welding, the heat from an electric arc is used to melt the hardfacing material onto the workpiece surface. The electric arc is developed by impressing a voltage between an electrode and the workpiece. The voltage required to sustain the arc varies with distance between the electrode and workpiece and the arc welding process. The filler material can derive directly from the electrode used to form the arc or can be externally introduced into the arc. Arc welding processes in which the filler derives directly from the electrode are sometimes referred to as consumable electrode processes. Arc welding processes in which the filler is externally introduced into the arc are sometimes referred to as nonconsumable electrode processes. As a rule, nonconsumable electrode processes are accomplished at lower power requirements than are consumable electrode processes, resulting in less dilution and lower deposition rates. Consumable and nonconsumable electrode hardfacing processes both require that the filler materials be protected from oxidation as they are melted and joined to the workpiece surfaces.

Filler materials in consumable electrode hardfacing processes are sometimes protected by fluxes as in SMAW, SAW, and open arc welding, or by inert gases as in GMAW. The selection of a particular consumable or nonconsumable electrode hardfacing process depends on consideration of the factors discussed in the section on hardfacing process selection of this article.

Shielded Metal Arc Welding. The electrodes used in the SMAW consumable electrode hardfacing process generally consist of discrete lengths of filler material rod covered with specially formulated coatings. The filler material rod generally ranges in diameter from $^1/_8$ to $^5/_{16}$ in. and may be cast or of tubular construction. The electrode coating protects the filler material during welding by chemical reaction as the electrode melts. The coating can be fortified to increase alloy content during melting, as well as to simply provide molten metal protection. Coating ingredients also determine or affect the shape of the weld, amount of penetration, cooling rate of the weld, stabilization of the arc, and refinement of the hardfacing deposit.

The selection of the proper power supply for SMAW consumable electrode hardfacing is not a serious problem. Motor-generators, direct current rectifiers, and alternating current transformers can be used. Direct current electrode positive (DCEP) normally is recommended for hardfacing. Direct current electrode negative (DCEN) can be used to increase deposition rates slightly, but there is a tendency for the weld beads to be high and narrow. Although direct current is recommended, alternating current has been used in many applications with reasonably good success. Arc stability and burnoff characteristics are not quite as good by this method.

Speed of travel, position of the electrode in relation to the weld pool, arc voltage, and amperage are modified to minimize penetration and the consequent dilution of the deposit. Typical welding currents for small $^1/_8$-in.-diam hardfacing electrodes range from 60 to 130 A, depending on the alloy. Welding currents for larger $^1/_4$-in.-diam hardfacing electrodes range from 150 to 330 A.

Depending on the surface area to be hardfaced, the bead-deposit pattern may be of the stringer or weave type, with or without staggering of beads, as dictated by service requirements. Selection of the bead pattern and sequence also depends on the accessibility of the surface to be hardfaced and the distortion limits of the workpiece.

Deposition rates vary considerably with welding conditions. Normally from 3 to 4 lb of metal per hour of arc time can be applied. The recovery of alloy in the deposit generally ranges from 65 to 75% of the electrode weight, excluding stub loss. The average stub loss is approximately 3 in., which corresponds to about 20% of a 14-in.-long electrode.

There are many advantages to the SMAW consumable electrode process. The intense heat of the arc permits the hardfacing of large parts without preheat. Small areas of parts machined or fabricated to close tolerances can be hardfaced without distorting the entire part. Difficult-to-reach areas are best accessed with coated electrodes. The overall process is relatively fast, as preheat and high interpass temperatures do not have to be maintained. Another attractive feature of this process is the portability of the equipment. Gasoline engine-driven generators can be readily taken into the field.

There are also some disadvantages to the SMAW consumable electrode process that must be considered. Penetration tends to be high, and dilution rates of 20% are not uncommon. This results in lower hardness and less resistance to abrasion. Two or three layers of weld metal normally are required to obtain maximum wear properties. Coated electrodes also leave a slag covering on the surface of the weld deposit which must be removed prior to each succeeding pass. Because coated electrodes usually are deposited with no more than a small amount of preheat, cracking or cross checking may occur, especially with the higher alloy electrodes that produce hard overlays.

The hardfacing material is at its maximum temperature, and hence its maximum thermal expansion, as it is being deposited. The workpiece, by contrast, may be near ambient temperature during the welding process and considerably less expanded than the hardfacing material. Severe strains develop in the hardfacing deposit as it cools after welding. These strains are accommodated by plastic deformation in the case of a ductile hardfacing material or by cracking and cross checking in the case of more brittle hardfacing materials. Cracking or cross checking does not affect the wear resistance of the overlay in many applications.

Submerged arc welding is a consumable electrode hardfacing process in which the area between the electrode and workpiece is shielded by a blanket of granular, fusible flux material. Electrodes in the SAW hardfacing process consist of continuous lengths, or coils, of filler materials—generally in tubular wire form.

The fluxes used for hardfacing by SAW are mineral compositions specially formulated to protect the molten filler metal without appreciable gas evolution. Submerged arc fluxes are generally finely divided, free-flowing granulars in the solid state, which are placed by gravity along the surface to be hardfaced in advance of the electrode. The flux becomes molten and highly conductive under the arc, creating localized conditions conducive to high deposition rates.

Hardfacing filler materials can be deposited over a wide range of welding currents, voltages, and travel speeds, each of which can be controlled independently. Generally, however, the current, voltage, and workpiece travel speed are adjusted for hardfacing applications to produce a bead approximately $^5/_8$ in. wide with a slightly convex surface and sides tapering into the base metal. Welding currents usually range from 300 to 500 A for most SAW hardfacing applications. However, it is not uncommon to increase current to as much as 900 A to increase deposition rate.

A number of features make SAW particularly attractive for hardfacing. Because it is an automatic process, SAW requires little skill on the part of the operator for efficient operation. High deposition rates are possible, usually ranging as high as 10 to 12 lb/h, depending on the job. The process is capable of making porosity-free deposits that are smooth and even on the surface and that can be used either as welded or that require only a minimum of grinding to clean up. The filler metal is protected from the atmosphere by flux, which also completely surrounds the arc, requiring no shielding for the operator.

The major disadvantages of hardfacing by SAW are that: (1) it is generally limited to hardfacing simple cylindrical or flat workpieces; and (2) it is not particularly adaptable to the surfacing of small parts. Hardfacing welding in other than the flat position requires special flux retainers. Preheating the workpiece in excess of 600 °F usually makes slag removal extremely difficult. Occasionally, small portable units are used, but normally SAW units are fairly large and are mounted on stationary tracks, thus limiting their use for field operations. It is imperative that the flux be kept dry prior to use; therefore, storage can be a problem. Dilution is relatively high. It is not uncommon to find interalloying with the base metal, reaching levels of 30%. Consequently, two or three layers of weld metal may be required for maximum wear resistance.

Open arc welding is a semiautomatic or automatic process, in which the arc between the electrode and workpiece is shielded by a self-generated or self-contained gas or flux. Open arc electrodes are continuous lengths, or coils, of filler material in tubular wire form. Open arc tubular filler materials are generally of low-carbon steel sheath construction with alloying elements, deoxidants, arc stabilizers, and a shielding gas source and/or flux contained in the core as powder or granules. High-alloy filler materials are generally shielded by a self-generated CO_2 gas deriving from the reaction between carboneous materials contained in the core and the atmosphere. Arc shielding is often supplemented in low-alloy filler materials with flux and/or aluminum or magnesium additions which serve as deoxidizers.

Low equipment cost is one of the primary advantages of the open arc hardfacing process. Neither flux-handling nor gas-regulating equipment is required. Thus, the open arc hardfacing process is a simple semiautomatic welding process requiring only a torch and a device to feed the continuous electrode wire. Conventional power supplies with drooping voltage-amperage characteristics are well adapted to open arc hardfacing equipment. Motor-generator units are frequently used for field applications because of their portability. In-plant operations normally make use of direct current rectifiers. Although open arc hardfacing is a semiautomatic process, it is not necessary to have a constant-potential power supply for its operation.

Open arc equipment generally is designed for small-diameter welding wires, usually $^{7}/_{64}$ in., although other diameters are manufactured. Most ferrous metals can be hardfaced with ease. Cast iron and chilled iron should be handled cautiously. Open arc is a high-energy process capable of much higher deposition rates than SMAW. The $^{7}/_{64}$-in.-diam rod can be deposited with welding currents as high as 375 A. The open arc hardfacing process is closely related to the submerged arc process, except that it does not require a flux burden over the bead; consequently, there is little or no slag to be removed from the overlay. Most hardfacing alloys can be deposited by the open arc process without preheat, and there is good recovery of elements across the arc. Hardfacing by open arc welding is an easy process to use, and only a short time is required to train an operator for the average application.

Arc stability and melting rates are excellent. The best hardfacing characteristics are obtained from DCEP. However, good results can be obtained with an alternating current supply at 26 to 30 V. Hardfacing deposits are comparable in soundness to deposits made with covered electrodes at deposition rates three to five times faster than with covered electrodes.

Hardfacing by open arc welding is not without disadvantages. Because the arc is not shielded with inert gas or a blanket of granular flux, considerable spatter and some porosity can be expected with this process. In addition, because of high welding currents, it is not particularly well adapted to the hardfacing of small parts. The recovery of weld metal from the tube wire to the deposit normally averages near 80 to 85% or higher. The amount of cross cracking and distortion is comparable to that of the other arc welding processes.

Gas metal arc welding is a consumable electrode hardfacing process in which the filler material and workpiece surfaces are protected by a flowing shielding gas—carbon dioxide, argon, or helium either singly or in combination with a small amount of oxygen. The GMAW process is suitable for semiautomatic hardfacing, and the versatility of the process lends itself to the hardfacing of complex shapes. Hardfacing using GMAW can be fully mechanized. The hardfacing deposit is visible at all times in GMAW, thereby enabling high-quality deposits.

Hardfacing by GMAW utilizes small-diameter (generally $^{1}/_{16}$ in. or less) hardfacing filler wires or electrodes. Gas metal arc hardfacing wires can be deposited in either the spray arc or short arc modes. The spray arc mode produces a continuous stream of droplets about the same diameter as the hardfacing wire, which are projected by electromagnetic force from the tip of the wire. Deposition rates are high, as is dilution. In the short arc mode, lower voltages are used, and transfer of the molten hardfacing wire is more globular, usually resulting in more spatter.

Hardfacing by GMAW, unlike SMAW, requires a constant-potential power source with voltage, slope, and wire-feed rate controls. Best results for $^{1}/_{16}$-in.-diam cobalt-based hardfacing wires in the spray arc mode are generally obtained at 18 to 26 V, with wire-feed rates ranging from 110 to 150 in./min. Best results in the short arc mode are generally obtained at 16 to 18 V, with wire-feed rates ranging between 80 to 100 in./min. The use of auxiliary shielding gases adds to the cost of hardfacing by GMAW. However, the higher cost is generally offset by the higher deposit quality associated with GMAW.

Gas tungsten arc welding is a nonconsumable electrode arc process in which the heated area of the workpiece, molten hardfacing alloy, and the nonconsumable electrode are shielded from the atmosphere by a flow of shielding gas fed through the torch. Thoriated tungsten is the preferred nonconsumable electrode material. Conventional power supplies with drooping voltage-amperage characteristics are normally used for hardfacing. Motor-driven generators can be used, as can alternating current transformers with continuous high-frequency current. Direct current electrode negative is always used to minimize tungsten contamination in the hardfacing deposit. Argon generally is used as the shielding gas, but helium is also suitable.

Hardfacing by manual GTAW is a useful alternative to hardfacing by OAW techniques, particularly when large components or reactive base metals are involved. Gas tungsten arc welding is preferred for hardfacing on reactive base metals such as titanium-stabilized stainless steels or aluminum-bearing nickel-based alloys. Gas tungsten arc welding also is preferred over OAW for hardfacing applications in which carbon pickup is unacceptable or in applications involving filler materials that tend to boil under the influence of an oxyacetylene flame. The localized, intense heat of GTAW generally results in more base-metal dilution than OAW. Base-metal dilution can be minimized by torch oscillation, by using no more amperage than is necessary, and/or by concentrating the arc on the hardfacing deposit rather than on the workpiece. Hardfacing deposition rates generally are comparable in manual GTAW and OAW, despite the difference in base-metal dilution.

Hardfacing by GTAW can be accomplished automatically by simply attaching the torch to an oscillating mechanism and using a mechanized device to feed the hardfacing filler material into the arc region. This, of course, requires long-length filler rod or continuous wire. Long-length rods are becoming increasingly more available with the emergence of the continuous casting process for producing hardfacing rod. High-quality, reproducible hardfacing deposits can be produced by automatic GTAW by controlling filler-metal feed rates, torch oscillation, and travel speed. Current-delay controls can be used to control the final stages of solidification in the hardfacing deposit to minimize shrinkage and crater cracking.

The arc action in GTAW is generally smoother, quieter, and freer of weld spatter than in other consumable electrode arc welding processes. The use of a separate filler rod or wire enables greater operator control of deposit shape. Gas tungsten arc welding is more adaptable to hardfacing small intricate parts and generally produces higher quality deposits than other hardfacing processes.

Plasma Arc Welding. Hardfacing by PAW is similar to hardfacing by GTAW in that both processes employ a gas-shielded arc between a nonconsumable tungsten electrode and the workpiece as the primary heat source for hardfacing. The processes differ in that the PAW process maximizes the use of plasma as a secondary heat source. The plasma is formed by ionizing the gas flowing in a nozzle surrounding the electrode. The electrode generally is recessed into the nozzle. The plasma gas generally emerges from a constricting orifice arrangement. The PAW process can use bare rod or wire as a hardfacing consumable, but more often powder is used as the consumable. When powder is used, the process is often referred to as the plasma transferred arc process.

A typical plasma transferred arc torch for powder consumables is illustrated in Fig. 8. The hardfacing filler material is carried from a powder feeder to the plasma torch in an inert gas stream. The powder is directed from the torch into the arc effluent, where it is melted and fusion welded to the workpiece. A direct current power supply connected between the tungsten electrode and the workpiece provides the energy for the transferred arc. A second direct current power supply connected between the tungsten electrode and the nozzle supports a nontransferred arc, which

Fig. 8 Schematic of the plasma transferred arc hardfacing process

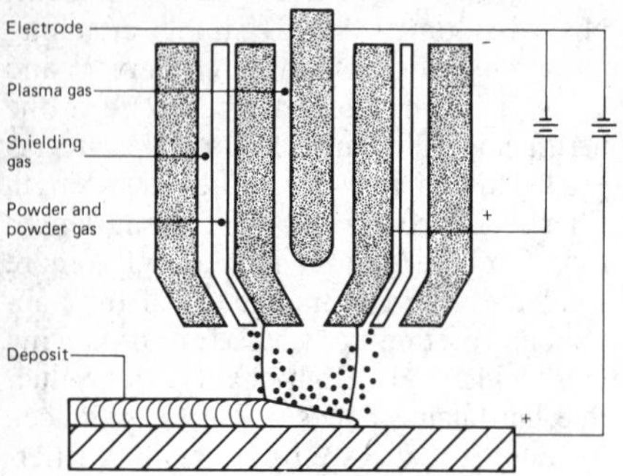

supplements the heat of the transferred arc and serves as a pilot arc to initiate the main transferred arc. The shielding gas is passed through a diffuser and forms a blanket in and around the arc zone.

Hardfacing by PAW using rod and/or wire filler materials is similar to hardfacing by GTAW. However, the stiffer arc intrinsic to PAW enables greater control of deposit profile and position than is possible with conventional GTAW.

Hardfacing by PAW is a high-energy process. Deposition rates up to 8 lb/h are common. Still higher deposition rates ranging up to 40 lb/h are possible using the plasma arc hot wire process—a process which, as its name implies, utilizes wire hardfacing consumables and two independently controlled systems operating in one weld pool to melt and fuse a hardfacing alloy to a substrate. One system raises the hardfacing wire to its melting point using resistance heating. The other system fuses the hardfacing filler to the workpiece using arc energy. A typical plasma arc hot wire system is shown in Fig. 9. In this case, two filler-metal wires are fed at a constant speed through individual hot wire nozzles to intersect in the weld pool beneath the plasma arc. The filler metals are electrically connected in series and energized by an alternating current constant-potential power supply. Resistance heating occurs in each electrode between the wire contact tip and the weld pool. Input power is selected for a given filler-metal feed rate so that maximum resistance heating is induced into both electrodes without resulting in an open arc between them.

Hardfacing with the plasma transferred arc process has many advantages. Hardfacing deposits can range in size from approximately 0.009 in. thick by $^3/_{16}$ in. wide to approximately $^1/_4$ in. thick by $1^1/_2$ in. wide by simply varying the welding current, powder-feed rate, oscillation, and travel speed. The process is amenable to automation, which makes it well suited for high production involving a large number of parts. Powder recoveries as high as 95%, with deposition rates up to 10 lb/h, are possible, depending on the size and shape of the part being hardfaced. The plasma transferred arc process is comparable to GTAW with respect to its ability to accommodate reactive substrate materials.

Hardfacing by the plasma transferred arc process has some disadvantages. The equipment is relatively expensive and is limited to straight-line or cylindrical parts except when special tooling is provided. Usually, several parts are needed for setup, which makes the process somewhat unattractive to the low-volume job shop. The torch is mounted on a carriage and track which require permanent installation. There are two circuits employed in this process; therefore, either a special power supply or two separate conventional power units are required. Argon consumption is somewhat higher than with GTAW. Argon has to be supplied to the center of the torch to protect the electrode. Shielding gas is necessary to protect the weld metal, and argon is also used as a carrier for the powder. Large parts being faced with hard alloys normally require preheat. This is a serious potential problem with the PAW process, if excessive preheats and prolonged hardfacing times cause overheating of the torch.

Comparison of Welding Processes

In rotary oil-well drilling, there is considerable abrasive wear on the shoulder areas of the joint used to connect sections of drill pipe as the pipe rotates and moves up and down in the well. Replacement costs for the tool joint justify the expense of hardfacing, particularly when abrasive rock formations are being drilled. In directional drilling, especially in offshore operations, drilling is done through a casing. When a hardfacing layer consists of fine particles of tungsten carbide embedded in a matrix of low-carbon steel, the deposits on the joints are less abrasive to the casing and retain high wear resistance.

A comparison was made of four welding processes used in hardfacing of drill joints with this carbide steel composite: manual OAW and automatic GMAW, GTAW, and PAW. The working details of each process, and the relevant advantages of the three arc welding processes over OAW for this application, are discussed in the example that follows.

Example 2. Comparison of Welding Processes. The design of a typical tool joint for rotary oil-well drilling is shown in Fig. 10. Joints varied in size as shown in accompanying table.

Tool joints were made of 4137H steel, oil quenched and tempered to 311 to 341 HB. Descriptions of the deposition of a wear-resistant coating of tungsten carbide particles in a matrix of low-carbon steel onto the tool joint by each of the four welding processes follow.

Oxyacetylene Welding. The tool joint was first positioned on a turning mechanism actuated by a foot pedal, with the axis of the joint horizontal, the threaded end toward the welder. The part was cleaned

Fig. 9 Schematic of the plasma arc hot wire hardfacing process

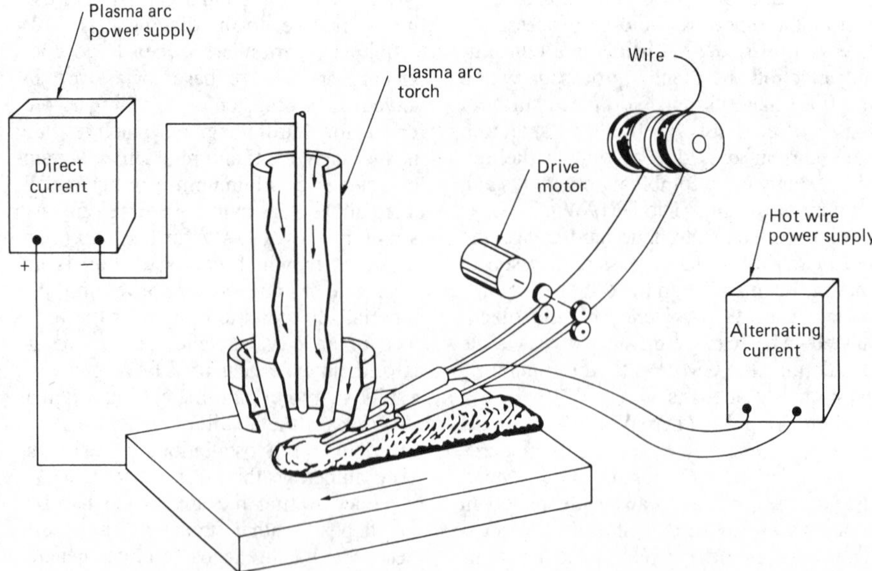

Fig. 10 Tool joint for oil-well drilling, showing the area that was hardfaced, and sections through the weld groove before and after welding. Positions of torch or electrode holder and carbide feed tube are shown at the right. Values of *a*, *m*, and *w* were determined by trial and error. See Table 10 for a comparison of data for four welding processes.

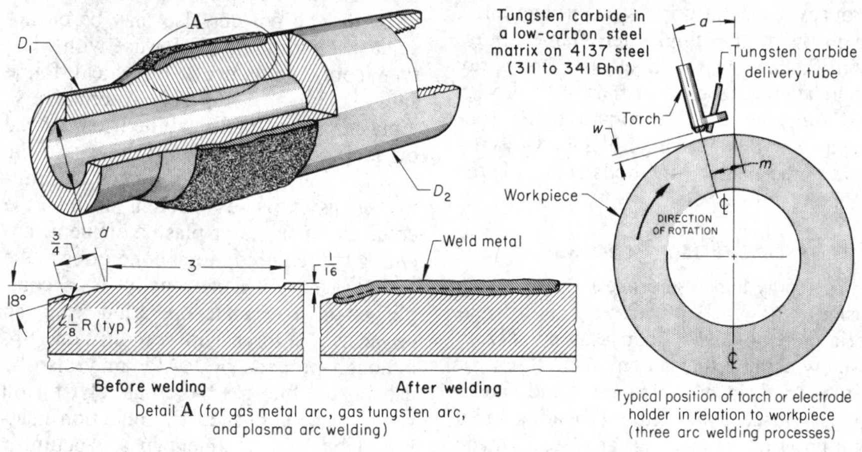

Size, in.	$2^7/_8$	$3^1/_2$	$4^1/_2$	5
Weight, lb	34	41	74	122
Length, in.	14	16	17	17
Dimension D_1, in.	3	$3^5/_8$	$4^5/_8$	$5^1/_4$
Dimension d, in.	$1^7/_8$	$2^7/_{16}$	$3^1/_4$	$3^3/_4$
Dimension D_2, in.	$4^1/_2$	$4^3/_4$	6	$6^1/_4$

and preheated to 400 to 600 °F, using an indicating crayon to check the temperature. By SMAW, stringer beads intended to serve as guide ribs for hardfacing were deposited circumferentially along the area to be hardfaced. The ribs were about $^1/_{16}$ in. high and were spaced about $^7/_8$ in. apart. They were interrupted every 120° by a gap of about $^1/_2$ in.; the interruptions in successive ribs were staggered. Where previous hardfacing was still visible, new ribs were deposited directly on the old ones.

After depositing the guide ribs, flux and weld spatter were removed from the surface. Hardfacing alloy was deposited manually, starting in the groove nearest the threaded end. Deposition continued in the direction of the tapered elevator shoulder, ending in a blend at the shoulder. The backhand technique was used with a slight excess of acetylene. The depth of tungsten carbide deposited was approximately $^1/_{16}$ in. Every effort was made to obtain a good bond with minimum pooling. The base-metal temperature, measured at a distance of 4 in. from the tool joint shoulder, was not permitted to exceed 700 °F. After hardfacing, the joint was cooled in still air.

Gas Metal Arc Welding. The tool joint was grooved as shown in detail A of Fig. 10 and preheated to 600 to 800 °F. It was positioned so that the 18° tapered shoulder was horizontal, and the positions of the electrode holder and carbide feed were approximately as shown in Fig. 10. Angle *a* and distances *m* and *w* were determined by trial and error. First, hardfacing was applied to the tapered elevator shoulder; tungsten carbide was fed into the molten hardfacing pool immediately behind the arc. The part was then repositioned so that its centerline was horizontal. The filler metal and tungsten carbide were deposited, smoothly overlapping the initial deposit at the shoulder. The welding head was then indexed automatically so that each successive bead overlapped the preceding bead by about $^1/_8$ in. After hardfacing, the tool joint was cooled in air.

Gas Tungsten Arc Welding. The procedure for GTAW hardfacing, including the location of the torch and carbide feed nozzle, was similar to that used for GMAW, except that the outside of the tool joint was not grooved for hardfacing and the tungsten carbide was introduced differently. The surface of the base metal was brought to fusion temperature by the heat of the arc, and the tungsten carbide particles were fed onto the molten surface. As the weld metal

Table 10 Operating conditions for hardfacing joints on 4137H steel tools

	OAW(a)	GMAW	GTAW	PAW
Mode	Manual	Automatic, oscillating	Automatic	Automatic, oscillating
Hardfacing alloy	$^3/_8$- to $^1/_4$-in.-diam steel tube binder, 20-mesh tungsten carbide filler(b)	$^1/_{16}$-in.-diam low-carbon steel wire, 20- or 50-mesh tungsten carbide granules	20 to 50 mesh tungsten carbide granules	Iron-based powder; 20 or 50-mesh tungsten carbide granules
Layer depth, in.	$^1/_{16}$-$^3/_{32}$	$^1/_{16}$-$^3/_{32}$	$^1/_{16}$-$^3/_{32}$	$^1/_{16}$-$^3/_{32}$
Power supply	...	500-A direct current	300-A direct current	500-A transferred arc; 75-A nontransferred arc
Hardfacing alloy feed	Manual	Variable carbide feed	Variable carbide feed	Variable carbide feed
Fuel-gas flow	Acetylene at 49-62 ft³/h; oxygen at 53-65 ft³/h	...	...	...
Shielding	Flux	Argon at 30-25 ft³/h	Helium at 70 ft³/h	Argon at 70 ft³/h(c)
Current, A	...	340-360 DCEP	240-260 DCEN	300-360, direct current transferred
Voltage, V	...	29-32	40	29-32
Number of passes	...	5	15	5
Bead width, in.	...	$^7/_8$	$^1/_4$	$^7/_8$
Surface speed	2.25 to 2.50 in.²/min	18-20 in./min	36 in./min	18-20 in./min
Tungsten carbide flow	...	600 g/min	250 g/min	600 g/min(d)
Oscillations per minute	...	80-110, $^3/_4$ in. wide	...	70-90, $^3/_4$ in. wide
Arc time, min	...	$4^1/_2$	$7^1/_2$	$4^1/_2$
Preheat, °F	400-600(e)	600-800(f)	600-800(f)	600-800(f)
Hardfacing time, min	2500	450	750	450

(a) Shielded metal arc welding was used to deposit strings of beads that served as guide ribs for hardfacing. Power for SMAW was supplied by a 200-A unit: electrodes were $^3/_{32}$- to $^1/_8$-in.-diam E6010 or E6012. (b) 40% binder: remainder filler. (c) Included argon for powder, shielding gas, and orifice flow. (d) Iron powder flow, 100 g/min. (e) Gas burner, furnace, or induction coil. (f) Induction coil

solidified, a coating of carbide particles dispersed in a matrix of the base metal was obtained.

Plasma Arc Welding. The PAW hardfacing process was an automatic operation similar to GMAW, except that a plasma arc torch was employed and iron powder was substituted for the $^{1}/_{16}$-in.-diam consumable electrode wire.

Details of welding conditions for the four processes are given in Table 10. A comparison of the three arc welding methods with OAW led to the following conclusions:

- Arc welding processes could be readily automated and, hence, were three to five times faster than OAW.
- Equipment cost for automatic arc welding was much higher than for OAW. Initial investment in equipment for PAW was about twice that for GTAW or GMAW and ten times that for OAW.
- Labor cost (indicated by hardfacing time in Table 10) was much higher for manual OAW than for automatic arc welding.

Hardfacing by Spraying

A variety of spraying techniques are available for applying a hardfacing material. However, the three basic methods of hardfacing by spraying are spray and fuse, plasma spraying, and detonation gun. The primary considerations involved in spraying process selection include (1) substrate composition, (2) coating wear and bond strength requirements, and (3) service environment. These spraying methods are line-of-sight processes in which powder is heated to a plastic or molten state and accelerated by either a gas stream or detonation wave. The powder is directed at the workpiece surface and, on impact, forms a coating consisting of many layers of overlapping particles or splats. The spray-and-fuse process includes an additional step of fusing the coating using either a heating torch or a furnace.

A single-step version of the spray-and-fuse process is to simply spray powdered materials onto a substrate, such as copper, nickel, or stainless steel, to provide protection against corrosive environments. A variation of this single-step process is to feed a continuous ductile wire into the flame, rather than powder. This method of flame spraying is known as metallizing or wire-flame spraying. Metallizing using wire also can be accomplished by the electric arc spraying process, in which two wires are used to create an electric arc, and the resulting molten metal droplets are propelled to the workpiece with a continuous blast of compressed air. While these methods are occasionally used to spray bronzes, especially for wear protection, a majority of the single-step flame spray, wire-flame spray, and electric arc spray applications are used primarily to protect the substrate from a corrosive environment. For this reason, a detailed discussion of these THSP methods is outside the scope of this article.

Spray-and-Fuse Process

The spray-and-fuse process is a two-step process in which powdered coating material is deposited by conventional THSP, usually using either a combustion gun (or torch) or a plasma spray gun, and subsequently fused using either a heating torch or a furnace. The coatings are usually made of nickel or cobalt self-fluxing alloys to which hard particles, such as tungsten carbide, may be added for increased wear resistance. Coatings ranging from 0.020 to 0.080 in. thick can be made by building up several layers at a rate of 0.005 to 0.030 in. per pass. Typical deposition rates are 9 to 12 lb/h.

There are several types of combustion powder spraying systems used for the initial deposition step, differing primarily in the method used to feed the powder into the flame. One system utilizes compressed air as a propellant for the powder, which is injected directly into the flame through a central orifice in the nozzle. Another system feeds the powder into an aspirating gas ahead of the nozzle and prior to ignition. In still another system, gravity feeds the powder directly into the flame; the powder is then conveyed to the workpiece by the force of the combustion gases.

Usually, acetylene is used as the fuel gas in conjunction with oxygen, although other fuel gases such as propane and MAPP are utilized. The powder also may be encapsulated in plastic tubing for use with wire-type spray equipment. A typical flame spray is shown in Fig. 11. Alternatively, a plasma spray torch may be used instead of a combustion gun. In all cases, the powdered alloy becomes molten or semimolten as it passes through or into the combustion flame or plasma effluent, enabling the required initial bond prior to the fusion step. Fusion usually is accomplished in a separate operation using an oxyacetylene torch with a multiflame tip, although propane or MAPP may also be used as the fuel gas. Alternatively, fusion may be accomplished by induction heating or by heat treatment in a vacuum or reducing atmosphere furnace.

The spray-and-fuse process usually employs self-fluxing hardfacing alloys that contain silicon and boron. These elements act as fluxing agents, which during fusion permit the coating to react with any oxide film on the workpiece or individual powder particle surfaces, allowing the coating to wet and interdiffuse with the substrate and result in virtually full densification of the hardfacing. Silicon and boron also depress the melting point of hardfacing alloys approximately 700 to 800 °F below the melting point of most steels.

Some spray-and-fuse hardfacing alloy compositions are listed in Table 11. Abrasive wear resistance of the alloys listed may be increased by adding up to 50% tungsten carbide particles to these alloys.

Fig. 11 Schematic of spray-and-fuse hardfacing process

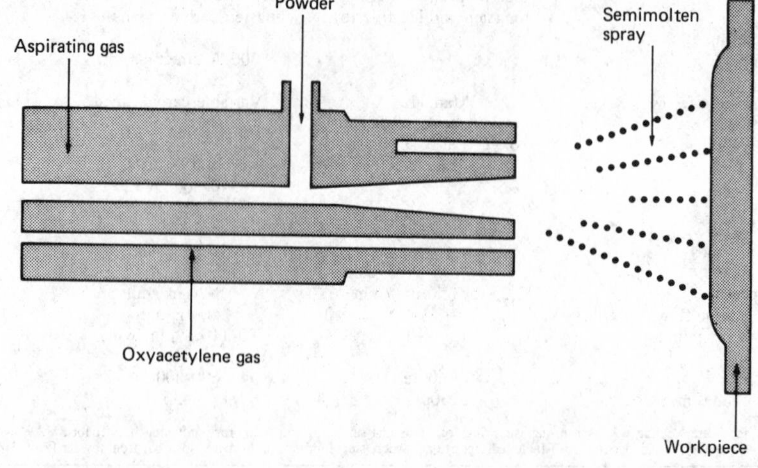

Table 11 Nominal spray-and-fuse powder alloy compositions

Alloy designation	Composition, wt% Ni	Co	Cr	C	Si	B	Fe	W
Deloro Alloy 40	rem	...	7.5	0.35	3.5	1.7	1.5	...
Deloro Alloy 50	rem	...	11.0	0.45	4.0	2.4	3.0	...
Deloro Alloy 60	rem	...	15.5	0.8	4.3	3.5	4.0	...
Stellite SF6	13	rem	19	0.7	2.3	1.7	3.0	7.5
Stellite Alloy 157	2 max	rem	22	0.1	1.6	2.4	2 max	4.5

Most workpiece substrate metals can be hardfaced using the spray-and-fuse process without special precautions, while others require special preheating or cooling procedures to prevent cracking of the hardfacing. Most plain carbon (SAE IXXX), manganese (SAE 12XX), molybdenum (SAE 4XXX), chromium (5XXX), chromium-vanadium (SAE 61XX), and nickel-chromium-molybdenum (SAE 8XXX) steels can be hardfaced by the spray-and-fuse process without special precautions, provided the carbon content is below approximately 0.25%. When the carbon content of the steel is above 0.25%, the workpiece requires a 500 to 700 °F preheat prior to fusing the sprayed coating and slow cooling after fusion.

Irons (such as gray cast iron, meehanite iron, malleable iron, ingot iron, and wrought iron) and nonferrous metals and alloys (such as copper, nickel, Monel 400, Inconel 600, Nichrome, and most high-temperature alloys) are also amenable to hardfacing by the spray-and-fuse process without special precautions.

The 300 series austenitic stainless steels may be hardfaced by the spray-and-fuse process, but because they have high coefficients of thermal expansion, it is advisable to preheat them to 600 to 700 °F prior to spraying. Preheat expands the workpiece sufficiently to prevent the subsequent expansion or contraction from breaking the mechanical bond prior to fusing. The fusion of these steels is best done immediately after spraying.

Air-hardening tool steels and the SAE 400 series martensitic stainless steels (except 414 and 421) can be hardfaced easily by the spray-and-fuse process; however, these metals require preheating prior to fusing the overlay and cooling with an isothermal hold after fusion to prevent the overlay from cracking. Use of 303 stainless steel or other free-machining metals containing sulfur, manganese, phosphorus, lead, or selenium can result in porous overlays due to the emanation of gas during fusion, and these materials usually are not considered for this process. Titanium- or aluminum-bearing alloys (321 stainless steel) are also not considered because of the highly stable oxide layers on these base metals, which interfere with proper bonding of the overlay to the base metal during the fusion operation.

In some cases, it is desirable to apply a coating. Spray-and-fuse hardfacings may be deposited directly over an existing deposit if the workpiece surface is clean, free of oxide, and preheated to about 1500 °F before the initial deposition. A thin coating layer is applied first, followed by additional layers applied by conventional spraying; fusing is done immediately after deposition.

Workpiece Surface Preparation. Undercutting is frequently necessary during workpiece surface preparation to produce an even deposit thickness on an unevenly worn surface or on parts not originally built to include a hardfacing, or to provide room for a sufficiently thick hardfacing to accommodate the expected wear. The depth of the undercut should be determined by the amount of wear permitted in service plus approximately 0.010 in.

When undercutting to a shoulder, it is advisable to feather up to the shoulder at a 30° angle from the surface. If an external corner is going to be sprayed, it should have a radius of at least $^1/_{32}$ in. If rough threading is to be used in preparing the surface, the shoulder angle should be reduced to 15°.

Workpiece surfaces must have all plating, carburizing, nitriding, or surface residues from other surface treatments removed. Surfaces should be degreased, if necessary. Workpiece surfaces to be sprayed usually must be roughened; grit blasting is the preferred method. When the workpiece surface hardness is less than HRC 30, it can be grit blasted with crushed angular chilled iron grit. Grit size can range from Society of Automotive Engineers (SAE) No. 12 to 16; some manufacturers also recommend mixtures of 25 and 40 grit.

Silicon carbide or aluminum oxide grit can be used as well, particularly for harder substrates. It is important to grit blast more area than will be sprayed and to blast surfaces around external corners and beyond the shoulder of undercut areas. Rough threading sometimes may be used to prepare the surface, but it is not the preferred method. Threading should be done with 32 to 40 threads per inch, not over 0.008 in. deep. Metal slivers should be removed by running a clean new file lightly over the surface.

Often there are surface areas adjacent to the area to be hardfaced that must be kept free of any sprayed deposit. An appropriate inhibitor or stop-off material (a liquid chemical formula that prevents the adhesion of metallic particles) may be applied to these areas. The stop-off material can be removed subsequently by wire brushing and polishing. If there are holes, keyways, or slots in the workpiece that are to remain uncoated, they should be masked with carbon. The top surface of the carbon plug should be level with the height of the finished overlay.

Spraying. During spraying with a combustion gun, the tip of the outer flame envelope is usually held about 1 in. from the workpiece surface. However, the base metal, size, or shape of the workpiece may require that this distance be modified. It is best to follow the manufacturer's recommendations. The distance from the work surface to gun tip generally falls in the range of 7 to 12 in. The relative gun-to-part surface speed is usually about 80 to 120 sfm. Multiple passes are preferred, with the application of 0.005 to 0.030 in. per pass. When a plasma gun is used for deposition, the manufacturer's suggested operating parameters should be used.

In calculating the required thickness of the sprayed overlay, about 20% must be allowed for shrinkage on fusing. In addition, an allowance of at least 0.010 in. per side for finishing should be made. The workpiece should be preheated to about 400 to 500 °F when spraying internal diameters or heavy sections, or when spraying heavy deposits. This helps prevent cracking of the as-sprayed mechanical bond.

Fusion. Fusing should be done immediately after spraying. It can be accomplished in one of several ways, but regardless of the method used, the main principle is to bring the overlay and base metal to the proper temperature so that the hardfacing material will wet and bond without losing its shape or running. Depending on the specific hardfacing alloy, fusing temperatures range from 1850 to 2050 °F.

Heating with an oxyacetylene torch is often the preferred fusion method, using a multiflame tip with a soft, bushy flame adjusted to a neutral flame. The torch is held close to a 90° angle to the workpiece and at the proper distance so that the outer flame envelope is close to but not touching the part. The oxyacetylene flame is played on the base metal adjacent to the sprayed overlay until it reaches a dull red, a minimum of 1300 °F. If the overlay does not extend to the end of the workpiece, heating should begin about 2 in. from the overlay. If the overlay extends to both ends of the workpiece, heating should begin at the center of one end, or on the inside if the part is a hollow cylinder. The heat should then be concentrated on the sprayed deposit. As the sprayed deposit reaches its bonding temperature, the rough sprayed surface becomes molten and reflects the torch flame (glazing), which is then moved along at a constant rate that is slow enough to heat the overlay to its bonding temperature, but fast enough to prevent sagging of the sprayed deposit.

Alternatively, a vacuum furnace or a furnace having a neutral or reducing atmosphere can be used. The exact temperature and degree of control required depend on the atmosphere being used, the melting point of the alloy selected, and the position of the sprayed surface. The furnace fusion method is best suited for parts with irregular cross sections, where the torch method is not practical. Fusing can also be accomplished with induction heating, which is particularly well suited for automated high-production applications.

Spray-and-fuse deposits of alloys of hardnesses up to 55 HRC can be machined. Alloys with higher hardnesses should only be ground. Machining should be done with a carbide-tipped tool. Most spray-and-fuse alloy deposits can be ground with green silicon carbide grinding wheels of H, I, or J hardness. Wet grinding is recommended. For roughing, 24-grit wheels should be used; for finishing, 60 grit or finer. Typical applications include arbors, cams, fan blades, glass molds, pump plungers, shafts, valve seats, valve stems, and calendar rolls.

Advantages and Disadvantages. The major advantages of the spray-and-fuse process are the virtually fully dense microstructure of the coating and the metallurgical bond that is created between the coating and the substrate. The major disadvantage is the high temperature to which the substrate is exposed during the fusion step. This frequently causes undesirable changes in the structure of steels and other substrates. Some dilution of the coating with substrate materials may also occur as a result of interdiffusion during the fusion step of the process.

Plasma Spraying

In PSP, a plasma gas stream heats and propels powder particles onto the workpiece surface. Because plasma temperatures are extremely high, PSP can be used to apply refractory coatings that cannot be applied by the spray-and-fuse process. Although plasma spray coatings have inherent porosity and are principally mechanically bonded to the substrate, they generally have higher density and better adhesion than is achieved with single-step flame spraying. In most applications, coating thickness is in the range of 0.005 to 0.015 in., but depending on the coating and the application, much thicker coatings can be applied, such as thermal barrier coatings. Usually, the workpiece is maintained below 300 °F, resulting in little or no distortion of the component and no dilution of the coating by the substrate. Very smooth deposits can be obtained, and finishing may not be required.

Figure 12 depicts a typical plasma spray gun, showing alternative powder inlet positions. A gas, usually argon or nitrogen (or a mixture of these) with hydrogen or helium, flows through a water-cooled copper anode which serves as a constricting nozzle. An arc is maintained internally from an axial rear tungsten electrode. The arc operates on direct current from a rectifier power supply, fed from an alternating current main supply at power levels that may exceed 100 kW. The gas plasma generated by the arc consists of free electrons, ionized atoms, and some neutral atoms and undissociated diatomic molecules if nitrogen or hydrogen is used. Controllable temperatures, to a range well in excess of the melting point of any known substance, can be obtained.

Powder introduced into the plasma stream is heated to a molten or semimolten state and propelled at high velocity onto the substrate. The powder is fed at a precisely regulated rate into an inert carrier gas stream and is usually introduced into the plasma stream either in the diverging portion of the nozzle or just beyond the exit. Usually, the carrier gas is the same composition as the primary plasma gas (argon and/or nitrogen). The choice of powder-feed rate is important. If the powder is fed faster than it can be properly heated, deposition efficiency decreases rapidly, and the coating contains trapped, unmelted powder. In contrast, if the feed rate is too low, operation costs significantly increase.

Powder velocity is strongly influenced by torch configuration. With most of the conventional commercial torches available up to the mid-970's, velocities varied from about 400 to 1000 ft/s. Higher velocity torches now result in coatings with higher densities and with some bond strengths in excess of 10 ksi.

Material for PSP must be in powder form and sized appropriately to achieve melting. For the highest coating density, powders are sized between about 44 μm (325 mesh) and 10 μm. The lower limit ensures free flow. Coarser powder may be sized up to 100 μm, with typical distribution falling between 88 μm (170 mesh) and 44 μm (325 mesh). Powder shape ranges from spherical to acicular; spherical is best because of reduced surface area and con-

Fig. 12 Schematic of PSP hardfacing process

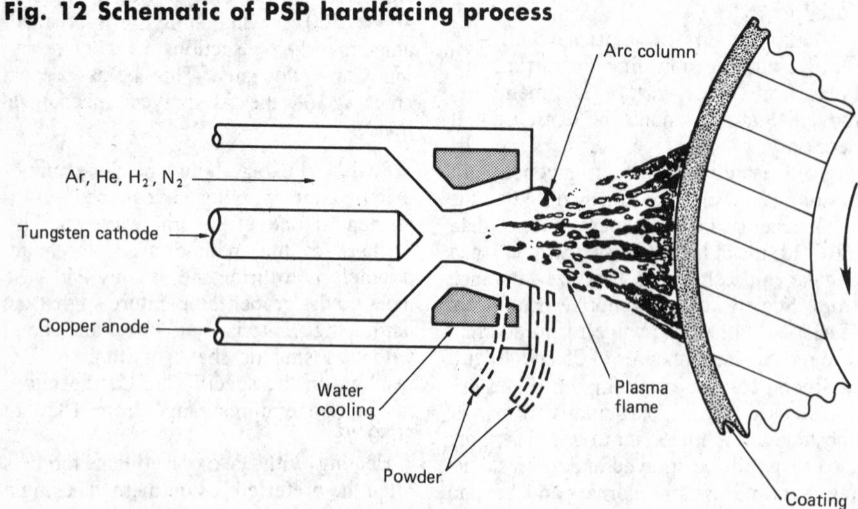

sistent flow. Any material that melts without subliming and is available as a properly sized powder has potential as a PSP coating. Plasma spray coatings may be elemental (aluminum, molybdenum, nickel, or chromium), alloyed (nickel-, iron-, or cobalt-based), compounds (Al_2O_3, Cr_2O_3, or Cr_3C_2), composites (nickel-clad aluminum, cobalt-bonded tungsten carbide TiO_2, or clad Al_2O_3), or mechanical blends.

Almost any substrate that can be adequately roughened can be plasma coated. The substrate surface roughness should normally exceed 150 μin. root mean square (rms). In addition, the coating should be adequately supported, so the substrate must have sufficient strength to resist deformation in service.

The size of the torch and the required standoff distance from the nozzle or the front face of the torch to the workpiece can be a limitation when the inside of a cylinder must be coated. Plasma torches are available that allow coatings to be deposited on the inside surfaces of cylinders as small as $1^1/_4$ in. in diameter. An additional limitation in coating complex parts, particularly those with narrow grooves or sharp angles, is that coating quality is a function of the angle of deposition (i.e., the angle between the axis of the plasma jet effluent and the surface being coated). Usually, a 90° angle of deposition is optimal. With some torches, angles as low as 60° can be tolerated.

Workpiece Surface Preparation. The surface to be coated must be free of oils, lubricants, oxide scales, or other foreign matter. After the surfaces are cleaned, they must be roughened. The preferred method is grit blasting. Chilled steel grit is often satisfactory for relatively soft substrates, while alumina or silicon carbide has better cutting action on harder substrates. As mentioned previously, the surface roughness should normally exceed 150 μin. rms, with the surface topology being sharply peaked.

A wide variety of masking techniques are used to limit the deposition to the required area on the part. In most cases, masking is less expensive than subsequent removal by grinding. Many types of tape and oxide-loaded paints or stop-off lacquers are satisfactory for low-velocity, long standoff plasma torches. For high-velocity, short standoff torches, more substantial masking is required, such as glass fiber-reinforced high-temperature tape, adhesive-backed steel or aluminum foil, or sheet metal masking.

Coating Deposition. The choice of torch parameters should be based on the manufacturer's recommendations. The use of automated torch and work-handling equipment is recommended to ensure uniform coating thickness and to minimize residual stress within the coating. It is always advisable to coat a quality-control specimen to verify the coating deposition rate and microstructure before coating any parts. Metallographic examination of this specimen should include general phase content, amount of oxidation occurring during deposition, apparent porosity, and microhardness. It is also advisable to check the grit inclusion level and/or amount of substrate surface contamination, but this is meaningful only if the quality-control specimen is of the same composition and condition as the parts to be coated.

In many applications, plasma coatings are used as-coated, but a variety of finishing techniques are also available. These include brush finishing to produce a nodular surface, machining (suitable for some metallic coatings), grinding (usually with silicon carbide or diamond), and lapping to produce surface roughnesses of less than 2 μin. rms. The best surface finish that can be obtained is a function not only of the finishing technique, but also of the coating composition and the deposition parameters. Recommendations for the machining, grinding, and lapping techniques to be used with a specific coating can be obtained from coatings service organizations or coating equipment manufacturers. Great care should be exercised in finishing operations to avoid damaging the coating through heat checking, pullout, or edge chipping.

Particle cooling rates measured in excess of 10^5 °C/s create structures typical of splat quenching in plasma coatings. Many of the major attributes of plasma coatings can be traced to rapid quenching—the retention of high-temperature metastable phases that provide high hardness and wear resistance and, in some compositions, quenched-in microcracks that provide some oxide ceramic coatings with impact resistance. Bridging between particles accounts for the porosity always present in coatings sprayed in air.

Metallographically apparent porosity can be reduced to less than $^1/_2$% with proper selection of powder size, type, and spray parameters. Bulk oxides or the effects of oxidation are readily seen in some coatings sprayed in air. The extent of oxidation is a function of torch design and the material being sprayed. Oxides typically assume a flattened shape and are called oxide stringers. In the case of tungsten carbide-cobalt powder, oxygen causes decarburization due to oxidation. Therefore, the coating shows little or no oxide inclusion, but a loss of carbon is evident.

Oxides may be reduced by using an argon or argon and helium plasma combined with short spray distances. Several means of shielding the plasma stream are also available. One method involves PSP in a vacuum chamber. However, a patented argon shroud provides greater versatility and is equally effective in producing clean plasma coatings. Plasma coatings are used throughout industry for applications including antifretting and wear resistance, thermal barriers, clearance control, abradable and abrasive coatings to control gas path clearance, high-temperature oxidation, corrosion resistance, and general buildup for reclamation of parts.

Advantages and Disadvantages. The plasma spray process has several advantages. Plasma spray coatings have higher density and better adhesion than is achieved with single-step flame spraying. Usually, the substrate is maintained below 300 °F, resulting in little or no distortion of the component and no dilution of the coating by the substrate. Also, very smooth coatings can be obtained, and finishing may not be necessary.

Limitations of PSP include: (1) it is a line-of-sight process, so there may be difficulty in coating some complex parts; (2) the bond to the substrate is primarily mechanical, not metallurgical; and (3) plasma coatings are inherently porous. For high-temperature applications, densification may be promoted by mechanically working the coating surface, by shot peening, and by a high-temperature sintering treatment.

Plasma spray under low pressure is a variation of PSP, in which the plasma gun and workpiece are enclosed in a vacuum tank. Preheating the workpiece, spraying the coating, and subsequent heat treating are carried out in an inert atmosphere at low pressure (50 torr). A typical installation is shown in Fig. 13.

Very fine powder generally is used for low-pressure PSP to achieve maximum density. If high-purity coatings are also desired, then the powder must be manufactured under inert cover and be atomized in an inert gas to preserve purity. Substrate size is limited by the size of the vacuum tank. Substrate requirements may also limit the preheat and postheat treatment temperatures. Typical applications include coatings on turbine air foils, blade tips, and shroud segments, and certain chemical environments that require impermeable coatings.

Fig. 13 Typical low-pressure plasma flame spray system installation. Pyrometer (infrared thermometer) located at rear of vacuum tank, not shown. Cables and hoses omitted for clarity.

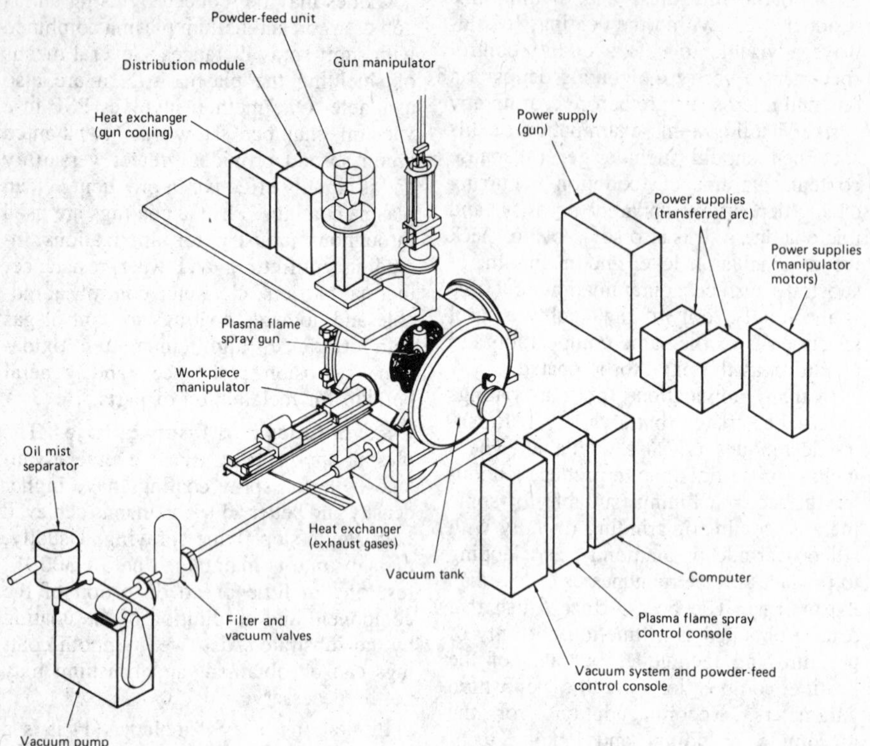

The advantages of PSP in a low-pressure, inert environment include:

- Higher bond strength may be achieved because higher substrate preheat temperatures and higher operating temperatures can be used without detrimental oxidation of the substrate or coating.
- There is excellent dimensional thickness control on substrates with irregular surfaces because longer torch standoffs from the substrate can be tolerated.
- Coating porosity is minimized.
- Deposition efficiency improves; more of the powder is actually deposited as coating.
- There are minimal changes in composition and metallurgy between the powder and coating.
- Noise and dust are eliminated.

The obvious disadvantages are higher initial capital investment, higher operating costs, and increased maintenance as compared to regular plasma or flame spray processes.

Detonation Gun

The detonation gun spray process is markedly different from other flame spraying processes and was initially developed for the deposition of hard, wear-resistant materials such as oxides and carbides. The extremely high particle velocities achieved in the detonation gun result in coatings with higher density, greater internal strength, and superior bond strength than can be achieved with conventional PSP or single-step flame spraying. Detonation gun coatings have been successfully applied to critical areas of precision components made from virtually all commercial alloys. The typical coating thickness is less than 0.015 in., but many applications require only 0.003 in. or less of finished coating.

The detonation gun, shown schematically in Fig. 14, consists of a water-cooled barrel several feet long with an inside diameter of about 1 in. and associated gas and powder metering equipment. In operation, a mixture of oxygen and acetylene is fed into the barrel along with a charge of powder. The gas is then ignited, and the detonation wave accelerates the powder to about 2400 ft/s, while heating it close to, or above, its melting point. The distance that the powder is entrained in the high-velocity gas is much longer than in a plasma or flame spray device, which accounts for the high particle velocity. After the powder has exited the barrel, a pulse of nitrogen purges the barrel.

The cycle is repeated many times a second. Because of the gases used in the detonation gun, the powder can be exposed to either an oxidizing, carburizing, or essentially inert environment. Coating deposition is closely controlled by fully automated equipment in order to achieve uniform coating thickness and to minimize substrate heating and coating residual stress.

Suitable materials for detonation gun coatings include pure metals, alloys, oxides, carbides, composites, and mechanical blends of two or more components. Historically, the detonation gun has proven particularly advantageous in applying hard wear-resistant carbide and oxide coatings. Detonation gun coatings have been successfully applied to critical areas of precision parts made from virtually all commercial alloys. The substrate surface is seldom heated above 300 °F during the coating process. As a result, a part can be fabricated and fully heat treated prior to coating. Usually, there is little or no dis-

Fig. 14 Schematic of detonation gun hardfacing process

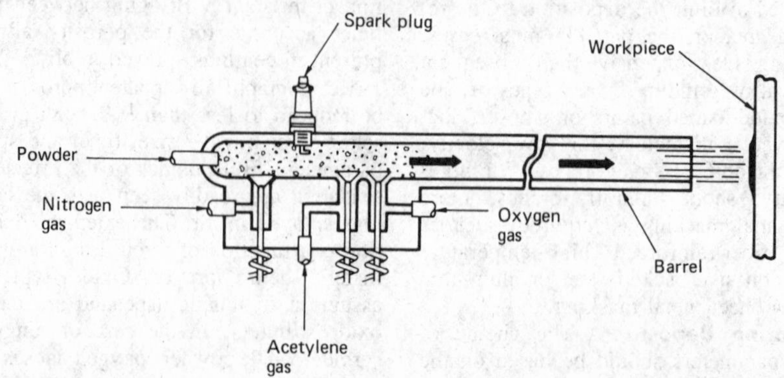

tortion, and the substrate microstructure and strength are not changed by the coating process.

The detonation gun is a line-of-sight process in which the structure of the coating is a function of the angle of deposition—the angle between the axis of the gun effluent and the surface of the part being coated. Normally, coatings with the highest density and bond strength are achieved at a 90° angle of deposition. The detonation gun has sufficiently high particle velocities so that, in many cases, coatings can be made at as low as a 45° angle of deposition without significant degradation of coating properties. This feature is useful in coating complex parts. The detonation gun, of course, cannot fit into a cylinder or other cavity. It can be used, however, to coat the inside of a cylinder to a depth about equal to the diameter, to an angle of deposition of about 45°.

Workpiece Surface Preparation. Cleaning the substrate surface is required as in all flame spray processes. However, grit blasting is not always necessary prior to the application of detonation gun coatings. Because the detonation gun accelerates the powder particles to such a high velocity, they are actually driven into the surface of some metallic substrates, such as titanium, and do not require grit blasting to achieve adequate bonding. With harder substrates, grit blasting is usually used to facilitate surface roughening. Sheet metal masking is often used with detonation gun coatings to limit the deposition to the required area on the part.

Coating Deposition. The detonation gun deposits a circle of coating with each detonation. The circles of coating are about 1 in. in diameter and a few thousandths of an inch thick. Each circle of coating is composed of many overlapping splats corresponding to the individual powder particles. The placement of the circles of coating is closely controlled to build up a smooth coating and to minimize substrate heating and residual stress in the coating.

In the detonation gun process, the extremely high kinetic energy of the impinging particles causes them to conform and adhere exceptionally well to the substrate surface. Thus, detonation gun coatings have higher density and bond strength than conventional plasma coatings.

The surface roughnesses of most as-deposited detonation gun coatings range from 125 to 250 μin. rms. In some applications, the coating is used as coated, and in at least one application, a tungsten carbide-cobalt coating is grit blasted to further roughen the surface for better gripping action. Frequently the coating surface is finished before being placed in service. Finishing techniques include brush finishing to produce a nodular surface, machining (used on some metallic coatings), grinding (usually with diamond), and lapping to produce surfaces with roughnesses as low as 2 μin. rms.

Detonation gun coatings are used in abrasive, adhesive, and erosive wear applications in virtually all industries. Many of these applications are in corrosive environments. They are particularly effective where close dimensional control must be maintained, such as on valve components, turbine blade Z-notches, compressor blade midspans, pump plungers, and compressor rods.

Advantages and Disadvantages. The chief advantage of the detonation gun is the extremely high particle velocities it develops. This results in coatings with higher densities, greater internal strengths, and superior bond strengths than achieved with conventional PSP or single-step flame spraying. Coating thicknesses less than 0.010 in. suffice for most applications. The substrate is seldom heated above 300 °F. Therefore, there is usually no distortion of components, no change in substrate microstructure or strength, and no dilution of the coating by the substrate.

A major disadvantage is that detonation guns are not commercially available. An additional limitation is that detonation coatings, while very dense, do contain a small percentage of interconnected porosity. For applications in corrosive environments up to about 350 °F, it is possible to seal the coating by the same methods used for plasma coatings. In high-temperature applications, detonation coatings are often sufficiently dense as deposited so that they seal themselves against internal oxidation and prevent attack of the substrate.

Appendix: Hardening of Metal Surfaces by Laser Processing

By J.D. Ayers
Metallurgist
Naval Research Laboratory
Department of the Navy
and
Daniel S. Gnanamuthu
Member of Technical Staff
Rockwell International

LASER PROCESSING techniques, used to improve the wear resistance of metals, became feasible in the early 1970's with the advent of high-power, continuous wave, carbon dioxide lasers. Process techniques include laser alloying, laser cladding, laser melt/particle injection, transformation hardening, and laser glazing. Some of these processes are carried out using heat sources other than lasers, but have become the subject of additional interest with the availability of selective heating attainable by using high-power lasers. Transformation hardening is a prime example of this point, because surface hardening of steel components has been employed on an industrial scale for many years using induction and flame heating. The principal advantage of using laser heating is that the surface can be heated with minimal heat transport into the bulk of the component, thus permitting conduction quenching to harden the surface to only the desired depth. Avoiding unnecessary heating of the bulk component minimizes the distortion produced during quenching. For more information on transformation hardening, see the article "Laser Surface Transformation Hardening" in Volume 4 of the 9th edition of *Metals Handbook*.

Laser glazing rapidly melts and resolidifies metal surfaces rather than inducing solid-state phase transformations, as in transformation hardening. Structural refinement caused by rapid solidification hardens many alloys, an effect that possibly could improve their wear resistance.

Laser Alloying

Controlled melting of a workpiece surface to a desired depth using a laser beam, with simultaneous addition of powdered alloying elements, leads to a rapid method of localized alloy synthesis. Because alloying elements diffuse in a thin liquid layer at the surface, the required depth of alloying can be obtained in small time intervals (typically 0.1 to 10 s). Thus, a desired alloy composition can be generated in situ on given substrate surfaces; alloy composition governs microstructure, and laser processing conditions (cooling rates) govern the degree of microstructural refinement. Depending on the choice of alloy design at the surface, a less expensive base material such as AISI 1018 steel can be locally modified to increase resistance to wear, erosion, corrosion, and high-temperature oxidation; only those surfaces locally modified possess properties characteristic of high-performance alloys.

Experimental Results. For this study, the surface of AISI 1018 steel was alloyed by using either one elemental powder or a mixture of powders. Carbon and chromium were chosen to generate stable car-

Table 1 Experimental conditions and results for laser alloying AISI 1018 steel

Alloying elements	Composition, wt%	Powder coating: Depth, in.	Powder coating: Width, in.	Application method	Processing conditions: Laser beam size, in. × in.	Type of laser beam	Laser power, W	Speed, in./s	Shielding gas
Chromium	100 Cr	0.02	0.65	Slurry	0.70 × 0.70	Stationary	12 500	0.07	Helium
Chromium, carbon	85 Cr-15 C	0.03	1.0	Slurry	0.25 × 0.75	Oscillating (690 Hz)	5 800	0.85	Helium and argon
Chromium, carbon, manganese	25 Cr-50 C-25 Mn	0.001	1.0	Spray	0.25 × 0.75	Oscillating (690 Hz)	3 400	0.35	None
Chromium, carbon, manganese, aluminum	24 Cr-48 C-24 Mn-4 Al	0.005	1.0	Spray	0.25 × 0.75	Oscillating (690 Hz)	5 000	0.35	None

(continued)

bides such as M_7C_3 and M_3C in an austenitic, pearlitic, or martensitic matrix. For example, for abrasion resistance, about 30% Cr and 4.5% C are required to produce M_7C_3 carbides in a pearlitic or martensitic matrix (Ref 1). For heat and corrosion resistance, about 25 to 35% Cr and 1 to 3% C are required for generating finely divided $Cr_{23}C_6$ carbides in a matrix of ferrite containing a sufficiently high concentration of chromium in solid solution (Ref 2). Manganese was used because it is a strong carbide stabilizer. Because aluminum is an excellent deoxidizer, it was used in certain alloying experiments when an inert gas shielding was not implemented.

Appropriate mixtures of metal powders, with an average particle size ranging between 0.4 and 2 mils, were applied directly on the specimen surface. This was done by spraying powders suspended in isopropyl alcohol to obtain a loosely packed powder coating of nominal thickness (up to 0.01 in.), or by coating a powder slurry suspended in organic binding compounds to a nominal thickness up to 0.08 in. If carbon contamination from binding compounds is detrimental to alloy properties, a mixture composed of 0.14 oz of cellulose dissolved in 3.4 fluid oz of acetone can be used for binding the powders. Powdered materials alone facilitate absorption of the laser beam and, consequently, alloying generally does not require additional application of an energy-absorbing coating. Typical experimental conditions are summarized in Table 1. Although metal powders were used in the alloying experiments (this generally is less expensive), appropriate experimental modifications can extend this method to use prefabricated rods, wires, thin ribbons, or sheets.

Alloying Procedure. From Table 1, the ratio of the amount of alloying elements added is not exactly the same as that determined in the alloyed casing (trace amounts of sulfur, phosphorus, and other elements were not analyzed). This can be associated with (1) nonautomated application of the required quantity of powder mix, either by spraying or as a slurry, which can be inconsistent; and (2) possible loss during laser processing. However, powders can be applied in controlled quantities by utilizing powder feeders with an electronic metering system.

Measurements of alloyed case depth were made at the midpoint of alloyed case width. The alloyed case width to be generated by a single pass of the workpiece under a laser beam dictates beam size. Views of an alloyed surface after surface grinding are shown in Fig. 1; this alloyed casing contains 16% Cr and was obtained by using a stationary beam.

Microstructures. Figure 2 shows the feasibility of obtaining desired microstructures based on alloy addition. These microstructures were generated under experimental conditions listed in Table 1. The microstructures can be interpreted by phase diagrams; however, quantitative evaluation must be based on metallographic techniques that use transmission electron microscopy to identify the various microstructural constituents.

The carbon-chromium-iron phase diagram was used in identifying microstructural constituents of the alloyed casing for (1) a 16% Cr addition, and (2) a 43% Cr and 4.4% C addition. With a 16% Cr addition, the matrix is essentially martensitic; islands in the microstructure are probably ferrite. Small quantities of carbides

Fig. 1 Top and cross-sectional views of alloyed casing on AISI 1018 steel. Casing contains 16% Cr. Etchant: nital. Base material: AISI 1018. Power: 12 500 W. Beam size: 0.7 × 0.7 in. Speed: 0.07 in./s

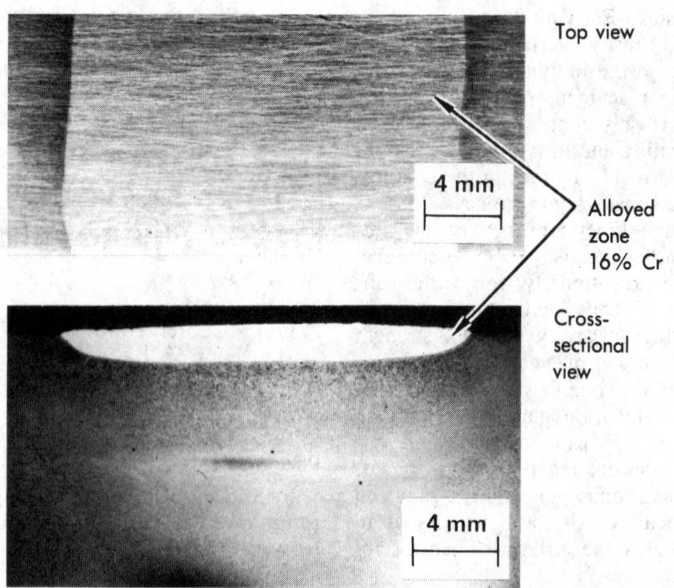

Table 1 continued

Alloying elements	Alloyed casing: Depth, in.	Width, in.	Composition, wt%	Hardness, HRC	Major microstructural constituents	Micro-structure
Chromium	0.08	0.85	16.0 Cr-0.7 Mn	53	Martensite	Fig. 2(a)
Chromium, carbon	0.01	0.6	43.0 Cr-4.4 C-0.5 Mn	64	Carbide M_7C_3	Fig. 2(b)
Chromium, carbon, manganese	0.005	0.6	3.5 Cr-1.9 C-1.3 Mn	64	Martensite and cementite	Fig. 2(c)
Chromium, carbon, manganese, aluminum	0.03	0.6	0.9 Cr-1.4 C-1.0 Mn-0.5 Al	56	Martensite and austenite	Fig. 2(d)

are present along grain boundaries because of small quantities of carbon alloyed from the base material. A structure similar to this has been observed in CA-15 stainless steel castings containing approximately 13% Cr, 1% Ni, and 0.15% C. With the addition of 43% Cr and 4.4% C, M_7C_3 carbides, which appear as irregular hexagons in the microstructure, are the primary phase to precipitate. These carbides are of uniform size and morphology. Formation of large quantities of this carbide in the alloyed zone leads to partial coalescence during solidification. The Knoop hardness (HK) at 50 g of these carbide particles is 1100 to 1200 kg/mm^2. The uniform dispersion of these hard carbides is useful for applications where resistance to abrasive and adhesive wear is needed.

Fig. 2 Microstructures of laser alloyed casings on AISI 1018 steel. (a) 16% Cr etched in 15% HCl + 5 mL H_2O_2. (b) 43% Cr and 4.4% C etched in Murakami's reagent. (c) 3.5% Cr, 1.9% C, and 1.3% Mn etched in nital, followed by Murakami's reagent. (d) 0.9% Cr, 1.4% C, 1.0% Mn, and 0.5% Al etched in super picral. Processing conditions are listed in Table 1.

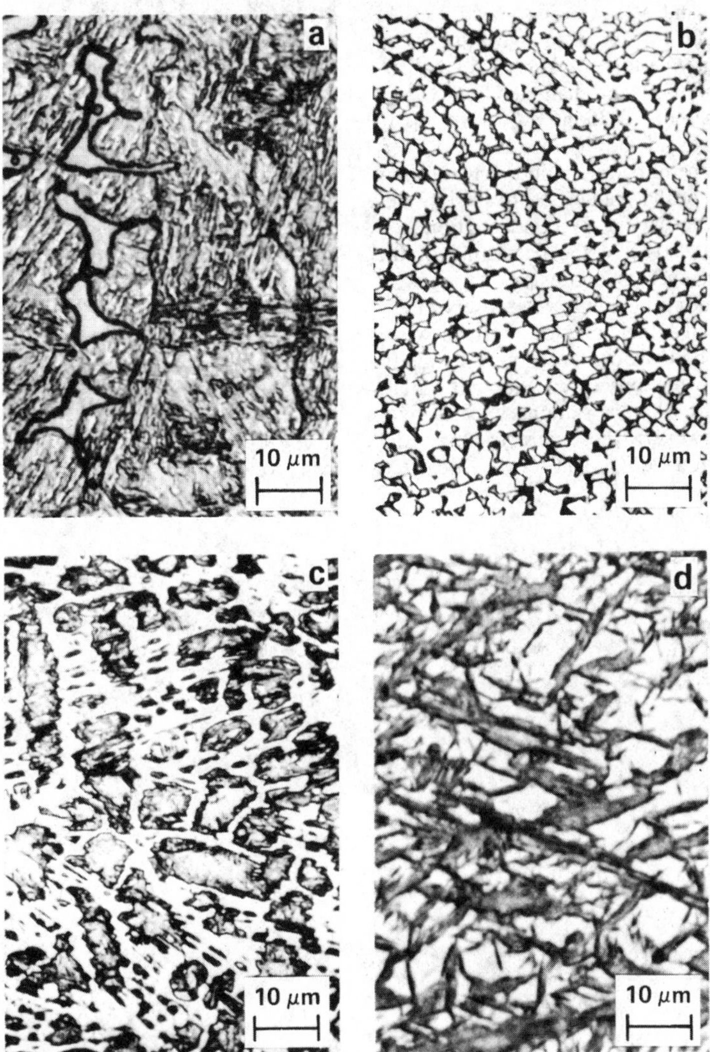

The iron-iron carbide phase diagram is used in identifying microstructures of the alloyed casing when strong carbide stabilizers, such as chromium and manganese, are present in small quantities (less than 5% total). The addition of 3.5% Cr, 1.9% C, and 1.3% Mn shows a microstructure composed of martensite, which results from transformation of primary austenite dendrites, surrounded by a eutectic network composed of cementite. Figure 3 shows a microstructure with Knoop hardness impressions and composition (using electron microprobe analysis) of the alloyed casing. The decrease in the amount of manganese near the surface may be the result of its high vapor pressure and of the small depth of the melt zone. In spite of this loss, the concentration of manganese is above that for the base material. The alloyed casing is uniform in dendrite morphology and in dendrite arm spacing. This observation indicates that cooling rates during solidification, at least for thin liquid layers, are uniform over the entire melt zone. The addition of 0.9% Cr, 1.4% C, 1.0% Mn, and 0.5% Al forms a microstructure composed of martensite (and possibly bainite), along with retained austenite.

Liquid Flow. During alloying, a thin liquid layer on the surface of a powder-coated workpiece is formed when it is irradiated with an adequately powerful laser beam (stationary or oscillating). Visual observations during surface alloying reveal a vigorous liquid flow along the surface where the laser beam is focused; the

Fig. 3 Microstructure of laser alloyed casing. Knoop hardness impressions at 500-g load, and concentration profile of added alloying elements. Etchant: nital, followed by Murakami's reagent. Base material: AISI 1018. Power: 3400 W. Beam size: 0.25 × 0.75 in. Speed: 0.33 in./s

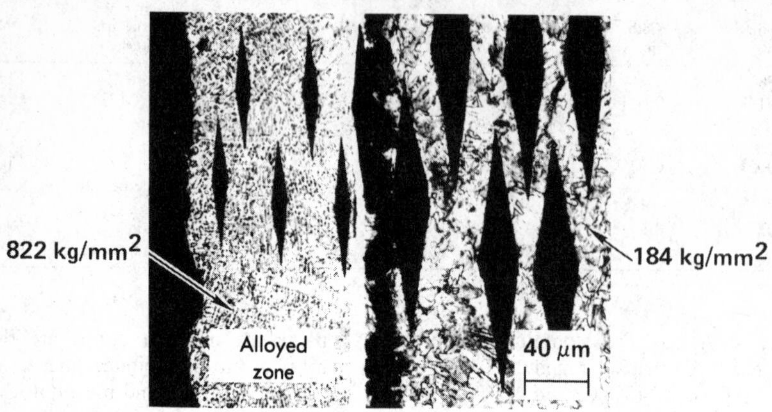

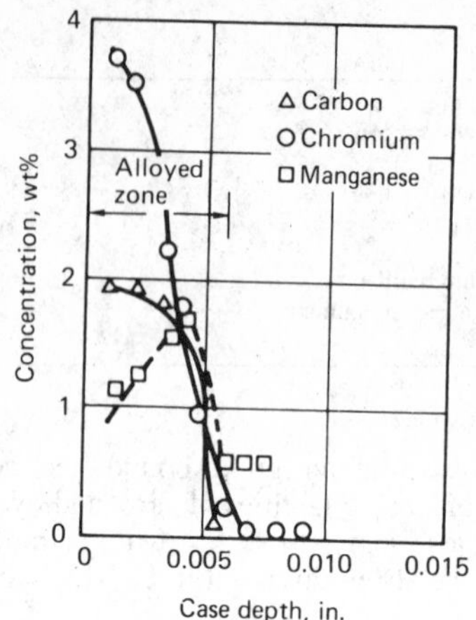

flow is momentary and quickly subsides when the laser beam is withdrawn. When liquid on a surface is directly radiated by a laser beam, the temperature of the liquid reaches a maximum with a consequent decrease in surface tension; the liquid near the beam (not directly radiated), however, has a temperature lower than the maximum value, and therefore has a higher surface tension. This establishes melt depressions and convection currents. During alloying, surface tension varies along the locally melted liquid because of temperature and concentration gradients associated with the addition of alloying elements.

A stationary laser beam induces liquid flow only in the area where it is focused; on the other hand, if the same laser beam is locally oscillated at rapid rates, localized liquid flow is spread out over the wider region radiated by the beam and, as the beam sweeps the same incremental area of the liquid surface several times, a potentially beneficial mixing action can occur. For example, at an oscillation rate, R = 690 Hz, and a processing speed of 0.15 in./s, each increment of the liquid is intermittently perturbed by as much as 1160 times; that is, beam diameter × R ÷ processing speed = 0.25 in. × 690 Hz ÷ 0.15 in./s. Thus, if the liquid surface is intermittently perturbed, a multidirectional flow may be locally induced. Physically, this implies that any alloy concentration gradient that exists at the beginning of the process is decreased after processing with an oscillating beam. It was observed experimentally that alloyed casings were uniform both in composition and microstructure when processed with an oscillating beam; when a stationary beam was used, however, the microstructure and composition were nonuniform in localized regions (especially along the edges of alloyed case width). Therefore, an oscillating beam is preferable for laser alloying.

Laser Cladding

The laser cladding process differs from surface alloying in that only enough substrate material is melted to ensure good bonding of the applied coating. Laser cladding differs little in principle from traditional forms of weld cladding, but advantages are evident when thin claddings are desired or when access to the clad surface can be achieved more readily by a laser beam than with an electrode or torch.

Laser beams offer potential in applying cladding alloys with high melting points on low-melting workpieces. Cladding or hardfacing alloys are usually cobalt-, nickel-, or iron-based and are used extensively for applications involving metal-to-metal wear, impact, erosion, corrosion, and

Table 2 Experimental conditions used for laser cladding

Cladding	Form of cladding material	Base material	Powder depth, in.	Powder width, in.	Powder application method	Preheat temperature, °F
Tribaloy T-800	Powder (plasma spray grade)	ASTM A387	0.25	1.0	Slurry	68
Haynes Stellite alloy No. 1	Cast rod (0.12-in.-diam)	AISI 4815	...	...	...	480
Silicon	Powder (44 μm size)	AA 390 aluminum alloy	0.04	0.20	Slurry	68
Tungsten carbide and iron	WC granules (0.02 in. size) and iron powder (44 μm size)	AISI 1018	0.04	0.75	Loose powder	68
Alumina	Powder (0.3 μm size)	2219 aluminum alloy	0.03	1.0	Loose powder	68

(continued)

Table 2 continued

Cladding	Laser beam size, in. × in.	Type of laser beam	Laser power, W	Processing speed, in./s	Shielding gas	Micro-structure
Tribaloy T-800	0.55 × 0.55	Stationary	12 500	0.05	Helium	Fig. 4
Haynes Stellite alloy No. 1	0.25 (diam)	Stationary	3 500	0.17	Hydrogen and argon	Fig. 5
Silicon	0.20 (diam)	Stationary	4 300	0.35	Helium and argon	Fig. 6
Tungsten carbide and iron	0.50 × 0.50	Stationary	12 500	0.25	Helium	Fig. 7
Alumina	0.25 × 0.75	Oscillating (960 Hz)	12 500	0.35	Oxygen	Fig. 8

abrasion. A single cladding alloy cannot satisfy all the applications mentioned above; consequently, selection of cladding alloys is based on factors such as service conditions, base materials, cladding processes, and cost. Cladding alloys that are available commercially as cast rods, wires, or powders can be melted under controlled conditions by using a laser beam, to allow the molten alloy to freely spread and freeze over a workpiece surface. Under controlled cladding conditions, a laser beam melts a very thin surface layer of the workpiece; this thin liquid layer mixes with the liquid cladding alloy and subsequently freezes to form a metallurgical bond between the cladding and substrate.

Experimental Results. Cladding experiments were performed with commercial cladding alloys as cast rods and powders. Prealloyed powders were applied to workpiece surfaces before laser processing with or without suitable binding compounds. Self-fluxing alloy powders (nickel- or cobalt-based alloys with a high chromium concentration, along with a sufficient quantity of silicon and boron) provide a rapid method of hardfacing when flame sprayed on workpieces and subsequently laser clad. Plasma spraying also can be used to deposit selected cladding or hardfacing alloys for subsequent laser cladding. When prealloyed cladding powders or rods were used, both oscillating and stationary laser beams provided dense and pore-free claddings that possessed uniform chemical composition and microstructure. Table 2 summarizes experimental conditions used for laser cladding.

Tribaloy Alloys. Tribaloy cobalt-based alloys (such as T-800) exhibit wear and corrosion resistance. Wear resistance of these alloys is attributed to the presence of a hard Laves intermetallic phase as primary dendrites surrounded by a solid solution and eutectic matrix. The composition of Laves phase typically is between cobalt-molybdenum-silicon and Co_3Mo_2Si, possessing a $MgZn_2$-type (close-packed hexagonal) crystal structure. The partitioning of chromium is about $^1/_3$ in Laves phase and $^2/_3$ in solid solution, thereby enhancing the corrosion resistance of both phases. Total quantity of carbon is kept below 0.08% to prevent formation of carbides. In Tribaloy T-800 alloy, volume fraction of Laves phase is nearly 0.5. This alloy was laser clad to ASTM A387 steel (2.25% Cr and 1% Mo). Cladding thickness of 0.24 to 0.28 in. and width of nearly 0.4 in. was achieved in a single sweep of the powder-coated workpiece under a laser beam. Dilution of the cladding alloy with the substrate was less than 5 wt%. Nominal deposition rates of 6 lb/h were obtained by using 12 500 W of laser power; by adapting superior powder application methods and optimized laser processing parameters, deposition rates can be improved. Microstructures (Fig. 4) indicate adequate fusion between substrate and cladding. Average hardness (Knoop hard ness) at 500 g of the cladding was 690 kg/mm^2 or 58 HRC.

Haynes Stellite Alloys. Using 0.12-in.-diam cast rods of Haynes Stellite alloy No. 1, cladding has been accomplished with minimal substrate dilution (substrate was AISI 4815 steel). Cast rods of Stellite al-

Fig. 4 Laser cladding Tribaloy T-800 alloy to ASTM A387 steel. (a) Cladding alloy, etched in Marble's reagent (electrolytically etched at 3 V for 15 s). (b) Interface between cladding and substrate, etched in nital. Processing conditions are listed in Table 2.

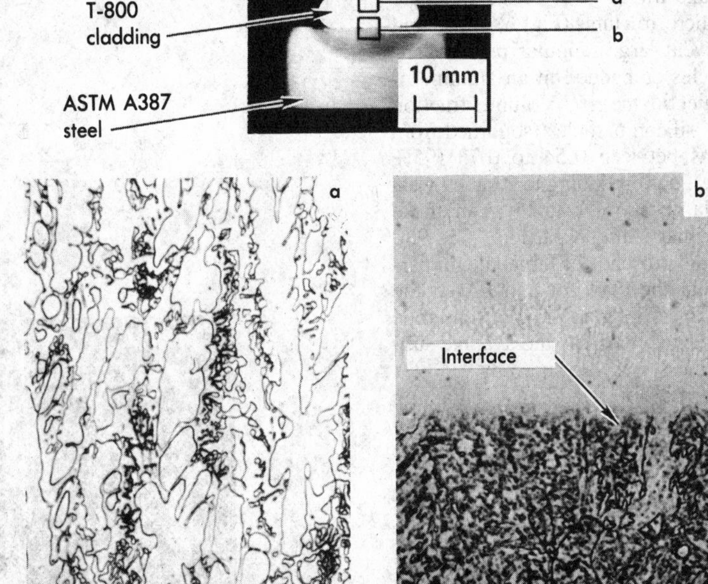

loy No. 1 contain M_7C_3 and M_6C carbides as a fine dispersion in an austenitic cobalt matrix. Appropriate lengths of Stellite rods were laid and subsequently clad to the workpiece surface using a laser beam; after cladding, the carbide structure was refined further, as shown in Fig. 5(a) and 5(b). Stellite alloy No. 1 can be kept from cracking by cladding only a small area or using a high-temperature preheat. Deposition rate was nominally 2 lb/h, using 3000 W of laser power. Electron microprobe analysis revealed the composition of the cladding to be 51% Co, 31% Cr, and 13% W. Composition and microstructure were uniformly constant across the cladding thickness. From this analysis, introduction of substrate into the cladding was determined to be less than 5 wt%. Hardness (HK at 500 g) of the cladding was 730 kg/mm^2, or 60 HRC. For applications involving fatigue, wear, and corrosion, an optimized microstructure of carbides dispersed in a matrix with uniform chemical composition is desirable. Carbides should be sufficiently refined to prevent crack nucleation, yet should possess adequate wear resistance.

Silicon. Claddings containing a high volume fraction of silicon particles in an aluminum-silicon eutectic matrix were generated by applying silicon (melting point 2610 °F) on low-melting substrates such as AA 390 aluminum alloy, nominally containing 17% Si (melting range of 950 to 1200 °F). During laser processing, locally generated heat at the aluminum surface melted the applied silicon powder (1.7 mil size) and a thin layer of the substrate surface. Microstructures (Fig. 6) of the clad surface reveal large, angular primary silicon particles embedded in an aluminum-silicon eutectic matrix. Volume fraction of primary silicon particles (solidified from liquid) was between 0.54 to 0.73 (95% confidence), corresponding to 53 to 69 wt% silicon. Hardness (HK at 25 g) of these angular primary silicon particles was 980 kg/mm^2; the presence of large silicon particles within the clad zone increased the overall hardness (HK at 500 g) of this zone to 400 kg/mm^2, whereas the average substrate hardness was 125 kg/mm^2.

Dense matrix cladding containing tungsten carbide particles was applied to the surface of AISI 1018 steel. A mixture of coarse (nominal particle size of 0.02 in.) tungsten carbide particles and iron powder (1.7 mil size) was applied to the steel surface without a binder. The carbide particles were completely immersed in the iron powder to minimize possible disintegration when directly irradiated by the laser beam (during processing, tungsten carbide particles were not melted). After laser processing, the average hardness (HK at 500 g) of tungsten carbide particles was 1100 kg/mm^2, and the hardness of the surrounding matrix was 870 kg/mm^2. Microstructure (Fig. 7) reveals a distinct region on each tungsten carbide particle that possibly is associated with interaction be-

Fig. 5 Microstructures of Haynes Stellite alloy No. 1. (a) As-cast rod. (b) As-laser clad, showing refinement of carbides. Processing conditions are listed in Table 2. Etchant: Murakami's reagent

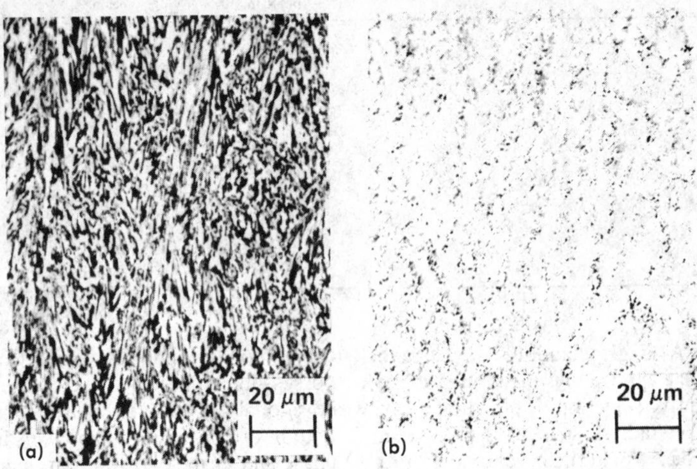

Fig. 6 Silicon-enriched cladding on AA 390 aluminum alloy. Microstructure of cladding containing angular primary silicon particles of HK 980 kg/mm^2. Processing conditions are listed in Table 2. Etchant: Keller's reagent

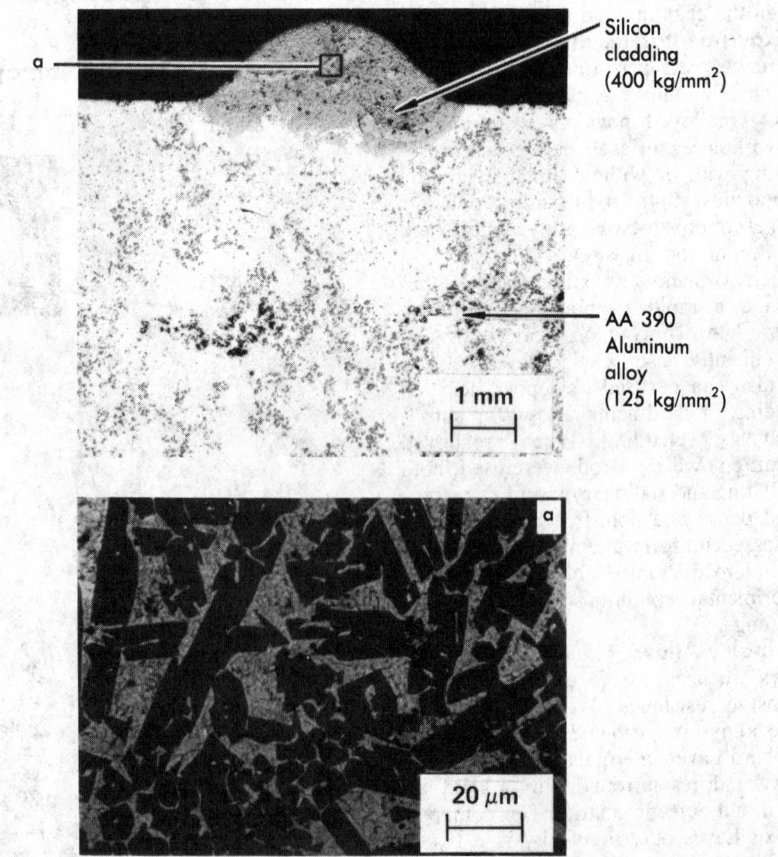

Fig. 7 Cladding of tungsten carbide particles embedded in a hard dense matrix on AISI 1018 steel. Processing conditions are listed in Table 2. Etchant: nital, followed by Murakami's reagent

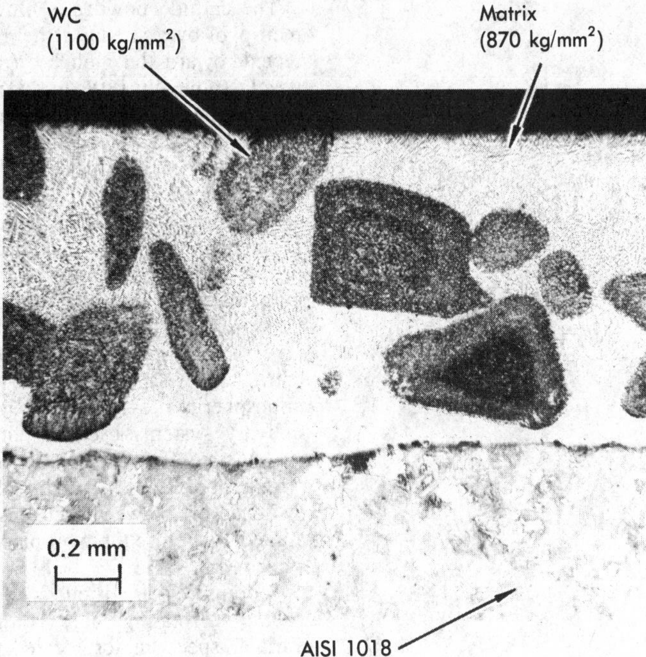

tween liquid iron and tungsten carbide during processing.

Alumina. High melting and stable ceramic materials such as alumina (melting point of 3750 °F) can be applied over aluminum alloys. By adapting the experimental conditions listed in Table 2, it was possible to melt the alumina powder (0.01 mil size) and form a dense alumina cladding over a 2219 aluminum alloy. Figure 8 shows Knoop hardness indentations at 100-g load; hardness readings in the cladding ranged between 2000 and 2300 kg/mm^2. It is not clear how well the alumina cladding is bonded to the substrate, or how the cladding behaves during thermal cycling. Angular-shaped voids are probably caused by chunks of alumina cladding that fell out during specimen preparation, because of formation of cracks

Fig. 8 Cladding of alumina on 2219 aluminum alloy. Microstructure shows Knoop hardness impressions at 100-g load. Processing conditions are listed in Table 2. Unetched

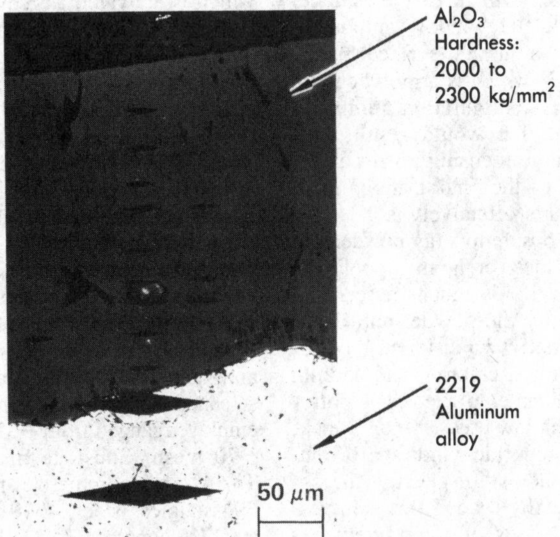

during processing. Although these experiments were only partially successful, they indicate promise in applying dense ceramic claddings on metallic substrates.

Laser Melt/Particle Injection

The laser melt/particle injection process bears a resemblance to the industrial practice of pouring tungsten carbide particles into a moving weld pool (a practice of secondary importance compared to weld cladding using carbide loaded rods), but the laser process offers far greater versatility. It is applicable to a wide range of material combinations because of the way in which the carbide particles are introduced into the melt.

Process Description. The method for injecting particles into surfaces is shown schematically in Fig. 9, where a converging laser beam melts a liquid pool on the surface of a metal substrate moved relative to the beam. Powder particles are blown from a fixed nozzle into the melt pool and are incorporated into the surface as the trailing edge of the pool solidifies.

Processing generally is done in the vacuum chamber illustrated in Fig. 10. The chamber is mechanically pumped initially to a pressure of 8 to 38 mtorr, but entry of the carrier gas elevates the pressure during processing to 115 mtorr or more. Processing is not restricted to a low-pressure environment, as demonstrated by success in producing titanium carbide injected surfaces of good quality on aluminum alloys in air. Properly designed gas shielding permits the processing of all metals under ambient conditions.

A continuous wave carbon dioxide laser rated at beam power levels of up to 15 kW was used. Although injection has been ac-

Fig. 9 Particle injection process. Injection of particles into a melt zone established by a high-power laser. The particles are carried by a stream of helium.

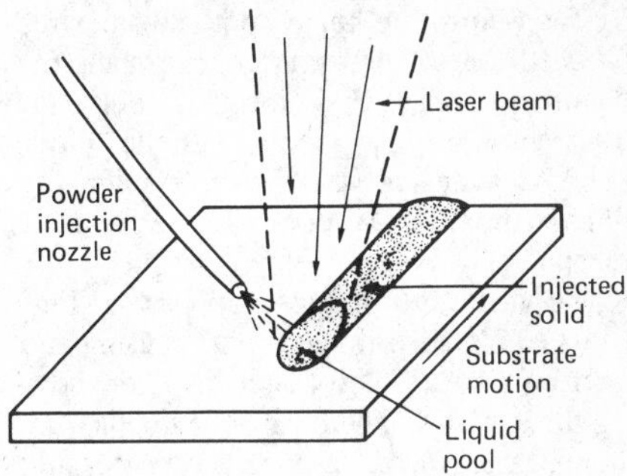

Fig. 10 Vacuum processing chamber with optical setup. The salt flat is tilted to prevent the partially reflected beam from entering the laser. *M*1: flat mirror. *M*2: focusing mirror. *M*3: final mirror

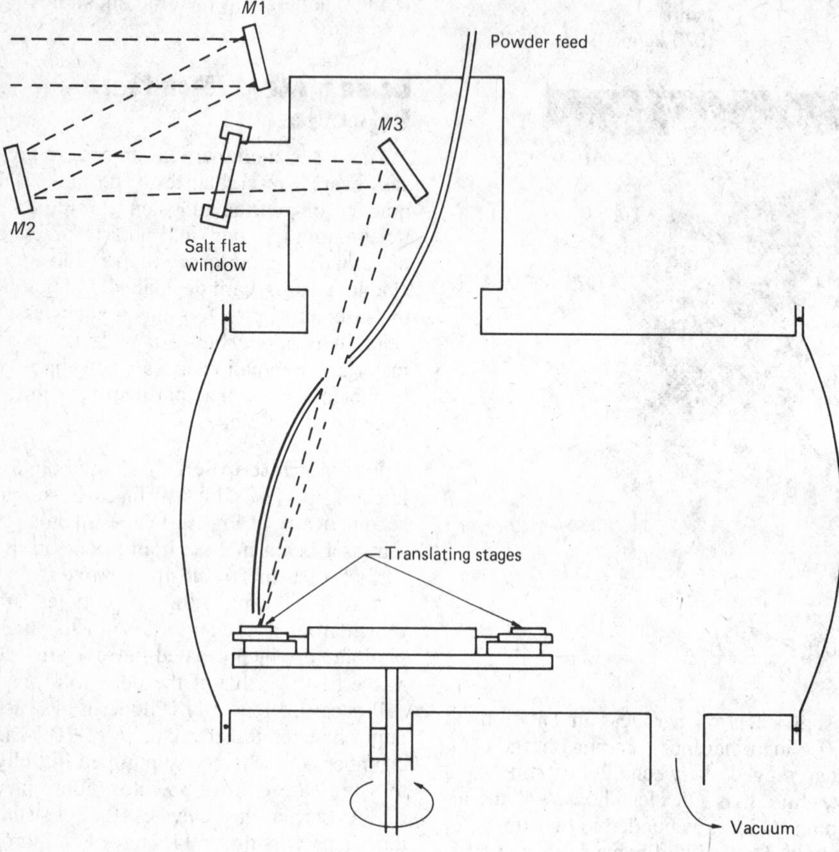

complished at powers ranging from 2 to 16 kW, most good structures are produced at power levels between 4 and 7 kW. The laser was configured to run with different energy density profiles in the output beam, using both the "tophat" and the unstable resonator mode. A "tophat" beam is produced by inducing multiple standing waves within the lasing cavity so that the output beam has a nearly uniform density profile. The beam from an unstable resonating cavity is nearly single mode, but it has a minimum in energy density at the center of the beam. The unstable resonator mode is preferable, because it can be more easily focused to the beam spot sizes desired (about 0.08 in.). A single-mode gaussian beam, produced by inducing only the transverse electric and magnetic mode $(TEM)_{00}$ standing wave, gives results similar to those achieved with the unstable resonator.

The vacuum processing apparatus (Fig. 10) is equipped with (1) beam focusing and steering mirrors, and (2) a stage for moving the samples. The beam is brought from the exit port of the laser to the processing apparatus by a series of metal mirrors that produce a 3-to-4-in.-diam beam of moderate divergence. At the apparatus, the beam is deflected by the flat mirror (*M*1 of Fig. 10), at a low angle of incidence to the focusing mirror, *M*2, which has a 3-ft focal length. The beam enters the chamber through the salt flat window and is directed toward the sample surface by the final mirror, *M*3, which is a molybdenum flat. This optical configuration minimizes the astigmatism that would result from employing a single focusing mirror at position *M*3. Placing the final turning mirror inside the chamber effectively isolates the salt flat from exposure to stray powder and metal vapor produced at the melt pool. The molybdenum mirror is much more resistant to damage by the powder and vapor than is the salt flat. Two different types of salt flats, potassium chloride and sodium chloride, have been employed. The potassium chloride window used is 0.3 in. thick, and the sodium chloride window is 0.6 in. thick. These windows, especially the sodium chloride, exhibit good service lives, surviving many days of intermittent experimentation. The molybdenum mirror inside the chamber is cleaned every few days.

The carbide powder is blown into the melt pool by a stream of helium. It is directed toward the melt by a fine copper nozzle (typically 0.06 in. ID) positioned approximately 0.4 in. from the melt. The carbide powder is fed into the carrier gas by an experimental metering device or by a commercial powder feeding device designed for use with plasma spray torches. When operating in a vacuum, the pressure of the carrier gas is determined by a vacuum regulator that generally is set at a pressure of 38 to 150 torr, absolute. When processing is done at ambient pressure, the carrier gas is set at 115 to 450 torr above atmospheric pressure. With the powder delivery systems used for this work, a steady-state flow of powder is achieved within about 1 s. The powder flow is switched off between melt passes and restarted 1 s before the workpiece enters the laser beam.

Workpieces are rotated under a laser beam. A rotating table is driven by a variable-speed motor located outside the chamber. Table speed is measured by counting pulses from a light chopping wheel mounted on the rotary motion feedthrough. Two sample stages on the wheel are driven by a leadscrew mechanism which drives both stages inward or outward in unison. This arrangement maintains the balance of the table, permitting the high-speed rotations necessary for rapid solidification experiments. A leadscrew device advances the workpieces an incremental amount between melt passes, the amount of advance determined by the number of pins engaged by the drive gear. The pins are mounted on a solenoid-driven arm, which is activated as required.

The setup used to do the particle injection processing described above can be used, without modification, to do laser cladding and surface alloying, making possible the formation of a wide range of surface structures with a single facility. By introducing the cladding material or the alloying elements in powder form through the nozzle shown in Fig. 9, these operations can be accomplished without precoating the surface of the workpiece.

Alloying Materials. The processes described above employed tungsten carbide and titanium carbide powders that were screened into sized fractions. Fraction sizes employed ranged from −325 mesh to +140 −70 mesh. Substrate materials used are Ti-6Al-4V, commercial-purity titanium, 304 stainless steel, 4340 tool steel, 1018 steel, Inconel X-750, aluminum alloys

5052, 6064, and 2024, and an aluminum bronze. Most samples were 2.5 by 4 by 0.2 in. The only surface preparation of the metals normally employed was solvent cleaning, though some materials were smoothed by surface grinding.

Many other metal and particle combinations are possible, because the only fundamental limitation on the process is that the metal must not exhibit excessive vapor pressure under the conditions of laser melting. Despite the wide range of alloy compositions used in the processes, the conditions necessary to achieve good volume fractions of carbide are surprisingly constant. This similarity in processing conditions results in part from carrying out carbide injection under soft vacuum conditions, thus eliminating oxidation problems without introducing problems associated with the use of shielding gases.

Microstructures. The isolated melt pass shown in Fig. 11 was produced by injecting titanium carbide into Ti-6Al-4V. This sectional view clearly demonstrates that the injection process produces not a coating, but a particulate composite surface in which the base metal serves as the matrix phase. The injected particles displace part of the base metal upward to produce the bulge evident in the sectioned surface.

On the cut surfaces in Fig. 11, the metal matrix between the injected carbide particles appears to be substantially darker than either the unmelted base metal or the injected titanium carbide particles. This contrast arises from the presence of fine carbide particles produced in the metal matrix by laser processing, not from an alteration in alloy chemistry. These fine carbide particles are evident in Fig. 12, which shows higher magnification views of the melt pass in Fig. 11. The fine carbide particles in these photographs are shown standing in relief above the metal matrix, which was deeply etched. These fine dendritic titanium carbide particles were formed during solidification of the molten metal matrix—a melt that had partially dissolved the injected carbide particles. Carbide dissolution and regrowth frequently exhibits cracking of the injected layer.

Fig. 11 Laser melt pass in a sample of Ti-6Al-4V. Injected with −170 +200 mesh titanium carbide. This experiment employed a laser power of 6 kW and a sample translation speed of 2 in./s

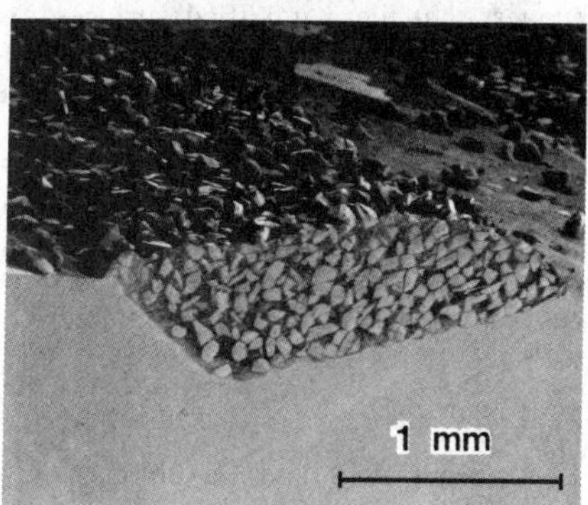

Fig. 12 Higher magnification views of the sample in Fig. 11. The fine titanium carbide particles partitioned from the metal matrix during solidification.

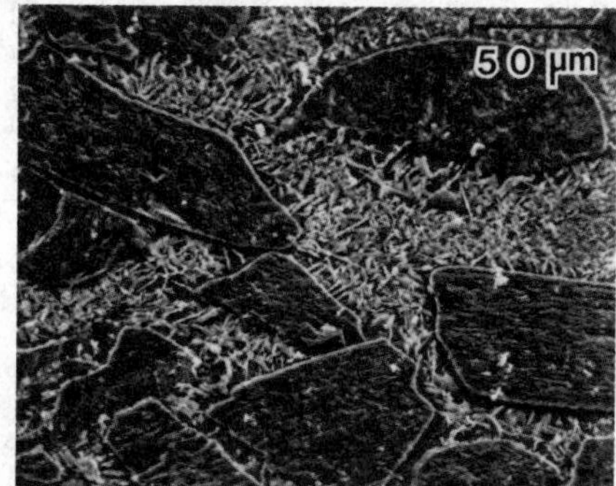

In titanium-based alloys, the titanium carbide particles grown from the melt tend to be relatively uncomplicated. They assume the somewhat feathery shape evident in Fig. 12 when grown from a melt only moderately enriched in carbon, and they assume the traditional ⟨100⟩ growth direction morphology expected of cubic crystals when the melt is highly supersaturated in carbon. An example of the latter morphology is shown in Fig. 13. Note that the higher carbon concentration in the melt also results in a greater tendency for the resolidified titanium carbide to grow epitaxially from the injected carbide particles. This microstructural difference accounts for the tendency to crack in melt passes that are rich in dissolved carbides.

Iron-based alloys injected with titanium carbide exhibit a wider range of microstructures than do titanium-based alloys. This greater complexity is evident in Fig. 14, which shows a portion of a 304 stainless steel workpiece that was injected with titanium carbide. In this melt pass, only slight dissolution of the titanium carbide occurred, as evidenced by the small volume fraction of titanium carbide that partitions from the melt upon solidification. Within this one small area, however, there are four distinct morphologies of carbide grown from the melt. One of these morphologies is evident in Fig. 14(a) as

Fig. 13 Dendritic titanium carbide. Formed from a melt that had dissolved a substantially higher fraction of the injected carbide than the melt pictured in Fig. 11 and 12.

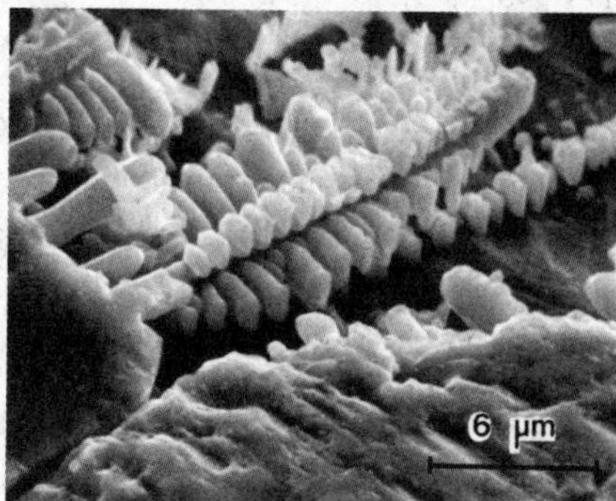

Fig. 14 Injected and resolidified titanium carbide in a matrix of 304 stainless steel

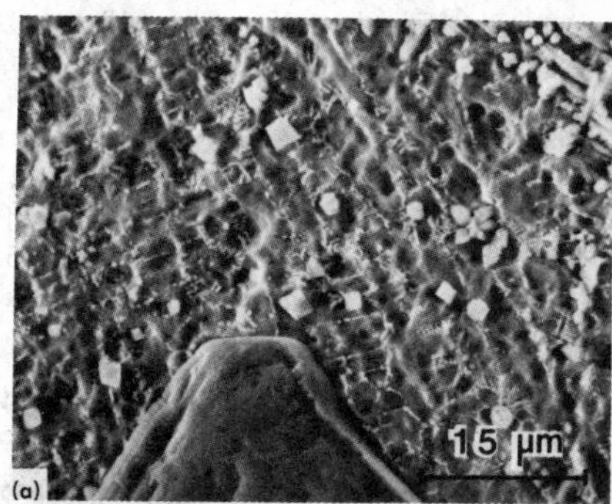

an epitaxial layer on the surface of the large injected titanium carbide particle at the bottom of the photo. The other forms present are the dendritic particles in the upper right corner of Fig. 14(a), the cuboidal particles evident in both photographs, and the fine, somewhat lacy particles that are most evident in Fig. 14(b).

The first three of these growth forms were produced by solidification within a titanium carbide plus liquid phase field in the multicomponent phase diagram, whereas the fine, lacy carbides were partitioned from the austenitic steel as it solidified at a lower temperature. The cuboidal carbide particles differ from the dendrites only in that they formed in a portion of the melt that was locally less saturated in carbon and titanium, so that the tips of the cuboids did not grow rapidly enough to form dendrite branches.

In 304 stainless steel melts injected with titanium carbide under conditions that result in a greater dissolution of titanium carbide, far more dendritic carbides are formed during solidification of the melt. Also, the titanium carbide that is regrown on the injected carbide particles has a dendritic morphology similar to that seen in Fig. 13. As in the titanium-based alloys, these differences in morphology create a tendency for surfaces with a greater degree of carbide dissolution to form cracks during laser processing.

Aluminum. Aluminum alloy 5052 shows no microstructural evidence of titanium carbide dissolution during laser processing. For this reason, the alloy retains good ductility, and a wide range of processing conditions can be employed without producing any cracks or other observable structural defects. As a result, very uniform injected layers can be produced by overlapping the melt passes. Figure 15 demonstrates this point with a workpiece that was injected with +200 −170 mesh titanium carbide in a series of 0.12-in.-wide melt passes that were consecutively overlapped by 0.08 in. The upper surface (more darkly shaded in this scanning electron microscope photo) was ground and lightly polished, steps necessary before using treated surfaces in many applications, because as-processed surfaces are relatively rough.

Fig. 15 Sample of 5052 aluminum alloy. Injected with −170 +200 mesh titanium carbide. In this scanning electron microscope view, the upper surface was ground and lightly polished.

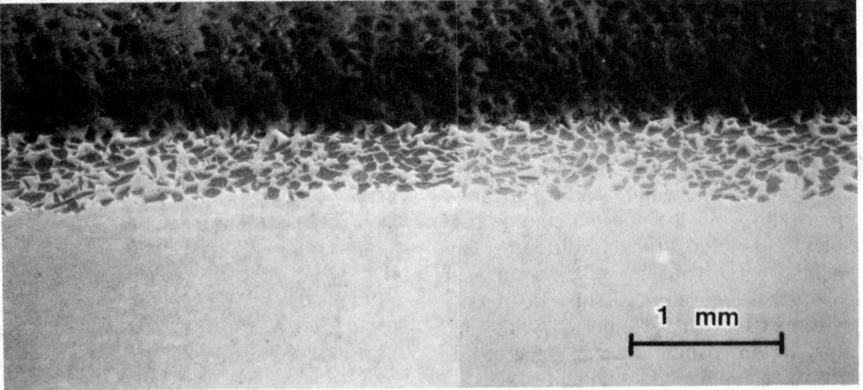

Mechanical Characterization

Surface layers produced by the particle injection process can greatly improve the wear resistance of metals. Three types of testing have been done to date: (1) hardness testing (micro and macro), (2) coefficient of friction measurements, and (3) dry sand/rubber wheel wear testing.

Both microhardness and macrohardness testing is necessary to characterize these surfaces because of their composite nature. Surface hardness can be altered by this process by at least three mechanisms, in addition to those of transformation hardening and microstructural refinement. In alloys that partially dissolve the injected carbide particles, the resolidified metal can be dispersion hardened by fine resolidified carbides and through solid solution strengthening, if part of the dissolved carbide is retained in solution. In all alloys, macrohardness is increased because of the impediments to plastic flow presented by large injected carbide particles. Micro- and macrohardness testing of Ti-6Al-4V and of 304 stainless steel, two alloys that should not exhibit transformation hardening, demonstrates that the partial dissolution and repartitioning of titanium-carbon in the metal matrix contributes to hardening of the injected surface layer, but that this effect is small compared to the transformation hardening produced in a 4340 tool steel by the same processing. On the other hand, the aluminum alloy 5052 metal matrix is softened by laser processing (from a Vickers hardness of 78 to 71), but the injected carbide particles still increase the macrohardness from 50 to 61 HRB.

Carbide injection processing is valuable in applications involving dry sliding wear because metals such as titanium alloys, which experience galling, can have their coefficient of friction reduced by the injection of titanium carbide. In the as-received condition, polished samples of Ti-6Al-4V and 304 stainless steel both showed kinetic coefficients of friction (μ_k) between 0.40 and 0.45 when tested in sliding contact with a 52100 steel ball, but injection of titanium carbide reduced the μ_k for both. It was found that μ_k drops with increasing volume fraction of titanium carbide, achieving a minimum value of 0.18 at a volume fraction of approximately 0.5. At this volume fraction of titanium carbide, the steel slider contacts only titanium carbide in the polished surfaces, so the measured μ_k is independent of the nature of the base metal.

The only wear testing on these materials employed the dry sand/rubber wheel test. Under standard test conditions, the laser-processed samples performed substantially better than did the same alloys in the as-received condition. The processed surfaces exhibit wear rates only a few percent of those of the untreated alloys, and they compare favorably with commercial wear-resistant alloys. Even aluminum alloy 5052 injected with titanium carbide has a wear volume loss of only 0.00055 in.^{3} after 1000 revolutions of the rubber wheel, a volume removal rate that is in the middle of the reported range of values for cast cobalt- and iron-based wear-resistant alloys.

Applications. The laser melt/particle injection process is capable of producing particulate-hardened surfaces on a wide range of metallic alloys. Microstructural characterization and mechanical test data suggest that these modified surfaces exhibit a significant potential for use in applications involving abrasive or adhesive wear. They also may be applicable to situations involving other sorts of wear. Preliminary results suggest that, in addition to applications involving the production of wear-resistant metal surfaces, the process may be suited for the manufacture of carbide-hardened cutting tools.

REFERENCES

1. Arnoldy, R.F. and Reynolds, G.H., Iron-Chromium-Carbon Hardfacing

With the Bulk Weld Process, *Metal Progress,* Vol 112 (No. 6), 1977, p 31
2. Hume-Rothery, W., Ed., *The Structures of Alloys of Iron,* 1st ed., Pergamon Press, Oxford, 1966, p 345

SELECTED REFERENCES

- Anthony, T.R. and Cline, H.E., Surface Rippling Induced by Surface-Tension Gradients During Laser Surface Melting and Alloying, *Journal of Applied Physics,* Vol 48 (No. 9), 1977, p 3888
- Ayers, J.D., Modification of Metal Surfaces by the Laser Melt/Particle Injection Process, *Journal of Thin Solid Films,* Vol 84, 1981, p 323
- Ayers, J.D., Particulate Composite Surfaces by Laser Processing, *Lasers in Metallurgy,* K. Mukherjee and J. Mazunder, Ed., Metallurgical Society of AIME, Warrendale, PA, 1982
- Ayers, J.D., Tucker, T.R., and Bowers, R.C., A Reduction in the Coefficient of Friction for Ti-6Al-4V, *Scripta Metallurgica,*Vol 14, 1980, p 549
- Ayers, J.D., Tucker, T.R., and Schaefer, R.J., Wear Resisting Surfaces by Carbide Particle Injection, *Rapid Solidification Processing—Principles and Technologies, II,* R. Mehrabian, B.H. Kear, and M. Cohen, Ed., Claitor's Publishing Division, Baton Rouge, 1980, p 212
- Beer, S.Z., Ed., *Liquid Metals Chemistry and Physics,* 1st ed., Marcel Dekker, Inc., New York, 1972, p 171-184
- Bhansali, K.J. and Silence, W.L., Metallurgical Factors Affecting Wear Resistance of Metals, *Metal Progress,* 1977, p 34
- Breinan, E.M., Kear, B.H., and Banas, C.M., Processing Materials With Lasers, *Physics Today,* Vol 29 (No. 11), 1976, p 44
- Cunningham, F.E., The Use of Lasers for the Production of Surface Alloys, M.S. Thesis, Massachusetts Institute of Technology, Cambridge, MA, 1964
- Gnanamuthu, D.S., Cladding, United States Patent No. Re 29815, Oct 24, 1978
- Irons, G.C., Laser Fusing of Flame Sprayed Coatings, *Welding Journal,* Vol 57 (No. 12), 1978, p 29
- Schaefer, R.J., Tucker, T.R., and Ayers, J.D., Laser Surface Melting With Carbide Particle Injection, *Laser and Electron Beam Processing of Materials,* C.A. White and P.S. Peercy, Ed., Academic Press, New York, 1980, p 749
- Schmidt, R.D. and Ferriss, D.P., New Materials Resistant to Wear and Corrosion to 1000 °C, *Wear,* Vol 32 (No. 3), 1975, p 279
- Seaman, F.D. and Gnanamuthu, D.S., Using the Industrial Laser to Surface Harden and Alloy, *Metal Progress,* Aug 1975, p 67
- Silence, W.L., Studies of Cast-Wear Resistant Alloys, Technology Dept. Report No. 8165, Stellite Division, Cabot Corp., Kokomo, IN, 1972

Weld Overlays

By the ASM Committee on Weld Overlays*

MANY METALLIC assemblies are constructed with coatings of different metallic or nonmetallic materials. The purpose of the coating is to impart some desired property to the surface of the article that is not intrinsic to the underlying base metal. Paint on steel is a common form of nonmetallic coating that imparts rust resistance to steel surfaces. Galvanize and terne are metallic coatings that protect steel from rusting. A weld overlay is defined as the deposit of a dissimilar weld metal laid on the surface of a metal part to obtain a desired property inadequate in the base metal. A weld deposit of stainless steel laid on the interior surface of a low-alloy steel chemical plant vessel for improved corrosion resistance is an example of a weld overlay.

Weld overlays fall into the general category of weld surfacing, which includes hardfacing and buttering. Hardfacing by arc welding, thermal spraying, and laser alloying methods are covered in the article "Hardfacing" in this Volume; buttering is a special technique involving weld surfacing the edges of a weld groove or plate surface prior to joining. The term weld overlay, also known as weld cladding, usually denotes the application of a relatively thick layer ($^1/_8$ in. or more) of weld metal to impart a corrosion-, erosion-, or wear-resistant surface. Hardfacing produces a thinner surface coating than a weld overlay and is normally applied where hardness is the primary criteria for wear or impact resistance. Metallic materials produced by weld surfacing are classified as composites.

In addition to weld overlays that develop a composite structure by a fusion welding process, there are many processes by which composite structures may be produced. A short description of a number of these processes follows.

Roll Cladding. In this process, a steel plate and a sheet of cladding material are welded together around the edges. The sandwiched material is heated to a high temperature and rolled until the cladding and base plate join together. The initial weld along the edges excludes air from the interface and prevents slippage of the two parts during rolling.

Explosion cladding is generally used to clad flat plates, but it can be used to clad the inside or outside of cylinders. In this process, the clad plate is separated from the base plate by an established standoff distance. The clad plate is covered with a suitable explosive material. Detonation of the explosive produces a progressive collision between the clad and base plate at an angle and velocity that breaks up and expels the surface films and produces a wave-like interface bond. This process is uniquely suited to cladding of steel with titanium or aluminum, metals that do not clad by the fusion welding processes.

Sheet and Strip Liner Cladding. Sheets or strips of a metal of the desired properties may be used to line the interior of a vessel or pipe or to cover a structural base plate. The sheets or strips may be fastened to the base metal with few or with many attachment welds. The attachment welds, if they do not have the desired properties, may be covered under overlapping zones of the sheets or strips, which are welded together to produce a continuous, impenetrable liner or cover.

Braze Cladding. In this process, a sheet of clad metal and a foil sheet of brazing alloy are sandwiched with the base metal under the platens of a press. Brazing heat is applied to the sandwich while in a suitable atmosphere or vacuum. Another variation includes placing the sandwich in a steel sheet envelope that can be evacuated and brought to brazing heat in a furnace or salt bath. In these processes, the base plate may be clad on both sides and two or more plates may be clad at one time.

Thermal Spraying. The following paragraphs are a definition and description of thermal spraying, abstracted from the American Welding Society's *Welding Handbook,* Volume 3, 7th edition.

"Thermal spraying (THSP) is a process in which a metallic or nonmetallic material is heated and then propelled in atomized form onto a substrate. The material may be initially in the form of wire, rod, or powder. It is heated to the plastic or molten state by an oxyfuel gas flame, by an electric or plasma arc, or by detonation of an explosive gas mixture. The hot material is propelled from the spray gun to the substrate by a gas jet. Most metals, cermets, oxides, and hard metallic compounds can be deposited by one or more of the process variations. The process can also be used to produce free-standing objects using a disposable substrate. It is sometimes called 'metallizing' or 'metal spraying.'

"Thermal spraying is widely used for surfacing applications to attain or restore desired dimensions; to improve resistance to abrasion, wear, corrosion, oxidation, or a combination of these; and to provide specific electrical or thermal properties. Frequently, thermal sprayed deposits are applied to new machine elements to provide surfaces with desired characteristics for the application." For more information

*R. David Thomas, Jr., *Chairman*, President, R.D. Thomas & Co., Inc.; J.J. Barger, Supervisor, Prototype Equipment, Combustion Engineering, Inc.; Howard N. Farmer, Director, Metallurgical Services, Stoody Co.; Jerald E. Jones, Assistant Professor, Department of Metallurgical Engineering, Colorado School of Mines; William E. Layo, Manager of Technical Services, Sandvik, Inc.; John J. Meyer, Chief Welding Engineer, Nooter Corp.; Louis E. Stark, Welding Engineer, Babcock & Wilcox

on thermal spraying, see the article "Hardfacing" in this Volume.

The economic value of composites is clearly seen in cases where they provide inexpensive low-alloy steel with a high degree of resistance to a wide range of corrosive environments. Composites make significant savings in the quantity of the more expensive corrosion-resistant alloys used in the construction of large or thick-walled structures. Low-strength alloys that may be ruled out for certain high-strength applications may be useful when joined to a higher strength steel.

Applicability

A wide variety of alloys are commercially available for weld overlay applications. Conditions that affect the selection of a particular composition of filler include: (1) expected service conditions; (2) base metal and possible dilution effects; (3) availability of the particular filler material desired in a form compatible with the available welding process; and (4) matching welding filler metal with the other welding consumables, such as flux or shielding gas, to be used. Filler material for weld overlay application can be divided into four categories: corrosion-resistant materials, wear-resistant materials, high-temperature materials, and finally, alloys for wear resistance in corrosive environments.

Purposes of the Overlay

Most weld overlays are made to impart to the surface of one material the properties of another material. The particular service conditions dictate the category of material to be selected. One exception, the buildup of worn or corroded parts to the original dimensions, is often done with a filler metal of essentially identical composition to the base metal.

Service conditions must be carefully defined so that the specific purposes of the overlay can be determined and material selection accomplished. Often, the requirements for the overlay conflict, so that a compromise must be made in material selection. For example, the surfacing of valve seats in a nuclear reactor might best be done with a cobalt alloy for wear resistance; however, because cobalt can become radioactive, the problem of residual radioactivity may be severe enough that a compromise material is selected. The basic types of service conditions that can be overcome by weld overlays are discussed below.

Corrosion can be defined as the removal of material or degradation of the mechanical properties of a metal by chemical or electrochemical processes, or by a combination of chemical, electrochemical, and mechanical processes. The principal forms of corrosion include:

- *Simple dissolution* of material, often associated with the formation of a corrosive product on the surface. If carefully selected for the service environment, the overlay material may form a corrosion product that protects the surface from further corrosion.
- *Selective leaching* occurs when one component of an alloy is soluble in the corrosion medium. Brass alloys are susceptible to this form of corrosion, with the zinc typically being selectively removed from the surface of the overlay, leaving a porous copper layer that has poor mechanical properties and continues to grow until leaking or mechanical failure results.
- *Galvanic corrosion* is the result of two dissimilar metals in the same corrosion medium being in electrical contact with each other. The galvanic series can be used as a guide to the dissimilarity of two metals in a galvanic couple. Two metals, either of which might be corrosion resistant in a given environment, can be electrically coupled in the same environment with disastrous results. The anode in the galvanic couple corrodes, with the resultant evolution of hydrogen at the cathode. The corrosion rate is determined by the galvanic dissimilarity of the two metals and the ratio of cathodic to anodic surface area exposed. Consequently, a base metal should be surfaced with a metal that is anodic to it so that small pores in the overlay expose only small areas of cathodic base material, resulting in a small cathode-to-anode surface area ratio.
- *Pitting corrosion* is the result of a localized corrosion occurring on the surface of a metal. Often, the pit begins with the breakdown of a small area of a protective oxide film on the surface of a material that depends on such a film for corrosion resistance. The local environment in the pit may exhibit a more concentrated corrosive material; thus, the pit growth usually accelerates with time, making a prediction of remaining life impossible.
- *Intergranular corrosion* occurs when the chemistry of the grain boundaries of a material is different from the bulk chemistry of the alloy. The corrosive environment can attack the grain boundaries even though the nominal composition of the metal should be expected to withstand attack. This form of corrosion may be the most dangerous because very little indication of the corrosion is observable. Almost no weight loss occurs and, other than some surface roughening, no visible corrosion product or material loss may be present. Such corrosion is observed in many alloy systems, notably stainless steel or precipitation-hardening aluminum alloys that have been improperly heat treated or welded, including weld overlays.
- *Stress corrosion cracking* is the result of a combination of applied tensile stress and a corrosive environment. A material that would, in the absence of stress, be impervious to attack in a given corrosive medium can often be subject to rapid stress corrosion cracking when an applied tensile load exists. The residual stress associated with a welding operation, including a weld overlay application, may be sufficient tensile stress to cause stress corrosion cracking.
- *Hydrogen* damage may occur in the form of hydrogen embrittlement, hydrogen blistering, or hydrogen sulfide cracking. The materials most susceptible are the high-strength iron alloys, so the selection of a high-strength steel for buildup or wear resistance should be reconsidered if significant hydrogen concentrations are present (see Volume 10 of the 8th edition of *Metals Handbook* for more information on hydrogen damage).

Wear is defined as the removal of material from a surface due to mechanical processes. Wear can be divided into several major categories, depending on the process that causes the problem. The weld overlay (buildup or hardfacing) material and process should be selected for the particular wear situation. The most common wear types include:

- *Abrasive wear* is caused by the sliding of an abrasive material across the surface being worn.
- *Impact wear,* often associated with either low- or high-stress abrasion, occurs when the material being worn experiences high-energy blows from the wearing material.
- *Adhesive wear* is generally characterized by the rubbing of two metal surfaces against each other, or possibly against a third material between them.
- *Erosion* is the removal of material due to the impingement of a flowing fluid medium.

- *Cavitation* is a process that also occurs in flowing liquid mediums, but is significantly different from erosion. If the flow is disrupted, turbulence results. Turbulent regions are characterized by regions of extremely low pressure in which voids can form. When those voids move to regions of higher pressure, implosion occurs, and the resulting shock wave causes removal of material from the surface of the adjacent metal.

This article deals mainly with weld overlay applications for corrosive and high-temperature environments. For more detailed information on welding to resist wear, see the article "Hardfacing" in this Volume.

High-temperature service may constitute a combination of wear, corrosion processes, and stress-related processes at elevated temperature. Oxidation of materials at elevated temperatures may not follow the expected trends developed from low-temperature information. Diagrams of oxide stability have been developed for many of the metals that are typically used in engineering alloys for elevated-temperature service. The composition of the atmosphere of service and the interaction of various components of the alloy chosen for the particular application must be known before the material/atmosphere interaction can be predicted. In addition, strength of the material may be an important consideration because mechanical properties of materials degrade with increased temperature within specific phase boundaries. Thus, unless elevated-temperature properties of a specific material are known, it may not provide the expected performance at elevated temperatures.

Types of Materials

Nearly any metal that can be used as a welding filler metal can be used as a weld overlay. With the exception of simple buildup situations, where the overlay is used to restore the original dimensions of a worn or corroded part, the composition and properties of the filler metal are often quite different from the base metal. The problems associated with dissimilar metal welds are commonly found in overlay operations and must be considered when selecting the type of material to be used. A discussion of the various materials that may be used for weld overlay applications follows; however, this discussion may not include all of the materials that could be useful in certain specialized applications, only the ones that are typical.

Stainless steels are commonly used for their corrosion and oxidation resistance. Table 1 lists some of the filler metals for stainless steel weld overlays. In certain moderately elevated-temperature applications, stainless steels have provided good service. Stainless steels are corrosion resistant due to chromium oxide films that form on the surface, which protect the surface from further oxidation and many corrosive media. However, use of stainless steels in environments where the chromium oxide film is not stable, or in the sensitized condition, results in pitting, intergranular corrosion, or possibly general attack of the surface. In addition, in certain (particularly chloride-containing) environments, stainless steels are susceptible to stress corrosion cracking. More detailed discussions of stainless steels, their applications, and their limitations can be found in Volume 3 of the 9th edition of *Metals Handbook* and in the article "Arc Welding of Stainless Steels" in this Volume. For information on dilution and composition control of stainless steels, see the section on composition control of stainless steels in this article.

Table 1 Filler metals for stainless steel weld overlays

Weld overlay type	First layer: Covered electrode(a)	First layer: Bare rod or electrode(b)	Subsequent layers: Covered electrode(a)	Subsequent layers: Bare rod or electrode(b)
304	E309	ER309	E308	ER308
304L	E309L E309Cb	ER309L	E308L	ER308L
321	E309Cb	ER309Cb	E347	ER347
347	E309Cb	ER309Cb	E347	ER347
309	E309	ER309	E309	ER309
310	E310	ER310	E310	ER310
316	E309Mo	ER309Mo	E316	ER316
316L	E309MoL E317L	ER309MoL ER317L	E316L	ER316L
317	E309Mo E317	ER309Mo ER317	E317	ER317
317L	E309MoL E317L	ER309MoL ER317L	E317L	ER317L
20 Cb	E320	ER320	E320	ER320

Note: Columbium (Cb) is also referred to as niobium (Nb).
(a) Refer to AWS specification A5.4. (b) Refer to AWS specification A5.9.

Nickel and cobalt alloys comprise a large number of materials, including most of the superalloys. These materials are generally used for their high-temperature oxidation resistance and ability to retain good mechanical properties at elevated temperatures. Cobalt alloys are also used where high-temperature corrosion resistance combined with wear resistance is required. Certain alloys of the cobalt system avoid the thermal expansion mismatch problems associated with deposits on ferritic base metal, due to their relatively low coefficient of expansion. Additional information concerning nickel and cobalt alloys can be found in Volume 2 of the 9th edition of *Metals Handbook* and in the articles "Arc Welding of Heat-Resistant Alloys" and "Brazing of Heat-Resistant Alloys" in this Volume.

Copper alloys combine the properties of good corrosion resistance in certain environments with good adhesive wear characteristics, particularly in contact with iron alloys. Both silicon and aluminum bronze have good resistance to seawater and are often used in naval applications to provide sealing and wear surfaces in seawater. In addition, because copper alloys cannot become radioactive, the usefulness of these materials in nuclear applications is apparent. Both brass and bronze alloys have shown good performance in metal-to-metal wear applications and are used extensively to repair bearing surfaces on many applications. Volume 2 of the 9th edition of *Metals Handbook* and the articles "Arc Welding of Copper and Copper Alloys" and "Brazing of Copper and Copper Alloys" in this Volume contain additional information concerning the uses of copper alloys.

Irons, usually thought of as casting alloys due to the eutectic liquidus temperature depression, have many properties that make them useful as weld overlay materials. The addition of chromium to white iron causes the carbides formed upon solidification to be of the M_7C_3 type. These complex (Fe, Cr)$_7$C$_3$ carbides have hardnesses greater than those of many minerals and metal oxides and give good abrasive wear characteristics. In addition, in erosion, cavitation, and combination wear and corrosive environments, the wear resistance of the carbides coupled with the corrosion resistance of the chromium makes chromium white iron a good choice of overlay material. The addition of other hard carbides by inoculation of the weld pool or inclusion from the core of a tubular electrode can enhance the wear resistance of iron overlays. The use of iron overlay deposits for wear resistance is discussed in more detail in the article "Hardfacing" in this Volume.

Alloy steels can be used for overlay, particularly to build up to the original dimensions material that is worn or corroded. The buildup of car and locomotive wheels with steel has been used extensively in the mining industry; however, the choice of a relatively soft plain carbon steel versus a harder high-alloy steel may be influenced by the particular application and nature of the mineral content of the material that is trapped between the wheel and rail. Alloy steels are also used to rebuild metal mill rolls and may be used in applications for rolling and finishing of other products. See the article "Hardfacing" in this Volume for more information on the uses of alloy steels for wear-resistant applications.

Manganese alloys are used to provide wear resistance in a low stress abrasion situation. The work-hardening characteristics of these 14% Mn materials provide the resistance to low stress abrasion with a hardened surface layer. However, gouging the surface removes the work-hardened layer and makes this material unsuitable for gouging-type wear. The manganese alloys, sometimes known as Hadfield alloys, have an austenitic structure and give excellent impact resistance. In an impact wear situation, a manganese alloy is usually an excellent choice of material. The addition of 14% Cr to this material imparts good corrosion resistance in many environments. The use and application of high-manganese materials for wear-resistant applications are discussed in more detail in the article "Hardfacing" in this Volume.

Ceramics such as aluminum oxide are deposited as a weld overlay in applications requiring extremely high-temperature wear and corrosion resistance. Although the application of ceramics is relatively limited, they represent a material that is able to provide unique properties to alleviate problems not otherwise solvable. The use and application of ceramic materials is discussed in more detail in the article "Hardfacing" in this Volume.

Overlay Processes

With very few exceptions, virtually any welding process can be used for weld overlay; however, certain processes are more applicable. The selection of a process should include consideration of the purpose of the overlay, the required thickness of the overlay, and the cost and availability of the welding equipment and consumables: filler, flux, and shielding gas. The increase in available automated processes and the increasing cost of labor have caused a shift away from labor-intensive manual welding processes; however, applications where labor cost is a smaller percentage of the total welding cost, or where small areas or repaired areas are to be overlaid, manual processes are still used extensively. The following discussion includes those processes that are commonly used for overlay applications. More detailed information on these processes for stainless steel weld overlays can be found in the section of this article on procedures for stainless steel weld overlays. Table 2 gives some relative values of deposition rates, deposit thicknesses, and characteristics for four of the processes.

Submerged arc welding (SAW) is used extensively to clad vessels with corrosion-resistant materials or to surface rolls with a wear-resistant overlay. The SAW process is particularly useful in depositing layers of material with excellent surface finish in the as-welded condition, which requires very little, if any, machining. Submerged arc welding is a high-deposition-rate process that is capable of depositing a wide variety of alloys, because the alloying can be accomplished in the filler metal, in the flux, or by a combination of both. The use of strip electrodes rather than wire electrodes allows deposition of beads up to 10 in. in width, with 1.2, 2.4, and 3.5 in. (30, 60, and 90 mm) the most popular commercially available strip widths. The absence of a visible welding arc and the substantial reduction of fumes gives SAW a substantial advantage over many other methods, particularly in welder appeal and productivity. The SAW process is limited to use as an automatic process and has the substantial limitation that it can be employed only in the flat position. The use of work positioners and rotating the objects to be overlaid can partially overcome this limitation.

Gas metal arc welding (GMAW) and flux cored arc welding (FCAW) are also used for weld overlay operation. Both can be used as automatic processes and result in a moderate deposition rate that is better than many processes, but not as great as SAW unless used as multiple wires. When used in the semiautomatic mode, or when positioned by multi-axial robot systems, these processes are much more productive than shielded metal arc welding (SMAW), can be used to weld in all positions, and are very useful to produce final touch-up or repairs. Both processes have the disadvantages that they produce large quantities of welding fumes and require the welder and surrounding workers to have shielding from the visible arc.

Shielded metal arc welding (SMAW) is used extensively for weld overlay application. The process is versatile, allowing the deposition of material in areas where automatic equipment cannot fit, and requires very little capital investment in comparison to the automatic processes. A wide variety of filler materials are available for SMAW, and the process can be used economically for small-area overlay, for one-of-a-kind or unusually shaped surfaces, or for repair overlay. In addition, field application of overlay material is often done by SMAW.

Gas tungsten arc welding (GTAW) is used only in specialized applications for overlays. The GTAW process allows extremely close control of heat input and dilution; however, surfacing of large areas is not usually economical with this process. Fully automated GTAW systems are sometimes used for overlays of alloys sensitive to microfissuring, for example, high nickel-chromium alloys or fully austenitic stainless steels.

Plasma arc welding (PAW) is becoming more widely used for overlay applications. The addition of a powder spray to the plasma allows the deposition of a wide variety of overlay materials. Materials such as cobalt-tungsten carbide or ceramics, which cannot be produced in wire form, can be easily deposited by PAW. Typically, deposition rates are low, and the deposit thicknesses achieved are not very

Table 2 Process and deposit characteristics of four overlay processes

Welding process	Potential deposition rate, lb/h^{-1}	Usual deposit thickness, in.	Deposit character	Dilution in first layer, %	Bond type
Spray	>44	>0.15	Single layer(a) Multilayer	Essentially zero	Mechanical
Gas metal arc	>22(b)	0.4-0.8	Multilayer	8-50	Metallurgical
Submerged arc	>66	>3.9	Multilayer	8-50	Metallurgical
Electroslag	440/880	0.6-3.9	Single layer	>50	Metallurgical

(a) Spray and fuse. (b) Plasma GMAW >66 lb/h^{-1}
Source: *Weld Surfacing and Hardfacing*, The Welding Institute, Abington Hall, Abington, Cambridge, 1980

deep; consequently, the process is limited to applications requiring only relatively thin layers of deposit and usually only over limited surface areas. The plasma arc process has one additional limitation, which is that the equipment to produce PAW spray coatings is somewhat complex, expensive, and requires a relatively skilled operator.

Electroslag welding (ESW) is often overlooked as an overlay method. By the addition of powder material to the flux bath, or in a coating on a consumable guide, or as an addition to cored wire, a wide variety of overlays can be produced. The use of wide strip electrodes or multiple wires can produce virtually unlimited width passes, and the use of specially designed dams with a cutout on the side at the bottom allows surfacing to be accomplished in a horizontal orientation. However, the process is best suited to surfaces that can be oriented within 45 to 60° from vertical, producing overlays in thicknesses of 1/2 to 2 in.

Thermal spray processes can be effectively used to produce overlays with very good properties. For information on various thermal spray processes for wear-resistant applications, see the article "Hardfacing" in this Volume.

Other processes that can be used to produce overlay include laser, electron beam, and spark emission. Electron beam and laser beam welding processes can be used, just as with GTAW, with a cold or hot wire feed to produce overlays. In the flat position, both of these processes can be used to fuse a layer of powder into an overlay, and the laser is used experimentally to produce implantation of hard particles into the surface of a substrate for wear resistance, particularly in aluminum alloys. Spark emission is used experimentally to produce hard surfaces by fusing a powder into the surface of a base metal with very rapid solidification. Such processes are used primarily in very specialized applications and have a low deposition rate. However, for those specialized applications, no other process may be applicable in many cases.

Application Considerations

The technique of weld overlay is an excellent method to impart properties to the surface of a substrate that are not available from that base metal, or to conserve expensive or difficult-to-obtain materials by only using a thin surface layer on a less expensive or abundant base material. This technique has several inherent limitations or possible problems that must be considered when planning for overlay welding. The thickness of the required surface must be less than the maximum thickness of the overlay that can be obtained with the particular process and filler metal selected.

Fig. 1 Constitution diagram for stainless steel weld metal. Point 100 is A387A. Example: Point *X* on the diagram indicates the equivalent composition of a type 318 (316 Cb) weld deposit containing 0.07% C, 1.55% Mn, 0.57% Si, 18.02% Cr, 11.87% Ni, 2.16% Mo, and 0.80% Nb. Each of these percentages was multiplied by the "potency factor" indicated for the element in question along the axes of the diagram to determine the chromium and the nickel equivalent. When these were plotted as point *X*, the constitution of the weld was indicated as austenite plus from 0 to 5% ferrite; magnetic analysis of the actual sample revealed an average ferrite content of 2%. For austenite-plus-ferrite structures, the diagram predicts the percentage of ferrite within 4% for the following stainless steels: 308, 309, 309 Cb, 310, 312, 316, 317, 318 (316 Cb), and 347. Dashed line is the martensite/*M* + *F* boundary modification by Eberhard Leinhos, Mechanische Eigenschaften und GefüMgeausbildung von mit Chrom und nickel ligiertum Schweissgut, *VEB Duetscher Verlag fürMr Grundstoffindustrie*, Leipzig, 1966.

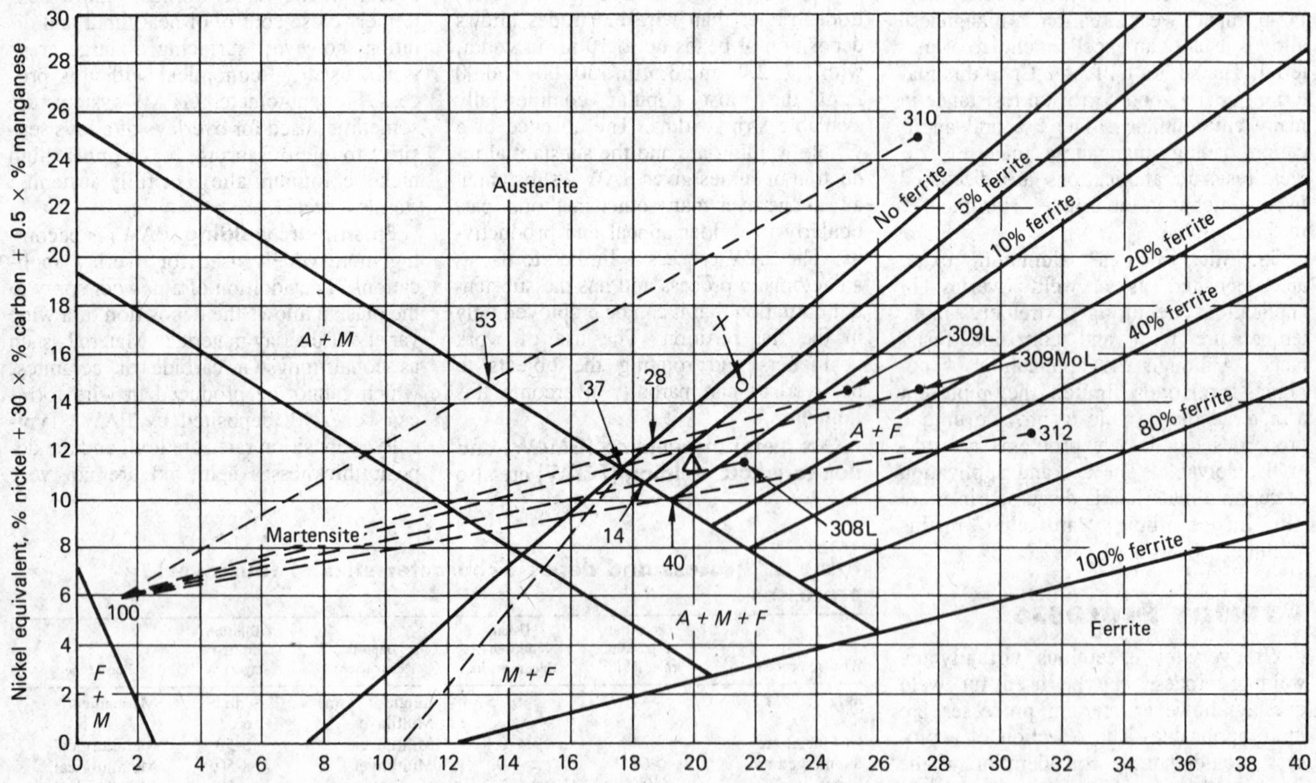

Fig. 2 Percentage of carbon in the deposit after base-metal carbon and cladding dilution factors have been established

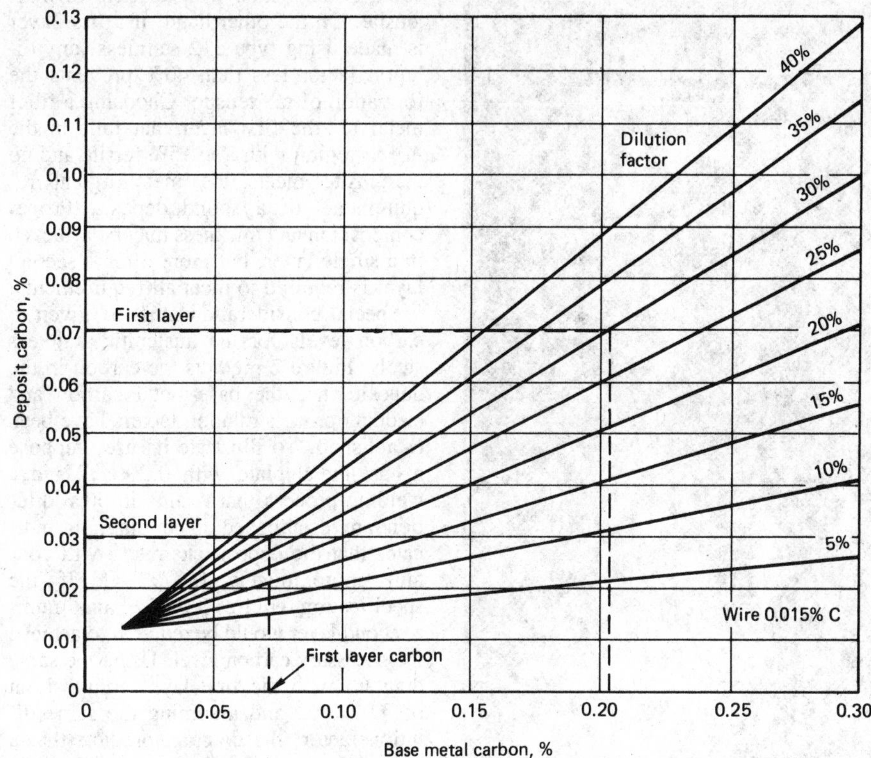

Many of the hardfacing materials are limited in thickness to only two passes due to spalling tendencies. In addition, when using a material that can only be applied by a spray process (such as a ceramic), the coating thickness is limited by the process itself, with the possible problem of poor bonding or disbonding between the layers if applied in excessive thicknesses.

Welding position also must be considered when selecting an overlay material and process. Certain processes are limited in their available welding positions; for example, SAW can only be used in the flat position. In addition, when using a high-deposition-rate process that exhibits a large liquid pool, welding vertically or overhead may be difficult or impossible. Some alloys exhibit eutectic solidification, which leads to large molten pools that solidify instantly rather than with a "mushy" (liquid plus solid) transition. Such materials are also difficult to weld except in the flat position.

Composition Control of Stainless Steel Weld Overlays

The economics of stainless steel weld cladding are dependent on achieving the specific chemistry at the highest practical deposition rate in a minimum number of layers. Selection of filler wire and welding process is the choice of the fabricator, whereas the purchaser specifies the surface chemistry and thickness, along with the base metal. The most outstanding difference between welding a joint and depositing an overlay is in the area of dilution. The percentage of dilution equals the amount of base metal melted (x) divided by the sum of base metal melted and filler metal added ($x + y$), the quotient of which is multiplied by 100.

$$\% \text{ dilution} = \frac{x}{x + y} \times 100$$

For stainless steel cladding, a fabricator must understand how the dilution of the filler metal with the base metal affects the

Table 3 Typical welding parameters for stainless steel weld overlays by SAW

Form of electrode	Size of electrode, in.	Power source	Current, A	Voltage, V	Travel speed, in./min	Oscillation, in.	Productivity, lb/h	Dilution(a), %
Single wire	1/16	DCEN(b)	240	34	5	3/4	9	15-20
	3/32	DCEN(b)	350	42	6	1 1/4	20	15-20
	3/32	ac	350	30	11	None	11	20-25
	1/8	ac	450	32	11	None	12	20-25
	5/32	ac	500	34	11	None	13	20-25
Series arc								
Two wires	5/32	ac(b)	400	26	10	None	41	20-25
Three wires	5/32	ac(b)	480	28	14	None	43	15-20
Six wire								
First wire	1/16	DCEP	255	27	5	5/8	83	15-20
Others	1/16	DCEN	1125	30	5	5/8	83	15-20
Strip	0.020 by 1.2	DCEP(b)	400	25	7	None	16	15-20
	0.020 by 2.4	DCEP(b)	700	27	7	None	32	15-20
	0.020 by 3.5	DCEP(b)	1250	27	5	None	60	15-20
	0.020 by 4.7	DCEP(b)	1550	27	5	None	78	15-20
	0.025 by 1	DCEP(b)	525	27	12	None	30	10-15
	0.025 by 2	DCEP(b)	1050	27	12	None	60	10-15
	0.025 by 4	DCEP(b)	2100	27	12	None	120	10-15

(a) Variations in current, voltage, and travel speed affect percentage of dilution. (b) Constant potential

Fig. 3 Strip weld overlay for a pressure vessel. Courtesy of Combustion Engineering, Inc.

composition and metallurgical balance, such as the proper ferrite level to minimize hot cracking, absence of martensite at the interface for bond integrity, and carbon at a low level to ensure corrosion resistance. The Schaeffler phase diagram shown in Fig. 1 is a useful tool in predicting deposit composition and ferrite levels for different filler metals for an established welding process. As an example of its use, suppose a type 304L clad surface is specified (shown as a triangle on the diagram). A desirable composition is approximately 20% Cr equivalent and 11% Ni equivalent. If an American Society for Testing and Materials (ASTM) A387A plate (point 100 on the diagram) is specified, all combinations of weld deposit lie on a line connecting point 100 with a proposed filler metal. As an example, note the line connecting point 100 with type 310 stainless steel. The diagram shows that weld-deposit compositions with up to 53% dilution (53% of the length of the line connecting the base metal and the filler metal) will be fully austenitic. Although sufficient alloy might be present, at least for a first layer, the absence of ferrite may make the deposit sensitive to fissuring. The same procedure can be used to estimate the usefulness of other filler metals.

Good metallurgical practice dictates that the first layer contain no martensite, shown as *M* in the diagram. This limits the dilution. If a process can be found that has less than 14% dilution, then a type 308L could be deposited with absence of martensite. On the other hand, if a first layer is made using type 312 stainless, any dilution factor less than 40% prevents the formation of martensite. Choosing a filler metal for the first layer that falls in the $A + F$ region with 3 to 15% ferrite and no martensite meets the metallurgical requirements of a sound deposit. Proper composition and thickness may be achieved in a single layer, but more often a second layer is required to meet all requirements.

Special consideration must be given to carbon levels for the austenitic stainless steels. Figure 2 predicts the carbon in the deposit once the base-metal carbon and cladding process dilution factors have been established. To illustrate its use, suppose a base-metal plate with 0.20% C and a welding process that results in 30% dilution are established. The diagram indicates that the deposit chemistry will contain approximately 0.07% C. If the specification requires 0.04% C maximum, a second layer would be required to achieve the necessary carbon level. Using the same diagram with the first layer composition of 0.07% C, and assuming the same dilution factor, the diagram predicts that a second layer will contain approximately 0.03% C. This would meet the specification of 0.04% C maximum.

Control of dilution plays an important part in the economics of the weld cladding process. Although each process has an expected dilution factor, experimenting with the welding parameters can minimize dilution. A value between 10 and 15% is generally considered optimum. Less than 10% raises the question of bond integrity, and greater than 15% increases the cost of the filler metal. Unfortunately, most welding processes have considerably greater dilution.

Because of the importance of dilution in overlay applications, each welding parameter must be carefully evaluated and recorded. Many of the parameters that affect dilution in overlay applications are not so closely controlled when arc welding a joint. These parameters include:

- *Amperage:* Increased amperage (current density) increases dilution. The arc becomes hotter, penetrates more deeply, and more base-metal melting occurs.
- *Polarity:* Direct current electrode negative (DCEN) gives less penetration and resulting lower dilution than direct current electrode positive (DCEP). Alternating current results in a dilution that lies between DCEN and DCEP.

- *Electrode size:* The smaller the electrode, the lower the amperage, which results in less dilution.
- *Electrode extension:* A long electrode extension for consumable electrode processes decreases dilution. A short electrode extension increases dilution.
- *Travel speed:* A decrease in travel speed decreases the amount of base metal melted and increases proportionally the amount of filler metal melted, thus decreasing dilution.
- *Oscillation:* Greater width of electrode oscillation reduces dilution. The frequency of oscillation also affects dilution—the higher the frequency of oscillation, the lower the dilution.
- *Welding position:* Depending on the welding position or work inclination, gravity causes the weld pool to run ahead of, remain under, or run behind the arc. If the weld pool stays ahead of or under the arc, less base-metal penetration and resulting dilution will occur. If the pool is too far ahead of the arc, there will be insufficient melting of the surface of the base metal and coalescence will not occur.
- *Arc shielding:* The shielding medium, gas or flux, also affects dilution. The following list ranks various shielding mediums in order of decreasing dilution: (1) helium (highest), (2) carbon dioxide, (3) argon, (4) granular flux without alloy addition, and (5) granular flux with alloy addition (lowest).
- *Additional filler metal:* Extra metal (not including the electrode) added to the weld pool as powder, wire, strip, or with flux reduces dilution by increasing the total amount of filler metal and reducing the amount of base metal that is melted.

Procedures for Stainless Steel Weld Overlays

Submerged Arc Welding. This is by far the most commonly used process for weld overlays. The process is adaptable for use with single wires as the electrode filler metal, multiple wires, or strip. Alternating current is often used, but direct current, with either reverse or straight polarity, has been found preferable in numerous instances. Oscillation of the filler metal is frequently found desirable. Typical welding parameters for stainless steel weld overlays by SAW are given in Table 3.

The single-wire process is largely used for overlaying limited areas or where it is necessary to limit the heat input. Wire sizes from $^1/_{16}$ to $^3/_{16}$ in. diam are practical. The small-size wires are generally employed with oscillation to keep dilution low, provide a wide (up to about 1-in.) weld bead, and give uniformly good tie-ins with the adjacent beads. Either alternating or direct current may be employed for single-wire overlays.

Productivity in pounds of deposited metal per hour is greatly increased by the use of multiple wires (Table 3). Two-wire series arc (each wire connected to the poles of an alternating current power source) is a frequently used method for producing type 347 stainless steel weld overlays of petrochemical pressure vessels. The wire size is usually either $^1/_8$ or $^5/_{32}$ in. in diameter. Alternating current is essential to achieve uniform fusion of both wires. The height of the welding head over the base metal greatly influences the penetration and hence the dilution. If the base metal has a nonuniform surface, or if the head cannot be accurately controlled above the base metal, the process may not be acceptable because of variations in dilution, or even lack of penetration in spots.

"Shingling" is a technique used to minimize dilution in single- or two-wire weld overlays. By providing an overlap of as much as one half a bead width, it is possible to reduce the dilution from 30 to 35% to 15 to 20%. In some applications, this allows the overlay to be deposited in a single layer at considerable cost savings as compared with two-layer overlays.

To further improve productivity, fabricators have increased the number of wires deposited by a single welding head. Up to six wires are not uncommon, which with oscillation deposit beads 6 in. or more in width. Usually, the wire sizes are $^1/_{16}$ to $^5/_{32}$ in. in diameter. Direct current electrode negative has been found best for relatively low dilution (around 15%). Sometimes the wire that overlaps the previously deposited bead is connected to the positive tap of the power source (DCEP), which gives somewhat greater penetration and ensures good tie-in with the previous bead.

For weld overlays on the inside surfaces of large pressure vessels, as shown in Fig. 3, wide beads produced by oscillated multiple-wire systems or strip electrodes have become the means to improve productivity and minimize dilution while offering a uniformly smooth surface. Welding parameters for stainless steel strip weld overlays are illustrated in Fig. 4 to 6. In Fig. 4, the effect of increased current on dilution, penetration, and bead thickness of 2.4-in. strip is shown. Figure 5 illustrates the effect of travel speed on dilution, penetration, and bead dimensions using 2.4-in. strip. Productivity of different strip sizes given in lb/arc h (100% duty cycle) is shown in Fig. 6.

The fluxes for submerged arc overlays should have low bulk density, especially for wide weld beads produced with multiple wire or strip electrodes. Standard fused fluxes, therefore, are usually unsuitable. Agglomerated fluxes, which have much

Fig. 4 Dilution, penetration, and bead thickness using 2.4-in. strip at various currents

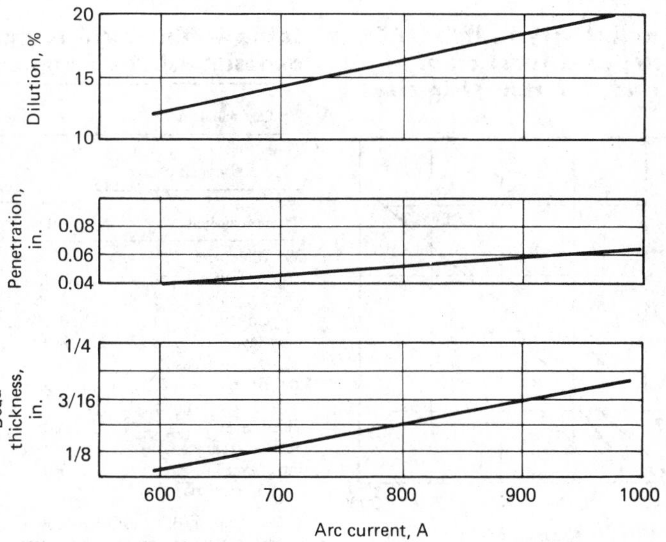

Fig. 5 Influence of travel speed on dilution, penetration, and bead thickness using 2.4-in. strip

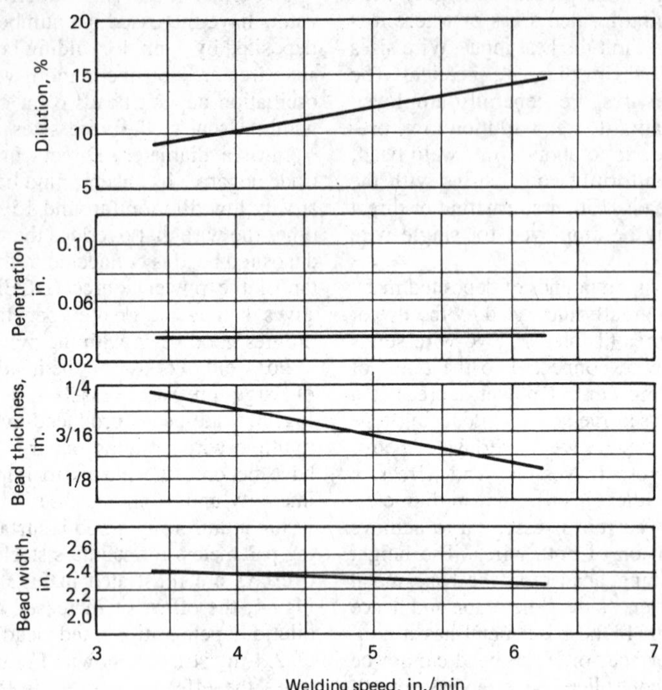

lower bulk density, are commonly used for this purpose. Low bulk density fused fluxes have been produced and found successful by a manufacturing process know as foaming. Such foamed fluxes have even lower bulk density than agglomerated fluxes.

Agglomerated fluxes (or bonded fluxes) have a distinct advantage over fused fluxes in that it is possible to introduce alloying elements by means of the flux. When this is done, they are known as reinforced, or alloyed, fluxes. For stainless steel weld overlays, chromium is commonly added to the fluxes to overcome the oxidation of chromium, which is characteristic of SAW of stainless alloys, and to compensate for the dilution with the nonchromium or low-chromium base metal. Other alloying elements often added to reinforced fluxes are manganese, nickel, molybdenum, and niobium. With reinforced fluxes, it is possible to produce various modified stainless weld overlays using a single filler metal as an electrode, for example, a 347 overlay with a 308 or 309 filler metal using a niobium alloyed flux. Tables 4 and 5 list variations in chemical compositions of single-wire and series-arc weld deposits using an agglomerated flux and type 308 or 309 wire.

Another method for providing the desired stainless steel overlay composition is to introduce a layer of powdered metal under the submerged arc flux. This procedure is sometimes known as bulk welding. It is suitable for use either with single-wire or with strip electrodes. The powdered metal not only provides the alloying elements needed in the overlay, but also greatly reduces the dilution.

Either method of adding alloying elements by using powdered alloys, either as an ingredient of the flux or separately underneath the flux, is subject to variability of the deposit composition. When reinforced fluxes are employed, the recovery of the alloying ingredient is dependent on the arc voltage, with a measurable difference in recovery by a change of as little as 1 or 2 V. This is not the case when the powdered alloys are separately fed below the flux, but in this case variability results if the height of the powder layer varies. Thus, it is unwise to rely on these methods of alloy adjustment of the weld overlay composition for more than three percentage points.

Still another method of providing the desired composition for the overlay is to use tubular wire for the filler metal. Whereas the steel producer is limited in the compositions that can be produced and still provide hot workability and cold

Fig. 6 Productivity in lb/h (100% duty cycle) as a function of current and different strip sizes

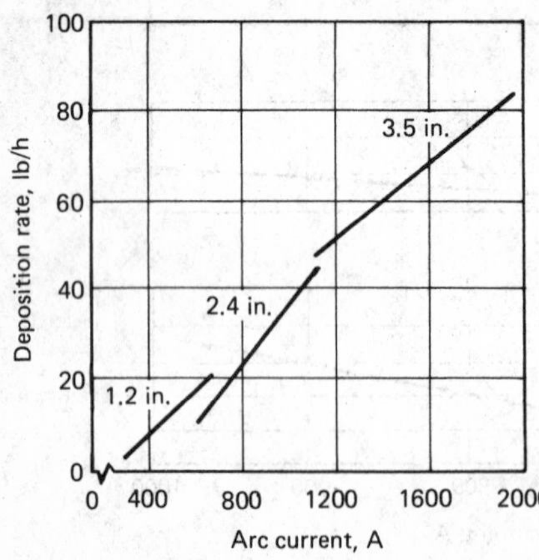

Table 4 Chemical composition variations in single-wire weld deposits using an agglomerated flux and type 308 or 309 wire

Welding conditions	Wire size, in.	Amperes, ac	Average voltage, V	Travel speed, in./min
Type 308 overlay	$^5/_{32}$	500	28	12
Type 309 overlay	$^3/_{16}$	600	30	12

Chemical analysis	Carbon	Manganese	Silicon	Sulfur	Phosphorus	Chromium	Nickel
	Element, %						
308 filler wire	0.04	1.67	0.29	0.013	0.023	20.13	9.94
Top or fourth layer	0.05	1.51	0.81	0.009	0.039	18.59	9.08
Third layer	0.06	1.45	0.71	0.008	0.038	18.16	8.80
Second layer	0.06	1.35	0.76	0.007	0.036	18.53	8.21
First layer	0.07	1.06	0.52	0.008	0.034	14.34	7.11
Carbon steel base	0.07	0.51	...	0.013	0.016	...	...
309 filler wire	0.05	1.81	0.32	0.016	0.022	24.68	13.36
Top or fourth layer	0.06	1.42	0.83	0.010	0.042	19.78	11.23
Third layer	0.06	1.37	0.76	0.009	0.041	19.00	10.49
Second layer	0.07	1.24	0.67	0.010	0.038	18.20	9.43
First layer	0.07	1.22	0.48	0.009	0.036	14.42	6.54
Carbon steel base	0.07	0.51	...	0.013	0.016	...	...

Source: Campbell, H. C. and Johnson, W. C., Welding Alloy Steels Under Bonded Fluxes, *The Welding Journal*, Nov 1958

Table 5 Chemical composition variations in series-arc weld deposits using an agglomerated flux and type 308 or 309 wire

Welding conditions	Wire size, in.	Angle between wires, °	Amperes, ac	Voltage, V	Travel speed, in./min
Type 308 overlay	5/32	45	500	20-25	10
Type 309 overlay	5/32	45	540	26-30	13

Chemical analysis	Element, % Carbon	Manganese	Silicon	Sulfur	Phosphorus	Chromium	Nickel
308 filler wire	0.04	1.74	0.30	0.011	0.022	21.30	9.83
Top or third layer	0.05	1.39	0.61	0.007	0.028	20.95	9.57
Second layer	0.06	1.36	0.63	0.010	0.029	20.29	9.02
First layer	0.07	1.20	0.58	0.006	0.026	18.08	8.16
Carbon steel base	0.07	0.51	...	0.013	0.016	...	...
309 filler wire	0.05	1.81	0.32	0.016	0.022	24.68	13.36
Top or third layer	0.06	1.49	0.93	0.005	0.043	24.43	13.14
Second layer	0.06	1.55	0.84	0.006	0.040	24.37	11.08
First layer	0.07	1.36	0.84	0.008	0.039	21.79	11.43
Carbon steel base	0.07	0.51	...	0.013	0.016	...	...

Source: Campbell, H. C. and Johnson, W. C., Welding Alloy Steels Under Bonded Fluxes, *The Welding Journal*, Nov 1958

drawing capability, the producer of tubular wire is virtually unlimited as to the compositions that can be produced. For example, the strip can be a chromium or chromium-nickel steel to which the core ingredients are added to give a weld deposit of an austenitic stainless steel, rich enough in the essential alloys to compensate for both oxidation and dilution losses. Another distinct advantage of tubular wires is for relatively small production applications of compositions that are not available as a "shelf item" in solid wire form. Small quantities of tubular wires can also be prepared for experiments to determine the optimum composition for a particular application. The effective cross section of tubular wire for current-carrying ability is substantially less than solid wire of the same diameter. Welding parameters that have been established for solid wires will have to be adjusted to take this and other differences into account.

Self-Shielded Flux Cored Wire. Tubular wires are produced that contain not only the alloying ingredients, but also the fluxing and gas shielding ingredients. This permits the deposition of weld overlays without the need of an external flux, as is needed with SAW, or an external gas, as is needed with gas shielded solid wire welding. Weld overlays are produced by manually directed welding guns for limited areas using a single tubular filler wire.

High-productivity procedures have been developed using multiple wires in a manner similar to six-wire SAW. These wires have inherently low penetration characteristics, making them highly suitable for cladding. Deposition rates of 80 to 100 lb/h are often achieved. In addition, improved wire feed roll mechanisms have lessened the problems of kinking and crushing of these thin-walled electrodes during the wire-feeding process, thereby increasing the productivity of multiple-wire applications. Some fabricators who use self-shielded flux cored wires prefer to limit the automatic overlay process to two or three wires, particularly because the gain in productivity with increasing numbers of wires diminishes, as shown in Fig. 7. Both six-wire and three-wire welding heads for overlay applications are shown in Fig. 8 and 9.

Metallurgically, flux cored wires offer potentially higher alloyed overlays, permitting a considerable cost saving to produce the desired composition in a single layer. Also, dilution is minimized with this process. Furthermore, carbon content can be kept to much lower levels than is pos-

Fig. 7 Effect of deposition rate on cost

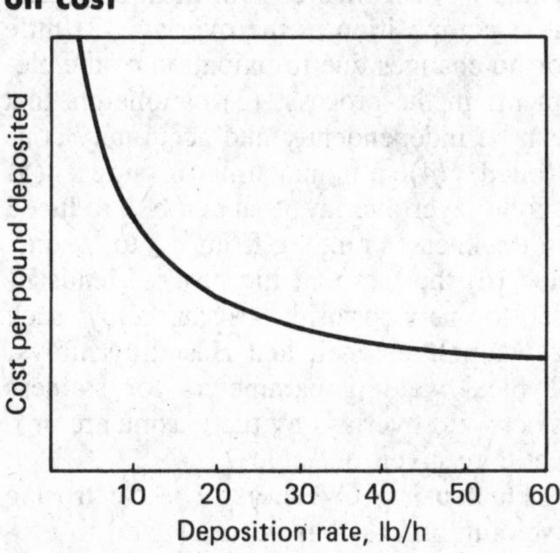

Fig. 8 Six-wire welding head for overlay applications. Courtesy of Stoody Co.

Fig. 9 Three-wire welding head for overlay applications. Courtesy of Stoody Co.

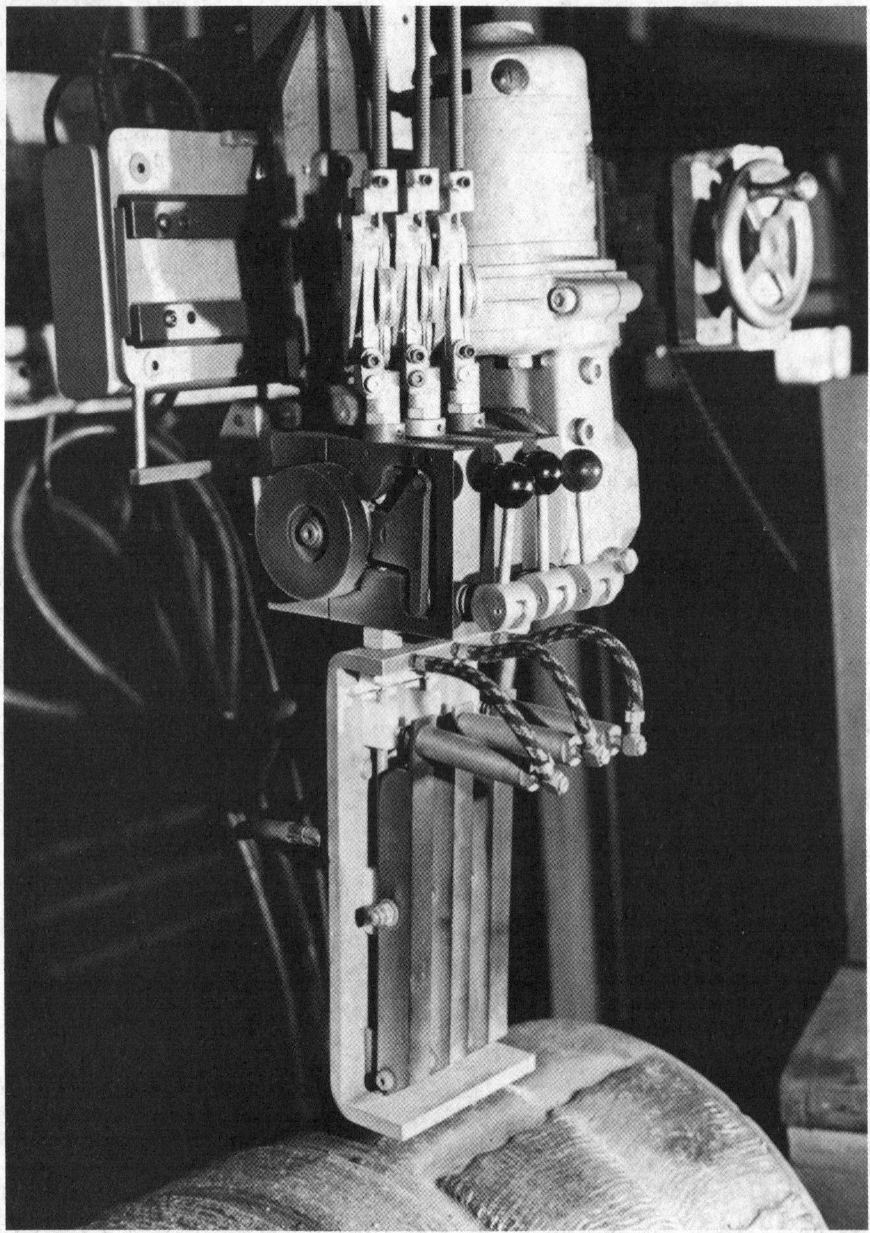

Table 6 Typical welding parameters for stainless steel weld overlays using self-shielded flux cored wires

No. of $^{3}/_{32}$-in.-diam wires	Constant potential power source capacity(a)	Current, A	Voltage, V	Travel speed, in./min	Oscillation, cycles/min	Productivity, lb/h	Dilution, %
1	400	300	27	20	None	12	20
2	800	600	27	4.5	20	30	12
3	1200	900	27	4.0	20	45	12
6	1200 each (2-in. parallel)	1800	27	3.5	20	85	12

(a) 100% duty cycle

sible with solid wires, partly due to the ability to have very low-carbon ingredients in the flux cored wire and partly due to the low dilution. Typical welding parameters for stainless steel weld overlays using self-shielded flux cored wires are given in Table 6.

Plasma Arc Processes. For overlaying relatively small surfaces, the plasma arc process offers many advantages. The control of the heat sources is independent of the feed of the filler metal as in the metal arc welding processes. This gives considerable flexibility to control the dilution and surface characteristics of the resulting overlay.

Corrosion-resistant weld overlays can be produced using stainless steel powders in a plasma arc gun in the same way as hard surfacing overlays are obtained. For large surfaces, however, this process is too slow to compete with other methods.

The filler metal used with the plasma process can be wire of the same types that are used in the metal arc processes. They are generally fed into the plasma arc stream roughly parallel and slightly above the surface of the base metal. Using a single wire under a plasma gun, the deposition rate is generally far below that of the metal arc processes and therefore is not competitive except in smaller areas.

The plasma arc hot wire process is used for depositing stainless steel overlays on pressure vessels subjected to corrosive, high-temperature, and high-pressure hydrogen environments. The set-up for this process is illustrated in Fig. 10. While a single hot wire increases the deposition rate considerably over the cold wire or powder methods, a dual hot wire process permits deposition rates as high as 40 to 70 lb/h, thus making the hot wire process commercially competitive with other weld overlay processes. A continuously made stainless steel weld overlay made by the hot wire process is shown in Fig. 11.

Advantages of the hot wire process include: (1) accurate control of dilution and thus composition of the overlay, (2) little or no changes due to oxidation of the elements in the process, (3) parameters that can be independently and accurately controlled, (4) minimal microfissures, (5) single-layer overlays that can be produced in thicknesses ranging from $^{5}/_{32}$ to $^{5}/_{16}$ in., and (6) the fact that the process lends itself to many corrosion-resistant alloys such as Monel, Inconel, and Hastelloy alloys. Typical welding parameters for stainess steel weld overlays by the plasma arc process are given in Table 7.

Electroslag Overlays. The electroslag welding process has been adapted to pro-

Fig. 10 Setup for plasma arc hot wire surfacing process

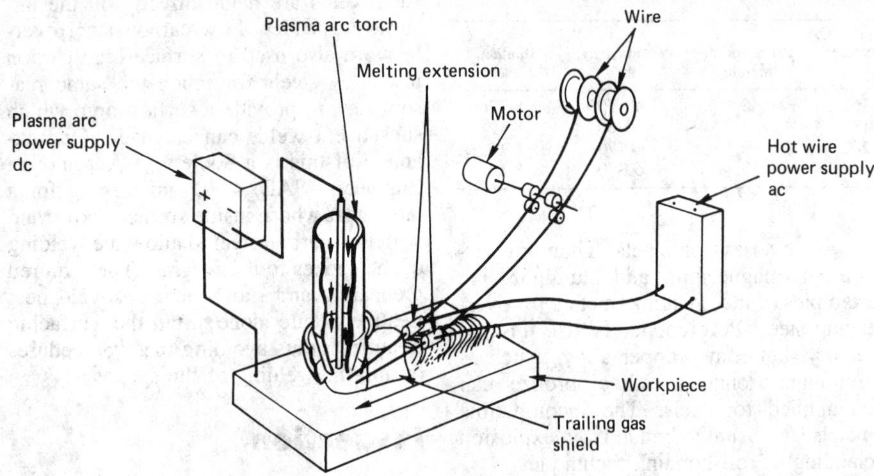

Fig. 11 Stainless steel weld overlay made continuously by the hot wire plasma arc process on a circular disc by spiraling toward the center. Courtesy of Nooter Corp.

duce stainless steel weld overlays. This process offers the ability to use wider strip (up to 12 in.) than the submerged arc process and more uniform penetration into and lower dilution with the base metal. Its principal disadvantages are operator exposure to the radiant heat from the molten slag and a sharper compositional change at the fusion line that may give unsatisfactory service in certain high-temperature pressure vessels, for example, disbonding in hydrocracker or hydrodesulfurizer service.

For large heavy-walled pressure vessels, the electroslag overlay process typically uses 0.16-in.-thick by 6-in.-wide strip. The effect of ESW parameters on dilution using 6-in. strip is shown in Fig. 12. As in ESW, energy is created by ohmic resistance as the electric current flows from the strip to the work through the molten slag; no arc exists. Unlike ESW, however, the slag bath is relatively thin. To provide the required electrical resistance, viscosity, and other properties, the fluxes must be significantly higher in fluorides than for SAW (Ref 1). A typical fused flux composition comprises 49% CaF_2, 21% CaO, 21% SiO_2, and 9% Al_2O_3.

One method of providing a uniform surface of the weld bead is to superimpose Lorentz forces by adding supplementary electrical coils at the edges of the strip (Ref 1). This changes the shape of the molten pool, as shown in Fig. 13. Another method that produces a good surface finish and good interbead penetration is to adjust the flux composition and weld parameters so that an arc exists at the edges of the strip and electroslag mode transfer for most of the strip width. This method does not require magnetic probes.

Because of disbonding of weld overlays deposited on pressure vessels in a hydrogen environment, electroslag weld overlays are not used for single-layer overlays; but where two-layer overlays are specified, the electroslag process is used for the second layer. Higher productivity and a smoother surface are thus obtained, the first layer having been deposited by the submerged arc process.

Extremely low carbon levels are possible because of low dilution with the base metal and a removal of carbon through oxidation in the process. Using a 0.18% C base metal and a 0.01% C strip, one investigator (Ref 2) produced weld-metal carbon levels of 0.02% in the first layer and 0.01% in the second. The electroslag overlay process also can be applied for weld overlays of nickel-chromium (Inconel) alloys in the same manner as for stainless steel.

Processes for Special Applications

The shielded metal arc process is commonly used for weld overlays of small areas, such as nozzles or other penetrants, through the walls of weld overlayed pressure vessels. The ease of application, the ability to deposit on vertical as well as downhand surfaces, and the availability of virtually any stainless alloy make this an adaptable method for providing a corrosion-resistant surface. Manual covered electrodes are produced with some of the alloying ingredients in the covering along with all of the slag-forming and gas

Table 7 Typical welding parameters for stainless steel weld overlays by the plasma arc process

Torch conditions			Filler-wire conditions					
Current, A	Voltage, V	Gas flow rate, in.^{3}/h	No. of wires	Size, in.	Current, A	Travel speed, in./min	Productivity, lb/h	Dilution, %
440	38	85	2	1/16	160	8	40-50	8-12
480	38	85	2	1/16	180	9	50-60	8-12
500	39	85	2	1/16	200	9	60-70	8-15
500	39	85	2	3/32	240	10	60-70	8-15

shielding components. Electrode sizes from 3/32 to 3/16 in. are most often employed, the smaller sizes used where heat input must be limited and the larger sizes used for massive articles.

Care must be taken to avoid excessive dilution, which varies with the current, travel speed, and arc length (voltage). Because these parameters may vary with the skill of the operator, it is important to establish reasonable limits and require operator qualification tests. The shingling technique may be employed to reduce dilution from about 25% to approximately 15%. Two or more layers are generally specified for manually deposited weld overlays because of the difficulty of controlling all of the variables affecting the composition of a single layer.

The gas metal arc and gas tungsten arc processes are also applied both manually directed and automatically controlled. Most of the more commonly used stainless steels are available for filler wires for these processes. For GMAW, wire sizes of 0.045 and 1/16 in. are generally used. For GTAW, the filler rod for manual use is generally 1/8 in. in diameter by 36 in. long.

Fully automatic procedures have been developed using GTAW in extremely confined areas, such as the inside surface of tubes as small as 3/4 in. in diameter. Filler wires of 0.035-in. and smaller diameters are used in such applications. The gun and wire-feed nozzle are miniaturized to accommodate the confined areas to be surfaced. Precise control of the welding parameters is essential for successful weld overlays.

Weld Overlays Other Than Stainless Steel

Many of the procedures described for stainless steel weld overlays are applicable for other alloys. Generally, nickel-based alloys can be produced successfully using essentially the same procedures as for stainless steels. Some of the common alloy overlays and the filler metals used for such are given in Table 8.

Metals and alloys that do not tolerate significant amounts of iron cannot be used in weld overlays on steels. Titanium, zirconium, magnesium, and aluminum are examples of incompatible metals for overlaying steel. Pure copper is also unsatisfactory, but some copper alloys, such as aluminum-bronze and silicon-bronze, can be applied to steels. The incompatible metals are usually bonded by explosion cladding or roll bonding techniques.

Low-carbon steel is often applied as weld overlay for specific purposes. A tube header, for example, is sometimes overlaid with low-carbon steel weld metal so that subsequent welds of the tubes to the header (usually by GTAW) are free from porosity or other defects that arise when such welds are made directly on the unsurfaced header. Low-carbon steel overlays are also used to surface high-carbon low-alloy steels for much the same reason, i.e., to provide a surface onto which subsequent welds can be made. One example of this is a Society of Automotive Engineers (SAE) 4340 rim forging for a large gear whose inside surface is overlaid with low-carbon steel to allow the welding of the spokes to the forging. The required preheating and subsequent postweld heat treatments are done after the surfacing treatment, thus avoiding these procedures during the welding of the spokes.

Inspection

Types of Discontinuities. The inspection procedures for weld overlays are directed toward identifying those flaws that would affect the integrity of the overlaid surface or would become initiating points for brittle fracture of the base metal in ser-

Fig. 12 Effect of welding parameters on dilution using 6-in. strip in electroslag weld overlay process

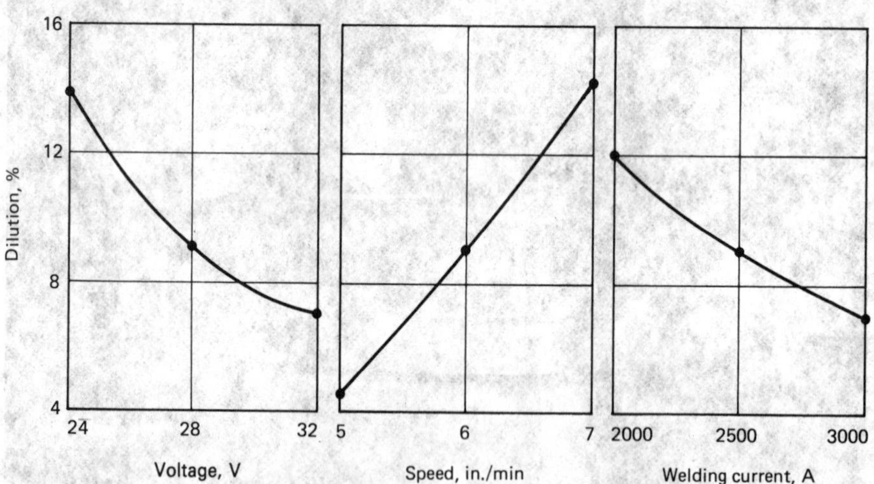

Fig. 13 Control of liquid slag flow by the Lorentz force

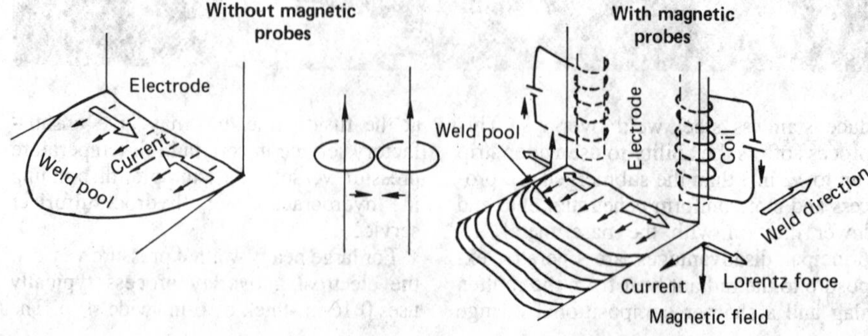

Table 8 Filler metals used for weld overlays on steels by metals and alloys other than stainless steels

Weld overlay type	Covered electrode: AWS specification No.	Covered electrode: Classification No.	Bare rod or electrode: AWS specification No.	Bare rod or electrode: Classification No.
Low-carbon steel	A5.1	E7016	A5.17	EL8
		E7018	A5.18	ER70S-2
Aluminum-bronze	A5.6	ECuAl-A2	A5.7	ERCuAl-A1
Silicon-bronze	A5.6	ECuSi	A5.7	ERCuSi
Copper-nickel	A5.6	ECuNi(a)	A5.7	ERCuNi(a)
Nickel	A5.11	ENi-1	A5.14	ERNi-1
Monel	A5.11	ENiCu-7	A5.14	ENiCu-7
Inconel	A5.11	ENiCrFe-3	A5.14	ERNiCr-3
Inconel 625	A5.11	ENiCrMo-3	A5.14	ERNiCrMo-3
Hastelloy B	A5.11	ENiMo-7	A5.14	ERNiMo-7
Hastelloy C	A5.11	ENiCrMo-4	A5.14	ERNiCrMo-4

(a) First layer must be either nickel or Monel.

Fig. 14 Laboratory test to ascertain that welding procedure and filler metal meet compositional requirements

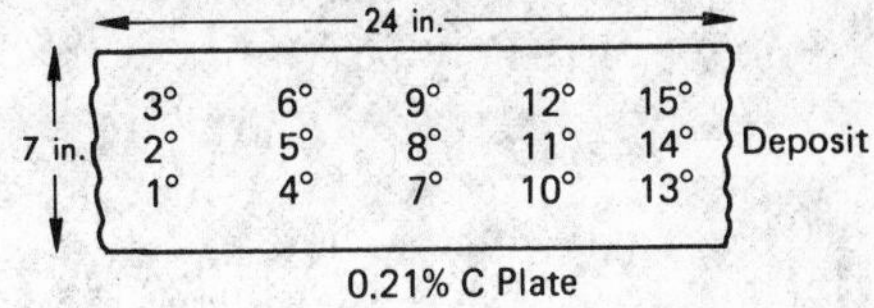

Specimen No.	Carbon, %	Chromium, %	Nickel, %	Ferrite (magne gage), %
1	0.022	20.49	10.28	11.6
2	0.032	20.38	10.12	11.7
3	0.032	20.72	10.35	12.8
4	0.030	20.31	10.12	10.2
5	0.024	20.43	10.26	10.2
6	0.026	20.60	10.15	11.8
7	0.026	20.45	10.30	10.2
8	0.023	20.87	10.32	12.0
9	0.026	20.61	10.42	10.4
10	0.034	20.67	10.30	11.6
11	0.025	20.49	10.30	12.0
12	0.021	20.68	10.32	10.6
13	0.023	20.49	10.42	11.6
14	0.023	20.84	10.32	12.2
15	0.022	20.49	10.34	11.6

vice. Cracks are the principal type of discontinuity to be identified.

The types of cracks most commonly encountered in weld overlays are underclad cracks, fusion-line cracks, and microfissures in the overlay itself. Underclad cracks are attributed to the cleanliness of the base metal and to the presence of hydrogen, often from moisture in the flux (SAW) or the covering (SMAW). Steels of poor weldability—for example, those with high-carbon equivalent—are more susceptible to underclad cracking.

Fusion-line cracks result from the use of filler metals or welding procedures that produce a substantial amount of martensite in the first layer. This is usually identified during qualification tests and is corrected by altering the composition of the filler metal or the welding procedure to minimize the dilution (see the section on composition control of stainless steels in this article).

Another flaw that may occur at the fusion line is lack of bonding. In welding processes where dilution is minimized, there is a danger that fusion will not occur at all points under the deposit. This occurs most often at the line where the bead overlaps the previously deposited bead.

Microfissures usually are the most troublesome type of discontinuity. Some alloys are more prone to microfissures than others. In stainless steels, control is established by providing a microstructure that has at least 3% ferrite in the austenitic matrix. Fully austenitic stainless steel overlays require filler metals having extremely low amounts of residual elements, sulfur, phosphorus, silicon, and other elements. Nickel-based and some copper-based alloys are also subject to microfissuring; these alloys require controlled amounts of deoxidizers, usually titanium and aluminum, in the filler metals to eliminate the microfissuring tendency. These deoxidizers also are needed to minimize porosity in these alloys.

Many overlay processes or consumables result in undercut where the bead does not fuse smoothly with the previously deposited bead. This is a surface condition that is readily identifiable visually. Some overlay alloys are subject to embrittlement when a postweld heat treatment is applied. This is particularly troublesome in austenitic stainless steels having an as-deposited microstructure high in ferrite (over 10%). When a steel pressure vessel, for example, requires a stress-relieving heat treatment following the overlay operation, the ferrite constituent may transform to the brittle sigma phase. This brittleness shows up in qualification tests when the weld overlay fails to elongate in the bend test.

Qualification Tests. Most fabrication codes or customer specifications require tests to demonstrate that the selected welding process and consumables meet the requirements. Because the main purpose of a weld overlay is to provide a surface that will resist a more aggressive environment than the base metal, one of the most important tests is the chemical composition of the overlay surface. An example of a qualification test for weld composition is shown in Fig. 14. A 7-in.-wide weld bead, which was to meet a type 308L deposit analysis, was applied to a low-carbon steel plate in a single layer using six self-shielding, open-arc, $^{3}/_{32}$-in.-diam wires that were oscillated at 7.8 in./min. Fifteen separate chemical analyses were taken to evaluate the uniformity of the deposit composition. Values are as shown.

The tie-ins between adjacent beads is tested by means of a laboratory sample on which the second pass is deposited at a slight angle with respect to the initial pass. Visually observing the surface for smoothness and undercut and sectioning the sample for lack of fusion in the overlap area provide evidence of the optimum overlap of adjacent beads (Fig. 15).

The ability of the overlay to meet the ductility criteria is best evaluated by the side bend test. Types of cracks that may be encountered are microfissures in the overlay, martensite at the weld interface, and formation of the sigma phase following stress-relief heat treatment.

Inspection During Fabrication. Surface inspection is relied upon by inspectors concerned with maintaining the quality of the overlaying operation once the qualification procedures have been established. In stainless steel weld overlays depending on close tolerances of the ferrite for control of microfissures and possible

Fig. 15 Method of establishing optimum bead overlap. By angling second pass, all practical overlaps are obtained in one test. Observation followed by sectioning quickly determines the optimum overlap.

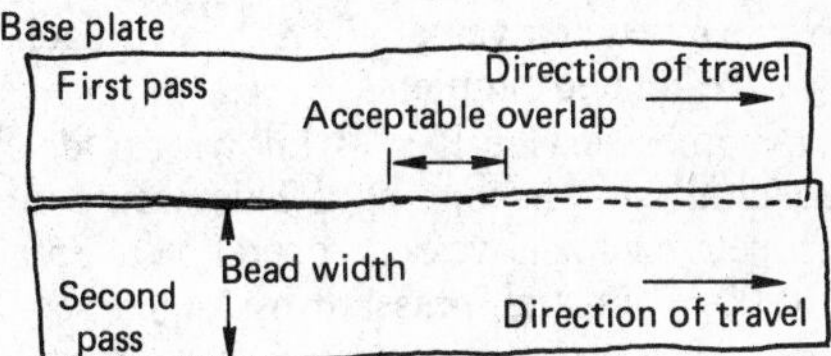

Fig. 16 Weld overlay for aggressive environment. Courtesy of Combustion Engineering, Inc.

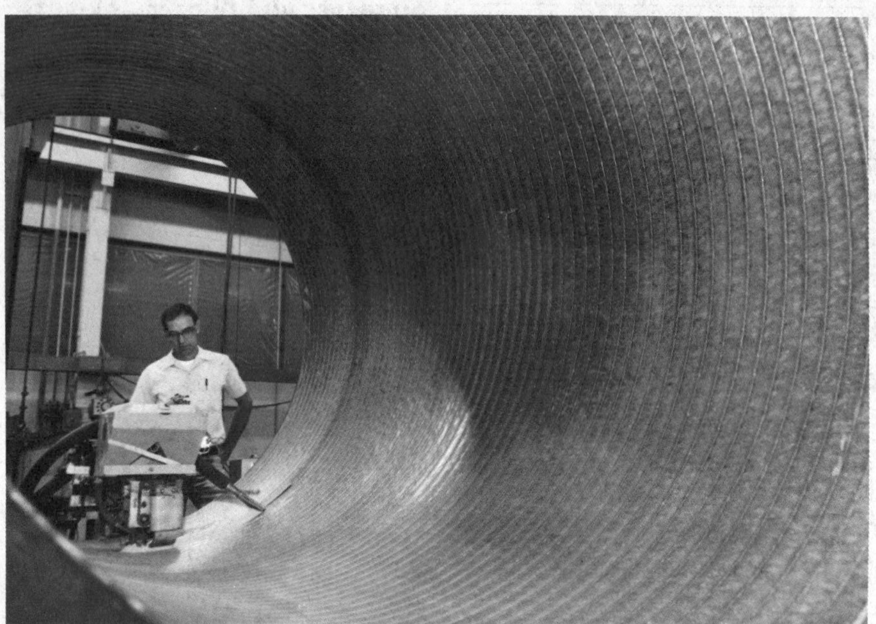

sigma formation, a portable device for measuring ferrite is employed periodically after the overlay has cooled to room temperature. Liquid-penetrant inspection methods are used to locate microfissures and other cracks, as well as undercut areas. It should be understood that this identifies only those defects that reach the surface. If the surface is to be machined, ground, or polished, surface inspection of this type is made after the finishing operation.

Radiographic tests are often required of the finished weldment, but this technique rarely uncovers the types of discontinuities that occur in or adjacent to weld overlays. Underclad cracks and fusion-line flaws also can be identified with ultrasonic tests.

Performance of Overlays in Service. In pressure vessels having weld overlays of stainless steels, two types of difficulties have been observed. Underclad cracks, which may have existed but were not identified during routine fabrication inspection, have been found associated with certain low-alloy steel-based metals. Temperature cycling in service and other strains have opened up small cracks at locations near the overlap areas of the overlay beads. Nuclear reactor vessels have shown such service-related problems.

Another service flaw is known as "debonding" or "disbonding." This occurs in the petrochemical vessel's operation at 850 to 900 °F in high-pressure hydrogen service. The cracks appear in the fused metal immediately above the line of fusion. This results when the vessel is cooled relatively rapidly during shutdowns when the base metal and overlay are highly charged with hydrogen and time has not been allowed for the hydrogen to diffuse into the atmosphere.

Weld overlays similar to the example in Fig. 16 have been increasingly employed in aggressive environments. Except on rare occasions such as those mentioned in previous paragraphs, the service life of weld overlays has been high.

For more information on types of discontinuities and inspection methods, see the article "Weld Discontinuities" in this Volume.

Economics

One of the first considerations in planning a weld overlay procedure is the selection of a welding process and the available deposition rate for that process. Generally, the higher the deposition rate, the lower the cost per pound of metal deposited. A comparison of deposition rates with various fusion welding processes is shown in Fig. 17.

The form of filler metal is either wire, rod, or strip, depending on the welding process selected. Wire in coils may be solid, flux cored, or metal powder cored. Rods may be solid straight cut lengths or flux-coated electrodes. Coils of strip are available in a wide range of widths and thicknesses. Generally, the larger filler-metal sizes in each type are less expensive than smaller sizes. Coils of solid wire usu-

Fig. 17 Deposition rates with various fusion processes

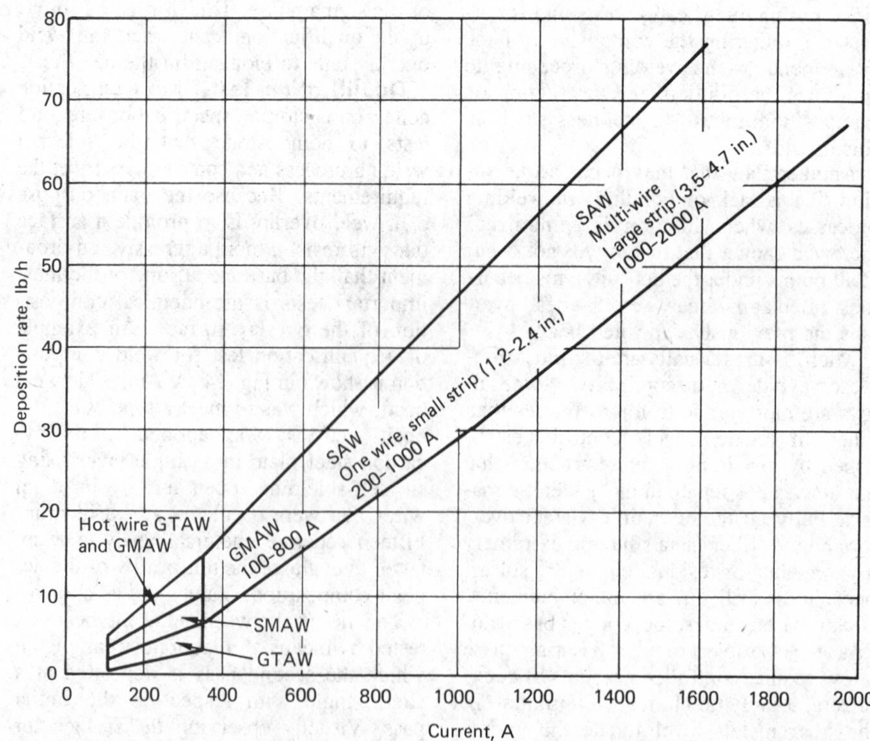

ally represent the lowest cost per pound of filler metal.

Many factors are involved in establishing the cost of producing a composite structure by the weld overlay process compared with other processes. One method of estimating the weld overlay cost per unit weight or per unit area is to use the following equations:

$$C_a = \left(\frac{P_s}{N} + KP_f + \frac{L}{HD}\right)(1 + R)$$

$$C_f = C_a\, t\, A\, d$$

where C_a is the cost of weld overlay per unit of weight, C_f is the cost of weld overlay per unit of area, P_s is the filler-metal cost, N is the deposition efficiency, P_f is the flux cost or shielding gas cost, K is the flux or gas consumption, L is the labor rate, H is the deposition rate, D is the duty cycle, R is the repair factor, t is the thickness of weld overlay, A is the area, and d is the density of the weld overlay alloy. Having determined this cost, the other processes previously suggested could be reviewed for suitability and cost effectiveness.

REFERENCES

1. Nakano, S., Hiro, T., Nishiyama, N., and Tsuboi, J., *The New Strip Electroslag Overlay Technique with Electro-Magnetic Control—The Maglay,* IIW Document XII-A-4-81
2. *M-C (Magnetic-Controlled) Overlay—The New Overlay Method with Electroslag Welding for Wide Strip Electrode,* Mitsubishi Heavy Industries, Ltd., Kobe Shipyard and Engine Works Report, Oct 1980

Other Welding Topics

Codes, Standards, and Inspection

CODES, STANDARDS, AND INSPECTION govern all facets of the metalworking process. An important factor in the development and use of materials and processes is a concern for those people who are exposed to possible hazards. It is essential that the materials used and the capabilities of the process meet predetermined levels of acceptance. This implies that detailed guidelines need be established to govern various materials and processes.

There are many different codes, standards, and specifications that involve welding. With the advent of modern welding technology, failures in welded structures have forced engineering societies and manufacturers to develop proper guidelines for the welding industry. Products that are governed by well-known welding guidelines include:

- Boilers and pressure vessels
- Nuclear reactors
- Pressure piping
- Industrial pipelines and piping
- Buildings, bridges, and similar structures
- Ships
- Field-welded storage tanks
- Railroad rolling stock
- Aircraft and spacecraft
- Construction equipment
- Industrial machinery

Generally, the codes, standards, and specifications issued by various organizations that apply to welding are similar. Welding guidelines for similar product forms that are issued by the American Society for Testing and Materials (ASTM), the American National Standards Institute (ANSI), and the American Welding Society (AWS) do not conflict.

Standards-issuing agencies, however, have no provisions for enforcing their recommended codes. Standards do not become law unless adopted by a local, state, or federal agency. Adoption may be made by reference or citation of the particular code and date, requiring that all operations be performed accordingly.

This article examines codes, standards, and specifications as they apply to three areas: (1) specific products, (2) filler metals, and (3) nondestructive inspection techniques. For a list of codes, standards, and specifications that are pertinent to the welding industry, the reader should refer to the list of Selected References at the end of this article.

Boilers and Pressure Vessels

The manufacture of boilers and pressure vessels falls under the general specification of the American Society of Mechanical Engineers (ASME) "Boiler and Pressure Vessel Code." This code is published in several parts:

- *Section I*: Power Boilers
- *Section II*: Material Specifications
  Part A, Ferrous
  Part B, Nonferrous
  Part C, Welding Rods, Electrodes, and Filler Metals
- *Section III*: Nuclear Power Plant Components
- *Section IV*: Heating Boilers
- *Section V*: Nondestructive Examination
- *Section VI*: Recommended Rules for Care and Operation of Heating Boilers
- *Section VII*: Recommended Rules for Care of Power Boilers
- *Section VIII*: Pressure Vessels, Divisions 1 and 2 (two sections)
- *Section IX*: Welding Qualifications
- *Section X*: Fiberglass-Reinforced Plastic Vessels
- *Section XI*: Rules for In-Service Inspection of Nuclear Reactor Coolant Systems

All products manufactured under the requirements of any of these guidelines may also be manufactured under laws that cite or reprint various sections of this code. To be able to display the ASME logo on a vessel, the manufacturer must have his shop inspected and approved by ASME. Welding procedures must be qualified as well. Inspectors commissioned by the National Board of Boiler and Pressure Vessel Inspectors, as well as independent testing laboratories, insurance companies, or state inspectors, must perform this certification.

National Board of Boiler and Pressure Vessel inspectors are required to have certain minimum levels of experience in the operation of high-pressure boilers or as designers and fabricators of pressure vessels. Additional guidelines from other organizations include:

- "Marine Engineering Regulations and Materials Specifications," U.S. Coast Guard
- "American Bureau of Shipping Rules for Building and Classing Steel Vessels Standards," Tubular Exchanger Manufacturers' Association, Inc.
- "Lloyd's Rules and Regulations," Lloyd's Register of Shipping
- CSA Standard B51, "Code for Construction of Boilers and Pressure Vessels," Canadian Standards Association

Nuclear Reactors

Nuclear reactors and related components and materials used in nuclear power plants are covered by the provisions of Section III (Nuclear Power Plant Components) of the ASME "Boiler and Pressure Vessel Code." Any part used in a nuclear plant should be manufactured under the guidelines of this specification. Components for naval ship use are covered by NAVSHIPS 250-1500-1, "Standard for Welding of Reactor Coolant and Associated Systems and Components for Naval Nuclear Power Plants," issued by the Department of Defense, Naval Ship Division. This code includes additional restrictions: (1) it requires the certification of materials; (2) it requires traceability of all materials, including welding filler metals, to the point of origin; and (3) it outlines specific inspection procedures required during the manufacture of nuclear power plant components.

Pressure Piping

Guidelines for pressure piping are covered by the American National Standards Institute document ANSI B31.1, "Code for Pressure Piping." This code is similar, as far as welding requirements are concerned, to Section IX (Welding Qualifications) of the ASME "Boiler and Pressure Vessel Code." Pressure piping also is regulated by many states. In some cases, welders are tested and certified by the individual states.

Industrial Pipelines and Piping

Generally, the welding of cross-country pipelines is governed by American Petroleum Institute (API) standard 1104, "Standard for Welding Pipe Lines and Related Facilities." It provides welding procedure information and welder qualification information for work on cross-country pipelines. This has become a worldwide specification used for transmission pipelines, particularly for the petroleum industry. It also is the accepted code for other types of pipe welding. Recently, additional codes have been issued for other types of pipe welding, most of which are in substantial agreement with API 1104.

ANSI Standard B31, "Code for Pressure Piping," Sections 1 through 8, also applies to industrial pipelines and piping. The sections of the code issued by ANSI include:

- *B31.1*: Power Piping
- *B31.2*: Fuel Gas Piping
- *B31.3*: Petroleum Refinery Piping
- *B31.4*: Liquid Petroleum Transportation Piping Systems
- *B31.5*: Refrigeration Piping
- *B31.6*: Chemical Plant Piping
- *B31.7*: Nuclear Power Piping
- *B31.8*: Gas Transmission and Distribution Piping Systems

Additional guidelines for pipelines and piping include:

- ASME "Boiler and Pressure Vessel Code," Sections I, III, VIII, and IX
- American Water Works Association (AWWA) standards covering fabricated piping
- Pipe Fabrication Institute standards
- U.S. Navy Bureau of Ships specifications, NAVSHIPS 250-582 and 250-1500-1
- Procedures by the Heating, Piping and Air Conditioning Contractors National Association

Bridges, Buildings, and Similar Structures

Bridges, buildings, and similar structures are typical structural welding applications that are governed by applicable codes and standards. The basis for these codes is AWS D1.1, "Structural Welding Code."

This code provides for the design of welded connections, filler metal and flux, welder qualification and techniques, and the strengthening and repair of buildings. Design considerations such as stresses and the proportioning of load-carrying members are not covered. Provision is made, however, for the qualification of processes, procedures, and joints not covered by prequalification.

Most states publish their own welding codes. Welding on highway bridges is under the jurisdiction of the state highway departments, and in many states, welders are examined yearly and certified by the state to work on bridges. Many state highway departments also require certification of welding electrodes and filler metals. "Standard Specifications for Highway Bridges," adopted by the American Association of State Highway Transportation Officials (AASHTO), and the "Specifications for Steel Railroad Bridges," published by the American Railway Engineering Association (AREA), are in substantial agreement with the AWS structural welding codes.

Large steel buildings welded in the major cities in North America are covered by city codes and specifications. These specifications and codes are also in substantial agreement with the AWS "Structural Welding Code." Some cities require certification of filler metals for welded structures.

Shipbuilding

Welding on ships is covered comprehensively by different specifications and codes. In the United States, all government vessels are covered by the U.S. Coast Guard "Marine Engineering Regulations," Subchapter F, Part 57, "Welding and Brazing," or the Naval Ships Division of the Department of Defense, NAVSHIPS 0900-000-1000, "Fabrication, Welding, and Inspection of Ship Hulls." These requirements are identical in the areas of welding qualification and welder qualification. They also are similar to the Maritime Administration, which issues the "Standard Specification for Merchant Ship Construction." Qualification of welders usually is transferable among these three organizations. The American Bureau of Shipping, which issues "Rules for Building and Classing Steel Vessels," has similar requirements for welding on ships. Lloyd's Register of Shipping and other classification societies also publish specifications that cover welding. Certification of weld metal usually is required by all of these codes.

Field-Welded Storage Tanks

Welding of storage tanks is covered by two major codes: AWWA D-100, "Standard for Welded Steel Elevated Tanks, Standpipes, and Reservoirs for Water Storage"; and API standard 650, "Standard for Welded Steel Tanks for Oil Storage." Both of these codes reference Section IX of the ASME boiler code for welding qualification.

Railroad Rolling Stock

The Department of Transportation issues specifications for the manufacture of rolling stock in the United States. Welding qualification and welding design requirements are issued by the Association of American Railroads (AAR). Various specifications exist, including "Specifications for Tank Cars" and "Specifications for Design, Fabrication, and Construction of Freight Cars." These specifications provide information concerning the design of welds and the qualification of welders. These specifications generally comply with requirements of the AWS "Structural Welding Code."

The Department of Transportation also covers the manufacture of tanks for transporting gas under high pressure (Code of Federal Regulations, Title 49, Transportation Section 178.340, Part D, "General Design and Construction Requirements") and for tanks carrying liquid petroleum and similar products (Code of Federal Regulations, Title 49, Transportation Section 178.337, "Specification for Cargo Tanks").

Aircraft and Spacecraft

Assemblies intended for use in aircraft and space vehicles are welded according to government specifications. Standards-issuing organizations include the Aerospace Materials Division of the Society of Automotive Engineers (SAE), which issues "Aerospace Material Specifications" or, as they are popularly referred to, AMS specifications.

The AMS committees issue specifications in the following areas: nonmetallics, refractory materials, electronic materials and processes, finishes, processes and fluids, corrosion- and heat-resistant alloys, carbon and low-alloy steels, and nonferrous alloys. Currently, there are over 10 000 AMS specifications. In addition to AMS specifications, there are those issued by the Aerospace Industries Association of America, "National Aerospace Standards."

Welding codes or requirements are covered by specifications of the National Aeronautics and Space Administration (NASA) and of the Department of Defense, which issues military standards and specifications, such as MIL-T-5021D, "Test, Aircraft and Missile Welding Operators Qualification." This standard covers many welding processes, filler metals, and welder qualifications and must be adhered to strictly when welding aircraft. Qualification under this standard is done under the jurisdiction of government inspectors.

In military specifications, the title is followed by a letter-number designation and the date of issuance—for example, MIL-T-18068(NAVY), entitled "Torches, Welding; and Torches, Cutting-Hand, Commercial, Oxyacetylene and Oxyhydrogen Gases." The first three letters (MIL) identify this as a military specification. The letter "T" indicates "torches," the significant word in the title. Following the second dash is the serial number of the specification. The enclosure (NAVY) indicates the branch of service. If no branch of service is designated, the specification is used in all branches.

If a specification is revised, the serial number is followed by a capital letter. Immediately following, and usually printed beneath the designation, is the date of issuance. Amendments are numbered sequentially and printed below the specification identification.

Construction Equipment

A wide variety of construction equipment is made to company standards, which are based on field applications. Most manufacturers of construction equipment follow their own specifications. The American Welding Society has issued specifications that establish acceptance standards for weld performance and process application for this industry (AWS D14.3, "Welding on Earth Moving and Construction Equipment"). Qualification of welders is not covered in this standard.

Industrial Machinery

Most industrial machinery containing weldments is not covered by code or specification. However, AWS has issued some specifications to establish common accepted standards for weld performance and process application, including AWS D14.1, "Welding Industrial and Mill Cranes," and AWS D14.2, "Metal Cutting Machine Tool Weldments." The welder qualification requirements for fabricating industrial machinery are similar to AWS D1.1, "Structural Welding Code" and the ASME "Boiler and Pressure Vessel Code," Section IX, Welding Qualifications.

Filler-Metal Specifications

In the United States, AWS filler-metal specifications are the primary guidelines used by industry. The AWS specifications are written to provide the specific chemical analysis of the filler material and the mechanical properties (minimum tensile and yield strengths) of the deposited weld metal. The mechanical properties of the weld-metal deposits are based on a standardized welding procedure for a specified weld-joint design to produce weld samples for testing. Some specifications may require additional property data, such as impact toughness, quality standards, or porosity standards. Most specifications cover useability factors, including welding position of the electrode or filler metal, type of welding current and polarity to be used, and type of coating to be used. Size and packaging information is also provided.

The American Welding Society does not test or approve filler metals, but provides specifications only as voluntary requirements. The manufacturer of the filler metal guarantees that his product conforms to a specific AWS specification and classification.

Table 1 lists filler-metal specifications that have been issued by AWS. These specifications are updated periodically, and a two-digit suffix indicating the year issued is attached to the specification number. For example, current specifications for flux cored stainless steel electrodes is

Table 1 AWS filler-metal specifications

AWS specification	Specifications title
A5.1	Carbon Steel Covered Arc Welding Electrodes
A5.2	Iron and Steel Gas Welding Rods
A5.3	Aluminum and Aluminum Alloy Arc Welding Electrodes
A5.4	Corrosion-Resisting Chromium and Chromium-Nickel Steel Covered Welding Electrodes
A5.5	Low-Alloy Steel Covered Arc Welding Electrodes
A5.6	Copper and Copper Alloy Covered Electrodes
A5.7	Copper and Copper Alloy Welding Rods
A5.8	Brazing Filler Metal
A5.9	Corrosion-Resisting Chromium and Chromium-Nickel Steel Bare and Composite Metal Cored and Standard Arc Welding Electrodes and Rods
A5.10	Aluminum and Aluminum Alloy Welding Rods and Bare Electrodes
A5.11	Nickel and Nickel Alloy Covered Welding Electrodes
A5.12	Tungsten Arc Welding Electrodes
A5.13	Surfacing Welding Rods and Electrodes
A5.14	Nickel and Nickel Alloy Bare Welding Rods and Electrodes
A5.15	Welding Rods and Covered Electrodes for Welding Cast Iron
A5.16	Titanium and Titanium Alloy Bare Welding Rods and Electrodes
A5.17	Bare Carbon Steel Electrodes and Fluxes for Submerged Arc Welding
A5.18	Carbon Steel Filler Metals for Gas Shielded Arc Welding
A5.19	Magnesium Alloy Welding Rods and Bare Electrodes
A5.20	Carbon Steel Electrodes for Flux Cored Arc Welding
A5.21	Composite Surfacing Welding Rods and Electrodes
A5.22	Flux Cored Corrosion-Resisting Chromium and Chromium-Nickel Steel Electrodes
A5.23	Bare Low-Alloy Steel Electrodes and Fluxes for Submerged Arc Welding
A5.24	Zirconium and Zirconium Alloy Bare Welding Rods and Electrodes
A5.25	Consumables Used for Electroslag Welding of Carbon and High-Strength Low-Alloy Steels
A5.26	Consumables Used for Electrogas Welding of Carbon and High-Strength Low-Alloy Steels
A5.27	Copper and Copper Alloy Gas Welding Rods
A5.28	Low-Alloy Steel Filler Metals for Gas Shielded Arc Welding
A5.29	Low-Alloy Steel Electrodes for Flux Cored Arc Welding
A5.30	Consumable Inserts
A5.31(a)	Brazing Fluxes

(a) To be issued in 1983

AWS A5.22, issued in 1980. The suffix 80, therefore, is attached to this specification number (A5.22-80) to indicate the issue date. Additional specifications are added when the need arises.

In addition to AWS, filler-metal specifications are also issued by other organizations. The American Society of Mechanical Engineers in the "Boiler and Pressure Vessel Code" issues filler-metal specifications that are identical with the AWS specifications. The numbers are changed slightly; ASME adds the prefix letters "SF" to the specification number. These are given in Section 2, Material Specification, of the ASME code. The Department of Defense also issues standards for welding filler metals—MIL specifications for procurement by the government. These specifications usually are in agreement with AWS specifications; however, if a military specification is referenced, it should be consulted.

Federal specifications for filler metals are issued under the supervision of the United States Federal Supply Service of the General Services Administration. Specification titles are composed so that the most significant word appears first, followed by modifiers and descriptive words—for example, "Rods, Welding, Copper and Nickel Alloys." The initial letter appears in the identification symbol of the specification. In the specification QQ-R-571, "R" represents rods, "QQ" indicates metals, and "571" is the serial number.

Inspection

It is the option of the welding inspector to call for any of the nondestructive techniques if there is reason to be suspicious of a specific joint, welder, or welded assembly. Of course, structures used in critical applications, such as nuclear reactor components, automatically are subjected to nondestructive testing. The growth of nondestructive testing has accelerated with the need for higher quality and reliability of manufactured products, due in part to the need for subjecting materials and welds to environmental conditions never encountered before. Nondestructive testing used properly provides assurance that each weld is acceptable.

In addition to visual inspection, four nondestructive tests are widely used in welding application: liquid-penetrant and fluorescent-penetrant testing, magnetic-particle testing, ultrasonic testing, and radiographic testing. Table 2 lists these nondestructive testing techniques, along with the equipment used, the defects detected, and the advantages and disadvantages of each process. This section briefly describes the applications of each of these processes and lists applicable codes, standards, and specifications for each of the techniques.

Visual Inspection

For many noncritical welds, integrity is verified principally by visual inspection. Even when other nondestructive methods are used, visual inspection still constitutes an important part of practical quality control. Widely used to detect flaws, visual inspection is simple, quick, and relatively inexpensive. The only aids that might be used to determine conformity of a weld are a low-power magnifier, a borescope, a dental mirror, or a gage. Visual inspection can and should be done before, during, and after welding.

Visual inspection is useful for checking the following:

- Dimensional accuracy of weldments
- Conformity of welds to size and contour requirements
- Acceptability of weld appearance with regard to surface roughness, weld spatter, and cleanness
- Presence of surface flaws such as unfilled craters, pockmarks, undercuts, overlaps, and cracks

Although visual inspection is an invaluable method, it is unreliable for detecting subsurface flaws. Therefore, judgment of weld quality must be based on information in addition to that afforded by surface indications. For a detailed description of this inspection method, see the article "Nondestructive Inspection of Weldments" in Volume 11 of the 8th edition of *Metals Handbook*.

Liquid-Penetrant Inspection

Liquid-penetrant inspection is a highly sensitive, nondestructive method for detecting surface discontinuities (flaws) such as cracks and porosity. It is applicable to

Table 2 Guide to nondestructive testing techniques

Technique	Equipment	Defects detected	Advantages	Disadvantages	Other considerations
Liquid-penetrant or fluorescent-penetrant	Fluorescent or visible penetrating liquids and developers; ultraviolet light for fluorescent dyes	Defects open to the surface only; good for leak detection	Detects small, surface imperfections; easy application; inexpensive; use on magnetic or nonmagnetic materials; low cost	Time consuming; not permanent	Used on root pass of highly critical pipe welds; indications may be misleading on poorly prepared surfaces
Magnetic-particle inspection	Wet or dry iron particles, or fluorescent; special power source; ultraviolet light for fluorescent dyes	Surface and near-surface discontinuities: cracks, porosity, slag	Indicates discontinuities not visible to the naked eye; useful for checking edges before welding; no size limitations	For magnetic materials; surface roughness may distort magnetic field; not permanent	Test from two perpendicular directions to detect any indications parallel to one set of magnetic lines
Radiographic inspection	X-ray or gamma ray; film processing and viewing equipment	Most internal discontinuities and flaws; limited by direction of discontinuity	Provides permanent record of surface and internal flaws; applicable to any alloy	Usually not suitable for fillet weld inspection; film exposure and processing critical; slow and expensive	Popular technique for subsurface inspection
Ultrasonic inspection	Ultrasonic units and probes; reference patterns	Can locate all internal flaws located by other methods, as well as small flaws	Extremely sensitive; complex weldments restrict usage; can be used on all materials	High interpretation skills; not permanent	Required by some specifications and codes

many materials, such as ferrous and nonferrous metals, glass, and plastics.

Liquid-penetrant inspection is useful for locating leaks in all types of welds. Welds in pressure and storage vessels, as well as piping for the petroleum industry, can be inspected for surface cracks and porosity. For a detailed description of this nondestructive testing method, see the articles "Liquid-Penetrant Inspection" and "Nondestructive Inspection of Weldments" in Volume 11 of the 8th edition of *Metals Handbook*.

Codes, standards, and specifications for this type of inspection include:

- "Nondestructive Testing Personnel Qualification and Certification Supplement D, Liquid Penetrant Testing Method," American Society for Nondestructive Testing (ASNT) Recommended Practice No. ASNT-TC-1A
- "Standard Method for Liquid Penetrant Inspection," ASTM E-165

Fluorescent-penetrant inspection is similar to liquid-penetrant inspection. The penetrant is fluorescent when exposed to ultraviolet or black light. A glowing fluorescent image is produced. Greater contrast is produced with greater sensitivity compared to the visible dye penetrants.

One of the most useful applications of fluorescent-penetrant testing is for leak detection in magnetic and nonmagnetic weldments. A fluorescent penetrant is applied to one side of the joint and a portable ultraviolet light (black light) is used on the reverse side of the joint to examine the weld for leaks. Fluorescent-penetrant inspection is also widely used to inspect the root pass of highly critical pipe welds. The codes, standards, and specifications listed under liquid-penetrant testing also are applicable to fluorescent-penetrant inspection.

Magnetic-Particle Inspection

Magnetic-particle testing is a nondestructive means of detecting cracks, porosity, seams, inclusions, lack of fusion, and other surface or shallow subsurface discontinuities in ferromagnetic alloys. Parts of any size or shape can be inspected.

Magnetic-particle inspection may be applied to all types of weldments. It is sometimes used to inspect each pass immediately after it has been deposited in multipass welds. Most steel weldments in aircraft components are reviewed by magnetic-particle inspection. In thin weldments, subsurface defects may be detected with this type of inspection process. For a detailed description of this nondestructive testing method, see the articles "Magnetic-Particle Inspection" and "Nondestructive Inspection of Weldments" in Volume 11 of the 8th edition of *Metals Handbook*.

Codes, standards, and specifications applicable to magnetic-particle inspection include:

- "Nondestructive Testing Personnel Qualification and Certification, Supplement B, Magnetic Particle Method," ASNT Recommended Practice No. ASNT-TC-1A
- "Standard Method for Wet Magnetic Particle Inspection," ASTM E-138
- "Standard Method for Dry Powder Magnetic Particle Inspection," ASTM E-109

Radiographic Inspection

Radiography is a nondestructive test method that uses x-ray or gamma radiation to examine the interior of materials. Radiographic examination provides a permanent record of defects. Although slow and expensive, radiographic testing is a reliable method of detecting inclusions, porosity, cracks, and voids in the interior of castings, welds, and other structures.

Radiography is also suitable for locating subsurface defects in all types of alloys. Radiography frequently is used in the pipeline industry to ensure proper weld quality. For a detailed description of this nondestructive testing method, see the articles "Radiographic Inspection" and "Nondestructive Inspection of Weldments" in Volume 11 of the 8th edition of *Metals Handbook*.

Codes, standards, and specifications applicable to radiographic testing include:

- "Nondestructive Testing Personnel Qualification and Certification, Supplement A, Radiographic Testing Method," ASNT Recommended Practice No. ASNT-TC-1A
- "Recommended Practice for Radiographic Testing," ASTM E-94
- "Industrial Radiographic Terminology," ASTM E-52
- "Standard Reference Radiographs for Steel Welds," ASTM E-390
- "IIW Collection of Reference Radiographs of Welds," available from AWS
- *Interpretation of Radiographs of Pipeline Welding Defects*, Hobart Brothers Co., Troy, Ohio

Ultrasonic Inspection

Ultrasonic inspection is a nondestructive method of analyzing weldments for internal quality. High-frequency mechanical vibrations similar to sound waves are used in the testing procedure. A beam of ultrasonic energy is directed into the specimen to be tested, which travels through the material. The beam is diverted when it encounters a discontinuity or a change in alloy composition. Ultrasonic testing can be used to test practically any metal or material. It is not suitable for testing complex weldments. For a detailed description of this nondestructive testing method, see the articles "Ultrasonic Inspection" and "Nondestructive Inspection of Weldments" in Volume 11 of the 8th edition of *Metals Handbook*.

Codes, standards, and specifications applicable to ultrasonic testing include:

- "Nondestructive Testing Personnel Qualification and Certification, Supplement C, Ultrasonic Testing Method," ASNT Recommended Practice No. ASNT-TC-1A
- "Standard Method for Ultrasonic Contact Inspection of Weldments," ASTM E-164
- IIS/IIW-340-69 (ex doc V-360-67), "Classification of Defects in Metallic Fusion Welds with Explanation," *Metal Construction and British Welding Journal*, February 1970, London, England
- "Welding Inspection," AWS

SELECTED REFERENCES

- "Boiler and Pressure Vessel Code," Section III, Nuclear Power Plant Components, American Society of Mechanical Engineers, New York, 1980
- "Quality Assurance Criteria for Nuclear Power Plants and Fuel Reprocessing Plants," Code of Federal Regulations, Section 10, Energy, Part 50, Appendix B (10CFR50-B)
- "Standard for Welding of Reactor Coolant and Associated Systems and Components for Naval Nuclear Power Plants," NAVSHIPS 250-1500-1
- "Code for Pressure Piping," ANSI B 31, American National Standards Institute, New York
- "Standard for Welding Pipe Lines and Related Facilities," API Standard 1104, 13th ed., American Petroleum Institute, Washington, DC, July 1973
- "Regulations for the Transportation of Natural and Other Gas by Pipeline," Part 192, Title 49 Code of Federal Regulations, Department of Transportation, Office of Pipeline Safety, Washington, DC, Oct 1973
- "Transportation of Liquids by Pipeline," Part 195, Department of Transportation, Hazardous Materials Regulations Board, Washington, DC

- "Structural Welding Code," AWS D1.175, 1975, American Welding Society, Miami
- "Standard Specifications for High Bridges," American Association of State Highway Transportation Officials, Washington, DC
- "Specification for Steel Railway Bridges," American Railway Engineering Association, Chicago, 1969
- Marine Engineering Regulations, Sub Chapter F, Part 57, "Welding and Brazing," U.S. Coast Guard, Department of Transportation, Code of Federal Regulations, Washington, DC
- "Fabrication, Welding and Inspection of Ship Hulls," NAVSHIPS 0900-000-1000, Department of the Navy, Naval Ship Systems Command, Washington, DC, Oct 1968
- "Standard Specification for Merchant Ship Construction," Reference ABS, U.S. Maritime Administration
- "Rules for Building and Classing Steel Vessels," American Bureau of Shipping, New York, 1974
- "Standard for Welded Steel Elevated Tanks, Standpipes and Reservoirs for Water Storage," D100-73, American Waterworks Association, New York
- "Welded Steel Tanks for Oil Storage," API Standard 650, American Petroleum Institute, Washington, DC
- "Specifications for Design, Fabrication and Construction of Freight Cars," Standard 1964, Association of American Railroads, Chicago
- "General Design and Construction Requirements," Code of Federal Regulations, Title 49, Transportation Section 178.340, Part D, Superintendent of Documents, Washington, DC
- "Specification for Cargo Tanks," Code of Federal Regulations, Title 49, Transportation Section 178.337 (ML 331), Superintendent of Documents, Washington, DC
- "Aerospace Material Specifications," Society of Automotive Engineers, Inc., New York
- "National Aerospace Standards," Aerospace Industries Association of America, Inc., available from National Standards Association, Washington, DC
- "Test, Aircraft and Missile Welding Operators Qualification," MIL-T-5021D, Department of Defense, Washington, DC
- "Welding on Earthmoving and Construction Equipment," AWS D14.3, American Welding Society, Miami
- "Welding Industrial and Mill Cranes," AWS D14.1, American Welding Society, Miami
- "Metal Cutting Machine Tool Weldments," AWS D14.2, American Welding Society, Miami

Weld Discontinuities

By the ASM Committee on Weld Discontinuities*

DISCONTINUITIES are interruptions in the desirable physical structure of a weld. A discontinuity constituting a danger to the fitness-for-purpose of a weld is a defect. By definition, a defect is a condition that must be removed or corrected (Ref 1). The word "defect" should therefore be carefully used, as it implies that a weld is defective and requires corrective measures or rejection. Thus, repairs may be made unnecessarily and solely by implication, without a critical engineering assessment. Consequently, the engineering community has recently begun to use the word "discontinuity" or "flaw" instead of "defect."

The significance of a weld discontinuity should be viewed in the context of the fitness-for-purpose of the welded construction. Fitness-for-purpose is a concept of weld evaluation that seeks a balance among quality, reliability, and economy of welding procedure. Fitness-for-purpose is not a constant. It varies depending on the service requirements of a particular welded structure, as well as on the properties of the material involved.

Neither construction materials nor engineered structures are free from imperfections. Welds and weld repairs are not exceptions. Weld acceptance standards are used when a discontinuity has been clearly located, identified, sized, its orientation determined, and its structural significance questioned. Critical engineering assessments of weld discontinuities are performed to define acceptable, harmless discontinuities in a structure that will not sacrifice weldment reliability. One of the major reasons for understanding the engineering meaning of weld discontinuities is to decrease the cost of welded structures by avoiding unnecessary repairs of harmless weld discontinuities (Ref 2). Welders, of course, must constantly be encouraged to make sound (perfect) welds independent of prevailing acceptance standards.

Classification of Weld Discontinuities

Discontinuities may be divided into three broad classifications: (1) design related, (2) welding process related, and (3) metallurgical. Design-related discontinuities include problems with design or structural details, choice of the wrong type of weld joint for a given application, or undesirable changes in cross section.

Discontinuities resulting from the welding process include:

- *Undercut*: A groove melted into the base metal adjacent to the toe or root of a weld and left unfilled by weld metal
- *Slag inclusions*: Nonmetallic solid material entrapped in weld metal or between weld metal and base metal
- *Porosity*: Cavity-type discontinuities formed by gas entrapment during solidification
- *Overlap*: The protrusion of weld metal beyond the toe, face, or root of the weld
- *Tungsten inclusions*: Particles from tungsten electrodes which result from improper gas tungsten arc welding procedures
- *Backing piece left on*: Failure to remove material placed at the root of a weld joint to support molten weld metal
- *Shrinkage voids*: Cavity-type discontinuities normally formed by shrinkage during solidification
- *Oxide inclusions*: Particles of surface oxides which have not melted and are mixed into the weld metal
- *Lack of fusion (LOF)*: A condition in which fusion is less than complete
- *Lack of penetration (LOP)*: A condition in which joint penetration is less than that specified
- *Craters*: Depressions at the termination of a weld bead or in the molten weld pool
- *Melt-through*: A condition resulting when the arc melts through the bottom of a joint welded from one side
- *Spatter*: Metal particles expelled during welding which do not form a part of the weld
- *Arc strikes (arc burns)*: Discontinuities consisting of any localized remelted metal, heat-affected metal, or change in the surface profile of any part of a weld or base metal resulting from an arc
- *Underfill*: A depression on the face of the weld or root surface extending below the surface of the adjacent base metal

Metallurgical discontinuities include:

- *Cracks*: Fracture-type discontinuities characterized by a sharp tip and high ratio of length and width to opening displacement
- *Fissures*: Small crack-like discontinuities with only a slight separation (opening displacement) of the fracture surfaces
- *Fisheye*: A discontinuity found on the fracture surface of a weld in steel that consists of a small pore or inclusion surrounded by a bright, round area
- *Segregation*: The nonuniform distribu-

*L.W. Sandor, *Chairman,* Manager of Materials Technology, Franklin Research Center, and Professor of Materials Engineering, Widener University; Michael F. Ahern, Engineer, Component Engineering Group, Northeast Utilities; Donald A. Bolstad, Section Chief, Materials Engineering, Martin Marietta Aerospace; Spencer H. Bush, Senior Staff Consultant, Battelle Pacific Northwest Laboratory; John C. Duke, Jr., Associate Professor of Engineering Science and Mechanics, Virginia Polytechnic Institute and State University; Maurice B. Kasen, Metallurgist, National Bureau of Standards; Koichi Masubuchi, Professor of Ocean Engineering and Materials Science, Massachusetts Institute of Technology; William H. Munse, Professor Emeritus of Civil Engineering, University of Illinois; Alan W. Pense, Professor of Metallurgy and Materials Engineering, Lehigh University

tion or concentration of impurities or alloying elements which arises during the solidification of the weld
- *Lamellar tearing*: A type of cracking that occurs in the base metal or heat-affected zone (HAZ) of restrained weld joints that is the result of inadequate ductility in the through-thickness direction of steel plate

In decreasing order of harmfulness, weld discontinuities may be classified as: (1) crack or crack-like, (2) geometric, (3) LOF and LOP, (4) solid inclusions, and (5) porosity. From a fracture mechanics point of view, weld discontinuities may be classified as planar or nonplanar (volumetric). A planar discontinuity is one which has two dimensions, such as a crack. Nonplanar discontinuities have three dimensions, such as porosity. Discontinuities may appear on the surface of weldments, they may be buried within the welded joint, or they may exist in the through-thickness of a section.

Occurrence of Weld Discontinuities

The observed occurrence of discontinuities and their relative amounts largely depend on (1) the welding process used, (2) the inspection method applied, (3) the type of weld made, (4) the joint design and fit-up obtained, (5) the material utilized, and (6) the working and environmental conditions. The most frequent weld discontinuities found during manufacture, ranked in order of decreasing occurrence on the basis of arc welding processes, are:

Shielded metal arc welding (SMAW)

- Slag inclusions
- Porosity
- LOF/LOP
- Undercut

Submerged arc welding (SAW)

- LOF/LOP
- Slag inclusions
- Porosity

Flux cored arc welding (FCAW)

- Slag inclusions
- Porosity
- LOF/LOP

Gas metal arc welding (GMAW)

- Porosity
- LOF/LOP

Gas tungsten arc welding (GTAW)

- Porosity
- Tungsten inclusions

Ranking of discontinuities and their relative amounts for manual and automatic welding, when established on the basis of nondestructive examination (NDE) methods, are illustrated in Table 1.

Table 1 Ranking of weld discontinuities for manual and automatic welding observed by NDE methods
Discontinuities are ranked in decreasing order of occurrence

Manual welding: Type of discontinuity	Amount, %	Automatic welding: Type of discontinuity	Amount, %
Detection by radiographic inspection(a)			
Slag	35-80	LOF/LOP	30-60
Porosity	10-20	Cracking	19-25
LOF/LOP	8-20	Slag	5-25
Cracks	1-10	Porosity	5-15
Detection by ultrasonic inspection(b)			
Slag	50-60	LOF/LOP	50-60
LOF/LOP	20-30	Slag	20-35
Porosity	5-30	Porosity	5-20
Detection by visual methods(c)			
Undercut	15-80		
Surface porosity	5-20		
Slag	13 (avg)		
Undesirable weld profile	2-15		
LOP/LOF	7 (avg)		
Cracks at craters	1-10		

(a) Typical range 5-20%. (b) Typical range 1-14%. (c) Typical range 2-25%. Data available for manual welding only.

Crack or Crack-Like Discontinuities

Cracks are the most serious type of weld discontinuity because they can greatly reduce the strength of welds or weldments. Types of cracking include hydrogen-induced cold cracking, microfissures, lamellar tearing, hot cracking, stress corrosion cracking, and graphitization.

Hydrogen-Induced Cold Cracking

Cold cracking is the term used for cracks that occur after the weld has solidified and cooled; it can occur in either the HAZ or the weld metal of low-alloy and other hardenable steels. Because these cracks occur under conditions of restraint, they are often referred to as restraint cracks. Cracking may occur several hours or days after the weld has cooled; consequently, the term "delayed cracking" is also used (Ref 3). On the basis of location, cracks are often described as toe cracking (Fig. 1), root cracking (Fig. 2), or underbead cracking.

For cold cracks to occur in steels, three principal factors must be present: (1) atomic hydrogen, (2) HAZ or weld metal that is susceptible to hydrogen embrittlement, and (3) a high tensile stress resulting from restraint. Controlling one or more of these factors may reduce the occurrence of cold cracking. The basic relationships among the variables responsible for cold cracking and the methods of controlling these variables are summarized in Fig. 3 and 4.

In steels, cracking in the base metal is often attributed to high carbon, alloy, or sulfur content. Control of this cracking requires the use of low-hydrogen electrodes, high preheat, sufficient interpass temperature, and less penetration through the use of lower currents and smaller electrodes. Lower heat input reduces the amount of alloy added to the weld from the base metal (less dilution). The process of HAZ crack-

Fig. 1 Photomacrograph of a fatigue crack that has initiated at the weld toe in hardenable steel. Etched, ×2

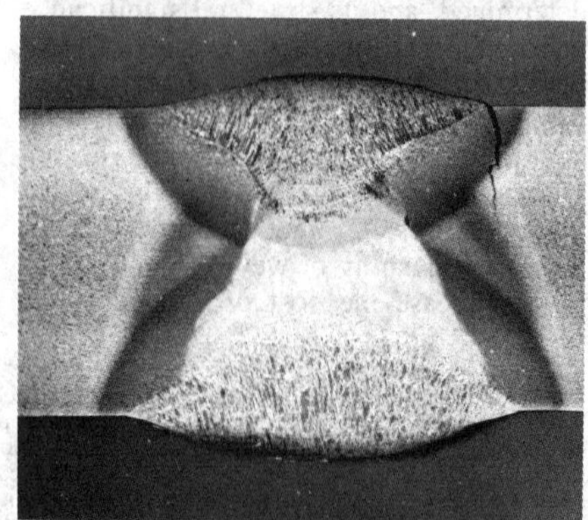

Fig. 2 (a) Root cracks, LOP, and (b) zinc dilution in an evaporator weldment consisting of a T fillet weld (309L) joining 1/2-in. A36 steel (coated with zinc-rich paint prior to welding) to 1/4-in. 316L stainless steel. Problem was discovered by observing excessive crater cracking in many of the intermittent fillet welds (SEM); ×400, as-polished, electron diffraction analysis. Courtesy of K.G. Wold, Aqua-Chem, Inc.

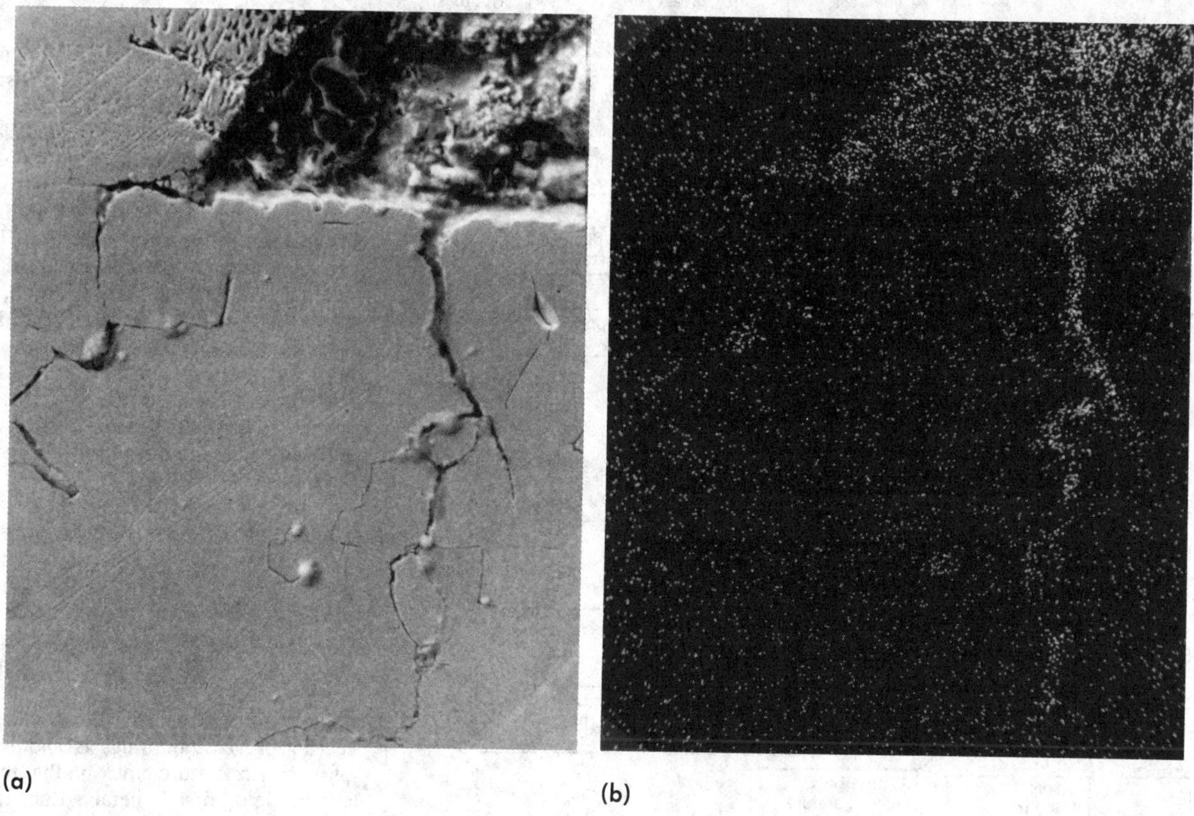

ing in steels is usually described as follows (Ref 4):

- Hydrogen is introduced into the hot weld metal and dissolves into the weld pool as atomic or ionized hydrogen.
- As the weld metal cools, it becomes supersaturated with hydrogen, and some hydrogen diffuses into the HAZ.
- Under rapid cooling, hydrogen has insufficient time to diffuse from the HAZ. Also, rapid cooling increases the chance for the HAZ to transform from austenite to martensite.
- Atomic hydrogen is insoluble in martensite and seeks rifts and discontinuities in the lattice, where it collects.
- External stress caused by thermal contraction, combined with the effects of the trapped hydrogen, causes discontinuities to enlarge to crack size.
- In time, hydrogen diffuses to the new crack root and causes the crack to enlarge, thus the occurrence of delayed cracking.

This process repeats itself until the stress in the weld has been relieved. Hydrogen and stress also play an important role in weld-metal cracking. Microcracks in the weld metal are often associated with small nonmetallic inclusions that provide points of initiation. More information on hydrogen-induced cold cracking can be found in the articles "Principles of Joining Metallurgy" and "Arc Welding of Hardenable Carbon and Alloy Steels" in this Volume.

Fisheye

Fisheyes are cracks surrounding porosity or slag in the microstructure of a weld. Fisheyes are a form of hydrogen embrittlement cracking caused by the presence of hydrogen that collects at these locations and that embrittles the surrounding metal in the same way cold cracking does. What distinguishes fisheye is a characteristically shiny appearance when viewed optically (Fig. 5 and 6).

Fisheyes, also termed "flakes," "blisters," and "halos," are predominantly found in ferritic steels welded with high-hydrogen electrodes. Complete understanding of the conditions for the formation of this type of crack is still evolving. Since they were first observed during bend testing, it is frequently assumed that stressing of the weldment beyond its tensile yield strength is required to form fisheyes, but this has not been confirmed. Fisheyes may be potentially harmful, if they form in service, as their presence will not be detected by the usual nondestructive examination procedures. Therefore, precautions should be taken to assess the probability of fisheye formation when structures welded with high-hydrogen electrodes are subject to high service stresses.

The overall fracture appearance of fisheyes usually indicates a mixed or quasicleavage fracture mode, as seen in Fig. 5(b). When a halo is examined at higher magnification, a fracture surface with a pattern radiating from a nucleating discontinuity is revealed. Small, buried microfissures also have been noted to nucleate fisheyes in as-welded E6010 weld metal. Elevated-temperature aging or postweld heat treatment can effectively eliminate fisheyes. Other factors that help prevent the occurrence of this type of discontinuity include high preheat, good joint design, low-hydrogen

Fig. 3 Causes and cures of cold cracking in weld metal. Thermal Severity Number (TSN), which is four times the total plate thickness capable of removing heat from the joint, is thus a measure of the member's ability to serve as a heat sink.

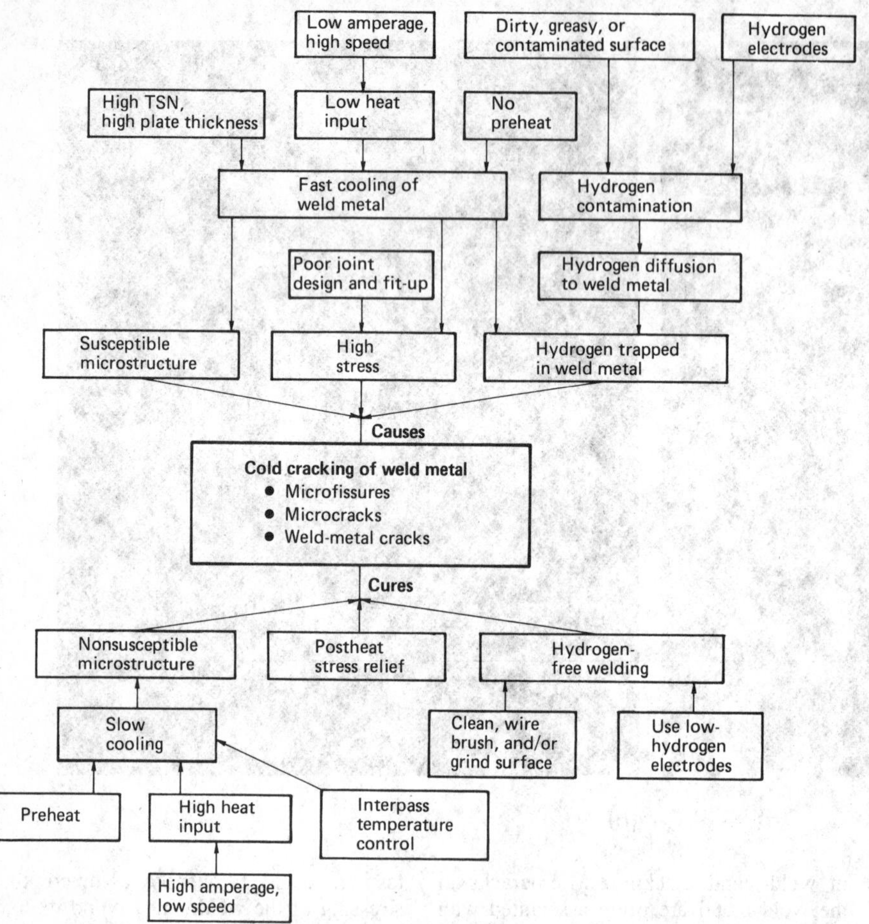

electrodes, and reduction in restraint. Note that use of low-hydrogen electrodes alone is apparently insufficient to eliminate fisheye formation.

Fisheyes do not form during conventional Charpy V-notch testing, apparently because the deformation rate is too high to permit the required diffusion of hydrogen (Fig. 5c).

Microfissures

Microfissures are metallurgically related flaws formed during weld-metal solidification. They are frequently found in austenitic stainless steel welds where the weld metal has an inadequate ferrite content. Microfissures can be of the hot or cold type of cracking, induced by small strain. Their detection is difficult by conventional nondestructive techniques; usually, metallographic preparation is necessary. Applied strain is needed to produce cracking; the amount needed to initiate cracking depends on material toughness. Because material is always subject to some type of strain, either from external or internal sources, the occurrence of cracking is therefore controlled by the conditions of welding, during which material toughness may be decreased. Examples of microcracking (microfissures) are shown in Fig. 7 and 8.

Lamellar Tearing

Lamellar tearing is a form of base-metal cold cracking that occurs in steels parallel to the plate surface adjacent to the welds. Lamellar tearing is associated with nonmetallic inclusions such as oxides, sulfides, and silicates that are elongated in the direction of rolling. The net result of these inclusions is a decrease in the through-thickness ductility. Other factors that affect lamellar tearing are design details, plate thickness, magnitude, sign and distribution of induced stresses, type of edge preparation and properties, as well as base-metal fabrication. More information on lamellar tearing can be found in the articles "Principles of Joining Metallurgy" and "Arc Welding of Hardenable Carbon and Alloy Steels" in this Volume.

Hot Cracking

Hot cracking is the most common form of weld-metal cracking in steels. The term encompasses cracks that form while a weld is solidifying or when a weldment is reheated. Hot cracking is often caused by excessive sulfur, phosphorus, or lead content in the base metal. It can also be caused by an improper method of breaking the arc or in a root pass when the cross-sectional area of the weld is small, compared to the mass of the base metal. Solidification cracking, liquation cracking, ductility-dip cracking, and reheat cracking are types of hot cracking. Such cracks are normally intergranular in nature and can occur in either the weld-metal or the base-metal HAZ.

Solidification cracking occurs within a few hundred degrees of the nominal liquidus temperature of the weld metal and is induced by welding stresses and the presence of low-melting-point constituents that form as a result of segregation during the liquid-to-solid-phase transformation process.

Liquation cracking (often called hot tearing or HAZ burning) is another type of high-temperature cracking that may occur in welds of all metals. Localized liquation of low-melting-point constituents such as foreign inclusions, carbide phases, or local segregations of certain alloying elements in the solid may form liquid grain-boundary films. The attendant welding strain may, in turn, be sufficiently high to open liquation cracks.

Ductility-dip cracking may occur in either the weld metal or the HAZ of austenitic stainless steels and nickel-based superalloys. This type of crack forms while the weld cools through a range of temperatures where the ductility of the particular metal is inherently low. The ductility-dip temperature for austenitic stainless steels falls just below the recrystallization temperature. If sufficient strain is present as the metal cools through the ductility-dip temperature range, cracking occurs. Additional information on ductility-dip cracking can be found in the article "Principles of Joining Metallurgy" in this Volume.

Reheat cracking or stress-relief cracking may form in the HAZ of welds in ferritic and austenitic steels. Postweld heat treatment may be specified for weldments to reduce (or eliminate) residual stresses and/or to restore toughness in the

Fig. 4 Causes and cures of cold cracking in base metal

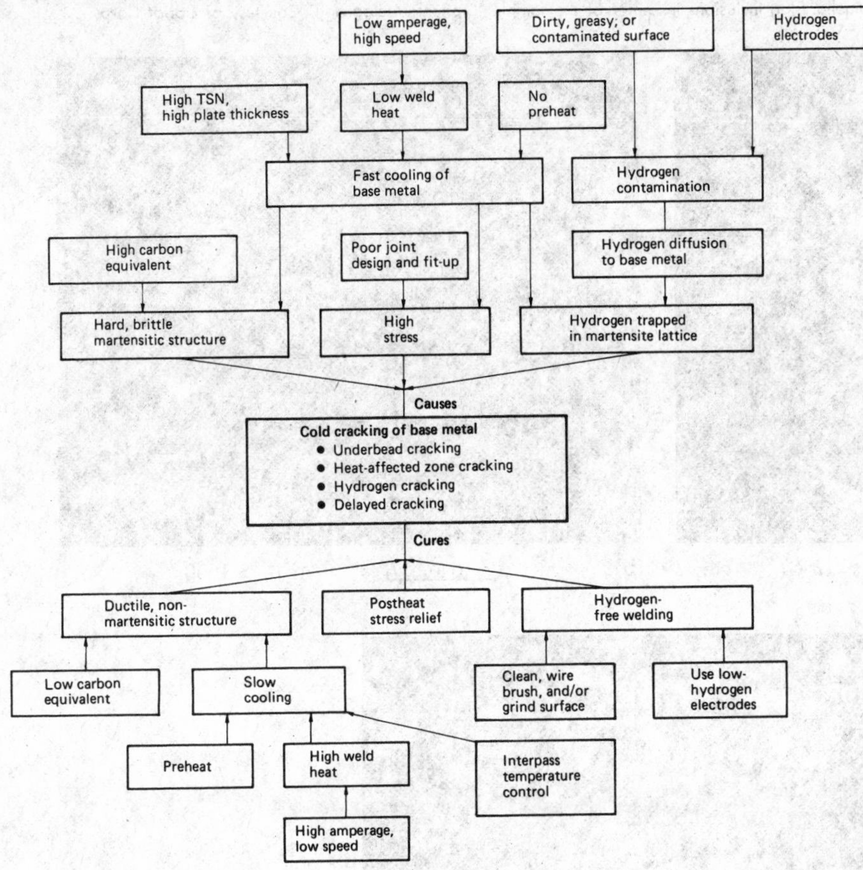

HAZ and weld metal. Reheat cracking may take place during this postweld heat treatment as a result of pre-postweld heat treatment microstructure of the HAZ brought about by welding. Welding heat input induces an elevated temperature solution treatment in the HAZ, which leads to dissolution of metallic carbides. Interstitial carbon may strengthen the grains sufficiently so that relaxation of internal stresses may occur only by grain-boundary sliding, with little or no grain deformation; hence, reheat cracking occurs. Reheat cracking has also been linked to temper embrittlement and the presence of trace elements, such as boron, arsenic, antimony, and tin, in steels. Remedies for reheat cracking include changes in metal composition (use of less susceptible steel), reduction in thickness, decrease in sulfur content, slower heating rate to avoid thermal shock, controlled cooling rate to prevent precipitation of carbides, higher creep ductility, good joint design, decrease in constraint, high preheat, elimination of stress concentration at welds, and use of weld metal with low hot strength. Additional information on reheat and stress-relief cracking can be found in the articles "Principles of Joining Metallurgy" and "Arc Welding of Hardenable Carbon and Alloy Steels" in this Volume.

Causes and Prevention of Hot Cracking. The factors affecting the formation of hot cracks include metallurgical properties, welding variables, and mechanical properties. The basic relationships among these factors and methods for controlling hot cracking in steels are shown in Fig. 9 and 10.

During solidification of weld metal, several compounds form whose presence, in their liquid phase, at grain boundaries contributes to hot cracking. Sulfur is considered the most detrimental element because it reacts to form many low-freezing-temperature compounds such as iron sulfide.

Sulfur compounds have the greatest effect on hot cracking in weldments stressed in the through-thickness of the base metal. Therefore, sulfur content in the base and filler metals should be kept low. Carbon is another detrimental element because it affects the liquid phase of the weld metal and tends to decrease high-temperature ductility of the weld metal. The level of carbon present in the base metal cannot be altered to any great extent, but the effects of carbon can be offset by high manganese-to-sulfur ratios. With a carbon content of 0.06 to 0.11 wt% in the weld metal, hot cracks can be eliminated completely when the manganese-to-sulfur ratio is >22. For carbon contents between 0.11 to 0.13%, a ratio of over 30 is required to avoid hot cracking. When the carbon content exceeds 0.13%, the effect of the manganese-to-sulfur ratio is nullified.

Silicon and phosphorous do not affect the liquid phase of the weld metal directly; however, they promote segregation of the sulfur and thereby aid sulfur reactions. Nickel has a more direct effect on the liquid phase, but its effects on crack formation are not completely understood.

No matter how pronounced the liquid phase in the weld metal is, hot cracks will not form unless tensile stresses are imposed on the weld. Unfortunately, stresses are impossible to avoid. Consequently, the greater the stresses imposed during solidification or reheating, the more severe the cracking.

The size and thickness of the base metal, the joint design, and the size and shape of the weld bead influence the mechanical stresses in a weld. Furthermore, different welding processes produce different amounts of heat input, thus producing variations in microstructural changes and levels of residual stresses. Joint design should provide a good fit-up. Excessive amounts of weld metal should be avoided. Welding procedures that minimize weld restraint should be specified.

Stress Corrosion Cracking

This type of cracking takes place when hot concentrated caustic solutions are in contact with steel that is stressed in tension to a relatively high level. The high level of tension stresses can be created by loading or by high residual stresses. Stress corrosion cracking will occur if the concentration of the caustic solution in contact with the steel is sufficiently high and if the stress level in the weldment is sufficiently high. This situation can be reduced by reducing stress level and reducing the concentration of the caustic solution. Various inhibitors can be added to the solution to reduce the concentration. Another solution is to maintain close inspection on highly stressed areas.

Stress corrosion often results in complete failure of a part. An example of stress corrosion failure in type 304 stainless steel fuel element cladding is shown in Fig. 11.

Fig. 5 (a) Fisheyes in as-welded E7018 tensile specimen tested at room temperature. Optical photomacrograph. (b) Mixed-mode fracture in as-welded E7018 three-point bend test specimen at room temperature (SEM). (c) Absence of fisheyes in as-welded E7018 Charpy V-notch specimen at −20 °F (11 ft · lb). Note the two lamellar slag inclusions in center of the sample. Optical fractograph. Courtesy of Cabot Corp.

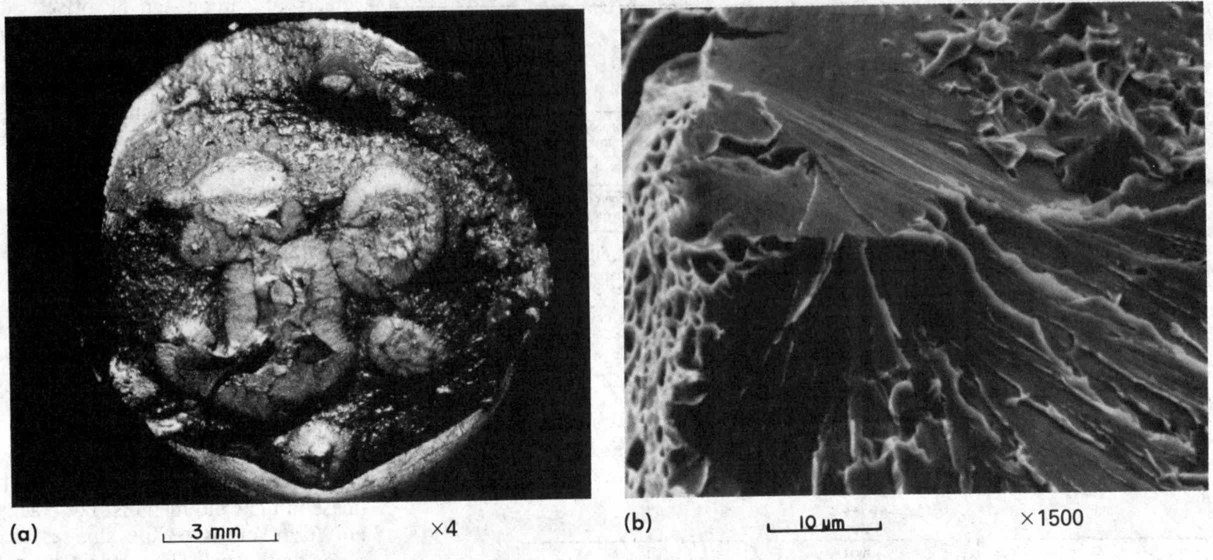

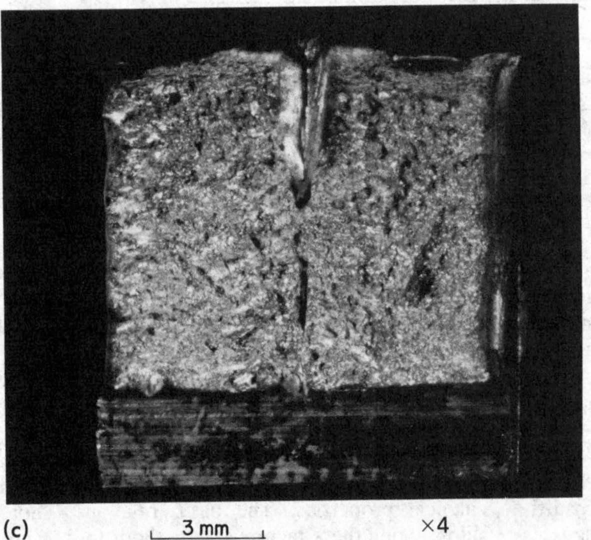

Graphitization

This type of cracking is caused by long service life exposed to thermal cycling, or repeated heating and cooling. This may cause a breakdown of carbides in the steel into small areas of graphite and iron. When these graphite formations are exposed to thermal cycling, cracking occurs. It will more often occur in carbon steels deoxidized with aluminum. The addition of molybdenum to the steel tends to restrict graphitization; for this reason, carbon-molybdenum steels are normally used in high-temperature power plant service. These steels must be welded with filler metals of the same composition.

Causes and Prevention of Cracking

The three principal types of weld-metal cracking are longitudinal, transverse, and crater cracks. Occurrence of weld-metal cracking in steel is likely if combinations of the following conditions are present:

- Incompatible composition and microstructure (high hardenability or carbon equivalent)
- High restraint
- High hydrogen content
- High tensile stress
- High weld travel speed
- Absence of preheat
- Low interpass temperature
- Poor weld profile
- Poor electrode manipulation
- Thin weld-bead cross section (particularly in the first pass)
- Improper fit-up
- Low weld current
- Small electrode diameter
- Improper polarity

Fig. 6 Formation of fisheyes. (a) At pore, in as-welded E6010 tensile specimen tested at room temperature. (b) Slag inclusions in as-welded E7018 tensile specimen at room temperature. (c) Slag/LOF in as-welded E11018 tensile specimen strained at 1.0% at room temperature (SEM). (Courtesy of C.R. Patriarca, M.S. thesis, 1981)

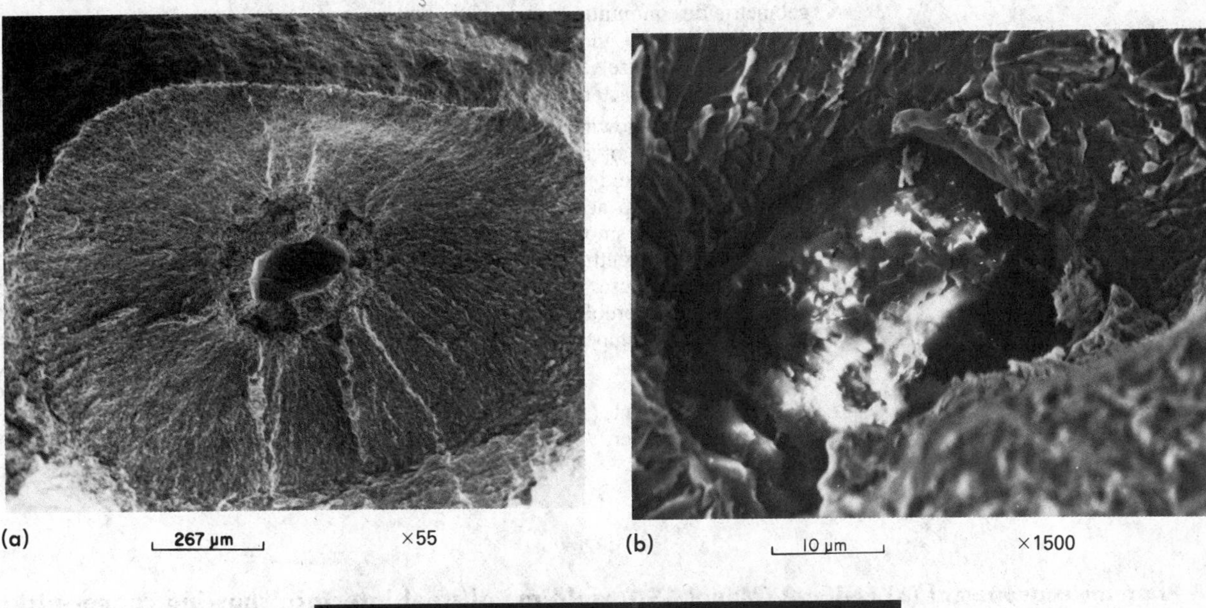

(a) 267 μm ×55 (b) 10 μm ×1500

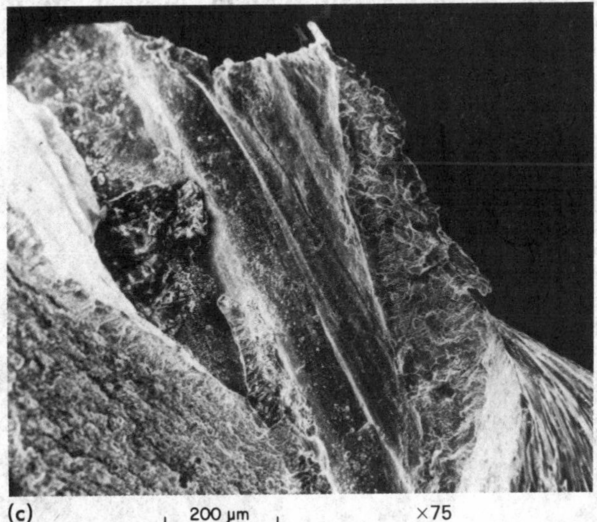

(c) 200 μm ×75

- Improper joint geometry (included angle or root opening too small
- Improper backgouging (narrow or shallow)
- Unbalanced heat input

To prevent cracking in multiple-pass groove or fillet welds:

- Increase bead size of the first pass by using lower travel speed, a short arc, and an uphill (~5°) welding mode.
- Change weld profile to flat or convex.
- Control interpass temperature and weld while the base metal is hot.
- Use filler metal and/or slag compositions that promote good wettability with the base metal.

To prevent cracking in craters:

- Fill crater sufficiently before breaking the arc.
- Employ a backstepping technique (terminate each weld on the crater of the previous weld).

To prevent cracking under restraint conditions:

- Eliminate or decrease the degree of restraint.
- Leave a gap between plates to permit shrinkage.
- Weld in the direction of no restraint.
- Peen the passes while hot. Caution: Avoid peening first and final passes due to the danger of either causing or covering cracks or of interfering with inspection.

Geometric Discontinuities

Geometric weld discontinuities are those associated with imperfect shape or unacceptable weld contour. Undercut, under-

Fig. 7 Photomicrograph of weld-metal microcrack. Etchant: 2% nital; 215×

fill, overlap, excessive reinforcement, fillet shape, and melt-through are included in this grouping. Morphology of these geometric discontinuities varies. One way of controlling the morphology of fillet welds in steel is by selecting electrodes with good wetting characteristics (Ref 5).

Undercut, as shown in Fig. 12, is a gap located at the toe or root of a weld that occurs when the weld metal does not completely fill the gap at the surface of the groove to form a smooth junction at the weld toe. Undercutting is particularly troublesome because it produces stress risers that create problems under impact, fatigue, or low-temperature service. To minimize undercut:

- Decrease current, travel speed, and electrode diameter, which controls weld-pool size.
- Change electrode angle so that the arc force holds the weld metal in the corners.
- Avoid weaving.
- Maintain constant travel speed.
- Use proper backing.
- Change the surface tension and viscosity of the molten weld metal where possible.

Undercut actually refers more to the base metal adjacent to the weld, whereas imperfect shape is a defect of the weld itself. This can include such things as excessive reinforcement on the face of a weld, which can occur on groove welds as well as fillet welds. There is also the problem of excessive reinforcement on the root of the weld, primarily open root groove welds. Excessive reinforcement is economically unsound. It can also be a stress riser and

Fig. 8 Photomicrographs of (a) ENiCu-2 (Monel 190) weld metal/steel interface showing copper-nickel penetration along prior austenite grain boundaries. Zinc and other trace elements such as titanium were also present. Voids and microcracks formed during cooling (optical photo). Etchant: picral; ×650. (b) Solidification voids in Monel weld metal (SEM), ×650. (c) Microfissures and shrinkage voids in the steel near Monel/steel interface (SEM). Etchant: 50% picral-50% nital; ×135. Courtesy of S.M. Fisher, Virginia Polytechnic Institute and State University, Doctoral Program, 1983.

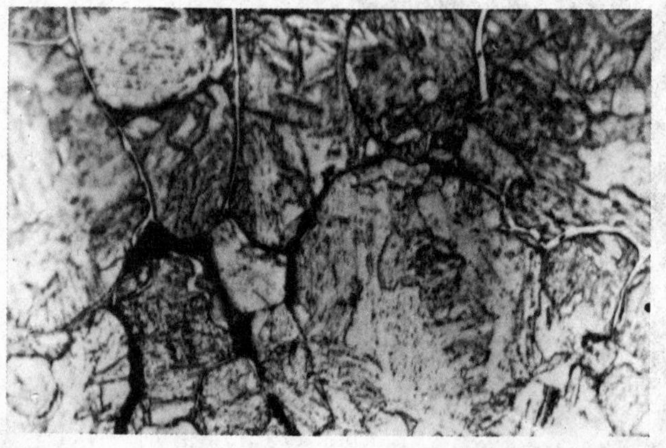

(a)

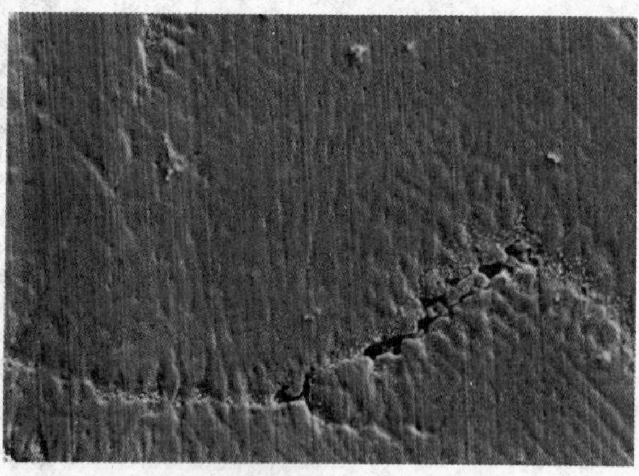

(b)

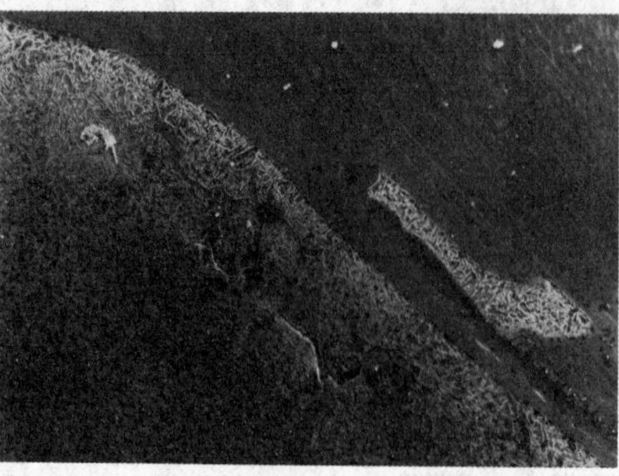

(c)

Fig. 9 Factors affecting hot cracking in weld metal

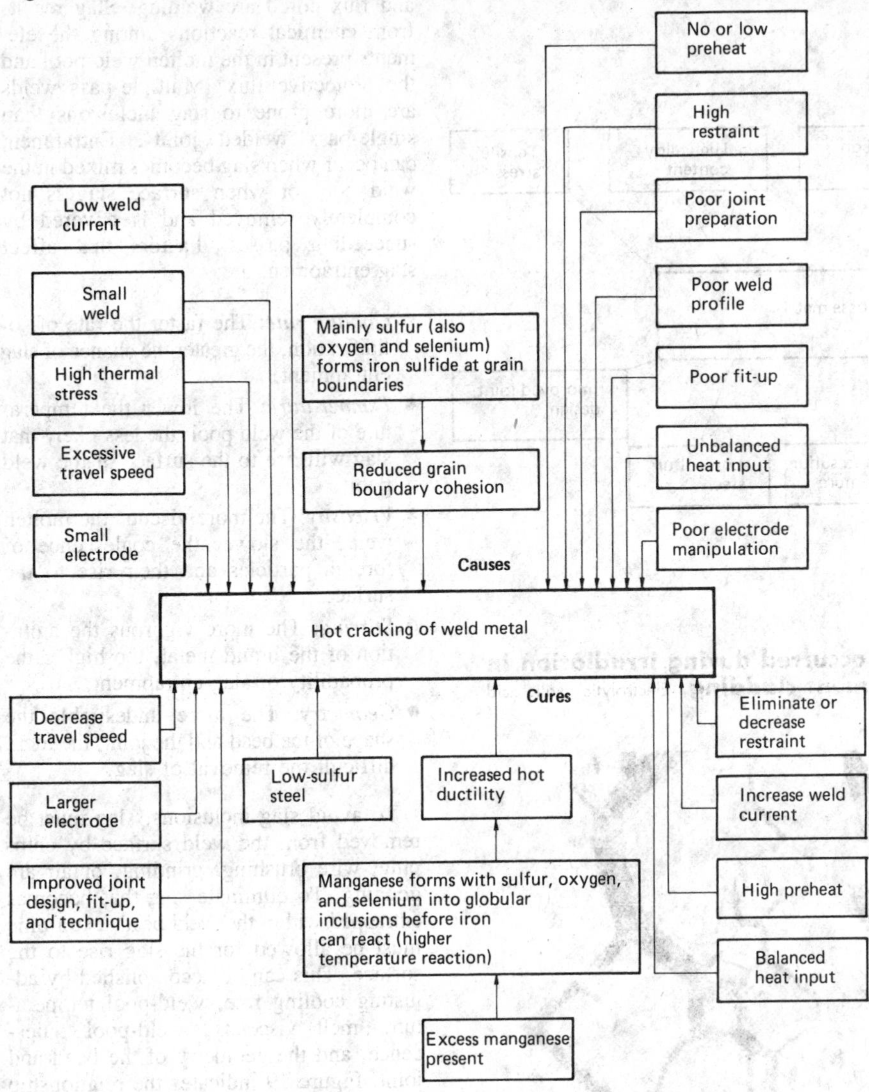

is objectionable from an appearance point of view. It is normally a factor involved with fit-up, welder technique, welding current, and type of electrode.

Underfill is defined as a depression on the face of a weld or root surface extending below the surface of the adjacent base metal. Underfill, as shown in Fig. 12, reduces the cross-sectional area of the weld below the designed amount and therefore is a point of weakness and potentially a stress riser where failure may initiate. To minimize underfill, the voltage, travel speed, and root opening should be reduced. Figure 13 shows underfill in a pulsed gas metal arc spot weld joining carbon steel to a 90Cu-10Ni alloy using ERNiCu-7 filler metal.

Overlapping, as shown in Fig. 12, is the protrusion of the weld metal over the edge or toe of the weld bead. This defect can cause an area of incomplete fusion, which creates a notch and can lead to crack initiation. To minimize overlapping:

- Use a higher travel speed.
- Use a higher welding current.
- Reduce the amount of filler metal added.
- Change the electrode angle so that the force of the arc will not push molten weld metal over unfused sections of the base metal.

Melt-through occurs when the arc melts through the bottom of the weld. To prevent melt-through, the welding current and the width of the root opening should be reduced, and travel speed should be increased.

Lack of Fusion and Lack of Penetration

Lack of fusion denotes a discontinuity caused by incomplete coalescence of some portion of the filler metal with the base material (Fig. 14). It can also form between weld beads in the case of multiple-pass welds. These essentially two-dimensional flaws occur when insufficient heat is absorbed by the underlying metal from the weld, causing incomplete melting at the interfaces of the weld and the base metal of successive passes. Lack of fusion usually is elongated in the direction of welding, with either sharp or rounded edges, depending on the conditions of formation.

When the weld metal has not penetrated to the bottom of the weld joint, the result is LOP (Fig. 15). Lack of penetration is caused by incorrect welding technique or by improper root gap. Incorrect welding techniques include low amperage, oversize electrode diameter, high travel speed, wrong electrode angle, incorrect joint cleaning, improper work position, misalignment of back side weld, excessive inductance in the dip transfer mode of gas metal arc welding, insufficient backgouging, too large a flat land in a groove weld, and undesirable arc manipulation by the welder, particularly in the vertical position.

To prevent LOF and LOP, heat input must be increased to allow complete fusion between the weld deposits and the workpiece, as well as between successive passes in multipass joints. Joint geometry must permit proper arc control and access for the electrode. The weldment surface must be cleaned of all contaminants. A welding technique must be chosen that will allow complete fusion and/or penetration. If the gap is excessive, weave manipulation should be adjusted accordingly.

Figure 16 illustrates many of the causes of and methods of prevention for LOF. Typical causes and methods of prevention for LOP are illustrated in Fig. 17.

Slag Inclusions

Slag inclusions are nonmetallic materials formed by the slag reaction that are trapped in the weld (Fig. 18). They frequently are nonplanar in shape and can range in size from very small globules to large, long bands along the axis of the weld. This type of discontinuity can occur randomly as isolated foreign particles or as continuous or intermittent lines of inclusions parallel to the axis of the weld. Slag inclusions occur in steel welds made by flux-shielded welding processes—

Fig. 10 Factors affecting hot cracking in the base-metal HAZ

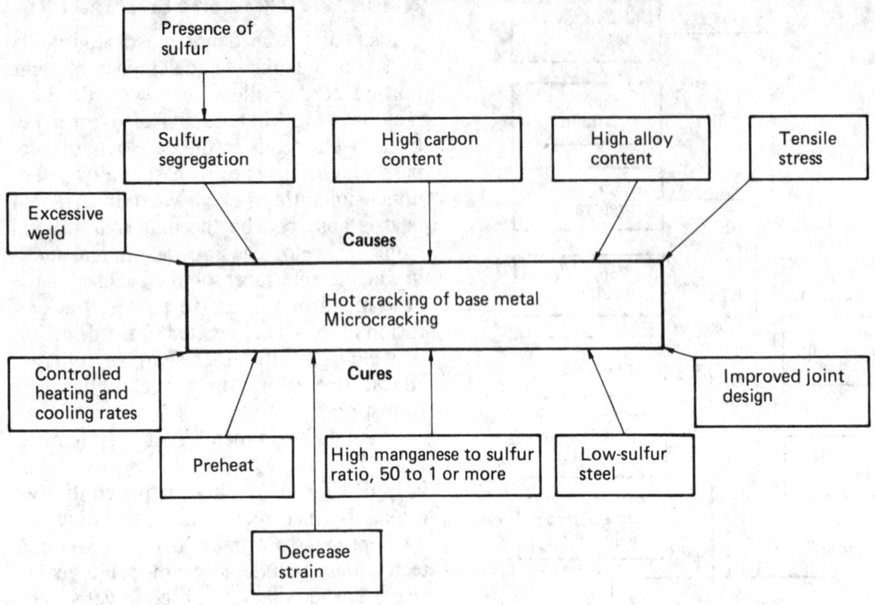

Fig. 11 Stress corrosion failure that occurred during irradiation in water in 304 stainless steel fuel element cladding. Electrolytic oxalic acid etch; ×425

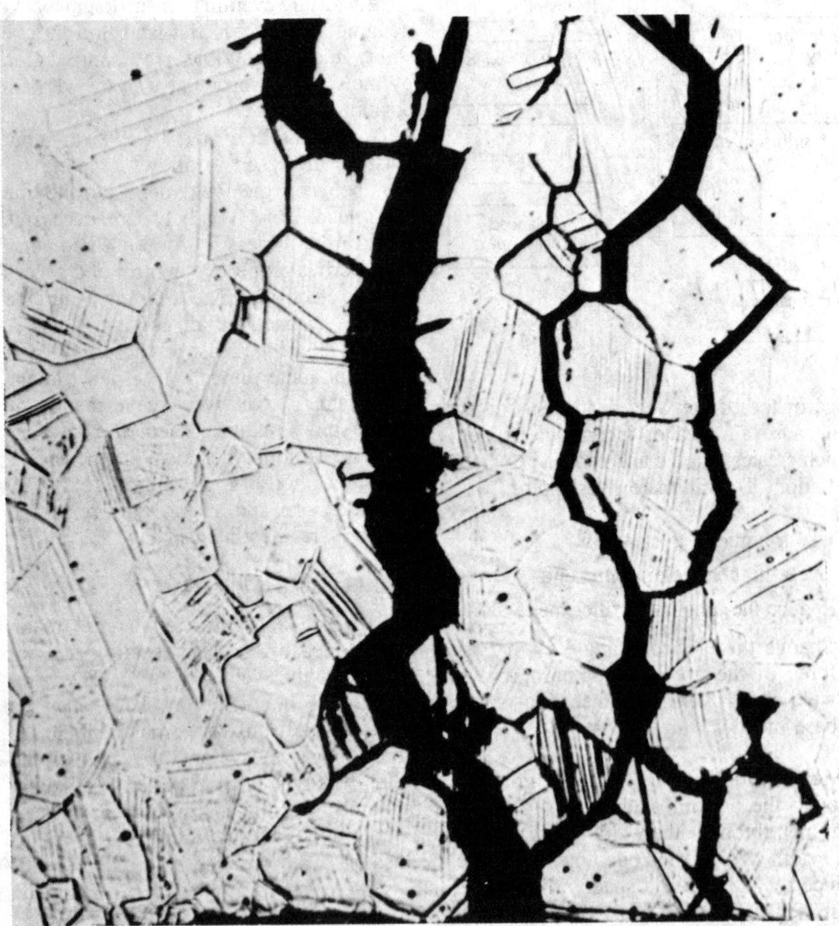

namely, shielded metal arc, submerged arc, and flux cored arc welding. Slag results from chemical reactions among the elements present in the molten weld pool and the protective flux. Multiple-pass welds are more prone to slag inclusions than single-pass welded joints. Entrapment can occur when slag becomes mixed in the weld pool or when surface slag is not completely removed and is covered by succeeding passes. Factors that affect slag entrapment are:

- *Cooling rate*: The faster the rate of solidification, the greater the chance of slag entrapment.
- *Temperature*: The lower the temperature of the weld pool, the less likely that slag will rise to the surface of the weld pool.
- *Viscosity*: The more viscous the molten weld, the slower the coalescence of foreign particles and their rise to the surface.
- *Stirring*: The more vigorous the agitation of the liquid metal, the higher the probability of slag entrapment.
- *Geometry*: The more undesirable the shape of the bead and the joint, the more difficult the removal of slag.

To avoid slag inclusions, slag must be removed from the weld surface by chipping, wire brushing, grinding, or air arc gouging. To eliminate slag that becomes entrapped within the weld bead, extra time must be allowed for the slag rise to the surface. This can be accomplished by adjusting cooling rate, weld-pool temperature, melt viscosity, weld-pool quiescence, and the geometry of the bead and joint. Figure 19 indicates the relationship among many of the factors that cause or prevent slag inclusions.

Fluxes exert a pronounced influence on the properties and soundness of the resultant weld. The type of flux used influences the quantity and type of foreign particles that may form in the weld bead. Basic fluxes promote low oxygen and/or sulfur contents in the weld metal, thereby minimizing the formation of slag inclusions.

Oxide Inclusions

Oxide inclusions are particles of surface oxides which have not melted and are mixed into the weld metal. These inclusions occur when welding metals that have surface oxides with very high melting points, such as aluminum and magnesium. Oxide inclusions weaken the weld and can serve as initiation points for cracking. The best method of prevention

Fig. 12 Weld discontinuities affecting weld shape and contour. (a) Undercut and overlapping in a fillet weld. (b) Undercut and overlapping in a groove weld. (c) and (d) Underfill in groove welds

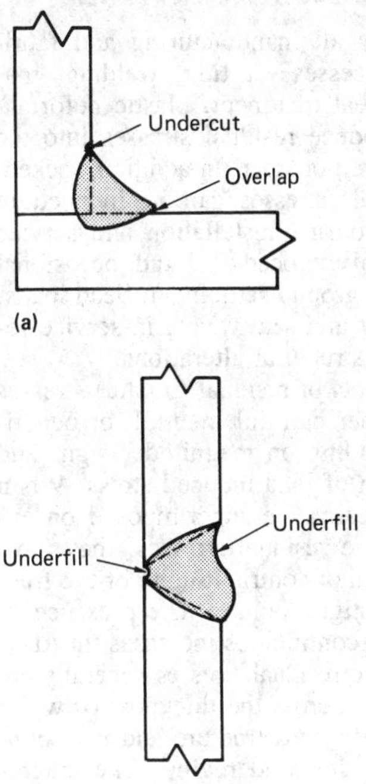

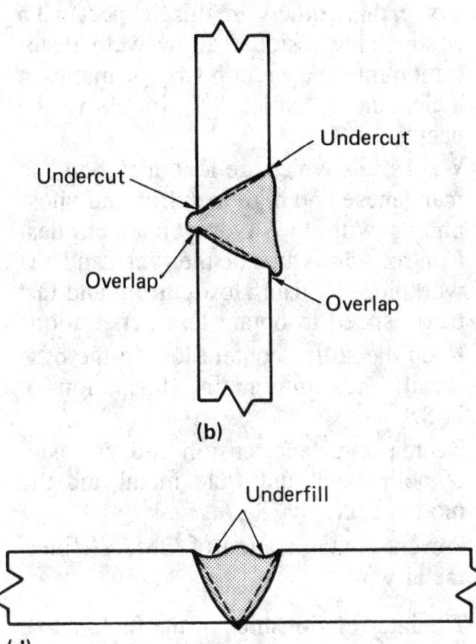

Fig. 13 Pulsed GMAW spot weld showing a lack of fill-in. Etchant: 50% nitric-50% acetic acid; ×4

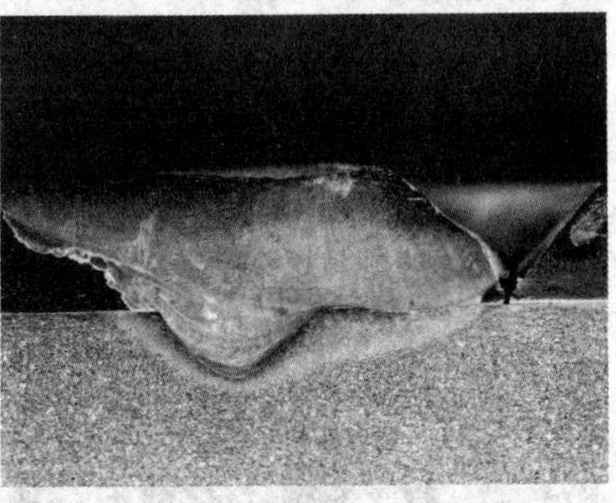

Fig. 14 Lack of fusion in a pulsed gas metal arc spot weld involving ERNiCu-7 (Monel 60), 0.035-in.-diam filler metal, copper-nickel to steel weldment. Etchant: 50% nitric-50% acetic acid. (a) ×30. (b) Note that LOF was eliminated by tapering the circular joint; ×4

(a)

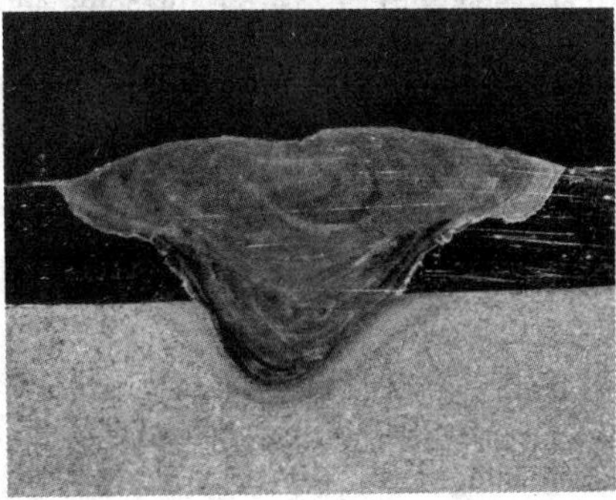

(b)

is to clean the joint and weld area thoroughly before welding.

Tungsten Inclusions

Tungsten inclusions are particles found in the weld metal from the nonconsumable tungsten electrode used in GTAW. These inclusions are the result of:

- Exceeding the maximum current for a given electrode size or type
- Letting the tip of the electrode make contact with the molten weld pool
- Letting the filler metal come in contact with the hot tip of the electrode
- Using an excessive electrode extension
- Inadequate gas shielding or excessive wind drafts, which result in oxidation
- Using improper shielding gases such as argon-oxygen or argon-CO_2 mixtures, which are used for GMAW

Tungsten inclusions, which are not acceptable for high-quality work, can only be found by internal inspection techniques, particularly radiographic testing.

Porosity

Porosity consists of cavities or pores that form in the weld metal as a result of entrapment of gases evolved or air occluded during the welding process, as evidenced by Fig. 20 and 21. Elements such as sulfur, lead, and selenium in the base metal, and foreign contaminants such as oil, grease, paint, rust, and moisture in the welding area, can increase porosity in the weld metal. Pores can take many shapes and sizes—from pinholes to large voids, from spherical to elongated or pear-shaped, with constrictions or expansions. Distribution of porosity in welds may be aligned, clustered, or uniformly scattered. Two common types of porosity are wormholes and blowholes. Wormholes are elongated voids with a definite worm-type shape and texture (Fig. 22). Blowholes may form on the surface of the weld if porosity cannot fully escape before the weld metal solidifies. "Herringbone" porosity (Fig. 34a) can form when the gas shield in an automated GMAW process in interrupted. Its appearance is due to the formation of highly elongated pores inclined to the direction of welding.

Porosity in aluminum is primarily caused by the entrapment of hydrogen during the solidification process from sources such as filler metal, base metal, surface contamination, and shielding gas. Weld-pool fluidity (or viscosity) is a major controlling factor. For example, pores form most readily in aluminum when it is welded with filler metal such as Al-Mg that produces a weld pool of a low fluidity. It is much less of a problem in welds made with an

Fig. 15 Photomacrographs of LOP. (a) In core-plated silicon steel laminates after the first GMAW pass in a U-groove joint sectioned longitudinally. As-polished, ~4×. (b) Butt welded joint in steel. Original plate thickness, 3/4 in. As polished, 1.4×

(a)

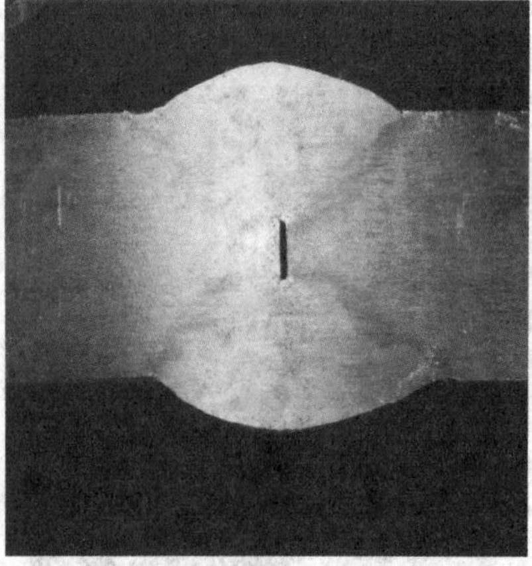

(b)

Al-Si filler metal that produces a highly fluid weld pool. In GMAW of wrought aluminum alloys, contamination of the filler-metal surface is the predominant source of porosity because of the large surface area. When powder metallurgy parts are gas metal arc welded, the primary controlling factor in the occurrence of pores is the base metal. To minimize porosity:

- Remove scale, rust, paint, grease, oil, and moisture from the surface of the joint.
- Eliminate moisture from the flux or the shielding gas.
- Use short arc length (particularly with low-hydrogen electrodes).
- Keep the weld molten long enough to allow the gas to escape.
- Provide sufficient shielding (flux or gas).
- Improve fit-up.
- In steel, remove slag residue that may cause porosity, particularly in butt welds (fluxes on certain types of electrodes are worse than others in this respect). To control slag residue, allow weld penetration into the backup strip or maintain a clearance of at least 5/32 in. above the backup.
- Weld steels which are low in carbon and manganese and high in sulfur and phosphorus with low-hydrogen electrodes. Control admixture of the weld and the workpiece by using low current and fast travel speed to obtain less penetration.
- Keep the sulfur content low in the base metal, thus minimizing formation of H_2S.
- Decrease surface tension and viscosity of both the liquid weld metal and the molten flux.
- Prevent contamination of GMAW filler-metal wire.

The interrelationships of the factors that cause or control porosity are illustrated in Fig. 23. Additional information on porosity can be found in the article "Principles of Joining Metallurgy" in this Volume.

Weld Repair

Welds are repaired to eliminate harmful defects. However, although weld repair may qualify a weld to a workmanship criterion, it does not necessarily ensure an improvement in the structural reliability of the weldment. The working conditions under which a repair is made are usually not conducive to good workmanship. Restraint is inherently greater, increasing the probability of cracking. The chances for creating new discontinuities, particularly metallurgical ones, are increased. The harmful effects of an unsatisfactory weld repair may include:

- Increase in residual stress
- Greater distortion
- Introduction of new discontinuities not detectable by nondestructive examination
- Degradation of microstructure and fracture toughness (e.g., grain growth, precipitation of carbides, hydrogen embrittlement, thermal straining, and sensitization of stainless steels such as type 304)
- Aggravation or extension of preexisting discontinuities that were undetected during original inspection

Effect of Residual Stresses on Weld Discontinuities

Virtually all manufacturing and fabrication processes—casting, welding, machining, heat treatment, plastic deformation—introduce residual stresses into the manufactured product. In addition, locked-in (residual) stresses can be induced in structures during installation and service by assembly procedures and occasional overloads, ground settlement, dead loads, high winds and sea waves, in-service repairs, and structural alterations.

The effects of residual (in situ) stresses can be either harmful, neutral, or beneficial, depending on magnitude, sign, and distribution of load-induced stress. When service stresses are superimposed on existing tensile residual stresses, they have the potential of contributing to brittle fracture and fatigue failure. The presence of existing discontinuities increases this danger, because residual stresses generally are not uniform across the thickness or width of a member in a structure and may concentrate in the weld region. The magnitude of residual stress is difficult to determine. It is therefore prudent to assess the significance of a weld flaw as if the weld in which it is contained is at yield.

Localized heating and contraction of the solidifying weld metal generate complex thermal stresses that are the source of the attendant residual stress and distortion in the completed weld. The largest residual stress usually occurs in regions near the weld parallel to the weld axis (longitudinal stress). The resulting tensile stress may be as high as the yield stress of the material and will increase the significance of planar flaws transverse to the weld. Usually, the amount of transverse stress is not as high, unless lateral movement of the joint is severely restrained. However, transverse stress can be of significance because the orientation of many common planar discontinuities (LOP, LOF) is normal to this stress. Residual stresses are therefore important because they can have a significant effect on the initiation of fractures from planar discontinuities. The deleterious effect of residual stress on the performance of welded constructions can be significant, even when the external load is rather low, if the weldment is of low toughness or subject to stress corrosion and contains significant discontinuities. For more information on residual stresses, see the article "Residual Stresses and Distortion" in this Volume.

Brittle fracture of a welded structure can occur at low applied stresses, if the

Fig. 16 Causes and cures of LOF

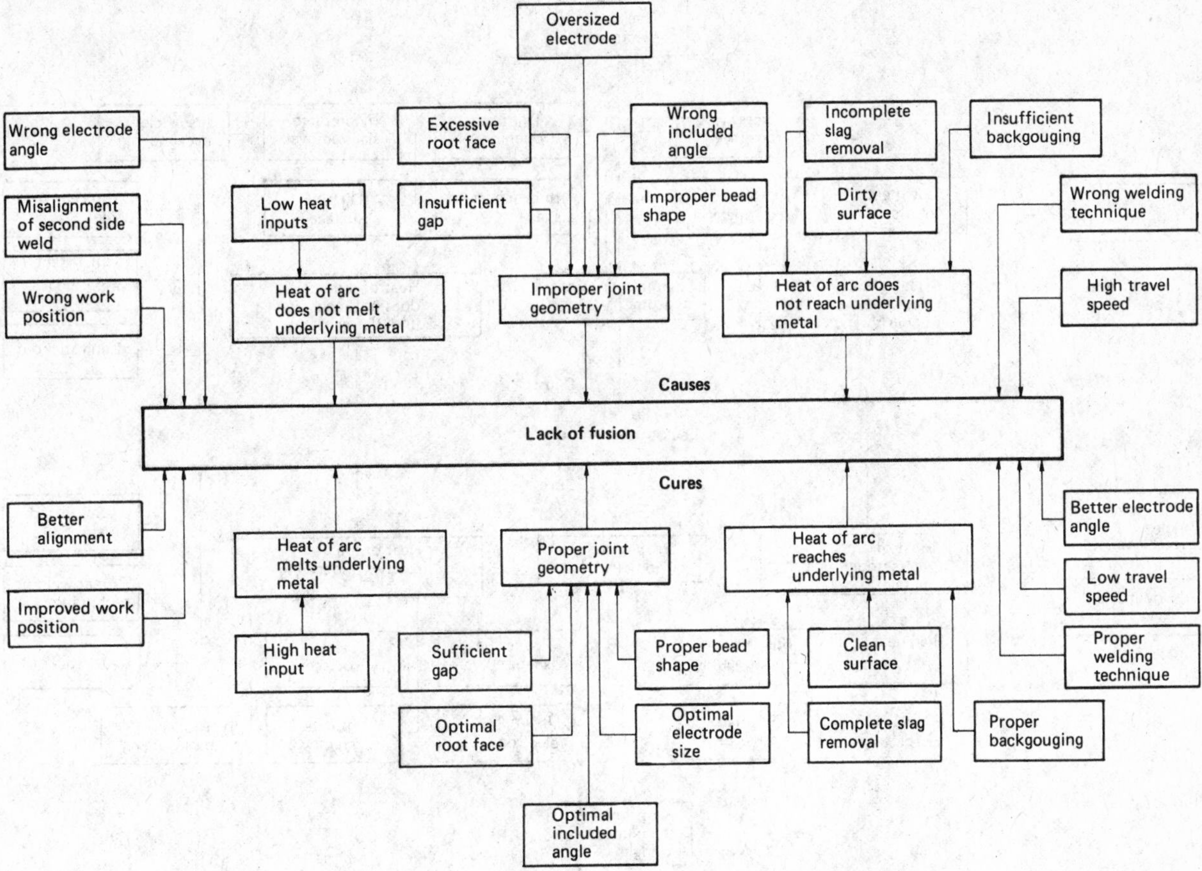

assembly is made of a material of low toughness and a sharp discontinuity exists in regions of high tensile residual stress. Unfortunately, welding tends to accentuate some of the undesirable characteristics that should be avoided in order to avoid brittle fracture. The thermal treatment resulting from welding tends to reduce the toughness of the steel and may raise its transition temperature in the HAZ. The monolithic structure of a weldment creates locked-up energy; residual stresses may exist at yield point levels. The monolithic structure also causes stresses and strains to be transmitted throughout the entire weldment, and defects in weld joints can be the nucleus for a notch or crack that will cause fracture initiation.

Brittle fracture can be greatly reduced by selecting steels that have sufficient toughness at service temperatures. The transition temperature should be below the service temperature to which the weldment will be subjected. Heat treatment, normalizing, or any method of reducing locked-up stresses will reduce the triaxial yield strength stresses within the weldment. Design notches must be eliminated and notches resulting from poor workmanship must not occur.

Fatigue Failure. The fatigue strength, expressed in terms of the number of cycles to fracture under a given load and environment (the endurance limit), increases when a specimen contains compressive residual stresses, especially on the specimen surface, but is reduced by tensile stresses. This is because fatigue failure usually originates from a discontinuity on the specimen surface (or sometimes from internal discontinuities, if severe enough). Surface smoothness is an important contribution to fatigue strength. Unfortunately, most welds are fabricated with a reinforcement that creates a stress concentration at the weld edge. This discontinuity is frequently found to be of more significance in limiting fatigue life than high levels of internal discontinuity. Fatigue initiation in fillet welded components, notably rotating machine parts, can occur at low stress if structural or weld discontinuities are present, as shown in Fig. 24.

The Significance of Weld Discontinuities

The significance of weld discontinuities is evaluated on the basis of their effect on the service life and integrity of a welded structure, often called fitness-for-purpose. Fracture initiation in a given metallic material from which a structure is fabricated is a function of material toughness properties, size and type of existing discontinuities, and the level of stress acting on the discontinuity. Toughness and applied stress control the rate of crack extension. Fracture mechanics provides a means of establishing an acceptable size of a given discontinuity in a material or a structure given these parameters. The result is an accept/reject criterion for flaw length versus the flaw through-wall dimension.

For the fracture mechanics approach to be effective, significant flaws must therefore be detected and sized by nondestructive examination. Fracture toughness of the base metal and the weld metal must be determined. Fracture toughness is a measure of the resistance of a material to crack ini-

Fig. 17 Factors influencing LOP

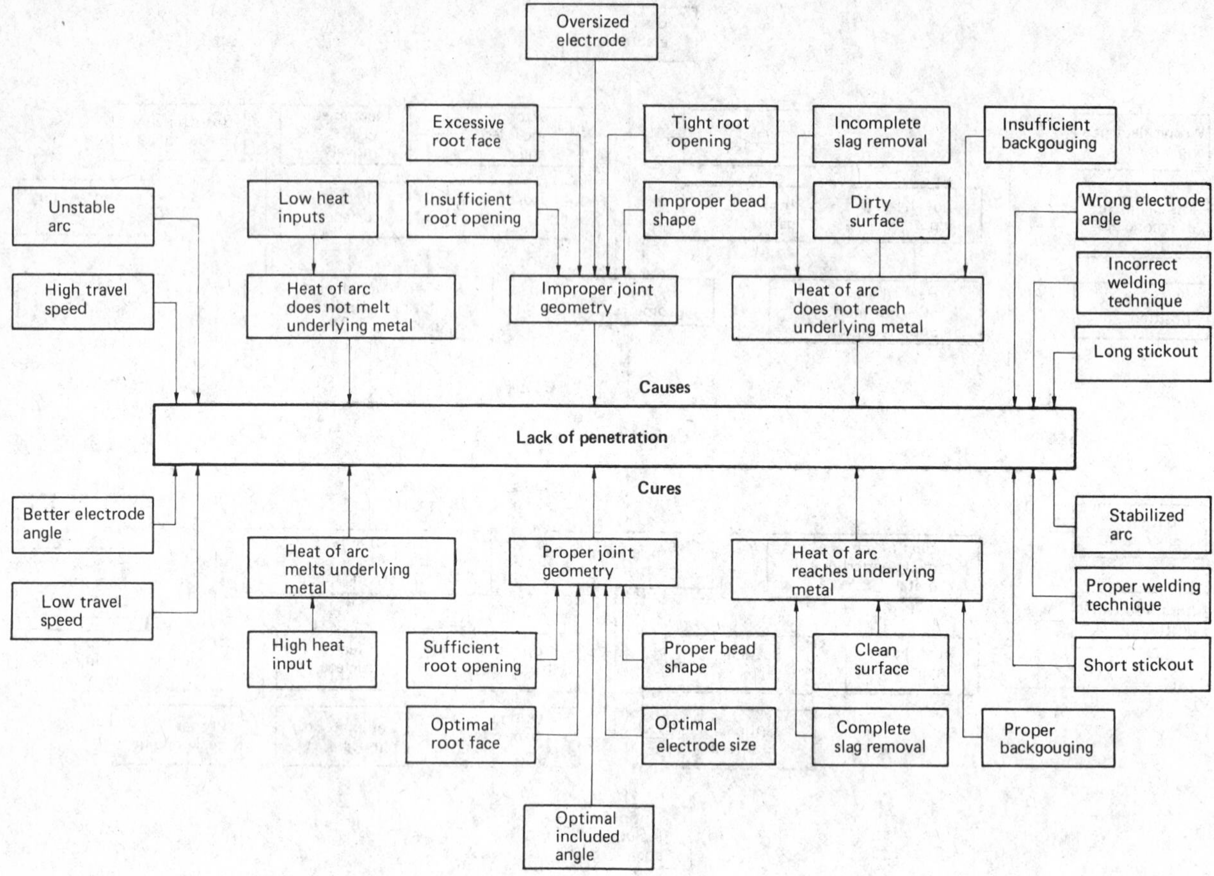

tiation and propagation and can be evaluated by established test procedures. Environmental factors (particularly temperature) must be considered in determining toughness. The primary advantage of the fracture toughness approach is that it permits judgment of the significance of discontinuities on a quantitative, objective basis related to the service of a given structure. For more detailed information on fracture toughness testing, the fracture mechanics principles involved, and their relationship to established fitness-for-purpose criteria, see Volume 10 of the 8th edition of *Metals Handbook*.

The greater the height, the longer the length, and the closer to the surface or to another discontinuity, the greater the severity of a discontinuity. When many of the same or different types of discontinuities are present in a weld, such as in Fig. 25, evaluating and correcting their interaction is difficult. The British Standards Institution (Ref 6) states that "where multiple slag inclusions occur on the same cross-section and the distance between the defects is less than 1.25 times the height of the larger defect, they should be treated as a single planar defect with an overall height equal to the distance between the outer extremities."

Modes of Failure

The significance of a discontinuity cannot be established fully until the modes of failure affected by the presence of a given type of discontinuity are determined. Common failure modes include:

- Brittle fracture
- Fatigue failure
- Stress corrosion
- Leakage
- Instability (buckling)
- Creep
- Yielding due to overload
- Corrosion
- Erosion-corrosion
- Corrosion fatigue

Weld discontinuities are not considered influential in general yielding, creep, and corrosion failure conditions. A decrease in tensile strength is commensurate with the fraction of the load-bearing cross-sectional area occupied by weld discontinuities. Large reductions in strength generally correspond to extremely large quantities of discontinuities in the weld. Weld discontinuities do, however, have a significant role in brittle fracture and fatigue modes of failure, as can be seen in the following sections.

Brittle Fracture. The initiation of brittle fracture at weld discontinuities is of concern in weldments that deform by elastic or by elastic-plastic mechanisms. The occurrence of unstable fractures initiating at weld discontinuities is governed primarily by material toughness, level of tensile stress, rate of loading, and severity of stress concentration for a discontinuity type. Therefore, the significance of discontinuities in fracture initiation depends on the material used and the structural application. As stated earlier in this article, the problem of brittle fracture can be greatly reduced in weldments by selecting steels that have sufficient toughness at service temperatures. The transition temperature

Fig. 18 Slag in steel metal. (a) Oblong. Etchant: 2% nital; 165×. (b) Lamellar. Etchant: 2% nital; 165×. (c) Jagged irregular. Fatigue fractograph as-is; ~15×

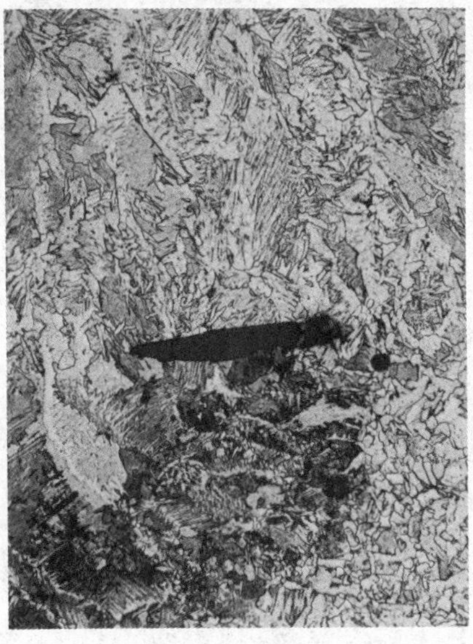

(a)

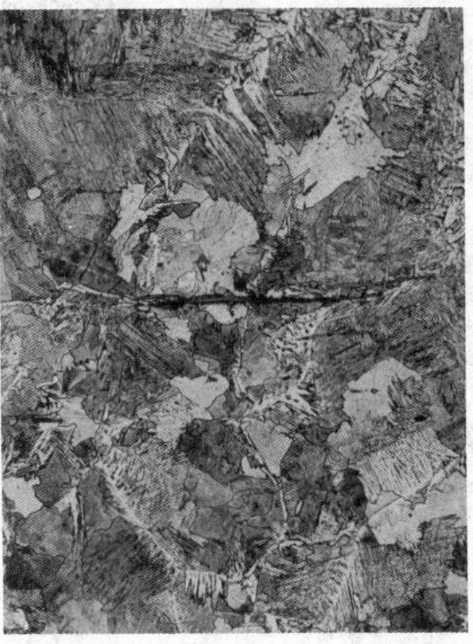

(b)

(c)

should be below the service temperature to which the weldment will be subjected.

Fatigue Failure. The fatigue strength of a weld that contains a discontinuity which contributes to the stress concentration factor can be rated or evaluated by comparison with the fatigue strength of a sound weld containing no discontinuities. Best-fit mean fatigue curves (Ref 7), based on more than 2100 fatigue tests of low-carbon steel and quenched and tempered steel butt welded specimens with the reinforcement either intact or removed, have been used to make such comparisons with butt welds containing various types of discontinuities (Ref 8). An example of a comparison between a group of fatigue test results with weld discontinuities and the mean fatigue curve is shown in Fig. 26. The curve at one standard deviation above the mean fatigue curve exceeds nearly all of the fatigue data for weldments that contained even small weld discontinuities and is a good representation of the fatigue strength for sound welds.

A weld in steel, such as a transverse butt weld, subjected to a static load often can contain a limited number of discontinuities without reduction in member strength, if the strength of the weld metal overmatches that of the base metal. For example, low-carbon steel welds can contain uniform porosity up to about 7% of the cross-sectional area without altering the tensile strength, ductility, or impact-energy absorption of the weld. Similar behavior also exists for other internal discontinuities.

A discontinuity that does not affect the strength of a weld member subjected to load can, however, significantly affect the fatigue strength of the member as a result of a concentration of stress that occurs at the discontinuity. In a statically loaded member, the increased stress is relieved by plastic strain of the material near the discontinuity. However, in a member subjected to a fluctuating load, the range of stress at the discontinuity remains much higher than in the surrounding material and may cause a reduction in fatigue strength.

Brittle Fracture Versus Fatigue Failure. The distinction between brittle fracture and fatigue failure can be made on the basis of the respective consequences. Brittle fracture is usually more serious, at times catastrophic. Thus, in terms of nature of the consequences of brittle fracture versus fatigue failure, unstable fracture is normally most severe—because it occurs without warning—although fatigue failure occurs more frequently (up to 70 to 90% of all failures). A through-thickness sudden failure is indicative of brittle fracture, which can progress the full length of the weldment. For plane-strain conditions, the critical size discontinuity is much smaller than for fatigue or elastic-plastic yield. With unstable fracture, it is best to ensure that it does not occur by selecting materials that are ductile under service conditions.

Types of Discontinuities

The different types of weld discontinuities vary in significance for given failure modes. For instance, planar discontinuities, such as undercut, LOF, and LOP, are more likely to cause fast fracture initiation than nonplanar ones, such as slag inclusions and porosity.

Cracks are the most detrimental of all weld discontinuities because of their sharp extremities. They have a high probability of initiating brittle fracture. Cracks frequently form at or near the fusion line, where stress concentration created by weld reinforcement increases their chances of becoming initiation sites for cleavage.

In terms of location, surface cracks are most harmful, partly because they are exposed to environmental effects. This exposure amplifies the adverse influence of

Fig. 19 Causes and remedies of slag inclusions

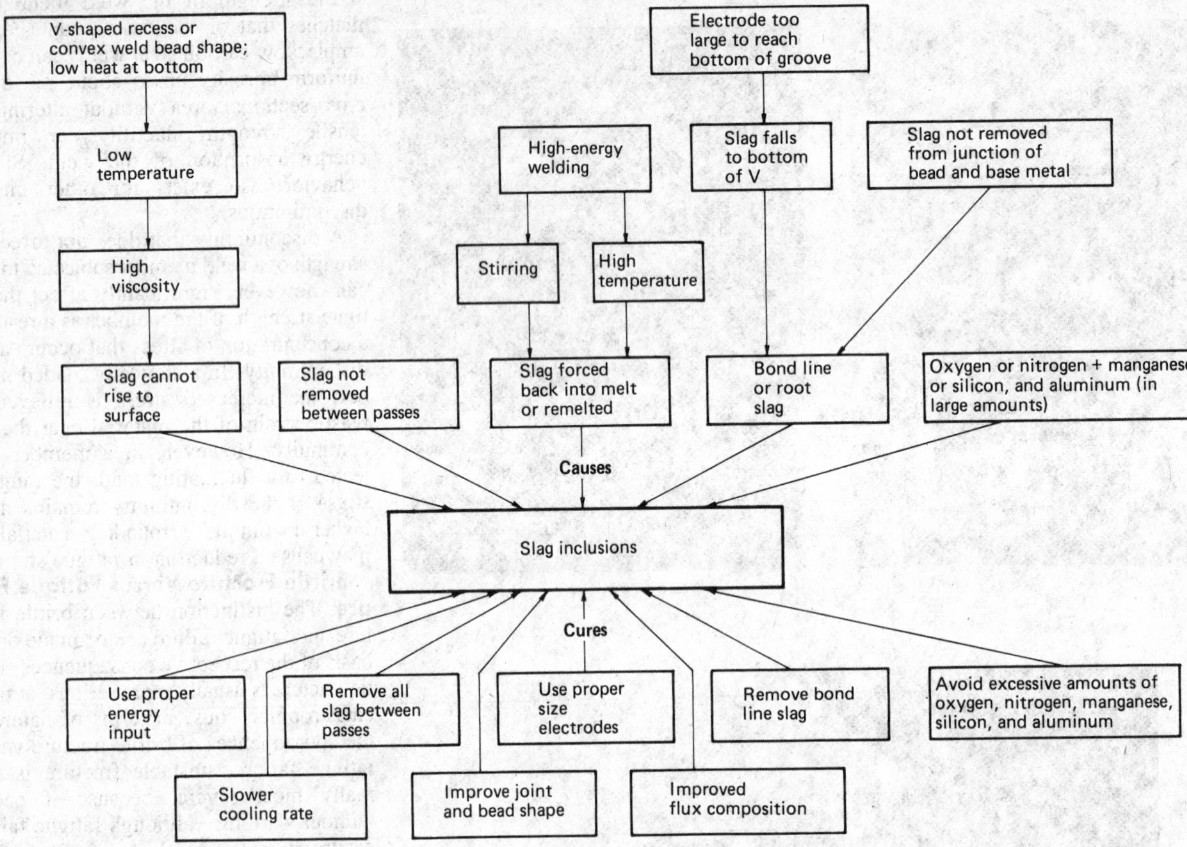

cracks on the structure, particularly in low-toughness materials.

Generally, cracks should not be tolerated in welded constructions, especially during manufacture. However, if a crack is detected in the welded product during service, the fitness-for-purpose concept based on fracture mechanics principles can be used to determine the most appropriate course of action.

Weld undercut is another undesirable type of weld discontinuity. Despite their negative influence on fatigue strength, welds are produced routinely with undesirable contours, such as large reinforcement angles. The critical size of a geometric discontinuity such as undercut in fillet welds for a given leg length decreases as the plate thickness increases (Ref 9).

The computed fatigue strengths for welds in $^3/_4$-in. plate with various undercut depths are compared at 100 000 and 2 000 000 cycles to the fatigue strength of a sound weldment in Fig. 27. Fatigue strength is markedly reduced by even small undercut depths.

LOF and LOP. The fitness-for-purpose analysis of LOF and LOP, when present in carbon steels, is typically related to the ductile-to-brittle transition temperature, because the significance of these discontinuities can change. When material is below its transition temperature, LOF and LOP are potentially harmful and can trigger brittle fracture. If, on the other hand, the steel is above its transition temperature, these discontinuities may be only slightly worse than slag inclusions or porosity. Consequently, it is essential to define the exact transition temperature and conditions that will be encountered in service.

The effect of LOF and LOP can be masked by weld-metal overmatching and reinforcement. The direction of the applied alternating stress with respect to LOF and LOP is also important. When the applied load is parallel to LOF and LOP, the effect is minimal; the reverse is true when the direction of stress is perpendicular. The significance of LOF and LOP, however, also depends on their "aspect ratio," which indicates the sharpness of the discontinuity (ratio of width-to-depth), and location.

Few experimental studies have been conducted to examine the effect of LOF on fatigue strength. However, fatigue behavior of weldments with LOF discontinuities an be estimated using the mathematical models for fatigue crack propagation.

Fatigue lives were computed for LOF at a $^1/_4$-in. depth for $^3/_4$-in. plate by adding the computed initiation and propagation lives at selected values for the stress range and LOF depth. The computed fatigue strengths at 100 000 and 2 000 000 cycles are compared to the mean fatigue strengths of sound weldments in Fig. 28.

Fatigue strengths of welds in $^3/_4$-in. plate containing LOP discontinuities are compared at 100 000 and 2 000 000 cycles to the mean fatigue strength of sound welds in Fig. 29, demonstrating that the fatigue strength of a weldment containing significant LOP can be greatly reduced below that of a sound weld.

In an investigation of the significance of LOF and LOP in an Al-4.4%Mg alloy welded by GMAW, it was found that LOF and LOP were more serious than undercut, surface irregularity, and macro- and microporosity (Ref 10). The fatigue strength of this aluminum alloy in a 0.47-in.-thick test sample was reduced 60% by the presence of LOF and/or LOP.

Fig. 20 Porosity in GMAW core-plated silicon steel laminations. (a) 100×. (b) 250×

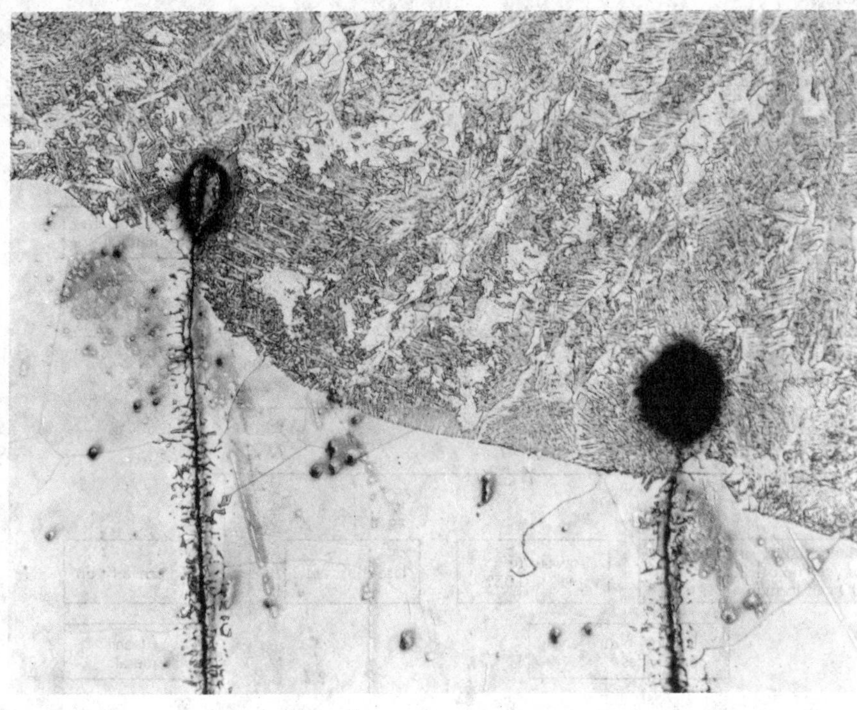

(a)

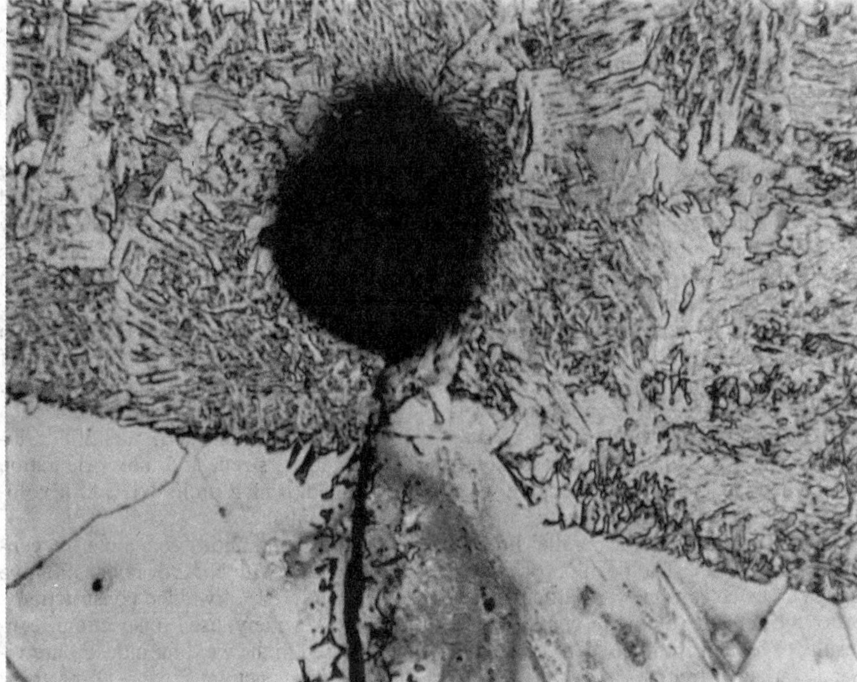

(b)

Slag Inclusions. The structural importance of the presence of slag inclusions in a weld is evaluated in terms of their size, shape, amount, distribution, and location. Tensile strength of the weld is reduced in proportion to the projected area of the slag; yield strength is affected to a lesser extent. Tensile ductility is reduced appreciably. Slag inclusions occurring along the fusion line, which is the weakest part of most welds, are more significant than porosity.

Quality band fatigue curves are usually plotted on an *SN* diagram and are structured in order of increasing length of slag inclusions, where *S* denotes the stress range and *N* signifies endurance (number of cycles). Each band corresponds to a specific size range of discontinuity, which is permitted in conformance with the design criteria for a given structure, as shown in Fig. 30. For a given length of slag, the number of cycles decreases as the stress range increases. Quality categories can be established on the basis of as-welded and

Fig. 21 Pulsed GMAW spot weld showing porosity in dissimilar metal weldment; a copper-nickel alloy to a carbon-manganese steel using an ERNiCu-7 (Monel 60) electrode. Etchant: 50% nitric-50% acetic acid; ×4

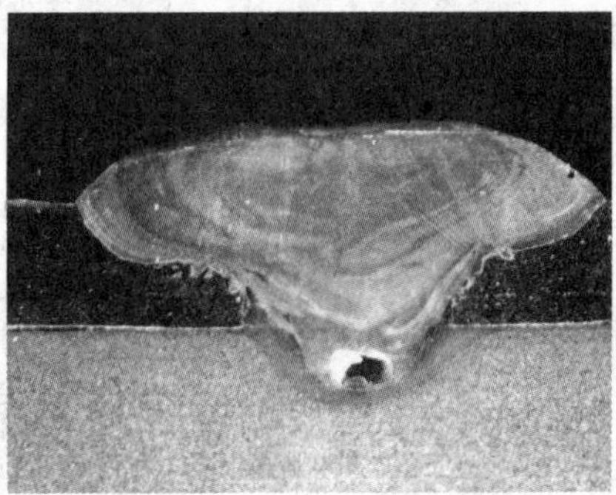

Fig. 22 Wormhole porosity in a weld bead. Longitudinal cut; ~20×

Fig. 23 Factors influencing porosity

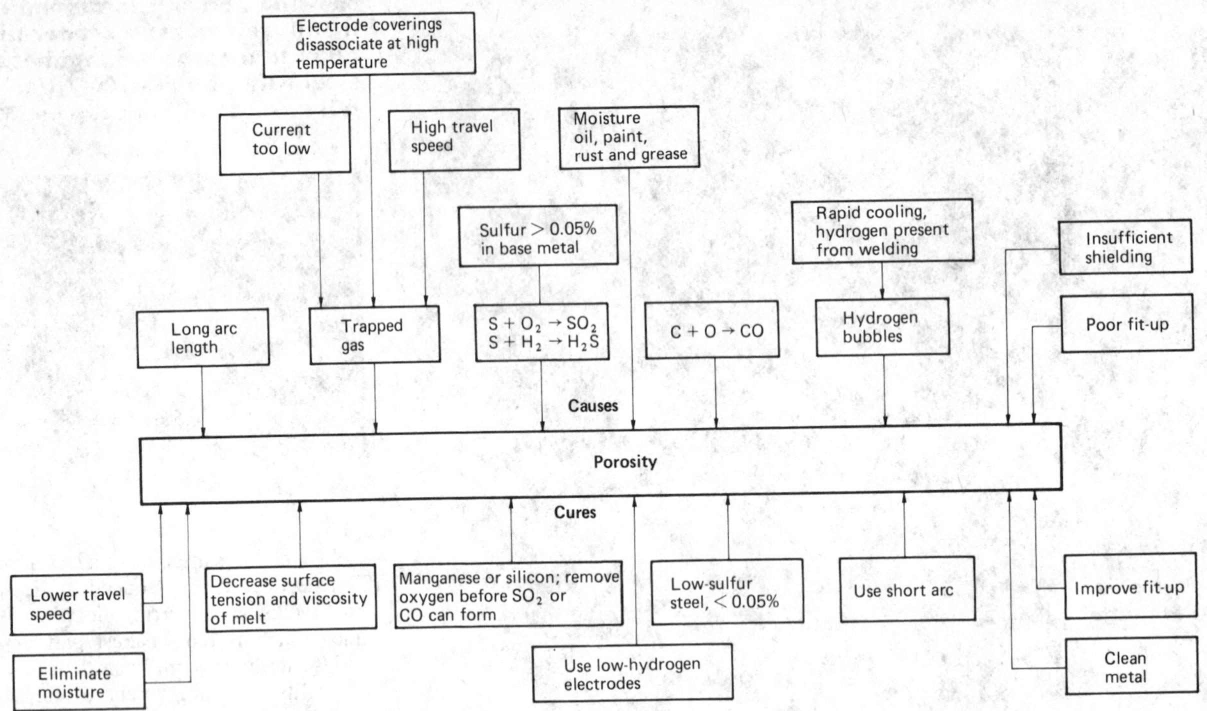

stress-relieved structures (Ref 6). Comparisons between the fatigue strengths from the quality band and sound weldment curves at 100 000 and 2 000 000 cycles are presented in Fig. 31. These curves relate the mean fatigue strength of a weld containing a particular slag inclusion length to that of a sound weld.

Porosity, from the fitness-for-purpose point of view, is regarded as the least harmful type of discontinuity. Most pores are very small; through-thickness depth lies inherently within the weld pass. Extensive porosity can interfere with detection of the more harmful planar discontinuities, such as undercut, LOF, and LOP, by obscuring their presence.

The influence of porosity is evaluated according to its location, i.e., surface or internal. Surface porosity has a detrimental effect on fatigue strength, especially when weld reinforcement is removed or the weld is undermatched. In high-cycle fatigue, porosity is least harmful, as long as weld buildup is not removed. If it is removed, the rate of decrease in fatigue strength initially is appreciable (up to 50% at 8% porosity); after reaching about 8% porosity, fatigue strength reduction rate decreases. To improve fatigue life of welds with surface porosity, the weld root should be dressed to minimize stress concentration. Fatigue failure originates at surface porosity rather than at buried slag inclusions. Pores in the surface are more harmful than internal pores of the same size and amount over a given weld length. Fatigue strength is highly susceptible to the presence of stress risers. Therefore, surface porosity causes a reduction in fatigue strength. Shape of the porosity is relatively unimportant. The presence of porosity in butt welds is structurally more significant than in fillet welds, because butt welds usually see more critical applications than fillet welds, and the stress concentrations at the toes of fillet welds are generally quite large.

A log-log plot of stress range (S) and number of cycles (N) for porosity levels of 0, 3, 8, 20, and over 20% can be seen in Fig. 32. This diagram is based on a comprehensive analysis of a literature study (Ref 11). These quality bands are the result of an arbitrary set of parallel lines at a slope of −0.25. The position of the fatigue curves corresponds to particular levels of the percent volume of porosity. The results of comparison between the quality band fatigue strength and the fatigue strength for sound weldment behavior at 100 000 and 2 000 000 cycles are given in Fig. 33. These curves relate the percent volume of porosity present in a weldment directly to the percent reduction in mean fatigue strength of a sound weldment with the reinforcement either ground off or left intact.

Nondestructive Examination

Nondestructive examination (NDE) of weldments has two functions: (1) quality control—the monitoring of welder and equipment performance and of the quality of the consumables and the base materials used, and (2) acceptance or rejection of a weld on the basis of its fitness-for-purpose under the service conditions imposed on the structure. The appropriate method of inspection is different for each function. If evaluation is a viable option, discontinuities must be detected, identified, located exactly, sized, and their orientation established, which limits NDE to a volumetric technique.

Weld discontinuities constitute the center of activity with NDE or nondestructive inspection (NDI) of welded constructions. The most widely used inspection techniques used in the welding industry are visual, liquid-penetrant, magnetic-particle, radiographic, ultrasonic, eddy current, and acoustic emission methods. Each of these techniques has specific advantages and limitations. Existing codes and standards that provide guidelines for these various techniques are based on the capabilities and/

(a)

(b)

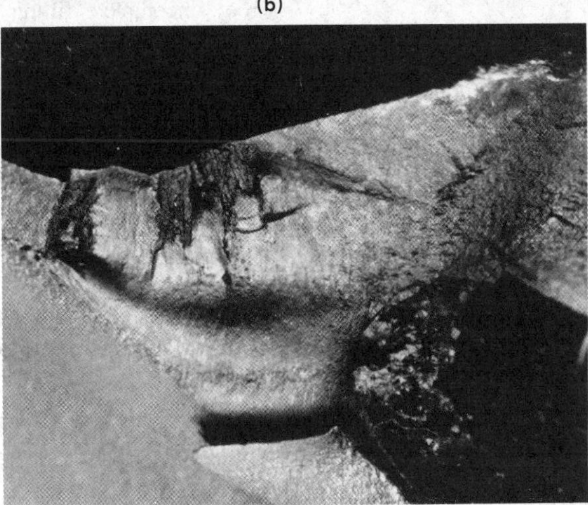

(c)

Fig. 24 Different initiation sites for a rotating bending type of fatigue failure. (a) Steel shaft intentionally broken to reveal extent of failure. (b) Sharp milled corner. (c) Slag inclusions in fillet weld and LOP

or limitations of these NDE methods. Additional information regarding these codes and standards can be found in the article "Codes, Standards, and Inspection" in this Volume.

Nondestructive Inspection Methods

Visual Inspection. For many noncritical welds, integrity is verified principally by visual inspection. Even when other nondestructive methods are used, visual inspection still constitutes an important part of practical quality control. Widely used to detect discontinuities, visual inspection is simple, quick, and relatively inexpensive. The only aids that might be used to determine conformity of a weld are a low-power magnifier, a boroscope, a dental mirror, or a gage. Visual inspection can and should be done before, during, and after welding. Although visual inspection is the simplest inspection method to use, a definite procedure should be established to ensure that it is carried out accurately and uniformly.

Visual inspection is useful for checking the following:

- Dimensional accuracy of weldments
- Conformity of welds to size and contour requirements

Fig. 25 Photomacrograph showing discontinuities in multipass joint of low-carbon steel. Etchant: 2% nital; 3×

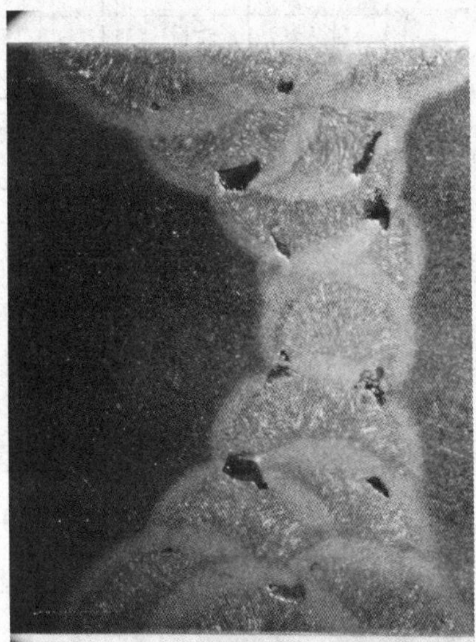

- Acceptability of weld appearance with regard to surface roughness, weld spatter, and cleanness
- Presence of surface flaws such as unfilled craters, pockmarks, undercuts, overlaps, and cracks

Although visual inspection is an invaluable method, it is unreliable for detecting subsurface discontinuities. Therefore, judgment of weld quality must be based on information in addition to that afforded by surface indications.

Liquid-penetrant inspection is capable of detecting discontinuities open to the surface in weldments made of either ferromagnetic or nonferromagnetic alloys, even when the discontinuities are generally not visible to the unaided eye. Liquid penetrant is applied to the surface of the part, where it remains for a period of time and penetrates into the flaws. For correct usage of liquid-penetrant inspection, it is essential that the surface of the part be thoroughly clean, leaving the openings free to receive the penetrant. Operating temperatures of 70 to 90 °F produce optimum results. If the part is cold, the penetrant may become chilled and thickened so that it cannot enter very fine openings. If the part or the penetrant is too hot, the volatile components of the penetrant may evaporate, reducing the sensitivity.

After the penetrating period, the excess penetrant remaining on the surface is removed. Then an absorbent, light-colored developer is applied to the surface. This

Fig. 26 Comparison of weld discontinuity fatigue data to mean fatigue curve

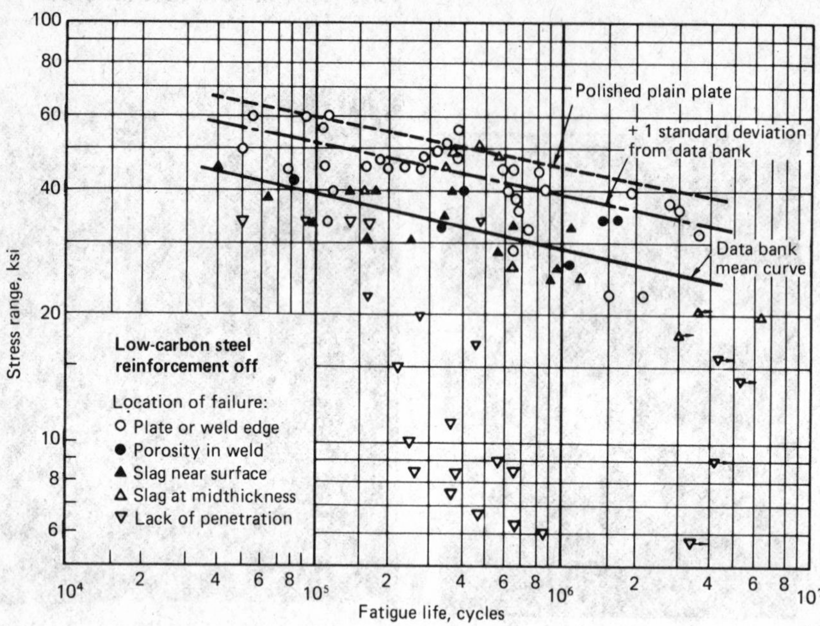

developer acts as a blotter, drawing out a portion of the penetrant that had previously seeped into the surface openings. As the penetrant is drawn out, it diffuses into the developer, forming indications that are wider than the surface openings. The inspector looks for these colored or fluorescent indications against the background of the developer.

Liquid-penetrant inspection is used to detect surface defects in aluminum, magnesium, and stainless steel weldments when the magnetic-particle inspection method cannot be used. It is very useful for locating leaks in all types of welds. Welds in pressure and storage vessels and in piping for the petroleum industry are inspected for surface cracks and for porosity using this method.

Magnetic-particle inspection is a nondestructive method of detecting surface and near-surface discontinuities in ferromagnetic materials. It consists of three basic operations:

- Establishing a suitable magnetic field in the material being inspected
- Applying magnetic particles to the surface of the material
- Examining the surface of the material for accumulations of the particles (indications), and evaluating the serviceability of the material

Magnetic-particle inspection is particularly suitable for the detection of surface discontinuities in highly ferromagnetic metals. Under favorable conditions, those flaws that lie immediately under the surface are also detectable. Nonferromagnetic and weakly ferromagnetic metals, which cannot be strongly magnetized, cannot be inspected by this method. With suitable ferromagnetic metals, magnetic-particle inspection is highly sensitive and produces readily discernible indications at flaws in the surface of the material being inspected.

The types of weld discontinuities normally detected by magnetic-particle inspection include cracks, LOP, LOF, and porosity open to the surface. Linear porosity, slag inclusions, and gas pockets may be detected if large or extensive, or if smaller and near the surface.

Radiographic inspection is a nondestructive method that uses a beam of penetrating radiation such as x-rays and gamma rays. When the beam passes through a weldment, some of the radiation energy is absorbed and the intensity of the beam is reduced. Variations in beam intensity are recorded on film, or on a screen when a fluoroscope or an image amplifier is used. The variations are seen as differences in shading that are typical of the types and sizes of any discontinuities present.

Surface discontinuities that are detectable by radiography include undercuts, longitudinal grooves, concavity at the weld root, incomplete filling of grooves, excessive reinforcement, overlap, irregularities at electrode-change points, grinding marks, and electrode spatter. Surface irregularities may cause density variations on a radiograph. When possible, they should be removed before a weld is radiographed. When impossible to remove, they must be considered during interpretation.

Fig. 27 Fatigue strength of a weldment containing undercut as a percentage of the mean fatigue strength of a sound low-carbon steel weld

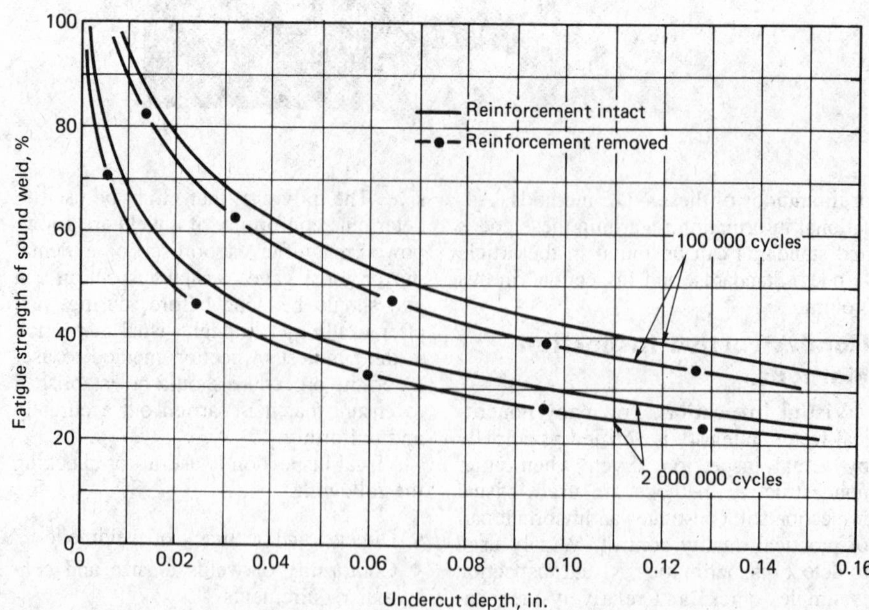

Fig. 28 Computed fatigue strength of a weldment containing lack of fusion as a percentage of the mean fatigue strength of a sound low-carbon steel weld

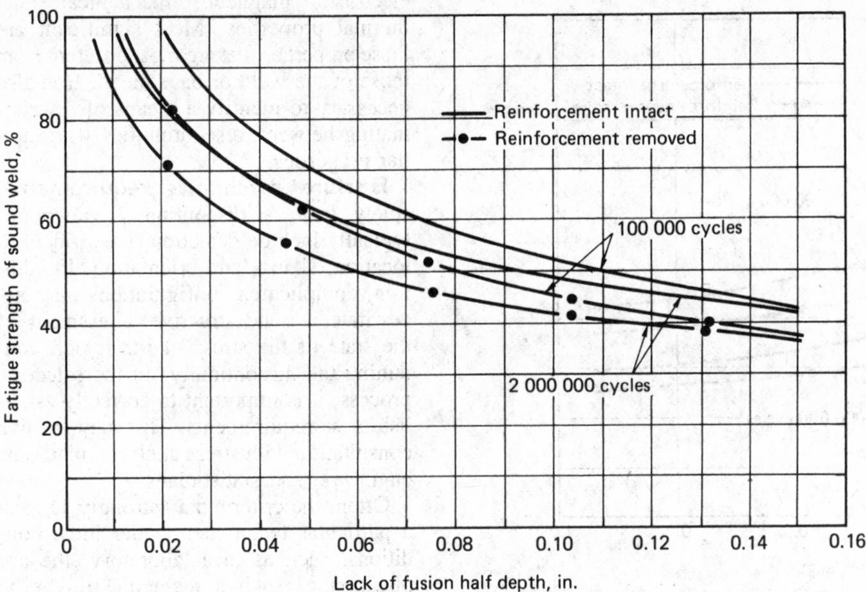

Subsurface discontinuities detectable by radiography include gas porosity, slag inclusions, cracks, LOP, LOF, and tungsten inclusions. Figure 34 shows radiographs of herringbone porosity, wagon-track slag, and lack of side-wall fusion.

Ultrasonic inspection is a nondestructive method which employs mechanical vibrations similar to sound waves, but of a higher frequency. A beam of ultrasonic energy is directed into the specimen to be tested. This beam travels through a material with only a small loss, except when it is intercepted and reflected by a discontinuity or by a change in material.

Ultrasonic testing is capable of finding surface and subsurface discontinuities. The ultrasonic contact pulse reflection tech-

Fig. 29 Fatigue strength of a weldment containing lack of penetration as a percentage of the mean fatigue strength of a sound low-carbon steel weld

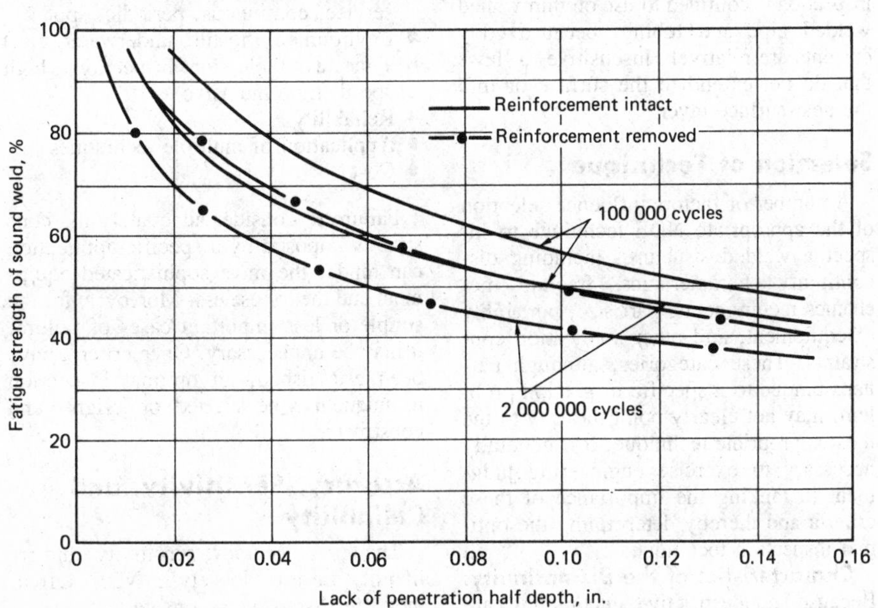

Fig. 30 Quality bands for slag inclusions representing as-welded carbon-manganese steel

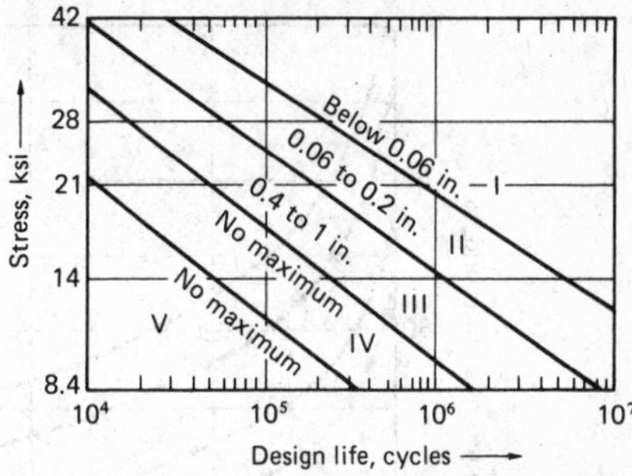

nique is used. This system uses a transducer, which changes electrical energy into mechanical energy. The transducer is excited by a high-frequency voltage, which causes a crystal to vibrate mechanically. The crystal probe becomes the source of ultrasonic mechanical vibrations. These vibrations are transmitted into the test piece through a coupling fluid, usually a film of oil, called a couplant. When the pulse of ultrasonic waves strikes a discontinuity in the test piece, it is reflected back to its point of origin. Thus the energy returns to the transducer. The transducer now serves as a receiver for the reflected energy. The initial signal, the returned echoes from the discontinuities, and the echo of the rear surface of the test material are all displayed by a trace on the screen of a cathodery oscilloscope.

Ultrasonic testing methods, which can be used to test most materials, are more sensitive to planar weld discontinuities than to nonplanar ones. Variation of test results with operator skill is significant with most commonly used ultrasonic systems (Ref 12 and 13). Effective examination of welds with coarse grain structures by ultrasonic testing is cumbersome, because small discontinuities (tight fissures) can be masked by the large grains.

Conventional pulse-echo ultrasonics suffer from the subjectivity inherent in signal interpretation. To some extent, this can be compensated for by well-developed calibration, examination procedure, and highly trained personnel.

Acoustic emissions are impulsively generated small-amplitude elastic stress waves created by deformations in a material. The rapid release of kinetic energy from the deformation mechanism propagates elastic waves from the source, and these are detected as small displacements on the surface of the specimen. The emissions indicate the onset and continuation of deformation, and can be used to locate

Fig. 31 Fatigue strength of a weldment containing slag inclusions as a percentage of the mean fatigue strength of a sound low-carbon steel weld

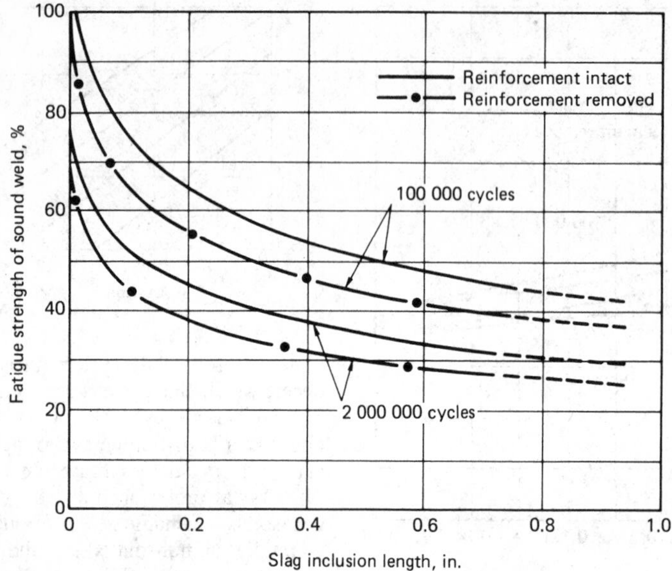

the source of deformation through triangulation techniques.

Acoustic emissions can be used to assess weld quality by monitoring during or after welding. In weldments, regions having incomplete penetration, cracking, porosity, inclusions, or other discontinuities can be identified by detecting the acoustic emissions originating at these regions. During welding processes, acoustic emissions are caused by many things, including plastic deformation, melting, friction, solidification, solid-solid phase transformations, and cracking. Monitoring of acoustic emission during welding can even include, in some instances, automatic feedback control of the welding process. In large-scale automatic welding, the readout equipment can be conveniently located near the welder controls or in a quality-monitoring area.

Eddy-current inspection is based on the principles of electromagnetic induction and is used to identify or to differentiate between a wide variety of physical, structural, and metallurgical conditions in electrically conductive ferromagnetic and nonferromagnetic metals.

Eddy-current inspection, like ultrasonic inspection, can be used for detecting subsurface porosity. Normally, eddy-current inspection is confined to use on thin-walled welded pipe and tubing, because eddy currents are relatively insensitive to flaws that do not extend to the surface or into the near-surface layer.

Fig. 32 Quality categories for different levels of porosity in as-welded carbon-manganese steel. Source: Ref 11

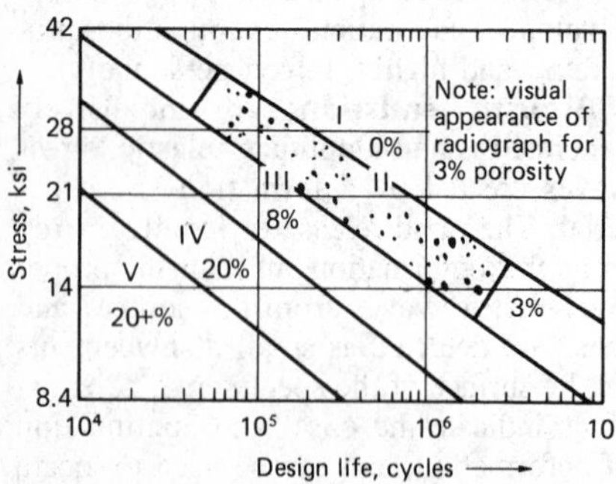

Selection of Technique

A number of factors influence selection of the appropriate NDE technique to inspect a welded structure, including discontinuity characteristics, fracture mechanics requirements, part size, portability of equipment, and other application constraints. These categories, although perhaps unique to a specific inspection problem, may not clearly point the way to the most appropriate technique. It is generally necessary to exercise engineering judgment in ranking the importance of these criteria and thereby determining the optimal inspection technique.

Characteristics of the Discontinuity. Because nondestructive techniques are based on physical phenomena, it is useful to describe the properties of the discontinuity of interest, such as composition and electrical, magnetic, mechanical, and thermal properties. Most significant are those properties that are most different from those of the weld or base metal. It is also necessary to identify a means of discriminating between discontinuities with similar properties.

Fracture mechanics requirements, solely from a discontinuity viewpoint, typically include detection, identification, location, sizing, and orientation. In addition, complicated configurations may necessitate a nondestructive assessment of the state of the stress of the region containing the discontinuity. In the selection process, it is important to correctly establish these requirements. This may involve consultation with stress analysts, materials engineers, and statisticians.

Often, the criteria may strongly suggest a particular technique. Under ideal conditions, such as in a laboratory, the application of such a technique might be routine. In the field, however, other factors may force a different choice of technique.

Constraints tend to be unique to a given application and may be completely different even when the welding process and metals are the same. Some of these constraints include:

- Access to the region under inspection
- Geometry of the structure (flat, curved, thick, thin)
- Condition of the surface (smooth, irregular)
- Mode of inspection (preservice, inservice, continuous, periodic, spot)
- Environment (hostile, underwater, etc.)
- Time available for inspection (high speed, time intensive)
- Reliability
- Application of multiple techniques
- Cost

Failure to consider adequately the constraints imposed by a specific application can render the most sophisticated equipment and theory useless. Moreover, for the simple or less important cases of failure, it may be unnecessary. Once criteria have been established, an optimal inspection technique may be selected, or designed and constructed.

Accuracy, Sensitivity, and Reliability

The terms accuracy, sensitivity, and reliability are used loosely in NDE. Often, they are discussed as one term to avoid

Fig. 33 Fatigue strength of a weldment containing porosity as a percentage of the fatigue strength of a sound low-carbon steel weld

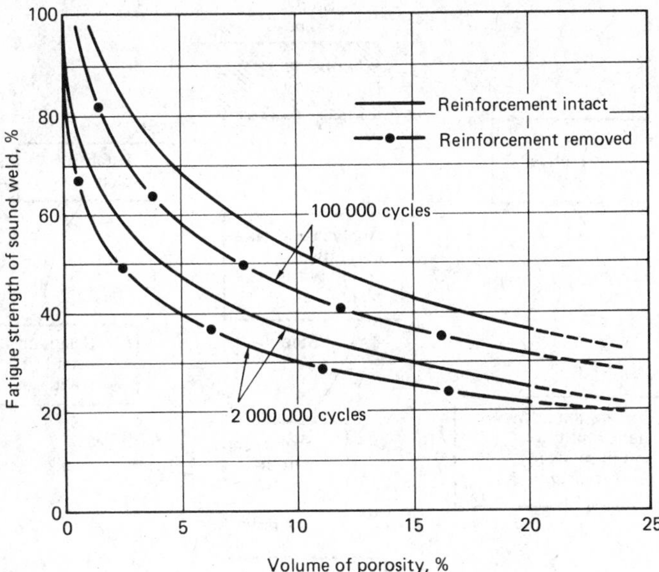

Fig. 34 Radiographs showing (a) herringbone porosity in automatic weld due to disruption of gas shield, (b) wagon-track slag in manual shielded metal arc weld, (c) LOF in automatic weld

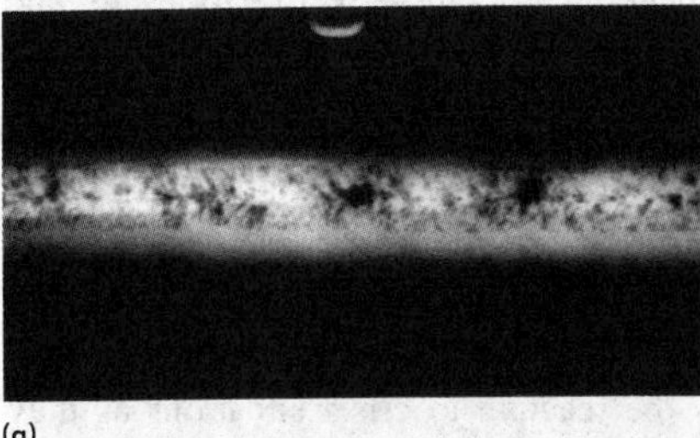

(a)

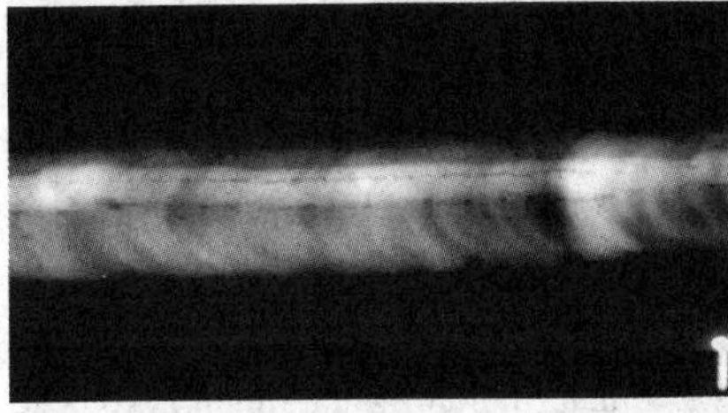

(b)

(c)

distinguishing among the specific aspects of these terminologies.

Accuracy is the attribute of an inspection method that describes the correctness of the technique within the limits of its precision. In other words, the technique is highly accurate if the indications resulting from the technique are correct. This does not mean that the technique was able to detect all discontinuities present, but rather that those indicated actually exist.

Sensitivity, on the other hand, refers to the capability of a technique to detect discontinuities that are small or that have properties only slightly different from the material in which they reside. Figure 35 schematically illustrates the concepts of accuracy and sensitivity in the context of detection probability. In general, sensitivity is gained at the expense of accuracy.

Reliability is a combination of both accuracy and sensitivity. Three factors influence reliability: (1) inspection procedure, including the instrumentation; (2) human factors (inspector motivation, experience, training, education, etc.); and (3) data analysis. Uncalibrated equipment, improper application of technique, and inconsistent quality of accessory equipment (transducers, couplant, film, chemicals, etc.) may affect accuracy and, in some instances, sensitivity. Poor inspector technique, unfamiliar response, lack of concentration, and other human factors can combine to reduce reliability. Data analysis, or the lack thereof, can influence reliability as well; generally, inspection is performed under conditions in which detection probability is less than 100% and is not constant with discontinuity severity. Consequently, statistics must be employed to establish the level of confidence that may be attached to the inspection results.

High sensitivity with low accuracy may be far worse, from the viewpoint of reliability, than low sensitivity with high accuracy, especially if the sensitivity level is adequate for detecting the weld discontinuities in question. As a rule of thumb, the transition region of the detection probability curve indicates the degree of reliability. If this region occurs with the limits encompassed by NDE capabilities, which are smaller than the values required for evaluating the fitness-for-purpose of the welds being inspected, reliability is satisfactory. If, on the other hand, the region occurs at values higher than those required, reliability is unsatisfactory. The transition region can be viewed as the "reliability threshold."

Fatigue and Fracture Control

Fatigue and fracture control of welded structures invariably entails methodologies that reduce stress concentrations (which are usually design related), reduce tensile residual stresses (fabrication, installation, or service related), introduce compressive stresses, avoid critical levels of applied stress (operation control related), eliminate critical size discontinuities (welding process and quality control related), and apply materials of adequate toughness for the service intended (design related). This approach serves the basis of the fitness-for-purpose concept. In most cases, failures result from a combination of more than one cause, during manufac-

Fig. 35 Detection probability. Indicates the contrary nature of sensitivity and accuracy. The minimum NDE parameter size required to establish fitness-for-purpose must lie to the right of the transition region, or reliability threshold, to achieve satisfactory reliability.

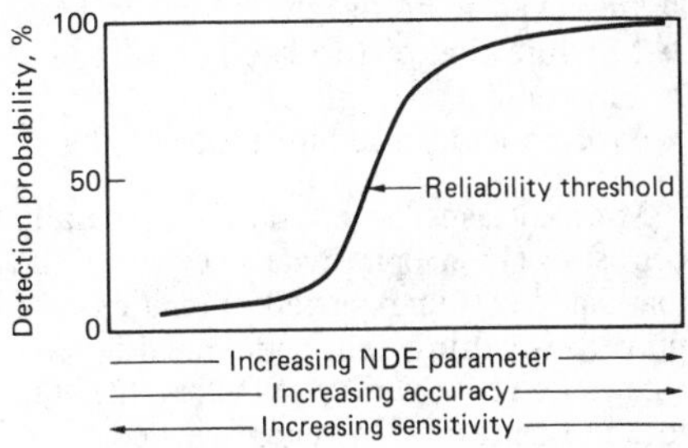

ture, installation, or in service. Therefore, quality control must be incorporated into evaluation for fitness-for-purpose. A good quality control system extends beyond construction to include installation and in-service monitoring of the structure, information feedback, data analysis, establishment of cause-and-effect relationship, and dissemination of conclusions and recommendations to those concerned (designers, builders, owner/operators, users, and code-writing bodies). Such an approach is called a Quality Control Systems Loop (QCSL), schematically shown in Fig. 36.

The conventional practice of quality control is open ended, isolated, and tends to be reactive to crisis situations as they arise. Usually, there are no follow-through provisions to check the installation and field performance of the inspected or non-inspected welds. Consequently, there is no reliable way to validate the action taken at the time of original inspection to satisfy code compliance or owner/operator requirements. Typically, the only time feedback from the field is obtained is if there was a major failure which might have been averted by a systematic and comprehensive approach. Properties and conditions of the material or the construction can change, at times unexpectedly, during fabrication, installation, or service. Corrective action may include:

- Stress relieving (postweld heat treatment)
- Spot heating
- Control of the weld profile
- Peening
- Tungsten inert gas or plasma dressing
- Grinding
- Quenching
- Cold working
- Improved procedures
- Application of ductile materials in selected areas of a welded structure (crack arresters)
- Plastic coating or corrosion protection
- Local pressing
- Drilling crack arrester holes
- Prior overloading
- Bolted splice plates
- Decreasing the applied stress (or elimination of in-service overload)
- Improved joint design
- Minimization of residual stress
- Environmental control
- Process and procedure control

Material aspects of failure generally consist of (1) inappropriate selection of the material, (2) unexpected alterations of properties during fabrication and/or service caused by a variety of factors, (3) high locked-in stresses, (4) unanticipated overload, and (5) presence of significant discontinuities. Poor design detail is almost always one of the principal causes of failure. Others include inadequate control of processes and procedures, and sometimes the dictates of the economics.

Fig. 36 Quality control systems loop. Loop A, in-plant set-up for short-range benefits; Loop B, in-plant and in-service set-ups for long-range benefits

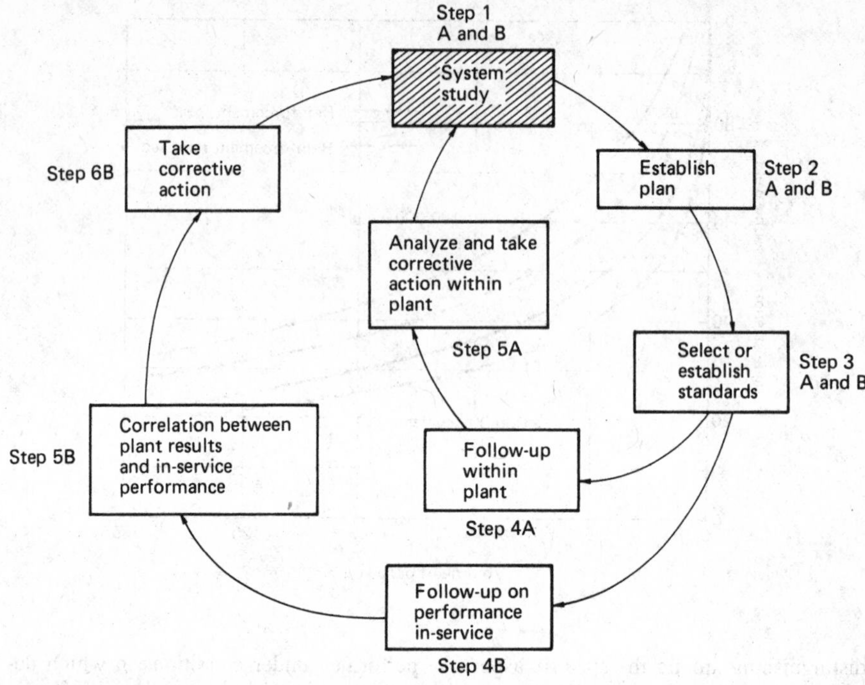

Production Examples

Example 1. Fracture Control of Space Shuttle External Tank Welds. The external tank of the Space Shuttle (composed of two tanks, a large hydrogen tank and a smaller oxygen tank, joined together by a collar-like intertank to form one large propellant storage container 154.2 ft long and 27.5 ft in diameter) was fabricated from 2219-T87 aluminum, a relatively tough alloy; consequently, it was recognized that expressions relating fracture strength to discontinuity size would be difficult to obtain and subject to criticism, as would any calculation of a toughness value (e.g., K_{Ic}). A semiempirical approach to the problem was taken. Specimens were generated at

Fig. 37 Discontinuity depth versus fracture strength for 2219-T87 welds. A leak mode failure occurs when the discontinuity depth exceeds the thickness at the critical stress. *a*/2*c* ratio describes the shape of the discontinuity, where *a* is crack depth and 2*c* is crack length.

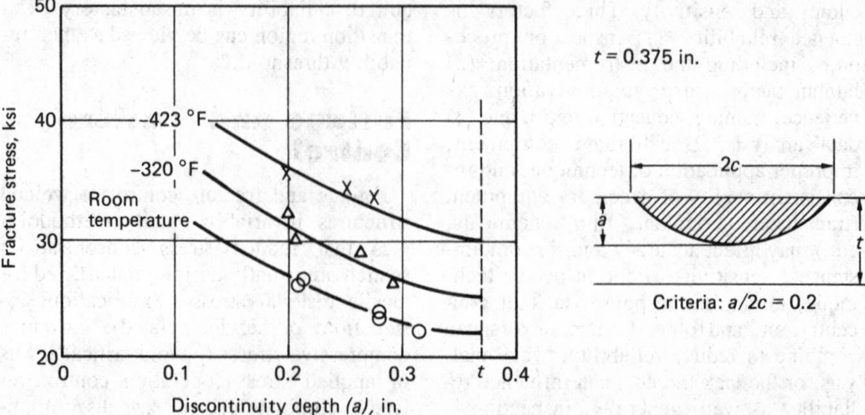

Fig. 38 Determining critical discontinuity depths in aluminum alloy 2219-T87 welds used in space shuttle external tanks

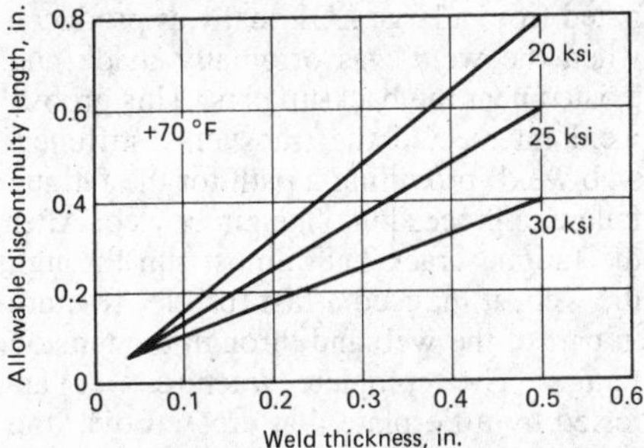

the production facility with production welding machines, using production weld schedules, specimen thicknesses equivalent to tank wall thicknesses, and test temperatures that were the same as flight temperatures to determine critical discontinuities at the anticipated proof and flight stresses. The resulting curves of discontinuity size, as a function of fracture strength, were used for design and analysis.

Test specimens were prepared with known cracks. The cracks were introduced by electrical discharge machining a slot and extending the slot by fatigue cycles. By trial and error, slot shape and fatigue loading were adjusted to produce a preexisting crack of known configuration. The specimen was exposed to a known load environment or load cycles, and crack growth or fracture strength was determined.

Figure 37 plots fracture strength for typical 2219 welds with a discontinuity in the weld centerline. Using these data, the designer could predict the failure stress for various sizes of discontinuities.

Using ambient temperature data similar to Fig. 37 for the various external tank weld thicknesses, weld acceptance criteria for 2219 were established by:

- Determining critical discontinuity depths at 20, 25, and 30 ksi stress, the most critical stress range in external tank welds.
- Converting discontinuity depths into lengths by multiplying the depth by 5 ($a/2c$ 0.2, see Fig. 37). This conversion was necessary because x-ray and liquid-penetrant inspection can only determine discontinuity lengths.
- Dividing critical lengths by 2. This provided a safety margin, allowing for uncertainties in stress analysis and inaccuracies in inspection techniques.
- Considering a semicircular discontinuity where the depth of the discontinuity was equal to the wall thickness—a discontinuity that will leak. This discontinuity length was compared to the length determined above and the smaller length was used.
- Rounding off the lengths to the nearest 0.05 in. to help the inspection.

The result of this approach is shown in Fig. 38.

In line with the practice of using grades to denote inspection level, the 2219 specification included a table of weld grades as a function of acceptable discontinuity length. The design drawing specified the

Fig. 39 Space shuttle external tank: hydrogen and oxygen tanks in production. Insert shows space shuttle launch. Courtesy of Martin Marietta Aerospace, Michoud Division.

weld grade for a particular weld. The inspector used the grade to determine the discontinuity length for radiographic and liquid-penetrant acceptance of the weld. Figure 39 shows the external tank under construction and during lift-off.

Example 2. Trans-Alaska Oil Pipeline. The original construction code applied to the Alaskan crude oil pipeline was API-1104. Discontinuity acceptance levels were established to maintain a required norm of workmanship. An audit of about 30 000 welds revealed discontinuities (LOF/LOP, slag, porosity, hollow weld beads, melt-through, cracks, and gas pockets). The largest number of repairs required by code involved gas pockets located at the bottom of the pipe where welders ended their welds. The approach taken to assess the structural significance of these discontinuities (planar and nonplanar) assumed worst-case conditions:

- All discontinuities as surface cracks
- Minimum material toughness
- Maximum stresses (high hoop and tensile residual stresses, pipeline loading, and pressure above yield)
- Earthquake possibility
- Worst-case fatigue
- Most adverse service environment (corrosivity and lowest temperature on a historical basis in Alaska)
- Crack growth rate under cyclic and sustained load conditions

The net conclusion according to the fitness-for-purpose criteria was that larger discontinuities than allowable under the API 1104 code could be permitted and that arc burns could be permitted under certain conditions. The fitness-for-purpose evaluation contributed to the Department of Transportation's decision to grant a waiver which permitted use of the alternative standards in place of API 1104 requirements. Applying the waiver procedures to critical welds located under a river crossing initially saved several millions of dollars in remedial construction cost. More importantly, a legal precedent was established. A subsequent audit of a statistical sample of girth weld radiographs (1500 welds) indicated that 7.9% of the welds did not meet API 1104 requirements. The fitness-for-purpose standards and the legal precedent provided the basis for deciding that the weld quality of the entire pipeline was acceptable, thus avoiding a potential delay in pipeline startup.

Example 3. Failure Analysis of a Large Structure. In May of 1975, one of the main girders of the Lafayette Street bridge over the Mississippi River in St. Paul, Minnesota, was discovered to be cracked in the 362-ft main span of the three-span structure. The crack extended completely through the bottom flange of the girder and up the web for a distance of about 30 in. out of 138 in. total girder depth.

In the vicinity of the crack in the web close to the bottom flange, a gusset plate and vertical stiffener were welded to the web. The crack also appeared to extend along the gusset-to-stiffener weld. The flange and the gusset were made of ASTM A441 and A36 steel, respectively. Chemical analysis and tension tests performed on the materials showed compliance with ASTM requirements.

Charpy V-notch testing at 15 ft · lb resulted in a transition temperature of about 50 °F, which did not meet existing toughness requirements for the geographic location of the bridge. However, no such specification existed at the time of bridge construction (between 1964 and 1968). The web proved to be tougher and met the current specification (1974 AASHTO) with a 15 ft · lb transition temperature of 15 °F. Based on Charpy impact and compact tests, the web K_{Ic} at the temperature of fracture (estimated to be 0 °F) was about 70 ksi $\sqrt{\text{in.}}$

Several stages and different modes (fatigue and cleavage) of cracking were noted. The fatigue crack in the lateral bracing gusset-to-transverse stiffener weld originated from a large LOF that was produced when the weld was originally made and existed near the back-up bars. This groove weld intersected the transverse stiffener-web weld, providing a path for the fatigue failure to proceed into the girder web. After the fatigue crack had almost run through the web, it induced a fast (brittle) fracture in part of the web and through the tension flange. The complete fracture was arrested by a tie plate that crossed over the crack. Subsequent cyclic loading produced further crack propagation, resulting in an appreciable web crack extension. This weld detail was not unique to just one location or this span; other design details of this type were examined, and fatigue cracks were found. Figure 40 shows a cracked specimen girder.

The traffic on this bridge was estimated to be 1500 trucks per day, or 3.3 million trucks at the time the fracture was discov-

Fig. 40 Cracking of bridge girder assembly

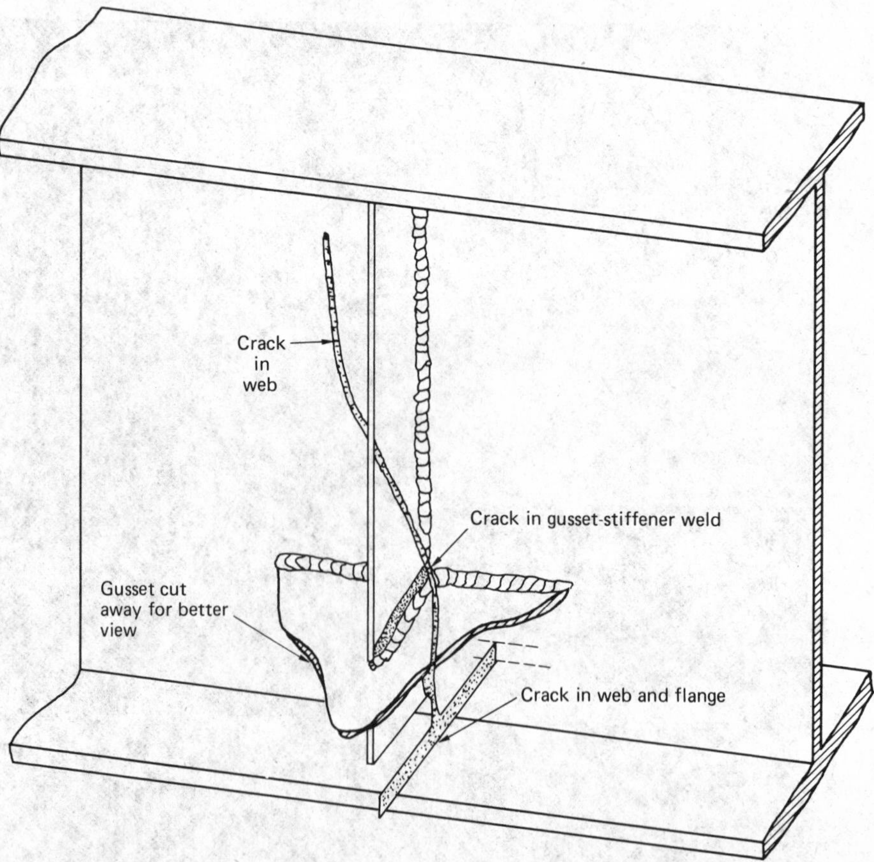

ered. Stress calculations showed that the dynamic crack driving force (ΔK) was small, but sufficient to cause growth of the LOF. Using the 3 million cycles applied to the bridge over 7 years of service, the weld discontinuity in combination with the operating stresses would have cracked the gusset-stiffener weld and penetrated the web by about $^3/_8$ in. From the high residual stress known to exist in the web at this location from the intersecting welds, the $^3/_8$-in. crack would have had a high enough stress ($K_{applied}$ = 70 ksi $\sqrt{\text{in.}}$) to become unstable and propagate. Because the flange was less tough than the web, rapid crack propagation down the web readily caused the flange to fail.

Repair consisted of drilling crack arrester holes in the web and gusset. The detail configuration was also changed to prevent gusset-plate welds from coming too close to the transverse stiffener welds, thus avoiding a direct path to the web in case a crack were to develop in the transverse stiffener-gusset welds. The gap also reduced restraint in the detail area. Furthermore, groove welds with back-up bars perpendicular to the bending stress were found to be unsatisfactory; this type of edge preparation tends to lead to LOF of critical size.

Example 4. Induction Motor Shaft Failures. The shaft of an eight-pole induction motor failed after 9 months of service. The shaft and spider section consisted of an 8-in.-diam 4140 steel shaft with eight $1^3/_4$-in.-thick C1020 steel ribs, which were fillet welded to the shaft by submerged arc welding and equally spaced around the circumference of the shaft. The synchronous motor drove an induced draft fan at a new electric utility plant.

Extensive examinations showed that materials conformed to required specifications. Fatigue failure was of the rotational bending type. Fatigue initiated at several locations of high stress concentrations—at the ends of the fillet welds on the fan side, at the roots of the fillet welds, and at slag inclusions. The ends of the fillet welds coincided with poor design details (sharp corners). Originally, a stiff coupler was placed between the motor and the fan. The two units were misaligned; ground settlement was also a possible contributing factor. The insufficiently balanced fan induced resonant frequency vibration in the motor, causing rapid failure.

A soft coupling was installed along with a new induction motor. The shaft failed again in the same manner and at the same location. The final repair consisted of: (1) balancing the fan, (2) better alignment, (3) soft coupling, (4) larger size shaft, (5) better design detail of the shaft and spider (large radii), (6) double groove welds, and (7) careful control of welding, manufacturing, and installation procedures.

REFERENCES

1. ASME Boiler and Pressure Vessel Code, Appendix IV, Draft 1, Oct 21, 1981
2. Sandor, L.W., The Meaning of Weld Discontinuities in Shipbuilding, Proceedings of First National Conference on Fitness-for-Service in Shipbuilding, U.S. MARAD, SNAME, Boulder, CO, Oct 1980
3. Munse, W.H. *et al.*, Effect of Weld Procedures on Weld Quality, Civil Engineering Studies, Structural Research Series No. 196, University of Illinois, July 1982
4. Stout, R.D. *et al.*, *Weldability of Steels*, 3rd ed., Welding Research Council, 1978
5. Sandor, L.W., The Influence of Different Welding Wires upon Fillet Weld Morphology, unpublished report, April 1977
6. British Standards Institution, Draft PD for Guidance on Some Methods for the Derivation of Acceptance Levels for Defects in Fusion Welded Joints, London, 1978
7. Munse, W.H., Fatigue Data Bank, University of Illinois, 1980. Also see, "Fatigue Data Bank and Data Analysis Investigation," Structural Research Series No. 405, University of Illinois, June 1973
8. Bowman, M.D. *et al.*, The Effect of Discontinuities on the Fatigue Behavior of Transverse Butt Welds in Steel, Civil Engineering Studies, Structural Research Series No. 491, University of Illinois, April 1981
9. Gurney, T.R., Theoretical Analysis of the Influence of Toe Defects on the Fatigue Strength of Fillet Welded Joints, TWI Research Report, 32/1977/E, March 1977
10. Screm, G. *et al.*, The Effect of Weld Defects in Reducing the Fatigue Strength of Welded Joints, *Alluminio*, Vol 44 (No. 3), March 1975
11. Harrison, J.D., Basis for a Proposed Acceptance Standard for Weld Defects, Part I: Porosity, *Metal Construction and British Welding Journal*, Vol 4 (No. 3), March 1972
12. Packman, P.F., Status of Non-Destructive Inspection Techniques with Special References to Welding Defects, Proc. Japan—U.S. Seminar, University of Tokyo Press, 1973
13. Rummel, W.D. *et al.*, Detection and Measurement of Fatigue Cracks in Aluminum Alloy Sheet by Non-Destructive Evaluation Techniques, in *Prevention of Structural Failure: The Role of Qualitative Non-Destructive Evaluation*, American Society for Metals, 1975

Residual Stresses and Distortion

By Koichi Masubuchi
Professor of Ocean Engineering
and Materials Science
Massachusetts Institute of Technology

COMPLEX THERMAL STRESSES occur in parts during welding due to the localized application of heat. Residual stresses and distortion remain after the welding process is completed. Thermal stresses, residual stresses, and distortion sometimes cause cracking and mismatching of joints. High tensile residual stresses in areas near the weld can cause premature failure of welded structures under certain conditions. They can also promote fatigue fracture and stress corrosion cracking in weldments. Distortion and compressive residual stresses in the base plate can reduce the buckling strength of structural members subjected to compressive loading. Correcting unacceptable distortion is costly and, in some cases, impossible.

Residual stresses and distortion in welded structures have been recognized and studied since 1930. Analyses of these subjects require complex computations; therefore, most past studies were primarily empirical or limited to analyses of simple cases. With the advancement of modern computers and techniques such as the finite element method, effort has been renewed in recent years to study these stresses and related phenomena.

In 1980, Masubuchi (Ref 1) published a book covering many subjects related to residual stress and distortion in welded structures. He also prepared a chapter on "Residual Stresses and Distortion" in Volume 1 of the 7th edition of the American Welding Society (AWS) *Welding Handbook* published in 1976 (Ref 2). This article has been excerpted from these publications. Efforts also have been made to update the information.

Many diverse notations have been used to express a variety of physical and material properties, including English, metric, and SI units. Although efforts have been made to use English units of measure throughout this article, some metric notations, as used in original documents, are retained.

Formation of Residual Stresses and Distortion

Residual stresses, also referred to as internal stresses, initial stresses, inherent stresses, reaction stresses, and locked-in stresses, are stresses that continue to exist in a body if all external loads are removed. Residual stresses also occur when a body is subjected to nonuniform temperature changes; these stresses are called thermal stresses. The intensity of stress is usually expressed in loads or forces per unit area, such as pounds per square inch (psi), kilogram per square millimetre (kg/mm^2), Newton per square metre (N/m^2), or Pascal (Pa).

Causes of Residual Stresses. Residual stresses in metal structures occur for many reasons during various manufacturing stages. Residual stresses may be produced in many structural elements, including plate, bar, and sections, during rolling, casting, or forging. They may occur during forming and shaping of metal parts by such processes as shearing, bending, machining, and grinding. They also occur during fabrication processes such as welding.

Heat treatments at various stages also can influence residual stresses. For example, quenching produces residual stresses, while stress-relieving heat treatments reduce residual stresses.

Macroscopic and Microscopic Residual Stresses. Areas in which residual stresses exist vary greatly from a large portion of a metal structure down to a microscopic area. For example, when a ship is heated by the sun on one side of the structure, thermal stresses and distortion are produced. In all welded structures, residual stresses are produced in areas around the weld. Grinding operations are known to produce localized residual stresses in thin layers below the ground surface.

Residual stresses also occur on a macroscopic scale. For example, residual stresses are produced in areas near the martensitic structures, because martensitic transformation that occurs at relatively low temperatures results in expansion of the metal. This article primarily discusses macroscopic residual stresses. Residual stresses can be classified into two groups according to the mechanisms that produce them—structural mismatch, and the uneven distribution of nonelastic strains, including plastic and thermal strains.

Residual Stresses Produced by Mismatch

Figure 1 illustrates residual stresses produced when bars of different lengths are forcibly connected. Tensile stresses are produced in the shorter bar, Q, and compressive stresses are produced in longer bars, P and P'.

Figure 2 shows a typical heating and cooling cycle that caused mismatch resulting in residual stresses. Three carbon steel bars of equal length and cross-sectional area are connected by two rigid blocks at the ends. The middle bar is heated to 1100 °F and then cooled to room temperature, while the side bars are kept at room temperature. Figure 2 plots the stress in the middle bar against the temperature and shows the amount of residual stress produced. Because the two side bars resist the deformation of the middle bar, the stress in each side bar is equal to half the stress in the middle bar and opposite in sign from the equilibrium condition:

$$\sigma_s = -\frac{1}{2}\sigma_m \qquad \text{(Eq 1)}$$

Fig. 1 Residual stresses produced when bars of different lengths are forcibly connected. (a) Free state. (b) Stressed state. (Source: Ref 1 and 2)

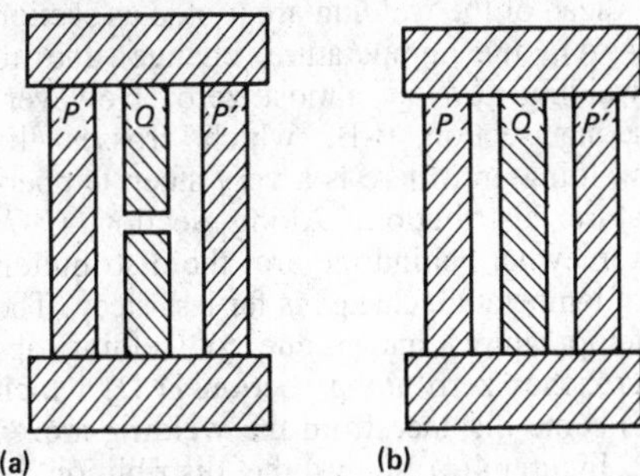

Fig. 2 Stress-temperature curve for middle bar of a three-bar frame

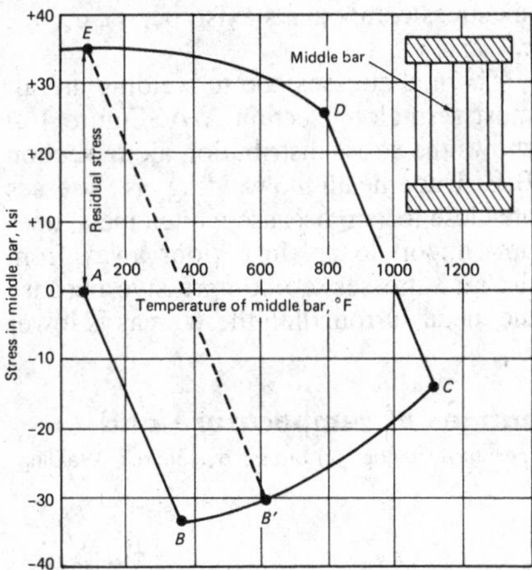

Because the length of the middle and side bars must be the same at any temperature, the following equation must be satisfied:

$$\frac{\sigma_m}{E_t} + \alpha\Delta T = \frac{\sigma_s}{E} \quad \text{(Eq 2)}$$

where E is the modulus of elasticity at room temperature (T_O); E_t is the modulus of elasticity at temperature (T); ΔT equals $T - T_O$, which equals the increment of temperature; and α is the coefficient of linear thermal coefficient. With these conditions,

$$\sigma_m = -\alpha\Delta T \frac{2E}{1 + 2E/E_t} \quad \text{(Eq 3)}$$

The stress in the middle bar (σ_m) at various stages of the thermal cycle can be calculated from Eq 3. As the temperature of the middle bar increases, the stress changes when the temperature is approximately 340 °F, as indicated by B. As the temperature rises above point B, the stress in the middle bar is limited to the yield stress at each corresponding temperature, as shown by curve BC. When the temperature falls below 1100 °F on cooling, the middle bar experiences elastic stress. Compressive stress in the middle bar also drops rapidly, changes to tension, and soon reaches the yield stress in tension as indicated by point D. As the temperature decreases further, the stress in the middle bar is limited to yield stress at each corresponding temperature, as shown by curve DE. Thus, a residual tension stress equal to the yield stress at room temperature is set up in the middle bar. Residual stresses in the side bars are compressive stresses and are equal to one half of the tensile stress in the middle bar. Line $B'E$ indicates that residual stress of the same magnitude, which is equal to the yield stress at room temperature, is produced by heating the middle bar at any temperature exceeding 600 °F.

Dislocations in a Multiply Connected Body. Mathematically, residual stresses caused by a mismatch can be regarded as stresses due to an elastic dislocation that exists in a multiply connected body (Ref 1 and 2). A solid body occupies a singly connected region, while a body containing holes occupies a multiply connected region. Figures 3(a) and (b) show dislocations in a circular ring, in the x and y directions, respectively, which is a simple form of a multiply connected body. Figure 3 shows two types of dislocations out of a total six possible basic types that can exist in a circular ring. The six types are caused by translations in the x, y, and z directions and by rotations around the x-, y-, and z-axes. In the 1900's, Volterra (Ref 3) studied residual stresses in multiply connected regions and called this mismatched deformation "distortioni." Love (Ref 4) termed such stresses "dislocations." Since the 1930's, Taylor (Ref 5), Orowan (Ref 6), and others have used the concept of dislocation to explain atomic mechanisms of plastic deformation; this concept also has been used to explain many phenomena in physical metallurgy. As a result, some readers may regard the term "dislocation" as solely related to atomic dislocations. Displacement is multivalued when dislocation occurs.

Residual Stresses Produced by Uneven Distribution of Nonelastic Strains

When materials are heated uniformly, thermal stress is not produced, as they also expand uniformly. However, when materials are not heated uniformly, one must expect stress to result. Also, residual stresses are produced when unevenly distributed nonelastic strains, or plastic strains, exist. Fundamental relationships for a two-dimensional plane stress ($\sigma_z = 0$) residual stress field are listed below. Generally, in a three-dimensional stress field, six stress components exist: σ_x, σ_y, σ_z, τ_{xy}, τ_{yz}, and τ_{zx}. Strains are composed of elastic strain and nonelastic strain:

$$\epsilon_x = \epsilon_x' + \epsilon_x''$$

$$\epsilon_y = \epsilon_y' + \epsilon_y''$$

$$\gamma_{xy} = \gamma_{xy}' + \gamma_{xy}'' \quad \text{(Eq 4)}$$

where ϵ_x, ϵ_y, and γ_{xy} are components of the total strain; ϵ_x', ϵ_y', and γ_{xy}' are components of the elastic strain; and ϵ_x'', ϵ_y'', and γ_{xy}'' are components of the nonelastic strain. The nonelastic strain can be plastic strain or thermal strain. In the case of thermal stress, $\epsilon_x'' = \epsilon_y'' = \alpha \cdot \Delta T$, and $\gamma_{xy}'' = 0$, where α is the coefficient of linear thermal expansion and ΔT is the change of temperature from the initial temperature.

A Hooke's Law relationship exists between stress and elastic strain; therefore:

$$\epsilon_x' = \frac{1}{E}(\sigma_x - \nu\sigma_y)$$

Fig. 3 Dislocations in a circular ring. (a) Translation in the x direction (ξ). (b) Translation in the y direction (η)

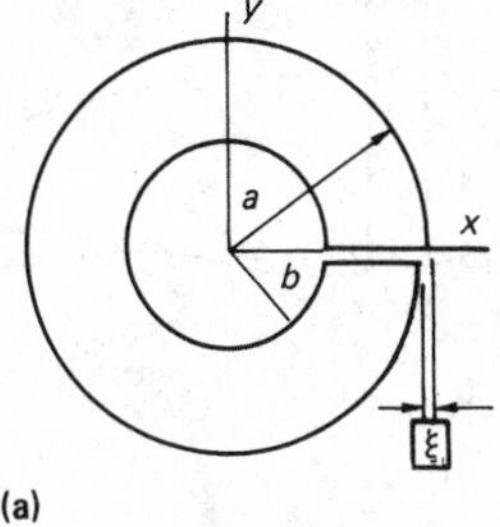

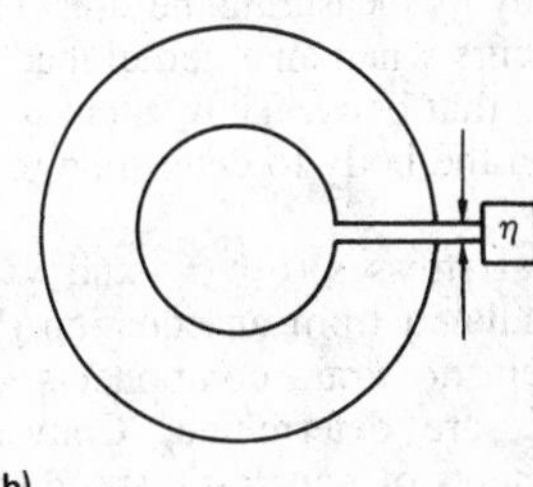

$$\epsilon'_y = \frac{1}{E}(\sigma_y - \nu\sigma_x)$$

$$\gamma'_{xy} = \frac{1}{G}\tau_{xy} \quad \text{(Eq 5)}$$

where E is modulus of elasticity; ν is Poisson's ratio; and G is shear modulus.

The stress must satisfy the equilibrium conditions:

$$\frac{\partial \sigma_x}{\partial x} + \frac{\partial \tau_{xy}}{\partial y} = 0$$

$$\frac{\partial \tau_{xy}}{\partial x} + \frac{\partial \sigma_y}{\partial y} = 0 \quad \text{(Eq 6)}$$

The total strain must satisfy the condition of compatibility:

$$\frac{\partial^2 \epsilon'_x}{\partial y^2} + \frac{\partial^2 \epsilon'_y}{\partial x^2} - \frac{\partial^2 \gamma'_{xy}}{\partial x \cdot \partial y} + \frac{\partial^2 \epsilon''_x}{\partial y^2} + \frac{\partial^2 \epsilon''_y}{\partial x^2} - \frac{\partial^2 \gamma''_{xy}}{\partial x \cdot \partial y} = 0 \quad \text{(Eq 7)}$$

Equations 6 and 7 indicate that residual stresses exist when the value of R, which is determined by the nonelastic strain as follows, is not zero:

$$R = -\frac{\partial^2 \epsilon''_x}{\partial y^2} + \frac{\partial^2 \epsilon''_y}{\partial x^2} - \frac{\partial^2 \gamma''_{xy}}{\partial x \cdot \partial y} \quad \text{(Eq 8)}$$

R, which has been called "incompatibility," can be considered the cause of residual stresses. If the nonelastic strain components can be expressed by linear functions of the position:

$$\epsilon''_x = a + bx + cy$$

$$\epsilon''_y = e + fx + gy$$

$$\gamma''_{xy} = k + mx + ny \quad \text{(Eq 9)}$$

then $R = 0$. Consequently, residual stresses will not occur.

Mathematical analyses of residual stresses have been made by many investigators. Several equations have been proposed to calculate stress components, σ_x, σ_y, and γ_{xy} for given values of nonelastic strain (ϵ''_x, ϵ''_y, and γ''_{xy}). From these mathematical analyses, many important findings have been obtained, including:

- One cannot determine residual stresses in a body by measuring the stress change that occurs when an external load is applied to that body. Thus, a cut must be made in the body to determine residual stresses.
- Residual stresses (σ_x, σ_y, and γ_{xy}) can be calculated from an equation (Eq 5) when elastic strain components ϵ'_x, ϵ'_y, and γ'_{xy} are determined. Conversely, components of nonelastic strain (ϵ''_x, ϵ''_y, and γ''_{xy}), which have caused residual stresses, cannot be ascertained unless the history of residual stress formation is known.

Thermal Stresses and Metal Movement During Welding

Because a weldment is heated locally by the welding heat source, temperature distribution in the weldment is not uniform and changes as the welding procedure progresses. During the welding thermal cycle, complex transient thermal stresses are produced in the weldment. The weldment also undergoes shrinkage and deformation during welding.

Thermal Stresses During Welding. Figure 4 schematically shows changes in temperature and resulting stresses that occur during welding, by examining a bead-on-plate weld made along the x-axis. The welding arc, which is moving at a speed (v), is presently located at the origin (O), as shown in Fig. 4(a).

The area where plastic deformation occurs during the welding thermal cycle is shown by the shaded area, M-M' in Fig. 4(a). The region where the metal is melted is indicated by the elipse near the origin (O). The region outside the shaded area remains elastic throughout the entire welding thermal cycle.

Temperature distributions along several cross sections are indicated in Fig. 4(b). Ahead of the welding arc (noted as section A-A), the temperature change due to welding (ΔT) is almost zero. However, along section B-B, which crosses the welding arc, there is a very steep temperature distribution. Along section C-C, somewhat behind the arc, the distribution of temperature change is far less steep. The temperature change due to welding approaches zero along section D-D, which is some distance from the welding arc.

Figure 4(c) shows the distribution of stresses along these sections in the x direction (σ_x). Stress in the y direction (σ_y) and shearing stress (τ_{xy}) also exist in a two-dimensional stress field. As stated previously, in a three-dimensional stress field, six stress components exist: σ_x, σ_y, σ_z, τ_{xy}, τ_{yz}, and τ_{zx}.

Thermal stresses due to welding are almost zero along section A-A. Figure 4(c) shows the stress distribution along section B-B. Underneath the welding arc, stresses are close to zero because molten metal does not support loads. In regions away from the arc, stresses are compressive because the metal surrounding these areas is lower

Fig. 4 Schematic representation of changes of temperature and stresses during welding. (a) Weld. (b) Temperature change. (c) Stress, σ_x. Source: Welding Research Council

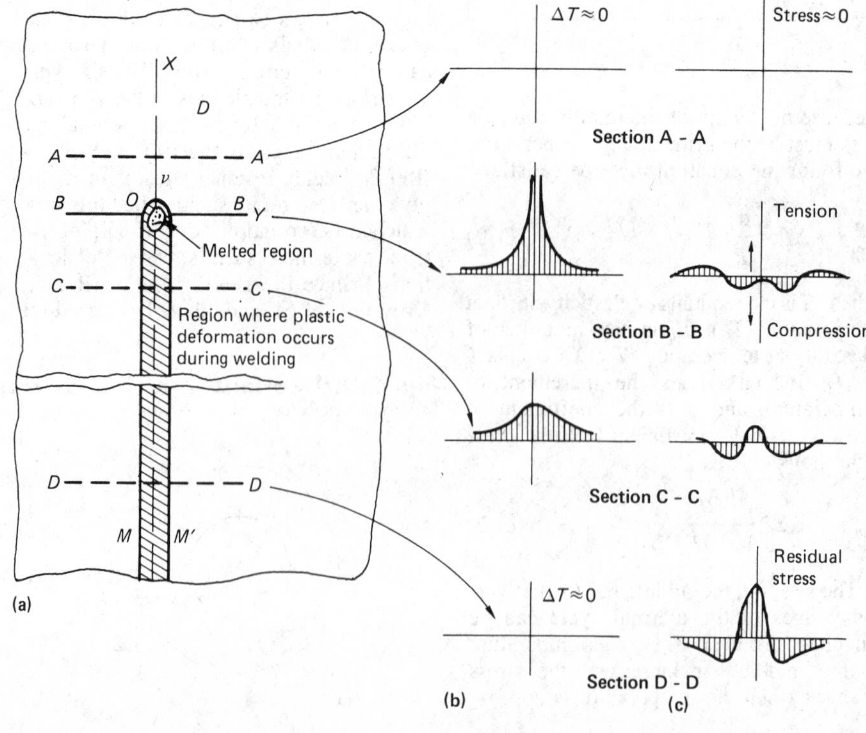

in temperature and expansion is restrained. Because the yield strength of the material is low and the temperatures of these areas are quite high, stresses in these areas are as high as the yield strength of the base metal at corresponding temperatures. The magnitude of compressive stress passes through a maximum with increasing distance from the weld or with decreasing temperature. However, stresses occurring in regions away from the weld are tensile in nature and balance with compressive stresses in areas near the weld.

Figure 4(c) shows that along section C-C the weld-metal and base-metal regions near the weld have cooled. As they shrink, tensile stresses are caused in regions in and adjacent to the weld. As the distance from the weld increases, stresses change to compressive and then become tensile. High tensile stresses are produced along section D-D in and adjacent to the weld. Compressive stresses are produced in areas away from the weld.

Equilibrium Condition of Residual Stresses. Because residual stresses exist without external forces, the resultant force and the resultant moment produced by the residual stresses must vanish:

$$\int \sigma \cdot dA = 0 \text{ on any plane section and} \quad \text{(Eq 10)}$$

$$\int dM = 0 \quad \text{(Eq 11)}$$

where dA is area, and dM is the resultant moment. Residual stress data must be checked in any experiment to ensure that it satisfies the above conditions.

Metal Movement During Welding. During welding, the weldment undergoes shrinkage and deformation. The transient deformation, or metal movement, is most evident when the weld line is away from the neutral axis of the weldment, causing a large amount of bending moment. Figure 5 schematically shows how a rectangular plate deforms when its longitudinal edge is heated by a moving heat source, such as a welding arc or an oxyacetylene torch (for cutting, welding, or flame heating). Areas near the heat source, or the upper regions of the rectangular plate, are heated to higher temperatures and thus expand more than areas away from the heat source or more than lower regions of the plate. Therefore, the plate first deforms as shown by curve *AB*.

If all parts of the material remained completely elastic during the entire thermal cycle, thermal stresses produced during the heating and cooling cycle would disappear when the temperature returned to the initial temperature. Deformation of the plate would be as indicated in curve *AB'C'D'*, indicating no deformation after the thermal cycle.

Fig. 5 Changes of deflection at the center of the lower edge of a rectangular plate due to heating by a heat source moving along the upper edge and subsequent cooling

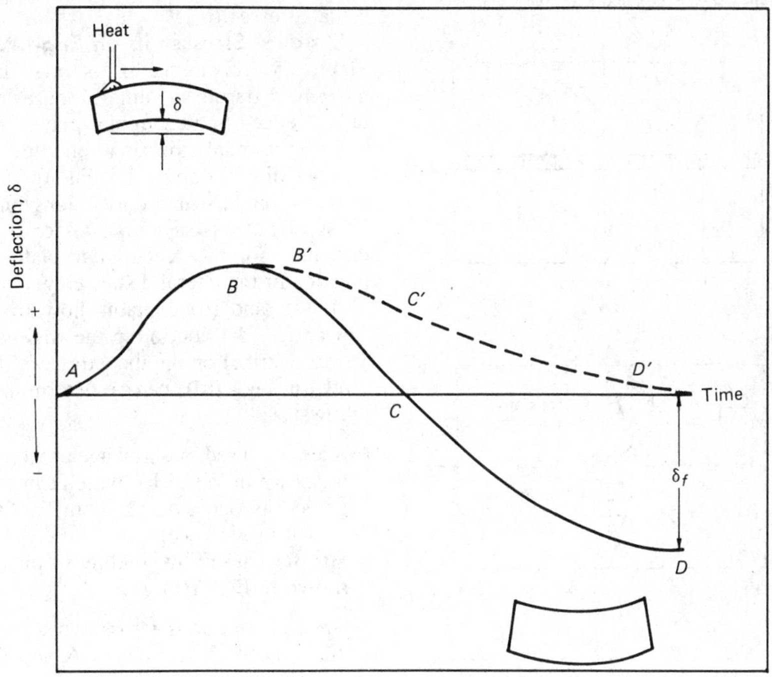

However, in most practical materials, plastic strains are produced in areas heated to elevated temperatures, causing plastic upsetting. The thermal stresses do not disappear when the temperature returns to the initial temperature, causing residual stresses. Transient deformation of the plate during heating and cooling is shown by curve *ABCD*. After the plate cools to the initial temperature, final deformation in the amount of δ_f remains, which also is called distortion. The metal movement during welding, and the distortion after welding is completed, are in opposite directions, generally on the same order of magnitude.

Residual Stresses and Distortion in Weldments

During the heating and cooling cycle of welding, nonelastic strains are produced in areas near the weld. As a result, when the weldment cools to the initial temperature, residual stresses and distortion remain. Welding also may cause elastic dislocation in the structure being fabricated, especially during the first pass welding in which two separate plates are joined together.

Residual Welding Stresses and Reaction Stresses. As stated earlier, residual stresses can be classified into two groups—stresses produced by structural mismatch or dislocation, and stresses produced by uneven distribution of nonelastic strains. The same method of classification is applicable to residual stresses in weldments. They are classified as reaction stresses that are caused when the weldment is restrained externally, and residual welding stresses that are produced in an unrestrained weldment. Distributions of these two types of residual stresses in a simple butt weld are discussed below.

Residual Stresses in a Butt Weld. Figure 6 shows a typical distribution of residual stresses in a butt weld. The significant stresses are those parallel to the weld direction, designated σ_x, and those transverse to it, designated σ_y. The distribution of longitudinal residual stress (σ_x) is shown in Fig. 6(b). Near the weld, tensile stresses of high magnitude are produced. These taper off quickly and become compressive at a distance equal to several times the width of the weld metal.

Fig. 6 Typical distributions of residual stresses in a butt weld. (a) Butt weld. (b) Distribution of σ_x along *YY*. (c) Distribution of σ_y along *XX*. Reprinted by permission of the Welding Research Council

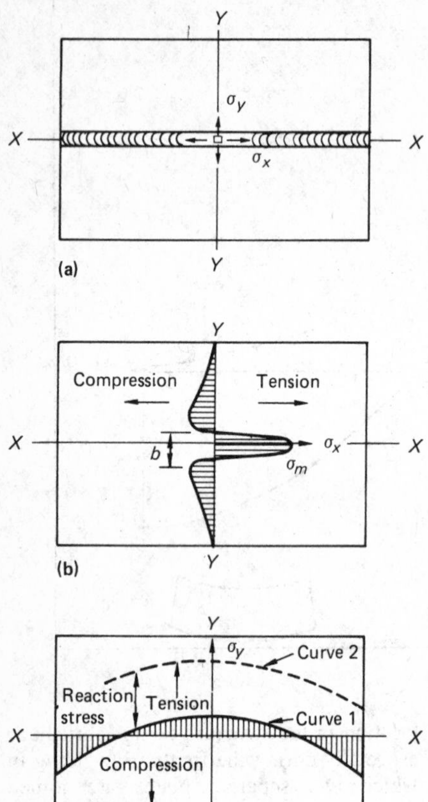

Stress distribution is characterized by two parameters: (1) the maximum stress at weld region (σ_m), and (2) the width of the tension zone of residual stress (b). In weldments made of low-carbon steel, the maximum residual stress (σ_m) is usually as high as the yield stress of the weld metal. The distribution of longitudinal residual stress (σ_x) may be approximated by the following equation:

$$\sigma_x(y) = \sigma_m \left[1 - \left(\frac{y}{b}\right)^2\right] e - \frac{1}{2}(y/b)^2 \quad \text{(Eq 12)}$$

Curve 1 of Fig. 6(c) shows the distribution of transverse residual stress (σ_y) along the length of the weld. In the center of the joint, tensile stresses of relatively low magnitude are produced. Compressive stresses are produced at the ends of the joint.

If an external constraint is used to restrain the lateral contraction of the joint, approximately uniform tensile stresses along the weld are added as the reaction stress. However, an external constraint has little influence on the distribution of residual stresses (σ_x).

Residual Stresses in an Edge Weld. Figure 5 schematically shows metal movement during welding along a longitudinal edge of a rectangular plate. Figure 7 shows residual stresses in an edge weld after welding is completed. Figure 7(a) illustrates the distribution of longitudinal stress (σ_x) across sections A-A or B-B in the center of the rectangular plate as a function of the lateral distance (y). Figure 7(b), (c), and (d) explains how to construct stress distribution in the edge weld. Stress distribution in the edge weld, as shown in Fig. 7(d), can be determined as follows:

- Stresses caused by shrinkage of the weld, as shown in Fig. 7(b), which can be regarded as one side of a butt weld, as shown in Fig. 6(b)
- Stresses caused by bending of plate, as shown in Fig. 7(c)

Stress distributions across two adjoining sections, such as sections A-A and B-B, are essentially the same. In other words, changes of stresses in the x-direction, except in areas near both ends of the rectangular plate, are far less than those in the y-direction. Mathematically, this means that:

$$\frac{\partial\sigma}{\partial x} \ll \frac{\partial\sigma}{\partial y} \quad \text{(Eq 13)}$$

The stress distribution shown above is an ideal for one-dimensional stress analysis. From the equilibrium condition of stresses as shown in Eq 6:

$$\frac{\partial\tau_{xy}}{\partial y} = 0$$

$$\frac{\partial\sigma_y}{\partial y} = 0 \quad \text{(Eq 14)}$$

This means that τ_{xy} and σ_y are constants. Because these stress components must be zero at the top and bottom edges, both τ_{xy} and σ_y are zero. In other words:

$$\sigma_x = f(y)$$

$$\tau_{xy} = \sigma_y = 0 \quad \text{(Eq 15)}$$

In analyzing the stress distribution, only the longitudinal stress (σ_x) is considered a function of the lateral distance; y, τ_{xy}, and σ_y are neglected. This type of analysis is called one-dimensional analysis.

Fundamental Types of Distortion in Weldments. Distortions found in fabricated structures are caused by three fundamental dimensional changes that occur during welding: (1) transverse shrinkage occurring perpendicular to the weld line, (2) longitudinal shrinkage occurring parallel to the weld line, and (3) an angular change consisting of rotation around the weld line. All of these dimensional changes are shown in Fig. 8.

Shrinkage and distortion that occur during fabrication of actual structures are far more complex than those shown in Fig. 8. For example, when longitudinal shrinkage occurs in a fillet welded joint, the joint bends longitudinally unless the weld line is located along the neutral axis of the joint.

Techniques for Analyzing Residual Stress and Distortion

Studies on transient thermal stresses during welding started in the 1930's. However, because the computations required for analyzing transient phenomena involving plastic deformation at elevated temperatures are complex, analyses using manual computations were limited to simple applications, such as spot welding pro-

Fig. 7 Residual stresses in an edge weld. (a) Edge weld. (b), (c), and (d) Distribution of residual stress

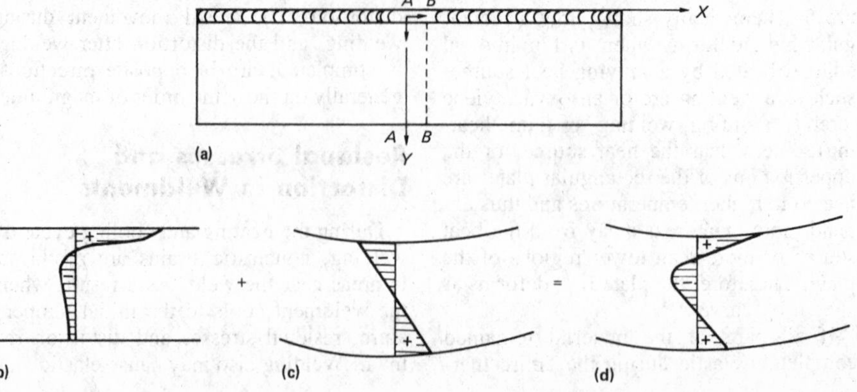

Fig. 8 Dimensional changes occurring in weldments. (a) Transverse shrinkage in a butt weld. (b) Longitudinal shrinkage in a butt weld. (Distribution of longitudinal residual stress, σ_x, is also shown.) (c) Angular change in a butt weld. (d) Angular change in a fillet weld. Source: Welding Research Council

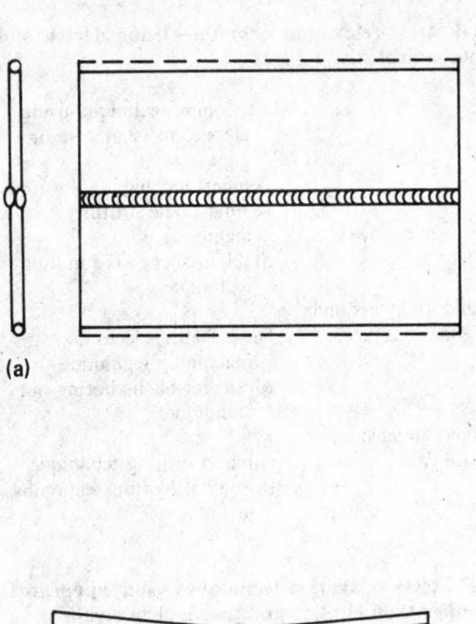

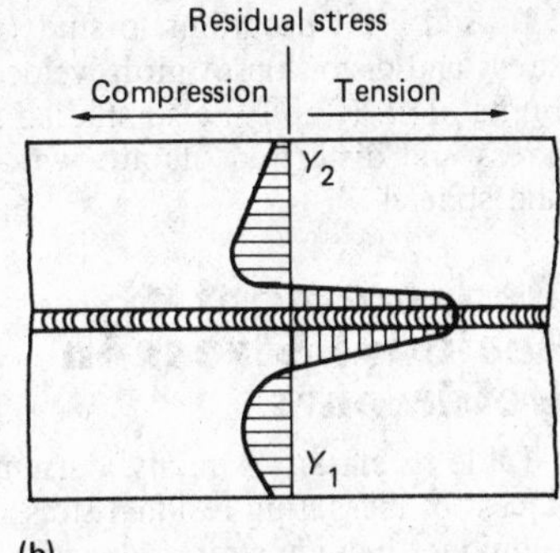

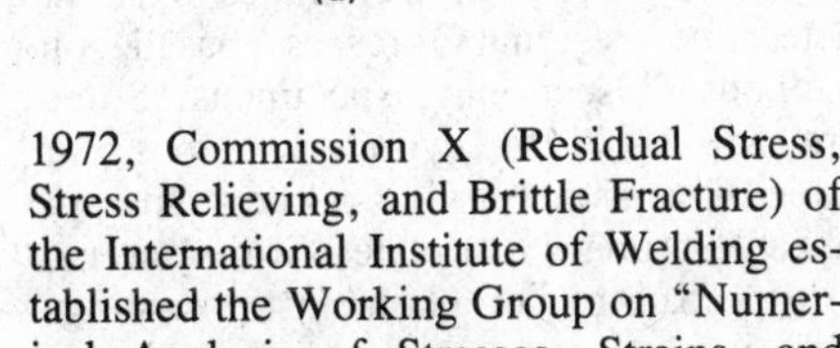

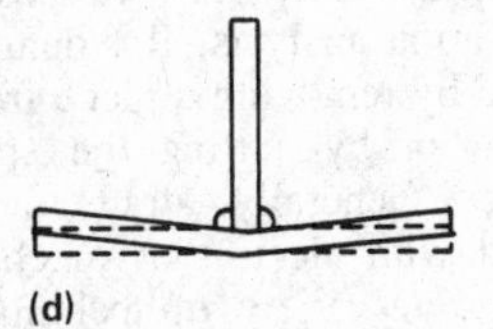

cedures in which temperature and stress changes are axially symmetric, or instantaneous heating along the edge of a strip in which temperature and stress changes are functions of only one axis. However, these analyses are far too simple to represent actual weld conditions.

The first significant attempt to use computers to analyze thermal stresses during welding was done by Tall (Ref 7 and 8) in 1961. He developed a simple program on thermal stresses during bead welding along the center line of a strip. The temperature distribution was treated as two-dimensional; however, in analyzing stresses it was assumed that longitudinal stress (σ_x) was a function of the lateral distance y only and that σ_y and γ_{xy} were zero. Such an analysis is called one-dimensional analysis. In 1968, Masubuchi *et al.* (Ref 9) developed, based on Tall's analysis, a FORTRAN computer program for one-dimensional analysis of transient thermal stresses during welding. One-dimensional computer programs have been further improved (Ref 10).

Since 1970, computer analyses of residual stresses and distortion in weldments have become more common. Investigators have developed computer programs for analyzing transient thermal stresses, residual stresses, and distortion in various types of weldments. The finite element method has been used widely for analyzing stresses under complex boundary conditions. In 1972, Commission X (Residual Stress, Stress Relieving, and Brittle Fracture) of the International Institute of Welding established the Working Group on "Numerical Analysis of Stresses, Strains, and Other Effects Produced by Welding." The Working Group has prepared reports detailing numerous stress analysis studies (Ref 11).

Methods for Analyzing Residual Stresses in Weldments. Figure 9 shows various ways to analyze residual stress and distortion in weldments. The "analytical simulation" examines what actually happens during welding. Analysis involves the following steps:

- *Step 1*: Analysis of heat flow
- *Step 2*: Analysis of transient thermal strain
- *Step 3*: Determination of incompatible strain
- *Step 4*: Analysis of residual stress and distortion

When analyzing residual stresses and distortion, step 3 is the most important. If transient thermal stress was completely elastic and no incompatible strain was produced, no residual stress would remain. When the material undergoes plastic deformation, the stress-strain relationship is not linear, and plastic properties of the material change with temperature. In addition to the plastic strain, incompatible strains that occur during welding may include dimensional changes associated with solidification of the weld metal, solid-state transformation, etc. It is extremely difficult to determine the exact distribution of incompatible strains that occur in many practical welds.

Step 4 is rather simple. Once the distribution of incompatible strains is deter-

Fig. 9 Methods of analyzing residual stresses in weldments

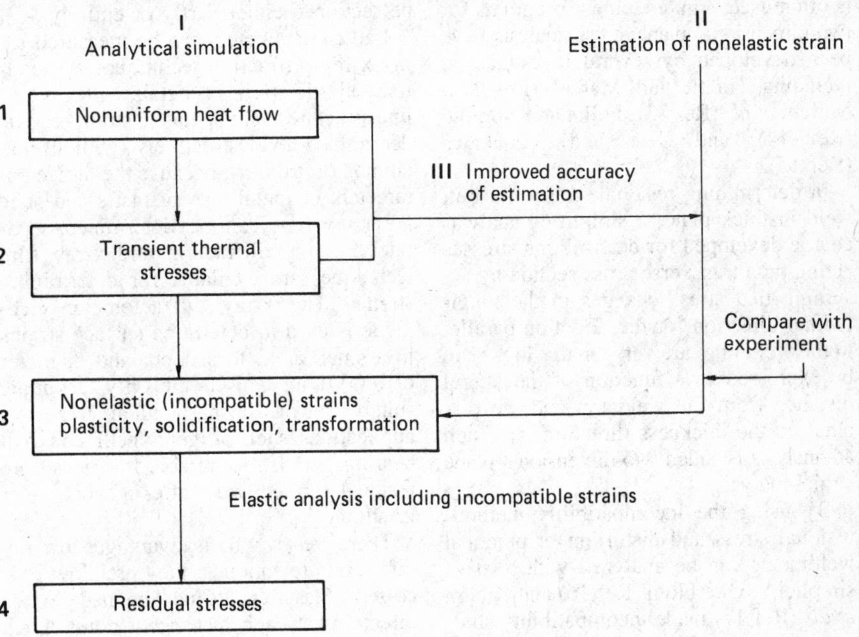

mined (either analytically, experimentally, or by estimation), residual stress and distortion can be determined through elastic calculations.

Because residual stress and distortion can be determined by knowing the distribution of incompatible strains, a second method that by-passes steps 1 and 2 and avoids complex plasticity analyses has been developed. This method, called the "incompatible strain method," has several advantages. It can be used for analyzing residual stress and distortion in complex weldments. However, there is no sure way to accurately estimate the distribution of incompatible strains.

The third method combines the two methods described above. An analytical simulation is used to refine the estimation that is made of the distribution of incompatible strains.

Of course, analytical simulation is the conventional method of analyzing residual stress and distortion in weldments. As stated earlier, however, analyses using manual computations were limited to very simple cases such as spot welding and instantaneous heating along the edge of a strip. Even in the simple case of welding along the edge of a rectangular plate, as shown in Fig. 5, a computer is needed to determine curve *ABCD* rapidly. The one-dimensional computer program is rather simple, and computations can be done at low cost.

To calculate two-dimensional distributions of transient thermal stress during welding a butt weld (similar to Fig. 6), required computations are complex enough so that the finite element method (FEM) is often used. Finite element programs for two-dimensional plane stress analysis have been developed by several investigators, including Hibbit and Marcel (Ref 12), Muraki *et al.* (Ref 13), Fujita and Nomoto (Ref 14), and Ueda and Yamakawa (Ref 15).

In determining residual stress in a butt weld in thick plate, a simplified analysis can be developed for determining stresses acting on a transverse cross section by assuming that stress changes in the longitudinal direction, or the direction parallel to the weld line, are very small. Stress can be expressed as a function of the lateral distance from the weld (y) and the distance in the thickness direction (z). Such an analysis is called two-dimensional plane strain analysis.

By using the incompatibility method, residual stress and distortion of practical weldments can be analyzed with relative simplicity. Okerblom (Ref 16) and Kihara *et al.* (Ref 17) made incompatibility studies before computers were available. With computers, it is now possible to extend the analysis to various types of practical weldments. For example, Rybicki *et al.* (Ref 18) used FEM programs to study residual stress and distortion of girth-welded pipes. Fujita *et al.* (Ref 19) also studied residual stress and distortion of butt welded pipe and spheres.

Measurement of Residual Stress in Weldments

Table 1 classifies many current techniques for measuring residual stress. These techniques include stress-relaxation, x-ray diffraction, stress-sensitive property analysis, and cracking techniques.

Elastic strain release is measured to determine residual stresses in stress-relaxation analysis. Residual stress is relaxed by removing a piece from the specimen, or by cutting the specimen into pieces. Generally, strain release is measured with electric or mechanical strain gages. A variety of techniques exist to determine residual stresses based on methods of sectioning specimens. Some are applicable to cylinders, tubes, or three-dimensional solids. Others are applicable primarily to plate. Grid systems, brittle coatings, or photoelastic coatings also can be used to determine strain release during stress relaxation. Stress-relaxation techniques are widely used to measure residual stress in weldments, because they produce reliable and quantitative data. However, stress-relaxation techniques are destructive; the specimen must be sectioned either partly or entirely.

Lattice parameters can be measured using x-ray diffraction techniques to determine elastic strain in metals with crystalline structures. Thus, elastic strains can be determined nondestructively (without machining or drilling), because the lattice parameters of metals in an unstressed state are known or can be determined separately. X-ray diffraction and x-ray film techniques are available for determining strains. The x-ray diffractometer technique is used to determine surface strains in a small area, to a depth and diameter of 0.0001 in. This is the only technique suitable for measuring residual stresses in applications such as gear teeth and ball bearings. It also is suitable for measuring residual surface stress after machining or grinding.

There are several disadvantages to x-ray diffraction techniques; they are slow processes. Measurements must be made in two directions at each measuring point. Each measurement requires 15 to 30 min of exposure for the film technique. In addition, measurements are not very accurate, particularly with heat treated materials in which atomic structures are distorted. Ultrasonic testing and hardness measurements methods are best suited to determine residual stresses in metals by measuring stress-sensitive properties.

Ultrasonic testing techniques may use polarized ultrasonic waves and may make use of stress-induced changes in the ab-

Table 1 Classification of techniques for measuring residual stresses

	Technique
A-1 Stress-relaxation techniques using electric and mechanical strain gages	
Plate	Sectioning technique using electric resistance strain gages
	Gunnert technique
	Mathar-Soete drilling technique
	Stäblein successive milling technique
Solid cylinders and tubes	Heyn-Bauer successive machining technique
	Mesnager-Sachs boring-out technique
Three-dimensional solids	Gunnert drilling technique
	Rosenthal-Norton sectioning technique
A-2 Stress-relaxation techniques using apparatus other than electric and mechanical strain gages	
	Grid system-dividing technique
	Brittle coating-drilling technique
	Photoelastic coating-drilling technique
B X-ray diffraction techniques	
	X-ray film technique
	X-ray diffractometer technique
C Techniques using stress-sensitive properties	
Ultrasonic techniques	Polarized ultrasonic wave technique
	Ultrasonic attenuation technique
Other	Hardness techniques
D Cracking techniques	
	Hydrogen-induced cracking technique
	Stress corrosion cracking technique

Source: Ref 2

sorption of ultrasonic waves (ultrasonic attenuation) or make use of stress-induced change in the angle of polarization of polarized ultrasonic waves (similar to the photoelastic technique).

Hardness testing techniques use stress-induced changes in hardness. These techniques have not been developed past the laboratory stage, and none has been used with success for measuring residual stresses in weldments.

Residual stresses can be studied by observing cracks in specimens produced by residual stresses. These cracks may be induced by the presence of hydrogen or stress corrosion. In studying complex structural models that have complicated residual stress distributions, these cracking techniques have been useful. However, these techniques result in qualitative rather than quantitative data.

Measurement of Residual Stresses by Stress-Relaxation Techniques

Using stress-relaxation techniques, it is possible to determine residual stresses without knowing the history of the material. Stress-relaxation techniques are based on the fact that strains occurring during unloading are elastic, even when the material has undergone plastic deformation.

Five techniques for measuring residual stresses, based on stress-relaxation techniques that can be used for weldments, are described in the following paragraphs. The first four techniques employ electric or mechanical strain gages. The first and second techniques apply primarily to plate, while the third and fourth techniques are used for three-dimensional solids. The last technique employs a photoelastic coating. For additional information, see Ref 1, which discusses 11 techniques.

Table 2 summarizes the range of application and the advantages and disadvantages of each of the five techniques described below.

Sectioning Technique Using Strain Gages. Residual stresses in plate can be determined by using electric-resistance strain gages mounted on the specimen surfaces. After the gage is mounted, a small piece of metal containing the gage is removed from the structure, as shown in Fig. 10. With resistance-bonded strain gage techniques, metallic wire or foil gage materials are bonded on the specimen. As the specimen is strained, the resistance of the gage changes, and the magnitude of strain is determined by measuring the resistance change. Information on electric strain gages is available from Ref 21 and 22. Measurement is made of strain changes ($\bar{\epsilon}_x$, $\bar{\epsilon}_y$, and $\bar{\gamma}_{xy}$) that take place during the removal of the piece. It can be assumed, if the piece is small enough, that residual stress no longer exists in the piece after removal, and therefore, the following can be stated:

Fig. 10 Complete stress-relaxation technique

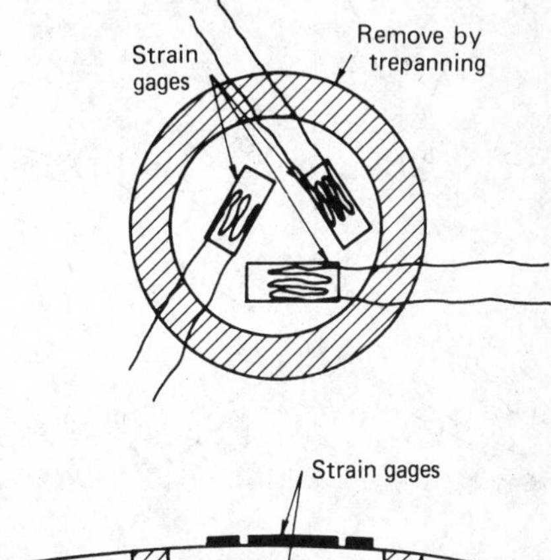

Table 2 Summary of stress-relaxation techniques for measuring residual stresses

Technique	Application	Advantages	Disadvantages
Sectioning technique for a plate using electric-resistance strain gages	Relatively all-around use, with the measuring surface placed in any position	Reliable method; simple principle; high measuring accuracy	Destructive; gives average stresses over the area of the piece removed from the specimen; not suitable for measuring locally concentrated stresses; machining is sometimes expensive and time consuming
Mathar-Soete drilling technique	Can be used for laboratory and field work and on horizontal, vertical, and overhead surfaces	Simple principle; causes little damage to the test piece; convenient to use on welds and adjoining material	Drilling causes plastic strains at the periphery of the hole which may displace the measured results; must be used with care
Gunnert drilling technique	Can be used for laboratory and field work; the surface of the plate must be substantially horizontal	Robust and simple apparatus; semi-nondestructive; damage to the object tested can be easily repaired	Relatively large margin of error for the stresses measured in a perpendicular direction; the underside of the plate must be accessible for the attachment of a fixture; entails manual training
Rosenthal-Norton sectioning technique	For laboratory measurements	Fairly accurate data can be obtained when measurements are carried out carefully	A troublesome, time-consuming, and completely destructive method
Photoelastic coating-drilling technique	Primarily a laboratory method, but it can also be used for field measurements under certain circumstances	Permits the measurement of local stress peaks; causes little damage to the material	Sensitive to plastic strains which sometimes occur at the edge of the drilled hole

Source: Ref 2

$$\bar{\epsilon}_x = -\epsilon'_x$$

$$\bar{\epsilon}_y = -\epsilon'_y$$

$$\bar{\gamma}_{xy} = -\gamma'_{xy} \quad \text{(Eq 16)}$$

where ϵ'_x, ϵ'_y, and γ'_{xy} are elastic strain components of the residual stress. The minus signs in Eq 16 indicate that, when tensile residual stress exists, shrinkage (not elongation) takes place during stress relaxation. Then residual stresses are:

$$\sigma_x = -\frac{E}{1-\nu^2}(\bar{\epsilon}_x + \nu\bar{\epsilon}_y)$$

$$\sigma_y = -\frac{E}{1-\nu^2}(\bar{\epsilon}_y + \nu\bar{\epsilon}_x)$$

$$\tau_{xy} = -G\,\bar{\gamma}_{xy} \quad \text{(Eq 17)}$$

where E is modulus of elasticity; ν is Poisson's ratio; and G is shear modulus.

Strain measurements should be made on both surfaces of the plate, as residual stresses may be caused by bending. The difference between the strains on both surfaces represents the stress component caused by bending. The mean value of strains measured on both surfaces represents the plate stress component.

Mathar-Soete Drilling Technique. When a small circular hole is drilled in a plate containing residual stresses, the residual stresses in areas outside the hole are

Fig. 11 120° star arrangement of strain gages for the Mathar-Soete drilling technique

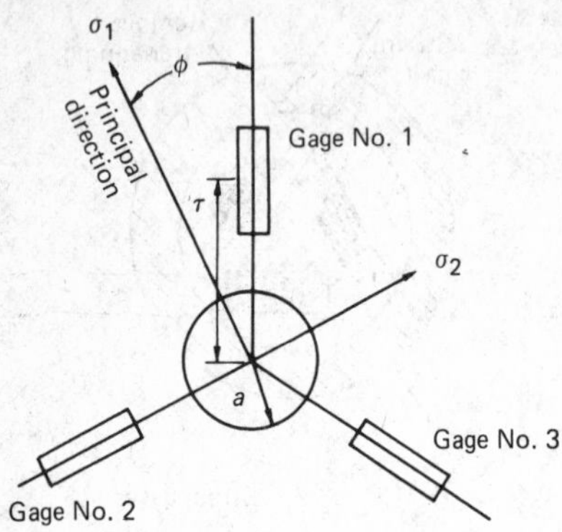

partially relaxed. By measuring stress relaxation in areas outside the drilled hole, it is possible to determine the amount of residual stress that exists in the drilled area. Mathar first proposed and used the hole method of measuring stress; it was further developed by Soete (Ref 23).

A common method of determining residual stress is to place strain gages in a star form 120° from each other and drill a hole in the center, as shown in Fig. 11. By measuring strain changes at the three gages, one can calculate the magnitude and direction of the principal stresses.

The Gunnert Drilling Technique. Four 0.12-in. parallel holes located on a circle with a 0.36-in. diameter are drilled through the plate at the measuring point, as shown in Fig. 12. Using a specially designed mechanical gage, one can measure the diametrical distances between these holes at

Fig. 12 Gunnert drilling method. (a) Plan. (b) Side view

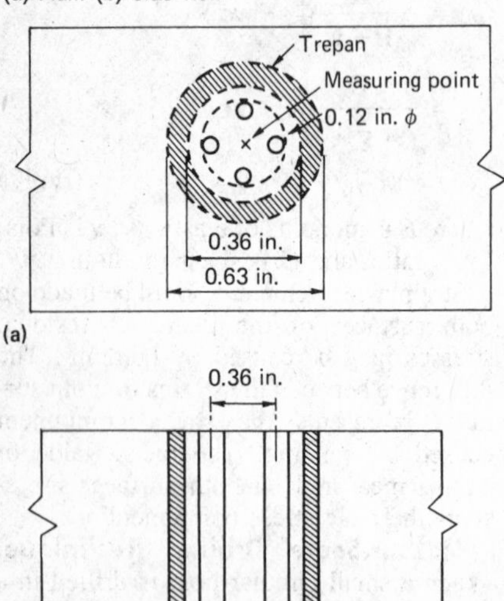

different levels below the surface of the plate. Also, the perpendicular distance between the plate surface and the gage location at different levels below the surface can be measured. A groove 0.63 in. from the measuring point is then trepanned around the holes in steps, and the same measurements are made. Formulas are proposed to calculate residual stress in the interior of the specimen from the measurement data.

Sectioning Techniques. Rosenthal and Norton (Ref 24) proposed a technique for determining residual stresses in heavy weldments. Two narrow blocks with full thickness of the plate, one parallel to the weld and the other transverse to the weld, are cut from the weld. Residual stress remaining in the narrow blocks is then measured. Formulas have been proposed to estimate residual stresses in the interior of the weldment from strain changes that occur while cutting the narrow blocks and residual stresses that are left in the blocks.

Recently, Ueda *et al.* (Ref 25) have developed a method of measuring residual stress in heavy weldments. Longitudinal and transverse blocks are cut from the weldment, which are further sliced into thin sections. Strain changes during cutting and slicing of the blocks are measured. Finite element computer programs have been developed to determine residual stress from measured data.

Photoelastic Coating-Drilling Technique. The photoelastic coating technique is a method of stress analysis in which the specimen to be stress analyzed is coated with a photoelastic plastic. When strain occurs in the specimen, the plastic becomes birefringent. Birefringence is the refraction of light in two slightly different directions to form two rays using a reflection polaroscope. Strains are determined by measuring birefringence.

Using this method, a hole is drilled to a certain depth (for instance, equal to the diameter) at the measuring point through the photoelastic coating and a portion of the specimen. Birefringence occurs in areas near the drilled hole if residual stresses exist. By analysis, the birefringence strain release that took place during drilling is determined. Then the residual stresses that existed in the drilled area are calculated.

Measurement of Residual Stresses by X-ray Diffraction

When internal or external forces are applied to a structure made of metallic crystals, the crystalline lattice is distorted, therefore changing the interatomic distances. Plastic deformation occurs as a result of slip along the lattice planes when deformation exceeds the elastic limit. Change in the interatomic spacing is directly proportional to the stress.

If a monochromatic plane wave is introduced into the atomic planes in the direction *AB*, as shown in Fig. 13, the reflected beams from successive parallel planes of atoms are reinforced in one direction (*BC*), the diffraction direction. Bragg's law defines the condition for diffraction as follows:

$$n\lambda = 2d \sin \theta \qquad \text{(Eq 18)}$$

where λ is the wavelength of incident beam; θ is the angle between incident or reflected beams and surface of reflecting planes; d is the interplanar spacing; and n is the order of reflection ($n = 1, 2, 3 \ldots$).

Equation 18 shows that, if the wavelength of the x-ray is known, the interplanar spacing (d) can be determined by measuring the angle θ.

Figure 14 illustrates schematic setups for x-ray diffraction techniques. Two general techniques are employed in the recording of diffraction patterns—the photographic or x-ray film technique, as shown in Fig. 14(a), and the x-ray diffractometer or counter tube technique, as shown in Fig. 14(b).

Figure 14(a) shows a portable setup that employs the film technique. The apparatus consists essentially of a film in a light-tight cassette mounted perpendicularly to the incoming x-ray beam. The beam is collimated by the pinhole system inserted through a hole in the film. The film records the rays diffracted by the specimen. On development, near-circular rings are exposed. The diameter of a diffraction ring divided by the distance from the film to the specimen gives 2 tan $(180 - 2\theta)$, from which θ is obtained for insertion in Eq 18.

Generally, the x-ray diffraction method and the film method differ only in the detector and in the angle made by the specimen with the x-ray beam. The angle between the x-ray beam and the specimen

Fig. 13 Diffraction produced by reflections from adjacent atomic planes of a monochromatic plane wave

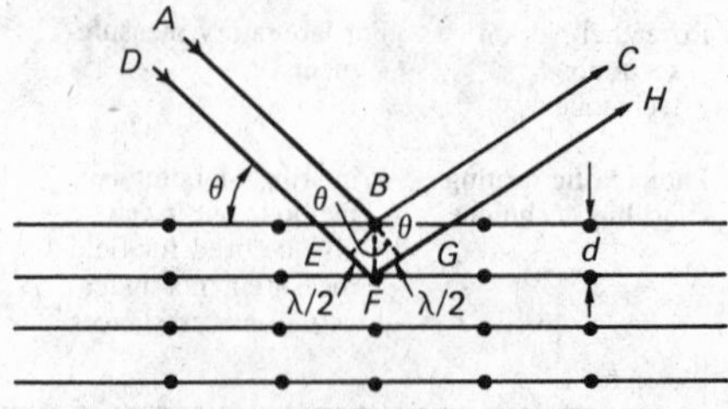

Fig. 14 Setups for x-ray diffraction techniques. (a) Portable x-ray film technique. (b) X-ray diffractometer technique

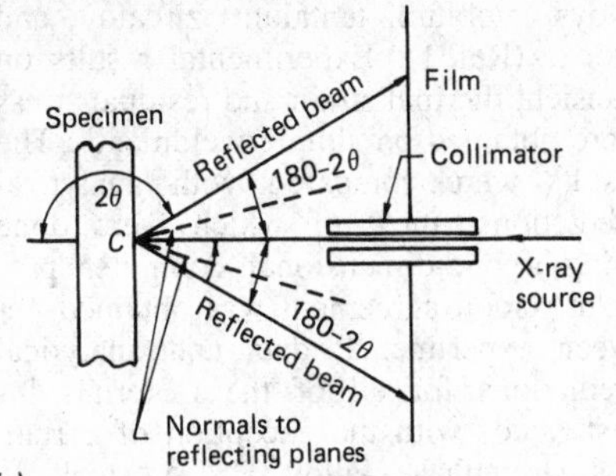

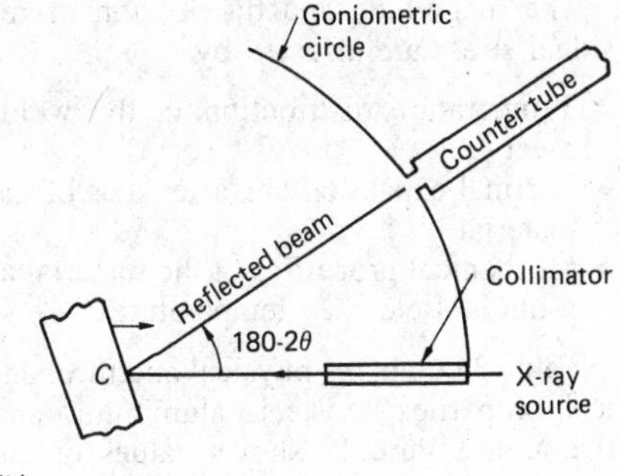

Fig. 15 Theoretical and experimental distribution of residual stresses in a plug weld

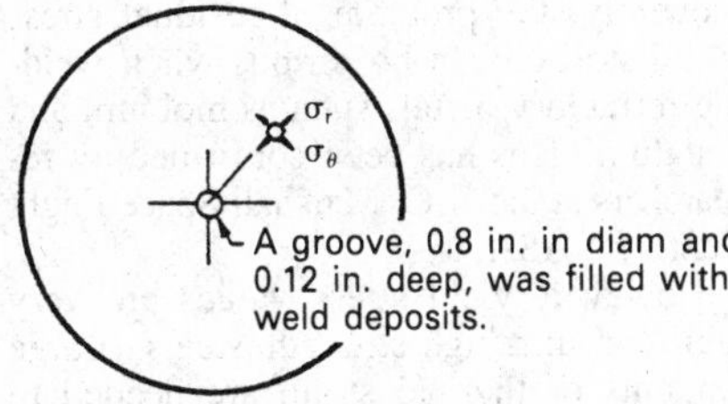

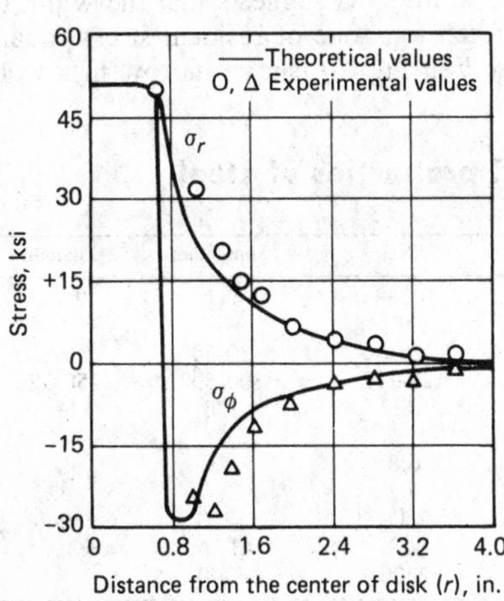

surface is 90° in the film method, but it is an angle of θ° for the diffraction method. A counter and receiving slit are moved along a goniometric circle to record the intensity of the reflected beam, while the diffraction angle is determined as the angle of maximum intensity.

More accurate results usually are provided by the x-ray diffractometer technique than with the film technique. Also, the size specimen that can be tested is limited by the geometry of the instrument.

Magnitude and Distribution of Residual Stress in Weldments

A number of investigators have studied distributions of residual stress in various types of weldments. Distributions of residual stress in a butt weld and an edge weld were discussed in Fig. 6 and 7, respectively. Distribution of residual stress in a plug weld, welded shapes and columns, and in welded pipe are discussed below.

Plug Weld. Figure 15 shows the distribution of residual stresses in a circular plug weld (Ref 17). Tensile stresses, as high as the yield stress of the material, are produced in both radial and tangential directions in the weld and adjacent areas. In areas away from the weld, radial stresses (σ_r) are tensile, and tangential stresses (σ_θ) are compressive; both stresses decrease with increasing distance from the weld (r).

Welded Shapes and Columns. Figure 16 shows typical distortion and distributions of residual stresses in welded shapes (Ref 26) produced in a welded T-shape. As shown in section xx, high tensile residual stresses are produced in areas near the weld, in the direction parallel to the axis. In the flange, stresses are compressive in areas away from the weld and

Fig. 16 Typical residual stresses in welded shapes. (a) T-shapes. (b) H-shapes. (c) Box shapes

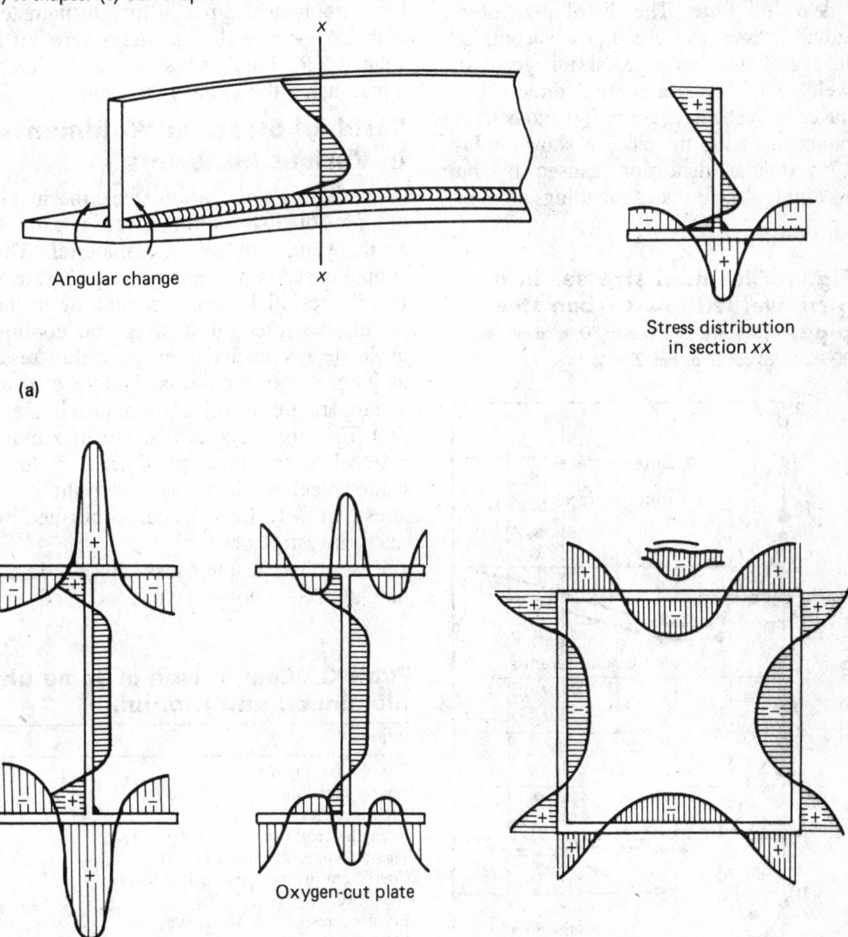

Fig. 17 Residual stresses in a welded pipe

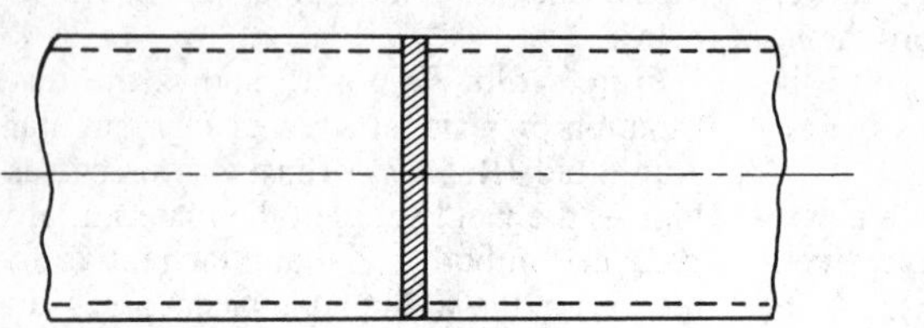

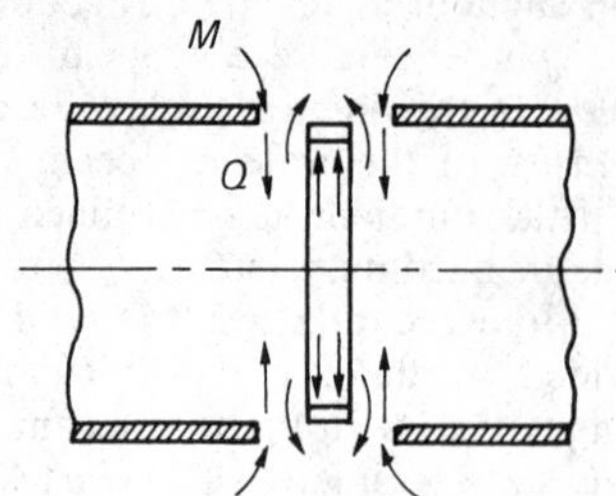

tensile in areas near the weld. Tensile stresses in areas near the upper edge of the web are caused by longitudinal bending distortion of the shape caused by the longitudinal shrinkage of the weld. Angular distortion also is produced.

Figures 16(b) and (c) show typical distributions of residual stresses in H-shapes (I-beams) and in a welded box shape, respectively. Residual stresses are shown parallel to the axis; they are compressive in areas away from the welds, and tensile in areas near the welds.

Welded Pipe. The distribution of residual stresses in welded pipe is complex. In a girth welded pipe, shrinkage of the weld in the circumferential direction induces both shearing force (Q) and bending moments (M) to the pipe, as shown in Fig. 17. Angular distortion caused by butt welding also induces bending moment.

Fig. 18 Residual stresses in a girth-welded low-carbon steel pipe. 7/16-in. wall thickness, 30 in. diam, and 30 in. long. (Source: Ref 27)

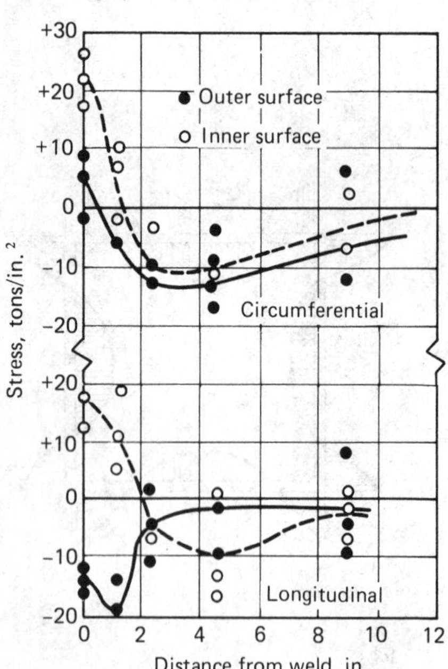

Distribution of residual stresses is affected by:

- Diameter and wall thickness of the pipe
- Joint design (square butt, V, X, etc.)
- Welding procedure and sequence (welded on outside only, welded on both sides, outside first, or welded on both sides, inside first)

Residual stresses in welded pipe have been studied by a number of investigators (Ref 18, 19, and 27). Figure 18 shows the results obtained by Burdekin (Ref 27). Two low-carbon steel pipes 30 in. in diameter with a 7/16-in. wall thickness were girth welded. Residual stresses were determined using the Gunnert technique.

Residual Stress in Weldments in Various Materials

In welding low-carbon steel, the maximum residual stress in the weld is as high as the yield stress of the material. The simple analysis presented in Fig. 2 shows that the residual stresses remaining in the middle bar after the heating and cooling cycle always reach the tensile yield stress, as long as the middle bar is subjected to a temperature increase of approximately 340 °F. This suggests that the maximum residual stress in a weldment in low-carbon steel would always reach the yield stress. In fact, this has been confirmed by many investigators.

Researchers at the Massachusetts Institute of Technology have conducted extensive studies on residual stresses in weldments in various materials, including low-carbon steel, high-strength steel, stainless steel, aluminum alloys, titanium alloys, niobium, tantalum, zircaloy, and bronze (Ref 1). Experimental results on transient thermal strain and residual stress were obtained on simple weldments. The results were compared with analytical predictions, most of which were done using a one-dimensional computer program. Good agreement was obtained between experimental data and analytical predictions for all of the materials investigated, with the exception of certain high-strength low-alloy (HSLA) steels.

The magnitude and distribution of residual stress are affected by:

- Temperature distribution in the weldment
- Thermal expansion characteristics of the material
- Mechanical properties of the material at room and elevated temperatures

Table 3 compares physical and mechanical properties of steel, aluminum, and titanium. Figure 19 shows values of the yield stress of several materials at various temperatures.

In Fig. 19, niobium and tantalum are shown to have relatively low values of yield stress over a wide temperature range (Ref 10). This suggests that areas of the plastic zone near the weld tend to be large in weldments made of these materials. Consequently, the problem of residual stress and distortion can be serious when welding refractory metals such as niobium and tantalum. This has been confirmed by researchers at the G.C. Marshall Space Flight Center, NASA.

However, yield stress values are very high for ultrahigh-strength steels. Large amounts of thermal strain are needed to produce plastic deformation in areas near the weld. This suggests that the width of the tension zone of residual stress (quantity b in Fig. 5) can be narrow in a weld

Table 3 Comparison of some physical properties of steel, aluminum, and titanium

Properties	Steel	Aluminium	Titanium
Density (σ), lb/in.^{3}	0.283	0.1	0.163
Young's modulus, $E \times 10^6$ psi	30	10	17
Yield strength, $\sigma_{ys} \times 10^3$ psi	35-150	30-50	40-150
Strength/weight, $\sigma_{ys}/\rho \times 10^3$	123-150	300-500	250-920
Thermal conductivity(λ), Btu/ft · h · °F	26.2	130	9
Coefficient of linear thermal expansion, $\alpha \times 10^{-6}$/°F	6.8	13	4.7
Electrical resistivity, $10^{-6}\Omega \cdot$ cm	9.7	2.7	42
Melting point, °F	2800	1220	3040
Melting point of oxide, °F	FeO, 2400	Al_2O_3, 3700	...

Fig. 19 Yield stresses of structural materials at elevated temperatures

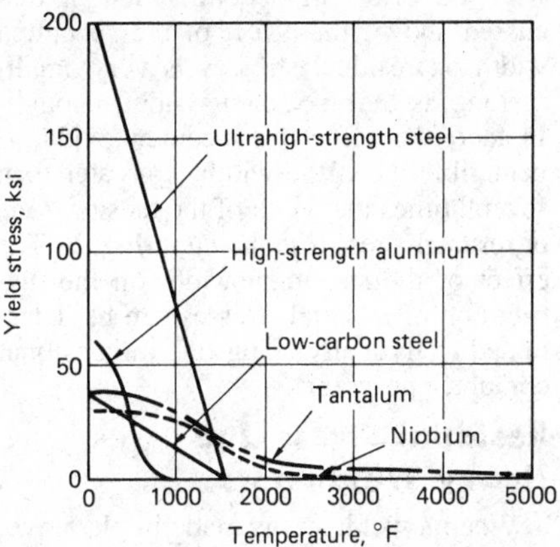

Fig. 20 Maximum mechanical strains observed at locations 1 in. from weld line (μin./in.) versus yield strength (ksi) of the base plate. Source: Ref 28

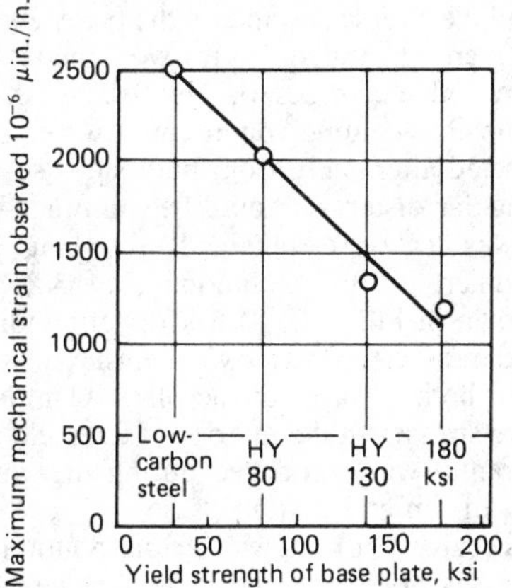

made in ultrahigh-strength steel. Figure 20 shows experimental data obtained on butt welds made in steels with different strength levels (Ref 1 and 28). In Fig. 20, the maximum mechanical strain observed at the location 1 in. from the weld center decreases as the yield strength of the base metal increases. Strain response is roughly proportional to the inverse of the plate strength.

Welds in High-Strength Steels. As stated earlier, the maximum residual stress approaches the yield stress value in weldments made with low-carbon steel with the yield strength of 35 to 40 ksi. Curve 0 in Fig. 21 schematically shows a typical distribution of longitudinal stress in a butt weld in low-carbon steel. However, information on the magnitude and distribution of residual stress in weldments in high-strength steels is limited and inconclusive. It has been well established that residual stresses generally are tensile in regions near the weld, and they become compressive in regions away from the weld. Critical considerations include:

- Are peak values of residual stresses as high as the yield stresses of the weld metal and base metal?
- How wide are the areas with high tensile residual stresses?

Curves 1, 2, and 3 in Fig. 21 schematically show three typical distributions of longitudinal residual stresses in a butt weld in high-strength steels. If one assumes that the maximum residual stress is as high as the yield stress, this distribution would be given by curve 1. If this were true, residual stress and distortion may cause severe problems in the fabrication of welded structures using high-strength steels. In curve 2, high tensile residual stresses are confined to small areas. In this case, distortion would be less, but cracking due to high tensile residual stresses may be a problem. Stress distributions illustrated by curve 3 would cause few problems.

No existing experimental and analytical data support curve 1. Fabrication experience with high-strength steel welded structures also indicates the possibility that curve 1 can be eliminated.

Experimental data obtained with the use of strain gages and sectioning techniques tend to support curve 3 (Ref 29). Experimental data exist that indicate residual stresses could be considerably low in some regions of the heat-affected zone (HAZ) of weldments, in certain high-strength steel (Ref 30 and 31). A possible explanation is the effect of solid-state transformation during cooling. However, experimental data on heavy-section HY-130 steel weldment have shown that the peak residual stress can be as high as the yield stress value (Ref 32).

Fig. 21 Distribution of longitudinal residual stresses in a butt weld in high-strength steel

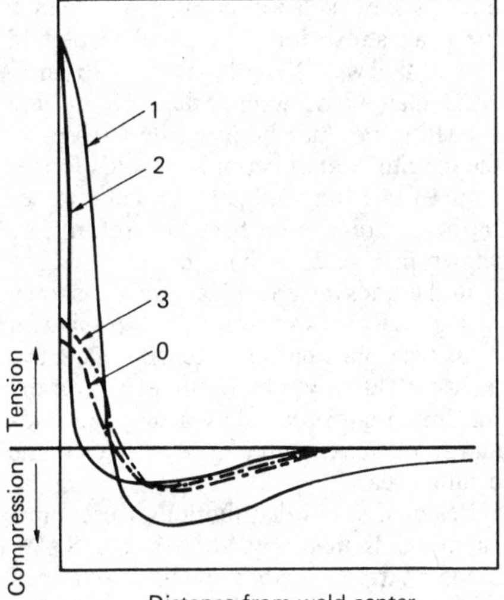

However, analytical data neglecting the effect of phase transformation tend to support curve 2 (Ref 1). In an analytical study that included the effect of solid-state transformation, it has been found that residual stresses in some regions near the weld can be reduced considerably (Ref 33).

On the basis of the experimental and analytical information obtained thus far, it appears that residual stress distributions in actual structures are mixtures of curves 2 and 3. Peak values of residual stresses in some areas near the weld can be as high as the yield stress. However, average values of residual stresses could be close to curve 3. If x-ray diffraction is used for measuring residual stresses in the weld metal and HAZ, it is likely that widely scattered results ranging from the yield stress to much lower values will be observed.

Welds in Aluminum Alloys. A considerable amount of information has been obtained on residual stresses in weldments in aluminum alloys. For example, Hill (Ref 34) investigated the residual stresses that occur when butt joints in 5456-H321 aluminum alloy plates are joined by inert gas shielded arc welding using 5556 alloy consumable electrodes. Figure 22 shows a typical distribution of longitudinal residual stresses in a $^1/_2 \times 36 \times 48$ in. panel that has been formed by welding two $^1/_2 \times 18 \times 48$ in. plates. These tensile stresses are confined to the region in which the heat of welding has lowered the yield strength of the material.

Welds in Titanium Alloys. Figure 23 shows results obtained by Hwang (Ref 35) on a weldment in Ti-6Al-2Nb-1Ta-1Mo made by gas metal arc welding (GMAW), using Ti-6Al-2Nb-1Ta-0.8Mo filler wire. Welding was done along a longitudinal edge of a strip 56 in. long, 7.5 in. wide, and $^1/_2$ in. thick. Experimental results agreed reasonably well with analytical predictions by the one-dimensional program.

Robelotto *et al.* (Ref 36) investigated residual stresses in butt welds made in titanium alloy plates 0.2 in. thick. Tests were made of three titanium alloys, Ti-5Al-2.5Sn, Ti-6Al-4V, and Ti-8Al-1Mo-1V.

Fig. 22 Distribution of yield strength and longitudinal residual stresses in a welded 5456-H321 plate 36 in. wide and 1/2 in. thick. Source: Ref 34

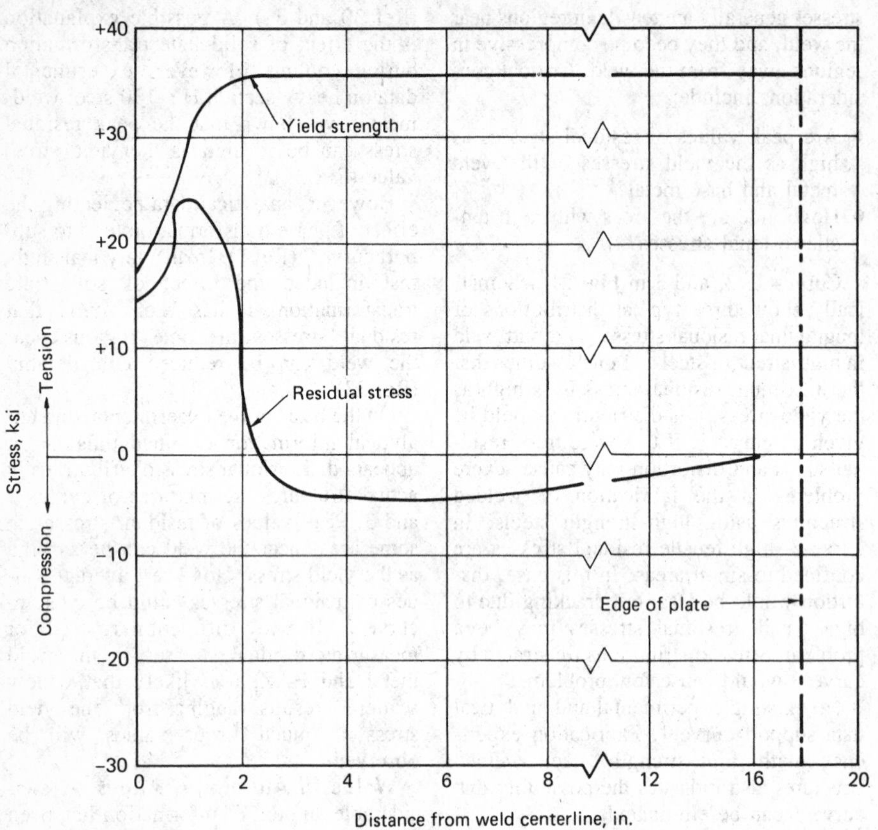

Effects of Specimen Size on Residual Stress

When measuring residual stress in a weldment, welded specimens must be large enough to contain residual stresses as high as those that exist in actual structures.

Effect of Specimen Length. DeGarmo *et al.* (Ref 37) investigated the effect of weld length on residual stress in unrestrained butt welds made in low-carbon steel. Two series of weldments were prepared using submerged arc welding (SAW) and shielded metal arc welding (SMAW), as shown in Fig. 24. In each series, the only variable was the length of the weldment:

Welding process	Weld length, in.
SAW	3, 4, 5, 8, 12, 18, 24, 36
SMAW	5, 7, 10, 18, 48

The width of each specimen was sufficient to ensure that full restraint could be applied.

Figure 25(a) and (b) show the distribution of residual stress in weldments made by SAW and SMAW, respectively. Distributions of longitudinal and transverse stress along the weld are shown.

Longitudinal residual stress values must be zero at both ends of the weld, while high tensile stresses exist in the central region. The peak stress in the central region increases with increasing weld length. This effect is shown clearly in Fig. 26, in which the peak stress for each panel is plotted versus the weld length. Figure 26 indicates that welds longer than 18 in. are needed to produce high tensile stresses in the longitudinal direction. In welds longer than 18 in., longitudinal residual stresses in the central region became uniform, as shown in Fig. 25(a) and (b).

In the transverse residual stresses shown in Fig. 25, stresses were compressive in areas near plate ends and tensile in central areas. Weld length had little effect on the maximum stress in areas near plate ends and on the maximum tensile stresses in the central area.

Residual stress distributions were similar in welds made by SMAW and SAW. Smooth stress distributions were obtained in welds made by SAW, while stress distributions in welds made by SMAW were somewhat uneven.

Effect of Specimen Width. Compared with the effect of specimen length discussed above, the effect of the specimen width on residual stresses is very small, as long as the specimen is long enough. In fact, the effect of specimen width is negligible when the width is greater than several times the width of the tension zone of residual stress (*b*), shown in Fig. 6. The effect of the specimen width on the distribution of residual stresses can be determined analytically using one-dimensional computer programs.

Residual Stress and Heavy Weldments

When a weldment is made in plate over 1 in. thick, residual stresses in the thickness direction (σ_z) can become significant. Figure 27 shows distributions of residual stress in the thickness direction in the weld metal of a butt joint 1 in. thick, 20 in. long, and 20 in. wide in low-carbon steel. As shown in Fig. 27, the longitudinal stresses (σ_x), the transverse stresses (σ_y), and the stresses normal to the plate surface (σ_z) are illustrated. Welds were made with covered electrodes 0.1 to 0.2 in. in diameter; welding operations were conducted alternately from both sides so that angular distortion could be minimized.

Results were obtained by using the Gunnert drilling technique (Ref 38). As shown in Fig. 27(a) and (b), longitudinal and transverse stresses were tensile in areas near both surfaces of the plate. Compressive stresses in the interior of the weld apparently were produced during the welding of top and bottom passes.

Figure 27(c) shows the distribution of stresses normal to the plate surface (σ_z). At both surfaces, σ_z must be zero. Residual stresses were compressive in the case shown here. However, many investigators believe that σ_z in the interior of a thick weld can be tensile.

Other investigators also have reported experimental results on residual stresses in heavy weldments. However, the information is still far from complete, and it is not possible at present to describe general tendencies of distributions of residual stresses in heavy weldments.

The major difficulty comes from the lack of adequate measuring techniques. As described earlier, in all of the stress-measuring techniques in use today, including strain gage and x-ray diffraction techniques, stresses are determined by measuring strains on the surface. There is no method, other than ultrasonic testing, that is capable of measuring stresses in-

Fig. 23 Distribution of residual stresses in weldment of titanium alloy

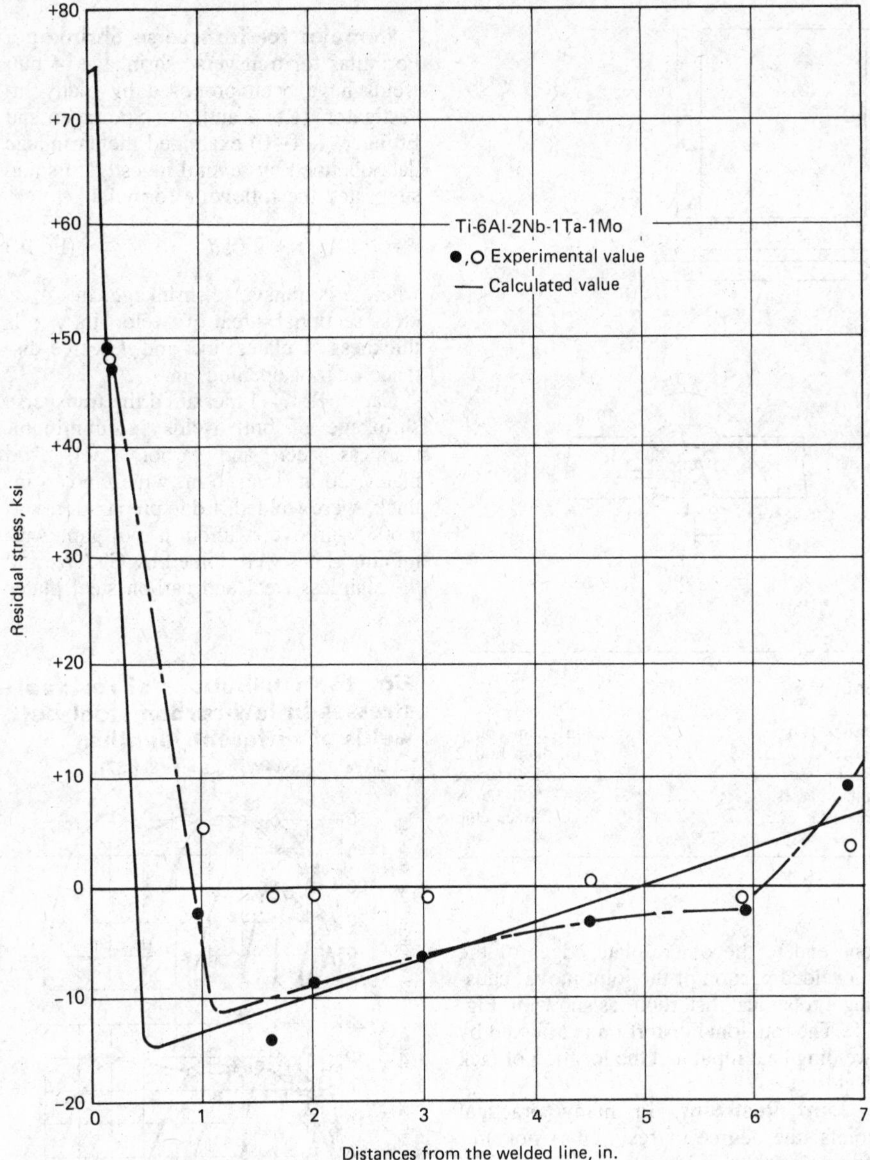

side a body. When residual stresses are measured in a heavy weldment, the surface on which the measurements are to be made must be exposed through drilling or sectioning. These methods are time consuming and costly and provide stress values only at the drilled holes or along the sectioned blocks.

On the other hand, engineers desire complete knowledge of the distribution of residual stresses throughout the entire weldment and, if possible, how residual stresses may be changed, or how residual stresses may be reduced by proper selection of welding procedures. In other words, there is a big gap between the information that current measuring techniques can offer and the information engineers would like. The recent development of analytical techniques that are now available may be able to close these gaps.

Residual Stress of Weldments Made by Different Welding Processes

Generally, it is believed that similar residual stresses are created in welds made by various processes, including SAW, SMAW, GMAW, and gas tungsten arc welding (GTAW). For example, Fig. 25(a) and (b) show similarities in the residual stress produced by SMAW and SAW.

Effects of Welding Sequence on Residual Stress

To reduce residual stress and distortion in welding a long butt joint, various types of welding sequences, such as backstep, block, built-up, or cascade, may be used. In welding joints with high restraint, such as joints involved in making patches, selection of a proper welding sequence is an important consideration.

The effects of welding sequence on residual stress have been widely studied (Ref 1 and 2). In a comprehensive investigation of the effects of welding sequence on residual stresses and shrinkage in restrained butt welds and circular patch welds, welding sequences were classified (Ref 1 and 2) as:

- Multilayer sequences in which the first layer is completed along the entire weld length; there may be various ways to complete the first layer, such as straight forward, backstep, skip, etc. Welding of the second layer is then completed, and so on.
- Block welding sequence in which some length of the joint, or a block, is completely welded, the next block is welded, and so on.

The effect of welding sequence on residual stress along the weld was minor. In all welds tested, high tensile longitudinal stresses were found. Differences in welding sequence caused considerable differences in transverse shrinkage, in the amount of total strain energy produced in restrained joints, and in the amount of reaction stresses in the inner plates of circular patch welds. Block welding sequences generally produced less shrinkage, less strain energy, and less reaction stress than multilayer sequences.

Distortion in Weldments

Distortions in fabricated structures are caused by three fundamental dimensional changes—transverse shrinkage, longitudinal shrinkage, and angular changes, as shown in Fig. 8. However, distortion that occurs in actual welded structures is far more complex than fundamental dimensional changes shown in Fig. 8.

Types and magnitudes of distortion that occur in weldments are affected by many parameters, including plate thickness, joint design, and welding process. When a weldment is made with a thin plate, for

Fig. 24 Low-carbon steel specimens for studying the effect of weld length on residual stresses. (a) SAW. (b) SMAW (double-V). (Source: Ref 37)

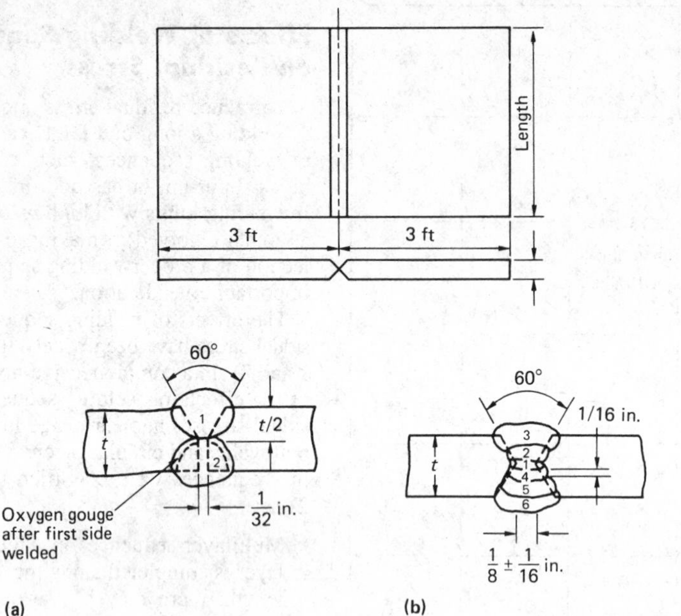

SAW		SMAW	
Electrode	$^1/_4$-in.-diam EH14 F62-EH14-200	Passes 1, 4	$^5/_{32}$-in.-diam E6010
Voltage	32	Passes 2, 3, 5, 6	$^1/_4$-in.-diam E6012
Current	1050 A	Root passes	150-165 A
Speed of arc travel	~12.5 in./min	Other passes	300-320 A
Pass 1	Oxygen-gouge back side	Passes 1, 2, 3	Back chip

example, the plate may buckle due to compressive residual stresses without applying external loads.

Over the years, a considerable amount of information has been accumulated on different types of distortions in various types of weldments. Most of the information has been on steel weldments. Some information has been obtained on aluminum weldments; however, information on weldments in other materials is limited. Most of the information presently available is empirical.

Transverse Shrinkage of Butt Welds

In Fig. 8(a), shrinkage is uniform along the weld. However, transverse shrinkage that occurs in butt welds, especially long welds, usually is not uniform and much more complex in actual structures. Major factors causing nonuniform transverse shrinkage in butt welds are discussed below.

Rotational Distortion. When two separate plates are joined progressively from one end to the other, plate edges of the unwelded portion of the joint move, causing a rotational distortion, as shown in Fig. 28. The rotational distortion is affected by welding heat input and the location of tack welds.

Joint Restraint. In many practical joints, the degree of restraint is not uniform along the weld. Because the amount of transverse shrinkage is affected by the degree of restraint of the joint, nonuniform distribution of degree of restraint of a joint results in uneven distribution of transverse shrinkage. For a given amount of the weld metal, the amount of transverse shrinkage decreased when the joint was more severely restrained. For example, Fig. 29 schematically shows the distribution of transverse shrinkage along a slit weld. Welds similar to the slit weld frequently are made in repairs. The amount of shrinkage is greater in regions near the center of the joint and very slight near the ends of the joint. This is because portions of the joint near the ends are more highly restrained than central portions of the joint. When a long butt joint is welded by a step-back sequence, the transverse shrinkage is not uniform.

Formulas for Transverse Shrinkage. Formulas for transverse shrinkage of butt welds have been proposed by many investigators (Ref 1 and 39). Spraragen and Ettinger (Ref 40) examined the shrinkage data obtained by several investigators and suggested the following formula:

$$S = 0.2A_w/t + 0.05d \qquad \text{(Eq 19)}$$

where S is transverse shrinkage, in.; A_w is cross-sectional area of weld, in.^{2}; t is thickness of plates, in.; and d is free distance or root opening, in.

Capel (Ref 41) measured the transverse shrinkage of butt welds in aluminum, stainless steel, and carbon steel. Two plates, 20 in. long, 6 in. wide, and $^1/_4$ in. thick, were welded. Edge preparation was a 60° V-groove, without a root gap. Aluminum plates were joined by GTAW, and the stainless steel and carbon steel plates

Fig. 25 Distributions of residual stresses in low-carbon steel butt welds of different lengths. (a) SAW. (b) SMAW. (Source: Ref 37)

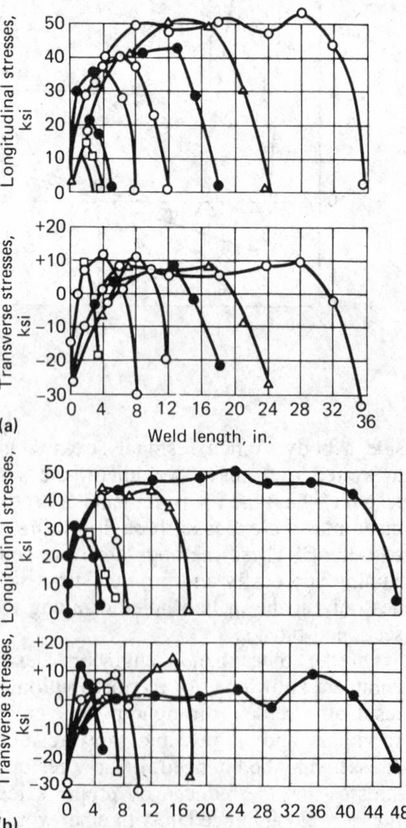

Fig. 26 Effect of length of weld on longitudinal residual stress in low-carbon steel. Source: Ref 37

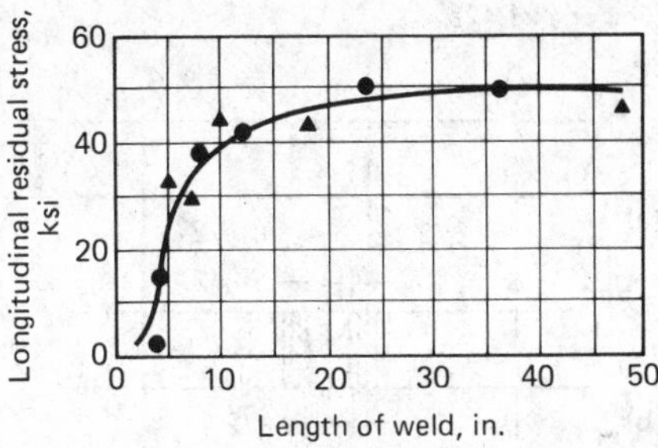

were welded with covered electrodes. The following formulas were proposed (Ref 41):

$$\Delta 1 \text{ (aluminum)} = \frac{20.4 \times W \times 10^3}{s \times u}$$

$$\Delta 1 \text{ (stainless steel)} = \frac{22.7 \times W \times 10^3}{s \times u}$$

$$\Delta 1 \text{ (carbon steel)} = \frac{17.4 \times W \times 10^3}{s \times u} \quad \text{(Eq 20)}$$

where $\Delta 1$ is transverse shrinkage, in.; s is thickness of layer of weld metal, in.; u is welding speed, in./min; W equals $I \times V$, which equals the electric power of welding arc; I is welding current, A; and V is arc voltage, V.

Table 4 provides the welding data used in Capel's experiments (Ref 41). Considerably higher heat inputs were used for welding aluminum. As a result, actual values of shrinkage were greater for aluminum welds than for stainless steel and carbon steel. Comparison of the actual measurements and the calculated shrinkage illustrates:

Welding pass	$\Delta 1_1$ measured, in.	$\Delta 1_2$ calculated, in.
Aluminum		
First layer	0.088	0.10
Second layer	0.044	0.064
Stainless steel		
First layer	0.032	0.032
Second layer	0.064	0.048
Carbon steel		
First layer	0.016	0.020
Second layer	0.036	0.028

Total shrinkage after two-layer welding was completed was 0.13 in. for aluminum, but only 0.051 in. for carbon steel.

Mechanisms of transverse shrinkage have been studied by several investigators. The most important finding of these mathematical analyses is that the major portion of transverse shrinkage of a butt weld is due to contraction of the base plate. The base plate expands during welding. When the weld metal solidifies, the expanded base metal must shrink; this shrinkage accounts for the major part of transverse shrinkage. Shrinkage of the weld metal is only about 10% of the actual shrinkage.

Figure 30 schematically presents the changes of transverse shrinkage in a single-pass butt weld in a free joint after welding (Ref 42). Shortly after welding, the heat of the weld metal is transmitted into the base metal, which causes the base metal to expand, with a consequent contraction of the weld metal. During this period the points of sections A and A' do not move (Fig. 30).

When the weld metal begins to resist the additional thermal deformation of the base metal, parts of sections A and A' begin to move in response. The starting time of movement of A and A' is indicated by t_s.

The various thermal deformations of both the weld and base metals are defined as:

- δ_s: Thermal expansion of the base metal at $t = t_s$
- δ: Additional thermal deformation of the base metal caused in $\overline{AA'}$ at $t > t_s$

Fig. 28 Rotational distortion. (a) Unwelded portion of the joint closes (covered electrode). (b) Unwelded portion of the joint opens (submerged arc process). Source: Welding Research Council

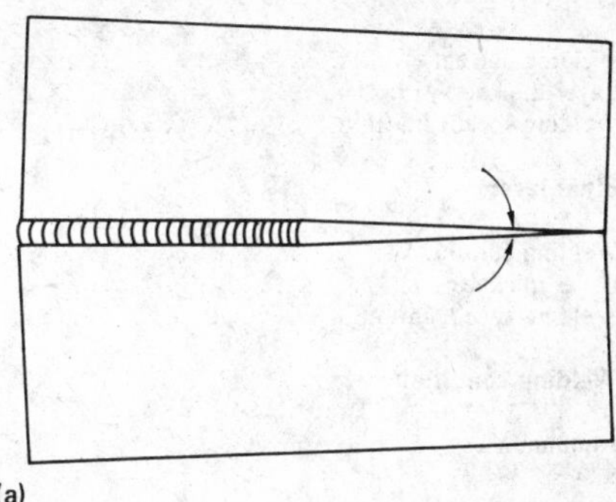

Fig. 29 Transverse shrinkage in a butt weld under restraint. Slit-weld specimen and typical distribution of shrinkage. Source: Welding Research Council

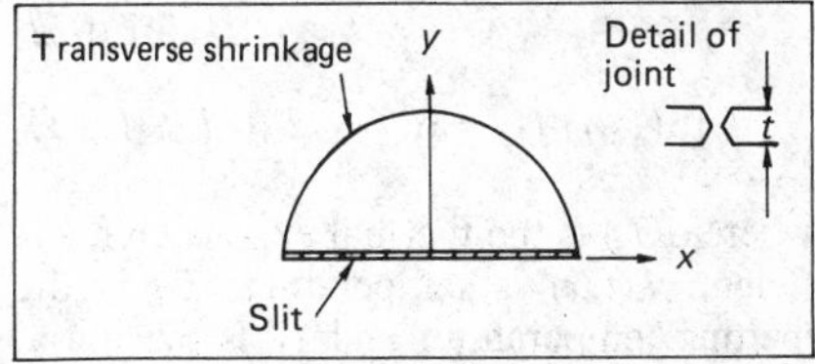

Fig. 27 Distributions of residual stresses in the thickness direction of the weld metal of a low-carbon steel butt joint. (a) Longitudinal stresses, σ_x. (b) Transverse stresses, σ_y. (c) Normal to surface stresses, σ_z. (Source: Ref 38)

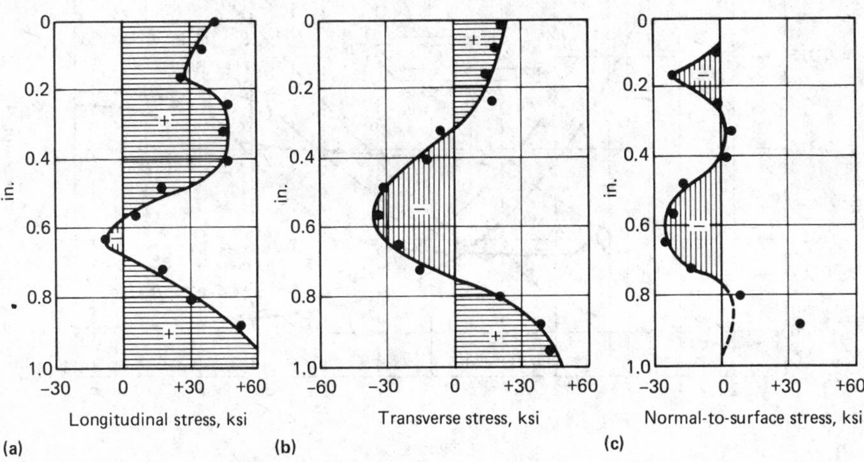

Table 4 Welding data used in Capel's experiments

Process variable	Aluminum	Stainless steel	Carbon steel
First layer			
Welding current, W	4050	2320	2100
Layer thickness, in.	0.140	0.133	0.145
Welding speed, in./min	3.5	7.5	7.8
Final layer			
Welding current, W	4050	2550	2100
Layer thickness, in.	0.2	0.1	0.09
Welding speed, in./min	3.6	7.0	8.6
Welding conditions			
Aluminum	First layer: 5/32-in.-diam filler rod, 5/32-in.-diam electrode; gas flow, 16.9 ft^3/h Second layer: 5/32-in.-diam filler rod, 5/32-in.-diam electrode; gas flow, 16.9 ft^3/h		
Stainless steel	First layer: 0.1-in.-diam electrode Second layer: 0.13-in.-diam electrode		
Carbon steel	First layer: 0.1-in.-diam electrode Second layer: 0.13-in.-diam electrode		

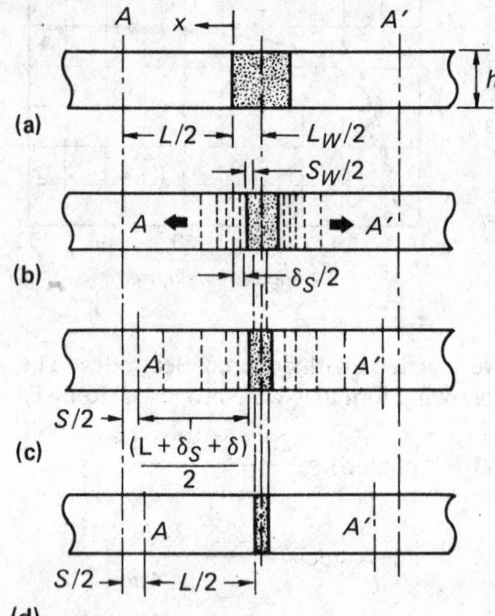

Fig. 30 Schematic presentation of a transverse shrinkage in a butt weld in a single pass. (a) $t = 0$. (b) $t = t_s$. (c) $t > t_s$. (d) $t = \infty$. (Source: Ref 42)

- S_W: Thermal contraction of the weld metal at $t > t_s$

These deformations can be calculated by:

$$\delta_s = 2\int_0^{L/2} [\alpha(T) \cdot T(t_s,x) - \alpha(T_0) \cdot T_0]\, dx \quad \text{(Eq 21)}$$

$$\delta = 2\int_0^{L/2} [\alpha(T) \cdot T(t,x) - \alpha(T) \cdot T(t_s,x)]\, dx \quad \text{(Eq 22)}$$

$$S_w = [\alpha(T_M) \cdot T_M - \alpha(T_0) \cdot T_0] \cdot L_w \quad \text{(Eq 23)}$$

where $\alpha(T)$ is the thermal expansion coefficient; $T(t,x)$ is temperature; T_M is the melting temperature; and T_0 is the initial and final (room) temperature.

Using these results, the transverse shrinkage can be calculated from:

$$S = \begin{cases} 0 & \text{for } 0 \le t \le t_s \\ -\delta + S_w & \text{for } t > t_s \\ \delta_s + S_w & \text{for } t = \infty \end{cases} \quad \text{(Eq 24)}$$

Thus, final transverse shrinkage depends on the thermal expansion of the base metal at $t = t_s$ and the thermal contraction of the weld metal.

Figure 31 shows experimental results obtained by Matsui (Ref 42) on butt welds in low-carbon steel. Curves labeled "T" indicate temperature changes, while curves labeled "S" indicate changes in transverse shrinkage. After the weldment has cooled to a relatively low temperature, most of the shrinkage occurs. Figure 31 shows that in a thicker plate transverse shrinkage starts earlier, but the final value of the shrinkage is smaller. However, this is true only if the same amount of heat input is always used, regardless of joint thickness. Welding thicker plate may require more than one pass.

Transverse Shrinkage During Multipass Welding. Extensive study (Ref 43 and 44) of transverse shrinkage during multipass welding of constrained butt joints in carbon steel shows that transverse shrinkage increases during multipass welding (Fig. 32). Shrinkage was relatively pronounced during the early weld passes, but diminished during later passes (Fig. 32b), due to the resistance against shrinkage that increases as the weld becomes larger. A linear relationship exists between transverse shrinkage (u) and the logarithm of the weight of the weld metal deposited (w):

$$u = u_o + b(\log w - \log w_o) \quad \text{(Eq 25)}$$

Fig. 31 Transverse shrinkage during welding and cooling for various plate thicknesses (*h*) and measuring distances (*L*) in free butt joints. Curves labeled *T* show the temperature changes, while the curves labeled *S* show the changes of transverse shrinkage. (Source: Ref 42)

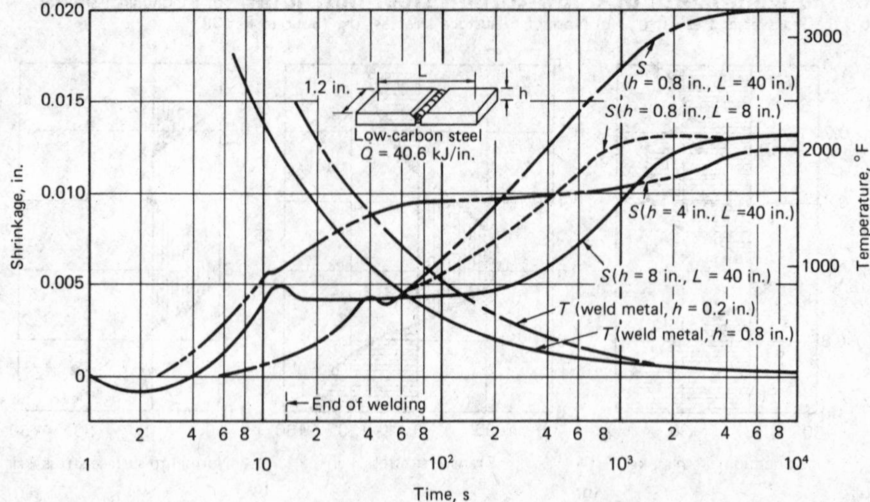

Fig. 32 Increase of transverse shrinkage during multipass welding of a butt joint. (a) Increase of transverse shrinkage in multipass welding. (b) Relationship between log w and u

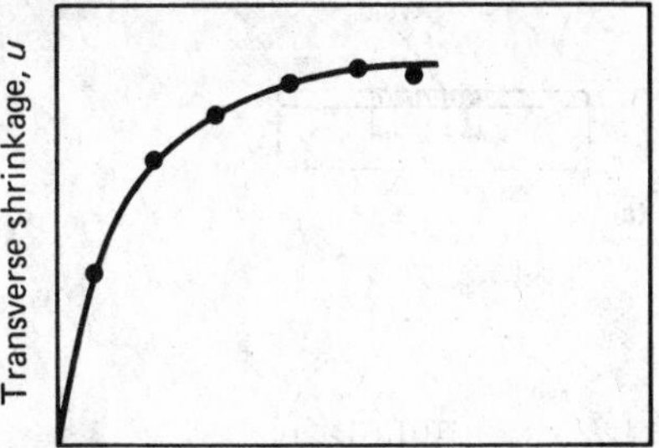

(a)

Transverse shrinkage, u

B (last pass)

μ_o

A (first pass)

log w_o

(b) log w

where u_o and w_o are transverse shrinkage and the weight of the weld metal deposited, respectively, after the first pass is welded; b is a coefficient.

Figure 33 demonstrates three methods to reduce transverse shrinkage:

- Decrease the total weight of weld metal, as shown by arrow 1; the amount of shrinkage changes from B to C.
- Decrease the tangent b, as shown by arrow 2; the amount of shrinkage changes from B to D.
- Move the shrinkage after the first pass from A to A', as shown by arrow 3; the amount of shrinkage after the completion of the weld changes from B to E.

The effects of various factors on transverse shrinkage, including joint design, root opening, type and size of electrodes, degree of constraint, peening, and oxygen gouging, were studied using ring-type specimens (Ref 43 and 44). The results are summarized in Table 5. Among the factors investigated, root gap and joint design produced the greatest effects.

Figure 34 shows the effect of root opening on transverse shrinkage. As the root opening increases, shrinkage increases and the total amount of weld metal increases (method 1 in Fig. 33). The results show that method 2 (Fig. 33) was also significant. A single-V-joint produced more shrinkage than a double-V-joint. This was partly due to method 1, the larger joint sectional area. Method 2, however, also was significant.

Fig. 33 Schematic diagram showing the methods to reduce transverse shrinkage of butt welds. Reprinted by permission of the Society of Naval Architects of Japan. (Source: Ref 43 and 44)

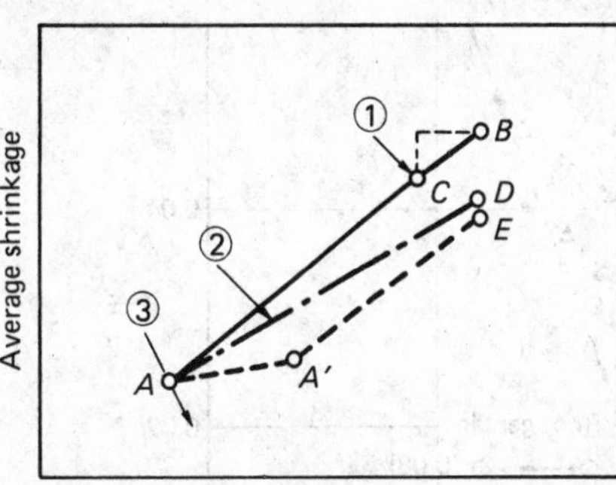

Logarithm of weight of deposited metal or consumed rod

Figure 34(b) shows the effect of electrode size on transverse shrinkage. Shrinkage decreases as the electrode size increases, and method 3 (Fig. 33) is most significant. Unless the electrodes are used in the first pass, reduction of shrinkage cannot be obtained with large size electrodes. This was also confirmed experimentally.

In considering the effect of welding heat input on transverse shrinkage, shrinkage decreases as total heat input required to weld a certain joint decreases (arrow 1 in Fig. 33). When a weld is completed in several passes, shrinkage decreases when the first pass is welded with greater heat input. In this case, an increase in shrinkage is produced after welding the first pass by using a greater heat input (compare points A' and A in Fig. 33). However, the amount of shrinkage after the completion of the weld decreases from B to E.

In considering the effect of chipping and gouging on transverse shrinkage, chipping a part of the weld metal produces little effect on shrinkage, and shrinkage increases due to rewelding. Because heat is applied to the workpiece during oxygen gouging, shrinkage increases by gouging; shrinkage increases further during repair welding (Ref 43 and 44).

Rotational Distortion of Butt Welds. Rotational distortion is affected by both heat input and welding speed (Ref 1). When $^1/_2$-in.-thick low-carbon steel plate are welded with covered electrodes at a low welding speed, the unwelded portion of the joint tends to close, as shown in Fig. 28(a). When steel plate are welded with the SAW process, the unwelded portion of the joint tends to open, as shown in Fig. 28(b). Tack welds of a submerged arc welded joint must be large enough to withstand stresses caused by the rotational distortion.

Effect of Restraint on Transverse Shrinkage. In the original studies made in Japan, investigators used K to express the degree of restraint. Because K is now widely used to express the stress intensity factor in the fracture mechanics theories, k_s is used to indicate degree of restraint. Japanese investigators have expanded the concept of degree of restraint to include not only transverse shrinkage but also bending and other types of deformation (see Ref 1 for more information).

To quantitatively study the effect of restraint on the transverse shrinkage of a butt weld, the degree of restraint of a joint must be defined analytically, as shown in Fig. 35(a) (Ref 1, 44, and 45). When welding is done along part of a slit from $x = x_1$ to x_2, the degree of restraint is defined as:

$$k_s = \frac{\pi}{2}\frac{E}{L}\frac{\ell}{L}\frac{1}{F} \qquad \text{(Eq 26)}$$

Table 5 Effects of various procedures on transverse shrinkage in butt welds

Methods 1, 2, and 3 are explained in Fig. 33 and related text.

Procedure	Effects
Root opening	Shrinkage increases as root opening increases (see Fig. 34); effect is great (methods 1 and 2)
Joint design	A single-V-joint produces more shrinkage than a double-V-joint; effect is great (methods 1 and 2)
Electrode diameter	Shrinkage decreases by using larger sized electrodes (see Fig. 34); effect is medium (method 3)
Degree of constraint	Shrinkage decreases as the degree of constraint increases; effect is medium (method 2)
Electrode type	Effect is minor (method 2)
Peening	Shrinkage decreases by peening; effect is minor (method 2)
Chipping and gouging	See text for explanation of process and effects

Fig. 34 Effect of root opening and electrode size on transverse shrinkage of butt welds. Specimens were ring-shaped, as shown in Fig. 37. Outer diameter, 24 in.; inner diameter, 12 in.; plate thickness, 0.7 in.; double-V-groove; w, weight of electrode deposited per unit weld length, g/cm; u, transverse shrinkage (mean value along the weld line). (a) Effect of root opening. (b) Effect of electrode size. Source: Welding Research Council

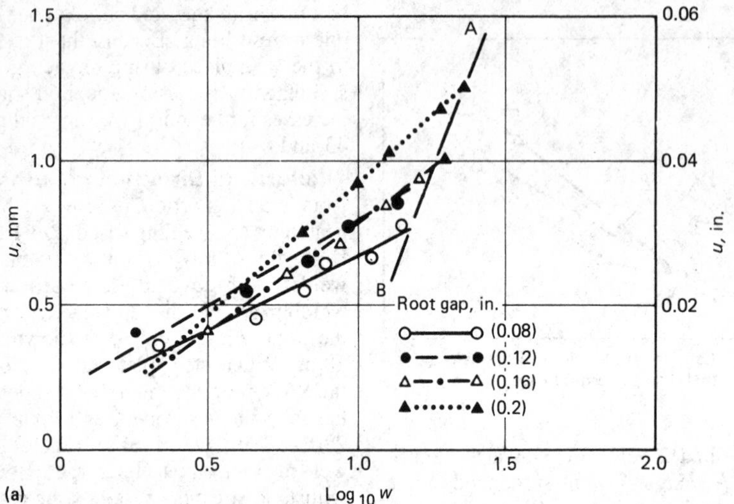

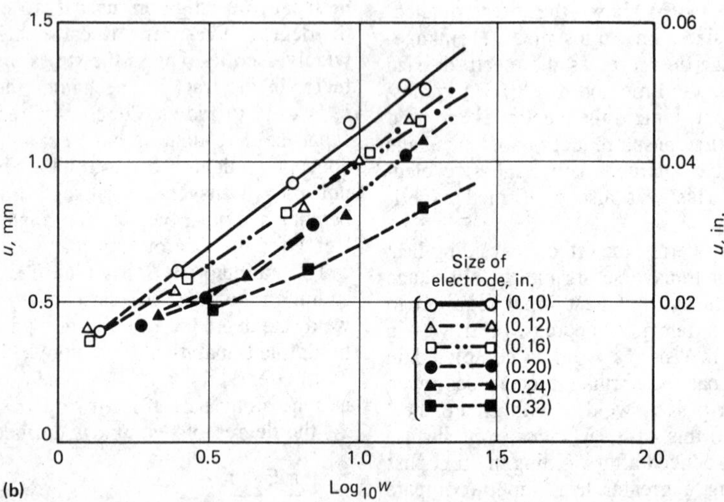

Fig. 35 Analysis of degree of restraint (k_s) of a slit-type weld. (a) Slit-type specimen. (b) Assumed stress distribution. (c) Displacement transverse

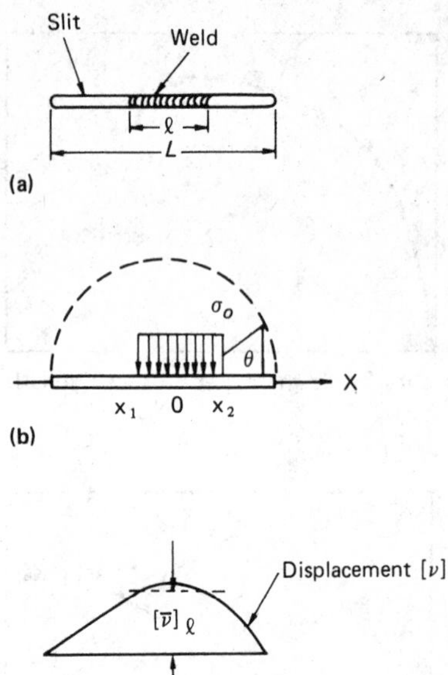

Fig. 36 Relationship between degree of restraint (k_s) and transverse shrinkage in a slit-type specimen

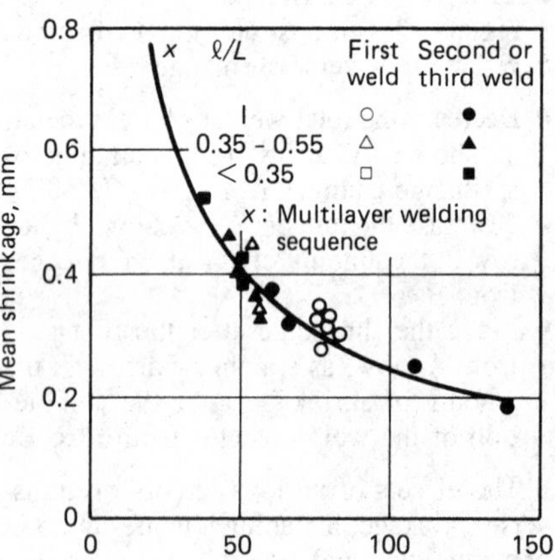

where k_s is the degree of restraint, kg/mm/mm^2 (lb/in./in.^{2}); L is slit length; $\ell = x_2 - x_1$, which equals weld length; E is Young's modulus of the material; x_1 is the abscissa of the starting point of welding measured from the center of the slit, 0; and x_2 is the abscissa of the end point of welding.

The parameter θ is used to express the abscissa x as follows (Fig. 35b):

$$x = \frac{L}{2} \cos \theta$$

$$x_1 = \frac{L}{2} \cos \theta_1$$

$$x_2 = \frac{L}{2} \cos \theta_2 \qquad \text{(Eq 27)}$$

F is a function determined by θ_1 and θ_2 as follows:

$$F = \sum_{n=1}^{\infty} \left[\int_{\theta_1}^{\theta_2} \sin \theta \cdot \sin n\theta \cdot d\theta \right]^2 \qquad \text{(Eq 28)}$$

The physical meaning of k_s is explained as follows. When uniform compressive stresses in the amount of σ_o are applied along a portion of the slit from x_1 to x_2, as shown in Fig. 35(b), both edges of the slit deform to come close together, or dislocation, $[v]$, occurs. The amount of $[\bar{v}]_\ell$ is the average value of $[v]$ in the range of length ℓ from x_1 to x_2, as shown in Fig.

35(c). The relationship between σ_o and $[\bar{v}]_\ell$ is:

$$\sigma_o = k_s [\bar{v}]_\ell \quad \text{(Eq 29)}$$

Stress in the amount of k_s is needed to produce a unit amount of shrinkage or dislocation along the weld length (ℓ).

Studies have been conducted on the effect of the degree of restraint on the transverse shrinkage of a slit-type weld (Ref 45). Figure 36 shows the relationship between k_s values of joints and values of transverse shrinkage determined experimentally. Experiments were conducted on joints with various lengths (L) ranging from 75 to 450 mm (3 to 17.7 in.) and weld lengths (ℓ) ranging from 75 to 150 mm (3 to 6 in.). Results obtained with a block sequence and multilayer sequence also are plotted. Figure 36 indicates that transverse shrinkage decreases as the degree of restraint increases.

A study also has been made of the degree of restraint of a ring-type joint, as shown in Fig. 37(a). Results obtained by different investigators using different types of specimens can be compared using the degree of restraint (Ref 46). Figure 38 shows the relationship between the degree of restraint (k_s) and the ratio between the shrinkage of free welds (S_{tf}) and the shrinkage of restrained welds (S_t). Figure 37 gives formulas to calculate the degree of restraint of different specimens.

Effect of Welding Sequence on Transverse Shrinkage. The effects of welding sequence on the magnitude and distribution of transverse shrinkage in slit weld specimens and patch weld specimens were the subject of another study (Ref 45). Figure 39 shows the experimental results obtained with slit weld specimens. Specimens 1-1, 1-2, and 1-3 were welded using a block welding sequence. The circled numbers under the specimen designation indicate the sequence in which blocks were welded. Specimen 1-4 was welded using a multilayer procedure. Figure 39 shows that shrinkage was reduced when the center block was welded first, followed by the blades on each side.

Transverse Shrinkage of Fillet Welds

More transverse shrinkage occurs across a butt weld than across a fillet weld. However, only limited studies have been made on the transverse shrinkage of fillet welds. Nevertheless, a simple formula has been proposed (Ref 40). For T-joints with two continuous fillets:

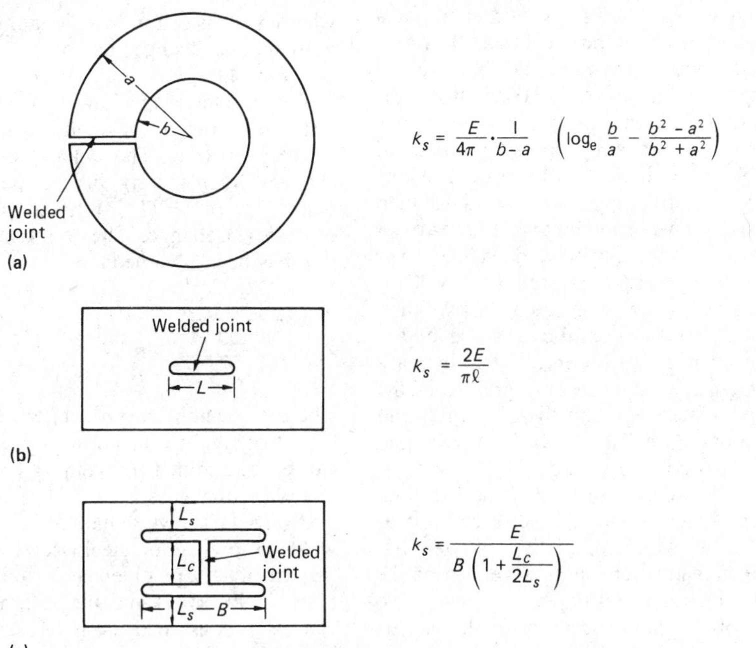

Fig. 37 Specimen types included in the analysis by Watanabe and Satoh (Ref 46). (a) Circular-ring type specimen. (b) Slit-type specimen. (c) H-type constrained specimen

Fig. 38 Effect of external constraint on the transverse shrinkage of butt-welded joints. S_t/S_{tf} is the ratio between the shrinkage of free welds and the shrinkage of restrained welds. (Source: Ref 46)

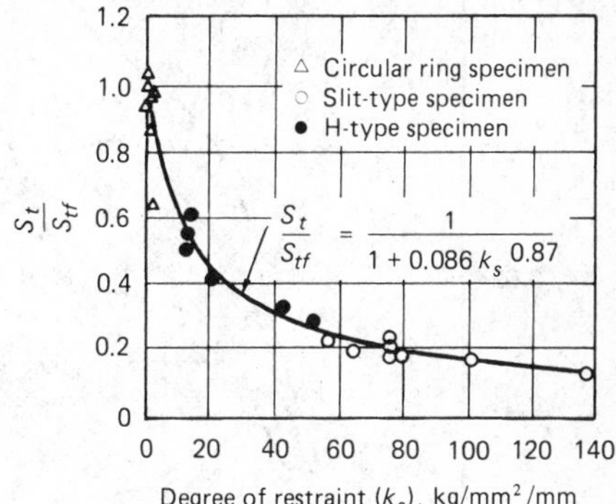

$$\text{Shrinkage} = \frac{\text{leg of fillet}}{\text{thickness of plate}} \times 0.04 \quad \text{(Eq 30)}$$

For shrinkage in millimetres, use fillet leg and plate thickness in millimetres. For shrinkage in inches, use fillet leg and plate thickness in inches.

For intermittent welds, a correcting factor of proportional length of fillet to total length should be used. For fillets in a lap joint (two fillet welds):

$$\text{Shrinkage} = \frac{\text{leg of fillet}}{\text{thickness of plate}} \times 0.06 \quad \text{(Eq 31)}$$

For shrinkage in millimetres, use fillet leg and plate thickness in millimetres. For shrinkage in inches, use fillet leg and plate thickness in inches.

Longitudinal Shrinkage of Butt Welds

The amount of longitudinal shrinkage in a butt weld is on the order of $^1/_{1000}$ of the weld length, much less than the transverse shrinkage. Only limited studies have been made of longitudinal shrinkage in butt welds, but the following formula has been proposed (Ref 47):

$$\Delta L = \frac{0.12 \times I \times L}{100\,000 \times t} \quad \text{(Eq 32)}$$

where I is the welding current, A; L is the length of weld, in.; and t is the plate thickness, in. If $t = {}^1/_4$ in. and $I = 200$ A:

Fig. 39 Distribution of transverse shrinkage obtained in a slit weld specimen with different welding sequence. Plates were about $^3/_4 \times 32 \times 47$ in. Welds were about 20 in. long. (Source: Ref 1, 2, and 45)

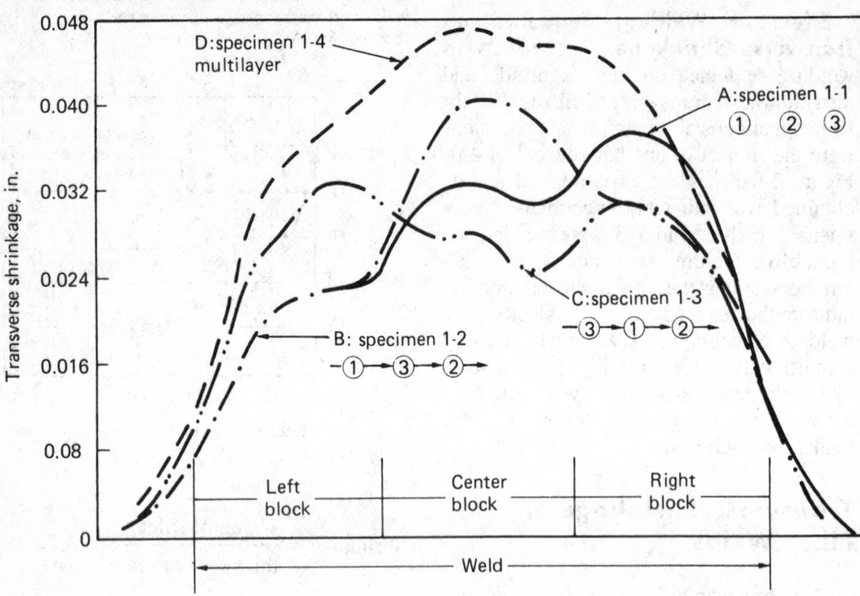

$$\Delta L = \frac{0.12 \times 200}{100\,000 \times 0.25} L \doteqdot 0.001 \cdot L \qquad \text{(Eq 33)}$$

Longitudinal Shrinkage of Fillet Welds

In an extensive study of longitudinal shrinkage in carbon steel fillet welds, longitudinal shrinkage was shown to be primarily a function of the total cross section of the joints involved. As plate width and thickness increase, restraint is more effective (Ref 48). Therefore, total cross section of the welded plate in the transverse section is called the "resisting cross section."

Figure 40 shows the results obtained by Guyot (Ref 48). As shown in Fig. 40, cross-shaped assemblies were used to maintain symmetry in the transverse section, thereby keeping longitudinal deflection to a minimum. Shrinkage values are expressed as a function of the resisting cross-sectional area (A_p) and the cross-sectional area of the weld metal (A_w). When the ratio of A_p to A_w is less than 20, the following formula may be used:

$$\delta = \frac{A_w}{A_p} \times 25 \qquad \text{(Eq 34)}$$

where δ is the longitudinal shrinkage in thousandths of an inch per inch of weld.

Angular Change in Butt Welds

In a butt weld, when transverse shrinkage is not uniform in the thickness direction, angular change often occurs. Many studies have been made of the angular change in butt welds and the effects of various welding procedure parameters, including the degree of constraint and the shape of the groove.

Figure 41 shows experimental data gathered on ring-type specimens (Ref 43). A radial groove was cut and then welded using covered electrodes 0.25 in. in diameter. Five specimens with different types of grooves ranging from symmetrical double-V to single-V were tested. Welding was first completed on one side; then the specimen was turned over and the other side was back-chipped and welded. Angular change was measured after welding each pass. In the earliest stage of welding on the first side, a mild increase of angular change was observed. The increase of angular change became greater in the intermediate stage and then became mild again in the final stage. Back chipping did not affect angular change.

During welding of the second side, angular change in the reverse direction was produced. After completion of welding, angular change remaining depended on the ratio of weld metal deposited on the two sides of the plates. Because angular change increased more rapidly during welding of the second side, minimum angular change was obtained in the specimen having a slightly larger groove in the first side welded. In this example (Ref 43), angular change could be minimized (near zero) for a butt joint having a ($t_1 + {}^1/_2\, t_3$) to t ratio of 0.6.

The Shipbuilding Research Association of Japan conducted an extensive research program on angular change in butt welds (Ref 49). Figure 42 shows the groove shape most suitable to minimizing angular changes in butt welds of various thicknesses. Curves are shown for two conditions, with and without strongbacks.

Distortion Caused by Angular Changes of Fillet Welds

Many research programs have studied angular changes during fillet welding. If a fillet joint is free from outer restraint, the joint simply bends to a polygonal form having a knuckle at the weld (see Fig. 43a). However, if the joint is restrained, a different type of distortion is produced. For example, when the movement of stiffeners welded to a plate is prevented, wavy distortion of the plate, as shown in Fig. 43(b), results. Analysis of wavy distortion and associated stresses can be handled in much the same manner as stress in a rigid frame (Ref 50). In the simplest case of a uniform distortion, the relationship between angular change and distortion at the weld is given as follows:

$$\frac{\delta}{\ell} = {}^1/_4\, \phi - \left[\frac{x}{\ell} - \frac{1}{2}\right]^2 \phi \qquad \text{(Eq 35)}$$

where δ is distortion; ℓ is length of span; ϕ is angular change; and x is the center (see Fig. 43).

The amount of angular change (ϕ) in a restrained structure is smaller than that in a free joint (ϕ_o). The amount of ϕ also changes when the rigidity of the bottom plate, $D = Et^3/12(1 - \nu^2)$, and the length of span (ℓ) change. The following equation has been obtained:

$$\phi = \frac{\phi_o}{1 + \dfrac{2D}{\ell} \cdot \dfrac{1}{C}} \qquad \text{(Eq 36)}$$

where C, which may be called the coefficient of rigidity for the angular change, can be determined by welding conditions and plate thickness.

Figure 44 shows values of ϕ_o for fillet welds in low-carbon steel (Ref 1 and 51). Values of ϕ_o are given as a function of plate thickness (t) and the weight of the electrode consumed is given per weld length (w). To convert from w to the size

Fig. 40 Variation of longitudinal shrinkage as a function of the resisting cross-section area (A_p) and transverse cross section (A_w). Source: Ref 48

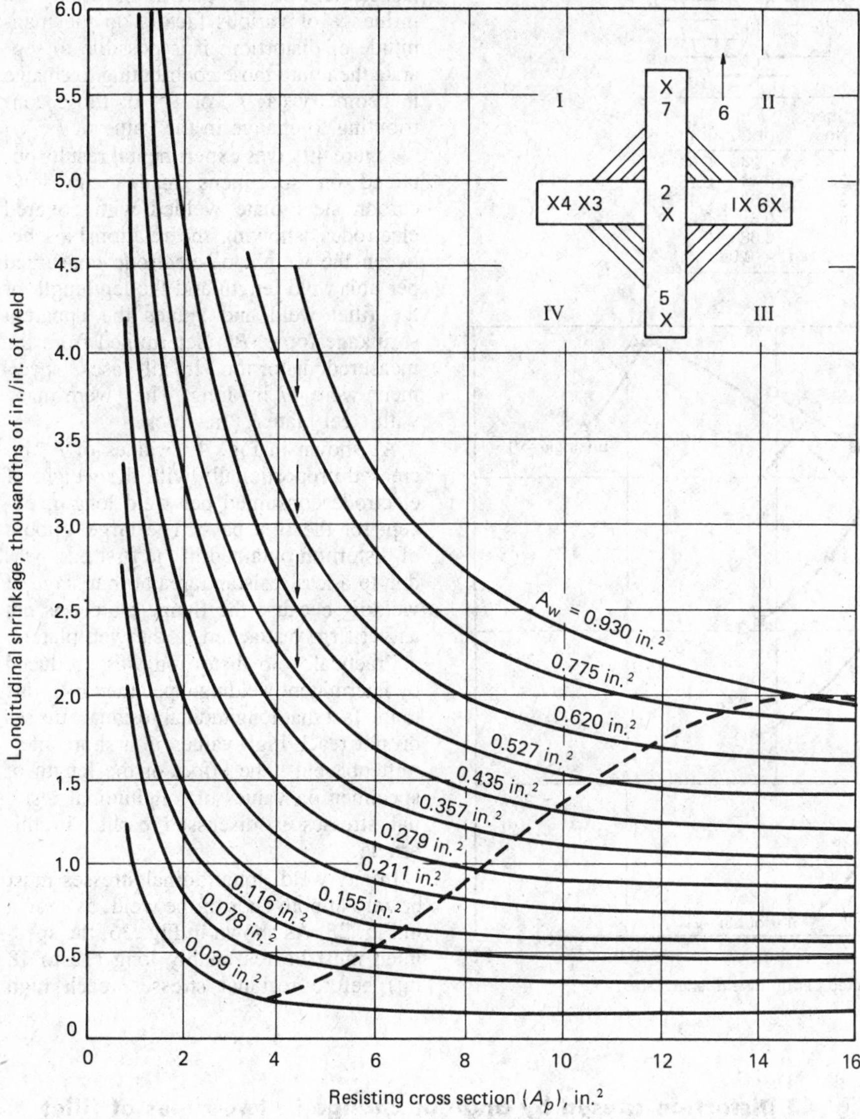

of the fillet weld (D_f), the following formula may be used (see Fig. 45):*

$$w = (D_f^2) \times 10^{-2} \times (\rho/\eta_d) \qquad \text{(Eq 37)}$$

where ρ is the density of weld metal, and η_d is the deposition efficiency. Fillet size (D_f) commonly is used in design work, while w is easy to determine in a welding experiment.

Experimental results shown in Fig. 44 were obtained using covered electrodes 0.2 in. in diameter. Maximum angular changes were obtained when the plate thickness was around 0.35 in. When the plate was thinner than 0.35 in., the amount of angular change was reduced with the plate thickness, because the plate was heated more evenly in the thickness direction, thus reducing the bending moment. When the plate was thicker than 0.35 in., the amount of angular change was reduced as the plate thickness increased (because of the increase of rigidity).

Table 6 lists values of C for steel weldments. Figure 46 shows values of ϕ_o for fillet welds in aluminum (Ref 1 and 52). Strain-hardened, aluminum magnesium structural alloy 5086-H32 was gas metal arc welded with alloy 5356 filler wire. Maximum distortion in aluminum welds occurred in thickness of approximately 0.28 in. Table 7 lists values of C.

Because aluminum is lighter than steel, values of w in Fig. 46 are considerably smaller than those in Fig. 44. Figure 47 compares values of distortion at the panel center (δ_m) for steel and aluminum welded structures on the basis of the same fillet size (D_f). Distortion in aluminum structures is less than in steel structures.

Bending Distortion Produced by Longitudinal Shrinkage

When the weld line does not coincide with the neutral axis of a weldment, longitudinal shrinkage of the weld induces bending moment, resulting in longitudinal distortion. The longitudinal distortion is the major distortion during welding fabrication of long built-up beams, as shown in Fig. 48. It is technically possible to analytically determine transient thermal and residual stresses and distortion during welding fabrication of long beams by extending the one-dimensional analysis (Ref 1). However, distortion after welding is the prime consideration; a much simpler analysis similar to the beam theory can be used.

In analyzing the longitudinal distortion of a long, slender weldment (Fig. 48), longitudinal residual stresses (σ_x) and the curvature of longitudinal distortion ($1/R$) are given as (Ref 53):

$$\sigma_x = -E\epsilon_x'' + \frac{M_y^*}{I_y} + \frac{P_x^*}{A}$$

$$\frac{1}{R} = \frac{M_y^*}{EI_y} = \frac{P_x^*\ell^*}{EI_y} \qquad \text{(Eq 38)}$$

where ϵ_x'' is incompatible strain; A is sectional area of the joint; I_y is moment of inertia of the joint around the neutral axis; P_x^* is apparent shrinkage force, $P_x^* = \iint E\epsilon_x'' dydz$; M_y^* is apparent shrinkage moment, $M_y^* = \iint E\epsilon_x'' zdydz = P_x^*\ell^*$; and ℓ^* is distance between the neutral axis and the acting axis of apparent shrinkage force.

Equation 38 shows that it is necessary to know the distribution of incompatible strain (ϵ_x'') in order to establish the distribution of residual stress (σ_x). However, information about moment (M_y^*) is only sufficient for determining the amount of distortion ($1/R$). Moment (M_y^*) is determined when the magnitude of apparent shrinkage force (P_x^*) and the location of its acting axis are known. Through experiments, it was found that the acting axis of P_x^* is located somewhere in the weld metal. The apparent shrinkage force is the origin of residual stress. Distortion is produced as the result of the existence of P_x^*. More

*There are two fillet welds, as shown in Fig. 45. There is an error in Ref 1; calculations in Ref 2 are correct.

Fig. 41 Effect of shape of groove on angular change. Source: Ref 28

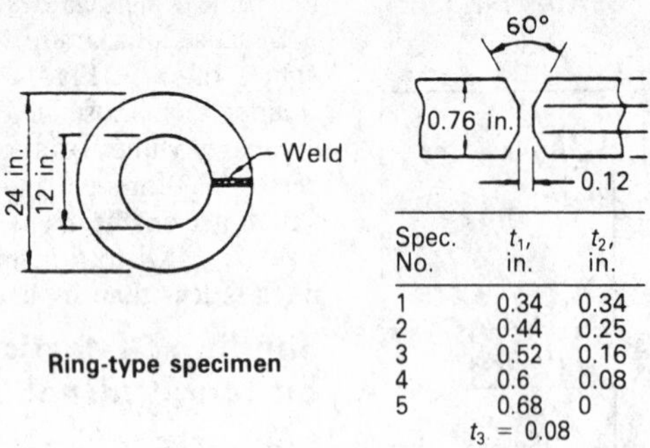

Spec. No.	t_1, in.	t_2, in.
1	0.34	0.34
2	0.44	0.25
3	0.52	0.16
4	0.6	0.08
5	0.68	0
t_3 = 0.08		

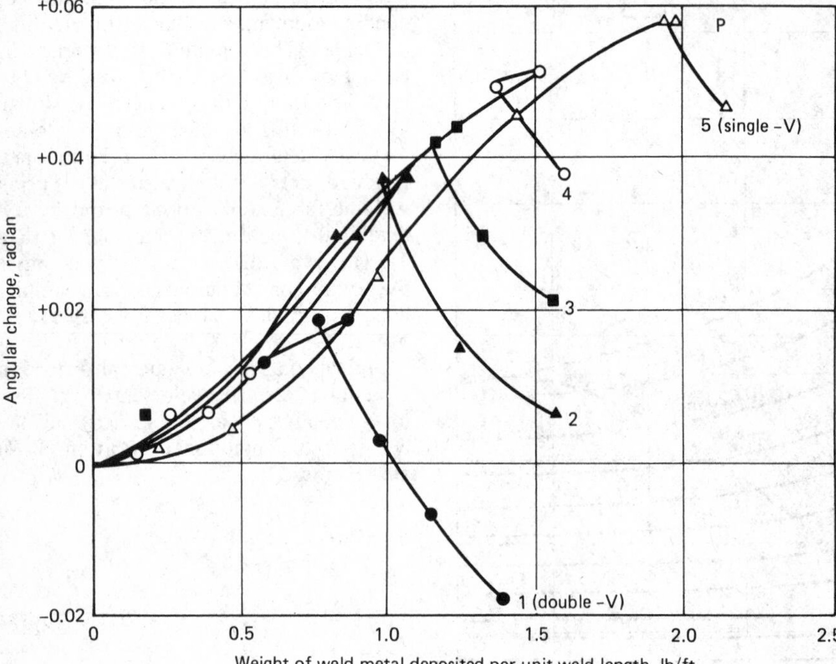

information can be obtained when the value of P_x^* rather than the value of distortion itself is used in the analysis of experimental results. For example, in discussing the influence of various factors on the magnitude of distortion, it is possible to separate them into those contributing to change in geometry (A, I_y, or ℓ^*) or those contributing to change in the value of P_x^*.

Figure 49 gives experimental results obtained on specimens made with low-carbon steel plate welded with covered electrodes, showing the relationships between the weight of electrode consumed per unit weld length and the leg length of the fillet weld and values the apparent shrinkage force (P_x^*) determined from the measured distortion. In all cases, specimens were 47 in. long. They were made with steel plate $^1/_2$ in. thick.

As shown in Fig. 49, values of P_x^* increased proportionally with the weight of electrode consumed per weld length, except for the first pass. The large amount of distortion obtained in the first pass was due to a less resistant area at that stage of welding because the flange plate was not as yet firmly attached to the web plate.

Practically no distortion was produced by intermittent welding (specimen I-4), due to the fact that longitudinal residual stresses do not reach high values in a short intermittent weld. The effect of the length of specimen on values of longitudinal residual stresses is discussed earlier in this article.

In any weld, longitudinal stresses must be zero at both ends of the weld, as shown in Fig. 25. As shown in Fig. 26, the specimen must be reasonably long (12 to 18 in.) before residual stresses reach high

Fig. 42 Suitable groove shape that gives zero angular change in butt welds. Source: Ref 49

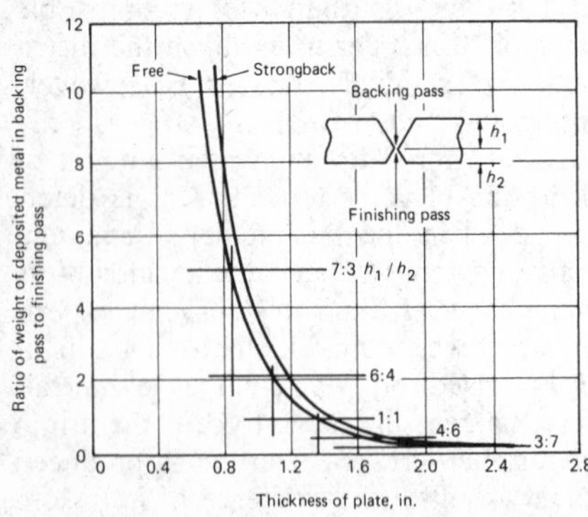

Fig. 43 Distortion caused by angular change in two types of fillet welded structures. (a) Free joint. (b) Restrained joint. Reprinted by permission of the Welding Research Council

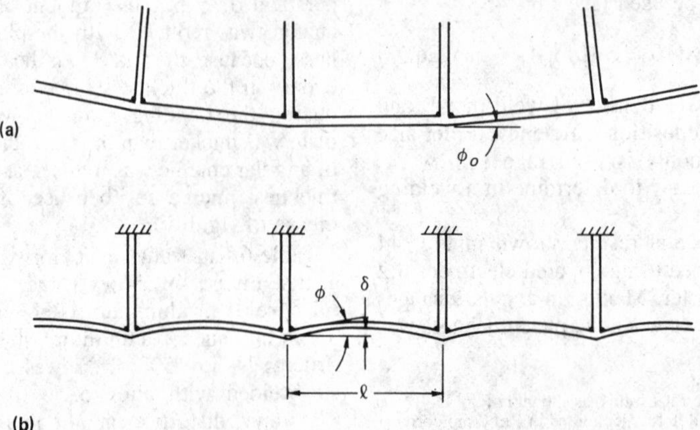

Fig. 44 Angular change of a free fillet weld (ϕ_o) in steel

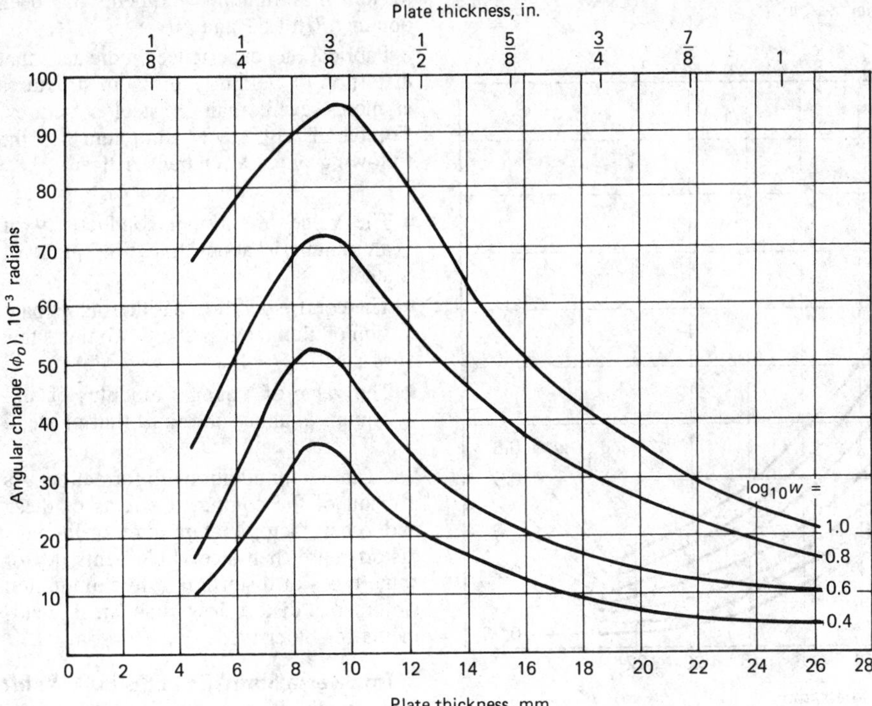

values. Because intermittent welds are short, longitudinal residual stresses are very low, resulting in a small amount of longitudinal bending distortion.

Comparison of Steel and Aluminum Welds. A study also was made at the Massachusetts Institute of Technology of longitudinal distortion due to welding fabrication of built-up beams in an aluminum alloy (Ref 54). T-beams were fabricated by fillet welding a web plate 4 or 6 in. high to a flange plate 4 in. wide. The 5052-H32 alloy plates were $^1/_2$ in. thick and 48 in. long; they were welded by GMAW using 4043 and 2319 filler wires.

Figure 50 shows the experimental results, including the relationships between the length of the fillet weld and the longitudinal distortion, expressed in terms of the radius of curvature ($1/R$). Figure 50 suggests that steel weldments distort more than aluminum weldments with equivalent dimensions.

Fig. 45 Angular change in a free fillet weld

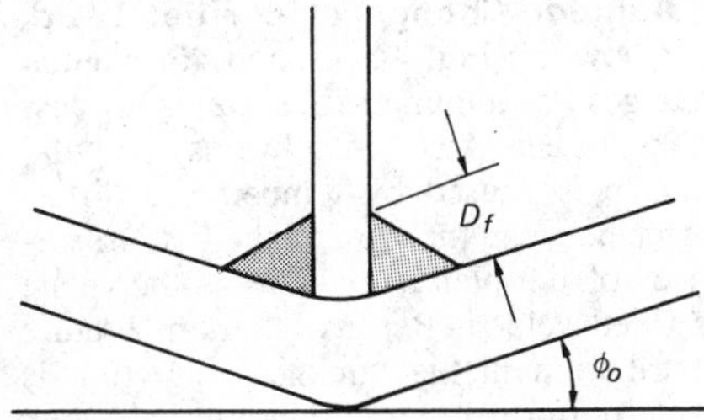

Buckling Distortion

In any weldment, longitudinal residual stresses are compressive in areas away from the weld, as shown in Fig. 6(b). When plate is thin, it may buckle due to the compressive residual stresses. Buckling distortion differs from bending distortion in that the amount of distortion in buckling distortion is much greater than that of bending distortion. Also, there may be more than one stable deformed shape in the buckling distortion. Buckling distortion can be prevented through proper selections of plate thickness, length of free span, and welding heat input.

Buckling distortion of welded plate due to residual stresses has been studied by several investigators (Ref 55 and 56). Efforts have been made to determine critical sizes (length and width) of welded plate in different thicknesses. Buckling distortion occurs when the specimen length exceeds a certain limit.

When stiffeners are fillet welded to thin plate, the plate may buckle due to the compressive residual stresses that occur in the plate. This type of distortion is important in the welding fabrication of stiffened thin-walled structures. Engineers at Kawasaki Heavy Industries conducted extensive experimental and analytical studies of out-of-plane distortion during welding fabrication of panel structures in thin steel plates (Ref 57 and 58). Experiments were conducted under the following conditions:

- Material: Low-carbon steel
- Plate thickness (t): 0.18 to 0.4 in.
- Panel dimension (b): 20 × 20 in. and 40 × 40 in.
- Heat input (Q): 19 050 J/in. and 34 300 J/in.
- Welding processes: GTAW and SMAW

Figure 51 shows how distortion at the center of a panel increases after welding. Results on a square panel 20 in. long and 0.24 in. thick are given. Figure 52 shows relationships between deflection and heat input for square panels 20 in. long made with plates 0.18 to 0.40 in. thick.

Figure 51 shows that it takes about 30 min before distortion is fully developed. After welding is completed, temperatures

Table 6 Values of angular rigidity coefficient *C* for low-carbon steel

Fillet size (D_f), mm	Weight of consumed electrode per unit weld length (w), g/cm	$\log_{10} w$	Rigidity coefficient (C), kg · mm/mm t = 10 mm	t = 13 mm	t = 18 mm	t = 25.4 mm
6.58	2.51	0.4	5400	19 900	76 100	170 100
7.38	3.16	0.5	4700	18 000	65 200	142 400
8.29	3.98	0.6	4100	16 300	56 100	130 200
9.30	5.01	0.7	3800	15 000	48 800	125 000
10.45	6.31	0.8	3500	13 600	43 000	116 800
12.20	7.95	0.9	3300	12 200	38 900	112 000
13.15	10.00	1.0	3100	11 000	36 100	108 200
14.80	12.60	1.1	3000	9 800	35 200	105 000
16.55	15.85	1.2	2950	8 800	34 800	102 000

Note: $w = (D_f^2) \times 10^{-2} \times 7.85/0.657 = 0.1195\, D_f^2$. According to Hirai and Nakamura (Ref 53), C can be expressed as:

$$C = \frac{1 + \frac{w}{5}}{t^4}$$, where t is plate thickness.

Fig. 46 Angular change in a free fillet weld (ϕ_o) in aluminum

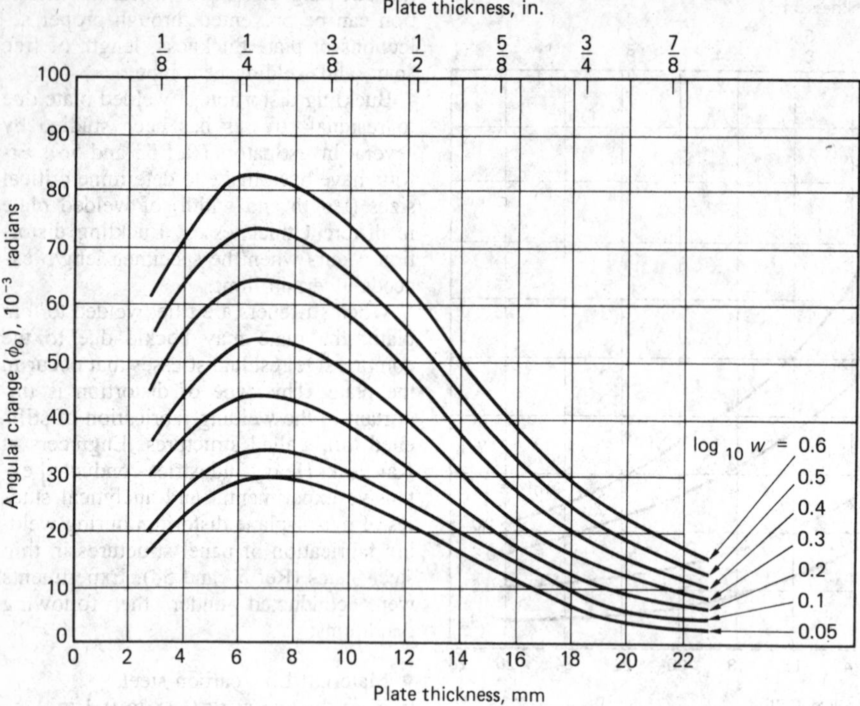

in regions near the weld cool quickly, in less than a minute. Heat is gradually distributed in greater regions in the weldment, and distortion of the entire panel gradually develops. As welding heat input increases, the structure starts to deform, and the amount of the final distortion increases. Figure 51 also shows that the amount of distortion increases significantly when the heat input increases from 21 340 to 23 470 J/in. This sudden increase of distortion is clearly shown in Fig. 52.

Figure 52 shows that the critical heat input (Q/t) for a 20 × 20 in. panel with a plate 0.24 in. thick is about 3700 cal/cm². When plate thickness is 0.18 in., the critical heat input is only about 2000 cal/cm².

In comparing experimental results on specimens made under various conditions, Kawasaki engineers developed Fig. 53, which indicates that buckling distortion occurs when:

$$H_{cr} = \frac{Q}{t^3} \cdot b \geqq \text{approx } 4 \times 10^5 \text{ cal/cm}^3 \quad \text{(Eq 39)}$$

The above parameter may be called the critical heat input index.

Comparison of Distortion in Aluminum and Steel Weldments

Although the majority of information on weld distortion available today pertains to steel weldments, some information on distortion in aluminum weldments has been obtained (Ref 59 and 60).

Fabrication experience indicates that distortion in welding aluminum structures is more severe than in steel structures. Compared with steel, aluminum has the following physical characteristics:

- The value of thermal conductivity of aluminum is about five times that of steel.
- The coefficient of linear thermal expansion of aluminum is about two times that of steel.
- The value of Young's modulus of aluminum is about one third that of steel.

See Table 3 for additional information. As a result of the combined effects of these and other factors, aluminum weldments distort more than steel weldments do for some types of distortion, while aluminum weldments distort less than steel weldments for other types of distortion.

Transverse Shrinkage of a Butt Weld. The amount of transverse shrinkage of a butt weld in aluminum is considerably larger than that of a steel weld of a similar dimension. Equation 20 shows simple empirical formulas for transverse shrinkage of butt welds in low-carbon steel, aluminum, and stainless steel. The reason why aluminum welds shrink more than steel welds can be explained on the basis of Eq 21 through 24, which analyze mechanisms of transverse shrinkage in a butt weld. As shown in these equations, the major portion of transverse shrinkage in a butt weld is due to contraction of the base plate. Because the value of thermal conductivity of aluminum is much greater than that of steel, heat generated by the welding arc dissipates faster in aluminum than in steel, resulting in a broader HAZ. Thus, values of $T(t_x, x)$ in Eq 21 in an aluminum butt weld are higher than those in a steel butt weld. In addition, the value of α for aluminum is larger than that of steel. Consequently, the value of transverse shrinkage (S) in Eq 24 for an aluminum weld is larger than that for a steel weld.

Angular Change of a Fillet Weld. As shown in Fig. 44 through 46, angular changes of aluminum fillet welds are less than those of steel fillet welds. Angular change is caused by temperature differences between the top and the bottom surfaces of the plate to which another plate is fillet welded. Because the temperature distribution in the thickness direction is more uniform in an aluminum weld than

Table 7 Values of angular rigidity coefficient for aluminum

Fillet size, (D_f), mm	Weight of consumed electrode per unit weld length (w), g/cm	$\log_{10} w$	Rigidity coefficient (C), kg · mm/mm t = 3.18 mm	t = 6.4 mm	t = 9.5 mm	t = 12.7 mm	t = 15.9 mm	t = 19.1 mm
8.969	1.122	0.05	57	782	14 390	22 800	31 000	78 400
9.567	1.259	0.1	55	762	13 600	20 800	25 300	72 500
10.660	1.585	0.2	52	725	7 900	17 000	18 000	31 800
11.960	1.995	0.3	49	686	5 600	13 800	12 900	22 200
13.420	2.512	0.4	46	645	4 300	11 000	9 200	17 000
15.057	3.162	0.5	43	608	3 600	8 900	6 900	13 500

Note: $w = (D_f^2) \times 10^{-2} \times 2.65/0.95 = 0.02789 D_f^2$

Fig. 47 Out-of-plane distortion (δ_o) as a function of plate thickness (T), span length (l), and the size of fillet weld (D_f) for steel and aluminum

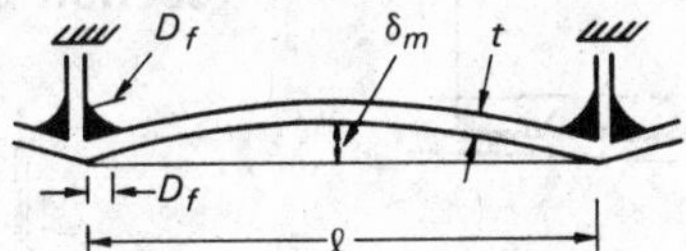

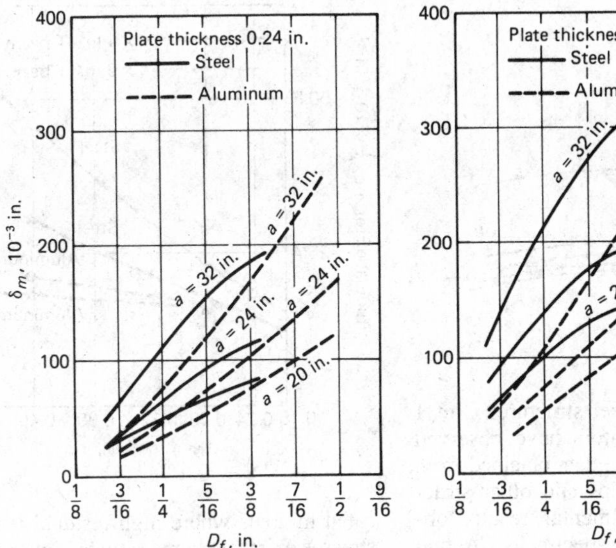

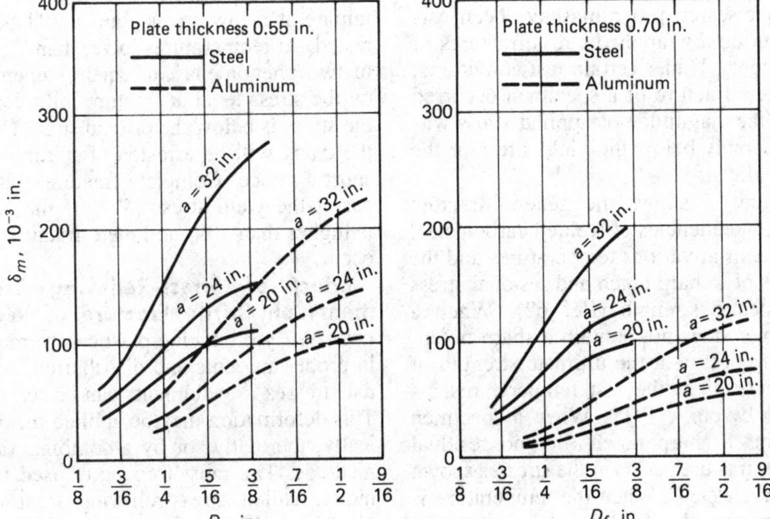

in a steel weld, the amount of angular change of an aluminum fillet weld is less than that of a steel fillet weld.

Longitudinal Distortion. Figure 50 compares values of longitudinal distortion expressed in terms of the radius of curvature of built-up beams in steel and aluminum. Aluminum welds distorted less than steel welds, perhaps due to the fact that the temperature distribution in the z-direction is more uniform in an aluminum weld than in a steel weld.

Effects of Residual Stresses and Distortion on Service Behavior of Welded Structures

Only when certain conditions exist do residual stresses decrease the fracture strength of welded structures; however, loss of strength can be drastic under these conditions. Generally, the effects of residual stresses are significant when fractures occur under low applied stress.

Changes of Residual Stresses in Weldments Subjected to Tensile Loading

Figure 54 shows how residual stresses change when a butt weld is subjected to tensile loading. Curve 0 shows the lateral distribution of longitudinal residual stress in the as-welded condition. When uniform tensile stress ($\sigma = \sigma_1$) is applied, the stress distribution is shown by curve 1. Stresses in areas near the weld reach yield stress, and most of the stress increase occurs in areas away from the weld. When the tensile applied stress increases to σ_2, the stress distribution is shown by curve 2. The stress distribution across the weld becomes more even as the level of applied stress increases; that is, the effect of welding residual stress on the stress distribution decreases.

General yielding occurs when the level of applied stress is further increased; that is, yielding occurs across the entire cross section. The stress distribution at general yielding is shown by curve 3. Beyond general yielding, the effect of residual stresses on the stress distribution virtually disappears.

Curve 1′ shows the residual stress that remains after unloading when the tensile stress ($\sigma = \sigma_1$) is applied to the weld and then released. Curve 2′ shows the residual stress distribution when the tensile stress ($\sigma = \sigma_2$) is applied and then released.

Compared to the original residual stress distribution (curve 0), residual stress distributions after loading and unloading are more even. As the level of loading increases, the residual stress distribution after unloading becomes more even; that is, the effect of welding residual stress on the stress distribution decreases.

The effect of residual welding stresses on the performance of welded structures is significant only on phenomena that occur under low applied stress, such as brittle fracture and stress corrosion cracking. As the level of applied stress increases, the effect of residual stress decreases.

Fig. 48 Analysis of longitudinal distortion in a fillet joint. (a) General view. (b) Incompatible strain (ϵ_x''). (c) Residual stress (σ_x). (Source: Ref 53)

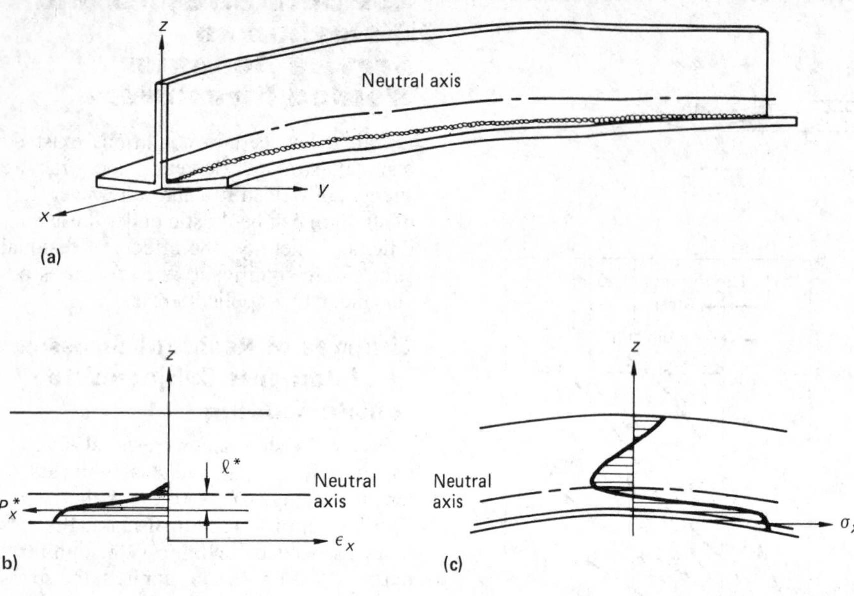

Fig. 50 Relationship between length of leg and curvature of longitudinal deflection in T-section beam. Source: Ref 1 and 54

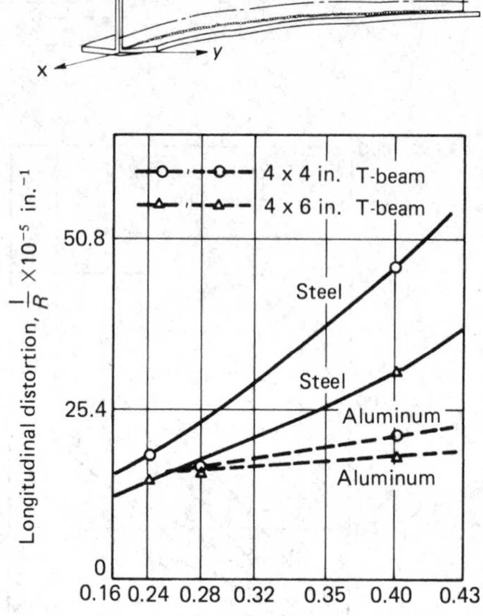

The effect of residual stress is negligible on the performance of welded structures under applied stresses beyond yielding. Also, the effect of residual stress tends to decrease after repeated loading.

Effects of Residual Stresses on Brittle Fractures of Welded Structures

Extensive studies have been conducted on the effects of residual stresses on brittle fracture of welded steel structures (Ref 1 and 61-64). Investigators have observed differences between the data obtained from brittle fractures in ships and other structures and the experimental results obtained with notched specimens. Actual fractures occurred at stresses far below the yield stress of the material; however, even when specimens contain very sharp cracks, nominal fracture stress of a notched specimen is as high as the yield stress. A number of research programs have been carried out on low applied stress fractures of weldments. Under certain test conditions, complete fracture of a specimen occurred while the magnitude of applied stress was considerably below the yield stress of the material.

Fig. 49 Increase of longitudinal distortion during multipass welding. Source: Ref 57

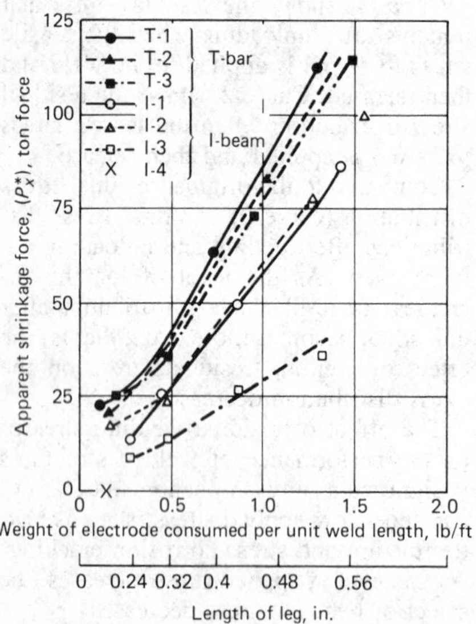

Figure 55 shows the general fracture strength tendencies of welded carbon steel specimens at various temperatures and the effects of a sharp notch and residual stress on fracture strength (Ref 62). When a specimen does not contain a sharp notch, fracture occurs at the ultimate strength of the material at the test temperatures, as shown by curve *PQR*. When a specimen contains a sharp notch (but no residual stress), fracture occurs at the stresses shown by curve *PQST*. When the temperature is higher than the fracture transition temperature (T_f), a high-energy (shear-type) fracture occurs at high stress. When the temperature is below T_f, fracture appearance changes to a low-energy (cleavage) type, and the stress at fracture decreases to near the yield stress. These various fractures can occur when a notch is located in areas where high residual tensile stresses exist. At temperatures higher than T_f, fracture stress is the ultimate strength (curve *PQR*). Residual stress has no effect on fracture stress. At room temperatures lower than T_f but higher than the crack-arresting temperature (T_a), a crack may initiate at a low stress, but it will be arrested. At temperatures lower than T_a, one of two phenomena can occur, depending on the stress level at fracture initiation. If the stress is below the critical stress (*VW*), the crack will be arrested after running a short distance. Complete fracture will occur at the yield stress (*ST*). If the stress is higher than *VW*, complete fracture will occur.

Effect of Stress-Relieving Treatments on Brittle Fracture of Weldments. By producing plastic deformation in proper amounts and distribution, residual stresses in weldments can be reduced. This deformation may be applied mechanically, thermally, or by a combination of methods. The most frequently used technique, called stress-relieving treatment, places a weldment in a furnace for a certain period at a specific temperature, depending on the type of material and thickness, and then cooled slowly.

When a load is applied to a weldment, residual stresses are redistributed due to local plastic deformation. Stresses are reduced when the load is removed. This ef-

Fig. 51 Formation of out-of-plane distortion during welding fabrication of stiffened panel structures. Welding was done between a low-carbon steel plate (1/4 in. thick) and frames.

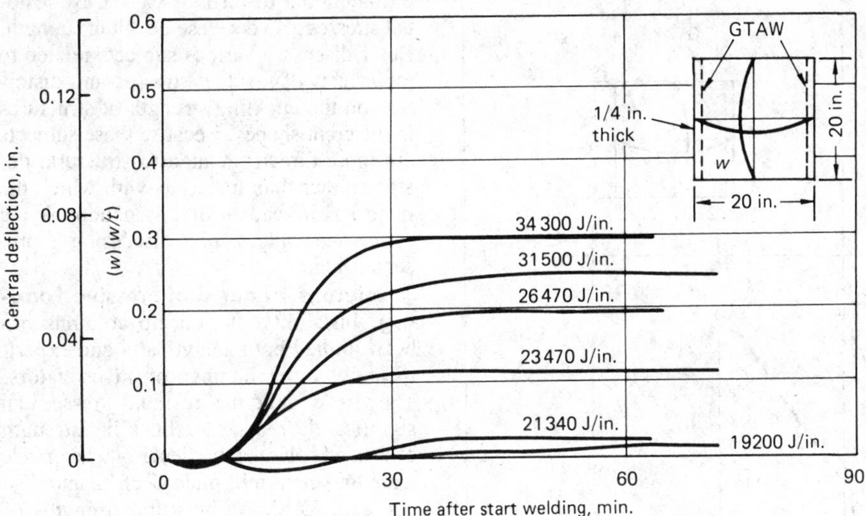

Fig. 52 Relationship between deflection and heat input for low-carbon steel panels 20 × 20 in.

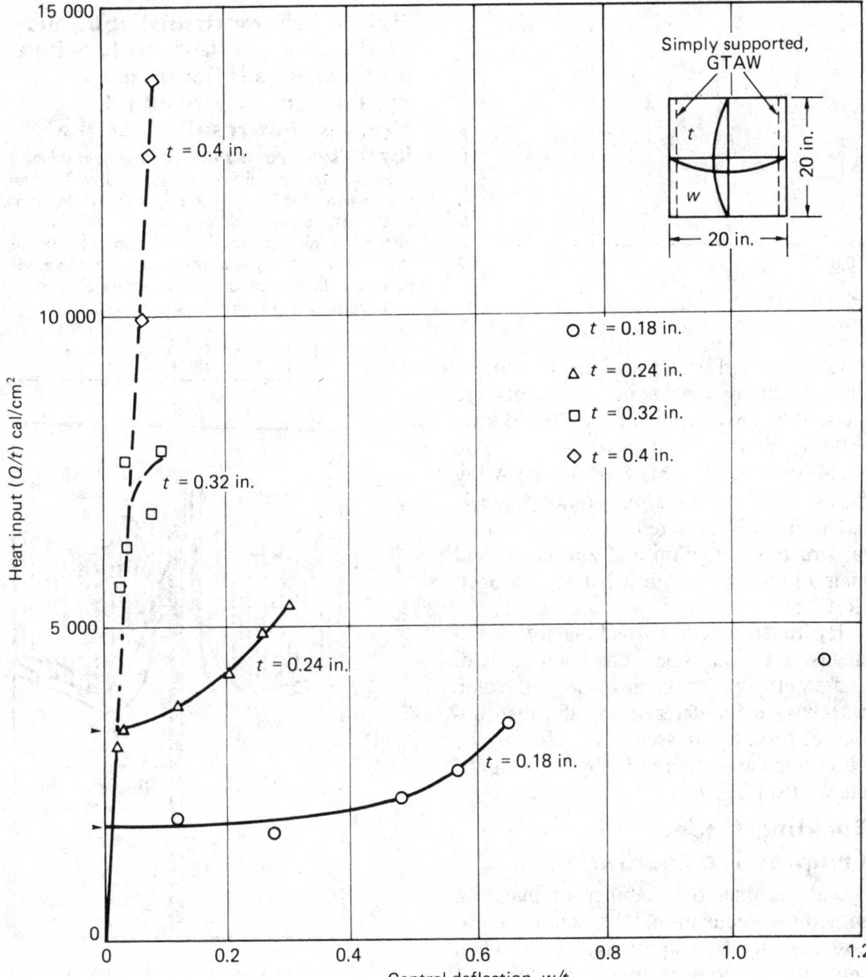

fect, called mechanical stress relieving, was demonstrated in a series of experiments on welded and notched wide plate specimens in carbon steel (Ref 65). Results are shown in Fig. 56.

In a series of tests, external loads were applied to different stress levels, 50, 100, 150, 200, and 230 MPa (7250, 14 500, 21 760, 29 010, and 33 360 psi) at 20 °C (68 °F), which was above the critical temperature for crack initiation (T_c). The load was then reduced. After mechanical stress-relieving treatments, specimens and applied tensile loads were cooled at temperatures below −30 °C (−22 °F). For these specimens, fractures occurred only after the preloaded stresses were exceeded, even at this low temperature, as shown in Fig. 56(a).

In another series of tests, specimens were thermally stress relieved by placing them in a furnace for 1 h at 320, 420, 520, and 620 °C (610, 790, 970, and 1150 °F). Fracture stresses for these specimens were higher when welds were heat treated at higher temperatures, indicating that more stresses were relieved by heating at higher temperatures (see Fig. 56b).

Effects of Residual Stresses on Fatigue Fracture of Welded Structures

Although the effects of residual stresses on fatigue strength of weldments have been studied extensively, the subject is still a matter of debate (Ref 66-68). Fatigue strength increases when a specimen has compressive residual stresses, especially on the specimen surface. Many investigators have reported that the fatigue strength (the number of cycles at fracture under a given load or the endurance limit) increased when the specimens had compressive residual stresses.

In experiments, local spot heating of certain types of welded specimens produced increases in fatigue strength. However, a number of investigators believe that residual stresses are relieved during cyclic loading; therefore, the effects of residual stress on fatigue strength of weldments are negligible.

Most fatigue cracks originate at the surface. Smoothness of surface is very important in obtaining high fatigue strength. For example, removing weld reinforcement, grinding surface irregularities, and other treatments are effective in reducing stress concentrations and increasing fatigue strength.

In conducting experiments to determine the effects of residual stresses on fatigue strength of weldments, the surface con-

Fig. 53 Relationship between deflection and Qb/t^3

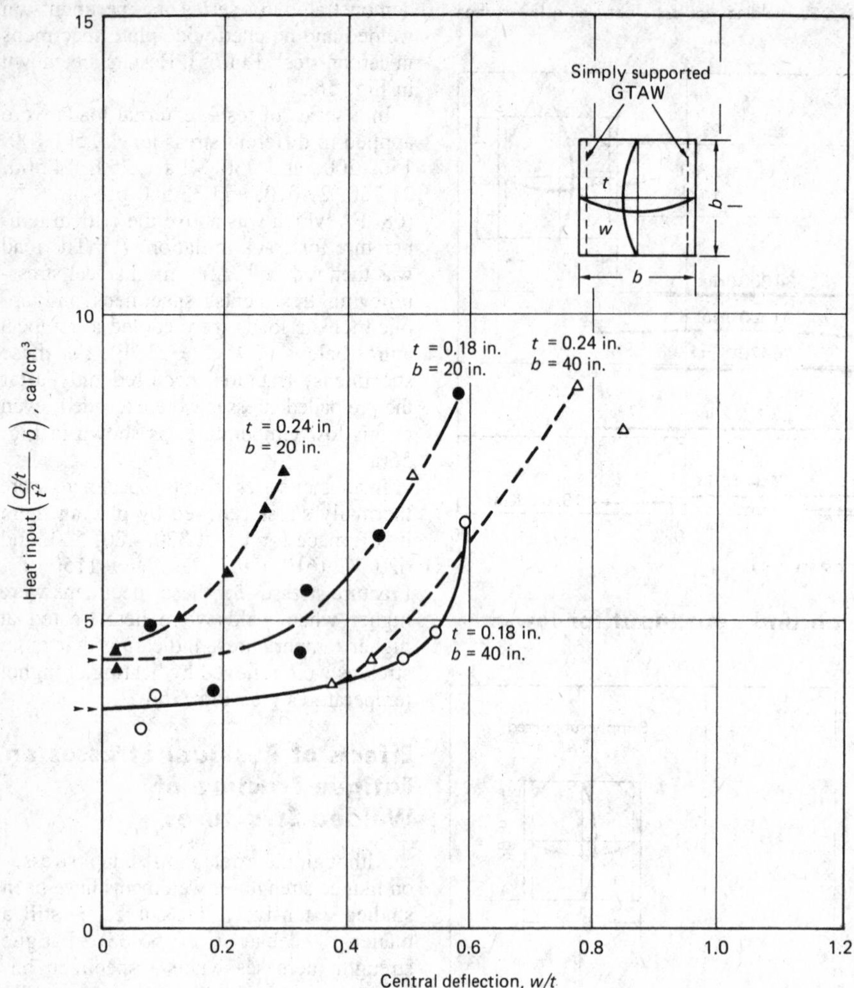

ditions of specimens must be carefully considered. Thus, a specimen that has compressive residual stresses on the surface is expected to have high fatigue strength. However, the specimen may yield poor results if it contains a sharp notch that causes fatigue cracks to initiate from the notch.

Effects of Environment

Cracking can occur in weldments, even without external loading, when the material is embrittled by exposure to certain environments and residual stresses are present. Stress corrosion cracking is a brittle-type fracture in a material exposed to a certain environment; it should not be confused with other types of localized attack, such as pitting, galvanic attack, intergranular corrosion, or cavitation (Ref 69-71). Stress corrosion cracking has been observed in a number of ferrous and nonferrous alloys exposed to certain environments.

High-strength steels are sensitive to hydrogen. Many cracks in weldments are caused by hydrogen introduced during welding, as discussed in the article "Arc Welding of Hardenable Carbon and Alloy Steels" in this Volume. In addition to steels having a body-centered cubic (bcc) crystal structure, titanium and zirconium and their alloys can be embrittled by hydrogen (Ref 73).

Hydrogen-induced cracking of weldments in various steels has been studied, and an attempt has been made to develop a technique for determining the distribution of residual stresses in a weldment by observing the pattern of hydrogen-induced cracks (Ref 29).

Buckling Under Compressive Loading

Failures due to instability or buckling sometimes occur in metal structures composed of slender bars or thin plate, when subjected to compressive axial loading, bending, or torsional loading. Residual compressive stresses decrease the buckling strength of a metal structure. In addition, initial distortions caused by residual stresses also decrease buckling strength. Ref 1 discusses various subjects related to the effects of residual stresses and distortion on the buckling strength of structures in different shapes. Because these subjects are more directly related to structural design, rather than materials with which the majority of readers of this handbook are concerned, only a short discussion is presented here.

Columns Under Compressive Loading. Instability of built-up columns has been studied both analytically and experimentally by a number of investigators. They have found that residual stresses can significantly reduce the buckling strength of welded columns, particularly when made from universal mill plate (Ref 72 and 73).

Figure 57 shows buckling strengths of steel I-beams built up by welding universal mill plate under compressive axial

Fig. 54 Schematic distributions of stresses in a butt weld when uniform tensile loads are applied and the residual stresses that result after the loads are released. Curve 0, residual stresses in the as-welded condition; curve 1, stress distribution at $\sigma = \sigma_1$; curve 2, stress distribution at $\sigma = \sigma_2$; curve 3, distribution of stresses at general yielding; curve 1′, distribution of residual stress after $\sigma = \sigma_1$ is applied and then released; curve 2′, distribution of residual stress after $\sigma = \sigma_2$ is applied and then released

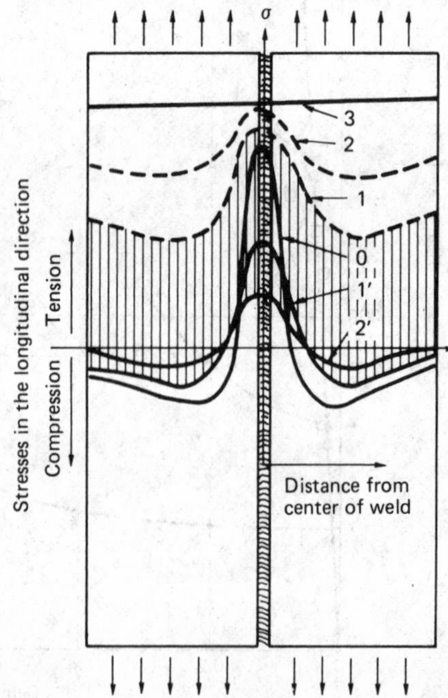

Fig. 55 Effects of sharp notch and residual stress on fracture strength. Source: Ref 62

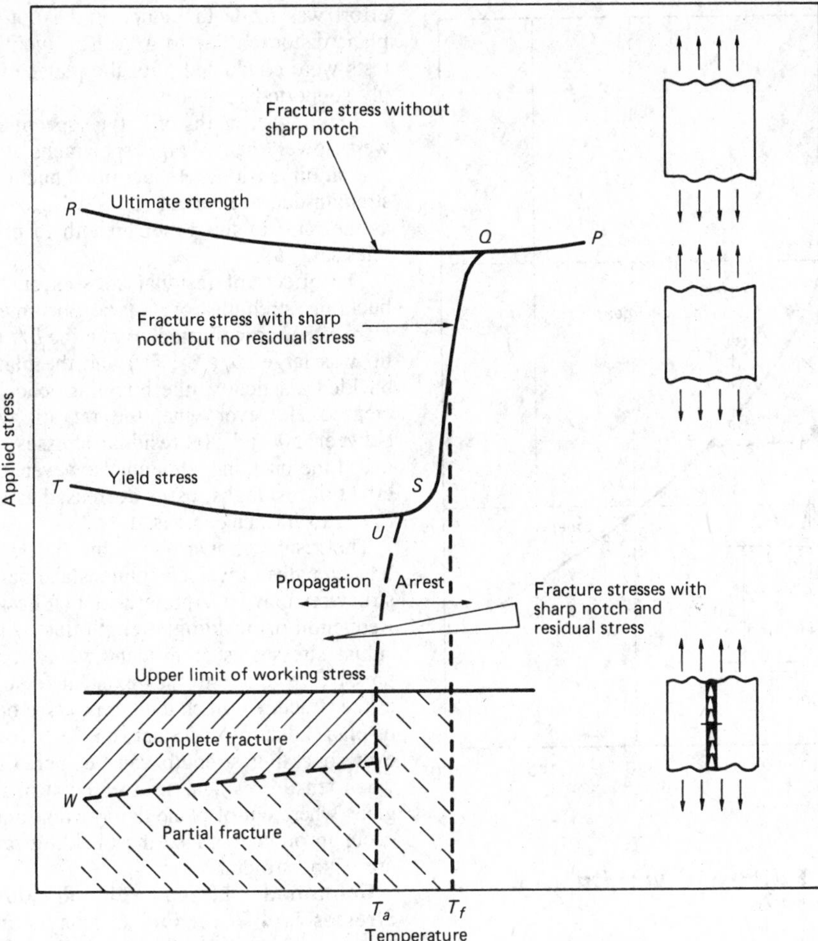

Fig. 56 Effects of stress relieving treatments on brittle-fracture characteristics of welded and notched wide plate specimens. (a) Effect of mechanical stress relieving. (b) Effect of thermal stress relieving. See Fig. 55 for the explanations of Curves *QST* and *UVW*. (Source: Ref 65)

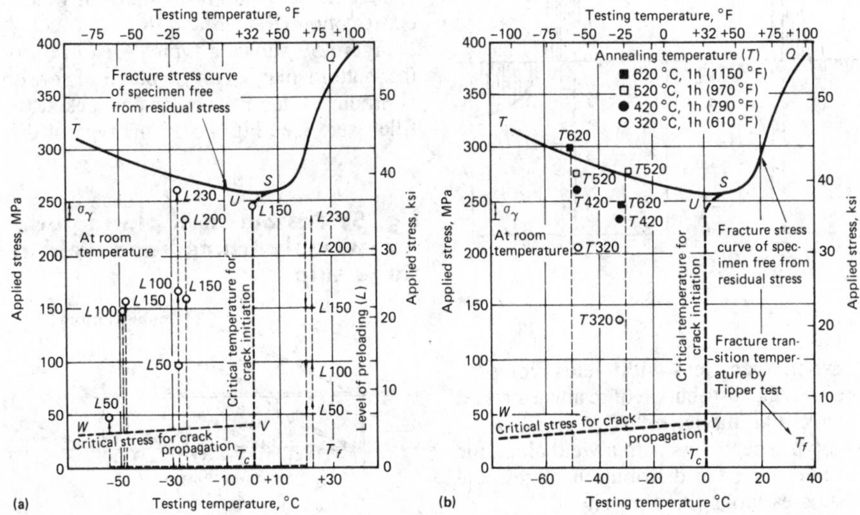

loading (Ref 72 and 73). The relationship between the slenderness ratio of specimen, L/r (L is the specimen length and r is the radius of gyration), and the ratio of the critical stress and yield strength of the material is given. Curves *PQR* and *ST* show distributions of residual stresses in the longitudinal direction of an as-welded column in the flange and web plate, respectively. Welded joints between the web and flange plates cause residual tensile stresses in areas near the weld and compressive stresses in outer areas of the flanges and in the web plate. Curve *AB* shows the buckling strengths of as-welded columns, while curve *CDE* shows the buckling strengths of stress-relieved specimens and specimens with tensile residual stresses in the outer areas of the flange. Curves *AB* and *CDE* are obtained by theoretical analysis; *DE* is the so-called Euler curve, and *AB* is obtained by considering the effect of residual stress on buckling strength. Buckling occurred in the weak axis of the columns. As-welded universal mill specimens had considerably lower critical buckling strength than did the other types of specimens, indicating that an unfavorable residual stress distribution can cause a substantial decrease in the buckling strength. Close agreement was obtained between the experimental data and theoretical values. If the column also has initial distortion, its buckling strength will be reduced further.

However, columns made from oxygen-cut plates normally have residual tensile stresses in the outer areas of the flange. From Fig. 57, it may be seen that in these columns the residual stresses assume a more favorable pattern and are similar to the stress-relieved columns. In this case, residual stresses that result from preparing plate by oxygen cutting and from subsequent welding into the column section will combine to counterbalance each other from the buckling strength viewpoint. The performance of these columns should be about equivalent to hot rolled columns.

Plate and Plate Structures Under Compressive Loading. Residual stresses in the direction parallel to the weld line are compressive in regions away from the weld, while they are tensile in regions near the weld, as shown in Fig. 6. Consequently, the effects of residual stresses on the buckling strengths of welded plate and plate structures are complex. Residual stresses may reduce the buckling strengths under some conditions, but they may increase the buckling strengths under some other conditions.

Fujita and Yoshida (Ref 74) conducted an analytical and experimental study of the

Fig. 57 Effects of residual stresses on buckling strength of columns. Stress relieved specimens (○) and specimens with tensile residual stresses in the outer areas of the flanges (□). (Source: Ref 66)

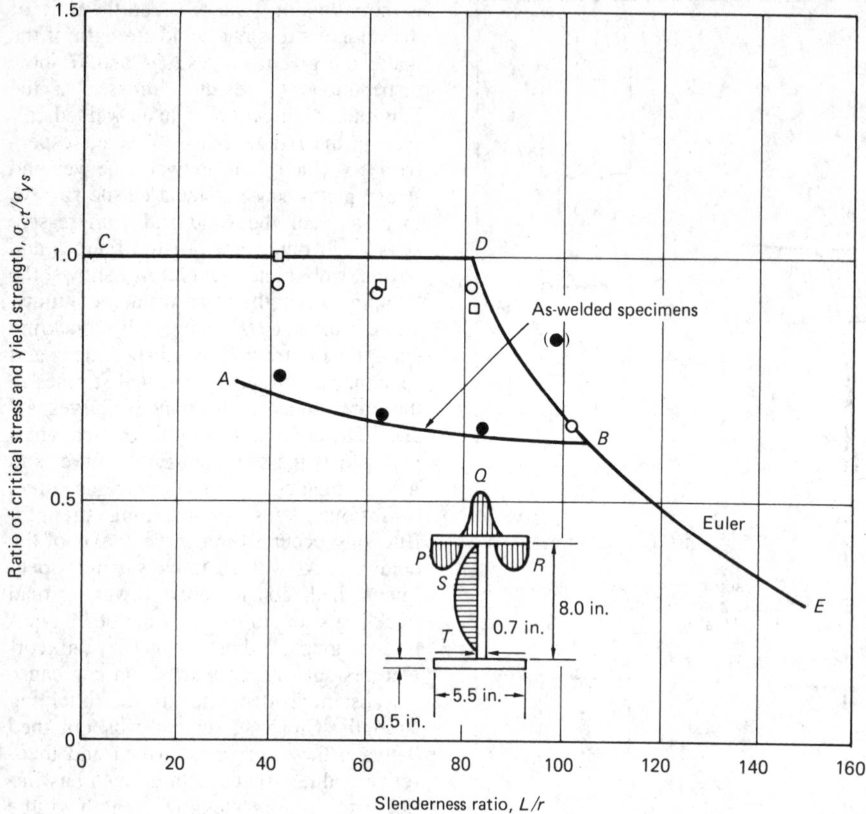

Fig. 58 Steel panel specimens used by Fujita and Yoshita. (a) Supported span, V-specimen. (b) E-specimen. (c) C-specimen. (Source: Ref 74)

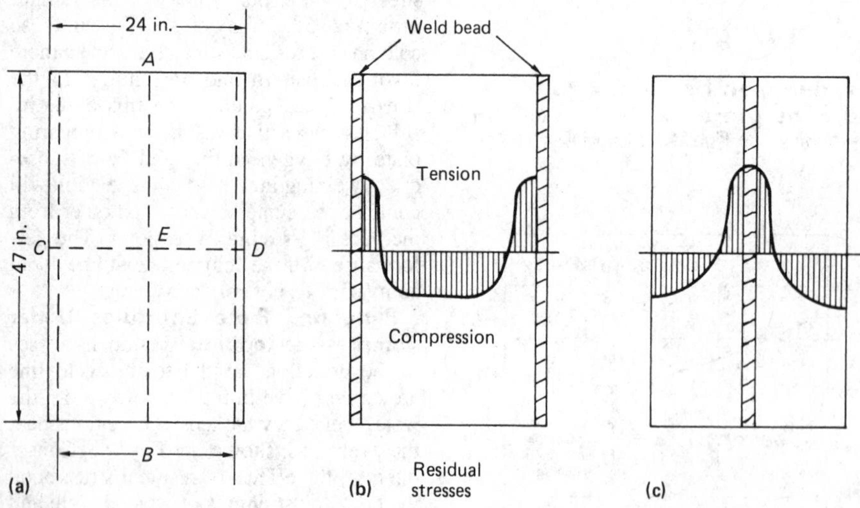

effects of residual stresses on the buckling strengths of welded panel structures. Figure 58 shows the three types of specimens used in the experiment:

- V-type specimens with no residual stress
- E-type specimens with welds along the edges; the distribution of residual stresses looks like the letter "E"
- C-type specimens with a weld along the center line; the distribution of residual stresses looks like the letter "C"

The specimens were 47 in. long, 24 in. wide, and 0.32 to 0.56 in. thick, and they were made of low-carbon steel. Special effort was made to reduce initial out-of-plane distortion due to welding. Buckling tests were conducted with the plates simply supported.

Buckling strengths of E-type specimens were lower than V-type specimens with similar dimensions. Reduction of buckling strength due to residual stresses increased as the ratio of the panel breadth to plate thickness (b/t).

The effects of residual stresses on the buckling strengths of C-type specimens were rather complicated. When the b/t ratio was large ($b/t > 70$) and the plates buckled elastically, the buckling load increased. However, when the b/t ratio was between 50 and 70, residual stresses reduced the buckling strength. However, the effect of residual stresses decreased as the b/t ratio further decreased.

The results shown above on the buckling strengths of welded columns and panel structures may be summarized as follows. Reduction of buckling strength due to residual stresses is significant when compressive residual stresses exist in regions of the structure, which tend to produce out-of-plane distortion rather easily. However, buckling strength can be increased when tensile residual stresses exist in regions where out-of-plane deformation must occur in order to cause the buckling failure of the structure.

Combined Effects of Residual Stresses and Distortion. In many structures, both residual stress and distortion exist, and they create complex effects on the buckling strength of welded structures. For example, corrugation damage of the bottom shell plating of a welded ship occurred in some ships having transversely framed bottom structures of welded construction (Ref 75 and 76).

Figure 59 shows a typical distortion of the bottom plate. Out-of-plane distortion is produced due to angular changes at the fillet weld (see Fig. 43). This type of dis-

Fig. 59 Bottom shell plating of a transversely framed and welded cargo ship

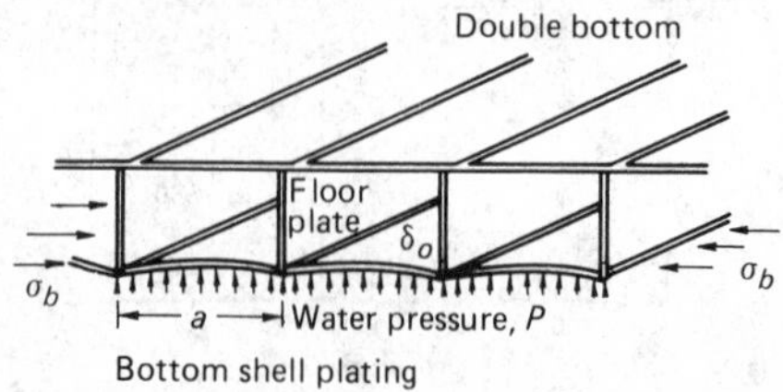

Fig. 60 Change in skin stress distribution and deflection curve of bottom shell

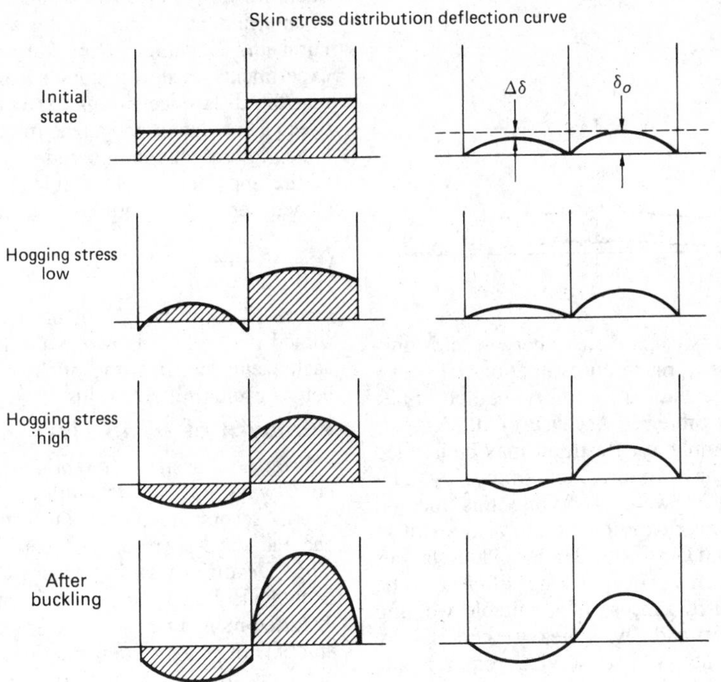

tortion is often called "hungry horse," because the outside appearance of the ship hull resembles the body of a hungry horse. Initial distortion and residual stresses decrease the buckling strength of the bottom plate, which is subjected to water pressure (P) and the axial compressive stress (σ_b). Compressive stresses are produced in the bottom structure when the ship is on the crest of a long wave, which is referred to as the hogging condition.

Figure 60 illustrates how stresses and plate deformation change when compressive stresses are applied to a structure with irregular distortion. Structures with irregular distortion buckle easier than structures with regular distortion. When compressive stresses are applied to a structure with irregular distortion, the following phenomena occur (Ref 77). While external compressive stresses are low, distortion increases in all spans. As external compressive stresses increase, the span with small distortion tends to buckle outward, while the span with large distortion tends to bend further inward. This results in an S-shaped buckling of the bottom plate.

Extensive analytical and experimental studies have been conducted on (1) the mechanisms producing distortion and residual stresses during welding, and (2) how these mechanisms affect the buckling strength of the bottom plate. Using results of the second type of research, proposals have been made of the maximum distortion allowable in the ship bottom plating (Ref 78).

Design and Procedures for Reducing Residual Stresses and Distortion

It is essential to understand how residual stress and distortion form and how they are affected by design and welding procedures in order to develop effective means for reducing residual stress and distortion. Thus, some of the discussion in earlier sections of this article should be useful in selecting appropriate procedures and designs. The following sections describe additional practical approaches to reducing residual stress and controlling distortion.

Reduction of Residual Stresses

Although the extent of the effects of residual stresses on the service behavior of welded structures varies significantly, depending on a number of conditions, it generally is preferable to have low residual stresses. Therefore, various precautions should be taken to reduce residual stress. This is particularly important when welding heavy sections.

Effects of Amount of Weld and Welding Sequence. Because residual stress and distortion are caused by thermal strains produced by welding, a reduction in the amount of weld metal usually results in reduction of residual stresses and distortion. For example, the use of a double-V-groove instead of a single-V-groove results in a reduction of weld metal. Bevel angles and the smallest root openings possible that will not restrict accessibility should be used. The effect of welding sequences, particularly block welding, was discussed previously.

Design for Control of Distortion

The most economic design for a welded fabrication is one that requires the fewest number of parts and a minimum of welding. Such a design also assists in reducing distortion. Type of joint preparation is particularly important for butt welds. Joint design that requires a minimum amount of weld metal is recommended for reducing transverse shrinkage. The amount of angular change is strongly influenced by the ratio of the weld metal in the top and the bottom side of the plate.

With fillet welded joints, effort should be made to use minimum sizes of welds as dictated by strength considerations. Fillet welds are widely used for fabricating stiffened panel structures. The degree of out-of-plane distortion is determined not only by the amount of the weld metal, but also by plate thickness and frame spacing. Reduction in frame spacing results in a reduction in the amount of distortion and an increase in the buckling strength of the panel; however, it also results in an increase in the weight of the structure. Careful analysis, considering both the amount of distortion and its effect on buckling strength, is needed in the initial design stage.

Careful analysis during the design stage is essential in preventing buckling distortion that often occurs in welded structures using thin plate. Because the amount of buckling distortion is significant, buckling distortion is best controlled through prevention by proper selection of plate thickness, size of the free panel, and the amount of weld heat input.

Assembly procedures also have various effects on the amount of distortion, especially in the fabrication of large, complex structures. Experience has shown that success may be achieved by adopting one of these assembly methods:

- Estimating the amount of distortion likely to occur during welding and then assembling the job with the members preset to compensate for the distortion

Fig. 61 Presetting for fillet and butt welds

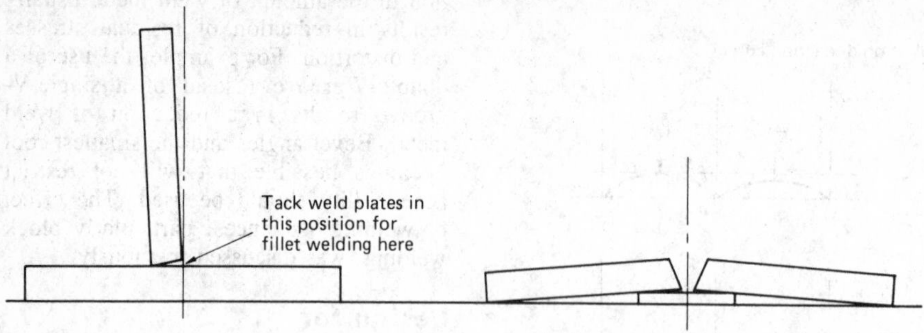

- Assembling the job so that it is nominally correct before welding and by employing some form of restraint to minimize the welding distortion after welding

In the first method, parts have almost complete freedom to move during welding, and there is less residual stress than with the second method. On the other hand, it is a difficult method to apply, except on relatively simple fabrications. Therefore, work is divided into subassemblies that can be welded without restraint for fabrications comprising many parts. Subassemblies are then assembled and welded to complete the job; this final welding often has to be carried out under conditions of restraint.

Figure 61 demonstrates the use of the presetting method as applied to fillet and butt welds. The amount of preset required varies somewhat, according to plate thickness, plate width, and welding procedure. Therefore, it is advisable to establish the amount by experiment; otherwise, it is only possible to make an estimation.

The restrained assembly method is frequently preferred because of its comparative simplicity. Restraint may be applied by clamps, fixtures, or simply by adequate tack welding. While this method minimizes distortion, it can also result in high residual stresses. High residual stresses and the risk of cracking often can be reduced by the use of a suitable welding sequence and, with heavy sections, by preheating. Where service requirements demand removal of residual stress, a stress-relieving heat treatment must be applied after welding. To impose complete restraint may be undesirable, but by restraining movement in one direction and allowing freedom in another, the overall effect generally can be controlled (see Fig. 62). The clamping arrangement is designed to prevent angular distortion in a V-butt joint while permitting transverse shrinkage.

Fig. 62 Arrangement of clamp to prevent angular distortion while permitting transverse shrinkage

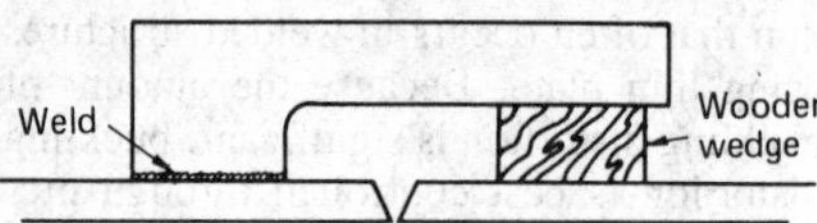

Fig. 63 Apparatus for welding T-joints subjected to plastic prestrain by bolting down both free ends. Round bar stock was used for the spacer.

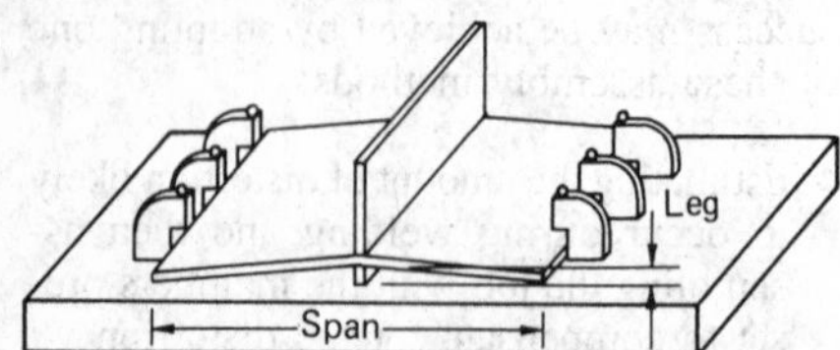

Elastic prespring is used to prevent angular change. The bottom plate is elastically bent, while stiffeners are fillet welded to the plate, as shown in Fig. 63. Angular changes after the removal of the restraint can be reduced significantly by this method. Figure 64 shows results obtained on steel weldments (Ref 79). Values of stresses on the plate surfaces required to produce zero angular change as functions of the thickness of the plate (flange) and the leg length of fillet welds are shown. Ref 1 presents data on the amounts of elastic prespring needed to minimize angular changes in aluminum weldments.

Preheating may be used to reduce angular change in a fillet weld. Figure 65 shows the effect of preheating on angular change in a steel fillet weld under different welding conditions and plate thicknesses (Ref 80).

Intermittent Fillet Welding. A fillet joint may be welded by use of continuous fillet welding or intermittent fillet welding. Multiple-pass welding is an effective method of reducing longitudinal distortion, as shown in Fig. 49. However, reduction of angular change due to the use of intermittent welding is not as drastic. Hirai and Nakamura (Ref 51) conducted experiments on steel weldments and found that the difference in angular change between a continuous and an intermittent weld is negligible when compared on the basis of the apparent weight of electrode consumed per weld length (w_a) as follows:

$$w_a = w \frac{l}{l + p} \qquad \text{(Eq 40)}$$

where w is the weight of electrode consumed per weld length; l is the length of each weld; and p is the pitch or interval between intermittent welds.

Removal of Distortion

With new designs or fabrications, it is not always possible to control distortion within acceptable limits. Distortion may become excessive, despite careful planning. Distortion can be removed by producing adequate plastic deformation in the distorted member or section. The required amount of plastic deformation can be obtained by the thermal or mechanical methods.

Thermal straightening, or flame straightening, has been used successfully in the shipbuilding industry to remove distortion. The area to be straightened is heated to between 1100 to 1200 °F and then quenched with a water spray; repeated applications of heat in specific areas in a selected sequence or pattern normally are needed to straighten a distorted member or structure. The patterns are usually spot or linear heating techniques.

Pressing. Distorted members can be straightened in a press if the members can

Fig. 64 Skin stress versus flange plate thickness required for obtaining zero angular change due to elastic prespringing. Source: Ref 79

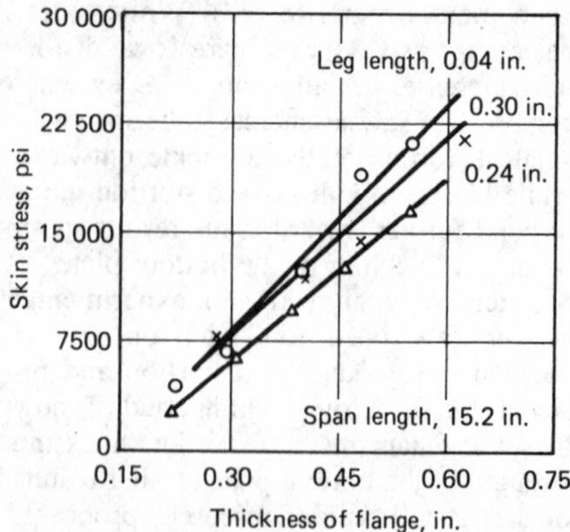

Fig. 65 Effect of preheating on angular change of T-fillet welded joint. *I*, welding current; *h*, thickness of horizontal plate (mm); *v*, welding speed

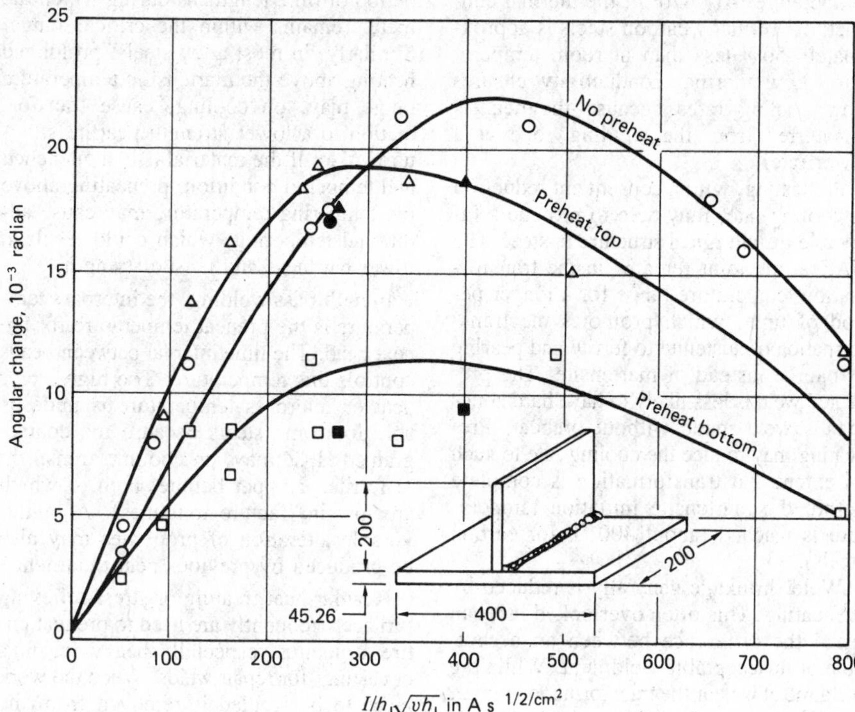

be moved and if the press is large enough to handle the parts. Heat may or may not be required for straightening.

Other Methods. Jacking is similar to pressing in that distortion is removed through the application of pressure with or without added heat. Although generally not approved, local hammering of heated areas is sometimes used for straightening.

Flame straightening, which is a relatively simple method and requires a minimum of equipment, has been widely used, especially for low-carbon steel structures. However, material degradation is possible with flame straightening of high-strength steel structures.

Thermal and Mechanical Treatments of Weldments

To reduce residual stresses and distortion, various thermal and mechanical treatments often are performed on weldments. They include preheat, peening, postweld thermal treatments, proofstressing, etc. Not only do these treatments reduce residual stresses and distortion, but they also may change the metallurgical properties of weldments.

Thermal and mechanical treatments are necessary to maintain or restore the properties of base metal affected by the heat of welding (Ref 2). Weld-deposit properties also may be improved or modified by thermal treatment. Additionally, these treatments may be necessary to relieve stresses and to produce the desired microstructure in the base metal and filler metal. The extent of the changes in the base metal, which determines the subsequent thermal treatment, depends on the temperature to which the metal is subjected, the period of exposure, the composition of the metal, and the rate of cooling. Other factors include joint design, welding process, welding procedure, and welding variables.

Included among the many properties controlled or improved by suitable thermal or mechanical treatments are distortion during welding, reduction of stresses that could seriously affect the service performance of a weldment, weldability (which may be improved considerably by preheat treatment), and improvement of dimensional stability and machinability. Optimum properties in the weldment can be expected when preweld and postweld treatments are correctly integrated with approved procedure, skillful performance, and good joint design.

In most metals, reduction of the metallurgical notch effect, resulting from abrupt changes in hardness or other microstructural discontinuities around welds, can be accomplished readily by correct postweld heat treatment. Sometimes, mechanical treatments are beneficial in superimposing counterstresses on normal welding stresses, which can result in the reduction of stresses to a level compatible with the service conditions required of the components. When welding carbon and alloy steels in the heat treated condition, complete residual stress control is not practical in some cases because preheating and postheating must be below the heat treating temperature used in producing the base metals.

On quenched and tempered steel where a limited stress-relieving temperature can be used, machinability can be improved by using a tempering bead or weld. This weld bead is applied to the last weld bead at the base metal. The application of this arc energy provides for a tempering treatment of the last weld bead and the HAZ of the base metal.

The resistance of the components to propagation of cracks, especially in the HAZ adjacent to the weld, often is directly related to correct welding procedures, which include preweld and postweld heat treatments. Improved mechanical properties in the weld and HAZ adjacent to the weld, which are due to thermal and mechanical treatments, result in a greatly increased resistance to crack propagation. Postweld heat treatment often greatly improves corrosion resistance of the weldment.

Code Requirements

Thermal treatments are specified for certain types of weldments when a weldment must be constructed in accordance with the recommendations of a code, such as the American Society of Mechanical Engineers (ASME) Boiler and Pressure Vessel Code. Codes prepared by technical organizations or regulatory bodies are based on existing evidence that indicates when thermal treatment is necessary or that the weldment can withstand service conditions in the as-welded condition. These are codes of minimum requirements. The fabricator should employ other treatments as experience dictates to produce a quality weldment, in addition to the thermal treatment specified.

Pertinent codes and standards are listed in the Selected References at the end of this article. Others are listed in the article "Codes, Standards, and Inspection" in this Volume. It should be noted however, that these documents are constantly revised and updated; the issuing organization should be consulted for the latest editions.

Most states have adopted one or more of these codes or standards for control and regulation of new construction installed under their jurisdiction. The fabricator must be knowledgeable in all laws and regulations that apply to the fabrication. For pressure vessel installation, the State Industrial Commission should be contacted for applicable rules concerning weld repairs on pressure vessels.

Preheating

Preheating involves raising the temperature of the base metal or a section of the base metal above the ambient temperature before welding. Preheat temperatures may be as low as 79 °F when welding outdoors in winter to as high as 1200 °F when welding ductile cast iron, or 600 °F when welding highly hardenable steels. Recommendations for preheating and postheating various grades of steel are included in publications such as *Weldability of Steels,* published by the Welding Research Council, and in the article "Arc Welding of Hardenable Carbon and Alloy Steels" in this Volume.

In many operations, the temperature to which the base metal is heated must be carefully controlled. The best means of control is to heat the part in a furnace held at the desired temperature, or to use electric induction coils or electric resistance heating blankets. On thin-walled materials, radiant lamps or a hot air blast can be used. In these methods, temperature indicators are attached to the part being preheated. If these methods are not practical, there are many other methods that may be used for measuring the temperature, such as surface thermocouples, magnetically attached bimetal surface thermometers, colored chalks that change color at known temperatures, and pellets that melt at a predetermined temperature. When using torches for preheating, localized overheating must be prevented, as well as deposits of incomplete combustion on the surfaces of the joints or areas to be welded.

Preheating is an effective means of reducing weld-metal and base-metal cracking. Preheating generally improves weldability, with two major beneficial effects. It retards the cooling rates in the weld metal and heat-affected base metal and reduces the magnitude of shrinkage stresses. However, when welding quenched or age-hardened materials, the effects of preheating can be detrimental unless controlled within allowable limits.

Cooling rates usually are faster for a weld made without preheat. The higher the preheating temperature, the slower the cooling rates after the weld is completed. The temperature gradient is reduced, and in the case of iron, the thermal conductivity is decreased. At 1100 °F, the thermal conductivity of iron is 50% less than at room temperature. At 1470 °F, the thermal conductivity of many carbon steels is approximately 50% less than at room temperature. Low thermal conductivity ensures slow cooling rates because the heat is transferred from the welding zone at a lower rate.

Preheating, with a consequent reduction in cooling rate, may help to provide a favorable metallurgical structure in steel. The HAZ at the joint remains in the transformation temperature range for a longer period of time, which promotes the transformation of austenite to ferrite and pearlite or bainite instead of martensite. The preheated weld is less likely to have hard zones than a weld made without preheat. Preheating may reduce the cooling rate to such an extent that transformation is complete before the martensite initiation temperature is reached (about 400 °F for carbon steels).

Weld shrinkage generally is reduced by preheating. This often overlooked fact can mean the difference between an acceptable or unacceptable weldment. While the weld metal is near the transformation range, it has little strength. Consequently, it cannot exercise any appreciable force on the joint members. Plastic flow experienced by the weld metal adjusting to the changing dimensions is similar to hot working with no loss of ductility. The weld metal and heat-affected base metal become stronger as the temperature decreases, and the shrinkage forces naturally become greater. Plastic flow becomes similar to cold working, which results in a progressive loss of ductility. Actual measurements across welds during cooling have shown that 30% less total contraction occurred in joints preheated to 400 °F compared with similar joints welded at room temperature.

However, when an area being welded is under severe restraint, localized preheat may increase the amount of shrinkage and cause cracking. Preheating to the higher temperatures, above 600 °F, reduces the yield strength of the material, which relieves shrinkage stresses. Hot cracks in the weld metal or in the base metal close to the weld may result if these stresses are forced to occur only in the highly heated zone close to the weld without preheat.

Because detrimental effects may result under certain conditions, preheat must be used with caution. Welding of stainless steel containing 16% Cr is improved by preheating, which prevents shrinkage stresses acting on the coarse-grained weld metal and HAZ at low temperatures causing embrittlement. When preheat is too high, cooling rates become slower and the period of time lengthens during which the metal remains within the critical range. Similarly, in most alloy steels, prolonged heating above the martensitic temperature range may, on cooling, cause transformation to a lower strength pearlite structure. Also, if the material is in a quenched and tempered condition, preheating above the tempering temperature may cause additional tempering, which could result in lower hardness and tensile strength.

In multipass welding, the interpass temperature is the preheat temperature for the next bead. The time interval between beads controls this temperature. Too high a preheat or interpass temperature is undesirable for some steels because the coarse-grained HAZ may, on cooling, transform to ferrite or upper bainite, both of which are low in fracture toughness. An unfavorable alteration of properties may also be produced by previous heat treatment.

Regular heat treating or stress-relieving furnaces frequently are used to preheat entire structures, especially heavy forgings or castings for repair welds. When the work piece to be welded is removed from the furnace, protection against rapid cooling should be provided by insulating with paper or blankets. Plate fabricating shops often use natural or manufactured gas as a fuel for the local preheating of weldments, as the gas is relatively clean and convenient.

Electrical strip heaters may be used for general and local preheating, for repairs in castings and pipe, and for pressure vessel joints, especially for field welding. To avoid the danger of shock to workers, strip heaters must be insulated properly at the terminals.

The most common method of preheating pipe joints for welding is induction heating, using 60-Hz transformers. These transformers may be operated manually or with automatic controls that maintain any temperature range or follow a predetermined time-temperature program.

Postweld Thermal Treatments

Stress-relief heat treatment is defined as the uniform heating of a structure to a suitable temperature, holding at this temperature for a predetermined period of time, followed by uniform cooling. Heat treatments that involve changes in grain structure and dimensional changes may be detrimental to a part; therefore, stress-relief heat treatment usually is performed below the critical range.

In the ASME Boiler and Pressure Vessel Code, minimum temperatures and times are specified at which postweld heat treatment is to be performed for welded power boilers and unfired pressure vessels. These vary with the type or grade of steel involved.

Thermal Stress Relief. Residual stresses resulting from welding may be reduced by postweld thermal stress-relief heat treatment. Stresses are reduced to a level just below the yield point of the material, at the temperature of the stress-relief treatment. Residual stress remaining in a material after thermal stress relief depends on the rate of cooling. Uneven cooling from stress relief to ambient temperatures may undo much of the value of the heat treatment and result in additional stresses within the weldment.

The percentage relief of internal stresses depends on steel type, composition, or yield strength. The effects of varying time and temperature are shown in Fig. 66. The temperature reached during stress-relief treatment has a far greater effect in relieving stresses than the length of time the specimen is held at that temperature. The closer the temperature is to the critical or recrystallization temperature, the more effective it is in removing residual stresses, if heating and cooling cycles are employed.

When a thermal stress-relief treatment is used to reduce residual stresses, other important properties must be taken into consideration. Microstructure and tensile and impact strength are among the properties affected by stress-relief treatment. Therefore, a temperature must be selected that will develop the desirable properties in the steel while providing the maximum stress relief (Fig. 66).

Controlled Low-Temperature Stress Relief. Originally, this process was developed to reduce residual stresses in weldments too large to be stress relieved in a furnace. The metal on either side of a welded joint is heated to a temperature of 350 to 400 °F, while the weld itself is kept relatively cool. The progression of these heated bands of metal, parallel to the weld and adjacent to each side, results in a traveling zone of thermal expansion in the base metal, and a reciprocal tensile stress in the weld. The theory of this procedure is that two zones of compression are expanded thermally, thereby increasing the tensile stress in the weld beyond the yield. The process operates in the temperature range associated with possible strain aging, but available data indicate that the medium-strength ductile materials are not adversely affected.

Fig. 66 Effect of temperature and time on stress relief.
Time and temperature, 4 h

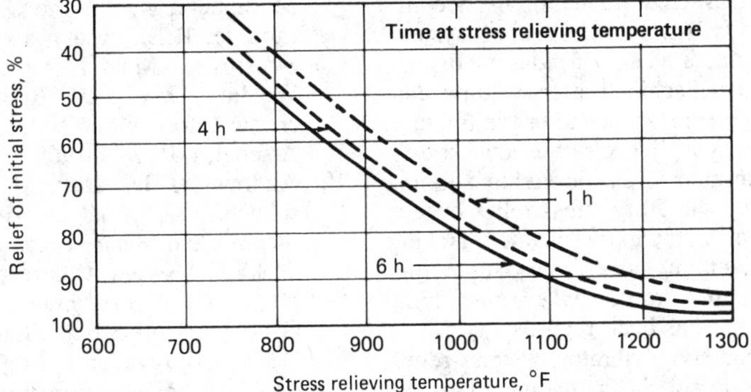

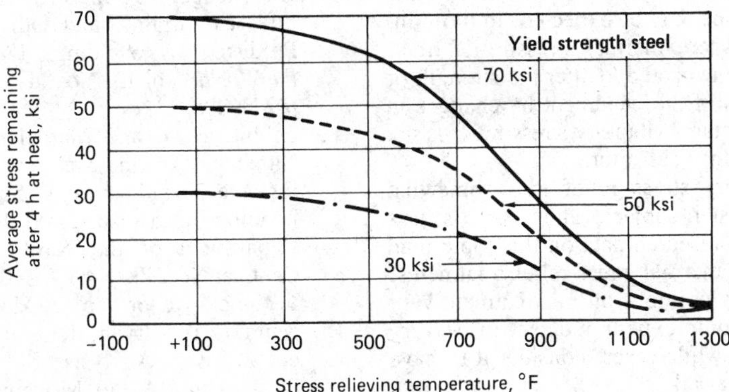

Peening

Peening has been used by the welding industry for many years, but code requirements and regulations governing this procedure are based on opinion rather than on scientific data; there is no practical method for measuring the effect of peening.

Various specifications and codes require that the first and last layers of a weld should not be peened. Vague statements such as "thoroughly peened" and "the work shall not be overpeened" do little to clarify a situation that is already confusing. There is a firm foundation for the requirement that the first layer shall not be peened. Peening of the first layer may actually pierce the weld or displace the member sideways. Peening of the last layer is prohibited largely because it is believed that cold working may injure the weld metal, as there is no subsequent application of heat to anneal it. In peening applications, the following special precautions may be necessary:

- Work hardening should be considered when certain AISI 300 series stainless steels are involved.
- Hot shortness may preclude hot peening of certain bronze alloys.
- AISI 400 series stainless steels have relatively poor notch ductility in the as-welded condition; utmost care should be exercised if peening is attempted.
- Relative elongation values or ductility of welds and base metals should be considered before employing the peening process.

Peening equipment should be selected with care. The hammer, pneumatic tools, and other related equipment should be sufficiently heavy in striking force to be effective without producing excessive work hardening. However, they should not be so heavy as to involve bending moments or as to produce cracks in the weld.

To be effective, peening should be employed on each weld bead or layer (except the first and last). Effectiveness decreases as the thickness of the bead or layer increases. Peening is not suitable for deposits of $^1/_4$ in. or more, except in special applications where the rigidity or weight of the weldment permits the use of heavy blows. When the metal subsequently cools and contracts, the stress falls below the yield.

When peening is used correctly, a partial reduction in the longitudinal stress of butt welds is achieved (Ref 81). Longitudinal weld stresses are considered to be of primary importance, because they attain values of up to the yield strength in tension. Others question the validity of low-temperature stress relief, claiming that, in restrained plate, both longitudinal and transverse residual stresses approach the yield point of the material. Reductions in transverse residual stresses ranging up to 60% have been reported, as well as considerable reduction in the longitudinal stresses. Consequently, low-temperature stress relief is an inexpensive means of obtaining a substantial reduction of shrinkage stress in ductile materials.

In most materials, low-temperature stress-relief treatment does not improve metallurgical properties of the weld metal and the HAZ, and treatment should not be regarded as a substitute for postweld thermal treatment when such treatment is required to provide ductility and notch toughness.

Peening of resistance welds to reduce distortion of sheet metal parts usually is done with a mechanically operated trip hammer, using flat, wide dies for both the hammer and supporting anvil. Because the welds must be backed by rigid anvils placed immediately under the point of impact of the hammer blow, parts usually are moved through a stationary hammer. Metal adjacent to the welds is usually peened, as well as the weld itself.

Proofstressing

Investigations indicate that uniform heavy loading of weldments may tend to decrease longitudinal residual stresses by the amount of loading (see Fig. 54). Using hydrostatic loading, cylindrical and spherical pressure vessels can be proofstressed readily. In cylindrical vessels, hydrostatic loading produces circumferential stress approximately double the longitudinal stress, so that relief of residual stress in circumferential welds is only half the relief in longitudinal welds.

Proofstressing of weldments of more complex geometry usually presents considerable difficulty and therefore is seldom attempted. Proofstressing often serves as a final fabricating operation by smoothing out irregularities formed by plastic straining, so that under subsequent service conditions, the strains will be completely elastic and within safe limits.

Proofstressing is based on the plastic straining of the material. Therefore, a potential cracking hazard may exist where the materials involved are notch sensitive or lack notch ductility at the proofstressing temperature.

Vibratory Stress Relief

Mechanical energy in the form of low- and high-frequency vibrations has been used to relieve residual stresses in weldments. The procedures used vary, but generally consist of an oscillating or rotating wave generator that is mechanically coupled to the part to be stress relieved. The part may be vibrated at one of its natural (resonant) frequencies, although this approach may be difficult to use in a part of complex shape. For such a system to be effective, it must produce plastic yielding in the region to be stress relieved; however, the yielded region may be quite local and may already be at a residual stress level at or near the yield point prior to treatment. The yielded region also may be moved progressively along the part to achieve full coverage. It may not be possible or even necessary to eliminate all residual stresses by this method, but it may be possible to reduce the magnitude of peak stresses. If such a reduction is accomplished, a vibratory stress-relief system could provide a means of reducing distortion in parts machined after welding and could improve resistance to brittle fracture if triggered by high levels of residual stress. In some instances, plastic yielding is not achieved by vibratory stress relief due to the size or complexity of the part involved. Even in these cases, some improvement in distortion when machining after welding has been reported.

Unfortunately, vibratory stress-relief treatments do not change the metallurgical structure of welds or HAZ to which they are applied, and thus no improvement or change in mechanical properties of the treated zone may be expected. In those instances where thermal stress-relief treatments are expected to alter the strength or toughness of the weldment by changes in microstructure, vibratory stress relief is not a satisfactory substitute.

Vibratory stress relief is a somewhat controversial subject, and there is less published research data on this topic than exists for thermal stress-relief treatments. Moreover, opinions of this treatment vary greatly; some experts judge it to be very effective, while others consider it to have no effect at all.

REFERENCES

1. Masubuchi, K., *Analysis of Welded Structures—Residual Stresses, Distortion, and Their Consequences,* Pergamon Press, Oxford/New York, 1980
2. Chapter 6, Residual Stresses and Distortion, in *Welding Handbook,* Vol 1, 7th ed., American Welding Society, Miami, 1976
3. Volterra, V., Sur L'equilibre des corps elastiques multiplement connexes, *Ann. Ec. norm.,* Vol 3, 1907, p 401-517
4. Love, A.E.H., *Mathematical Theory of Elasticity,* 4th ed., Cambridge University Press, London, 1934
5. Taylor, G.I., The Mechanisms of Plastic Deformation of Crystals—Part I, Theoretical, *Proceedings of the Royal Society London A,* Vol 145, 1934, p 362-387
6. Orowan, E., Problems of Plastic Gliding, *Physical Society London,* Vol 52, 1940, p 8-22
7. Tall, L., The Strength of Welded Built-up Columns, Ph.D. dissertation, Lehigh University, 1961
8. Tall, L., Residual Stresses in Welded Plates—A Theoretical Study, *Welding Journal,* Vol 43 (No. 1), Research Supplement, 1964, p 10s-23s
9. Masubuchi, K., Simmons, F.B., and Monroe, R.E., Analysis of Thermal Stresses and Metal Movement During Welding, RSIC-820, Redstone Scientific Information Center, Redstone Arsenal, AL, July 1968
10. Andrews, J.B., Arita, M., and Masubuchi, K., Analysis of Thermal Stresses and Metal Movement During Welding, NASA Contractor Report NASA CR-61351, prepared for the G.C. Marshall Space Flight Center, Dec 1970 (available from the National Technical Information Service, Springfield, VA 22151)
11. Masubuchi, K., Report on Current Knowledge of Numerical Analysis of Stresses, Strains, and Other Effects Produced by Welding, *Welding in the World,* Vol 13 (No. 11/12), 1975, p 271-288
12. Hibbit, H.D. and Marcel, P.V., A Numerical Thermo-Mechanical Model for the Welding and Subsequent Loading of a Fabricated Structure, Department of the Navy, NSRDC Contract No. N00014-67-A-019-0006, Technical Report No. 2, March 1972
13. Muraki, T., Bryan, J.J., and Masubuchi, K., Analysis of Thermal Stresses and Metal Movement Dur-

ing Welding. Part I: Analytical Study and Part II: Comparison of Experimental Data and Analytical Results, *Journal of Engineering Materials and Technology,* ASME, Jan 1975, p 81-84, 85-91

14. Fujita, Y. and Nomoto, T., Studies on Thermal Elastic-Plastic Problems, First Report, *Journal of the Society of Naval Architects of Japan,* Vol 130, 1971, p 183-191
15. Ueda, Y. and Yamakawa, T., Analysis of Thermal Elastic-Plastic Stress and Strain During Welding, Document X-616-71, Commission X of the International Institute of Welding, 1971 (distributed by the American Welding Society)
16. Okerblom, N.O., *The Calculation of Deformations of Welded Metal Structures,* Her Majesty's Stationary Office, London, 1958
17. Kihara, H., Watanabe, M., Masubuchi, K., and Satoh, K., *Researches on Welding Stress and Shrinkage Distortion in Japan,* Vol 4 of the 60th Anniversary Series of the Society of Naval Architects of Japan, Tokyo, 1959
18. Rybicki, E.F., Schmueser, D.R., Stonesifer, R.B., Groom, J.J., and Mishler, H.W., A Finite Element Model for Residual Stresses and Deflections in Girth-Butt Welded Pipes, *Journal of Pressure Vessel Technology,* Vol 100, Aug 1978, p 256-262
19. Fujita, Y., Nomoto, T., and Hasegawa, H., Deformation and Residual Stresses in Butt-Welded Pipes and Spheres, Document X-963-80, Commission X of the International Institute of Welding, 1980 (distributed by the American Welding Society)
20. Masubuchi, K., Nondestructive Measurement of Residual Stresses in Metals and Metal Structures, RSIC-410, April 1965, Redstone Scientific Information Center, U.S. Army Missile Command, Redstone Arsenal, AL
21. Dally, J.W. and Riley, W.F., *Experimental Stress Analysis,* McGraw-Hill, New York, 1978
22. Perry, C.C. and Lissner, H.R., *The Strain Gage Primer,* McGraw-Hill, New York, 1962
23. Soete, W., Measurement and Relaxation of Residual Stresses, *Welding Journal,* Vol 28 (No. 8), Research Supplement, 1949, p 354s-364s
24. Rosenthal, D. and Norton, T., A Method for Measuring Residual Stresses in Plates, *Welding Journal,* Vol 24 (No. 5), Research Supplement, 1945, p 295s-307s
25. Ueda, Y., Fukuda, K., and Tanigawa, M., New Measuring Method of Three Dimensional Residual Stresses Based on Theory of Inherent Strain, *Transactions of JWRI,* Vol 8 (No. 2), 1979, p 89-96, Welding Research Institute of Osaka University, Osaka, Japan
26. Nagaraja Rao, N.R., Esatuar, F.R., and Tall, K., Residual Stresses in Welded Shapes, *Welding Journal,* Vol 39 (No. 3), Research Supplement, 1964, p 295s-306s
27. Burdekin, F.M., Local Stress Relief of Circumferential Butt Welds in Cylinders, *British Welding Journal,* Vol 10 (No. 9), 1963, p 483-490
28. Klein, K.M., Investigation of Welding Thermal Strains in Marine Steels, M.S. thesis, Massachusetts Institute of Technology, May 1971
29. Masubuchi, K. and Martin, D.C., Investigation of Residual Stresses by Use of Hydrogen Cracking, *Welding Journal,* Part I, Vol 40 (No. 12), Research Supplement, 1961, p 553s-563s, and Part II, Vol 45 (No. 9), Research Supplement, 1966, p 401s-418s
30. Adams, C.M., Jr. and Corrigan, D.A., Mechanical and Metallurgical Behavior of Restrained Welds in Submarine Steels, Final Report, Massachusetts Institute of Technology (AD-634 747)
31. Wahlfart, H., Schweisseigenspannungen, HTM 31, 1976
32. Papazoglou, V.J. and Masubuchi, K., Study of Residual Stresses and Distortion in Structural Weldments in High-Strength Steels, Second Technical Progress Report of Contract N00014-75-0469 from the Massachusetts Institute of Technology to the Office of Naval Research, Nov 30, 1980
33. Papazoglou, V.J., Analytical Techniques for Determining Temperatures, Thermal Strains, and Residual Stresses During Welding, Ph.D. thesis, Massachusetts Institute of Technology, May 1981
34. Hill, H.N., Residual Welding Stresses in Aluminum Alloys, *Metal Progress,* Vol 80 (No. 2), 1961, p 92-96
35. Hwang, J.S., Residual Stresses in Weldments in High-Strength Steels, M.S. thesis, Massachusetts Institute of Technology, Jan 1976
36. Robelotto, R., Lambase, J.M., and Toy, A., Residual Stresses in Welded Titanium and Their Effects on Mechanical Behavior, *Welding Journal,* Vol 47 (No. 7), Research Supplement, 1968, p 289s-398s
37. DeGarmo, E.P., Meriam, J.L., and Jonassen, F., The Effect of Weld Length upon the Residual Stresses of Unstrained Butt Welds, *Welding Journal,* Vol 25 (No. 8), Research Supplement, 1946, p 485s-486s
38. Gunnert, R., Method for Measuring Tri-Axial Residual Stresses, *Welding Research Abroad,* Vol 4 (No. 10), 1958, p 17-25
39. Masubuchi, K., Control of Distortion and Shrinkage in Welding, Welding Research Council Bulletin, No. 149, April 1970
40. Spraragen, W. and Ettinger, W.G., Shrinkage Distortion in Welding, *Welding Journal,* Vol 29 (No. 6 and 7), Research Supplement, 1950, p 292s-294s and 323s-335s
41. Capel, L., Aluminum Welding Practice, *British Welding Journal,* Vol 8 (No. 5), 1961, p 245-248
42. Matsui, S., Investigation of Shrinkage, Restraint Stresses, and Cracking in Arc Welding, Ph.D. thesis at Osaka University, 1964 (Japanese).
43. Kihara, H. and Masubuchi, K., Studies on the Shrinkage and Residual Welding Stress on Constrained Fundamental Joint, *Journal of the Society of Naval Architects of Japan,* Part I, Vol 95, 1954, p 181-195; Part II, Vol 96, 1955, p 99-108; Part III, Vol 97, 1955, p 95-104 (Japanese)
44. Masubuchi, K., Analytical Investigation of Residual Stresses and Distortions due to Welding, *Welding Journal,* Vol 39 (No. 12), Research Supplement, 1960, p 525s-537s
45. Kihara, H., Masubuchi, K., and Matsuyama, Y., Effect of Welding Sequence on Transverse Shrinkage and Residual Stresses, Report 24, Jan 1957, Tokyo: Transportation Technical Research Institute
46. Watanabe, M. and Satoh, K., Effect of Welding Conditions on the Shrinkage and Distortion in Welded Structures, *Welding Journal,* Vol 40 (No. 8), Research Supplement, 1961, p 377s-384s
47. King, C.W.R., in Transactions of the Institute of Engineers and Shipbuilders in Scotland, Vol 87, 1944, p 238-255
48. Guyot, F., A Note on the Shrinkage and Distortion of Welded Joints, *Welding Journal,* Vol 26 (No. 9), Research Supplement, 1947, p 519s-529s
49. Researches on Welding Procedures of Thick Steel Plates Used in the Construction of Large Size Ships, Report of the Shipbuilding Research Association of Japan, Sept 1959
50. Masubuchi, K., Ogura, Y., Ishihara,

Y., and Hoshino, J., Studies on the Mechanism of the Origin and the Method of Reducing the Deformation of Shell Plating in Welded Ships, *International Shipbuilding Progress,* Vol 3 (No. 19), 1956, p 123-133
51. Hirai, S. and Nakamura, I., Research on Angular Change in Fillet Welds, Ishikawajima Review, April 1956, p 59-68 (Japanese)
52. Taniguchi, C., Out-of-Plane Distortion Caused by Fillet Welds in Aluminum, M.S. thesis, Massachusetts Institute of Technology, 1972
53. Sasayama, T., Masubuchi, K., and Moriguchi, S., Longitudinal Deformation of a Long Beam Due to Fillet Welding, Document X-88-55, London: International Institute of Welding (distributed in the United States by the American Welding Society), 1955
54. Yamamoto, G., Study of Longitudinal Distortion of Welded Beams, M.S. thesis, Massachusetts Institute of Technology, May 1975
55. Masubuchi, K., Buckling-type Deformation of Thin Plate Due to Welding, in *Proceedings of Third International Congress for Applied Mechanics of Japan,* 1954, p 107-111
56. Watanabe, M. and Satoh, K., Fundamental Studies of Buckling of Thin Steel Plate Due to Bead-Welding, *Journal of the Japan Welding Society,* Vol 27 (No. 6), 1958, p 313-320
57. Terai, K., Matsui, S., and Kinoshita, T., Study on Prevention of Welding Deformation in Thin Plate Structures, Kawasaki Technical Review, Kawasaki Heavy Industries, Ltd., Iobe, Japan, No. 61, Aug 1978, p 61-66
58. Terai *et al.*, unpublished document
59. Masubuchi, K., Residual Stresses and Distortion in Welded Aluminum Structures and Their Effects on Service Performance, Welding Research Council Bulletin 174, July 1972
60. Masubuchi, K. and Papazoglou, V.J., Analysis and Control of Distortion in Welded Aluminum Structures, *Transactions of the Society of Naval Architects and Marine Engineers,* Vol 86, 1978, p 77-100
61. Wells, A.A., The Brittle Fracture Strength of Welded Steel Plates, Quarterly Transactions of Institute of Naval Architects, Vol 48 (No. 3), 1956, p 296-326
62. Kihara, H. and Masubuchi, K., Effect of Residual Stress on Brittle Fracture, *Welding Journal,* Vol 38 (No. 4), Research Supplement, 1959, p 159s-168s
63. Hall, W.J., Kihara, H., Soete, W., and Wells, A.A., *Brittle Fracture of Welded Plates,* Prentice-Hall, Englewood Cliffs, NJ, 1967
64. Kammer, P.A., Masubuchi, K., and Monroe, R.E., Cracking in High-strength Steel Weldments—A Critical Review, DMIC Report 197, Defense Metals Information Center, Battelle Memorial Institute, Columbus, OH, Feb 1964
65. Kihara, H., Masubuchi, K., Iida, K., and Ohba, H., Effect of Stress Relieving on Brittle Fracture Strength of Welded Steel Plate, Document X-218-59, 1959, London: International Institute of Welding (distributed in the United States by the American Welding Society)
66. Gurney, T.R., *Fatigue of Welded Structures,* Cambridge University Press, London, 1968
67. Munse, W.H., *Fatigue of Welded Steel Structures,* Welding Research Council, 1964
68. Masubuchi, K., *Materials for Ocean Engineering,* Massachusetts Institute of Technology Press, Cambridge, 1970
69. Tetelman, A.S. and McEvily, A.J., Jr., *Fracture of Structural Materials,* John Wiley & Sons, New York, 1967
70. Berry, W.E., Stress-Corrosion Cracking—a Nontechnical Introduction to the Problem, DMIC Report 144, Battelle Memorial Institute, Columbus, OH, Jan 1961
71. Uhlig, H.H., *Corrosion and Corrosion Control, and Introduction to Corrosion Science and Engineering,* John Wiley & Sons, New York, 1963
72. Kihara, H. and Fujita, Y., The Influence of Residual Stresses on the Instability Problem, in colloquium on the *Influence of Residual Stresses on Stability of Welded Structures and Structural Members,* International Institute of Welding, London (distributed in the United States by the American Welding Society), 1960
73. Tall, L., Huber, A.W., and Beedle, L.S., Residual Stress and the Instability of Axially Loaded Columns, in colloquium on the *Influence of Residual Stresses on Stability of Welded Structures and Structural Members,* International Institute of Welding, London (distributed in the United States by the American Welding Society), 1960
74. Fujita, Y. and Yoshida, K., Plastic Design in Steel Structures (4th Report), Influence of Residual Stresses on the Plate Instability, *Journal of the Society of Naval Architects of Japan,* Vol 115, 1964, p 106-115
75. Murray, J.M., Corrugation of Bottom Shell Plating, Transactions of the Institute of Naval Architects, London, Vol 94, 1954, p 229-250
76. Investigation on the Corrugation Failure of Bottom Plating of Ships, Report No. 19 of the Shipbuilding Research Association of Japan, Tokyo, June 1967
77. Akita, Y. and Yoshimoto, K., Effect of Bottom Unfairness on Bottom Plate Buckling, *Journal of the Society of Naval Architects of Japan,* Vol 95, 1954
78. Yoshiki, M., Kanazawa, T., and Ando, N., A Study on the Strength of Ship's Bottom Platings, Proceedings of the Second Congress on Theoretical and Applied Mechanics, New Delhi, Oct 1956
79. Kumose, T., Yoshida, T., Abe, T., and Onoue, H., Prediction of Angular Distortion Caused by One-Pass Fillet Welding, *Welding Journal,* Vol 33 (No. 10), 1954, p 945-956
80. Watanabe, M., Satoh, K., Morii, H., and Ichikawa, I., Distortion in Web Plate of Welded Built-up Girder Due to Welding of Stiffeners and the Methods for Decreasing it, *Journal of the Japan Welding Society,* Vol 26, 1957, p 591-596
81. McKinsey, C.R., Effect of Low-Temperature Stress Relieving on Stress-Corrosion Cracking, *Welding Journal*, Vol 33 (No. 4), 1954, p 161s-166s

SELECTED REFERENCES

- ASME Boiler and Pressure Vessel Code, Section I, III, and VIII Div. 1 and 2, American Society of Mechanical Engineers
- Code for Pressure Piping, ANSI B31.1 to B31.8, American National Standards Institute
- Fabrication Welding and Inspection; and Casting Inspection and Repair for Machinery, Piping and Pressure Vessels in Ships of the United States Navy, MIL-STD-278 (Ships), Navy Department, Washington, DC
- General Specification for Ships of the United States Navy, Spec. S9-1, Navy Department, Washington, DC
- Gurney, T.R., *Fatigue of Welded Steel Structures,* Cambridge University Press, London, 1968
- Hall, W.J., Kihara, H., Soete, W., and Wells, A.A., *Brittle Fracture of Welded Plates,* Prentice-Hall, Englewood Cliffs, NJ, 1967

- Masubuchi, K., *Analysis of Welded Structures—Residual Stresses, Distortion, and Their Consequences,* Pergamon Press, Oxford/New York, 1980
- Masubuchi, K., Control of Distortion and Shrinkage in Welding, Welding Research Council Bulletin 149, 1970
- Masubuchi, K., Residual Stresses and Distortion in Welded Aluminum Structures and Their Effects on Service Performance, Welding Research Council Bulletin 174, 1974
- Munse, W.H., *Fatigue of Welded Steel Structures,* Welding Research Council, 1964
- Osgood, W.R., *Residual Stresses in Metal and Metal Construction,* Reinhold Publishing, New York, 1954
- Rules for Building and Classing Steel Vessels, American Bureau of Shipping
- Structural Welding Code, AWS D1.1, American Welding Society
- Tall, L., Ed., *Structural Steel Design,* 2nd ed., Ronald Press Co., New York, 1974
- Treuting, R.G., Lynch, J.J., Wishart, H.B., and Richards, D.G., *Residual Stress Measurements,* American Society for Metals, 1952
- United States Coast Guard Marine Engineering Regulations and Materials, Spec CG-115, United States Coast Guard, Washington, DC

Thermal Cutting

By Rosalie Brosilow
Editor
Welding Design & Fabrication

THERMAL CUTTING processes differ from mechanical cutting (machining) in that the cutting action is initiated by chemical reaction (oxidation) or by melting (heat from arc). All cutting processes result in the severing of or removal of metals.

Oxygen cutting is accomplished through a chemical reaction in which preheated metal is cut, or removed, by rapid oxidation in a stream of pure oxygen. Typical oxygen cutting processes are oxyfuel gas, oxygen lance, chemical flux, and metal powder cutting. Oxyfuel gas cutting (OFC) and its modifications, chemical flux cutting (FOC) and metal powder cutting (POC), which are used to cut oxidation-resistant materials, are discussed in this article.

Arc cutting melts metal by heat generated from an electric arc. Because extremely high temperatures are developed, arc cutting can be used to cut almost any metal. Modifications of the process include the use of compressed gases to cause rapid oxidation (or to prevent oxidation) of the workpiece, thus incorporating aspects of the gas cutting process. Arc cutting methods include air carbon arc, gas metal arc, gas tungsten arc, shielded metal arc, plasma arc, and oxygen arc cutting. The methods of industrial importance that are covered in this article include air carbon arc cutting, plasma arc cutting, and oxygen arc cutting.

Oxyfuel Gas Cutting

Oxyfuel gas cutting includes a group of cutting processes that use controlled chemical reactions to remove preheated metal by rapid oxidation in a stream of pure oxygen. A fuel gas/oxygen flame heats the workpiece to ignition temperature, and a stream of pure oxygen feeds the cutting (oxidizing) action. The OFC process, which is also referred to as burning or flame cutting, can cut carbon and low-alloy plate of virtually any thickness. Castings more than 30 in. thick commonly are cut by OFC processes. With oxidation-resistant materials, such as stainless steels, either a chemical flux or metal powder is added to the oxygen stream to promote the exothermic reaction.

The simplest oxyfuel gas cutting equipment consists of two cylinders (one for oxygen and one for the fuel gas), gas flow regulators and gages, gas supply hoses, and a cutting torch with a set of exchangeable cutting tips. Such manually operated equipment is portable and inexpensive. Cutting machines, employing one or several cutting torches guided by solid template pantographs, optical line tracers, numerical controls, or computers, improve production rates and provide superior cut quality. Machine cutting is important for profile cutting—the cutting of regular and irregular shapes from flat stock.

Principles of Operation

Oxyfuel gas cutting begins by heating a small area on the surface of the metal to the ignition temperature of 1400 to 1600 °F with an oxyfuel gas flame. Upon reaching this temperature, the surface of the metal will appear bright red. A cutting oxygen stream is then directed at the preheated spot, causing rapid oxidation of the heated metal and generating large amounts of heat. This heat supports continued oxidation of the metal as the cut progresses. Combusted gas and the pressurized oxygen jet flush the molten oxide away, exposing fresh surfaces for cutting. The metal in the path of the oxygen jet burns. The cut progresses, making a narrow slot, or kerf, through the metal.

To start a cut at the edge of a plate, the edge of the preheat flame is placed just over the edge to heat the material. When the plate heats to red, the cutting oxygen is turned on, and the torch moves over the plate to start the cut.

During cutting, oxygen and fuel gas flow through separate lines to the cutting torch at pressures controlled by pressure regulators, adjusted by the operator. The cutting torch contains ducts, a mixing chamber, and valves to supply an oxyfuel gas mixture of the proper ratio for preheat and a pure oxygen stream for cutting to the torch tip. By adjusting the control valves on the torch handle or at the cutting machine controller, the operator sets the precise oxyfuel gas mixture desired. Depressing the cutting oxygen lever on the torch during manual operation initiates the cutting oxygen flow. For machine cutting, oxygen is normally controlled by the operator at a remote station or by numerical control. Cutting tips have a single cutting oxygen orifice centered within a ring of smaller oxyfuel gas exit ports. The operator changes the cutting capacity of the torch by changing the cutting tip size and by resetting pressure regulators and control valves. Because different fuel gases have different combustion and flow characteristics, the construction of cutting tips, and sometimes of mixing chambers, varies according to the type of gas.

Oxyfuel gas flames initiate the oxidation action and sustain the reaction by continuously heating the metal at the line of the cut. The flame also removes scale and dirt that may impede or distort the cut.

The rate of heat transfer in the workpiece influences the heat balance for cutting. As the thickness of the metal to be cut increases, more heat is needed to keep the metal at its ignition temperature. Increasing the preheat gas flow and reducing the cutting speed maintains the necessary heat balance.

Oxygen flow also must increase as the thickness of the metal to be cut increases.

The jet of cutting oxygen must have sufficient volume and velocity to penetrate the depth of the cut and still maintain its shape and effective oxygen content.

Quality of Cut. Oxyfuel gas cutting operations combine more than 20 variables. Suppliers of cutting equipment provide tables that give approximate gas pressures for various sizes and styles of cutting torches and tips and recommended cutting speeds; these variables are operator controlled. Other variables include type and condition (scale, oil, dirt, flatness) of material, thickness of cut, type of fuel gas, and quality and angle of cut.

Where dimensional accuracy and squareness of the cut edge are important, the operator must adjust the process to minimize the kerf, the width of metal removed by cutting, and to increase smoothness of the cut edge. Careful balancing of all cutting variables helps attain a narrow kerf and smooth edge. The thicker the work material, the greater the oxygen volume required and, therefore, the wider the cutting nozzle and kerf.

Process Capabilities

Oxyfuel gas cutting processes are used primarily for severing carbon and low-alloy steels. Other iron-based alloys and some nonferrous metals can be oxyfuel gas cut, although process modification may be required, and cut quality may not be as high as is obtained in cutting the more widely used grades of steel. High-alloy steels, stainless steels, cast iron, and nickel alloys do not readily oxidize and therefore do not provide enough heat for a continuous reaction. As the carbon and alloy contents of the steel to be cut increase, preheating or postheating, or both, often are necessary to overcome the effect of the heat cycle, particularly the quench effect of cooling.

Some of the high-alloy steels, such as stainless steel, and cast iron can be cut successfully by injecting metal powder (usually iron) or a chemical additive into the oxygen jet. The metal powder supplies combustion heat and breaks up oxide films. Chemical additives combine with oxides to form lower temperature melting products that flush away.

Applications. Large-scale applications of oxyfuel cutting are found in shipbuilding, structural fabrication, manufacture of earthmoving equipment, machinery construction, and in the fabrication of pressure vessels and storage tanks. Many machine structures, originally made from forgings and castings, can be made at less cost by redesigning them for OFC and welding with the advantages of quick delivery of plate material from steel suppliers, low cost of oxyfuel gas cutting equipment, and flexibility of design.

Structural shapes, pipe, rod, and similar materials can be cut to length for construction or cut up in scrap and salvage operations. In steel mills and foundries, projections such as caps, gates, and risers can be severed from billets and castings. Mechanical fasteners can be quickly cut for disassembly using OFC. Holes can be made in steel components by piercing and cutting. Machine OFC is used to cut steel plate to size, to cut various shapes from plate, and to prepare plate edges (bevel cutting) for welding.

Gears, sprockets, handwheels, clevises and frames, and tools such as wrenches can be cut out by oxyfuel gas torches. Often, these oxyfuel cut products can be used without further finishing. However, when cutting medium- or high-carbon steel or other metal that hardens by rapid cooling, the hardening effect must be considered, especially if the workpiece is to be subsequently machined.

Thickness Limits. Gas can cut steel less than $^1/_8$ in. thick to over 60 in. thick, though some sacrifice in quality occurs near both ends of this range. With very thin material, operators may have some difficulty in keeping heat input low to avoid melting the kerf edges and to minimize distortion. Steel under $^1/_4$ in. thick often is stacked for cutting of several parts in a single torch pass.

There are a number of advantages and disadvantages when OFC is compared to other cutting operations such as arc cutting, milling, shearing, or sawing. The advantages of OFC are:

- Metal can be cut faster by OFC. Setup is generally simpler and faster than for machining and about equal to that of mechanical severing (sawing and shearing).
- Oxyfuel gas cutting patterns are not confined to straight lines as in sawing and shearing, or to fixed patterns as in die cutting processes. Cutting direction can be changed rapidly on a small radius during operation.
- Manual OFC equipment costs are low compared to machine tools. Such equipment is portable and self-contained, requiring no outside power and well suited for field use.
- When properties and dimensional accuracy of gas cut plate are acceptable, OFC can replace costly machining operations. It offers reduced labor and overhead costs, reduced material costs, reduced tooling costs, and faster delivery.
- With advanced machinery, OFC lends itself to high-volume parts production.
- Large plates can be cut quickly in place by moving the gas torch rather than the plate.
- Two or more pieces can be cut simultaneously using stack cutting methods and multiple-torch cutting machines.

The disadvantages of the OFC process include:

- Dimensional tolerances are poorer than for machining and shearing.
- Because OFC relies on oxidation of iron, it is limited to cutting steels and cast iron.
- Heat generated by OFC can degrade the metallurgical properties of the work material adjacent to the cut edges. Hardenable steels may require preheat and/or postheat to control microstructure and mechanical properties.
- Preheat flames and the expelled red hot slag pose a fire hazard to plant and personnel.

Chemistry of Cutting

Carbon or low-alloy steel heated to ignition temperature (1600 °F) and supplied with pure (99.5%) oxygen oxidizes rapidly. The principal oxidation product is iron oxide (Fe_3O_4). The following can be used as a guide for oxygen requirements and heat emission:

$$3Fe + 2O_2 \rightarrow Fe_3O_4 + 267\,000 \text{ cal}$$

By weight of solid material, the reaction can be expressed as:

$$1 \text{ lb Fe} + 4.6 \text{ ft}^3\ O_2 \rightarrow$$

$$1.38 \text{ lb } Fe_3O_4 + 2870 \text{ Btu}$$

For the complete oxidation of iron (with oxygen volumes at 68 °F and at 1 atm):

- 1 lb of iron requires 0.38 lb or 4.6 ft^3 of oxygen.
- 1 in.^{3} of iron requires 0.109 lb or 1.31 ft^3 of oxygen.
- 1 lb of oxygen requires 2.62 lb or 9.19 in.^{3} of iron.
- 1 ft^3 of oxygen requires 0.22 lb or 0.76 in.^{3} of iron.

In OFC, all of the iron removed from the cut is not oxidized. Some of it (up to 30 or 40%) blows out of the cut as molten iron, along with the oxide, and becomes part of the slag. Process users should note that OFC is accomplished by a chain reaction that requires just enough heat to maintain cutting action; excessive preheat after ignition and pierce is wasteful and may damage the base material and reduce the quality of the cut edge.

Oxygen consumption and flow rates vary depending on whether economy, speed, or accuracy of cut is desired. For average straight-line cutting of low-carbon steel, consumption of cutting oxygen per pound of metal removed varies with thickness of the metal and is lowest at a thickness of 4 to 5 in.

By assuming that for every unit mass of iron oxidized an equal mass of iron melts, one can calculate the amount of heat generated by the cutting reaction—heat emitted is 2870 Btu/lb of iron oxidized. Melting of 1 lb of iron takes 680 Btu, based on a melting point of 2800 °F, 0.2 Btu/lb · °F as the specific heat, and 117 Btu/lb as the heat of fusion. Only a small amount of the heat melts the iron; most of it, about 2100 Btu/lb, goes into the reaction. Some of this superheats the molten metal, some soaks into the workpiece, and some leaves by radiation and convection. Most of it leaves with the slag and hot exhaust gases.

As cutting oxygen flows down through the cut, the quantity available for reaction decreases. If the flow of oxygen is large and well collimated, the rate of cutting through the depth of the cut is approximately constant. The cutting face remains vertical if the oxygen is in excess and if cutting speed is not excessive.

If oxygen flow is insufficient, or cutting speed too high, the lower portions of the cut react more slowly, and the cutting face curves behind the torch. The horizontal distance between point of entry and exit is called drag (see Fig. 1).

Drag influences edge quality. Optimum edge quality results from zero drag—the oxygen stream enters and leaves the cut in a straight line along the cutting tip axis. This is called a "drop-cut," designating a clean, fully severed edge. Increasing cutting speed or reducing oxygen flow makes less oxygen available at the bottom of the cut, causing the bottom of the cut to "drag" behind the top of the cut. A drag of 20% means that the bottom of the cut edge lags the top surface by 20% of the material thickness. Drag lines appear as curved ripples on the cut edge. For fast, rough cuts, some drag is acceptable.

Fig. 1 Cross section of work metal during OFC showing drag on cutting face

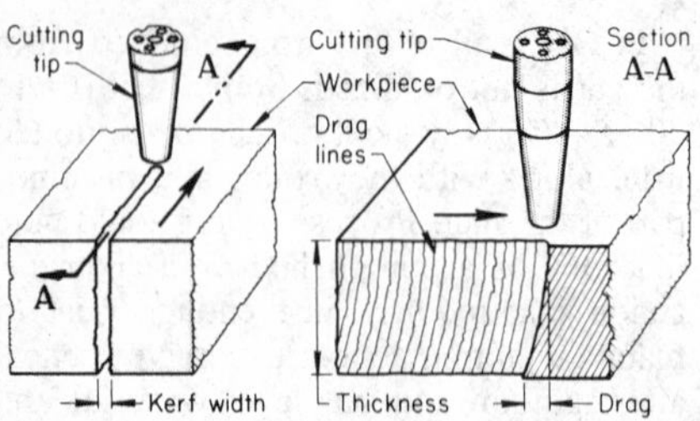

Table 1 Effects of alloying elements on resistance of steel to oxyfuel cutting

Element	Effect on oxyfuel cutting
Aluminum	Extensively used as a deoxidizer in steelmaking; has no appreciable effect on oxygen cutting unless present in amounts above 8-10%; above this percentage, metal powder cutting should be used.
Carbon	Steels containing up to 0.25% C can be flame cut readily; higher carbon steels should be preheated to prevent hardening and cracking; graphitic carbon makes flame cutting of cast iron difficult; cast iron containing up to 4% C can be flame cut when a powder, flux, or filler rod is used as a supplemental oxidizing agent.
Chromium	Steels containing up to 5% Cr can be flame cut without difficulty; steels with chromium content of 10% or more require metal powder or chemical flux cutting.
Cobalt	When present in the amounts normally used in steelmaking, cobalt has no noticeable effect on flame cutting.
Copper	Up to 3% Cu has no effect on flame cutting.
Manganese	Has no effect on flame cutting of carbon steels; steel containing 14% Mn and 1.5% C are difficult to cut and must be preheated.
Molybdenum	Steels with up to 5% Mo can be cut easily; this is true of AISI 41XX steels; high molybdenum-tungsten steels require metal powder cutting.
Nickel	Steels with up to 3% Ni and less than 0.25% C may be readily cut by OFC; up to 7% Ni requires flux additions to the oxygen stream; stainless steels, from 18-8 to 35-15 types, require chemical flux or metal powder cutting techniques.
Phosphorus	The amount usually found in steel has no effect on flame cutting.
Silicon	No effect results in steels with up to 4% Si; in higher silicon steels with high carbon and manganese contents, preheating and post annealing usually are needed to avoid hardening and cracking.
Sulfur	Amounts usually found in steel have no effect; higher sulfur content slows cutting speed and emits sulfur dioxide fumes.
Tungsten	Steels containing up to 14% W are readily flame cut, but cutting is more difficult with a higher percentage; high red-hardness tungsten steels are difficult to flame cut and require preheating.
Vanadium	The amounts normally found in steel do not interfere with flame cutting.

Drag is a rough measure of cut quality and of economy in oxygen consumption. In metal thicknesses up to 2 or 3 in., 10 to 15% drag indicates good quality of cut and economy. The key to quality cuts is control of heat input. Higher quality demands less drag; more drag indicates poorer quality and low oxygen consumption. Excessive drag may lead to incomplete cutting.

In very thin sections, drag has little significance. In very thick sections, the goal is to avoid excessive drag.

Oxygen purity and the alloy content of the steel being cut affect oxyfuel gas cutting quality and speed. Oxygen supplied from cylinders for OFC usually is at least 99.5% pure. A 0.5% departure from this purity (99% oxygen) decreases cutting efficiency. At 90% purity, cutting is very difficult, and at lower purities, it is impossible. The effective purity of oxygen can be reduced during cutting as it mixes with gaseous combustion products from the preheat flames and from the metal being cut, usually a result of insufficient oxygen flow or excessively slow travel speed.

Alloying of iron affects OFC, usually by reducing the rate of oxidation. The total alloy content in low-alloy steel usually does not exceed 5%, and the effect on cutting speed is slight. Alloying elements affect OFC of steel in two ways. They may make the steel more difficult to cut, or they harden the cut edge, or both. In highly alloyed steel, the oxidizing characteristics of alloying elements and the constituents formed in alloying may make sustained oxidation difficult or even impossible. The effects of alloying elements on cutting are evaluated in Table 1. In any steel, preheat accelerates the chemical reaction; higher alloy steels, therefore, may need preheating beyond that provided by the preheat flames of the gas torch to promote cutting.

Preheating may consist of merely warming a cold workpiece with a torch or may require furnace heating of the work beyond 1000 °F. For some alloy steels, preheat temperatures are 400 to 600 °F. Carbon steel billets and other sections occasionally are cut at 1600 °F and higher.

In OFC, preheating is accomplished by means of the oxyfuel gas flame, which surrounds the cutting oxygen stream. At cut initiation, the preheat flame, the result of oxygen and fuel gas combustion, brings a small amount of material to ignition temperature so that combustion can proceed. After cutting begins, the preheat

Table 2 Properties of common fuel gases

Properties	Fuel gases: Acetylene (C_2H_2)	MPS(a) (C_3H_4)	Propane (C_3H_8)	Propylene (C_3H_6)	Natural gas(b)
Neutral flame temperature, °F	5600	5200	4580	5200	4600
Primary flame heat emission, Btu/ft^3	507	517	255	433	11
Secondary flame heat emission, Btu/ft^3	963	1889	2243	1938	989
Total heat value(c) Btu/ft^3	1470	2406	2498	2371	1000
Combustion ratio (neutral flame), vol oxygen/vol fuel	2.5 to 1	4 to 1	5 to 1	4.5 to 1	2 to 1
Oxygen supplied through torch (neutral flame), ft^3 oxygen/lb fuel (60 °F)	18.9	22.1	37.2	31.0	44.9
Maximum allowable regulator pressure, psi	15	Cylinder	Cylinder	Cylinder	Line
Volume-to-weight ratio, ft^3/lb (60 °F)	14.6	8.85	8.66	8.25	23.6
Specific gravity(d)	0.906	1.48	1.52	1.48	0.62

(a) Stabilized methylacetylene propadiene. (b) Principally methane (CH_4). (c) Total heat value after vaporization. (d) At 60 °F; air = 1.

flame merely adds heat to compensate for heat lost by convection and radiation or through gas exhausted during cutting. The flame also helps to remove or burn off scale and dirt on the plate surface; the hot combusted gases protect the stream of cutting oxygen from the atmosphere.

Preheating may also be applied over a broader area of the work. It may include soaking the entire workpiece in a furnace to bring it up to 200 to 400 °F, or a simple overall warm-up with a torch to bring cold plate to room temperature. A preheat improves cutting speed significantly, allowing faster torch travel for greater productivity and reduced consumption of fuel gas. Broader preheat smooths the temperature gradient between the base metal and the cut edge, possibly reducing thermal stress and minimizing hardening effects in some steels.

Chemical reactions are used in FOC and POC to overcome the effects of some alloying elements, such as chromium, on flame cutting. Fluxes combine with refractory oxides to form lower melting temperature compounds. Metal powders, such as iron, or iron and aluminum mixtures, burn with a high evolution of heat. Addition of these powders to the combustion stream aids in the rapid cutting of steels that normally resist oxidation.

Properties of Fuel Gases

Each cutting job entails a different type or volume of work to be completed. Consequently, the best gas for all cutting in a fabricating plant is found through experimentation. Evaluating a gas for a single job requires a test run that monitors fuel gas and oxygen flow rate, labor costs, overhead, and the amount of work performed. If plant production varies from week to week, gas performance should be measured over a long enough period to achieve an accurate cost analysis. Any of the fuel gases may perform well over a range of flow rates. When comparing gases, performance should be rated at the lowest flow rate that gives acceptable results for each gas.

Gas manufacturers provide comprehensive data on flame temperature, heat of combustion, oxygen-to-gas consumption ratio, heat transfer, and heat distribution in the flame. Table 2 lists properties of common fuel gases. Gas properties can be rated on a weight or volume basis. Flow rates, for example, commonly are given in terms of volume. Gas may be sold, however, by weight or by volume, and specifications may state properties in terms of one or the other to make the product seem as attractive as possible. For example, acetylene occupies more volume per pound than does propane. Consequently, cost based on weight makes propane liquid seem inexpensive. Cost based on volume makes the lighter fuels, such as acetylene, seem less costly. The user must understand these ratings and comparisons. Figure 2 shows a form for cost analysis of operations using various cutting fuels.

Flame temperature is theoretical temperature calculated from the chemical reactions that occur when a fuel burns. This

Fig. 2 Convenient method for comparing cutting fuels. Keeps a complete record of cutting conditions and helps analyze direct cost of operation

How to compare cutting fuels

Variables	Acetylene	Natural gas	Propane	MPS	Propylene
Plate thickness, in.					
Type of cutting tip					
Cutting oxygen orifice: Tip No. Drill size					
Optimum cutting speed, in./min					
Fuel gas pressure, psi					
Fuel gas flow, ft^3/h					
Preheat oxygen pressure, psi					
Preheat oxygen flow, ft^3/h					
Ratio of preheat oxygen to fuel gas					
Cutting oxygen pressure, psi					
Cutting oxygen flow, ft^3/h					
Time per 100 ft cut, h					
Fuel gas cost per 100 ft^3					
Oxygen cost per 100 ft^3					
Direct labor cost per hour					
Fuel gas cost per 100 ft cut					
Oxygen cost per 100 ft cut					
Direct labor cost per 100 ft cut					
Total cost per 100 ft cut					

is the highest possible temperature a flame can reach when burning at 100% efficiency. This flame is capable of melting materials that have a melting point slightly below the flame temperature. Any of the common fuel gases easily can cut steel, which melts at 2800 °F, if flame temperature is the only consideration.

Heat of combustion is the amount of heat emission (in Btu) given off when the fuel burns. Heat of combustion can be given in heat per unit weight or heat per unit volume—an important difference to consider when evaluating fuel gases. Fuel gases have different densities, thus the heat of combustion per unit volume varies widely. On a weight basis, all hydrocarbon fuels release 20 000 to 24 000 Btu/lb. For an accurate comparison of the cost of a gas (not the cost of the whole cutting operation), it is necessary to convert all combustion values to either a weight or volume basis, then compare heat per weight, or per volume, per unit cost.

Heat Distribution in the Flame. Combustion of fuel gas produces the preheat flame, which initiates cutting action and helps sustain the operation. These flames heat the surface of the work to ignition temperature to (1) initiate cutting, (2) descale and clean the work surface, (3) supply heat to the work and the cutting oxygen stream to maintain the heat needed for continuous cutting, and (4) shield the cutting oxygen stream from surrounding air. Preheat gases consist of hydrocarbons, which produce water vapor and carbon dioxide as products of chain reactions. These reactions occur in cones within the flame, usually visible as an inner and outer flame. A gas whose inner flame has a high temperature and high heat release provides the most concentrated heat. These gases are superior for fast starts in flame cutting of high-alloy steels that are difficult to cut.

When low heat release is accompanied by high flame temperature, even though it is above the melting point of steel, heat is diffused and gives slow starts in flame cutting. Gases, such as natural gas, that release most of their heat in the outer flame are well suited to heating and heavy cutting. Heat distribution is a good indication of the potential performance of a particular gas.

Coupling distance (standoff distance) is the distance between the torch tip and the work surface. Heat distribution within the gas flame determines the required torch tip-to-work distance and the allowable variation in that distance. The heat intensity of acetylene is greatest (507 Btu/ft^3) at the tip of the inner flame cone and decreases drastically in the outer flame envelope. Coupling distance must be small enough to take advantage of the high-intensity inner cone. Some fuels exhibit the opposite effect—short coupling distance wastes the high heat of the outer envelope. The most "forgiving" fuel is one whose coupling distance has the greatest tolerance. A less skilled operator can achieve good results with such a gas. With the proper cutting tip, allowable variation in coupling distance increases, even with a less "forgiving" gas. A tip that maintains smooth (laminar) and well-defined oxygen flow directs the heat flow farther out from the point of combustion than a tip that allows turbulence in the oxygen stream.

Heat transfer is described by gas manufacturers as the amount of heat passing from the flame into the work. It is affected by the temperature difference between the flame and the work, the heat conductivity of the work, and the size of the workpiece. Assuming that coupling distance is well suited to the particular application and that the workpiece is fairly large, the ability of the flame to transfer heat to the workpiece depends primarily on flame temperature. Fuel gases force a trade-off between fast, concentrated heating (acetylene) and slower heating over a wider area (natural gas).

Combustion Ratio. Oxygen consumption varies with fuel gas. To compare gases, the amount of oxygen consumed per unit weight or per unit volume must be measured. Weight and volume measurements should not be compared to each other. Combustion ratio is the volume of oxygen needed to burn a unit volume of fuel gas. If the equation of burning shows that 2 parts of acetylene will burn 5 parts of oxygen, the combustion ratio (the ratio of the volume of line and atmosphere oxygen to the volume of acetylene burned) is 2.5 to 1. This ratio is different for different gases and includes the total oxygen consumption for burning, some of which comes from the air. Line oxygen accounts for the main cost of OFC. With complete understanding of the chemistry of combustion, coupled with manufacturer's data, one can predetermine line oxygen consumption for any gas.

Acetylene

Acetylene is the most widely used cutting gas. It burns hotter than any of the other common fuel gases, making it indispensable for certain jobs. The overall reaction for complete combustion of acetylene is:

$$2C_2H_2 + 5O_2 \rightarrow 4CO_2 + 2H_2O$$

The volume of oxygen consumed per volume of acetylene is $2^1/_2$. Combustion occurs in stages. In the small inner cone at the tip of the torch, acetylene burns with feed-line oxygen:

$$2C_2H_2 + 2O_2 \rightarrow 4CO + 2H_2$$

This reaction gives off a blistering amount of heat; the tip of the inner cone is the hottest part. Burning continues in the outer envelope of the flame, in a cooler, blue-colored region:

$$4CO + 2H_2 + 3O_2 \rightarrow 4CO_2 + 2H_2O$$

Most oxygen for this reaction comes from the air surrounding the flame. Of the $2^1/_2$ parts of oxygen needed for acetylene to burn completely, about $1^1/_2$ comes from the air and 1 part from line oxygen. A neutral flame, recommended for manual cutting, consumes equal volumes of line oxygen and acetylene. In practice, operators may raise the line oxygen ratio to $1^1/_2$, a proportion that produces an oxidizing flame. The inner cone contracts and becomes sharply defined. This is the hottest flame attainable with acetylene or with any other raw fuel gas. The triple bond in acetylene, shown in Fig. 3, makes it the hottest burning gas. In cutting of steel, operators should adjust the acetylene-to-

Fig. 3 Chemical bonding of common fuel gases. When a gas burns, the bonds between the atoms break. Breaking of single bonds absorbs heat. Breaking of double and triple bonds releases heat. Most of the heat of burning is generated when the atoms recombine with oxygen.

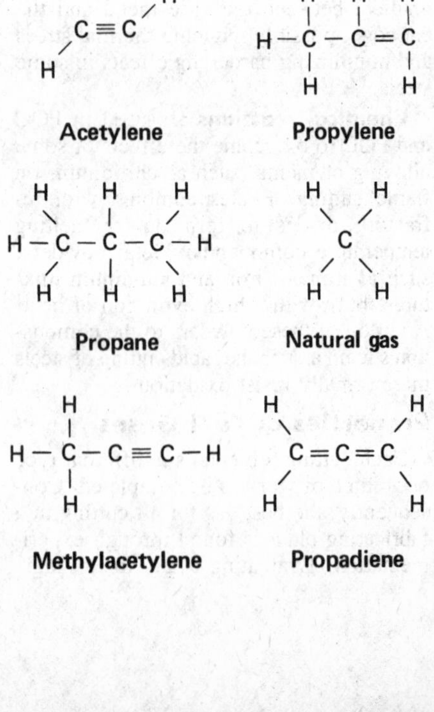

```
      H            H   H   H
     /             |   |   |
  C≡C           H—C — C = C
 /                 |       |
H                  H       H

 Acetylene        Propylene

 H   H   H          H   H
  \  |  /            \ /
H—C—C—C—H             C
  /  |  \            / \
 H   H   H          H   H

  Propane         Natural gas

     H           H       H
     |            \     /
H—C—C≡C—H          C=C=C
     |            /     \
     H           H       H

Methylacetylene    Propadiene
```

oxygen ratio to produce a neutral flame, even though it is below the highest possible temperature. A neutral flame minimizes oxidation and carburization of the base metal. Users sometimes use the hotter oxidizing flame for piercing at the start of a cut.

Acetylene is unstable at room temperature and moderate pressure and can explode under a blow even without the presence of air or oxygen. For safe storage and handling, it is dissolved in acetone. A typical cylinder is filled with heavy porous clay, which holds the acetone. Acetylene, forced into the cylinder, saturates the acetone. When the cylinder valve opens, acetylene vaporizes above the acetone solution. The regulator limits outlet pressure to 15 psig or less.

Cylinders for acetylene should be handled with care. Safety requires them to be placed upright to prevent acetone solvent from entering the regulator. The ratio of line oxygen consumption to acetylene consumption is lower than that for the other fuel gases. Acetylene uses less oxygen at its maximum flame temperature, so that handling of oxygen cylinders may be reduced, thus offsetting the inconvenience of the acetylene cylinder.

Cost of acetylene as a fuel gas changes with consumption. Large-volume users of acetylene often install bulk trailers that hook up to manifold piping systems. When volume justifies, users may install acetylene-generating plants, thus reducing handling costs.

Heat content of acetylene (Btu/ft^3) is lower than all gases except natural gas. Although acetylene burns at a high temperature, it has to burn at higher flow rates than other gases to deliver the same amount of heat. Acetylene releases heat rapidly in a small, concentrated area. Acetylene is a poor choice for cutting a large block of metal because of high fuel cost and the need for large volumes of the fuel to obtain required total heat. To concentrate heat on a limited area, as in cutting thin plate, acetylene is a good choice. It burns faster than other gases, and it burns close to the torch tip. Heat energy concentrates at the tip of the tiny cone.

Acetylene gives a narrow heat-affected zone (HAZ) and possibly less distortion (depending on the volume and shape of the work) than a wider spreading heat. It is particularly well suited to fast cutting of plate under $^1/_2$ in. thick. The hot flame cuts through heavy mill scale or rust and can make a bevel cut quickly with good edge quality. Acetylene facilitates short, stop-and-start cutting jobs, such as cutting structural members and reinforcing bars, because of its short preheat time. For stack cutting and for cutting heavy sections—operations requiring large heat input into a deep kerf—acetylene does not perform as well as other gases. This type of work calls for a fuel gas with high heat output in the secondary flame, such as propane or methylacetylene-propadiene-stabilized (MPS or MAPP) gas.

However, manual cutting is cooler with acetylene than with other gases; less heat rises into the operator's face. It is one of the most versatile and convenient fuels for shops that perform a wide range of fuel gas operations and for shops that cannot justify handling more than one gas (see Table 2 for properties of acetylene). Data for both machine and manual oxyacetylene cutting (OFC-A) of carbon steel plate of varying thickness are given in Tables 3 and 4.

Natural Gas

The composition of natural gas varies, depending on its composition at the well, but its main component is methane (CH_4). When natural gas burns with oxygen, the reaction is:

$$CH_4 + 2O_2 \rightarrow CO_2 + 2H_2O$$

One volume of methane requires two volumes of oxygen. The combustion ratio is 2 to 1.

Although available in pressurized cylinders, natural gas usually is supplied through low-pressure lines from a local utility. Torchmakers overcome low line pressure through torch design. A siphon stream in the mixing chamber pulls the gas in at the required rate for complete combustion. Torches operate at pressures as low as 4 oz/in.^{2} gage line pressure.

A natural gas flame is more diffuse than an acetylene flame—heat intensity is lower, and adjustment for reducing (carburizing), neutral, and oxidizing flames is less clearly defined. Preheating time is longer.

Because heat content is low and heat emission is diffuse, natural gas cannot be used to weld steel. Consequently, if cutting and welding must be performed with a single fuel gas, natural gas cannot be used.

Despite these disadvantages, use of natural gas for cutting has increased. It is the lowest cost commercial fuel gas and, with careful torch adjustment, produces excellent cuts in light- to heavy-gage material (see Table 2 for properties of natural gas). Data for manual oxynatural gas straight-line cutting of carbon steel plate of varying thickness are given in Table 5. Table 6 gives data for oxynatural gas shape cutting of carbon steel plate. Table 7 gives data for machine oxynatural gas drop cutting of shapes from thick carbon steel plate.

Methylacetylene-Propadiene-Stabilized Gas

The common trade name of methylacetylene-propadiene-stabilized gas is MAPP gas. Methylacetylene-propadiene-stabilized mixtures are by-products of the manufacture of chemicals such as ethylene. These mixtures combine the qualities of an acetylene flame with a more even heat distribution in a fuel that is less prone to explosion and less costly than acetylene. It is supplied as a liquid, in large tanks or in portable cylinders.

Table 3 Recommended conditions for manual OFC-A of carbon steels

Plate thickness, in.	Diameter of cutting orifice, in.	Oxygen pressure, psi	Cutting speed(a), in./min	Gas consumption, ft^3(b)			
				Per hour		Per linear foot	
				Oxygen	Acetylene	Oxygen	Acetylene
$^1/_8$	0.0380-0.0400	15-23	20-30	45-55	7-9	0.37-0.45	0.06-0.07
$^1/_4$	0.0380-0.0595	11-20	16-26	50-93	9-11	0.63-0.72	0.08-0.11
$^3/_8$	0.0380-0.0595	17-25	15-24	60-115	10-12	0.80-0.96	0.10-0.13
$^1/_2$	0.0465-0.0595	20-30	12-22	66-125	10-13	1.10-1.14	0.12-0.17
$^3/_4$	0.0465-0.0595	24-35	12-20	117-143	12-15	1.43-1.95	0.15-0.20
1	0.0465-0.0595	28-40	9-18	130-160	13-16	1.78-2.89	0.18-0.29
1$^1/_2$	0.0595-0.0810	35-48	6-14	143-178	15-18	1.96-3.18	0.21-0.33
2	0.0670-0.0810	22-50	6-13	185-231	16-20	3.55-6.16	0.31-0.53
3	0.0670-0.0810	33-55	4-10	240-290	19-23	5.80-12.00	0.46-0.95
4	0.0810-0.0860	42-60	4-8	293-388	21-26	9.70-14.64	0.65-1.05
5	0.0810-0.0860	53-70	3.5-6.4	347-437	24-29	13.66-19.83	0.91-1.37
6	0.0980-0.0995	45-80	3.0-5.4	400-567	27-32	21.00-26.70	1.19-1.80
8	0.0995	60-77	2.6-4.2	505-615	31.5-38.5	29.30-38.84	1.83-2.42
10	0.0995	75-96	1.9-3.2	610-750	36.9-45.1	46.90-64.20	2.57-3.84
12	0.1200	69-86	1.4-2.6	720-880	42.3-51.7	67.70-103.00	3.98-6.05

Note: Values do not necessarily vary in exact proportion to plate thickness, because straight-line relations do not exist among pressure, speed, and orifice sizes.
(a) Lowest speeds and highest gas consumptions are for inexperienced operators, short cuts, dirty or nonuniform material. Highest speeds and lowest gas consumptions are for experienced operators, long cuts, clean and uniform material. (b) Pressure of acetylene for the preheating flames is more a function of torch design than of the thickness of the part being cut. For acetylene pressure data, see charts of manufacturers of apparatus.

Table 4 Recommended conditions for machine OFC-A of carbon steels

Plate thickness, in.	Diameter of cutting orifice, in.	Oxygen pressure, psi	Cutting speed(a), in./min	Gas consumption, ft^3(b) Per hour Oxygen	Per hour Acetylene	Per linear foot Oxygen	Per linear foot Acetylene
1/8	0.0250-0.0400	15-23	22-32	40-55	7-9	0.34-0.36	0.05-0.06
1/4	0.0310-0.0595	11-35	20-28	45-93	8-11	0.34-0.66	0.07-0.08
3/8	0.0310-0.0595	17-40	19-26	82-115	9-12	0.86-0.89	0.08-0.09
1/2	0.0310-0.0595	20-55	17-24	105-125	10-13	1.04-1.24	0.11-0.12
3/4	0.0380-0.0595	24-50	15-22	117-159	12-15	1.45-1.56	0.14-0.16
1	0.0465-0.0595	28-55	14-19	130-174	13-16	1.83-1.86	0.17-0.19
1 1/2	0.0670-0.0810	25-55	12-15	185-240	14-18	3.20	0.23-0.24
2	0.0670-0.0810	22-60	10-14	185-260	16-20	3.70-3.72	0.29-0.32
3	0.0810-0.0860	33-50	8-11	240-332	18-23	6.00-6.04	0.42-0.45
4	0.0810-0.0860	42-60	6.5-9	293-384	21-26	8.53-9.02	0.58-0.65
5	0.0810-0.0860	53-65	5.5-7.5	347-411	23-29	10.97-12.62	0.77-0.84
6	0.0980-0.0995	45-65	4.5-6.5	400-490	26-32	15.10-17.78	0.98-1.16
8	0.0980-0.0995	60-90	3.7-4.9	505-625	31-39	25.52-27.30	1.59-1.68
10	0.0995-0.1100	75-90	2.9-4.0	610-750	37-45	37.50-42.10	2.25-2.55
12	0.1100-0.1200	69-105	2.4-3.5	720-880	42-52	49.70-60.00	2.97-3.50

Note: Values do not necessarily vary in exact proportion to plate thickness, because straight-line relations do not exist among pressure, speed, and orifice sizes.
(a) Lowest speeds and highest gas consumptions are for inexperienced operators, short cuts, dirty or nonuniform material. Highest speeds and lowest gas consumptions are for experienced operators, long cuts, clean and uniform material. (b) Pressure of acetylene for the preheating flames is more a function of torch design than of the thickness of the part being cut. For acetylene pressure data, see charts of manufacturers of apparatus.

Table 5 Recommended conditions for manual oxynatural gas straight-line cutting of carbon steels

Plate thickness, in.	Diameter of cutting orifice(a), in.	Oxygen pressure, psi	Minimum natural gas pressure, psi	Cutting speed(b), in./min	Approximate gas consumption, ft^3/h Oxygen	Natural gas
1/8	0.046	15	3	20-28	25	8
1/4	0.046	18	3	18-28	35	10
3/8	0.046	20	3	16-20	45	14
1/2	0.059	30	3	13-17	50	18
3/4	0.059	35	3	10-15	75	20
1	0.059	40	3	9-13	100	24
1 1/2	0.067	40	3	7-12	145	26
2	0.067	45	3	6-10	190	28
2 1/2	0.067	50	3	6-9	245	30
3	0.093	50	3	5-8	270	32
4	0.093	55	3	5-7	320	36
5	0.093	60	3	4-6	400	40
6	0.110	60	3	4-6	470	48
7	0.110	70	3	3-5	520	52
8	0.110	80	3	3-4	580	56
10	0.110	90	3	3-4	850	60
12	0.110	100	3	2-3	1000	64

(a) Using injector-type torch and two-piece tips. (b) Variations in cutting speeds may be caused by mill scale on plate, variation in oxygen purity, flame adjustment, condition of equipment, impurities in steel, and variation in heat content of natural gas.

Methylacetylene-propadiene-stabilized gases contain a mixture of several hydrocarbons, including propadiene (allene), propane, butane, butadiene, and methylacetylene. Methylacetylene, like acetylene, is a high-energy triple bond compound (see Fig. 3). It is unstable, but other compounds in the mixture dilute it sufficiently to enable safe handling.

Compositions of methylacetylene-propadiene-stabilized mixtures are proprietary and may vary; consequently, an exact combustion equation cannot be specified. The mixture burns hotter than propane or propylene. It also affords a high release of heat energy in the primary flame cone, characteristic of acetylene. The outer flame gives relatively high heat release, similar to propane and propylene. The overall heat distribution in the flame is the most even of any of the other gases. The inner cone releases 517 Btu/ft^3, and the outer flame releases 1889 Btu/ft^3 (see Table 2). The coupling distance is therefore less exacting than for acetylene. The best coupling distance for MPS fuel places the outer cone on the plate; however, a shorter coupling distance also delivers considerable heat.

The neutral MPS gas-oxygen flame generates 2406 Btu/ft^3 with a 5300 °F flame with a ratio of 3.5 to 4 parts oxygen to 1 part fuel. Values vary with the composition of the gas. The carburizing flame, 2.2 parts or less oxygen to 1 part fuel, can weld alloys, such as aluminum, which oxidize readily. A neutral flame, with a ratio of 2.3 parts line oxygen to 1 part fuel gas, can weld steel. At 2.8 line oxygen ratio, the flame becomes oxidizing, unsuitable for welding. Methylacetylene-propadiene-stabilized gas produces its hottest flame at 3.3 line oxygen-to-gas ratio.

In comparing the cost of MPS gas with the cost of acetylene, differences in cylinder yield and consumption rate must be considered. A 120-lb cylinder of MPS gas yields 620 ft^3 of gas; a 240-lb cylinder of acetylene yields only 260 ft^3 of gas. In addition, acetylene burns faster. Thus, storage, transportation, and time and labor for changing cylinders become important cost factors.

Methylacetylene-propadiene-stabilized gas competes with acetylene for almost every job that uses fuel gas. Its most unusual use perhaps is in deep water cutting. Because acetylene outlet pressure is limited to 15 psig, it cannot be used below 30 ft of water. The Navy and several shipyards use MPS gas for underwater work (see Table 2 for properties of MPS). Table 8 provides data for oxy/MPS gas cutting of carbon steel plate.

Propane

Propane (C_3H_8) is a liquefied petroleum fuel that comes from oil refineries or as a gas off the top of oil wells. It may vary in composition, containing traces of other refinery products such as ethane, butane, and pentane, which burn with the propane but do not change its heating properties significantly. Supply and price of propane are related to oil production. Producers liquefy the gas and store it in high-pressure tanks. Users draw it off as a gas. The combustion reaction for propane is:

$$C_3H_8 + 5O_2 \rightarrow 3CO_3 + 4H_2O$$

The combustion ratio is 5 to 1. Half the oxygen comes from the surrounding air.

The maximum flame temperature of propane in oxygen is 5100 °F. This is obtained when the oxygen-to-propane ratio is 4.5 to 1. See Table 2 for pertinent heating values. Propane is low on the scale of fuel gas quality. The combustion equation shows that burning gives off large volumes of gas, a disadvantage because this gas carries off considerable heat.

The gas releases low heat in the primary flame (only natural gas provides lower heat input). Flame cutting and beveling using propane are slower than for other gases at the same flow rate. Propane is widely used

Table 6 Recommended conditions for machine oxynatural gas shape cutting of carbon steels

Plate thickness, in.	Diameter of cutting orifice(a), in.	High preheat: Oxygen, psi	High preheat: Natural gas, psi	Low preheat: Oxygen, psi	Low preheat: Natural gas, oz/in.^{2}	Natural gas, psi	Cutting speed(b), in./min	Width of kerf (approx), in.
1/4	0.036	28	3	7-12	1	70-75	18-28	0.08
3/8	0.037	28	3	7-12	1	70-80	18-26	0.08
1/2	0.039	34	3	7-12	1	75-80	16-24	0.09
5/8	0.039	34	3	7-12	1	75-80	16-23	0.09
3/4	0.046	46	3	8-12	1	80-85	16-22	0.10
1	0.046	48	3	8-12	1	80-85	14-20	0.10
1 1/4	0.054	44	3	8-12	1	80-85	14-18	0.12
1 1/2	0.054	44	3	8-12	1	90-95	13-18	0.12
1 3/4	0.054	45	3	8-12	1	90-95	12-17	0.12
2	0.054	45	3	8-12	1	100	10-15	0.12
2 1/4	0.055	46	3	8-12	1	100	9-15	0.13
2 1/2	0.055	46	3	8-12	1	100	8-14	0.13
2 3/4	0.067	46	3	8-12	1	100	8-13	0.14
3	0.067	46	3	8-12	1	100	7-13	0.14
3 1/2	0.073	50	3	10-14	1	105	6-12	0.15
4	0.073	50	3	10-14	1	110	6-11	0.15
4 1/2	0.082	50	3	10-14	1	110	5-10	0.17
5	0.082	50	3	10-14	1	115	5-10	0.17
5 1/2	0.096	50	3	10-14	1	115	5-9	0.18
6	0.096	50	3	10-14	1	120	5-9	0.18
6 1/2	0.096	50	3	10-14	1	120	4-8	0.18
7 1/2	0.096	50	3	10-14	1	120	4-8	0.18
8	0.096	50	3	10-14	1	120	3-7	0.18

(a) Two-piece tips for high-speed machine cutting. (b) Variations in cutting speed may be caused by mill scale on plate, variation in oxygen purity, flame adjustment, condition of equipment, impurities in steel, and variation in heat content of natural gas.

Table 7 Typical data for machine oxynatural gas drop cutting of shapes from low-carbon steel plate 10 to 21 1/2 in. thick(a)

Plate thickness, in.	Cutting orifice diam, in.	Preheat: Oxygen, psi	Preheat: Natural gas, psi	Cutting oxygen, psi(b)	Cutting speed, in./min	Cutting oxygen, ft^3/h
10	0.250	35-45	5	27	3.25	1400
12 3/4	0.281	35-45	5	28	3.25	1750
15 1/2	0.281	35-45	5	30	3.0	1900
18	0.312	35-45	5	28	4.25	2245
21	0.312	35-45	5	28	3.0	2245
21 1/2	0.312	35-45	5	30	3.75	2365

(a) Two-piece recessed tips with milled preheat flutes (heavy preheat) and straight-bore cutting-oxygen orifices. (b) Pressure measured at torch inlet

because it offers high heat input energy at a lower cost than most of the other gases. Properly used, it gives a good cut. The low-temperature flame gives smooth cut edges, with sharp tops and bottoms. After a cut is started, the burning metal supplies almost enough heat to maintain cutting at a low speed. Excessive heat melts the edges, increasing kerf width. Because the temperature is lower, propane causes less slag adhesion and less edge hardening than other gases.

Cutting characteristics of propane flames are similar to those of natural gas flames. Both gases use the same cutting torches and tips. Flow rates for propane are half those of natural gas. Table 2 lists properties of propane. Table 9 gives data on machine straight-line cutting of carbon steel plate using oxypropane.

Propylene

Propylene, a liquefied gas similar to propane, has a higher flame temperature than propane (it has one double bond, and propane has none, as shown in Fig. 3). It has a flame temperature about equal to MPS mixtures, although its heat content is less. On a volume basis, propylene is about 20% less expensive than acetylene. The combustion equation for propylene is:

$$2C_3H_6 + 9O_2 \rightarrow 6CO_2 + 6H_2O$$

The combustion ratio for propylene is 4.5 to 1. Line oxygen for a neutral flame is about 3.5 to 1. Distributors sell propylene under various trade names, either pure or as improved mixtures with propane and other hydrocarbon additives. Properties of propylene are given in Table 2. Tables 10 and 11 give data for both manual and machine oxypropylene cutting of carbon steel plate.

Effect of Oxyfuel Cutting on Base Metal

During cutting of steel, the temperature of a narrow zone adjacent to the cut face is raised considerably above the transformation range. As the cut progresses, the steel cools through this range. Cooling rate depends on the heat conductivity and the mass of the surrounding material, on loss of heat by radiation and convection, and on speed of cutting. When steel is at room temperature, the rate of cooling at the cut is sufficient to produce a quenching effect on the cut edges, particularly in heavier cuts in large masses of cold metal. Depending on the amount of carbon and alloying elements present, and on the rate of cooling, pearlitic steel transforms into structures ranging from spheroidized carbides in ferrite to harder constituents. The HAZ may be 1/32 to 1/4 in. deep for steels 3/8 to 6 in. thick. Approximate depths of the HAZ in oxyfuel gas cut carbon steels are given in Table 12. Some increase in hardness usually occurs at the outer margin of the HAZ of nearly all steels.

Stainless steels do not support oxyfuel combustion and, therefore, require POC, FOC, or PAC processes. Except for stabilized types, stainless steels degrade under the heat of POC or FOC processes. Carbide precipitation occurs in the HAZ about 1/8 in. from the edge, where the metal has been heated to 800 to 1600 °F long enough for dissolved carbon to migrate to the grain boundaries and combine with the chromium to form chromium carbide. The chromium-poor (sensitized) regions near grain boundaries are subject to corrosion in service. This type of corrosion can be prevented by a stabilizing anneal, which puts the carbon back into solution. However, the required quench through the sensitizing temperature range may distort the material.

Water quenching of the cut edge directly behind the cutting torch may avert sensitization. Because it takes about 2 min at sensitizing temperature for carbide precipitation to occur, water quenching must be done immediately. Distortion is more likely with this method than with the stabilizing anneal. Still another procedure is to remove the sensitized zone entirely by chipping or machining.

Low-Carbon Steel. For steels containing 0.25% C or less cut at room temperature, the hardening effect usually is negligible, though at the upper carbon limit it may be significant if subsequent machining is required. Short of preheating or annealing the workpiece, hardening may be lessened by ensuring that (1) the cutting flame is neutral to slightly oxidizing, (2) the flame is burning cleanly, and (3) that the inner cones of the flame are at the correct height. By increasing the machining allowance slightly, the first cut usually can be made deep enough to penetrate below the hardened zone in most steels. Mechanical properties of low-carbon steels generally are not adversely affected by OFC. Typical data for OFC of low-carbon steel plate are given in Tables 3 to 11.

Medium-Carbon Steels. Steels having carbon contents of 0.25 to 0.45% are affected only slightly by hardening caused by OFC. Up to 0.30% C, steels with very low alloy content show some hardening of the cut edges, but generally not enough to cause cracking. Over 0.35% C, preheating to 500 to 600 °F is needed to avoid cracking. All medium-carbon steels should be preheated if the gas cut edges are to be machined.

High-Carbon and Alloy Steels. Oxyfuel gas cutting of steels containing over 0.45% C and of hardenable alloy steels at room temperature may produce a thin layer of hard, brittle material on the cut surface that may crack from the stress of cooling. Preheating and annealing may alleviate hardening and the formation of residual stress.

Preheating to 500 to 600 °F is sufficient for high-carbon steels; alloy steels may require preheating as high as 1000 °F. Preheat temperature should be maintained during cutting. Thick preheated sections should be cut as soon as possible after the piece has been withdrawn from the furnace.

Local preheating heats the volume of the workpiece enclosing the HAZ of the cut. If the volume of material to be heated is small, the flame of a cutting torch can be used for preheating. When the workpiece is thick and broad, a special heating torch, as shown in Fig. 4, may be necessary. Workpieces must be heated uniformly through the section to be cut, without excessive temperature gradient.

Annealing restores the original structure of the steel and provides stress relief. The preheating flame of the cutting torch or a special heating torch may be used, depending on the mass of the workpiece and the area to be covered. The HAZ of the workpiece should be heated uniformly; the temperature gradient at the boundary of the heated mass should be gradual enough to avoid distortion of the workpiece.

Table 8 Recommended conditions for oxy/MPS gas cutting of carbon steels

			Oxygen				MPS gas		
Plate thickness, in.	Cutting tip No.(a)	Cutting speed, in./min	Cutting pressure(b), psi	Cutting rate of flow, ft^3/h	Preheat pressure, psi	Preheat rate of flow, ft^3/h	Cutting pressure, psi	Cutting rate of flow, ft^3/h	Kerf width, in.
Cutting with standard-pressure tips									
1/8	75	30-36	40-50	12-15	5-10	7-25	2-10	2-10	0.025
3/16	72	26-32	40-50	20-30	5-10	7-25	2-10	2-10	0.03
1/4	68	24-30	40-50	30-40	5-10	7-25	2-10	2-10	0.04
1/2	61	22-28	40-50	55-65	5-10	12-25	2-10	5-10	0.05
3/4	56	16-22	40-50	60-75	5-10	12-25	2-10	5-10	0.06
1	56	14-20	40-50	60-75	5-10	12-25	2-10	5-10	0.06
1 1/4	54	13-17	50-60	105-120	10-20	20-35	2-10	8-15	0.08
1 1/2	54	12-16	50-60	105-120	10-20	20-35	2-10	8-15	0.08
2	52	10-14	50-60	145-190	10-20	20-35	2-10	8-15	0.09
2 1/2	48	9-13	50-60	210-265	10-30	20-50	6-10	8-20	0.10
3	48	8-13	50-60	210-265	10-30	20-50	6-10	8-20	0.10
4	46	7-12	60-70	290-330	10-30	25-50	6-10	10-20	0.15
5	46	6-10	70-80	330-405	10-30	25-50	6-10	10-20	0.15
6	42	5-8	60-70	375-470	10-30	25-50	6-15	10-20	0.16
8	35	4-7	60-70	485-590	30-50	40-100	10-15	20-45	0.19
10	30	3-6	40-70	500-625	30-50	40-100	10-15	20-45	0.20
12	30	3-5	50-85	645-865	30-50	60-150	10-15	30-60	0.21
Cutting with high-speed tips									
1/8	75	32-38	60-70	20-25	5-10	7-25	2-10	3-10	0.025
3/16	72	28-32	70-80	30-40	5-10	7-25	2-10	3-10	0.03
1/4	68	26-32	70-80	55-65	5-10	7-25	2-10	3-10	0.05
1/2	61	24-30	80-90	75-95	5-10	12-25	2-10	5-10	0.06
3/4	56	20-26	80-90	115-130	5-10	12-25	2-10	5-10	0.07
1	56	18-24	80-90	115-130	5-10	12-25	2-10	5-10	0.07
1 1/4	54	16-20	70-80	155-170	10-20	20-35	2-10	8-15	0.08
1 1/2	54	15-19	80-90	170-180	10-20	20-35	2-10	8-15	0.08
2	52	14-18	80-90	215-255	10-20	20-35	2-10	8-15	0.09
2 1/2	52	12-17	80-90	215-255	10-20	20-35	2-10	8-15	0.09
3	48	10-15	80-90	335-400	10-20	20-35	6-10	10-15	0.10
4	46	9-14	80-90	375-425	10-20	20-35	6-10	10-15	0.12

Note: All recommendations are for straight-line cutting with a three-hose torch perpendicular to work.
(a) All tips are of design recommended by the supplier. (b) Pressure of cutting oxygen measured at the torch

Table 9 Recommended conditions for machine oxypropane straight-line cutting of carbon steels

		Operating pressure, psi				Gas consumption, ft^3/h		
Plate thickness, in.	Diameter of cutting orifice, in.	Cutting oxygen	Preheat oxygen	Propane	Cutting speed, in./min	Cutting oxygen	Preheat oxygen	Propane
8-12	0.1562	40-65	20-40	10-15	3-6	715-1210	200-375	55-105
10-16	0.1875	30-60	20-40	10-15	3-5	840-1500	200-375	55-105
15-25	0.250	30-60	20-40	10-15	2-3 1/2	1500-2510	200-375	55-105
20-32	0.2812	30-60	20-60	10-20	2-3	1900-3150	220-425	60-115
24-36	0.3125	25-55	20-65	10-25	1 1/2-2 1/2	2080-3650	250-450	60-125
28-44	0.3437	25-55	20-65	10-25	1 1/2-2 1/2	2570-4400	250-450	60-125

Note: Two-piece recessed tips with milled preheat flutes (heavy preheat) and straight-bore cutting oxygen orifices

Local annealing is not a substitute for preheating; it cannot correct damage done during cutting, such as upsetting of the metal or cracking at the cut edges. Local annealing is limited to steel plate up to 1 1/2 in. thick. From 1 1/2 to 3 in. thick, heat should be applied to both sides of the plate. This method is not suitable for thicknesses over 3 in. If local annealing cannot be done simultaneously with cutting, as shown in Fig. 4, the cut edges should be tempered after cutting with a suitable heating torch.

Distortion, which is the result of heating by the gas flame, can cause considerable damage during (1) cutting of thin plate (less than 5/16 in. thick), (2) cutting of long narrow widths, (3) close-tolerance profile cutting, and (4) cutting of plates that contain high residual stresses. The heat may release some of the restraint to locked-in stress, or may add new stress. In either case, deformation (warpage) may occur, thus causing inaccurate finished cuts. Plates in the annealed condition have little or no residual stress.

Deformation. In cuts made from large plates, the cutting thermal cycle changes the shape of narrow sections and leaves residual stress in the large section (see Fig. 5). The temperature gradient near the cut is steep, ranging from melting point at the cut to room temperature a short distance from it. Plate does not return to its original shape unless the entire plate is uniformly heated and cooled.

As the metal heats, it expands and its yield strength decreases; the weakened heated material is compressed by the surrounding cooler, stronger metal. The hotter metal continues to expand elastically in all directions until its compressive yield strength is reached, at which point it yields plastically in directions not under restraint. The portion of this upset metal above 1600 °F is virtually stress-free; the remainder is under compressive stress that is equal to its yield strength. Metal that expands but does not upset is under compressive strength below yield. The net stress on the heated side of the neutral axis causes bowing of a narrow plate during cutting, as shown in Fig. 5.

As the heated metal begins to cool, it contracts, and its strength increases. First, the contraction reduces the compressive stress in the still-expanded metal. When the compressive stress reaches zero and the plate regains its original shape, previously upset metal also has regained strength. This metal is now in tension as it cools, and its tensile yield strength increases. Tension increases until the metal reaches room temperature. Residual tensile stress in the cooling side of the neutral axis causes the bowing of narrow plates after cooling (see Fig. 5). Controlled upsetting is the basis of flame straightening.

Control of Distortion. Preheating the workpiece can reduce distortion by reducing differential expansion, thereby decreasing stress gradients. Careful planning

Table 12 Approximate depths of HAZ in gas cut carbon steels

Plate thickness, in.	HAZ depth, in.
Low-carbon steels	
Less than 1/2	Less than 1/32
1/2	1/32
6	1/18
High-carbon steels	
Less than 1/2	Less than 1/32
1/2	1/32-1/16
6	1/18-1/4

Note: The depth of the fully hardened zone is considerably less than the depth of the HAZ. For most applications of gas cutting, the affected metal does not have to be removed.

Table 10 Recommended conditions for manual oxypropylene cutting of carbon steels

Plate thickness, in.	Tip size(a)	Cutting orifice drill size(b)	Kerf width, in.	Propylene Pressure, psi	Propylene Flow, ft³/h	Preheat oxygen Pressure, psi	Preheat oxygen Flow, ft³/h	Cutting oxygen Pressure, psi	Cutting oxygen Flow, ft³/h	Cutting speed, in./min
To 1/4	0	68	0.03-0.05	2-5	3-8	5-10	8-24	25-45	20-35	20-30
3/8	0	68	0.04-0.06	2-5	5-10	5-10	12-30	35-50	35-55	20-28
1/2	1	60	0.05-0.07	3-10	5-10	5-10	12-30	40-50	50-60	18-24
3/4	1	60	0.06-0.08	3-10	5-10	5-10	12-30	40-50	55-70	16-22
1	2	57	0.06-0.08	3-10	6-12	10-20	15-35	40-55	60-80	14-20
1 1/4	2	57	0.08-0.10	3-10	8-12	10-20	20-35	40-60	90-120	12-18
1 1/2	2	57	0.08-0.12	5-10	8-15	10-20	20-45	40-65	90-120	12-18
2	3	53	0.09-0.12	5-10	8-15	10-20	20-45	40-70	140-190	10-14
2 1/2	3	53	0.09-0.12	5-10	8-15	10-30	20-45	40-70	160-220	9-12
3	5	46	0.10-0.14	5-10	8-20	10-30	20-50	40-70	200-270	8-11
4	5	46	0.14-0.16	5-10	8-20	10-30	20-50	40-75	300-380	7-10
6	7	39	0.14-0.19	5-15	10-25	20-30	25-70	50-75	320-400	5-8
8	7	39	0.17-0.24	10-15	15-35	30-50	40-100	50-90	400-500	4-6
10	8	30	0.19-0.26	10-15	15-40	30-50	40-120	50-100	550-700	3-5
12	8	30	0.22-0.30	10-15	18-40	30-50	50-120	50-100	700-900	3-5
14-16	10	18	0.25-0.34	10-15	20-45	30-50	50-130	60-120	900-1100	2-4

(a) APCI style 700. (b) Supplier's recommendations

Table 11 Recommended conditions for machine oxypropylene cutting of carbon steels

Plate thickness, in.	Tip size(a)	Cutting orifice drill size(b)	Kerf width, in.	Propylene Pressure, psi	Propylene Flow, ft³/h	Preheat oxygen Pressure, lb/in.²	Preheat oxygen Flow, ft³/h	Cutting oxygen Pressure, psi	Cutting oxygen Flow, ft³/h	Cutting speed, in./min
1/4	1	68	0.06	3-10	4-10	5-10	9-25	70-90	50-70	25-30
3/8	1	68	0.06	3-10	4-10	5-10	12-30	70-90	60-80	24-30
1/2	2	64	0.07	3-10	5-10	5-10	12-30	80-100	80-100	22-28
3/4	2	64	0.07	3-10	5-12	5-10	15-35	80-100	110-130	18-24
1	2	64	0.07	3-10	8-12	10-20	20-35	80-100	120-160	18-22
1 1/4	3	57	0.08	3-10	8-12	10-20	20-35	80-100	150-180	16-22
1 1/2	3	57	0.08	3-10	8-15	10-20	20-45	80-100	170-200	14-20
2	4	55	0.10	3-10	8-15	10-30	20-45	80-100	200-250	12-18
2 1/2	4	55	0.10	3-10	8-15	10-30	20-45	80-100	230-280	11-16
3	5	53	0.12	5-10	10-15	10-30	30-45	80-100	280-340	10-15
4	5	53	0.12	5-10	10-15	10-30	30-45	80-100	320-380	8-12

Note: Pressures are measured at the regulators and assume 25-ft maximum of 1/4-in.-ID hose or larger. For low-pressure, injector-type torches, set fuel pressure at 1 to 5 psi and increase preheat oxygen pressure to 30 lb/in² for thin materials; increase to 90 lb/in² for thick materials.
(a) APCI style 700. (b) Supplier's recommendations
Source: Oxyacetylene, oxynatural gas, oxymethylacetylene-propadiene gas, and oxypropane data from Vol 4, 8th Edition, *Metals Handbook*. Oxypropylene data courtesy Air Products and Chemicals, Inc.

Fig. 4 Local preheat-anneal torch head. Provides multi-ported, diffused, slow-burning flame to maintain relatively low temperatures over a wide work area. To preheat, it precedes the cutting torch; to anneal, it follows.

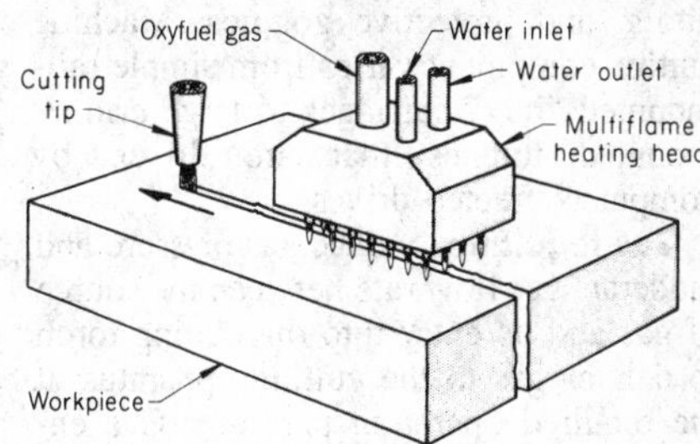

Fig. 5 Effects of OFC thermal cycle on shape of sections. (a) Plate with large restraint on one side of kerf, little restraint on the other side. Phantom lines indicate direction of residual stress that would cause deformation except for restraint. (b) Plate with little restraint on either side

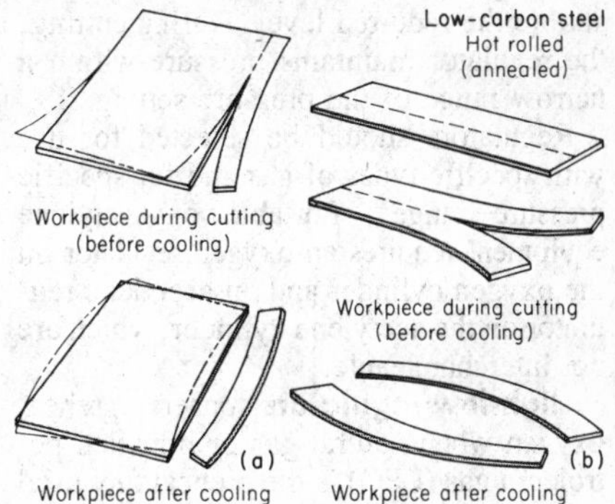

of the cutting sequence also may help. For example, when trimming opposite sides of a plate, both sides should be cut in the same direction at the same time. When cutting rings, the inside diameter should be cut first; the remaining plate restrains the material for the outside-diameter cut. In general, the larger portion of material should be used to retain a shape for as long as possible; the cutting sequence should be balanced to maintain even heat input and resultant residual stresses about the neutral axis of plate or part.

Equipment

Commercial gases usually are stored in high-pressure cylinders. Natural gas—primarily methane—is supplied by pipeline from gas wells. The user taps into local gas lines. Acetylene, dissolved in acetone, is available in clay-filled cylinders. High-volume users often have acetylene generators on site. For heavy consumption or where many welding and cutting stations use fuel gas, banks of gas cylinders are maintained at a central location in the plant, and the gas is manifolded and piped to the point of use.

Manual gas cutting equipment consists of gas regulators, gas hoses, cutting torches, cutting tips, and multipurpose wrenches. Auxiliary equipment may include a hand truck, tip cleaners, torch ignitors, and protective goggles. Machine cutting equipment varies from simple rail-mounted "bug" carriages to large bridge-mounted torches that are driven by computer-directed drives.

Gas regulators reduce gas pressure and moderate gas flow rate between the source of gas and its entry into the cutting torch to deliver gas to the cutting apparatus at the required operating pressure. Gas enters the regulating device at a wide range of pressures. Gas flows through the regulator and is delivered to the hose-torch-tip system at the operating pressure, which is preset by manual adjustment at the regulator and at the torch. When pressure at the regulator drops below the preset pressure, regulator valves open to restore pressure to the required level. During cutting, the regulator maintains pressure within a narrow range of the pressure setting.

Regulators should be selected for use with specific types of gas and for specific pressure ranges. Portable oxyacetylene equipment requires an oxygen regulator on the oxygen cylinder and an acetylene regulator on the acetylene cylinder, which are not interchangeable.

High-low regulators conserve preheat oxygen when natural gas or liquefied petroleum gas (LPG) is the preheat fuel used in OFC. These gases require a longer time to start a cut than acetylene or MPS. High-low regulators reduce preheat flow to a predetermined level when the flow of cutting oxygen is initiated. When the regulator switches from high to low, preheat cutback may range from 75 to 25% as plate thicknesses increase from $^3/_8$ to 8 in. High-low regulators are used for manual and automatic cutting with natural gas and with LPG.

Hose. Flexible hose, usually $^1/_8$ to $^1/_2$ in. in diameter, rated at 200 psig maximum, carries gas from the regulator to the cutting torch. Oxygen hoses are green; the fittings have right-hand threads. Fuel gas hoses are red; the fittings have left-hand threads and a groove cut around the fitting. For heavy cutting, two oxygen hoses may be necessary—one for preheat and one for cutting oxygen. Multi-torch cutting machines often have three-hose torches.

Cutting torches, such as the one shown in Fig. 6, control the mixture and flow of preheat oxygen and fuel gas and the flow of cutting oxygen. The cutting torch discharges these gases through a cutting tip at the proper velocity and flow rate. Pressure of the gases at the torch inlets, as well as size and design of the cutting tip, limits these functions, which are operator controlled.

Oxygen inlet control valves and fuel gas inlet control valves permit operator adjustment of gas flow. Fuel gas flows through a duct and mixes with the preheat oxygen; the mixed gases then flow to the preheating flame orifices in the cutting tip. The oxygen flow is divided—a portion of the flow mixes with the fuel gas, and the remainder flows through the cutting oxygen orifice in the cutting tip. A lever-actuated valve on the manual torch starts the flow of cutting oxygen; machine cutting starts the oxygen from a panel control.

Fuel gases supplied at low pressure, such as natural gas tapped from a city line, require an injector-mixer (Fig. 6b) to increase fuel gas flow above normal operating pressures. Optimum torch performance relies on proper matching of the mixer to the available fuel gas pressure.

Cutting tips are precision-machined nozzles, produced in a range of sizes and types. Figure 7(a) shows a single-piece acetylene cutting tip. A two-piece tip used for natural gas (methane) or LPG is shown in Fig. 7(b). A tip nut holds the tip in the torch. For a given type of cutting tip, the diameters of the central hole, the cutting oxygen orifice, and the preheat ports increase with the thickness of the metal to be cut. Cutting tip selection should match the fuel gas; hole diameters must be balanced to ensure an adequate preheat-to-cutting oxygen ratio. Preheat gas flows through ports that surround the cutting oxygen orifice. Smoothness of bore and accuracy of size and shape of the oxygen orifice are important to efficiency. Worn, dirty bores reduce cut quality by causing turbulence in the cutting oxygen stream.

The size of the cutting tip orifice determines the rate of flow and velocity of the preheat gases and cutting oxygen. Flow to the cutting tip can be varied by adjustment at the torch inlet valve or at the regulator, or both.

Increasing cutting oxygen flow solely by increasing the oxygen pressure results in turbulence and reduces cutting efficiency. Turbulence in the cutting oxygen causes wide kerfs, slows cutting, increases oxygen consumption, and lowers quality of cut. Consequently, larger cutting tips are required for making heavier cuts.

Fig. 6 Typical manual cutting torch in which preheat gases are mixed before entering torch head. (b) and (c) Sections through preheat gas duct showing two types of mixers commonly used with the torch shown. After the workpiece is sufficiently preheated, the operator depresses the lever to start the flow of cutting oxygen. Valves control the flow of oxygen and fuel gas to achieve required flow and mixture at the cutting tip.

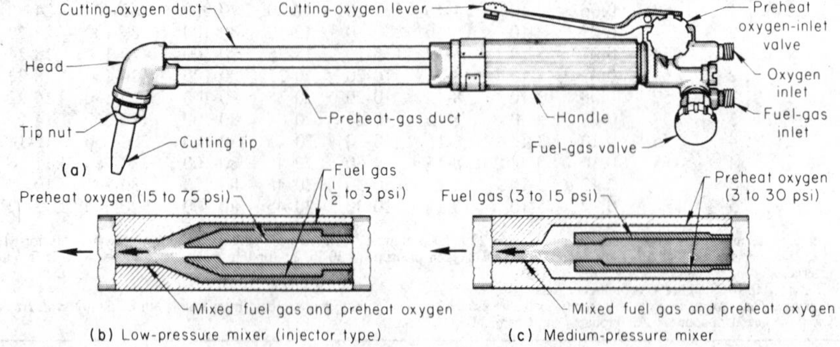

Fig. 7 Types of cutting tips.
(a) Single-piece acetylene cutting tip. (b) Two-piece tip for natural gas or LPG. Fuel gas and preheat oxygen mix in tip. Recessed bore helps promote laminar flow of gas.

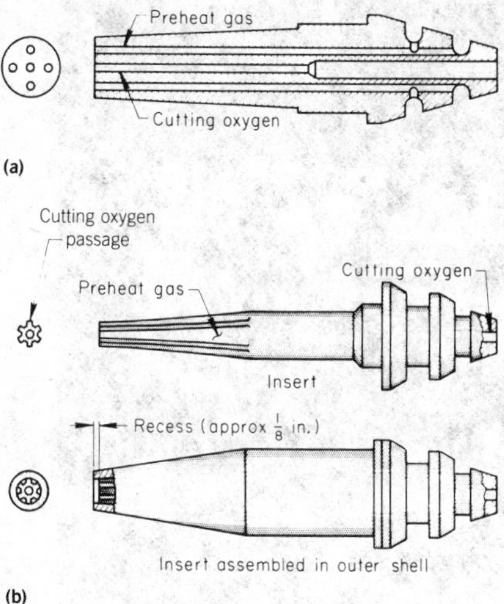

Standard tips, as shown in Fig. 8(a), have a straight-bore oxygen port. Oxygen pressures range from 30 to 60 psi and are used for manual cutting. High-speed tips, or divergent cutting tips (Fig. 8b), use a converging, diverging orifice to achieve high gas velocities. The oxygen orifice flares outward. High-speed tips operate at cutting oxygen pressures of about 100 psi and provide cutting jets of supersonic velocity. These tips are precision made and are more costly than straight-drilled tips, but they produce superior results—improved edge quality and cutting speeds 20% higher than standard tips. Best suited to machine cutting, high-speed tips produce superior cuts in plate up to about 6 in. thick. Above this thickness, advantages of their use decrease; they are not recommended for cutting metal more than 10 in. thick.

Fig. 8 Oxyfuel cutting tips.
(a) Standard cutting tip with straight-bore oxygen orifice. (b) High-speed cutting tip with divergent-bore oxygen orifice

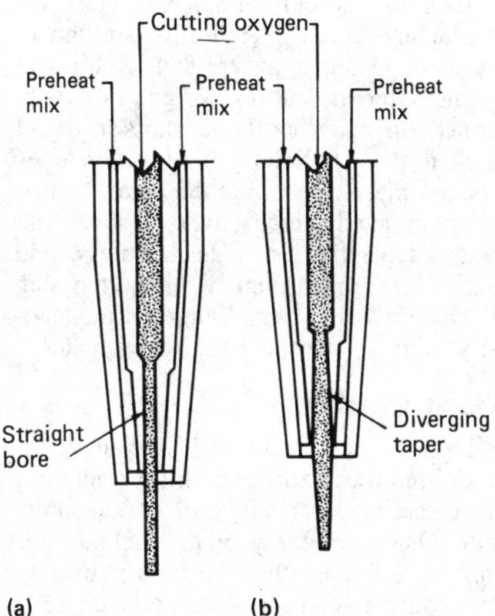

Equipment Selection Factors. Natural gas and liquefied petroleum gases operate most efficiently with high-low gas regulators, injector-type cutting torches, and two-piece, divergent, recessed cutting tips. Acetylene cutting is most efficient with divergent single-piece tips. If acetylene is supplied by low-pressure generators, an injector-type torch is best suited to most cutting applications.

Two-piece, divergent cutting tips are best suited for use with MPS gas; the tip recess should be less than for use of natural gas or propane. Injector-type torches and high-low regulators are not required with MPS gas.

Guidance Equipment. In freehand cutting, the operator usually can follow a layout accurately at low speeds, but the cut edges may be ragged. For accurate manual cutting at speeds over 10 in./min, the torch tip should be guided with a metal straightedge or template. Circles and arcs are cut smoothly with the aid of a radius bar—a light rod adjustably clamped to the torch at one end while the other end is held at the circle center.

Machine guidance equipment includes magnetic tracing of a metal template, manual spindle tracing, optical tracing of a line drawing, guidance by numerically controlled tape or by programmable controllers, and computer-programmed guidance equipment.

Portable cutting machines are used primarily for straight-line and circular cutting. Components include a torch mounted on a motor-driven carriage that travels on a track or other torch guidance device. The operator adjusts travel speed and monitors the operation.

Machine cutting torches are of heavy construction of in-line design. The torch casing has a rack, which fits into a gear on the torch holder, for raising and lowering the torch over the work. Ducts and valves are encased in a single tube. The cutting tip is mounted axially with the tube. A valve knob or a lever-operated poppet valve replaces the spring-loaded cutting oxygen lever of the manual torch.

On some portable machines, gases are supplied to connections on the carriage, rather than directly to the torch, to avoid hose-drag on the torch. Short hoses are used from machine connections to the torch. Some carriages can accommodate two or more torches operating simultaneously, for such operations as squaring and beveling.

The operator follows the carriage to make adjustments. When plates are wavy or distorted, the operator may need to adjust torch height to avoid losing the cut. When carefully operated, track-guided torches can produce cuts at speeds and quality approaching those obtainable with stationary cutting machines.

Stationary cutting machines, as shown in Fig. 9, are used for straight-line and circular cuts, but their primary use is for cutting complex parts—shape cutting. Plate to be cut is moved to the machine.

On shape cutting machines, cutting torches move left and right on a bridge mounted over the cutting table. The bridge moves back and forth on supports that ride on floor-mounted tracks. The combined movement of the torches on the bridge and the bridge on the track allows the torch to cut any shape in the *x-y* plane. Bridges are either cantilever or gantry design. Suppliers classify cutting machine capacity by the maximum width of plate that can be cut.

Machine Directions. Methods for directing motion of shape cutting machines have become increasingly sophisticated and include manual, magnetic, or electronic means of control. The simplest machines have one or two torches and use manual or magnetic tracing.

For manual tracing, the operator either steers an idler wheel or a spindle around a template, or he guides a wheel or a focused light beam around an outline on paper. Cutting speed is controlled by setting the speed of the tracing head (pantograph director) or by setting the speed of the torch carriage (coordinate drive). Cutting speed in manual tracing is about 14 in./min, depending on operator skill.

Magnetic tracing is done with a knurled magnetized spindle that rotates against the edge of a steel template. The spindle is linked to a pantograph. Direct-reading tachometers, showing cutting speed in inches per minute, assist in adjusting cutting speed. These control methods are relatively slow.

Faster electronic tracers use a photoelectric cell that scans the reflection of a beam of light directed on the outline of a template. Templates are line drawings on paper, white-on-black paper cutouts, or photonegatives of a part outline. To continuously hold tolerances closer than 1/16 in., templates of plastic film, glass cloth, or some other durable, dimensionally stable material should be used.

In scanning the edge of a white-on-black template, the circuit through the photo-

Fig. 9 Stationary OFC machine

electric cell balances when the cell "sees" an equal amount of black and white. A change in this balance sends an impulse to a motor that moves the tracing head back to balance. In line tracing, the photoelectric cell scans the line from side to side. As long as the light reflects equally from both sides of the line, the steering signals balance. When the photocell scans more light on one side of the line than on the other, the scanner rotates to balance.

Some machines adjust to permit parts to be cut about $^1/_{32}$ in. larger or smaller than the template. This feature, called kerf compensation, is useful for cutting to close tolerances, especially when the template has insufficient kerf allowance.

Coordinate drive machines translate motion 1-to-1 or in other ratios. Such ratio cutting permits the use of templates in any proportion, from full-scale to one tenth of part size.

Tape Control. Cutting machine movement may be controlled by electronic signals from punched tape (numerical control). These machines do not require templates, and the tape may be easily stored and used many times.

Some cutting machines receive direction from a microprocessor, programmed directly or from punched tape. The most sophisticated machines take directions from a computer (computer numerical control).

Starting the Cut

To start a cut at the outer edge of the workpiece, the operator should place the cutting torch so that the ends of the inner cones in the preheating flame just clear the work metal. When a spot of metal at the top of the edge is heated to bright red, the stream of cutting oxygen is turned on. The reaction forms a slot in the plate edge, and the torch can move along the desired line of cut. Piercing starts are necessary to cut holes, slots, and shapes.

In manual piercing, the spot should be heated to bright red. Then the torch should be raised to $^1/_2$ in. above the normal cutting distance, and cutting oxygen should be turned on slowly. As soon as the metal is penetrated, the torch is lowered to cutting height and cutting action is started.

Machine torch piercing is similar to manual operations, except that torch travel begins after the cutting oxygen is slowly turned on and the flame starts to eject molten slag. Initial piercing action does not completely penetrate the plate, so that the completed pierce covers a short distance (depending on plate thickness) and forms a sloping trough with molten slag ejected from the torch. Length of the lead-in should be increased for heavy plate.

Light Cutting

Light cutting (material less than $^3/_8$ in. thick) requires extra care in tip selection, tip cleanliness, and control of gas pressure. Quality of cut when oxyfuel gas cutting steel thinner than $^1/_8$ in. is virtually impossible to control. Wide, uneven kerf, ragged edge, and distortion result from heat buildup.

For straight or large-radius cuts in thicknesses $^1/_8$ in. and over, angling the cutting torch forward increases the distance through the cut and may help to deflect some of the heat input. When cutting shapes, however, the torch must be vertical.

One of the most important techniques for cutting of thin sections is stack cutting (see the section of this article on stack cutting for more detailed information on this process). This technique avoids the difficulties inherent in light cutting and also increases production. Sections stacked for cutting must be flat and tightly clamped together.

Medium Cutting

In cutting plate $^3/_8$ to 10 in. thick, OFC achieves its highest efficiency and produces the most satisfactory cuts. Conditions for medium cutting of carbon steel plate using a variety of oxyfuel combinations are presented in Tables 3 to 11. Commercial cuts must meet these requirements:

- Drag is short, markings on the face of the cut approach vertical.
- Side of the cut is smooth, not fluted, grooved, or ragged.
- Slag should not adhere to the bottom of the cut.
- Upper and lower edges should be sharp.
- Cost is moderate.

The following variables require adjustment to obtain satisfactory commercial cuts:

- Suitable cutting tip, with cutting orifice of correct type and size, and proper degree of preheat
- Suitable oxygen and fuel gas pressures
- Correct cutting speed
- Uniform torch movement
- Clean, smooth-bore cutting orifice and preheat holes
- High-purity oxygen
- Proper angle of the cutting jet in relation to the upper edge of the cut

Surface finish for gas cut production parts is compared with sample cuts. One such group of machine cut samples is shown in Fig. 10. Sample A represents high quality, with drag varying about 2 to 7%. Samples B, C, D, and E show typical defects resulting from improper cutting speed or torch settings. The American Welding Society (AWS) provides standard plastic samples of cut edges.

Fig. 10 Comparison of surface finish of gas cut specimens. (A) Proper speed, preheat, and cutting-oxygen pressure. Note clean face and nearly straight drag lines. (B) Proper speed and cutting-oxygen pressure, too much preheat. Note excessive slag and rounding of top edge. (C) Proper preheat and cutting-oxygen pressure, but too much speed. Note increase in drag and uncut corner at lower left. (D) Proper speed and preheat, but too much cutting-oxygen pressure. Surface is rough and top edge is melted. (E) Proper preheat and cutting-oxygen pressure, but speed too low. Burned slag adheres to cut surface.

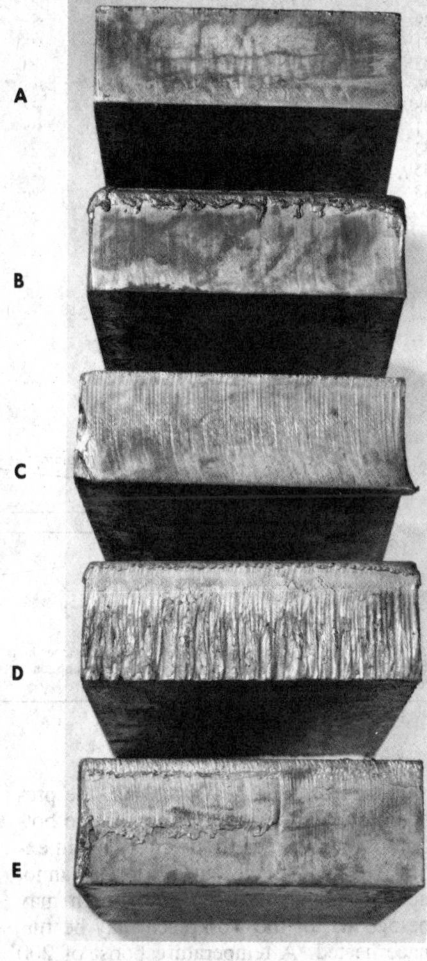

Torch settings for bar A(a)

Preheat oxygen pressure:	
High (edge starting)	45 psi
Low (cutting)	8 to 12 psi
Preheat natural gas pressure:	
High (edge cutting)	3 psi
Low (cutting)	Under 1 oz/in.^{2}
Cutting oxygen pressure	100 psi
Cutting speed	10 to 15 in./min

(a) Settings for bars B, C, D, and E were varied as noted in the legend above. All bars were cut with two-piece divergent-nozzle (high-speed) tips with 0.054-in.-diam cutting oxygen orifice. All cutting was done with low preheat settings, except for bar B, which was cut with both preheat gases set higher than for starting.

Tip Design. General-purpose cylindrical-bore tips operate best with cutting oxygen pressures of 30 to 60 psi; a pressure of 40 psi is a good initial setting. Divergent-orifice tips operating with cutting oxygen pressures of about 100 psi, with variations of ±20 psi, may provide better results. Because tip size controls kerf width, a smaller tip should be used for a narrower kerf; the part may not drop free, and slower speed must be used. Larger-than-normal tips may be used when the metal is covered with thick scale.

Cutting Speed. Within recommended speed ranges, slow cutting produces the best results; higher speeds are used for maximum output. Optimum cutting speed may be determined by observing the material that is expelled from the underside of the workpiece. When speed is too slow, the cutting stream forms slag drippings. At excessive speed, a stream trails at a sharp angle to the bottom of the workpiece. Highest quality cutting occurs when a single, continuous stream is expelled from the underside for about $^1/_4$ in., then breaks into a uniform spray. This condition is accompanied by a sound similar to that of canvas ripping.

Preheat. Less preheat is needed to continue a cut than to start it. When optimum quality is desired, the amount of preheat should be reduced after starting the cut. Commercially available devices automatically reduce preheat after cutting begins.

In thicknesses up to 2 in., excessive preheat may cause the top corner to melt and roll over. Insufficient preheat results in loss of cut. Correct preheat provides a continuous reaction with a sharp, clean top edge of the cut.

Kerf compensation, or the allowance of half the kerf width on the outline of the template, is the amount added for outside cuts and subtracted for inside cuts. In straight-line cutting or in following a layout, the operator may compensate for this measurement; in machine cutting, kerf compensation should be incorporated in the template. Many automatic machines incorporate kerf compensation in their controllers.

Machine Accuracy. When a machine is properly maintained, the guidance system controls accuracy. Mechanical followers are very accurate when properly adjusted. Compensation for the diameter of the follower wheel or spindle must be made in the same manner as for kerf width when making a template.

Kerf angle, the small deviation of the kerf wall from a right angle, may be caused by failure to set the torch perpendicular to the work. It may also be the result of widening of the cutting oxygen stream. In the latter case, the width of the kerf increases from the top to the bottom of the cut; this may be corrected in straight cutting by angling the torch slightly to make one wall of the kerf perpendicular. In shape cutting, where this cannot be done, kerf angle can be corrected by slightly reducing cutting oxygen pressure.

Plate movement may be vertical, caused by warp, or lateral, caused by planar expansion and contraction of the part. Vertical movement requires either manual or automatic control of torch height. A very slight allowance can be made in the initial setting of the tip-to-work distance. Lateral movement can be controlled by fixing the part to be cut on the plate support, wedging the cut within the stock plate, or allowing only a small amount of scrap trim around the part so that the scrap will move instead of the workpiece.

Heavy Cutting

Cutting of metal 10 in. or more in thickness requires attention to gas flow, preheat flame setting, drag, and starting technique. Standard cutting torches may be used to cut steel up to 18 or 20 in. thick; heavy-duty torches are necessary for sections up to 60 in. thick.

Gas Flow Requirements. Heavy cutting requires a uniform supply of high-purity oxygen and fuel gas at constant pressure, and an adequate volume of cutting oxygen. Torches should have three hoses—one for the cutting oxygen, one for the preheat oxygen, and one for the fuel gas. This allows setting of proper pressures independently for preheating and cutting. Cutting oxygen hose should be at least 1/2 in. in diameter to ensure an adequate volume of oxygen at low pressure settings. Regulators should be used that are capable of providing the large volumes of gas required. Cutting oxygen flow requirements (ft^3/h) vary from 80 to 120 times ($80t$ to $120t$) the thickness to be cut measured in inches. Table 13 gives typical cutting oxygen flow rates, operating pressures, and oxygen-orifice diameters for cutting heavy steel plate from 8 to 48 in. thick. Cutting oxygen consumption for shape cutting approaches $120t$ or higher; for straight-line cutting, it is between $80t$ and $100t$. Pressures are 20 to 60 psi at the torch.

Table 13 Relationship among metal thickness, cutting oxygen flow rates, operating pressures, and oxygen orifice diameters for cutting heavy steel sections

Data based on gas cutting of plate 8 to 48 in. thick at 2 to 6 in./min, using a straight-drilled torch tip of the type and with the dimensions shown in accompanying diagram

Work-metal thickness(a), in., for thickness-range of:			Flow rate, ft^3/h	Cutting oxygen — Operating pressure, psi, at torch entry, for orifice diameter, in., of:									
$80t$	$100t$	$120t$		0.147	0.1695	0.1935	0.221	0.250	0.290	0.332	0.375	0.422	0.468
12	10	8	1000	56	39	28	...	...	...	...	...	...	...
14 1/2	12	9 1/2	1200	...	49	35	26	...	...	...	...	...	...
17	14	11	1400	...	59	42	31	...	...	...	...	...	...
19	16	13	1600	...	...	49	36	27	...	...	...	...	...
21 1/2	18	14 1/2	1800	...	...	57	42	31	...	...	...	...	...
24	20	16	2000	...	...	...	48	35	26	...	...	...	...
27	22	17 1/2	2200	...	...	...	53	39	28	...	...	...	...
29	24	19	2400	...	...	...	...	44	31	...	...	...	...
31 1/2	26	21	2600	...	...	...	...	49	34	24	...	...	...
33 1/2	28	22 1/2	2800	...	...	...	...	54	37	26	...	...	...
36	30	24	3000	...	...	...	...	...	40	29	...	...	...
38 1/2	32	25 1/2	3200	...	...	...	...	...	43	31	...	...	...
41	34	27	3400	...	...	...	...	...	47	33	25	...	...
43	36	29	3600	...	...	...	...	...	50	36	27	...	...
45 1/2	38	30 1/2	3800	...	...	...	...	...	54	38	29	...	...
48	40	32	4000	...	...	...	...	...	...	40	30	...	...
...	42	33 1/2	4200	...	...	...	...	...	...	43	32	24	...
...	44	35	4400	...	...	...	...	...	...	45	34	25	...
...	46	37	4600	...	...	...	...	...	...	48	36	27	...
...	48	38 1/2	4800	...	...	...	...	...	...	50	38	28	...
...	...	40	5000	...	...	...	...	...	...	53	40	30	...
...	...	41 1/2	5200	...	...	...	...	...	...	...	42	31	24
...	...	43	5400	...	...	...	...	...	...	...	44	32	25
...	...	45	5600	...	...	...	...	...	...	...	46	34	26
...	...	46 1/2	5800	...	...	...	...	...	...	...	48	35	27
...	...	48	6000	...	...	...	...	...	...	...	50	37	28

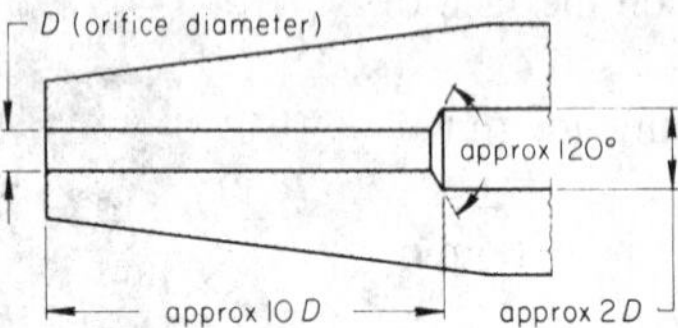

(a) Based on the formula that oxygen flow in cubic feet per hour will range from 80 times metal thickness ($80t$) to 120 times metal thickness ($120t$) in heavy cutting. Cutting oxygen consumption for shape cutting is generally near, or slightly above, $120t$; straight-line cutting will use $80t$ to $100t$.

Preheating. In heavy cutting, the preheat flame must extend almost to the bottom of the thickness to be cut to avoid excessive drag. If the cutting torch cannot deliver enough heat, an extra torch may be needed, or the workpiece may be furnace heated. A temperature boost of 200 or 300 °F assists cutting action considerably. Tip height above the work is important; if the tip is too close, excessive melting and rounding of the top edge occur. If too high, preheat will be insufficient and cutting will be slowed. Generally, 1 to 2 in. of separation is satisfactory.

The starting edge should be thoroughly preheated, with the preheated zone extending far down the face as shown in Fig. 11(a). As soon as the metal begins to melt, the flow of cutting oxygen and the cutting motion should proceed simultaneously, cutting at normal speed. The cut should progress down the face at a constant rate until it breaks though the bottom. Drag will be long at first, but it will shorten as soon as the cut is confined. Too slow a speed causes a shelf part way through the cut edge (Fig 11e). Too high a speed results in incomplete penetration or in extremely long drag, as shown in Fig. 11(f).

Drag. To successfully complete the cut (drop cutting), minimum drag conditions must exist, as shown in Fig. 12(a). Too much drag leaves an uncut corner at the end of the cut (Fig. 12b and c). Angled cutting may be used in straight-line work to counter drag (Fig. 12d and e), but for shape cutting, the torch must be perpendicular to the work.

Starting. Extra care is needed in starting a heavy cut to avoid leaving an uncut corner or pocketing the flame in the lower portion of the cut. Starting sometimes is

Fig. 11 Proper and improper techniques for heavy cutting. (a) Proper torch position; preheat is primarily on starting face. (b) Improper start; oxygen stream is too far onto work, which results in action of cut as shown in (c) and in uncut corner as shown in (d). (e) Excessive oxygen pressure or action of cut at too low a speed. (f) Insufficient oxygen pressure or action of cut at too high a speed

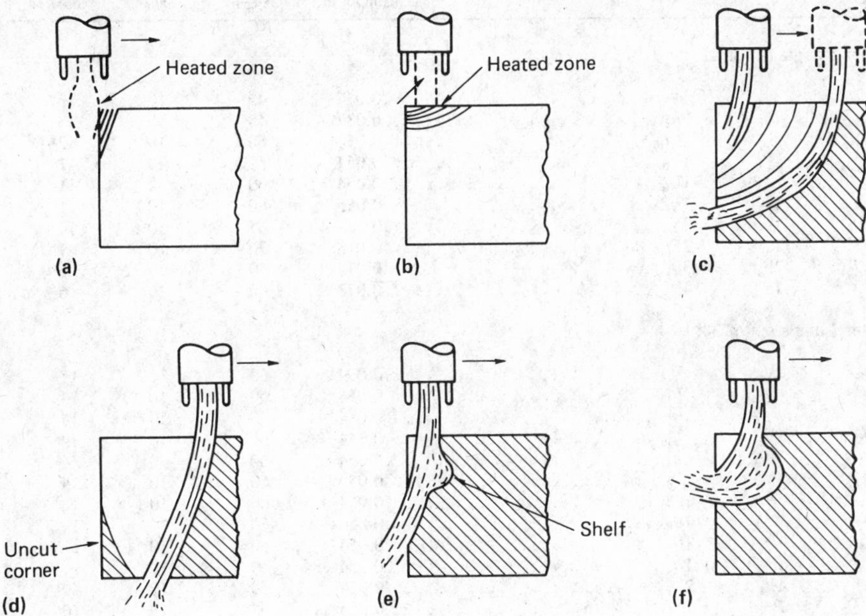

Fig. 12 Proper and improper techniques for a heavy cut. (a) Minimum drag (typical at balanced conditions) permits flame to break through cutting face uniformly at all points. (b) Excessive drag, typically caused by insufficient oxygen or excessive speed, results in undercut. (c) Forward drag resulting from excessive oxygen pressure or too little speed. (d) and (e) Forward angling of torch to minimize drag and thus avoid an uncut corner. (f) Excessive forward angling of torch, resulting in undercut

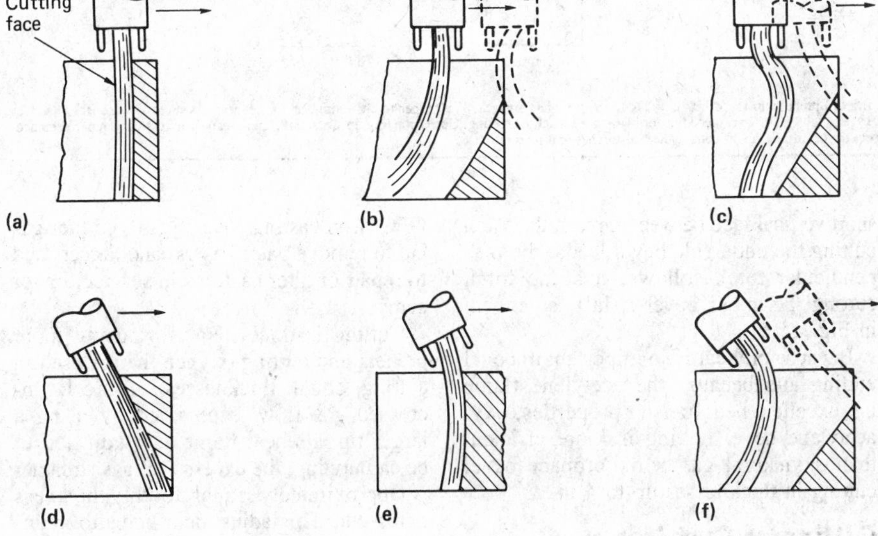

facilitated by first undercutting the forward edge of the material with a hand torch.

Stack Cutting

Stack cutting is a practical approach to cutting thin sheet; it also reduces the cost of cutting thick material. Savings are multiplied by using multi-torch machines for stack cutting. Several torches can be used, each cutting up to 30 pieces of $^1/_8$-in.-thick material simultaneously. Thicknesses most practical for this technique are 20 gage (0.0359 in.) to $^1/_2$ in., although material $^3/_4$ in. thick has been stack cut.

Material for stack cutting should be clean and free of scale. The sheets or plates must be clamped tightly or even welded together in a pile with edges aligned where the cut is to start. Air gaps between pieces interfere with the cutting flame. The most practical total thickness for the stack is 5 or 6 in., although 10-in. stacks have been cut. The risk of a costly mishap increases with the thickness of the metal to be cut and the number of pieces in the process.

When stack cutting pieces less than $^3/_{16}$ in. thick, a waster plate $^1/_4$ in. thick is clamped on top of the stack to ensure better starting and a sharper edge for the top. This precaution also prevents buckling.

Preparation of Weld Edges

Steel plate to be welded is squared, trimmed, cut to shape, and beveled by OFC, often in one operation. Cutting accuracy can be held to $^1/_{32}$ in. with proper care. Adherence of slag to the underside of the cut can be minimized with careful adjustment of cutting parameters. Light cleaning, if required, can be done with a grinding wheel or a wire brush.

Bevel cutting severs the plate at an angle and prepares plate edges for welding. Single- and double-bevel cuts vary from 30 to 45°. Low-quality cut surfaces may result for bevels of less than 10° or bevels greater than 60°.

Standard cutting equipment is used to cut bevels in straight-line cutting applications and in circular cutting applications. Most standard machines do not have provisions for rotating the torch with a change in direction. Some tape-controlled machines have a rotating-torch mechanism for cutting beveled shapes. Circular bevel cutting can be done with portable equipment that uses a radius bar. In some applications, the workpiece may be rotated past a stationary cutting torch. Oxyfuel cutting torches, angled and mounted on circular band tracks, travel around pipes to prepare pipe ends for welding.

Cutting tip selection for beveling requires special attention. When the cutting tip is inclined to the metal surface, heat transfer to the plate is reduced. Because of the angle, the actual cutting thickness is greater than the plate thickness.

Preheating for Bevel Cutting. Because torch inclination does not affect the oxidizing action of the cutting oxygen stream if preheat is adequate, tip size (size of cutting oxygen orifice) and cutting oxygen flow should be selected on the basis of the cut thickness. As the angle of inclination increases, heat transfer becomes less efficient and more preheat gas flow is required, as shown in Fig. 13. In contrast

Fig. 13 Preheating for bevel cutting. Preheat flames become less effective as bevel angle increases.

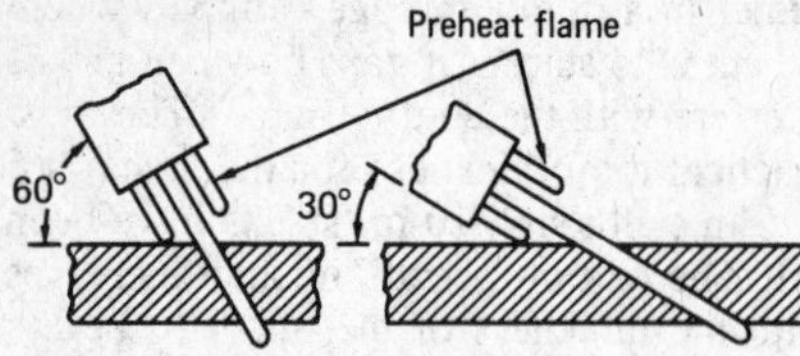

Fig. 14 Cutting a bevel with an extra preheat tip or bevel adapter

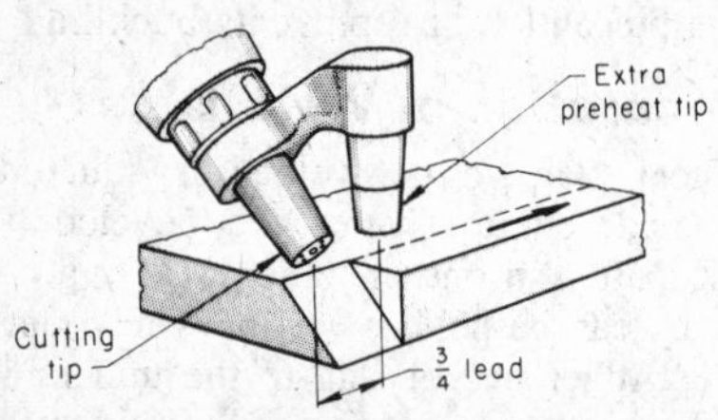

to perpendicular cutting, more preheating is required for thinner materials.

Preheating is an important step in the bevel cutting process. The best position for the oxyacetylene cutting flame is one in which the uppermost cones of flame touch the plate surface while the lower cones extend into the cut. A second torch may precede the cutting torch for additional preheat. Extra preheat tips, called bevel adapters, have a special heating nozzle offset from the cutting tip so that when the cutting tip is inclined at 45° the heating tip is normal to the plate, as shown in Fig. 14. Bevel adapters are right- or left-handed; therefore, two may be needed. Much beveling is done, however, without special equipment.

Data for bevel cutting with adapters, using natural gas preheat, are given in Table 14. The table provides torch settings for cutting bevels without the use of extra preheat tips. The next larger tip should be selected to provide the extra preheat needed due to torch inclination. Cutting tip height for oxynatural gas beveling should be about 1/8 to 1/4 in. at the closest point above the work.

When a plate must be trimmed to size as well as beveled, both operations can be done at the same time, using two torches close together, with the perpendicular torch leading, as shown in Fig. 15. Multi-tip bevel cutting heads are available for this purpose. Two torches may similarly be used for cutting a double bevel, with the leading torch cutting the underside bevel, as shown in Fig. 15(b). Three torches can also be used to cut a double bevel with a short vertical face between bevels; the torch cutting the underside bevel leads, the perpendicular torch follows, and the torch cutting the upper bevel is last, as shown in Fig. 15(c).

Because preheat is so important in bevel cutting and because the acetylene flame has excellent heat transfer properties, oxyacetylene bevel cutting is more efficient than oxynatural gas or oxypropane bevel cutting in thicknesses up to 4 in.

Table 14 Torch settings for various bevels, using extra preheat tip (bevel adapter) in oxynatural gas cutting(a)

Bevel dimensions, in. A	B	C	Cutting oxygen orifice diam(b), in.	Pressure, psi Oxygen	Natural gas	Speed, in./min
Bevel angle (α) of 45°						
1/4	1/4	3/8	0.037	45	10	13
3/8	3/8	1/2	0.037	45	10	13
1/2	1/2	11/16	0.054	60	10	12 1/2
3/4	3/4	1 1/16	0.054	60	10	11
1	1	1 7/16	0.054	60	10	11
1 1/4	1 1/4	1 3/4	0.055	60	10	11
1 1/2	1 1/2	2 1/8	0.055	65	10	10
2	2	2 13/16	0.073	70	10	9
2 1/2	2 1/2	3 1/2	0.073	70	10	6 1/2
3	3	4 1/4	0.073	80	10	6
Bevel angle (α) of 30°						
5/32	1/4	5/16	0.054	60	10	15
7/32	3/8	7/16	0.054	60	10	14
1/4	7/16	1/2	0.054	60	10	14
9/32	1/2	9/16	0.054	60	10	14
3/8	21/32	3/4	0.054	60	10	14
7/16	3/4	7/8	0.054	60	10	14
1/2	27/32	1	0.054	60	10	13
9/16	31/32	1 1/8	0.054	60	10	13
3/4	1 1/4	1 1/2	0.054	60	10	12
7/8	1 1/2	1 3/4	0.054	60	10	11
1	1 3/4	2	0.054	60	10	10 1/2
1 5/32	2	2 5/16	0.055	65	10	10
1 1/4	2 5/32	2 1/2	0.055	65	10	9
1 7/16	2 1/2	2 7/8	0.055	70	10	9
1 1/2	2 19/32	3	0.055	70	10	7
1 3/4	3 1/32	3 1/2	0.073	75	10	7
2	3 15/32	4	0.073	75	10	6 1/2

Partial bevel   Full bevel

(a) Extra preheat tip to be set 1/4 to 3/8 in. above plate surface. For proper torch setting, adjust preheat for slight whistle, then reduce oxygen slightly. (b) Two-piece tips for high-speed machine cutting. Corresponding tip sizes of different manufacturers vary in performance because of differences in construction of torches and mixers.

Cutting of Cast Iron

Because of its high carbon content, cast iron resists ordinary OFC. Cast iron contains carbon in the form of graphite flakes and nodules and iron carbide, which hinder oxidation of the iron and interrupt cutting action. For this reason, cast iron is considered oxidation-resistant with respect to OFC. Iron castings usually are oxyfuel gas cut to remove gates, risers, and defects and to repair or alter castings in service, or for scrap.

Cutting is done manually, using more preheat and cutting oxygen than is used in cutting equal thicknesses of steel. Increased gas flow is obtained by using a larger tip. Preheat flames are adjusted to be carburizing; the excess fuel gas streamer is approximately equal to the thickness being cut. This adjustment helps to maintain preheat in the cut, because the excess fuel gas combines with cutting oxygen beyond the tip.

To start the cut, the area of the initial cut is preheated, the point of starting is heated to melting, and then the cutting oxygen is turned on. In cutting, the torch advances in a weave pattern—the size of the

Fig. 15 Guidelines for edge preparation of steel by OFC. (a) Simultaneous trimming and beveling used two torches with torch 1 (perpendicular torch) leading. (b) Two torches used to cut a double bevel with torch 1 cutting the underside bevel and torch 2 cutting the tip bevel. (c) Three torches used to cut a double bevel with a vertical land. Torch 1 cuts the underside bevel; torch 2 (perpendicular torch) cuts the land face, and torch 3 cuts the upper bevel.

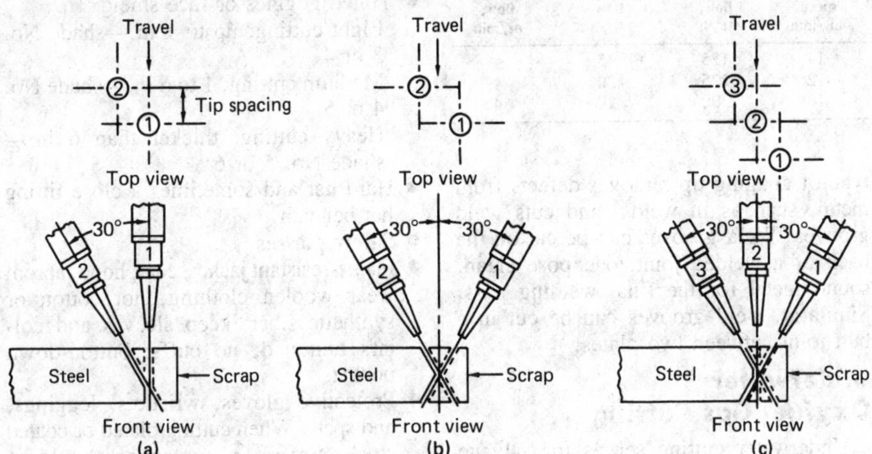

half circles and the speed of advance depend on the thickness of the cut, as shown in Fig. 16. This oscillating technique helps the cutting jet to get behind the cut and blow out the slag and molten metal at the cut. The kerf is wider, and the cut edges are considerably rougher than in cutting steel. Oxygen and acetylene consumption is greater.

Other methods for cutting cast iron, more effective than OFC, include POC and FOC processes. These methods are used in foundries for removing gates, risers, and sprues, or for breaking up ladle skulls.

Cutting of Bars and Structural Shapes

To cut round steel bars, gas pressure should be adjusted for the maximum thickness (diameter), and the cut should be started at the outside. As the cut progresses, the torch should be raised and lowered to follow the circumference. The bar should be nicked with a chisel to make a burr at the point where the cut is to begin. Oxyfuel gas cutting of round bars is usually manual, so the cut surface is rough. The chief advantage of OFC over sawing is that the torch can be brought to the work. The same advantage applies to OFC of various structural shapes.

Fig. 16 Movement of torch when cutting cast iron

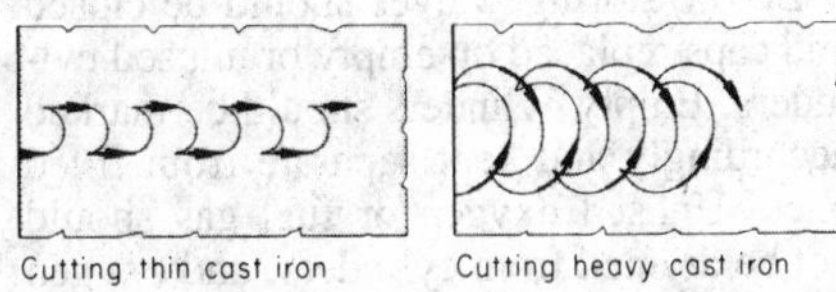

Nesting of Shapes

Savings in material, labor, and gas consumption can be gained by nesting parts in the stock layout for single-torch or multi-torch operation. Savings can be realized whenever one cut can be made instead of two. Sometimes a shape can be modified for better nesting. The advent of computer graphics allows cutting machine programmers to create layouts of part patterns on cathode ray tube screens, manipulating cutting patterns for greatest plate utilization. Several firms offer programs that closely optimize parts nesting.

Close-Tolerance Cutting

Shapes can be oxyfuel gas cut to tolerances of $^1/_{32}$ in. in plate up to $2^1/_2$ in. thick with multi-torch pantograph or coordinate-drive machines. Conditions must be optimum:

- Operators must be well trained and experienced.
- Machines must be properly adjusted, so that they can trace and retrace a pattern with accuracy of ± 0.005 in. in a given direction.
- Workpiece must be flat and free of dirt, scale, grease, and oil.
- Workpiece should be supported or restrained to prevent movement of the shape during cutting.
- Cutting tip must be clean (selected for high-quality, narrow-kerf cutting).
- Cutting speed and gas pressures should be tuned to specific production conditions.

Optical tracing devices can follow an ink line 0.025 in. wide on paper. If cutting speed does not exceed 30 in./min, a template with an outside corner radius of $^3/_{32}$ in. will produce a sharp corner. A $^3/_{32}$-in. inside radius on the template produces a somewhat larger radius in the workpiece, depending on thickness and cutting tip. Numerical control tracing, whether punched tape or computer assisted, offers the greatest accuracy. Inaccuracies introduced by distortion of templates are avoided, and electronic directions are more exacting than those generated by mechanical cutting machines, which set limits on cutting speed.

Metal Powder Cutting

Finely divided iron-rich powder suspended in a jet of moving air or dispensed by a vibratory device is directed into the gas flame in POC. The iron powder passes through and is heated by the preheat flame so that it burns in the oxygen stream. Heat generated by the burning iron particles improves cutting action. Cuts can be made in stainless steel and cast iron at speeds only slightly lower than those for equal thicknesses of carbon steel. By adding a small amount of aluminum powder, cuts can be made through copper and brass.

Equipment. Metal powder cutting uses a gas torch with an external powder attachment, or a torch with built-in powder passages. A vibratory or pneumatic powder dispenser, air supply, and powder hose are required, in addition to fuel and oxygen lines. The equipment may be used manually for removing metal, such as risers from castings, or mechanized for straight-line or shape cutting by machine. Powder cutting torches mounted on gas cutting machines are capable of cutting stainless steel. Table 15 gives normal operating conditions for metal powder cutting of type 302 stainless steel.

Thickness. In plate 1 to 4 in. thick, powder cuts can be produced by machine to an accuracy of $^1/_{32}$ to $^1/_{16}$ in. Heavier sections are seldom cut except for the trimming of castings; in this application, hand cutting requires greater allowances to avoid damage. Typical POC applications include removal of risers; cutting of bars, plates, and slabs to size; and scrapping.

Quality of Cut. The kerf has a layer of scale which, on stainless steel, flakes off as the workpiece cools. The surface exposed after scale removal has the texture of sandpaper. Light grinding normally is sufficient to smooth high spots and remove iron particles and oxide. Unstabilized austenitic stainless steel may become sensitized by the heat of cutting. Powder cut cast iron develops a hardened

Table 15 Metal powder cutting parameters for type 302 stainless steel

Plate thickness, in.	Cutting oxygen orifice diameter, in.	Cutting oxygen pressure, psi	Cutting speed, in./min	Oxygen flow, ft^3/h	Acetylene flow, ft^3/h	Powder flow, oz/min
1/2	0.040	50	14	125	15	4
1	0.060	50	12	225	20	4
2	0.060	50	10	300	20	4

case at the surface, which may require annealing or removal by grinding.

Chemical Flux Cutting

Chemical flux cutting processes are well suited to materials that form refractory oxides. Finely pulverized flux is injected into the cutting oxygen before it enters the cutting torch. The torch has separate ducts for preheat oxygen, fuel gas, and cutting oxygen. When the flux strikes the refractory oxides that are formed when the cutting oxygen is turned on, it reacts with them to form a slag of lower melting temperature compounds. This slag is driven out, enabling oxidation of the metal to proceed.

Chemical fluxing methods are used to cut stainless steel. The operator should have an approved respirator for protection from toxic fumes generated by the process.

Oxyfuel Gas Gouging

Oxyfuel gas gouging differs from OFC in that, instead of cutting through the material in a single pass, the process makes grooves or surface cuts in material. This process uses special cutting torches or standard torches and special tips. Tips for gouging vary to suit the size and shape of the desired groove or surface cut, as shown in Fig. 17. Torches may include attachments for dispensing iron powder to increase the speed of cutting or to permit the scarfing of stainless steel. Gas consumption, especially of oxygen, is much greater than in ordinary OFC.

A bent tip, as shown in Fig. 17(b), makes it possible to hold the torch so that the cutting flame will strike at a low angle. This type of gouging tip removes defects from metal, such as in welds, and cuts weld grooves. Weld grooves can be cut on the back of a welded joint to expose clean, sound metal for the final welding pass. Similarly, weld grooves can be cut in a butt joint between two plates.

Fig. 17 Gouging tips. (a) Tip for cutting flat grooves. (b) Tip for cutting round grooves

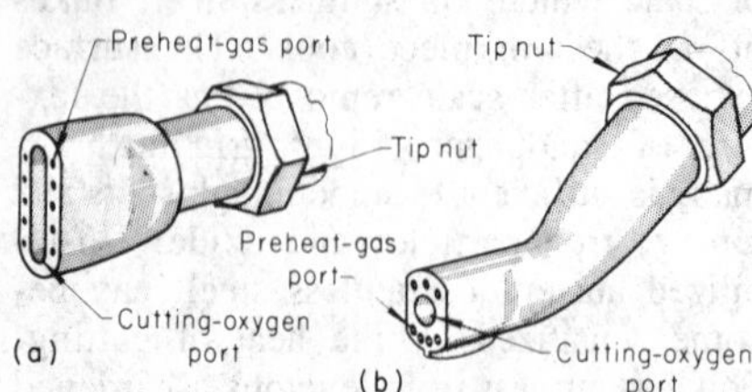

Underwater Oxyfuel Gas Cutting

Underwater cutting serves for salvage of ships, piers, and other submerged structures and for cutout and replacement of sections of offshore structures and pipelines. Underwater OFC uses a special torch, similar to the standard manual torch with an added passage for a stream of compressed air to shield the flame and a spacer to help the operator maintain standoff distance. The shield of compressed air protects the preheat flame and keeps water out of the cutting area. The spacer is an aid to the operator in murky conditions where he may not be able to see well. Hoses connect to fuel gas and oxygen lines and to the line of compressed air.

Hydrogen, methylacetylene-propadiene, and natural gas are used under water. Acetylene is safe for use only down to 20 ft; it should not be used at pressures greater than 15 psi. As depth of water increases, so must pressure of the gases. A rule of thumb is to add 1 psi for every 2 ft of depth to the settings used at 15 psi.

For more information on underwater cutting, see the article "Underwater Welding and Cutting" in this Volume.

Safety*

The hazards of combustion or possible explosion associated with the gases used in OFC, as well as the presence of toxic gases and dust, make it necessary for the user to follow established safety precautions. The three areas of most concern, dealt with in this section, are protective clothing, handling and storage of gas cylinders, and the working environment.

Protective Clothing. Size, nature, and location of the work to be cut dictate the necessary protective clothing. The operator may require some or all of the following:

- Tinted goggles or face shield:
  Light cutting, up to 1 in.—shade No. 3 or 4
  Medium cutting, 1 to 6 in.—shade No. 4 or 5
  Heavy cutting, thicker than 6 in.—shade No. 5 or 6
- Hard hat and sometimes a close-fitting hat beneath
- Safety glasses
- Flame-resistant jacket, coat, hood, apron; wear woolen clothing, not cotton or synthetic fiber; keep sleeves and collars buttoned, no cuffs; button-down pockets
- Protective gloves, wristlets, leggings, and spats. When cutting plated or coated stock, respiratory protection should be used if necessary.
- Clothing should be free of grease and oil, and free of ragged edges.

Cylinders. Fuel gas cylinders should be placed in a location where they cannot be knocked over. Cylinders should be kept upright with valve ends up to avoid hazardous liquid withdrawal; the protective caps should never be removed except when the cylinders are connected for use. Cylinders should never be exposed to high heat or open flame, which could cause them to explode.

Cylinders should not be placed where they might complete an electric circuit, such as against a welding bench. They should not be used as an electrical ground or as a surface on which to strike an arc.

Damage to cylinders by impact should be avoided. Regulators should be removed before cylinders are moved, except on portable outfits. Slings or magnets should not be used to move cylinders. Cylinders should not be used as supports or rollers.

Damaged or defective cylinders should be removed from service immediately, tagged with a statement of the problem, and returned to the supplier. Leaking cylinders should be transported to an isolated location outdoors, the supplier notified, and his instructions for disposal followed.

No attempt should be made to fill a small cylinder from a large one. Filling requires special equipment and training.

Before storing, valves should be closed and caps replaced on empty or unused cylinders. Empty cylinders should be marked accordingly and kept separate from filled ones. Unused oxygen or fuel gas should not be drained from cylinders, and oxygen

*From Welding Engineering Data Sheet No. 491, "Safety Checklist for Oxyfuel Gas Cutting"

Fig. 18 Components of PAC torch

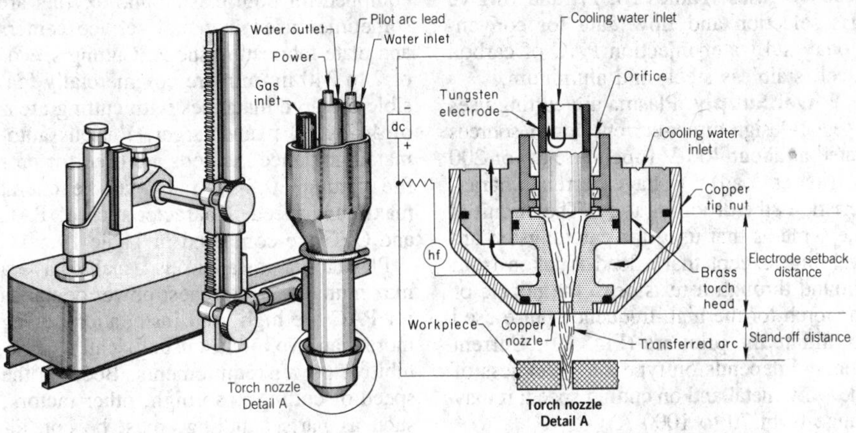

cylinders should be kept away from fuel gas cylinders.

Working Environment. Torches should not be used around chlorinated solvent vapors; heat forms phosgene and other corrosive, toxic products from them. Adequate ventilation must be provided, especially when alloys containing lead, cadmium, zinc, mercury, beryllium, or other toxic elements are being cut. A flame should not be used in a closed vessel or pipeline that has held flammable or explosive material, such as gasoline. AWS A6.0, "Safe Practices for Welding and Cutting Containers That Have Held Combustibles," should be followed. Adequate ventilation should also be provided when work is being performed in a confined area. Under such conditions, the person doing the cutting should have an assistant in case trouble occurs.

Fire prevention warnings in American National Standards Institute (ANSI) Z49.1, "Safety in Welding and Cutting," which have legal status as Occupational Safety and Health Administration (OSHA) regulations, should be heeded. Workers should know the locations and operating procedures of the nearest fire extinguishers.

If cutting is to be done over concrete, it should be protected with metal or another suitable material; concrete spalls explosively when overheated. Hoses should be protected from sparks, hot slag, hot objects, sharp edges, and open flame.

Plasma Arc Cutting

Plasma arc cutting (PAC) uses a high-velocity jet of high-temperature ionized gas to sever carbon steel, stainless steel, aluminum, copper, and other metals. The plasma jet melts and displaces the workpiece material in its path. Because PAC does not depend on a chemical reaction between the gas and the work metal, because the process relies on heat generated from an arc between the torch electrode and the workpiece, and because it generates very high temperatures (50 000 °F compared to 5500 °F for oxyfuel), it can be used on almost any material that conducts electricity, including those that are resistant to OFC. The past decade has seen great increase in use of PAC, due to its high cutting speed (100 in./min for 1-in.-thick low-carbon steel). The process increases productivity of cutting machines over OFC without increasing space or machinery requirements.

Principles of Operation

At temperatures above about 10 000 °F (as in a welding arc), gases partially ionize and exist as a plasma—a mixture of free electrons, positively charged ions, and neutral atoms. The plasma torch confines the plasma in an arc chamber. The central zone of the plasma reaches a temperature of 20 000 to 50 000 °F and is completely ionized. In the arc, the plasma expands and accelerates as it moves through the torch orifice, leaving with enormous energy.

The PAC torch, as shown in Fig. 18, is similar to a gas tungsten arc welding (GTAW) torch, but the tungsten electrode is recessed into a nozzle with a small opening. It is more rugged, built to withstand long work cycles at high temperatures. A high-frequency pulsing potential initiates a pilot arc between the tungsten electrode (cathode, negative) and the copper nozzle (anode, positive), both of which are water cooled. This pilot or nontransferred arc initiates an external transferred arc between the torch electrode and the workpiece, which is connected as the anode (positive). The pilot arc shuts off, and the external arc supplies the energy to sustain cutting. The PAC circuit is shown in Fig. 19.

Plasma arc torches vary in design, as shown in Fig. 20 and 21. The conventional PAC process (Fig. 20) relies on a copper nozzle to constrict the arc and attain high plasma temperatures. Reducing the size of the orifice raises plasma temperature, desirable for cutting. The orifice can be constricted only so far before double-arcing starts and destroys the nozzle. Typical operating conditions for conventional PAC of aluminum, stainless steel, and carbon steel are given in Table 16.

Water-injection PAC (Fig. 21) injects water inside the nozzle, constricts the arc, and increases current density. It extends the hot region of the arc—the arc is longer and narrower—to produce cuts square within $^1/_2$° from the vertical. Water-injection PAC allows higher cutting speeds and reduces dross. Operating conditions for water-injection PAC are given in Tables 17 and 18.

In the torch illustrated in Fig. 18, gas is introduced around the electrode at the top. Passage through the lower half of the electrode chamber causes the gas to swirl around the walls of the nozzle, helping to stabilize and constrict the arc. The layer of cooler gas next to the nozzle acts as a thermal and electrical insulator and prevents melting of the nozzle.

The plasma jet heats the workpiece by bombardment with electrons and by transfer of energy from the high-temperature, high-energy gas. The plasma arc constricting orifice in the torch nozzle concentrates the arc energy on a small area of the workpiece and heats the area to melting temperature. The plasma jet forces the melted material from the kerf.

Cutting power of the torch depends on the intensity and velocity of the plasma,

Fig. 19 Plasma arc cutting circuit. The process operates on DCEN (straight polarity). The arc is initiated by a pilot arc between the electrode and torch nozzle. Pilot arc is initiated by the high-frequency generator, which is connected to the electrode nozzle.

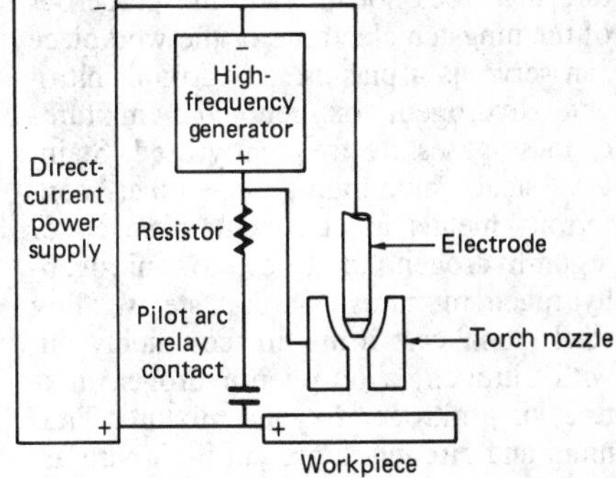

Fig. 20 Conventional PAC torch, transferred arc

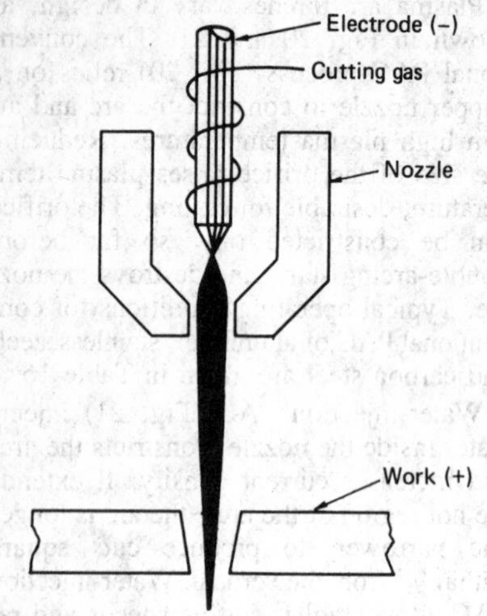

Fig. 21 Water-injection PAC torch

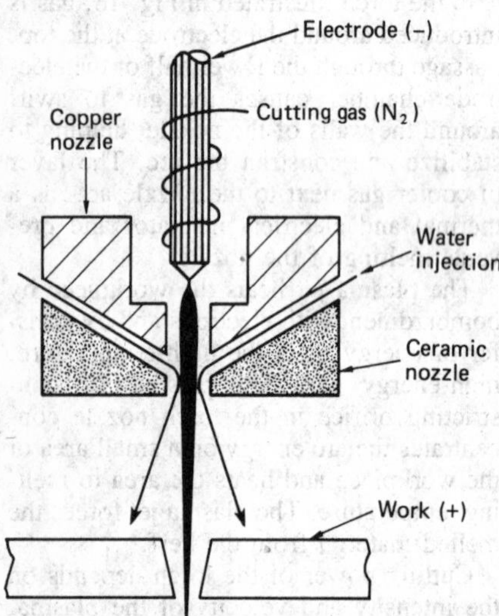

which depends in turn on gas composition, its inlet pressure, and the shape and size of the nozzle orifice.

Selection of Gas. Any gas or gas mixture that does not degrade the properties of the tungsten electrode or the workpiece can serve as a plasma gas. Argon, nitrogen, hydrogen, oxygen, and mixtures of these gases are frequently used. Stainless steel, aluminum, and other nonferrous metals are cut with nitrogen or argon-hydrogen mixtures, or nitrogen-hydrogen mixtures. Carbon steels, alloy steels, and cast irons are commonly cut with nitrogen, a nitrogen-hydrogen mixture, or a nitrogen-oxygen mixture. Titanium and zirconium are cut with pure argon due to their embrittlement caused by reactive gases. Tables 16, 17, and 18 give gas selection and flow data for conventional and water-injection PAC of carbon steel, stainless steel, and aluminum.

Power Supply. Plasma arc cutting uses special-design direct-current power sources rated at about 400 V (open circuit) or 200 V (under load). Voltage-current characteristic is the drooping type. The circuit is the same as that used for gas tungsten arc welding, except that a lead is taken from ground through a resistor to the nozzle of the torch for the high-frequency pulse used to initiate the pilot arc (Fig. 19). Current demand depends on type and thickness of the work metal and on cutting speed; it may range from 70 to 1000 A.

Work Metals. Plasma arc cutting produces clean, smooth edges in stainless steel, aluminum alloys, and carbon steel. It can cut stainless steel up to 6 in. thick, aluminum alloys up to about 8 in. thick, and carbon steel up to 3 in. thick. It also works well on magnesium, copper, nickel, and copper-nickel alloys. Titanium can be cut by either PAC or OFC, but is cut more rapidly by OFC.

Reduction in cutting time compared with OFC is considerable on long continuous cuts, such as those made in sections for bridge construction, shipbuilding, and tank fabrication. The relative economy of shorter cuts with frequent starts and stops may be less. Figure 22 compares cutting speeds for various thicknesses of carbon steel using PAC and OFC methods. Machines equipped for both plasma and oxyfuel are sometimes used in metal service centers and plate fabricating shops. Cutting speeds of 2 to 240 in./min are commercially feasible on large machines with cutting areas of 44 by 80 ft and larger. Various automated guidance methods are used for profile cutting—0.8-in. radii can be cut at maximum speed. Characteristics of PAC and OFC are compared in Table 19.

Plasma arc installations usually have a maximum of two torches; power demands for PAC are high, and installations using more than two torches usually call for prohibitive power requirements. Because the speed of cutting is so high, other factors, such as parts handling, must be considered when determining the number of torches. High-production setups have two cutting beds, one for loading and unloading plate, while cutting proceeds on the second bed.

Technique. With a machine-operated plasma arc torch, standoff distance from the work metal is 1/4 to 5/8 in. In manual operation, a less common procedure, the current and rate of gas flow are set, and the arc is struck by pressing a button on the torch, which is guided manually over the work. At the end of the cut, the arc is automatically extinguished, and the control opens the contactor and closes the gas valves. The operator can extinguish the arc at any time by moving the torch away from the work metal.

Table 16 Operating conditions for conventional PAC of aluminum, stainless steel, and carbon steel

Thickness, in.	Speed, in./min	Orifice diameter, in.	Power, kW	Gas	Flow rate, ft^3/h
Aluminum					
1/4	300	1/8	60	65% Ar-35% H_2	175
1/2	100	1/8	70	65% Ar-35% H_2	175
1	50	5/32	80	65% Ar-35% H_2	200
2	20	5/32	80	65% Ar-35% H_2	200
3	15	3/16	90	65% Ar-35% H_2	200
4	12	3/16	90	65% Ar-35% H_2	200
Stainless steel					
1/4	200	1/8	60	N_2	150
1/2	100	1/8	60	N_2	150
1	50	5/32	80	N_2	170
2	20	5/32	100	65% Ar-35% H_2	200
3	16	3/16	100	65% Ar-35% H_2	200
4	8	3/16	100	65% Ar-35% H_2	200
Carbon steel					
1/4	200	1/8	55	N_2	170
1/2	100	1/8	55	N_2	170
1	50	5/32	85	N_2	170
2	25	3/16	110	N_2	200

Table 17 Operating conditions for water-injection PAC of low-carbon steel, stainless steel, and aluminum(a)

Thickness, in.	Nozzle and swirling size, in.	Gas flow, ft^3/h	Injection water flow, gal/min	Torch-to-work distance, in.	Arc settings V	Arc settings A	Travel speed, in./min
Low-carbon steel							
0.035	0.120	110	0.43	1/8	125	250	450
0.075	0.120	110	0.43	1/8	135	250	300
1/8	0.120	110	0.43	3/16	145	260	200
1/4	0.120	110	0.43	1/4	160	260	150
1/8	0.166	165	0.38	1/4	140	300	200
1/4	0.166	165	0.38	1/4	145	380	150
3/8	0.166	165	0.38	1/4	150	400	125
1/2	0.166	165	0.38	1/4	155	400	100
1/2	0.187	165	0.38	3/8	160	550	115
3/4				3/8	165	575	75
1				3/8	165	600	60
1 1/2				3/8	170	600	30
2				3/8	170	600	20
2	0.220	260	0.48	1/2	190	700	25
3				5/8	200	750	12
Stainless steel							
0.035	0.120	110	0.43	1/8	125	250	450
0.075				1/8	135	250	300
1/8				3/16	145	250	300
1/4				1/4	160	260	150
1/8	0.166	165	0.38	1/4	140	300	200
1/4				1/4	145	380	150
3/8				1/4	150	400	125
1/2				1/4	155	400	100
3/4				5/16	160	400	50
1				3/8	165	400	30
3/4	0.187	165	0.38	3/8	165	575	75
1				3/8	165	600	60
1 1/2				3/8	170	600	30
2				3/8	170	600	20
2	0.220	260	0.48	1/2	190	700	25
3				5/8	200	750	12
Aluminum							
0.035	0.120	110	0.43	1/8	125	250	560
0.075				1/8	135	250	375
1/8				3/16	145	260	225
1/4				1/4	160	260	200
1/8	0.166	165	0.38	1/4	140	300	225
1/4				1/4	145	325	200
3/8				1/4	150	350	140
1/2				1/4	155	375	120
3/4				5/16	160	400	70
1				3/8	165	400	45
1	0.187	165	0.38	3/8	165	500	80
1 1/2				3/8	170	600	45
2				3/8	170	600	30
2	0.220	260	0.43	1/2	190	700	40
3				5/8	200	750	25

(a) Plasma gas, nitrogen
Source: Metallurgical Industries, Inc.

Quality of Cut. As a result of the swirl of the plasma gas, walls of plasma arc cuts have a V-shaped included angle of 2 to 4° on one of the cut edges. Where a straight edge is required on the cut part, the operator must operate the torch carefully so that the bevel is on the scrap side of the cut. When facing the direction of torch travel, if the gas swirls clockwise, the bevel will be on the left side of the cut. In many cases, a small bevel is acceptable; it may even be used as a weld preparation. The relationship of torch travel direction to the part with clockwise swirl of the orifice gas is illustrated in Fig. 23.

Quality of cut takes into consideration surface smoothness, kerf width, degree of parallelism of the cut faces, dross adhesion on the bottom of the cut, and sharpness of top and bottom faces. Table 20 provides data on the causes of imperfections in PAC of low-carbon steel, stainless steel, and aluminum.

Width of kerf is 1 1/2 to 2 times the kerf of conventional OFC. The range is usually 3/16 to 3/8 in., although some users achieve 1/32 in. For thick work metal, width of kerf may exceed 1/2 in.

Heat-Affected Zone. The high speeds possible with PAC result in relatively low heat input to the workpiece. Heat-affected zones are therefore narrow. The HAZ on stainless steel plate 1 in. thick cut at 50 in./min is 0.003 to 0.005 in. Sensitization is usually avoided.

Bevel cutting for weld preparation is an important application of PAC. The intense heat of the process makes it suitable for all types of beveling at a higher efficiency than OFC. The PAC process produces surfaces of acceptable quality.

Safety Precautions. Operators require protection from electric shock, arc radiation, spatter, fume, and noise. The plasma arc emits more radiation than does the gas cutting flame. Emission is similar to that of the arcs of gas tungsten arc and gas metal arc welding processes. Helmets and eye protection used for these processes must also be employed when PAC processes are used.

Shades No. 11 and 12 protective lenses should be worn. Ultraviolet radiation from the plasma arc requires that the skin be protected by clothing to prevent burns. Cutting is usually performed over a water-filled table. The water traps dross and fumes and cools the work. During some PAC, noise levels may warrant the use of ear protectors. Some torches come with water mufflers, a spray of water that surrounds the torch nozzle to shield the arc and muffle noise.

The best available equipment uses the underwater process. Cutting proceeds under 1 to 2 in. of water, which may contain radiation-absorbing dye. This process is clean and quiet. Operators need no eye, ear, or fume protection.

Air Carbon Arc Cutting and Gouging

Air carbon arc cutting (AAC) and gouging severs or removes metal by melting it with the heat of an arc struck between a carbon-graphite electrode and the base metal. A stream of compressed air blows the molten metal from the kerf or groove. Its most common uses are (1) weld joint preparation; (2) removal of defective welds;

Table 18 Operating conditions for high-current water-injection PAC of stainless steel and aluminum(a)

Thickness, in.	Nozzle swirl ring size, in.	Gas flow, ft^3/h	Injection water flow, gal/min	Torch-to-work distance, in.	Arc settings V	Arc settings A	Travel speed, in./min
Stainless steel							
3	0.250	270-300	0.48	1	210	900	15
4	0.250	270-300	0.48	1	210	1000	10
5	0.250	270-300	0.48	1	220	1000	6
Aluminum							
3	0.250	270	0.48	1	210	900	30
4	0.250	270	0.48	1	210	900	15
5	0.250	270	0.48	1	210	1000	10
6	0.250	270	0.48	1	210	1000	7

(a) Plasma gas, 65% Ar-35% H_2
Source: Metallurgical Industries, Inc.

Fig. 22 Comparison of cutting times for PAC and OFC of a 12-by-12-in. square in carbon steel

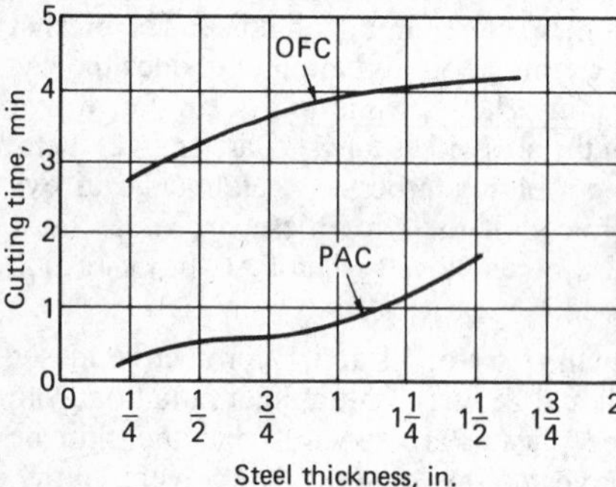

(3) removal of welds and attachments when dismantling tanks and steel structures; and (4) removal of gates, risers, and defects from castings. The process cuts almost any metal, because it does not depend on oxidation to keep the cut going. A holder clamps the carbon-graphite electrode in position parallel to an air stream, which issues from orifices in the electrode holder to strike the molten metal immediately behind the arc. The electrode holder contains an air flow control valve, an air hose, and a cable. The cable connects to the welding machine; the air hose connects to a source of compressed air. Cutting action in the AAC process is illustrated in Fig. 24.

The low heat input of air carbon arc gouging makes this process ideal for joint preparation and for weld removal on high-strength steels. Base-metal temperatures rise very little, about 150 °F in most applications.

Rough cutting is done manually. Accurate work calls for electrode holders mounted on motor-driven carriages.

Pipe Fabrication. Fabricators of structural steel, pressure vessels, tanks, and pipe use hand torches, semiautomatic torches, and fully automatic torches. A typical pipe fabrication plant uses two automatic air carbon arc torches. One, mounted on a large traveling manipulator, works with several sets of turning rolls and in tandem with submerged arc welding units. Longitudinal and circumferential seams are square-butted, welded on the inside, backgouged to sound weld metal, then welded on the outside. The second torch, mounted on a pedestal, backgouges circumferential seams at another station.

Power Supply. Constant-voltage direct current with a flat to slightly rising voltage characteristic is best for most AAC applications. Direct current is preferred; copper alloys, however, cut better with alternating current. Table 21 provides data on power sources for AAC and gouging.

Air Supply. Compressed air from a shop line or a compressor at 80 to 100 psi should be used; pressure as low as 40 psi is suitable for light work. Deep grooves in thick metal require pressures up to 125 psi. Air hoses should have a minimum inside diameter of 1/4 in. with no constrictions. Air pressure is not critical in AAC; the process requires a sufficient volume of air to ensure a clean, slag-free surface. The amount of air required depends on the type of work—3 to 33 ft^3/min for manual operations, 25 to 50 ft^3/min for mechanized operations.

Air carbon arc cutting electrodes are made from mixtures of carbon and graphite. The three basic types of AAC electrodes are:

- Direct current copper-coated electrodes, which are used most frequently because of long electrode life, stable arc characteristics, and groove uniformity.

Fig. 23 Relationship of torch travel direction to the part with clockwise swirl of the orifice gas. With the clockwise swirling plasma gas, the bevel side of the cut is on the left, looking in the direction of the torch travel. To achieve straight cuts on the inner diameter and the outer diameter of the ring, torch directions must reverse to keep the right side of the cut on the part edge.

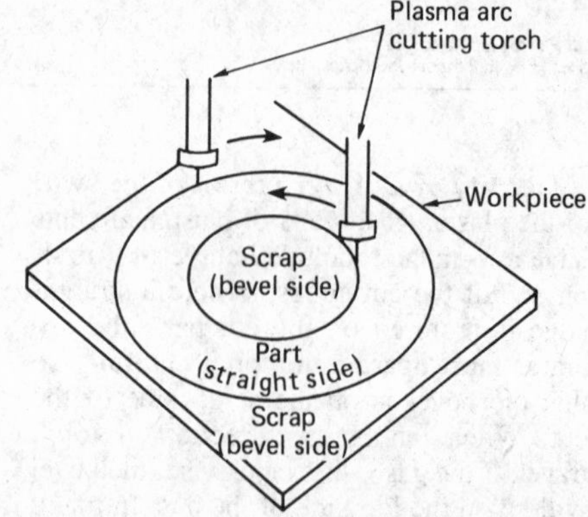

Table 19 Comparison of OFC and PAC processes

	Oxyfuel	Plasma arc
Flame temperature	5500 °F	50 000 °F
Action	Oxidation, melting, expulsion	Melting, expulsion
Preheat	Yes	No
Kerf	Narrow	Wide
Cut	Both sides square	One side square
Speed	Moderate	High
Heat-affected zone	Moderate	Narrow
Cutting ability:		
Carbon steel	Yes	Yes
Stainless steel	Requires special process	Yes
Aluminum	No	Yes
Copper	No	Yes
Special alloys	Some	Yes
Nonmetallics	No	Yes

Table 20 Causes of imperfections in plasma arc cuts

Type of imperfection	Cause of imperfection: Low-carbon steel	Stainless steel	Aluminum
Top edge rounding	Excessive speed, excessive standoff	Excessive speed, excessive standoff	Seldom occurs
Top edge dross	Excessive standoff, dross easily removed	Excessive standoff, excessive hydrogen	Excessive standoff, dross easily removed
Top side roughness	Seldom occurs	Excessive hydrogen or standoff, insufficient speed	Insufficient hydrogen
Side bevel-positive	Excessive speed, excessive standoff	Excessive speed, excessive standoff	Excessive speed, insufficient hydrogen
Side bevel-negative	Seldom occurs	Seldom occurs	Excessive hydrogen
Top side undercut	Excessive hydrogen	Excessive hydrogen	Insufficient speed, insufficient hydrogen
Bottom side undercut	Seldom occurs	Slight effect at near optimum conditions	Seldom occurs
Concave surface	Seldom occurs	Excessive hydrogen	Excessive hydrogen, insufficient speed
Convex surface	Excessive speed	Insufficient hydrogen, excessive speed	Seldom occurs
Bottom edge rounding	Excessive speed	Seldom occurs	Seldom occurs
Bottom dross	Excessive hydrogen or speed, insufficient standoff	Insufficient speed, excessive hydrogen	Excessive speed
Bottom side roughness	Insufficient standoff	Seldom occurs	Insufficient hydrogen

Table 22 Recommended conditions for air carbon arc gouging of carbon steel

Electrode diameter, in.	Groove depth, in.	Current, A(a)	Travel speed, in./min
5/16	1/8	400	65
5/16	3/16	400	45
5/16	1/4	450	36
5/16	5/16	450	33
3/8	1/8	500	65
3/8	3/16	500	57
3/8	1/4	500	46
3/8	3/8	500	25
3/8	1/2	500	46
1/2	1/8	850	72
1/2	1/4	850	57
1/2	3/8	850	35
1/2	1/2	850	24
1/2(b)	5/8	850	40
1/2(b)	3/4	850	35
5/8	1/8	1250	72
5/8	1/4	1250	50
5/8	3/8	1250	30
5/8	1/2	1250	28
5/8	5/8	1250	22
5/8	3/4	1250	30
5/8	1	1250	21

Note: Data obtained with Arcair jointed electrodes, 100 psi air pressure, 45° torch-to-work angle, 3 1/2-in. electrode stickout. Use this table as a starting point and adjust for specific conditions. (a) Direct current electrode positive. (b) Two passes

Fig. 24 Cutting action in AAC. Maximum electrode stickout, 7 in.; air stream always under electrode between electrode and work; cut and gouge in direction of air flow only; best torch-to-work angle, 35°; maintain arc length to allow air to remove molten metal. (Source: Airco Co.)

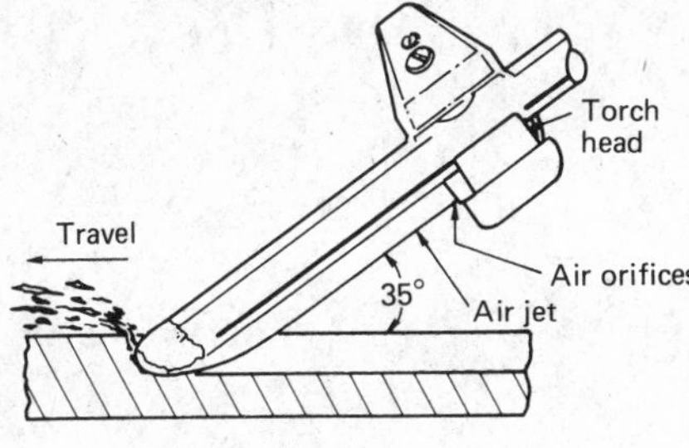

These electrodes are produced in diameters from 5/32 to 3/4 in.

- Direct current uncoated electrodes, which have limited use. These electrodes, although generally restricted to diameters of less than 3/8 in., are available from 1/8 to 1 in. in diameter.
- Alternating current copper-coated electrodes, which have additions of rare-earth metals to provide arc stabilization with alternating current. These electrodes are produced in 3/16-, 1/4-, 3/8-, and 1/2-in. diameters.

Cross sections vary; round electrode rods are most common. Electrodes also come in flat, half-round, and special shapes to produce specially designed groove shapes.

Technique. Angle of the electrode, speed of cut, and amount of current determine depth and contour of the cut or groove. The electrode is held at an angle, and an arc is struck between the end of the electrode and the work metal. The electrode is then pushed forward. Table 22 provides data on groove depth, electrode size, current, and travel speed for air carbon arc gouging.

For through-cutting, the electrode is placed at a steeper angle, almost vertically inclined. Plate thicknesses greater than 1/2 in. may require multiple passes.

Grooves as deep as 1 in. can be made in a single pass. A steep angle, approaching that used for through-cutting, and rapid advance produce a deep, narrow groove; a flatter angle and slower advance produce a wide, shallow groove. Electrode diameter directly influences groove width. Operators should use a wash or weave action to remove excess metal such as risers and pad stubs, or in surfacing. Smoothness of the gouged or cut surface depends on the stability of electrode positioning, as well as on the maintained steadiness of the electrode as it advances during the cutting operation. Mechanized gouging, with the electrode and holder traveling in a car-

Fig. 25 Components of an oxygen arc electrode

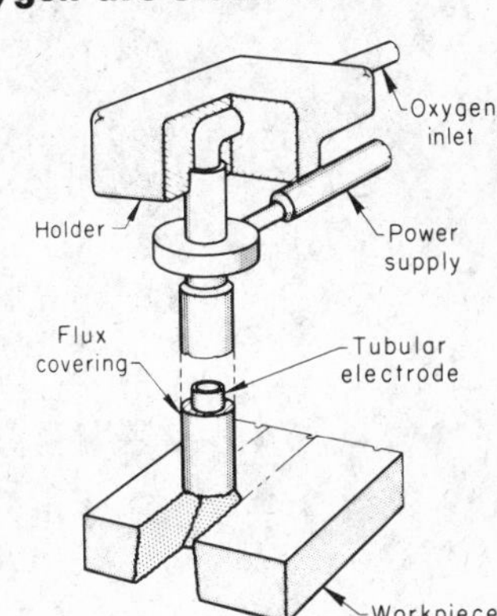

Table 21 Power sources for AAC and gouging

Equipment	Polarity	Use
Variable-voltage motor-generator, resistor, and resistor grid	Direct current	All electrode sizes
Constant-voltage motor-generator, rectifier	Direct current	Electrodes over 1/4 in. in diameter
Transformer	Alternating current	Alternating current electrodes only
Rectifier	Alternating current, direct current	Direct current from 3-phase transformer only; single-phase source not recommended. Use alternating current with alternating current electrodes only

riage on a track, produces surfaces smoother than does manual work.

Absorption of Carbon. Reverse polarity AAC removes metal faster than straight polarity. However, the current carries carbon from the electrode to the base metal, increasing its carbon content. To minimize hardenability, adjust the air stream to ensure removal of all molten metal.

Oxygen Arc Cutting

Oxygen arc cutting uses a flux-covered tubular steel electrode. The covering insulates the electrode from arcing between it and the sides of the cut. The arc raises the work material to kindling temperature; the oxygen stream burns the material away. Oxidation, or combustion, liberates additional heat to support continuing combustion of sidewall material as the cut progresses. The electric arc supplies the preheat necessary to obtain and maintain ignition at the point where the oxygen jet strikes the surface of the work. The process finds greatest use in underwater cutting.

When cutting oxidation-resistant metals, melting action occurs. The covering on the electrode acts as a flux. The electrode covering functions in a manner similar to that of powdered flux or powdered metal injected into the gas flame in the flux-injection method of OFC of stainless steel. Components of an oxygen arc electrode are shown in Fig. 25.

Equipment. Oxygen arc cutting uses direct or alternating current, although direct current electrode negative (DCEN) is preferred. The electrode and the electrode holder convey the electric current and oxygen to the arc. Electrode holders must be fully insulated; underwater cutting requires a flashback arrester, and the electrode must have a watertight plastic coating.

Underwater Welding and Cutting

By E.A. Silva
Leader, Ocean Technology Division
Office of Naval Research

UNDERWATER WELDING AND CUTTING are processes that are unique because they are used in an underwater environment. There has been increased interest in recent years in underwater welding processes due to construction and repair requirements associated with offshore exploration, drilling, and recovery of oil and natural gas. Underwater welding and cutting procedures exist for virtually any required underwater task. While some of the approaches currently being used are complex and expensive to implement, the ability to make sound joints with appropriate properties and to cut material at an underwater work site offers ocean engineers, salvors, and others who work at sea a valuable capability.

Past stigmas attached to underwater welding and cutting that related to poor joint quality and questionable repeatability have been overcome during the past decade due to the great interest in offshore work prompted by the petroleum industry. The capability to make sound welds exists, proper inspection techniques are available, and underwater welding and joining approaches are viable options.

Underwater Welding Processes

As new applications for underwater welding are discovered, new processes are developed or existing ones are modified to suit specialized needs. Therefore, it is not possible to provide exhaustive descriptions of all the processes in use at any given time. Nevertheless, underwater welding processes can be divided into the following generic classes:

- *Wet or open-water welding*: The workpiece and the welder (diver) are located in the water; special efforts are not made to separate the materials being welded from the water. The welder/diver uses electrodes with waterproof coatings and holders, guns, or torches that have been specially designed or modified for underwater use. High-quality welding may also involve the use of gas-filled containers to protect electrodes or flux from water absorption before use.
- *Dry-spot welding*: The volume containing the arc and all or a portion of the work is separated from the water within a small gas-filled enclosure that is at ambient pressure and has unsealed openings for access. This chamber provides a protective region between the workpiece and the wet environment and is large enough to accommodate only an electrode or the welder's torch. The backside of the workpiece may remain in contact with the water, and the fixture may or may not be attached to the part to be welded.
- *Dry-box welding*: Welding is done in an open-bottomed gas-filled chamber that is large enough to accommodate the work area and at least the head and shoulders of the welder/diver in full diving dress. This approach removes water from a much larger region than the dry-spot method, and sometimes the box can be designed so that the entire assembly area (weld face and backside) is protected from water.
- *Habitat welding*: Water is displaced from a large chamber with a background gas. The weld area is isolated completely from the water, but is at ambient pressure. The welder/diver does not wear diving equipment in the habitat, and there is room for elaborate weld preparation and positioning equipment, such as jacks or hoists.
- *Chamber welding*: Welding is accomplished at a pressure of 1 atm in a pressure chamber that is placed or erected around the work site. Under these conditions, welding is virtually unaffected by water depth or properties. It is important to note that this is the only class of underwater welding that is not a hyperbaric approach. This means that it is the only class of underwater welding that is free from the pressure effects of depth.
- *Remote underwater welding*: This classification includes processes such as explosion welding, friction joining, and chemical mixture applications that do not require continuous welder/diver participation during the joining operation.

Although these classifications illustrate major approaches to underwater welding, they are not definitive. However, they do illustrate the present capability to vary the interaction of the welder/diver and the weld with water from complete immersion during wet welding to total isolation from the environment during the chamber processes.

Virtually all welding processes have been attempted under water in the laboratory or in natural waters; consequently, underwater welding technology has benefited, as improved techniques are developed. Most of these improvements entail modification to familiar procedures that help isolate the welding process from the surrounding water. Nevertheless, the majority of underwater welding applications are done with conventional procedures that are optimized by judicious electrode selection and attempts to reduce environmental effects.

Underwater welding in a dry environment, such as that provided by habitat or chamber welding, usually is done using gas metal arc welding (GMAW), gas tungsten arc welding (GTAW), or plasma arc welding (PAW). Gas metal arc and flux cored arc welding (FCAW) approaches are the most common processes selected for dry-box or dry-spot welding. These procedures work well in tailored environments when the processes have been properly adapted and qualified for underwater work.

Wet or open-water welding usually is accomplished with shielded metal arc welding (SMAW) methods. Selection of proper electrodes for the base material and protection of the electrodes from water absorption before and during welding are primary concerns for successful welding. All position and multipass welds can be made with all of these processes if proper adaptations are made for underwater use, in addition to the use of well-trained welders/divers.

Underwater Cutting Processes

Underwater cutting processes can be grouped into the following categories:

- *Underwater gas cutting*: With this process, the reaction initiating cutting action comes from the burning of a gas mixture.
- *Underwater arc cutting*: The initiating heat for melting and oxidation of the work material depends on the existence of an arc.
- *Remote underwater cutting*: Explosives or other chemical mixtures are used to cut materials without the continuous involvement of a welder/diver.
- *Mechanical cutting*: Mechanical cutting or grinding devices are used to make a cut.

Although mechanical cutting produces high-quality edges, such as during preparation for welding, it is usually slower than thermal cutting methods.

The most common underwater arc cutting procedure is the oxygen arc method, which (1) offers high cutting rates, (2) requires minimal operator skill, (3) can be used under conditions of poor visibility, and (4) can cut a wide variety of materials by utilizing oxidizing reactions and/or melting and erosion. Typically, this method is limited to materials that are no thicker than $1^1/_2$ in. The major disadvantages are the need to replace electrodes during cutting, the hazard of shock to the welder, the danger of igniting trapped explosive gas mixtures, and the considerable amount of equipment required to support the procedure.

Straight polarity welding (the workpiece is the anode) usually is used for underwater cutting and other open-water welding activities to minimize electrolytic damage to the equipment. In addition, increased thermal benefits may be derived due to electron bombardment of the positive workpiece. Alternating current is seldom used under water because of its greater hazard of shock and its corrosive effect on equipment.

Where accessibility and logistics are severe limitations, such as at great depths or for work in remote locations, underwater gas cutting can be used if reduced cutting speeds are acceptable. Hydrogen usually is used as the fuel gas. Oxygen-hydrogen underwater cutting systems date back to 1925, and the procedures are well established. More recently, oxygen/methylacetylene-propadiene stabilized gas (MAPP gas) and oxygen/gasoline have been used to minimize logistics requirements. The MAPP method is useful to 115-ft depths in cold water, and oxygen-gasoline has been used to 325 ft. Underwater gas cutting procedures require more operator skill than underwater arc methods, but they are particularly useful where nonconducting material such as paint or oxides is present and where welding machines are not available.

Since the early 1960's, shaped explosive charges have been developed for a wide variety of remote underwater cutting applications. This procedure involves placing specially designed explosive packages on the material to be cut. When these charges are fired, the blast effects are concentrated on this material, and the cut is produced. These procedures are being used routinely for underwater work, and straight, curved, and hole cutting can be done. The only factor that must be considered from structural and safety standpoints is the increased shock propagation that occurs in water as opposed to that in less dense air.

Depth Limitations

The depth at which underwater welding and cutting processes can be used is a frequent concern. Many processes have been tried in laboratory and field applications with varying success. Of these, a large number are available that allow underwater welding and cutting at virtually any depth that divers can reach. Underwater cutting and welding procedures involving gases are limited by liquefication and decomposition caused by high pressure and low temperatures. Arc processes are limited by the power required to produce arcs of substantial length; however, some form of underwater welding and cutting can be used successfully at all depths presently attainable by divers.

Environmental Effects on Underwater Welds

One of the most important effects found when welding under water is the increased quenching rate experienced when the workpiece is exposed to the aqueous environment. Although it is common practice to regard welds made in air as cast structures when considering their properties, exposed underwater welds possess properties similar to quenched as well as cast materials.

Highly stressed martensite can form during open-water welding, and the rapid quenching sometimes leads to an increase in porosity and slag inclusions. These inhomogeneities result from the limited time available for gas and slag particles to reach the weld surface before solidification is complete.

Hydrogen is often present in underwater welding as a product of the dissociation of water in an arc. The water can come from direct exposure, as in open-water welding, or from humidity effects in all other processes. The combination of highly stressed martensite and hydrogen can lead to underbead cracking. Medium- and deep-hardening steels are most susceptible to hydrogen embrittlement, although martensitic regions of underwater welded low-carbon steels also can be affected adversely. Other hydrogen-related imperfections include the appearance of flakes or small fissures on a fracture surface that are parallel to the direction of maximum stress. Similarly, slowly deformed hydrogen embrittled materials exhibit fisheyes (bright spots with cracks at their centers) on ruptured surfaces; delayed cracking may also occur.

The quantity of absorbed hydrogen required to cause serious damage is difficult to quantify; there is agreement in the literature that no single correlation exists. Hydrogen embrittlement is a function of steel type, thermal treatment, physical properties, and existing welding conditions. However, rapid quenching and/or potentially high hydrogen concentrations should be guarded against when selecting a procedure for open-water welding. If proper care is taken, open-water or any of the other underwater processes can be used to produce welds that give excellent service.

Processes that involve the use of partial- or full-protecting chambers also require special consideration of process conditions such as pick-up from shielding or background gases, humidity, and changes in solidification processes due to hyperbaric effects. Thus, the mechanical and physical properties of underwater welds are affected most by the quenching ability of the medium, chemicals that come from natural waters, pressure effects, and the potential for gas absorption from gas shielding.

Environmental Effects on Underwater Welding Processes

Welding processes are affected by the aqueous environment in two significant ways. First, arc processes tend to have smoother arcs in seawater than in fresh water. This is attributed to the stabilizing influence of ions from the salts in seawater that are produced by dissociation of the water in the arc. These extra charge carriers provide more arc stability than when the same arcs are operated in fresh water. However, this is a minor consequence of welding in seawater, and no welding process is dramatically affected by the difference between fresh water or seawater operation.

Secondly, depth, or pressure, does have a significant effect on underwater welding processes. All underwater welding is done using arcs, which are constricted by hydrostatic pressure and the tremendous cooling capacity of the large bodies of water. This results in higher than surface arc core temperatures, increased penetration, and more rapid metal transfer. It also results in increased current requirements for underwater work. However, there is disagreement in the literature on the magnitude of current or voltage increases required to sustain a given welding process. There are no reports of a need to decrease machine settings as welding or cutting processes are moved to greater water depths. Surface-supplied electrical processes require additional power to compensate for losses occurring in supply cables. Similarly, surface-supplied gas operations require increased operating pressures to compensate for hose losses with increasing depth.

The effect of pressure also can lead to decomposition of some fuel gases and liquefication of others. Acetylene seldom is used at depths exceeding $16^1/_2$ ft because it becomes unstable at greater pressures. Other gases can be used at greater depths, depending on the pressure and temperature of the work site. For example, it has been calculated that oxygen-hydrogen cutting processes can be used at depths of 4595 ft, and oxygen arc processes would be suitable at depths greater than 13 000 ft in the ocean. Similarly, MAPP gas liquefies at 115 ft in water at 32 °F, and the density of an argon shielding gas approaches that of water at approximately 11 800 ft at the same temperature. Consequently, selection of a shielding, oxidizing, or fuel gas must carefully take into consideration working depth and conditions.

Environmental Effects on Design

Designers must consider the strengths and limitations of underwater welding and cutting when designing facilities that may necessitate the use of these processes during new construction or in repair applications. Materials that require very complex welding processes on land may require elaborate habitat or chamber environments if welding must be done under water. Similarly, unless conditions at the work site are known, it cannot be assumed that working conditions will afford the welder/diver good visibility and ideal working conditions.

While underwater welding can produce welds exactly like those made in air, designers should take into account the specific application as it relates to the unique nature of underwater welding processes. For example, accessibility may be more difficult in underwater applications than land applications because of bulky diving equipment. Conversely, the welder/diver is freed from many gravitational constraints by buoyancy effects and may have ready access to work at locations that could not be reached if the welding were being done on land. Designers must understand that underwater processes offer opportunities and have constraints that are not the same as those for welding practices on land.

Underwater welding processes tend to have less latitude in procedural variation than land welding. Designers of new construction or repair procedures must have thorough knowledge of the chemical properties and heat treatment of materials to be welded under water. Complete knowledge of a relatively simple parameter, such as carbon equivalent, can be a key factor in the selection or proper application of an underwater welding procedure.

Environmental Effects on Cutting

The same constraints on fuel and oxidizing gases brought about by the pressure and temperature effect of water apply to cutting processes. Similarly, processes that include arcs experience benefits from constriction effects.

Most underwater thermal cutting processes rely on the impingement of an oxygen stream on a preheated metal surface to accomplish cutting action. Metal is removed as a reaction product or by the erosive effects of the jet. The only factor that differs under water is the increased heat loss to the surrounding medium. However, this does not handicap underwater cutting processes, and excellent work can be done at all depths that divers are capable of reaching.

Environmental Effects on the Welder

With the exception of 1-atm chamber welding, welders/divers are placed in an environment that is very different from surface welding. The welder/diver is frequently in a diving suit and wearing a helmet that tends to restrict movement and visibility. Of course, welding leathers and helmets used with surface welding also have these effects. The welder/diver should be protected from the heat-absorbing properties of water, and poor visibility and instability due to wave and current water motions may be experienced.

The wet underwater welder should be provided with a stable work platform. This may be of simpler design than the staging and platforms used on land because of the buoyancy control available to welders/divers. Although the diver requires eye protection, the shades used are frequently lighter than those required in air because of the light absorption properties of water. Infrared hazards are greatly minimized in water; the main hazard is eye damage from ultraviolet radiation. For welding applications in dark or murky water, lights are often provided, and clear water injection is sometimes used to help improve visibility.

The wet welder/diver also must avoid becoming part of the welding circuit. Diving dress minimizes this hazard, and surface control of the welding power supply is used as the major electrical safety factor. Power is only provided when a weld or cut is actually being made.

Welders/divers in habitats must guard against getting burned and must wear eye protection similar to that used in conventional welding. Diver/welder breathing gas is generally independent of the habitat background gas; consequently, fire hazards are minimized in hyperbaric environments.

Process Selection

The principal welding processes that are being used under water are SMAW, GMAW, GTAW, FCAW, and PAW. Each can be used for a variety of underwater joining tasks depending on the work to be done, the skill and experience of the welder/diver, and the underwater environment that is to be confronted. Given this wide variation in available processes, an equally wide variation in underwater weld quality and capability is possible. No single process can be recommended as best for a given job without a thorough understanding of the work environment, joint requirements, and individual capabilities. An effective means for process selection has been developed by the offshore construction community. It calls for designers to specify performance or service requirements of welds and for underwater welding engineers to prove that proposed joining approaches will meet these requirements. The designers' performance requirements are developed from the structural and regulatory constraints on proposed bonds. Proof of joint performance is obtained from procedure and welder/diver qualification tests. This emphasis on weld performance allows the constantly improving underwater welding community to respond to bonding requirements with the best available process.

Procedure Qualification

Underwater welding procedure qualification is best done under actual or simulated work-site conditions. Approaches that do not duplicate underwater conditions accurately can produce very misleading results. A sufficient amount of welding must be done to allow the effectiveness of the proposed welding procedure to be established. Materials, weld types (fillet or groove), position, number of passes, depth range, weld-metal properties, electrode types, and machine settings frequently are investigated.

Test welds are examined by both destructive and nondestructive methods to establish the usefulness of a procedure. The amount of testing needed to establish qualification is directly related to the importance of the work to be done with the process and the regulatory constraints imposed. Underwater welding differs from air welding in that microhardness surveys are recommended for some process qualifications, such as critical pipeline or structural welds, to ensure that proper weld-metal and heat-affected zone properties are produced. Because typical underwater welding operations are costly, thorough procedure qualification is clearly not a task to ignore or minimize.

Welder/Diver Qualification

Underwater welding and cutting usually are done by specially trained divers, and personnel experience is often a major factor in achieving acceptable results. Welder/diver qualification is used to establish a given welder's ability to make sound welds with a qualified welding procedure. Welder qualifications usually are granted for lengthy periods following completion of testing—not on a daily or weekly basis.

Welder/diver qualification involves making welds under actual or closely simulated conditions where the essential variables of the welding process are satisfied. As with procedure testing, a sufficient number of welds are made to allow assessment of the welder's skill with various weld types, positions, and diving equipment variations. These welds are examined by destructive and nondestructive methods, depending on the type of qualification being sought.

On-site "confirmation welds" also are recommended by experienced underwater welders. Such a weld is made at the actual site with the production welding equipment to ensure that the welder and the equipment are performing as expected before production welding is started.

Visual Inspection

Underwater welds must be subjected to the same levels of inspection as surface welds if equivalent service requirements are to be obtained. Visual inspection is the most common method of weld assessment. This may be done with or without support from hand-held or remotely operated film or television cameras. Color imaging can be particularly helpful in assessing visual weld properties.

A key factor in visual inspection, or any other inspection method, is establishing a reference system so that all participants in a project can locate areas on a weld precisely. This is particularly important when the inspection is being viewed at the surface, because a given section of weld may appear differently to the diver than to those viewing a television or remote image. Lighting and perspective effects can enhance or mask defects on underwater weld surfaces. The criteria for visual acceptance of underwater welds are the same as those for surface welds, and standard gages can be used to help establish contour compliance.

Magnetic-particle inspection can be used effectively in both gaseous and wet underwater environments. Fluorescent magnetic particles and high-intensity ultraviolet lights often are used to optimize the procedure. Permanent, direct, or alternating current magnets may be used with underwater procedures.

Magnetic-Particle Inspection

Radiographic inspection techniques are well adapted to underwater welding procedures and are similar to those required for weld evaluation in air. However, longer exposure times or increased source strength may be required to allow for the increase in radiation absorption and scatter in the water if gaseous radiation paths cannot be provided. The inspector and the user must have a clear understanding of the inspection process before work is begun; redoing work involving diving can be very costly.

Ultrasonic Inspection

Ultrasonic inspection methods are well adapted to underwater welding when properly prepared (wire brushed) surfaces are present. Experience has shown that this approach is most effective when the inspector/diver can see the ultrasonic signal during surveying. Existing techniques allow both the inspector/diver and individuals on the water's surface to view signals simultaneously, which is particularly useful when the welds being inspected are of utmost importance or controversial in nature. As with ultrasonic techniques in air, system calibration and operator proficiency are key factors in the successful identification of weld flaws.

SELECTED GENERAL REFERENCES

- Avilov, T.I., Properties of Underwater Arcs, *Welding Production,* Vol 7 (No. 2), 1960, p 30-33
- Kandel, C., Underwater Cutting and Welding, *Welding Journal,* Vol 25 (No. 3), 1946, p 209-212
- Madatov, N.M., Some Peculiarities of an Underwater Arc, *Welding Production,* Vol 9 (No. 3), 1962, p 72-76
- Mills, L., Arc Welding and Cutting Underwater, *Transaction Institute of Welding,* Vol 9 (No. 4), 1946, p 128-130, 156-158, 167
- Mishler, H.W. and Randall, M.D., Underwater Joining and Cutting—Present and Future, *Proc. Offshore Tech. Conf.*, No. 1251, Houston, 1970, p II-235 to II-242
- Silva, E.A., Welding Processes in the Deep Ocean, *Nav. Eng. Jour.*, Vol 80 (No. 4), 1968, p 561-568

- U.S.N., *Underwater Welding and Cutting,* Tech. Man. NAVSHIPS 0929-000-8010, Naval Ship Systems Command, Washington, DC, 1969, 146 p

SELECTED REFERENCES: UNDERWATER WELDING

- Brown, R.T. and Masubuchi, K., Fundamental Research on Underwater Welding, *Welding Journal,* Vol 54 (No. 6), 1975, p 178s-188s
- Chandiramani, D., Some Aspects of Bead Deposition in Underwater Gas Metal Arc Welding, *Welding Journal,* Vol 61 (No. 5), 1982, p 35-38
- Ellis, J.B., Arc Welding in the Ocean, *Met. Const. & Brit. Weld. Journal,* Vol 1 (No. 3), 1969, p 151-154
- Gibson, D.E. and Lythall, D.J., Report on the Effects of Depth on Underwater Code Quality Gas Metal Arc Welding, *Proc. Offshore Tech. Conf.,* No. 2643, Houston, 1976
- Gilman, B.G., The Application of Hyperbaric Welding for the Offshore Pipeline Industry, *Proc. Offshore Tech. Conf.,* No. 1252, Houston, 1970, p II-243 to II-248
- Grubbs, C.E., Wet Welding's Role in Underwater Repair, *Ocean Resources Engineering,* Vol 11 (No. 2), 1977, p 7-10
- Grubbs, C.E. and Seth, O.W., Multipass All-Position Wet Welding—A New Underwater Tool, *Proc. Offshore Tech. Conf.,* No. 1620, Houston, 1970, p II-41 to II-54
- Hipperson, A.J., Under-Water Arc Welding, *Journal Am. Soc. Naval Eng.,* Vol 55 (No. 4), 1943, p 767-771
- Holden, W.T., Code Quality Hyperbaric Welding of Offshore Pipelines, *Proc. ASME Pet. M. E. Conf.,* Sept 21-25, Tulsa, 1969, p 1-5
- Hrenoff, K. and Livshitz, M., Electric Arc Welding Under Water, *Journal Am. Welding Soc.,* Vol 13 (No. 4), 1934, p 15-18
- Ivanenko, V.M. and Budnik, N.M., Welding with the Arc Shielded by Vapours and Gases Evolved from the Weld Pool, *Welding Production,* No. 1, Jan, 1963, p 14-18
- Kemp, W.N., Underwater Arc Welding, *Trans. Inst. of Welding,* Vol 8 (No. 4), 1945, p 152-156
- Kirkley, D.W. and Lythall, D.J., Underwater Repair and Construction Using Advanced Wet-Welding Techniques, *R.I.N.A. Symp. on O. E.,* Paper No. 12, 1974
- Levin, M.L. and Kirkley, D.W., Welding Underwater, *Met. Const. & Brit. Weld. Journal,* Vol 40 (No. 5), 1972, p 167-170
- Loper, C.R. and Nagarajan, V., Underwater Welding of Mild Steel: A Metallurgical Investigation of Critical Factors, *Proc. Offshore Tech. Conf.,* No. 2668, Houston, 1978
- Madatov, N.M., Concerning the Effect of the Salinity of Sea Water on the Process of Underwater Welding by Consumable Electrodes, *Welding Production,* Vol 8 (No. 4), 1961, p 26-41
- Madatov, N.M., The Properties of the Bubble of Steam or Gas Around the Arc in Underwater Welding, *Automatic Welding,* Vol 18 (No. 12), 1965, p 29-34
- Madatov, N.M., Energy Characteristics of the Underwater Welding Arc, *Welding Production,* Vol 13 (No. 3), 1966, p 19-27
- Madatov, N.M., The Static Volt-Ampere Characteristics of Underwater Arcs, *Automatic Welding,* Vol 19 (No. 4), 1966, p 49-53
- Madatov, N.M., Use of CO_2 for Underwater Welding, *Welding Production,* Vol 14 (No. 12), 1967, p 21-26
- Masubuchi, K. and Anderssen, A.H., Underwater Applications of Exothermic Welding, *Proc. Offshore Tech. Conf.,* No. 1910, Houston, 1973
- Morrissey, G., Hyperbaric Welding Grows Up, *Oceanology,* Vol 7 (No. 3), 1972, p 28-29
- Perlman, M., *et al.,* Ambient Pressure Effects on Gas Metal-Arc Welding of Mild Steel, *Welding Journal,* Vol 48 (No. 6), 1969, p 231s-238s
- Sadowski, E.P., Underwater Wet Welding Mild Steel with Nickel Base Stainless Steel Electrodes, *Welding Journal,* Vol 79 (No. 7), 1980, p 30-38
- Silva, E.A. and Hazlett, T.H., Shielded Metal-Arc Welding Underwater with Iron Powder Electrodes, *Welding Journal,* Vol 50 (No. 6), 1971, p 406-415
- Stepath, M., Underwater Welding: Where Do We Go From Here, *Pipe Line Industry,* Oct, 1975, p 44-46

SELECTED REFERENCES: UNDERWATER CUTTING

- Goldberg, F., Survey of Underwater Cutting of Metals, *Welding in the World,* Vol 16 (No. 9/10), 1978, p 178-198
- Rodman, M.F., Underwater Cutting of Metal, *Journal of Am. Weld. Soc.,* Vol 23 (No. 7), 1944, p 603-609
- Slottman, G.V., *Oxygen Cutting,* McGraw-Hill, New York, 1951, p 351-367
- Spraragen, W. and Claussen, G.E., Underwater Cutting, Arc Cutting, the Oxygen Lance, and Oxygen Deseaming and Machining, *Journal of Am. Weld. Soc.,* Vol 19 (No. 3), 1940, p 81s-96s
- Wodtke, C.H., *et al.,* Underwater Plasma Arc Cutting, *Welding Journal,* Vol 55 (No. 1), 1976, p 15-24

Brazing

Furnace Brazing of Steels*

FURNACE BRAZING is a mass-production process for joining the components of small assemblies by a metallurgical bond, using a nonferrous filler metal as the bonding material and a furnace as the heat source. Furnace brazing is only practical if the filler metal can be placed on the joint before brazing and retained in position during brazing. This article describes the application of furnace brazing to the joining of low-carbon (less than 0.30% C) and low-alloy steels, usually with a copper filler metal and occasionally with a silver alloy filler metal. The low-alloy steel category includes the American Iron and Steel Institute (AISI) 23XX nickel steels, 31XX nickel-chromium steels, 41XX chromium-molybdenum steels, 43XX nickel-chromium-molybdenum steels, and other types containing less than 5% total alloy content.

Furnace brazing requires the use of a suitable atmosphere to protect the steel assemblies against oxidation, or oxidation and decarburization, during brazing and during cooling, which is accomplished in chambers adjacent to the brazing furnace. The proper brazing atmosphere facilitates proper wetting of the joint surfaces by the molten copper filler metal, usually without use of a brazing flux.

Although filler metals other than copper can be used in furnace brazing carbon and low-alloy steels, copper generally is preferred because of its low cost and the high strength of the joints produced. The high brazing temperature necessary when copper filler metals are used (2000 to 2100 °F) is also advantageous when steel assemblies are to be heat treated after brazing.

Applicability

Generally, steel assemblies that are brazed most efficiently and economically are small and weigh less than 5 lb. Much larger assemblies can be brazed in specially built furnaces; the size of assemblies is limited by the heat required to bring them to the brazing temperature. Most steel assemblies are brazed at 2000 to 2100 °F. This temperature, which is considerably higher than those employed in the heat treatment of steel, imposes limitations on furnace design and operation, including the maximum feasible size of the heating chamber, the degree of tightness and temperature uniformity that can be maintained, the time required to heat the workpieces to the brazing temperature, and the weight of loads that can be supported at 2000 °F without sagging of furnace fixtures.

Steel components that are commonly furnace brazed include machined parts, light stampings, deep drawn sheet metal parts, small forgings, and some castings. Usually, components are designed to be self-jigging—that is, capable of being assembled for brazing without the use of fixtures. Fixtures are sometimes required, but are avoided whenever possible; they add weight and change in dimensions after repeated exposure to elevated temperature. Staking, expanding, spinning, swaging, knurling, crimping, press fitting, and tack welding ensure adequate assembly for brazing.

Advantages

The principal advantage of furnace brazing over other brazing processes is that it permits the use of a variety of prepared protective atmospheres, notably the rich exothermic-based, endothermic-based, and some prepared and commercial nitrogen-based atmospheres. These atmospheres are among the least expensive; they can be generated in the plant in large volume, or in the case of commercial nitrogen-based atmospheres, they can be stored in liquid form outside the plant. They provide excellent protection against oxidation, and they can be prepared with any carbon potential in the range of about 0.2% to more than 1.0% C, depending on the atmosphere. This range of carbon potential is sufficient to accommodate all carbon and low-alloy steels, including those carburized before brazing. By selecting an atmosphere with a carbon potential that matches the carbon content of the work metal, brazing can be accomplished without carburizing or decarburizing the work metal.

Because the protective atmospheres used for furnace brazing are sufficiently reducing to iron oxide, they usually eliminate the need for fluxes when brazing carbon steel with copper filler metal. These atmospheres can reduce light oxide films present on the surfaces of workpieces at the time of entry into the furnace, and they can prevent any further oxidation of the surfaces during the brazing cycle. An oxide-free surface normally promotes wetting of the workpiece by the molten filler metal. However, some low-alloy steels that contain a total of more than 2 or 3% of chromium, manganese, aluminum, and silicon form more stable surface oxides, and they require highly reducing atmospheres (such as dry hydrogen or dissociated ammonia), a flux, or nickel plating to obtain adequate wetting action.

Another major advantage of furnace brazing is its ability to process large quantities of assemblies at low unit cost on either a batch or continuous basis. Furnace brazing is most efficient and economical when used in mass production, but it is sufficiently flexible to handle occasional small loads and low-production items, although at a higher unit cost.

Furnace brazing sometimes replaces another brazing process to increase the production rate. For instance, masonry drills of the type shown in Fig. 1 were initially assembled by induction brazing the sintered carbide tip to the low-alloy steel body, using a silver alloy filler metal. A slot was

*Revised by Gary W. Gaines, Project Engineer, Linde Division, Union Carbide Corp.

milled in the cutting end of the drill body to receive the sintered carbide tip with a light press fit. When it became necessary to produce these drills in larger quantities, furnace brazing was selected to replace induction brazing. The press fit used for induction brazing was found to be adequate to hold the drill body and carbide tip in position during furnace brazing, but the filler metal was changed to a copper paste slurry into which the tips were dipped before assembly. The drills were carried through the brazing furnace on a mesh-belt conveyor at a speed adjusted to give maximum penetration of the filler metal. Brazing was at 2030 to 2050 °F, and production rate was 1000 to 1100 $^{3}/_{16}$-in.-diam drills per hour and 500 to 600 $^{5}/_{8}$-in.-diam drills per hour.

Furnace brazing can provide close temperature control and uniformity at all stages of the brazing cycle, including cooling. It can provide atmospheric protection during both heating and cooling. It can provide a different protective atmosphere in different chambers or compartments of the furnace. This is common with commercial nitrogen-based atmospheres. Also, furnaces of special design make it possible to braze in a controlled vacuum, although this is seldom done in brazing carbon and low-alloy steels. Vacuum brazing is used for brazing stainless steels, superalloys, aluminum alloys, titanium alloys, and refractory metals.

Furnace brazing provides a uniform distribution of heat in the mating parts at the brazing temperature. However, if the difference in sectional thickness of the parts to be brazed is great, it is sometimes necessary to preheat them to just below the melting point of the filler metal, soak them until the heat is equalized, and then increase the temperature to the brazing range. If joint design and fit are satisfactory, and the correct amount and form of filler metal are provided, brazed joints are uniformly strong and sound. Several joints on the same assembly can be brazed in a single operation. With proper atmospheric protection, the brazed assemblies that leave the furnace cooling chamber (at about 300 °F) are clean and bright, requiring no cleaning.

Fig. 1 Furnace brazing of masonry drill. Carbide tip was furnace brazed to drill body in high-production application.

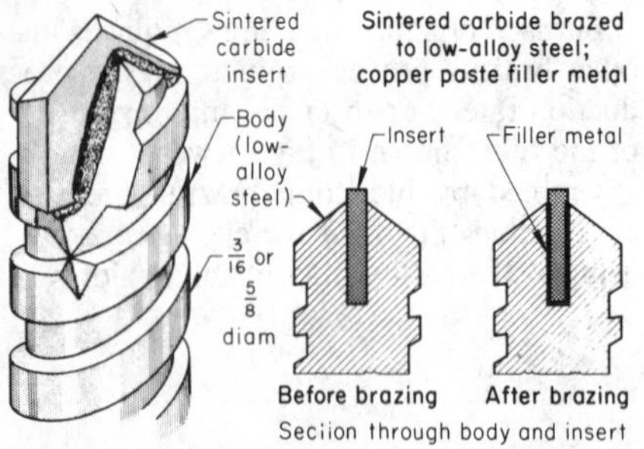

Limitations

Most of the limitations of furnace brazing are directly related to the high temperatures required to braze steels with copper filler metal. These temperatures exceed the average brazing temperature required to braze with silver alloy filler metals by 500 °F or more. They are high enough to cause grain coarsening in medium-carbon, high-carbon, and low-alloy steels; however, grain refinement can be obtained by subsequent heat treatment. The high brazing temperatures adversely affect the life of furnace components, especially those components that are exposed to the maximum temperature, such as furnace linings, electrical heating elements, muffles, rails, trays, and conveyor belts.

The initial cost of a furnace and atmosphere generator is high, compared with that of most other types of brazing equipment, although the use of commercial nitrogen-based atmosphere systems can lower initial capital requirements. For this reason, the purchase of furnace brazing equipment cannot be justified for a new project if production quantities are low, and another brazing method may have to be considered. However, if a brazing furnace is already available, a few assemblies can be brazed economically.

Commercial and generator-prepared atmospheres may contain toxic compounds. Those that have a total of 5% or more of combustible gases (H_2, CO, and CH_4) constitute a potential fire and explosion hazard. Safe operating practice and preventive maintenance of furnaces, generators, and venting systems are mandatory (see the Appendix on safe operation of brazing furnaces). Most of the disadvantages related to the life of furnace components have been lessened by improvements in furnace design and materials.

Sequence of Operations

Furnace brazing entails four processing operations: cleaning, assembling and fixturing, brazing, and cooling.

Cleaning generally is limited to the removal of oils used in machining operations. The preferred cleaning methods are alkaline cleaning, solvent cleaning, and vapor degreasing. When alkaline cleaning is used, it is important that all alkaline compounds be removed from workpieces before they enter the brazing furnace. Pigmented drawing compounds containing lead are generally removed by mechanical cleaning methods, such as dry grit blasting or wet blasting with an abrasive slurry. If they are not completely removed, drawing compounds containing lead are extremely detrimental to the quality of the brazed joint and to the life of furnace components.

Assembling and Fixturing. Components to be furnace brazed are generally designed for assembling by press fitting, expanding, swaging, or other means that eliminate the need for fixtures. However, fixtures (jigs) are occasionally required for holding parts in proper relationship or for positioning an assembly in the brazing furnace so that the molten filler metal flows in the required direction.

The cleaned components are assembled with filler metal preplaced within or adjacent to the joints to be brazed. Assemblies are then loaded onto trays for batch-type or roller-hearth continuous furnace brazing or are transferred directly to the conveyor for mesh-belt conveyorized furnace brazing.

Brazing. The assemblies are moved into the brazing chamber of the furnace, where they are heated under a suitable protective atmosphere. When the assembly reaches a temperature higher than the melting point of the filler metal, the filler metal wets and flows over the steel surfaces and is drawn into the joints by capillary action. In making the bond, the filler metal forms a solid solution with, but does not melt, the steel surface. Heating time for furnace brazing most steel assemblies is from 10 to 15 min.

Cooling. The assemblies are moved to the cooling chamber of the furnace, where they are cooled under a protective atmosphere (usually the same atmosphere as was used in the brazing chamber). They remain in the cooling chamber until they have cooled enough so that they will not discolor when exposed to air, usually to about 300 °F.

Brazing Furnaces

Furnaces used for brazing are classified into four groups: (1) batch type, with either air or controlled atmospheres, in which workpieces are loaded and unloaded manually; (2) continuous type, with either air or controlled atmospheres, which feature an automatic conveying system; (3) retort type, with controlled atmospheres; and (4) vacuum type. The batch- and continuous-type furnaces are used most frequently for brazing of carbon and low-alloy steel assemblies. The method of heating varies with the application. Some furnaces are heated by gas or oil, but most are electrically resistance heated. Heating elements in electric furnaces are made of nickel-chromium, silicon carbide, or refractory metal heating elements. In all types of brazing furnaces, accurate control of the heating-zone temperature is essential. Most furnaces have a temperature control of the potentiometer type connected to thermocouples and gas control valves or contractors. More information on temperature instrumentation and control systems used in furnace applications can be found on pages 345 to 366 in Volume 4 of the 9th edition of *Metals Handbook*.

Furnace Atmospheres. Brazing furnaces utilizing an air atmosphere can be used for brazing parts that can be protected with a paste flux. A controlled or vacuum atmosphere, however, must be used when the brazing cycle exceeds the protective capability of available fluxes. With controlled-atmosphere furnaces, a continuous flow of the atmosphere gas is maintained in the heating zone to avoid any contamination from outgassing of the metal parts and dissociation of oxides. Because many furnace atmospheres are flammable and a slightly positive gas pressure must be maintained in the furnace, some gas escapes into the work area. Adequate venting of the work area, therefore, is necessary to avoid toxicity and explosion hazards from the flammable gases. Effective venting practices are discussed later in this article.

Furnace Ratings. The capacity of a brazing furnace is generally expressed in terms of gross weight of workload processed per hour. Trays, fixtures, and conveyor belts are included in gross weight and are estimated to be about one third of the total workload. Thus, a furnace with a gross-weight capacity of 150 lb/h can be expected to process about 100 lb of assemblies per hour.

Batch-type furnaces, which heat each workload separately, normally consist of an insulated chamber with an external reinforced steel shell, a heating system for the chamber, and one or more access doors to the heated chamber. Standard batch furnaces may be box type (Fig. 2), top loading (pit type), side loading, or bottom loading. Gas- or oil-fired batch furnaces without retorts require that flux be used on the parts for brazing. Electrically heated batch-type furnaces are often equipped for controlled-atmosphere brazing, because the heating elements usually can be operated in the controlled atmosphere.

An electrically heated batch box-type furnace commonly used for copper brazing of carbon and low-alloy steel assemblies is shown schematically in Fig. 2. It consists of a heating chamber and a water-jacketed cooling chamber that is at least three times as long as the working length of the heating chamber. The brazing atmosphere, which is supplied by a generator (not shown in Fig. 2), is maintained within both the heating and cooling chambers. Throat baffle doors, which normally are closed, serve to increase effective heating length and to reduce end losses. Sliding doors at the entrance and exit of the furnace also are normally closed and help to maintain the protective atmosphere inside the furnace at high purity and to reduce the loss of atmosphere. The end doors are equipped with automatic gas-flame curtains.

The refractory brick walls and roof of the heating chamber are backed up with thermal insulation. The load-carrying hearth plates are made of a heat-resistant alloy, silicon carbide, or alumina. Side rails (not shown in Fig. 2) serve to guide trays and to protect sidewall heating elements. The heating elements under the hearth plates and on the sidewall and roof are spaced to ensure temperature uniformity and are regulated by thermocouples.

The brazing atmosphere is fed to both heating and cooling chambers through inlets located outside the exit baffle door. Gas flow is manually set and is measured by a flowmeter. An inlet for purging gas is located inside the heating chamber to provide gas for burnout. A protective atmosphere venting system is not usually required with batch box-type furnaces, because the entrance and exit doors are never opened at the same time, thus avoiding contamination of the atmosphere by a through draft of air.

The cooling chamber is constructed in one or more segments, which are joined together and to the heating chamber by welding or by bolting with gaskets at interconnecting flanges, to provide a gas-tight seal. Each segment is equipped with its own water-cooling system, thereby providing maximum control of cooling from zone to zone. The temperature of the cooling water is thermostatically controlled to minimize water consumption and to avoid condensation in liners during furnace idling. When work is resumed, condensed moisture could turn to steam and cause oxidation of steel parts.

Typically, such a furnace accommodates four trays at a time—one in the heating chamber and three in the cooling chamber. As soon as the tray in the heating chamber reaches brazing temperature, the operator pulls the end tray out of the cooling chamber and pulls the other two trays closer to the end. The operator then pushes the hot tray of brazed assemblies

Fig. 2 Batch box-type brazing furnace

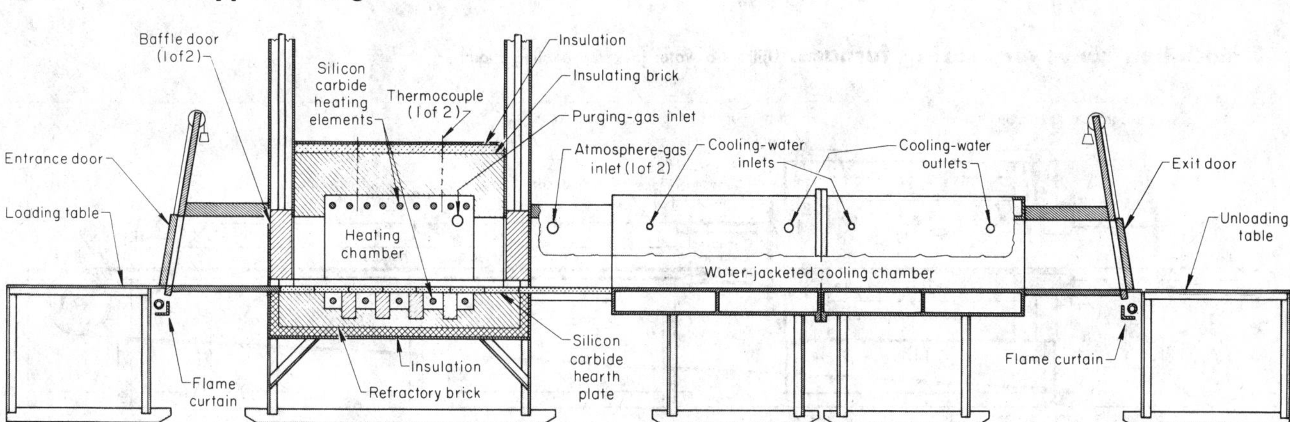

into the empty space in the cooling chamber and pushes a new tray of unbrazed assemblies into the heating chamber.

Continuous-type furnaces receive a steady flow of incoming assemblies. The heat source may be gas or oil flames, or electrical heating elements. The parts, either singly or in trays or baskets, are moved continuously through the furnace. Three common types of continuous furnaces are pusher, mesh-belt conveyor, and roller-hearth conveyor. Continuous furnaces are employed when high volumes are processed in continuous cycles. In addition to providing more consistent part quality, continuous furnaces greatly reduce the manual labor inherent in batch operations. With the continuous flow of work through the furnace, the equipment can be integrated readily into the material flow required in a manufacturing process. Continuous-type furnaces are also readily adaptable to automation.

Continuous furnaces usually contain a preheat or purging area through which the parts enter first. The parts are slowly brought to a temperature below the brazing temperature. The furnace atmosphere gas, if used, surrounds the parts under a positive pressure. The gas flow removes any entrapped air and starts the reduction of surface oxides.

Two common furnace types used for copper brazing of steels are the mesh-belt and roller-hearth conveyor furnaces. Mesh-belt conveyor furnaces offer the advantages of continuous operation at high capacity and accurate, automatic cycle timing in both the heating and cooling chambers. An electrically heated mesh-belt conveyor furnace, incorporating several special features, is shown in Fig. 3.

The construction of the heating chamber, including the brickwork, is similar to that of a batch box-type furnace. Refractory-lined baffle doors, such as those shown in the batch box-type furnace of Fig. 2, can also be installed. These are manually adjusted to the precise height of the workload or to close the throats completely during standby. The height of the end doors can also be adjusted to reduce the consumption of protective atmosphere and to aid in directing atmosphere flow. The exit door is equipped with a flame curtain. As with the batch box-type furnace, brazing atmosphere is fed to both the heating and cooling chambers through inlets located outside the heating chamber exit baffle. The control of brazing atmosphere flow is described in the subsequent section on venting for mesh-belt conveyor furnaces. The cooling chamber is similar to that of a box furnace. Heating chambers of mesh-belt conveyor furnaces commonly range in length from about 2 to 12 ft; cooling chambers are at least three times as long.

Roller-hearth conveyor furnaces are continuous furnaces in which the conveyor consists of driven rolls made of a heat-resistant alloy. The ends of the rolls extend through the sidewalls of the heating and cooling chambers, where they are held in suitable bearings and are rotated by sprockets and an endless chain controlled by a variable-speed driving mechanism.

Roller-hearth furnaces are generally used for high-production applications, particularly when loading density or height of assemblies is considerably above average. Some furnaces require the use of purging chambers preceding the heating chambers to prevent contamination of the brazing atmosphere.

Assemblies to be brazed can be loaded directly on the roll table, if they are long enough, but more commonly they are loaded on flexible trays made of a wrought or cast heat-resistant alloy. There is virtually no limit to the length of the heating and cooling chambers in roller-hearth furnaces. Typical production rates range from 350 to 2000 lb of brazed assemblies per hour.

Retort-type furnaces are batch-type furnaces in which the assemblies are placed in a sealed retort for brazing. After the air in the retort is purged by the furnace atmosphere gas, the retort is placed in the furnace and heated by conventional furnace heat (electric, oil, or gas). After the parts have been brazed, the retort is removed from the furnace and cooled. The furnace atmosphere gas is then purged from the retort. The retort is then opened, and the brazed parts removed. Because the retorts are exposed to high temperatures, they are usually made of heat-resistant alloys to avoid oxidation. Carbon and low-alloy steels generally are not brazed in a retort-type furnace. Most brazing using a BNi filler metal or chromium-bearing base metal, however, is carried out in a retort to preserve a low dew point environment.

Vacuum-type furnaces are divided into two major types of equipment—hot and cold wall furnaces. Both types of furnaces, which can be gas-fired or electrically heated, are designed with side-loading, bottom-loading, or top-loading (pit-type) configurations. In hot wall equipment, the workload is placed into the retort; the retort is sealed, evacuated, and heated externally by a furnace. Most hot wall vacuum brazing applications require that vacuum pumping be continuous throughout the heat cycle to remove the gases given off by the workload. The hot wall type of vacuum furnace is limited in size and maximum heating temperature by the ability of the retort to withstand the high pressure at brazing temperature. Vacuum brazing furnaces of this type can operate at temperatures up to 2100 °F, but most are limited to 1600 °F or lower.

In some larger installations, the hot wall vacuum furnaces have a separate vacuum in a space between the outside of the work chamber and the heating elements of the furnace. This design is referred to as the double-pumped or double hot wall retort vacuum furnace, which allows the use of

Fig. 3 Mesh-belt conveyor brazing furnace. Utilizes a water-jacketed cooling chamber

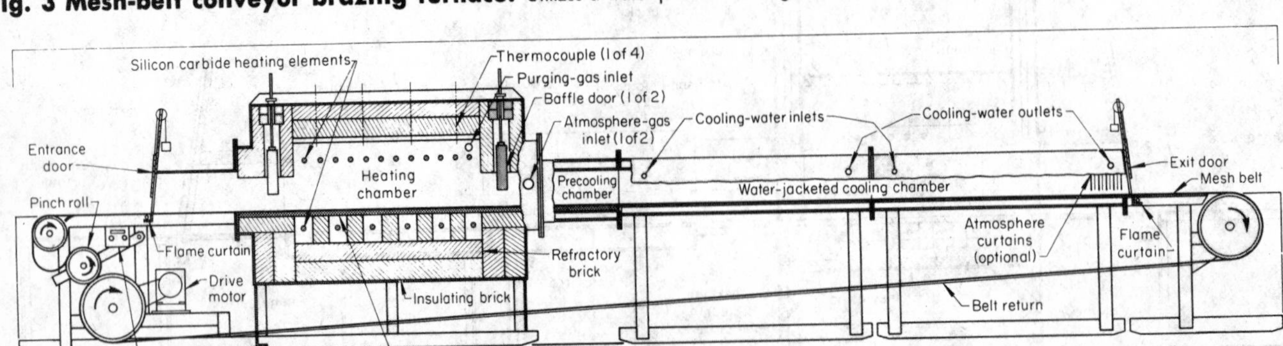

thinner retort wall materials because the work chamber is not subjected to high pressure at brazing temperature. An advantage of this furnace type is that the heating elements and thermal insulation are not subjected to high pressures, which results in extended service life of those components.

The second type of vacuum furnace, the cold wall furnace, consists of a single vacuum chamber, with the thermal insulation and electrical heating elements located inside the chamber. The vacuum chamber is usually water cooled. Suitable heat shields, typically made of multiple layers of molybdenum, tungsten, tantalum, graphite, or other high-temperature materials, are placed between the heating elements and the furnace wall to concentrate the radiant heat on the work and prevent heat losses to the furnace walls. Temperatures up to 4000 °F and pressures as low as 10^{-6} torr are possible.

Vacuum brazing equipment generally is not used for brazing of carbon and low-alloy steels. It is used mainly for brazing stainless steels, nickel-, iron-, or cobalt-based superalloys, aluminum alloys, titanium alloys, and refractory and reactive metals.

Venting for Mesh-Belt Conveyor Furnaces

Proper venting of mesh-belt brazing furnaces is of prime importance in avoiding discoloration or decarburization of assemblies because of contamination from infiltrating air, air or moisture contained in assemblies, gas flame curtains, and volatilized oil from unclean components, and in preventing the contaminants from drifting in the wrong direction through the heating and cooling chambers. A venting system of poor design can counteract the beneficial effects of the brazing atmosphere. In general, the brazing atmosphere should flow toward the ends of the furnace. With this flow pattern, the atmosphere flows counter to the work in the heating chamber, sweeping air and contaminants from the entrance, and it flows with the work in the cooling chamber, providing protection up to the point of exit.

Drafts. The direction of room drafts often corresponds to the direction of the wind outside, particularly if doors and windows are open. Sometimes large baffles are installed across the ends of furnaces to block or deflect room drafts. Exhaust hoods with stacks are mounted at the furnace ends over the door openings to reduce drafts further and to carry away the hot products of combustion. Also, the input flow of brazing atmosphere at various points in the furnace can be adjusted to meet the demands of the moment. Because all of these techniques for draft control are subject to failure, properly designed exhaust hoods and ductwork systems are needed.

Poor Design for Venting. Typical of poor venting design is the system shown in Fig. 4(a). Stacks from exhaust hoods at the entrance and exit ends of the furnace go directly out through the plant roof and have no dampers for adjustment of draft. Because winds blowing over the top of one stack do not necessarily travel with the same velocity over the top of the other stack, there is a difference in the amount of suction in the two stacks. Also, the differential velocities can reverse from hour to hour, and this can affect the direction or velocity of flow of brazing atmosphere through the furnace. Because the stacks are connected directly to the hoods, downdrafts can blow air into the heating and cooling chambers, contaminating the brazing atmosphere and disrupting temperature uniformity. In addition, the hoods are located over the tops of door openings only, leaving the sides and bottoms unprotected; thus they are ineffective in blocking room drafts, although this is their intended function.

Effective Venting. A successful venting system is shown in Fig. 4(b). With this system, the exhaust hoods feed into open collectors with dampers, not directly into stacks. The collectors draw in room air along with the hot products of combustion, which helps to cool the stacks and to minimize downdrafts. Ducts from the collectors lead to a single stack through the roof, which, with or without an exhaust fan and depending on the room height and the amount of natural draft, prevents variations in stack draft. The dampers permit adjustment of stack draft to control the direction of flow of brazing atmosphere through the entire furnace, ensuring that this flow is outward toward the end doors. This provides maximum protection and minimum contamination in the heating and cooling chambers.

Adjustable end doors are another feature that contributes to successful operation of the venting system shown in Fig. 4(b). The end doors are lowered so that they are as close to the tops of the workpieces as possible to reduce air infiltration and to minimize atmosphere consumption. Raising the door at one end of the furnace

Fig. 4 Venting systems for mesh-belt conveyor furnaces

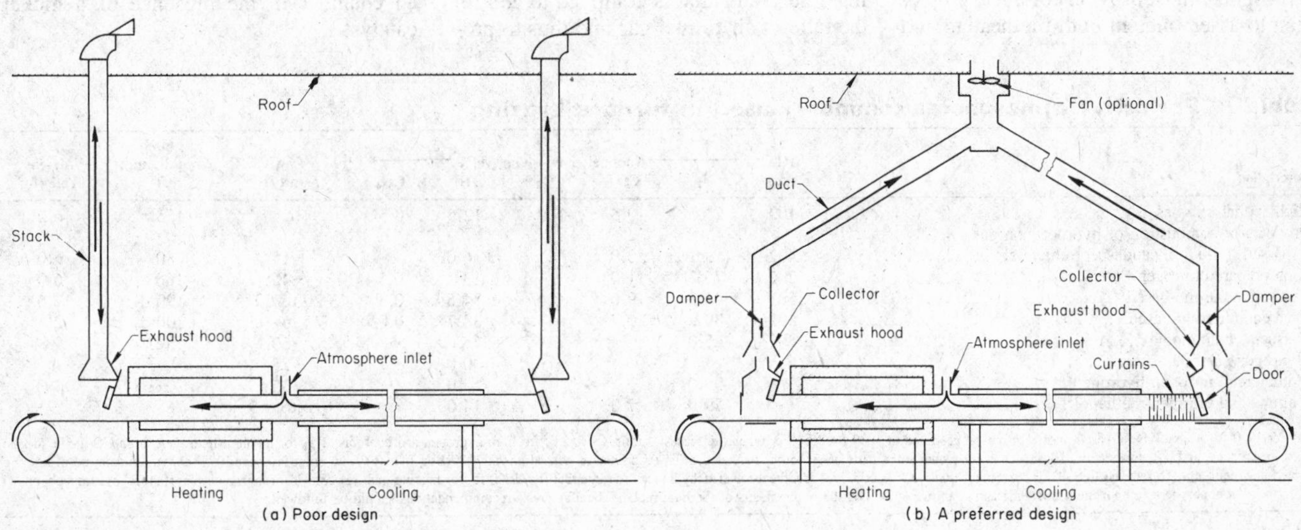

slightly higher than that at the other end creates a draft or chimney action, which also counterbalances room drafts.

As shown in Fig. 4(b), the door at the entrance end (left) is set higher than that at the discharge end (right), and the damper in the collector at left is farther open than that at the right. This encourages maximum counterflow of the brazing atmosphere, which enters the furnace at the hot end of the cooling chamber. The atmosphere is preheated by the outgoing hot workpieces and, after passing through the heating chamber, forces air and other contaminants out through the entrance door. If a room draft is flowing mildly in this same counterflow direction, air can infiltrate the cooling chamber with adverse effects. To counteract this condition, the damper and end door at the exit end are opened and those at the entrance end are partially closed. This counterbalances the room draft and provides a rate of flow of brazing atmosphere out the entrance end. Unless the venting system is subjected to extremely strong room drafts, the design in Fig. 4(b) affords complete control of the flow of atmosphere through the furnace in either direction.

Protective Furnace Atmospheres

The gas atmospheres used in furnace brazing serve primarily to protect the steel assemblies from oxidation or scaling and to assist the flow of filler metal by promoting wetting of steel surfaces. Both functions require a gas atmosphere that is reducing. When required, the atmosphere may also serve to maintain the carbon content of the steel by preventing carburization or decarburization at elevated temperatures. To satisfy all requirements, the atmosphere must provide complete protection to assemblies in both the heating and the cooling chambers of the brazing furnace. The American Gas Association (AGA) classifications of prepared atmospheres are given on pages 393 to 394 in Volume 4 of the 9th edition of *Metals Handbook*.

Rich Exothermic-Based Atmosphere. In theory, almost any reducing atmosphere can be used in furnace brazing of low-carbon steel with a copper filler metal. In practice, a rich exothermic atmosphere (AGA class 102) is usually selected, because it (1) is the least expensive of the generated atmospheres, (2) is adequately reducing, (3) has relatively low sooting potential compared with drier atmospheres containing more carbon monoxide, and (4) requires a minimum of generator maintenance.

The content of reducing gases in a rich exothermic atmosphere (12.5 to 15% H_2 and about 11% CO) usually is sufficient to promote good wetting and to maintain bright surfaces on steel assemblies. When hydrogen is low and moisture is high, refrigeration of the gas to reduce its dew point is required to ensure bright surfaces.

The carbon potential of the atmosphere is normally low (about 0.10% C), but it can be increased to about 0.40% C by decreasing the dew point and removing the carbon dioxide. Such a modification is seldom required, because the initial carbon content of the steel assemblies to be brazed is low (less than 0.30% C) or because superficial decarburization can be tolerated. When control of carbon potential is required, a rich endothermic atmosphere usually is preferred.

The nominal composition of a rich exothermic-based atmosphere is given in Table 1. This atmosphere is produced by partial combustion of a hydrocarbon fuel, such as natural gas or propane, in air using a generator that is equipped to control the ratio of air to fuel gas and thus to produce the desired composition. Detailed information regarding the generation of a rich exothermic atmosphere and the generator equipment employed is given on pages 395 to 396 in Volume 4 of the 9th edition of *Metals Handbook*.

Rich Prepared Nitrogen-Based Atmosphere. Although a rich exothermic-based atmosphere is most widely used in copper brazing of low-carbon steel, it is much less versatile than a rich prepared nitrogen-based atmosphere (AGA class 202). Because of its low dew point (−40 °F) and the absence of carbon dioxide, this nitrogen-based atmosphere is more reducing than and does not exhibit the decarburizing effects of an unpurified exothermic atmosphere. It can be used whenever decarburization of steel assemblies is not permissible; hence, it is used to protect carbon and low-alloy steels of medium carbon content. It is likely to cause partial decarburization of high-carbon steel and to carburize low-carbon steel. It is effective in protecting low-carbon steel from oxidation.

The nominal composition of a rich prepared nitrogen-based atmosphere is given in Table 1. A nitrogen-based atmosphere generator incorporating a scrubbing system is described on pages 400 to 401 in Volume 4 of the 9th edition of *Metals Handbook*. The main economic disadvantage of prepared nitrogen-based atmospheres lies in the high initial cost of generating and scrubbing equipment.

Endothermic-Based Atmosphere. The nominal compositions of a lean (AGA class 301) and of a rich (AGA class 302) endothermic-based atmosphere are given in Table 1. A lean atmosphere is generated by reacting lean mixtures of hydrocarbon gas and air in an externally heated chamber in the presence of a nickel catalyst.

Table 1 Protective atmospheres commonly used in furnace brazing

Description	AGA class	Nominal composition, vol% N_2	CO	CO_2	H_2	CH_4	Dew point, °F	Fuel required(a), ft^3	Air-gas ratio(b)
Rich exothermic-based	102	71.5	10.5	5.0	12.5	0.5	(c)	155	6.0
Products of combustion of hydrocarbon gas passed through incandescent charcoal	402	rem	30.0	...	16.0	...	−15	80	6.0
Rich prepared nitrogen-based	202	75.3	11.0	...	13.2	0.5	−40	160	6.0
Lean endothermic-based	301	45.1	19.6	0.4	34.6	0.3	+20 to 50	190(d)	2.6
Rich endothermic-based	302	39.8	20.7	...	38.7	0.8	+25 to −5	200(d)	2.5
Dissociated ammonia	601	25.0	...	...	75.0	...	−60	(e)	...
Hydrogen purified	...	...	...	...	100.0	...	−75	...	...
Commercial nitrogen-hydrogen(f)	...	95.0	...	...	5.0	...	−35	...	...
Commercial nitrogen-methanol(f)	...	79.0	7.0	...	14.0	...	+10 to +55	...	...

(a) Per 1000 ft^3 of atmosphere; based on use of natural gas rated at 1000 Btu/ft^3. For other fuel gases, multiply by: 2.0, for high-hydrogen artificial gas; 2.5, for medium-hydrogen, high-CO artificial gas; 0.4, for propane; and 0.3, for butane. (b) Values indicate number of parts of air to one part of gas (based on use of natural gas at 1000 Btu/ft^3). (c) Dew point is about 10 °F above temperature of cooling water; dew point may be reduced to +40 °F by refrigeration, or to −50 °F by adsorbent-tower dehydration. (d) Plus 250 ft^3/1000 ft^3 for heating gas. (e) 23.5 lb of ammonia per 1000 ft^3 of atmosphere. (f) Percentage of atmosphere components can significantly vary depending on the requirements of the base metal. Manufacturer's recommendations should be followed.

By controlling the dew point of endothermic-based atmospheres, their carbon potential can be accurately controlled in the range from 0.20 to 1.30% C, making it possible to ensure equilibrium conditions for carbon and low-alloy steels of low, medium, and high carbon content, thereby avoiding carburization and decarburization. The atmospheres are reducing and thus protect against oxidation and promote wetting during brazing.

Generators for producing these atmospheres are considerably less complicated and less expensive than those required for producing prepared nitrogen-based atmospheres; the operating costs for generating the gases are roughly equivalent, depending on fuel costs. New processes such as the Pressure Swing Absorption system should be evaluated. Endothermic-atmosphere generators and the reactions obtained in them with several different hydrocarbon fuels are described on pages 397 to 398 in Volume 4 of the 9th edition of *Metals Handbook*.

The principal disadvantages of endothermic atmospheres are their potential explosiveness when mixed with air and their sooting potential when dry, at the temperatures encountered in the cooling chambers of brazing furnaces.

Commercial nitrogen-based atmospheres provide a viable substitute for generated atmospheres, including exothermic (AGA class 102) atmospheres. Although the desired results are often the same with either a commercial nitrogen-based system or a conventional generated atmosphere, there are some differences in equipment, operation, and function among these systems. A generated atmosphere has a relatively fixed composition, which is determined by the input ratio of air to hydrocarbon, the scrubbing and purification process equipment, and the condition of the equipment and catalysts. The commercial nitrogen-based system uses volatilized pure fluid blended to the desired atmosphere proportions.

Commercial nitrogen is pure (99.9995%), dry, and inert. While it does not oxidize metals, it also does not remove oxides or alter the carbon content. Reactive components, such as hydrogen, methane, or methanol, are added to provide oxide-reducing or carbon potential qualities needed for specific applications. Because both the nitrogen and the reactive components are piped and precisely blended into the furnace, the atmosphere composition can be set or altered to meet almost any requirement.

Many of the advantages offered by commercial nitrogen-based atmospheres are related to their purity and controllability. They provide lower impurity levels than generated atmospheres—and have a lower dew point and lower oxygen and carbon dioxide impurity levels (less than 10 ppm). The dryness of the commercial system also facilitates filler-metal flow and wetting, which can result in filler-metal savings, a lowering of the furnace temperature, and improved appearance of brazed assemblies.

Because commercial nitrogen-based atmospheres are inherently dry, their use can lead to soot formation when used with some brazing pastes. Exothermic atmospheres typically contain enough water and carbon dioxide to react with the paste and transform it into gases that escape the furnace before soot can form. Special pastes are formulated for use with dry atmospheres and should be tried if sooting occurs. Water or air also can be introduced into the furnace atmosphere to create the conditions that exist in an exothermic atmosphere. Suppliers can create the desired wet conditions in the heat-up zone where the paste becomes reactive, while maintaining the desired low dew point conditions in the brazing and cooling sections (zone separation).

The composition of commercial nitrogen-based atmospheres is formulated in accordance with the requirements of the base metals to be brazed. Because the additives to the nitrogen are more expensive than the nitrogen, the proportion of additive is kept to a minimum. Consequently, commercial nitrogen-based atmospheres vary considerably in composition, as is footnoted in Table 1. Commercial nitrogen-based atmospheres are discussed on pages 402 to 408 in Volume 4 of the 9th edition of *Metals Handbook*.

Other more expensive supplied atmospheres and vacuum are also suitable for the furnace brazing of steel. Because of the stability of surface oxides containing chromium, manganese, titanium, vanadium, aluminum, and silicon, alloy steels containing a total of more than 2 or 3% of these elements can be brazed without flux in either a vacuum or one of the strongly reducing atmospheres, such as dissociated ammonia or purified dry hydrogen (see Table 1 for compositions). These more expensive supplied atmospheres are rarely used without special justification, although in some plants that require them for brazing nonferrous alloys or for use in heat treating, they may be used for brazing steel simply because they are available and because justification for installing another atmosphere generator is lacking.

Vacuum Brazing

There are two types of vacuum brazing—high vacuum, and medium or partial vacuum. High vacuum is well suited for brazing base metals containing hard-to-dissociate oxides such as nickel-based superalloys. Partial vacuums are used when the base or filler metal volatilizes at its brazing temperature under high vacuum conditions.

The following advantages are associated with vacuum brazing when compared with other high-purity brazing atmospheres. Vacuum removes essentially all gases from the brazing area, which eliminates the necessity of purifying a supplied atmosphere. A vacuum system evacuated to 10^{-5} torr contains about 0.000001% residual gases. Certain base-metal oxides dissociate in vacuum at brazing temperatures. Chromium-bearing base metals such as stainless steels are often brazed in vacuum. Problems experienced from contamination of brazing interfaces due to expulsion of gases by the base metal are not encountered in vacuum. Occluded gases are removed from the interfaces immediately upon evolution from the base metal. The low pressure existing around the base and filler metals at elevated temperatures removes volatile impurities and gases from the metals.

Vacuum purging prior to high-purity dry-hydrogen brazing is frequently employed where extra precautions must be taken to ensure optimum freedom of the atmosphere from even small amounts of foreign or contaminating gases. Similarly, dry-hydrogen or inert-gas purging prior to evacuation is sometimes helpful in obtaining improved brazing results in a high-vacuum atmosphere.

Brazing With Copper Filler Metals

Copper is the preferred filler metal for furnace brazing of carbon and low-alloy steel assemblies without flux in reducing protective atmospheres. Significant amounts of two trace elements, arsenic and phosphorus, should be avoided because they form brittle compounds in the brazed joint. The copper should be essentially free of arsenic, and if it was deoxidized with phosphorus, the residual phosphorus content should be low.

Filler Metals. There are three standard copper brazing filler metals, bearing the American Welding Society (AWS) designations BCu-1, BCu-1a, and BCu-2 (Table 2).

BCu-1 filler metal contains a minimum of 99.90% Cu and a maximum of 0.10%

Table 2 Copper filler metals commonly used in furnace brazing (AWS A5.8)

AWS classification	Minimum copper, %	Brazing temperature, °F
BCu-1	99.90	2000-2100
BCu-1a	99.0	2000-2100
BCu-2	86.5	2000-2100

of other elements. It is available in strip, rod, and wire on spools.

BCu-1a filler metal contains a minimum of 99.0% Cu and a maximum of 0.30% of other metallic elements. It is available as a powder in two standard sieve analyses, medium-1 and medium-2. It is applied as a powder in some applications, but is frequently mixed with a liquid vehicle and applied as a paste. In most applications, BCu-1 and BCu-1a are interchangeable.

BCu-2 filler metal is available in the form of a paste and contains a minimum of 86.5% Cu and a maximum of 0.50% other metallics and 1.3% nonmetallic contaminants, including chlorides, sulfates, and matter insoluble in nitric acid or soluble in acetone. The remainder is oxide. The paste is a suspension of particles of copper and cuprous oxide in a volatile vehicle.

Various proprietary pastes, identified by trade name and not covered by an AWS specification, are available commercially. These pastes are prepared with several different types of hygroscopic and nonhygroscopic vehicles, with different thinning, drying, and spattering characteristics. In addition to the vehicle, some contain commercially pure copper powder only, and others contain powdered cuprous oxide only. Some pastes contain mixtures of copper powder and cuprous oxide; of copper powder, cuprous oxide, and iron oxide; or of copper powder, cuprous oxide, and iron powder. Generally, pastes containing oxides require strongly reducing atmospheres to promote flow. Pastes containing iron or iron oxide are intended to fill joint clearances up to about 0.003 in. at brazing temperature.

The unalloyed copper filler metals (BCu types) should not be confused with the copper-phosphorus (BCuP) filler metals, which are never used for joining steel, or with the copper-zinc (RBCuZn) filler metals, which generally require the use of a borax-boric acid flux.

Joint Strength. A principal advantage of the copper filler metals used in the furnace brazing of steel is the high strength they impart to the brazed joint. The shear strength of copper joints in low-carbon steel generally ranges from 22 to about 31 ksi, while the tensile strength ranges from 25 to almost 50 ksi. The rotating-beam fatigue strength of these joints is also high; in one series of tests, copper brazed joints in low-carbon steel withstood 10 million cycles without fracture at stresses of about 12 ksi. In all tests of joint strength, it is apparent that fit affects strength. When other variables are constant, a diametral interference fit of 0.001 in. is slightly stronger than a slight diametral clearance fit (such as 0.0005 in.), and a diametral interference fit of 0.002 in. generally is even stronger. Joint clearances and joint strength are discussed in greater detail in the section of this article on joint fit and design.

Carburizing of Copper Brazed Assemblies. The liquidus temperature of the copper filler metals is about 1980 °F, and the recommended temperature range for brazing is 2000 to 2100 °F. This range is safely above the temperature ranges for austenitizing and carburizing of carbon and low-alloy steels. Gas carburizing temperatures seldom exceed 1725 °F. Carburizing a steel assembly after copper brazing has no adverse effect on the brazed joint.

Brazing Carburized Components. A distinct advantage of furnace brazing with copper filler metal is that it permits carburizing before brazing, as well as after brazing. This means that only those components that require hardening need to be carburized. This is advantageous in some applications, such as the one described in the following example.

Example 1. Copper Brazing of an Assembly Containing Carburized Components. Three of the five components of the cam assembly shown in Fig. 5 were carburized prior to assembling for copper brazing. The two 1010 steel cams were selectively carburized to produce a 0.020-in. case along the peripheral working surfaces, and the 1215 steel stud was carburized on all surfaces except the tenon portion to the same case depth. Surface carbon content of the case was about 1.1% after carburizing. This initially high surface carbon content compensated for subsequent diffusion during brazing and reheating for hardening. Surface carbon content after austenitizing in a neutral salt and oil quenching was not reported, but was satisfactory for the intended service. Resulting hardness was at least 79 HR-30N.

After carburizing, the cams and stud were thoroughly degreased and flash copper plated with a 0.0002-in.-thick coating. The spacer and hub of 1215 steel were also degreased, the five components were assembled for brazing, and copper filler metal in the form of preformed wire rings was preplaced as shown in Fig. 5. Diametral clearance ranged from 0.000 to 0.002 in. The assemblies were brazed at 2070 °F in a 100-kW electrically heated mesh-belt conveyor furnace (12-in.-wide belt) under a lean endothermic atmosphere. The carbon potential of the atmosphere was maintained between 0.3 and 0.4% C. Brazing time was 11 min per piece and production rate was 1000 assemblies per hour.

Originally, the copper plating had been applied only to the components that had not been carburized (the hub and spacer) to protect them from acquiring a hard, difficult-to-machine surface during brazing in an atmosphere of high carbon potential, which was required to keep the carburized components from decarburizing. Copper plate approximately 0.0005 in. thick was needed to protect the surfaces, and it ran during brazing, forming puddles that were difficult to remove. Therefore, it was decided to use an atmosphere that was compatible with the bare noncarburized components and to copper plate only the carburized components. With the atmosphere of lower carbon potential, a 0.0002-in.-thick copper plate was sufficient to prevent decarburizing of the components to be hardened. The desired results were obtained; use of the thinner copper plate virtually eliminated puddling of the copper, and the atmosphere with the

Fig. 5 Selective plating of cam assembly for protection of carburized surfaces during brazing

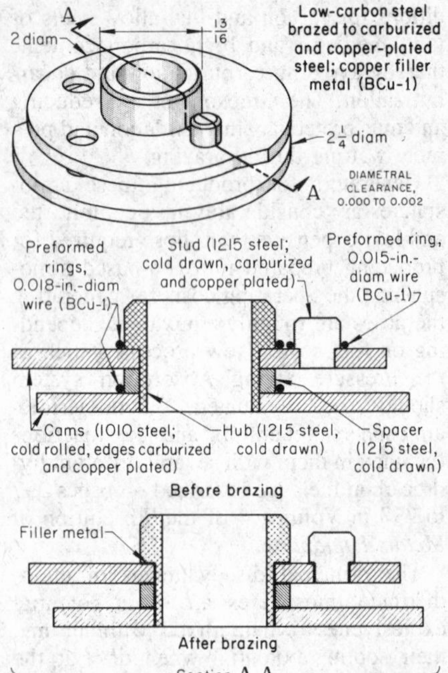

Fig. 6 Use of copper filler metal in paste form during brazing

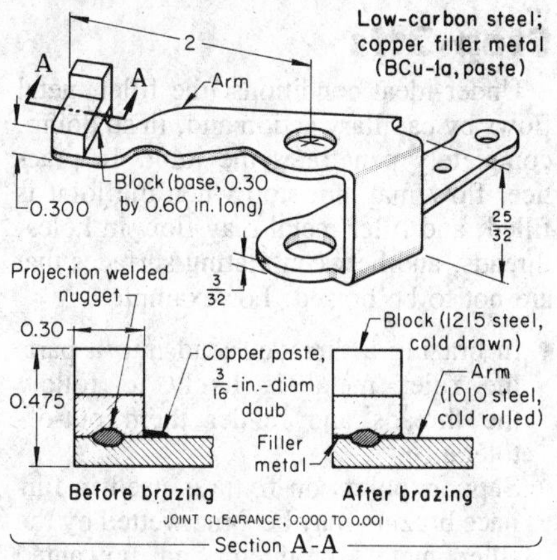

lower carbon potential effectively eliminated sooting in the furnace and in the generator.

Elimination of Flux. Copper filler metals, when used in conjunction with a suitable protective atmosphere in the brazing of steel, are self-fluxing by virtue of the ability of the atmosphere to reduce surface oxides and thereby to promote the flow of filler metal. Fluxes increase brazing costs and contaminate and corrode heating elements and other furnace components. The relative importance of flux contamination varies considerably among applications.

Selection of Filler-Metal Form. Copper filler metal is available in several forms, including wire, strip, special preforms, powder, and a variety of pastes. Occasionally, copper that is to serve as filler metal is deposited on the base metal by electroplating. Selection of the preferred form of filler metal is usually based on joint design, ease of placement in assembly, ability to retain a fixed location during transit through the furnace, and production quantities required. The next two examples discuss the reasons for selecting two forms of filler metal—paste and electroplating—on the basis of the application requirements.

Example 2. Use of a Paste Form of Copper Filler Metal. In the business machine assembly shown in Fig. 6, the 1215 steel block was projection welded to the 1010 steel arm, and then the joint area surrounding the projection weld nugget was furnace copper brazed to provide the strength required in service. In selecting the form of copper filler metal, several factors had to be considered. Production volume was relatively low and, therefore, the cost of preforming special copper wires to the desired shape could not be justified. Placing a piece of wire at the joint would have been unsatisfactory because the wire could be moved out of position by vibration during transit through the furnace. Hand crimping the wire would have prevented displacement, but crimping would have been time consuming and difficult. Mechanical precrimping could not be justified on the basis of production quantity. Under the circumstances and because fillets in the joint area were not objectionable, filler metal in the form of a copper paste proved most economical and easiest to apply. The paste was applied to one side of the joint in the form of a daub about $^{3}/_{16}$ in. in diameter.

The assemblies were brazed at 2070 °F in a 100-kW electrically heated mesh-belt conveyor furnace (12-in.-wide belt), in an endothermic atmosphere. Time in the heating chamber was 9 min, and the production rate was 900 assemblies per hour.

Example 3. Use of Copper Plating as a Filler Metal to Control the Volume of Copper in Brazed Joints. The small size of the two studs and adjacent joint areas in the plate-and-stud assembly shown in Fig. 7 required close control of the amount of copper in the brazed joint. Copper paste applicators could not deliver the required volume with enough accuracy. Manual placement of wire rings was not feasible because the small diameters of the wire and ring would make placement slow and tedious. Automatic ring placement was feasible, but could not be justified because of low production rate. The stud tenons could have been dipped in copper paste before assembly by press fitting, but dipping would have been slow and messy unless mechanized, an expense that was not justifiable.

Fig. 7 Application of filler metal by plating to small brazed joints

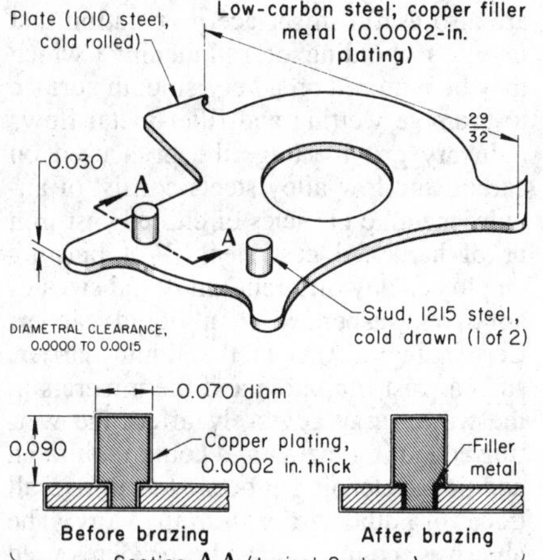

Fig. 8 Service life of woven belts in copper brazing furnaces. Operating temperature was 2050 °F. See text for comparison of data.

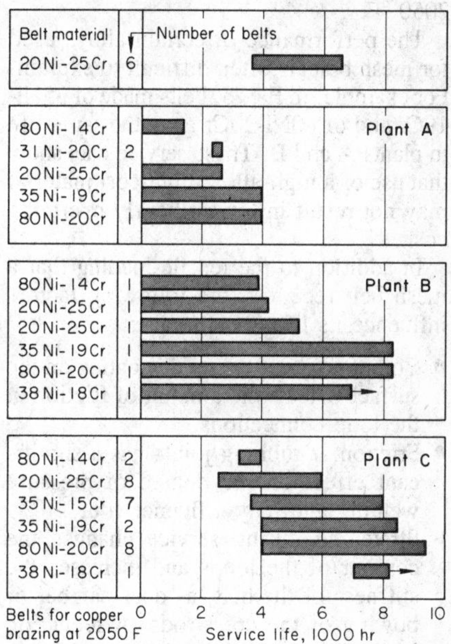

Barrel plating the studs with copper proved to be feasible and economical. Plating time and operating conditions had to be accurately controlled to ensure an optimum thickness of copper—enough to fill the joint without developing an excessive fillet at the joint or copper puddles in critical areas. Plating thickness was controlled by a drop test to 0.00012 to 0.00023 in. Diametral clearance between stud and plate was held to 0.0000 to 0.0015 in.

The degreased components (1010 steel plate and 1215 steel studs) were assembled and then brazed in a 100-kW electric mesh-belt conveyor furnace (12-in.-wide belt), in an endothermic atmosphere. The brazing time was 8 min at 2070 °F. Production rate was 2200 assemblies per hour.

Mesh belts are used in furnaces of various capacities and for loads of various sizes. Figure 8 shows service life for rod-reinforced mesh belts made from several different alloys and used in copper brazing furnaces at 2050 °F. Load-density limits for mesh belts are established mainly by operating temperature, length from the loading position to the discharge end of the heating chamber, belt alloy, cross-sectional area of the belt, the brazing atmosphere used, and an estimate of acceptable life. It is essential that the mesh does not become crushed, either from excessive loading or from parts dropping into the mesh. For an operating temperature of

1450 to 1650 °F, a load range of 16 to 140 lb/ft^2 is usually permissible, but the allowable loading decreases to about 4 to 30 lb/ft^2 when the temperature is increased to 2050 °F.

The performance of certain alloys used for mesh belts is often difficult to explain. For example, in Fig. 8, belts made of 35Ni-19Cr and of 80Ni-20Cr gave the same life in plants A and B. These service data show that use of a high-alloy, high-cost material may not result in lowest hourly operating cost.

In addition to the tensile loading that a mesh belt receives, the following factors influence its life in a furnace:

- Composition of the oxide coating on the surface affects the amount of friction at the loop connections.
- Friction at rubbing joints has a significant effect on the degree of pressure welding and may influence life.
- Stretching during service changes the contour of the loops and increases the stiffness of the belt, as does camber or bowing of the cross-rods. A reduction in flexibility, such as is generally encountered when the alloy wire becomes old and brittle, can hasten failure during the return trip of the belt over drive rolls, even though this repeated stress takes place at room temperature.
- Use of fluxes reduces belt life because fluxes spill on the belt and promote deterioration.
- The contour of the loops and the method of joining at the edges of the belt are equally important. Certain patterns or weaves perform better on a specific furnace and yield longer belt life.

Surface Preparation

Cleaning is almost always advisable before brazing because the presence of oil, grease, pigmented drawing lubricants, excessive amounts of oxide, and other surface contaminants in or near the brazed joint have a deleterious effect on the soundness and strength of the joint. Some contaminants interfere with the wetting action so that normal flow of molten filler metal in the joint is prevented.

Both chemical and mechanical cleaning methods are used to clean steel components and assemblies for brazing, but chemical methods are the more widely used. Chemical methods include alkaline cleaning, solvent cleaning, vapor degreasing, and sometimes acid pickling. The mechanical methods most commonly used are dry and wet abrasive blast cleaning. Other cleaning methods, if they prove satisfactory, are employed largely because the necessary equipment is available. If warranted, machining or grinding may be used to obtain the necessary joint cleanness and to ensure satisfactory wetting.

Chemical Cleaning. Alkaline cleaning, including soak, spray, and barrel cleaning, is widely used for removing oily, semisolid, or solid soils from steel components before furnace brazing. It is generally satisfactory for removing most cutting and grinding fluids, grinding and polishing abrasives, and some pigmented drawing compounds. The solutions, methods, equipment requirements, advantages, and limitations are described on pages 1 to 68 in Volume 5 of the 9th edition of *Metals Handbook*.

Solvent cleaning is capable of removing oil, grease, loose metal chips, and other contaminants from steel components. Parts are immersed and soaked in a common organic solvent. Spray methods can also be employed. The solvents, process variables, equipment, and limitations of solvent cleaning are described on pages 40 to 58 in Volume 5 of the 9th edition of *Metals Handbook*.

Vapor degreasing employs the hot vapors from a boiling chlorinated hydrocarbon solvent to remove surface contaminants such as oils, greases, and waxes. To supplement the vapor, some degreasing units are equipped with facilities for immersing the work in the hot solvent or for spraying with clean solvent. Solvents, procedures, equipment, costs, and other aspects of the cleaning process are described on pages 44 to 58 in Volume 5 of the 9th edition of *Metals Handbook*.

Mechanical Cleaning. As previously noted, mechanical methods are less widely used than chemical methods in cleaning for brazing. However, they are usually preferred for removing heavy scale and may be indispensable in removing the more tenacious lubricants, such as a pigmented drawing compound. Mechanical methods are also useful in surface preparation that involves abrading or roughening, which may be required on a very smooth surface to promote wetting and filler-metal flow.

In dry grit blasting, the grits used on carbon and low-alloy steels consist of angular metallic particles of chilled cast iron or of hardened cast steel. Wet blasting employs many different kinds and sizes of abrasives suspended in a liquid carrier. Certain ingredients in the liquid carrier, such as rust inhibitors and the minerals in the water, may adversely affect the wetting action in brazing and could require an additional cleaning process to remove all traces of liquid carrier from the work. The abrasives, equipment, and procedures used in both dry and wet blasting are described on pages 83 to 96 in Volume 5 of the 9th edition of *Metals Handbook*.

Stop-Offs

Under ideal conditions, the filler metal flows by capillary action and, in so doing, completely penetrates the joint. In practice, flow may not stop when the joint is filled, and filler metal may flow in holes, threads, and between mating surfaces that are not to be brazed. For example:

- In brazing a threaded stud into a part, the filler metal is likely to follow the threads and render them out-of-tolerance.
- Support points on fixtures used in furnace brazing may become wetted by the filler metal, producing an unwanted braze and perhaps resulting in loss of the fixture and assembly, because it may be impossible to separate them without damage.
- Some parts, such as turbine and compressor brazements, are designed to close tolerances and excess filler metal may be dimensionally objectionable.
- Tubular assemblies, particularly small capillary tubes ($^1/_{16}$-in. ID or less), can easily become partly or completely blocked with filler metal.
- Excess filler metal may be unacceptable because of appearance.
- In production brazing, where it may be necessary to use more filler metal than called for to allow for variations in fit-up between parts, some joints have excess filler metal that flows away from the joint area.

Stop-Off Materials. For brazing carbon and low-alloy steels in the more commonly used atmospheres, such as exothermic-based atmospheres, milk of magnesia painted on the appropriate areas is an effective stop-off. Also, painting fixtures with a water solution of chromic acid and then heating them to the brazing temperature renders them resistant to wetting by the filler metal, because a thin layer of chromium oxide forms.

For brazing in a hydrogen atmosphere or in vacuum, commercial materials are used that are composed of graphite or oxides of aluminum, titanium, and magnesium prepared in the form of a water slurry or organic binder mixture.

Application of Stop-Off. For large areas, the use of a paint brush or roller is satisfactory. This method is used to protect touch points of metal fixtures. Often it is advisable to repaint the fixture prior to each use, because some stop-off may crack off during each heating cycle. The use of an artist's brush is ideal for precision application of fine areas of stop-off, although the operation is time consuming and requires considerable skill.

Use of a medical syringe makes it possible to obtain extremely fine detail in stop-off. Needles of 0.010-in. ID are often used. With a small needle, a drop of stop-off can be applied at a precise point. Hypodermic syringes are not suitable for use with fast-drying stop-offs because the needles soon become clogged.

With fast-drying stop-off, there is always danger that some of it will inadvertently run into the joint area. If this happens, the assembly must be taken apart, and all stop-off must be removed.

A nonwicking stop-off can be applied by conventional equipment designed for application of liquid plastics, paste brazing alloys, and other organic compounds. It remains stable over long periods of time and does not clog the valves and tubing in the system.

Stop-Off Removal. Brazing stop-off materials of the "parting compound" type can be removed by wire brushing, air blowing, or water flushing. The "surface reaction" type can best be removed by a hot nitric acid-hydrofluoric acid pickle, except when the brazed assemblies contain copper and silver. Solutions of sodium hydroxide or ammonium bifluoride can be used in all applications, including copper and silver. Other stop-off materials can be removed by dipping in a 5 to 10% solution of either nitric or hydrochloric acid.

Assembly for Brazing

The component parts of an assembly to be furnace brazed must be assembled in an essentially fixed position before entering the furnace, and they must be capable of maintaining this position throughout brazing and cooling. The filler metal, a part of the assembly, must be preplaced in the proper location, and it must maintain this location. Filler metal should be applied in the most convenient product form to facilitate assembly and subsequent brazing.

Self-jigging is the method of assembly in which the component parts incorporate design features that ensure that the components, when assembled, will remain in proper relationship throughout the brazing cycle without the aid of auxiliary fixtures. This is the preferred method of assembly, as it eliminates the initial and replacement cost of auxiliary fixtures and the cost of heating them during brazing. It usually is a more reliable method of holding the components. Self-jigging can be accomplished by several methods, including gravity locating, interference or press fitting, knurling, staking, expanding, spinning, swaging, crimping, thread joining, riveting, folding, peening, and tack welding, as shown in Fig. 9.

Gravity Locating. Perhaps the simplest method of assembling two components is to rest one on top of the other with the brazing filler metal either wrapped around one component near the joint (Fig. 9a) or placed between components (Fig. 9b). The principal disadvantage of gravity locating may be the lack of a dependable means of orienting the components or keeping them from moving in relation to one another. Nevertheless, some production components are assembled in this manner, especially those in which the upper component is relatively heavy.

Interference or press fitting, which requires expansion or contraction of mating component surfaces, provides a very tight fit—sometimes called a "tight press fit." Interference fitting is illustrated in Fig. 9(c) and (d). In Fig. 9(c), the cup has an inside diameter that is smaller than the mating projection of the underlying plate. The extent of interference seldom exceeds about 0.001 in./in. of diameter, up to about 3-in. diameters. Nevertheless, most interference fits require considerable force to achieve assembly, a force generally provided by an arbor press or similar tool. Thus, an interference fit is a press fit.

Lighter interference fits, such as that shown in Fig. 9(d), may provide zero clearance or a very slight gap between the mating surfaces of components. These also require some external force, such as that provided by an arbor press, to achieve assembly. Fits with zero clearance are referred to as "size-to-size" fits. Some method is used to prevent slippage when the components are heated in the furnace, particularly if the joint has a vertical axis. As shown in Fig. 9(d), a shoulder on one of the components can be used to ensure stability.

Knurling. In high-production manufacturing, considerable variation exists in joint clearance among the assemblies being brazed. Typical brazed assemblies in which

Fig. 9 Typical methods of self-jigging

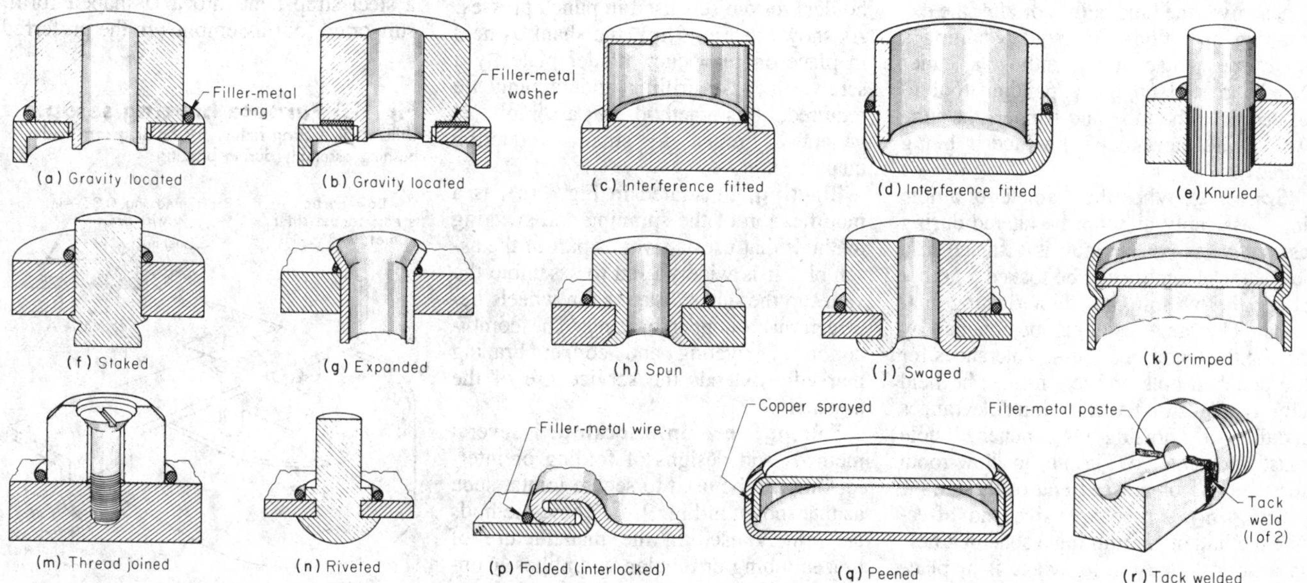

a round male member is fitted to a female member are subject to either of two conditions: (1) the male part is off-center, thus allowing all of the diametral clearance on one side; or (2) the male member is out-of-round so that all of the clearance is on two opposite sides with no clearance (or even interference) on the other two sides.

Knurling the end of the male member (Fig. 9e) may correct the conditions described above, and uniformity among brazed joints can thus be obtained. Often knurling can be done during machining of the part, thus adding very little to the cost. If knurling must be done in a secondary operation, the extra cost often can be balanced against the cost of the rejects that would be encountered if knurling were not done.

When the male member is tubular, prick punching may be substituted for knurling. Usually two rows of prick punch marks near the end of the male member are sufficient. Prick punching is easily done, but because it involves a secondary operation, the cost must be justified.

Staking. Figure 9(f) shows how staking locks two components in position. Burrs are turned up on the shaft by driving a punch into it. This method, which may be modified to suit the brazing application, is commonly used to retain the orientation of such assemblies as cams, levers, and gears on shafts or on common hubs. It is sometimes a substitute for tack welding, knurling, or interference fitting.

Expanding. This method of assembly is commonly used for joining tubes to tube sheets. The tubular component is pressed into a header sheet and expanded in the hole to lock the assembly, as shown in Fig. 9(g). Rings of brazing filler metal can be placed over the tube before or after the expanding operation. To avoid obtaining a mere line contact in expanding, a leader can be placed on the expanding tool to project into the tube and support the tube wall while the end of the tube is being flared.

Spinning. When the diameter of a hole in an assembly may not be altered during assembly, as when a hub is fastened to a lever, the assembly can be locked together (Fig. 9h) by spinning in a riveting machine. The same result can be obtained by flaring the tenon in a press. Tolerances for the punched hole and tenon must be held closely to ensure the close joint clearances required for brazing. The punched hole must be chamfered (Fig. 9h) to allow room for the spun or pressed end of the tenon.

Swaging. An inexpensive and effective method of assembling a spud in a hole in a hollow body is to swage it in place (Fig. 9j). This is acceptable when it is not necessary to maintain accuracy of the diameter of the hole in the hub and when the projection on the flange can be tolerated. The principal advantage of swaging is that close tolerances do not have to be held on the tenon or the punched hole because the swaging operation forces the components into intimate contact. In addition to other applications, swaging has been used to assemble a valve body in a float chamber for refrigerators. The resulting bond after furnace copper brazing is strong, tight, leakproof, and capable of withstanding high pressure.

Crimping. Figure 9(k) shows the assembly of a disk, shell, and copper filler-metal ring in which the disk and ring are held in place by crimping the end of the shell. Also shown is an inexpensive method of forming stoppers against which the disk is located; these stoppers consist of three or four indentations around the shell.

It generally is preferable to set an assembly of this type on end in the furnace so that the filler metal flows downward through the joints. However, if the tubular component is long, the assembly must be laid on its side to clear the furnace interior; consequently, an oversize ring of hard copper wire filler metal that can be sprung in place close to the joint is used. If the diameter of the tube is 2 in. or more, the filler-metal wire and adjoining steel surfaces should be coated with copper powder paste, which hardens and prevents the wire from sagging away from the joint at the top as the assembly is heated. The paste also provides an auxiliary supply of filler metal.

Thread joining has been used for assembling components of replacement punch holders for die sets used in punch presses. As shown in Fig. 9(m), the shank is held in place on the punch holder plate by a screw. Because drilling and tapping are required, this method of assembly is generally limited to small production quantities.

Riveting, illustrated in Fig. 9(n), is a modification of the spinning and swaging methods that uses a rivet as part of the assembly. It is widely used to assemble the vanes to the outer disks of fan wheels before furnace copper brazing. The combination of riveting and copper brazing markedly extends the service life of the assembly.

Folding or Interlocking. Several methods and designs of folding or interlocking can be used to secure joints, such as that shown in Fig. 9(p). These methods are widely used in the manufacture of brazed tubing or tubular assemblies. Copper filler metal is supplied either in the form of copper plating or by wedging a copper wire at the base of the joint. Capillary action draws the filler metal to all areas of the joint.

Peening. Assembly of two hollow shells by the peening method is shown in Fig. 9(q). The stamped components are pressed together, and the outer shell is peened with an air hammer along the periphery. To apply filler metal to an assembly of this type, copper can be sprayed on the joint interfaces before assembly, using an oxyacetylene spray gun.

Tack Welding. Prior to copper brazing, the tip and shank of the electrode holder for a welding torch were assembled by tack welding, as shown in Fig. 9(r). Filler metal consisted of a small amount of copper powder paste daubed around the joint. Any oxide formed during tack welding was reduced by the protective atmosphere during furnace brazing. The tack welding method of assembly usually requires careful investigation to determine the most strategic point or points for placing the weld. For economy, the number of tack welds per assembly should be held to a minimum.

Auxiliary Fixtures. In assembling some components for furnace brazing, self-jigging may not be feasible, or the assembly may require additional positioning or support that cannot be provided by self-jigging alone. Consequently, the use of auxiliary fixtures is unavoidable. These fixtures can take the form of a simple bracket or wire stand, machined graphite blocks, clamps, or cast supports.

An example of an extremely simple fixture for supporting arm-and-bushing assemblies is shown in Fig. 10. The fixture, a steel strap bent into a U-shaped form, supported four assemblies in the preferred

Fig. 10 Furnace brazing setup. Utilizes steel strap fixture for holding arm and bushing assembly during brazing

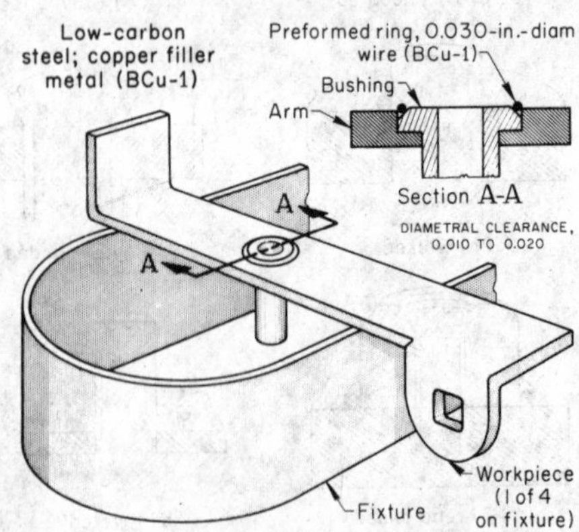

position for brazing. It provided clearance for the lugs at the end of the arms and supported the bushing in the vertical position (Fig. 10) so that the copper filler metal would flow downward into the joint. These assemblies, made of low-carbon steel, were brazed in a conveyor furnace with a 12-in.-wide mesh belt and an exothermic atmosphere.

Low-carbon steel is commonly used for fixtures for short runs; although it is economical, it has the disadvantage of producing low strength at brazing temperatures. For long production runs, stainless steel and wrought and cast heat-resistant alloys are preferred.

Fixture design should adhere to the following principles. Sections of fixtures should be as thin as possible, consistent with required rigidity and durability. Fixtures should be designed for minimum contact with the assembly. Point or line contact is preferable to overall surface contact. In general, external fixtures should expand faster, and internal fixtures slower, than the assembly; in applications where tight clamping is required, the reverse is true.

To equalize pressure on the assembly from shrinkage during cooling, systems of levers, cams, and weights can be used. Wedges and weights often provide good follow-up. The use of bolts or screws should be avoided in fixtures as they tend to relax upon heating and can pressure weld in place. Springs or clamps must be designed to withstand brazing temperatures if used in the heated area; otherwise, relaxation may occur during the brazing operation.

The use of dissimilar metals for fixturing should be avoided when differences in thermal expansion might affect assembly dimensions. Fixtures should be subjected to the brazing environment and temperatures prior to the actual brazing operation to ensure stability and relieve stresses, and they should be designed for easy inspection and should be inspected often.

Fixtures are closed or clamped by driving wedges into slots and lugs. The wedges can be easily removed after the assembly is brazed. Spring clamps are seldom used because they undergo relaxation at the brazing temperatures. However, spring clamps made of Inconel, which do not relax as much as those made of carbon steel, are sometimes used. Nickel-based alloy fixtures, however, should never be used for brazing titanium, because the nickel titanium eutectic forms at 1729 °F, which is lower than brazing temperatures normally used for titanium- or nickel-based alloys. Threaded fasteners should be avoided; if used, they should be made to loose fits and should be coated with a magnesium hydroxide and alcohol mixture to prevent sticking.

Wetting of Fixtures. Even with a minimum number of contact points between the fixture and the components to be brazed, it is sometimes difficult to prevent the fixture from being wetted by filler metal and sticking or being brazed to the assembly. If the contact surface between fixture and components must be extensive, selection of a fixture material that resists wetting becomes critical.

Joint Fit and Design

In brazing, the molten filler metal is drawn by capillary action between closely adjacent, substantially symmetrical surfaces. The distance the filler metal flows through a joint depends on the clearance between the mating surfaces and on the filler metal used. Molten copper flows freely and for greater distances than other filler metals in joints with size-to-size fit (zero clearance) or an interference fit (negative clearance). The distance of flow of copper increases as joint interference increases, up to the point where the seizing and galling of mating surfaces interferes with capillarity. Conversely, as joint clearance (gap) is increased, a clearance is reached at which filler-metal flow stops completely. Uniform fit throughout the joint is important, because nonuniform fit results in nonuniform strength.

Gap Principle. When joint fit is nonuniform, the areas of maximum clearance constitute gaps that interfere with capillary action and impede the flow of filler metal. The effects of uniform joint fit, and of fits containing gaps, on filler-metal flow are shown in Fig. 11. Figure 11(a) shows a joint designed with uniform fit throughout the joint, including the fit at the square internal and external corners. As shown in Fig. 11(b), filler metal from an externally placed wire ring flows uniformly to all parts of the joint, providing a good bond. The joint shown in Fig. 11(c) has uniform fit except at a rounded internal corner. As shown in Fig. 11(d), the gap created by the rounded corner prevents the filler metal from flowing any farther. The joint in Fig. 11(e) has nonuniform fit. The rounded corner and the loose fit at the bottom cause a snug fit at the top. Filler metal flows to most parts of the joint but is blocked at the rounded internal corner (Fig. 11f). The result is an incomplete braze. The same general principle applies to the design of a joint between a cap and a shell. A joint with uniform fit, as shown in the top view of Fig. 12, results in a uniform bond throughout the joint, whereas a joint with nonuniform fit (or line contact), as shown in the bottom view of Fig. 12, results in

Fig. 11 Effects of uniform and nonuniform fit on filler-metal flow

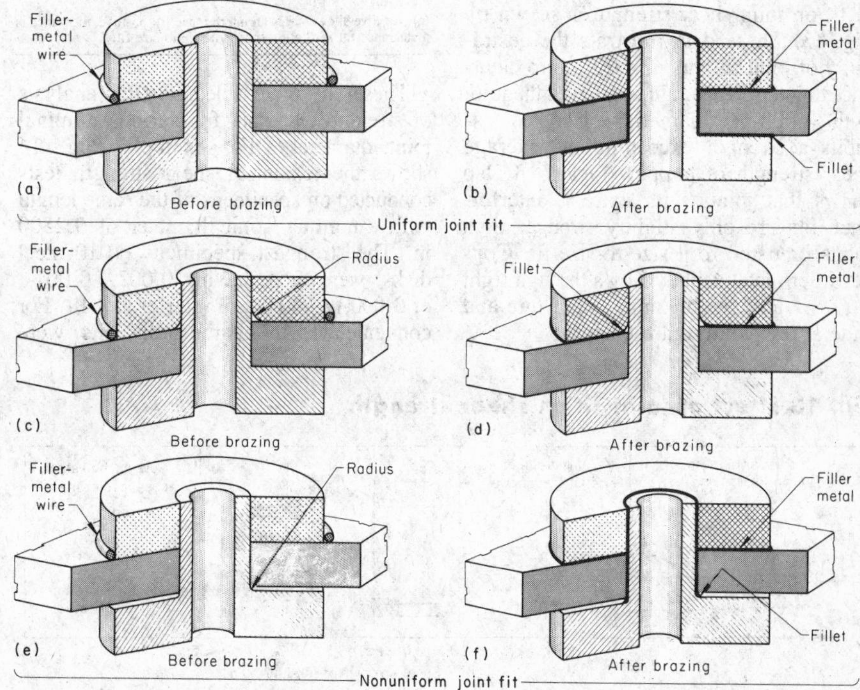

Fig. 12 Effect of joint fit on bonding

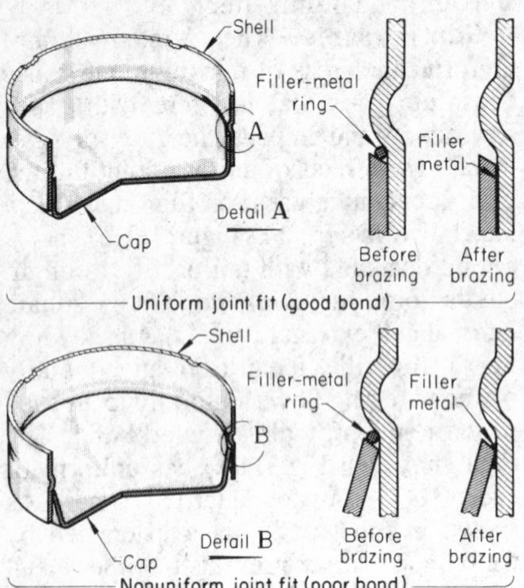

an incomplete bond and unreliable strength and tightness.

Joint Fit. The amount of clearance or interference provided in a brazed joint directly affects the strength of the joint, although other factors, such as cleanness of the joint, length of the joint, composition of base and filler metals, time at brazing temperature, and subsequent heat treatment, if any, also affect strength. When low-carbon steel components are joined by a copper filler metal, the strength of the joint almost always exceeds the strength of the copper filler metal and, under favorable conditions, is almost equal to the strength of the steel base metal. The effect of fit on joint shear strength is shown by Fig. 13. These data illustrate the desirability of joint fits ranging from zero clearance to interference, although even the joint with 0.003-in. diametral clearance exhibits a strength exceeding the average shear strength of copper. Figure 13 also shows that joints with extreme interference fits are only slightly stronger than joints prepared to a size-to-size fit. Copper filler metal usually flows through tight fits, provided the brazing temperature and time at temperature are adequate.

For most applications, the recommended diametral fit for copper brazing of low-carbon steel is 0.000- to 0.003-in. interference. High-carbon steels require a slightly looser diametral fit, usually 0.001-in. clearance to 0.002-in. interference. The usefulness of interference fits is limited to assemblies in which the mating surfaces can expand or contract, as required, when pressed together. When the expansion or contraction approaches zero, as is usual with massive components, an interference fit may result in seizing and galling, making it difficult or impossible for copper to flow through the joint.

When maximum joint shear strength is required in brazed assemblies involving low-carbon carbon steels or low-carbon low-alloy steels, one manufacturer of business machines specifies the following diametral allowances on joints between hole walls and tenons to be furnace brazed, using a copper filler metal (wire, foil, or paste), and using an endothermic-exothermic gas mixture (71% nitrogen, 15% carbon monoxide, 12% hydrogen, and 2% carbon dioxide, by volume) or dissociated ammonia:

Diameter, in. — Hole (+0.0010, −0.0000 in.)	Diameter, in. — Tenon (+0.0000, −0.0015 in.)	Diametral allowance on joint fit(a), in.
0.0625	0.0635	−0.0010 to +0.0015
0.1250	0.1265	−0.0015 to +0.0010
0.2500	0.2520	−0.0020 to +0.0005
0.5000	0.5025	−0.0025 to 0.0000
1.0000	1.0030	−0.0030 to −0.0005
2.0000	2.0030	−0.0030 to −0.0005

(a) Negative allowance is an interference fit; positive allowance is a clearance fit; zero allowance is a size-to-size fit.

These fits were selected after analysis of strength test data for various nominal joint diameters. For example, Fig. 14 shows the results of shear strength tests conducted on specimens of the same length with a nominal joint diameter of 0.2500 in. The strongest specimens (1010 steel disks) were those having 0.002-, 0.001-, or 0.000-in. diametral interference fit. For convenience, the same joint fits were

Fig. 14 Effect of joint fit and composition on shear strength of copper brazed joints

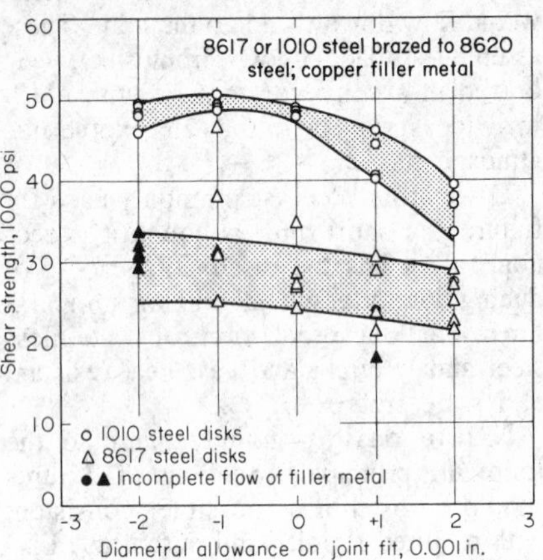

Fig. 15 Effect of joint fit on fatigue strength in copper brazed joints

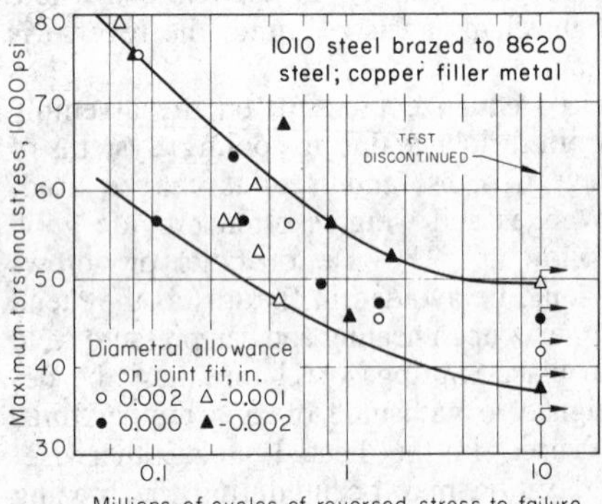

specified for critical assemblies that were subject to alternating torsional loading (torsional fatigue), but test results like those shown in Fig. 15 showed that torsional fatigue strength was essentially independent of joint fit, at least for the range of fits studied.

Low-carbon steels 1010, 8617, and 8620 were used for these tests because they were easier to machine and form than medium- or high-carbon steels and because they were suitable for carburizing after brazing. It was determined that one of the components tested should be a plain carbon steel. As shown in Fig. 14, the specimens with 1010 steel disks had shear strengths as much as 40% greater than those with 8617 steel disks. The reason for this difference was evident when the microstructure of the joints was examined. The joint using 1010 steel disks showed a much deeper penetration of copper on the 1010 steel side of the joint than on the 8620 steel side of

Fig. 13 Effect of joint fit on shear strength

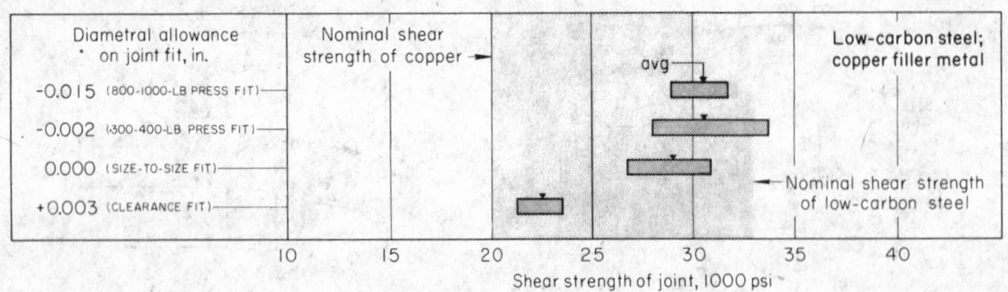

the joint. The presence of a chromium-containing oxide film, resulting from incomplete reduction by the furnace atmosphere, probably restricted the diffusion of copper into the 8620 steel. When both components were made of a low-alloy steel, both sides of the joint were affected.

Factors Affecting Fillet Size. Provided the volume of filler metal is controlled, joint clearance or interference within limits can serve to control the size of fillets. With a fixed volume of filler metal available to the joint, an increase in joint clearance decreases top fillet size and increases bottom fillet size. A decrease in joint clearance has the opposite effect. Also, when joint clearance or interference is fixed, an increase or decrease in the volume of filler metal has a similar effect on fillet size.

With brazed joints, fillet size should be indicated in drawing notes only if it is significant to the function of the brazement—large fillets do not necessarily increase joint strength.

Joint Design. Lap joints generally are preferred for brazing. These joints generally depend on penetration between close, conforming surfaces for their strength, rather than on external fillets; the joints are usually intended to be stressed in shear. A rule of thumb is to make the length of the joint at least three times the thickness of the thinnest section.

Joints should be designed especially for brazing. Six joint designs commonly used for arc welding are compared with their counterparts for brazing in Fig. 16. Any of the joints shown for arc welding could be joined by brazing, but at great sacrifice in strength. Actually, each of the joint designs shown in Fig. 16 for brazing applications is a lap joint.

Butt joints are used when the thickness of the lap joint is objectionable and when the strength of the completed joint will meet the service requirements of the brazement. The strength of a properly made butt joint must be sufficiently high that failures will occur in the base metal away from the joint. The joint strength depends on the filler-metal strength in the joint as compared to the base-metal strength, the degree of filler metal-base metal interaction during the brazing operation, and the service requirements. Maximum strength may not be obtained in the butt joint when the filler metal in the joint is much weaker than the base metal. A means to improve the joint strength of a butt joint is to utilize joint clearances that are as small as possible. An assembly where a butt joint has been used must be designed to ensure that there will be no deflection or bending stresses at the brazed joint. Concentration of bending stresses at the joint will cause the brazed filler metal to tear, especially if the base metal has a higher modulus of elasticity than the filler metal.

Change in Section Thickness. When a brazed assembly is subjected to bending stresses or fatigue, an abrupt change in section thickness in the joint area promotes failure at the joint, just as in a welded joint. This type of failure is caused by the relative flexibility of the thin section and rigidity of the heavy section. Failure can be avoided by tapering the heavy section in the joint area, thereby providing the two components with approximately equal stiffness at the point where failure would otherwise occur. This reduction in section also facilitates brazing, because it provides more uniform heating, uniform clearance, and uniform filler-metal flow.

Surface Condition. The importance of cleaning as a preparatory step before brazing is discussed in the section of this article on surface preparation. Satisfactory flow of filler metal is highly dependent on the cleanness of the joint area.

Surface finish also affects capillarity. The relatively smooth surfaces that are obtained by polishing or fine grinding are less readily wetted by molten filler metal than are rough surfaces. Therefore, it is sometimes advisable to roughen the mating surfaces, such as by rubbing with 60- to 100-grit silicon carbide papers, to prepare them for brazing. A grit-blasted or shot-blasted surface is most readily wetted. Sand blasting is not recommended, because embedded sand particles interfere with wetting action. Any surface treatment that involves the use of aluminum oxide or titanium oxide abrasives should be avoided, because these abrasives deposit a surface contaminant a few microns thick on the parts, which generally inhibits wetting.

Brazing With Silver Alloy Filler Metal

Because heat input to the work is localized in torch and induction brazing of steel, the lower melting and brazing temperatures of a silver alloy filler metal are distinctly advantageous; less often are these advantages a significant factor in furnace brazing of steel. Consequently, the use of silver alloy filler metals is generally limited in furnace brazing to applications for which the copper filler metals or the high brazing temperatures (2000 to 2100 °F) they require are not suited. These applications include brazing:

Fig. 16 Comparison of joint designs for arc welding and brazing

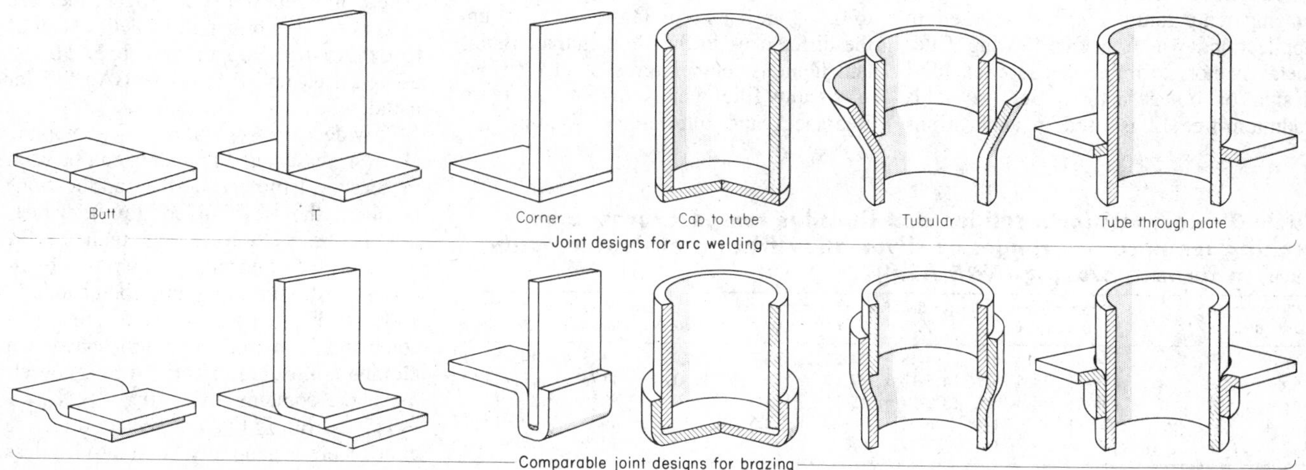

- A carbon or low-alloy steel to a stainless steel or to a nonferrous metal or alloy such as copper or brass
- With joint clearances that are larger than those permitted in copper brazing
- An assembly that is likely to distort excessively or to suffer a loss of desirable properties at the temperatures used in copper brazing
- Subsequent joints in a copper brazed assembly, a practice referred to as step brazing

Types and Forms of Filler Metal. Silver alloy filler metals generally used in furnace brazing are identified in Table 3, including chemical compositions, and solidus, liquidus, and brazing temperatures.

BAg-1 and BAg-1a filler metals are used in furnace brazing carbon steel to carbon steel, to austenitic stainless steel, and to copper or copper alloys. These filler metals are quaternary silver-based alloys containing copper, zinc, and cadmium. They are free-flowing, with a relatively low brazing temperature range. In furnace brazing, BAg-1a is often preferred to the less costly BAg-1 because of its higher silver and lower cadmium contents, which render it less susceptible to joint porosity.

Silver alloy filler metals that contain substantial percentages of cadmium were originally intended for use in torch or induction brazing, not furnace brazing. There is a large loss of cadmium during furnace brazing, and the resulting joint metal is considerably different in cadmium content than that in a joint brazed by torch or induction. However, the cadmium-bearing filler metals are widely used for furnace brazing.

The same combinations of metals and alloys can be furnace brazed with BAg-5 filler metal as with BAg-1 and BAg-1a filler metals. BAg-5 is a ternary silver-based alloy with a considerably higher brazing temperature range. It is used in applications where a free-flowing filler metal is not desirable because of joint design or other factors. Because it is cadmium-free, it is used in applications where cadmium is prohibited, as in food handling and pharmaceutical equipment.

Filler metals BAg-3 and BAg-4 contain nickel and are used in furnace brazing steel to ferritic or martensitic stainless steels containing chromium as the principal alloying element and little or no nickel. Joints in these steels are susceptible to interface corrosion in water or moist air unless they are brazed without flux, using a nickel-containing filler metal. BAg-4 is used when a wider range between solidus and liquidus is required than that of BAg-3. Both alloys are also used in brazing carbide tips to steel tools.

Fluxes. When filler metals BAg-1, BAg-1a, BAg-3, BAg-4, and BAg-5 are used, a flux is required in furnace brazing steel to steel or steel to a dissimilar metal. Without corroding the base metals, the flux serves to reduce surface oxides, ensuring that filler metal will flow and that base metals will be satisfactorily wetted by it.

Flux of AWS type 3A satisfies these requirements and is widely used in furnace brazing with silver alloy filler metals. It contains boric acid, borates, fluorides, fluoborate, and a wetting agent; it is effective in furnace brazing (with protective atmosphere) within the temperature range of 1050 to 1600 °F. It is available as a paste or powder and can be mixed with water, alcohol, or monochlorobenzene for thinning to a liquid consistency. Depending on consistency, flux is applied by brushing or spraying, or by metering from a gun. Flux residues must be completely removed from assemblies after brazing. The fluoride-type fluxes generally used in silver brazing develop residues that are hygroscopic and corrosive. Type 4 flux is used when brazing steel to alloys containing aluminum.

Joint Clearance. Recommended diametral joint clearances for brazing with silver alloy filler metals range from 0.002 to 0.005 in. This range takes into account the differences in the flow characteristics and liquidus temperatures of the various silver alloy filler metals. Size-to-size (zero clearance) and interference fits are not recommended, although joints with interference fits have been successfully silver brazed.

Furnaces. The batch box-type furnaces and mesh-belt conveyor furnaces used in copper brazing are generally suitable for brazing with silver alloy filler metals, provided that the furnaces can maintain satisfactory temperature uniformity in the lower ranges.

Because of their chemical activity, brazing fluxes are potentially damaging to furnace components such as heating elements, mesh belts, rolls, hearth plates, and refractories. They temporarily reduce protective oxides, until their activity is exhausted, and then the oxides re-form. The most severe damage occurs when fluxes are permitted to make direct contact with furnace components. It is common practice to equip silver brazing furnaces with alloy muffles to protect furnace components against flux attack, but some plants report that the use of muffles is unnecessary. Assemblies are usually placed on catch trays that collect flux drips throughout the brazing cycle. The damaging effects of flux constituents in volatilized form are apparently modified to a considerable extent by dilution with the protective furnace atmosphere and by diversion to a venting system.

Protective Atmospheres. The protective atmospheres most commonly used in furnace brazing with silver alloy filler metals are rich exothermic gas, endothermic gas, dissociated ammonia, dry hydrogen, and commercial nitrogen-based atmosphere blends, principally with hydrogen. Even when a flux is used, an atmosphere is usually employed to minimize or prevent oxidation and discoloration of the base metals and to ensure that the flux performs its functions.

Exothermic and endothermic atmospheres are less expensive than dissociated ammonia or dry hydrogen; they are used in furnace brazing of steel to steel or to oxygen-free copper or copper alloys, using a flux and BAg-1a or BAg-5 filler metal.

Provided dissociation is complete (100%), dissociated ammonia can be used in nearly all furnace brazing applications involving the use of silver alloy filler metals. If dissociation is less than 100%, however, the atmosphere may promote nitrogen pickup (nitriding) in carbon and low-alloy steels. In such cases, the use of a commercial nitrogen-based atmosphere can eliminate nitriding effects, even if the atmosphere contains up to 70 vol% N_2.

The use of dry hydrogen as a protective atmosphere is generally limited to fluxless

Table 3 Compositions, solidus and liquidus temperatures, and brazing temperature ranges of silver alloy filler metals commonly used in furnace brazing (AWS A5.8)

AWS classification	Composition, %					Temperature, °F		
	Ag	Cu	Zn	Cd	Ni	Solidus	Liquidus	Brazing
BAg-1	44-46	14-16	14-18	23-25	...	1125	1145	1145-1400
BAg-1a	49-51	14.5-16.5	14.5-18.5	17-19	...	1160	1175	1175-1400
BAg-3	49-51	14.5-16.5	13.5-17.5	15-17	2.5-3.5	1170	1270	1270-1500
BAg-4	39-41	29-31	26-30	...	1.5-2.5	1240	1435	1435-1650
BAg-5	44-46	29-31	23-27	...	...	1250	1370	1370-1550

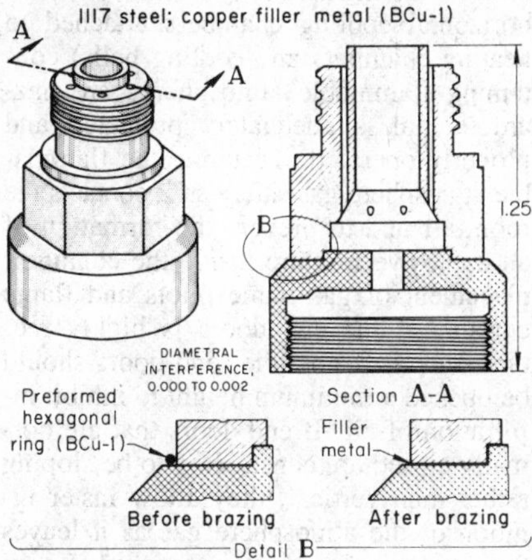

Fig. 17 Use of furnace brazing to reduce machining. Fuel injection pump nozzle holder was produced as a brazed assembly of two components to simplify production.

brazing applications in which the most strongly reducing atmosphere available is required to ensure removal of surface oxide and the satisfactory flow of filler metal. The hydrogen must be extremely dry (dew point, −60 °F or less) for maximum effectiveness.

Furnace Brazing to Supplement, Modify, or Replace Other Fabricating Processes

From a design standpoint, furnace brazed assemblies are of two general types—those designed initially for brazing, and those redesigned for brazing for any of several reasons, including product improvement, increased production rate, ease of fabrication, and cost. Examples in which assemblies were redesigned for brazing are considered in this section; each describes a brazing application that was adopted to supplement, modify, or replace another fabricating process.

Machining and Brazing. Furnace brazing can sometimes be used in combination with machining to conserve material and machining time and to simplify the fabrication of complex shapes. Usually, this entails the subdivision of a single, complex part into two or more relatively simple elements that can be assembled and joined by furnace brazing. An application in which machining and furnace brazing were judiciously combined to simplify complex parts is described in the example that follows.

Example 4. Simplifying Fabrication by Combining Machining and Furnace Brazing. Several features in the design of the fuel injection pump nozzle holder shown in Fig. 17 made it difficult and expensive to fabricate in one piece. It required an internally threaded round base to adjoin a smaller hexagonal section—a combination requiring several different machining operations and different machines. In the upper portion, it also required six $^{1}/_{32}$-in.-diam holes extending from an outer shoulder to an inner recess.

To simplify fabrication and reduce cost, the nozzle holder was machined as two separate pieces, both of 1117 steel bar stock, in automatic bar machines. The threaded base was machined from 1-in.-diam round bar, and the upper portion from $^{3}/_{4}$-in. hexagonal bar. The internal cavity was made by counterboring the hexagonal stock, and the six holes were drilled through the hexagonal body to connect with the counterbored recess.

A shoulder was machined on the base, and a matching counterbore on the upper portion, to nest the upper portion on the base. The nesting surfaces were machined to obtain a light interference fit (0.000 to 0.002 in. in diameter) for positive location and good flow of copper filler metal. A preformed hexagonal ring of BCu-1 copper wire was slipped over the hexagonal section to complete the assembly (Fig. 17). Assemblies were brazed in an electric box furnace at 2050 °F in an exothermic atmosphere. Production rate was about 1500 assemblies per hour.

Substitution for Forging or Casting. Furnace brazed assemblies comprising two or more separately machined components may be more economical to fabricate than a complex forged or cast component that requires extensive machining. Cost savings result from the elimination of expensive punch and die sets (with limited life), elimination of one or more difficult machining operations, or overcoming a high rejection rate resulting from porosity and other defects in the cast product, as illustrated by the following example.

Example 5. Substitution of a Furnace Brazed Assembly for a Casting. Originally, the leakoff-valve assembly shown in Fig. 18 had been produced as a single machined brass casting, but porosity of the casting resulted in leakage in service. The method of producing the leakoff valve was changed to assembly by furnace brazing of two pieces machined from 1113 steel bar stock.

The components of the assembly were turned and partly drilled in an automatic bar machine; the transverse hole was drilled in a drill press. The components were assembled, and filler metal in the form of a preformed ring of copper wire (BCu-1) was slipped over the inserted component and located, as shown in Fig. 18. Assemblies were then loaded on furnace trays, with the inserted component upright. The assemblies were brazed at 2050 °F in an electric box furnace in an exothermic atmosphere. After brazing, drilling was completed, and the hole in the inserted component was threaded.

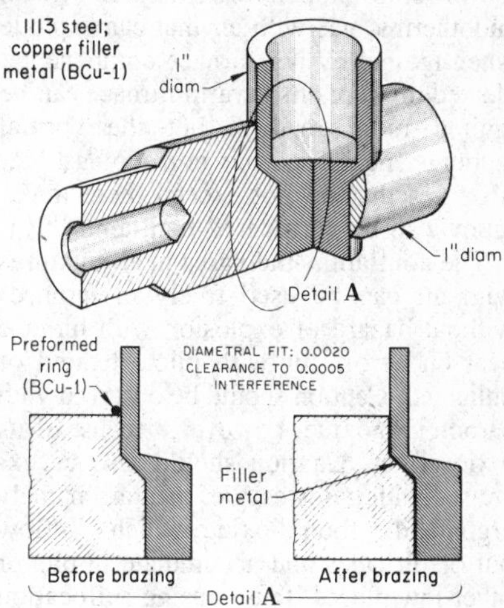

Fig. 18 Use of brazing versus casting. A leak-off valve assembly was changed from a brass casting to a two-piece steel brazement to avoid leakage due to porosity in the casting.

Appendix: Safe Operation of Brazing Furnaces

In starting, operating, shutting down, and maintaining brazing furnaces that contain protective atmospheres, special procedures must be followed for the safety of personnel and equipment. Safety features are incorporated in modern equipment to help guard against making unfortunate mistakes, but the best safety feature is a well-trained operator who thoroughly understands the equipment, the nature of the gases utilized, and the importance of correct procedures.

Nonflammable gases will not burn in air, or explode if mixed with air in any proportions, when exposed to temperatures that can ignite flammable mixtures. Typical nonflammable gases are nitrogen, argon, helium, and carbon dioxide, and mixtures of these gases containing flammable gases, such as hydrogen and carbon monoxide in quantities below flammable concentration.

Lean exothermic gas is a mixture of nitrogen and carbon dioxide that contains low percentages of hydrogen and carbon monoxide. The hydrogen and carbon monoxide content is safely below the minimum amounts of hydrogen and carbon monoxide in room-temperature mixtures of lean endothermic gas with air that can explode when ignited by hot furnace components. Heated mixtures in a warm furnace can be ignited more readily, but the normal amounts mentioned are not combustible. Mixtures that do not exceed these maximums can be considered nonflammable.

The nonflammable gases in all mixtures with air can be used freely in furnaces without hazard of explosion with furnace heat on or off, or with pilots lighted or unlighted. Caution should be observed with carbon monoxide, however, because of its toxic effect. Caution should also be exercised with gases heavier than air, namely argon and carbon dioxide, which can flow out of furnaces and accumulate in pits or other low places, thus causing suffocation to personnel in those places.

Flammable Gases. The flammable gases commonly encountered in furnace atmospheres are hydrogen, carbon monoxide, and methane (CH_4). Flammable prepared mixtures containing hydrogen, carbon monoxide, and a small amount of methane, together with percentages of nonflammable gases, include rich exothermic gas, purified rich exothermic gas, and endothermic gas. Dissociated ammonia (75% H_2, 25% N_2) is highly flammable.

When a brazing furnace chamber contains 3% or less of hydrogen and a similar amount of carbon monoxide, or more than 75% of hydrogen and carbon monoxide, it is safe from explosion if mixed with air at ignition temperatures. However, all in-between percentages of flammable gas mixed with air can explode, producing pressure, heat, and flames that can destroy or damage equipment and injure or kill personnel. The furnace operator must understand these basic principles, and he should take precautions to see that ignition, such as may result from turning on the heat or lighting the pilots, does not take place except under known safe conditions.

Test results indicate that hydrogen-air mixtures can be explosive at room temperature and can be ignited by a hot spark in a glass tube when the hydrogen content is as low as 4 vol%, or as high as 75 vol%. If the mixture is heated before ignition, the low limit decreases and the high limit increases as the temperature increases. Therefore, it is suggested that all concentrations of hydrogen in excess of 2% be considered flammable. Although the room temperature upper flammable limit of a mixture of hydrogen in air is 75% hydrogen, the remaining 25% is air made up of 21% oxygen and 79% nitrogen. At room temperature, a hydrogen-air mixture containing oxygen in amounts of 5% or more is explosive. At elevated temperatures, the dangerous oxygen level decreases to 3%. In mixed gases containing air, a carbon monoxide content of 12 vol% or more is explosive at room temperature; the explosive level decreases to 9% carbon monoxide at elevated temperatures.

In laboratory tests to determine explosive limits, electric hot sparks have been used as sources of ignition, as mentioned. In brazing furnaces, however, the ignition source can be a hot heating element, hot muffle, or some other component. Also, static electricity may be present where there is gas flow, which presents the possibility of an electric static discharge.

Ignition Temperatures. The minimum ignition temperature of hydrogen in air or oxygen is about 1065 °F; for carbon monoxide, the ignition temperature is about 1202 °F. Thus, an explosive mixture will ignite if exposed to these temperatures or to higher temperatures. If a furnace is to be purged by the burnout method, these combustible gases should not be fed to the furnace until the furnace temperature is 1400 °F or above, thereby ensuring ignition of the combustible gases as they enter the air-filled furnace.

Purging is the replacement of one atmosphere in a furnace or retort with another atmosphere, in such a manner as to prevent the formation of an explosive mixture. At start-up in brazing furnace operation, the air in the furnace or retort is replaced with a nonflammable gas, followed by a flammable protective atmosphere gas, or the burnout method is used. At shutdown, the flammable furnace atmosphere gas is replaced by a nonflammable gas (or the burnout method is used), followed by filling the furnace or retort with air.

Critical Periods

In operating a furnace containing a flammable protective atmosphere, potentially critical periods exist when:

- The flammable protective atmosphere gas is introduced into the furnace chamber.
- A cold chamber filled with flammable gas is opened.
- The flammable atmosphere gas is removed and air is allowed to re-enter the furnace.
- The flow of flammable gas to the furnace is accidentally interrupted, allowing air to enter, during otherwise normal operation.

Cold Chambers

Cold chambers (such as water-jacketed horizontal cooling chambers attached to heating chambers and cooling bells) containing flammable atmospheres are hazardous unless adequately protected and properly operated. To ignite the flammable atmosphere gas safely as it contacts the room air and to prevent the formation of an explosive mixture inside the chamber, placement of gas flame pilots and flame curtains at the end doors is highly recommended. If possible, end doors should be opened a minimum height to inhibit infiltration of air. If end doors that are normally closed are constructed to be sloping rather than vertical, they allow faster ignition of the atmosphere gas as it leaves the furnace when they are opened. Room drafts arising from open doors and windows should be kept to a minimum. If possible, the furnace should be situated so that the normal room drafts are crosswise to rather than parallel with the length of the furnace. Ventilating fans and room inlets for makeup air should be located with this in mind. Baffles may have to be erected at the ends of the furnace to block normal room drafts or to divert unexpected drafts.

Emergency Procedure When Flow of Flammable Gas Is Interrupted

Whenever flow of flammable gas to the furnace is interrupted, the operator should make an immediate attempt to re-establish flow. If this is not possible and flow cannot be restored immediately, the furnace should be purged, by either the nonflammable gas method or the burnout method to restore pressure, thereby preventing entry of air. The type of furnace, its locations, drafts, temperature and the degree of flammability of the gas may affect the time delay permissible before purging, but the safest procedure is to immediately and automatically start purging with nonflammable gas. An alarm should be sounded and the area should be cleared of all personnel.

If an emergency supply of nonflammable gas is not available and the flow of flammable atmosphere gas cannot be restored promptly, the burnout method of purging should be used. For batch box-type and other straight-through furnaces, continuously burning gas pilots supply a source of ignition. If these are not available or are not operative, portable torches should be used, or twisted newspapers or oily waste should be laid across the full width of outside door openings and lighted. Then the doors should be fully opened. The pilots, torches, newspapers, or waste should be kept burning until the residual gases are burned out.

For a bell or elevator type of brazing furnace with retorts, it is recommended that nonflammable gas be made available and that an automatic system be provided to purge the retorts with nonflammable gas in the event of failure of flammable gas supply; an alarm system should also be provided.

Leaky Gas Valves

Safety shutoff valves and manual shutoff valves in flammable gas lines have been known to leak when closed. Such leaks can, of course, allow flammable gas to seep into the furnace chamber, muffle, or retort during standby periods, when it is assumed that the furnace is free of flammable gas. Gas valve leaks are potentially dangerous. For example, if a small leak should supply flammable gas to an air-filled furnace or cooling chamber over an extended shutdown period while the furnace is cold, an explosive mixture could develop. This mixture could be ignited when the furnace heat is turned on at start-up, prior to purging. Ignition could trigger a dangerous explosion. Several simple precautions can be taken to minimize or prevent explosions caused by leaky valves, the best of which is to provide "block-and-vent" valves. First, there should be a safety shutoff valve in the flammable gas line, followed by a normally closed solenoid-operated (or safety shutoff) blocking valve. The interconnecting pipe between these valves should be joined to a venting line running to the outdoors at a safe location, and the venting line should be equipped with a solenoid-operated vent valve that is normally open. When the two mainline valves are open, the vent valve should be closed. When the mainline valves are closed, the vent valve should be open to permit drain-off of any flammable gas that might leak through the closed safety shutoff valve. A gas safety expert should approve all installations.

Leaky Retorts or Muffles

Leaks in furnace retorts and muffles invariably provide their own warning signals. Deterioration in the appearance and quality of the work and an appreciable increase in the dew point of the atmosphere within the retort or muffle are prime indications of leakage. Nevertheless, all retorts and muffles should be pressure or vacuum tested periodically, and whenever a leak is detected, it should be repaired promptly.

When a muffle can be removed from the furnace, it can be tested more conveniently without the encumbrance of the surrounding furnace. However, many muffles are built into the furnace and are difficult to remove. Testing a built-in muffle often can be accomplished by removing the mesh belt and by sealing the extreme ends of the furnace for pressure or vacuum testing. The sealing is generally accomplished by clamping steel plates with gaskets to the entrance and exit ends of the furnace.

When it is suspected that a muffle is leaking and the furnace is in operation at high temperature, the hot chamber should be permitted to cool down to 1400 °F and should then be purged with nonflammable gas or by the burnout method.

Carbon Monoxide Poisoning and Suffocation Hazards

The possibility of poisoning is always present in the vicinity of a protective atmosphere furnace or gas generator when the atmosphere gas contains carbon monoxide. Usually, the concentrations of the atmosphere gases coming from furnaces are kept within safe health limits by adequate room ventilation or by exhaust systems. Room atmosphere can also be kept safe by burning the escaping gases at flame curtains.

Because carbon monoxide is colorless, odorless, tasteless, and nonirritating, a person inhaling air contaminated with carbon monoxide receives no warning of its presence. He should, therefore, be warned of the toxic effects of carbon monoxide and be acquainted with the symptoms of carbon monoxide poisoning.

A comparatively small concentration of carbon monoxide can have a marked effect on a person. Safe concentrations of carbon monoxide in air are considered to be less than 0.01%. Exposure to 0.04% for about $1^1/_2$ h may produce the characteristic primary symptoms of headache, mental dullness, and physical loginess. Greater concentrations or longer exposure times may prove fatal. Accordingly, when symptoms are noted, personnel should leave the area at once and should report the condition to the proper authorities. Simple and inexpensive testing equipment for detecting carbon monoxide and warning of its presence is available. If safe limits of carbon monoxide are to be established for given working environments, medical authorities should be consulted.

If personnel are to enter a furnace that has been shut down for repairs, particular care should be taken to ensure that all gases, particularly (but not exclusively) those containing carbon monoxide, are first flushed out of the furnace with air. In addition, a continuous flow of fresh air from a fan or blower should be provided to the furnace during the repair periods. For safety, some users require that atmosphere gas supply lines be blanked or disconnected from the furnace gas inlet lines during repair periods. This eliminates the hazard of leaking shutoff valves.

Suffocation and death can occur from lack of oxygen, either within or outside the furnace. As gases that are heavier than air, namely argon and carbon dioxide, flow from the furnace, they settle near the floor, cutting off the available oxygen supply.

Purging With Nonflammable Gas

These recommendations apply to all types of furnaces with protective atmosphere heating chambers, with or without muffles or retorts, rated at any temperature, and heated by all types of heat sources. Where practical, it is preferable to purge an air-filled furnace with nonflammable gas prior to introducing a flammable gas atmosphere. Likewise, during shutdown, it is preferable to purge the flammable gas with nonflammable gas. This applies particularly to heated furnaces, or sections of furnaces, that are below the generally reliable gas ignition temperature of 1400 °F. Flow rate indicators and a time clock should always be used when nonflammable gas purging, and a method for checking the extent of purging is desirable. The cooling zone should receive particular attention.

When air is purged from a furnace chamber using nonflammable gas, a volume of nonflammable gas equal to about five volumes of the chamber may be needed to reduce the oxygen content to about 0%. For maximum safety, most operators observe this rule. Typical nonflammable gases used for purging are listed in Table 4. The extent of purging by removing air with nonflammable gas can be determined with the use of an oxygen analyzer or a specific gravity indicator. The same specific gravity indicator can also be used to determine the extent of purging flammable gas with nonflammable gas when the furnace is being shut down.

The density of the purging gas has some bearing on the purging procedure. When

Table 4 Nonflammable gases used for furnace purging

Gas	Specific gravity(a)
Argon	1.379
Carbon dioxide	1.527
Helium	0.137
Nitrogen	0.972
Lean burned dissociated ammonia (99% N_2, 1% H_2)	0.963
Lean exothermic gas	1.030
Lean purified exothermic gas	0.966

(a) Specific gravity of air is 1.000.

a chamber is purged by replacing one gas with another, the least amount of mixing is generally desirable, which permits the most rapid and economical purging. If the density of the two gases is approximately the same, as for example air (sp gr 1.0) and nitrogen (sp gr 0.972) or lean exothermic gas (sp gr 1.03), the manner in which the two gases are mixed together is of little consequence. Therefore, if the air in a chamber is to be displaced by nitrogen, the nitrogen can be introduced at the top or at the bottom, whichever is more convenient, assuming that there is an opening at the opposite extremity for allowing the gases to escape.

If, however, it is desired to replace air (sp gr 1.0) with helium (sp gr 0.137), the helium should be added at the highest level, so that it can collect at the top and force the heavier air out at the bottom, with a minimum amount of mixing of the two gases. If the purging gas is argon (sp gr 1.379) or carbon dioxide (sp gr 1.527), the purging gas should be introduced at the bottom, so that the lighter air can escape at the top. The same principle applies when shutting down a furnace or taking a retort out of service when it is necessary to replace the flammable gas with a nonflammable purging gas.

In some furnaces, it may be impractical to employ the density consideration, and the gases may have to be diffused with turbulence and mixed, regardless of specific gravity, in order to replace one with the other. In general, this can be done satisfactorily, but it is likely to involve more time and greater expense. Manufacturers' purging instructions should be secured and followed.

Follow-Up Purging With Flammable Gas. After purging with nonflammable gas and before starting up the furnace, the next step is to purge with flammable gas. About five volume changes generally ensure complete purging. Purging with flammable gas also should be measured with flow-rate indicators and a time clock to ensure complete replacement.

Some typical flammable gases and their specific gravities are shown in Table 5. All of the flammable gases in Table 5 are lighter than the nonflammable gases in Table 4, except for helium.

When it is assumed that the chamber is completely purged with flammable gas, it is highly recommended that the composition of the effluent from the furnace be checked, using one of the methods listed below to make sure that it is not explosive, before turning on the heat, lighting pilots, or providing a source of ignition in any other way. Such routine checks may seem to be of little value when other prescribed procedures are adhered to, but they can show up a rare mistake by the operator or a fault in the equipment and can help to prevent an explosion. For example, the furnace operator may forget to purge with nonflammable gas prior to injecting the flammable gas; the flow rate or time interval for either of the purging gases may be inadequate; or an unnoticed crack in a retort or muffle, or an open door or pipeline, may provide a source of unwanted air.

Several methods are available for determining whether the flammable gas in the cold chamber is safe or explosive. A specific gravity indicator is effective for this purpose. An oxygen analyzer of the proper type can also be used. Some oxygen analyzers operate on a combustion principle, which supplies a hazardous source of ignition to an explosive atmosphere. Safe instruments are the chemical absorption type and the magnetic susceptibility type of oxygen analyzers, which provide no source of ignition. An atmosphere sample can be collected in a test tube and ignited at a safe distance from the furnace. Quiet burning or a pop or whistle indicates the type of combustion. After it has been determined that the furnace is safely filled with flammable gas and there is no chance for the atmosphere to explode, it is safe to turn on the heat, to light pilots, or to otherwise subject the furnace atmosphere to ignition sources.

Table 5 Flammable gases used for furnace purging

Gas	Specific gravity(a)
Hydrogen	0.069
Dissociated ammonia (75% H_2, 25% N_2)	0.295
Dissociated ammonia burned with air, rich (24% H_2, 76% N_2)	0.755
Rich exothermic gas:	
Unpurified (15% H_2)	0.858
Purified (15.5% H_2)	0.825
Endothermic gas:	
Dry (38% H_2)	0.622
Wet (28% H_2)	0.798

(a) Specific gravity of air is 1.000.

Purging by the Burnout Method

The burnout method of purging is utilized where it is impractical to purge an air-filled furnace with nonflammable gas before a flammable gas atmosphere is introduced during start-up or to purge out the flammable gas with nonflammable gas during shutdown. The following procedures apply to horizontal furnaces with heating chambers rated at 1400 °F or above, with attached relatively cool chambers that may or may not be heated for preheating, burn-off, purging, and cooling. These procedures do not apply to furnaces with molybdenum or tungsten heating elements, because such elements are seriously damaged in an air atmosphere at 1400 °F or above. Furnace types covered include box, pusher, mesh-belt conveyor, and roller-hearth conveyor.

In the burnout method of start-up, a flammable gas is injected directly into the air-filled heating chamber after the chamber has been heated to above 1400 °F. Care must be taken not to inject the gas into adjoining air-filled cool chambers, where explosive mixtures could form. If the main heating chamber has a muffle, these instructions apply only if a gas inlet leads directly into the muffle.

Start-up safety can be ensured by having a safety shutoff valve in the flammable gas line to the furnace interlocked with a 1400 °F contact in a furnace temperature control instrument. This arrangement ensures that only when the furnace is 1400 °F or above can flammable gas be supplied to it. If the furnace temperature drops below 1400 °F, the safety shutoff valve closes and stops the flammable gas flow, sounding an alarm. An auxiliary timer and solenoid valve in the atmosphere gas line to the cooling chamber allows flow only after adequate time has elapsed for complete burnout purging by the gas entering the heating chamber.

If the furnace operating temperature is to be below 1400 °F, such as might be required for silver brazing, the foregoing procedure needs to be modified slightly. First the furnace should be heated to 1400 °F and purged as described above. When that temperature has been reached, the auxiliary contact in the temperature control instrument can be bypassed with a holding relay, and the temperature can be lowered safely without the contact closing the safety shutoff valve and cutting off the atmosphere gas flow to the furnace. To shut down the furnace after use at reduced temperature, the furnace should be reheated to 1400 °F and the burnout procedure followed as outlined below. Or, if the alternate nonflammable gas procedure is preferred, shutdown can be instituted directly from the low operating temperature.

During start-up, as the flammable gas enters the brazing chamber, which has been heated to 1400 °F or above, the gas burns oxygen out of the air. It continues to burn and to spread into adjoining chambers, eventually flooding the entire furnace and making it ready to operate with protective atmosphere.

During shutdown, the procedure is reversed. With furnace temperature above 1400 °F the flammable gas is ignited at an opened end door, and gas flow is shut off. As the gas burns, air sweeps in and the flammable constituents are entirely consumed, leaving the furnace filled with air and nonflammable products of combustion.

Burnout Method Procedures for Start-Up and Shutdown

For convenience, the following lists present, in chronological sequence, the typical steps to be taken in starting up and shutting down a horizontal furnace of the box, mesh-belt, or roller-hearth type, using the burnout method. For a particular furnace, the manufacturer's operating instructions should be consulted and followed because a typical procedure cannot be employed for all types and makes of equipment.

Procedure for Atmosphere Introduction During Start-Up

1 Open wide all furnace doors.
2 Make sure all inlets for flammable atmosphere gas are shut off tight and have been shut for some time. If there is any question that there has been some leakage, do not turn on the heat source or light any pilots or curtains. First blow or flow air or nonflammable gas into the furnace, taking particular care to purge high spots such as the arched roof and refractory door cavities. Vent the nonflammable gas outdoors, so that it cannot accumulate in areas where there are personnel.
3 Turn on heat and bring main heating chamber up to 1400 °F or above.
4 If there is a heated preheat or precool chamber attached, with muffle, bring it up to 1400 °F also, preparatory to burning any flammable gas that might leak through pores or cracks in the muffle. If the maximum temperature rating is below 1400 °F, or if there is no muffle, heating of this chamber is optional.
5 Close all purge and vent pipes.
6 Light all pilots at flame curtains, purge pipes, and vent pipes.
7 If the hot chamber (or chambers) is double-ended and has an attached cool chamber (or chambers) at only one end, close the outer entrance door to the hot chamber, leave all intermediate throat-baffle doors wide open, and lower the outer exit door of the cool chamber to an opening of 2 or 3 in. For this purpose, gas flame curtains should be adjusted to extend only slightly above hearth level to facilitate seeing the atmosphere gas flames described in step 9.
8 If attached cool chambers are at both ends of the hot chamber (or chambers), leave all intermediate doors wide open and lower both outer entrance and exit doors to openings of 2 or 3 in. Adjust gas flame curtains in accordance with step 7.
9 Start flow of flammable gas into hot (1400 °F or above) main heating chamber, using a flow rate equivalent to normal consumption. Continue until atmosphere gas burns at the outer entrance and exit doors. The burning of gas at the furnace doors is an indication that all air has been burned out and that the furnace is filled with protective atmosphere.
10 Lower and raise intermediate refractory doors. Piston action in upper door cavities may create vacuum or pressure, causing air to rush in or gas to be pushed out forcefully at end openings. Slower door-operating speed will usually rectify this objectionable effect. Reduce speed of air-operated doors by closing down the small needle valves in vent lines of the pneumatic cylinders.
11 Open the outer end doors wide and adjust gas flame curtains to full height to cover the openings completely with nearly transparent flame.
12 Close all outer end doors and intermediate doors, if so desired.
13 Open purge and vent pipes as desired. Atmosphere gas will be ignited by the pilots.
14 Bring temperature of heating chamber (or chambers) up or down to desired operating temperature.
15 As a drying-out procedure, allow time for heat to soak into refractory walls and moisture to be purged out of protective atmosphere.
16 The furnace should now be ready for operation.

Procedure for Withdrawing Atmosphere

1 Adjust temperature of hot chamber (or chambers) to 1400 °F or above.
2 Open all interior doors wide.
3 Provide continuous, reliable ignition sources at both outer entrance and exit doors. These may be pilot lights, low-burning gas flame curtains, or portable torches. The gas flame curtains for this purpose should be adjusted to extend only slightly above hearth level to allow fresh air to sweep over them into the furnace after the outer end doors of the furnace are opened and the atmosphere gas is burning.
4 Partly open outer entrance and exit doors. Be sure atmosphere gas is ignited.
5 Shut off flow of flammable atmosphere gas to all inlets of hot and cool chambers. The ignited atmosphere gas will burn with insweeping air throughout the furnace, and the flame will extinguish itself in a few minutes.
6 Caution: Make sure that all atmosphere-gas shutoff valves are closed tightly and do not leak. Use of the block-and-vent valves previously described will ensure against leakage.
7 With the inner flame extinguished and all chambers filled with air and nonflammable products of combustion, all pilots and flame curtains may be extinguished and the heat source (or sources) shut off.
8 Leave all furnace doors open for self-venting in the event of leakage of flammable gas.
9 At this stage the furnace can be considered completely shut down.

Torch Brazing of Steels*

TORCH BRAZING is a brazing process in which the heat is obtained from a gas flame or flames impinging on or near the joint to be brazed. Torches used in this process may be of the hand-held type or may consist of fixed burners with one or many flames. Several types of fuel gas are available for combustion with oxygen or air. Torch brazing can be performed as a completely manual, a partly mechanized, or a completely automatic process.

This article primarily discusses the torch brazing of carbon and low-alloy steels. Torch brazing, however, is also used on stainless steels, cast irons, copper and copper alloys, and carbides. Highly alloyed steels, heat-resistant alloys, aluminum alloys, and reactive metals are usually brazed by other methods, because they require special atmospheres or closer control of the thermal cycle, or both.

Manual torch brazing is most suitable in applications with small production quantities, because equipment cost is low. It is also useful when physical size, joint configuration, or other considerations make brazing by other methods difficult or impossible. The main drawbacks are the high labor content and the relative skill needed for efficient production. Where production quantities are larger, automatic torch brazing is used, often producing brazed assemblies at the rate of 400 to 1400/h.

Fuel Gases

Acetylene, natural gas, propane, and proprietary gas mixtures are the types of fuel gas most often used in the torch brazing of steel. Hydrogen, butane, and producer (city) gas are seldom employed. In manual torch brazing, pure oxygen is chiefly used as the combustion agent because of its fast heating rate. As a cheaper source of lower grade oxygen, compressed air or a high-volume low-pressure blower is also suitable, if lower flame temperatures and heating rates are acceptable. Either oxygen or compressed air is used in automatic torch brazing.

The economics of gas consumption plays an important role in the selection of gases. The increased production of oxygen for steelmaking and other purposes has resulted in larger supplies and lower prices. This trend has favored the use of natural gas and propane, which requires larger oxygen-to-fuel ratios than acetylene, but is less costly than acetylene.

The use of compressed air entails only the cost of compression, which can be advantageous in automatic operations where the investment in pumping and mixing equipment can be amortized with the rest of the equipment. Another factor favoring compressed air/natural gas combustion in automatic applications is that the workpieces are less likely to be destroyed by overheating if conveyor or rotary indexing equipment malfunction occurs.

Principles and Techniques

The principles of torch brazing differ from those of other brazing processes only in technique. The braze bond is the same; a suitable filler metal is melted in a closely fitted joint between heated but unmelted base-metal parts and fills the joint by capillary action.

Brazing Filler Metals. The two types of filler metal established as suitable for torch brazing of steel are the silver alloys and the copper-zinc alloys that are described in the section on filler metals in this article. Although the silver alloys are more costly, they generally melt at lower temperatures than the copper-zinc alloys. The tensile strength of the silver alloy filler metals is about half that of the low-carbon steels, but considerably greater strength is developed when they are used in correctly designed joints. Strengths exceeding 100 ksi are possible using 0.0015-in. joint clearance. Tensile strength of the copper-zinc filler metals is about the same as that of the low-carbon steels.

Joint Design. In torch brazing of steel, the frequently used filler metals need a joint clearance (at brazing temperature) of 0.001 to 0.005 in. for capillary flow. Where thermal expansion is significant, an allowance is made on room-temperature measurements. Lap joints designed for shear loads are generally preferred to butt joints designed for tensile loads. For maximum joint efficiency, the length of the overlap should measure at least three times the thickness of the thinnest member to be joined.

Joints are designed for either face feeding or preplacement of filler metal. Preplacements are used in automated brazing or in joints of large area or a configuration difficult to penetrate by face feeding. For additional discussion of joint fit and design, see the article "Furnace Brazing of Steels" in this Volume.

Prebraze Cleaning. Filler metals will not wet and adhere to contaminated surfaces. Before torch brazing, the joint surfaces must be cleaned to remove all dirt, oil, grease, rust, and scale. Prebraze cleaning is discussed in the article "Furnace Brazing of Steels" in this Volume.

Fluxing. A flux is applied to the joint before heating to promote the flow and bonding of filler metal throughout the joint to be torch brazed. Flux is used to keep parts clean; it is not used to clean dirty parts. The three principal functions of a flux are to prevent the oxidation of metal surfaces during heating by excluding oxygen, to absorb and dissolve residual surface oxides that may form during heating, and to assist in the flow of the brazing filler metal by providing a clean surface over which the molten filler may flow. Flux may be applied as a powder, paste, slurry, or liquid, or as a mixed paste of flux and filler metal. It may also be applied as a vapor through the gas flame. The types of fluxes are discussed later in this article.

*Revised by C. Philp, Product Engineer—Brazing, Handy & Harman; C. Van Dyke, Brazing Consultant, Lucas-Milhaupt, Inc.

To prevent oxidation of the joint surfaces, flux must be applied so that both surfaces to be joined are completely protected as the heating operation proceeds. Flux is effectively applied before assembly, especially if the joints are more than about $^{3}/_{16}$ in. deep, by brushing a thin layer of paste on the cold or slightly warmed joint surfaces or by dipping the joint in paste flux heated to about 150 °F or in liquid flux. Paste, slurry, or liquid flux may also be applied by brushing or spraying on the joint after assembly, if sufficient distribution can be obtained before heating begins. The latter method is frequently used in automatic torch brazing.

Flux is often applied on the area around the joint to minimize discoloration and scaling of the workpiece surface. Although this increases flux cost, postbraze cleaning time may be reduced or eliminated. When flux is used for this purpose, it should be applied sparingly to the entire surface that will be subjected to heating.

Flux inclusions can be caused by entrapment (not providing an escape path), by too loose a joint fit, or by overheating the workpiece. These conditions can be minimized by moving one member in relation to the other, while at brazing temperature, before the filler metal begins to solidify. The movement aids the filler metal in penetrating the joint area and displacing the flux.

The filler metal also should be protected by flux. Preformed filler metal can be dipped in flux, or flux can be brushed onto it. Either face-fed rod, wire, or strip can be brushed with flux, or the end can be dipped into the flux. Filler-metal powder is either dusted on wet flux or mixed with it (see the section on filler metals in this article).

Steels readily oxidize during torch heating, resulting in a surface that resists wetting by the filler metal. The flux protects the steel surface and absorbs oxides that form, thus providing a joint interface that is readily wet by the filler metal. The flux must be heated to its active state, and it must flow over all joint surfaces before the filler metal starts to flow. The filler metal will flow around areas not adequately protected by flux, resulting in voids.

Heating. Brazing should be done away from drafts to avoid uneven heating and cooling. Torches adjusted for a slightly reducing or a neutral flame are used in bringing steel parts to brazing temperature. To prevent overheating and melting of the base metal, the torch tip is held away from the surface of the work so that the inner cone of the flame does not approach too closely. Heating is accomplished by the outer flame envelope. In manual torch brazing, the torch should be kept in motion to avoid localized overheating. The flame should be applied by the following procedure to bring both members of the joint to uniform brazing temperature:

- With joint members of unequal mass, heat the more massive member first, until flux melting indicates that brazing temperature has been reached.
- Heat the less massive member next, until it reaches brazing temperature.
- Heat the joint last, being careful not to overheat. If preplaced, the filler metal should flow all around the joint, appearing at the joint edges, and the flame should be removed. If face fed, the filler metal should be placed on the joint before or behind the point where the flame strikes the part.
- Avoid applying the torch heat directly on the filler metal, if possible.

In some applications, a massive member may receive all of the heating, the less massive part being heated entirely by conduction. When joining steel to another metal, the metal having the higher thermal conductivity is treated as the more massive member.

When heating a tubular or socket assembly for brazing, the flame is applied first to the inner member, where it emerges from the outer member, to expand the inner member and tighten the fit. Then heating is concentrated on the outer, or more massive, member; thus, further heating of the inner member is mainly by conduction. Conduction is poor if the fit is not tightened previously by heating as described above. Heating should continue until the workpieces are slightly hotter than the flow temperature of the filler metal. Overheating should be avoided; excess heat will break down the flux and interfere with its action.

Brazing fluxes melt at somewhat lower temperatures than the filler metals with which they are used. Their melting serves as an indication that brazing temperature is being reached. When flux is heated, the water content first boils off, leaving a powdery deposit. Upon further heating, the flux melts as a clear, thin liquid that flows through the joint and actively protects the surfaces from oxidation. If the procedures described above have been followed, the joint is at brazing temperature, and the filler metal (if face-fed) melts and very quickly flows into the joint after contact. If the filler metal does not melt and flow after contact with the base metal, the joint is not at brazing temperature and heating must be continued.

Good brazing practice requires either that heat from the properly preheated workpiece is used to melt and flow the face-fed filler metal, or that the workpiece and preplaced filler metal are simultaneously heated to the proper brazing temperature to fill the joint. Heating of the filler metal alone, or ahead of the base-metal members, will restrict capillary flow and wetting and can result in defective joints. Overheating of the filler metal is also likely with direct exposure to the flame and can result in the evolution of excessive brazing fumes such as cadmium and zinc oxides.

Feeding Filler Metal. The two primary types of filler metals used in torch brazing of steel are silver and copper-zinc alloys. They are usually available in any of the following forms: rod and wire of various diameters, strip (shim stock) of various thicknesses, and powder. Characteristics of these filler metals are discussed in the section on filler metals in this article.

The quantity of filler metal required per joint depends on the volume of the joint space to be filled plus the volume of the desired fillets. Usually, the joint is slightly overfilled to allow for filler-metal shrinkage. In manual face feeding with rod or wire, the size of the filler metal used should be large enough to avoid interruption of brazing to start a new length before completing the joint. For thin materials, the diameter of the rod or wire should be comparably thinner. Thicker filler metal melts slowly and can result in poor flow or overheating, causing damage to the thin member. Automatic face feeding of wire and powdered filler metals is described in the section on equipment for automatic torch brazing in this article; the preplacement of wire, strip, and powder is discussed in the section on filler metals in this article.

Quality Control and Inspection. Joint quality can be controlled by carefully checking the adequacy of each of the procedural steps involved. Inadequate cleaning, improper joint clearance, underheating, and overheating frequently are sources of difficulty. Defects that are likely to recur are usually in the form of voids caused by: (1) unreduced oxide or foreign matter that prevents wetting of the joint surface, (2) gas pockets resulting from gas evolution or vaporization of lower melting constituents of the filler metal, (3) entrapped flux, (4) incomplete filler-metal flow, and (5) too short a heating time.

Small and widely distributed voids seldom affect the serviceability of the part adversely. Most inspection standards require no more than 80% coverage between faying surfaces, as revealed by peel testing, ultrasonic inspection, or radiography. Visual inspection is frequently used; joints are accepted if a continuous fillet is formed all around the joint, especially on the side opposite to that of filler-metal placement or feeding.

Pressure tests and ultrasonic and radiographic examination are used to reveal defects nondestructively. In addition, destructive testing can be used to set up both the original procedure and, on a sampling basis, to check whether the procedure is under control. One form of destructive test is the peel testing of lap joints, frequently used in resistance spot welding. To make this test, one member of the joint is held in a vise while the joint is opened with a chisel; the other member is then peeled or twisted away to reveal the condition of the bonded area.

Brazing Versus Braze Welding. One of the joining methods often confused with torch brazing is braze welding. The equipment and some of the filler metals used in this type of joining may be the same as those used in manual torch brazing. In some applications, even the torch and filler-metal manipulations may appear the same. Generally, however, the two joining methods differ considerably. Some of the characteristics of braze welding are discussed in the article "Oxyacetylene Braze Welding of Steel and Cast Irons" in this Volume.

Equipment for Manual Torch Brazing

Manual torch brazing equipment is the same as, or very similar to, the equipment used for manual oxyfuel gas welding. A description of the design and use of this equipment, which consists of a gas supply, regulators, hoses, and a gas torch, is given in the article "Oxyfuel Gas Welding Processes and Their Application to Steel" in this Volume. Although most of the equipment described in that article can be used for torch brazing, tip selection and the technique of applying the flame to the workpieces differ. The safety precautions stressed in that article apply with equal emphasis to torch brazing.

Because the high heat intensity of the oxyacetylene flame, which is usually necessary for oxyfuel gas welding of steel, is not necessary for brazing, other gases such as natural gas, propane, and various proprietary gases may be used in combination with oxygen or air. The high heat of oxyacetylene flames is often used to advantage in high-production applications because of shorter brazing times and reduced labor costs. Because of its relatively high cost, oxyacetylene heating is sometimes combined with air/propane or air/natural gas heating. For example, in production brazing of massive steel assemblies, air/propane (or air/natural gas) is used to preheat the workpiece to about 900 °F before torch brazing with an oxyacetylene flame.

The choice of oxyfuel gas combination determines the type of gas supply equipment and regulator design and also affects the size and color of the hoses used and the design of the torch components. The sizes of the torch components, hoses, regulators, and gas supply system also must be appropriate to the general range of heating capacity required.

Equipment for manual torch brazing usually includes attachments for general welding and cutting operations. When fuel gas in cylinders is employed, manual equipment is usually portable; otherwise, torch mobility is limited by proximity of the fuel gas to the work. Accessories may include heating torches and tips, gas savers, flashback arresters, and gas fluxing equipment.

Manual torch brazing is usually performed on a work table covered with fire-resistant cement board overlaid with refractory brick. A supply of these bricks should be available. They can be used to aid in the positioning of workpieces and to reflect torch heat on the part being brazed. Reflected heat can also be used to preheat parts waiting to be brazed. Fiberglass, plastic, and ceramic screens and heat shields may be used for protecting nonbrazing personnel and surrounding equipment from the heat of the torches and for conserving torch heat that might otherwise be dissipated. Such screens also protect the assemblies from drafts.

All torch brazing, whether manual, mechanized, or automatic, requires adequate ventilation in accordance with American National Standards Institute (ANSI) Standard Z49.1, "Safety in Welding and Cutting." The sources of the most harmful fumes in torch brazing are the cadmium- and zinc-bearing filler metals and the fluorine-containing fluxes. Ventilation must be provided in the manner and to the extent prescribed in the standard cited above. The level of the contaminant must not exceed established threshold limit values. For the filler material used, "Supplier's Material Safety Data Sheet, OSHA-20" should be consulted. Furthermore, to minimize the evolution of harmful fumes, the heating of filler metals above their recommended brazing temperatures should be avoided (see Table 1). Cadmium boils at 1409 °F; zinc boils at 1665 °F. Exceeding these temperatures during brazing with cadmium- and zinc-bearing alloys can result in excessive fuming.

The filler metals that contain cadmium and zinc are identified in the section on filler metals in this article, and the fluxes that contain fluorine compounds are discussed in the section on fluxes in this article. Major sources of safety rules and recommendations are listed in the article "Oxyfuel Gas Welding Processes and Their Application to Steel" in this Volume.

Manual brazing torches designed for oxyfuel gas consumption have three principal components: the torch body, which serves as the handle and is equipped with needle valves to control the flow of oxygen and fuel gas; a mixing head; and a set of torch tips, which may be supplied with or without extension tubes. Essential details of these parts are shown in the article "Oxyfuel Gas Welding Processes and Their Application to Steel" in this Volume.

Mixing-chamber design includes two basic types: equal-pressure and injector. Equal-pressure mixing chambers (also called medium-pressure and positive-pressure mixers) are used chiefly with acetylene, which can be supplied at pressures of about 1 to 15 psi; this type of mixing chamber cannot be used with fuel gas supplied at a pressure less than about 1 psi.

Injector mixing chambers are used with fuel gases that are supplied at pressures less than about 1 psi or require large quantities of oxygen. Injector mixing chambers, however, can also be used with higher pressure fuel gases by throttling the fuel gas and oxygen using the torch inlet control valves. Individual adjustment is necessary for each tip size.

Because oxygen is readily available at high pressure, the injector mixing chamber is able to utilize the venturi principle by which a high-velocity jet of oxygen is made to draw or aspirate a low-pressure fuel gas. Injector mixing chambers are essential to the use of low-pressure natural gas, which is often supplied at a pressure of about 1 psi or less. When using propane and certain proprietary gases that can be supplied at higher pressures, the injector mixing chamber is a convenient means of obtaining the higher oxygen-to-fuel gas ratios required. In torch brazing, volumetric oxygen-to-fuel gas ratios are approximately 1 to 1 for acetylene, 2 to 1 for natural gas, and 4.5 to 1 for propane.

The heating capacity of a torch is determined by the fuel-gas consumption rate,

Table 1 Filler metals for torch brazing low-carbon and low-alloy steels(a)

AWS classification	Product form	Nominal composition, % Ag	Cu	Zn	Cd	Ni	Sn	Fe	Mn	Si	P	Temperature, °F Solidus	Liquidus	Brazing
Silver alloys														
BAg-1	Strip, wire, powder	45	15	16	24	...	...	...	...	...	...	1125	1145	1145-1400
BAg-1a	Strip, wire, powder	50	15.5	16.5	18	...	...	...	...	...	...	1160	1175	1175-1400
BAg-2	Strip, wire, powder	35	26	21	18	...	...	...	...	...	...	1125	1295	1295-1400
BAg-2a	Strip, wire, powder	30	27	23	20	...	...	...	...	...	...	1125	1310	1310-1400
BAg-3	Strip, wire, powder	50	15.5	15.5	16	3.0	...	...	...	...	...	1170	1270	1270-1400
BAg-4	Strip, wire, powder	40	30	28	...	2.0	...	...	...	...	...	1220	1435	1435-1650
BAg-5	Strip, wire, powder	45	30	25	...	...	...	...	...	...	...	1225	1370	1370-1550
BAg-6	Strip, wire, powder	50	34	16	...	...	...	...	...	...	...	1250	1425	1425-1600
BAg-7	Strip, wire, powder	56	22	17	...	...	5.0	...	...	...	...	1145	1205	1205-1400
BAg-20	Strip, wire, powder	30	38	32	...	...	...	...	...	...	...	1250	1410	1410-1600
BAg-27	Strip, wire, powder	25	35	26.5	13.5	...	...	...	...	...	...	1125	1375	1300-1400
BAg-28	Strip, wire, powder	40	30	23	...	...	2	...	...	...	...	1200	1310	1310-1500
Copper-zinc alloys														
RBCuZn-A(a)	Strip, rod, wire, powder	...	59	40	...	...	0.6	...	...	...	...	1630	1650	1670-1750
RBCuZn-D(a)	Strip, rod, wire, powder	...	48	41	...	10.0	...	...	...	0.15	0.25	1690	1715	1720-1800
RCuZn-B(b)	Rod	...	58	38	...	0.5	0.95	0.7	0.25	0.08	...	1590	1620	...
RCuZn-C(b)	Rod	...	58	39	...	...	0.95	0.7	0.25	0.08	...	1595	1620	...

(a) Classified for braze welding and brazing. (b) Classified for braze welding
Source: Abstracted from the mandatory and nonmandatory sections of AWS A5.7, AWS A5.8, and other sources

which in turn depends on the size of the tip orifice and the mixing chamber. As the diameter of the orifice increases, the gas flow required to produce a proper flame and the heating capacity increase also. A given size of mixing chamber can usually supply the proper gas flow for a specific range of tip sizes. Selection of a tip size outside this range necessitates replacement of the mixing chamber with one of the appropriate size. The interchangeability of mixing chambers and of tips (including tip tube extensions and adapters) is limited and is prescribed by the manufacturer of the torch.

Torch tips for manual brazing differ according to the kind of fuel gas used, the tip-orifice diameter (heating capacity), and the style. Tips used with acetylene are usually faced off square across the orifice. Tips used with natural gas and propane are recessed or cupped at the orifice; otherwise, the lower flame-propagation rates of these gases may allow the flame to be blown away by the gas jet. In general, tips of both types are available in a range of orifice diameters designated by code numbers of manufacturers, on the basis of which recommended gas settings are supplied. Also available are multiple-flame tips that provide a broader flame coverage, useful in speeding the brazing operation. Figure 1 shows several tip designs used in manual torch brazing.

Multiple-flame tips of the type shown in Fig. 1(c) and (d) are used on heating torches for brazing with the lower melting silver-alloy filler metals. Because the flames are broader, heat transfer is more diffuse than with single-flame tips. These torches are frequently used in soldering. They are less desirable for brazing with copper-alloy filler metals and are not generally used for braze welding because of the time needed to reach operating temperatures. Torches equipped with multiple-flame tips are usually of a heavy-duty type and require larger capacity mixers than those used with single-flame

Fig. 1 Tip designs used in manual torch brazing. (a) One-piece tip and tube extension. (b) Two-piece tip and tube assembly that permits quick replacement of tips. (c) and (d) Basic screw-in tips. (e) Dual tip holder. Designs are available in various sizes, styles, and capacities, used with complementary torch systems and gas settings.

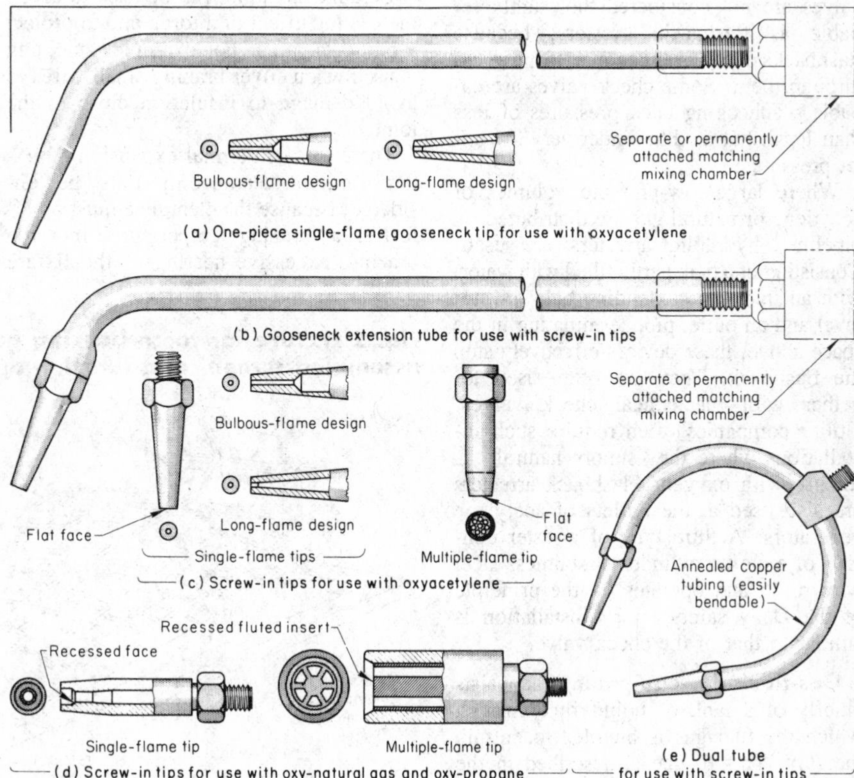

tips to handle higher gas flows. They may need larger gas supply capacity.

Gas savers are used at brazing stations to conserve gas between brazing operations. The hoses are connected to dual valves actuated by a lever that also serves as a torch rest. In one type, the torch rest shuts off the gas completely, but an adjacent pilot flame permits instant re-ignition at the previous setting when the torch is lifted. Another type reduces the flame to a small size; when the torch is lifted from the rest, the flame reverts to its former size.

Flashback arresters are used to prevent combustion or explosion from taking place in the hose if the flame, as a result of some malfunction, should burn back through the mixer and torch handle. If not promptly stopped, flashbacks can result in injury to personnel and damage to equipment and property. The use and installation of flashback arresters are covered in ANSI Standard Z49.1.

The two principal types of flashback arrester are the mechanical check valve and the water-seal, or hydraulic back-pressure, valve. Check valves for oxygen and fuel gases supplied at about 1 psi or more consist of small threaded fittings that are inserted between the torch and the hoses to check back flow. They may also be installed between the hoses and the regulators or line gages. When used alone, check valves are not considered thoroughly reliable as flashback arresters, because flashbacks have been known to progress through them. Some check valves are capable of checking back pressures of less than 1 psi, but tend to hinder very low inlet pressures.

Where large low-pressure volumes of aceytlene or natural gas are distributed by pipeline, hydraulic arresters are used. Consisting of a tank partly filled with water, with an inlet pipe extending below water level and an outlet pipe terminating in the space above, these devices effectively stop the flashback. They are often used together with mechanical check valves. Utility companies often require such installations where they supply natural gas for use with oxygen. Flashback arresters are also used at the outlets of acetylene generators. A third type of arrester consists of a closed cylinder of stainless steel wire mesh and operates on the principle of the Davy safety lamp. Installation is similar to that of the check valve.

Gas-fluxing equipment consists chiefly of a tank of liquid flux through which the fuel gas is bubbled to entrain the flux. This device is described in the section on fluxes used for torch brazing in this article.

Miscellaneous standard equipment includes tip cleaners, which are necessary for defouling tips, maintaining proper flame shape, and preventing tip overheating (one of the causes of backfires and flashbacks); goggles, which are needed for eye protection; and a set of wrenches, for proper tightening of tip, mixing-chamber hose, and other connections.

Fixtures for Manual Torch Brazing

Holding devices include vises, clamps, pliers, tongs, and special fixtures. Sometimes, a simple holding fixture can be made to support a number of small assemblies. Ordinarily, self-jigging assemblies are preferred when possible. Where a fixture must be used, it should be constructed to avoid interference with the torch flame, the application of filler metal, or the view of the operator. The fixture should be as lightweight as practical and should contact the assembly only where necessary.

Where a low heat sink is desired, point or line contact is preferable to contact over an area of surface. Ceramic points or contacts are available to minimize heat loss to fixtures. Conversely, where heat-sensitive components must be protected, larger surface contact and a conductive material, such as copper, can be used.

Ceramic shielding can be used to divert the heating effect of a torch and to protect a particular area. For example, shielding is used when silver brazing small wires to avoid damage to insulation close to the joint.

The effect of thermal expansion of fixtures on joint clearance must also be considered, because the clearance must be effective at brazing temperature. In torch brazing, excessive heating of the fixture usually can and should be avoided, especially where complicated assemblies or high temperatures are involved. If parts are tightly clamped and rigid fixtures are overheated, the effect of expansion and contraction can result in overstressing and cracking of the joint and/or distortion of the part. In the examples that follow, simplicity of fixturing was a primary reason for selecting torch brazing.

Example 1. Use of Torch Brazing To Simplify Fixturing. An important reason for selecting torch brazing for assembling an air-inlet screen for a gas turbine was the simple fixture that could be used for holding the numerous parts in position (Fig. 2). All of the parts to be assembled—frame members, trunnion, and ribs—were made of 1020 steel. Before brazing, they were cleaned in a hot alkaline solution, rinsed, acid pickled, hot rinsed, and dried. The clean parts were then assembled in the fixture. Each joint was manually fluxed and torch brazed, using filler metal in wire form. After brazing, flux residue was removed by immersing the assembly in boiling water, the joints were visually inspected, and the fabricated screen was cadmium plated. Brazing filler metal was silver alloy BAg-1a, which was well suited to torch brazing at a relatively low temperature (about 1200 °F). Flux was type 3A in paste form.

Example 2. Meeting Airtightness and Dimensional Requirements of a Tubing Assembly. The low-carbon steel tube assembly shown in Fig. 3 was manufactured in quantities of about 15 000 units annually. Essential requirements were:

- The two short pieces of tubing should be parallel within ±0.005 in.
- The cross-sectional area on the inside of the long tube should not be decreased by more than 25% by excess filler metal.

Fig. 2 Fixture for torch brazing gas turbine air-inlet screen, the assembled screen, and location of brazed joints

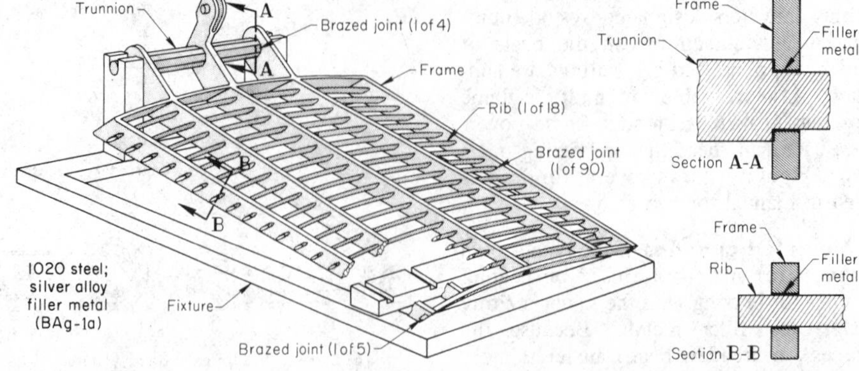

Fig. 3 Torch brazed tubing assembly

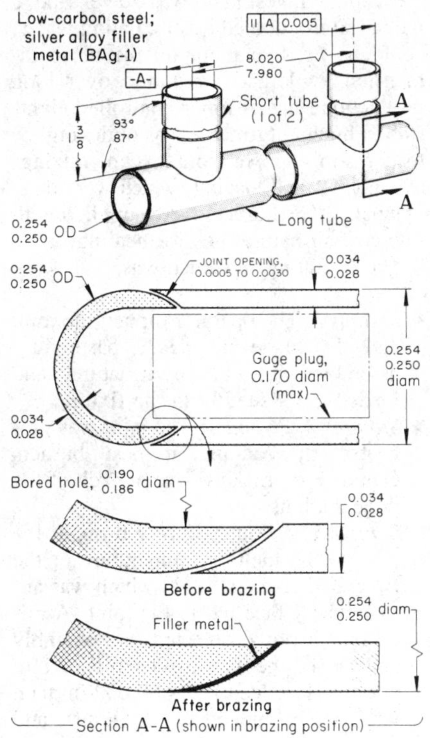

Torch type	Oxyacetylene
Flame type	Neutral
Tip-orifice diameter	0.0380 in.
Filler metal	0.031-in. wire, BAg-1
Flux	AWS type 4
Production rate	30 assemblies (60 joints) per hour

- All joints should be airtight (withstand 10 psi minimum).
- A 0.170-in.-diam gage plug should fit freely through the short tubes into the long tube after brazing.

The brazing fixture consisted of a V-grooved bar for the long tube with two V-grooved bars at 90° for the short tubes, two locating plugs that were inserted through the short tubes into the holes in the long tube, and toggle clamps to hold the assembly in alignment. The locating plugs were removed after the parts were clamped in place.

Joint preparation consisted of drilling two holes, 0.188 ± 0.002 in. in diameter, in the long tube, milling the mating ends of the short tubes to fit saddlewise over the holes in the long tube (Fig. 3), degreasing the parts in a water-soluble cleaner, and rinsing the parts in cold water and then hot water to speed drying. The joints were brushed with a paste flux that corresponded to an American Welding Society (AWS) type 4A and which proved to be an acceptable alternate to type 3A in this application.

Joint clearances, after fixturing, varied from 0.0005 to 0.003 in. The joints were heated by manually passing the neutral flame of the torch around the joint to heat the area to the brazing temperature, approximately 1200 °F. The wire of BAg-1 silver-alloy filler metal was face fed at the top surface of the joint, with the assembly positioned as shown in Fig. 3 (section A-A). The filler metal flowed around to complete the joint.

Equipment for Machine Torch Brazing

Various mechanical devices are used to perform repetitive torch brazing operations. Machines have been built for fixturing, fluxing, heating, face feeding of filler metal, forced cooling, hardening, and other operations. Flowmeters and manometers are often used to monitor and control gas ratios where large gas volumes are involved. Mechanization of all torch brazing operations, however, is not always possible or desirable. Workpiece configuration, joint design, and accessibility or other factors determine whether some operations are done best by hand or by machine, even though large production quantities are involved. Machine torch brazing is the term used for operations that are partly manual and partly mechanical.

In principle, motion can be imparted to the workpiece, the brazing equipment, or both. Usually, however, the workpiece at a certain stage of the operation is moved past one or more stations, where a brazing function is performed manually or by machine. The two most frequently used means of movement are the conveyor belt and the turntable. Brazing operations can be set up for either mode of travel. Heating is the operation most often mechanized, because moving workpieces past fixed torches or burners is relatively simple with controlled timing. Manual operations may include fluxing, filler-metal preplacement, and fit-up.

Conveyor belts are often used to carry the workpieces under a series of fixed burners and then through an air blast or water jet for cooling. In this type of setup, joint assembly, fluxing, and filler-metal addition are usually done manually before the part is placed on the conveyor.

Turntables provide a means for any degree of automation. A circular table is fitted with a series of duplicate holding fixtures equally spaced on the rim. The number of fixtures equals the number of operating stations that are located in fixed positions around the outside of the table. At any station, the operation may be manual, partly manual and partly automatic, or completely automatic; the choice is determined by the nature of the workpiece. For instance, devices are built to face feed liquid, paste, or powder flux at one station and wire filler metal at another, or an applicator can be used to face feed a paste mixture of powder filler metal and flux. Preform rings can be automatically positioned at the joint either on or off the machine.

The joint design of the workpiece must be amenable to face feeding from one or more fixed-point sources. Even if the joint area is not too large or inaccessible to require prefluxing or filler-metal preplacement, there must be at least a small area on the joint where the filler metal can rest momentarily. Complex joint configurations usually necessitate hand feeding.

Heating from fixed stations presents few difficulties, because single or multiple burners can be mounted at one or more stations and connected to a common gas supply. In more complicated torch brazing machines, the dwell time at each station is determined by the assembly of piece parts, time to heat the mass, or time taken for the filler-metal application; the total heating time is obtained by adding the number of heating stations needed to complete the particular thermal cycle. Air-cooling and water-cooling stations are added to obtain desired cooling rates. Turntable indexing and rotation are controlled by electronic timers.

The three examples that follow describe applications in which simple turntables were used to rotate the assemblies past several heating burners to achieve required production rates at relatively low capital investment. In two applications, assembly and fluxing were done manually. The third application incorporates automatic preform ring loading and automatic fluxing.

Example 3. Mechanized Heating and Cooling of a Joint. The contact assembly shown in Fig. 4, which consisted of a spring steel arm and a sintered carbide contact, was brazed on a turntable using burners supplied with premixed air/natural gas. Fluxing and loading were done by hand; the remaining steps were automatic, including the cooling operation. The production rate was 300 assemblies per hour, and the normal manufacturing lot was 100 000 assemblies.

The spring steel arm was supplied in the hardened condition and with a bright finish. It was vapor degreased before assembly. The parts were held by tweezers and

Fig. 4 Contact assembly and turntable setup used for brazing and quenching

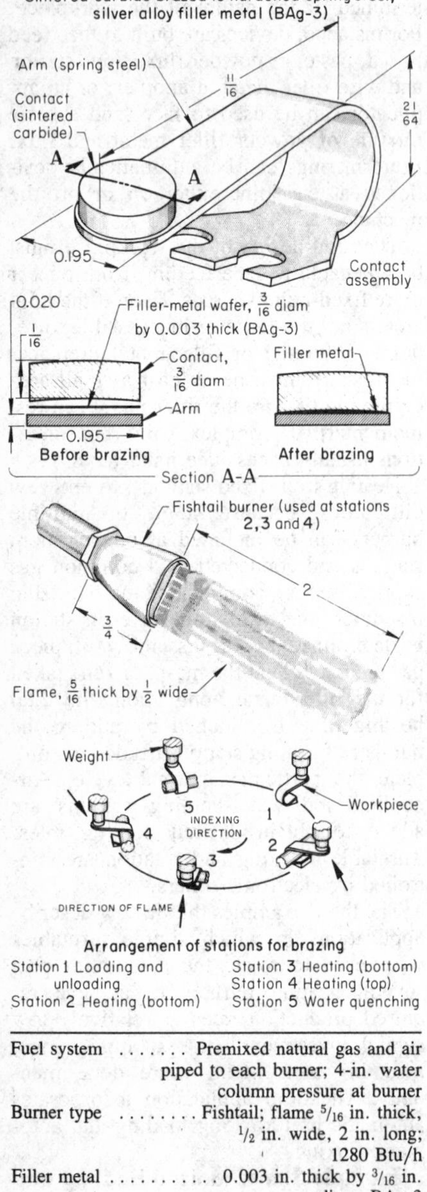

Fuel system	Premixed natural gas and air piped to each burner; 4-in. water column pressure at burner
Burner type	Fishtail; flame 5/16 in. thick, 1/2 in. wide, 2 in. long; 1280 Btu/h
Filler metal	0.003 in. thick by 3/16 in. diam, BAg-3
Flux	AWS type 3A
Production rate	300 assemblies per hour

dipped in AWS type 3A liquid flux and loaded on the turntable with a wafer of BAg-3 silver-alloy filler metal, 3/16 in. in diameter (the diameter of the carbide tip) and 0.003 in. thick, between them. A weight on the top of the carbide tip held the parts in position during brazing.

Three fishtail burners, with the flat "tail" in the horizontal position, were used to heat the assemblies. Burners at stations 2 and 3 heated the assembly on the underside of the joint, and the burner at station 4 applied heat to the joint on the top side to melt and flow the filler metal. Waterspray cooling followed at station 5. The hardness of the arm was not changed in a manner to affect its function as a spring.

Example 4. Machine Torch Brazing. Magnet armatures of the type shown in Fig. 5 were used as the striking members of a printing machine. The 0.040-in.-thick striker blade was made of high-carbon (0.65 to 0.85% C) spring steel; thus, the outer end, which engaged a linkage at 250 times a minute, could be locally hardened to resist wear. A brazed joint was required between the blade and the 5/32-in.-thick armature, made of 2½% Si steel. Conventional furnace brazing with copper filler metal was not suitable for this operation because (1) the slotted joint design did not lend itself to the tight fit required for copper brazing; (2) the high brazing temperature (2050 °F) would coarsen the grain structure of the high-carbon steel blade; and (3) the furnace atmosphere needed to prevent decarburization of the blade was expensive.

Brazing the assemblies on a turntable using a silver-alloy filler metal and either induction or flame heating would provide the desired production rate, while avoiding the objections to furnace brazing. Machine torch brazing was selected, because the capital investment was lower and a metal fixture, cooled by circulating water, could be used to maintain the close dimensional tolerances required over long production runs. A timer-controlled eight-station brazing turntable was built (Fig. 5) to perform all operations except fluxing, assembling, and loading, which were done manually. The burners were supplied with low-cost air/natural gas for heating.

The operating sequence was:

- *Station 1*: The operator dipped the joint end of a blade in type 3C paste flux, assembled it with an armature, and loaded the assembly in the fixture.
- *Stations 2, 3, and 4*: The joint was preheated by rotating it past burners clamped to brackets, one at each of the three stations.
- *Station 5*: A fourth burner brought the joint to the melting temperature of the BAg-1 filler-metal wire, which was automatically face fed to the joint. A microswitch probe detected the assembly and caused the wire feeder to slide into position, face feed 1 in. of 0.032-in-diam wire (which melted on the joint), and retract.

Fig. 5 Armature assembly and the turntable used for torch brazing. Assemblies were brazed at the rate of 230/h.

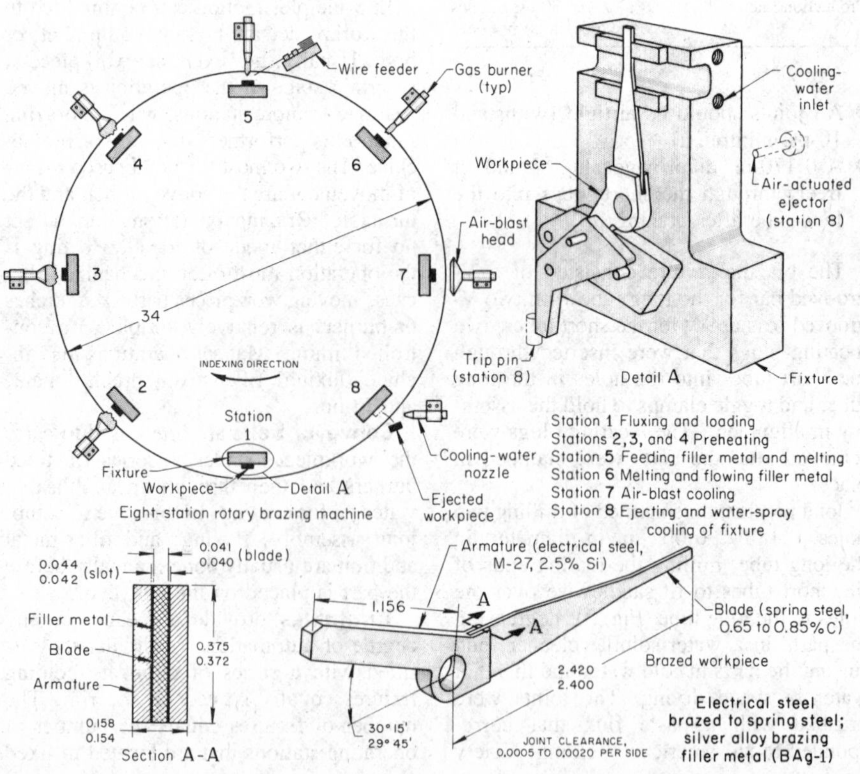

- *Station 6*: Filler metal was brought to brazing temperature by a fifth burner and flowed through the joint.
- *Station 7*: The assembly and fixture were cooled by an air blast.
- *Station 8*: A stationary strip pin unclamped the fixture, releasing the assembly, which was then ejected by an air-actuated plunger. The fixture was cooled by a water spray.

Production rate averaged 230 assemblies per hour. After brazing, the striker ends of the blades were hardened in a separate operation. Induction heating was used for this operation because it provided closer control over the dimension being hardened and allowed an oil quench.

Example 5. Automatic Torch Brazing of a Cutting Torch Attachment. Positive alignment of three stainless steel tubes in relation to a brass head and base with strong, leakproof, silver brazed joints was the main criterion in developing this high-production brazing machine. Although the press fit previously used aided in alignment, it created a "blind hole" condition with a very severe gas entrapment that resulted in minimum alloy penetration into the joint area, as well as a leaker condition (Fig. 6).

Stainless steel fixturing (Fig. 7b) eliminated the troublesome press-step system of alignment. In the fixture, spring buttons and a guide block were used to ensure vertical alignment. The cylindrical weight allowed the head and base to settle and bottom against the stainless steel tubes. Perfect alignment was achieved using a 0.005-in. clearance between the stainless steel tube and the brass bodies and heads

Fig. 6 Previously used press-fit setup that produced a blind hole condition

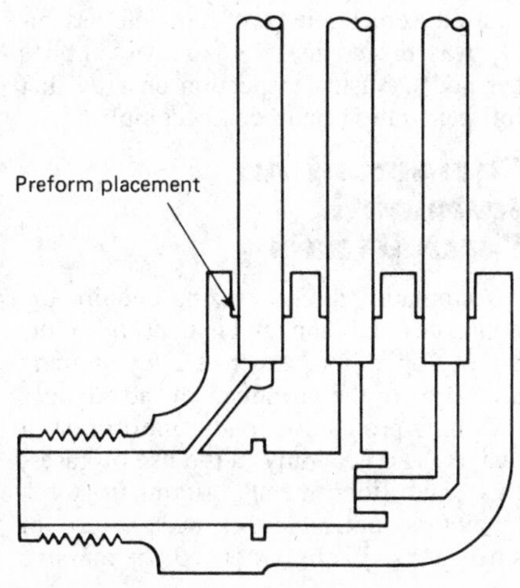

Fig. 7 Cutting torch attachment brazed in a turntable setup. (a) Twelve-station rotary brazing machine. (b) Fixturing of cutting torch attachment. (c) Preform ring (detail A) made of $1^2/_3$ turns of type BAg-1 filler metal. (d) Setup at base of assembly (detail B) achieved perfect alignment using 0.005-in. clearance between stainless steel tube and brass base and head.

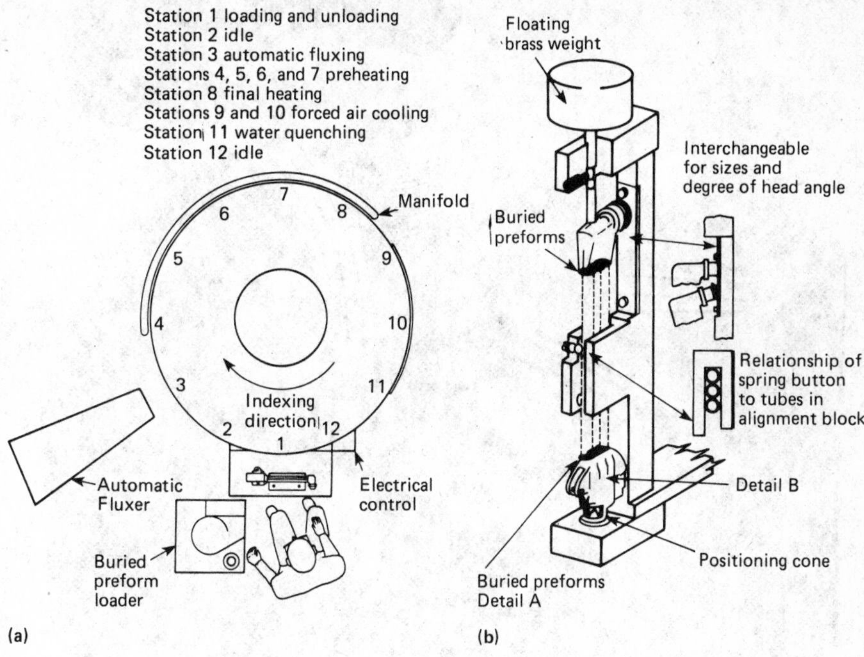

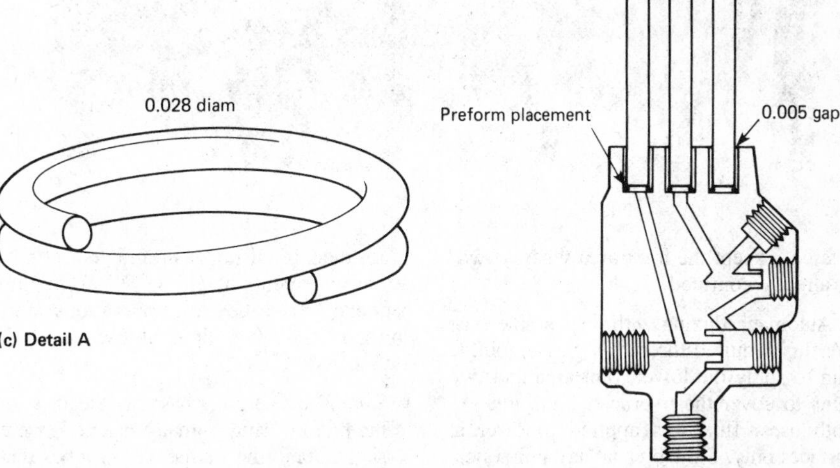

through to the shoulders (Fig. 7d). A semiautomatic preform ring loader, shown in Fig. 8, was used for alloy placement. The operator held the heads and bases over three pins. By stepping on a foot control, the pins were retracted, allowing three preform rings to drop into place. The rings were then pressed into the head or base by the same pins. The BAg-1 preforms (Fig. 7c) were $1^2/_3$ turns of 0.028-in.-diam wire made 0.005 in. oversize to fit snugly in the brass holes.

The buried preforms offered the following benefits:

- Use of the least amount of silver brazing material
- Visual joint inspection with a fine line of fillet (no excess) at the top of the joint
- Maximum strength and leaktightness from the brazing alloy
- Assemblies brought to heat faster without fear of the alloy running in any other

Fig. 8 Semiautomatic buried preform ring loader

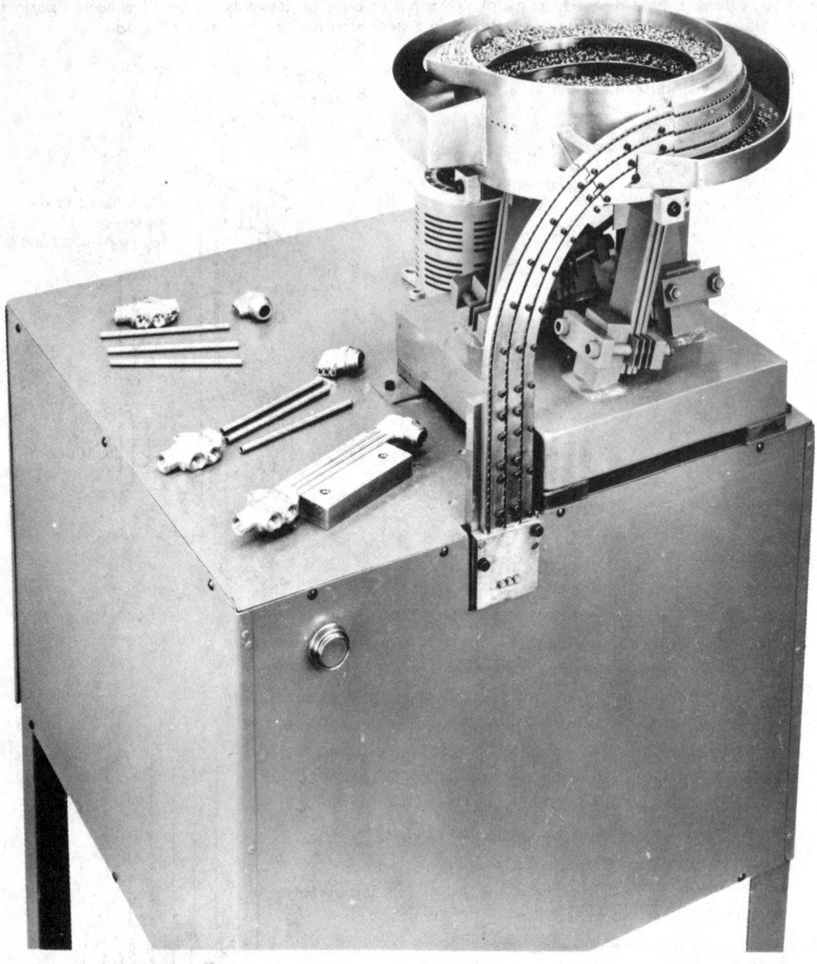

area, except the joint area where it was already confined

Automatic fluxing with dispensable type 3A flux required three flux guns: a single gun to spray the lower joint area and two guns to cover the overhead joint area. In both uses, flux was applied to external surfaces only. Slow preheat at the first heat station drove the water out of the flux, making it tacky, and allowed it to penetrate the joint area by capillary flow during progressive heating.

Oxynatural gas burners provided the heat at five progressive heat stations. The burners were secured to heavy-wall stainless steel and a 2-in.-diam manifold covering 220 to 230° of the stainless steel tabletop.

Figure 7a shows the layout of the tabletop with the stations and functions of the stations identified. A right-angle roller gear drive provided smooth transfer. An adjustable timer controlled the variable dwell time. Indexing time was regulated by pulley arrangements at 0.5 s. The following sequence describes the operations performed and the equipment used at each station:

- *Station 1*: The brass base and head, with the preform rings already inserted, were slipped onto the stainless steel tubes that were secured in a separate holding fixture on a shelf directly in front of the operator at station 1 (Fig. 7a). The operator then placed the assembly in the fixture (Fig. 7b) on the machine. A heavy brass weight secured the assembly and added weight to push the unit squarely into place after the assembly was progressively heated and the filler metal was melted and flowed.
- *Station 2*: This idle station provided space between the operator and the automatic fluxing operation.
- *Station 3*: Automatic fluxing, which was activated by a whisker sensor at station 2, sprayed flux through three individual guns, two overhead and one downward, only if an assembly was in the fixture.
- *Station 4*: Preheating was performed with the burners backed away; thus, the heat drove the liquid out of the flux, making it tacky.
- *Stations 5, 6, 7, and 8*: Stainless steel tubes supported the burners, which were positioned at slightly different angles to produce a full, even heat over the entire joint area. Three burners heated the head and four heated the base. Oxynatural gas was employed through a simple venturi system with no moving parts. Air was supplied by a high-volume, low-pressure blower. The burners operated at low through high heat. A gas flowmeter aided in repeatability.
- *Stations 9 and 10*: Still and forced air cooling allowed the alloy to set and solidify, using excess air from the high-volume burner.
- *Station 11*: Final cooling of the assembly and the stainless steel fixtures was accomplished by water quenching. The majority of the flux was removed by water cooling with the aid of thermal shock.
- *Station 12*: This idle station provided working space for the operator.

The machine met all of the design criteria; alignment was perfect, and the alloy penetrated the full joint area, producing strengths two and a half times greater than previously obtained. Automatic ring loading, operating three at a time, pressed the rings snugly in place. Six brazed joints were made simultaneously—three in the overhead and three in the downhand position. Using two sets of stainless steel fixtures that were adjustable to three heights with inserts to handle 45°, 75°, and 90° cutting-head angles, and using two tube sizes, 12 styles were produced. Production increased tremendously; a complete assembly was brazed in 15 s (four assemblies per hour). Visual inspection ensured that full penetration had been accomplished.

Equipment for Automatic Torch Brazing

Automatic torch brazing equipment eliminates the human element from the brazing operation, except for loading and unloading of the machine. The advantages are high production rate, uniform joint quality, and economy in the use of gases, flux, and filler metal. Automatic torch brazing is applicable to a wide variety of small parts. Large, extended, or massive workpieces present the usual problems of heat control, filler-metal flow, and distortion. For such pieces (a filler pipe joined to an automotive gasoline tank is an example), soldering is usually selected for the joining method rather than brazing, if the strength is satisfactory.

The machines and other equipment used in automatic torch brazing are basically the same as those described in the previous section on machine torch brazing in this article. The use of burners rather than torches usually requires the installation of flowmeters and manometers for accurate flame control. An automatic setup lends itself more easily to multiple-joint or multiple-part brazing. Production rates are much higher.

Two types of face-feeding machines have made automatic torch brazing possible. They are the wire feeder and the paste feeder. Both machines are the fixed-station type and are used in conjunction with a work-fixtured turntable or conveyor belt. Both are controlled by timers and are activated by a probe or limit microswitch that determines whether the approaching fixture is loaded. Various combinations of fluxing, filler-metal feeding, and heating sequences are possible.

Wire Feeders. One type of wire feeder incorporates a double slide. The first slide advances the gas torches or burners into position. After a preset interval, the second slide advances wire-guide tips to the joint, and filler metal is fed, melting off on contact. When the correct amount of filler metal has been melted off, the wire is quickly retracted to prevent balling at the end. Heating is continued until the filler metal is distributed through the joint, and the torches or burners are then retracted.

In the double slide system, flux is applied in either liquid, slurry, or paste form at a previous station. Gas fluxing is performed through the torch or burner flames. The assembly is cooled by air or water at following stations.

The most critical part of the operation is the setting of the wire-feed rate to coincide with the wire-melting rate; lack of coordination results in either an underfed joint or waste of filler metal (usually silver alloy). Cycle timing, slide motion, and wire-feed rate are individually controlled by an electronic timing circuit.

Paste Feeders. A mixture of flux and powder filler metal is applied to the joint or joints to be brazed when using paste feeders. The paste is contained in a cylinder equipped with a piston to eject the paste through a nozzle. One or more cylinders are mounted on a single slide that enables the applicator to be positioned at the desired points of deposition.

The operating sequence, which is triggered by a microswitch striking the fixtured workpiece, consists of sliding the applicator into position, depositing a predetermined quantity of paste, and retracting. The workpiece is then moved through a series of burner stations where the flux is dried, melted, and activated, and the filler metal is finally melted and distributed through the joint. Cycle timing, slide motion, and paste feeding are controlled by electronic timing circuits. As in other automatic machines, stations are set up for loading, cooling, and unloading. This system can also use gas fluxing in applications where the workpiece must remain clean.

Example 6. Automatic Torch Brazing of an Appliance Part. A machine for high-production torch brazing of the joint between the low-carbon steel beater blades and shaft of an electric food beater required only manual loading. All other operations were controlled automatically, including ejection from the brazing fixture. The assembly, together with joint details, is shown in Fig. 9. The formed blades were fixed at one end of the shaft by riveting over a through-pin integral with the shaft. The four ends of the blades were notched to form a mitered cross, which fit precisely into the 0.045-in.-wide by 1/16-in.-deep annular groove cut in the shaft. Squeezing the blades together produced the desired joint configuration.

The machine consisted of a 12-station turntable with a station dwell time of 3.55 s and a between-station indexing time of 1.25 s, producing 750 assemblies per hour. The braze was made using a face-fed paste of BAg-1 silver-alloy filler metal and flux. Heating was done with air/natural gas burners, with added gas fluxing to keep the assembly clean. Figure 9 shows the general arrangement of stations around the turntable and identifies the operations performed at each station. Turntable rotation was controlled by a timer that was set for station dwell time; indexing time was a machine constant. The following sequence describes the operations performed and the equipment used at each station:

- *Station 1*: An external loading device slipped the beater blades into the circular opening in the locator arm of the holding fixture on the turntable. The loading device consisted of a cup-shaped die with four equally spaced slots to hold the beater blades in position for brazing and a hole in the center for the beater shaft. The die was fixed on a crossbar supported by two spring-loaded shafts. Mounted on the shafts was a sliding crossbar that was attached to a loading pin. An air cylinder actuated the pin (Fig. 9). As the assembly was manually loaded into the die, the beater blades were squeezed together, positioning the ends in the joint. By pressing a button-switch, the operator caused the entire loading device to move upward. The spring-loaded shafts were end-stopped part way through the loading operation, but the sliding crossbar and loading pin continued to the end of the stroke, pushing the beater shaft until the four blades were engaged in the locator arm. When the loading device retracted, the machine indexed to the next station. The turntable fixture, in addition to supporting the locator arm, had a bolt-and-locknut vertical stop with a hole through the center to permit entry of an ejector pin (station 11). A cemented, fiber-insulating board over the fixture support served as a heat reflector.
- *Station 2*: This station was idle to give the operator freedom from interference with the sizable apparatus at station 3.
- *Station 3*: This station was equipped with a device consisting of a paste applicator and a microswitch post-mounted on a slide, an air-pressurized paste reservoir connected to the applicator by a plastic tube, and an air-supply tube to actuate the applicator piston. Figure 9 shows the general arrangement of these parts. This station also had a separate control panel with a paste-pressure control knob set for 17 psi, a meter indicating applicator operating pressure (40 psi), a timer that controlled the paste-feed cycle, and a cycle counter. When the slide moved the applicator nozzle into position over the joint to be brazed, the probe finger of the microswitch was deflected by the workpiece. This closed the switch and started the timed cycle in which the piston ejected the correct amount of paste onto the joint. Approximately 1 oz of paste supplied 200 parts.
- *Station 4*: This station was idle to separate the paste-feeding equipment from the heat of the torches.
- *Stations 5, 6, 7, and 8*: Each of these four stations had two air/natural gas burners with tips mounted on bendable copper tubing. In stations 5, 6, and 7, the tips directed the gas flames onto the beater shaft above and below the workpiece joint, for progressive heating. At station 8, the flames were directed to converge on the joint, completing the melting and causing flow of the filler

Fig. 9 Torch brazed food beater and paste-feeding type of automatic torch brazing machine. Station 1 (loading, position A): The workpiece is manually fitted in the joint-forming die preparatory to loading. Station 1 (loading, position B): The workpiece is loaded into the locator arm of the holding fixture, and the loading device has begun to retract. Station 3 (applying paste): Microswitch probe detects the presence of the workpiece and initiates paste-feeding cycle.

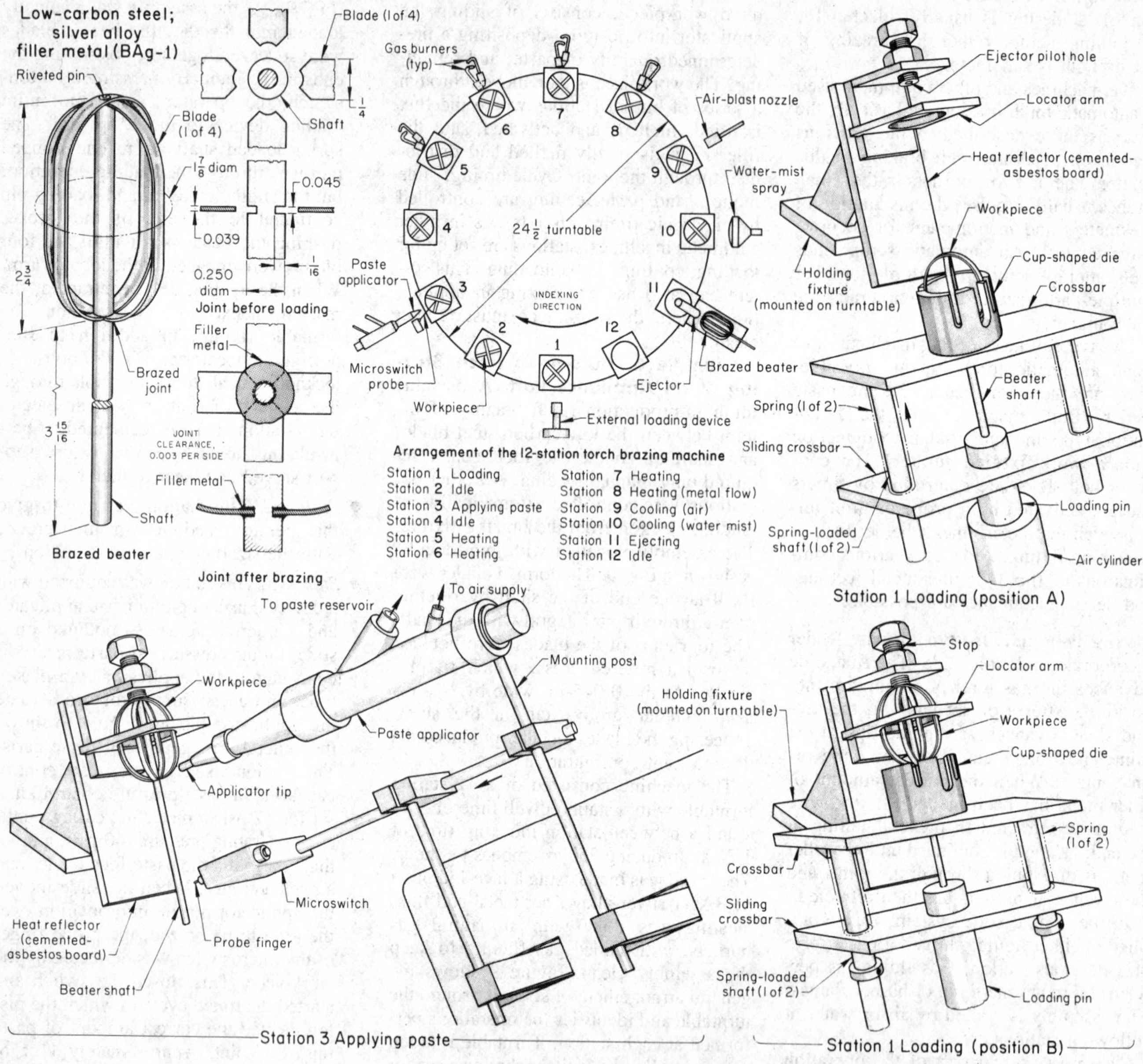

metal. Air and natural gas were pumped, metered, and mixed in a combustion control unit mounted in the base of the machine. The ratio of air to natural gas was 9 to 1. The supply of natural gas was monitored by a flowmeter; a manometer indicated the pressure of the gas mixture in the burners.

- *Station 9*: This station was fitted with an air-blast nozzle for cooling the workpiece.
- *Station 10*: The workpiece was further cooled with a water mist spray.
- *Station 11*: When a fixture was indexed at this station, an air piston moved a pin down through the hole in the vertical stop, ejecting the part into a chute and out the side of the machine into a bin.
- *Station 12*: This station was idle to give the operator freedom from interference with the collection bin.

After flux removal, the assemblies were chromium plated. The main control panel for the machine had the following controls: emergency stop button, start button, a light signaling that paste was applied, a turntable-cycle on-off switch, an applicator on-off switch, and an on-off switch for the gas-combustion control unit.

Success with this machine indicated that the operator could easily load two beaters at one time, using both hands. Accordingly, a machine with 12 duplex stations was installed. Although larger and more complex than the machine described above, the equipment and operation were essentially the same. The new machine brazed 1400 to 1500 assemblies per hour.

Filler Metals

Silver and copper-zinc brazing alloys are the filler metals used in torch brazing of low-carbon and low-alloy steels. The product forms, nominal compositions, and melting and brazing temperature ranges of the filler metals most frequently used are given in Table 1.

Choice of filler-metal form depends primarily on joint design, method of assembly, and heating method. The amount of filler metal used can be controlled accurately by the use of preformed shapes bent from wire or blanked from strip. These can be preplaced in and around joints of various designs, in a manner that enhances flow and completes filling of the joint. Typical shapes of filler-metal preforms are illustrated in the article "Induction Brazing of Steels" in this Volume.

Filler metal in powder form is often used. If applied on the exterior of the joint, however, some provision must be made to hold it in place. One method is first to brush or automatically spray the joint with flux and then to dust the filler-metal powder lightly onto the flux while it is still wet. Another method is to mix the powder with flux and a binder to form a paste, which is then applied by brushing or face feeding through an applicator. Rod and wire can be manually face fed; automatic wire face-feeding equipment and paste-applying equipment are also available, as described earlier in this article.

Silver-alloy filler metals BAg-1 through 7, 20, 27, and 28 are used for torch brazing most types of steel to themselves or to other metals, except aluminum and magnesium, and are available in several product forms. Alloys BAg-8 through 19 and BAg-21 through 26 are used chiefly in furnace or induction brazing, and are, for the most part, used in joining base metals other than low-carbon and low-alloy steels.

The most frequently used alloys of the group shown in Table 1 are BAg-1, 1a, and 3; the first two are outstanding for high fluidity, low melting temperature, and narrow melting range. BAg-1 is much preferred for fast-cycle torch brazing, especially for automatic torch brazing with low-cost air/natural gas heating. The BAg-3 alloy contains nickel for improved wettability when brazing stainless steel and tungsten carbide (tool tips) and has lower fluidity for bridging larger joint clearances. BAg-2, 2a, and 27 are similar to 1 and 1a, but contain less silver and are therefore cheaper; they also have a somewhat wider melting range and are less fluid.

These filler metals contain cadmium, which is added to depress the melting range and to improve flow properties. Because cadmium fumes are toxic, industrial brazing installations must be provided with adequate ventilation as described in the section on equipment for manual torch brazing in this article. Prolonged heating and excessive brazing temperatures must be avoided to minimize fuming (see Table 1 for brazing temperature limits). Federal regulations stipulate that cadmium-bearing filler metals and their toxic properties be identified by a suitable warning label.

BAg-4 through 7, 20, and 28 are cadmium-free filler metals, a requirement for some food-processing applications. BAg-4, which contains some nickel, is used for brazing carbide tool tips to steel. BAg-5, 6, 20, and 28 are widely used on heat exchanger, electrical, and food equipment. The melting range of BAg-7 is the lowest and narrowest in temperature range of the cadmium-free alloys; because it contains tin, BAg-7 produces a white color that matches steel, in contrast to the somewhat yellowish color produced by other silver alloys.

Joint clearances of 0.001 to 0.005 in. are often recommended for silver-alloy filler metals. The clearances used in torch brazing, however, vary considerably. Because joint clearance is a factor in capillary flow, too little or too much clearance affects filler-metal distribution, which thus affects joint strength.

Maintaining consistently proper clearances is particularly important in automatic face-feeding operations. Some latitude is possible in face-fed manual torch brazing, especially if joint strength is not critical. Where joint strength is critical, a poor fit, even if not apparent in the completed joint, is likely to show up in service.

Copper-zinc filler metals are used extensively in manual torch brazing and braze welding of low-carbon and low-alloy steels. They can also be used to join nickel-based and copper-nickel alloys to themselves or to steel, where corrosion resistance is not required. Table 1 gives the nominal composition and melting characteristics of four copper-zinc alloys used as filler metals for torch brazing steels. The two RB types are classified for braze welding and brazing; the two R types are for braze welding only. The R types are included in Table 1 to clarify a distinction among the four standard copper-zinc filler metals that is often overlooked; only the two RB types are used in torch brazing, but all four copper-zinc filler metals can be used in braze welding.

Copper-zinc filler metals have higher melting and brazing temperatures than silver brazing alloys. Overheating must be avoided, however, because of their high zinc content. When the alloys are overheated, zinc vaporizes (fumes), causing voids in the joint. RBCuZn-A, often called naval brass or bronze, produces some zinc fuming and is brassy in color. RBCuZn-D, containing 10% Ni, is often called nickel silver, because of its whitish color, although it contains no silver. RBCuZn-D is often selected for its low fuming and high strength, as well as for its good color match with steel. Because of the high brazing temperature of this filler metal, when brazing some heat treatable steels, brazing can be combined with heating for hardening. Chromium-molybdenum steels have been brazed and heat treated in this manner. Both RB types of filler metal have a relatively narrow melting range. When used for torch brazing, joint clearances of 0.002 to 0.005 in. are recommended. These filler metals are also available as pastes and preforms for automatic torch brazing.

Fluxes

Surface oxide films inhibit the wetting of the base metal by the filler metal and, therefore, the capillary flow of the filler metal in the joint. They also prevent the formation of a true metal-to-metal braze bond. Fluxes must have sufficient chemical and physical activity to reduce or dissolve the thin surface oxide films without attacking the base metal severely. They are not made to dissolve heavy oxides, greases, or dirt. For these reasons, fluxes are not intended to serve as a substitute for prebraze cleaning or for removal of heavy oxide films. The chemical reactivity of a flux with the oxides and other compounds encountered on the surface of the base metal varies according to the stability of the compounds.

Because the heat of brazing accelerates the formation of oxide films, fluxes must also melt and flow freely over joint surfaces to shield the metal from the atmosphere under the torch brazing conditions. In addition, because different filler metals vary in melting range and brazing temperature, fluxes must maintain their stability and effectiveness throughout the brazing temperature range of the filler metal being used.

Finally, the density, viscosity, and surface tension of the molten flux must be low enough to enable the molten filler metal to replace the flux on the joint surfaces by capillary flow. In addition, fluxes must be easy to remove after brazing, preferably by washing in hot water.

Flux Constituents. Brazing fluxes are available chiefly as proprietary formulations; there are no standard composition limits, such as those for filler metals. A general list of the compounds used in brazing fluxes follows; of these, the boron and fluorine compounds are the active deoxidizing constituents used in fluxes for the torch brazing of steel:

- *Fused borax* is a high-melting material that is active at high temperatures. It is seldom used in low-melting brazing fluxes.
- *Borates* melt at 1400 °F or higher. They have moderately high viscosity and must be used with other constituents. They are potent oxide solvents and protect against oxidation for long periods of time.
- *Fluorides* react readily with most metallic oxides at high temperatures and are used in fluxes as cleaning agents. They are especially useful in counteracting refractory oxides, such as chromium oxide and aluminum oxide. They also increase the fluidity of molten borates.
- *Chlorides* function in much the same manner as fluorides, but at a lower temperature. They must be used carefully, because at high temperatures they may oxidize the work metal. Chlorides are used to depress the melting temperature of fluoride-based fluxes and are useful in brazing aluminum-containing alloys.
- *Fluoborates* react in somewhat the same way as borates, but they do not give as long-lasting protection. They flow better in the molten state than the straight borates and have superior oxide-dissolving properties. Fluoborates are used in combination with borates and alkaline compounds such as carbonates.
- *Fluosilicaborates* have a somewhat higher melting range than fluoborates. They cover and adhere to surfaces well, but they are more limited in use than fluoborates because of their high melting point.
- *Hydroxides* of sodium and potassium absorb moisture from the air and, therefore, are used in fluxes for special applications only. Even small amounts in other fluxing agents can cause difficulty and limit storage life in humid environments. They raise the useful working temperature of fluxes and are used in fluxes for brazing molybdenum-bearing tool steels.
- *Boric acid* is used in the conventional and in the calcined form. The calcined form has a somewhat higher melting point. Both forms promote friability to help facilitate the removal of the glass-like flux residue that remains after brazing. The melting range of boric acid is lower than that of the borates, but higher than that of the fluorides.
- *Wetting agents* are used in paste, slurry, and liquid fluxes to promote the flow and spread of the flux on the work metal. The agents that are used must not interfere with the normal functions of the flux.
- *Water* is present in brazing fluxes either as water of hydration in the various chemicals, or as an addition to make the flux into a paste or a liquid. Water of hydration is removed by calcining the chemical that contains it. Water that is used to dilute fluxes must be tested for mineral content. Excessively hard water should be avoided. Sometimes alcohol is substituted for water, or demineralized water is used.

Flux Types. For torch brazing of low-carbon and low-alloy steels, fluxes can be grouped roughly into three general types, as shown in Table 2. The flux constituents are formulated to ensure that useful temperature ranges meet the applicable filler-metal temperature requirements. Because of their stability, boron compounds are present in all three types. For compatibility with the silver-alloy filler metals, fluorine compounds are added to lower the useful temperature range and the viscosity of the flux, as indicated by type 3A. Boron compounds are the main ingredients in type 3D flux, because the higher useful temperature range and viscosity are compatible with the copper-zinc filler metals. Type 3C flux falls between type 3A and type 3D in useful temperature range and represents a second choice for either filler metal in the torch brazing of steel.

Gas Fluxing. In some applications of manual and automatic torch brazing, more efficient production can be realized by applying flux directly through the gas flame. This is accomplished by bubbling the fuel gas through a small tank containing the flux in the form of a volatile liquid. A typical gas-fluxing assembly is shown in Fig. 10. Vaporized flux is entrained with the fuel gas, which flows to a conventional torch or burner, mixes with oxygen (or air), and burns in the flame, depositing a thin, uniform film on the workpiece. In liquid form, the flux is a type of methyl borate produced from methanol, boric acid, and diluents, under moisture-free conditions. Its temperature range corresponds to flux type 3D (Table 2).

Any fuel gas that is free of water vapor can be used with gas fluxing. In the presence of water vapor, the fluxing agent (boric acid) is precipitated in the form of a whitish, crystalline solid. This condition is prevented by installing a chemical drier (calcium chloride) in the line and substituting ethylene glycol for the water used in flashback arresters. For best results, a neutral flame is used. Brazing is done using goggles with number 4, 5, or 6 green lenses, because the flux produces a brilliant green flame that masks the inner flame cone from view by the naked eye.

The flux can be deposited only on areas that the flame can reach. Gas fluxing has two principal uses. In manual torch brazing and especially in braze welding of low-carbon steel, where rods of copper-zinc filler metal are used in joints with varying or relatively large clearances, the flux is used for brazing the joint and for protecting the base metal from oxidation. In such applications, gas fluxing eliminates the need for continually dipping the filler-metal rod in flux or use of the more costly flux-covered rods. For postbraze cleaning, the flux can be removed by washing in hot water (150 °F). Cost savings are most significant in production braze welding operations.

The second use of gas fluxing is in applications where joint clearances are closely held or where silver-alloy filler metals are used. In these applications, a suitable flux (type 3A or 3C, Table 2) must be applied to the joint before heating, and the gas

Table 2 Types of flux ordinarily used in torch brazing of low-carbon and low-alloy steels

AWS type	Useful temperature range, °F	Principal constituents	Available forms	Applicable filler metals
3A	1050-1600	Boric acid, borates, fluorides, fluoborates, wetting agent	Paste, liquid, slurry, powder	BAg
3C	1050-1800	Boric acid, borates, boron, fluorides, fluoborates, wetting agent	Paste, slurry, powder	BAg, RBCuZn
3D	1400-2200	Boric acid, borates, borax, fluorides, fluoborates, wetting agent	Paste, slurry, powder	RBCuZn

Fig. 10 Gas fluxing hookup for a manual brazing station. Fuel gas flows through the inlet valve A into the liquid-flux tank, where it bubbles through the flux, entrains some of it and carries it through the outlet valve B to the torch. The bypass valve C controls the proportion of fuel gas entering the bubble chamber to provide rich or lean fluxing mixtures. Valves D and E operate to refill the liquid-flux tank and remove the reserve tank for remote filling without stopping the operation. Joints must be sealed to prevent moisture pickup from air, which causes precipitation of flux and clogging of lines.

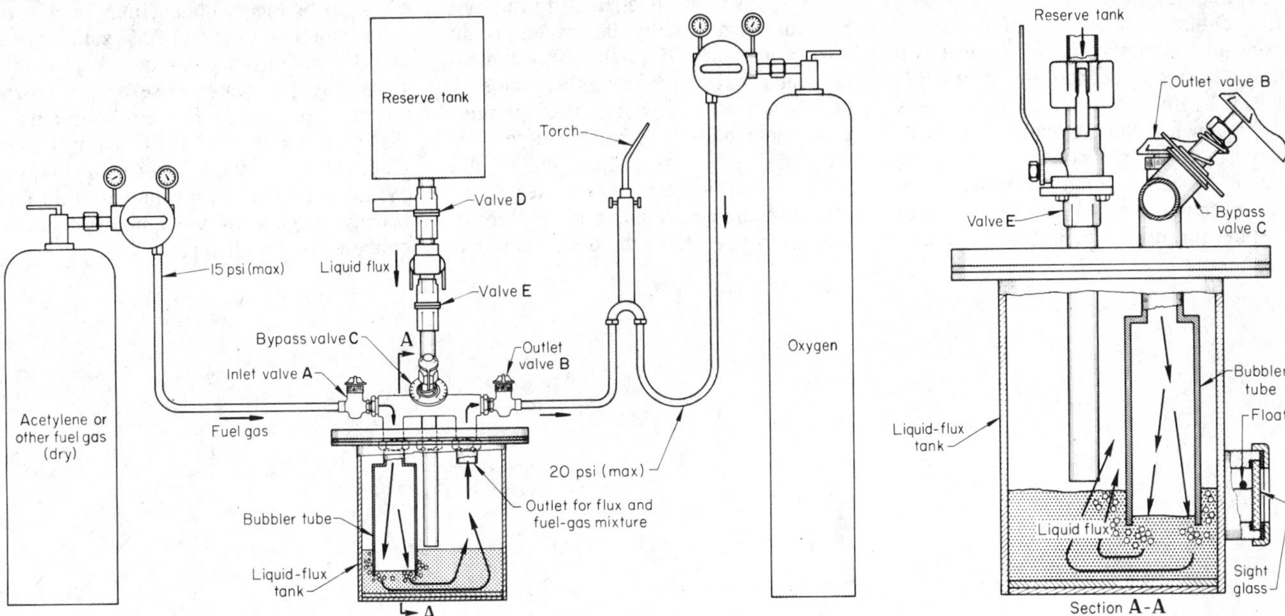

fluxing process is used primarily to protect the part from oxidation when surface appearance is important or when parts are to be subsequently plated. As an aid in fluxing, gas fluxing also reduces the amount of paste flux required and, in hand-fed operations, the amount of silver-alloy filler metal used.

Gas fluxing is also used in a variety of semiautomatic and automatic torch brazing equipment. In addition to its use in joining low-carbon steel, it is employed in brazing alloy steels, stainless steels, copper and copper alloys, and nickel alloys and in joining these metals to each other. Applications of gas fluxing in the torch brazing and braze welding of steel include automotive and machinery parts, bicycle and motorcycle frames, metal furniture, doors and partitions, and many kinds of special fixtures.

Removal of Flux After Torch Brazing

There are five major reasons for removing residual flux after brazing: (1) the joint cannot be inspected for soundness until the cover of flux residue is removed; (2) the joint may be bound together by the flux in the semblance of a brazed joint, only to break apart later in service; (3) in fluid or pressure service, the flux may block pinholes that might withstand a pressure test, but would leak soon after being placed in service; (4) if left on the joint, the flux attracts available water, resulting in oxidation and corrosion; and (5) painting, coating, or plating cannot be done satisfactorily on areas covered with flux residue.

If parts have been well cleaned before brazing and not overheated during brazing, flux residue can usually be removed by a hot water rinse followed by thorough drying. To aid in flux removal and avoid corrosion, postbraze cleaning should be delayed no more than 48 h.

A quick method of removing glasslike residues is to quench the joint in water after brazing, thus cracking off the deposit by thermal shock. In some applications, however, such treatment may cause distortion of the brazed assembly. Scrubbing, applying a steam jet, and most of the standard abrasive techniques, such as wire brushing and abrasive blasting, are also used to dislodge stubborn flux residues, provided that the operation does not impair the function of the assembly.

When flux cannot be removed from steel assemblies by rinsing in hot water, a 5% solution of sulfuric acid works more effectively. The solution may be warmed to accelerate the action, but caution must be used to prevent excessive attack on the assembly. A small addition of sodium dichromate to the solution speeds up action, but the time of immersion must be carefully controlled to avoid the greater risk of etching the steel.

Phosphate solutions similar to those used for cleaning steel are effective flux removers and have the added advantage of giving carbon steel assemblies a temporary protective coating. This coating can, however, hamper subsequent brazing operations.

Boric acid, as applied in gas fluxing, can be removed by washing in clean water heated to at least 150 °F. Boric acid is only slightly soluble in cold water.

Mixed borax and boric acid fluxes are more difficult to remove than other types. Fortunately, moisture absorption and corrosion are minimal with borax fluxes. In fact, rather than risk damage to delicate assemblies, such as electronic components, when mixed borax and boric acid fluxes are used, the flux is sometimes allowed to remain after brazing. The manufacturer accepts the possibility of some corrosion occurring and some imperfect joints being hidden under the flux.

These fluxes can be removed by quenching, shot blasting, sand blasting, chipping, filing, scraping, and wire brushing. The rate of solution in water is slow, and even if a dilute sulfuric acid solution is used, the necessary period of immersion may be inconveniently long for production work.

Fluoride fluxes are soluble in water and are much easier to remove than borax fluxes. Holding assemblies under running water while brushing with a wire or bristle brush is usually sufficient. Alternatively, the assemblies may be boiled in water for a few minutes, and then rinsed in cold water. Dilute sulfuric acid solutions and phosphate solutions can also be used for quicker results. The residue of fluoride fluxes is hygroscopic, and if the assembly is not quenched after brazing, it is often advantageous to postpone the postbraze cleaning for 24 h. Under normal atmospheric conditions, the residue absorbs moisture during this period and becomes more readily soluble in any of the solvents previously mentioned. Flux removal should not be delayed for more than 48 h because many types of steel may begin to corrode.

If a fluoride flux is difficult to remove, the cause can usually be traced to the brazing operation. If too little flux is used, the residue is a hard, cokelike oxide. If the flux is heated above its rated operating temperature or for too long a time, it is likely to leave a hard, glasslike residue similar to that left by borax fluxes.

Care should be taken to prevent fluoride residues from entering the body via mouth or skin openings. Gloves should be worn, and the hands should be washed well before handling food.

Preparation for plating, if the assembly is to be electroplated, may consist of immersion in a solution of 5% sulfuric acid and 3% sodium or potassium dichromate, followed by a water rinse and a hydrochloric acid pickle. The anodic treatment that is normally used for cleaning before plating removes the last traces of flux, but should not be used for removal of all flux because the electrolyte quickly becomes contaminated with flux.

Induction Brazing of Steels*

INDUCTION BRAZING is a process in which the surfaces of components to be joined are selectively heated to brazing temperature by electrical energy transmitted to the workpiece by induction, rather than by a direct electrical connection, using an inductor or work coil. Heating is the result of eddy currents or I^2R losses in the workpiece; because of the electrical resistivity of the workpiece and the flow of induced alternating current through it, heat is generated. When the work metal being heated is ferromagnetic, as most steels are, some slight additional heating results from hysteresis. All heating due to hysteresis, however, ceases when the temperature of the work metal is raised to the Curie point (about 1420 °F). Above this temperature, heating by electrical resistance continues at a reduced rate as the temperature rises.

The depth of heating by induction depends primarily on the frequency of the alternating current. As the frequency is increased, both the theoretical depth of current penetration and the depth of the heated zone in the workpiece decrease. For example, the theoretical depth of current penetration is about 0.035 in. at a frequency of 3 kHz, but decreases to about 0.003 in. at 500 kHz.

Distribution of heat to other areas of the workpiece depends on conduction. In general, heat flow by conduction, although fairly rapid, is minimized by the rapidity at which induction heating takes place.

Frequency Range. Frequencies used for induction brazing can range from the powerline frequency of 60 Hz to approximately 450 kHz. The higher frequencies should be selected when shallow heat penetration is desired. In the induction heating of steel for surface hardening, shallow heat penetration at high power inputs is generally desirable for rapid heating and control of depth of heating. In the brazing of steel components, however, deeper and more uniform heating of the matching surfaces that will form the joint may be preferable. The advantages of a high operating frequency, such as 450 kHz, may be more suitable for the brazing of nonferrous metals or the brazing of steel to a nonferrous metal or nonmagnetic (austenitic) steel. The role of frequency in induction brazing is more fully discussed in the section on selection of frequency in this article. For the majority of brazing applications, the frequencies used are seldom below 10 kHz, the peak frequency obtainable from motor-generator power-supply units.

Process Capabilities

The primary advantage of induction brazing over other brazing processes is high-speed localized heating that minimizes oxidation and thus reduces cleaning after brazing. Because the heating is localized, warpage is often less than when the entire assembly is heated, and the nature and extent of metallurgical changes, such as the softening of cold worked or heat treated metal, are also minimized.

Most of the variables in induction brazing are machine controlled. Therefore, operators require a minimum of training and skill. Induction brazing is capable of making clean, neat joints without excessive spatter or excessive flow of filler metal to areas where it is neither needed nor wanted. This latter capability accounts, in part, for the extensive use of the process in the electronics and electrical industries.

Metals Brazed. With the exception of aluminum and magnesium, most of the common metals and alloys that can be joined by other brazing processes can be brazed satisfactorily by induction in air. Sometimes, provision of an inert atmosphere enclosure for brazing is required; however, the use of vacuum chambers is unsuitable for mass production. Because most induction brazing is done without the protection of a prepared atmosphere, the heating and cooling must be rapid and the temperature attained must be relatively low (seldom higher than 1550 °F); otherwise, the workpieces oxidize excessively beyond the flux-protected joint.

Dissimilar metals can be induction brazed, although special techniques must be employed to equalize differences in heating rates when a magnetic metal is induction brazed to a nonmagnetic metal. Special techniques are also required to equalize differences in coefficients of thermal expansion of dissimilar metals being induction brazed (see the section on brazing of dissimilar metals in this article).

Size Limitations. Induction brazing is applied most conveniently to small- and medium-sized assemblies. Brazing large assemblies, such as cylindrical bodies several feet in diameter, entails major design and installation problems, even with an adequate power supply. Brazing temperature also may impose a limit on the size of assembly that can be brazed by induction. Power requirements rise with increases in both the area to be heated and the required brazing temperature.

Nevertheless, the practical limit on the maximum size of an assembly for induction brazing normally exceeds that for furnace brazing. In furnace brazing, the entire assembly must be heated to the brazing temperature; whereas in induction brazing, the inductor is designed to restrict heating to the joint area.

Shape Limitations. Assemblies of almost any shape can be heated for brazing by induction, depending primarily on limitations imposed by construction of a suitable inductor, matching the impedance of the inductor and setup with output characteristics of the power supply, efficiency in heating, and cost. Although many inductors of intricate and unusual design have been constructed, practical limits are entailed in the bending and forming of copper tubing of a given diameter and wall thickness, as well as in providing the inductor with enough cooling water to pre-

*Revised by Pat Capolongo, Manager, Technical Sales, Lepel Corp.

vent overheating and in matching impedances.

Some of the problems associated with large assemblies have been overcome by rotating or otherwise moving the workpiece in relation to the inductor to develop an intricate heating pattern with a relatively simple inductor. In this manner, for example, a cylindrical object can be heated evenly while it is rotating within a simple, nonconforming "hairpin" inductor.

Maximum efficiency in heating is generally obtained when heating the external surface of a cylinder using a simple ring inductor of one or more turns that surrounds the external surface. Such an inductor is also one of the easiest to construct and to cool. Heating the internal surface of a hollow cylinder by locating the inductor inside the cylinder is feasible and often advantageous for reasons such as heat transfer and thermal expansion. Internal coils may, at times, prove less efficient.

At the higher operating frequencies, such as those provided by vacuum-tube power supplies, workpieces that have sharp corners or projections (such as threads) may be susceptible to damage in induction brazing, because the crests or points may overheat before the joint reaches brazing temperature. There are techniques, however, that limit damage to thread profiles.

Although induction brazing is by no means limited to assemblies that are self-jigging, this design feature, as in furnace brazing, often eliminates the need for special fixturing. Assemblies that are not self-jigging by virtue of design may often be temporarily held together by staking or tack welding. The use of fixtures in induction brazing is not uncommon, but may impose special requirements regarding the selection of materials from which the fixtures are to be made.

For fixtures that are to be located close (2 in. or closer) to any portion of the induction or connections leading to the inductor, heat-resistant nonmetallic materials, such as insulating fiberboard, ceramics, and quartz, are preferred. Aluminum, copper, and titanium are useful fixture materials, provided they are far enough from the inductor not to be heated by it. Because titanium does not wet readily (or braze readily), it may be heated in proximity to a workpiece. Fixtures that are allowed to become heated deteriorate and detract from the efficiency of the brazing operation.

Matching Impedance. An important limitation in all induction heating operations is the requirement for matching impedance. All characteristics of the inductor (or work coil), including diameter, number of turns, length, coupling with the work load, and operating frequency, together with work load variables such as the electrical resistivity of the workpiece, combine to constitute a factor known as impedance. Heating the workpiece depends on obtaining sufficient power in the inductor, which in turn depends on matching the impedance of the setup (inductor and workpiece) with the output characteristics of the power supply (generator or oscillator).

In essence, the circuit that constitutes the inductor and setup must be in preferred balance or resonance with the circuit of the power supply, or the transfer of energy from the power supply to the inductor and workpiece is ineffective. Thus, when a multiple-turn inductor is substituted for a single-turn inductor, or any other major variable of the work-load circuit is changed, some readjustment usually must be made in the power-supply circuit to compensate for the resulting change in impedance. Depending on the type of power supply employed, matching transformers or variable capacitors are ordinarily used for this purpose. With any power supply, however, the ability to match impedance is limited; inductor design and work load are thereby affected.

Quantity Limitations. Although induction heating equipment is relatively expensive, and most induction brazing setups (inductors and fixtures) are designed for one assembly only, the induction brazing process is not limited to mass repetitive production. Because most inductors are simple and inexpensive to construct and many brazing fixtures are neither elaborate nor expensive, the process is also well suited for handling small quantities of assemblies in a variety of shapes and sizes. Also, because the same induction heating machines are often capable of performing a variety of heat treating operations, induction brazing has been widely adopted by companies that perform both brazing and heat treating.

For high-volume production of assemblies of relatively few designs, the compactness of induction brazing equipment enables a plant with limited space to set up for induction brazing. When production quantities warrant the investment, the induction brazing setup can be partly or fully automated.

Brazing Combined With Heat Treating. Because the temperatures used in induction brazing are frequently close to those used in heat treating, the two operations can often be performed simultaneously or in sequence. In some applications, significant cost savings can be realized by brazing, austenitizing for hardening, quenching (usually with a spray), and tempering in an automated sequence of operations.

Principles of Operation

In induction brazing, the distribution of heat to sections of the joint depends largely on the contour of the inductor and the proximity of the inductor to the surfaces to be heated. The rate of heating varies inversely with the distance between the inductor and the workpiece surface. This relationship is not linear; as the distance is increased, the heating rate drops very rapidly. Increasing the distance between the inductor and the joint reduces the thermal gradient across the joint, but at the same time rapidly reduces heating efficiency. At some distance from the inductor, coupling between the inductor and workpiece is broken, and all heating ceases.

In addition to the inductor and the energy transmitted by it to the workpiece, the heating of a metal by induction for any purpose is also influenced by the mass, electrical resistivity, and magnetic permeability of the metal. Under identical conditions, a thin section heats to a given temperature faster than a thick section; a metal with high electrical resistivity heats faster than one with low resistivity; and magnetic metals heat faster than nonmagnetic metals.

Brazing of Steel. Carbon and low-alloy steels and ferritic and martensitic (400 series) stainless steels are magnetic and have comparatively high electrical resistivity. Therefore, they heat very rapidly with less power input than is required for heating a nonmagnetic, low-resistivity metal such as copper. Hysteresis effects play a minor role in the heating of magnetic metals; rapid heating is primarily the result of high electrical resistivity and magnetic linkages.

When brazing steel to steel, differences in the mass of the components are often significant. Much more energy must be introduced into the heavier component than into the lighter component to achieve a similar temperature rise in each. If the mass of one component is much greater than that of the other, generation of most of the heat energy in the heavy component may be advisable, and the light component may be heated solely or largely by thermal conduction.

Brazing of Steel to Copper. In brazing steel to copper, the steel heats much more rapidly than the copper, unless provision is made to equalize the heating rates. In practice, this is accomplished by coupling the inductor more closely to the copper than to the steel or by adding more turns to that portion of the inductor heating the copper. Similar provision must be made in brazing carbon steel to brass or to austenitic (nonmagnetic) stainless steel. Carbon steel heats faster than either of these materials, although the differential in heating rates is less than that for carbon steel and copper.

Power Supply

Induction brazing is usually done at frequencies of 10 kHz and higher. A variety of types of commercial power supplies are available, ranging in rating from about 1/2 to several hundred kilowatts, thus providing a wide selection that can be used for single or multiple inductors.

Motor-Generator Units. Power output ratings of motor-generator units range from 5 to 500 kW or more, and operating frequencies vary from 960 Hz to about 10 kHz. A typical motor-generator unit consists of a high-frequency generator driven by a motor, capacitors connected across the generator output terminals to create a high power factor, a transformer between inductor and generator with a turn ratio to a desired work-load impedance, and electrical controls such as voltage regulators, switches, meters, and automatic timers.

Most motor-generators are used with a 60-Hz power input and are designed to deliver full power at a particular voltage, current, and power factor. To transfer power from the generator to the workpiece, the generating source must be adjusted to the work load. This function is performed at the transformer and, when required, by altering the turns of the inductor or work coil.

Solid-State Power Supplies. Solid-state, high-frequency power supplies can be substituted for motor-generator units to provide power outputs of up to 50 kW or more and output frequencies up to 50 kHz. The solid-state unit provides high-frequency power through the use of silicon-controlled rectifiers, together with automatic power factor control, to compensate for changes that occur at the inductor when the resistivity of the work metal increases with rising temperature or when a magnetic workpiece passes the Curie point. The unit consists of a rectifier section operating on 60-Hz alternating current and producing a substantially constant level of direct current voltage, an inverter section producing variable-frequency alternating voltage, a series reactor, an output transformer, and power-factor corrective capacitors.

Vacuum-Tube Units. Rated capacities of vacuum-tube units vary from 1/2 to 600 kW, and although higher frequencies are available, output frequencies usually range from 100 to 450 kHz. Vacuum-tube units are often referred to as RF, or radio-frequency, units. Vacuum-tube units consist of a power-supply section and an oscillator section. The power-supply section provides the high voltage for the oscillator tube after rectification to a pulsating direct current, usually solid-state diodes.

The oscillator tube and a tank circuit, consisting of a matched inductor and capacitor, comprise the oscillator section. The oscillator tube controls the amount of electrical energy delivered to the tank circuit, from which the energy is removed by the coupled load. A small and proportionate amount of the power in the tank circuit is fed back into the grid of the oscillator tube to control the current that is delivered to the tube, and the tube in turn controls the amount of electrical energy entering the tank circuit. The frequency developed in the converter is determined by the inductance of the tank coil and the capacitor, which form a parallel tuned circuit. A load-matching network electrically coupled to the tank circuit is used to transmit tank circuit energy to the work.

Selection of Frequency. Power for most induction brazing operations is supplied by high-frequency induction heating equipment, usually with operating frequencies of 10 kHz or higher. The range of suitable frequencies higher than 10 kHz, however, is very broad and suggests that, in brazing, the factors of power input and impedance outweigh frequency in importance. Solid-state and motor-generator units with frequencies of about 10 kHz and vacuum-tube units with frequencies as high as 460 kHz can be used for induction brazing.

In general, solid-state power supplies are more efficient for heating large and heavy components at high power inputs. Vacuum-tube units are generally more versatile, permitting processing of a wider variety of brazing and other heating applications with one power supply. These units often present fewer problems in matching impedances and permit the use of inductors with close or loose coupling, or combinations of both, in the same multiple-turn inductor, which is often advantageous in induction brazing of dissimilar metals. Vacuum-tube units also permit the combination of brazing and thin-case hardening in a single operation.

The lower frequencies of the motor-generator can more often be tolerated for heating steel than for heating copper and may be advantageous. Because for the same power input, steel is only a fair conductor compared to copper, the steel surface heats faster, and heat penetration is much slower.

Inductors

Variables that affect the pattern of heating obtained by induction and are therefore pertinent to induction brazing include: (1) the shape of the inductor that produces the magnetic field, (2) the number of turns in the inductor, (3) the spacing between turns of the inductor, (4) the distance (air gap) between the turns of the inductor and the workpiece, (5) the presence of sharp corners on that portion of the workpiece within the magnetic field, (6) the presence of metallic shields within or near the inductor, (7) the operating frequency, and (8) the alternating current power input.

Magnetic Fields and Heating Patterns. Figure 1 shows examples of magnetic fields and heating patterns produced by induction. The patterns of magnetic flux for a single-turn and a multiple-turn inductor and the heating patterns developed by these inductors are shown in Fig. 1(a). Figures 1(b), (c), and (d), respectively, illustrate heating patterns of inductor pitch, or the distance between turns in a multiple-turn inductor, showing that finer pitch windings develop a deeper heat pattern than loose windings; coupling, or the air gap between inductor and workpiece, showing that deepest heat patterns occur with loose couplings; and sharp corners in a multiple-turn inductor, such as corners in a keyway, showing that heat builds excessively at sharp corners. Figure 1(e) shows alterations in heating pattern that result from modification of coil contour in a multiple-turn inductor designed for heating internally; contour of the inductor tube in a single-turn inductor; and coupling, coil contour, and pitch in a multiple-turn inductor.

The use of an external copper shield to dissipate some of the power input of the inductor, and thereby reduce heating rate and intensity, is illustrated in Fig. 1(f). Although use of a dissipator shield is rare, it is sometimes more effective than a change in coupling in the control of heat input. Finally, the development and cancellation of flux fields is shown in Fig. 1(g). When the turns of an inductor are made to carry the current in the same direction, a flux field is developed; whereas when

Fig. 1 Magnetic fields and heating patterns produced by various inductors

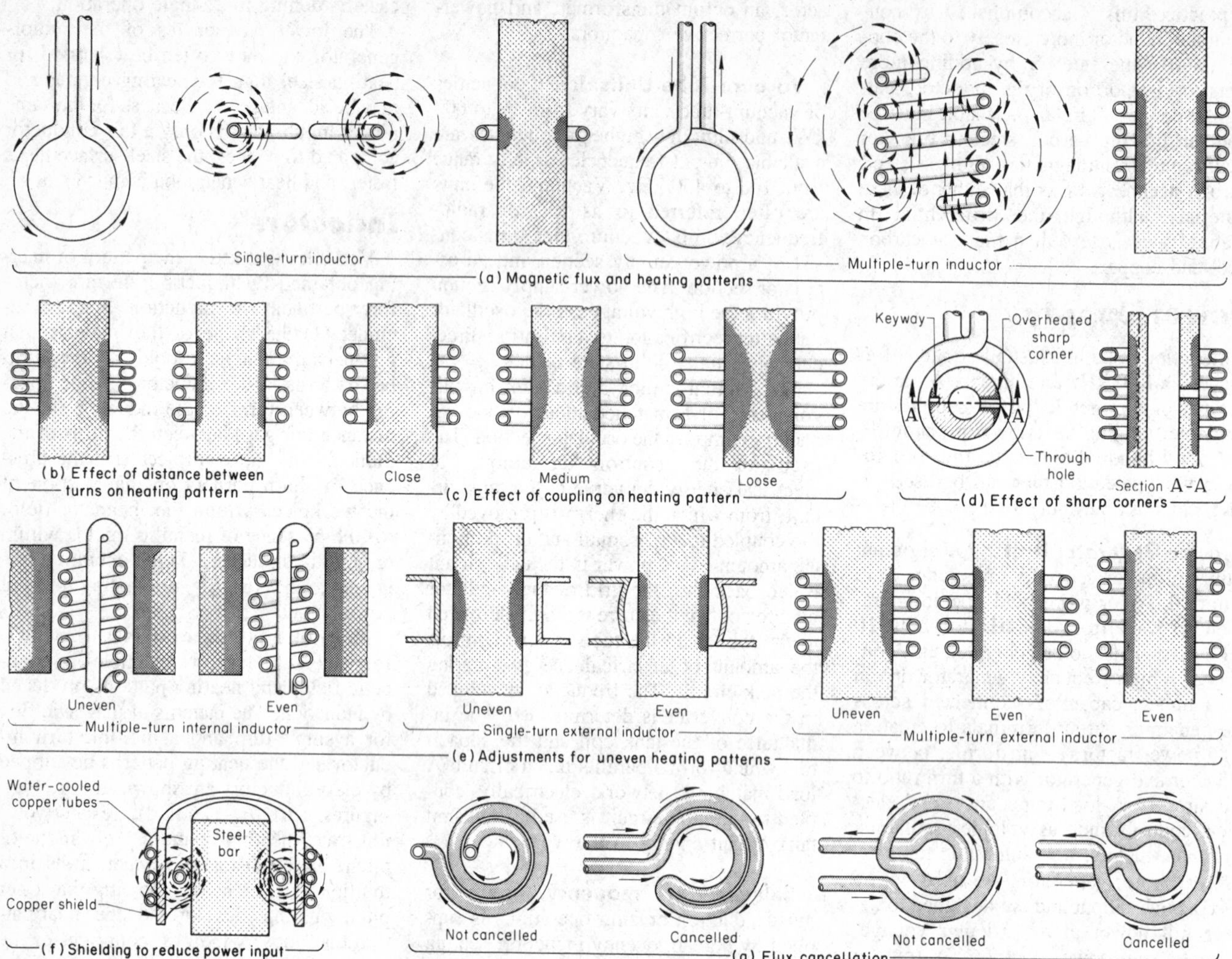

the turns are made to carry current in opposing directions, magnetic flux is cancelled. Although occasionally useful, cancellation of the flux field generally should be avoided.

Heating patterns can be altered in many ways. The rate of heating obtained by induction, however, depends on the resistance of the work metal to the flow of current induced in it by the inductor. This, in turn, depends on the strength of the magnetic field to which the workpiece is exposed.

Inductor Design for Brazing

The success of an induction brazing application relies largely on the design of the inductor, which must be related to the dimensions and configuration of the assembly to be brazed, the heat pattern desired, the heating time that produces minimal work-metal discoloration and oxidation, and the amount of power available. The size of the production run also influences inductor design. Although a single multiple-turn inductor may suffice for brazing small to medium quantities of assemblies intermittently, a series-type inductor, consisting of two or more single-turn or multiple-turn inductors of identical design, can braze larger quantities of the same assembly intermittently. The increase in production rate usually depends on the number of individual inductors in the series. In still another design, assemblies can be induction brazed continuously as they are carried on a conveyor belt or turntable to hairpin or pancake-type inductors that permit entrance to and exit from the heating zone without obstruction.

For the most rapid heating rates, inductors are designed to provide the maximum flow of current in the inductor and the closest permissible coupling between inductor and workpiece, after consideration of heat distribution, work-handling features, and high-voltage arcing between the turns of the inductor or between the inductor and the workpiece. In practice, considerable variation exists in the design of inductors. For the range of frequencies suitable for induction brazing, however, the inductors are made of copper, because of the high electrical conductivity and wide availability at relatively low cost characteristic of this metal.

Basic Designs. A variety of basic designs of inductors for use primarily with vacuum-tube power supplies with or without matching transformers is shown in Fig. 2. The inductor designs illustrated include: (1) a modified single-turn inductor capable of simultaneous brazing of two different joints on a single assembly (Fig. 2a); (2) multiple-turn inductors that have been formed to different geometric shapes to accommodate the contours of specific assemblies (Fig. 2b through e); (3) two types of multiple-turn inductors that are widely used to braze tungsten carbide cutting tips to steel shanks (Fig. 2f and g); (4) an inductor that differs from the preceding inductors that were made from copper tubing, because it is machined from a solid copper bar into which holes have been drilled for water cooling (Fig. 2h); and (5) two-station and four-station versions of the solid copper inductor (Fig. 2j and k).

The pancake type of inductor represents a different design category. Double-pancake inductors, such as those shown in Fig. 2(m) and 2(n), are sometimes substituted for the multiple-turn inductors in Fig. 2(f) and 2(g) to braze carbide tips to steel tools. Assuming that both types of work coils are equally satisfactory for brazing, the selection of a pancake-type coil may be based on matching impedance. Because flat pancake inductors permit unobstructed passage of workpieces either above or below them, they are widely used for continuous brazing on conveyor belts (Fig. 2p) and turntables (Fig. 2q). A variety of nonmetallic insulating products is suitable for conveyor belting and tabletop coverings.

Another type of work coil used in induction brazing, especially in conveyor-type operations, is the hairpin inductor. Examples of hairpin inductors suitable for use with conveyor belts or turntables are shown in Fig. 2(r) and 2(s). Because the heating of steel in a hairpin inductor is often restricted to a very limited area, rotation of the workpiece is usually necessary to expose all of the joint area that requires heating to the flux path of the inductor. Rotation is often achieved by forced frictional contact of the workpiece with a stationary guide as it is being conveyed forward in a horizontal plane.

The inductor shown in Fig. 2(t) is typical of a class of rather elaborately contoured work coils that require considerable ingenuity to design and fabricate. This inductor is essentially a double pancake,

Fig. 2 Basic designs of inductors for use primarily with vacuum-tube power supplies

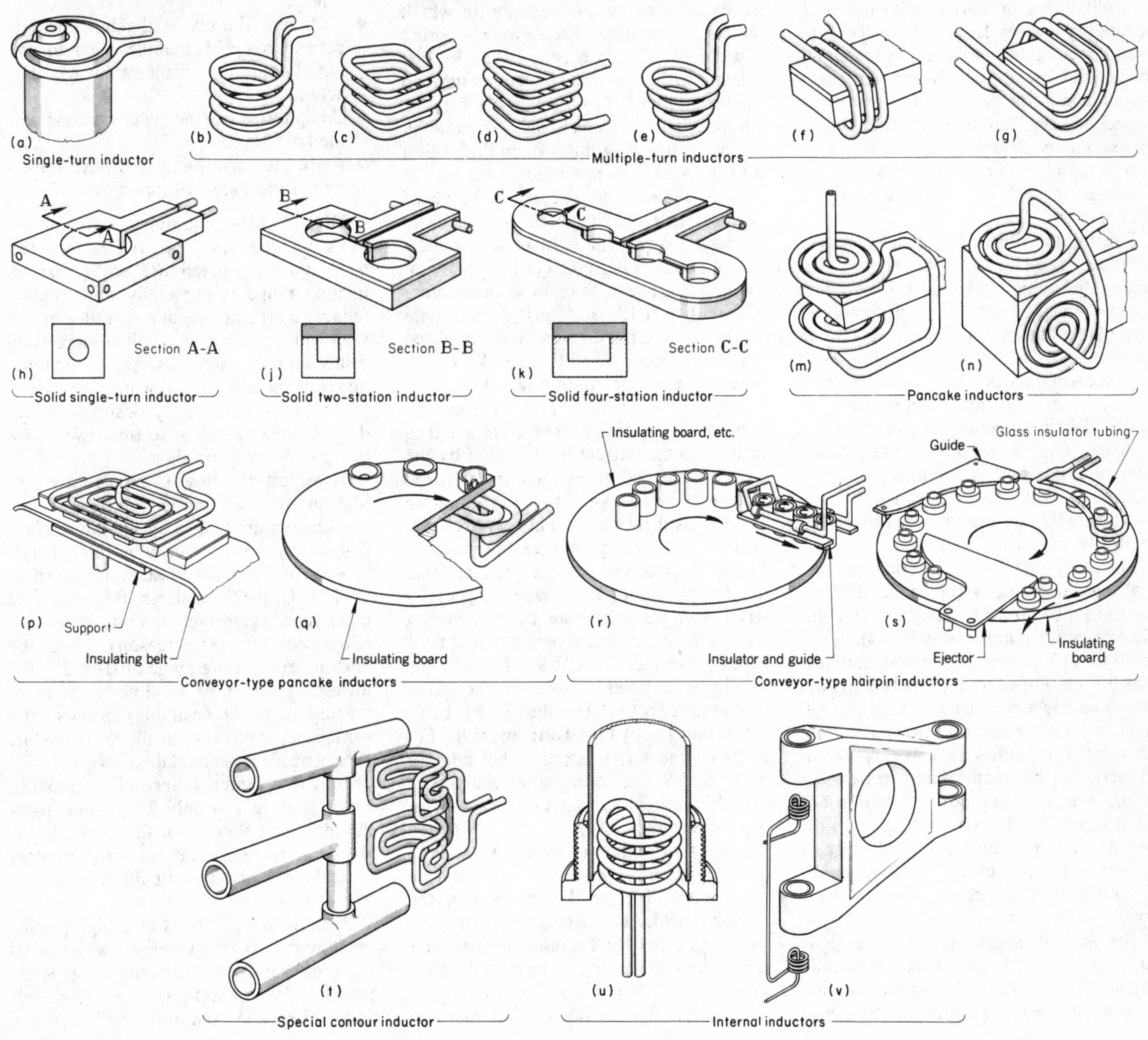

with each pancake formed to a U-shaped contour. The coil surrounds a sufficient area of the joints to ensure uniform heating by induction and conduction. If more time is required to allow for conduction of heat to certain remote areas of the joints, infrared or other automatic temperature controls can be employed to ensure uniform heating. Two-level power controls are also available.

Inductors designed for heating internally are useful in brazing dissimilar metals and in special applications where the use of an external inductor is not feasible. A typical work coil for heating internally is shown in Fig. 2(u). The application requires brazing a portion of the external surface of a copper tube to the internal surface of an externally threaded steel bushing. Heating the assembly with an external inductor primarily heats the steel, expanding it away from cold copper, and thus affecting heat transfer by enlarging the annulus. In contrast, heating internally favors heating of the copper, so that both copper and steel can be brought to brazing temperature at about the same time. A two-coil internal inductor for brazing two bushings in a hinge simultaneously is shown in Fig. 2(v).

Tubing for Inductors. Most inductors are made of commercial copper tubing. The size of the tubing selected for a specific inductor must be large enough to accommodate the current input and to permit an adequate flow of water for cooling. With machines of low power, the tubing may be only $^1/_8$ in. in diameter. For units of 20 to 50 kW, it is usually $^3/_{16}$ or $^1/_4$ in. in diameter. Only when the flow of pressurized cooling water is adequate, a $^1/_8$-in.-OD tubing with a wall thickness of only 0.018 to 0.020 in. may be used in the 25- to 50-kW range.

Inductors for frequencies obtained from motor-generator units (up to 10 kHz) are generally of two basic designs: (1) a single-turn coil, for high current densities (20 to 50 times the generator output current), which is employed to confine heating to a comparatively narrow band or segment; and (2) a multiple-turn coil for low current densities (1 to 5 times the generator output current), which is employed to heat a wider or larger area. A step-down transformer is ordinarily utilized between single-turn inductors and the generator; multiple-turn inductors are connected to the generator either directly or indirectly through a step-down transformer.

The wall thickness of the copper conductor used with motor-generator units is important. For efficient operation at various frequencies, the following minimum wall thicknesses may be used as a guide in constructing multiple-turn coils:

Frequency, kHz	Minimum wall thickness, in.
1	0.120
3	0.070
10	0.040

In the construction of single-turn coils, with which current density may be considerably higher, the values of minimum wall thicknesses listed above should, whenever possible, be multiplied by a factor of three or four.

Cooling. Single-turn and multiple-turn inductors may be formed from tubing or a combination of tubing and solid bus bar. Because of the high current density and the extremely thin cross section in which the current confines itself, water cooling is required. This is accomplished by circulating water through the tubes or through channels provided in the bus bar for that purpose. These channels may be made by drilling connecting holes or milling out a path to make a complete loop around the bore of the inductor. The exposed ends of the holes are then plugged or a copper sheet is brazed over the milled passage to make a continuous watertight cooling channel. Cooling passages with cross-sectional areas of 0.050 to 0.125 in.^{2} provide adequate cooling at water pressures of 40 to 50 psi for power inputs of 30 to 150 kW.

Water used for cooling should have a hardness of less than 12 grains per gallon. If the water-cooling passages are small in relation to the current load carried by the inductor, the use of distilled or deionized water may be necessary to avoid a buildup of deposits that could eventually stop circulation. Preferably, the water should be filtered to remove foreign particles that might clog small passageways, especially when inductors of intricate design are being used. The water should have an inlet temperature between 70 and 95 °F, and flow should be sufficient to prevent the outlet temperature from rising above 150 °F.

Coupling and Coil Turn Spacing. For efficient transfer of energy, the inductor bore may be $^1/_8$ to $^1/_4$ in. larger in diameter than the workpiece, which results in an air gap, or coupling, of $^1/_{16}$ to $^1/_8$ in. between the inductor and the workpiece. As the bore of the inductor is increased, the workpiece is subjected to a weaker portion of the magnetic field, heating time increases, and the heating pattern becomes deeper, unless the power density is increased correspondingly. Although loose coupling is seldom desirable for induction hardening, it is often employed in brazing to provide more uniform heating.

Inductor coil turns are normally spaced $^1/_{16}$ to $^1/_8$ in. apart. Considerable variation in the heat pattern can be obtained by adjustment of either the spacing between turns or the dimension of the air gap for individual turns. Some heat patterns are shown in Fig. 1.

Filler Metals

Requirements of a filler metal for induction brazing are:

- Melting temperature lower than temperatures at which the metals being brazed are adversely affected
- Ability to wet the metals being brazed
- Narrow melting range (difference between solidus and liquidus temperatures)
- Sufficient fluidity at the brazing temperature to enable the filler metal to flow rapidly through the joint by capillary action
- Composition chemically compatible with the base metal
- Ability to form joints that have the required mechanical properties

Most of these requirements also apply, to varying degrees, in many torch and furnace brazing operations, but a narrow melting range is especially important in induction brazing operations. Filler metals with wide melting ranges flow more sluggishly and are more susceptible to liquation than alloys with narrow melting ranges. Fast flow is particularly desirable in induction brazing, because time cycles for this process may be short.

Selection of Alloy. Compositions, solidus and liquidus temperatures, and brazing temperature ranges for the three alloys that are most widely used as filler metals in induction brazing of steel are given in Table 1. Of the three alloys, BAg-1 is used much more extensively than the other two alloys, for two main reasons. First, the brazing temperature range of BAg-1 is the lowest of the three. A low-temperature brazing range is desirable, because the workpiece, which is normally heated in air, experiences less oxidation with lower brazing temperatures. Second, the melting range of BAg-1 is only 20 °F, making it the most free-flowing of the three alloys. BAg-1 is recommended for brazing steel to steel and is also suitable for brazing steel to copper alloys.

Because BAg-2 melts at a higher temperature than BAg-1 and has a wider melting range, it is less suitable for general-purpose use. This property, however, makes BAg-2 more useful than BAg-1 for

Table 1 Compositions, solidus and liquidus temperatures, and brazing temperature ranges for filler metals frequently used in induction brazing

AWS classification	Composition, %					Temperature, °F		
	Ag	Cu	Zn	Cd	Ni	Solidus	Liquidus	Brazing
BAg-1	44-46	14-16	14-18	23-25	...	1125	1145	1145-1400
BAg-2	34-36	25-27	19-23	17-19	...	1125	1295	1295-1550
BAg-3	49-51	14.5-16.5	13.5-17.5	15-17	2.5-3.5	1170	1270	1270-1500
BAg-1a	49-51	14.5-16.5	15.5-17.5	17-19	1160-1175	...	...	1145-1400
BAg-5	44-46	29-31	24-26	...	1225-1370	...	...	1370-1550
BAg-7(a)	55-57	21-23	16-18	...	1145-1205	...	...	1205-1400

(a) Contains 5.0% Sn

large-clearance joints. BAg-3 is often preferred for joining carbide tips or inserts to steel because it has good wetting action on carbide. In addition, this filler metal is preferred for joining stainless steel to itself or to carbon steel.

Selection of Form. The filler metals listed in Table 1 may be purchased as spools of wire of various diameters and as strips of various thicknesses that can be made into preformed shapes, if desired. Filler metal is also available in powder form.

Joint design, method of assembly, and cost are the main factors that govern the choice among these forms. Preforms are usually preferred because they can be used with minimum waste and lend themselves to preplacement in or around joints of various designs. Preforms of almost any shape can be made from strip with a simple punch and die or from wire by forming. Some typical preforms are shown in Fig. 3.

Rings often serve as preforms. They can be made by winding wire on an arbor and then slitting the springlike coil lengthwise to produce the rings. Often, experimentation is needed to determine the arbor diameter that produces rings of the correct diameter after springback.

Many standard types and sizes of rings (and washers) also are commercially available. The sizes are ordinarily specified by their inside diameters, the rings being meant to fit around joints. The butt-end ring shown in Fig. 3 is a design that allows for expansion during preplacement. The $1^1/_{16}$-turn ring (Fig. 3) is designed to allow compression during preplacement inside a joint, and it is specified by outside diameter.

For some joints, preforms made from strip are more suitable than shapes formed from wire. When large numbers of blanked or formed preforms are needed, their cost may be reduced by buying them from a metal producer who can reprocess the waste punchings, whereas a user must sell the waste as scrap.

Filler metal in powder form is often used, although if it is used on the exterior of the joint, some provision must be made to hold it in place. One method is first to brush the joint with paste flux and then to dust the filler-metal powder lightly onto the flux while it is still wet. Another method is to mix the powder with flux and a binder to form a paste, which is then applied by brushing.

Fluxes

Flux or another means of oxygen exclusion is required for induction brazing. The flux used should decompose oxides without corroding the base metal or the filler metal, should be extremely active because of the short brazing times employed, and should be easy to remove after brazing.

Type 3A flux meets these requirements and is used for an estimated 95% of the induction brazing applications that involve steel. This flux is composed of a mixture of a wetting agent and one or more of the following materials: boric acid, borates, fluorides, and fluoborates. It is effective within the temperature range of 1050 to 1600 °F. Type 3A flux is available in the form of paste or powder and can be mixed to a liquid consistency with water, alcohol, or monochlorobenzene. Paste and liquid fluxes are most often applied to the joint by brushing. When more convenient, flux powder can be sprinkled on or in the joint, or liquid flux can be sprayed.

Fig. 3 Typical filler-metal preforms used in induction brazing

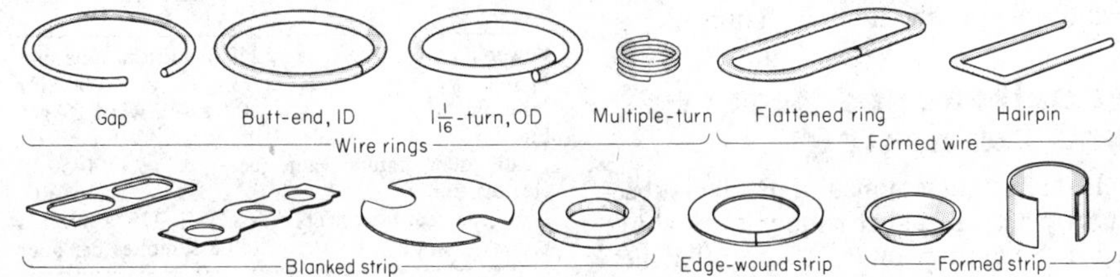

Assembly

Methods of preparing joints and assembling the components for induction brazing are generally the same as those used prior to furnace brazing (see the article "Furnace Brazing of Steels" in this Volume). The recommended diametral clearance for joints brazed with the filler metals listed in Table 1 is generally 0.002 to 0.005 in. Based on experience with similar joints, however, a clearance less than or greater than the recommended value is often selected for a specific joint. If the distance the filler metal travels while filling the joint is greater than normal, the clearance (or the brazing temperature, or both) may have to be increased.

Fixturing

Most assemblies that are brazed by induction require some fixturing, even though the components may be held securely together mechanically before being positioned in the inductor. Fixtures can range from a simple locating pin to hold the assembly in the center of the inductor to an elaborate clamping and holding arrangement. Every assembly must be considered separately in relation to its fixturing requirements. Fixture design is influenced not only by the size and shape of the assembly, but also by the rate at which assemblies are to be produced. Functions of a fixture may include supporting the assembly, positioning the assembly in the inductor, and holding the assembly securely together during the brazing cycle until the filler metal has solidified.

Selection of materials for fixtures is especially important. If all components of a fixture are sufficiently remote from the inductor (several inches away) to be unaffected by its magnetic field, materials used for fixtures for other brazing processes are suitable. If all or part of a fixture is within about 2 in. from the inductor or the leads to the inductor, however, the fixture must be made of a nonmetallic and, preferably, heat-resistant material. Several such materials are commercially available, including heat-resistant glass, ceramics, quartz, and plaster. When use of a metal near the inductor or the leads is unavoidable, aluminum, copper, and brass are the preferred metals.

Brazing of Tight Joints

For joints brazed with the high-silver alloys, the recommended diametral clearance (positive allowance) is 0.002 to 0.005 in. In some applications, however, this much clearance results in an unacceptable amount of eccentricity in the brazed assembly. A common method of eliminating eccentricity in most joints is to use a clearance range of 0.001 to 0.003 in.; with an active, free-flowing filler metal such as BAg-1, acceptable joint penetration can usually be obtained. For some joints, however, even a diametral clearance of 0.001 to 0.003 in. is unacceptable because of the resulting eccentricity; therefore, a press fit, requiring special provisions, becomes necessary (Example 1).

Example 1. Induction Brazing With an Interference-Fit Joint. Requirements for the medium-carbon steel cam-and-shaft assembly shown in Fig. 4 demanded special consideration in brazing. The specified concentricity necessitated an interference fit of the joint at room temperature, and the hardness of the bearing area of the shaft had to be retained. The cam and shaft were assembled with a press fit (slight interference) to ensure concentricity; the end of the shaft was chamfered slightly to facilitate assembly and to form a V-groove for a preformed ring of BAg-3 filler metal (Fig. 4).

During development of the technique for making the joint, an excessive amount of heat was conducted away from the joint area, which resulted in too long a heating time and softening of the bearing area of the shaft. This difficulty was overcome by machining a cup-shaped recess in the end of the shaft, by employing vacuum-tube power-supply units with a frequency of 450 kHz, and by using flat inductors that faced the V-groove.

The production requirement of 2400 assemblies per hour was met by the use of two 25-kW machines, each with two inductors. The assemblies were dipped in liquid type 3A flux and manually placed in holding fixtures mounted on a conveyor traveling at 30 in./min. Preformed 370° rings, 0.480 in. OD and made from 0.031-in.-diam wire, were dropped in place in the grooves.

The assemblies were passed under a 36-in.-long inductor that preheated the joint area and dried the flux and then under a 30-in.-long inductor where brazing was completed. The rate of heating by the inductors was controlled by adjusting the air gap between the inductor and the work-

Fig. 4 Induction brazed cam and shaft. Assembly required a press fit to maintain concentricity.

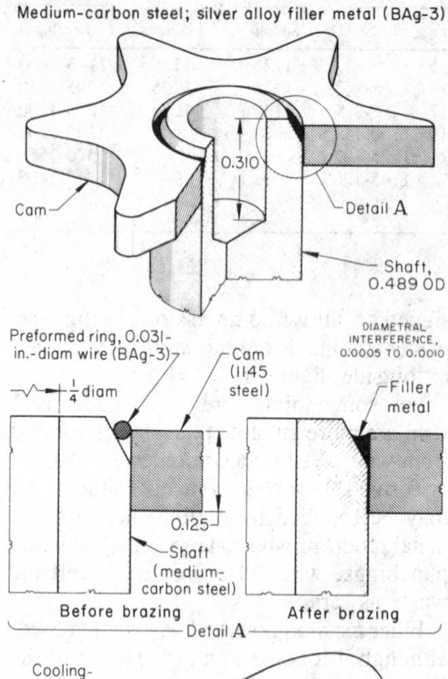

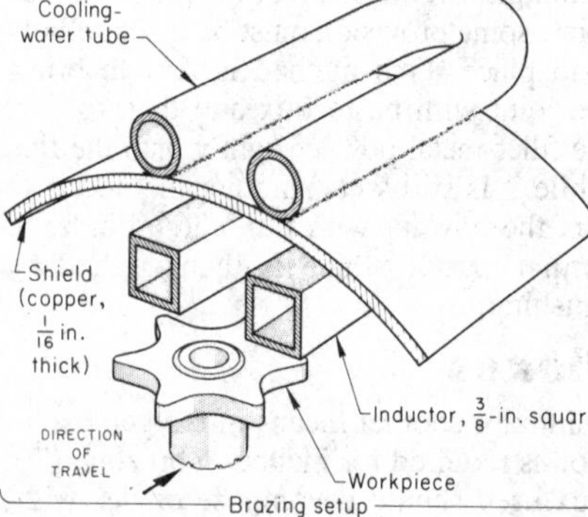

piece. The preheat inductor was inclined to give decreasing air gap; the brazing inductor was horizontal, giving a constant air gap. A water-cooled dissipator shield above the brazing inductor dispersed the flux field to help reduce the heating rate to the desired level.

After brazing, the heated portions of the assemblies cooled in air to about 800 °F during the next 120 in. of conveyor travel and were then sprayed with water during the following 20 in. of travel to remove flux and further cool the assemblies. Finally, the assemblies were ejected automatically from the holding fixtures.

Simultaneous Brazing and Hardening

If brazed assemblies must be subsequently heat treated at temperatures above the melting temperature of the filler metals, the use of induction brazing is generally precluded for joining steels, because those temperatures destroy brazed joints. This potential disadvantage of induction brazing can often be overcome, however, by brazing and heat treating in a single operation. Therefore, in the planning of brazing and heat treating operations, the possibility of combining the two operations or of using the induction equipment for heat treating local areas on the workpiece in a separate induction heating operation should always be considered.

Example 2. Brazing and Surface Hardening in One Operation. Originally, the valve-and-tube assembly shown in Fig. 5, which was used in grease-pumping equipment, was furnace brazed with BCu-1 filler metal. The tube was made of 1010 steel and the valve of 1040 steel. The valve face was subsequently hardened by induction heating and water quenching. Quenching the entire assembly after furnace brazing was unacceptable because it produced distortion. Furnace brazing, which was subcontracted because the manufacturer did not have suitable equipment, resulted in difficulties. First, the valves, having been decarburized in the brazing furnace atmosphere, did not harden

Fig. 5 Pump valve-and-tube assembly. Components were induction brazed and surface hardened in one operation.

Power supply	25kW vacuum-tube unit
Frequency	400 kHz
Filler metal	0.040-in. wire, BAg-1
Flux	AWS type 3A
Heating temperature (valve face)	1450 °F
Heating time	30 s
Postbraze heat treatment	375 °F for 1 h
Production rate	80 assemblies per hour

correctly. In addition, the cost per assembly with separate furnace brazing and induction hardening was uneconomical. To eliminate these problems, processing was changed to in-plant induction brazing performed simultaneously with heating for quenching.

Filler metal was 0.040-in.-diam BAg-1 wire formed into $^{9}/_{16}$-in.-ID rings. A groove was machined in the valve (Fig. 5) to make a pocket for the preformed ring. Assembled in this manner, the filler-metal ring just reached the brazing temperature of 1145 to 1250 °F when the valve face reached the hardening temperature of 1450 °F, because of the characteristic steep gradient caused by the rapid heating. If the filler metal had reached the hardening temperature, it would have been excessively oxidized.

A press-fit joint had been employed when BCu-1 filler metal was used with furnace brazing. On changing to BAg-1 filler metal and induction brazing, the diametral clearance between the two components was established at 0.004 to 0.008 in. In practice, clearance was consistently in the middle of this range.

The inductor was made from $^{3}/_{16}$-in.-OD copper tubing and had two turns on a 1-in. diameter. The brazing fixture comprised a holder for the tube and a quench ring of low-carbon steel tubing, which surrounded the inductor with a $^{1}/_{2}$-in. clearance to avoid induction heating of the quench ring. The spray of tap water from the quench ring was directed at the face of the valve.

Timing for the brazing cycle was determined visually using the following procedure. An assembly was placed in the machine and the inductor energized. When the operator saw the filler metal emerge and form a slight fillet at the edge of the joint, he noted the time interval, shut off the power, and opened the quench-water valve. After the heating time was established, it was controlled by a timer, but the operator continued to operate the quench-water valve manually. The sequence of operations was:

- Degrease tube and valve.
- Place preformed ring of filler-metal wire in groove.
- Flux mating surfaces of tube and valve with type 3A flux.
- Assemble components and place assembly in fixture.
- Heat for 30 s.
- Spray quench.
- Remove from fixture.

After brazing and hardening, the assemblies were washed and pickled to remove flux. Then they were tempered for 1 h at 375 °F to obtain the required minimum hardness of 48 HRC on the valve face. Other inspection requirements were a strong joint and squareness of the top of the valve with the tube (to ensure correct final grinding). In addition to an increased production rate and improved cost per assembly, several years of field experience demonstrated the reliability of the simultaneous operation; there were no failures.

Brazing of Dissimilar Metals

In addition to the differences in characteristics of dissimilar base metals that affect their brazeability by any process—namely, differences in thermal expansion, thermal conductivity, and compatibility with specific filler metals—differences in magnetic characteristics also have an effect on induction brazing. Despite these variables, induction brazing is used successfully for joining a number of other base metals to steel. One method for dealing with the difference in heating rates is to design the assembly so that the component that is heated more easily has the larger mass, which acts as the large heat sink and reduces the heating rate. Another method is to design the inductor so that the component that heats more easily is positioned in a weaker part of the magnetic field.

Sometimes, when other base metals are to be brazed to steel, a compromise must be made in selection of filler metal. For instance, BAg-1 is generally preferred for brazing one ferritic steel to another, but when austenitic stainless steel is one of the base metals, BAg-3 is a better choice, because it is suitable for both base metals.

Difference in coefficient of thermal expansion (and contraction) of the base metals produces shear stress in the cooling joint, which sometimes must be accommodated to avoid cracking of the joint. One means of adaptation is placement of a relatively thick, ductile metallic interlayer between the base metals, such as that obtained with a sandwich consisting of a fine iron gauze between strips of filler metal. Another method is to use a ductile copper strip clad on both sides with filler metal (Example 3).

Example 3. Use of a Filler-Metal Sandwich. To avoid cracking of brazed joints between tungsten carbide tips and tool shanks of low-carbon steel or tool steel, a sandwich brazing technique was used. Brazing was done with a 10-kHz, 25-kW machine. The assembly rested on a nonmetallic insulating board through which

Fig. 6 Induction brazing of carbide tip to a steel tool shank. A filler-metal sandwich was used to prevent shear-stress cracking of the carbide.

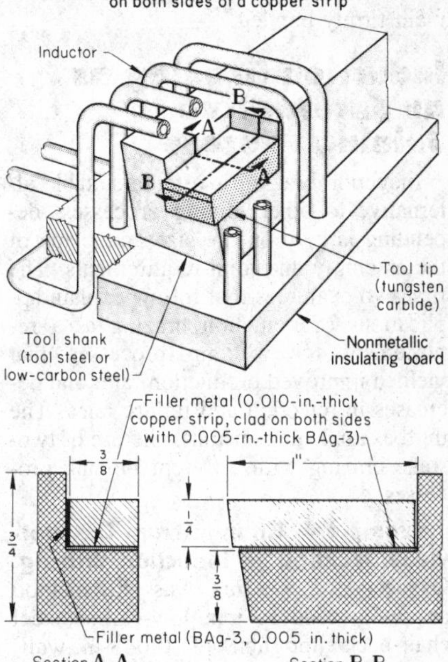

the inductor penetrated, as shown in Fig. 6. Surfaces of the carbide tip to be brazed were prepared by grinding, and surfaces of the steel shank were prepared by machining and vapor degreasing.

Filler metal for the portion of the joint between the bottom of the tip and the recess in the shank was a sandwich of 0.010-in.-thick copper strip, clad on both sides with 0.005 in. of BAg-3; filler metal for the vertical portion of the joint was a 0.005-in.-thick strip of BAg-3. The tip and shank were coated with flux and then assembled so that the tip rested on the filler-metal sandwich and also held the strip of solid filler metal against the sidewall of the recess.

The tip end of the assembly was inserted into the inductor only as far as the base of the tip to minimize the effect of the heat of brazing on the main portion of the shank. When the filler metal was melted, the tip was moved back and forth with a nonmetallic rod to pool the filler metal, release gas bubbles, and ensure good wetting. After brazing, the assembly was removed from the inductor, allowed to cool in still air, and then cleaned of flux.

The coefficients of thermal expansion of carbide and steel are different, causing the tip and shank to expand or shrink different amounts during heating to and cooling from the brazing temperature, which in turn produces a shear stress along the joint, sometimes large enough to crack it. The advantage of having the filler-metal sandwich beneath the tip was that the ductile copper yielded under the shear stress, allowing all components of the joint to remain firmly bonded.

Induction Brazing as an Alternative Joining Process

Induction brazing is often a suitable alternative to other joining processes, depending largely on the size and shape of the assembly and joint requirements. The first two examples that follow explain applications of induction brazing as a replacement for welding processes that yielded improved production rates and decreases in rejection and failure rates. The third example demonstrates the use of two-stage brazing with different brazing processes.

Example 4. Change From Oxyacetylene Welding to Induction Brazing. Oxyacetylene welding was the method originally used to join low-carbon steel chair-back uprights, of 0.048-in.-wall, 27¹/₄-in.-long, square, welded tubing to low-carbon steel end caps ⁷/₈ in. square by 0.060 in. thick. The welded joints had to be ground and polished, and many were rejected because of undercutting and porosity. Reject and rework rate was 10%. Hourly production per operator was 110 assemblies.

Induction brazing of five assemblies simultaneously in one inductor was selected to replace welding. Figure 7 shows the setup used. The power supply was a vacuum-tube unit, which operated on a 440-V, three-phase, 60-Hz line. It had an output of 15 kW at a frequency of 450 kHz. The inductor, made of copper tubing and arranged to accommodate five assemblies in a single turn, was shaped to heat at the joint surface and tilted to allow access to the fixture. Optimum air gap to the workpiece was about ¹/₈ in. The inductor and the power supply were cooled by water at a minimum rate of 7 gal/min under pressure of 40 to 80 psi. A heat exchanger was used to keep the water temperature at about 70 °F to prevent condensation that might cause short circuits in the inductor and power supply.

The fixture consisted of a nonmetallic insulating board on an aluminum support. It had ⁷/₈-in. slots spaced about ¹/₂ in. apart to accommodate and locate five assemblies. Five end caps were placed in the slots of the fixture, and then a single strip of BAg-1 filler metal 0.005 in. thick by ¹/₄ in. wide was placed over them in the position shown in Fig. 7. Each of the five uprights, which had been dipped first in a suitable degreaser to remove mill oil and then in type 3A flux, was placed on the filler metal so that the filler metal was pressed between it and the end cap. The upper ends of the uprights were held by magnets.

Fig. 7 Setup for simultaneous induction brazing of five chair-back uprights to five end caps

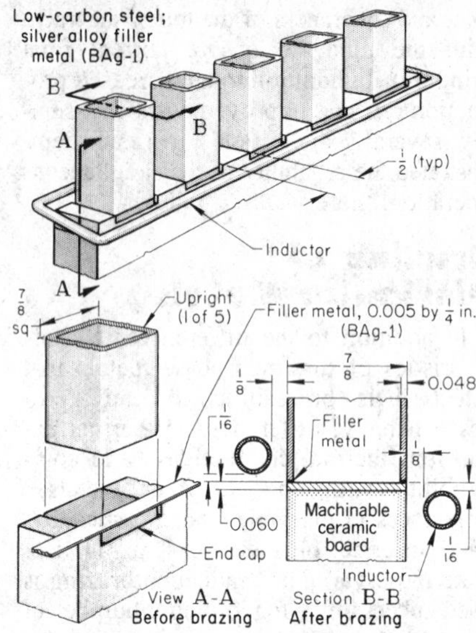

When all five assemblies were in place, an induction heating cycle of 28 s was started. Heating was greatest next to the upper (back) half turn of the inductor; thus, when the filler metal melted (at approximately 1150 °F), it flowed by capillary action toward the hottest part of the joint and filled the entire joint. While the filler metal was still molten, the end caps and the uprights were aligned manually with a short paddle made from insulating ceramic rod. The inductor was lower in front of the work to allow access for the paddle.

The brazed assemblies were washed in a tank containing a 1-to-6 emulsion of soap and water at 190 °F. This emulsion removed 80 to 90% of the flux residue and also acted as a rust inhibitor. The brazed joints were then lightly belt-sanded.

Quality was controlled by visual inspection. The rejection rate was less than 1%, as compared with 10% for oxyacetylene welding. Average production rate for one operator was 232 assemblies per hour—more than twice the rate for welding.

Example 5. Use of Induction Brazing Instead of Shielded Metal Arc Welding (SMAW). Originally, a ⁵/₃₂-in.-wall resistance welded and drawn low-carbon steel tube and a hot rolled 1045 steel collar were joined by SMAW. The joint design for welding is shown in Fig. 8. During welding, unintentional melt-through of the tube occasionally occurred. Moreover, the weld beads had to be ground for satisfactory appearance, and some of the welded joints cracked in service.

To improve the appearance of the joint and increase strength, induction brazing was substituted for welding. Two joint designs were used for brazing (Fig. 8). In the first design, a single groove for holding a preformed ring of filler-metal wire was machined in the collar, and the collar and tube were then threaded together. Before brazing, both parts were cleaned with a suitable degreaser, and type 3A flux was brushed on the tube. Then a preformed ring of ¹/₁₆-in.-diam BAg-1 wire was inserted in the groove provided for it, and the parts

Fig. 8 Induction brazed tube-and-collar assembly. Brazing replaced welding with appropriate changes in joint design.

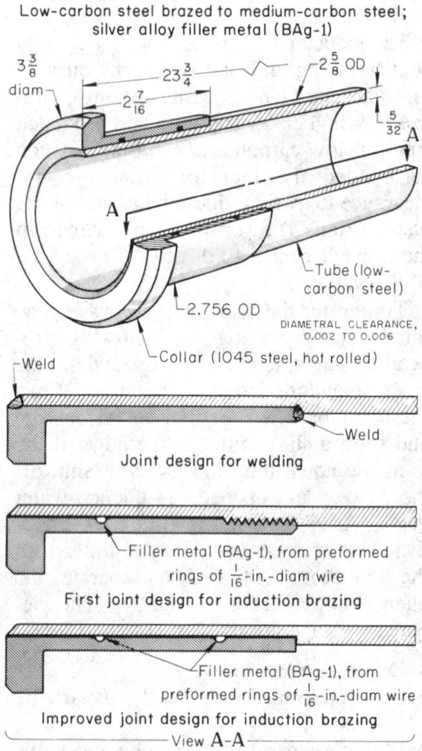

Operation	Time, h: Welding	Brazing(a)
Time for joining 30 assemblies		
Setup	0.20	0.80
Joining	3.66(b)	2.37(c)
Setup and joining	3.86	3.17

(a) Using improved design shown. (b) 0.122 h per assembly. (c) 0.079 h per assembly

Fig. 9 Housing-cover assembly. One joint was furnace brazed; the other was induction brazed.

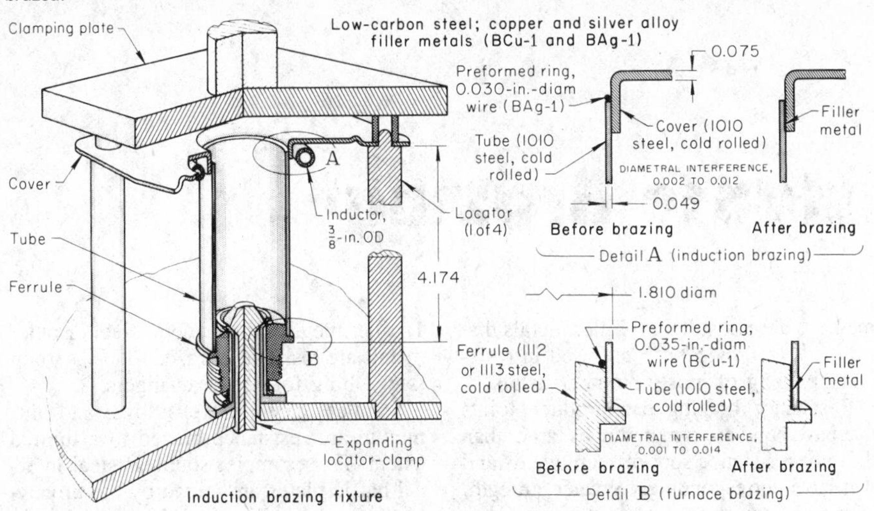

Furnace brazing (ferrule to tube)	
Furnace type	Electric roller hearth(a)
Atmosphere	Exothermic, 2400 ft^3/h
Brazing temperature	2050 °F
Time in heat zone	39 min
Total time in furnace	2 h

Induction brazing (cover to tube)	
Power supply	20-kW vacuum-tube unit
Frequency	450 kHz
Heating time	6 s
Air-blast (cooling) time	5 s
Total cycle time(b)	20 s

(a) Preheat zone, 15 ft; heating, 30 ft; fan cooling, 45 ft. (b) For clamping, heating, cooling

were screwed together, mounted on centers, and brazed using a 100-kW, 10-kHz power supply. The joint seemed strong, but in the field it failed by fatigue through the innermost thread.

In an improved joint design, the threads were eliminated and a second groove for filler metal was added. Filler metal for both grooves was $^1/_{16}$-in.-diam BAg-1 wire in preformed rings. Visual inspection after brazing showed a full perimeter of filler metal at both ends of the joint. The brazed assembly withstood an axial pull of 60 000 lb (equivalent to a shear stress of 3.4 ksi), which was above service requirements. None of the assemblies that incorporated the improved joint design failed in service.

Brazing, employing the improved joint design, had the advantage of a stronger assembly. The higher strength was attributed to the fact that, in the brazed joint, a large area of the joint between the collar and the tube was bonded; whereas in the welded joint, bonding was restricted to the collar ends. Additional advantages included elimination of melt-through, improved appearance of the joint and elimination of the grinding operation, and increased production rate.

Example 6. Use of Two-Process Brazing. A housing-cover assembly for an automotive brake booster was fabricated by fastening a machined ferrule to one end of a welded tube and a press-formed cover to the other end, as shown in Fig. 9. The ferrule was made of cold rolled 1112 or 1113 steel; the tube and cover were made of 1010 steel.

For some assemblies that incorporate two joints, the use of induction brazing for one joint and furnace brazing for the other joint may be advantageous. The ferrule-to-tube joint (Fig. 9, detail B) lent itself to brazing in a controlled-atmosphere furnace, which was available in the plant. The cover-to-tube joint (Fig. 9, detail A) was subsequently brazed by induction, because the close final dimensions necessary could not be held in furnace brazing, and brazing in a furnace would distort the press-formed cover.

For induction brazing, the furnace brazed subassembly was loaded over a locator-clamp in the induction brazing fixture. In automatic sequence, this locator-clamp expanded and was drawn down to clamp the subassembly firmly in position. The cover was spray cleaned with a hot alkaline solution, type 3A flux was brushed inside the hole, and a preformed ring of BAg-1 filler-metal wire was set in place inside the hole. The cover was then put into the fixture and over the tube, and four locator posts on the fixture were inserted into four mounting holes in the cover.

The joint was designed for a press fit to accommodate the tolerance of the tubing as-purchased; thus, the 1.808- to 1.813-in.-OD tube and the 1.801- to 1.806-in.-ID cover at the joint produced a room-temperature interference of 0.002 to 0.012 in. Differential expansion of tube and cover during heating allowed the filler metal to flow through the joint.

Pressing two palm buttons started the following automatic sequence. The piston of a bottom air cylinder drew down the locator-clamp, expanding it and clamping the ferrule. The piston of a top air cylinder pushed the clamping plate down, which pressed the cover down onto the tube and the four locator posts, thus holding the 4.174-in. dimension between the cover and the flange on the ferrule. Heating time was 6 s, followed by a 5-s cooling air blast. The piston of the bottom cylinder pushed up the locator-clamp, releasing the ferrule, and then the piston of the top cylinder and the clamping plate retracted, releasing the cover. The assembly was unloaded by hand and cleaned of flux. The single-turn inductor was positioned $^1/_8$ in. below the cover; the gap between cover lip and inductor was 0.090 in.

Resistance Brazing*

RESISTANCE BRAZING is a resistance joining process in which the workpieces are heated locally and filler metal that is preplaced between the workpieces is melted by the heat obtained from resistance to the flow of electric current through the electrodes and the work. In the usual application of resistance brazing, the heating current is passed through the joint itself. Equipment is the same as that used for resistance welding, and the pressure needed for establishing electrical contact across the joint is ordinarily applied through the electrodes. The electrode pressure also is the usual means for providing the tight fit needed for capillary behavior in the joint. The heat for resistance brazing can be generated mainly in the workpieces themselves, in the electrodes, or in both, depending on the electrical resistivity and dimensions.

Applicability

Parts of many different shapes can be resistance brazed, provided that the surfaces to be joined are either flat or conform over a sufficient contact area and that they can be held together under pressure to permit the heating current to flow through the joint and the filler metal to be distributed throughout the joint by capillary action. Workpieces that can be joined by resistance brazing range from 0.001-in.-diam wire to assemblies with joint areas of about 10 to 15 in.^{2} Joint area in most high-production resistance brazing is small, usually not more than 0.1 to 0.6 in.^{2}

The use of portable welding heads or tongs permits resistance brazing to massive parts or structures and in locations inaccessible to standard resistance welding machines. Resistance brazing also is used where the workpieces must be heated locally and maximum local temperatures are below the melting point of the workpieces. In some applications, total heat input can be sufficiently small and localized to make joints for which conventional arc or resistance welding or other brazing methods are not suitable. Filler metals that flow at 1100 to 1500 °F are used in resistance brazing of most common metals.

Resistance brazing can produce joints that have bond areas many times larger than those of resistance spot or seam welds and that have correspondingly greater strength. When resistance brazing is done in resistance welding machines, joining can be done at high production rates, at lower labor cost, and with low operator skill.

Metals Joined. The work metal most frequently joined by resistance brazing is copper. Resistance brazing with high-resistivity electrodes or electrode facings is an efficient method of providing localized heating at the joint in this highly conductive metal, but avoiding fusion of the copper base metal. In addition, copper is the only frequently used metal that can be brazed in air with self-fluxing filler metals (copper-phosphorus alloys, BCuP type) and, thus, without the use of a flux.

The metals that rank next in frequency of joining by resistance brazing are copper alloys, such as 184, 360, and 353, which are used in subsequent examples. Copper alloys (and copper) have been resistance brazed in producing a variety of assemblies employed in circuit breakers, electrical switchgear, and power-distribution equipment. One important application of resistance brazing is electrical contacts made of silver, silver-graphite, silver-molybdenum, silver-tungsten, and copper-tungsten, which have been resistance brazed to copper in making heavy-duty circuit breakers and other types of electrical equipment.

In plants where copper and copper alloy assemblies are resistance brazed, the process occasionally is applied to assemblies made of steel or other metals. Typical resistance brazed low-carbon steel assemblies are transformer brackets made by joining hat-shaped strips $^1/_8$ in. thick by 1 in. wide to flat strips $^1/_8$ by 1 by 6 in., making a 1-in.-square joint at each end of the flat strip using preplaced foil of BAg-1a filler metal. Fins made of steel or other metals are resistance brazed to low-carbon steel tubing for heat exchangers.

Stainless steel, nickel alloys, and aluminum are resistance brazed to a limited extent. For example, stainless steel internal baffle plates are joined by the process to the inner walls of 1020 steel tubes in heat-exchanger applications. Additional metals, not mentioned above, are also occasionally resistance brazed.

Equipment

Resistance brazing ordinarily is done with conventional resistance welding equipment, as described in the article "Resistance Spot Welding" (RSW) in this Volume. This equipment can be used for some resistance brazing applications without modification. Generally, heating and cooling times are longer and electrode force is lower for resistance brazing than for RSW.

Resistance spot welding machines may be modified to provide ranges of operating conditions suitable for resistance brazing, or machines may be designed especially for resistance brazing. Other changes often needed to adapt RSW equipment for resistance brazing are in electrode holders and electrodes. Electrode holders that provide low inertia and fast follow-up are sometimes needed. Adaptation for high-volume production is generally the same as in resistance spot or projection welding, except for the additional need in most resistance brazing applications of providing for the preplacement of filler metal and, sometimes, flux in the joint.

Power sources for most resistance brazing are conventional resistance welding transformers, which deliver low-voltage, high-amperage power (see the article "Resistance Spot Welding" in this Volume). A higher voltage is usually needed in the secondary circuit for resistance brazing than for resistance welding, especially when using high-resistance

*Revised by Austin Dixon, Consultant, Process Design, Metals Joining Technology

electrodes; hence, the transformer should have a relatively high kV · A rating.

Capacitors or capacitor banks can be used as power sources for resistance brazing of wires or other small parts in joints for which the heat input must be extremely small and the heating time very short. Before each operation, the capacitors are charged with direct current from either a rectifier or a motor-generator.

Controls for resistance brazing are similar to those for resistance spot and projection welding (see the article "Resistance Spot Welding" in this Volume). Brazing current, electrode force, and timing usually are controlled automatically. Slope control and pulsing, where needed, can be programmed into the machine. Electrode advance and retraction, and sequencing of some or all machine functions, can be done manually or automatically.

Fixed timing is sometimes unsatisfactory, because it does not provide compensation for variations in heat input to the joint from the electrodes or in the heat-sink capacity of workpieces, fixtures, and other equipment during continuous production runs. Automatic termination of heating by a controller that senses temperature at or near the joint can provide uniform heating of the joint and filler metal, regardless of variation in heat input or heat loss. Manually pulsing the heat on and off using a hand or foot switch during the brazing cycle is frequently employed to promote uniform heat distribution on larger parts where brazing time may exceed 5.0 s.

Example 1. Use of an Infrared Sensor for Heat Control. The assembly shown in Fig. 1, a connector for an electric controller, was composed of a U-shaped copper-strip contact finger brazed to a lug consisting of a short length of copper tubing that was flattened at the joint. No cleaning was necessary before brazing, because both members of the assembly had been tin plated.

Erratic joint strength was observed in the first production runs, in which the assemblies were joined by induction brazing. When the joining process was changed to resistance brazing to improve consistency of joint strength, adjustment of the heating time during continuous production runs was necessary to compensate for progressive heat buildup in the electrodes. Timing was accurate to within one cycle ($^1/_{60}$ s).

Fig. 1 Electrical-connector assembly. Automatic resistance brazing of the assembly used an infrared sensor to control heating time.

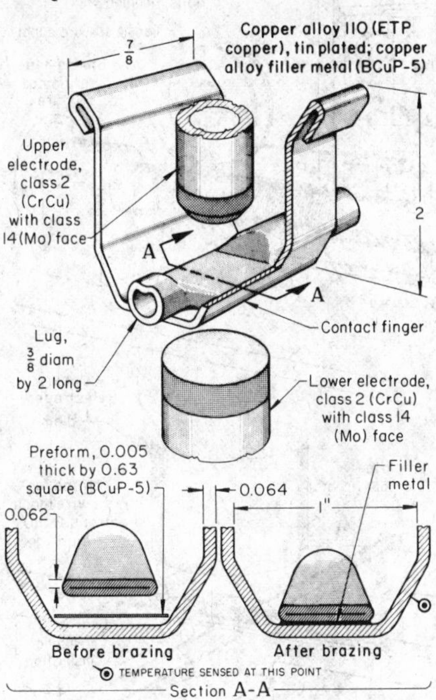

Machine	Press-type, air-operated, 75-kV · A automatic resistance welding machine
Loading and unloading	Manual
Filler metal	BCuP-5, 0.005-in.-thick preforms
Flux	None
Electrodes	RWMA class 2 (chromium copper) with class 14 (molybdenum) facings
Electrode force	350 lb
Brazing current	18 000 A
Heating time	Approx 50 cycles (controlled by infrared sensor)
Hold time	40 cycles
Fixtures	Insulating, heat-resistant material for alignment of workpieces

The need for periodic adjustment of the heating time to obtain consistent joint strength was eliminated by changing from a time-control method of regulating heating to a method in which heating was automatically terminated by a controller actuated by an infrared photocell that sensed the work temperature near the joint. The sensor was sighted on the contact finger at the point shown in Fig. 1.

The filler-metal preform (section A-A in Fig. 1) was a 0.63-in. square of 0.005-in.-thick BCuP-5. This filler metal, self-fluxing on copper, was selected to avoid the need for using a separate flux. In spite of the relatively low electrical conductivity (10% IACS) of BCuP-5, electrical resistance across the joint area (about 0.4 in.^{2}) was low enough in relation to the small current load (100 A maximum) carried by the connector in service. The machine used for brazing was a 75-kV · A, air-operated, press-type resistance welding machine. The electrodes had Resistance Welder Manufacturers Association (RWMA) class 2 (chromium-copper) shanks and class 14 (molybdenum) faces. The force applied to the electrodes was 350 lb. The brazing current was 18 000 A, and the brazing time averaged 50 cycles, or slightly less than 1 s. Alignment of the workpieces was maintained by insulating, heat-resistant fixtures (not shown in Fig. 1). Recommended materials include heat-resistant products such as pressed mica board, ceramic fiber, silicon oxide, and aluminum oxide.

The operator loaded the contact finger into the fixture, placed the filler-metal preform in position, placed the lug on top of the filler metal, and pushed the start button to close the electrodes and actuate the brazing sequence, which was completed without further attention by the operator. Quality was controlled by random checks on the operation and the brazed assemblies. In addition to visual examination of the completed joints for presence of a fillet, randomly selected brazed assemblies were tested to destruction in a pull test. The brazed joint was required to remain intact in the destructive test.

Machine construction of the machines used for resistance brazing, which generally are conventional resistance welding machines, is ordinarily press-type construction (see the article "Resistance Spot Welding" in this Volume). In these machines, the upper electrode and welding head move in a straight line and thus assist in maintaining alignment of the workpieces during brazing. Machines for intermediate to high production are usually floor-standing or bench-mounted air-operated types.

Portable machines, although ordinarily rated for lighter duty than floor-standing or bench-mounted machines, may have the same capabilities in all other respects (see the article "Resistance Spot Welding" in this Volume). The portable head can be of a size suitable for use in close quarters and for carriage mounting in applications where the brazing head must travel.

Example 2. Incorporation of Cross Wire Resistance Brazing Into a Mechanized Coil-Winding Operation. The 36-to-48-in.-long coil shown under construction in Fig. 2 was used to read out positions of machine-tool components on a numerically controlled machine tool. For this purpose, precisely located tap leads were brazed to the coil wire at predetermined intervals in a straight line along the coil. As many as 48 taps were made on a single coil, and to ensure the precision of the readout, the taps had to be located within ±0.003 in. of true position. Resistance brazing with a carriage-mounted

Fig. 2 Arrangement for cross wire resistance brazing. Stranded copper wire tap leads were joined to solid copper wire while the solid wire was being wound into a coil.

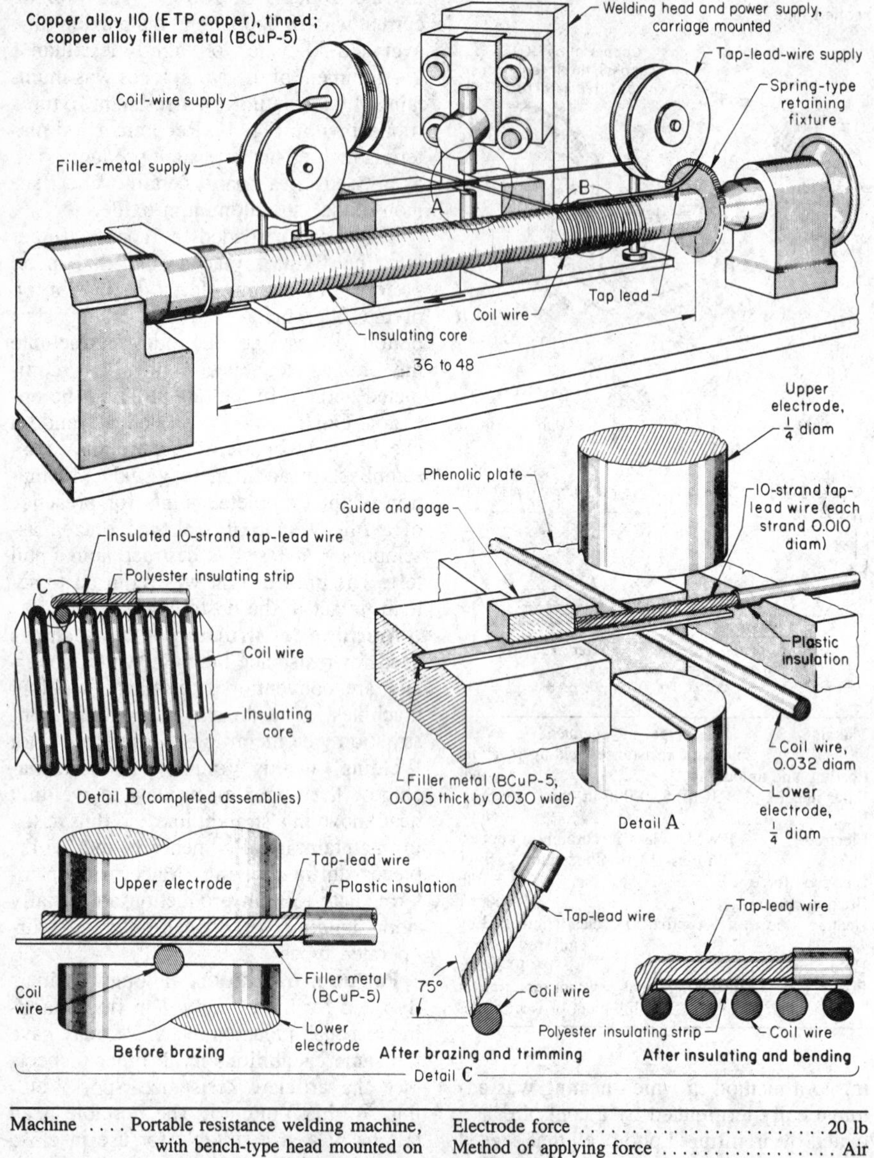

Machine	Portable resistance welding machine, with bench-type head mounted on traveling carriage, incorporated into coil-winding machine	Electrode force	20 lb
		Method of applying force	Air
		Brazing current	2450 A
		Heating time	60 cycles
Loading and unloading	Manual	Hold time	30 cycles
Filler metal	BCuP-5	Timing method	Solid-state timer
Flux	None	Joint strength(c):	
Precleaning	None(a)	Mean	24.3 lb, min
Electrodes	RWMA class 14 (Mo), $^1/_4$ in. diam(b)	Range	5.75 lb or less

(a) Insulation manually stripped from tap leads. (b) With 0.010-in.-deep (0.022-in. radius) locating grooves on faces. (c) Determined by performing a pull test on five sample joints

welding head was combined with other assembly operations in the manufacturing line.

Flexible wires for the tap leads were resistance brazed to the coil wire in cross-wire joints while the coil was being wound. For this purpose, a portable resistance welding head was mounted to the winding machine so that it could move along the 36-to-48-in.-long coil into position for attaching each tap lead. The electrodes were $^1/_4$-in.-diam molybdenum rods that were grooved—the lower one to take the coil wire, and the upper one at right angles to the lower groove to accommodate the tap lead (see detail A in Fig. 2).

Both wires were made of tinned copper corresponding to American Society for Testing and Materials (ASTM) B3—soft or annealed copper wire for electrical purposes, which ordinarily is alloy 110 (ETP copper) or low-resistance lake copper. The thickness of the tin coating was about 30 to 50 μin. The coil wire was a single strand 0.032 in. in diameter; the tap-lead wire consisted of ten strands of copper, each 0.010 in. in diameter, and was covered with a plastic insulating coating. The insulation was manually stripped from the joint area at the end of the stranded tap-lead wire before brazing.

The BCuP-5 filler metal was furnished in strip 0.030 in. wide by 0.005 in. thick. Because the workpieces were tinned, the joint surfaces did not need cleaning, and because of the phosphorus in the filler metal, they did not need fluxing. The wires and the filler metal were furnished on spools from which they were fed during production of the coils.

In production, the operator pushed a button, and the coil-winding machine automatically wound the correct number of turns of coil wire onto the threaded insulating core. The coil wire from the spool was guided between the electrodes by a grooved phenolic plate. The operator drew the filler-metal strip, which was guided by a groove in the filler-metal guide, across the coil wire, stripped the insulation from the end of the tap-lead wire, and placed the bare end over the filler-metal strip so that the tap-lead wire lay in its guide groove. A gage at the end of the guide for the tap-lead wire indicated the proper cutoff length (detail A in Fig. 2).

When the operator pressed a button, the electrodes closed and the joint was brazed. Timing of the brazing sequence was controlled by a solid-state timing device. Each joint was made at a point on the coil wire exactly one turn away from its final position. The filler metal and the tap-lead wires were trimmed by hand, and the final turn was made manually to bring the tap lead into final position on the coil. The operator inspected the joint, trimmed the excess tap-lead wire at the joint at the proper angle (75° from horizontal), placed a polyester insulating strip between the tap lead and the adjacent turn of the coil to avoid shorting that would give false readouts, and bent the tap lead to fit closely against the body of the coil. Detail B in Fig. 2 shows completed assemblies of tap leads to a coil wire; detail C in Fig. 2 shows the setup for brazing and the assembly after brazing and trimming and after insulating and bending. Next, the operator placed the end portion of the tap lead in a spring-type retaining fixture mounted on a disk (shown at right in the top view in Fig. 2) that was temporarily attached to the end of the coil core to hold the brazed tap leads during the remainder of the operation.

Joints had to be 100% sound to be electrically efficient. For machine qualification, five sample joints were made by the operator before the start of each coil, at the beginning of each shift, and after any interruption of about 1 h in production. The sample joints were checked visually for the presence of fillets, and joint strength was measured on each in a pull test. The mean breaking load for the five specimens had to be 24.3 lb minimum, and the range could not exceed 5.75 lb.

Production was not started (or resumed after an interruption) until acceptable test results were obtained. Making and evaluating the sample joints took about 5 min. During production, brazed joints were spot checked visually for the presence of fillets. Rejects for quality of brazed joints averaged about 1%. Rejects for faulty positioning were negligible.

Resistance brazing was preferred to a mechanical attachment technique, which had been tried on prototype coils, because of the greater reliability of brazed connections and because of production problems anticipated for the mechanical technique. Attaching the tap leads by welding would have required higher temperatures, with the likelihood of burning the small strands of the tap-lead wire; soldering would have produced weaker joints with a lower operating-temperature limit in service.

Hand-held tongs are used to equip a simple type of portable resistance brazing machine that can be used conveniently to make brazed connections on massive assemblies that cannot be brought to a conventional resistance welding machine. In the most rudimentary form, the hand-held current-carrying tongs, with electrodes attached at the end of each arm, are squeezed together by the operator to exert pressure on the electrodes and the work, and the current is turned on and off and adjusted manually by the operator. The amount of current passed through the joint, the duration of heating, and the repetition rate may require water cooling of the flexible current-carrying cables that connect the tongs to the power source (usually a conventional resistance welding transformer). The tongs also may be water cooled.

Except for the very lightest units, the tongs or portable welding gun is suspended from an adjustable counterbalancing unit for convenient manipulation by the operator. The electrode force on most units is applied by air pressure, and a hydraulic booster is sometimes used. The tongs or welding gun can be constructed so that the electrode force is applied in a straight line, to avoid disturbing the alignment of the workpieces during the brazing operation.

Metal Electrodes

Selection of electrode material for resistance brazing depends on the electrical conductivity of the work metal or work metals to be joined, the design and dimensions of the joint, appearance requirements for the brazed product, susceptibility of the work metal to damage or marking, pressure needed on the work, production quantity of brazed joints, and cost of reconditioning or replacing an electrode.

Electrode materials for resistance brazing usually are made either of the standard electrode materials used in RSW or of carbon. Low-resistivity metallic electrodes can be used for resistance brazing metals that have high to moderate resistivity; electrodes or facings made of highly resistive materials must be used for brazing low-resistivity metals. In the former instance, most of the heat for brazing is generated within the workpieces; in the latter, most of the heat is generated within the electrodes and is conducted to the work.

The materials used most frequently in resistance brazing are RWMA class 2 (chromium copper), RWMA class 14 (molybdenum), and various grades of carbon-graphite and graphite. Other standard and special electrode materials are sometimes used for special applications. Properties of the standard metallic electrode materials are given in the article "Resistance Spot Welding" in this Volume.

Class 2 (chromium copper), which is a general-purpose resistance welding electrode material with a minimum electrical conductivity of about 75% IACS, is usually used to make electrode holders and shanks of faced electrodes for resistance brazing. In addition, RWMA class 2 electrodes frequently are employed for resistance brazing of work metals that have high to moderate electrical resistivity, because sufficient heat for brazing can be generated in the work metal itself.

Refractory metal electrodes include class 14 (molybdenum), which is intermediate in mechanical and electrical properties in that classification, having a nominal electrical conductivity of 30% IACS. This material and RWMA class 13 (tungsten), because of resistance to high temperatures and nonsticking characteristics, are the only common electrode materials that have long life in resistance brazing of copper and other highly conductive nonferrous metals. They are about equal in electrical conductivity.

Class 14 (molybdenum) is generally preferred because class 13 (tungsten) is harder to machine and is likely to develop radial cracks in service. For reasons of economy, common practice is to use facings, buttons, or inserts of the refractory metal instead of making the entire electrode of the refractory metal.

Special electrode materials are sometimes needed to meet the unusual requirements of specific resistance brazing applications. In Example 6, an electrode made of 60%Pt-40%Rh was used to provide long life in high-speed mass-production resistance brazing of insulated copper wires to copper terminal pads clad with BCuP-5. In this application, the electrode had to withstand the heating that was used for melting the polyurethane insulation on the copper wire, contact with molten BCuP-5 filler metal, and heating in air that was used for burning off insulation residues after each brazed connection was made. The electrode reached a temperature of 1700 °F.

Carbon Electrodes

Ordinarily, two general types of carbon electrodes are used in resistance brazing: carbon-graphite and electrographite (artificial graphite). These electrode materials are made by simultaneously heating and blending the finely divided raw materials with coal tar pitch, which serves as a binder.

The stock from which resistance brazing electrodes are made comprises a few of the many grades of carbon stock that are manufactured primarily for use in melting ferrous and nonferrous metals. When these grades are used as resistance brazing electrodes, their very high electrical resistivity, which is much higher than that of the frequently used metallic resistance brazing electrode materials (classes 2 and 14), permits the generation of a larger quantity of resistive heat than metallic electrode materials. Accordingly, carbon electrodes are used chiefly in resistance brazing of copper and other highly conductive (low-resistivity) work metals. They are less expensive than metallic electrodes in raw-material cost and fabrication cost, but wear more rapidly, oxidize in air at operating temperatures (red heat), and cannot withstand pressures as high as those used on metallic electrodes. Thus, the use of carbon electrodes in resistance brazing

Table 1 Properties of five carbon electrode materials used in resistance brazing

Electrode material	Electrical resistivity, Ω · in.	Scleroscope hardness	Flexural strength, psi (min)	Apparent density, g/cm³
Carbon-graphite, hard(a)	0.00080	70	3500	1.74
Carbon-graphite, hard, oxidation resistant(b)	0.00080	70	3500	1.75
Carbon-graphite, soft	0.00075	40	2400	1.57
Electrographite(a)	0.00042	50	2500	1.73
Electrographite, oxidation resistant(b)	0.00042	50	2500	1.75

(a) This type of carbon electrode material is also frequently used in air carbon arc cutting. (b) Similar to the electrode material listed immediately above, but impregnated with a small percentage of an oxidation retardant, usually an inorganic compound containing boron or phosphorus, for longer life

is restricted to brazing of small quantities of parts for which low electrode pressure is satisfactory.

Composition and properties of the commercial carbon electrode materials vary; no generally accepted industry standards and terminology exist. The properties of five grades that are generally typical of the materials used in carbon electrodes for resistance brazing are given in Table 1.

Carbon-Graphite. The first three types of electrode materials listed in Table 1 are called carbon-graphite. They are made from mixtures of finely ground petroleum coke (carbon) and natural or artificial (electro) graphite with coal tar pitch. These electrode materials are extruded in suitable shapes and heated in an electric furnace at about 1500 °F to develop the desired properties shown in Table 1. The heating converts about half of the pitch to carbon; the remainder is driven off as gases.

The final properties can be varied from those shown by changing the proportions of the raw materials, the particle size, and the time and temperature of heating. Increasing the ratio of graphite to carbon in the mixture, prolonging the heating time, or increasing the temperature lowers resistivity, hardness, and strength. Density is influenced chiefly by particle size.

Oxidation resistance for some grades of carbon-graphite is improved, for longer electrode life, by impregnating the cured material with a small percentage of an oxidation retardant, usually an inorganic compound that contains boron or phosphorus. In Table 1, oxidation-resistant grades 2 and 5 are the same as grades 1 and 4, respectively, except for the presence of an additive of this type.

Electrographite Electrode Material. Carbon-graphite grades and electrographite electrode materials are produced in the same way, except that the extruded mixture of raw materials is heated at about 4500 to 4900 °F, thus converting the mixture to graphite (called artificial graphite or electrographite). Properties of a representative material of this type and of an oxidation-resistant or long-life impregnated form of the same material are given in Table 1. Resistivity, hardness, and flexural strength of electrographite electrode materials are substantially lower than for the hard carbon-graphite electrode materials.

Selection of Carbon Electrode Material. The general types of carbon electrode material shown in Table 1 (carbon-graphite and electrographite) have almost completely replaced the formerly used "straight-carbon" types, which are made by heating finely ground petroleum coke and a binder at about 1500 °F. The straight-carbon grades typically have an electrical resistivity of about 0.0020 Ω · in. and a Scleroscope hardness of about 100. Machining of these materials is much more difficult and costly than that of carbon-graphite and electrographite.

The hard carbon-graphite types are general-purpose materials and are preferred for most resistance brazing with carbon electrodes because brazing temperatures can be reached with less current than when the less resistive electrographite types are used. Electrographite is preferred for resistance brazing of metals that have a high surface resistance, particularly when one of the metals being joined is steel or another iron-based alloy. Soft carbon-graphite material combines high heating capacity with a low tendency to produce local hot spots on the work metal, but has comparatively low wear resistance. The first and fourth grades in Table 1 are the same types of material ordinarily used in air carbon arc cutting.

In resistance brazing with carbon electrodes, nonuniform current flow and resultant local overheating of the electrodes can shorten electrode life excessively. To prevent this, carbon electrodes must be tightly fitted into matching tapered adapters or clamped securely to the electrode holder, making contact with the holder over as large an area as possible. Provision also must be made for adequate flow of cooling water in the electrode holder. In addition, some carbon brazing electrodes are electroplated or sprayed with a copper coating about 0.002 in. in thickness to reduce contact resistance against the electrode holder and minimize internal temperature buildup in the carbon electrode.

Design of Electrodes

Commercially available RSW electrodes are used where the design and dimensions of workpiece and joint permit. The electrode tip is machined, where necessary, to provide a tip shape and face dimensions suitable for the work. Carbon electrodes, depending on the size and design of the work, are in the shape of either standard metallic electrodes or flat blocks that may be several inches in length and width (size is usually limited by ability to obtain uniform heating and water cooling or by the current capacity of the power source).

When a high-resistance or long-wearing electrode material is required, it is often used in the form of facings, inserts, or buttons that are attached to electrode shanks or bodies made of a less-expensive material. Electrode design is developed to work with workpiece and joint design in eliminating the need for holding and locating fixtures or clamps wherever possible and in permitting rapid and easy loading and unloading of workpieces, filler metal, and flux. Provision for water cooling of electrode holders and electrodes generally is the same as in resistance welding (see the article "Resistance Spot Welding" in this Volume).

Arrangement of Electrodes

In most resistance brazing, the electrodes apply the brazing force and are arranged in line with the workpieces between them. This arrangement of electrodes, workpieces, and filler metal is shown in Fig. 3. In making some butt joints or where space limitations or work configurations do not permit the use of opposed electrodes, the electrodes are merely connected on either side of the joint to provide the brazing current; other means are used to apply the brazing force to the joint.

Press-fitted internal or external members can be resistance brazed to tubes or other workpieces of cylindrical shape without the application of force. Electrodes are attached to the ends of the tube, and the brazing current is not passed through the joint, rather, only through the tube. A filler-metal ring is preplaced at the joint, and the internal or external members to be attached are fluxed before they are press fitted. This technique has been used

Fig. 3 Arrangements for resistance brazing. (a) For small flat parts or small flat portions of larger components, using opposed water-cooled metal electrodes of the conventional resistance welding type. (b) For large flat parts, typically of a highly conductive metal such as copper, using opposed carbon block electrodes attached to water-cooled copper alloy electrode holders. (c) Fir flanged fins to a tube, using circular clamping electrodes

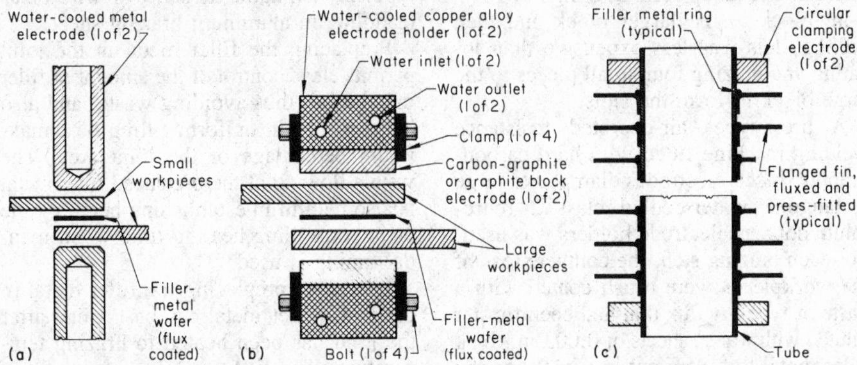

to a limited extent in resistance brazing of internal stainless steel baffle plates and external steel or copper fins to 1020 steel tubes in heat exchangers (Fig. 3c).

Filler Metals

Of the large number of filler metals available, only a few are used extensively in resistance brazing. Selection of filler metal for resistance brazing is similar to that for other brazing processes and is discussed in the articles on those processes and on the brazing of cast irons, stainless steels, aluminum alloys, and copper alloys in this Volume.

More attention is given in resistance brazing to selecting compatible filler metals having the lowest brazing temperature, because in resistance brazing, the maximum local temperature reached by the work must be kept as low as possible, while providing uniform heating of the abutting joint surfaces and the filler metal. Fluidity of the filler metal is not critical in most resistance brazing, because the filler metal is usually preplaced and the bond area is relatively large. The general types of filler metal usually selected for resistance brazing various classes of work metals are:

Work metal	Filler-metal alloys
Steel, stainless steel, heat-resistant alloys, copper, copper alloys, nickel alloys	Silver (BAg type)
Aluminum alloys	Al-Si
Copper and copper alloys	Cu-P

These types of filler metal all have relatively low brazing temperatures.

Silver Alloys. Of the silver alloy (BAg type) group of filler metals, the two most often used are BAg-1 and BAg-1a, which are free-flowing alloys that permit the use of low brazing temperatures. In addition, their narrow melting ranges (20 and 15 °F, respectively) prevent liquation, making them insensitive to variations in the rate of heating or cooling. Their narrow melting ranges are also advantageous in step brazing. For making corrosion-resistant brazed joints in stainless steel, BAg-3 and BAg-18 are preferred.

Aluminum-Silicon Alloys. The article "Brazing of Aluminum Alloys" in this Volume lists the aluminum-silicon alloys (BAlSi type) used in resistance brazing of aluminum alloys. These low-melting alloys are available in sheet or wire form; BAlSi-2 and BAlSi-5 are also available as cladding on aluminum brazing sheet. Temperature must be controlled with special care in brazing of the lower melting brazeable aluminum alloys, such as wrought alloy 6151 and cast alloys 443.0 and 356.0, because the melting range of these filler metals approaches closely or overlaps the melting range of the work metal.

Copper-Phosphorus Alloys. Filler metals of copper-phosphorus alloys (BCuP type) are widely used for resistance brazing of copper and copper alloys. They are low-cost, general-purpose filler metals and have the special advantage of being self-fluxing on copper, although flux is ordinarily needed when using them on copper alloys because phosphorus does not reduce metallic oxides other than those of copper. BCuP-5 is by far the most frequently used of the group.

In some applications on copper or copper alloys, the wider melting range, higher working temperature, and lower ductility of the BCuP filler metals make them less suitable than the BAg types. For example, in uses requiring higher ductility and corrosion resistance, silver brazing alloys sometimes are preferred to the BCuP filler metals for brazing of copper, in spite of the resulting need for fluxing. Difficulty in restricting flow of BCuP filler metals to joint surfaces may be a disadvantage.

Filler Metals for Step Brazing. When two or more joints are to be resistance brazed in sequence at points not widely separated on the same assembly, filler metals that differ in working temperature are selected, and the higher melting filler metal is used first. Some overlap of the brazing temperature ranges can be tolerated, depending on the work metal, the size of the parts, proximity of the joints, the presence or absence of heat sinks, and the closeness of control of temperature.

Three general-purpose filler metals were used for all resistance brazing in one plant, in which the work metals were copper, copper alloys, carbon steel, and electrical contact materials. The filler metals and their important characteristics were:

Filler metal	Temperature, °F Solidus	Liquidus	Brazing
BCuP-2	1310	1460	1350-1550
BCuP-5	1190	1475	1300-1500
BAg-1a	1160	1175	1175-1400

Step brazing of copper and copper alloys was done on a variety of assemblies using these filler metals in sequence in any combination of two, as needed. Brazing was done first with the filler-metal composition having the higher brazing temperature range.

Example 3. Use of Resistance Step Brazing. The lower terminal for a 1200-A circuit breaker (Fig. 4) was originally made by machining the main portion of the terminal from a solid $1^1/_2$-in.-thick bar of copper and then attaching the end plates by torch brazing. Nearly half of the copper was removed by milling when this method of manufacture was used.

In an improved method, two copper-phosphorus filler metals (BCuP-2 and BCuP-5) were used in resistance step brazing a simple but fairly massive assembly of four copper parts, for which the areas of the three joints were 12, $4^1/_2$, and $4^1/_2$ in.^{2} The terminal was made by resistance brazing and machining in three sequential operations:

- Brazing a small (3-by-4-in.) block to a large block (section A-A in Fig. 4)
- Milling three slots $^1/_4$ in. wide by $^3/_4$ in. deep across the full width of the subassembly
- Brazing two end plates to the subassembly simultaneously (section B-B in Fig. 4), using a lower temperature brazing filler metal than in the initial brazing

Fig. 4 Circuit-breaker subassembly

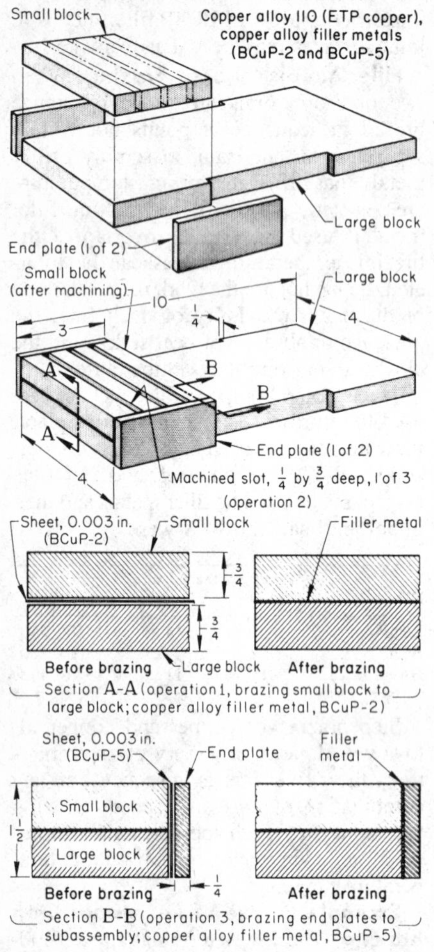

Machine	Press-type, air-operated, 50-kV · A resistance welding machine
Loading and unloading	Manual
Timing	Manual
Electrodes	Hard carbon-graphite blocks, 3/4 in. thick, clamped to RWMA class 2 (chromium copper) electrode holder
Precleaning	Solvent degreasing
Production quantity	30 assemblies (approx)

	Operation 1	Operation 3
Filler metal(a):		
AWS classification	BCuP-2	BCuP-5
Flow temperature	1350 °F	1300 °F
Working temperature	1500 °F	1325 °F
Flux(b)	Type 3A	Type 3A
Electrode size, in.	3 by 4	2 by 3
Electrode force, lb	30	30
Brazing current, A	40	40
Heating power, kV·A	40	25
Heating time, s(c)	50	40
Hold time, s(d)	15	15

(a) Foil 0.003 in. thick. (b) Water-diluted paste was brushed on contact area of workpieces just before brazing. (c) Current was turned off when filler metal was seen to melt. (d) Electrodes were released and work was removed when filler metal was seen to solidify.

Changing to the improved method reduced the amount of copper machined away to about one eighth of the amount originally removed and substantially lowered material and labor cost. Brazing the 3-by-4-in. block to the larger block and then milling slots was less expensive than locating and brazing four small pieces to the large block to form the slots.

A press-type, air-operated resistance welding machine fitted with hard carbon-graphite block electrodes clamped to large rectangular water-cooled class 2 (chromium copper) electrode holders was used. For each brazing step, the contact areas of the workpieces were brush coated with a paste of type 3A flux that had been further diluted with water, sheets of 0.003-in.-thick filler-metal foil were put in place, and the assembled parts were placed between the carbon electrodes. The flux was used to prevent oxidation of the work surfaces during the heating, although BCuP-2 and BCuP-5 are usually considered self-fluxing in brazing copper.

Filler metal for the initial brazing operation was BCuP-2, which flows freely at 1350 °F (flow temperature) and was used at a working temperature of 1500 °F; filler metal for the final brazing was BCuP-5, which flows freely at 1300 °F and was used at 1325 °F. The operator turned off the heating current when filler-metal flow was observed, and released the electrode pressure and removed the work when the filler metal was seen to solidify. A total of about 30 lower terminal assemblies were resistance brazed as replacement units for circuit breakers that had failed in service.

Form and Application of Filler Metal. Filler metal for resistance brazing is usually preplaced in the joint in the form of foil, wire, ribbon, or a washer about 0.003 to 0.005 in. thick. The shape and size generally cover the approximate contact area of the joint. Filler metals not available in these preforms can be preplaced in the joint in the desired amount as a powder, paste, or preform that may also contain flux. High-speed injection can be used in mechanized operations, if precautions are taken to prevent separation and settling and to ensure the addition of a sufficient quantity of the filler-metal paste.

One technique for preplacing powdered filler metal is to dip the prefluxed parts into the powder, using flux of a viscosity selected to pick up the desired thickness of filler-metal powder. In some resistance brazing applications, one or both workpieces may be plated, clad, or coated with a metal or alloy that serves as a filler metal.

The silver brazing alloys (BAg type) and the copper-phosphorus alloys (BCuP type), as described in the preceding sections, are commercially available as sheet, wire, powder, or custom-made preforms. The aluminum-silicon brazing alloys (BAlSi type) are available as sheet or wire or as cladding on aluminum brazing sheet.

Preplacing the filler metal in the joint permits close control of the amount of filler metal used, thus avoiding waste, and also helps to obtain uniform filling of a maximum percentage of the joint area. The visible flowing of the preplaced filler metal is also helpful in establishing heating time or in controlling heating time when manual timing is used.

Instead of preplacing the filler metal in the joint, filler metal can be applied after the joint has been heated to brazing temperature. The filler metal in the form of ribbon, wire, or rod is fed to the joint, being drawn into the interface by capillary action. This method is used only rarely in resistance brazing.

Fluxes and Cleaning

A flux is used in almost all resistance brazing. It serves the same purposes in resistance brazing as in other brazing processes: providing a coating to prevent or minimize oxidation of the work metal during heating; dissolving oxides that are present or that may form during heating; and assisting the molten filler metal in wetting the work metal to promote capillary flow. The flux in resistance brazing, however, has the additional function of serving as an electrical conductor to permit passage of the brazing current through the joint; most dry fluxes are nonconductors and must be mixed with water in order to conduct current.

Application. The flux is usually applied as a dilute water-based paste shortly before the parts and filler metal are assembled for brazing. Arcing and an explosion may occur if the paste is not a thin, uniform layer and free from lumps. If the flux should dry before brazing is started, it may be possible to restore electrical conductivity by moistening it, but results are not always consistent. Once melted, the flux remains conductive. If the filler metal is in powder form, flux can be combined with it in a fine-particle paste.

Selection. The same fluxes are used for resistance brazing as for other brazing processes on the same work metal. Selection and properties of fluxes are described in this Volume in the article "Torch Brazing of Steels" and articles describing brazing of specific types of alloys. Type 3A fluxes are general-purpose fluxes suitable for most metals that are commonly resistance brazed (although type 4 flux is needed for copper alloys that contain tin, aluminum, or silicon); type 1 fluxes are used on aluminum alloy work metals.

Brazing Without Flux. The two general situations in which a flux is not used in resistance brazing are brazing in a vacuum or protective reducing gas or inert atmosphere (see the article "Furnace Brazing of Steels" in this Volume) and brazing of copper with a BCuP filler metal. A flux is not ordinarily needed in resistance brazing of copper when a BCuP filler metal is used, because these filler metals are self-fluxing on copper by virtue of their phosphorus content. Flux was used, however, in step brazing of copper with BCuP-2 and BCuP-5 filler metals in Example 3.

In some special situations, brazing can be done in air without the use of flux when brazing is done immediately after cleaning or mechanically abrading the joint surfaces. Special cleanliness of the workpieces made it possible to avoid the use of flux in the automatic resistance brazing of leaded copper alloy parts in Example 5, in which the filler metal was BAg-1.

Cleaning. As in other brazing methods, unless the work metal at the joint is free from grease, oil, dirt, and interfering oxide coatings, chemical or mechanical cleaning must be done before resistance brazing to permit wetting by the molten filler metal and capillary flow. Removal of flux residues after resistance brazing, which is normally necessary (as in other brazing processes), is accomplished by washing in hot water or, if this is not effective, by chemical or mechanical means.

A production technique frequently used to remove flux residue from brazed copper parts is to immerse or flood them in cold water while the parts are still hot (above 500 °F) but below the solidus temperature of the braze metal. Removal of flux residues from braided cable and other parts where crevices are present is especially difficult and not always completely effective.

Joint Design

Joints for resistance brazing are usually lap joints, although other joint arrangements are used where lap joints are not suitable. Workpieces can have a wide variety of configurations, but the joint design must permit contact surfaces, usually flat or conforming, to be pressed together, ordinarily between electrodes, to make the joint.

Special-shaped workpieces that are joined by resistance brazing include crossed wires, wires or tubes laid flat against the surface of a second part, stranded or braided wires or cables, shaft ends or collars butt-joined to flanges or rings, armature leads of various shapes inserted into slots in commutator bars, overlapping-spot-brazed flat stock, attachments to the interior of cylindrical shapes or in other types of recesses, and solidified joints of stranded or braided conductor cable to terminals.

Lap joints are preferred to butt joints because they can be made with greater joint strength; brazed joints are stronger in shear than in tension or bending. In lap joints, overlap should be at least three times the thickness of the thinner member for full joint strength or at least $1^1/_2$ times this thickness to avoid a significant loss in electrical conductivity across the joint.

Where practical, joint design is coordinated with electrode design to make the workpieces self-aligning (or, ideally, self-nesting), thus minimizing the need for special fixtures. Joint design must permit the workpieces to move when the filler metal melts and flows and should, at the same time, maintain proper workpiece alignment until the filler metal has solidified completely.

The contacting surfaces of the workpieces must be designed to fit closely together during brazing to avoid local overheating, permit capillary action, and minimize voids in the brazed joint. Forming or machining of the areas to be joined is sometimes necessary for proper fit. The design also must permit the application of the electrode force without distortion of the workpieces. In addition, joint design, in conjunction with electrode design and arrangement, should permit easy assembly of workpieces and filler metal, convenient fluxing (if a flux is used), and easy loading and unloading of the machine.

Example 4. Design of Workpieces and Electrodes for Self-Fixtured Resistance Brazing. In resistance brazing a 0.094-in.-thick 1025 steel hub to a $^3/_8$-in.-diam low-carbon steel shaft in a press-type resistance welding machine, the workpieces and the electrodes were designed for self-fixturing, as shown in Fig. 5. Resistance brazing was the only available method suitable for joining parts of the exact design shown in Fig. 5, because of the need to avoid heat damage to the fiber gear. If redesign of the shaft for annular projection welding had been permissible, that method would also have been suitable.

For joint strength, the shaft had a collar to provide a flat annular contact surface with a $^1/_4$-in. ID and a $^3/_4$-in. OD. The $^1/_4$-in.-diam chamfered locating tip on the bottom of the shaft extended through the clearance hole (diametral clearance of $^1/_{32}$ in.) in the hub and projected about 0.015 in. into a clearance recess (diametral clearance of $^1/_{32}$ in.) in the lower electrode. The flat upper surface of the collar on the shaft provided an annular contact surface for the upper electrode, and the upper portion of the shaft fit into a clearance hole (diametral clearance of $^1/_{32}$ in.) in the upper electrode.

In preparing the assembly for brazing, the operator first applied a water-based paste of type 3A flux to the shaft by dipping and then slipped a washer of BAg-1 filler metal and the hub over the locating lower tip of the shaft, fitting them against the lower surface of the collar. Then, holding the assembly in an upright position by the hub, the operator inserted the shaft into the recess in the upper electrode and lowered the upper electrode at a controlled rate until the electrodes were closed on the assembly at low pressure. Pushing

Fig. 5 Shaft-and-hub assembly. Workpieces and electrodes were designed for self-fixtured resistance brazing.

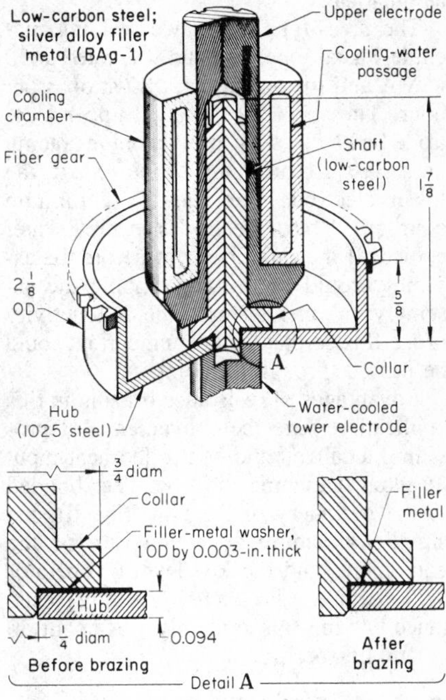

Machine	Press-type, air-operated, 50-kV · A resistance welding machine, rated at 80 strokes per hour
Loading and unloading	Manual
Filler metal	BAg-1 preform washer, 0.003 in. thick by 1 in. OD
Flux	Type 3A water-base paste
Electrodes:	
Upper	RWMA class 2 (chromium copper), $1^3/_{16}$ in. diam, class 14 (molybdenum) facing, water cooled, recessed to hold shaft
Lower	RWMA class 2 (chromium copper), $^9/_{16}$ in. diam, water cooled, recessed to hold pilot tip of shaft
Timing method	Electronic
Production rate	80 assemblies per hour

the start switch on the welding machine actuated the brazing force and current; the brazing sequence was completed automatically.

The press-type resistance welding machine was air operated and was rated at 50 kV · A and for operation at 80 strokes per hour. The specially designed upper electrode had a class 14 (molybdenum) facing on a class 2 (chromium copper) shaft; the lower electrode was class 2 (chromium copper). Both electrodes were water cooled. If the flux dried out before the assembly could be brazed, dipping the assembly in water restored the conductivity of the flux so that the brazing current could be passed through the joint.

Advantages of resistance brazing in this application were the convenience of providing localized and controlled heat input to avoid damaging the fiber gear bonded to the periphery of the hub (Fig. 5), the simplicity and ready availability of the equipment, and the low level of operator skill required. Production rate for resistance brazing this assembly was about 80 assemblies per hour.

Special Techniques Used in Resistance Brazing

Characteristics common to nearly all resistance brazing operations include localized heat input; passage of the heating current through the joint, which is held together under pressure; and use of standard or modified resistance welding machines. In most other respects, resistance brazing applications vary widely, ranging from the simplest manual operations on small quantities of noncritical parts to fully mechanized high-speed mass production of brazed assemblies to rigid standards. Workpieces vary widely in size, shape, and type and condition of material, and often special techniques and procedures are needed.

Step Brazing. This process entails two or more brazed joints made in succession on the same assembly, using filler metals that have progressively lower working-temperature ranges. It is conveniently accomplished by resistance brazing, in which the heat input is localized and can be controlled accurately. Step brazing is used extensively in making circuit breakers and other types of electrical switchgear and power distribution equipment; Example 3 describes step brazing of a circuit-breaker subassembly.

Use of Fast Follow-Up. The inertia of the moving mass of the electrode holder and electrode can prevent electrodes and workpieces from responding quickly enough to the melting of the filler metal, especially when small parts are being brazed. The delayed response and rebound of the electrode holder and the workpieces can then cause expulsion of too much of the filler metal, producing a weak brazed joint bonded over only a small percentage of its area.

Fast follow-up is especially important in resistance brazing with filler metals (such as BAg-1 and BAg-1a) that flow at a temperature only 15 to 20° above the temperature at which they start to melt. For applications using such filler metals, resistance welding machines are equipped with special low-inertia, low-friction electrode holders that permit rapid follow-up by the electrode when the filler metal flows.

Example 5. Use of a Low-Inertia Upper Electrode Holder for Fast Follow-Up in Automatic Resistance Brazing. A spring-loaded fast-follow-up electrode holder was used in an air-operated automatic resistance welding machine in brazing a post-and-flange assembly. The binding post and flange shown in Fig. 6 were resistance brazed with BAg-1 filler metal (solidus of 1125 °F, liquidus of 1145 °F) used as a preformed ring. Because of the short time interval between melting and flowing of this filler metal, a special low-inertia, low-friction upper electrode holder was used to provide fast follow-up and to minimize rebound. The spring-controlled action of this electrode holder kept the two workpieces in intimate contact with the filler metal throughout the melting-and-flowing sequence and resulted in the proper capillary action and maximum filling of the brazed joint.

Brazing was done on a 75 kV · A automatic resistance welding machine with a 440-V input, a 4- to 7.33-V output, and synchronous heat controls. A hollow lower electrode was shaped to hold the post. The filler-metal preform and the flange rested on the collar of the binding post. The upper electrode was 5/8 in. in diameter and flat. Both electrodes were of RWMA class 2 material (chromium copper).

First, the two parts were cleaned by acid dipping and were assembled with a preform of 0.032-in.-diam BAg-1 wire and flux. The assembly, which was self-fixturing, was mounted by hand in the lower electrode. Force, voltage, and time were controlled automatically. The assembly was clamped with a force of 190 lb and heated with a secondary voltage of 6.9 V for 27 cycles, or a little less than half a second. Electrode cost per assembly was negligible.

Fig. 6 Post-and-flange assembly

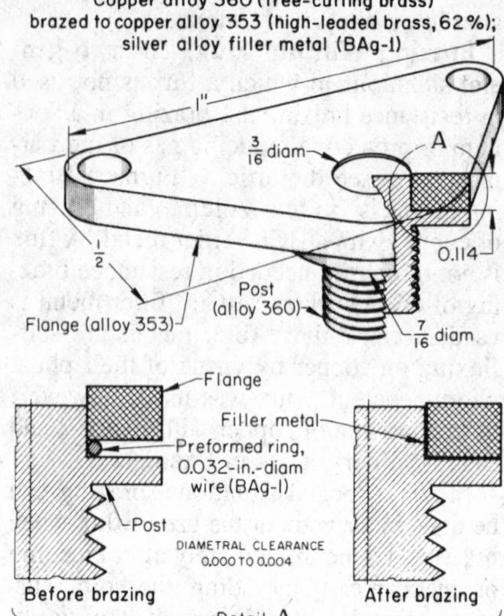

Machine	Press-type, air-operated, 75-kV · A automatic resistance welding machine, synchronous controls, equipped for fast follow-up
Loading and unloading	Manual
Filler metal	Preformed ring of 0.032-in.-diam BAg-1 wire
Flux	None
Precleaning	Acid dip both parts
Electrodes	RWMA class 2 (chromium copper), 5/8 in. diam(a)
Electrode force	190 lb
Brazing voltage	6.9 V
Heating time	27 cycles
Production rate	278 assemblies per hour

(a) Upper electrode had a flat face; lower electrode was recessed to hold post.

The flange had to bottom on the shoulder of the binding post and be perpendicular to it within 1°. The filler metal was required to show on the lower side of the flange all the way around the joint. The post was made of copper alloy 360, and the flange was made of copper alloy 353. Arc or resistance welding processes are not ordinarily suitable for joining leaded brasses of such high lead content (nominally 2% for alloy 353 and 3% for alloy 360), because the low-melting lead concentrates in the grain boundaries during exposure to welding temperatures, causing cracking and weakening of the joints. With the use of resistance brazing and a small and closely controlled heat input, these difficulties were avoided.

The heat for brazing was generated chiefly in the filler metal and the regions of the workpieces in the immediate vicinity of the joint. Electrical conductivities of the materials involved were, in % IACS, RWMA class 2 electrodes, 75 minimum; BAg-1 filler metal, 28; copper alloy 360 (post), 26; and copper alloy 353 (flange),

26. The use of class 2 electrodes for this operation allowed localization of heating in the joint and use of a low total heat input.

Resistance brazing was preferred to torch brazing because of better controllability of heat input, greater speed, and lower operator-skill requirements. This process was preferred to induction brazing because of the higher cost of equipment for induction brazing and the availability of resistance welding equipment in the plant.

Brazing Plastic-Coated Wire. To permit wetting of the entire joint area by the molten filler metal, organic coatings, grease, oil, oxides, and other types of nonmetallic materials must ordinarily be removed from the contact surfaces of workpieces to be resistance brazed. Some types of organic coatings, however, can be removed readily and completely by electrode heat and pressure as part of the brazing sequence.

Example 6. Use of Capacitor-Discharge Energy Pulses To Remove Insulation and Resistance Braze. A special energy-supply system and a dual electrical circuit were used to remove polyurethane insulation from copper wire with the electrode, braze the stripped wire to a terminal, and burn off residues from the electrode after brazing each joint. The brazing electrode was made of a precious metal alloy to withstand this heating sequence and to resist alloying with copper or the BCuP-5 filler metal that was used. In addition, fast follow-up was provided by a specially designed spring-loaded decoupling system in a gravity-operated brazing head.

Resistance heating of the brazing electrode by a preliminary capacitor-discharge energy pulse was used to melt insulation from the contact surfaces of polyurethane-coated copper wires in resistance brazing them to terminal pads in a ferrite-core storage unit for a computer (Fig. 7). An energy pulse from a capacitor was also

Fig. 7 Setup for capacitor-discharge resistance brazing of a ferrite-core memory frame for a computer. Total of 392 connections. Pulsed energy from capacitors was used to melt the polyurethane insulation from the wire during brazing and to burn residues off the 60%Pt-40%Rh electrode tip between brazing operations.

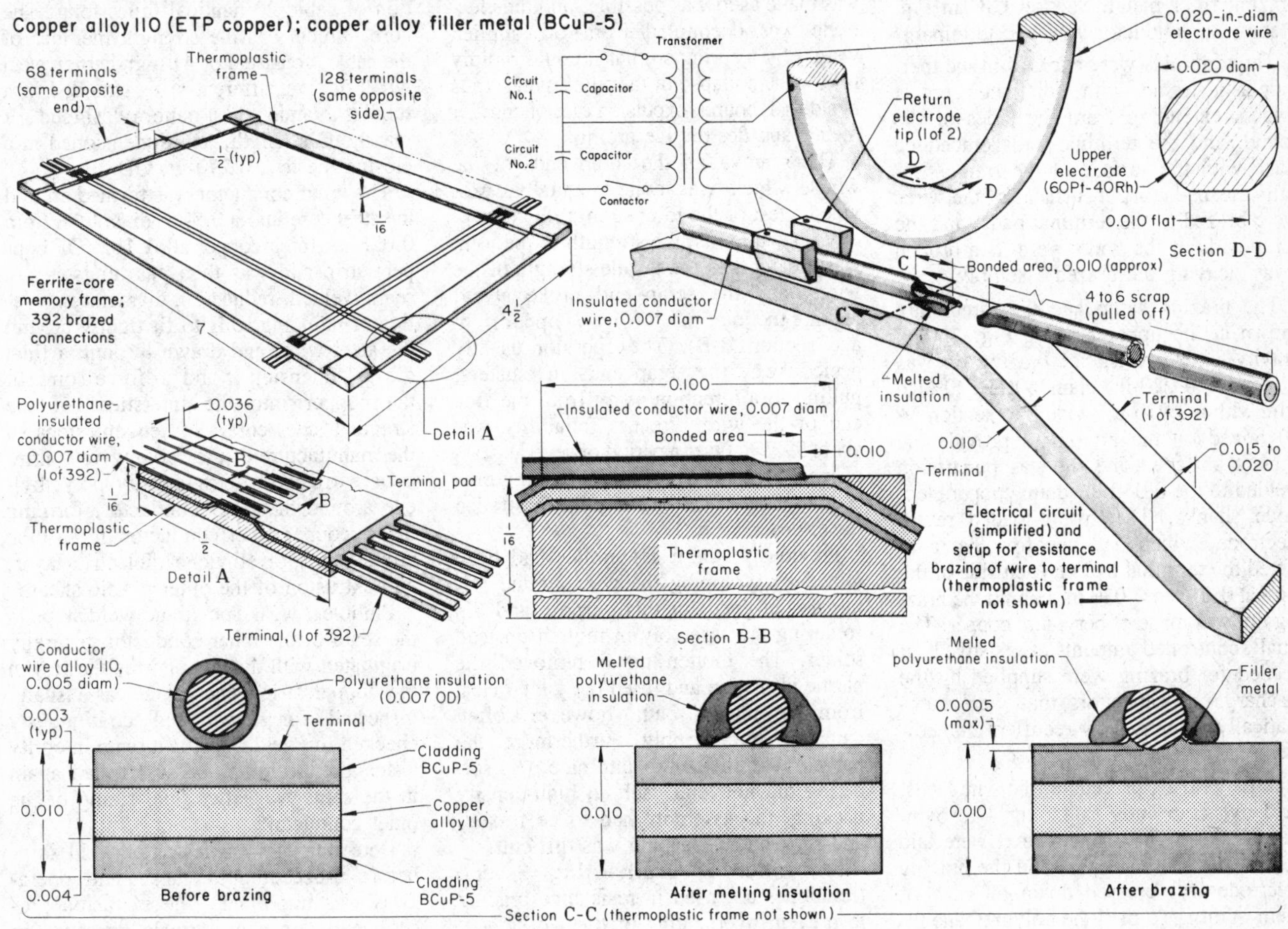

Machine	Bench-mounted capacitor-discharge resistance welding machine
Capacitance and voltage of power supplies (capacitors):	
Circuit No. 1	735 μF, approx 128 V
Circuit No. 2	735 μF, approx 82 V
Flux	None
Filler metal	0.003-in.-thick BCuP-5 cladding on terminals
Precleaning	None
Brazing electrode	0.120-in.-radius loop of 0.020-in.-diam 60Pt-40Rh wire with a 0.010-in. flat honed on bottom surface
Return electrodes	Beryllium copper shoes
Electrode force	100 g (weight)
Electrode-head mass	20 g
Energy and duration of pulses(a) for:	
Wire stripping	5-7 W·s, 2-3 ms
Brazing	2-3 W·s, 2-3 ms
Electrode cleaning	5-7 W·s, 2-3 ms
Postbraze cleaning	None

(a) Data are approximate values for wire with conductor diameter of 0.002 to 0.005 in.

passed through the retracted electrode between brazing operations to burn residual insulation material from the electrode tip. The brazing electrode was made of 60%Pt-40%Rh to withstand a temperature of 1700 °F in continuous high-speed production and to resist alloying with copper or filler metal. Figure 7 shows the general construction of a single two-dimensional ferrite-core memory frame, and detail A is an enlarged view of a portion of the frame.

The copper conductor of the wires was 0.005 in. in diameter (other conductor sizes, down to 0.002 in. in diameter, were used in other frames of this type), and the overall diameter for the wire shown, including the insulation, was about 0.007 in. The terminals were punched from 0.004-in.-thick copper strip clad on both sides with a 0.003-in.-thick layer of BCuP-5 brazing filler metal. Section C-C in Fig. 7 shows a conductor wire and terminal.

The terminals were encased in the thermoplastic frame, with their outer contact fingers extending from the sides of the frame, and the terminal pads embedded flush with the surface of the frame. Locally cleaned contact surfaces of the wires were brazed to the terminal pads, and the scrap ends of the wires were then broken away, leaving the brazed assembly.

The brazing setup and the circuit are shown at the upper right in Fig. 7. The gravity-loaded resistance brazing tip was a loop of 0.020-in.-diam 60%Pt-40%Rh wire with a 0.010-in.-wide flat (section D-D) honed on the bottom to provide the contact surface. The tip was percussion welded to the 0.040-in.-diam copper electrode shanks. The spring-loaded return electrode, which had two tips that contacted the terminal pad on each side of the wire at 0.010 to 0.030 in. behind the brazing tip, was made of beryllium copper. The small, controlled amounts of resistive heat needed for brazing were supplied by the discharge of capacitors that were automatically recharged by rectifiers immediately before each energy pulse.

In the brazing procedure, the wire leads, held by scrap ends extending 4 to 6 in. beyond the edges of the frame, were laid across the terminal pads, and the brazing electrode was pressed down on each of them in turn. To melt the polyurethane insulation, the switch was closed in the circuit identified in Fig. 7 (top center) as circuit No. 1. This passed a current through the loop-shaped brazing electrode, producing enough heat to break down and strip the insulation (section C-C). Electrical contact was thus firmly established between the brazing electrode and the wire lead and between the lead and the terminal.

To initiate brazing, the switch in circuit No. 2 was closed, causing a pulse of current to flow from the electrode through the wire and terminal and back through the return electrode. This pulse generated enough heat at the wire-terminal interface to melt the filler metal at the contact area to a depth of about 0.0005 in. and produce a fillet on each side of the wire (Fig. 7, section C-C). The brazing electrode was cleaned before each braze by pulsing current through the tip (circuit No. 1 in Fig. 7) to vaporize polyurethane residues. Timing of the entire sequence of operations was done with a solid-state timing device.

The sudden melting of the filler metal was too rapid for a conventionally mounted welding head to follow. The electrode mass was made as low as possible, and the electrode was decoupled from the applied weight by a spring so that it could rapidly follow the collapse of the filler metal, thus avoiding rebound, expulsion of molten filler metal, and destructive arcing.

The removal of the scrap ends of the wires, which was done manually, provided a test of the joint strength. On properly brazed joints, the strength of the joint was greater than the tensile strength of the wires, and the scrap end broke away, leaving the joint intact (Fig. 7, upper right and section B-B). The operator usually peeled away the scrap ends in clusters, pulling in a direction away from the free end of the wire. Process reliability was about 99.7%. Over a period of several years of mass production using this joining procedure, none of the brazed joints has failed in service.

Previously, these assemblies had been joined by wrapping the wire around terminals designed for this purpose and dip soldering the assembly in molten tin-lead solder. The molten metal removed the plastic insulation and fused the joints. Heat from the molten bath, however, often damaged the assembly. Furthermore, the process was difficult to automate. Dip soldering did not lend itself to high-density packing, the assembly had to be fluxed, and subsequent cleaning was difficult.

Prevention of Overheating. Several techniques are used in resistance brazing to prevent overheating of the workpiece. A single energy pulse from a capacitor was used in Example 6 to provide the small, closely controlled rapid heat input needed to resistance braze copper wires 0.002 to 0.005 in. in diameter to flat terminals. In Example 2, the cross-wire joint and precision timing and current control were the major factors in limiting the heat input to the workpiece. In Example 5, joint design and the relative electrical conductivities of the electrodes, workpieces, and filler metals made highly localized heating and the use of low total heat input possible. In the following example, small overlapping spot brazes were used to avoid annealing of adjacent high-carbon steel wires in resistance brazing of repair patches to the copper inner-conductor tube of ocean cable.

Example 7. Use of Overlapping Spot Resistance Brazing. Weld defects or unwelded sections (weld skips) in the seam welded inner-conductor tubes of broadband-transmission ocean cable were repaired by resistance brazing patches over the defects or unwelded sections, as shown in Fig. 8. The ocean cable was manufactured in a process that was operated continuously for over 80 h to produce a section of cable 20 nautical miles long. The core, which was the strength member of the cable, consisted of 41 high-carbon steel wires of five different sizes stranded in a tubular strander, in a pattern designed for maximum strength, and dimensioned in a closing die to a 0.290-in. OD.

The inner conductor was formed around the steel core into a 0.5-in.-diam tube from 0.023-in.-thick copper alloy 102 (OF copper) strip and was then gas tungsten arc seam welded, reduced in diameter in a series of reducing rolls to fit tightly around the steel wire, and drawn through a final die to dimension it and to force some of the copper into the interstices of the stranded steel core. Subsequent steps in the manufacture of the cable were continuous extrusion of low-density polyethylene around the inner conductor, forming of the copper outer conductor into a tube around the polyethylene dielectric layer, and extrusion of the outer plastic sheath.

Pinholes, weak spots, and weld skips in the seam of the inner conductor generally originated with defects or irregularities in the copper strip. The seam was visually inspected for flaws and continuously checked for weld skips by a seam integrity tester after welding and was tested again in the same way after final sizing of the inner conductor.

Detection of weaknesses or weld skips before proceeding to subsequent operations was important. The extrusion process built up considerable pressure in trapped air inside the welded inner conductor because of the heat generated and the choking effect of the 20-nautical-mile length. Where flaws or skips existed in the welding of the inner conductor, this compressed air blew bubbles in the extruded polyethylene, creating voids in the dielectric and impairing the electrical function

Fig. 8 Ocean cable made in continuous production and copper alloy patch. To repair gas tungsten arc seam welding defects in the copper alloy inner conductor tube, the copper alloy patch was overlapping spot resistance brazed to the tube.

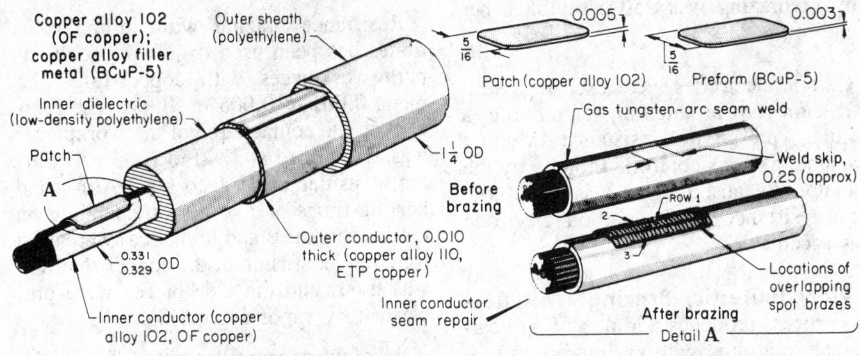

Machine	Portable resistance welding machine, 50 kV · A	Electrode force	100 lb
Welding head	(a)	Method of applying force	Air
Filler metal	BCuP-5, 0.003 by 5/16 wide, length to suit defect length	Brazing current	50 000 A (typical)
Flux	None	Brazing voltage (open circuit)	30 V
Precleaning	Wipe with trichlorethylene	Heating time	2 cycles
Electrodes	(b)	Hold time	30 cycles (typical)
		Timing method	Electronic
		Joint reliability	98%

(a) Mounted in fixed ways; upper electrode holder, cantilever mounted; lower electrode holder, stationary. (b) Class 1 (cadmium copper), 5/8 in. diam, with class 12 (copper-tungsten) conforming tips

of the cable. Defects that were not detected and repaired before extrusion of the inner dielectric could necessitate repair or removal of the defective section of cable later in the manufacturing process, with greater difficulty and cost. Only four repairs of any kind were permitted in a 20-nautical-mile length of cable. The average number of inner-conductor repairs per cable length was 2.3. Reliability of these resistance brazed patch repairs was 98%.

Finding a technique for repairing defects in the inner-conductor seam was complicated by the need for a ductile patch and by the fact that excessive heat might anneal the steel core wires and thus weaken the cable unacceptably. Techniques that were tried and found unsatisfactory used tin-lead solder patches, epoxy patches, and resistance welded gold-plated copper patches; the last method overheated the steel core wires.

The technique that was finally adopted consisted of resistance brazing 0.005-in.-thick copper alloy 102 patches over the defects, using 0.003-in.-thick strips of BCuP-5 filler metal and a resistance welding machine. A strip of filler metal and a copper patch, each 5/16 in. wide and of a suitable length, were placed over each defect (Fig. 8). Then the patch was stitched all over with overlapping spot brazes, one braze at a time. The individual spot brazes were about 1/8 by 3/16 in. and were spaced on 0.030- to 0.040-in. centers (Fig. 8, detail A). Afterward, the patch was smoothed with emery paper to keep the diameter of the inner conductor from exceeding 0.333 in. and was polished with crocus cloth.

The electrode tips were machined to conform to the curvature of the inner conductor. Both electrodes were made of 5/8-in.-diam class 1 material (cadmium copper) faced with class 12 material (copper tungsten). The upper electrode had a groove with a 0.167-in. radius, and the face was reduced to 1/8 in. wide. The lower electrode also had a groove with a 0.167-in. radius, but the face was not reduced in width; thus, the lower electrode provided maximum support for the conductor and a large contact area. Because the lower electrode had a much larger contact area than the upper electrode, it functioned only as a return electrode. Brazing force was 100 lb. The current and heating time were selected to provide just enough heat to melt the brazing filler metal completely.

Solidified Joints. Resistance brazing is frequently used to join braided or stranded electrical conductors, especially in the larger sizes (rated to carry currents of about 60 A or more), to each other or to other types of conductors. Braided conductors of smaller sizes are conveniently joined by resistance welding, thus avoiding the use of filler metal and flux. The currents needed for resistance welding of heavy braided or stranded conductors are prohibitively high, however, and conductor cables rated for 60 A or more are more conveniently joined by resistance brazing of solidified joints.

This type of joint has better resistance to corrosion than mechanically crimped connectors and, in addition, can be machined and formed much like a single continuous piece of metal. As shown in Example 8, the solidified end of the braided or stranded cable provides the filler metal for resistance brazing. Where the shapes of the cable and the other workpiece permit, solidification and brazing can be done simultaneously to minimize exposure to heating. Carbon electrodes, simpler to prepare and less expensive than high-resistivity metallic electrodes, often are used in short-run resistance brazing of solidified joint applications, where their low resistance to wear is not important.

Example 8. Use of Molybdenum-Faced Electrodes To Permit Short Heating Time. In resistance brazing the solidified end of a 60-A braided copper

Fig. 9 Solidified end of a 60-A braided DLP copper conductor cable and ETP copper terminal.

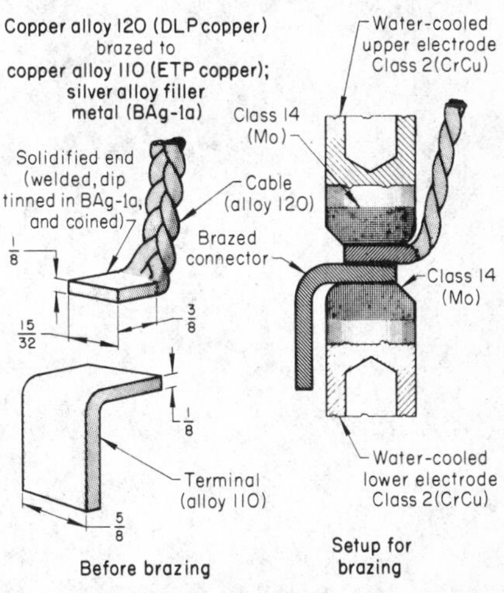

Machine	Press-type 75 kV · A resistance welding machine, bench mounted, air operated, with automatic timing
Loading and unloading	Manual
Filler metal	BAg-1a(a)
Flux	AWS type 3A
Precleaning	Terminals degreased and bright dipped
Electrodes	RWMA class 2 (chromium copper) with 1/2- by 3/8-in. class 14 (molybdenum) facing, water cooled
Electrode force	350 lb
Brazing current	15 000 A
Heating time	5 s
Hold time	20 cycles
Post treatment	Water quenching(b)
Production rate	250 assemblies per hour

(a) Filler metal was applied to the end of the braided cable by dip tinning to produce a solidified end. (b) Assemblies were dropped into cold running water immediately after brazing.

conductor cable to a copper terminal, the heating time had to be minimized to avoid excessive annealing of the terminal and excessive penetration of filler metal into the braided cable. The end of the braided cable first was resistance welded using class 14 (molybdenum) electrodes to fuse the strands partially and to consolidate the braid. It was then dip tinned with BAg-1a brazing filler metal, coined to the desired shape and size, and finally resistance brazed to a solid terminal or other workpiece.

By using water-cooled class 2 (chromium copper) electrodes with a class 14 (molybdenum) facing, the joint was made with a heating time of 5 s. The braided cable, with its resistance welded, dip tinned, and coined solidified end, and the terminal are shown at the left in Fig. 9. The completed assembly, ready for unloading, is shown between the electrodes at the right in Fig. 9.

A bench-mounted resistance welding machine was used. Timing of the brazing sequence was automatic. The production rate for brazing was 250 assemblies per hour.

Carbon electrodes were used in a similar short-run production, employing a manual, press-type resistance welding machine that was controlled totally by the operator. Because the run was short, low wear resistance of the carbon electrode was acceptable.

Alloy (Eutectic) Brazing. This fluxless process produces a joint by first heating the joint area with an interposed thin piece of metal (shim) to form a liquid phase and then extruding this liquid along with surface oxides from the joint cross section. Because resistance brazing in a resistance welding machine offers localized, controlled heat input to the joint and controlled application of brazing force, it is well suited for alloy, or eutectic, brazing.

Resistance brazing with carbon electrodes has been used to join 1350 aluminum workpieces with copper-foil filler metal 0.001 to 0.005 in. thick. The pressure on the contact area of the workpieces was maintained at 1200 to 2000 psi. Current densities of 2500 to 4000 A/in.^{2} and heating times of 3 to 60 s (depending on the foil thickness and joint area) were used for complete fusion of the copper. No flux was used, and the workpieces were precleaned by vapor degreasing.

All of the copper filler metal alloyed with the aluminum work metal to form a molten phase, most of which was squeezed out of the joint. The brazed joints, including butt joints, had electrical conductivity essentially the same as that of the work metal.

Dip Brazing of Steels in Molten Salt*

DIP BRAZING in molten salt is also referred to as salt-bath dip brazing and molten chemical-bath dip brazing. In this process, the assembly to be brazed is immersed in a bath of molten salt, which provides the heat and may supply the fluxing action for brazing as well. The bath temperature is maintained above the liquidus of the filler metal, but below the melting range of the base metal. This article describes the application of dip brazing to carbon and low-alloy steels with silver alloy, copper-zinc alloy, and copper filler metals.

Advantages and Limitations

The advantages of salt-bath dip brazing are:

- Time for heating is about one fourth of that required in a controlled-atmosphere furnace.
- A protruding joint can be selectively brazed by partially immersing the assembly.
- A cocoon of frozen salt forms instantly around the cold assembly when it is immersed in the molten salt, which usually prevents premature melting of the brazing filler metal by providing a temporary insulator.
- By selection of an appropriate salt composition, heating and fluxing of the work can often be combined in a single step, although flux can be applied to the joint and dried before brazing.
- Brazing can usually be combined with carburizing or hardening, without the necessity for a separate reheating operation.
- More than one assembly or joint in an assembly can be brazed at the same time, because production is limited only by the size and heating capacity of the furnace.
- The workpiece is protected from scaling or decarburization by a thin film of salt that adheres to the surface of the assembly when it is removed from the salt bath.
- Removal of the salt film is accomplished by dissolving during quenching or washing operations. When flux is used, there is no removal problem; flux is either dissipated during the brazing operation or dissolved simultaneously with the salt film during washing.
- Because the density of the molten salt supports a considerable portion of the weight of the workpiece, the assembly weighs less than when immersed, which can reduce the likelihood of distortion during heating.

Limitations of salt-bath dip brazing are:

- The process is not generally used for intermittent operation, being better suited for work that requires daily production.
- Joints that do not protrude from the assembly cannot be selectively brazed by partial immersion; most or all of the assembly must be heated to the brazing temperature in order to braze such joints.
- The workpieces must be completely dry, because the molten salt reacts violently with moisture, splattering and possibly even exploding. If moisture is present, all work requires preheating.
- The shape of the part must be designed to avoid trapping air or salt and to drain completely after removal from the salt bath.
- The assemblies should not require large, complicated fixtures.
- Part cleaning may be difficult.
- Assemblies can be corroded by salt residues, especially chloride.
- Proper maintenance of a salt-bath furnace is difficult and has special problems, such as solidification during power outages.
- Salt vapors may present health hazards unless properly ventilated.

Furnaces

A salt-bath furnace consists essentially of a metal or ceramic (refractory) pot that serves as a container for the molten salt. Some salt-bath furnaces are externally heated by gas, oil, or electrical resistance; this type of furnace lends itself more readily to intermittent operation and is not widely used for high-volume production. On the other hand, furnaces that are internally heated by immersed or submerged electrodes are not well suited to intermittent operation; therefore, they are used for high-volume production. Figure 1 shows the typical construction of the four principal types of furnaces. Detailed descriptions of internally and externally heated salt-bath furnaces can be found in the article "Liquid Carburizing and Cyaniding" in Volume 4 of the 9th edition of *Metals Handbook*.

Externally heated furnaces (Fig. 1a and b) are usually gas-fired or oil-fired and, less frequently, are heated by means of electrical resistance elements. When using electrical resistance heating, pot failure may result in total destruction of the heating elements. The waste heat of flue gases from fuel-fired furnaces may be fed to an adjacent chamber and used to preheat workpieces.

Internally heated furnaces (Fig. 1c and d) are energized with alternating cur-

*Revised by Quentin D. Mehrkam, Senior Vice President, Ajax Electric Co.

Fig. 1 Principal types of furnaces used for salt-bath dip brazing

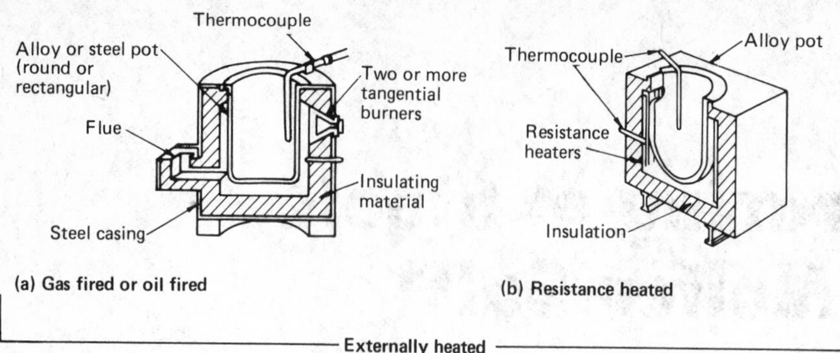

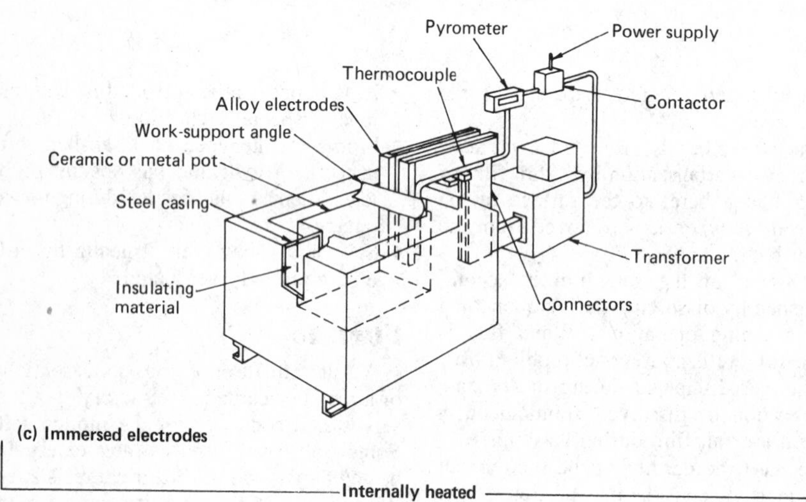

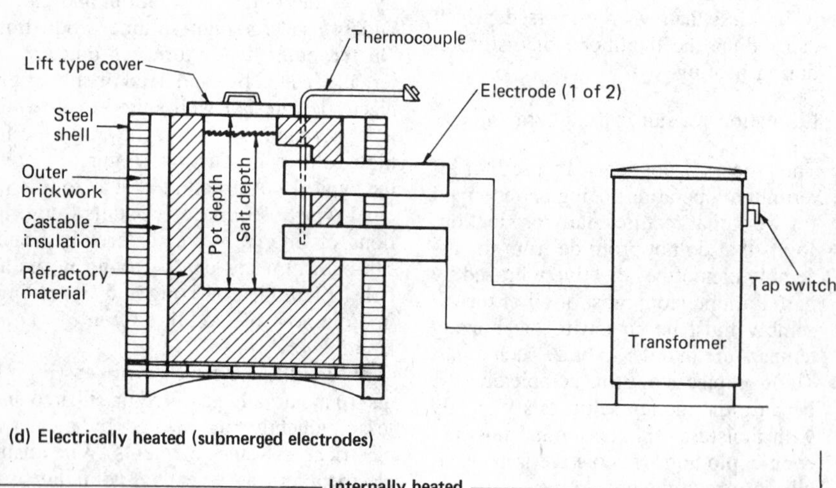

rent at 10 to 15 V, supplied from the multiple-tap secondary side of a step-down transformer. The molten salt is an electrical conductor, and heat is generated within the salt between the electrodes from resistance to the passage of current. By closely spacing the electrodes, an electromagnetic stirring action of the salt is obtained that assists in maintaining temperature uniformity and a control of ±5 °F.

Pot materials, those mediums used for construction of the pot in salt-bath furnaces, depend on the type of salt to be contained. The lining of a submerged-electrode furnace is generally made of high-fired fireclay refractory brick. For use with cyanide salts and noncyanide carburizing salts that contain sodium carbonate, the pot is constructed of magnesia-chrome refractory brick.

Because the salt pot of an externally heated furnace is ordinarily supported from a flange (Fig. 1a and b), the pot size is limited by the strength of the material used. Externally heated pots for use with all types of brazing salts are made of metals ranging in composition from low-carbon aluminized steel to high-nickel-chromium alloys. A small furnace with a pot 10 in. in diameter and 12 in. deep contains about 43 lb of salt; a fairly large furnace with a pot 24 in. in diameter and 30 in. deep has a capacity of 700 lb of salt.

With internally heated furnaces, a ceramic (refractory) pot is usually preferred for neutral chloride salts and fluxing salts that consist of neutral chloride salts plus a fluxing agent such as borax or cryolite. Slight modifications can be made in the pot material when neutral salts and salts containing flux are used. When carburizing or cyaniding is to be performed in addition to brazing, a steel or heat-resistant alloy can be used; ceramic linings are also applicable.

Salts

The types of salts used in dip brazing of carbon and low-alloy steels are neutral chloride salts, neutral chloride salts plus a fluxing agent such as borax or cryolite, and carburizing and cyaniding salts, which are also fluxing types of salts. Types and compositions of brazing salts and temperatures used for brazing of carbon and low-alloy steels with various filler metals are given in Table 1.

Neutral salts, so called because normally they do not add or subtract anything from the surface of the steel being treated, protect the surface from attack by oxygen in the air. Oxide on the workpiece, however, cannot be reduced by the salt, and a flux must generally be provided.

The neutral salts are mildly oxidizing to steel when they are used at recommended austenitizing temperatures. The oxides produced by heating steel in molten salt are largely soluble; hence, the steel is scale-free after heating. The accumulation of oxide in the molten salt, however, progressively makes the salt more decarburizing, and for this reason baths may require periodic rectification, as discussed in Volume 4 of the 9th edition of *Metals Handbook*.

Flux that is applied to the surface of the assembly and dried before the assembly is

Table 1 Typical salts used for salt-bath dip brazing of carbon and low-alloy steels with various filler metals

Filler metal(a)	Type of salt	Nominal composition, %	Brazing temperature range(b), °F
BAg-1 through BAg-8, and BAg-18	Neutral	55 $BaCl_2$, 25 NaCl, 20 KCl	1150-1600
	Cyaniding-fluxing	20-30 Na_2CO_3, 20-30 KCl, 30-40 NaCN	1200-1600
	Neutral	50 NaCl, 50 KCl	1350-1600
RBCuZn-A	Neutral	80 $BaCl_2$, 20 NaCl	1675-1725
	Fluxing	79 $BaCl_2$, 20 NaCl, 1 borax	1675-1725
	Carburizing-fluxing (water soluble)	30 NaCl, 30 KCl, 20 carbonate, 15-20 NaCN, activator (proprietary)	1675-1725
	Carburizing and self-fluxing	50 carbonate, 50 chloride with graphite addition(c)	1500-1700
RBCuZn-D	Neutral	90 $BaCl_2$, 10 NaCl	1900-1925
BCu-1 and 1a	Neutral	95 $BaCl_2$, 5 NaCl	2000-2100
	Neutral	100 $BaCl_2$	2000-2100

(a) Nominal compositions and brazing temperature ranges are given for silver alloys and copper-zinc alloys in the article "Torch Brazing of Steels" in this Volume and for copper filler metals in the article "Furnace Brazing of Steels" in this Volume. (b) Temperatures shown are those of the salt bath. (c) Used with mechanical agitation

immersed in the neutral salt will be quickly dissipated by dissolving in the salt or escaping from the surface of the bath as a volatized salt or gas. For this reason, there is generally no difficulty in removing flux from an assembly that has been brazed in a salt bath.

Fig. 2 Self-clamping pliers partially brazed in a salt bath

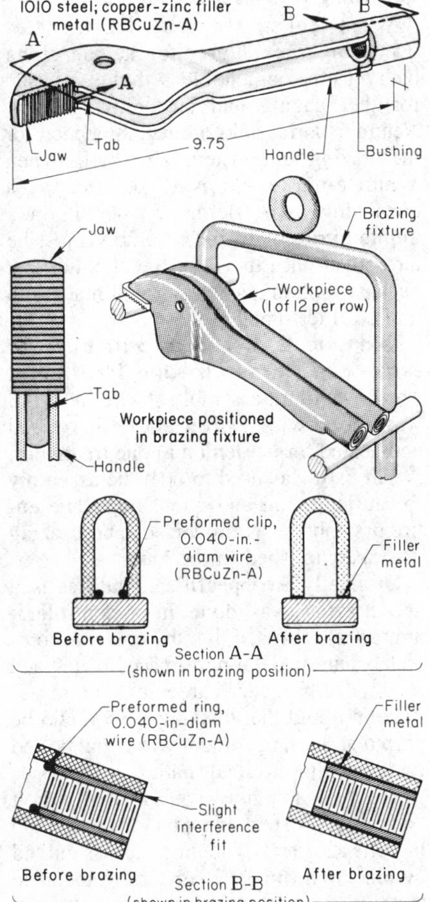

Fluxing agents such as borax and cryolite are added to neutral chloride salts to produce a fluxing environment in the bath. When these fluxing agents are used with silver alloy or copper-zinc filler metals, periodic flux additions are required to maintain the fluxing potential of the bath. Above 1200 °F, the fluxing potential can decrease rapidly because of oxidation from contact with air or the parts being brazed; therefore, the fluxing agent must be replenished more frequently.

Carburizing and cyaniding salts provide their own fluxing action. In addition, they supply carbon or carbon and nitrogen to the surface of the steel assembly as it is being brazed. Although silver brazing alloys have been used successfully, RBCuZn-A filler metal is generally preferred. A case depth up to 0.012 in. can be obtained without adversely affecting the quality of a joint brazed with this copper-zinc filler metal. Cyaniding and carburizing in salt baths (without brazing) are discussed in Volume 4 of the 9th edition of *Metals Handbook*.

Example 1. Combining Brazing and Carburizing in a Salt Bath. The 1010 steel assembly shown in Fig. 2 is part of a pair of adjustable self-clamping pliers. The jaw and a threaded bushing for the adjustment mechanism were brazed to the handle in a salt bath that, simultaneously with brazing, also carburized the bushing and handle.

The jaw was first separately carburized for 30 min in a water-soluble carburizing salt at 1675 °F to obtain a deeper case than would be obtained in the time allowed for brazing. After being air cooled to room temperature, the jaw was washed and dried.

A hairpin-clip preform of RBCuZn-A copper-zinc filler metal and the previously carburized jaw and tab were pressed in place in the handle, and the bushing with a preform ring of the same filler metal was pressed into the opposite end of the handle. Twenty-four assemblies were then loaded on a rack so that the jaws were held in place by the weight of the handles (Fig. 2). An interference fit held the bushing and handle together.

The rackload of assemblies was brazed and carburized by immersing for 20 min in a water-soluble carburizing salt bath at 1675 °F, which also provided the fluxing action. The brazed assemblies were quenched directly in oil and then tempered until the jaw hardness was 48 to 51 HRC. The total case depth obtained on the jaw in the two carburizing operations was 0.010 to 0.012 in. A file-hard case about 0.004 to 0.005 in. deep was also developed on the threaded bushing and handle, which provided a hard wear-resistant surface to serve as a base for hard chromium plating.

Fluxes

An adequate fluxing environment is needed to ensure good flow and penetration of the brazing alloy in salt-bath dip brazing. When brazing is done in a neutral chloride salt bath, a flux is usually applied to the assemblies before brazing. Generally, the application of flux to the assembly is not necessary when using a cyanide bath or other fluxing bath.

Flux can be applied by brushing, dipping, or spraying the parts to be brazed before, during, or after assembly. After flux application, if any moisture is present, the assemblies must be preheated to dry them before immersion in the salt bath. Typical fluxes employed for prefluxing carbon steels and low-alloy steels that are to be brazed in a salt bath are American Welding Society (AWS) type 3A and 3B.

Filler Metals

The brazing filler metals shown in Table 1 are the most widely used for salt-bath dip brazing of carbon and low-alloy steels. Although silver alloys BAg-13 and 13A (not shown in Table 1) can be used for brazing in a salt bath, they have been supplanted in most applications by copper-zinc alloys, which are less costly and have similar brazing temperature ranges. The rapid heating rate and nonoxidizing environment in a salt bath minimize dezincification of copper-zinc alloys, thereby facilitating the use of these alloys.

Although a temporary insulating cocoon of frozen salt forms instantly around a piece of cold metal when it is immersed in a salt bath, externally located filler metal can reach the melting temperature range before the steel workpiece has reached a temperature high enough for proper wetting to take place. When this occurs, the molten filler metal flows away from the joint with no brazing. This can be avoided either by preheating the assembly to a temperature below the melting range or by relocating the filler metal inside the joint in grooves, recesses, or drilled holes. Many assemblies of thin-section steel, however, have been brazed successfully with the filler metal placed on the outside of the joint.

The brazing filler metal must be in contact with the joint. Filler metals in the form of wire, strip, powder, paste, and cladding are available. Special preform rings or other shapes can be obtained for specific joint requirements.

Joint Design

The filler metal providing the bond in brazed joints is drawn by capillary action between closely adjacent, matching surfaces. A diametral clearance of 0.001 to 0.003 in. is considered necessary for good flow and penetration of silver or copper-zinc filler metals in most joints. For copper brazing, joint clearance can range from a slight interference fit to a positive diametral clearance of about 0.002 in. When brazing dissimilar metals that have differing coefficients of thermal expansion or dissimilar masses of the same metal, the designer must be aware of the differing rates of expansion to ensure that the required joint clearance is obtained between the components at the brazing temperature.

For brazing, lap joints designed for shear loads are preferred to butt joints designed for tensile loads. For additional information, see the article "Furnace Brazing of Steels" in this Volume.

Preparation for Brazing

Burrs can interfere with the capillary action and should be removed. In addition, the joint surfaces must be free of grease, oil, paint, oxide, and scale that would prevent the filler metal from wetting the workpiece surfaces. For cleaning methods, see the article "Furnace Brazing of Steels" in this Volume. Fluxing, as mentioned previously in this article, can be done before, during, or after assembly, as convenient, but the assembly must be warmed before brazing to eliminate any moisture.

Assemblies of self-jigging joints minimize fixturing; a hook, rod, basket, or rack may be all that is required to support the assembly during the brazing process. Fixtures should not touch the joint being brazed in order to avoid the risk of brazing the assembly to its fixture. Methods of self-jigging, including gravity, locating, press fitting, thread joining, and swaging, are also discussed in the article "Furnace Brazing of Steels" in this Volume.

Although the fixturing of workpieces that cannot be self-jigged creates problems such as distortion, difficulty in maintaining dimensional tolerances, and the necessity of heating an added mass of material, these difficulties can be overcome by careful fixture design, use of fixture material having thermal expansion compatible with that of the work metal, and attention to maintenance of tolerances during brazing. The fixture material should also be compatible with the salt and flux used and have reasonably long life. Furthermore, the design of the fixture should facilitate draining after removal from the bath.

Preheating an assembly before brazing serves several purposes. If prefluxing is used, preheating dries the flux and vaporizes all moisture from the assembly and fixture. Even a slight amount of moisture can cause spattering in contact with molten salt.

Drying in an oven at 400 to 600 °F before the assembly is immersed in the molten bath is recommended. Assemblies may also be warmed to about 200 °F by placing them on top of the salt-bath furnace.

Preheating assemblies not only decreases the temperature drop of the salt bath and reduces brazing time, but also minimizes the premature melting of externally placed filler metal. In joining an assembly consisting of both heavy and light sections, preheating reduces thermal gradients and subsequent distortion, and improves the wetting action on the heavier parts as well. To be effective, preheating temperature must be at least 100 °F lower than the melting range of the filler metal. If an oven preheat is used, oxidation must be avoided by using temperatures below 900 °F. Otherwise, an inert atmosphere is desirable.

Brazing

General ranges of brazing temperature that are used with various salts are listed in Table 1; specific brazing ranges for the individual filler metals are given in the articles "Torch Brazing of Steels" and "Furnace Brazing of Steels" in this Volume. The time in the molten salt bath differs from one job to another. For thin-section parts to be brazed only, the holding time may be as short as 1 min. For assemblies that are to be superficially case hardened (0.005- to 0.012-in. case depth) as well as brazed, the holding time is the time required to produce the desired depth of case.

After the workpieces have been in the bath for the required time, they are carefully lifted from the salt bath. A uniform motion is necessary during removal from the bath; jerky movements can cause the liquid filler metal to be displaced from the joint.

Brazing and Heat Treating. Brazing can be combined with a heat treating operation such as neutral hardening or carburizing. For this procedure, the assembly can be air cooled to room temperature or quenched in a suitable liquid medium, as required. If the assembly is to be quenched, it should cool in air until the brazing filler metal has fully solidified, before quenching. Otherwise, the filler metal is usually blown out of the joint during quenching. This is more likely to occur with water quenching than with oil quenching.

When molten salt baths are used for quenching, the salt used for the quenching bath must be compatible with the salt used for the brazing and austenitizing bath. Neutral chloride salts are recommended for the brazing and austenitizing baths when a nitrate-nitrite salt is to be used as a quenching bath. Before quenching in a nitrate-nitrite bath, the assembly must be air cooled until the filler metal solidifies, because molten quench salt will react with molten filler metal.

Example 2. Use of a Salt Bath To Braze and Harden in One Treatment. The breech-bolt assembly shown in Fig. 3 was brazed with a silver alloy filler metal and heated for hardening in one treatment. A salt bath was used to heat the assembly for austenitizing and quenching. The entire assembly was immersed, and brazing was accomplished in the bath.

Originally, a copper filler metal was used and brazing was done in a controlled-atmosphere furnace, but the bolt was completely annealed during furnace brazing and required subsequent hardening. An investigation found that this assembly could be brazed at the same time it was being heated for hardening in a salt bath.

The bolt assembly consisted of a bolt body of 1137 steel and a handle and guide lug of 1120 steel. The unit was assembled by placing a ring of BAg-1 filler metal in the bottom of the guide lug adjacent to the

Fig. 3 Rifle breech-bolt assembly that was simultaneously heated for hardening and brazed in a salt bath

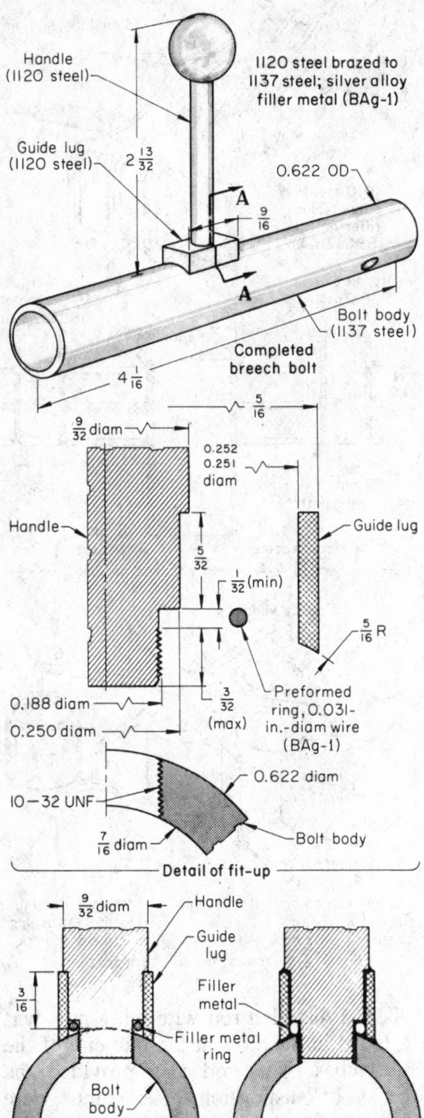

bolt body and then screwing the handle through the lug snugly into the threaded hole in the bolt body. This method of assembly held the filler metal in place so that, after brazing, the handle, lug, and bolt body would all be joined by brazing. The joints were coated with type 3A flux and dried by heating to 400 to 500 °F.

The bolt assembly was then immersed in a salt bath for 7 min at 1550 °F. The bath was a neutral chloride salt mixture containing 50% NaCl and 50% KCl. The bolt body was austenitized at the same time the filler metal penetrated the joints to be brazed. The assembly was immediately transferred to a nitrate-nitrite salt bath at 500 °F for 7 min.

In this application, it was not necessary to permit the filler metal to solidify before the quench; any exposed molten filler metal that was oxidized by the quench was removed in a later machining operation. The parts were then air cooled, tempered in a nitrate-nitrite bath at 820 °F for 20 min, air cooled, and washed in hot water. The final hardness of the bolt body was 30 to 34 HRC, and the strength and quality of the brazed joint were completely satisfactory.

Example 3. Joining Carbide Tips to Tool Bits While Heating the Tool Bits for Hardening. Carbide-tipped rock bit drills, shown in Fig. 4, were joined using a silver brazing filler metal and heated for hardening in one operation in a salt bath. The bits ranged in diameter from 6 to 9 in. and weighed 100 to 300 lb. The shanks were made of air-hardening or oil-hardening tool steels.

Shanks and tips were vapor degreased and the surfaces to be joined were grit blasted and coated with flux. The carbide tips were inserted into the slots in the ends of the shanks with BAg-3 alloy shims placed under and on each side of each tip. Metal plates welded to the outer end of each slot, as shown in Fig. 4, kept flux and molten filler metal from escaping and kept the molten salt from entering when the assemblies were heated. The plates were removed later by machining.

The assemblied bits were placed in a fixture with the carbide-tipped ends up, and were preheated to about 1000 °F to dry the flux and assembly and to reduce the time required to reach austenitizing temperature in the salt bath. Each assembly was then partially immersed in the salt bath to a level at which the metal plates welded on the ends of the slots would safely prevent molten salt from entering the joint area. The assemblies were held in the bath for 20 to 30 min, depending on the size of the bit. This time period was needed to austenitize the shank and was not necessarily required for the joining operation. The tips were not submerged to:

- Avoid thermal shock that might crack the carbide
- Avoid washing away the flux
- Keep the joints visible and accessible so that dry powdered flux and filler metal could be added to fill the joint

The bath, a neutral chloride salt (50% NaCl and 50% KCl), was maintained at selected temperatures between 1600 and 1650 °F, and held within ±5 °F. The exposed ends of the assemblies, containing the joints being brazed, were heated only by conduction from the immersed portion. This lower heat input, and the heat losses by radiation from the exposed metal, served to keep the temperature in the joint area within the range (1270 to 1500 °F) required to prevent overheating of the high-cadmium filler metal.

The assembly was air cooled or oil quenched, depending on the type of tool steel in the shank. During the cooling period, additional filler-metal rod and dry powdered flux were fed into the molten pool of BAg-3 alloy to fill the voids left by the approximately 20% shrinkage of the filler metal during solidification. After cooling, any salt remaining was washed from the bit, which then was tempered to the hardness specified for the shank.

The larger than normal clearances for filler metal (0.015 in. on each side and 0.030 in. under the carbide inserts) were found by field experience to provide a shock-impact cushion between the carbide and steel, which reduced the likelihood of premature failure.

Salt-bath heating was selected because the rapid heating minimized oxidation and prevented breakdown of flux. In addition, several assemblies could be processed at one time.

Induction heating was rejected as a means of performing the operation because: (1) capacity would have been limited to one rock bit per station; (2) high heat input would have demanded diffusion cycles to allow heat to be conducted from the surface to the interior of the heavy bits; (3) temperature would have varied considerably in the carbide area, causing erratic melting and too many voids; (4) overheating of carbide would have caused

Fig. 4 Carbide-tipped rock bit that was simultaneously heated for hardening and brazed in a salt bath

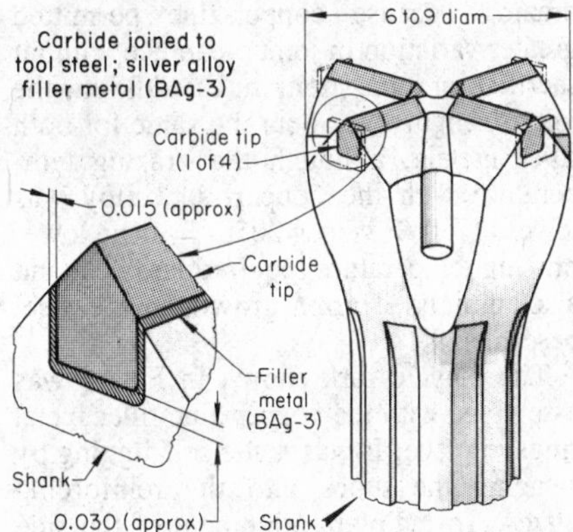

cracking; and (5) subsequent overall heating would have been necessary to harden the shank.

Torch brazing would have been unsatisfactory because: (1) uneven heating would have caused erratic melting of filler metal; (2) with localized heating, the whole bit would have required reheating to harden the shank; (3) flames from the various torches would have interfered with the operator's ability to monitor the joint and to face feed the filler metal after heating was completed; and (4) the torch flames would have caused excessive scale and spent flux, which would have had to be removed.

Cyanide Salts. The use of cyanide salts in brazing baths allows carburizing, cyaniding, or austenitizing to take place during brazing. The correct quenching medium must be chosen, however, since assemblies brazed in salts containing cyanide must not be quenched in nitrate-nitrite salts because of the explosion hazard, and the filler metal must solidify before quenching. A suitable procedure for using cyanide salts in brazing baths follows.

First, a brazing filler metal that solidifies above the transformation temperature range of the steel being brazed, such as RBCu-Zn-A, is selected. The brazed assembly is transferred from the cyanide-containing bath into a neutral chloride salt bath rinse that is maintained below the solidus temperature of the filler metal but within the austenitizing range of the steel. The assembly is then transferred from the neutral chloride rinse to the nitrate-nitrite bath for the quenching operation. Control of the amount of cyanide buildup in the neutral chloride rinse bath is essential. When tests indicate more than 5% cyanide in the chloride rinse, part of the chloride salt should be discarded, and the remainder should be diluted with the addition of new neutral chloride salt.

After quenching in the nitrate-nitrite bath, the assemblies are air cooled, washed, and then tempered, if required. All fixtures must be thoroughly cleaned and dried after the quenching operation to prevent transfer of nitrate-nitrite salt to cyanide baths or neutral chloride baths. Nitrate-nitrite salts cause an explosion if mixed with cyanide, and a chloride bath contaminated with nitrate-nitrite salts causes pitting and decarburization of steel parts immersed in it. The use of cyanide-free carburizing salt eliminates the nitrate-nitrite explosion hazard.

Mechanized Brazing

The salt-bath dip brazing process can be mechanized to increase production rates, using semiautomatic or fully automatic handling mechanisms. The article "Liquid Carburizing" in Volume 4 of the 9th edition of *Metals Handbook* describes some of those mechanisms. One simple automatic arrangement is the "merry-go-round" (Example 4), in which the time at each station is the same.

Programmable automatic arrangements have been used with hoists to lift the workpieces in and out of the various furnaces and tanks in the line. Another machine is the "jackrabbit" mechanism, which has synchronized continuous-chain conveyors that carry the work through various operations with fixed time cycles. The programmable work-handling system is more advantageous because time cycles and sequences of each operation are easily changed to suit a variety of applications.

Example 4. Partial Immersion of Assemblies During Brazing To Preserve Work-Hardened Areas. Bicycle-fork assemblies made of 1010 steel were brazed in a salt bath using RBCuZn-A filler metal. Selective heating by partial immersion was necessary to retain, as nearly as possible, the strength developed in the side tubes by cold working. This was accomplished by suspending the fork so that only the stem tube, reinforcing plates, and a portion (2$^3/_4$ in.) of the side tubes were immersed in salt (Fig. 5).

Originally, the forks had been brazed in two gas-fired pots containing molten brass. The forks were preheated by placing them on top of the furnace and then were dipped in the molten brass, one fork at a time. A high degree of operator skill and rapid manipulation were required. Production rate was 75 forks per hour. Because loose fits (up to $^1/_{32}$-in. clearance) were used, the assembly had to be tack welded before brazing to make it self-supporting. Rejection rate was 2%, and rejects were not reworkable.

Salt-bath dip brazing was tried next. A copper-zinc alloy, rather than copper, was selected as filler metal for salt-bath dip brazing because copper-zinc permitted greater variation in joint clearance without sacrifice in joint strength. In addition, the joint strength was about the same for both filler metals, although the brazing temperature with the copper-zinc alloy was lower (1700 °F versus 2050 °F). The lower brazing temperature required less heat and also minimized grain growth in the steel base metal.

The bicycle fork shown in Fig. 5 was assembled with the copper-zinc filler metal rings in place. It was made self-jigging by swaging the tubes into the reinforcing plates. To maintain alignment of the side tubes, a threaded rod with wing nuts was fastened in the slots at the hub end of the side tubes. This rod also provided the means of suspension. The joints were fluxed before brazing.

Fig. 5 Mechanically brazed bicycle-fork assembly

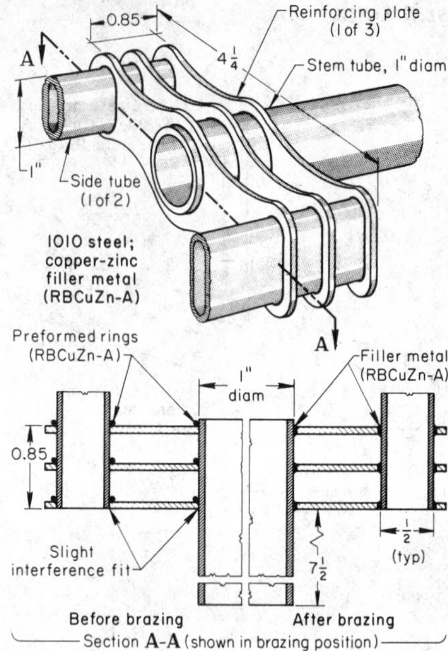

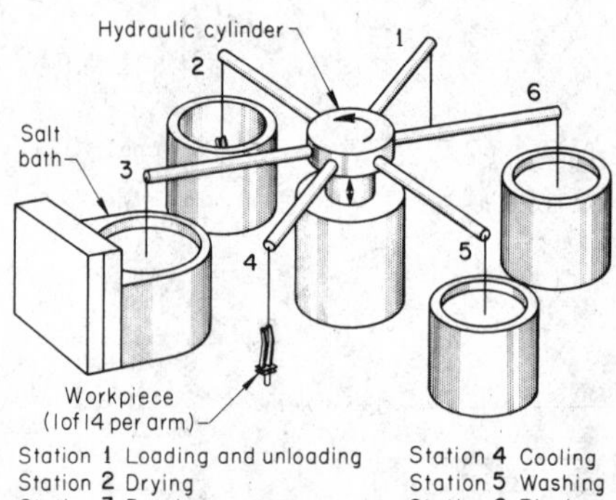

"Merry-go-round" for salt bath brazing

Because of the production requirements, brazing was done in a merry-go-round brazing machine (Fig. 5) that consisted of six stations. Each station accommodated a fixture from which up to 14 fork assemblies were suspended. The fixtures were raised, lowered, and indexed by arms that connected them to a central hydraulic cylinder. Time at each station was 3 min and transfer time between stations was 30 s.

Forks were hand-loaded and unloaded at the first station. Assemblies were dried at station 2 in a 5-kW electric drier. The salt-bath furnace at station 3 was an internally heated immersed-electrode furnace that operated at 1700 °F and used a neutral chloride salt composed of 80% $BaCl_2$ and 20% NaCl. Air cooling to solidify the filler metal followed at station 4. Stations 5 and 6 were a hot water wash and a hot water rinse. Unloading completed the cycle. The new method increased production from 75 to 240 forks per hour. Less than 1% of the forks showed defects, and these were salvaged by reworking.

Safety Precautions

Cautions and references for the safe use of nitrate-nitrite salt baths and oil quench tanks, as well as the need for all parts to be free of moisture before immersion in a salt bath, have been discussed earlier in this article. In addition, salt-bath dip brazing may produce dust, fumes, and gases that are hazardous to health. Therefore, adequate exhaust systems and ventilation are necessary. For example, cadmium is contained in some silver alloy filler metals, and inhalation of cadmium oxide fumes, which are toxic, can be fatal. Other metals, salts, and materials present various degrees of hazard.

The fluorides in fluxes pose a dual problem; it is essential not only to provide adequate ventilation to carry away fumes, but also to avoid skin contact with these fluxes. Thus, procedures must include means to make any handling in the flux unnecessary. Brazing personnel should thoroughly wash their hands before making any body contacts or touching food.

Table 2 Threshold limit values (TLV) for substances encountered in salt-bath dip brazing

Substance	TLV(a), mg/m^3
Barium (soluble compounds)	0.5
Cadmium oxide fume	0.1
Copper dusts and mists	1.0
Cyanide (as CN), skin(b)	5.0
Fluoride (as F)	2.5
Hydrogen cyanide, skin(b)	11.0
Hydrogen fluoride	2.0
Silver metal and soluble compounds	0.01
Zinc oxide fume	5.0

(a) Approximate milligrams of particulate per cubic meter of air. (b) "Skin" indicates that the substance can penetrate the skin to cause systemic effects

Source of data and discussion: "Threshold Limit Values of Airborne Contaminants and Intended Changes Adopted by ACGIH for 1970," American Conference of Governmental Industrial Hygienists, Cincinnati, OH 45202. "This is not a definitive list of exposure levels but serves only as a relative comparison."

Two brazing salts listed in Table 1 contain cyanide. If taken internally, cyanides are fatally poisonous; if allowed to come in contact with scratches or wounds, they are highly toxic. In addition, fatally poisonous fumes are evolved when cyanides are brought into contact with acids. To avoid possible toxic effects, the provision of exhaust systems is recommended to remove the fumes from salt baths. Precautions regarding the use of cyanide salts and the disposal of cyanide wastes are discussed further on pages 248 and 249 in Volume 4 of the 9th edition of *Metals Handbook*. Information is also available from Occupational Safety and Health Administration (OSHA) and Environmental Protection Agency (EPA) publications.

Threshold limit values for airborne concentrations of substances commonly encountered in salt-bath dip brazing are given in Table 2. A summary of some of the limitations of these values follows, but the reader is directed to the reference in Table 2 for complete information. Threshold limit values represent conditions under which it is believed that nearly all workers may be repeatedly exposed day after day without adverse effect. These values are time-weighted concentrations for a 7- or 8-h workday and a 40-h week. They should be used as guides in the control of health hazards, but not as fine dividing lines between safe and dangerous concentrations.

Threshold limit values are intended for use in the field of industrial hygiene and should be interpreted and applied only by persons trained in that field. They are not intended for use or for modification for use (1) as a relative index of toxicity or hazard; (2) in the evaluation or control of community air pollution or air pollution nuisances; (3) in estimating the toxic potential of continuous, uninterrupted exposures; or (4) as proof or disproof of an existing disease or physical condition.

Brazing of Cast Irons*

BRAZING gray, ductile, and malleable cast irons differs from the brazing of steel in two principal respects: special precleaning methods are necessary to remove graphite from the surface of the iron, and the brazing temperature is kept as low as feasible to avoid reduction in the hardness and strength of the iron.

The processes used for brazing cast irons are the same as those used for brazing steel—furnace, torch, induction, and dip brazing. As with other metals, selection of the brazing process depends largely on the size and shape of the assembly, the quantity of assemblies to be brazed, and the equipment available.

Filler Metal and Flux

Because most cast irons are brazed at relatively low temperatures, the filler metals used are almost exclusively silver brazing alloys. Compositions and other information concerning the more common silver alloy filler metals are listed in the articles "Furnace Brazing of Steels" and "Brazing of Stainless Steels" in this Volume. Of these silver alloys, BAg-1 is most often used for brazing of cast iron, principally because it has the lowest brazing-temperature range. A fluoride-type flux such as American Welding Society (AWS) type 3A is usually used with BAg-1 filler metal.

Brazeability

Relatively high silicon content and sand inclusions on as-cast surfaces have some adverse effects on the brazeability of cast iron. These effects, however, are less significant than the adverse effect of graphite, which is present in all gray, ductile, and malleable cast irons. Graphite has essentially the same effect on machined joint surfaces as on as-cast surfaces. Although gray, ductile, and malleable irons all have lower brazeability than carbon or low-alloy steels, the three types of iron are not equal in brazeability.

Malleable iron is generally considered the most brazeable of the three types of cast iron, largely because the total carbon content is somewhat lower (seldom over 2.70%) and, therefore, graphitization is lower. Brazeability is also enhanced because the graphite occurs in the form of approximately round nodules and thus is easier to remove or cover up (as by abrasive blasting). Also, malleable iron is lower in silicon than the other types of cast iron and thus is less graphitized, which makes it better suited for brazing.

Ductile iron can have a composition nearly the same as gray iron, but the graphite particles are spheroidal rather than flake-shaped. The spheroidal shape is more favorable for brazing. Shot or grit blasting is effective in rolling metal over graphite particles that are exposed at the surface.

Gray iron, which is characterized by large flakes of graphite, is the most difficult type of cast iron to braze. Until the development of electrolytic salt-bath cleaning, brazing of gray iron was considered impractical.

Applicability

Brazing is sometimes used to repair defective or damaged castings, although braze welding is more often used for this purpose (see the article "Oxyacetylene Braze Welding of Steel and Cast Irons" in this Volume).

Most brazing of cast iron is performed to join assemblies at lower cost than is possible by another process or to fabricate parts that are difficult to produce, for example, one-piece castings (Example 1). In some applications, two or more cast iron components are brazed together; in other applications, one or more components of a brazed assembly are made of another metal, most often steel (Examples 2 and 3). Copper alloys and cast iron can also be joined by brazing with silver alloy filler metal.

Preparation of Castings for Brazing

Preferred joint designs for brazing cast irons are generally the same as for steel. Best results are obtained using diametral clearances in the range of 0.002 to 0.005 in. Diametral clearances up to 0.010 in. may be used, but this much clearance will result in lower joint strength and added filler-metal cost.

Methods of Surface Preparation. A number of methods have been tried for preparing cast iron surfaces for brazing; most of them have been only partly successful. Abrasive blasting with steel shot or grit has proved reasonably successful for preparing the surfaces of ductile and malleable iron castings, but is seldom suitable for preparing surfaces of gray iron castings. Electrolytic treatment in a molten salt bath, alternately reducing and oxidizing, has been the most successful method for surface preparation and is applicable to all graphitic cast irons. Ordinary chemical cleaning methods such as degreasing, detergent washing, or acid pickling have the distinct disadvantage of not removing surface carbon, which interferes with bonding.

Before any procedure for cleaning is adopted, tests should be made by cleaning samples of the iron intended for use in the castings to be brazed, fluxing the samples, and applying filler metal (preferably on a smooth, flat surface). The samples are then heated to the pre-established brazing temperature, cooled, and examined visually. If the samples show indication that the filler metal has not uniformly wetted the test piece, the surface is not sufficiently clean.

Fused Salt Cleaning

Alkaline-based fused salts operating at 750 to 900 °F are extremely effective in removing surface oxides and sand from iron castings. At these temperatures, salt baths exhibit the required high chemical activ-

*Revised by William G. Wood, Vice President—Technology/Research & Development, Kolene Corp.

ity. This action is further enhanced in the cleaning operations by the introduction of electrical energy.

Work polarity in fused salt-bath cleaning is fundamentally important for removal of specific contaminants. Current density has less influence, providing that amperage is sufficient to maintain bath activity. In the cleaning process, current enters the system through a reversing switch. A positive or negative pole is developed at the workpiece, depending on the switch position. The interior surface of the salt pot becomes the opposite pole. Frequently, a sacrificial liner is used for this pole to protect the salt-bath container. Dwell time at either polarity may vary from 1 to 20 min.

Charging the work negatively generates a chemically reducing medium in the immediate vicinity, removing oxide and sand according to the following:

$$SiO_2 + 2Na^+ + 2OH^- = Na_2SiO_3 + H_2O \quad \text{(Eq 1)}$$

$$Fe_2O_3 + 2Na^+ + 2OH^- = Na_2Fe_2O_4 + H_2O \quad \text{(Eq 2)}$$

The products of these reactions, sodium silicate and sodium ferrite, are removed in a sludge collection system. The removal of this sludge and the addition of makeup salt keep the bath regenerative and at maximum efficiency without change.

Graphitic surface carbon must be removed to obtain good adhesion of silver braze. The work polarity must be reversed for this operation. With the work positive, a chemical oxidizing condition results in the following reactions:

$$C + 4OH^- = CO_2 + 2H_2O \quad \text{(Eq 3)}$$

$$2NaOH + CO_2 = Na_2CO_3 + H_2O \quad \text{(Eq 4)}$$

$$2Fe + 6OH^- = Fe_2O_3 + 3H_2O \quad \text{(Eq 5)}$$

The bath product of these reactions, sodium carbonate, is removed in a sludge collection system. The iron oxide formed in Eq 5 is deposited on the work surface and must be removed in a final reduction cycle, which has a chemical reaction identical to that in Eq 2. The completely cleaned casting is thoroughly rinsed in cold water, then immersed in hot water and air dried.

The excellent metallurgical interface obtainable with this cycle is shown in Fig. 1. Bonds of this quality have permitted direct babbitting of cast iron bearings and silver brazing of fittings to cast iron surfaces. With proper cleaning and preparation of cast iron, it is possible to produce a brazed bond between metal surfaces that exceeds the strength of the parent metal.

Fig. 1 Interface of brazed cast iron and steel after fused salt cleaning

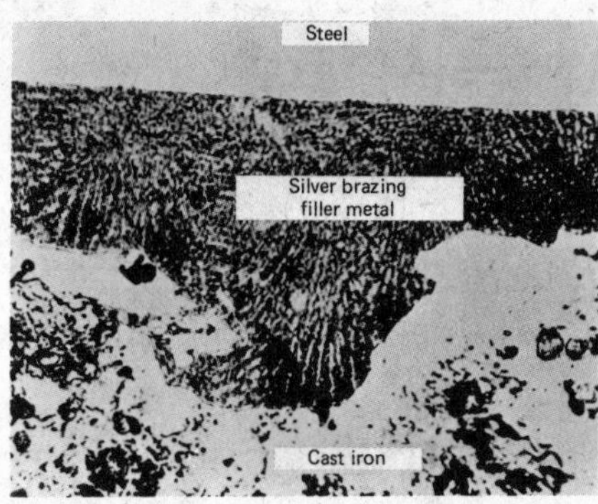

Preheating assemblies that contain one or more cast iron components minimizes brazing time and dries the flux when brazing in a furnace or salt bath. Because cast irons have moderate to high thermal expansion coefficients, coupled with relatively low thermal conductivities, temperature differences may be reduced by preheating the entire assembly to a temperature between 400 and 800 °F before torch or induction brazing.

Production Applications

Foundry problems encountered in the production of complex castings can sometimes be overcome by producing two or more simpler castings and brazing them together. Molds for castings with complex internal passages are difficult to make, and removal of the cores and burned-in sand from the castings may be even more cumbersome.

Example 1. Brazing Two Castings for an Engine Cylinder Block. The gray iron cylinder block (Fig. 2) was originally machined from a single casting. Not only were the cores for the internal passages difficult to mold, but removing the core sand from the casting was an additional problem.

Both difficulties were overcome by changing from the one-piece design to two castings brazed together (Fig. 2b). Mating surfaces of the block and cylinder liner were finish machined so that diametral clearance was 0.002 to 0.007 in. Two grooves were machined in the block to receive the preformed filler-metal wires.

Both of the cast components were then cleaned electrolytically in a fused salt bath. Finally, the two castings were assembled, with the preformed rings of filler-metal wire (BAg-1) and flux (type 3A) trapped in the grooves, and brazed in a furnace at 1325 °F under a protective atmosphere.

Example 2. Brazing Four Gray Iron Cylinder Liners to a Steel Deck Plate. Brazing cast iron to steel may be done to attain a specific set of properties, to simplify the casting process, or to reduce cost. The brazed assembly shown in Fig. 3, consisting of four centrifugally cast cylinder liners of class 30 or 40 gray iron and deck plate of 1020 steel, served as a component of a gasoline engine.

The five parts of the assembly shown in Fig. 3 were machined to the sizes given, permitting a diametral clearance of 0.002 to 0.008 in. between each gray iron liner and the steel plate. In preparation for brazing, the steel plate was cleaned by degreasing in trichlorethylene. The gray iron liners were cleaned electrolytically in a bath of fused salt at 860 °F for a total time of

Fig. 2 Cross-section comparison of single- and double-cast gray iron cylinder blocks

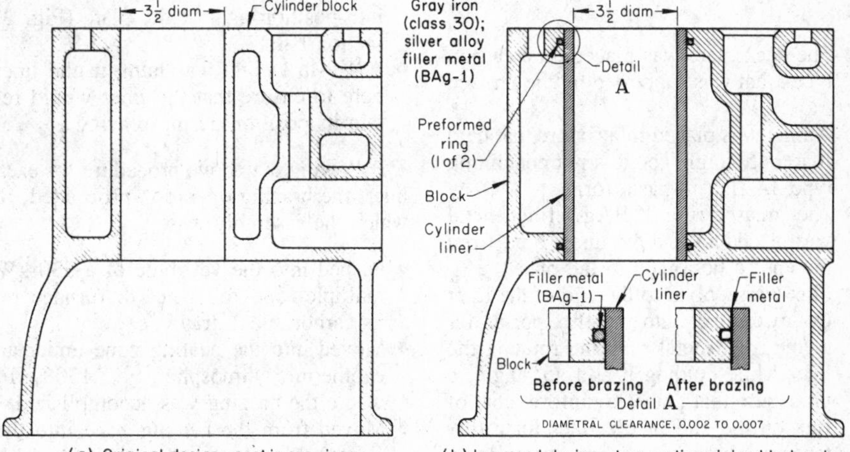

(a) Original design—cast in one piece (b) Improved design—two castings joined by brazing

Fig. 3 Gray iron cylinder liners furnace brazed to a steel deck plate for a gasoline engine. Liners were cleaned in an electrolytic salt bath.

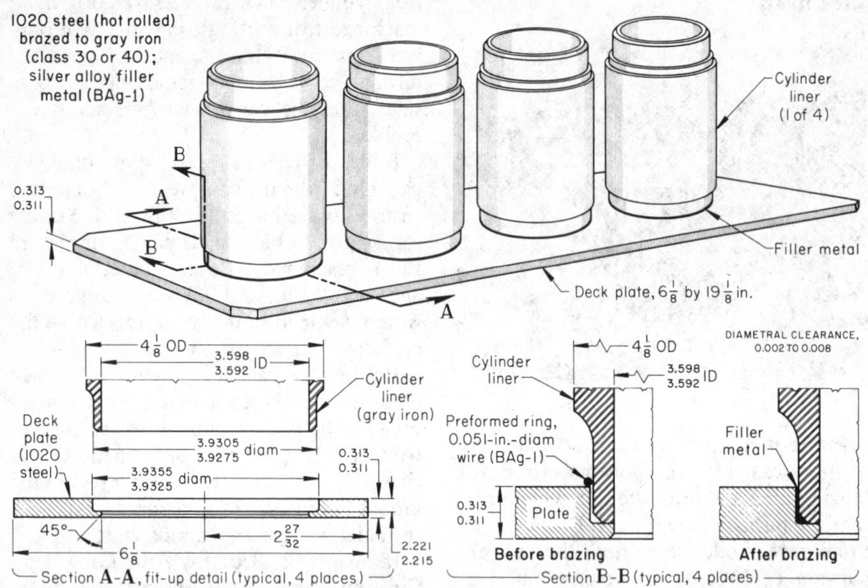

Electrolytic salt-bath prebraze cleaning

Tank size	4 by $3^1/_2$ by 4 ft high (usable volume)
Tank capacity	13 000 lb of molten salt
Salt composition	Principally sodium hydroxide
Current	1800 A (DCEP)
Bath temperature	860 °F
Reduction-oxidation cycle:	
Reduction (workpiece negative), agitator on	5 min
Oxidation (workpiece positive), agitator off	20 min
Reduction (workpiece negative), agitator on	10 min
Total time in bath	35 min
Salt removal:	
Dip hot castings	Cold water
Dip cold castings	Water at 160 °F

Conditions for furnace brazing

Furnace	Roller hearth, multiple zone
Electrical heating power	330 kW
Furnace zone length:	
Vestibule	4 ft, 2 in.
Heating zone	22 ft, 6 in.
Slow-cooling zone	10 ft
First water-cooled zone	6 ft, 10 in.
Second water-cooled zone	5 ft, 4 in.
Third water-cooled zone	6 ft, 11 in.
Atmosphere	Exothermic, 6.5-to-1 air-to-gas ratio
Furnace temperature	1300 °F
Filler metal	BAg-1
Flux	Type 3A
Brazing time, door-to-door	144 min
Production rate	$62^1/_2$ assemblies per hour

35 min, before being immersed first in cold water and then in water at 160 °F.

Immediately after prebraze cleaning, the following fluxing and assembly procedure was used:

- The deck plate was placed on a carbon block that was supported by the furnace tray
- A liner was placed in a fixture that held it at a 45° angle above a pot containing type 3A flux in paste form
- A degreased ring of BAg-1 filler metal was placed around the liner $^1/_4$ in. from the end to be joined to the plate
- Flux was applied to the end of the liner by lowering it into the flux pot as far as the filler-metal ring and rotating the liner while holding it at a 45° angle, a procedure that places a uniform coat of flux on the end of the liner for a distance of about $^1/_4$ in. from the end
- The liner was removed from the fluxing fixture and placed in one of the holes in the plate
- The filler-metal ring was pushed down into position against the plate (Fig. 3, section B-B)
- A $4^1/_2$-lb weight was hung in the liner bore to ensure that the liner would retain its position during brazing

Following the above procedure for each liner, the brazing operation proceeded, in which the assembly was:

- Pushed into the vestibule of a 330-kW multiple-zone roller-hearth furnace on its carbon block tray
- Moved into the heating zone under an exothermic atmosphere at 1300 °F, where the brazing was accomplished
- Moved from the heating zone into the slow-cooling zone, then through three consecutive water-cooled zones, and finally removed at a temperature of 300 °F maximum

Total time in the heating zone for each assembly was $60^1/_2$ min. Soaking at brazing temperature was approximately 15 min. Total time in the furnace for each assembly was 144 min (door-to-door). By following the procedure outlined here, the assemblies were brazed at an average rate of $62^1/_2$ per hour.

After brazing, each assembly was cleaned ultrasonically in a solution of 10 oz/gal of rust stripper, rinsed in cold water, and dipped in a solution of 2 oz/gal of alkaline rust preventive. Quality of the brazed joints was checked visually by the furnace operator and verified by liquid-penetrant inspection.

The application described in Example 2 is one of numerous similar applications performed in various plants. Details of processing procedure vary considerably among plants, but the use of electrolytic salt-bath cleaning for surface preparation is widespread.

Example 3. Replacement of One-Piece Castings with Silver Brazed Sections. One-piece castings requiring complicated coring can be designed in two or more open sections. These open-cored sections are salt bath cleaned and silver brazed to form the finished structure, as with the three-piece cylinder liner construction in Fig. 4.

The two steel sleeves shown to the right and left of the gray iron casting in Fig. 4

Fig. 4 Three-piece cylinder liner construction. (a) and (c) Steel sleeves. (b) Gray iron casting

were electrolytically cleaned in a fused salt bath. After cleaning, the three sections were silver brazed to form the water jacket of the liner. This design and construction was selected to avoid a complicated casting and the requirement for a watertight jacket. Because production items such as cylinder liners are seldom permitted to remain in batch cleaning operations, a completely automated system was required to meet the demand for 36 liners per hour; Fig. 5 shows the stations required for automatic processing of the 36 castings and 72 sleeves to meet the hourly production quota.

Because the surfaces to be silver brazed had been machine finished, an initial reduction cycle in the electrolytic salt was not required. Twelve cast liners and 24 expanded steel sleeves were cycled through the fused salt using an 8-min oxidation cycle followed by a 14-min reduction cycle. After the rinse and drying stations, the liners were transferred to an assembly point where the BAg-1 silver brazing ring, type 3 brazing flux, plain carbon steel sleeves, and cast iron liner were automatically assembled. The assembly then passed into a preheat furnace at 900 °F and through an induction coil where the brazing was completed. The brazed liner was slow cooled to 600 °F, washed in a vibratory cleaning system, and finally tested for leakproof joints.

Retention of Strength

Cast irons that contain pearlite or free carbide graphitize and decrease in strength at elevated temperatures. Because graphitization is a function of both time and temperature, some experimentation is usually necessary to develop a brazing cycle that produces acceptable joints without excessive graphitization and decrease in strength.

In Example 2, the assemblies were heated for an hour at 1300 °F maximum, with no significant decrease in strength. In another plant, the assemblies were heated to about 1450 °F without significant decrease in strength. In the latter application, however, heating was by induction instead of in a furnace, and the heating time was much shorter than would have been required in a furnace.

In some plants using furnace brazing, the furnace is operated at a considerably higher temperature than is desired for brazing (sometimes as high as 1600 °F), but using time cycles so short that the assemblies never reach furnace temperature.

For applications in which little or no decrease in strength of the cast iron can be tolerated, it is mandatory to use a filler metal with as low a flow-temperature range as possible, thus permitting a low brazing temperature, and to keep the time at brazing temperature to the minimum.

The temperature required for wetting the base metal and for flow of filler metal having a melting range of 1145 to 1195 °F may vary from about 1275 to 1550 °F, depending on the complexity of the joint design, especially the distance the filler metal must flow. Simple joints with short flow distances can be brazed at lower temperatures than more complex joints.

In Example 4, the salt bath was operated at 1600 °F. This high temperature severely reduces the strength of many cast irons. If this decrease in strength cannot be tolerated, either of two methods can be used for brazing in a salt bath: employment of a salt that can be used without excessive dragout at a temperature lower than 1600 °F, or development of a time cycle short enough that the temperature of the casting never exceeds about 1425 °F.

Fig. 6 Malleable-iron-to-steel fitting assembly dip brazed in molten salt

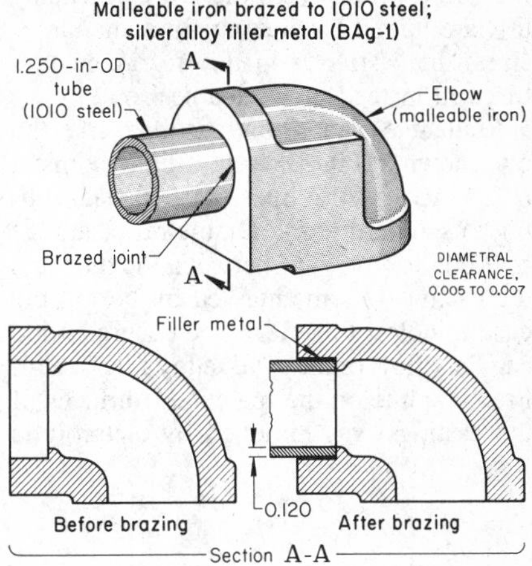

Conditions for salt-bath brazing

Furnace	Pit-type salt bath(a)
Salt	Sodium chloride and potassium chloride(b)
Fixtures	Type 304L stainless steel
Prebraze cleaning:	
Steel tube	Degrease
Cast fitting	Electrolytic salt bath clean(c)
Filler metal	BAg-1
Flux	Type 3A
Brazing bath temperature	1600 °F(d)
Time in salt bath	Varied with load(d)
Postbraze cleaning	Water quench

(a) Brick lined, 48 in. long, 13½ in. wide, and 25 in. deep. Heated by three-phase, 60-cycle, 460-V, 110-kW, submerged electrodes; with on-and-off temperature control and over-temperature alarm. (b) Rectified with methyl chloride to maintain neutrality. (c) See conditions given in table with Fig. 3. (d) Temperature of parts probably did not exceed 1425 °F.

Dip Brazing in a Fused Salt Bath

One of the more common applications for brazing of cast iron is the joining of steel tubing to headers, special fittings, and stanchions. Various mechanical assemblies are brazed to cast iron because different metal properties are needed in different parts of the assembly or, more often, because the fitting portion of such an assembly can be made most economically from cast iron. These mechanical assemblies can be satisfactorily brazed by torch, induction, atmosphere furnace, or salt-bath processes. Selection of process depends primarily on the quantity to be brazed and the equipment available.

Example 4. High-Production Dip Brazing. Dip brazing in fused salt was selected for joining a cast fitting of malleable iron to steel tubing because it was

Fig. 5 Salt-bath cleaning system for diesel locomotive cylinder liners

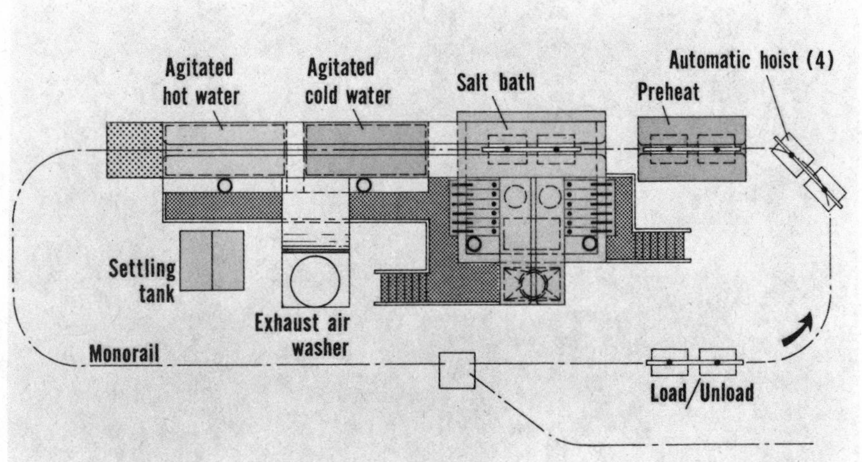

possible to achieve a higher production rate than by other available processes. The particular type and size of fitting assembly discussed in this example is one of many that are joined by brazing. In some larger sizes, the casting is made of gray or ductile iron instead of malleable iron.

Malleable iron elbow castings, like the one shown in Fig. 6, were used to connect hydraulic pressure lines of 1010 steel tubing to standard ports. Diametral clearance of 0.005 to 0.007 in. between the tube and the casting was maintained by boring the casting holes; this clearance ensured good capillary flow of the silver alloy filler metal. Free graphite on the machined surfaces of the castings was removed by electrolytic salt-bath cleaning. The filler metal was BAg-1, and the flux was type 3A.

Three methods of heating for brazing were available: with gas burners, by induction, and in a salt bath. Heating with gas burners was too slow to be economical. Induction heating was satisfactory, but the production rate obtained with the available induction equipment was lower than with the available salt-bath equipment. Therefore, the salt bath was selected.

The salt-bath system included a semiautomatic conveyor (manually loaded and unloaded) that carried the assemblies through the bath, which was maintained at 1600 °F, and through a subsequent water quench, which removed the salt. The time in the salt bath was such that the assembly did not reach bath temperature. Fixtures were of type 304L stainless steel and were designed so that assemblies of various sizes could be accommodated.

In Example 4, the assemblies containing the malleable iron castings were water quenched to remove the brazing salt—a practice that is common with steel assemblies. To avoid warpage and cracking, however, brazed assemblies containing gray and ductile iron castings should never be water quenched from the brazing temperature. Instead, the assemblies should be cooled slowly to about 300 °F and then immersed in water to remove the salt.

Brazing of Stainless Steels*

EXCELLENT RESULTS can be achieved in brazing standard wrought stainless steels. Brazeability, however, as with weldability of these steels, varies with composition. The quality of brazed joints is dependent on judicious selection of brazing process, brazing temperature, filler-metal composition, and protective atmosphere or activating flux. These choices must be compatible with performance in service.

Applicability

Brazing of stainless steels is often used to join these steels to dissimilar metals, including stainless steels of dissimilar composition, carbon steels, low-alloy steels, and copper alloys. The ability of brazing to join a stainless steel to a variety of dissimilar metals in combinations that would not otherwise weld satisfactorily is a principal advantage of the brazing process. A wide variety of brazing filler-metal compositions are available to achieve compatibility, strength, corrosion resistance, and other desirable properties when dissimilar metals are to be joined.

When the most appropriate heating techniques are employed, brazing provides a means to obtain strong, corrosion-resistant, leaktight joints in small or thin-walled components with a minimum of buckling or warpage. Brazing can produce joints in delicate assemblies and in very thin-gage metals that are difficult or impossible to obtain with conventional welding processes. Brazing is also suitable for the mass production of small- and medium-size assemblies on various types of continuous equipment. Finally, brazing enables the production of joints in inaccessible locations by the preplacement of filler metal and subsequent heating of the workpiece; often, such joints cannot be made by other joining processes.

Brazeability

As a class, stainless steels are no more difficult to braze than carbon and low-alloy steels. The high quantities of chromium present in stainless steels cause the chromium oxide films on the surface of all stainless steels, as well as films of titanium oxide that form on the surface of titanium-stabilized stainless steels such as 321. These oxides, which are refractory and strongly adherent, prevent wetting of the base metal by the molten filler metal.

The formation of chromium oxide is accelerated when stainless steels are heated in air. Therefore, although the oxide may have been removed from the surface by chemical cleaning at room temperature, a new oxide layer that seriously interferes with wetting forms rapidly when the steel is heated in air to the brazing temperature. The adverse effect of oxides on wetting can be alleviated by:

- Chemical cleaning of surface oxide from the steel at room temperature, quickly followed by heating to the brazing temperature in a chemically inert gaseous atmosphere, such as argon
- Heating the steel directly to the brazing temperature in a strongly reducing atmosphere, such as dry hydrogen, after less intensive cleaning, which chemically reduces the oxide and thereby promotes wetting action
- Coating the area at the joint with a chemically active flux that dissolves the oxide during heating to the brazing temperature
- Heating and brazing in a vacuum (after cleaning), which thus prevents exposure to oxygen or water vapor and enables effective wetting
- Degreasing alone, if the proper flux or controlled atmosphere is used, prior to brazing stainless steels with clean surfaces (free of black or green oxide coatings)
- Brazing with filler metals which melt at about 1100 °F to reduce oxidation

Inclusions and Surface Contaminants. In brazing stainless steel, base-metal inclusions and surface contaminants are even more deleterious than in brazing carbon steel. Base-metal inclusions, such as oxides, sulfides, and nitrides, interfere with the flow of filler metal. Flow is also impeded by surface contaminants, which may include lubricants, such as oil, graphite, molybdenum, disulfide, and lead, that are applied during machining, forming, and grinding or by aluminum oxide particles produced by grit blasting, or by grinding with aluminum oxide wheels or belts.

Some filler-metal powders (in paste form) used with stainless steel contain organic binders. Acrylics and other plastics are often used for this purpose. Although some binders form a soot residue, this residue does not usually interfere with filler-metal flow.

The brazing characteristics of stainless steel can also be seriously impaired by unsuitable fixturing materials, such as graphite, or by a protective atmosphere with a nitriding potential. Carbon in graphite fixtures unites with hydrogen to form methane (CH_4), which carburizes stainless steel and impairs its corrosion resistance. Dissociated ammonia, unless sufficiently dry and completely (100%) dissociated, nitrides stainless steel. Both carburizing and nitriding interfere with brazing quality.

Brazing Processes. Stainless steels can be brazed by all conventional brazing processes, including furnace, torch, induction, resistance, and salt-bath dip brazing. Furnace brazing is most widely used, because applications generally require brazing in a prepared atmosphere or vacuum. Furnace brazing is used in most of the brazing applications presented in this article.

Filler Metals

Most stainless steels can be brazed with any one of several different filler metals, including silver alloys, nickel alloys, copper, and gold alloys. In most applications, filler metals are selected for mechanical

*Revised by A.J. Moorhead, Group Leader, Oak Ridge National Laboratory; S.J. Whalen, President, Aerobraze Corp.

properties, corrosion resistance, service temperature, and compatibility, rather than for brazeability. Table 1 lists composition requirements for the filler metals most often used in brazing of stainless steels.

Silver Alloys. The most widely used filler metals for brazing stainless steel are the silver alloys (the BAg group). Alloy BAg-3, which contains 3% Ni, is probably the silver alloy selected most frequently, although several other silver alloys can also be used successfully.

Silver brazed joints cannot be used for high-temperature service; the recommended maximum service temperature is 700 °F (BAg-13). Recommended allowances on joint fit for silver brazing are relatively loose—generally, 0.002- to 0.004-in. diametral clearance.

Of the silver alloy filler metals shown in Table 1, all except BAg-19, and possibly BAg-13, are used at brazing temperatures that fall within the effective range of sensitizing temperatures (1000 to 1600 °F) for austenitic stainless steels. Chromium carbide precipitation occurs in the sensitizing temperature range, resulting in impairment of the corrosion resistance of the base metal. Carbide precipitation, however, depends on time as well as temperature, and exposure to the sensitizing-temperature range for only a few minutes is unlikely to result in a significant amount of precipitate. Nevertheless, the lower melting temperatures of the silver alloys prohibit re-solution treatment of the base metal after brazing, and if corrosion resistance in service is sufficiently critical, an extra-low-carbon, titanium-stabilized, or niobium-tantalum-stabilized type should be selected instead of a nonstabilized type of austenitic steel.

Ferritic and martensitic stainless steels that contain little or no nickel are susceptible to interface corrosion in plain water or moist atmospheres, when brazed with nickel-free silver alloy filler metals using a liquid or paste flux. Filler metal containing nickel helps to prevent interface corrosion. However, for complete protection, special brazing alloys containing nickel and tin should be used, and brazing should be done in a protective atmosphere without flux.

Most silver alloy filler metals contain appreciable amounts of copper and zinc, singly or in combination. Overheating or heating for an excessive period of time may result in extensive penetration of grain boundaries by copper and zinc, thereby embrittling the brazed joint.

Cadmium, which is added to some silver alloy filler metals to lower the melting temperature, also penetrates grain boundaries. The effect is accelerated by brazing parts under tensile stress. Cadmium-containing fumes are extremely toxic, and operators must take every precaution to avoid inhaling them. Filler metals containing cadmium should not be used in a protective atmosphere, including vacuum, because cadmium gases off readily. Sim-

Table 1 Typical compositions and properties of standard brazing filler metals for brazing stainless steels

Filler metal	Composition(a), % Ag	Cu	Zn	Cd	Ni	Sn	Li	Mn	Other elements total	Solidus temperature, °F	Liquidus temperature, °F	Brazing temperature range, °F
Silver alloys												
BAg-1	44.0-46.0	14.0-16.0	14.0-18.0	23.0-25.0	...	...	...	...	0.15	1125	1145	1145-1400
BAg-1a	49.0-51.0	14.5-16.5	14.5-18.5	17.0-19.0	...	...	...	...	0.15	1160	1175	1175-1400
BAg-2	34.0-36.0	25.0-27.0	19.0-23.0	17.0-19.0	...	...	...	...	0.15	1125	1295	1295-1550
BAg-2a	29.0-31.0	26.0-28.0	21.0-25.0	19.0-21.0	...	...	...	...	0.15	1125	1310	1310-1550
BAg-3	49.0-51.0	14.5-16.5	13.5-17.5	15.0-17.0	2.5-3.5	...	...	...	0.15	1170	1270	1270-1500
BAg-4	39.0-41.0	29.0-31.0	26.0-30.0	...	1.5-2.5	...	...	...	0.15	1240	1435	1435-1650
BAg-5	44.0-46.0	29.0-31.0	23.0-27.0	...	...	...	...	...	0.15	1250	1370	1370-1550
BAg-6	49.0-51.0	33.0-35.0	14.0-18.0	...	...	...	...	...	0.15	1270	1425	1425-1600
BAg-7	55.0-57.0	21.0-23.0	15.0-19.0	...	...	4.5-5.5	...	...	0.15	1145	1205	1205-1400
BAg-8	71.0-73.0	rem	...	...	...	...	...	...	0.15	1435	1435	1435-1650
BAg-8a	71.0-73.0	rem	...	...	...	...	0.25-0.50	...	0.15	1410	1410	1410-1600
BAg-9	64.0-66.0	19.0-21.0	13.0-17.0	...	...	...	...	...	0.15	1240	1325	1325-1550
BAg-10	69.0-71.0	19.0-21.0	8.0-12.0	...	...	...	...	...	0.15	1275	1360	1360-1550
BAg-13	53.0-55.0	rem	4.0-6.0	...	0.5-1.5	...	...	...	0.15	1325	1575	1575-1775
BAg-13a	55.0-57.0	rem	...	...	1.5-2.5	...	...	...	0.15	1420	1640	1600-1800
BAg-18	59.0-61.0	rem	...	...	...	9.5-10.5	...	...	0.15	1115	1325	1325-1550
BAg-19	92.0-93.0	rem	...	...	...	...	0.15-0.30	...	0.15	1400	1635	1610-1800
BAg-20	29.0-31.0	37.0-39.0	30.0-34.0	...	...	...	...	...	0.15	1250	1410	1410-1600
BAg-21	62.0-64.0	27.5-29.5	...	...	2.0-3.0	5.0-7.0	...	...	0.15	1275	1475	1475-1650
BAg-22	48.0-50.0	15.0-17.0	21.0-25.0	...	4.0-5.0	...	...	7.0-8.0	0.15	1260	1290	1290-1525
BAg-23	84.0-86.0	...	...	...	...	...	...	rem	0.15	1760	1780	1780-1900
BAg-24	49.0-51.0	19.0-21.0	26.0-30.0	...	1.5-2.5	...	...	...	0.15	1220	1305	1305-1550
BAg-25	19.0-21.0	39.0-41.0	33.0-37.0	...	...	...	...	4.5-5.5	0.15	1360	1455	1455-1555
BAg-26	24.0-26.0	37.0-39.0	31.0-35.0	...	1.5-2.5	...	...	1.5-2.5	0.15	1305	1475	1475-1600
BAg-27	24.0-26.0	34.0-36.0	24.5-28.5	12.5-14.5	...	...	...	...	0.15	1125	1375	1375-1575
BAg-28	39.0-41.0	29.0-31.0	26.0-30.0	...	...	1.5-2.5	...	...	0.15	1200	1310	1310-1550

Filler metal	Composition(a), % Cu	Zn	Sn	Fe	Mn	Ni	P	Pb	Al	Si	Other elements total	Solidus temperature, °F	Liquidus temperature, °F	Brazing temperature range, °F
Copper alloys													
BCu-1	99.90 min	...	...	...	...	...	0.075	0.02	0.01	...	0.10	1981	1981	2000-2100
BCu-1a	99.0 min	...	...	...	...	...	...	...	...	...	0.30	1981	1981	2000-2100
BCu-2	86.5 min	...	...	...	...	...	...	...	...	...	0.50	1981	1981	2000-2100

(continued)

Table 1 (continued)

Filler metal	Cr	B	Si	Fe	C	P	S	Al	Ti	Mn	Cu	Zr	Ni	Other elements total	Solidus temperature, °F	Liquidus temperature, °F	Brazing temperature range, °F
	Composition(a), %																
Nickel alloys																	
BNi-1	13.0-15.0	2.75-3.50	4.0-5.0	4.0-5.0	0.6-0.9	0.02	0.02	0.05	0.05	...	...	0.05	rem	0.50	1790	1900	1950-2200
BNi-1a	13.0-15.0	2.75-3.50	4.0-5.0	4.0-5.0	0.06	0.02	0.02	0.05	0.05	...	...	0.05	rem	0.50	1790	1970	1970-2200
BNi-2	6.0-8.0	2.75-3.50	4.0-5.0	2.5-3.5	0.06	0.02	0.02	0.05	0.05	...	...	0.05	rem	0.50	1780	1830	1850-2150
BNi-3	...	2.75-3.50	4.0-5.0	0.5	0.06	0.02	0.02	0.05	0.05	...	...	0.05	rem	0.50	1800	1900	1850-2150
BNi-4	...	1.5-2.2	3.0-4.0	1.5	0.06	0.02	0.02	0.05	0.05	...	...	0.05	rem	0.50	1800	1950	1850-2150
BNi-5	18.5-19.5	0.03	9.75-10.50	...	0.10	0.02	0.02	0.05	0.05	...	...	0.05	rem	0.50	1975	2075	2100-2200
BNi-6	...	...	...	...	0.10	10.0-12.0	0.02	0.05	0.05	...	...	0.05	rem	0.50	1610	1610	1700-2000
BNi-7	13.0-15.0	0.01	0.10	0.2	0.08	9.7-10.5	0.02	0.05	0.05	0.04	...	0.05	rem	0.50	1630	1630	1700-2000
BNi-8	...	...	6.0-8.0	...	0.10	0.02	0.02	0.05	0.05	21.5-24.5	4.0-5.0	0.05	rem	0.50	1800	1850	1850-2000

Filler metal	Au	Cu	Pd	Ni	Other elements total	Solidus temperature, °F	Liquidus temperature, °F	Brazing temperature range, °F
	Composition(a), %							
Precious metal alloys								
BAu-1	37.0-38.0	rem	...	...	0.15	1815	1860	1860-2000
BAu-2	79.5-80.5	rem	...	...	0.15	1635	1635	1635-1850
BAu-3	34.5-35.5	rem	...	2.5-3.5	0.15	1785	1885	1885-1995
BAu-4	81.5-82.5	...	...	rem	0.15	1740	1740	1740-1840
BAu-5	29.5-30.5	...	33.5-34.5	35.5-36.5	0.15	2075	2130	2130-2250
BAu-6	69.5-70.5	...	7.5-8.5	21.5-22.5	0.15	1845	1915	1915-2050

Filler metal	Cr	Ni	Si	W	Fe	B	C	P	S	Al	Ti	Zr	Co	Other elements total	Solidus temperature, °F	Liquidus temperature, °F	Brazing temperature range, °F
	Composition(a), %																
Cobalt alloys																	
BCo-1	18.0-20.0	16.0-18.0	7.5-8.5	3.5-4.5	1.0	0.7-0.9	0.35-0.45	0.02	0.02	0.05	0.05	0.05	rem	0.50	2050	2100	2100-2250

(a) Single values are maximum percentages, unless otherwise indicated.
Source: AWS A5.8-81, "Specification for Brazing Filler Metals"

ilarly, zinc gases in a protective atmosphere. Because the loss of zinc and cadmium raises the melting point of the filler metal, the brazeability is adversely affected.

Virtually all of the silver alloy filler metals are suitable for brazements used in fabricating vacuum chambers and pumps where pressures that are as low as 10^{-5} torr are encountered. For high-vacuum work (pressures of less than 10^{-6} torr), however, the filler metal must not contain cadmium or zinc. Vaporization of these metals interferes with the production of the vacuum and can contaminate the vacuum chamber and pumps. Selection of a suitable silver alloy filler metal for service in a vacuum at pressure below 10^{-5} torr is considered in the following example.

Example 1. Selection of Filler Metal for Use in High Vacuum. The sleeve-and-tube assembly shown in Fig. 1 is typical of brazements used in vacuum systems. The tubes and the sleeve were of type 304L austenitic stainless steel. Filler metal BAg-18 (60%Ag-10%Sn-30%Cu), which has proved satisfactory for brazing of assemblies used in high-vacuum applications, was used with type 3B flux. This silver alloy filler metal contains no zinc or cadmium.

Brazing was done with a manually manipulated oxyacetylene torch, employing a strongly reducing flame. The assembly shown in Fig. 1 could have been brazed in a furnace or by induction, but production was small and did not justify the investment for such equipment.

Nickel Alloys. After the silver alloys, nickel alloys usually rank next in frequency of use as brazing filler metals for stainless steels. Nickel alloy filler metals provide joints that have excellent corrosion resistance and high-temperature strength. These filler metals alloy with stainless steel, however, and form phases with two undesirable characteristics: the phases are considerably less ductile than either the base metal or the filler metal, even at elevated temperatures and, thus, are a potential source of rupture; and the alloys formed with stainless steel are higher melting alloys that are likely to freeze and block further flow into the joint during brazing.

To achieve flow in deep joints, diametral clearances of as much as 0.004 to 0.008 in. are necessary. Knurling of male members sometimes helps in centering loosely

Fig. 1 Assembly, torch brazed with a silver alloy filler metal, for use in a high-vacuum system

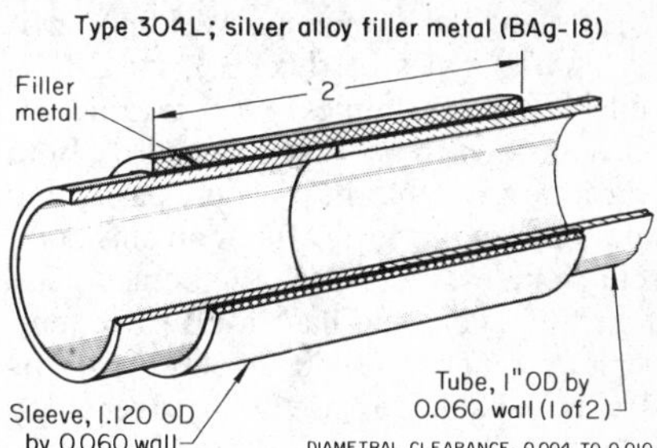

fitting components. With such large clearances, joints brazed with nickel alloy filler metals do not develop their greatest strength. With 0.001- to 0.003-in. clearance, the brazed joint is stronger.

Because of the relatively high brazing temperatures required for the nickel alloy filler metals, their use is generally restricted to furnace brazing in a controlled atmosphere (including vacuum), although there are occasional exceptions (Fig. 12).

Copper Filler Metals. The high brazing temperature and the need for a protective atmosphere generally restrict the use of copper filler metal to furnace brazing. These filler metals (the BCu group) melt at about 1980 °F and flow freely at 2050 °F.

Copper is not recommended for exposure to certain corrosive substances, such as the sulfur in jet fuel and in sulfur-bearing atmospheres. Furthermore, copper filler metals exhibit poor oxidation resistance at elevated temperatures and should not be exposed to service temperatures above 800 °F. When a copper filler metal is used, recommended diametral allowances on joint fit range from 0.004-in. clearance to 0.002-in. interference.

Gold alloys (the BAu group) are sometimes used for brazing stainless steel, although their high cost restricts the use of these filler metals to specialized applications, such as fabrication of aerospace equipment (Example 3). When a gold alloy is used, alloying with the stainless steel base metal is minimized, and as a result, joints exhibit good ductility.

Cobalt Alloys. Cobalt-based filler metals are very rarely used for brazing stainless steel. The alloy is included in Table 1, however, is available for that purpose.

Fluxes

For furnace brazing in strongly reducing or inert atmospheres, flux usually is not required. In some furnace brazing applications, however, a flux is necessary. A flux is always required for torch brazing and is usually required for induction and resistance brazing, unless atmospheric protection is provided.

Either of two American Welding Society (AWS) types of flux (3A and 3B) is suitable for all stainless steel brazing applications where a flux is needed. Both types are available in powder, paste, and liquid forms. Type 3A flux contains boric acid, borates, fluorides, fluoborates, and a wetting agent and has an effective temperature range of 1050 to 1600 °F. This flux is suitable for use with silver alloy filler metals.

Type 3B flux contains the same ingredients as type 3A, but not in the same proportions, and has a higher effective temperature range, 1350 to 2100 °F. This type of flux is often selected for use with silver alloy filler metals if the brazing temperature is above 1350 °F and is well suited for use with copper, nickel alloy, and gold alloy filler metals.

Furnace Atmospheres

Almost all furnace brazing of stainless steel is done in a protective atmosphere, or vacuum. One exception is the application described in Example 9, in which air was the furnace atmosphere. In this application, a liquid flux was used, and the time at brazing temperature was short, which helped to prevent excessive oxidation.

The protective atmospheres most often used in furnace brazing of stainless steel are dry hydrogen and dissociated ammonia. These atmospheres are effective in reducing oxides, protecting the base metal, and promoting the flow of filler metal. The low-cost exothermic atmospheres that are widely used in furnace brazing of low-carbon steel are not suitable for stainless steel. An inert gas such as argon or vacuum may be used to satisfy special requirements and provide protection in applications for which hydrogen or hydrogen-bearing gases are unsatisfactory.

Selection of furnace atmosphere depends on the degree of protection that must be given to the base metal or metals, the flow characteristics of the filler metal, the brazing temperature, and cost. Special requirements that arise from the brazing of dissimilar metals are often a major factor in atmosphere selection. Availability of equipment may also be important.

Furnace Brazing in Dry Hydrogen

A dry hydrogen atmosphere is preferred for many applications of brazing stainless steel. Hydrogen, the most strongly reducing of protective atmospheres, reduces chromium oxide and provides for excellent wetting by some filler metals without the need for flux. The principal disadvantages of hydrogen are high cost, difficulty in drying sufficiently, need for special furnace equipment, and danger involved in storing and handling hydrogen. The following examples describe applications in which a specific type of stainless steel was joined to the same type or to another stainless steel.

Example 2. Selection of BAg-13 Silver Alloy Filler Metal for Brazing at 1700 °F. The type 347 stainless steel retainer assembly shown in Fig. 2 was furnace brazed in dry hydrogen, using BAg-13 silver alloy filler metal. This filler metal was selected in preference to lower melting silver alloys because the upper limit of the brazing temperature range of BAg-13 (1575 to 1775 °F) permitted brazing at 1700 °F. At furnace temperatures above 1600 °F, dry hydrogen is strongly reducing, and the use of a brazing flux is not required for satisfactory wetting action. Thus, by judicious selection of filler metal and furnace atmosphere, the extra cost of applying a flux and removing flux residue after brazing was avoided.

The components were vapor degreased, assembled with outside diameters concentric, and (to make them self-jigging) were spot welded at four locations 90° apart (Fig. 2). A ring of 0.040-in.-diam filler-metal wire was preplaced at the joint, and assemblies were loaded two-across on the mesh belt of a conveyor-type furnace. The heating chamber of the furnace was elevated from the entrance and discharge level to conserve the lighter-than-air hydrogen

Fig. 2 Retainer assembly furnace brazed with BAg-13 filler metal

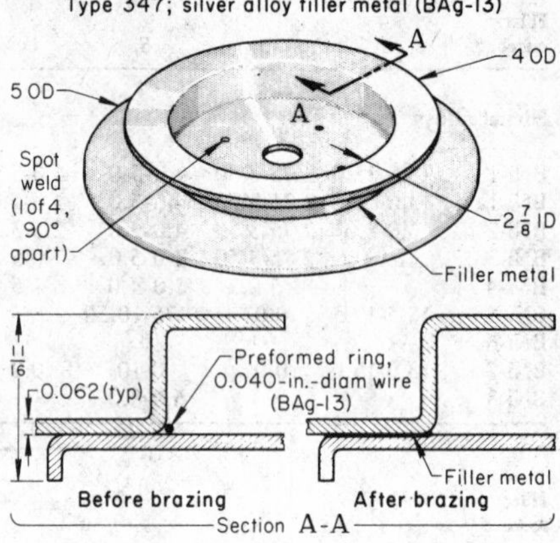

Furnace brazing in dry hydrogen

Furnace	Continuous conveyor(a)
Fixtures	None
Furnace temperature	1800 ± 10 °F
Brazing temperature	1700 ± 10 °F
Hydrogen dew points	−100 °F (incoming); −70 °F (exhaust)
Hydrogen flow rate	400 ft^3/h
Filler metal	BAg-13, 0.040-in.-diam wire(b)
Joint position during brazing	Horizontal
Conveyor travel speed	30 ft/h
Time at brazing temperature	5 min(c)
Production rate	120 assemblies per hour

(a) Electrically heated (60 kW), constructed with heating chamber (6 in. high, 12 in. wide, 36 in. long) higher than entrance and discharge ends. (b) Preformed full rings. (c) Cooled in hydrogen atmosphere to room temperature and prevent oxygen in the atmosphere from mixing with the hydrogen to either raise its dew point or cause an explosion.

Quality standards for brazed assemblies, which were checked by 100% visual inspection, required that the joint exhibit full braze penetration (360° fillets on both sides of the joint) and be pressure-tight.

Example 3. Use of a Gold Alloy Filler Metal for Brazing in an Aerospace Heat Exchanger. In the fabrication of a high-reliability heat exchanger for manned space flights, 2552 fins of type 347 stainless steel 0.004 in. thick were brazed to 0.025-in.-thick type 347 stainless steel side panels, as shown in Fig. 3. The 5104 fin-to-panel joints had to be strong and corrosion resistant.

Silver alloy and copper brazing filler metals could not be used because of incompatibility with sulfur-bearing rocket fuel. The BNi series of nickel alloys had the necessary compatibility, but made brittle joints that were unreliable under tension peel stress. Therefore, gold alloy filler metals were used. The necessary brazing characteristics for the fin-to-panel joints were found in BAu-4 (nominal composition: 82%Au-18%Ni). The strength and toughness of the brazed joints made with this alloy justified its high cost.

The fins and side panels were cleaned by vapor degreasing. The side panels were pickled, rinsed in clean water, and dried. The filler metal was coated on the panels in the form of a powder suspended in an organic binder. Multiple lap joints were made between the flat-crown-hairpin ends of the fins and the flat side panels. The assembly was placed in a fixture (Fig. 3), and the entire assembly and fixture were placed in the retort of a bell-type furnace and sealed. The sealed retort was purged with a volume of hydrogen equivalent to five times that of the retort. The retort was then heated to the brazing temperature of 1860 °F and held for 7 to 10 min. The joint gaps at brazing temperature were 0.000 to 0.010 in. After brazing, the retort was purged with argon while being cooled to 300 °F and was not opened until after purging and cooling.

The joints brazed by this procedure were the final brazed joints in the assembly. In a prior brazing operation, tubes had been joined to the fins (Fig. 3) by brazing at 1970 °F using a higher melting gold alloy filler metal composed of 70%Au-22%Ni-8%Pd.

Completed assemblies were visually inspected and pressure tested at pressures that far exceeded the pressures to be encountered in service. Typical service pressures were 270 psi on the outside of the tubes and 1710 psi on the inside. Acceptance pressure tests were at 540 psi on the outside of the tubes and 2275 psi on the inside. Selected brazed assemblies were tested to bursting. These samples were required to withstand at least three times the service pressures before bursting. The assemblies brazed with gold alloy filler metals passed all tests and had three times the bursting strength of assemblies brazed with nickel alloys.

Fig. 3 Heat-exchanger assembly in brazing fixture and detail of joints brazed with gold alloy filler metal

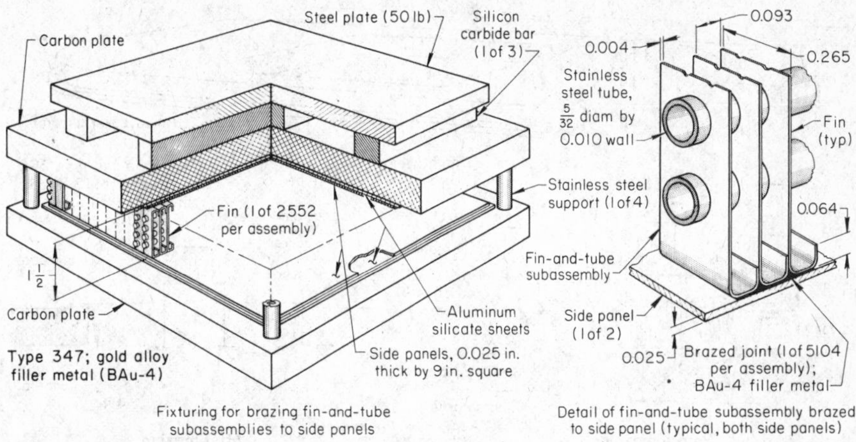

Furnace brazing in dry hydrogen		Processing time per assembly	
Furnace	Bell(a)	Clean components	45 min
Fixtures	(See illustration)	Preplace filler metal	$1^1/_4$ h
Brazing temperature	1860 °F	Assemble components in fixture	4 h
Hydrogen dew point (max)	−80 °F(b)	Time at brazing temperature	7-10 min
Purging	5 volume changes in retort	Total time in furnace(d)	4 h
Filler metal	BAu-4(c)	Inspect	1 h
Number of assemblies per load	1	Pressure test	40 h

(a) Electrically heated, with 36-in.-diam retort with water-cooled rubber seals. (b) Hydrogen was purchased as cylinder hydrogen, then passed through an electrolytic drier. (c) 200-mesh powder suspended in an organic binder. (d) Including cooling to 300 °F in retort, which was purged with argon before being opened

Example 4. Combination Brazing and Solution Heat Treatment of an Assembly of Three Types of Stainless Steel. Three different stainless steels were selected to make the cover for a hermetically sealed switch. The switching action had to be transmitted through the cover without breaking the seal. This was accomplished by providing a diaphragm through which a shouldered pin was inserted, as shown in Fig. 4. The switch was actuated by depressing the pin, which in turn deflected the diaphragm. The pin (type 303), the diaphragm (PH 15-7 Mo), and the cover (type 305) were assembled as shown in Fig. 4 and furnace silver brazed in dry hydrogen.

Silver alloy filler metal BAg-19 was chosen because it flowed at a temperature that coincided with the solution heat treating temperature for the PH 15-7 Mo diaphragm (1750 °F). A holding fixture was needed to keep the PH 15-7 Mo diaphragm in position during the brazing cycle; to avoid carburizing the diaphragm, the material selected for the fixture was stainless steel, rather than graphite. The furnace was a batch-type tube furnace with a 5-in.-diam high-heat zone 18 in. long. Moisture content of the hydrogen atmosphere was carefully controlled, because with too dry an atmosphere, the lithium-containing filler metal flowed too freely and did not seal the joints.

After being cleaned, the components were assembled with two preformed rings of BAg-19 wire; tweezers were used to avoid contamination of the cleaned surfaces. Each assembly was held in a stainless steel fixture, which in turn was placed on a stainless steel furnace sled. The sled was pushed into the high-heat zone of the furnace and held at 1750 °F for 10 min, pulled into an intermediate cooling zone at 1000 °F and held for 5 min, and finally pulled to the water-cooled zone, where it cooled to room temperature. Brazing of the two joints and solution treating of the PH 15-7 Mo diaphragm were accomplished simultaneously at the brazing temperature (1750 °F). To complete the heat

Fig. 4 Three-steel switch-cover assembly which combined brazing temperature as part of solution heat treatment

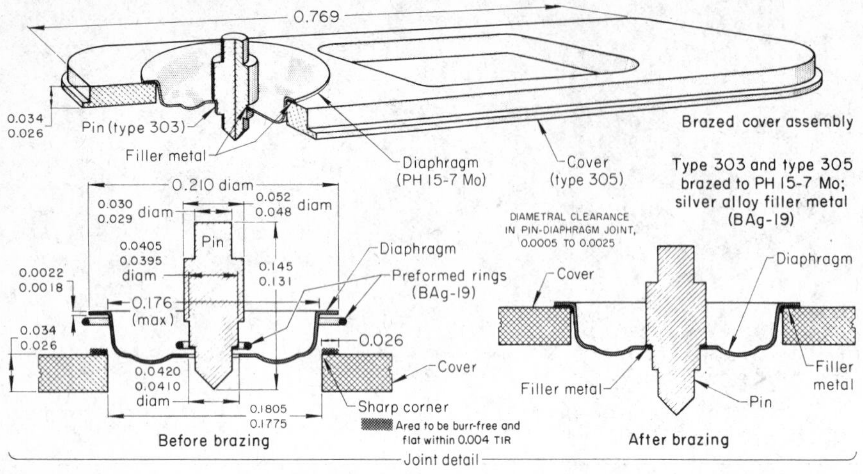

Furnace brazing in dry hydrogen

Furnace	Batch-type tube(a)	Time at brazing temperature	10 min
Fixture material	Stainless steel(b)	Time in first cooling zone (1000 °F)	5 min
Brazing temperature	1750 ± 15 °F	Time in final cooling zone(d)	5 min
Filler metal	BAg-19 wire(c)	Production in 8 h	1000 assemblies

(a) Three-zone furnace, with a high-heat zone 5 in. in diameter by 18 in. long. (b) Fixture located and held components of assembly together and was placed on a stainless steel sled for transport through the furnace. (c) Preformed rings. (d) Water-cooled zone, in which assembly was cooled to room temperature. Then, to complete heat treatment of the PH 15-7 Mo diaphragm, assembly was cooled to −100 °F and held for 8 h, then aged at 950 °F for 1 h.

treating process, the assembly was cooled to −100 °F, held for 8 h, and then aged at 950 °F for 1 h.

A 1-in.-square piece of PH 15-7 Mo was processed with each batch of cover assemblies and was used as a hardness test specimen to verify that the diaphragms had been correctly heat treated. Brazed assemblies were inspected by the brazing operator. The joints were required to be fully sealed and have no voids. The pins were required to be perpendicular within 4°. Perpendicularity was measured on a comparator. Randomly selected samples were given a push-out test, in which joints had to withstand a push of 14 lb. All assemblies were given 100% visual inspection at a magnification of 13 diameters.

Example 5. Simultaneous Brazing in a Heat-Exchanger Assembly. An air-to-air heat-exchanger assembly, shown in Fig. 5, consisted of 185 thin-walled (0.008-in.) tubes and two $^1/_{16}$-in.-thick headers. All components were made of type 347 stainless steel. The tubes were assembled with the headers by flaring the tube ends to lock them in place and provide metal-to-metal contact for brazing filler metal. All 370 joints were brazed during a single pass through a continuous conveyor-type electric furnace.

Although a nickel alloy filler metal was preferred for this high-temperature application, because of the resistance to heat and corrosion that nickel alloys provide, selection among the nickel alloys presented a problem. Higher melting, boron-containing nickel filler metals, such as BNi-1 and BNi-3, alloy with the base metal and therefore are likely to erode thin materials. Fortunately, the extent of erosion can be modified by control of the brazing temperature and time, and the amount of filler metal.

Although there are nickel filler-metal alloys that contain silicon in place of boron, these alloys generally require much higher brazing temperatures, which can result in grain coarsening in the base metal. Therefore, after numerous tests, BNi-3 filler metal was selected on the basis of brazing temperature and excellent fluidity, and the problem of applying the correct amount of filler metal to avoid erosion was solved by preparing a slurry from an accurately controlled mixture of filler-metal powder, acrylic-resin binder, and xylene thinner.

Before the filler metal was applied, the heat-exchanger assembly was cleaned ultrasonically in acetone and was carefully weighed to determine the proportionate weight of filler metal required. One half of the total amount of filler metal was then applied to one end of the assembly by spraying. The assembly was reweighed, and the remainder of the filler metal was applied to the opposite end. At all stages of processing, the assembly was handled by operators wearing clean, lint-free white cotton gloves.

The assembly was placed on a holding fixture made of stainless steel sheet, to which a stop-off compound had been applied to prevent the assembly from brazing to the fixture if the brazing filler metal flowed excessively. Assemblies were placed 12 in. apart on the conveyor belt as they traveled through the furnace at 30 ft/h under protection of dry hydrogen.

After brazing, both sides of each joint were subjected to 100% visual inspection for the presence of fillets; assemblies were pressure tested in accordance with customer requirements. Because of the thin-walled (0.008-in.) tubing, brazing this assembly provided more consistent and less costly results than could have been achieved by other joining processes.

Example 6. Combined Brazing and Hardening of a Shaft Assembly. The shaft assembly shown in Fig. 6 consisted of three bars or screw machine products (a shaft, a drive pin, and a guide pin) and

Fig. 5 Heat-exchanger assembly with tube-to-header joints brazed in one pass through a furnace

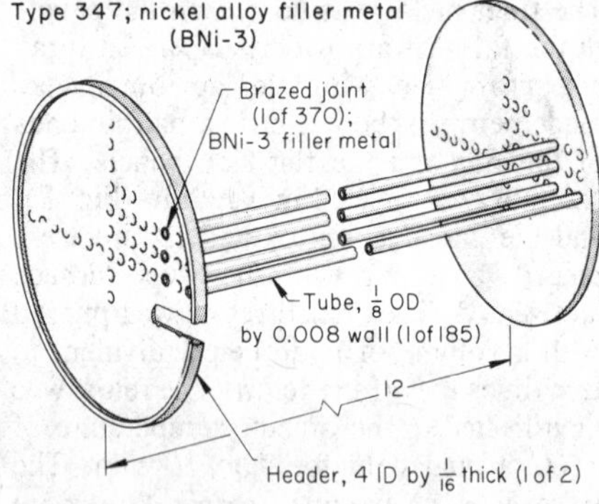

Furnace brazing in dry hydrogen

Furnace	Continuous conveyor(a)
Fixture material	Type 347 stainless steel(b)
Furnace temperature	2050 ± 10 °F
Brazing temperature	1950 ± 10 °F
Hydrogen dew points	−100 °F (incoming); −70 °F (exhaust)
Hydrogen flow rate	600 ft³/h
Filler metal	BNi-3 powder(c)
Conveyor travel speed	30 ft/h
Time at brazing temperature	5 min
Cooling	In hydrogen atmosphere
Production rate	15 assemblies per hour

(a) Electrically heated (60 kW), constructed with heating chamber (6 in. high, 12 in. wide, 36 in. long) higher than entrance and discharge ends. (b) Holding fixture fabricated from $^1/_8$-in.-thick sheet. (c) Mixed to a slurry with acrylic resin and xylene thinner; powder-to-vehicle ratio, 70 to 30

Fig. 6 Four-joint shaft assembly that was simultaneously furnace brazed and heated for hardening

Type 410; copper filler metals (BCu-1 wire, BCu-2 paste); Rockwell C 40 (min) after brazing

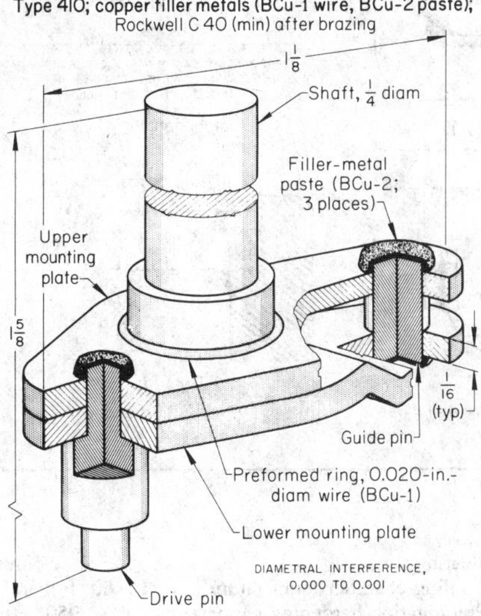

Furnace brazing in dry hydrogen

Furnace	Continuous conveyor(a)
Fixtures	None(b)
Furnace temperature	2150 ± 10 °F
Brazing temperature	2050 ± 10 °F
Hydrogen dew points	−100 °F (incoming); −70 °F (exhaust)
Hydrogen flow rate	400 ft^3/h
Filler metal	BCu-1 wire, BCu-2 paste(c)
Conveyor travel speed	20 ft/h
Time at brazing temperature	8 min(d)
Cooling	In hydrogen atmosphere
Production rate	800 assemblies per hour

(a) Electrically heated (60 kW), constructed with heating chamber (6 in. high, 12 in. wide, 36 in. long) higher than entrance and discharge ends. (b) Components were staked, for self-fixturing. Assemblies, supported by ceramic spacers to keep shaft end up, were brazed on trays. (c) 0.020-in.-diam wire in a preformed ring; paste was applied at one end of drive pin, both ends of guide pin. (d) Assemblies were in high-heat zone for about 10 min.

two stampings (upper and lower mounting plates), all made from type 410 stainless steel, that were furnace brazed together with four joints. By brazing with copper filler metal at 2050 °F, it was possible to austenitize and harden the assembly to the required minimum hardness of 40 HRC during the brazing and cooling operations, thereby avoiding a separate hardening operation after brazing.

Because the joints were all relatively short, an interference fit of 0.000 to 0.001 in. was satisfactory. Normally, with longer joints in stainless steel, a clearance fit between mating parts is required. Automatic staking of components was used to make the assembly self-fixturing.

As shown in Fig. 6, a full ring of 0.020-in.-diam BCu-1 copper wire was preplaced around the 1/4-in.-diam shaft to braze the shaft to the upper and lower mounting plates, and a small amount of BCu-2 copper paste was applied at one end of the drive pin to braze it to the two mounting plates. Because of the separation between the two plates on the guide-pin side, a small amount of BCu-2 copper paste was placed on each end of the guide pin. The BCu-2 copper paste was applied manually. The assemblies were placed in brazing trays, with the shaft in a vertical position, and were supported in this position by ceramic spacers.

The brazing trays were placed on the mesh belt of a continuous conveyor furnace containing a dry hydrogen atmosphere and were transported up an incline to the horizontal preheat and high-heat chambers at a speed of 20 ft/h. Because the assemblies were small, they heated to brazing temperature in about 2 min. After 8 min at brazing temperature, the assemblies were conveyed into water-jacketed cooling chambers, where they cooled rapidly in the hydrogen atmosphere to room temperature. Brazed assemblies emerged from the exit end of the furnace bright and free of oxidation.

The brazed assemblies were visually inspected 100% for complete joint coverage. Hardness tests on a sampling basis were used to determine whether the assemblies had responded properly to hardening. Tempering to the desired final hardness followed the simultaneous brazing and hardening operation.

Furnace Brazing in Dissociated Ammonia

Dissociated ammonia, when it is free of moisture and 100% dissociated, is a suitable atmosphere for brazing stainless steel with some filler metals without need for a flux. Dissociated ammonia is strongly reducing, but less so than pure dry hydrogen. Consequently, although it will promote wetting action by reducing chromium oxide on the surface of stainless steel, dissociated ammonia may not be sufficiently reducing to promote the flow of some filler metals, such as copper oxide powders. Because of its high (75%) hydrogen content, dissociated ammonia forms explosive mixtures with air and must be handled with the same precautions as those required for the handling of hydrogen.

A dissociated-ammonia atmosphere is prepared by heating anhydrous liquid ammonia in the presence of an iron or nickel catalyst. The decomposition of ammonia to form hydrogen and nitrogen begins at 600 °F, and the rate of decomposition increases with temperature. Unless the atmosphere used in brazing stainless steel is completely decomposed (100% dissociated), even minute amounts of raw ammonia (NH_3) in the atmosphere will nitride stainless steel, especially steels containing little or no nickel. In addition, because of the solubility of ammonia in water, the atmosphere coming from the dissociator must be extremely dry (preferably having a dew point of −80 °F or lower). To ensure a very low dew point, the atmosphere coming from the dissociator is commonly processed by passing through a molecular-sieve drier. To avoid oxidation of base metal and filler metal, the atmosphere must be kept pure and dry while it is inside the furnace. In the following examples of production practice, dissociated ammonia was used successfully in the furnace brazing of austenitic and precipitation-hardening stainless steels.

Example 7. Brazing in Dissociated Ammonia Without Flux. The pressure-gage subassembly shown in Fig. 7 was composed of five diaphragms of 17-7 PH stainless steel, a deep drawn cup and a top fitting of type 304 stainless steel, and a connector of copper alloy 145 (tellurium-bearing copper). Originally, these subassemblies were furnace brazed with a silver alloy filler metal that required a flux. Because applying flux and assembling the fluxed components with gloved hands was time consuming, the decision was made to change to fluxless brazing in an atmosphere of dissociated ammonia. Although this necessitated using a more expensive filler metal (BAg-19), higher cost was offset by the greater productivity of each operator. In addition, subassemblies brazed with BAg-19 in dissociated ammonia exhibited fewer leaks and had improved corrosion resistance and better appearance than those brazed with the original filler metal and with flux.

Prior to brazing, the deep drawn type 304 cups were fully annealed at 2000 °F. Annealing served to avoid erosive penetration of filler metal in zones of high residual stress. All components were chemically cleaned and then were assembled by hand, together with seven preplaced rings of filler metal. The assemblies were placed on holding fixtures, and the fixtures were loaded on the belt of a conveyor furnace heated to 1800 °F. The cooling chamber of the furnace was cooled to below 60 °F to ensure rapid cooling of the 17-7 PH diaphragms from the solution treating temperature, thereby combining solution treating with the brazing operation.

After brazing, the assemblies were cooled to −40 °F, dried, and then heated to 950 °F in dry dissociated ammonia to

Fig. 7 Pressure-gage subassembly that combined furnace brazing with solution heat treatment

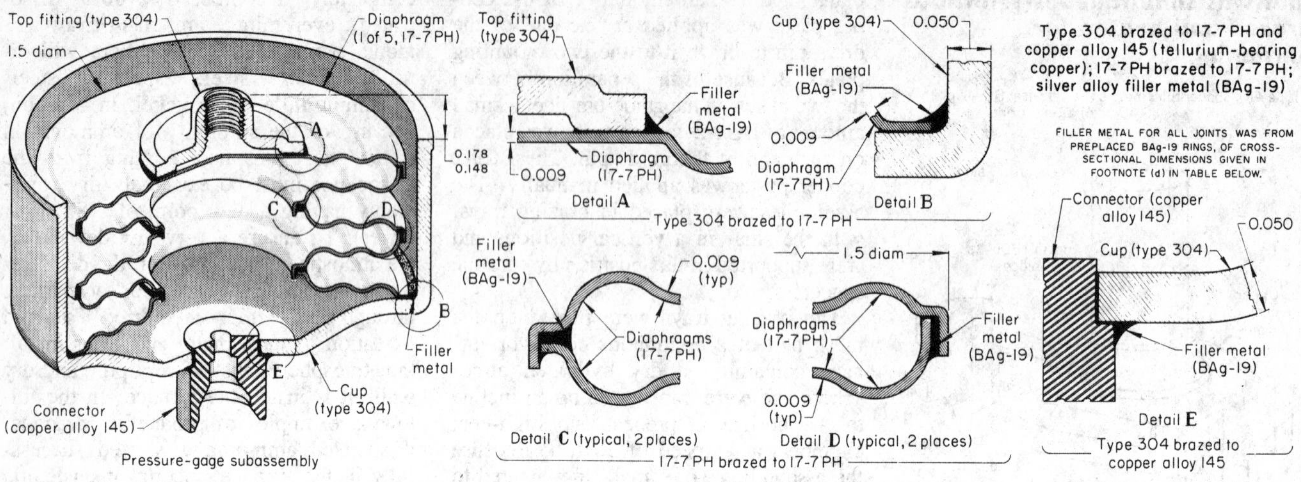

Furnace brazing in dissociated ammonia

Furnace	Chain-belt conveyor(a)	Filler metal	BAg-19, preplaced rings(d)	Heating time	5 min
Furnace temperature(b)	1800 °F	Flux	None(e)	Cooling-chamber temperature	60 °F max(f)
Dissociated-ammonia dew point	−80 °F(c)	Furnace belt speed	10 in./min	Precipitation-hardening temperature	950 °F(g)

(a) Electrically heated, with elevated high-heat zone. (b) For brazing the subassembly and simultaneously solution heat treating the 17-7 PH diaphragms. (c) Achieved by running the dissociated ammonia through a molecular-sieve drier after cracking. (d) Cross-sectional dimensions (and product forms) of filler-metal rings were: for joint between diaphragm and top fitting (detail A), 0.050 in. wide by 0.004 to 0.005 in. thick (stamping); for outside joints between diaphragm segments (detail D) and joint between diaphragm and cup (detail B), 0.040 in. wide by 0.004 to 0.005 in. thick (ribbon); for inside joints between diaphragm segments (detail C), 0.030 in. wide by 0.005 in. thick (ribbon); and for joint between cup and connector (detail E), 0.060 by 0.010 in. (wire). (e) Use of BAg-19 eliminated the need for flux, which had been required with the silver alloy filler metal originally used. (f) To cool rapidly from 1800 °F and ensure solution treatment of the 17-7 PH diaphragms. (g) In dry dissociated ammonia, after subassembly had been cooled to −40 °F and dried. The 17-7 PH diaphragms were hardened to 44 to 48 HRC.

harden the diaphragms to 44 to 48 HRC. Brazed assemblies were pressure tested in a bellows halogen leak detector by applying freon at 75 psi and adding compressed air to bring the total pressure up to 300 psi. Leakage of freon in the gas-air mixture would have been detected by the halogen leak detector. The requirement was for no leakage at the most sensitive setting of the leak detector.

The rejection rate for leakage, based on 750 000 bellows produced, dropped from 2.8% with the original silver alloy filler metal to 1.0% with the BAg-19 filler metal. Field corrosion returns dropped 96%. By eliminating the stains caused by flux, it was no longer necessary to paint the assemblies.

Furnace Brazing in Argon

Argon is occasionally used as a furnace atmosphere in brazing stainless steels to other stainless steels or to reactive metals such as titanium (Example 8). Argon has the advantage of being chemically inert in relation to all metals; thus, it is a useful protective atmosphere for metals that can combine with or absorb reactive atmospheres, such as hydrogen. An argon atmosphere had the disadvantage of being unable to reduce oxides; consequently, the surface of stainless steel components must be exceptionally clean and free of oxides when brazed in argon.

Example 8. Brazing in an Argon Atmosphere. A manufacturer of jet engines designed a gear-reduction box made of commercially pure titanium. This complicated fabrication was made from assemblies of stampings and machined forgings, most of which were joined by GTAW in argon-filled welding chambers. For joining some assemblies, however, brazing was more appropriate.

A typical assembly that was furnace brazed in argon is shown in Fig. 8. This assembly consisted of a machined forging of commercially pure titanium (Aerospace Material Specification [AMS] 4921) and a length of seamless type 347 stainless steel tubing that was flared or expanded for a distance of approximately $^5/_{16}$ in. to accept the titanium forging.

The outside diameter of the titanium forging was held to 0.500 +0.000, −0.001 in. The inside diameter of the stainless steel tube was held to 0.501 +0.001, −0.000 in. This allowed for a diametral clearance of 0.001 to 0.003 in. between components at room temperature. From 32 to 1650 °F, the mean coefficient of expansion of commercially pure titanium is 5.7 μin./in. · °F; from 32 to 1600 °F, the mean coefficient of expansion of type 347 stainless steel is 11.1 μin./in. · °F. Calculation of the expansion that would occur during heating both components to 1650 °F indicated a 0.004-in. diametral clearance between the titanium and the stainless steel. Adding the diametral clearance at room temperature (0.001 to 0.003 in.) to the 0.004-in. clearance gave a total diametral clearance at brazing temperature of 0.005 to 0.007 in., which is within the clearance range commonly recommended for silver brazing.

Silver alloy BAg-19 was selected as the brazing filler metal, because of its high fluidity in an argon atmosphere and because its brazing temperature is lower than that of pure silver. Most alloy elements in silver brazing alloys form brittle intermetallic compounds with titanium, which result in unreliable joints. With the exception of a minute amount of lithium, the only alloying element contained in BAg-19 is 7.5% Cu. By limiting the time at brazing temperature, sound ductile joints were made and formation of the titanium-copper intermetallic phase was minimized.

Prior to assembly, the titanium forging was degreased and cleaned in a solution containing 40% nitric acid plus 2% hydrofluoric acid. The stainless steel tubing

Fig. 8 Stainless-and-titanium assembly furnace brazed in an argon atmosphere

Type 347 brazed to commercially pure titanium (AMS 4921); silver alloy filler metal (BAg-19)

Furnace brazing in argon

Furnace	Pit, gas-fired(a)
Retort	Inconel, 24 in. diam and length
Fixture material	Titanium sheet(b)
Furnace temperature	1700 ± 10 °F
Brazing temperature	1650 ± 10 °F
Argon dew points	−85 °F (incoming); −70 °F (exhaust)
Argon flow rate (purging to cooling)	150 ft^3/h
Filler metal	BAg-19, 0.040-in.-diam wire(c)
Time at brazing temperature	Less than 1 min
Cooling	In argon, to room temperature
Production lot per cycle	1 to 20 assemblies

(a) 72 in. diam, 72 in. deep. (b) Holding fixture, to keep assembly upright (forging down) during brazing. (c) Preformed full ring

and the filler metal (preformed rings of 0.040-in.-diam BAg-19 wire) were cleaned by washing in acetone. In all subsequent handling of the components during assembly, and until after brazing, operators wore clean, lint-free white cotton gloves.

The titanium forging was inserted into the expanded end of the stainless steel tube until it was completely seated. A filler-metal ring was placed around the outside diameter of the titanium tube at the joint intersection. The assembly was placed upright (forging down) in a titanium sheet metal holding fixture and loaded into an Inconel retort. The retort was designed for displacement purging with an inlet and exit manifold for the argon atmosphere. After loading, the retort cover was seal welded to its base, using GTAW.

The retort was purged for 30 min with pure, dry argon at 150 ft^3/h and then was placed in a gas-fired pit-type furnace. The retort was heated to 600 °F and held at that temperature for an additional 30 min, or until the dew point of the exiting argon was −70 °F, as recorded on an electrolytic water analyzer. The furnace temperature was then raised until the assembly temperature reached 1650 °F, as indicated by a Chromel-Alumel thermocouple attached to the titanium holding fixture within the retort. As soon as this temperature was reached, the retort was removed from the furnace and fan-cooled to room temperature.

The retort cover was opened by grinding away the seal weld, and the assemblies were removed. The titanium and stainless steel components emerged bright and clean, with evidence of excellent filler-metal flow. Radiographic inspection showed over 95% joint coverage. All joints were visually inspected on both sides.

Furnace Brazing in Air Atmosphere

The principal advantages of furnace brazing are high production rates and means for using controlled protective atmospheres at controlled dew points, which often make use of a flux to obtain satisfactory wetting action unnecessary. In most furnace brazing applications, both of these advantages are exploited. Occasionally, however, furnace brazing is selected solely on the basis of production rate, and brazing is performed without a protective atmosphere, but with a suitable flux. The lower melting filler metals are generally selected for brazing under these conditions, as in the following application.

Example 9. Substitution of Furnace Brazing in Air Atmosphere for Torch Brazing. The gas-valve bobbin assembly shown in Fig. 9 was satisfactorily brazed by both the torch and furnace processes, the choice of process depending primarily on the production rate required. Cost data proved that furnace brazing increased the production rate per hour, reduced the direct labor rate per hour, and reduced the direct labor cost per assembly.

As Fig. 9 shows, the bobbin assembly consisted of four parts: a screw made of type 446 stainless steel, a base made of type 303 stainless steel, a tube made of a copper alloy closely related to nickel silver (74%Cu-22%Ni-4%Zn), and a plug made of copper alloy 187 (99%Cu-1%Pb) which held the screw in place and blocked gas passage through the tube.

All components were thoroughly vapor degreased before brazing. They were assembled with two preformed rings of 0.031-in.-diam BAg-3 filler-metal wire; the diametral clearance on the joints was 0.003 to 0.005 in. One preform, with a $^3/_8$-in. internal diameter, was placed over the neck of the plug; the other, with a $^1/_2$-in. internal diameter, was placed over the tube adjacent to the base joint. The joint areas were coated with type 3A brazing flux, and the assemblies were brazed in a continuous-belt conveyor furnace. Filler metal BAg-3 was chosen in preference to BAg-1 or BAg-1a to avoid the risk of interface corrosion.

Furnace Brazing in Vacuum

The majority of vacuum brazing is performed in two types of equipment utilizing either hot or cold walled furnaces. The hot walled structure utilizes a retort that is evacuated and placed into a furnace which provides the heat source. The retort can be single-pumped, providing vacuum to temperatures up to 1800 °F, or double-pumped for temperatures above 1800 °F to prevent the retort from collapsing. Limitations of the hot walled furnace (retort) include longer cycle times due to heating the retort externally, a 2200 °F temperature limit, and slower cooling rates. Some advantages of hot walled furnaces are less cost in initial capital expenditure, reduced contamination from the retort, and easy upkeep and maintenance.

The cold walled vacuum furnace is typically designed with a water-cooled outer jacket protected by radiation shielding adjacent to the inner wall. The heating elements are exposed directly to the workload. A braze cycle may have temperatures

Fig. 9 Stainless-and-copper gas-valve bobbin furnace brazed in an air atmosphere

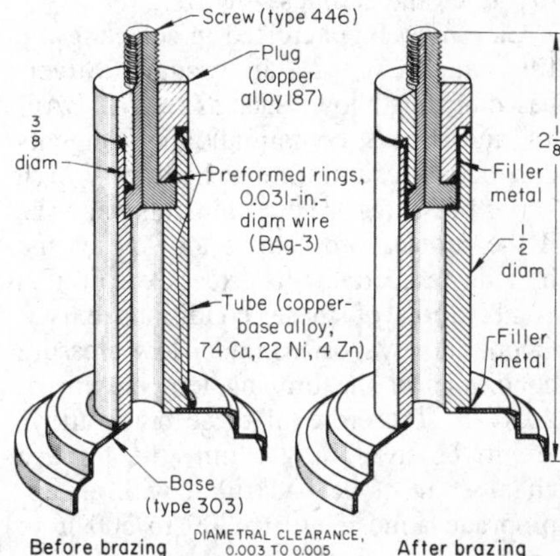

Furnace brazing in air atmosphere

Furnace	Continuous belt, air atmosphere
Furnace temperature	1370-1450 °F
Filler metal	BAg-3, 0.031-in.-diam wire(a)
Flux	AMS 3410D (AWS type 3A)
Time at brazing temperature	1 min (max)
Production rate	140 assemblies per hour

(a) Preformed rings, preplaced as shown in illustration

in excess of 4000 °F, depending on heating element material and load support structure. Heating and cooling rates for cold walled vacuum furnaces are substantially less than for the hot walled vacuum retort. Higher braze temperatures, in excess of 2300 °F, are obtainable in vacuum furnaces and are frequently employed.

Effect of Filler-Metal Composition. Vacuum brazing of many structural configurations made of austentic stainless steels offers excellent heat and corrosion resistance for high-temperature service applications. Filler metals, such as gold, gold-palladium, and the nickel brazing alloys offer greater high-temperature strength and oxidation resistance. Problems can occur when brazing the 300 series stainless steels because of the carbide precipitation and loss of corrosion resistance when brazing in the temperature range of 900 to 1500 °F. Brazing at temperatures in excess of 1500 °F with filler metals that have melting points higher than 1500 °F followed by rapid cooling reduces the carbide precipitation and improves corrosion properties.

Occasionally, wetting does not occur with a particular lot of stainless steel; generally, contaminants such as titanium and aluminum oxides contribute to the poor wettability.

Martensitic stainless steels, such as the 400 series, can be brazed successfully in vacuum. The use of a suitable filler metal can result in austenitizing and brazing at the same time, followed by rapid cooling to harden the stainless steel.

Care must be exercised in selecting the filler metals for use in vacuum. Silver-based brazing alloys, such as BAg-1, BAg-1a, and BAg-3, contain alloying elements of zinc and cadmium that have very high vapor pressures. These elements vaporize if the furnace pressure is too low or the brazing temperature is excessive, or if a combination of these conditions exists. Copper also vaporizes under low pressure conditions at the brazing temperature of 2050 °F. Therefore, if these braze alloys are to be used in vacuum, the furnace chamber must be backfilled with an appropriate atmosphere to 300 to 500 μ to prevent the loss of these elements.

Torch Brazing

For stainless steel, the fundamentals of torch brazing, as well as advantages and limitations, are basically the same as for carbon steels (see the article "Torch Brazing of Steels" in this Volume). Because of the metallurgical characteristics of stainless steel and its requirements for corrosion resistance, however, best results are obtained when special consideration is given to type of flame at the torch and filler-metal composition.

Flame Adjustment. To aid in reducing the oxide already present, as well as to prevent further oxidation of the work-metal surfaces, a strongly reducing flame should be used for torch brazing stainless steel to itself. A reducing flame is also satisfactory for brazing stainless steel to nickel alloys or carbon steels. When brazing stainless steel to copper alloys, however, some compromise is necessary. Although a slightly oxidizing flame is normally best for brazing copper, for brazing stainless steel to copper, a slightly reducing flame usually provides a satisfactory compromise in flame adjustment.

Filler Metals. The silver alloy filler metals that flow at relatively low temperatures are used almost exclusively for torch brazing of stainless steels. BAg-3 is most often used, because it flows well in the temperature range of 1300 to 1400 °F and provides joints that have greater resistance to corrosion than those brazed with filler metals such as BAg-1 or BAg-1a (although these filler metals are also used). The use of filler metals that require temperatures higher than about 1400 °F results in excessive oxidation. Sometimes, for special applications, higher melting filler metals must be used (Example 1).

Flux. Type 3A flux is used almost exclusively for torch brazing of stainless steel. It has a working range of 1050 to 1600 °F and is well suited for use with the lower melting silver alloy filler metals. In some applications, however, type 3B flux is preferred (Example 1).

Example 10. Torch Brazing of Stainless Steel to Nickel. The brazed assembly shown in Fig. 10, which consists of a type 304 stainless steel tube and a pure nickel tube, was resistance heated in service. Requirements for this assembly were:

- Transmission of electricity without developing hot spots
- Straightness and smoothness, because the assembly had to slide into a larger assembly
- Joining at minimum temperature, because numerous small insulated wires were in the assembly at the time of joining and were subject to damage if the joining temperature was too high

Torch brazing with silver alloy filler metal proved to be a desirable way to make the joint, because of the relatively low brazing temperature, high electrical conductivity of the joint brazed with the silver alloy filler metal, minimum distortion, ease of removing excess filler metal, and ease of radiographic inspection.

Fig. 10 Torch brazed assembly of a stainless steel tube and a pure nickel tube

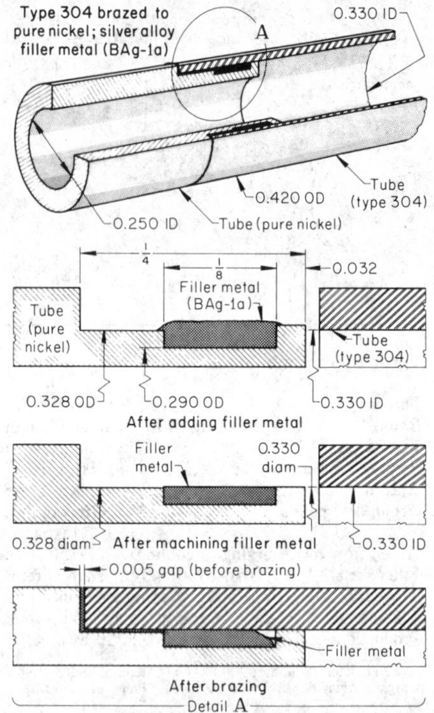

The joint design (Fig. 10) provided for preplacement of the filler metal (BAg-1a) in a 1/8-in.-wide groove machined into the shoulder of the nickel tube. The filler metal was then melted with a torch, and both filler metal and shoulder were machined to match the inside diameter of the stainless steel tube (zero-clearance fit).

For brazing the nickel tube to the stainless steel tube, several heating methods were tried, including induction heating in an inert gas and multiple-torch heating. The most successful method, however, proved to be the use of a single oxyacetylene torch by a skilled technician. The sequence of operations for single-torch brazing was:

- Components were cleaned with acetone.
- Flux paste was placed on the nickel tube in the area to be brazed.
- The two tubes were assembled in a fixture with a 0.005-in. gap showing at the surface (detail A in Fig. 10).
- The assembly was heated with a torch until the filler metal flowed to the outside surface (flow temperature of BAg-1a is 1175 °F).

- Excess flux that flowed to the surface was removed manually.

It was possible to inspect the entire joint by making two radiographs. Only scattered porosity was detected in routine radiographic inspection of the brazed joints.

Induction Brazing

Depending on the metallurgical and physical properties of particular stainless steels, their behavior in heating by electrical induction may differ considerably from that of carbon and low-alloy steels and from that of the more widely used nonferrous metals. In addition, depending on whether a stainless steel is magnetic or nonmagnetic at room temperature, the response of the steel to induction heating varies considerably. Differences in specific heat and electrical conductivity markedly affect response to heating by induction.

Ferritic and martensitic (400 series) stainless steels are ferromagnetic at all temperatures up to the Curie temperature. Thus, given the same power input, these steels generally heat faster than austenitic stainless steels, which are nonmagnetic in the annealed condition. Cold working may induce slight magnetism in the austenitic chromium-nickel steels, whereas the 400 series chromium steels are strongly magnetic. Rate of heating to the temperature at which the filler metal flows usually affects induction-coil design and coupling and may also influence selection of power-output frequency and other processing variables (see the article "Induction Brazing of Steels" in this Volume).

Stainless steels may be induction brazed in an air atmosphere, using a suitable flux, although for critical applications induction brazing is sometimes done in a protective atmosphere, or in vacuum (see the subsequent section in this article on induction brazing in vacuum and Fig. 12). In other applications, an inert gas such as argon may be used as a backing gas (as is often done in arc welding) to minimize oxidation.

Example 11. Brazing a Tube to an End Blank. The assembly shown in Fig. 11, part of a solenoid, consisted of a type 321 stainless steel tube brazed to a type 416 end blank. Type 321 is a nonmagnetic, austenitic stainless steel, whereas type 416 is martensitic and ferromagnetic. Consequently, although both metals were easily brazed, achieving proper joint clearance between the two components was complicated by the marked differences in coefficients of thermal expansion of the two steels. Thus, calculations were needed to determine the room-temperature clearance required to provide suitable clearance at the brazing temperature.

Fig. 11 Induction brazed assembly

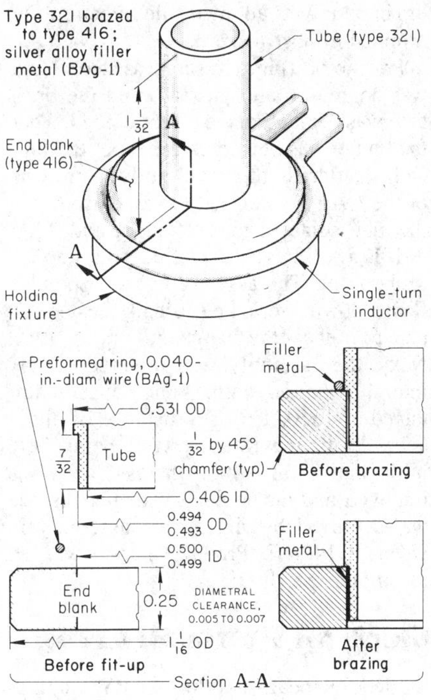

Induction brazing

Power supply	Vacuum tube, 20 kW, 450 kHz
Inductor	Single-turn, copper tube
Brazing temperature	1200 °F
Filler metal	BAg-1, 0.040-in.-diam wire
Flux	AWS type 3A
Time at brazing temperature	10 s
Cooling time	10 s in fixture
Production rate	140 assemblies per hour

Because the assembly was not intended for high-temperature service, the selection of a low-melting silver alloy filler metal (BAg-1) and a relatively low brazing temperature (1200 °F) was utilized to minimize heating and oxidation of the stainless steel components. For brazing at 1200 °F, calculations based on coefficients of thermal expansion showed that the following dimensions and tolerances in the joint area would be satisfactory: for the tube diameter, 0.494 +0.000, −0.001 in.; and for the inside diameter of the end blank, 0.500 +0.000, −0.001 in. Thus, diametral clearance at room temperature was 0.005 to 0.007 in.

The shape of the assembly and the low brazing temperature both favored brazing by induction. The end blank was in the hardened-and-tempered condition prior to brazing, and the short induction heating cycle (10 s) did not reduce the hardness to less than the required minimum.

Prior to brazing, the components were vapor degreased. The end of the tube was dipped in flux and inserted in the end blank. A preformed ring of filler-metal wire was slipped over the tube and located at the top of the joint. Then the end blank was placed on a holding fixture, positioned in a single-turn inductor (Fig. 11), and heated for 10 s. After brazing, the assembly was cooled in air for 10 s before it was removed from the holding fixture. The assembly was then washed in hot water to remove the flux residue.

Induction Brazing in Vacuum. A distinctive advantage of induction brazing, as applied to stainless steel, is suitability for simple setups that permit brazing in vacuum. Closed, nonmetallic containers with reasonably good strength and dielectric properties can provide enclosure for the assembly to be brazed and can be evacuated prior to brazing. Because the inductor can be placed outside the container, it can heat the assembly efficiently without being part of the vacuum system.

Stainless steel collar-and-tube assemblies (Fig. 12) were brazed in a simple setup that combined induction heating and the protection afforded by heating in vacuum. The vacuum container consisted of a high-silica, low-expansion glass tube with copper end fittings connected to a vacuum

Fig. 12 Collar-and-tube assembly that was induction brazed in vacuum

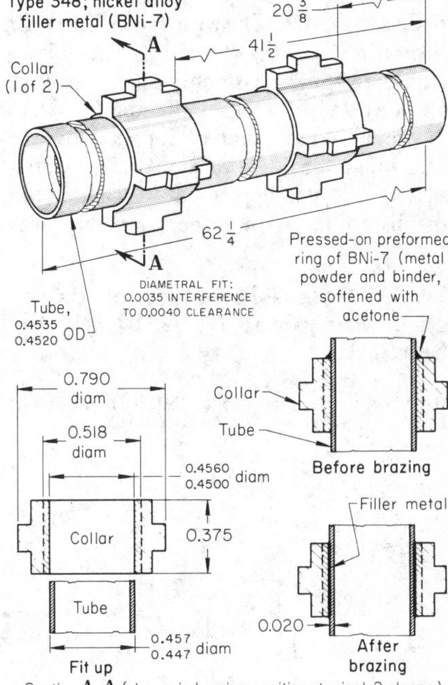

system. The collar-and-tube components, with preformed BNi-7 filler-metal rings pressed into place on the shoulder of each collar, were positioned and held inside the glass tube by means of a simple holding fixture. The tube was sealed and evacuated with a multiple-turn inductor, outside the tube, in position to heat one of the collars. When the vacuum reached 10^{-3} torr, heating was started. The collar was heated slowly to 1775 °F, and after 4 min the power was shut off. The tube was then repositioned to bring the second collar into the field of the inductor, and the heating sequence was repeated.

When the second collar had cooled to a black heat (no glow visible in normal light), the tube was backfilled for 5 min with argon. Brazed joints were inspected visually and using liquid-pentrant and metallographic methods. They were found to be sound and acceptable in all respects. The induction heating source was an 8-kV · A spark-gap converter with an operating frequency of 175 to 200 kHz. The water-cooled external inductor coil was made of $^1/_4$-in.-diam copper tubing. Production rate was 22 assemblies per day.

Dip Brazing in a Salt Bath

Brazing of stainless steel by immersion of all or a portion of the assembly in molten salt offers essentially the same advantages that would apply to brazing similar assemblies made of carbon steel. Similarly, the same limitations are applicable. See the article "Dip Brazing of Steels in Molten Salt" in this Volume.

Example 12. Change From Torch or Induction to Dip Brazing. The television wave-guide assembly shown in Fig. 13 consisted of a type 304 stainless steel flange brazed to a tube of copper alloy 230 (red brass, 85% Cu and 15% Zn). Satisfactory end use depended on a minimum of distortion. When the assembly was brazed by torch or induction, rejection rate sometimes reached 70% because of distortion caused by uneven heating. When dip brazing was adopted, the rejection rate dropped almost to zero.

Prior to brazing, the stainless steel flange was degreased and pickled, and the brass tube was degreased and bright dipped. Then the flange was placed on the tube, the tube end was flared outward slightly, a preform of the BAg-3 filler metal was placed over the tube adjacent to the flange, and AMS 3410D (AWS type 3A) flux was applied to the joint. The assembly was suspended flange-down over an electrically heated salt bath to preheat the flange and dry the flux. Next, the assembly was lowered slowly into the molten bath, which was maintained at 1350 °F, for a distance of about 1 in. above the flange. After being held in the bath for $1^1/_2$ min, the assembly was removed and air cooled. The flux residue was removed by rinsing the assembly in water at 140 °F. Production rate was 30 assemblies per hour.

Fig. 13 Television wave-guide assembly joined by salt-bath dip brazing

Type 304 brazed to copper alloy 230 (red brass, 85%); silver alloy filler metal (BAg-3)

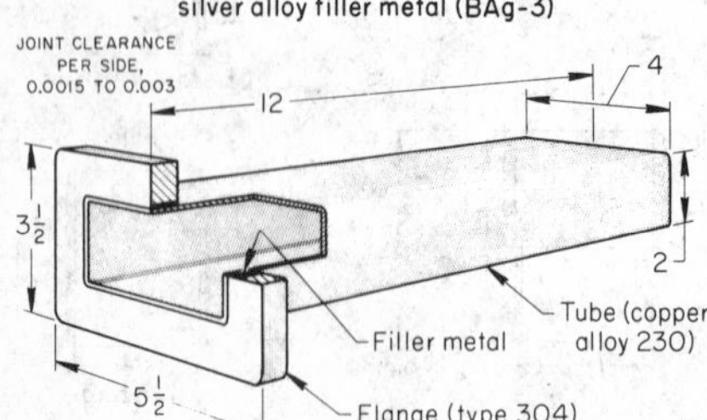

Electron Beam Brazing

Electron beam welding (EBW) equipment and techniques have been used to a limited extent for brazing. The high vacuum used in the work chamber (10^{-4} to 10^{-5} torr) permits adequate flow of brazing filler metal on properly cleaned joints without using a reducing atmosphere or flux. Thus, flux entrapment does not occur, and the work does not require cleaning after brazing. The high vacuum and absence of flux provide a brazing environment that avoids the problems associated with prepared atmospheres encountered when brazing some stainless steels, as well as the more reactive metals (such as titanium).

Electron beam brazing is performed similarly to EBW, except the beam is defocused to provide a larger beam spot and reduce the power density (kW/in.^{2}) or the heating effect on the work (see the article "Electron Beam Welding" in this Volume). If necessary, the beam spot diameter can be enlarged substantially, depending on the type of equipment (defocused beam, Example 13), while providing an adequate amount of heat input for brazing. Work movement can be used if an area substantially larger than the beam spot size is to be heated, and the work can be rotated under the beam for uniform heating.

Brazing temperatures are reached quickly, and heat can be localized to minimize grain growth, softening of cold worked metal and, in austenitic stainless steels, sensitizing of the material by carbide precipitation.

Applications. Electron beam brazing is a convenient method for brazing small assemblies, such as instrument packages, and combines the versatility and close controllability of electron beam heating with the advantages of vacuum brazing. Packaged devices can be encapsulated with an internal vacuum without damaging the basic package.

Tube-to-header joints in small heat-transfer equipment made of heat-resistant alloys and refractory metals are sometimes electron beam brazed. In one technique, the tube-to-header joint is electron beam welded on the top side of the header, and the heat of the beam causes the brazing filler metal preplaced on the reverse side of the header at the joint to melt and flow. Small-diameter, thin-walled stainless steel tubes are readily joined by electron beam brazing, as in the following example.

Example 13. Use of Defocused Beam for Electron Beam Brazing Small Tubes. Capillary and other small-diameter tubes used in instrument packages required leaktight joints to be made without over-

Fig. 14 Joint between two capillary tubes of an instrument package. Joined by electron beam brazing using a low-power defocused beam in a high vacuum

Type 304; copper alloy filler metal (BCu-1a)

Preformed rings, $\frac{1}{32}$-in.-diam wire (1 of 2, BCu-1a)
0.140 diam
$\frac{3}{4}$
$\frac{1}{4}$ (typ)
Socket coupling
0.010
Tube (1 of 2)
Before brazing
0.100 diam
DIAMETRAL CLEARANCE, 0.003 TO 0.005
DEFOCUSED ELECTRON BEAM ($\frac{3}{16}$ - DIAM SPOT)
Filler metal
After brazing

Electron beam brazing

Joint type	Cylindrical sleeve
Filler metal	$^1/_{32}$-in.-diam wire, BCu-1a
Machine capacity	3 kW
Gun type	Fixed diode
Vacuum chamber	24 in. diam
Maximum vacuum	10^{-5} torr
Fixtures	Holding jig
Pumpdown time	30 min
Brazing power	18 kV, 20-30 mA
Beam spot size	$^3/_{16}$ in. diam (approx)
Brazing vacuum	10^{-5} torr
Brazing time	30 s

heating other portions of the assembly. The use of flux also had to be avoided, because entrapped flux was difficult or impossible to remove. These conditions were met by electron beam brazing.

Figure 14 shows a typical joint in the 0.100-in.-OD by 0.010-in.-wall type 304 tubing that was brazed by the electron beam process. Joint design was based on the use of a $^{3}/_{4}$-in.-long socket coupling counterbored with a total clearance of 0.003 to 0.005 in. over the tube diameter and to a depth of $^{1}/_{4}$ in. Average joint clearance (per side) was therefore 0.002 in. Tubes and socket coupling were deburred and solvent cleaned. They were then assembled with two wire-ring preplacements of BCu-1a filler metal as shown in Fig. 14. The tubes were held in position with a small clamping fixture, and the assembly was mounted in a fixed position on a table in the vacuum chamber.

After pumpdown, the joint was brought to brazing temperature by moving the table back and forth under the defocused electron beam, causing the heat of the $^{3}/_{16}$-in.-diam beam spot to be applied chiefly to the central portion of the coupling. Heated by conduction at relatively low beam power, the filler metal melted and flowed through the joint in approximately 30 s.

About ten assemblies were brazed, one per pumpdown, using the machine settings and other brazing conditions given with Fig. 14. Sensitizing of the austenitic stainless steel was not a problem in this application, because the service environment was not significantly corrosive. The relatively short-time brazing cycle minimized grain growth and dilution of the thin-walled tubing with copper filler metal.

Brazing of Heat-Resistant Alloys

By the ASM Committee on Brazing of Heat-Resistant Alloys*

HEAT-RESISTANT ALLOYS are frequently referred to as superalloys, because of their strength, oxidation resistance, and corrosion resistance at elevated temperatures (1200 to 2200 °F). This article discusses primarily nickel- and cobalt-based alloys. Superalloys can be further subdivided into two categories: conventional cast and wrought alloys and powder metallurgy (P/M) products. Powder metallurgy products may be produced in conventional alloy compositions and oxide dispersion-strengthened (ODS) alloys. Almost any metal, as well as nonmetallics, can be brazed to these heat-resistant alloys, if it can withstand the heat of brazing. For additional information on these alloys, see Volume 3 of the 9th edition of *Metals Handbook*.

Brazing Filler Metals

The American Welding Society (AWS) has classified several gold-, nickel-, and cobalt-based brazing filler metals which can be used for elevated-temperature service (Table 1). In addition to these brazing filler metals, there are many that are not classified by AWS. The AWS classified brazing filler metals are suitable for high-temperature service; however, if the application is for temperatures above 1800 °F or in severe environments, the required brazing filler metal may not be in Table 1. It should be noted that for lower service temperatures, copper (BCu) and silver (BAg) brazing filler metals have been used for many successful applications.

Generally, heat-resistant alloys are brazed with nickel- or cobalt-based alloys containing boron and/or silicon, which serve as melting-point depressants. In many commercial brazing filler metals, the levels are 2 to 3.5% B and 3 to 10% Si. Phosphorous is another effective melting-point depressant for nickel and is used in filler metals from 0.02 to 10%. It is also used where good flow is important in applications of low stress, where temperatures do not exceed 1400 °F.

In addition to boron, silicon, and phosphorous, chromium is often present to provide oxidation and corrosion resistance. The amount may be as high as 20%, depending on the service conditions. Higher amounts, however, tend to lower brazement strength.

Cobalt-based filler metals are used mainly for brazing cobalt-based components, such as first-stage turbine vanes for jet engines. Most cobalt-based filler metals are proprietary. In addition to containing boron and silicon, these alloys usually contain chromium, nickel, and tungsten to provide corrosion and oxidation resistance and to improve strength.

Product Forms

Available forms of AWS classified and proprietary brazing filler metals include wire, foil, tape, paste, and powder. The form used is dictated frequently by the application. If the filler metal required for a specific application is only available as a dry powder, then brazing aids such as cements and pastes are available to help position the brazing filler metal.

Brazing filler-metal powders usually are atomized and sold in a range of specified particle sizes, which ensures uniform heating and melting of the brazing filler metal during the brazing cycle. These powders can be mixed with water, plastisizers, or organic cements to facilitate positioning. If the mixture must support its own weight until the brazing cycle begins, an organic binder or cement is required. These binders burn off in atmosphere brazing, and little or no residue results. When the brazing filler metal is supplied as a paste, it is simply a premixed powder and binder.

Brazing Tapes and Foils. Brazing filler metals in the form of tapes and foils appear similar, but the foils are usually made by melt spinning operations and tend to be very homogeneous microcrystalline structures. Brazing tapes are made of powder that is held together by a binder and formed into a sheet. Most foils have a high metalloid (phosphorus, silicon, boron) content, while tapes can be made from brazing filler metals that have no metalloid content. The metalloids usually are melting-point depressants and frequently form brittle phases. In some cases, where the composition is workable, such as BAu, foils can be made by cold rolling. Most of the nickel- or cobalt-based brazing filler metals require melt spinning to form foils. Foil also can be produced by rolling an alloy of suitable composition into a foil before adding the metalloids. Tapes and foils are best suited for applications requiring a large area joint, good fit-up, or where brazing flow and wetting may be a problem.

Brazing wires of nonfabricable alloys usually are made by P/M processes from atomized powder, which is held together by a binder or by extruding powder into

*Thomas J. Kelly, *Chairman*, Senior Metallurgist, The International Nickel Co.; G.A. Andreano, Manufacturing Technology Engineer, Garrett Turbine Engine Co.; Edward J. Cove, Senior Joining & Coatings Engineer, General Electric Co.; Jeff Flynn, Materials Project Engineer, Pratt & Whitney Aircraft; John M. Gerken, Staff Engineer, TRW Inc.; Edward J. Ryan, Senior Metallurgical Engineer, Pratt & Whitney Aircraft; Charles J. Sponaugle, Fabrication Manager—Technical and Product Services, Cabot Corp.

Table 1 AWS brazing alloys for elevated-temperature service

AWS classification	Composition, % Cr	B	Si	Fe	C	P	S	Al	Ti	Mn	Cu	Zr	Ni	Other elements total	Solidus, °F	Liquidus, °F	Brazing range, °F
Nickel-based alloy filler metals(a)																	
BNi-1	13.0-15.0	2.75-3.50	4.0-5.0	4.0-5.0	0.6-0.9	0.02	0.02	0.05	0.05	···	···	0.05	rem	0.50	1790	1900	1950-2200
BNi-1a	13.0-15.0	2.75-3.50	4.0-5.0	4.0-5.0	0.06	0.02	0.02	0.05	0.05	···	···	0.05	rem	0.50	1790	1970	1970-2200
BNi-2	6.0-8.0	2.75-3.50	4.0-5.0	2.5-3.5	0.06	0.02	0.02	0.05	0.05	···	···	0.05	rem	0.50	1780	1830	1850-2150
BNi-3	···	2.75-3.50	4.0-5.0	0.5	0.06	0.02	0.02	0.05	0.05	···	···	0.05	rem	0.50	1800	1900	1850-2150
BNi-4	···	1.5-2.2	3.0-4.0	1.5	0.06	0.02	0.02	0.05	0.05	···	···	0.05	rem	0.50	1800	1950	1850-2150
BNi-5	18.5-19.5	0.03	9.75-10.50	···	0.10	0.02	0.02	0.05	0.05	···	···	0.05	rem	0.50	1975	2075	2100-2200
BNi-6	···	···	···	···	0.10	10.0-12.0	0.02	0.05	0.05	···	···	0.05	rem	0.50	1610	1610	1700-2000
BNi-7	13.0-15.0	0.01	0.10	0.2	0.08	9.7-10.5	0.02	0.05	0.05	0.04	···	0.05	rem	0.50	1630	1630	1700-2000
BNi-8	···	···	6.0-8.0	···	0.10	0.02	0.02	0.05	0.05	21.5-24.5	4.0-5.0	0.05	rem	0.50	1800	1850	1850-2000

AWS classification	Composition, % Au	Cu	Pd	Ni	Other elements total	Solidus, °F	Liquidus, °F	Brazing range, °F
Precious metals								
BAu-1	37.0-38.0	rem	···	···	0.15	1815	1860	1860-2000
BAu-2	79.5-80.5	rem	···	···	0.15	1635	1635	1635-1850
BAu-3	34.5-35.5	rem	···	2.5-3.5	0.15	1785	1885	1885-1995
BAu-4	81.5-82.5	···	···	rem	0.15	1740	1740	1740-1840
BAu-5	29.5-30.5	···	33.5-34.5	35.5-36.5	0.15	2075	2130	2130-2250

AWS classification	Composition, % Cr	Ni	Si	W	Fe	B	C	P	S	Al	Ti	Zr	Co	Other elements total	Solidus, °F	Liquidus, °F	Brazing range, °F
Cobalt-based alloy filler metals																	
BCo-1	18.0-20.0	16.0-18.0	7.5-8.5	3.5-4.5	1.0	0.7-0.9	0.35-0.45	0.02	0.02	0.05	0.05	0.05	rem	0.50	2050	2100	2100-2250

(a) If determined, cobalt is 0.1% maximum unless otherwise specified.
Source: AWS 5.8-81

wire and sintering. This form of brazing filler metal is better able to support itself than are pastes and powders of filler metal, but is not used to replace tapes or foils where precision is needed in placing the filler metal, such as joint gaps less than 0.005 in.

Surface Cleaning and Preparation

Cleaning of all surfaces that are involved in the formation of the desired brazed joint is necessary to achieve successful and repeatable brazed joints. All obstruction to wetting, flow, and diffusivity of the thermally induced molten brazing filler metal must be removed from both surfaces to be brazed prior to fit-up assembly. The presence of contaminants on one or both surfaces to be brazed may result in void formation, restricted or misdirected filler-metal flow, and contaminants included within the solidified brazed area, which reduces mechanical properties of the resulting brazed joint. Common contaminants are oils, greases, residual zyglo fluids, pigmented markings, residual casting or coring materials, and oxides formed either through previous thermal exposures or by exposures to contaminating environments.

Chemical cleaning methods are most widely used. As part of any chemical cleaning procedure for preprocessing assemblies for brazing, a degreasing solvent to remove all oils and greases should be the first operation. This is necessary to ensure wettability of the chemicals used for cleaning. Oils and greases form a very thin film on metals, which prevents wettability by both the subsequent chemical cleaning and/or the molten filler metals. Oils and grease removers that are widely used include degreasing solutions such as stabilized perchloroethylene or stabilized trichloroethylenes. These may be used as simple manual soaks, sprays, or by suspending the parts in a hot vapor of the aforementioned chemicals, commonly referred to as a vapor degreasing process. In conjunction with these processes, anodic and cathodic electrolytic cleaning can also be used.

A chemical cleaning procedure can be a simple single-step process or may involve multistep operations. If surfaces of the braze joint are in the machined condition, vapor degreasing may be sufficient to remove machining oils, handling oils, and zyglo penetrants to yield a sound, clean surface for brazing. If, on the other hand, one or both of the surfaces to be brazed is not a machined surface, then additional chemical cleanings should be employed. Once vapor degreasing is accomplished, care must be taken to maintain the surface integrity of the brazed components by handling in an environmentally clean atmosphere. Additional methods of chemical cleaning to remove oxides and other adherent metallic contaminants include immersion in phosphate acid cleaners or acid pickling solutions, which are comprised of nitric, hydrochloric, or sulfuric acids or combinations of these.

Care must be taken in time of exposure for both acid cleaning and acid pickling of heat-resistant base metals. Overexposure during chemical cleaning can lead to excessive metal loss, grain-boundary attack, and selective phase structure attack. As a last step in chemical cleaning, an ultrasonic cleaning in alcohol or clean hot water is recommended to ensure removal of all traces of previous cleaning solutions.

If no subsequent mechanical cleaning is used, the components to be brazed should be stored and transported to the braze preparation areas in dry, clean containers, such as plastic bags. The time between cleaning and braze application to the assembled joints should be kept as short as manufacturing processes allow.

Mechanical cleaning usually is confined to those metals with heavy tenacious oxide films or to repair brazing on components exposed to service. Mechanical methods are standard machining processes—abrasive grinding, grit blasting, filing, or wire brushing (stainless steel bristles must be used). These are used not only to remove surface contaminants, but to slightly roughen or fray the surfaces to be brazed.

Care must be taken that the surfaces are not burnished and that mechanical cleaning materials are not embedded in the metals to be brazed. In grit blasting, medium choice is critical. Wet and dry grit blasting commonly are used, but wet mediums are subject to additional cleaning requirements to remove the embedded grit. The mediums used are iron grits, silicon carbide grits, and grits comprised of brazing filler metals. Grit sizes as coarse as No. 30 (0.0232 in.) are recommended for cleaning forgings and castings. Finer grits (No. 90 and No. 100, 0.0065 and 0.0059 in., respectively) are used for general blasting. All grit mediums should be changed frequently, as extensive reuse of the same medium results in loss of sharp angles or facets. Once these configurations are lost or markedly reduced in the grit medium, burnishing rather than cleaning occurs. Overused medium results in the entrapment of oxides in the metal. If possible, the angle of grit blasting should be less than 90° to the surfaces to be cleaned. This also reduces the chances of embedding the oxides or medium in the surfaces to be brazed.

Care must be taken to remove all blasting medium from the surface after mechanical cleaning, as these mediums will contaminate the braze. Iron grit may impart an iron film which oxidizes as a rust. Aluminum oxides, if not removed, prevent wetting and flow of the brazing filler metal; thus, use of aluminum oxides are not recommended. Silicon carbide is extremely hard and has sharp facets. Consequently, it becomes embedded if an improper blasting angle is used. Blasting with a nickel brazing filler metal or similar alloy gives the best results; stainless steel blasting medium is also acceptable.

After mechanical cleaning, air blasting or ultrasonic cleaning should be used to remove all traces of loosened oxides or cleaning medium. Care must be taken to ensure maintenance of the clean surfaces and components; once cleaned, they should be assembled and brazed as soon as possible.

Nickel Flashing. Certain heat-resistant base-metal alloys, particularly nickel-based alloys containing high percentages of aluminum and titanium (Inconel 718), may require a surface pretreatment to ensure maintenance of the cleaned surfaces. This surface pretreatment after cleaning is generally an electroplate of nickel, commonly referred to as nickel flashing. Thickness of the plate flashing is kept under 0.0006 in. for alloys with less than 4% Ti plus aluminum and 0.0008 to 0.0012 in. for alloys with greater than 4% Ti plus aluminum. This promotes wettability in the braze joint without seriously affecting the braze strength and other mechanical properties of the braze. The thickness of nickel plating may have to be increased as the brazing temperature is increased and as the time above 1800 °F is increased. Titanium and aluminum will diffuse to the surface of the nickel plating upon heating.

Fixturing

One prerequisite to successful brazing that is often neglected is proper fixturing, when required for assembly and brazing of various components. One type of fixturing is classified as cold fixturing and is used primarily for assembly purposes.

In most cases, cold fixtures are made of hot and cold rolled iron, stainless alloys, nonferrous alloys, and nonmetals, such as phenolics and micarta. These fixtures are used for assembly and tack welding details, they need not be massive or heavy, but should be sturdy enough to assemble components as required by design.

Hot fixtures (fixtures used in the furnace for brazing) must have good stability at elevated temperatures and the ability to cool rapidly; metals are not stable enough to maintain tolerances during the brazing cycle. Therefore, ceramics, carbon, or graphite is used for hot fixturing. Ceramics, due to their high processing cost, are used for small fixtures and for spacer blocks to maintain gaps during brazing of small components. Graphite has been found to be the most suitable material for maintaining flatness in a high vacuum or argon atmosphere, and it provides faster cooling because of its porosity. Graphite should be coated with an Al_2O_3 slurry to prevent carburization of parts during the brazing cycle. It should not be used in a pure dry hydrogen atmosphere as it will cause carburization of the base metals by gaseous transfer. Molybdenum and tungsten may be used, but they are generally avoided because of their cost.

Controlled Atmospheres

Controlled atmospheres (including vacuum) are used to prevent the formation of oxides during brazing and to reduce the oxides present so that the brazing filler metal can wet and flow on clean base metal. Controlled-atmosphere brazing is widely used for the production of high-quality joints. Large tonnages of assemblies of a wide variety of base metals are mass produced by this process.

Controlled atmospheres are not intended to perform the primary cleaning operation for the removal of oxides, coatings, grease, oil, dirt, or other foreign materials from the parts to be brazed. All parts for brazing must be subjected to appropriate prebraze cleaning operations as dictated by the particular metals. Controlled atmospheres commonly are employed in furnace brazing; however, they may also be used with induction, resistance, infrared, laser, and electron beam brazing. In applications where a controlled atmosphere is used, postbraze cleaning is generally not necessary. In special cases, flux may be used with a controlled atmosphere (1) to prevent the formation of oxides of titanium and aluminum when brazing in a gaseous atmosphere, (2) to extend the useful life of the flux, and (3) to minimize postbraze cleaning. Fluxes should not be used in a vacuum environment.

The use of controlled atmospheres inhibits the formation of oxides and scale over the whole part and permits finish machining to be done before brazing in many applications. In some applications, such as the manufacture of electronic tubes, eliminating flux is tremendously important. Some types of equipment, such as metallic muffle furnaces and vacuum systems, may be damaged or contaminated by the use of flux.

Pure dry hydrogen is used as a protective atmosphere because it dissociates the oxides of many elements. See the section on metal-metal oxide equilibria in hydrogen in this article for more information. Hydrogen with a dew point of −60 °F dissociates the oxides of most elements found in heat-resistant alloys, with the exception of the oxides of elements shown in the lower corner of Fig. 1. The most notable of these are aluminum and titanium, which are found in most of the high-strength heat-resistant base metals.

Inert gases, such as helium and argon, do not form compounds with metals. In equipment designed for brazing at ambient pressure, inert gases reduce the evaporation rate of volatile elements, in contrast to brazing in a vacuum. Inert gases permit the use of weaker retorts than required for vacuum brazing. Elements such as zinc and cadmium, however, vaporize in pure dry inert atmospheres.

Vacuum. An increasing amount of brazing of heat-resistant alloys, particularly precipitation-hardenable alloys that contain titanium and aluminum, is done in a vacuum. Vacuum brazing in the range of 10^{-4} torr has proved adequate for brazing most of the nickel-based superalloys. By removal of gases to a suitably low pressure, including gases that are evolved during heating to brazing temperature, very clean surfaces are obtainable. A vacuum is particularly useful in the aerospace, electronic tube, and nuclear fields, where metals that react chemically with a hydrogen atmosphere are used or where entrapped fluxes or gases are intolerable. The maximum tolerable pressure for successful brazing depends on a number of factors that are primarily determined by the composition of the base metals, the brazing filler metal, and the gas that remains in the evacuated chamber.

Vacuum brazing is economical for fluxless brazing of many similar and dissimilar base-metal combinations. Vacuums are especially suited for brazing very large, continuous areas where (1) solid or liquid fluxes cannot be removed adequately from the interfaces during brazing, and (2) gaseous atmospheres are not completely efficient because of their inability to purge occluded gases evolved at close-fitting brazing interfaces. It is interesting to note that a vacuum system evacuated to 10^{-5} torr contains only 0.00000132% residual gases based on a starting pressure of 1 atm (760 torr).

Vacuum brazing has the following advantages and disadvantages compared with other high-purity brazing atmospheres:

- Vacuum removes essentially all gases from the brazing area, thereby eliminating the necessity for purifying the supplied atmosphere. Commercial vacuum brazing generally is done at pressure varying from 10^{-5} to 10^{-1} torr, depending on the materials brazed, the filler metals being used, the area of the brazing interfaces, and the degree to which gases are expelled by the base metals during the brazing cycle.
- Certain oxides of the base metal dissociate in vacuum at brazing temperatures.
- Difficulties sometimes experienced with contamination of brazing interfaces, due to base-metal expulsion of gases, are negligible in vacuum brazing.
- Low pressure existing around the base and filler metals at elevated temperatures removes volatile impurities and gases from the metals. Frequently, the properties of base metals are improved. This characteristic is also a disadvantage when elements in the filler metal or base metals volatilize at brazing temperatures, thus changing the melting point of the filler metal or properties of the base metal. This tendency may, however, be corrected by employing partial-pressure vacuum brazing techniques.

Metal-Metal Oxide Equilibria in Hydrogen and Vacuum.* Beginning with the thermochemical data for the formation of metal oxides, Fig. 1 shows the formation of metal-metal oxide equilibria in hydrogen. In experiments with pure metals and filler metals, it was found that vacuum pressure could be substituted for hydrogen dew point with similar results in vacuum. The vacuum scale used in Fig. 1 represents the vapor pressure of water and ice at the temperatures shown on the left scale for hydrogen dew point. At brazing temperatures around 1700 °F and above, experiments with pure metals follow the curves.

In all metallurgical processing where metal oxides are reduced or dissociated in an atmosphere, a plot of the metal-metal oxide equilibria shows the varying degree of difficulty in dissociating the metal oxides of various elements with varying temperature and dew point or vacuum pressure.

The left triangle in the top center of Fig. 1 indicates the elements whose oxides are so readily dissociated that they are off the graph on the far left. Thus, elements such as gold, copper, nickel, and cobalt can be dissociated readily in almost any controlled atmosphere. The right triangle indicates an element whose metal-metal oxide curve would fall off the right side of the graph, thus indicating that it is very difficult to dissociate its oxide.

In working with chromium-containing alloys of approximately 1% and above, the line for chromium on the graph indicates equilibrium. To provide an atmosphere that will result in adequate brazing, assuming the base metal and filler metal contain little or no elements shown on the graph to be to the right of chromium, the dew point and vacuum pressure versus temperature point coordinate point must be sufficiently to the right of the chromium curve. The further the coordinate point is to the right of the chromium curve, the higher the dissociation pressure, which increases the wetting and flow of a given filler metal on the surface of the base metal.

Aluminum and titanium, when contained in a base metal, form oxides in pure dry hydrogen unless the percentage contained is very low (approximately 0.3%) or is tied up with carbon, or as another stable compound. In a laboratory, using a vacuum atmosphere in a very clean furnace, aluminum and titanium (TiO) evaporate from the surface of the part to be brazed, leaving a clean surface for the wetting and flow of the filler metal, and thus deviating from the equilibria graph. In production brazing, however, furnace conditions are not as suitable and a series of colors may be seen on the surface of the base metal, depending on the percent of titanium and/or aluminum in the base metal. Nickel-based alloys containing very low amounts of titanium (less than 0.2%) will remain bright when processed in a good-quality atmosphere. As the percentage of titanium in the base metal is increased, the surface color of the cleanly machined base metal will vary from gray to light gold, to gold, to brown, to light purple, and finally, with high titanium contents (3 to 4%), to dark purple. With aluminum, the color fringes are gray, ranging from a very light tint to a dark gray as the percentage of aluminum in the base metal is increased.

The metal-metal oxide equilibria graph is a useful tool to obtain an advance indication of the atmosphere required for a given base metal or filler metal. It is also useful for other atmospheres, such as helium, argon, and nitrogen, as well as mixed gases and partially combusted gases. At temperatures of 1200 °F and above, the mechanism for the dissociation of oxide appears to be a partial pressure phenomenon and not an oxidation-reduction reaction. For this reason, inert atmospheres of helium, argon, and nitrogen dissociate metal oxides under the proper conditions of temperature, and partial pressure of oxygen dissociates oxides of metal to produce a clean base-metal surface with good wetting and flow of the filler metal. While no definitive data have been obtained on brazing of various heat-resistant base met-

*Excerpted from Bredzs, N. and Tennenhouse, C.C., Metal-Metal Oxide-Hydrogen Atmosphere Chart for Brazing or Bright Metal Processing, *Welding Journal*, May 1970

Fig. 1 Metal-metal oxide equilibria in pure hydrogen atmospheres. Source: Bredzs, N. and Tennenhouse, C.C., *Welding Journal*, May 1970

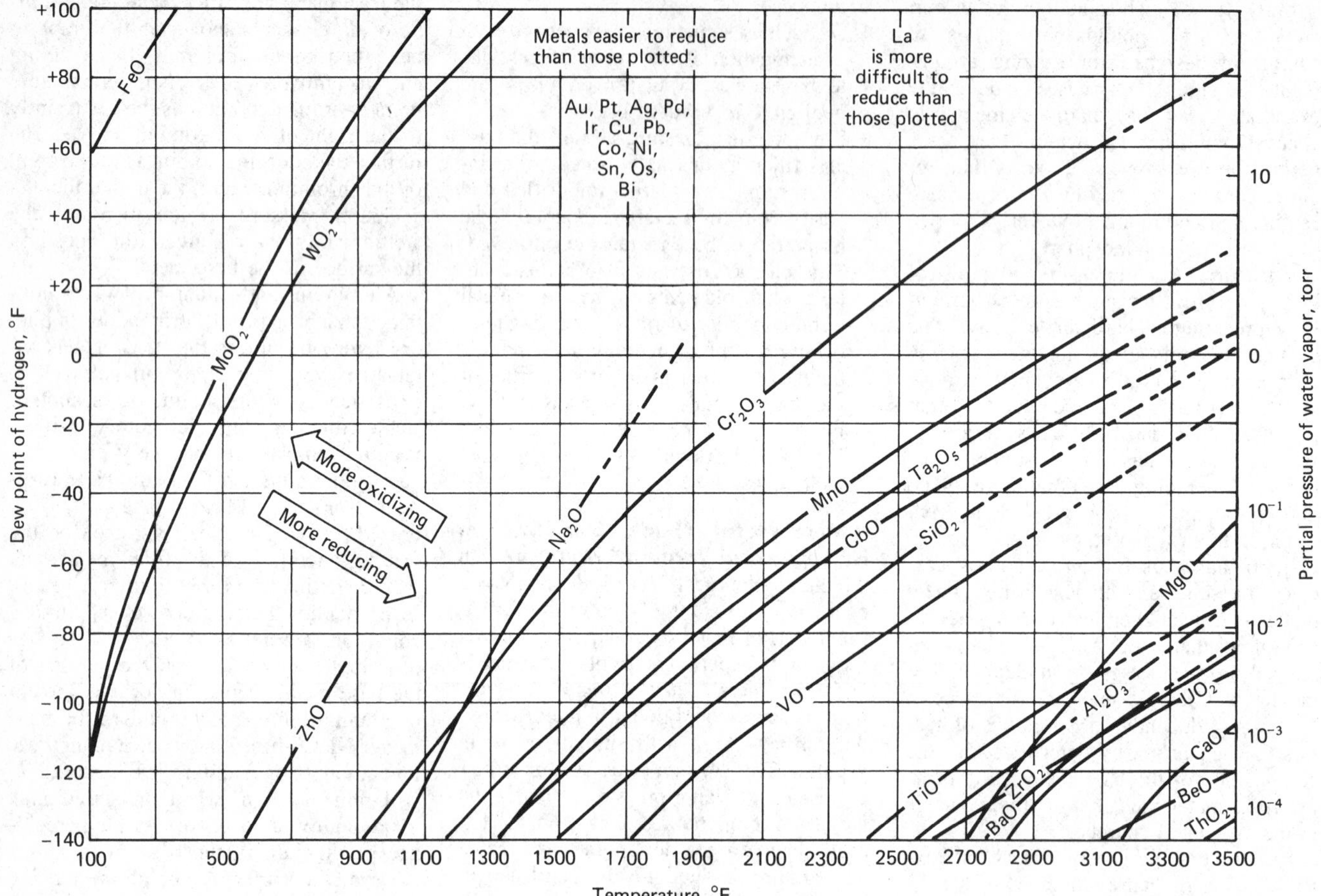

als in these inert atmospheres, it has been shown that they also fit the metal-metal oxide equilibria graph. The dew point of hydrogen can be readily converted to partial pressure of oxygen, as can the vacuum pressure; therefore, relating the dissociation of oxides to the partial pressure of oxygen in all atmospheres follows the observed brazing conditions in these atmospheres.

Nickel-Based Alloys

In the selection of a brazing process for a nickel-based alloy, the characterisitics of the alloy must be carefully considered. The nickel-based alloy family includes alloys that differ significantly in physical metallurgy, such as precipitation strengthened versus solid-solution strengthened, and in process history, cast versus wrought. These characteristics can have a profound effect on their brazeability.

Precipitation-hardenable alloys present several difficulties not normally encountered with solid-solution alloys. Precipitation-hardenable alloys often contain appreciable (greater than 1%) quantities of aluminum and titanium. The oxides of these elements are almost impossible to reduce in a controlled atmosphere (vacuum, hydrogen). Therefore, nickel plating or the use of a flux is necessary to obtain a surface that allows wetting by the filler metal.

Because these alloys are hardened at temperatures of 1000 to 1500 °F, brazing at or above these temperatures may alter the alloy properties. This frequently occurs when using silver-copper (BAg) filler metals, which occasionally are used on heat-resistant alloys.

Liquid metal embrittlement is another difficulty encountered in brazing of precipitation-hardenable alloys. Many nickel-, iron-, and cobalt-based alloys crack when subjected to tensile stresses in the presence of molten metals. This is usually confined to the silver-copper (BAg) filler metals. If precipitation-hardenable alloys are brazed in the hardened condition, residual stresses are often high enough to initiate cracking.

Cleanliness, as in all metallurgical joining operations, is important when brazing nickel and nickel-based alloys. Cleanliness of base metal, filler metal, flux (when used for induction brazing), and purity of atmosphere should be as high as practical to achieve the required joint integrity. Elements that cause surface contamination or interfere with braze wetting or flow should be avoided in prebraze processing. All forms of surface contamination such as oils, chemical residues, scale, or other oxide products should be removed by using suitable cleaning procedures. The use of nickel-based filler metals offers some cost effectiveness in this regard, as nickel-based brazes are known to be self-fluxing and thus more forgiving to slight imperfections in cleanliness.

Attempting to braze over the refractory oxides of titanium and aluminum that may be present on precipitation-hardenable nickel-based alloys must be avoided. Procedures to prevent or inhibit the formation of these oxides before and/or during brazing include special treatments of the sur-

Fig. 2 Braze specimen assembly used for evaluating stress-rupture properties of brazed Inconel MA 754

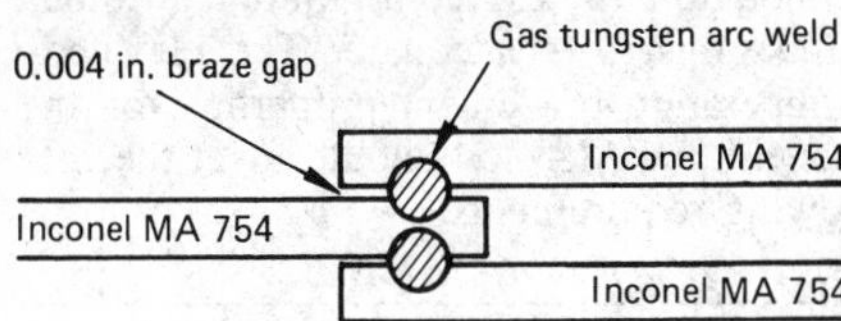

faces to be joined or brazing in a highly controlled atmosphere. Surface treatments include electrolytic nickel plating and reducing the oxides to metallic form. As stated earlier in this section, a typical practice is to nickel plate the joint surface of any alloy that contains aluminum and/or titanium. For vacuum brazing, when aluminum and titanium are present in trace amounts, use of 0.0001- to 0.0003-in. plate is considered optional. Alloys with up to 4% Al and/or titanium require 0.0004- to 0.0006-in. plate, while alloys with aluminum and/or titanium contents greater than 4% require 0.0008- to 0.0012-in. plate. When brazing in a pure dry hydrogen atmosphere, thicker plating (0.001 to 0.0015 in.) is desirable for alloys with high (>4%) aluminum and/or titanium contents.

Atmospheres. Dry, oxygen-free atmospheres that are frequently used include inert gases, reducing gases, and vacuum. The brazing atmosphere, whether gaseous or vacuum, should be free from harmful constituents such as sulfur, oxygen, and water vapor. When brazing in a gaseous atmosphere, monitoring of the water vapor content of the atmosphere as a function of dew point is common practice. A dew point of −60 °F is average; −80 °F or below produces a better quality braze.

Stresses. During brazing, residual or applied tensile stress should be eliminated or minimized as much as possible.

Table 2 Typical compositions of several cobalt-based alloys

Alloy(a)	Nominal composition, %															Characteristics and typical applications
	C	Mn	Si	Cr	Ni	Co	Mo	W	Nb	Ti	Al	B	Zr	Fe	Other	
AiResist 13(b)	0.45	0.5(c)	...	21	1.0(c)	rem	...	11	2.0	...	3.5	...	...	2.5(c)	0.1 Y	High-temperature parts
AiResist 213	0.18	...	...	19	...	rem	...	4.7	...	...	3.5	...	0.15	...	6.5 Ta, 0.1 Y	Sheets, tubing; resistant to hot corrosion
AiResist 215(b)	0.35	...	...	19	...	rem	...	4.5	...	...	4.3	...	0.13	...	7.5 Ta, 0.17 Y	Nozzle vanes; resistant to hot corrosion
Elgiloy	0.15	2.0	...	20.0	15.0	40.0	7.0	...	...	...	...	...	...	rem	0.04 Be	Springs; corrosion resistant, high strength
FSX-414(b)	0.25	1.0(c)	1.0(c)	29.5	10.5	rem	...	7.0	...	...	...	0.012	...	2.0(c)	...	Gas turbine vanes
FSX-418(b)	0.25	1.0(c)	1.0(c)	29.5	10.5	rem	...	7.0	...	...	...	0.012	...	2.0(c)	0.15 Y	Gas turbine vanes; improved oxidation resistance
FSX-430(b)	0.40	...	...	29.5	10.0	rem	...	7.5	...	...	...	0.027	0.9	...	0.5 Y	Gas turbine vanes; improved strength and ductility
X-40(b)	0.50	0.50	0.50	25	10	rem	...	7.5	...	...	...	...	...	1.5	...	Gas turbine parts, nozzle vanes
Haynes 150	0.08	0.65	0.35	28	3.0(b)	rem	1.5(b)	...	...	...	...	...	...	20.0	...	Resists thermal shock, high-temperature corrosion (air and air-SO_2)
Haynes 188 (sheet)	0.10	1.25(b)	0.3	22	22	rem	...	14	...	...	...	...	...	3.0(b)	0.04 La	Better oxidation resistance than Hastelloy X
MAR-M302(b)	0.85	0.10	0.20	21.5	...	rem	...	10.0	...	...	...	0.005	0.15	...	9.0 Ta	Jet engine blades, vanes
MAR-M322(b)	1.00	0.10	0.10	21.5	...	rem	...	9.0	...	0.75	...	...	2.25	...	4.5 Ta	Jet engine blades, vanes
MAR-M509(b)	0.60	0.10(c)	0.10(c)	21.5	10	rem	...	7.0	...	0.2	...	0.010(c)	0.50	1.0	3.5 Ta	Jet engine blades, vanes
MAR-M918	0.05	0.2(c)	0.2(c)	20	20	rem	...	...	...	...	...	...	0.16	0.5(c)	7.5 Ta	High-temperature sheets
MP35N	...	...	...	20.0	35.0	35.0	10.0	...	...	...	...	...	...	...	...	Stress corrosion resistant, high-strength fasteners
NASA Co-W-Re(b)	0.40	...	...	3	...	rem	...	25	...	1.0	...	...	1.0	...	2.0 Re	High-temperature space applications
S-816	0.38	1.20	0.40	20	20	rem	4.0	4.0	4.0	...	...	...	...	4	...	Gas turbine blades, bolts, springs
V-36	0.27	1.00	0.40	25	20	rem	4.0	2.0	2.0	...	...	...	...	3	...	High-temperature sheets
WF-11, L605, Haynes 25	0.10	1.50	0.50	20	10	rem	...	15	...	...	...	...	...	3.0(b)	...	Jet engine parts, sheets
WF-31	0.15	1.42	0.42	20	10	rem	2.6	10.7	...	1.0	...	...	...	...	...	High-temperature sheets
WI-52(b)	0.45	0.50(c)	0.50(c)	21	1.0(c)	rem	...	11	2.0	...	...	...	...	2.0	...	Gas turbine parts, nozzle vanes
X-45(b)	0.25	1.0(c)	...	25.5	10.5	rem	...	7.0	...	...	...	0.010	...	2.0(c)	...	Nozzle vanes

(a) Some superalloys are made by more than one manufacturer. The proprietary designation for such alloys has been used in this compilation. (b) Cast alloy. (c) Maximum composition

Also, inherent stresses present in the precipitation-hardenable alloys may lead to stress corrosion cracking. Stress relieving or annealing prior to brazing is also recommended for all furnace, induction, or torch brazing. Brazing filler metals that melt below the annealing temperature are likely to cause stress corrosion cracking of the base metal.

Thermal Cycles. Consideration must be given to the effect of the brazing thermal cycle on the base metal. Filler metals that are suitable for brazing nickel-based alloys may require relatively high thermal cycles. This is particularly true for the filler-metal alloy systems most frequently used in brazing of nickel-based alloys—the nickel-chromium-silicon or nickel-chromium-boron systems.

Solid solution-strengthened nickel-based alloys such as Inconel 600 may not be adversely affected by nickel braze filler-metal brazing temperatures of 1850 to 2250 °F. Precipitation-strengthened alloys such as Inconel 718 may, however, display adverse property effects when exposed to brazing cycles higher than their normal solution heat treatment temperatures. Inconel 718, for example, is solution heat treated at 1750 °F for optimum stress rupture life and ductility. Braze temperatures of 1850 °F or above result in grain growth and an attendant decrease in stress rupture properties, which cannot be recovered by subsequent heat treatment.

Consideration of base-metal property requirements for service enables selection of an appropriate braze alloy. Lower melting temperature (below 1900 °F) braze filler metals are available within the nickel-based alloy family and within other braze filler-metal systems (see Table 1).

Inconel 718 is often used in the fabrication of air diffusers for aerospace turbine engines. One manufacturer found that vacuum brazing of diffuser components at 10^{-4} torr in a cold-walled vacuum furnace provided the best results. Prior to brazing, all joint surfaces were nickel plated to 0.0002- to 0.0006-in. thicknesses. Plating was done in accordance with specification AMS 2424 or equivalent. Prior to assembly, application of BNi-2 braze filler metal tape (approximately 0.0045 in. thick) was placed between all joint surfaces. After assembly, a braze slurry of BNi-2 filler metal was applied to all joints to ensure soundness.

Brazing of ODS Alloys

Oxide dispersion-strengthened alloys are P/M alloys that contain stable oxide evenly distributed throughout the matrix. The oxide does not go into solution in the alloy even at the liquidus temperature of the matrix. However, the oxide is usually rejected from the matrix on melting of the matrix, which occurs during fusion welding, and cannot be redistributed in the matrix on solidification; therefore, these alloys are usually joined by brazing. There are two commercial alloy classes of ODS alloys: the dispersion-strengthened nickel-chromium and dispersion-strengthened nickel, and mechanically alloyed Inconel MA 754, Inconel MA 6000, and Incoloy MA 956.

Inconel MA 754, dispersion-strengthened nickel alloys, and dispersion-strengthened nickel-chromium alloys are the easiest to braze of the family of ODS alloys. Vacuum, hydrogen, or inert atmospheres can be used for brazing. Prebraze cleaning consists of grinding or machining the faying surfaces and washing with a solvent that evaporates without leaving a residue. Generally, brazing temperatures should not exceed 2400 °F unless demanded by a specific application that has been well examined and tested. The brazing filler metals for use with these ODS alloys usually are not classified by AWS. In most cases, the brazing filler metals used with these alloys have brazing temperatures in excess of 2250 °F. These include proprietary alloys that are nickel-, cobalt-, gold-, or palladium-based.

Brazements made of ODS alloys to be used at elevated temperature must be tested at elevated temperature to prove fitness-for-purpose. In the case of stress rupture testing, AWS specification C3.2 may be used as it represents the actual joint configuration. The configuration shown in Fig. 2 is preferred by some as a test model, although any test configuration without stress raisers is adequate. The elevated-temperature brazement properties for Inconel alloy MA 754 should meet the following requirements:

Shear stress, ksi	Temperature, °F	Service life, h
3.8	1800	1000+
1.3	2000	1000+

Inconel MA 6000 is a nickel-based ODS alloy that is also γ' strengthened. The amount of alloying elements plus the γ' precipitation cause an interesting problem in the brazing of this alloy. Inconel MA 6000 has a solidus temperature of 2372 °F; therefore, the brazing temperature should be no higher than 2282 °F. Additionally, because 2250 °F is the γ' solution treatment temperature, it becomes important to carefully select the brazing filler metal and to heat treat the assembly after brazing. The BNi, BCo, and specially formulated filler metals have been used for this alloy.

Inconel MA 6000 is used for its high-temperature strength and corrosion resistance; unfortunately, the passive oxide scale that provides good corrosion resistance also prevents wetting and flow of brazing filler metal. Therefore, correct cleaning procedures are very important. Surfaces to be brazed should be mechanically cleaned with

Table 3 Effect of brazing on mechanical properties of Haynes 25

Condition	Test temperature, °F	Ultimate strength, ksi	0.2% offset yield strength, ksi	Elongation, %
Tensile testing				
Mill anneal	Room	147.9	69.2	56
After braze cycle	Room	108.3	69.2	12
Mill anneal	1500	57.4	30.5	17
After braze cycle	1500	57.0	33.2	24
Mill anneal	1800	21.0	18.1	35
After braze cycle	1800	21.5	18.8	35

Condition	Test temperature, °F	Stress, ksi	Hours to rupture
Stress-rupture testing			
Mill anneal	1500	24.5	82
After braze cycle	1500	24.5	72.3
Mill anneal	1650	15.0	56
After braze cycle	1650	15.0	36
Mill anneal	1800	6.5	110
After braze cycle	1800	6.5	120

a water-cooled, low-speed belt or wheel of approximately 320-grit and stored in a solvent, such as methanol, until immediately before the beginning of the brazing cycle.

Cobalt-Based Alloys

The brazing of cobalt-based alloys is readily accomplished with the same techniques used for nickel-based alloys. Because most of the popular cobalt-based alloys do not contain appreciable amounts of aluminum or titanium, brazing atmosphere requirements are less stringent. Table 2 gives typical compositions of several cobalt-based alloys. These materials can be brazed in either a hydrogen atmosphere or a vacuum. Filler metals are usually nickel- or cobalt-based alloys or gold-palladium compositions. Silver or copper braze filler metals may not have sufficient strength and oxidation resistance in many high-temperature applications. Although cobalt-based alloys do not contain appreciable amounts of aluminum or titanium, an electroplate or flash of nickel is often used to promote better wetting of the brazing filler metal.

Nickel-based brazing alloys such as AWS BNi-3 have been used successfully on Haynes 25 for honeycomb structures. It has been reported that after brazing, a diffusion cycle is used to raise the braze joint remelt temperature to 2300 to 2400 °F. Table 3 presents the effects of a high-temperature braze (2240 °F for 15 min) on the mechanical properties of Haynes 25. One cobalt-based brazing filler metal (AWS BCo-1, see Table 1) appears to offer a good combination of strength, oxidation resistance, and remelt temperature for use on Haynes 25 foil.

Cobalt alloys, much like nickel alloys, can be subject to liquid metal embrittlement or stress corrosion cracking when brazed under residual or dynamic stresses. This frequently is observed when using silver or silver-copper (BAg) filler metals. Liquid metal embrittlement of cobalt-based alloys by copper (BCu) filler metals occurs with or without the application of stress; therefore, BCu filler metals should be avoided when brazing cobalt.

Brazing of Aluminum Alloys*

BRAZING of aluminum alloys was made possible by the development of fluxes that disrupt the oxide film on aluminum without harming the underlying metal and filler metals (aluminum alloys) that have suitable melting ranges and other desirable properties. The aluminum-based filler metals used for brazing aluminum alloys have liquidus temperatures much closer to the solidus temperature of the base metal than those for brazing most other metals. For this reason, close temperature control is required in brazing aluminum. The brazing temperature should be approximately 70 °F below the solidus temperature of the base metal, but if temperature is accurately controlled and the brazing cycle is short, it can be as close as 10 °F. Aluminum alloys, depending on composition, can be brazed with commercial filler metals from 1020 to 1180 °F. Most brazing is done at temperatures between 1040 and 1140 °F.

Much of the equipment and many of the techniques used to prepare, braze, and inspect aluminum alloys are the same as those used for other metals. For general information, see the articles "Furnace Brazing of Steels," "Torch Brazing of Steels," "Resistance Brazing," and "Dip Brazing of Steels in Molten Salt" in this Volume.

Base Metals

The non-heat-treatable wrought alloys that have been brazed most successfully are the 1*xxx* and 3*xxx* series and low-magnesium members of the 5*xxx* series. The alloys containing a higher magnesium content are more difficult to braze by the usual flux methods, because of poor wetting and excessive penetration by the filler metal. Filler metals are available that melt below the solidus temperatures of most commercial, non-heat-treatable wrought alloys.

The commonly brazed, heat treatable wrought alloys are the 6*xxx* series. The 2*xxx* and 7*xxx* series of aluminum alloys are low melting and, therefore, not normally brazeable, with the exception of 7072 (used as a cladding material only) and 7005.

Alloys that have a solidus above 1100 °F are easily brazed with commercial binary aluminum-silicon filler metals. Stronger, lower melting alloys can be brazed with proper attention to filler-metal selection and temperature control, but the brazing cycle must be short to minimize penetration by the molten filler metal. Sand and permanent mold casting alloys with a high solidus temperature are brazeable; the most frequently brazed alloys are 443.0, 356.0, and 710.0. Formerly, aluminum die castings were not brazed because of blistering due to high gas content, but advances in casting technique have resulted in improved quality.

Some common wrought and cast aluminum alloys are listed in Table 1 with their melting temperature ranges and brazeability ratings. Brazing of aluminum is generally limited to parts more than 0.015 in. thick, but dip brazing and fluxless vacuum brazing have been accomplished successfully on aluminum fin stock as thin as 0.005 in.

Filler Metals

Commercial filler metals for brazing aluminum are aluminum-silicon alloys containing 7 to 12% Si. Lower melting points are attained, with some sacrifice in resistance to corrosion, by adding copper and zinc. Filler metals for vacuum brazing of aluminum usually contain magnesium. The compositions and solidus, liquidus, and brazing temperatures of the most frequently used brazing filler metals for aluminum are given in Table 2.

The optimum brazing-temperature range for an aluminum-based filler metal is determined not only by the melting range of the filler metal and the amount of molten filler metal needed to fill the joint, but also is limited by the mutual solubility between the filler metal and the base metal being brazed. The brazing-temperature ranges of some filler metals are related to those of some base metals in Fig. 1.

Filler metals for separate application from the base metal to be brazed are available as wire and sheet (thin-gage shim stock). The manufacture of filler metal in sheet and wire forms becomes more difficult as the silicon content increases. Only filler metals BAlSi-2 (alloy 4343), BAlSi-4 (alloy 4047), and alloy 4004 are available as sheet.

Most filler metals are used for any of the common brazing processes and methods, but two (alloys 4004 and 4104, which contain additions of magnesium and magnesium-bismuth, respectively, and have a brazing-temperature range of approximately 1090 to 1120 °F) have been de-

Table 1 Melting ranges and brazeability of some common aluminum alloys

Alloy	Melting range, °F	Brazeability(a)
Non-heat-treatable wrought alloys		
1350	1195-1215	A
1100	1190-1215	A
3003(b)	1190-1210	A
3004	1165-1205	B
5005	1170-1205	B
5050	1160-1205	B
5052	1100-1200	C
Heat treatable wrought alloys		
6053	1100-1205	A
6061	1100-1200	A
6063	1140-1210	A
6951(c)	1140-1210	A
7005	1125-1200	B
Casting alloys(d)		
443.0	1065-1170	B
356.0	1035-1135	B
710.0	1105-1195	B
711.0	1120-1190	A

(a) A, generally brazeable by all commercial procedures; B, brazeable with special techniques or in specific applications that justify preliminary trials or testing to develop the procedure and to check the performance of brazed joints; C, limited brazeability. (b) Used both plain and as the core of brazing sheet. (c) Used only as the core of brazing sheet. (d) Sand and permanent mold castings only

*Revised by Arthur H. Lentz, Welding Engineer, Reynolds Metals Co.

Table 2 Compositions and solidus, liquidus, and brazing temperature ranges of brazing filler metals for use on aluminum alloys

AWS classification	Composition(a), % Si	Cu	Mg	Zn	Mn	Fe	Temperature, °F Solidus	Liquidus	Brazing
BAlSi-2	6.8-8.2	0.25	...	0.20	0.10	0.8	1070	1135	1110-1150
BAlSi-3(b)	9.3-10.7	3.3-4.7	0.15	0.20	0.15	0.8	970	1085	1060-1120
BAlSi-4	11.0-13.0	0.30	0.10	0.20	0.15	0.8	1070	1080	1080-1120
BAlSi-5(c)	9.0-11.0	0.30	0.05	0.10	0.05	0.8	1070	1095	1090-1120
BAlSi-6(d)	6.8-8.2	0.25	2.0-3.0	0.20	0.10	0.8	1038	1125	1110-1150
BAlSi-7(d)	9.0-11.0	0.25	1.0-2.0	0.20	0.10	0.8	1038	1105	1090-1120
BAlSi-8(d)	11.0-13.0	0.25	1.0-2.0	0.20	0.10	0.8	1038	1075	1080-1120
BAlSi-9(d)	11.0-13.0	0.25	0.10-0.5	0.20	0.10	0.8	1044	1080	1080-1120
BAlSi-10(d)	10.0-12.0	0.25	2.0-3.0	0.20	0.10	0.8	1038	1080	1080-1120
BAlSi-11(d,e)	9.0-11.0	0.25	1.0-2.0	0.20	0.10	0.8	1038	1105	1080-1120

(a) Principal alloying elements. (b) Contains 0.15% Cr. (c) Contains 0.20% Ti. (d) Solidus and liquidus temperature ranges vary when used in vacuum. (e) Contains 0.02-0.20% Bi

Fig. 1 Comparison of brazing-temperature ranges of aluminum alloy base metals and aluminum alloy brazing filler metals

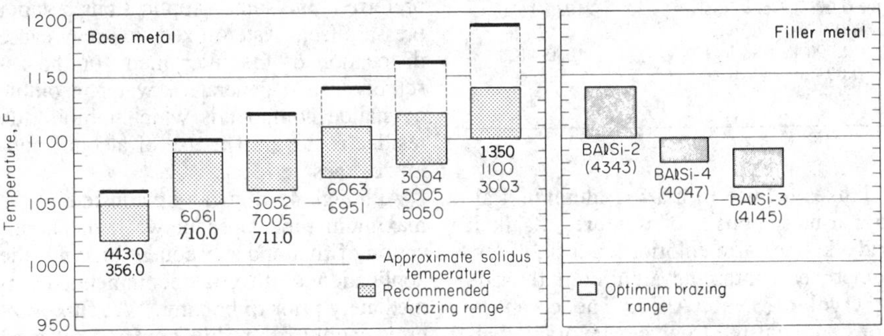

veloped exclusively for use in fluxless vacuum brazing. Similarly, a proprietary mixture of filler metal BAlSi-4 (alloy 4047) in powder form and a chemical compound is used exclusively with dip brazing. This mixture can be brushed or extruded onto the joints and can be applied to surfaces in all positions. The mixture stays in place because it is baked onto the metal surface during preheating. It may be employed in brazing overhead joints in wave-guide assemblies and in applications where a small, controlled flow of aluminum alloy filler metal is desired.

Brazing Sheet

Brazing sheet consists of a brazeable aluminum alloy roll bonded or clad with an aluminum brazing alloy. Brazing sheet, which is available with one or both sides clad, provides a more convenient method of supplying the filler metal than wire, shims, or powder and is particularly convenient for mass-produced, complex assemblies. The selection of brazing sheet instead of filler metal in other forms, however, is based on cost in a given application. Brazing sheet can be subjected to drawing, bending, or any other forming process.

Three types of brazing sheet are shown in Fig. 2. The most common type (Fig. 2a) has filler metal on one or both sides. The compositions and brazing-temperature ranges of commercially available brazing sheets of this type are listed in Table 3. The other types of brazing sheet shown in Fig. 2 are available only by special order. The brazing sheet shown in Fig. 2(b) has an interlayer of aluminum alloy to act as a diffusion barrier between the high-silicon filler metal and the core sheet of structural alloy. The brazing sheet shown in Fig. 2(c) has filler metal on the joint side of the structural alloy and is alclad for corrosion resistance on the opposite side.

The structural alloys frequently used in brazing sheet are 3003, which is resistant to sagging at brazing temperatures, and 6951, which is heat treatable after brazing and is used where higher strength is desired. Some of the commercially available filler metals are used as cladding of brazing sheet.

Silicon Diffusion. Diffusion between the core metal and the coating of filler metal limits the general applications of brazing sheet. Long heating times above 900 °F increase diffusion of silicon from the coating into the core, which can lower the mechanical properties of the core and reduce the amount of filler metal available for flow.

To restrict diffusion, the brazing cycle should be as short and at as low a temperature as possible. Core alloys must be selected to prevent formation of harmful intermetallic compounds. Under conditions where the core material is aggressively penetrated by the coating alloy, an intermediate protective layer of commercial-purity aluminum or of an alloy not easily penetrated by the coating can be used.

Fluxes

Conventional brazing, performed in air or other oxygen-containing atmospheres,

Fig. 2 Three types of aluminum brazing sheet. Shown in joints with a vertical member

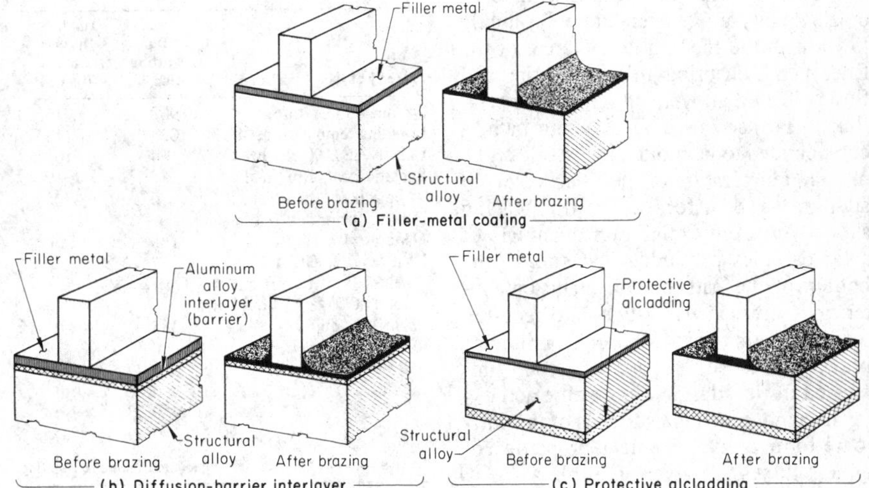

Table 3 Compositions and brazing-temperature ranges of aluminum brazing sheet

Brazing sheet(a)	Sides clad	Core alloy	Cladding alloy	Cladding on each side, % of sheet thickness	Optimum brazing range, °F
7	1	3003	4004	15% for 0.024 in. and less	1120-1130
				10% from 0.025 in. to 0.062 in.	...
				7½% for 0.063 in. and over	...
8	2	3003	4004	15% for 0.024 in. and less	1120-1130
				10% from 0.025 in. to 0.062 in.	...
				7½% for 0.063 in. and over	...
11	1	3003	4343	10% for 0.063 in. and less	1100-1140
				5% for 0.064 in. and over	...
12	2	3003	4343	10% for 0.063 in. and less	1100-1140
				5% for 0.064 in. and over	...
13	1	6951	4004	15% for 0.024 in. and less	1110-1115
				10% from 0.025 in. to 0.062 in.	...
				7½% for 0.063 in. and over	...
14	2	6951	4004	15% for 0.024 in. and less	1110-1115
				10% from 0.025 in. to 0.062 in.	...
				7½% for 0.063 in. and over	
21	1	6951	4343	10% for 0.090 in. and less	1100-1120
				5% for 0.091 in. and over	...
22	2	6951	4343	10% for 0.090 in. and less	1100-1120
				5% for 0.091 in. and over	...
23	1	6951	4045	10% for 0.090 in. and less	1080-1120
				5% for 0.091 in. and over	...
24	2	6951	4045	10% for 0.090 in. and less	1080-1120
				5% for 0.091 in. and over	...

(a) Designations registered with the Aluminum Association.

requires the use of a chemical flux. Fluxes, which become active before brazing temperature is reached and are molten over the entire brazing range, penetrate the film of oxide, exclude air, and promote wetting of the base metal by the filler metal. A satisfactory flux must: (1) begin to melt at a temperature low enough to minimize oxidation of the parts, (2) be essentially molten at the time the filler metal melts, (3) flow over the joint and the filler metal to shield them from oxidizing gases, (4) penetrate the oxide films, (5) lower the surface tension between the solid and liquid metals to encourage wetting, (6) remain liquid until the filler metal has solidified, and (7) be relatively easy to remove after brazing is complete.

A superior flux for furnace and torch brazing melts at a temperature only slightly lower than the melting temperature of the filler metal, ensuring uniform wetting and flow of filler alloy in minimum time. A flux to be used as a dip brazing bath is compounded to be molten and stable at the melting temperature of the filler metal. In addition, a flux for use in dip brazing should form only minimum quantities of solid particles or sludge that sink to the bottom of the bath or collect in joint interstices. Less active fluxes are required for dip brazing than for torch or furnace brazing, because the parts are totally immersed in flux during dip brazing and oxygen cannot reach the surfaces of the parts to re-form oxide. Physical properties of typical fluxes are given in Table 4.

Fluxes for use in brazing aluminum alloys usually consist of mixtures of alkali and alkaline earth chlorides and fluorides, sometimes containing aluminum fluoride or cryolite ($3NaF \cdot AlF_3$). The compositions are adjusted to give a favorable balance among melting range, density, chemical activity, etching characteristics, and cost. Small amounts of one or more of the chlorides of antimony, cadmium, chromium, cobalt, copper, iron, lead, manganese, nickel, silicon, tin, zinc, precious metals, or rare earths improve the performance of fluxes. Reduction of fluoride can reduce effective oxide removal, but too high a concentration results in an undesirably high melting range.

Table 4 Physical properties of typical fluxes for brazing aluminum alloys

Property of flux	Dip brazing (flux 33)	Torch and furnace brazing (flux 34)
Solidus temperature, °F	900	915
Liquidus temperature, °F	1035	1115
Density at 1100 °F, lb/ft^3	104	107
Specific heat, Btu/lb °F (approx)	0.2	(a)
Heat of fusion, Btu/lb (approx)	168	(a)
Heat requirement, Btu to heat 1 lb of flux from solid at 70 °F to liquid at 1150 °F (approx)	385	(a)
Resistivity, Ω · cm at:		
1080 °F	0.43	(a)
1130 °F	0.36	(a)
1150 °F	0.33	(a)
1180 °F	0.29	(a)

(a) These properties are not pertinent to torch and furnace brazing.

Flux usually is received in dry powder form in sealed, moistureproof containers. It can be stored for long periods if the seal is maintained. Once a flux container is opened, stringent precautions must be followed to prevent contamination of the flux by atmospheric moisture. Flux containers should be of perfectly clean aluminum, glass, or earthenware—never of steel.

Technique. Aluminum brazing fluxes can be applied dry, or they can be mixed with tap water or alcohol and applied by painting, spraying, or dipping. Dry flux can be sprinkled on the work, or a heated filler rod can be dipped into the dry flux. Although flux can be mixed with tap water to form a paste, the use of alcohol may be preferred in some applications; vapor pressure from water-mixed flux may cause dislocation of the filler metal or the assembly and generates water-insoluble oxyhalide compounds which inhibit flux residue removal. The use of alcohol minimizes these effects.

Although 45 min can be considered the maximum time lapse between the application of flux and subsequent brazing, the application of flux is recommended immediately prior to brazing. Wet flux mixtures should be freshly prepared (at least once in each shift).

The wetting action of a flux can be improved considerably by the use of a wetting agent. A mixture of two thirds flux and one third water by weight usually is satisfactory for painting. Spraying or dipping requires a thinner consistency using more water. The amount of water needed for spray gun operation is determined best by trial.

A proprietary furnace brazing process that utilizes a noncorrosive flux and an inert gas is available. The $K_3AlF_6-KAlF_4$ flux is inactive at temperatures both above and below the aluminum brazing temperature. Satisfactory production has been accomplished on the 1*xxx* and 3*xxx* series alloys, as well as the nonvacuum, commercially available brazing sheet alloys.

Flux Stop-Offs. Sometimes positive action to prevent filler metal from flowing beyond a certain area is desirable. Stop-offs suitable for this purpose sometimes consist of a mixture of equal parts by weight of a medium-heavy engine oil (SAE 30), finely powdered graphite, and benzene or naphtha (mineral spirits) or slurries of refractory oxides. Often, a mark made by a soft graphite pencil is an effective stop-off. Proprietary, commercial stop-off compounds are also available.

Some of these may be applied in paste form without being baked and later may be removed by brushing.

Furnace and dip brazing frequently require the use of a stop-off to prevent the jigs and fixtures from being brazed to the work. The mixture is brushed or sprayed on the areas to be stopped off and then baked at 400 to 600 °F to carbonize the oil. One application usually lasts for several brazing cycles. Stop-offs for fluxless vacuum brazing are usually refractory oxides, which are sprayed on the jigs and fixtures, or are formed on the jigs and fixtures by heating to high temperature in an air atmosphere. Such coatings generally last until mechanically damaged. In torch brazing, the use of stop-offs usually is not required, because the operator has adequate control over the flow of the filler metal.

Joint Design

Joints to be brazed with the use of flux must be designed to permit application of the flux to the joint surfaces before assembly or to permit entry of the flux between the components after assembly. In addition, provision must be made for the flux to be displaced by the filler metal, because entrapped flux is a potential source of corrosion. Joint design must also allow the escape of gas and subsequent penetration by the filler metal to ensure the complete distribution of filler metal in the joint (see the article "Furnace Brazing of Steels" in this Volume).

Assemblies to be brazed may be designed with any of the several types of joints. For brazing processes that require a flux, joint strength equal to the strength of the base metal can be obtained with lap joints. The designs of six lap joints for brazing are compared with butt, T-, and corner joints used for welding in the article "Furnace Brazing of Steels" in this Volume. Lap joints that require the filler metal to flow long distances should be designed for flow in one direction only; otherwise, filler metal flowing from both edges of such joints may entrap flux. The need for flow in joints having wide laps is nullified by using brazing sheet as one of the members.

Butt and scarf joints are not usually as strong as the base metal, but when correctly designed, such joints may give satisfactory service. For fluxless brazing processes, joints having narrow or line contact are preferred to joints having wide contact because line contact provides column strength and prevents flux entrapment. Joints of short length or line contacts are highly desirable during any aluminum brazing process, whether torch, furnace, or dip. For long lap joints, corrugations can be used to provide an outlet for molten flux, and the final result is the same as when line contacts are used.

In joints for furnace and torch brazing, capillary rise is limited to about $^1/_4$ in. and must be considered in the design of the joint. In joints for dip brazing, capillary rise is seldom a limiting factor in design. During brazing of aluminum, base metal and filler metal are mutually soluble. This causes the filler metal to change progressively in composition as it flows in joints, which progressively raises the liquidus temperature of the filler metal and reduces the ability of the filler metal to wet and flow. Clearances must be sufficient to prevent premature solidification of the filler metal in small capillary spaces which entrap flux and cause porosity. As the distance the filler metal must flow increases, clearance requirements increase also.

In dip brazing with preplaced filler metal, joint clearance at room temperature ranging from 0.002 to 0.004 in. may suffice for narrow laps ($^1/_4$ in. or less). Wider laps may require clearances up to 0.010 in. For furnace and torch brazing, clearance ranging from 0.004 to 0.010 in. is required for narrow laps ($^1/_4$ in. or less); as much as 0.025 in. is required for wider laps. With brazing sheet, clearances may be smaller. To ensure formation of a continuous fillet in fluxless vacuum brazing with brazing sheet, clearances normally should not exceed 0.003 in. In some applications, however, continuous fillets have been formed in joints with clearances as great as 0.009 in.

Tube-to-tube joints to be torch or furnace brazed require that the outer tube be flared 10 to 12° to produce sound joints. In joining fittings to a tube, knurling of the tube or fitting permits complete penetration through the joint to be achieved.

The correct preplacement of the filler metal is extremely important. Gravity is usually sufficient to keep the filler metal in place for fixed-position furnace brazing. In dip brazing, the filler may have to be held in place because of the buoyancy of the molten flux.

Prebraze Cleaning

Oil and grease must be removed from components of assemblies to be brazed to eliminate stop-off effects. For non-heat-treatable alloys, vapor or solvent cleaning is usually adequate, although chemical cleaning may be required for components that have been severely formed, as by spinning. For the heat treatable alloys, chemical cleaning is usually necessary to reduce the amount of tenacious oxide film. Chemical cleaning is not recommended for fluxless vacuum brazing. Scrubbing with steel wool, abrasive cloth, or a power-driven wire brush (preferably with stainless steel bristles) can also be used. Burrs should be removed before brazing.

Chemical cleaning methods used prior to brazing include nitric acid, hydrofluoric acid, or nitric-hydrofluoric acid mixtures at room temperature. A widely used method is immersion for about 30 s in a solution containing equal parts of commercial nitric acid and water, followed by rinsing in clean water (preferably hot) and drying in hot air.

Aluminum-silicon alloys require a special etchant, because the silicon constituent is not attacked readily by many alkaline or acid solutions. For these alloys, a room-temperature solution of three parts concentrated nitric acid and one part concentrated hydrofluoric acid is employed. This solution requires a tank lined with an inert material such as carbon brick or certain types of plastic. The presence of fluorides necessitates caution in handling and special waste disposal procedures. For thick and resistant oxide coatings, immersion for about 30 s in a warm (150 °F) aqueous solution of 5% sodium hydroxide is recommended. To remove the surface smut produced by this treatment, the treatment should be followed by a cold water rinse, immersion in a room-temperature solution containing equal parts of commercial nitric acid and water, a final water rinse (preferably hot), and hot air drying.

Chemical cleaning of aluminum provides excellent surfaces for brazing. Cleaning in a caustic solution is particularly effective, although residues from caustic solutions can interfere with brazing, probably because of the large amounts of aluminum oxide formed. Nitric, sulfuric, and phosphoric acid residues may prevent brazing entirely. To eliminate the possibility of harmful residues, chemical treatments should be followed by a hot distilled water rinse, after which the components are dried. Hydrofluoric acid residues are not detrimental to brazing, but hydrofluoric acid is ineffectual for removing oil and grease and, thus, is useful only on components that have been degreased by a solvent or emulsion cleaner.

For best results, brazing should be done immediately after cleaning or within 48 h. If precautionary measures are taken to prevent their contamination, however, adequately cleaned components do not lose brazing qualities even in several weeks (see the article "Cleaning and Finishing of Aluminum and Aluminum Alloys" in

Volume 5 of the 9th edition of *Metals Handbook*).

Assembly

Components to be brazed must be correctly located during assembly and must be held in position by some type of jigging. Correct jigging is particularly important in dip brazing, where the displacement of air and the buoyancy of the flux must be considered. Self-jigging, when possible, offers an economic advantage over the cost of design, maintenance, and replacement of jigs. In some applications, spot welding is an effective means for jigging. Several methods of self-jigging are described and illustrated in the article "Furnace Brazing of Steels" in this Volume. Another method, the use of tabs and mating slots, is described in Example 1.

Mating surfaces of brazing sheet should not be spot welded for self-jigging because they can separate in the flux bath. Jigs that may be required to hold the parts in correct alignment must be resistant to the highly reactive molten flux. Low-carbon steel fixtures may have sufficiently long life for a specific job, but aluminum-coated steel has longer useful life.

Because of differential thermal expansion, the aligning jig should be designed with spring relief. Stainless steel type 304 and Inconel X-750 are both good materials for such springs; Inconel X-750 retains more of its spring characteristics at brazing temperatures. Inconel X-750 also has better resistance to corrosion, not only during brazing, but also in postcleaning operations.

Dip Brazing

The best method of heating and fluxing aluminum joints simultaneously is to immerse the entire assembly in a bath of molten flux. Because of the low specific heat of flux, assemblies usually are heated to about 1000 °F prior to flux immersion. This is known as dip brazing, or flux-bath brazing. Dip brazing has been used successfully in the manufacture of complex, multiple-joint heat exchangers.

Immersing the entire assembly into molten flux has many advantages. Heat is applied to all parts simultaneously and uniformly, and air is replaced by a buoyant and surface-active environment, promoting brazing filler-metal flow. In addition, the uniform temperature permits production assembly of parts with dimensional tolerances as low as ±0.002 in. or even less.

Heat-transfer units assembled from alternate corrugated and flat aluminum brazing sheets or from various crimped and formed pieces are examples of the type of work that dip brazing can handle advantageously. Units weighing up to 20 000 lb have been joined by dip brazing. Certain designs have to withstand a service pressure of 650 psi. Brazing sheet is essential to this type of work, reducing assembly and brazing costs. The rapid and even heating and flux buoyancy minimize distortion.

For assemblies designed with components in close proximity, flux removal can be tedious and expensive. For instance, when components such as those of a heat-exchanger matrix are spaced closer than $^1/_8$ in., the flux holds to the surfaces by surface tension and capillary action; it will not drain from the components freely. This is not as great a problem with spacings greater than $^1/_8$ in. between components of normal length; wider spacing of long components is desirable.

Equipment. Dip brazing equipment may be as simple as a heat-resistant glass beaker inside a resistance-heated furnace or as complex as a large steel vessel lined with high-alumina, acid-proof brick. For adequate resistance to flux, the alumina content of lining brick should be at least 40%. The molten bath is usually heated by low-voltage alternating current passing through the flux between wrought nickel, Inconel 600, or carbon electrodes. These show less attack than copper or copper-bearing electrodes and cause minimum contamination of the flux. Attack at the electrode-flux-air interface has led to preference for submerged electrodes. The bath temperature should be controlled within ±5 °F.

Technique. The amount of flux required to fill the bath is about 100 lb/ft^3. Approximately 385 Btu/lb is required to melt and heat the flux. Thermocouples enclosed in protective tubes should be used to determine the bath temperature. Flux quantity should be sufficient to prevent the temperature from dropping more than 5 to 10 °F when parts are immersed. The specific heat of dip brazing flux, approximately 0.2 Btu/lb · °F is about the same as that of aluminum. For assemblies that have been preheated to 1000 to 1050 °F, about 16 lb of molten flux should be used per pound of aluminum. With this ratio, the bath temperature may drop 2 to 5 °F, thereby facilitating brazing of four to six loads per hour.

The composition of the flux should be adjusted periodically by fluoride and chloride additions. Proprietary additive mixtures are available for this purpose. Even when the molten bath is idle, side reactions reduce its activity. Because molten flux may contain water vapor, it should be dehydrated periodically with aluminum to minimize the formation of hydrogen when the assemblies are dipped. The aluminum used for dehydrating the bath may be a conveniently loose coil of alloy 1100 or alloy 3003. When hydrogen stops burning at the surface of the bath, dehydration is essentially complete. Initial dehydration should be conducted for 4 to 48 h, depending on bath size. Insertion of aluminum into the bath before the brazing operation is begun also removes heavy-metal impurities such as nickel, copper, iron, and zinc. The heavy-metal deposit on the aluminum coil is removed by quenching the coil in water, dipping it in nitric acid, and thoroughly rinsing it with water. The coil should be dried before being reused.

These operations, as well as the actual brazing, produce a sludge containing oxides from the brazed parts and the brickwork and insoluble fluoride complexes. After settling, the sludge should be ladled out at periodic intervals with a perforated tool.

Flux is removed by dragout on the parts being brazed and must be replaced. When an assembly is dip brazed, approximately 0.5 oz of flux dragout occurs per square foot of surface. With heat exchangers having complex and devious passages, this amount may be larger, because of capillary forces holding the flux. For a specific unit, dragout may vary as much as threefold, depending on the melting point and viscosity of the flux.

Before immersion in the flux bath, all moisture must be removed from the assembly and from any fixture used with it. Even a slight amount of moisture in contact with molten flux can cause spattering. Drying by preheating is recommended. The use of preheated assemblies decreases the drop in temperature of the salt bath, shortening processing time. Large or complex assemblies should be preheated to 1000 to 1050 °F, usually in an air-recirculating furnace, at a rate that provides a suitable compromise between distortion from fast heating and diffusion between the layers in brazing sheet during slow heating. Methods and advantages of preheating are discussed further in the article "Dip Brazing of Steels in Molten Salt" in this Volume.

After being preheated, parts are immediately immersed in the flux bath for the scheduled period. This period depends somewhat on the mass of the assembly, but immersion is usually only $^1/_2$ to 3 min in duration. For large assemblies, such as a cryogenic heat exchanger that may weigh

Fig. 3 Dip brazing a complex chassis assembly. Slot-and-tab arrangement was used for self-jigging.

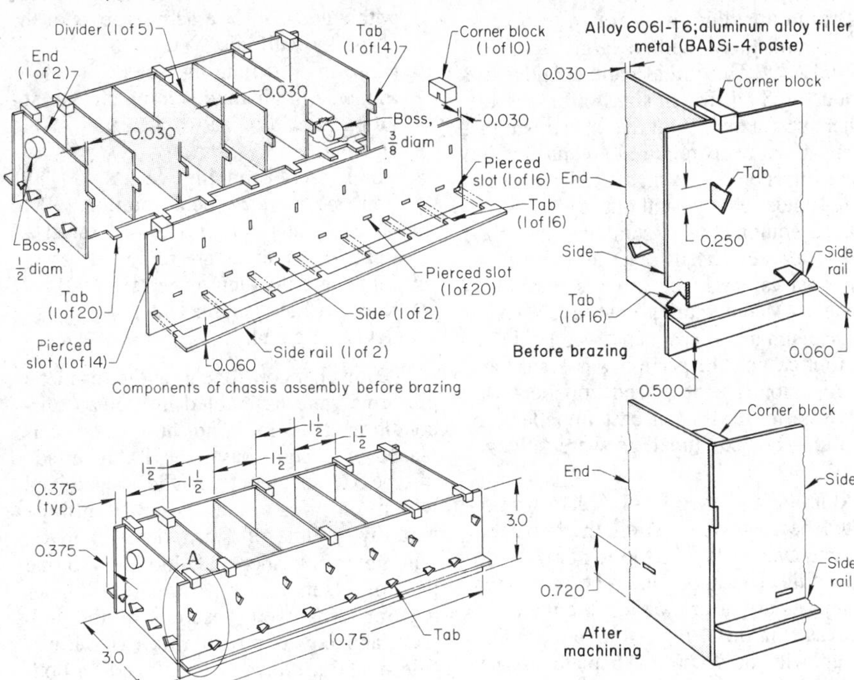

more than 1000 lb, immersion time may be as long as 20 min. Time in the bath should be no longer than is required to obtain melting and complete flow of the filler metal. After being preheated, the self-jigged assembly in the following example was dip brazed in only 20 s.

Example 1. Use of Dip Brazing To Join a Complex Assembly. The chassis assembly shown in Fig. 3 was used for airborne electronic equipment. The large, thin-walled structure of complex design and small production quantities made casting impractical. Resistance spot welding, arc welding, and torch brazing were ruled out by the need for a continuous metal surface, close dimensional tolerances, and the presence of hidden joints in the design. Thus, dip brazing was selected as the joining method best fulfilling the design requirements.

The dip brazed assembly was made of alloy 6061-T6. The two ends, two sides, five dividers, two side rails, and the bottom were blanked and pierced in a numerically controlled press. The ten corner blocks and the two round bosses were machined from bar stock. Self-jigging of the 24 components was achieved by providing the sheet metal components with slots and mating tabs; the tabs were inserted into the slots and hand twisted after assembly to hold the components in place. After brazing, the tabs and other extraneous material were machined off (Fig. 3, detail A). This type of assembly allowed brazing with no external fixturing and permitted the use of assembly-line techniques, thus reducing costs.

After burrs larger than 0.005 in. were removed from the sheet metal components, the chassis was assembled and then cleaned by degreasing followed by immersion in either a caustic bath or a phosphoric acid etching solution. Brazing paste containing BAlSi-4 filler metal was applied to one side of each joint. The assembly was preheated for 20 to 30 min in an electric furnace operating at 1000 °F and then was dipped for 20 s in a bath of proprietary brazing flux electrically heated to 1400 °F.

After brazing, the assembly was cooled to about 500 °F by an air blast (to minimize distortion) before being quenched in hot water and washed to remove flux. The heating for brazing, followed by the air blast cooling and the hot water quench, put the 6061 alloy in a condition that yielded a T4 temper after natural aging for several days.

After the self-jigging tabs and other extraneous metal had been removed by machining, the brazed chassis assembly was inspected visually. Parts were held within 0.015-in. total tolerance of warp over the 10-in. dimension and the angularity of 1° in 3 in. Fillet radii were held within 0.030 to 0.060 in. at junctures of two parts and within 0.030 to 0.090 in. at junctures of three parts. Fluorescent-penetrant or radiographic inspection was used in special cases. The rejection rate for this chassis was only 2% from all causes.

Furnace Brazing

Brazing in an atmosphere furnace is a high-production process that requires minimum training and skill of operators. Production rates can be considerably higher and costs can be lower than for torch brazing. Aircraft hot air ducts are furnace brazed in as little as 1/20th of the time required to torch braze similar parts. Mass-production assemblies, such as refrigerator evaporators, are brazed at about 500 pieces per hour in continuous furnaces.

Equipment. Heating methods for furnace brazing include electrical heating elements and direct combustion and radiant tubes. Direct combustion furnaces are inexpensive, but the furnace gases may cause undesirable metallurgical effects during brazing of the heat treatable alloys—for example, the 6*xxx* series. Furnaces are generally refractory lined, although such linings become saturated with flux components. Heat-resistant steels, which are normally not recommended because of flux attack, are satisfactory for furnace linings if kept clean, particularly if aluminum coated.

Whatever the type of furnace, the temperature in the brazing zone must be uniform within ±10 °F and preferably within ±5 °F. Circulation of atmosphere, preferably with appropriate baffles, is required to prevent local heat variations and to obtain the maximum rate of temperature rise.

Technique. Flux slurry may be applied to the parts by dipping, brushing, or spraying. Tap, distilled, or deionized water can serve as a vehicle; tap water should be free of heavy metals, because these can cause subsequent corrosion. Because hydrogen may be evolved when wet flux is heated on aluminum parts, closed assemblies must be vented. Gas generation can be reduced by drying the flux on the part prior to brazing. Mixing the flux with alcohol instead of water speeds the drying, but the explosive fumes from the alcohol must be dissipated.

Ambient air or chemically inert gas is normally used as the furnace atmosphere. A dry atmosphere (dew point of −40 °F) consisting of the products of combustion of fuel can sometimes reduce the amount of flux needed. For brazing aluminum to other metals, an inert, dry atmosphere is particularly beneficial.

Continuous furnace brazing requires furnaces divided into several progressive heating zones to improve heating rate and joint quality and to reduce warpage. A furnace cycle of 15 min or less is desirable. In automated operations, the brazing zone usually requires a travel time of 2 to 3 min for assemblies of moderate size. Because the flux loses its activity in about 30 min, aluminum assemblies large enough and heavy enough to require heating times exceeding this limit should not be furnace brazed.

Beyond the heating portion of the furnace, from 1 to 5 min of conveyor travel should be in an unheated zone to allow the filler metal to solidify. Directly following should be an air blast, a hot water spray (180 to 212 °F), or a boiling water quench, which begins the flux removal process. For the heat treatable alloys, a water quench after brazing permits improvement in the mechanical properties, especially if the parts are subsequently given an aging treatment.

Example 2. Use of Furnace Brazing. Impellers like the one shown in Fig. 4 were used in radial-flow turbo-compressors to increase the recoverable energy content of the gas passing through the compressor. Speeds of various models of impellers ranged from 3000 to 30 000 rpm. More than 80 different sizes of impellers were made. Aluminum was used for impellers that, depending on size and speed, were to run at temperatures up to about 400 °F.

The casting of impellers, which was done for some other applications, would have been expensive because of the large investment required for a multiplicity of patterns, the need to make new patterns for design changes and to store the old patterns for replacement orders, the high cost of developing the casting techniques, and expensive inspection requirements.

Brazed impellers were found to be easily manufactured with a large saving in capital cost. The initial capital outlay for producing 80 different sizes of brazed impellers was about 5% of the investment that would have been required for production of the impellers as castings. Design changes were made easily without the need to change equipment. A standard hub plate and cover were used for a number of impeller sizes; only the blade size was changed. Manufacturing cost (exclusive of savings in initial investment) was 1% lower than for casting. Inspection was easier, and the rejection rate of brazed impellers was only one fourth that of cast impellers. In addition, brazed impellers were salvageable.

Aluminum alloy 6061-T6 plate was selected because of its strength, brazeability, and availability. Although brazing destroyed the T6 temper and reheat treatment was necessary after brazing, the plate was purchased in this temper to avoid possible mixup with other 6061-T6 plate used in the plant. In addition, no cost premium was needed for the temper. Blades were cut from a cylinder made by curving the plate on a three-roll former, gas metal arc welding the longitudinal seam, and stress relieving. The blades were then finish machined to size and shape, and holes for locating pins were drilled on one edge.

The blades, the hub plate (machined and stress relieved), and the cover, together with 0.006-in.-thick BAlSi-4 filler-metal preforms, were cleaned in the following sequence of operations:

- Immerse in emulsion cleaner at 150 °F. The emulsion was a mixture (2 to 3 oz/gal of water) of alkaline salts combining complex phosphates and sulfates with anionic surface-active agents of the aryl-alkyl sulfonic type.
- Rinse in cold running water.
- Immerse in alkaline etching cleaner solution (2 to 3 oz/gal of water) at 150 to 160 °F.
- Rinse in cold running water.
- Immerse in nitric acid solution (30 to 40 wt%) at room temperature until all stains or smudges are removed.
- Rinse in cold running water.
- Rinse in hot running water.
- Dry with air blast.

The impeller components and filler-metal preforms were assembled in a clean, air-conditioned room. The hub plate was placed on a rigid baseplate fixture made of type 316 stainless steel. Flux was mixed within 4 h prior to being used from three parts by volume of a proprietary powder (aluminum fluorides and chlorides) to one part of clean water. Aluminum blade-locating pins were inserted in the hub plate and covered with flux. The lower filler-metal preforms were fluxed on both surfaces and put in position over the pins. The blades were fluxed and put in position (located by the pins). The upper filler-metal preform strips were fluxed and placed on the blades.

Tabs were bent down over the blades to hold the filler-metal strips in position. Flux was brushed onto the joint side of the cover. The cover was placed in position over the blades, concentric with the hub plate, and held there by four locating clamps fastened to the cover at 90° intervals and extending down to the bottom of the outside edge of the hub plate.

The cover had been drilled with four 1/8-in.-deep holes on the top surface to ac-

Fig. 4 Furnace brazed impeller assembly. Locating pins were used to hold the blades in position during brazing.

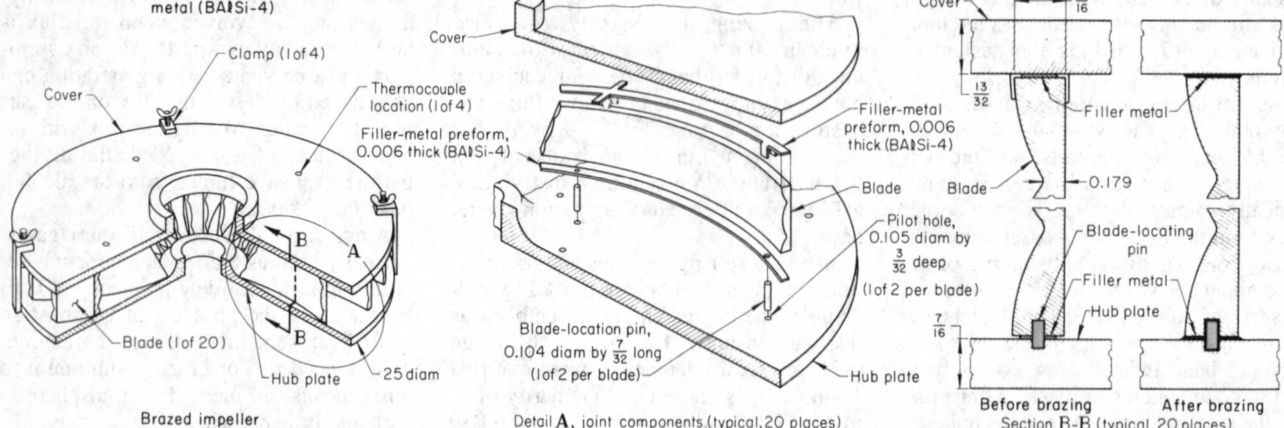

commodate thermocouple beads. For small assemblies, only the holes at the front and back positions during brazing were used. For large assemblies, the holes at the side positions in the furnace also had thermocouple beads peened into them. Care was taken to avoid pickup of foreign metal when making the thermocouple joints and also to avoid contaminating the thermocouple wires with flux.

A box-type furnace, electrically heated on all interior surfaces, was used. The assembly was placed in the furnace, which had been preheated to 1200 °F (120 to 130 °F above the brazing temperature), and the temperature of the assembly was checked with the thermocouples at regular intervals. When any thermocouple reading was within 50 °F of the brazing temperature of 1070 to 1080 °F, readings were taken every minute. When the assembly reached brazing temperature, the furnace was opened, and the assembly was removed to cool in air to room temperature. The brazed assembly was washed in hot water to remove flux and then cleaned by the same procedure used before brazing.

After cleaning, the impeller was solution heat treated, quenched and aged to regain the T6 temper, and again given the same sequence of cleaning operations as before to remove the slight surface oxidation that formed during heat treatment. Final inspection included visual examination, liquid-penetrant inspection, and a hardness check. In addition, the brazed assemblies were inspected for deformation and brazing failure after the impeller was test run 20% over design speed for 2 min in vacuum. Assemblies that showed porosity in a brazed joint were salvaged by recleaning, refluxing, positioning small clips of additional filler metal where needed, and rebrazing. If cracks were found in a joint, the hub plate and cover were salvaged for reuse by cutting apart the assembly. The used blades were discarded.

Fluxless Vacuum Brazing

Furnace brazing in a vacuum with the use of no flux offers several advantages: the possibility of flux inclusions is eliminated, and blind cavities, tortuous paths, and small passageways can be designed into the assembly without regard to flux removal or entrapment after brazing. Fluxless brazing also eliminates the cost of flux and its application, the need for cleaning the assembly after brazing, and potential corrosion of equipment and pollution of air and water by flux residues or flux reaction products.

With correct techniques, alloys of the *1xxx*, *3xxx*, *5xxx*, *6xxx*, and *7xxx* series can be vacuum brazed using No. 7, 8, 13, or 14 brazing sheet, which are clad with 4004 filler metal. When additional filler metal is required, 4004 in wire and sheet form also can be introduced. The joint designs used for brazing with flux can be used for fluxless vacuum brazing.

Equipment. Cold wall vacuum furnaces with electrical resistance radiant heaters are recommended for aluminum vacuum brazing. Both batch-type and semicontinuous furnaces are used. The vacuum pumping system should be capable of evacuating a conditioned chamber to a high vacuum (about 10^{-5} torr) made in 5 min. For most applications, rectangular chambers made of hot rolled steel are suitable. The temperature distribution within the work being brazed should be reasonably uniform, ideally within ±5 °F. For many applications, however, wider ranges are used.

Technique. Components are cleaned, usually by vapor degreasing with a common solvent such as perchlorethylene (ordinarily, chemical cleaning is avoided), assembled, and clamped in a suitable fixture made of stainless steel. The use of dry cotton gloves is recommended for assembling by hand, because fingerprints can inhibit filler-metal flow. Heating of the assembly is started simultaneously with pumpdown of batch-type furnaces. Average time for heating to brazing temperature is about 15 min. The assembly is then held at brazing temperature (see optimum brazing ranges, Table 3) for about 1 min. If the chamber is backfilled with chemically inert gas, the assembly can be removed at temperatures above 900 °F. Then, heat treatable alloys can be quenched; non-heat-treatable alloys are air cooled. The clean, dry, brazed assembly is ready for use or further processing, such as conversion coating, as soon as it is cool.

Torch Brazing

Torch brazing is used for either manual or automatic fabrication and for repair operations. The uses of torch brazing range from rather simple tube-to-tube joints to more complex and mechanized assemblies. Some of the more common commercial applications are tubular joints in refrigerator coils, miter joints in extruded window frames, and joints between electric heating elements and structures.

Equipment. Oxyacetylene, oxyhydrogen, and oxynatural gas are employed commercially for torch brazing. Gasoline blow torches and all types of gas burners can also be used.

Torch brazing is similar to oxyfuel gas welding in that the heat to effect the joint is applied locally. The torch tip sizes used are similar to those for gas welding. The choice of tip size and gas pressures depends on the thickness of the parts and should be determined by trial, using the values in Table 5 as starting points.

With the generally employed fillers, BAlSi-3 and BAlSi-4, close temperature control is needed, especially for torch brazing of alloys that have low solidus (wrought alloys 5052, 6053, 6061, and 7005 and the casting alloys listed in Table 1). Because aluminum alloys show no color when hot, even melting without a color change, some means for determining when the parts are reaching brazing temperature is necessary. The flux used should be one that melts at a slightly lower temperature than the filler metal and, thus, serves as a temperature indicator. Aligning jigs should be insulated to avoid excessive heat conduction.

Technique. After the components of an assembly to be torch brazed have been suitably cleaned, the joint areas and the filler metal are painted with a slurry of brazing flux, and the components are assembled and (if required) jigged. The assembly is then brazed by directing a soft, slightly reducing flame over the entire joint

Table 5 Typical conditions for oxyacetylene and oxyhydrogen torch brazing of aluminum alloys

	Oxyacetylene brazing			Oxyhydrogen brazing		
Metal thickness, in.	Orifice diameter, in.	Oxygen pressure, psi	Acetylene pressure, psi	Orifice diameter, in.	Oxygen pressure, psi	Hydrogen pressure, psi
0.020	0.025	0.5	1	0.035	0.5	1
0.025	0.025	0.5	1	0.045	0.5	1
0.032	0.035	0.5	1	0.055	0.5	1
0.040	0.035	0.5	1	0.065	1	2
0.051	0.045	1	2	0.075	1	2
0.064	0.055	1	2	0.085	1	2
0.081	0.065	1.5	3	0.095	1.5	3
0.102	0.075	1.5	3	0.105	1.5	3
0.125	0.085	2.0	4	0.115	1.5	3

area. The filler metal can be preplaced, or it can be face fed (flowed into the joint when touched against the heated work). The brazed joint should have a smooth fillet, usually requiring little or no finishing. Residual flux should be removed.

Specialized Brazing Processes

The various processes described in this section are not currently in wide use. For some of these processes, the basic art, knowledge, and materials are available, but the applications to date are meager; for others, the technology is not fully developed.

Modifications of Dip Brazing. In one modification of dip brazing, mixtures of filler-metal powder and one of the active components of the flux, such as a fluoride, are applied to the areas to be brazed and dried. Then the assembly is dipped into a less-active molten flux.

Another modification is useful when, because of joint design, supply of the filler metal from a molten bath is desirable. This is done in either of two ways: (1) assemblies that are coated with flux (by spraying or dipping) and thoroughly dried are placed into the molten filler metal, or (2) assemblies are dipped into a molten flux bath and then dipped into the molten filler metal. In either method, the molten filler metal flows into the capillary spaces at the joints to effect a braze.

Resistance Brazing. Aluminum alloys that are joined by resistance brazing are most often small parts. Connections in electric motor windings are a typical application. Usually, the work is clamped for heating between two carbon blocks held in a tong arrangement. For some jobs, it is clamped in a resistance welding machine.

Alloy Brazing. This fluxless process achieves a braze by first heating the joint area with an interposed shim to form a liquid phase, then extruding this liquid along with surface oxides from the joint cross section. Because of considerations such as the rate of alloying, fluidity, melting temperature, and availability as foil, the preferred shim material is copper. The extreme brittleness of the aluminum-copper intermetallic compounds is no deterrent, because the compounds are entirely displaced by extrusion.

Many types of heating are used for alloy brazing; resistance heating is the most widely used method. Clamping pressure between the carbon blocks is generally 1200 to 2000 psi, based on the overlap area. Current densities of 2500 to 4000 A/in.^{2} with an operating potential of 9 to 13 V are satisfactory.

Alloy brazing is used in the electrical industry, because the electrical conductivity of the brazed joints is essentially the same as the parent aluminum conductor, and the process is low in cost. In addition to simple overlap and butt joints, multiple-ply overlap, thin-to-thick, and round-to-flat joints are brazed. The simple, inexpensive, fluxless features also make this process attractive for brazing narrow sheet and plate members, for butt joining round and square aluminum rod, and for sealing the ends of thin-walled tubing. Alloys such as 1350, 1060, 1100, and 3003, with solidus temperatures considerably above 1018 °F (aluminum-copper eutectic temperature), are the most readily brazed. The 5*xxx* and 6*xxx* alloys with lower solidus temperatures can be brazed if close temperature control is provided.

Motion Brazing. Both vibration brazing and flow brazing are types of motion brazing. As might be expected, vibration of low or ultrasonic frequency has a pronounced beneficial effect in brazing aluminum, particularly when brazing sheet is used. Brazed joints can be made with brazing sheet parts in the absence of flux and in air. The brazing sheet surfaces, typically of types 7, 11, 13, and 23, are held together at a temperature that is, preferably, above the liquidus of the brazing alloy. A relative movement between the brazing sheets displaces the oxide film on the semiliquid contact surfaces.

In flow brazing, simple joints can be brazed between brazing sheets or even plain aluminum parts. The part to be brazed or the molten filler metal is moved rapidly with respect to the other, causing a mechanical removal of oxide film and the mating of liquid-liquid or solid-liquid interfaces. For simple shapes, this can be done in air; for more difficult shapes, in an inert gas. The total time must be short, and no flux is used. Vibration is helpful.

The motion brazing concept has certain obvious limitations; parts are restricted to rather elementary structures because of the requirements of directional vibration, rate and type of motion, supply of premelted filler alloy to the joint, shape of part, precleaning, and other complicating factors.

Brazing to Other Metals

Aluminum can be brazed to many other metals. In specific applications, painting or other suitable coating may be required after brazing to minimize subsequent galvanic corrosion of the joint area. Stresses from nonuniform expansion must also be considered.

Aluminum to Ferrous Alloys. Steel should be protected from oxidation during preheating and brazing to aluminum. In dip brazing, oxidation can be prevented by dipping unheated parts into molten flux, but this procedure has limited application because it is likely to cause warping and misalignment of the components.

Plated or coated steel can be brazed to aluminum more readily than bare steel. Copper, nickel, or zinc electroplates and aluminum, silver, tin, or hot dip zinc coatings are used to promote wetting of the steel and to minimize formation of brittle aluminum-iron compounds, thus producing a more ductile joint. The furnace brazing of plated steel liners or sleeves in aluminum alloy cylinder blocks, as well as steel valve seats in aluminum alloy cylinder heads, has been done experimentally.

Aluminum-coated steels can be torch brazed readily to aluminum, using aluminum filler metals and fluxes. The procedure is the same as in brazing aluminum to aluminum, except that preheating should be rapid, and brazing time must be minimized to avoid the formation of brittle aluminum-iron phases at the interface. Tube-to-tube joints, with a nominal clearance of about 0.010 in. and laps varying from 0.50 to 2.50 in., have shown shear strengths of 10 000 to 15 000 psi.

In certain complex applications, a multiple-step joining procedure must be employed to permit flux removal. For instance, a section of steel tube was hot dip coated with aluminum at one end, and the aluminum-coated end was dip brazed to a section of aluminum tube. After thoroughly cleaning the brazed subassembly to remove residual brazing flux, the aluminum tube portion was welded to an aluminum container that had been furnace brazed and cleaned in separate operations. The completed assembly was vacuum-tight.

Aluminum to Copper. Brazing of aluminum to copper is difficult, because of the low melting temperature (1018 °F) of the aluminum-copper eutectic and its extreme brittleness. By heating and cooling rapidly, however, reasonably ductile joints are made for applications such as copper inserts in aluminum castings for electrical conductor use. The usual filler metals and fluxes for brazing aluminum to aluminum can be used, or the silver alloy filler metals BAg-1 and BAg-1a can be used if heating and cooling are rapid (to minimize diffusion). Pretinning the copper surfaces with solder or silver alloy filler metal improves wetting and permits shorter time at brazing temperature. A more practical way to braze aluminum to copper is to braze one end of a short length of aluminum-coated steel tube to the aluminum, and then silver braze the other end of the tube to the copper.

Aluminum to Other Nonferrous Metals. Aluminum-silicon filler metals are unsuitable for brazing aluminum to uncoated titanium because of the formation of brittle intermetallic compounds. Titanium can be hot dip coated with aluminum, however, after which it can be brazed to aluminum with the usual aluminum filler metals.

Under correct conditions, nickel and nickel alloys are no more difficult to braze to aluminum than ferrous alloys. They can be brazed directly or precoated with aluminum. Although Monel alloys can be wetted directly, brazed joints are likely to be brittle; thus, Monel alloys are preferably precoated with aluminum.

Beryllium can be wetted directly by aluminum brazing alloys. Magnesium alloys can be brazed to aluminum, but the brazed joints have limited usefulness because of the extremely brittle aluminum-magnesium phases that form at the interface.

Flux Removal

Fluxes used in brazing aluminum alloys can cause corrosion if allowed to remain on the parts. Therefore, cleaning of joints after brazing is essential. A thorough water rinse followed by a chemical treatment is the most effective means of complete flux removal.

As much flux as possible should be removed by immersing the parts in an overflowing bath of boiling water just after the filler metal has solidified. If such a quench produces distortion, the parts should be allowed to cool in air before immersion to decrease the thermal shock. When both sides of a brazed joint are accessible, scrubbing with a fiber brush in boiling water removes most of the flux. For parts too large for water baths, the joints should be scrubbed with hot water and rinsed with cold water. A pressure spray washer may be the best first step. A stream jet is also effective in opening passages plugged by flux.

Any of several acid solutions (Table 6) remove flux that remains after washing. The choice depends largely on the thickness of the brazed parts, accessibility of fluxed areas, and the adequacy of flux removal in the initial water treatment. A pitting or intergranular type of attack on parts can result as chlorides from the flux build up in the acid solution. Some solutions have a greater tolerance for these chlorides than others before parts are attacked. The degree of flux contamination tolerable for the five typical flux removal solutions listed in Table 6 are given in the footnotes of the table.

The two chromium-containing solutions in Table 6 have a greater tolerance for chlorides and are preferred for thin-walled assemblies. In areas where disposal of chromates presents a problem, the nitric acid solution can be used if inhibitors such as 1% thiourea or triethanolamine salt of sulfolaurylalkylbenzoate are added. As a corrosion inhibitor, about 0.5% sodium or potassium dichromate is sometimes added to the final rinse water.

Agitation and turbulence improve the efficiency of any flux removal treatment. Ultrasonic cleaning works effectively for cleaning inaccessible areas, decreases the immersion time, and reduces the possibility of attack on the aluminum.

Checking for complete flux removal should be a routine inspection procedure. To detect the presence of flux, a few drops of distilled water are put on the surface to be tested and left there for a few seconds. The water is then picked off with an eyedropper and placed in an acidified solution of 5% silver nitrate. If the solution stays clear, the metal is clean. If a white precipitate clouds the solution, chloride residues were present on the surface. Flux removal procedures must then be repeated until the brazed assembly tests clean. Complete removal of the flux is essential, because it is corrosive to aluminum in the presence of moisture.

Table 6 Solutions for removing brazing flux from aluminum parts

Type of solution	Concentration: Amount	Concentration: Component(a)	Operating temperature, °F	Procedure(b)
Nitric acid	5 gal	58-62% HNO_3	Room temperature	Immerse for 10-20 min; rinse in hot or cold water(c)
	34 gal	Water		
Nitric-hydrofluoric acid	4 gal	58-62% HNO_3	Room temperature	Immerse for 10-15 min; rinse in cold water, rinse in hot water; dry(d)
	1 qt	48% HF (1.15 sp gr)		
	36 gal	Water		
Hydrofluoric acid	10 pt	48% HF	Room temperature	Immerse for 5-10 min; rinse in cold water; dip in nitric acid solution shown at top of table; rinse in hot or cold water(d)
	40 gal	Water		
Phosphoric acid-chromium trioxide	$1^1/_2$ gal	85% H_3PO_4	180	Immerse for 10-15 min; rinse in hot or cold water(e)
	$7^1/_4$ lb	CrO_3		
	40 gal	Water		
Nitric acid-sodium dichromate	$4^1/_4$ gal	58-62% HNO_3	140	Immerse for 5-30 min; rinse in hot water(f)
	32 lb	$Na_2Cr_2O_7 \cdot 2H_2O$		
	36 gal	Water		

(a) All compositions in weight percent. (b) Before using any of the above solutions, it is recommended that the assembly first be immersed in boiling water to remove the major portion of the flux. (c) Flux contamination in acid should not exceed 5 g/L of chloride expressed as sodium chloride. Solution is not recommended for use on base metals less than 0.020 in. thick. (d) Flux contamination in acid should not exceed 3 g/L of chloride expressed as hydrochloric acid. Solution is aggressive and not recommended for base metals less than 0.020 in. thick. (e) Tolerance for flux contamination is in excess of 100 g/L and permissible limit is probably governed by cleaning ability. If large pockets of flux are present, solution promotes intergranular attack at the pocket. Recommended for final cleaning of thin-gage parts, when most of the flux can be removed easily in water. (f) Exceptionally high flux tolerance. Recommended for cleaning thin-gage assemblies, if adequacy of water cleaning is doubtful. License required.

Postbraze Heat Treatment

Brazing temperatures exceed the solution heat treatment temperatures used for aluminum alloys. Heat treatable aluminum alloys can attain full strength by aging after being quenched from the brazing temperature. After brazing of alloy 7005, a normal air quench for small parts of 5 °F/s, or even a cooling rate as slow as 1 °F/s, is adequate for precipitation hardening to occur at room temperature. Except as dictated by distortion problems, postbraze quenching and aging treatments are the same as for the base alloy. Typical treatment for artificially aging a heat treatable alloy brazed assembly is 16 to 20 h at 320 °F or 6 to 10 h at 350 °F.

Finishing

Because of the smooth, uniform fillets resulting from the brazing operation, little if any mechanical treatment is required before final finishing. If flux has been completely removed (or if fluxless brazing has been used), all chemical and electrochemical finishing treatments are effective when the brazed structures are aluminum throughout. Because of the high silicon content of the filler-metal fillets, any treatment that thickens the oxide or preferentially etches aluminum, leaving a residue of silicon, may cause the fillets to be a darker color than the remainder of the product.

Brazing fluxes containing chlorides of zinc or other heavy metals deposit that metal on the surface of aluminum parts. These fluxes, as well as fluxes that cause

Table 7 Tensile properties of 0.063-in.-thick alloy 7005 heated as in brazing

Heated 10 min at 1090 °F, air cooled, aged as designated

Room temperature aging treatment	Tensile strength, ksi	Yield strength, ksi	Elongation in 2 in., %
None	28	12	26
3 days	42	21	22
1 week	45	24	22
1 month	49	27	21
3 months	52	30	21
6 months	54	32	21
T63(a)	52	44	13

(a) Artificially aged (after solution heat treatment)

Table 8 Results of microscopic examination of furnace brazed specimens exposed 2 yr to 3.5% sodium chloride intermittent spray

Specimens were small inverted T-joints of 0.064-in. sheet; filler metal used was BAlSi-3.

Sheet alloy	Sheet (base metal) Type of attack	Depth of attack, in. max	Depth of attack, in. avg	Joint (filler metal) Type of attack	Depth of attack, in. max	Depth of attack, in. avg
3003	Pitting	0.0098	0.0022	Pitting	0.0014	0.0011
5052	Pitting	0.0182	0.0042	Pitting	0.0042	0.0014
6053	Pitting, intergranular	0.0126	0.0028	Pitting	0.0012	0.0008
6061	Pitting, slight intergranular	0.0126	0.0033	Pitting	0.0042	0.0014

severe etching of aluminum, should be avoided for highest quality in chemical finishing.

Mechanical Properties

At least a part of the base metal is heated above its annealing temperature during the brazing cycle. Torch brazing may anneal only a small region near the joint, whereas dip or furnace brazing anneals the entire assembly. Unless the completed part is quenched and aged, heat treated, or cold worked, the metal that was heated has mechanical properties typical of the annealed alloy.

When the alloy is heat treatable, improved strength can be imparted by quenching directly from the brazing furnace or dip pot, then artificially or naturally aging according to regular procedures for the alloy involved. Another alternative is solution heat treating and aging as separate operations after brazing. Heat treating is not always possible, because the rapid quenching required for most heat treatable alloys can cause distortion. Alloy 7005 age hardens at room temperature to T6 properties after normal air cooling (1 to 2 °F/s) from brazing temperature. Table 7 lists typical properties of alloy 7005 air cooled from brazing temperature.

Resistance to Corrosion

The aluminum alloys best suited for brazing are also among those most resistant to corrosion. Corrosion resistance of aluminum alloys generally is unimpaired by brazing if a fluxless brazing process is used or if flux is completely removed after brazing. If flux removal is inadequate, the presence of moisture can lead to interdendritic attack on the filler metal at joint faces and to intergranular attack on the base metal.

When two aluminum alloys are brazed together, exposure to salt water or some other electrolyte may result in attack on the more anodic alloy. This condition is aggravated if the anodic part is relatively small compared with the other piece; therefore, the anodic aluminum alloy should be the larger of the two members.

Torch brazed alclad 3003 and alclad 3004 show excellent resistance to corrosion. Furnace or dip brazing, however, causes a certain amount of silicon diffusion from the clad surface, which limits application of these methods with conventional alclad products. A brazing sheet with filler metal on one side and alclad with a special alloy on the other (Fig. 2c) performs well in furnace or dip brazing.

Commercial filler metals of the aluminum-silicon type have high resistance to corrosion, comparable to that of the base metals usually brazed. Filler metals containing substantial amounts of copper or zinc are less resistant to corrosion, but they are usually adequate, except for service in severe environments.

Joints brazed with aluminum-silicon filler metals (BAlSi-2, BAlSi-4, and BAlSi-5) show a potential of −0.82 V with respect to a 0.1*N* calomel reference electrode in an aqueous solution of 53 g/L of sodium chloride and 3 g/L of hydrogen peroxide. This potential is barely cathodic to the frequently brazed base metals, for which the value is −0.83 V for 1100, 3003, 6061, and 6063. Therefore, little electrolytic action occurs in assemblies of these base metals that are brazed with the usual filler metals.

The potential of joints brazed with filler metal BAlSi-3 (alloy 4145), which contains copper in addition to aluminum and silicon, depends on the cooling rate after brazing. For slow cooling, these joints have about the same potential as joints brazed with the aluminum-silicon filler metals (−0.82 V). If the cooling is rapid enough to retain a substantial amount of copper in solid solution, the potential is lower; a potential of −0.73 V has been found for T-joints in 0.064-in. sheet brazed with BAlSi-3 filler metal and rapidly cooled.

Although considerable undissolved silicon-containing constituent is evident in brazed joints, it polarizes strongly (except in acid chloride environments) and has little influence on the potential of the brazed joint and its corrosion resistance. Table 8 shows the results of long-time exposure in a highly corrosive environment of various sheet alloys that were furnace brazed with filler metal BAlSi-3. The good performance can be considered typical of a variety of brazing combinations.

Safety

Many of the safety considerations discussed in the articles in this Volume on brazing processes are applicable to the brazing of aluminum alloys by the same processes. The principal hazard in brazing aluminum alloys that is not present in brazing steel or copper alloys arises from the use of molten fluorine-containing fluxes in dip brazing. Toxic effects may be produced by the inhalation of fumes from the fluorine compounds; thus, exhaust facilities are required for dip brazing. Furnace brazing with fluorine-bearing fluxes requires exhaust of the fumes generated to prevent attack on exposed metals; the air changes necessary for this reason are adequate from the standpoint of health protection. Applicable local, state, and Occupational Safety and Health Administration regulations should be enforced.

Brazing of Copper and Copper Alloys*

MOST COPPERS AND COPPER ALLOYS can be brazed satisfactorily, using one or more of the conventional brazing processes. These processes include furnace, torch, induction, resistance, and dip brazing.

Brazeability

Brazeability is generally rated from good to excellent. With some alloys, however, difficulties may be encountered. For example, some lead-containing alloys can form a dross that interferes with wetting, and tin-containing alloys, if not stress relieved before brazing, may crack when subjected to rapid localized heating.

Coppers. Included in this group are oxygen-containing or tough pitch, phosphorus-deoxidized, and oxygen-free coppers, together with coppers containing not more than about 1% of an additive element, such as silver, zirconium, chromium, lead, selenium, tellurium, or sulfur.

Tough pitch coppers are subject to embrittlement when heated in reducing atmospheres containing hydrogen. At temperatures above 900 °F even relatively small amounts of hydrogen lead to embrittlement and internal cavitation caused by the reduction of copper oxide and the formation of high-pressure steam within the solid metal. Consequently, although tough pitch coppers are generally rated as having good to excellent brazeability, they should not be brazed in a furnace that contains hydrogen at a reducing potential such as dissociated ammonia or in an exothermic-based or endothermic-based atmosphere. Heating by open flame or by torch also may result in hydrogen diffusion and embrittlement.

Phosphorus-deoxidized and oxygen-free coppers can be brazed without flux in hydrogen-containing atmospheres without risk of embrittlement, provided self-fluxing filler metals (BCuP series) are used. The use of flux is required, however, when the silver alloy filler metals that contain additives such as zinc, cadmium, or lithium (BAg series) are used to braze these coppers or to braze these coppers to copper alloys or other metals.

The coppers, including those that contain small additions of silver, lead, tellurium, selenium, or sulfur (generally no more than 1%), are readily brazed with the self-fluxing BCuP filler metals, but wetting action is improved when a flux is used and when a sliding motion between components is provided while the filler metal is molten. Precipitation-hardenable (PH) copper alloys that contain beryllium, chromium, or zirconium form oxide films that impede the flow of filler metal. To ensure proper wetting action of the joint surface by the filler metal, beryllium copper parts, for example, should be freshly machined or mechanically abraded before being brazed. Removal of beryllium oxide from joint surfaces requires the use of a high-fluoride-content flux.

Brazing PH coppers in the aged condition reduces their mechanical properties. Properties can sometimes be partly restored by a subsequent aging treatment. Beryllium copper, e.g., C17200 which contains 1.8 to 2.0% Be, can be brazed and solution heat treated by heating to 1450 °F, followed by rapid quenching. Subsequent aging at about 600 °F develops adequate hardness. When the sections to be brazed are thin and can be cooled very rapidly, solution heat treated beryllium copper may be brazed in the temperature range of 1150 to 1200 °F.

When chromium coppers are brazed at a temperature within the solution treatment range (1800 °F, for example), brazing and solution treatment can be combined. Brazed chromium copper parts can then be aged in a subsequent operation to develop improved mechanical properties and electrical conductivity.

Zirconium coppers do not precipitation harden without the benefit of prior cold working, a sequence that is incompatible with brazing. In the absence of cold working, the strength of zirconium coppers is not improved by aging treatments, although electrical conductivity is improved.

Red brasses (copper-zinc alloys that contain up to 20% Zn) are readily brazed with a variety of filler metals. Flux is normally required for best results, especially when the zinc content is above 15%.

Yellow brasses (copper-zinc alloys that contain 25 to 40% Zn) are readily brazed, but low-melting filler metals should be used to avoid dezincification of the base metal.

Leaded Brasses. If added to red brass or yellow brass, lead forms a dross on heating that can seriously impede wetting and the flow of filler metal. Consequently, in brazing leaded brasses, the use of a flux is mandatory to prevent dross formation in the joint area.

The susceptibility of leaded brasses to hot cracking varies directly with lead content. Therefore, these alloys must have low residual stresses before brazing. Heating to the brazing temperature should be uniform to minimize thermal stressing. Brazing results are poor at a lead content of 3% and above because of liquation and the brittleness of phases formed in the joint by lead and the filler metal. Alloys containing more than 5% Pb are usually not brazed.

Tin-containing brasses, which include admiralty brass, naval brass, and leaded naval brass, contain up to 1% Sn and may contain other alloying elements such as lead, manganese, arsenic, nickel,

*Revised by Martin Prager, Consultant

and aluminum. Except for the aluminum-containing alloys, these brasses are readily brazed; they have greater resistance to thermal shock and are less susceptible to hot cracking than the high-lead brasses. For proper wetting, brasses that contain aluminum require a special flux, such as American Welding Society (AWS) type 4.

Phosphor Bronzes. These copper-tin alloys contain small amounts of phosphorus, up to about 0.25%, added as a deoxidizer. Although susceptible to hot cracking in the cold worked condition, alloys in this group have good brazeability and are adaptable to brazing with any of the common filler metals that have melting temperatures lower than that of the base metal. The use of a flux is generally preferred. To avoid cracking, parts made from phosphor bronzes should be stress relieved at approximately 550 to 650 °F before brazing.

Silicon bronzes, which contain up to about 3.25% Si and are in a highly stressed condition, are susceptible to hot shortness and stress cracking by molten filler metal. To avoid cracking, the alloys should be stress relieved at about 550 to 650 °F before brazing. For best brazing results, joint surfaces should be freshly machined or mechanically cleaned. Silicon bronzes containing aluminum require the use of AWS type 4 flux (Table 2).

Aluminum Bronzes. Because of their aluminum content, which results in the formation of aluminum oxide on the surface, aluminum bronzes are generally considered difficult to braze. However, alloys containing 8% Al or less are brazeable, provided AWS type 4 flux is used to dissolve the aluminum oxide. The oxide, which inhibits the flow of filler metal, cannot be reduced in dry hydrogen. Use of the low-melting high-silver filler metals is recommended for these bronzes.

Copper nickels, which may contain from about 5 to 40% Ni, are susceptible both to hot cracking and to stress cracking by molten filler metal. The silver alloy filler metals (BAg series) are preferred for brazing these alloys. In general, the use of filler metals containing phosphorus should be avoided, because the copper nickels are susceptible to the formation of brittle nickel phosphides at the interface, which lower joint strength and ductility.

Nickel silvers (brasses that contain up to about 20% Ni but that do not contain silver) are highly susceptible to hot cracking and should be stress relieved at about 550 °F before being brazed. They have low thermal conductivity and should be heated and cooled uniformly.

Brazing Dissimilar Alloys. Most of the alloys belonging to any one of the above groups can be brazed to an alloy of another group. However, to achieve compatibility, some compromise may be required in the selection of brazing temperature, filler metal, and flux. For example, if a component made of copper is to be brazed to a component made of aluminum bronze, the brazing temperature should be predicated on the lower melting temperature of the bronze, and a suitable flux should be selected to accommodate the bronze. It is not uncommon for three or more different copper alloys to be brazed together in a single assembly.

Filler Metals

Table 1 presents the nominal compositions, solidus and liquidus temperatures, and electrical conductivities of some filler metals used in brazing of copper and copper alloys, as well as the joint clearance used with each. The filler metals listed in Table 1 represent four series: (1) copper-zinc (RBCuZn) alloys, (2) copper-phosphorus and copper-silver-phosphorus (BCuP) alloys, (3) silver (BAg) alloys, and (4) gold (BAu) alloys. The copper (BCu) filler metals are omitted. Because of their high liquidus temperature (1980 °F), these filler metals are restricted to use with the copper-nickel alloys only. Of the filler metals listed in Table 1, the BCuP and BAg alloys are by far the most widely used in brazing copper and its alloys.

Copper-Zinc (RBCuZn) Filler Metals. Because of their high liquidus temperature, poorer corrosion resistance, anodic position relative to copper, and sensitivity to overheating, copper-zinc filler metals are seldom used in brazing copper and copper alloys. Overheating causes the zinc to vaporize and form voids in the joint. In applications where corrosion resistance is not important, these filler metals can be used for joining copper, copper nickel, or silicon bronze. They are sometimes used in joining copper to steel, stainless steel, and nickel alloys. These filler metals generally require the use of a flux.

Copper-Phosphorus and Copper-Phosphorus-Silver (BCuP) Filler Metals. The BCuP alloys are self-fluxing when used for brazing unalloyed coppers, but the use of a flux is generally recommended when these filler metals are used in brazing special coppers and copper alloys. In general, BCuP filler metals are not suitable for use in brazing copper-nickel alloys containing 20% or more nickel. The BCuP filler metals have good corrosion resistance, but they are severely attacked in sulfur-bearing atmospheres at elevated temperature. Among the advantages of these filler metals is their relatively low cost.

Silver Alloy (BAg) Filler Metals. The silver alloy filler metals listed in Table 1 are suitable for use with all brazeable coppers and copper alloys as well as most dissimilar alloy combinations. The lower brazing temperatures indicated in Table 1 for some of the silver alloy filler metals make them particularly well suited for brazing copper alloys with an appreciable zinc content. These include yellow brasses, which are subject to dezincification. The

Table 1 Nominal compositions, solidus and liquidus temperatures, and electrical conductivities of filler metals used in brazing of copper and copper alloys, and joint clearances used with these filler metals

AWS filler metal	Nominal composition, %							Solidus temperature, °F	Liquidus temperature, °F	Conductivity(a), % IACS	Typical diametral joint clearance, in.
	Ag	Cu	P	Zn	Cd	Ni	Other				
RBCuZn-A	...	59.25	...	40	...	...	0.75Sn	1630	1650	26	0.002-0.005
RBCuZn-D	...	48	...	42	...	10	...	1690	1715	...	...
BCuP-1	...	95	5	...	...	...	...	1310	1650	...	0.002-0.005
BCuP-2	...	92.75	7.25	...	...	...	...	1310	1460	...	0.001-0.003
BCuP-4	6	86.75	7.25	...	...	...	...	1190	1335	...	0.001-0.003
BCuP-5	15	80	5	...	...	...	...	1190	1475	10	0.001-0.005
BAg-1	45	15	...	16	24	...	...	1125	1145	28	0.002-0.005
BAg-1a	50	15.5	...	16.5	18	...	...	1160	1175	24	0.002-0.005
BAg-2	35	26	...	21	18	...	...	1125	1295	29	0.002-0.005
BAg-3	50	15.5	...	15.5	16	3	...	1170	1270	18	0.002-0.005
BAg-5	45	30	...	25	...	...	...	1250	1370	19	0.002-0.005
(b)	75	22	...	3	...	...	...	...	1365	...	0.002-0.005
BAg-8	77	23	...	...	...	...	...	1435	1435	...	0.002-0.005
BAg-8a	72	27.8	...	...	...	...	0.2Li	1410	1410	89(c)	0.002-0.005
BAg-19	92.5	7.3	...	...	...	...	0.2Li	1435	1635	88(c)	0.002-0.005
BAu-4	...	...	...	...	...	18	82Au	1740	1740	6	0.002-0.005

(a) Ratio of the resistivity of the material at 68 °F to that of IACS, expressed as a percentage and calculated on a volume basis. (b) Special filler metal used in brazing nickel silver knife handles. (c) Conductivity of filler metal after volatilization of lithium in brazing

Table 2 Fluxes used in brazing of copper and copper alloys

AWS flux type	Working temperature, °F	Constituents	Available forms	Base metals	Filler metals
3A	1050-1600	Boric acid Borates Fluorides Fluoborates Wetting agent	Powder Paste Liquid	All coppers and copper alloys except aluminum bronzes	BCuP and BAg series, except for BCuP-1 and BAg-19
3B	1350-2100	Boric acid Borates Fluorides Fluoborates Wetting agent	Powder Paste Liquid	All coppers and copper alloys except aluminum bronzes	All listed in Table 1
4	150-1600	Chlorides Fluorides Borates Wetting agent	Powder Paste	Aluminum bronzes	BAg series (principally)
5	1400-2200	Borax Boric acid Borates Wetting agent	Powder Paste Liquid	All coppers and copper alloys except aluminum bronzes	RBCuZn series (principally)

corrosion resistance of silver alloy filler metals ranges from good to excellent, although the cadmium-containing silver alloys are avoided when brazing equipment for the dairy, food, and pharmaceutical industries, because of the high toxicity of cadmium. The principal disadvantage of silver alloy filler metals is high cost, but this disadvantage can be largely offset by correct joint design. Flux is generally required for all but the lithium-containing filler metals (BAg-8a and BAg-19), which are self-fluxing in dry, nonoxidizing protective atmospheres. This is because of the ability of lithium to reduce refractory oxides on the base-metal surfaces at brazing temperature.

Gold alloy (BAu) filler metals, such as BAu-4 in Table 1, are high-cost compositions that are generally restricted to highly specialized applications such as joining vacuum tube components that are hermetically sealed. In this application, the low vapor pressure of gold is advantageous. The high liquidus temperatures of gold alloy filler metals further limit their use to brazing coppers and a few high melting temperature copper-nickel alloys.

Brazing Fluxes

The types of fluxes used in brazing of coppers and copper alloys are listed in Table 2. All are marketed as proprietary compositions with no standard composition ranges.

Type 3A is a general-purpose, low-temperature flux suitable for use with all copper and copper alloy base metals except those containing substantial amounts of aluminum. The filler metals that are compatible with this flux include most of the copper-phosphorus alloys (with the exception of BCuP-1) and most of the silver alloys (including all those with liquidus temperatures below 1600 °F).

Type 3B flux is a modification of type 3A for use at higher temperatures. The active temperature range of type 3B flux is 1350 to 2100 °F. This flux can be used for brazing with any of the filler metals listed in Table 1, provided the brazing temperature is above 1350 °F.

Type 4 flux is specifically prepared for brazing the aluminum-bearing copper alloys, and has the same working temperature range as type 3A flux, 1050 to 1600 °F. This flux is generally used with the silver alloy filler metals.

Type 5 flux, which has a working temperature range of 1400 to 2200 °F, is used with the copper-zinc filler metals. It is less active than type 3B flux, but it remains active longer (as required by the longer heating cycles used to braze heavy components) and costs less.

Joint Clearance

Joint clearance is a principal factor in determining the mechanical strength of brazed joints. It is also a factor in minimizing voids in the joint area and establishing the capillary force required to fill a joint.

Typical diametral joint clearances used with the filler metals commonly used in joining copper and copper alloys are given in Table 1. These are clearances at room temperature and are applicable to brazing components of about the same mass made from the same copper or copper alloy. Adjustments may be required for brazing dissimilar metals to compensate for different coefficients of thermal expansion.

Selection of Brazing Process

Often the size and shape of an assembly suggest a preferred brazing process to the exclusion of other processes. When two or more brazing processes have approximately equal suitability for brazing a given assembly, the quantity of assemblies to be brazed, because of the direct bearing on cost, is likely to be the deciding factor in process selection. Comparative production rates for brazing each of three different assemblies by different brazing processes are given in Table 3.

Furnace Brazing

Furnace brazing is a mass-production process. Its primary advantage is that it can be used to process a large number of assemblies on a batch or continuous basis at low unit cost. Furnace brazing can be used

Table 3 Comparison of production rates for brazing

Assembly brazed(a)	Brazing process	Production rate, assemblies/h	Assembly brazed(a)	Brazing process	Production rate, assemblies/h
A	Induction	720	B	Torch (manual)	60(c)
A	Torch (manual)	120(b)	C	Induction	60(c)
B	Furnace	140	C	Furnace	12(d)

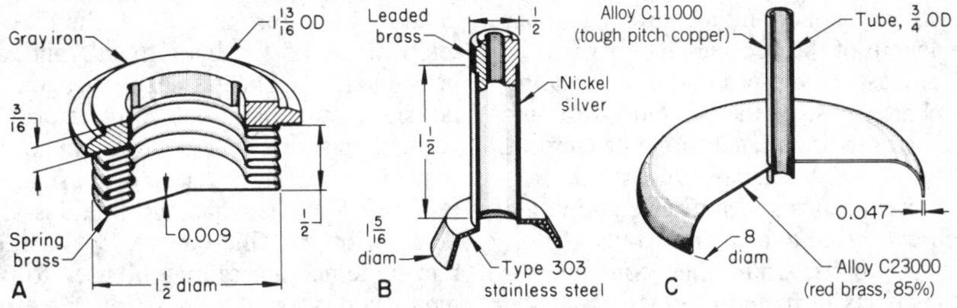

(a) Refer to labeled figures. (b) 30 s for preparation and heating; one assembly at a time. (c) 60 s for preparation and heating; one assembly at a time. (d) 5 min for preparation and heating; one assembly at a time to braze a number of joints on the same assembly simultaneously or to braze a variety of different assemblies simultaneously. Furnace brazing also provides an enclosed container for atmospheres that can protect assemblies against surface oxidation and other undesirable effects encountered when heating in air.

Advantages of furnace brazing that are more specifically applicable to the joining of copper and copper alloys relate to the furnace as a source of heat and to the cooling chamber that is provided on conveyor belt furnaces as a means for cooling assemblies from the brazing temperature to 300 °F or below. To a lesser degree, the prepared protective atmospheres that can be used most conveniently in furnace brazing, notably the exothermic-based and endothermic-based atmospheres, constitute another advantage when brazing deoxidized coppers and copper alloys in a furnace.

The rate of heating assemblies in a brazing furnace is low when compared with rates normally used in torch, induction, and resistance brazing. Few furnace heating cycles are less than 5 min in duration. For heating copper alloys susceptible to hot cracking, however, the slower, more uniform heating of a furnace is desirable.

In conveyor furnaces with multiple-zone heating chambers, the heating rate can be controlled with great accuracy. Depending on furnace capacity and the size of the assemblies to be brazed, the production rate in furnace brazing may equal or exceed that obtainable in induction brazing for the same amount of energy expended (assemblies per kilowatt-hour of input, for example).

The cooling rates that can be obtained in the cooling chambers of brazing furnaces can be closely controlled to ensure slow, uniform cooling. For copper alloys susceptible to hot cracking, the control of cooling rate from the brazing temperature is as important as control of the rate of heating to the brazing temperature.

Limitations. Apart from the high initial equipment costs and the floor space required to accommodate a furnace with both a heating chamber and a cooling chamber (the length of the cooling chamber is usually at least three times that of the heating chamber), most of the limitations of furnace brazing are related to the deleterious effects of furnace temperatures and brazing fluxes on furnace muffles and linings, electrical heating elements, rails, trays, conveyor belts, and other components. These effects increase in seriousness as the operating temperature of the furnace increases. A furnace used for brazing copper and copper alloys with silver alloy filler metals and operating at a temperature below 1500 °F could be expected to require considerably less maintenance, repair, and replacement of components than a similar furnace used to braze carbon steel with copper filler metals and operating at a temperature of 2000 °F. The use of flux, required for brazing with most silver alloy filler metals, could offset the advantage of a lower operating temperature, however, by introducing corrosion and corrosive deterioration that are not encountered in a flux-free furnace chamber.

A furnace operating at 1500 °F requires idling at elevated temperature when not in use, as does a furnace operating at 2000 °F, adding to power costs. Idling prevents thermal cycling that can cause serious damage to components such as furnace brickwork. Lower operating temperatures usually permit the use of lower idling temperatures. However, for furnaces operating with combustible protective atmospheres, the hazard of explosion is greatest when the furnace is operating at temperatures below about 1400 °F, and special precautions must be scrupulously observed in this range.

Furnaces. In design and equipment, the furnaces used for brazing copper and copper alloys are essentially the same as those used for brazing steel (see the article "Furnace Brazing of Steels" in this Volume).

Because of the damaging effects of sulfur and sulfur-bearing compounds on copper and copper alloys, the products of combustion in gas-fired furnaces operating without a muffle must be completely free of sulfur. Electric furnaces may be operated without a muffle unless damage to brickwork and other furnace components caused by volatilized flux and flux droppings warrants use of a muffle. Muffles may also be used to ensure the purity and to maintain the dew point of a protective atmosphere, particularly when the atmosphere is serving in place of a flux and its effectiveness depends on freedom from contamination.

Temperatures for furnace brazing depend on the filler metal used, the melting temperature of the base metal, and any harmful effects, such as dezincification and hot-short cracking, that result from exceeding prescribed temperature limits. The brazing temperature selected is usually at least 50 °F higher than the liquidus temperature of the filler metal; even higher brazing temperatures may be used to promote fluidity or to achieve other desired results (such as combining brazing and solution heat treatment in the same furnace operation). Some filler metals can be used at a temperature below the liquidus; e.g., BCuP-3, with a liquidus of 1485 °F, and BCuP-5, with a liquidus of 1475 °F, flow freely and make good joints at 1300 °F. In the following example, furnace brazing was done at 75 °F below the liquidus of the filler metal (a self-fluxing silver-containing BCuP alloy), without using a protective atmosphere.

Fig. 1 Furnace brazed contact arm assembly. Brazed without protective atmosphere, 75 °F below the liquidus of the filler metal

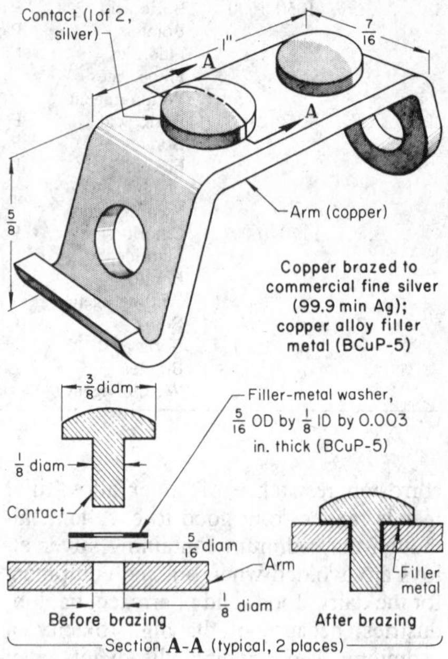

Example 1. Brazing at 75 °F Below the Liquidus of the Filler Metal—Without Protective Atmosphere. The contact arm assembly (Fig. 1) was a component of a line voltage thermostat. The two contacts were made of fine silver and, when brazed to the copper arm, required a highly conductive joint to accommodate high service currents without overheating. Neither a flux nor a protective atmosphere was used because the brazing temperature was considerably below the liquidus of the filler metal. The copper-silver-phosphorus (BCuP-5) filler metal is self-fluxing on both copper and silver, and its unusually wide temperature range between solidus and liquidus (1190 and 1475 °F) permits brazing at a temperature below the liquidus. The assembly was brazed at 1400 °F, 75 °F below the liquidus temperature, in a mesh belt conveyor furnace at a production rate of 377 assemblies per hour.

Accelerated Heating. The furnace temperature setting used in a given brazing operation may exceed that attained by the workload during the heating cycle to accelerate the heating of the work. Temperature differentials between the thermocouple and the work being processed must be closely monitored and controlled. Experimentation is required to establish the time spent by the workload in the heating zone when using higher furnace temperatures. Above 1750 °F, the substitution of a reducing protective atmosphere for a brazing flux helps extend the life of heating elements and other furnace components by eliminating both oxidation and the corrosive reactions of volatized flux.

Furnace Atmospheres. Protective atmospheres are used in furnace brazing of copper and copper alloys, although numerous exceptions exist. Exothermic-based, endothermic-based, dissociated ammonia, and other suitable prepared atmospheres are widely used to protect copper and copper alloys that are not adversely affected by hydrogen at elevated temperature. Atmospheres and maximum dew points recommended for brazing copper and copper alloys are given in Table 4. Additional information on these atmospheres is given in the article "Furnace Brazing of Steels" in this Volume. Detailed information regarding the generation of exothermic-based, endothermic-based, and dissociated ammonia atmospheres is given in the article "Furnace Atmospheres" in Volume 4 of the 9th edition of *Metals Handbook*.

Table 4 Recommended protective atmospheres for furnace brazing of copper and copper alloys

Base metal	Suitable atmosphere	Maximum dew point, °F
Coppers, phosphor bronzes, and copper nickels	Lean or rich exothermic	20
	Reacted endothermic	20
	Dissociated ammonia	20
Red brasses(a)	Purified lean exothermic	10
	Reacted endothermic	10
	Dissociated ammonia	20
Yellow brasses(b), leaded brasses, tin brasses(b), and nickel silvers	Purified lean exothermic	−40
	Reacted endothermic	−20
	Dissociated ammonia	20
Silicon and aluminum bronzes	Purified lean exothermic	−40
	Dissociated ammonia	−40

(a) Low zinc. (b) High zinc

Depending on base-metal and filler-metal combinations, the use of a brazing flux may be avoided by brazing in either a dry hydrogen or a prepared nitrogen-based atmosphere. These are among the more expensive furnace atmospheres; however, at low dew points, they have the advantage of being highly reducing. Prepared exothermic-based atmospheres are considerably less expensive and are effective in preventing oxidation at elevated temperature, but the reducing potential of these atmospheres is limited because hydrogen content does not exceed about 13%, and consequently they cannot be used as a substitute for a chemical flux.

Assembly for Brazing. The component parts to be furnace brazed must be assembled in an essentially fixed position with filler metal preplaced before entering the furnace, and they must maintain this position throughout brazing and cooling. Self-jigging is the preferred method of assembly. An assembly is self-jigging when its component parts incorporate design features that ensure that each component, when assembled, remains in proper relationship throughout the brazing cycle without the aid of auxiliary fixtures. Self-jigging can be accomplished by numerous methods (see the article "Furnace Brazing of Steels" in this Volume). When self-jigging is not feasible or when the assembly requires positioning or support in addition to that provided by self-jigging, auxiliary fixtures are used. These fixtures may take the form of a simple bracket or wire stand, ceramic blocks, clamps, or cast supports.

Venting Fully Enclosed Assemblies. Fully enclosed assemblies, whether assembled by self-jigging or supported by auxiliary fixturing, must be vented to permit the escape of entrapped air. When heated, entrapped air expands and, unless adequately vented, escapes from a sealed assembly in the area of the joint, resulting in flux spatter, joint porosity, and misalignment of brazed components. A small pinhole or slot, located safely away from the area of the joint, is usually sufficient to provide adequate venting and to avoid damage to the joint.

Torch Brazing

Torch brazing of copper and copper alloys follows the same basic principles and uses the same equipment as the torch brazing of steel (see the article "Torch Brazing of Steels" in this Volume). The properties of copper and copper alloys introduce certain specific considerations in torch brazing.

Applications. Joining components of various types of heat exchangers probably represents the largest field of application for torch brazing of copper and copper alloys; products include condensers, evaporators, air conditioners, radiators, and refrigerators, all of which depend on the high thermal conductivity of copper. Torch brazed products that depend on the high electrical conductivity of copper include motor windings, reactor coils, switches, contactors, and terminal leads.

Process Selection. The selection of torch brazing in preference to another brazing process, such as furnace or induction brazing, depends largely on feasibility and cost. Torch brazing is often used when workpiece limitations preclude the use of alternative brazing processes.

Manual Torch Brazing. Low equipment cost is a major advantage in manual torch brazing, which is particularly useful for assemblies involving unequal masses. A brazer with moderate skill can adjust and apply the heating flame so that unequal masses are brought uniformly to brazing temperature. The brazer can also apply the heat selectively to the joints of assemblies involving both large and small areas. The size of brazing flames may range from those of extremely small torches, the size of a hypodermic needle used for electronic leads, up to those of large torches used in brazing assemblies weighing hundreds of pounds.

Several precautions should be taken when torch brazing certain coppers and copper alloys. Where it is necessary to braze oxide-containing (tough pitch) coppers, a reducing atmosphere in the flame must be avoided because it can promote hydrogen embrittlement. For these coppers, a neutral or slightly oxidizing flame and a short brazing cycle are necessary. Brasses are subject to volatilization of zinc when overheated or when held too long at brazing temperature. Application of flux suppresses zinc volatilization.

Alloys containing elements that readily form refractory oxides (aluminum, beryllium, chromium, and silicon) must be protected by flux and should not be exposed to an oxidizing flame.

One of the notable differences between the torch brazing of steel and the torch brazing of copper alloys is that steel can withstand very rapid heating rates, whereas some of the copper alloys (the phosphor

Fig. 2 Return bend assembly of a heat exchanger. Includes a dual-tip torch used for rapid heating of the self-fluxing filler metal

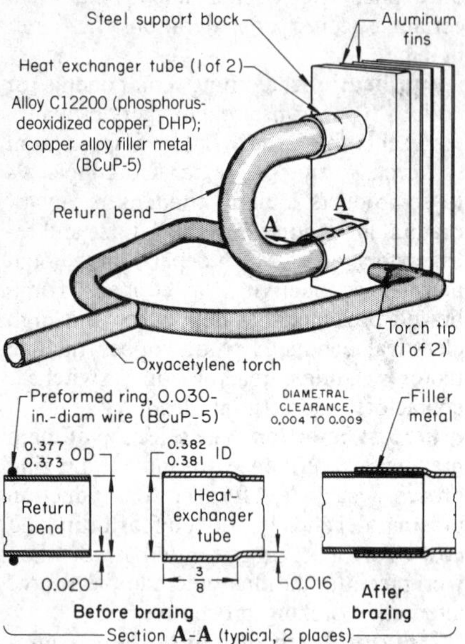

bronzes, for example) are subject to cracking if heated too rapidly when the metal is under high restraint during brazing.

Filler metals used for torch brazing of copper and copper alloys include the lower melting BAg silver alloys, the lower melting BCuP copper-phosphorus alloys, and the copper-zinc filler metal RBCuZn-A.

The BAg-1 and BAg-1a alloys have the lowest melting temperatures of all the filler metals used in torch brazing of copper and copper alloys and possess excellent flow characteristics. These filler metals contain more silver than BAg-2 (Table 1) and are therefore more costly. For this reason, BAg-2 is sometimes preferred, although it has a higher liquidus temperature than either BAg-1 or BAg-1a.

The color of silver alloy filler metals does not match that of most copper alloys. Joints that are visible are plated or painted if appearance is important. Face-fed filler metal in the form of wire, powder, or a paste with flux is less neat in appearance than filler metal deposited from preplaced preforms. The latter, made from either wire or strip, are used in deep joints or in joints having a change in direction of the mating surfaces.

The BCuP filler metals contain phosphorus, which makes them self-fluxing on copper; however, when these fillers are used with copper alloys, an AWS type 3A flux is advised. These filler metals are not used in steel joints to avoid formation of iron phosphide. Rapid heating and short brazing time at low temperature are recommended to avoid reduced joint strength that results from filler-metal liquation.

Of the types of filler metal used in torch brazing, BCuP-2 and 4 are the most fluid and fill joints having diametral clearances of 0.001 to 0.003 in. BCuP-5 can tolerate looser fits; acceptable diametral clearance is 0.001 to 0.005 in. BCuP alloys are lower in cost than silver alloy filler metals and also provide a better color match on copper after the oxide film that forms during cooling has been removed. The next two examples describe the use of BCuP-5 in torch brazing.

Example 2. Use of Self-Fluxing Filler Metal in Torch Brazing Return Bends for a Heat Exchanger. Return bends were brazed to heat exchanger tubes (Fig. 2). The tubes and return bends were made of alloy C12200 (phosphorus-deoxidized copper, DHP), $^{3}/_{8}$-in. OD, with 0.016-in. and 0.020-in. wall thicknesses, respectively. Ends of the tubes were expanded to fit over the ends of the return bends with diametral clearance of 0.004 to 0.009 in. Brazed joints had to be leaktight at 400-psi pressure.

All parts were vapor degreased before brazing. Each leg of the return bends was mechanically fitted with a preformed ring of 0.030-in.-diam BCuP-5 filler wire, which does not require a flux when used for brazing copper. The wire ring had to fit tightly on the tube, so that the filler metal would be heated by conduction as the base metal was heated to brazing temperature. If heated directly, the filler metal might have melted before the base metal reached brazing temperature.

The joints were assembled and manually brazed using a Y-shaped oxyacetylene torch with two tips. The opposing tips surrounded the joint with flame, making heating faster and more uniform than it would have been with only one flame. The joint was brought up to temperature as rapidly as possible to prevent liquation of the filler metal. Each joint took 5 s to braze. Because no flux was used, brazed joints did not have to be cleaned.

The manufacturer also used multiple-flame gas-air burners for this product. Such equipment was used for high production on a given size of return bend. Multiple burners heated both joints at one time while the assemblies were on a conveyor line. Natural gas was used in preference to acetylene for economy and heating control; some joints were overheated and others underheated when using the rapid heating of acetylene. Radiant gas burners were also used for this work. Furnace brazing could not be used, because the remainder of the assembly, including the aluminum fins, could not be heated.

Example 3. Use of Self-Fluxing Filler Metal in Torch Brazing a Fin Tube Assembly. Circular cooling coils were made of two nested counterflowing segments coupled by a return bend (Fig. 3). Each coil segment consisted of $5^{1}/_{2}$ turns of alloy C12200 (phosphorus-deoxidized copper) tubing 0.629 ± 0.005-in. ID by 0.042-in. wall, fitted with closely spaced fins about $1^{9}/_{16}$-in. OD. Four brazed joints were required, two at the return bend and one each at the inlet and outlet elbows. The return bend and elbows, also of alloy C12200, were formed from $^{5}/_{8}$-in.-OD by 0.042-in.-wall tubing. The diametral clearance for brazing varied from 0.002 to 0.005 in.; because of tubing tolerances, the return bends and the elbows were dressed as required to obtain clearance within this range. The assembly had to be

Fig. 3 Fin tube assembly. Shows four joints manually torch brazed using self-fluxing filler metal

Alloy C12200 (phosphorus-deoxidized copper, DHP); copper alloy filler metal (BCuP-5)

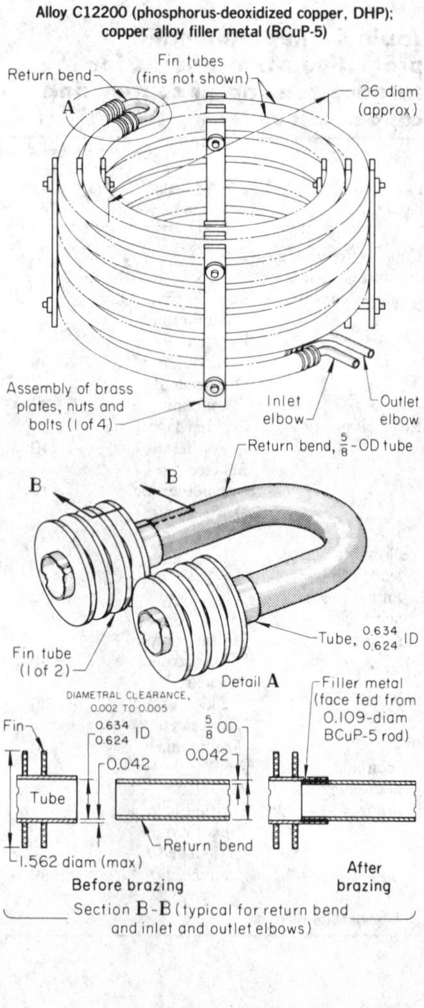

leaktight while holding 250-psi internal air pressure.

Manual torch brazing was selected as the most practical method of brazing the assemblies because (1) annual production was not high, (2) the workpiece was large compared with the joints to be brazed, (3) some manipulation of the assembly was needed, and (4) localized torch heating offered flexibility and freedom from general distortion.

The components were vapor degreased after forming. The two coil segments were then positioned in a rack consisting of four steel uprights (1-in. by $^1/_4$-in. steel flats) mounted on a circular base. A 0.109-in.-diam rod of self-fluxing brazing alloy BCuP-5 was selected for filler metal to eliminate the need for flux application and removal. An available protective atmosphere (a furnace gas containing 1 to 2% CO, 10 to 12% CO_2, remainder nitrogen) was piped through the tubing during brazing to protect the internal surfaces from oxidation.

The joints were heated to brazing temperature with a manual oxynatural gas torch, the natural gas being supplied at approximately 1 psi. The filler metal was face fed. The return bend joints were brazed first; then the inlet and outlet elbows were correctly positioned and brazed. After brazing, the coil was strapped with flat brass bars and bolts and removed from the holding rack. Production rate averaged about $1^1/_2$ assemblies per hour.

BCuP filler metals are supplied as wire, rod, and powder. BCuP-1 and BCuP-5 alloys are also available as strip. Like several other filler metals available in strip or sheet form, BCuP-5 is also produced by some manufacturers as a clad sheet (brazing sheet) having a copper core and a cladding of BCuP-5 alloy on both sides. Clad brazing sheet is cut into strips and placed in joints or on surfaces that cannot be filled with other forms of filler metal.

The copper-zinc filler metal used in torch brazing is RBCuZn-A, which has the highest brazing temperature range of the filler metals considered in this article. Overheating must be avoided, both to protect the base metal and to prevent voids in the braze caused by vaporization of the zinc.

Corrosion resistance and galvanic potential are considerations in selecting filler metal for applications where exposure to reactive environments is encountered. Filler metal of inferior resistance can be selectively attacked. For example, RBCuZn-A is anodic and generally inferior in corrosion resistance to copper, copper-silicon, and copper-nickel alloys, and it should not be used in seawater with those metals. It is acceptable for other applications where air is the only environment. The silver alloy brazing filler metals are cathodic, have good corrosion resistance, and are suitable for all corrosive environments where copper alloys may be used.

Fluxes. Selection of flux for torch brazing of copper and copper alloys depends not only on the filler metal used (as it does for brazing of steel), but also on the base metal. Some copper alloys form refractory oxides, and this influences choice of flux.

When BAg or BCuP filler metal is used in brazing copper or in brazing a copper alloy that contains no elements that cause it to form refractory oxides (aluminum, beryllium, chromium, and silicon), a type 3A flux is used. When brazing copper alloys that form refractory oxides, a more active flux is usually required. For instance, aluminum bronzes require a type 4 flux to inhibit the oxidation of aluminum.

Gas fluxing, by entraining volatilized flux in the gas flame, is used in torch brazing copper and copper alloys. The equipment for and use of this fluxing method are described in the article "Torch Brazing of Steels" in this Volume. When used on copper and its alloys, gas fluxing protects the surface of the base metal from becoming oxidized or discolored, eliminating postbraze cleaning. The joint surfaces are also protected with an appropriate flux. This secondary use of a gas flux is illustrated in the following example.

Example 4. Use of Gas Fluxing To Prevent Oxidation During Brazing. Two types of flux were used when torch brazing a capillary tube system consisting of a bulb and two capillary tubes (Fig. 4). One type was necessary for brazing. The other, a gas flux fed through the flame, was used to prevent oxidation and discoloration of the copper assembly. The torch used oxyacetylene and had a standard brazing tip and conventional gas-fluxing equipment. The gas flux was composed of $71^1/_2$% methyl borate and $28^1/_2$% acetone. BAg-1 filler metal was used, although other filler metals, such as BAg-1a and BCuP-2 to 5, would have been satisfactory.

After five assemblies had been loaded on a five-station manually operated fixture, the joint surface of each tube was brushed with a small amount of type 3A flux to ensure penetration of flux and filler metal into the joint. The joint was heated with the oxyacetylene flame containing the gas flux. The torch was passed over the assembly to give it a flux coating and to prevent oxidation during brazing. The torch was then held close to the bulb. The filler

Fig. 4 Bulb and capillary tubes joined by torch brazing. Used gas fluxing in addition to type 3A brazing flux to prevent oxidation

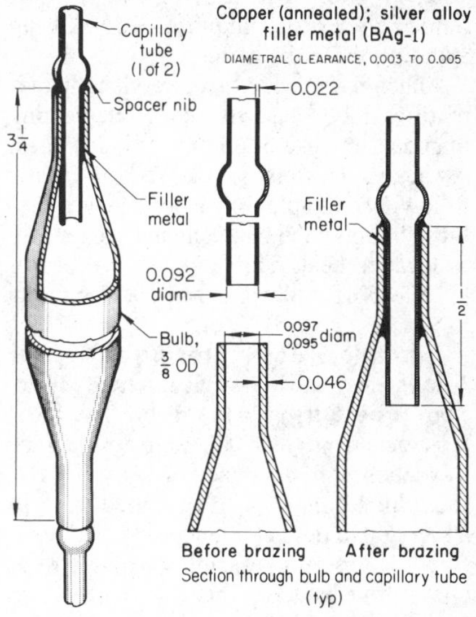

metal was hand fed to complete the joint. The brazed assembly was removed and quenched with water to cool it and remove the flux. This sequence of operations was repeated for each of the five assemblies in the fixture. The finished assembly was bright and free of deposits.

Fuel Gases. The fuel gas mixtures generally used in torch brazing of copper and copper alloys are oxyacetylene, oxynatural gas, oxypropane, oxyhydrogen, and air/natural gas. Oxyfuel gases are highest in cost, flame temperature, and heating rate, with oxyacetylene the highest in the group in each respect. Oxyfuel gases are widely used in manual torch brazing where temperature can be controlled by torch manipulation. The cost advantage of air/natural gas is exploited particularly in mechanized and automatic high-production torch brazing, where the lower flame temperature offers protection from damage caused by overheating. Desired heating rates are obtained economically by using high flow rates together with multiple torches or radiant burners. Neutral or slightly reducing flames are used to help the fluxes prevent oxidation of base-metal surfaces.

Although high heat input is necessary to overcome high thermal conductivity, many copper alloys cannot withstand the rapid heating rates that can be used on steel. Under too rapid heating, high thermal expansion can cause local stresses, resulting in distortion and cracking in some alloys. Accidental overheating can cause damage to copper and copper alloy assemblies more readily than to those made of steel. Particular care should be taken with phosphor bronzes, leaded brasses, nickel silvers, and silicon bronzes. The use of gases of lower flame temperature requires less skill in avoiding these difficulties.

Although the use of oxyacetylene flames requires more care to avoid overheating than does the use of other types of flames, oxyacetylene flames heat faster. In the following example, oxyacetylene was useful in bringing joints to temperature quickly to localize heat. This avoided the loss of the effects of cold work and kept labor costs low.

Example 5. Torch Brazing of Copper Alloys and Stainless Steel in a Bourdon Tube Spring Assembly. The measuring element of a 5000-psi pressure gage was a bourdon tube made of alloy C17200 (beryllium copper, 1.9%), coiled to form a spring that deflected under internal pressure. The spring assembly (Fig. 5) required four brazed joints. Section A-A in Fig. 5 shows two joints that were brazed simultaneously: a type 304 stainless steel capillary tube brazed to a copper bushing, which was brazed to the inside of the beryllium copper bourdon tube. At section B-B in Fig. 5, the bourdon tube was brazed to a takeoff arm of alloy C36000 (free-cutting brass), which in turn was brazed to a takeoff lug of alloy C26000 (cartridge brass, 70%), shown in section C-C in Fig. 5.

Torch brazing was used because it was simpler to control the amount of heat input required for the different joints by torch brazing than by induction brazing. Furnace brazing was considered unsatisfactory because the entire spring would have been exposed to the brazing temperature, which would have removed the effects of cold work and reduced the potential mechanical properties obtainable by the heat treatment after brazing. A fixture was used to position the components accurately for brazing. Parts were fluxed with type 3A flux, assembled in the fixture, and heated by an oxyacetylene torch. Silver alloy filler metal (BAg-1a) was hand fed to the joints. Immediately after the filler metal had solidified and while the assembly was still hot, it was plunged into cold water to remove the flux residue and limit the annealing action on the beryllium copper bourdon tube. After brazing, springs were age hardened and then bright dipped. Each assembly was made, quenched and dried, and ready for heat treatment in $2^1/_4$ min.

Mechanized and automatic torch brazing equipment and operations are the same as those applied on steel assemblies, except for the special considerations discussed earlier in this section. Mechanized torch brazing is widely used in manufacturing air conditioners, radiators, and other types of heat exchangers that use hairpin coils of copper tubing with aluminum fins. Conveyor belts and turntables are used extensively for both mechanized and automatic torch brazing. The following example illustrates the use of a turntable in an operation that was wholly automatic except for assembly.

Example 6. Use of an Eight-Station Turntable for Automatic Brazing of Bulb and Tube Assemblies. An eight-station automatic setup (Fig. 6) replaced a manual torch brazing setup that required two operators and used a rotating fixture for brazing copper bulb and tube assemblies. Production rate (about 700 joints per hour, gross) was not increased with the automatic method, and improvement in quality was negligible. However, only one unskilled operator was needed. Therefore, the cost of direct labor was reduced more than 50%. The sequence of operations is given in Fig. 6. The air/natural gas torches at stations 4, 5, and 6 were adjusted to produce neutral or slightly reducing flames. Filler metal was BAg-1 in flux-paste form.

Fig. 6 Bulb and tube assembly. Brazed by mechanically held torches on an eight-station turntable

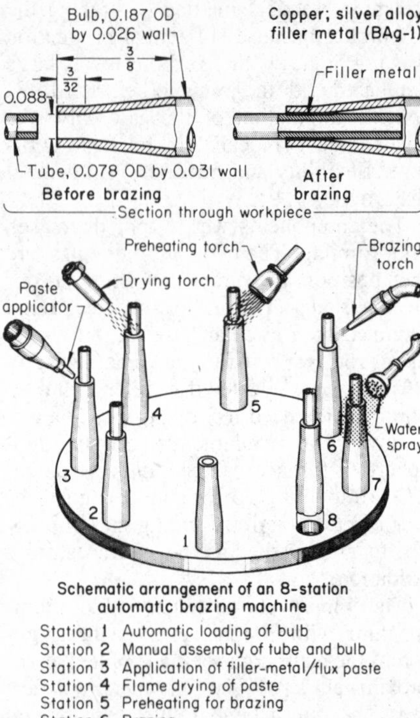

Fig. 5 Torch brazed bourdon tube spring assembly. Components of copper alloys and stainless steel

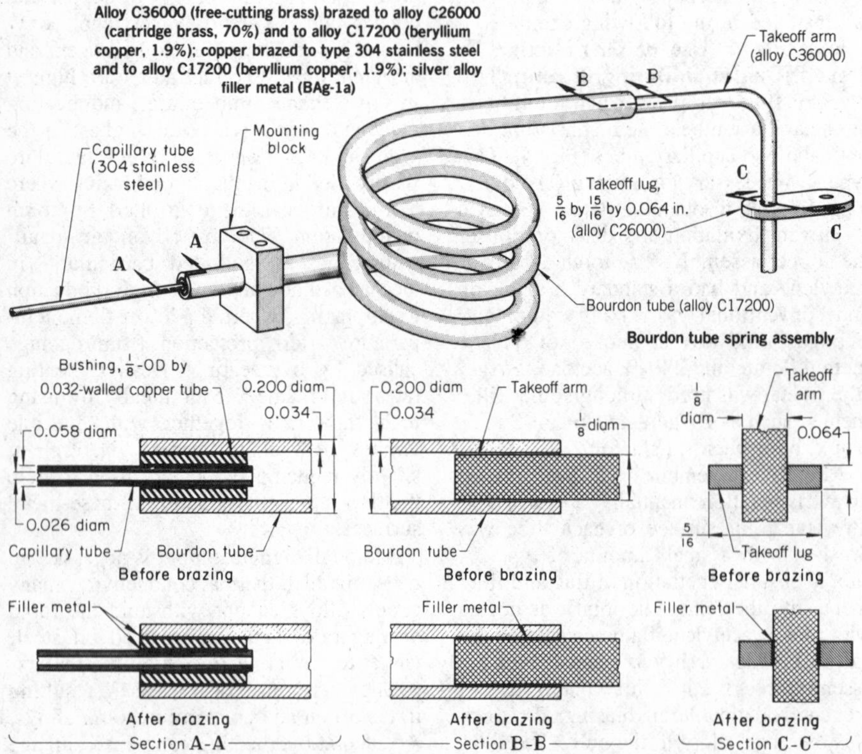

Fig. 7 Torch brazed component of a cooler for an electronic system. Change in joint design eliminated leakage

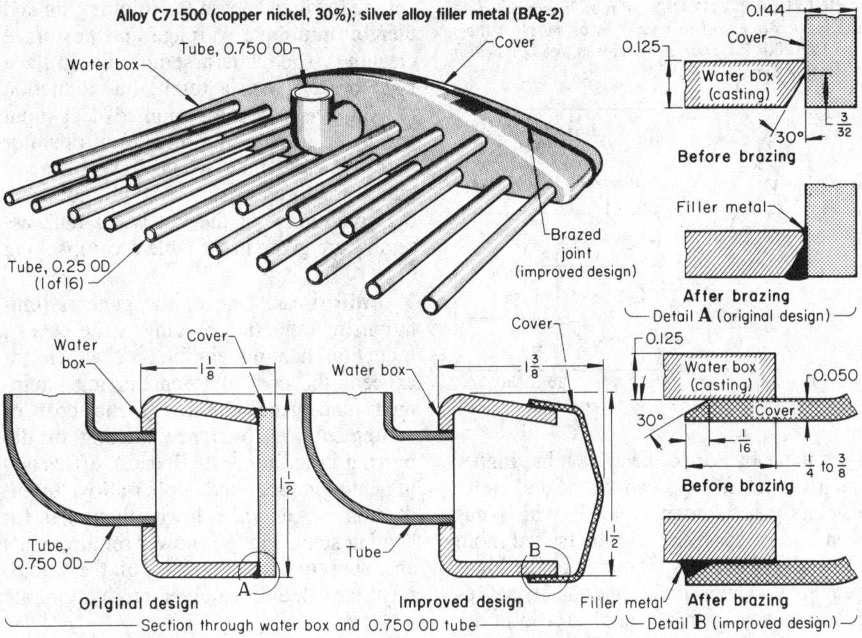

Alloy C71500 (copper nickel, 30%); silver alloy filler metal (BAg-2)

The completed assemblies were leak tested with air at 60 psi. Rejection rate was 1%, which was considered satisfactory. The joints were clean and ductile.

Joint Design. For widely used silver alloy and copper-zinc alloy filler metals, diametral clearances of 0.002 to 0.005 in. are suitable. For BCuP filler metals, diametral clearances differ; clearances of 0.001 to 0.003 in. are suitable for BCuP-2 and BCuP-4, and 0.001 to 0.005 in. for BCuP-5. Much torch brazing is done by face feeding of filler metal, and when this is desired, joints should be limited to a depth that can be quickly and adequately fed by the filler metal used. Deep joints and those that change direction sharply require the use of preplaced filler metal. The following example involves a failure that was corrected by changing from a corner joint that placed the braze in tension, due to bending, to a lap joint that placed the braze in shear and enlarged the area of contact of the joint.

Example 7. Change in Joint Design That Eliminated Failures Caused by Tensile Stresses. A torch brazed assembly (Fig. 7) was the heat-dissipation section of a cooling unit for electronic equipment. Ethylene glycol solution circulated through the tubes and the water box. All components shown were made of alloy C71500 (copper nickel, 30%). The tube fins were made of copper, mechanically attached. The assembly was brazed with a manually operated oxynatural gas torch, using BAg-2 filler metal.

The view at lower left and detail A in Fig. 7 show the joint between the water box and the cover as originally designed. A high percentage of these joints failed in service on the inside corner of the joint because of tensile loading. Varying pressure loads imposed bending moments on the box walls, causing the joint to act like a hinge. The braze itself was far from perfect. Only about $^1/_{16}$ in. of the joint depth was a true braze; the remainder had only the inherent strength of the silver alloy filler metal. Small defects nucleated leaks that caused loss of cooling and failure of the system.

The cover plate was redesigned as a formed cup, which provided a lap joint that placed the braze in shear and increased the joint area considerably. The improved design of the cover assembly and the joint are shown in the lower middle view and detail B. No field failures were reported for several thousand of the redesigned units.

Precision Torch Brazing. Although conventional torch brazing is capable of joining parts only a fraction of an inch in size, some small assemblies require special equipment. When joining copper and copper alloy parts of this type, heat control is the principal difficulty, but other problems may be encountered. For example, many electrical and electronic joints are not only small but also are located close to components that can be functionally disabled by heat, spatter, and the removal of flux residues. If allowed to remain, an active flux can, in time, corrode the joint.

One method of meeting these problems is the use of a miniature oxyhydrogen torch that uses a special gas generator. Torch tips are made from several sizes of hypodermic needle tubing. Oxygen and hydrogen are generated by electrolysis of water in the exact ratio required for complete combustion. Alcohol vapor can be added to the fuel line to impart a blue cast to the otherwise less visible flame. This technique, when used with a self-fluxing BCuP-5-clad brazing sheet, proved useful in joining copper leads, as described in the next example.

Example 8. Technique for Brazing 0.008-in.-Diam Wires to 0.010-in.-Thick Terminals. Terminals to be attached to 0.008-in.-diam wire leads (Fig. 8) were made of a 0.004-in.-thick core of copper clad on both sides with 0.003-in.-thick layers of BCuP-5 brazing filler metal. Several experimental assemblies were

Fig. 8 Assembly of a lead wire and terminals. Brazed with a special oxyhydrogen torch and self-fluxing filler metals

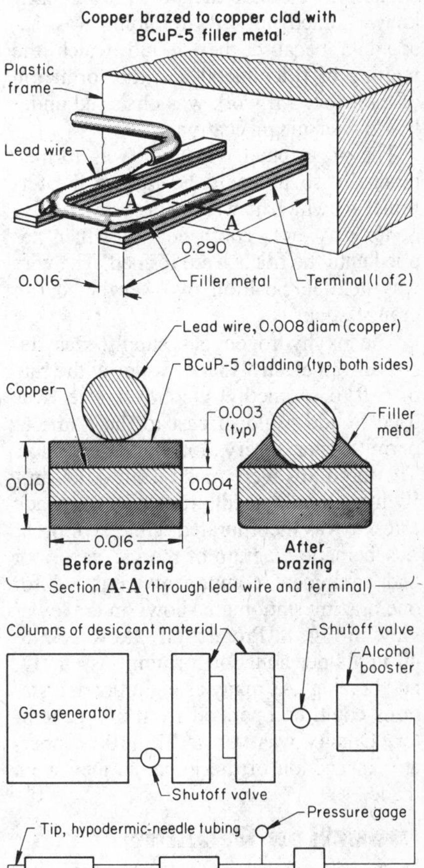

originally made by gas tungsten arc welding. This method was unsatisfactory because the heat could not be controlled. Arc starts occurred in a burst that made observation by the operator difficult.

Production joints were made initially by capacitor discharge resistance brazing. Expulsion from the joint caused spatter to be lodged on surfaces or concealed in crevices, where it changed the electrical properties of the device by induction or short circuiting. Inspection of the joint was difficult and time consuming, and repair time was excessive. By changing to torch brazing with a special oxyhydrogen torch, the procedure was brought under control. Quality and reliability of the joint improved greatly.

During the time operators were learning to use this process, production rates were lower than before. After approximately 3 months, operator skill increased enough that production time was equal to or slightly less than that achieved with the resistance brazing process. An overall saving in time was realized because rework for broken or improperly brazed joints was almost eliminated. Operators came to prefer this method of brazing because they were better able to monitor their own work; they knew immediately if the joint was acceptable because they could watch the melting of filler metal and the formation of the fillet. All work was observed under low-power magnification.

The prestripped wire lead was formed manually so it would lie flat on the terminal to which it was to be brazed (Fig. 8, upper view). Torch heat was then applied until the fillets were formed. The wire was held in position by tweezers or an orange stick.

The oxyhydrogen gas supply was derived from electrolysis of water at the rate of 6 ft^3/h. A methyl alcohol booster was used to add a bluish cast to the flame to permit easier observation by the operator. The torch tips were made from 0.009-in.-ID hypodermic needle tubes. A flashback arrester was incorporated with each torch. A schematic diagram of the gas generator and equipment components required for one brazing station are shown in the lower view in Fig. 8. Production rate was 35 to 40 joints per hour for forming, assembly, and brazing. As many as eight brazing stations could be operated from one generator. Quality was verified by 100% operator inspection of the joints as they were made.

Induction Brazing

The efficiency of heating by induction varies directly with the electrical resistivity of the alloy. Brass, because it has higher electrical resistivity, can be heated more efficiently than copper; steel, which has even higher resistivity, can be heated more efficiently than brass. In terms of the high-frequency power input and the time required to heat 1 lb of metal in a joint assembly to a brazing temperature of about 1300 °F, a power input of 15 kW (at 450 kHz) can heat a steel joint to this temperature in about 16 s, whereas brass requires about 30 s and copper about 55 s (Fig. 9).

Fig. 9 Power input and heating time required for high-frequency induction brazing. Power is expressed as kilowatts required to heat 1 lb of metal at the joint to 1300 °F. (Source: Wilkinson, W.D., *British Welding Journal,* Oct 1965)

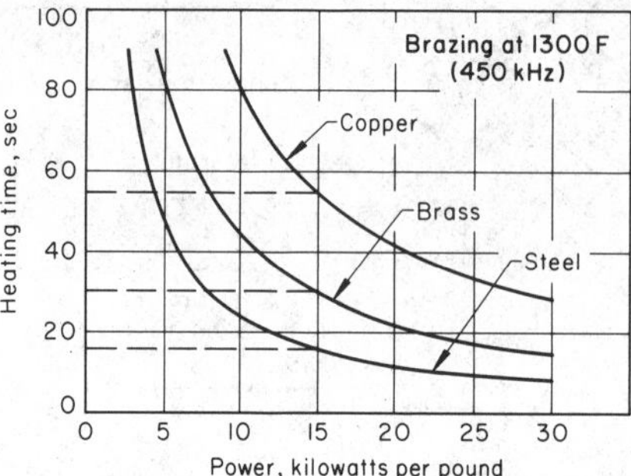

Advantages. The general advantages of induction brazing are applicable to copper and its alloys. In the next example, induction brazing minimized warpage and postbraze cleaning. Induction brazing also required less operator skill than would have been needed for torch brazing.

Example 9. Use of Induction Brazing To Minimize Warpage and Postbraze Cleaning. Alloy C11000 (tough pitch copper) tubes $^3/_4$ in. in outside diameter were induction brazed to hot water tank covers, formed from alloy C23000 (red brass, 85%), using the setup shown in Fig. 10. The resulting joints had to be leaktight at a pressure of 300 psi. The holding fixture (Fig. 10), which was made of heat-resistant board, had a retaining shoulder at the edge and a stop at the center to support the pipe. The fixture was raised and lowered manually on a rack to move the joint into and out of the inductor. Tubes and covers were degreased in trichlorethylene and bright dipped before being assembled for brazing. The joint surfaces were brushed with flux before the parts were assembled. A preformed ring of 0.050-in.-diam BCuP-5 filler-metal wire was placed over the tube, as shown in detail A. The assembly was then raised to position in the single-turn inductor, where it was heated for 28 s and then lowered and unloaded from the fixture. Selection of induction heating for this application was based primarily on the ability of the process to localize heat in the joint region and thereby minimize warpage and postbraze cleaning. The same assembly could have been brazed with a torch, but induction brazing provided programmed heat input and made possible the use of an operator with less skill than would have been required for torch brazing. Additional brazing conditions for the hot water tank assembly are given in the table accompanying Fig. 10.

Limitations. One of the general limitations of induction brazing is the cost of induction heating equipment, which far exceeds the cost of torch brazing equipment and usually exceeds the cost of equipment for resistance brazing or dip brazing in molten salt. Because efficiency in heating copper and copper alloys by induction is generally lower than that for heating steel (Fig. 9), power requirements and consequently the cost of the equipment for achieving a given production rate

Fig. 10 Induction brazed tank cover assembly. Induction brazing minimized warpage and postbraze cleaning.

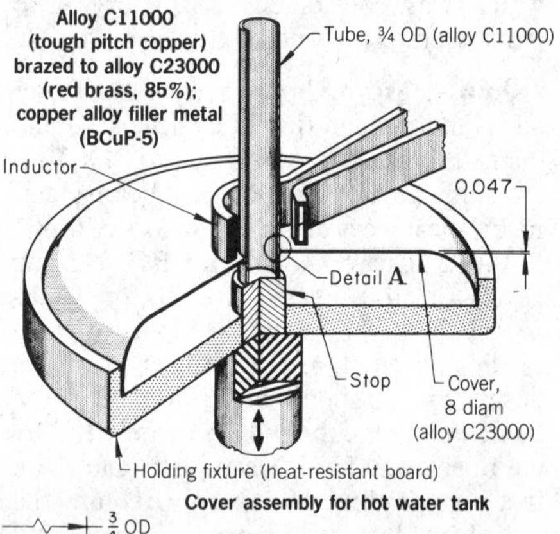

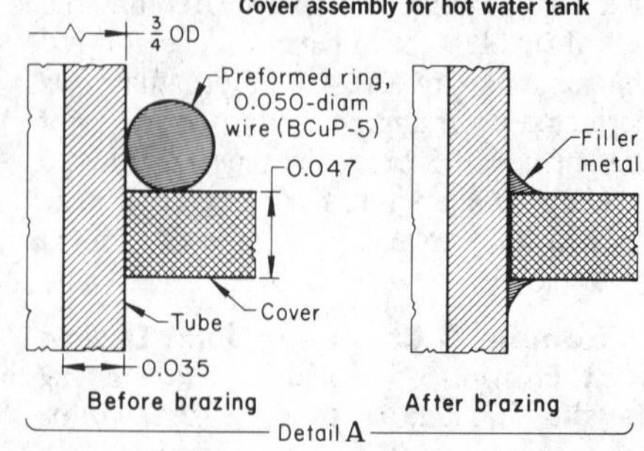

Power supply	Vacuum tube unit, 10 kW, 450 kHz
Inductor	Single-turn
Filler metal	BCuP-5 wire, 0.050 in. diam, preformed to $^3/_4$-in.-ID ring
Flux	Type 3A (AMS 3410D)
Brazing time	28 s per joint
Production rate	60 joints/h

are proportionately higher. Long-term operating and maintenance costs of induction units, however, may outweigh the higher initial costs.

Other general limitations of induction brazing that relate to the size and shape of assemblies that can be brazed, the design of inductors, and requirements for matching impedances apply to the brazing of copper and copper alloys as well as to the brazing of steel and are discussed in the article "Induction Brazing of Steels" in this Volume. Limitations that relate more specifically to the brazing of copper and copper alloys by induction include heating rate and the avoidance of overheating, both of which are less amenable to control in induction brazing than in furnace brazing or dip brazing in molten salt. Consequently, induction brazing of copper alloys that are susceptible to hot cracking or to dezincification is more problematic.

Power supplies used in the induction brazing of copper and copper alloys, including motor-generator sets, solid-state units, and vacuum tube units, are discussed at length in the article "Induction Brazing of Steels" in this Volume. Geometrical considerations may require loose coupling between the inductor and the load (workpieces). Under conditions of loose coupling, two methods for improving heating efficiency can be used. If a variable output transformer is used, an additional tuning capacitor may be connected directly across the inductor to increase inductor current. As an alternative, a high-impedance output circuit is designed to yield a relatively high voltage and current in the work circuit.

Design of Inductors. Guidance in the design of inductors is presented in the article "Induction Brazing of Steels" in this Volume, which contains examples of single-turn and multiple-turn inductors used in brazing applications. Although the use of single-turn inductors is not uncommon in brazing copper and copper alloys (Example 9), the efficiency in heating and, to some extent, the control of heating rate are greatly improved with the use of multiple-turn inductors. In the following example, the control of heat input and distribution was critical, and excellent results were obtained at a relatively low operating frequency (10 kHz) using a three-turn inductor.

Example 10. Use of a Multiple-Turn Inductor To Control Heat Input and Minimize Overaging of Chromium Copper. Chromium copper (C18200, 0.6% Cr) has high electrical conductivity (78% IACS) and develops improved mechanical properties in the precipitation-hardened condition. After solution heat treatment, it is aged at 750 to 930 °F. Heating for brazing is critical, because the copper can overage or partly anneal at temperatures above 930 °F, the overaging effect increasing with increasing temperature and time at temperature. Consequently, in brazing a tungsten-silver alloy contact to a precipitation-hardened chromium copper post (Fig. 11), heat input to the post required careful control to minimize overaging. The joint required high electrical conductivity and impact strength, and the chromium copper post had to retain its hardness and strength to resist distortion when subjected to repeated impact. Brazing the tungsten-silver alloy contact required the use of a nickel-containing silver-alloy filler metal for adequate wetting. BAg-3 was chosen rather than BAg-4 or BAg-13, because BAg-3 has a lower brazing range (1270 to 1500 °F); induction brazing was selected to control heat input and minimize the duration of the brazing cycle. Optimum control of heat input to the chromium copper post was obtained using the square three-turn inductor shown in Fig. 11.

Before brazing, the chromium copper base was bright dipped, and the tungsten-silver alloy contact was cleaned with an abrasive cloth. Type 3A flux paste was applied to both components with a brush. They were then assembled with a disk of filler metal placed between them (Fig. 11). Power supply was a 50-kW, 10-kHz motor-generator set. When the brazing alloy melted and flowed, the contact was rotated to release any entrapped gases and improve wetting. Production rate was ten joints per hour.

Fig. 11 Tungsten-silver alloy contact and chromium post in position for induction brazing. Used three-turn inductor

Tungsten-silver alloy brazed to C18200 chromium copper (0.6% Cr); silver alloy filler metal (BAg-3)

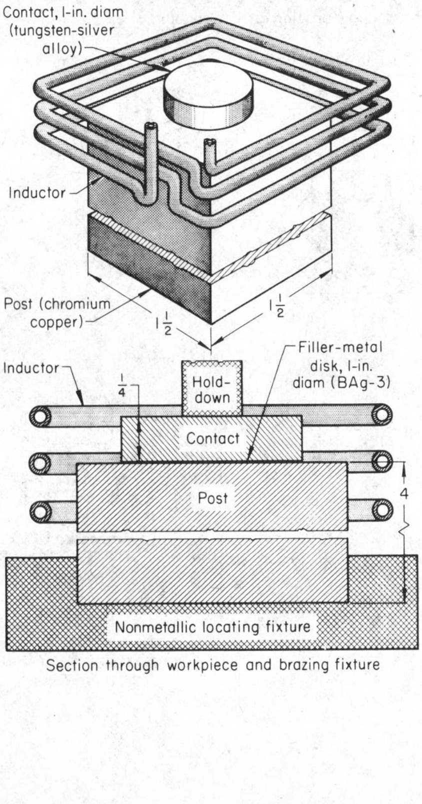

Section through workpiece and brazing fixture

Inductors for Mass Production. Induction heating is suited for mass producing brazed assemblies, primarily because inductors can be designed that heat a line of assemblies as they are carried through the induction field by conveyor belt or turntable. Hairpin and pancake inductors are most widely used for conveyorized applications, because they do not obstruct passage of the assemblies as they travel through the induction field; inductors for conveyorized setups are described in the article "Induction Brazing of Steels" in this Volume.

Because steel assemblies heat more efficiently, they are generally more adaptable to brazing in a conveyorized setup than are assemblies made of copper and copper alloys. An alternative design for use with copper alloys is the multiple-station inductor. With this type of inductor, two or more (often as many as six) assemblies can be brazed simultaneously or in a selective sequence that permits loading at one or more stations while other stations are heating.

Examples 11 and 12 which follow describe the application of a conveyorized setup and a multiple-station inductor to the brazing of copper and copper alloy assemblies at high production rates.

Example 11. Use of a Conveyorized Setup for Induction Brazing of Connector Assemblies. Small connector assemblies, consisting of two alloy C36000 (free-cutting brass) studs and an alloy C17200 (beryllium copper, 1.9%) spring clamp (Fig. 12), were induction brazed on a 5-ft-long conveyor at a rate of 1900 assemblies per hour, using a modified pancake inductor and a $2\frac{1}{2}$-kW, 300-kHz vacuum tube power supply.

The brass studs, which had a tapered bore, were brazed to the platform of the spring clamp. The critical portions of the assembly that could not be overheated without causing serious damage were the thin (0.015-in.) wall at the top of the studs and the receptacle portion of the beryllium copper spring clamp, which was heavily cold worked and precipitation hardened before assembly for brazing. Heating, therefore, had to be largely restricted to

Fig. 12 Connector assembly and section through inductor used for conveyorized brazing

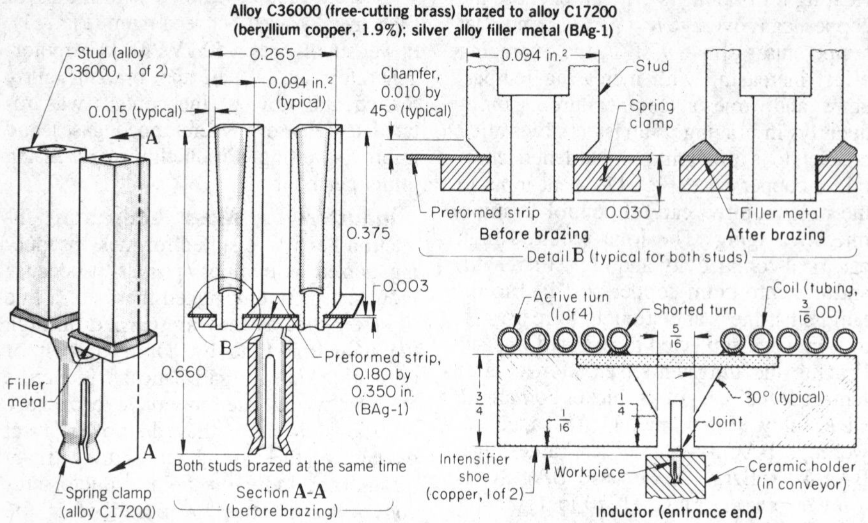

the area of the joint without disrupting the flow of filler metal or interfering with the seating of the studs on the clamp platform.

Proper heating was obtained with the modified pancake inductor shown in cross section at lower right in Fig. 12. The inductor, which was made of copper tubing, consisted of four active turns and a shorted central turn to which was soldered a plate with a pair of 6-in.-long copper intensifier shoes. The upper part of the gap between the shoes was tapered to direct energy toward the joint but not toward the thin walls of the studs. At the inductor entrance, the gap between the shoes was $^5/_{16}$ in.; from the midpoint of the inductor, the gap width flared to $^1/_2$ in. at the inductor exit. The assemblies were held in place by inserting the receptacle end of the spring clamp into ceramic holders; the holders were located $^1/_2$ in. apart and traveled on the conveyor chain at a speed of 16 in./min.

Stud and clamp components were bright dipped before assembly. The studs were hopper fed into clamping cavities in a rotary indexing table. A preform of BAg-1 strip 0.180 by 0.350 by 0.003 in. thick was pierced, cut off, and placed automatically on each pair of studs. The beryllium copper spring clamp was placed on each assembly by hand. The ends of the studs were staked to the clamp to make the assembly self-jigging. The staked assemblies were dipped in a 20-to-1 solution of hot water and type 3A flux and air dried. Then the assemblies were placed, spring clamp down, into the ceramic holders, as shown at lower right in Fig. 12, and passed through the inductor. Time at temperature was approximately 22 s. After being brazed, the assemblies were washed in hot water.

Example 12. Use of Two Four-Station Inductors for High-Production Brazing of Motor Armatures. In the production of motor armatures, copper end rings were induction brazed to both ends of 14 copper rivets that held the armature laminations (Fig. 13) at a rate of 433 assemblies per hour (two joints per assembly) in two four-station, single-turn inductors, using a 5-kW, 450-kHz vacuum tube power supply. Induction brazing was selected for this application partly because the rapid rate of heating left the magnetic properties of the silicon steel armature laminations relatively unaffected.

Each of the eight inductor stations contained a graphite bushing into which an armature assembly could be placed. The inductors and bushings were attached to graphite blocks that supported the armatures during the brazing cycle.

Silver-containing filler metal BCuP-5 was selected for its brazing range and because it is self-fluxing on copper. Because the filler metal would not wet the graphite holders, there was no difficulty in brazing rings to both ends of the armatures simultaneously. Conductivity was adequate.

Before brazing, the copper rings were degreased and bright dipped. Each inductor station was loaded by first placing a copper end ring at the bottom of the graphite bushing, followed by a preform of brazing alloy. The armature was then inserted in the hole in the graphite block (Fig. 13, section A-A), and a second ring of filler metal was placed on top of the armature, followed by another end ring of copper. When one of the four-station inductors was loaded in this manner, power was applied and the brazing cycle began. The operator then began loading the second four-station inductor.

Use of Fluxes. The proximity of a moist or liquid flux to an uninsulated high-voltage, high-amperage inductor is often a source of difficulty unique to induction

Fig. 13 Motor armature and a four-station inductor. Used for high-production brazing of copper end rings to rivets holding armature laminations

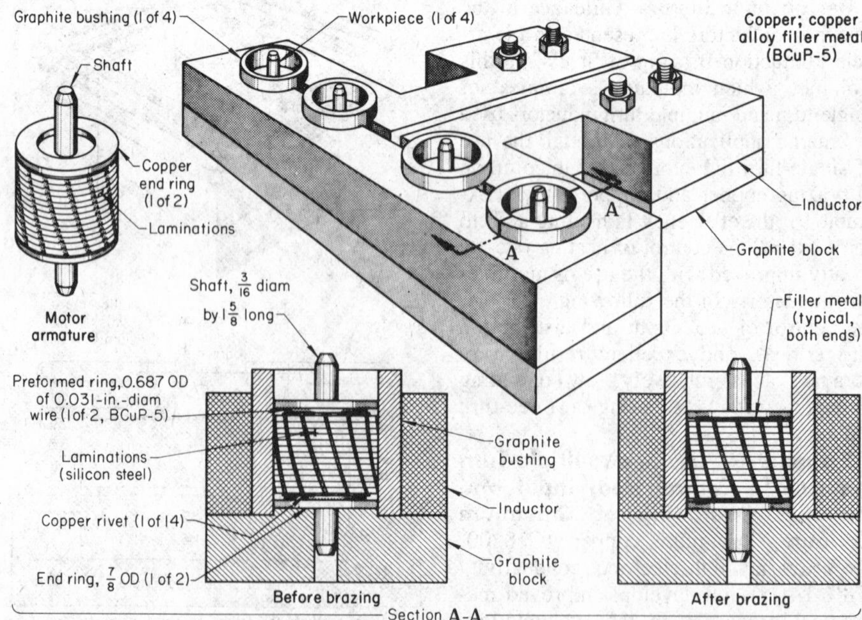

brazing. The difficulty increases with the tightness of coupling (the narrow distance across the air gap between inductor and workpiece) that is often required to heat copper and copper alloys efficiently. When the base metal heats, and the water in a moist flux is driven off in the form of steam, a violent sputtering often occurs, followed by high-voltage arcing caused by shorting between inductor and workpiece. If the arcing is of sufficiently high intensity, it punctures the thin wall of the inductor tubing at one or more points, releasing a high-pressure spray of cooling water from the inductor. Damage to expensive inductors, fixtures, and assemblies from flux-induced arcing can be considerable. Although relatively uncommon, a molten flux, because of its electrical conductivity, may sputter when driven by field forces in a field of high intensity.

The hazard of arcing can be substantially reduced by protecting the inductor with an insulated cover, such as woven fiberglass sleeving, grounding the inductor, allowing the water in moist fluxes to dry before they are heated, using an insulating lacquer, or using a low-sputtering flux.

Avoiding the Use of Flux. Some filler metals, notably those in the BCuP series, have self-fluxing properties when used on commercial coppers. Because of their alloying constituents, however, most copper alloys require the use of a flux, even when BCuP filler metals are used. For example, lead-containing alloys require the protective action of flux to prevent formation of a dross that can seriously impede wetting and the flow of filler metal.

In some applications, however, the use of a flux is unacceptable because of special service requirements or because flux residues cannot be removed completely after brazing. In the example that follows, copper plating eliminated the need for flux in induction brazing cartridge brass to a leaded commercial bronze.

Example 13. Avoiding the Use of Flux by Copper Plating Copper Alloy Components. A reversing valve assembly for an air conditioner (Fig. 14) required leak-resistant brazed joints between the body, made from alloy C26000 (cartridge brass, 70%), tubing, and the cover, made from alloy C31400 (leaded commercial bronze). Two specific restrictions were imposed in brazing one of the cover joints (section A-A): (1) during brazing a nylon needle located ½ in. from the brazed joint could not be heated above 450 °F; and (2) the joint had to be brazed without the use of flux, not only to minimize the possibility of leakage, but also because the inside of the joint could not be cleaned after it was brazed.

Fig. 14 Induction brazed valve assembly. Induction brazed to avoid overheating the nylon needle and selectively copper plated to avoid using flux

Alloy C26000 (cartridge brass, 70%) brazed to alloy C31400 (leaded commercial bronze); copper alloy filler metal (BCuP-4)

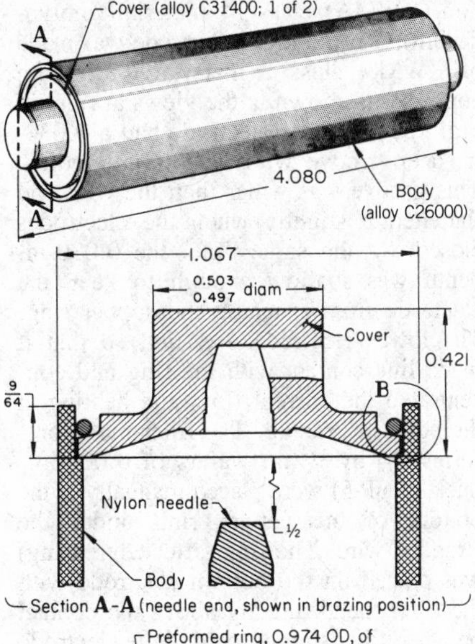

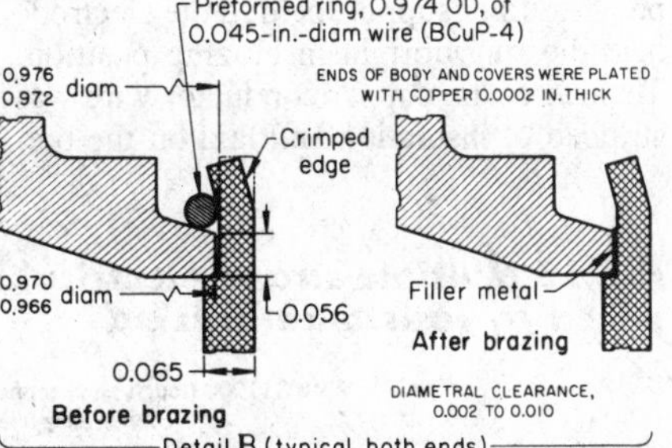

Power supply	Vacuum tube unit, 20 kW, 450 kHz
Inductor	Two-turn(a)
Fixture	For accurate centering and lengthwise positioning
Filler metal	BCuP-4, 0.045-in.-diam wire(b)
Flux	None
Heating time	15 s

(a) Made of 3/16-in.-OD copper tubing. Inside diameter of the inductor was 1 3/16 in. (b) Formed into a 0.974-in.-OD ring to fit the inside diameter of the body

By heating the joint in 15 s and quenching it with water as soon as the filler metal solidified, heat was sufficiently localized so that the nylon needle was not overheated. By plating both the end of the body and the cover with a 0.0002-in. thickness of copper before brazing, leaktight joints were made without the use of flux.

The diametral clearance between the cover and the body was 0.002 to 0.010 in. When the two components were assembled, a preformed ring of BCuP-4 filler wire was pushed in place by hand, and then the end of the tube was crimped slightly (detail B). Then the assembly was induction brazed using a two-turn inductor. Because of the short heating cycle (15 s), surfaces did not require postbraze cleaning and, following inspection, were immediately spray coated with black lacquer.

Joints were inspected 100% using a halogen leak detector. The internal pressure at the time of testing was 400 psi. In addition, these assemblies were required to withstand 2600-psi internal pressure without rupture. The percentage of rejections because of leakage was extremely low (0.15%).

Cost of Brazing. The labor costs for induction brazing are usually lower than those for torch brazing, the process with which induction brazing is most often competitive. At high production rates, the total brazing cost per assembly generally favors induction brazing over torch brazing. However, the direct savings that accrue from the use of induction brazing will vary considerably among different plants because of differences in the time periods established for amortizing capital equipment.

Resistance Brazing

Resistance brazing is often used in the joining of copper conductors, terminals, and other parts in lap joints for electrical connections where heating must be localized and closely controlled while brazing the joint, and the brazed joint must have low electrical resistance. Generation of heat in the filler metal and nearly complete filling with a thin layer of the filler metal in the joint help in meeting both of these objectives.

The filler metal most frequently used is BCuP-5, which is used without a flux when brazing copper in most applications. In spite of the comparatively low electrical conductivity of BCuP-5 (approximately 10% IACS, Table 1), producing brazed joints having a low voltage drop acceptable for nearly all applications is not difficult, because the layer of filler metal in the brazed joint is very thin and the joints are designed to provide a conducting area larger than the cross section of the smaller member.

Brazing of Multiple-Strand Copper Wire. Resistance brazing is preferred for making connections of terminals or assemblies to stranded or braided copper electrical conductors. Torch brazing does not provide sufficiently localized or controlled heating for most applications of this type. Induction brazing, the only other brazing process that permits selective heating, provides well-controlled but somewhat less localized heating than resistance brazing, and the equipment for induction brazing is more costly. Resistance brazing, like other brazing processes, is often selected over soldering, because of the higher strength of the filler metal.

The major arc welding processes are usually ruled out on the same basis as torch brazing, and arc processes that involve impact, such as stud and percussion welding, generally cannot be used because the stranded or braided copper conductors do not provide a rigid workpiece for the lap connections in use. T-joints of multiple-strand wires to flat surfaces have been made by percussion welding in some instances.

Resistance welding cannot be used on multiple-strand copper electrical conductors rated to carry a current higher than about 60 A, because the amount of current needed to make the weld is prohibitively large. Such connections are made by resistance brazing using the solidified joint technique.

Connections to smaller multiple-strand conductors are made by either resistance welding or resistance brazing. Resistance welding has the advantage of requiring no filler metal and is often faster than resistance brazing. However, when individual strands are small in diameter, melting of strands in resistance welding is difficult to prevent. In the example that follows, resistance brazing was selected instead of resistance welding to join a 19-strand silver-plated copper wire (strand diameter, 0.014 in.) to a copper ring, because the separate strands were melted by resistance welding of this smaller strand size. The use of BCuP-5 filler metal in the resistance brazing procedure enabled melting to begin at a temperature of 1190 °F and provided enough molten metal for effective heat transfer within the multiple-strand wire and from the wire to the copper ring, so that the strands did not overheat. The original joining procedure, based on manual tin-lead soldering, had been too slow and had produced joints of inconsistent quality with excessive spreading of the solder.

Example 14. Use of Resistance Brazing for Stranded Copper Wire. An electrical connection (Fig. 15) was originally made by manual soldering with a tin-lead alloy. Production rate was slow (about 60 connections per hour), the solder spread over an undesirably large area, and joint quality was unpredictable. Resistance welding failed because the temperature required to produce a bond invariably melted the thin (0.014-in.-diam) strands. The method was changed to resistance brazing using the setup illustrated in Fig. 15. The upper brazing electrode was made of Resistance Welder Manufacturers Association (RWMA) class 14 material (molybdenum), and the lower electrode was made of RWMA class 2 material (chromium copper). As shown in the views at right in Fig. 15, the upper electrode had a 0.030-in.-deep groove with a 0.060-in. radius. This groove was wider than the wire and flattened it slightly when the electrodes closed. At the same time, the 0.030-in. depth was shallow enough to keep the electrode from touching the copper ring. The lower electrode was flat, so that it made line contact with the ring and concentrated the current flow and heating at the point of brazing. The filler-metal preforms ($^1/_4$ by $^1/_8$ in. wafers of 0.005-in.-thick BCuP-5) were placed manually at the bottom of the copper ring under the stranded wire. The ring (after degreasing) was placed on the bottom electrode with the filler metal directly above the contact point. A fiber support behind the electrode held the ring upright in brazing position. The end of the copper conductor wire was stripped of insulation and laid on the preform. No flux was used. The wire was hand held, and the brazing cycle was initiated with a foot switch. Additional brazing conditions are given in the table with Fig. 15.

Brazing of Leads to Commutator Bars. The success with resistance brazing for joining small copper electrical conductors to massive copper assemblies has led to the use of the process for attaching armature leads to commutator bars on large electrical motors and generators. Joints made by this method provide a large conducting cross section that prevents significant resistive heating at the connections in service. The joints are made at much lower temperature than would be possible by resistance welding, which could be done only with excessively high current and would produce weaker joints with too small a conducting area at the joint.

The use of conventional resistance welding equipment and high-resistivity electrodes with specially contoured tips makes it possible to concentrate the heating at the joint, to keep the heating time to a minimum, and to obtain efficient handling and comparatively high production rates. In the example that follows, joint quality, localization and control of heat input, and initial cost of equipment were major factors in the selection of resistance

Fig. 15 Multiple-strand electrical conductor wire and copper ring joined by resistance brazing

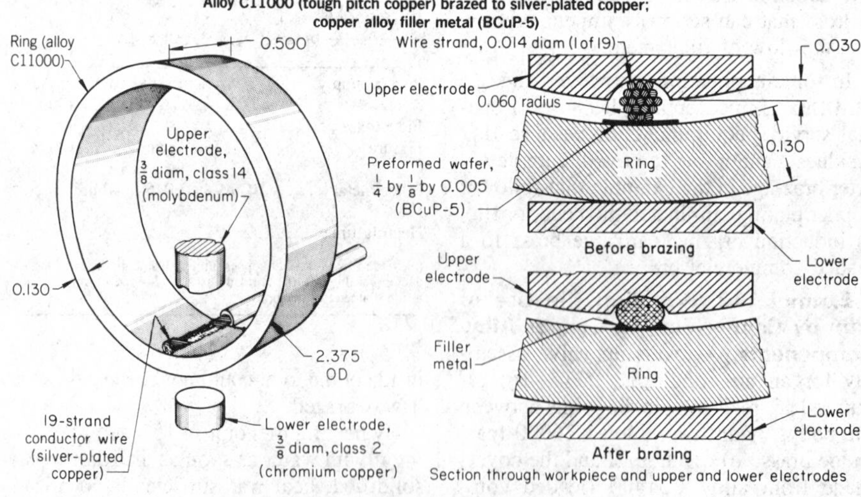

Machine	Press, air operated, 10 kV·A resistance welding machine	Filler metal	BCuP-5 preformed wafer(c)
Upper electrode	$^3/_8$-in.-diam RWMA class 14 (molybdenum)(a)	Brazing current	80 A
		Voltage	3.7 V
Lower electrode	$^3/_8$-in.-diam RWMA class 2 (chromium copper)(b)	Squeeze time	5 cycles
		Heating time	17 cycles
		Hold time	30 cycles
Electrode force	150 lb	Production rate	180 assemblies/h

(a) Contoured to a half circle, water cooled. (b) Flat end, water cooled. (c) Flux not used

brazing for attaching dual armature leads to commutator bars.

Example 15. Resistance Brazing for Joining Armature Leads to Commutators. When carbon arc brazing was used for joining alloy C10200 armature leads to alloy C11000 commutator bars (Fig. 16), the joints were porous and high in electrical resistance. This caused excessive temperature rise in the motors in which they were used. Changing the method of making the connections to resistance brazing, using BCuP-5 filler metal and the conditions listed with Fig. 16, eliminated the porosity and reduced the electrical resistance across the joint. A temperature of 1300 to 1350 °F was required to ensure an acceptable joint. Torch brazing was ruled out because of excessive width of heating and inadequate heat control. Induction heating was rejected because the initial investment for equipment would have been too great.

Fig. 16 Resistance brazing of armature leads to commutator bars

Alloy C10200 (OF copper) brazed to alloy C11000 (tough pitch copper); copper alloy filler metal (BCuP-5)

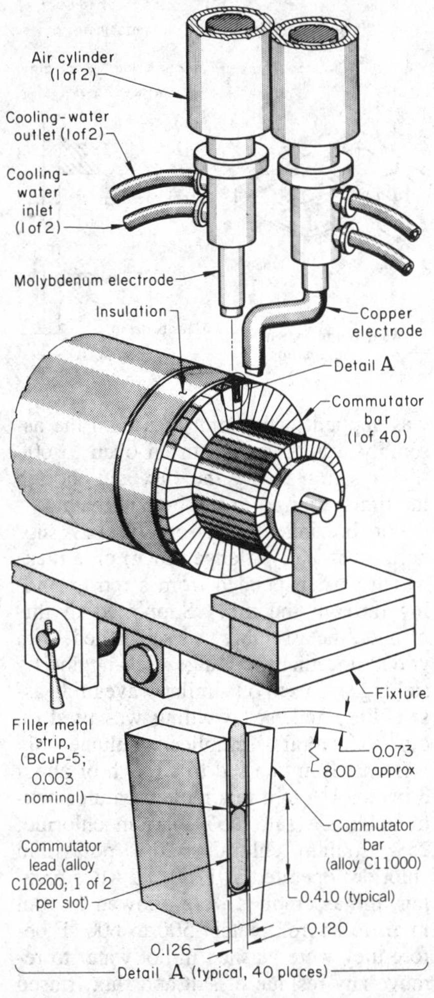

Machine	Press, air operated 50 kV·A resistance welding machine
Electrodes contacting commutator bar	RWMA class 1 (cadmium copper)
Electrodes contacting armature lead	RWMA class 14 (molybdenum)
Electrode force	80 lb(a)
Voltage	5.0 V(a)
Filler metal	BCuP-5, 0.003-in.-thick strip
Flux	None
Production per hour	Four commutators(b)

(a) Squeeze time, 10 s; five heating pulses, each 2 s long. (b) Forty-slot commutators

The setup used for resistance brazing is shown in Fig. 16. The width of the slots in the riser portions of the commutator bars was increased by 0.006 in. to accommodate a U-shaped strip of filler metal. Detail A shows the work arranged for brazing with the upper lead extending about 0.073 in. above the commutator bar to be contacted by the molybdenum electrode. Neither fluxing nor prebraze cleaning was required.

To initiate the brazing sequence, a valve was manually actuated, and the air cylinders positioned and applied pressure to the electrodes. The contactor switch was then activated by a push button to energize the primary of the transformer. High secondary current heated the commutator bar in the area between the electrode tips. A red color was visible, starting at the tips and progressing to a point between them. The filler metal melted, the leads in the slot partially melted, and the leads were compressed approximately flush with the bar surface. Use of a controlled amount of filler metal made final grinding unnecessary.

Observation by the operator permitted termination of the heating sequence as soon as the leads in the commutator slot reseated themselves. The transformer circuit was then opened by a push button and cooling began. The electrodes were then retracted by releasing the air valve. As a safety measure, a timer in the circuit limited brazing time to a maximum of 1 min. If required time exceeded this setting, the operator had to release and repush the operating button.

Brazing With Portable Machines. One common use of portable resistance welding machines for resistance brazing is attaching bus bar terminals or similar strip connectors to large electrical equipment that cannot be brought to or positioned for brazing in a conventional fixed-position resistance welding machine. Electrical connections to such equipment can often be made more economically by resistance brazing than by mechanical means and are made more readily by resistance brazing than by arc welding.

Resistance brazing done with portable resistance welding machines is a convenient and economical way of interconnecting large copper electrical bus bars or of attaching either large or small copper bus bars to motor-generators, transformers, and other electrical equipment. Lap joints made in this way have a bonding area that provides adequate strength and current-carrying capacity. Joints that have a large area of contact are brazed by making a series of spot brazes that overlap to provide full-joint or nearly full-joint bonding.

The low melting temperature of the filler metal helps to avoid overheating and excessive annealing of the work, and the usual selection of self-fluxing BCuP-5 alloy for the filler metal avoids corrosion problems and the need for flux removal. In the example that follows, production time was cut in half and waste of filler metal was eliminated by changing from torch brazing to a portable machine, multiple-spot

Fig. 17 Transformer terminal assembled by manual multiple-spot resistance brazing. Used carbon-graphite electrodes

Alloy C11000 (tough pitch copper); copper alloy filler metal (BCuP-5)

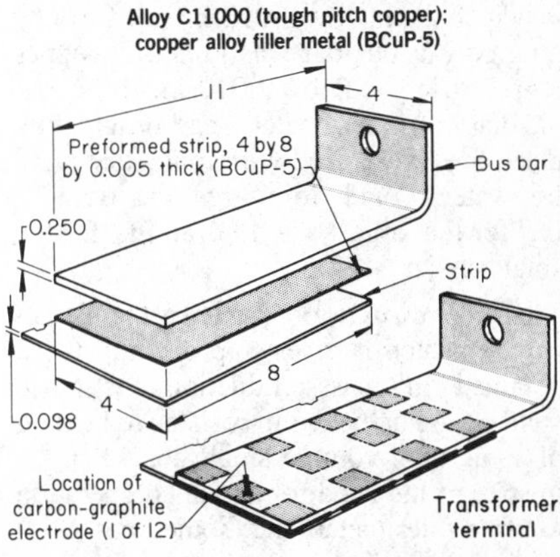

Machine	Portable 22-kV·A resistance welding machine with manual, water-cooled tongs
Fixtures	None
Electrodes	Carbon-graphite, 1 in.^{2}
Electrode force	Hand pressure by operator
Filler metal	BCuP-5 strip (see figure)
Flux	None
Brazing current	2000 A (max)
Heating time	Controlled manually by observation of filler-metal flow
Joint area	32 in.^{2}
Number of braze of braze spots	12(a)

(a) The filler metal was melted at each braze spot in an area larger than the electrode area, thus providing bonding on 80% or more of the total joint area.

resistance brazing for making a bus bar connection of large area.

Example 16. Resistance Brazing of a 32-in.² Lap Joint on Bus Bar. A bus bar and strip assembly (Fig. 17) served as a terminal for a transformer. Originally, the two alloy C11000 components were joined by torch brazing using a filler metal of the BAg series. Torch brazing required the operator to use both hands, one to hold the torch and the other to feed the filler metal. Consequently, the assemblies had to be clamped for brazing. In addition, some of the silver alloy filler metal was dropped on the floor and wasted.

With the improved method, components were joined by multiple-spot resistance brazing in a portable resistance welding machine, using a 4-by-8-in. preformed strip of 0.005-in.-thick BCuP-5 filler metal between the bus bar and strip (Fig. 17). This method reduced production time by 50% and eliminated waste of filler metal and the need for clamping. The material being joined was received in good condition, and no precleaning was necessary. Because no flux was required with BCuP-5, no cleaning after brazing was needed. The filler metal was melted at each braze spot in an area larger than that of the electrode, providing bonding on 80% or more of the total joint area.

In making small connections of this type in which $^1/_2$-in. by 0.012-in. by $2^1/_2$-in.-long copper bus bar was joined to copper strip $1^3/_4$ in. wide by 0.010 in. thick, $^3/_8$-in.-diam copper-coated carbon-graphite electrodes were used with tongs that were not water cooled. Joint area was from 32 in.² for the largest size to 0.88 in.² for the smallest size.

High-Production Resistance Brazing. Production rates approaching those obtained in mass-production resistance welding are achieved in resistance brazing of some high-volume small copper parts, in spite of the need for an added operation to place filler metal (and, sometimes, flux) in the joint. When resistance brazing copper parts in large quantities, the self-fluxing filler metal BCuP-5 is usually selected, eliminating flux application and removal.

In some applications, filler metal for resistance brazing can be provided in the form of a coating already present on one or both members to be joined, eliminating not only the use of flux but also the operation of placing filler metal at the joint. This is done when copper wire is joined by high-speed resistance brazing to copper terminals clad with BCuP-5 filler metal.

Dip Brazing

Salt-bath furnaces used for the brazing of steel assemblies with silver alloy filler metals can be used also for brazing of copper and copper alloys with silver alloy filler metals. The same neutral salts, operating temperatures, and brazing procedures are used for dip brazing of most copper alloys as are used for steel (see the article "Dip Brazing of Steels in Molten Salt" in this Volume). Applications of dip brazing of copper alloys in molten salt include: waveguides and waveguide hardware, flowmeter hardware, and capillary tube and bellows assemblies.

The support given immersed workpieces by the buoyancy of the molten salt and the rapid and even heating afforded by dip brazing make this process especially useful for joining assemblies requiring minimum distortion, such as waveguides. When a waveguide assembly has a large number of brazed joints, both external and internal, a salt bath is particularly efficient for simultaneous brazing of all joints in a single immersion in the molten salt. With dip brazing, a flange fitting and the end of a waveguide tube to be joined to it can be brazed by immersing only the joint portion in the molten salt bath, as in the example that follows.

Example 17. Use of Dip Brazing to Minimize Distortion. A waveguide assembly (Fig. 18), which consisted of a straight tube of alloy C22000 (commercial bronze, 90%) and a flange of cast bronze, was used in electronic equipment. The dimensions of the thin-walled tube ($1^1/_4$ by $2^3/_8$ in. by $^1/_{16}$-in. wall) had to be maintained to close tolerances, and dip brazing was chosen as the joining process to avoid unacceptable amounts of warpage. Torch brazing had been tried but resulted in excessive distortion.

After burrs were removed from the joint area, tube and flange were degreased, bright dipped, rinsed in water, and dried. The parts were assembled so that the tube extended through the flange. Two accurately located sets of holes were drilled through the longer sides of the rectangular tube, and a pin was inserted through each set of holes to support the flange in the vertical position during brazing.

Filler metal in the form of a rectangular preform of $^1/_{32}$-in.-diam BAg-1 wire was placed around the tube and adjacent to the flange (opposite the face). Type 3A flux was applied to the joint area, and the assembly was preheated in an oven to 600 °F for 5 min to dry the flux and shorten the time required in the brazing bath.

Fig. 18 Waveguide assembly brazed by partial immersion in a molten salt bath

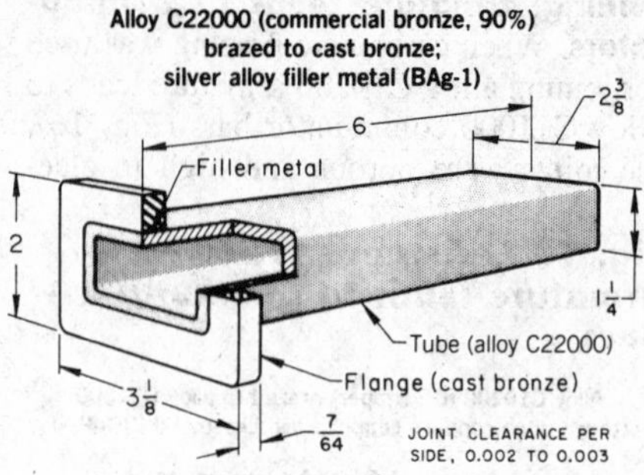

Wave guide after brazing and machining

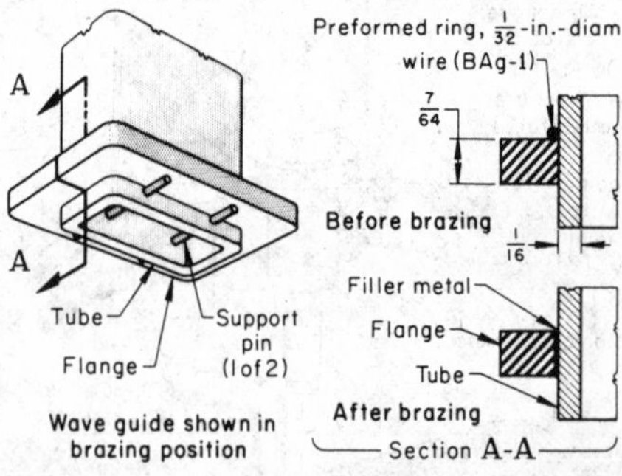

Wave guide shown in brazing position

For brazing, the assembly was supported vertically (flange down) on a rack, which was suspended from a rod extending through the tube. Supported in this manner, the assembly was self-jigged, with gravity locating the flange. Self-jigging by staking was used on similar waveguide assemblies, and tack welding was used on others. Several assemblies simultaneously were partly immersed to a depth of about 3 in. for $1^1/_4$ min in a molten bath of neutral chloride salt (55% barium chloride, 25% sodium chloride, 20% potassium chloride) operated at 1350 °F. After brazing, the assemblies were allowed to cool in air to approximately 500 to 600 °F before they were washed in hot water to remove any residue of salt and flux, rinsed in hot water, bright dipped to remove slight oxidation that formed during cooling, rinsed in cold water, and dried. Uniform fillets were formed and satisfactory joint penetration was obtained. The flange face was subsequently machined, which removed the portion of the tube that extended beyond the flange. The final dimensions of the assembly were within tolerance.

Brazing of Reactive Metals and Refractory Metals

By Mel Schwartz
Manager—Manufacturing Technology
Sikorsky Aircraft
Division of United Technologies

REACTIVE AND REFRACTORY METALS, because of their unique physical and mechanical properties, are finding increased use in the nuclear, aerospace, electronic, and chemical industries. The reactive metals, including titanium, zirconium, and beryllium, combine readily with oxygen at elevated temperatures to form very stable oxides. Reactive metals can also become embrittled by the interstitial absorption of oxygen, hydrogen, and nitrogen. Because these metals have a high affinity for oxygen and other gases, they must be brazed in a vacuum or dry inert-gas atmosphere.

The refractory metals include those metals that have melting points in excess of 4000 °F. The characteristics that affect the brazeability of refractory metals include ductile-to-brittle transition behavior, recrystallization temperature, and reactivity with oxygen, nitrogen, hydrogen, and carbon. The most important refractory metals, from a structural standpoint, are niobium, molybdenum, tantalum, and tungsten.

This article addresses the factors that affect the applicability and brazeability of reactive and refractory metals. Additional information on the physical and mechanical properties of these metals can be found in Volume 2 of the 9th edition of *Metals Handbook*.

Reactive Metals: Titanium, Zirconium, and Beryllium

Titanium, zirconium, and beryllium and their alloys are engineering materials used in a variety of industries. Commercial titanium alloys are frequently used in aircraft and numerous corrosion-resistant applications. Zirconium and beryllium are used primarily in nuclear applications, although beryllium possesses significant potential as a high-efficiency structural material for aerospace applications and has been used in limited applications on such vehicles as the Space Shuttle Orbiter.

Base Metals

These elements are considered highly reactive. They combine readily with oxygen at elevated temperature and react to form brittle intermetallic compounds with many other metals. These metals have a surface oxide film that requires careful cleaning before brazing; surfaces should be maintained in a clean condition until brazed. Titanium and zirconium can become embrittled by the interstitial absorption of hydrogen, nitrogen, and oxygen gases, whereas beryllium is only superficially affected.

Brazing should be done in a vacuum at a pressure of 10^{-5} to 10^{-4} torr or in dry inert-gas atmosphere. Titanium and zirconium are embrittled by the absorption of minute quantities of hydrogen and nitrogen. Nitrogen reacts with beryllium, titanium, and zirconium to form a surface film that presents problems in wetting. Filler-metal selection in brazing reactive metals is also critical, because they react with many of the constituents of brazing filler metals to form undesirable intermetallic compounds.

Titanium

The three types of titanium alloys commercially available are alpha, beta, and alpha-beta. Impurity limits and nominal compositions of titanium grades and alloys are given in Table 1. Pure titanium and a few of its alloys are classified as alpha. Material properties of alpha alloys are not affected by brazing and are not heat treatable. Several beta base metals are available commercially. In the annealed condition, these metals are unaffected by brazing; however, if heat treated, the brazing temperature may have an effect on beta alloy properties. Ductility in base metal is obtained by brazing at the solution treatment temperature, while the ductility of the base metal decreases as braze temperature increases.

The mechanical properties of alpha-beta titanium alloys may be altered by heat treatment and variations in microstructure. Wrought alpha-beta titanium alloys generally are fabricated to obtain a fine-grain equiaxed duplex microstructure to produce maximum ductility. It is desirable to maintain this microstructure by requiring that the brazing temperature not exceed the beta-phase transformation temperature (beta transus), which varies from 1650 to 1900 °F. Alpha-beta alloys are used in annealed, solution-treated, and aged conditions.

Selection of filler metals and brazing cycles that are compatible with the heat treatment required for alpha-beta and beta-titanium base metals may present some difficulty. Ideally, brazing should be conducted 100 to 150 °F below the beta transus; the ductility of alpha-beta base metals may be impaired if this temperature is exceeded. The beta transus can be exceeded when beta-titanium base metals are brazed; however, if the brazing temperature is too high, base-metal ductility after heat treatment may be impaired and considerable interaction between the filler metal and base metal may occur. The tensile properties of heat treatable titanium alloys also may be adversely affected by brazing, unless the

Table 1 Summary of commercial and semicommercial grades and alloys of titanium

Designation	Impurity limits, wt% N (max)	C (max)	H (max)	Fe (max)	O (max)	Nominal composition, wt% Al	Sn	Zr	Mo	Others
Unalloyed grades										
ASTM grade 1	0.03	0.10	0.015	0.20	0.18	...	...	...	...	...
ASTM grade 2	0.03	0.10	0.015	0.30	0.25	...	...	...	...	...
ASTM grade 3	0.05	0.10	0.015	0.30	0.35	...	...	...	...	...
ASTM grade 4	0.05	0.10	0.015	0.50	0.40	...	...	...	...	...
ASTM grade 7	0.03	0.10	0.015	0.30	0.25	...	...	...	...	0.2 Pd
Alpha and near-alpha alloys										
Ti code 12	0.03	0.10	0.015	0.30	0.25	...	...	...	0.3	0.8 Ni
Ti-5Al-2.5Sn	0.05	0.08	0.02	0.50	0.20	5	2.5	...	...	...
Ti-5Al-2.5Sn-ELI	0.07	0.08	0.0125	0.25	0.12	5	2.5	...	...	...
Ti-8Al-1Mo-1V	0.05	0.08	0.015	0.30	0.12	8	...	...	1	1 V
Ti-6Al-2Sn-4Zr-2Mo	0.05	0.05	0.0125	0.25	0.15	6	2	4	2	...
Ti-6Al-2Nb-1Ta-0.8Mo	0.02	0.03	0.0125	0.12	0.10	6	...	...	0.8	2 Nb, 1 Ta
Ti-2.25Al-11Sn-5Zr-1Mo	0.04	0.04	0.008	0.12	0.17	2.25	11.0	5.0	1.0	0.2 Si
Ti-5Al-5Sn-2Zr-2Mo(a)	0.03	0.05	0.0125	0.15	0.13	5	5	2	2	0.25 Si
Alpha-beta alloys										
Ti-6Al-4V(b)	0.05	0.10	0.0125	0.30	0.20	6.0	...	...	...	4.0 V
Ti-6Al-4V-ELI(b)	0.05	0.08	0.0125	0.25	0.13	6.0	...	...	...	4.0 V
Ti-6Al-6V-2Sn(b)	0.04	0.05	0.015	1.0	0.20	6.0	2.0	...	...	0.75 Cu, 6.0 V
Ti-8Mn(b)	0.05	0.08	0.015	0.50	0.20	...	...	...	...	8.0 Mn
Ti-7Al-4Mo(b)	0.05	0.10	0.013	0.30	0.20	7.0	...	...	4.0	...
Ti-6Al-2Sn-4Zr-6Mo(c)	0.04	0.04	0.0125	0.15	0.15	6.0	2.0	4.0	6.0	...
Ti-5Al-2Sn-2Zr-4Mo-4Cr(a)(c)	0.04	0.05	0.0125	0.30	0.13	5.0	2.0	2.0	4.0	4.0 Cr
Ti-6Al-2Sn-2Zr-2Mo-2Cr(a)(b)	0.03	0.05	0.0125	0.25	0.14	5.7	2.0	2.0	2.0	2.0 Cr, 0.25 Si
Ti-10V-2Fe-3Al(a)(c)	0.05	0.05	0.015	2.5	0.16	3.0	...	...	...	10.0 V
Ti-3Al-2.5V(d)	0.015	0.05	0.015	0.30	0.12	3.0	...	...	...	2.5 V
Beta alloys										
Ti-13V-11Cr-3Al(c)	0.05	0.05	0.025	0.35	0.17	3.0	...	...	...	11.0 Cr, 13.0 V
Ti-8Mo-8V-2Fe-3Al(a)(c)	0.05	0.05	0.015	2.5	0.17	3.0	...	...	8.0	8.0 V
Ti-3Al-8V-6Cr-4Mo-4Zr(a)(b)	0.03	0.05	0.020	0.25	0.12	3.0	...	4.0	4.0	6.0 Cr, 8.0 V
Ti-11.5Mo-6Zr-4.5Sn(b)	0.05	0.10	0.020	0.35	0.18	...	4.5	6.0	11.5	...

(a) Semicommercial alloy; mechanical properties and composition limits subject to negotiation with suppliers. (b) Mechanical properties given for annealed condition; may be solution treated and aged to increase strength. (c) Mechanical properties given for solution treated and aged condition; alloy not normally applied in annealed condition. Properties may be sensitive to section size and processing. (d) Primarily a tubing alloy; may be cold drawn to increase strength

assembly can be heat treated after brazing. For example, alpha-beta titanium alloys must be solution treated, quenched, and aged to develop optimum properties. It is difficult to select a filler metal that is suitable for brazing and solution treating in a single operation. Similarly, it is not always possible to quench a brazed assembly at the desired cooling rate, and certain configurations, such as honeycomb sandwich structures, cannot be quenched rapidly without distortion. Brazing at the aging temperature is impractical, because few filler metals melt and flow at these temperatures.

The possibility of galvanic corrosion must be considered when filler metals are selected for brazing titanium-based metals. While titanium is an active metal, its activity tends to decrease in an oxidizing environment, because its surface undergoes anodic polarization similar to that of aluminum. Thus, in many environments, titanium becomes more chemically inactive than most structural alloys. The corrosion resistance of titanium is generally not affected by contact with structural steels, but other metals, such as copper, corrode rapidly in contact with titanium under oxidizing conditions. Thus, filler metals must be chosen carefully to avoid preferential corrosion of the brazed joint.

When titanium is brazed, precautions must be taken to ensure that the brazing retort or chamber is free of contaminants from previous brazing operations. Mechanical properties of titanium may deteriorate because of gaseous contamination from the brazing furnace. Also, the choice of materials to be used in fixtures must be carefully considered. Nickel or materials containing high amounts of nickel generally should be avoided; nickel and titanium form a low-melting eutectic at about 1728 °F (28.4 wt% Ni). If titanium workpieces come in contact with fixtures or a retort made of a nickel-based alloy, the parts may fuse together if the brazing temperature is in excess of 1728 °F. If a fixture material, such as stainless steel, which may contain a high nickel content is used, it should be oxide coated. In most applications, coated graphite or carbon steel fixture materials are used.

Filler Metals. Braze filler metals initially used for brazing titanium and its alloys were silver with additions of lithium, copper, aluminum, or tin. Most of these braze filler metals were used in low-temperature applications (1000 to 1100 °F). Commercial braze filler metals, including silver-palladium, titanium-nickel, titanium-nickel-copper, and titanium-zirconium-beryllium, are now available that can be used in the 1600 and 1700 °F range. Higher strengths and improved resistance to crevice-type corrosion are desirable characteristics that current braze filler metals enjoy. For joining applications requiring a high degree of corrosion resistance, the 48Ti-48Zr-4Be and 43Ti-43Zr-12Ni-2Be braze filler metals were developed. A silver-palladium-gallium braze filler metal (Ag-9Pd-9Ga), which flows at 1650 to 1675 °F, is another excellent filler metal with which to fill large gaps.

During the recently concluded supersonic transport materials and process technology program (Ref 1, 2, 3, and 4), methods to braze titanium honeycomb sandwich assemblies with aluminum braze filler metal were developed. Aircraft structures up to 23 ft in length were brazed successfully using 3003 brazing foils. Use of aluminum brazing filler metal (3003) provided satisfactory strength up to about 600 °F. If temperatures of 1000 to 1100 °F are required, high-strength, corrosion-resistant titanium-zirconium-beryllium or titanium-zirconium-nickel-beryllium braze filler metals should be used. A typical titanium honeycomb assembly used for aircraft structures is shown in Fig. 1.

Zirconium

Zircaloy, the most commonly used zirconium alloy, contains small percentages of tin, iron, chromium, and nickel. These structural alloys are used for corrosion resistance in nuclear applications, especially pressurized water nuclear power reactors. Zirconium is useful in constructing various components in nuclear reactors because of its low-absorption cross section for thermal neutrons; its excellent corrosion resistance in water, liquid sodium, and liquid potassium; and its good mechanical properties.

Zirconium frequently is used as a cladding material for fuel elements, because brittle intermetallic compounds are not formed in the zirconium-uranium alloy system. Zircaloy-2 and Zircaloy-4 were developed specifically for nuclear applications. Table 2 lists chemical compositions of nuclear-grade zirconium alloys. Like titanium and beryllium, zirconium reacts readily with oxygen, hydrogen, and nitrogen and is thus embrittled. It also reacts with many metals and alloys to form intermetallic compounds. Oxidation occurs at temperatures as low as about 400 °F, and its rate increases at higher temperatures. The reaction of zirconium and nitrogen occurs slowly at about 750 °F. Above 1500 °F, reaction rate increases rapidly. Zirconium is embrittled by the absorption of hydrogen, which occurs rapidly at temperatures between 600 and 1800 °F. As a result of its reactivity with gaseous contaminants, zirconium must be brazed in a vacuum or a dry atmosphere of argon or helium. Zirconium joint members must be carefully cleaned before brazing to remove oxides and other surface contaminants, and brazing should be done immediately after cleaning.

Filler Metals. Compared to other reactive metals, development of filler metals and brazing methods to join zirconium and zirconium alloys is limited. Much of the research done to date is classified by the government. Also, zirconium does not enjoy wide commercial usage. To a degree, the lack of research on brazing can be attributed to the availability of other joining methods. Most commercial filler metals do not wet or flow well on zirconium base metals, and they are not metallurgically compatible. Additionally, many such filler metals do not possess the required corrosion resistance in a reactor environment. Much of the development of braze filler metals for zirconium has been directed toward zirconium alloy tubing for pressurized water nuclear reactors.

Research (Ref 5) was conducted to develop filler metals for brazing Zircaloy-2 joints that possessed good resistance to corrosion in pressurized water at 680 °F. Single-lap joints with zero-clearance were induction brazed in a dry helium (or argon) atmosphere with 30 commercial and experimental filler metals. After brazing, portions of each joint were exposed in pressurized 680 °F water for up to 1450 h. Additional joints using spacers that produced a clearance of 0.003 in. between joint members also were brazed and corrosion tested in 680 °F water; the data from these studies and from metallographic examinations of the brazed joints indicated that the following filler metals most nearly met the service requirements:

- Zr-5Be
- Cu-20Pd-3In
- Ni-20Pd-10Si
- Ni-30Ge-13Cr
- Ni-6P

Joints brazed in high vacuum with 48Ti-48Zr-4Be filler metal have also exhibited excellent corrosion resistance in 680 °F water at pressures up to 3 ksi.

Additional studies to develop improved filler metals for brazing Zircaloy base metals for use in water-cooled reactors were completed recently (Ref 6). Candidate zirconium and non-zirconium base (nickel and copper base materials) filler metals were formulated, produced by arc melting, and screened by wetting tests and corrosion tests in pressurized, high-temperature water (600 °F). Zircaloy tubing joints made with suitable filler metals were evaluated in test loops, through which the following 600 °F solutions flowed under pressure: (1) lithium hydroxide pH-controlled borated water, or (2) ammonium hydroxide pH-controlled water. Following long-term corrosion tests (2 to 4 months), tensile properties of the joints were determined at room temperature and at 600 °F. The following filler metals had acceptable corrosion resistance and mechanical strength:

- Zr-50Ag at a brazing temperature of 2768 °F
- Zr-29Mn at a brazing temperature of 2516 °F

Fig. 1 Brazed titanium honeycomb sandwich aerospace assembly

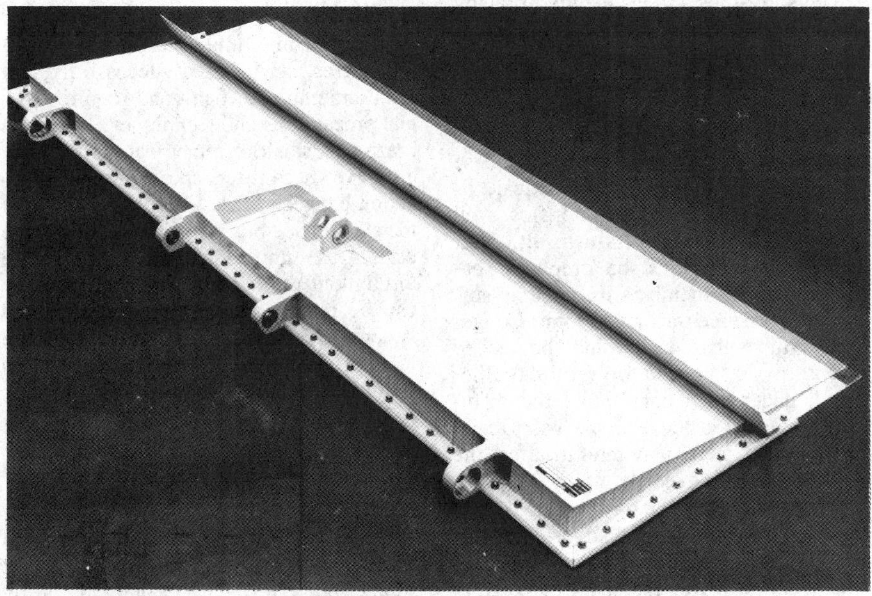

Table 2 Nominal composition of nuclear-grade zirconium-based alloys

Alloy	Composition, %									
	Sn	Fe	Cr	Ni	Nb	O	Fe + Cr + Ni	Fe + Cr	Hf	Zr
UNS R60862	1.02-1.70	0.07-0.20	0.05-0.15	0.03-0.08	...	...	0.18-0.38	...	0.010	rem
UNS R60804	1.20-1.70	0.18-0.24	0.07-0.13	...	...	...	...	0.28-0.37	0.010	rem
UNS R60901	...	...	...	...	2.40-2.80	0.09-0.13	...	...	0.010	rem

- Zr-24Sn at a brazing temperature of 3146 °F

The test joints evaluated during this investigation were vacuum brazed; the pressure within the furnace was 10^{-6} to 10^{-5} torr.

The Zr-5Be filler metal has been used extensively to braze zirconium base metals to themselves and to other metals such as stainless steel. For example, zirconium sheet stock was brazed with Zr-5Be filler metal using the following cycle: 10 min at 1840 °F, followed by 4 to 6 h at 1470 °F. Because of its ability to wet ceramic surfaces, Zr-5Be has been used to braze zirconium to uranium oxide and beryllium oxide.

Beryllium

Much of the early research on beryllium was for nuclear applications; therefore, high purity was vital. Currently, beryllium is available as a nuclear grade and as hot pressed block, large wrought plate, and sheet stock. Due to its high elastic modulus to density ratio, beryllium is about four to six times as efficient as conventional metal alloy structures (aluminum and magnesium alloys, for example). Beryllium consequently finds increasing use in space hardware and certain low-inertia instruments.

Beryllium is assuming an increasingly important role as a structural metal as its advantages become more widely known and fabrication technology becomes available. Because of its high permeability to x-rays, beryllium has been long used for windows in x-ray tubes and similar devices. It is a natural choice for nuclear reactor moderators and reflectors, because of its low thermal neutron-absorption cross section and its large scattering cross section. Recently, beryllium has been used for aerospace structures because of its low density, high stiffness-to-weight ratio, and excellent dimensional stability. However, beryllium has disadvantages as well. Its limited ductility, which is partly due to its hexagonal close-packed crystal structure, presents problems in the design and fabrication of beryllium structures. The toxicity of beryllium and its compounds are well known, and extreme precautions should be observed in fabricating this metal. (See the section on safety in this article.)

Like titanium, beryllium reacts with oxygen at conventional brazing temperatures. Beryllium also reacts with atmospheric nitrogen. Because the presence of an oxidized or nitrided surface impairs the wetting and flow properties of brazing filler metals, high-temperature brazing usually is done in an argon atmosphere or in a vacuum after thorough cleaning of base-metal surfaces. These surface contaminants cannot be removed in a controlled atmosphere. Fluxes may be used to prevent oxidation during low-temperature brazing of beryllium in air. Also, beryllium surfaces may be plated with silver to improve wetting and flow of filler metals.

Joint design, filler-metal selection, and choice of brazing cycle are complicated by the low ductility of beryllium, as well as its reactivity with many of the filler metals used for brazing. The low ductility of beryllium is aggravated by the presence of structural discontinuities, such as surface scratches, notches, and asymmetrical stress patterns produced by single-lapped joints. Beryllium parts must be handled carefully, and joint configurations such as butt, scarf, step, and double-lapped joints should be considered when beryllium structures are designed.

Because beryllium reacts with the constituents of most filler metals, brazing should be done under conditions that minimize the formation of intermetallic compounds—rapid heating and cooling cycles, low brazing temperatures, minimum time at brazing temperature, and minimum amounts of filler metal. However, these procedures are not always practical, as many filler metals do not wet or flow well on beryllium surfaces. As a result, filler metal frequently must be preplaced between the joint members in amounts sufficient to produce the brazed joint. Longer brazing times than desired may be needed to ensure wetting and flow of the filler metal; under such conditions, the diffusion that occurs between the base metal and the filler metal may tend to affect the joint properties adversely.

Filler metals used for brazing beryllium include pure silver, silver-based alloys, pure aluminum, and aluminum-silicon alloys. Silver and silver-based filler metals, such as BAg-19 (Ag-7Cu-0.2Li), are used for high-temperature applications. Lithium is added to improve wettability. The aluminum-silicon filler metals, such as BAlSi-2 or BAlSi-4, that contain 7 to 12% Si can provide high-strength brazed joints for service in temperatures up to 300 °F. Both the silver-based and aluminum-silicon fillers, however, exhibit poor capillary flow, and preplacement is recommended.

Safety. Beryllium has limited ductility at room temperature, which requires that components be handled with care to avoid damage. Beryllium and its compounds are toxic in dust, vapor, or liquid form. Proper handling and safety precautions are mandatory, and governmental regulations have been issued to protect personnel handling these materials. Safe exposure limits for beryllium and its compounds have been determined, and the metal may be safely handled by observing those limits. The use of ventilation and filtering systems to control airborne contaminants is recommended to meet these requirements. Brazing or any other fabrication operation should not be attempted without first instituting proper safety standards.

Process and Equipment

Brazing processes should be used that do not allow joint surfaces to come in contact with air during heating. Induction and furnace brazing in inert-gas or vacuum atmospheres can be used successfully, but torch brazing is difficult and requires special precautions and techniques. Induction brazing of small, symmetrical parts is very effective, because its speed minimizes reaction between braze filler metal and base metal. Furnace brazing is favored for large parts, as uniformity of temperature throughout the heating and cooling cycle can be readily controlled. Titanium and zirconium assemblies frequently are brazed in high-vacuum, cold-wall furnaces; beryllium assemblies braze better in argon or helium gas.

Precleaning and Surface Preparation

Titanium. If heavy scale is formed on titanium when heated above 1000 °F, it should be removed mechanically or by molten salt-bath dipping, as described in Volume 5 of the 9th edition of *Metals Handbook*. Light oxide film is removed by pickling. Pickling solutions for titanium should contain 20 to 45% nitric acid, plus 2% hydrofluoric acid. When degreasing is required, nonchlorinated solvents such as acetone, methylethyl ketone, or ethyl alcohol should be used. Methyl alcohol has been known to cause stress corrosion cracking and should therefore be avoided.

Zirconium parts are cleaned in solutions similar to titanium.

Beryllium components should be degreased and pickled in 10% hydrofluoric and nitric-hydrofluoric acid mixtures prior to brazing. Ultrasonic rinsing with deionized wáter is also required. Precoating the joint faying surfaces by special vacuum-metallizing techniques or electroplating with silver, titanium, copper, aluminum, or BAlSi-4 prior to brazing is effective for subsequent wetting by various braze filler metals.

Utilization of vacuum deposition of titanium vapor is effective in promoting wetting on beryllium. Titanium vapor deposition treatments drastically change the surface of the beryllium and promote extensive spreading and wetting when pure aluminum filler metal is used. Ultimate strength obtained at room temperature for joints butt brazed with commercially pure aluminum compares favorably with that for beryllium base metal.

Brazing of beryllium generally is carried out on lap joints, because they are easiest to make. Joints made in beryllium may be undesirable because of the inherent notch present in this type of joint. Therefore, joints that contain no notches are required to achieve maximum usefulness of brazed joints for beryllium. Step, scarf, or butt joints also have been used when possible.

Fluxes and Atmospheres

Argon, helium, and vacuum atmospheres are satisfactory for brazing titanium, zirconium, and beryllium. For torch brazing, special fluxes must be used on titanium and zirconium. Beryllium should not be torch brazed. Fluxes for titanium are primarily mixtures of fluorides and chlorides of the alkali metals, sodium, potassium, and lithium.

Vacuum and inert-gas atmospheres protect titanium during furnace and induction brazing operations. A vacuum of 10^{-3} torr or more is required to braze titanium, whereas a dew point of −70 °F or less is necessary to prevent discoloration of the titanium in a brazing temperature range of 1400 to 1700 °F. Generally, brazing techniques used for titanium are applicable to zirconium.

Beryllium usually is brazed in purified argon, although helium and vacuum of less than 10^{-3} torr are also suitable. For use in furnace or induction brazing, 60% LiF-40% LiCl and $SnCl_2$ flux have produced good results.

Applications

Titanium

Titanium and its alloys are used as brazed assemblies for aerospace hardware and jet engine inlet vanes, hydraulic tubing or fittings, and honeycomb sandwich panels. For example, anti-icing airfoils have been made with wrap-around skins and corrugated core members, which are brazed in vacuum with 70Ti-15Cu-15Ni braze filler metal.

Titanium plumbing systems providing additional weight savings are an outgrowth from aircraft and aerospace brazed stainless steel high-pressure fluid line joints and fittings. Application of hinged induction heater tools and protective argon atmosphere permit rapid, reliable brazes to be made in open lines. Close-out joints or dead-end lines present difficulties in ensuring adequate backside shielding and should be avoided.

In one example, a wide-body jetliner utilized over 250 brazed joints involving titanium tubing and Ti-6Al-4V fittings. The braze filler metal was 90Ag-10Pd. The supersonic transport design relied heavily on high-efficiency honeycomb structures. Metallurgically bonded titanium sandwich panels were considered for fuselage, wing, and control surfaces. The honeycomb core was 0.002-in.-thick titanium foil; facings were Ti-6Al-4V, 0.010 in. thick, and aluminum alloy 3003 braze filler metal was successfully used with ribbon heated ceramic tooling and argon atmosphere at rarefied pressure to produce aluminum brazed titanium alloy honeycomb sandwich panels. Engine cowls were made successfully in production for other aircraft systems.

Successful joining of dissimilar metal combinations has been achieved with titanium and stainless steel and copper. A brazed transition section located between a titanium tank and stainless steel feed lines was evaluated during the course of the space program. Titanium alloy Ti-6Al-4V was vacuum induction brazed to type 304L stainless steel with Au-18Ni braze filler metal. The presence of a brittle intermetallic compound and an indication of cracking led to extended joint evaluations. It was concluded that the brazed joint could sustain loads in excess of the yield strength of the stainless steel. The successful performance of this joint has been attributed to the rigid control of all the brazing process variables. The formation of brittle intermetallic compounds was minimized because the joint was brazed rapidly and holding time at temperature was kept to a minimum. The same procedures have been used to make joints between titanium and low-carbon steel, chromium-molybdenum-vanadium (such as Vasco-jet-1000), and other metals.

Alloy development studies have also been conducted to produce a braze filler metal for joining titanium to stainless steel tubing for other uses in the space program. The braze filler metal Pd-9.0Ag-4.2Si was developed and successfully brazed steel and titanium in vacuum at 1360 °F. Excellent flow properties were exhibited by this filler metal. Recently, titanium and its alloys have been successfully brazed to carbon and austenitic stainless steels with the 48Ti-48Zr-4Be filler metal.

Titanium alloy Ti-3Al-1.5Mn has been brazed to Cu-0.8Cr alloy using silver-based braze filler metals. Three braze filler metals were evaluated during this program: Ag-28Cu, Ag-40Cu-35Zn, and Ag-27Cu-5Sn. The joints were brazed in a vacuum at a temperature of 1520 °F for 5 min. Shear strengths of 28.4 to 38.3 ksi were obtained at room temperatures. The heating rate and brazing temperature were critical. For maximum joint strength, the brazing temperature was maintained between 1518 and 1526 °F.

A process for vacuum brazing copper-plated titanium has also been developed. Copper was electroplated on the surface of titanium alloy Ti-3Al-1.5Mn after the surface had been etched in a sulfuric acid solution. The etching of the titanium surface removed hydrogen or hydrogen compounds, thereby creating an effective surface for the copper plate. Joints were made between the titanium alloy and commercially pure copper, as well as between Ti-3Al-1.5Mn and stainless steel or nickel-based alloy using Ag-27Cu-5Sn braze filler metal. The joints were brazed at 1400 to 1500 °F for 15 to 20 min. Under ideal conditions, the average shear strength was 28 ksi at room temperature.

Zirconium

Applications for brazed zirconium and zirconium alloys usually relate to nuclear requirements. In one production application, 500 Zircaloy-2 19-tube clusters were brazed in vacuum. The fuel tube cap assemblies were 3 in. in diameter by 2 in. long.

A recent application of electron beam brazing was for a Zircaloy-2 in-pile tube-burst specimen. A stainless steel capillary tube was brazed into a molybdenum adapter with copper, and the adapter was brazed into the Zircaloy-2 specimen with 48Ti-48Zr-4Be. Both brazes were made simultaneously by impinging the defocused beam from a conventional low-voltage electron gun on the molybdenum adapter. In this application, visual control of the brazing cycle was used.

Beryllium

Brazing of beryllium has been used in fabrication of complex, high-performance spacecraft structures and instrument payloads. These aerospace applications include:

- An electronic camera pressure housing 4 in. in diameter by 8 in. high with Kovar sleeves has been brazed to beryllium using BAg-19; gas tungsten arc welding of the alumina/Kovar electrical lead-through completes the mass spectrometer tight assembly.
- Another structural brazement was the main support ring for VISSR satellites. The joints were furnace brazed with BAlSi-4 (88Al-12Si) braze filler metal.
- Brazing of beryllium to itself and other metals has included the following applications: honeycomb sandwiches, truss structures, built-up I-beams, laminated spars, lapped tubes, and radiation windows.
- Several satellites are powered by nuclear generators, using a critical beryllium housing brazement. Beryllium sheet fins were brazed to a beryllium case with BAg-19. A brazed stainless steel transition ring at each end of the case was also brazed to produce a vacuum-tight joint.

Several dissimilar metals in combination with beryllium have been successfully brazed. Stainless steel tubes have been successfully brazed to beryllium end caps with 49Ti-49Cu-2Be (Fig. 2). Tests have indicated that the braze filler metal readily wets beryllium and that the joints exhibit adequate strength for the applications, provided brazing time is sufficiently short to minimize the formation of intermetallic compounds. Induction brazing in vacuum provided a ready means for performing this operation.

The eutectic diffusion brazing of beryllium and B120VCA titanium alloy (Ti-3Al-13V-11Cr) has been achieved with pure silver in a vacuum at 1650 °F. These joints had shear strengths of about 20 ksi.

X-ray-tube beryllium window assemblies have been vacuum brazed into Monel retainers with silver braze filler metal. All assemblies are vacuum tight and suitable for use in x-ray tubes.

Type 303 stainless steel pressure fittings were joined to a 0.080-in.-thick sheet of QMV beryllium using pure silver as the braze filler metal. The joint was eutectic diffusion brazed at 1690 °F for 20 to 30 min in a vacuum; a pressure of 5 psi was applied to hold the joint members together. Other researchers have found that brazing temperatures and times at temperature must be kept to a minimum to obtain highest joint strengths when only silver braze filler metal is used. Also, other studies have reported that beryllium can be brazed to titanium or stainless steel at temperatures as low as 1650 °F.

QMV beryllium was brazed to type 304L stainless steel to demonstrate the feasibility of joining these metals for nuclear applications. The steel surface was coated with a thin film of the Ag-28Cu braze filler metal by vapor deposition techniques, and brazing was conducted at 1510 °F. Lap shear strengths ranged from 14.2 to 20.5 ksi at room temperature.

Fig. 2 Stainless steel/beryllium brazement using 49Ti-49Cu-2Be filler metal

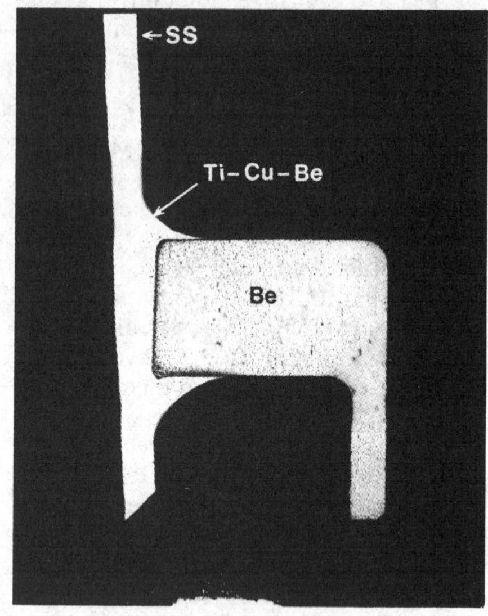

Refractory Metals: Niobium, Molybdenum, Tantalum, and Tungsten

Refractory metals have a history of use in applications where structural integrity at high temperatures is required. For example, filaments for incandescent lamps are made from tungsten, which is used extensively in vacuum tubes for cathode structures. Similarly, molybdenum is used for grids and supporting members in vacuum tubes and for lead wires in incandescent lamps. However, joining requirements for such applications are relatively unsophisticated, and tack welding, spot welding, and mechanical fastening methods are usually adequate. Such is not the case in current applications of the refractory metals in the aircraft, space, nuclear, and electronic industries. Joints of significant dimensions with properties that are adequate to meet exacting requirements of strength, ductility, and corrosion resistance are needed. Because of its limited effect on base-metal properties, brazing has been widely accepted as a method to join assemblies fabricated from the refractory metals.

The refractory metal group usually is defined to include those metals that melt at temperatures above 4000 °F—ruthenium, niobium, iridium, molybdenum, tantalum, osmium, rhenium, and tungsten. Chromium, vanadium, rhodium, and hafnium sometimes are included in this classification. Regardless of definition, niobium, molybdenum, tantalum, and tungsten are the most important of the refractory metals from a structural standpoint.

Niobium, molybdenum, tantalum, and tungsten have much in common; however, there are differences that have a bearing on the manner in which these metals are used and joined. All of the refractory metals have a body-centered cubic crystal structure, with very high melting temperatures, high to very high densities, low specific heats, and low coefficients of thermal expansion. Mechanical strength and structural integrity at high temperatures are excellent. In most acids, corrosion resistance ranges from good to excellent; the resistance of niobium, molybdenum, and tantalum is also good in liquid alkali metals. Niobium and molybdenum have very low-capture cross sections for thermal neutrons; however, cross sections for tantalum and tungsten are much higher.

The mechanical properties of the refractory metals are affected markedly by ductile-to-brittle transition behavior, recrystallization temperature, and reactions with carbon and various gases. These characteristics must be considered when procedures to braze refractory metals are established.

Ductile-to-Brittle Transition Behavior

Refractory metals, because of their body-centered cubic structure, have a well-defined transition from ductile-to-brittle behavior. With these metals, a large decrease in energy absorption occurs over a narrow temperature range, and an associated change in the type of fracture from tough to brittle is also evident. This temperature range is not a fixed property of the metal, as it varies with strain rate, alloying additions, impurities, heat treatment, and method of fabrication. The transition temperature ranges for the pure refractory metals are:

Niobium	−250 to −150 °F
Molybdenum	300 to 500 °F
Tantalum	<−320 °F
Tungsten	500 to 700 °F

Thus, molybdenum and tungsten are brittle at room temperature and must be handled carefully to avoid damage. Also, these metals must be brazed in a stress-free condition.

Recrystallization Temperature

The strength and ductility of the refractory metals are adversely affected by microstructural changes that occur when the recrystallization temperatures of these metals are exceeded. Recrystallization temperature range also varies with alloying additions, interstitial content, fabrication method (including degree of cold working), and time at temperature. Recrystallization temperature ranges for unalloyed refractory metals are:

Niobium	1790 to 2100 °F
Molybdenum	2100 to 2190 °F
Tantalum	2010 to 2500 °F
Tungsten	2190 to 3000 °F

Some applications permit brazing with braze filler metals that melt below the recrystallization temperature range. Other applications require the use of braze filler metals that melt above this temperature range. The joint also must be designed to accommodate the loss in mechanical properties associated with recrystallization.

Research has improved the high-temperature mechanical properties of the refractory metals and has led to an increase in recrystallization temperatures through alloying. For example, titanium, zirconium, and hafnium can be used to strengthen molybdenum and increase its recrystallization temperature. The recrystallization temperature of unalloyed molybdenum is about 2100 to 2200 °F. In contrast, the recrystallization temperature of Mo-0.5Ti-0.7Zr molybdenum alloy is about 2700 °F, and the stress-rupture strength of this alloy at elevated temperatures (1800 to 2000 °F) is several times that of unalloyed molybdenum.

Reactions With Gases and Carbon

The environment in which the refractory metals are brazed is determined by the reactivity of these metals with oxygen, hydrogen, and nitrogen and the effect of these elements on the mechanical properties of the refractory metals. All of the refractory metals react with oxygen at moderately elevated temperatures, but they form different types of oxides. Niobium and tantalum form hard adherent oxides at temperatures above 400 and 750 °F, respectively. On the other hand, molybdenum and tungsten form volatile oxides at temperatures above 750 and 950 °F, respectively. In either case, the surfaces of the refractory metals must be protected during brazing to ensure wetting by the braze filler metal. Also, these metals must be coated with an oxidation-resistant material—a nickel or silver plating, for example—if they are exposed in air at elevated temperatures. For such service conditions, the brazing filler metal must be compatible with both the base metal and the coating.

Niobium and tantalum are embrittled by the presence of hydrogen at relatively low temperatures. In contrast, molybdenum and tungsten can be brazed in a hydrogen atmosphere. Niobium, molybdenum, and tantalum are embrittled by nitrogen at high temperatures; however, reactions with nitrogen begin at relatively low temperatures. For example, nitrogen is dissolved in molybdenum at temperatures as low as 1100 °F, but severe embrittlement does not occur until the temperature exceeds about 2000 °F. Tantalum behaves in much the same manner. All of the refractory metals form carbides in the presence of minute quantities of carbon and its compounds.

Brazing Process

If maximum joint strength is required, refractory metals should be brazed at temperatures below those at which recrystallization occurs. However, brazing at much higher temperatures may be necessitated by service requirements, and some decrease in joint properties must be anticipated. All of the refractory metals can be brazed in a vacuum or in an argon or helium atmosphere with a very low dew point; molybdenum and tungsten also can be brazed in a dry hydrogen atmosphere. Graphite fixturing should not be used to position refractory metal parts during brazing, because these metals form carbides readily. However, graphite tooling may be acceptable if coated with a refractory ceramic material. Ceramics also can be used for fixturing, but care in their selection must be exercised. Some ceramics cannot be used in a vacuum because of outgassing; others react with refractory metals at high temperatures. Refractory metals with higher melting temperatures than the metal being brazed can be used for fixturing also.

Braze filler metals for brazing refractory metals are selected on the basis of service conditions, specific application, and compatibility with the base metal and coating, if one is used. When brazing is done in a vacuum, the braze filler-metal vapor pressure must be also considered.

Refractory metals can be brazed with many of the silver and copper-zinc braze filler metals. Joints made with these braze filler metals usually are designed for low-temperature service. Refractory metals can be brazed for high-temperature service with braze filler metals based on nickel chromium, the other noble metals (gold, palladium, and platinum), and the reactive metals. High-melting refractory metals can also be brazed with refractory metals that melt at lower temperatures. Tantalum alloys, for example, can be brazed with niobium and molybdenum. Table 3 lists applicable filler metals for brazing refractory metals.

Research on brazing refractory metals has concentrated on the development and evaluation of braze filler metals for high-temperature applications. The results of various programs provide data on the braze filler metals and procedures that have been used to braze each of the refractory metals. Many of the braze filler metals may not be available commercially because of limited applicability.

Niobium and Niobium Alloys

Braze Filler Metals and Properties. Niobium is used mainly for nuclear and aerospace applications. As a result, research has been directed toward the development of braze filler metals with char-

Table 3 Brazing filler metals for refractory metals

Filler metal(a)	Liquidus temperature(b), °F	Filler metal(a)	Liquidus temperature(b), °F
Nb	4380	Co-Cr-W-Ni	2600
Ta	5425	Mo-Ru	3450
Ag	1760	Mo-B	3450
Cu	1980	Cu-Mn	1600
Ni	2650	Nb-Ni	2175
Ti	3300	Pd-Ag-Mo	2400
Pd-Mo	2860	Pd-Al	2150
Pt-Mo	3225	Pd-Ni	2200
Pt-30W	4170	Pd-Cu	2200
Pt-50Rh	3720	Pd-Ag	2400
Ag-Cu-Zn-Cd-Mo	1145-1295	Pd-Fe	2400
Ag-Cu-Zn-Mo	1325-1450	Au-Cu	1625
Ag-Cu-Mo	1435	Au-Ni	1740
Ag-Mn	1780	Au-Ni-Cr	1900
Ni-Cr-B	1950	Ta-Ti-Zr	3800
Ni-Cr-Fe-Si-C	1950	Ti-V-Cr-Al	3000
Ni-Cr-Mo-Mn-Si	2100	Ti-Cr	2700
Ni-Ti	2350	Ti-Si	2600
Ni-Cr-Mo-Fe-W	2380	Ti-Zr-Be(c)	1830
Ni-Cu	2460	Zr-Nb-Be(c)	1920
Ni-Cr-Fe	2600	Ti-V-Be(c)	2280
Ni-Cr-Si	2050	Ta-V-Nb(c)	3300-3500
Mn-Ni-Co	1870	Ta-V-Ti(c)	3200-3350
Co-Cr-Si-Ni	3450		

(a) Not all of the filler metals listed are commercially available. (b) Liquidus temperature may vary depending on percentage of alloying elements. (c) Depends on specific composition

acteristics that are compatible with the base metal and its intended use. For example, braze filler metals to produce joints that are resistant to liquid alkali metals, such as sodium, and that possess useful properties at 1300 to 1500 °F have been developed. Ti-48Zr-4Be and Zr-19Nb-6Be braze filler metals produced as-brazed shear strengths that ranged from 20 to 33 ksi at room temperature and from 12 to 20 ksi at 1300 °F. Some decrease in room-temperature shear strength occurred with specimens that were aged for 100 h at 1500 °F before testing. Braze filler metals in the Ta-V-Nb and Ta-V-Ti alloy systems were also investigated; the compositions and flow temperatures of suitable braze filler metals are shown in Table 4. Ta-V-Nb and Ta-V-Ti braze filler metals are compatible with niobium and the other refractory metals and

Table 4 Shear strengths of brazed TZM (Mo-0.5Ti-0.08Zr) joints in vacuum at 2000 °F

Filler-metal composition, wt%	Brazing temperature, °F	Shear strength(a), ksi
25Ta-50V-25Nb	3400	24.4
30Ta-40V-30Nb	3500	22.4
5Ta-65V-30Nb	3300	24.6
30Ta-65V-5Nb	3400	23.4
30Ta-65V-5Ti	3350	25.6
25Ta-55V-20Ti	3350	18.9
20Ta-50V-30Ti	3200	30.2
10Ta-40V-50Ti	3200	20.5

(a) Each data point for shear strength is an average of two tests.

are resistant to corrosion in alkali metals. The base metal Nb-1Zr was vacuum brazed at 2280 °F with the Ti-28V-4Be braze filler metal. These joints were unaffected by exposure in potassium liquid and vapor for 1000 h at 1500 °F.

Manufacturing procedures for fabricating flat and curved niobium alloy sandwich panels and heat-shield panels have been established. Niobium alloy D36 (Nb-10Ti-5Zr) was used for the honeycomb core and for face sheets that were brazed to the resistance welded core. Among the braze filler metals evaluated for brazing were palladium, Pd-30Cu, titanium, and Ti-11Cr-13V-3Al (B120VCA). Palladium and palladium copper produced ductile joints with excellent filleting; however, considerable interaction between the base metal and braze filler metal occurred. Titanium and the titanium-based alloy also produced joints with excellent ductility, and there was less reaction with the base metal. The flow characteristics of the B120VCA braze filler metal were more sluggish than those of pure titanium; as a result, the braze filler metal could be used to bridge wide joint clearances. Honeycomb sandwich panels were brazed with pure titanium at 3200 °F; heat-shield panels were brazed with B120VCA braze filler metal at 3000 °F in a vacuum environment. Tests in a simulated re-entry environment indicated that the heat-shield panel could be used at temperatures up to 2400 °F. Room-temperature and elevated-temperature tests of the structural panel indicated that it possessed useful properties up to 2300 °F.

As a result of high-remelt-temperature brazing techniques based on diffusion sink and reactive brazing concepts, two new braze filler metals have been developed for D-36 niobium. The diffusion-sink technique involves either permitting the braze filler metal to react with the base metal under controlled conditions or adding base-metal powder to the braze filler-metal powder. One diffusion-sink braze filler metal is Ti-33Cr, requiring a brazing temperature of 2650 to 2700 °F. The other alloy is a Ti-30V-4Be reactive braze filler metal that requires a brazing temperature of 2350 to 2400 °F.

Test results of Ti-33Cr brazed joints indicate an increase in lap-shear strength from approximately 2.5 ksi to greater than 4.5 ksi at 2500 °F and from 0 to 1 ksi at 3000 °F. A diffusion treatment 300 °F below the original brazing temperature actually increased the joint remelt temperature by 600 °F. It essentially doubled the lap shear strength at 2500 °F and produced excellent strength at 3000 °F.

Lap joints reactive brazed with Ti-30V-4Be exhibited substantially lower strength than similar joints diffusion sink brazed with Ti-33Cr, even though remelt temperatures were similar for both systems. Lap-shear strength of Ti-30V-4Be joints was approximately one half and one third the strength of Ti-33Cr braze filler-metal joints at 2500 and 3000 °F, respectively. Nevertheless, the Ti-30V-4Be braze filler-metal joints exhibited attractive potential for lower stress-level applications.

Other niobium alloys, D-43 (Nb-10W-1Zr-0.1C), Cb-752 (Nb-10W-2.5Zr), and C-129Y (Nb-10W-11Hf-0.07Y), have been successfully brazed with two braze filler metals (B120VCA and Ti-8.5Si) at 2650 °F. A further demonstration of the variety of braze filler metals is the use of copper- or zirconium-based braze filler metals for brazing Nb-1Zr fins to Nb-1Zr tubes. On the basis of the test results, the Zr-28V-16Ti-0.1Be braze filler metal was used to braze full-scale heat receivers.

Precleaning and Surface Preparation. One method of preparing niobium as well as tantalum prior to brazing is to electroplate either copper or nickel onto an acid-cleaned surface. The deposits are bonded to the tantalum or niobium by diffusion, and in the case of copper, melting actually occurs. Brazing is subsequently accomplished by using the plated surfaces as a base. Niobium can be cleaned by both mechanical and chemical methods. Table 5 lists cleaning methods for refractory metals.

Table 5 Cleaning methods for refractory metals

Method	Procedure
Tungsten	
Method 1	Immerse in 20% potassium hydroxide solution (boiling)
Method 2	Electrolytic etching in a 20% potassium hydroxide solution
Method 3	Chemical etching in a 50 vol% HNO_3, 50 vol% HF solution
Method 4	Immerse in molten sodium hydroxide
Method 5	Immerse in molten sodium hydride
Tantalum	
Method 1	Shot blast tantalum surface
	Follow by immersion into HCl solution to remove embedded iron particles
	Immerse in glass cleaning solution composed of 95% H_2SO_4, 4.5% HNO_3, 0.5% HF, and Cr_2O_3
Molybdenum	
Method 1	Immerse in a glass cleaning etch composed of 95% H_2SO_4, 4.5% HNO_3, 0.5% HF, and Cr_2O_3
Method 2	Immerse in alkaline bath of 10% NaOH, 5% $KMnO_4$ and 85% H_2O by weight operating at 150 to 180 °F for 5 to 10 min
	Immerse 5 to 10 min in second bath to remove smut formed during first treatment, bath consists of 15% H_2SO_4, 15% HCl, 70% H_2O, and 6 to 10 wt% per unit volume of chromic acid
Method 3	Mo-0.5 Ti alloy
	Degrease 10 min in trichloroethylene
	Immerse in commercial alkaline cleaner 2 to 3 min
	Rinse in cold water
	Buff and vapor blast
	Immerse in commercial alkaline cleaner 2 to 3 min
	Rinse in cold water
	Electropolish in 80% H_2SO_4 at 120 °F with 8 to 12 A
	Immerse in commercial alkaline cleaner 2 to 3 min
Niobium	
Method 1	Degrease
	Immerse in commercial alkaline cleaner 5 to 10 min
	Rinse with water
	Immerse in 35 to 40% HNO_3 for 2 to 5 min at room temperature
	Rinse with tap water, followed by distilled water
	Force air dry

Fluxes and Atmospheres. Only inert-gas (argon and helium), controlled atmospheres, and a vacuum are used in joining niobium. Brazing may be performed in air using fluxes that are suited to the particular braze filler metal being used. However, niobium normally requires a protective coating, such as nickel or copper electroplate.

Special techniques are necessary to satisfactorily braze niobium. All gases that have any reactivity must be removed—oxygen, carbon monoxide, ammonia, hydrogen, nitrogen, and carbon dioxide. Above 750 °F, niobium should be protected from oxidation during brazing. One method, as stated above, is to electroplate the surfaces to be brazed with copper or nickel; consequently, the braze filler metal must be compatible with the plating.

Molybdenum and Molybdenum Alloys

Braze Filler Metals and Properties. In recent years, there has been an emphasis on the use of molybdenum as a structural material for high-temperature applications in the electronic, aircraft, missile, and nuclear reactor fields. However, problems have been encountered in making ductile molybdenum joints for high-temperature service, because unalloyed molybdenum recrystallizes at about 2100 °F. Thus, when brazing is conducted above the recrystallization temperature, brazing time must be kept to a minimum to avoid excessive grain growth and a subsequent increase in the transition temperature of molybdenum. This problem has been alleviated somewhat by the development of base metals with increased recrystallization temperatures.

Copper- and silver-based braze filler metals can be used to braze molybdenum for low-temperature service. For high-temperature applications, molybdenum can be brazed with gold, palladium, and platinum filler metals, nickel-based filler metals, reactive metals, and refractory metals that melt at lower temperatures than molybdenum. It should be noted, however, that nickel-based alloys have limited applicability for high-temperature service, because nickel and molybdenum form a low-melting eutectic at about 2400 °F.

There are three basic limits to brazing molybdenum and its alloys for use above 1800 °F:

- Recrystallization of molybdenum
- Formation of intermetallics between the refractory metal and braze filler metals
- Relative weakness of braze filler materials at elevated temperatures

The formation of intermetallics between molybdenum and braze filler metals is detrimental to joint soundness because the intermetallics become brittle and may fracture at relatively low loads when the joint is stressed.

The relative weakness of braze filler materials at elevated temperatures limits the use of brazed molybdenum assemblies. Most of the nickel-based elevated-temperature braze filler materials melt between 1800 to 2100 °F, where the superior elevated-temperature strength of molybdenum begins to manifest itself. Special high-melting-point braze filler metals for molybdenum and its alloys are required—provided, of course, that they do not cause recrystallization or form intermetallics with the metal.

Two binary braze filler metals, V-35Nb and Ti-30V, have been evaluated for use with the Mo-0.5Ti molybdenum base metal. T-joints were brazed in a vacuum for 5 min at 3000 °F for the Ti-30V braze filler metal and 3400 °F for the V-Nb braze filler metal. The braze filler metals had excellent metallurgical compatibility with the molybdenum base metal, and minimum erosion of the base metal occurred during brazing.

Brazements using Ti-8.5Si diffusion-sink braze filler metal, which melts at approximately 2425 °F, exhibited excellent filleting and wetting that produced crack-free ductile joints. Specimens with molybdenum powder added to the braze filler-metal powder were brazed at 2550 °F.

Because of its excellent high-temperature properties and compatibility with certain environments, molybdenum is a prime candidate for use in isotropic power systems, certain nuclear reactor components, and chemical processing systems. As a result of these stringent criteria (liquid alkali metals effects on braze filler metal), a series of iron-based brazing filler metals have been developed. The best filler metal based on overall performance is Fe-15Mo-5Ge-4C-1B.

Molybdenum alloy TZM (0.5Ti-0.08Zr-Mo) has also been successfully vacuum brazed at 2550 °F with molybdenum powder added to the Ti-8.5Si braze filler metal. Other research in this field has shown that other braze filler metals can be used. Ti-25Cr-13Ni braze filler metal, with a brazing temperature of 2300 °F, has produced the highest remelt temperatures on TZM. T- and lap joint remelt temperatures were about 3100 °F.

The results of shear tests at 2000 °F are presented in Table 4. The room-temperature strengths possessed by these joints are in the range of 20 to 30 ksi and are considered excellent.

Recent investigations conducted to develop vacuum brazes for molybdenum and tungsten that can be used in seal joint applications up to 2907 °F were completed with the following braze filler metals:

- Ti-65V wire
- V wire
- MoB_2-50MoC powder mixture
- V-50Mo powder mixture
- Mo-15MoB_2 powder mixture
- Mo-49V-15MoB_2 powder mixture

Braze temperatures ranged from 2960 to 4095 °F, and leaktight joints were successfully made with all the braze filler metals. Molybdenum joints made with Ti-65V, pure vanadium, V-50Mo, and MoB_2-50MoC are as strong or stronger than the base metal at elevated temperature. The most resistant joints among those tested were brazed with Ti-65V, V-50Mo, and MoB_2-50MoC. Without further testing, the maximum service temperature of the seals should be limited to 4095 °F. Excellent results have been obtained with vanadium as a vacuum furnace braze filler metal for molybdenum and tungsten.

Molybdenum has an extremely low coefficient of thermal expansion, which should be considered in the design of brazed joints, particularly when molybdenum is joined to other metals. Usually, joint clearances in the range of 0.002 to 0.005 in. at the brazing temperature are suitable. The selection of a fixture material also should be made with care and should be based on this property. Tantalum and tungsten fixture materials also have been used successfully for brazing molybdenum.

Precleaning and Surface Preparation. Surface oxide removal before brazing is mandatory. The cleaning operation should be performed immediately before brazing to prevent contamination from occurring. Degreasing should be used to remove oil, fingerprints, and grease. Mechanical and chemical cleaning methods are satisfactory. Sandblasting, liquid abrasive cleaning, or abrasion may be used on simple parts to remove oxide films, but chemical cleaning is preferred, especially for complex assemblies. For heavy oxide films, molten salt baths, such as 70% sodium hydroxide and 30% sodium nitrite at 500 to 700 °F or commercial martempering salt (mixture of sodium and potassium nitrates) operating at 700 °F, have achieved good results. Procedures are described in Volume 5 of the 9th edition of *Metals Handbook*. The first cleaning bath should be controlled closely because it attacks molybdenum. Gross attack has not been noted with the second bath. Light surface oxide films should be removed after the salt bath by an appropriate treatment. Electrolytic etchants may be used to remove surface oxides on simple parts; however, grain-boundary attack of molybdenum can be severe. Chemical etchants are the most popular cleaning method; three successful methods are given in Table 5.

Fluxes and Atmospheres. When utilizing an oxyacetylene torch, sufficient protection may be obtained by using a combination of fluxes, a commercial borate-based or silver brazing flux, plus a high-temperature flux containing calcium fluoride. Fluxes are active between 1050 and 2600 °F. The molybdenum is first coated with the commercial silver brazing flux, and then the high-temperature flux is applied. The silver brazing flux is active at the lower end of the active temperature range; the high-temperature flux then takes over and is active up to 2600 °F. Purified dry hydrogen and inert-gas (helium and argon) atmospheres have been found suitable for brazing molybdenum. For brazing pure molybdenum, the purity of the hydrogen atmosphere is not critical. A dew point of 80 °F can be tolerated in hydrogen when reducing molybdenum oxide. A low dew point (−50 °F or lower) at 2200 °F is required when brazing titanium-bearing molybdenum alloys. Vacuum furnaces have been used to braze molybdenum; however, precautions should be taken in selecting a proper brazing filler metal that will not volatilize during the brazing cycle. Pressures of less then 10^{-4} torr are desirable when brazing molybdenum alloys containing titanium.

Processing and Equipment. Torches, controlled-atmosphere furnaces, vacuum furnaces, and induction and resistance heating equipment can be used to braze molybdenum. Oxyacetylene torch brazing can be accomplished with silver- and copper-based braze filler metals and an appropriate flux.

Tantalum and Tantalum Alloys

Braze Filler Metals and Properties. Braze filler metals are chosen on the basis of the intended application. Nickel-based braze filler metals (such as the nickel-chromium-silicon filler metals) have been used to braze tantalum. Tantalum forms a homogeneous solid solution with nickel up to a tantalum content of 36%; the liquidus is reduced from 2640 to 2460 °F. These braze filler metals are satisfactory for service temperatures below 1800 °F. Copper-gold alloys having less than 40% Au can be used as braze filler metals, but gold in amounts between 40 to 90% tends to form age-hardening compounds that are also brittle. Although silver-based braze filler metals have been used to join tantalum, they are not recommended because of a tendency to embrittle the base metal. Copper-tin, gold-nickel, gold-copper, and copper-titanium braze filler metals also have been used in brazing tantalum.

Because tantalum and its alloys are usually used for elevated-temperature applications at 3000 °F and above, only a few braze filler metals have been developed. Investigations have examined conventional brazing, reactive brazing, and diffusion-sink brazing concepts for new tantalum braze filler metals. The reactive brazing concept uses a braze filler metal containing a strong melting temperature depressant. The depressant is selected to react with the base material or powder additions to form a high-melting-point intermetallic compound during a post-braze diffusion treatment. By removing the depressant in this manner, the joint remelt temperature is increased. Successful application of this concept appears highly dependent on controlling the intermetallic compound reaction to form discrete particles.

Diffusion-sink brazing with titanium and Ti-30V (braze temperature of 3050 to 3200 °F) has produced remelt temperatures exceeding 3800 °F on T- and lap joints. Diffusion-sink brazing with 33Zr-34Ti-33V (braze temperature 2600 °F) has produced remelt temperatures exceeding 3200 °F. The remelt temperature indicates that service temperatures for the titanium and Ti-30V alloys could be 3500 °F, and for the 33Zr-34Ti-33V braze filler metal could be 3000 °F.

Most of the braze filler metals currently available for tantalum are in powder form, which is difficult to work with at elevated temperatures. New powder braze filler metals, such as Hf-7Mo, Hf-40Ta, and Hf-19Ta-2.5Mo, are being developed for tantalum as foil-type braze filler metals. This form is produced either by direct rolling of the braze filler metal or pack-diffusion heat treatment of hafnium foil.

Precleaning and Surface Preparation. Tantalum can be cleaned by mechanical and chemical methods. Hot chromic acid (glass cleaning solution) is satisfactory, but hot caustic cleaning solutions attack the metal and should not be used. Prior to chromic acid cleaning, tantalum can be blast cleaned. This procedure, however, should be followed by immersion in hydrochloric acid solution to dissolve the blasting media, which renders the glass cleaning solution more effective. Abrasion and other usual mechanical cleaning methods are also suitable. Tantalum has a tenacious oxide film that reforms immediately on exposure to air after any cleaning treatment. One method of preparing tantalum is given in Table 5. Additional processing is detailed in Volume 5 of the 9th edition of *Metals Handbook*.

Fluxes and Atmospheres. Only inert-gas (argon and helium) controlled atmospheres and a vacuum are used in joining tantalum. Atmospheres containing constituents that embrittle tantalum, such as oxygen, nitrogen, and hydrogen, should be avoided. Brazing may be performed in air using fluxes normally used for aluminum filler metals. However, tantalum requires a protective coating, such as nickel or copper electroplate, to induce wetting during brazing.

Processes and Equipment. Special techniques are necessary to satisfactorily braze tantalum. All reactive gases must be removed; oxygen, as well as carbon monoxide, must be eliminated. Tantalum readily forms oxides, carbides, nitrides, and hydrides with these gases, with a subsequent loss of ductility. At high temperatures, tantalum should be protected from oxidation. Because this is often accomplished by electroplating the surfaces with copper or nickel, it is thus necessary for the braze filler metal to be compatible with the plating. Controlled atmosphere and vacuum (both hot- and cold-walled) furnaces are used for brazing tantalum. Induction, resistance, and torch brazing are not recommended.

Tungsten and Tungsten Alloys

Braze Filler Metals and Properties. Brazing of tungsten structures has been successfully accomplished in vacuum. A wide variety of braze filler metals and pure metals with liquidus temperatures ranging from 1200 to 3500 °F can be used for brazing. Table 3 shows typical braze filler metals that have been used to braze not only tungsten, but also other refractory metals and alloys.

Tungsten can be brazed in much the same manner as molybdenum and its alloys, using many of the same braze filler metals. Brazing can be accomplished in a vacuum or in a dry argon, helium, or hydrogen atmosphere. To some extent, the selection of the brazing atmosphere depends on the braze filler metal used. For example, braze filler metals that contain elements with high vapor pressures at the brazing temperature cannot be used effectively in a high vacuum.

Care must be exercised in handling and fixturing tungsten parts because of their inherent brittleness; these parts should be assembled in a stress-free condition. Contact between graphite fixtures and tungsten must be avoided to prevent the formation of brittle tungsten carbides. While nickel-based braze filler metals have been used successfully to braze tungsten, a reaction between nickel and tungsten that results in base-metal recrystallization can occur; this reaction can be minimized by short brazing cycles, minimum brazing temperatures, and the use of small quantities of braze filler metal.

Data on brazing tungsten are more limited than data on the brazing of other refractory metals. During research on the development of braze filler metals for brazing niobium alloys, unalloyed tungsten was vacuum brazed with two experimental braze filler metals—Nb-2.2B and Nb-20Ti. The lap-shear strength of joints brazed with Nb-2.2B was about 5 ksi at 3000 °F and 8 ksi at 2500 °F. The strength of joints brazed with Nb-20Ti was somewhat lower, about 3 ksi at both test temperatures; the flow of this braze filler metal was more sluggish than that obtained with Nb-2.2B.

Braze filler metals based on the platinum-boron and iridium-boron systems were developed to braze tungsten for service at 3500 °F. They contained up to 4.5% B and could be used for brazing below the recrystallization temperature of tungsten. Tungsten lap joints were vacuum brazed and diffusion treated in a vacuum at 2000 °F for 3 h. This cycle resulted in the production of joints with remelt temperatures of about 3700 °F. A slight increase in joint remelt temperature was noted when tungsten powder was added to the braze filler metal. The highest remelt temperature, 3940 °F, was obtained when joints were brazed with Pt-3.6B, to which 11 wt% tungsten powder had been added.

Studies were conducted to develop and evaluate braze filler metals that could be used to braze tungsten for nuclear reactor service at 4530 °F in hydrogen. Butt joints were brazed using a gas tungsten arc as the heat source, using braze filler metals W-25Os, W-50Mo-3Re, and Mo-5Os.

The joint clearances required in many instances for joining tungsten are governed by the brazing filler metal being used. Frequently, the engineering design, base metal, and manufacturing operation may automatically predetermine the conditions of the assembly prior to brazing.

Precleaning and Surface Preparation. Thorough mechanical or chemical cleaning prior to brazing is essential. Cleaning methods found acceptable for tungsten are given in Table 5. The most effective cleaning procedure depends on the tenacity of the oxide film. In cases where wrought tungsten sheet has been mill cleaned, degreasing is sometimes the only cleaning operation necessary prior to brazing. Cleaning processes are detailed in Volume 5 of the 9th edition of *Metals Handbook*. However, optimum conditions for preparation of tungsten should be determined for each particular application. In some cases, electroplating of tungsten with nickel has been satisfactory to stop diffusion of elements that form brittle intermetallic compounds with the base metal. A hydrogen atmosphere furnace cleaning operation at 1950 °F for 15 min has been effective in reducing light oxide films. Tumble polishing in a slurry is commonly used for tungsten contact material.

Fluxes and Atmospheres. Tungsten can be brazed in either an inert-gas atmosphere (helium or argon), reducing atmosphere (hydrogen), or a vacuum. Two precautions should be observed when vacuum brazing: (1) the vapor pressures of the compositional elements of the brazing filler metals should be compatible with the temperatures and pressures involved, and (2) the effect of outgassing on the soundness of the base metals and the deposited filler metal should be evaluated. Conventional low-temperature fluxes are used when brazing tungsten for electrical contact applications when silver- and copper-based braze filler metals are used.

Process and Equipment. Tungsten can be brazed by furnace (inert-gas or reducing atmosphere and vacuum), torch, resistance, or induction brazing. Furnace brazing operations usually are used when the parts are larger than those considered practical for the induction and resistance processes.

Applications. Tungsten-copper electrode tips are materials made by powder metallurgy processes. A normal analysis of one of the materials is 80W-20Cu. These tips may be used for resistance welding electrodes that are usually brazed to the copper-chromium alloys, which constitute the main current-carrying portion of the electrode. Any of the BAg series of braze filler metals are suitable for this application. Tungsten-copper tips usually are applied by induction or torch heating.

Tungsten-copper tips are also used for electrical discharge machining (EDM), in which the same or similar composition of tungsten-copper is used for the finish machining electrode. Only the end of the electrode is made from the tungsten-copper alloy. A base of either phosphorized or OFHC copper is used to attach the electrode to the current supply. The large mass and high thermal conductivity of the brazement present problems during joining; in most cases, brazing is accomplished in a neutral or reducing atmosphere. The most suitable braze filler metal is BAg-8. To hold parts in alignment, pins are incorporated in the design, and the braze filler metal is preplaced in the form of shim stock located between the mating parts.

When brazing resistance welding electrodes with pure tungsten tips, precoating of the tungsten surface aids in a uniform flow of the braze filler metal between the faying surface of the joint. Without pretreatment, poor wetting of the tungsten by the braze filler metal often results, causing premature joint failure under the high pressures and temperatures of resistance welding.

Joining of Dissimilar Refractory Metals

Braze Filler Metals. Molybdenum base metals that have been brazed to other metals include:

- Silver-palladium braze filler metals used to braze TZM face sheets to honeycomb core fabricated from nickel-based superalloy Inconel 702. Brazing was done below 2200 °F to prevent recrystallization of the molybdenum base metal.
- TZM foil brazed to Hastelloy X, René 41, and L605 sheet stock. Each dissimilar metal combination was brazed with the following filler metals: Ni-7Si-5Fe-5Cr-4Co-3W-0.7B, Ni-15.5Cr-16Mo-5Fe-4W-2.5Co-1Mn-1Si, and Pd-40Ni.
- A tungsten nozzle brazed to a molybdenum plenum for a propulsion engine application. These components were brazed in argon or hydrogen with pure iron for service at temperatures up to 1800 °F and with Cr-25V for service at 2700 to 3000 °F. Molybdenum-to-tungsten joints have also been brazed with Cu-35Mn.

Fig. 3 Niobium/cobalt alloy vacuum brazement using 82Au-18Ni filler metal

Niobium and its alloys have been successfully joined to cobalt-based alloys with an 82Au-18N braze filler metal (Fig. 3), type 316 stainless steel to Nb-1Zr alloy using Co-21Ni-21Cr-8Si braze filler metal at 2150 °F, and niobium to alumina using zirconium nickel as braze filler metal. Niobium alloys B66 and D43 have been joined to Hastelloy X, L605, and René 41 with the following braze filler metals: NX-77 (Ni-7Si-5Fe-5Cr-4Co-3W-0.7B-0.1Mn), Hastelloy C (Ni-15.5Cr-16Mo-5Fe-4W-2.5Co-1Mn-1Si), and Pd-40Ni.

Vacuum Brazing of Tantalum Sheet to OFHC Copper Plate for Missile Applications. This joint combined the high-temperature strength and abrasion resistance of tantalum and the heat-dissipation capabilities of copper. The bimetallic material was produced in plate form. The plate was then cold drawn into a nose-cone configuration for evaluation in rapidly flowing oxidizing gases at 3600 °F. Joints withstanding this environment for a 1-min period were produced with the following braze filler metals: Cu-8Sn, Au-18Ni, and Ag-15Mn.

Tantalum has been brazed to molybdenum in vacuum at 3244 °F using pure zirconium as the braze filler metal. Gold-nickel and pure copper braze filler metals have produced excellent tensile test results in brazing type 304 stainless steel to tantalum and Ta-10W.

REFERENCES

1. Elrod, S.D., Lovell, D.T., and Davis, R.A., Aluminum Brazed Titanium Honeycomb Sandwich System, *Weld. J.*, Vol 52 (No. 10), Oct 1973, p 425s-432s
2. Lovell, D.T. and Lindh, D.V., Titanium Structures Technical Summary, DOT/SST Ph. I and Ph. II, Rep. FFA-SS-73-27, Oct 1974
3. Taylor, R.Q., Elrod, S.D., and Lovell, D.T., Development and Evaluation of the Aluminum-Brazed Titanium System, Vol III, Rep. FAA-SS-73-5-3, D6-60277-3, May 1974
4. Hurwitz, D., Manufacturing Methods for Brazed Titanium Hybrid Structures, AFML-TR-76-119, Contr. F33615-74-C-5047, Final Report, June 1976
5. Beal, R.E. and Saperstein, Z.P., Development of Brazing Filler Metals for Zircaloy, *Weld. J.*, Vol 50 (No. 7), 1971, p 2755-2915
6. Amato, I. and Ravizza, M., Some Developments in Zircaloy Brazing Technology, *Energ. Nucl.* (Milan), Vol 16 (No. 2), 1969, p 35-39

Brazing of Carbon and Graphite*

By A.J. Moorhead
Group Leader
Oak Ridge National Laboratory
and
C.R. Kennedy
Research Staff Member
Oak Ridge National Laboratory

CARBON AND GRAPHITE, brazed both to themselves and to metals, are the subject of this article. These materials vary widely in the degree of crystallinity, in the degree of orientation of the crystals, and in the size, quantity, and distribution of porosity in the microstructure. These factors are strongly dependent on the starting materials and on processing and, in turn, govern the physical and mechanical properties of this product. Accordingly, the first section of this article discusses how these materials are produced.

Material Production

Carbons and graphites can be manufactured by several processes that yield materials with a wide range of crystalline perfection and properties. In the most widely used process, polycrystalline graphites are made from cokes produced as a by-product of the residuum from the manufacture of petroleum or from natural pitch sources; the coke is the final product that remains after all volatile materials have been removed after heating to about 930 °F. The coke product is broken up and then calcined at temperatures from 1650 to 2550 °F to reduce the volatile content and create a dimensionally stable filler coke. This stabilization prevents the occurrence of excessive shrinkage during subsequent heat treatments. Calcination is omitted in the production of some special high-strength graphites.

The calcined coke is crushed, milled, and sized through screens into various fractions. The coarse fraction may exceed 0.04 in. in the fabrication of large bodies and may be as small as 0.004 in. in smaller blocks, depending on the desired properties. The shape and properties of the crushed coke particles depend largely on the coke source. Some cokes naturally break up into highly anisotropic particles with platelike or needlelike shapes, while others produce rounder isotropic particles that have a strong influence on the properties of the final body.

The size fractions are mixed according to properties desired in the final material and blended with a hot coal tar pitch. The pitch creates a plastic mix that can be shaped by extrusion or molding. The shape of particles is important in that forming processes tend to preferentially align the anisotropic particles. Particles that are aligned in this manner as the result of extrusion or molding processes produce bodies with highly anisotropic properties. Extruded bodies and molded bodies yield rotational symmetry; however, the axes of the graphite crystals are aligned parallel to the extrusion direction and normal to the molding direction.

The formed body is baked to pyrolize the binder pitch at temperatures from 1470 to 1650 °F, usually in large gas-fired floor furnaces. During baking, the binder experiences approximately 50 to 60% weight loss and an even greater volume loss. The effect is to reduce overall density and subsequently increase porosity. Density can be increased by rebaking following impregnation with low-melting-point, high-viscosity pitches. Special impregnants, such as thermal setting resins or mixtures of resins and pitches, can be used to control porosity.

Final graphitization of carbon artifacts is achieved at temperatures ranging from 4170 to 5430 °F in an Acheson furnace, an electric furnace similar to ones used primarily for the production of silicon carbide. The baked carbon bodies are stacked in a conducting bed that is buried under insulating material. A large electric current (6000 A at 230 V) is passed through the bed using large, water-cooled electrodes at both ends of the furnace until the graphitization temperature is obtained, and then the mass is allowed to cool. The heating cycle generally lasts about 15 days, after which the furnace is disassembled and the graphite blocks are removed.

Two alternate processes may be used to produce carbon bodies. Carbon blacks may be used as the filler material in place of petroleum coke. Carbon blacks, in this case, are mixed with pitch and briquetted. They are then baked, reground, and remixed with pitch. A reforming step completes production. These grades primarily are used in brushes for electrical motors where good wear resistance and conduc-

*Research sponsored by the Gas-Cooled Reactor Programs Division, U.S. Department of Energy, under contract W-7505-eng-26 with the Union Carbide Corp.

tivity are ensured through adjustment of the final heat treatment temperature, which is usually lower than 4530 °F. Carbon fibers also can be made into yarns that are woven into cloth or the desired shape. The product is pitch infiltrated, baked to reduce porosity, and graphitized. These bodies generally have very high tensile strength, as well as low coefficients of thermal expansion, which provides for excellent resistance to thermal shock or stresses. However, carbon bodies fabricated from carbon fibers are very anisotropic, with lower shear and flexure strengths.

Applications

Carbon and graphite find widespread use as electrodes in metallurgical applications and as moderator materials in nuclear applications. Specialized uses include rocket nozzles, guide vanes, nose cones, electric motor brushes and switches, bushings and bearings, high-temperature heat-exchangers, and plumbing, as well as heart valves, synthetic teeth posts, air frame composites, and high-performance brake linings. The physical-property requirements of these products vary considerably, thus illustrating the numerous carbon and graphite structures commercially available. Electrical or thermal conductivity, thermal expansion, and strength requirements may also vary considerably, depending on the application. Also, because of the highly anisotropic crystal structure of carbon and graphite, materials can be produced with physical properties that are capable of being highly anisotropic to structures that are virtually isotropic.

Brazing Characteristics

Wettability. The wetting characteristics of all the carbons and graphites are strongly influenced by impurities, such as oxygen or water, that are either absorbed on the surface or absorbed in the bulk material. Moisture absorption always occurs to some extent, with levels as high as 0.25 wt%. Brazeability also depends on the size and distribution of pores, which can vary significantly from one grade to another. For example, some graphites are so porous that all available filler metal is drawn into them, resulting in alloy-starved joints. Others are so dense and impervious that adherence of filler metal is poor.

Thermal Expansion. A major consideration when brazing carbon and graphite is the effect of the coefficient of thermal expansion of these materials. This can range from about 2×10^{-6}/°C up to 8×10^{-6}/°C between 25 and 1000 °C (75 and 1830 °F), depending on the type and grade of product, as well as within a given piece, depending on the degree of anisotropy. In these materials, expansion coefficients may be less than, equal to, or greater than those of the reactive or refractory metals. However, they are always less than the more common structural materials such as iron and nickel. Before brazing graphite, the type and grade of carbon or graphite must be established to ascertain the expansion characteristics of the particular material. This information is also important when brazing carbon or graphite to itself. Joint failure, particularly during thermal cycling, may occur if too great a difference exists between the coefficients of thermal expansion of the graphite and the brazing filler metal.

Brazing to a Dissimilar Material. If the braze gap increases significantly on heating because of a large mismatch in coefficient, the brazing filler metal may not be drawn into the joint by capillary flow. However, if the materials and joint design cause the gap to become too small, the alloy may not be able to penetrate the joint. In conventional brazing of dissimilar materials, the material having the greater coefficient of expansion is made the outer member of the joint. Joint tolerances are used that do not allow the gap between the surfaces to become too great for capillary flow.

Additional problems occur in brazing dissimilar materials when one part of the joint is a carbon or graphite. Carbons and graphites have little or no ductility and are relatively weak under tensile loading. These adverse conditions are usually compensated for in graphite-to-metal joints by brazing the graphite to a transition piece of a metal, such as molybdenum, tantalum, or zirconium, with a coefficient of expansion near that of the graphite. This transition piece can be subsequently brazed to a structural metal if required. This minimizes shear cracking in the graphite by transposing the stresses resulting from the large difference in thermal expansion to the metallic components. Thin sections of metals, such as copper or nickel, that deform easily when stressed have also been successfully used for brazing dissimilar metals.

A special graded transition piece was developed by Hammond and Slaughter (Ref 1) to accommodate the mismatch between the coefficients of expansion of graphite and of a nickel-based structural alloy (Hastelloy N). Using powder metallurgy techniques, a series of seven W-Ni-Fe rings having different tungsten contents were fabricated. By varying the tungsten content from 97.5% in the first ring down to 40% in the seventh, they were able to fabricate a graded seal with a thermal coefficient of expansion near that of graphite on one end and of Hastelloy N on the other. All of the W-Ni-Fe compacts and the graphite and Hastelloy N terminal pieces were then brazed together in a single operation using pure copper as the filler metal. Brazing of the heavy alloys to the graphite was made possible by a prior metallization of the graphite with chromium, as discussed in the following section of this article.

Filler Metals

Graphite is inherently difficult to wet with the more common brazing filler metals. Most merely ball up at the joint, with little or no wetting action. Two techniques are used to overcome this wetting deficiency: the graphite is coated with a more readily wettable layer, or brazing filler metals containing strong carbide-forming elements are used. Several researchers have developed techniques for coating graphite with either a metallic or intermetallic layer so that brazing can be accomplished with a conventional filler metal.

Example 1. Chemical Vapor Deposition of Graphite With Molybdenum or Tungsten. Graphite was coated with a thin film (0.008 to 0.31 mils thick) of molybdenum or tungsten by a chemical vapor deposition (CVD) process (Ref 2). This was accomplished by passing a mixture of hydrogen and the appropriate hexafluoride gas (molybdenum or tungsten) over the graphite at a temperature of 1470 or 1110 °F, respectively. These metals were selected because of their low coefficients of thermal expansion, which are approximately that of the particular graphite being used, not because of carbide formation at the interface. Carbide formation does not occur at these low temperatures, so the coating-to-graphite bond is essentially mechanical. In this instance, the coated graphite parts were brazed to molybdenum with copper (BCu-1), but other filler metals are equally acceptable. Although the CVD process is not complicated, it is not widely used. There are commercial companies, however, that do utilize this process to apply coatings.

Example 2. Formation of a Chromium Carbide Substrate on Graphite by Chromium Vapor Plating. Hammond and Slaughter (Ref 1) developed a process for treating graphite that produces a metallurgically attached chromium carbide substrate. This treatment is applied by a novel procedure involving vapor plating chromium on the graphite in a partial vacuum at 2550 °F. The chromium carbide forms by chemical reaction as chromium deposits on the graphite. The chromium vapor is supplied by reacting a mixture of fine carbon and chromium oxide powder, spread over the hearth of a graphite crucible in which the plating is carried out.

Push-pin shear tests were conducted on graphite specimens coated by this technique and then brazed with pure copper filler metal. Reported shear strengths were 20 ksi both at room temperature and at 1290 °F. Metallographic evaluation of failed test pieces showed the specimens coated with chromium carbide to have failed by shear in the graphite pin just inside the bond region.

Filler-Metal Compositions. A number of experimental brazing filler metals have been developed for brazing of graphite either to itself or to refractory metals. These filler metals typically contain one or more of the strong carbide-forming elements such as titanium, zirconium, silicon, or chromium. For example, Donnelly and Slaughter (Ref 3) reported on the successful brazing of graphite using filler metals of composition 48%Ti-48%Zr-4%Be, 35%Au-35%Ni-30%Mo, and 70%Au-20%Ni-10%Ta. In addition, Fox and Slaughter (Ref 4) recommended the use of a filler metal with composition 49%Ti-49%Cu-2%Be for brazing of graphite as well as oxide ceramics. These alloys wet graphite and most metals well in either a vacuum or inert atmosphere (pure argon or helium) and span a fairly wide range in brazing temperatures (from 1830 °F for 49%Ti-49%Cu-2%Be to 2460 °F for 35%Au-35%Ni-30%Mo). However, they have not been evaluated for oxidation resistance or mechanical properties. These materials are not available commercially, and this presents a problem for a potential user who does not have access to arc-melting services.

At least two commercially available brazing filler metals reportedly wet carbon or graphite, as well as a number of metals. One is a modified version of the silver-copper alloy with a small titanium addition to promote wetting of oxide ceramics and graphite. This alloy has the composition of 68.8%Ag-26.7%Cu-4.5%Ti, with a solidus of 1525 °F and a liquidus of 1560 °F. This alloy is suitable for low- to medium-temperature applications but appears to have only moderate oxidation resistance. The second commercially available filler metal for graphite brazing has the composition of 70%Ti-15%Cu-15%Ni. It has a somewhat higher melting range (1670 °F solidus and 1760 °F liquidus) than the first and, by virtue of its greater titanium content, has better oxidation resistance than the silver-bearing alloy.

A considerable amount of work has been done by researchers in Russia on joining of graphite. For example, graphite was brazed to steel at 2100 °F using a filler metal of 80%Cu-10%Ti-10%Sn (Ref 5). In another technique, known as diffusion brazing, a metallic interlayer was placed between the graphite components; the components were pressed together with a specific pressure and heated to the temperature of formation of a carbon-bearing melt or a eutectic (Ref 6 and 7). On heating to higher temperatures, the melt dissociated with the precipitation of finely divided crystalline deposits of carbon that interacted with the graphite base material to form a strong joint. Depending on the physical nature of the metal of the interlayer and on the type of carbon-metal phase diagram, carbon is formed in the joint either during the thermal dissociation of a carbon-bearing melt or a carbide-carbon eutectic. Iron, nickel, and aluminum are typical metals that form carbon-bearing melts when heated at high temperature in contact with graphite. Molybdenum is capable of forming a thermally dissociating carbide-carbon eutectic.

For in situ formation of a liquid film of brazing material that is subsequently dissociated at a graphite interface, a specific compressive force of 0.5 kgf/mm^2 was used, and argon pressure of 0.3 to 0.5 atm was supplied. The optimum temperature range for joining of graphite using an intermediate nickel layer was 3810 to 3990 °F; for iron, 3990 to 4350 °F; for aluminum, 3990 to 4170 °F; and for molybdenum, 4350 to 4710 °F. Metallographic examination of joints made using this technique showed that with increasing temperature the amount of metallic or carbide phase in the joint decreased, but the amount of the graphite phase increased. This increase in graphite content resulted in a marked increase in joint strength as compared to those samples brazed at lower temperatures and having significant metal- or carbide-containing microstructures.

Amato *et al.* (Ref 8) developed a procedure for brazing a special grade of graphite to a ferritic stainless steel for a seal in a rotary heat exchanger. It seems apparent that the selection of type 430 stainless steel was based at least partly on its lower coefficient of thermal expansion (7.3×10^{-6}/°F) as compared to that of a typical austenitic stainless steel (11×10^{-6}/°F). In addition, a joint geometry was developed that minimized the area of the braze joint, thereby reducing thermally induced stresses to acceptable levels. Specimens of graphite brazed in a vacuum furnace to thin type 430 stainless steel sheet with either Ni-20%Cr-10%Si or Ni-18%Cr-8%Si-9%Ti at 2060 to 2150 °F performed well in tests at 1200 °F.

Applicable Heating Methods

As graphite begins to oxidize at about 840 °F (depending on the grade), brazing operations must be conducted in environments that exclude oxygen. This can be accomplished either in a vacuum of $\sim 1 \times 10^{-4}$ torr or less or through the use of high-purity inert gas (argon or helium) protection. Heating typically has been done in a furnace, but induction heating also has been used.

REFERENCES

1. Hammond, J.P. and Slaughter, G.M., Bonding Graphite to Metals with Transition Pieces, *Welding Journal,* Vol 50 (No. 1), 1971, p 33-40
2. Werner, W.J. and Slaughter, G.M., Brazing Graphite to Hastelloy N for Nuclear Reactors, *Welding Engineer,* March 1968, p 65
3. Donnelly, R.G. and Slaughter, G.M., The Brazing of Graphite, *Welding Journal,* Vol 41 (No. 5), 1962, p 461-469
4. Fox, C.W. and Slaughter, G.M., Brazing of Ceramics, *Welding Journal,* Vol 43 (No. 7), 1964, p 591-597
5. Kochetov, D.V., *et al.*, Investigation of Heat Conditions in the Brazing of Graphite to Steel, *Weld. Prod.*, Vol 21 (No. 3), 1974, p 15-18 (English translation)
6. Anikin, L.T., *et al.*, The High Temperature Brazing of Graphite, *Weld. Prod.*, Vol 24 (No. 1), 1977, p 39-41 (English translation)
7. Anikin, L.T., *et al.*, The High Temperature Brazing of Graphite Using An Aluminum Brazing Alloy, *Weld. Prod.*, Vol 24 (No. 7), 1977, p 23-25 (English translation)
8. Amato, I., *et al.*, Brazing of Special Grade Graphite to Metal Substrates, *Welding Journal,* Vol 53 (No. 10), Oct 1974, p 623-628

Laser Brazing

By Charles E. Witherell
Leader, Metal Fabrication Group
Lawrence Livermore National Laboratory

LASER BRAZING is a metals-joining process that uses the thermal energy developed by laser beams to perform localized brazing on thin-walled precision parts. Because of the higher costs of brazing with lasers, the method should be considered only when conventional brazing methods are not adequate. A list of "problem" applications that are candidates for laser brazing appear below. These metals-joining applications frequently are difficult, if not impossible, to do by ordinary brazing methods, but have been accomplished with laser brazing:

- Miniature precision parts that require minimal heat input during joining operations to maintain dimensional tolerances
- Thin base metals, 0.004 in. and less, that tend to become eroded and sometimes perforated during brazing operations using filler metals or fluxes
- Joints on assemblies containing heat-sensitive materials or parts that cannot be removed during joining operations
- Brazed joints near glass-to-metal seals, adhesively bonded joints, or other thermally sensitive connections
- Connections inside evacuated or pressurized vessels or containers (e.g., within sealed glass vacuum tubes)

The main advantage laser brazing offers over more conventional brazing processes is its ability to produce a brazed connection locally without heating the entire part or component to the flow point of the brazing filler metal (Ref 1). Another advantage is the high degree of control of the thermal energy of laser beams, including intensity, spot size, duration, and ability to be located or positioned precisely. Also, because a laser beam is capable of being transmitted through solids that are transparent to its wavelength, brazing can be accomplished within hermetically sealed vacuum or high-pressure atmosphere enclosures.

While other laser brazing modes are possible and are likely avenues of future development, most laser brazing is confined to line-contact types of joints where fillets are made to bridge across the intersection of two surfaces, or of a surface and an edge. Infrequently, a bead is made to connect two abutting edges (butt joint). Because the flow of the filler metal is confined to the region heated by the laser beam (generally a circular spot), a seam is brazed by producing a series of overlapping spots.

Laser Brazing Process

In most applications, the laser beam is aimed directly at the preplaced filler metal in the joint, and the weld is executed (Ref 2). The technique of heating the region of the joint adjacent to the filler metal is not the dominant mode of brazing with lasers. Although such an approach does work, it offers little advantage for using lasers over conventional brazing techniques. Relatively extensive areas of the part must be heated, and the time to reach the flow temperature of the filler metal is relatively long because the thermal coupling efficiency of lasers is poor. Also, distortion from attempts to locally heat a part to the brazing temperature using an intensive heat source, while taking care not to melt the surface, can cause as much distortion as a fusion weld (and sometimes more), eliminating one of the advantages of choosing a laser brazing process.

Generally, laser brazing involves positioning the workpiece under the fixed laser beam. While it is possible to position laser beams upon a fixed part or workpiece, it is simply more convenient and requires less costly equipment to fix the laser and move the workpiece, which is usually very light and small. Both continuous wave (cw) CO_2 and pulsed lasers can be used, but solid-state pulsed lasers, such as neodymium-doped yttrium-aluminum-garnet (YAG) pulsed lasers, appear to be more adaptable to the progressive overlapping spot mode of brazing. Their shorter wavelength also seems to couple more efficiently to brazing alloys of high-conductivity and high-reflectivity metals, such as copper, silver, and aluminum.

The workpiece is positioned above the focal point of the laser beam to obtain the proper balance between energy density (a somewhat diffused thermal pattern is desired for brazing) and beam width (at its point of intersection with the surface to be brazed). In lasers that have an adjustable beam output mode (i.e., transverse electromagnetic mode, or TEM), a favorable pattern for brazing has been the TEM_{01} mode, which provides a toroidally shaped thermal energy pattern. Brazing is accomplished with one pulse per spot, with considerable overlap (more than 50%) from one spot to the next.

Brazing filler metal is preplaced in the joint, either as a powdered alloy or a shim sandwiched between the members. The powdered alloy is easier to use and handle, and most alloys remain in place during the series of laser pulses. However, increased beam power levels needed for filler metals of high reflectivity or thermal conductivity may require the filler metal to be used in the form of foil or shims sandwiched between the joint members to retain it during application of the necessarily higher beam power levels. Attempts to use cements or binders to hold the powdered alloys in the joint during laser brazing have been unsuccessful due to contamination of the joint and the surrounding region.

Table 1 Typical laser brazing conditions for T-fillet joints

Joint members: Side A	Joint members: Side B	Filler-metal classification and form	Laser characteristics: Pulse energy, J	Beam spot diam, in.	Pulse width, s	Pulse peak power, W
Type 304 stainless steel (0.005 in.)	Same	BAg-8 powder	7.3	0.044	0.010	860
Type 302 stainless steel (0.001 in.)	Same	BAg-1 powder	0.2	0.012	0.008	30
Type 316 stainless steel (0.005 in.)	Same	BNi-2 powder	5.2	0.044	0.010	610
Monel 400 (0.001 in.)	Type 316 stainless steel (0.004 in.)	BAg-6 powder	4.5	0.028	0.010	520
Copper (0.001 in.)	Nickel 200 (0.001 in.)	BAg-18 powder	14.7	0.044	0.010	1730
Nickel 200 (0.001 in.)	Iron (0.001 in.)	BCu-1 foil	6.8	0.032	0.010	800

Note: For filler-metal classification information, see AWS A5.8-76, "Specification for Brazing Filler Metal," American Welding Society, 1976. Powders used were −325 mesh. Fluoride-based silver brazing flux used with all filler metals except BNi-2 and BCu-1, which were 0.002-in.-thick foils and were brazed without flux. Protective argon shielding gas used for all brazed joints. All joints were brazed with neodymium YAG pulsed laser with 80-J maximum pulse energy and 50-W maximum output power, equipped with 100-mm focal length lens.

Adequate atmospheric protection is required for all laser brazing, whether or not a flux is used, even if precautions are taken to keep the joint region free of contaminants. Depending on the composition of the materials, the purity of the filler alloy, and other variables, joints may be brazed without a flux, but a protective atmosphere (argon shielding or vacuum) should always be used. When a flux is used, it may be mixed with the powder filler alloy into a paste using water or alcohol and applied to the joint. The mixture must be thoroughly dried before attempting to braze.

Best results have been obtained by firing the laser beam directly over the joint containing the preplaced filler, with the beam axis coinciding with the joint axis. Where both joint members are of equal thickness, the joint should be oriented so that the beam bisects the angle between the members. For example, for a 90° T-fillet joint between materials of equal thickness and conductivity, the joint should be tilted so that the beam is positioned 45° from both members and normal to the longitudinal axis of the joint. For members of different thicknesses, the joint should be tilted so that the laser beam is directed toward the thicker or more conductive member. The objective is to obtain the same thermal energy input in both members and thereby obtain a uniformly contoured fillet braze profile.

Table 1 lists typical conditions for laser brazing fillet joints between various metals and alloys. Actual values for specific laser installations can be expected to vary from the given values because of differences in pulse duration and pulse shape from laser to laser and variations in thermal energy coupling effects with differing surface finishes of the material and compositions of the brazing filler metals. It is important to note that base-metal thicknesses, as well as thermal conductivity and optical reflectivity of both the base metal and filler metal, influence the level of laser pulse power required to melt the brazing alloy and heat the base metal sufficiently for flow across the joint to occur.

Laser Brazing Versus Conventional Brazing

In most, if not all, other brazing processes, the filler metal is distributed between the closely fitting surfaces of the joint by capillary action. Because the flow of the filler metal is limited to the region heated by the laser beam in laser brazing, capillary action plays a much less dominant role. The area heated by the laser beam is often less than a millimeter in width.

Conventional brazing frequently is used in lap seams or to make joints between two closely fitting parallel surfaces, as in socket joints or sleeve connections, where preplaced or manually added filler metal fed from an adjacent location flows during the heating process by capillary action. While lap seams conceivably could be brazed using a laser in a rastering (scanning) mode, conventional brazing would be much simpler and less costly.

Limitations of Laser Brazing

Laser brazing generally should be reserved for precision jobs that require the strength of a brazed joint but for some reason cannot tolerate the heat, distortion, or other consequences of more conventional techniques. Another fundamental consideration is the availability of a laser of adequate power and the availability of accessory workpiece-positioning equipment. Currently, metalworking laser installations are expensive in comparison with the costs of conventional brazing facilities. In time, and with wider use, the costs of metalworking laser systems will be reduced. However, laser brazing will never replace conventional brazing techniques, except for specific applications such as thin-walled precision parts that must be produced in large quantities. For such applications, lasers would be cost-effective competitors of conventional brazing processes for production joining of large numbers of precision parts that facilitate computer-controlled automation.

Studies of laser brazing have revealed a number of effects of the very rapid rate of solidification associated with pulsed lasers that are not encountered in conventional brazing. Some of these are advantageous, while others can be detrimental. For example, laser brazed deposits in some alloys have exhibited unusually high levels of as-deposited hardness and strength. The strength of laser brazed joints has, in some cases, exceeded that of fusion welds made in the same base materials. Table 2 compares microhardnesses of conventionally brazed and laser brazed deposits.

At the same time, the rapid rate of solidification characteristic of laser-melted deposits has led to levels of brittleness that would be unsuitable in joints for many applications made with some fillers (the BNi series, for example). This can be overcome by varying laser parameters, such as use of greater beam width, longer pulse

Table 2 Comparative microhardness of conventionally brazed and laser brazed deposits

AWS brazing filler-metal classification	Deposit microhardness(a), DPH: Oxyfuel gas brazed(b)	Laser brazed(c)	Hardness increase ratio
BAg-1(d)	120(e)	232(e)	1.93
BAg-6(d)	106(e)	176(e)	1.66
BAg-8(d)	102(e)	200(e)	1.96
BAg-18(d)	123(e)	148(e)	1.20
BNi-2(d)	692(g)	920(g)	1.33
BCu-1(f)	61(e)	72(e)	1.18
BAu-4(d)	268(e)	340(e)	1.27
BAlSi-2(f)	53(e)	62(e)	1.17

(a) Averages of five determinations. (b) Brazing filler metal placed on surface of 1020 carbon steel strip (0.062 in. thick); surface protected from atmospheric contamination by argon shield. Underside of steel strip heated by oxyfuel gas torch until flow of brazing filler metal occurred. (c) Determinations made on T-joint fillet brazed deposits. (d) Powder (−325 mesh). (e) With flux. (f) Foil. (g) Without flux

Fig. 1 Laser brazing of capillary tubing to pressure sensor fitting

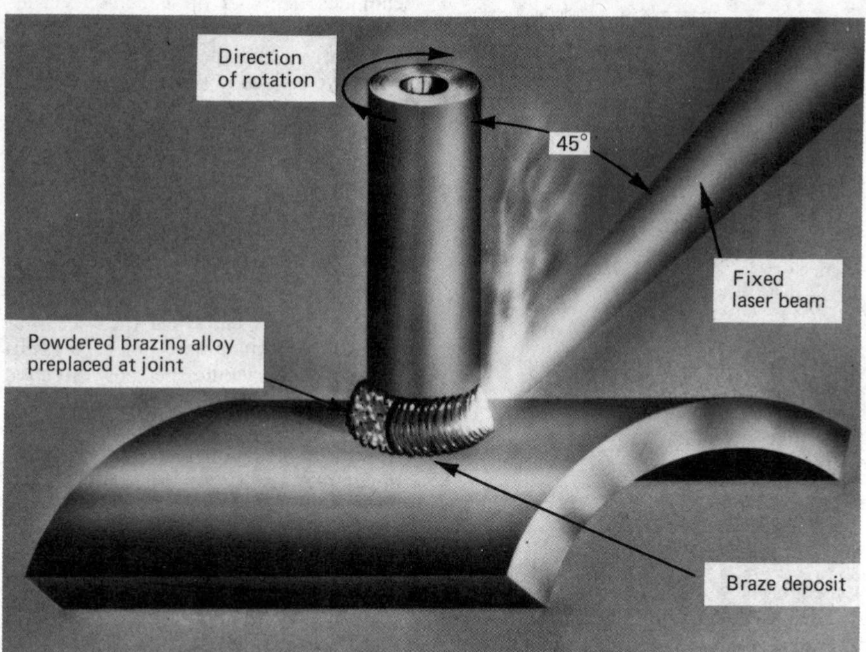

duration, and increased braze spot overlap, in alloys that are hard and experience a further increase in hardness when solidified rapidly. The same considerations also apply when brazing base materials that are hardenable under conditions of rapid quench.

The effects of high levels of solidification stress combined with high deposit hardness can cause a liquid-metal-embrittlement type of cracking with some brazing filler metals in some alloys in a highly cold worked condition. These isolated occurrences are not unexpected with such radically different modes and rates of heating and solidification, but they do indicate the wisdom of conducting a well-planned and comprehensive series of tests to determine the appropriateness of use before committing to any new joining procedure like the laser brazing process.

Example 1. Laser Brazing Capillary Tubing to a Pressure Sensor Fitting. A miniature pressure sensor assembly required a strong, pressure-tight (several thousand psi) joint between 70Cu-30Ni capillary tubing, with an outside diameter of 0.007 in. and a wall thickness of 0.002 in., and a type 316 stainless steel fitting having a wall thickness of 0.005 in. Previous attempts by conventional brazing techniques invariably resulted in flow of the brazing alloy into the 0.003-in.-diam bore of the capillary tubing and the closing of it. Attempts to fusion weld the copper-nickel to stainless steel were similarly unsuccessful. The parts were successfully laser brazed using the procedure described below:

- *Laser characteristics*: Neodymium YAG 50-W pulsed laser; 4-in. focal length lens; TEM_{01} mode; pulse interval, 3 s; laser pulse energy, 0.87 J; measured beam spot diameter at workpiece, 0.036 in.; total pulse width, 6 ms; effective pulse width, 5.1 ms; peak pulse power, 0.2 kW (energy density = 1.4×10^2 J/cm^2; power density = 2.7×10^4 W/cm^2
- *Flux*: Standard fluoride-based silver brazing flux mixed with powdered brazing filler metal in a flux-to-filler-metal proportion of 1-to-20 by volume
- *Filler metal*: Prealloyed powdered −325 mesh BAg-1 class (silver-cadmium-copper-zinc) filler and flux mixture (water added to make paste consistency) was preplaced in joint region and a controlled volume provided in the joint by drawing a 0.04-in.-diam metering rod through the mixture, allowing the surface of the rod to run along the sides of the joint members. The resulting meniscus-shaped volume of the flux and filler mixture was dried thoroughly under an infrared heat lamp before brazing.
- *Shielding gas*: The entire joint region was blanketed with pure argon using a shaped porous bronze gas-distribution shielding fixture with a hole to allow access for the laser beam.
- *Laser/joint configuration*: The length of capillary tubing was first assembled to the stainless steel fitting by inserting the end of the tubing into a hole drilled through the fitting. This positioned the tubing in the desired location and retained it in alignment during brazing. As shown in Fig. 1, the entire joint assembly was rotated incrementally with each laser pulse, allowing about an 80% overlap of each previously brazed spot. The laser head was fixed above the rotation fixture holding the part. A series of tacks equally spaced around the periphery of the joint were first made, followed by a full rotation of the joint for the brazing cycle.

REFERENCES

1. *Brazing Manual,* 3rd ed., American Welding Society, Miami, 1976
2. Witherell, C.E. and Ramos, T.J., Laser Brazing, *Welding Journal,* Vol 59 (No. 10), Research Supplement 267s-277s, Oct 1980

Soldering

Soldering

By the ASM Committee on Soldering*

SOLDER AND FLUXES have existed for 2000 or 3000 years, but soldering technology began its serious development in the early 19th century when the industrial revolution demanded better means of fabrication. Early industrial soldering applications were followed by the radio, the telephone, light bulbs, car radiators, plumbing, and an endless variety of business, commercial, and industrial products. The conquest of space required reliable joining and service of many soldered joints. Modern computer and data processing equipment require soldering techniques. This ancient craft has therefore survived and become a modern technology.

The American Welding Society (AWS) defines soldering as metal coalescence below 800 °F. Soldering facilitates joining parts without heat damage and provides a system for rapid multiple-part joining of products such as printed circuits.

Principal soldered alloys are a combination or alloy of tin and lead. The tin component in the alloy reacts with the metals to be joined to form a metallurgical bond. Variations in tin and lead content and the addition of various alloying elements result in different melting ranges and joining characteristics of the alloy. Figure 1 compares solder temperature ranges with some base-metal melting points. Solders may contain antimony, silver, zinc, indium, and bismuth. These alloys allow specific melting resistance of the completed joints. Several important steps in achieving a good soldered joint are:

- Design the joint specifically for soldering.
- Select the correct solder alloy for the application.
- Select the proper flux for joining.
- Clean or prepare surfaces to be joined.

Fig. 1 Solder temperature ranges compared with base metal melting points. Source: Lead Industries Association, Inc.

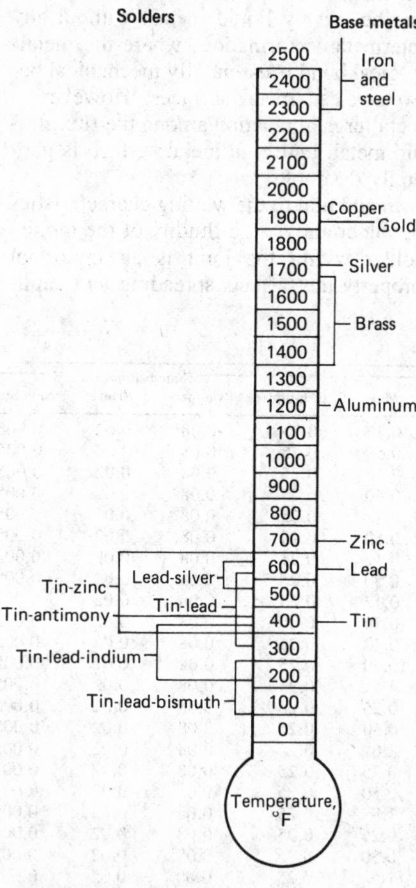

- Apply the optimum heat to produce the required temperature and time for soldering.
- Remove flux residues when necessary after joining.

Soldering is selected over alternative joining methods such as adhesive bonding, welding, brazing, or mechanical joining, because it offers the following advantages:

- A low energy input is required for soldering.
- Precise control over the amount of solder used is possible.
- A variety of heating methods can be used.
- Solders with various melting ranges can be selected to fit the application.
- Soldering can be easily and economically automated with a low capital expense outlay.
- Sequential assembly is possible.
- Joint reliability is high.
- Joints are easily repaired and reworked.
- Solder alloys can be selected for service in differing environments.
- Solder joints have good thermal and electrical conductivity.
- Solder joints are impermeable to gas and liquid.
- A long shelf life for soldered assemblies is common.

Principles of Soldering

Soldering involves metallurgy, physics, and chemistry in the interaction of elements, the constitution of fluxes, the thermal chemistry involved in heating both fluxes and metals to the molten state, and the underlying thermodynamics and fluid dynamics in promoting formation of a soldered joint. When the selected solder is heated with the appropriate flux to make a joint with a particular base metal, the liquid materials formed flow over the surfaces to be joined. Relative surface tensions of the materials dictate these flow

*Roy E. Beal, *Chairman*, President, Amalgamated Technologies, Inc.; Paul J. Bud, President, Electrovert Consulting Services; William B. Hampshire, Assistant Manager, Tin Research Institute, Inc.; Jerome F. Smith, Vice President, Lead Industries Association, Inc.; C.E.T. White, Executive Vice President, The Indium Corporation of America

characteristics and define the capability for work adhesion in making the soldered joint.

The cleanliness and chemical composition of the surfaces to be joined are critical to the process. One function of the flux is to ensure that the base metal is sufficiently cleaned to provide adequate spread and flow of the soldering alloy to promote joint formation. Under most soldering conditions, a very low contact angle at the edge of the solder and good wetting is necessary for joint formation. Under certain conditions, the contact angle required is higher, and the capability to bridge or fill a gap between two metal surfaces is equally important.

At the soldering temperature, the liquid metal displaces the flux from the joint surfaces and is in intimate contact with the base metal to be joined, making possible a metallurgical reaction between the liquid solder and the base metal. This reaction varies in direct relation to the proximity and compatibility of the liquid solder and solid base metal. In many soldering systems, an intermetallic compound is developed at the interface between the solder and the base metal, producing an essentially complete metallurgical joint. Reference to the constitutional diagrams of the metals involved is essential to understanding the soldering process. During cooling of the soldered joints, reaction products modify the metallurgical properties of the solder alloy, which is reflected in its melting range and solidification pattern. The intermetallic compound formed at the interface continues to grow at a substantial rate until solidification takes place. When the joint is completely solidified, diffusion between the base metal and soldered joint continues until the completed item is cooled to room temperature. Mechanical properties of soldered joints, therefore, are generally related to, but not equivalent to, the mechanical properties of the soldering alloy.

Most soldering operations are carried out in air, with the flux acting as a barrier to surface oxidation and interaction with the atmosphere. In some cases, however, a protective atmosphere can enhance the formation of the desired soldered joint and also serve to reduce the amount of flux used in particular products. Surfaces to be joined can be wet by liquid metals without any intermetallic formation, where the metallurgical bond is essentially mechanical between clean metal surfaces. However, a metallurgical reaction among the flux, liquid metal, and base metal surface is generally desirable.

In addition to the wetting characteristics of solder alloys, the fluidity of the molten solder within the joint is an important property that affects spreading and capillary action between gapped surfaces, particularly in such joints as pipe and tube joints. When conditions of fluxing or metal surface and solder alloy compatibility are inadequate, two other actions can occur that are essential principles of soldering. The molten solder may simply not wet the surface to be joined, for example, tin on chromium. In this case, wetting and spreading will not occur, and the molten material will simply sit in a globule on a surface with a wetting angle of greater than 90°. A second, more insidious phenomenon is dewetting, which is a function of solder composition, flux selection, thermal conditions, and cleanliness of the base-metal surfaces. Dewetting occurs when a solder alloy has initially spread over a given surface area and then retracts, or pulls back into globules, as a result of incomplete wetting adhesion or the formation of oxide films between the molten metal and the material to be joined. Dewetting sometimes can leave a film of partially metallurgically reacted material, generally an indication of material incompatibility, poor flux selection, or inadequate cleaning of the base materials.

Tin-Lead Solders

Care should be taken in specifying the correct solder for the job, because each alloy is unique with regard to its composi-

Table 1 Tin-lead solder compositions

Alloy grade	Tin desired	Lead, nominal	Antimony, % Min	Antimony, % Desired	Antimony, % Max	Composition, max %: Bismuth	Copper	Iron	Aluminum	Zinc	Arsenic	Melting range, °F: Solidus	Liquidus
70A	70	30	...	...	0.12	0.25	0.08	0.02	0.005	0.005	0.03	361	378
70B	70	30	0.20	...	0.50	0.25	0.08	0.02	0.005	0.005	0.03	361	378
63A	63	37	...	...	0.12	0.25	0.08	0.02	0.005	0.005	0.03	361	361
63B	63	37	0.20	...	0.50	0.25	0.08	0.02	0.005	0.005	0.03	361	361
60A	60	40	...	...	0.12	0.25	0.08	0.02	0.005	0.005	0.03	361	374
60B	60	40	0.20	...	0.50	0.25	0.08	0.02	0.005	0.005	0.03	361	374
50A	50	50	...	...	0.12	0.25	0.08	0.02	0.005	0.005	0.03	361	421
50B	50	50	0.20	...	0.50	0.25	0.08	0.02	0.005	0.005	0.03	361	421
45A	45	55	...	...	0.12	0.25	0.08	0.02	0.005	0.005	0.03	361	441
45B	45	55	0.20	...	0.50	0.25	0.08	0.02	0.005	0.005	0.03	361	441
40A	40	60	...	...	0.12	0.25	0.08	0.02	0.005	0.005	0.02	361	460
40B	40	60	0.20	...	0.50	0.25	0.08	0.02	0.005	0.005	0.02	361	460
40C	40	58	1.8	2.0	2.4	0.25	0.08	0.02	0.005	0.005	0.02	365	448
35A	35	65	...	...	0.25	0.25	0.08	0.02	0.005	0.005	0.02	361	477
35B	35	65	0.20	...	0.50	0.25	0.08	0.02	0.005	0.005	0.02	361	477
35C	35	63.2	1.6	1.8	2.0	0.25	0.08	0.02	0.005	0.005	0.02	365	470
30A	30	70	...	...	0.25	0.25	0.08	0.02	0.005	0.005	0.02	361	491
30B	30	70	0.20	...	0.50	0.25	0.08	0.02	0.005	0.005	0.02	361	491
30C	30	68.4	1.4	1.6	1.8	0.25	0.08	0.02	0.005	0.005	0.02	364	482
25A	25	75	...	...	0.25	0.25	0.08	0.02	0.005	0.005	0.02	361	511
25B	25	75	0.20	...	0.50	0.25	0.08	0.02	0.005	0.005	0.02	361	511
25C	25	73.7	1.1	1.3	1.5	0.25	0.08	0.02	0.005	0.005	0.02	364	504
20B	20	80	0.20	...	0.50	0.25	0.08	0.02	0.005	0.005	0.02	361	531
20C	20	79	0.8	1.0	1.2	0.25	0.08	0.02	0.005	0.005	0.02	363	517
15B	15	85	0.20	...	0.50	0.25	0.08	0.02	0.005	0.005	0.02	440	550
10B	10	90	0.20	...	0.50	0.25	0.08	0.02	0.005	0.005	0.02	514	570
5A	5(a)	95	...	...	0.12	0.25	0.08	0.02	0.005	0.005	0.02	518	594
5B	5(a)	95	0.20	...	0.50	0.25	0.08	0.02	0.005	0.005	0.02	518	594
2A	2(b)	98	...	...	0.12	0.25	0.08	0.02	0.005	0.005	0.02	518	594
2B	2(c)	98	0.20	...	0.50	0.25	0.08	0.02	0.005	0.005	0.02	518	594

Note: Solder metal specification ASTM B-32-76 (Table 1) is printed in full in the *1981 Annual Book of Standards Part 8*, American Society for Testing and Materials
(a) Permissible tin range, 4.5 to 5.5%. (b) Permissible tin range, 1.5 to 2.5%. (c) Federal specifications are similar to the above alloy grade No. 70B, 63B, 60B, 50B, 40B, 35C, 30C, and 20C.

tion and its properties. Table 1 gives the composition and melting characteristics of some tin-lead solders. When referring to tin-lead solders, the tin content is customarily given first, for example, 40%Sn-60%Pb. Table 2 presents information relating to tin-lead solder analysis and use.

Solders in the tin-lead system are the most widely used of all joining materials. Industrial solder alloys are in use that contain a combination of materials from 100% Pb to 100% Sn, as demanded by the particular application. The utility of the tin-lead combination is highlighted by examination of the constitution diagram between these two materials, shown in Fig. 2. Solder alloys can be obtained with melting temperatures as low as 360 °F and as high as 600 °F within this system. Except for the pure metals and the eutectic solder at 63%Sn-37%Pb, all solder alloys melt within a temperature range that varies according to the alloy composition. Each alloy has unique characteristics. In general, properties are influenced by the melting characteristics of the alloys, which in some measure are related to their load-carrying and temperature capabilities.

Applications. Solder alloys containing less than 5% Sn are used for joining tin-plated containers and for automobile radiator manufacture. For automobiles, a small additional amount of silver is usually added to provide extra joint strength at automobile radiator operating temperatures. Solder alloys of 10%Sn-90%Pb and 20%Sn-80%Pb are also used in radiator joints. With compositions between 10%Sn-90%Pb and 25%Sn-75%Pb, care must be taken to avoid any kind of movement during the solidification phase to prevent hot tearing in solders with a wide freezing range, as indicated by the constitution diagram (Fig. 2).

Higher tin content solders at the 25%Sn-75%Pb and 30%Sn-70%Pb compositions have lower liquidus temperatures and can be used for joining materials with sensitivity to high temperature, or where the wetting characteristics of the tin are important to providing sound soldered joints. Solder alloys in the composition range described above usually are applicable to industrial products and generally are used in conjunction with inorganic fluxing materials.

The widely used general-purpose solder alloys contain 40 to 50% Sn. These solders are used for plumbing applications, electrical connections, and general soldering of domestic items. The 60%Sn-40%Pb and 63%Sn-37%Pb alloys are used most extensively in the electronic industries for both hand soldering and wave or dip applications. Sometimes silver additions are made to alloys used in the electronics industries to reduce dissolution of silver-based coatings.

Table 2 Tin-lead solders

Composition, %		Temperature, °F			
Tin	Lead	Solidus	Liquidus	Pasty range	Uses
2	98	518	594	76	Side seams for can manufacturing
5	95	518	594	76	Coating and joining metals
10	90	514	570	56	
15	85	440	550	110	
20	80	361	531	170	Coating and joining metals, or filling dents or seams in automobile bodies
25	75	361	511	150	Machine and torch soldering
30	70	361	491	130	
35	65	361	477	116	General purpose and wiping solder
40	60	361	460	99	Wiping solder for joining lead pipes and cable sheaths. For automobile radiator cores and heating units
45	55	361	441	80	Automobile radiator cores and roofing seams
50	50	361	421	60	General purpose. Most popular of all
60	40	361	374	13	Primarily used in electronic soldering applications where low soldering temperatures are required
63	37	361	361	0	Lowest melting (eutectic) solder for electronic applications

Impurities in Tin-Lead Solders

Impurities in solders can affect their performance and must be kept to a minimum. American Society for Testing and Materials (ASTM) standards for solder alloys set maximum tolerable impurities in alloys as provided by the supplier or refinery. Impurities can be inadvertently picked up during normal usage of the alloys, especially when solder pots with recirculation systems and passage of components through the molten materials are used. The purity of solders supplied by reputable manufacturers usually is adequate for most applications. Particular soldering operations may require the use of super-purity materials that can be supplied upon request. Impurities present in sufficient quantities can affect wetting properties, flow within the joint, melting temperature of the solder, strength capabilities of joints, and oxidation characteristics of the solder alloys. The most common im-

Fig. 2 Tin-lead phase diagram. Source: Lead Industries Association, Inc.

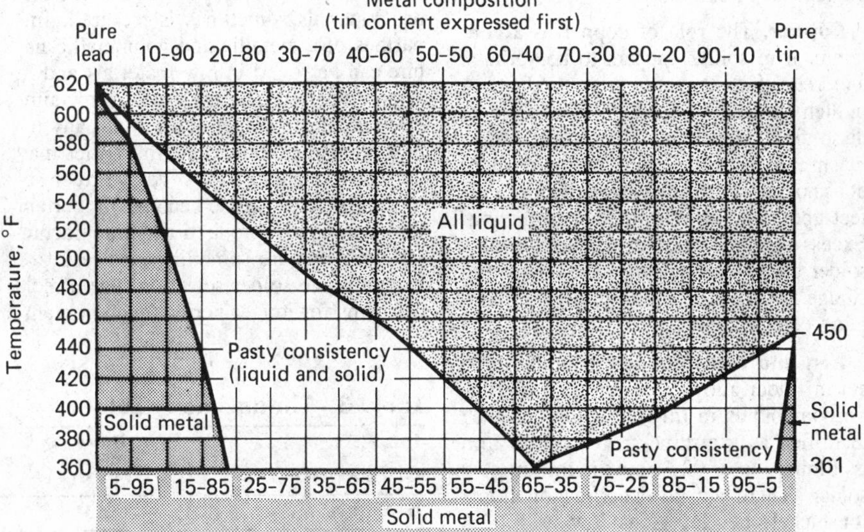

purity elements are listed below with their principal levels and effects.

Aluminum. Traces of aluminum in a tin-lead solder bath can seriously affect soldering qualities. More than 0.005% of the metal can cause grittiness, lack of adhesion, and surface oxidation of the solder alloy. A deterioration in surface brightness of a molten bath sometimes is an indication of the presence of aluminum.

Antimony is often used in solders as a deliberate addition. Some specifications even require a minimum antimony content. As an impurity, antimony tends to reduce the effective spread of a solder alloy. High-lead solder specifications usually require a 0.5% Sb maximum limit. The general rule is that antimony should not exceed 6% of the tin content, although in some applications this rule can be invalid.

Arsenic. A progressive deterioration in the quality of the solder is observed with increases in arsenic content. As little as 0.005% As induces some dewetting, which becomes more severe as the percentage of arsenic is increased to 0.02%. Arsenic levels should be kept within this range.

Bismuth. Low levels of bismuth in the solder alloy generally do not cause any difficulties, although some discoloration of soldered surfaces occurs at levels above 0.5%.

Cadmium. A progressive decrease in wetting capability occurs with additions of cadmium to tin-lead solders. While there is no significant change in the molten solder appearance, small amounts of cadmium can increase the risk of bridging and icicle formation in printed circuits. For this and health reasons, cadmium levels should be kept to a minimum.

Copper. The role of copper as a contaminant in solder appears to be variable and related to the particular product. A molten tin-lead solder bath is capable of dissolving copper at a high rate, easily reaching 0.3% Cu. Copper in liquid solder does not appear to have any deleterious effect upon wetting rate or joint formation. Excess copper settles to the bottom of a solder bath as an intermetallic compound sludge. New solder alloy allows a maximum copper content of 0.08%.

Iron and nickel are not naturally present in solder alloy. The presence of iron-tin compounds in tin-lead solders can be identified as a grittiness. Generally, iron is limited to 0.02% maximum in new solder. There are no specification limits for nickel, but levels as low as 0.02% can produce some reduction in wetting characteristics.

Phosphorus and Sulfur. Phosphorus at a 0.01% level is capable of producing dewetting and some grittiness. At higher levels, surface oxidation occurs, and some identifiable problems such as grittiness and dewetting become readily discernible. Sulfur causes grittiness in solders at a very low level, and should be held to 0.001%. Discrete particles of tin-sulfide can be formed. Both of these elements are detrimental to good soldering.

Zinc. The ASTM new solder alloy specification states that zinc content must be kept to 0.005% maximum in tin-lead solders. At this maximum limit, even with new solders in a molten bath, some surface oxidation can be observed and oxide skins may form, encouraging icicles and bridging. Up to 0.01% Zn has been identified as the cause of dewetting on copper surfaces.

The combined effects of the above impurity elements can be significant. Excessive contamination in solder baths or dip pots generally can be identified through surface oxidation, changes in the product quality, and the appearance of grittiness or frostiness in joints made in this bath. A general sluggishness of the solder also is observed. In addition to analysis, experience with solder bath operation is helpful in determining the point at which the material should be renewed for good solder joint production. American Society for Testing and Materials specifications, which specify maximum allowable concentrations, are useful in purchasing solder for general use. In particular applications, specific contaminants or a combination of elements may be detrimental to a particular soldered product. A determination of a revised or limited specification for solder materials sometimes is required. Impurities of a metallic and nonmetallic nature can be found in raw materials and in scrap solder sometimes used by reclaimers. Reclaimed solder is used in many industrial applications where impurities may not be detrimental. However, correct selection of solder grade becomes important for economical production. Manufacturing problems are sometimes the result of inappropriate solder selection, use of solder baths for longer periods than contamination buildup will tolerate, or processing methods that rapidly contaminate a solder bath. Determination of suitable specifications, allowable impurities in new materials, and allowable impurities through deterioration to the point at which a solder bath is discarded should be included in any soldering quality-control program.

Other Solder Alloys

A wide range of alternate alloys are available. At low temperatures, the ternary gallium-indium-tin eutectic (62.5%Ga-16%Sn-21.5%In), with a melting point of 50 °F, is useful. Bismuth-based fusible alloys with melting points ranging from 110 to 485 °F are manufactured. Alloys based on indium with lead, tin, and silver additions are available to cover the temperature range of 200 to 600 °F. Solders available in the temperature range between 600 to 750 °F are limited, but combinations available include cadmium-zinc and cadmium-silver alloys (liquidus temperatures from 510 to 750 °F), the zinc-aluminum eutectic (liquidus 720 °F), and three gold-based solders—the gold-tin eutectic at 80%Au-20%Sn, with a melting point of 535 °F; the gold-germanium eutectic at 88%Au-12%Ge, with a melting point of 675 °F; and the gold-silicon eutectic at 96.4%Au-3.6%Si, with a melting point of 700 °F. The temperature range from 750 to 840 °F is limited to the only available alloy, the aluminum-germanium eutectic (45%Al-55%Ge), with a melting point of 795 °F. The gold-indium composition (82%Au-18%In), with a liquidus of 905 °F and a solidus of 845 °F, is occasionally used.

Tin-Antimony Solder. The tin-antimony solder listed in Table 3 has excellent soldering and strength characteristics. It is used where a slightly higher temperature range is needed and in applications for joining stainless steels where lead contamination must be avoided. The alloy also is used in plumbing and refrigeration because of its good creep strength and fatigue resistance.

Tin-Silver Solders. Because of their comparatively high cost, tin-silver solders, shown in Table 4, are used in fine instrument work and in food applications. They have good wetting characteristics and

Table 3 Tin-antimony solder

ASTM alloy grade	Federal specification QQS-571	Composition, %		Temperature, °F	
		Tin	Antimony	Solidus	Liquidus
95TA	Sb 5	95	5	452	464

Table 4 Tin-silver solders

ASTM alloy grade	Federal specification QQS-571	Composition, % Tin	Silver	Temperature, °F Solidus	Liquidus
96.5 TS	Sn 96	96.5	3.5	430	430
(a)	...	97.5	2.5	430	438
(a)	...	95.0	5.0	430	465
(a)	...	90.0	10.0	295	563

(a) No ASTM or federal designation is available.

superior joint strength, compared to conventional solders.

Tin-zinc solders were developed primarily for joining aluminum. These alloys resist galvanic corrosion of soldered joints in aluminum. Soldering of aluminum is more difficult and has important differences with the joining of other commonly used materials. Aluminum forms a tenacious surface oxide that is very difficult to remove, necessitating the use of very active flux. The wetting properties of solders on aluminum surfaces are generally not as good as on other common metals and have resulted in the development of special processes such as rub soldering and ultrasonic soldering to alleviate these difficulties.

Although aluminum can be joined using tin-lead, cadmium-zinc, and aluminum-zinc solder alloys, low-temperature soldering generally is performed with tin-zinc alloys, which have melting temperatures of about 400 °F. These solders wet aluminum readily and provide some joint strength. Higher temperature solders, utilizing a substantial portion of zinc, can be very useful for aluminum joining. Wetting angles are generally higher in these solders, resulting in larger fillets. The joints generally are strong and more corrosion resistant than lower temperature soldered joints. The highest temperature solders for aluminum contain 90 to 100% Zn, with small amounts of high-melting materials such as silver, copper, nickel, and aluminum. Flux selection is an important and integral part of the aluminum soldering system and reference should be made to commercially made materials for successful joining.

Typical tin-zinc compositions are:

Composition, % Tin	Zinc	Temperature, °F Solidus	Liquidus
91	9	390	390
80	20	390	520
70	30	390	590
60	40	390	645
30	70	390	710

Lead-Silver and Lead-Silver-Tin Solders. The lead-silver solders listed in Table 5 have solidus temperatures high enough to make them useful where strength at moderately elevated temperatures is required. Flow characteristics are, however, not satisfactory. The addition of 1% Sn improves wetting and flow characteristics and reduces susceptibility to humid atmosphere corrosion. The listed alloys with tin contents over 1% are used primarily in electronic applications where high wettability is required. The 97.5%Pb-1%Sn-1.5%Ag alloy is useful in cryogenic duty, where the high lead content is advantageous, and also for fine copper wire soldering to reduce dissolving tendencies of the wire in the solder.

Cadmium-silver solders are used primarily in applications where high service temperatures are required. High tensile strength joints can be produced that retain a substantial portion of their strength at service temperatures up to 425 °F. A 95%Cd-5%Ag solder has a solidus temperature of 640 °F and a liquidus temperature of 740 °F.

Cadmium-zinc solders are useful for soldering aluminum. Joints of intermediate strength and corrosion resistance are attainable with proper techniques. Several cadmium-zinc solders are:

Composition, % Cadmium	Zinc	Temperature, °F Solidus	Liquidus
82.5	17.5	509	509
40	60	509	635
10	90	509	750

Zinc-Aluminum Solder. The 95%Zn-5%Al solder was developed specifically for use on aluminum. It develops joints with good strength and corrosion resistance where the soldering temperature is high. It has been used in certain specialized electronic applications where the high solidus and liquidus temperature (720 °F) is advantageous.

Indium-based solders, listed in Table 6, are generally specialty solders. They possess properties that make them valuable for certain applications. The 52%In-48%Sn and 97%In-3%Ag alloys will wet glass, quartz, and many ceramics, making them useful in glass-to-metal seals. Because of their low vapor pressure, they also are useful for seals in vacuum systems. The indium-lead and indium-lead-silver sol-

Table 5 Lead-silver and lead-silver-tin solders

ASTM alloy grade	Federal specification QQS-571	Composition, % Lead	Silver	Tin	Temperature, °F Solidus	Liquidus
2.5S	Ag 2.5	97.5	2.5	...	579	579
5.5S	Ag 5.5	94.5	5.5	...	579	579
1.5S	Ag 1.5	97.5	1.5	1.0	588	588
(a)	...	92.5	2.5	5.0	549	565
(a)	...	88	10.0	2.0	514	576
(a)	...	36	1.5	62.5	354	372

(a) No ASTM or federal designation is available.

Table 6 Indium-based solders

Composition, % Indium	Lead	Tin	Other	Temperature, °F Solidus	Liquidus
44	...	42	(a)	200	200
52	...	48	...	244	244
70	9.6	15	(b)	257	(c)
97	...	...	(d)	290	290
80	15	...	(e)	301	(c)
12	18	70	...	324	(c)
25	37.5	37.5	...	274	358
50	50	...	...	356	408
90	...	...	(f)	290	446
75	25	...	...	264	508
5	92.5	...	(g)	572	(c)
5	90	...	(h)	554	590
5	95	...	...	558	598

(a) 14% Cd. (b) 5.4% Cd. (c) Melting point only determined. (d) 3% Ag. (e) 5% Ag. (f) 10% Ag. (g) 2.5% Ag. (h) 5% Ag

ders have improved resistance to thermal fatigue compared to the conventional lead-tin solders. This, coupled with their marked reduction in the scavenging and leaching of gold surfaces, has led to their use in electronic assemblies. Indium solders have good resistance to alkaline corrosion, but corrosion resistance in the presence of traces of chloride ions is not satisfactory, necessitating the use of hermetic packages or conformal coatings. The lead-indium system is a solid-solution system permitting a choice of alloys with temperature differentials large enough for step soldering.

Fusible Solders. The bismuth-based solders, often called fusible solders, are useful for soldering operations where temperatures below 360 °F are required. A representative group is listed in Table 7.

Alloys rich in bismuth generally are not good solders. They do not wet base metals readily, and either corrosive fluxes or pretinning is required. Applications include:

- Soldering heat treated surfaces where high soldering temperatures would result in softening of the part.
- Soldering joints where the adjacent material is temperature-sensitive and would deteriorate at a high soldering temperature.
- Step soldering operations where a low soldering temperature is necessary to protect a nearby solder joint.
- Soldering temperature-sensitive devices (safety links, fuses, and safety plugs) where the failure of the solder joint is required at a predetermined low temperature for safety purposes.

Precious Metal Solders. Gold-based solders find application primarily in semiconductor device assembly and package sealing. Although the base cost of the solders is high, their corrosion resistance, good wettability and strength, and compatibility with silicon justify their use. Compositions of several precious metal solders are:

Composition, %		Temperature, °F	
Gold	Other	Solidus	Liquidus
80	(a)	536	536
88	(b)	673	673
96.4	(c)	698	698
82	(d)	843	905

(a) 20% Sn. (b) 12% Ge. (c) 3.6% Si. (d) 18% In

Miscellaneous Solders. The composition 65%Sn-25%Ag-10%Sb is a special high-strength solder with a tensile strength of 18 ksi recently developed for electronic applications. The germanium-aluminum eutectic is the highest temperature solder available, and although solderability is not particularly good, it is used in special electronic applications. Compositions for miscellaneous solders are:

Composition, %	Temperature, °F	
	Solidus	Liquidus
65Sn-25Ag-10Sb	451	(a)
55Ge-45Al	795	795

(a) Melting point only determined

Forms. In addition to being available in a wide variety of alloys, solders are commercially available in various sizes, shapes, and forms that can be classified as follows:

- *Pig:* Available in 20-, 40-, 50-, and 100-lb sizes
- *Cakes or ingots:* Rectangular or circular in shape, weighing 3, 5, or 10 lb
- *Bars:* Available in weights from 1/2 to 2 lb
- *Paste:* Available as a mixture of solder and flux in paste form in quantities of 1 lb or more
- *Segment or drop:* Wire or triangular bar cut into pieces or lengths of any desired number
- *Foil, sheet, and ribbon:* Supplied in various thicknesses and widths
- *Wire—solder:* Diameters of 0.010 to 0.250 in. on spools weighing 1, 5, 20, 25, or 50 lb or in bulk packs
- *Wire—flux cored:* Solder can be cored with organic, inorganic, or rosin fluxes, 0.010 to 0.250 in. in diameter, on spools weighing 1, 5, 20, 25, or 50 lb or in bulk packs
- *Preforms:* A wide range of custom-designed preform shapes are available. Each shape is a derivative of one or more of the following four most common shapes: wire, punched parts, spheres, and flux-coated metal forms.

Table 7 Fusible solders

Composition, %					Temperature, °F	
Bismuth	Lead	Tin	Cadmium	Other	Solidus	Liquidus
44.7	22.6	8.3	5.3	(a)	117	117
49	18	12	...	(b)	136	136
48	25.63	12.77	9.60	(c)	142	149
50	26.7	13.3	10.0	...	158	158
42.5	37.7	11.3	8.5	...	160	190
55.5	44.5	...	...	...	255	255
58	...	42	...	...	281	281
40	...	60	...	...	281	338
48	28.5	14.5	...	(d)	217	440

(a) 19.1% In. (b) 21% In. (c) 4.0% In. (d) 9.0% Sb

Base and Coated Metal Solderability

The concept of solderability is fundamental to any discussion of metals to be joined (base metals). Because soldering is a relatively low-temperature joining method, little thermal activation is available to speed chemical and metallurgical reactions. Therefore, the base metals must be properly designed and prepared before assembly. Solderability testing evaluates the effectiveness of these preparatory stages before actual soldering.

Various test schemes have been devised to evaluate solderability or, more specifically, the ability of the solder to wet the cleaned and fluxed base metal. The simplest of these is the vertical dip test, which relies on a visual assessment of wetting after the test specimen has been dipped into a solder pot under specified conditions. A more sophisticated version suspends the test sample from a load cell arrangement. While wetting of a sample is attempted, the forces acting on the sample are recorded as a function of time. This test quantifies solderability, but the physical significance of some of the parts of the load versus time measurements is not clear.

One deficiency of all present solderability tests is that they give no indication of future solderability. Storage of components for future assembly is an inevitable consequence of production scheduling, so there is an urgent need for accelerated aging procedures to complement solderability testing. In a partial attempt to allow for the effects of storage and materials variability, solderability tests are carried out under conditions more difficult than those actually expected. For example, lower temperatures and a less active flux may be specified in the solderability test.

To put solderability into perspective, the soldering designer nearly always finds that the base metals have been selected for some property other than solderability, such as electrical or thermal conductivity, coefficient of linear expansion, or strength-to-weight ratio. Once the choice of base metals is fixed, the designer must understand their inherent solderabilities, how to improve solderabilities by coating, and how to preserve solderability until the assembly procedure is completed.

Base Metals

General ratings of a variety of materials with regard to ease of soldering are:

Material	Ease of soldering
Tin, gold, cadmium (unpassivated), silver, palladium, and rhodium	Excellent
Copper, bronze, brass, lead, nickel silver, and beryllium copper	Good
Steel (plain carbon and low-alloy types), zinc, and nickel	Fair
Aluminum, aluminum bronze, and high-alloy steels (especially high chromium)	Difficult
Cast iron, chromium, titanium, tantalum, magnesium, and ceramics	Require preplating with a solderable metal

The precious metals, which show excellent solderability, dissolve rapidly in liquid tin-lead solders and quickly form a range of intermetallic compounds with tin. Because the tarnish films on such metals are very thin (if they exist at all), even a weak flux will provide rapid wetting and soldering action. On the basis of cost, these metals are usually encountered as soluble coatings.

Copper joining is practically synonymous with soldering. Copper and tin react readily to form two intermetallic compounds, Cu_6Sn_5 and Cu_3Sn. As soon as the flux removes any tarnish films on the copper, wetting proceeds rapidly, presuming that proper precleaning has been performed. The precleaner, flux, solder, and postcleaner form a system of interrelated materials and processes. Soldering success requires careful consideration of each as a part of the whole process.

Iron and tin also form intermetallic compounds (FeSn and Fe_2Sn), but at a slower rate than for copper. Low-alloy steels can be soldered, but the cleaners and fluxes should be slightly more aggressive. Zinc and nickel also require stronger fluxes to dissolve the thicker oxide layers that form on these metals.

Aluminum forms a tenacious oxide layer upon exposure to air and is therefore difficult to solder, but is still solderable. Chromium, as contained in high-alloy steels, creates a similar situation. These oxide films interfere with soldering unless particularly aggressive cleaning and fluxing actions are used, perhaps with a special solder composition as well.

The most difficult-to-solder materials in the table above cannot be joined directly by normal soldering techniques. However, these materials can be plated with a solderable metal such as copper or nickel to effect joining. Excellent examples are the varieties of ceramic chips used for electronics applications. To render these solderable, even after storage, they may be given a zincate treatment to promote metallizing, followed by an electroless nickel plating to build up a solderable layer, and finally a tin or gold plating to preserve the solderable nickel surface during storage.

Actually, some ceramics can be wet and soldered directly by special indium-containing solder compositions. There are no immutable laws in soldering; there is great flexibility. The solderability of nearly any material can be altered by coating it with another material.

Coated Metals

As stated in the preceding section, certain materials require plating before they can be soldered. The wetting speed requirements of mass soldering, as used in electronics soldering, may necessitate plating to improve the solderability of metals such as nickel and beryllium copper which, under less rigorous conditions, have fairly good solderability. A properly designed coating will also allow long-term storage of a metal, while preserving its solderability for future use.

The first criterion for classifying coatings is whether they melt upon exposure to normal soldering temperatures. If so, they are termed fusible coatings. Metal coatings that do not melt are further subdivided according to their solubility in liquid solder as soluble or insoluble. There are also nonmetallic coatings used to preserve solderability. The need to store electronic components for use in repair, and as a natural consequence of production scheduling, has brought about much development work on solderable coatings and their applications.

Fusible Coatings. Two common fusible coatings are pure tin and near-eutectic tin-lead alloy compositions. In neither case is the coating anodic to the base metal. Therefore, the coating thickness must be sufficient to ensure an essentially pore-free layer; otherwise, corrosion products may appear at the pores and spread across the surface to later inhibit soldering. The minimum thickness required for this pore-free condition is about 200 μin. for both materials.

During normal storage, tin or tin-lead will slowly oxidize, but the oxide film remains thin. A weak flux dissolves these films or else the coating melts and moves, thereby disrupting the films. In some cases, the method of soldering may contribute to improved solderability, as in wave soldering, where wave turbulence helps to sweep away and replace a fusible coating.

Fusible coatings normally are applied by hot dipping or by a plating-type process. Successful hot dipping requires a wettable surface on the base metal to form a metallurgical bond during dipping. By centrifuging or air knife leveling of a hot dipped coating, a thickness of 200 to 300 μin. can be applied, thereby providing good protection and a pleasing appearance. Electroplating probably is used more often, especially for electronics soldering, because, until recently, it offered better control of coating thickness. Care must be taken, for tin or tin-lead can be easily electroplated over an unwettable base metal surface. The plating may appear sound, but when soldering is attempted, the surface will not wet properly.

A compromise coating widely used in the electronics industry is called a reflowed or flow-melted coating. It is applied by electroplating. The plating is then quickly heated to just above its melting point by immersion in hot oil or exposure to infrared radiation. This gives the coating a pleasing bright appearance and, more importantly, if wetting is accomplished without difficulty, the surface should remain solderable after storage, provided the coating thickness is sufficient. Reflowing thus adds to the processing an advantage inherent to hot dipping, i.e., the metallurgical bond.

It is possible to plate tin or tin-lead in a bright condition through the use of bath additions. These platings have the staining resistance of reflowed or hot dipped coatings without the need for subjecting the base metal to soldering temperatures. There is, unfortunately, still no assurance that the base-metal surface is solderable. Soldering can, in some instances, be impaired by the presence of the organic brightening materials included in the coating.

One type of base metal that should not be plated directly with tin or tin-lead is brass or similar zinc-containing alloys. Zinc from the base metal will diffuse quite rapidly through the tin or tin-lead and react with the atmosphere to form zinc corrosion products on the surface that impair solderability. This difficulty is avoided by applying a barrier layer of copper or nickel electroplated to at least 100 μin. thick before the tin or tin-lead plating.

Soluble and Insoluble Coatings. The distinction between soluble and insoluble coatings is sometimes a matter of coating thickness. Common soluble coatings include thin gold, zinc, and cadmium, of which gold is used most often. All soluble coatings present the risk of solder-joint contamination, but the problem is most acute with gold. The gold-tin intermetallic compound, $AuSn_4$, is hard and brittle and, if the gold content of a soldered joint exceeds about 5%, the joint will almost certainly be unacceptably brittle. Therefore, gold thicknesses above about 60 μin. are generally avoided. The problem with thin gold is the lack of corrosion protection.

One system that has been successful for printed circuit boards is an insoluble coating of 65%Sn-35%Ni, followed by a soluble thin gold layer. Tin-nickel about 300 μin. thick provides corrosion protection for the underlying copper conductor, while the thin gold, about 20 μin. thick, keeps the tin-nickel surface active and more receptive to soldering. Other insoluble coatings are nickel and copper, which can be applied to a wide variety of materials to improve and maintain their solderability. It is often desirable to protect the nickel or copper with a topcoat of tin or tin-lead. In nearly all cases, an investment in improved solderability is fully recouped by fewer flaws after soldering.

Nonmetallic coatings include rosin lacquers, which give only limited protection due to reaction with the atmosphere. Some evidence indicates that the reaction with the atmosphere may actually make the surface less solderable than if no lacquer had been used. If the storage time is short, however, the coatings will perform well.

Imidiazole is sometimes used to confer reproducible solderability on a component that may face long-term storage. Some evidence, however, indicates that this compound may react chemically with some materials (e.g., copper) and actually decrease their solderability on a short-term basis.

Solderability Testing

Solderability testing should be used to evaluate materials prior to actual assembly and soldering, particularly in mass soldering where thousands of joints may be made simultaneously in one assembly. One defective joint may cause rejection of an entire assembly, representing a very large investment. Also, some forms of accelerated aging may need to be included in the test to simulate the effects of storage on solderability, i.e., to determine the shelf life of base-metal surfaces. To assess solderability, one or more tests may be used, with the specific choice depending in part on the size and shape of the component. The five most commonly used tests are the vertical dip test, the rotary dip test, the surface tension balance test, the globule test, and the wave soldering test. These and several other tests are described in the paragraphs that follow.

Vertical Dip Test. A cleaned and fluxed test sample is suspended from a mechanical arm. A motor-cam arrangement causes the arm to lower the sample at a specified rate to a specified depth in a solder pot. The sample is immersed for a specified time of about 5 s, then removed from the bath, and the solder on it is allowed to solidify. The sample is cleaned and visually inspected for evidence of wetting, using criteria set out in the particular standard being followed.

Such a test has been specified in Military Standard 202 Method 208, in S-801 from the Institute for Interconnecting and Packaging Electronic Circuits (IPC), in Specifications B545 and B579 from ASTM, in the specification of the Electronic Industries Association (EIA), and in various other national and international specifications.

The rotary dip test is similar to the vertical dip test and is used mostly for printed circuit board testing. In this procedure, the test sample is lowered in an arc so that it skims the surface of the solder pot at the bottom of the arc of travel. This motion better simulates typical mass soldering processes. The sample is then removed, cleaned, and visually inspected for wetting. By adjusting the speed of rotation of the arm, a "minimum wetting time" can be determined. This procedure can be used to assess the solderability of plated through-holes and component leads, as well.

More widely specified outside the U.S., this test is nevertheless a part of ASTM B545 and has been proposed in IPC specifications work. Devices to perform the test are built ruggedly to withstand the rigors of day-to-day testing.

Surface Tension Balance Test. This is one of the more recently devised test systems and can be considered as a sophisticated version of the dip test. The test sample is suspended from an arm that is attached to a load cell. The solder pot typically is raised to control the immersion depth of the sample. A strip-chart recorder attached to the tester produces a record of the forces acting on the sample as a function of immersion time. An initial upward thrust caused by buoying will give way, if wetting occurs, to a downward-acting force called the wetting force. The time span between immersion and the emergence of the downward force gives a measure of the wetting time.

In general, the surface tension balance test is applicable to all types of solderable components, but it is most successful with those whose entire surface is meant to be solderable. If a fair proportion of the surface is nonsolderable, a buoyant force remains throughout the test, which tends to mask the effect of the wetting force. This test has been incorporated in Military Standard 883 and several company specifications. Although it requires more expensive, less rugged test devices, this test method is certain to be more widely specified in the future.

The globule test is used for round component leads. The lead is put in the test fixture, which lowers it so as to bisect almost completely a globule of molten solder of specified size on a heated aluminum block. When wetting occurs, the two parts of the globule flow together on top of the lead and trip a timer, giving a readout of the time elapsed since the lead entered the globule. This result is the wetting time.

The globule test is used widely in Europe, where it is specified for leads. It has also found some use in the U.S., although there are relatively few U.S. specifications covering its use. A variation of the test can be used for plated through-hole testing. Here, the molten globule is brought into contact with the bottom of the test hole and the timer measures the elapsed time until solder appears at the top of the hole. This variation is a recent development and has not yet been widely written into specifications.

Wave Soldering Test. More closely simulating actual production conditions is the wave soldering test of IPC specification S-803. A flat surface such as a printed circuit board is processed on a wave soldering machine as it would be during actual soldering, except that no components are loaded onto it. Besides the disadvantages of using a production soldering machine for testing or else buying a separate machine for testing, this test relies on visual inspection of the surface to evaluate wetting.

Area-of-spread tests have been applied in the past to evaluate large surfaces. A measured quantity of solder is placed on the fluxed surface of the specimen, which is placed on a heat source. The solder melts and flows outward to cover an area, the size of which is one measure of the solderability of the material. This test has been largely supplanted by the previously mentioned tests, which give a better indication of the kinetics of the important reactions.

Capillary penetration tests have also been devised, but these are used more in the mechanical joining of larger components than for electronic applications. The capillary penetration tests lack sensitivity, particularly for coated metals, and show only small differences between acceptable and unacceptable results. Like the area-of-spread test, this type of test is no longer widely used.

Applications of the Tests. Most solderability tests have been developed for use by the electronics industry, where mass soldering is the usual processing mode. In such applications it is not usually practical to use longer times, higher temperatures, or stronger fluxes in an attempt to improve solderability. Any of these adjustments could prove detrimental to long-term reliability. Other solder-using industries are beginning to appreciate the value of solderability testing as a quality-control technique. Altering a well-conceived soldering operation to accommodate less-than-adequate solderability is likely to lead to additional problems.

Accelerated Aging Tests. Storage of a solderable surface can result in two forms of deterioration. One arises from the reaction of the environment with the solder surface to cause tarnish films. These consist of oxides and, depending on the exact environment, may also include chlorides, sulfides, and carbonates. These films almost universally reduce solderability, and their effect can be aggravated by the presence of moisture. The other form of deterioration results from reaction at the interface between the tin or tin-lead coating and the base metal, if it forms intermetallic compounds with the tin. These compounds form and grow slowly, even at room temperature. Eventually, by a mechanism not well understood, their presence adversely affects solderability. Time and temperature of storage are the only important variables.

An accelerated aging test usually includes two parameters, moisture and elevated-temperature exposure. Beyond this, there seems to be little agreement at present. Military standards specify that a sample be given 1 h of exposure suspended above boiling water. This procedure has been used with other vertical dip tests. Recent work suggests that 1 h is too short to prove shelf life, and that boiling water tests may provide an insufficiently controlled condition. At the other extreme is a specification calling for 16 h of dry heat at a temperature of 310 °F. This exposure is regarded by some as too severe, but one accelerated aging procedure will not be universally applicable. Nevertheless, considerable effort is ongoing to generate the data needed to establish the desired correlation between accelerated aging and actual long-term storage.

Solder Joint Design

Solder alloys generally have lower strength properties than the materials to which they will be joined. Overall design of a product involving soldered joints must therefore be evaluated to ensure that the joints are capable of carrying the supplied loads for the expected life of the product. Stress-rupture and creep properties are therefore important to solder joints under load in service. Care must be taken also not to use bulk solder properties for this evaluation because these do not take into account the effect of joint formation, interfacial solder joint reactions, and stress transfer capabilities across soldered joints. In designing a product, several solder/base metal selections can be made that will adequately perform the task. In addition to design aspects, overall costs in materials and in manufacturing the product are usually taken into account. The lap joint provides a capability of conservative design by allowing larger areas of joints to be utilized at lower unit stress.

Most data available in the literature on joint strengths are not directly applicable to the design of a soldered joint. It is often necessary to fabricate sample parts and test the joints to ensure their producibility and capability regarding strength properties. Commonly used soldered joints are illustrated in Fig. 3, and self-jigging joints are illustrated in Fig. 4. The most definitive work available on loading of soldered joints is that used for load-carrying capabilities of copper tube with sleeve-type joints or fittings. Conservative joint design requires that only 50% of the joint be considered filled for design purposes. Under normal good practice for soldering, 70% or more of the joint can be expected to be of sound soldered joint material.

Automotive radiators are among the most common applications of soldering where the strength of the joint for both short- and long-term use is very important. Radiator tubes are normally made by fabricating a lock seam and joining this seam continuously in a soldering machine. Joints between the tubes and header plates are made by piercing a flat plate and inserting the ends of the tubes. These joints are essentially shear loaded in both directions as the radiator is heated, pressurized, and subsequently cooled. The tank-to-header joint of radiators is a trough-type joint with a peel-type joint load. The radiator is a good example of the utilization of joint design in a manner controlled by a need for rapid assembly and manufacture. Because several joints are involved,

Fig. 3 Joint designs frequently used in soldering. Solder joints terminology has not been standardized. (Source: American Welding Society Soldering Manual)

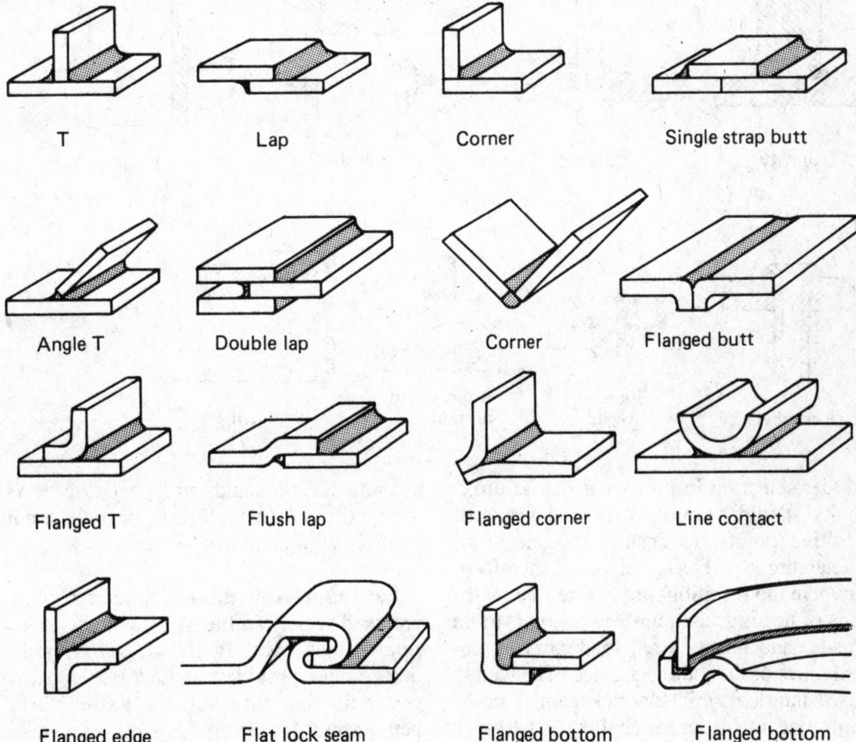

Fig. 4 Twenty-one methods that can be used to make solder joints self-jigging. Source: American Welding Society Soldering Manual

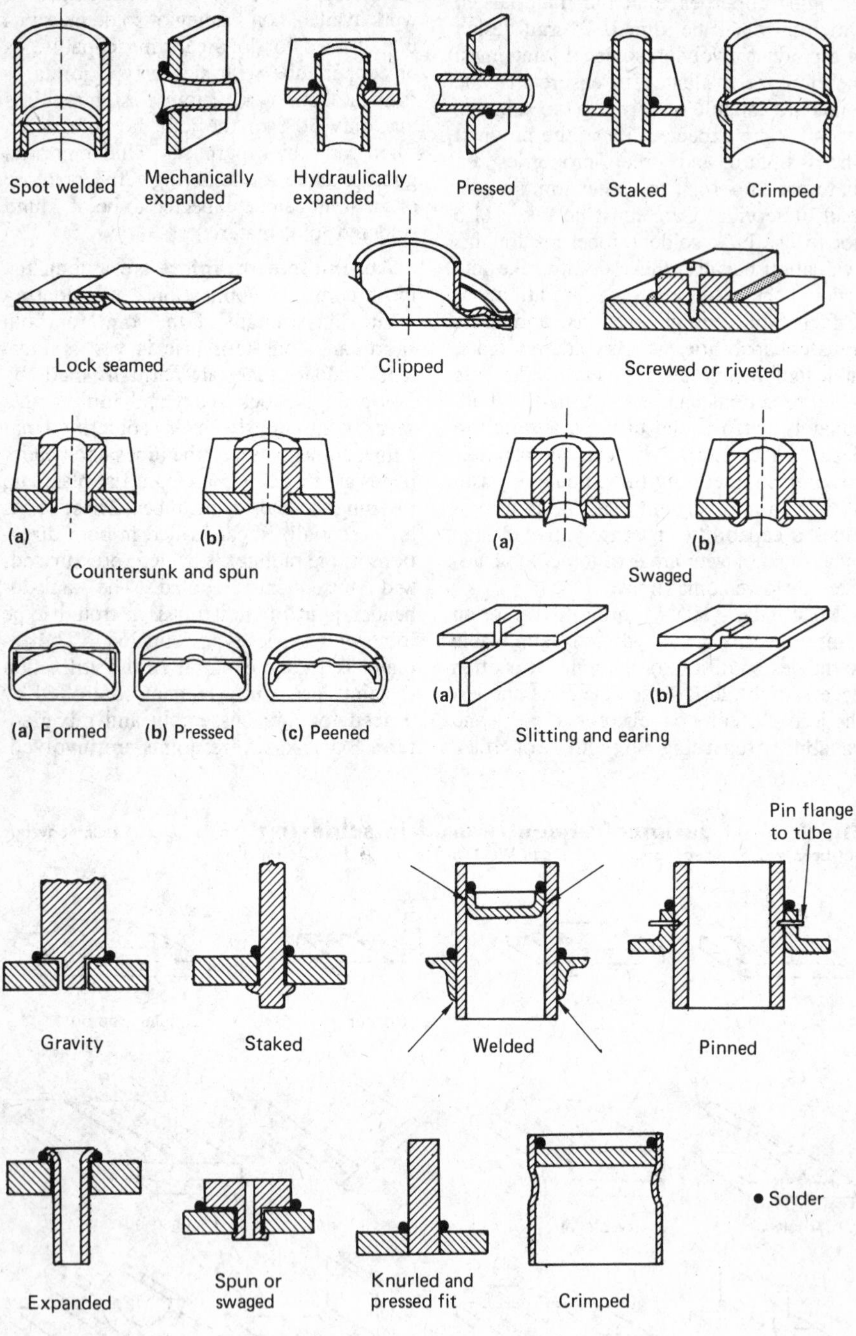

solder selection in joints with close proximity should be progressive, in terms of melting points, to ensure a sound completed product. Designed joints, therefore, involve the assembly procedure, the methods of heating, the numbers of joints to be made, and the required load-carrying capabilities of the joint in service. The widely used interlock joint or lockseam is probably one of the more challenging from a manufacturing standpoint, because it is more difficult to uniformly heat the joint prior to application of solder.

Electronic Assemblies. Soldered joints are used widely in the electrical and electronics industries. In this function, joint design must provide satisfactory electrical continuity and have sufficient strength to permanently fix the components and leads.

Joining leads to printed circuits is performed with different types of attachments. The shape of joint chosen is dictated by the proximity of other components, a need for efficient electrical transfer, and mechanical strength requirements. Clinched wires provide a strong mechanical attachment where some heating of a printed circuit board may occur, causing higher stresses on the soldered joint. Electrical connections can be made with strain-relieved leads to reduce the chances of thermal fatigue, and many alternate methods exist to ensure the required longevity and service of a soldered joint. The essence of any design, and this applies to electrical joining also, is that the solder alloy and joint material should not be stressed beyond their capabilities in terms of short-term or long-term strength or corrosion characteristics. Printed circuit board lands and wire insertions are illustrated in Table 8 and Fig. 5 and 6.

Soldering of printed circuit boards is carried out continuously on wave soldering machines. The designs of individual joints and heat sinks provided at each section of the board are important to overall capability of joining many components simultaneously. This has led to a wide variation in the areas of lands on boards at the joint area, the requirement for leads in terms of solderability, and thermal heat sinks. The compatibility among solder alloys, fluxes being used, and the required, resultant, sound soldered joint is important. As electronic packages become smaller and components are added in higher density configurations, the importance of joint design to suit the soldering process becomes more critical. Joint design is becoming more involved in the overall soldering techniques and metallurgy of a soldered interface. Critical component design and miniature sizing require more attention to this detail to ensure satisfaction and long service from the soldered joint. Verification trials for manufacturing procedures, testing of joint designs, and inspection for production capabilities are very important to the success of a soldering program. Constant review should be made of these factors in the design process to reduce, rework, and carefully control the costs and time necessary to produce a correctly joined package. Some typical data for electrical connections in design are shown in Table 8. Many joint configurations are illustrated in Tables 9, 10, and 11 as a guide to the flexibility of soldering as a method of joining. The capabilities of a soldering system are limited only by the imagination of the designer, as long as basic material properties are taken into account.

Table 8 Configurations of printed circuit lands

Type		Preferred direction for component lead	Solder fillet contour	Remarks
	Teardrop	Toward long end	Even, and almost round	Good design. Enlarged contact area
	Round	Any	Even and round	Universal pattern
	"D"	Toward tip	Uneven	Not widely used
	Rectangular	Toward a corner or long end	Uneven	Not widely used
	Delta	Toward base	Uneven	Used if space very limited

Precleaning and Surface Preparation

Oil, film, grease, tarnish, paint, pencil markings, cutting lubricants, and general atmospheric dirt interfere with the soldering process. A clean surface is imperative to ensure a sound and uniform quality soldered joint. Fluxing alone cannot substitute for adequate precleaning. Therefore, a variety of techniques are used to clean and prepare the surface of metal to be soldered. The importance of cleanliness and surface preparation cannot be overemphasized. These steps help ensure sound soldered joints, as well as a rapid production rate. Precleaning can also greatly reduce repair work due to defective soldered joints. Two general methods of cleaning are chemical and mechanical. The most common of these are degreasing, acid cleaning, mechanical cleaning with abrasives, and chemical etching.

Fig. 5 Methods of making wire-to-tag joints. Source: American Welding Society Soldering Manual

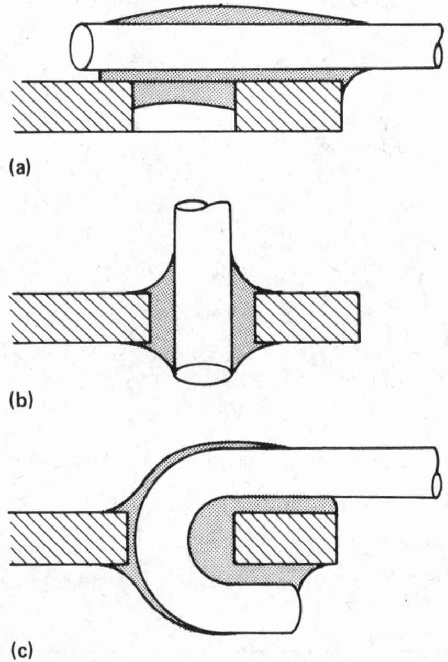

Degreasing. Either solvent or alkaline degreasing prior to soldering is recommended for cleaning oily or greasy surfaces. Of the solvent degreasing methods, the vapor condensation of halogenated hydrocarbon-type solvents probably leaves the least residual film on the surface. The cold articles to be degreased are suspended above the boiling solvent, causing the vapor to condense on the articles and drain back into the boiling liquid. Only clean, freshly distilled solvent contacts the material to be cleaned, so there is no recontamination to hinder the degreasing.

The least satisfactory method of degreasing is to rub the articles with a cloth saturated with solvent. In the absence of vapor degreasing apparatus, immersion in liquid solvents or in detergent solutions is often a suitable procedure. The efficiency of this method of cleaning can be considerably enhanced by incorporating ultrasonic cleaning. This method employs vibrational waves which, through cavitation, promote removal of soils, grit, or grease.

Alkali detergents are also used for degreasing. In general, a 1 to 3% solution of trisodium phosphate and a wetting agent is satisfactory. All cleaning solutions must be thoroughly washed from base-metal surfaces by steam or water before soldering. Whenever water is used, soft water is preferable, as residues from hard water may interfere with the soldering. These cleaning methods are especially designed for substantial volume and should be used according to proper safety precautions and suitability for the application.

Acid cleaning, or pickling, is used to remove rust and oxide scale from metal, which provides a chemically clean surface. Hydrochloric, sulfuric, orthophosphoric, nitric, and hydrofluoric acids can be used, either singly or mixed, for acid cleaning. Hydrochloric and sulfuric acids are the most often used. An inhibitor is sometimes added to prevent pitting of the base metal once the scale has been removed.

After pickling, if droplets of water show on the metal surfaces, traces of grease or other contaminants may still be on the surface and should be removed before proceeding. The articles should be thoroughly washed in hot water after pickling and dried as quickly as possible. For many electronic applications, such as printed circuit boards and component leads, spe-

Fig. 6 Methods of improving joint strength. (a) For single-sided boards, larger pads and longer leads increase solder mass. (b) A thinner double-sided board provides a longer path for crack propagation. (Source: American Welding Society Soldering Manual)

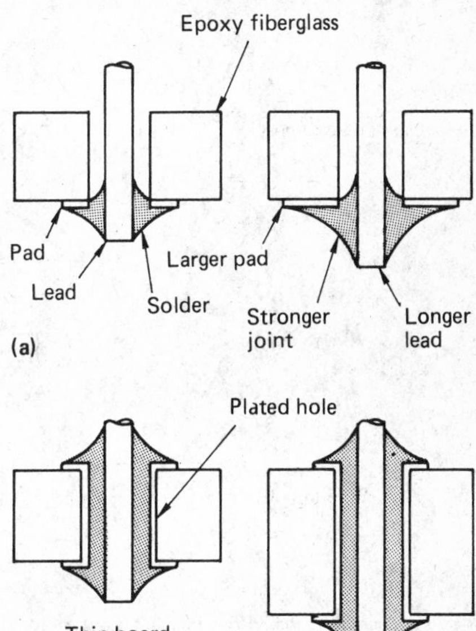

Table 9 Data for electrical connections design

Group 1—No mechanical security prior to soldering

No.	Type	Diagram	Controlling formula	Conditions	Fixtures	Current
Butt connections						
1	Round to round	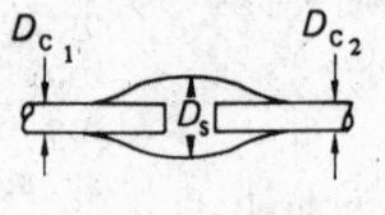	$D_S = \sqrt{\delta D_{C_1}}$	$\rho_{C_1} \geqslant \rho_{C_2}$ $D_{C_1} \leqslant D_{C_2}$	Yes	Small
2	Square to square	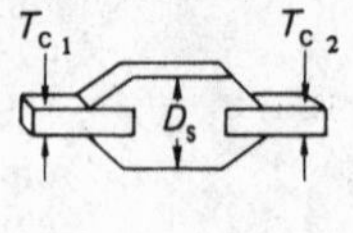	$D_S = \sqrt{\frac{4}{\pi}\delta T_{C_1}}$	$\rho_{C_1} \geqslant \rho_{C_2}$ $T_{C_1} \leqslant T_{C_2}$	Yes	Small
3	Rectangle to rectangle	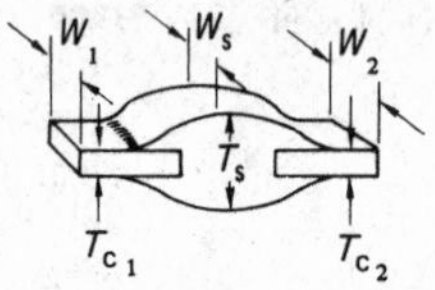	$T_S = \delta T_{C_1}$	$\rho_{C_1} \geqslant \rho_{C_2}$ $W_1 = W_2 = W_S$ $T_{C_1} \leqslant T_{C_2} \neq T_S$	Yes	Small
Lap connections						
1	Round to round(a)	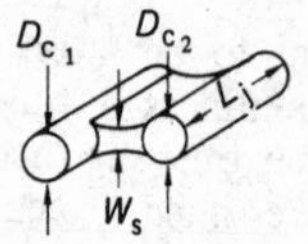	$L_j = \frac{\pi}{2}\delta D_{C_1}$	$\rho_{C_1} \geqslant \rho_{C_2}$ $D_{C_1} \leqslant D_{C_2}$ $W_S \geqslant \frac{D_{C_1}}{2}$	Yes	Large
2	Round to flat	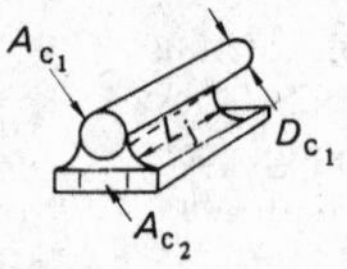	$L_j = \frac{\pi}{4}\delta D_{C_1}$	$\rho_{C_1} \geqslant \rho_{C_2}$ $A_{C_1} \leqslant A_{C_2}$	Optional	Large
3	Flat to flat	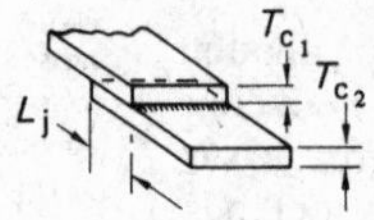	$L_j = \delta T_{C_1}$	$\rho_{C_1} \geqslant \rho_{C_2}$ $W_1 = W_2 = W_S$ $T_{C_1} \leqslant T_{C_2}$	Optional	Large
4	Wire to post	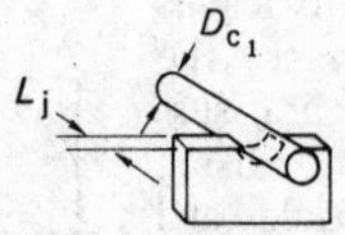	$L_j = \frac{1}{2}\delta D_{C_1}$	$\rho_{C_1} \geqslant \rho_{C_2}$ Solder fillet $\geqslant \frac{D_{C_1}}{2}$	No	Medium
5	Wire to cup	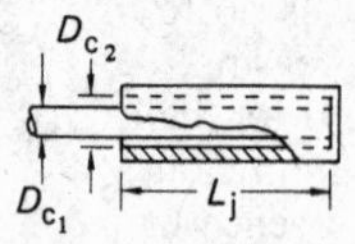	$L_j = \frac{1}{4}(\delta - 1)D_{C_1}$	$\rho_{C_1} \geqslant \rho_{C_2}$	No	Large
6	Wire to hole	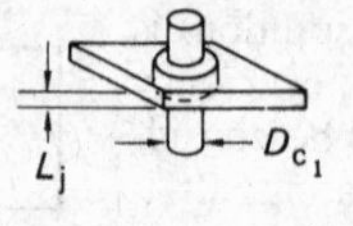	$L_j = \frac{1}{4}\delta D_{C_1}$	$\rho_{C_1} \geqslant \rho_{C_2}$	Optional	Medium

Note: D_{C_1} is diameter of smaller conductor; A_{C_1} is area of smaller conductor; S is solder; W is width; L_j is length of joint; T is thickness; N is number of turns; δ is resistivity ratio: $\frac{\rho_S}{\rho_{C_1}}$; ρ is resistivity in $\mu\Omega \cdot$cm.

(a) Use only when large conductor diameter is 3 to 4 times larger than small diameter; otherwise, use round-to-flat lap joint formula.

Source: Manko, H.H., How to Design the Soldered Electrical Connection, *Prod. Eng.*, 12 June 1961, p 57

Table 10 Data for electrical connections design
Group II—Partial mechanical security prior to soldering

No.	Type	Diagram	Controlling formula	Conditions	Fixtures	Current
Hook connections						
1	.......Round to round	D_{c_1} D_{c_2}	$D_{C_1} = \frac{2}{\delta} D_{C_2}$	$\rho_{C_1} \geqslant \rho_{C_2}$ $D_{C_1} \leqslant D_{C_2}$ Hook $\geqslant$ 180°	No	Large
2	.......Round to flat	D_{c_1} T_{c_2} L_j	$D_{C_1} = \frac{1}{\pi\delta}(8\,L_j + 4T_{C_2})$	$\rho_{C_1} \geqslant \rho_{C_2}$ $A_{C_1} \leqslant A_{C_2}$ Hook $\geqslant$ 180°	No	Medium

Note: D_{C_1} is diameter of smaller conductor; A_{C_1} is area of smaller conductor; L_j is length of joint; T is thickness; δ is resistivity ratio: $\frac{\rho_S}{\rho_{C_1}}$; ρ is resistivity in $\mu\Omega \cdot$cm.
Source: Manko, H.H., How to Design the Soldered Electrical Connection, *Prod. Eng.*, 12 June 1961, p 57

cial mild proprietary surface cleaners and solutions are available.

Mechanical Preparation with Abrasives. A commonly used method of cleaning is abrasion, which consists of grit or shotblasting, mechanical sanding or grinding, filing or hand sanding, cleaning with steel wool, wire brushing, or scraping with a knife or shave hook. For best soldering results, cleaning should extend beyond the joint area. A simple solderability test should be performed following abrasive cleaning. Care should be taken to avoid embedding abrasive grit in the surface, because this will affect solderability.

Fluxes

Flux technology is multifaceted and complex. Published data generally do not suggest any relationship among flux composition, strength properties of joints, and optimized processing, but experimentation has clearly proved that these factors are inextricably intertwined. Earlier work on contact angles, where flux composition was determined to influence the tin-lead content at which best spread values are achieved, and work which showed that precise chemical additive levels optimize spread values are very important. These observations demonstrate how specific the flux-solder-processing parameter relationship is.

Inorganic salts, organic salts, solvents, and acids form their own binary eutectics (and other complex physical relation-

Table 11 Data for electrical connections design
Group III—Full mechanical security prior to soldering(a)

No.	Type	Diagram	Controlling formula	Conditions	Fixtures	Current
Wrap connections						
1	.......Round to round	D_{c_1} L_j D_{c_2}	$L_j = \frac{\pi}{2}\,\delta D_{C_1}$	$\rho_{C_1} \geqslant \rho_{C_2}$ $D_{C_1} \leqslant D_{C_2}$ $N > 1$	No	Large
2	.......Round to flat	T_{c_2} L_j D_{c_1}	$D_{C_1} = \frac{8}{\pi\delta}(L_j + T_{C_2})$	$\rho_{C_1} \geqslant \rho_{C_2}$ $A_{C_1} \leqslant A_{C_2}$ $N = 1$	No	Medium
3	......Round to post	D_{c_1} D_{c_2}	$D_{C_1} = \frac{4N}{\delta} D_{C_2}$	$\rho_{C_1} \geqslant \rho_{C_2}$ $D_{C_1} < D_{C_2}$ $N \geqslant 1$	No	Large

Note: D_{C_1} is diameter of smaller conductor; A_{C_1} is area of smaller conductor; L_j is length of joint; T is thickness; N is number of turns; δ is the resistivity ratio: $\frac{\rho_S}{\rho_{C_1}}$; ρ is resistivity in $\mu\Omega \cdot$cm.
(a) In cases where loosening or breaking of the joint would result in a hazardous condition, mechanical security should be specified.
Source: Manko, H.H., How to Design the Soldered Electrical Connection, *Prod. Eng.*, 12 June 1961, p 57

ships), which influence flux melting characteristics and fluxing speed. Additions are not necessarily additive in effect, and benefits gained by small additions of a salt or acid may be destroyed by higher concentrations. Substances used in fluxes are classified as follows:

Inorganic

- Zinc chloride
- Ammonium chloride
- Stannous chloride
- Hydrochloric acid
- Orthophosphoric acid
- Hydrofluoric acid
- Fluoboric acid
- Boron trifluoride
- Sodium chloride
- Potassium chloride
- Lithium chloride
- Cupric chloride
- Hydrogen
- Dry hydrochloride

Organic

- Glutamic acid
- Aniline hydrochloride or aniline phosphate
- Hydrazine hydrobromide
- Lactic acid
- Oleic acid
- Stearic acid
- Urea
- Rosins
- Abietic acid
- Phthalic acid
- Ethylene diamine
- Palmitic acid

Vehicles

- Water
- Glycerine
- Petroleum jelly
- Methylated spirit
- Isopropyl alcohol
- Polyethylene glycol
- Turpentine
- Wetting agents

The equilibrium diagrams shown in Fig. 7 illustrate solid and liquid states of four salt mixtures.

The need for specific fluxes for specific tasks has not been extensively commercialized, except in the electronics industry. The need for better flux control by the user, with quality tests on as-received flux, simulated process tests, and corrosion evaluation, is clear.

Physical and chemical methods for measuring flux effectiveness, which should be an integral part of any well-managed processing quality-control program, are available. One way to ensure good joints is pH control of fluxes, because a definite relationship exists among pH, chloride ion, and solder spread. Fluxes of low-adjusted pH and chloride content have lower corrosive aftereffects. Resin-based fluxes are used without any halides in some applications for best corrosion resistance.

Fig. 7 Typical flux salt equilibrium diagrams. Source: Copper Development Association

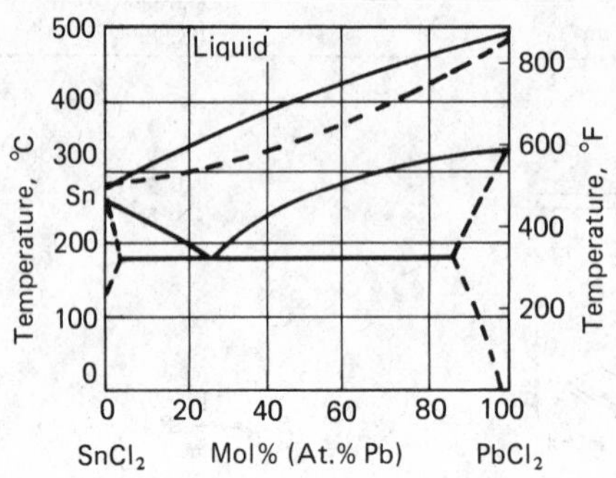

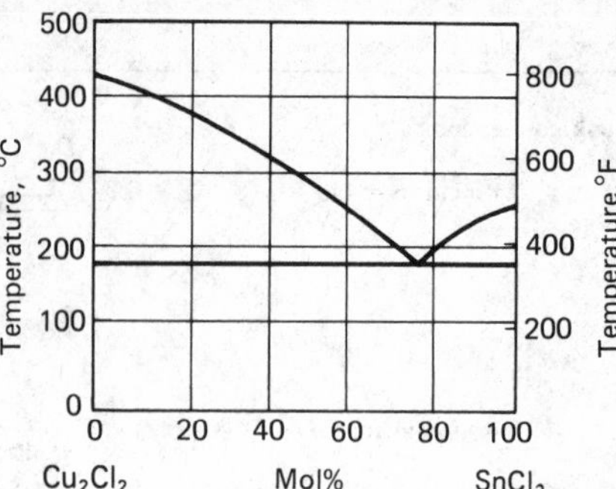

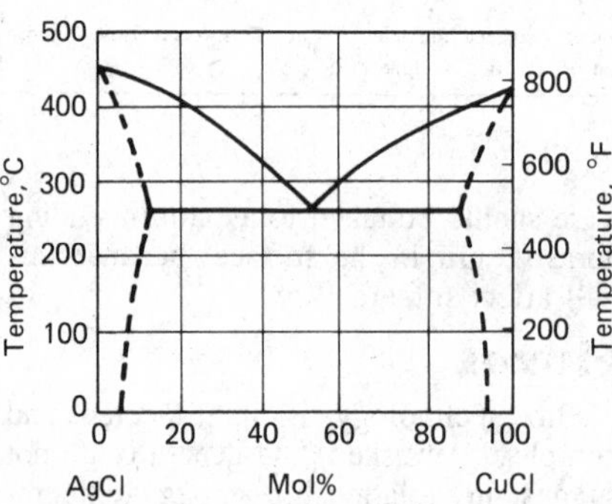

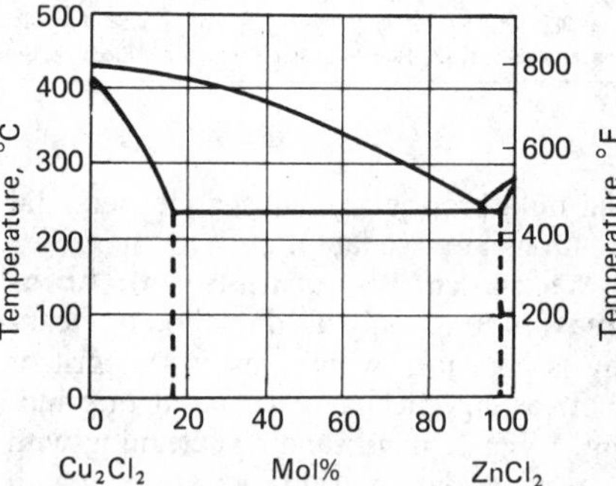

Fluxes with a highly active chloride content produce low contact angles at higher lead contents than rosin-only fluxes. Addition of a tin salt to the flux accelerates this tendency. Electrochemical action in which the solder is the anode and tin from the flux is deposited on the copper surface in advance of the moving solder pool is thought to be the reason for better results on spreading when a tin salt is included in the flux.

Fluxes include a wide variety of inorganic and organic substances, as the list above shows. Typical acids and salts used are zinc chloride, ammonium chloride, stannous (tin) chloride, sodium or potassium chloride, lithium chloride, aluminum chloride, hydrochloric acid, orthophosphoric acid, and sometimes, for difficult cases, fluorides. Zinc chloride is most common, with a melting temperature well above the solidus of most tin-lead solders. For a lower melting flux, this salt is combined with ammonium chloride, with which it forms a eutectic at 350 °F. Dilute hydrochloric acid sometimes is present in the flux to serve as an activator.

Molten solder reacts with metallic salts in fluxes. This action can influence solder wetting of the base metal. Active surface adsorption of heavy metal ions promotes wetting; for example, when a copper salt is used in the flux, an improvement in wetting and spreading on copper results. Similarly, when a nickel salt is included, improvements are obtained on a nickel surface. Copper stearate causes rapid wetting and spreading on copper substrates.

A basic requirement is that the metal salt be more noble than the base metal in the particular solvent at the fluxing conditions to be used in soldering for it to activate and produce the improved spreading characteristics. Elements potentially useful as additions are nickel, silver, lead, palladium, copper, and tin. A very specific metal salt addition is required for optimum fluxing effectiveness, as illustrated in Fig. 8. The percentage of tin and lead in the solder composition and the optimum value of salt addition for maximum spreading on a surface is specific to the salt and solder alloy.

Fluxes that are easily removed from the surface of the soldered metal are desirable. Organic acids and compounds are used as mild activators. Chlorides and

Fig. 8 Spreading of tin solder on copper plate. With aniline-based flux at 480 °F, with added organic salts. (Source: Copper Development Association)

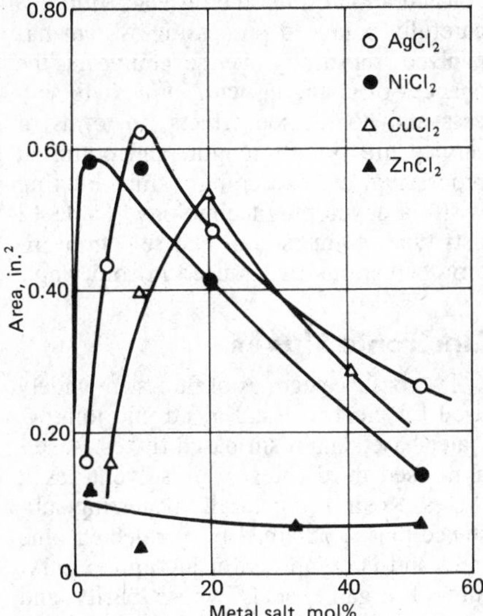

bromides are used at temperatures of about 400 to 500 °F. These salts decompose slowly. Naphthalene tetrachloride and naphthalene tetrabromide, upon heating, release hydrogen chloride without forming reaction products and permit soldering with minimum residue. Ammonium chloride works similarly in breaking down to ammonia, then to nitrogen and hydrogen, during soldering to act as a reducing and protective agent.

Less active fluxes are especially useful when minimum quantities of flux are allowable. When they are used, care in heating ensures that decomposition does not occur too soon. Typical constituents in this less active group are lactic acid, citric acid, oleic acid, stearic acid, and glutamic acid. Totally noncorrosive fluxes are claimed by some manufacturers.

Rosin-based and solvent-based fluxes are used for lower melting point solders. Water-based fluxes predominate for high-temperature soldering operations. The desirable properties of a soldering flux can be defined as:

- A melting point slightly below the soldering temperature
- Ability to remove oxides
- Ability to protect metal surfaces from further oxidation
- Electrically nonconducting after soldering, continuous, and protective

The function of a flux in soldering is to remove the oxide film from the base metal by reacting with or otherwise loosening that film from the base-metal surface. The molten flux then forms a protective blanket which prevents re-formation of the oxide film until molten solder displaces the flux and reacts with the base metal to form an intermetallic bond. Many good fluxes dissolve oxides poorly but remove them by "tunneling" beneath the oxide adjacent to the clean metal surface.

The effectiveness of a flux is often measured by a spreading test and the determination of the contact angle between solder and base material when the flux is used. Low contact angles mean good solder alloy spread in a joint.

Each flux has an optimum processing temperature and solder joint strength relationship. As soldering temperature increases for a given flux, flux activity increases, producing better-shaped and stronger joints until an optimum fluxing temperature range is reached. Then, at higher temperatures, the flux begins to break down, or to react too quickly, and the soldered joint appearance worsens as the flux burns, chars, or is otherwise degraded into a vapor and leaves the joint surfaces. Fluxes can be designed for optimum processing at any temperature above the liquidus of any given solder alloy. A family of fluxes can be generated that are useful for a particular solder alloy from about 450 to 1100 °F.

The relationship among solder, base metal, and flux is such that a flux that is optimum for one solder composition is not necessarily the best flux for a different composition. Fluxing action is a separate, independent function and depends on a combination of metallurgical, chemical, and thermal considerations. Few attempts have been made to produce fluxes specifically matched to particular solders. No fluxes are ideal for all jobs, even within a single plant. At present, there are no standard tests that adequately determine the suitability of a flux for a particular job.

Where fluxes are used in bulk, a useful measure of the concentration uniformity of a flux can be achieved by monitoring the electrical conductivity of the flux solution and providing alarm limits to signal when the solution is either too weak or has become overly concentrated by evaporation.

Corrosivity of residual flux deposits is important. A water extract method of measuring corrosivity through conductivity has been developed. The conductivity test was readily established because work experience showed a relationship between general corrosion tendencies and conductivity values. Rapid quantitative evaluation of flux corrosiveness can be defined and may be relied upon to give a good indication of the general corrosion properties of a flux. Extracts from flux residues can also be evaluated.

The three basic methods of flux residue detection are radioactive tracers, standard chemical analysis, and fluorescent dye evaluation. The fluorescent dye technique provides the best means of determining the efficiency of a flux removal process. Minute amounts of dye in the remaining flux readily fluoresce under ultraviolet light. The method can be used to test the efficiency of washing techniques.

Flux activity studies show that certain active ingredients control the operating rate of a flux. Equally important, totally inert ingredients can dilute the active ingredient to provide a stable carrying medium for it.

The minimum temperature of activity of a flux is determined by either its melting temperature or component breakdown temperature, where initially inactive materials become active. A second activity level is determined by chemical interactions of the flux component materials, the base metal, and the solder being used. The rate of fluxing action increases with respect to temperature unless some specific reaction interferes. Fluxing duration (at temperature) depends upon chemical breakdown, usage of all active ingredients, and completion of interactive effects with base and solder metals. Mechanical property relationships (with fluxing action) can be identified in terms of temperature, time, joint strength, and specific flux activity.

Corrosive general-purpose fluxes are effective on low-carbon steel, copper, brass, and bronze. Applications are in the production of auto radiators, air conditioning and refrigerating equipment, and sheet metal assembly. Compositions of these fluxes include:

1	Zinc chloride	40 oz
	Ammonium chloride	4 oz
	Water to make	1 gal
2	Zinc chloride	36 oz
	Sodium chloride	10 oz
	Ammonium chloride	1/2 oz
	Hydrochloric acid	1 oz
	Water to make	1 gal
3	Zinc chloride	21 oz
	Sodium chloride	6 oz
4	Zinc chloride	25 oz
	Ammonium chloride	3 1/2 oz
	Petroleum jelly	65 oz
	Water	6 1/2 oz

Flux composition 3 is a dry flux for molten solder cover in dip soldering.

Additional fluxes for limited purposes are:

For stainless steel and galvanized iron	
1 Zinc chloride	85 oz
Ammonium chloride	6½ oz
Stannous chloride	9 oz
Hydrochloric acid	2 oz
Water to make	1 gal
Wetting agent (optional)	0.1 wt%
For stainless steel	
2 Zinc chloride	48 oz
Ammonium chloride	5 oz
Hydrochloric acid	3 oz
Water to make	1 gal
Wetting agent (optional)	0.1 wt%
For Monel	
3 Zinc chloride	16 oz
Ammonium chloride	16 oz
Glycerin	16 oz
Water	1 pint
For high tensile manganese, bronze, copper, or brass	
4 Orthophosphoric acid (85%)	34 oz
Water	16 oz
For cast iron	
5 Zinc chloride	32 oz
Ammonium chloride	4 oz
Sodium chloride	8 oz
Hydrochloric acid	8 oz
Water to make	1 gal
For cast iron	
6 Zinc chloride	40 oz
Ammonium chloride	4 oz
Hydrofluoric acid	1¼ oz
Water to make	1 gal
Paste flux for soldering aluminum	
7 Stannous chloride	83 oz
Zinc dihydrazinium chloride	7 oz
Hydrazine hydrobromide	10 oz
Water	10 oz
For soldering aluminum	
8 Cadmium fluoboride	5 oz
Zinc fluoboride	5 oz
Fluoboric acid	6 oz
Diethanol amine	20 oz
Diethanol diamine	4 oz
Diethanol triamine	10 oz
9 Potassium chloride	45 oz
Sodium chloride	30 oz
Lithium chloride	15 oz
Potassium fluoride	7 oz
Sodium pyrophosphate	3 oz
10 Triethanolamine	25 oz
Fluoboric acid	3 oz
Cadmium fluoborate	2 oz
11 Stannous chloride	44 oz
Ammonium chloride	5 oz
Sodium fluoride	1 oz
12 Zinc chloride	44 oz
Ammonium chloride	5 oz
Sodium fluoride	1 oz

Composition 10 is a chloride-free organic flux for soldering aluminum, with a fluxing range of ≈350 to 525 °F. The viscous liquid can be dissolved with water or alcohol to any desired concentration.

Composition 11 is a reaction-type flux for soldering aluminum, with a fluxing range of ≈540 to 720 °F or higher. It may be used as a dry powder mixture or it may be suspended in alcohol.

Composition 12 is a reaction flux for soldering aluminum. It may be used as a dry powder or mixed with water or alcohol.

Intermediate fluxes contain organic compounds that decompose at soldering temperatures. When properly used, the mildly corrosive elements in the flux volatilize, leaving a residue relatively inert and easily removed with water. They are effective on all materials that are solderable with mild fluxes. Typical compositions are as follows:

1 Glutamic acid hydrochloride	19 oz
Urea	11 oz
Water	1 gal
Wetting agent	0.2 wt%
2 Hydrazine monohydrobromide	10 oz
Water	90 oz
Nonionic wetting agent	1/20 oz
3 Lactic acid (85%)	9 oz
Water	42 oz
Wetting agent	1/10 oz

Noncorrosive fluxes are the rosin-based fluxes—nonactivated, mildly activated, and activated. For all electronic and critical soldering applications, water-white rosin dissolved in an organic solvent (item 1 below) is the safest known flux. Activators added to the rosin increase activity, but the flux residue from these fluxes should pass tests for noncorrosivity and nonconductance when used on electronic applications. These fluxes are effective on clean copper, brass, bronze, tinplate, terneplate, electrodeposited tin, and in alloy coatings, cadmium, nickel, and silver.

1 Water-white rosin	10 to 25 wt%
Alcohol, turpentine, or petroleum	rem
2 Water-white rosin	40 wt%
Glutamic acid hydrochloride	2 wt%
Alcohol	rem
3 Water-white rosin	40 wt%
Cetyl pyridinium bromide	4 wt%
Alcohol	rem
4 Water-white rosin	40 wt%
Stearine	4 wt%
Alcohol	rem
5 Water-white rosin	40 wt%
Hydrazine hydrobromide	2 wt%
Alcohol	rem

The knowledge available today makes it possible, by careful work, to optimize fluxing, thereby improving processing. Fluxes should not be changed without due consideration of all other effects. Where a carefully balanced processing system has evolved empirically over several years, the chances of changing with immediate success and no ill side effects, in terms of service life, joint strength, and optimum processing, are exceedingly slim. Fluxing is still a developing technology. Table 12 lists type, composition, and selection criteria of various fluxes used in soldering.

Electronic Fluxes

Two basic categories of fluxes are widely used for electrical and electronic joining: water-based and rosin-based fluxes. A seldom used third category is solvent-based fluxes. Rosin is a naturally occurring substance that is obtained from selected pine trees and is composed of a complex mixture of organic acids whose purity and composition depend upon the type of tree and its growth history. The basic material is usually purified for flux usage to a water-white condition. More recently, chemically synthesized materials are being used for fluxing operations. Rosin alone will work as a soldering flux because of the organic acids it contains. Chemical compounds called activators are utilized to increase the cleaning capability of these fluxes. Activity levels have been divided into three levels: rosin (labeled R), rosin mildly activated (labeled RMA), and rosin activated (labeled RA). The R-type fluxes are the weakest of the three. The RMA fluxes contain small amounts of activators and are widely utilized in electrical applications. The RA fluxes contain greater amounts of active fluxing materials and are used in hard-to-solder situations.

A range of water-soluble fluxes is available for use in the electrical industries. These materials tend to be used most widely in the appliance industries, where corrosion of residues may not be as difficult a problem as with instrumentation and fine electronic applications. Water-soluble fluxes may use water or an alcohol base.

Organic Fluxes. Organic moderately active fluxes are comprised largely of glutamic acid and stearic acid dissolved in water or alcohols. These fluxes are less active than inorganic fluxes but more active than rosin fluxes. They can be used for some electrical soldering.

Rosin Fluxes. Rosin is the base of least-active fluxes used for electronic soldering. It is unique because it becomes active as

Table 12 Flux selection

Type	Composition	Carrier	Uses	Temperature stability	Ability to remove tarnish	Corrosiveness	Recommended cleaning after soldering
Inorganic							
Acids	Hydrochloric, hydrofluoric, orthophosphoric	Water, petrolatum paste	Structural	Good	Very good	High	Hot water rinse and neutralize; organic solvents
Salts	Zinc chloride, ammonium chloride, tin chloride	Water, petrolatum paste, polyethylene glycol	Structural	Excellent	Very good	High	Hot water rinse and neutralize; 2% HCl solution; hot water rinse and neutralize; organic solvents
Organic							
Acids	Lactic, oleic, stearic glutamic, phthalic	Water, organic solvents, petrolatum paste, polyethylene glycol	Structural, electrical	Fairly good	Fairly good	Moderate	Hot water rinse and neutralize; organic solvents
Halogens	Aniline hydrochloride, glutamic hydrochloride, bromide derivatives of palmitic acid, hydrazine hydrochloride (or hydrobromide)	Same as organic acids	Structural, electrical	Fairly good	Fairly good	Moderate	Same as organic acids
Amines and amides	Urea, ethylene diamine	Water, organic solvents, petrolatum paste, polyethylene glycol	Structural, electrical	Fair	Fair	Noncorrosive normally	Hot water rinse and neutralize; organic solvents
Activated rosin	Water-white rosin	Isopropyl alcohol, organic solvents, polyethylene glycol	Electrical	Poor	Fair	Noncorrosive normally	Water-based detergents; isopropyl alcohol; organic solvents
Water-white rosin	Rosin only	Same as activated	Electrical	Poor	Poor	None	Same as activated water white rosin but does not normally require post-cleaning

a flux when heated and returns to an inactive state when cooled. Within the family of rosin fluxes, there are three recognized classifications:

- *Activated rosin*: Rosin fluxes to which small amounts of strong activating agents have been added to improve the fluxing action
- *Mildly activated rosin*: Rosin fluxes to which have been added mild activating agents to improve the fluxing action
- *Non-activated rosin:* Rosin fluxes with no activating agents

Rosin fluxes find their greatest applications in high-reliability electronic soldering because the residues are the least corrosive and conductive of all of the flux types.

Fluxes can be supplied as a liquid, paste, or solid or combined with solder, in core solder, or as paste solder. Table 13 may be used in selecting fluxes for soldering base metals. Where more than one flux is effective on a given base metal, the mildest flux should be used. Soldering of aluminum is a difficult process that requires special fluxes and alloys.

The Soldering Process

In addition to surface preparation, solder selection, and fluxing, another important part of the soldering process is the

Table 13 Metal solderability chart and flux selector guide

Metals	Solderability	Rosin fluxes: Non-activated	Rosin fluxes: Mildly activated	Rosin fluxes: Activated	Organic fluxes (water soluble)	Inorganic fluxes (water soluble)	Special flux and/or solder
Platinum, gold, copper, silver, cadmium plate, tin (hot dipped), tin plate, solder plate	Easy to solder	Suitable	Suitable	Suitable	Suitable	Not recommended for electrical soldering	...
Lead, nickel plate, brass, bronze, rhodium, beryllium copper	Less easy to solder	Not suitable	Not suitable	Suitable	Suitable	Suitable	...
Galvanized iron, tin-nickel, nickel-iron, low-carbon steel	Difficult to solder	Not suitable	Not suitable	Not suitable	Suitable	Suitable	...
Chromium, nickel-chromium, nickel-copper, stainless steel	Very difficult to solder	Not suitable	Not suitable	Not suitable	Not suitable	Suitable	...
Aluminum, aluminum-bronze	Most difficult to solder	Not suitable	Not suitable	Not suitable	Not suitable	...	Suitable
Beryllium, titanium	Not solderable	...	...	...	...	...	...

choice of best heating method. Available methods are discussed below.

The Soldering Iron or Bit

A common method of soldering is by means of a soldering iron or bit, which may be heated electrically, by direct flame, or by oven. Because soldering is a heat transfer process, the maximum surface area of the heated tip should contact the base metal during soldering. In making a joint, the solder itself should not be melted upon the tip of the iron. If flux cored wire is used, such a practice destroys the effectiveness of the flux.

A large variety of size and design of irons are available, from a small pencil size to special irons or bits that weigh 5 lb or more. The selection of the iron depends on the task and how much heat is needed at the joint, with the heat recovery time of the iron being fast enough to keep up with the job.

Plain copper bits and iron-coated bits are commonly used. Both types must be kept well tinned (coated with molten solder) during use. Oxidized iron tips should never be filed, because the thin electroplated layer will be removed and the tip ruined. Regardless of the way in which they are heated, bits perform the following functions:

- Transfer heat from the heat source to the parts being soldered
- Bring the workpiece joint area to the soldering temperature
- Store molten solder
- Convey molten solder
- Withdraw surplus molten solder

Flame or Torch Soldering

Where fast soldering is necessary, flame is frequently used. The flame temperature is controlled by the fuel mixture used. Fuel gas burned with oxygen gives the highest flame temperature possible with that gas. The highest flame temperatures are attained with acetylene, and lower temperatures are obtained with propane, butane, natural gas, and manufactured gas.

Hot Dip Soldering

When conducted properly, dip soldering may be very useful and economical, because an entire assembly, comprising any number of joints, can be soldered merely by dipping in a bath of molten solder. However, jigs or fixtures must be used to contain the unit and keep the proper clearance at the joint until solidification of the solder takes place.

The molten bath supplies both the heat and the solder necessary for completing the joint. The soldering pot capacity should be large enough so that at a given rate of production the units being dipped will not appreciably lower the temperature of the solder bath. A preliminary treatment of the assembly to be soldered, such as degreasing, cleaning, and fluxing, is required before dipping.

Electronic components are commonly soldered to printed circuit boards by a process called wave soldering, which is described in detail in a later section of this article. This approach lends itself to automatic mass soldering and is a continuously pumped crest of solder that peaks at the printed circuit board. Another method is "drag" soldering, which solders by use of a bath into which the board is automatically dipped with controlled time.

Both of these techniques can be highly automated, incorporating a fluxing station, preheating or drying station, soldering station, and cleaning area. Automatic mass soldering speeds can range from 2 to 16 ft/min. Modern mass soldering machines offer adjustable tracks, variable carriers, and a variety of options to accommodate a large variety of workpieces.

Induction Soldering

The only requirement for a material that is to be induction soldered is that it be an electrical conductor. The rate of heating is dependent on the induced current flow. The distribution of heat obtained with induction heating is a function of the induced current frequency. The higher frequencies concentrate the heat at the surface of the workpiece.

Several types of equipment are available for induction heating: the vacuum tube oscillator, the resonant spark gap, the motor-generator unit, and solid-state units. Induction heating is generally applicable for soldering operations, with the following requirements:

- Large-scale production
- Application of heat to a localized area
- Minimum oxidation of surface adjacent to the joint
- Good appearance and consistently high joint quality
- Simple joint design that lends itself to mechanization

Resistance Soldering

In resistance soldering, the workpiece is placed either between a ground and a movable electrode or between two movable electrodes as part of an electrical circuit. Heat is applied to the joint both by the electrical resistance of the metal being soldered, and by conduction from the electrode, which is usually carbon. Production assemblies may utilize multiple electrodes, rolling electrodes, or special electrodes, depending on which will be advantageous with regard to soldering speed, localized heating, and power consumption. Resistance soldering electrode bits generally cannot be tinned, and the solder must be fed directly into the joint.

Furnace Soldering

Furnace soldering should be considered when entire assemblies can be brought to the soldering temperature without damage to any of the components; when production is sufficiently great to allow expenditure for jigs and fixtures to hold the parts during soldering; and when the assembly is complicated in nature, making other heating methods impractical.

Infrared Soldering

Heating by radiative transfer is commonly called infrared heating. The most common sources of infrared heating for soldering applications are heated filaments. The quartz-iodine, tungsten filament lamp is widely used because it is very stable and reliable over a wide range of temperatures.

In general, infrared soldering systems are simple and inexpensive to operate. Variations in the condition of the solder surface can be compensated for to some extent by adjusting the heating power, but one of the most critical operating parameters is surface condition. Advantages are process repeatability, ability to concentrate or focus the energy with reflectors and lenses, economy of operation, and absence of contact with the workpiece.

Ultrasonic Soldering

Ultrasonic soldering can be considered a form of dip soldering, in which the workpiece is immersed in a tank of molten solder. When a properly designed transducer (source of ultrasonic energy) is energized in a bath of molten solder, it produces cavitation in the molten solder around the end of the transducer and at the joint surfaces. Aided by the solvent action of the molten tin, the cavitating solder actually scrubs, attacks, and removes the surface compounds and soils found on the workpiece, much as a flux would do in the usual solder operation.

After cleaning, which takes only a fraction of a second, solder wets the clean metal and is deposited on it. The thickness of the deposit is a function of immersion time

Fig. 9 Principle of flow soldering. (a) Flat dipping illustrates the curved path the panel must take across the solder surface. (b) Flow solder illustrates the straight path of the panel through the solder wave.

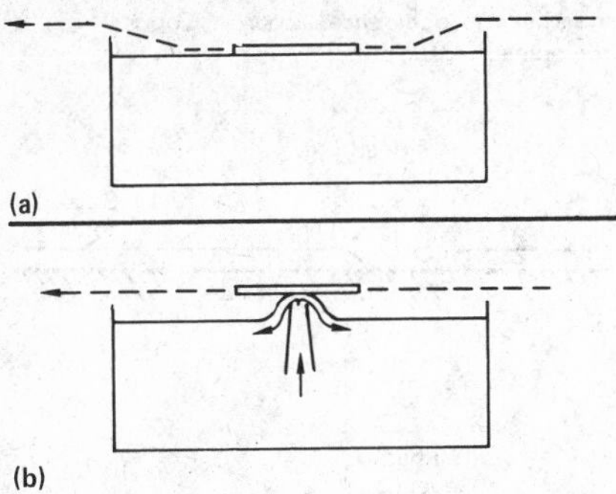

and solder temperature. Ultrasonic soldering is fluxless and enables the soldering of metals otherwise considered difficult to solder, such as Kovar, nickel alloys, and aluminum.

Wave Soldering

Wave soldering became possible by the introduction of the liquid wave principle, which replaces the stagnating surface of a static solder bath with a continuously maintained, gently overflowing liquid wave or fountain (Fig. 9) of solder. This wave is created and maintained by the continuous circulation of hot, liquid solder driven through a narrow nozzle. The solder ejected from the nozzle creates a solder head or crest which overflows the nozzle edges and drains back into the solder reservoir of the machine. The solder wave has a perfect horizontal crest and a mirror-like, steady, continuously replenished wave surface characterized by laminar flow characteristics.

Wave soldering is performed by passing a printed circuit board across the crest of the wave, whereby the bottom side of the board touches the top portion of the hot solder wave, which supplies both the heat and the solder for the solder joints to be formed. The combined effects of capillary action, complemented by slight wave pressure, result in the formation of sound, reliable solder joints.

The practical and reliable performance of the soldering operation requires a conveyor to move the boards and auxiliary operational stages to perform certain supplementary functions required to make sound solder joints, such as fluxing and preheating.

Three operational stages—fluxing, preheating, and soldering—are the functional elements of wave solder production lines or systems. Integrated systems range in size from miniature to very large.

Wave solder systems are available in different levels of design sophistication; the use of any particular one depends on the specific application and the requirements of productivity, economics, and quality and reliability of the soldered assembly. Manual controls have been replaced by electromechanic or partially electronic controls. Systems are now available that can be used with the owner's present computer or microprocessor setup. The latest development is a fully microprocessor-controlled, integrated wave solder system.

The basic liquid wave technology has undergone significant improvements in wave generation concepts and in the shape, size, and flow pattern of the solder wave. As an intermediary step, efforts were made to improve the process by applying liquid additives such as oil into or onto the solder stream. By utilizing wave geometry, the fluid dynamics of the liquid solder, and their coordination with the thermodynamics of the soldering process, wave solder machines have been developed with the capability of processing printed circuit boards and assemblies of complex, fine-line, multilayer, or multiwire design at conveyor speeds approaching 20 ft/min and reject rates of less than a fraction of a tenth of a percent.

Satisfactory results require appropriate printed circuit board and component design, board and component lead preparation, cleanliness, and the assurance of perfect solderability. Proper selection of soldering materials, flux, solder, and the correct process and apparatus in their application is also required.

The Wave Solder Unit. A variety of design concepts are used in the agitation, movement, or circulation of solder to contribute to optimum conditions in the formation of the solder wave. There is wide diversification in the internal design of the pump and sump system submerged in the solder pot reservoir and in the way the solder stream is guided to the nozzle assembly. Pumping the solder is generally implemented by rotating mechanical pumps; electrodynamic drive systems for solder circulation were also developed for this purpose. Figure 10 shows the general outline of the dual tank pumping principle, which intermixes oil in the solder stream. Intermixing oil in the solder stream and certain related processes like forced oil in-

Fig. 10 Solder wave generation—dual tank pumping principle

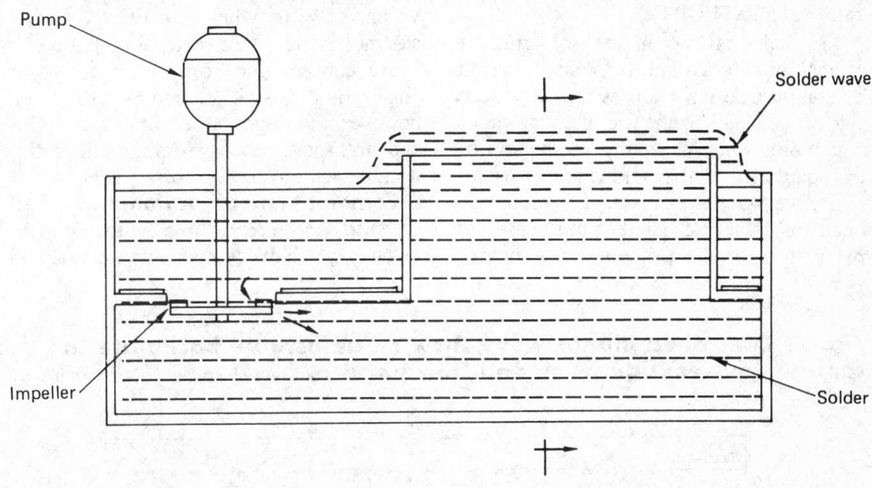

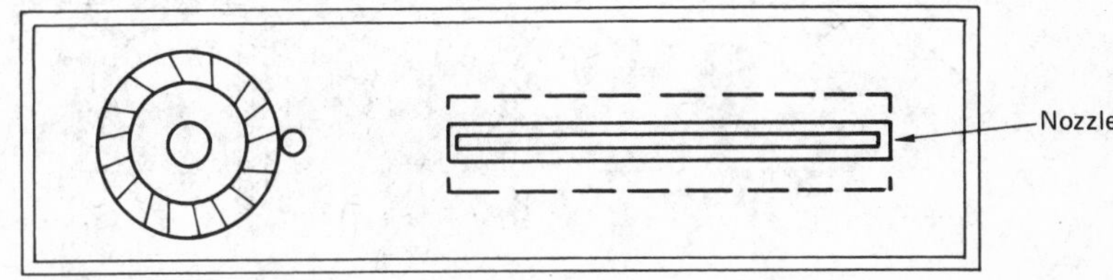

jection and lamination of a thin film of oil to the solder surface are considered temporary advancements in the improvement of soldering results. The general trend, however, is to use dry waves, which are solder waves without any liquid additive.

Research and development has improved wave soldering results, keeping pace with the ever-increasing sophistication of board and assembly design by improving the shape and flow characteristics of the solder wave through improved circulating and nozzle systems. These efforts essentially centered around the elimination of shortcomings of the early, narrow, parabolically shaped, double-sided or single-sided jet-type wave, which offered only a very narrow contact area between board and solder. This wave thermodynamically represented a heat source of moderate heat transfer capacity. By its flow characteristics and flow pattern, the wave provided an abrupt entry and, worse, an abrupt exit for the board, causing rapid cooling, thereby retarding drainage of excess solder from the board.

One development that deserves attention is solder waves that offer a substantially planar, elongated wave surface area to engage with the incoming board, optimum entry (high wave flow velocities and fast heat transfer), and good exit conditions. Progressive separation between board and wave is implemented by a low separation angle (Fig. 11).

An additional aid in the soldering of printed circuit assemblies, characterized by complex design features on the bottom side of the board, is the use of an air knife, which blows hot air against the bottom of the board as it exits from the solder wave.

Double-wave systems have been developed for soldering leadless mini-chips to the bottom of the printed circuit board. Most of these machines can be described as two independent soldering units applied in combination, with the first one producing a relatively narrow wave and second, a large surface wave. Some of them are tandem wave solder systems embodying two reservoirs, two pumps, two funnels, and two wave and nozzle systems. The structural material of soldering machines is preferably high-grade chromium-nickel alloy stainless steel.

Flux is applied either as liquid or foam by means of a liquid wave, foamhead, spray, or rotating brush. Some sophisticated designs make minutely controlled flux applications possible. Because fluxes can be rosin based or organic acid based, fluxers generally are built of stainless steel or plastic.

Preheat, properly applied and controlled, is a factor in wave soldering. Infrared radiation of wavelengths used in wave soldering may range from the black part of the spectrum, through off-red to white radiation. Combination heatbanks utilizing a preliminary drying stage operated by a combination of radiation and convective heat, the latter being implemented by a forced airflow and followed by a more intensive black radiating heatbank section, is used in industry.

Conveyance generally is performed at an incline to the horizontal, usually not exceeding 8° (Fig. 11). Horizontal conveyance also may be used. Printed circuit assemblies are either secured by pallets or board carriers that are put on conveyor chains or bands. The modern design is, however, the finger conveyor, which holds and transports the assemblies directly, usually without the use of a pallet.

Combination of Automatic Lead Cutting and Wave Soldering. A technique in which the components are mounted on the printed circuit board with long uncut leads has been developed. In the process, the assemblies with long leads are passed to a preliminary component securing stage that provides means to temporarily secure (lock) the component lead wires in the holes of the board while they are passing the automatic cutting module, which cuts the leads to predetermined lengths by fast, rotating cutting disks. The assembly is subsequently passed through a standard wave soldering operation, which makes the final, permanent solder joint. The preliminary securing of the leads is performed either by solder in the solder-cut-wave solder system, or by a low melting wave of wax-flux stabilizer in a wax-cut-wave solder system.

Supplementary Operations. Cleaning the printed wiring assembly in line—an

Fig. 12 Typical appearances of soldered joints. Appearance is related to joint quality. (a) Good mechanical properties. (b) Moderate fracture initiation strength. (c) Poor mechanical properties, no tinned layer. (d) Poor fracture initiation strength. (e) Poor mechanical properties, no adherence. (Source: Copper Development Association)

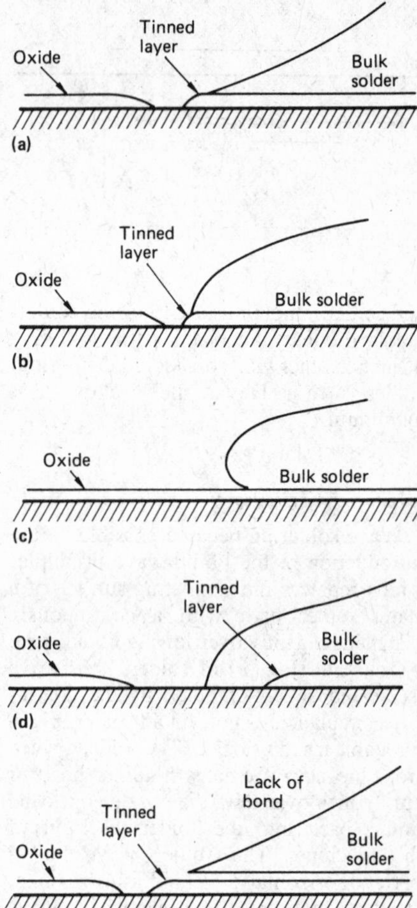

Fig. 11 Controlled planar wave showing adjustable backplate to control wave configuration and flow pattern. Courtesy of Electrovert Consulting Services

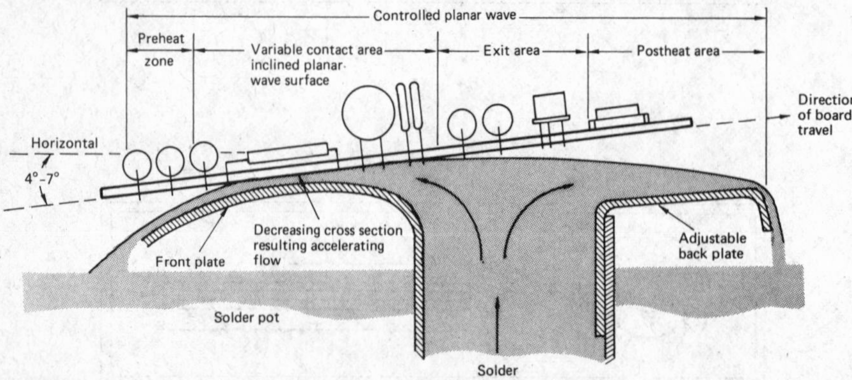

optional step prior to soldering, but usually done after soldering—has become an integral part of wave soldering sequences, except where cleaning is performed as batch cleaning in vapor degreasers or other appropriate cleaners.

In-line cleaning, mostly performed in the past by means of solvent cleaning or in combination with conveyorized vapor degreasing processes, is performed by aqueous cleaning if water-soluble fluxes are used in the soldering operation or by aqueous-detergent cleaning if rosin fluxes are used. The integrated aqueous cleaners may utilize a variety of spray techniques.

Other Applications. The application of the liquid wave principle has not been limited to printed circuit assembly and joining. It has entered the domain of integrated circuit surface coating and joining, the reflowing and fusing of hot air leveling of printed wiring boards, and tinning (solder coating) of the leadwires of discrete circuits and the terminals of integrated circuits. Most of these operations are performed using 63%Sn-37%Pb or 60%Sn-40%Pb alloys at soldering temperatures of 450 to 550 °F.

Certain special operations like the soldering of motor armatures or radiator cores may require much higher solder temperatures of 650 to 750 °F or more. Copper wire has been tinned by wave soldering at 750 °F and 500 ft/min. For fusible alloys, special-purpose and small-size wave solder machines have been applied sporadically.

Fluxing Effects on Joint Formation

Examination of hundreds of joints has led to a classification system based on appearance; under this system, joints are classified as having good, intermediate, or poor mechanical properties. After some practice, a simple visual examination, or at most examination at 20× magnification, provides evidence of the efficiency of the fluxing operation. The variations in soldered joint appearance shown in Fig. 12 reflect the quality and effectiveness of the fluxing action.

When all conditions are favorable, fluxing completely removes the oxide film (or other surface film) and protects the newly exposed metal. A tinning layer of bulk solder material then advances over the fluxed area to give primary wetting action. The tinned layer is followed by the bulk solder, and enough flux must remain to allow secondary wetting action to occur between the already solidified tinned layer and the bulk solder. A strong bond is produced, and the remaining flux ensures a good joint profile with low stress concentrations. This favorable joint profile will not be obtained if the volume of flux is insufficient, if the flux cannot cope with surface impurities, or if fluxing action is too slow. These situations result in a wide variance in mechanical properties and physical appearance. In some cases, even where tinning action has occurred, a poor bond results because of flux inactivity or inability to remelt the tinned layer or bond to it.

Inspection and Testing

Visual Inspection

Nearly all soldered joints are visually inspected for defects. The hand solderer visually inspects the joint as it is made and cleaned. The operator of a mass soldering machine usually gives at least a quick glance to each printed circuit board to ensure that the machine is still working properly. The only joints that need not be visually inspected are those for which a convenient nondestructive test is available.

A soldered joint surface should be shiny, smooth, and free of cracks, porosity, or holes. Ordinarily, no flux residues should remain. The transition from the edge of the solder surface to the exposed base metal should show a smooth profile. There should be no extraneous solder beyond the intended amount. Any area of base metal that has contacted the solder should show evidence of wetting.

The reasons for conditions different from those above include the heat sink effect of a nearby large conductive mass or a restraint of solder flow caused by the design of the joint. The inspector must question any condition that cannot be explained, for it may indicate deeper problems. For example, slight dewetting around the pad of a printed circuit board is sometimes considered acceptable, but if the dewetting were to extend over the solder-land interface, the strength and reliability of the joint could be seriously impaired.

The degree to which any flaw is cause for rejection must be determined for each application. It is not possible to generalize. For example, some military specifications require that all holes on a soldered printed circuit board be filled with solder, while a commercial application, such as a small radio, would not require such perfection. Often the differences are small, and good quality joints are found in all types of applications.

Normally, inspection is facilitated by the use of a magnifying device that should provide no more than 10× magnification factor, because too much magnification may make very minor flaws seem more important than they really are. However, very small joints, such as in microelectronics, require additional magnification to bring the image up to "normal" size. Visual inspection of soldered joints requires training and experience for the inspectors. Training assistance, ranging from simple training aids up to full courses of instruction, is commercially available. Despite this requirement, no other method of testing is currently competitive with visual inspection for a wide range of soldered joints. Common flaws in soldered assemblies that are detectable by visual inspection are covered in the following sections.

Dewetting is a phenomenon where solder flows over a surface and wets it, but before the solder solidifies, a change in the relative surface tension forces takes place so that the solder withdraws into ridges and globules similar to water on a greasy surface. The areas between solder globules retain the color of solder, but they have poor solderability, and this surface can be made to wet properly only with great effort, if at all. The exact mechanism of dewetting is not well understood, but dry abrasive cleaning that leaves particles imbedded in copper is one cause.

Nonwetting is more obvious than dewetting because, in most cases, the base metal retains its original color. Like dewetting, nonwetting is a major flaw because the solder joint is not continuous and is therefore weak. In most cases, nonwetting in very small amounts scattered across the surface can be tolerated. Typically, nonwetting can be traced to insufficient flux activity or inadequate time and/or temperature during soldering.

A flaw related to dewetting or nonwetting is the poor filling of capillary spaces, such as the holes on printed circuit boards. Nonwetting or dewetting may cause the poor fill, or excessive joint clearances may prevent joint completion. If the latter is the cause, the joint may still be acceptable, depending on the required soldering quality level.

Dull or rough solder surfaces can result from overheating or underheating. Overheating is not so damaging to the solder itself, but may damage surrounding heat-sensitive materials. Underheating produces poor contact of the solder with the base metal, but can probably be repaired. A dull, not necessarily rough, surface may also result from vibration or movement during solidification or from certain solder contaminants. Contamination is often inconsequential, but vibration could mean that a degree of weakness has been introduced into the joint structure.

Bridging. Bridges of solder spanning the space between solder joints cannot be tolerated, especially in electrical connections. Assuming that the spacing of the connections is adequate (a design parameter that depends on many factors), the most likely cause of bridging is contamination of the solder by an element that promotes oxides in the solder, such as excess cadmium or zinc.

A related phenomenon is icicling, which resembles incomplete bridges. These spikes of material attached to the joints are normally associated with wave or drag soldering. The same impurity that causes bridging can also cause icicles by interfering with the drainage of the solder as the printed circuit board is removed. Another possible cause is improper alignment of the path of the board when exiting the bath.

Porosity in the solder joint, if small and scattered, poses no threat to reliability. Large pores can significantly decrease the volume of solder in the joint area and therefore the strength of the joint. Pores that break the solder surface can trap corrosive materials or can extend down to the base metal, allowing local corrosion. Such gross porosity is formed when a bubble of air, flux vapor, or water vapor is trapped in the solidifying solder. In some cases, such defects can only be found through sectioning of the soldered joint, but usually there is some other evidence of a problem to alert the inspector.

Excessive Solder. Another flaw is excess solder in the joint areas, which may mask other flaws. Obviously, where visual inspection is the only possibility, any condition that interferes with the inspection cannot be tolerated.

Nondestructive Testing

For a few soldering applications, especially those for which the solder forms a fluid seal, pressure or vacuum testing can be used to check for leakage. The choice is usually determined by the intended application. For example, food cans that will be vacuum-packed are tested by sealing and evacuating the can. The special test machine then monitors for any loss of vacuum. Vehicle radiators and other heat exchangers that operate under pressure are leak tested by creating an air pressure inside the part being tested while it is immersed in a water tank. The test pressure normally is selected to be at least as great as the pressure level to be encountered in use. Plumbing joints usually are checked visually for leaks by test operation of the system.

For large, flat areas, such as the mounting of photovoltaic cells for solar arrays, x-radiography has been used. Unfortunately, the design of the soldered joint rarely allows this test method.

New developments include acoustic emission and laser inspection techniques. Acoustic emission requires slight deformation of the solder to generate a signal, so there is some risk of damaging the joint during testing. The laser inspection technique is most easily applied to instances where many joints of the same geometry are to be tested. The basic concept of the technique is a burst of heat generated in the joint by the laser. The dissipation of this heat is monitored and related to joint quality. A drawback of this method is that laser radiation only penetrates the solder surface a very small distance and therefore cannot indicate joint quality deeper within the joint.

Destructive Testing

As mentioned above, sectioning of soldered joints may be the only means for determining conditions in the interior. Porosity, cracking, and excessive intermetallic growth may be revealed by this procedure when other methods are inadequate. Care is required in preparing the sample for examination to avoid inadvertently cracking the joint during cutting or altering the structure by heating during sawing, polishing, or metallographic mounting.

Metallographic preparation of soldered joints also requires some experience. Because solder usually is much softer than base metals, it is easy to introduce a step at the interface during polishing. This step may be interpreted as, or may mask, a crack at the interface. It is also very easy to smear solder over the surface, obliterating pertinent features of the joint.

Similar precautions in regard to extracting the sample apply to pull testing. Various test schemes have been devised for different joint designs and intended applications. Simple butt joints can be pulled apart in tension, but lap joints tend to twist and tear apart when pulled, giving results with some scatter. Peel testing is appropriate in some cases. Application of a test method should be determined by examination of the parts and evaluation of intended service loads and mode of stressing.

A corrosion test also may be employed. Failures of electronic assemblies due to corrosion are rare, due to the good corrosion resistance of both tin and lead. A reduction in the number of service failures is achieved by applying the body of knowledge on the subject and testing parts in the appropriate environment. Corrosion inhibitors, for example, are routinely added

Table 14 Physical properties of tin-lead solders

Tin content, %	Density, g/cm^3	Electrical conductivity of copper (IACS), %	Thermal conductivity, Btu · in./s · ft^2 · °F	Coefficient of linear thermal expansion per °F × 10^{-6}
0	11.34	7.9	0.067	16.3
5	10.80	8.1	0.068	15.8
10	10.50	8.2	0.069	15.5
20	10.04	8.7	0.072	14.7
30	9.66	9.3	0.078	14.2
40	9.28	10.1	0.084	13.7
50	8.90	10.9	0.090	13.1
60	8.52	11.5	0.096	12.0
63	8.34	11.8	0.098	11.8
70	8.17	12.5	...	11.5

Table 15 Mechanical properties of bulk tin-lead solders

Tin content, %	Tensile strength, psi	Shear strength, psi	Elongation, %	Modulus of elasticity, × 10^6 psi	Hardness, HB	Impact strength (Izod), ft · lbs	Stress to produce creep rate of 0.0001 in./in./day, psi
0	1800	1800	55	2.61	4	6	250
5	4000	2100	45	...	8	7	200
10	4400	2400	30	2.76	10	8	...
20	4800	3000	20	2.90	11	11	...
30	5000	4000	18	3.05	12	12	115
40	5400	4600	25	3.34	12	14	...
50	6000	5200	35	...	14	15	125
60	7600	5600	40	4.35	16	15	...
63	7800	5400	37	...	17	15	335
70	7800	5200	30	5.08	17	14	...

to coolant mixtures to prevent solder joint failures in automotive radiators. Where galvanic corrosion is a known possibility, a protective material can be used to cover the entire joint area and reduce or eliminate possible corrosion effects on soldered joints.

Repair

Once a flaw has been found, it can be marked and prepared for repair. Depending on the type and severity of the flaw, simple remelting of the joint may be sufficient, or desoldering may be required. Desoldering involves remelting the solder and removing it by a vacuum device or by using a copper braid to wick the solder from the area. The joint is then resoldered using fresh flux and solder.

In most cases, the likelihood of soldering success is greatest when first attempted. Repair, therefore, often requires more drastic measures, such as the use of higher soldering temperatures, longer times, or the use of more active fluxes. Because most repair operations are done by hand, they can be inspected at the same time.

The basic philosophy of any soldering operation, however, should be to avoid repair. This objective can be accomplished by careful design and control of the various operations leading up to and including the actual soldering. A considered analysis of repair costs, including the possible degradation of equipment by the repair process, justifies proper preparation.

Properties of Solders and Solder Joints

Joining by soldering is utilized in a wide range of industries. The bulk of these industries use tin-lead solders for their products. Some physical and mechanical properties of tin-lead solders and their joints with base materials are known, but there is a clear need for much further property value development to enable the designer and metallurgist to approach soldering technology from a more scientific basis. A few physical properties of tin-lead solders are listed in Table 14. Densities, electrical conductivity, thermal conductivity, and coefficients of linear thermal expansion are available for a range of compositions from 0 to 70% Sn. Their properties vary in an expected manner.

The mechanical properties of bulk tin-lead solders are presented in Table 15 for tin contents up to 70%. Tensile strength generally increases with tin content, together with modulus and hardness. Creep properties do not vary in a linear manner and are higher at each end of the tin-lead system. In performing mechanical testing for the strengths of soldered joints and bulk solder alloys, the rate of testing is an important factor. Solder alloys have good ductility but are sensitive to the rate of strain during testing. This effect is illustrated in Fig. 13, which shows that at higher testing rates, higher strength measurements will be made. This fact sometimes makes data obtained from various sources difficult to interpret.

Fig. 13 Influence of rate of testing on tensile strength of soldered joints and bulk solder alloys. Source: International Tin Research Institute

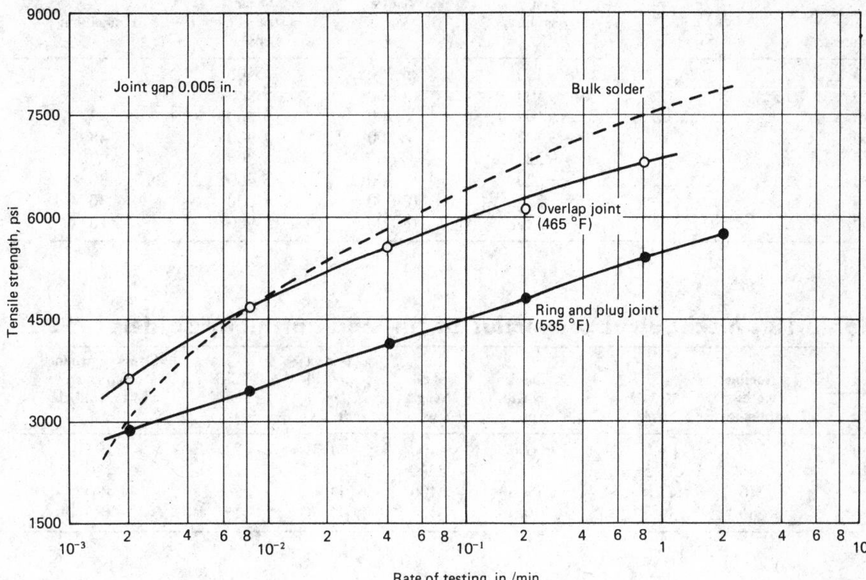

For correct bulk solder or joint strength evaluation, the rate of testing information must be included. The tensile properties of some bulk solder alloys, including tin-lead, tin-lead-antimony, and tin-lead-silver alloys are shown in Table 16 at 68 and 212 °F testing temperatures. Although more information is available on the properties of bulk solders, the important figure for service is the strength obtained at a completed joint. Bulk tin-lead solders at cryogenic temperatures are presented in Table 17, which shows that significant increases in strength occur at these very low temperatures. Retention of good elongation values to temperatures as low as −320 °F illustrates the versatility of the solder alloy material.

Improvements in the mechanical properties of bulk alloys can be obtained to some extent by the addition of antimony. Table 18 illustrates four tin-lead compositions with the addition of antimony. The creep rate is significantly improved, with moderate improvements in both tensile and shear strength. A range of alloys tested at temperatures from −328 to 400 °F are presented in Fig. 14. Strengths of all alloys increase with decreases in temperature,

Table 16 Tensile properties of bulk solders
Tested at 0.002 in./min

Composition, %			Tensile strength, psi		Elongation, %		Loss in tensile strength (68 to 212 °F), %
Tin	Lead	Other	68 °F	212 °F	68 °F	212 °F	
60	40	...	2700	580	135	>100	79
40	60	...	2420	...	130	...	...
10	90	...	2850	1160	56	>100	60
62	36	(a)	6120	2700	7	...	70
40	58	(b)	3270	960	78	>100	56
95	...	(c)	4410	2900	25	31	35
96.5	...	(d)	5260	...	31	...	...
5	93.5	(e)	4270	2750	20	25	27
1	97.5	(e)	4980	...	28	...	...
...	95	(f)	4550	...	25	...	...

(a) 2% Ag. (b) 2% Sb. (c) 5% Sb. (d) 3.5% Ag. (e) 1.5% Ag. (f) 5% Ag

Table 17 Mechanical properties of bulk tin-lead solders at cryogenic temperatures

Tin content, %	Test temperature, °F	Tensile strength, psi	Shear strength, psi	Elongation, %
10	≈−100	5 900	4 500	34
20	≈−100	7 000	5 300	32
40	≈−100	6 900	5 800	43
60	≈−100	8 500	7 900	48
10	≈−320	8 600	6 300	27
20	≈−320	12 400	8 400	30
40	≈−320	12 600	11 200	30
60	≈−320	18 800	15 900	10

Table 18 Mechanical properties of tin-lead-antimony solders

Nominal composition, %			Tensile strength, psi	Shear strength, psi	Impact strength (Izod), ft·lbs	Elongation, %	Stress to produce creep rate of 0.04 mil/mil/day, psi
Tin	Antimony	Lead					
30	2.0	68.0	6600	4400	11.3	21	295
40	2.5	57.5	7100	5300	10.4	34	420
50	3.0	47.0	7500	6100	11.1	29	480
60	3.6	36.4	8800	6100	11.5	18	480

except for high-tin alloys, which show a significant drop in tensile strength at approximately −148 °F. For very low-temperature cryogenic applications, higher lead alloys should be used. Service at even moderately elevated temperatures for solders requires some attention to loss of strength. As shown in Table 19, which

Fig. 14 Variation in tensile properties of bulk solder alloys with temperature of testing. Note the loss in ductility of tin-rich alloys below about −150 °F. (Source: International Tin Research Institute)

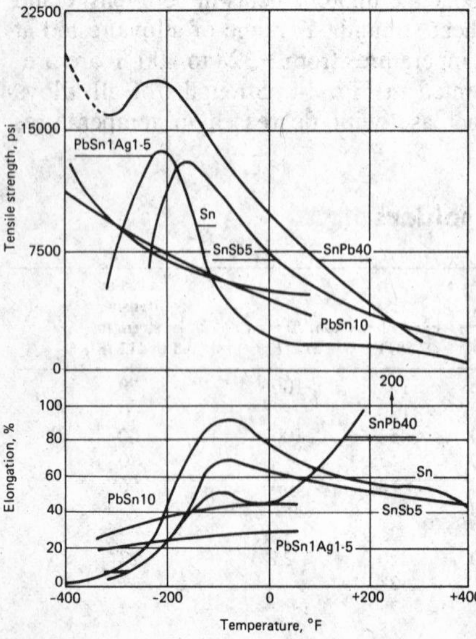

gives data for creep stress and life for bulk solder alloys at two test temperatures, the strength values dropped significantly between room temperature and 212 °F for all alloys. In most applications concerning soldered joints, the creep resistance of the material is an important parameter that should be evaluated in any proposed application.

The Soldered Joint. Physical and mechanical property values of bulk solders give an indication of their capabilities in providing the desired joint properties for service. However, the specific nature of the solder joint/base metal relationship requires that for practical purposes, measurement of soldered joint properties directly is advisable for successful application. The most widely reported joint strengths for soldered joints are shear strengths in lap joints as presented in Table 20, where data are given for a wide range of tin-lead and antimony-containing solders. Generally, the higher tin content solders provide a higher shear strength.

The relationship of shear strength to tin content is more clearly illustrated in Fig. 15, which shows that the maximum shear strength is obtained at approximately 60% Sn content. A limited number of short-term shear strength data for soldered joints at 68 and 212 °F are presented in Table 21. Differences between these data and those shown in Table 20 are due to different sources of information. Note the loss in strength as the temperature of testing is increased to 212 °F.

As stated earlier, an important measure of the capability of the soldered joint is its ability to sustain stress for long periods of time, which is good creep resistance. Table 22 shows a number of tin-lead solders and strength properties at three separate test temperatures. Note that the higher lead content solders perform much better in a creep situation than the higher tin content solder alloys. Additional data are presented graphically in Fig. 16 for three separate solder alloys. The 95%Sn-5%Sb solder alloy has the best creep strength of the three materials shown. Tin-antimony solders are therefore utilized at higher temperatures. However, the costs of the alloy must be taken into account in overall design of any system using soldered joints. Good high-temperature properties are also obtained with a 95%Sn-5%Ag alloy, for

Table 19 Creep stress—life data for bulk solder alloys

Composition, %			Initial creep stress for life of 1000 h, psi	
Tin	Lead	Other	68 °F	212 °F
60	40	...	4206	653
40	60	...	3046	609
10	90	...	5076	159
95	...	(a)	3046	1305
95	...	(b)	3191	...
62	36	(c)	...	3916
40	58	(d)	7107	856
5	93.5	(e)	2321	...
1	97.5	(e)	2828	...

(a) 5% Sb. (b) 5% Ag. (c) 2% Ag. (d) 2% Sb. (e) 1.5% Ag

Table 20 Shear strength of soldered lap joints

Tin content, %	Joint between low-carbon steel members, psi	Joint between copper members, psi	Joint between brass members, psi
ASTM grade A tin-lead solders			
10	2700	2100	1800
20	4000	3000	2800
30	4700	4000	3300
40	5000	5000	4000
50	5000	5600	4500
60	4800	5700	4300
ASTM grade C tin-lead-antimony solders			
10	1800	2100	1800
20	3100	3100	2800
30	4000	4200	3300
40	4600	5000	4000
50	4900	5700	4000
60	4500	6100	4000

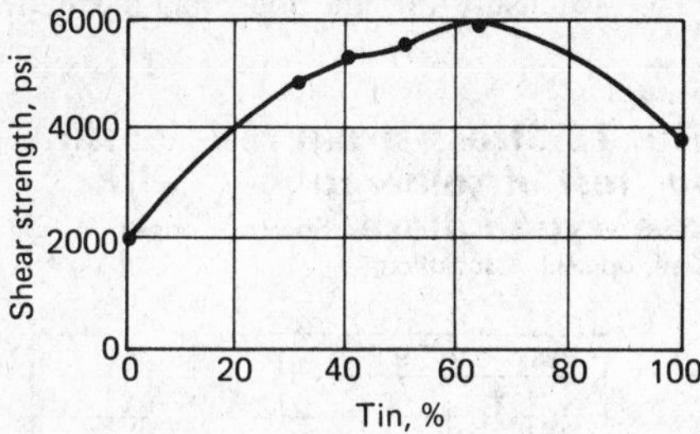

Fig. 15 Shear strengths of copper joints soldered with tin-lead solders. Source: International Tin Research Institute

which curves for creep tests under shear are presented in Fig. 17.

For elevated-temperature usage, tin-lead-silver solders frequently are selected. Table 23 shows results obtained with a specific loading program comparing several tin-lead-silver alloys and tin-lead solders. The advantages of the silver-containing materials are clearly evident in their ability to withstand much higher stresses on creep loading. These tests were carried out initially at a low load, which was increased daily to obtain the values presented. Although this is not a creep test, the results are comparable to creep tests and can be used on a relative basis for design purposes. Sufficient data are presented, however, to demonstrate to the designer that specific material selection can be made in relation to a product, depending upon loading requirements in service. From an economic standpoint, it is essential to use the advantages of a silver addition for its inherent benefits. Conversely, sometimes savings can be made where silver-bearing solders are used without properly assessing the need for them.

There are a number of applications involving soldered joints where the applied stresses tend to open or peel the joint apart, rather than pull or shear the joint. Tearing a joint apart produces a fracture curve similar to that presented in Fig. 18. The load increases until fracture initiation occurs, at which point the load value drops to a lower level and proceeds at a constant value as the joint is peeled apart. Peel strength, or fracture initiation strength, is a specific mechanical property of a soldered joint and should be determined for each particular system and design. An example of data obtained in this way at various processing temperatures and with various fluxes is shown in Fig. 19. The importance of the total solder system in determining strength properties is clearly evident. The solder alloy has an inherent strength property that it can give to the joint. The realization of this property depends upon the application of an appropriate flux to obtain maximum adhesion at the joint and also the performance of soldering at a precise temperature to obtain optimum strength values. This method of testing is appropriate to lead attachments in electronics, solder joints in containers under pressure, and in the tanks of automotive radiators. A comparison of the peel property or fracture initiation strength with various solder alloys and alloy C26000 brass under this type of load at various test temperatures from room to 300 °F is presented in Fig. 20. The superiority of a tin-lead-silver solder alloy over various tin-lead materials can be observed. The particular alloy selected for joining should depend upon the economics of the materials and the loading required upon the joint.

The fatigue strength of soldered joints is a complex and difficult subject to examine. Because solder alloys are strain-rate sensitive and have large elongation capabilities, the performance of fatigue tests under constant stress causes progressive and rapid relaxation of the joint, and conversely, tests under constant strain do not reflect a practical application situation. The influence of the rate of stress cycling in terms of rate of straining on the fatigue life of copper soldered joints with 60%Sn-40%Pb alloy is presented in Fig. 21. Fatigue of soldered joints has been obtained from a thermal cycling of printed circuit joints where the only stresses implied are those imposed by the joint itself during the thermal cycle. Metallurgical change during thermal cycling and straining occurs by internal diffusion, complicating analysis of fatigue in soldered joints. The fatigue life of different solder alloys at two test temperatures at a strain rate of 5 reversals/min is shown in Table 24. A lead-indium solder alloy composition gives best results under these conditions.

Table 22 Maximum sustained stress at various temperatures
Values will not cause failure of soldered lap joints in 10 years (in air)

Tin content, %	Sustained stress, psi 68 °F	212 °F	300 °F
5	500	250	150
10	470	200	100
20	360	120	50
30	300	90	30
40	260	75	30
50	250	75	30
60	250	75	30

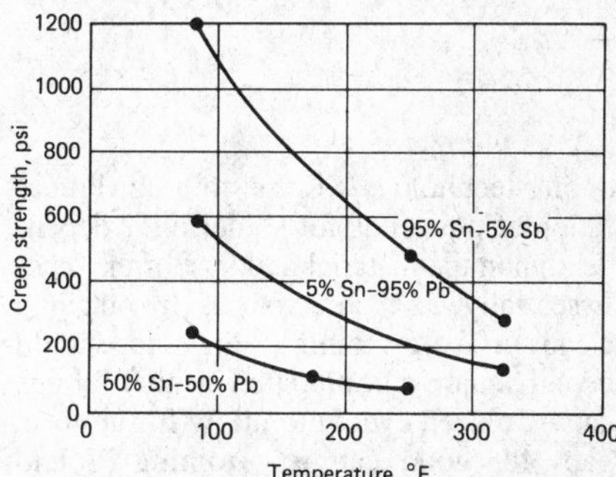

Fig. 16 Creep strengths at elevated temperatures for copper joints soldered with several alloys. Source: International Tin Research Institute

Pipe and Tube Soldering

Solder joints are widely used in the plumbing industry. The universal application of soldered copper tubing attests to the relative simplicity of performing the task and the reliability of the joints that are made. There are several important rules to ensure the correct application of sol-

Table 21 Short-term shear strength of soldered joints
Tested at 0.002 in./min

Composition, % Tin	Lead	Other	Shear strength, psi 68 °F	212 °F	Loss in strength (68 to 212 °F), %
60	40	...	2840	1850	35
10	90	...	2420	1560	35
62	36	(a)	3980	1710	57
40	58	(b)	3410	1560	54
95	...	(c)	3980	1990	50
5	93.5	(d)	2560	1710	33

(a) 2% Ag. (b) 2% Sb. (c) 5% Sb. (d) 1.5% Ag

Fig. 17 Stress-time to failure curves at various temperatures for overlap joints between brass components soldered with 95%Sn-5%Ag alloy. Source: International Tin Research Institute

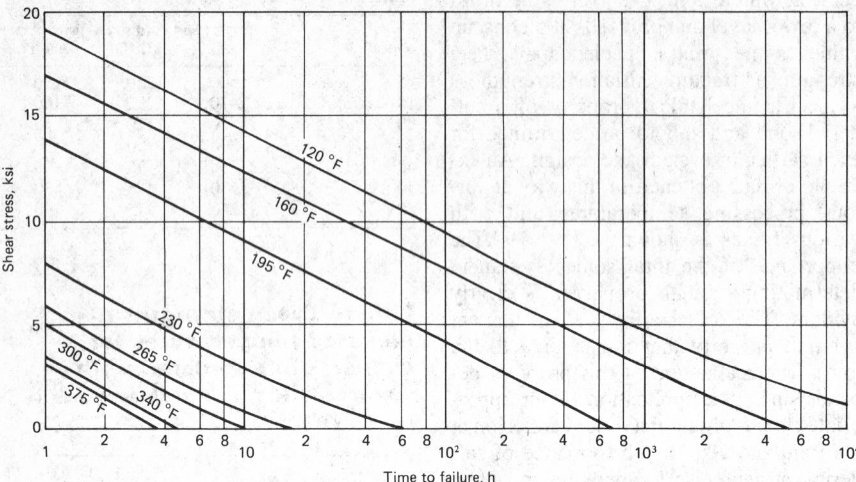

dering technology to pipe joining. Preparation of pipe ends for soldering is dependent upon the material and size of the pipe. Essentially, the objective is to obtain a clean surface for joining. Pipe ends should be cut square and prepared so that fittings can be placed evenly to allow for uniform gaps for good capillary joining. Sound soldered joints usually can be made with a 0.004-in. joint clearance. Tube end cleaning prior to solder joining can be achieved by light abrasion with fine grades of abrasive paper.

A wide range of fluxes is available for tube end pipe joining. These vary in quality and effectiveness, and care should be taken to select a flux known to provide good joint solder spread with low residual corrosion tendencies. Paste fluxes usually are applied by a small brush or clean cloth to the pipe end and internally in the fitting. The parts then are pushed together and the joint assembly heated to make the solder joint.

Pipe and tube soldering usually is accomplished with gas torches fueled with propane or natural gas. For larger sized copper piping, acetylene torches sometimes are used because of the high heat conductivity of the material. Generally, neutral to very slightly reducing flames should be used and the pipe and fitting heated as uniformly as possible until the solder temperature is reached. Soldering temperature can be determined by the ability to melt the end of a solder wire when applied to the surface of the pipe or fitting. When the solder begins to melt, the normal practice is to feed additional solder into the capillary space while moving the torch around the fitting on the opposite side to that at which the solder is being applied. This ensures that the joint is heated sufficiently so that cold soldered joints are not made. Penetration of the solder around the joint can be visually observed, and usually some fillet should form between the pipe and fitting.

Other methods of joint heating are in use, including induction furnace heating, radiant heating, and direct resistance heating. Portable equipment for resistance heating using tongs with graphite electrodes can be directly applied to the tube in applications where the presence of a flame may be undesirable. Care must be taken to ensure that the joint has reached

Fig. 18 Stress-strain relationship for test of soldered joint with peel-type failure. Source: Copper Development Association

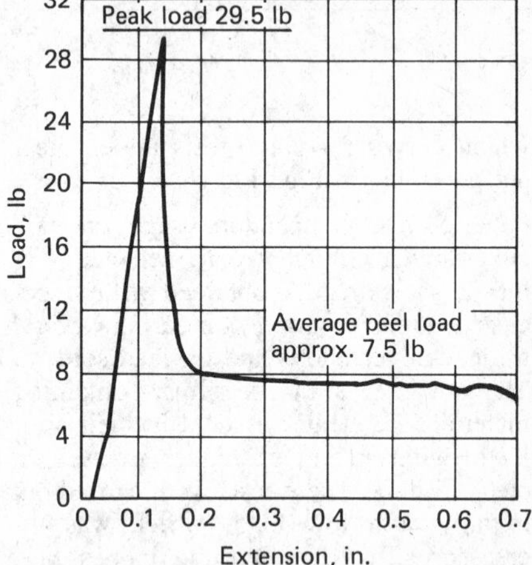

Fig. 19 Fracture initiation strength for copper joints soldered with four fluxes over a range of temperatures. Source: Copper Development Association

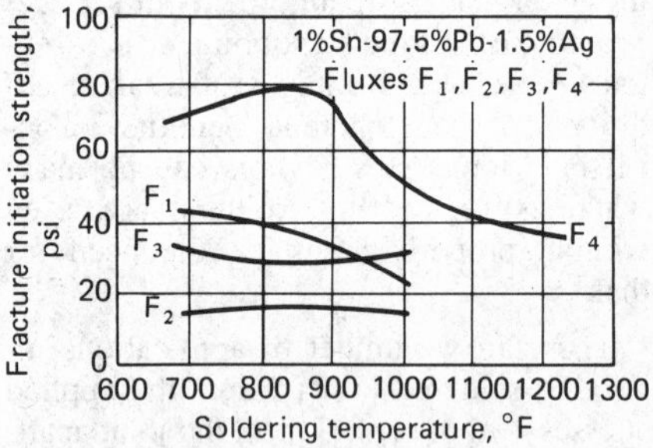

Fig. 20 Fracture initiation strength for several solders. Source: Copper Development Association

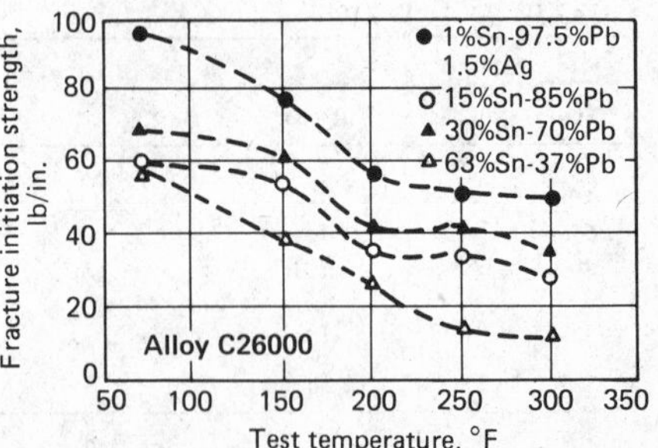

Table 23 Stepped loading creep tests on nominal $^1/_2$ by $^1/_8$ in. overlap joints on copper

Composition, %			Temperature, °F		Breaking stress, psi		
Lead	Tin	Silver	Liquidus	Solidus	68 °F	212 °F	300 °F
98	2	...	608	580	...	600	<300
95	5	...	594	518	...	540	<300
97.5	1	1.5	595	573	1640	900	700
96.5	2	1.5	583	573	1500	880	700
93.5	5	1.5	579	573	2150	800	560
70	30	...	491	361	900	375	...
60	40	...	466	361	850	300	...
50	50	...	421	361	875	300	...

Fig. 21 Influence of rate of stress cycling. In terms of straining on the fatigue life of ring-and-plug joints of copper soldered with 60%Sn-40%Pb alloy. The minimum stress imposed was nominally zero and the maximum stress in the loading cycle is the ordinate. (Source: International Tin Research Institute)

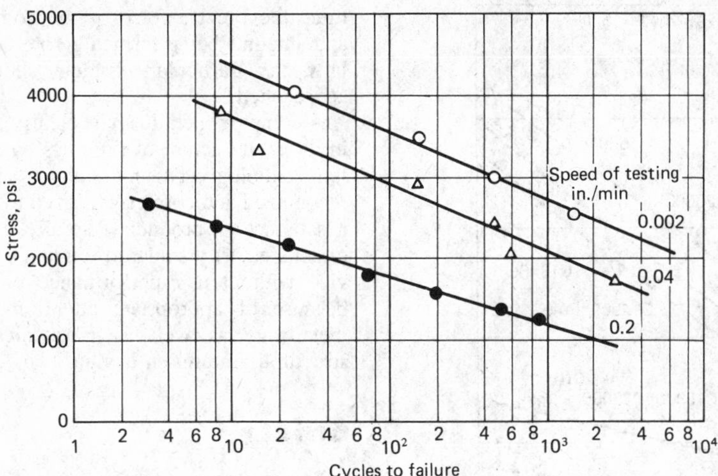

the soldering temperature, which can be tested by the uniformity of the observed heat and oxidation pattern and the capability of melting the solder wire into the joint.

Solder joints are made on pipe and tubing rapidly and effectively. The solder alloys used depend on the strength requirements of the piping system, but solders from 40 to 60% Sn with the balance lead are often used. Where creep conditions are known to exist, or are required in the load-carrying capability of the tube, a higher lead solder is desirable. Changes to the higher lead solder generally require a change in flux. Tests should be made to ensure that the appropriate solder alloy and flux combination has been selected. Pipe and tubing used for elevated temperatures at higher pressure usually are manufactured with 95%Sn-5%Sb solder. Soldering techniques with this alloy are different, and care should be taken to ensure that the techniques used are capable of producing the desired sound joints.

Soldering of stainless steel pipes can be accomplished with 50%Sn-50%Pb solders. Fluxes utilized for stainless steel joining are specifically formulated for the removal of tenacious oxide films found on stainless alloys.

Aluminum alloy pipe and tube generally is joined with tin-zinc or zinc-aluminum alloys. Joint clearances for aluminum pipe joining are greater than for most other materials, and heating must be uniform, especially when working with higher melting point solders. A socketed tube joint is common when joining aluminum alloys. Fluxes used for aluminum alloy joining have aggressive flux residues and must be thoroughly cleaned after joining to prevent subsequent corrosion.

Most soldered pipe joints are examined after manufacture by visual means and pressure testing of the piping systems. Occasionally, in systems where high pressure or elevated temperature is a known service duty requirement, radiographic and/or ultrasonic inspections may be made. Visual criteria for examination of pipe and tube joints are the presence of adequately joined solder material completely around the periphery of the fittings at both sides. Radiographic examination usually requires at least a 70% joint fill to satisfy load-carrying capabilities. A number of small disconnected voids are preferable to occasional large single-void areas. Density differences between solder alloys and commonly used piping materials make radiographic interpretation easy and void areas easily identifiable.

Table 24 Fatigue life of solder alloys subjected to 5 strain reversals/min

Composition, %			Life at 3% shear strain imposed at	
Tin	Lead	Other	68°F	212 °F
63	37	...	16 000	3000
40	60	...	13 000	5100
95	...	(a)	6 300	3600
42	...	(b)	8 500	6300
50	...	(c)	17 000	6700
95	...	(d)	20 000	4900
...	50	(c)	52 000	5400

(a) 5% Sb. (b) 58% Bi. (c) 50% In. (d) 3.5% Ag, 0.5% Sb, and 1% Cd

Solder Joint Metallurgy

When a solder joint is produced by the spread of a liquid metal or alloy over a solid base metal or alloy, a metallurgical reaction between the two frequently occurs, resulting in the formation of intermetallic compounds. Effective soldering is possible without this intermetallic compound formation when the metal systems involved do not contain reaction products. In this latter instance, the soldering action essentially produces diffusion, metallic adhesion, or perhaps even a mechanical bond. Metallurgical joining can be attained with no intermetallic compound. The most widely used systems for soldering, however, tend to be between alloy systems that have a tendency toward compound formation.

There are two main intermetallics in the copper-tin system, Cu_6Sn_5 and Cu_3Sn, both of which are found at joint interfaces between copper or brass and tin-lead solders. In this particular joint, lead shows little tendency to react with the base metal. Tin diffuses quite deeply into the brass, in addition to forming an interfacial compound. An important effect is the dissolution of the base metal into the solder alloy. The effect of time and temperature on the reaction of copper and copper alloys with two separate solders is shown in Fig. 22. The temperature effect is dominant, followed by the effect of alloy addition or modification to the base metal. The effect of tin content on the time required to react with a base-metal strip is presented in Fig. 23 for the entire tin-lead system. This figure gives an indication of the relative reaction rates of different solder alloys at various processing temperatures. These effects are summarized in a nomograph presented as Fig. 24.

The dissolution of base metal into the solder alloy and formation of an intermetallic compound, should this occur naturally in the system, results simultaneously during a soldering operation. Examination of soldered joints in brass with one particular solder is shown in Fig. 25, which demonstrates that as the intermetallic compound thickness increases, the strength properties of the joint are decreased. After a solder joint is complete, intermetallic compound formation can occur by diffusion of solder alloy elements and base-metal elements. The extent of diffusion depends on the service temperature of the joint. Figure 26 shows various curves with a 60%Sn-40%Pb solder after exposure at various temperatures for extended periods of time. Intermetallic com-

Fig. 22 Effect of time and temperature on the reaction of copper and copper alloys with two solders. (a) 30%Sn-70%Pb solder. (b) 1%Sn-97.5%Pb-1/5%Ag solder. (Source: Copper Development Association)

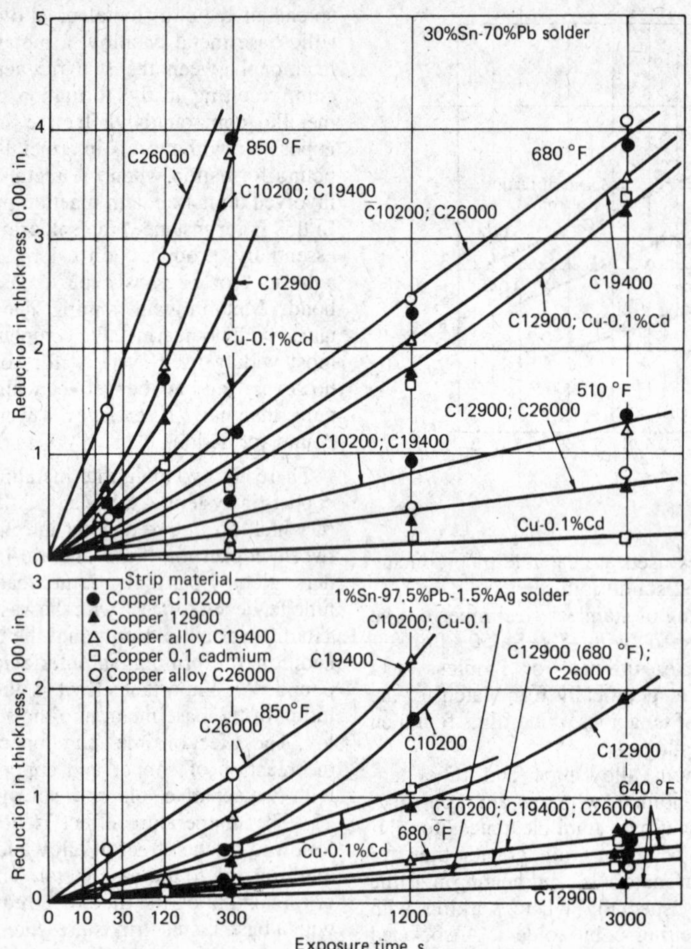

pounds form in the copper-tin system through solid-state diffusion, resulting in a reduction in strength properties of the joints, as shown in Fig. 27.

Intermetallic compounds are very important to soldering technology. Potential problems with rapid compound formation sometimes can be overcome by changes in base-metal composition or by plating of surfaces prior to the joining operation. Dissolution rate during soldering, reaction rate during soldering, and solid-state diffusion after soldering must be taken into account in any concerned manufacturing program. Intermetallic compounds can be influential in soldered joint reliability over a long period of time. In addition, these compounds also can affect electrical properties in solder joints and may become very significant in microelectronic applications.

The formation of gold-tin intermetallic compounds when soldering gold-plated leads and components is the cause of problems in service. Palladium and platinum, two precious metals used for plating important component leads, are also capable of forming several intermetallic compounds. Nickel is often used as a diffusion barrier on copper where gold plating has been used to prevent the problems associated with gold-tin intermetallic formation. However, it is possible that the nickel itself will form a nickel-tin Ni_3Sn_4 intermetallic compound, although this reaction rate is somewhat less than in materials previously discussed. Some problems ascribed to intermetallic compound formation with gold-plated components have been determined to be caused by poor plating practices that leave organic materials, or barriers, on the surfaces of components to be joined. Where possible, therefore, mechanical property determinations and metallographic examination of samples should be performed to determine the precise form of the compound and amounts being formed before deciding how any particular problem should be approached.

A complex metallurgy is clearly present in the manufacture of products by soldering technology. This metallurgy should be recognized and correctly utilized to ensure a satisfactory product. Literally millions of satisfactorily soldered joints are in service without any performance problems because of appropriate attention to the metallurgy as well as mechanical and structural features in design.

Safety

Rosin Fluxes. Rosin dissolved in organic solvent, such as alcohol or turpentine, is one of the mildest fluxes. In addition to rosin, these fluxes may contain organic halogens. Rosin is an organic-type flux, primarily abietic acid, and is not water soluble. A large portion of the evolving fumes during soldering is the organic solvent used to dissolve the rosin. Both liquid fluxes and core solder fluxes contain these solvents. The thermal decomposition products from rosin fluxes may be irritating to the membranes of the eyes, nose,

Fig. 23 Effect of tin content on the time required for tin-lead solders to react with 1 mil of base metal strip thickness. At four temperatures. (Source: Copper Development Association)

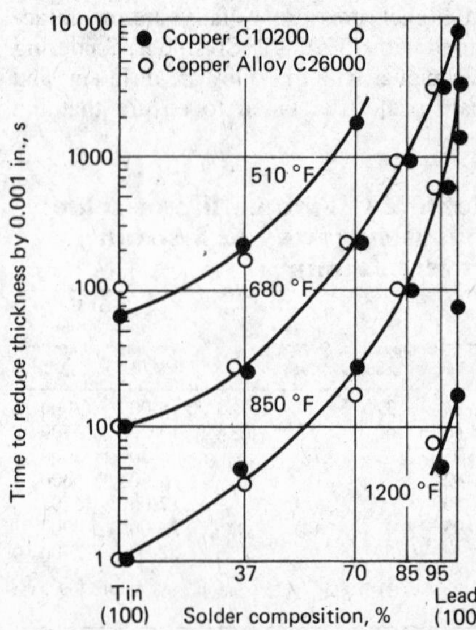

Fig. 24 Reaction of tin-lead solder with copper alloys C10200 and C26000 under static infinite-volume conditions. Select the exposure conditions on scales A and B. With a straight edge between the conditions, find the intersection with base line C. Using a straight edge, connect this point on line C with the tin content of the solder on scale D. Read the reaction-zone thickness on scale E. The nomograph may also be used in reversed sequence to select operating conditions given a permissible reaction zone thickness and solder composition. (Source: Copper Development Association)

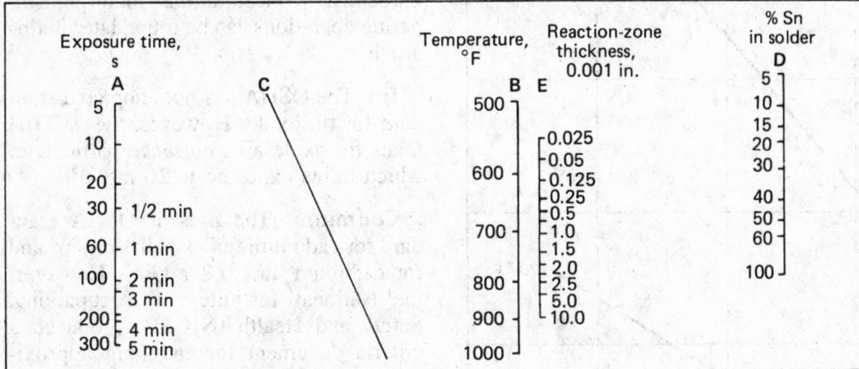

and throat. The American Conference of Governmental and Industrial Hygienists (ACGIH) has recommended a threshold limit value (TLV) for pyrolysis products of rosin core solder of 0.1 mg/m^3,* measured as formaldehyde. The thermal decomposition products of such rosin fluxes include:

- Acetone
- Methyl alcohol
- Aliphatic aldehydes
- Carbon dioxide

Fig. 25 Effect of intermetallic compound at soldered joint on peel strength. Source: Copper Development Association

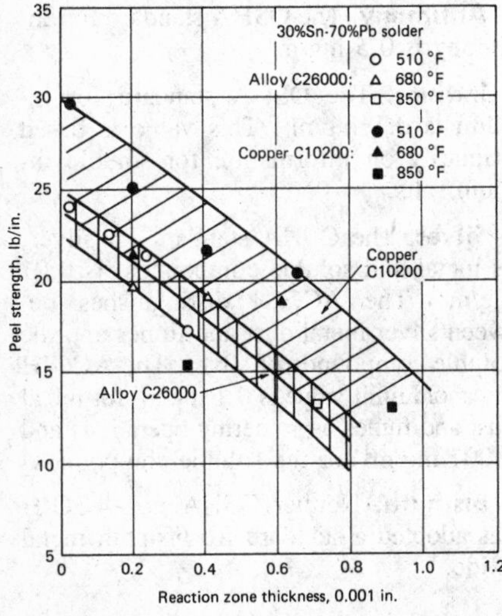

*Units mg/m^3 refer to milligrams (10^{-3} g) of material per cubic metre of air.

- Carbon monoxide
- Methane
- Ethane
- Abietic acid
- Diterpene acids

The products vary with temperature both qualitatively and quantitatively. Aliphatic aldehydes, measured as formaldehyde, have been selected as the best indicator of rosin pyrolysis products. In a study of human volunteers exposed to rosin pyrolysis products, it was found that 80% of the volunteers reported moderate to severe irritation of the eyes, nose, or throat when the concentration exceeded 0.12 mg/m^3 and that only 10% of the subjects reported irritation at a concentration below 0.05 mg/m^3.

Thirty percent of the subjects reported irritation at a concentration of aliphatic aldehydes of 0.07 mg/m^3. Adaption to the exposure occurred quickly, and after 1 to 2 min of exposure, most subjects reported that the odor was no longer detectable; therefore, the TLV was set at 0.1 mg/m^3 and the short-term exposure limit (STEL) at 0.3 mg/m^3 aliphatic aldehyde for "seasoned" workers. The Occupational Safety and Health Administration (OSHA) does not have an airborne standard for rosin pyrolysis products, although there is a standard for formaldehyde of 3 mg/m^3 maximum.

Organic water-soluble fluxes are intermediate fluxes that are more active than the rosin fluxes, but that do not exhibit the high corrosive property of inorganic fluxes. These fluxes contain organic salts such as amine hydrohalides and organic acids such as citric, lactic, benzoic, and glutamic. Ammonium chloride, the simplest hydrochloride, is at worst a very mild irritant to the skin and respiratory tract with a TLV of 10 mg/m^3. The organic acids may produce mild allergic irritation of the skin and respiratory tract, while contact with concentrated solutions can cause severe burns to the skin or eyes. The Occupational Safety and Health Administration has no published standards for these materials as airborne contaminants.

Inorganic water-soluble fluxes are strong, corrosive fluxes that contain zinc chloride, stannous chloride, hydrochloric acid, and phosphoric acid dissolved in water in various concentrations. Zinc chloride can cause irritation of the nose, throat, and respiratory tract. To prevent respiratory irritation, the ACGIH threshold limit value is 1 mg/m^3. This level has also been adopted by OSHA. The ACGIH has not published a TLV for stannous chloride.

Cleaning Agents

Cleaning agents may be used to clean surfaces prior to soldering or for removal of remaining unwanted solder or flux residues after soldering.

Acids. Contact with acids can result in severe burns, and acid mists and vapors are highly irritating to the nose, throat, and lungs. Airborne standards for hydrochloric, sulfuric, and nitric acids, set by OSHA, are as follows:

Acid	Standard
Hydrochloric	7 mg/m^3 (5 ppm)
Sulfuric	1 mg/m^3
Nitric	5 mg/m^3 (2 ppm)

Alkalies. Alkaline baths may contain caustic soda, soda ash, trisodium phosphate, and other soaps, wetting agents, or emulsifiers. The splash hazard of an alkaline bath is mainly dependent upon the temperature of the bath and its degree of alkalinity. Caustic soda in sustained contact with the skin may produce deep and painful destruction. Even weak alkalies can cause severe skin irritation if not quickly removed. These materials represent a serious hazard if splashed into the eyes.

Organic solvents such as chlorinated or fluorinated hydrocarbons can exert effects ranging from irritation of the eyes, nose, and throat, to an anesthetic effect, headaches, and dizziness. Alcohols also may be constituents of a blend of solvents. The most general effect of alcohols at high exposure is as an acute narcotic. By removing normal protective skin oils, the solvents can expose skin to chemical irritation. Organic solvents should be used

Fig. 26 Rate of growth of layers of intermetallic compound between a 60%Sn-40%Pb solder and copper at various temperatures. Source: International Tin Research Association

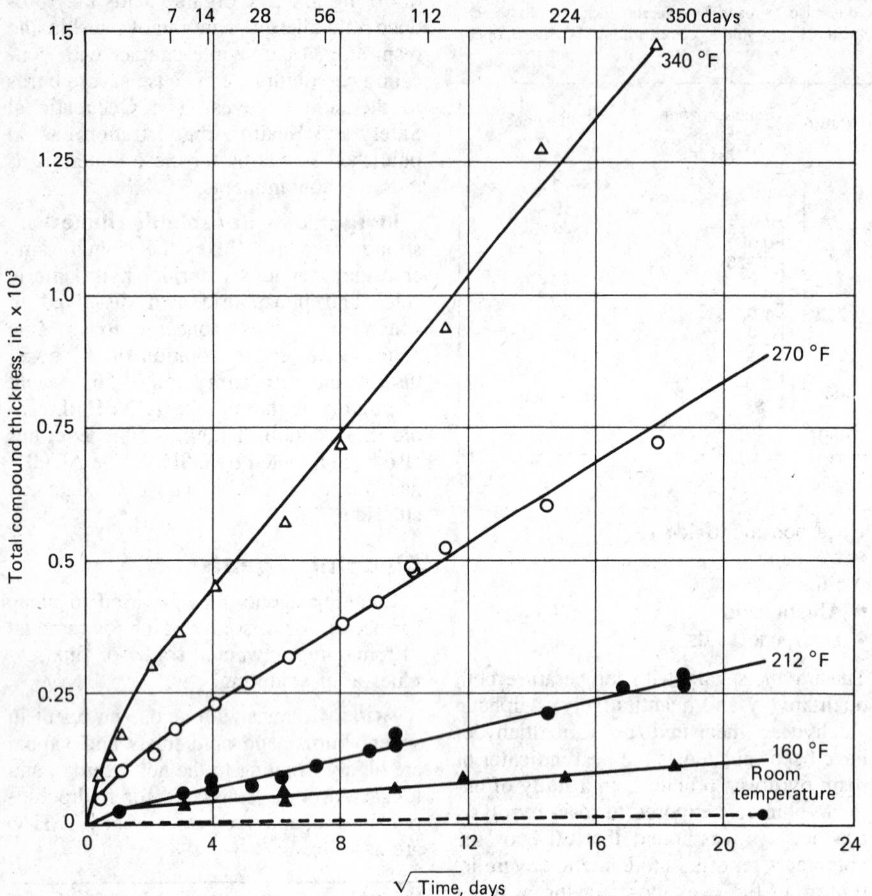

only with health and safety department approvals.

Solder Constituents

Air concentration limits for some of the more common base and filler materials used

Fig. 27 Effect of thickness of intermetallic compound in soldered joints on tensile strength at room temperature

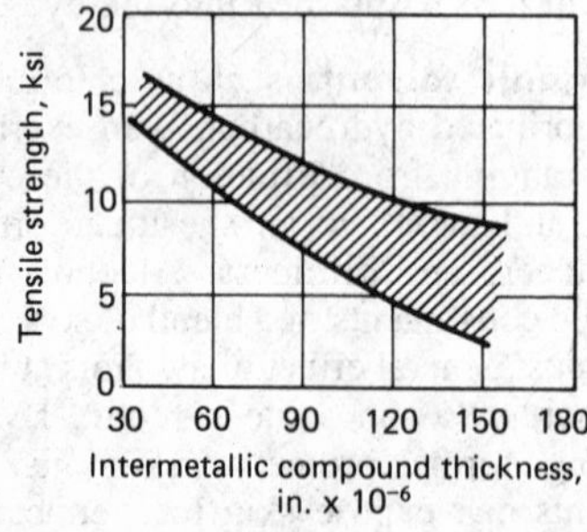

in solder which release fumes upon heating, usually as oxides, are:

Fumes as oxide	OSHA levels, mg/m^3
Cadmium	0.100
Beryllium	0.002
Zinc	5.000
Antimony	0.500
Arsenic	0.010
Tin(a)	...
Lead	0.050
Indium	0.100
Silver(a)	0.010

(a) The American Conference of Governmental and Industrial Hygienists (ACGIH) sets a level of 10.000 mg/m^3 for tin and 0.010 mg/m^3 for soluble silver compound and 0.100 mg/m^3 for silver metal dust and fumes.

Lead. A standard for lead was set by OSHA in 1978 that reduces the permissible exposure limit (PEL*) to 0.05 mg/m^3 from the previous level of 0.2 mg/m^3, with the present action level** set at 0.030 mg/m^3. The lead standard has been the subject of much controversy, and OSHA is presently reviewing certain aspects of the standard. A detailed discussion of airborne lead concentrations found in soldering operations can be found later in this article.

Tin. The OSHA has not adopted a standard for tin oxide. However, the ACGIH treats tin oxide as a nuisance particulate, which limits exposure to 10 mg/m^3.

Cadmium. The present OSHA standard for cadmium fumes is 0.1 mg/m^3 and for cadmium dust 0.2 mg/m^3. However, the National Institute for Occupational Safety and Health (NIOSH) produced a criteria document for cadmium approximately 3 years ago with a recommendation for an airborne level of 0.04 mg/m^3. This value was recommended as a ceiling value to protect against both chronic and acute effects of cadmium fumes.

Zinc. The OSHA standard for zinc oxide is 5 mg/m^3. The most serious effect reported for zinc exposure has been the fully reversible, rather innocuous metal fume fever.

Arsenic. The OSHA standard for arsenic is 0.01 mg/m^3, down from a previous level of 0.5 mg/m^3. The ACGIH threshold limit value was reduced from a previous level of 0.5 mg/m^3 to 0.2 mg/m^3.

Beryllium. The OSHA standard for beryllium is 0.002 mg/m^3. However, OSHA has proposed that this level be reduced to 0.001 mg/m^3.

Antimony. The OSHA standard for antimony is 0.5 mg/m^3.

Indium. The OSHA standard for indium is 0.1 mg/m^3. This value is based primarily on information for soluble indium salts.

Silver. The OSHA standard for silver, as metal and soluble compounds, is 0.01 mg/m^3. The ACGIH distinguishes between silver metal dust and fumes and the soluble compounds of silver. The ACGIH threshold limit value is 0.1 mg/m^3 for metal dust and fumes as soldering operations and 0.001 mg/m^3 for the soluble compounds.

Bismuth. Neither OSHA nor ACGIH has adopted a standard for bismuth metal oxide.

*Permissible exposure limit (PEL) refers to the airborne concentration level averaged over an 8-h period, above which no employee shall be exposed.

**Action level refers to the airborne concentration level averaged over an 8-h period, which triggers certain requirements of OSHA lead standards.

Assessing Lead Exposure in Soldering Operations

While the OSHA lead standard is complex and extensive, the majority of employee exposures to lead in soldering operations may be lower than the lead standard action level of 0.030 mg/m^3. Operations where exposures to lead are below the action level of 0.030 mg/m^3 are exempt from most of the requirements of the OSHA lead standard. However, such operations must notify employees of any individual sampling results, document the sampling results, meet training requirements, and resample if there are production, process control, or personnel changes.

Lead standard hearings held by OSHA in 1977 indicated that in many, if not all, soldering operations, air lead levels do not exceed the OSHA lead standard action level of 0.03 mg/m^3 as shown:

Type of solder(a)	Amount of solder	Air lead concentration, mg/m^3
Conduction	Medium(b)	0.0013
Dip	Heavy(c)	0.0014
Conduction	Medium(b)	0.0010

(a) Type of soldering by 60%Sn-40%Pb solder with a melting range of 360 to 375 °F. (b) Up to 3 or 4 h/day. (c) Most of the time or constantly, up to 8 h/day

The data were for conduction and dip soldering operations with medium to continuous soldering. The reported air lead concentrations were within normal levels for general ambient air in urban areas and only slightly above air lead levels reported in rural areas (0.0005 mg/m^3). In addition to air lead data, blood lead levels reported on 37 individuals representative of the solder-exposed population are:

Employee(a)	Blood lead level, μg/100 mL	Job description
1	15	Mass soldering
2	14	Hand iron soldering
3	14	Hand iron soldering
4	14	Hand iron soldering
5	13	Occasional hand soldering
6	11	Layout operator
7	17	Mass soldering
37	19	Mass soldering

Note: Mean blood level for 37 employees, 11.7 μg/100 mL
(a) Values cited are typical for all 37 employees.

The mean blood lead level was 11.7 μg/100 mL,* which is within the range of blood lead values found in the normal population with no occupational exposure to lead, i.e., 10 μg/100 mL to 30 μg/100 mL. Data on lead exposures during soldering operations were compiled. Surveys were conducted during a 5-year period (1976-1981) by certified industrial hygienists in accordance with accredited procedures. The detection sensitivity for air lead measurements was in the 0.002 to 0.005 mg/m^3 range. The search included data from a variety of soldering methods that were categorized as:

- Automatic wave soldering
- Manual gun and iron soldering
- Open dip pot tinning
- Open flame joining

These data were further categorized according to the primary industrial classification. This included the following subcategories:

- Arts and crafts
- Battery manufacture
- Electronics assembly
- Electromechanical assembly
- Radiator manufacture and repair
- Tool and die repair

Automatic Wave Soldering. Airborne lead levels as a result of automatic wave soldering activities were found to average 10 μg/m^3. The highest sample was 20 μg/m^3. The absence of any significant airborne lead levels during automatic wave soldering operations was attributed to local exhaust ventilation provided by enclosing hoods and operating bath temperatures (575 to 600 °F) only slightly above the melting point of the solder.

Manual Gun Soldering. In the electronics industry, airborne lead level data from soldering averaged less than 10 μg/m^3. These results were attributed to good general and local exhaust ventilation that was provided for these operations. Also, the relatively low percentage of sample time actually spent soldering (50% or less) contributed to lower average levels.

Iron Soldering. Airborne lead levels during iron soldering on radiators averaged 28 μg/m^3, with one sample in excess of 50 μg/m^3. Adequate general ventilation was usually provided, but often there was no local exhaust ventilation in this industry. In addition, a larger volume of solder per piece and higher temperatures (up to 1000 °F) were required for radiator soldering operations. Recent improvements have been made.

Open Dip Pot Tinning. Results for open dip pots for tinning of metal were available for four industries. Evaluation of data for electronics assembly, electromechanical assembly, and the tool and die repair industry revealed average airborne lead levels of less than 5 μg/m^3 for a total of 25 samples. These low results again were attributed to good local and general exhaust ventilation and the moderate temperature of the molten materials. The average lead level during tinning operations in the radiator manufacture and repair industry was 46 μg/m^3.

Open Flame Joining. Data were available for four industries utilizing open flame soldering. In the arts and crafts industry, open flame soldering techniques are used to fill cracks and crevices between pieces of cut glass, either by copper foil or channel joining techniques. Average airborne lead levels for the arts and crafts industry were less than 15 μg/m^3. These lower levels were attributed to good general ventilation, moderate lead content of the filler material (about 50%), and the limited duration of soldering activities. In the radiator manufacture and repair industry, average airborne lead levels during soldering were about 38 μg/m^3.

Another open flame soldering technique used in the radiator manufacture and repair industry was "sweating" joints. The average airborne lead level found for this activity was about 80 μg/m^3. Of the five samples evaluated, two were greater than 50 μg/m^3. These high airborne lead levels were attributed to the lack of local exhaust ventilation for these operations and the use of moderate open flame temperatures up to 1500 °F.

Airborne lead results for the various soldering operations and industries are:

Type of soldering	Typical airborne lead concentrations (mean), mg/m^3
Automatic wave:	
Electronics assembly	0.010
Manual gun and iron:	
Electronics and electromechanical assembly	0.010
Radiator manufacture and repair	0.028
Open dip pot tinning:	
Electronics, electromechanical assembly, and tool and die repair	0.004
Radiator manufacture and repair	0.046
Open flame joining:	
Arts and crafts	0.015
Radiator manufacture and repair	0.038

*Units refer to micrograms of lead per 100 mL of blood.

A review of data of soldering operations involving lead indicates that airborne lead concentrations are below the OSHA lead standard action level of 0.030 mg/m^3 for automatic wave soldering operations, manual gun and iron soldering, open dip pot tinning, and open pot joining for electronics assembly, electromechanical assembly, tool and die repair, and arts and crafts. These operations would therefore be excluded from the bulk of the requirements of the OSHA occupational standard for lead.

Radiator repair operations without local exhaust frequently exceed the action level and, occasionally, the permissible exposure level. These operations should be evaluated on a shop-by-shop basis. The contribution of lead from lead-related adjacent operations such as filing, grinding, and chipping tended to confound the measurement of lead levels in soldering operations involving radiator repair.

By keeping soldering temperatures as low as possible to do the job, generation of fumes can be kept to a minimum. In addition, local exhaust ventilation and good general ventilation may be required to minimize exposure to the potentially hazardous substances involved in soldering operations. Representative air sampling may be required in some cases to ensure that the soldering process does not represent any undue exposure. In addition, employee education and training concerning exposure and the importance of good personal hygiene habits are required parts of safe and healthy soldering operations. Smoking, eating, and drinking in lead exposure areas, particularly in areas where the airborne lead level exceeds the OSHA action level, should be prohibited. Smoking materials and foodstuffs should not be brought into lead-processing areas. Lead could be transferred from contaminated hands to food products and smoking materials and ingested. As a further precaution, hands should be washed prior to handling food.

Soldering Case Histories

Air Rifles. One of the leading producers of high-compression air rifles and pistols consistently produces trouble-free products by soldering the principal components. Due to the inaccessibility of the air chamber after it has been manually inserted in the barrel, a solder preform is used for this soldering operation. The preshaped 40%Sn-60%Pb solder ring melts around the chamber when heat is applied to the barrel, automatically joining the components.

When soldering the air chamber of the rifle has been accomplished, the assembly stud and other components to be hand soldered are positioned on the barrel. After brush coating with flux, the next four soldering steps are performed. This involves torch heating 60%Sn-40%Pb solder to join the front sight, rear sight mount, shot retainer (breech end), and shot tube assembly and assembly stud to the barrel. One operator completes an entire gun.

Trumpet Valves. Solder plays an important role in the production of quality trumpet valves at one leading manufacturer of musical instruments. The company solders over 200 000 joints annually in assembling the valves. Soldering is considered to be the most reliable method of securing the brass spring retainer tube into the nickel-plated brass piston of the valve.

Originally, the piston assembly required 80% of the time of two experienced solderers. Using 60%Sn-40%Pb wire solder and a torch, the two craftsmen were able to produce about 125 hand-soldered joints per hour. Although the company was obtaining the desired quality and reliability, it decided to switch to specially designed solder preform and a semiautomatic soldering system run by a single operator to increase productivity and cut costs.

In this case, a 60%Sn-40%Pb preform produced alloy savings of 50% over the manual method, because a more precise amount could be applied to each joint. In addition, the new system was able to turn out 250 soldered joints per hour, reducing assembly costs by 50%.

The four equipment stations with workholding fixtures have preset dwell times to permit loading, fluxing, and optimum heating of the valve tube and brass piston. Dwell is also timed for thorough melting and penetration of the preform filler by capillary action around the entire circumference of the brass piston shoulder, mating the retainer tube to the piston. With the benefits gained by switching to the preform soldering method, the company has realized a return on its investment of over 70% annually after taxes.

Solar conversion for hot water and space heating is basically a heat transfer process in which energy from the sun heats a solar collector panel. The heat generated is absorbed by tubes containing a liquid heat transfer medium and routed by pipes or ducts to meet a current demand, or for future use.

Many manufacturers of solar systems have discovered the benefits of using tin-lead solder for joining heat transfer tubes to the flat solar collector plate. In most cases, solar equipment manufacturers specify a solder with metal ratios ranging from 5%Sn-95%Pb to 3%Sn-97%Pb because a high melting point is required. These high-temperature solders have a solidus or melting point between 580 and 600 °F. Usually, soldering is accomplished in ovens. Solder with a 5%Sn-95%Pb ratio also is employed to ensure a reliable bond at the collector joints of the copper tubing which transports the heat transfer medium through the system.

Copper tubing is soldered to a 24-gage cold rolled steel absorber panel that is corrugated in a V-shape. The solder used has a 3%Sn-97%Pb ratio. The high lead content increases the capacity of the solder to withstand the high temperature necessary for the oven soldering process and the subsequent application of an additional

Fig. 28 Superconducting cable composed of an outer copper jacket. Wrapped around and soldered to a superconducting core. (Source: Lead Industries Association, Inc.)

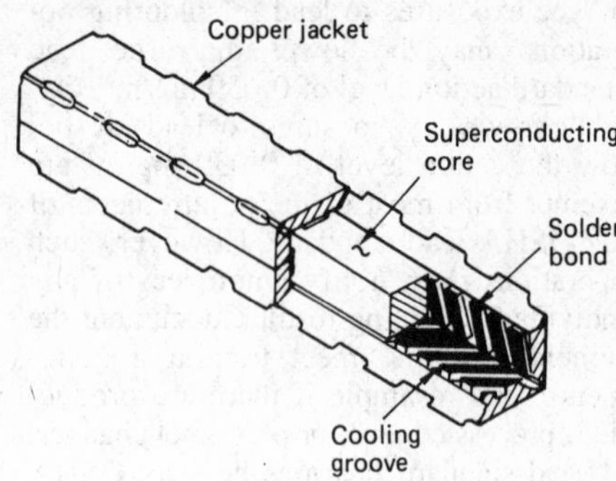

Fig. 29 Solder bonds and blocked cooling grooves. Between the outer copper jacket and the superconducting core. All are ultrasonically evaluated (one side). (Source: Lead Industries Association, Inc.)

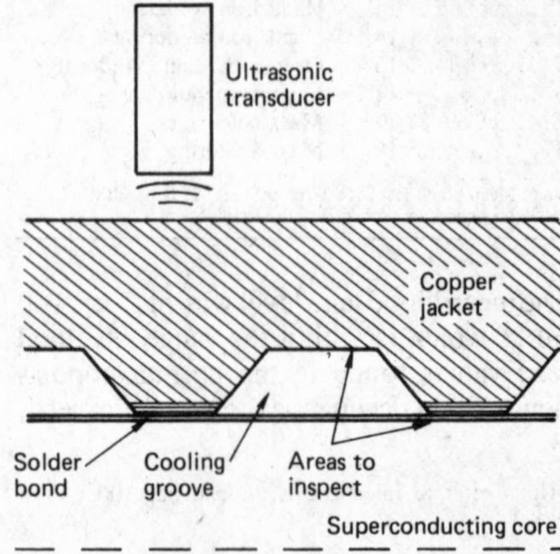

Fig. 30 Ultrasonic inspection system for soldered superconductors.
Source: Lead Industries Association, Inc.

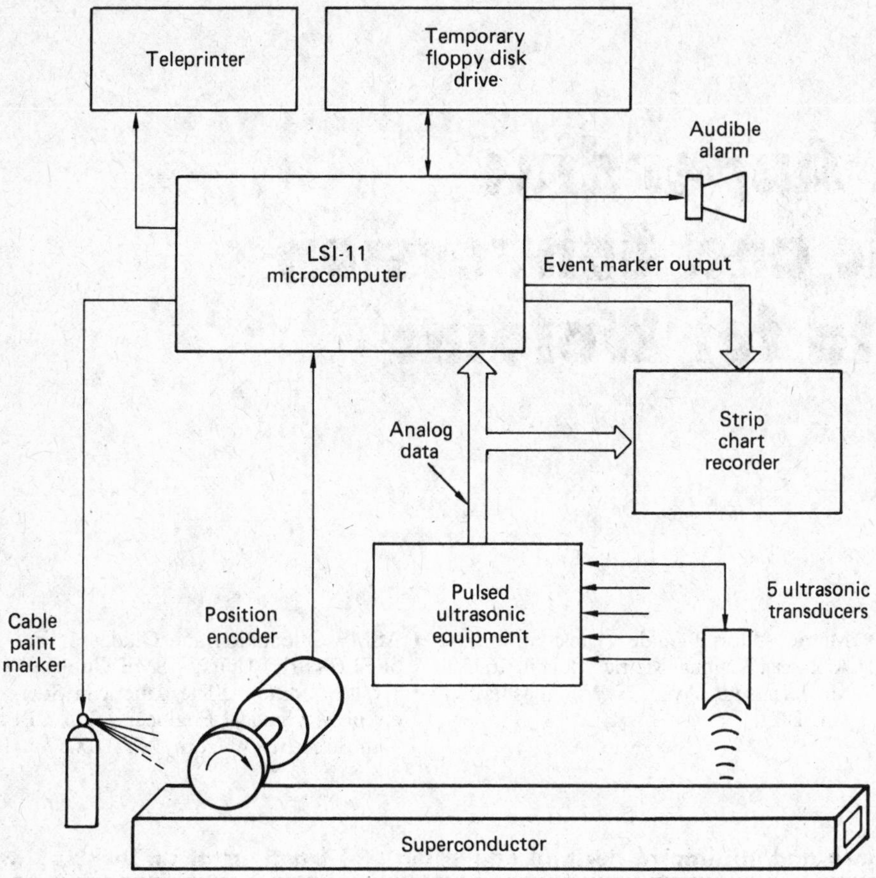

overcoat onto the panel. Solder is applied in wire form and is compressed between the steel and the copper. Each absorber requires 2 lb of solder. The absorber panels are then stacked and placed in a large oven for 1 h, during which time the tubing is soldered in place. The directional selectivity of the panels, combined with their oxide coating, reportedly yields a high-performance, low-cost solar collector. Solder provides the necessary heat transfer reliability.

Paste Solder. The potential for cost savings through the use of tin-lead paste solder for certain joining applications is becoming increasingly recognized in industry. Paste solder can provide material and labor savings, productivity gains, and increased joint integrity. Newer pastes can be applied with low-cost automated machinery that closely controls the amount dispensed.

In the manufacture of paste, tin and lead (or other necessary metals) are alloyed and the molten metal is atomized. The powder is then screened to isolate the required particle sizes. This controlled powder is then blended with a flux and suspension binder system. Typical paste solder consists of 70 to 80% metal, 2 to 5% flux, and the balance a liquid binder.

If the parts to be soldered are small enough, automatic paste soldering can often be profitably substituted for manual soldering. One company assembles a $1^1/_2$ V zinc carbon dry cell battery using 50%Sn-50%Pb paste solder to automatically solder any one of three different terminals directly to the battery casing.

A 16-station rotary indexing machine is used to solder the terminals to the casing. Each battery moves in front of the solder applicator, where a precise, measured amount of paste solder is applied near the top of the battery can. The battery is then indexed to the operator system, where any one of three different terminal types is placed in the paste. A fixture is lowered to hold the terminal and battery firmly in place until the solder has been flame heated to achieve the proper bond.

Soldering in High Technology. New applications of soldering technology include sophisticated joining. Soldering was selected as the best method of manufacturing superconductor cables. The superconductor consists of a composite core of stranded niobium-titanium fibers clad in a copper matrix. An outer jacket of copper is soldered around the outside of the superconducting core. Solder bond quality had to be evaluated on a total of 27 miles of cable. Ultrasonic testing using a new microcomputer-based system was selected for bond testing. The basic cable design is shown in Fig. 28, with the cable test arrangement shown in Fig. 29 and 30.

Some Applicable Standards for Soldering

- ASTM B32-76: Specification for Solder Metal
- EIA-RS-319: Solderability of Printed Wiring Boards
- MIL-Q-9858: Quality Assurance
- MIL-STD-105: Sampling Procedure and Tables for Inspection by Attributes
- MIL-STD-202: Test Methods for Electronic and Electrical Component Parts
- MIL-S-46844: Solder Bath Soldering of Printed Wire Assemblies
- MIL-F-14256: Flux, Soldering, Liquid (Rosin Base)
- IPC-TM-650: Test Method Manual
- QQ-S-571: Solder, Tin, Alloy, Lead-Tin Alloy, and Lead Alloy
- FED-STD-151: Metals-Test Methods

Appendix: Metric and Conversion Data for Welding

This Appendix is intended as a guide for expressing weights and measures in the Système International d'Unités (SI) for use in the welding industry. The purpose of SI units, developed and maintained by the General Conference of Weights and Measures, is to provide a basis for world-wide standardization of units and measure. For more information on metric conversions, the reader should consult the following references.

Selected References

- Standard for Metric Practice, ASTM E 380-82, American Society for Testing and Materials, 1916 Race Street, Philadelphia, PA, 19103
- Metric Practice Guide for the Welding Industry, AWS A1.1 (latest issue), American Welding Society, 2501 N.W. 7th Street, Miami, FL 33125
- SI Units and Conversion Factors for the Steel Industry, 1978, American Iron and Steel Institute, 1000 16th Street, N.W., Washington, D.C., 20036
- SP 330 The International System of Units, National Bureau of Standards (SD Catalog No. 13 C13.10:330/2), Order from Superintendent of Documents, U.S. Government Printing Office, Washington, D.C., 20402
- Metric Editorial Guide, (latest edition), American National Metric Council, 1625 Massachusetts Ave., N.W., Washington, DC
- ASME Orientation and Guide for Use of SI (Metric) Units, ASME Guide SI-1 (latest edition), The American Society of Mechanical Engineers, 345 East 47th Street, New York, NY, 10017

Inch and millimetre decimal equivalents of fractions of an inch

in.			in.		
Fraction	Decimal	mm	Fraction	Decimal	mm
1/64	0.015 625	0.396 875	33/64	0.515 625	13.096 875
1/32	0.031 250	0.793 750	17/32	0.531 250	13.493 750
3/64	0.046 875	1.190 625	35/64	0.546 875	13.890 625
1/16	0.062 500	1.587 500	9/16	0.562 500	14.287 500
5/64	0.078 125	1.984 375	37/64	0.578 125	14.684 375
3/32	0.093 750	2.381 250	19/32	0.593 750	15.081 250
7/64	0.109 375	2.778 125	39/64	0.609 375	15.478 125
1/8	0.125 000	3.175 000	5/8	0.625 000	15.875 000
9/64	0.140 625	3.571 875	41/64	0.640 625	16.271 875
5/32	0.156 250	3.968 750	21/32	0.656 250	16.668 750
11/64	0.171 875	4.365 625	43/64	0.671 875	17.065 625
3/16	0.187 500	4.762 500	11/16	0.687 500	17.462 500
13/64	0.203 125	5.159 375	45/64	0.703 125	17.859 375
7/32	0.218 750	5.556 250	23/32	0.718 750	18.256 250
15/64	0.234 375	5.953 125	47/64	0.734 375	18.653 125
1/4	0.250 000	6.350 000	3/4	0.750 000	19.050 000
17/64	0.265 625	6.746 875	49/64	0.765 625	19.446 875
9/32	0.281 250	7.143 750	25/32	0.781 250	19.843 750
19/64	0.296 875	7.540 625	51/64	0.796 875	20.240 625
5/16	0.312 500	7.937 500	13/16	0.812 500	20.637 500
21/64	0.328 125	8.334 375	53/64	0.828 125	21.034 375
11/32	0.343 750	8.731 250	27/32	0.843 750	21.431 250
23/64	0.359 375	9.128 125	55/64	0.859 375	21.828 125
3/8	0.375 000	9.525 000	7/8	0.875 000	22.225 000
25/64	0.390 625	9.921 875	57/64	0.890 625	22.621 875
13/32	0.406 250	10.318 750	29/32	0.906 250	23.018 750
27/64	0.421 875	10.715 625	59/64	0.921 875	23.415 625
7/16	0.437 500	11.112 500	15/16	0.937 500	23.812 500
29/64	0.453 125	11.509 375	61/64	0.953 125	24.209 375
15/32	0.468 750	11.906 250	31/32	0.968 750	24.606 250
31/64	0.484 375	12.303 125	63/64	0.984 375	25.003 125
1/2	0.500 000	12.700 000	1	1.000 000	25.400 000

Base, supplementary, and derived SI units pertaining to welding

Measure	Unit	SI symbol formula
Base units		
Length	metre	m
Mass	kilogram	kg
Time	second	s
Electric current	ampere	A
Thermodynamic temperature	kelvin	K
Amount of substance	mole	mol
Luminous intensity	candela	cd
Supplementary units		
Plane angle	radian	rad
Solid angle	steradian	sr
Derived units		
Area dimensions	square millimetre	mm^2
Current density	ampere per square millimetre	A/mm^2
Deposition rate	kilogram per hour	kg/h
Electrical resistivity	ohm metre	$\Omega \cdot m$
Electrode force (upset squeeze, hold)	newton	N
Flow rate (gas and liquid)	liter per minute	L/min
Fracture toughness	meganewton metre$^{-3/2}$	$MN \cdot m^{-3/2}$
Impact strength	joules	J
Linear dimensions	millimetre	mm
Power density	watt per square metre	W/m^2
Pressure	kilopascal	kPa
Tensile strength	megapascal	MPa
Thermal conductivity	watt per metre kelvin	$W/m \cdot K$
Travel speed	millimetre per second	mm/s
Volume dimension	cubic millimetre	mm^3
Wire feed rate	millimetre per second	mm/s

SI prefixes—names and symbols

Exponential expression	Multiplication factor	Prefix	SI symbol
10^{18}	1 000 000 000 000 000 000	exa	E
10^{15}	1 000 000 000 000 000	peta	P
10^{12}	1 000 000 000 000	tera	T
10^{9}	1 000 000 000	giga	G
10^{6}	1 000 000	mega	M
10^{3}	1 000	kilo	K
10^{2}	100	hecto(a)	h
10^{1}	10	deka(a)	da
10^{0}	1	BASE UNIT	
10^{-1}	0.1	deci(a)	d
10^{-2}	0.01	centi(a)	c
10^{-3}	0.001	milli	m
10^{-6}	0.000 001	micro	μ
10^{-9}	0.000 000 001	nano	n
10^{-12}	0.000 000 000 001	pico	p
10^{-15}	0.000 000 000 000 001	femto	f
10^{-18}	0.000 000 000 000 000 001	atto	a

(a) Nonpreferred. Prefixes should be selected in steps of 10^3 so that the resultant number before the prefix is between 0.1 and 1000. These prefixes should not be used for units of linear measurement, but may be used for higher order units. For example, the linear measurement, decimeter, is nonpreferred, but square decimeter is acceptable.

Wire gage diameter conversions

U.S. steel wire gage No.	in.	mm
7/0's	0.4900	12.447
6/0's	0.4615	11.7221
5/0's	0.4305	10.9347
4/0's	0.3938	10.0025
3/0's	0.3625	9.2075
2/0's	0.3310	8.4074
0	0.3065	7.7851
1	0.2830	7.1882
2	0.2625	6.6675
3	0.2437	6.1899
4	0.2253	5.7226
5	0.2070	5.2578
6	0.1920	4.8768
7	0.1770	4.4958
8	0.1620	4.1148
9	0.1483	3.7668
10	0.1350	3.429
11	0.1205	3.0607
12	0.1055	2.6797
13	0.0915	2.3241
14	0.0800	2.032
15	0.0720	1.8389
16	0.0625	1.5875
17	0.0540	1.3716
18	0.0475	1.2065
19	0.0410	1.0414
20	0.0348	0.8839
21	0.0317	0.8052
22	0.0286	0.7264
23	0.0258	0.6553
24	0.0230	0.5842
25	0.0204	0.5182
26	0.0181	0.4597
27	0.0173	0.4394
28	0.0162	0.4115
29	0.0150	0.381
30	0.0140	0.3556
31	0.0132	0.3353
32	0.0128	0.3251
33	0.0118	0.2997
34	0.0104	0.2642
35	0.0095	0.2413
36	0.0090	0.2286
37	0.0085	0.2159
38	0.0080	0.2032
39	0.0075	0.1905
40	0.0070	0.1778
41	0.0066	0.1678
42	0.0062	0.1575
43	0.0060	0.1524
44	0.0058	0.1473

Conversion factors

To convert from	to	multiply by
Angle		
degree	rad	1.745 329 E − 02
Area		
in.^{2}	mm^2	6.451 600 E + 02
in.^{2}	cm^2	6.451 600 E + 00
in.^{2}	m^2	6.451 600 E − 04
ft^2	m^2	9.290 304 E − 02
Bending moment or torque		
lbf · in.	N · m	1.129 848 E − 01
lbf · ft	N · m	1.355 818 E + 00
kgf · m	N · m	9.806 650 E + 00
ozf · in.	N · m	7.061 552 E − 03
Bending moment or torque per unit length		
lbf · in./in.	N · m/m	4.448 222 E + 00
lbf · ft/in.	N · m/m	5.337 866 E + 01
Current density		
A/in.^{2}	A/mm^2	1.550 003 E − 03
A/ft^2	A/m^2	1.076 400 E + 01
Deposition rate		
lb/h	kg/h	4.536 000 E − 01
lb/min	kg/h	2.721 600 E + 01
Electrode force		
lbf	N	4.448 222 E + 00
kgf	N	9.806 650 E + 00
Energy (impact, other)		
ft · lbf	J	1.355 818 E + 00
ft · poundal	J	4.214 011 E − 02
Btu (thermochemical)	J	1.054 350 E + 03
cal (thermochemical)	J	4.184 000 E + 00
kW · h	J	3.600 000 E + 06
W · h	J	3.600 000 E + 03
Flow rate		
ft^3/h	L/min	4.719 475 E − 01
ft^3/min	L/min	2.831 000 E + 01
gal/h	L/min	6.309 020 E − 02
gal/min	L/min	3.785 412 E + 00
Force per unit length		
lbf/ft	N/m	1.459 390 E + 01
lbf/in.	N/m	1.751 268 E + 02
Fracture toughness		
ksi$\sqrt{in.}$	MPa$\sqrt{m}$	1.098 800 E + 00
Heat content		
Btu/lb	kJ/kg	2.326 000 E + 00
cal/g	kJ/kg	4.186 800 E + 00
Heat input		
J/in.	J/m	3.937 008 E + 01
kJ/in.	kJ/m	3.937 008 E + 01
Length		
Å	nm	1.000 000 E − 01
μin.	μm	2.540 000 E − 02
mil	μm	2.540 000 E + 01
in.	mm	2.540 000 E + 01
ft	m	3.048 000 E − 01
yd	m	9.144 000 E − 01
mile	km	1.609 300 E + 00
Magnetic core loss		
W/lb	W/kg	2.204 600 E + 00
Magnetic field strength		
Oersted	A/m	7.957 700 E + 01
Magnetic flux		
maxwell	μWb	1.000 000 E − 02
Magnetic flux density		
gauss	T	1.000 000 E − 04
Mass		
oz	kg	2.834 952 E − 02
lb	kg	4.535 924 E − 01
ton (short, 2000 lb)	kg	9.071 847 E + 02
ton (long, 2240 lb)	kg	1.016 047 E + 03
Mass per unit area		
oz/in.^{2}	kg/m^2	4.395 000 E + 01
oz/ft^2	kg/m^2	3.051 517 E − 01
oz/yd^2	kg/m^2	3.390 575 E − 02
lb/ft^2	kg/m^2	4.882 428 E + 00
Mass per unit length		
lb/ft	kg/m	1.488 164 E + 00
lb/in.	kg/m	1.785 797 E + 01
Mass per unit time		
lb/h	kg/s	1.259 979 E − 04
lb/min	kg/s	7.559 873 E − 03
lb/s	kg/s	4.535 924 E − 01
Mass per unit volume (includes density)		
g/cm^3	kg/m^3	1.000 000 E + 03
lb/ft^3	kg/m^3	1.601 846 E + 01
lb/in.^{3}	kg/m^3	2.767 990 E + 04
Power density		
W/in.^{2}	W/m^2	1.550 003 E + 03
Pressure or stress (force per unit area)		
atm (standard)	Pa	1.013 250 E + 05
bar	Pa	1.000 000 E + 05
in.Hg (32 °F)	Pa	3.386 380 E + 03
in.Hg (60 °F)	Pa	3.376 850 E + 03
kgf/mm^2	Pa	9.806 650 E + 06
lbf/ft^2	Pa	4.788 026 E + 01
lbf/in.^{2} (psi)	Pa	6.894 757 E + 03
torr (mmHg, 0 °C)	Pa	1.333 220 E + 02
Specific heat		
Btu/lb · °F	J/kg · K	4.186 800 E + 03
cal/g · °C	J/kg · K	4.186 800 E + 03
Stress		
tonf/in.^{2}	MPa	1.378 951 E + 01
kgf/mm^2	MPa	9.806 650 E + 00
ksi	MPa	6.894 757 E + 00
lbf/in.^{2} (psi)	MPa	6.894 757 E − 03
Temperature		
°F	°C	5/9 · (°F − 32)
°R	°K	5/9
Temperature interval		
°F	°C	5/9
Thermal conductivity		
Btu · in./s · ft^2 · °F	W/m · K	5.192 204 E + 02
Btu/ft · h · °F	W/m · K	1.730 735 E + 00
Btu · in./ft^2 · h · °F	W/m · K	1.442 279 E − 01
cal/cm · s · °C	W/m · K	4.184 000 E + 02
Thermal expansion		
in./in. · °C	m/m · K	1.000 000 E + 00
in./in. · °F	m/m · K	1.800 000 E + 00
Travel speed		
in./min	mm/s	4.233 333 E − 01
Velocity		
ft/h	m/s	8.466 667 E − 05
ft/min	m/s	5.080 000 E − 03
ft/s	m/s	3.048 000 E − 01
in./s	m/s	2.540 000 E − 02
km/h	m/s	2.777 778 E − 01
mph	km/h	1.609 344 E + 00
Viscosity		
poise	Pa · s	1.000 000 E − 01
stokes	m^2/s	1.000 000 E − 04
ft^2/s	m^2/s	9.290 304 E − 02
in.^{2}/s	mm^2/s	6.451 600 E + 02
Volume		
in.^{3}	m^3	1.638 706 E − 05
ft^3	m^3	2.831 685 E − 02
fluid oz	m^3	2.957 353 E − 05
gal (U.S. liquid)	m^3	3.785 412 E − 03
Wavelength		
Å	nm	1.000 000 E − 01

Sheet metal gage thickness conversions

Gage	in.	mm	Gage	in.	mm
30	0.0120	0.3048	16	0.0598	1.5189
29	0.0135	0.3429	15	0.0673	1.7094
28	0.0149	0.3785	14	0.0747	1.8974
27	0.0164	0.4166	13	0.0897	2.2784
26	0.0179	0.4547	12	0.1046	2.6568
25	0.0109	0.5309	11	0.1196	3.0378
24	0.0239	0.6071	10	0.1345	3.4163
23	0.0269	0.6833	9	0.1495	3.7973
22	0.0299	0.7595	8	0.1644	4.1758
21	0.0329	0.8357	7	0.1793	4.5542
20	0.0359	0.9119	6	0.1943	4.9352
19	0.0418	1.0617	5	0.2092	5.3137
18	0.0478	1.2141	4	0.2242	5.6947
17	0.0538	1.3665	3	0.2391	6.0731

Pressure/tensile strength conversion, ksi to MPa

1 ksi (1000 psi) = 6.894 757 MPa

ksi	MPa	ksi	MPa	ksi	MPa	ksi	MPa	ksi	MPa	ksi	MPa
1	6.894 757	43	296.474 551	85	586.054 345	127	875.634 139	169	1165.213 933	210	1447.898 970
2	13.789 514	44	303.369 308	86	592.949 102	128	882.528 896	170	1172.108 690	211	1454.793 727
3	20.684 271	45	310.264 065	87	599.843 859	129	889.423 653	171	1179.003 447	212	1461.688 484
4	27.579 028	46	317.158 822	88	606.738 616	130	896.318 410	172	1185.898 204	213	1468.583 241
5	34.473 785	47	324.053 579	89	613.633 373	131	903.213 167	173	1192.792 961	214	1475.477 998
6	41.368 542	48	330.948 336	90	620.528 130	132	910.107 924	174	1199.687 718	215	1482.372 755
7	48.263 299	49	337.843 093	91	627.422 887	133	917.002 681	175	1206.582 475	216	1489.267 512
8	55.158 056	50	344.737 850	92	634.317 644	134	923.897 438	176	1213.477 232	217	1496.162 269
9	62.052 813	51	351.632 607	93	641.212 401	135	930.792 195	177	1220.371 989	218	1503.057 026
10	68.947 570	52	358.527 364	94	648.107 158	136	937.686 952	178	1227.266 746	219	1509.951 783
11	75.842 327	53	365.422 121	95	655.001 915	137	944.581 709	179	1234.161 503	220	1516.846 540
12	82.737 084	54	372.316 878	96	661.896 672	138	951.476 466	180	1241.056 260	221	1523.741 297
13	89.631 841	55	379.211 635	97	668.791 429	139	958.371 223	181	1247.951 017	222	1530.636 054
14	96.526 598	56	386.106 392	98	675.686 186	140	965.265 980	182	1254.845 774	223	1537.530 811
15	103.421 355	57	393.001 149	99	682.580 943	141	972.160 737	183	1261.740 531	224	1544.425 568
16	110.316 112	58	399.895 906	100	689.475 700	142	979.055 494	184	1268.635 288	225	1551.320 325
17	117.210 869	59	406.790 663	101	696.370 457	143	985.950 251	185	1275.530 045	226	1558.215 082
18	124.105 626	60	413.685 420	102	703.265 214	144	992.845 008	186	1282.424 802	227	1565.109 839
19	131.000 383	61	420.580 177	103	710.159 971	145	999.739 765	187	1289.319 559	228	1572.004 596
20	137.895 140	62	427.474 934	104	717.054 728	146	1006.634 522	188	1296.214 316	229	1578.899 353
21	144.789 897	63	434.369 691	105	723.949 485	147	1013.529 279	189	1303.109 073	230	1585.794 110
22	151.684 654	64	441.264 448	106	730.844 242	148	1020.424 036	190	1310.003 830	231	1592.688 867
23	158.579 411	65	448.159 205	107	737.738 999	149	1027.318 793	191	1316.898 587	232	1599.583 624
24	165.474 168	66	455.053 962	108	744.633 756	150	1034.213 350	192	1323.793 344	233	1606.478 381
25	172.368 925	67	461.948 719	109	751.528 513	151	1041.108 307	193	1330.688 101	234	1613.373 138
26	179.263 682	68	468.843 476	110	758.423 270	152	1048.003 064	194	1337.582 858	235	1620.267 895
27	186.158 439	69	475.738 233	111	765.318 027	153	1054.897 821	195	1344.477 615	236	1627.162 652
28	193.053 196	70	482.632 990	112	772.212 784	154	1061.792 578	196	1351.372 372	237	1634.057 409
29	199.947 953	71	489.527 747	113	779.107 541	155	1068.687 335	197	1358.267 129	238	1640.952 166
30	206.842 710	72	496.422 504	114	786.002 298	156	1075.582 092	198	1365.161 886	239	1647.846 923
31	213.737 467	73	503.317 261	115	792.897 055	157	1082.476 849	199	1372.056 643	240	1654.741 680
32	220.632 224	74	510.212 018	116	799.791 812	158	1089.371 606	200	1378.951 400	241	1661.636 437
33	227.526 981	75	517.106 775	117	806.686 569	159	1096.266 363	201	1385.846 157	242	1668.531 194
34	234.421 738	76	524.001 532	118	813.581 326	160	1103.161 120	202	1392.740 914	243	1675.425 951
35	241.316 495	77	530.896 289	119	820.476 083	161	1110.055 877	203	1399.635 671	244	1682.320 708
36	248.211 252	78	537.791 046	120	827.370 840	162	1116.950 634	204	1406.530 428	245	1689.215 465
37	255.106 009	79	544.685 803	121	834.265 597	163	1123.845 391	205	1413.425 185	246	1696.110 222
38	262.000 766	80	551.580 560	122	841.160 354	164	1130.740 148	206	1420.319 942	247	1703.004 979
39	268.895 523	81	558.475 317	123	848.055 111	165	1137.634 905	207	1427.214 699	248	1709.899 736
40	275.790 280	82	565.370 074	124	854.949 868	166	1144.529 662	208	1434.109 456	249	1716.794 493
41	282.685 037	83	572.264 831	125	861.844 625	167	1151.424 419	209	1441.004 213	250	1723.689 250
42	289.579 794	84	579.159 588	126	868.739 382	168	1158.319 176				

Impact energy conversions, ft · lb to J

1 ft · lb = 1.355818 J

ft · lb		J	ft · lb		J	ft · lb		J
0.7376	1	1.3558	37.6157	51	69.1467	77.4440	105	142.3609
1.4751	2	2.7116	38.3532	52	70.5025	81.1318	110	149.1400
2.2127	3	4.0675	39.0908	53	71.8583	84.8196	115	155.9191
2.9502	4	5.4233	39.8284	54	73.2142	88.5075	120	162.6982
3.6878	5	6.7791	40.5659	55	74.5700	92.1953	125	169.4772
4.4254	6	8.1349	41.3035	56	75.9258	95.8831	130	176.2563
5.1629	7	9.4907	42.0410	57	77.2816	99.5709	135	183.0354
5.9005	8	10.8465	42.7786	58	78.6374	103.2587	140	189.8145
6.6381	9	12.2024	43.5162	59	79.9933	106.9465	145	196.5936
7.3756	10	13.5582	44.2537	60	81.3491	110.6343	150	203.3727
8.1132	11	14.9140	44.9913	61	82.7049	114.3221	155	210.1518
8.8507	12	16.2698	45.7288	62	84.0607	118.0099	160	216.9308
9.5883	13	17.6256	46.4664	63	85.4165	121.6977	165	223.7099
10.3259	14	18.9815	47.2040	64	86.7723	125.3856	170	230.4890
11.0634	15	20.3373	47.9415	65	88.1282	129.0734	175	237.2681
11.8010	16	21.6931	48.6791	66	89.4840	132.7612	180	244.0472
12.5386	17	23.0489	49.4167	67	90.8398	136.4490	185	250.8263
13.2761	18	24.4047	50.1542	68	92.1956	140.1368	190	257.6054
14.0137	19	25.7605	50.8918	69	93.5514	143.8246	195	264.3845
14.7512	20	27.1164	51.6293	70	94.9073	147.5124	200	271.1636
15.4888	21	28.4722	52.3669	71	96.2631	154.8880	210	284.7218
16.2264	22	29.8280	53.1045	72	97.6189	162.2637	220	298.2799
16.9639	23	31.1838	53.8420	73	98.9747	169.6393	230	311.8381
17.7015	24	32.5396	54.5796	74	100.3305	177.0149	240	325.3963
18.4390	25	33.8954	55.3172	75	101.6863	184.3905	250	338.9545
19.1766	26	35.2513	56.0547	76	103.0422	191.7661	260	352.5126
19.9142	27	36.6071	56.7923	77	104.3980	199.1418	270	366.0708
20.6517	28	37.9629	57.5298	78	105.7538	206.5174	280	379.6290
21.3893	29	39.3187	58.2674	79	107.1096	213.8930	290	393.1872
22.1269	30	40.6745	59.0050	80	108.4654	221.2686	300	406.7454
22.8644	31	42.0304	59.7425	81	109.8212	228.6442	310	420.3036
23.6020	32	43.3862	60.4801	82	111.1771	236.0199	320	433.8617
24.3395	33	44.7420	61.2177	83	112.5329	243.3955	330	447.4199
25.0771	34	46.0978	61.9552	84	113.8887	250.7711	340	460.9781
25.8147	35	47.4536	62.6928	85	115.2445	258.1467	350	474.5363
26.5522	36	48.8094	63.4303	86	116.6003	265.5224	360	488.0944
27.2898	37	50.1653	64.1679	87	117.9562	272.8980	370	501.6526
28.0274	38	51.5211	64.9055	88	119.3120	280.2736	380	515.2108
28.7649	39	52.8769	65.6430	89	120.6678	287.6492	390	528.7690
29.5025	40	54.2327	66.3806	90	122.0236	295.0248	400	542.3272
30.2400	41	55.5885	67.1182	91	123.3794	302.4005	410	555.8854
30.9776	42	56.9444	67.8557	92	124.7452	309.7761	420	569.4435
31.7152	43	58.3002	68.5933	93	126.0911	317.1517	430	583.0017
32.4527	44	59.6560	69.3308	94	127.4469	324.5273	440	596.5599
33.1903	45	61.0118	70.0684	95	128.8027	331.9029	450	610.1181
33.9279	46	62.3676	70.8060	96	130.1585	339.2786	460	623.6762
34.6654	47	63.7234	71.5435	97	131.5143	346.6542	470	637.2344
35.4030	48	65.0793	72.2811	98	132.8702	354.0298	480	650.7926
36.1405	49	66.4351	73.0186	99	134.2260	361.4054	490	664.3508
36.8781	50	67.7909	73.7562	100	135.5818	368.7811	500	677.9090

Temperature conversions

The general arrangement of this conversion table was devised by Sauveur and Boylston. The middle columns of numbers (in boldface type) contain the temperature readings (°F or °C) to be converted. When converting from degrees Fahrenheit to degrees Celsius, read the Celsius equivalent in the column headed °C. When converting from Celsius to Fahrenheit, read the Fahrenheit equivalent in the column headed °F.

°F		°C	°F		°C	°F		°C	°F		°C	°F		°C
···	**−458**	−272.22	···	**−328**	−200.00	−324.4	**−198**	−127.78	−90.4	**−68**	−55.56	143.6	**62**	16.67
···	**−456**	−271.11	···	**−326**	−198.89	−320.8	**−196**	−126.67	−86.8	**−66**	−54.44	147.2	**64**	17.78
···	**−454**	−270.00	···	**−324**	−197.78	−317.2	**−194**	−125.56	−83.2	**−64**	−53.33	150.8	**66**	18.89
···	**−452**	−268.89	···	**−322**	−196.67	−313.6	**−192**	−124.44	−79.6	**−62**	−52.22	154.4	**68**	20.00
···	**−450**	−267.78	···	**−320**	−195.56	−310.0	**−190**	−123.33	−76.0	**−60**	−51.11	158.0	**70**	21.11
···	**−448**	−266.67	···	**−318**	−194.44	−306.4	**−188**	−122.22	−72.4	**−58**	−50.00	161.6	**72**	22.22
···	**−446**	−265.56	···	**−316**	−193.33	−302.8	**−186**	−121.11	−68.8	**−56**	−48.89	165.2	**74**	23.33
···	**−444**	−264.44	···	**−314**	−192.22	−299.2	**−184**	−120.00	−65.2	**−54**	−47.78	168.8	**76**	24.44
···	**−442**	−263.33	···	**−312**	−191.11	−295.6	**−182**	−118.89	−61.6	**−52**	−46.67	172.4	**78**	25.56
···	**−440**	−262.22	···	**−310**	−190.00	−292.0	**−180**	−117.78	−58.0	**−50**	−45.56	176.0	**80**	26.67
···	**−433**	−261.11	···	**−308**	−188.89	−288.4	**−178**	−116.67	−54.4	**−48**	−44.44	179.6	**82**	27.78
···	**−436**	−260.00	···	**−306**	−187.78	−284.8	**−176**	−115.56	−50.8	**−46**	−43.33	183.2	**84**	28.89
···	**−434**	−258.89	···	**−304**	−186.67	−281.2	**−174**	−114.44	−47.2	**−44**	−42.22	186.8	**86**	30.00
···	**−432**	−257.78	···	**−302**	−185.56	−277.6	**−172**	−113.33	−43.6	**−42**	−41.11	190.4	**88**	31.11
···	**−430**	−256.67	···	**−300**	−184.44	−274.0	**−170**	−112.22	−40.0	**−40**	−40.00	194.0	**90**	32.22
···	**−428**	−255.56	···	**−298**	−183.33	−270.4	**−168**	−111.11	−36.4	**−38**	−38.89	197.6	**92**	33.33
···	**−426**	−254.44	···	**−296**	−182.22	−266.8	**−166**	−110.00	−32.8	**−36**	−37.78	201.2	**94**	34.44
···	**−424**	−253.33	···	**−294**	−181.11	−263.2	**−164**	−108.89	−29.2	**−34**	−36.67	204.8	**96**	35.56
···	**−422**	−252.22	···	**−292**	−180.00	−259.6	**−162**	−107.78	−25.6	**−32**	−35.56	208.4	**98**	36.67
···	**−420**	−251.11	···	**−290**	−178.89	−256.0	**−160**	−106.67	−22.0	**−30**	−34.44	212.0	**100**	37.78
···	**−418**	−250.00	···	**−288**	−177.78	−252.4	**−158**	−105.56	−18.4	**−28**	−33.33	215.6	**102**	38.89
···	**−416**	−248.89	···	**−286**	−176.67	−248.8	**−156**	−104.44	−14.8	**−26**	−32.22	219.2	**104**	40.00
···	**−414**	−247.78	···	**−284**	−175.56	−245.2	**−154**	−103.33	−11.2	**−24**	−31.11	222.8	**106**	41.11
···	**−412**	−246.67	···	**−282**	−174.44	−241.6	**−152**	−102.22	−7.6	**−22**	−30.00	226.4	**108**	42.22
···	**−410**	−245.56	···	**−280**	−173.33	−238.0	**−150**	−101.11	−4.0	**−20**	−28.89	230.0	**110**	43.33
···	**−408**	−244.44	···	**−278**	−172.22	−234.4	**−148**	−100.00	−0.4	**−18**	−27.78	233.6	**112**	44.44
···	**−406**	−243.33	···	**−276**	−171.11	−230.8	**−146**	−98.89	+3.2	**−16**	−26.67	237.2	**114**	45.56
···	**−404**	−242.22	···	**−274**	−170.00	−227.2	**−144**	−97.78	+6.8	**−14**	−25.56	240.8	**116**	46.67
···	**−402**	−241.11	−457.6	**−272**	−168.89	−223.6	**−142**	−96.67	+10.4	**−12**	−24.44	244.4	**118**	47.78
···	**−400**	−240.00	−454.0	**−270**	−167.78	−220.0	**−140**	−95.56	+14.0	**−10**	−23.33	248.0	**120**	48.89
···	**−398**	−238.89	−450.4	**−268**	−166.67	−216.4	**−138**	−94.44	+17.6	**−8**	−22.22	251.6	**122**	50.00
···	**−396**	−237.78	−446.8	**−266**	−165.56	−212.8	**−136**	−93.33	+21.2	**−6**	−21.11	255.2	**124**	51.11
···	**−394**	−236.67	−443.2	**−264**	−164.44	−209.2	**−134**	−92.22	+24.8	**−4**	−20.00	258.8	**126**	52.22
···	**−392**	−235.56	−439.6	**−262**	−163.33	−205.6	**−132**	−91.11	+28.4	**−2**	−18.89	262.4	**128**	53.33
···	**−390**	−234.44	−436.0	**−260**	−162.22	−202.0	**−130**	−90.00	+32.0	**±0**	−17.78	266.0	**130**	54.44
···	**−388**	−233.33	−432.4	**−258**	−161.11	−198.4	**−128**	−88.89	+35.6	**+2**	−16.67	269.6	**132**	55.56
···	**−386**	−232.22	−428.8	**−256**	−160.00	−194.8	**−126**	−87.78	+39.2	**+4**	−15.56	273.2	**134**	56.67
···	**−384**	−231.11	−425.2	**−254**	−158.89	−191.2	**−124**	−86.67	+42.8	**+6**	−14.44	276.8	**136**	57.78
···	**−382**	−230.00	−421.6	**−252**	−157.78	−187.6	**−122**	−85.56	+46.4	**+8**	−13.33	280.4	**138**	58.89
···	**−380**	−228.89	−418.0	**−250**	−156.67	−184.0	**−120**	−84.44	+50.0	**+10**	−12.22	284.0	**140**	60.00
···	**−378**	−227.78	−414.4	**−248**	−155.56	−180.4	**−113**	−83.33	+53.6	**+12**	−11.11	287.6	**142**	61.11
···	**−376**	−226.67	−410.8	**−246**	−154.44	−176.8	**−116**	−82.22	+57.2	**+14**	−10.00	291.2	**144**	62.22
···	**−374**	−225.56	−407.2	**−244**	−153.33	−173.2	**−114**	−81.11	+60.8	**+16**	−8.89	294.8	**146**	63.33
···	**−372**	−224.44	−403.6	**−242**	−152.22	−169.6	**−112**	−80.00	+64.4	**+18**	−7.78	298.4	**148**	64.44
···	**−370**	−223.33	−400.0	**−240**	−151.11	−166.0	**−110**	−78.89	+68.0	**+20**	−6.67	302.0	**150**	65.56
···	**−368**	−222.22	−396.4	**−238**	−150.00	−162.4	**−108**	−77.78	+71.6	**+22**	−5.56	305.6	**152**	66.67
···	**−366**	−221.11	−392.8	**−236**	−148.89	−158.8	**−106**	−76.67	+75.2	**+24**	−4.44	309.2	**154**	67.73
···	**−364**	−220.00	−389.2	**−234**	−147.78	−155.2	**−104**	−75.56	+78.8	**+26**	−3.33	312.8	**156**	68.83
···	**−362**	−218.89	−385.6	**−232**	−146.67	−151.6	**−102**	−74.44	+82.4	**+28**	−2.22	316.4	**158**	70.00
···	**−360**	−217.73	−382.0	**−230**	−145.56	−148.0	**−100**	−73.33	+86.0	**+30**	−1.11	320.0	**160**	71.11
···	**−358**	−216.67	−378.4	**−228**	−144.44	−144.4	**−98**	−72.22	+89.6	**+32**	±0.00	323.6	**162**	72.22
···	**−356**	−215.56	−374.8	**−226**	−143.33	−140.8	**−96**	−71.11	+93.2	**+34**	+1.11	327.2	**164**	73.33
···	**−354**	−214.44	−371.2	**−224**	−142.22	−137.2	**−94**	−70.00	+96.8	**+36**	+2.22	330.8	**166**	74.44
···	**−352**	−213.33	−367.6	**−222**	−141.11	−133.6	**−92**	−68.89	+100.4	**+38**	+3.33	334.4	**168**	75.56
···	**−350**	−212.22	−364.0	**−220**	−140.00	−130.0	**−90**	−67.78	+104.0	**+40**	+4.44	338.0	**170**	76.67
···	**−348**	−211.11	−360.4	**−218**	−138.89	−126.4	**−88**	−66.67	+107.6	**+42**	+5.56	341.6	**172**	77.78
···	**−346**	−210.00	−356.8	**−216**	−137.78	−122.8	**−86**	−65.56	+111.2	**+44**	+6.67	345.2	**174**	78.89
···	**−344**	−208.89	−353.2	**−214**	−136.67	−119.2	**−84**	−64.44	+114.8	**+46**	+7.78	348.8	**176**	80.00
···	**−342**	−207.78	−349.6	**−212**	−135.56	−115.6	**−82**	−63.33	+118.4	**+48**	+8.89	352.4	**178**	81.11
···	**−340**	−206.67	−346.0	**−210**	−134.44	−112.0	**−80**	−62.22	+122.0	**+50**	+10.00	356.0	**180**	82.22
···	**−338**	−205.56	−342.4	**−208**	−133.33	−108.4	**−78**	−61.11	+125.6	**+52**	+11.11	359.6	**182**	83.33
···	**−336**	−204.44	−338.8	**−206**	−132.22	−104.8	**−76**	−60.00	+129.2	**+54**	+12.22	363.2	**184**	84.44
···	**−334**	−203.33	−335.2	**−204**	−131.11	−101.2	**−74**	−58.89	+132.8	**+56**	+13.33	366.8	**186**	85.56
···	**−332**	−202.22	−331.6	**−202**	−130.00	−97.6	**−72**	−57.78	+136.4	**+58**	+14.44	370.4	**188**	86.67
···	**−330**	−201.11	−328.0	**−200**	−128.89	−94.0	**−70**	−56.67	+140.0	**+60**	+15.56	374.0	**190**	87.78

(continued)

Temperature conversions (continued)

°F		°C	°F		°C	°F		°C	°F		°C	°F		°C
377.6	**192**	88.89	629.6	**332**	166.67	881.6	**472**	244.44	1580.0	**860**	460.00	2840.0	**1560**	848.89
381.2	**194**	90.00	633.2	**334**	167.78	885.2	**474**	245.56	1598.0	**870**	465.56	2858.0	**1570**	854.44
384.8	**196**	91.11	636.8	**336**	168.89	888.8	**476**	246.67	1616.0	**880**	471.11	2876.0	**1580**	860.00
388.4	**198**	92.22	640.4	**338**	170.00	892.4	**478**	247.78	1634.0	**890**	476.67	2894.0	**1590**	865.56
392.0	**200**	93.33	644.0	**340**	171.11	896.0	**480**	248.89	1652.0	**900**	482.22	2912.0	**1600**	871.11
395.6	**202**	94.44	647.6	**342**	172.22	899.6	**482**	250.00	1670.0	**910**	487.78	2930.0	**1610**	876.67
399.2	**204**	95.56	651.2	**344**	173.33	903.2	**484**	251.11	1688.0	**920**	493.33	2948.0	**1620**	882.22
402.8	**206**	96.67	654.8	**346**	174.44	906.8	**486**	252.22	1706.0	**930**	498.89	2966.0	**1630**	887.78
406.4	**208**	97.73	658.4	**348**	175.56	910.4	**488**	253.33	1724.0	**940**	504.44	2984.0	**1640**	893.33
410.0	**210**	98.89	662.0	**350**	176.67	914.0	**490**	254.44	1742.0	**950**	510.00	3002.0	**1650**	898.89
413.6	**212**	100.00	665.6	**352**	177.78	917.6	**492**	255.56	1760.0	**960**	515.56	3020.0	**1660**	904.44
417.2	**214**	101.11	669.2	**354**	178.89	921.2	**494**	256.67	1778.0	**970**	521.11	3038.0	**1670**	910.00
420.8	**216**	102.22	672.8	**356**	180.00	924.8	**496**	257.78	1796.0	**980**	526.67	3056.0	**1680**	915.56
424.4	**218**	103.33	676.4	**358**	181.11	928.4	**498**	258.89	1814.0	**990**	532.22	3074.0	**1690**	921.11
428.0	**220**	104.44	680.0	**360**	182.22	932.0	**500**	260.00	1832.0	**1000**	537.78	3092.0	**1700**	926.67
431.6	**222**	105.56	683.6	**362**	183.33	935.6	**502**	261.11	1850.0	**1010**	543.33	3110.0	**1710**	932.22
435.2	**224**	106.67	687.2	**364**	184.44	939.2	**504**	262.22	1868.0	**1020**	548.89	3128.0	**1720**	937.78
438.8	**226**	107.78	690.8	**366**	185.56	942.8	**506**	263.33	1886.0	**1030**	554.44	3146.0	**1730**	943.33
442.4	**228**	108.89	694.4	**368**	186.67	946.4	**508**	264.44	1904.0	**1040**	560.00	3164.0	**1740**	948.89
446.0	**230**	110.00	698.0	**370**	187.78	950.0	**510**	265.56	1922.0	**1050**	565.56	3182.0	**1750**	954.44
449.6	**232**	111.11	701.6	**372**	188.89	953.6	**512**	266.67	1940.0	**1060**	571.11	3200.0	**1760**	960.00
453.2	**234**	112.22	705.2	**374**	190.00	957.2	**514**	267.78	1958.0	**1070**	576.67	3218.0	**1770**	965.56
456.8	**236**	113.33	708.8	**376**	191.11	960.8	**516**	268.89	1976.0	**1080**	582.22	3236.0	**1780**	971.11
460.4	**238**	114.44	712.4	**378**	192.22	964.4	**518**	270.00	1994.0	**1090**	587.78	3254.0	**1790**	976.67
464.0	**240**	115.56	716.0	**380**	193.33	968.0	**520**	271.11	2012.0	**1100**	593.33	3272.0	**1800**	982.22
467.6	**242**	116.67	719.6	**382**	194.44	971.6	**522**	272.22	2030.0	**1110**	598.89	3290.0	**1810**	987.78
471.2	**244**	117.78	723.2	**384**	195.56	975.2	**524**	273.33	2048.0	**1120**	604.44	3308.0	**1820**	993.33
474.8	**246**	118.89	726.8	**386**	196.67	978.8	**526**	274.44	2066.0	**1130**	610.00	3326.0	**1830**	998.89
478.4	**248**	120.00	730.4	**388**	197.78	982.4	**528**	275.56	2084.0	**1140**	615.56	3344.0	**1840**	1004.4
482.0	**250**	121.11	734.0	**390**	198.89	986.0	**530**	276.67	2102.0	**1150**	621.11	3362.0	**1850**	1010.0
485.6	**252**	122.22	737.6	**392**	200.00	989.6	**532**	277.78	2120.0	**1160**	626.67	3380.0	**1860**	1015.6
489.2	**254**	123.33	741.2	**394**	201.11	993.2	**534**	278.89	2138.0	**1170**	632.22	3398.0	**1870**	1021.1
492.8	**256**	124.44	744.8	**396**	202.22	996.8	**536**	280.00	2156.0	**1180**	637.78	3416.0	**1880**	1026.7
496.4	**258**	125.56	748.4	**398**	203.33	1000.4	**538**	281.11	2174.0	**1190**	643.33	3434.0	**1890**	1032.2
500.0	**260**	126.67	752.0	**400**	204.44	1004.0	**540**	282.22	2192.0	**1200**	648.89	3452.0	**1900**	1037.8
503.6	**262**	127.78	755.6	**402**	205.56	1007.6	**542**	283.22	2210.0	**1210**	654.44	3470.0	**1910**	1043.3
507.2	**264**	128.89	759.2	**404**	206.67	1011.2	**544**	284.44	2228.0	**1220**	660.00	3488.0	**1920**	1048.9
510.8	**266**	130.00	762.8	**406**	207.78	1014.8	**546**	285.56	2246.0	**1230**	665.56	3506.0	**1930**	1054.4
514.4	**268**	131.11	766.4	**408**	208.89	1018.4	**548**	286.67	2264.0	**1240**	671.11	3524.0	**1940**	1060.0
518.0	**270**	132.22	770.0	**410**	210.00	1022.0	**550**	287.78	2282.0	**1250**	676.67	3542.0	**1950**	1065.6
521.6	**272**	133.33	773.6	**412**	211.11	1040.0	**560**	293.33	2300.0	**1260**	682.22	3560.0	**1960**	1071.1
525.2	**274**	134.44	777.2	**414**	212.22	1058.0	**570**	298.89	2318.0	**1270**	687.78	3578.0	**1970**	1076.7
528.8	**276**	135.56	780.8	**416**	213.33	1076.0	**580**	304.44	2336.0	**1280**	693.33	3596.0	**1980**	1082.2
532.4	**278**	136.67	784.4	**418**	214.44	1094.0	**590**	310.00	2354.0	**1290**	698.89	3614.0	**1990**	1087.8
536.0	**280**	137.78	788.0	**420**	215.56	1112.0	**600**	315.56	2372.0	**1300**	704.44	3632.0	**2000**	1093.3
539.6	**282**	138.89	791.6	**422**	216.67	1130.0	**610**	321.11	2390.0	**1310**	710.00	3650.0	**2010**	1098.9
543.2	**284**	140.00	795.2	**424**	217.78	1148.0	**620**	326.67	2408.0	**1320**	715.56	3668.0	**2020**	1104.4
546.8	**286**	141.11	798.8	**426**	218.89	1166.0	**630**	332.22	2426.0	**1330**	721.11	3686.0	**2030**	1110.0
550.4	**288**	142.22	802.4	**428**	220.00	1184.0	**640**	337.78	2444.0	**1340**	726.67	3704.0	**2040**	1115.6
554.0	**290**	143.33	806.0	**430**	221.11	1202.0	**650**	343.33	2462.0	**1350**	732.22	3722.0	**2050**	1121.1
557.6	**292**	144.44	809.6	**432**	222.22	1220.0	**660**	348.89	2480.0	**1360**	737.78	3740.0	**2060**	1126.7
561.2	**294**	145.56	813.2	**434**	223.33	1238.0	**670**	354.44	2498.0	**1370**	743.33	3758.0	**2070**	1132.2
564.8	**296**	146.67	816.8	**436**	224.44	1256.0	**680**	360.00	2516.0	**1380**	748.89	3776.0	**2080**	1137.8
568.4	**298**	147.78	820.4	**438**	225.56	1274.0	**690**	365.56	2534.0	**1390**	754.44	3794.0	**2090**	1143.3
572.0	**300**	148.89	824.0	**440**	226.67	1292.0	**700**	371.11	2552.0	**1400**	760.00	3812.0	**2100**	1148.9
575.6	**302**	150.00	827.6	**442**	227.78	1310.0	**710**	376.67	2570.0	**1410**	765.56	3830.0	**2110**	1154.4
579.2	**304**	151.11	831.2	**444**	228.89	1328.0	**720**	382.22	2588.0	**1420**	771.11	3848.0	**2120**	1160.0
582.8	**306**	152.22	834.8	**446**	230.00	1346.0	**730**	387.78	2606.0	**1430**	776.67	3866.0	**2130**	1165.6
586.4	**308**	153.33	838.4	**448**	231.11	1364.0	**740**	393.33	2624.0	**1440**	782.22	3884.0	**2140**	1171.1
590.0	**310**	154.44	842.0	**450**	232.22	1382.0	**750**	398.89	2642.0	**1450**	787.78	3902.0	**2150**	1176.7
593.6	**312**	155.56	845.6	**452**	233.33	1400.0	**760**	404.44	2660.0	**1460**	793.33	3920.0	**2160**	1182.2
597.2	**314**	156.67	849.2	**454**	234.44	1418.0	**770**	410.00	2678.0	**1470**	798.89	3938.0	**2170**	1187.8
600.8	**316**	157.78	852.8	**456**	235.56	1436.0	**780**	415.56	2696.0	**1480**	804.44	3956.0	**2180**	1193.3
604.4	**318**	158.89	856.4	**458**	236.67	1454.0	**790**	421.11	2714.0	**1490**	810.00	3974.0	**2190**	1198.9
608.0	**320**	160.00	860.0	**460**	237.78	1472.0	**800**	426.67	2732.0	**1500**	815.56	3992.0	**2200**	1204.4
611.6	**322**	161.11	863.6	**462**	238.89	1490.0	**810**	432.22	2750.0	**1510**	821.11	4010.0	**2210**	1210.0
615.2	**324**	162.22	867.2	**464**	240.00	1508.0	**820**	437.78	2768.0	**1520**	826.67	4028.0	**2220**	1215.6
618.8	**326**	163.33	870.8	**466**	241.11	1526.0	**830**	443.33	2786.0	**1530**	832.22	4046.0	**2230**	1221.1
622.4	**328**	164.44	874.4	**468**	242.22	1544.0	**840**	448.89	2804.0	**1540**	837.78	4064.0	**2240**	1226.7
626.0	**330**	165.56	878.0	**470**	243.33	1562.0	**850**	454.44	2822.0	**1550**	843.33	4082.0	**2250**	1232.2

(continued)

Temperature conversions (continued)

°F		°C	°F		°C	°F		°C	°F		°C	°F		°C
4100.0	**2260**	1237.8	4586.0	**2530**	1387.8	5072.0	**2800**	1537.8	5558.0	**3070**	1687.8	7772.0	**4300**	2371.1
4118.0	**2270**	1243.3	4604.0	**2540**	1393.3	5090.0	**2810**	1543.3	5576.0	**3080**	1693.3	7862.0	**4350**	2398.8
4136.0	**2280**	1248.9	4622.0	**2550**	1398.9	5108.0	**2820**	1548.9	5594.0	**3090**	1698.9	7952.0	**4400**	2426.6
4154.0	**2290**	1254.4	4640.0	**2560**	1404.4	5126.0	**2830**	1554.4	5612.0	**3100**	1704.4	8042.0	**4450**	2454.4
4172.0	**2300**	1260.0	4658.0	**2570**	1410.0	5144.0	**2840**	1560.0	5702.0	**3150**	1732.2	8132.0	**4500**	2482.2
4190.0	**2310**	1265.6	4676.0	**2580**	1415.6	5162.0	**2850**	1565.6	5792.0	**3200**	1760.0	8222.0	**4550**	2510.0
4208.0	**2320**	1271.1	4694.0	**2590**	1421.1	5180.0	**2860**	1571.1	5882.0	**3250**	1787.7	8312.0	**4600**	2537.7
4226.0	**2330**	1276.7	4712.0	**2600**	1426.7	5198.0	**2870**	1576.7	5972.0	**3300**	1815.5	8402.0	**4650**	2565.5
4244.0	**2340**	1282.2	4730.0	**2610**	1432.2	5216.0	**2880**	1582.2	6062.0	**3350**	1843.3	8492.0	**4700**	2593.3
4262.0	**2350**	1287.8	4748.0	**2620**	1437.8	5234.0	**2890**	1587.8	6152.0	**3400**	1871.1	8582.0	**4750**	2621.1
4280.0	**2360**	1293.3	4766.0	**2630**	1443.3	5252.0	**2900**	1593.3	6242.0	**3450**	1898.8	8672.0	**4800**	2648.8
4298.0	**2370**	1298.9	4784.0	**2640**	1448.9	5270.0	**2910**	1598.9	6332.0	**3500**	1926.6	8762.0	**4850**	2676.6
4316.0	**2380**	1304.4	4802.0	**2650**	1454.4	5288.0	**2920**	1604.4	6422.0	**3550**	1954.4	8852.0	**4900**	2704.4
4334.0	**2390**	1310.0	4820.0	**2660**	1460.0	5306.0	**2930**	1610.0	6512.0	**3600**	1982.2	8942.0	**4950**	2732.2
4352.0	**2400**	1315.6	4838.0	**2670**	1465.6	5324.0	**2940**	1615.6	6602.0	**3650**	2010.0	9032.0	**5000**	2760.0
4370.0	**2410**	1321.1	4856.0	**2680**	1471.1	5342.0	**2950**	1621.1	6692.0	**3700**	2037.7	9122.0	**5050**	2787.7
4388.0	**2420**	1326.7	4874.0	**2690**	1476.7	5360.0	**2960**	1626.7	6782.0	**3750**	2065.5	9212.0	**5100**	2815.5
4406.0	**2430**	1332.2	4892.0	**2700**	1482.2	5378.0	**2970**	1632.2	6872.0	**3800**	2093.3	9302.0	**5150**	2843.3
4424.0	**2440**	1337.8	4910.0	**2710**	1487.8	5396.0	**2980**	1637.8	6962.0	**3850**	2121.1	9392.0	**5200**	2871.1
4442.0	**2450**	1343.3	4928.0	**2720**	1493.3	5414.0	**2990**	1643.3	7052.0	**3900**	2148.8	9482.0	**5250**	2898.8
4460.0	**2460**	1348.9	4946.0	**2730**	1498.9	5432.0	**3000**	1648.9	7142.0	**3950**	2176.6	9572.0	**5300**	2926.6
4478.0	**2470**	1354.4	4964.0	**2740**	1504.4	5450.0	**3010**	1654.4	7232.0	**4000**	2204.4	9662.0	**5350**	2954.4
4496.0	**2480**	1360.0	4982.0	**2750**	1510.0	5468.0	**3020**	1660.0	7322.0	**4050**	2232.2	9752.0	**5400**	2982.2
4514.0	**2490**	1365.6	5000.0	**2760**	1515.6	5486.0	**3030**	1665.5	7412.0	**4100**	2260.0	9842.0	**5450**	3010.0
4532.0	**2500**	1371.1	5018.0	**2770**	1521.1	5504.0	**3040**	1671.1	7502.0	**4150**	2287.7	9932.0	**5500**	3037.7
4550.0	**2510**	1376.7	5036.0	**2780**	1526.7	5522.0	**3050**	1676.7	7592.0	**4200**	2315.5	10022.0	**5550**	3065.5
4568.0	**2520**	1382.2	5054.0	**2790**	1532.2	5540.0	**3060**	1682.2	7682.0	**4250**	2343.3	10112.0	**5600**	3093.3

Abbreviations and Symbols

a crystal lattice edge length along *a* axis

A ampere

Å angstrom

AA Aluminum Association

AAC air carbon arc cutting

AAR Association of American Railroads

AASHTO American Association of State Highway Transportation Officials

AAW air acetylene welding

AB arc brazing

ac alternating current

Ac_1 temperature at which austenite begins to form upon heating

Ac_3 temperature at which transformation of ferrite to austenite is completed upon heating

Ac_{cm} in hypereutectoid steel, temperature at which cementite completes solution in austenite

ACGIH American Conference of Governmental Industrial Hygienists

Ae_{cm}, Ae_1, Ae_3 equilibrium transformation temperatures in steel

AFS American Foundrymen's Society

AHW atomic hydrogen welding

AISI American Iron and Steel Institute

AMS Aerospace Material Specification (of SAE)

ANSI American National Standards Institute, Inc.

AOC oxygen arc cutting

API American Petroleum Institute

Ar_1 temperature at which transformation to ferrite or to ferrite plus cementite is completed upon cooling

Ar_3 temperature at which transformation of austenite to ferrite begins upon cooling

Ar_{cm} temperature at which cementite begins to precipitate from austenite on cooling

AREA American Railway Engineering Association

ASM American Society for Metals

ASME American Society of Mechanical Engineers

ASTM American Society for Testing and Materials

atm atmosphere (pressure)

AWG American wire gage

AWS American Welding Society

AWWA American Water Works Association

bal balance

bcc body-centered cubic

BI basicity index

BMAW bare metal arc welding

Btu British thermal unit

CAC carbon arc cutting

cal calorie

CAW carbon arc welding

CAW-G gas carbon arc welding

CAW-S shielded carbon arc welding

CAW-T twin carbon arc welding

CCT continuous cooling transformation

CDA Copper Development Association

CE carbon equivalent

CGA Compressed Gas Association

cm centimetre

CO_2 carbon dioxide

CSA Canadian Standards Association

CVN Charpy V-notch

CW cold welding or continuous wave

DB dip brazing

DBTT ductile brittle transition temperature

dc direct current

DCEN direct current electrode negative

DCEP direct current electrode positive

DFB diffusion brazing

DFW diffusion welding

diam diameter

D_o diffusion coefficient

DPH diamond pyramid hardness

DS dip soldering

E modulus of elasticity

EBW electron beam welding

EBW-HV electron beam welding—high vacuum

EBW-MV electron beam welding—medium vacuum

EBW-NV electron beam welding—nonvacuum

EGW electrogas welding

EIA Electronics Industries Association

EPA Environmental Protection Agency

Eq equation

ESW electroslag welding

ETP electrolytic tough pitch

eV electron volt

EXW explosion welding

F farad

FB furnace brazing

FCAW flux cored arc welding

fcc face-centered cubic

FN ferrite number

FOC chemical flux cutting

FOW forge welding

FRW friction welding

FS furnace soldering

ft foot

FW flash welding

G shear modulus

g gram

gf gram force

GMAW gas metal arc welding

GMAW-P pulsed gas metal arc welding

GMAW-S short circuiting gas metal arc welding

GPa gigapascal

GTAW gas tungsten arc welding

GTAW-P pulsed gas tungsten arc welding

h hour

HAZ heat-affected zone

HB Brinell hardness

hcp hexagonal close-packed

HF hardenability factor

HFIW high frequency induction welding

HFRW high frequency resistance welding

HK Knoop hardness

hp horsepower

HPW hot pressure welding

HRB Rockwell "B" hardness

HRC Rockwell "C" hardness

HSLA high-strength low-alloy

HV Vickers hardness

Hz hertz

IACS International Annealed Copper Standard

IB induction brazing

ID inside diameter

IIW International Institute of Welding

INS iron soldering

in. inch

IPC Institute for Interconnecting and Packaging Electronic Circuits

I^2R resistance heating

IRB infrared brazing

IRS infrared soldering

IS induction soldering

ISO International Standards Organization

IW induction welding

J joule

K kelvin

kg kilogram

kHz kilocycles per second (kilohertz)

K_{Ic} plane-strain fracture toughness

K_{Iscc} threshold stress intensity for stress corrosion cracking

ksi 1000 pounds per square inch

kV kilovolt

kW kilowatt

L litre

lb pound

LB laser brazing

LBW laser beam welding

log common logarithm (base 10)

LPG liquefied petroleum gas

LSC laser supported combustion

LSD laser supported detonation

m metre

MAPP or MPS methyacetylene-propadiene stabilized

max maximum

M_f temperature at which martensite formation finishes during cooling

mg milligram

MIL military

min minimum, minute

mL millilitre

mm millimetre

MPa megapascal

M_s temperature at which martensite starts to form from austenite upon cooling

N newton

NBS National Bureau of Standards

NDTT nil ductility transition temperature

NEMA National Electrical Manufacturers Association

NFPA National Fire Protection Association

No. number

OAW oxyacetylene welding

OD outside diameter

OFC oxyfuel gas cutting

OFHC oxygen-free high conductivity

OFW oxyfuel gas welding

OHW oxyhydrogen welding

OSHA Occupational Safety and Health Administration

oz ounce

Pa pascal

PAC plasma arc cutting

PAW plasma arc welding

PEW percussion welding

pH negative logarithm of hydrogen-ion activity

PH precipitation hardenable

P/M powder metallurgy

ppm parts per million

pps pulses per second

POC metal powder cutting

psi pounds per square inch

PSP plasma spraying

RA reduction in area, or rosin activated

RB resistance brazing

Ref reference

rem remainder

RF radio frequency

RMA rosin mildly activated

rms root mean square

rpm revolutions per minute

RPW projection welding

RS resistance soldering

RSEW resistance seam welding

RSW resistance spot welding

RWMA Resistance Welder Manufacturers Association

s second

SAE Society of Automotive Engineers

SAW submerged arc welding

SAW-S series submerged arc welding

SCR silicon controlled rectifier

SEM scanning electron microscopy

sfm surface feet per minute

SMAW shielded metal arc welding

SSW solid-state welding

SW stud arc welding

TB torch brazing

TCAB twin carbon arc brazing

TEA transverse excited atmospheric (lasers)

TEM transmission electron microscopy or transverse electric and magnetic mode

THSP thermal spraying

TLV threshold limit value

T_m melting temperature

TS torch soldering

TTT time-temperature transformation

UNS Unified Numbering System (ASTM-SAE)

USW ultrasonic welding

UW upset welding

V volt

vol volume

vol% volume percent

W watt

WRC Welding Research Council

WS wave soldering

wt% weight percent

YAG yttrium aluminum garnet

yr year

YS yield strength

° degree, angular measure

°C degree Celsius (centigrade)

°F degree Fahrenheit

°R degree rankine

÷ divided by

= equals

> greater than

< less than

± maximum deviation

μin. micro-inch

μm micron

− minus, negative ion charge

× multiplied by, diameters (magnification)

· multiplied by

/ per

% percent

+ plus, in addition to, including, positive ion charge

√ surface roughness

ε strain

ν Poisson's ratio

π pi (3.141592)

ρ density

σ stress

Ω ohm

Index

Numbered Alloys

A

C

D

E

F

G

H

I

J

K

L

M

N

O

P

Q

R

T

U

V